W9-BVX-746

The Desk Encyclopedia of Microbiology

The Desk Encyclopedia of Microbiology

Editor

Moselio Schaechter

Consulting Editor

Joshua Lederberg

Amsterdam • Boston • Heidelberg • London • New York • Oxford
Paris • San Diego • San Francisco • Singapore • Sydney • Tokyo

This book is printed on acid-free paper

Elsevier Academic Press
525 B Street, Suite 1900, San Diego, California 92101-4495, USA
http://www.elsevier.com

Elsevier Academic Press
84 Theobald's Road, London WC1X 8RR, UK
http://www.elsevier.com

British Library Cataloguing in Publication Data
A catalogue record for this book is available from the British Library

Library of Congress Catalog Number: 2002114100

ISBN 0-12-621361-5

Printed and bound in China
03 04 05 06 07 08 9 8 7 6 5 4 3 2 1

Contents

Contributors

David W. K. Acheson
Center for Food Safety and Applied Nutrition,
Food and Drug Administration,
Rockville, MD 20740, USA

H.-W. Ackermann
Department of Medical Biology, Laval University,
Local 2332, Pav. Ferdinand Vandry, Laval, Quebec,
Canada G1K 7P4

George N. Agrios
Department of Plant Pathology, University of
Florida, 1453 Fifield Hall, P.O. Box 110680,
Gainesville, FL 32611, USA

Adriano Aguzzi
Institute of Neuropathology, University of Zurich,
University Hospital of Zürich, Schmelzbergstrasse
12, Zurich CH-8091, Switzerland

Shin-Ichi Aizawa
Soft Nano-Machine Project, CREST,
Japan Science and Technology Agency,
1064-18 Takahori, Hirata, Takanezawa,
Shioya-gun, Tochigi 329-1206, Japan

Orna Amster-Choder
Department of Molecular Biology,
Hebrew University School of Medicine,
P.O. Box 12272, Bldg. 3, 2nd Floor, Room 34,
Jerusalem 91120, Israel

Thomas M. Anderson
Microbiology Manager, Archer Daniels Midland
BioProducts, P.O. Box 1470, Decatur, IL 62525, USA

Ann M. Arvin
Department of Pediatrics, Stanford University
School of Medicine, Mailcode 5208, 300 Pasteur
Drive, G-312A Stanford, CA 94305, USA

Mariette R. Atkinson
Department of Biological Chemistry,
University of Michigan Medical School,
4310 Med Sci I, 1301 Catherine, Ann Arbor,
MI 48109-0606, USA

Farah K. Bahrani-Mougeot
Department of Medicine, University of Maryland
School of Medicine, MSTF 90, 656 W Baltimore St.,
Baltimore, MD 21201-1559, USA

Joseph T. Barbieri
Department of Microbiology, Medical College of
Wisconsin, P.O. Box 26509, 8701 Watertown Plank
Rd., Milwaukee, WI 53226-0509, USA

Douglas H. Bartlett
Center for Marine Biotechnology and
Biomedicine, University of California,
San Diego, Scripps Institution of Oceanography,
9500 Gilman Drive, Dept 0202, La Jolla,
CA 92093-0202, USA

Arnold J. Bendich
Professor of Botany and Genetics, Department of
Biology, 522 Hitchcock Hall, University of
Washington, Box 351800, Seattle, WA 98195, USA

Peter M. Bennett
Department of Pathology and Microbiology,
University of Bristol, School of Medical Sciences,
University Walk, Bristol BS8 1TD, UK

Mary K.B. Berlyn
Department of Biology, Yale University, 355-OML,
165 Prospect St., New Haven, CT 06520-8104, USA

Paul Blum
George Beadle Center for Genetics,
Nebraska University-Lincoln, P.O. Box 880666,
Lincoln, NE 68588-066, USA

Andrea D. Branch
Division of Liver Diseases, Department of Medicine, Mount Sinai Medical Center, Recanati/Miller Transplantation Institute, One Gustave L. Levy Place, Box 1633, New York, NY 10029-6574, USA

Yves V. Brun
Department of Biology, Indiana University, Jordan Hall, Bloomington, IN 47405, USA

Trevor N. Bryant
Medical Statistics and Computing, University of Southampton, Southampton General Hospital, Tremona Rd, Southampton SO16 6YD, UK

George H. Bowden
Department of Oral Microbiology, University of Manitoba, Faculty of Dentistry, 780 Bannatyne Avenue, Winnipeg, Manitoba, Canada R3E 0W2

Arturo Casadevall
Department of Medicine, Infectious Diseases, Albert Einstein College of Medicine, Golding Bldg. Rm. 701, 1300 Morris Park Avenue, Bronx, NY 10461, USA

Ricardo Cavicchioli
Department of Microbiology and Immunology, University of New South Wales, Sydney, NSW 2052, Australia

Jane A. Cecil
The Johns Hopkins University, Ross Research Building 1159, 720 Rutland Avenue, Baltimore, MD 21205-2196, USA

Peter J. Christie
Department of Microbiology and Molecular Genetics, University of Texas Health Science Center, 6431 Fannin St., Houston, TX 77030-1501, USA

Laurie E. Comstock
Channing Laboratory, Harvard Medical School, 181 Longwood Avenue, Boston, MA 02115-5899, USA

Sandra Da Re
Department of Molecular Biology, Princeton University, 330 Lewis Thomas Lab, Princeton, NJ 08544, USA

Robert B. Dadson
University of Maryland, Eastern Shore, Princess Anne, Maryland, USA

Julian Davies
Department of Microbiology and Immunology, University of British Columbia, Vancouver, British Columbia, Canada

Bruce Demple
Department of Cancer Cell Biology, Harvard School of Public Health, Bldg. 1 Floor 6, 665 Huntington Avenue, Boston, MA 02115-6021, USA

Kevin M. Devine
Department of Genetics, Trinity College, Dublin, Lincoln Place Gate, Dublin 2, UK

Vidula Dixit
George Beadle Center for Genetics, Nebraska University-Lincoln, P.O. Box 880666, Lincoln, NE 68588-066, USA

Karen W. Dodson
Department of Molecular Microbiology, Washington University School of Medicine, Campus Box 8230, 660 S Euclid Avenue, 8230, St. Louis, MO 63110-1010, USA

Michael S. Donnenberg
Department of Medicine, University of Maryland School of Medicine, MSTF 90, 656 W Baltimore St., Baltimore, MD 21201-1559, USA

Brian A. Dougherty
Department of Applied Genomics, Bristol-Myers Squibb Company, Pharmaceutical Research Institute, 5 Research Parkway, Wallingford, CT 06492-7660, USA

Angela E. Douglas
Department of Biology, University of York, P.O. Box 373, York YO1 5YW, UK

Karl Drlica
Public Health Research Institute, 225 Warren Street, Newark, NJ 07103

Charles F. Earhart
Department of Microbiology, University of Texas, ESB 226, A5000, 24th St. & Speedway, Austin, TX 78712-1095, USA

Thomas Egli
Department of Microbiology, Swiss Federal Institute for Environmental Science and Technology, BU-D04, Ueberlandstrasse 133, Dübendorf CH-8600, Switzerland

Gerald H. Elkan
Department of Microbiology, North Carolina State University, Raleigh, NC 27605, USA

Karen T. Elvers
Hatherly Labs, University of Exeter, Exeter, Prince of Wales Road, Devon EX4 4PS, UK

Larry E. Erickson
Department of Chemical Engineering, Kansas State University, 105 Durland Hall, Manhattan, KS 66506-5102, USA

Ana A. Espinel-Ingroff
Department of Medicine, Division of Infectious Diseases, Medical College of Virginia, Sanger Hall, Room 7049, 1101 E. Marshall Street, Richmond, VA 23298, USA

Stuart J. Ferguson
Department of Biochemistry, University of Oxford, South Parks Road, Oxford OX1 3QU, UK

Laura S. Frost
Department of Biological Sciences, University of Alberta, CW 405 Biological Sciences Bldg., Edmonton, Alberta T6G 2E9, Canada

Clay Fuqua
Department of Biology, 1001 E. 3rd Street, Jordan Hall 418, Indiana University Bloomington, IN 47405

Jorge Galan
Department of Microbial Pathogenesis, Yale University, New Haven, CT 06520, USA

Emil C. Gotschlich
Laboratory of Bacterial Pathogenesis and Immunology, Rockefeller University, 1230 York Avenue, New York, NY 10021-6399, USA

Peter H. Graham
Department of Soil, Water and Climate, University of Minnesota, 256 Borlaug Hall, 1991 Upper Buford Circle, St. Paul, MN 55108, USA

Carol A. Gross
Department of Microbiology, University of California, San Francisco, Medical Sciences Rm 534, #0512, 513 Parnassus Avenue, San Francisco, CA 94143-0512, USA

Lawrence Grossman
Department of Biochemistry, Johns Hopkins University School of Hygiene and Public Health, 615 N Wolfe Street, Baltimore, MD 21205-2179, USA

Janine Guespin-Michel
Laboratoire de Microbiologie du Froid, IFR CNRS – Université de Rouen, Faculté de Sciences et Techniques, Place Emile Blondel, Mont-Saint-Aignan 76821, France

Ian R. Hamilton
Department of Oral Microbiology, University of Manitoba, Faculty of Dentistry, 780 Bannatyne Avenue, Winnipeg, Manitoba R3E 0W2, Canada

Fawzy Hashem
University of Maryland, Eastern Shore, Princess Anne, Maryland, USA

J. Woodland Hastings
Department of Biology, Harvard University, Biological Laboratories, 16 Divinity Avenue, Cambridge, MA 02138-2020, USA

Jack A. Heinemann
Norweigian Institute of Gene Ecology, Tromso, Norway

James A. Hejna
Department of Molecular and Medical Genetics, Oregon Health Sciences University, M/C L103, 3181 SW Sam Jackson Park Rd., Portland, OR 97201-3098, USA

Christophe Herman
Department of Microbiology, University of California, San Francisco, Medical Sciences Rm 534, #0512, 513 Parnassus Avenue, San Francisco, CA 94143-0512, USA

David L. Heymann
Executive Director, Communicable Diseases, World Health Organization, Geneva 27 CH-1211, Switzerland

Susan K. Hoiseth
Wyeth Vaccines Research, Pearl River, New York, USA

Joachim-V. Höltje
Max-Planck-Institut für Entwicklungsbiologie, Abteilung Biochemie, Spemannstrasse 35, Tübingen D-72076, Germany

Joseph B. Hughes
Energy and Environmental Systems Institute, Rice University, 6100 S. Main, MS-316, Houston, TX 77005, USA

Scott James Hultgren
Department of Molecular Microbiology, Washington University School of Medicine, Campus Box 8230, 660 S Euclid Avenue, 8230, St. Louis, MO 63110-1010, USA

Francoise Joset
Laboratoire de Chimie Bactérienne,
CNRS 13412 Marseille, France

Robert J. Kadner
Department of Microbiology, University
of Virginia School of Medicine, Box 441,
Health Sciences Center, Charlottesville,
VA 22908, USA

Isao Karube
School of Bionics, Tokyo University of Technology,
1404-1 Katakura-cho, Hachioji,
Tokyo 192-0982, Japan

Dennis L. Kasper
Channing Laboratory, Harvard Medical School,
181 Longwood Avenue, Boston,
MA 02115-5899, USA

Michael J. Klug
Department of Microbiology, Michigan State
University, W.K. Kellogg Biological Station, 3700
East Gull Lake Drive, Hickory Corners,
MI 49060, USA

Roger Knowles
Department of Natural Resource Sciences,
McGill University, MacDonald Campus, 21,111
Lakeshore Road, Ste-Anne de-Bellevue, Quebec,
Canada H9X 3V9

L. David Kuykendall
Agricultural Research Service, Beltsville, US
Department of Agriculture, Bldg. 011A, Barc West
Rm. 252, Plant Molecular Pathology Laboratory, PSI,
Beltsville, MD 20705, USA

Hilary M. Lappin-Scott
Hatherly Labs, University of Exeter, Exeter,
Prince of Wales Road, Devon EX4 4PS, UK

Piet Lens
Department of Environmental Technology,
Wageningen Agricultural University, P.O. Box 8129,
Wageningen 6700 EV, The Netherlands

Charles R. Lovell
Department of Biological Sciences, University of
South Carolina, Coker Life Sciences 408, Columbia,
SC 29208, USA

K. Brooks Low
Department of Therapeutic Radiology, Yale
University, Hunter Radiation Therapy, M353, 333
Cedar Street, New Haven, CT 06520, USA

Millicent Masters
Institute of Cell & Molecular Biology, University of
Edinburgh, Darwin Bldg. King's Buildings,
Mayfield Road, Edinburgh EH9 3JR, UK

A. C. Matin
Department of Microbiology and Immunology,
Stanford University, Sherman Fairchild Science
Bldg. D317, Stanford, CA 94305-5402, USA

Peter A. Meacock
Department of Genetics, University of Leicester,
Adrian Building, University Road,
Leicester LE1 7RH, UK

Ian R. McDonald
Department of Biological Sciences, University of
Warwick, Coventry CV4 7AL, UK

Paul Messner
Center for Ultrastructure Research and Institute
for Molecular Nanotechnology, University of
Agricultural Sciences, Vienna, Gregor-Mendel-Str.
33, Vienna A-1180, Austria

Linda A. Miller
Department of Automicrobial Profiling/
Clinical Microbiology, SmithKline Beecham
Pharmaceuticals, P.O. Box 5089, 1250 S.
Collegeville Rd., Mail Code UP1340,
Collegeville, PA 19426-0989, USA

Saroj K. Mishra
Division of Life Sciences, NASA/Johnson Space
Center, Houston, 3600 Bay Area Blvd.,
Houston, TX 77058, USA

Luc Montagnier
Institut Pasteur, 25–28 rue du Dr Roux,
75015 Paris, France and Department of Biology,
Queens College, NY 11367, USA

Stephen S. Morse
DARPA – Defense Advanced Research Project
Agency, Columbia University, 3701 N Fairfax Drive,
Room 838, Arlington, VA 22203-1714, USA

Robb E. Moses
Department of Molecular and Medical Genetics,
Oregon Health Sciences University,
M/C L103, 3181 SW Sam Jackson Park Rd.,
Portland, OR 97201-3098, USA

Matthew A. Mulvey
Department of Molecular Microbiology,
Washington University School of Medicine,
Campus Box 8230, 660 S Euclid Avenue, 8230,
St. Louis, MO 63110-1010, USA

Noreen E. Murray
Institute of Cell and Molecular Biology, University of Edinburgh, King's Buildings, Mayfield Road, Edinburgh, EH9 3JR, UK

J. Colin Murrell
Department of Biological Sciences, University of Warwick, Coventry CV4 7AL, UK

Christine Musahl
Institute of Neuropathology, University of Zurich, University Hospital of Zürich, Schmelzbergstrasse 12, Zurich CH-8091, Switzerland

C. Nelson Neale
Energy and Environmental Systems Institute, Rice University, 6100 S. Main, MS-316, Houston, TX 77005, USA

T. G. Nagaraja
Department of Diagnostic Medicine/ Pathobiology, Kansas State University, College of Veterinary Medicine, Manhattan, KS 66506-5606, USA

Alexander J. Ninfa
Department of Biological Chemistry, University of Michigan Medical School, 4310 Med Sci I, 1301 Catherine, Ann Arbor, MI 48109-0606, USA

Yoko Nomura
School of Bionics, Tokyo University of Technology, 1404-1 Katakura-cho, Hachioji, Tokyo 192-0982, Japan

David A. Odelson
Invitrogen Corp. Carlsbad, CA, USA

Donald B. Oliver
Department of Molecular Biology and Biochemistry, Wesleyan University, Hall-Atwater and Shanklin Labs, Middletown, CT 06459-0175, USA

Mary J. Osborn
Department of Microbiology, University of Connecticut Health Center, Farmington, CT 06032, USA

Carol J. Palmer
Department of Pathobiology, University of Florida, USA

Sarad R. Parekh
Dow AgroSciences, 9330 Zionsville Road, Indianapolis, IN 46268-1054, USA

Christine Paszko-Kolva
Accelerated Technology Laboratories, Inc., Belmont, CA, 496 Holly Grove School Road, West End, NC 27376, USA

D. L. Pierson
Division of Life Sciences, NASA/Johnson Space Center, Houston, 3600 Bay Area Blvd., Houston, TX 77058, USA

Patrick J. Piggot
Department of Microbiology and Immunology, Temple University School of Medicine, 3400 N Broad Street, Philadelphia, PA 19140-5104, USA

Look Hulshoff Pol
Department of Environmental Technology, Wageningen Agricultural University, P.O. Box 8129, Wageningen 6700 EV, The Netherlands

Pablo J. Pomposiello
Department of Microbiology, University of Massachusetts, Amherst, MA 01003, USA

James A. Poupard
Department of Automicrobial Profiling/Clinical Microbiology, SmithKline Beecham Pharmaceuticals, P.O. Box 5089, 1250 S. Collegeville Rd., Mail Code UP1340, Collegeville, PA 19426-0989, USA

Thomas C. Quinn
The Johns Hopkins University, Ross Research Building 1159, 720 Rutland Avenue, Baltimore, MD 21205-2196, USA

Mary F. Roberts
Chemistry Department, 140 Commonwealth Avenue, Boston College, Chestnut Hill, MA 02467, USA

Milton J. Schlesinger
Department of Molecular Microbiology, Washington University School of Medicine, Box 8230, 4566 Scott Avenue, St. Louis, MO 63110-1093, USA

Sondra Schlesinger
Department of Molecular Microbiology, Washington University School of Medicine, Box 8230, 4566 Scott Avenue, St. Louis, MO 63110-1093, USA

Kate M. Scow
Department of Land, Air, Water Resources, University of California, Hoagland Hall, Davis, CA 95616-5224, USA

Uwe B. Sleytr
Center for Ultrastructure Research and Institute for Molecular Nanotechnology, University of Agricultural Sciences, Vienna, Gregor-Mendel-Str. 33, Vienna A-1180, Austria

Joan L. Slonczewski
Department of Biology,
Kenyon College, 100 College Road,
Gambier, OH 43022, USA

Peter G. Sohnle
Department of Infectious Diseases,
Medical College of Wisconsin,
Research Service/151, VA Medical Center,
Milwaukee, WI 53295, USA

Mohammad Sondossi
Department of Microbiology,
Weber State University, 2506 University Circle,
Ogden, UT 84408-2506, USA

Gabriel E. Soto
Department of Molecular Microbiology,
Washington University School of Medicine,
Campus Box 8230, 660 S Euclid Avenue, 8230,
St. Louis, MO 63110-1010, USA

Kevin R. Sowers
Center of Marine Biotechnology, University of
Maryland Biotechnology Institute, Suite 236,
Columbus Center, 701 E Pratt Street, Baltimore,
MD 21202-4031, USA

Jeff B. Stock
Department of Molecular Biology,
Princeton University, 330 Lewis Thomas Lab,
Princeton, NJ 08544, USA

Morton N. Swartz
Infectious Disease Unit, Massachusetts General
Hospital and Harvard Medical School, 70 Blossom
Street, Boston, MA 02114-2696, USA

Christopher M. Thomas
School of Biological Sciences,
University of Birmingham, Edgbaston,
Birmingham B15 2TT, UK

Torsten Thomas
Department of Microbiology and Immunology,
University of New South Wales, Sydney, NSW 2052,
Australia

Sue Tolin
Department of Plant Pathology, Virginia
Polytechnic Institute and State University,
Blacksburg, VA 24061, USA

Arthur O. Tzianabos
Channing Laboratory, Harvard Medical School,
181 Longwood Avenue, Boston,
MA 02115-5899, USA

Marcus Vallero
Department of Environmental Technology,
Wageningen Agricultural University, P.O. Box 8129,
Wageningen 6700 EV, The Netherlands

Costantino Vetriani
Department of Environmental Biology,
Portland State University, Science Building 2,
1719 S.W. 10th Avenue, Portland,
OR 97201, USA

Anne Vidaver
Department of Plant Pathology,
University of Nebraska-Lincoln,
Lincoln, NE 68588, USA

David K. Wagner
Department of Infectious Diseases,
Medical College of Wisconsin,
Research Service/151, VA Medical Center,
Milwaukee, WI 53295, USA

Graeme M. Walker
School of Molecular and Life Sciences,
University of Albertay Dundee,
Kydd Building, Bell Street,
Dundee, DD1 1HG, UK

C. Herb Ward
Energy and Environmental Systems Institute,
Rice University, 6100 S. Main, MS-316, Houston,
TX 77005, USA

Vera Webb
Lookfar Solutions Inc.,
P.O. Box 811, Tofino V0R2Z0,
British Columbia, Canada

Chris Whitfield
Department of Microbiology, University of Guelph,
172 Chemistry-Microbiology, Guelph,
Canada ON N1G 2W1

Richard J. Whitley
Department of Pediatrics, University of Alabama
School of Medicine, Ambulatory Care Center 616,
1600 7th Avenue S, Birmingham,
AL 35294-0011, USA

Brian M. Wilkins†
Department of Genetics, University of Leicester,
Adrian Building, University Road,
Leicester LE1 7RH, UK

† Deceased.

Kevin W. Winterling
Department of Biology, Emory and Henry College,
P.O. Box 75, One Garnand Drive, Emory,
VA 24327, USA

Bernard S. Wostmann
Lobund Laboratory, University of Notre Dame,
16977 Adams Road, Granger,
IN 46530, USA

Charles Yanofsky
Department of Biological Science, Stanford University,
Gilbert Bldg., 371 Serra Street, Stanford,
CA 94305, USA

Judith W. Zyskind
San Diego State University, Elitra Pharmaceuticals,
3510 Dunhill St, Suite A, San Diego,
CA 92121, USA

Preface

The field of Microbiology encompasses highly diverse life forms—bacteria, archaea, fungi, protists, and viruses. They have a profound influence on all life on Earth: they play an essential role in the cycles of matter in nature, affect all biological environments, interact in countless ways with other living beings, and play a crucial role in agriculture and industry. The literature associated with Microbiology, of necessity, tends to be specialized and focused. For that reason, it is difficult to find sources that provide a broad perspective on a wide range of microbiological topics. That is the aim of *The Desk Encyclopedia of Microbiology*.

The concept behind this venture is to provide a single reference volume with appeal to microbiologists on all levels and fields, including those working in research, teaching, industry, and government. We believe that this book will be helpful, especially for accessing material in areas in which the reader is not a specialist. It is intended to facilitate preparing lectures, grant applications and reports, and to satisfy curiosity regarding microbiological topics.

The Desk Encyclopedia of Microbiology is principally a synthesis from the comprehensive and multivolumed *Encyclopedia of Microbiology*. Our intention is to provide affordable and ready access to a large variety of topics within one set of covers. To this end we have chosen subjects that, in our opinion, will be of greatest interest to the largest number of readers. Included are the most general chapters from *The Encyclopedia of Microbiology*, brought up to date and augmented with current references and related URLs. We have emphasized topics that are currently "hot" in the field of Microbiology, including additional chapters from other sources.

The result is a volume where coverage is extensive but not overly long in specific details. We believe this will be a most appropriate reference for anyone with an interest in the intriguing field of Microbiology.

Moselio Schaechter, 2003

General websites

American Society for Microbiology. An extensive list of links is in "Search Microbiology Sites" (members only)
http://www.asm.org/

Society for General Microbiology links page
http://www.socgenmicrobiol.org.uk/links.htm

Links to microbiology courses at various universities
http://www.geocities.com/CapeCanaveral/3504/courses.htm

Microbiology clinical cases
http://www.medinfo.ufl.edu/year2/mmid/bms5300/cases/index.html

List of bacterial names with standing in nomenclature (J. P. Euzéby)
http://www.bacterio.cict.fr/index.html

Access to online resources on Bacterial Infections and Mycoses. Karolinska Institutet
http://www.mic.ki.se/Diseases/c1.html

1

Adhesion, Bacterial

Matthew A. Mulvey and Scott J. Hultgren
Washington University School of Medicine

GLOSSARY

adhesin A molecule, typically a protein, that mediates bacterial attachment by interacting with specific receptors.

extracellular matrix A complex network of proteins and polysaccharides secreted by eukaryotic cells. Functions as a structural element in tissues, in addition to modulating tissue development and physiology.

invasin An adhesin that can mediate bacterial invasion into host eukaryotic cells.

isoreceptors Eukaryotic cell membrane components which contain identical receptor determinants recognized by a bacterial adhesin.

lectins Proteins that bind carbohydrate motifs.

Adhesion is a principal step in the colonization of inanimate surfaces and living tissues by bacteria. It is estimated that the majority of bacterial populations in nature live and multiply attached to a substratum. Bacteria have evolved numerous, and often redundant, mechanisms to facilitate their adherence to other organisms and surfaces within their environment. A vast number of structurally and functionally diverse bacterial adhesive molecules, called adhesins, have been identified. The adhesins expressed by different bacterial species can directly influence bacterial tropism and mediate molecular crosstalk among organisms.

I. MECHANISMS OF BACTERIAL ADHESION

Bacterial adhesion to living cells and to inanimate surfaces is governed by nonspecific electrostatic and hydrophobic interactions and by more specific adhesin–receptor binding events. Studies of bacterial adherence indicate that initial bacterial interactions with a surface are governed by long-range forces, primarily van der Waals and electrostatic interactions. The surface of most gram-negative and many gram-positive bacteria is negatively charged. Thus, bacteria will often readily adhere nonspecifically to positively charged surfaces. In some cases, bacterial proteins possessing hydrophobic surfaces, including many adhesins, can also mediate nonspecific bacterial interactions with exposed host cell membrane lipids and with other hydrophobic surfaces encountered in nature. If the approach of bacteria to a surface, such as a negatively charged host cell membrane, is unfavorable, bacteria must overcome an energy barrier to establish contact. Protein–ligand binding events mediated by bacterial adhesins can often overcome or bypass repulsive forces and promote specific and intimate microbial interactions with host tissues and other surfaces.

Bacteria can produce a multitude of different adhesins, usually proteins, with varying specificities for a wide range of receptor molecules. Adhesins are presented on bacterial surfaces as components of

The Desk Encyclopedia of Microbiology
ISBN: 0-12-621361-5

filamentous, nonflagellar structures, known as pili or fimbriae, or as afimbrial (or nonfimbrial) monomeric or multimeric proteins anchored within the bacterial membrane. Other nonprotein components of bacterial membranes, including lipopolysaccharides (LPS) synthesized by gram-negative bacteria, and lipoteichoic acid in some gram-positive bacteria, can also function as adhesive molecules. Adhesins are often only minor subunits intercalated within pilus rods or located at the distal tips of pili, but they can also constitute the major structural subunits of adhesive pili. The molecular machinery required for the synthesis of many different adhesive pili and afimbrial adhesins is conserved, although the receptor specificities of the different adhesins can vary widely. Many bacterial adhesins function as lectins, mediating bacterial interactions with carbohydrate moieties on glycoproteins or glycolipids. Other adhesins mediate direct contact with specific amino acid motifs present in receptor proteins. Plant and animal cell surfaces present a large array of membrane proteins, glycoproteins, glycolipids, and other components that can potentially serve as receptors for bacterial adhesins. Protein constituents of the extracellular matrix (ECM) are also often used as bacterial receptors. In some cases, ECM proteins can function as bridges, linking bacterial and host eukaryotic cells. In addition, organic and inorganic material that coats inanimate surfaces, such as medical implants, pipes, and rocks, can act as receptors for bacterial adhesins, allowing for the establishment of microbial communities or biofilms. Adhesins also mediate interbacterial associations, facilitating the transfer of genetic material between bacteria and promoting the coaggregation of bacterial species in sites such as the oral cavity.

A single bacterium can often express multiple adhesins with varying receptor specificities. These adhesins can function synergistically and, thus, enhance bacterial adherence. Alternately, adhesins may be regulated and expressed differentially, allowing bacteria to alter their adhesive repertoire as they enter different environmental situations. To date, a large number of bacterial adhesins have been described, but relatively few receptors have been conclusively identified. Bacterial adhesins can show exquisite specificity and are able to distinguish between very closely related receptor structures. The ability of bacterial adhesins to recognize specific receptor molecules is dependent upon the three-dimensional architecture of the receptor in addition to its accessibility and spatial orientation. Most studies to date of bacterial adhesion have focused on host–pathogen interactions. Numerous investigations have indicated that bacterial adhesion is an essential step in the successful colonization of host tissues and the production of disease by bacterial pathogens. Examples of adhesins expressed by bacterial pathogens and their known receptors are presented in Table 1.1. To illustrate some of the key concepts of bacterial adhesion, the modes of adhesion of a few well-characterized pathogens are discussed in the following sections.

A. Adhesins of uropathogenic *Escherichia coli*

Uropathogenic strains of *E. coli* are the primary causative agents of urinary tract infections among humans. These bacteria can express two of the best characterized adhesive structures, P and type 1 pili. These pili are composite organelles, consisting of a thin fibrillar tip structure joined end-to-end to a right-handed helical rod. Chromosomally located gene clusters, that are organizationally as well as functionally homologous, encode P and type 1 pili. The P pilus tip fibrillum contains a distally located adhesin, PapG, in association with three other tip subunits, PapE, PapF, and PapK. The adhesive tip fibrillum is attached to the distal end of a thicker pilus rod composed of repeating PapA subunits. An additional subunit, PapH, anchors the PapA rod to the outer membrane.

The P pilus PapG adhesin binds to the α-D-galactopyranosyl-(1–4)-β-D-galactopyranoside (Galα(1–4)Gal) moiety present in the globoseries of glycolipids, which are expressed by erythrocytes and host cells present in the kidney. Consistent with this binding specificity, P pili have been shown to be major virulence factors associated with pyelonephritis caused by uropathogenic *E. coli*. Three distinct variants of the PapG adhesin (G-I, G-II, and G-III) have been identified that recognize three different Galα(1–4)Gal-containing isoreceptors: globotriaosylceramide, globotetraosylceramide (globoside), and globopentaosylceramide (the Forssman antigen). The different PapG adhesins significantly affect the tropism of pyelonephritic *E. coli*. For example, urinary tract *E. coli* isolates from dogs often encode the G-III adhesin that recognizes the Forssman antigen, the dominant Galα(1–4)Gal-containing isoreceptor in the dog kidney. In contrast, the majority of urinary tract isolates from humans express the G-II adhesin that preferentially recognizes globoside, the primary Galα(1–4)Gal-containing isoreceptor in the human kidney.

In comparison with P pili, type 1 pili are more widely distributed and are encoded by more than 95% of all *E. coli* isolates, including uropathogenic and

TABLE 1.1 Selected examples of bacterial adhesins and their receptors

Organism	Adhesin	Receptor	Form of receptor[a]	Associated disease(s)
Escherichia coli	P pili (PapG)	Galα(1–4)Gal	GL	Pyelonephritis/cystitis
	Type 1 pili (FimH)	D-mannose (uroplakin 1a and 1b, CD11, CD18, uromodulin)	GP	Cystitis
	Curli (CsgA)	Fibronectin/laminin/ plasminogen	ECM	Sepsis
	Prs pili	Galα(1–4)Gal	GL	Cystitis
	S pili	α-sialyl-2,3-β-galactose	GP	UTI, newborn meningitis
	K88 pili (K88ad)	IGLad (nLc$_4$Cer)	GL	Diarrhea in piglets
	K99 pili (FanC)	NeuGc(α2–3)Galβ4Glc	GL	Neonatal diarrhea in piglets, calves, and lambs
	DR family			
	DR	Decay accelerating factor (SCR-3 domain)	P	UTI
	DR-II			UTI
	AFA-I			UTI
	AFA-III			UTI, diarrhea
	F1845			diarrhea
	Nonfimbrial adhesions 1–6	Glycophorin A	GP	UTI, newborn meningitis
	M hemagglutinin	A^M determinant of glycophorin A	GP	Pyelonephritis
	Intimin	Tir (EPEC encoded phosphoprotein)	P	Diarrhea
Neisseria	Type 4a pili	CD46	GP	Gonorrhea/meningitis
	Opa proteins	CD66 receptor family/HSPG	P, GL	
	Opa$_{50}$	Vitonectin/fibronectin	ECM	
	Opc	HSPG/Vitronectin	GL, ECM	
	LOS	ASGP-R	GP	
	Inducible adhesin	Lutropin receptor	GP	
Listeria monocytogenes	Internalin	E-cadherin	GP	Listeriosis (meningitis, septicemia, abortions, gastroenteritis)
Haemophilus influenzae	Hemagglutinating pili	AnWj antigen/lactosylceramide	GP, GL	Respiratory tract infections
	Hsp-70-related proteins	Sulfoglycolipids	GL	
	HMW1, HMW2	Negatively charged glycoconjugates	GP	
Campylobacter jejuni	CadF	Fibronectin	ECM	Gastroenteritis
Yersinia	Invasin	β_1 integrins	P	Plague, Enterocolitis
	YadA	Cellular fibronectin/collagen/laminin	ECM	
Bordetella pertussis	FHA	CR3 integrin	P	Whooping cough
	Pertactin, BrkA	Integrins	P	
	Pertussis toxin	Lactosylceramides/gangliosides	GP/GL	
Mycobacterium	BCG85 complex, FAP proteins	Fibronectin	ECM	Tuberculosis, leprosy
Streptococcus	Protein F family	Fibronectin	ECM	Pharyngitis, scarlet fever, erysipelas, impetigo, rheumatic fever, UTI, dental caries, neonatal sepsis, glomerulonephritis, endocarditis, pneumonia, meningitis
	Polysaccharide capsule	CD44	GP	
	ZOP, FBP4, GAPDH	Fibronectin	ECM	
	Lipoteichoic acid (LTA)	Fibronectin/macrophage scavenger receptor	ECM/GP	
	M protein	CD46/fucosylated glycoconjugates/fibronectin	GP/ECM	
Staphylococcus	FnbA, FnbB	Fibronectin	ECM	Skin lesions, pharyngitis, pneumonia, endocarditis, toxic shock syndrome, food poisoning
	Can	Collagen	ECM	
	Protein A (Spa)	von Willebrand factor	GP	
	ClfA	Fibrinogen	ECM	
	EbpS	Elastin	ECM	

[a]P, protein–protein interactions; GP, interaction with glycoproteins; GL, glycolipids; ECM, extracellular matrix proteins.

commensal intestinal strains. The type 1 pilus tip fibrillum is comprised of two subunits, FimF and FimG, in addition to the adhesin, FimH. The adhesive tip is connected to the distal end of a thicker pilus rod composed of repeating FimA subunits. In addition to its localization within the pilus tip, the FimH adhesin also appears to be occasionally intercalated along the length of the type 1 pilus rod. FimH binds to mannose containing host receptors expressed by a wide variety of host cell types and has been shown to be a significant virulence determinant for the development of bladder infections. Natural phenotypic variants of the FimH adhesin have been identified by Sokurenko *et al.* (1998), which differentially bind to mono-mannose structures. Interestingly, most uropathogenic isolates express FimH variants that bind well to mono-mannose residues, whereas most isolates from the large intestine of healthy humans express FimH variants that interact poorly with mono-mannose structures. Mono-mannose residues are abundant in the oligosaccharide moieties of host proteins, known as uroplakins, that coat the luminal surface of the bladder epithelium. *In vitro* binding assays by Wu *et al.* (1996) have demonstrated that type 1-piliated *E. coli* can specifically bind two of the uroplakins, UP1a and UP1b. Scanning and high-resolution electron microscopy have shown that type 1 pili can mediate direct and intimate bacterial contact with the uroplakin-coated bladder epithelium (Fig. 1.1).

The assembly of P pili and type 1 pili requires two specialized assembly proteins: a periplasmic chaperone and an outer membrane usher. Periplasmic chaperones facilitate the import of pilus subunits across the inner membrane and mediate their delivery to outer membrane usher complexes, where subunits are assembled into pili. Homologous chaperone/usher pathways modulate the assembly of over 30 different adhesive organelles, expressed by uropathogenic *E. coli* and many other gram-negative pathogens. Among the adhesive structures assembled via a chaperone/usher pathway by uropathogenic *E. coli* are S pili, nonfimbrial adhesin I, and members of the Dr adhesin family. This family includes the uropathogenic-associated afimbrial adhesins AFA-I and AFA-III and the fimbrial adhesin Dr, in addition to the diarrhea-associated fimbrial adhesin F1845. These adhesins recognize the Dr^a blood group antigen present on decay accelerating factor (DAF), a complement regulatory factor expressed on erythrocytes and other tissues, including the uroepithelium. These four members of the Dr adhesin family appear to recognize different epitopes of the Dr^a antigen. The Dr adhesin, but not the other three, also recognizes type IV collagen. Members of the Dr adhesin family are proposed to facilitate ascending colonization and chronic interstitial infection of the urinary tract. It is unclear why the Dr and F1845 adhesins assemble into fimbria while AFA-I and AFA-III are assembled as nonfimbrial adhesins on the bacterial surface. It has been suggested that afimbrial adhesins, such as AFA-I and AFA-III, are derived from related fimbrial adhesins, but have been altered such that the structural attributes required for polymerization into a pilus are missing while the adhesin domain remains functional and anchored on the bacterial surface.

B. Neisserial adhesins

Neisseria gonorrhoeae and *N. meningitidis* are exclusively human pathogens that have developed several adhesive mechanisms to colonize mucosal surfaces. Initial contact with mucosal epithelia by *Neisseria* species is mediated by type 4a pili. These adhesive organelles are related to a group of multifunctional structures expressed by a wide diversity of bacterial species, including *Pseudomonas aeruginosa, Moraxella* species, *Dichelobacter nodus*, and others. Type-4a pili are assembled by a type II secretion system that is distinct from the chaperone/usher pathway. They are comprised primarily of a small subunit, pilin, that is packaged into a helical arrangement within pili. The type 4a pilin can mediate bacterial adherence, but in *Neisseria* species, a separate, minor tip protein, PilC, has also been implicated as an adhesin. A eukaryotic membrane protein, CD46, is proposed to be a host receptor for type 4a pili expressed by *N. gonorrhoeae*, although it is currently unclear which pilus component binds this host molecule.

Following primary attachment mediated by type-4a pili, more intimate contact with mucosal surfaces is apparently established by the colony opacity-associated (Opa) proteins of *Neisseria* species. These proteins constitute a family of closely related but size-variable outer membrane proteins that are expressed in a phase variable fashion. Opa proteins mediate not only adherence, but they also modulate bacterial invasion into host cells. A single neisserial strain can encode from 3 to 11 distinct Opa variants, with each Opa protein being expressed alternately of the others. The differential expression of Opa variants can alter bacterial antigenicity and possibly modify bacterial tropism for different receptors and host cell types. Some Opa variants recognize carbohydrate moieties of cell surface-associated heparin sulfate proteoglycans (HSPGs), which are common constituents of mammalian cell membranes. The majority of Opa variants, however, bind via protein–protein interactions to CD66 transmembrane glycoproteins, which comprise

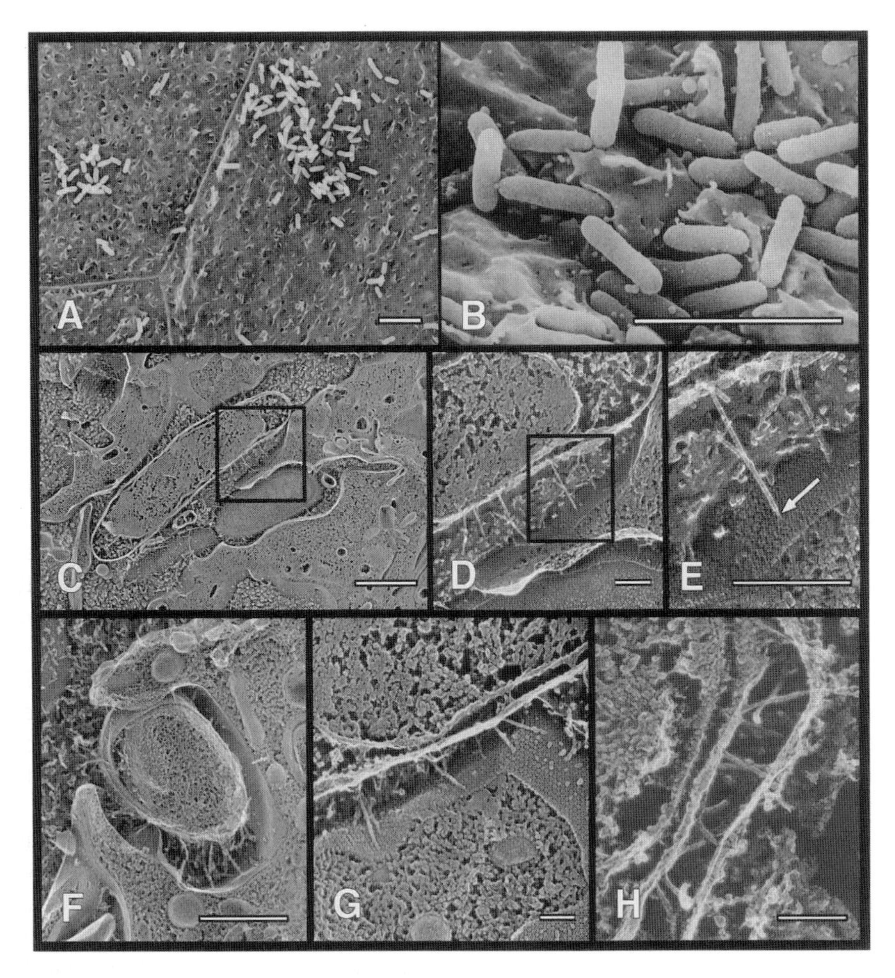

FIGURE 1.1 Type 1 pilus-mediated bacterial adherence to the mouse bladder epithelium was visualized by (A and B) scanning and (C–H) high-resolution freeze–fracture, deep-etch electron microscopy. Mice were infected via transurethral inoculation with type 1-piliated uropathogenic *E. coli*. Bladders were collected and processed for microscopy at 2 h. after infection. Bacteria adhered randomly across the bladder lumenal surface, both singly and in large, biofilmlike microcolonies, some of which contained several hundred bacteria (A and B). The type 1 pili-mediating bacterial adherence were resolved by high-resolution electron microscopy techniques. The adhesive tips of type 1 pili make direct contact with the uroplakin-coated surface of the bladder epithelium (D–G). Hexagonal arrays of uroplakin complexes are visible. The boxed areas in (C) and (D) are shown magnified, respectively, in (D) and (E). In (H), type 1 pili span from the host cell membrane on the right to the bacterium on the left. These images demonstrate that type 1 pili can mediate intimate bacterial attachment to host bladder epithelial cells. Scale bars indicate 5 μm (A and B), 0.5 μm (C and F), and 0.1 μm (D, E, G, H) (Plate 1). (Reprinted with permission from Mulvey, M. A., *et al.* (1998). Induction and evasion of host defenses by type 1-piliated uropathogenic *Escherichia coli*. *Science* **282**, 1494–1497. Copyright 1998 American Association for the Advancement of Science.).

a subset of the carcinoembryonic antigen (CEA) receptor family of the immunoglobulin super-family. Individual Opa variants specifically recognize distinct CD66 receptors and this likely influences both the tissue tropism of *Neisseria* and the host cell responses to neisserial attachment. In addition to pili and Opa proteins, the lipopolysaccharide (lipooligosaccharide, LOS) and a distinct outer membrane protein, Opc, expressed by *Neisseria* can also influence bacterial adhesion and invasion. Deconvoluting the various roles of the different adhesive components of *Neisseria* during the infection process remains a major challenge.

C. Adhesins of *Haemophilus influenzae*

Haemophilus influenzae is a common pathogen of the human respiratory tract. Isolates of *H. influenzae* can be divided into encapsulated and nonencapsulated, or nontypable, forms. Prior to the use of *H. influenzae* conjugate vaccines, capsulated strains of *H. influenzae* were the primary cause of childhood bacterial meningitis and a major cause of other bacteremic diseases in children. Vaccines effective against nontypable strains have not yet been developed and these strains remain important human pathogens, causing pneumonia, otitis media, sinusitis, and bronchitis. Several adhesins have been identified which facilitate the colonization of the respiratory epithelium by both encapsulated and nontypable *H. influenzae*.

During the initial stages of the infection process, nontypable *H. influenzae* associates with respiratory mucus, apparently through interactions between bacterial outer membrane proteins (OMPs P2 and P5) and sialic acid-containing oligosaccharides within the mucus. Both nontypable and encapsulated strains of *H. influenzae* can initiate direct contact with the respiratory epithelium via adhesive pili. Over 14 serological types of adhesive pili have been indentified in *H. influenzae*. These pili are composite structures assembled by chaperone/usher pathways similar to those used by uropathogenic *E. coli* to assemble P and type 1 pili. Piliated strains of *H. influenzae* preferentially bind to nonciliated cells or damaged epithelium. The pili of *H. influenzae* can recognize the AnWj antigen, in addition to gangliosides and other compounds containing siallyllactoceramide. Following initial attachment mediated by pili, the polysaccharide capsule of encapsulated strains is reduced, enabling a second adhesin, Hsf, to establish more intimate bacterial contact with host epithelial cells. Hsf assembles into short, thin fibrils on the bacterial surface. While Hsf expression is restricted to encapsulated strains of *H. influenzae*, a subpopulation of nontypable strains expresses a Hsf homolog called Hia. Both Hsf and Hia share homology with other bacterial adhesins including AIDA-1, an adherence factor produced by diarrheagenic *E. coli*.

Instead of adhesive pili and Hia, the majority of nontypable *H. influenzae* isolates produce two alternate adhesins: high molecular weight surface-exposed proteins called HMW1 and HMW2. These two adhesins share significant sequence identity with each other and are similar to filamentous hemagglutinin (FHA), an adhesin and colonization factor expressed by *Bordetella pertussis*. HMW1 and HMW2 have distinct adhesive specificities and may function at different steps in the infection process. The receptors for the HMW adhesins appear to be negatively charged glycoconjugates that have not yet been completely defined. Nontypable *H. influenzae* encodes several other adhesive factors, including two Hsp-70-related proteins, which can mediate bacterial binding to sulfoglycolipids. Interestingly, other heat shock proteins have been implicated in the adherence of other microbial pathogens including *Helicobacter pylori*, *Mycoplasma*, and *Chlamydia trachomatis*.

Work by St. Geme and coworkers (1998) has highlighted an additional adhesin, Hap, which is expressed by virtually all nontypable *H. influenzae* isolates. Hap mediates low-level adherence to epithelial cells, complementing the binding activities of pili and Hia or HMW1 and HMW2. Hap also promotes interbacterial associations leading to bacterial aggregation and microcolony formation on the epithelial surface. The mature Hap adhesin consists of a C-terminal outer membrane protein domain, designated Hap_{β}, and a larger extracellular domain designated Hap_{s}. The Hap_{s} domain, which is responsible for mediating adherence, has serine protease activity and can be autoproteolytically cleaved, releasing itself from the bacterial surface. Interestingly, secretory leukocyte protease inhibitor (SLPI), a natural host component of respiratory-tract secretions, which possibly protects the respiratory epithelium from proteolytic damage during acute inflammation, has been shown to inhibit Hap autoproteolysis and enhance bacterial adherence. Despite the presence of SLPI, Hap_{s}-mediated adherence *in vivo* is likely transient. Over time, the eventual autoproteolysis and release of the Hap_{s} adhesin domain from the bacterial surface may allow bacterial spread from microcolonies on the respiratory epithelium and aid the bacteria in evading the host immune response. Identification of the receptor molecules recognized by Hap awaits further studies.

D. Adherence to components of the extracellular matrix

One of the principal functions of the ECM is to serve as substrate for the adherence of eukaryotic cells within animal tissues. The ECM is composed of polysaccharides and numerous proteins including fibronectin, vitronectin, laminin. elastin, collagen, fibrinogen, tenascin, entactin, and others. Thin flexible mats of specialized ECM, known as basal laminae or basement membranes, underlie all epithelial cells and surround individual fat cells, muscle cells, and Schwann cells. Binding of ECM proteins is one of the primary mechanisms used by many pathogenic bacteria to adhere to host tissues. Bacterial adhesins have been identified which recognize specific components of the ECM and a few adhesins, such as the Opa_{50} protein of *Neisseria* and the YadA adhesin of *Yersinia enterolitica*, are able to recognize multiple ECM components. Some bacterial adhesins preferentially recognize immobilized, cell-bound ECM components over soluble forms. The YadA adhesin expressed by *Y. enterolitica*, for example, mediates adherence to cell-bound fibronectin, but not to soluble fibronectin within plasma. This may allow *Y. enterolitica* to more efficiently bind tissue rather than circulating molecules.

The tissue distribution of ECM components can directly influence the tropism of a bacterial pathogen.

For example, *Mycobacterium leprae*, the causative agent of leprosy, binds LN-2, an isoform of the ECM component laminin. This ECM component recognizes a host cell-surface receptor, α-dystroglycan, and serves as a bridge linking host and bacterial cells. *M. leprae* targets the Schwann cells of the peripheral nervous system and can also invade the placenta and striated muscle of leprosy patients. The tissue distribution of LN-2, which is restricted to the basal laminae of Schwann cells, striated muscles, and trophoblasts of the placenta, directly correlates with sites of natural infection by *M. leprae*.

In contrast to the restricted tissue distribution of LN-2, most components of the ECM are more widely apportioned and can interact with receptor molecules expressed by a broad range of cell types present within a variety of different tissues. By interacting with widely distributed components of the ECM, bacteria greatly enhance their adhesive potential. Numerous bacteria are able to bind fibronectin, an ECM component present in most tissues and body fluids and a prominent constituent of wounds. The bacterial adhesins that bind fibronectin are diverse. For example, *E. coli* and *Salmonella* species express thin, irregular, and highly aggregated surface fibers, known as curli, that bind fibronectin in addition to other receptor molecules. *Mycobacterium* species produce at least five fibronectin-binding molecules, three of which are related and collectively known as the BCG85 complex. *Streptococcus* expresses an even larger number of different fibronectin-binding adhesins, including ZOP, lipoteichoic acid, GAPDH, FBP54, M protein, and several related molecules represented by Protein F. Binding of Protein F and related adhesins to fibronectin is specific and essentially irreversible. Members of the Protein F family of adhesins have similar domain architectures, although they appear to interact with fibronectin differently. Protein F possesses two distinct domains, composed of repeated sequence motifs, which bind independently of each other to different sites at the N-terminus of fibronectin. Additional fibronectin-binding proteins related to the Protein F family of adhesins have also been identified in *Staphylococcus*. These gram-positive bacteria, in addition to producing fibronectin-binding proteins, can also express an array of other adhesive molecules, which bind other widely distributed ECM components, including collagen, fibrinogen, and elastin. By encoding a large repertoire of adhesins able to recognize ECM components, *Streptococcus, Staphylococcus*, and other pathogens, presumably, increase their capacity to effectively bind and colonize sites within host tissues.

II. CONSEQUENCES OF BACTERIAL ADHESION

Research in recent years has demonstrated that interactions between bacterial adhesins and receptor molecules can act as trigger mechanisms, activating signal transduction cascades and altering gene expression in both bacterial and host cells. Zhang and Normark showed in 1996 that the binding of host cell receptors by P pili activated the transcription of a sensor–regulator protein, AirS, which regulates the bacterial iron acquisition system of uropathogenic *E. coli*. This response may enable uropathogens to more efficiently obtain iron and survive in the iron-poor environment of the urinary tract. Around the same time, Wolf-Watz and colleagues showed, using *Y. pseudotuberculosis*, that bacterial contact with host cells could increase the rate of transcription of virulence determinants called Yop effector proteins. More recently, Taha and coworkers (1998) demonstrated that transcription of the PilC1 adhesin of *N. meningitidis* was transiently induced by bacterial contact with host epithelial cells. The PilC1 adhesin can be incorporated into the tips of type-4a pili, but it can also remain associated with the bacterial outer membrane, where it can, presumably, facilitate pilus assembly. The up-regulation of the PilC1 adhesin may enhance bacterial adherence to host cells by promoting the localization of PilC1 into the tips of type 4a pili.

Signal transduction pathways are activated within host eukaryotic cells in response to attachment mediated by many different bacterial adhesins. For example, the binding of type-4a pili expressed by *Neisseria* to host cell receptors (presumably, CD46) can stimulate the release of Ca^{++} stores within target epithelial cells. Fluxes in intracellular Ca^{++} concentrations are known to modulate a multitude of eukaryotic cellular responses. Similarly, the binding of P pili to Galα(1–4)Gal-containing host receptors on uroepithelial cells can induce the release of ceramides, important second messenger molecules that can influence a number of signal transduction processes. Signals induced within urepithelial cells upon binding P-piliated bacteria result in the up-regulation and eventual secretion of several immunoregulatory cytokines. The binding of type 1-piliated and other adherent bacteria to a variety of host epithelial and immune cells has also been shown to induce the release of cytokines, although the signaling pathways involved have not yet been well defined. In some cases, bacteria may co-opt host signal transduction pathways to enhance their own attachment. For example, binding of the FHA adhesin of *B. pertussis* to a monocyte integrin receptor complex activates host signal pathways that lead to the up-regulation of another integrin, complement receptor 3 (CR3). FHA can bind CR3 through a separate domain and, thus, enhance the adhesion of *B. pertussis*.

The activation of host signal pathways following bacterial attachment can result in dramatic rearrangements of the eukaryotic cytoskeleton, which can lead to the internalization of adherent bacteria. Many pathogenic bacteria invade host eukaryotic cells to evade immune responses or to pass through cellular barriers, such as the intestinal epithelium. In some cases, bacteria introduce effector molecules into their target host cells to trigger cytoskeletal rearrangements and intense ruffling of the host cell membrane that results in bacterial uptake. In other situations, bacterial adhesins (which are sometimes referred to as invasins) more directly mediate bacterial invasion by interacting with host cell membrane receptors that sequentially encircle and envelope the attached bacterium. This type of invasion is referred to as the "zipper" mechanism and requires the stimulation of host signaling cascades, including the activation of protein tyrosine kinases. The invasin protein of *Yersinia* and internalin expressed by *Listeria* can both mediate bacterial internalization into host cells by such a zipper mechanism by interacting with β_1-integrin and E-cadherin, respectively. The Opa proteins of *Neisseria* can also mediate bacterial internalization into host cells by a zipperlike mechanism. Recent work by several labs has indicated that fimbrial adhesins, such as FimH within type 1 pili, can also function as invasins.

III. TARGETING ADHESINS FOR ANTIMICROBIAL THERAPY

Bacterial adhesin–receptor binding events are critical in the pathogenesis of virtually every bacterial disease. In some cases, the knockout of a specific adhesin can greatly attenuate bacterial virulence. Uropathogenic *E. coli* strains, for example, which have been engineered to express type 1 pili lacking the FimH adhesin, are unable to effectively colonize the bladder. Similarly, a P-piliated pyelonephritic strain of *E. coli* lacking a functional PapG adhesin is unable to infect the kidney. For many other bacteria, attachment is a multifaceted process involving several adhesins that may have complementing and overlapping functions and receptor specificities. In these cases, it has been more difficult to discern the roles of individual adhesins in disease processes. The construction of mutants with knockouts in more than one adhesin is beginning to shed light on the interrelationships between multiple bacterial adhesins.

The central role of bacterial adhesins at the host–pathogen interface during the infection process has made them attractive targets for the development of new antimicrobial therapies. Vaccines directed against individual adhesins and adhesive pili have had some success in the past. However, antigenic variation of the major immunodominant domains of some adhesive organelles and the immunorecessive nature of others have frustrated progress in this area. Fortunately, by unraveling the molecular details of adhesin structure and biogenesis, substantial progress is being made. For example, the identification of FimH as the adhesive subunit of type 1 pili and the elucidation of the chaperone/usher pathway used to assemble these adhesive organelles has made it possible to purify large quantities of native FimH and to test its efficacy as a vaccine. Unlike the major type 1 pilus subunit, FimA, there is relatively little heterogeneity among the FimH adhesins expressed by diverse *E. coli* strains. The use of purified FimH as a vaccine, rather than whole type 1 pili in which FimH is present only in low numbers, has proven to significantly enhance the host immune response against the FimH adhesin. In early trials, FimH-vaccinated animals showed substantial resistance to infection by a wide variety type 1-piliated uropathogenic *E. coli* strains.

In addition to the prophylactic approach of generating vaccines to inhibit bacterial adhesion, other anti-adhesin strategies are being explored. With increased knowledge of the mechanisms used to assemble adhesins on the bacterial surface, it may be possible to design specific inhibitors of adhesin biogenesis. For example, synthetic compounds that specifically bind and inactivate periplasmic chaperones could potentially inhibit the biogenesis of a wide range of bacterial adhesive organelles. The use of soluble synthetic receptor analogs that bind bacterial adhesins substantially better than the natural monomeric ligands represents an additional strategy for inhibiting bacterial attachment and colonization. Recent advances in the synthesis of multimeric carbohydrate polymers have highlighted the possibility of creating high affinity receptor analogs that could potentially work at pharmacological concentrations within patients. Such compounds could also be used to competitively remove adherent bacteria from medical implants, industrial pipes, and other surfaces. Furthermore, it may be possible to inhibit multiple bacterial adhesins with a single compound by incorporating several receptor analogs within a single carbohydrate polymer. Continued research into the structure, function, and biogenesis of bacterial adhesins promises not only to enhance our knowledge of pathogenic processes, but may also help augment our current arsenal of antimicrobial agents.

BIBLIOGRAPHY

Dalton, H. M., and March, P. E. (1998). Molecular genetics of bacterial attachment and biofouling. *Curr. Op. Biotech.* **9**, 252–255.

Davey, M. E., and O'Toole, G. (2000). Microbial biofilms: from ecology to molecular genetics. *Microbiol. Mol. Biol. Rev.* **64**, 847–867.

Dehio, C., Gray-Owen, S. D., and Meyer, T. F. (1998). The role of neisserial Opa proteins in interactions with host cells. *Trends Microbiol.* **6**, 489–495.

Finlay, B. B., and Falkow, S. (1997). Common themes in microbial pathogenicity revisited. *Microbiol. Mol. Biol. Rev.* **61**, 136–169.

Foster, T. J., and Höök, M. (1998). Surface adhesins of *Staphylococcus aureus*. *Trends Microbiol.* **6**, 484–488.

Goldhar, J. (1996). Nonfimbrial adhesins of *Escherichia coli*. *In* "Toward Anti-Adhesion Therapy for Microbial Diseases" (Kahane and Ofek, eds.), pp. 63–72. Plenum Press, New York.

Hultgren, S. J., Jones, C. H., and Normark, S. (1996). Bacterial adhesins and their assembly. *In "Escherichia coli* and *Salmonella*," Vol. 2 (F. C. Neidhardt, ed.), pp. 2730–2756. ASM Press, Washington, DC.

Jacques, M., and Paradis, S. E. (1998). Adhesin–receptor interactions in *Pasteurellaceae*. *FEMS Microbiol. Rev.* **22**, 45–59.

Jenkinson, H. F., and Lamont, R. J. (1997). Streprococcal adhesion and colonization. *Crit. Rev. Oral Biol. Med.* **8**, 175–200.

Kerr, J. R. (1999). Cell adhesion molecules in the pathogenesis of and host defence against microbial infection. *Mol. Pathol.* **52**, 220–230.

Kolenbrander, P. E. (2000). Oral microbial communities: biofilms, interactions, and genetic systems. *Annu. Rev. Microbiol.* **54**, 413–437.

Lingwood, C. A. (1998). Oligosaccharide receptors for bacteria: A view to a kill. *Curr. Op. Chem. Biol.* **2**, 695–700.

O'Toole, G., Kaplan, H. B., and Kolter, R. (2000). Biofilm formation as microbial development. *Annu. Rev. Microbiol.* **54**, 49–79.

Schilling, J. D., Mulvey, M. A., and Hultgren, S. J. (2001). Structure and function of *Escherichia coli* type 1 pili: new insight into the pathogenesis of urinary tract infections. *J. Infect. Dis.* **183**, (Suppl 1), S36–S40.

Sharon, N. (1996). Carbohydrate–lectin interactions in infectious disease. *In* "Toward Anti-Adhesion Therapy for Microbial Diseases" (Kahane and Ofek, eds.), pp. 1–8. Plenum Press, New York.

Soto, G. E., and Hultgren, S. J. (1999). Bacterial adhesins: Common themes and variations in architecture and assembly. *J. Bacteriol.* **181**, 1059–1071.

Whittaker, C. J., Klier, C. M., and Kolenbrander, P. E. (1996). Mechanisms of adhesion by oral bacteria. *Annu. Rev. Microbiol.* **50**, 513–552.

Wilson, M. (2002). "Bacterial Adhesion to Host Tissues." Cambridge University Press, Cambridge.

Wizemann, T. M., Adamou, J. E., and Langermann, S. (1999). Adhesins as targets for vaccine development. *Emerg. Infect. Dis.* **5**, 395–403.

WEBSITE

The *E. coli* Cell Envelope Protein Data Collection includes many proteins involved in adhesion
http://www.cf.ac.uk/biosi/staff/ehrmann/tools/ecce/ecce.htm

2

Agrobacterium and plant cell transformation

Peter J. Christie
University of Texas Health Science Center at Houston

GLOSSARY

autoinducer An acyl homoserine lactone secreted from bacteria which, under conditions of high cell density, passively diffuses across the bacterial envelope and activates transcription.

border sequences 25-bp direct, imperfect repeats that delineate the boundaries of T-DNA.

conjugal pilus An extracellular filament encoded by a conjugative plasmid involved in establishing contact between plasmid-carrying donor cells and recipient cells.

conjugation Transfer of DNA between bacteria by a process requiring cell-to-cell contact.

mobilizable plasmid Conjugal plasmid that carries an origin of transfer (*oriT*) but lacks genes coding for its own transfer across the bacterial envelope.

T-DNA Segment of the *Agrobacterium* genome transferred to plant cells.

transconjugant A cell that has received a plasmid from another cell as a result of conjugation.

transfer intermediate A nucleoprotein particle composed of a single strand of the DNA destined for export and one or more proteins that facilitate DNA delivery to recipient cells.

type IV transporters A conserved family of macromolecular transporters evolved from ancestral conjugation systems for the purpose of exporting DNA or protein virulence factors between prokaryotic cells or to eukaryotic hosts.

Agrobacterium tumefaciens is a gramnegative soil bacterium with the unique ability to infect plants through a process that involves delivery of a specific segment of its genome to the nuclei of susceptible plant cells. The transferred DNA (T-DNA) is a discrete region of the bacterial genome defined by directly repeated border sequences. The T-DNA is important for infection because it codes for genes which, when expressed in the plant cell, disrupt plant cell growth and division events.

Approximately 20 years ago, it was discovered that oncogenic DNA could be excised from the T-DNA and in its place virtually any gene of interest could be inserted. *Agrobacterium tumefaciens* could then efficiently deliver the engineered T-DNA to a wide array of plant species and cell types. Transformed plant cells could be selected by cotransfer of an antibiotic resistance marker and regenerated into fertile, transgenic plants. The discovery that *A. tumefaciens* is a natural and efficient DNA delivery vector for transforming plants is largely responsible for the burgeoning industry of plant genetic engineering, which today has many diverse goals ranging from crop improvement to the use of plants as "pharmaceutical factories" for high-level production of biomedically important proteins.

Because of the dual importance of *Agrobacterium* as a plant pathogen and as a DNA delivery system, an extensive literature has emerged describing numerous aspects of the infection process and the myriad of ways this organism has been exploited for plant genetic

The Desk Encyclopedia of Microbiology
ISBN: 0-12-621361-5

engineering. The aim of this article is to summarize recent advances in our knowledge of this system, with particular emphasis on chemical signaling events, the T-DNA processing and transport reactions, and exciting novel applications of *Agrobacterium*-mediated gene delivery to eukaryotic cells.

I. OVERVIEW OF INFECTION PROCESS

Agrobacterium species are commonly found in a variety of environments including cultivated and nonagricultural soils, plant roots, and even plant vascular systems. Despite the ubiquity of *Agrobacterium* species in soil and plant environments, only a small percentage of isolates are pathogenic. Two species are known to infect plants by delivering DNA to susceptible plant cells. *Agrobacterium tumefaciens* is the causative agent of crown gall disease, a neoplastic disease characterized by uncontrolled cell proliferation and formation of unorganized tumors. *Agrobacterium rhizogenes* induces formation of hypertrophies with a hairy root appearance referred to as "hairy root" disease. The pathogenic strains of both species possess large plasmids that encode most of the genetic information required for DNA transfer to susceptible plant cells. The basic infection process is similar for both species, although the gene composition of the transferred DNA (T-DNA) differs, and therefore, so does the outcome of the infection. This article focuses on recent advances in our understanding of the *A. tumefaciens* infection process.

The basic infection cycle can be described as follows (Fig. 2.1). Pathogenic *A. tumefaciens* strains carry large, ~180-kb tumor-inducing (Ti) plasmids. The Ti plasmid harbors the T-DNA and virulence (*vir*) genes involved in T-DNA delivery to susceptible plant cells. As with many bacterial pathogens of plants and mammals, *A. tumefaciens* infects only at wound sites. As part of the plant wound response, various plant cell wall precursors, including defined classes of phenolic compounds and monosaccharide sugars, are released into the extracellular milieu. These molecules play an important role in the infection process as inducers of the *vir* genes. On *vir* gene activation, T-DNA is processed into a nucleoprotein particle termed the T-complex. The T-complex contains information for (i) export across the *A. tumefaciens* cell envelope via a dedicated transport system, (ii) movement through the plant plasma membrane and cytosol, (iii) delivery to the plant nuclear pore, and (iv) integration into the plant genome. Once integrated into the plant genome, T-DNA genes are expressed and the resulting gene products ultimately disrupt the balance of two endogenous plant hormones that synergistically coordinate plant cell growth and division events. The imbalance of these hormones contributes to loss of cell growth control

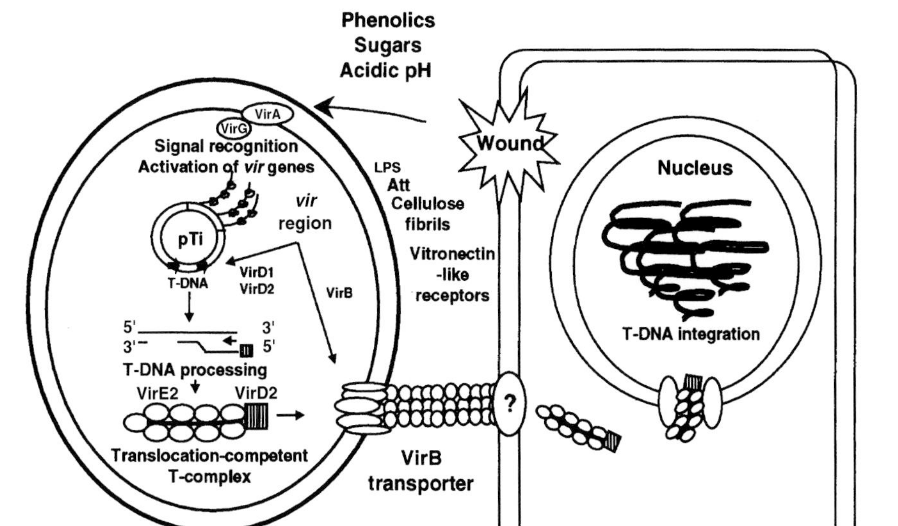

FIGURE 2.1 Overview of the *Agrobacterium tumefaciens* infection process. Upon activation of the VirA/VirG two-component signal transduction system by signals released from wounded plant cells, a single strand of T-DNA is processed from the Ti plasmid and delivered as a nucleoprotein complex (T-complex) to plant nuclei. Expression of T-DNA genes in the plant results in loss of cell growth control and tumor formation (see text for details).

and, ultimately, the proliferation of crown gall tumors.

II. Ti PLASMID

Genetic and molecular analyses have resulted in the identification of two regions of the Ti plasmid that contribute directly to infection (Fig. 2.2). The first is the T-DNA, typically a segment of 20–35 kb in size delimited by 25-bp directly repeated border sequences. The T-DNA harbors genes that are expressed exclusively in the plant cell. Transcription of T-DNA in the plant cell produces 3′ polyadenylated RNA typical of eukaryotic RNA message that is translated in the cytoplasm. The translated proteins ultimately disrupt plant cell growth and division processes resulting in the characteristic tumorous phenotype. The second region of the Ti plasmid involved in infection harbors the genes responsible for processing the T-DNA into a transfer-competent nucleoprotein particle and exporting this particle across the bacterial envelope. Two additional regions of the Ti plasmid code for functions that are not essential for the T-DNA transfer process per se but are nevertheless intimately associated with the overall infection process. One of these regions harbors genes involved in catabolism of novel amino acid derivatives termed opines that *A. tumefaciens* induces plants to synthesize as a result of T-DNA transfer. The second region encodes Ti plasmid transfer functions for distributing copies of the Ti plasmid and its associated virulence factors to other *A. tumefaciens* cells by a process termed conjugation. Intriguing recent work has described a novel regulatory cascade involving chemical signals released both from the transformed plant cells and from the infecting bacterium that activates conjugal transfer of the Ti plasmid among *A. tumefaciens* cells residing in the vicinity of the plant tumor.

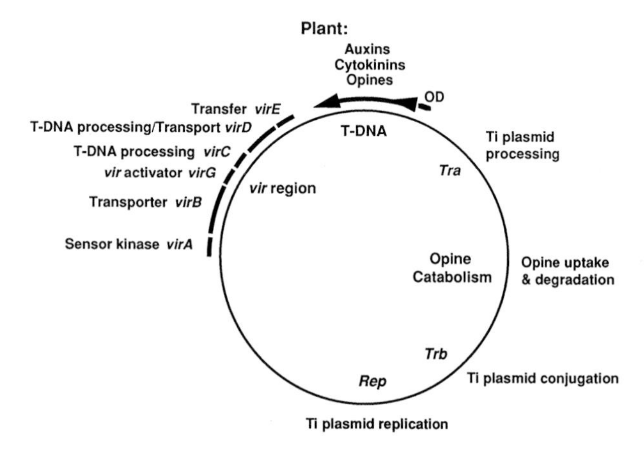

FIGURE 2.2 Regions of the Ti plasmid that contribute to infection (*vir* region and T-DNA), cell survival in the tumor environment (opine catabolism), and conjugal transfer of the Ti plasmid to recipient agrobacteria (*tra* and *trb*). The various contributions of the *vir* gene products to T-DNA transfer are listed. T-DNA, delimited by 25-bp border sequences (black arrows), codes for biosynthesis of auxins, cytokinins, and opines in the plant. OD, overdrive sequence that enhances VirD2-dependent processing at the T-DNA border sequences.

A. T-DNA

The T-DNA is delimited by 25-bp direct, imperfect repeats termed border sequences (Fig. 2.2). Flanking one border is a sequence termed overdrive that functions to stimulate the T-DNA processing reaction. All DNA between the border sequences can be excised and replaced with genes of interest, and *A. tumefaciens* will still efficiently transfer the engineered T-DNA to plant cells. This shows that the border sequences are the only *cis* elements required for T-DNA transfer to plant cells and that genes encoded on the T-DNA play no role in movement of T-DNA to plant cells. Instead, the T-DNA genes code for synthesis of two main types of enzymes within transformed plant cells. Oncogenes synthesize enzymes involved in the synthesis of two plant growth regulators, auxins and cytokinins. Production of these plant hormones results in a stimulation of cell division and a loss of cell growth control leading to the formation of characteristic crown gall tumors. The second class of enzymes code for the synthesis of novel amino acid derivatives termed opines. For example, the pTiA6 plasmid carries two T-DNAs that code for genes involved in synthesis of octopines, a reductive condensation product of pyruvate and arginine. Other Ti plasmids carry T-DNAs that code for nopalines, derived from α-ketoglutarate and arginine, and still others code for different classes of opines.

Plants cannot metabolize opines. However, as described later, the Ti plasmid carries opine catabolism genes that are responsible for the active transport of opines and their degradation, thus providing a source of carbon and nitrogen for the bacterium. The "opine concept" was developed to rationalize the finding that *A. tumefaciens* evolved as a pathogen by acquiring the ability to transfer DNA to plant cells. According to this concept, *A. tumefaciens* adapted a DNA conjugation system for interkingdom DNA transport to incite opine synthesis in its plant host. The cotransfer of oncogenes ensures that transformed plant cells proliferate, resulting in enhanced opine synthesis. The environment of the tumor, therefore, is a rich chemical environment favorable for growth and propagation of the infecting *A. tumefaciens*. It is also notable that a given *A. tumefaciens* strain catabolizes only those opines that it incites plant cells to synthesize. This ensures a selective advantage of the infecting bacterium over other

A. tumefaciens strains that are present in the vicinity of the tumor.

B. Opine catabolism

The regions of two Ti plasmids coding for opine catabolism have been sequenced and shown to code for three functions related to opine catabolism (Fig. 2.2). The first is a regulatory function that controls expression of the opine transport and catabolism genes. The regulatory protein is OccR for the octopine catabolism region of plasmid pTiA6. Recent studies have shown that OccR positively regulates expression of the *occ* genes involved in octopine uptake and catabolism by inducing a bend in the DNA at the OccR binding site. Interestingly, octopine alters both the affinity of OccR for its target site and the angle of the DNA bend, suggesting that octopine modulates OccR regulatory activity. The regulatory protein is AccR for the nopaline catabolism region of plasmid pTiC58. In contrast to OccR, AccR functions as a negative regulator of *acc* genes involved in nopaline catabolism.

The second and third functions, opine transport and catabolism, are encoded by several genes that are transcribed from a single promoter. At the proximal end of the operon is a set of genes that code for one or more transport systems conferring opine-specific binding and uptake. Typically, one or more of these genes encode proteins homologous to energy-coupling proteins found associated with the so-called ATP-binding cassette (ABC) superfamily of transporters. The ABC transporters are ubiquitous among bacterial and eukaryotic cells, and they provide a wide variety of transport functions utilizing the energy of ATP hydrolysis to drive the transport reaction. At the distal end of the operon are genes involved in cleaving the opines to their parent compounds for use as carbon and nitrogen sources for the bacterium.

C. Ti plasmid conjugation

The Ti plasmid transfer (*tra* and *trb*) functions direct the conjugal transmission of the Ti plasmid to bacterial recipient cells. The transfer genes of conjugative plasmids code for DNA processing and transport system that assembles at the bacterial envelope for the purpose of delivering conjugal DNA transfer intermediates to recipient cells. DNA sequence studies have shown that one set of transfer genes codes for many proteins that are related to components of other plasmid and protein toxin transport systems. As described later in more detail, this evolutionarily conserved family of transporters is referred to as a type IV secretion system.

1. Autoinduction-dependent Ti plasmid transfer

Recent work has demonstrated that a regulatory cascade exists to activate Ti plasmid transfer under conditions of high cell density (Fig. 2.3). This regulatory cascade initiates when *A. tumefaciens* imports opines released from plant cells. For the octopine pTiA6 plasmid, OccR acts in conjunction with octopine to activate transcription of the *occ* operon. Although the

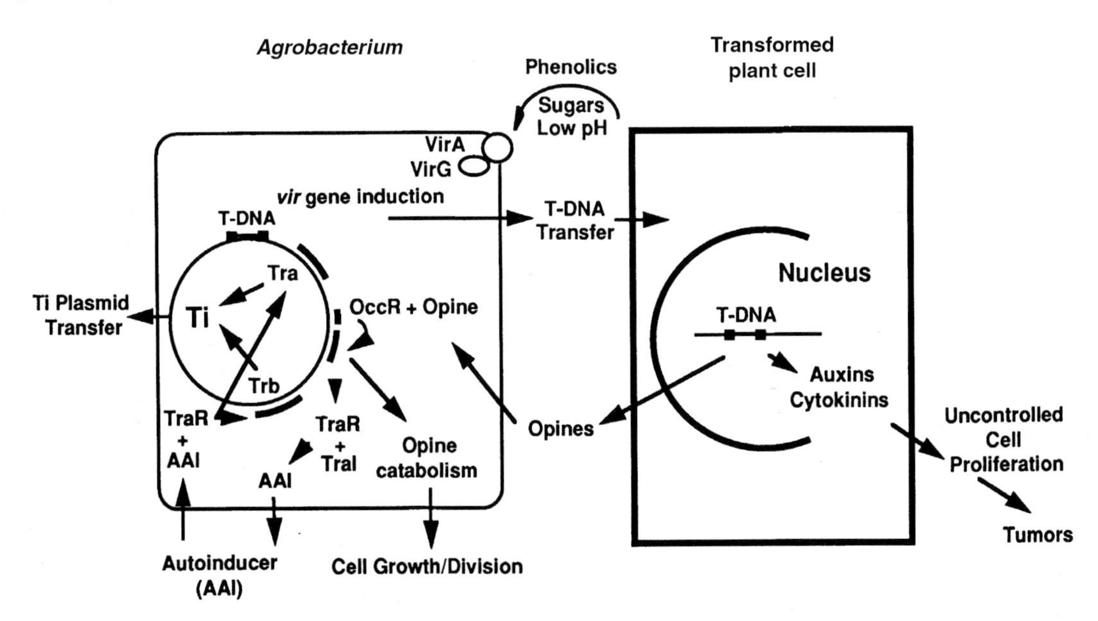

FIGURE 2.3 A schematic of chemical signaling events between *Agrobacterium* and the transformed plant cell. Signals released from wounded plant cells initiate the infection process leading to tumor formation. Opines released from wounded plant cells activate opine catabolism functions for growth of infecting bacteria. Opines also activate synthesis of TraR for autoinducer (AAI) synthesis. TraR and AAI at a critical concentration activate the Ti plasmid conjugation functions (see text for details).

majority of the *occ* operon codes for octopine transport and catabolism functions, the distal end of the occR operon encodes a gene for a transcriptional activator termed TraR. TraR is related to LuxR, an activator shown nearly 20 years ago to regulate synthesis of an acyl homoserine lactone termed autoinducer. Cells that synthesize autoinducer molecules secrete these molecules into the environment. At low cell densities, autoinducer is in low concentration, whereas at high cell densities this substance accumulates in the surrounding environment and passively diffuses back into the bacterial cell to activate transcription of a defined set of genes. In the case of *A. tumefaciens*, the autoinducer is an *N*-3-(oxo-octonoyl)-L-homoserine lactone termed *Agrobacterium* autoinducer (AAI). AAI acts in conjunction with TraR to activate transcription of the Ti plasmid *tra* genes and *traI*, whose product mediates synthesis of AAI. Therefore, synthesis of TraR under conditions of high cell density creates a positive-feedback loop whereby a TraR–AAI complex induces transcription of TraI, which in turn results in enhanced synthesis of more AAI. It must be noted that this regulatory cascade, involving opine-mediated expression of *traR* and TraR–AAI-mediated expression of Ti plasmid transfer genes under conditions of high cell density, has the net effect of enhancing Ti plasmid transfer in the environment of the plant tumor. Given that the Ti plasmid encodes essential virulence proteins for stimulating T-DNA transfer, *A. tumefaciens* might have evolved this complex regulatory system to maximize the number of bacterial cells in the vicinity of the plant wound site that are competent for delivery of opine-encoding T-DNA to plant cells.

D. *vir* genes

The Ti plasmid carries an ~35-kb region that harbors at least six operons involved in T-DNA transfer. Two of these operons have a single open reading frame, whereas the remaining operons code for 2–11 open reading frames. The products of the *vir* region direct events within the bacterium that must precede export of a copy of the T-DNA to plant cells. These events include (i) elaboration of the VirA/VirG sensory transduction system for perception of plant-derived signals and transcriptional activation of the *vir* genes, (ii) T-DNA processing into a nucleoprotein particle for delivery to plant nuclei by the VirC, VirD, and VirE proteins, and (iii) assembly of a transenvelope transporter composed of VirB proteins for exporting the T-DNA transfer intermediate across the bacterial envelope.

1. vir *gene activation*

Infection is initiated when bacteria sense and respond to an array of signals, including specific classes of plant phenolic compounds, monosaccharides, and an acidic pH that are present at a plant wound site (Fig. 2.1). Signal perception is mediated by the VirA/VirG signal transduction system together with ChvE, a periplasmic sugar-binding protein, and possibly other phenolic-binding proteins. VirA was one of the first described of what is recognized as a very large family of sensor kinases identified in bacteria and recently in eukaryotic cells. The members of this protein family typically are integral membrane proteins with an N-terminal extracytoplasmic domain. Upon sensory perception, the kinase autophosphorylates at a conserved histidine residue and then transfers the phosphate group to a conserved aspartate residue on the second component of this transduction pathway, the response regulator. The phosphorylated response regulator coordinately activates transcription of several operons, whose products mediate a specific response to the inducing environmental signal. For the *A. tumefaciens vir* system, the response regulator is VirG, and phosphorylated VirG activates transcription of the six essential *vir* operons and many other Ti plasmid-encoded operons that are dispensable for virulence.

VirA senses all three of the plant-derived signals discussed previously. The most important signal molecules are phenols that carry an *ortho*-methoxy group. The type of substitution at the *para* position distinguishes strong inducers such as acetosyringone from weaker inducers such as ferulic acid and acetovanillone. A variety of monosaccharides, including glucose, galactose, arabinose, and the acidic sugars D-galacturonic acid and D-glucuronic acid, strongly enhance *vir* gene induction. The inducing phenolic compounds and the monosaccharides are secreted intermediates of biosynthetic pathways involved in cell wall repair. Therefore, the presence of these compounds is a general feature of most plant wounds and likely contributes to the extremely broad host range of *A. tumefaciens*. VirA functions as a homodimer, and recent genetic studies support a model indicating that VirA interacts directly with inducing molecules that diffuse across the outer membrane into the periplasm. Sugar-mediated inducing activity occurs via an interaction between sugars and the periplasmic sugar-binding protein ChvE. In turn, ChvE–sugar interacts with the periplasmic domain of VirA to induce a conformational change that increases the sensitivity of VirA to phenolic inducer molecules. The periplasmic domain of VirA also senses the third environmental signal, acidic pH, required for maximal induction of the *vir* genes; however, the underlying mechanism responsible for stimulation of VirA activity is unknown.

On the basis of recent crystallographic analysis of CheY, a homolog of VirG, phosphorylation of this family of response regulators is thought to induce a conformational change. Phospho-VirG activates transcription of the *vir* genes by interacting with a *cis*-acting regulatory sequence (TNCAATTGAAAPy) called the *vir* box located upstream of each of the *vir* promoters. Interestingly, both nonphosphorylated and phosphorylated VirG bind to the *vir* box, indicating that a phosphorylation-dependent conformation is necessary for a productive interaction with components of the transcription machinery.

III. CHROMOSOMALLY ENCODED VIRULENCE GENES

Most studies of the *A. tumefaciens* infection process have focused on the roles of Ti plasmid genes in T-DNA transfer and opine response. Several essential and ancillary chromosomal genes also have been shown to contribute to *A. tumefaciens* pathogenicity. Although mutations in these genes are often pleiotropic, they generally function to regulate *vir* gene expression or mediate attachment to plant cells.

A. Regulators of *vir* gene expression

At least three groups of chromosomal genes have been identified that activate or repress *vir* gene expression. As described previously, the periplasmic sugar-binding protein ChvE complexed with any of a wide variety of monosaccharides induces conformational changes in VirA, allowing it to interact with phenolic inducers. Interestingly, *chvE* mutants are not only severely compromised for T-DNA transfer but also show defects in chemotaxis toward sugars, suggesting that ChvE interacts both with VirA and with another membrane protein(s) involved in chemotaxis. ChvE therefore plays a dual role in the physiology of *A. tumefaciens* by promoting chemotaxis toward nutrients and by enhancing the transfer efficiency of opine-encoding T-DNA to plant cells.

A second locus codes for Ros, a transcriptional repressor of certain *vir* operons. As described later, the VirC and VirD operons contribute to the T-DNA processing reaction. Although the promoters for these operons are subject to positive regulation by the VirA/VirG transduction system in response to phenolics and sugars, they are also negatively regulated by the Ros repressor. A mutation in *ros* leads to constitutive expression of *virC* and *virD* in the complete absence of VirG protein. Ros binds to a 9-bp inverted repeat, the *ros* box residing upstream of these promoters. In the absence of plant signals, Ros binding to the *virC* and *virD* promoters prevents the T-DNA processing reaction, whereas in the presence of plant signals Ros repression is counteracted by the VirA/VirG induction system. Interestingly, Ros was recently shown to be a novel prokaryotic zinc finger protein that functions to repress not only the expression of T-DNA processing genes in the absence of a suitable plant host but also the expression of the T-DNA oncogenes in the bacterium.

A second two-component regulatory system has been identified that, like the VirA/VirG transducer pair, senses environmental signals and mounts a behavioral response by modulating gene expression. ChvG is the sensor kinase and ChvI is the response regulator. Null mutations in genes for these proteins result in cells which cannot induce the *vir* genes or grow at an acidic pH of 5.5. The molecular basis underlying the effect of the ChvG and ChvI proteins on *vir* gene expression is unknown.

B. Attachment to plant cells

Binding of *A. tumefaciens* to plant cells is required for T-DNA transfer. Recent evidence indicates there are at least two binding events that may act sequentially or in tandem. The first is encoded by chromosomal loci and occurs even in the absence of the Ti plasmid genes. This binding event directs bacterial binding to many plant cells independently of whether or not the bacterium is competent for exporting T-DNA or the given plant cell is competent for receipt of T-DNA. The second binding event is mediated by a pilus that is elaborated by the *virB* genes (see Section V.B.1).

Binding via the chromosomally encoded attachment loci is a two-step process in which bacteria first attach loosely to the plant cell surface, often in a polar fashion. A series of genes termed *att* are required for this binding reaction. The second step involves a transition resulting in the tight binding of the bacteria to plant cells. The *cel* genes that mediate this form of binding direct the synthesis of cellulose fibrils that emanate from the bacterial cell surface. Recent studies indicate that binding due to these chromosomal functions occurs at specific sites on the plant cell surface. Binding is saturable, suggestive of a limited number of attachment sites on the plant cell, and binding of virulent strains can also be prevented by attachment of avirulent strains. Although the identity of a plant cell receptor(s) has not been definitively established, a good candidate is a vitronectin-like protein found in detergent extracts of plant cell walls. Attachment-proficient *A. tumefaciens* cells bind radioactive vitronection, whereas attachment-deficient cells do not bind this molecule. Intriguingly, human vitronectin and antivitronectin antibodies both inhibit the binding of *A. tumefaciens* to plant cells.

Efficient attachment of bacteria to plant cells also requires the products of three chromosomal loci: *chvA*, *chvB*, and *exoC* (*pscA*). All three loci are involved in the synthesis of transport of a cyclic β-1,2 glucan molecule. Mutations in these genes are pleiotropic, suggesting that β-1,2 glucan synthesis is important for the overall physiology of *A. tumefaciens*. Periplasmic β-1,2 glucan plays a role in equalizing the osmotic pressure between the inside and outside of the cell. It has been proposed that loss of this form of glucan may indirectly disrupt virulence by reducing the activity or function of cell surface proteins. Interestingly, *chv* mutants accumulate low levels of VirB10, one of the proposed components of the T-complex transport system (see Section V), suggesting that β-1,2 glucan might influence T-DNA export across the bacterial envelope by contributing to transporter assembly.

IV. T-DNA PROCESSING

One of the early events following attachment to plant cells and activation of *vir* gene expression in response to plant signals involves the processing of T-DNA into a form which is competent for transfer across the bacterial cell envelope and translocation through the plant plasma membrane, cytosol, and nuclear membrane. The prevailing view, strongly supported by molecular data, is that T-DNA is transferred as a single-stranded molecule that is associated both covalently and noncovalently with Vir proteins. Two proteins identified to date are components of the transfer intermediate: VirD2, an endonuclease that participates in the T-DNA processing reaction, and VirE2, a single-stranded DNA-binding protein which is proposed to associate noncovalently along the length of the single-stranded transfer intermediate (Fig. 2.1). Intriguingly, recent studies have provided strong evidence that *A. tumefaciens* can export the VirE2 SSB to plant cells independently of T-DNA (see Section IV.B).

A. Formation of the transfer intermediate

More than a decade ago, investigators determined that the T-DNA border repeats are cleaved by a strand-specific endonuclease and that the right T-DNA border sequence is essential for and determines the direction of DNA transfer from *A. tumefaciens* to plant cells. The predominant product of this nicking reaction was shown to be a free single-stranded T-DNA molecule that corresponds to one strand of T-DNA. It was noted that these features of the T-DNA processing reaction are reminiscent of early processing events involved in the conjugative transfer of plasmids between bacterial cells. In the past 10 years, a large body of evidence has accumulated supporting the notion that DNA processing reactions associated with T-DNA transfer and bacterial conjugation are equivalent. Extensive studies have shown that two systems in particular, the T-DNA transfer system and the conjugation system of the broad host-range plasmid RP4, are highly similar. The substrates for the nicking enzymes of both systems, T-DNA border sequences and the RP4 origin of transfer (*oriT*), exhibit a high degree of sequence similarity. Furthermore, the nicking enzymes VirD2 of pTi and TraI of RP4 possess conserved active-site motifs that are located within the N-terminal halves of these proteins. Purified forms of both proteins cleave at the nick sites within T-DNA borders and the RP4 *oriT*, respectively. In the presence of Mg^{2+}, purified VirD2 will catalyze cleavage of oligonucleotides bearing a T-DNA nick site. However, VirD1 is essential for nicking when the nick site is present on a supercoiled, double-stranded plasmid. Both VirD2 and TraI remain covalently bound to the 5′ phosphoryl end of the nicked DNA via conserved tyrosine residues Tyr-29 and Tyr-22. Finally, both proteins catalyze a joining activity reminiscent of type I topoisomerases. VirD1 was reported to possess a topoisomerase I activity, but recent work suggests instead that VirD1 supplies a function analogous to TraJ of RP4, which is thought to interact with *oriT* as a prerequisite for TraI binding to an *oriT* DNA–protein complex.

The current model describing the T-DNA and plasmid conjugation processing reactions is that sequence and strand-specific endonucleases initiate processing by cleaving at T-DNA borders and *oriT* sequences, respectively. This reaction is followed by a strand displacement reaction, which generates a free single-stranded transfer intermediate. Concomitantly, the remaining segment of T-DNA or plasmid serves as a template for replacement synthesis of the displaced strand. It is important to note that the single-stranded transfer intermediates of the T-DNA and RP4 transfer systems remain covalently bound to their cognate endonucleases. Considerable evidence suggests that these protein components play essential roles in delivering the respective transfer intermediates across the bacterial envelope.

B. The role of VirE2 SSB in T-DNA transfer

The *virE2* gene codes for a single-stranded DNA-binding protein that binds cooperatively to single-stranded DNA (ssDNA). Early studies supplied evidence that VirE2 binds with high affinity to any ssDNA *in vitro* and that it binds T-DNA in *A. tumefaciens*. By analogy to other ssDNA-binding proteins (SSBs) that play important roles in DNA replication, VirE2 was proposed to participate in the T-DNA processing reaction by binding to the liberated T-strand and preventing it from reannealing to its complementary strand on the Ti plasmid. The translocation-competent form of DNA therefore has been depicted as a ssDNA molecule covalently bound at the 5′ end by VirD2 and coated along its length with an SSB. The single-stranded form of T-DNA delivered to plants is termed the T-strand, and the VirD2–VirE2-T-strand nucleoprotein particle is termed the T-complex (Fig. 2.1).

Considerable evidence indicates that the T-complex represents the biologically active transfer intermediate. The T-complex, composed of a 20-kb T-strand capped at its 5′ end with a 60-kDa endonuclease and approximately 600 VirE2 molecules along its length, is a large nucleoprotein complex of an estimated size of 50×10^6 Da. This size approaches that of some bacteriophages, and it has been questioned whether such a complex could be exported intact across the *A. tumefaciens* envelope without lysing the bacterial cell. Although this is still unknown, several recent discoveries support an alternative model that assembly of the T-complex initiates within the bacterium but is completed within the plant cell.

Approximately 15 years ago, it was discovered that two avirulent *A. tumefaciens* mutants, one with a deletion of T-DNA and a second with a *virE2* mutation, could induce the formation of tumors when inoculated as a mixture on plant wound sites. To explain this observation, it was postulated that *A. tumefaciens* separately exports VirE2 and VirD2 T-strands to the same plant cell. The *virE2* mutant was proposed to export the VirD2 T-strands (T-DNA donor), and the T-DNA deletion mutant could export the VirE2 protein only (VirE2 donor). Once exported, these molecules could then assemble into a nucleoprotein particle, the T-complex, for transmission to the plant nucleus. In strong support of this model, recent genetic analyses have shown that both the proposed T-DNA donor strain and the VirE2 mutant in the mixed infection experiment must possess an intact transport machinery and intact genes mediating bacterial attachment to the plant cell. Furthermore, current genetic data argue against the possible movement of T-DNA or VirE2 between bacterial cells by conjugation as an alternative explanation for complementation by mixed infection. Finally, a *virE* mutant was shown to incite the formation of wild-type tumors on transgenic plants expressing *virE2*. This finding indicates that VirE2 participates in *A. tumefaciens* pathogenesis by supplying essential functions within the plant.

C. Role of cotransported proteins in T-DNA transfer and plasmid conjugation

As discussed previously, processing of T-DNA and conjugative plasmids results in the formation of a ssDNA transfer intermediate covalently bound at its 5′ end to the nicking enzyme. Recent studies have shown that the protein component(s) of these conjugal transfer intermediates participates in the delivery of the DNA to the recipient cell. In the case of T-DNA, the transferred proteins facilitate movement of the T-DNA transfer intermediate to plant nuclei by (i) piloting the T-DNA transfer intermediate across the bacterial envelope and protecting it from nucleases and/or (ii) directing T-DNA movement and integration in plant cells. In the case of the IncP plasmid RP4, TraI relaxase is thought to promote plasmid recircularization, and a primase activity associated with the TraC SSB is considered to be important for second-strand synthesis in the recipient cell.

1. *Piloting and protection*

A piloting function for VirD2 is suggested by the fact that VirD2 is covalently associated at the 5′ end of the T-strand and also from the finding that the T-strand is transferred to the plant cell in a 5′–3′ unidirectional manner. A dedicated transporter functions to export substrates to plant cells (see Section V). VirD2 might guide T-DNA export by providing the molecular basis for recognition of the transfer intermediate by the transport machinery. By analogy to other protein substrates exported across the bacterial envelope by dedicated transport machines, VirD2 might have a linear peptide sequence or a protein motif in its tertiary structure that marks this molecule as a substrate for the T-DNA transporter.

Studies of T-DNA integrity in transformed plant cells have shown that the 5′ end of the transferred molecule generally is intact, suffering little or no loss of nucleotides as a result of exonuclease attack during transit. By contrast, the 3′ end of the transferred molecule typically is often extensively deleted. These findings suggest that a second role of the VirD2 endonuclease is to protect the 5′ end of the transfer intermediate from nucleases. Recent molecular studies have also shown that T-DNA transferred to plant cells by an *A. tumefaciens virE2* mutant is even more extensively degraded than T-DNA transferred by wild-type cells, suggesting that VirE2 SSB also functions to protect the DNA transfer intermediate from nucleases during transfer.

2. *T-DNA movement and integration*

DNA sequence analyses revealed the presence of a bipartite type of nuclear localization sequence (NLS) near the C terminus of VirD2. The nuclear localizing function of this NLS was confirmed by fusing the *virD2* coding sequence to a reporter gene and demonstrating the nuclear localization of the reporter protein activity in tobacco cells transiently expressing the gene fusion. As predicted, *A. tumefaciens* strains expressing mutant forms of VirD2 with defects in the NLS sequence are very inefficient in delivering T-DNA to plant nuclei. Similar lines of investigation showed that VirE2 also possesses two NLS sequences that both contribute to its delivery to the nuclear pore. Therefore, both VirD2 and VirE2 are proposed to promote T-DNA delivery to and across the plant nuclear membrane. In this context, VirD2 has been shown to interact with a plant NLS receptor localized at the nuclear pore. Of further interest, VirD2 has also been shown to interact with several members of a family of proteins termed cyclophilins. The postulated role for cyclophilins in this interaction is to supply a chaperone function at some stage during T-complex trafficking to the nucleus. *Agrobacterium tumefaciens* has been demonstrated to transport DNA to representatives of prokaryotes, yeasts, and plants. Cyclophilins are ubiquitous proteins found in all these cell types and therefore may be of general importance for *A. tumefaciens*-mediated DNA transfer.

T-DNA integrates into the plant nuclear genome by a process termed "illegitimate" recombination. According to this model, T-DNA invades at nicks or gaps in the plant genome possibly generated as a consequence of active DNA replication. The invading ends of the single-stranded T-DNA are proposed to anneal via short regions of homology to the unnicked strand of the plant DNA. Once the ends of T-DNA are ligated to the target ends of plant DNA, the second strand of the T-DNA is replicated and annealed to the opposite strand of the plant DNA. Recent mutational analysis of VirD2 showed that a C-terminal sequence termed Ω appears to play a role in promoting T-DNA integration. A recent study also supports a model that VirE2 also participates in the T-DNA integration step, but the precise functions of VirD2, VirE2, and possible host proteins in this reaction have not been defined.

V. THE T-DNA TRANSPORT SYSTEM

A. The essential components of the T-complex transporter

Exciting progress has been made during the past 6 years on defining the structure and function of the transporter at the *A. tumefaciens* cell surface that is dedicated to exporting the T-DNA transfer intermediate to plant cells. Early genetic studies suggested that products of the ~9.5-kb *virB* operon are the most likely candidates for assembling into a cell surface structure for translocation of T-DNA across the *A. tumefaciens* envelope. Sequence analyses of the *virB* operon have supported this prediction by showing that the deduced products have hydropathy patterns characteristic of membrane-associated proteins. Recently, a systematic approach was taken to delete each of the 11 *virB* genes from the *virB* operon without altering expression of the downstream genes. Analyses of this set of nonpolar null mutants showed that *virB2–virB11* are essential for T-DNA transfer, whereas *virB1* is dispensable. As described in more detail later, the VirB proteins, along with the VirD4 protein, are thought to assemble at the cell envelope as a channel dedicated to the export of T-complexes.

B. The T-complex transporter

1. *Type IV transporters: DNA conjugation systems adapted for export of virulence factors*

DNA sequence studies within the past 4 years have identified extensive similarities between products of the *virB* genes and components of two types of transporters dedicated to movement of macromolecules from or between cells (Fig. 2.4). The first type, encoded by *tra* operons of conjugative plasmids, functions to deliver conjugative plasmids to bacterial recipient cells. The IncN plasmid, pKM101, and the IncW plasmid, R388, code for Tra protein homologs of each of the VirB proteins. Furthermore, the genes coding for related proteins are often colinear in these respective *virB* and *tra* operons, supporting the view that these DNA transfer systems share a common ancestral origin. Other broad host-range plasmids such as RP4 (IncPα) and the narrow host-range plasmid F (IncF) code for proteins homologous to a subset of the VirB proteins.

DNA sequence studies also identified a related group of transporters in several bacterial pathogens of humans that function not to export DNA but rather to secrete protein toxins (Fig. 2.4). *Bordetella pertussis*, the causative agent of whooping cough, uses the Ptl transporter to export the six-subunit pertussis toxin across the bacterial envelope. All nine Ptl proteins have been shown to be related to VirB proteins, and the *ptl* genes and the corresponding *virB* genes are colinear in their respective operons. Type I strains of *Helicobacter pylori*, the causative agent of peptic ulcer disease and a risk factor for development of gastric adenocarcinoma, contain a 40-kb cag pathogenicity island (PAI) that codes for several virulence factors, of which several are related to Vir proteins. These Cag proteins are thought to assemble into a transporter for

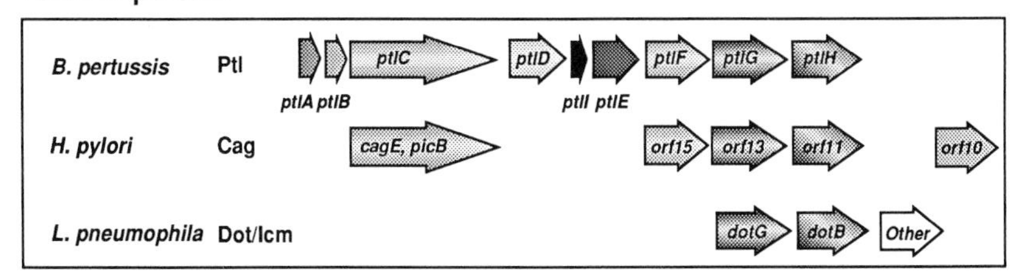

FIGURE 2.4 Alignment of genes encoding related components of the type IV transport systems. Of the 11 VirB proteins, those encoded by *virB2–virB11*, as well as *virD4*, are essential for T-complex transport to plant cells. The broad host-range (BHR) plasmid pKM101 encodes a conjugation apparatus composed of the products of the *tra* genes shown. Other BHR plasmids and the narrow host-range (NHR) F plasmid code for Tra proteins related to most or all the VirB genes. A second subfamily of type IV transporters found in bacterial pathogens of humans export toxins or other protein effectors to human cells.

exporting an unidentified protein toxin(s) that induces synthesis of the proinflammatory cytokine IL-8 in gastric epithelial cells. Finally, *Legionella pneumophila*, the causative agent of Legionnaire's disease and Pontiac fever, possesses the *icm/dot* genes, of which *dotG* and *dotB* code for proteins related to VirB10 and VirB11 and others code for homologs of transfer proteins encoded by other bacterial conjugation systems. The Icm/Dot proteins are proposed to assemble into a transporter that exports a virulence factor(s) that promotes intracellular survival of *L. pneumophila* and macrophage killing.

The transporters described previously are grouped on the basis of evolutionary relatedness as a distinct transport family. Designated as the type IV secretion family, this classification distinguishes these transporters from other conserved bacterial protein targeting mechanisms that have been identified in bacteria. Although this is a functionally diverse family, the unifying theme of the type IV transporters is that each system has evolved by adapting an ancestral DNA conjugation apparatus or a part of this apparatus for the novel purpose of exporting DNA or proteins that function as virulence factors.

2. *Functional similarities among type IV transporters*

Functional studies have supplied compelling evidence that the type IV transporters are mechanistically related. The non-self-transmissible plasmid RSF1010 of the IncQ incompatibility group possesses an *oriT* sequence and mobilization (*mob*) functions for generating a ssDNA transfer intermediate. This transfer intermediate can be delivered to recipient bacteria by the type IV transporters of the IncN, IncW, IncP, and F plasmids. In addition, approximately 10 years ago it was shown using an *A. tumefaciens* strain harboring a disarmed Ti plasmid (with *vir* genes but lacking the T-DNA or its borders) and an RSF1010 derivative that the T-complex transporter could deliver the IncQ transfer intermediate to plant cells. This discovery was followed soon afterwards by the demonstration that the T-complex transporter also functions to conjugally deliver the IncQ plasmid to *A. tumefaciens* recipient cells. Interestingly, *A. tumefaciens* strains carrying both an IncQ plasmid and an intact T-DNA efficiently deliver the IncQ plasmid to plant cells but do not transfer the T-DNA. Preferential transfer of the IncQ plasmid over the T-DNA transfer intermediate could result from the transporter having a higher affinity for the IncQ plasmid or the IncQ plasmid being more abundant than the T-DNA. Of further interest, the coordinate overexpression of *virB9*, *virB10*, and *virB11* relieved the IncQ suppression and restored efficient T-DNA transfer to plant cells. These findings suggest that the T-complex and the IncQ transfer intermediate compete for the same transport apparatus. Furthermore, the data suggest that VirB9–VirB11 stoichiometries determine the number of transporters a given cell can assemble or influence the selection of substrates destined for export.

Although the toxin substrates have not been identified for the *H. pylori* Cag and *L. pneumophila* Dot/Icm transporters, it is intriguing to note that the Dot/Icm system also has been shown to deliver the non-self-transmissible IncQ plasmid RSF1010 to bacterial recipient cells by a process requiring cell-to-cell contact. Also, as observed for T-complex export, the presence of an IncQ plasmid suppresses export of the natural substrate of the Dot/Icm transporter of *L. pneumophila*, resulting in inhibition of intracellular multiplication and human macrophage killing. These parallel findings show that the type IV DNA and protein export systems are highly mechanistically related.

C. Architecture of the T-complex transporter

The T-complex transporter, like other DNA conjugation machines, is proposed to be configured as a transenvelope channel through which the T-DNA transfer intermediate passes and as an extracellular pilus termed the T-pilus for making contact with recipient cells. Most of the VirB proteins fractionate with both membranes, consistent with the view that these proteins assemble as a membrane-spanning protein channel. All the VirB proteins except VirB11 possess periplasmic domains, as shown by protease susceptibility and reporter protein fusion experiments. Although detailed structural information is not available for the T-complex transporter, important progress has been made in the characterization of the VirB proteins, especially in the following areas: (i) characterization of the *virB*-encoded pilus termed the T-pilus, (ii) structure–function studies of the VirB4 and VirB11 ATPases, and (iii) identification of a nucleation activity of a disulfide cross-linked VirB7/VirB9 heterodimer during transporter assembly (Table 2.1).

TABLE 2.1 Properties of the VirB proteins

VirB	Localization	Proposed function
B1	Periplasm	Transglycosylase
B1*	Cell exterior	Cell contact/pilin subunit?
B2	Exported/cell exterior	Cell contact/pilin subunit
B3	Exported	Unknown
B4	Transmembrane	ATPase/transport activation
B5	Exported	Cell contact/pilin subunit?
B6	Transmembrane	Candidate pore former
B7	Outer membrane	Lipoprotein/transporter assembly
B8	Periplasmic face of inner membrane	Unknown
B9	Outer membrane	Lipoprotein/transporter assembly
B10	Transmembrane	Coupler of inner and outer membrane subcomplexes?
B11	Cytoplasm/ inner membrane	ATPase/transport activation
D4	Transmembrane	ATPase/coupler of DNA processing and transport systems

1. *The T-pilus*

The type IV systems involved in conjugation elaborate pili for establishing contact between plasmid-bearing donor cells and recipient cells. Recent studies have demonstrated that VirB proteins direct the assembly of a pilus which is essential for T-DNA transfer. Electron microscopy studies have demonstrated the presence of long filaments (~10 nm in diameter) on the surfaces of *A. tumefaciens* cells induced for expression of the *virB* genes. These filaments are absent from the surfaces of mutant strains defective in the expression of one or more of the *virB* genes. Furthermore, an interesting observation was made that cells grown at room temperature rarely possess pili, whereas cells grown at ~19 °C possess these structures in abundance. This finding correlates well with previous findings that low temperature stimulates the *virB*-dependent transfer of IncQ plasmids to bacterial recipients and T-DNA transfer to plants.

Recently, compelling evidence demonstrated that VirB2 is the major pilin subunit. Early studies showed that VirB2 bears both sequence and structural similarity to the TraA pilin subunit of the F plasmid of *E. coli*. Recent work demonstrated that VirB2, like TraA, is processed from an ~12-kDa propilin to a 7.2-kDa mature protein that accumulates in the inner membrane. During F plasmid conjugation, TraA is mobilized to the surface of the donor cell where it polymerizes to form the pilus. Similarly, the appearance of pili on the surface of *A. tumefaciens* cells induced for expression of the *vir* genes is correlated with the presence of VirB2 on the cell exterior. Finally, VirB2 is a major component of pili that have been sheared from the cell surface and purified.

Many adhesive and conjugative pili possess one or more minor pilin subunits in addition to the major pilin structural protein. Interestingly, VirB1, a periplasmic protein with transglycosylase activity, is processed such that the C-terminal two-thirds of the protein, termed VirB1*, is secreted to the outer surface of the cell. This localization is consistent with a proposed function for VirB1* as a minor pilus subunit. VirB5 might also assemble as a pilus subunit based on its homology to a possible pilin subunit encoded by the IncN plasmid pKM101 transfer system.

2. *Studies of the VirB ATPases*

Two VirB proteins, VirB4 and VirB11, possess conserved mononucleotide-binding motifs. Mutational analyses established the importance of these motifs for the function of both proteins. In addition, purified forms of both proteins exhibit weak ATPase activities, suggesting that VirB4 and VirB11 couple the energy of ATP hydrolysis to transport. Both of these putative ATPases appear to contribute functions of general importance for macromolecular transport since homologs have been identified among many DNA and protein transport systems. Of further possible significance, VirB11 and two homologs, TrbB of IncP RP4 and EpsE, of *Vibrio cholerae* have been reported to autophosphorylate. VirB4 and VirB11 might activate substrate transport by using the energy of ATP hydrolysis or a kinase activity to facilitate assembly of the transport apparatus at the cell envelope. Alternatively, by analogy to the SecA ATPase of *E. coli* which uses the energy of ATP hydrolysis to drive translocation of exported proteins, one or both of the VirB ATPases may contribute directly to export of the DNA transfer intermediate. Recent studies have shown that both VirB4 and VirB11 assemble as homodimers. Dimerization is postulated to be critical both for protein stability and for catalytic activity. Accumulation of these ATPases to wild-type levels depends on the presence of other VirB proteins, suggesting that complex formation with other components of the T-complex transporter contributes to protein stability. Specific contacts between these ATPases and other transporter components have not been identified.

3. *The VirB7 lipoprotein and formation of stabilizing intermolecular disulfide bridges*

Detailed studies have shown that VirB7 is critical for assembly of a functional T-complex transport system. VirB7 possesses a characteristic signal sequence that ends with a consensus peptidase II cleavage site characteristic of bacterial lipoproteins. Biochemical studies have confirmed that VirB7 is processed as a lipoprotein. Furthermore, maturation of VirB7 as a lipoprotein is critical for its proposed role in T-complex transporter biogenesis. Recent studies have shown that the VirB7 lipoprotein interacts directly with the outer membrane protein VirB9. The first hint of a possible interaction between these proteins was provided by the demonstration that VirB9 accumulation is strongly dependent on co-synthesis of VirB7, suggesting that VirB7 stabilizes VirB9. Interestingly, this stabilizing effect has been shown to be mediated by formation of a disulfide bridge between these two proteins. VirB7 assembles not only as VirB7/VirB9 heterodimers but also as covalently cross-linked homodimers, and there is evidence that VirB9 assembles into higher order multimeric complexes. These dimers and higher order multimers might correspond to stable subcomplexes of the larger transport system. In the case of the VirB7/VirB9 heterodimer, considerable evidence indicates that this heterodimer plays a critical role early during transporter biogenesis by recruiting and stabilizing newly synthesized VirB proteins. The heterodimer has been shown to interact with VirB1*. The heterodimer also interacts with VirB10, a cytoplasmic membrane protein with a large C-terminal periplasmic domain. VirB10 has been postulated to join the VirB7/VirB9 heterodimer at the outer membrane with a VirB protein subcomplex located at the inner membrane.

4. *VirB protein stimulation of IncQ plasmid uptake by bacterial recipient cells*

The T-complex transport system seems designed to function unidirectionally to export substrates to recipient cells. However, a recent discovery indicates that VirB proteins can also assemble as a transenvelope structure that stimulates DNA uptake during conjugation. The fundamental observation is that *A. tumefaciens* cells harboring an IncQ plasmid conjugally transfer the IncQ plasmid to recipient cells expressing the *virB* genes at a frequency of ~1000 times that observed for transfer to recipient cells lacking the *virB* genes. Furthermore, only a subset of *virB* genes, including *virB3, virB4*, and *virB7–virB10*, was required for enhanced DNA uptake by recipient cells. These findings suggest that a subset of the VirB proteins might assemble as a core translocation channel at the bacterial envelope that accommodates the bidirectional transfer of DNA substrates. Such a channel might correspond to an early assembly intermediate that, upon complex formation with additional VirB proteins, is converted to a dedicated T-complex export system.

VI. AGROBACTERIUM HOST RANGE

One of the most appealing features of the *A. tumefaciens* DNA transfer system for genetic engineering is its extremely broad host range. Pathogenic strains of *Agrobacterium* infect a wide range of gymnosperms and dicotyledonous plant species of agricultural importance. Crown gall disease can cause devastating reductions in yields of woody crops such as apples, peaches, and pears and vine crops such as grapes. Various host range determinants present in different *A. tumefaciens* strains determine whether a given bacterial strain is virulent for a given plant species.

A. Transformation of monocots

In the past 5 years, dramatic progress has been made toward the development of protocols for stably transforming agriculturally important monocotyledonous plant species. The first indication of gene transfer involved the introduction of a plant viral genome into a plant host via *A. tumefaciens*-mediated transfer of T-DNA carrying the viral genome. Once inside the plant host, the viral DNA excises from the T-DNA and infects the host, inciting disease symptoms that are characteristic of the virus. This process, termed agroinfection, supplied compelling evidence that *A. tumefaciens* transfers T-DNA to monocot plants such as maize. A notable feature of agroinfection is that the introduced viral DNA incites disease without incorporating into the plant genome. Early efforts to obtain stable transformation of monocot species were unsuccessful. The demonstration of agroinfection and the inability to demonstrate T-DNA integration together led to the suggestion that the T-DNA integration step was somehow blocked in monocots. However, protocols have been developed for the efficient and reproducible stable transformation of rice, corn, wheat, and other monocot species. Key to the success of these protocols was the use of actively dividing cells such as immature embryos. In addition, preinduction of *A. tumefaciens* with phenolic inducers appears to enhance T-DNA transfer efficiencies. Additional factors, such as plant genotype, the type and age of plant tissue, the kinds of vectors and bacterial strains, and the types of selectable genes delivered to plant cells, influence the transformation efficiencies. For rice and corn, most of these parameters have been optimized, so that the delivery of foreign DNA to these crop plants is a routine technique.

B. Gene transfer to yeast and fungi

Intriguing recent work has extended the host range of *A. tumefaciens* beyond the plant kingdom to include budding and fission yeast and many species of filamentous fungi. The successful transfer of DNA to yeast depends on the presence of stabilizing sequences, such as a yeast origin of replication sequence or a telomere, or regions of homology between the transferred DNA and the yeast genome for integration by homologous recombination. When the T-DNA lacks any extensive regions of homology with the *Saccharomyces cerevisiae* genome, it integrates at random positions by illegitimate recombination reminiscent of T-DNA integration in plants (see Section IV.C.2). The transformation of filamentous fungi with *A. tumefaciens* is an exciting advancement. *Agrobacterium tumefaciens* was shown to efficiently deliver DNA to fungal protoplasts and fungal conidia and hyphal tissue. This discovery extends well beyond academic interest because the simplicity and high efficiency make this gene delivery system an extremely useful tool for the genetic manipulation and characterization of fungi. This DNA transfer system is especially valuable for species such as the mushroom *Agaricus bisporus* which are recalcitrant to transformation by other methods. It is also of interest to consider that both *A. tumefaciens* and many fungal species exist in the same soil environment, raising the possibility that *A. tumefaciens*-mediated gene transfer to fungi may not be restricted solely to the laboratory bench.

VII. GENETIC ENGINEERING OF PLANTS AND OTHER ORGANISMS

The extent to which any biological system is understood is reflected by our ability to manipulate that system to achieve novel ends. For *A. tumefaciens* transformation, the holy grail has been monocot transformation. As described previously, exciting progress has been made toward attaining this objective for several agriculturally important monocot species. Currently, plant genetic engineers are developing the *A. tumefaciens* gene delivery to achieve equally challenging goals such as (i) designing T-DNA tagging methods for isolating and characterizing novel plant genes, (ii) designing strategies to deliver foreign DNA to specific sites in the plant genome, and (iii) characterizing and genetically engineering other organisms such as agriculturally or medically important fungi.

A. Overcoming barriers to transformation

It is remarkable that all progress in *A. tumefaciens*-mediated monocot transformation has been achieved in the intervening period between the publication of the first and second editions of this encyclopedia. In fact, currently *A. tumefaciens* is the biological DNA delivery system of choice for transformation of most dicot and monocot plant species. The reasons are twofold. First, *A. tumefaciens* is readily manipulated such that plasmids carrying foreign genes of interest are easily introduced into appropriate bacterial strains for delivery to plants. Typically, strains used for gene delivery are "disarmed," that is, deleted of oncogenic T-DNA but still harboring intact Ti plasmid and chromosomal *vir* genes. Foreign genes destined for delivery to plants generally are cloned onto a plasmid that carries a single T-DNA border sequence or two T-DNA border sequences that flank various restriction sites for cloning as well as an antibiotic resistance gene to select for transformed plant cells. If the plasmid carries a single border sequence, the entire plasmid is delivered to plants, and recent work indicates that *A. tumefaciens* can deliver as much as 180-kb of DNA to plants. If the plasmid carries two border sequences, only the DNA bounded by T-DNA borders is delivered to plants. Second, the frequency of stable transformation is often very high, far exceeding frequencies achieved by other gene delivery methods. For example, co-cultivation of *A. tumefaciens* with regenerating protoplasts of certain plant species can result in transformation of up to one-half of the protoplasts.

However, with protoplast transformation there is often a significant reduction in the number of transgenic, fertile plants recovered during selective regeneration of transformed protoplasts. For certain species, protoplasts can be transformed but are recalcitrant to regeneration into intact plants. Consequently, other transformation methods have relied on transformation of plant tissues such as excised leaves or root sections. In the case of monocot species such as maize, immature embryos are the preferred starting material for *A. tumefaciens*-mediated DNA transfer. For rice, success has been achieved with callus tissue induced from immature embryos.

In addition to the need to identify transformable and regenerable plant tissues, many varieties of a given species often need to be screened to identify the susceptible varieties. A large variation in transformation efficiencies is often observed depending on which cell line is being tested. This underscores the notion that interkingdom DNA transfer is a complex process dependent on a genetic interplay between *A. tumefaciens* and host cells. Fortunately, many of the agonomically important species are readily transformable, but additional efforts are needed to overcome the current obstacles impeding efficient transformation of other species of interest.

B. Other applications

Agrobacterium tumefaciens is increasingly used to characterize and isolate novel plant genes by an approach termed T-DNA tagging. Several variations to this methodology exist depending on the desired goals. For example, because insertions are generally randomly distributed throughout the plant genome, T-DNA is widely used as a mutagen for isolating plant genes with novel phenotypes. If the mutagenic T-DNA carries a bacterial origin of replication, the mutated gene of interest can easily be recovered in bacteria by suitable molecular techniques. Furthermore, if the T-DNA is engineered to carry a selectable or scorable gene near one of its ends, insertion downstream of a plant promoter will permit characterization of promoter activity. Conversely, if the T-DNA is engineered to carry an outward reading promoter, insertion can result in a modulation of gene expression with potentially interesting phenotypic consequences. Finally, the discovery that *A. tumefaciens* can transform fungal species of interest means that all approaches developed for plants can be applied to the characterization of fungi.

Although random T-DNA insertion is a boon to investigators interested in characterizing plant or fungal genes, it is an undesired event for plant genetic engineering. In addition to the potential result that T-DNA will insert into an essential gene, insertion often is accompanied by rearrangements of flanking sequences, thus further increasing the chances that the insertion will have undesired consequences. Ideally, T-DNA could be delivered to a restricted number of sites in the plant genome. Recent progress toward this goal has involved the use of the bacteriophage P1 Cre/lox system for site-specific integration in the plant genome. The Cre site-specific recombinase catalyzes strand exchange between two lox sites which, for P1, results in circularization of the P1 genome upon infection of bacterial cells. For directed T-DNA insertion, both the plant and the T-DNA are engineered to carry lox sequences and the plant is also engineered to express the Cre protein. Upon entry of T-DNA into the plant cell, Cre was shown to catalyze the site-specific integration of T-DNA at the plant lox site. The frequency of directed insertion events is low compared to random insertion events, but additional manipulation of this system should enhance its general applicability.

VIII. CONCLUSIONS

The early discovery that oncogenes can be excised from T-DNA and replaced with genes of interest paved the way for the fast growing industry of plant genetic engineering. Currently, much information has been assembled on the *A. tumefaciens* infection process. This information has been used to successfully manipulate the T-DNA transfer system both to enhance its efficiency and to broaden the range of transformable plants and other organisms. Furthermore, it must be noted that this information has also often established a conceptual framework for initiating or extending the characterization of other pathogenic and symbiotic relationships. The discovery that secreted chemical signals initiate a complex dialogue between *A. tumefaciens* and plant cells as well as other *A. tumefaciens* cells has stimulated a global effort to identify extracellular signals and characterize the cognate signal transduction systems in many bacterial systems. The discovery of T-DNA transport provided a mechanistic explanation for how horizontal gene transfer impacts the evolution of genomes of higher organisms. This discovery also established a precedent for interkingdom transport of virulence factors by bacterial pathogens. Indeed, in only the past 6 years, studies have revealed that numerous pathogens employ interkingdom transport to deliver a wide array of effector proteins to plant and animal hosts. These so-called type III transport systems, like the *A. tumefaciens* T-complex transporter and related type IV transporters, deliver substrates via a process dependent on cell-to-cell contact and, in some cases, elaboration of an extracellular filament or pilus. It is clear that, in the future, studies of all the various aspects of the *A. tumefaciens* infection process will continue to spawn new applications for this novel DNA transfer system and yield new insights about the evolution and function of pathogenic mechanisms that are broadly distributed in nature.

ACKNOWLEDGMENTS

I thank members of my laboratory for helpful and stimulating discussions. Studies in this laboratory are funded by the National Institutes of Health.

BIBLIOGRAPHY

Binns, A. N., and Howitz, V. R. (1994). The genetic and chemical basis of recognition in the *Agrobacterium:* plant interaction. *Curr. Topics Microbiol. Immunol.* **192**, 119–138.

Binns, A. N., Joerger, R. D., and Ward, J. E., Jr. (1992). *Agrobacterium* and plant cell transformation. *In* "Encyclopedia of Microbiology" (J. Lederberg, ed.), pp. 37–51. Academic Press, San Diego, CA.

Christie, P. J. (1997). The *Agrobacterium tumefaciens* T-complex transport apparatus: A paradigm for a new family of multifunctional transporters in eubacteria. *J. Bacteriol.* **179**, 3085–3094.

Christie, P. J., and Covacci, A. (1998). Bacterial type IV secretion systems: Systems utilizing components of DNA conjugation machines for export of virulence factors. *In* "Cellular Microbiology" (P. Cossart, P. Boquet, S. Normark, and R. Rappuoli, eds.). ASM Press, Washington, DC.

Christie, P. J., and Vogel, J. P. (2000). Bacterial type IV secretion: conjugation systems adapted to deliver effector molecules to host cells. *Trends Microbiol.* **8**, 354–360.

Citovsky, V., and Zambryski, P. (1993). Transport of nucleic acids through membrane channels: Snaking through small holes. *Annu. Rev. Microbiol.* **47**, 167–197.

Das, A. (1998). DNA transfer from *Agrobacterium* to plant cells in crown gall tumor disease. *Subcell. Biochem.* **29**, 343–363.

Fernandez, D., Spudich, G. M., Dang, T. A., Zhou, X.-R., Rashkova, S., and Christie, P. J. (1996). Biogenesis of the *Agrobacterium tumefaciens* T-complex transport apparatus. *In* "Biology of Plant– Microbe Interactions" (G. Stacey, B. Mullin, and P. Gresshof, eds.), pp. 121–126. ISMPMI, St. Paul, MN.

Firth, N., Ippen-Ihler, K., and Skurray, R. A. (1996). Structure and function of the F factor and mechanism of conjugation. *In "Escherichia coli* and *Salmonella typhimurium,"* 2nd ed., pp. 2377–2401. American Society for Microbiology, Washington, DC.

Fuqua, W. C., Winans, S. C., and Greenberg, E. P. (1996). Census and consensus in bacterial ecosystems: The LuxR–LuxI family of quorum-sensing transcriptional regulators. *Annu. Rev. Microbiol.* **50**, 727–751.

Lai, E. M., and Kado, C. I. (2000). The T-pilus of *Agrobacterium tumefaciens. Trends Microbiol.* **8**, 361–369.

Moriguchi, K., Maeda, Y., Satou, M., Hardayani, N. S., Kataoka, M., Tanaka, N., and Yoshida, K. (2001). The complete nucleotide sequence of a plant root-inducing (Ri) plasmid indicates its chimeric structure and evolutionary relationship between tumor-inducing (Ti) and symbiotic (Sym) plasmids in Rhizobiaceae. *J. Mol. Biol.* **307**, 771–784.

Nester, E. W., Kemner, J., Deng, W., Lee, Y.-W., Fullner, K., Liang, X., Pan, S., and Heath, J. D. (1996). *Agrobacterium:* A natural genetic engineer exploited for plant biotechnology. *In* "Biology of Plant–Microbe Interactions" (G. Stacey, B. Mullin, and P. Gresshof, eds.), pp. 111–144. ISMPMI, St. Paul, MN.

Ream, W. (1998). Import of *Agrobacterium tumefaciens* virulence proteins and transferred DNA into plant cell nuclei. *Subcell. Biochem.* **29**, 365–384.

Sheng, J., and Citovsky, V. (1996). *Agrobacterium*–plant cell DNA transport: Have virulence proteins, will travel. *Plant Cell* **8**, 1699–1710.

Spudich, G. M., Dang, T. A. T., Fernandez, D., Zhou, X.-R., and Christie, P. J. (1996). Organization and assembly of the *Agrobacterium tumefaciens* T-complex transport apparatus. *In* "Crown Gall: Advances in Understanding Interkingdom Gene Transfer" (W. Ream, and S. Gelvin, eds.), pp. 75–98. APS Press, St. Paul, MN.

Winans, S. C., Burns, D. L., and Christie, P. J. (1996). Adaptation of a conjugal transfer system for the export of pathogenic macromolecules. *Trends Microbiol.* **4**, 1616–1622.

Zupan, J. R., and Zambryski, P. (1995). Transfer of T-DNA from *Agrobacterium* to the plant cell. *Plant Physiol.* **107**, 1041–1047.

WEBSITES

An overview of crow gall and its control
http://helios.bto.ed.ac.uk/bto/microbes/crown.htm

Website for Comprehensive Microbial Resource of The Institute for Genomic Research. Links to many other microbial genomic sites
http://www.tigr.org/tigr-scripts/CMR2/CMRHomePage.spl

Website of the University of Washington Crown Hall Group
http://depts.washington.edu/agro/

3

Antibiotic resistance in bacteria

Julian Davies
University of British Columbia

Vera Webb
Lookfar Solutions Inc., British Columbia

We must swim with the microbes and study their survival and adaptation to new habitats.

Richard M. Krause (1994).

GLOSSARY

Broad-spectrum antibiotic A drug that affects a wide range of bacteria.

Cell wall, bacterial A rigid outer layer of bacterial cells containing peptidoglycan.

Integron A mobile genetic element that serves as the site of integration for DNA. It encodes a site-specific recombinase called integrase.

Multidrug Efflux Pumps Mechanisms for exporting a number of compounds across the cell membrane.

Peptidoglycan or Murein A long chain of disaccharides consisting of N-acetylglucosamine and muramic acid with short peptides that are often cross linked to give a mesh-like two-dimensional structure. Its synthetic machinery is the target for β-lactam antibiotics.

Protein synthesis machinery A large complex of proteins and RNAs, including ribosome, transfer RNAs (tRNA), messenger RNA (mRNA), proteins, and small molecules.

Resistance determinants (or R determinants) Genes that confer antibiotic resistance.

Target for antibiotics Ideally, a constituent present in the target cell (e.g., bacterial pathogen, cancer cell) and not present in the host.

I. INTRODUCTION

The development of antibiotic resistance can be viewed as a global problem in microbial genetic ecology. It is a very complex problem to contemplate, let alone solve, due to the geographic scale, the variety of environmental factors, and the enormous number and diversity of microbial participants. In addition, the situation can only be viewed retrospectively, and what has been done was uncontrolled and largely unrecorded. Simply put, since the introduction of antibiotics for the treatment of infectious diseases in the late 1940s, human and animal microbial ecology has been drastically disturbed. The response of microbes to the threat of extinction has been to find genetic and biochemical evolutionary routes that led to the development of resistance to every antimicrobial agent used. The result is a large pool of resistance determinants in the environment. The origins, evolution, and dissemination of these resistance genes is the subject of this review.

II. MECHANISMS OF RESISTANCE: BIOCHEMISTRY AND GENETICS

The use of antibiotics should have created a catastrophic situation for microbial populations; however, their genetic flexibility allowed bacteria to survive (and even thrive) in hostile environments. The alternatives for survival for threatened microbial populations were either mutation of target sites or acquisition of novel

The Desk Encyclopedia of Microbiology
ISBN: 0-12-621361-5

biochemical functions (also known as resistance determinants or R determinants).[1] Table 3.1 lists antibiotic resistance mechanisms that are transferable among bacteria. Mutation and the acquisition of R determinants are not mutually exclusive resistance strategies. Under the selective pressure of the antibiotic, mutation can lead to "protein engineering" of the acquired resistance determinant which may expand its substrate range to include semisynthetic molecules designed to be refractory to the wild-type enzyme (see Section II.A.2.a).

Theoretically, the ideal target for a chemotherapeutic agent is a constituent which is present in the target cell (e.g. bacterial pathogen, cancer cell) and not present in the host cell. The first antibiotics to be employed generally, the penicillins, targeted the synthesis of peptidoglycan, a component unique to the bacterial cell wall. Antibiotics have been found that inhibit the synthesis or interfere with the function of essentially all cellular macromolecules. Table 3.2 lists some of the common antimicrobial drugs used clinically and their targets within the bacterial cell. Our discussion of the mechanisms of resistance will be organized according to the targets of antibiotic activity.

TABLE 3.1 Transferable antibiotic resistance mechanisms in bacteria

Mechanism	Antibiotic
Reduced uptake into cell	Chloramphenicol
Active efflux from cell	Tetracyclines
Modification of target to eliminate or reduce binding of antibiotic	β-Lactams
	Erythromycin
	Lincomycin
	Mupirocin
Inactivation of antibiotic by enzymatic modification	
Hydrolysis	β-Lactams
	Erythromycin
Derivatization	Aminoglycosides
	Chloramphenicol
	Lincomycin
Sequestration of antibiotic by protein binding	β-Lactams
	Fusidic acid
Metabolic bypass of inhibited reaction	Trimethoprim
	Sulfonamides
Binding of specific immunity protein	Bleomycin
Overproduction of antibiotic target (titration)	Trimethoprim
	Sulfonamides

TABLE 3.2 Antimicrobial drugs: mechanisms of action

Target	Drug
Protein synthesis	
30 S subunit	Tetracyclines
	Aminoglycosides: streptomycin, amikacin, apramycin, gentamicin, kanamycin, tobramycin, netilmicin, isepamicin
50 S subunit	Chloramphenicol
	Fusidic acid
	Macrolides: erythromycin, streptogramin B
	Lincosamides
tRNA synthetase	Mupirocin
Cell wall synthesis	
Penicillin binding proteins	β-Lactams
	Penicillins: ampicillin, methicillin, oxacillin
	Cephalosporins: cefoxitin, cefotaxime
	Carbapenems: imipenen
	β-Lactamase inhibitors: clavulanic acid, sulbactam
D-Ala-D-Ala binding	Cyclic glycopeptides: vancomycin, teicoplanin, avoparcin
Muramic acid biosynthesis	Fosfomycin
Nucleic acid synthesis	
DNA	Quinolones and fluoroquinolones: ciprofloxicin, sparfloxacin
RNA	Rifampicin
Folic acid metabolism	Trimethoprim
	Sulfonamides

A. Targets and specific mechanisms

1. *Protein synthesis*

Protein synthesis involves a number of components: the ribosbme, transfer RNA (tRNA), messenger RNA (mRNA), numerous ancillary proteins, and other small molecules. When protein synthesis is inhibited by the action of an antibiotic, the ribosome is usually the target. The bacterial ribosome is composed of two riboprotein subunits. The small, 30 S subunit consists of approximately 21 ribosomal proteins (rprotein) and the 16 S ribosomal RNA (rRNA) molecule (about 1500 nucleotides), whereas the large, 50 S subunit contains approximately 34 rproteins and the 23 S and 5 S rRNA molecules (about 3000 and 120 nucleotides, respectively). The complexity of the ribosome structure and the redundancy of many of the genes encoding ribosomal components in most bacterial genera makes resistance due to point mutation an unlikely event. Generally, resistance to antibiotics that inhibit protein synthesis is mediated by R determinants.

a. Aminoglycosides

Aminoglycosides, which are broad-spectrum antibiotics, are composed of three or more aminocyclitol units.

[1]A note concerning terminology: "determinant" refers to the genetic element which encodes a "mechanism" or biochemical activity which confers resistance.

They bind to the 30 S subunit and prevent the transition from the initiated complex to the elongation complex; they also interfere with the decoding process. As noted above, target site mutations of ribosomal components resulting in antibiotic resistance are rare; however, they do occur. For example, in the slow-growing *Mycobacterium tuberculosis*, mutants resistant to streptomycin appear more frequently than they do in *Escherichia coli*. Mutations leading to resistance result from an altered S12 rprotein or 16 S rRNA such that the ribosome has reduced affinity for the antibiotic. Most fast-growing bacteria have multiple copies of the rRNA genes, and because resistance is genetically recessive to antibiotic sensitivity, only rare mutations in the gene for protein S12 are isolated under normal situations. However, because the slow-growing mycobacteria possess only single copies of the rRNA genes, streptomycin resistance can arise by mutational alteration of either 16 S rRNA or ribosomal protein S12. Both types of mutations have been identified in *M. tuberculosis* (Finken *et al.*, 1993).

The introduction and therapeutic use of a series of naturally occurring and semisynthetic aminoglycosides over a 20-year period (1968–1988) led to the appearance of multiresistant strains resulting from selection and dissemination of a variety of aminoglycoside resistance determinants. For example, in 1994 the Aminoglycoside Resistance Study Group examined of the occurrence of aminoglycoside-resistance mechanisms in almost 2000 aminoglycoside-resistant *Pseudomonas* isolates from seven different geographic regions. In their study 37% of the isolates overall had at least two different mechanisms of resistance (Aminoglycoside Resistance Study Group, 1994).

A dozen different types of modifications are known to be responsible for resistance to the aminoglycosides. When one considers that each of these enzymes has a number of isozymic forms, there are at least 30 different genes implicated in bacterial resistance to this class of antibiotics. Different proteins in the same functional class may show as little as 44% amino acid similarity. The phylogenic relationships between different aminoglycoside-modifying enzymes were the subject of an excellent review by Shaw and co-workers, who have collated the nucleotide and protein sequences of the known aminoglycoside acetyltransferases, phosphotransferases, and adenylyltransferases responsible for resistance in both pathogenic bacteria and antibiotic-producing strains (Shaw *et al.*, 1993).

Rarely do a few point mutations in the aminoglycoside resistance gene arise sufficient to generate a modified enzyme with an altered substrate range that would lead to a significant change in the antibiotic resistance spectrum (Rather *et al.*, 1992; Kocabiyik and Perlin, 1992); in other words, these genes do not appear to undergo facile mutational changes that generate enzymes with altered substrate activity. *In vitro* mutagenesis studies have failed to generate extended-spectrum resistance to this class of antibiotics, and, so far, such changes have not been identified in clinical isolates, in contrast to the situation with the β-lactamases (see Section II.B.2). The bacterial response to the introduction of a new aminoglycoside antibiotic is to acquire a different resistance gene. For example, when the use of kanamycin was superseded by a new but structurally related aminoglycoside (gentamicin), a previously unknown class of antibiotic-inactivating enzymes was detected in the gentamicin-resistant strains that appeared in hospitals.

Many gram-positive pathogens possess an unusual bifunctional aminoglycoside-modifying enzyme, the only reported instance of fused resistance genes. The enzyme encodes acetyl- and phosphotransferase activities based on protein domains acquired by the fusion of two independent resistance genes (Ferretti *et al.*, 1986; Rouch *et al.*, 1987). The hybrid gene is widely distributed among hospital isolates of staphylococci and enterococci and can be assumed to have evolved as a fortuitous gene fusion during the process of insertion of the two resistance genes into the "cloning" site of a transposon. The therapeutic use of aminoglycoside antibiotics has decreased in recent years, largely because of the introduction of the less toxic broad-spectrum β-lactams. As a result, there has been a relatively limited effort to seek specific inhibitors of the aminoglycoside-modifying enzymes, which might have offered therapeutic potential to extend the effective range of this class of antibiotics. Although some active inhibitors have been identified, none of them has been deemed fit for introduction into clinical practice. The situation may change as a result of structural studies by Puglisi and co-workers (Fourmy *et al.*, 1996). Employing three-dimensional (3-D) nuclear magnetic resonance techniques, they were able to model the aminoglycoside paromomycin into its binding site (the decoding site) in a 16 S RNA fragment. The logical extension of this important information, which identifies the key binding functions on the antibiotic, would be to design aminoglycoside molecules that: (a) could bind to a resistant ribosome modified by base substitution or methylation; or (b) might be refractory to modification by one or more of the aminoglycoside-modifying enzymes yet retain high affinity to the rRNA receptor site. This aminoglycoside/rRNA binding is the first to be characterized at this level of resolution and may permit the design and production of not only aminoglycoside analogs refractory to modification but also those which have reduced affinity for mammalian ribosomes; the latter is an important component of aminoglycoside toxicity.

b. Tetracycline

Tetracyclines are perhaps the prime example of the development of antibiotic resistance; introduced in the late 1940s, they were the first group of broad-spectrum antibiotics (penicillin use was limited mostly to infections caused by gram-positive bacteria). Tetracyclines act by blocking the binding of aminoacyl-tRNAs to the A-site of the ribosome. They are active against most types of bacteria including gram-positives, gram-negatives, mycoplasmas, chlamydiae, and rickettsiae. This broad spectrum of activity and their relative safety, and low cost, have made tetracyclines the second most widely used group of antimicrobial agents after the β-lactams. It is probable that larger quantities of the tetracyclines have been produced and used than any other antibiotic. From a point of view of antibiotic resistance, almost everything that could have happened to the tetracyclines has done so! In 1953, just 6 years after their discovery, the first tetracycline-resistant *Shigella dysenteriae* were isolated in Japan. They were the earliest of the antibiotic substances to be used as feed additives in agriculture; enormous amounts have been dispensed in the last 40 years. In addition, because of their spectrum of activity, the tetracyclines became widely dispensed in both hospitals and the community. In many countries tetracycline has been virtually an over-the-counter drug for many years. As a result resistance to the tetracyclines is widely disseminated in many forms (see Tables 3.3 and 3.4). This class of antibiotics has become a paradigm for antibiotic resistance studies, and a number of useful reviews of tetracycline resistance have appeared (Hillen and Berens, 1994; Roberts, 1994).

Because tetracyclines are extensively used, one might expect that resistance is also widespread (Table 3.3). At least 16 different tetracycline-resistance (Tet) determinants from "target organisms" and three different oxytetracycline-resistance (Otr) determinants from the producing organisms, *Streptomyces* species, have been described and characterized (Table 3.4). The majority of these determinants (13), are plasmid encoded. For example, the genes encoding TetA–E are found on plasmids in gram-negative bacteria, while those for TetK and TetL are found only in gram-positive microbes. The TetM determinant is generally found on the chromosome of both gram-negative and gram-positive bacteria, while TetO in gram-negative and TetK genes in gram-positive organisms have been detected in either location.

Three different mechanisms for tetracycline resistance have been described: (a) energy-dependent efflux of tetracycline from the cell; (b) ribosome protection (summarized in Table 3.4); and (c) enzymatic inactivation of tetracycline; only the first two mechanisms have been shown to have clinical significance. The majority of the Tet determinants code for a membrane-bound efflux protein of about 46 kilodaltons (kDa). Nucleotide sequence analysis of the genes encoding these proteins indicate that they are members of the so-called major facilitator (MF) family of efflux proteins (Sheridan and Chopra, 1991; Marger and Saier, 1993) (see Section II.B.2 and Table 3.5). The

TABLE 3.3 Distribution of tetracycline-resistance determinants among various bacterial genera

Gram-negatives	Tet determinant	Gram-positives	Tet determinant
Actinobacillus	TetB	*Actinomyces*	TetL
Aeromonas	TetA, B, D, E	*Aerococcus*	TetM, O
Bacteroides	TetM, Q, X	*Bacillus*	TetK, L
Campylobacter	TetO	*Clostridium*	TetK, L, M, P
Citrobacter	TetA, B, C, D	*Corynebacterium*	TetM
Edwardsiella	TetA, D	*Enterococcus*	TetK, L, M, O
Eikenella	TetM	*Eubacterium*	TetK, M
Enterobacter	TetB, C, D	*Gardnerella*	TetM
Escherichia	TetA, B, C, D, E	*Gemella*	TetM
Fusobacterium	TetM	*Lactobacillus*	TetO
Haemophilus	TetB, M	*Listeria*	TetK, L, M, S
Kingella	TetM	*Mobiluncus*	TetO
Klebsiella	TetA, D	*Mycobacterium*	TetK, L, OtrA, B
Moraxella	TetB	*Mycoplasma*	TetM
Neisseria	TetM	*Peptostreptococcus*	TetK, L, M, O
Pasteurella	TetB, D, H	*Staphylococcus*	TetK, L, M, O
Plesiomonas	TetA, B, D	*Streptococcus*	TetK, L, M, O
Prevotella	TetQ	*Streptomyces*	TetK, L, OtrA, B, C
Proteus	TetA, B, C	*Ureaplasma*	TetM
Pseudomonas	TetA, C		
Salmonella	TetA, B, C, D, E		
Serratia	TetA, B, C		
Shigella	TetA, B, C, D		
Veillonella	TetM		
Yersinia	TetB		
Vibrio	TetA, B, C, D, E, G		

Based on Roberts, 1994.

TABLE 3.4 Tetracycline-resistance determinants: location and mechanism of resistance

Mechanism	Location
Efflux	
TetA, C, D, E	Plasmid
TetB	Primarily plasmid
TetG, H	Plasmid
TetK	Chromosome and plasmid
TetL	Primarily chromosome
TetA(P)	Plasmid
OtrB	Chromosome
Ribosomal	
TetM	Primarily chromosome
TetO	Chromosome and plasmid
TetS	Plasmid
TetQ	Chromosome
TetB(P)	Plasmid
OtrA	Chromosome

TABLE 3.5 Multidrug efflux pumps[a]

Protein	Organism	Substrate
MFS		
QacA	*Staphylococcus aureus*	Mono- and divalent organic cations
NorA	*S. aureus*	Acriflavin, cetyltrimethyl-ammonium bromide, fluoroquinolones, chloramphenicol, rhodamine 6G
EmrB	*Escherichia coli*	CCCP, nalidixic acid, organomercurials, tetrachlorosalicylanilide, thiolactomycin
Bcr	*E. coli*	Bicyclomycin, sulfathiazole
Blt	*Bacillus subtilis*	Similar range as NorA
Bmr	*B. subtilis*	Similar range as NorA
SMR		
Smr	*S. aureus*	Monovalent cations, e.g., cetyltrimethylammonium bromide, crystal violet, ethidium bromide
EmrE	*E. coli*	Monovalent cations, e.g., cetyltrimethylammonium bromide, crystal violet, ethidium bromide, methyl viologen, tetracycline
QacE	*Klebsiella pneumonia*	Similar range to Smr
RND		
AcrAB TolC	*E. coli*	Acriflavin, crystal violet, detergents, decanoate, ethidium bromide, erythromycin, tetracycline, chloramphenicol, β-lactams, nalidixic acid
AcrEF	*E. coli*	Acriflavin, actinomycin D, vancomycin
MexAB OprM	*Pseudomonas aeruginosa*	Chloramphenicol, β-lactams, fluoroquinolones, tetracycline

[a]Summarized from Paulsen *et al.* (1996).

efflux proteins exchange a proton for a tetracycline–cation complex against a concentration gradient. The other major mechanism for tetracycline resistance is target protection. Ribosome protection is conferred by a 72.5-kDa cytoplasmic protein. Sequence analysis has shown regions with a high degree of homology to elongation factors Tu and G (Taylor and Chau, 1996; Sanchez-Pescador *et al.*, 1988).

Every cloud has its silver lining, and the analysis of the *tetA* operon and its regulation by a modulation of a repressor–operator interaction has provided an exquisitely controllable gene expression system for eukaryotic cells. A number of genes in a variety of cell lines have been placed under the control of the tetracycline-regulated *tetA* promoter with the result that transcription can be activated on the addition of the drug.

c. *Chloramphenicol*

Chloramphenicol acts on the 50 S ribosomal subunit to inhibit the peptidyl transferase reaction in protein synthesis. It is an effective broad-spectrum antibiotic, is inexpensive to produce, and is employed extensively in the Third World for the treatment of a variety of gram-negative pathogens (*Salmonella, Vibrio,* and *Rickettsia*). Liberal over-the-counter availability in many countries ensures strong selection pressure for the maintenance of chloramphenicol resistance. The number of indications for which chloramphenicol is the drug of choice has declined rapidly because of chronic toxicity, namely, depression of bone marrow function leading to blood disorders such as aplastic anemia. Thus, though chloramphenicol resistance determinants are widespread, their presence is not of major consequence from a therapeutic standpoint in Europe and North America.

Chloramphenicol acetyltransferases (CATs) are widely distributed among bacterial pathogens of all genera. This group of enzymes has been analyzed in great detail by Shaw and collaborators (Shaw, 1984; Murray and Shaw, 1997); at least a dozen breeds of CAT genes encoding similar but not identical acetyltransferases have been identified (Bannam and Rood, 1991; Parent and Roy, 1992). As with the aminoglycoside-modifying enzymes, the CAT genes are (presumably) of independent derivation, because they cannot be linked by a small number of point mutations to a single ancestral gene. Potential origins for the CAT family have yet to be clearly identified. The type I CAT (encoded by transposon Tn*9*) has two activities: in addition to catalyzing the acylation of chloramphenicol, the protein forms a tight stochiometric complex with the steroidal antibiotic fusidic acid (Bennett and Chopra, 1993); thus sequestered, the latter antibiotic is ineffective. This is the only plasmid-determined resistance to fusidic acid that has been characterized. Notwithstanding their different structures, the two antibiotics bind competitively to the enzyme; interestingly, both chloramphenicol and fusidic acid exert their antibiotic action through inhibition of bacteria protein synthesis (fusidic acid blocks the translocation of the ribosome along the mRNA). Hence, the type I CAT is bifunctional, determining resistance to two structurally related antibiotics by distinct mechanisms.

Given our understanding of the basic biochemistry of the drug and knowledge of the mechanism of chloramphenicol resistance (Day and Shaw, 1992), it is surprising that more effort has not been put into rational drug design of chloramphenicol analogs of lower toxicity that would be active against resistant strains. Shaw's group has studied the CAT protein extensively; the 3-D structure of the protein is known, and site-directed mutagenesis has been used to establish the kinetics of the reaction mechanism (Murray and Shaw, 1997). Fluorinated analogs of chloramphenicol with reduced substrate activity for CAT have been produced but have not been afforded extensive clinical study (Cannon *et al.*, 1990). The antibiotic has an excellent inhibitory spectrum: might it be possible to use fusidic acid analogs to inhibit chloramphenicol acylation?

Although the CAT mechanism is the most common form of chloramphenicol resistance, a second type causing the active efflux of chloramphenicol from *Pseudomonas* cells has been described (Bissonnette *et al.*, 1991). The *cml* determinant is encoded on integrons (see Section IV.A) and, like the tetracycline efflux pump, is a member of the MF family of efflux systems (see Section II.B.2, and Table 3.5). The presence of the cloned *cml* gene in *E. coli* leads to, in addition to active pumping of the drug from the cell, a reduction in outer membrane permeability to chloramphenicol by repressing the synthesis of a major porin protein (Bissonnette *et al.*, 1991). This effect has also been reported for the homolog of the *cml* gene in*Haemophilus influenzae*, a major causative agent of meningitis.

d. Macrolides, lincosamides, and streptogramins

The macrolide–lincosamide–streptogramin (MLS) group of antibiotics have been used principally for the treatment of infections caused by gram-positive bacteria. The macrolides, especially erythromycin and its derivatives, have been used extensively and may be employed for the treatment of methicillin-resistant *Staphylococcus aureus* (MRSA), although the multiple drug resistance of the latter often includes the MLS class. Extensive studies on the mechanism of action of erythromycin identify the peptidyltransferase center as the target of the drug although there are clearly some subtleties in mechanism that remain to be resolved.

The principal mechanism of resistance to macrolide antibiotics involves methylation of the 23 S rRNA of the host giving the erm^R phenotype. In clinical isolates, enzymatic modification of rRNA, rendering the ribosome refractory to inhibition, is the most prevalent mechanism and, worldwide, compromises the use of this class of antibiotic in the treatment of gram-positive infections (Leclercq and Courvalin, 1991). Mutation of rRNA has also been shown to be important in some clinical situations: nucleotide sequence comparisons of clarithromycin-resistant clinical isolates of *Helicobacter pylori* revealed that all resistant isolates had a single base pair mutation in the 23 S rRNA (Debets-Ossenkopp *et al.*, 1996). In addition, a number of different mechanisms for the covalent modification of the MLS group have been described. For example, O-phosphorylation of erythromycin has been identified in a number of bacterial isolates (O'Hara *et al.*, 1989), and hydrolytic cleavage of the lactone ring of this class of antibiotics has also been described. The lincosamides (lincomycin and clindamycin) have been shown to be inactivated by enzymatic O-nucleotidylation in gram-positive bacteria. Macrolides are also inactivated by esterases and acetyltransferases. For macrolides and lincosamides and the related streptogramins (the MLS group) (Arthur *et al.*, 1987; Brisson-Noël *et al.*, 1988), the latter forms of antibiotic inactivation seem (for the moment) to be a relatively minor mechanism of resistance.

The macrolide antibiotics, especially the erythromycin group (14-membered lactones), contain a number of semisynthetic derivatives that have improved pharmacological characteristics. This includes activity against some resistant strains: however, no derivative has been found with effective potency against erm^R strains. The 23 S rRNA methyltransferases from pathogenic gram-positive cocci remain the most significant problem. To our knowledge, no useful inhibitor of these enzymes has been identified; with the availability of ample quantities of the purified enzymes it would not be surprising if rationally designed, specific inhibitors of some of the *erm* methyltransferases may eventually be developed.

The control of expression of the *erm* methyltransferases has been studied extensively by Weisblum and by Dubnau and their colleagues (Weisblum, 1995; Monod *et al.*, 1987). The majority of the resistant strains possess inducible resistance which is due to a novel posttranslational process in which the ribosome stalls on a 5′-leader sequence in the presence of low concentrations of antibiotic. The biochemistry of this process has been analyzed extensively. Point mutations or deletions that disrupt the secondary structure of the leader protein sequence generate constitutive expression, and such mutants have been identified clinically. These strains are resistant to the majority of MLS antibiotics. The regulation of antibiotic resistance gene expression takes many forms (see tetracycline and vancomycin for other examples); the leader-control process seen with the *erm* methylases is reminiscent of the control of amino acid biosynthesis by attenuation, which has been extensively studied. One cannot help but marvel at the simplicity and "cheapness" of this form of control of gene expression—all that is required is a few extra bases flanking the 5′ end of the gene with no requirement of additional regulatory genes.

The streptogramins (especially virginiamycins) contain two components, a macrocyclic lactone and a depsipeptide which have synergistic activity (one such combination has been named Synercid®). These compounds have long been used as animal feed additives with the result that bacteria resistant to the MLS class of antibiotics are commonly found in the flora of farm animals. This nontherapeutic application is likely to compromise the newer MLS antibiotics being developed for human therapy. For example, derivatives of the pristinamycins (quinupristin and dalfopristin) will shortly be recommended for the treatment of vancomycin-resistant enterococci, against which they have potent activity. Regrettably, due to animal applications of MLS antibiotics a significant gene pool of resistance determinants is already widely disseminated.

e. Mupirocin

Mupirocin is an example of how quickly resistance can develop to a new antibiotic. Introduced in 1985, it has been used solely as a topical treatment for staphylococcal skin infections and as a nasal spray against commensal MRSA. Mupirocin, also known as pseudomonic acid A, is produced by the gram-negative bacterium *Pseudomonas fluorescens*. It acts by inhibiting isoleucyl-tRNA synthetase, which results in the depletion of charged tRNA Ile, amino acid starvation, and ultimately the stringent response. The first reports of mupirocin resistance (MuR) appeared in 1987; while not yet a widespread problem, the Mu^R phenotype in coagulase-negative staphylococci and MRSA among others is emerging in many hospitals all over the world (Cookson, 1995; Zakrzewska-Bode *et al.*, 1995; Udo *et al.*, 1994). Resistance is due to the plasmid-encoded *mup*A gene, a mupirocin-resistant isoleucyl-rRNA synthetase (Noble *et al.*, 1988). Transfer has been demonstrated by filter mating and probably accounts for the rapid spread of resistance (Rahman *et al.*, 1993). An analysis of plasmids from clinical isolates of *S. aureus* has shown that mupirocin resistance can be found on multiple resistance, high copy number plasmids (Needham *et al.*, 1994).

2. *Cell wall synthesis*

The synthesis and integrity of the bacterial cell wall have been the focus of much attention as targets for antimicrobial agents. This is principally because this structure and its biosynthesis is unique to bacteria and also because inhibitors of cell well synthesis are usually bacteriocidal.

a. β-Lactams

In the years following the introduction of the β-lactam antibiotics (penicillins and cephalosporins) for the treatment of gram-negative and gram-positive infections, there has been a constant tug-of-war between the pharmaceutical industry and the bacterial population: the one to produce a novel β-lactam effective against the current epidemic of resistant bugs in hospitals, and the other to develop resistance to the newest "wonder" drug.

The role of mutation is especially important in the evolution or expansion of resistance in the case of β-lactams. In a 1992 review Neu showed the "phylogeny" of development of β-lactam antibiotics in response to the evolution of bacterial resistance to this class of antibiotics (Neu, 1992). Of the several known mechanisms of resistance to the β-lactam antibiotics (Table 3.1), the most elusive target is hydrolytic inactivation by β-lactamases. A single base change in the gene for a β-lactamase can change the substrate specificity of the enzyme (Jacoby and Archer, 1991). Such changes occur frequently, especially in the Enterobacteriaceae, and it is frightening to realize that one single base change in a gene encoding a bacterial β-lactamase may render useless $100 million worth of pharmaceutical research effort.

The cycle of natural protein engineering in response to changing antibiotic-selection pressure has been demonstrated especially for the TEM β-lactamase (penicillinase and cephalosporinase) genes. The parental genes appear to originate from a variety of different (and unknown) sources (Couture *et al.*, 1992). The β-lactamase families differ by a substantial number of amino acids, as is the case for other antibiotic resistance genes. Sequential expansion of their substrate range to accommodate newly introduced β-lactam antibiotics is a special case and occurs by a series of point mutations at different sites within the gene that change the functional interactions between the enzyme and its β-lactam substrate. More than 30 of the so-called extended-spectrum β-lactamases have been identified (and more will come). The fact that the β-lactamase genes so readily undergo mutational alterations in substrate recognition could have several explanations, one being that the β-lactamases, like the related proteases, have a single active site that does not require interaction with any cofactors. Other antibiotic-modifying enzymes often have two active binding sites. The pharmacokinetic characteristics of the different classes of antibiotics (e.g. dose regimen, active concentration, and route of excretion) also may favor the pathway of mutational alteration in the development of resistance. One aspect of the mutational variations of the β-lactamase genes might involve the presence of mutator genes on the R plasmids in the bacterial hosts (LeClerc *et al.*, 1996). These genes increase mutation frequency several-fold and could explain the facile evolution of the TEM-based enzymes. This characteristic of bacterial pathogens has received comparatively little study.

As one approach to counteracting the destructive activity of β-lactamases, a series of effective inhibitors of these enzymes has been employed. The inhibitors are structural analogs of β-lactams that are, in most cases, dead-end irreversible inhibitors of the enzyme. Several have been used in combination with a β-lactam antibiotic for the treatment of infections from resistant microbes, for example, the successful combination of amoxicillin (antibiotic) and clavulanic acid (inhibitor). However, the wily microbes are gaining the upper hand once again by producing mutant β-lactamases that not only are capable of hydrolyzing the antibiotic but concomitantly become refractory to inhibition (Blazquez *et al.*, 1993).

In addition to active site mutation, other changes in β-lactamase genes have evolved in response to continued β-lactam use. In some cases, increased resistance results from increased expression of the gene through an upregulating promoter mutation (Chen and Clowes, 1987; Mabilat *et al.*, 1990); alternatively, chromosomal β-lactamase genes can be overexpressed in highly resistant strains as a result of other changes in transcriptional regulation (Honoré *et al.*, 1986).

Drug inactivation is not the only mechanism of resistance to the β-lactams. In fact, mutations that alter access to the target (penicillin binding proteins, pbp) of the drug through porin channels have been widely reported (Nikaido, 1994). Methicillin resistance in *S. aureus* (MRSA) is due to an unusual genetic complex which replaces the normal pbp2 with the penicillin refractory pbp2A. The origin of this resistance determinant is unknown, but the consequences of the clonal distribution of MRSA is well documented. It would appear that most MRSA are close relatives of a small number of parental derivatives that have been disseminated by international travel. Alterations of other pbp's in different bacterial pathogens have been responsible for widespread resistance to β-lactam antibiotics; in the case of *Streptococcus pneumoniae* and *Neisseria gonorrhoeae* this has occurred by interspecific recombination leading to the formation of mosaic genes that produce pbps with markedly reduced affinity for the drugs.

b. *Glycopeptides*

The glycopeptide antibiotics, vancomycin and teichoplanin, were first discovered in the 1950s but have only come into prominence since the late 1980s, being the only available class of antibiotic effective for the treatment of MRSA and methicillin-resistant enterococci (MRE). However, numerous outbreaks of vancomycin-resistant enterococci (VRE) have been reported in hospitals around the world (VRE are essentially untreatable by any approved antibiotic), and there is great concern that vancomycin-resistance determinants will be transferred to pathogenic staphylococci; the resulting methicillin, vancomycin resistant *S. aureus* (MVRSA) will be the "Superbug," the "Andromeda" strain that infectious disease experts fear most (at least according to the newspapers). The glycopeptides block cell wall synthesis by binding to the peptidoglycan precursor dipeptide D-alanyl-D-alanine (D-Ala-D-Ala) and preventing its incorporation into the macromolecular structure of the cell wall. The most common type of resistance is the vanA type found principally in *Enterococcus faecalis* (Walsh *et al.*, 1996; Arthur *et al.*, 1996). Resistance due to the VanA phenotype results from the substitution of the depsipeptide D-alanyl-D-lactate (D-Ala-D-Lac) for D-Ala-D-Ala residues, thereby reducing the binding of the antibiotic by eliminating a key hydrogen bond in the D-Ala-D-Ala complex. The introduction of D-Lac is encoded by the nine genes of the VanA cluster and is associated with a mechanism to prevent the formation of native D-Ala-D-Ala containing peptide in the same host; thus, the resistance is dominant. Vancomycin resistance is inducible by a two-component regulatory system; the inducers are not the antibiotics, but rather the accumulated peptidoglycan fragments produced by the initial inhibitory action of the glycopeptides. The VanA cluster is normally found on a conjugative transposon related to Tn*1546*, and it is probably responsible for the widespread dissemination of glycopeptide resistance among the enterococci. A plasmid carrying the VanA resistance determinants has been transferred from enterococci to staphylococci under laboratory conditions (Noble *et al.*, 1992), and there is apprehension that this will occur in clinical circumstances. Regrettably a glycopeptide antibiotic, avoparcin, has been employed extensively as a feed additive for chickens and pigs in certain European countries. This has led to the appearance of a high proportion glycopeptide-resistant enterococci in natural populations. This feeding practice has now been banned by the European Union (EU), but is it too late? Only time will tell.

c. *Fosfomycin*

A widely used mechanism for the detoxification of cell poisons in eukaryotes is the formation of glutathione adducts; for example, this mechanism is commonly used for herbicide detoxification in plants (Timmermann, 1988). However, in spite of the fact that many microbes generate large quantities of this important thiol, only one example of an antibiotic resistance mechanism of this type has so far been identified in bacteria, namely, that of fosfomycin. Fosfomycin, produced by a streptomycete, is an analog of phosphoenol pyruvate and interferes with bacterial cell wall synthesis by inhibiting the formation of *N*-acetylmuramic acid, a unique component of bacterial cell walls. It is employed in the treatment of sepsis, both alone and in combination with other antimicrobial agents. In gram-negative bacteria transmissible resistance is due to a plasmid-encoded glutathione *S*-transferase that catalyzes the formation of an inactive fosfomycin–glutathione adduct (Arca *et al.*, 1990). Two independent genes for fosfomycin resistance have been cloned and sequenced, one from *Serratia marcescens* (Suárez and Mendoza, 1991) and the other from *Staphylococcus aureus* (Zilhao and Courvalin, 1990); the two genes are unrelated at the sequence level. It is unlikely that the gram-positive gene encodes an enzyme involved in the production of a glutathione adduct, because *S. aureus* does not contain glutathione!

3. *DNA and RNA synthesis*

Nucleic acid metabolism has attracted much attention as a potential target for antimicrobial drugs; the strategy of hitting at the "heart" of the microbe seemed the most likely to lead to effective bactericidal agents. Unfortunately, the ubiquity of DNA and RNA and the failure to identify discriminating target differences in the biosynthetic enzymes made this search unproductive until relatively recently when the fluoroquinolones (a class of synthetic drugs with no known natural analogs) were introduced. However, resistance mechanisms were not long in appearing, and the fluoroquinolones instead of being the "superdrugs" needed, are already limited by resistance.

a. Fluoroquinolones

Nalidixic acid, the prototype quinolone antibiotic discovered in 1962, had limited use, principally for urinary tract infections caused by gram-negative bacteria. Resistance developed by mutation, and the nal^R phenotype proved to be the first useful marker for *gyrA* the gene encoding the A subunit of DNA gyrase (topoisomerase I). The development of resistance to nalidixic acid occurred solely by this type of mutation, and plasmid-determined resistance has never been reliably identified. This is not surprising, given that nalidixic acid is a purely synthetic chemical and no natural analog has been identified.

In the late 1970s, the first fluoroquinolone antimicrobials were introduced; these proved vastly superior to nalidixic acid and are among the most potent antimicrobial agents known. A number of fluoroquinolone antibiotics have now been introduced as antiinfectives and most show good, broad-spectrum activity. In laboratory studies, mutations to high-level resistance occurred at relatively low frequency, and genetic studies identified a number of different DNA replication-associated targets. As with nalidixic acid, topoisomerase I is the principal target, but other targets associated with bacterial DNA replication which give the FQ^R phenotype have been identified in different bacterial species. In clinical use, resistance to the newer fluoroquinolones has been found to develop quite rapidly as a result of one or more mutations. In addition to target mutation, active efflux of the drug is also an important mechanism of resistance; for example, the *norA* mutation identifies a multiple drug resistance (mdr) system in *S. aureus* (see Section II.B.2 and Table 3.5). Strains resistant to high levels of the drug have been identified frequently during the course of treatment especially with *P. aeruginosa* and *S. aureus* infections. Resistance is due to multiple mutations leading to increased efflux and alteration of components of the DNA synthetic apparatus. Clonal dissemination of FQ^R strains appears to be quite common in nosocomial *P. aeruginosa* infections. No plasmid-mediated resistance to fluoroquinolones has been identified to date, although possible mechanisms leading to dominant resistance genes can be envisaged.

b. Rifampicin

Rifampicin is the only inhibitor of bacterial (and mitochondrial) RNA polymerases that has ever been used in the treatment of infectious disease. Mutations in the gene encoding the β subunit of RNA polymerase (*rpoB*) give high level resistance to the drug; the study of these mutations has provided important information on the structure and function of the RNA polymerase proteins in bacteria. Rifampicin is used in the treatment of gram-positive bacteria and is effective against mycobacterial and staphylococcal infections. Because it is lipophilic and thus diffuses rapidly across the hydrophobic cell envelop of mycobacteria, rifampicin is one of the frontline drugs for the treatment of tuberculosis and leprosy. However, the appearance of resistant strains as a result of *rpoB* mutations is quite common in multiple drug-resistant strains (Cole, 1994). In addition, inactivation of rifampicin in fast growing mycobacterial strains by phosphorylation, glucosylation, and ribosylation has been reported (Dabbs *et al.*, 1995).

4. Folic acid biosynthesis

As the above discussion illustrates, the biosynthesis and function of cellular macromolecules are the principal targets for the majority of antimicrobial agents. However, interference with the activity of enzymes involved in intermediary metabolism is also an effective strategy. Although many potential antimicrobial targets have been identified and tested, to date only folic acid biosynthesis has been successfully exploited in the development of useful drugs.

Folic acid is involved in the transfer of one-carbon groups utilized in the synthesis and metabolism of amino acids such as methionine and glycine and in the nucleotide precursors adenine, guanine, and thymine. Folic acid is converted in two reduction steps to tetrahydrofolate (FH_4), which serves as the intermediate carrier of hydroxymethyl, formyl, or methyl groups in a large number of enzymatic reactions in which "one-carbon" groups are transferred from one metabolite to another or are interconverted. The synthetic antibacterial agents sulfonamides and trimethoprim inhibit specific steps in the biosynthesis of FH_4. The current state of resistance to sulfonamides and trimethoprim in major bacterial pathogens and the mechanisms of sulfonamide and trimethoprim resistance have been reviewed (Huovinen *et al.*, 1995).

a. Sulfonamides

The sulfonamides were first discovered in 1932 and introduced into clinical practice in the late 1930s. They have a wide spectrum of activity and have been used in urinary tract infections due to the Enterobacteriaceae, in respiratory tract infections due to *Streptococcus pneumoniae* and *Haemophilus influenzae*, in skin infections due to *S. aureus*, and in gastrointestinal tract infections due to *E. coli* and *Shigella* spp. The wide range of clinical indications and low production costs maintain the popularity of the sulfonamides in Third World countries. The enzyme dihydropteroate synthase (DHPS) catalyzes the formation of dihydropteric acid, the immediate precursor of dihydrofolic acid. DHPS found in bacteria and some protozoan parasites, but not in human cells, is the target of sulfonamides. These drugs are structural analogs of *p*-aminobenzoic acid, the normal substrate of DHPS, and act as competitive inhibitors for the enzyme, thus blocking folic acid biosynthesis in bacterial cells. A large number of sulfonamides have been synthesized that show wide variations in therapeutic activity! One class, the dapsones, remains an effective anti-leprosy drug.

b. Trimethoprim

Trimethoprim was first introduced in 1962, and since 1968 it has been used (often in combination with sulfonamides due to a supposed synergistic effect) for numerous clinical indications. Like the sulfonamides, trimethoprim has a wide spectrum of activity and low cost. The target is the enzyme dihydrofolate reductase (DHFR), which is essential in all living cells. Trimethoprim is a structural analog of dihydrofolic acid and acts as a competitive inhibitor of the reductase. The human DHFR is naturally resistant to trimethoprim, which is the basis for its use.

Resistance has been reported to both trimethoprim and sulfonamides since their respective introductions into clinical practice. Although the principal form of resistance is plasmid mediated, a clinical isolate of *E. coli* was described in which the chromosomal DFHR was overproduced several hundredfold, leading to very high trimethoprim resistance (minimum inhibitory concentration $>1\,g/l$). Sul^R and Tmp^R strains carry plasmid-encoded *dhps* and *dfhr* genes that may be up to 100 times less susceptible to the inhibitors. Extensive genetic and enzymatic studies, principally by Sköld and collaborators, have characterized resistance in the Enterobacteriaceae. The *sul* gene is part of the 3′ conserved region of integron structures and may have been the first resistance determinant acquired (see Section IV.A). Less information is available for resistant gram-positive pathogens. Nonetheless, the origins of Sul^R and Tmp^R genes remains a mystery.

B. Broad-spectrum resistance systems

In the first part of this section, we examined bacterial targets for antibiotic activity and mechanisms that specifically provide resistance to those antibiotics. The mechanisms included modification of the target (e.g. vancomycin resistance), modification of the antibiotic (e.g. aminoglycoside methyltransferases), overproduction of the target (e.g. trimethoprim resistance), and extrusion of the drug from the cell (e.g. TetA-type tetracycline resistance). However, even as the first antimicrobial agents were being tested, researchers noticed "intrinsic" differences in sensitivity among the target organisms to a wide range of compounds.

1. Membranes and cell surfaces

Initially, intrinsic differences in resistance to antibiotics and other chemotherapeutic agents were attributed to structures such as the gram-negative outer membrane and the mycobacterial cell surface. Intrinsic drug resistance in mycobacteria has been reviewed by

Nikaido and collaborators (Jarlier and Nikaido, 1994). They suggest that lipophilic molecules are slowed by the low fluidity of the lipid bilayer surrounding the cell wall and that hydrophobic molecules enter the mycobacterial cell slowly because the porins are inefficient and few in number. They note that although these surface features strongly contribute the high level of natural resistance in mycobacteria, other factors are also involved such as pbps with low affinity for penicillin and the presence of β-lactamases.

2. *Multidrug efflux pumps*

A cell has a number of different ways to export material across its membrane. All of them involve the expenditure of energy. The most well-characterized multiple drug resistance pumps in mammalian cells are the ABC (ATP binding cassette) transporters. In bacteria, the ABC transporters are primarily seen in the translocation of virulence factors such as hemolysin in *E. coli* and cyclolysin in *Bordetella pertussis*. A second type of active export system has been characterized in which the energy to drive the pump comes from the proton motive force (PMF) of the transmembrane electrochemical proton gradient. These multidrug efflux systems have been the subject of a number of reviews (Lewis, 1994; Nikaido, 1996; Paulsen *et al.*, 1996). There are three families of PMF multidrug efflux pumps: the major facilitator superfamily (MFS), the staphylococcal (or small) multidrug resistance (SMR) family, and the resistance/nodulation/cell division (RND) family. Table 3.5 lists examples from each of these families and the types of compounds they pump out of the cell. In addition to these multisubstrate pumps, proton motive force efflux pumps for specific antibiotic resistance, such as the TetK and TetL (MFS-type) pumps, have also been described (see Section II.A.1.b).

The Acr multidrug efflux pump found in *E. coli* is one of the most well characterized of the RND-type pumps. Expression of the Acr efflux pump is controlled by the marA protein. The multiple antibiotic resistance (mar) phenotype was first described in 1983 by Levy and co-workers when they plated *E. coli* on medium containing either tetracycline or chloramphenicol and obtained mutants that were also resistant to the other antibiotic. Further analysis showed that these mutants had additional resistances to β-lactams, puromycin, rifampicin, and nalidixic acid. The most striking aspect of this multiple drug resistance phenotype was the wide range of structurally unrelated compounds with which it was observed. Subsequent studies have found that the mar phenotype is part of a complex stress response system which includes the superoxide response locus *soxRS* (Miller and Sulavik, 1996). Rather than encoding the structural components responsible for the mar phenotype, the *mar* locus encodes a regulatory system. The marA protein is a positive regulator which controls expression of at least two loci, the *acr* locus, which encodes the genes for a multidrug efflux pump, and the *micF* locus, which encodes an antisense repressor of the outer membrane protein ompF. Expression from the *mar* locus is tightly controlled by the first gene in the operon, *marR*, which encodes a represser protein (Miller and Sulavik, 1996).

As noted above, the *mar* locus is part of a complex stress response system. Levy and co-workers (Goldman *et al.*, 1996) have reported that mutations of the marR repressor protein protected *E. coli* from rapid killing by fluoroquinolones. They hypothesize that such protection may allow cells time to develop mutations that lead to higher levels of fluoroquinolone resistance, and may thus explain the increasing frequency of occurrence of resistant clinical isolates. Such mutations have been found among clinical strains of fluoroquinolone-resistant *E. coli*, suggesting that mutations at the *mar* locus may be the first step in clinically significant fluoroquinolone resistance. The latter may have been the case for all clinically significant antibiotic resistance: a mutational event leading to a low-level increase in drug efflux, followed by the acquisition of a heterologous resistance determinant, leading to high-level antibiotic resistance.

3. *Other types of natural resistance*

Other types of natural resistance in bacteria will depend on the organism and the drug in question. For example, three species of enterococci, *Enterococcus gallinarum*, *E. casseliflavus*, and *E. flavescens*, produce peptidoglycan precursors which end in D-serine residues and are intrinsically resistant to low levels of vancomycin. Similarly, genera from the lactic acid bacteria *Lactobacillus*, *Leuconostoc*, and *Pediococcus* are resistant to high levels of glycopeptides because their cell wall precursors end with D-lactate. As noted above, mycobacteria species have pbps which have a low affinity for penicillin. In addition, some mycobacteria have been reported to have ribosomes that have a lower affinity for macrolides than the ribosomes of *S. aureus* (Jarlier and Nikaido, 1994).

III. GENE TRANSFER

All available evidence suggests that the acquisition and dissemination of antibiotic resistance genes in bacterial pathogens has occurred since the late 1940s.

The best support for this notion comes from studies of Hughes and Datta (Datta and Hughes, 1983; Hughes and Datta, 1983) who examined the "Murray Collection," a collection of (mostly) gram-negative pathogens obtained from clinical specimens in the pre-antibiotic era. None of these strains show evidence of resistance to antibiotics in current use. Given the critical role of gene exchange in bacterial evolution, it is self-evident that extensive interspecific and intergeneric gene transfer must have occurred during the golden age of antibiotics. This subject has been reviewed many times, and a variety of mechanisms of gene transfer have been invoked in the process of resistance determinant acquisition and dispersion (see Table 3.6) and characterized in laboratory studies. It can be assumed that these (and probably other processes) all occur in nature.

A. Transduction

Transduction is the exchange of bacterial genes mediated by bacteriophage, or phage. When a phage infects a cell, the phage genes direct the takeover of the host DNA and protein synthesizing machinery so that new phage particles can be made. Transducing particles are formed when plasmid DNA or fragments of host chromosomal DNA are erroneously packaged into phage particles during the replication process. Transducing particles (those carrying nonphage DNA) are included when phage are liberated from the infected cell to encounter another host and begin the next round of infection. Although there are many laboratory studies of transduction of antibiotic resistance, this mechanism has been considered less important in the dissemination of antibiotic resistance genes because phage generally have limited host ranges; they can infect only members of the same or closely related species, and the size of DNA transferred does not usually exceed 50 kb. However, phage of extraordinarily broad host specificity have been described. For example, phage PRR1 and PRD1 will infect any gram-negative bacterium containing the resistance plasmid RP1 (Olsen and Shipley, 1973; Olsen *et al.*, 1974) and could, in principle, transfer genetic information between unrelated bacterial species. Transduction has been documented in at least 60 species of bacteria found in a wide variety of environments (Kokjohn, 1989). Although the actual level of intraspecies and interspecies transduction are unknown, the potential for transductional gene exchange is likely to be universal among the eubacteria. A study of the presence of bacteriophage in aquatic environments has shown that there may be as many as 108 phage particles per milliliter (Bergh *et al.*, 1989); the authors calculate that at such concentrations one-third of the total bacterial population is subject to phage attack every 24 h. In transduction the DNA is protected from degradation within the phage particle, and Stotzky has suggested that transduction may be as important as conjugation or transformation as a mechanism of gene transfer in natural habitats (Stotzky, 1989).

B. Conjugation

Conjugation is the process in which DNA is transferred during cell-to-cell contact. It has long been considered the most important mechanism for the dissemination of antibiotic resistance genes. During an epidemic of dysentery in Japan in the late 1950s increasing numbers of *Shigella dysenteriae* strains were isolated that were resistant to up to four antibiotics simultaneously. It soon became clear that the emergence of multiply resistant strains could not be attributed to mutation. Furthermore, both sensitive and resistant *Shigella* could be isolated from a single patient, and the *Shigella* sp. and *E. coli* obtained from the same patients often exhibited the same multiple resistance patterns. These finding led to the discovery of resistance transfer factors and were also an early indication of the contribution of conjugative transfer to the natural evolution of new bacterial phenotypes.

In addition to plasmid-mediated conjugal transfer, another form of conjugation has been reported to take place in gram-positive organisms. Conjugative transposons were first reported in *S. pneumoniae* when the transfer of antibiotic resistance determinants occurred in the absence of plasmids (Shoemaker *et al.*, 1980). Salyers has suggested that conjugative transposons may be more important that conjugative plasmids in broad host-range gene transfer between some species of bacteria (Salyers, 1993). Conjugative transposons are not considered typical transposons (Scott, 1992). As of 1993, three different families of conjugative transposons had been found: (a) the Tn*916* family (originally found in streptococci but now known also to occur in gram-negative bacteria such as *Campylobacter*) (Salyers, 1993); (b) the *S. pneumoniae* family; and (c) *Bacteroides* family. Conjugative transposons range in size from 15 to 150 kb, and, in addition to other resistance genes, most encode tetracycline-resistance determinants (e.g. Tn*916* encodes the TetM determinant, and TetQ (Salyers, 1993) is found on the conjugative transposon from the *Bacteroides* group).

Does conjugation take place in the environment? The ideal site for gene transfer is the warm, wet, nutrient-abundant environment of the mammalian intestinal tract with its associated high concentration of bacteria.

TABLE 3.6 Gene transfer processes

		Required genetic determinants		Observed in					
Process	Components	Donor	Recipient[a]	Laboratory	Nature	DNA form[b]	DNA size	Host range	Comments
Conjugation	Cell/cell	Transfer genes	May require receptor	+	+	ss	≥4 Mb	Very broad interspecific, intergeneric	Can be of very high efficiency; can reduce problem of restriction in recipient
Fusion	Cell/cell	?	?	+ (rare)	–	ds	Unlimited	?	Likely to involve partners with damaged wall or membranes (protoplasts or spheroplasts)
Transduction	Bacteriophage /cell	Phage receptor	Phage receptor	+	+	ds	≤50 kb	Limited to closely replated species	Very efficient
Transformation[c]	DNA/cell	–	a.Competence determinants b.Chemical or physical changes	+	+	ss/ds	≤50 kb	Very broad	Chemically and electrically induced competence required; efficiency variable

[a]All DNA exchange processes are subject to the negative effects of restriction (nuclease action) in the recipient.
[b]All processes, in principle, may take place with transfer of intact plasmids, chromosomes, or linear fragments. The requirement for recombination in the recipient depends on the nature of transferred DNA and its properties. ss, Single stranded; ds, double stranded.
[c]Many bacterial species are genetically non-competent, but may be converted to a competent state by laboratory processes. Electrotransformation is a good example.

The resident microflora is believed to serve as a reservoir for genes encoding antibiotic resistance which could be transferred not only to other members of this diverse bacterial population, but also to transient colonizers of the intestine, such as soil or water microbes or human pathogens. Using oligonucleotide probes having DNA sequence similarity to the hypervariable regions of the TetQ determinant, Salyers and co-workers provided evidence that gene transfer between species of *Bacteroides*, one of the predominant genera of the human intestine, and *Prevotella* sp., one of the predominant genera of livestock rumen, has taken place under physiological conditions (Nikolich *et al.*, 1994).

C. Transformation

Natural transformation is a physiological process characteristic of many bacterial species in which the cell takes up and expresses exogenous DNA. Although natural transformation has been reported to occur only in a limited number of genera, these include many pathogenic taxa such as *Haemophilus, Mycobacterium, Streptococcus, Neisseria, Pseudomonas*, and *Vibrio* (Stewart, 1989). Initial studies suggested that natural transformation in some of these genera was limited to DNA from that particular species. For example, an 11-bb recognition sequence permits *Haemophilus influenzae* to take up its own DNA preferentially compared to heterologous DNAs (Kahn and Smith, 1986). Given such specificity one could ask if natural transformation is an important mechanism in the transfer of antibiotic resistance genes. Spratt and co-workers have reported the transfer of penicillin resistance between *S. pneumoniae* and *N. gonorrhoeae* by transformation. Further, Roberts reported that when the TetB determinant (which is conferred by conjugation in other gram-negative groups) is present in *Haemophilus* species and highly tetracycline-resistant *Moraxella catarrhalis*, it is disseminated by transformation (Roberts *et al.*, 1991).

D. General considerations

There is a world of difference between laboratory and environmental studies. While the isolation of pure cultures is an important component of bacterial strain identification, the use of purified bacterial species in gene transfer studies does little to identify the process and probably bears no relationship to the processes of genetic exchange that take place in the complex microbial populations of the gastrointestinal tract (for example). When antibiotic-resistant bacteria are isolated from diseased tissue and identified as the responsible pathogen, this is the identification of the final product of a complex and poorly understood environmental system. Although gene exchange may under normal circumstances be rare in stable microbial microcosms, the intense selective pressure of antibiotic usage is likely to have provoked cascades of antibiotic resistance gene transfer between unrelated microbes. These transfers must involve different biochemical mechanisms during which efficiency is not a critical factor since the survival and multiplication of a small number of resistant progeny suffices to create a clinically problematic situation.

It should be apparent that a great deal of additional study using modern molecular and amplification methods with complex microbial communities is necessary before the parameters of natural antibiotic resistance gene transfer can be defined properly.

IV. EVOLUTION OF RESISTANCE DETERMINANT PLASMIDS

A. The integron model

Studies by Hughes and Datta of plasmids they isolated from the Murray collection (Hughes and Datta, 1983; Datta and Hughes, 1983) suggest that the appearance of resistance genes is a recent event, that is, the multiresistance plasmids found in pathogens must have been created since the 1940s. What really takes place when a new antimicrobial agent is introduced and plasmid-determined resistance develops within a few years? The most significant component of the process of antibiotic resistance flux in the microbial population is gene pickup, which has now been emulated in the laboratory. Largely due to the studies of Hall and co-workers (Stokes and Hall, 1989; Collis *et al.*, 1993; Recchia and Hall, 1995), we have a good idea of the way in which transposable elements carrying multiple antibiotic resistance genes might be formed. From their studies of the organization of transposable elements, these researchers have identified a key structural constituent of one class of transposon that they named an "integron." The integron is a mobile DNA element with a specific structure consisting of two conserved segments flanking a central region in which "cassettes" that encode functions such antibiotic resistance can be inserted. The 5′ conserved region encodes a site-specific recombinase (integrase) and a strong promoter or promoters that ensure expression of the integrated cassettes. The integrase is responsible for the insertion of antibiotic resistance gene cassettes downstream of the promoter; ribosome binding sites are conveniently provided. More than one promoter sequence exists on the element (Lévesque *et al.*, 1994); transcription initiation is very efficient and functions in

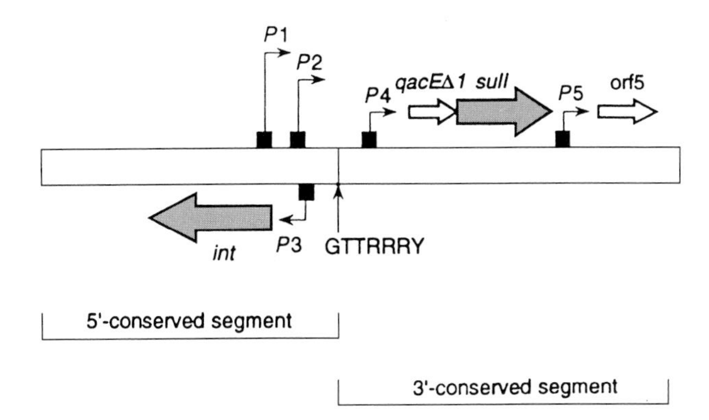

FIGURE 3.1 The general structure of an integron. Integrons consist of a 5′ conserved sequence that encodes an integrase and contains promoter sequences (P) responsible for the transcription of inserted gene cassettes. The 3′ conserved sequence encodes resistance to sulfonamide drugs (sul). The insertion site for gene cassettes (GTTRRRY) is indicated.

both gram-negative and gram-positive bacteria. The 3′ conserved region carries a gene for sulfonamide resistance (*sul*) and two open reading frames of unknown function. Probably the ancestral integron, encoded no antibiotic resistance (Fig. 3.1). The ubiquitous presence of *sul* in an element of this type might be surprising, although sulfonamides (see above) have been employed since the mid-1930s and (apart from mercury salts) are the longest used agents for the treatment of infectious diseases. The resistance gene cassettes are integrated into a specific insertion site in the integron. Typically, in the case of Tn*21*-related transposons, each antibiotic resistance cassette is associated with one of a functional family of closely related, palindromic 59-b elements (or recombination hot spots) located to the 3′ side of the resistance gene (Fig. 3.1). Integrase-catalyzed insertion of resistance gene cassettes into resident integrons has been demonstrated (Collis *et al.*, 1993; Martinez and de la Cruz, 1990). In addition, site-specific deletion and rearrangement of the inserted resistance gene cassettes can result from integrase-catalyzed events (Collis and Hall, 1992).

Francia *et al.* (1993) have expanded our understanding of the role of integrons in gene mobilization by showing that the Tn*21* integrase can act on secondary target sites at significant frequencies and so permit the fusion of two R plasmids by interaction between the recombination hot spot of one plasmid and a secondary integrase target site on a second plasmid. The secondary sites are characterized by the degenerate pentanucleotide sequence sequence Ga/tTNa/t. Though the details of the mechanism by which new integrons are then generated from the fusion structure are not established, the use of secondary integration sites could explain how new genes may be inserted into integrons without the necessity for a 59-bp element, as the authors point out.

B. Other multiple resistance plasmids

Analyses of the integron-type transposons provide a good model for the way in which antibiotic resistance genes from various (unknown) sources may be incorporated into an integron by recombination events into mobile elements and hence into bacterial replicons, providing the R plasmids that we know today (Fig. 3.2) (Bissonnette and Roy, 1992). However, in bacterial pathogens a variety of transposable elements have been found that undergo different processes of recombinational excision and insertion. It is not known what evolutionary mechanisms are implicated or whether some form of integron-related structure is present in all cases. For the type of integron found in the Tn*21* family, we have plausible models, supported by *in vivo* and *in vitro* studies, to provide a modus operandi by which antibiotic resistance genes were (and are) molecularly cloned in the evolution of R plasmids. A large number of transposable elements carrying virtually all possible combinations of antibiotic resistance genes have been identified (Berg, 1989), and nucleotide sequence analysis of multiresistant integrons shows that the inserted resistance gene cassettes differ markedly in codon usage, indicating that the antibiotic resistance determinants are of diverse origins. Microbes are masters at genetic engineering, and heterologous expression vectors of broad host range in the form of integrons were present in bacteria long before they became vogue for biotechnology companies in the 1980s.

V. ORIGINS

We have described how the majority of antibiotic resistance genes found in microbes have been acquired by their hosts. The important question is, From where did they acquire these genetic determinants? The integron model defines a mechanism by which antibiotic genes can be procured by members of the Enterobacteriaceae and pseudomonads. This mechanism requires the participation of extrinsic resistance genes (or cassettes); however, the origins of these open reading frames are a mystery. The same questions can be applied to any of the resistance genes found in pathogenic bacteria. Available evidence concerning some of the origins is discussed below.

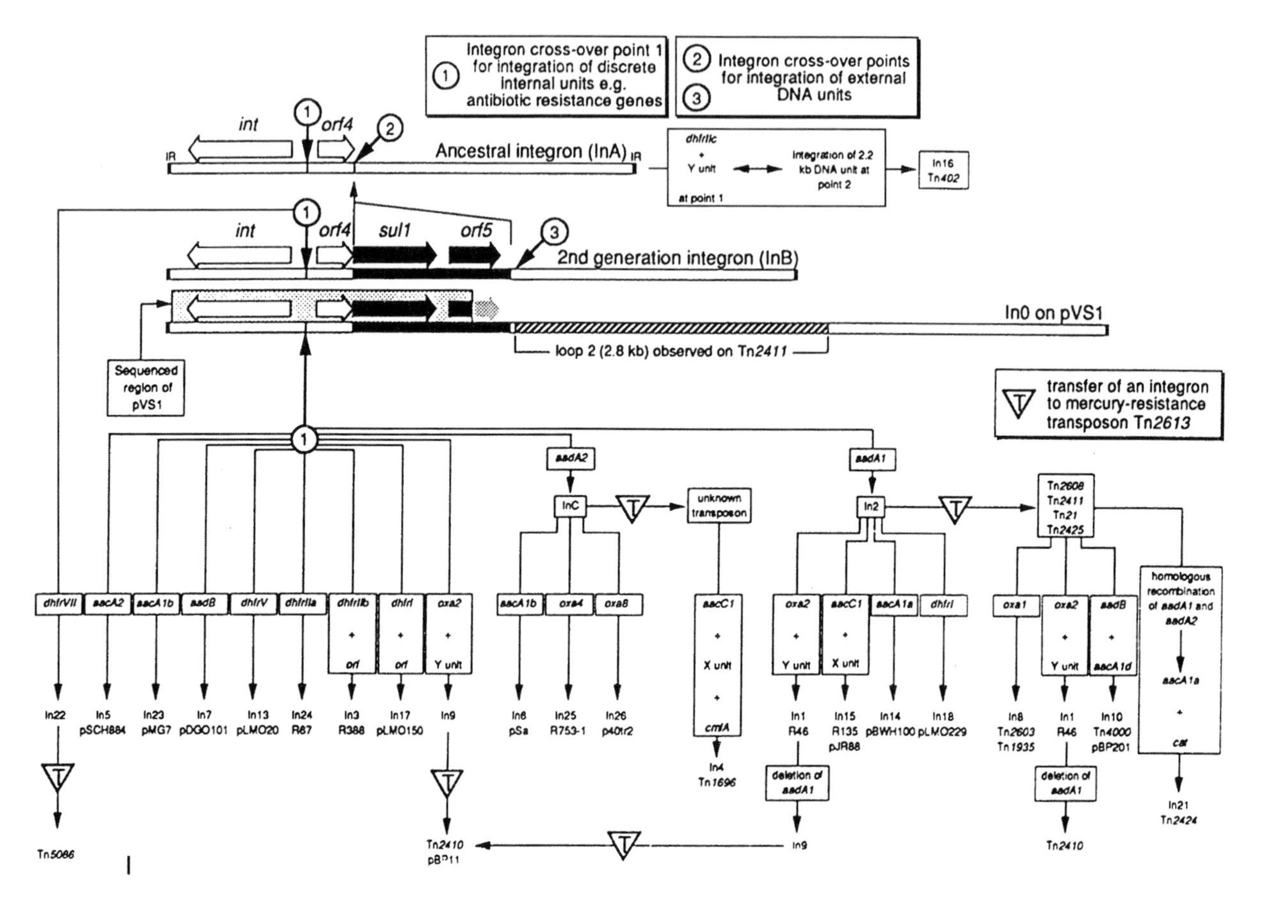

FIGURE 3.2 The diversity of antibiotic resistance integrons. The diagram illustrates the insertion of antibiotic resistance gene cassettes into the basic integron structure (Fig. 3.1). It should be noted that the antibiotic resistance gene cassettes are usually inserted in tandem array and more than five genes may be found in a single integron structure.

A. Antibiotic producers

Antibiotic-producing microbes are the prime suspects for the maintenance of a pool of resistance genes in nature. Any organism producing a toxic molecule must, by definition, possess a mechanism(s) to survive this potentially suicidal situation. Because the majority of antibiotics are produced by bacteria (principally the actinomycete group), one would expect these organisms to have mechanisms of protection against the antibiotics they make. These mechanisms take various forms, and it is significant that the mechanisms of resistance for the known antibiotics in producing organisms and clinical isolates are biochemically identical (see Table 3.7). The gene clusters for antibiotic biosynthesis in producing organisms almost invariably comprise one or more genes that encode resistance to the antibiotic produced. Many of these genes have been cloned and sequenced. Sequence comparisons of the genes from the producer and the clinical isolate often show very high degrees of similarity. Thus, the homologous biochemical mechanisms and the relatedness of the gene sequences support the hypothesis that producing organisms are likely to be the source of most resistance genes. However, other sources of resistance genes are not excluded, and some of these are discussed below.

TABLE 3.7 Resistance determinants with homologs in antibiotic producing organisms

Antibiotic	Mechanism
Penicillins	β-Lactamases
Cephalosporins	Penicillin binding proteins
Aminoglycosides	Phosphotransferases, acetyltransferases, nucleotidyltransferases
Chloramphenicol	Acetyltransferases
Tetracyclines	Ribosomal protection, efflux
Macrolides	rRNA methylation
Streptogramins	Esterases
Lincosamides	Phosphotransferases, acetyltransferases
Phosphonates	Phosphorylation, glutathionylation (?)
Bleomycin	Acetyltransferases, immunity protein

1. *Tetracyclines*

The first tangible evidence of resistance gene transfer involving antibiotic-producing streptomycetes in a clinical setting has come from Pang *et al.* (1994). These researchers analyzed human infections of nontuberculous mycobacteria and *Streptomyces* spp.; the infections did not respond to treatment with tetracyclines. Both microbial species contained resistance genes (*tetK* and *tetL*) known to be the basis of tetracycline resistance in gram-positive bacteria (see Section II.A.1.b on tetracycline resistance determinants). These resistance determinants promote efflux of the drug and are typically transposon-associated. Surprisingly, the mycobacteria and the streptomycetes both had the tetracycline resistance genes (*otrA* and *otrB*) previously identified in the tetracycline-producing strain *Streptomyces rimosis* (Davies, 1992; Doyle *et al.*, 1991). Reciprocally, the streptomycetes had acquired the *tetK* and *tetL* genes. Because the latter are clearly "foreign" genes (*tetK* and *tetL* have a G + C content different from those of streptomycete and mycobacterial chromosomal DNA), they must have been acquired as the result of a recent gene transfer. Although this evidence is consistent with resistance gene transfer between the streptomycetes and other bacteria, it is not known which is donor and which is recipient, nor whether the newly acquired tetracycline resistance genes are plasmid or chromosomally encoded.

2. *Aminoglycosides*

Covalent modification of their inhibitory biochemical products is very common in antibiotic-producing bacteria. It was the discovery of antibiotic modification as a means of self-protection in the streptomycetes that led to the proposal that antibiotic-producing microbes were the origins of the antibiotic resistance determinants found in other bacteria (Benveniste and Davies, 1973; Walker and Skorvaga, 1973). Support for this hypothesis has been provided by nucleic acid and protein sequence comparisons of aminoglycoside resistance determinants from producing organisms and clinical isolate sources (Shaw, 1984; Davies, 1992). As mentioned above, producing organisms are not the only potential source of antibiotic resistance mechanisms. The proposal that the enzymes that modify aminoglycosides evolved from such "housekeeping" genes as the sugar kinases and acyltransferases has been made by a number of groups (Udou *et al.*, 1989; Shaw *et al.*, 1992; Rather *et al.*, 1993).

3. *Macrolides*

The principal mechanism of resistance to the macrolide antibiotics involves methylation of the 23 S rRNA of the host. Methylation occurs on a specific adenine residue in the rRNA, and mono- and dimethylation has been described. The *N*-methyltransferase genes responsible for encoding this protective modification have been studied in detail from both the producing organisms and clinical isolates (Rather *et al.*, 1993). Although no direct transfer of resistance between producer and clinical isolates has been found, the identity of the biochemical mechanism and its regulation make for a compelling evolutionary relationship.

4. *Other antibiotics*

Perhaps the most straightforward path toward the development of a drug resistance mechanism is via the major facilitator superfamily (MFS; see Section II.B.2) of transporters. The MFS is found in all organisms involved and consists of membrane transport systems involved in the symport, antiport, or uniport of various substrates. Other examples of MFS pumps include sugar uptake systems, phosphate ester/phosphate antiport, and oligosaccharide uptake. One might expect that, in addition to methods for the import of nutrients, a cell would have methods for the removal of harmful substances. The β-lactamases are an interesting case because they are widely distributed among the bacterial kingdom. The ubiquity of β-lactamases suggests that these enzymes may play a part in normal metabolic or synthetic processes, but an essential role has not been shown. Examination of the sequences of β-lactamases by Bush *et al.* (1995) permitted the establishment of extensive phylogenetic relationships, which includes those from microbes employed in the commercial production of β-lactam antibiotics.

In the case of the glycopeptide antibiotics (vancomycin and teichoplanin), resistance in the enterococci is due to the acquisition of a cluster of genes that encode a novel cell wall precursor (see Section II.A.3). A comparison of D-Ala-D-Ala ligases from different sources has shown that the *vanA* gene is dissimilar to the other known genes, suggesting a divergent origin (Arthur *et al.*, 1996). The *vanA* homolog from the producing organism, *Streptomyces toyocaensis*, has been cloned and sequenced. The predicted amino acid sequence was compared with the D-Ala-D-Ala ligase from *Enterococcus* and is greater than 60% similar (G. D. Wright *et al.*, 1997).

B. Unknown origin

There are gaps in our understanding of the origin of resistance determinants. For example, the aminoglycoside nucleotidyltransferases have been found only in clinical isolates and have no known relatives; the potential origins of the chloramphenicol acetyltransferase genes are still unclear (see Section II.A.1.c); and the sources for sulfonamide and trimethoprim resistance are still not known (see Section II.A.4). There may be examples of "housekeeping" functions of resistance genes. For example, *Providencia stuartii* has a chromosomally encoded aminoglycoside acetyltransferase that may play a role in cell wall peptidoglycan formation (Payie *et al.*, 1995).

VI. MAINTENANCE OF ANTIBIOTIC RESISTANCE

A variety of surveys have indicated that normal healthy humans (who are not pursuing a course of antibiotic therapy) carry antibiotic-resistant enteric species in their intestinal tract; a substantial proportion are found to contain transmissible antibiotic resistance plasmids. Studies have demonstrated that a lack of antibiotic selective pressure, for example, removing antimicrobials from cattle feed, can lead to a gradual decrease in the percentage of resistance genes and resistance bacteria found in a population (Langlois *et al.*, 1986; Hintone *et al.*, 1985).

Although the results of these studies are encouraging, other studies indicate that once resistance cassettes have been developed it is unlikely that they will disappear completely from an environment where antibiotics are routinely used. In a sense "the cat is out of the bag." Chemostat studies (Chao *et al.*, 1983; Hartl *et al.*, 1983) suggest that insertion elements may themselves provide their hosts with a selective advantage independent of the resistance determinants they carry. Roberts has postulated that commensal bacteria can be reservoirs for tetracycline resistance determinants. When bacteria from the urogenital tracts of females who had not taken antibiotics for 2 weeks previously were examined, 82% of the viridans-type streptococci hybridized with at least one Tet determinant (Roberts *et al.*, 1991).

A. Multiple antibiotic resistance and mercury resistance

Although the cooccurrence of antibiotic resistance and resistance to heavy metals such as mercury has long been known, its implications for public health are only now becoming clear. In 1964, the cotransduction of genes encoding resistance to penicillin and mercury by a staphylococcal phage was reported (Richmond and John, 1964). Ten years later it was found that 25% of the antibiotic resistance plasmids isolated from enteric bacteria in Hammersmith Hospital also carried mercury resistance (Schottel *et al.*, 1974). DNA sequence analyses has shown that the Tn*21*-type transposons carry both a copy of the *mer* locus and an integron (see above) (Stokes and Hall, 1989; Grinsted *et al.*, 1990). Summers and co-workers (1993) have observed that resistance to mercury occurs frequently in human fecal flora and is correlated with the occurrence of multiple antibiotic resistance. How does this phenomenon become a public health concern? In the same report Summers' group found that the mercury released from the amalgam in "silver" dental fillings in monkeys led to the rapid enrichment of many different mercury-resistant bacteria in the oral and fecal flora. They suggest that in humans this chronic and biologically significant exposure to mercury may foster the persistence of multidrug-resistant microbes through selection of linked markers.

B. Other examples

There may well be other examples of this phenomenon, where subclinical concentrations of an antibiotic could serve as selection for the maintenance (and propagation) of genetic elements and their resident resistance genes. There is also maintenance by constant selection pressure for other phenotypes, a type of "linked" selection (we have mentioned the role of mercury in this respect). How else does one explain the fact that streptomycin and chloramphenicol resistance can be still found on plasmids in hospitals, even though these antibiotics are no longer used? Are there other positive selective functions carried by plasmids? Resistance to ultraviolet light or other physical or chemical toxins (e.g. detergents, disinfectants) would be a possibility. Is it also conceivable that plasmids improve the fitness of their hosts under "normal" conditions? Evidence for this comes from the work of Lenski (personal communication, 1996).

VII. CONCLUDING REMARKS—FOR NOW AND THE FUTURE

It should be apparent from the foregoing discussion of antibiotic modification that there must be a substantial pool of antibiotic resistance genes (or close relatives of these genes) in nature. Gene flux between bacterial replicons and their hosts is likely to be the rule rather than the exception, and it appears to respond quickly to environmental changes (Levy and Novick, 1986; Levy and Miller, 1989; Hughes and Datta, 1983). This gene pool is readily accessible to bacteria when they are exposed to the strong selective pressure of antibiotic usage—in hospitals, for veterinary and agricultural purposes, and as growth promoters in animal and poultry husbandry. It is a life-or-death situation for microbes, and they have survived. A better knowledge of the components of this gene pool, particularly with respect to what might happen on the introduction of a new chemical entity such as an antimicrobial agent, might, on the one hand, permit early warning and subsequent chemical modification of antibiotics to permit them to elude potential resistance mechanisms and, on the other, lead to more prudent use of antibiotics under circumstances where the presence of specific resistance determinants can be predicted.

The development of resistance to antimicrobial agents is inevitable, in response to the strong selective pressure and extensive use of antibiotics. Resistance may develop as the result of mutation or acquisition, or a combination of the two. The use of antibiotics should be such as to delay the inevitable, and knowledge gained over the past 50 years if correctly interpreted and used to modify current practices appropriately should permit this. Unfortunately, for all of the antimicrobial agents in current use, we have reached a state of no return. The American Society for Microbiology Task Force on Antibiotic Resistance has made a number of recommendations to deal with the current crisis of antibiotic resistant microbes (see Table 3.8; ASM, 1995). It has been suggested that such measures can, at the least, maintain and improve the status quo.

TABLE 3.8 Recommendations from the American Society for Microbiology Task Force on Antibiotic Resistance

Establish a national surveillance system immediately.
- Lead agency should be the National Center for Infectious Diseases of the U.S. Centers for Disease Control (CDC) and should involve the National Institute of Allergy and Infectious Disease of the National Institutes of Health (NIH), the Environmental Protection Agency, and the Food and Drug Administration.

Strengthen professional and public education in the area of infectious disease and antibiotics to reduce inappropriate usage of antibiotics.
- The curriculum for health professionals should include the appropriate handling, diagnosis, and treatment of infectious disease and antibiotic resistance.
- Reduce the spread of infectious agents and antibiotic resistance in hospitals, nursing homes, day care facilities, and food production industries.
- Educate patients and food producers.
- Improve antimicrobial use for cost-effective treatment and preservation of effectiveness.

Increase basic research directed toward development of new antimicrobial compounds, effective vaccines, and other prevention measures.
- Fund areas directly related to new and emerging infections and antibiotic resistance.
- Fund basic research in bacterial genetics and metabolic pathways.
- Establish a culture collection containing representative antibiotic-resistance biotypes of pathogens.
- Sequence genomes of microbial pathogens.
- Develop better diagnostic techniques.
- Develop vaccines and other preventative measures.

Two of the options that would permit continued success of antibiotic therapy in the face of increasing resistance are: (a) the discovery of new antibiotics (by "new" this implies novel chemical structures) and (b) the development of agents, that might be used in combination with existing antibiotics, to interfere with the biochemical resistance mechanisms. Such a strategy has already been partially successful in the development of inhibitors of β-lactamases to permit "old" antibiotics to be used. However, there appears to be little success (or effort) in applying this approach to other antibiotic classes. In our discussion of biochemical mechanisms (Section II) we have noted a number of cases where this approach could be taken (e.g. fluorinated analogs of chloramphenicol). With respect to novel antibiotics, several valid approaches exist: (a) natural product screening, especially directed at products of the 99.9% of microbes that cannot be grown in the laboratory; (b) combinatorial chemistry that can be used as a means of discovery of new active molecules or to new substitutions on known ring structures; and (c) rational chemical design based on identification of specific biochemical targets. The success of these methods will depend on the availability of cell-based and biochemical assays that will detect low concentrations of active molecules by high flux screening methods. It seems redundant to insist on the requirement for early identification of natural resistance mechanisms for any compounds of interest (thus permitting the design of analogs) and the study of structure–toxicity relationships at the earliset stage possible.

Last, but not least, the introduction of a novel antimicrobial agent into clinical practice must be accompanied by strict limitations on its use. No novel therapeutic should be used for other than human use under prescription, and no structural analog should be employed for "other" purposes.

ACKNOWLEDGMENTS

We thank the Canadian Bacterial Diseases Network and the Natural Sciences and Engineering Council of Canada for support.

REFERENCES

Aminoglycoside Resistance Study Group. (1994). Resistance to aminoglycosides in *Pseudomonas*. *Trends Microbiol.* **2** (10), 347–353.

Arca, P., Hardisson, C., and Suárez, J. E. (1990). Purification of a glutathione *S*-transferase that mediates fosfomycin resistance in bacteria. *Antimicrob. Agents Chemother.* **34**, 844–848.

Arthur, M., Brisson-Noël, A., and Courvalin, P. (1987). Origin and evolution of genes specifying resistance to macrolide, lincosamide and streptogramin antibiotics: Data and hypothesis. *J. Antimicrob. Chemother.* **20**, 783–802.

Arthur, M., Reynolds, P., and Courvalin, P. (1996). Glycopeptide resistance in gram-positive bacteria. *Trends Microbiol.* **4**, 401–407.

ASM. (1995). Report of the ASM Task Force on antibiotic resistance. *Antimicrob. Agents Chemother.* (Suppl.), 123.

Bannam, T. L., and Rood, J. I. (1991). Relationship between the *Clostridium perfringens catQ* gene product and chloramphenicol acetyltransferases from other bacteria. *Antimicrob. Agents Chemother.* **35**, 471–476.

Bennett, P. M., and Chopra, I. (1993). Molecular basis of β-lactamase induction in bacteria. *Antimicrob. Agents Chemother.* **37**, 153–158.

Benveniste, R., and Davies, J. (1973). Aminoglycoside antibiotic-inactivating enzymes in actinomycetes similar to those present in clinical isolates of antibiotic-resistant bacteria. *Proc. Natl. Acad. Sci. USA* **70**, 2276–2280.

Berg, D. E. (1989). Transposable elements in prokaryotes. *In* "Gene Transfer in the Environment" (S. B. Levy and R. V. Miller, Eds.), pp. 99–137. McGraw-Hill, New York.

Bergh, O., Borsheim, K. Y., Bratbak, G., and Heldal, M. (1989). High abundance of viruses found in aquatic environments. *Nature (London)* **340**, 467–468.

Bissonnette, L., and Roy, P. H. (1992). Characterization of In*0* of *Pseudomonas aeruginosa* plasmid pVS1, an ancestor of integrons of multiresistance plasmids and transposons of gram-negative bacteria. *J. Bacteriol.* **174**, 1248–1257.

Bissonnette, L., Champetier, S., Buisson, J.-P., and Roy, P. H. (1991). Characterization of the nonenzymatic chloramphenicol resistance (*cmlA*) gene of the In*4* integron of Tn*1696:* Similarity of the product to transmembrane transport proteins. *J. Bacteriol.* **173**, 4493–4502.

Blazquez, J., Baquero, M.-R., Canton, R., Alos, I., and Baquero, F. (1993). Characterization of a new TEM-type β-lactamase resistant to clavulanate, sulbactam, and tazobactam in a clinical isolate of *Escherichia coli. Antimicrob. Agents Chemother.* **37**, 2059–2063.

Brisson-Noël, A., Delrieu, P., Samain, D., and Courvalin, P. (1988). Inactivation of lincosaminide antibiotics in *Staphylococcus. J. Biol. Chem.* **263**, 15880–15887.

Bush, K., Jacoby, G. A., and Medeiros, A. A. (1995). A functional classification scheme for β-lactamases and its correlation with molecular structure. *Antimicrob. Agents Chemother.* **39**, 1211–1233.

Cannon, M., Harford, S., and Davies, J. (1990). A comparative study on the inhibitory actions of chloramphenicol, thiamphenicol and some fluorinated derivatives. *J. Antimicrob. Chemother.* **26**, 307–317.

Chao, L., Vargas, C., Spear, B. B., and Cox, E. C. (1983). Transposable elements as mutator genes in evolution. *Nature (London)* **303**, 633–635.

Chen, S.-T., and Clowes, R. C. (1987). Variations between the nucleotide sequences of Tn*1*, Tn*2*, and Tn*3* and expression of β-lactamase in *Pseudomonas aeruginosa* and *Escherichia coli. J. Bacteriol.* **169**, 913–916.

Cole, S. T. (1994). *Mycobacterium tuberculosis:* Drug-resistance mechanisms. *Trends Microbiol.* **10** (2), 411–415.

Collis, C. M., and Hall, R. M. (1992). Site-specific deletion and rearrangement of integron insert genes catalysed by the integron DNA integrase. *J. Bacteriol.* **174**, 1574–1585.

Collis, C. M., Grammaticopoulos, G., Briton, J., Stokes, H. W., and Hall, R. M. (1993). Site-specific insertion of gene cassettes into integrons. *Mol. Microbiol.* **9**, 41–52.

Cookson, B. (1995). Aspects of the epidemiology of MRSA in Europe. *J. Chemother.* **7** (Suppl. 3), 93–98.

Couture, F., Lachapelle, J., and Levesque, R. C. (1992). Phylogeny of LCR-1 and OXA-5 with class A and class D β-lactamases. *Mol. Microbiol.* **6**, 1693–1705.

Dabbs, E. R., Yazawa, K., Mikami, Y., Miyaji, M., Morisaki, N., Iwasaki, S., and Furihata, K. (1995). Ribosylation by mycobacterial strains as a new mechanism of rifampin inactivation. *Antimicrob. Agents Chemother.* **39**, 1007–1009.

Datta, N., and Hughes, V. M. (1983). Plasmids of the same Inc groups in enterobacteria before and after the medical use of antibiotics. *Nature (London)* **306**, 616–617.

Davies, J. (1992). Another look at antibiotic resistance. *J. Gen. Microbiol.* **138**, 1553–1559.

Day, P. J., and Shaw, W. V. (1992). Acetyl coenzyme A binding by chloramphenicol acetyltransferase: Hydrophobic determinants of recognition and catalysis. *J. Biol. Chem.* **267**, 5122–5127.

Debets-Ossenkopp, Y., Sparrius, M., Kusters, J., Kolkman, J., and Vandenbroucke-Grauls, C. (1996). Mechanism of clarithromycin resistance in clinical isolates of *Helicobacter pylori. FEMS Microbiol. Lett.* **142**, 37–42.

Doyle, D., McDowall, K. J., Butler, M. J., and Hunter, I. S. (1991). Characterization of an oxytetracycline-resistance gene, *otrA*, of *Streptomyces rimosus. Mol. Microbiol.* **5**, 2923–2933.

Ferretti, J. J., Gilmore, K. S., and Courvalin, P. (1986). Nucleotide sequence analysis of the gene specifying the bifunctional 6′-aminoglycoside acetyltransferase 2″-aminoglycoside phosphotransferase enzyme in *Streptococcus faecalis* and identification and cloning of gene regions specifying the two activities. *J. Bacteriol.* **167**, 631–638.

Finken, M., Kirschner, P., Meier, A., Wrede, A., and Böttger, E. C. (1993). Molecular basis of streptomycin resistance in *Mycobacterium tuberculosis:* Alterations of the ribosomal protein S12 gene and point mutations within a functional 16S ribosomal RNA pseudoknot. *Mol. Microbiol.* **9**, 1239–1246.

Fourmy, D., Recht, M. I., Blanchard, S. C., and Puglisi, J. D. (1996). Structure of the A site of *Escherichia coli* 16S ribosomal RNA complexed with an aminoglycoside. *Science* **274**, 1367–1371.

Francia, M. V., de la Cruz, F., and García Lobo, J. M. (1993). Secondary sites for integration mediated by the Tn*21* integrase. *Mol. Microbiol.* **10**, 823–828.

George, A., and Levy, S. B. (1983) Amplifiable resistance to tetracycline, chloramphenicol, and other antibiotics in *Escherichia coli*: involvement of a non-plasmid-determined efflux of tetracycline. *J. Bacteriol.* **155**, 531–540.

Goldman, J. D., White, D. G., and Levy, S. B. (1996). Multiple antibiotic resistance (*mar*) locus protects *Escherichia coli* from rapid cell killing by fluoroquinolones. *Antimicrob. Agents Chemother.* **40**, 1266–1269.

Grinsted, J., de la Cruz, F., and Schmitt R. (1990). The Tn*21* subgroup of bacterial transposable elements. *Plasmid*, **24**, 163–189.

Hartl, D. L., Dykhuizen, D. E., Miller, R. D., Green, L., and de Framond, J. (1983). Transposable element IS*50* improves growth rate of *E. coli* cells without transposition. *Cell (Cambridge, Mass.)* **35**, 503–510.

Hillen, W., and Berens, C. (1994). Mechanisms underlying expression of Tn*10* encoded tetracycline resistance. *Annu. Rev. Microbiol.* **48**, 345–369.

Hinton, M., Linton, A. H., and Hedges, A. J. (1985). The ecology of *Escherichia coli* in calves reared as dairy-cow replacements. *J. Appl. Bacteriol.* **85**, 131–138.

Honoré, N., Nicolas, M.-H., and Cole, S. T. (1986). Inducible cephalosporinase production in clinical isolates of *Enterobacter cloacae* is controlled by a regulatory gene that has been deleted from *Escherichia coli. EMBO J.* **5**, 3709–3714.

Hughes, V. M., and Datta, N. (1983). Conjugative plasmids in bacteria of the 'pre-antibiotic' era. *Nature (London)* **302**, 725–726.

Huovinen, P., Sundström, L., Swedberg, G., and Sköld, O. (1995). Trimethoprim and sulfonamide resistance. *Antimicrob. Agents Chemother.* **39**, 279–289.

Jacoby, G. A., and Archer, G. L. (1991). New mechanisms of bacterial resistance to antimicrobial agents. *N. Engl. J. Med.* **324**, 601–612.

Jarlier, V., and Nikaido, H. (1994). Mycobacterial cell wall: Structure and role in natural resistance to antibiotics. *FEMS Microbiol. Lett.* **123**, 11–18.

Kahn, M., and Smith, H. O. (1986). Role of transformazomes in *Haemophilus influenzae* Rd transformation. *In* "Antibiotic Resistance Genes: Ecology, Transfer, and Expression" (S. B. Levy and R. P. Novick, Eds.), pp. 143–152. Cold Spring Harbor Laboratory, Cold Spring Harbor, New York.

Kocabiyik, S., and Perlin, M. H. (1992). Altered substrate specificity by substitutions at Tyr218 in bacterial aminoglycoside 3′-phosphotransferase-II. *FEMS Microbiol. Lett.* **93**, 199–202.

Kokjohn, T. A. (1989). Transduction: Mechanism and potential for gene transfer in the environment. *In* "Gene Transfer in the Environment" (S. B. Levy and R. V. Miller, Eds.), pp. 73–97. McGraw-Hill, New York.

Krause, R. M. (1994). Dynamics of emergence. *J. Infect. Dis.* **170**, 265–271.

Langlois, B. E., Dawson, K. A., Cromwell, G. L., and Stahly, T. S. (1986). Antibiotic resistance in pigs following a 13 year ban. *J. Anim. Sci.* **62**, 18–32.

LeClerc, J. E., Li, B., Payne, W. L., and Cebula, T. A. (1996). High mutation frequencies among *Escherichia coli* and *Salmonella* pathogens. *Science* **274**, 1208–1211.

Leclercq, R., and Courvalin, P. (1991). Bacterial resistance to macrolide, lincosamide, and streptogramin antibiotics by target modification. *Antimicrob. Agents Chemother.* **35**, 1267–1272.

Lévesque, C., Brassard, S., Lapointe, J., and Roy, P. H. (1994). Diversity and relative strength of tandem promoters for the antibiotic-resistance genes of several integrons. *Gene* **142**, 49–54.

Levy, S. B., and Miller, R. V., eds. (1989). Gene transfer in the environment. *In* "Environmental Biotechnology." McGraw-Hill, New York.

Levy, S. B., and Novick, R. P., eds. (1986). "Antibiotic Resistance Genes: Ecology, Transfer, and Expression." Banbury Report 24. Cold Spring Harbor Laboratory, Cold Spring Harbor, New York.

Lewis, K. (1994). Multidrug resistance pumps in bacteria: Variations on a theme. *Trends Biochem. Sci.* **19**, 119–123.

Mabilat, C., Goussard, S., Sougakoff, W., Spencer, R. C., and Courvalin, P. (1990). Direct sequencing of the amplified structural gene and promoter for the extended-broad-spectrum β-lactamase TEM-9 (RHH-1) of *Klebsiella pneumoniae*. *Plasmid* **23**, 27–34.

Marger, M. D., and Saier, M. H. (1993). A major superfamily of transmembrane facilitators that catalyse uniport, symport and antiport. *Trends Biochem. Sci.* **18**, 13–20.

Martinez, E., and de la Cruz, F. (1990). Genetic elements involved in Tn*21* site-specific integration, a novel mechanism for the dissemination of antibiotic resistance genes. *EMBO J.* **9**, 1275–1281.

Miller, P. F., and Sulavik, M. C. (1996). Overlaps and parallels in the regulation of intrinsic multiple-antibiotic resistance in *Escherichia coli*. *Mol. Microbiol.* **21**, 441–448.

Monod, M., Mohan, S., and Dubnau, D. (1987). Cloning and analysis of *ermG*, a new macrolide-lincosamide-streptogramin B resistance element from *Bacillus sphaericus*. *J. Bacteriol.* **169**, 340–350.

Murray, I. A., and Shaw, W. V. (1997). *O*-Acetyltransferases for chloramphenicol and other natural products. *Antimicrob. Agents Chemother.* **41**, 1–6.

Needham, C., Rahman, M., Dyke, K. G. H., and C., N. W., Noble, W. C. (1994). An investigation of plasmids from *Staphylococcus aureus* that mediate resistance to mupirocin and tetracycline. *Microbiology* **140**, 2577–2583.

Neu, H. C. (1992). The crisis in antibiotic resistance. *Science* **257**, 1064–1073.

Nikaido, H. (1994). Prevention of drug access to bacterial targets: Permeability barriers and active efflux. *Science* **264**, 382–387.

Nikaido, H. (1996). Multidrug efflux pumps of gram-negative bacteria. *J. Bacteriol.* **178**, 5853–5859.

Nikolich, M. P., Hong, G., Shoemaker, N. B., and Salyers, A. A. (1994). Evidence for natural horizontal transfer of *tetQ* between bacteria that normally colonize humans and bacteria that normally colonize livestock. *Appl. Environ. Microbiol.* **60**, 3255–3260.

Noble, W. C., Rahman, M., and Cookson, B. (1988). Transferable mupirocin resistance. *J. Antimicrob. Chemother.* **22**, 771.

Noble, W. C., Virani, Z., and Cree, R. G. A. (1992). Co-transfer of vancomycin and other resistance genes from *Enterococcus faecalis* NCTC 12201 to *Staphylococcus aureus*. *FEMS Microbiol. Lett.* **93**, 195–198.

O'Hara, K., Kanda, T., Ohmiya, K., Ebisu, T., and Kono, M. (1989). Purification and characterization of macrolide 2′-phosphotransferase from a strain of *Escherichia coli* that is highly resistant to erythromycin. *Antimicrob. Agents Chemother.* **33**, 1354–1357.

Olsen, R. H., and Shipley, P. (1973). Host range and properties of the *Pseudomonas aeruginosa* R factor R 1822. *J. Bacteriol.* **113**, 772–780.

Olsen, R. H., Siak, J., and Gray, R. H. (1974). Characteristics of PrD1, a plasmid dependent broad host range DNA bacteriophage. *J. Virol.* **14**, 689–699.

Pang, Y., Brown, B. A., Steingrube, V. A., Wallace, R. J., Jr., and Roberts, M. C. (1994). Tetracycline resistance determinants in *Mycobacterium* and *Streptomyces* species. *Antimicrob. Agents Chemother.* **38**, 1408–1412.

Parent, R., and Roy, P. H. (1992). The chloramphenicol acetyltransferase gene of Tn*2424:* A new breed of *cat*. *J. Bacteriol.* **174**, 2891–2897.

Paulsen, I. T., Brown, M. H., and Skurray, R. A. (1996). Proton-dependent multidrug efflux systems. *Microbiol Rev.* **60**, 575–608.

Payie, K. G., Rather, P. N., and Clarke, A. J. (1995). Contribution of gentamicin 2′-*N*-acetyltransferase to the O-acetylation of peptidoglycan in *Providencia stuartii*. *J. Bacteriol.* **177**, 4303–4310.

Rahman, M., Noble, W. C., and Dyke, K. G. H. (1993). Probes for the study of mupirocin resistance in staphylococci. *J. Med. Microbiol.* **39**, 446–449.

Rather, P. N., Munayyer, H., Mann, P. A., Hare, R. S., Miller, G. H., and Shaw, K. J. (1992). Genetic analysis of bacterial acetyltransferases: Identification of amino acids determining the specificities of the aminoglycoside 6′-*N*-acetyltransferase Ib and IIa proteins. *J. Bacteriol.* **174**, 3196–3203.

Rather, P. N., Orosz, E., Shaw, K. J., Hare, R., and Miller, G. (1993). Characterization and transcriptional regulation of the 2′-*N*-acetyltransferase gene from *Providencia stuartii*. *J. Bacteriol.* **175**, 6492–6498.

Recchia, G. D., and Hall, R. M. (1995). Gene cassettes: A new class of mobile element. *Microbiology* **141**, 3015–3027.

Richmond, M. H., and John, M. (1964). Co-transduction by a staphylococcal phage of the genes responsible for penicillinase synthesis and resistance to mercury salts. *Nature (London)* **202**, 1360–1361.

Roberts, M. C. (1994). Epidemiology of tetracycline-resistance determinants. *Trends Microbiol.* **2** (10), 353–357.

Roberts, M. C., Pang, Y. J., Spencer, R. C., Winstanley, T. G., Brown, B. A., and Wallance, R. J. (1991). Tetracycline resistance in *Moraxella* (*Branhamella*) *catarrhalis:* Demonstration of two clonal out breaks by using pulsed-field gel electrophoresis. *Antimicrob. Agents Chemother.* **35**, 2453–2455.

Rouch, D. A., Byrne, M. E., Kong, Y. C., and Skurray, R. A. (1987). The *aacA–aphD* gentamicin and kanamycin resistance determinant of Tn*4001* from *Staphylococcus aureus:* Expression and nucleotide sequence analysis. *J. Gen. Microbiol.* **133**, 3039–3052.

Salyers, A. A. (1993). Gene transfer in the mammalian intestinal tract. *Curr. Opin. Biotechnol.* **4**, 294–298.

Sanchez-Pescador, R., Brown, J. T., Roberts, M., and Urdea, M. S. (1988). Homology of the TetM with translational elongation factors: Implications for potential modes of tetM conferred tetracycline resistance. *Nucleic Acids. Res.* **16**, 12–18.

Schottel, J., Mandal, A., Clark, D., Silver, S., and Hedges, R. W. (1974). Volatilisation of mercury and organomercurials determined by inducible R-factor systems in enteric bacteria. *Nature (London)* **251**, 335–337.

Scott, J. R. (1992). Sex and the single circle: Conjugative transposition. *J. Bacteriol.* **174**, 6005–6010.

Shaw, K. J., Rather, P. N., Sabatelli, F. J., Mann, P., Munayyer, H., Mierzwa, R., Petrikkos, G. L., Hare, R. S., Miller, G. H., Bennett, P., and Downey, P. (1992). Characterization of the chromosomal *aac(6′)-Ic* gene from *Serratia marcescens. Antimicrob. Agents Chemother.* **36**, 1447–1455.

Shaw, K. J., Rather, P. N., Hare, R. S., and Miller, G. H. (1993). Molecular genetics of aminoglycoside resistance genes and familial relationships of the aminoglycoside-modifying enzymes. *Microbiol. Rev.* **57**, 138–163.

Shaw, W. V. (1984). Bacterial resistance to chloramphenicol. *Br. Med. Bull.* **40**, 36–41.

Sheridan, R. P., and Chopra, I. (1991). Origin of tetracycline efflux proteins: Conclusions from nucleotide sequence analysis. *Mol. Microbiol.* **5**, 895–900.

Shoemaker, N. B., Smith, M. D., and Guild, W. R. (1980). DNase resistant transfer of chromosomal *cat* and *tet* insertions by filter mating in pneumococcus. *Plasmid* **3**, 80–87.

Stewart, G. J. (1989). The mechanism of natural transformation. *In* "Gene Transfer in the Environment" (S. B. Levy and R. V. Miller, Eds.), pp. 139–163. McGraw-Hill, New York.

Stokes, H. W., and Hall, R. M. (1989). A novel family of potentially mobile DNA elements encoding site-specific gene-integration functions: Integrons. *Mol. Microbiol.* **3**, 1669–1683.

Stotzky, G. (1989). Gene transfer among bacteria in soil. *In* "Gene Transfer in the Environment" (S. B. Levy and R. V. Miller, Eds.), pp. 165–221. McGraw-Hill, New York.

Suárez, J. E., and Mendoza, M. C. (1991). Plasmid-encoded fosfomycin resistance. *Antimicrob. Agents Chemother.* **35**, 791–795.

Summers, A. O., Wireman, J., Vimy, M. J., Lorscheider, F. L., Marshall, B., Levy, S. B., Bennet, S., and Billard, L. (1993). Mercury released from dental "silver" fillings provokes an increase in mercury- and antibiotic-resistant bacteria in oral and intestinal floras of primates. *Antimicrob. Agents Chemother.* **37**, 825–834.

Taylor, D. E., and Chau, A. (1996). Tetracycline resistance mediated by ribosomal protection. *Antimicrob. Agents Chemother.* **40**, 15.

Timmermann, K. P. (1988). *Physiol. Plant.* **77**, 465–471.

Udo, E. E., Pearman, J. W., and Grubb, W. B. (1994). Emergence of high-level mupirocin resistance in methicillin-resistant *Staphylococcus aureus* in western Australia. *J. Hospital Infect.* **26**, 157–165.

Udou, T., Mizuguchi, Y., and Wallace, R. J., Jr. (1989). Does aminoglycoside-acetyltransferase in rapidly growing mycobacteria have a metabolic function in addition to aminoglycoside inactivation? *FEMS Microbiol. Lett.* **57**, 227–230.

Walker, J. B., and Skorvaga, M. (1973). Phosphorylation of streptomycin and dihydrostreptomycin by *Streptomyces*. Enzymatic synthesis of different diphosphorylated derivatives. *J. Biol. Chem.* **248**, 2435–2440.

Walsh, C. T., Fisher, S. L., Park, I.-S., Prahalad, M., and Wu, Z. (1996). Bacterial resistance to vancomycin: Five genes and one missing hydrogen bond tell the story. *Curr. Biol.* **3**, 21–28.

Weisblum, B. (1995). Insights into erythromycin action from studies of its activity as inducer of resistance. *Antimicrob. Agents Chemother.* **39**, 797–805.

Zakrzewska-Bode, A., Muytjens, H. L., Liem, K. D., and Hoogkamp-Korstanje, J. A. (1995). Mupirocin resistance in coagulase-negative staphylococci, after topical prophylaxis for the reduction of colonization of central venous catherters. *J. Hospital Infect.* **31**, 189–193.

Zilhao, R., and Courvalin, P. (1990). Nucleotide sequence of the *fosB* gene conferring fosfomycin resistance in *Staphylococcus epidermidis. FEMS Microbiol. Lett.* **68**, 267–272.

WEBSITE

Website of the CDC National Center for Infectious diseases
http://www.cdc.gov/drugresistance/

4

Antifungal agents

Ana Espinel-Ingroff
Medical College of Virginia Commonwealth University

GLOSSARY

emerging fungal infections Fungal infections caused by new or uncommon fungi.

granulocytopenia/neutropenia Acquired or chemically induced immunosuppression caused by low white blood cell counts.

immunocompromised Having a defect in the immune system.

in vitro *and* in vivo Describing or referring to studies carried out in the test tube and in animals, respectively.

mycoses and mycotic infections Diseases caused by yeasts or molds.

nephrotoxicity Damage to the kidney cells.

opportunistic infections Infections caused by saprophytic fungi or not true parasites.

Antifungal Agents are naturally occurring or synthetically produced compounds that have *in vitro* or *in vivo* activity against yeasts, molds, or each. Fungi and mammalian cells are eukaryotes, and antifungal agents that inhibit synthesis of proteins, RNA, and DNA are potentially toxic to mammalian cells.

Fungi can be unicellular (yeasts) and multicellular or filamentous (molds) microorganisms. Some medically important fungi can exist in both of these morphologic forms and are called dimorphic fungi. Of the estimated 250 000 fungal species described, fewer than 150 are known to be etiologic agents of disease in humans. Most fungi associated with disease are considered opportunistic pathogens (especially the yeasts) because they live as normal flora in humans, lower animals, and plants and rarely cause disease in otherwise healthy individuals. Many fungi, on the other hand, are important plant and lower animal parasites and can cause damage to crops (wheat rust, corn smut, etc.) and to fruit (banana wilt), forest (Dutch elm disease), and ornamental trees and other plants. Historically, the potato famine, which was the reason for the great migration from Ireland to the Americas, was caused by a fungal infection (potato blight). At the same time, fungi and their products play an important economic role in the production of alcohol, certain acids, steroids, antibiotics, etc.

Due to the high incidence of toxicity among antifungal agents and the perception before the 1970s that the number of severe and invasive infections was low, only 12 antifungal agents are currently licensed for the treatment of systemic fungal infections: the polyene amphotericin B and its three lipid formulations, the pyrimidine synthesis inhibitor 5-fluorocytosine (flucytosine), the imidazoles miconazole and ketoconazole, the triazoles fluconazole, itraconazole and voriconazole, and the echinocandin caspofungin. However, the number of fungal diseases caused by both yeasts and molds has significantly increased during the past 20 years, especially among the increased number of immunocompromised patients, who are at high risk for life-threatening mycoses.

The Desk Encyclopedia of Microbiology
ISBN: 0-12-621361-5

There are more antifungal agents for topical treatment and agriculture and veterinary use, and several agents are under investigation for the management of severe and refractory fungal infections in humans (Table 4.1).

This article summarizes the most relevant facts regarding the chemical structure, mechanisms of action and resistance, pharmacokinetics, safety, adverse interactions with other drugs and applications of the established systemic and topical antifungal

TABLE 4.1 Antifungal agents, mechanisms of action, and their use[a]

Antifungal class	Antifungal target of action	Agent	Use
Polyenes	Membranes containing ergosterol	Amphotericin B (AMB)	Systemic mycoses[b,c]
		Nystatin (NYS)	Superficial mycoses[b,c]
		AMB lipid complex	Systemic mycoses intolerant or refractory to AMB
		AMB colloidal dispersion	
		Liposomal AMB	
		Liposomal NYS	Under investigation
		Pimaricin	Topical keratitis[b,c]
Phenolic benzyfuran cyclo-hexane	Microtubule aggregation and DNA inhibition	Griseofulvin	Dermatophytic infections[b]
Natural glutarimide	Protein synthesis inhibition	Cycloheximide	Laboratory and agriculture
Phenylpyrroles	Unknown	Fenpiclonil	Agriculture
		Fludioxonil	
Synthetic pyrimidines	Fungal cytosine permeae and deaminase	Flucytosine	Systemic (yeasts) in combination with AMB[b,c]
	Ergosterol inhibition	Triarimol	Agriculture
		Fenarimol	
Anilinopyrimidines	Enzyme secretion	Pyrimethanil	
		Cyprodinil	
Azoles	Ergosterol biosynthesis inhibition	Imidazoles	
		Clotrimazole	Topical, oral troche[b,c]
		Econazole	
		Isoconazole	
		Oxiconazole	
		Tioconazole	
		Miconazole	*P. boydii* infections only and veterinary[c]
		Ketoconazole	Secondary alternative to other agents and veterinary[b,c]
		Enilconazole	Veterinary
		Epoxiconazole	Agriculture
		Fluquinconazole	
		Triticonazole	
		Prochoraz	
		Triazoles	
		Fluconazole	Certain systemic and superficial diseases[b,c]
		Itraconazole	
		Terconazole	Intravaginal
		Voriconazole	Treatment of acute aspergillosis and salvage therapy for serious fungal infections
		Posaconazole	Under Investigation
		Ravuconazole	Under Investigation
Allylamines		Terbinafine	Superficial infections
		Naftifine	Topical
Benzylamines		Butenafine	Topical
Thiocarbamates		Tolnaftate	Topical
		Tolciclate	
		Piritetrade	
Dithiocarbamates	Non-specific	Mancozeb	Agriculture
		Thiram	

TABLE 4.1 *(Continued)*

Benzimidazoles and methylbenz-imidazole carbamates	Nuclear division	Carbendazim Benomyl Thiophanate	Agriculture
Morpholines	Ergosterol biosynthesis inhibition	Amorolfine	Topical
		Fenpropimorph Tridemorph	Agriculture
Pyridines		Buthiobate Pyrifenox	Agriculture
Echinocandins	Fungal β(1,3)-glucan synthesis inhibition	Papulocandins	None
		Caspofungin	Treatment of candidemia and refractory aspergillosis
		Anidulafungin	Under investigation
Pradimicins	Fungal sacharide (mannoproteins)	Pradamicin FA-2 (BMY 28864)	Under investigation
Benanomycins		Benanomycin A	Under investigation
Polyoxins	Fungal chitin synthase inhibition	Polyoxin D	None
Nikkomycins		Nikkomycin Z	Under investigation
Sordarins	Protein synthesis inhibition	GM 222712 GM 237354 GM 211676 GM 193663	Under investigation
Cinnamic acid	Cell wall	Dimethomorph	Agriculture
Oomycete fungicide	Oxidative phosphorylation	Fluazynam	Agriculture
Phthalimides	Non-specific	Captan Captafol Folpet	Agriculture
Cationic peptides	Lipid bilayer of biological membranes	Natural peptides	
		Cecropin Indolicidin	Under investigation
		Synthetic peptides	Under investigation
Amino acid analogs	Amino acid synthesis interference	RI 331 Azoxybacillins Cispentacin	Under investigation

[a]Only licensed, commonly used, and antifungals under clinical investigation are listed; see text for other antifungals.
[b]Clinical and veterinary use; other applications for use in humans only.
[c]A human product used in veterinary practice.

agents currently licensed for clinical, veterinary, or agricultural uses. A shorter description is provided for antifungal compounds that are in the last phases of clinical development, under clinical trials in humans, or that have been discontinued from additional clinical evaluation. The former compounds have potential use as therapeutic agents. More detailed data regarding these agents are found in the references.

I. THE POLYENES

The polyenes are macrolide molecules that target membranes containing ergosterol, which is an important sterol in the fungal cell membranes. Traces of ergosterol are also involved in the overall cell cycle of fungi.

A. Amphotericin B

Amphotericin B is the most important of the 200 polyenes. Amphotericin B replaced 2-hydroxystilbamidine in the treatment of blastomycosis in the mid-1960s. Two amphotericins (A and B) were isolated in the 1950s from *Streptomyces nodosus*, an aerobic bacterium, from a soil sample from Venezuela's Orinoco River Valley. Amphotericin B (the most active molecule) has seven conjugated double bounds, an internal ester, a free carboxyl group, and a glycoside side-chain with a primary amino group (Fig. 4.1A). It is unstable to

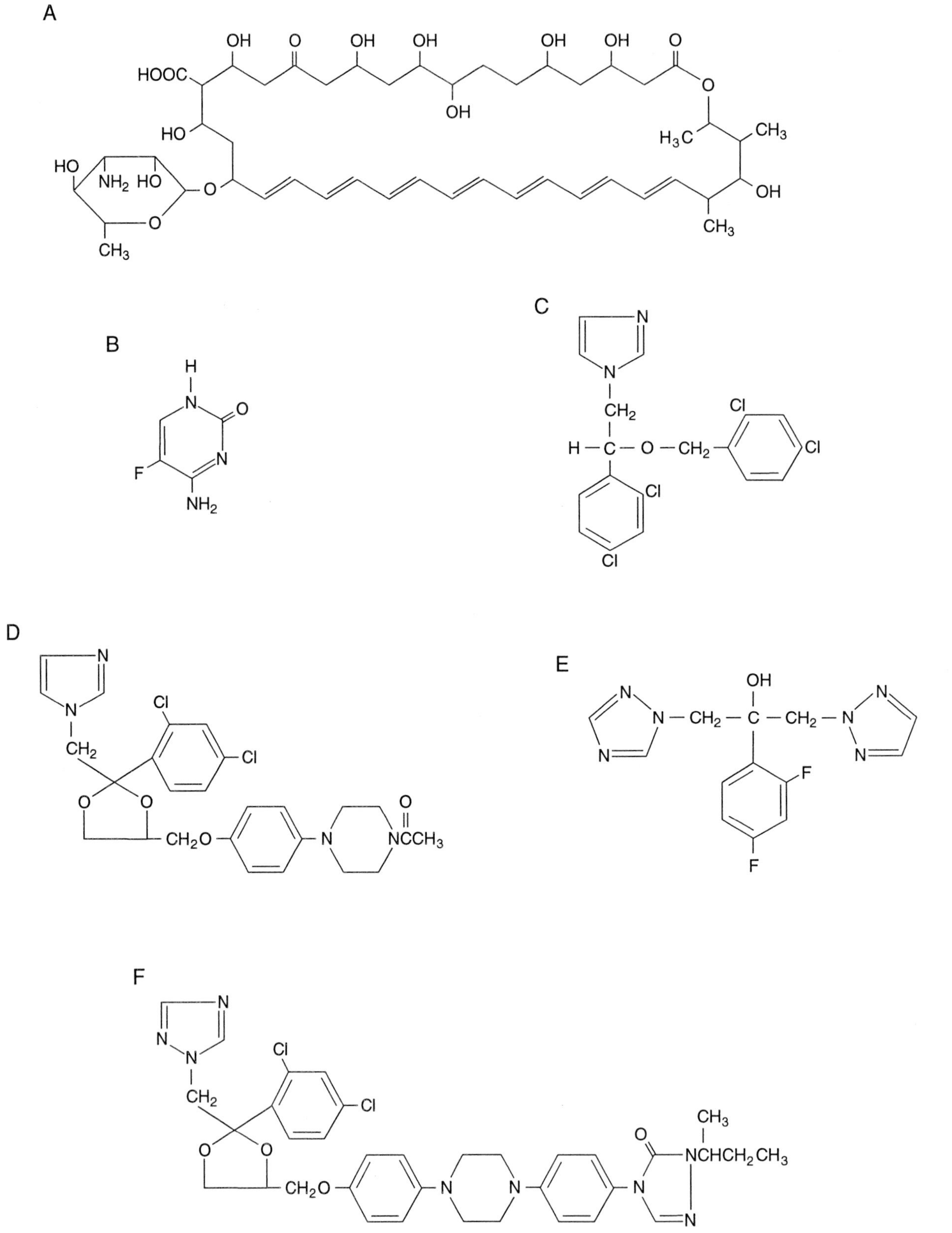

FIGURE 4.1 Chemical structures of some systemic licensed antifungal agents: (A) amphotericin B, (B) 5-fluorocytosine, (C) miconazole, (D) ketoconazole, (E) fluconazole, and (F) itraconazole.

heat, light, and acid pH. The fungistatic (inhibition of fungal growth) and fungicidal (lethal) activity of amphotericin B is due to its ability to combine with ergosterol in the cell membranes of susceptible fungi. Pores or channels are formed causing osmotic instability and loss of membrane integrity. This effect is not specific; it extends to mammalian cells. The drug binds to cholesterol, creating the high toxicity associated with all conventional polyene agents. A second mechanism of antifungal action has been proposed for amphotericin B, which is oxidation dependent. Amphotericin B is highly protein bound (91–95%). Peak serum of 1–3 μg/ml and trough concentrations of 0.5–1.1 μg/ml are usually measured after the intravenous (i.v.) administration of 0.6 mg/kg doses. Its half-life of elimination is 24–48 μh, with a long terminal half-life of up to 15 days.

Although resistance to amphotericin B is rare, quantitative and qualitative changes in the cell membrane sterols have been associated with the development of microbiological resistance both *in vitro* and *in vivo*. Clinically, resistance to amphotericin B has become an important problem, particularly with certain yeast and mold species, such as *Candida lusitaniae, C. krusei, C. glabrata, Aspergillus terreus, Fusarium* spp., *Malassezia furfur, Pseudallescheria boydii, Scedosporium prolificans, Trichosporon beigelii*, and other emerging fungal pathogens.

The *in vitro* spectrum of activity of amphotericin B includes yeasts, dimorphic fungi, and most of the opportunistic filamentous fungi. Clinically, amphotericin B is considered the gold standard antifungal agent for the management of most systemic and disseminated fungal infections caused by both yeasts and molds, including endemic (infections caused by the dimorphic fungi, *Cocccidioides immitis, Histoplasma capsulatum*, and *Blastomyces dermatitidis*) and opportunistic mycoses. Although it penetrates poorly into the cerebrospinal fluid (CSF), amphotericin B is effective in the treatment of both *Candida* and *Cryptococcus* meningitis alone and/or in combination with 5-fluorocytosine. Current recommendations regarding daily dosage, total dosage, duration, and its use in combination with other antifungal agents are based on the type of infection and the status of the host. Since severe fungal infections in the granulocytopenic host are difficult to diagnose and cause much mortality, empirical antifungal therapy with amphotericin B and other agents has improved patient care. Systemic prophylaxis for patients at high risk for invasive mycoses has also evolved. Toxicity is the limiting factor during amphotericin B therapy and has been classified as acute or delayed (Table 4.2). Nephrotoxicity is the most significant delayed adverse effect. Therefore, close monitoring of renal function tests, bicarbonate, electrolytes including magnesium, diuresis, and hydration status is recommended during amphotericin B therapy. Adverse drug interactions can occur with the administration of electrolytes and other concomitant drugs. This drug is also used for the treatment of systemic infections in small animals, especially blastomycosis in dogs, but it is not effective against aspergillosis. Side effects (especially in cats) and drug interactions are similar to those in humans.

TABLE 4.2 Adverse Side effects of the Licensed Systemic Antifungal Agents[a]

Side effect	Drug
Fever, chills	A, K, C, V
Rash	FC, K, I, FL, C, V
Nausea, vomiting	A, FC, K, I, FL, V
Abdominal pain	FC, K, V
Anorexia	A, K
Diarrhea	FC, V
Elevation of transaminases	FC, K, I, FL, C, V
Hepatitis (rare)	FC, K, I, FL
Anemia	A, FC
Leukopenia, thrombocytopenia	FC
Decreased renal function (azotemia, acidosis, hypokalemia, etc.)	A, C
Decreased testosterone synthesis	K (I, rare)
Adrenal insufficiency, menstrual irregularities, female alopecia	K
Syndrome of mineralocorticoid excess, pedal edema	I
Headache	A, FC, K, I, FL, V
Photophobia	K, V
Dizziness	I, V
Seizures	FL
Confusion	FC, V
Arthralgia, myalgia, thrombophlebitis	A
Abnormal vision	V
Cardiovascular (tachycardia and others)	C, V
Hypokalemia	C, V

[a]See Groll *et al.* (1998) for more detailed information. A, amphotericin B; FC, flucytosine; K, ketoconazole; FL, fluconazole; I, itraconazole; C, caspofungin, and V, voriconazole. (% of side effects for C and V are usually lower than those for other licensed agents.)

B. Nystatin

Nystatin was the first of the polyenes to be discovered when it was isolated from *S. noursei* in the early 1950s. It is an amphoteric tetrane macrolide that has a similar structure (Fig. 4.2A) and identical mechanism of action to those of conventional amphotericin B. Although it has an *in vitro* spectrum of activity similar to that of amphotericin B, this antifungal is used mostly for the therapy of gastrointestinal (orally) and mucocutaneous candidiasis (topically). This is not only due to its toxicity after parenteral administration to humans and lower animals but also to its lack of effectiveness when given i.v. to experimental animals.

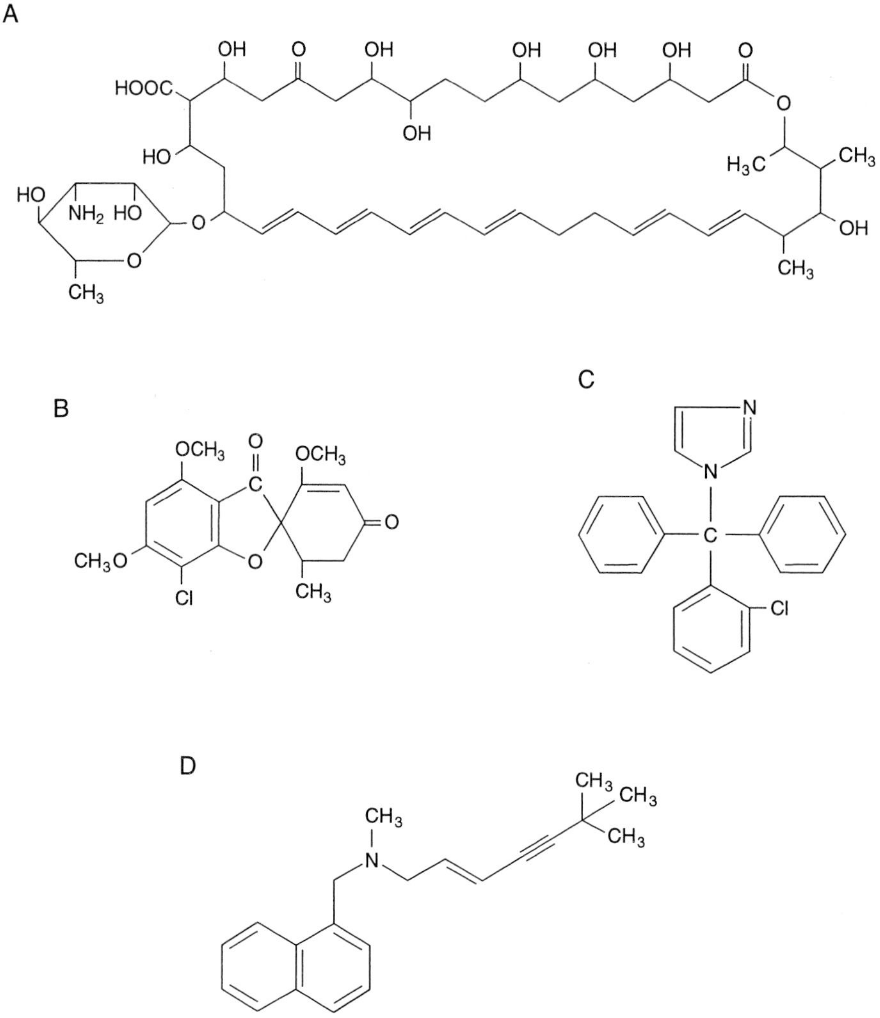

FIGURE 4.2 Chemical structures of the most commonly used topical antifungal agents: (A) nystatin, (B) griseofulvin, (C) clotrimazole, and (D) terbinafine.

It is used for candidiasis in small animals and birds and for otitis caused by *Microsporum canis*.

C. Lipid formulations

1. *Amphotericin B lipid formulations*

In an attempt to decrease the toxicity and increase the efficacy of amphotericin B in patients with deep-seated fungal infections refractory to conventional therapy, several lipid formulations of this antifungal have been developed since the 1980s. These preparations have selective toxicity or affinity for fungal cell membranes and theoretically promote the delivery of the drug to the site of infection while avoiding the toxicity of supramaximal doses of conventional amphotericin B. Because lysis of human erythrocytes is reduced, higher doses of amphotericin B can be safely used. Three lipid formulations of amphotericin B have been evaluated in clinical trials: an amphotercin B lipid complex, an amphotericin B colloidal dispersion, and a liposomal amphotericin B. However, despite evidence of nephrotoxicity reduction, a significant improvement in their efficacy compared to conventional amphotericin B has not been clearly demonstrated. Although these three formulations have been approved for the treatment of invasive fungal infections that have failed conventional amphotericin B therapy, enough information is not available regarding their pharmacokinetics, drug interactions, long-term toxicities, and the differences in both efficacy and tolerance among the three formulations. Also, the most cost-effective clinical role of these agents as first-line therapies has not been elucidated.

a. Liposomal amphotericin B

In the only commercially available liposomal formulation (ambisome), amphotericin B is incorporated into small unilamellar, spherical vesicles (60–70-nm liposomes). These liposomes contain hydrogenated soy phospatidylcholine and disteaoryl phosphatidylglycerol stabilized by cholesterol and amphotericin B in a 2 : 0.8 : 1 : 0.4 molar ratio. In the first liposomes, amphotericin B was incorporated into large, multilamellar liposomes that contained two phospholipids, dimyristoyl phosphatidylcholine (DMPC) and dimyristoyl phosphatidylglycerol (DMPG), in a 7:3 molar ratio (5–10% mole ratio of amphotericin B to lipid). This formulation is not commercially available, but it led to the development of commercial formulations.

b. Amphotericin B lipid complex

Amphotericin B lipid complex (abelcet) contains a DMPC/DMPG lipid formulation in a 7:3 ratio and a 50% molar ratio of amphotericin B to lipid complexes that form ribbon-like structures.

c. Amphotericin B colloidal dispersion

Amphotericin B colloidal dispersion (amphotec) contains cholesteryl sulfate and amphotericin B in a 1:1 molar ratio. This formulation is a stable complex of disk-like structures (122-nm in diameter and 4-nm thickness).

2. Liposomal nystatin

In order to protect human erythrocytes from nystatin toxicity and thus make this drug available as a systemic therapeutic agent, nystatin has been incorporated into stable, multilamellar liposomes, which contain DMPC and DMPG in a 7:3 ratio. It has been demonstrated that the efficacy of liposomal nystatin is significantly superior to that of conventional nystatin and is well tolerated in experimental murine models of systemic candidiasis and aspergillosis (fungal infections caused by *Candida* spp. and *Apergillus* spp.). In patients with hematological malignancies and refractory febrile neutropenia, dose-limiting nephrotoxicity has not been observed at high dosages. A 37% response to therapy has been documented in a small group of patients. Ongoing clinical trials would confirm the potential value of liposomal nystatin.

D. Candicidin

Candicidin is a conjugated heptaene complex produced by *S. griseus* that is selectively and highly active *in vitro* against yeasts. It is more toxic for mammalian cells than either amphotericin B or nystatin; therefore, its use was restricted to topical applications for the treatment of vaginal candidiasis (infections by *Candida albicans* and other *Candida* spp.).

E. Pimaricin

Pimaricin is a tetraene polyene produced by *S. natalensis*. It has a higher binding specificity for cholesterol than for ergosterol and, therefore, it is highly toxic for mammalian cells. The therapeutic use of pimaricin is limited to the topical treatment of keratitis (eye infections; also in horses) caused by the molds, *Fusarium* spp., *Acremonium* spp., and other species.

II. GRISEOFULVIN

Griseofulvin is a phenolic, benzyfuran cyclohexane agent (Fig. 4.2B) that binds to RNA. It is a product of *Penicillium janczewskii* and was the first antifungal agent to be developed as a systemic plant protectant. It acts as a potent inhibitor of thymidylate synthetase and interferes with the synthesis of DNA. It also inhibits microtubule formation and the synthesis of apical hyphal cell wall material. With the advent of terbinafine and itraconazole, the clinical use of griseofulvin as an oral agent for treatment of dermatophytic infections has become limited. However, it is frequently used for these infections in small animals, horses, and calves (skin only) as well as for equine sporotrichosis. Abdominal adverse side effects have been noted, especially in cats.

III. CYCLOHEXIMIDE

This is a glutaramide agent produced by *S. griseus*. This agent was among the three antifungals that were reported between 1944 and 1947. Although cycloheximide had clinical use in the past, it is currently used as a plant fungicide and in the preparation of laboratory media.

IV. PYRROLNITRIN, FENPICLONIL, AND FLUDIOXONIL

Pyrrolnitrin is the fermentation product of *Pseudomonas* spp. It was used in the past as a topical agent.

Fenpiclonil and fludioxonil (related to pyrrolnitrin) were the first of the phenylpyrrols to be introduced as cereal seed fungicides.

V. THE SYNTHETIC PYRIMIDINES

A. 5-Fluorocytosine (5-FC, flucytosine)

The synthetic 5-fluorocytosine is an antifungal metabolite that was first developed as an antitumor agent, but it is not effective against tumors. It is an oral, low-molecular-weight, fluorinated pyrimidine related to 5-fluorouracil and floxuridine (Fig. 4.1B). It acts as a competitive antimetabolite for uracil in the synthesis of yeast RNA; it also interferes with thymidylate synthetase. Several enzymes are involved in the mode of action of 5-fluorocytosine. The first step is initiated by the uptake of the drug by a cell membrane-bound permease. Inside the cell, the drug is deaminated to 5-fluorouracil, which is the main active form of the drug. These activities can be antagonized *in vitro* by a variety of purines and pyrimidine bases and nucleosides. At least two metabolic sites are responsible for resistance to this compound: one involves the enzyme cytosine permease, which is responsible for the uptake of the drug into the fungal cell, and the other involves the enzyme cytosine deaminase, which is responsible for the deamination of the drug to 5-fluorouracil. Alterations of the genetic regions encoding these enzymes may result in fungal resistance to this drug by either decreasing the cell wall permeability or synthesizing molecules that compete with the drug or its metabolites. Development of flucytosine resistance during therapy against *Candida* spp. and *C. neoformans* has been documented since the early 1970s.

5-Fluorocytosine has fungistatic but not fungicidal activity mostly against yeasts; its activity against molds is inoculum dependent. Clinically, the major therapeutic role of 5-fluorocytosine is its use in combination with amphotericin B in the treatment of meningitis caused by the yeast *C. neoformans*. The synergistic antifungal activity of these two agents has been demonstrated in clinical trials in non-HIV-infected and AIDS patients. 5-Fluorocytosine should not be used alone for the treatment of any fungal infections. Therapeutic combinations of 5-fluorocytosine with several azoles are under investigation. The most serious toxicity associated with 5-fluorocytosine therapy is bone marrow suppression (6% of patients), which leads to neutropenia, thrombocytopenia, or pancytopenia (Table 4.2). Therefore, monitoring of the drug concentration in the patient's serum (serial 2-h levels post-oral administration) is highly recommended to adjust dosage and maintain serum levels between 40 and 60 μg/ml. Since the drug is administered in combination with amphotericin B, a decrease in glomerular filtration rate, a side effect of the latter compound, can induce increased toxicity to 5-fluorocytosine. Adverse drug interactions can occur with other antimicrobial and anticancer drugs, cyclosporine, and other therapeutic agents. Because of its toxic potential, 5-fluorocytosine should not be administered to pregnant women or animals. This drug has been used in combination with ketoconazole for cryptococcosis in small animals (very toxic for cats) and also for respiratory apergillosis and severe candidiasis in birds.

B. Triarimol, fenarimol, pyrimethanil, and cyprodinil

Triarimol and fenarimol are pyrimidines with a different mechanism of action than that of 5-fluorocytosine. They inhibit lanosterol demethylase, an enzyme involved in the synthesis of ergosterol, which leads to the inhibition of this biosynthetic pathway. Triarimol and fenarimol are not used in medicine but are used extensively as antifungal agents in agriculture.

The anilino-pyrimidines, pyrimethanil and cyprodinil, inhibit the secretion of the fungal enzymes that cause plant cell lysis. Pyrimethanil has activity (without cross-resistance) against *Botrytis cinerea* (vines, fruits, vegetables, and ornamental plants infections) and *Venturia* spp. (apples and pears), whereas cyprodinil has systemic activity against *Botrytis* spp., but only a preventive effect against *Venturia* spp.

VI. THE AZOLES

The azoles are the largest single source of synthetic antifungal agents; the first azole was discovered in 1944. As a group, they are broad-spectrum in nature and mostly fungistatic. The broad spectrum of activity involves fungi (yeasts and molds), bacteria, and parasites. This group includes fused ring and N-substituted imidazoles and the N-substituted triazoles. The mode of action of these compounds is the inhibition of lanosterol demethylase, a cytochrome P-450 enzyme.

A. Fused-ring imidazoles

The basic imidazole structure is a cyclic five-member ring containing three carbon and two nitrogen molecules. In the fused-ring imidazoles, two carbon molecules are shared in common with a fused benzene ring. Most of these compounds have parasitic activity (anthelmintic) and two have limited antifungal activity: 1-chlorobenzyl-2-methylbenzimidazole and thiabendazole.

1. *1-Chlorobenzyl-2-methylbenzimidazole*

The azole 1-chlorobenzyl-2-methylbenzimidazole was developed specifically as an anti-*Candida* agent. It has been used in the past in the treatment of superficial yeast and dermatophytic infections.

2. *Thiabendazole*

Thiabendazole was developed as an anthelmintic agent and has a limited activity against dermatophytes. It was also used in the past in the treatment of superficial yeast and dermatophytic infections. Thiabendazole has been used for aspergillosis and penicillosis in dogs.

B. N-substituted (mono) imidazoles

In this group, the imidazole ring is intact and substitutions are made at one of the two nitrogen molecules. At least three series of such compounds have emerged for clinical and agricultural use. In the triphenylmethane series, substitutions are made at the nonsymmetrical carbon atom attached to one nitrogen molecule of the imidazole ring. In the second series, the substitutions are made at a phenethyl configuration attached to the nitrogen molecule. The dioxolane series is based on a 1,3-dioxolane molecule rather than on the 1-phenethyl molecule. These series vary in spectrum, specific level of antifungal activity, routes of administration, and potential uses.

1. *Clotrimazole*

Clotrimazole is the first member of the triphenylmethane series of clinical importance (Fig. 4.2C). It has good *in vitro* activity at very low concentrations against a large variety of fungi (yeasts and molds). However, hepatic enzymatic inactivation of this compound, after systemic administration, has limited its use to topical applications (1% cream, lotion, solution, tincture, and vaginal cream) for superficial mycoses (nail, scalp, and skin infections) caused by the dermatophytes and *M. furfur*, for initial and/or mild oropharyngeal candidiasis (OPC; 10-mg oral troche), and for the intravaginal therapy (single application of 500-mg intravaginal tablet) of vulvovaginal candidiasis. Other intravaginal drugs require 3–7-day applications. This drug is also used for candidal stomatitis, dermatophytic infections, and nasal aspergillosis (infused through tubes) in dogs.

2. *Bifonazole*

Bifonazole is a halogen-free biphenylphenyl methane derivative. Bifonazole is seldom utilized as a topical agent for superficial infections, despite its broad spectrum of activity. Its limited use is the result of its toxic side effects for mammalian cells. Bifonazole is retained in the dermis for a longer time than clotrimazole.

3. *Econazole, isoconazole, oxiconazole, and tioconazole*

Other frequently used topical imidazoles include econazole (1% cream), isoconazole (1% cream), oxiconazole (1% cream and lotion), and tioconazole (6.5% vaginal ointment) (Table 4.1). As with clotrimazole, a single application of tioconazole is effective in the management of vulvovaginal candidiasis and as a nail lacquer for fungal onychomycosis (nail infections). Mild to moderate vulvovaginal burning has been associated with intravaginal therapy. Oxiconazole and econazole are less effective than terbinafine and itraconazole in the treatment of onychomycosis and other infections caused by the dermatophytes. Although topical agents do not cure onychomycosis as oral drugs do, they may slow down the spread of this infection. However, the recommended drugs for the treatment of onychomycosis are terbinafine (by dermatophytes) and itraconazole.

4. *Lanoconazole*

In recent years, lanoconazole has been introduced for topical treatment of dermatomycoses. It appears to have superior activity *in vitro* and in experimental infections in guinea pigs than those of earlier compound.

5. *Miconazole*

Miconazole was the first azole derivative to be administered intravenously for the therapy of systemic fungal infections. Its use is limited, due to toxicity and high relapse rates, to certain cases of invasive infections caused by the opportunistic mold, *P. boydii*. Since this compound is insoluble in water, it was dissolved in a polyethoxylated castor oil for its systemic administration. This solvent appears to be the cause of the majority of miconazole side effects (pruritus, headache, phlebitis, and hepatitis). On the other hand, miconazole is used for dermatophytic infections in large animals, fungal keratitis and pneumonia in horses, resistant yeast infections to nystatin in birds, and aspergillosis in raptors. However, safety and efficacy data are not available (veterinary use).

6. *Ketoconazole*

Ketoconazole was the first representative of the dioxolane series (Fig. 4.1D) to be introduced into clinical

TABLE 4.3 Adverse interactions of the licensed systemic azoles with other drugs during concomitant therapy[a]

Azole	Concomitant drug	Adverse side effect of interaction
K, Fl, I	Nonsedating antihistamines, cisapride, terfenadine, astemizole	Fetal arrhythmia
K, Fl, I, V	Rifampin, isoniazid, phenobarbital, rifabutin, carbamazepine, and phenytoin	Reduce azole plasma concentrations
K, Fl, I, V	Phenytoin, benzodiazepines, rifampin	Induces the potential toxicity levels of cocompounds
K, I	Antacids, H_2 antagonists, omeprazole, sucralfate, didanosine	Reduces azole absorption
K, Fl, I	Lovastin, simvastatin	Rhabdomyolysis
I	Indinavir, vincristine, quinidine, digoxin, cyclosporine, tacrolimus, methylprednisolone, and ritonavir	Induces potential toxicity cocompounds
Fl, I, V	Warfarin, rifabutin, sulfonylurea	Induces potential toxicity of cocompounds
K	Saquinavir, chlordiazepoxide, methylprednisone	Induces potential toxicity of these compounds
K	Protein-binding drugs	Increases the release of fractions of free drug
K, C, V	Cyclosporine A	Nephrotoxicity (concomitant use with C is not recommended)
C	Tacrolimus	Tacrolimus concentration can be decreased
C	Efavirenz, nevirapine, phenytoin, dexamethasone, carbamazepine, and rifampin	Can significantly reduce C concentrations (use of daily dose of 70 mg of C should be considered when C is co-administered with some of these compounds.

[a]See Groll *et al.* (1998) for more detailed information. K, ketoconazole; Fl, fluconazole; I, itraconazole; C, caspofungin; V, voriconazole.

use and was the first orally active azole. Ketoconazole requires a normal intragastric pH for absorption. Its bioavailability is highly dependent on the pH of the gastric contents; an increase in pH will decrease its absorption, for example, in patients with gastric achlorhydria or treated with antacids or H_2-receptor antagonists (Table 4.3). This drug should be taken with either orange juice or a carbonated beverage.

Ketoconazole pharmocokinetics corresponds to a dual model with an initial half-life of 1–4 h and a terminal half-life of 6–10 h, depending on the dose. This drug highly binds to plasma proteins and penetrates poorly into the CSF, urine, and saliva. Peak plasma concentrations of approximately 2, 8, and 20 μg/ml are measured 1–4 h after corresponding oral doses of 200, 400, and 800 mg. The most common and dose-dependent adverse effects of ketonazole are nausea, anorexia, and vomiting (Table 4.2). They occur in 10% of the patients receiving a 400-mg dose and in approximately 50% of the patients taking 800-mg or higher doses. Another limiting factor of ketoconazole therapy is its numerous and significant adverse interactions with other concomitant drugs (see Table 4.3 for a summary of the interactions of the azoles with other drugs administered to patients during azole therapy).

In vitro, ketoconazole has a broad spectrum of activity comparable to that of miconazole and the triazoles. However, due to its adverse side effects, its adverse interaction with other drugs, and the high rate of relapses, ketoconazole has been replaced by itraconazole as an alternative to amphotericin B for the treatment of immunocompetent individuals with non-life-threatening, non-central nervous system, localized or disseminated histoplasmosis, blastomycosis, mucocutaneous candidiasis, paracoccidioidomycosis, and selected forms of coccidioidomycosis. In non-cancer patients, this drug can be effective in the treatment of superficial *Candida* and dermatophytic infections when the latter are refractory to griseofulvin therapy. Therapeutic failure with ketoconazole has been associated with low serum levels; monitoring of these levels is recommended in such failures. Ketoconazole also has been used for a variety of systemic and superficial fungal infections in cats and dogs.

7. *Enilconazole*

This is the azole most widely used in veterinary practice for the intranasal treatment of aspergillosis and penicillosis as well as for dermatophytic infections. The side effects are few.

8. *Epoxiconazole, fluquinconazole, triticonazole, and prochloraz*

Epoxiconazole, fluquinconazole, and triticonazole are important agricultural fungicides which have a wider spectrum of activity than that of the earlier triazoles, triadimefon and propiconazole, and the imidazole, prochloraz, as systemic cereal fungicides. However, development of resistance to these compounds has been documented.

C. The triazoles

The triazoles are characterized by a more specific binding to fungal cell cytochromes than to mammalian cells due to the substitution of the imidazole ring by the triazole ring. Other beneficial effects of this substitution are: (i) an improved resistance to metabolic degradation; (ii) an increased potency; and (iii) a superior antifungal activity. Although fluconazole, itraconazole and voriconazole are the only three triazoles currently licensed for antifungal systemic therapy, several other triazoles are at different levels of clinical evaluation. Voriconazole has been licensed in Europe (Table 4.1).

1. *Fluconazole*

Fluconazole is a relatively small molecule (Fig. 4.1E) that is partially water soluble, minimally protein bound, and excreted largely as an active drug in the urine. It penetrates well into the CSF and parenchyma of the brain and the eye, and it has a prolonged half-life (up to 25 h in humans). The pharmacokinetics are independent of the route of administration and of the drug formulation and are linear. Fluconazole is well absorbed orally (its total bioavailability exceeds 90%), and its absorption is not affected by food or gastric pH. Plasma concentrations of 2–7 μg/ml are usually measured in healthy subjects after corresponding single doses of 100 and 400 mg. After multiple doses, the peak plasma levels are 2.5 times higher than those of single doses. The CSF to serum fluconazole concentrations are between 0.5 and 0.9% in both healthy human subjects and laboratory animals.

Fluconazole does not have *in vitro* or *in vivo* activity against most molds. Both oral and i.v. formulations of fluconazole are available for the treatment of candidemia in nonneutropenic and other nonimmunosuppressed patients, mucosal candidiasis (oral, vaginal, and esophageal), and chronic mucocutaneous candidiasis in patients of all ages. Fluconazole is the current drug of choice for maintenance therapy of AIDS-associated cryptococcal and coccidioidal meningitis. It is also effective as prophylactic therapy for immunocompromised patients to prevent both superficial and life-threatening fungal infections. However, since the cost of fluconazole is high and resistance to this drug can develop during therapy, fluconazole prophylaxis should be reserved for HIV-infected individuals or AIDS patients, who are refractory and intolerant to topical agents, or for patients with prolonged (+2 weeks) and profound neutropenia (+1500 cells). Although the recommended dosage of fluconazole for adults is 100–400 mg qd, higher doses (+800 mg qd) are required for the treatment of severe invasive infections and for infections caused by a *Candida* spp. that exhibit a minimum inhibitory concentration (MIC) of +8 μg/ml. However, despite the fluconazole MIC obtained when the infecting yeast is either *Candida krusei* or *C. glabrata*, intrinsic resistance to these yeasts precludes its use for the treatment of such infections. In contrast to the imidazoles and itraconazole, fluconazole does not exhibit major toxicity side effects (2.8–16%). However, when the dosage is increased above 1200 mg, adverse side effects are more frequent (Table 4.2). Fluconazole interactions with other concomitant drugs are similar to those reported with other azoles, but they are less frequent than those exhibited by ketoconazole and itraconazole (Table 4.3). Fluconazole has been used to treat nasal aspergillosis and penicillosis in small animals and birds when topical enilconazole is not feasible.

2. *Itraconazole*

Itraconazole is another commercially available oral triazole for the treatment of certain systemic mycoses. In contrast to fluconazole, itraconazole is insoluble in aqueous fluids; it penetrates poorly into the CSF and urine but well into skin and soft tissues; and it is highly protein bound (+90%). Its structure is closely related to that of ketoconazole (Fig. 4.1F), but itraconazole has a broader spectrum of *in vitro* and *in vivo* antifungal activity than those of both ketonazole and fluconazole. Similar to ketoconazole, itraconazole is soluble only at low pH and is better absorbed when the patient is not fasting. Absorption is erratic in cancer patients or when the patient is taking concomitant H_2-receptor antagonists, omeprazole, or antacids. Therefore, this drug should be taken with food and/or acidic fluids. Plasma peak (1.5–4 h) and trough concentrations between 1 and 2.2 and 0.4 and 1.8 μg/ml, respectively, are usually obtained after 200-mg dosages (capsule) as either single daily dosages (po or bid) or after i.v. administration (bid) for 2 days and qd for more days; these concentrations are also obtained in cancer patients receiving 5 mg/kg divided into two oral solution dosages.

Clinically, itraconazole (200–400 mg/day) has supplanted ketoconazole as first-line therapy for endemic, non-life-threatening mycoses caused by *B. dermatitidis*, *C. immitis*, and *H. capsulatum* as well as by *Sporothrix schenckii*. For more severe mycoses, higher doses are recommended and clinical resistance may emerge. It can also be effective as a second-line agent for refractory or intolerant infections to conventional amphotericin B therapy, for example, infections by the phaeoid (dematiaceous or black molds or yeasts) fungi and *Aspergillus* spp. Itraconazole is commercially available as oral solution, tablet, and i.v. suspension. The oral solution is better absorbed than the tablet and has become useful for the treatment of HIV-associated oral and esophageal candidiasis, especially for those cases that are resistant to fluconazole. However, monitoring of itraconazole plasma concentrations is recommended during treatment of both superficial and invasive diseases: Drug concentration

+0.5 μg/ml by high-performance liquid chromatography and +12 μg/ml by bioassay appear to be critical for favorable clinical response. Treatment with itraconazole has been associated with less adverse and mostly transient side effects (+110%) than that with ketoconazole (Table 4.2), and these effects are usually observed when the patient takes up to 400 mg during several periods of time. Itraconazole has been used for the treatment of endemic mycoses, aspergillosis, and crytococcosis in dogs (especially blastomycosis), equine sporotrichosis, and osteomyelitis caused by *C. immitis* in large animals, but its use is minimal. No data are available regarding its side effects or drug interactions in animals.

3. *Voriconazole (UK-109496)*

Voriconazole is a novel fluconazole derivative obtained by replacement of one triazole moiety by fluoropyrimidine and α-methylation groups (Fig. 4.3A).

In contrast to fluconazole and similar to itraconazole, voriconazole is non-water soluble. As do the other azoles, voriconazole acts by inhibiting fungal cytochrome P450-dependent, 14-α-sterol

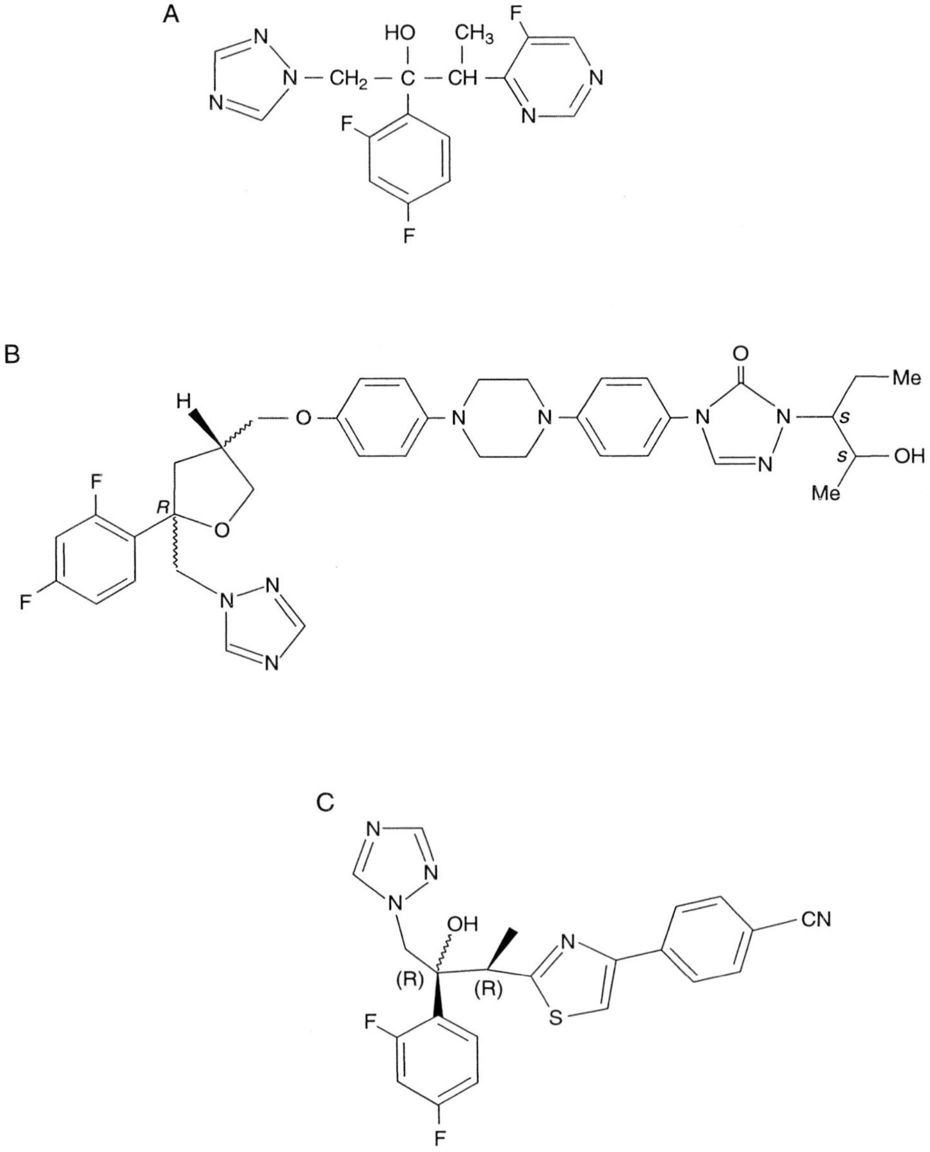

FIGURE 4.3 Chemical structures of three new triazoles: (A) voriconazole, (b) posaconazole, and (C) ravuconazole.

demethylase-mediated synthesis of ergosterol. Voriconazole pharmacokinetics in humans are nonlinear. Following single oral doses, peak plasma concentrations were achieved after 2 h and multiple doses resulted in a higher (eight times) accumulation. The mean half-life of elimination is about 6 h. Voriconazole binds to proteins (65%), is extensively metabolized in the liver, and is found in the urine (78–88%) practically unchanged after a single dose. Voriconazole has an improved *in vitro* fungistatic activity and an increased potency against most fungi compared to those of fluconazole. It is fungicidal against some fungi, especially *Aspergillus* spp. However, less *in vitro* activity has been demonstrated for the opportunistic molds *Fusarium* spp., *Rhizopus arrhizus*, *S. schenckii*, and other less common emerging fungi. Studies in neutropenic animal models have demonstrated that voriconazole is superior to both amphotericin B and itraconazole for the treatment of certain opportunistic (especially aspergillosis) and endemic mycoses. This compound has undergone phase III evaluation for the treatment of invasive aspergillosis and infections refractory to established antifungal agents in humans. In the United States, voriconazole was approved May 24, 2002 for primary treatment of acute aspergillosis and as salvage therapy for serious fungal infections caused by *S. apiospermum* and *Fusarium* spp. However, the European label is for treatment of invasive aspergillosis, fluconazole-resistant serious invasive *Candida* infections (including *C. krusei*) and treatment of serious fungal infections caused by *Scedosporium* spp. and *Fusarium* spp. The drug has been well tolerated with only reversible side effects. In patients, hepatic (10–15%), transient visual (10–15%) and skin rash (1–5%) side effects have been observed.

4. *Terconazole*

Terconazole was the first triazole marketed for the topical treatment of vaginal candidiasis and superficial dermatophyte infections. Currently, it is only used for vulvovaginal candidiasis (0.4 and 0.8% vaginal creams and 80-mg vaginal suppositories).

D. Investigational triazoles

As fungal infections became an important health problem and resistance to established agents began to emerge, new triazoles were developed with a broader spectrum of antifungal activity. Early investigational triazoles, such as R 66905 (saperconazole), BAY R 8783, SCH 39304, and SCH 51048, were discontinued from further development due a variety of adverse side effects. Two triazoles (posaconazole and ravuconazole) are currently under clinical investigation (Table 4.1) and others are at earlier stages of development.

1. *SCH 39304, SCH 51048, and SCH 56592*

a. SCH 39304

SCH 39304 is an N-substituted difluorophenyl triazole with both *in vitro* and *in vivo* (oral and parenteral) activity for both yeasts and molds. Although preliminary clinical trials demonstrated that this compound was well tolerated by humans and had good pharmacokinetic properties, additional clinical development was precluded by the incidence of hepatocellular carcinomas in laboratory animals during prolonged treatment.

b. SCH 51048

SCH 51048 is a tetrahydrofurane-based triazole that has superior potency (orally) than that of SCH 39304 toward the target enzyme and good *in vitro* activity against a variety of fungi. Although animal studies demonstrated that this drug is also orally effective for the treatment of systemic and superficial yeast and mold infections, the slow absorption rate from the intestinal track due to its poor water solubility precluded its further clinical development.

c. Posaconazole

Posaconazole (SCH 56592) is the product of a modification of the *n*-alkyl side chain of SCH 51048 which included a variety of chiral substituents (Fig. 4.3*b*). The *in vitro* fungistatic and fungicidal activities of posaconazole are similar to those of voriconazole and ravuconazole and superior or comparable to those of the established agents against yeasts, the dimorphic fungi, most opportunistic molds including *Aspergillus* spp., the Zygomycetes, certain phaeoid fungi, and the dermatophytes. It has been demonstrated that posaconazole is superior to itraconazole for the treatment of experimental invasive aspergillosis in animals infected with strains of *Aspergillus fumigatus* with high and low itraconazole MICs. Posaconazole has been effective in the treatment of patients with non-meningeal coccidioidomycosis, oropharangeal infections and refractory and invasive mold infections including these caused by *Fusarium* spp. and the Zygomycetes. Similar results have been obtained for a variety of superficial and invasive infections in other animal models. The pharmacokinetics of posaconazole have been studied in laboratory animals and although drug concentrations above both MIC and MFC (fungicidal) have been determined after a single po dose at 24 h, it has been demonstrated that plasma concentrations should be 5–10 times higher than the MIC. Also, its absorption from the intestinal tract is slow and peak serum concentrations are achieved 11–24 h after the actual dose. The clinical utility of this compound has yet to be determined in clinical trials in humans.

2. *Ravuconazole (BMS-207147; ER-30346)*

Ravuconazole (BMS-207147) is a novel oral thiazole-containing triazole (Fig. 4.3c) with a broad spectrum of activity against the majority of opportunistic pathogenic fungi. The antifungal activity of this triazole against *A. fumigatus* appears to be enhanced by the introduction of one carbon chain between the benzylic *tert* carbon and thiazole substituents and the cyano group on the aromatic ring attached to the thiazole. Ravuconazole has a similar or superior *in vitro* activity compared to those of the other investigational and established drugs against most pathogenic yeasts, with the exceptions of *C. tropicalis* and *C. glabrata*. Ravuconazole also has good *in vivo* antifungal activity in murine models for the treatment of invasive aspergillosis, candidiasis, and cryptococcosis. Ravuconazole shows good pharmacokinetics in animals that is similar to that of itraconazole. This indicates that ravuconazole is absorbed at levels comparable to those of itraconazole. However, the half-life of ravuconazole (4 h) is longer than that of itraconazole (1.4 h) and similar to that of fluconazole. The potential use of ravuconazole has yet to be determined in clinical trials in humans.

3. *Saperconazole (R 66905)*

Saperconazole is a lipophilic and poorly water-soluble fluorinated triazole; its chemical structure resembles that of itraconazole. Although both *in vitro* and *in vivo* antifungal activities were demonstrated against yeasts and molds and it was well tolerated during three clinical trials, this triazole was discontinued due to the incidence of malignant adrenal tumors in laboratory animals (long-term toxicity experiments).

4. *BAY R 3783*

This metabolite triazole was also discontinued from further clinical development due to the potential toxic effect during prolonged therapy.

5. *SDZ 89-485*

The antifungal activity of the D-enantiomer SDZ 89–485 antifungal triazole was demonstrated only in a few laboratory animal studies, and additional studies were not conducted with this compound.

6. *D 0870*

Although more *in vitro* and *in vivo* studies were conducted with D 0870 than with SDZ-89-485, and D 0870 showed good antifungal activity, this drug was also discontinued by its original developers. The *in vitro* activity of D 0870 is lower than that of itraconazole against *Aspergillus* spp., but higher for the common *Candida* spp. Therefore, evaluation of this compound has been continued by another pharmaceutical company for the treatment of OPC in HIV-infected individuals. It has also shown activity against the parasite *Trypanosoma cruzei*.

7. *T-8581*

T-8581 is a water-soluble 2-fluorobutanamide triazole derivative. High peak concentrations (7.14–12 μg/ml) of T-8581 were determined in the sera of laboratory animals following the administration of single oral doses of 10 mg/kg, and the drug was detected in the animals sera after 24 h. The half-life of T-8581 varies in the different animal models from 3.2 h in mice to 9.9 h in dogs. Animal studies suggest that the absorption of this compound is almost complete after po dosages. The maximum solubility of T-8581 is superior (41.8 mg/ml) to that of fluconazole (2.6 mg/ml), which suggests the potential use of this compound as an alternative to fluconazole for high-dose therapy.

T-8581 has shown potent *in vitro* antifungal activity against *Candida* spp., *C. neoformans*, and *A. fumigatus*. The activity of T-8581 is similar to that of fluconazole for the treatment of murine systemic candidiasis and superior to itraconazole for aspergillosis in rabbits. The safety of T-8581 is under evaluation.

8. *UR-9746 and UR-9751*

UR-9746 and UR-9751 are similar and recently introduced fluoridated triazoles that contain an *N*-morpholine ring, but UR-9746 has an extra hydroxyl group. The pharmacokinetics of these two compounds in laboratory animals has demonstrated peak concentrations (biological activity) of 184 (UR-9746) and 34 μg/ml (UR-9751) after 8 and 8–24 h, respectively. A slow decline of these levels was seen after 48 h. Chronic (19 days) doses of 100 mg/kg produced higher peak levels than single doses; two peaks were observed after 1 and 8 h. However, the rate of decline of the drug after 24 h was faster after multiple than after single doses. Superior *in vitro* and *in vivo* activity than that of fluconazole has been demonstrated with these compounds against *Candida* spp., *C. neoformans*, *H. capsulatum*, and *C. imitis*. Both antifungals lacked detectable toxicity in experimental animal infections. Although UR-9751 MICs were fourfold higher than those of UR-9746, the *in vivo* activity in the animal model of systemic murine coccidioidomycosis was similar. Additional studies will determine the potential use of these compounds as systemic therapeutic agents in humans.

9. TAK 187 and SSY 726

Some *in vitro* and very little *in vivo* data are available for these new triazoles.

10. KP-103

KP-103 is another novel triazole that is being developed for the local treatment of dermatomycoses. Because this azole has low affinity for keratin, its antifungal activity is not lost as it penetrates skin tissue. The clinical values of this agent is to be determined in clinical trials.

VII. THE ALLYLAMINES

The allylamines are synthetic compounds that were introduced in the 1970s. They act by inhibiting squalene epoxidase, which results in a decrease of the ergosterol content and an accumulation of squalene affecting membrane structure and function (e.g. nutrient uptake).

A. Terbinafine

Terbinafine is the most active derivative of this class of antifungals. It has an excellent *in vitro* activity against the dermatophytes and other filamentous fungi, but its *in vitro* activity against the yeasts is controversial. It follows linear pharmacokinetics over a dose range of 125–750 mg; drug concentrations of 0.5–2.7 μg/ml are detected 1 or 2 h after a single oral dose. Terbinafine has replaced griseofulvin and ketoconazole for the treatment of onychomycosis and other infections caused by dermatophytes (oral and topical). It is also effective for the treatment of vulvovaginal candidiasis. It is usually well tolerated at oral doses of 250 and 500 mg/day and the side effects (~10%) are gastrointestinal and cutaneous. The metabolism of terbinafine may be decreased by cimetidine and increased by rifampin. Resistance has been reported for *Ustilago maydis*, a corn pathogen; resistance involved a decreased affinity for the target enzyme as well as a decreased accumulation of drug inside the fungal cell.

B. Naftifine

Pharmacokinetics and poor activity have limited the use of naftifine to topical treatment of dermatophytic infections.

VIII. THE BENZYLAMINES, THIOCARBAMATES, AND DITHIOCARBAMATES

The benzylamine, butenafine, and the thiocarbamates, tolnaftate, tolciclate, and piritetrade, also inhibit the synthesis of ergosterol at the level of squalene. Their clinical use is limited to the topical treatment of superficial dermatophytic infections.

The Bordeaux mixture (reaction product of copper sulfate and lime) was the only fungicide used until the discovery of the dithiocarbamate fungicides in the mid-1930s. Of those, mancozeb and thiram are widely used in agriculture, but because they are only surface-acting materials frequent spray applications are required. Ferbam, maneb, and zineb are not used as much.

THE BENZIMIDAZOLES AND METHYLBENZIMIDAZOLE CARBAMATES

A great impact on crop protection was evident with the introduction of the benzimidazoles and other systemic (penetrate the plant) fungicides. These compounds increased spray intervals to 14 days or more. The methylbenzimidazole carbamates (MBCs; carbendazim, benomyl, and thiophanate) inhibit nuclear division and are also systemic agricultural fungicides. However, since MBC-resistant strains of *B. cinerea* and *Penicillium expansum* have been isolated, these compounds should be used in combination with *N*-phenylcarbamate or agents that have a different mode of action.

THE MORPHOLINES

The morpholines interfere with δ14 reductase and δ7 and δ8 isomerase enzymes in the ergosterol biosynthetic pathway, which leads to an increase of toxic sterols and an increase in the ergosterol content of the fungal cell.

A. Amorolfine

Amorolfine, a derivative of fenpropimorph, is the only morpholine that has a clinical application for the topical treatment of dermatophytic infections and candidal vaginitis.

B. Fenpropimorph, tridemorph, and other morpholines

Protein binding and side effects have precluded the clinical use of these morpholines, but they are important agricultural fungicides.

XI. THE PYRIDINES

The pyridines are another class of antifungal agents that inhibit lanosterol demethylase.

A. Buthiobate and pyrifenox

These agents are important agricultural fungicides.

XII. THE ECHINOCANDINS, PNEUMOCANDINS, AND PAPULOCANDINS

The echinocandins and papulocandins are naturally occurring metabolites of *Aspergillus nidulans* var. *echinulatus* (echinocandin B), *A. aculeatus* (aculeacin A), and *Papularia sphaerosperma* (papulocandin). They act specifically by inhibiting the synthesis of fungal β(1,3)-glucan synthesis, which results in the depletion of glucan, an essential component of the fungal cell wall.

A. The papulocandins

The papulocandins A–D, L687781, BU4794F, and chaetiacandin have *in vitro* activity only against *Candida* spp., but poor *in vivo* activity, which precluded clinical development.

B. The echinocandins

The echinocandins include echinocandins, pneumocandins, aculeacins, mulundo- and deoxymulundocandin, sporiofungin, vWF 11899 A–C, and FR 901379. The echinocandins have better *in vitro* and *in vivo* antifungal activity than the papulocandins. Pharmaceutical development has resulted in several semisynthetic echinocandins with an improved antifungal activity compared to those of the naturally occurring molecules described previously.

1. *Cilofungin (LY 121019)*

Cilofungin is a biosemisynthetic analog of the naturally occurring and toxic (erythrocytes lysis) 4-*n*-octyloxybenzoyl-echinocandin B. Although it showed good *in vitro* activity against *Candida* spp., this drug was discontinued due to the incidence of metabolic acidosis associated with its intravenous carrier, polyethylene glycol.

2. *Anidulafungin (V-echinocandin, LY 303366)*

This is another semisynthetic cyclic lipopeptide, which resulted from an increase of aromatic groups in the cilofungin side-chain (Fig. 4.4A). It has high potency and oral and parenteral bioavailability. In laboratory animals, peak levels in plasma (5 or 6 h) of 0.5–2.9 μg/ml have been measured after single doses of 50–250 mg/kg. In humans, peak levels of 105–1624 ng/ml are measured after oral administrations of 100–1000 mg/kg; its pharmacokinetics is linear and the half-life is about 30 h and is dose independent. Tissue concentrations are usually higher than those in plasma in animals.

Anidulafungin has good *in vitro* activity against a variety of yeasts, including isolates resistant to itraconazole and fluconazole, and molds. This compound is not active against *C. neoformans, T. beigelii,* and *B. dermatitidis;* its MICs for certain molds are higher than those of the three new investigational azoles. However, its fungicidal activity against some species of *Candida* is superior to those of the azoles, which are mostly fungistatic drugs. Although the drug is well tolerated up to 700 mg/kg doses, gastrointestinal adverse effects have been observed with 100 mg/kg doses in human subjects. Potentially peak plasma concentrations in excess of MIC values have been demonstrated in rabbits and tissue levels are above MIC endpoints in major organs. In human volunteers, it exhibits linear pharmacokinetics after single oral doses of 100–1000 mg. Anidulafungin peak plasma levels occurred after 6–7 h after ingestion with an elimination half-life of approximately 30 h. Anidulafungin has *in vivo* activity in experimental murine (normal and immunocompromised animals) aspergillosis and candidiasis and *P. carinii* pneumonia. Clinical trials are being conducted to assess the efficacy of anidulafungin against *Candida* infections.

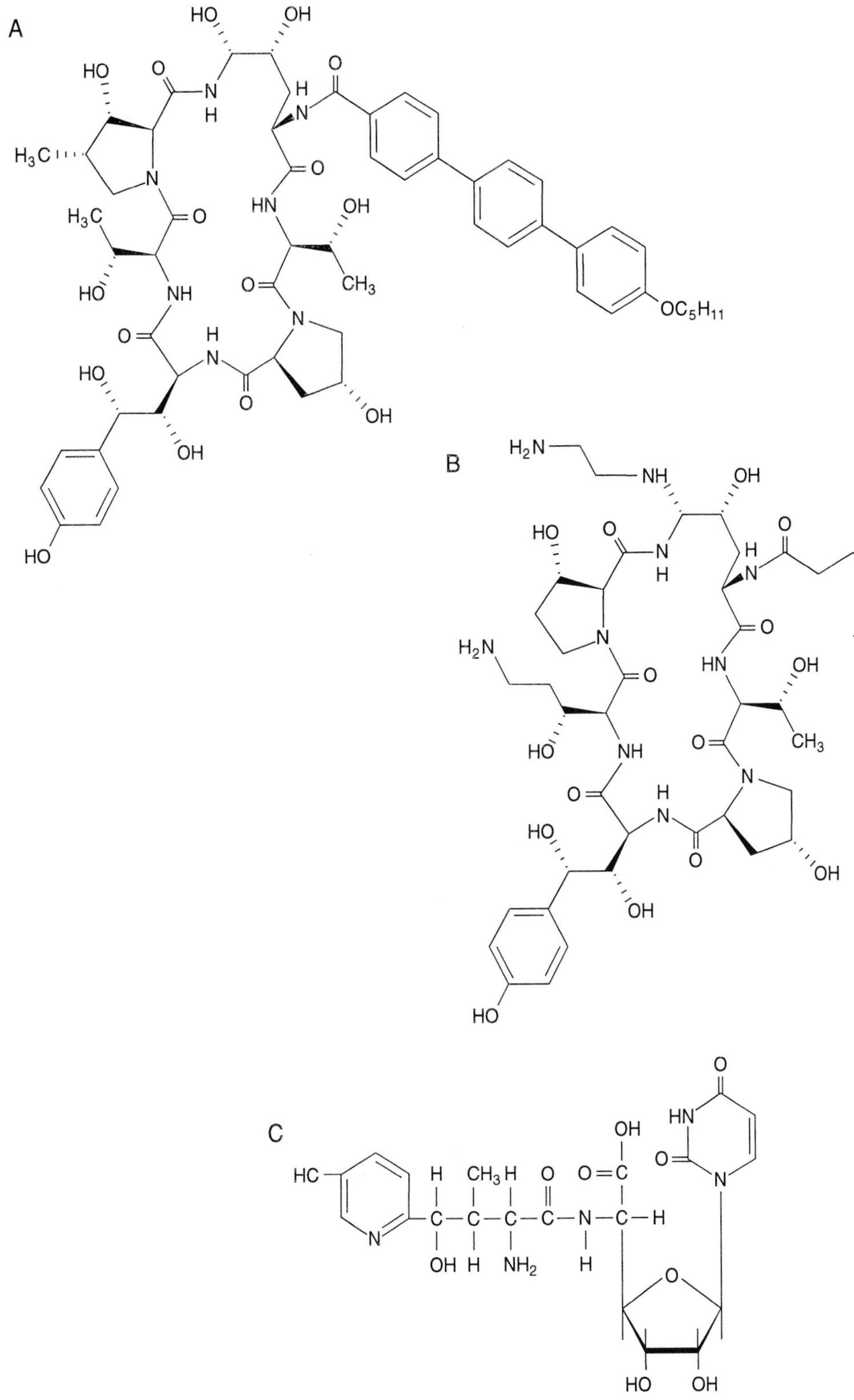

FIGURE 4.4 Chemical structures of: (A) anidulafungin (LY-303366), (B) caspofungin (L-743872 or MK-0991), and (C) nikkomycin Z.

3. *Anidulafungin derivatives*

Several derivatives of anidulafungin have been synthesized including the phospate derivative LY-307853 and phosphate ester derivatives LY-329960 and LY-333006, which had improved water solubility. Good activity has been demonstrated in a murine model of candidiasis with the two latter compounds.

Other echinocandin derivative is mulundocandin, which was obtained from a variant of *A. sydowii*. It has *in vitro* antifungal activity against itraconazole-resistant *Candida* spp. and *A. fumigatus*.

4. *Cyclopeptamines*

A-192411.29 is a novel cyclopeptamine antifungal lipopeptide derived by total synthesis from the structural template of the natural product echinocandin. It has similar *in vitro* activity to that of amphotericin B against *Candida* spp. and *C. neoformans* and partial activity against *A. fumigatus*.

5. *Non-echinocandin macrocyclic lipopeptidolactones (FR-901469)*

FR-901469 is a water-soluble, non-echinocandin-type lipoprotein fungal derivative. It has good *in vitro* activity against *C. albicans* and *A. fumigatus*.

C. Pneumocandin derivatives

The pneumocandins have similar structures to those of the echinocandins, but they possess a hexapeptide core with a β-hydroxyglutamine instead of the threonine residue, a branched-chain $_{14}C$ fatty acid acyl group at the N-terminal, and variable substituents at the C-terminal proline residue. The pneumocandins are fermentation products of the mold *Zalerion arbolicola*.

Of the three naturally occurring pnemocandins (A–C), only A and B have certain antifungal activity *in vitro* and *in vivo* against *Candida* spp. and *Pneumocystis carinii* (in rodents), but they are non-water-soluble; this group has pneumocandin Ao (L-671329) and pneumocandin Bo (L-688786), which are fermentation products produced by *Zalerion arboricola* ATCC 20868 and related semysinthetic derivatives.

1. *L-639989, L-733560, L-705589, and L-731373*

Modification of the original pneumocandin B by phosphorylation of the free phenolic hydroxyl group led to the improved, water-soluble pneumocandin B phosphate (L-639989). Further modifications of pneumocandin B led to the water-soluble semisynthetic molecules L-733560, L-705589, and L-731373. Although studies were conduced in laboratory animals, these molecules were not evaluated in humans.

D. Caspofungin (MK-0991 or L-743872)

Caspofungin acetate (Fig. 4.4B) is the product of a modification of L-733560 and was selected for further evaluation in clinical trials in humans. As are the other semisynthetic pneumocandins, caspofungin is water soluble. Caspofungin is highly protein bound (97%) with a half-life that ranges from 5 to 7.5 h and drug concentrations are usually higher in tissue than in plasma.

Caspofungin has fungistatic and fungicidal activities similar to those of anidulafungin against most *Candida* spp. and lower activity against the dimorphic fungi. It also has fungistatic *in vitro* activity against some of the other molds, especially *Aspergillus* spp. However, both anidulafungin and caspofungin pose difficulties regarding their *in vitro* laboratory evaluation and the data are controversial regarding their MICs for the molds. Animal studies have demonstrated that this compound has good *in vivo* activity not only against yeast infections but also in murine models of disseminated aspergillosis and pulmonary pneumocystosis and histoplasmosis. The drug is not effective for the treatment of disseminated experimental infections caused by *C. neoformans*. In laboratory animals, the drug is mostly well tolerated, but histamine release and mild hepatotoxicity have been reported. Caspofungin is generally well tolerated in humans, but side effects include hypokalemia, nephrotoxicity, chills, fever, intestinal, tachycardia, rash, and sweating (Table 4.2).

Data from 69 patients with either invasive refractory aspergillosis or intolerant infection to standard therapies demonstrated a 41% favorable response in patients receiving at least one dose of caspofungin and a 50% favorable response in patients receiving more than 7 days of therapy. Caspofungin has been licensed in the United States for candidemia and other *Candida* infections and for refractory aspergillosis.

E. Micafungin (FK 463)

FK 463 is a semisynthetic derivative of naturally occurring lipopeptide that was synthesized by a chemical modification from a product of the mould *Coleophoma empedri*. As the other related compounds, it has good *in vitro* activity against *Candida* and *Aspergillus* species, but it is inactive against *C. neoformans, T. beigelii*, and *F. solani*. It also has good activity *in vivo* in experimental invasive candidiasis and pulmonary aspergillosis in neutropenic mice; the activity was similar to that of amphotericin B. The drug was well tolerated (up to 2.5–50 mg single, or 25 mg multiple doses, 7 days) in healthy adult male volunteers and adult bone marrow or peripheral stem cell transplant patients (200 mg/day, 10 days dosing average). The serum concentrations in the latter two groups of patients were higher than the serum levels obtained in experimental candidiasis and aspergillosis. Micafungin is protein binding (99%) and plasma concentrations attain a steady state by day 4 with repeated doses. Clearing of esophageal candidiasis symptoms has been demonstrated among 74 HIV-positive patients treated with micafungin at 50, 25, and 12.5 mg/day as 1-h infusion for a mean of 12 days. Diarrhea was the only side effect reported. Micafungin is undergoing Phase III clinical trials.

F. Other fungal cell inhibitors

Various aromatic natural products have been shown to have antifungal activity against *S. cerevisiae* (xanthofulvin) and *C. albicans* (Ro-41-0986 and its derivative Ro-09-3024).

XIII. THE PRADIMICINS AND BENANOMYCINS

The pradimicins and benanomycins are fungicidal metabolites (benzonaphthacene quinones) of *Actinomadura* spp., which were introduced in the 1980s. Several semisynthetic molecules have also been produced. They act by disrupting the cell membrane through a calcium-dependent binding with the saccharide component of mannoproteins, which results in disruption of the plasma membrane and leakage.

A. Pradimicin A (BMY 28567) and FA-2 (BMY 28864)

The poor solubility of pradimicin A led to the development of BMY 28864, which is a water-soluble derivative of pradimicin FA-2. BMY 28864 appears to have good *in vitro* and *in vivo* activity against most common yeasts and *A. fumigatus*. Clinical trials in humans have not been conducted.

B. BMS 181184

This compound is either a semisynthetic or biosynthetic derivative of BMY 28864. Although it was selected for further clinical evaluation due its promising *in vitro* and *in vivo* data, elevation of liver transaminases in humans led to the discontinuation of this drug.

C. Benanomycin A

This compound has shown the best antifungal activity among the various benanomycins. Its great advantage compared to other new antifungals is its good *in vivo* activity in animals against *P. carinii*.

XIV. THE POLYOXINS AND NIKKOMYCINS

The polyoxins are produced by *S. cacaoi* and the nikkomycins by *S. tendae*. The former compounds were discovered during a search for new agricultural fungicides and pesticides. Both polyoxins and nikkomycins are pyrimidine nucleosides that inhibit the enzyme chitin synthase, which leads to the depletion of chitin in the fungal cell wall; they were introduced in the 1960s and 1970s. These molecules are transported into the cell via peptide permeases.

A. Polyoxin D

Although polyoxin D has *in vitro* antifungal activity against *C. immitis* (parasitic phase), *C. albicans*, and *C. neoformans*, it was not effective in the treatment of systemic candidiasis in mice.

B. Nikkomycin Z

This compound appears to have both *in vitro* and *in vivo* activity against *C. immitis, B. dermatitidis*, and *H. capsulatum*, which are highly chitinous fungi. It also has *in vitro* modest activity against *C. albicans* and *C. neoformans*. Studies to evaluate its safety have been conducted and clinical trials have been designed for the treatment of human coccidioidomycosis. These studies will determine its role as a therapeutic agent in humans.

XV. THE SORDARINS

The natural sordarin GR 135402 is an antifungal fermentation product of *Graphium putredinis*. The compounds GM 103663, GM 211676, GM 222712, and GM 237354 are synthetic derivatives of GR 135402. *In vitro*, GM 222712 and GM 237354 have shown broad-spectrum antifungal activity for a variety of yeasts and molds. Development of these two sordarines has been discontinued.

XVI. DIMETHOMORPH AND FLUAZINAM

Dimethomorph is a cinnamic acid derivative for use against *Plasmopara viticola* on vines and *Phytophthora infestans* on tomatoes and potatoes; it is not cross-resistant to phenylamides (systemic controllers of *Phycomycetes* plant infections). Fluazinam is used in vines and potatoes but also acts against *B. cinera* as an uncoupler of oxidative phosphorylation.

XVII. THE PHTHALIMIDES

The discovery of captan in 1952 and later of the related captafol and folpet initiated the proper protection of crops by the application of specific fungicides. Captan is also used to treat dermatophytic infections in horses and cattle, but it causes skin sensitization in horses.

XVIII. THE SPHINGOLIPID SYNTHESIS INHIBITORS

The sphingofungins, lipoxamycins, myriocin (ISP-1) and viridiofungins selectively inhibit the fatty acid-like natural products. Fumonisins (produced by the corn and human pathogen *F. moniliforme*) and the structually related *Alternaria* toxin are inhibitors of ceramide synthase; resemble the structurally unrelated australifungins. Because these compounds inhibit the mammalian sphingolipid pathway and cause accumulation of sphinganine (sphingolipid depletion of both mammalian and fungal cells), they are toxic. The fumonisins have been associated with cancer in humans and also could cause disease in animals.

Aureobasidin A is an inhibitor of the IPC (inositolphosphorylceramide) synthase that is produced by the black yeast *Aureobasidium pullulans*; it is the less toxic of this class of compouds. The oral fungicidal activity of aureobasidin has been demonstrated in experimental murine candidiasis. Khafrefungin and rustmicin (galbonolide A) also inhibit IPC synthase and have fungicidal activity against some yeasts and moulds by causing ceramide accumulation in the cell membrane. Due to their toxic effects, only aureobasidin A had both preclinical and Phase I clinical trials. However, its development was discontinued owing to its limited activity against *Candida* spp.

XIX. OTHER ANTIFUNGAL APPROACHES

A. Natural and synthetic cationic peptides

Cationic peptides provide a novel approach to antifungal therapy that warrants further investigation.

1. *Cecropin*

Cecropin is a natural lytic peptide that is not lethal to mammalian cells and binds to ergosterol. Its antifungal activity varies according to the fungal species being challenged.

2. *Indolicidin*

Indolicidin is a tridecapeptide that has good *in vitro* antifungal activity and when incorporated into liposomes has activity against experimental aspergillosis in animals.

3. *Synthetic peptides*

Synthetic peptides have been derived from the natural bactericidal-permeability increasing factor. They appear to have *in vitro* activity against *C. albicans*, *C. neoformans*, and *A. fumigatus* and also show synergistic activity with fluconazole *in vitro*.

B. Amino acid analogs

RI 331, the azoxybacillins, and cispentacin are amino acid analogs with good *in vitro* antifungal activity against *Aspergillus* spp. and the dermatophytes (RI 331 and azoxybacillins) and also good *in vivo* activity (cispentacin). RI 331 and the azoxybacillins inhibit homoserine dehydrogenase and the biosynthesis of sulfur-containing amino acids, respectively. The derivative of histatin 5 called P-113 has antifungal *in vitro* activity against Candida species.

See also the following articles:

ANTIVIRAL AGENTS, BACTERIOCINS, FUNGAL INFECTIONS, FUNGI, FILAMENTOUS

BIBLIOGRAPHY

Allen, D. G., Pringle, J. K., Smith, D. A., Conlon, P. D., and Burgmann, P. M. (1993). *Handbook of Veterinary Drugs*. Lippincott, Philadelphia, PA.

Chiou, C.C., Groll, A. H., and Walsh, T. J. (2000). New drugs and novel targets for treatment of invasive fungal infections in patients with cancer. *The Oncologist* **5**, 120–135.

Clemons, K. V. and Stevens, D. A. (1997). Efficacies of two novel azole derivatives each containing a morpholine ring, UR-9746 and UR-9751, against systemic murine coccidioidomycosis. *Antimicrob. Agents Chemother.* **41**, 200–203.

Espinel-Ingroff, A. (1996). History of medical mycology in the United States. *Clin. Microbiol. Rev.* **9**, 235–272.

Espinel-Ingroff, A. (1998). Comparison of *in vitro* activity of the new triazole SCH 56592 and the echinocandins MK-0991 (L-743,872) and LY303366 against opportunistic filamentous and dimorphic fungi. *J. Clin. Microbiol.* **36**, 2950–2956.

Espinel-Ingroff, A., and Shadomy, S. (1989). *In vitro* and *in vivo* evaluation of antifungal agents. *Eur. Clin. Microbiol. Infect. Dis.* **8**, 352–361.

Espinel-Ingroff, and Pfaller, M. A. (2003). Susceptibility methods: yeasts and filmentous fungi. In *Manual of Clinical Microbiology* (P. R. Murray, E. J. Baron, M. A. Pfaller, F. C. Tenover, and R. H. Yolken, Eds.), 8th edn. ASM, Washington, DC.

Espinel-Ingroff, A., Boyle, K., and Sheehan, D. J. (2001). In vitro antifungal activities of voriconazole and reference agenst as determined by NCLLS methods: review of the literature. *Mycopathologia* **150**, 101–115.

Georgopapadakou, N. H. (2000). Antifungals targeted to sphingolipid synthesis: focus on inositol phosphorylceramide synthase. *Exp. Opin. Invest. Drugs* **9**, 1787–1796.

Georgopapadakou, N. H. (2001). Update on antifungals targeted to the cell wall: focus on β-1,3-glucan synthase inhibitors. *Exp. Opin. Invest. Drugs* **10**, 269–280.

Groll, A. H., Piscitelli, S. C., and Walsh, T. J. (1998). Clinical pharmacology of systemic antifungal agents: A comprehensive review of agents in clinical use, current investigational compounds, and putative targets for antifungal drug development. *Adv. Pharmacol.* **44**, 343–500.

Russell, P. E., Milling, R. J., and Wright, K. (1995). Control of fungi pathogenic to plants. In *Fifty Years of Antimicrobials: Past Perspectives and Future Trends* (P. A. Hunter, G. K. Darby, and N. J. Russell, Eds.). Cambridge University Press, New York.

Sheehan, D. J., Hitchcock, C. A., and Sibley, C. M. (1999). Current and emerging azole antifungal agents. *Clin. Microbiol. Rev.* **12**, 40–79.

St Georgiev, V. (2000). Membrane transporters and antifungal drug resistance. *Curr. Drug Targets* **1**, 261–268.

Yang, Y. L. and Lo, H. J. (2001). Mechanisms of antifungal agent resistance. *J Microbiol Immunol. Infect.* **34**, 79–86.

WEBSITES

Practice guidelines of the Infectious Diseases Society of America
http://www.journals.uchicago.edu/IDSA/guidelines/

Dr. Fungus: Antifungal drugs. With links. (Merck & Co.)
http://www.doctorfungus.org/

5

Antisense RNAs

Andrea Denise Branch
The Mount Sinai School of Medicine

GLOSSARY

artificial RNAs RNA molecules expressed from genes that have been introduced into cells (transgenes) or RNA molecules synthesized in cell-free systems. The mode of action of artificial antisense RNAs is under active investigation. In some biological systems, artificial RNAs may themselves form double-stranded RNAs that mediate target-gene inhibition through novel mechanisms.

complementarity A measure of the percentage of nucleotides in two sequences that are theoretically able to form Watson–Crick base pairs.

cosuppression A type of posttranscriptional gene silencing in which transcripts of both an endogenous gene and an homologous transgene are synthesized and then degraded.

homology-dependent viral resistance A form of posttranscriptional gene silencing in which viral RNAs are degraded in transgenic plants expressing RNAs homologous to viral RNAs, resulting in inhibition of viral replication and attenuation of virus symptoms.

perfect double-stranded RNA (dsRNA) duplex A helical structure in which two segments from a single RNA molecule (an intramolecular duplex), or segments of two separate RNA molecules (an intermolecular duplex) in an anti-parallel orientation to each other form an uninterrupted series of Watson–Crick base pairs (C pairing with G; A pairing with U).

posttranscriptional gene silencing A process through which specific RNAs are degraded posttranscriptionally, resulting in loss of expression of associated genes.

RNA interference (RNAi) An efficient process that allows small (21–23 nucleotide-long) RNAs derived from double-stranded RNAs to inhibit (silence) specific target genes. The small interfering RNAs (siRNAs) that mediate RNAi may be produced through the cleavage of longer double-stranded RNAs, by an enzyme called Dicer, or they may be derived from artificial siRNA duplexes.

Antisense RNAs are RNA molecules that bind to a second, sense, RNA through complementary Watson–Crick base pairing of anti-parallel strands; RNA molecules that are at least 70% complementary to a second RNA for at least 30 nucleotides and thus have the potential for binding; or RNA molecules that are transcribed from the DNA strand opposite that of a second RNA. Antisense RNAs in gene therapy are complementary to target RNAs and are intended to eliminate the expression of specific genes; target RNAs may be either associated with diseases or with normal cellular functions. Naturally occurring antisense RNAs comprise a structurally and functionally diverse group that includes RNAs known to bind to their target RNAs and RNAs that simply contain sequences complementary to other previously identified RNAs.

In 1984, Izant and Weintraub thrust antisense RNA into the center stage of molecular research by proposing

The Desk Encyclopedia of Microbiology
ISBN: 0-12-621361-5

that artificial antisense RNAs could be used to eliminate the expression of specific target genes, offering an alternative to the labors of classical mutational analysis. Rather than producing random mutations and then screening for those affecting genes of interest, they suggested that mutants could be created at will by introducing antisense RNAs complementary to sense transcripts of selected genes. They envisioned antisense RNAs binding to messenger RNAs (mRNAs) or their precursors, forming duplexes, and thereby inhibiting gene expression. The promise of streamlined genetic analysis, and improved pharmaceutical agents, livestock, and crops stimulated tremendous interest in antisense technology in members of the research community and on Wall Street. However, artificial antisense RNAs have not always performed as intended. The molecular events responsible for their unexpected behavior are not yet known, but enough information has emerged to indicate that these events merit thorough investigation. To understand the properties of artificial antisense RNA and to gain a more complete understanding of RNA's regulatory functions, it is essential to study both naturally occurring and artificial antisense RNA. This collection of molecules includes RNAs known to alter the expression of their sense RNA counterparts and RNAs whose sequences appear to equip them to interact with their sense counterparts (i.e., RNAs that are at least 70% complementary to a second RNA for at least 30 nucleotides). Many natural and artificial antisense RNAs exist—far too many for each to be discussed here. Therefore, this article focuses on the principles governing their behavior. (Information about antisense oligomers composed of DNA is not included, but has been reviewed by the author.)

I. INTRODUCTION

A. Naturally occurring antisense RNAs are extremely versatile

Antisense RNAs are best known for their ability to eliminate the expression of target RNAs by binding to complementary sequences. However, antisense RNAs do much more than turn off other genes. For example, in virus-infected mammalian cells, antisense RNA combines with sense RNA to form biologically active double-stranded RNA (dsRNA), which triggers the interferon (IFN) response. Other antisense RNAs are involved in RNA maturation. These molecules are often omitted from lists of antisense RNAs because they promote expression of their target RNAs, rather than inhibit it. However, there has never been a requirement for antisense RNAs to function as negative regulators of gene expression. The antisense RNAs involved in RNA maturation illustrate how complementary RNA sequences contribute to essential cellular functions. The guide RNAs of certain parasites bind to mitochondrial mRNA precursors through short complementary regions and direct upstream editing of the pre-mRNA. (RNA editing is any process leading to an alteration in the coding capacity of an mRNA, other than splicing or 3′-end processing). Similarly, small nucleolar RNAs (snoRNAs) bind to complementary regions of ribosomal RNA (rRNA) precursors, leading to methylasemediated site-specific modification of the precursor.

Of the antisense RNAs on the frontiers of research, those transcribed from mammalian genes are among the most intriguing and in greatest need of further investigation. Based on evidence showing that certain of these RNAs down-regulate their targets—diminishing synthesis of sense RNA, interfering with pre-mRNA processing, and inhibiting sense RNA translation—it has generally been assumed that any newly discovered antisense RNA would also function as a negative regulator. However, recent data indicate that each RNA must be individually investigated. An antisense transcript of the Wilms's tumor gene (a gene imprinted under certain circumstances) appears to enhance expression of the sense RNA (Moorwood *et al.*, 1998). Concerning the range of possible antisense RNA functions, it is interesting to note that an antisense RNA to basic fibroblast growth factor mRNA is thought to serve in two capacities: to act as the mRNA for a highly conserved protein of its own and to down-regulate growth factor expression. Several additional antisense RNAs contain open reading frames and may specify proteins.

When interpreting a report of a newly discovered antisense RNA, particularly one detected in eukaryotic cells, it is important to remember that terminology in this part of the field permits a molecule to be designated an "antisense RNA" on the basis of sequence information alone. There is no requirement that the RNA bind to its sense counterpart or alter expression of the sense RNA in any way. Furthermore, throughout the entire antisense field, there is no requirement that sense and antisense RNAs be transcribed from opposite strands of the same DNA, and thus they are not necessarily exact complements of each other.

The looseness in antisense terminology could be problematic. However, it serves a useful purpose, increasing the chances that meaningful similarities will be recognized. Such similarities illustrate the principles governing the behavior of antisense RNAs. Examples are selected from three areas: prokaryotic systems, virus-infected mammalian cells, and artificial inhibitory RNAs. Antisense RNAs involved in RNA maturation, such as guide RNAs and snoRNAs, are not discussed further due to space limitations. However, the ability of guide RNAs to transfer genetic information is reflected in the function of the minus-strand viral RNAs, which are included; and the ability of snoRNAs to induce site-specific methylation is echoed in the gene-specific DNA methylation associated with some of the artificial inhibitory RNAs.

B. Artificial RNAs expanded the antisense field in unexpected directions

Some naturally occurring antisense RNAs are highly effective gene regulators. Their efficacy, and the conceptual simplicity of antisense-mediated gene ablation stimulated efforts to develop artificial antisense RNAs that could be used to inhibit specific genes in higher organisms and to confer resistance to micro-organisms. These efforts have already yielded commercial agricultural products, such as the transgenic Flavr Savr tomato. They have also revealed that it is sometimes possible to substitute a sense transcript for an antisense transcript and achieve the same level of target-gene inhibition. Because it is usually impossible for sense transcripts and their targets to form a perfect duplex containing more than about 7–12 bp, sense inhibition appears to be a manifestation of a novel regulatory pathway. It is important to learn the details of this pathway in order to gain insight into RNA function and to facilitate the development of more effective artificial RNAs for use in biotechnology and basic research.

C. Despite the diversity of antisense RNAs, four general principles account for most antisense effects

The first principle is that, above all else, antisense RNAs are ribonucleic acids. As such, they are endowed with a unique combination of properties. RNAs can store and transmit genetic information, just as DNA can. Moreover, naturally occurring RNAs readily form intricate three-dimensional structures and can produce catalytic active sites. RNAs are directly involved in protein synthesis at a variety of levels. Most RNAs are transcribed from DNA through a complex process involving *cis*-active promoter elements and many proteins. Nascent transcripts are converted into mature RNAs through an equally complex set of reactions. Regulation can occur at any of a number of points during transcription and subsequent processing. RNAs can be stable, or they can turn over rapidly. RNAs can readily move from the nucleus to the cytoplasm, and shuttle back and forth. They can form structural signals recognized by proteins, and they can interact with other nucleic acids through complementary base paring. These properties allow antisense RNAs to weave their way in and out of an enormous variety of cellular processes.

The second principle is that complementarity between an antisense RNA and a second nucleic acid is no guarantee that the two molecules will bind to each other. The tendency of antisense and target RNAs to form complexes, or to remain as separate molecules, is strongly influenced by their individual intramolecular structures. Potential nucleation sites can be prominently displayed, or virtually inaccessible. Complex formation is a bimolecular reaction whose rate is sensitive to concentration; the rate increases with increasing RNA concentration. The relationship between antisense RNA structure and function is illustrated most clearly by the antisense RNAs of prokaryotic systems, which are described in Sections II.A and II.B. Subcellular location also affects the probability that two RNAs will interact. Sense and antisense RNAs transcribed from the same genetic locus are more likely to encounter each other than RNAs transcribed from distant sites in the DNA. Similarly, two RNAs that accumulate in the same membrane-bound compartment are more likely to interact than those in separate compartments.

The third principle is that antisense activity is often mediated by proteins; these proteins must be identified and their modes of action characterized for antisense RNA function to be understood. Many different proteins bind to antisense RNAs and to the RNA–RNA duplexes they create. Depending on the protein and the nature of the duplex, binding can have a variety of effects. The same protein may catalyze a range of reactions, with the outcome determined by information encoded in the structure of the RNA–RNA duplex. Interactions between antisense RNAs and proteins are described in Sections II.D and III.C.

The fourth principle is that dsRNA can act as a signal. Mammalian cells recognize dsRNA as a sign that they, or their neighbors, are infected by a virus. Double-stranded RNA causes mammalian cells to enter an antiviral state. This response is mediated by a group of dsRNA-binding proteins, which make up a very sensitive dsRNA biosensor. There is growing evidence that dsRNA has symbolic value to cells from a variety of plants and animals. The potential of dsRNA to act as a signal is described in Sections III.A and IV.B–D and should be kept in mind when considering the possible biological effects of an antisense RNA.

D. Summary

Antisense RNAs have many roles. They can act as negative regulators of gene expression, induce interferon, or direct RNA maturation. Antisense effects are strongly influenced by internal RNA structure and are often mediated by proteins. Artificial antisense RNAs have allowed new agricultural products to be developed and revealed unexpected roles of RNA in gene regulation.

II. ANTISENSE RNAs IN PROKARYOTIC SYSTEMS: INHIBITION BY DIRECT BINDING TO TARGET RNAs

A. The copy number of plasmid ColE1 is regulated by RNA I, an antisense transcript

RNA I of the *Escherichia coli* plasmid ColE1 was the first regulatory antisense RNA to be discovered. In 1981, Lacatena and Cesareni reported that base pairing between complementary regions of RNA I and RNA II inhibits plasmid replication. Because this system illustrates many principles of antisense RNA action it is discussed in detail.

As is typical of antisense reactions, binding between RNA I and RNA II is a bimolecular process. Its rate is concentration dependent. The concentration dependence of the RNA I–RNA II binding reaction is harnessed to achieve the desired biological effect—maintenance of plasmid copy number at a stable 10–20 copies per cell. Formation of the RNA I and RNA II complex inhibits plasmid DNA replication by preventing RNA II from maturing into the RNA primer required for DNA synthesis (for an excellent review of this system by a leading research group, see Eguchi *et al.*, 1991). RNA I is constitutively synthesized at a high rate and has a short half-life. Its concentration reflects the number of template DNA molecules. When copies of the plasmid are numerous, RNA I concentration is high, binding to RNA II is favored, and plasmid DNA synthesis is inhibited. Conversely, when the plasmid DNA concentration is low, RNA I concentration falls, and plasmid replication is stimulated. Because RNA I contains regions complementary to the RNA II molecules produced by related plasmids, it provides the basis for plasmid compatibility and incompatibility. RNA I has no coding capacity.

Synthesis of RNA II is initiated 555 bases upstream from the origin of DNA replication. RNA I is perfectly complementary to 108 bases at the 5′-end of RNA II and is transcribed from the same region of the genome, but in the opposite direction. RNA I must bind RNA II shortly after RNA II synthesis is initiated. If binding is delayed, the nascent RNA II transcript forms structures that render it resistant to inhibition by RNA I. This competition between formation of the RNA I–RNA II complex and the RNA II self-structure means that antisense activity requires rapid association.

RNA I and RNA II interact through an intricate process whose individual steps are predetermined by the structures of the two RNAs. As illustrated in Fig. 5.1. RNA I has three stem–loop structures and a short tail at its 5′-end. The loops contain seven bases. Figure 5.1 also depicts RNA II, in a conformation that may exist in nascent RNA II transcripts. The secondary structures of RNA I and RNA II maximize the chances that they will form a bimolecular complex. Three sets of complementary bases are exposed in single-stranded loops. Bases making up these potential nucleation sites are displayed in structures somewhat similar to those that project the bases of tRNA anti-codons toward the

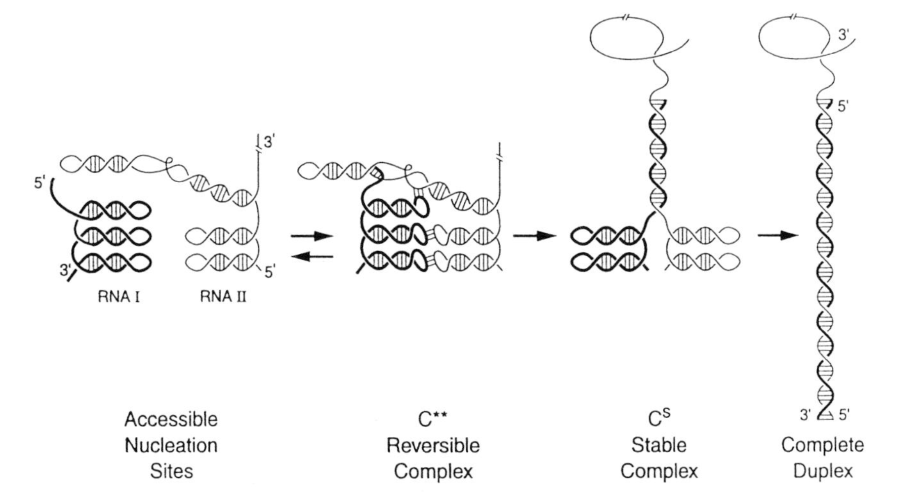

FIGURE 5.1 Binding of ColEI RNA I to RNA II is a stepwise process. RNA I and RNA II interact through complementary sequences present in loops to form C**. Pairing between the 5′-end of RNA I and RNA II is followed by a series of structural changes that culminate in the formation of a stable complex, C^s. Finally, RNA I hybridizes to RNA II throughout its entire length. Adapted from Eguchi *et al.* (1991).

mRNA codons. During nucleation, bases in corresponding loops of RNA I and RNA II interact weakly, evidently forming a limited number of bonds between bases in only one or two loops. The very unstable early intermediates can dissociate rapidly, or establish a "kissing complex," which can, in turn, produce the C** complex, a structure in which all three corresponding loops (reversibly) interact with each other.

The formation of C** can be facilitated by the plasmid-encoded protein, Rom (*R*NA-*o*ne-*m*odulator), which binds to the kissing complex and reduces its equilibrium dissociation constant. The solution structure of a helix modeled after the RNA I–RNA II loop–loop helix was recently solved by nuclear magnetic resonance spectroscopy (Lee and Crothers, 1998). As expected from previous biochemical studies, all seven bases in the loops form complementary base pairs. The loop–loop helix partially stacks on the stem helices, producing a nearly linear structure. The loop–loop helix is bent toward the major groove, which is thereby narrowed. This bend, as well as phosphate clusters flanking the major groove, distinguish this helix from standard A-form RNA, perhaps accounting for the ability of Rom to recognize it. Conversion of C** to stable structures, such as C^s, begins with events occurring at the 5′-end of RNA I. If enough time elapses, the stable complex convert into full-length dsRNA molecules. However, inhibition can result from stable complex formation itself and does not require the generation of a complete duplex. Any lengthy dsRNA regions that form are likely to be rapidly degraded by the endonuclease RNase III.

B. Antisense RNAs in prokaryotic systems have many features in common

1. *Antisense RNA structure*

RNA I is only the first example of the many well-characterized antisense RNAs in prokaryotic systems. Four other representative antisense RNAs are depicted in Table 5.1 and their modes of action are presented. Prokaryotic antisense RNAs are typically small (65–100 bases), stable, noncoding RNAs, whose secondary structures contain one or more stem-loops (see Table 5.1). The loops usually contain 5–8 bases and often have sequences that are similar to each other. The stems of antisense RNA stemloop structures often contain unpaired nucleotides at precise locations. These "imperfections" protect the RNAs from cleavage by dsRNA-specific ribonucleases and also reduce the stability of the stems, allowing them to open up during binding to their target RNAs (Hjalt and Wagner, 1995).

2. *Antisense and target RNAs associate through a stepwise binding process*

Antisense and target RNAs interact through a stepwise process that proceeds from nucleation to stable complex formation. This process has been studied extensively in three systems: RNA I and RNA II of the plasmid ColE1, CopA and CopT of the plasmid R1 (CopT occurs in repA mRNA; see Table 5.1), and RNA-OUT and RNA-IN from the mobile genetic element IS*10*. The apparent second-order rate constants for pairing between these RNAs are in the range of 0.3–1.0 $\times 10^6 M^{-1} s^{-1}$. In the binding reactions between RNA I and RNA II and between CopA and CopT, loop–loop interactions between antisense and target RNAs nucleate binding and formation of the kissing complex. In the binding reaction between RNA-OUT and RNA-IN, the first bonds form between bases in the loop of RNA-OUT and the 5′-end of RNA-IN, an RNA thought to have a relatively open structure. Rapid association appears to be a general requirement for efficient antisense activity. Because they mediate this early and rapid association, the initial base pairs are far more important to the overall binding process than their thermodynamic contribution to the final complex would suggest. Early intermediates are rapidly replaced by complexes containing more intermolecular base pairs. The final outcome of antisense binding depends on the system. Effects include termination of transcription, destabilization, and inhibition of translation. All known prokaryotic antisense RNAs are negative regulators of gene expression, although according to Wagner and Simons, "mechanisms for positive control are quite plausible" (Wagner and Simons, 1994).

Most, but not all, antisense RNAs are transcribed from overlapping gene sequences. As indicated in Table 5.1, mRNA-interfering complementary (micF) RNA is not closely linked to its target, ompF RNA. The duplex they form contains looped-out regions and noncanonical base pairs, in addition to conventional bonds (Delihas *et al.*, 1997). This duplex helps to establish the lower limit of complementarity for an antisense–target RNA pair. Within the duplex, only 24 of 33 bases (73%) of micF RNA are Watson–Crick base-paired to nucleotides in ompF RNA.

C. Bioengineers hope to use the special features of naturally occurring antisense RNAs to develop effective artificial antisense RNAs

Engdahl and colleagues attempted to apply their knowledge of naturally occurring antisense RNAs to develop bioengineered antisense RNAs capable of

TABLE 5.1 Representative antisense RNAs of prokaryotic accessory DNA elements and bacterial DNA

System	Antisense RNA structure	Mode of action
Plasmid Plasmid R1 Antisense: CopA (91 bases) Target: CopT	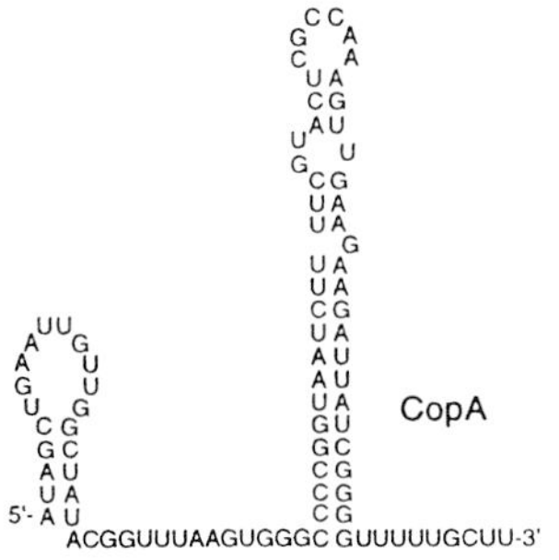 CopA 5'-AU ... ACGGUUUAAGUGGGC GUUUUUGCUU-3'	CopA indirectly prevents synthesis of RepA, a protein required for plasmid replication, by pairing with CopT sequences in the polycistronic RNA that encodes RepA. Binding blocks the ribosome binding site of the tap gene, and prevents its translation, which is coupled to translation of the repA gene. Binding also yields duplexes which are RNase III substrates.
Phage Bacteriophage λ Antisense: oop RNA (77 bases) Target: cII mRNA	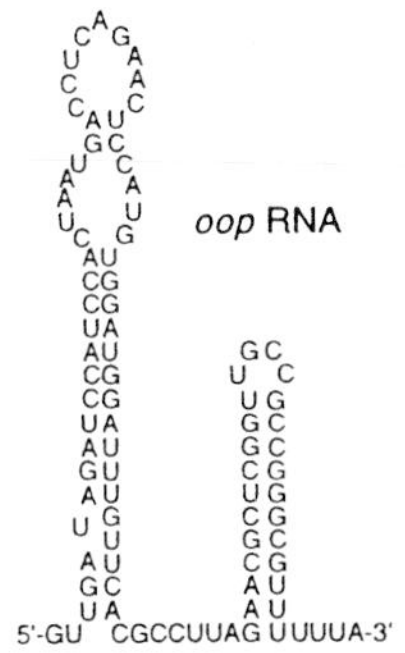 *oop* RNA 5'-GU CGCCUUAG UUUUA-3'	oop RNA binds to (55) bases at the 3′-end of the cII portion of the cII-O mRNA, creating an RNase III cleavage site, destabilizing the cII message, and thereby enhancing the burst size during induction.
Transposable element Insertion sequence IS10 Antisense: RNA OUT (70 bases) Target: RNA-IN	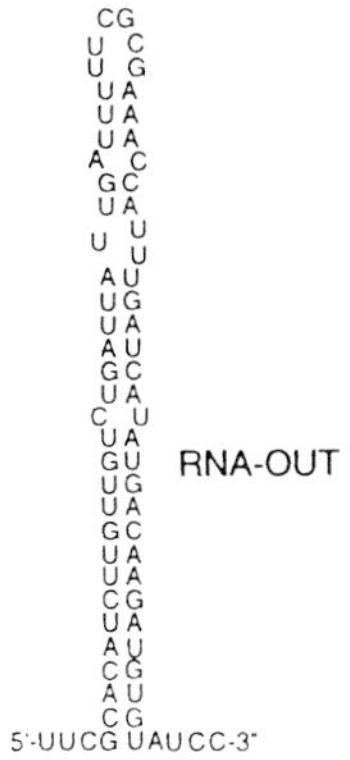 RNA-OUT 5'-UUCG UAUCC-3'	RNA-OUT binds to (35) nucleotides at the 5′-end of RNA-IN, transposase mRNA, preventing translation by blocking the ribosome binding site or by creating an RNase III cleavage site. Antisense inhibition increases as IS10 copy number increases, producing multicopy inhibition. IS10 is the mobile element of Tn10 (a tetracycline-resistance transposon).
Bacterial chromosome *Escherichia coli* Antisense: micF RNA (93 bases) Target: ompF mRNA	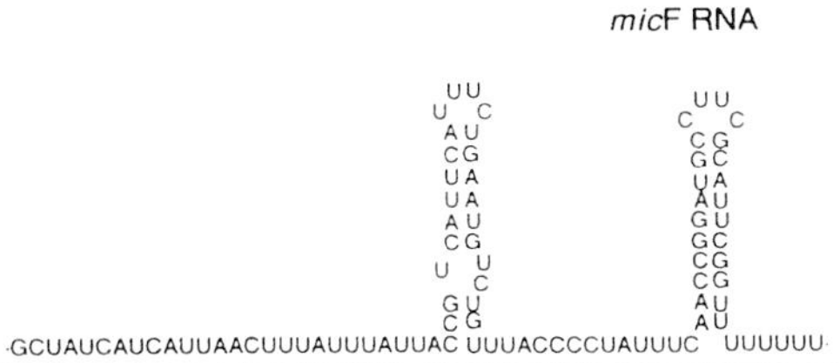 *mic*F RNA GCUAUCAUCAUUAACUUUAUUUAUUAC UUUACCCCUAUUUC UUUUUU	micF RNA is about 70% complementary to the 5′-end of ompF mRNA in the region of the ribosome binding site. Binding modulates production of OmpF, a major component of the *E. coli* outer membrane.

inhibiting selected target genes (Engdahl *et al.*, 1997). They tested antisense RNAs containing a recognition element resembling the major stem-loop of CopA and either a segment complementary to the ribosome binding site of the target RNA or a ribozyme. None of their antisense RNAs inhibited the target genes by more than 50%. They concluded, "we still have too little insight into the factors that determine this property [the ability to rapidly associate] and, hence cannot yet tailor such structures to any chosen target sequence." However, their study and similar studies by other investigators are helping to identify the structural features needed to produce effective artificial antisense RNAs for use in bacterial systems.

D. Proteins often mediate antisense RNA activity: RNase III is a prototypic dsRNA-binding protein

Important steps in several antisense systems are carried out by RNase III. Purified in 1968, RNase III was the first enzyme to be discovered that recognizes RNA–RNA duplexes as substrates. Like many other dsRNA-binding proteins, RNase III does not act exclusively on perfect duplexes. The varied interactions between RNase III and its substrates illustrate the range of functions that can be included in the repertoire of a single enzyme.

As illustrated in Fig. 5.2A, RNase III cleaves perfect RNA–RNA duplexes into short fragments averaging about 15 bases in length. Cuts are made across both strands of the duplex, at sites that are usually offset by one or two bases. This reaction shows no sequence specificity. Certain complexes of antisense and target RNAs are degraded by such randomly placed double-cleavages. In contrast, the 30S rRNA precursor is cut at four precise bonds. The rRNA precursor folds into a structure with two large stems, one topped by the sequence of 16S rRNA, the other by 23S rRNA. RNase III cleaves each of these stems exactly twice at specific nucleotides, releasing rRNA intermediates. The stem flanking 23S rRNA is shown in Fig. 5.2B. The cleavage reactions carried out on the nearly perfect stems in the rRNA precursor resemble those carried out on perfect duplexes in certain respects, but not others. They are double-cleavages, but at predetermined locations. Surprisingly, RNase III also makes a series of precise cuts in the early mRNA precursor of bacteriophage T7 (Fig. 5.2c), even though this molecule contains no structures that bear an obvious similarity to perfect duplexes. Although the bacteriophage T7 cleavage sites lack the hallmark feature of dsRNA—a consecutive series of Watson–Crick bonds—they almost certainly contain noncanonical bonds that confer stability and allow them to fold into three-dimensional structures with features similar to dsRNA.

As a group, the RNase III cleavage sites demonstrate the subtlety of the interactions between dsRNA-binding proteins and RNA molecules. RNase III acts as a rampant random nuclease on dsRNA substrates. Duplexes are cleaved at multiple sites regardless of

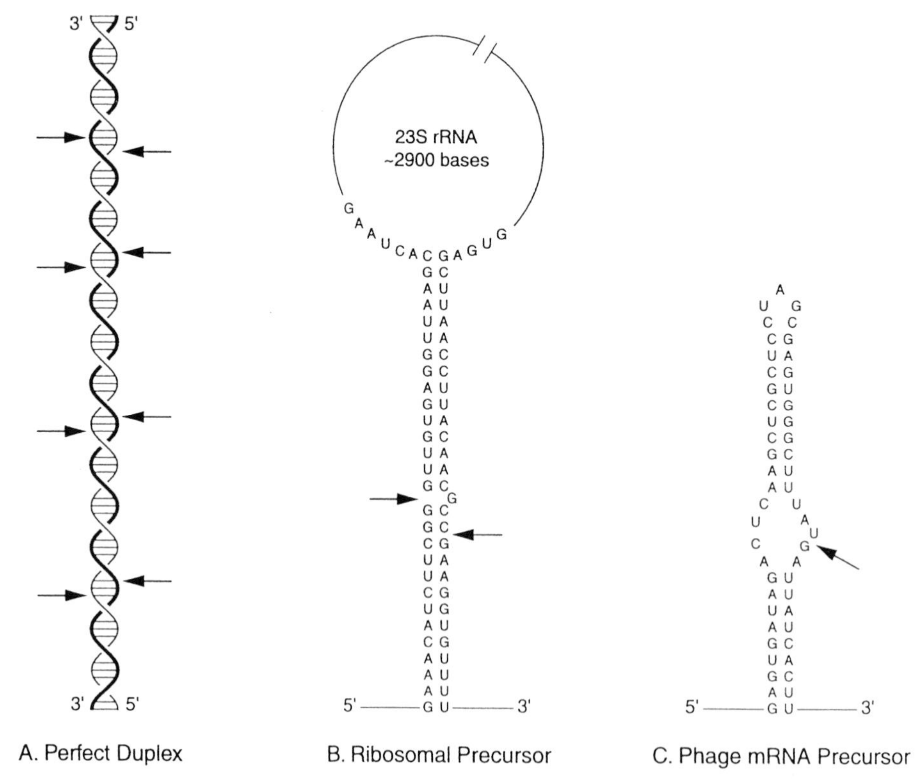

FIGURE 5.2 Structural features of dsRNA regions direct RNase III cleavage. RNase III cleaves perfect RNA–RNA duplexes into fragments averaging 15 base pairs in length (A). In addition, it cuts the 30S ribosomal RNA precursor at four positions, releasing intermediates that will become the mature ribosomal RNAs. The double-cleavage site in the region 23S region (Bram *et al.*, 1980) is shown (B). Finally, RNase III cleaves certain complex RNA structures, such as those in the bacteriophage T7 early mRNA precursor at single sites. These complex structures are typically depicted as "bubbles" because they are devoid of Watson–Crick pairs; however, it is likely that they contain a precise array of non-canonical bonds and are not open loops (C).

their sequence; yet RNase III makes precise cleavages at sequence-specific sites in certain RNA precursor molecules. The structure of the RNA dictates the outcome of encounters with RNase III. The ability of single-stranded RNAs to exploit and sometimes to thwart dsRNA-specific enzymes is a significant survival factor for viruses, as illustrated by bacteriophage T7, and by the mammalian viruses discussed in the next section.

E. Summary

Naturally occurring antisense RNAs in prokaryotic systems are small, noncoding molecules whose structures facilitate rapid association with target RNAs. These antisense RNAs are much more than RNA molecules that happen to have sequences complementary to those of other RNAs. They are highly evolved machines designed to snare and entwine their targets. Inhibition can be a direct consequence of binding, or it can involve cellular proteins, especially endonucleases. RNase III is an endonuclease that cleaves RNA– RNA duplexes, such as those produced by certain antisense RNAs and their targets. Degradation is not the inevitable fate of an RNase III substrate. Depending on the structure of the duplex, RNase III can also introduce specific cleavages and promote RNA maturation.

III. ANTISENSE RNAs IN VIRUS-INFECTED MAMMALIAN CELLS: SIGNALS OF DANGER

A. dsRNAs are potent inducers of interferon and activators of antiviral defenses

Because the interferon response is central to mammalian antiviral defense, few antisense molecules can compete for significance with those producing the dsRNAs that induce interferon and otherwise contribute to the antiviral state. Despite the technical obstacles that make these dsRNAs difficult to study, they are analyzed here because of their importance to mammalian survival and because they illustrate the ability of dsRNA to act as a signal. The protective power of an intact interferon response is demonstrated by studies of transgenic mice deficient in either the type I or the type II interferon receptor. These mice rapidly succumb to viral infections that they would otherwise readily clear. For example, while the median lethal dose of intravenously administered vesicular stomatitis virus is normally in the range of 10^8 plaque-forming units, mutant mice lacking the type I IFN receptor die within 3–6 days after receiving 30–50 units (Muller *et al.*, 1994). A functional interferon system raises the lethal dose of this virus 2 million-fold.

Mammalian cells have an extraordinarily sensitive mechanism for detecting dsRNA and for responding to dsRNA by synthesizing interferons. In 1977, Marcus and Sekellick reported that primary chick embryo cells are capable of responding to a single molecule of dsRNA. They exposed cells to defective interfering particles of vesicular stomatitis virus containing a covalently linked antisense–sense RNA molecule in a ribonucleoprotein complex. A single dsRNA molecule was presumed to form upon entry into the cell. Peak interferon titers were obtained in cultures incubated with 0.3 particles per cell. Both higher and lower doses of particles resulted in very marked reductions in interferon production, producing a bell-shaped dose– response curve. The data indicated that cells that attached two or more particles produced little or no interferon. Cells in the cultures exposed to the defective interfering particles entered a general antiviral state, as indicated by their resistance to a subsequent challenge with Sindbis virus. Control experiments ruled out the possibility that interferon induction was due to proteins in the particles.

In these experiments, cells responded to intracellular dsRNA. Cells also respond to exogenously applied dsRNA. This ability may allow cells to respond to dsRNA released from their moribund neighbors. A pharmaceutical form of dsRNA has been developed for intravenous delivery to virus-infected patients. When dsRNA acts as a danger signal, it is performing a function similar to that of unmethylated CpG-containing DNA. Such DNA is a strong immune stimulant and is interpreted by B-cells and cells of the innate immune system as a sign of microbial attack. Thus, two types of nucleic acid, dsRNA and unmethylated CpG-containing DNA, are central to mammalian defenses against infectious agents.

B. Viruses are thought to be the source of interferon-inducing antisense RNA and dsRNA, but hard evidence is very difficult to obtain

Conventional wisdom holds that the dsRNAs responsible for interferon induction are entirely of viral origin. However, as stated by Jacobs and Langland, "Actually identifying the potential sources of dsRNA in infected cells has in fact been problematic over the years" (Jacobs and Langland, 1996). The replication intermediates of RNA viruses contain both plus and minus RNAs, and are obvious potential sources of dsRNA. DNA viruses could generate dsRNA by aberrant transcription or by transcription of overlapping genes encoded on opposite strands.

The lack of direct evidence concerning the identity of the interferon-inducing dsRNA reflects technical difficulties that make it nearly impossible to analyze the critical dsRNA molecules. Ironically, the cell's extreme sensitivity is a major problem; it sets a standard for dsRNA detection that existing molecular techniques cannot equal. RNA viruses generate relatively large quantities of complementary (minus) strands during the course of replication. Minus strands create a potential background problem because they can associate with sense strands during the extraction process. Particularly if the extraction is carried out in the presence of phenol, which catalyzes nucleic acid hybridization, the presence of viral dsRNA in extracts does not prove that it existed in cells.

Even when dsRNA can be shown to be present in cells, for example by the use of dsRNA-specific antibodies, questions remain about whether this dsRNA was accessible to the cell's dsRNA-sensing machinery. Magliano and colleagues demonstrated that the membrane-bound cytoplasmic vacuoles, called rubella virus "replication complexes," are virus-modified lysosomes (Magliano *et al.*, 1998). This compartmentalization may effectively hide the rubella virus dsRNA. Similar membranous structures have been described in other virus-infected cells. The sequestration of dsRNA may be a common viral defense strategy. Vaccinia virus, a DNA virus that produces large quantities of complementary transcripts, also produces proteins encoded by the E3L gene that bind dsRNA. This countermeasure is effective. Although vaccinia virus produces dsRNA, it is relatively resistant to interferon unless the E3L gene is mutated. Moreover, interferon resistance can be restored to mutant vaccinia by supplying RNase III (Shors and Jacobs, 1997), the bacterial endonuclease which destroys dsRNA. The ability of vaccinia to replicate in cells expressing RNase III suggests that the viral dsRNA is not required by the virus, but rather is a side product. It would be interesting to know whether RNase III-producing cells support the replication of mammalian RNA viruses, or if the RNA to RNA replication cycle of such viruses render them sensitive to this endonuclease. The replication intermediates of RNA bacteriophage contain little if any exposed dsRNA and are not sensitive to RNase III *in vivo*.

The studies of rubella virus and vaccinia virus illustrate the general point that the stronger the evidence that viral dsRNA exists inside infected mammalian cells, the stronger the evidence that the dsRNA is obscured in some way. These experiments indicate that viruses produce very little dsRNA that is exposed to the interferon sensing machinery, and suggest that the RNAs reaching the dsRNA biosensor may not be mainstream viral RNAs required for replication. They also raise the possibility that viral dsRNA may not be the only sign cells use to detect viral infection. The high concentration of viral mRNA is a potential additional stigma. It has been proposed that higher plants have a surveillance system that recognizes and eliminates RNAs whose concentrations rise above threshold limits. This system has been associated with posttranscriptional gene silencing (see Section IV).

C. Double-stranded RNA-binding proteins contribute to the antiviral state

Double-stranded RNAs play two different roles in the interferon response. First, they stimulate interferon production and thus induce expression of at least 30 genes. Second, they interact with interferoninduced enzymes and thereby promote the antiviral state of the cell. Despite technical difficulties that impede direct analysis, it is possible to deduce some characteristics of these dsRNAs by studying the properties—length preferences, affinities, and concentrations—of the interferon-induced enzymes. Like dsRNA itself, at least one of these proteins, the interferon-induced RNA-dependent protein kinase (PKR), plays a dual role in the interferon response. PKR transduces the dsRNA signal, communicating to the nucleus that a virus is present. In addition, PKR and two other dsRNA-binding enzymes catalyze reactions that inhibit virus production.

1. *The interferon-induced RNA-dependent protein kinase, PKR*

PKR levels have been measured in human Daudi cells. Each cell contains about 5×10^5 molecules in the cytoplasm (mostly associated with ribosomes) and 1×10^5 in the nucleus (mostly in the nucleolus). Following interferon treatment, the PKR concentration rises three- to fourfold, with almost all of the increase occurring in the cytoplasm. IFN-treated cells contain approximately one molecule of PKR for each ribosome (Jeffrey *et al.*, 1995). Ribosomes compete with dsRNA for binding to PKR (Raine *et al.*, 1998). Ribosome-bound PKR may constitute a reserve supply that can be rapidly deployed without the need for new RNA or protein synthesis.

PKR binds to and is activated by long dsRNA molecules. As is true for many interactions between dsRNA-specific enzymes and perfect duplexes, there is no discernible sequence specificity for the reactions between PKR and dsRNA. Duplexes shorter than 30 base pairs do not bind stably to PKR and do not activate the enzyme. Those longer than 30 base pairs bind with increasing efficiency, reaching a maximum at about 85 base pairs. The lack of sequence specificity allows the PKR to recognize dsRNAs regardless of their origins, satisfying a prerequisite for any broadbased antiviral response. The rather long optimal length of dsRNA makes PKR resistant to activation by the short and imperfect duplexes present in many cellular RNAs, including rRNAs. However, PKR's requirement for perfect duplexes is not absolute. Certain RNAs lacking extensive perfect duplexes activate PKR, including a cellular RNA recently identified by Petryshyn *et al.* (1997).

Binding to long dsRNA causes PKR to undergo an autophosphorylation reaction and become activated. This process displays a marked concentration dependence: PKR is activated by low concentrations of dsRNA (in the range of 10–100 ng/ml), but higher concentrations are less and less effective, giving rise to a bell-shaped activation curve. The shape of this curve indicates that PKR is optimally activated by a particular ratio of dsRNA to PKR. It also suggests that virus-infected cells do not contain high concentrations of accessible dsRNA. If they did, the PKR defense system would not function efficiently.

Once it has been activated, PKR phosphorylates a number of proteins, most notably the translation initiation factor eIF-2. This phosphorylation inhibits protein synthesis, thereby diminishing virus production. In addition, PKR appears to transduce the dsRNA signal, at least in part, by phosphorylating I-κB, releasing and activating the transcription factor NF-κB. Analysis of the interferon response is progressing rapidly, producing a picture of an intricate combinatorial cascade in which PKR activation by dsRNA is an early event in a series of reactions that culminates in the antiviral state.

PKR is a highly effective antiviral agent. Accordingly, several viruses have evolved strategies for neutralizing it. One of the best characterized of the anti-PKR viral products, and the one with the closest ties to dsRNA, is VAI RNA of adenovirus. In cells infected with mutant viruses deficient in VAI RNA synthesis, PKR is activated and protein synthesis comes to a halt. VAI RNA is an abundant RNA (10^8 copies per infected cell) that is about 160 bases long. It has enough similarity to *bona fide* dsRNA to bind to PKR, but lacks the structural features required for activation. Thus, it competitively inhibits activation by dsRNA.

2. *The 2′,5′-oligo(A) synthetases*

The 2′,5′-oligo(A) synthetases are activated upon binding to dsRNA molecules. They then polymerize ATP into 2′,5′-oligo(A), which in turn activates RNase L (a normally latent endonuclease), leading to mRNA degradation. Constituitive expression of 2′,5′-oligo(A) synthetase has been shown experimentally to confer resistance to picorna virus infection.

In an attempt to determine the form of picorna virus RNA physically associated with the enzyme, Gribaudo and colleagues analyzed RNAs coprecipitating with 2′,5′-oligo(A) synthetase extracted from encephalomyocarditis virus (EMCV)-infected HeLa cells (Gribaudo *et al.*, 1991). Precipitates contained both plus and minus EMCV RNAs. About 10% of the viral RNA was resistant to single-stranded RNA-specific ribonucleases, and thus potentially representative of preexisting dsRNA. However, the authors commented that this value might be "an overestimate resulting from the annealing of regions of complementary strands upon removing proteins which blocked the annealing." Providing further evidence that only a small percentage of the viral RNA was double-stranded *in vivo*, the synthetase prepared from these cells was not fully activated. Activation could be enhanced 20-fold by incubation with artificial dsRNA [poly(I)-poly(C)]. Because both adenovirus VAI RNA (Desai *et al.*, 1995) and heterogeneous nuclear RNA (Nilsen *et al.*, 1982) activate 2′,5′-oligo(A) synthetases *in vitro*, extensive regions of perfect duplex structure are clearly not required for synthetase activation. Plus and minus EMCV RNAs may have stable structural elements capable of partially activating the synthetase, particularly in cells primed to respond. Further study is needed to determine whether both duplexes (composed of plus and minus viral RNAs) and free viral RNAs contribute to synthetase activation in picorna virus-infected cells.

3. *The dsRNA adenosine deaminase, dsRAD*

The dsRNA adenosine deaminase, dsRAD, catalyzes the C-6 deamination of adenosine to yield inosine. Deamination reduces the stability of dsRNAs. In conjunction with a ribonuclease specific for inosine-containing RNA (Scadden and Smith, 1997), dsRAD may play a role in viral defense. However, its contribution to viral defense is not as clearly established as that of the PKR kinase and the 2′,5′-oligo(A) synthetases. Primarily a nuclear enzyme that is expressed in virtually all mammalian cells, dsRAD contributes to normal metabolism. It has also been implicated in the production of hypermutated measles virus RNAs during chronic infection of the central nervous system. Such chronic infection can lead to a fatal degenerative neurological disease, subacute sclerosing panencephalitis.

The substrate preferences of dsRAD are rapidly coming to light, and will clarify its biological functions when they are fully known. The deaminase has no clear *in vitro* sequence specificity for action on perfect dsRNA; however, some adenosines may be preferred. For maximum modification, intermolecular or intramolecular duplexes need to contain a minimum of 100 base pairs. From the standpoint of molecular structure, it is remarkable that this enzyme acts upon continuous duplex RNA. It is able to deaminate adenosines even though the C-6 amino group of adenosine lies in the deep and narrow major groove of standard A-form RNA, a space that is usually inaccessible to amino acid side-chains.

In addition to perfect duplexes, dsRAD acts on RNAs that lack extensive duplex structure. For example, it edits certain cellular pre-mRNAs, such as that of the glutamate-gated ion channel GluR. Like PKR and the 2′,5′-oligo(A) synthetases, dsRAD binds to adenovirus VAI RNA *in vitro* (Lei *et al.*, 1998). In a manner similar to the assistance RNase III lends to bacteriophage T7, dsRAD aids the human hepatitis delta agent by deaminating a specific adenosine residue and carrying out an RNA editing event essential for the survival of this viroid-like pathogen. Exhibiting an additional similarity to RNase III, dsRAD behaves like one type of enzyme when interacting with long dsRNA molecules, in this case acting as a very robust and vigorous deaminase while it functions as a highly selective editing enzyme when interacting with single-stranded RNAs that have specific structural elements.

The dsRNA adenosine deaminase is clearly an enzyme with important cellular functions, and it is sometimes exploited by pathogens. Its role in antiviral defense is less clearly established, but the fact that its level rises following interferon treatment suggests that dsRAD contributes to the antiviral state.

D. Summary

Antisense RNAs play an important role in antiviral defenses by forming dsRNA. dsRNA is recognized by mammalian cells as a danger signal indicating that viral infection has occurred. A single molecule of dsRNA can induce interferon. Viruses often produce detectable amounts of dsRNA. However, in many cases, this dsRNA is obscured by other viral products. As a result, it has been difficult to pinpoint the actual dsRNA responsible for inducing the interferon response. Several cellular dsRNA-specific enzymes are induced by interferon and are thought to contribute to the antiviral state. Two of these enzymes degrade RNA, a third blocks protein synthesis. This enzyme, the PKR, also contributes to the antiviral state by setting off a cascade that activates interferon synthesis.

IV. ARTIFICIAL RNAs, dsRNA, AND POSTTRANSCRIPTIONAL GENE SILENCING

A. Artificial sense, antisense, and dsRNAs place a spotlight on gene silencing

Just as Rutherford was unprepared for what happened when he shot alpha particles at a thin sheet of gold, biologists were unprepared for what occurred when they engineered plants to express sense transcripts of the *chalcone synthetase A* (*chA*) gene (van der Krol *et al.*, 1990; Napoli *et al.*, 1990). In many plants, both the transgene and the endogenous homolog of the transgene were silent, or co-suppressed. Thus, an attempt to overexpress genes involved in pigment formation resulted in plants with white flowers. These plants, and many studied subsequently, exhibit posttranscriptional gene silencing (PTGS), a condition in which specific RNA molecules are degraded.

The examples of PTGS in plants involve many different genes, plant species, and DNA constructs. It is believed that PTGS could be produced in all plant species with most endogenous genes. Furthermore, PTGS can be used to confer virus resistance to transgenic plants expressing sense transcripts of viral genes. The RNA affected by PTGS may be the product of a transgene, an endogenous plant mRNA, or a viral RNA. According to Balcombe, who reviewed hypotheses concerning the pathway leading to PTGS, all mechanisms require the production of an antisense RNA (Baulcombe, 1996). Evidence reveals that PTGS also occurs in nematodes (Fire *et al.*, 1998). At least in some cases, PTGS involves dsRNA molecules functioning as signals. In order to gain a full understanding of gene regulatory pathways and the biological role of RNA, it is critical to identify the molecular events leading to PTGS. To this end, PTGS is described here in three experimental settings.

B. Homology-dependent virus resistance occurs in transgenic plants

Viral genes have been transformed into a wide range of plant species to obtain viral protection. Results of these experiments support a two-phase model of homology-dependent resistance. In the first phase, viral and transgene RNA concentrations rise. After their combined concentration exceeds a threshold (or for some other reason), a switch is triggered in a surveillance system and a factor capable of causing gene-specific RNA turnover is produced. Appearance of this factor—which is probably either a perfect or an imperfect RNA duplex—initiates the second (maintenance) phase. During the second phase, the factor moves from the inoculated leaf to other portions of the plant, causing systemic resistance (Voinnet and Baulcombe, 1997).

Like other forms of PTGS in plants, virus resistance is usually seen in only a fraction of the plants carrying a particular transgene. This variability suggests that PTGS results from a rare event. The level of virus resistance is also variable. In the most extreme cases, there is no detectable accumulation of the virus anywhere in the inoculated plant. In others, virus accumulates at least in the inoculated leaves. A plant showing intermediate resistance to tobacco etch virus (TEV) allowed the sequential events that take place during PTGS to be analyzed. This plant was initially susceptible to TEV and the transgene expressed high levels of viral sense transcripts. However, symptoms were attenuated in the (upper) leaves, which developed after inoculation. These new leaves contained little TEV RNA or RNA of the transgene. Such plants are said to have "recovered" (although the tissues showing "recovery" never developed symptoms in the first place) (Lindbo *et al.*, 1993; Goodwin *et al.*, 1996). This plant was resistant to challenge inoculation with a related virus, but was susceptible to infection by unrelated viruses, indicating that resistance was homology dependent.

Very similar events take place during certain natural viral infections, suggesting that homology-dependent virus resistance and natural resistance have common features. Ratcliff and colleagues demonstrated that *Nicotiana clevelandii* inoculated with tomato black ring nepovirus (strain W22) initially showed symptoms, and then recovered. Inoculated plants were resistant to challenge with a second inoculation of the same strain and were partially resistant to challenge with a related strain, BUK, which is 68% identical to W22. However, they were sensitive to an unrelated virus, potato virus X (Ratcliff *et al.*, 1997). Thus, in both homology-dependent virus resistance in transgenic plants and in certain natural viral infections, a factor is generated in highly infected tissues that moves to other parts of the plant, rendering them resistant. The specificity of these events strongly implicates a nucleic acid, almost certainly an RNA.

PTGS can be a two-way street, as illustrated by RNA viruses that have been engineered to contain inserts of nuclear genes. When such viruses replicate, they produce RNAs with sequences of the nuclear genes and they inhibit expression of the nuclear gene. For example, the upper leaves of *N. benthamiana* inoculated with a recombinant tobacco mosiac virus containing part of the *phytoene desaturase* gene displayed a photobleaching effect, indicating that the desaturate gene had been inhibited (resulting in the loss of carotenoid-mediated protection) (Kumagai *et al.*, 1995). Because the RNAs of plant viruses are thought to have a strictly cytoplasmic location, these results indicate that either all steps of PTGS can take place in the cytoplasm or that the RNAs interact with DNA during cell division following breakdown of the nuclear envelope.

Waterhouse and colleagues produced strong evidence that dsRNA can trigger PTGS. They showed that *N. tabaccum cv* W38 engineered to express dsRNAs of the potato virus Y (PVY) protease gene were much more likely to acquire virus immunity than plants expressing either sense or antisense transcripts of this gene (Waterhouse *et al.*, 1998). As they pointed out, the efficacy of their constructs, which were designed to produce dsRNA in which both strands of the duplex originated from transgenes (rather than duplexes in which one strand originated from the transgene and the other strand was of viral origin), argues against conventional models of antisense action. Conventional models require pairing to take place between transcripts of transgenes and their targets, while the data of Waterhouse and colleagues indicate that constructs were much more effective when their transcripts were capable of forming dsRNA on their own. It will be interesting to learn what makes such constructs more effective than those expressing antisense RNA and to identify the cellular factors interacting with the dsRNA. Proteins will almost certainly be involved in this process. Changes at the DNA level, such as altered methylation, may also occur.

There are many reports of DNA methylation of transgenes associated with PTGS. For example, in tobacco plants displaying homology-dependent resistance and PTGS of a PVY transgene, the DNA of the transgene was methylated (Smith *et al.*, 1995). At the moment, the significance of DNA methylation is not clear. It may be a cause of homology-dependent PTGS, or it may be a consequence of it.

When one considers the impact RNA may have on DNA methylation patterns, it is interesting to consider the results of studies carried out on transgenic plants carrying defective viroid cDNAs. Viroids are circular RNAs that replicate through an RNA-to-RNA rolling-circle process in the nucleus. Inoculation of the transgenic plants with infectious viroid led to viroid replication and to specific methylation of the viroid cDNA sequences, even though this DNA was not the template for the viroid RNA (Wassenegger *et al.*, 1994). Unfortunately, because viroid-infected nuclei contain both free viroid RNA and viroid replication intermediates (which may have dsRNA segments of unknown length), these experiments do not reveal whether single-stranded viroid RNA or double-stranded viroid RNA induced the DNA methylation.

C. Co-suppression of nuclear Genes in plants involves RNA–RNA duplexes

In studies paralleling those described above, Waterhouse and colleagues also compared the ability of various constructs to inhibit a Δβ-glucuronidase (GUS) reporter gene in rice. Some constructs were designed to produce single-stranded RNA transcripts, others to produce dsRNA. Based on their results, they concluded that co-suppression, like homology-dependent virus resistance, is triggered by dsRNA (Waterhouse *et al.*, 1998). Their conclusion is consistent with observations made by many earlier investigators who found that PTGS is more likely to occur in plants containing two tandem copies of a transgene arranged as inverted repeats, an organization favoring the production of transcripts capable of forming dsRNA.

Despite the strength of the evidence implicating a perfect duplex as the PTGS trigger, Metzlaff and colleagues have developed a model that involves an imperfect duplex. Their model is based on the profiles of calcone synthetase-specific RNAs present in wild-type *Petunia* and in transgenic plants manifesting PTGS. They reported that white flowers of transgenic plants had little full-length poly$(A)^+$ *chsA* RNA, but instead had characteristic mRNA fragments. They proposed that a self-sustaining degradation cycle is set in motion when sequences from the 3′-portion of an "aberrant" *chsA* transcript bind to partially complementary sequences in a second *chsA* RNA molecule, causing it to be cleaved and to release a new 3′ fragment (Metzlaff *et al.*, 1997). Significantly, fragments similar to those in the white-flowering transgenic plants accumulate in the white portions of a nontransgenic *Petunia* (Red Star) whose flowers have purple-white patterned flowers. This result shows that co-suppression uses steps of a preexisting control pathway. Eventually, the model proposed by Metzlaff and colleagues will need to be reconciled with the evidence that perfect dsRNA can trigger PTGS. A consistent model may emerge when the events leading to the production of the "aberrant" transcript are understood in greater detail.

Further studies are also needed to define the role of DNA methylation in gene silencing. DNA methylation has been studied in *Petunia* and *Arabidopsis* manifesting PTGS of the calcone synthetase gene. In purple *Petunia* flowers, an EcoRII site in the 3′-end of the endogenous genes only rarely contains a methylated cytosine, whereas in leaf DNA, these sites are frequently methylated. There is, therefore, a developmentally regulated loss of methylation at these sites. In transgenic plants with white flowers, this developmental change does not occur, and these sites are frequently methylated (Flavell *et al.*, 1998). To determine whether methylation is required for PTGS, Furner and colleagues studied transgenic *Arabidposis* plants that carry a mutation in a gene required for DNA methylation. PTGS was reversed and expression of the *chs* transgene was restored in plants defective in DNA methylation, leading these investigators to conclude that "methylation is absolutely necessary" for PTGS (Furner *et al.*, 1998). Confirmation of this observation in another system will shed further light on the significance of DNA methylation.

D. dsRNA-induced homology-dependent posttranscriptional gene silencing takes place in *Caenorhabditis elegans*

RNA interference in nematodes was discovered by Guo and Kemphues, who found that antisense and sense transcripts yielded identical results when used to block gene expression in the maternal germ line (Guo and Kemphues, 1995). Fire and colleagues later demonstrated that dsRNA mediates PTGS in nematodes and that these worms have a transport system that allows dsRNA-mediated interference to move across cell membranes. On a mole-per-mole basis, they found that dsRNA transcripts were about 100 times more potent than either antisense or sense transcripts. Their initial experiments involved the *unc-22* gene, which encodes an abundant but nonessential myofilament protein. They injected a mixture of sense and antisense transcripts covering a 742-nucleotide segment of *unc-22* into the body cavity of adults and observed robust interference in the somatic tissues of the recipients and in their progeny broods. Only a few molecules of dsRNA were required per affected cell, suggesting that the dsRNA signal may have been amplified.

Several genes have been tested for susceptibility to dsRNA-mediated interference in addition to *unc-22*. The following observations have been made (Montgomery *et al.*, 1998; Tabara *et al.*, 1998; Fire *et al.*, 1998). First, PTGS effects are usually similar to those of null mutants, indicating that the target gene has been fully and selectively inhibited. However, some expression of the target gene can occur in dsRNA-treated worms. In some cases, suppression occurs in only some cells. Second, mRNAs appear to be the targets of PTGS, rather than mRNA precursors. dsRNA segments corresponding to introns and promoter segments are not effective, as might be expected if precursors or the transcription process were the target. Furthermore, dsRNA covering upstream genes in polar operons have no effect on downstream genes, underscoring the conclusion that mRNAs, rather than precursors, are the target. In addition, cytoplasmic levels of target RNAs drop precipitously. Third, dsRNA-mediated interference is able to cross cellular boundaries. In fact, the dsRNA crosses cellular membranes so readily that it is possible to induce PTGS by feeding worms transgenic *Escherichia coli* expressing dsRNA covering the target gene (Timmons and Fire, 1998) or by soaking them in dsRNA—although neither of these approaches is as effective as microinjection. Fourth, PTGS passes into the F_1 generation; remarkably, Tabara and colleagues report that for certain genes, "interference can be observed to transmit in the germ line apparently indefinitely" (Tabara *et al.*, 1998).

E. The discovery of the importance of RNA interference is a major breakthrough

Growing evidence indicates that dsRNA is involved in gene regulation in higher organisms. In mammalian cells, dsRNA (whether applied exogenously or synthesized endogenously) induces interferon; in plants, dsRNA has been linked to homologydependent virus resistance and to PTGS of both endogenous and transgenes; and in nematodes, dsRNA moves across cell boundaries and selectively inhibits gene expression. It was once thought that bioengineered antisense RNA would function by establishing Watson–Crick base pairs with target RNAs, thereby eliminating their function. It now appears that dsRNA mediates the effects of artificial RNAs in nematodes and in some cases of PTGS in higher plants. This dsRNA is evidently recognized by cellular factors that work in conjunction with it to destroy other RNAs of similar sequence. Most of the molecular intermediates of PTGS remain a mystery. However, DNA methylation frequently occurs during PTGS in plants and is thought to play an essential role by some investigators. When one considers the possible significance of PTGS for organisms other than vascular plants and worms, it is interesting to note that some of the mammalian genes that are subject to imprinting, a process associated with DNA methylation, express antisense transcripts (Ward and Dutton, 1998; Moorwood *et al.*, 1998; Rougeulle *et al.*, 1998; Reik and Constancia, 1997). As more studies are carried out on mammalian genes producing natural antisense RNA and on mammalian cells synthesizing artificial antisense RNAs, mechanistic ties to PTGS may become apparent.

Continued exploration of phenomena such as PTGS, homology-dependent virus resistance, and co-suppression in plants; and RNA interference in nematodes have yielded important insights into the molecular biology of eukaryotes at the most basic level. It is now clear that many eukaryotic organisms—including humans, fruit flies, plants, and fungi—have the capacity to use short dsRNAs to (down) regulate expression of homologous target sequences. Several genes and gene products that contribute to this process have been identified. Their roles in a variety of biological processes, such as embryonic development, inhibition of transposon movement, and transcriptional silencing of chromosomal genes, remain areas of very active investigation. Gene regulation that is mediated by short dsRNA molecules is now most commonly referred to as "RNA silencing". The short dsRNAs involved in RNA silencing are part of a much larger group of noncoding RNAs that perform both structural and regulatory functions—often through complementary base pair (antisense) interactions. A five-part "Special Section"—RNA Silencing and Noncoding RNA, *Science* (**296**, 1259–1273, 2002) provides an up-to-date and well-referenced review of RNA silencing, non-coding RNAs, and the biological functions they perform.

Several proteins involved in RNA silencing have been identified. In the "classical" RNA silencing degradative pathway, long dsRNA molecules are cleaved to short duplexes (21–23 nucleotides) by an RNase III-like enzyme called the Dicer, which leaves 2- to 3-nucleotide long 3′ overhangs. These short dsRNAs may bind to 250–500 kD nuclease complex called RISC (RNA-induced silencing complex). The individual strands of the short dsRNAs associate with target RNAs through complementary base pairing, perhaps using an RNA helicase to separate from each other and thus gain assess to sequences in the target mRNAs. The target RNAs are then cleaved by RISC. Should one of the individual strands of the short dsRNA bind to the target RNA in the absence of RISC, a structure results that may be amplified by an RNA dependent RNA polymerase, yielding additional long dsRNA that is a substrate for Dicer.

In the fields of genomics and biotechnology, short artificial dsRNAs are being introduced into cells to knock out target genes and thereby determine its function. Short dsRNAs can be more effective inhibitors than single-standard antisense oligonucleotides, and are being considered as potential therapeutic agents, as well as research tools. While initial experiments have produced impressive results and demonstrate the RNA silencing can be obtained in mammalian tissue culture cells, the extent of target gene inhibition is variable. If RNA silencing molecules can be optimized, they may provide a shortcut for deletion mutagenesis and for large scale efforts to assign functions to individual genes. RNA silencing is one intriguing component of the larger and perennially eye-opening field of RNA biochemistry. Many more exciting discoveries can be expected as the full range of RNA's role in biology comes to light.

ACKNOWLEDGMENTS

I thank Dr. Jose Walewski, Mr. Decherd Stump, and Ms. Toby Keller for help with the manuscript. This work was supported in part by NIDDK (grants R01-DK52071 and P01-DK50795, project 2), the Liver Transplantation Research Fund, and the Division of Liver Diseases research funds.

BIBLIOGRAPHY

Altuvia, S., and Wagner, E. G. (2000). Switching on and off with RNA. *Proc. Natl. Acad. Sci. USA*. **97**, 9824–9826.

Bass, B. L. (1997). RNA editing and hypermutation by adenosine deamination. *Trends Biochem. Sci.* **22**, 157–162.

Baulcombe, D. C. (1996). RNA as a target and an initiator of post-transcriptional gene silencing in transgenic plants. *Plant Mol. Biol.* **32**, 79–88.

Bram, R. J., Young, R. A., and Steitz, J. A. (1980). The ribonuclease III site flanking 23S sequences in the 30S ribosomal precursor RNA of *E. coli*. *Cell* **19**, 393–401.

Branch, A. D. (1998). A good antisense molecule is hard to find. *Trends Biochem. Sci.* **23**, 45–50.

Delihas, N., Rokita, S. E., and Zheng, P. (1997). Natural antisense RNA/target RNA interactions: Possible models for antisense oligonucleotide drug design. *Nat. Biotechnol.* **15**, 751–753.

Desai, S. Y., Patel, R. C., Sen, G. C., Malhotra, P., Ghadge, G. D., and Thimmapaya, B. (1995). Activation of interferon-inducible 2′–5′ oligoadenylate synthetase by adenoviral VAI RNA. *J. Biol. Chem.* **270**, 3454–3461.

Eguchi, Y., Itoh, T., and Tomizawa, J. (1991). Antisense RNA. *Annu. Rev. Biochem.* **60**, 631–652.

Engdahl, H. M., Hjalt, T. A., and Wagner, E. G. (1997). A two unit antisense RNA cassette test system for silencing of target genes. *Nucleic Acids. Res.* **25**, 3218–3227.

Fire, A., Xu, S., Montgomery, M. K., Kostas, S. A., Driver, S. E., and Mello, C. C. (1998). Potent and specific genetic interference by double-stranded RNA in Caenorhabditis elegans. *Nature* **391**, 806–811.

Flavell, R. B., O'Dell, M., and Metzlaff, M. (1998). Transgene-promoted epigenetic switches of chalcone synthase activity in petunia plants. *Novartis Found. Symp.* **214**, 144–154.

Franch, T., and Gerdes, K. (2000). U-turns and regulatory RNAs. *Curr. Opin. Microbiol.*, **3**, 159–164.

Furner, I. J., Sheikh, M. A., and Collett, C. E. (1998). Gene silencing and homology-dependent gene silencing in arabidopsis: Genetic modifiers and DNA methylation. *Genetics* **149**, 651–662.

Goodwin, J., Chapman, K., Swaney, S., Parks, T. D., Wernsman, E. A., and Dougherty, W. G. (1996). Genetic and biochemical dissection of transgenic RNA-mediated virus resistance. *Plant Cell* **8**, 95–105.

Gribaudo, G., Lembo, D., Cavallo, G., Landolfo, S., and Lengyel, P. (1991). Interferon action: Binding of viral RNA to the 40-kilodalton 2′-5′-oligoadenylate synthetase in interferon-treated HeLa cells infected with encephalomyocarditis virus. *J. Virol.* **65**, 1748–1757.

Guo, S., and Kemphues, K. J. (1995). par-1, a gene required for establishing polarity in C. elegans embryos, encodes a putative Ser/Thr kinase that is asymmetrically distributed. *Cell* **81**, 611–620.

Hjalt, T. A., and Wagner, E. G. (1995). Bulged-out nucleotides in an antisense RNA are required for rapid target RNA binding in vitro and inhibition in vivo. *Nucleic. Acids Res.* **23**, 580–587.

Izant, J. G., and Weintraub, H. (1984). Inhibition of thymidine kinase gene expression by anti-sense RNA: A molecular approach to genetic analysis. *Cell* **36**, 1007–1015.

Jacobs, B. L., and Langland, J. O. (1996). When two strands are better than one: The mediators and modulators of the cellular responses to double-stranded RNA. *Virology* **219**, 339–349.

Jeffrey, I. W., Kadereit, S., Meurs, E. F., Metzger, T., Bachmann, M., Schwemmle, M., Hovanessian, A. G., and Clemens, M. J. (1995). Nuclear localization of the interferoninducible protein kinase PKR in human cells and transfected mouse cells. *Exp. Cell Res.* **218**, 17–27.

Katze, M. G. (1992). The war against the interferon-induced dsRNA-activated protein kinase: Can viruses win? *J. Interferon Res.* **12**, 241–248.

Kumagai, M. H., Donson, J., della Cioppa, G., Harvey, D., Hanley, K., and Grill, L. K. (1995). Cytoplasmic inhibition of carotenoid biosynthesis with virus-derived RNA. *Proc. Natl. Acad. Sci. USA* **92**, 1679–1683.

Lee, A. J., and Crothers, D. M. (1998). The solution structure on an RNA loop-loop complex: The Col*E1* inverted loop sequence. *Structure* **6**, 993–1005.

Lei, M., Liu, Y., and Samuel, C. E. (1998). Adenovirus VAI RNA antagonizes the RNA-editing activity of the ADAR adenosine deaminase. *Virology* **245**, 188–196.

Lindbo, J., Silva-Rosales, L., Proebsting, W. M., and Dougherty, W. G. (1993). Induction of a highly specific antiviral state in transgenic plants: Implications for regulation of gene expression and virus resistance. *Plant Cell* **5**, 1749–1759.

Magliano, D., Marshall, J. A., Bowden, D. S., Vardaxis, N., Meanger, J., and Lee, J. Y. (1998). Rubella virus replication complexes are virus-modified lysosomes. *Virology* **240**, 57–63.

Malmgren, C., Wagner, E. G. H., Ehresmann, C., Ehresmann, B., and Romby, P. (1997). Antisense RNA control of plasmid R1 replication. The dominant product of the antisense RNA-mRNA binding is not a full RNA duplex. *J. Biol. Chem.* **272**, 12508–12512.

Marcus, P. I., and Sekellick, M. J. (1977). Defective interfering particles with covalently linked [+/−] RNA induce interferon. *Nature* **266**, 815–819.

Mathews, M. B., and Shenk, T. (1991). Adenovirus virus-associated RNA and translation control. *J. Virol.* **65**, 5657–5662.

Matzke, M. A., and Matzke, A. J. (1995). Homology-dependent gene silencing in transgenic plants: What does it really tell us? *Trends Genet.* **11**, 1–3.

Metzlaff, M., O'Dell, M., Cluster, P. D., and Flavell, R. B. (1997). RNA-mediated RNA degradation and chalcone synthase A silencing in petunia. *Cell* **88**, 845–854.

Mizuno, K., Chou, M. Y., and Inouye, M. (1984). A unique mechanism regulating gene expression: Translational inhibition by a complementary RNA transcript (micRNA). *Proc. Natl. Acad. Sci. USA* **81**, 1966–1970.

Montgomery, M. K., Xu, S., and Fire, A. (1998). RNA as a target of double-stranded RNA-mediated genetic interference in *Caenorhabditis elegans*. *Proc. Natl. Acad. Sci. USA* **95**, 15502–15507.

Moorwood, K., Charles, A. K., Salpekar, A., Wallace, J. I., Brown, K. W., and Malik, K. (1998). Antisense WT1 transcription parallels sense mRNA and protein expression in fetal kidney and can elevate protein levels *in vitro*. *J Pathol* **185**, 352–359.

Muller, U., Steinhoff, U., Reis, L. F., Hemmi, S., Pavlovic, J., Zinkernagel, R. M., and Aguet, M. (1994). Functional role of type I and type II interferons in antiviral defense. *Science* **264**, 1918–1921.

Napoli, C., Lemieux, C., and Jorgensen, R. (1990). Introduction of a chimeric chalcone synthase gene into petunia results in reversible co-suppression of homologous genes in *trans*. *Plant Cell* **2**, 279–289.

Nilsen, T. W., Maroney, P. A., Robertson, H. D., and Baglioni, C. (1982). Heterogeneous nuclear RNA promotes synthesis of (2′,5′)oligoadenylate and is cleaved by the (2′,5′)oligoadenlate-activated endoribonuclease. *Mol. Cell Biol.* **2**, 154–160.

Petryshyn, R. A., Ferrenz, A. G., and Li, J. (1997). Characterization and mapping of the double-stranded regions involved in activation of PKR within a cellular RNA from 3T3-F442A cells. *Nucleic Acids Res.* **25**, 2672–2678.

Proud, C. G. (1995). PKR: A new name and new roles. *Trends Biochem. Sci.* **20**, 241–246.

Raine, D. A., Jeffrey, I. W., and Clemens, M. J. (1998). Inhibition of the double-stranded RNA-dependent protein kinase PKR by mammalian ribosomes. *FEBS Lett.* **436**, 343–348.

Ratcliff, F., Harrison, B. D., and Baulcombe, D. C. (1997). A similarity between viral defense and gene silencing in plants. *Science* **276**, 1558–1560.

Reik, W., and Constancia, M. (1997). Genomic imprinting. Making sense or antisense? *Nature* **389**, 669–671.

Robertson, H. D. (1982). Escherichia coli ribonuclease III cleavage sites. *Cell* **30**, 669–672.

Rougeulle, C., Cardoso, C., Fontes, M., Colleaux, L., and Lalande, M. (1998). An imprinted antisense RNA overlaps UBE3A and a second maternally expressed transcript. *Nat. Genet.* **19**, 15–16.

Samuel, C. E. (1994). Interferon-induced proteins and their mechanisms of action. *Hokkaido Igaku. Zasshi.* **69**, 1339–1347.

Scadden, A. D., and Smith, C. W. (1997). A ribonuclease specific for inosine-containing RNA: A potential role in antiviral defence? *EMBO J.* **16**, 2140–2149.

Shors, T., and Jacobs, B. L. (1997). Complementation of deletion of the vaccinia virus E3L gene by the Escherichia coli RNase III gene. *Virology* **227**, 77–87.

Simons, R. W. (1988). Naturally occurring antisense RNA control—a brief review. *Gene* **72**, 35–44.

Smith, H. A., Powers, H., Swaney, S., Brown, C., and Dougherty, W. G. (1995). Transgenic potato virus Y resistance in potato: Evidence for an RNA-mediated cellular response. *Phytopathol* **85**, 864–870.

Smith, H. A., Swaney, S. L., Parks, T. D., Wernsman, E. A., and Dougherty, W. G. (1994). Transgenic plant virus resistance mediated by untranslatable sense RNAs: Expression, regulation, and fate of nonessential RNAs. *Plant Cell* **6**, 1441–1453.

Tabara, H., Grishok, A., and Mello, C. C. (1998). RNAi in *C. elegans:* Soaking in the genome sequencing. *Science* **282**, 430.

Timmons, L., and Fire, A. (1998). Specific interference by ingested dsRNA. *Nature* **395**, 854.

van der Krol, A. R., Mur, L. A., Beld, M., Mol, J. N., and Stuitje, A. R. (1990). Flavonoid genes in petunia: Addition of a limited number of gene copies may lead to a suppression of gene expression. *Plant Cell* **2**, 291–299.

Vanhée-Brossollet, C., and Vaquero, C (1998). Do antisense transcripts make sense in eukaryotes? *Gene* **211**, 1–9.

Voinnet, O., and Baulcombe, D. C. (1997). Systemic signalling in gene silencing. *Nature* **389**, 553.

Wagner, E. G., Blomberg, P., and Nordstrom, K. (1992). Replication control in plasmid R1: Duplex formation between the antisense RNA, CopA, and its target, CopT, is not required for inhibition of RepA synthesis. *EMBO J.* **11**, 1195–1203.

Wagner, E. G., and Simons, R. W. (1994). Antisense RNA control in bacteria, phages, and plasmids. *Annu. Rev. Microbiol.* **48**, 713–742.

Ward, A., and Dutton, J. R. (1998). Regulation of the Wilm's Tumour suppressor (WT1) gene by an antisense RNA: A link with genomic imprinting? *J. Pathol.* **185**, 342–344.

Wassenegger, M., Heimes, S., Riedel, L., and Sänger, H. L. (1994). RNA-directed de novo methylation of genomic sequences in plants. *Cell* **76**, 567–576.

Waterhouse, P. M., Graham, M. W., and Wang, M. B. (1998). Virus resistance and gene silencing in plants can be induced by simultaneous expression of sense and antisense RNA. *Proc. Natl. Acad. Sci. USA* **95**, 13959–13964.

WEBSITES

List of websites giving information about antisense RNA
http://www.ambion.com/techlib/hottopics/rnai
http://www.imb-jena.de/RNA.html

6

Antiviral agents

Richard J. Whitley
The University of Alabama at Birmingham

GLOSSARY

acyclic purine nucleoside analog A molecule with the structure of the normal purine components of DNA or RNA but with the sugar ring cleaved open (acyclic).

alanine amino transferase An enzyme found in the liver and blood serum, the concentration of which is often elevated in cases of liver damage.

antiretroviral agent Any drug used in treating patients with human immunodeficiency virus (HIV) infection.

bioavailability The property of a drug to be absorbed and distributed within the body in a way that preserves its useful characteristics; for example, it is not broken down, inactivated, or made insoluble.

condyloma acuminatum Venereal warts.

conjunctivitis Inflammation of the conjuctiva or white of the eye.

EC_{50} Concentration of a drug which produces a 50% effect, e.g., in virus yield.

enterovirus One of a group of viruses which infect the intestinal track.

Epstein–Barr virus A member of the herpesvirus family.

hantavirus pulmonary syndrome A pneumonia-like illness resulting from infection with hantavirus, a virus normally carried in rodents.

hepatotoxicity Liver toxicity.

hyperemia Literally excess blood; flushed, reddened, and engorged with blood.

interferon Any group of glycoproteins with antiviral activity.

interstitial nephritis An inflammation of the substance of kidney exclusive of the structure called the glomerulus.

leukopenia Deficiency in circulating white blood cells.

monotherapy Treatment with a single drug, contrasted with combination therapies with more than one drug at the same time.

mucocutaneous Refers to the skin where there is both exterior skin and mucus membranes, such as the borders of the mouth.

nephrotoxicity Kidney toxicity.

neuraminidase An enzyme, present on the surface of some viruses, which catalyzes the cleavage of a sugar derivative called neuraminic acid.

neuropathy Pathological changes in the nervous system.

papillomavirus A group of viruses causing warts of various kinds.

peptidomimetic A molecule having properties similar to those of a peptide or short protein.

pharmacokinetic Refers to the rates and efficiency of uptake, distribution, and disposition of a drug in the body.

phase III The final stage in testing of a new drug, after determination of its safety and effectiveness, in which it is tested on a broad range,

The Desk Encyclopedia of Microbiology
ISBN: 0-12-621361-5

and large population of patients for comparison to existing treatments and to test for rare complications.

picornavirus A group of viruses with small RNA genomes, such as poliovirus.

prodrug A drug that is given in a form that is inactive and must be metabolized in the body to the active form.

stromal keratitis Inflammation of the deep layers of the cornea of the eye.

superficial punctate keratopathy Fine, spot-like pathological changes in the superficial layer of the cornea of the eye.

$t_{1/2}$ The time for reduction of some observed quantity, for example, the blood concentration of a drug, by 50%.

thrombocytopenia Deficiency of platelets, the blood-clotting agents, in the blood.

tubular necrosis Death of the tubule cells in the kidney.

uveitis Inflammation of the iris or related structures in the eye.

zoster Infection with varicella-zoster virus which leads to skin lesions on the trunk (usually) following the distribution of the sensory nerves; commonly called shingles.

Antiviral Agents are drugs that are administered for therapeutic purposes to humans with viral diseases. Importantly, many people are infected by viruses but only some develop disease attributed to these microbes. Antiviral agents used to treat these diseases are currently limited and only exist for the management of herpes simplex virus, varicella-zoster virus, cytomegalovirus, hepatitis B, hepatitis C, human immunodeficiency virus, respiratory syncytial virus, human papillomavirus, and influenza virus-related diseases.

Only a few antiviral agents of proven value are available for a limited number of clinical indications. Unique problems are associated with the development of antiviral agents. First, viruses are obligate intracellular parasites that utilize biochemical pathways of the infected host cell. Second, early diagnosis of viral infection is crucial for effective antiviral therapy because by the time symptoms appear, several cycles of viral multiplication usually have occurred and replication is waning. Third, since many of the disease syndromes caused by viruses are relatively benign and self-limiting, the therapeutic index, or ratio of efficacy to toxicity, must be extremely high in order for therapy to be acceptable.

Fortunately, molecular biology research is helping solve two of these problems. Enzymes unique to viral replication have been identified and, therefore, distinguish between virus and host cell functions. Unique events in viral replication are sites which serve as ideal targets for antiviral agents; examples include the thymidine kinase (TK) of herpes simplex virus (HSV) or protease of human immunodeficiency virus (HIV). Second, several sensitive and specific viral diagnostic methods are possible because of recombinant DNA technology [e.g., monoclonal antibodies, DNA hybridization techniques, and polymerase chain reaction (PCR)]. This article will synthesize knowledge of the existing antiviral agents as it relates to both pharmacologic and clinical properties.

I. THERAPEUTICS FOR HERPESVIRUS INFECTIONS

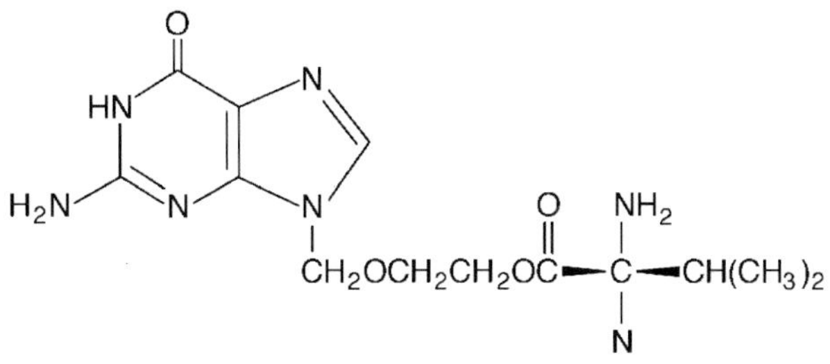

A. Acyclovir and valaciclovir

Acyclovir has become the most widely prescribed and clinically effective antiviral drug available to date. Valaciclovir, the L-valine ester oral prodrug of acyclovir, was developed to improve the oral bioavailability of acyclovir. Valaciclovir is cleaved to acyclovir by valine hydrolase which then is metabolized in infected cells to the active triphosphate of acyclovir.

1. *Chemistry, mechanism of action, and antiviral activity*

Acyclovir [9-(2-hydroxyethoxymethyl) guanine], a synthetic acyclic purine nucleoside analog, is a selective inhibitor of HSV-1 and -2 and varicellazoster virus (VZV) replication. Acyclovir is converted by

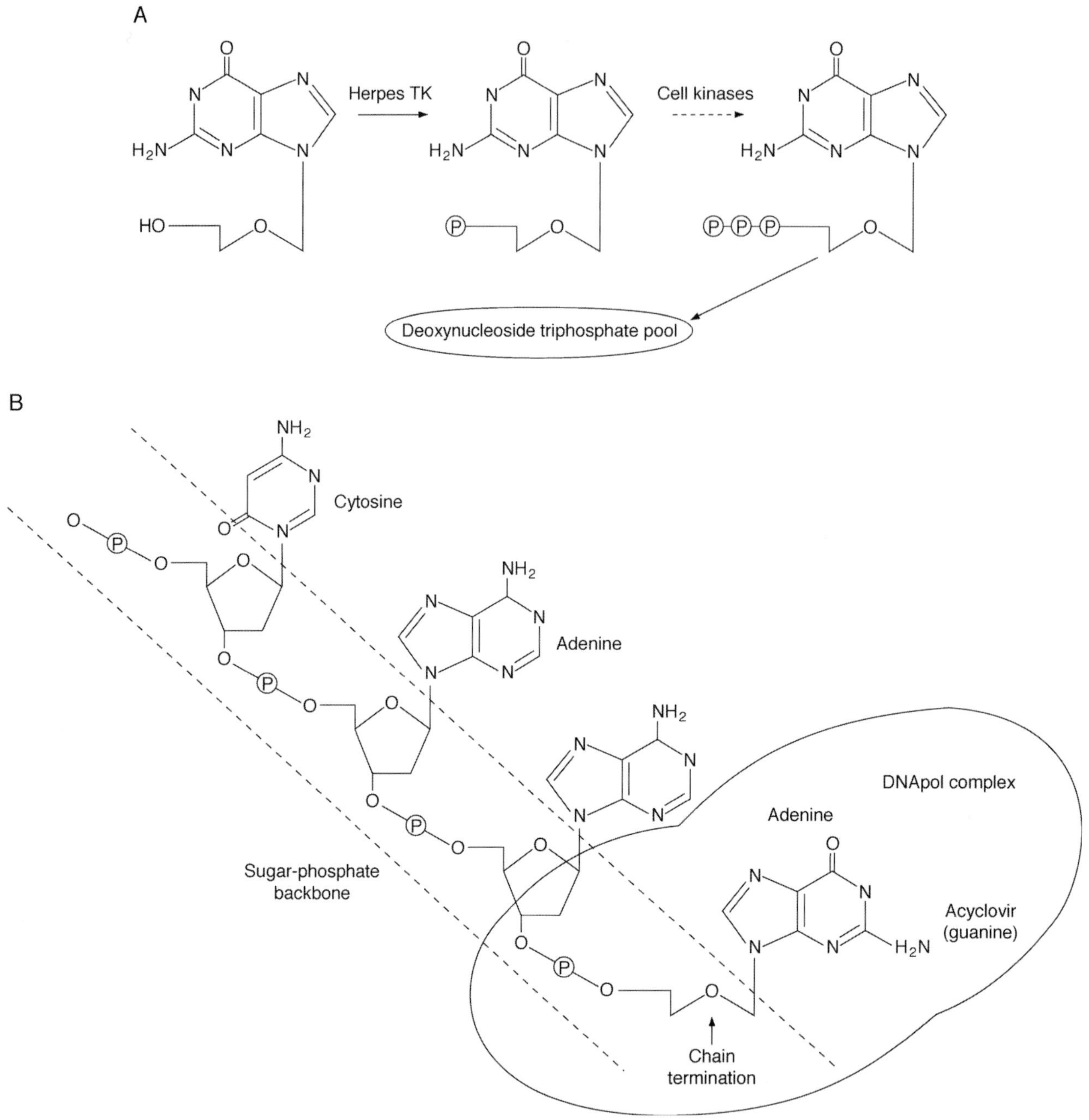

FIGURE 6.1 The mechanism of action of acyclovir. (A) activation and (B) Inhibition of DNA synthesis and chain termination.

virus-encoded TK to its monophosphate derivative, an event that does not occur to any significant extent in uninfected cells. Subsequent di- and triphosphorylation is catalyzed by cellular enzymes, resulting in acyclovir triphosphate concentrations 40–100 times higher in HSV-infected than in uninfected cells. Acyclovir triphosphate inhibits viral DNA synthesis by competing with deoxyguanosine triphosphate as a substrate for viral DNA polymerase, as illustrated in Fig. 6.1. Because acyclovir triphosphate lacks the 3′ hydroxyl group required for DNA chain elongation, viral DNA synthesis is terminated. Viral DNA polymerase is tightly associated with the terminated DNA chain and is functionally inactivated. Also, the viral polymerase has greater affinity for acyclovir triphosphate than does cellular DNA polymerase, resulting in little incorporation of acyclovir into cellular DNA. *In vitro*, acyclovir is most active against

HSV-1 (average EC_{50} = 0.04 μmg/ml), HSV-2 (0.10 μg/ml), and VZV (0.50 μg/ml). Epstein–Barr virus (EBV) requires higher acyclovir concentrations for inhibition, and cytomegalovirus (CMV), which lacks a virus-specific TK, is relatively resistant.

Acyclovir is available in topical, oral, and intravenous preparations. Oral formulations include a 200-mg capsule, a 800-mg tablet, and suspension (200 mg/5 ml) and absorption of acyclovir results in 15–30% bioavailability. After multidose oral administration of 200 or 800 mg of acyclovir, the mean steady-state peak levels are approximately 0.57 and 1.57 μg/ml, respectively. Higher plasma acyclovir levels are achieved with intravenous administration. Steady-state peak acyclovir concentrations following intravenous doses of 5 or 10 mg/kg every 8 hr are approximately 9.9 and 20.0 μg/ml, respectively. The terminal plasma time for a 50% decrease in drug concentration ($t_{1/2}$) is 2 or 3 hr in adults with normal renal function. Acyclovir is minimally metabolized, and approximately 85% is excreted unchanged in the urine via renal tubular secretion and glomerular filtration.

Valaciclovir is only available as a tablet formulation and is metabolized nearly completely to acyclovir within minutes after absorption. Plasma levels of acyclovir, following 2 g of valaciclovir given three times a day by mouth, approximate 5 mg/kg administered every 8 h intravenously. Both acyclovir and valaciclovir must be dose adjusted if renal impairment exists.

2. *Clinical indications*

a. Genital herpes

First episode genital HSV infection can be treated with topical, oral, or intravenous acyclovir. Topical application is less effective than oral or intravenous therapy. Intravenous acyclovir is the most effective treatment for first-episode genital herpes and results in a significant reduction in the median duration of viral shedding, pain, and time to complete healing (8 versus 14 days) but is reserved for patients with systemic complications. Oral therapy (200 mg five times daily) is the standard treatment. Neither intravenous nor oral acyclovir treatment alter the frequency of recurrences.

While neither valaciclovir nor famciclovir have been evaluated in patients with primary genital herpes, their pharmacokinetic properties would predict efficacy. Many experienced physicians would preferentially use these drugs over acyclovir. The dose of valaciclovir is one gram t.i.d. for 7–10 days.

Recurrent genital herpes is less severe and resolves more rapidly than primary infection, offering a shorter time interval for successfully antiviral chemotherapy. Topically applied acyclovir has no clinically beneficial effect. Orally administered acyclovir (200 mg five times daily or 400 mg three times daily) shortens the duration of virus shedding and time to healing (6 versus 7 days) when initiated within 24 h of onset, but the duration of pain and itching is not affected.

Oral acyclovir administration effectively suppresses frequently recurring genital herpes. Daily administration of acyclovir reduces the frequency of recurrences by up to 80%, and 25–30% of patients have no further recurrences while taking the drug. Successful suppression for as long as 3 years has been reported, with no evidence of significant adverse effects. Titration of acyclovir (400 mg twice daily or 200 mg two to five times daily) may be required to establish the minimum dose that is most effective and economical. Asymptomatic virus shedding can continue despite clinically effective acyclovir suppression, resulting in the possibility of person-to person transmission.

Valaciclovir therapy of recurrent genital herpes (either 1 g or 500 mg twice a day) is clinically equivalent to acyclovir administered at either 200 mg three times daily or five times daily. It is also effective for suppression of recurrences when 1 g per day is administered.

b. Herpes labialis

Topical therapy for HSV-1 mouth or lip infections is of no clinical benefit. Orally administered acyclovir (200 or 400 mg five times daily for 5 days) reduces the time to loss of crust by approximately 1 day (7 versus 8 days) but does not alter the duration of pain or time to complete healing. Oral acyclovir therapy has modest clinical benefit but only if initiated very early after a recurrence. Valaciclovir is easier to administer and is given at 500 mg 2 × day every 12 hours.

c. Immunocompromised host

HSV infections of the lip, mouth, skin, perianal area, or genitals may be more severe in immunocompromised patients. Clinical benefit from intravenous or oral acyclovir therapy is documented as evidenced by a significantly shorter duration of viral shedding and accelerated lesion healing. Acyclovir prophylaxis of HSV infections is of significant clinical value in severely immunocompromised patients, especially those undergoing induction chemotherapy or organ transplantation. Intravenous or oral acyclovir administration reduces the incidence of symptomatic HSV infection from 70 to 5–20%. A variety of oral dosing regimens, ranging from 200 mg three times daily to 800 mg twice daily, have been used successfully.

d. Herpes simplex encephalitis

Acyclovir therapy (10 mg/kg every 8 h for 14–21 days) reduces mortality overall from 70 to 19%. Furthermore, 38% of acyclovir recipients returned to normal neurologic function.

e. Neonatal HSV infections

Acyclovir treatment of babies with disease localized to the skin, eye, or mouth yielded 100% survival, whereas 18 and 55% of babies with central nervous system (CNS) or disseminated infection died, respectively. For babies with HSV localized to the skin, eye, and mouth, 98% of acyclovir recipients developed normally 2 years after infection. For babies surviving encephalitis and disseminated disease, 43 and 57% of acyclovir recipients developed normally. The currently recommended intravenous dose is 20 mg/kg every 8 h for 14–21 days.

f. Varicella

Oral acyclovir therapy in normal children and adolescents with chicken pox shortens the duration of new lesion formation by about 1 day, reduces total lesion count, and improves constitutional symptoms. Therapy of older patients with chicken pox (who may have more severe manifestations) is indicated, whereas treatment of younger children must be decided on a case-by-case basis. The oral dose of acyclovir is 20 mg/kg/t.i.d. upto 800 mg p.o. t.i.d.

Acyclovir therapy of chicken pox in immunocompromised children substantially reduces morbidity and mortality. Intravenous acyclovir treatment (500 mg/m^2 of body surface area every 8 h for 7–10 days) improved the outcome, as evidenced by a reduction of VZV pneumonitis from 45 to $<5\%$. Oral acyclovir therapy is not indicated for immunocompromised children with chicken pox; instead, treatment is with intravenous drug.

g. Herpes zoster

Intravenous acyclovir therapy of herpes zoster in the normal host produces some acceleration of cutaneous healing and resolution of pain—both acute neuritis and zoster-associated pain. Oral acyclovir (800 mg five times a day) administration results in accelerated cutaneous healing and reduction in the severity of acute neuritis. Oral acyclovir treatment of zoster ophthalmicus reduces the incidence of serious ocular complications such as keratitis and uveitis. Valaciclovir (1 g three times daily for 7–10 days) is superior to acyclovir for the reduction of pain associated with shingles. Similar data established efficacy for famciclovir as shown below.

The increased frequency of significant morbidity in immunocompromised patients with herpes zoster highlights the need for effective antiviral chemotherapy. Intravenous acyclovir therapy significantly reduces the frequency of cutaneous dissemination and visceral complications of herpes zoster in immunocompromised adults. Acyclovir is the standard therapy at a dose of 10 mg/kg (body weight) or 500 mg/m^2 (body surface area) every 8 h for 7–10 days. Oral acyclovir therapy in immunocompromised patients with herpes zoster likely is effective, but valaciclovir is presumably superior.

3. *Antiviral resistance*

Resistance of HSV to acyclovir develops through mutations in the viral gene encoding TK via generation of TK-deficient mutants or the selection of mutants possessing a TK which is unable to phosphorylate acyclovir.

DNA polymerase mutants also have been recovered from HSV-infected patients. Acyclovir-resistant HSV isolates have been identified as the cause of pneumonia, encephalitis, esophagitis, and mucocutaneous infections, all occurring in immunompromised patients.

Acyclovir-resistant mutants have been described in the normal host but are uncommon. Acyclovir-resistant isolates of VZV have been identified much less frequently than acyclovir-resistant HSV but have been recovered from marrow transplant recipients and AIDS patients. The acyclovir-resistant VZV isolates all had altered or absent TK function but remained susceptible to vidarabine and foscarnet.

4. *Adverse effects*

Acyclovir and valaciclovir therapies are associated with few adverse effects. Renal dysfunction can occur but is relatively uncommon and usually reversible. A few reports have linked intravenous acyclovir use with CNS disturbances, including agitation, hallucinations, disorientation, tremors, and myoclonus.

An Acyclovir in Pregnancy Registry has gathered data on prenatal acyclovir exposures. Though no significant risk to the mother or fetus has been documented, the total number of monitored pregnancies remains too small to detect any low-frequency teratogenic events.

B. Cidofovir

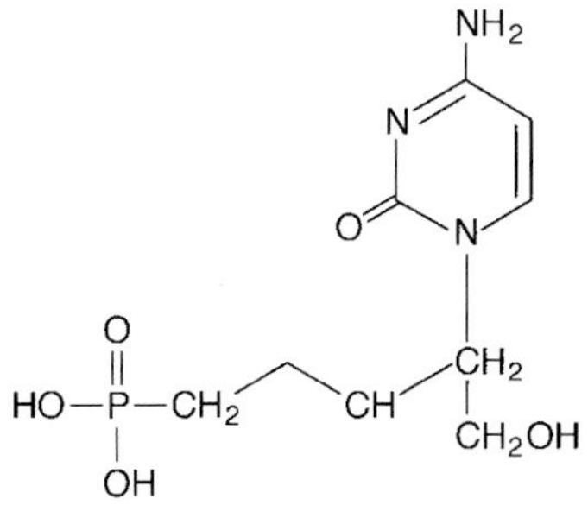

1. Chemistry, mechanism of action, and antiviral activity

Cidofovir, (S)-1-(3-hydroxy-2-phosphonomethoxypropyl) cytosine (HPMPC), is a novel acyclic phosphonate nucleoside analog and is used to reat acyclovir- and foscarnet-resistant HSV infections as well as CMV retinitis. The drug has a similar mechanism of action as the other nucleoside analog but employs cellular kinases to produce the active triphosphate metabolite. Activated HPMPC has a higher affinity for viral DNA polymerase, and therefore it selectively inhibits viral replication. The drug is less potent than ACV *in vitro*; however, *in vivo* HPMPC persists in cells for prolonged periods, increasing drug activity. In addition, HPMPC produces active metabolites with long half-lives (17–48 h), permitting once-weekly dosing. Unfortunately, HPMPC concentrates in kidney cells 100 times greater than in other tissues and produces severe proximal convoluted tubule nephrotoxicity when administered systemically. Attempts to limit nephrotoxicity include coadministration of probenecid with intravenous hydration, synthesis of cyclic congener prodrugs of HPMPC, and use of topical formulations. HPMPC has limited and variable oral bioavailability (2–26%) when tested in rats and, therefore, is administered intravenously.

2. Clinical indications

Cidofovir is licensed for treatment of CMV retinitis and has been used to treat acyclovir-resistant HSV infection. A treatment regimen of 5 mg/kg per week for 2 weeks followed by the same dose once weekly provides superior benefit over lower maintenance doses. Probenecid and liberal intravenous hydration have been added to prevent significant nephrotoxicity. Because of nephrotoxicity, this regimen is less attractive than oral valganciclovir therapy.

3. Resistance

The development of resistance with clinical use is uncommon; however, mutations in CMV DNA polymerase can mediate altered susceptibility.

4. Adverse events

Nephrotoxicity is associated with the cidofovir administration, occurring in up to 30% of patients. Oral probenecid administration accompanies intravenously administered HPMPC in order to prevent significant nephrotoxicity.

C. Fomivirsen

Fomivirsen is the first antisense oligonucleotide licensed for the treatment of a viral disease.

1. Chemistry, mechanism of action, and antiviral activity

Fomivirsen (5′-GCG TTT GCT CTT CTT-3′0) is approved for the treatment of CMV retinitis. The IC_{50} against laboratory strains of CMV is about 0.37 μM. Drug binds to the mrRNA of the immediate early 2 gene of CMV. Fomivirsen can only be administered by intravitreal injection. The pharmacokinetics of drug administration to the rabbit eye indicates a half-life of 62 hours.

2. Clinical indications

Fomivirsen delays progression of CMV retinitis when administered at a dosage of 330 μg every other weeks on three occasions, followed by the same dose monthly. Drug is approved for patients intolerant to other medications.

3. Resistance

No isolates from humans have been reported as resistant to fomivirsen.

4. Adverse events

Increased intraocular pressure and inflammation have been reported as a major side effect in as many as 20% of patients.

D. Foscarnet

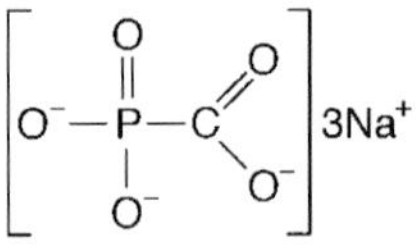

1. Chemistry, mechanism of action, and antiviral activity

Foscarnet, a pyrophosphate analog of phosphonoacetic acid has potent *in vitro* and *in vivo* activity against all herpesviruses and inhibits the DNA polymerase by blocking the pyrophosphate binding site, inhibiting the formation of the 3′,5′ phosphodiester bond between primer and substrate and preventing chain elongation. Unlike acyclovir, which requires activation by a virus-specific TK, foscarnet acts directly on the virus DNA polymerase. Thus, TK-defi- cient, acyclovir-resistant herpesviruses remain sensitive to foscarnet.

The oral bioavailability of foscarnet is poor; thus, administration is by the intravenous route. An intravenous infusion of 60 mg/kg every 8 h results in peak and trough plasma concentrations which are approximately 450–575 and 80–150 μ*M*, respectively. The cerebrospinal fluid concentration of foscarnet is approximately two-thirds of the plasma level.

Renal excretion is the primary route of clearance of foscarnet, with >80% of the dose appearing in the urine. Bone sequestration also occurs, resulting in complex plasma elimination.

2. Clinical indications

Foscarnet is licensed for the treatment of CMV retinitis as well as HSV and VZV disease caused by acyclovir- or penciclovir-resistant viruses. Administration of foscarnet at 60 mg/kg every 8 h for 14–21 days followed by maintenance therapy at 90–120 mg/kg per day is associated with stabilization of retinal disease in approximately 90% of patients. However, as is with the case with ganciclovir therapy of CMV retinitis, relapse occurs.

Mucocutaneous HSV infections and those caused by VZV in immunocompromised hosts can be treated with foscarnet at dosages lower than that for the management of CMV retinitis. Foscarnet dosages of 40 mg/kg administered every 8 hr for 7 days or longer will result in cessation of viral shedding and healing of lesions in the majority of patients. However, relapses will occur which may or may not be amenable to acyclovir therapy.

3. Resistance

Isolates of HSV, CMV, and VZV have all been demonstrated to develop resistance to foscarnet both in the laboratory and in the clinical setting. Isolates of HSV which are resistant to foscarnet have EC_{50} 100 μg/ml. These isolates are all DNA polymerase mutants.

4. Adverse effects

Although foscarnet has significant activity in the management of herpesvirus infections, nephrotoxicity, including acute tubular necrosis and interstitial nephritis, can occur. Metabolic aberrations of calcium, magnesium, phosphate, and other electrolytes are associated with foscarnet administration and warrant careful monitoring. Symptomatic hypocalcemia and resultant seizures are the most common metabolic abnormality. Increases in serum creatinine will develop in one-half of patients who receive medication but usually are reversible after cessation. Other CNS side effects include headache (25%), tremor, irritability, and hallucinations.

E. Ganciclovir and valganciclovir

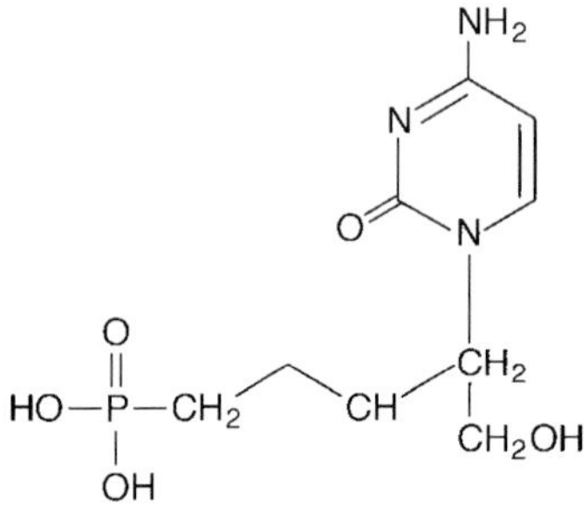

1. Chemistry, mechanism of action, and antiviral activity

Ganciclovir [9-(1,3-dihydroxy-2-propoxymethyl) guanine] (Cytovene) has enhanced *in vitro* activity against all herpesviruses as compared to acyclovir, including an 8–20 times greater antiviral activity against CMV. Like acyclovir, the activity of ganciclovir in herpesvirus-infected cells depends on phosphorylation by virus-induced TK. Also like acyclovir, ganciclovir monophosphate is further converted to its di- and triphosphate derivatives by cellular kinases. In cells infected by HSV-1 or -2, ganciclovir triphosphate competitively inhibits the incorporation of guanosine-triphosphate into viral DNA. Ganciclovir triphosphate is incorporated at internal and terminal sites of viral DNA, inhibiting DNA synthesis. The mode of action of ganciclovir against CMV is mediated by a protein kinase, UL-97, that efficiently promotes the obligatory initial phosphorylation of ganciclovir to its monophosphate.

The oral bioavailability of ganciclovir is poor (5–7%). Peak plasma levels are approximately 1.0 μg/ml after administration of 1 g every 6 h. Intravenous administration of a standard dose of 5 mg/kg will result in peak and trough plasma concentrations of 8–11 and 0.5–1.2 μg/ml, respectively. Concentrations of ganciclovir in biologic fluids, including aqueous humor and cerebrospinal fluid (CSF), are less than plasma levels. The plasma elimination $t_{1/2}$ is 2–4 h for individuals with normal renal function. The kidney is the major route of clearance of the drug, and therefore, impaired renal function requires adjustment of dosage. Valganciclovir is orally bioavailable (approximately 60%) and is rapidly converted to ganciclovir after absorption. It is currently in clinical development.

Valganciclovir, L-valine, 2-[2-amino-1,6-dihydro-6-oxo-9H-purin-9-yl)methoxy]-3-hydroxypropyl ester, is metabolized completely to ganciclovir; thus, it has the same spectrum of activity and mechanism of activity as the parent compound.

2. *Clinical indications*

a. *HIV-infected patients*

Ganciclovir has been administered to large numbers of patients with AIDS having CMV retinitis. Most patients (78%) experience either improvement or stabilization of their retinitis based on fundoscopic exams compared to historical controls. Induction therapy is usually at a dosage of 5.0 mg/kg twice a day given intravenously for 14–21 days. Maintenance therapy is essential. Median time to relapse for patients receiving no maintenance therapy averages 47 days. Maintenance therapy of 25–35 mg/kg per week significantly lengthens median time to relapse to 105 days. Virtually every patient treated will experience either a cessation or reduction of plasma viremia. Visual acuity usually stabilizes at pretreatment levels but rarely improves dramatically. Relapse occurs quickly in the absence of maintenance therapy but usually occurs eventually, even in patients receiving maintenance therapy (5 mg/kg for 5–7 days per week). The significance of bone marrow toxicity must be taken into consideration since 30–40% of patients develop neutropenia. Benefit has been reported with the use of ganciclovir for the treatment of other CMV infections, particularly in those involving the gastrointestinal tract.

Ganciclovir can be administered orally for prevention of CMV disease and retinitis in patients with AIDS. The utilization of ganciclovir at dosages of 1 g three to six times daily, following intravenous induction therapy, provides a sustained period prior to the next episode of reactivated retinitis at similar, albeit less (but not significantly less) intervals as when drug is given intravenously.

Valganciclovir is comparable to ganciclovir for the treatment of CMV retinitis. The dose is 900 mg twice daily for 3 weeks followed by 900 mg once daily.

b. *Transplant recipients*

Prophylaxis and preemptive therapy of CMV infections in high-risk transplant recipients is common. Both prevention and therapy of CMV infection of the lung are amenable to ganciclovir therapy. Ganciclovir of CMV pneumonia in conjunction with CMV immune globulin is therapeutically beneficial. Ganciclovir has been administered in anticipation of CMV disease to bone marrow transplant recipients (preemptive therapy). Several clinical trials utilizing different designs (e.g., initiation of ganciclovir after engraftment versus at the time of documentation of infection by bronchial alveolar lavage but in the absence of clinical symptomatology) have established the effectiveness of ganciclovir in preventing CMV pneumonia and reducing mortality during the treatment period. The utilization of ganciclovir in these circumstances has support among transplant physicians; however, long-term survival benefit (>120 days) is not apparent.

Valganciclovir is under investigation in organ transplant recipients.

3. *Resistance*

Resistance to CMV is associated with a deteriorating clinical course. Two mechanisms of resistance to ganciclovir have been documented: (i) the alteration of protein kinase gene, UL-97, reduces intracellular phosphorylation of ganciclovir, and (ii) point mutations in the viral DNA polymerase gene. Resistance is associated with decreased sensitivity up to 20-fold.

4. *Adverse effects*

The most important side effects of ganciclovir therapy are the development of neutropenia and that of thrombocytopenia. Neutropenia occurs in approximately 24–38% of patients. The neutropenia is usually reversible with dosage adjustment of ganciclovir, including withholding of treatment. Thrombocytopenia occurs in 6–19% of patients.

Ganciclovir has gonadal toxicity in animal models, most notably as a potent inhibitor of spermatogenesis. It causes an increased incidence of tumors in the preputial gland of male mice, a finding of unknown significance. As an agent affecting DNA synthesis, ganciclovir has carcinogenic potential.

F. Idoxuridine and trifluorothymidine

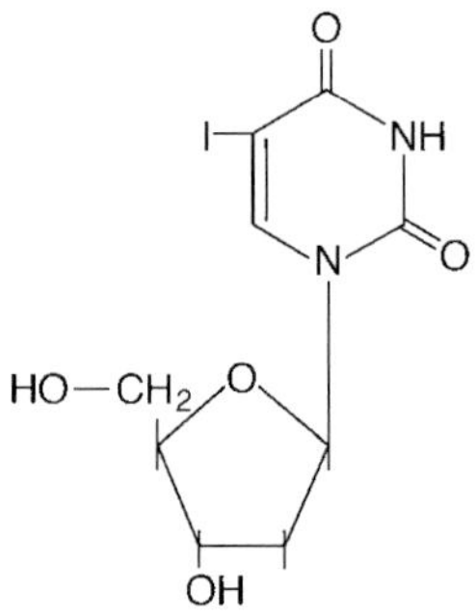

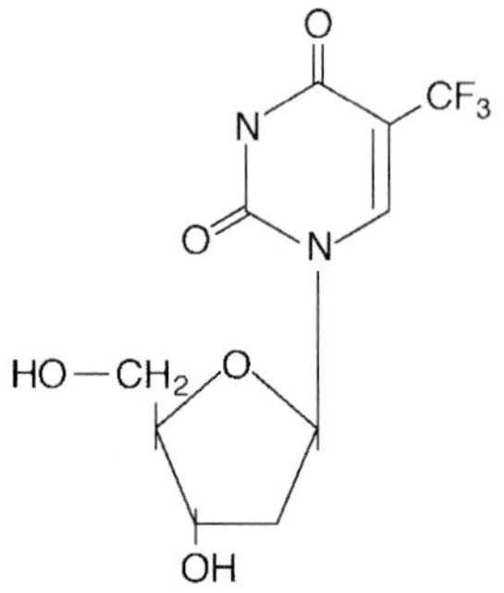

1. *Chemistry, mechanism of action, and antiviral activity*

Idoxuridine (5-iodo-2′-deoxyuridine) and trifluorothymidine (trifluridine, Viroptic) are analogs of thymidine. When administered systemically, these nucleosides are phosphorylated by both viral and cellular TK to active triphosphate derivatives which inhibit both viral and cellular DNA synthesis. The result is antiviral activity but also sufficient host cytotoxicity to prevent the systemic use of these drugs. The toxicity of these compounds is not significant when applied topically to the eye in the treatment of HSV keratitis. Both idoxuridine and trifluorothymidine are effective and licensed for treatment of HSV keratitis. Topically applied idoxuridine or trifluorothymidine will penetrate cells of the cornea. Low levels of drugs can be detected in the aqueous humor.

2. *Clinical indications*

Trifluorothymidine is the most efficacious of these compounds. These agents are not of proven value in the treatment of stromal keratitis or uveitis, although trifluridine is more likely to penetrate the cornea and, ultimately, may prove beneficial for these conditions.

3. *Resistance*

Little effort has been directed to evaluating HSV isolates obtained from the eye, in large part because of the difficulty in accomplishing this task.

4. *Adverse effects*

The ophthalmic preparation of idoxuridine and trifluridine causes local irritation, photophobia, edema of the eyelids, and superficial punctate keratopathy.

G. Penciclovir and famciclovir

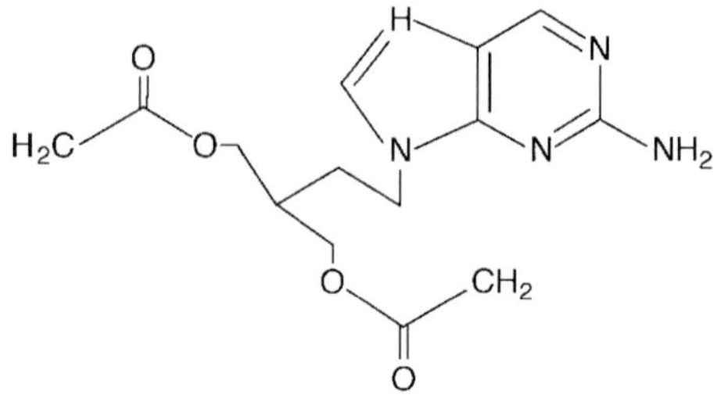

1. *Chemistry, mechanism of action, and antiviral activity*

A new member of the guanine nucleoside family of drugs is famciclovir [9-(4-hydroxy-3-hydroxymethyl-but-1-yl) guanine; Famvir], the prodrug of penciclovir. Penciclovir does not have significant oral bioavailability (<5%), but famciclovir is orally bioavailable (approximately 77%) and has a good therapeutic index for the therapy of both HSV and VZV infections. Famciclovir is the diacetyl ester of 6-deoxy penciclovir. When administered orally, it is rapidly converted to penciclovir. The spectrum of activity of penciclovir is similar to that of acyclovir. Penciclovir is phosphorylated more efficiently than acyclovir in HSV- and VZV-infected cells. Host cell kinases phosphorylate both penciclovir and acyclovir to a small but comparable extent. The preferential metabolism in HSV and VZV-infected cells is the major determinant of its antiviral activity. Penciclovir triphosphate has, on average, a 10-fold longer intracellular half-life than acyclovir triphosphate in HSV-1, HSV-2, and VZV-infected cells after drug removal. Penciclovir triphosphate is formed at sufficient concentrations to be an effective inhibitor of viral DNA polymerase, albeit at a lower K_i than that of acyclovir triphosphate. Both compounds have good activity against HSV-1, HSV-2, and VZV. The activity of penciclovir *in vitro*, like acyclovir, is dependent on both the host cell and the assay (plaque reduction, virus yield, and viral DNA inhibition). The mean penciclovir $EC_{50} \pm$ standard deviation for HSV-1 in MRC-5, HEL, WISH, and W138 cells is 0.4 ± 0.2, 0.6 ± 0.4, 0.2 ± 0.2, and 1.8 ± 0.8 μg/ml, respectively. For HSV-2, similar levels of activity in the identical cell lines are 1.8 ± 0.6, 2.4 ± 2.5, 0.8 ± 0.1, and 0.3 ± 0.2 μg/ml, respectively. These assays utilize a plaque reduction procedure. In virus yield reduction assays, inhibition of VZV replication in MRC-5 cells is between 3.0 and 5.1 μg/ml, values virtually identical to those of acyclovir. Penciclovir, like acyclovir, is relatively inactive against CMV and EBV. Penciclovir is also active against hepatitis B.

Conversion of famciclovir to penciclovir occurs at two levels. The major metabolic route of famciclovir is de-acetylation of one ester group as the prodrug crosses the duodenal barrier of the gastrointestinal tract. The drug is transported to the liver via the portal vein where the remaining ester group is removed and oxidation occurs at the sixth position of the side chain, resulting in penciclovir, the active drug. The first metabolite which appears in the plasma is almost entirely the de-acetylated compound, with little or no parent drug detected. Thus, the major metabolite of famciclovir is penciclovir. Pharmacokinetic parameters for penciclovir are linear over famciclovir oral dose ranges of 125–750 mg. Penciclovir is eliminated rapidly and almost unchanged by active tubular secretion and glomerular filtration by the kidneys. The elimination $t_{1/2}$ in healthy subjects is approximately 2 h.

2. *Clinical indications*

Famciclovir is available in an oral preparation. Penciclovir is available for topical therapy (Denavir).

a. Herpes zoster

Famciclovir (250, 500, or 750 mg three times a day) therapy is equivalent to the standard acyclovir treatment and superior to no therapy of herpes zoster for cutaneous healing, and in a subgroup analysis it accelerated resolution of pain (zoster-associated pain).

b. Genital HSV infection

Studies of patients with recurrent gential HSV infection (either intravenous penciclovir or oral famciclovir therapy) indicate beneficial effects in acceleration of all clinical parameter (e.g., pain, virus shedding, and duration). Famciclovir is given twice daily (125, 250 or 500 mg twice daily for 5 days). Famciclovir therapy on recurring HSV infections of immunocompromised hosts also is effective as suppressive therapy.

c. Herpes labialis

Topical application of penciclovir (Denavir) accelerates lesion healing (1 day) and resolution of pain. It is available over-the-counter in many countries.

3. *Resistance*

Herpes simplex virus and VZV isolates resistant to penciclovir have been identified in the laboratory. These isolates have similar patterns of resistance as those of acyclovir. Namely, resistance variants can be attributed to alterations or deficiencies of TK and DNA polymerase.

4. *Adverse effects*

Therapy with oral famciclovir is well tolerated, being associated only with headache, diarrhea, and nausea—common findings with other orally bioavailable antiviral agents. Preclinical studies of famciclovir indicated that chronic administration was tumorigenic (murine mammary tumors) and causes testicular toxicity in other rodents.

H. Vidarabine

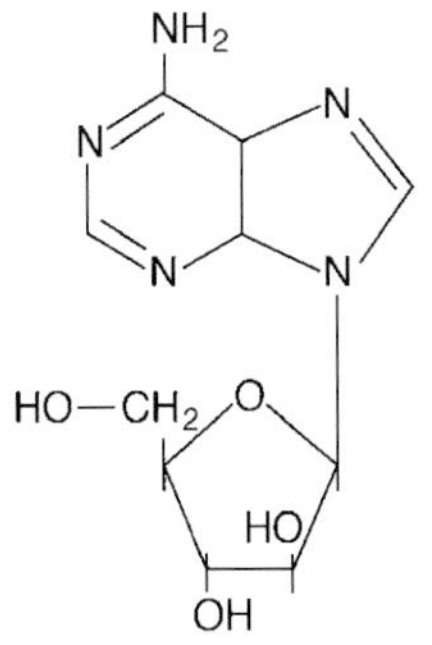

NH2, N, N, N, N, HO—CH2, O, HO, OH

1. *Chemistry, mechanism of action, and antiviral activity*

Vidarabine (vira-A, adenine arabinoside, and 9D-arabinofuranosyl adenine) is active against HSV, VZV, and CMV. Vidarabine is a purine nucleoside analog that is phosphorylated intracellularly to its mono-, di, and triphosphate derivatives. The triphosphate derivative competitively inhibits DNA dependent DNA polymerases of some DNA viruses approximately 40 times more than those of host cells. In addition, vira-A is incorporated into terminal positions of both cellular and viral DNA, thus inhibiting elongation. Viral DNA synthesis is blocked at lower doses of drug than is host cell DNA synthesis, resulting in a relatively selective antiviral effect. However, large doses of vira-A are cytotoxic to dividing host cells.

The benefit demonstrated in initial placebo-controlled clinical trials of this drug was a major impetus for the development of antiviral therapies. However, because of poor solubility and some toxicity, it was quickly replaced by acyclovir in the physician's armamentarium. Today, it is no longer available as an intravenous formulation. Vidarabine should be recognized historically as the first drug licensed for systemic use in the treatment of a viral infection.

2. *Clinical indications*

Vidarabine is only available as a topical formulation for ophthalmic administration.

3. *Resistance*

Studies of resistance to vidarabine have not been pursued.

4. *Adverse effects*

Ocular toxicity consists of occasional hyperemia and increased tearing, both of low incidence.

II. THERAPEUTICS FOR RESPIRATORY VIRUS INFECTIONS

A. Amantadine and rimantadine

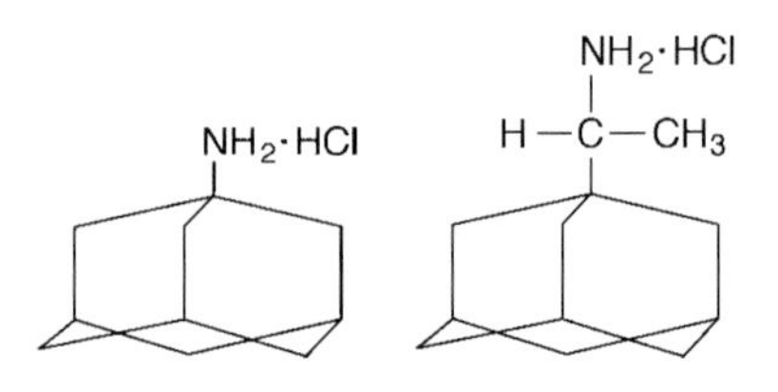

1. *Chemistry, mechanism of action, and antiviral activity*

Amantadine (1-adamantane amine hydrochloride; Symmetrel) is a tricyclic amine which is effective against all influenza A variants. Amantadine has a narrow spectrum of activity, being useful only against influenza A infections. Rimantidine is the α-methyl derivative of amantadine (α-methyl-1-adamantane methylamine hydrochloride). Rimantidine is 5- to 10-fold more active than amantadine and has the same spectrum of activity, mechanism of action, and clinical indications. Rimantadine is slightly more effective against type A viruses at equal concentrations. The mechanism of action of these drugs relates to the influenza A virus M2 protein, a membrane protein which is the ion channel for this virus. By interfering with the function of the M2 protein, amantadine and rimantidine inhibit the acid-mediated association of the matrix protein from the ribonuclear protein complex within endosomes. This event occurs early in the viral replicate cycle. The consequences of this drug are the potentiation of acidic pH-induced conformational changes in the viral hemagglutinin during its intracellular transport.

Absorption of rimantadine is delayed compared to that of amantadine, and equivalent doses of rimantadine produce lower plasma levels compared to amantadine, presumably because of a larger volume of distribution. Both amantadine and rimantadine are absorbed after oral administration. Amantadine is excreted in the urine by glomerular filtration and, likely, tubular secretion. It is unmetabolized. The plasma elimination, $t_{1/2}$, is approximately 12–18 h in individuals with normal renal function. However, the elimination, $t_{1/2}$, increases in the elderly with impaired creatinine clearance. Rimantadine is extensively metabolized following oral administration, with an elimination $t_{1/2}$ which averages 24–36 h. Approximately 15% of the dose is excreted unchanged in the urine.

2. *Clinical indications*

Amantadine and rimantadine are licensed both for the chemoprophylaxis and treatment of influenza A infections. The efficacy of amantadine and rimantadine when used prophylactically for influenza A infections averages 70–80% (range, 0–100%), which is approximately the same as with influenza vaccines. Effectiveness has been demonstrated for prevention of both experimental (i.e., artificial challenge) and naturally occurring infections for all three major subtypes of influenza A. Because of a lower incidence of side effects associated with rimantadine, it is used preferentially. Rimantadine can be given to any unimmunized member of the general population who wishes to avoid influenza A, but prophylaxis is especially recommended for control of presumed influenza outbreaks in institutions housing high-risk persons. High-risk individuals include adults and children with chronic disorders of the cardiovascular or pulmonary system requiring regular follow-up or hospitalization during the preceding year as well as residents of nursing homes and other chronic-care facilities housing patients of any age with chronic medical conditions.

These drugs are also effective for the treatment of influenza A. All studies showed a beneficial effect on the signs and symptoms of acute influenza as well as a significant reduction in the quantity of virus in respiratory secretions at some time during the course of infection. Because of the short duration of disease, therapy must be administered within 48 h of symptom onset to show benefit.

3. *Resistance*

Rimantadine-resistant strains of influenza have been isolated from children treated for 5 days. There have been subsequent reports of rimantadine-resistant strains being transmitted from person to person and producing clinical influenza. Development of resistance of influenza A viruses is mediated by single nucleotide changes in RNA segment 7, which results in amino acid substitutions in the transmembrane of the M2 protein. Obviously, amantadine and rimantadine share cross-resistance.

4. *Adverse effects*

Amantadine is reported to cause side effects in 5–10% of healthy young adults taking the standard adult dose of 200 mg/day. These side effects are usually mild and cease soon after amantadine is discontinued, although they often disappear with continued use of the drug. Central nervous system side effects, which occur in 5–33% of patients, are most common and include difficulty in thinking, confusion, lightheadedness, hallucinations, anxiety, and insomnia. More severe adverse effects (e.g., mental depression and psychosis) are usually associated with doses exceeding 200 mg daily. About 5% of patients complain of nausea, vomiting, or anorexia. Rimantadine appears better tolerated. Side effects associated with rimantadine administration are significantly less than those encountered with amantadine, particularly of the CNS. Rimantadine has been associated with exacerbations of underlying seizure disorders.

B. Oseltamivir

1. *Chemistry, mechanism of action, and antiviral activity*

Oseltamivir [ethyl (3R,a4R,5S)-4-acetomido-5-amino-3-(1-ethylpropoxy)-1cyclohexene-1-carboxylate] is a selective neuroaminidase inhibitor. It inhibits both influenza A and B virus at concentrations of 2 nM. Drug inhibits viral replication by targeting the neuraminidase protein via binding in a competitive fashion to the enzyme, rendering the virus incapable of reproducing. Because it has activity against influenza B, it has an advantage over the adamantadines. It has no activity against any other virus.

2. *Clinical indications*

Oseltamivir is licensed for the treatment and prevention of influenza A and B infections for individuals 2 years of age and older. Clinical trials indicate 30% acceleration in resolution of clinical symptoms. In pediatric studies, treatment accelerates disease resolution and is associated with a significantly decreased incidence of otitis media and antibiotic usage by 30% to 40%. Prophylactic efficacy is reported to be 75% to 85%.

3. *Resistance*

Mutations in the neuraminidase have been detected rarely in patients exposed to medication. In clinical studies, 1.3% to 8.6% of posttreatment isolates have altered susceptibility to oseltamivir. In vitro, the emergence of a resistant variant occurs with the substitution of a lysine for the conserved arginine at amino acid 292 of the neuraminidase.

4. *Adverse events*

Oseltamivir is generally well tolerated. Nausea with or without vomiting occurs in about 10% of the patients. Food alleviates this side effect.

C. Zanamivir

1. *Chemistry, mechanism of action, and antiviral activity*

Zanamivir (5-acetylamino-4-[aminoiminomethyl-amino]-2,6-anhydro-3,4,5-trideoxy-D-glycero-D-galacto-non-2-enonic acid) is a neuraminidase inhibitor. It binds competitively to influenza neuraminidase, habiting both influenza A and B. Influenza neuraminidase catalyzes the cleavage of the terminal sialic acid attached to glycolipids and glycoproteins. The oral bioavailability of zanamivir is poor, about 2%, thus it is only available as an inhaled medication.

2. *Clinical indications*

Zanamivir is licensed for the treatment and prevention of influenza A and B infections in patients over 7 years of age. In clinical trials, treatment reduced the duration of symptoms from 6 to 5 days and symptom scores by about 44%. The prophylactic efficacy of zanamivir is about 80%.

3. *Resistance*

Resistance is an uncommon occurrence in clinical trials, occurring no more frequently than in 1% of exposed patients. The site of mutation is that where drug binds the neuraminidase.

4. *Adverse events*

Most adverse effects are related to the respiratory tree. These include rhinorrhea and, rarely, bronchospasm.

Nausea and vomiting have been reported in less than 3%.

D. Ribavirin

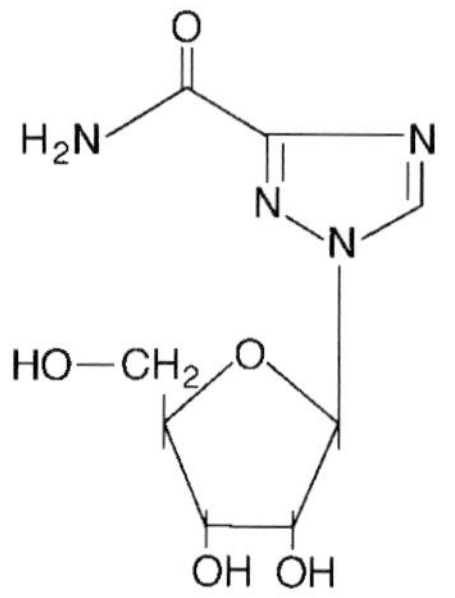

1. *Chemistry, mechanism of action, and antiviral activity*

Ribavirin (α-methyl-1-adamantane methylamine hydrochloride) has antiviral activity against a variety of RNA and DNA viruses. Ribavirin is a nucleoside analog whose mechanisms of action are poorly understood and probably not the same for all viruses; however, its ability to alter nucleotide pools and the packaging of mRNA appears important. This process is not virus specific, but there is a certain selectivity in that infected cells produce more mRNA than noninfected cells. A major action is the inhibition by ribavirin-5′-monophosphate of inosine monophosphate dehydrogenase, an enzyme essential for DNA synthesis. This inhibition may have direct effects on the intracellular level of GMP; other nucleotide levels may be altered, but the mechanisms are unknown. The 5′-triphosphate of ribavirin inhibits the formation of the 5′-guanylation capping on the mRNA of vaccinia and Venezuelan equine encephalitis viruses. In addition, the triphosphate is a potent inhibitor of viral mRNA (guanine-7) methyltransferase of vaccinia virus. The capacity of viral mRNA to support protein synthesis is markedly reduced by ribavirin. Of note, high concentrations of ribavirin also inhibit cellular protein synthesis. Ribavirin may inhibit influenza A RNA-dependent RNA polymerase.

Ribavirin can be administered orally (bioavailability of approximately 40–45%) or intravenously. Aerosol administration has become standard for the treatment of respiratory synctial virus (RSV) infections in children. Oral doses of 600 and 1200 mg result in peak plasma concentrations of 1.3 and 2.5 μg/ml, respectively. Intravenous dosages of 500 and 1000 mg result in 17 and 24 μg/ml plasma concentrations, respectively. Aerosol administration of ribavirin results in plasma levels which are a function of the duration of exposure. Although respiratory secretions will contain milligram quantities of drug, only microgram quantities (0.5–3.5 μg/ml) can be detected in the plasma.

The kidney is the major route of clearance of drug, accounting for approximately 40%. Hepatic metabolism also contributes to the clearance of ribavirin. Notably, ribavirin triphosphate concentrates in erythrocytes and persists for a month or longer. Likely, the persistence of ribavirin in erythrocytes contributes to its hematopoietic toxicity.

2. *Clinical indications*

a. Respiratory syncytial virus

While ribavirin is licensed for the treatment of carefully selected, hospitalized infants and young children with severe lower respiratory tract infections caused by RSV, it is no longer used. Use of aerosolized ribavirin in adults and children with RSV infections reduced the severity of illness and virus shedding. In patients receiving 8 or more hours of continuous therapy, the mean peak level in tracheal secretions may be 100 times greater than the minimum inhibitory concentration preventing RSV replication *in vitro*. The use of ribavirin for the treatment of RSV infections is controversial and remains discretionary. It is under study for prevention of RSV pneumonia in bone marrow transplant recipients. Combination ribavirin and pegylated IFN therapy is licensed for the treatment of hepatitis C.

3. *Resistance*

Emergence of viruses resistant to ribavirin has not been documented.

4. *Adverse effects*

Adverse effects attributable to aerosol therapy with ribavirin of infants with RSV include bronchospasm, pneumothorax in ventilated patients, apnea, cardiac arrest, hypotension, and concomitant digitalis toxicity. Pulmonary function test changes after ribavirin therapy in adults with chronic obstructive pulmonary disease have been noted. Reticulocytosis, rash, and conjunctivitis have been associated with the use of ribavirin aerosol. When given orally or intravenously, transient elevations of serum bilirubin and the occurrence of mild anemia have been reported. Ribavirin has been found to be teratogenic and mutagenic in preclinical testings. This drug is therefore contraindicated in women who are or may become pregnant during exposure to the drug.

Concern has been expressed about the risk to persons in the room of infants being treated with ribavirin aerosol, particularly females of childbearing age. Although this risk seems to be minimal with limited exposure, awareness and caution are warranted. Furthermore, the use of a "drug salvage" hood is mandatory.

5. *Hepatitis C*

With interferon-α, ribavirin is approved for combination therapy of chronic hepatitis C (see Section III.A).

III. HEPATITIS AND PAPILLOMAVIRUS

A. Interferons

1. *Chemistry, mechanism of action, and antiviral activity*

Interferons (IFNs) are glycoprotein cytokines (intracellular messengers) with a complex array of immunomodulating, antineoplastic, and antiviral properties. Interferons are currently classified as α, β, or γ, the natural sources of which, in general, are leukocytes, fibroblasts, and lymphocytes, respectively. Each type of IFN can be produced via recombinant DNA technology. Binding of IFN to the intact cell membrane is the first step in establishing an antiviral effect. Interferon binds to specific cell surface receptors; IFN-γ appears to have a different receptor from those of IFN-α and -β which may explain the purported synergistic antiviral and antitumor effects sometimes observed when IFN-γ is given with either of the other two IFN species.

A prevalent view of IFN action is that, following binding, there is synthesis of new cellular RNAs and proteins, particularly protein kinase R, which mediate the antiviral effect. Chromosome 21 is required for this antiviral state in humans, no matter which species of IFN is employed. At least three of the newly synthesized proteins in IFN-treated cells appear to be associated with the development of an antiviral state: (i) 2′5′-oligoadenylate synthetase, (ii) a protein kinase, and (iii) an endonuclease. The antiviral state is not fully expressed until these primed cells are infected with virus.

Interferon must be administered intramuscularly or subcutaneously (including into a lesion such as a wart). Plasma levels are dose dependent, peaking 4–8 h after intramuscular administration and returning to baseline between 18 and 36 h. There appears to be some variability in absorption between each of the three classes of IFN and, importantly, resultant plasma levels. Leukocyte and IFN-α appear to have elimination $t_{1/2}$ values of 2–4 h. Interferon is inactivated by various organs of the body in an as yet undefined method.

2. *Clinical indications*

a. *Condyloma acuminatum*

Several large, controlled trials have demonstrated the clinical benefit of IFN-α therapy of condyloma acuminatum which was refractory to cytodestructive therapies. Administration of 1.0×10^6 International Units (IU) of recombinant IFN-α led to significant benefit as evidenced by enhancing clearing of treated lesions (36 vs 17% placebo recipients) and by reduction in mean wart area (40% reduction vs 46% increase). In other well-controlled studies, either a similar rate (46%) or higher rates (62%) of clearance were reported. Notably, clearing responses of placebo recipients averaged 21 or 22%.

b. *Hepatitis B*

Hepatitis B DNA polymerase level, a marker of replication, is reduced with IFN therapy. Treatment with IFN-α in chronic hepatitis B has subsequently been investigated in several large, randomized, controlled trials. Clearance serum HBeAg and hepatitis B virus (HBV)–DNA polymerase occurs with treatment (30–40%).

c. *Hepatitis C*

The activity of IFN as a treatment of hepatitis C has undergone extensive evaluation. Interferon dosages have ranged from 1×10^6 to 10×10^6 IU three times weekly for 1–18 months. Of the placebo controls, only 2.6% normalized serum alanine amino transferase (ALT). In contrast, treatment led to serum ALT normalization in 33–45% of patients. Unfortunately, 50–80% of patients relapsed. Recently, IFN-α has been administered with ribavirin. Concomitant therapy for 40 weeks resulted in sustained responses in more than 60% of patients.

3. *Resistance*

Resistance to administered interferon has not been documented although neutralizing antibodies to recombinant interferons have been reported. The clinical importance of this latter observation is unknown.

4. *Adverse effects*

Side effects are frequent with IFN administration and are usually dose limiting. Influenza-like symptoms (i.e., fever, chills, headache, and malaise) commonly occur, but these symptoms usually become less severe with repeated treatments. At doses used in the treatment of condyloma acuminatum, these side effects rarely cause termination of treatment. For local treatment (intralesional administration), pain at the injection site does not differ significantly from that for placebo-treated patients and is short-lived. Leukopenia is the most common hematologic abnormality, occurring in up to 26% of treated patients. Leukopenia is usually modest, not clinically relevant, and reversible upon discontinuation of therapy. Increased alanine aminotransferase levels may also occur as well as nausea, vomiting, and diarrhea.

At higher doses of IFN, neurotoxicity is encountered, as manifested by personality changes, confusion, loss of attention, disorientation, and paranoid ideation. Early studies with IFN-γ show similar side effects as those of treatment with and IFN-α and -β but with the additional side effects of dose-limiting hypotension and a marked increase in triglyceride levels.

B. Adefovir dipivoxil

1. *Chemistry, mechanism of action, and antiviral activity*

Adefovir dipivoxil, bis-pivaloyloxymethyl-9-(2-phosphonyl-methoxyethyl)adenine, is the orally bioavailable prodrug of adefovir.

2. *Clinical indications*

Adefovir has activity against both herpes and hepadnavirus. It is in the nucleotide class of medications. Treatment of chronic hepatitis B at 10 mg daily significantly decreases HBV DNA polymerase (3.56 logs compared with 0.55 logs in placebo recipients), improves hepatitic hitopathology scores, and induces loss of HBeAg.

3. *Resistance*

Mutations within the HBV DNA polymerase which confer resistance to adefovir have not been identified in clinical trials. HBV isolates resistant to lamivudine or hepatitis B hyperimmune globulin retain susceptibility to adefovir.

4. *Adverse events*

Severe acute exacerbation of hepatitis has been reported who have discontinued anti-HBV therapy. Lactic acidosis and severe hepatomegaly with steatosis have also been reported.

C. Entecavir

[1S-(1α,3α,4β)]-2-amino-1,9-dihydro-9[4-hydroxymethyl-2-methyllenecyclopentyl]-6H-purin-6-one, is a nucleoside analog that is orally bioavailable for the treatment of chronic hepatitis B. Phase III trials are in progress.

IV. PROSPECTS FOR ENTEROVIRAL THERAPIES

Pleconaril, a compound with activity against many rhinoviruses and enteroviruses, is the first compound for which data exist to define anti-viral drug interaction with a virion at the atomic level. This compound is one of a class of compounds which resembles arildone, a drug known to inhibit uncoating of poliovirus. X-ray diffraction studies of bound to rhinovirus 14 show that the compound adheres tightly to a hydropic pocket formed by VP1, one of the structural proteins of rhinovirus 14. These hydrophobic pockets were found in the VP1 proteins of poliovirus and meningovirus and may be common to all picomaviruses. These compounds may lock into the conformation of the VP1 so that the virus cannot disassemble.

Pleconaril is under investigation for chronic enterovirus infections of the CNS in the immune deficient patients. It was shown to have an inadequate therapeutic index for therapy of rhinovirus cold.

V. ANTIRETROVIRAL AGENTS

A. Reverse transcriptase inhibitors

1. *Zidovudine*

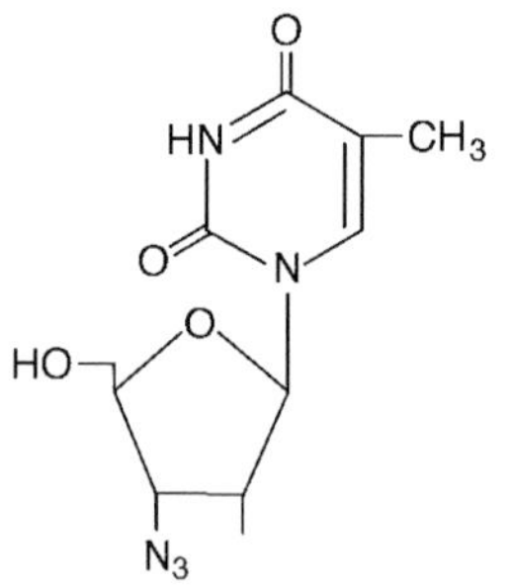

a. Chemistry, mechanism of action, and antiviral activity

Zidovudine (3′-azido-2′,3′-dideoxythymidine; azidothymidine and Retrovir) is a pyrimidine analog with an azido group substituting for the 3′ hydroxyl group on the ribose ring. The drug is initially phosphorylated by cellular TK and then to its diphosphate by cellular thymidylate kinase. The triphosphate derivative competitively inhibits HIV reverse transcriptase and also functions as a chain terminator. Zidovudine inhibits HIV-1 at concentrations of approximately 0.013 μg/ml.

In addition, it inhibits a variety of other retroviruses. Synergy has been demonstrated against HIV-1 when zidovudine is combined with didanosine, zalcitabine, lamiviudine, nevirapine, delavirdine, saquinavir, indinavir, ritonavir, and other compounds. It was the first drug to be licensed for the treatment of HIV infection and still is used in combination with other drugs as initial therapy for some patients.

Zidovudine is available in capsule, syrup, and intravenous formulations. Oral bioavailability is approximately 65%. Peak plasma levels are achieved approximately 0.5–1.5 h after treatment. Zidovudine is extensively distributed, with a steady-state volume of distribution of approximately of 1.6 liters/kg. The drug penetrates cerebrospinal fluid, saliva, semen, and breast milk, and it crosses the placenta. The drug is predominately metabolized by the liver through the enzyme uridine diphosphoglucuronosyltransferase to its major inactive metabolite 3′-azido- 3′-deoxy-5′-*O*-*B*-D-glucopyranuronosylthymidine. The elimination $t_{1/2}$ is approximately 1 h; however, it is extended in individuals who have altered hepatic function.

b. Clinical applications

Zidovudine was the first approved antiretroviral agent, and as a consequence, has been the most widely used antiretroviral drug in clinical practice. In monotherapy studies, zidovudine improves survival and decreases the incidence of opportunistic infections in patients with advanced HIV disease. Importantly, zidovudine decreased the incidence of transmission of HIV infection from pregnant women to their fetuses. However, its usefulness as monotherapy has been outlived.

Recently, zidovudine has been incorporated into multidrug regimens, including combinations with didanosine or zalcitabine which demonstrate a delay in disease progression and improved survival compared to zidovudine monotherapy; zidovudine plus didanosine and zidovudine plus lamivudine have also been shown to improve both outcome and important markers of disease, including CD4 counts and plasma HIV RNA levels.

Currently, three-drug combinations include the use of zidovudine with other reverse transcriptase inhibitors and nonnucleoside reverse transcriptase inhibitors and protease inhibitors. Triple-drug combinations offer enhanced therapeutic benefits, particularly as noted by survival and restoration of normal immune function.

c. Resistance

Zidovudine resistance occurs rapidly after the onset of therapy. Numerous sites of resistance have been identified, with the degree of resistance being proportional to the number of mutations. The development of resistant HIV strains correlates with disease progression. The utilization of combination drug therapies delays the onset of resistance.

d. Adverse events

The predominant adverse effect of zidovudine is myelosuppression, as evidenced by neutropenia and anemia, occurring in 16 and 24% of patients, respectively. Zidovudine has been associated with skeletal and cardiac muscle toxicity, including polymyositis. Nausea, headache, malaise, insomnia, and fatigue are common side effects.

2. *Didanosine*

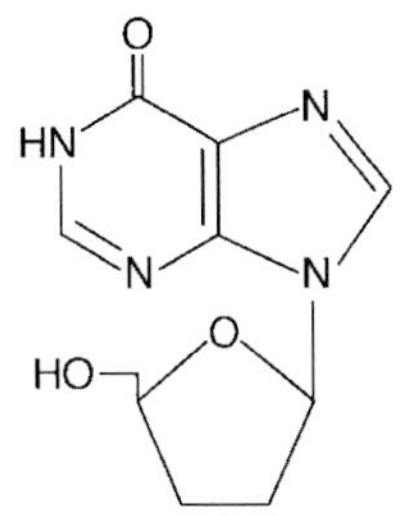

O
HN
N
N
N
HO
O

a. Chemistry, mechanism of action, and antiviral activity

Didanosine (2′,3′-dideoxyinosine; ddl and Videx) is a purine nuceloside with inhibitory activity against both HIV-1 and HIV-2. Didanosine is activated by intracellular phosphorylation. The conversion of 2′,3′-dideoxyinsine-5′-monophosphate to its triphosphate derivative is more complicated than that with other nucleoside analogs because it requires additional enzymes, including a 5′ nucleotidase and subsequently, adenylosuccinate synthetase and adenylosccinate lyase. The triphosphate metabolite is a competitive inhibitor of HIV reverse transcriptase and is also a chain terminator. The spectrum of activity of didanosine is enhanced by synergism with zidovudine and stavudine as well as the protease inhibitors.

Didanosine is available in an oral formulation; however, it is acid labile and has poor solubility. A buffered tablet results in 20–25% bioavailability. A 300-mg oral dose achieves peak plasma concentrations of 0.5–2.6 μg/ml with a $t_{1/2}$ of approximately 1.5 h. Drug is metabolized to hypoxanthine and is cleared primarily by the kidney.

b. Clinical indications

Didanosine is used in combination with other nucleoside analogs and protease inhibitors. In combination with zidovudine, improvement in both clinical outcome and immunologic markers of disease has been reported (CD4 lymphocyte counts).

c. Resistance

As with zidovudine, mutations and reverse transcriptase appear promptly after administration of didanosine therapy, resulting in a 3- to 10-fold decrease in susceptibility to therapy.

d. Adverse effects

The most significant adverse effect associated with didanosine therapy is the development of peripheral neuropathy (30%) and pancreatitis (10%). Adverse effects of note include diarrhea (likely attributed to the phosphate buffer), headache, rash, nausea, vomiting, and hepatotoxicty. Myelosuppression is not a component of toxicity associated with didanosine administration.

3. *Zalcitabine*

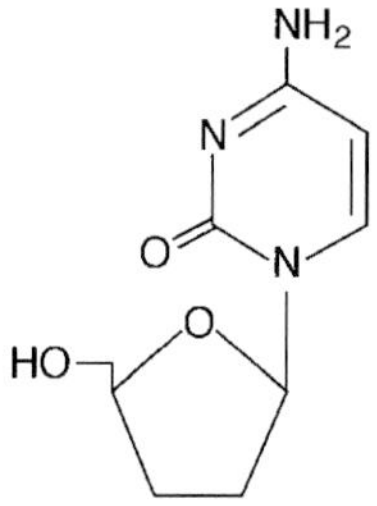

a. Chemistry, mechanism of action, and antiviral activity

Zalcitabine (2′,3′-dideoxycytidine; ddC and Hivid) is a pyrimidine analog which is activated by cellular enzymes to its triphosphate derivative. The enzymes responsible for activation of zalcitabine are cell cycle independent, and therefore this offers a theoretical advantage for nondividing cells, specifically dendritic and monocyte/macrophage cells. Zalcitabine inhibits both HIV-1 and HIV-2 at concentrations of approximately 0.03 μM. Synergy has been described between zidovudine and zalcitabine as well as with saquinavir.

The oral bioavailability following zalcitabine administration is more than 80%. The peak plasma concentrations following an oral dose of 0.03 mg/kg range from 0.1 to 0.2 μM, and the $t_{1/2}$ is short (approximately 20 min). The drug is cleared mainly by the kidney, and therefore, in the presence of renal insufficiency a prolonged plasma $t_{1/2}$ is documented.

b. Clinical indications

Zalcitabine is used in combination with other reverse transcriptase and protease inhibitors. As with other nucleoside combinations, zidovudine and zalcitabine do not benefit patients to the same extent as combinations of zidovudine and didanosine. Currently, it is used as part of a two- or three-drug regimen in combination with zidovudine and saquinavir.

c. Resistance

Zalcitabine-resistant HIV-1 variance has been documented both *in vitro* and *in vivo*.

d. Adverse effects

Peripheral neuropathy is the major toxicity associated with zalcitabine administration, occurring in approximately 35% of individuals. Pancreatitis can occur but does so infrequently. Thrombocytopenia and neutropenia are uncommon (5 and 10%, respectively). Other zalcitabine-related side effects include nausea, vomiting, headache, hepatotoxicity, and cardiomyopathy.

4. *Stavudine*

a. Chemistry, mechanism of action, and antiviral activity

Stavudine (2′,3′-didehydro, 3′-deoxythymidine; d4T and Zerit) is a thymidine analog with significant activity against HIV-1, having inhibitory concentrations which range from 0.01 to 4.1 μM. Its mechanism of action is similar to that of zidovudine. It is either additive or synergistic *in vitro* with other combinations of both nucleoside and nonnucleoside reverse-transcriptase inhibitors.

The oral bioavailability of stavudine is more than 85%. Peak plasma concentrations of approximately 1.2 μg/ml are reached within 1 h of dosing at 0.67 mg/kg per dose. The drug penetrates CSF and breast milk. The drug is excreted by the kidney unchanged and, in part, by renal tubular secretion.

b. Clinical indications

Stavudine has been studied both as monotherapy and in combination with other antiretroviral drugs. It is gaining increasing use as front-line therapy for HIV infection. Stavudine's clinical benefit is superior to that of zidovudine, particularly as it relates to increasing CD4 cell counts, slowing progression to AIDS or mortality.

c. Resistance

The development of resistance on serial passage in the laboratory can be achieved. Cross-resistance with didanosine and zalcitabine has been identified by specific mutations for stavudine. The development of resistance in clinical trials has not been identified.

d. Adverse effects

The principal adverse effect of stavudine therapy is the development of peripheral neuropathy. The development of this complication is related to both dose and duration of therapy. Neuropathy tends to appear after 3 months of therapy and resolves slowly with medication discontinuation. Other side effects are uncommon.

5. *Lamivudine*

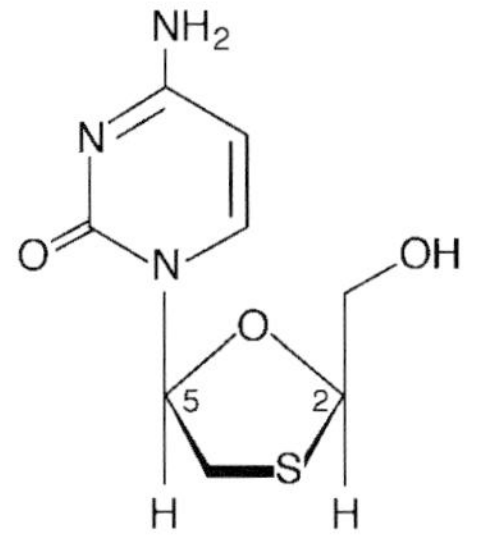

a. Chemistry, mechanism of action, and antiviral activity

Lamivudine is the (−) enantiomer of a cytidine analog, with sulfur substituted for the 3′ carbon atom in the furanose ring [(−) 2′,3′-dideoxy, 3′-thiacytidine; 3TC, Epivir]. It has significant activity *in vitro* against both HIV-1 and HIV-2 as well as HBV. Lamivudine is phosphorylated to the triphosphate metabolite by cellular kinases. The triphosphate derivat-ive is a competitive inhibitor of the viral reverse transcriptase.

Lamivudine's oral bioavailability in adults is in excess of 80% for doses between 0.25 and 8.0 mg/kg. Peak serum concentrations of 1.5 μg/ml are achieved in 1–1.5 h and the plasma $t_{1/2}$ is approximately 2–4 h. The drug is cleared by the kidney unchanged by both glomerular filtration and tubular excretion.

b. Clinical indications

Lamivudine is used in combination with other reverse transcriptase inhibitors and protease inhibitors. In combination with zidovudine, enhanced CD4 responses and suppression of HIV RNA levels occur to a greater extent than with zidovudine monotherapy. The combination of zidovudine and lamivudine is without significant adverse event. Because of this degree of tolerability, it is widely used in clinical practice.

In addition, lamivudine is licensed for the treatment of chronic hepatitis B. However, resistance appears soon after administration in many patients.

c. Resistance

With clinical therapy, resistance to lamivudine monotherapy develops rapidly. In large part, resistance is mediated by amino acid change at codon 184, resulting in a 100- to 1000-fold decrease in susceptibility. The 184 mutation site, which is of importance, also occurs with didanosine and zalcitabine and appears to increase sensitivity to zidovudine, providing a logical basis for its combination with this agent.

d. Adverse effects

Lamivudine has an extremely favorable toxicity profile. This may largely be attributed to the low affinity of lamivudine for DNA polymerase. At the highest doses of 20 mg/kg/day, neutropenia is encountered but at a low frequency. In pediatric studies, pancreatitis and peripheral neuropathies have been reported.

6. *Abacavir*

1. Chemistry, mechanism of action, and antiviral activity

Abacavir is a carbocyclic synthetic nucleoside analogue. Intracellularly, abacavir is phosphorylated by cellular enzymes to its active metabolite, carbovir triphosphate, which is an analogue of deoxyguanosine-5′-TP. Carbovir TP then inhibits the activity of HIV reverse transcriptase both by competing with the natural substrate dGTP and by its incorporation into viral DNA. The lack of a 3′-OH group in the incorporated nucleoside analogue prevents the formation of the 5′ to 3′ phosphodiester linkage essential for DNA chain elongation, producing chain termination.

2. Clinical indications

Abacavir often is used in combination with lamivudine and zidovudine, as well as with either a non-nucleoside reverse transcriptase inhibitor or protease inhibitor. It is licensed for the treatment of HIV infections of humans. The proposed dosage is 300 mg daily.

3. Resistance

Abacavir resistance is conferred by mutations in the HIV reverse transcriptase gene that resulted in amino acid substitutions at positions K65R, L74V, Y115F, and M184V. M184V and L74V are the most frequently observed mutations among clinical isolates. Multiple reverse transcriptase mutations conferring abacavir resistance exhibit cross-resistance to lamivudine, didanosine, and zalcitabine *in vitro*.

4. Adverse events

Adverse reactions associated with abacavir therapy include nausea, headache, stomach pain, diarrhea, insomnia, rash, fever, and dizziness. Importantly, fatal hypersensitivity reactions have been associated with abacavir use.

B. Non-nucleoside reverse transcriptase inhibitors

1. *Nevirapine*

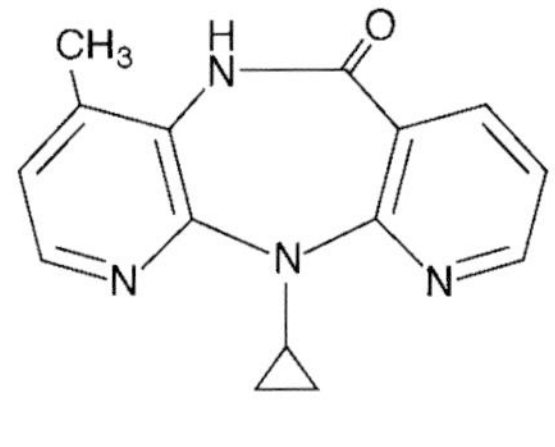

CH_3 H N O N N N

a. Chemistry, mechanism of action, and antiviral activity

Nevirapine (11-cyclopropyl-5,11-dihydro-4-methyl-6*H*-dipyrido[3,2-b:2′, 3′-e]; [1,4]diazepin-6-one and Viramune) is a reverse transcriptase inhibitor of HIV-1. Nevirapine is rapidly absorbed with a bioavailability of approximately 65%. Peak serum concentration is achieved approximately 4 h after a 400-mg oral dose of 3.4 μg/ml. Nevirapine is metabolized by liver microsomes to hydroxymethyl-nevirapine. *In vitro*, synergy has been demonstrated when administered with nucleoside reverse transcriptase inhibitors.

b. Clinical indications

Nevirapine monotherapy is associated with a non-sustained antiviral effects at a dosage of 200 mg/day. Concomitant with this minimal effect is the rapid emergence of resistant virus, such that by 8 weeks all patients had evidence of viral resistance. Thus, drug can only be administered in combination with other antiretroviral agents. In combination with nucleoside reverse-transcriptase inhibitors, there is evidence of reduction in viral HIV RNA load as well as increasing CD4 counts.

c. Adverse effects

The most common adverse effects include the development of a nonpruritic rash in as many as 50% of patients who received 400 mg/day. In addition, fever, myalgias, headache, nausea, vomiting, fatigue, and diarrhea have also been associated with administration of drug.

d. Resistance

Nevirapine resistance has been identified according to its binding site on viral polymerase. Specifically, two sets of amino acid residues (100–110 and 180–190) represent sites at which resistant mutations have occurred. Nevirapine monotherapy is associated with resistance, most frequently appearing at codon 181. Because of the rapid appearance of resistance, nevirapine must be administered with other antiretroviral agents.

2. *Delavirdine*

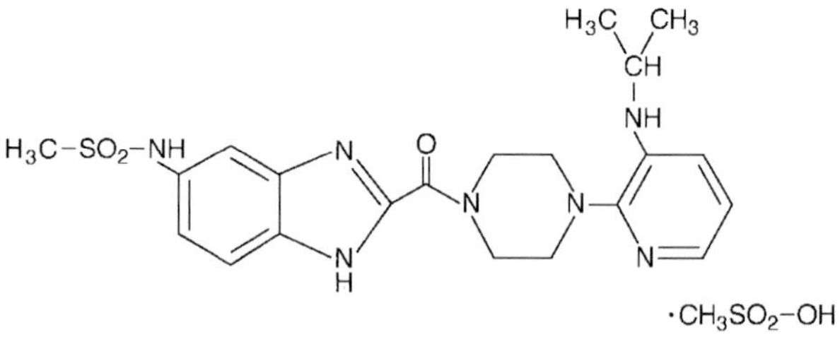

H_3C CH_3 CH NH H_3C-SO_2-NH N O N N N N H $\cdot CH_3SO_2-OH$

a. Chemistry, mechanism of action, and antiviral activity

Delavirdine (1-[5-methanesulfonamido-1*H*-indol-2-yl-carbonyl]-4-[3-(1-methylethylamino) pyridinyl] piperazine; Rescriptor) is a second-generation bis (heteroaryl) piperazine licensed for the treatment of HIV infection. It is absorbed rapidly when given orally to >60%. Delavirdine is metabolized by the liver with an elimination $t_{1/2}$ of approximately 1.4 h. It has an inhibitory concentration against

HIV-1 of approximately 0.25 μM. Inhibitory concentrations for human DNA polymerases are significantly higher.

b. Clinical indications

Reductions in plasma HIV RNA of more than 90% have been documented when delavirdine is administered such that trough levels exceed 50 μM. However, there is a rapid return to baseline over 8 weeks as resistance develops. As a consequence, delavirdine must be administered with either zidovudine or didanosine to have a more protracted effect.

c. Adverse effects

Delavirdine administration is associated with a maculopapular rash. Other side effects are less common.

d. Resistance

Delavirdine resistance can be generated rapidly both *in vitro* and *in vivo* with the codon change identified at 236, resulting in an increase and susceptibility to >60 μM. Delavirdine resistance can also occur at codons 181 and 188, as noted for nevirapine administration.

3. *Efavirenz*

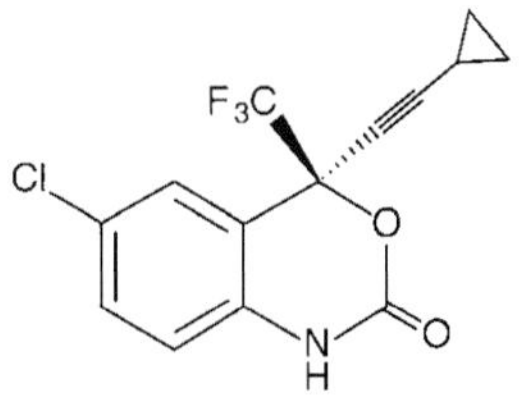

a. Chemistry, mechanism of action, and antiviral activity

Efavirenz Sustiva [(*S*)-6-chloro-4-(cyclopropylethynyl)-1,4-dihydro-4-(trifluoromethyl)-2*H*-3,1-benzoxazin-2-one; Sustiva and DMP266] is a nonnucleoside reverse-transcriptase inhibitor which can be administered once daily. Activity is mediated predominately by noncompetitive inhibition of HIV-1 reverse transcriptase. HIV-2 reverse transcriptase in human cellular DNA polymerases α, β, γ, and δ are not inhibited by efavirenz. The 90–95% inhibitory concentration of efavirenz is approximately 1.7–25 nM. In combination with other anti-HIV agents, particularly zidovudine, didanosine, and indinaver, synergy is demonstrated.

b. Clinical indications

Efavirenz is employed in combination with other antiretroviral agents indicated in the treatment of HIV-1 infection. Efficacy has been documented in the demonstration of plasma HIV negativity (<400 HIV RNA copies/ml) in approximately 80% of patients. Combination therapy has resulted in a 150-fold or greater decrease in HIV-1 RNA levels. Importantly, data have shown efficacy in children for both virologic and immunologic end points.

c. Adverse effects

The most common adverse events are skin rash (25%), which is associated with blistering, moist desquamation, or ulceration (1%). In addition, delusions and inappropriate behavior have been reported in 1 or 2 patients per 1000.

d. Resistance

As with other non-nucleoside reverse-transcriptase inhibitors, resistance appears rapidly and is mediated by similar enzymes.

4. *Future Prospect*

Capravirine

Capravirine is a non-nucleoside reverse transcriptase inhibitor currently under investigation. Need more It has potent in vitro activity against HIV variants with RT substitutions, including K103N that confer broad cross resistance to the other drugs in this class.

C. Protease inhibitors

1. *Saquinavir*

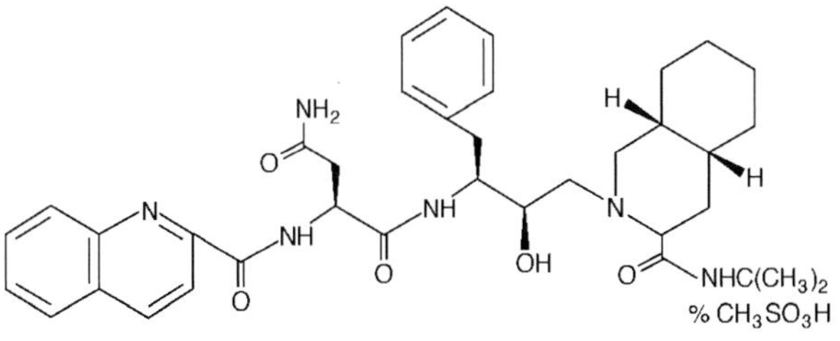

a. Chemistry, mechanism of action, and antiviral activity

Saquinavir (*cis*-*N*-tert-butyl-decahydro-2[2(*R*)-hydroxy-4-phenyl-3-(*S*)-([*N*-(2-quinolycarbonyl)-L-asparginyl] amino butyl)-4a*S*, 8a*S*]-isoquinoline-3[S]-carboxyamide methanesulfonate; Invirase) is a hydroxyethylamine-derived peptidomimetic HIV protease inhibitor.

Saquinavir inhibits HIV-1 and HIV-2 at concentrations at 10 nM and is synergistic with other nucleoside analogs as well as selected protease inhibitors.

Oral bioavailability is approximately 30% with extensive hepatic metabolism. Peak plasma concentrations of 35 mg/μl are obtained following a 600-mg dose.

b. Clinical indications

The clinical efficacy of saquinavir is limited by poor oral bioavailability, but improved formulation (soft-gel capsule) will likely enhance efficacy. Currently, it is used in combination therapy with other nucleoside analogs, particularly zidovudine, lamivudine, zalcitabine, and stavudine.

c. Adverse effects

Adverse effects are minimal, with no dose-limiting toxicities. Abdominal discomfort, including diarrhea, nausea, and photo sensitization has been reported infrequently.

d. Resistance

Resistance to saquinavir develops rapidly when it is administered as monotherapy. By 1 year, 45% of patients develop resistance at codon sites 90 and 48, resulting in, an approximately 30-fold decrease in susceptibility.

2. *Indinavir*

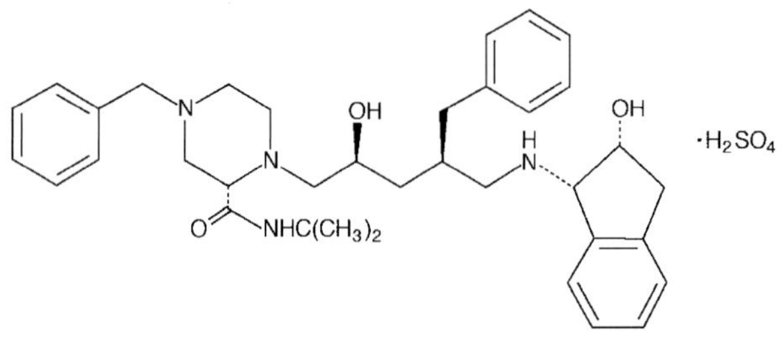

OH OH $\cdot H_2SO_4$ N N H N $NHC(CH_3)_2$ O

a. Chemistry, mechanism of action, and antiviral activity

{*N*-[2(*R*)-hydroxy-1(*S*)-indanyl]-5-[2(*S*)-(1,1-dimethylethlaminocarbonyl)-4-(pyridin-3-yl) methylpiperazin+++-1-yl]-4[*s*]-hydroxy-2[2]-phenylmethyl entanamide; Crixivan} is a peptidomimetic HIV-1 and HIV-2 protease inhibitor. At concentrations of 100 nM, indinavir inhibits 90% of HIV isolates. Indinavir is rapidly absorbed with a bioavailability of 60% and achieves peak plasma concentrations of 12 μM after an 800-mg oral dose.

b. Clinical indications

Indinavir has been established as effective therapy for the treatment of HIV infection, particularly in combination with nucleoside analogs. At a dose of 800 mg per 8 h, 80% of patients experience at least a 100-fold reduction in HIV-RNA levels, and in 50% of patients there is up to a 2 log reduction. In approximately 30% of patients plasma HIV RNA levels are reduced below 500 copies/ml, with an associated increase in CD4 cell counts over baseline. In combination with zidovudine and lamivudine, a >2 log decrease in plasma RNA levels can be achieved for a majority of patients (more than 80%).

c. Adverse effects

Although indinavir is well tolerated, commonly encountered adverse effects include indirect hyperbilirubinemia (10%) and nephrolithiasis (5%).

d. Resistance

Indinavir resistance develops rapidly with monotherapy and occurs at multiple sites. The extent of resistance is directly related to the number of codon changes in the HIV protease gene. Codon 82 is a common mutation in indinavir-resistant HIV isolates.

3. *Ritonavir*

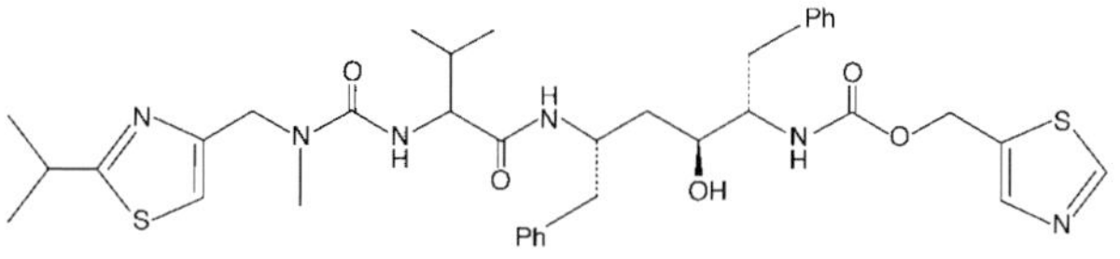

Ph O H O N N N N O S N O O S H H OH N Ph

a. Chemistry, mechanism of action, and antiviral activity

Ritonavir (10-hydroxy-2-methyl-5-[1-methylethyl]-1[2-(1-methylethyl)-4-thiazo lyl]-3,6,dioxo-8,11-bis [phenylmethyl]-2,4,7,12-tetra azatridecan-13-oic-acid, 5-thiazolylmethylester, [5*S*-(5*R*, 8*R*, 10*R*, 11*R*)]; Norvir) is a symmetric HIV protease inhibitor which has exquisite activity *in vitro* against HIV-1 laboratory strains (0.02–0.15 μM). It is synergistic when administered with nucleoside analogs.

Oral bioavailability is approximately 80%, with peak plasma levels of approximately 1.8 μM after 400 mg administered every 12 h. The plasma halflife is approximately 3 h.

b. Clinical indications

Ritonavir is used for treatment of HIV infection in combination with nucleoside analogs. As monotherapy, a 10- to 100-fold decrease in plasma HIV RNA is achieved with a concomitant increase in CD4 cell counts of approximately 100 cells/mm^3. Combination therapy results in a more significant decrease in HIV RNA plasma levels.

c. Adverse effects

Adverse effects include nausea, diarrhea, and headache, but all occur at a low frequency.

d. Resistance

Resistance to ritonavir resembles that to indinavir. Mutations at codon 82 are the most common.

4. *Nelfinavir*

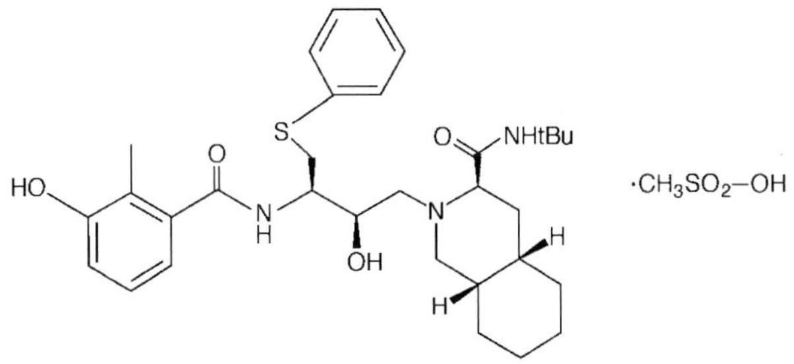

HO, NH, OH, S, O, NHtBu, N, H, H, $\cdot CH_3SO_2-OH$

a. Chemistry, mechanism of action, and antiviral activity

Nelfinavir [3*S*-(3*R*, 4a*R*, 8a*R*, 22′*S*, 3′*S*)]-2- [2′-hydroxy-3′-phenylthiomethyl-4′-aza-5′-ox-o-5′-(2″ methyl-3′-hydroxyphenyl)pentyl]-decahydroisoquinoline-3-*N*-(tert-butyl-carboxamide methanesulfonic acid salt) is another peptidomimetic HIV protease inhibitor. Inhibitory concentrations of HIV-1 are in the range of 20–50 nM. Nalfinavir is orally bioavailable at approximately 40%, achieving peak plasma concentrations of 2 or 3 mg following a 800-mg dose every 24 h. The drug is metabolized by hepatic microsomes.

b. Clinical indications

Nalfinavir is utilized in combination with nucleoside analogs. Monotherapy will achieve significant decreases in HIV RNA plasma levels up to 100-fold. Currently, the drug is used in combination with nuceloside analogs, particularly zidovudine, lamivudine, or stavudine, which results in 100- to 1000-fold reductions of HIV plasma RNA levels.

c. Adverse effects

Nelfinavir is well tolerated, with mild gastrointestinal complication reported.

d. Resistance

Cross-resistance to other protease inhibitors, particularly saquinavir, indinavir, or ritonavir, is not common. The most frequently demonstrated site of mutation is at codon 30.

5. *Amprenavir*

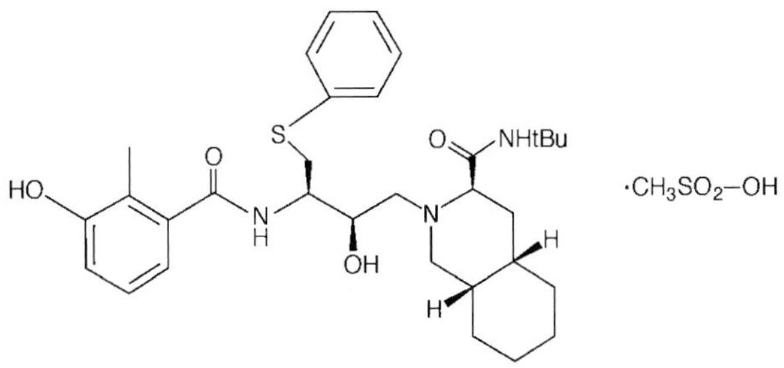

HO, NH, OH, S, O, NHtBu, N, H, H, $\cdot CH_3SO_2-OH$

a. Chemistry, mechanism of action, and antiviral activity

Amprenavir is a hydroxyethylamine sulfonamide peptidomimetric with a structure identified as (3*S*)-tetrahydro- 3-furyl *N*-(1*S*,2*R*)-3-(4-amino-*N*-isobutyl-benzenesulfonamido)-1-benzyl-2-hydroxypropyl carbamate. It is active at a concentration of 10–20 nM. The oral bioavailability is >70%, and peak plasma concentrations of 6.2–10 μg/ml are achieved after dosages of 600–1200 mg. The plasma half-life is 7–10 h. Cerebrospinal fluid concentrations are significant.

Amprenavir acts by binding to the active site of HIV-1 protease, preventing the processing of viral gag and gag-pol polyprotein precursors and resulting in the formation of immature non-infectious viral particles. *In vitro*, amprenavir has synergistic anti-HIV-1 activity in combination with abacavir, zidovudine, didanosine, or saquinavir, and additive anti-HIV-1 activity in combination with indinavir, nelfinavir, and ritonavir.

2. Clinical indications

Amprenavir is licensed for the treatment of HIV infections. The recommended dosage for adults is 1200 mg twice daily.

3. Adverse effects

The most serious adverse effect is a rash. Other side effects include nausea, vomiting, diarrhea, abdominal pain, and perioral paresthesias.

4. Resistance

Resistance to amprenavir is conferred by amino acid substitutions primarily at positions M46I/L, I47V, I50V, I54L/V, and I84V, as well as mutations in the viral protease p1/p6 cleavage site. Cross-resistance between amprenavir and the other protease inhibitors is possible.

6. Lopinavir/Ritonavir

Lopinavir/ritonavir combination (marketed as Kaletra) interferes with processing of viral polyprotein precursors, resulting in non-infectious progeny virions. The addition of ritonavir enhances the concentrations of lopinavir that which can be achieved following oral administration. It is given in combination with nucleoside and/or non-nucleoside reverse transcriptase inhibitors. Side effects of lopinavir/ritonavir include diarrhea, nausea, abdominal pain, and headache.

As lopinavir/ritonavir is a new addition to the protease inhibitors, a complete understanding of resistance profiles will await its widespread utilization.

D. Nucleotide Analogues

Viread; Tenofovir (tenofovir disoproxil fumarate)

Tenofovir or disproxil fumarate salt is an acyclic nucleoside phosphonate diester analog of adenosine monophosphate with an in vitro the 50% inhibitor concentration for HIV is 0.04–8.5 μmol. After diester hydrolysis, tenofovir is phosphorylated to the DP that then inhibits HIV reverse transcriptase by competing with the natural substrate deoxyadenosine 5-TP and, after incorporation into DNA, by DNA chain termination. Tenofovir DP is a weak inhibitor of mammalian DNA polymerases alpha, beta and mitochondrial DNA polymerase gamma. Additive or synergic anti-HIV activity with nucleoside analog, non-nucleoside analog and protease inhibitors has been demonstrated *in vitro*. Side effects include lactic acidosis, hepatomegaly with steatosis, and diarrhea.

Resistance is uncommon and occurs at codon 65.

E. HIV Fusion Inhibitors

Fuseon; Enfuvirtide

The recent licensure of a fusion inhibitor introduces a new class of antiviral compounds for the treatment of HTV. T-20 is an inhibitor of fusion of HIV-1 with CD4 cells that consists of a 36 amino acid synthetic peptide with the N-terminus acetylated and the C-terminus is a carboxamide. Medication is administered subcutaneously.

VI. SUMMARY

It is anticipated that many new compounds will be licensed for the treatment of viral disease because many are currently under development.

ACKNOWLEDGMENTS

Work performed and reported by the author was supported by Contracts NO1-Al-15113, NO1-Al-62554, NO1-Al-12667, and NO1-A1–65306 from the Antiviral Research Branch of the National Institute of Allergy and Infectious Diseases, a grant from the Division of Research Resources (RR-032) from the National Institutes of Health, and a grant from the state of Alabama.

BIBLIOGRAPHY

Abu-ata, O., Slim, J., Perez, G., and Smith, S. M. (2000). HIV therapeutics: past, present, and future. *Adv. Pharmacol.* **49**, 1–40.

Balfour, H. H., Jr. (1999). Antivirals (non-AIDS). *N. Engl. J. Med.* **340**, 1255–1268.

Beutner, K. R., Friedman, D. J., Forszpaniak, C., *et al.* (1995). Valaciclovir compared with acyclovir for improved therapy for herpes zoster in immunocompetent adults. *Antimicrob. Agents Chemother.* **39**, 1547–1553.

Crumpacker, C. S. (1996). Ganciclovir. *N. Engl. J. Med.* **335**, 721–728.

DeClercq E. (2002). New anti-HIV agents and targets. *Med. Res. Rev.* **22**, 531–565.

Douglas, J. M., Critchlow, C., Benedetti, J., *et al.* (1984). Double-blind study of oral acyclovir for suppression of recurrences of genital herpes simplex virus infection. *N. Engl. J. Med.* **310**, 1551–1556.

Dunkle, L. M., Arvin, A. M., Whitley, R. J., *et al.* (1991). A controlled trial of acyclovir for chicken pox in normal children. *N. Engl. J. Med.* **325**, 1539–1555.

Emery, S., and Cooper, D. A. (2003). Antiviral agents. *In* "Antibiotic and Chemotherapy", R. G. Finch, D. Greenwood, S. R. Norrby, and R. J. Whtley (Eds.), pp. 473–493, Churchill Livingstone, Philadelphia.

Galasso, G., Whitley, R. J., and Merigan, T. C. (Eds.) (1997). "Antiviral Agents and Viral Diseases of Man." Lippincott-Raven, New York.

Inouye, R. T., Panther, L. A., Hay, M. H., and Hammer, S. M. (2002). Antiviral Agents. *In* "Clinical Virology", D. D. Richman, R. J. Whitley, and F. G. Hayden (Eds.), 2nd edn., pp. 171–242, ASM Press, Washington, DC.

Lalezard, J. P., Henry, K., O'Hearn, M., *et al.* (2003). *N. Engl. J. Med.* **348**, 2175–2185.

Richman, R., Whitley, F., and Hayden. (Eds.) (2002). "Clinical Virology." ASM Press, Washinton.

Tyring, S., Barbarash, R. A., Nahlik, J. E., *et al.* (1995). Famciclovir for the treatment of acute herpes zoster. Effects on acute disease and postherpetic neuralgia: A randomized, double-blind, placebo-controlled trial. *Ann. Intern. Med.* **123**, 89–96.

Whitley, R. J. (in press). Antiviral therapy. *In*: "Infections Diseases", S. L. Gorbach, J. G. Bartlett, and N. R. Blacklow (Eds.), 3rd edn., pp. 330–350, Saunders, Philadelphia.

Whitley, R. J. (2003). Other antiviral agents. *In*: "Antibiotic and Chemotherapy", R. G. Finch, D. Greenwood, S. R. Norrby, and R. J. Whtley (Eds.), pp. 495–509, Churchill Livingstone, Philadelphia.

Whitley, R. J., and Gnann, J. (1992). Acyclovir: A decade later. *N. Engl. J. Med.* **327**, 782–789.

Whitley, R. J., Alford, C. A., Jr., Hirsch, M. S., *et al.* (1986). Vidarabine versus acyclovir therapy in herpes simplex encephalitis. *N. Engl. J. Med.* **314**, 144–149.

WEBSITES

Center for Disease Control (USA) website on antiviral drugs for influenza
http://www.cdc.gov/ncidod/diseases/flu/fluviral.htm

Antiviral Agents FactFile by International Medical Press (with search capabilities)
http://www.mediscover.net/antiviral.cfm

7

Archaea

Paul Blum and Vidula Dixit
George Beadle Center for Genetics, University of Nebraska-Lincoln

GLOSSARY

Archaea One of three domains of life. From the Greek *archaios* (ancient, primitive). Prokaryotic cells; membrane lipids predominantly isoprenoid glycerol diethers or diglycerols tetraethers. Formerly called archaebacteria.

Bacteria One of three domains of life. From the Greek *bacterion* (staff, rod). Prokaryotic cells; membrane lipids predominantly diacyl glycerol diesters. Formerly called eubacteria.

Crenarchaeota One of two kingdoms of organisms of the domain Archaea. From the Greek *crene-* (spring, fountain) for the resemblance of these organisms to the ancestor of the Archaea, and *archaios* (ancient). Include sulfur-metabolizing, extreme thermophiles.

Eukarya One of three domains of life. From the Greek *eu-* (good, true) and *karion,* (nut; refers to the nucleus). Eukaryotic cells; cell membrane lipids predominantly glycerol fatty acyl diesters.

Euryarchaeota One of two kingdoms within the domain Archaea. From the Greek *eurys-* (broad, wide), for the relatively broad patterns of metabolism of these organisms, and *archaios* (ancient). Include halophiles, methanogens, and some anaerobic, sulfur-metabolizing, extreme thermophiles.

halophiles From the Greek *halos-* (salt) and *philos* (loving). Includes organisms that grow optimally at high salt concentrations.

hyperthermophiles From the Greek *hyper-* (over), *therme-* (heat), and *philos* (loving). Includes organisms that grow optimally at temperatures warmer than 80°C.

mesophiles From the Greek *mesos-* (middle) and *philos* (loving). Includes organisms that grow optimally at temperatures between 20 and 50°C.

methanogens Strictly anaerobic Archaea that produce (Greek *gen,* to produce) methane.

psychrophiles From the Greek *psychros-* (cold) and *philos* (loving). Includes organisms that grow optimally at temperatures between 0 and 20°C.

thermophiles From the Greek *therme-* (heat) and *philos* (loving). Includes organisms that grow optimally at temperatures between 50 and 80°C.

I. INTRODUCTION

In an effort to accommodate molecular signatures evident in ribosomal small subunit RNAs, Woese and Fox (1977) proposed that prokaryotes are not a monophyletic group (single root). Instead, they argued for two distinct evolutionary lineages of organisms represented by the Bacteria and those now called Archaea (Fig. 7.1). Archaea, Bacteria and Eukarya are placed in separate taxonomic groups called Domains. The distinction between Archaea and Bacteria has since received impressive support from many sources.

The Desk Encyclopedia of Microbiology
ISBN: 0-12-621361-5

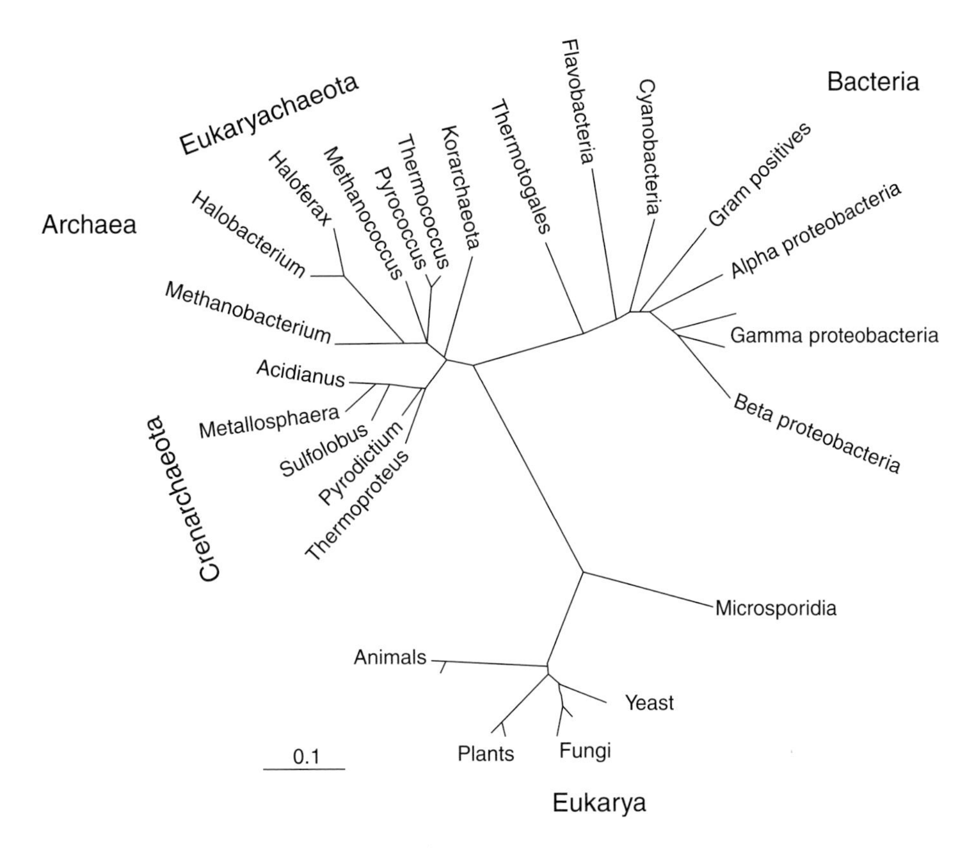

FIGURE 7.1 Universal (distance) phylogenetic tree based on ribosomal small subunit RNA sequences.

Perhaps the most compelling data comes from whole genome sequencing studies which reveal extensive gene and protein sequence conservation between members of the Archaea but not Bacteria. Comparative genomics also suggest that Archaea are a distinct group and share a common origin with Eukarya. This is further supported by the finding that Archaea employ a broad range of eukaryotic-like genes for conducting subcellular processes including the synthesis and processing of DNA, RNA, and protein (Blum, 2001). In contrast, Archaea use bacterial-like mechanisms for much of central metabolism including key biosynthetic (anabolic) and degradative (catabolic) pathways.

Archaea derived from extreme environments are studied intensively leading to the incorrect interpretation that all Archaea are extremophiles. Since extreme environments comprise only a small portion of inhabitable earth, Archaea are often thought to comprise only a small proportion of total prokaryotes. In contrast to this notion, studies on marine prokaryotic abundance indicate that global oceans harbor approximately equal numbers of archaeal and bacterial cells (Karner *et al.*, 2001). The major fraction of these Archaea are assigned to the subdivision called Crenarchaeota and these represent one of the ocean's single most abundant cell types. Archaea are also evident in soil and fresh water, and are associated with plant roots. It may be that Archaea are as abundant a life form as Bacteria.

Archaea are readily distinguished from Bacteria by several universal and unique chemical features. In contrast, the ecology and physiology of Archaea often overlaps that of Bacteria with the notable absence of differentiated cell forms or developmental cycles. Investigations of archaeal genomics and their molecular biology, however, reveals the true magnitude of the similarities between Archaea and Eukarya and differences from Bacteria. Perhaps this latter finding explains the utter lack of pathogenic Archaea despite the occurrence of host-adpated species such as the methanogens.

II. CELL STRUCTURE OF THE ARCHAEA

Archaea, like Bacteria, are prokaryotes (Table 7.1). They lack membrane-bound organelles such as a nucleus, or mitochondria, and are devoid of a cytoskeleton. Their chromosomal DNA is typically a single circular molecule and their ribosomes are of the 70S type. Their cell membranes and surface layers

TABLE 7.1 Major features distinguishing archaea and bacteria

Characteristic	Archaea[a]	Bacteria
Cell envelope		
(a) Wall	Mostly S-layer	Peptidoglycan
Membrane		
(a) Chirality of glycerol	L-glycerol	D-glycerol
(b) Hydrocarbon glycerol linkage	Ether-linkage	Ester-linkages
(c) Side-chains	Isoprene	Fatty acids
(d) Side-chain branching	Highly branched	Linear
Bacterial antibiotic sensitivity		
(a) Cell wall inhibitors	No	Yes
(b) Protein synthesis inhibitors	No[a]	Yes
(c) Transcription inhibitors	No	Yes

[a]Some methanogenic Archaea are sensitive to puromycin and chloramphenicol.

(envelope) are structurally and chemically distinct from those of Bacteria.

Archaeal cell envelope. Bacterial envelopes usually comprise a peptidoglycan wall with a single lipid bilayer internal to the wall (gram-positive Bacteria) or two lipid bilayers one internal and the other external to the wall (gram-negative Bacteria). In contrast, Archaea have neither peptidoglycan nor a wall. Instead the most common outer layer of their envelope consists of a paracrystalline S-layer, composed of noncovalently linked hexagonally or tetragonally arranged protein or glycoprotein subunits. Methanogens show particular diversity in envelope composition. Methanobacteria have a structure referred to as pseudomurein which resembles peptidoglycan, Methanosarcina have a structure referred to as methanochondroitin and Methanococcus and Methanoplanus have protein or glycoprotein layers. Pseudomurein is distinguished from peptidoglycan by its use of L-isomeric amino sugars with α-1,3 linkages rather than D-isomeric amino sugars using α-1,4 linkages. Haloarchaea have a protein layer containing a great excess of acidic amino acids whose negative charges counterbalance the high concentration of positively charged sodium ions in their salty environment. The envelope of Natronobacteria contains a novel layer composed of a glutamine polymer with *N*-acetylglucosamine, glucose and other sugars linked via the amide group of glutamine.

Archaeal cell membranes. Archaeal cell membranes are chemically unique from those of Bacteria or Eukarya. Differences include: (a) chirality of glycerol; (b) hydrocarbon–glycerol linkage; (c) isoprenoid chains; and (d) side-chain branching. The basic unit from which cell membranes are built is the phospholipid. The glycerol in archaeal phospholipids is a stereoisomer (L-glycerol) of that found in bacterial and eukaryal membranes (D-glycerol). The side chains in the phospholipids of Bacteria and Eukarya are fatty acids usually 16–18 carbon atoms in length and are coupled to glycerol via an ester linkage. In Archaea, side chains are coupled via an ether linkage. Archaea have side chains built from isoprene. Isoprene is the simplest member of a class of chemicals called terpenes. The ester linked fatty acids of Bacteria are linear, whereas the ether linked hydrocarbons of Archaea are highly branched. Archaeal glycerol diether lipids form a true bilayer membrane, whereas archaeal glycerol tetraether lipids form lipid monolayers. Lipid monolayer membranes occur in certain methanogens, are widespread among hyperthermophilic Archaea and are thought to confer additional thermal stability. Archaeol (diphytanylglycerol diether) is the predominant membrane core lipid in most methanogens and all extreme halophiles. In contrast, cell membrane of hyperthermophilic archaea and a few methanogens contain caldarchaeol, a dibiphytanyldiglycerol tetraether. The production of ether-linked lipids is so distinctive in Archaea that it is used as a biomarker for detecting fossilized Archaea in micropaleontological studies of rocks, sediment cores and other ancient materials. The chemical difference between archaeal and bacterial lipids provides additional support for the evolutionary distance between the Archaea and Bacteria.

III. ECOLOGY, PHYSIOLOGY AND SYSTEMATICS OF THE ARCHAEA

The cultivated Archaea are distributed into two kingdoms, the Euryarchaeota and the Crenarchaeota (Table 7.2). A third kingdom called Korarchaeota consists of uncultivated members. Cultivated Archaea are divided into 12 orders, 20 families, and 69 genera (Boone *et al.*, 2001). Cultivated Euryarchaeota include all of the known methanogens and extreme halophiles as well as some extreme thermophiles. Cultivated members of the Crenarchaeota include other hyperthermophiles and most thermoacidophiles. In addition, many other archaeal taxa have been detected in a variety of environments using molecular phylogenetic methods. While as yet uncultivated, these Archaea also are distributed across both the euryarchaeotal and the crenarchaeotal kingdoms. Cultivated Archaea include terrestrial and aquatic members, and occur in diverse locations from anaerobic sediments to hypersaline pools and geothermally heated environments. Some also occur as symbionts in animal digestive tracts. Archaea

TABLE 7.2 Systematics of the Archaea

Orde	Family	Habitat	Features
Kingdom: Euryarchaeota			
Archaeoglobales	Archaeaglobaceae	Geothermally heated sites	Strict anaerobes, facultative chemolithoautotrophs, hyperthermophiles, neutrophiles
Halobacteriales	Halobacteriaceae	Ubiquitous in areas of high salt concentration like salt lakes, salterns	Aerobes or facultative anaerobes, require 3.5–4.5 M NaCl, chemoorganotrophs, usually mesophiles
Methanobacteriales	Methanobacteriaceae	Aquatic sediment, sewage digestor, GI tract of animals	Strict anaerobes, chemolithoautotrophs, mesophiles to thermophiles
	Methanothermaceae	Hot solfataric fields	Strict anaerobes, chemolithotrophs, hyperthermophiles
Methanococcales	Methanococcaceae	Marine environments	Strict anaerobes, chemolithoautotrophs, mesophiles to thermophiles, selenium required
	Methanocaldococcaceae	Deep sea hydrothermal vents	Strict anaerobes, chemolithoautotrophs, hyperthermophiles
Methanomicrobiales	Methanocorpusculaceae	Lake sediments, digestors	Strict anaerobes, chemoorganotrophs, mesophiles
	Methanomicrobiaceae	Marine sediments, sewage digestors	Strict anaerobes, acetate required, chemolithoautotrophs, mesophiles to thermophiles
	Methanospirillaceae	Sewage sludge, waste digestors	Strict anaerobes, fix N_2, autotrophs or chemoorganotrophs, mesophiles
Methanopyrales	Methanopyraceae	Hot vents	Strictly anaerobic, chemolithoautotrophs, hyperthermophiles
Methanosarcinales	Methanosaetaceae	Sewage sludge and sediments	Strictly anaerobic, chemolithotrophs, mesophiles to thermophiles
	Methanosarcinaceae	Aquatic sediment, sewage digestors, GI tract of animals	Strictly anaerobic, N_2 may be fixed, chemoautotophs, mesophiles to thermophiles
Thermococcales	Thermococcaceae	Marine and terrestrial thermal environments	Strict anaerobes, heterotrophs, hyperthermophiles
Thermoplasmatales	Thermoplasmataceae	Self heating coal refuse piles, acidic solfatara fields	Wall-less, facultative aerobes, heterotrophs, thermoacidophiles
	Picrophilaceae	Hot geothermal solfotara soils and springs	Obligate aerobes, heterotrophs, hyperacidophiles
Kingdom: Crenarchaeota			
Desulfurococcales	Desulfurococcaceae	Marine environments, hot solfataric areas	Mostly anaerobic, chemolithotrophs or heteroptrophs, hyperthermophiles
	Pyrodictiaceae	Hot sea floors, sediments, black smokers	Anaerobic, facultative chemolithoautotrophs, hyperthermophiles
Sulfolobales	Sulfolobaceae	Hot acidic solfataric springs	Aerobic or anaerobic, chemolithoautotrophs, S^0 metabolizers, extreme thermoacidophiles
Thermoproteales	Thermofilaceae	Solfataric hot springs	Strict anaerobes, organotrophs, thermoacidophiles
	Thermoproteaceae	Acidic hot springs	Anaerobes to facultative anaerobes, chemolithoautotrophs or organotrophs, hyperthermophiles
Kingdom: Korarchaeota			
	Uncultivated Archaea delineated on the basis of 16 rRNA sequences		

Key: M, Molar; NaCl, sodium chloride; GI, gastrointestinal; N_2, nitrogen; S^0, sulfur.

include aerobes, anaerobes and facultative anaerobes, chemolithotrophs, organotrophs and facultative organotrophs. Archaea can be mesophilic or thermophilic with some species growing at temperatures up to 110 °C. Psychrophilic members have also been detected but not cultured. The Archaea often are divided into three key biotypes; methanoarchaea, haloarchaea and hyperthermophilic Archaea (Madigan *et al.*, 2000; also see *http://www. ncbi.nlm. nih.gov:80/entrez/query. fegi? db=Taxonomy*).

The methanogenic Archaea. The methanoarchaea belong to the Euryarchaeota and constitute a phylogenetically unique biotype. Methane (CH_4) production is an integral part of their energy metabolism and this form of metabolism is still unique to the Archaea. Such organisms are called methanogens and the process of methane formation is called methanogenesis, which is the terminal step in biodegradation of organic matter in many anaerobic environments. Habitats of methanogens range from anoxic sediments such as swamps, and wastewater treatment facilities to ruminant digestive tracts. They can also be found as endosymbionts of various anaerobic protozoa. Most known methanogens are mesophilic although "extremophilic" species growing optimally at very high or very low temperatures or at very high salt concentrations have also been isolated. These are strict anaerobes that are able to form methane as the principal metabolic end product using various oxidized forms of one- and two-carbon compounds as terminal electron acceptors. Key genera of this group include; *Methanobacterium, Methanococcus,* and *Methanosarcina.*

The halophilic Archaea. This diverse group of Euryarchaeota inhabit highly saline environments such as solar evaporation ponds, natural salt lakes and surfaces of salted foods. Moderately halophilic Archaea require 2–4 M sodium chloride (12–23%) for optimal growth. Virtually all extremely halophilic Archaea grow at 5.5 M sodium chloride (32%) which is the limit of saturation of sodium chloride. Haloarchaea can be aerobic or facultatively anaerobic and are chemoorganotrophic. They are generally mesophilic but can be moderately thermophilic (growth up to 55°C). Some species contain a protein called bacteriorhodopsin and the carotenoid pigments, bacterioruberins, which are used for the light mediated synthesis of ATP. Presence of these proteins results in a distinct pigmentation leading to their red-purple coloration. *Halobacterium* employs potassium as a compatible solute to withstand high external sodium concentrations to ensure a positive water balance. Extreme halophilicity was thought to occur only in Archaea; recently, however, molecular phylogenetic evidence indicates Bacteria may also have this property. Key genera of the Haloarchaea include; *Halobacterium, Haloferax,* and *Natronobacterium.*

The thermophilic Archaea. A number of Archaea thrive in thermal environments. Those with growth optima at or above 80°C are called hyperthermophiles. Thermophilic Archaea inhabit terrestrial and marine regions with geothermal or hydrothermal intrusion. A growth requirement for exterme acid (acidophily) can accompany the thermophilic lifestyle. The sulfur- and sulfate-reducing thermophiles are split between members of the Euryarcheota and the Crenarchaeota. These are strict anaerobes using elemental sulfur as a terminal electron acceptor, which is reduced to hydrogen sulfide. These organisms are extremely thermophilic with growth up to 100°C. Members of the *Thermococci* and *Pyrococci* are Euryarchaeota. *Methanopyrus,* another Euryarchaeote, is a hyperthermophilic methanogen. Only *Archaeoglobus* is a true sulfate reducer, capable of reducing sulfate to hydrogen sulfide. Sulfur reducing Crenarchaeota include *Thermoproteus, Pyrodictium,* and *Pyrolobus. Pyrolobus fumarii* holds the current record for the most thermophilic of all known organisms, its growth temperature maximum is 113°C. Crenarchaeotal sulfur oxidizers include the *Sulfolobales,* which are found in acidic sulfur containing geothermal pools such as those found in Yellowstone National Park, USA (Fig. 7.2). These are obligate aerobes that use oxygen as a terminal electron acceptor and grow in the temperature range 70–90°C and at pH values of between 2 and 5. Acidophily is also found among members of two euryarchaeotal genera, *Thermoplasma* and *Picrophilus,* which are among the most acidophilic of all known prokaryotes. Most species of *Thermoplasma* have been obtained from self-heating coal refuse piles. *Thermoplasma* resemble the mycoplasma in having an envelope comprising of only a cytoplasmic membrane. *Picrophilus* is capable of growth even below pH 0. These organisms are both chemoorganotrophic. Again, members of the Bacteria share at least the thermophilic and hyperthermophilic biotypes of Archaea while extreme acidophily remains unique to the Archaea. Key genera of the thermophilic Archaea include; *Thermococcus, Pyrococcus, Pyrobaculum,* and *Sulfolobus.*

Archaea as non-extremophiles. The known phenotypes of cultivated Archaea are still largely represented by extreme halophiles, sulfur metabolizing thermophiles, and methanogens. This picture has altered as new methods have been applied to the study of uncultivated microbes. The presence of new types of uncultivated Archaea was first suggested during surveys of marine plankton. Surveys of polymerase chain reaction (PCR) amplified small subunit rRNA genes revealed archaeal rRNA sequences in deep ocean water samples. The discovery of high numbers of Archaea in a wide number of oceans and the association of a novel crenarchaeotal isolate with a marine sponge living at 10°C, provide further evidence of new archaeal biotypes native to cold seawater. Since their initial detection, evidence for widespread distribution of new uncultivated Archaea has been extended to include forest and agricultural soils, deep subsurface paleosols, freshwater lakes and various sediments.

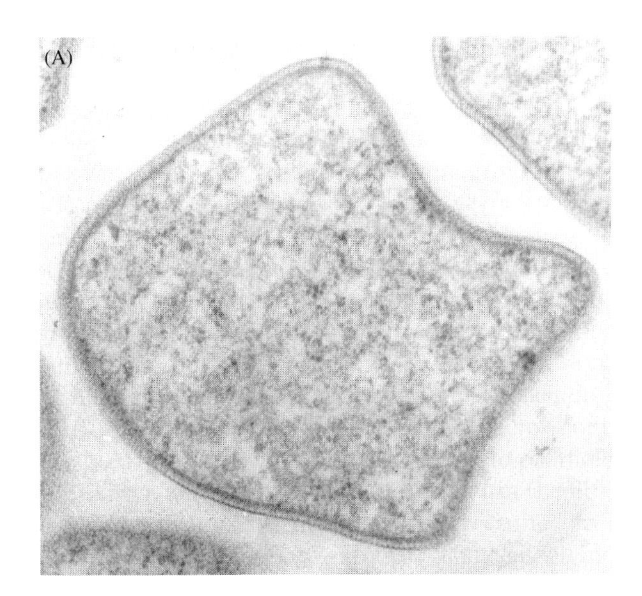

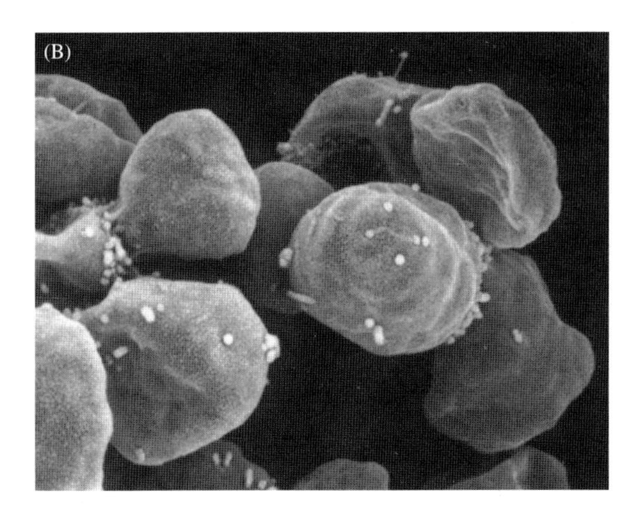

FIGURE 7.2 Electron micrographs of Sulfolobus solfataricus. (A) transmission electron micrograph; (B) scanning electron micrograph. Both images show the irregular cell shape of the organism.

TABLE 7.3 Ecology and physiology of the model archaeal species

Genus	Biotype	Habitat	Energy metabolism	Energy source
Methanococcus maripaludis	Methanogen	Salt marsh sediment	Obligate anaerobes chemolithotrophs/	$H_2 + CO_2$, pyruvate + CO_2, formate
Methanosarcina acetivorans	Methanogen	Marine sediment	Obligate anaerobes chemolithotrophs/	$H_2 + CO_2$, methanol, methylamines, acetate
Haloferax volcanii	Halophile	Dead Sea	Usually obligate aerobe chemoorganotroph,	Amino acids
Halobacterium salinarium	Halophile	Hypersaline lakes and salted foods	Usually obligate aerobe chemoorganotroph,	Amino acids, organic acids
Sulfolobus solfataricus	Hyperhermophile	Sulfur-rich hot springs	Aerobic chemolithotrophs/ chemoorganotrophs	S^0, H_2S, sugars, amino acids
Sulfolobus acidocaldarius	Hyperhermophile	Sulfur-rich hot springs	Aerobic chemolithotrophs/ chemoorganotrophs	S^0, H_2S, sugars, amino acids

Key: H_2, hydrogen; CO_2, carbon dioxide; S^0, sulfur; H_2S, hydrogen sulfide.

Archaeal model systems. As is the case with Bacteria, particular archaeal taxa are employed as model experimental systems to address mechanistic questions about this group of prokaryotes. The key feature distinguishing these species is their genetic systems. Among the Haloarchaea *Haloferax volcanii* and *Halobacterium salinarum* (including NRC-1) are employed. *Methanococcus maripaludis* and *Methanosarcina acetivorans*, are popular among the methanogenic Archaea, while among the thermophilic Archaea *Sulfolobus solfataricus* and *Sulfolobus acidocaldarius* are best studied (Table 7.3).

IV. MOLECULAR BIOLOGY OF THE ARCHAEA

The distinguishing feature of Archaea is their use of eukaryotic-like genes rather than bacterial-like genes for the synthesis, repair, and turnover of DNA, RNA, and protein (Table 7.4). Key aspects of this relationship have been recently described in greater detail (Blum, 2001). This striking example of gene conservation supports the idea of a shared or common evolutionary origin for Archaea and Eukarya. An understanding of these subcellular processes is therefore a prerequisite for appreciating their place in

The key mechanism employed in Archaea for protein degradation involves a multisubunit hollow barrel-like structure called the 20S proteasome. The same structure is universally present in Eukarya but only rarely present in Bacteria. The proteasome is readily observed by electron microscopy, being 15 nm in length and containing 28 subunits. When present in Bacteria, proteasomes are thought to have been acquired by horizontal transmission. The proteasome associates with ATPase regulatory components in the form of a cap increasing the size of the structure considerably. The eukaryotic ubiquitin protein targeting system is not evident in Archaea and though other energy dependent proteases exist in Archaea additional studies are necessary to determine their distribution and relationship to bacterial proteases.

Protein secretion. The translocation (secretion) of proteins through membranes in all organisms employs a ribonucleoprotein complex called signal recognition particle (SRP). To date, this appears to be the sole mechanism in Eukarya and Archaea, while Bacteria have the additional (Sec) system. SRP comprises RNA and associated proteins, which bind nascent polypeptides to promote their interaction and transfer across the cytoplasmic membrane. Archaea like Eukarya have a 7.5S SRP RNA while Bacteria have a smaller 4.5S RNA. The SRP protein components of Archaea (SRP19 and SRP54) are more similar to those of Eukarya though present in fewer numbers. Other components of the archaeal protein secretion system including the leader peptidase that removes a portion of the N-terminal end of secreted proteins, exhibit greater similarity to eukaryotic enzymes.

V. CONCLUSIONS

Archaea like Bacteria are prokaryotes with simple cell structures. They can be readily distinguished from Bacteria by differences in the structure and chemistry of their envelopes. With some exceptions they cannot generally be distinguished by biotype, physiology, or metabolism. The most striking difference which appears to unite the Archaea is their use of eukaryal-type genes often with unique functions, to make and process DNA, RNA, and protein. This latter feature has major implications for the origin of life and the relationship between Archaea and the last common ancestral cell.

BIBLIOGRAPHY

Amend, J. P., and Shock, E. L. (2001). Energetics of overall metabolic reactions of thermophilic and hyperthermophilic Archaea and bacteria. *FEMS Microbiol. Rev.* **25**, 175–243.

Bell, S. D., and Jackson, S. P. (2001). Mechanism and regulation of transcription in archaea. *Curr. Opin. Microbiol.* **4**, 208–213.

Bini, E., Dikshit, V., Dirksen, K., Drozda, M., and Blum, P. (2002). Stability of mRNA in hyperthermophilic archaea. *RNA* **8**, 1129–1136.

Blum, P. (2001). Ancient microbes, extreme environments, and the origin of life. *Adv. Appl. Microbiol.* **50**.

Boone, D. R., Castenholz, R. W., and Garrity, G. M. (2001). Procaryotic domains. *In* "Bergey's Manual of Systematic Bacteriology", Vol. 1, Springer-Verlag, Berlin.

Karner, M. B., DeLong, E. F., and Karl, D. M. (2001). Archaeal dominance in the mesopelagic zone of the Pacific Ocean. *Nature* **409**, 507–510.

Madigan, M. T., Martinko, J. M., and Parker, J. (2000). Prokaryotic diversity: the Archaea *In* "Brock Biology of Microbiology", Chap. 14, pp. 545–572, Prentice-Hall, Englewood Cliffs, NJ.

Reysenbach, A. L., and Cady, S. L. (2001). Microbiology of ancient and modern hydrothermal systems. *Trends Microbiol.* **9**, 79–86.

Sandman, K., Soares, D., and Reeve, J. N. (2001). Molecular components of the archaeal nucleosome. *Biochimie* **83**, 277–281.

Thomas, N. A., Bardy, S. L., and Jarrell, K. F. (2001). The archaeal flagellum: a different kind of prokaryotic motility structure. *FEMS Microbiol Rev.* **25**, 147–174.

Woese, C. R., and Fox, G. E. (1977). Phylogenetic structure of the prokaryotic domain: the primary kingdoms. *Proc. Natl. Acad. Sci. USA* **74**, 5088–5090.

WEBSITES

For more information on the systematics of Archaea see:
http://www.ncbi.nlm.nih.gov:80/entrez/query.fcgi?db=Taxonomy

For more information on archaeal histones see:
http://www.biosci.ohio-state.edu/~microbio/Archaealhistones/

For more information on RNA modification in Archaea see: *http://rna.wustl.edu/snoRNAdb/*

Archaea browser of the NCBI (National Center for Biotechnology Information, USA)
http://www.ncbi.nlm.nih.gov/htbin-post/Taxonomy/wgetorg?name=Archaea

Website for Comprehensive Microbial Resource of The Institute for Genomic Research and links to many other microbial genomic sites
http://www.tigr.org/tigr-scripts/CMR2/CMRHomePage.spl

List of bacterial names with standing in nomenclature (J. P. Euzéby)
http://www.bacterio.cict.fr/index.html

8

Attenuation, Transcriptional

Charles Yanofsky
Stanford University

GLOSSARY

antiterminator An RNA hairpin structure that generally contains several paired nucleotides that are essential for terminator formation. When these nucleotides are paired in the antiterminator they cannot participate in terminator formation.

attenuator A short DNA region that functions as a site of regulated transcription termination.

charged tRNA A transfer RNA bearing its cognate amino acid (e.g., Trp-tRNA^{Trp}).

leader peptide A peptide encoded by the leader segment of a transcript.

leader peptide coding region A short peptide coding region in the leader segment of a transcript.

RNA-binding attenuation regulatory protein An RNA-binding protein that binds to a specific RNA sequence and, by so doing, either promotes or prevents formation of a transcription terminator.

RNA hairpin structure A base-paired stem and loop structure that has sufficient stability to remain in the base-paired, hairpin configuration.

terminator (factor-dependent) An RNA sequence usually causing transcription pausing that serves as a site of factor-dependent transcription termination.

terminator (intrinsic) An RNA hairpin followed immediately by a sequence rich in U's. The terminator serves as a signal to RNA polymerase to terminate transcription.

transcription pausing A temporary pause or delay in RNA polymerase movement on its DNA template.

transcription pause structure An RNA hairpin that causes RNA polymerase to pause or stall during transcription.

transcription termination Cessation of RNA synthesis and release of transcript and DNA template from RNA polymerase.

transcriptional attenuation A mechanism used to regulate continuation vs termination of transcription.

Transcriptional Attenuation is the term used to describe a general transcription regulatory strategy that exploits various sensing events and molecular signals to alter the rate of transcription termination at a site or a site preceding one or more genes of an operon. Many mechanisms of transcriptional attenuation exist. Each regulates operon expression by responding to an appropriate molecule or event and determining whether transcription will or will not be terminated.

I. OBJECTIVES AND FEATURES OF REGULATION BY TRANSCRIPTIONAL ATTENUATION

It is evident that an appreciable fraction of the genetic material of each organism is dedicated to regulating gene expression. The ability to alter expression provides

The Desk Encyclopedia of Microbiology
ISBN: 0-12-621361-5

the variability that an organism needs in order to initiate or respond to the many changes that are associated with or responsible for each physiological and/or developmental event. Initiation of transcription is perhaps the single biological act that is most often subject to regulation. There are numerous examples of "negative-acting" repressor proteins—proteins that inhibit transcription initiation by binding to their respective DNA operator site(s) within or in the vicinity of the regulated promoter. Similarly, there are many examples of "positive-acting" regulatory proteins that activate transcription by binding at specific DNA elements in the vicinity of the affected promoter. Both negative- and positive-acting regulatory proteins are commonly activated or inactivated by small or large molecules as well as by reversible processes, i.e., phosphorylation and dephosphorylation. However, transcription initiation is only one of several common metabolic events that may be modulated to alter gene expression. The two subsequent stages in transcription, transcript elongation and transcription termination, are also common targets for regulatory change. The principal advantages achieved by regulating these events is that different classes of molecules and different metabolic processes can participate in regulatory decisions. Thus, once transcription has begun, the nascent transcript is a potential target for a regulatory event. In addition, in prokaryotes, in which most transcripts are initially translated as they are being synthesized, components of the translation machinery may participate in regulatory decisions. By exploiting these additional targets, organisms have greatly increased their regulatory options. A separate objective may have been to devote as little unique genetic information as possible to a regulatory process. Accordingly, some of the transcriptional attenuation regulatory mechanisms that will be described use less than 150 bp of DNA to achieve gene- or operon-specific control. Often attenuation regulation is achieved using only the common cell components that participate in RNA and protein synthesis. In this article, I review the features of several examples of regulation by transcriptional attenuation.

II. MECHANISMS OF TRANSCRIPTIONAL ATTENUATION

A. Regulation of termination at an intrinsic terminator

Many operons regulated by transcriptional attenuation contain a DNA region that specifies a RNA sequence that can fold to form a hairpin structure followed by a run of U's, a structure called an intrinsic terminator. Intrinsic terminators instruct RNA polymerase to terminate transcription. The region encoding the intrinsic terminator is located immediately preceding the gene or genes that are being regulated. The transcript segment before and including part of the terminator often contains a nucleotide sequence that can fold to form a competing, alternative hairpin structure called the antiterminator. The existence of this structure prevents formation of the terminator. Antiterminator and terminator structures generally share a short nucleotide sequence, which explains why prior formation of the antiterminator prevents formation of the terminator. Additional features of the nucleotide sequence preceding or following a terminator or antiterminator can influence whether these structures will form or act. The transcript segment preceding the terminator often contains sequences that allow the organism to sense a relevant metabolic signal and to respond to that signal by allowing or preventing antiterminator or terminator formation.

A variety of mechanisms are used to sense specific cell signals. In operons concerned with amino acid biosynthesis, ribosome translation of a peptide coding region rich in codons for a crucial amino acid is often used to sense the presence or absence of the corresponding charged tRNA. Depending on the location of the translating ribosome on the transcript, an antiterminator will or will not form. In another example, in an operon concerned with pyrimidine biosynthesis, coupling of RNA synthesis with translation is employed to sense the availability of a nucleotide needed for RNA synthesis. In some mechanisms, RNA-binding proteins regulate termination. These proteins bind to specific transcript sequences or structures and allow or prevent antiterminator or terminator formation. One common regulatory mechanism is designed to sense the relative concentrations of a charged and uncharged tRNA and, depending on which is in excess, induce formation of an antiterminator or terminator. It is evident from these and other examples that regulation of the formation of an intrinsic terminator is a common strategy used to alter operon expression.

1. *Ribosome-mediated attenuation*

Synthesis of most proteins requires not only the availability of all 20 amino acids but also these amino acids must be in their activated state, covalently attached to their respective tRNAs. As such, they are primed for participation in polypeptide synthesis. The intracellular concentration of each amino acid reflects a balance of several events, including rates of synthesis, utilization, import from the environment, and release from proteins by degradation. Occasionally, induction of a degradative pathway also affects the cellular level of an amino acid. The concentration of a specific charged tRNA also reflects several events, including the presence of the corresponding amino acid, its rate of charging onto tRNA, the availability of that tRNA, and use of that charged tRNA in protein synthesis. Other factors also affect the rate of protein synthesis, such as whether there are rare codons in the coding region being translated and whether all needed species of charged tRNA are available. The availability of free ribosomes and accessory molecules required for protein synthesis also has an impact on the rate of protein synthesis. Given these many variables, it is not surprising that so many attenuation mechanisms are used to sense and respond to specific cellular needs.

a. The trp *operon of* Escherichia coli

Transcription of the *trp* operon of *E. coli* is regulated by both repression and transcriptional attenuation. The initial event in regulation by attenuation is the formation of a RNA hairpin structure that directs the transcribing RNA polymerase molecule to pause after initiating transcription (Fig. 8.1, stage 1). This transcription pause provides sufficient time for a ribosome to bind to the ribosome binding site of a peptide coding region in the leader transcript and initiate translation (Fig. 8.1, stage 2). The moving ribosome in fact releases the paused polymerase, permitting resumption of transcription (Fig. 8.1, stage 3). Thereafter, transcription and translation proceed in unison. As the polymerase molecule transcribes the leader region, the translating ribosome moves along the transcript and reaches a segment that is capable of folding to form an antiterminator structure. However, whether or not this structure forms depends on the location of the translating ribosome. In a bacterium deficient in charged $tRNA^{Trp}$, the translating ribosome would stall over either of two adjacent Trp codons in the leader peptide coding region (Fig. 8.1, stage 4). Stalling would allow a downstream RNA segment to fold and form an antiterminator hairpin structure. As transcription proceeds, persistence of the antiterminator would prevent formation of the terminator since paired nucleotides at the base of the antiterminator must be free for terminator formation to occur. Under these conditions, transcription would continue into the structural genes of the operon. In a cell with adequate levels of charged $tRNA^{Trp}$ (Fig. 8.1, stage 4 alternate), the tandem Trp codons would be translated and the translating ribosome would proceed to the leader peptide stop codon. At this position, the ribosome would block formation of the antiterminator and allow the terminator to form; hence, transcription would be terminated.

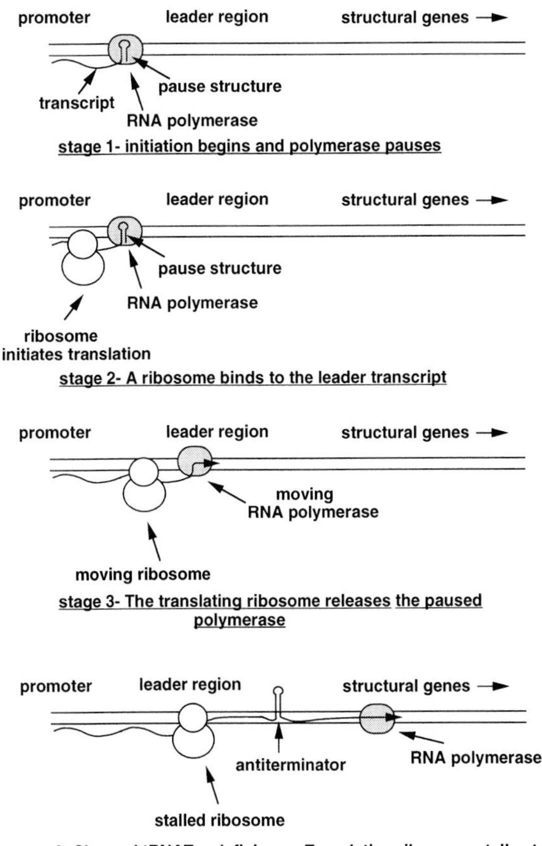

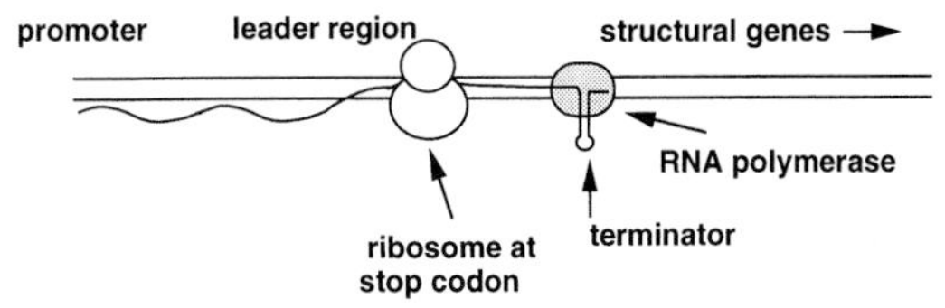

FIGURE 8.1 Stages in ribosome-mediated transcriptional attenuation regulation of the *trp* operon of *E. coli*. Transcription initiation and pausing (stage 1), ribosome loading (stage 2), and initiation of translation and release of the pause transcription complex (stage 3) occur under all conditions. When a cell is deficient in tryptophan-charged $tRNA^{Trp}$ (stage 4), the translating ribosome stalls at either of the two Trp codons in the leader peptide coding region. Stalling permits the antitterminator to form; this prevents terminator formation, allowing transcription to continue into the structural genes of the operon. When a cell has sufficient charged $tRNA^{Trp}$ to support ongoing protein synthesis (stage 4 alternate), translation proceeds to the leader peptide stop codon. A ribosome at this position blocks formation of the antiterminator structure and permits the terminator to form and cause termination.

The leader regions of many bacterial operons, such as the *his, phe, leu, thr, ilvGMEDA*, and *ilvBN* operons, are organized much like that of the *trp* operon of *E. coli*. These operons appear to be regulated by the same mechanism, with only minor variations tailored to each operon's needs. Generally, the leader region sequence and organization reflects differences in regulatory requirements. For example, transcription of the *his* operon of *S. typhimurium* is regulated only by attenuation, unlike transcription of the *trp* operon of *E. coli* which is regulated by both repression and attenuation. The *his* operon's leader peptide coding region contains seven consecutive His codons. This organization allows greater sensitivity to changes in the cellular level of charged $tRNA^{His}$; a slight reduction is sufficient to delay ribosome movement through the His codon region. Any delay promotes antiterminator formation. An operon with a leader region that is organized differently is the *ilvGMEDA* operon of *E. coli*. Here, attenuation is regulated in response to the availability of three charged tRNAs those for $tRNA^{Ile}$, $tRNA^{Val}$, and $tRNA^{Thr}$. Codons for these tRNAs are arranged in the leader peptide coding region so that a deficiency of any of these charged species would promote antiterminator formation. Another operon regulated similarly is *pheST* of *E. coli*. This operon specifies the two polypeptides of phenylalanyl-tRNA synthetase. Translation of its leader peptide coding region containing five Phe codons is used to regulate termination/antitermination. An interesting consideration that bears on *pheST* operon regulation is that the product of this operon, phenylalanyl-tRNA synthetase, is needed under all growth conditions.

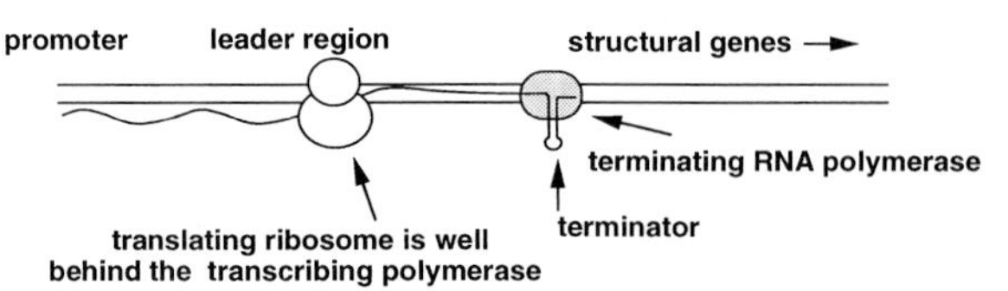

FIGURE 8.2 Stages in transcriptional attenuation in the *pyrBI* operon of *E. coli*. When a cell is deficient in UTP, the RNA polymerase molecule that is transcribing the *pyrBI* operon leader region pauses at one or more UTP deficiency-dependent pause sites (stage 1). While the polymerase is paused a ribosome binds to the leader transcript and initiates translation (stage 2). When the polymerase is released, the translating ribosome moves closely behind the transcribing polymerase. Continued translation by this ribosome prevents formation of the terminator structure; thus, transcription continues into the structural genes of the operon (stage 3). When there are adequate levels of UTP to support rapid RNA synthesis (bottom) the transcribing polymerase pauses very briefly in the leader region and then continues transcription. The terminator sequence is formed well before the translating ribosome can approach this segment of the transcript. Terminator formation results in termination.

b. The pyrBI *operon of* E. coli

Another well-studied example in which ribosome-mediated attenuation regulates transcription of an operon concerns the *pyrBI* operon of *E. coli*. When a cell has inadequate levels of UTP for RNA synthesis, the UTP deficiency triggers transcription antitermination in the leader region of this operon (Fig. 8.2). Continued transcription of the operon allows the cell to increase its rate of pyrimidine nucleotide synthesis. The *pyrBI* leader transcript has several features that explain its role in transcription regulation. It can fold to form alternative antiterminator and terminator structures. In addition, the leader segment contains the coding region for a leader peptide; this coding segment overlaps the antiterminator and terminator. The leader transcript also has several U-rich sequences which play a role in transcription pausing. When the UTP level is insufficient to sustain continued RNA synthesis, the polymerase transcribing the *pyrBI* operon stalls at these U-rich pause sites (Fig. 8.2, stage 1). Reduced polymerase migration allows sufficient time for a ribosome to bind to the transcript and move closely behind the polymerase (Fig. 8.2, stage 2). A translating ribosome at this position could prevent formation of the terminator structure; thus, transcription of the operon would continue (Fig. 8.2, stage 3). When a cell has an adequate level of UTP, the transcribing polymerase molecule moves through the pause sites rapidly and is positioned well ahead of the translating ribosome. This separation permits the terminator to form and cause transcription termination. Transcription of this operon is also regulated by an unrelated UTP-dependent mechanism.

c. Other examples

A related although different mechanism of ribosome-mediated transcription attenuation is used to regulate expression of the *ampC* operon of *E. coli*. The regulatory region of this operon, preceding *ampC*, encodes a leader transcript segment containing a ribosome binding site, adjacent start and stop codons, and a sequence that can form an intrinsic terminator. Expression of this operon is subject to growth rate regulation. During rapid growth, when the ribosome content per cell is high, a ribosome is likely to bind at the ribosome binding site in the leader segment and interfere with terminator formation. Under these conditions, transcription of the operon will continue. When the ribosome content per cell is low, the leader segment of the transcript is likely to be ribosome free for a period sufficiently long to allow the terminator to form and promote termination. An antiterminator is not used in this attenuation mechanism.

2. *Binding protein-mediated attenuation*

In several operons regulated by transcriptional attenuation, specific RNA-binding proteins determine whether or not transcription will be terminated. These proteins recognize specific sites or sequences in a transcript and, by binding, regulate formation of an antiterminator or terminator. Well-studied examples include the *bgl* operon of *E. coli* and the *sac* operon of *Bacillus subtilis*, which are regulated similarly, and the *trp* and *pyr* operons of *B. subtilis*, which are regulated differently. The RNA-binding regulatory proteins that regulate transcription of these operons function much like the stalled ribosome in amino acid biosynthetic operons, as described previously.

a. The bgl *operon of* E. coli

The *bgl* operon of *E. coli*, *bglG–bglF–bglB*, is a three-gene operon encoding proteins required for the utilization of β-glucosides as carbon sources. The operon contains two independent sites of regulated transcription termination, the first before *bglG* and the second between *bglG* and *bglF*. The products of the first two genes of the operon, BglG and BglF, are necessary for regulation of this operon by attenuation. BglG exists in two forms: a phosphorylated, monomeric, inactive species and a dephosphorylated, dimeric, active (RNA-binding) species (Fig. 8.3). BglF is a membrane-bound phosphoenolpyruvatedependent phosphotransferase. When BglF senses a β-glucoside, it phosphorylates the sugar and transports it into the cell. Substrate-activated BglF also dephosphorylates BglG, converting it into the active, RNA-binding dimeric form. In the absence of a β-glucoside, BglF phosphorylates BglG, rendering it monomeric and inactive.

The transcript segment preceding *bglG* and *blgF* can fold to form either an antiterminator or a terminator structure. When BglG is dephosphorylated and active, it binds to and stabilizes the antiterminator (Fig. 8.3). Since the stem of the antiterminator contains bases that are part of the terminator, the terminator does not form. Nucleotides in the single-stranded loop region of each antiterminator as well as paired bases in the antiterminator stem appear to be the sites of BglG binding. BglG is believed to act similarly at the two antiterminators. When BglG is inactive and the *bgl* operon is being transcribed, terminator structures form in the transcript and terminate transcription (Fig. 8.3).

The *sacB* and *sacPA* genes of *B. subtilis*, genes concerned with sucrose utilization, appear to be regulated by a very similar antitermination/termination mechanism. The protein products of genes *sacY* and

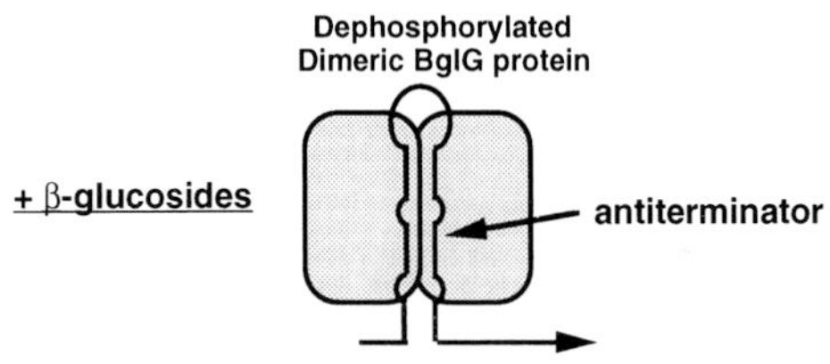

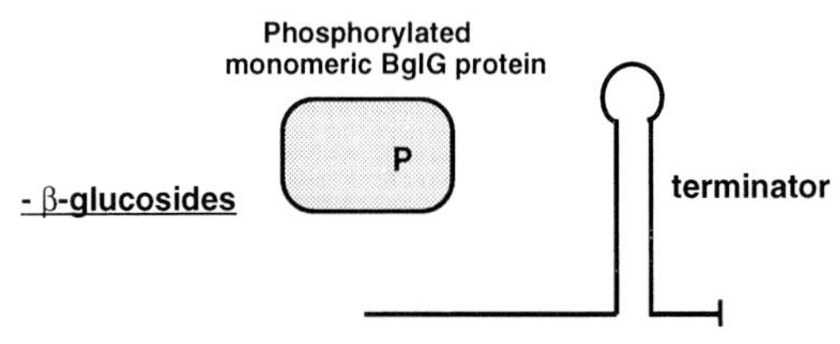

FIGURE 8.3 Protein-mediated transcriptional attenuation in the *bgl* operon of *E. coli*. In the presence of a β-glucoside carbon source the BglF protein phosphorylates the sugar and transports it into the cell (top). β-Glucoside-activated BglF also dephosphorylates the BglG protein. Dephosphorylated BglG dimerizes, and the dimer binds at one or both of the antiterminators in the transcript of the *bgl* operon, stabilizing the antiterminator structure. The existence of the antiterminator prevents formation of the terminator; thus, transcription proceeds. In the absence of a β-glucoside BglF phosphorylates BglG and the phosphorylated form remains as a monomer, incapable of binding to RNA (bottom). Under these conditions, the antiterminator is not stabilized, and the terminator forms, terminating transcription.

sacT regulate *sacB* and *sacPA* expression, respectively. The leader regions preceding *sacB* and *sacPA* specify RNA antiterminator structures that closely resemble those of the *bgl* operon. Dephosphorylation of SacY by SacX, in response to the presence of sucrose, leads to antitermination of transcription in the leader region preceding *sacB*. The proteins and sites involved in attenuation control in the *sac* and *bgl* systems are homologous.

b. The trp *operon of* B. subtilis

The leader segment of the transcript of the *trp* operon of *B. subtilis* can fold to form mutually exclusive antiterminator and terminator structures (Fig. 8.4). When a cell is deficient in tryptophan and the leader region of the operon is being transcribed, the antiterminator forms, preventing terminator formation and termination (Fig. 8.4). In the presence of excess tryptophan, an RNA-binding protein, TRAP (*trp* RNA-binding attenuation protein), encoded by the *mtrB* gene, binds tryptophan and becomes activated. Activated TRAP can bind to the *trp* operon leader transcript while it is being synthesized. The TRAP binding site consists of a series of U/GAG repeats located immediately preceding and including part of the antiterminator structure (Fig. 8.4). TRAP binding to the transcript prevents formation of the antiterminator, thereby promoting formation of the terminator. The 3D structure of TRAP has been described, and the residues in the protein principally responsible for RNA binding have been identified. The protein is doughnut shaped and consists of 11 identical subunits, each of which associates with a U/GAG sequence in the transcript. TRAP is believed to wrap the single-stranded leader transcript around its periphery and, by so doing, prevent formation of the antiterminator.

Tryptophan-activated TRAP also binds to a similar sequence of U/GAG repeats that precede the *trpG* coding region. *trpG* is the sole *trp* gene of *B. subtilis* that is not in the *trp* operon. *trpG* is located in a folate operon and specifies a bifunctional polypeptide that functions both in tryptophan and in folate biosynthesis. A TRAP binding site overlaps the *trpG* ribosome binding site; thus, TRAP binding inhibits translation of *trpG*. TRAP action therefore coordinates *trp* gene expression in the folate and tryptophan operons.

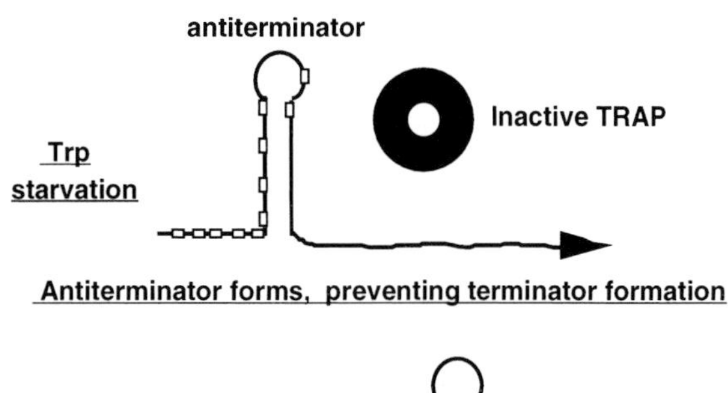

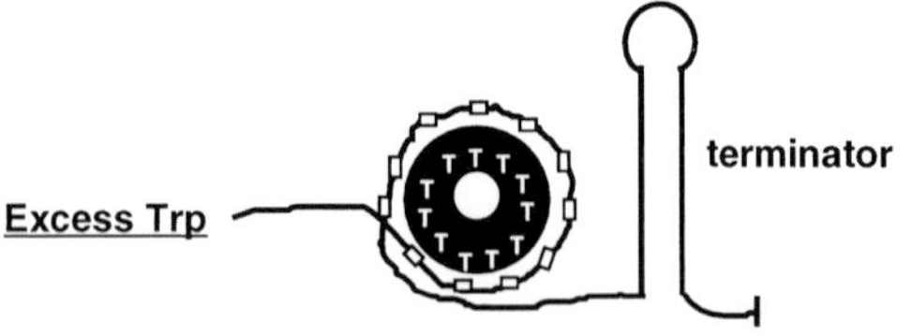

FIGURE 8.4 Protein-mediated transcriptional attenuation in the *trp* operon of *B. subtilis*. When a cell is deficient in tryptophan the TRAP protein is not active, the leader region of the *trp* operon is transcribed, the antiterminator forms, and transcription continues into the operon (top). When a cell has sufficient tryptophan to support rapid growth, the TRAP protein is activated by bound tryptophan (bottom). Activated TRAP binds at U/GAG repeat sequences (small boxes) in the transcript segments located before and within the antiterminator. Bound TRAP essentially melts the antiterminator, allowing a sequence at the base of the antiterminator to exist in an unpaired form. This unpaired sequence participates in the formation of the terminator hairpin structure, which promotes termination.

c. The pyr *operon of* B. subtilis

Another example of transcriptional attenuation mediated by a RNA-binding protein concerns regulated expression of the *pyr* operon of *B. subtilis*. This organism produces a novel uracil phosphoribosyltransferase, PyrR, that also functions as a RNA-binding transcription regulator. PyrR can bind at similar sites in three ~150-nt untranslated segments of the *pyr* transcript, each preceding a polypeptide coding segment and each containing a terminator. The first terminator precedes the first gene in the operon, *pyrR*, which in fact encodes this RNA-binding regulatory protein/enzyme. The second terminator is located between *pyrR* and *pyrP; pyrP* encodes a uracil permease. The third terminator is located between *pyrP* and *pyrB*. *pyrB* specifies aspartate transcarbamylase. Each of the three untranslated segments of the transcript can fold to form an alternative antiterminator structure that can prevent formation of an intrinsic terminator. In addition, each untranslated transcript segment can form a third structure, earlier in the transcript, termed an anti-antiterminator. This structure includes part of the antiterminator; thus, its formation prevents formation of the antiterminator. When pyrimidines are plentiful PyrR binds to the nascent *pyr* operon transcript and stabilizes the anti-antiterminator stem-loop structure. Stabilization of this structure blocks formation of the antiterminator structure, promoting formation of the terminator and thereby causing termination. When cells are deficient in pyrimidines and PyrR is inactive the antiterminator prevents formation of the terminator, allowing transcription to continue.

The three antiterminators are predicted to be the most stable of the several RNA structures. PyrR's RNA binding ability is responsive to the relative concentrations of UMP and PRPP, with bound UMP favoring RNA binding and bound PRPP preventing UMP binding and activation of the protein.

The 3D structure of PyrR of *B. subtilis* has been determined. The RNA sequences that are recognized have also been identified. Several bacterial species appear to produce homologs of PyrR and to regulate their *pyr* operons by the same or a similar mechanism.

The organization of the leader region of the *pur* operon of *B. subtilis* suggests that this operon is regulated by transcription termination/antitermination in response to changes in the availability of guanine nucleotides.

d. The S10 operon of E. coli

The 11-gene S10 ribosomal protein operon of *E. coli* contains a 172-base pair leader regulatory region which is used to achieve protein-mediated transcriptional attenuation. The S10 operon is regulated autogenously; that is, the product of one of its structural genes, protein L4, binds to the S10 leader transcript and regulates transcription termination. L4 binding also inhibits translation of coding regions of the operon. The transcript of the leader region forms multiple hairpin structures, two of which are essential for L4 activity. During transcription of the leader region RNA polymerase pauses after synthesizing one of these hairpins, a potential intrinsic terminator. Pausing at this site is enhanced *in vitro* by bound NusA protein and, most important, the pause complex is further stabilized by bound L4 protein. Enhanced stabilization of the terminator hairpin is believed to be responsible for efficient transcription termination. The leader RNA terminator structure, hairpin HE, participates in these events. An additional hairpin, HD, just preceding hairpin HE, also influences termination. How the structure HD is involved is not understood.

3. *tRNA-mediated attenuation*

a. tRNA synthetase operons of B. subtilis

In *B. subtilis* and other gram-positive bacteria, many operons encoding aminoacyl-tRNA synthetases, and some operons encoding amino acid biosynthetic enzymes, are regulated by tRNA-mediated transcriptional attenuation (Fig. 8.5). Each of these operons contains a leader region that specifies a transcript segment that can fold to form a complex set of structures, two of which are mutually exclusive and function as

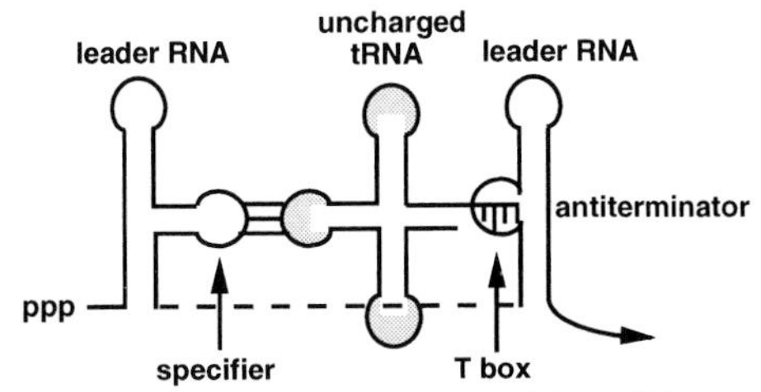

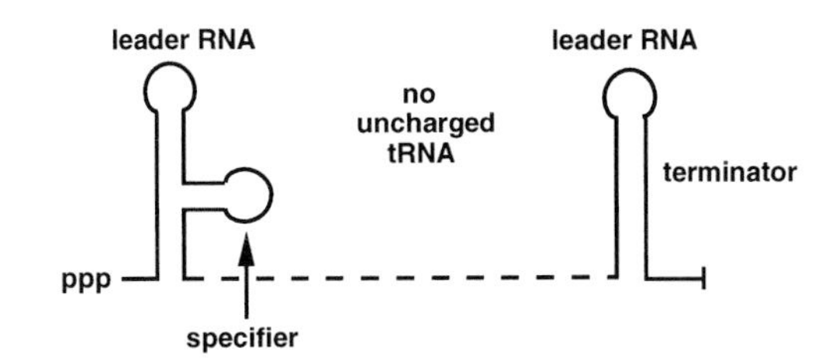

FIGURE 8.5 Uncharged tRNA-mediated transcriptional attenuation in tRNA synthetase and amino acid biosynthetic operons of *B. subtilis* and other gram-positive bacteria. When a bacterial cell is defective in charging a tRNA with the corresponding amino acid, the uncharged tRNA pairs with the leader transcript of the operon specifying the tRNA synthetase that charges that amino acid. Two segments of the tRNA are believed to be involved in RNA–RNA pairing. One segment, the anticodon of the tRNA, is thought to pair with a complementary sequence in a side bulge in the leader transcript, called the specifier. The acceptor end of the tRNA is also believed to pair with the leader transcript. Its target is a single-stranded bulge sequence in the antiterminator, called a T box. Pairing of the uncharged tRNA at these two sites is proposed to stabilize an antiterminator structure, thereby preventing formation of the terminator (top). When the relevant tRNA is mostly charged, it cannot pair with the T box sequence. The leader RNA then folds to form the terminator structure, which terminates transcription (bottom).

antiterminator and terminator. Translation is not used to choose between these alternative RNA structures. Rather, each leader RNA is designed to recognize the accumulation of an uncharged tRNA species as the signal to prevent termination. The crucial recognition sequence in leader RNA includes a single-stranded segment with a triplet codon, designated the specifier sequence (Fig. 8.5). The triplet specifier is located in a side bulge of a RNA hairpin structure. The specifier sequence is complementary to the anticodon of the tRNA that is a substrate of the tRNA synthetase that is being regulated. In amino acid biosynthetic operons regulated by this mechanism, this triplet codes for the amino acid that is synthesized by the proteins specified by the operon. A second tRNA binding site, termed a T box, located within a side bulge in the antiterminator, is complementary to nucleotides preceding the acceptor end of the tRNA (Fig. 8.3). The current regulatory model (Fig. 8.5) predicts that when an uncharged tRNA is plentiful, it binds to the specifier and T box of an appropriate leader RNA, stabilizing the antiterminator and thereby preventing terminator formation. When the tRNA is charged, its acceptor end is blocked by an amino acid and thus it cannot pair with the T box. Under these conditions, the terminator will form, resulting in transcription termination. The charged tRNA apparently is still recognized because it competes with uncharged tRNA. Switching the codon in a leader RNA can change the specificity of the response. Although the events described only concern interactions between tRNA and leader RNA, unidentified factors may also participate.

B. Regulation of termination at a factor-dependent terminator

In many bacterial species there is a second class of transcription termination sites—factor-dependent sites—at which a specific protein, Rho, interacts with RNA polymerase and causes termination. Rho-dependent termination requires an unstructured RNA segment as a site of Rho binding and a downstream RNA segment as a site of RNA polymerase pausing and termination. Accessory proteins that interact with RNA polymerase or with Rho also influence the termination process. Generally, once Rho binds to a transcript it migrates in the 3′ direction until it contacts a stalled polymerase. When it does, it can trigger the act of termination. Rho-dependent termination sites are not intrinsic terminators.

1. *N protein-mediated antitermination in phage λ*

The earliest studied and most thoroughly analyzed example of regulation by transcription termination/antitermination involves the action of the N protein of bacteriophage λ in mediating antitermination at sites of Rho-dependent termination in the phage genome. Regulation at these sites controls expression from major leftward and rightward phage promoters. N protein functions by interacting with RNA polymerase, forming an antitermination complex. This requires *cis*-acting transcript sites and sequences, called nut sites. These sites are composed of two elements, a BoxA sequence and a BoxB sequence. Box-B folds to form a hairpin loop structure. N associates with BoxB, and several host proteins associate with N and RNA polymerase in the formation of the antitermination complex. Other proteins in the complex either recognize the BoxA sequence or associate with N and the transcribing RNA polymerase complex. The fully formed N protein–RNA polymerase antitermination complex is resistant to the action of Rho. This antitermination complex can transcribe through intrinsic terminators as well as sites of Rho-dependent termination.

There are other examples, particularly in bacteriophage, in which specific viral proteins mediate antitermination events. In these instances, the mechanisms of antitermination vary somewhat from the mechanism attributed to the N protein. In addition, the ribosomal RNA operons of *E. coli* are known to be regulated by an antitermination mechanism that prevents Rho-dependent termination. This system has several features in common with N-mediated antitermination, including use of some of the same proteins and similar RNA binding sites.

2. *Translation-mediated antitermination in the* tna *operon*

Escherichia coli and other bacteria contain operons that encode enzymes that can degrade specific amino acids, making them available as carbon and/or nitrogen sources. The tryptophanase (*tna*) operon of *E. coli* is one example. This operon encodes two polypeptides—one that degrades tryptophan and another that transports tryptophan into the cell. Transcription of the structural genes of this operon is subject to regulation by transcriptional attenuation. Transcription initiation in the operon is regulated by catabolite repression. Attenuation is mediated by a mechanism that involves tryptophan-induced transcription antitermination. The anititermination process prevents Rho from terminating transcription at specific sites in the leader region of the operon. The transcript of the leader region contains a short peptide coding region, *tnaC*, which has a single Trp codon. Synthesis of the 24-residue TnaC peptide, with its crucial Trp residue and certain other key residues, is essential for antitermination. In the presence of inducing levels of tryptophan the nascent TnaC peptide is believed to act in *cis* on the ribosome engaged in synthesizing the peptide, inhibiting its release at the leader peptide stop codon. Ribosome release at this stop codon is thought to be essential for termination since release is required to expose a presumed Rho entry/binding site in the vicinity of the *tnaC* stop codon. How the inducer tryptophan is recognized is not known, nor is it known how the leader peptide interacts with the translating ribosome to block its release at the *tnaC* stop codon.

III. CONCLUSIONS

The transcriptional attenuation mechanisms described previously achieve operon-specific regulation by modifying one or more biological events that influence transcription termination. Use of these mechanisms greatly expands the regulatory capacity of each organism. An additional advantage is that these mechanisms permit a facile adjustment of the basal level of expression of an operon—expression in the absence of signals that regulate termination. Thus, variations in RNA structure, stability, or arrangement can establish an appropriate basal level of operon expression. In addition, as mentioned previously, some transcriptional attenuation mechanisms are economical because they require little unique genetic information. In eukaryotes there are several examples of regulated transcription delay with features resembling those of some of the attenuation mechanisms that were described.

BIBLIOGRAPHY

Grunberg-Manago, M. (1996). Regulation of the expression of aminoacyl-tRNA synthetases and translation factors. *In* "Transcription Attenuation in *Escherichia coli* and *Salmonella:* Cellular and Molecular Biology" (F. Neidhardt *et al.*, Eds.), pp. 1432–1457. ASM Press, Washington, DC.

Hatfield, G. W. (1996). Codon context, translational step—Times and attenuation. *In* "Regulation of Gene Expression in *E. coli*" (E. C. C. Lin and A. S. Lynch, Eds.), pp. 47–65. Landes/Chapman & Hall, Austin, TX.

Henkin, T. M. (1996). Control of transcription termination in prokaryotes. *Annu. Rev. Genet.* **30**, 35–57.

Landick, R., Turnbough, C. L., Jr., and Yanofsky, C. (1996). Transcription attenuation. *In* "*Escherichia coli* and *Salmonella*: Cellular and Molecular Biology" (F. Neidhardt *et al.*, Eds.), pp. 1263–1286. ASM Press, Washington, DC.

Platt, T. (1998). RNA structure in transcription elongation, termination, and antitermination. *In* "RNA Structure and Function," pp. 541–574. Cold Spring Harbor Laboratory Press, Cold Spring Harbor, NY.

Roberts, J. W. (1996). Transcription termination and its control. *In* "Regulation of Gene Expression in *E. coli*" (E. C. C. Lin and A. S. Lynch, Eds.), pp. 27–45. Landes/Chapman & Hall, Austin, TX.

Switzer, R. L., Turner, R. J., and Lu, Y. (1999). Regulation of the *Bacillus subtilis* pyrimidine biosynthetic operon by transcriptional attenuation: control of gene expression by an mRNA-binding protein. *Proc. Nucleic Acid Res. Mol. Biol.* **62**, 329–67.

Yanofsky, C. (2000). Transcription attenuation: once viewed as a novel regulatory strategy. *J. Bacteriol.* **182**(1), 1–8.

WEBSITE

Website for Comprehensive Microbial Resource of The Institute for Genomic Research and links to many other microbial genomic sites *http://www.tigr.org/tigr-scripts/CMR2/CMRHomePage.spl.*

9

Bacillus subtilis, Genetics

Kevin M. Devine
Trinity College, Dublin

GLOSSARY

competence Development of the ability to bind and internalize DNA from the medium.

endospore A metabolically quiescent cell that is resistant to desiccation, ultraviolet light, and other environmental insults.

forespore The cell compartment of the sporangium destined to become the spore.

integrating plasmid A plasmid that cannot replicate autonomously in a host bacterium. It can, however, establish itself by integration into the chromosome through recombination between homologous plasmid and chromosomal sequences.

mother cell The compartment of the sporangium which engulfs the forespore, synthesizes spore coat proteins, and lyses when the mature endospore is formed.

polymerase chain reaction Amplification of specific DNA sequences *in vitro* using oligonucleotide primers and thermostable DNA polymerase.

sigma factor A transcription factor which recognizes specific DNA sequences and directs RNA polymerase to initiate transcription at these sites.

SOS response A regulon that is induced to protect cells against DNA damage.

sporangium The developing bacterial cell.

sporulation The developmental process whereby the bacterial cell forms a quiescent spore.

two-component system A signal transduction system composed of a sensor kinase and a response regulator. The kinase is activated when it senses some environmental or nutritional parameter. It then activates the response regulator, which alters gene expression in a manner that allows the bacterium to respond to the prevailing conditions.

Bacillus subtilis is an endospore-forming, gram-positive, rod-shaped bacterium. Several characteristics of *B. subtilis* have attracted intense interest and therefore it has become a model system for bacterial research. It produces enzymes that are widely used in the brewing, baking, and washing powder industries.

Because its products have traditionally been used in the food industry, *B. subtilis* is classified as a GRAS organism (generally regarded as safe) and is therefore a natural choice of host for the production of heterologous proteins using recombinant DNA methodology. *Bacillus subtilis* cells become naturally competent during the transition between exponential growth and the stationary phase of the growth cycle. Competent cells have the ability to bind and internalize DNA present in the medium. Therefore, although the regulation of competence development has attracted research interest, competence development has provided the means through which *B. subtilis* can be readily genetically manipulated. This has led to the development of sophisticated molecular techniques, primarily based on integrating plasmids and transposons, for genetic analysis.

Spore formation is a developmental process whereby a vegetative cell undergoes a series of morphological

The Desk Encyclopedia of Microbiology
ISBN: 0-12-621361-5

events to become a metabolically quiescent spore. This process involves temporal and spatial regulation of gene expression and communication between the forespore and mother cell of the sporangium. These features make spore formation in *B. subtilis* an attractive model system to study development. The complete genome sequence of *B. subtilis* was published in November 1997. This knowledge has greatly expedited research efforts in this bacterium. It has also revealed that the genome encodes many genes that cannot be assigned a function. The challenge now is to determine how these genes contribute to the cellular metabolism and physiology.

I. CHARACTERISTICS OF *BACILLUS SUBTILIS*

A. Taxonomy and habitat

The genus *Bacillus* consists of gram-positive, endospore-forming, rod-shaped bacteria. There are more than 70 species, which display wide morphological and physiological diversity. Only 2 (*B. anthracis* and *B. cereus*) are known to be human pathogens. The defining feature of the genus is endospore formation. The genus is subdivided into six groups using a variety of morphological (particularly sporangial) and metabolic criteria. *Bacillus subtilis* belongs to group II, whose distinguishing features are (i) the formation of an ellipsoidal spore which does not swell the mother cell, (ii) the ability of cells to grow anaerobically with glucose as the carbon source in the presence of nitrate, and (iii) the production of acid from a variety of sugars.

The natural habitat of *B. subtilis* is the soil, but it is also found in fresh water, coastal waters, and oceans. The ubiquity of the bacterium is probably a consequence of endospore formation, which allows survival after exposure to even the most hostile environments. *Bacillus subtilis* is also associated with plants, animals, and foods and is found in animal feces. The significance of these associations is not clear. It is thought that its presence in feces is merely the result of ingestion and passage through the gut, whereas a synergistic relationship may exist with plants in which the bacterium enhances the supply of nutrients.

B. Development of competence

Bacillus subtilis cells become competent naturally. Competence is the ability to bind and internalize exogenous DNA from the medium. This capability develops during nutrient limitation when cells are in transition between exponential growth and the stationary phases of the growth cycle. Only 10% of the cell population becomes competent. The competent and non-competent fractions can be separated using renograffin gradients indicating that they are morphologically distinguishable. Competent cells can also be distinguished because they do not engage in either macromolecule or nucleotide synthesis and the SOS response is induced.

The mechanics of DNA binding and internalization have been established. DNA fragments of heterogeneous size adhere noncovalently to approximately 50 binding sites on the cell surface. DNA binding is not sequence specific. The DNA is then fragmented randomly. During internalization, one strand (chosen randomly) is degraded while the other is transported into the cytoplasm. Internalized DNA fragments are approximately 10 kilobases in size. The nature of the transforming DNA determines its fate: DNA that is homologous to the bacterial chromosome will form a heteroduplex with the chromosome leading to homologous recombination. Plasmid DNA will be established as autonomously replicating molecules.

C. Enzyme and antibiotic production

Bacillus species produce a range of enzymes and antibiotics in response to nutrient limitation. The enzymes include proteases, amylases, cellulases and lipases. Production is maximal when cells are in the stationary phase of the growth cycle. Production of these enzymes is presumably a survival strategy to scavenge macromolecular energy sources when nutrient levels are low. Many of these enzymes are widely used in the food, brewing, and biological washing powder industries. Enzymes with useful properties, such as thermostability, activity over a wide pH range, activity in detergents and oxidizing environments, have been identified in many *Bacillus* species. The role of *B. subtilis* in the enzyme industry is twofold: (i) Many *Bacillus* species are refractory to genetic analysis and *B. subtilis* is therefore the organism of choice to study the regulation of enzyme production and (ii) heterologous genes encoding enzymes with desirable properties can be cloned into *B. subtilis* strains which have been manipulated to give high product yields.

Bacillus species also produce antibiotics when cells enter the stationary phase of the growth cycle. This is probably a strategy to limit bacterial competition for the energy sources liberated through macromolecular degradation by the scavenging enzymes. *Bacillus subtilis* produces a range of peptide antibiotics, including subtilin, surfactin, bacillomycin, bacilysin, and fengycin, that display a range of antibacterial and antifungal activities. Although the synthesis of these antibiotics and their role(s) in bacterial cell physiology and survival are academically interesting, they are not of great medical importance. They are synthesized by a variety of mechanisms: For example, subtilin is a lantibiotic (contains the modified amino acid lanthionine) which is produced ribosomally, whereas surfactin is produced by the multienzyme thiotemplate mechanism. The complete genome sequence (see Section II) has revealed many of the loci encoding enzymes for antibiotic synthesis: for example, *pks* encodes a polyketide synthase, *srf* encodes surfactin synthetase, and *pps* encodes a peptide synthetase. These three loci comprise 4% of the total genome length.

D. Sporulation

Bacillus subtilis undergoes spore formation in response to carbon, nitrogen, or phosphate limitation. This process results in the formation of a metabolically quiescent cell that is resistant to desiccation, ultraviolet (UV) light, and other environmental insults. The process of sporulation involves temporal and cell type-specific regulation of gene expression, intercellular communication (between mother cell and forespore), morphological differentiation and programmed cell death (bacterial apoptosis). Such features are characteristic of more complex developmental systems. Sporulation in *B. subtilis* is therefore a simple developmental system amenable to genetic and biochemical analysis. The process requires 6–8 hr for completion and can be divided into seven stages (Fig. 9.1). At stage 0, the cell senses its environment and makes the decision to initiate sporulation. At stage II an asymmetric cell division has occurred, with the larger cell becoming the mother cell and the smaller cell the forespore. At stage III, the mother cell has completely engulfed the forespore to produce a cell within a cell. A cell type-specific program of gene expression has been established in each compartment at this stage. A series of morphological changes occur between stages IV and VI that lead to the formation of the spore cortex and spore coat. At stage VI, the developing endospore becomes resistant to heat, UV light, and desiccation, and at stage VII the mother cell lyses and releases the mature dormant spore.

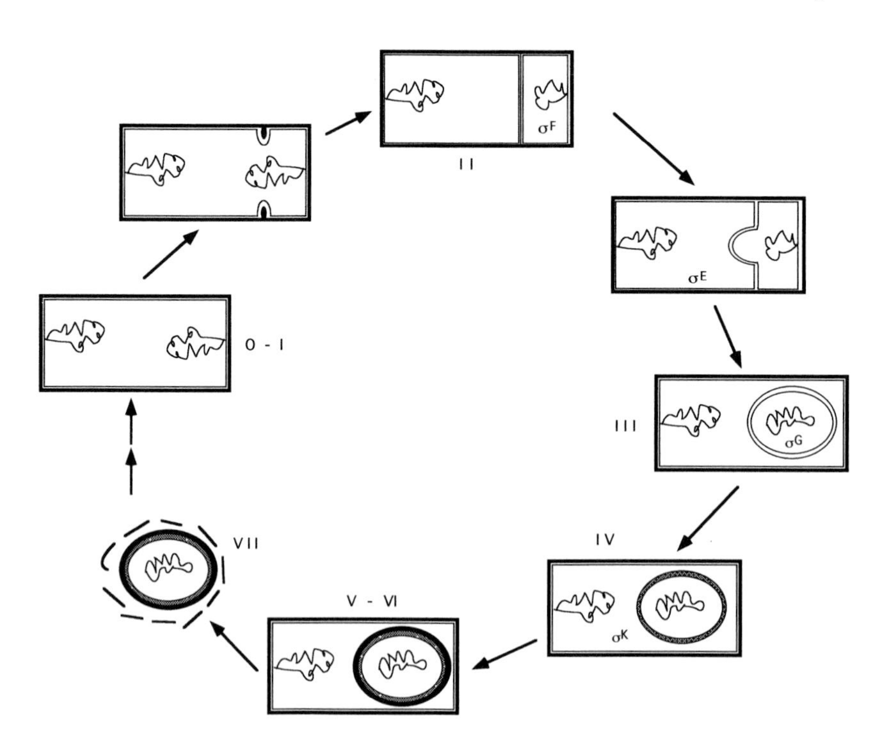

FIGURE 9.1 The morphological stages of sporulation in *Bacillus subtilis*. The decision to sporulate has occurred (stages 0 and 1) with two chromosomes (wavy circles) located at opposite poles of the cell. An asymmetric cell division occurs (stage II) with a single chromosome positioned in each compartment. At this stage, SigmaF is activated only in the forespore (smaller) compartment. During engulfment of the forespore (stages II and III), SigmaE becomes active in the mother cell (larger) compartment. SigmaG becomes active on completion of engulfment (stage III). At stage IV, SigmaK is activated in the mother cell and a layer of cortex (stippled ellipse) surrounds the developing spore. Further morphological changes occur during stages IV–VI that include deposition of a coat (dark ellipse) outside the cortex. The mother cell lyses (stage VII), releasing the mature ellipsoid spore [from Stragier and Losick (1996) with permission, from the *Annual Review of Genetics*, Volume 30, © 1996, by Annual Reviews].

II. THE COMPLETE GENOME SEQUENCE OF *B. SUBTILIS*

A. Genome organization

The complete nucleotide sequence of the *B. subtilis* genome was published in 1997. The circular genome is 4,214 kilobases in size and has an average G + C content of 43.5%. There are 10 regions which have a G + C content significantly lower than average and that correspond to known bacteriophage and bacteriophage-like elements. The origin and terminus of replication are almost perfectly diametrically opposed on the genome. The *B. subtilis* genome displays significant GT skew at third codon positions in common with many other bacteria. The leading strands have an excess of G (9%) and T (4%) over the lagging strands, and the position at which skew reversal occurs corresponds to the positions of the origin and terminus of replication. Approximately 87% of the genome is coding. More than 74% of all open reading frames and 94% of ribosomal genes are transcribed co-directionally with replication.

B. Gene composition

Fifty-three percent of genes are present in single copy. The remainder are present in multigene families, which range in size from those with 2 gene copies (568 genes are duplicated) to the ABC family of transporters that has 77 members. Multigene families present the opportunity for individual member genes to diverge and fulfill different functions and roles within the cell. In addition, individual members can have different regulatory signals so that they can be expressed under different environmental and nutritional conditions. Approximately 220 transcriptional regulators have been identified, including a family of 18 sigma factors (18 different types of promoter), a family of 34 two-component systems, 20 members of the GntR family, 19 members of the LysR family, and 12 members of the LacI family. It is evident, therefore, that *B. subtilis* has the potential to sense and respond to nutritional and environmental signals in a complex manner. This may be a reflection of the varied habitats in which *B. subtilis* can survive.

C. Gene identity

Approximately 58% of genes can be assigned an identity based either on functional analysis or extensive homology to a gene of known function. Therefore, the function of 42% of genes is unknown. This is a feature common to all genomes sequenced to date. Twelve percent of the unknown genes have homologs in other organisms. The large number of genes with unknown functions represents a formidable challenge to understanding the metabolism and physiology of *B. subtilis*. It is not clear why such a large number of genes were refractory to discovery by classical genetic analysis. It is probable that among this group are essential genes, redundant genes, and genes which participate in metabolic and physiological processes not yet discovered. Some of these questions will be resolved during the ongoing joint European–Japanese functional analysis project, the objective of which is to examine the expression of all genes of unknown function in *B. subtilis* by systematically inactivating each gene and testing the resultant mutant strain for a wide variety of phenotypes.

The intermediary metabolic pathways and the metabolic potential of a bacterium can be constructed from knowledge of the complete genome sequence. Analysis of this type shows that both the glycolytic and TCA cycles are complete and functional in *B. subtilis*, and the enzymes and regulator genes required for anaerobic growth with glucose as carbon source and nitrate as electron acceptor are also present. Anaerobic growth under these conditions has been experimentally verified.

III. GENETIC METHODOLOGY IN *B. SUBTILIS*

The knowledge of the complete genomic sequence has had a profound effect on research on *B. subtilis*. This is manifest most clearly in the accelerated pace at which research is now done. The information in the complete genomic sequence is enhanced by three additional features: (i) Polymerase chain reaction techniques allow any chromosomal fragment to be rapidly amplified, (ii) the transformation frequency of *B. subtilis* is high, and (iii) there is a sophisticated range of integrating plasmids and transposons available for use in *B. subtilis*.

Integrating plasmids are the predominant and most versatile tool for genetic manipulation of the *B. subtilis* chromosome. The essential features of an integrating plasmid are (i) the inability to replicate autonomously in *B. subtilis*, (ii) the presence of a gene for selecting plasmid establishment in *B. subtilis*, and (iii) a segment of *B. subtilis* chromosomal DNA through which the plasmid can integrate into the chromosome by homologous recombination. There are basically two types of integration events. When the transforming plasmid is circular, integration occurs by a single crossover event. When the transforming plasmid is linear, and contains two regions of homology with the chromosome, integration occurs by a double-crossover event that results in gene replacement. Incorporating additional genetic functions into the integrating plasmid can extend the repertoire of genetic manipulation. Such functions include reporter genes to generate transcriptional and translational fusions (e.g., β-galactosidase and chloramphenicol acetyl transferase to detect protein accumulation and green fluorescence protein to determine intracellular location), inducible promoters such as P_{spac} (an IPTG-inducible system based on the *lac* operon of *Escherichia coli*), and site-specific recombination functions.

The details of these systems and the mechanisms through which specific genetic manipulations can be achieved using integrating plasmids are beyond the scope of this article. However, it is useful to illustrate the range of genetic analysis that can be performed using integrating plasmids. Any gene can be mutated through either insertional inactivation or deletion. Complementation analysis and the dominance or recessivity of specific mutations can be tested. Genetic loci can be inserted into heterologous sites to test whether they function in *cis* or in *trans*. The phenotype caused by overproduction of a gene product can be assessed by gene amplification. Similarly, amplification of a control region can be used to test for titration of repressors. Large chromosomal fragments can be deleted or inverted. Strains can be constructed with multiple deletions in non-contiguous chromosomal regions. Genes and/or their control regions can be mutated *in vitro* and reinserted into homologous or heterologous sites of the chromosome in single or multiple copy. The expression profile of a gene/operon can be established by generating transcriptional and translational fusions to reporter genes. Regulation at the transcriptional and posttranscriptional levels can be distinguished. Conditional expression of any gene can be effected by placing it under the control of an inducible promoter. This is particularly useful for analysis of essential genes.

IV. GENETIC ANALYSIS OF *B. SUBTILIS*

An objective of bacterial research is to understand how individual processes are regulated and integrated within the cell. Two themes have emerged from the study of how post-exponential phenomena, such as competence development, enzyme production, and sporulation, are regulated: (i) Multiple signals detected by the cell are integrated by a signal transduction cascade which converges on a central regulator and (ii) the regulation of these processes overlaps so that entering one of these physiological states precludes activation of the other states.

A. Two-component signal transduction systems

It is imperative that bacteria adapt their gene expression and metabolism to the prevailing conditions. Two-component systems comprise a family of proteins, found ubiquitously in bacteria, which sense environmental and nutritional conditions and effect appropriate metabolic and physiological responses. They are generally (but not always) composed of two proteins: a sensor kinase and a response regulator. The kinase detects a parameter(s) of the environment that results in enzyme activation. The active kinase autophosphorylates and then transfers the phosphate to the response regulator. Phosphorylation of the response regulator activates (or alters) its transcriptional activity. Thirty-four two-component systems have been identified in *B. subtilis*, suggesting great versatility and flexibility in its response to changing environmental and nutritional conditions. Three such systems, ComP–ComA, DegS–DegU, and the unusual phosphorelay KinABC–Spo0F–Spo0B–Spo0A, are involved in regulating the post-exponential phase phenomena of competence development, enzyme synthesis, and sporulation, respectively, in *B. subtilis*.

B. Competence development

The regulation of competence development can be divided into three stages: (i) sensing the environmental and nutritional conditions which trigger the process, (ii) the signal transduction pathway that integrates the signals, and (iii) activation of the transcription factor ComK. The composition of the growth medium is an important parameter in competence development. Cells do not become competent in rich medium. In defined medium supplemented with amino acids, competence develops when cells enter the stationary phase of the growth cycle. In defined glucose-minimal medium, cells become competent during exponential growth. Cell density is a second parameter to which competence development responds. This signal is mediated by peptide factors that accumulate in the medium as cells grow to high density. Two such peptides have been identified. Competence stimulating factor (CSF) is a small peptide that is secreted from the cell after signal sequence cleavage. The secreted peptide is further proteolytically processed and a pentapeptide is reimported into the cell through the oligopeptide transport system. The second peptide, ComX, is secreted from the cell by an unknown mechanism. Accumulation of both these peptides causes an increase in the level of phosphorylated ComA (ComA~P). ComX does this by stimulating the ComP kinase that specifically phosphorylates ComA, whereas CSF is thought to inhibit the activity of a phosphatase which dephosphorylates ComA~P. Phosphorylated ComA then activates expression of *srf*, the surfactin synthetase operon, leading to increased levels of ComS, the next regulator in the signal transduction cascade. ComS is encoded by a small gene (46 codons) located entirely within the much larger *srfA* gene. The reason for this unusual gene organization is not known, but it provides a link between the post-exponential growth phase phenomena of competence development and antibiotic production. ComS destabilizes a ternary protein complex composed of MecA, ClpC, and ComK leading to release of free ComK, which can then function as a transcription factor. ComK also activates its own expression leading to very high levels of the protein, thereby further committing cells to the competent state. The ComK regulon comprises the group of genes and operons encoding the proteins required for binding, fragmentation, and uptake of DNA. The *comF* operon encodes a helicase that is involved in unwinding transforming DNA. The *comG* operon encodes proteins homologous to (i) the pilin protein and proteins involved in pilin assembly, (ii) proteins involved in pullulanase secretion in *Klebsiella pneumoniae*, and (iii) proteins encoded by the *virB* operon of *Agrobacterium tumefaciens* which function to transfer T-DNA from the bacterium to the plant. It is interesting that the transfer of DNA into *B. subtilis* cells shares features with other systems designed to transfer both DNA and proteins across cell walls and membranes.

C. Regulation of enzyme production

Production of extracellular enzymes occurs in response to nutrient limitation, and accumulation is observed when cells enter the stationary phase of the growth cycle. This is approximately the same growth period during which the cells become competent (see Sections I,B and IV,B). Although the regulatory pathways of these two physiological states overlap, it appears that enzyme production and competence development are alternate physiological states. The signals which trigger enzyme production (the nature of these signals is not precisely known) are sensed by the DegS kinase. Activation of the kinase leads to accumulation of phosphorylated DegU (DegU~P). DegU~P is a transcriptional activator which stimulates transcription of genes encoding the amylases, proteases, and glucanases produced when cells enter the stationary phase of the growth cycle. Two additional regulators, DegQ and DegR, are also required for enzyme production. DegU~P has an additional role in that it inhibits production of ComS, the regulator required for activation of the competence transcription factor ComK. Therefore, accumulation of Deg~P leads to stimulation of enzyme production and inhibition of competence development. In contrast, the nonphosphorylated form of DegU stimulates competence development. Phosphorylation of DegU therefore acts as a switch mechanism allowing cells to become competent (high levels of DegU) or to produce extracellular enzymes (high levels of DegU~P). The equilibrium between the phosphorylated and nonphosphorylated states will depend on the extent to which the kinase (which is responsive to nutritional and environmental conditions) is activated.

D. Regulation of sporulation

1. *Initiation of sporulation*

The conditions that trigger sporulation include limitation of carbon, nitrogen, and phosphorous and high cell density. These signals, and perhaps others, are sensed and integrated by a signal transduction pathway which converges on the transcriptional regulator Spo0A. The critical parameter in the decision to sporulate is the level of phosphorylated Spo0A, the form of the protein required for transcriptional activation. The nonphosphorylated form of Spo0A has no known transcriptional activity, whereas high Spo0A~P levels are required for initiation of sporulation. The commitment to sporulate is reinforced by a positive autoregulatory loop whereby Spo0A~P activates transcription of the *spo0A* gene. Spo0A is unusual among two-component transcriptional activators in that it is phosphorylated indirectly by a so-called phosphorelay (Fig. 9.2). Spo0F is phosphorylated by sensor kinases in response to nutritional and environmental conditions. The phosphate is then transferred from Spo0F to Spo0A via the Spo0B phosphotransferase. The relative cellular levels of Spo0A and Spo0A~P are the result of competing kinase and phosphatase activities. There are at least three kinases which phosphorylate Spo0F that lead to a buildup of Spo0A~P in the cell. The precise nature of the nutritional and/or environmental signals that activate the kinases is not firmly established. There are also four phosphatases that function to lower the cellular level of Spo0A~P: RapA, RapB, and RapE specifically dephosphorylate Spo0F~P, whereas the Spo0E phosphatase

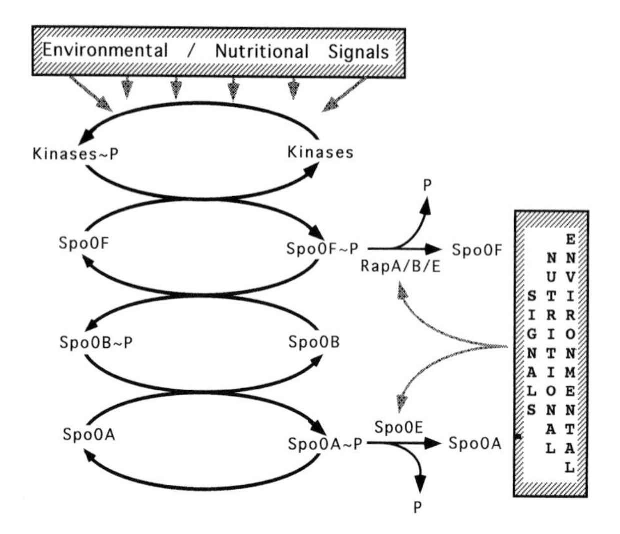

FIGURE 9.2 The phosphorelay leading to formation of Spo0A~P. A variety of environmental and nutritional conditions (stippled arrows) are sensed by sensor kinases leading to their autophosphorylation. The phosphate group is then transferred from the kinase to Spo0F (to give Spo0F~P). It is subsequently transferred from Spo0F~P to Spo0A (to give Spo0A~P) by the phosphotransferase Spo0B. Phosphate groups can be drained from the phosphorelay at two points: (i) by dephosphorylation of Spo0F~P by any of three response regulator aspartate phosphatases (Rap A/B/E) and (ii) by dephosphorylation of Spo0A~P by Spo0E phosphatase. The phosphatase activities are also responsive to a distinct group of environmental and nutritional conditions (stippled arrows). The competing actions of the kinase and phosphatase activities determine the relative cellular levels of Spo0A and Spo0A~P.

specifically dephosphorylates Spo0A~P. It appears that cell density signals can be detected through these phosphatases by a quorum-sensing mechanism. There is a gene encoding a small peptide juxtaposed to the RapA and RapE phosphatase genes called *phrA* and *phrE*, respectively. The PhrA and PhrE peptides are secreted from the medium, processed, and reimported into the cell. This results in inhibition of Spo0F~P dephosphorylation by RapA and RapE and leads to an increase in the cellular level of Spo0A~P. The genetic evidence indicates that those conditions which favor competence development signal an inhibition of sporulation. For example, the high levels of Com~P that direct competence development also lead to increased levels of RapA, which results in dephosphorylation of Spo0F. This leads to a decrease in cellular levels of Spo0A~P, thereby inhibiting sporulation. The AbrB regulator also provides a link between competence development, enzyme and antibiotic production, and sporulation. The level of AbrB varies throughout the growth cycle to ensure that cells can become competent, produce enzymes, or sporulate but cannot enter all three physiological states at the same time.

2. *Regulation of endospore development*

a. Activation of SigmaF in the forespore

Asymmetric septum formation in the sporangium is one of the first morphological events of endospore development (Fig. 9.1). At this stage (stage II) there are two complete chromosomes in the sporangium, each having been directed into one of the two compartments by a chromosome partitioning mechanism. When septum formation is complete, the fates of the two cells differ. The smaller compartment becomes the spore and the larger becomes the mother cell. Therefore, it is necessary to establish a separate program of gene expression in each compartment. The first step in this process is activation of expression of the operon encoding the transcription factor SigmaF by high levels of Spo0A~P. This operon is expressed before completion of septum formation and the SigmaF protein is therefore present in both compartments. However, it becomes active only in the forespore compartment. Three additional proteins, SpoIIAA, SpoIIAB, and SpoIIE, effect asymmetric activation of SigmaF. SpoIIAB is an anti-sigma factor that can bind either to SigmaF (making SigmaF inactive) or to SpoIIAA (allowing SigmaF to be transcriptionally active). The phosphorylation state of SpoIIAA determines whether SpoIIAB binds to SigmaF or to SpoIIAA. When SpoIIAA is phosphorylated, SpoIIAB binds to SigmaF preventing it from engaging in transcription; when SpoIIAA is not phosphorylated, it binds to SpoIIAB and SigmaF can now engage in transcription. The dephosphorylation of SpoIIAA is effected by SpoIIE, a phosphatase that is located in the asymmetric septum and dephosphorylates SpoIIAA only in the forespore (Fig. 9.3). This is a very clear example of morphological differentiation coupled with regulation of gene expression.

b. Activation of SigmaE in the mother cell

A cell-type-specific pattern of gene expression, mediated by the SigmaE transcription factor, is established in the mother cell after SigmaF has been activated in the forespore (Fig. 9.1). SigmaE protein is also synthesized in the predivisional sporangium and is therefore present in both the forespore and the mother cell compartments. However, it is activated only in the mother cell. SigmaE is activated by cleavage of a small peptide from the amino terminus of the protein. The proteolytic cleavage is effected by SpoIIGA, a membrane-localized protease (Fig. 9.3). Genetic analysis has revealed that SigmaF must be activated in the forespore before SigmaE can be activated in the mother cell. The basis of this requirement is that

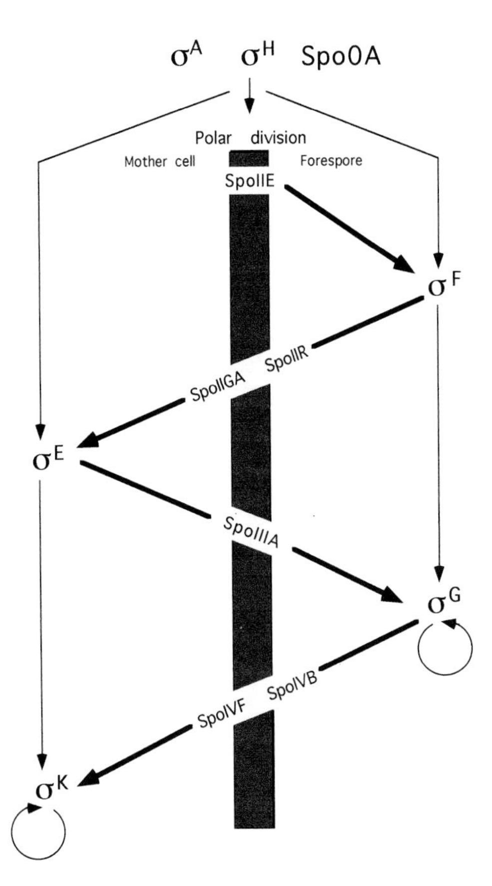

FIGURE 9.3. Crisscross regulation of compartmental gene expression during sporulation. Transcriptional dependency is indicated by thin arrows: transcription of both the sigmaE- and sigmaF- encoding genes requires SigmaA, SigmaH, and Spo0A~P. The formation of the septum (shaded rectangle) between the forespore and mother cell compartments is also dependent on these three transcription factors. SigmaF is required for transcription of the sigmaG-encoding gene in the forespore and SigmaG positively autoregulates its own expression (arrowed circle). Likewise, SigmaE is required for transcription of the sigmaK-encoding gene, which also positively regulates its own expression. A second level of control operates at the level of the activities of these factors (thick arrows). The activity of SigmaF in the forespore is dependent on septum formation and the septum-linked SpoIIE protein. The activity of SigmaE is dependent on the activity of SigmaF through the septumlinked SpoIIGA and SpoIIR proteins. The activity of SigmaG is dependent on the activities of SigmaE in the mother cell and the activity of SpoIIIA, whereas the activity of SigmaK is dependent on the activity of SigmaG in the forespore and the activities of SpoIVF and SpoIVB [from Stragier and Losick (1996) with permission, from the *Annual Review of Genetics*, Volume 30, © 1996, by Annual Reviews].

SigmaF is required to produce SpoIIR in the forespore (Fig. 9.3). SpoIIR is then secreted from the forespore into the intercompartmental space between forespore and mother cell where it binds to, and activates, the membrane-localized protease SpoIIGA. This protease then activates SigmaE, which effects the mother cell-specific program of gene expression.

c. *Activation of SigmaG and SigmaK in the forespore and mother cell, respectively*

Separate programs of gene expression are first established in the two compartments by activation of SigmaE and SigmaF. These sigma factors are then replaced by two new compartment-specific sigma factors, SigmaG in the forespore and SigmaK in the mother cell. Production of the SigmaK protein is directed by the mother cell-specific SigmaE, whereas production of SigmaG is directed by the foresporespecific SigmaF. However, activation of SigmaG and SigmaK requires a signal from the other compartment (Fig. 9.3). Activation of SigmaG in the forespore requires gene products encoded by the *spoIIIA* operon which is transcribed by SigmaE in the mother cell. Activation of SigmaK in the mother cell is similar to activation of SigmaE. SigmaK must be proteolytically processed to become active. The protease is produced in the mother cell. However, the forespore produces a product (under SigmaG control) which is secreted into the space between the forespore and mother cell that is required for activation of the protease in the mother cell.

d. *Features of the regulation of endospore formation*

The establishment of temporal and cell-type-specific programs of gene expression in the forespore and mother cell compartments of the sporangium displays many interesting features. Activation of both SigmaE and SigmaF is coupled to the morphological event of asymmetric septum formation by locating SpoIIE and SpoIIGA in the septum. Temporal regulation of gene expression is effected by sequential activation of sigma factors. The timing of sigma factor activation and the coordination of gene expression in the forespore and mother-cell compartments are controlled by so-called crisscross regulation (Fig. 9.3). Activation of SigmaF in the forespore is required before SigmaE can be activated in the mother cell; SigmaE must be activated in the mother cell before SigmaG can be activated in the forespore, and SigmaG must be activated in the forespore before SigmaK can be activated in the mother cell (Fig. 9.3). Both SigmaE and SigmaK are activated in the mother cell by two similar (but not identical) signal transduction systems. In both cases, a signal is produced and secreted from the forespore to effect sigma factor activation in the mother cell.

V. CONCLUSION

Bacillus subtilis is a very useful model organism for bacterial research. The complete genome sequence is known, it is amenable to genetic manipulation, and it exhibits many fundamental biological processes. Whole bacterial genomes can be sequenced with relative ease. However, only a small number of the bacteria are amenable to genetic manipulation. Therefore, the metabolic and physiological capabilities of these bacteria will have to be deduced from knowledge of their gene content coupled with research performed in model organisms such as *B. subtilis*. The large number of genes to which we cannot assign a function suggests that there is still much to be discovered in *B. subtilis*. It is likely, therefore, that *B. subtilis* will remain a primary focus of bacterial research.

BIBLIOGRAPHY

Fabret, C., Feher, V. A., and Hoch, J. A. (1999). Two-component signal transduction in *Bacillus subtilis*: how one organism sees its world. *J. Bacteriol.* **181**(7), 1975–1983.

Fisher, S. H. (1999). Regulation of nitrogen metabolism in *Bacillus subtilis*: vive la difference! *Mol. Microbiol.* **32**(2), 223–232.

Harwood, C., and Cutting, S. M. (Eds.) (1990). "Molecular Biological Methods for *Bacillus*." Wiley, Chichester, UK.

Kunst, F., *et al.* (1997). The complete genome sequence of *Bacillus subtilis*. *Nature* **390**, 249–256.

Sonenshein, A. L. (2000). Control of sporulation initiation in *Bacillus subtilis*. *Curr. Opin. Microbiol.*

Sonenshein, A. L., Hoch, J. A., and Losick, R. (Eds.) (2001). "*Bacillus subtilis* and Its Closest Relatives: From Genes to Cells." ASM Press.

Stragier, P., and Losick, R. (1996). Sporulation in *Bacillus subtilis*. *Annu. Rev. Genet.* **30**, 297–341.

WEBSITES

SubtiList World-Wide Web Server
http://genolist.pasteur.fr/SubtiList/

The Non-redundant *Bacillus subtilis* database
http://pbil.univ-lyon1.fr/nrsub/nrsub.html

Bacillus subtilis Japan Functional Analysis Network
http://bacillus.genome.ad.jp/

Bacillus subtilis Genetics at the University of London
http://web.rhul.ac.uk/Biological-Sciences/cutting/index.html

Micado (formerly Mad Base). A relational database on *B. subtilis* genetics (V. Biaudet, F. Samson & Ph. Bessieres)
http://locus.jouy.inra.fr/cgi-bin/genmic/madbase_home.pl

Website for Comprehensive Microbial Resource of the Institute for Genomic Research and links to many other microbial genomic sites
http://www.tigr.org/tigr-scripts/CMR2/CMRHomePage.spl.

10

Bacteriophages

Hans-Wolfgang Ackermann
Laval University

GLOSSARY

bacteriophage Virus that replicates in a bacterium; literally "eater of bacteria".
capsid Protein coat surrounding the nucleic acid of a virus.
envelope Lipoprotein membrane surrounding a virus capsid.
genome Complete set of genes in a virus or a cell; in viruses, it consists of either DNA or RNA.
host range Number and nature of organisms in which a virus or group of viruses replicate.
integrase Viral enzyme mediating the integration of viral DNA into host DNA.
prokaryote Type of cell whose DNA is not enclosed in a membrane.
restriction endonuclease Enzyme that recognizes a specific base sequence in double-stranded DNA and cuts the DNA strand at this site.
superinfection Infection of a virus-infected host by a second virus.
virion Complete infectious virus particle.

Bacteriophages, or "phages", are viruses of prokaryotes including eubacteria and archaebacteria. They were discovered and described twice, first in 1915 by the British pathologist Frederick William Twort and then in 1917 by the Canadian bacteriologist Félix Hubert d'Herelle working at the Pasteur Institute of Paris. With about 5150 isolates of known morphology, phages constitute the largest of all virus groups. Phages are tailed, cubic, filamentous, or pleomorphic and contain single-stranded or double-stranded DNA or RNA. They are classified into 13 families. Tailed phages are far more numerous than other types, are enormously diversified, and seem to be the oldest of all phage groups.

Bacteriophages occur in over 140 bacterial genera and many different habitats. Infection results in phage multiplication or the establishment of lysogenic or carrier states. Bacterial genes may be transmitted in the process. Some phages (e.g., T4, T7, λ, MS2, fd, and φX174) are famous experimental models. Phage research has led to major advances in virology, genetics, and molecular biology (concepts of lysogeny, provirus, induction, transduction, eclipse; DNA and RNA as carriers of genetic information; discovery of restriction endonucleases). Phages are used in phage typing and genetic engineering, but the high hopes set on phage therapy have generally been disappointed. In destroying valuable bacterial cultures, some phages are nuisances in the fermentation industry.

I. ISOLATION AND IDENTIFICATION OF PHAGES

A. Propagation and maintenance

1. Propagation

On solid media, phages produce clear, lysed areas in bacterial lawns or, if sufficiently diluted, small holes

The Desk Encyclopedia of Microbiology
ISBN: 0-12-621361-5

called "plaques," each of them corresponding to a single viable phage. In liquid media, phages sometimes cause complete clearing of bacterial cultures. Phages are grown on young bacteria in their logarithmic phase of growth, usually in conditions that are optimal for their host. Some phases require divalent cations (Ca^{2+}, Mg^{2+}) or other cofactors. Phages are propagated by three types of techniques: (i) in liquid media inoculated with host bacteria; (ii) on agar surfaces with a monolayer of bacteria; and (iii) in agar double layers consisting of normal bottom agar covered with a mixture of soft agar (0.3–0.9%), phages, and bacteria. Phages are harvested after a suitable incubation time, generally 3 h for liquid cultures and 18 h for solid media. Phages from agar cultures are extracted with buffer or nutrient broth. Phage suspensions, or lysates, are sterilized, best by filtration through membrane filters (0.45 μm pore size) and then titrated. Sterilization by chloroform or other chemicals is of questionable value.

2. *Storage*

No single technique is suitable for all phages. Many phages can be kept as lystes at +4°C or in lyophile, but others are quickly inactivated under these conditions. Lystes should be kept without additives such as thymol or chloroform. The best procedures seems to be preservation at −70°C in 50% glycerol. Phages may also be preserved in liquid nitrogen, by drying on filter paper, and, in the case of endospore-forming bacteria, by trapping phage genomes in spores. Ideally, any phage should be preserved by several techniques.

B. Isolation of phages

1. *Isolation from nature*

All samples must be liquid. Soil and other solid material are homogenized and suspended in an appropriate medium. Solids and bacteria are removed, usually by filtration preceded or not by centrifugation. Very rich samples can be assayed directly on indicator bacteria. In most cases, phages must be enriched by incubating the sample in a liquid medium inoculated with indicator bacteria. The culture is then filtered and titrated and phages are purified by repeated cloning of single plaques. Large samples must be concentrated before enrichment. This is done by centrifugation, filter adsorption and elution, flocculation, or precipitation by polyethylene glycol 6000. Adsorption–elution techniques may involve strongly acidic or alkaline conditions that inactivate phages.

2. *Isolation from lysogenic bacteria*

Many bacteria produce phages spontaneously. These phages may be detected by testing culture filtrates on indicator bacteria. It is generally preferable to induce phage production by mitomycin C, ultraviolet (UV) light, or other agents. A suspension of growing bacteria is exposed to the agent (e.g., 1 μg/ml of mitomycin C for 10 min or UV light for 1 min), incubated again, and then filtered. After mitomycin C induction, the bacteria should be separated from the agent by centrifugation and transferred into a fresh medium. Bacteriocins (see Section II.C), which are a source of error, are easily identified because they cannot be propagated and do not produce plaques when diluted.

C. Concentration and purification

Small samples of <100 ml are usually concentrated by ultracentrifugation (about 60 000*g* in swinging-bucket rotors), followed by several washes in buffer. Fixed-angle rotors allow considerable reduction of the *g* force because large phages sediment at as little as 10 000*g* for 1 h. Further purification may be achieved by centrifugation in a CsCl or sucrose density gradient. Large samples raise problems of contamination, aeration, and foaming. Preparation schedules are often complex: (a) pretreatment by low-speed centrifugation and/or filtration; (b) concentration, mostly by precipitation with polyethylene glycol; and (c) final purification in a density gradient or by ultracentrifugation.

D. Identification

Phase identification relies greatly on the observation that most phages are specific for their host genus; however, enterobacteria, in which polyvalent phages are common, are considered in this context as a single "genus." Phages are first examined in the electron microscope. This usually provides the family diagnosis and often indicates relationships on the species level. If no phages are known for a given host genus or only phages of different morphology, the new isolate may be considered as a new phage. If the same host genus has phages of identical morphology, they must be compared to the isolate by DNA–DNA hybridization and/or serology. Further identification may be achieved by determining restriction endonuclease cleavage patterns or constitutive proteins.

II. PHAGE TAXONOMY

A. General

D'Herelle thought that there was only one phage with many races, the *Bacteriophagum intestinale*. Early attempts at classification by serology, host range, and inactivation tests showed that phages were highly diversified, but these attempts proved premature. Modern taxonomy started in 1962 when a system of viruses based on the properties of the virion and its nucleic acid was introduced by Lwoff, Horne, and Tournier. In 1967, phages were grouped into six basic types on the basis of morphology and nature of nucleic acid. Other types were later established, and this process is likely to continue if more archaebacteria and other "unusual" microbes are investigated for the presence of phages. The International Committee on Taxonomy of Viruses presently recognizes one order, 13 families, and 31 genera in phages. Their morphology is illustrated in Fig. 10.1 and their basic characteristics and hosts are listed in Tables 10.1–10.5. The most important family criteria are type of nucleic acid, particle shape, and presence or absence of an envelope. As in other viruses, family names end in *-viridae* and genus names in *-virus*. Species are designated by the vernacular names of their best-known (or only) members (e.g., T4 or λ).

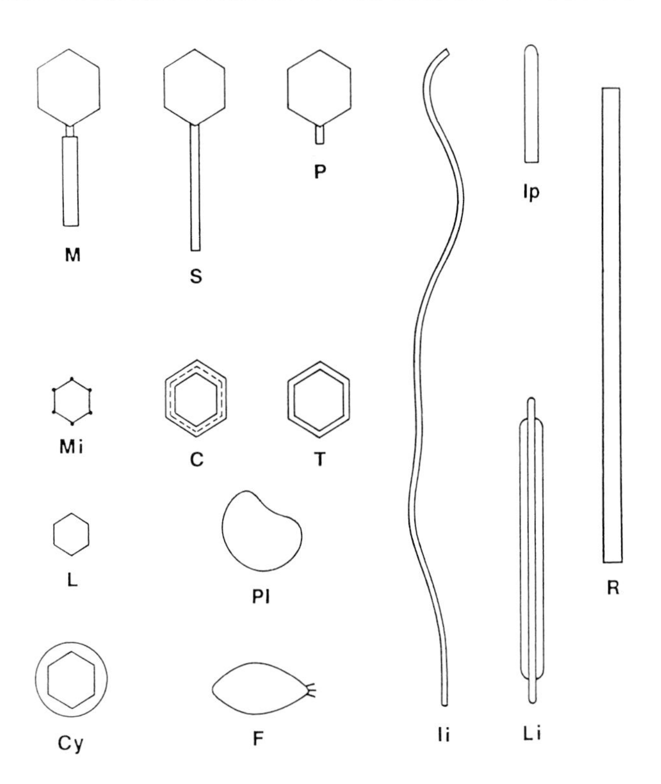

FIGURE 10.1 Morphology of phage families. C, Corticoviridae; Cy, Cystoviridae; F, Fuselloviridae; Ii, Inoviridae, *Inovirus* genus; Ip, Inoviridae, *Plectrovirus* genus; L, Leviviridae; Li, Lipothrixviridae; M, Myoviridae; Mi, Microviridae; P, Podoviridae; Pl, Plasmaviridae; R, Rudiviridae; S, Siphoviridae; T, Tectiviridae. [Modified from Ackermann, H.-W. (1987). *Microbiol. Sci.* **4**, 241–218. With permission of Blackwell Scientific Publications Ltd., Oxford, England]

B. Phage families and genera

1. *Tailed phages*

With approximately 4950 observations, tailed phages comprise 96% of phages and are the largest virus group known. They contain a single molecule of dsDNA and are characterized by a tubular protein

TABLE 10.1 Main properties and frequency of phage families[a]

Shape	Nucleic acid	Family	Genera	Particulars	Example	No. of Members[b]
Tailed	DNA, ds, L	Myoviridae	6, see text	Tail contractile	T4	1143
		Siphoviridae	6, see text	Tail long, noncontractile	λ	3011
		Podoviridae	3, see text	Tail short	T7	698
Cubic	DNA, ss, C	Microviridae	*Microvirus* *Bdellomicrovirus* *Chlamydiomicrovirus* *Spiromicrovirus*	Conspicuous capsomers	φX174	40
	DNA, ds, C, S	Corticoviridae	*Corticovirus*	Complex capsid, lipids	PM2	3?
	DNA, ds, L	Tectiviridae	*Tectivirus*	Double capsid, pseudo-tail, lipids	PRD1	18
	RNA, ss, L	Leviviridae	*Levivirus* *Allolevirirus*		MS2	39
	RNA, ds, L, M	Cystoviridae	*Cystovirus*	Envelope, lipids	φ6	1
Filamentous	DNA, ss, C	Inoviridae	*Inovirus* *Plectrovirus*	Long filaments Short rods	fd L51	57
	DNA, ds, L	Lipothrixviridae	*Lipothrixvirus*	Envelope, lipids	TTV1	6?
	DNA, ds, L	Rudiviridae	*Rudivirus*	Stiff rods, no envelope, no lipids	SIRV1	2
Pleomorphic	DNA, ds, C, S	Plasmaviridae	*Plasmavirus*	Envelope, no capsid, lipids	MVL2	6?
	DNA, ds, C, S	Fuselloviridae	*Fusellovirus*	Lemon-shaped, envelope, lipids	SSV1	8?

[a]Modified from Ackermann (1987) with permission of Blackwell Scientific Publications Ltd. C, circular; L, linear; M, multipartite; S, supercoiled; ss, single-stranded; ds, double-stranded.
[b]Exluding phage-like bacteriocins and known defective phages. Computed October 31, 2000.

TABLE 10.2 Dimensions and physicochemical properties[a]

	Virion					Nucleic acid		
Phage group or family	**Particle size (nm)**	**Tail length (nm)**	**Weight (MDa)**	**Buoyant density**	**Lipids (%)**	**Content (%)**	**Size (kbp or kb)**	**G + C (%)**
Tailed phages								
Average	63[b]	153	100	1.49	—	46	79	48
Range	38–160[b]	3–825	29–470	1.4–1.54	—	30–62	17–745	27–72
Microviridae	27	—	7	1.39	—	26	4.4–6.1	44
Corticoviridae	60	—	49	1.28	13	14	9.0	43
Tectiviridae	63	—	70	1.29	15	14	15.2	51
Leviviridae	23	—	4	1.46		30	3.5–4.3	51
Cystoviridae	75–80	—	99	1.27	20	10	13.4	56
Inoviridae								
Inovirus	760–1950 × 7	—	12–34	1.30	—	6?–21	5.8–7.3	40–60
Plectrovirus	85–250 × 15	—		1.37	—		4.4–8.3	
Lipothrixviridae	400–2400 × 38	—	33	1.25	22	3	16–42	
Rudiviridae	780–950 × 20–40	—		1.36	—		33–36	
Plasmaviridae	80	—			11		11.7	32
Fuselloviridae	85 × 60	—		1.24	10		15.5	

[a]Modified from Ackermann (1987) with permission of Blackwell Scientific Publications Ltd. Buoyant density is g/ml in Cscl; G + C, guanine–cytosine content; —, absent.
[b]Isometric heads only.

TABLE 10.3 Comparative biological properties

		Infection		Adsorption		Assembly		
Shape	**Phage group**	**Nature**	**By**	**By**	**To**	**Site**	**Start**	**Release**
Tailed	Caudovirales	V or T	DNA	Tail end	Cell wall, pili, capsule, flagella	Nucleoplasm, cell periphery	Capsid	Lysis
Isometric	Microviridae[a]	V	DNA	Spikes	Cell wall	Nucleoplasm	Capsid	Lysis
	Corticoviridae	V	DNA	Spikes	Pili	PM	Capsid	Lysis
	Tectiviridae	V	DNA	Pseudo-tail	Pili, cell wall	Nuceloplasm	Capsid	Lysis
	Leviviridae	V	RNA	A protein	Pili	Cytoplasm	RNA	Lysis
	Cystoviridae	V	Capsid	Envelope	Pili	Nuceloplasm	Capsid	Lysis
Filamentous	Inoviridae							
	Inovirus	S or T	Virion	Virus tip	Pili	PM	DNA	Extrusion
	Plectrovirus	S	Virion?	Virus tip	PM	PM	DNA?	Extrusion
	Lipothrixviridae	V or T		Virus tip	Pili			Lysis
	Rudiviridae	S	Virion	Virus tip	Pili			
Pleomorphic	Plasmaviridae	T	DNA?	Envelope	PM	PM	DNA	Budding
	Fuselloviridae	T		Spikes				Extrusion

[a]Data are for *Microvirus* genus only.
Abbreviations used: PM, plasma membrane; S, steady state; T, temperate; V, virulent.

tail, a specialized structure for the transfer of phage DNA into host bacteria. Tailed phages have recently been given order rank and the name *Caudovirales.* They fall into three families:

1. Myoviridae: phages with long complex tails consisting of a core and a contractile sheath (25% of tailed phages, six genera named after phages T4, P1, P2, Mu, SPO1, and ΦH);
2. Siphoviridae: phages with long noncontractile, more or less flexible tails (61%, six genera named after phages λ, T1, T5, L1, c2, ψM); and
3. Podoviridae: phages with short tails (15%, three genera named after T7, P22, and ϕ29).

Classification of tailed phages into genera is still in its infancy. Phage capsids, usually named heads, are icosahedra or derivatives thereof. Capsomers are rarely visible. Elongated heads are relatively rare but occur in all three families. Heads and tails vary enormously in size and may have facultative structures such as head or tail fibers, collars, base plates, or terminal spikes (Fig. 10.2). The DNA is coiled inside the head. Its composition generally reflects that of the

TABLE 10.4 Occurrence and frequency of tailed and cubic, filamentous, and pleomorphic (CFP) phages[a]

Volume and section	Bacterial group according to Bergey's Manual (Holt, 1990)	Phages		
		Tailed	*CFP*	*Total*
I				
1	Spirochetes	11		11
2	Spirilla and vibrioids	40	9	49
4	Gram-negative aerobic rods and cocci	856	22	878
5	Gram-negative facultatively anerobic rods and cocci	1080	93	1173
6	Gram-negative anaerobic rods	30		30
7	Gram-negative sulfate and sulfur reducers	2		2
8	Gram-negative anaerobic cocci	4		4
9	Rickettsias and chlamydias	2	2	4
10	Mycoplasmas	17	21?	38
11	Endosymbionts	2		2
II				
12	Gram-positive cocci	1217		1217
13	Endospore formers	625	10	635
14	Gram-positive nonsporing regular rods	286		286
15	Gram-positive nonsporing pleomorphic rods	196	1	197
16	Mycobacteria	78		78
17, 26	Nocardioforms	97		97
III				
18	Anoxygenic phototrophs	12		12
19	Cyanobacteria	44		44
20	Chemolithotrophs	2		2
21	Budding and/or appendaged bacteria	112	8	120
22	Sheathed bacteria	1		1
23	Nonfruiting gliding bacteria	32	2	34
24	Myxobacteria	16		16
25	Archaebacteria	14	18	32
IV				
28	Actinoplanetes	5		5
29	Streptomycetes	131		131
30	Maduromycetes	3		3
31	Thermomonosporae	27		27
32	Thermoactinomycetes	4		4
33	Other actinomycete genera	6		6
Total		4950	186	5139

[a]Excluding phage-like baceriocins and known defective phages; computed October 31, 2000. Based on a detailed computation published in 2001 (Ackermann, 2001).

host bacterium, but it may contain unusual bases such as 5-hydroxymethylcytosine. Genetic maps are complex and include 271 genes in phage T4 (possibly more in larger phages). Genes for related functions cluster together. Up to 40 structural proteins have been found in some phages (T4). Lipids are generally absent, but have been reported in a few exceptional cases. Response to inactivating agents is variable and no generalization is possible here. Despite the absence of lipids, about one-third of tailed phages are chloroform-sensitive, making chloroform use in phage isolation a dangerous procedure. Most properties of tailed phages appear as individual or species characteristics. Accordingly, genera have not yet been established, but about 250 species are currently recognizable, mostly on the basis of morphology, DNA–DNA hybridization, and serology.

2. *Cubic, filamentous, and pleomorphic phages*

This group includes 10 small phage families that correspond to approximately 4% of phages, differ greatly in nucleic acid nature and particle structure, and sometimes have a single member. Host ranges are mostly narrow (Table 10.4). Capsids with cubic symmetry are, with one exception, icosahedra or related bodies. Filamentous phages have, according to present knowledge, helical symmetry. Particles may or may not be enveloped. As in other viruses, the

TABLE 10.5 Host range

Bacterial division	Phage group	Bacterial group or genus
Eubacteria	Caudovirales	Any
	Microviridae	Enterobacteria, *Bdellovibrio, Chlamydia, Spiroplasma*
	Corticoviridae	*Alteromonas*
	Tectiviridae	a. Enterics, *Acinetobacter, Pseudomonas, Thermus, Vibrio*
		b. *Bacillus, Alicyclobacillus*
	Leviviridae	Enterics, *Acinetobacter, Caulobacter, Pseudomonas*
	Cystoviridae	*Pseudomonas*
	Inoviridae:	
	Inovirus	Enterics, *Pseudomonas, Thermus, Vibrio, Xanthomonas*
	Plectrovirus	*Acholeplasma, Spiroplasma*
	Plasmaviridae	*Acholeplasma, Spiroplasma*
Archaea	Caudovirales	Extreme halophiles and methanogens
	Lipothrixviridae	*Acidianus, Sulfolobus, Thermoproteus*
	Rudiviridae	*Sulfolobus*
	Fuselloviridae	*Acidianus, Sulfolobus (Methanococcus, Pyrococcus?)*

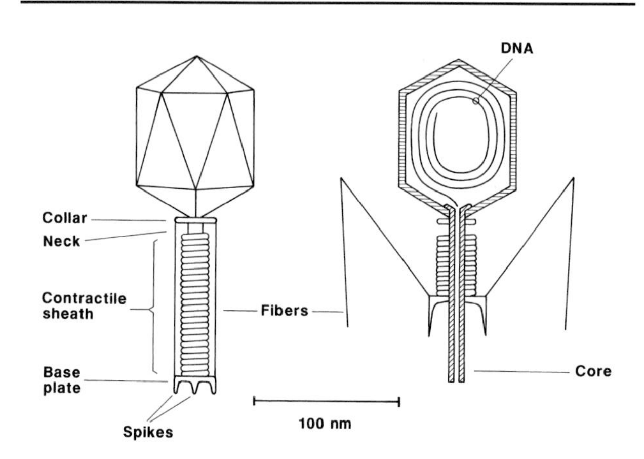

FIGURE 10.2 Schematic representation of phage T4 with extended tail and folded tail fibers (left) and sectioned with contracted tail (right). [Modified from Ackermann, H.-W. (1985). Les virus des bactéries. *In* "Virlogie médicale" (J. Maurin, ed.), p. 200. With permission of Flammarion Médecine-Sciences, Paris.]

presence of lipids is accompanied by low buoyant density and high sensitivity to chloroform and ether.

a. Cubic DNA phages

i. Microviridae The genus *Microvirus* includes the phage ϕX174 and related phages of enterobacteria and is characterized by large capsomers. Similar phages occur in so taxonomically distinct bacteria as *Bdellovibrio, Chlamydia,* and *Spiroplasma.*

ii. Corticoviridae The only certain member of the family Corticoviridae is a maritime phage, PM2. Its capsid consists of two protein shells and a lipid bilayer sandwiched in between. Two similar, little-known phages were isolated from seawater.

iii. Tectiviridae Phages are characterized by a double capsid and a unique mode of infection. The outer capsid, which is rigid and apparently proteinic, surrounds a thick, flexible lipoprotein membrane. Upon adsorption to bacteria or chloroform treatment, this inner coat becomes a tail-like tube of about 60 nm in length, obviously a nucleic acid ejection device. Tectiviruses of bacilli have apical spikes. Despite their small number, tectiviruses are found in widely different bacteria.

b. Cubic RNA phages

i. Leviviridae Leviviruses resemble enteroviruses and have no morphological features. Most of them are plasmid-specific coliphages that adsorb to F or sex pili and have been divided, by serology and other criteria, into two genera. Several not yet classified leviviruses are specific for other plasmid types (C, H, M, etc.) or occur outside of the enterobacteria family.

ii. Cystoviridae The single officially recognized member of the family Cystoviridae is unique in several ways. It is the only phage to contain dsRNA and RNA polymerase. The RNA is multipartite and consists of three molecules.

c. Filamentous phages

i. Inoviridae The Inoviridae family includes two genera with very different host ranges and similarities in replication and morphogenesis that seem to derive from the single-stranded nature of phage DNA rather than a common origin of these phages. Despite the absence of lipids, viruses are chloroform-sensitive. The *Inovirus* genus includes 42 phages that are long, rigid, or flexible filaments of variable length. They are restricted to a few related Gram-negative bacteria, sensitive to sonication, and resistant to heat. Many of them are plasmid-specific. The *Plectrovirus* genus includes 15 isolates. Phages are short, straight rods and occur in mycoplasmas only.

ii. Lipothrixviridae This family includes four viruses of the archaebacterial genus *Thermoproteus.* Particles are characterized by the combination of a lipoprotein envelope and rodlike shape.

iii. Rudiviridae This family includes two viruses of different length, isolated from the archaebacteria *Acidianus, Sulfolobus,* and *Thermoproteus.* Particles are straight rods without envelopes and closely resemble the tobacco mosaic virus.

d. Pleomorphic phages

i. Plasmaviridae Only one certain member is known, *Acholeplasma* virus MVL2 or L2. It contains dsDNA, has no capsid, and may be called a nucleprotein condensation with a lipoprotein envelope. Four similar isolates are known, but one them has been described as containing single-stranded DNA and their taxonomic status is uncertain.

ii. Fuselloviridae This family has only one certain member, SSV1, which is produced upon induction by the archaebacterium *Sulfolobus shibatae*. Particles are lemon-shaped with short spikes at one end. The coat consists of two hydrophobic proteins and host lipids and is disrupted by chloroform. SSV1 has not been propagated for absence of a suitable host. It persists in bacterial cells as a plasmid and as an integrated prophage (see Section IV.B.1). Possibly related spindle- or droplet-shaped viruses have been found in *Acidianus* and *Sulfolobus*.

C. Plasmids, episomes, and bacteriocins

Plasmids are extrachromosomal genetic elements that consist of circular or linear dsDNA and replicate independently of the host chromosome. Certain prophages behave as plasmids, but phages and plasmids are sharply differentiated: contrary to plasmids, phages have a coat and genomes of uniform size, occur free in nature, and generally lyse their hosts. The term *episome* designates both plasmids and prophages that can integrate reversibly into host DNA. Bacteriocins are antibacterial agents that are produced by bacteria, require specific receptors, and kill other bacteria. High-molecular weight or "particulate" bacteriocins are defective phages (e.g., contractile or non-contractile tails without heads). Low-molecular weight or "true" bacteriocins are a mixed group of entities, including enzymes and phage tail spikes.

D. Origin and evolution of Bacteriophages

Phages are probably polyphyletic in nature and originated at different times. This is indicated by seemingly unbridgeable fundamental differences between most phage families and by their host ranges. Phages may have derived from cell constituents that acquired a coat and became independent (e.g., leviviruses from messenger RNA (mRNA) and filamentous inoviruses from plasmids). Tailed phages are obviously phylogenetically related and may be the oldest phage group of all. Their occurrence in eubacteria and archaebacteria suggests that they appeared before their hosts diverged, thus at least 3 billion years ago. Phage groups linked to aerobic bacteria may have emerged at the same time or after the atmosphere became oxygenated by the activity of cyanobacteria. In some cases, nature repeated itself. Convergent evolution is evident in the pseudo-tails of tectiviruses and perhaps in the general resemblance of *Inovirus* and *Plectrovirus* phages.

Microviruses, tectiviruses, and leviviruses show little or no morphological differentiation, possibly because of constraints imposed by capsid size or of the relatively young geological age of these phages. Inoviruses differentiated by elongation. By contrast, tailed phages are extremely diversified and must have an eventful evolutionary history. In terms of structural simplicity and present-day frequency, the archetypal tailed phage from which the other types evolved is a *Siphovirus* with an isometric head. The diversification of tailed phages is attributed to point mutation and uniparental reproduction, which are found in all viruses, and two principal factors: modular evolution by exchange of genes or gene blocks, and the frequency of lysogeny (see Section IV.B.1), which perpetuates prophages and makes them available for recombination with superinfecting phages. Other avenues are gene rearrangement (deletions, duplications, inversions, and transpositions) and recombination with plasmids or the host genome. On the other hand, morphological properties may be highly conserved and some phages appear as living fossils, indicating phylogenetic relationships of their hosts.

III. PHAGE OCCURRENCE AND ECOLOGY

A. Distribution of phages in bacteria

Phages have been found in over 140 bacterial genera distributed all over the bacterial world: in aerobes and anaerobes, actinomycetes, archaebacteria, cyanobacteria and other phototrophs, endospore formers, appendaged, budding, gliding, and sheathed bacteria, spirochetes, mycoplasmas, and chlamydias (Table 10.4). Phage-like particles of the podovirus type have even been found in endosymbionts of paramecia. However, tailed phages reported in cultures of green algae and filamentous fungi are probably contaminants.

Most phages have been found in a few bacterial groups: enterobacteria (approximately 900 phages), bacilli, clostridia, lactococci, pseudomonads, staphylococci, and streptococci. This largely reflects the availability and ease of cultivation of these bacteria and the amount of work invested. About half of phages have been found in cultures of lysogenic bacteria. Tailed phages predominate everywhere except in mycoplasmas. In archaebacteria, they have been found in halobacteria and methanogens, but not yet in extreme thermophiles. Siphoviridae are particularly frequent in actinomycetes, coryneforms, lactococci, and streptococci. Myoviruses and podoviruses are relatively frequent in enterobacteria, pseudomonads, bacilli, and clostridia. There must be phylogenetic reasons for this particular distribution.

B. Phage ecology

1. *Habitats*

Phages have essentially the same habitats as their hosts; indeed, their most important habitat is the lysogenic bacterium because it protects prophages from the environment and frees them from the need to find new bacteria for propagation. In nature, phages occur in an extraordinary variety of habitats ranging from Icelandic solfataras to fish sauce, fetal calf serum, and cooling towers of thermal power stations. They are found on the surfaces and in normal and pathological products of humans and animals, on plants, and in food, soil, air, and water, especially sewage. Body cavities with large bacterial populations, such as intestines and rumen, are extremely rich in phages. According to their habitat, phages may be acido-, alkali-, halo-, psychro-, or thermophilic. These properties are not linked to particular phage groups but, rather, appear as individual adaptations. Psychrophilic phages are often temperature-sensitive and occur frequently in spoiled, refrigerated meat or fish.

2. *Geographical distribution*

Except for phages from extreme environments, phage species generally seem to be distributed throughout the whole earth. This is suggested by: (a) electron microscopical observations of rare and characteristical phage morphotypes in different countries; and (b) worldwide distribution of certain lactococcal phage species in dairy plants and of RNA coliphages in sewage. Unfortunately, most data are from developed countries.

3. *Frequency of phages in nature*

Sizes of phage populations are difficult to estimate because plaque assays and enrichment and (most) concentration techniques depend on bacterial hosts; they therefore only detect phages for specific bacteria and environmental conditions. Consequently, phage titers vary considerably—for example, for coliphages between 0 and 10^9 per gram in human feces and between 0 and 10^7 per milliliter in domestic sewage. Titers of actinophages in soil vary between 0 and 10^5 per gram. Purely electron microscopic phage counts, which do not allow phage identification, indicate that total phage titers are between 10^4 and 10^7 per milliliter in seawater and may attain 10^{10} per milliliter in sewage and 10^9 per milliliter in the rumen.

4. *Persistence of phages in the environment*

Phage survival in nature is frequently studied with the aim of using phages as indicators of contamination. The principal experimental models are cubic RNA phages (MS2, f2) because of their resemblance to enteroviruses, other coliphages (ϕX174, T4), and cyanophages. The indicator value of phages has not been conclusively proven and still lacks a solid statistical basis, but considerable data on phage ecology have been obtained. Phages appear as parts of complex ecosystems including various competing bacteria. Their numbers are affected by factors governing bacterial growth, notably nutrient supply and, in cyanobacteria, sunlight. The lowest bacterial concentration compatible with phage multiplication seems to be 10^4 cells/ml. In addition, phage counts are affected by association of phages with solids and colloids (e.g., clay), presence of organic matter, concentration and type of ions, pH, temperature, UV and visible light, the type of water (e.g., seawater), and nature and phage sensitivity of bacteria. Finally, phage titers depend on intrinsic phage properties such as burst size (see Section IV.A.4) and host range. No generalization is possible and each phage seems to have its own ecology.

IV. PHAGE PHYSIOLOGY

A. The lytic cycle

The lytic cycle, also called vegetative or productive, results in the production of new phages. Phages undergoing lytic cycles only are virulent. Lytic cycles consist of several steps and show considerable variation according to the type of phage (Table 10.3).

1. *Adsorption*

Phages encounter bacteria by chance and adsorb to specific receptors, generally located on the cell wall, but also on flagella, pili, capsules, or the plasma membrane. Adsorption sometimes consists of a reversible and an irreversible stage and may require cofactors (see Section I.A.1).

2. *Infection*

In most phage groups, only the viral nucleic acid enters the host and the shell remains outside. The mechanism of this step is generally poorly understood. In the *Inovirus* genus and in cystovirus φ6, the capsid penetrates the cell wall but not the plasma membrane. In phages with contractile tails, the cell wall is degraded by phage enzymes located on the tail tip. The sheath then contracts (Fig. 10.2) and the tail core is brought in contact with the plasma membrane.

3. *Multiplication*

The interval from infection to the release of new phages is called the latent period. It depends largely on the nature and physiological state of the host and varies between 20 min and 30–50 h. After infection, normal bacterial syntheses are shut off or modified. Phage nucleic acid is transcribed into mRNA using host and/or phage RNA polymerases. The RNA of leviviruses acts as mRNA and needs no transcription. In tailed phages, gene expression is largely sequential. Host syntheses are shut off first and structural genes are expressed last. According to present knowledge, replication of phage DNA and RNA is semiconservative, each strand of a double helix acting as a template for the synthesis of a complementary strand. In phages with single-stranded nucleid acid, double-stranded replicative forms are produced. In tailed phages, replication generally starts at fixed sites of the DNA molecule, is bidirectional, and generates giant DNA molecules, or concatemers, which are then cut to fit into phage heads. Translation is generally poorly know in phages. Microviridae, Leviviridae, the *Inovirus* genus, and Fuselloviridae have overlapping genes that are translated in different reading frames, allowing the synthesis of different proteins from the same DNA or RNA segment. Lipids, if present at all, are of variable origin. Phospholipids are specified or regulated by phages and fatty acids seem to derive from the host.

The assembly of new phages is called maturation. Phage constituents assemble spontaneously or with the help of specific enzymes. In most phage families, the nucleic acid enters a preformed capsid; in others, the capsid is constructed around or co-assembled with the nucleic acid. In tailed phages, assembly is a highly regulated process with sequentially acting proteins and separate pathways for heads and tails, which are finally joined together. The envelope of plasmaviruses is acquired by budding, but that of cystovirus φ6 is of cellular origin. The assembly of tailed phages often results in aberrant particles including giant or multi-tailed phages and structures consisting of polymerized head or tail protein, called polyheads, polytails, or polysheaths. Inoviruses produce particles of abnormal length. Leviviruses and some tailed phages produce intracellular crystalline inclusion bodies.

4. *Release*

Phages are liberated by lysis, extrusion, or budding. Lysis occurs in tailed and cubic phages and in the Lipothrixviridae. Bacterial cells are weakened from the inside and burst, liberating some 20–1000 phages (often 50–100). Exceptional burst sizes (up to 20 000) have been recorded in leviviruses. Extrusion, with phages being secreted through the membranes of their surviving hosts, is observed in inoviruses and fuselloviruses. Budding is found in plasmaviruses. Cells are not lysed and produce phages for hours. Progeny sizes for budding and extruded phages have been estimated at 130–1000 per cell.

B. The temperate cycle

1. *Lysogeny*

In phages called temperate, infection results in a special equilibrium between phage and host. The phage genome persists in a latent state in the host cell, replicates more or less in synchrony with it, and may be perpetuated indefinitely in this way. It behaves as a part of the bacterium. If this equilibrium breaks down, either spontaneously or after induction, phages are produced as in a lytic cycle. A bacterium harboring a latent phage genome or prophage is called lysogenic because it has acquired the ability to produce phages. Polylysogenic bacteria may carry up to five different prophages. Defective lysogeny is the perpetuation of temperate phages that are unable to replicate and often consist of single heads or tails.

Most temperate phages are tailed, but some members of the *Inovirus* genus and the Lipothrixviridae, Plasmaviridae, and Fuselloviridae can also lysogenize (Table 10.3). Lysogeny is nearly ubiquitous and occurs in eubacteria, including cyanobacteria, and in archaebacteria. Its frequency in a given bacterial species varies between 0 and 100% (often approximately 40%) according to the species, induction techniques, and number of indicator strains. Mitomycin C and UV light are the principal inducing agents (see Section I.B.2). Many others are known, notably antitumor agents, carcinogens, and mutagens. They often act by damaging host DNA or inhibiting its synthesis.

The λ type of lysogeny is particularly well understood. After infection, the genome of coliphage λ forms a circle and some λ proteins are immediately synthesized. They direct the bacterial cell to make a choice between the lytic and the temperate cycle. If a certain λ protein prevails, the λ genome integrates via a cross-over, mediated by an enzyme called "integrase," at a specific site of the host DNA. It is then replicated at every bacterial division and makes the bacterium immune against superinfection by related phages. Spontaneous or induced excision of the λ prophage leads to normal phage replication. Some phages have several integration sites.

In the P1 type of lysogeny, the phage genome, though able to integrate, usually persists as a plasmid, perhaps in association with the plasma membrane. In the Mu type, the infecting DNA does not form circles and integrates at random at any site of the bacterial genome. The core domain of Mu integrase (transposase) resembles retrovirus integrase.

2. *Pseudolysogeny and steady-state infections*

In pseudolysogenic bacteria, only part of a culture is infected with phages and an equilibrium exists between free phages and noninfected, phage-sensitive bacteria. Phage-free strains can be obtained by simple cloning or by cultivating the bacteria in antiphage serum. In steady-state infections, the whole culture is infected, but cells are not lysed and produce phages continuously (see Section IV.A.4).

3. *Transduction and conversion*

Transduction is transfer of host DNA by viruses and is normally a rare event. In generalized transduction, fragments of bacterial DNA are packaged by accident into phage heads and transferred to a new bacterium. Any host gene may be transferred and the implicated phages may be virulent or temperate. Specialized transduction is carried out by temperate phages that can integrate into host DNA (e.g., λ). If the phage DNA is not properly excised, bacterial genes adjacent to the prophage site may be packaged into phage heads along with normal genes. The resulting particle has a defective genome and may be nonviable. In conversion, bacteria acquire new properties through lysogenization by normal temperate phages. Conversion is a frequent event, affecting the whole bacterial population that has been lysogenized. The new properties are specified by phage genes and include new antigens, antibiotic resistance, colony characteristics, or toxin production (e.g., of diphtheria or botulinus toxin). They will disappear if the bacterium loses its prophage. Transduction and conversion are common in tailed phages; conversion to cholera toxin production has recently been found in the *Inovirus* genus.

V. PHAGES IN APPLIED MICROBIOLOGY

A. Therapeutic agents, reagents, and tools

1. *Therapy and prophylaxis of infectious diseases*

Phage therapy started with high hopes and was strongly advocated by d'Herelle. Phages were, enthusiastically and uncritically, applied in many human and animal diseases and spectacular results were reported as well as failures. When antibiotics became available, phage therapy was practically abandoned. The main reasons were host specificity of phages and rapid appearance of resistant bacteria. However, unbeknownst to the West, phage therapy was widely practiced in the former USSR until 1990 and surprisingly good results were reported in the 1980s from Poland. They suggest that phage therapy is a viable therapeutic alternative in antibiotic-resistant pyogenic infections (wounds, abscesses, furunculosis, osteomyelitis, septicemia). More basic research is needed on, for example, inactivation of phages by body fluids. Phage prophylaxis of infectious diseases was also attempted. Despite encouraging results in cholera prevention, it is of historic interest only. Phage control of plant diseases has not been recommended.

2. *Identification and classification of bacteria*

Early attempts to use phages for bacterial identification were abandoned because no phage lyses all strains of a bacterial species and no others. A few diagnostic phages are still used as screening agents in specialized laboratories, for example, for rapid identification of *Bacillus anthracis*, members of the genus *Salmonella*, or the biotype E1 of *Vibrio cholerae*. By contrast, phage typing is an important technique in epidemiology. In analogy with the antibiogram, bacteria are tested against a set of phages and subdivided into resistance patterns or phage types. Briefly, a continuous layer of bacteria is created on an agar surface, phage suspensions are deposited on it, and results are read the next day. Phage typing is invaluable for subdividing biochemically and serologically homogeneous bacterial species. Besides international typing schemes for *Salmonella typhi* and *Salmonella paratyphi*

B, there are typing sets for most human, animal, and plant pathogenic bacteria. Because of their host specificity, phages are also valuable tools in bacterial taxonomy. Phage host ranges were a major argument in reclassifying *Pasteurella pestis* as an enterobacterium of the genus *Yersinia*.

3. *Genetic engineering*

Phages have made many contributions to recombinant DNA technology. Restriction endonucleases were first identified in a phage-host system and the DNA ligase of phage T4 is used to insert foreign DNA into viral or plasmid vectors. In addition, phages have several major applications:

1. Phage λ, derivatives of it, cosmids (hybrids between λ DNA and plasmids), phage P1, and "mini-Mu" phages (derivatives of Mu containing the left and and right ends of the Mu genome) are used as cloning vectors. Recombinant DNA (vector plus foreign DNA) is introduced into phage proheads. After completion of phage assembly, it can be injected into bacteria.
2. Filamentous coliphages of the *Inovirus* genus are used for DNA sequencing. Foreign DNA is introduced into the double-stranded replicative form of these phages. The same coliphages are also used to express proteins or peptides at their surface. The technique, called "phage display," is a powerful tool for the selection and cloning of antibody fragments.
3. Phage Mu DNA, able to integrate at random into any gene, is used to create mutations and to displace genes to other locations.

4. *Other applications*

1. Destruction of unwanted bacteria in bacterial and cell cultures, milk, meat, and freshwater (e.g., of cyanobacteria in "algal blooms").
2. Assay of antivirals, disinfectants, air filters, and aerosol samplers.
3. Detection of fecal pollution in water and of carcinogens, mutagens, and antitumor agents.
4. Detection of *Listeria* bacteria in food using luciferase-marked phages.
5. Tracing of water movements (surface water and aquifers).

B. Phages as pests

In industrial microbiology, phages may destroy valuable starter cultures or disrupt fermentation processes. Phage interference has been reported in various branches of the fermentation industry, notably in the production of antibiotics, organic solvents, and cheese. In the dairy industry, phage infection is considered as the largest single cause of abnormal fermentations and a great source of economic losses. Phages derive from raw material, plant environment, or phage-carrying starter cultures. They are disseminated mechanically by air and may persist for months in a plant. Phage control is attempted by: (i) preventing contamination by cleanliness, sterilization of raw material, sterile maintenance of starter cultures, and use of phage-free starters; (ii) disinfection by heat, hypochlorites, UV light, and other agents; or (iii) impeding phage development by starter rotation, use of genetically heterogeneous starters, and phage-inhibiting media. A recently developed approach is to construct phage-resistant starters by genetic engineering.

BIBLIOGRAPHY

Ackermann, H.-W. (1985). Les virus des bactéries. *In* "Virologie Médicale" (J. Maurin, ed.). Flammarion Médecine-Sciences, Paris).

Ackermann, H.-W. (1987). Bacteriophage taxonomy in 1987. *Microbiol. Sci.* **4**, 214–218.

Ackermann, H.-W. (1998). Tailed bacteriophages—The order *Caudovirales*. *Adv. Virus Res.* **51**, 135–201.

Ackermann, H.-W. (2001). Frequency of morphological phage descriptions in the year 2000. *Arch. Virol.* **146**, 843–857.

Ackermann, H.-W., and DuBow, M. S. (1987). "Viruses of Prokaryotes," Vols **1** and **2**. CRC Press, Boca Raton, FL.

Calendar, R. (ed.) (1988). "The Bacteriophages," Vols **1** and **2**, Plenum Press, New York.

Casjens, S., Hatfull, G., and Hendrix, R. (1992). Evolution of the dsDNA tailed-bacteriophage genomes. *Sem. Virol.* **3**, 383–397.

Friedman, D. I., and Court, D. L. (2001). Bacteriophage Lambda: alive and well and still doing its thing. *Curr. Opin. Microbiol.* **4**, 201–207.

Goyal, S. M., Gerba, C. P., and Bitton, G. (eds.) (1987). "Phage Ecology." John Wiley & Sons, New York.

Holt, J. G. (editor-in-chief) (1984, 1986, 1989, 1990). "Bergey's Manual of Systematic Bacteriology," Vols **1–4**. Williams & Wilkins, Baltimore, MD.

Klaus, S., Krüger, D., and Meyer, J. (1992). "Bakterienviren." Gustav Fischer, Jena, 1992.

Maniloff, J., and Ackermann, H.-W. (1998). Taxonomy of bacterial viruses: establishment of tailed virus genera and the order *Caudovirales*. *Arch. Virol.* **143**, 2051–2063.

Smith, M. C. M., and Rees, C. E. D. (1999). Exploitation of bacteriophages and their components. *In* "Methods in Microbiology" (M.C.M. Smith, and R.E. Sockett, eds.) Academic Press, San Diego, Vol. **29**, 97–132.

Tidona, C. A., and Darai, G. (eds.) (2001). "The Springer Index of Viruses." Springer Verlag, Berlin.

Van Regenmortel, M. H. V., Fauquet, C. M., Bishop, D. H. L., Carstens, E., Estes, M. K., Lemon, S., Maniloff, J., Mayo, M. A., McGeoch, D. J., Pringle, C. R., and Wickner, R. B. (eds.) (2000). "Virus Taxonomy. Classification and Nomenclature of Viruses." Seventh Report of the International Committee on Taxonomy of Viruses. Academic Press, San Diego, CA.

Wilson, D. R., and Finlay, B. B. (1998). Phage display: applications, innovations, and issues in phage and host biology. *Can. J. Microbiol.* **44**, 313–329.

Zillig, W., Arnold, H. P., Holz, I., Prangishvili, D., Schweier, A., Stedman, K., She, Q., Phan, H., Garrett, R., and Kristjansson, J. K. (1998). Genetic elements in the extremely thermophilic archeon *Sulfolobus. Extremophiles* **2**, 131–140.

WEBSITES

ATCC Bacteriophage Collection Catalog
http://www.atcc.org/SearchCatalogs/bacteriophages.cfm

Bacteriophage Division of the American Society for Microbiology
http://www.asmusa.org/division/m/M.html

Phage Ecology and Phage Catalog
http://www.phage.org/

Phage Therapy
http://www.evergreen.edu/phage/

The Institute for Genomic Research (TIGR) and links to many other microbial genomic sites
http://www.tigr.org/tigr-scripts/CMR2/CMRHomePage.spl

11

Biocides (Nonpublic health, Nonagricultural antimicrobials)

Mohammad Sondossi
Weber State University

GLOSSARY

biocidal agent An agent that kills all living organisms.

biocide Primarily a chemical substance or composition used to kill microorganisms considered to be undesirable (i.e. pest organisms).

biodegradation A chemical breakdown of a substance into smaller molecules caused by microorganisms or their enzymes.

biodeterioration A physical or chemical alteration of a product, directly or indirectly caused by living organisms, their enzymes, or by-products, thereby making the product less suitable for its intended use.

biostatic agent An agent that inhibits or halts growth and multiplication of organisms. This means that when the agent is removed, the organism resumes growth and multiplication.

deteriogenic organisms The organisms that cause biodeterioration.

minimal inhibitory concentration The concentration of a particular biocide/antimicrobial agent necessary to inhibit the growth of a particular microorganism.

selective toxicity The ability of a chemotherapeutic agent (antibiotic, etc.) to kill or inhibit a microbial pathogen with minimal damage to the host at concentrations used.

The Simplest Definition of a Biocide is evident from the terminology: *bio*, meaning life, and *cide*, referring to killing–an agent that destroys life. Therefore, any word with the suffix *cide* would be classified under the category of biocides. This broad literal meaning encompasses many other topics in this and other encyclopedias. The terms herbicide and insecticide indeed have biocidal activities. Therefore, the topic biocides has to be defined more narrowly in relation to specific subjects and applications.

Perhaps the most appropriate terminology for this topic drawn from common usage is "industrial biocides." It should be accepted that scientific terms can acquire new or multiple meanings according to common usage. It is of particular importance that terminologies used are consistent and descriptive of the activity of the agents involved. It is imperative to recognize that the terminology implies a specific use and the language used for labeling biocides may have legal consequences. However, the common usage changes the meaning of the terms and even the perception of the spectrum of their activities. Table 11.1 represents some of the common terms related to the control or suppression of microbial growth. According to the Environmental Protection Agency's (EPA) Office of Pesticide Programs, antimicrobial pesticide (a term used by the EPA) products contain approximately 300 different active ingredients and are marketed in several formulations: sprays, liquids, concentrated powders, and gases. Currently, more than 8000 antimicrobial pesticide products that are registered with the EPA are sold, constituting an approximately $1 billion market. The EPA estimates that the total amount of antimicrobial pesticide active ingredients used in 1995 was 3.3 billion

The Desk Encyclopedia of Microbiology
ISBN: 0-12-621361-5

TABLE 11.1 Terminology related to control of microbial growth

Term	Definition	Comments
Microbicide	An agent that kills microbes but not necessarily their spores	Usually a general term
Germicide	An agent that kills pathogens and many non-pathogens but not necessarily their spores	A general term
Bactericide	An agent that kills bacteria but not necessarily bacterial endospores	A general term
Fungicide	An agent that kills fungi (mold and yeast) but not necessarily their spores	A general term
Algicide	An agent that kills algae	A general term
Virucide	An agent that inactivates viruses	A general term
Sporocide	An agent that kills bacterial endospores and fungal spores	Sporocidal action is not to be equated to sporistatic action which results in inhibition of spore germination
Sanitizer	An agent that reduces the microbial contaminants to safe levels as determined by public health requirements	Usually used on inanimate objects and places where no specific pathogens are present or suspected and complete killing of all forms of microorganisms is not necessary
Disinfectant	An agent, usually chemical, commonly used on inanimate objects to destroy pathogenic and harmful organisms but not necessarily their spores	Widely used term, legal definition includes more details (relative to factors of time, temperature, percentage kill, concentration, etc.); may inactivate viruses
Antiseptic	Usually a chemical agent commonly applied to living tissue, skin, or mucous membrane to kill or inhibit microorganisms	Not considered safe for internal use
Preservative	An agent that prevents or preempts biodeterioration and spoilage of a product or material under storage conditions	A chemical or physical agent or process resulting in the act of preservation
Chemical sterilizer	A chemical agent that destroys all forms of life, including spores, and inactivates viruses	Sterilization is an absolute term and there are no degrees of sterilization
Bacteriostatic agent	An agent that inhibits growth and multiplication of bacteria but does not kill them	If the bacteriostatic agent is removed bacterial growth may resume
Fungistatic agent	An agent that inhibits growth and multiplication of fungi but does not kill them	If the fungistatic agent is removed fungal growth may resume
Antimicrobial pesticide	Substance or mixture of substances used to destroy or suppress the growth of undesirable (pest) microorganisms on inanimate objects and surfaces	This is the terminology used by the EPA and it encompasses all the antimicrobial agents fitting the definition under public health and nonpublic health categories
Antibiotic	A substance, usually produced by microorganisms, which in low concentrations inhibits the growth or kills disease-causing microorganisms	Synthetic and semisynthetic antibiotics are also available, usually used in treatment of disease; has selective toxicity

lbs, accounting for 75% of all pesticides' active ingredients used. Antimicrobial pesticides, from a regulatory standpoint, are divided into public health products and nonpublic health products. The latter products are used to control microorganisms which cause spoilage, deterioration, or fouling of materials. Industrial biocides are classified under this category. They are chemical compositions used to control and prevent microbial biodeterioration and contamination of industrial/commercial material, systems, and products and/or to improve the efficiency of operation. The application of industrial biocides varies greatly depending on major use categories. Although biocides used in the preservation of cosmetics and personal care products are relevant to this article, they will not be discussed. The emphasis will be on nonpublic health-related topics. In the United States, cosmetics and toiletries are regulated under the Federal Food, Drug, and Cosmetic Act.

I. HISTORICAL PERSPECTIVE

Current uses of biocides are aimed at the inhibition and control of undesirable microorganisms based on their antimicrobial action and the microbes' potential roles in biodeterioration, spoilage, and disease. However, even before the discovery of the microbial world by Antoni van Leeuwenhoek in 1674, many practices were used to preserve material, food, and animal and human bodies (mummification). Although early methods were effective, the scientific foundation of these practices was not understood until almost two centuries after van Leeuwenhoek's discovery. Drying and salting fish and meat was one of the earliest food preservation techniques developed. In cold climates, food could also be buried in snow or underground. An early version of surface sterilization was the practice of passing metal objects through fire to clean them. One of the first documented cases of chemical sanitation was practiced as early as 450 BC in the Persian Empire, where boiled water was stored in copper and silver containers. This allowed for a portable water supply that helped the Persian army in many military conquests. Medical applications of chemical sanitation include the use of mercuric chloride as a wound dressing by Arab physicians in the Middle Ages, the use of bleaching powder as a disinfectant by Alcock in 1827, the use of iodine by Davies in 1839, the use of chlorinated water for hand washing in hospitals by Semmelweis in the 1840s, and the use of phenol on surgical wounds by Lister in the 1860s.

It was not until after the mid-nineteenth century that quantitative and comparative antimicrobial efficacies of some compounds were established. For example, in 1875 Bucholtz determined the minimal inhibitory concentrations of phenol, creosote, salicylic acid, and benzoic acid against bacterial growth. These and other findings were followed by the introduction of hydrogen peroxide as a disinfectant by Traugott in 1893 and chlorine-releasing compounds in 1915 by Dakin. It is obvious that most of these findings and their applications were in the areas of medicine and public health. Here, the description of biocides is not in the context of the general definition based on antimicrobial activity but instead covers applications in more specific areas dealing with biodeterioration. The latter term has been in use for the past 30 years and is basically not present in most dictionaries, traditional printed material, or even recent electronic media. Hueck (1968) defined biodeterioration as any undesirable change in the properties of a material used by humans caused by the vital activities of organisms in which the material is any form of matter, with the exception of living organisms. The term biodeterioration is distinguished from biodegradation in having a more negative or harmful connotation. Where the material is known to be at risk, preventative measures can be taken. These measures could be based on many physical and chemical parameters affecting microbial growth and include the use of biocides. It should be mentioned that the term biocide is also a new term introduced in the past few decades. The range of materials used by man has changed dramatically in comparison to days when raw materials were used with minimal processing. Currently, complex and heavily processed materials, composites, and synthetic and semisynthetic materials are everywhere. Complex man-made environments and the materials used, combined with a wide range of biotic and abiotic parameters, provide abundant and ideal environments for the growth of microorganisms. The total cost of losses due to biodeterioration and spoilage is approximately $100 billion per year. When the costs of replacement of deteriorated material, remedial measures, and lost productivity are considered, the importance of preventative measures and the role of biocides is clear.

II. CURRENT APPLICATIONS

There are many arbitrary categories of biocide applications in published articles and reference books. They have been grouped from an application aspect, with some differences. It should be noted that the use of a particular biocide is not restricted to any one group and may be used in different areas. One such summary of application categories could be assembled from articles, reviews, and reference materials as presented in Table 11.2.

Another grouping based on areas of application is used by regulatory agencies (the EPA and the Food and Drug Administration). These groups can be divided into two categories based on the type of microorganism (pest) against which the biocide (antimicrobial pesticide) is used. First, public health antimicrobials are intended to control infectious microorganisms (to humans) in any inanimate environment. Disinfectants, sterilizers, sanitizers, and antiseptics are included in

TABLE 11.2 Selected application areas of biocides

Human drinking water dis-infectants and purifiers	Paper and pulp
Freshwater algicide	Metalworking lubricants and hydraulic fluids
Swimming pools	Oil field operations
Animal husbandry	Fuels
Animal feed preservatives	Textiles
Food and beverage processing hygiene	Paint and paint film preservation
Food preservation	Wood preservation
Crop protection	Plastics
Hospital disinfectants	Resins
Hospital and medical antiseptics	Polymer emulsion, latex, adhesives, slurries
Pharmaceuticals	Tannery
Cosmetics	Museum specimens
Personal care disinfectants	Construction
Process cooling water	

TABLE 11.3 Antimicrobial product use sites and categories under consideration by the EPA[a]

Agricultural premises and equipment[b]
Food handling/storage establishment premises and equipment[b]
Commercial, institutional, and industrial premises and equipment
Residential and public access premises[b]
Medical premises and equipment
Human drinking water systems[b]
Materials preservatives[b]
Industrial processes and water systems
Antifouling coatings
Wood preservatives
Swimming pools
Aquatic areas

[a]Major use categories are subdivided further based on exposure scenarios.
[b]Use of a biocide product on some sites in this category with direct or indirect food contact will be considered a food use and registration must be supported by data sufficient to support the establishment of a tolerance or exemption from the requirement of a tolerance under the Federal Food, Drug, and Cosmetic Act.

this category. Second, the nonpublic health antimicrobial products are the products used to control growth of microorganisms causing deterioration, spoilage, and fouling of material, including growth of algae and odor-causing bacteria. Human exposure, product chemistry, and toxicology are considered in assigning antimicrobial agents into these main application categories.

In order to meet registration and data requirements, the EPA is currently considering classifying the antimicrobial products into 12 major use categories which are further subdivided based on exposure scenarios. The major categories are shown in Table 11.3.

III. EXAMPLES OF INDUSTRIAL BIOCIDES

Providing a complete list of registered industrial biocides is beyond the scope of this article and is not practically possible for many reasons. For example, in the United States, most states require biocide registration after registration with the EPA. In addition to Canada, European countries, and Japan, many other countries have laws regulating biocide registration and use. Toxicological and environmental impact regulations vary worldwide and are reevaluated constantly. An extensive list of biocides with current and past use in North America and Europe is given in Table 11.4, although some have been discontinued, have use limitations, and may not be registered in some countries. Therefore, for updated information on industrial biocides, contacting biocide manufacturers and regulatory agencies is strongly recommended. Databases are available that indicate the name and location of the basic manufacturers of compounds as listed by CAS registration number. Most of the regulatory statuses can also be obtained from on-line databases and search engines.

IV. CLASSIFICATION OF BIOCIDES

Classifying industrial biocides based on their chemical structures is not an easy task. Many review articles present group classifications and include a miscellaneous group whose members do not fit in any major class. Table 11.5 is a representation of this type of classification.

Biocides are sometimes also grouped based on their mode of action. This can be organized based on the target region of the microorganism affected by biocide action. Terms and categories such as membrane-active biocides and permeabilizers, cell wall inhibitors, cytotoxic agents (affecting targets in cytoplasm and interfering with metabolism and total cell function), and genotoxic agents (affecting DNA biosynthesis and reacting with DNA) have occasionally been used.

The chemical reactivity of biocides provides another, less frequently used, classification. Terms such as oxidizing, non-oxidizing, and electrophilic biocides have been used to separate industrial biocides into smaller groups. Some have used terms such as chlorine-yielding, bromine-yielding, and formaldehyde-releasing compounds to designate specific groups of biocides based on the active moiety/mechanism of action. It is therefore understandable that, depending on the subject, audience, users, and presenters (regulatory, academic, and industry) of information pertaining to industrial biocides, any of these classifications could be used.

V. EVALUATION

There are numerous methods that have been and can be employed to evaluate biocide efficacy, including a variety of basic microbiological tests, simulation tests in the laboratory, practical tests, and field tests to demonstrate the effectiveness of a biocide. First, it has to be demonstrated that the chemical or preparation being evaluated has antimicrobial activity. In this stage, the spectrum of activity is determined against bacteria (gram-positive, gram-negative, Mycobacteria, etc.), fungi (mold and yeast), and spores (bacterial and fungal) and dose-response relationships are established.

TABLE 11.4 Selected industrial biocides[a,b]

Active chemical	Applications
6-Acetoxy-2,4-dimethyl-*m*-dioxane	Metalworking fluids, textile lubricants, polymer emulsions, other aqueous emulsions
Acrolein (acrylaldehyde)	Paper and pulp
Alkenyl (C12–C18)dimethylethyl ammonium bromide	Paper and pulp, process cooling waters
Alkyldimethylbenzyl ammonium chloride	Wood, process cooling waters
Arsenic pentoxide	Wood
1-Aza-3,7-dioxa-5-ethylbicyclo-[3.3-0]octane	Metalworking fluids
1,2-Benzisothiazolin-3-one	Adhesive, latex, paper coatings, aqueous emulsions
Benzyl bromoacetate	Paint raw materials (cellulose and casein)
Benzyl-hemiformals mixture	Adhesives
Bis(1,4-bromoacetoxy)-2-butene	Paper and pulp
5,5-Bis(bromoacetoxymethyl)-*m*-dioxane	
2,6-Bis(dimethylaminomethyl) cyclohexanone	
1,2-Bis(monobromoacetoxy) ethane	
Bis(tributyltin)oxide	Wood, process cooling water
Bis(trichloromethyl)sulphone	Paper and pulp, process cooling water, adhesives, wet state protection concrete additives
Boric oxide	Wood
Brominated salicylanilides	Water-based latex pains and emulsions, joint cement, PVC plastic, acrylic and PVA water-based paints, polyvinyl acetate latex, adhesives
5,4′-Dibromosalicylanilide	
3,5,6′-Tribromosalicylanilide	
Other brominated salicylanilides	
Bromine-yielding chemicals	Process cooling waters
Sodium bromide, NaBr (must be activated by oxidizing agent, e.g. NaOCl, Cl_2, and potassium peroxymono-sulfate)	
1-Bromo-3-chloro-5,5-dimethylhydantoin	
4-Bromoacetoxymethyl-*m*-dioxolane	Paper and pulp
2-Bromo-4′-hydroxyacetophenone	Pulp and paper mills, paper making chemicals, felt
2-Bromo-2-nitro-1,3-propanediol	Metalworking fluids, textile, Process cooling waters
Bromo-nitrostyrene	Pulp and paper mills, water systems, lignosulphonates
1,1′-(2-Butenylene)bis(3,5,7-triaza-1-azoniaadamantane chloride)	Latex paints, resin emulsions, adhesives, dispersed colors
Chlorethylene bisthiocyanate	Water systems, emulsions
Chlorinated levulinic acids	Paper and pulp
Chlorine/chlorine-yielding chemicals:	Process cooling waters
Chlorine (gas), Cl	
NaOCl	
$Ca(OCl)_2{\cdot}4H_2O$	
Na dichloro-*s*-triazinetrione/trichloro-*s*-tri azinetrione	
1-Bromo-3-chloro-5,5-dimethylhydantoin	
1-Bromo-3-chloro-5-methyl-5-ethylhydantoin	
1,3-Dichloro-5,5-dimethylhydantoin	
Chlorine dioxide, ClO_2	
1-(3-Chloroallyl)-3,5,7-triaza-1-azoniaadamantane chloride + sodium bicarbonate	Adhesives, metalworking fluids, latex paints, textile, emulsions, water-based coating formulations
Chloromethyl butanethiolsulfonate	Paper and pulp
5-Chloro-2-methyl-4-isothiazolin-3-one + 2-methyl-4-isothiazolin-3-one	Wood veneer, cutting fluids and coolants, paste, slimes, cooling towers, paper, and paperboard
p-Chlorophenyl diiodomethyl sulfone	Paint
Chromic oxide	Wood
Coal tar creosote	Wood
Copper naphthenate	Wood
Copper sulfate	Wood
Cupric oxide	Wood
Copper-8-quinolinolate	Wood and wood products, glues and adhesives, paper products
Cresylic acids	Rubbers (synthetic and natural)
Cupric nitrate	Paper and pulp
Dialkyl methylbenzyl ammonium chloride	Paper and pulp, wood, process cooling waters
1,2-Dibromo-2,4-dicyanobutane	Metalworking fluids, aqueous paints, latex emulsions, joint cement adhesive

(*Continued overleaf*)

TABLE 11.4 *(Continued)*

Active chemical	Applications
2,2-Dibromo-3-nitrilopropionamide (20, 10, or 5%) in polyethylene glycol	Water cooling towers, pulp and paper mills, metalworking fluids, oil recovery
2,3-Dibromopropionaldehyde	Paper and pulp
2,4 Dichlorobenzyl alcohol	Textile
2,3-Dichloro-1,4-naphthoquinon	Toxic wash of construction material (interior use)
Didecyl dimethyl ammonium chloride	Wood
(2,2′-Dihydroxy-5,5′-dichloro)-diphenyl methane	Textile, cement additive, toxic wash (exterior use)
(2,2′-Dihydroxy-5,5′-dichloro-diphenyl monosulfide	Textile
Di-iodomethyl-*p*-tolyl sulfone	Paint
Di-isocyanate	Toxic wash construction material (interior use)
Ditmethyl aminomethyl phenol	Rubbers (synthetic and natural)
Dimethylbenzyl ammonium chloride	Construction toxic wash (exterior and interior use)
4,4-Dimethyloxazoldine + 3,4,4-trimethyloxazolidine	Metalworking fluids, in-can paint
3,5-Dimethyl-tetrahydro-1,3,5,-2*H*-thiadiazine-2-thion	Leather, paint, glue, casein, starch, paper mill
Dioctyl dimethyl ammonium chloride and ethanol	Cooling water systems
Diquat 1,1-ethylene-2,2-dipyridiylium	Construction toxic wash (exterior and interior use)
Disodium cyanodithioimidocarbamate	Paper and pulp, process cooling waters
Disodium ethylenebis(dithiocarbamate)	Paper and pulp, process cooling waters
Dithio-2,2-bis-benzmethylamide	Adhesives
Dithiocarbamates + benzimidazole derivatives	Adhesives, filters, stoppers, groutings, jointing compounds, sealants, putty
Dodecylamine salicylate	Construction toxic wash (exterior and interior use)
Dodecylguanidine hydrochloride, dodecylguanidine hydro-chloride	Paper and pulp, process cooling waters
Fatty acids of quaternary compounds	Textile
Fluorinated sulfonamide	Filters, stoppers, groutings
Formaldehyde	Toxic wash (exterior and interior use)
Glutaraldehyde (1,5-pentanedial)	Metalworking fluids, Process cooling waters
1,2,3,4,5,6-Hexachlorocyclohexane (lindane)	Wood
Hexachloro dimethyl sulfone	Industrial emulsions
Hexahydro-1,3,5-triethyl-*s*-triazine	Cutting oils, synthetic rubber latexes, adhesives, latex emulsions
Hexahydro-1,3,5-*tris*(2-hydroxyethyl)-*s*-triazine	Cutting oils and diluted coolants
Hydrogen peroxide	Process cooling waters
p-Hydroxybenzoic acid esters Ethyl *p*-hydroxybenzoate Methyl *p*-hydroxybenzoate Propyl *p*-hydroxybenzoate Butyl *p*-hydroxybenzoate	Adhesives, starch and gum solutions, inks, polishes, latexes, other emulsions
5-Hydroxymethoxymethyl-1-aza-3,7-dioxabicyclo(3.3.0)octane 5-Hydroxymethyl-L-aza-3,7-dioxabicyclo(3.3.0)octane 5-Hydroxypoly[methyleneoxy (74% C2, 21% C3, 4% C4, 1% C5)]methyl-L-aza-3,7-dioxabicyclo(3.3.0)octane	Latex paints
2-[(Hydroxymethyl) amino] ethanol	Paints, resin emulsions, in-can paint
2-[(Hydroxymethyl) amino]-2-methylpropanol	Latex paints, resin emulsions
2-(*p*-Hydroxyphenyl) glyoxylohydroximoyl chloride	Paper and pulp
2-Hydroxypropyl methanethiol sulphonate	Paper and pulp, paint films
3-Iodo-2-propynyl butyl carbamate	Interior and exterior coatings
3-Methyl-4-chlorophenol	Adhesives, filters, stoppers, groutings
Methyl-2,3-dibromopropionate	Process cooling waters
2,2′-Methylenebis(4-chlorophenol)	Textiles, rubber products, hoser
Methylene bis thiocyanate	Paper slimes, recirculating cooling water systems
N-[alpha-(nitroethyl)benzy] ethylenediamine	Paper and pulp
N-dimethyl-*N*′-fluorodichloromethylthio) sulfamide	Construction toxic wash (exterior use)
N-(fluordichloromethylthio) phthalmide	Construction toxic wash (exterior and interior use), jointing compounds, sealants, putty, plastic products
N-trichloromethylthio-4-cyclohexene-1,2-dicarboximide	Paper and pulp, polyethylene, paint, paste, rubber and rubber-coated products
N-(trichloromethylthio) phthalimide	Nonaqueous paints and caulking compounds

TABLE 11.4 *(Continued)*

Active chemical	Applications
N-(trimethylthio) phthalimide	Paint film
2-Nitrobutyl bromoacetate	Paper and pulp
4-(2-Nitrobutyl) morpholine + 4.4′-(2-ethyl-2-nitrotrimethylene) dimorpholine	Metalworking fluids, pulp and paper industry, petroleum production, jet fuels
2-*n*-Octyl-4-isothiazolin-3-one	Latex and oil-based paints, in-can paint preservative, fabrics, wet processing of hides
Organic mercurials	Paints
Organosulfur compound blends	Cooling towers, air washer systems
Organotin, quartemaries and amines	Cooling water systems
10,10′-Oxybisphenoxarsine 5% in a polymeric resin carrier	PVC, polyurethane, other polymeric compositions
10,10′-Oxybisphenoxarsine in various nonvolatile plasticizer carriers	Film and sheeting, extruded plastics, plastisols, molded goods, organosols, fabric coatings, etc.
2,2′-Oxybis-(4.4,6-trimethyl-1,3,2-dioxaborinane)-2,2′-(1-methyltrimethylenedioxy)-bis-(4-methyl-1,3,2-dioxaborinane)	Hydrocarbon fuels, boat and ship fuel and marine storage, home heating fuel
Oxyquinofine	Plastic products
Oxyquinoline sulphate	Plastic products
Ozone	Process cooling waters
Para-chloro-meta-cresol	Metalworking fluids
Para-chloro-meta xylenol	Metalworking fluids
Pentachlorophenol	Wood preservation, adhesives, cement additive, toxic wash (exterior use), rubbers (synthetic and natural)
Pentachlorophenyl laurate	Adhesives, wet state protection concrete additives, plastic products, rubbers (synthetic and natural), textile
Phenoxy fatty acid polyester	Bitumen products, jointing compounds, sealants, putty
Phenyl mercury acetate	Adhesives
Phenyl mercury nonane	Adhesives, wet state protection concrete additives
Phenyl mercury oleate	Adhesives, wet state protection concrete additives
o-Phenylphenol(sodium-*o*-phenylphenate tetrahydrate)	Protein-based paints, metalworking fluids, polishes, adhesives, gums, latexes, textiles
Polychlorophenates, alcohol, and amines	Cooling towers and evaporative condensers
Polychlorophenates, organosulfurs	Cooling towers
Poly[hydroxyethylene (dimethyliminio)-ethylene (dimethyliminio)]methylene dichloride	Industrial water systems, process cooling waters
Poly[oxyethylene (dimethyliminio)-ethylene (dimethyliminio)]-ethylene dichloride	Cooling water systems, cutting fluids
Potassium dichromate	Wood
Potassium dimethyl thiocarbamate	Metalworking fluids, process cooling water
Potassium *N*-hydroxymethyl-*N*-methyldithiocarbamate	Water-thinned colloids, emulsion reins, emulsion paints, waxes, cutting oils, adhesives
Potassium *N*-methyldithiocarbamate	Paper and pulp, process cooling waters
Quaternary phosphonium salt + surfactant	Textile
Rosin amine D-pentachlorophenate	Paper, textiles, rope, emulsion systems
Salicylamide	(Cable insulation) jointing compounds, sealants, putty, plastic products
Silver fluoride, silver nitrate	Paper and pulp
Sodium dimethyldithiocarbamate	Paper mills, cooling towers, paper and paperboard, cotton fabrics, paste, wood, veneer, cutting oils
Sodium fluoride	Wood
Sodium 2-mercaptobenzothiazole	Adhesives, wet state protection concrete additives, paper mills, cooling towers, paper and paperboard, cotton fabrics, paste, wood, veneer, cutting oil, water-thinned colloids, emulsion reins, emulsion paints, waxes, adhesives, textiles, rug backings
Sodium pentachlorophenate	Paper making, pulp, paper and paper products, leather, hides, drilling muds
Sodium 2-pyridinethiol-1-oxide	Aqueous-based metalworking fluid systems, vinyl, latex emulsions for short-term, in-can inhibition of bacterial growth
1,3,6,8-Tetraazatricyclo[6.2.1.1]dodecane	Paper and pulp
2,4,5,6-Tetrachloroisophthalonitrile	Adhesives, jointing compounds, sealants, putty, plastic products, latex paints

(Continued overleaf)

TABLE 11.4 *(Continued)*

Active chemical	Applications
2,3,4,6-Tetrachlorophenol	Wood preservation
3.3,4.4-Tetrachlorotetrahydrothiophene-1,1-dioxide	Paper and pulp
Tetrahydro-3,5-dimethyl-2*H*-1,3,5-thiadiazine-2-thione and blends	Slimicide for coatings, clay slurries, adhesives, glues, latex emulsions, casein, titanium slurries, cooling towers
2-(4-Thiazolyl)benzimidazole	Paint
2-(Thiocyanomethylthio)benzothiazole	Wood, paint films
Tributyltin acetate	Cement additive, paints
Tributyltin fluoride	Wood, antifouling paint
Tributyltin maleate	Textile
Tributyltin oxide	Construction toxic wash (exterior and interior use), wet state protection concrete additives, antifouling paints, adhesives, wood
Tributyltin oxide + nonionic emulsifier	
Trifluoromethyl thiophthalmide	Construction toxic wash (exterior and interior use), wet state protection concrete additives
3-(Trimethoxysilyl)-propyl-dimethyloctadecyl ammonium chloride	Construction toxic wash (exterior use)
Tris(hydroxymethyl) nitromethane	Oil in water emulsions, pulp and paper industry, water treatment, in-can paint
Zinc dimethyldithiocarbamate	Adhesives, cooling water, paper mill, paper and paperboard, textiles
Zinc 2-mercaptobenzothiazole	
Zinc naphthenate	Textile, wood
Zinc 2-pyridinethiol-1-oxide	Aqueous-based metalworking fluids, PVC plastics

[a]Not necessarily currently registered and some have been discontinued. This list was collected from many sources, including Sharpell (1980), Allsopp and Allsopp (1983), Bravery (1992), Rossmoore (1995a,b), Lutey (1995), Eagon (1995), McCarthy (1995), Downey (1995), and Leightley (1995).
[b]Current registration eligibility decision (RED) and fact sheets on biocides can be found on: *http://www.epa.gov/pesticides/reregistration/status.htm*.

TABLE 11.5 Biocides by chemical class[a]

Phenols	Alcohols
Organic and inorganic acids: esters and salts	Perooxygens
Aromatic diamidines	Chelating agents
Biguanides	Heavy metals and organometallic compounds
Surface-active agents: cationic and anionic agents	Anilides
Aldehydes: formaldehyde, glutaraldehyde, and others	Formaldehyde adducts
Dyes	Isothiazolones
Halogen compounds	Organosulfur compounds
Quinolines and isoquinoline derivatives	Essential oils
	Miscellaneous

[a]From Hugo and Russell (1992), Rossmoore (1995a), and others.

The nature of antimicrobial activity, biocidal or biostatic, may also be determined. The second stage includes suspension tests to determine MIC, establishing kill curves by plating, capacity tests (several reinoculations), and carrier tests (effects on organisms on the carrier). These may be followed by practical tests that, although done in the laboratory, demonstrate the efficacy under real-life conditions. Third, and most important, is evaluation in the field and under actual use conditions. For the regulatory agencies and registration purposes the tests should satisfy the label claims for specific applications. There are numerous test methods issued by federal and state governments or government-sanctioned publications, standards societies, and trade organizations as well as test methods developed by biocide manufacturers, users, and testing laboratories to demonstrate efficacy. Among governmental-sanctioned publications are those by the Association of Official Analytical Chemists, the American Public Health Association, and the United States Pharmacopeia. Voluntary consensus standards societies and groups include the American National Standards Institute, the American Society for Testing and Materials, and the International Standards Organization.

VI. MODE OF ACTION

Considering the heterogeneity of chemicals used as biocides, and the fact that they have been considered general cell poisons for a long time, one can understand the lack of detailed information on mode of action of industrial biocides. For most of the biocides, mode of action seems to be a concentration-dependent phenomenon by which individual effects can be identified and studied. Since biocides will act on organisms in an outside to inside direction, many have classified the target regions based on this directional impact. In other words, the cell wall, cell membrane and membrane-associated components, and the cytoplasmic regions will be sequentially affected by the biocide as it interacts with the intended target organism. Unlike antibiotics that have very specific targets, biocides may have more than one potential target. These could be located at any or all areas of the affected cell. The chemical structure of a biocide determines its affinity to specific targets and is the key to understanding its mode of action. Furthermore, the accumulated effects of sequentially affected regions of the cell may ultimately manifest as antimicrobial activity. Considering the structural and physiological differences of organisms and extrinsic parameters affecting the activity of biocides, the knowledge about mode of action is far from comprehensive. However, it is becoming increasingly evident that biocides indeed have a specific target(s) and cannot be labeled as general cell poisons.

Abnormal morphology of organisms exposed to biocides, studied by light and electron microscopy, has long been considered evidence of damage to the cell wall or its construction process. Lysis of cells due to initiation of autolysis has also been included in this category of mode of action. Early reports classified phenol, formaldehyde, mercurials, alcohols, and some quaternary ammonium compounds in this category. It should be noted that any of these events could also be a consequence of damage(s) exerted on cytoplasmic targets or initiated by trans-membrane signaling events. Since the cell wall composition of microorganisms (e.g., gram-positive and gram-negative bacteria, mycobacteria, fungi, and algae) differs significantly, one biocide may cause damage to the cell wall of one organism and may have no effect on the cell wall of another. Interaction with the cytoplasmic membrane, membrane-bound enzymes, electron transport, and substrate transport systems are the next group to be affected by the action of biocides. Among other biocides, chlorhexidine, 2-bromo-2-nitro-1,3-propanediol, and 1,2-benzisothiazolin-3-one have been reported to affect targets in the cytoplasmic membrane. Some membrane-active biocides may cause leakage of the intracellular material, whereas others have been reported to produce an increased permeability to ions acting as uncouplers of oxidative phosphorylation and inhibitors of active transport. Although there are many early reports that describe the mode of action of certain biocides in terms such as coagulation of cytoplasmic proteins and precipitation of cytoplasmic constituents, there are recent reports of more specific actions on selective inhibition of cytoplasmic enzymes and reactions with essential biomolecules. These more specific interactions result in inhibition of selected biosynthetic and energy-producing processes in the affected organisms.

Some recent reports describing more specific aspects of industrial biocides' modes of action include the following examples: 2-pyridinethiol-1-oxide has been suggested to act on cell membranes to eliminate important ion gradients used to store energy. In fungi, it eliminates the membrane charge gradient and interferes with nutrient transport. The collapse of delta, the pH component of the proton motive force, affects bacterial cells. It has been suggested that 2-pyridinethiol-1-oxide is not accumulated in cells and is not destroyed during action on cells but rather acts catalytically. 2-Hydroxybiphenyl ethers effectively inhibit fatty acid synthesis *in vivo* and the key enzyme of the fatty acid synthase system *in vitro*. This contradicts early reports on mode of action due to direct disruption of cell membranes. 5-Chloro-2-methyl-4-isothiazolin-3-one has been claimed to have multiple modes of action in the inhibition of microorganisms. These include lethal loss of protein thiols by covalent modification of protein molecules through direct electrophilic attack, generation of secondary electrophiles by disulfide exchange and tautomerization to a thioacyl chloride, and intracellular generation of free radicals as a result of the severe metabolic disruption.

Early studies on the mode of action of 2-bromo-2-nitro-1,3-propanediol using *Escherichia coli*, *Staphylococcus aureus*, and *Pseudomonas aeruginosa* indicated effects on cell membrane integrity and also on aerobic glucose metabolism. The major finding was the inhibition of thiol-containing enzymes such as glyceraldehyde-3-phosphate dehygrogenase. Studies on cysteine and glutathione showed the ability of 2-bromo-2-nitro-1,3-propanediol to oxidize the thiol group to form a disulfide bond and this was postulated as the inhibition mechanism. Later work confirmed this property of 2-bromo-2-nitro-1,3-propanediol, showing that it acted catalytically to oxidize thiol groups under aerobic conditions. In addition, there was evidence that 2-bromo-2-nitro-1,3-propanediol led to the formation of active oxygen species such as superoxide, suggesting interference with the electron transport mechanism within the cell.

There have also been reports which do not directly describe mode of action but rather clarify active moiety(s) involved in mode of action of certain groups of biocides. There are many industrial biocides that are synthesized with formaldehyde as one of the starting materials. The question about the role of formaldehyde in mode of action of formaldehyde-adduct

TABLE 11.6 Formaldehyde–adduct biocides with formaldehyde as active moiety in mode of action[a]

Hexahydro-1,3,5-tris(2-hydroxyethyl)-*s*-triazine
Hexahydro-1,3,5-triethyl-*s*-triazine
2-[(Hydroxymethyl) amino]-2-methylpropanol
2-[(Hydroxymethyl) amino] ethanol
4,4-Dimethyloxazolidine + 3,4,4-trimethyloxazolidine
1,3-Dihydroxymethyl)-5,5-dimethylhydantoin
5-Hydroxymethoxymethyl-L-aza-3,7-dioxabicyclo(3.3.0)octane
5-Hydroxymethyl-L-aza-3,7-dioxabicyclo(3.3.0)octane
5-Hydroxypoly[methyleneoxy(74% C2, 21% C3, 4% C4, 1% C5)] methyl-L-aza-3,7-dioxabicyclo(3.3.0)octane

[a]From Rossmoore and Sondossi (1988).

TABLE 11.7 Biocide mixtures used in metalworking and hydraulic fluids[a]

Hexahydro-1,3,5-tris(2-hydroxyethyl) triazine + 2-sodium-2-pyridinethiol-1-oxide
Sodium dimethyldithiocarbarnate + sodium 2-mercapto-benzothiazole
5-Chloro-2-methyl-4-isothiazolin-3-one + 2-methyl-4-isothiazolin-3-one + $CuSO_4$
1,3,5-Hexahydro-tris-(2-hydroxyethyl)-triazine + 5-chloro-2-methyl-4-isothiazolin-3-one
Bisoxazolidine + 5-chloro-2-methyl-4-isothiazolin-3-one
Formols + 5-chloro-2-methyl-4-isothiazolin-3-one
Dimethylolurea + formols + 5-chloro-2-methyl-4-isothiazolin-3-one
1,2-Dibromo-2,4-dicyanobutane + 5-chloro-2-methyl-4-isothiazolin-3-one

[a]From Rossmoore (1995b).

biocides was not always clearly addressed. Studies with bacterial strains resistant to formaldehyde and formaldehyde-adduct biocides derived separately from sensitive wild types and concurrent development of cross-resistance among all resistant strains established formaldehyde as the active moiety in mechanism of action of many formaldehyde-adduct biocides. Subsequently, it was shown that high levels of resistance were coupled to high levels of formaldehyde dehydrogenase in resistant cells. Although this does not resolve the problem of mode of action of formaldehyde, it consolidates the mechanism of action question and clearly demonstrates the involvement of formaldehyde as the active moiety in mode of action of many formaldehyde-adduct biocides. Table 11.6 shows some of these formaldehyde-adduct industrial biocides. There are many other adducts that are not included, although partial formaldehyde involvement in mode of action has been suggested.

VII. COMBINATION BIOCIDES (MIXTURES AND FORMULATIONS)

Biocides are used in combination for many reasons: (i) to broaden the antimicrobial spectrum; (ii) to minimize physical and chemical incompatibilities; (iii) to minimize toxicity; and (iv) to produce biochemical synergism. In its broadest sense, the subject of interactions among industrial biocides that alter their biological activities would include interactions in the extracellular environment and those within the target organisms. Gathering data on combined modes of action of biocide mixtures is not an easy task. Although many biocide mixtures have been used and are commercially available, a considerable amount of research will be required to define the biochemical nature of these interactions and their physiological effects on microorganisms. A select number of biocide mixtures primarily used in metalworking and hydraulic fluids are listed in Table 11.7. It should be noted that there are many more mixtures with a variety of applications and different active ingredients.

Traditionally, a toxicological interaction has been described as "a condition in which exposure to two or more chemicals results in a quantitatively or qualitatively altered biological response relative to that predicted from the action of a single chemical. Such multiple-chemical exposures may occur simultaneously or sequentially in time and the altered response may be greater or lesser in magnitude" (Murphy, 1980). It has been common practice to classify the quantitative joint action of chemicals including biocides using three general terms:

Addition: when the toxic effect produced by two or more biocides in combination is equivalent to that expected by simple summation of their individual effects.
Antagonism: when the effect of a combination is less than the sum of the individual effects.
Synergism: when the effect of the combination is greater than would be predicted by summation of the individual effects.

For biocides, there must be a quantitatively definable effect for each compound involved. Minimal inhibitory concentrations and other dose–response information could be used to construct a graphic representation (isobologram) to show additive, synergistic, and antagonistic interactions in biocide mixtures. This could easily be applied to combinations of any number of agents. For a combination of more than three agents, no graphic construction is possible; however, the interaction index could be calculated mathematically to describe additivity, antagonism, and synergism.

Because development costs of new biocides are estimated between $10 and $15 million, the introduction of new biocidal compounds to the market is difficult. This cost escalation is due to increasing legislative requirements and concerns regarding environmental impact. Therefore, the use of registered biocides in combinations that yield synergistic activity is an attractive alternative.

There is an extensive range of biocides from which to formulate mixtures, with substantially reduced initial screening costs and possibly an easier registration process if the active ingredients are well known and individually registered for particular end use(s). However, there are concerns regarding the toxicity and environmental impact of biocide mixtures.

VIII. RESTRICTIONS ON USE AND REGULATION

The fundamental requirements for industrial biocides suitable for protection of material in the early days of use were effective and aggressive antimicrobial activities, broad spectrum of activity, stability and persistence, and economical feasibility. With the constant expansion of biocide use and number, there has been increasing concern about their impact on human and environmental health. In approximately the past two decades, regulatory agencies have put in place restrictions for the application and selection of industrial biocides. The new requirements include spectrum of activity and effectiveness according to the category of application, stability relevant to application, very low human toxicity according to required toxicological data, very low environmental impact (ecotoxicity), and economical feasibility of use. This has produced a stream of new regulations.

The EPA registers and regulates antimicrobial pesticides including industrial biocides under the Federal Insecticide, Fungicide and Rodenticide Act (FIFRA). To register an industrial biocide, manufacturers of such products must meet EPA requirements to show that: (i) the product will not cause unreasonable adverse effects on human health and the environment; and (ii) the product labeling and composition comply with requirements of FIFRA. Since 1996, the antimicrobials division within the Office of Pesticide Programs (OPP) has been responsible for all activities related to regulating antimicrobial pesticides. The OPP reviews submitted detailed and specific information on the chemical composition of the product, efficacy data against specific intended microorganisms, support of directions for use on the label, appropriate labeling for safe and effective use, and extensive toxicological data and hazards associated with the product use.

Title 40 CFR, Part 158, explicitly outlines the data requirements for antimicrobial pesticides. Further amendments are being considered to Part 158. These data requirements are for tiered human health and exposure data requirements for nonfood uses, product chemistry, and toxicology. Explicit tiered testing approaches for environmental fate and effects have been developed for antimicrobials, wood preservatives, antifoulants, and algicides. Specifically, data are required for end-use antimicrobial products, including data on end-use formulation, active ingredient, product chemistry information, residue chemistry, efficacy, toxicity, environmental fate, and ecotoxicity. Data are required to assess acute toxicity, chronic and subchronic toxicity, developmental toxicity, reproductive toxicity, mutagenicity, neurotoxicity, metabolic effects, and immunological effects. Toxicology test requirements are set out in tiers based on general requirements and risk assessment. It should be noted that the EPA may require additional data on a case-by-case basis in order to conduct a risk assessment for the product.

To ensure that biocides meet current scientific and regulatory standards, EPA is reviewing older pesticides, including biocides, under FIFRA. This process is called reregistration and considers human health and ecological effects of pesticides. Biocides registered prior to November 1984 are being processed first. Numerous documents containing information on biocides could be found on EPA web sites.

IX. PROBLEMS ASSOCIATED WITH BIOCIDE USE

It should be kept in mind that biocides are all toxic by definition and most are also corrosive. For decades there have been concerns regarding toxicity issues of biocides, even though biocides must be registered for specific use and require a battery of data submissions which include human and environmental toxicological profiles. Recognition for the need to control risks from biocides has come from scientists in academia, regulatory agencies, and industry, and there is a need for comprehensive information and new data especially on the biocide mixtures and biocide-containing formulations. One of the most important problems based on toxicity concerns is the international recognition of the registration of biocides. Acceptance of toxicity data, including ecotoxicity of biocides, by the regulatory bodies in different countries has been of great concern to multinational producers of biocides.

In addition to the regulatory concerns associated with biocide use, there are problems beyond the obvious toxicity of biocides that have to be considered. Like antibiotics, when biocides interact with mixed populations of microorganisms they kill all susceptible organisms and promote the selection of resistant populations. The level of resistance may be intrinsic, developed by mutation, or could even be acquired by gene exchange. It should be very clear from the previous discussion that the ultimate result of biocide use is selection of resistant populations. Although resistance is a relative term, a resistant organism is one that is not affected by biocide concentrations used to control microorganisms regularly found in a system. There are many review articles on the development of resistance to antimicrobial agents (including antibiotics). These reviews and numerous other research articles have classified organisms according to their intrinsic levels of resistance to antimicrobials, mechanism(s) of resistance development, and concepts such as phenotypic and genotypic resistance to antimicrobial agents. Regarding intrinsic resistance to biocides, in general, it can be stated that gram-negative bacteria are more resistant to biocides than are gram-positive bacteria. In addition to this generalization, there are specific examples of intrinsic resistance of microorganisms, such as mycobacterial, peudomonad, and fungal species, to biocides.

Microorganisms may also gain the capacity to resist the biocide by the acquisition of gene function(s). These gene functions are mostly concerned with inactivation or modification of the biocide, efflux systems, specification of a new target, or enzymatic modification of the target. Microorganisms could also simply persist in the presence of the biocide. This phenomenon may result from mutation or temporary resistance due to gene regulatory events or phenotypic changes. It is accepted that the general resistance mechanisms producing biocide resistance in microorganisms are the same mechanisms found in antibiotic resistance. Excessive use of biocides may produce organisms with a non-specific mechanism(s) of cross-resistance to other biocides and, most important, to antibiotics (double resistance to biocides and to antibiotics).

Resistance development to industrial biocides has received much attention, and numerous published research articles and reviews have been devoted to this subject. There should be no doubt that the inappropriate and excessive use of biocides often results in selection of resistant populations. This usually includes unintentional under-dosing and, more important, misunderstanding of the kinetics of biocide effects and the dynamics of the system treated with biocides.

With regard to the resistance categories mentioned previously, there are some points worth noting. When the mechanism of resistance involves inactivation or modification of a biocide and the organism(s) becomes the dominant microbial population in the biocide-treated system, other less susceptible populations may also be protected. There have been reports indicating survival of biocide-sensitive organisms in the presence of resistant populations in metalworking fluids.

The addition of biocides (intermittent slug dosing or continuous addition with pumps) to a system may result in an unintended biocide buildup in which the extremely hostile toxic environment kills virtually all the microbial populations usually found in systems treated with recommended doses of biocide (gram-negative bacteria, especially *Pseudomonas* species). This results in an environment, although hostile, with little or no competition. Microbial populations which could tolerate these conditions will eventually colonize this environment and flourish. There are reports indicating the isolation of unusual organisms in the presence of biocide several times in excess of the recommended dose. The evidence strongly suggests that these organisms have an extremely low permeability to hydrophilic biocides used in the system.

This scenario may be involved in recent outbreaks of hypersensitivity pneumonitis associated with exposure to metalworking fluid aerosols. The evidence suggests that hypersensitivity pneumonitis has occurred where "atypical" flora have predominated in metalworking fluids. Although the microbiological origin of hypersensitivity pneumonitis is strongly suspected, the involvement of biocides and other constituents of the fluids has not been excluded. There have been many reports on allergic contact dermatitis among workers exposed to biocide-treated components of industrial systems and contact sensitization to products containing biocides at the consumer end. These types of reports include most of the frequently used industrial biocides.

X. RECENT DEVELOPMENTS AND CONCERNS OVER RESISTANCE DEVELOPMENT

In July 2001, the OPP of the EPA announced purely voluntary pesticide resistance management labeling guidelines based on mode/target site of action for agriculturally used herbicides, fungicides, bactericides, and insecticides. This, Pesticide Registration Notice to registrants, concerns pesticide products intended for general agricultural use at this time. One could envision similar notices for all biocides in the future. This possibility becomes even more realistic if new studies establish the existence of cross-resistance between biocides and antibiotics.

This aspect of cross-resistance could be approached considering two scenarios: (a) antibiotic resistant organism is also resistant to biocides; and (b) biocide resistant organism also exhibits cross-resistance to antibiotics. At this time, there is no evidence that antibiotic resistant microorganisms are more resistant to biocides, excluding the intrinsic resistance characteristics of some organisms. There have been reports studying the susceptibility of antibiotic resistant bacteria to disinfectants that suggest the above conclusion.

With the ever increasing use of antimicrobials in consumer products and biocides to control biodeterioration, there is a recent but growing and legitimate concern over cross-resistance of biocide resistant organisms to antibiotics. It is possible that selection of microorganisms with extremely low permeability as a result of biocide treatment may indeed produce populations less accessible to antibiotics. Although similarities between the mechanisms of resistance development to biocides and antibiotics exist, efflux pumps, thickening of the cell wall, outer membrane alterations in gram-negative bacteria, it has not yet been established that biocide use will result in selection of an antibiotic resistance trait in the same organisms.

XI. HANDLING OF BIOCIDES

It should be clear that "biocides" are poisons and are designed to kill living cells. Therefore, direct contact with concentrated biocides (undiluted product) should be avoided. The Material Safety Data Sheet (MSDS) for a biocide should be studied carefully and appropriate personal protective equipment (protective clothing, glasses or face shield, gloves, and other protective material) should be used when handling biocides. If there is any likelihood of exposure through inhalation, then respiratory protection should be included as protective equipment. An emergency action plan should be in place to deal with accidental spill. This should include the decontamination and deactivation of small spills based on recommendations of the biocide manufacturer and in case of major spills, local authorities should be contacted immediately.

Since biocides are considered hazardous materials, disposal of biocides should be done according to regulations and through licensed disposal contractors. Biocide concentrates should not be allowed to enter surface waters, ground waters, wastewater treatment systems or the environment in general.

BIBLIOGRAPHY

Allsopp, C., and Allsopp, D. (1983). An updated survey of commercial products used to protect material against biodeterioration. *Int. Biodeterioration Bull.* **19**, 99–145.

Block, S. (ed.) (1991). "Disinfection, Sterilization, and Preservation," 4th ed. Leo & Feiger, Philadelphia, PA.

Bravery, A. F. (1992). Preservation in construction industry. In "Principles and Practice of Disinfection, Preservation and Sterilization" (A. D. Russell, W. B. Hugo, and G. A. J. Ayliffe, eds.), 2nd ed., pp. 437–458. Blackwell, London.

Denyer, S. P. (1990). Mechanisms of action of biocides. *Int. Biodeterioration* **29**, 89–100.

Downey, A. (1995). The use of biocides in paint preservation. *In* "Handbook of Biocide and Preservative Use" (H. W. Rossmoore, ed.), pp. 254–266. Chapman & Hall, New York.

Eagon, R. G. (1995). Paper, pulp and food grade paper. *In* "Handbook of Biocide and Preservative Use" (H. W. Rossmoore, ed.), pp. 83–95. Chapman & Hall, New York.

Hueck, H. J. (1968). The biodeterioration of materials—an appraisal. *In* "Biodeterioration of Materials 6–12." Elsevier, London.

Hugo, W. B. (1992). Historical introduction. *In* "Principles and Practice of Disinfection, Preservation and Sterilization" (A. D. Russell, W. B. Hugo, and G. A. J. Ayliffe, eds.), 2nd ed., pp. 3–6. Blackwell, London.

Hugo, H. W., and Russell, A. D. (1992). Types of antimicrobial agents. *In* "Principles and Practice of Disinfection, Preservation and Sterilization" (A. D. Russell, W. B. Hugo, and G. A. J. Ayliffe, eds.), 2nd ed., pp. 7–88. Blackwell, London.

Leightley, L. (1995). Biocide use in wood preservation. *In* "Handbook of Biocide and Preservative Use" (H. W. Rossmoore, ed.), pp. 283–301. Chapman & Hall, New York.

Lutey, R. W. (1995). Process cooling water. *In* "Handbook of Biocide and Preservative Use" (H. W. Rossmoore, ed.), pp. 50–76. Chapman & Hall, New York.

McCarthy, B. J. (1995). Biocides for use in the textile industry. *In* "Handbook of Biocide and Preservative Use" (H. W. Rossmoore, ed.), pp. 238–253. Chapman & Hall, New York.

Murphy, S. D. (1980). Assessment of the potential for toxic interactions among environmental pollutants. *In* "The Principles and Methods in Modern Toxicology" (C. L. Galli, S. D. Murphy, and R. Paoletti, eds.), p. 277. Elsevier/North-Holland Biomedical Press.

Murtough, S. M., Hiom, S. J., Palmer, M., and Russell, A. D. (2001). Biocide rotation in the healthcare setting: is there a case for policy implementation? *J. Hosp. Infect.* **48**, 1–6

Rossmoore, H. W. (1995a). Introduction to biocide use. *In* "Handbook of Biocide and Preservative Use" (H. W. Rossmoore, ed.), pp. 1–17. Chapman & Hall, New York.

Rossmoore, H. W. (1995b). Biocides for metalworking lubricants and hydraulic fluids. *In* "Handbook of Biocide and Preservative

Use" (H. W. Rossmoore, ed.), pp. 133–156. Chapman & Hall, New York.

Rossmoore, H. W., and Sondossi, M. (1988). Application and mode of action of formaldehyde condensate biocides. *Adv. Appl. Microbiol.* **33**, 233–277.

Russell, A. D., Hugo, W. B., and Ayliffe, G. A. J. (eds.) (1992). "Principles and Practice of Disinfection, Preservation and Sterilization," 2nd ed. Blackwell, London.

Rutala, W. A., Stiegel, M. M., Sarabi, F. A., and Wber, D. J. (1997). Susceptibility of antibiotic-susceptible and antibiotic-resistant hospital bacteria to disinfectants. *Infect. Control Hosp. Epidemiol.* **18**, 417–421.

Sharpell, F. (1980). Industrial use of biocides in processes and products. *Dev. Ind. Microbiol.* **21**, 133–140.

WEBSITES

U.S. Environmental Protection Agency. Tolerance Reassessment and Registration
http://www.epa.gov/pesticides/reregistration/

U.S. Environmental Protection Agency, Office of Pesticide Programs
http://www.epa.gov/pesticides/about/index.htm

European Chemicals Bureau, Links to regulations, assessment procedures, documents
http://ecb.jrc.it/

OECD (Organisation for Economic Co-operation and Development) website, Links to regulations in various countries and international organizations on biocides
http://www1.oecd.org/ehs/biocides/

12

Biofilms and biofouling

Karen T. Elvers[1] *and Hilary M. Lappin-Scott*
University of Exeter

GLOSSARY

biofilm Complex association or matrix of microorganisms and microbial products attached to a surface.

biofouling Damage caused to a surface by microorganisms attached to a surface.

consortia Spatial grouping of bacterial cells within a biofilm in which different species are physiologically coordinated with each other, often to produce phenomenally efficient chemical transformations.

planktonic Free-floating bacteria living in the aqueous phase and not associated with a biofilm.

sessile Bacteria living within a biofilm.

Biofilms are generally described as consisting of the cells of microorganisms immobilized at a substratum, attached to a surface, and frequently embedded in an extracellular polymer matrix of microbial origin. In this context, studies have concentrated on bacterial cells rather than on other microorganisms. Bacteria attach firmly to almost any surface submerged in an aquatic environment or bulk liquid. Immobilized bacterial cells within a biofilm are called sessile, whereas those free floating in the aquatic environment are called planktonic. The immobilized cells grow and reproduce, with the newly formed cells attaching to each other as well as to the surface. They also produce extracellular polymers, which extend from the cells to form a matrix of fibers. This matrix entraps debris, nutrients, and other microorganisms establishing a biofilm that has a very heterogeneous structure and composition. Although bacteria may attach to surfaces in minutes, biofilms can take hours or days to develop. Biofouling refers to the damage caused by biofilms to surfaces. The combination of growth processes, the production of metabolites, and the physical presence of the biofilm can damage the surface and reduce its efficiency or effectiveness.

Early studies on the significance of bacterial adhesion to surfaces emerged from the work of Claude ZoBell in the 1930s. Since this initial research, it is now known that bacterial adhesion is widespread in the environment and the subject of biofilms constitutes an extensive field within microbiology. In addition, modern biofilm studies have shown that biofouling affects a surprisingly wide range of materials. This article contains examples of biofilms in medical and industrial situations and describes how particular biofilms cause damage and resist treatment.

I. BIOFILM FORMATION AND DETACHMENT

Many physical, chemical, and biological processes determine biofilm formation. A general description of biofilm formation on a surface begins with the transportation of molecules and small particles to the surface by molecular diffusion to form a conditioning film. This occurs very rapidly or almost instantaneously on exposure of a surface to an aqueous environment. Its effect is to cause changes in the surface properties,

[1]Present address: University of Wales Institute.

The Desk Encyclopedia of Microbiology
ISBN: 0-12-621361-5

including the acquisition of a small negative surface charge and a decrease in hydrophobicity. The composition of the conditioning film varies depending on the surface type but apparently it contains polysaccharides, glycoproteins, and proteins. It is generally uniform in composition and coverage and is an important influence on the subsequent adsorption of bacteria. The growth phase follows the initial phase of biofilm formation.

Fluid dynamics within the aquatic environment play an important role in determining the transport of bacteria to the surface during the growth phase. Under quiescent flow conditions, bacterial transport is affected by gravitational forces, Brownian motion, or motility. Under laminar flow conditions, bacteria are transported to the surface by diffusion with a significant increase in transport rate if the cells are motile. Under turbulent flow conditions, bacteria are transported to the surface by fluid dynamic forces (inertia, lift, drag, drainage, and downsweeps) which can be enhanced by further increasing turbulence and surface roughness. Furthermore, eddies that develop in turbulent flow are able to propel bacteria to the surface. The Reynolds number, which describes the relative magnitude of inertia to viscous forces, can be used to describe whether a system is laminar or turbulent.

Bacteria approaching the surface are subjected to repulsion forces which must be overcome if they are to adsorb to the surface. The outcome of the forces is described by the Derjaguin and Landau; Verwey and Overbeek theory of colloidal stability. The theory postulates two separation distances from the surface where adhesion can occur. Two- and three-step mechanisms have been proposed for the adhesion process, which results in irreversible adsorption usually by the production of exopolysaccharide (EPS). Development of the biofilm includes further attachment, cell growth, cell division, and EPS production resulting in the formation of distinct microcolonies. Mature biofilms then develop by the attraction of more planktonic bacteria and entrapment of inorganic and organic molecules and microbial products, developing a complex consortia within which there is physiological cooperation between different species. This results in increased heterogeneity and the development of chemical microgradients within the biofilm.

With time, portions of the biofilm detach and biofilm development reaches a plateau or steady state of development with accumulation equaling loss by detachment. Detachment is defined as the loss of components (biomass) from the biofilm matrix to the bulk liquid and is a means of interaction and cell turnover between the planktonic organisms in the liquid phase and the sessile organisms within the biofilm. This interaction can affect the overall species distribution. Detachment occurs by erosion, sloughing, and abrasion. It can be caused by several factors: the action of polymerases from the biofilm organisms, the result of grazing or predator harvesting by protozoa, the effects of substratum texture and surface chemistry, the production of unattached daughter cells through attached cell replication, and the availability of nutrients. Fluid dynamics is also thought to significantly influence detachment, in which an increase in fluid velocity causes an increase in detachment. There are also artificial methods of detachment which aim to control biofilm growth and biofouling, including chemical treatment (chelants, surfactants, and oxidants) and physical treatments (increased fluid velocity, ultrasound, and scrubbing).

II. EPS AND THE GLYCOCALYX

EPS and glycocalyx are terms used to describe the polysaccharide produced by bacterial cells. EPS refers to one of the major components of biofilms, and glycocalyx refers to the polysaccharide matrix surrounding individual cells. EPS has an important role in biofilm structure and function and has a complex physical and chemical nature. Its functions are mostly protective in nature and this is one of the benefits for bacteria in the sessile state. Because the glycocalyx is the outermost component of bacterial cells, this layer mediates virtually all bacterial associations with surfaces and other cells: it dictates location, juxtaposition, and the eventual success in the ecosystem.

EPS production may be a direct response to selective pressures in the environment and may protect against desiccation (by binding water molecules) and predation by feeding protozoa. It also provides protection against antimicrobial agents, including antibiotics, biocides, and host defense mechanisms. This defense may occur by means of a physical barrier or through aiding the bacteria to evade phagocytosis. Other advantages for bacteria of EPS derive from its polyanionic nature, which confers on the biofilm some ion-exchange properties that assist entrapment and the concentration of nutrients, the removal of toxins, and the exchange of metabolites within the consortia. Finally, the close proximity of cells within the biofilm allows plasmid transfer and an alteration in phenotypical characteristics as a response to changes in the environment.

III. BIOFILM STRUCTURE

Biofilm structure has been studied using many techniques, including transmission electron microscopy, scanning electron microscopy (SEM), and confocal scanning laser microscopy (CSLM). SEM has been used extensively to study the surface architecture of biofilms. The resulting images reveal an uneven outer surface topography, with the high resolution achieved by this method allowing individual cells to be clearly distinguished among a condensed matrix. Although this technique provides valuable information regarding the nature of biofilms, it is not entirely useful because it is well-known that the dehydration stages of sample preparation for SEM can destroy the EPS matrix. CSLM allows nondestructive *in situ* analysis of hydrated biofilms in combination with a wide range of fluorescent compounds. This technique can be used to form 3-D computer reconstructions of biofilms. These show a variable distribution of biomass with bacteria aggregating at different horizontal and vertical sites, with the highest cell densities at the biofilm base or at the top of the biofilm, forming "mushroom," "cone," or "stacks" shapes. Where biofilms have developed under turbulent conditions, they form additional structures termed "streamers." CSLM has also shown that biofilms are highly hydrated and that the total biofilm volume is made up of cell clusters, horizontal and vertical interstitial voids, and conduits beneath the clusters. These clusters and channels produce biofilms of varying depth and structure. Species composition has been shown to be an important determining factor in biofilm structure.

Recently, it was suggested that the structural complexity of biofilms is determined by the organisms through signaling molecules. It has been established that a family of diffusible chemical signals (*N*-acyl homoserine lactones) can regulate the production of virulence determinants and secondary metabolites, in suspended cultures, in a cell density-dependent manner. Also known as quorum sensing, it is thought that this may be important for the formation of biofilms which also contain densely packed cells. Evidence based on a pure culture *Pseudomonas aeruginosa* biofilm growing in laminar flow by Davies and coworkers (1998) supports this theory. They reported that *N*-(3-oxododecanoyl)-L-homoserine lactone (OdDHL) was required for the biofilm to develop a complex structure by comparing wild-type biofilms with a Lasl defective mutant (Lasl directs the synthesis of OdDHL). This work illustrates the interest in this field, which has a huge potential for biofilm control. However, since quorum sensing is a concentration-dependent phenomenon, it will be strongly influenced by mass transfer processes. It may be expected that quorum sensing will have a greater significance in diffusion-dominated regions such as those found in large cell clusters or channels when bulk liquid flow is very low.

IV. MIXED-SPECIES BIOFILMS

Much of the understanding of biofilm development, activity, and physiology is derived from studying single cultures of bacteria, with relatively few studies having been done on mixed cultures. There has been little discussion of the significance of other species within biofilms (e.g., bacterial and algal interactions and fungi), despite the fact that they are excellent colonizers of surfaces. Fungi are able to respond by growth at a surface and fungal hyphal slimes may have many of the functions attributed to bacterial EPS. These functions include the anchorage of mycelium to the substrate, retardation of desiccation, and service as a source of support and nutrition. Biofilms of filamentous fungi have been involved in industrial processes, such as the degradation of aromatic pollutants, biofouling of cooling tower timbers, voice prostheses, and photoprocessors.

V. TECHNIQUES FOR BIOFILM ANALYSIS

Techniques available to cultivate and study biofilms can be broadly categorized as disruptive and nondisruptive to the biofilm. These include fermentors and sampling devices such as the modified Robbins device (MRD). The MRD contains replaceable surfaces that can be examined (viewed by epifluorescence microscopy and SEM) for viable counts, total carbohydrate, total protein, and metabolic activity. The MRD is versatile in that colonization of different surfaces can be investigated, surface roughness can be controlled, and biocides and antibiotics can be tested.

Chemostats are widely used for studying microorganisms under constant environmental conditions over long periods of time. They can be used to cultivate well-defined two- or three-member mixed cultures or those with even more members. Other reports demonstrate the use of a two-stage chemostat system, with the inoculum grown in the first vessel before being passed to the second. This second or test vessel allows parameters to be changed, e.g., addition of biocide and insertion of coupons of differing materials on which the biofilm develops. Chemostats have also been used in combination with the MRD.

Other biofilm fermentors include the constant depth film fermentor, which allows the biofilm to accumulate to a preset depth which can be maintained, and the continuous perfused biofilm fermentor, which allows establishment of a biofilm on the underside of a cellulose membrane perfused with sterile fresh medium.

Microscopic techniques include light, epifluorescence and electron microscopy, CSLM and computer-enhanced microscopy with image analysis, and Nomarski differential contrast microscopy. Epifluorescence is particularly useful when cells are stained with fluorescent dyes such as acridine orange and propidium iodide, which mark nucleic acids, or those that determine metabolic activity, such as rhodamine and 5-cyano-2,3-ditolyl tetrazolium chloride. CSLM can be used in combination with microelectrodes which, depending on their construction, can measure oxygen, pH, and sulfide gradients within cell clusters. Voids can be visualized by following the movement of fluorescent latex beads.

Other techniques include the use of continuous-flow cell cultures with image analysis, Fourier transform infrared spectroscopy, nuclear magnetic resonance, atomic force microscopy, and cryosectioning. The data generated from all these techniques have greatly altered the understanding of biofilms in both pure and mixed cultures. They have shown biofilms as being spatially and temporally heterogeneous systems with microscale variations in architecture, chemistry, micro-gradients, and reactions to antimicrobial agents.

VI. BIOFILMS AND BIOFOULING IN DIFFERENT ENVIRONMENTS: THEIR CONTROL AND RESISTANCE

The phenotypic plasticity of bacteria allows colonization in a wide variety of environments. Adhering bacterial species are inherently different from their planktonic equivalents. In particular, biofilm bacteria are more resistant to medical and industrial control strategies than their planktonic counterparts. Biofilms may be either beneficial or detrimental for their host systems. The following sections discuss the variety and importance of biofilms and biofouling in the context of medical and industrial systems.

A. Biofilms in medical systems

Biofilms affecting human health can be divided into the following categories:

1. Biofilms formed on human tissue. These biofilms occur in the healthy body, for example, on teeth, in the digestive tract, and in the female genital tract. They may have a role in prevention of certain infections but can be overgrown by pathogenic microorganisms.

2. Biofilms formed on medical implants within the body.

3. Biofilms formed on surfaces outside the body that may harbor harmful pathogens. Examples of these surfaces include those in water systems that may harbor potentially pathogenic bacteria such as *Legionella* sp. and that consequently may be protected from chlorination. Biofilms have also been found on contact lenses and contact lens storage cases, for which bacteria induce severe eye irritation and inflammation and may play a role in persistence of the organisms.

In all cases, once established, the biofilm's resistance to phagocytosis and antibiotics allows the organism within it to continue living after planktonic organisms in the same environment have been killed by treatment.

Artificial implants are used for the replacement of diseased or damage body parts, e.g., joint or vascular prostheses. Many temporary devices, (e.g., urinary catheters, intravascular catheters, and endotracheal tubes) are inserted into patients for various lengths of time. Many inert materials are used for such devices, including vitallium, titanium, stainless steel, polyethylene, polymethyl methacrylate, silicone rubber polyttrafluorethylene (Teflon), and polyvinyl chloride. All of these can serve as substrata for bacterial biofilms.

A variety of bacteria are involved in the colonization of implants. These include gram-negative organisms (e.g., *Pseudomonas aeruginosa* and *Escherichia coli*) and gram-positive bacteria (e.g., *Staphylococcus aureus* and *S. epidermidis*). The latter, which are normally found on the skin, possess a high degree of adhesiveness to the prosthetic device surface. In the biofilm these species are protected from the effects of antibiotics and they can act as the disseminating center for infection. In addition, biofilm formation can also lead to malfunction of the device and destruction of adjacent tissue. Biofilms have been the cause of significant problems for patients receiving artificial hearts (Jarvik hearts).

In cystic fibrosis, patients' lungs become chronically infected by EPS-producing strains of *P. aeruginosa*. This bacterium has also been found frequently in catheterized patients. The production of large amounts of EPS and copious quantities of mucous by *P. aeruginosa* allows it to cause persistent infections. Isolation of mucoid strains and subsequent subculture results in reversion to non-mucoid colonies. This suggests that the host defense mechanisms must have a selective effect in favor of the mucoid strains. Treatment with antibiotics further selects for the mucoid strains. The mucous and EPS protects *P. aeruginosa* from attack by antibiotics, surfactants, and macrophages. It has been shown that EPS is a large anionic hydrated matrix that can partition charged molecules, preventing them from reaching the bacterial cell.

Silicone is a material that is widely used for tubing, catheters, mammary and testicular implants, and voice prostheses. Voice prostheses become colonized rapidly by mixed biofilms of bacteria and yeasts and these devices must be removed and replaced frequently before infection can be eradicated. It has been shown by SEM that voice prostheses become damaged by the yeast cells, which grow under the silicone surface. Treatment of infections for short-term devices, such as urinary and intravenous catheters, consists of their immediate removal followed by administration of antibiotics to the patient.

Urinary tract infections are most commonly caused by *E. coli*, *Proteus mirabilis*, *Enterococcus*, and *Streptococcus* spp., found in the gastrointestinal tract, and by pathogens directly transmitted through sexual activity. These infections include acute and chronic cystitis, struvite urolithiasis, chronic prostatitis, and catheter-associated infections. Once the microorganisms are established, they adopt the biofilm mode of growth. The bladder resists infection by the periodic passing of urine, which washes out unattached pathogens, and by sloughing of colonized uroepithelial cells on the glycosaminoglycan (GAG) mucous layer. The GAG layer is a very thin cover on the cell epithelium of the bladder that physically shields the bladder from surface pathogens. Catheter-associated infections increase by approximately 10% each day the catheter is in place. The organisms initially colonize the external surfaces, form a biofilm, and ascend into the bladder where the biofilm can act as a source of infection for the bladder and kidneys. Mineralization can also occur, which can reduce the diameter or block the catheter. Frequent replacement of catheters would reduce infections but is not always practical. Methods to control biofilm growth on catheters are being investigated and include the development of materials that block or kill adherent organisms. These methods include altering the hydrophobicity of polymers and the incorporation of disinfectants and antibiotics in the design of the implant.

Surfaces within the mouth become readily colonized with bacterial deposits, forming a biofilm, usually called dental plaque. By attaching to the teeth or dental implants, the biofilm helps to prevent colonization of the mouth by pathogenic bacteria. Although dental plaque forms naturally without good oral hygiene, it can be a source of dental caries or periodontal disease. The attached organisms obtain nutrients from the ingested food, saliva, and gingival crevice fluid found between the teeth and gums: It is thought that most of the nutrients are derived from the host rather than the from the host's diet.

Environmental factors that contribute to plaque formation are an optimum temperature of 35 or 36 °C and a neutral pH. The pH can become more acidic when carbohydrates are metabolized or more alkaline during an inflammatory host response. These local changes in pH can lead to shifts in the colonized species. The resident microflora of dental plaque is extremely diverse and consists of gram-positive and -negative bacteria. Few are truly aerobic; most are facultatively or obligately anaerobic. These include *Streptococcus*, *Neisseria*, *Actinomyces*, *Lactobacillus*, *Corynebacterium*, and *Fusobacterium* species, but there are many others and not all are culturable in the laboratory. Sometimes, a particular species will colonize a preferred habitat in the oral cavity, e.g., *Streptococcus mutans* colonizes the occlusal fissures.

Formation of dental plaque begins by the adsorption of a proteinaceous conditioning film or acquired pellicle which is composed of albumin, lysozyme, glycoproteins, and lipids from saliva and gingival cervicular fluid. The first colonizers are streptococci and actinomycetes, which rapidly divide to form microcolonies that quickly change into a confluent film of varying thickness. Species diversity increases with the attachment of rods and filaments and layering that is attributable to bacterial succession. Unusual combinations of bacteria such as "corn-cobs" (gram-positive filaments covered by cocci), "rosettes" (cocci covered by small rods), or "bristle brushes" (large filaments surrounded by rods) are seen under SEM.

If plaque formation continues undisturbed for weeks, its composition will vary with location on the teeth. Different environmental conditions exist on and between the teeth, resulting in different chemical gradients and shear forces. The bacterial communities are thought to form as a result of short-range specific molecular interactions between the bacterial cell adhesions of primary colonizers and host receptors in the conditioning film, the attachment of secondary colonizers to primary colonizers (coaggregation) and EPS synthesis and growth. Coaggregation has a major role in forming the distinct patterns in plaque.

Dental caries are formed as a result of the localized dissolution of the tooth enamel by acids produced by metabolism of carbohydrates, lowering the pH and favoring the growth of mutans streptococci and lactobacilli. Periodontal diseases occur when the supporting tissues of the teeth are attacked by obligately anaerobic gram-negative rods, filaments, or spiral-shaped bacteria. Prevention of dental plaque is by efficient oral hygiene (brushing and flossing can almost completely prevent plaque-mediated diseases), fluoridization of drinking water, and the addition of antiplaque or antimicrobial agents to toothpastes and mouthwashes.

Biofilms are central to the survival of bacteria when they are attacked by the normal host immune system or antibiotics. Gram-positive and -negative bacteria activate the complement system, with the major components causing this reaction being peptidoglycan and lipopolysaccharide, respectively. Activation of complement would eradicate the serum-sensitive bacteria but may also react to live or dead bacteria or bacterial fragments. Continual production of complement may destroy the tissues. Studies on the immune response to biofilms have concentrated on *P. aeruginosa* and infection in cystic fibrosis patients. It was shown that biofilm-grown *P. aeruginosa* was able to resist complement action. The biofilm bacteria activated the complement system to a lesser extent than did the planktonic bacteria. However, some fragments of the activated complement were deposited on the biofilm contributing to chronic inflammation.

B. Biofilms in industrial systems

Biofilm formation in industries involves the ability of the biofilm to act as a reservoir for potential pathogens and in instances in which the biofilm causes surface damage. The degree of contamination in a system is often measured by planktonic counts that fail to detect the presence of sessile bacteria and this leads to incorrect conclusions regarding the level of pathogens in the system. In the water industry, biofilms form on the pipe surfaces that connect the consumer to the supply. The level of biofilm formation is difficult to monitor in these situations, and levels of coliforms, pseudomonads, and *Flavobacterium* sp., detaching from biofilms, have been reported as being higher than permitted the levels. Treatment with chlorine does not control the problem because the biofilm bacteria are protected from the disinfectant. There is also evidence of accelerated material deterioration (corrosion) due to biofilm accumulation in water distribution pipes.

Excessive biofilm accumulation in porous media, on heat exchanger surfaces, and in storage tanks is responsible for reduced efficiency of heat transfer and reduction of flow rates. Transfer of heat is reduced because the thick surface growth physically prevents an efficient heat exchange between the liquid phase and the cooling surface.

Biofilms on ship hulls consist of diatoms, single-celled algae, and bacteria. This biofilm growth reduces speed in the water and increases fuel consumption. As the biofilm develops, the hull must be physically cleaned, which results in further expense. In an attempt to control this growth, antifouling paints have been used. These are not always effective; although good at preventing colonization of small animals, they do not stop bacterial growth.

Corrosion in marine environments on structures such as oil rigs is a result of biofilms that contain sulfate-reducing or acid-producing bacteria. These microorganisms create anodes and cathodes on metal surfaces. This unequal distribution of ions causes electrical currents, resulting in metal loss. Ideal anaerobic conditions for the biofouling action of sulfate-reducing bacteria are found around oil rig legs.

Biofilms in the food processing industry may form on food contact and non-contact surfaces. These biofilms may contain food spoilage and pathogenic micro-organisms which affect the quality and safety of the food product by reducing shelf life and increasing the probability of food poisoning. Stainless steel is commonly used as a food contact surface because it is chemically and physiologically stable at a variety of processing temperatures, easy to clean, and has a high resistance to corrosion. Food processing environments provide a variety of conditions that favor the formation of biofilms, e.g., flowing water, suitable attachment surfaces, ample nutrients (although possibly sporadic), and the raw materials or the natural flora providing the inocula; however, these conditions may be extremely varied. Time available for biofilm development is relatively short; for example, the production line may run for a few hours before cleaning. Various preventative and control strategies, such as hygienic plant lay-out and design of equipment, choice of materials, correct use and selection of detergents and disinfectants, coupled with physical methods, can be applied for controlling biofilm formation on food contact surfaces.

Biofilms have been shown to occur in many food environments. In the dairy industry, pasteurization ensures the destruction of pathogens and most vegetative organisms within raw milk. However, heat-resistant organisms and spores survive and may form biofilms that could result in post-pasteurization contamination. Biofilms have been found on gaskets and "O" rings from the pipes within the dairy industry. Pathogens, *Listeria* and *Bacillus* sp., have also been isolated from food contact and environmental surfaces in the dairy industry. Biofilms have been found in pipes of breweries, on rubber seals, conveyor belts, and in waste-water pipes, and in flour mills and malt houses. There is also evidence of microbial adherence to environmental surfaces during poultry processing. This could result in cross-contamination during the slaughter process, which may play an important role in product contamination with *Listeria*, *Campylobacter*, and *Staphylococcus aureus*.

BIBLIOGRAPHY

Costerton, J. W., Lewandowski, Z., Caldwell, D. E., Korber, D., and Lappin-Scott, H. M. (1995). Microbial biofilms. *Annu. Rev. Microbiol.* **49**, 711–745.

Costerton, J. W., Stewart, P. S., and Greenberg, E. P. (1999). Bacterial biofilms: a common cause of persistent infections. *Science* **284**(5418), 1318–1322.

Davey, M. E., and O'Toole, G. A. (2000). Microbial biofilms: from ecology to molecular genetics. *Microbiol. Mol. Biol. Rev.* **64**, 847–867.

Davies, D., Parsek, M. R., Pearson, J. P., Iglewski, B. H., Costerton, J. W., and Greenburg, E. P. (1998). The involvement of cell-to-cell signals in the development of a bacterial biofilm. *Science* 280, 295–298.

Lappin-Scott, H. M., and Costerton, J. W. (eds.) (1995). "Microbial Biofilms." Cambridge Univ. Press, Cambridge, UK.

O'Toole, G., Kaplan, H. B., and Kolter, R. (2000). Biofilm formation as microbial development. *Annu. Rev. Microbiol.* **54**, 49–79.

Stewart, P. S. (2001). Multicellular resistance: biofilms. *Trends Microbiol.* **9**(5), 204.

Stewart, P. S. and Costerton, J. W. (2001). Antibiotic resistance of bacteria in biofilms. *Lancet* **358**(9276), 135–138.

Wackett, L. P. (2001). Microbioal biofilms. *Environ. Microbiol.* **3**(2), 144.

ZoBell, C. E. (1937). The influence of solid surface upon the physiological activities of bacteria in sea water. *J. Bacteriol.* **33**, 86.

WEBSITES

The website for the Center for Biofilm Engineering, U. of Montana *http://www.erc.montana.edu/*

American Society for Microbiology, Biofilm image collection *http://www.asmusa.org/edusrc/biofilms/index.html*

13

Biological warfare

James A. Poupard and Linda A. Miller
SmithKline Beecham Pharmaceuticals

GLOSSARY

biological warfare Use of microorganisms, such as bacteria, fungi, viruses, and rickettsiae, to produce death or disease in humans, animals, or plants. The use of toxins to produce death or disease is often included under the heading of BWR (U.S. Army definition, included in U.S. Army report to the Senate Committee on Human Resources, 1977).

biological weapons Living organisms, whatever their nature, which are intended to cause disease or death in man, animals, or plants and which depend for their effects on their ability to multiply in the person, animal, or plant attacked [United Nations definition, included in the report of the secretary general titled "Chemical and Bacteriological (Biological) Weapons and the Effects of Their Possible Use," 1969].

genetic engineering Methods by which the genomes of plants, animals, and microorganisms are manipulated: includes but is not limited to recombinant DNA technology.

recombinant DNA technology Techniques in which different pieces of DNA are spliced together and inserted into vectors such as bacteria or yeast.

toxin weapon(s) Any poisonous substance, whatever its origin or method of production, which can be produced by a living organism, or any poisonous isomer, homolog, or derivative of such a substance (U.S. Arms Control and Disarmament Agency definition, proposed on August 20, 1980).

The Most General Concept of Biological Warfare involves the use of any biological agent as a weapon directed against humans, animals, or crops with the intent to kill, injure, or create a sense of havoc against a target population. This agent could be in the form of a viable organism or a metabolic product of that organism, such as a toxin. This article will focus on the use of viable biological agents because many of the concepts relating to the use of toxins are associated more with chemical warfare. The use of viable organisms or viruses involves complex issues that relate to containment. Once such agents are released, even in relatively small numbers, the focus of release has the potential to enlarge to a wider population due to the ability of the viable agent to proliferate while spreading from one susceptible host to another.

I. INTRODUCTION

As the twentieth century draws to a close three events mark significant alterations in the concept of biological warfare (BW): the end of the Cold War, the open threat of using BW agents in the Gulf War, and the realization that the developed world is quite susceptible to attack by radical terrorists employing BW agents. These events mark major changes in the concept of BW and transform the subject from one which was once limited to the realm of political and military policy makers to one that must be considered by a wide range of urban disaster planners, public health officials, and the general public. BW is a complex

The Desk Encyclopedia of Microbiology
ISBN: 0-12-621361-5

subject that is difficult to understand without a basic knowledge of a long and convoluted history. BW can be traced to ancient times and has evolved into more sophisticated forms with the maturation of the science of bacteriology and microbiology. It is important to understand the history of the subject because often there are preconceived notions of BW that are not based on facts or involve concepts related more to chemical rather than biological warfare. Many of the contemporary issues relating to BW deal with Third World conflicts, terrorist groups, or nonconventional warfare. An understanding of these issues is important because many of the long-standing international treaties and conventions on BW were formulated either in an atmosphere of international conflict or during the Cold War period of international relations. Many of the classic issues have undergone significant alteration by recent events. The issue of BW is intimately bound to such concepts as offensive versus defensive research or the need for secrecy and national security. It is obvious that BW will continue to demand the attention of contemporary students of microbiology and a wide range of specialists during the twenty-first century.

II. HISTORICAL REVIEW

A. 300 BC to 1925

Many early civilizations employed a crude method of warfare that could be considered BW as early as 300 BC when the Greeks polluted the wells and drinking water supplies of their enemies with the corpses of animals. Later, the Romans and Persians used these same tactics. All armies and centers of civilization need palatable water to function, and it is clear that well pollution was an effective and calculated method for gaining advantage in warfare. In 1155 at a battle in Tortona, Italy, Barbarossa broadened the scope of BW by using the bodies of dead soldiers as well as animals to pollute wells. Evidence indicates that well poisoning was a common tactic throughout the classical, medieval, and Renaissance periods. In modern times, this method has been employed as late as 1863 during the Civil War by General Johnson, who used the bodies of sheep and pigs to pollute drinking water at Vicksburg.

Catapults and siege machines in medieval warfare were a new technology for delivering biological entities. In 1422 at the siege of Carolstein, catapults were used to project diseased bodies over walled fortifications, creating fear and confusion among the people under siege. The use of catapults as weapons was well established by the medieval period, and projecting diseased bodies over walls was an effective strategy employed by besieging armies. The siege of a well-fortified position could last for months or years, and it was necessary for those outside the walls to use whatever means available to cause disease and chaos within the fortification. This technique became commonplace, and numerous classical tapestries and works of art depict diseased bodies or the heads of captured soldiers being catapulted over fortified structures.

In 1763, BW took a significant turn from the crude use of diseased corpses to the introduction of a specific disease, smallpox, as a weapon in the North American Indian wars. It was common knowledge at the time that the Native American population was particularly susceptible to smallpox, and the disease may have beeen used as a weapon in earlier conflicts between European settlers and Native Americans. In the spring of 1763, Sir Jeffrey Amherst, the British commander-in-chief in North America, believed the western frontier, which ran from Pennsylvania to Detroit, was secure, but the situation deteriorated rapidly during the next several months. The Indians in western Pennsylvania were becoming particularly aggressive in the area near Fort Pitt (Pittsburgh). It became apparent that unless the situation was resolved, western Pennsylvania would be deserted and Fort Pitt isolated. On June 23, 1763, Colonel Henry Bouquet, the ranking officer for the Pennsylvania frontier, wrote to Amherst, describing the difficulties Captain Ecuyer was having holding the besieged Fort Pitt. These difficulties included an outbreak of smallpox among Ecuyer's troops. In his reply to Bouquet, Amherst suggested that smallpox be sent among the Indians to reduce their numbers. This well-documented suggestion is significant because it clearly implies the intentional use of smallpox as a weapon. Bouquet responded to Amherst's suggestion by stating that he would use blankets to spread the disease.

Evidence indicates that Amherst and Bouquet were not alone in their plan to use BW against the Indians. While they were developing a plan of action, Captain Ecuyer reported in his journal that he had given two blankets and handkerchiefs from the garrison smallpox hospital to hostile chiefs with the hope that the disease would spread. It appears that Ecuyer was acting on his own and did not need persuasion to use whatever means necessary to preserve the Pennsylvania frontier. Evidence also shows that the French used smallpox as a weapon in their conflicts with the native population.

Smallpox also played a role in the American Revolutionary War, but the tactics were defensive rather than offensive: British troops were vaccinated against smallpox, but the rebelling American colonists were not. This protection from disease gave the British an advantage for several years, until Washington ordered vaccination of all American troops.

It is clear that by the eighteenth century BW had become disease oriented, even though the causative agents and mechanisms for preventing the spread of diseases were largely unknown. The development of the science of bacteriology in the nineteenth and early twentieth centuries considerably expanded the scope of potential BW agents. In 1915, Germany was accused of using cholera in Italy and plague in St. Petersburg. Evidence shows that Germany used glanders and anthrax to infect horses and cattle, respectively, in Bucharest in 1916 and employed similar tactics to infect 4500 mules in Mesopotamia the next year. Germany issued official denials of these accusations. Although there apparently was no large-scale battlefield use of BW in World War I, numerous allegations of German use of BW were made in the years following the war. Britain accused Germany of dropping plague bombs, and the French claimed the Germans had dropped disease-laden toys and candy in Romania. Germany denied the accusations.

Although chemical warfare was far more important than BW in World War I, the general awareness of the potential of biological weapons led the delegates to the Geneva Convention to include BW agents in the 1925 Protocol for the Prohibition of the Use in War of Asphyxiating, Poisonous or Other Gases, and of Bacteriological Methods of Warfare. The significance of the treaty will be discussed in Section III.

B. 1925–1990

The tense political atmosphere of the period following the 1925 Geneva Protocol and the lack of provisions to deter biological weapons research had the effect of undermining the treaty. The Soviet Union opened a BW research facility north of the Caspian Sea in 1929; the United Kingdom and Japan initiated BW research programs in 1934. The Japanese program was particularly ambitious and included experiments on human subjects prior to and during World War II.

Two factors were significant in mobilizing governments to initiate BW research programs: (i) continuing accusations regarding BW and (ii) the commitment of resources for BW research by several national adversaries, thus creating insecurity among governments. The presence of BW research laboratories in nations that were traditional or potential adversaries reinforced this insecurity. Thus, despite the Geneva Protocol, it was politically unwise for governments to ignore the threat of BW, and the result was increasingly sophisticated biological weapons.

In 1941, the United States and Canada joined other nations and formed national programs of BW research and development. Camp Detrick (now Fort Detrick) became operational as the center for U.S. BW research in 1943, and in 1947 President Truman withdrew the Geneva Protocol from Senate consideration, citing current issues such as the lack of verification mechanisms that invalidated the underlying principles of the treaty. However, there was no widespread use of BW in a battlefield setting during World War II. BW research, however, continued at an intense pace during and after the war. By the end of the decade, the United States, the United Kingdom, and Canada were conducting collaborative experiments involving the release of microorganisms from ships in the Caribbean. In 1950, the U.S. Navy conducted open-air experiments in Norfolk, Virginia, and the U.S. Army conducted a series of airborne microbial dispersals over San Francisco using *Bacillus globigii*, *Serratia marcescens*, and inert particles.

Not surprisingly, the intense pace of BW research led to new accusations of BW use, most notably by China and North Korea against the United States during the Korean War. In 1956, the United States changed its policy of "defensive use only" to include possible deployment of biological weapons in situations other than retaliation. During the 1960s, all branches of the U.S. military had active BW programs, and additional open-air dissemination experiments with stimulants were conducted in the New York City subway system. By 1969, however, the U.S. military concluded that BW had little tactical value in battlefield situations, and since it was believed that nuclear weapons dominated the strategic equation the United States would be unlikely to need or use BW. Thus, President Nixon announced that the United States would unilaterally renounce BW and eliminate stockpiles of biological weapons. This decision marked a turning point in the history of BW: Once the U.S. government made it clear it did not consider biological weapons a critical weapon system, the door was opened for negotiation of a strong international treaty against BW.

Once military strategists had discounted the value of BW, an attitude of openness and compromise on BW issues took hold, leading to the 1972 Convention on the Prohibition of the Development, Production and Stockpiling of Bacteriological (Biological) and Toxin Weapons and on Their Destruction (see Section III). The parties to the 1972 convention agreed to destroy or convert to peaceful use all organisms, toxins, equipment, and delivery systems. Following the signing of the 1972 treaty, the U.S. government generated much publicity about its compliance activities, inviting journalists to witness destruction of biological weapons stockpiles.

The problem of treaty verification beleaguered the 1972 convention. Press reports accusing the Soviet

Union of violating the treaty appeared as early as 1975. When an outbreak of anthrax was reported in Sverdlovsk, Soviet Union, in 1979, the United States claimed it was caused by an incident at a nearby Soviet biological defense laboratory that had released anthrax spores into the surrounding community. The Soviet government denied this allegation, claiming the outbreak was caused by contaminated black market meat.

BW continued to be discussed in the public media throughout the 1980s. In 1981, reports describing the American "cover-up" of Japanese BW experiments on prisoners of war began to surface in the public and scientific literature. In 1982, *The Wall Street Journal* published a series of articles on Soviet genetic engineering programs that raised many questions about the scope of Soviet BW activities. The environmental effects of testing biological agents at Dugway Proving Grounds in Utah received considerable press attention in 1988, leading to a debate over the need for such a facility.

The 1980s also were characterized by debate over larger issues relating to BW. A public debate in 1986 considered the possible role of biological weapons in terrorism. Scientific and professional societies, which had avoided discussing BW for many years, began considering both specific issues, such as Department of Defense support for biological research, and more general issues, such as adopting ethical codes or guidelines for their members.

C. 1990 and contemporary developments

The last decade of the twentieth century witnessed three significant events that will have long-term effects on developing policies relating to BW. The first event was the demise of the Soviet Union. Most U.S. defensive research was directed to counter potential use by the Soviet Union. As the wall of Soviet secrecy eroded during the 1990s the extent of the Soviet BW program became apparent. There is international concern that many unemployed BW researchers will find work as advisors for developing countries that view BW as a rational defense strategy, especially those countries without nuclear capability or those without restrictive laws against radical terrorist groups. This is an ongoing issue without readily apparent solutions.

The second major event was the Gulf War. The open threat by the Iraqi military to use BW agents raised serious concerns and changed attitudes about BW. The plans for Operation Desert Storm included provisions for protective equipment and prophylactic administration of antibiotics or vaccines to protect against potential biological weapons. Many of the critics of the U.S. Biological Defense Research Program (BDRP) were now asking why the country was not better prepared to protect its troops against biological attack. Fortunately, BW was not used during the Gulf War, but the threat of its use provided several significant lessons. Although there was considerable concern that genetic engineering would produce new, specialized biological weapons, most experts predicted that "classical" BW agents, such as anthrax and botulism, would pose the most serious threats to combat troops in Operation Desert Storm. Efforts by the United Nations after the war to initiate inspection programs demonstrated the difficulty of verifying the presence of production facilities for BW agents; these difficulties highlight the need for verification protocols for the BW convention. Verification and treaty compliance are major contemporary BW issues. Following the Gulf War the extent of the intense Iraq BW research programs demonstrated the inadequacy of all estimates and post war verification procedures of Iraq BW capacity.

The third significant contemporary development is the realization that urban centers and public facilities are vulnerable to attack by terrorists employing BW agents. Local and national governments are now realizing the extent of this vulnerability and are taking early measures to formulate policies to address these issues. Much work remains to be accomplished in this area.

III. INTERNATIONAL TREATIES

A. The 1925 Geneva protocol

The 1925 Geneva Protocol was the first international treaty to place restrictions on BW. The Geneva Protocol followed a series of international agreements that were designed to prohibit the use in war of weapons that inflict or prolong unnecessary suffering of combatants or civilians. The St. Petersburg Declaration of 1868 and the International Declarations Concerning the Laws and Customs of War, which was signed in Brussels in 1874, condemned the use of weapons that caused useless suffering. Two major international conferences were held at the Hague in 1899 and 1907. These conferences resulted in declarations regarding the humanitarian conduct of war. The conference regulations forbid nations from using poison, treacherously wounding enemies, or using munitions that would cause unnecessary suffering. The so-called Hague Conventions also prohibited the use of projectiles to diffuse asphyxiating or deleterious gases. The Hague Conventions still provide much of the definitive law of war as it exists today.

The Hague Conventions did not specifically mention BW, due in part to the lack of scientific understanding of the cause of infectious diseases at that time. The conventions have, however, been cited as an initial source of the customary international laws that prohibit unnecessary suffering of combatants and civilians in war. Although biological weapons have been defended as humanitarian weapons on the grounds that many biological weapons are incapacitating but not lethal, there are also biological weapons that cause a slow and painful death. It can be argued, therefore, that the Hague Conventions helped to set the tone of international agreements on laws of war that led to the 1925 Geneva Protocol.

The 1925 Geneva Protocol, formally called the Prohibition of the Use in War of Asphyxiating, Poisonous or Other Gases, and of Bacteriological Methods of Warfare, was opened for signature on June 17, 1925 in Geneva. More than 100 nations signed and ratified the protocol, including all members of the Warsaw Pact and North Atlantic Treaty Organization (NATO). The 1925 Geneva Protocol was initially designed to prevent the use in war of chemical weapons; however, the protocol was extended to include a prohibition on the use of bacteriological methods of warfare. The Geneva Protocol distinguishes between parties and nonparties by explicitly stating that the terms of the treaty apply only to confrontations in which all combatants are parties and when a given situation constitutes a "war." In addition, many nations ratified the Geneva Protocol with the reservation that they would use biological weapons in retaliation against a biological weapons attack. This resulted in the recognition of the Geneva Protocol as a "no first-use" treaty.

B. The 1972 biological warfare convention

International agreements governing BW have been strengthened by the 1972 BW convention, which is officially called the 1972 Convention on the Prohibition of the Development, Production and Stockpiling of Bacteriological (Biological) and Toxin Weapons and on Their Destruction. The convention was signed simultaneously in 1972 in Washington, London, and Moscow and entered into force in 1975. The preamble to the 1972 BW convention states the determination of the parties to the treaty to progress toward general and complete disarmament, including the prohibition and elimination of all types of weapons of mass destruction. This statement places the convention in the wider setting of international goals of complete disarmament. The 1972 BW convention is also seen as a first step toward chemical weapons disarmament.

The 1972 BW convention explicitly builds on the Geneva Protocol by reaffirming the prohibition of the use of BW in war. The preamble, although not legally binding, asserts that the goal of the convention is to completely exclude the possibility of biological agents and toxins being used as weapons and states that such use would be repugnant to the conscience of humankind. The authors of the 1972 convention, therefore, invoked societal attitudes as justification for the existence of the treaty.

The 1972 BW convention evolved, in part, from a process of constant reevaluation of the Geneva Protocol. From 1954 to the present, the United Nations has periodically considered the prohibition of chemical and biological weapons. The Eighteen-Nation Conference of the Committee on Disarmament, which in 1978 became the Forty-Nation Committee on Disarmament, began talks in 1968 to ban chemical weapons. At that time, chemical, toxin, and biological weapons were being considered together in an attempt to develop a comprehensive disarmament agreement. However, difficulties in reaching agreements on chemical warfare led to a series of separate negotiations that covered only BW and toxin weapons. The negotiations resulted in the drafting of the 1972 BW convention.

The 1972 BW convention consists of a preamble followed by 15 articles. Article I forms the basic treaty obligation. Parties agree never in any circumstance to develop, produce, stockpile, or otherwise acquire or retain the following:

1. Microbial or other biological agents, or toxins whatever their origin or method of production, of types and in quantities that have no justification for prophylactic, protective, or other peaceful purposes.
2. Weapons, equipment, or means of delivery designed to use such agents or toxins for hostile purposes or in armed conflict.

Article II requires each party to destroy, or divert to peaceful purposes, all agents, toxins, equipment, and delivery systems that are prohibited in Article I and are under the jurisdiction or control of the party. It also forbids nations from transferring, directly or indirectly, materials specified in Article I and prohibits nations from encouraging, assisting, or inducing any state, group of states, or international organizations from manufacturing or acquiring the material listed in Article I. There is no specific mention of subnational groups, such as terrorist organizations, in the treaty.

Articles IV requires each party to the convention to take any measures to ensure compliance with the terms of the treaty. Article IV has been interpreted by some states as the formulation of civil legislation or regulations to ensure adherence to the convention. This civil legislation could regulate activities by individuals, government agencies, universities, or corporate groups.

Articles V–VII specify procedures for pursuing allegations of noncompliance with the 1972 BW convention. The United Nations plays an integral part in all the procedures for investigating allegations of noncompliance. According to Article VI, parties may lodge a complaint with the Security Council of the United Nations if a breach of the treaty is suspected. All parties must cooperate with investigations that may be initiated by the Security Council. Article VII requires all parties to provide assistance or support to any party that the Security Council determines has been exposed to danger as a result of violation of the convention. Articles VII–IX are general statements for obligations of the parties signing the protocol. Article X gives the parties the right to participate in the fullest possible exchange of equipment, materials, and scientific or technological information of the use of bacteriological (biological) agents and toxins for peaceful purposes. Article XI allows parties to propose amendments to the convention. The amendments only apply to those states that accept them and enter into force after a majority of the states' parties to the convention have agreed to accept and be governed by the amendment.

Article XII requires that a conference be held 5 years after the entry into force of the BW convention. Article XIV states that the 1972 BW convention is of unlimited duration. A state party to the treaty is given the right to withdraw from the treaty if it decides that extraordinary events, related to the subject matter of the convention, have jeopardized the supreme interests of the country. This article also opens the convention to all nations for signature. Nations that did not sign the convention before its entry into force may accede to it at any time.

C. Review conferences

The 1972 convention contained a stipulation that a conference be held in Geneva 5 years after the terms of the convention entered into force. The purpose of the conference was to review the operation of the convention and to ensure that the purposes of the convention were being realized. The review was to take into account any new scientific and technological developments that were relevant to the convention. The first review conference was held in Geneva in 1980. Several points contained in the original convention were clarified at this conference. The second review conference was held in 1986, and a third was held in 1991. There is general agreement that these conferences and the one that followed serve a definite function in solving contemporary problems that need clarification based on changing events and have made significant contributions in keeping the 1972 convention relevant to the needs of a changing world situation.

TABLE 13.1 Representative organisms regulated by the CDC

Bacteria	Viruses
Bacillus anthracis	Crimean–Congo hemorrhagic fever virus
Brucella abortus, melitensis, suis	Eastern equine encephalitis virus
Burkholderia mallei, pseudomallei	Ebola virus
Clostridium botulinum	Equine morbillivirus
Francisella tularensis	Lassa fever virus
Yersinia pestis	Marburg virus
	Rift Valley fever virus
	South American hemorrhagic fever viruses
Rickettsiae	Tick-borne encephalitis complex virus
Coxiella burnetii	Variola (smallpox) major virus
Rickettsia prowazekii	Venezuelan equine encephalitis virus
Rickettsia rickettsii	Hantavirus
	Yellow fever virus

D. Additional U.S. laws and acts

The following U.S. laws have been enacted since 1989 that impact on BW:

- Biological Weapons and Anti-Terrorist Act (1989): established as a federal crime the development, manufacture, transfer, or possession of any biological agent, toxin, or delivery system for use as a weapon.
- Chemical and Biological Weapons Control Act (1991): places sanctions on companies that knowingly export goods or technologies relating to biological weapons to designated prohibited nations.
- The Defense Against Weapons of Mass Destruction Act (1996): designed to enhance federal, state and local emergency response capabilities to deal with terrorist incidents.
- Antiterrorism and Effective Death Penalty Act (1996): established as a criminal act any threat or attempt to develop BW or DNA technology to create new pathogens or make more virulent forms of existing organisms.
- Centers for Disease Control (CDC) Hazardous Biological Agent Regulation (1997): identification of infective agents that pose a significant risk to public health. Some of the organisms regulated by the CDC are listed in Table 13.1.

IV. CURRENT RESEARCH PROGRAMS

Biological weapons research in the United States is under the direction of the BDRP, headquartered at Fort Detrick, Maryland. In accordance with official U.S. policy, the BDRP is solely defensive in nature, with the goal of providing methods of detection for, and protective measures against, biological agents that could be used as weapons against U.S. forces by hostile states or individuals.

Current U.S. policy stems from the 1969 declaration made by President Nixon that confined the U.S. BW program to research on biological defense such as immunization and measures of controlling and preventing the spread of disease. Henry Kissinger further clarified the U.S. BW policy in 1970 by stating that the United States biological program will be confined to research and development for defensive purposes only. This did not preclude research into those offensive aspects of biological agents necessary to determine what defensive measures are required.

The BDRP expanded significantly in the 1980s in an apparent response to alleged treaty violations and perceived offensive BW capabilities of the Soviet Union. These perceptions were espoused primarily by representatives of the Reagan administration and the Department of State. At congressional hearings in May 1988, the U.S. government reported that at least 10 nations, including the Soviet Union, Libya, Iran, Cuba, Southern Yemen, Syria, and North Korea, were developing biological weapons. Critics of the U.S. program refuted the need for program expansion.

The BDRP is administered through two separate government organizations—the army and the CIA. Details of the program are described in the April 1989 Environmental Impact Statement published by the Department of the Army, U.S. Army Medical Research and Development Command.

The BDRP is located at three sites: the U.S. Army Medical Research Institute of Infectious Diseases (USAMRIID) at Fort Detrick, Maryland; the Aberdeen Proving Ground in Maryland; and the Dugway Proving Ground in Utah. USAMRIID is designated as the lead laboratory in medical defense against BW threats. Research conducted at the USAMRIID focuses on medical defense such as the development of vaccines and treatments for both natural diseases and potential BW agents. Work on the rapid detection of microorganisms and the diagnosis of infectious diseases is also conducted. The primary mission at the Aberdeen Proving Ground is nonmedical defense against BW threats including detection research, such as the development of sensors and chemiluminescent instruments to detect and identify bacteria and viruses, and development of methods for material and equipment decontamination. The U.S. Army Dugway Proving Ground is a Department of Defense major range and test facility responsible for development, testing, evaluation, and operation of chemical warfare equipment, obscurants and smoke munitions, and biological defense equipment. Its principle mission with respect to the BDRP is to perform developmental and operational testing for biological defense material, including the development and testing of sensors, equipment, and clothing needed for defense against a BW attack.

One hundred secondary sites have received contracts for biological defense research. Secondary sites include the Swiftwater Lab, operated by the Salk Institute in Swiftwater, Pennsylvania; the Naval Medical Research Institute in California; medical centers; universities; and private biotechnology firms in the United States, Scotland, and Israel.

The CIA also participates in the administration of the BDRP. In 1982, Thomas Dashiell of the office of the secretary of defense reported on a classified technology watch program related to BW that was operated by the intelligence community. The program was designed to monitor worldwide developments related to BW that could affect the vulnerability of U.S. and NATO forces to biological attack.

BDRP research focuses on five main areas:

1. Development of vaccines.
2. Development of protective clothing and decontamination methods.
3. Analysis of the mode of action of toxins and the development of antidotes.
4. Development of broad-spectrum antiviral drugs for detecting and diagnosing BW agents and toxins.
5. Utilization of genetic engineering methods to study and prepare defenses against BW and toxins.

The BDRP has often been a center of controversy in the United States. One BDRP facility, the Dugway Proving Ground, was the target of a lawsuit that resulted in the preparation of the environmental impact statement for the facility. A proposal for a high-level containment laboratory (designated P-4) was ultimately changed to a plan for a lower-level (P-3) facility.

The use of genetic engineering techniques in BDRP facilities has also been a focus of controversy. The BDRP takes the position that genetic engineering will be utilized if deemed necessary. The Department of Defense stated that testing of aerosols of pathogens derived from recombinant DNA methodology is not precluded if a need should arise in the interest of national defense.

One specific program requires special note. The Defense Advanced Research Project Agency is a Pentagon program that invests significantly in pathogen research through grants to qualified institutions. This project initially focused on engineering and electronics (computer) projects; however, starting in 1995 biology became a key focus, and several BW defensive research grants are now in operation at several academic and private institutions.

Very little is written in the unclassified literature on BW research conducted in countries other than the United States. Great Britain has maintained the Microbiological Research Establishment at Porton Down; however, military research is highly classified in Great Britain and details regarding the research conducted at Porton are unavailable.

During the 1970s and 1980s, much of the U.S. BW policy was based on the assumption of Soviet offensive BW capabilities. Most U.S. accounts of Soviet BW activities were unconfirmed accusations or claims about treaty violations. The Soviet Union was a party to both the 1925 Geneva Protocol and the 1972 BW convention. According to Pentagon sources, the Soviet Union operated at least seven top-security BW centers. These centers were reported to be under strict military control. Although the former Soviet Union proclaimed that their BW program was purely defensive, the United States consistently asserted that the Soviet Union was conducting offensive BW research.

V. CONTEMPORARY ISSUES

A. Genetic engineering

There has been considerable controversy regarding the potential for genetically engineered organisms to serve as effective BW agents. Recombinant DNA technology has been cited as a method for creating novel, pathogenic microorganisms. Theoretically, organisms could be developed that would possess predictable characteristics, including antibiotic resistance, altered modes of transmission, and altered pathogenic and immunogenic capabilities. This potential for genetic engineering to significantly affect the military usefulness of BW has been contested. It has been suggested that because many genes must work together to endow an organism with pathogenic characteristics, the alteration of a few genes with recombinant DNA technology is unlikely to yield a novel pathogen that is significantly more effective or usable than conventional BW agents.

The question of predictability of the behavior of genetically engineered organisms was addressed at an American Society for Microbiology symposium held in June 1985. Some symposium participants believed that the use of recombinant DNA increases predictability because the genetic change can be precisely characterized. Other participants, however, believed that the use of recombinant DNA decreases predictability because it widens the potential range of DNA sources. Other evidence supports the view that genetically engineered organisms do not offer substantial military advantage over conventional BW. Some studies have shown that in general, genetically engineered organisms do not survive well in the environment. This fact has been cited as evidence that these organisms would not make effective BW agents.

Despite the contentions that genetic engineering does not enhance the military usefulness of BW, a significant number of arguments support the contrary view. At the 1986 review conference of the BW convention, it was noted that genetic engineering advances since the convention entered into force may have made biological weapons a more attractive military option.

Several authors have contended that the question of the potential of genetic engineering to enhance the military usefulness of BW is rhetorical because the 1972 BW convention prohibits development of such organisms despite their origin or method of production. Nations participating in both the 1980 and 1986 review conferences of the BW convention accepted the view that the treaty prohibitions apply to genetically engineered BW agents. An amendment to the treaty, specifically mentioning genetically engineered organisms, was deemed to be unnecessary. In addition, the United States, Great Britain, and the Soviet Union concluded in a 1980 briefing paper that the 1972 BW convention fully covered all BW agents that could result from genetic manipulation.

Although the utility of genetic engineering for enhancing the military usefulness of BW agents has been questioned, the role of genetic engineering for strengthening defensive measures against BW has been clear. Genetic engineering has the potential to improve defenses against BW in two ways: (i) vaccine production and (ii) sensitive identification and detection systems. The issues of the new technologies in defensive research have been evident in the U.S. BW program. Since 1982, U.S. Army scientists have used genetic engineering to study and prepare defenses against BW agents. Military research utilizing recombinant DNA and hybridoma technology includes the development of vaccines against a variety of bacteria and viruses, methods of rapid detection and identification of BW agents, and basic research on protein structure and gene control. By improving defenses against BW, it is possible that genetic engineering may potentially reduce the risk of using BW.

The primary effect of BW on government regulations on genetic engineering is the tendency toward more stringent control of the technologies. The fear of genetically engineered BW agents has prompted proposals for government regulation of BW research utilizing genetic engineering research. The Department of Defense released a statement indicating that all government research was in compliance with the 1972 BW convention. The government has also prepared an environmental impact statement of research conducted at Fort Detrick.

Government regulations on genetic engineering also affect BW research through limitations on exports of biotechnology information, research products, and equipment. In addition to controls of exports due to competitive concerns of biotechnology companies, a substantial amount of information and equipment related to genetic engineering are prohibited from being exported from the United States. The Commerce Department maintains a "militarily critical technology" list, which serves as an overall guide to restricted exports. Included on the list are containment and decontamination equipment for large production facilities, high-capacity biological reactors, separators, extractors, dryers, and nozzles capable of disseminating biological agents in a fine mist.

Genetic engineering has altered the concept of BW. A current, comprehensive discussion of BW would include both naturally occurring and potential genetically engineered agents. Many current defenses against BW are developed with genetic engineering techniques. Government regulations on biotechnology have limited BW research, while fears of virulent genetically engineered BW agents have strengthened public support for stronger regulations. Future policies related to BW will need to be addressed in light of its altered status.

B. Mathematical epidemiology models

Although genetic engineering may potentially alter characteristics of BW agents, mathematical models of epidemiology may provide military planners with techniques for predicting the spread of a released BW agent. One of the hindrances that has prevented BW from being utilized or even seriously considered by military leaders has been the inability to predict the spread of a BW agent once it has been released into the environment. Without the capability to predict the spread of the released organisms, military planners would risk the accidental exposure of their own troops and civilians to their own weapons. The development of advanced epidemiology models may provide the necessary mechanisms for predicting the spread of organisms that would substantially decrease the deterrent factor of unpredictability.

C. Low-level conflict

Another important factor that has affected the current status of BW is the increase in low-level conflict or the spectrum of violent action below the level of small-scale conventional war, including terrorism and guerrilla warfare. In the 1980s, the low-intensity conflict doctrine, which was espoused by the Reagan administration, was a plan for U.S. aid to anti-Communist forces throughout the world as a way of confronting the Soviet Union without using U.S. combat troops. Despite the significant changes in the world since the inception of the low-intensity conflict doctrine, the probability of increasing numbers of small conflicts still exists. Although no evidence indicates that the United States would consider violating the 1972 BW convention and support biological warfare, the overall increase in low-level conflicts in the future may help create an environment conducive to the use of BW.

Although BW may not be assessed as an effective weapon in a full-scale conventional war, limited use of BW agents may be perceived as advantageous in a small-scale conflict. Although strong deterrents exist for nuclear weapons, including unavailability and, most formidably, the threat of uncontrolled worldwide "nuclear winter," BW may be perceived as less dangerous. In addition, the participants of low-level conflicts may not possess the finances for nuclear or conventional weapons. BW agents, like chemical weapons, are relatively inexpensive compared to other weapon systems and may be seen as an attractive alternative to the participants and leaders of low-level conflicts. Low-level conflict, therefore, increases the potential number of forums for the use of BW.

D. Terrorism

A final factor that could significantly affect BW is the worldwide increase in terrorism or the violent activities of subnational groups. Although there has not been an incident to date of the successful use of BW by a terrorist group, the possibility of such an event has increased in many forums.

The relationship of terrorism and BW can be divided into two possible events. The first is terrorist acts against laboratories conducting BW-related research. The level of security at Fort Detrick is high, the possibility of a terrorist attack has been anticipated, and contingency plans have been made. Complicating the problem of providing security against terrorist attack in the United States is the fact that although most BW research projects are conducted with the BW research program of the Army, an increasing number of projects are supported by the government that are conducted outside of the military establishment. These outside laboratories could be potential targets.

The second type of terrorist event related to BW is the potential use of BW by terrorists against urban areas or major public facilities. Biological weapons are relatively inexpensive and easy to develop and produce compared to conventional, nuclear, or chemical weapons. BW agents can be concealed and easily transported across borders or within countries. In addition, terrorists are not hampered by a fear of an uncontrolled spread of the BW agent into innocent civilian populations. On the contrary, innocent civilians are often the intended targets of terrorist activity and the greater chance for spread of the BW agent may be considered to be a positive characteristic (see Section II.C).

E. Offensive versus defensive biological warfare research

The distinctions between "offensive" and "defensive" BW research have been an issue since 1969, when the United States unilaterally pledged to conduct only defensive research. The stated purpose of the U.S. BDRP is to maintain and promote national defense from BW threats. Although neither the Geneva Convention nor the 1972 convention prohibits any type of research, the only research that nations have admitted to conducting is defensive. The problem is whether or not the two types of research can be differentiated by any observable elements.

Although production of large quantities of a virulent organism and testing of delivery systems have been cited as distinguishing characteristics of an offensive program, a substantial amount of research leading up to these activities, including isolating an organism and then using animal models to determine pathogenicity, could be conducted in the name of defense.

Vaccine research is usually considered defensive, whereas increasing the virulence of a pathogen and producing large quantities are deemed offensive. However, a critical component of a strategic plan to use biological weapons would be the production of vaccines to protect the antagonist's own personnel (unless self-annihilation was also a goal). This means that the intent of a vaccine program could be offensive BW use. Furthermore, research that increases the virulence of an organism is not necessarily part of an offensive strategy because one can argue that virulence needs to be studied in order to develop adequate defense.

The key element distinguishing offensive from defensive research is intent. If the intent of the researcher or the goals of the research program are the capability to develop and produce BW, then the research is offensive BW research. If the intent is to have the capability to develop and produce defenses against BW use, then the research is defensive BW research. Although it is true that nations may have policies of open disclosures (i.e., no secret research), "intent" is not observable.

Although the terms offensive BW research and defensive BW research may have some use in describing intent, it is more a philosophical than a practical distinction—one that is based on trust rather than fact.

F. Secrecy in biological warfare-related research

Neither the Geneva Protocol nor the 1972 BW convention prohibits any type of research, secret or nonsecret. Although the BDRP does not conduct secret or classified research, it is possible that secret BW research is being conducted in the United States outside of the structure of the BDRP. The classified nature of the resource material for this work makes it impossible to effectively determine if secret research is being conducted in the United States or any other nation.

It is not, however, unreasonable to assume that other nations conduct significant secret BW research. Therefore, regardless of the facts, one cannot deny the perception that such research exists in a variety of countries and that this perception will exist for the foreseeable future.

Secrecy has been cited as a cause of decreased quality of BW research. If secret research, whether offensive or defensive, is being conducted in the United States or other nations, it is unclear if the process of secrecy affects the quality of the research. If the secret research process consists of a core of highly trained, creative, and motivated individuals sharing information, the quality of the research may not suffer significantly. It must be stated, however that secrecy by its very nature will limit input from a variety of diverse observers.

Secrecy may increase the potential for violations of the 1972 BW convention; however, violations would probably occur regardless of the secrecy of the research. Secrecy in research can certainly lead to infractions against arbitrary rules established by individuals outside of the research group. The secret nature of the research may lure a researcher into forbidden areas. In addition, those outside of the research group, such as policy-makers, may push for prohibited activities if the sense of secrecy prevails. Secrecy also tends to bind those within the secret arena together and tends to enhance their perception of themselves as being above the law and knowing what is "right." As in the case of Oliver North and the Iran-Contra Affair, those within the group may believe fervently that the rules must be broken for a justified purpose and a mechanism of secrecy allows violations to occur without penalty.

The distrust between nations exacerbates the perceived need for secret research. The animosity between the United States and the Soviet Union during the 1980s fueled the beliefs that secret research leading to violations of the 1972 BW convention was being conducted in the Soviet Union. As the belligerence of the 1980s faded into the new world order of the 1990s, the questions focus less on the Soviet Union and more on the Middle East and Third-World countries. There are factions in the United States that believe strongly that other countries are conducting secret research that will lead to violations of the convention. There is also a tendency to believe that the secrecy in one's own country will not lead to treaty violations, whereas the same secret measures in an enemy nation will result in activities forbidden by international law.

The importance of the concept of secrecy in BW research is related to the perception of secrecy and arms control agreements. Regardless of the degree of secrecy in research, if an enemy believes that a nation is pursuing secret research, arms control measures are jeopardized. The reduction of secrecy has been suggested as a tool to decrease the potential for BW treaty violations. A trend toward reducing secrecy in BW research was exemplified by the 1986 review conference of the 1972 BW convention, which resulted in agreements to exchange more information and to publish more of the results of BW research. Whether or not these measures have any effect on strengthening the 1972 BW convention remains to be seen.

Organizations and individuals have urged a renunciation by scientists of all secret research and all security controls over microbiological, toxicological, and pharmacological research. This action has been suggested as a means of strengthening the 1972 BW convention. The belief that microbiologists should avoid secret research is based on the assumption that (i) secret research is of poor quality due to lack of peer review and (ii) secrecy perpetuates treaty violations.

Although it may be reasonable to expect microbiologists to avoid secret research, it is not realistic. Secrecy is practiced in almost every type of research including academic, military, and especially industrial. Furthermore, there will always be those within the military and intelligence structures who believe that at least some degree of secrecy is required for national security.

Secrecy in BW research is a complex issue. The degree to which it exists is unclear. Individuals are generally opposed to secrecy in BW research although other examples of secrecy in different types of research exist. The effect of secrecy on the quality of research, the need for the secrecy, and the choice of microbiologists to participate in secret BW research remain unanswered questions.

G. Problems relating to verification

One of the major weaknesses of the 1972 BW convention has been the lack of verification protocols. Problems with effectively monitoring compliance include the ease of developing BW agents in laboratories designed for other purposes and the futility of inspecting all technical facilities of all nations. Measures that have been implemented with the goal of monitoring compliance included (i) open-inspections, (ii) intelligence gathering, (iii) monitor-research, (iv) use of sampling stations to detect the presence of biological agents, and (v) international cooperation. The progress achieved with the Chemical Weapons Convention has renewed interest in strengthening mechanisms for verification of compliance with the 1972 BW convention. Although this renewed interest in verification along with the emergence of the Commonwealth of Independent States from the old Soviet Union has brought an optimism to the verification issue, the reticence of countries such as Iraq to cooperate with United Nations inspection teams is a reminder of the complexities of international agreements.

The examples discussed in this article are typical of the many issues attached to the concept of BW.

BIBLIOGRAPHY

Atlas, R. M. (1998). Biological weapons pose challenge for microbiology community. *ASM News* **64**, 383–389.

Buckingham, W. A., Jr. (Ed.) (1984). "Defense Planning for the 1990s." National Defense Univ. Press, Washington, DC.

Cole, L. (1996, December). The specter of biological weapons. *Sci. Am.* 60–65.

Frisna, M. E. (1990). The offensive–defensive distinction in military biological research. *Hastings Cent. Rep.* **20**(3), 19–22.

Gravett, C. (1990). "Medieval Siege Warfare." Osprey, London.

Harris, R., and Paxman, J. (1982). "A Higher Form of Killing." Hill & Wang, New York.

Khan, A. S., Morse, S., Lillibridge, S. (2000) Public-health preparedness for biological terrorism in the USA. *Lancet* **356**(9236), 1179–1182.

Livingstone, N. C. (1984). Fighting terrorism and "dirty little wars." *In* "Defense Planning for the 1990s" (W. A. Buckingham, Jr., Ed.), pp. 165–196. National Defense Univ. Press, Washington, DC.

Livingstone, N. C., and Douglass, J., Jr. (1984). "CBW: The Poor Man's Atomic Bomb." Tufts University, Institute of Foreign Policy Analysis, Medford, MA.

Louria, D. B. (1986). Recombinant DNA technology and biological warfare. *N. J. Med.* **83**(6), 399–400.

Meselson, M., Guillemin, J., Hugh-Jones, M., Langmuir, A., Popova, I., Shelokov, A., and Yampolskaya, O. (1994). The Sverdlovsk anthrax outbreak of 1979. *Science* **266**, 1202–1208.

Milewski, E. (1985). Discussion on a proposal to form a RAC working group on biological weapons. *Recombinant DNA Technol. Bull.* **8**(4), 173–175.

Miller, L. A. (1987). The use of philosophical analysis and Delphi survey to clarify subject matter for a future curriculum for microbiologists on the topic of biological weapons. Unpublished thesis, University of Pennsylvania, Philadelphia. (University Micro-films International, Ann Arbor, MI. 8714902).

Murphy, S., Hay, A., and Rose, S. (1984). "No Fire, No Thunder." Monthly Review Press, New York.

Poupard, J. A., Miller, L. A., and Granshaw, L. (1989). The use of smallpox as a biological weapon in the French and Indian War of 1763. *ASM News* **55**, 122–124.

Smith, R. J. (1984). The dark side of biotechnology. *Science* **224**, 1215–1216.

Stockholm International Peace Research Institute (1973). "The Problem of Chemical and Biological Warfare," Vol. 2. Humanities Press, New York.

Taubes, G. (1995). The defense initiative of the 1990s. *Science* **267**, 1096–1100.

Wright, S. (1985). The military and the new biology. *Bull. Atomic Sci.* **42**(5), 73.

Wright, S., and Sinsheimer, R. L. (1983). Recombinant DNA and biological warfare. *Bull. Atomic Sci.* **39**(9), 20–26.

Zilinskas, R., (Ed.) (1992). "The Microbiologist and Biological Defense Research: Ethics, Politics and Intermediate Security." New York Academy of Sciences, New York.

WEBSITES

A website of the Stockholm International Peace Research Institute, *http://projects.sipri.se/cbw/bw-mainpage.html*

DefenseLINK. A website of the U.S. Department of Defense *http://www.defenselink.mil/sites/*

14

Bioluminescence, Microbial

J. Woodland Hastings
Harvard University

GLOSSARY

autoinducer A homoserine lactone produced by bacteria which, after accumulating in the medium to a critical concentration, initiates transcription of specific genes by a mechanism referred to as autoinduction, recently dubbed **quorum sensing**.

bioluminescence Emission of light by living organisms that is visible to other organisms. It derives from an enzyme-catalyzed chemiluminescence, a highly exergonic reaction in which chemical energy is transformed into light energy.

bioluminescence quantum yield The number of photons produced per luciferin (substrate) molecule oxidized in a bioluminescent reaction.

blue and yellow fluorescent proteins Accessory proteins in the bioluminescence system in some bacteria, carrying lumazine and flavin chromophores, respectively, and serving as secondary emitters under some conditions.

luciferase The generic name for enzymes that catalyze bioluminescent reactions. Luciferases from different major groups of organisms are not homologous (e.g. firefly and jellyfish luciferases are unrelated to bacterial luciferase) so the organism must be specified in referring to a specific luciferase.

luciferin (light bearing) The generic name for a substrate that is oxidized to give light in a bioluminescent reaction; identified as a **flavin** in bacteria and a **tetrapyrrole** in dinoflagellates.

scintillons Bioluminescent organelles unique to dinoflagellates which emit brief bright flashes of light following stimulation.

Bioluminescence is defined as an enzyme-catalyzed chemiluminescence, a chemical reaction in which the energy released is used to produce an intermediate or product in an electronically excited state, which then emits a photon. It does not come from or depend on light absorbed, as in fluorescence or phosphorescence. However, the excited state produced in such a chemical reaction is indistinguishable from that produced in fluorescence after the absorption of a photon by the ground state of the molecule concerned.

All bioluminescent reactions involve the oxidation by molecular oxygen of a substrate by an enzyme, generically referred to as luciferin and luciferase, respectively, with the production of an electronically excited state, typically luciferase bound (Fig. 14.1A). The energy released from the oxidation of a luciferin in such reactions is about 10 times greater than that obtained from the hydrolysis of ATP.

There are numerous (20–30) extant bioluminescent systems, which mostly bear no evolutionary relationships with one another. The many different luciferases are thus considered to have arisen *de novo* and evolved independently, and the luciferins are likewise different. Thus, genes coding for luciferases from fireflies and jellyfish, for example, have no sequence similarities to bacterial or dinoflagellate luciferase, which themselves are unrelated. The luciferases discussed in this article

The Desk Encyclopedia of Microbiology
ISBN: 0-12-621361-5

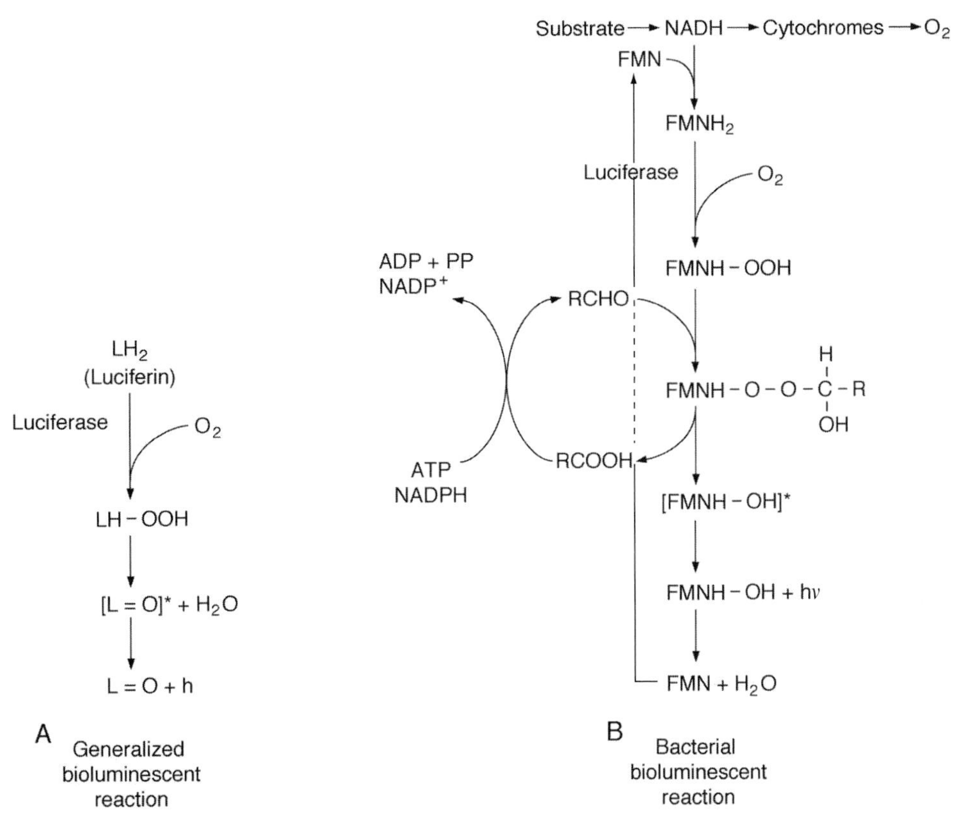

FIGURE 14.1 (A) Generalized reaction scheme for bioluminescent reactions. (B) The luciferase reaction showing the components involved in the bacterial system. Reduced flavin derived from the electron transport pathway reacts with luciferase and molecular oxygen to form an intermediate peroxide. In a mixed function oxidation with long-chain aldehyde (RCHO), hydroxy-FMN is formed in its excited state (*), from which emission of a photon (hν) occurs. The FMN and long-chain acid products are recycled.

should therefore be called bacterial luciferase and dinoflagellate luciferase.

I. BACTERIA

A. Occurrence, habitats, species, and functions

In the ocean luminous bacteria occur ubiquitously and can be isolated from most seawater samples from the surface to depths of ~1000 m, and they appear as bright colonies on plates (Fig. 14.2). They are very often found in some kind of symbiotic association with higher organisms (e.g. fish or squid), in which the light emission is evidently of functional importance to the host. In parasitic or saprophytic associations the advantage of light emission accrues more to the bacteria: The light attracts animals to feed, enhancing the

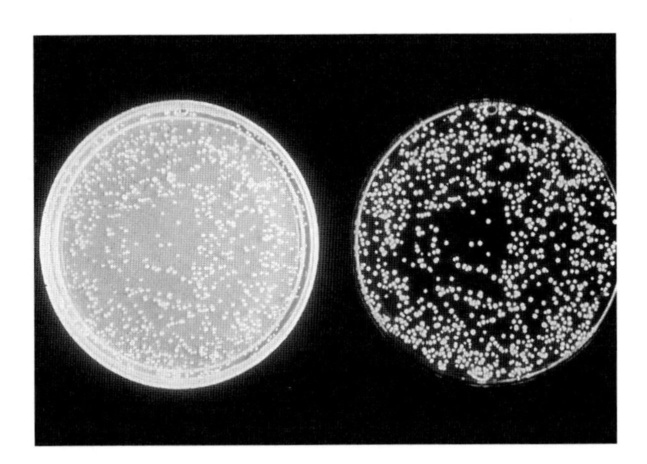

FIGURE 14.2 Colonies of luminous bacteria photographed by their own light (right) and in room light (left). Light is emitted continuously but is controlled by a quorum-sensing mechanism (autoinducer) and is thus not proportional to growth or cell density (Plate 2).

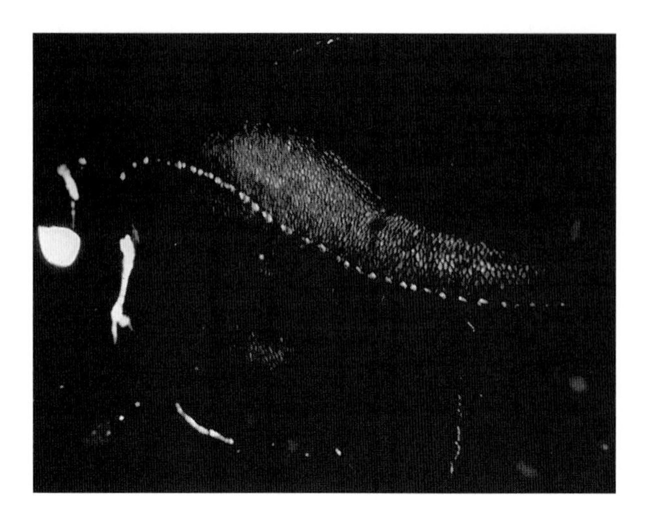

FIGURE 14.3 The flashlight fish (*Photoblepharon*) showing the exposed light organ, which harbors luminous bacteria and is located just below the eye. A special lid allows the fish to turn the light on and off (photograph by Dr. James Morin) (Plate 3).

dispersal of the bacteria. The maximum light emission of a bacterial cell is about $10^4 \mathrm{q\,s^{-1}}$, meaning that to be seen, the cell density must be high—about 10^9–10^{10} cells/ml.

Most luminous bacteria are classed under three major genera—*Vibrio, Photobacterium*, and *Photorhabdus*, the first two of which are almost exclusively marine, whereas the last is terrestrial. All are characterized as enteric bacteria and are notable for the symbioses in which they participate, most commonly in light organs in which the light is used by the host for some purpose. The flashlight fish, *Photoblepharon* (Fig. 14.3), maintains cultures of such bacteria in special organs located beneath the eyes. For all luminous bacterial species the primary habitat can be assumed to be in some association, either as a light organ or gut symbiont, or in a parasitic or saprophytic association.

If such associations are viewed as primary habitats, planktonic or "free-living" bacteria in the ocean may be considered secondary or reservoir habitats, produced as overflows or escapees into an environment in which luminescence may not be advantageous and thus not selected for. Thus, the failure, for whatever reason, of the luminescence system to be expressed under these conditions may be advantageous for the survival of the bacteria possessing the genes for luminescence and thus for their ability to compete favorably with heterotrophs not carrying the genes.

Different species occupy different specific habitats. *Vibrio harveyi* is the most cosmopolitan species and is not known to be involved in a light organ symbiosis. It occurs as a gut symbiont in many marine animals, and it is known to parasitize and/or infest saprophytically crustaceans and other species. Fish or squid having specific associations with *Vibrio fischeri, Photobacterium phosphoreum*, and *Photobacterium leiognathi* as symbionts have been identified, and all of these species have more restricted requirements for growth. Still other symbionts have not been cultured successfully, but affinities and relationships are known from their luciferase DNA sequences.

Photorhabdus luminescens is symbiotic with nematodes, which parasitize caterpillars, where they release the bacteria as an inoculum into the body cavity along with their own fertilized eggs. The bacteria grow, providing nutrient for the developing nematode larvae. The caterpillar does not survive but becomes brightly luminous, possibly to attract animals to feed on it and thereby disperse both nematodes and bacteria. Each young nematode then carries a fresh inoculum, estimated to be about 50–100 bacteria.

B. Biochemistry

1. Light-emitting reaction

Biochemically, light emission results from the luciferase-catalyzed mixed function oxidation of reduced flavin mononucleotide and long-chain aldehyde by molecular oxygen, populating the excited state of a luciferase-hydroxyflavin intermediate, which emits a blue-green light ($\lambda_{max} \sim 490$ nm).

Bacterial luciferase is an α–β heterodimer lacking metals, prosthetic groups, and non-amino acid residues. To date, no sequences in data bases exhibit any similarities to it, so the origin of the gene for the enzyme remains unknown. Although the two subunits are homologous, the active site and the detailed kinetics features of the reaction are properties of the α subunit (singular kinetic).

The reaction represents a biochemical shunt of the respiratory electron transport system, carrying electrons from the level of reduced flavin ($FMNH_2$) directly to oxygen (Fig. 14.1B). $FMNH_2$ reacts first with oxygen to form a linear hydroperoxide, which then reacts with long-chain fatty aldehyde to give the postulated peroxyhemiacetal intermediate. This breaks down to give long-chain acid and the intermediate hydroxyflavin in a high-energy electronically excited state. Although aldehydes with chain lengths from 7 to 18 carbon atoms give light in the reaction with isolated luciferase, tetradecanal (14C) has been identified as the naturally occurring molecule in the species studied. While its oxidation provides energy, the aldehyde is not the emitter and is thus not a luciferin, which means "light-bearing." The flavin is the luciferin in the bacterial system.

One photon is produced for about every four molecules of $FMNH_2$ oxidized; thus, the bioluminescence quantum yield is 25%. However, since the fluorescence quantum yield of FMN is about 30%, and the excited state produced as in Fig. 14.1B is equivalent to that which would be produced from light absorption by the hydroxyflavin, it may be concluded that the luciferase reaction is highly efficient.

In the living cell light is produced continuously; the oxidized FMN formed in the reaction is reduced again as indicated in Fig. 14.1B by pyridine nucleotide. Similarly, the myristic acid product is converted back to the corresponding aldehyde by enzymes of a specific fatty acid reductase complex with ATP and NADPH as cofactors.

2. *Luciferase structure*

The tertiary structure of luciferase (Fig. 14.4) was correctly predicted based on the X-ray crystal structure of a related and homologous protein expressed from *lux F*. The structures of the α and β subunits are similar, differing in a region in which substrate is presumed to bind to the α, although the structure of the site is not known because a determination of the structure with flavin bound has not been made. Both subunits exhibit the so-called $[\beta/\alpha]_8$ barrel form (β/α do not refer to subunits), in which stretches of beta sheet alternate with alpha-helical strands, all of which are parallel to one another, together forming a closed barrel with the eight α helices on the outside. The region of the α subunit where substrate is likely to bind was not seen in the electron density map, including a region known to be highly sensitive to protease attack that is absent in the β subunit.

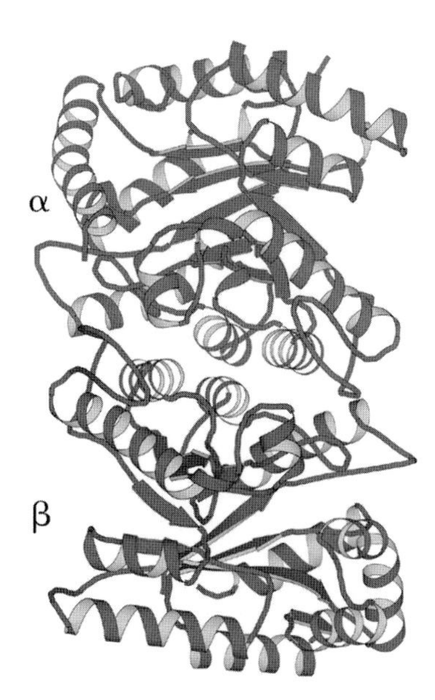

FIGURE 14.4 Ribbon representation of the structure of bacterial luciferase showing the α and β subunits and how they may associate (from Fisher *et al.*, 1995).

3. *Antenna proteins: blue and yellow fluorescent proteins as emitters*

The basic structure of the luciferase and the biochemistry of the reaction are the same in all luminous bacteria, which typically emit light peaking at about 490 nm. However, in some bacterial strains, the color of the light emitted by the living cell is blue or red shifted, even though the isolated luciferases still peaks at ~490 nm. In a strain of *P. phosphoreum*, the emission is blue shifted, peaking at about 480 nm, whereas in a strain of *V. fischeri* the light is yellow in color (λ_{max} ~ 540 nm). In both cases a second ("antenna") protein with its own chromophore is responsible (lumazine and flavin, respectively, for the two cases). The mechanisms involved in these cases have not been fully resolved, but they may be mechanistically similar because the proteins are homologous. Nonradiative energy transfer has been suggested, but evidence indicates that this alone cannot be responsible since the antenna protein appears to actually enter into the light-emitting reaction in the case of the yellow-emitting system. The functional importance for such spectral shifts has not been elucidated, although strains with a blue-shifted emission occur at depths of ~600 m in the ocean.

4. *Molecular biology: genes of the* lux *operon*

Lux genes cloned from several different species exhibit sequence similarities indicative of evolutionary relatedness and conservation. In *V. fischeri*, the species most extensively studied, the *lux* operon has five structural and two regulatory genes with established functions. As shown in Fig. 14.5, these include two that code for the α and β subunits of luciferase and three that code for the reductase, transferase, and synthetase components of the fatty acid reductase complex responsible for aldehyde synthesis and reduction and recycling of the acid product. Upstream, in the same operon, is the regulatory gene *lux I*, and immediately adjacent but transcribed in the opposite direction is *lux R*, whose product is responsible for the transcriptional activation of *lux A–E* and others. The latter include *lux F* and *G*, found in some species and located in the same region; these code for proteins whose functions are not well established. However, genes coding for the antenna proteins responsible for color shifting are located elsewhere on the genome but still subject to regulation by the autoinduction mechanism. In luminous Vibrios the

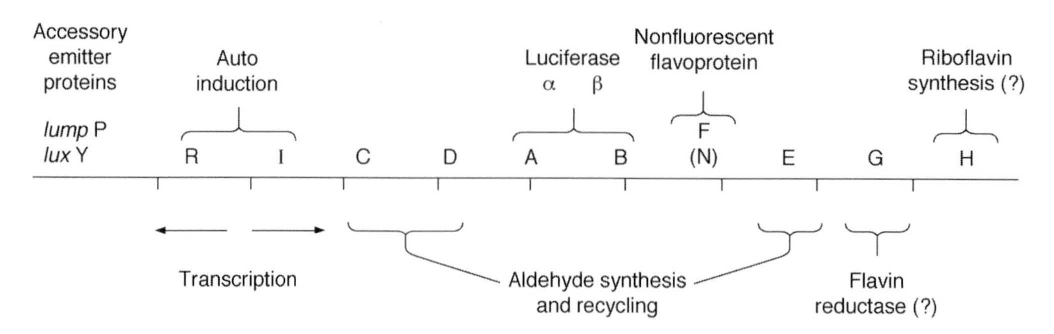

FIGURE 14.5 Organization of the *lux* genes in *Vibrio fischeri*. The operon on the right, transcribed from the 5′ to the 3′ end, carries genes for synthesis of autoinducer (*lux I*), for luciferase α and β subunits (*lux A* and *B*), and for aldehyde synthesis (*lux C–E*). The operon on the left carries the *lux R* gene, which encodes for a receptor molecule that binds autoinducer; the complex controls the transcription of the right operon. Other genes, *lux F (N)*. *G*, and *H* (right), are associated with the operon but with uncertain functions; genes for accessory emitter proteins also occur (left).

regulatory genes (*lux I* and *R* or their counterparts) are also located remotely from the *lux A–E* operon.

5. *Physiology: regulation of light emission*

a. *Autoinduction and quorum sensing*

Quorum sensing, which was first discovered and referred to as autoinduction in luminous bacteria, refers to a mechanism causing the *lux* genes to be transcribed only at higher cell densities. It is mediated by a homoserine lactone molecule produced by the cells that has been dubbed the autoinducer. In luminous bacteria in a confined environment, such as a light organ, autoinducer can accumulate and act, whereas in free-living bacteria, in which the light could not be seen at the low cell densities, luciferase is of no value and is not produced.

Autoinduction was discovered in the early 1970s and was proposed is as an explanation for the fact that luciferase and other components of the light-emitting system are not produced in cells growing at low cell densities but are produced, and rapidly so, above a critical cell concentration (Fig. 14.6). In laboratory cultures subjected to continuous (or repetitive) dilution (maintaining densities lower than ~10^7 cells/ml), such that autoinducer accumulation is not possible, no synthesis of luciferase or its messenger RNA occurs. The same is true of planktonic populations in the ocean, which are typically at densities of no greater than ~10^2 cells/ml.

The autoinducer in *V. fischeri* is the product of the *lux I* gene and acts as a positive regulator of the *lux* operon in the presence of a functional *lux R* gene. In the early 1990s it was found that a similar mechanism, also utilizing specific homoserine lactones, occurs in many other diverse groups of bacteria in which expression of certain specific genes is functionally important only at higher cell densities. The phenomenon was then appropriately dubbed "quorum sensing."

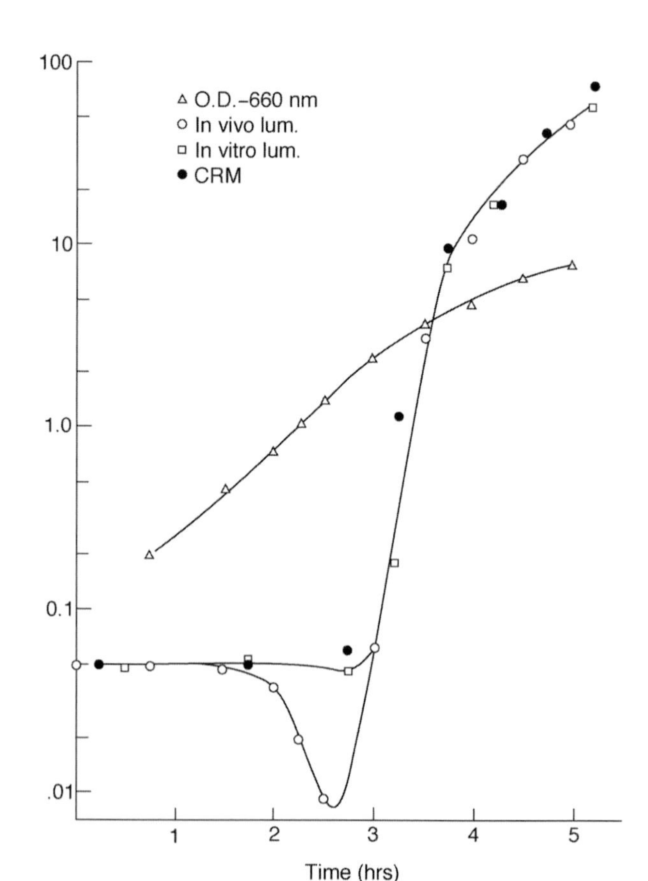

FIGURE 14.6 An experiment demonstrating autoinduction in luminous bacteria. Growth of the cells (measured by optical density at 660 nm) is exponential for the first few hours, during which time there is no change in the bioluminescence or the luciferase content of the culture (arbitrary units). After about 2.5 h, at a cell density of 1 or greater, the *lux* genes are rapidly transcribed and luciferase and other related proteins are synthesized, determined also by reaction with antiluciferase (CRM).

b. *Glucose, iron, and oxygen*

In both *V. harveyi* and *V. fischeri* luminescence is repressed by glucose, reversible by cyclic AMP. All species except *Photorhabdus luminescens* exhibit induction of luminescence by the addition of iron after growth under conditions of iron limitation, suggesting that eukaryotic hosts might use iron limitation to limit growth while maximizing luminescence of bacterial symbionts.

As a reactant in the luciferase reaction, oxygen can control light emission directly but only at extremely low oxygen concentrations (lower than ~0.1%), above which the luminescence is independent of oxygen. Growth, however, is reduced at concentrations lower than that of air (21%), so growth could be strongly inhibited without affecting the luciferase reaction. Indeed, in some species and strains, transcription of the *lux* operon is greatly favored over growth under microaerophilic conditions, where bright light emission may occur at low cell densities. Regulation by control of oxygen has thus been proposed as a mechanism whereby eukaryotic hosts might control growth of bacterial symbionts in light organs while maximizing luminescence.

In *P. luminescens* grown in pure (100%) oxygen, which is lethal for many bacteria including other luminous species, both luciferase synthesis and luminescence are enhanced in relation to cell mass. It has been proposed that in this case the luciferase system serves to detoxify damaging oxygen radicals.

c. *Dark variants*

In culture collections it has often been reported that bacteria may lose their luminosity over time in cases in which subculturing care has not been taken to reisolate single bright colonies each time. This can be attributed to the spontaneous occurrence and overgrowth of dark (e.g. very dim) variants, which is some ways appear to be similar to phase variants reported in other groups of bacteria. These dark variants do not produce luciferase or other luminescence components and are pleiotropic, being altered in several other properties such as cell morphology and phase sensitivity.

Dark variants could presumably provide a genetic mechanism whereby cells could respond to environmental conditions that select for or against the property of luminescence. Although such conditions have not been established, this would allow cells to compete better with other heterotrophic bacteria under conditions where luminescence is of no use and thereby become more widely dispersed but prepared to populate a niche where luminescence is functionally important.

II. DINOFLAGELLATES

A. Occurrence, habitats, species, and functions

Ocean "phosphorescence," commonly seen at night (especially in summer) when the water is disturbed, is due in large part to the bioluminescence of dinoflagellates. The organisms occur ubiquitously in the oceans as planktonic forms and respond to mechanical stimulation when the water is disturbed, such as by waves or fish swimming, by emitting brief (~0.1 s) bright flashes (~10^9 photons each). The wake of a large ship may be evident from such light emission for approximately 20 miles. Luminescent dinoflagellates occur primarily in surface waters and many species are photosynthetic. Only approximately 20–30% of marine species are bioluminescent.

The so-called red tides are transient blooms (usually for weeks) of individual dinoflagellate species. Cells typically migrate vertically during the night to deeper water where available nutrients are taken up, returning to the surface to photosynthesize during the day. Phosphorescent bays (e.g. in Puerto Rico and Jamaica) are persistent blooms of this type; in Puerto Rico, the dominant species is *Pyrodinium bahamense*.

As a group, dinoflagellates are important as symbionts, notably for contributing photosynthesis and carbon fixation in certain animals. Unlike bacteria, however, luminous dinoflagellates are not known to be harbored as symbionts on the basis of their light emission.

Since dinoflagellates are stimulated to emit light when predators (e.g. crustaceans) are active, predators might thereby be alerted to feed on crustaceans, resulting in a reduced predation on dinoflagellates generally. Predation on dinoflagellates may also be impeded more directly and help individual cells. The flash could startle or divert a predator, allowing that cell to escape predation. The response time to stimulation (milliseconds) is certainly fast enough to have this effect. This latter explanation, though supported by experiment, does not easily account for the fact that not all species are bioluminescent.

B. Cell biology

Luminescence in dinoflagellates is emitted from many small (~0.5 μm) cortical locations (Fig. 14.7). The structures have been identified as a new type of organelle termed scintillons (flashing units). They occur as outpocketings of the cytoplasm into the cell vacuole, like a balloon, with the neck remaining connected. Scintillons contain only dinoflagellate luciferase and luciferin and a protein that binds luciferin and keeps it from reacting with luciferase in between flashes. Other cytoplasmic components are somehow excluded from scintillons, which can be identified ultrastructurally by immunolabeling with antibodies raised against the luminescence proteins. They can also be visualized using image intensification by their bioluminescent flashing following stimulation as well as by the fluorescence of luciferin, which is present nowhere else in the cell.

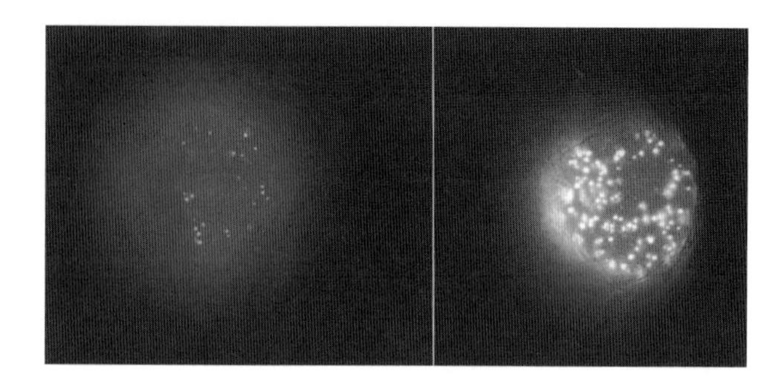

FIGURE 14.7 *Gonyaulax* cells viewed by fluorescence microscopy showing scintillons (bioluminescent organelles) visualized by the fluorescence of dinoflagellate luciferin (λ_{max} of emission, 475 nm), with chlorophyll fluorescence as the red background. Scintillons are structurally formed and destroyed on a daily basis, controlled by the circadian clock. (Right) Night phase cell with many scintillons; (left) day phase cell with few scintillons (Plate 4).

(a)

(b)

$$\underset{(pH\ 7.5)}{LBP\text{–}LH_2} \xrightarrow{H^+} LBP + \underset{(pH\ 6)}{LH_2} \xrightarrow[\text{luciferase}]{O_2} h\nu + L{=}O + H_2O$$

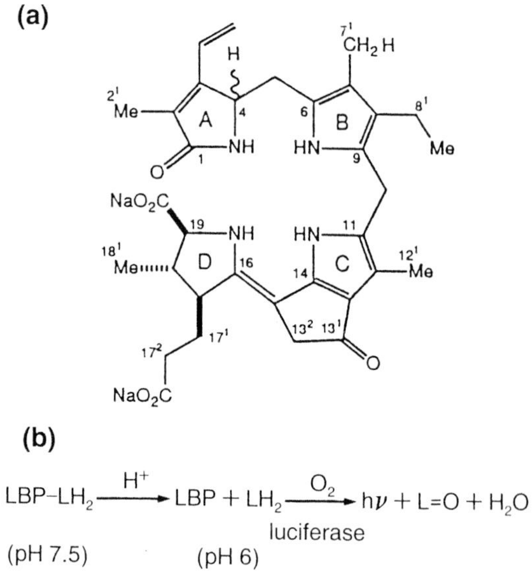

FIGURE 14.8 (A) Dinoflagellate luciferin, a tetrapyrrole, showing the location (13^2) of the oxygen addition and (B) the steps in the bioluminescent reaction.

C. Biochemistry

Dinoflagellate luciferin is a highly reduced novel tetrapyrrole related to chlorophyll (Fig. 14.8) and in extracts remains tightly bound to a ~75-kDa specialized protein at cytoplasmic pH (~8). The luciferase is also inactive at pH 8; it is a large single polypeptide chain of about 136 kDa with three contiguous and intramolecularly homologous domains, each having luciferase activity.

Activity can be obtained in soluble extracts made at pH 8 simply by shifting the pH to 6; the luciferin is released from its binding protein and the luciferase assumes an active conformation. The *pK* for both proteins is at pH 6.7. A similar activity can be found in the particulate (scintillon) fractions. Together, these results suggest that during extraction some scintillons are lysed with the proteins released into the soluble fraction, whereas others seal off at the neck and form closed vesicles. With the scintillon fraction, the *in vitro* activity is also triggered by a pH change and occurs as a flash (~100 ms), very close to that of the living cell, and the kinetics are independent of the dilution of the suspension. For the soluble fraction, the kinetics are dependent on dilution, as in an enzyme reaction.

D. Cellular flashing

The flashing of dinoflagellates *in vivo* is typically initiated by mechanical shear or cell stimulation, which has been shown to result in the generation of a conducted action potential in the vacuolar membrane. It is postulated that as this action potential traverses the vacuolar membrane it sweeps over the scintillons, opening voltage-gated ion channels, thus allowing protons from the acidic vacuole to enter, causing a transient pH change in the scintillons and thus a flash. Spontaneous flashes also occur (Fig. 14.9B).

E. Circadian clock control of dinoflagellate luminescence

Unlike bacteria, cell density and growth conditions have no effect on the development and expression of bioluminescence in dinoflagellates. However, in *G. polyedra* and some other dinoflagellates, luminescence is regulated by day–night light–dark cycles and an internal circadian biological clock mechanism. Spontaneous flashing (and also flashing in response to mechanical stimulation) is far greater during the night than during the day (and therefore flashes are more frequent), and a steady low-level emission (glow) exhibits a peak toward the end of the night phase. The regulation is attributed to an endogenous mechanism; cultures maintained under constant conditions (light and temperature)

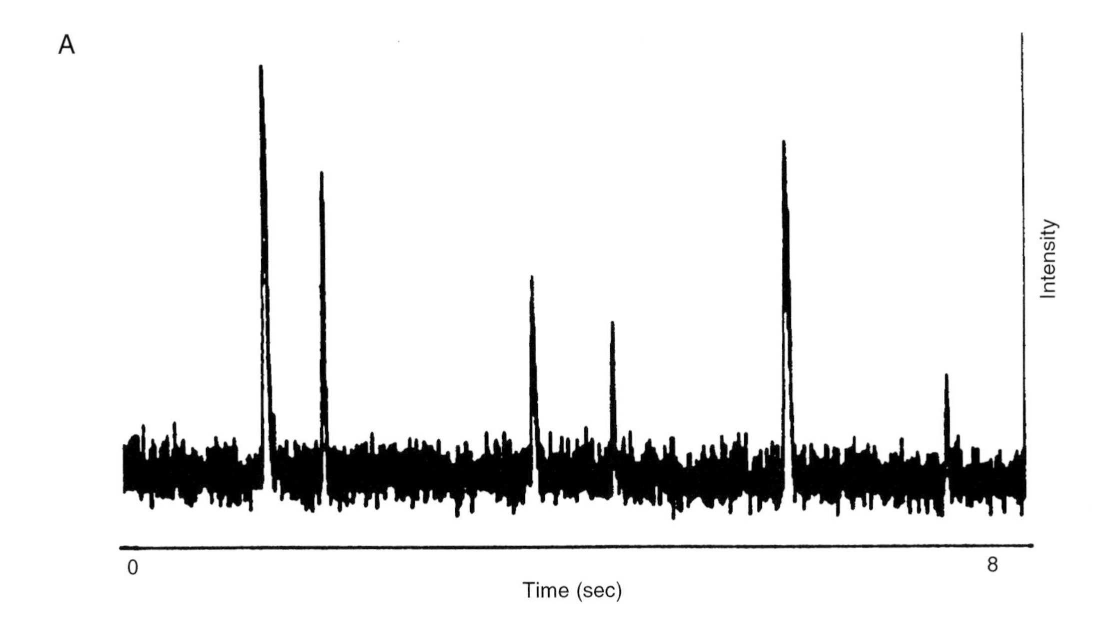

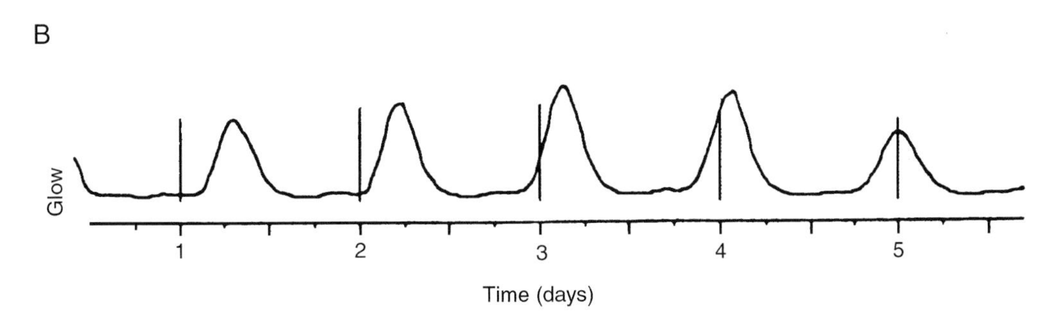

FIGURE 14.9 Bioluminescent flashes and glow of *Gonyaulax*. (A) Oscilloscope trace (ordinate, light intensity; abscissa, time, 8 s) recording from a vial with 32 000 cells. Six flashes having durations of about 100 ms are superimposed on a background glow. (B) Recording for 5 days from a similar vial kept in constant conditions showing the circadian rhythm of the background glow. The frequency of flashing also exhibits a circadian rhythm (not shown).

continue to exhibit rhythmicity (Fig. 14.9B), but with a period that is not exactly 24 h—it is only about (*circa*) 1 day (*diem*) (thus the origin of the term).

Genes coding for molecular components of the circadian clock have been identified and studied in several systems, and a mechanism involving negative feedback on the transcription of a gene by its protein product is postulated to be responsible for the rhythm. How such a mechanism might exert physiological control is not understood. In humans and other higher animals, in which it regulates the sleep–wake cycle and many other physiological processes, the mechanism involves the nervous system. However, it also occurs in plants and unicellular organisms, such as *G. polyedra*, in which daily changes in the cellular concentrations of luciferase, luciferin, and its binding protein occur. The two proteins are synthesized and destroyed each day, as are the scintillons in which they are located. Hence, the biological clock exerts control at a very basic level by controlling gene expression. This might explain the greater amount of luminescence at night, but a biochemical basis for the increased sensitivity to mechanical stimulation is not evident.

III. FUNGI

A. Occurrence, habitats, species, and functions

Light emission now known to be due to fungi has been observed since ancient times and was noted by both Aristotle and Pliny. Robert Boyle placed "shining wood" in his vacuum apparatus and showed that light emission was reversibly extinguished by the removal of air, anticipating the requirement for molecular oxygen by all bioluminescent systems. Definitive knowledge of the fungal origin of luminous wood emerged during the nineteenth century from extensive studies of timbers used for support in mines, and by the mid-twentieth century about 80 luminous species had been inventoried. As in bacteria, the emission is continuous and not affected by mechanical stimulation, but it is really quite dim. There is no indication that luminescence is regulated in relation to cell growth or density, but there is some evidence that nutrition may play a role. There have been reports that the luminescence is circadian regulated, as in dinoflagellates, but such results have not been confirmed or extended.

With one possible exception, all luminous fungi are basidiomycetes, and most are in the mushroom family; both the mycelium and fruiting body are luminous (Fig. 14.10). Such fungi occur in the many diverse habitats in which fungi occur, with the luminescence being visible most readily in dark forests—both tropical and temperate. The most striking reports describe luminescence from the interior or an infested tree split open by lightning.

The function of bioluminescence in fungi is not well understood. It has been suggested that the light serves as an attractant, which is consistent with the generalization that a continuous light emission acts in this way. If so, insects or other invertebrates might be attracted and enhance spore dispersal. However, this leaves the function of emission in the mycelium unexplained. The system might have evolved biochemically without constraints regarding its localization, and since it is probably not an energy-intensive function its value in any part of the lifecycle could be adequate to justify its retention.

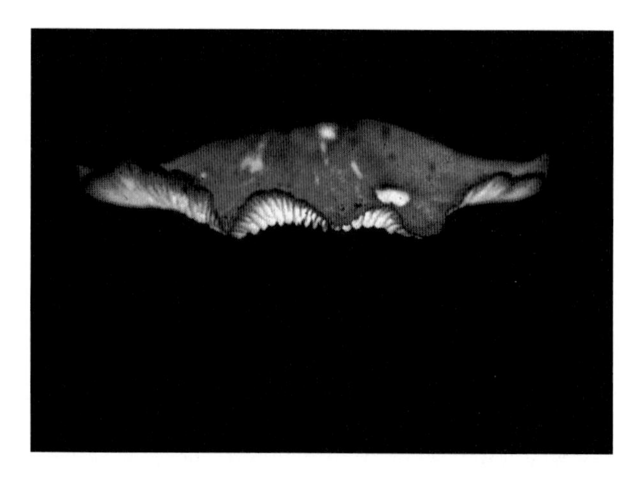

FIGURE 14.10 Bioluminescent mushroom photographed by its own light (photograph by Dr. Dan Perlman) (Plate 5).

B. Biochemistry

The spectrum of the light emitted has been determined from several species, all of which peak at about 525 nm, consistent with a flavin emitter. However, no biochemical evidence indicates this, and indeed no satisfactory understanding of the chemical basis for light emission has been obtained. Many years ago, a luciferin–luciferase-type system was reported with a link to reduced pyridine nucleotide, comparable to the bacterial system. However, this has not been confirmed in more recent studies, which suggest that the reaction may be a nonenzymatic chemiluminescence.

BIBLIOGRAPHY

Case, J. F., Herring, P. J., Robison, B. H., Haddock, S. H. D., Kricka, L., and Stanley, P. E. (2002). "Bioluminescence and Chemiluminescence: Proceedings of the 11th International lymporium". World Scientific, Singapore, 517 pp.

Fisher, A. J., Raushel, F. M., Baldwin, T. O., and Rayment, I. (1995). Three dimensional structure of bacterial luciferase from *Vibrio harveyi* at 2.4 Å resolution. *Biochemistry* **34**, 6581–6586.

Hastings, J. W. (1994). The bacterial luciferase reaction: model or maverick in flavin biochemistry? *In* "Flavins and Flavoproteins" (K. Yagi, ed.), pp. 813–822, De Gruyter, Amsterdam.

Hastings, J. W. (1996). Chemistries and colors of bioluminescent reactions—a review. *Gene* **173**, 5–11.

Hastings, J. W. (2001). Bioluminescence. *In* "Cell Physiology" (N. Sperelakis, ed.), 3rd edn., pp. 1115–1131, Academic Press, New York.

Hastings, J. W., and Johnson, C. (2003). Bioluminescence and chemi luminescence. Biophotonics. *Meth. Enz.* **360**, 76–104.

Hastings, J. W., and Wood, K. V. (2001). Luciferases did not all evolve from precursors having similar enzymatic properties. *In* "Photobiology 2002" (D. Valenzeno and T. Coohill, eds.). pp. 199–210, Valdenmar Publ. Co., Overland Park, KS.

Hastings, J. W., Kricka, L. J., and Stanley, P. E. (eds.) (1997). "Bioluminescence and Chemiluminescence: Molecular Reporting with Photons." Wiley, Chichester, UK.

Roda, A., Kricka, L. J., and Stanley, P. E. (eds.) (1999). "Bioluminescence and Chemiluminescence: Perspectives for the 21st Century." Wiley, Chichester, UK.

Taylor, F. J. R. (ed.) (1987). "The Biology of Dinoflagellates." Blackwell, Oxford.

Wilson, T., and Hastings, J. W. (1998). Bioluminescence. *Annu. Rev. Cell. Dev. Biol.* **14**, 197–230.

WEBSITES

These websites provide material and images on bioluminescence.

http://siobiolum.ucsd.edu/Biolum_intro.html
http://lifesci.ucsb.edu/~biolum/
http://mcb.harvard.edu/hastings/Images/bioluminescence.html
http://www.hboi.edu/marinesci/biolum.html
http://www.herper.com/Bioluminescence.html

Journal of Bioluminescence and Chemiluminescence; WebSite, Wiley Publ.
http://www3.interscience.wiley.com/cgi-bin/jtoc? ID=4087

15

Bioreactors

Larry E. Erickson
Kansas State University

GLOSSARY

airlift reactor Column with defined volumes for upflow and downflow of the culture broth; vertical circulation occurs because air is bubbled into the upflow volume.

batch bioreactor Culture broth is fed into the reactor at the start of the process; air may flow continuously.

bubble reactor Aerated column without mechanical agitation.

fed batch Liquid media is fed to the reactor continuously; the broth accumulates in the reactor because there is no outflow of liquid.

heterotrophs Microorganisms growing on an organic compound that provides carbon and energy.

insect cell culture Cultivation of insect cells in a bioreactor to produce a protein or other product.

photoautotrophs Microorganisms that use light for energy and carbon dioxide for their carbon source.

plant cell culture Production of plant cells in a bioreactor to produce useful products.

protein engineering The design, development, and production of new protein products with properties of commercial value.

tissue engineering The design, development, and production of tissue cells (biomaterials) for use on or in humans.

Bioreactors are vessels or tanks in which whole cells or cell-free enzymes transform raw materials into biochemical products and/or less undesirable byproducts. The microbial cell is a miniature bioreactor; other examples include shake flasks, petri dishes, and industrial fermentors. Diagnostic products based on enzymatic reactions, farm silos for silage fermentations, bread pans with fermenting yeast, and the soil in a Kansas wheat field may also be viewed as bioreactors. Although the bioreactor may be simple or highly instrumented, the important consideration is the ability to produce the desired product or result. The bioreactor is designed and operated to provide the environment for product formation selected by the scientist, baker, or winemaker. It is the heart of many biotechnological systems that are used for agricultural, environmental, industrial, and medical applications.

I. INTRODUCTION

The importance of the bioreactor is recorded in early history. The Babylonians apparently made beer before 5000 BC. Wine was produced in wineskins, which were carefully selected for their ability to produce a beverage that met the approval of the king and other members of his sensory analysis taste panel. Food and beverage product quality depended on art and craftsmanship rather than on science and engineering during the early years of bioreactor selection and utilization. Early recorded history shows that some understood the importance of the reactants and the environmental or operating conditions of the reactor. This allowed leavened bread and cheese to be produced in Egypt more than 3000 years ago.

The Desk Encyclopedia of Microbiology
ISBN: 0-12-621361-5

The process of cooking food to render it microbiologically safe for human consumption and to improve its sensory qualities is also an ancient tradition. The process of thermal inactivation of microorganisms through the canning of food to allow safe storage was an important early achievement in bioreactor design and operation.

As humans learned to live in cities, waste management including wastewater treatment became a necessity for control of disease. One of the first process engineering achievements was the biological treatment of wastes in bioreactors designed and built by humans for that purpose. Because a significant fraction of the population of a city could die from disease spread by unsanitary conditions, these early bioreactors represented important advancements.

After microorganisms were discovered, microbiologists and engineers increased their understanding of the biochemical transformations in bioreactors. Simple anaerobic fermentations for the production of ethyl alcohol, acetone, and butanol were developed. Aerobic and anaerobic treatment of waste-water became widely used. Sanitary engineering became a part of civil engineering education.

In the 1940s, the field of biochemical engineering emerged because of developments in the pharmaceutical industry that required large-scale bioreactors for the production of streptomycin and penicillin. Progress in bioreactor design and control resulted from research on oxygen transfer, air and media sterilization, and pH control. The central concern of the early biochemical engineers was the development of bioreactors that could achieve and maintain the chemical and physical environment for the organism that the biochemist/microbiologist recommended. The ability to scale-up from laboratory bioreactors to large fermentors required the development of instrumentation such as the sterilizable oxygen electrode. Early courses in biochemical engineering were concerned with the analysis, design, operation, and control of bioreactors. Although the field of biochemical engineering is less than 60 years old and some of the pioneers are still available to provide a first-person account of those exciting days, great progress has been made in bioreactor engineering. Some of the significant developments in bioreactor technology and their approximate dates are listed in Table 15.1.

TABLE 15.1 Significant developments in bioreactor technology

Development	Year[a]
Fermented beverages	5000 BC
Pasteur's discovery of yeast	1857
First medium designed for culturing bacteria	1860
Trickling filter for wastewater	1868
Anaerobic digester	1881
Production of citric acid using mold	1923
Production of penicillin in a petri dish	1928
Production of penicillin in small flasks	1942
Hixon and Gaden paper on oxygen transfer	1950
Air sterilization in fermentors	1950
Continuous media sterilization	1952
Aiba, Humphrey, and Millis biochemical engineering textbook on bioreactor design	1965
Continuous airlift reactor for production of yeast	1969
Advances in instrumentation and computer control	1970
Progress in airlift bioreactor design	1973
Recombinant DNA technology	1973
Insect cells grown in suspension culture	1975
Large-scale cell culture to produce interferon	1980
Insulin produced using bacteria	1982
Bioreactors for fragile cell cultures	1988
Textbook on plant cell biotechnology	1994
Textbook on protein engineering	1996
Textbook on tissue engineering	1997

[a]The dates are approximate and are indicative of periods of time when advances were progressing from initial studies to published works or commercial use.

II. CLASSIFICATIONS OF BIOREACTORS

Several methods have been used to classify bioreactors, including the feeding of media and gases and the withdrawal of products; the mode of operation may be batch, fed batch, or continuous. The classification may be based on the electron acceptor; the design may be for aerobic, anaerobic, or microaerobic conditions. In aerobic processes, the methods of providing oxygen have resulted in mechanically agitated bioreactors, airlift columns, bubble columns, and membrane reactors. The sterility requirements of pure culture processes with developed strains differ from those of environmental mixed-culture processes, which are based on natural selection. In some bioreactors the vessel is made by humans and there are also natural bioreactors, such as the microbial cell, the flowing river, and the field of native grass. In this article, the classification of bioreactors is based on the physical form of the reactants and products.

A. Gas phase reactants or products

Oxygen and carbon dioxide are the most common gas phase reactants and products; others include hydrogen, hydrogen sulfide, carbon monoxide, and methane. Oxygen is a reactant in aerobic heterotrophic growth processes, whereas it is a product in photoautotrophic growth. Generally, the concentration of the reactants and products in the liquid phase in the microenvironment of the cell influences the kinetics of the cellular reaction. Mass transfer to and from the gas phase affects bioreactor performance in most processes with gas phase reactants or products. The anaerobic reactor is designed to exclude oxygen. In some cases, inert gases are bubbled into the anaerobic reactor to provide a gas–liquid interfacial area to remove the product gases.

Because the solubility of oxygen in water is very low, the dissolved oxygen in the broth is rapidly depleted if oxygen transfer from the gas to the liquid phase is disrupted in aerobic processes. The distribution of dissolved oxygen throughout the reactor volume and the transient variation affect reactor performance. When mold pellets or biofilms are present, the diffusion of oxygen into the interior should be considered. A significant portion of the bioreactor literature is devoted to oxygen transfer and the methods recommended for the design and operation of aerobic bioreactors. The phase equilibrium relationship is based on thermodynamic data, whereas the rate of oxygen transfer depends on the gas–liquid interfacial area and the concentration driving force. Mechanical agitation increases the gas–liquid interfacial area. Aeration provides the supply of oxygen, and it affects the gas–liquid interfacial area.

Oxygen has been supplied by permeation through membranes in cultures in which bubbles may damage shear-sensitive cells. The membrane area and concentration driving force determine the oxygen transfer rate in these bioreactors.

Most large-scale bioreactors have either oxygen or carbon dioxide among the reactants or products. In many anaerobic fermentations the formation of carbon dioxide results in bubbling, and often no additional mixing is required for either mass transfer or suspension of the microbial cells. Methane is produced through anaerobic digestion of waste products. It is also a product of microbial action in landfills, bogs, and the stomach of the cow.

Packed-bed bioreactors are used to biodegrade volatile organic compounds in air pollution control applications. The rhizosphere provides a natural environment in which many volatile compounds in soil are transformed by microbial and plant enzymes.

B. Liquid phase reactants or products

Many bioreactors have liquid phase reactants and products. Ethanol, acetone, butanol, and lactic acid are liquid products that can be produced by fermentation. The kinetics of biochemical reactions depends on the liquid phase concentrations of the reactants and, in some cases, the products. The Monod kinetic model and the Michaelis–Menten kinetic model show that many biochemical reactions have first-order dependence on reactant (substrate) concentration at low concentrations and zero-order dependence at higher concentrations. Rates are directly proportional to concentrations lower than 10 mg/liter for many reactants under natural environmental conditions. At very high concentrations, inhibition may be observed.

Hydrocarbons that are relatively insoluble in the water phase, such as hexadecane, may also be reactants or substrates for biochemical reactions. Microbial growth on hydrocarbons has been observed to occur at the liquid–liquid interface and in the water phase. The oxygen requirements are greater when hydrocarbon substrates are used in place of carbohydrates. In the past, there was great interest in the production of microbial protein from petroleum hydrocarbons. The commercialization of the technology was most extensive in the former Soviet Union. The airlift bioreactor is uniquely suited for this four-phase process because of the tendency of the hydrocarbon phase to migrate to the top of the fermentor. The hydrocarbons are found suspended as drops in the water phase, adsorbed to cells, and at the gas–liquid interface. The cells are found adsorbed to hydrocarbon drops, suspended in the water phase, and at the gas–liquid interface. In the airlift fermentor, the vertical circulation mixes the hydrocarbons and cells that have migrated to the top of the fermentor with the broth that enters the downflow side of the column.

One of the oldest and most widely practiced fermentations is the microbial production of ethanol and alcoholic beverages such as beer and wine. Because ethanol inhibits fermentation at high concentrations, the process of inhibition has been extensively studied for this fermentation. Ethanol affects the cell membrane and the activities of enzymes. This inhibition limits the concentration of ethanol that can be obtained in a fermentor. Because ethanol is also produced for use as a motor fuel, there is still considerable research on ethanol production. Because the cost of the substrate is a major expense, inexpensive raw materials such as wastes containing cellulose have been investigated.

C. Solid phase reactants or products

There are many examples of bioreactors with solid phase reactants. The cow may be viewed as a mobile bioreactor system that converts solid substrates to methane, carbon dioxide, milk, and body protein. Although the cow is a commercial success, many efforts to transform low-cost cellulosic solid waste to commercial products in human-made bioreactors have not achieved the same level of success.

Solid substrates such as soybean meal are commonly fed into commercial fermentations. Through the action of enzymes in the fermentation broth, the biopolymers are hydrolyzed and more soluble reactants are obtained.

Many food fermentations involve the preservation of solid or semisolid foods such as in the conversion of cabbage to sauerkraut and meats to sausage products. Cereals, legumes, vegetables, tubers, fruits, meats, and fish products have been fermented. Some fermented milk processes result in solid products such as cheeses and yogurts.

Other examples include the composting of yard wastes, leaching of metals from ores, silage production, biodegradation of crop residues in soil, microbial action in landfills, and the remediation of contaminated soil.

In many of these fermentations, mixing is difficult or expensive. Transport of essential reactants may depend on diffusion; the concentrations of reactants and products vary with position. Rates may be limited by the transport of essential reactants to the microorganisms.

Most compounds that are present as solids in bioreactors are somewhat soluble in the water phase. For reactants that are relatively insoluble, biochemical reaction rates may be directly proportional to the available interfacial area. The surface of the solid may be the location of the biochemical transformation. An example of microorganisms growing on the surface of a solid substrate is mold on bread. To design bioreactors for solid substrates and solid products, the solubility and the transport processes should be considered as well as the kinetics of the process.

Recently, there has been considerable progress in tissue engineering. The rational design of living tissues and the production of these tissues by living cells in bioreactors are advancing rapidly because of progress in systems design and control for both *in vitro* flow reactors and *in vivo* maintenance of cell mass.

D. Microorganisms in bioreactors

The rate of reaction in bioreactors is often directly proportional to the concentration of microbial biomass. In biological waste treatment, the influent concentration of the organic substrate (waste) is relatively low, and the quantity of microbial biomass that can be produced from the waste is limited. The economy of the operation and the rate of biodegradation are enhanced by retaining the biomass in the bioreactor. In the activated sludge process, this is done by allowing the biomass to flocculate and settle; it is then recycled. The trickling filter retains biomass by allowing growth on the surfaces of the packing within the bioreactor.

A variety of immobilized cell reactors and immobilized enzyme reactors have been designed and operated because of the economy associated with reuse of cells and enzymes. In the anaerobic production of ethanol, lactic acid, and the other fermentation products, the product yield is greatest when the organisms are not growing and all the substrate is being converted to products. Continuous processes can be designed in which most of the cells are retained and the limiting maximum product yield is approached. Ultra-filtration membrane bioreactors have been used to retain cells, enzymes, and insoluble substrates.

In nature, cells are retained when biofilms form along flow pathways. The biofilms allow microorganisms to grow and survive in environments in which washout would be expected. The excellent quality of groundwater is the result of microbial biodegradation and purification under conditions in which microbial survival is enhanced by biofilm formation and cell retention on soil and rock surfaces. The ability of microorganisms to survive even after their food supply appears to be depleted is well established; this is the reason that there are microorganisms almost everywhere in nature. When spills occur, organic substances will often be degraded by microorganisms, if the nutritional environment is balanced. Nitrogen, phosphorous, and other inorganic nutrients often must be added.

The concentration of cells adsorbed to the surface and the concentration in the water phase depends on an adsorption phase equilibrium relationship and the operating conditions. In many environmental applications, most of the cells are adsorbed to surfaces. However, in large-scale fermentors with high cell concentrations and rich media feeds, only a small fraction of the cells are found on surfaces.

E. Photobioreactors

Light is the energy source that drives photoautotrophic growth processes. Because light is absorbed by the growing culture, the intensity decreases rapidly as the distance from the surface increases. Photobioreactors are designed to produce the quantity of product that is desired. Heat transfer is an important design aspect because any absorbed light energy that is not converted to chemical energy must be dissipated as heat.

III. PRINCIPLES OF BIOREACTOR ANALYSIS AND DESIGN

The basic principles of bioreactor analysis and design are similar to those for chemical reactors; however, many biochemical processes have very complex biochemistry. The chemical balance equations or stoichiometry of the process must be known or investigated. The yield of microbial biomass and products depends on the genetics of the strain and the operating conditions. The consistency of data from experimental measurements can be evaluated using mass balances such as the carbon balance and the available electron balance.

Microorganisms obey the laws of chemical thermodynamics; some heat is produced in heterotrophic growth processes. The free energy change is negative for the complete system of biochemical reactions associated with heterotrophic growth and product formation. Thus, the chemical energy available for growth and product formation decreases as a result of microbial assimilation of the reactants.

The rate of growth and product formation depends on the number of microorganisms and the concentrations of the nutrients. The kinetics of growth and product formation is often written in terms of the concentration of one rate-limiting substrate; however, in some cases, more than one nutrient may be rate limiting. The kinetics must be known for rational design of the bioreactor.

Heat is evolved in microbial bioreactors. For aerobic processes, the quantity of heat generated (heat of fermentation) is directly proportional to the oxygen utilized. Thus, the heat transfer and oxygen transfer requirements are linked by the energy regularity of approximately 450 kJ of heat evolved per mole of oxygen utilized by the microorganisms.

Transport phenomena is widely applied in bioreactor analysis and design. Many fermentation processes are designed to be transport limited. For example, the oxygen transfer rate may limit the rate of an aerobic process. Bioreactor design depends on the type of organism and the nutritional and environmental requirements. For example, in very viscous mycelial fermentations, mechanical agitation is often selected to provide the interfacial area for oxygen transfer. Likewise, animal cells that grow only on surfaces must be cultured in special bioreactors, which provide the necessary surface area and nutritional environment. In other cases, animals are used as the bioreactors because the desired biochemical transformations can best be achieved by competitively utilizing animals; cost and quality control are both important when food and pharmaceutical products are produced.

IV. SENSORS, INSTRUMENTATION, AND CONTROL

The ability to measure the physical and chemical environment in the fermentor is essential for control of the process. In the past 50 years, there has been significant progress in the development of sensors and computer control. Physical variables that can be measured include temperature, pressure, power input to mechanical agitators, rheological properties of the broth, gas and liquid flow rates, and interfacial tension. The chemical environment is characterized by means of electrodes for hydrogen ion concentration (pH), redox potential, carbon dioxide partial pressure, and oxygen partial pressure. Gas phase concentrations are measured with the mass spectrometer. Broth concentrations are measured with gas and liquid chromatography; mass spectrometers can be used as detectors with either gas or liquid chromatography. Enzyme thermistors have been developed to measure the concentrations of a variety of biochemicals. Microbial mass is commonly measured with the spectrophotometer (optical density) and cell numbers through plate counts and direct microscopic observation. Instruments are available to measure components of cells such as reduced pyridine nucleotides and cell nitrogen. On-line biomass measurements can be made using a flow cell and a laser by making multi-angle light scattering measurements. Multivariate calibration methods and neural network technology allow the data to be processed rapidly and continuously such that a predicted biomass concentration can be obtained every few seconds.

The basic objective of bioreactor design is to create and maintain the environment that is needed to enable the cells to make the desired biochemical transformations. Advances in instrumentation and control allow this to be done reliably.

V. METABOLIC AND PROTEIN ENGINEERING

Genetic modification has allowed many products to be produced economically. With the use of recombinant DNA technology and metabolic engineering, improved cellular activities may be obtained through manipulation of enzymatic, regulatory, and transport functions of the microorganism. The cellular modifications of metabolic engineering are carried out in bioreactors. Successful manipulation requires an understanding of the genetics, biochemistry, and physiology of the cell. Knowledge of the biochemical pathways involved, their regulation, and their kinetics is essential.

Living systems are bioreactors. Through metabolic engineering, man can modify these living bioreactors and alter their performance. Metabolic engineering is a field of reaction engineering that utilizes the concepts that provide the foundation for reactor design, including kinetics, thermodynamics, physical chemistry, process control, stability, catalysis, and transport phenomena. These concepts must be combined with an understanding of the biochemistry of the living system. Through metabolic engineering, improved versions of living bioreactors are designed and synthesized.

Although many products are produced in microbial cells, other cell lines, including insect cells, mammalian cells, and plant cells, are utilized for selected applications. The science to support these various living bioreactors is growing rapidly and the number of different applications is increasing steadily. The choice of which organism to select for a specific product must be made carefully, with consideration of biochemistry, biochemical engineering, safety, reliability, and cost. Both production and separation processes affect the cost of the product; however, the costs of product development, testing, regulatory approval, and marketing are also substantial.

Proteins with specific functional properties are being designed, developed, and produced through applications of protein engineering. Through molecular modeling and computer simulation, proteins with specific properties are designed. Protein production may involve applications of recombinant DNA technology in host cell expression bioreactors. An alternative is the production of a protein with the desired amino acid sequence through direct chemical synthesis.

VI. STABILITY AND STERILIZATION

Although beneficial genetic modification has led to many successful industrial products, contamination and genetic mutations during production operations have resulted in many batches of useless broth. Batch processes are common in bioreactors because of the need to maintain the desired genetic properties of a strain during storage and propagation. Continuous operation is selected for mixed-culture processes such as wastewater treatment, in which there is natural selection of effective organisms.

Bioreactors that are to operate with pure cultures or mixed cultures from selected strains must be free of contamination; i.e., the reactor and associated instrumentation must be sterilizable. The vessels that are to be used for propagation of the inoculum for the large-scale vessel must also be sterilizable. Methods to sterilize large vessels, instrumentation, and connecting pipes are well developed; however, there is a continuous need to implement a wide variety of good manufacturing practice principles to avoid contamination problems.

Steam sterilization has been widely applied to reduce the number of viable microorganisms in food and in fermentation media. As temperature increases, the rates of biochemical reactions increase exponentially until the temperature affects the stability of the enzyme or the viability of the cell. The Arrhenius activation energies, which have been reported for enzymatic reactions and rates of cell growth, are usually in the range of 20–80 kJ/g/mol, whereas activation energies for the thermal inactivation of microorganisms range from 200 to 400 kJ/g/mol. Many of the preceding principles also apply to the thermal inactivation of microorganisms in bioreactors. When solids are present in foods or fermentation media, heat transfer to the interior of the solids occurs by conduction. This must be considered in the design of the process because of the increase in the required sterilization time.

VII. CONCLUSIONS

Bioreactors are used for a variety of purposes. The knowledge base for their application has increased significantly because of the advances in chemical, biochemical, and environmental engineering during the past 60 years.

BIBLIOGRAPHY

Alleman, B. C., and Leeson, A. (Eds.) (1999). "Bioreactor and Ex Situ Biological Treatment Technologies: The Fifth International In Situ and On-Site Bioremediation Symposium: San Diego, California."Battelle Press: Columbus OH.

Asenjo, J. A., and Merchuk, J. C. (Eds.) (1995). "Bioreactor System Design." Dekker, New York.

Bailey, J. E., and Ollis, D. F. (1986). "Biochemical Engineering Fundamentals," 2nd ed. McGraw-Hill, New York.

Barford, J. P., Harbour, C., Phillips, P. J., Marquis, C. P., Mahler, S., and Malik, R. (1995). "Fundamental and Applied Aspects of Animal Cell Cultivation." Singapore Univ. Press, Singapore.

Blanch, H. W., and Clark, D. S. (1996). "Biochemical Engineering." Dekker, New York.

Carberry, J. J., and Varma, A. (Eds.) (1987). "Chemical Reaction and Reactor Engineering." Dekker, New York.

Characklis, W. G., and Marshall, K. C. (Eds.) (1990). "Biofilms." Wiley–Interscience, New York.

Christi, M. Y. (1989). "Airlift Bioreactors." Elsevier, New York.

Cleland, J. L., and Craik, C. S. (Eds.) (1996). "Protein Engineering: Principles and Practice." Wiley, New York.

Cookson, J. T. (1995). "Bioremediation Engineering." McGraw-Hill, New York.

Drioli, E., and Giorni, L. (1999). "Biocatalytic Membrane Reactors." Taylor & Francis, London, UK.

Endress, R. (1994). "Plant Cell Biotechnology." Springer-Verlag, Berlin.

Erickson, L. E., and Fung, D. Y. (Eds.) (1988). "Handbook on Anaerobic Fermentations." Dekker, New York.

Fan, L. T., Gharpuray, M. M., and Lee, Y. H. (1987). "Cellulose Hydrolysis." Springer-Verlag, Heidelberg.

Goosen, M. F. A., Daugulis, A. J., and Faulkner, P. (Eds.) (1993). "Insect Cell Culture Engineering." Dekker, New York.

Lanza, R. P., Langer, R. S., and Chick, W. L. (Eds.) (1997). "Principles of Tissue Engineering." Academic Press, San Diego.

Lubiniecki, A. S. (Ed.) (1990). "Large-Scale Mammalian Cell Culture Technology." Dekker, New York.

Mitchell, D. A., Berovic, M., and Krieger, N. (2000). Biochemical engineering aspects of solid state bioprocessing. *Adv. Biochem. Eng. Biotechnol.* **68**, 61–138.

Moo-Young, M. (Ed.) (1988). "Bioreactor Immobilized Enzymes and Cells: Fundamentals and Applications." Elsevier, New York.

Nielsen, J. H., and Villadsen, J. (1994). "Bioreaction Engineering Principles." Plenum, New York.

Schugerl, K. (2000). Development of bioreaction engineering. *Adv. Biochem. Eng. Biotechnol.* **70**, 41–76.

Shuler, M. L., and Kargi, F. (2002). "Bioprocess Engineering," 2nd edn. Prentice Hall, Englewood Cliffs, NJ.

Sikdar, S. K., and Irvine, R. L. (Eds.) (1998). "Bioremediation: Principles and Practice." Technomic, Lancaster, PA.

Twork, J. V., and Yacynych, A. M. (Eds.) (1990). "Sensors in Bioprocess Control." Dekker, New York.

Van't Riet, K., and Tramper, J. (1991). "Basic Bioreactor Design." Dekker, New York.

WEBSITES

The Electronic Journal of Biotechnology
http://www.ejbiotechnology.info/

List of biotechnology organizations and research institutes
http://www.ejb.org/feedback/borganizations.html

16

Bioremediation

Joseph B. Hughes, C. Nelson Neale, and C. Herb Ward
Rice University

GLOSSARY

bioattenuation The nonengineered, natural decomposition of organic contaminants in soil and groundwater systems.

biochemical markers Easily monitored (e.g., substrate-specific microbial population) or chemical (e.g., metabolic intermediates and end products) indicators of biodegradation or biotransformation.

biodegradation Metabolism of a substance by microorganisms that yields mineralized end products.

bioslurry treatment Accelerated biodegradation of contaminants by the suspension of contaminated soil or sediment in water through mixing energy.

biotransformation Microbially mediated process in which the original compound is converted to secondary or intermediate products.

bioventing Accelerated biodegradation of contaminants in contaminated subsurface materials by forcing and/or drawing air through the unsaturated zone.

cometabolism Fortuitous metabolism of a compound by a microorganism that neither yields energy directly nor produces a metabolic product that can subsequently be involved in energy metabolism.

composting Accelerated biodegradation of contaminants at high temperatures by aerating and adding bulking agents and possibly nutrients to waste in a compost pile.

ex situ *bioremediation* Biological treatment of excavated or removed contaminated media.

immobilization Chemical and/or physical processes by which contaminants become strongly associated or sorbed with a soil matrix or sludge and desorption is limited.

in situ *bioremediation* Treatment without physical disruption or removal of contaminated media.

land treatment Accelerated biodegradation of contaminated media through application to surface soils to enhance aeration and, in some cases, to allow for nutrient amendment.

microcosm Highly controlled laboratory-scale apparatus used to model or simulate the fate or transport of compounds under the biological, chemical, and physical conditions found in the natural environment.

plume Dissolved contaminants emanating from a source region due to groundwater transport processes.

unsaturated zone Region which spans the area located just beneath the surface and directly above the water table.

Bioremediation is defined in this article as the process by which microorganisms are stimulated to rapidly degrade hazardous organic contaminants to environmentally safe levels in soils, sediments, subsurface materials, and groundwater. Biological remediation processes have also recently been devised to either precipitate or effectively immobilize

The Desk Encyclopedia of Microbiology
ISBN: 0-12-621361-5

inorganic contaminants such as heavy metals; however, treatment of inorganic contaminants will not be included in this definition of bioremediation. Stimulation is achieved by the addition of growth substrates, nutrients, terminal electron acceptor, electron donors, or some combination therein, resulting in an increase in contaminant biodegradation and biotransformation. The microbes involved in bioremediation processes may obtain both energy and carbon through the metabolism of organic contaminants. In some cases, metabolism occurs via a cometabolic process or by a terminal electron-accepting process. Independent of the metabolic pathways, bioremediation systems are designed to degrade hazardous organic contaminants sorbed to soils and sediments or dissolved in water. Bioremediation of contaminants may occur *in situ* or within the contaminated soil, sediments, or groundwater. Alternatively, the contaminated media may be removed and treated using *ex situ* techniques.

I. INTRODUCTION

With the advent of petroleum refining and manufacture of synthetic chemicals, many potentially hazardous organic compounds have been introduced into the air, water, and soil. One method for removing these undesirable compounds from the environment is bioremediation, an extension of carbon cycling. Given the appropriate organism(s), time, and growth conditions, a variety of organic compounds, such as oil and petroleum products, creosote wastes, and a variety of synthetic organic chemicals, can be metabolized to innocuous materials, usually carbon dioxide (CO_2), water, inorganic salts, and biomass (mass of bacterial cells); however, metabolic byproducts, some of which are undesirable, may accumulate when biodegradation of compounds is incomplete. Bioremediation is normally achieved by stimulating the indigenous microflora (naturally occurring microorganisms) present in or associated with the material to be treated. In instances in which the indigenous microflora fails to degrade the target compounds or has been inhibited by the presence of toxicants, microorganisms with specialized metabolic capabilities may be added (bioaugmentation).

The technical basis for modern bioremediation technology has a very long history (e.g., composting of organic wastes into mulch and soil conditioners). Bioremediation technology has grown to include the biological treatment of sewage and wastewater, food processing wastes, agricultural wastes, and, recently, contaminants in soils and groundwater. In this article, bioremediation is defined and limited to the biological treatment of organic contaminants. First, we present important background information on the metabolic processes that drive bioremediation, the requirements for the stimulation of specific metabolic processes, and the influence of contaminant behavior and distribution on contaminant availability for microbial uptake. This discussion is followed by sections outlining favorable growth conditions for microorganisms that are capable of degrading common classes of organic contaminants found in soil, sediments, and groundwater. Engineered systems used for the treatment of contaminated media are then presented.

II. BACKGROUND SCIENCES

The following two sections describe the fundamental metabolic processes that govern bioremediation as well as biodegradation and biotransformation characteristics of selected classes of organic contaminants. A more thorough treatment of the subject matter may be found in many texts and monographs, and the reader is encouraged to consult these materials for further information. Some of the more notable and recent references include *Microbial Transformation and Degradation of Toxic Organic Chemicals* (Young and Cerniglia, 1994), *Biological Degradation and Bioremediation of Toxic Chemicals* (Chaudry, 1994), *Biodegradation and Bioremediation* (Alexander, 1994), *Biology of Microorganisms* (Madigan *et al.*, 1997), and *Biodegradation of Nitroaromatic Compounds* (Spain, 1995).

A. Metabolic processes

The metabolism of organic contaminants can be broadly differentiated by the ability of the organisms to gain energy for cell growth from the process. If the metabolism of a compound provides energy for cell maintenance and division, the contaminant is referred to as a primary substrate. In some cases, a compound is metabolized and provides the cell with energy but does not support growth. Contaminants of this type are referred to as secondary substrates. If a compound is transformed without benefit of the cell (no energy or carbon provided for use by the organism) while the cell is obtaining energy from another transformable compound, the biotransformation is referred to as cometabolic. Finally, an additional classification has been recently identified in which some contaminants are capable of serving as the terminal electron acceptor in the respiratory chain of certain anaerobic (without oxygen) bacteria. In this case, energy is not obtained from the contaminant itself, but its transformation is a component of metabolic processes that provide energy to the cell for growth.

B. Growth requirements

An essential element of bioremediation processes is the ability to sustain enhanced levels of metabolic activity for extended periods of time. To accomplish this objective, an assessment of conditions at contaminated sites is conducted to determine limiting factors that will be manipulated in an engineered process. A comprehensive list of considerations in this assessment is provided in Table 16.1. For naturally occurring organic compounds (i.e., petroleum hydrocarbons), the availability of oxygen as an electron acceptor is often the primary limiting factor. This can be demonstrated through an evaluation of the stoichiometry of hydrocarbon mineralization, as is shown here for benzene or C_6H_6 (a common contaminant of concern at sites at which gasoline has been spilled):

$$C_6H_6 + 7.5O_2 \rightarrow 6CO_2 + 3H_2O$$

For the complete mineralization of 1 mol of C_6H_6, 7.5 mol of oxygen will be consumed. Water, in equilibrium with the atmosphere, contains approximately 8 mg/liter dissolved oxygen, which can support the oxidation of 2.6 mg/liter C_6H_6. Since the solubility of C_6H_6 in water is approximately 1800 mg/liter, the availability of oxygen limits the extent to which hydrocarbons may be biodegraded.

C. Bioaugmentation

For contaminant metabolism to occur in a bioremediation system, organisms with the genetic capacity to transform compounds of interest must be present. Experience has demonstrated that this is often the case in media in which contamination has been present for even short time periods. In certain cases, the addition of organisms acclimated to specific contaminants, or bioaugmentation, may decrease the duration of lag phases. The ability to effectively bioaugment bioremediation systems is a function of the process used. Bioaugmentation is best suited for processes in which contaminated soil or sediments have been excavated and can be mixed or tilled. The bioaugmentation of *in situ* processes is more difficult because of difficulties in uniformly distributing cells throughout a porous medium. Few cases exist in which bioagumentation of contaminated groundwater aquifers has proven beneficial.

TABLE 16.1 Requirements for microbial growth in bioremediation processes

Requirement	Description
Carbon source	Carbon contained in many organic contaminants may serve as a carbon source for cell growth. If the organism involved is an autotroph, CO_2 or HCO_3^- in solution is required. In some cases, contaminant levels may be too low to supply adequate levels of cell carbon, or the contaminant is metabolized via cometabolism. In these cases the addition of carbon sources may be required.
Energy source	In the case of primary metabolism, the organic contaminant supplies energy required for growth. This is not the case when the contaminant is metabolized via secondary metabolism or cometabolism or as a terminal electron acceptor. If the contaminant does not serve as a source of energy, the addition of a primary substrate(s) is required.
Electron acceptor	All respiring bacteria require a terminal electron acceptor. In some cases, the organic contaminant may serve in this capacity. Dissolved oxygen is a common electron acceptor in aerobic bioremediation processes. Under anaerobic conditions, NO_3^-, SO_4^{2-}, SO^{2+} and CO_2 may serve as terminal electron acceptors. Certain cometabolic transformations are carried out by fermentative and other anaerobic organisms, in which terminal electron acceptors are not required.
Nutrients	Nitrogen (ammonia, nitrate, or organic nitrogen) and phosphorus (*ortho*-phosphate or organic phosphorus) are generally the limiting nutrients. In certain anaerobic systems, the availability of trace metals (e.g., Fe, Ni, Co, Mo, and Zn) can be of concern.
Temperature	Rates of growth and metabolic activity are strongly influenced by temperature. Surface soils are particularly prone to wide fluctuations in temperature. Mesophilic conditions are generally best suited for most applications (with composting being a notable exception).
pH	A pH ranging between 6.5 and 7.5 is generally considered optimal. The pH of most ground-water (8.0–8.5) is not considered inhibitory.
Absence of toxic materials	Many contaminated sites contain a mixture of chemicals, organic and inorganic, which may be inhibitory or toxic to microorganisms. Heavy metals and phenolic compounds are particular concerns.
Adequate contact between microorganisms and substrates	For a contaminant to be available for microbial uptake it must be present in the aqueous phase. Thus, contaminants that exist as nonaqueous phase liquids or are sequestered within a solid phase may not be readily metabolized.
Time	This is an important factor in the start-up of bioremediation systems. Even when the first eight considerations in this table are met, lag phases are often observed prior to the onset of activity. In some cases, the dramatic bacterial population shifts that are required for bioremediation will lengthen periods of slow activity.

III. ENHANCEMENT OF CONTAMINANT METABOLISM

The method by which the rate or extent of contaminant metabolism can be increased in a bioremediation system is governed largely by the substrate-specific metabolic processes that result in its transformation. An understanding of contaminant metabolism is essential. In the following sections, an overview of specific metabolic pathways for common contaminant classes is presented.

A. Monoaromatic hydrocarbons

As constituents of gasoline, diesel, and jet fuels, monoaromatic hydrocarbons enter the subsurface environment due to accidental spills and leaking underground storage tanks (UST). These contaminants are commonly found in the environment in the form of free product entrapped or sorbed to porous media or dissolved in water. Monoaromatics are typically referred to as light nonaqueous phase liquids because their specific gravity is less than that of water. Of particular interest in this class of pollutants are benzene, toluene, ethylbenzene, and xylene isomers (BTEX). The biodegradation of these compounds has been and continues to be studied extensively. Under aerobic (containing oxygen) conditions, all the constituents are rapidly biodegraded as primary substrates. Oxygen is important in this process in two ways. First, it is a substrate in the initial attack of the aromatic ring catalyzed by oxygenase enzymes. Second, oxygen serves as the terminal electron acceptor for respiratory chains. Figure 16. 1 illustrates the dual functionality of oxygen in the metabolism of benzene.

The biodegradation of BTEX compounds is not as well characterized under anaerobic conditions as it is under aerobic conditions. Certainly, the biodegradation of all BTEX compounds has been observed under a range of anaerobic electron acceptor conditions, but it does not occur in all cases. In particular, benzene can be recalcitrant under anaerobic conditions. In some cases, however, the metabolism of benzene has been observed in the absence of oxygen. Little is known about the pathways of these processes or the enzymes that may be involved in these reactions.

In any case, the anaerobic degradation of BTEX compounds is generally slower than aerobic processes. Thus, bioremediation systems targeted for BTEX remediation are typically operated under aerobic conditions. The basis of most BTEX bioremediation systems is the enhancement of the rate of aerobic metabolism by increasing the availabililty of oxygen in contaminated areas. Several methods for doing so are presented later.

B. Polynuclear aromatic hydrocabons

Polynuclear aromatic hydrocarbons (PAHs) typically result from activities including combustion of fossil fuels and coal gasification processes, and they may also be found in creosote wastes used in wood preservation. PAHs are generally found sorbed to soils and sediments in the natural environment. This class of chemicals contains many compounds with varying biodegradation and physicochemical characteristics. In general, PAH biodegradation is limited to aerobic metabolism and is initiated by oxygenase attack (similar to that depicted in Fig. 16.1). PAHs of three or fewer rings, including naphthalene, fluorene, and phenanthrene, are known to be primary substrates for bacterial growth. Larger PAHs (i.e., four rings and larger) tend to behave as secondary substrates in the presence of the smaller, more water-soluble PAHs.

C. Phenolic compounds

Phenol and chlorinated phenols have historically been used in the treatment or preservation of wood products and have served as bacterial disinfectants. These compounds are biodegraded as primary substrates under aerobic and anaerobic conditions. These compounds are often recalcitrant in the environment due to their toxicity and the low water solubility of certain chlorinated forms (e.g., pentachlorophenol). When present at concentrations lower than toxic thresholds, phenols can be rapidly mineralized by a wide range of microorganisms. As the degree of chlorine substituents increases, the rate of degradation often decreases, especially under aerobic conditions.

D. Chlorinated hydrocarbons

Chlorinated methanes, ethanes, and ethenes comprise a group of compounds commonly referred to as chlorinated hydrocarbons (also referred to as chlorinated solvents). These compounds have been used extensively as degreasers, dry cleaning agents, and paint removers, and they are widely present and persistent in the environment. They are common contaminants of subsurface soils and groundwater, and contamination has resulted from leaking storage facilities or improper disposal practices. Due to their high specific gravity and density, chlorinated hydrocarbon compounds may often be referred to as dense nonaqueous phase liquids (DNAPLs). DNAPLs will typically be found near the lower confining unit of an acquifer since their densities are greater than the density of water. Common chlorinated hydrocarbon contaminant compounds include trichloroethane, perchloroethene (PCE),

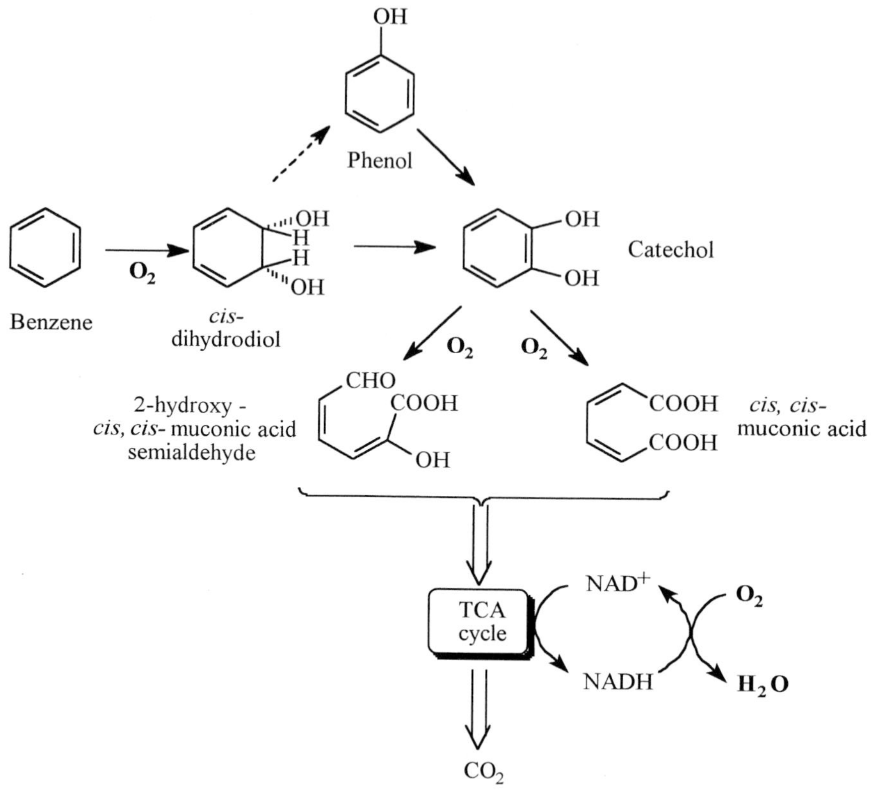

FIGURE 16.1 The metabolic pathways of benzene biodegradation under aerobic conditions.

TABLE 16.2 Summary of the biotransformation processes of chlorinated hydrocarbons

Compound	1° Substrate	2° Substrate	Cometabolic substrate	Terminal electron acceptor
Dichloromethane	Yes	Yes	Yes	No
Chloroform	No	No	Yes	No
Carbon tetrachloride	No	No	Yes	No
Perchloroethene	No	No	Yes	Yes
Trichloroethene	No	No	Yes	Yes
Dichloroethenes[a]	No[b]	No	Yes	Yes[c]
Vinyl chloride	Yes	Possible	Yes	Yes[c]
Hexachloroethane	No	No	Yes	No
1,1,1-Trichloroethane	No	No	Yes	No

[a]Three isomers of dichloroethenes exist: 1,1-dichloroethene, *cis*-dichloroethene, and *trans*-dichloroethene.
[b]Recent studies have identified oxidative pathways for *cis*-dichloroethene that may yield energy for growth.
[c]Vinyl chloride and *cis*-dichloroethene are intermediates during the respiration of perchloroethene and trichloroethene to ethene.

trichloroethene (TCE), dichloroethene (DCE), carbon tetrachloride, chloroform, and vinyl chloride.

The metabolism of chlorinated hydrocarbons is perhaps more diverse than that of any other group of environmental contaminants. Depending on the compound of interest, the electron acceptor condition, and the presence of inducing substrates, the metabolism of chlorinated hydrocarbons may occur through primary metabolism, secondary metabolism, cometabolism, or terminal electron acceptor processes. Table 16.2 lists common chlorinated hydrocarbon contaminants and the processes by which individual compounds are known to be transformed.

E. Nitroaromatic compounds

Nitroaromatics are common pollutants of water and soils as a result of their use in plastics, dyes, and explosives. Typical nitroaromatic contaminants used in explosives manufacture include trinitrotoluene

(TNT), hexahydro-1,3,5-trinitro-1,3,5-triazine (RDX), and octahydro-1,3,5,7-tetranitro-1,3,5,7,-tetrazoncine (HMX) (Spain, 1995). The nitro group has a strong electron withdrawing functionality, which previously was thought to reduce the potential for oxygenase attack toward the aromatic ring. Recent studies have demonstrated that certain organisms are capable of oxidizing selected nitroaromatic compounds to obtain energy for growth (i.e., primary metabolism). The activity is generally limited to nitroaromatics containing two or fewer nitro groups.

Under anaerobic conditions, nitroaromatic transformation generally yields reduced aromatic products. For example, the product of the complete reduction of nitrobenzene is aniline (aminobenzene). The formation of an aryl amine from an aryl nitro group requires that two intermediate forms be produced; the first is any aryl nitroso intermediate followed by the second intermediate, an aryl hydroxylamine. Recent work has demonstrated the importance of the aryl hydroxylamine intermediate in the ultimate fate of nitroaromatics under anaerobic conditions. The hydroxylamine can be reduced to the amine or undergo more complex reactions that can result in binding with natural organic matter or the formation of an aminophenol through rearrangement reactions.

IV. APPLICATION OF CONTAMINANT METABOLISM IN ENGINEERED SYSTEMS

The process of transforming an individual contaminant molecule into a nontoxic form occurs at the enzymatic level. Potential remediation sites contain kilograms to tons of contaminants distributed over large areas. Reconciling the difference in scale between molecular processes and the cleanup of tremendous volumes of contaminated media is a significant engineering challenge. Fundamentally, the application of microscale phenomena to the field-scale bioremediation of large, complex contaminated sites begins with a thorough analysis of site conditions. Key steps in site characterization may include (i) determination of the contaminants present and their concentration and distribution, (ii) delineation of the volume of material undergoing treatment, (iii) evaluation of the physical and chemical state of contaminants, (iv) analysis of the redox conditions at the site, and (v) establishment of site hydrogeologic conditions. Upon completion of this phase of the investigation, an analysis would be conducted to determine whether *in situ* or *ex situ* treatment or some other technology would be most appropriate given the site conditions. Regardless of the selected mode of treatment, systems would be designed to create the appropriate ecological conditions to select for organisms that possess the ability to degrade target contaminants. Furthermore, considerations outlined in Table 16.1 would be evaluated to identify potential limiting factors to bioremediation so that modifications or additions to the treatment scheme could be made to enhance the rates of contaminant biodegradation and biotransformation.

Specific bioremediation technologies are discussed in detail later. In all cases, these technologies are predicated on the stimulation of specific metabolic activities. The selection of a bioremediation process begins with the understanding of how specific contaminants may be metabolized. In some cases, metabolism may already be occurring, and application of a bioremediation system to those sites would focus on accelerating the rate of the naturally occurring processes. In other cases, contaminant metabolism may be negligible and conditions may require significant alteration through an engineered process. In all cases, bioremediation technologies intended to distribute metabolic activity throughout a region of contamination that is vastly larger than that of a bacterial cell. Thus, the coupling of microscale metabolic processes with macroscale mass transfer processes is one of the most significant challenges in the development of efficient bioremediation technologies. It should be noted that more detailed information on these technologies may be found in many texts as well as in various collections of monographs. Some of the more pertinent references include *In Situ Bioremediation: When Does It Work?* (Rittman *et al.*, 1993), *Handbook of Bioremediation* (Norris *et al.*, 1994), *Bioremediation: Field Experience* (Flathman *et al.*, 1994), *Innovative Site Remediation Technology: Bioremediation* (Ward *et al.*, 1995), *Bioremediation Engineering: Design and Application* (Cookson, 1995), *Innovations in Ground Water and Soil Cleanup* (Rao *et al.*, 1997), *Soil Bioventing: Principles and Practice* (Leeson and Hinchee, 1997), *Subsurface Restoration* (Ward *et al.*, 1997), and *Innovative Site Remediation Technology Design & Application: Bioremediation* (Dupont *et al.*, 1998).

V. NATURAL BIOATTENUATION

A. Overview

Natural bioattenuation, sometimes termed intrinsic bioremediation or natural bioremediation, refers to the biodegradation or biotransformation of both subsurface soil and groundwater contaminants through

TABLE 16.3 Application of bioremediation treatment options to various classes of contaminants[a]

		Contaminant class				
		Monoaromatic hydrocarbons	Chlorinated solvents	Nitroaromatics	Phenols	PAHs
In situ	Natural bioattenuation	Yes	Yes	?[b]	?	?
	Biostimulation	Yes	Yes	?	Yes	?
	Electron donor delivery	No	Yes	?	?	No
	Bioventing	Yes	No	No	No	Yes
	Permeable reactive barriers	Yes	Yes	?	?	?
Ex situ	Land treatment	Yes	No	Yes	Yes	Yes
	Composting	Yes	No	Yes	Yes	Yes
	Bioslurry processes	Yes	Yes	Yes	Yes	Yes

[a]Adapted in part from Rao *et al.* (1997).
[b]?, Undetermined or in developmental stages.

microbially mediated processes. This mechanism for contaminant mass reduction represents a key component of the broader remediation process of natural attenuation which focuses on the reduction of contaminant concentration, mass, mobility, and/or toxicity through natural processes, including dilution, dispersion, volatilization, adsorption to solid surfaces, and chemical and biological transformation reactions. Bioattenuation is a nonintrusive process (i.e., does not require a mechanical or engineered system for remediation) and is generally more cost-effective than other *in situ* and *ex situ* cleanup strategies. This remediation process has been successfully demonstrated in the mitigation of BTEX contamination resulting from leaking underground storage tanks and may also be applicable to the remediation of other compounds, including chlorinated solvents. The applicability of this treatment scheme has not been determined for PAHs, PCBs, explosives, and pesticides (as shown in Table 16.3; Rao *et al.*, 1997). For many UST sites (47%), natural attenuation is the chief mechanism for groundwater remediation, whereas nearly 67% of U.S. states recognize and implement natural attenuation as a viable alternative for soil and groundwater cleanup (USEPA, 1997). Table 16.4 compares the use of natural attenuation with other remediation technologies (some of which are described later) for both soil and groundwater cleanup. It is important to recognize that natural bioattenuation may not necessarily replace active remediation processes, but it does offer an attractive option to complement techniques that may be very costly to implement. For example, at UST sites, large pools of nonaqueous phase liquids may serve as a continual source for groundwater contamination. Typically, a more intensive strategy is required to remove most of the free-phase contaminant before natural bioattenuation is implemented as a cleanup method. Natural bioremediation should also not be characterized as a "no action" approach to cleanup; rather, it requires both long-term monitoring of the parent contaminant compound and secondary metabolites and monitoring of other chemical and biological markers that are indicative of the attenuation process (Ward *et al.*, 1997). However, the efficacy of natural attenuation for meeting remedial objectives and managing risk of groundwater contamination is controversial in research and regulatory communities. Its role in soil and groundwater cleanup continues to develop as we learn more about quantifying the processes involved.

TABLE 16.4 Use of remediation technologies at UST sites[a]

Remediation technology	Use at UST sites (% of sites)
Soil remediation	
Soil washing	0.2
Bioventing	0.8
Incineration	2
Thermal desorption	3
Landfarming	7
Soil vapor extraction	9
Biopiles	16
Natural attenuation	28
Landfilling	34
Groundwater remediation	
Biosparging	2
Dual-phase extraction	5
In situ bioremediation	5
Air sparging	13
Pump and treat	29
Natural attenuation	47

[a]*Source*: USEPA (1997).

The rate and extent of contaminant biodegradation are governed by many environmental factors, including contaminant and cell biomass concentrations, temperature, pH, supply of nutrients, adequacy of carbon and energy sources, the presence of toxins such as heavy metals, availability of contaminants to microorganisms (i.e., contact, contaminant solubility and hydrophobicity, and desorption from solids), time for acclimation, and availability of electron acceptors (Table 16.1). The supply and availability of electron acceptors is often cited as the controlling or limiting factor in the bioattenuation process. Aerobic biodegradation occurs in the presence of oxygen, whereas alternate electron acceptors, including nitrate, sulfate, trivalent iron (Fe^{3+}), and carbon dioxide, are utilized under anaerobic conditions. For relatively soluble petroleum hydrocarbons such as BTEX, the rate and extent of biological transformation of contaminants is typically much greater under aerobic conditions than under oxygen-limited conditions. For highly chlorinated solvents (e.g., PCE and TCE), the rate and extent of biological transformation, through reductive dechlorination, is greater under anaerobic conditions than under aerobic conditions, whereas less halogenated compounds (e.g., vinyl chloride) may be more amenable to aerobic biodegradation.

The oxygen demand exerted by the microorganisms during petroleum hydrocarbon biodegradation generally exceeds the rate of oxygen replenishment, especially in areas of high contaminant concentration (i.e., near the source zone). In fact, the limiting dissolved oxygen concentration for aerobic biodegradation is approximately 2 mg/liter (Rao *et al.*, 1997), although a study by J. Salanitro and coworkers published in *Ground Water Monitoring and Review* in 1997 suggests that this concentration may be as low as 0.2 mg/liter. The characteristic shape of a groundwater contaminant plume may be partially explained by the presence or absence of oxygen as a terminal electron acceptor. Zones of aerobic biodegradation generally occur on the outermost and leading edges of the plume where contaminated groundwater meets with uncontaminated, well-oxygenated groundwater. Mixing of the two waters via dispersion provides an adequate supply of oxygen for the aerobic biodegradation process. However, contaminant biodegradation in the central region of the plume is generally governed by contaminant-specific anaerobic processes due to the rapid depletion of oxygen in these areas of high metabolic activity. The diffusion rate of oxygen in water is four orders of magnitude less than the diffusion rate in air; thus, the rate of oxygen consumption easily exceeds its rate of transport in water, which results in anaerobic conditions (Norris *et al.*, 1994; Rao *et al.*, 1997). Figure 16.2 presents a typical UST spill and indicates the areas or zones of aerobic and anaerobic biodegradation.

The previous discussion illustrates the importance of oxygen as a driving force in the natural aerobic bioattenuation of subsurface contaminants. Oxygen may be naturally delivered to the contaminant plume either through mixing with uncontaminated groundwater or through reaeration from the overlying unsaturated zone. Reaeration may serve as a significant source of oxygen and is governed by many factors, including soil hydrogeologic properties, soil moisture, precipitation, and respiration rate of soil microorganisms. Macroporous sand materials can easily

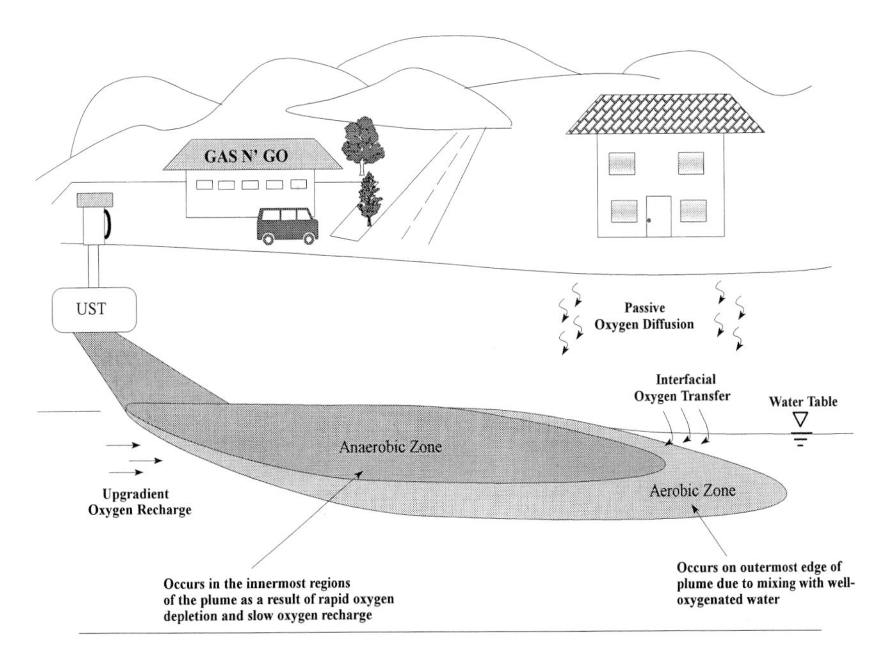

FIGURE 16.2 The role of oxygen flux in the bioattenuation of hydrocarbon contaminants (adapted in part from Norris *et al.*, 1994).

and rapidly contribute to the oxygen supply of a contaminant plume, whereas the diffusion of oxygen through clay materials is generally impeded, especially in the presence of moisture. Moisture may cause swelling of certain clay soils and reduces the number of air-filled pore spaces in the soil through which oxygen may easily diffuse. The respiration rate of soils may also serve as a significant oxygen sink and will restrict delivery of the gas to a contaminant plume.

B. Demonstration

Several demonstrations have been completed to determine the efficacy of natural bioattenuation of contaminants in groundwater. A study by C. Y. Chiang and coworkers, published in *Ground Water* in 1989, investigated the aerobic bioattenuation of a petroleum hydrocarbon plume at a gas manufacturing plant in Michigan using a large network of monitoring wells. The contaminated aquifer was located 10–25 ft below the surface and was overlain by coarse sand. The initial mass of benzene at the site was approximately 10 kg; after a period of 21 months, the mass had been reduced to approximately 1.3 kg of benzene. A first-order bioattenuation rate of 0.0095/day was determined for the site. Analysis of groundwater samples throughout the study period revealed an inverse relationship between dissolved oxygen and benzene concentration in the plume. Samples from locations in which the BTX concentration was high were coupled with low dissolved oxygen concentrations as a result of increased microbial activity or biodegradation. Likewise, samples with low BTX concentrations contained higher dissolved oxygen concentrations due to the reduction in oxygen demand. Other natural attenuation processes were found to have very little effect on the reduction of benzene contaminant mass. Only 5% of benzene contaminant loss was attributed to volatilization from the groundwater, and sorption was thought to have little impact on the attenuation process based on the low organic carbon content of the aquifer material. Subsequent modeling and microcosm studies confirmed that natural aerobic bioattenuation was a dominant mechanism in the reduction of contaminant mass at this particular site.

Bioattenuation of chlorinated solvents has also been demonstrated in field studies. In 1995, L. Semprini and coworkers investigated the biotransformation of chlorinated solvents in groundwater at a site in Michigan. A series of multilevel samplers bounding the width and length of the contaminant plume were used to determine the concentrations of TCE, isomers of DCE, vinyl chloride, and ethane with depth in the sandy aquifer. The results of the study indicated that the reductive dechlorination of TCE had occurred over time in the plume based on the presence of less chlorinated compounds as well as the presence or absence of biochemical markers, including methane (methanogenesis) and sulfate (sulfate reduction).

The transformation of TCE to isomers of DCE (predominantly *cis*-DCE) was associated with sulfate reducing conditions, whereas further transformation of *cis*-DCE to vinyl chloride and ethene was coupled with areas high in methanogenic activity. Although the mass of original TCE contaminant could not be established, Semprini and coworkers suggested that, on the basis of calculated flux rates for each of the contaminant compounds, TCE reduction to ethane was on the order of 20%.

VI. ENHANCED *IN SITU* BIOREMEDIATION/ BIOATTENUATION

Many *in situ* bioremediation processes have been designed or engineered to maintain or accelerate the biodegradation of organic contaminants in soil and groundwater by supplying those materials which may limit the breakdown of the contaminant compound. For example, in petroleum hydrocarbon-contaminated environments which are nutrient limited, nitrogen, phosphorus, and other minerals may be added to the environmental system (either through soil tilling or through injection into groundwater) to stimulate growth of the contaminant-degrading microbial population. In many cases, the natural supply of oxygen is rapidly extinguished in areas of high metabolic activity, and therefore limits the rate and extent of biodegradation. Oxygen replenishment may also be achieved by soil tilling or by the introduction of air, pure oxygen, hydrogen peroxide (H_2O_2), or other oxygen-releasing materials into soil and groundwater systems. Microorganisms that have been previously acclimated to the contaminant of interest may also be mixed into soil or groundwater environments to stimulate the rate of contaminant decomposition. A recent approach to enhancing *in situ* treatment of chlorinated solvents is the introduction of electron donors such as hydrogen, lactic acid, etc. Installation of reactive barrier walls into groundwater aquifers may also improve the bioremediation of a variety of organic contaminants.

A. Biostimulation

1. *Overview*

The earliest system of enhanced or engineered *in situ* bioremediation of groundwater was designed by R. L. Raymond and co-workers with Sun Research and Development Company in Philadelphia and described in a 1978 American Petroleum Institute report titled "Field Application of Subsurface Biodegradation of Gasoline in a Sand Formation." The "Raymond process" was a patented system [U.S. Patent No. 3,846,290 (1974)] in which biodegradation of petroleum hydrocarbons (namely, gasoline constituents) in groundwater by indigenous subsurface microorganisms was stimulated through the injection of nutrients and oxygen. Although subsurface microorganisms require many minerals, the two most common nutrient amendments are nitrogen (as ammonium or nitrate) and phosphorus (as phosphate). The necessary mass of each nutrient may be determined stoichiometrically, and the ratio between carbon, nitrogen, and phosphorus is typically 100:10:1 (C:N:P) (Norris *et al.*, 1994). Many compounds have been used to increase the dissolved oxygen concentration of contaminated groundwater aquifers. Initial attempts to increase dissolved oxygen concentration in groundwater aquifers relied on air sparging of water before injection into the subsurface or sparging air into wells; however, the low solubility of oxygen in water (8–12 mg/liter) precluded maximum contaminant biodegradation. Subsequent methods for oxygen delivery resulted in greater solubility of the gas. Dissolved oxygen concentrations increase to approximately 40–50 mg/liter when sparging with pure oxygen, whereas injection of H_2O_2 can easily result in dissolved oxygen concentrations in the range of 250–500 mg/liter and higher, although H_2O_2 may be toxic to the microorganisms at higher concentrations (Rao *et al.*, 1997).

Oxygen-releasing compounds (e.g., ORC®, marketed by Regenesis, Inc.) are a more recent innovation for meeting the oxygen demand of microorganisms in subsurface environments. These formulations promote the slow release of oxygen which is produced when magnesium peroxide (the typical active ingredient in these compounds) reacts with water. Oxygen-releasing compounds may be added as a slurry injection or may be placed in a series of wells in the aquifer to form a barrier through which the contaminated groundwater flows. Oxygen is therefore introduced into the central regions of the plume (usually anaerobic) to promote aerobic biodegradation of the contaminant compound (DuPont *et al.*, 1998).

Bioaugmentation, or the introduction of contaminant-acclimated or genetically engineered microorganisms, is another method for enhancing biodegradation of contaminants. Although bioaugmentation has been successfully implemented in a variety of laboratory-scale and larger *ex situ* reactor systems, the success of bioaugmentation in soils and groundwater is limited (Alexander, 1994). Some of the problems associated with adding an inoculum to soils or groundwater aquifers are the presence of environmental toxins and bacterial predators, adequate conditions for growth (i.e., nutrients, pH, temperature, etc.; see Table 16.1), contact with target contaminant, and cell movement or distribution (Alexander, 1994). The movement or distribution of the inoculum throughout the contaminated region of interest may be inhibited by sorption to solid surfaces, bacterial mobility, and the structure of the porous media. Microorganisms may be able to move freely through large macropores in the soil; however, their movement may be restricted through soil micropores in which larger pools of the contaminant may often reside (Alexander, 1994).

A schematic of a typical biostimulation *in situ* remediation system is presented in Figure 16.3. Groundwater is initially drawn from a series of recovery wells downgradient from the source of contamination. The groundwater is then pumped to an aboveground treatment facility at which both nutrient amendments and oxygen (through either sparging or H_2O_2 injection) are added after the water is initially passed through a contaminant treatment scheme. After addition of chemicals, the groundwater is returned to the aquifer either through an injection well or through an infiltration gallery above the zone of contamination. This process continues until the contaminant concentration is reduced to the target level (DuPont *et al.*, 1998).

2. *Demonstration*

J. T. Wilson, J. M. Armstrong, and H. S. Rifai reported the bioremediation of an aviation gasoline spill at a Coast Guard air station in Traverse City, Michigan,

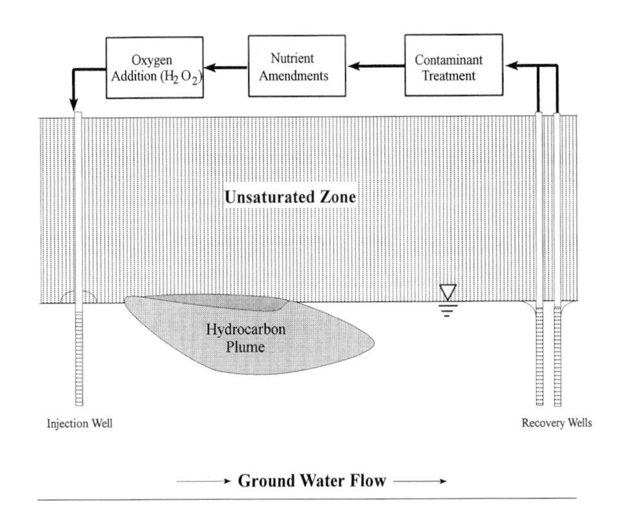

FIGURE 16.3 Simplified schematic of an *in situ* biostimulation remediation system (Raymond process).

using nutrient and H_2O_2 amendments (Flathman *et al.*, 1994). The shallow sandy aquifer underlying the site was contaminated by approximately 25,000 gallons of aviation fuel emanating from an UST. Analysis of soil cores taken from the site identified a vertical zone of petroleum hydrocarbon contamination, and the remediation process was aimed at reducing the concentration of BTEX dissolved in the groundwater and bound in the aquifer material. A series of six monitoring wells were placed downgradient of the infiltration wells to quantify BTEX loss due to biodegradation.

Nutrients introduced into the contaminated region included ammonium chloride, disodium phosphate, and potassium phosphate, whereas H_2O_2 was incrementally injected into the subsurface to allow for microbial acclimation (to a maximum of approximately 400 mg/liter oxygen). A high degree of BTEX removal was observed in each of the six monitoring wells. For example, BTEX was reduced from 1200 to 380 mg/liter at a monitoring well 83 ft from the infiltration wells. Benzene was most amenable to biodegradation in the groundwater, whereas the isomers of xylene were most resistant to biodegradation. Remediation of the aquifer material was also investigated by taking soil cores after completion of the experiment. The results indicated that the process generally led to a reduction in BTEX concentrations in the aquifer materials; however, concentrations of total petroleum hydrocarbons at each of the monitoring points remained fairly high. In areas in which oxygen concentration was low, nitrate was determined to be the terminal electron acceptor supporting most of the BTEX biodegradation.

B. Electron donor delivery

1. *Overview*

The delivery of electron donors to groundwater systems is another *in situ* biostimulation approach to promote or accelerate the biotransformation of chlorinated solvent compounds. Unlike many of the petroleum hydrocarbon contaminants that may be directly utilized by microorganisms to obtain cell energy and promote cell growth, chlorinated solvents are not often used as primary substrates. However, chlorinated solvents may be biodegraded or biotransformed under either aerobic or anaerobic conditions. Under aerobic conditions, biodegradation can be achieved through cometabolic processes in which enzymes capable of breaking down the target contaminant are fortuitously expressed during the biodegradation of other primary substrates. More frequently, chlorinated compounds are biotransformed to less chlorinated products under anaerobic conditions through the process of reductive dechlorination. Microorganisms require a sufficient supply of electron donors in order to carry out the reductive dechlorination process. A variety of potential electron donors have been evaluated both in field and in laboratory studies and include acetate, methane, hydrogen, ammonia, benzoate, lactate, and methanol. Cocontaminants such as petroleum hydrocarbons may also serve as electron donors in subsurface environments.

Many of the same difficulties experienced with the introduction of nutrients, microorganisms, and other materials into subsurface environments also occur with electron donor delivery (Norris *et al.*, 1994). Excessive microbial growth may result near the point of injection due to the high electron donor concentration and availability. This problem may be further exacerbated by the simultaneous introduction of other materials that are essential for bacterial growth (i.e., oxygen and nutrients).

2. *Demonstration*

In 1997, in a paper titled "Scale-up Issues for *in situ* Anaerobic Tetrachloroethene Bioremediation" M. D. Lee and co-workers reported laboratory and field experiments relating to electron donor selection and delivery to enhance anaerobic PCE biotransformation. Microcosm studies were used to determine the most efficient substrates to effect PCE dechlorination. The substrates included yeast extract [38% carbon (C)], wastewater (4.5% C), cheese whey permeate (26% C), molasses (29% C), corn steep liquor (17% C), manure tea (3% C), sodium benzoate (58% C), and acetate (29% C). The field site under investigation was a landfill associated with a chemical plant in Texas. The saturated zone of the aquifer underlying this site consisted primarily of sand. A series of injection, withdrawal, and monitoring wells were used in the field experiments which were designed to determine an optimal electron donor delivery system.

The initial microcosm study investigated dechlorination of chlorinated ethenes (23 μM PCE, 0.6 μM TCE, and 3.3 μM 1,2-DCE) using yeast extract, wastewater, molasses, corn steep liquor, and manure tea at varying initial carbon loadings. A high degree of carbon utilization was achieved in each microcosm (>60% TOC removal), and PCE was transformed to vinyl chloride at carbon concentrations >60 mg/liter. Cell counting experiments indicated a positive relationship between the amount of carbon in the microcosm and the number of microorganisms present. A second microcosm experiment, using sodium benzoate, sodium acetate, corn steep liquor, and molasses, studied the effect of very high organic loadings on both PCE dechlorination and microorganism concentration. In contrast to the results from the first study, higher organic loadings inhibited TOC removal and resulted in either no transformation or only partial transformation of PCE.

The field tracer experiments compared the injection and distribution of the substrate and nutrient amendments under three conditions: tracer (bromide) injection at the rate of groundwater flux, tracer (iodide) injection at a rate of 60 times groundwater flux, and cross-gradient injection [bromide, substrate, and nutrients (N, P)] at a rate of 3.8 liters/min for the withdrawal wells (two) and the injection well. Distribution or dispersion was enhanced when the injection rate was increased to 60 times the natural groundwater gradient. The third condition, employing the cross-gradient injection scheme, further increased dispersion of the bromide tracer. Measurements of TOC also indicated that sodium benzoate was successfully transported downgradient from the injection well. However, transport of both the injected nutrients and electron acceptor (sulfate) was retarded either by sorption to the aquifer solids or by biodegradation processes.

C. Bioventing

1. *Overview*

Bioventing is another *in situ* technique designed to stimulate aerobic biodegradation of contaminants by replenishing oxygen levels in oxygen-depleted, unsaturated zone environments. The process was first described by J. T. Wilson and C. H. Ward in an article published in 1986 in the *Journal of Industrial Microbiology*. R. E. Hinchee, R. R. DuPont, R. N. Miller, and others were responsible for developing and testing the process (Leeson and Hinchee, 1997). Bioventing has been successfully applied at several petroleum hydrocarbon-contaminated sites and may also be implemented to remediate sites contaminated with chlorinated solvents, although it typically requires the addition of substrates to promote cometabolism of the chlorinated hydrocarbons (Norris *et al.*, 1994). A recent report suggests that bioventing is being used at less than 1% of UST contaminated soil sites. However, this percentage translates into the application of bioventing at more than 800 UST sites (USEPA, 1997). Bioventing is designed to emphasize biodegradation over contaminant volatilization; however, the relationship between the two mechanisms may be determined by both the properties of the contaminant of interest (e.g., molecular weight and vapor pressure) and site conditions (Norris *et al.*, 1994; Leeson and Hinchee, 1997). Hydrocarbon contaminants with higher vapor pressures will tend to volatilize with injection of air and may not be suitable for bioventing, whereas biodegradation of heavy hydrocarbons with lower vapor pressures may be achieved using this treatment technology (Cookson, 1995).

In addition to contaminant characteristics, soil properties play a major role in the success of bioventing applications. These properties include physical characteristics, such as hydraulic conductivity, gas permeability, and moisture content, as well as properties that will affect the growth of microorganisms (in addition to oxygen supply), such as temperature, pH, and nutrient supply. Bioventing applications are best suited for soils with high gas permeabilities so that the injected air readily moves through the contaminated soil matrix (Leeson and Hinchee, 1997; Ward *et al.*, 1997). The flow of air may be severely restricted in low-permeability soils and will subsequently impact the extent of contaminant biodegradation. Gravel and sand may be considered highly gaspermeable materials, whereas silts and clays represent soils with lower gas permeabilities. Although moisture is required by soil microbes for contaminant metabolism, high percentages of water in the soil will negatively impact the flow of oxygen in the soil pores. The majority of pore spaces in high moisture content soils are filled with water, and the diffusion of oxygen through water is much slower than through air.

As with other treatment technologies, microbial growth factors (temperature, pH, and nutrients) may also impact the rate and extent of biodegradation in bioventing applications. Higher temperatures are generally associated with higher rates of contaminant biodegradation. However, Leeson and Hinchee (1997) pointed out that psychrophilic microorganisms at a field test site in Alaska were able to signifycantly biodegrade petroleum hydrocarbons in the subsurface. Nutrient amendment also may need to be considered in deficient subsurface soils. Delivery or transport of nutrients to the target zone may be difficult and will also depend on the soil physical properties. Soils generally contain sufficient nutrients to support the biodegradation of hydrocarbon contaminants, and the advantage of adding nutrients to subsurface soils has not been firmly established (Leeson and Hinchee, 1997).

Figure 16.4 presents a schematic of a bioventing system (Leeson and Hinchee, 1997). Bioventing is accomplished by advective flow of air through a series of vent wells into areas that have exhausted their supply of oxygen in the soil gas. Monitoring points are distributed throughout the zone of contamination to determine both the decrease in oxygen concentration

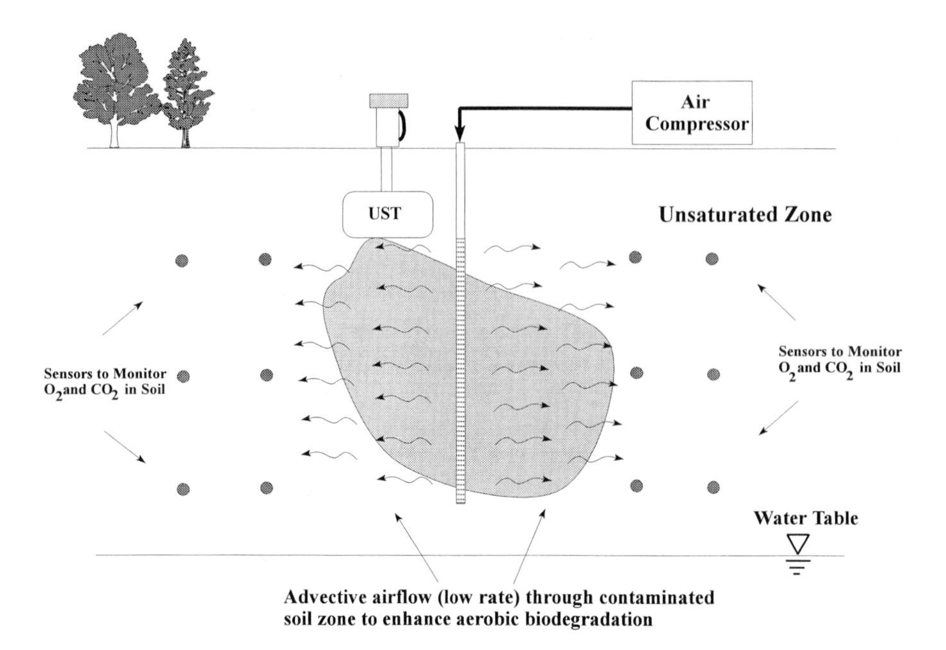

FIGURE 16.4 Schematic of a bioventing system (adapted from Leeson and Hinchee, 1997).

and the increase in the carbon dioxide concentration (primary product of biodegradation). Low airflow rates are usually incorporated into the system design to favor contaminant biodegradation over volatilization. In theory, contaminant biodegradation should be much greater than volatilization; therefore, management of gases that are transported to the soil surface is usually not required. If volatilization is high, a separate extraction and treatment system will have to be installed to handle the vapors being generated by the bioventing system. Bioventing is frequently applied in cooperation with and usually following soil vapor extraction (SVE) for vadose zone remediation.

2. *Demonstration*

One of the earliest and best known bioventing remediation systems was implemented at a site at Hill Air Force Base in Utah where 27,000 gallons of jet fuel were spilled (Leeson and Hinchee, 1997). Fuel contamination migrated to a depth of approximately 65 ft in the unsaturated zone, which consisted primarily of sand and gravel constituents. The average contaminant concentration in the soils was approximately 400 mg/kg. A series of vent wells were installed over the length of the contamination to evaluate the potential of bioventing. Enhanced bioventing using both moisture and nutrient addition was also investigated. The system was initially constructed to investigate contaminant volatilization and capture at higher air flow rates (SVE); however, after several months of testing, lower airflow rates were employed to favor biodegradation over volatilization.

A series of soil samples were taken with depth before and after implementation of the SVE and bioventing remediation system at the site. After application of SVE and bioventing, nearly all of the post-treatment hydrocarbon concentrations in the soils were < 5 mg/kg, indicating successful cleanup of the site. It was estimated that approximately 1500 lbs of hydrocarbon fuel was removed through volatilization, whereas 93,000 lbs of fuel was removed through biodegradation. Much of the biodegradation of the fuel was attributed to the bioventing phase of cleanup. Enhanced bioventing by adding moisture and nutrients yielded mixed results. The addition of moisture resulted in a significant increase in contaminant biodegradation, whereas addition of nutrients (N and P) did not enhance removal of the petroleum hydrocarbons.

D. Permeable reactive barriers

1. *Overview*

The use of *in situ* permeable reactive barriers is another method that introduces reactants which may stimulate the biotic or abiotic transformation of environmental contaminants. As opposed to other *in situ* remediation methods that rely on injection of reactants (i.e., oxygen and nutrients) and their transport with groundwater flow through the contaminated aquifer, stationary barrier walls containing the reactive porous media are placed into the subsurface using a variety of methods. The contaminated groundwater is directed to and passes through the reactive barrier, thus enabling reaction between the porous media and the contaminant of interest. Permeable reactive barriers are typically used to treat target contaminants, including petroleum hydrocarbons and chlorinated solvents (Table 16.3).

A schematic of an *in situ* permeable reactive barrier is presented in Figure 16.5. The barrier may span the width of the zone of contamination or a series of sheet pilings may be inserted into the subsurface which direct the contaminated water through the reactive barrier. This "funnel and gate" system was devised R. C. Starr, J. A. Cherry, and other researchers at the University of Waterloo and published in 1994 in *Ground Water*. In cases in which the reactants in the barrier may be rapidly extinguished due to biochemical reactions, these materials may be replenished by designing replacement cassettes that may be easily exchanged in the barrier. Cassettes can also be placed in series to target the biodegradation or biotransformation of a particular contaminant or to remediate groundwater with many contaminants, each of which has specific requirements for degradation. Permeable reactive barrier technologies may also incorporate a series of smaller reactive walls if the size of the plume is large, thus facilitating easier removal and replacement when compared to a single large wall.

Starr, Cherry, and others pointed out that the most important considerations for the design and application of this type of system include (i) time required to effect desired biochemical reaction (combination of groundwater flow rate, reaction rate, influent contaminant concentration, and target effluent contaminant concentration), (ii) the potential formation of hazardous products resulting from reactions within the reactive barrier, and (iii) costs associated with installation and media regeneration. Examples of reactants or amendments used in permeable reactive barrier technologies to enhance contaminant biodegradation and transformation include nutrient amendments (N and P), oxygen addition using oxygen-releasing compounds or biosparging (introduction of air), and introduction of zero valent iron to achieve abiotic reduction of chlorinated compounds.

2. *Demonstration*

J. F. Barker, J. F. Devlin, and co-workers demonstrated the use of permeable reactive barrier remediation technologies at the Canadian Forces Base Borden site in Ontario, Canada. The work was documented in a 1998 report sponsored by the Department of Defence Advanced Applied Technology Demonstration Facility (AATDF) at Rice University. Contaminants within the groundwater plume targeted for remediation using this *in situ* technology included carbon tetrachloride, PCE, and toluene. Remediation of the highly chlorinated aliphatics (i.e., carbon tetrachloride and PCE) was accomplished through both biotic

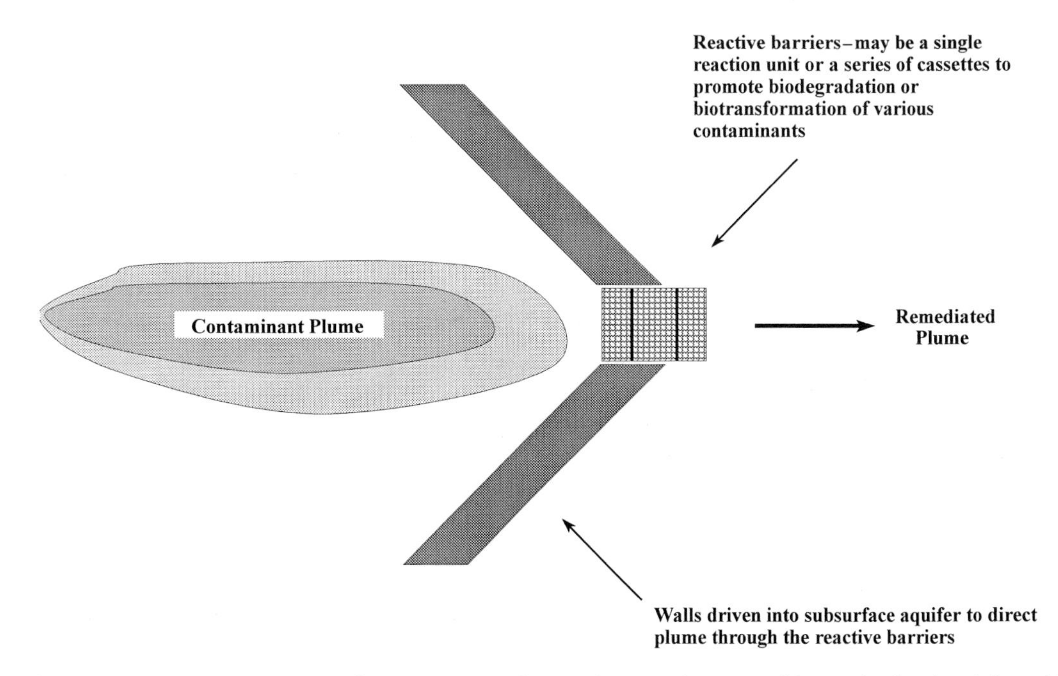

FIGURE 16.5 Funnel and gate system to remediate contaminated groundwater using permeable reactive barriers (adapted in part from Starr and Cherry, 1994).

and abiotic reduction processes, whereas biodegradation of the less chlorinated aliphatics and toluene was achieved in an aerobic treatment zone. Three test gates containing different sequences of reactive cassette barriers were installed at the site. Gate 1 consisted of two granular iron cassettes (to effect abiotic reductive dechlorination of chlorinated hydrocarbons) followed by a cassette containing oxygen-releasing compounds. Gate 2 did not contain reactive media and served as a control and/or natural attenuation gate. Gate 3 contained a benzoate injection well (to serve as substrate for microorganisms and to achieve anaerobic conditions for microbially mediated reductive dechlorination) followed by a biosparging wall. Initial spiked contaminant concentrations in the groundwater were approximately 1 or 2 mg/liter for carbon tetrachloride and PCE and 10 mg/liter for toluene.

Results from Gate 1 indicated that PCE was completely transformed to ethane and ethene (halflife = 0.5 days). Carbon tetrachloride was transformed to chloroform prior to reaching the gate and was rapidly dechlorinated in the iron barrier to dichloromethane. High pH next to the granular iron cassette required emplacement of the ORC downgradient at lower pH to stimulate oxygen release and toluene biodegradation. Gate 2 (natural biodegradation and biotransformation) results also indicated biotransformation of carbon tetrachloride and chloroform (each having a half-life of approximately 11 days). In Gate 3, PCE was successfully transformed to *cis*-DCE. Subsequent passage of the contaminated groundwater in Gate 3 through the biosparging wall (aerobic zone) resulted in biodegradation of toluene as well as the less chlorinated end products of reductive dechlorination produced in the anaerobic zone (e.g., *cis*-DCE).

VII. ENHANCED *EX SITU* BIOREMEDIATION

Like *in situ* processes, enhanced *ex situ* bioremediation processes are engineered to accelerate the biodegradation or biotransformation of organic contaminants in soils and solids. In many cases, *ex situ* technologies generally offer better control of the parameters that govern biodegradation than do *in situ* techniques. *Ex situ* bioremediation technologies include land farming, composting, and slurry reactors. Land treatment employs nutrient addition and aeration (by tilling) to stimulate biodegradation of land-applied wastes. Composting of contaminated soils and sludges is a high-temperature, exothermic process in which bulking agents (e.g., wood chips and mulch) and, in some cases, nutrients are added to encourage biodegradation. Finally, *ex situ* bioslurry reactors may be used to treat either contaminated liquids or solids through nutrient addition, aeration, and bioaugmentation.

A. Land treatment

1. *Overview*

Land treatment or land farming refers to the accelerated aerobic biodegradation of organic wastes in either near-surface or excavated soils through the addition of nutrients, lime (pH control), and moisture and through increased aeration by tilling or other mechanical mixing (Loehr and Malina, 1986). Biodegradation is generally carried out by indigenous soil microorganisms, although some form of bioaugmentation may enhance the rate of degradation. Typically, land farming is an *ex situ* process whereby the contaminated soils are excavated from a site and sent to an engineered treatment unit (also termed a prepared bed system or reactor); however, *in situ* methods (nutrient addition and tilling) may be adequate to enhance biodegradation of the soil contaminants near the surface of excavated wastes that are mixed or tilled into the top soil layer (to a depth of approximately 1 ft). A wide variety of organic contaminants have been successfully treated in land farming applications, including petroleum hydrocarbons, pesticides, PCPs, PCBs, and PAHs (Cookson, 1995).

Land farming typically requires the addition of nutrient amendments (namely, N and P) to the soils to enhance biodegradation of contaminants. The minerals may be introduced into the soil either as a solid or mixed with water and applied through a spraying system (Cookson, 1995). The spraying system may also provide needed moisture to soils and enhance contaminant biodegradation. Plowing, tilling, or other methods of mechanical mixing of the soils stimulates biodegradation through (i) mixing and distribution of soil amendments (nutrients, lime), (ii) distribution of contaminants (by breaking up soils) and increased contact between contaminants and microorganisms, and (iii) increased aeration of soils (increased oxygen supply) (DuPont *et al.*, 1998; Ward *et al.*, 1995; Cookson, 1995).

Figure 16.6 presents an engineered land treatment system. Much like conventional municipal solid waste treatment units or landfills, controls or collection systems may be placed in the land treatment unit. To prevent groundwater contamination, the base of a typical land treatment unit is covered with a highly impermeable clay or geosynthetic (plastic) liner. A series of leachate recovery pipes are then placed near the base of the system to collect wastes that may percolate through the soil. A thick layer of sand covers the collection system, and the contaminated soil is then placed on the

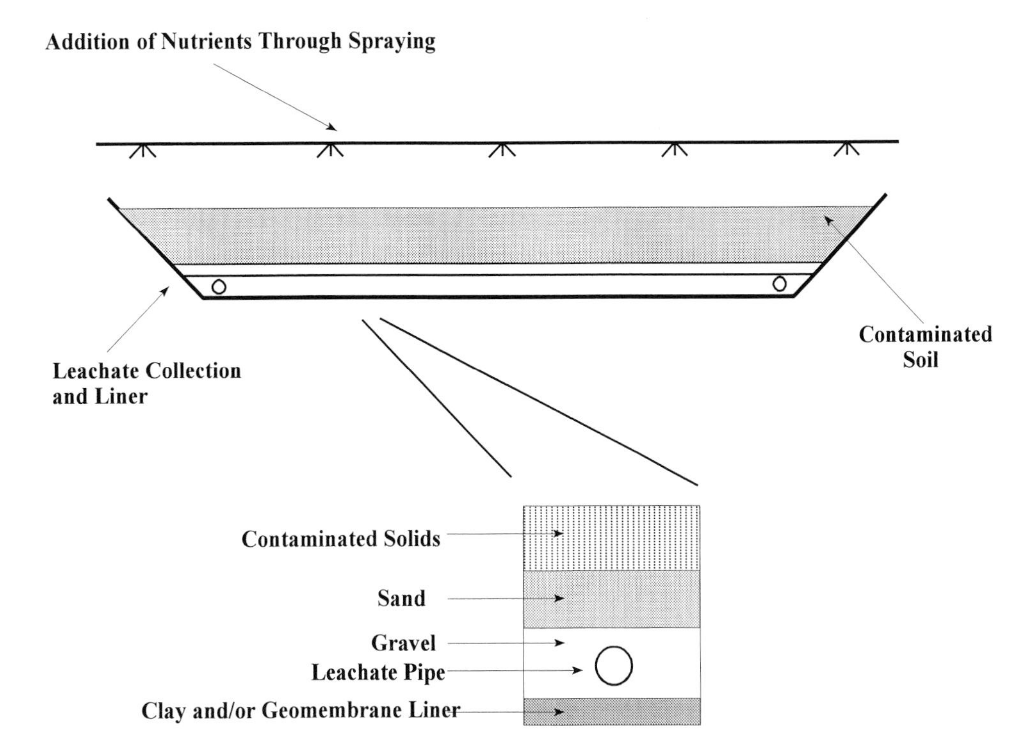

FIGURE 16.6 Typical land treatment unit (inset depicts liner configuration) [adapted in part from Cookson (1995) and Alexander (1994)].

sand layer. Collected leachate may be recycled or sent to another treatment system. A cover or enclosure may be added to the land treatment unit to eliminate rainfall percolation and off-site migration of contaminant vapors (Cookson, 1995; Alexander, 1994).

2. *Demonstration*

In 1979, J. T. Dibble and R. Bartha published work (*Soil Science*) on the land treatment of hydrocarbon-contaminated soil resulting from a pipeline leak in New Jersey in which approximately 1.9 million liters of kerosene was spilled onto 1.5 hectares (ha) of an agricultural plot. After removal of approximately 200 m^3 of heavily contaminated soil, both lime and nutrients (N, P, and potassium) were applied to the field soil (to a depth of 117 cm) to enhance biodegradation. Aerobic biodegradation of the kerosene contaminant was further stimulated by periodic tilling or mixing of the soil to promote distribution of oxygen. The applied nutrient concentrations were 200 kg nitrogen/ha, 20 kg phosphorus/ha, and 17 kg potassium/ha and were added at two different times after an initial loading of 6350 kg/ha of lime. The contaminated soil was characterized as a well-drained, sandy loam soil. Kerosene biodegradation was monitored during a 24-month period to determine the efficiency of the treatment system.

Over the course of the experiment, kerosene concentrations in the soil decreased from 8700 mg/kg to very low levels in the upper 30 cm of the soil. Likewise, kerosene concentration in the lower portion of the soil (30- to 45-cm depth) also decreased over the 24-month period to <3000 mg/kg. Biodegradation was determined to be the primary mechanism for removal (compared to volatilization) based on the disappearance patterns of compounds of varying molecular weight. The rate of biodegradation in the upper portion of the soil was initially greater than the rate of biodegradation in the lower portion. However, more rapid kerosene biodegradation was observed in the lower portion of the soil after 6 months of system operation and evaluation. Temperature also played a major role in the biodegradation of the kerosene contaminants. The greatest decreases in kerosene concentration due to biodegradation occurred during time periods when temperatures were at or higher than 20 °C; contaminant removal was diminished at lower temperatures.

B. Composting

1. *Overview*

Composting is another *ex situ* process in which organic compounds are biodegraded, biotransformed, or otherwise stabilized by mesophilic and thermophilic bacteria. Addition of other readily biodegradable materials, bulking agents, moisture, and possibly nutrients to contaminated soils enhances or stimulates the composting process (Ro *et al.*, 1998) Nitroaromatics, petroleum hydrocarbons, PAHs, chlorinated phenols, and pesticides are classes of contaminants that are amendable to biodegradation using a composting system (Table 16.3).

Composting at high temperatures typically accelerates the biodegradation or biotransformation process and has historically been required to kill pathogenic organisms remaining in wastewater treatment sludges; however, high-temperature composting of contaminated soils and other solids may not be required due to the absence of pathogens in these materials (Cookson, 1995). As a result, composting of contaminated soils may occur either under mesophilic (15–45 °C) conditions or under thermophilic (50–70 °C) conditions. The most suitable temperatures for composting range from 55 to 60 °C, whereas temperatures higher than 70 °C severely inhibit the ability of the microorganisms to metabolize or transform the contaminants (Ro *et al.*, 1998). Bulking agents are used to increase the pore space or porosity of contaminated soil which results in greater air movement through the compost pile and higher rates of aerobic biodegradation. Typical bulking agents used in composting include wood chips, straw, tree bark, and plant matter (Alexander, 1994). Moisture content may also be a controlling parameter in the composting of contaminated soils. Excess moisture will impede the diffusion of oxygen through the compost pile and therefore restrict aerobic biodegradation. Optimal percentage moisture contents in compost piles range from 50 to 65% (Ro *et al.*, 1998). Cookson (1995) reports that hydrocarbon biodegradation in a composting system is optimized at a moisture content of 60%, whereas slightly lower moisture contents are suitable for other hazardous constituents.

The three most common types of composting systems are the windrow, static pile, and in-vessel systems. In windrow composting systems, the composting materials (mixture of organic material, contaminated soil, and bulking agents) are placed in long rows and are turned or mixed using a mechanical device to provide aeration or oxygen replenishment. Static pile systems are similarly arranged in long rows of composting material, but oxygen is supplied through a series of pipes that are placed at the base of the piles. Finally, in-vessel systems are typically enclosed units in which composting materials are transported on a conveyor system through a reaction vessel that is responsible for mixing, forced aeration, and temperature control (Cookson, 1995). In-vessel systems allow for greater engineering control (i.e., capture of volatile compounds and temperature and aeration regulation) and may greatly accelerate the biodegradation process when compared to the other two types of composting systems.

2. *Demonstration*

In 1992, R. T. Williams, P. S. Ziegenfuss, and W. E. Sisk published work in the *Journal of Industrial Microbiology* on the static pile composting of explosives contaminated sediments from two U.S. Army facilities—Louisiana Army Ammunition Plant (LAAP) and Badger Army Ammunition Plant (BAAP). Target contaminants to be biodegraded using this remediation scheme included TNT, RDX, HMX, tetryl, and nitrocellulose (NC). Additives to the contaminated soils included alfalfa, straw, manure, wood chips, and horse feed in varying combinations along with nitrogen, phosphorus, potassium, and water. The initial concentration of contaminants in the LAAP soil included 56,800 mg/kg TNT, 17,900 mg/kg RDX, 2390 mg/kg HMX, and 650 mg/kg tetryl, whereas the initial concentration of NC in the BAAP soil was 18,800 mg/kg. Compost pile temperatures of the various configurations (two LAAP piles and four BAAP piles) were set at 35 and 55 °C to investigate composting of explosives under both mesophilic and thermophilic conditions.

The results indicated successful remediation of the contaminants using various combinations of bulking agents and temperatures. For the LAAP soil static piles, the combined concentration of explosives decreased from 17,870 to 74 mg/kg in the thermophilic pile, whereas mesophilic conditions effected a concentration decrease from 16,560 to 326 mg/kg during the 153-day test cycle. Half-lives for TNT, RDX, and HMX under thermophilic conditions were 12, 17, and 23 days, respectively. Half-lives of the contaminants under mesophilic conditions were nearly double those calculated under thermophilic conditions. Very low levels of amino transformation products (e.g., 2-amino-4,6-dinitrotoluene, 4-amino-2,6-dinitrotoluene) were detected at the end of the test cycle. Nitrocellulose concentrations were dramatically reduced in the BAAP soil static piles. Under thermophilic conditions, NC concentration in two of the test piles was reduced from an average of approximately 13,090 mg/kg to approximately 20 mg/kg.

R. Valo and M. Slkinoja-Salonen (1986) investigated windrow composting of chlorophenol-contaminated soils from a sawmill in Finland. Two 50-m^3 compost piles were constructed and contained contaminated soil along with bark and ash. Nitrogen, phosphorus, and potassium were also added to the compost pile in the aqueous phase to ensure adequate nutrient levels throughout the system. The initial concentration of chlorophenols in the soil ranged from 400 to 500 mg/kg soil, whereas the initial concentration in the compost ranged from 200 to 300 mg/kg. The composting unit was monitored over the course of 17 months with the majority of sampling and analysis being conducted during the first 5 months of operation (summer months, presumably when microbial activity was greatest).

The results of the study indicated a significant decrease in the concentration of chlorophenols in the composting unit. Most of the contaminant loss or destruction occurred during the first few months of operation. The concentration of chlorophenol decreased rapidly to approximately 30 mg/kg after a few months of operation, and the concentration decreased to approximately 15 mg/kg by the end of the test period. Bacterial identification and counts also indicated that the compost material contained a greater number of pentachlorophenol-degrading microorganisms when compared to an agricultural clay soil that was not previously contaminated by chlorophenols. Subsequent jar tests on the compost material with radiolabeled pentachlorophenol confirmed that biodegradation was a major removal mechanism because approximately 30% of the radiolabeled carbon evolved as $^{14}CO_2$ during a 40-day test period.

C. Bioslurry processes

1. *Overview*

Slurry reaction systems are another type of *ex situ* bioremediation process in which biodegradation of contaminants is effected either in a highly controlled bioslurry reactor or in a waste lagoon. Typically, contaminated soils, sediments, or other solids are added to the system and mixed with a host of amendments, including water, nutrients, oxygen (if an aerobic environment is desired), surfactants to enhance contaminant mobilization, and/or microorganisms that may specifically biodegrade or biotransform the contaminant of interest. Bioslurry systems, especially enclosed bioreactors, offer a high degree of engineering control. Most of the major parameters that impact contaminant biodegradation (e.g., temperature, pH, and nutrient addition) may be monitored and adjusted using this remediation approach. Many contaminant compound classes are amenable to biodegradation using bioslurry processes and include phenols, chlorinated phenols, PAHs, pesticides, chlorinated hydrocarbons, and petroleum hydrocarbons (Cookson, 1995; Alexander, 1994). Highly viscous contaminants such as tars and certain oils may not be appropriate for bioslurry treatment.

Treatment of the contaminated material may occur either in a bioslurry reactor or in a lagoon. The bioreactor system may consist of a single reactor or a series of reactors, and these reactors are typically closed to the atmosphere to prevent escape of volatile compounds and to maintain system control. Both the size and the volume of the bioslurry reactors can be highly variable. The percentage solids concentration in these reactors ranges from 5 to 50% and depends on both the physical properties of the soil and the characteristics of the treatment system. Reactors in series offer a potential method of creating both aerobic and anaerobic environments in sequence, and this type of system may be very efficient in coupling the reductive dechlorination of highly chlorinated solvents (anaerobic conditions) with the subsequent metabolism of less chlorinated compounds (aerobic conditions). Typically, engineered waste lagoons vary in size and may be equipped with a variety of mixers and aerators to provide oxygen to the system. As in land treatment applications, a highly impermeable layer of clay soil or synthetic liner should be situated at the base of waste lagoons to prevent contaminant percolation into groundwater.

The continuously mixed bioreactors or waste lagoons have significant engineering advantages compared to other *ex situ* technologies such as composting and land farming, although bioreactor systems may be costly to operate (Cookson, 1995). The mixing of the solids within the reactor enhances the distribution of both the solid materials containing the contaminants and the amendments that are added to the reactor system. Increased aeration and mass transfer of oxygen throughout the bioreactor as a result of mixing or sparging encourages aerobic biodegradation of the contaminants. If the bioslurry system contains volatile compounds, the rate of volatilization may be increased due to agitation and mixing. Both the accessibility of the contaminant to the microorganisms (contact) and the fraction of bioavailable contaminant are increased in this type of reactor system.

Bioslurry processes promote the breakup of aggregated soil particles that may sequester or contain high concentrations of the target contaminant. This breakup of larger particles into smaller ones leads to a greater soil surface area to volume ratio and thus may increase contaminant desorption by maximizing surface contact with the aqueous phase. Contaminated clay soils may be especially amenable to bioslurry treatment processes due to their decreased permeability and difficulty in treating *in situ*. Organic contaminants adsorb strongly to adhesive clay materials and may be easily entrapped in the intra- and interparticle pore spaces, and subsequent transport of materials (i.e., nutrients, oxygen or other electron acceptor, and microorganisms) through clay soils is very difficult.

2. *Demonstration*

In 1991, G. C. Compeau and coworkers reported results from laboratory-scale experiments used in the design and implementation of a full-scale bioslurry reactor treatment system. The full-scale system was designed to treat 3400 cubic yards of PCP-contaminated soil from a spill. PCP concentrations in the soil were variable, with a maximum of 9000 mg/kg. Initial testing of the contaminated soil indicated that the indigenous microflora were not capable of breaking down the PCP contaminant; therefore, subsequent bench and full-scale treatment systems incorporated specialized PCP-degrading microorganisms to augment the existing soil organisms and stimulate biodegradation of the contaminant.

The laboratory-scale bioslurry treatability studies indicated that for a variety of solids with concentrations ranging from 5 to 40%, PCP could be successfully biodegraded when the reactor was inoculated with the PCP-degrading culture. For example, in the 40% solids concentration bioslurry test with an initial concentration of approximately 275 mg/liter, very little of the PCP was removed from the system during the initial 13-day test period. After inoculation, nearly all the PCP in the reactor was biodegraded during a 10-day period. The laboratory studies also revealed that the majority of the PCP contaminant resided within the more coarse particles, which may not be easily suspended in a bioslurry reactor system. As a result, design of the full-scale treatment system was amended to include a washing step to desorb the PCP from the coarse particles. The wash solution containing PCP along with the finer unwashed soil particles represented the major influents to the bioslurry reactors.

A pair of 25,000-gallon bioslurry reactors were used in the full-scale cleanup operation. Nitrogen and phosphorus were added to the reactors to meet microbial growth requirements. Results from the full-scale experiment supported the laboratory findings that inoculation was necessary to stimulate PCP biodegradation. A testing period of 2 weeks was required to decrease PCP soil concentrations from 370 to <0.5 mg/kg in one of the inoculated reactors. This degree of treatment was also reached in the second reactor, but only after addition of the inoculum following a 7-day test period in which biodegradation did not occur. It is generally believed that bioaugmentation decreases the lag time in biodegradation studies but that, given time, selection processes will result in microbial populations capable of degrading target contaminants.

VIII. SUMMARY

Bioremediation refers to the transformation of organic wastes by microorganisms into biomass, carbon dioxide, water, and inorganic salts, depending on the structure of the compounds in the waste. These organic waste materials may also be biotransformed to less toxic compounds or compounds that may be more amenable to complete mineralization. Many *in situ* and *ex situ* technologies are currently available to address organic waste contamination in both soils and groundwater.

Natural bioattenuation is an *in situ* method for remediating contaminated subsurface soils and groundwater. It is a nonintrusive method that takes advantage of the abilities of natural microflora in subsurface environments to biodegrade organic contaminants. Although natural bioattenuation does not require complex engineered systems, monitoring for contaminant concentration and biochemical markers is necessary to validate the efficiency of this type of treatment option. Enhanced *in situ* bioremediation technologies include the Raymond process (biostimulation), electron donor delivery, bioventing, and permeable reactive barriers. Biostimulation technology involves the addition of nutrients (namely, N and P), oxygen, and perhaps microorganisms to groundwater aquifers to enhance biodegradation of organic contaminants. Oxygen may be added to the groundwater by injection of air, pure oxygen, hydrogen peroxide, or oxygen releasing compounds. Injection of microorganisms into the subsurface, or bioaugmentation, also may be an option for biodegradation enhancement; however, to date there has been little success in field-scale experiments.

In situ biodegradation and biotransformation of chlorinated organic compounds may be accelerated by the injection of electron donors, such as lactate, methanol, and hydrogen, to the contaminant region of interest in groundwater. In a cometabolic process, breakdown of the primary substrate (e.g., methane) leads to the fortuitous expression of enzymes that are capable of oxidizing highly chlorinated organics such as TCE. Biotransformation through reductive dechlorination also requires an adequate supply of electron donors to serve as a primary substrate for the microorganisms that mediate this process.

Bioventing of contaminated subsurface unsaturated zone soils focuses on delivering oxygen to areas in which oxygen has been previously depleted due to microbiological reactions in the soil. This treatment technology is typically applied to soils that have high permeabilities and thus a greater ability to transfer oxygen to the contaminant region. The treatment technology is designed to favor contaminant biodegradation over volatilization, but site hydrogeology and contaminant physical properties may also have an impact on the effectiveness of the two mechanisms. Permeable reactive barriers provide another *in situ* method in which groundwater is routed through a wall or barrier containing materials that will enhance either the biodegradation or biotransformation of organic contaminants. Materials contained in the reaction zone may include both nutrients and oxygen releasing compounds to enhance aerobic biodegradation of contaminants or other reactive agents such as zero-valent iron that lead to the abiotic reductive dechlorination of chlorinated solvents.

Land treatment or land farming is an *ex situ* method to treat contaminated soils and sludges. Excavated materials are placed either in an engineered treatment unit or on top of the natural soil surface. Biodegradation in these systems is enhanced through tilling and mixing to promote aeration, nutrient amendment, and possibly the addition of contaminant-degrading microorganisms. *Ex situ* composting of contaminated soils is achieved by addition of bulking agents such as wood chips to promote aeration, nutrients, and readily biodegradable materials to the contaminated soils. These mixtures are then placed in windrow, static pile, or in-vessel composting systems in which higher temperature, thermophilic conditions are initiated to accelerate the biodegradation process. Finally, bioslurry processes promote treatment of contaminated soils, solids, and sludges by placement in bioslurry reactors and mixing with water, nutrients, oxygen, microorganisms, or other materials that might enhance the biodegradation or biotransformation of the contaminant. Bioslurry processes offer a high degree of control over the parameters that influence biodegradation, including temperature, pH, and nutrient concentration, but they can be expensive when compared to land treatment and composting.

Bioremediation processes for the treatment of contaminated environmental media such as soils and the saturated (aquifer) and unsaturated (vadose) zones of the subsurface may prove to be the most cost-effective treatment options depending on site-specific remedial objectives. In common with most other remediation technologies, bioremediation processes are usually applied as part of a system or treatment in conjunction with other complementing processes to obtain optimal results.

BIBLIOGRAPHY

Alexander, M. (1994). "Biodegradation and Bioremediation." Academic Press, San Diego.

Alleman, C. B., Leeson, A. (Eds.) (1999). "Bioreactor and Ex Situ Biological Treatment Technologies: The Fifth International In Situ and On-Site Bioremediation Symposium: San Diego, California." Battelle.

Chaudry, G. R. (Ed.) (1994). "Biological Degradation and Bioremediation of Toxic Chemicals." Dioscorides Press, Portland, OR.

Cookson, J. T. (1995). "Bioremediation Engineering: Design and Application." McGraw-Hill, New York.

DuPont, R. R., Bruell, C. J., Downey, D. C., Huling, S. G., Marley, M. C., Norris, R. D., and Pivetz, B. (1998). "Innovative Site Remediation Technology Design & Application: Bioremediation" (W. C. Anderson, Ed.). American Academy of Environmental Engineers, Annapolis, MD.

Flathman, P. E., Jerger, D. E., and Exner, J. H. (Eds.) (1994). "Bioremediation: Field Experience." Lewis, Boca Raton, FL.

Leeson, A., and Hinchee, R. E. (1997). "Soil Bioventing: Principles and Practice." Lewis, Boca Raton, FL.

Loehr, R. C., and Malina, J. F. (Eds.) (1986). "Land Treatment—A Hazardous Waste Management Alternative." Center for Research in Water Resources, University of Texas at Austin/Van Nostrand Reinhold, Austin/New York.

Madigan, M. T., Martinko, J. M., and Parker, J. (1997). "Biology of Microorganisms," 8th ed. Prentice-Hall, Upper Saddle River, NJ.

Norris, R. D., Hinchee, R. E., Brown, R., McCarty, P. L., Semprini, L., Wilson, J. T., Kampbell, D. H., Reinhard, M., Bouwer, E. J., Borden, R. C., Vogel, T. M., Thomas, J. M., and Ward, C. H. (1994). "Handbook of Bioremediation." Lewis, Boca Raton, FL.

Pieper, D. H., and Reineke, W. (2000). Engineering bacteria for bioremediation. *Curr. Opin. Biotechnol.* 11, 262–70.

Rao, P. S., Brown, R. A., Allen-King, R. M., Cooper, W. J., Gardner, W. R., Gollin, M. A., Hellman, T. M., Heminway, D. F., Luthy, R. G., Olsen, R. L., Palmer, P. A., Pohland, F. G., Rappaport, A. B., Sara, M. N., Syrrist, D. M., and Wagner, B. J. (1997). "Innovations in Ground Water and Soil Cleanup: From Concept to Commercialization." National Academy Press, Washington, DC.

Rittman, B. E., Alvarez-Cohen, L., Bedient, P. B., Brown, R. A., Chappelle, F. H., Kitanidis, P. K., Mahaffey, W. R., Norris, R. D., Salanitro, J. P., Shauver, J. M., Tiedje, J. M., Wilson, J. T., and Wolfe, R. S. (1993). "*In Situ* Bioremediation: When Does It Work?" National Academy Press, Washington, DC.

Ro, K. S., Preston, K. T., Seiden, S., and Bergs, M. A. (1998). Remediation composting process principles: Focus on soils contaminated with explosive compounds. *Crit. Rev. Environ. Sci. Technol.* **28**(3), 253–282.

Spain, J. C. (Ed.) (1995). "Biodegradation of Nitroaromatic Compounds." Plenum, New York.

U.S. Environmental Protection Agency (USEPA) (1997). Cleaning up the nation's waste sites: Markets and technology trends, EPA 542-R-96-005. USEPA, Washington, DC.

Ward, C. H., Loehr, R. L., Norris, R., Nyer, E., Piotrowski, M., Spain, J., and Wilson, J. (1995). "Innovative Site Remediation Technology: Bioremediation" (W. C. Anderson, Ed.). American Academy of Environmental Engineers, Annapolis, MD.

Ward, C. H., Cherry, J. A., and Scalf, M. R. (Eds.) (1997). "Subsurface Restoration." Ann Arbor Press, Chelsea, MI.

Young, L. Y., and Cerniglia, C. E. (Eds.) (1994). "Microbial Transformation and Degradation of Toxic Organic Chemicals." Wiley–Liss, New York.

WEBSITES

The University of Minnesota Biocatalysis/Biodegradation Database
http://umbbd.ahc.umn.edu/

The Bangor (Wales) Biodegradation Group
http://biology.bangor.ac.uk/research/biodegradation/

The Electronic Journal of Biotechnology
http://www.ejb.org/

List of biotechnology organizations and research institutes
http://www.ejb.org/feedback/borganizations.html

The Biodegradative Strain Database
http://bsd.cme.msu.edu/bsd/index.html

Bioremediation Discussion Group
http://bioremediationgroup.org

17

Biosensors

Yoko Nomura and Isao Karube
Tokyo University of Technology

GLOSSARY

***biochemical oxygen demand* (*BOD*)** The amount of dissolved oxygen (DO) needed to biologically degrade the organic compounds in an aquatic environment. BOD measurements are conventionally carried out according to the BOD 5-day method which measures the DO by titration (modified Winkler method) before and after a 5-day sample incubation period during which biodegradation occurs.

biological sensing element A biomolecule or a biomaterial used in a biosensor for analyte recognition; sometimes referred to as a biological recognition element. It undergoes a specific biological reaction with the analyte so that the analyte can be selectively detected by the biosensor. Examples include enzymes, antibodies, DNA oligomers, and microorganisms.

***flow injection analysis* (*FIA*)** A technique sometimes used in flow-type biosensors; developed in the 1970s. The sample is injected directly into carrier solution running through fine tubes of manifolds which include an electrical detection apparatus. Reactions such as those resulting in fluorescence or chemiluminescence occur *in situ* and are detected as electronic signals. FIA systems afford precise control of the mixing ratio of injected samples and carrier solutions, which results in highly reproducible measurements. Because the detection by FIA is simple, rapid, precise, and continuous, it is used for various measurements, such as in medical diagnosis, food quality control, and environmental monitoring.

glucose oxidase An enzyme which catalyzes glucose oxidation. It is usually suitable for industrial use because of its high stability. It is one of the typical biological sensing elements and was used in the first biosensor.

transducer An electronic signal-transducing element that can convert a change in the concentration of a product of a biological reaction into an electronic signal. Examples are electrodes and optical apparatus.

A Biosensor is broadly defined as a sensing system which uses biological reactions such as enzymatic or immunological reactions. Most biosensors are composed of a biological sensing element and a transducer.

Many biosensors (Cass, 1990; Suzuki, 1990; Buerk, 1993) have been constructed since the first report describing the development of the first enzyme sensor for glucose measurement by Clark in 1962 (Buerk, 1993). This biosensor measured the product of an enzyme reaction using an electrode, which was a remarkable achievement even though the enzyme was not immobilized on the electrode.

Today's basic biosensors originated from the investigation by Updike and Hicks in 1967 (Buerk, 1993). Their sensor combined membrane-immobilized glucose oxidase with an oxygen electrode, and oxygen measurements were carried out before and after the enzyme reaction. Many other biological elements and

The Desk Encyclopedia of Microbiology
ISBN: 0-12-621361-5

transducers have been examined since this first glucose biosensor was developed. For example, in 1977 Karube reported the first microbial sensor which used the whole cell as a biological sensing element. Many immuno-sensors using antibodies were fabricated in 1980s and recently DNA, RNA, and even artificial recognition elements have been employed. Transducers have also been improved, and novel transducing elements such as optical detectors, including surface plasmon resonance (SPR) detectors, are widely used in modern biosensors.

I. PRINCIPLE OF BIOSENSORS

A biosensor is a detection system composed of biological sensing elements, such as enzymes, antibodies, microorganisms, and DNA, and an electronic signal-transducing element (Fig. 17. 1) (Buerk, 1993). A transducer can convert a change in the concentration of a product of a biological reaction into an electronic signal. For example, an oxygen electrode converts a change in the oxygen concentration of a sample caused by enzymes or biodegradation into a change in electric current. The transducers used in biosensors include electrodes, piezo-electric quartz crystals, and optics. Many biosensors can be applied to important fields such as diagnosis and environmental monitoring.

The principle of a simple glucose sensor using glucose oxidase (GOD) is shown in Fig. 17.2. The target analyte, glucose, is a substrate of GOD. Glucose diffuses into the membrane and a biological reaction occurs within the membrane. When GOD catalyzes the glucose oxidation, dissolved oxygen in the sample is consumed and gluconic lactone and hydrogen peroxide are produced. The oxygen electrode detects a change in dissolved oxygen concentration of the sample as a change in electric current. Since the change in dissolved oxygen concentration is a result of the glucose oxidation, the glucose concentration of the sample can be measured. The biosensor does not directly detect the target analyte (glucose). Instead, it measures the change in the concentration of a co-reactant (oxygen) or a co-product (hydrogen peroxide) of the reaction catalyzed by the immobilized biological-sensing material (i.e., GOD).

Biosensor configurations are categorized into two types; batch type (a) and flow type (b) as depicted in Fig. 17.3 (Buerk, 1993). In flow-type sensors, a bioreactor, a column stuffed with biomaterial-immobilized beads, can be incorporated separately from a transducer. Flow-type biosensors are very useful for continuous monitoring of target analytes. Immobilization of biomaterials is usually required for biosensor fabrication. Typical immobilization carriers are glass, alginate and artificial resin beads, and membranes.

Disposable biosensors such as amperometric glucose sensors for medical diagnosis are examples of a batch-type sensor. Flow injection analysis (FIA) is a technique sometimes used in flow-type biosensors. Because biosensors using FIA yield very rapid and accurate measurements, these have been applied to various fields, such as environmental monitoring and food quality control.

In a typical biosensor study, parameters such as the amount of immobilized biological-sensing elements (e.g., enzymes), pH, and temperature of the buffer are optimized, and then a calibration curve is generated using standard solutions containing known concentrations of the analyte. Figure 17.4 shows a typical calibration curve of a biosensor. Calibration curves are normally obtained by one of two methods: from reaction curves of the biological co-reaction measurement (Fig. 17.4a) or the biological co-product measurement

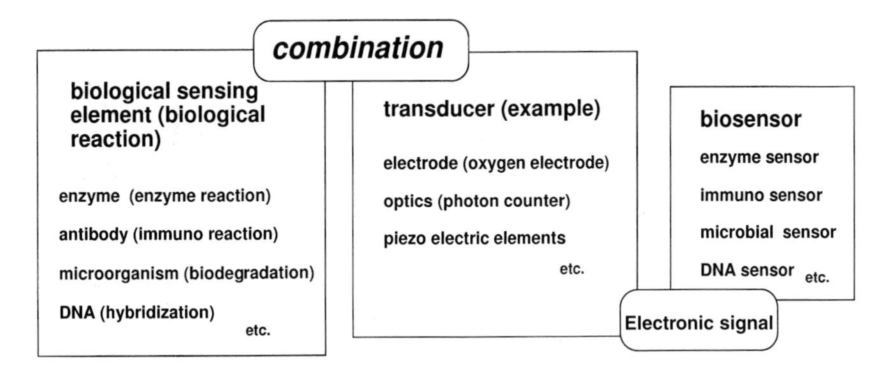

FIGURE 17.1 Biological elements are combined with transducers when biosensors are fabricated. Electrodes and optical devices are widely used as transducers in biosensors. A piezoelectric device such as a quartz crystal detects a weight change before and after the biological reaction as applied to immuno- and DNA sensors. Transducers convert a change in the concentration of the compound into an electronic signal.

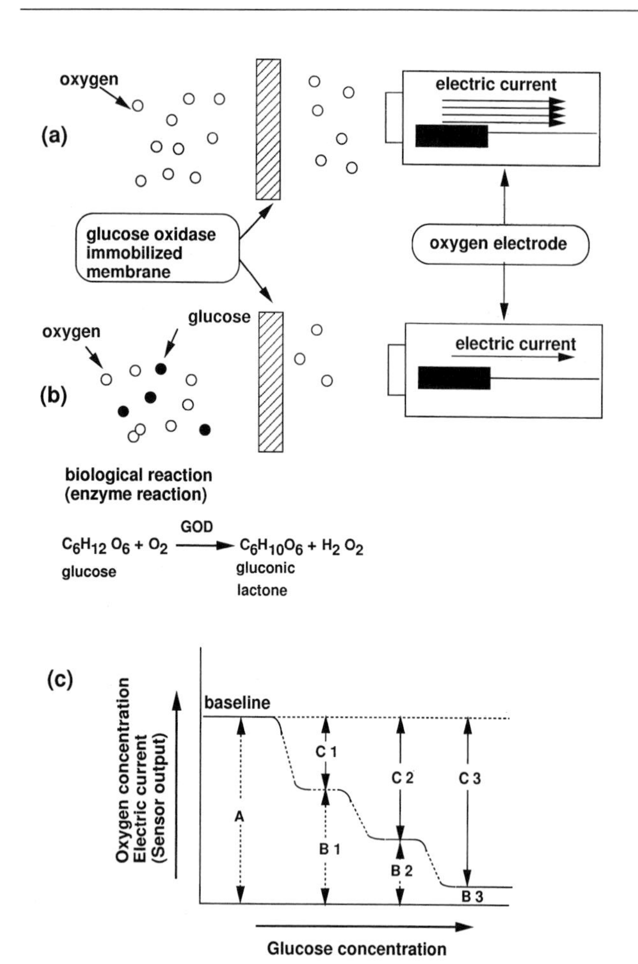

FIGURE 17.2 Principle of a glucose sensor. The analyte is glucose, and the biological-sensing element of the biosensors is glucose oxidase (GOD). The membrane-immobilized GOD is combined with an oxygen electrode as a transducer. An oxygen electrode generates electric current depending on the dissolved oxygen concentration. Glucose and oxygen permeate into the membrane-immobilized GOD. (a) Dissolved oxygen is not consumed when a sample which does not contain glucose such as pure buffer solution is measured by the glucose sensor. The sample is air-saturated and the oxygen electrode produces stable high current. This value is defined as the baseline current of the sensor. (b) A sample containing glucose is measured by the glucose sensor. GOD catalyzes the glucose oxidation using oxygen and dissolved oxygen is consumed. The oxygen concentration is lower than that in (a) and the electric current decreases from the baseline current. (c) Typical reaction curve of the sensor is illustrated. A, the baseline current: B1–B3, the electric currents obtained from samples containing glucose. The differences between A and B1–B3 are the sensor responses (C1–C3). As glucose concentration increases, the sensor response will also increase.

(Fig. 17.4b). In both cases the sensor responses are calculated from the difference between the values of pure buffer and the sample containing the analyte.

The sensitivity of a biosensor is defined as the slope of the linear range of the calibration curve. The linear range is between the lower detection limit (C1) and the nonlinear profile at higher concentration (C6, −7,

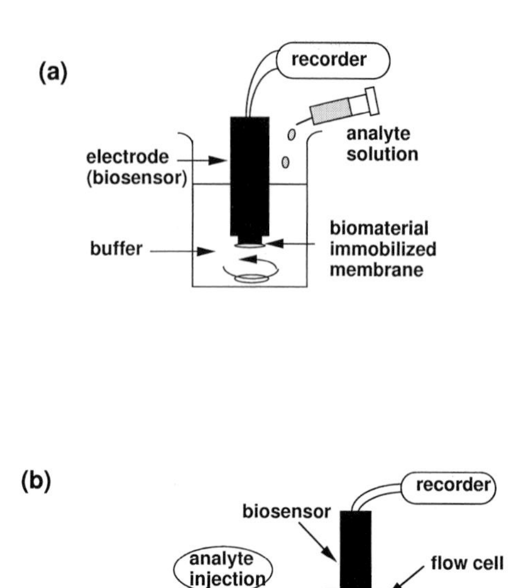

FIGURE 17.3 Two types of biosensor measurements. The illustrated biosensor is a combination of an electrode and a membrane-immobilized biomaterial. For example, adsorption or entrapment methods may be used to immobilize biomaterials. The membrane with immobilized biomaterial such as cellulose acetate membrane (0.45 μm) is attached to the electrode by an O-ring. (a) Batch-type measurement. The biosensor is immersed in the buffer solution and the analyte solution is directly injected into the buffer solution. (b) Flow-type measurement. The biosensor is attached to a flow cell and a flow-type biosensor is fabricated. The buffer solution is propelled by a peristaltic pump and the analyte solution is injected into the flow line. The buffer solution mixed with the analyte is introduced to the flow cell and is measured by the biosensor.

and −8). The biosensor response (biosensor output signal Rs) obtained from the sample measurement is substituted into the calibration curve in Fig. 17.4 and the analyte concentration of the sample can be calculated (C*x*).

Reproducibility and the life-time of a biosensor are usually examined, and selectivity must also be evaluated in some cases.

Biosensors have replaced conventional methods, which are often complicated, time-consuming, expensive, and require pretreatment or clean-up of real samples prior to analysis. Biosensors generally have the following advantages compared to other analytical methods:

1. Rapid and convenient detection.
2. Direct measurement of real samples.
3. Very specific detection.

On the other hand, stability and reproducibility have been problematic for biosensors due to the inherent instability of biomaterials used as sensing

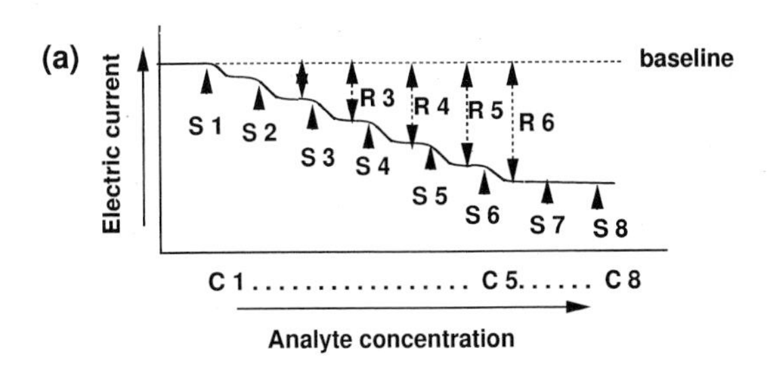

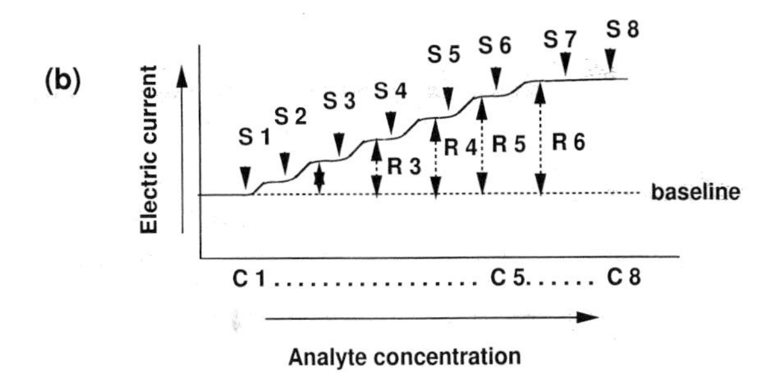

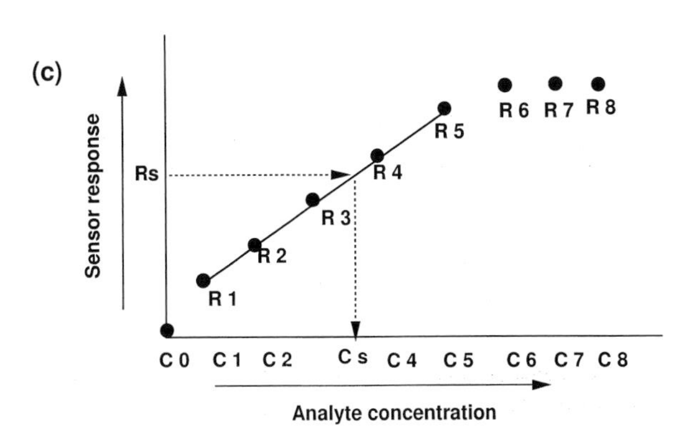

FIGURE 17.4 Typical calibration curve of a biosensor. The sample injections were carried out eight times (S*n*). (a) A reaction curve of a biosensor based on the measurement of co-reactants. Co-reactant is consumed during the measurement and the sensor output such as the electric current in the figure decreases as the analyte concentration increases. The baseline current is higher than that of the sample containing the analyte. The sensor response (R*n*) is calculated by subtracting the measurement value (e.g. the electric current) of the sample containing the analyte from the baseline value. (b) A reaction curve of a biosensor based on the measurement of co-products. Co-product or sensor output increases as the analyte concentration increases. The baseline value is less than that of the sample containing the analyte. The sensor response (R*n*) is obtained by subtracting the baseline value from the measurement value of the sample containing the analyte. (c) A calibration curve. The dynamic range of a biosensor is the linear correlation from the detection limit C1–C5. The sensitivity of the biosensor, for example, is calculated as (R4 − R2)/(C4 − C2). Rs is obtained when an unknown concentration sample is measured by the biosensor and it is substituted into the linear correlation, and the sample concentration Cs is obtained.

elements. Although biosensors have disadvantages, numerous investigations on biosensors to date have helped to overcome at least some of these difficulties, allowing practical application of many biosensors in the real world.

II. APPLICATIONS OF BIOSENSORS

Biosensors have been applied in many important fields, such as food quality control, environmental monitoring, and medical diagnosis. Recently, some groups reported biosensors for detecting chemical or toxic agents for use in military settings (Buerk, 1993). A few examples of biosensors in practical use in major fields are described in the following sections.

A. Food analysis

Many biosensors such as enzyme sensors have been developed for food analysis and food quality control. Enzyme sensors are used for the measurement of sugars, such as glucose, sucrose, and fructose. Vitamin C (ascorbic acid) and glutamate in food are also measured by biosensors. Some of these sensors are also being used in the medical field and for monitoring waste-water. Many of these sensors operate on the same principle as described previously for the glucose sensor shown in Fig.17. 2. The freshness of fish meat can also be measured by enzyme sensors. The freshness sensor that detects the degradation products of adenosine triphosphate (ATP) in fish meat, as well as many other enzyme sensors, achieves very high sensitivity by combining an FIA system coupled with chemiluminescence detection.

Microbial sensors are also applied to food quality analysis for measuring free fatty acids in milk (Schmidt *et al.*, 1996), alcohol, or acetic acid (Suzuki, 1990). The free fatty acid sensor uses *Arthrobacter nicotianase*, and both the alcohol and the acetic acid sensors use same bacterium, *Tricosposporon blassicae*.

B. Medical diagnosis

Enzyme sensors and microbial sensors for antibiotics and vitamins have been fabricated since the 1970s. Diabetes has been a particularly important target for many biosensors intended for medical use. Glucose sensors have been applied to diagnosis and monitoring of diabetes patients and to food quality analysis. Insulin biosensors have also been developed for use in diabetes treatment.

Since the first glucose sensor using an oxygen electrode was reported, numerous glucose sensors have been developed using techniques aimed at practical use. For example, screen-printing (Nagata *et al.*, 1995) and micromachining techniques (Hiratsuka *et al.*, 1998) have been applied to glucose sensors.

Disposable biosensors have very promising applications in medical diagnosis, such as monitoring blood glucose concentrations in diabetes patients.

Matsushita Company (Osaka, Japan) has commercialized a circular-type glucose sensor which utilizes semiconductor technology.

C. Environmental monitoring

Biosensors can rapidly detect and measure eutoriphicants, toxicants, and biochemical oxygen demand (BOD) in the environment. Enzyme immuno and microbial sensors have been mainly developed to detect pollutants in the environment. Biosensors for environmental monitoring have been reviewed (Karube *et al.*, 1995).

The phosphate sensor is a typical example of an enzyme sensor for environmental monitoring. In the past few years, dramatic improvements have been made in phosphate detection systems using enzyme sensors (Nakamura *et. al.*, 1997), and it is expected that some of these sensors will be used in the field in the near future.

Many biosensors have been developed to detect toxicants in the environment such as pesticides and cyanide. Jeanty and Marty (1998) reported organophosphates detection systems based on acetylcholine esterase inhibition. Immuno-sensors, in place of conventional enzyme-linked immuno sorbent assays are also frequently used to detect pesticides. Carl *et al.* (1997) fabricated an immuno-sensor which detects chemical endocrine disrupters (PCB) using a screen-printing technique, and recently Seifert and Hock (1998) investigated a novel sensor which uses a human estrogen receptor and an SPR-detection system. Microbial sensors have been constructed which detect cyanide and detergents. Karube *et al.* (1995) reviewed developments in microbial sensors for environmental monitoring.

The BOD sensor, which uses an omnivorous yeast *Trichosporon cutaneum*, is a well-known example of a microbial biosensor applied to environmental monitoring (Cass, 1990). The microbial BOD sensor measures the oxygen uptake by the respiratory system of the microorganism which changes as a result of the biodegradation of organic compounds in the sample. (Fig. 17.5). The microbial BOD sensor systems have been commercialized and marketed since 1983 by several companies (DKK, Nisshin Denki, and Central Kagaku Co., Tokyo).

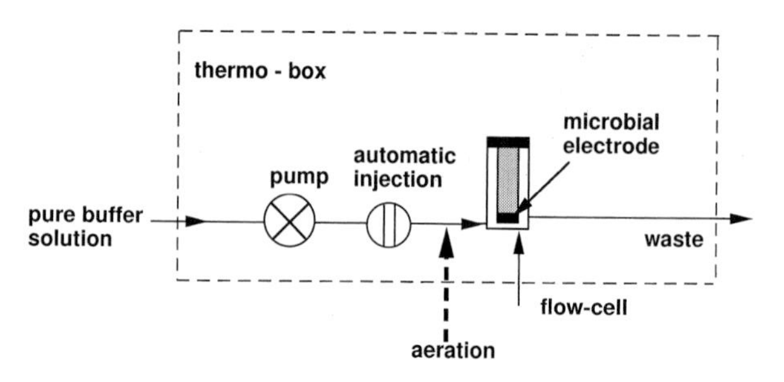

FIGURE 17.5 A schematic of a basic microbial BOD sensor. A microbial electrode consists of a membrane-immobilized *T. cutaneum* and an oxygen electrode. BOD measurement is performed aerobically using phosphate buffer at 30 °C. Glucose–glutamic acid solution is used as the standard solution for preparation of a calibration curve. After the standard solution measurement is obtained, real samples are examined and the BOD value is estimated. This system has been commercialized and the measurements are carried out automatically and continuously.

III. CURRENT TOPICS IN BIOSENSORS

Enzyme-, immuno-, and microbial sensors have become very popular. Sensors that take advantage of new transducing techniques such as SPR and a charge-coupled device have increasingly been reported (Buerk, 1993). Novel molecular recognition elements, such as DNA oligomers having specific affinity to various target molecules, have also been tested for use in biosensors.

Stability of a biosensor is very important, especially when the sensors must be used continuously in the field. Microorganisms and few stable enzymes are suitable for this purpose. However, many enzymes and antibodies used in biosensors are not stable enough for long-term practical use. Molecularly imprinted polymers (MIPs) have been employed in place of antibodies to construct biosensors with enhanced stability (Kriz *et al.*, 1995). These polymers are prepared by polymerization of various monomers in the presence of the target compound which acts as a template. After the polymer is removed from the template compound, the polymer retains the memory of the template and can selectively rebind the template compound. The use of MIPs to replace biomolecules in biosensors has advantages such as enhanced stability and inexpensive cost, which may expand the scope and applicability of the biosensors of tomorrow.

ACKNOWLEDGMENT

We thank Yohei Yokobayashi at The Scripps Research Institute for his help and advice in preparing the manuscript.

BIBLIOGRAPHY

Buerk, D. G. (1993). "Biosensors." Technomic, Lancaster, Pennsylvania.

Carl, M. D., Iionti, I., Taccini, M., Cagnini, A., and Mascini, M. (1997). Disposable screen-printed electrode for the immunochemical detection of polychlorinated biphenyls. *Anal. Chim. Acta* **324**, 189–197.

Cass, A. E. G. (1990). "Biosensor." IRL/Oxford University Press, New York.

Cosnier, S. (2000). Biosensors based on immobilization of biomolecules by electrogenerated polymer films. New perspectives. *Appl. Biochem. Biotechnol.* **89**, 127–138.

Hiratsuka, A., Sasaki, S., and Karube, I. (1998). A self-contained glucose sensor chip with arrayed capillaries. *Electro-analysis* **10**, 231–235.

Jeanty, G., and Marty, J. L. (1998). Detection of paraoxon by continuous flow system based enzyme sensor. *Biosensor Bioelectronics* **13**, 213–218.

Karube, I., Nomura, Y., and Arikawa, Y. (1995). Biosensors for environmental control. *Trends Anal. Chem.* **14**, 295–299.

Kriz, D., Ramstrom, O., Svensson, A., and Mosbach, K. (1995). Introducing biomimetic sensors based on molecularly imprinted polymers as recognition elements. *Anal. Chem.* **67**, 2142–2144.

Nagata, R., Yokoyama, K., Clark, S. A., and Karube, I. (1995). A glucose sensor fabricated by the screen printing technique. *Biosensor Bioelectronics* **10**, 261–267.

Nakamura, H., Ikebukuro, I., McNiven, S., Karube, I., Yamamoto, H., Hayashi, K., Suzuki, M., and Kubo, I. (1997). A chemiluminescent FIA biosensor for phosphate ion monitoring using pyruvate oxidase. *Biosensor Bioelectronics* **12**, 959–966.

Schmidt, A., Gabisch, C. S., and Bilitewski, U. (1996). Microbial biosensor for free fatty acids using an oxygen electrode based on thick film technology. *Biosensor Bioelectronics* **11**, 1139–1145.

Seifert, M., and Hock, B. (1998). Analytics of estrogens and xenoestrogens in the environmental using a SPR-Biosensor, *Proc. Biosensor* **98**, 47.

Stefan, R. I., van Staden, J. F., Aboul-Enein, H. Y. (2000). Immunosensors in clinical analysis. *Fresenius J. Anal. Chem.* **366**, 659–668.

Stephens, D. (2001). New genetically encoded 'biosensors'. *Trends Cell Biol.* **11**, 241.

Suzuki, S. (1990). "Biosensor," 5th ed. Kohdan-sha, Tokyo. (In Japanese)

Van Regenmortel, M. H. (2000). Analysing structure–function relationships with biosensors. *Cell Mol. Life Sci.* **58**, 794–800.

WEBSITES

Website by K. Bruce Jacobson of the Oak Ridge Ridge National Laboratory
http://www.ornl.gov/ORNLReview/rev29_3/text/biosens.htm

The Electronic Journal of Biotechnology
http://www.ejb.org/

List of biotechnology organizations and research institutes
http://www.ejb.org/feedback/borganizations.html

18

Cell membrane: structure and function

Robert J. Kadner
University of Virginia

GLOSSARY

ABC proteins Proteins that contain the widely conserved ATP-binding cassette, a motif that couples energy from ATP binding and hydrolysis to various transport processes.

detergent A molecule with polar and nonpolar portions that can disrupt membranes by stabilizing the dispersion of hydrophobic lipids and proteins in water.

hydrophobic Molecules or portions of molecules that cannot form hydrogen bonds or other polar interactions with water.

hydrophobic effect The tendency of hydrophobic regions of molecules to avoid contact with water.

osmolarity The tendency of water to flow across a membrane in the direction of the more concentrated solution.

proton motive force The electrochemical measure of the transmembrane gradient of protons, consisting of an electrical potential due to separation of charge and the pH gradient due to different concentration of protons.

symporter A transport system in which movement of the coupling ion moves in the same direction as the substrate molecule, in contrast to an antiporter or uniporter.

Every Cell Possesses a Surface Membrane that separates it from the environment or from other cells. Animal cells and other eukaryotic cells possess, in addition to the plasma membrane, numerous intracellular membranes which form the organelles that perform specialized metabolic functions. Bacterial and archaeal cells typically lack intracellular membrane organelles and contain only the single cytoplasmic membrane, perhaps surrounded by an outer membrane. The cytoplasmic membrane of bacteria is typically composed of simple phospholipids that form a membrane bilayer, into which are inserted a large number of different proteins. The phospholipid bilayer forms the osmotic barrier that prevents movement of most materials into or out of the cell. The various membrane proteins carry out numerous important functions, including the generation and storage of metabolic energy and the regulation of uptake and release of all nutrients and metabolic products. Membrane proteins recognize and transmit many signals that reflect changes in environmental conditions and trigger an appropriate cellular response. They also play key roles in the control of cell growth and division, bacterial movement, and the export of surface proteins and carbohydrates.

I. ULTRASTRUCTURE AND THE ROLE OF CYTOPLASMIC MEMBRANE

A. Ultrastructure of cell membranes

Cell membranes are readily visible when thin sections of cells are stained with heavy metals and viewed in

The Desk Encyclopedia of Microbiology
ISBN: 0-12-621361-5

the electron microscope. They appear as a characteristic triple-layered structure with two dark electron-dense layers, representing the region of the lipid head groups, surrounding a light layer which reflects the hydrophobic central portion of the membrane bilayer. All cellular membranes appear quite similar by this technique, regardless of the source of the membrane or their protein content (Fig. 18.1a). Most biological membranes are 4 or 5 nm in width. In the technique of freeze-fracture electron microscopy a knife blow is used to split a frozen sample of cells. The fracture plane often extends along or through the weakly connected central section of a membrane bilayer. This technique can reveal the presence and density of proteins embedded in and spanning through the membrane. Figure 18.1 supports the fluid mosaic model of membrane structure in which the polar membrane lipids form a lamellar, or leaf-like, bilayer in which their nonpolar portions face each other in the central region and their polar regions are on the outside. Integral membrane proteins span across the bilayer, but can diffuse within the plane of the bilayer and even associate into large complexes. Many peripheral membrane proteins do not span across the membrane and can be transiently bound through hydrophobic anchors or by association with other membrane proteins or the lipid head groups.

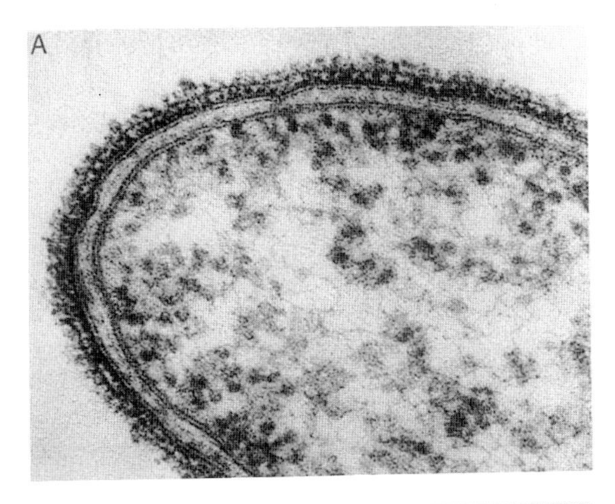

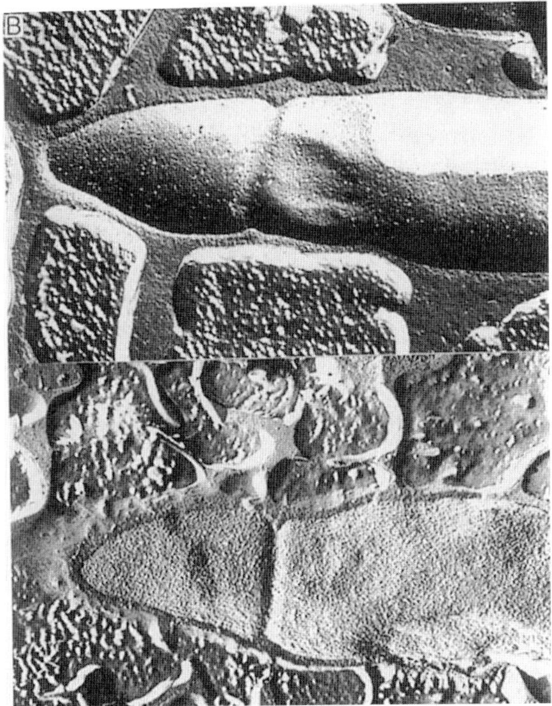

FIGURE 18.1 Structure of bacterial membranes. (a) Electron micrograph of a thin section of the fish pathogen, *Aeromonas salmonicida*. Cells were embedded in plastic, cut to a thin slice, stained with heavy metals (uranyl acetate and lead citrate), and visualized by transmission electron microscopy. The cytoplasmic membrane is seen as the triple-layered structure bounding the cytoplasm. Outside the 7.5-nm thick cytoplasmic membrane is the periplasmic space, the triple-layered outer membrane, and the thick surface S layer. (b) Electron micrograph of a freeze-fractured sample. A suspension of cells of *Bacillus licheniformis* was frozen, and the block was shattered by a sharp blow with a knife edge. The fracture plane ran occasionally though membrane bilayers. A carbon replica of the two faces thus revealed was made, the organic material was etched away with acid, and surface structures were enhanced by shadowing with a beam of platinum atoms. The image is viewed in transmission electron microscope. The two faces show the two halves of the cytoplasmic membrane layer, indicating the presence of particles that are embedded in and span the membranes (courtesy of Terrance Beveridge, University of Guelph, Canada).

B. Role as osmotic barrier

The cytoplasmic membrane is the osmotic barrier of the cell, owing to its ability to restrict the passage of salts and polar organic compounds. If a cell is placed in a medium in which the osmolarity is higher or lower than the osmolarity of the cytoplasm, water will flow across the cytoplasmic membrane out or into the cell, respectively. This osmotic flow of water occurs in response to the natural forces that seek to eliminate gradients or differences in the concentration of water on the two sides of the membrane. Hence, the cytoplasm either shrinks or swells under these two conditions as a result of the loss or gain of water. In most bacteria, the cell does not change size owing to the presence of its rigid cell wall.

C. Role as cell boundary

The cytoplasmic membrane is the boundary between the cell and its surroundings and thus must regulate the passage of nutrients and metabolic products. The presence of the hydrophobic layer formed by the membrane lipids greatly restricts the passage of any polar molecules and of macromolecules. It prevents the loss of cellular macromolecules and metabolic intermediates.

D. Regulation of transport

Transport systems allow the passage of specific molecules, such as ion, nutrients, and metabolic products, either into or out of the cell. Bacteria in general have a large number of very specific transport systems that carry out active transport. They consume some form of metabolic energy to be able to pump their substrate from a low concentration on one side of the membrane to a much higher concentration on the other side. Transport is an integral part of the universal process of bioenergetics, which refers to the formation and consumption of sources of cellular energy, most of which involve transmembrane ion gradients.

E. Role in cell growth and division

Expansion of the membrane surface is intimately related to growth rate of any cell, and all components must be inserted in a timely manner to allow cells to expand in size. Cell division requires a carefully controlled process whereby the membranes of the parental cell pinch together, fuse, and separate to create two progeny cells, without loss of internal material. The membrane contains export systems of control the release of structural components to the cell surface and other secreted factors beyond the cell. Other membrane-localized protein complexes regulate the process of initiation of DNA replication, separation of chromosomes into the dividing bacterial cells, and in-growth of the cell surface that occurs during cell division.

II. STRUCTURE AND PROPERTIES OF MEMBRANE LIPIDS

A. Lipid composition

The key ingredients of biological membranes are polar lipids, primarily phospholipids. In most bacteria, phospholipids consist of two fatty acids, usually with 16–18 carbon atoms in the hydrocarbon chain with zero or one *cis*-double bond. The fatty acid content changes in response to environmental conditions, particularly temperature. As described later, lower growth temperatures result in a higher degree of fatty acid unsaturation, which has dramatic effects on the membrane's fluidity and function. Some fatty acids are branched or contain cyclopropane rings. Fatty acids are joined in ester linkage to two of the hydroxyl groups of glycerol, usually with a saturated fatty acid at the 1-position and an unsaturated fatty acid at the 2-position. To the third hydroxyl group of glycerol is attached a phosphate moiety and to it the head group. In bacteria, the range of head groups is narrow, and the phospholipids in *Escherichia coli* are approximately 75% phosphatidyl ethanolamine (PE) and 20% phosphatidyl glycerol (PG), and the remainder is cardiolipin (diphosphatidyl glycerol), phosphatidyl serine, and trace amounts of other phospholipids (Fig. 18.2).

Other bacteria possess more complex types of membrane lipids, although these lipids are usually much less complex than those in the plasma membrane of animal cells. Some bacteria possess phosphatidyl choline, or lecithin, which is characteristic of higher organisms. Other bacteria produce glycolipids, such as monogalactosyl diglyceride. The membrane lipids from archaea are quite different from those in bacteria and eukarya. Their hydrocarbon chains are based on isoprenoid units and these are linked to the glycerol backbone in an ether, rather than ester, linkage. In some archaea, a glycerol backbone and head group are attached to both ends of a pair of isoprenoid units.

Sterols, such as cholesterol in mammalian cells or ergosterol in fungi, are invariant features of membranes in eukaryal cells, in which they appear to stiffen the membrane by increasing the degree of order of the hydrocarbon chains. Sterols are not commonly found in bacteria and archaea, except for the cell wall-less *Mycoplasma*. Very complex lipids, including the very long, branched mycolic acids, are common in *Mycobacterium* but occur in a very thick and rigid outer layer rather than in the cytoplasmic membrane.

B. Hydrophobic effect

Phospholipids spontaneously form membranous structures when suspended in aqueous solution. The forces that drive their assembly into a bilayer or more complex structure are called the hydrophobic effect. This organizing force depends on the ability of water molecules to donate and accept hydrogen bonds from one another to form an extensive network of water molecules transiently linked through hydrogen bonds and polar interactions. Polar molecules can participate in this interactive network of water molecules and are thereby able to dissolve in aqueous solution. In contrast, hydrophobic or nonpolar molecules, such as a long hydrocarbon or aromatic chain, are unable to participate in the hydrogen-bonded network of water molecules. For them to dissolve in water would result in loss of the energy resulting from the intrusion in the mobile water bonding and from the organization of

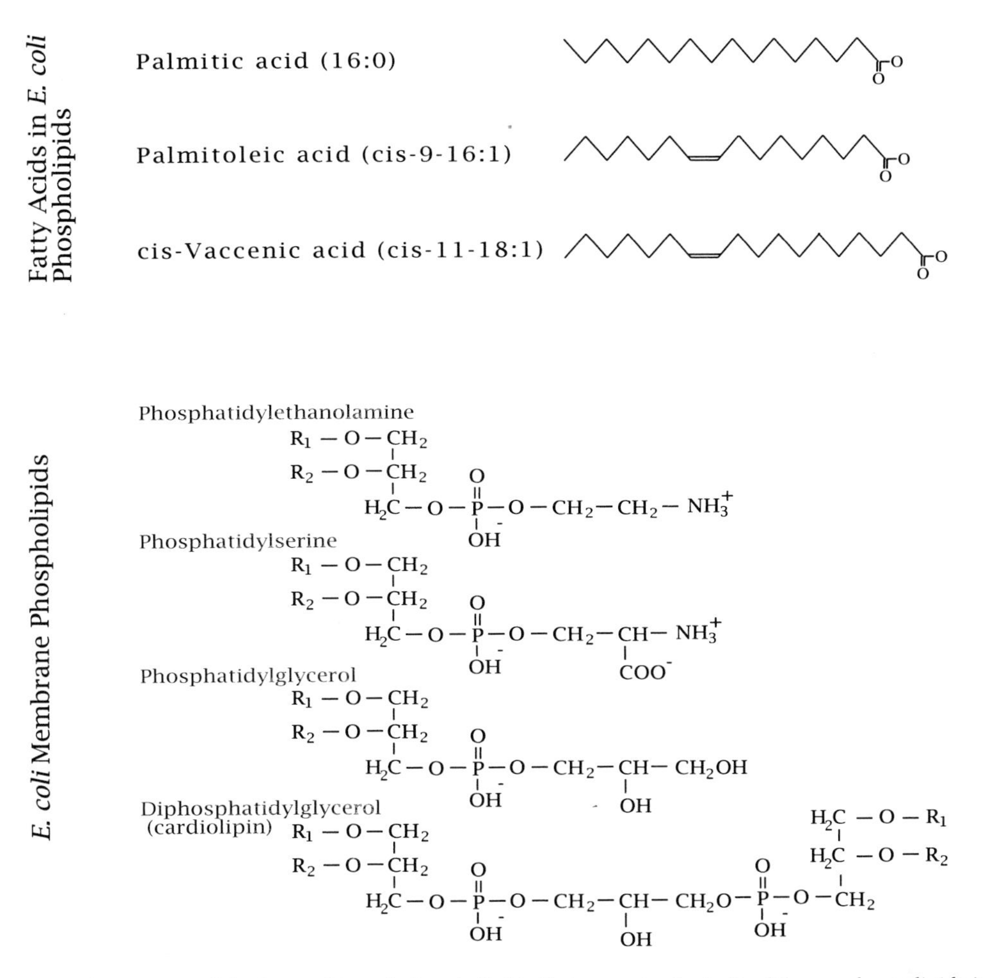

FIGURE 18.2 Chemical structures of the fatty acids and phospholipids that comprise the bulk of the membrane lipids in *E. coli*. For the phospholipids, the R groups represent a fatty acid.

water molecules into a cage-like structure around the intruding nonpolar molecule. It is the loss of the energy of interaction between the water molecules and the entropic cost of organizing them that drive the non-polar molecules to associate with one another, out of contact with the water. Hence, non-polar lipids tend to form oil droplets in water, so as to present the smallest possible hydrophobic surface to the water.

C. Membrane bilayer

Polar lipids, such as phospholipids, have chemical structures of two different natures. The hydrophobic acyl chains strive to be sequestered from contact with water, whereas the charged and polar head groups seek contact with aqueous ions to help dissipate their electrical charge and to form a hydrogen-bonded network with water. Thus, polar lipids are driven by basic physical characteristics to form aggregate structures in which the hydrophobic portions are segregated out of contact with water while the head groups face the water. There are numerous ways in which these requirements can be accommodated. Of greatest biological relevance is the lamellar bilayer, in which large flat surfaces of bilayer form with the acyl chains facing each other on the inside and the head groups facing the solution on the outside (Figs. 18.3 and 18.4). Other non-lamellar structures, such as hexagonal phases, can form under certain conditions of head group, temperature, and salt concentration. The propensity of different lipids to form nonlamellar bilayers is a function, in part, of the relative cross-sectional areas of the hydrophobic acyl chains in reference to the area of the head group. When the areas of the polar and nonpolar parts are similar, a lamellar bilayer is favored. If the nonpolar part is substantially smaller or larger than the head group, formation of a spherical micelle or an inverted structure, such as the HexII phase (hexagonal phase II), respectively, is favored.

When most biological phospholipids are dried into a film from a solution in an organic solvent and then water is added, the lipids spontaneously form a

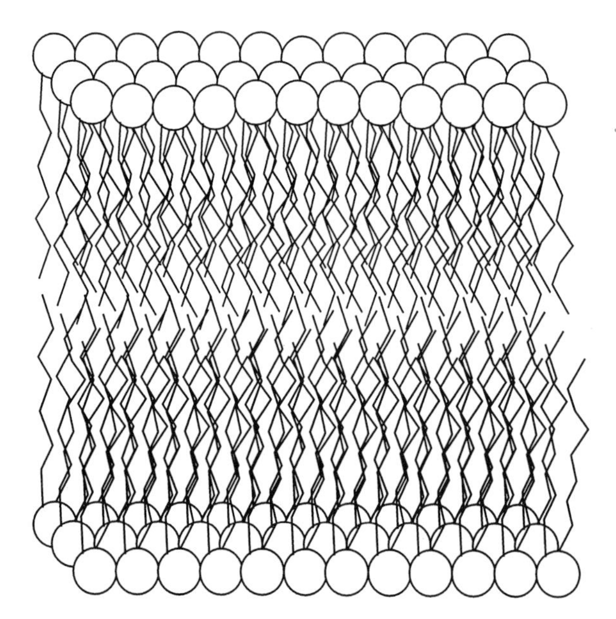

FIGURE 18.3 Schematic representation of the lipid bilayer or lamellar arrangement. The polar head groups of the lipids face the water, and the hydrocarbon acyl chains are segregated away from the water in the interior of the bilayer.

multilamellar liposome in which concentric bilayers assemble like an onion. When these liposomes are sonicated (disrupted by intense ultrasonic irradiation), they break down into small unilamellar vesicles. These are small spherical particles with a single membrane bilayer surrounding an aqueous cavity.

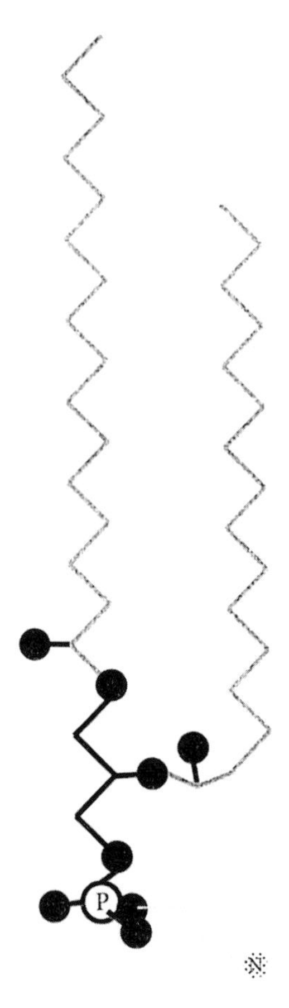

FIGURE 18.4 Representation of the orientation of the acyl chains and head groups of a phospholipid in a bilayer. This orientation of phosphatidylethanolamine is seen in crystal structures. The solid circles represent oxygen atoms; the heavy lines represent the glycerol; the thin lines are ethanolamine; and the gray lines are the hydrocarbon chains of the two fatty acids, all in their most extended configuration.

D. Lipid phase behavior

Above a certain temperature, the lipid molecules in a bilayer can move rapidly within the plane of the membrane but do not move very far in or out of the membrane owing to the hydrophobic effect. In this liquid crystalline state, the acyl chains are parallel to each other but undergo frequent rotations around the carbon bonds to produce kinks in the chain. These kinks provide transient discontinuities in the hydrophobic barrier that allow movement of other lipid molecules within the membrane and of water molecules across the membrane. As the suspension of membranes is cooled below a critical temperature (T_m), there is a transition of the lipids from the liquid crystalline phase to the rigid gel phase. In this phase, all the hydrocarbon chains form the all-*trans* configuration, which increases the bilayer thickness and greatly decreases diffusion of the lipids within the membrane and of permeants across the membrane. This transition reflects the motion and packing of the acyl chains.

The critical temperature at which the gel-to-liquid crystalline phase transition occurs is dependent on the lipid composition, including the nature of the head group and the lipid chains. The longer the hydrocarbon chains, the higher the T_m for transition to the liquid crystalline state. The presence of unsaturated fatty acids with one or more double bonds has a very dramatic effect reducing the T_m by as much as 60 °C. The effect of the double bond is greatest when it is in the middle of the acyl chain. The double bond introduces a permanent kink or bend in the chain that interferes with packing of the chains in the gel state. Most cells adjust their lipid composition to the growth temperature to ensure that their membrane remains in the liquid crystalline state. Most membrane proteins are excluded from or are inactive in the rigid gel phase membranes. Another lipid phase transition can occur at temperature higher than T_m. This corresponds to the change of certain lipids from a lamellar to a nonlamellar configuration.

III. STRUCTURE AND PROPERTIES OF MEMBRANE PROTEINS

Although the lipids are very important for determining the barrier functions of the cell membrane, the membrane proteins confer most of the important functions of biological membranes. There are many different types of membrane proteins, reflecting their very different functions. Integral membrane proteins are those that cross the membrane with one or more transmembrane segments and are usually exposed to both sides of the membrane. They cannot be easily removed from the membrane unless detergents are added to disrupt the bilayer structure. Some integral membrane proteins are stably anchored to the membrane, usually by covalent attachment to a lipid such as a fatty acid, isoprenoid, phosphatidyl inositol, or a lipoprotein derivative. The membrane-spanning proteins include most of the functionally important transporters and signal receptors. Peripheral membrane proteins are defined operationally as those that are readily removed from association with the membrane by procedures that do not disrupt the bilayer, such as washing with high salt, urea, or sodium carbonate at pH 11. These proteins generally lack transmembrane segments, although it is recognized that some peripheral proteins can transiently insert a segment across the membrane as part of their normal function.

Some integral membrane proteins possess one or two transmembrane segments only, and the bulk of these proteins is located in the aqueous solutions on one or both sides of the membrane. The nonmembranous segments of these proteins are similar in character to that of a normal soluble globular protein. Charged or polar amino acid residues line the surfaces exposed to the water, and nonpolar amino acid residues form the interior of the domain, driven into this sequestered state by the same hydrophobic effect that stabilizes the membrane lipid structure. The membrane-spanning portion of these proteins has a very limited range of composition and structures. Owing to the very hydrophobic environment of the interior of the membrane bilayer, the presence of charged or polar residues or of unpaired hydrogen bonds is energetically unfavorable. Thus, the amino acid residues that comprise single transmembrane segments are generally highly hydrophobic. In addition, an α helix is the only peptide conformation in which all the hydrogen bonding possibilities of the polypeptide backbone are satisfied. Thus, single or double transmembrane segments are most likely to be non-polar α-helical segments, with lengths of approximately 20 amino acids (approximately six turns of the helix), which is sufficient to span the width of the membrane.

Many membrane proteins cross the membrane multiple times and insert the bulk of their mass within the membrane. Transport proteins typically have 10–14 transmembrane segments, and a major class of signal receptors have seven transmembrane segments. These proteins have a very different character than that of the multidomain bitopic proteins described previously. The large surface of the protein that is imbedded in the membrane and exposed to the hydrophobic environment of the hydrocarbon lipid chains is highly hydrophobic and cannot possess charged amino acid residues. Other surfaces of these proteins are exposed to the solution on either side of the membrane. These surfaces, which are composed of the loops joining transmembrane segments, must possess mainly polar residues capable of remaining soluble in water. The presence of the very different surfaces of membrane proteins, two polar belts and one very nonpolar belt, holds the protein tightly within the membrane bilayer and restricts its movement out of the membrane.

The transmembrane segments of such a polytopic protein are α-helical in the few proteins for which detailed structural information is available (the photosynthetic reaction center, some electron transport complexes, bacteriorhodopsin, the vitamin B12 transporter, and lactose permease). The amino acid residues that comprise transmembrane segments need not be all nonpolar. These residues can be exposed to different environments, namely the lipid hydrocarbon chains, the neighboring α-helical transmembrane segments, and a potential water-filled channel. Thus, the residues of these transmembrane segments exhibit periodic variability, with very non-polar residues along one face and generally polar residues along the opposite face. In contrast, the transmembrane segments of most bacterial outer membrane proteins are composed of antiparallel β sheets of approximately eight amino acid residues in length.

Proteins are inserted in cellular membranes in a defined orientation. The orientation of the entire protein appears to be determined by the orientation of its individual transmembrane segments. The orientation of each transmembrane segment is determined mainly by the nature of the charged residues flanking the transmembrane segment rather than by the residues within the transmembrane segment. The "inside-positive" rule seems to govern the orientation of transmembrane segments in bacteria, and it states that relatively short, positively charged loops are retained on the cytoplasmic side of the membrane. The orientation of a protein can be affected by its association with other cellular or membrane proteins or by the presence of a signal sequence that directs the protein into the secretory pathway.

The orientation of a membrane protein can be determined experimentally by detecting the sites of action of proteases or other enzymes added to one side of the membrane. A simpler and more generally informative approach makes use of fusions of a part of the tested protein to a topological reporter, which is an enzyme whose activity depends on which side of the membrane it resides. It has thus been possible to gain a considerable degree of understanding of the structure of many membrane proteins, even without high-resolution crystallographic information.

IV. FUNCTIONS OF CYTOPLASMIC MEMBRANE

The cytoplasmic membrane of bacteria is a very busy site at which an almost bewildering number of important processes occur. It carries out most of the reactions that are handled by the many organelles of eukaryotic cells. Since bacteria are generally far more metabolically capable and diverse than eukaryotic cells, it is not surprising that an estimated one-fourth of the cell's proteins are membrane associated.

One of the key principles in biology is the basic universality of bioenergetics, which refers to the processes of energy generation, storage, and utilization in cells. The major feature of these processes is the use of ion gradients that are formed across cellular membranes, such as the membranes of mitochondria and chloroplasts in higher organisms or the cytoplasmic membrane of bacteria. In most systems, the gradient of protons is the central factor in bioenergetics, although gradients of sodium ions are used by some bacteria for energy generation and by most eukaryotic cells for cellular signaling and nutrient transport. The chemiosmotic proposal, initially made by Peter Mitchell, that ion gradients are the intermediate between the processes of electron transport and the formation of ATP has been overwhelmingly accepted. In bacteria, ion gradients are also used to drive several types of active transport systems, bacterial motility, and protein secretion.

A. Energy generation

Energy generation refers to the trapping in a metabolically useful form of the energy absorbed from sunlight or released by reduction of some inorganic molecule or the oxidation or breakdown of an organic molecule serving as energy source. Bacteria can generate energy by many different processes. In simple fermentative pathways, such as the glycolytic breakdown of glucose to lactate, ATP can be formed during several enzymatic steps by the process of substrate level phosphorylation. Important energy-producing processes use electron transport chains to pass electrons from a carrier of high negative redox potential to carriers of successively lower energy states. During respiration, electrons enter these chains following transfer from an organic molecule and are ultimately transferred to an inorganic electron acceptor. During photosynthesis electrons are excited to a higher energy state following absorption of light by a chlorophyll-related molecule. Figure 18.5 summarizes several key steps of microbial energy generation.

During the processes of electron transport to carriers of successively higher redox potential, there is a separation of charge across the membrane. This is ultimately coupled to the movement of protons from one side of the membrane to the other, resulting in the formation of a proton motive force (pmf) which has two aspects. Movement of the positively charged proton creates an electrical charge across the membrane, which is termed $\Delta\Psi$. In bacteria and mitochondria, in which electron transport results in the release of protons, the electrical charge is negative inside and can be 100–200 mV. Proton pumping also results in a difference in proton concentration, or ΔpH, across the membrane, such that the exterior is usually more acidic than the interior. Photosynthetic or respiratory electron transport thus results in proton pumping and creation of the pmf.

The proton gradient can be tapped to bring about the formation of ATP, which is the ultimate energy source for most energy-requiring processes in the cytoplasm. Synthesis of ATP is carried out by a family of protein complexes, which include the F_1F_0 proton-translocating ATPases of mitochondria and bacteria. Related complexes are found in chloroplasts and archaea. These protein complexes contain a membrane-embedded sector, the F_0 portion, which includes the pathway to allow the protons to flow back into the cell in response to the chemical and electrical forces acting on them. The F_1 portion of the complex contains the sites for conversion of ADP + Pi to ATP, and the energy for this process is coupled to the movement of protons. The stoichiometry of the process is such that entry of three or four protons results in the formation of one molecule of ATP. The individual steps of proton movement through the F_1F_0-ATPase and ATP synthesis or hydrolysis are tightly coupled under most conditions to prevent wasteful loss of the pmf or of the ATP pool in the cell and consequent heat generation. In bacteria, this ATP synthase can

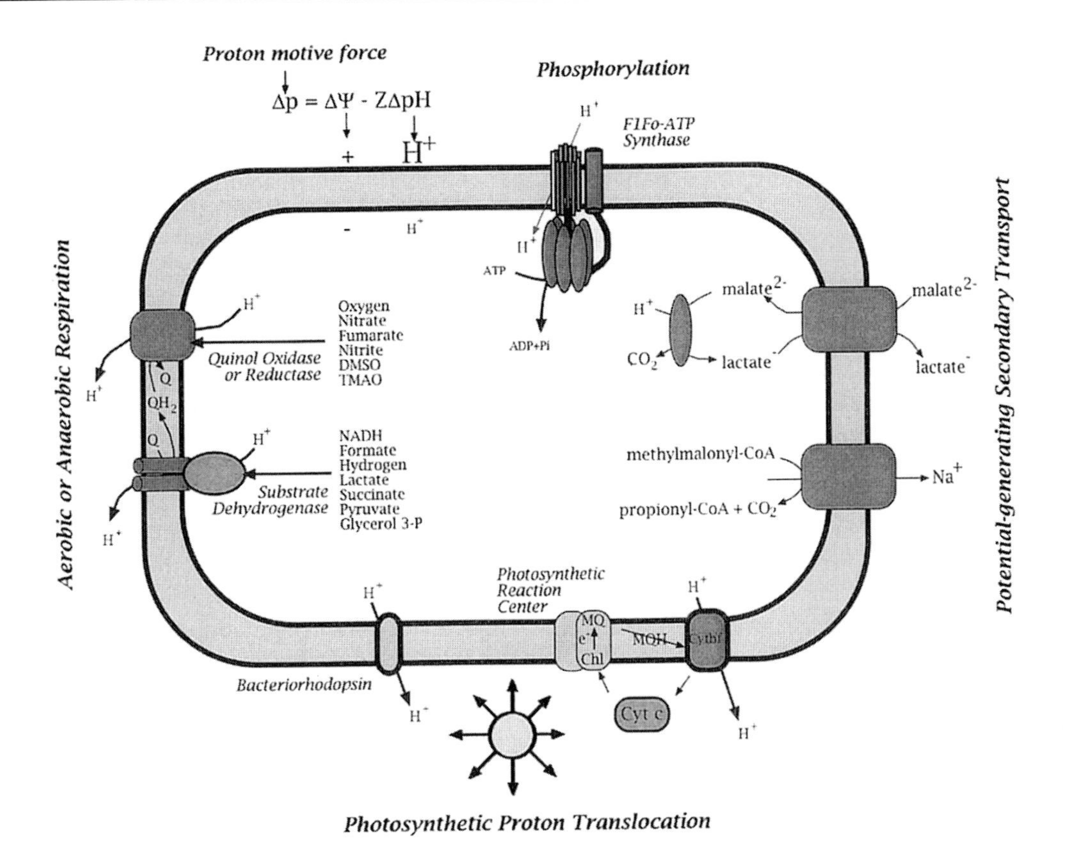

FIGURE 18.5 Summary of some processes of generation of metabolic energy in bacteria. (Left) Components of respiratory electron transport chains, in which specific substrate dehydrogenases oxidize their substrate, transfer the released electrons to membrane quinones, and in some cases extrude protons to the exterior. The electrons of the reduced quinones are transferred by the terminal oxidase or reductase to the terminal electron acceptor, such as oxygen, with the extrusion of additional protons. (Bottom) Photosynthetic systems whereby absorption of light is converted to a transmembrane gradient of protons. (Right) Two examples in which substrate/product exchange and metabolism result in generation of pmf. (Top) The electrical and chemical components of the proton motive force and a representation of the action of the F_1F_0-ATP synthase which interconverts the proton gradient and ATP synthesis or hydrolysis.

function in a reversible manner to allow ATP that was generated by substrate-level phosphorylation to drive formation of a proton gradient which can then be used to drive transport systems or motility. Movement of protons causes the F_0 sector to rotate within the membrane, like the action of a turbine. The rotation of the F_0 sector is coupled to changes in the conformation of the nucleotide-binding sites in the stationary F_1 sector, which is linked to interconversion of ADP + Pi to ATP.

1. *Photosynthesis*

Photosynthesis traps the energy of sunlight and converts it into metabolically useful forms such as ATP during cyclic electron transport and NADPH during noncyclic electron transport processes. In some organisms, photosynthetic electron transport is coupled to the formation of oxygen from water—the process that is essential for aerobic life. There are numerous pigments in cells that absorb light energy for use in photosynthesis, but the most important of these are the chlorophylls. Chlorophylls are porphyrin molecules, similar to heme, but containing a magnesium atom instead of iron. Some pigments are carried in protein molecules that serve as antennae or light-harvesting complexes, but the key processes occur in a protein complex called the photosynthetic reaction center, which typically consists of three proteins that spread across the membrane and contain bacteriochlorophyll, bacteriopheophytin, menaquinone, and nonheme iron as electron carriers. The light energy is ultimately absorbed by a chlorophyll molecule and this energy excites an electron to a higher energy level. This excited electron passes through a series of electron-carrying prosthetic groups within the reaction center and then out through the pool of membrane-bound quinones, which transfer the electron to a cytochrome-containing electron transport chain. Passage of the electron through the electron transport chain is coupled to pumping of protons across the specialized membranes containing the photosynthetic apparatus. In cyclic phosphorylation, as carried out in the photosynthetic bacteria, the electron ultimately returns to the chlorophyll molecules after transfer to the periplasmic heme protein, cytochrome *c*. In the process of noncyclic phosphorylation in plants and cyanobacteria, the electron can be transferred ultimately to pyridine nucleotides for use as a reductant in biosynthetic processes. In this process, the electron can be replaced on chlorophyll by another light-absorption process that removes electrons from water to create oxygen.

A completely different system for conversion of light into metabolic energy is present in *Halobacterium salinarum*, an extremely halophilic archaeon that thrives in very saline environments such as the Dead Sea or brine evaporation ponds. These bacteria produce patches of membrane that are densely packed with the membrane protein, bacteriorhodopsin, having seven transmembrane helices and covalently bound retinal as in the visual pigment in mammalian retina. Light absorption by the retinal causes the isomerization of one of the double bonds in the molecule, causing a change in the conformation of the protein which results in the change of the p*K* of several acidic groups on either side of the membrane. The consequence of these changes is that a proton is released from the bacteriorhodopsin on the outside of the membrane and replaced by one from the cytoplasm. In this way, light is directly converted into a transmembrane proton gradient without the requirement of an electron transport chain.

2. *Respiration*

Respiration is the process whereby electrons from the metabolism of an energy source are transferred through a proton-pumping electron transport chain to some inorganic molecule. The most familiar form of respiration is aerobic respiration, in which oxygen serves as the ultimate electron acceptor. Owing to the ability of oxygen to accept electrons, aerobic respiration is the most energetically favorable, but some partially reduced forms of oxygen, hydrogen peroxide, superoxide anion, and hydroxyl radical, are extremely reactive and thus toxic to the organism. Many bacteria are capable of carrying out anaerobic respiration, in which the electrons are transferred to alternative acceptors, such as nitrate, nitrite, sulfate, or sulfite. These processes yield less energy but can occur in anoxic environments.

All respiratory metabolism uses electron transport chains, whereby electron transfer from the donor to oxygen or other acceptor is coupled to proton movement across the membrane. In *E. coli*, electron donors for respiration include NADH, succinate, glycerol 3-P, formate, lactate, pyruvate, hydrogen, and glucose. Electron acceptors include oxygen, nitrate, nitrite, fumarate, dimethylsulfoxide, and trimethylamine-*N*-oxide. Typical respiratory systems contain two to four transmembrane protein complexes. These include substrate-specific dehydrogenases, which transfer electrons from the donor to quinones in the membrane. The reduced quinones migrate to another protein complex which accepts electrons from them and transfers the electrons to cytochromes and ultimately to the terminal electron acceptor. The transmembrane protein complexes often contain flavin and/or non-heme iron and are arranged in the membrane in such a way that the passage of electron results in the release of proton to the outside or its consumption from the cytoplasm, i.e. the formation of the pmf.

3. *Coupled processes*

Some bacteria couple the transport and metabolism of their energy source directly to the production of the pmf. An example of this very simple, but not very energy-rich, process is malo-lactate fermentation in *Leuconostoc*. The substrate malate is transported into the cell and converted to lactate, which leaves the cell in exchange for a new molecule of malate. The net result of this process is the movement of one negative change into the cell and the consumption of one proton inside the cell, which results in the creation of a pmf that is interior negative and alkaline.

B. Membrane transport

Biological membranes form the permeability barrier separating the cell from its environment. The hydrophobic barrier of the membrane bilayer greatly restricts passage of polar molecules, although non-polar molecules can pass. Transport mechanisms exist to move nutrients and precursors into the cell and metabolic products, surface components, and toxic materials out of the cell. Several types of transport mechanisms and families of transporters have been identified.

1. *Types of transport systems*

Transport can occur through energy-dependent and energy-independent processes. Several general classes of transport process have been identified. Passive diffusion occurs spontaneously without the involvement of metabolic energy or of transport proteins. It only allows the flow of material down a concentration gradient, and the rate of this process is a linear function of the concentration gradient. The rate of passive diffusion depends on the ability of the permeant to dissolve in the membrane bilayer and thus depends of the polarity of the permeant and its size. These factors are related to the ability of the permeant to fit into transient defects that form in the membrane bilayer. Only water and a few hydrophobic molecules use this mechanism for entry into bacteria.

Facilitated diffusion requires the operation of a membrane protein for passage of the permeant across the membrane. These transporters merely provide a route for diffusion of their substrate down its concentration gradient, and thus the concentration of the substrate on both sides of the membrane will become equal. Transport is not dependent on the polarity of the substrate and usually exhibits stereospecificity, in which isomeric forms of the same compound are transported at very different rates. Because of the involvement of the transporter as a catalyst for movement, the rate of transport can be saturated in the same manner as an enzyme-catalyzed reaction. A possible example is the glycerol facilitator GlpF of *E. coli*, a transmembrane protein that allows glycerol and other small molecules to diffuse across the cytoplasmic membrane at rates much faster than they cross lipid bilayers. However, this protein may act in the manner of a channel rather than a carrier.

The overwhelming majority of transport systems in bacteria catalyze active transport and expend metabolic energy to allow the accumulation of even very low external concentrations of a nutrient to a much greater concentration inside the cell. Active transport is carried out by a transport protein or complex and thus exhibits substrate stereospecificity and rate saturation. The difference compared to facilitated diffusion is that the substrate can be accumulated at concentrations as much as 1 million times higher than that outside. This accumulation requires the expenditure of energy, and in the absence of energy many but not all transport mechanisms can carry out facilitated diffusion. These active transport systems differ in their molecular complexity and in the mechanism by which metabolic energy is coupled to substrate accumulation. It is important to distinguish active transport, in which the substrate is accumulated in unaltered form, from group translocation, in which the substrate is converted into a different molecule during the process of transport. Some types of transport processes in bacterial cells are summarized in Fig. 18.6.

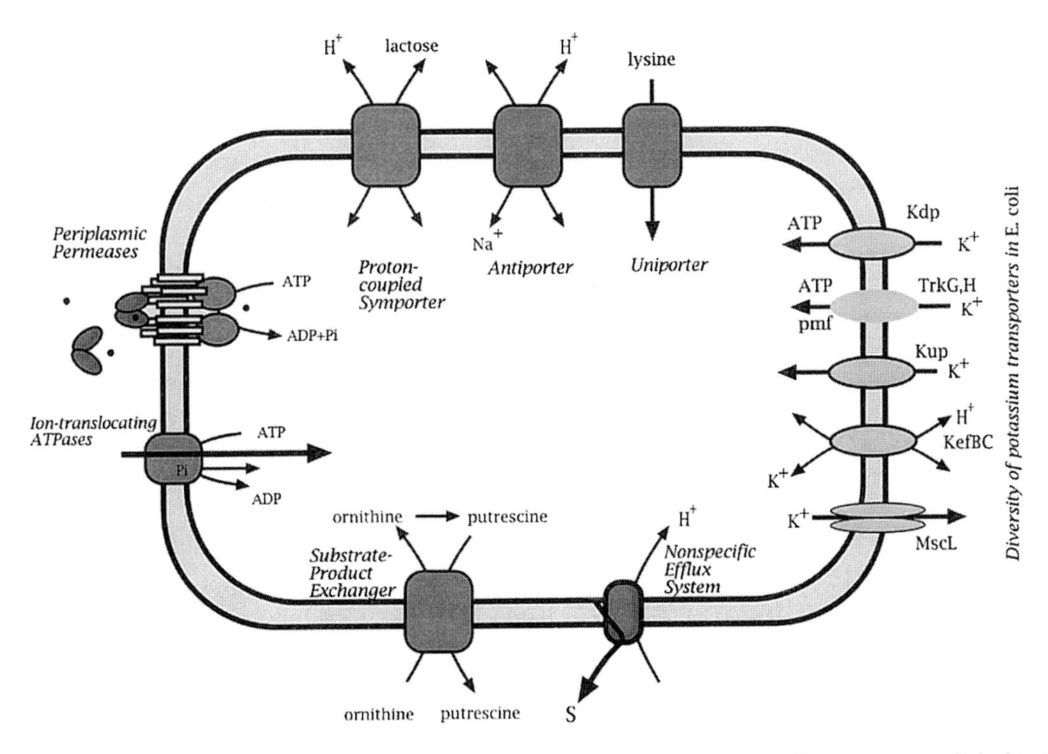

FIGURE 18.6 Schematic representation of several types of active transport systems in bacteria. Top, transporters linked to the pmf; left, ATP-driven transports; bottom, examples of systems that carry out release of metabolic products of toxic chemicals. (Right) A presentation of all the transport systems known to mediate uptake or release of potassium in *E. coli*. Some of these transporters are ATP driven; others are coupled to the pmf; and MscL is a channel activated by mechanical stretch of the membrane.

2. *ATP-driven active transport*

Several groups of transport mechanisms use the energy gained during ATP hydrolysis to drive active transport. One of these groups is called the P-type ion-translocating ATPases to indicate the fact that a phospho-enzyme is formed as an intermediate in their reaction cycle. Typically, these transport systems consist of a large polypeptide of approximately 100 kDa which spans the membrane and contains the site for ATP binding and the residue that is phosphorylated. A smaller subunit usually participates in the activity. The ATP is used to phosphorylate a specific acidic aspartate residue, and this phosphorylation causes a change in conformation of the transporter that is part of the ion pumping process. Several examples of this type of transport system have been extensively studied. The sodium/potassium ATPase in the plasma membrane of higher organisms is responsible for pumping sodium out and potassium into cells, thereby generating the ion gradients that are necessary for many steps of nutrient transport and for neural signal transmission. The electrical potential that exists across the plasma membrane of mammalian and some other cells is based mainly on the difference in sodium ion concentration that is maintained by the action of this transporter. Similarly, the calcium-translocating ATPase located in the sarcoplasmic reticulum acts to lower the intracellular calcium concentrations that accumulate following the processes that initiate muscle contraction. Other P-type ATPases include the proton-translocating ATPase in the plasma membranes of fungi and plants, which establishes the gradient of protons that is used to drive many of these cells' nutrient transport systems, and some of the transport systems for magnesium or potassium ions in bacteria.

Although P-type ATPases function only in the transport of ions, another family of ATP-driven systems transports a wide range of substrates and is involved in the uptake of numerous types of nutrients and in the efflux of both surface and secreted macromolecules (proteins, carbohydrates, and lipids) and of toxic chemicals. This family of transporters is usually the largest family of any set of related genes in those organisms whose genomes have been completely sequenced. Although the subunit composition of these transporters differs, they are called the ABC family to indicate the presence in at least one subunit of a highly conserved ATP-binding cassette, a protein domain that couples ATP binding and hydrolysis to the transport process. A more descriptive name for these proteins is traffic ATPases.

One subset of the family of ABC transporters includes a large group of nutrient uptake mechanisms present only in bacteria and archaea and called periplasmic permeases. These transport systems, such as those for histidine, maltose, oligopeptides, etc., consist of a heterotetramer in which two highly hydrophobic transmembrane proteins with usually five or six membrane-spanning segments are associated with two subunits that contain the ATP-binding cassette and are mainly exposed to the cytoplasm. A fifth protein subunit is responsible for the substrate specificity of these transport systems. This substrate-binding protein usually has very high affinity for the substrate and allows uptake of nutrients even in nanomolar-range concentrations. The structure of these substrate-binding proteins resembles a clam, with two large lobes hinged in the middle. The substrate binds to specific residues in both lobes, which close around the substrate molecule for carriage to the membrane-bound components and entry into the cell. In gram-negative bacteria, the substrate-binding protein floats freely in the periplasmic space between the cytoplasmic and outer membranes. Gram-positive bacteria lack the outer membrane and, to prevent its loss, the binding protein is tethered to the cytoplasmic membrane by a lipoprotein anchor. The mechanism by which ATP hydrolysis is coupled to the release of substrate from the binding protein and its movement across the membrane is currently being studied. A striking feature of these transporters is that they act in a unidirectional manner and only allow nutrient to enter the cell but not to be released.

The basic features of the ABC transport process and homologous transport components also operate in the opposite direction in many processes of macromolecular export, including proteins and surface carbohydrates.

3. *Transporters coupled to ion gradients*

In addition to ATP-driven active transport systems, bacteria possess many transporters in which the movement of their substrate is obligately coupled to the movement, in the same or opposite direction, of an ion. In this way, accumulation of a substrate is coupled to the expenditure of the gradient of the coupling ion, which is usually a proton or sodium ion. Transmembrane ion gradients are a very convenient source of energy for active transporters. Entry into the cell of at least three protons must occur for synthesis of one molecule of ATP. It is thus very economical for the cell if it can accumulate a molecule of substrate at the expenditure of one proton rather than having to expend one ATP molecule for the same purpose. Symport refers to the process in which the coupling ion moves in the same direction as the substrate, that is, when the downhill movement of a proton into the negative and alkaline interior of the cell is coupled to the uptake of substrate. Antiport is the reverse process, in which the two molecules move in opposite directions. If movement of substrate is coupled to movement of a proton in a 1:1 stoichiometry, then a pmf of −120 mV can achieve a 100-fold accumulation of substrate inside the cell. A pmf of −180 mV can allow a 1000-fold gradient of substrate. Uniporters allow coupling of the movement of a positively charged molecule to the pmf without movement of any other ion. The cationic molecule is drawn into the negatively charged cell interior simply by electrostatic attraction.

These types of transporters are referred to as secondary active transport systems since they use the pmf that was generated by other means and do not use an immediate source of energy, such as ATP. Ion-coupled transport systems are inhibited by conditions that dissipate or prevent formation of the pmf, such as ionophores which allow ions to distribute across the membrane in response to the electrical and chemical gradients that act on it. An uncoupler or protonophore, such as 2,4-dinitrophenol or carbonylcyanide *p*-trifluoro-methoxy phenylhydrazone, is a hydrophobic molecule that can cross the membrane in either its ionized or its neutral form. Its presence allows protons to equilibrate across the membrane, thereby dissipating both the electrical and the chemical gradients of protons and thus the entire pmf. The ionophore valinomycin carries potassium ions and allows them to distribute across the membrane in response to electrical or chemical gradients. The addition of valinomycin to cells or membrane vesicles that have a pmf allows potassium ions to accumulate inside the negatively charged interior. This accumulation results in dissipation of the electrical potential $\Delta\Psi$. The ionophore nigericin carries protons and sodium or potassium ions across the membrane, but only in the process of exchange. The action of nigericin thus does not result in any net gain or loss of charge and thus does not dissipate $\Delta\Psi$. However, if there is a concentration gradient of protons, ΔpH, nigericin allows the concentration gradient to dissipate, whereas the electrical potential is maintained or even increased.

Ion-coupled transporters typically consist of a single polypeptide chain with 12 transmembrane segments, although examples with 10–14 transmembrane segments are known. It has been proposed that these proteins arose as the result of tandem duplication of a precursor protein with six transmembrane segments.

In the well-studied *E. coli* lactose permease LacY, several major experiments demonstrated the coupling of lactose accumulation to the pmf. The magnitude of the pmf affects the magnitude of the accumulation ratio of lactose inside the cell in a direct manner indicative of a 1:1 stoichiometry of lactose and proton. When membrane vesicles are energized by provision of a substrate for the electron transport system, a pmf is generated and lactose is accumulated. Even when a final steady-state level of lactose accumulation is reached, the lactose is in continual movement in both directions across the membrane. At the steady state, the rates in and out of the vesicle are equal, although the internal concentration of lactose is much higher than the external concentration. This indicates that the energy has resulted in a decreased affinity of the carrier for lactose on the inside face of the membrane relative to its affinity on the outside. Instead of using the electron transport system, a transmembrane electrical potential, interior negative, can be generated experimentally by diluting vesicles loaded with a high concentration of potassium ions into a medium of low potassium concentration in the presence of valinomycin. The potassium ions flow out down their concentration gradient, carrying positive charge out of the vesicle and leaving behind an interior negative charge. This negative interior can attract protons into the vesicle through the lactose permease, thereby driving lactose accumulation. Finally, if unenergized vesicles are placed in an unbuffered solution that contains a high concentration of lactose, the lactose will flow into the vesicle, bringing along a proton and thereby causing a measurable decrease in the pH of the medium. All these results provide convincing evidence for the coupled movement of proton and lactose.

Transporters are designed to prevent uncoupled movement of substrate without protons or of protons without substrate. If the latter case occurred, it would result in the operation of an uncoupler and allow the futile dissipation of the pmf. How the binding of a proton affects the affinity or binding of the substrate remains an intriguing and central question. It has been shown that the proton must bind before the lactose. If saturating concentrations of lactose are present on both sides of the membrane, the lactose transporter carries out their very rapid exchange independent of the release or re-binding of the proton. This result indicates that the reorientation of the loaded substrate-binding site from facing the interior to facing the exterior does not require changes in proton binding by LacY. If lactose is present on only one side of the membrane, however, its downhill movement is much slower than in the case of exchange and is strongly influenced by the pH. This result indicates that the bound proton must be released from the carrier to allow the empty carrier to re-orient its substrate binding site to pick up another molecule of lactose. This result is in agreement with the model that the binding of a proton is needed to increase the affinity of the carrier for lactose, and that the low concentration of protons inside the cell or vesicle is the factor that slows the rate of release of lactose and hence drives lactose accumulation.

The lactose permease has been subjected to extensive genetic and biochemical analysis. Surprisingly few amino acid residues are essential for function, although amino acid substitutions that introduce or affect charged residues in the transmembrane region usually interfere with the stability of the protein or its ability to be stably inserted in the membrane. Models for the folding of the transmembrane segments relative to one another have been proposed from genetic and biophysical assays of regional proximity.

4. *Efflux systems*

The existence and clinical importance of non-specific drug efflux systems have recently been recognized. An important medical finding was the discovery of an ABC protein that mediates the efflux from cells of non-polar planar molecules, many of which are used in cancer chemotherapy. Overexpression of this protein, called the P-glycoprotein or multidrug resistance protein-1 (MDR-1), is often associated with failure of chemotherapy and recurrence of the disease after a previous round of treatment. This protein is a single polypeptide, but it resembles the periplasmic permeases in having two separate transmembrane domains, each with six membrane-spanning segments followed by a domain with the ABC consensus motifs. It is interesting that a homologous protein in mammalian cells, MDR-2, does not mediate drug efflux but is responsible for the translocation of lipids or their flipping from one side of the membrane to the other. Spontaneous flipping of a lipid across membrane bilayers is extremely slow, and hence an ABC transporter may be assigned this function in cells. Another related protein catalyzes bile acid transport out of liver cells. There are probably many more functions for ABC transporters.

It subsequently has become clear that non-specific efflux systems are widespread in bacteria and account for serious examples of multiple antibiotic resistance. Unlike the ATP-dependence of the MDR carrier, the bacterial efflux systems are coupled to the proton gradient. Most systems possess 4, 12, or 14 transmembrane segments; those with four appear to function as a trimer. In *E. coli*, the major non-specific efflux system is called the Acr system, indicating its initial discovery as a factor that conferred resistance to acriflavins. It comprises three proteins: AcrB, a proton-coupled cytoplasmic membrane transporter; AcrA, a lipoprotein than spans the periplasmic space; and TolC, an outer membrane pore-forming protein. Although not essential for growth, the Acr system can very effectively pump many amphiphilic or lipophilic molecules from the cytoplasmic membrane through the outer membrane directly into the medium. This systems is just one example of complex transport systems that act across multiple cellular compartments.

5. *Phosphotransferase system*

Many bacterial species, but not archaea or eukarya, possess sugar uptake systems that carry out the simultaneous transport and phosphorylation of the sugar. The phosphoenolpyruvate : sugar phosphotransferase system (PTS) uses phosphoenolpyruvate (PEP) as phosphate donor, which is transferred successively to two proteins that act in common in all PTS systems. The enzyme I protein transfers phosphate from PEP to HPr, a small protein that is phosphorylated on a histidine residue. HPr serves as phosphate donor to the sugar-specific components, which comprise three protein domains that can be linked together in various orders and combinations. The IIC domain spans the membrane and catalyzes transport of the sugar, but only under conditions in which it is phosphorylated during its passage. The IIA and IIB domains receive phosphate from HPr-P and transfer it to the sugar molecule that is carried by the corresponding IIC domain. This transport mechanism provides a very economical method to combine the phosphorylation that must occur during sugar metabolism with the transport process. Although the PTS allows substrate accumulation, it is not considered active transport because the substrate is modified during transport, and the metabolic energy has been expended in the substrate modification rather than during the transport.

It is interesting that the PTS system plays a major role in the regulation of cellular metabolism, and the transport of PTS sugars inhibits uptake of potential carbon sources through other types of transport systems. Part of this mechanism involves the inhibitory interaction of the IIA component of the glucose PTS with a variety of transporter proteins for other sugars.

C. Export of surface molecules

All bacteria have specialized processes for the secretion or export of proteins and polysaccharides across the cytoplasmic membrane to the periplasm, cell wall, outer membrane, or external medium. Exported components can comprise 10% or more of the weight of the cell, indicating the magnitude and variety of macromolecular transport activities that occur. For peptidoglycan, lipopolysaccharide, and capsular polysaccharide synthesis, the precursor subunits are translocated across the cytoplasmic membrane and assembled on its outer surface. Some of these precursors are flipped across the membrane after they are coupled to a lipid membrane carrier.

There are several specialized mechanisms for secretion of proteins. To be secreted, a protein must carry a suitable secretion signal. The most common of these is the signal peptide, an amino-terminal extension that is removed during secretion. The typical signal peptides in all cell types comprise 20–30 amino acids with positively charged residues at the amino terminus and then a stretch of hydrophobic residues, followed by a more polar stretch and a peptidase cleavage site. Variants of this structure are used to target proteins to specific cellular organelles in eukaryotic cells. The Sec system acts on the majority of translocated proteins in bacteria. It operates along with several cytoplasmic chaperone proteins, mainly one called SecB, that slow the folding of the precursor protein into its stable final structure, which would prevent its movement across the membrane. The SecYEG protein complex in the cytoplasmic membrane forms a channel through which the polypeptide chain can move. The SecA protein plays a crucial role by binding to the precursor in the cytoplasm, bringing it to the SecYEG translocation complex, and sequentially inserting approximately 30 amino acid segments to the other side of the membrane. The insertion process requires ATP hydrolysis. The SecA and SecB proteins are found only in bacteria and not eukaryotic cells. Once the signal peptide has appeared on the other side of the membrane, it is cleaved off by the action of a leader peptidase enzyme.

Other secretory systems operate in bacteria. The Tat system is named for the twin-arginine motif present in the signal sequences of its substrates, and can move even folded proteins across the cytoplasmic membrane. The SRP-dependent system is related to the process that predominates in eukaryotic cells and involves a signal recognition particle. This system also uses the SecYEG complex for transmembrane movement of its substrate proteins during their synthesis on the ribosome, and appears to be used mainly for proteins destined to the cytoplasmic membrane.

At least five types of transport systems allow export of specific proteins out of the cell. These proteins employ special targeting signals to specify their entry into the appropriate secretory pathway. One of these, the type II secretion system, uses the Sec pathway for initial movement of the precursor into the periplasmic space, where a very complex protein assemblage recognizes the exported protein and moves it across the outer membrane. The type I secretion system resembles the Acr multidrug efflux system in the simplicity and location of its protein components, except that the cytoplasmic membrane component uses the energy of ATP hydrolysis for its action. The type III and type IV secretion systems are of particular interest because they can secrete their substrate proteins, or DNA–protein complexes carried by some type IV systems, directly into a eukaryotic cell. Other specialized systems exist for the assembly of flagella and fimbrial adhesins (pili and fimbriae) on the cell surface. The number, mechanism, and regulation of the many specialized export systems have only recently begun to be understood.

D. Cell growth

1. *Cell division*

The cytoplasmic membrane plays numerous important roles in the processes of cell growth and division. The rate of growth of a cell is determined by the rates of synthesis of its macromolecules, RNA and protein. Their continued production produces pressure that drives the expansion of the cytoplasmic membrane, with the insertion of additional membrane proteins and of the phospholipids to maintain a set protein : lipid ratio. Somehow, the pressure of the expanding cytoplasm triggers insertion of additional cell wall material to allow the increase in cell volume. Once the cell volume has reached a critical point, or the concentration of a specific protein has reached a certain concentration, or a certain time has passed since the chromosome was replicated, the process of cell division is initiated. A critical step in cell division is the assembly of a ring of tubulin-like FtsZ proteins from the cytoplasm onto the membrane at the division site. This assembled complex contracts to pull the membrane together at the division septum to close off the two progeny cells.

A poorly understood process of cell division that may require membrane action is the separation and equal partitioning of the bacterial chromosomes into each progeny cell. In one model, the chromosomes attach to a membrane protein and are pulled apart by the growth of the membrane. This model is unlikely to account for the rapidity with which chromosome separation can occur under certain conditions, but no convincing evidence for cytoskeletal components that might pull the chromosomes apart has been presented.

2. *Signal transduction*

Bacterial gene expression is very responsive to changes in the cell's environment, and many types of regulatory systems allow specific genes to respond to the presence of or need for a wide range of pathway substrates or products. Some regulatory systems are controlled by the binding of the effector molecule to a specific DNA-binding protein that controls the level of expression of the controlled gene. Other systems are controlled by effector molecules that remain outside the cell. Many of these are controlled by two-component regulatory systems, which are widespread in bacteria and archaea and even occur in eukarya. One component of these systems is a transmembrane protein that recognizes its effector molecule in the medium and responds to its binding by a transmembrane signaling event that changes its ability to phosphorylate or to transfer that phosphate. Phosphorylation of the sensor kinase protein can result in transfer of that phosphate to the second component, which is a response regulator protein. The ability of the response regulator protein to bind to a target DNA sequence or to activate transcription of that gene is directly related to the level of its phosphorylation. This mode of transmembrane signaling by regulation of protein phosphorylation is reminiscent of the myriad signaling processes in eukaryotic cells, although the mechanisms and components are not related.

3. *Cell movement*

Bacteria exhibit several types of motility. Many rod-shaped bacteria and a few cocci can swim through liquid medium through the use of flagella. Flagella are long, helical filaments that extend from the poles or the periphery of the cell body and are assembled from a single protein subunit, called flagellin. Motility is initiated by the rotation of the flagellar filament, which in the case of bacteria with multiple filaments results in their coalescence into a bundle that acts similar to a propeller. Their rotation is driven by the downhill entry of protons through the flagellar basal body, a complex structure embedded in the cytoplasmic membrane and driven by the pmf.

Spirochetes are spiral-shaped bacteria in which the cell body is wrapped around the flagellar filaments which grow from the cell poles. Their flagella do not extend into the medium but are retained in the periplasmic space. These bacteria exhibit a characteristic corkscrew motility that is thought to result from rotation of the endo-flagella.

Several types of bacteria are capable of movement on solid surfaces or in very viscous solutions, in which flagella are ineffective. In some types of these bacteria the mechanism of movement is unknown, whereas in some enteric bacteria, such as *Proteus*, this swarming motility is related to differentiation of the cell into very long forms that are covered with a profusion of lateral flagellar filaments.

Motility is a regulated process that allows bacteria to swim toward or away from gradients of nutrients or repellents or physical conditions, such as oxygen or light. Response to chemical attractants or repellents is called chemotaxis and this senses whether a bacteria is moving in a direction which increases or decreases the concentrations of the chemical signal. In the absence of a signal, bacteria exhibit periods of swimming in a straight line followed by short periods in which they tumble aimlessly before setting off in a new direction. Straight-line swimming is associated with counterclockwise (CCW) rotation of the flagellum (viewed toward the cell), and tumbling is associated with clockwise (CW) flagellar rotation. The process of chemotaxis controls the direction of rotation so that when cells are swimming in a favorable direction, their period of straight swimming (CCW rotation) is extended. Movement in an unfavorable direction results in an increased frequency of tumbling (CW rotation). This process is controlled by several chemoreceptors in the cytoplasmic membrane, whose occupancy by substrates indicates the level of the chemical signals. Occupancy of these receptors by their ligands results in changes in the activity of a protein kinase that is related to the two-component regulatory systems described previously. Changes in the activity of the kinase result in changes in the level of phosphorylation of a small cytoplasmic protein, CheY, whose phosphorylated form signals CW rotation of the flagella. Another form of covalent modification is the methylation of certain glutamate residues on the chemoreceptors which serves to adjust their signaling properties and allow the receptors to respond to changes in the level of the chemicals rather than to their static concentration. This adaptive response allows the bacterium to resume its normal behavior once it finds itself in a steady supply of the chemical attractant and to prepare to set off in response to new signals. Chemotaxis thus provides a well-studied example of transmembrane signaling and communication between protein complexes in the cytoplasmic membrane. Of particular interest is the localization of the chemoreceptors and associated components at the poles of the cell.

V. ISOLATION OF MEMBRANES

Procedures have been developed for the isolation of cellular membranes from many bacteria. These procedures are modified owing to differences in the composition and content of peptidoglycan cell wall and of the outer membrane or other surface layers. For the gram-negative *E. coli*, membranes are most reliably isolated following disruption of the cell either by decompression in a French pressure cell or by osmotic lysis following disruption of the peptidoglycan cell wall with the degradative enzyme lysozyme. Cytoplasmic membrane vesicles that are used for transport studies are obtained by osmotic lysis of such spheroplasts (or operation of the French pressure cell at low pressure) and generally have only a somewhat smaller volume than that of the intact cell and are mainly in the right-side-out orientation. Membrane vesicles that are prepared by more vigorous cell disruption by sonication or operation of the French pressure cell at high operating pressure tend to be much smaller in volume and in the inside-out or everted orientation.

After unbroken cells are removed by low-speed centrifugation, the total membrane fraction can be collected by ultracentrifugation. Sucrose density gradient centrifugation can be used for removal of ribosomes, peptidoglycan fragments, and other large complexes and for separation of the cytoplasmic and outer membranes. Because of its content of lipopolysaccharide, the density of the outer membrane in *E. coli* is higher than that of the cytoplasmic membrane. In other gram-negative bacteria, the densities of the two membranes are similar, and their separation by this technique is not as effective.

An alternative approach for separation of the proteins from the cytoplasmic and outer membranes in enteric bacteria takes advantage of the resistance of most outer membrane proteins to solubilization by nonionic detergents in the presence of divalent cations. This rapid method cannot be relied on for localization of all membrane proteins since it has been tested only on a limited set of proteins and depends on features that are not necessarily universally maintained. Detergents are amphipathic molecules that are able to associate with and disrupt membrane structure. They allow the lipids to form a spherical micellar structure and provide the membrane proteins with a suitable environment for their non-polar surfaces in molecular aggregates much smaller than the membrane. Many nonionic detergents allow membrane proteins to retain their structure and function.

VI. MEMBRANE ASSEMBLY AND CONTROL OF COMPOSITION

The process of membrane assembly is becoming increasingly understood. Insertion of integral proteins into the cytoplasmic membrane is determined by the presence of hydrophobic or amphiphilic stretches of amino acid residues that represent potential transmembrane segments. Most cytoplasmic membrane proteins do not have a cleaved signal sequence that directs them to the membrane. The exceptions are proteins which are anchored in the membrane but are mainly exposed to the outside; these undergo processing and removal of a signal sequence. The orientation of the transmembrane segments of an integral membrane protein is determined primarily by the pattern of charged amino acid residues on the two polar loops on either side of the segment, which generally obey the positive-inside behavior. Although most integral membranes can insert spontaneously into lipid bilayers, their insertion in the cell appears to be facilitated by cytoplasmic chaperones and the Ffh and FtsY proteins that are related to the components of the eukaryotic signal recognition system, SRP. The components of the Sec system do not seem to be necessary, unless the protein contains a large (160 residues) external segment.

Membrane phospholipids are made by cytoplasmic and membrane-associated enzyme pathways. New phospholipids are inserted into the inner face of the cytoplasmic membrane and must flip across to the outer leaflet. This flipping process is extremely slow without the involvement of a protein catalyst. There is a considerable degree of membrane asymmetry in membranes of eukaryotic cells, with a substantially different lipid composition in the two leaflets of the bilayer. Such asymmetry appears less prominent in bacterial membranes but has not been studied in detail.

The lipid composition of cellular membranes changes in response to environmental conditions and growth state. Lower growth temperatures result in the incorporation of fatty acids with a higher degree of unsaturation than those that are present at higher growth temperatures so as to maintain a sufficient level of membrane fluidity. This sensing and response to ambient temperature involves selective synthesis and incorporation of different fatty acids by the appropriate enzymes of lipid biosynthesis rather than changes in levels of these enzymes. When cells of *E. coli* enter the stationary phase of growth, an enzyme is produced that adds a methyl group across double bonds in fatty acids to convert unsaturated fatty acids to the corresponding cyclopropane fatty acids. These cyclopropane fatty acids have similar effects on membrane dynamics as those with double bonds but are more stable to oxidative damage.

Phospholipid head groups determine certain properties of the membrane. The acidic phospholipids in

E. coli, PE and cardiolipin, are necessary for the incorporation and function of many membrane enzymes and transporters. Anionic phospholipids are essential for stimulating the ATPase activity of the SecA protein during its translocation of proteins across the membrane. These lipids are also necessary for the activation of ATP binding to the DnaA protein, which plays the central role in the initiation of bacterial chromosome replication. The use of mutants defective in the various lipid biosynthetic pathways revealed that a minimal quantity of acidic phospholipids was essential for cell growth, although cardiolipin was dispensable. The other major lipid, PE, is noted for its ability to form non-lamellar structures under certain conditions. Mutants that are completely blocked in the synthesis of PE can grow, but only in the presence of elevated concentrations of divalent cations, which promote non-lamellar structures in cardiolipin. In this mutant, cardiolipin synthesis is essential for growth. It is thus clear that the membrane can tolerate a wide range in the lipid head group composition, as long as there is a minimal level of acidic phospholipids and of lipids that can form nonlamellar structures. It has proven difficult to determine the effect of changes in lipid composition on individual membrane functions in the intact cell because of the activation of a stress response system under these conditions. The fluid state of the membrane certainly affects the activity of numerous membrane proteins, indicating that acyl chain or lipid mobility are necessary for their proper function or insertion. One expects substantial progress in the future in understanding the structures and interactions of lipids and proteins in the cytoplasmic membrane.

BIBLIOGRAPHY

deKruijff, B., Killian, J. A., Rietveld, A. G., and Kusters, R. (1997). Phospholipid structure and *Escherichia coli* membranes. *Curr. Topics Membranes* **44**, 477–515.

Driessen, A.J., Manting, E. H., and van der Does, C. (2001). The structural basis of protein targeting and translocation in bacteria. *Nat. Struct. Biol.* **8**, 492–498.

Fekkes, P., and Driessen, A. J. (1999). Protein targeting to the bacterial cytoplasmic membrane. *Microbiol. Mol. Biol. Rev.* **63**, 161–173.

Harold, F. M. (1986). "The Vital Force: A Study of Bioenergetics." Freeman, New York.

Koebnik, R., Locher, K. P., and Van Gelder, P. (2000). Structure and function of bacterial outer membrane proteins: barrels in a nutshell. *Mol. Microbiol.* **37**, 239–253.

Konings, W. N., Kaback, H. R., and Lolkema, J. S. (1996). "Transport Processes in Eukaryotic and Prokaryotic Organisms." Elsevier, Amsterdam.

Muller, M., Koch, H. G., Beck, K., and Schafer, U. (2000). Protein traffic in bacteria: multiple routes from the ribosome to and across the membrane. *Prog Nucleic Acid Res. Mol. Biol.* **66**, 107–157.

Neidhardt, F. C., Curtiss, R., III, Ingraham, J. L., Lin, E. C. C., Low, K. B., Magasanik, B., Reznikoff, W. S., Riley, M., Schaechter, M., and Umbarger, H. E. (1996). "*Escherichia coli* and *Salmonella*. Cellular and Molecular Biology." ASM Press, Washington, DC.

Pao, S. S., Paulsen, I. T., and Saier, M. J., Jr. (1998). Major facilitator superfamily. *Microbiol. Mol. Biol. Rev.* **62**, 1–3.

Walker, J. E., ed. (2000). See a series of review articles dealing with the structure and function of the F_1F_0-ATPase. *Biochim. Biophys. Acta* **1458**, issue 2.

Yeagle, P. L. (1993). "The Membranes of Cells," 2nd edn. Academic Press, San Diego.

WEBSITE

The *E. coli* Cell Envelope Protein Data Collection
http://www.cf.ac.uk/biosi/staff/ehrmann/tools/ecce/ecce.htm

19

Cell Walls, Bacterial

Joachim-Volker Höltje
Max-Planck-Institut für Entwicklungsbiologie

GLOSSARY

autolysins Endogenous murein hydrolases, which cleave bonds in the peptidoglycan sacculus that are critical for the mechanical strength of the structure and thus cause lysis (autolysis) of the bacterium.

cell envelope A multilayered structure that engulfs the cytoplasm. The innermost layer, the cytoplasmic membrane, is stabilized by an exoskeleton of peptidoglycan (murein). Gram-negative bacteria have a second lipid bilayer, called the outer membrane, making the envelope of gram-negative bacteria less permeable compared to that of gram-positive bacteria.

Gram stain Iodine–gentian violet complex that is retained by some bacteria (called gram-positive) but is released from the envelope by acetone or ethanol from another group of bacteria (called gram-negative). Whereas gram-positive bacteria have a multilayered shell of peptidoglycan, gram-negative bacteria have a thin (monolayered) peptidoglycan and a second membrane, the outer membrane.

lipoproteins Proteins carrying at their amino terminus a cysteine to which glycerol is linked that has its hydroxyl groups substituted by fatty acids. In addition, the amino group of Cys is also modified by a fatty acid. A murein lipoprotein is covalently attached to the cell wall of gram-negative bacteria with its carboxyl terminus and at the same time is inserted with its lipophilic amino terminus into the outer membrane, thereby connecting murein and the outer membrane.

lysozyme β-1,4-*N*-acetylmuramidase that hydrolyses the β-1,4 glycosidic bond between *N*-acetylmuramic acid and *N*-acetylglucosamine in peptidoglycan. It is found in many bacteria, where it is involved in the growth processes of the cell wall. It is also present in various tissues and secretions of higher organisms (e.g. hen egg white lysozyme), where it functions as a powerful antibacterial agent.

murein Synonym for peptidoglycan, a cross-linked biopolymer of poly-(*N*-acetylglucosamine-β-1,4-*N*-acetylmuramic acid) that is cross-linked by peptide bridges that are linked to the lactyl group of the muramic acid residues (from *Murus*, a Latin word meaning wall).

murein sacculus The bacterial exoskeleton that forms a bag-shaped macromolecule completely enclosing the cell. It endows the cell with mechanical strength and confers the specific shape to the bacterium.

penicillin-binding proteins Enzymes that covalently interact with penicillin by forming a penicilloyl–enzyme complex. The family of penicillin-binding proteins consists of bifunctional transglycosylase-DD-transpeptidases, DD-transpeptidases, DD-endopeptidases, and DD-carboxypeptidases.

periplasm A specific cellular compartment of the cell envelope of gram-negative bacteria confined by the cytoplasmic membrane and the outer membrane.

The Desk Encyclopedia of Microbiology
ISBN: 0-12-621361-5

teichoic acid Anionic polyol-phosphate polymer covalently bound to the muramic acid of peptidoglycan in gram-positive bacteria. Ribitol- and glycerol teichoic acids are known.

The Bacterial Cell Envelope is a complex structure consisting of different layers that have to fulfill many critical functions for the cell. Besides being a protective shield against the hostile environment, it must allow communication with the surroundings in order for the bacteria to find optimal conditions for growth. This highly developed cell organelle has to be mechanically stabilized to withstand the high intracellular osmotic pressure of approximately 2–25 atm. Eubacteria as a rule reinforce their cell envelope by an exoskeleton made of peptidoglycan (murein), a cross-linked biopolymer that is extremely well suited to endow the cell with sufficient strength. In order to achieve this, the murein forms a closed bag-shaped structure, called sacculus, completely wrapping up the cell. Importantly, the murein sacculus not only stabilizes the cytoplasmic membrane but also maintains the specific shape of the bacterium. Morphogenesis of bacteria can therefore be studied by analyzing the metabolism of a single macromolecule—the murein sacculus.

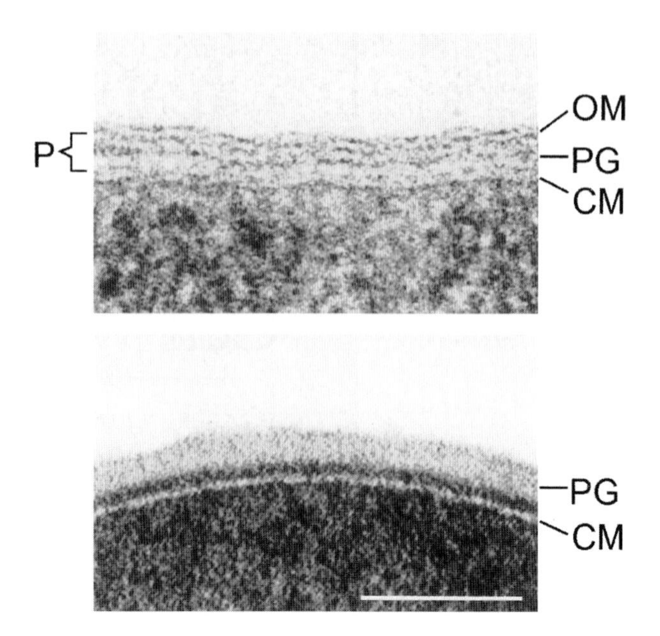

FIGURE 19.1 Electron micrographs of negatively stained sections of bacterial cell walls. (Top) Gram-negative bacterium *Escherichia coli*; (bottom) gram-positive bacterium *Staphylococcus aureus*. P, periplasmic space; OM, outer membrane; PG, peptidoglycan; CM, cytoplasmic membrane. Scale bar = 0.1 mm (courtesy of Drs. H. Frank and H. Schwarz).

I. CELL ENVELOPE STRUCTURE

On the basis of a special staining procedure bacteria can be subdivided in essentially two groups, the gram-positive and the gram-negative bacteria. Interestingly, this reflects a fundamental difference in the general construction of the cell envelope (Figs 19.1 and 19.2). The staining method by Christian Gram consists of two steps. First, the heat-fixed cells are stained with a dark-blue iodine gentian violet dye complex. In the second step, the cells are extracted with ethanol. If the cells are destained during this step they are referred to as gram-negative bacteria, and if the stain remains within the cell envelope the bacteria are called gram-positive.

A typical gram-positive envelope consists of a cytoplasmic membrane and a thick, multilayered murein shell (20–50 nm thick) which is decorated by teichoic acids (Fig. 19.2). The gram-negative bacterium is enclosed by two bilayers, a cytoplasmic membrane and an outer membrane confining a unique compartment called the periplasmic space (Fig. 19.2). The murein sacculus of gram-negative bacteria is embedded into this space and is characteristically extremely thin, forming only one continuous layer (approximately 3 nm thick). The outer membrane and the murein layer are connected by a lipoprotein that has its fatty acid-substituted amino terminus immersed in the outer membrane and its carboxyl terminus covalently linked to the murein (Figs 19.2 and 19.4). Unlike the cytolasmic membrane, the outer membrane is a highly asymmetric lipid bilayer structure with an inner leaflet consisting of phospholipids and an outer leaflet of lipopolysaccharides. Typically, the outer membrane contains pore-forming proteins, known as porins.

The most important structural distinction between gram-positive and gram-negative bacteria is the presence in the latter group of a second bilayer system, the outer membrane. Due to this additional permeability barrier, gram-negative bacteria show significantly higher minimal inhibitory concentration values for many antibiotics than do gram-positive bacteria. As a consequence, some otherwise quite powerful antibiotics are of no therapeutic use for the treatment of gram-negative infections. Besides a general hindrance to the uptake of hydrophilic compounds, the existence of a second permeability barrier has many additional consequences for cell wall metabolism. It allows the cell to reduce the thickness of its murein to just one single layer and makes it possible to efficiently recycle the valuable murein turnover products that accumulate during growth in the periplasmic space.

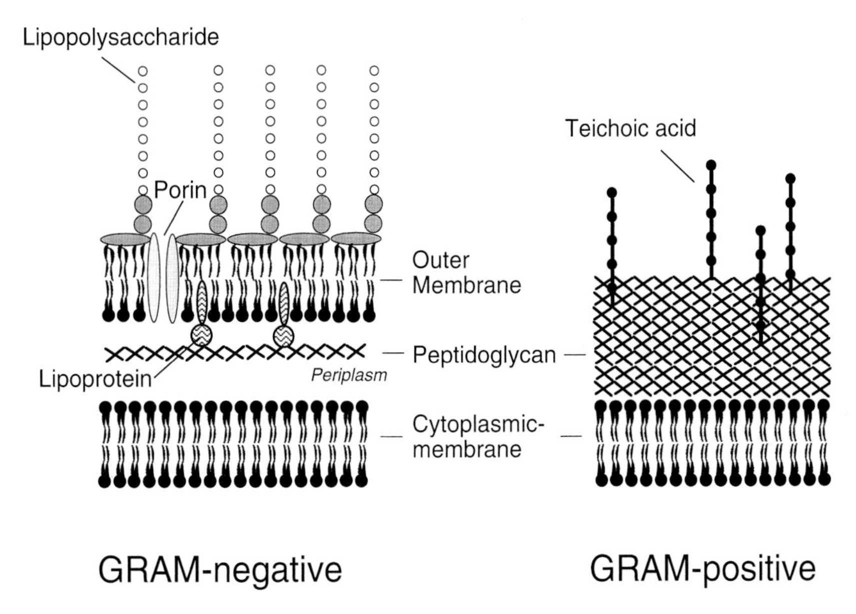

FIGURE 19.2 Schematic view of the construction of the cell envelope of gram-positive and gram-negative bacteria.

II. CHEMISTRY OF BACTERIAL CELL WALLS

The structural principle that glycan strands are cross-linked by peptides in order to form a strong latticework for a bacterial exoskeleton has been invented in nature twice, in Archaea and in Eubacteria. Archaea contain a compound called pseudomurein. Although the architectural style is the same in pseudomurein and in murein, the building materials are different.

A. Peptidoglycan (murein)

The murein of Eubacteria consists of glycan strands that are cross-linked by peptides (Figs 19.3 and 19.7). The amino sugars *N*-acetylglucosamine (GlcNAc) and *N*-acetylmuramic acid (MurNAc) are polymerized by β-1,-4 glycosidic bonds in an alternating order; thus, the glycans have one GlcNAc and one MurNAc terminus. The latter represents the reducing end of the polysaccharide. In some species the terminal MurNAc residue is modified to a non-reducing 1,6-anhydromuramic acid (Fig. 19.3). The lactyl group of the muramic acid is substituted by a short peptide consisting of L- and D-amino acids in an alternating sequence. The presence of LD and DD peptide bonds renders the peptides resistant to the majority of peptidases. Cross-linkage of two peptides is made possible by the presence of a diamino acid, such as L-lysine, *meso*-diaminopimelic acid, or L-ornithine. This allows the formation of a tail to tail bridging peptide bond between the terminal carboxyl group of a D-Ala in position 4 of one peptide moiety and the non-alpha amino group of the diamino acid at position 3 of another peptide side chain (Figs 19.3 and 19.7). In many gram-positive bacteria the cross-linkage is not a direct one but is mediated by an additional peptide called "Intervening peptide." A pentaglycine, for example, is inserted between the D-Ala and the L-Lys in the case of *Staphylococcus aureus*. In contrast to the types of cross-linkages referred to as group A, in which the cross-linkage is between positions 4 and 3 of two peptides, the linkage in group B is via an intervening peptide containing one diamino acid that allows for a cross-linkage between the alpha-carboxyl group of the D-Glu in position 2 of one peptide to the carboxyl group of D-Ala in position 4 of another peptide (Fig. 19.3). In general there are only a few modifications of the sugar part. The glycans can be O_6-acetylated, which makes the murein resistant towards the action of hen egg white lysozyme, or they can be de-N-acetylated, as is the case in *Bacillus cereus*.

B. Murein lipoprotein

The murein of gram-negative bacteria is substituted by a lipoprotein that on average is attached to every tenth muramic acid residue. The linkage is from the epsilon amino group of the lysine at the carboxyl terminus to the carboxyl group at the L-center of *meso*-diaminopimelic acid (Fig. 19.4). The murein lipoprotein is a member of the superfamily of lipoproteins that are characterized by a typical consensus processing site (LLLAGCSSNS). The unmodified prolipoprotein that carries a typical amino-terminal leader is exported to

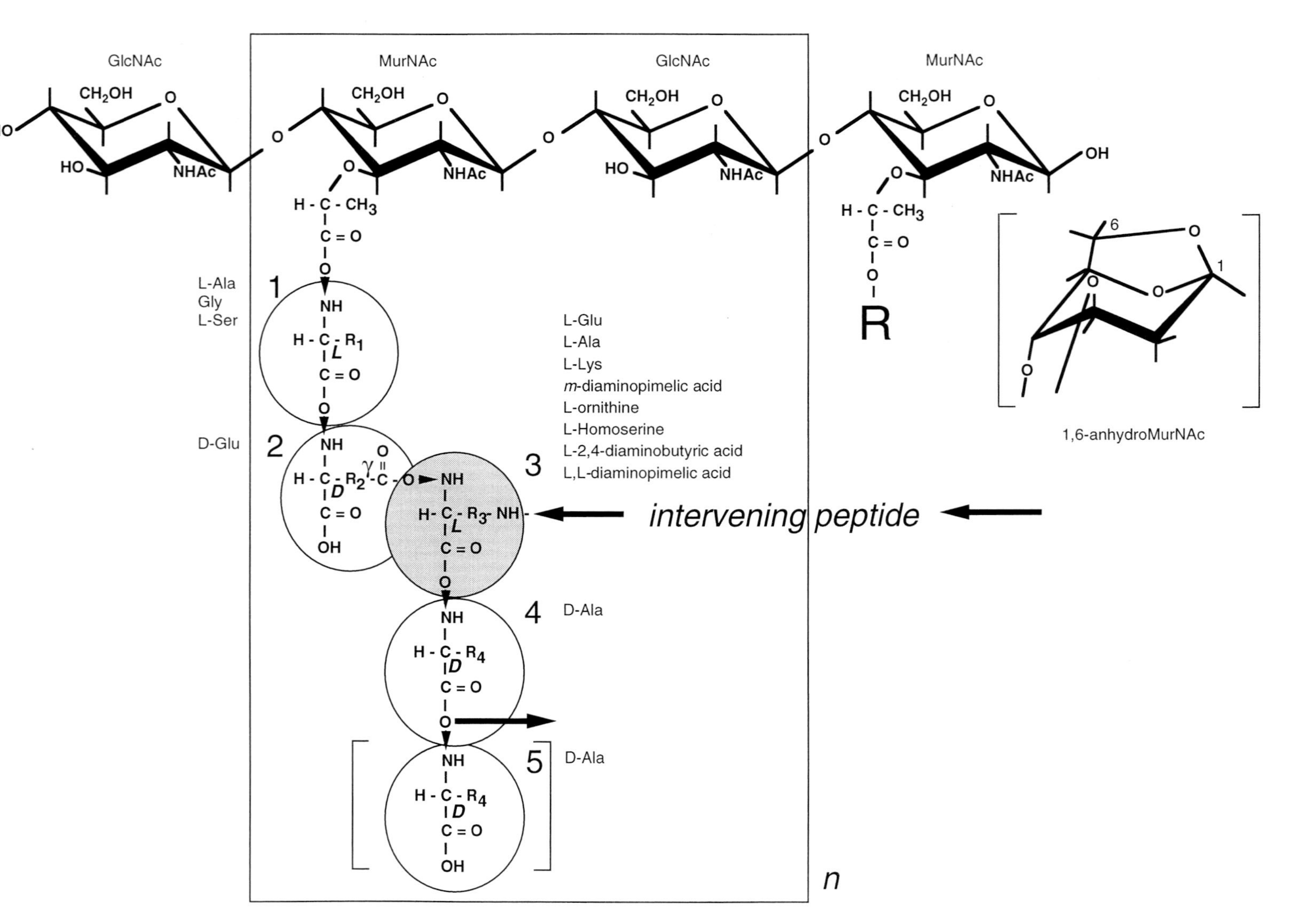

FIGURE 19.3 Generalized chemical structure of peptidoglycan (murein). The box indicates the subunit of the polymer. The different amino acids found at the indicated sites of the pentapeptide moiety in various peptidoglycans are listed. R, peptidyl-moiety; GlcNAc, *N*-acetylglucosamine; MurNAc, *N*-acetylmuramic acid. The arrows indicate the donor and acceptor sites for cross-linkage. The intervening peptide that occurs in many gram-positive bacteria is explained in the text.

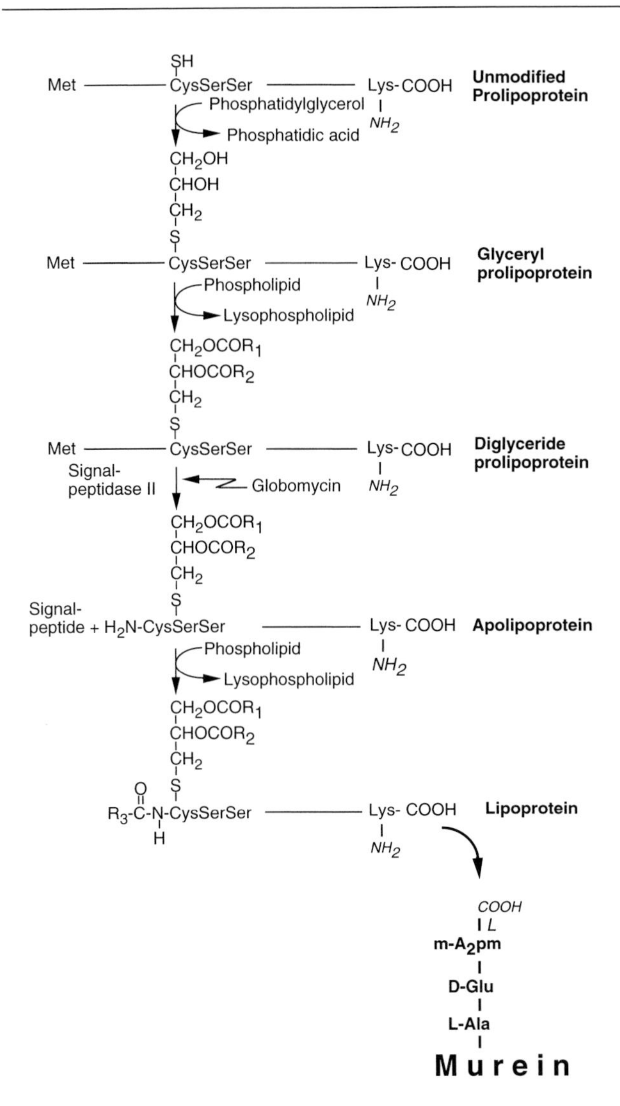

FIGURE 19.4 Biosynthetic pathway of the murein lipoprotein. R_1–R_3 indicate fatty acids with a composition similar to that of the phospholipids of the bacterium.

the periplasmic side of the cytoplasmic membrane where the SH group of the Cys is substituted by glycerol (Fig. 19.4). Two fatty acids are added to the hydroxyl groups of the glycerol moiety to yield the diglyceride proliprotein. A specific signal peptidase (signal peptidase II) cleaves the leader, releasing the amino group of the Cys to which a third fatty acid is linked. The free form of the mature lipoprotein is hooked to the murein by an unknown adding enzyme. Globomycin specifically inhibits the signal peptidase II.

The function of the lipoprotein is probably to connect the outer membrane with the murein sacculus (Fig. 19.2). Although mutants in Lpp are viable, they often show blebs of outer membrane vesicles at the site of cell division where all three layers of the envelope, the outer membrane, the murein, and the cytoplasmic membrane have to be contracted simultaneously. When the Lpp mutation is combined with a defect in the major outer membrane protein OmpA, *Escherichia coli* can no longer maintain its rod shape and grows as spheres, but only when sufficient Mg^{2+} is added.

C. Teichoic acid

An anionic polyol-phosphate polymer, called teichoic acid, constitutes the outermost part of the peptidoglycan shell of gram-positive bacteria. The polyol can be either ribitol or glycerol. Most glycerol teichoic acids are 1,3-linked. Many substitutions, such as D-alanine, L-serine, glycine, glucose, and GlcNAc can occur. Teichoic acids are linked via a special linkage unit by a phophodiester to the C_6 of the muramic acid. Because of the equally spaced phosphate groups, teichoic acids strongly bind magnesium ions. It has been speculated that teichoic acids function as a kind of ion exchanger. Thus, the teichoic acid layer forms a type of cell compartment that may have a function similar to that of the periplasmic space of gram-negative bacteria.

Upon transfer to phosphate-limiting growth conditions, teichoic acid synthesis rapidly decreases and synthesis of a teichuronic acid takes over. Unlike teichoic acids, teichuronic acids lack the phosphate group in the repeating units. The teichuronic acid of *Micrococcus luteus* is formed from the disaccharide D-*N*-acetylmannosaminuronic acid-β-1,6-D-glucose that is polymerized by α-1,4-glycosidic bonds up to 10–40 mers. The teichuronic acid of *Bacillus lichenformis* consists of equimolar amounts of *N*-acetygalactosamine and D-glucuronic acid.

D. Pseudomurein

Pseudomurein resembles murein in that it has the same structural engineering as murein. However, the glycan strands consist of GlcNAc or *N*-acetylgalactosamine and *N*-acetyltalosaminuronic acid, linked together by β-1,-3 glycosidic bonds. The peptide bridges that are hooked to the carboxyl group of the talosaminuronic acid contain L-amino acids, including Lys, Glu, Ala, Thr, and Ser. As a consequence of this different chemical setup, the synthesis of pseudomurein is not inhibited by D-cyloserine, vancomycin, or penicillin, which are typical inhibitors of murein synthesis.

III. BIOSYNTHESIS OF PEPTIDOGLYCAN (MUREIN)

A. Cytoplasmic reaction steps

The biosynthetic pathway of the murein sacculus involves three cellular compartments: synthesis of

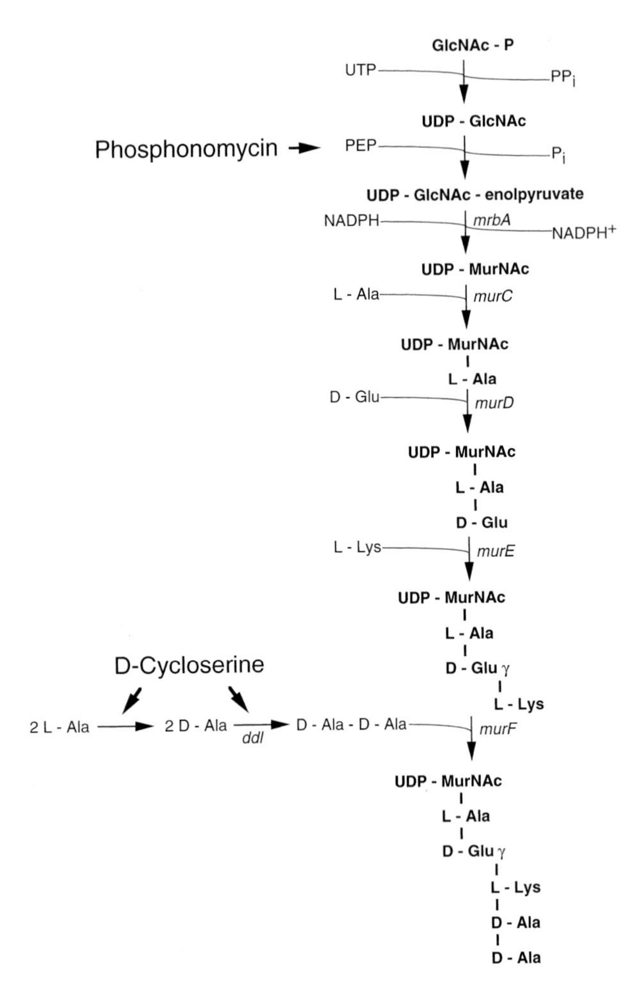

FIGURE 19.5 Biosynthetic pathway of the murein precursor UDP-*N*-acetylmuramyl pentapeptide.

soluble murein precursors in the cytoplasm (Fig. 19.5), transport of the lipid-linked murein precursors across the cytoplasmic membrane (Fig. 19.6), and insertion of the precursors into the pre-existing murein sacculus in the extracellular space or, in the case of gram-negative bacteria, in the periplasmic space (Figs 19.6 and 19.7).

The first biosynthetic intermediate that is specific for murein is UDP-*N*-acetylmuramic acid (UDP-MurNAc). It is formed by a transfer of enolpyruvate from phosphoenolpyruvate to UDP-*N*-acetylglucosamine (UDP-GlcNAc), followed by NADPH catalyzed reduction of the intermediate UDP-GlcNAc-enolpyruvate to the lactate-substituted glucosamine known as muramic acid (Fig. 19.5). The antibiotic fosfomycin (phosphonomycin, L-*cis*-1,2-enolpropylphosphoric acid) is a potent inhibitor of the UDP-GlcNAc-enolpyruvyl transferase. The lactate moiety of muramic acid functions as an acceptor for the first amino acid of the pentapeptide side chain. The additional amino acids are added sequentially by specific synthases that couple the cleavage of ATP to the formation of a peptide bond. The correct sequence of the pentapeptide, an alternating succession of L- and D-amino acids, is determined by the substrate specificity of the synthases. However, the L-alanine adding enzyme also accepts L-serine and glycine and the *meso*-diaminopimelic acid adding enzyme from *E. coli* also accepts the sulfur-containing diaminopimelic acid analog, lanthionine. Interestingly, the last two amino acids, both of which are D-alanine, are added in the form of the D-alanyl-D-alanine dipeptide pre-synthesized by a D-alanine ligase. The D-Ala-D-Ala adding enzyme completes the synthesis of the UDP-MurNAc-pentapeptide precursor. Other DD-dipeptides in addition to D-Ala-D-Ala can be added by the ligase; in particular, glycine and D-lactate can substitute for one of the D-Ala residues. Both the L-alanine racemace and the D-alanine racemace are inhibited by D-cycloserine and related compounds such as *O*-carbamoyl-D-serine, haloalanines, and alaphosphin (L-alanyl-L-l-aminoethyl phosphonic acid).

B. Membrane translocation of murein precursors

For insertion into the murein sacculus the activated precursors UDP-GlcNAc and UDP-MurNac-pentapeptide must be transported across the cytoplasmic membrane (Fig. 19.6). The details of this important step are poorly understood. An undecaprenylphosphate (C_{55}-isoprenoid) molecule, also called bactoprenol, functions as a vehicle (lipid carrier) to shuffle the hydrophilic precursors from the inner to the outer side of the cytoplasmic membrane. A translocase, phospho-*N*-acetylmuramyl pentapeptide translocase (MraY), transfers the phosphoryl-muramyl pentapeptide to undecaprenylphosphate to yield the so-called lipid intermediate I, undecaprenyl-diphosphoryl-*N*-acetylmuramyl pentapeptide. The translocase reaction is fully reversible and is inhibited by tunicamycin. This enzyme also catalyzes an exchange reaction between UMP and UDP-MurNAc pentapeptide. In a second step, catalyzed by the transferase *N*-acetylglucosamine transferase (MurG), GlcNAc is transferred from UDP-GlcNAc to lipid I to form the final murein precursor undecaprenyl-diphosphoryl-*N*-acetylmuramyl(pentapeptide)-*N*-acetylglucosamine, also called lipid intermediate II. Both translocase and transferase are bound to the inner side of the cytoplasmic membrane thus, the final murein precursor accumulates at the inner side of the membrane to which it is anchored via the undecaprenylphosphate moiety. To be available for insertion into the murein sacculus, the lipid-linked precursor has to be translocated across the cytoplasmic membrane. The antibiotic

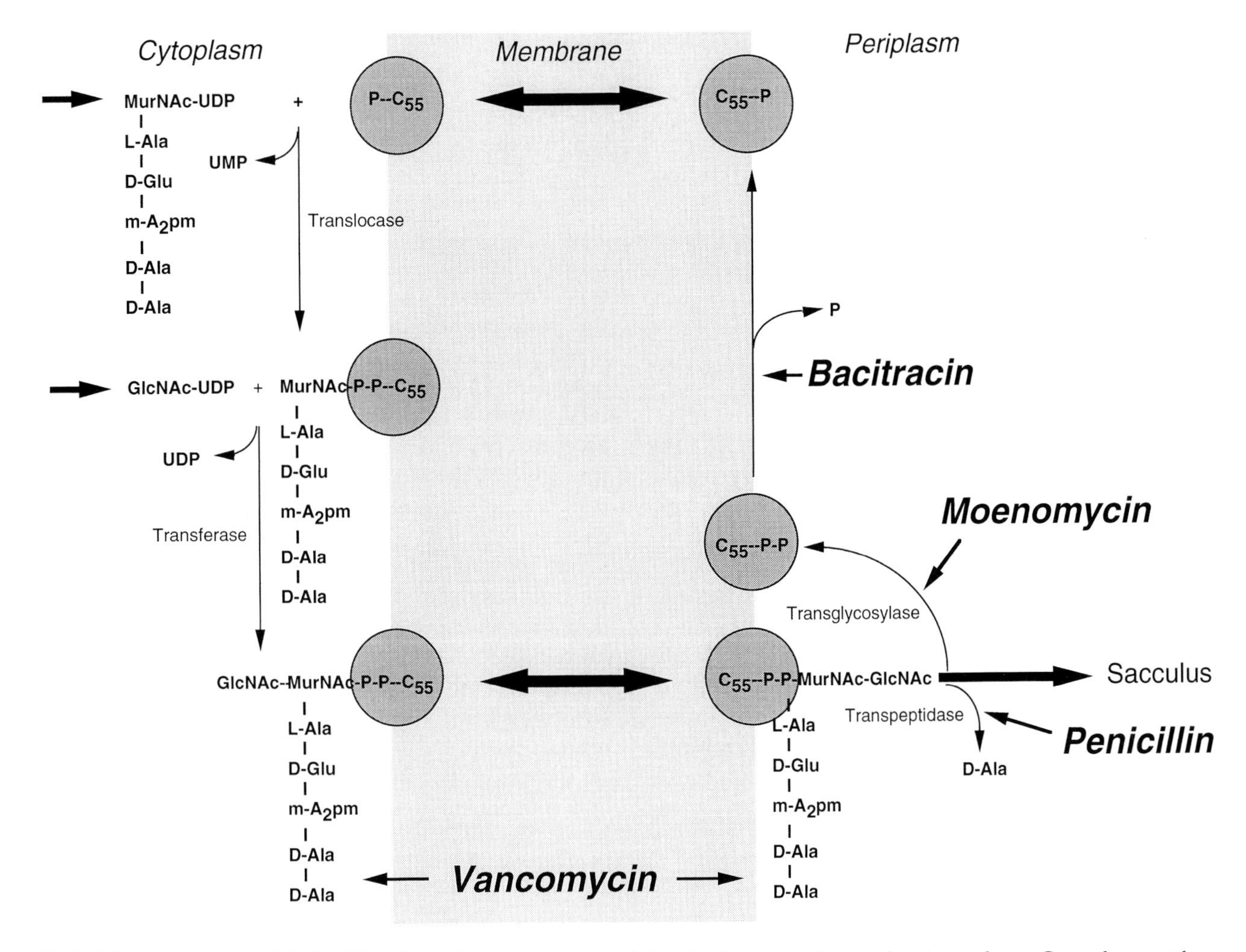

FIGURE 19.6 Formation of the lipid-linked murein precursors and their translocation across the cytoplasmic membrane. C_{55}, undecaprenol.

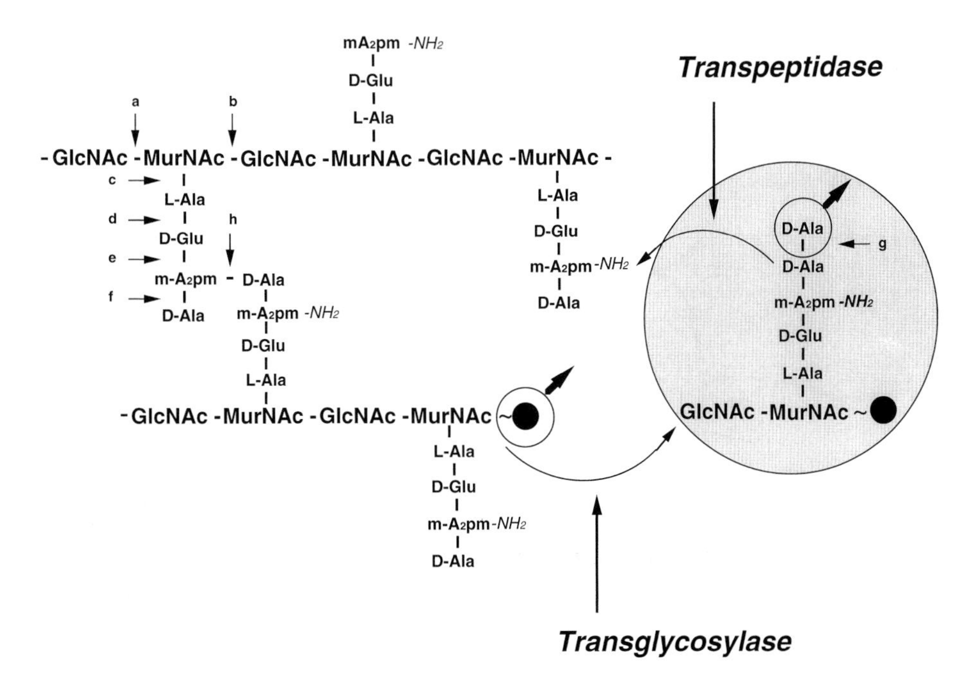

FIGURE 19.7 Polymerization of the lipid intermediate II of *E. coli* by transglycosylation and transpeptidation. The black circles represent the undecarenyl pyrophosphate moiety. The lipid intermediate II is highlighted by a gray circle and the leaving groups of the polymerization reactions are encircled and marked with an arrow. The sites of action of specific murein hydrolases are indicated by small letters: a, *N*-acetylglucosaminidase; b, muramidase (lysozyme); c, *N*-acetylmuramyl-L-alanine amidase; d, L-alanyl-D-glutamyl-endopeptidase; e, D-glutamyl-L-diaminopimely-endopeptidase; f, LD-carboxypeptidase; g, DD-carboxypeptidase; h, DD-endopeptidase; GlcNAc, *N*-acetylglucosamine; MurNAc, *N*-acetylmuramic acid; A_2pm, *meso*-diaminopimelic acid.

vancomycin forms complexes with the D-Ala-D-Ala terminus of the precursor and this blocks murein synthesis. It is unlikely that a spontaneous flip-flop of the lipid intermediate II would be fast enough to match the rate of murein. Therefore, it is expected that the process is facilitated by auxiliary proteins.

C. Insertion of murein precursors into the wall

The murein disaccharide pentapeptide is inserted into the existing murein net by the formation of glycosidic and peptide bonds (Fig. 19.7). A transglycosylase first catalyzes the cleavage of the phosphodiester bond by which a nascent glycan strand is still linked to the undecaprenyl pyrophosphate and then catalyzes the transfer of the glycan strand onto the C_4 hydroxyl group of the GlcNAc residue of a lipid-linked disaccharide pentapeptide precursor. The glycolipid antibiotic moenomycin specifically inhibits the transglycosylation reaction. The released bactoprenol pyrophosphate is then processed to bactoprenol monophosphate in order to be accepted again by the translocase (Fig. 19.6). Bacitracin, an antibiotic that binds strongly to the bactoprenol pyrophosphate, inhibits the recycling of the lipid carrier. The formation of the cross-linkage of the peptide side chains to acceptor peptide side chains in the sacculus is catalyzed by a transpeptidase. Like transglycosylation, transpeptidation is also a two-step transferase reaction. First, the terminal D-Ala-D-Ala peptide bond of the pentapeptide precursor is cleaved, the terminal D-Ala is released, and a substrate–enzyme intermediate is formed. In a second reaction step the murein peptidyl moiety is transferred to a free amino group on an acceptor peptide side chain. As a consequence of transglycosylation and transpeptidation, the precursors are polymerized in two directions yielding the characteristic net structure of murein. Interestingly, in some bacteria, bifunctional enzymes combining a transglycosylase domain with a transpeptidase domain are responsible for the polymerization of the murein precursors.

Penicillin inhibits murein transpeptidases because of its analogy to the D-alanyl-D-alanine terminus of the murein precursor (Fig. 19.8a). The inhibition depends on the enzymatic interaction of the enzyme with the antibiotic. In analogy to the cleavage of the D-Ala-D-Ala bond (Fig. 19.7), the β-lactam ring is cleaved by the

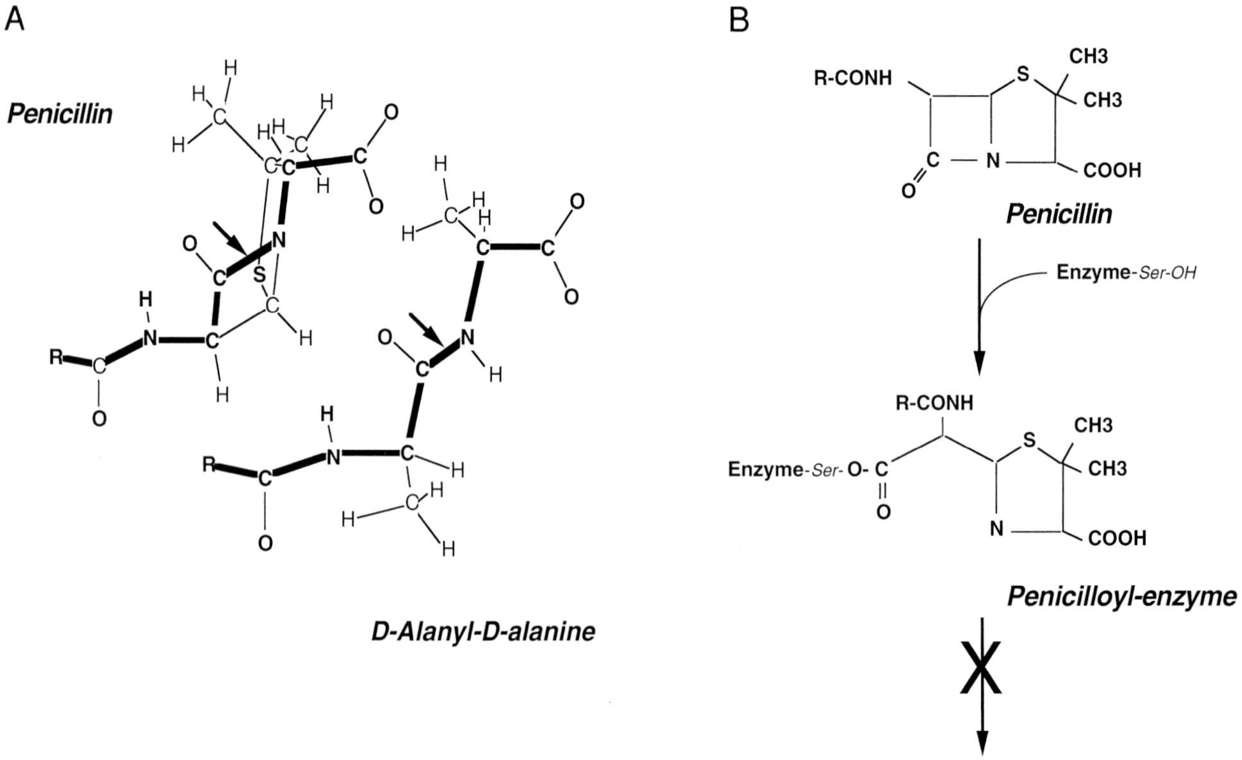

FIGURE 19.8 (A) Structural analogy between penicillin and D-alanyl–D-alanine. The arrows indicate the bonds cleaved by penicillin-sensitive DD-transpeptidases, DD-endopeptidases, and DD-carboxypeptidases. (B) Formation of the penicilloyl–enzyme intermediate.

enzyme and a covalent substrate (penicilloyl)-enzyme intermediate is formed that involves a Ser in the catalytic site (Fig. 19.8b). This intermediate is inert and does not react further. Thus, the enzyme is blocked by the covalently linked penicillin molecule, which may be considered a "suicide substrate" for the transpeptidases. Not only the DD-transpeptidases but also other enzymes specifically recognizing D–D peptide bonds, including DD-carboxypeptidases and DD-endopeptidases, are penicillin-sensitive enzymes. These proteins are collectively referred to as penicillin-binding proteins (PBPs). Since inhibition of the PBPs by penicillin is due to the formation of a covalent enzyme–penicillin intermediate (Fig. 19.8), a simple assay allows the identification of all PBPs of a given strain. Incubation of whole cells or cell fractions with a labeled (e.g. radioactively) penicillin followed by sodium dodecyl sulfate (SDS)–polyacryl-amide gel electrophoresis results in a species-specific PBP pattern.

IV. ARCHITECTURE OF THE MUREIN SACCULUS

A. Structure of murein sacculi

The cell wall can be isolated in the shape of the intact murein sacculus by boiling cells in 4% SDS. When inspected by electron microscopy the final sacculi preparation shows empty bag-shaped structures that reflect the shape of the cells from which they have been isolated. The murein structure (i.e. its composition of muropeptide subunits) can be analyzed by a complete hydrolysis of isolated murein sacculi using a muramidase, followed by separation of the products by reversed-phase high-pressure liquid chromatography.

Sterochemical calculations and X-ray diffraction data indicate that the disaccharide units in the strands are twisted in relation to one another, forming a four- to five-fold helix structure. As a result, the peptide moieties protrude alternatingly upwards, to the left, downwards, right, and so forth. This arrangement allows a strand to be cross-linked to both neighboring strands in one layer and strands in an upper and lower level, thus forming a perfect three-dimensional framework. The average length of the glycan strands can be calculated when the total number of disaccharide subunits and the number of reducing ends are known. For gram-positive bacteria, lengths between 40 and 80 units have been observed. The average length of the glycan strands in the gram-negative *E. coli* was found to be approximately 21 disaccharide units. For *E. coli*, the length distribution could be determined by separating the glycan strands that were released from the sacculus by amidase treatment. The majority (70%) of the glycans were found to be short, with a length of approximately seven to nine disaccharide units.

The orientation of the glycan strands in the murein sacculus is still a matter of debate. In the case of rod-shaped bacteria, the glycan strands could either be running along the shorter circumference or arranged parallel to the long axis of the cell. A rhombic, "Chinese finger puzzle-like pattern" can also be envisaged, and even a completely unordered structure cannot be discounted. Despite the lack of clear-cut experimental data, it is tempting to speculate that some kind of order must exist in order to facilitate (assist) the ordered growth of the shape-maintaining structure of the bacterium.

The murein structure is endowed with a great degree of flexibility. In particular, the peptide bridges can be stretched by a factor of four. By contrast, the glycan strands are quite stiff and show almost no elasticity. As a consequence, the sacculus can increase in surface when under stress and shrink when relaxing, as is the case for isolated murein sacculi.

B. Barrier function of murein

Because of the latticework of murein the sacculus represents a molecular sieve for larger compounds. The smallest mesh, also called a tessera, in a systematically constructed murein latticework would have a length of a peptide bridge and a width of approximately four disaccharide units. Thus, in the case of *E. coli* it measures approximately $1\text{–}4 \times 4\text{–}5$ nm. The meshes in the net may not just form rectangles; due to the tension in the wall, they may be stretched into a honeycomb-like hexagon. The exclusion limit for the passage of molecules across the net has been determined *in vitro* to be approximately 50–60 kDa and to be almost the same for the multilayered murein of gram-positive and the thin murein of gram-negative bacteria. For bulky proteins with greater mass, murein is an effective barrier and therefore a localized opening of the murein net is a prerequisite for such molecules to pass through the sacculus. Consequently, the participation of specific murein hydrolases has been proposed for processes such as the export of bulky proteins (i.e. pili and flagellar assembly) and transfer of DNA during conjugation.

V. BACTERIAL CELL WALL GROWTH

A. Cell wall metabolizing enzymes

The vast number of enzymes involved in murein metabolism reflects the complexity and importance of growth of the cell wall in bacteria. Different and distinct processes are involved. First, the biosynthetic pathway leads to the formation of activated murein precursor molecules. Second, the precursors are inserted into the preexisting murein sacculus. Due to the lattice structure of murein, a concerted action of murein hydrolases and synthases is needed in order to enlarge the surface of the sacculus. Depending on the mechanism employed, many bacteria release surprisingly large quantities of turnover products during growth (see Section III.D). Therefore, a third set of enzymes are specifically involved in degrading and trimming these valuable murein turnover products into structures suitable for recycling for *de novo* murein synthesis. Further complexity is added by the performance of two fundamental processes during growth of the bacterial cell wall: the general expansion of the surface (in the case of rod-shaped bacteria elongation of the cylindrical middle part) and subsequent formation of the septum (see Section V.B). Different enzyme systems seem to be responsible for these two processes. Due to the importance of the perfect execution of cell wall growth and division for cell viability, regulatory enzymes, backup enzymes, and repair enzymes are likely to be involved.

Specific enzymes exist that can cleave all covalent bonds in murein (indicated in Fig. 19.7). Glycosylases have only two specificities. *N*-acetylglucosaminidases cleave the β-1,-4 glycosidic bond between GlcNAc and MurNAc. Muramidases (lysozymes) split the β-1,-4 glycosidic bond between MurNAc and GlcNAc. A unique type of muramidase, called lytic transglycosylase, combines the cleavage of the glycosidic bond with a concomitant formation of a 1-6-anhydro ring at the muramic acid residue (see Fig. 19.3).

Because so many different peptide bonds are present in murein, the number of peptidases is correspondingly high. *N*-acetylmuramyl-L-alanine amidases cleave the amide bond between the lactyl group of the MurNAc and the L-Ala of the peptide side chain. DD-Endopeptidases specifically hydrolyze the bridging peptide bond between the D-Ala of one peptide and the D-center of the dibasic amino acid of another peptide. Whereas DD-endopeptidases are penicillin sensitive (see Section III.C), LD-endopeptidases that specifically hydrolyze the LD-peptide bonds are not inhibited by most β-lactams. DD-Carboxypeptidases remove the terminal D-Ala residue from pentapeptides and LD-carboxypeptidases split off the D-Ala residue in position 4. DD-Carboxypeptidases are sensitive towards β-lactam antibiotics however, LD-carboxypeptidases are inhibited only by β-lactams that carry a D-amino acid in their side chains, such as nocardicin A and cephalosprin C.

With the exception of the carboxypeptidases, murein hydrolases that cleave bonds in the murein sacculus are potentially autolytic enzymes and are thus also referred to as autolysins. Although potentially suicidal, these enzymes are essential for growth. Cleavage of bonds in the pre-existing murein net by hydrolases that allows the insertion of new subunits is a prerequisite for the enlargement of the latticework. In addition, cell separation depends on the splitting of the murein septum by murein hydrolases. It is this group of autolysins that is responsible for antibiotic-induced bacteriolysis. The mechanisms that control the murein hydrolases are not fully understood.

B. Growth and division of the murein sacculus

Growth and division of the murein sacculus are a risky enterprise since the stress-bearing structure that is essential for the cell's integrity has to be enlarged and split into two daughter sacculi. These operations have to be executed while maintaining both the mechanical strength of the wall and the specific shape of the cell. Whereas shape maintenance remains poorly understood, the mechanical stability of the wall during growth is thought to be preserved by a mechanism that enlarges the sacculus by an inside-to-outside growth mechanism (see animation in *http://www.eb.tuebingen.mpg.de/papers/hoeltje/*). Accordingly, new material is first hooked in a relaxed state underneath the existing stress-bearing layers of the wall and, in a second step, is exposed to stress by the cleavage of critical bonds in the old material; this is a strategy called "make-before-break." This mechanism may even result in the release of old material from the sacculus, a phenomenon known as murein turnover. Growth of the thin, mostly monolayered murein of gram-negative bacteria is a far more delicate process. Therefore, it very likely also follows the safe make-before-break strategy. One model proposes that first a murein triplet (i.e. three cross-linked glycan strands) is covalently attached to the cross bridges on both sides of a so-called docking strand of the murein layer under tension. Specific removal of the docking strand by the action of murein hydrolases provokes insertion of the murein triplet into the stress-bearing sacculus.

Growth of spherical bacteria (i.e. cocci) occurs exclusively at the equator of the cell, the future site of cell division (Fig. 19.9). The role of the FtsZ protein in placing the division site at the midpoint of the cell is discussed elsewhere. New material is added in a sharp growth zone to the leading edge of the nascent cross wall. Splitting and peeling apart of the newly

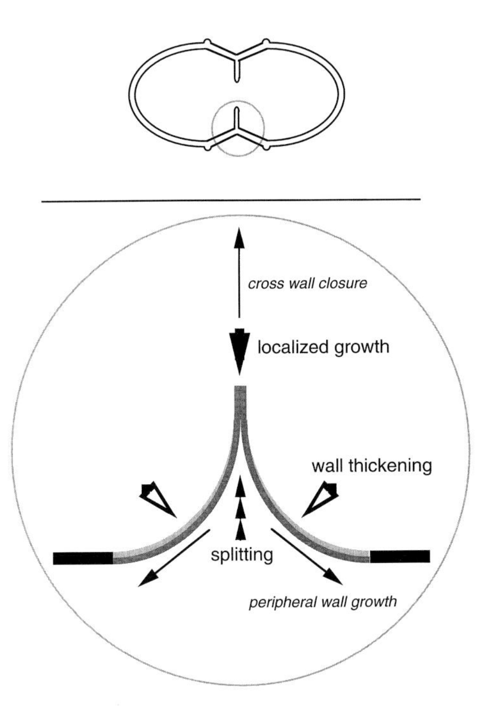

FIGURE 19.9 Schematic representation of the growth modus of cocci. The encircled area of a dividing coccus (top) is shown in detail (bottom). The large, solid arrowhead indicates the addition of new material at the leading edge of the septum. The group of three small arrowheads marks the site of action of the murein hydrolases that split the newly added material. As a result, the material is pushed outwards, thereby increasing the surface of the two daughter hemispheres. The two open arrowheads point to the thickening process of the newly formed peripheral wall. Arrows indicate the two processes of peripheral wall enlarging and centripetal cross-wall extension. Old peptidoglycan is shown in black, newly synthesized material in gray, and the murein added during the thickening process in light gray (modified from Higgins and Shockman, 1976).

added murein results in the material being pushed outwards, thereby causing an increase in the surface of the coccus. The newly synthesized wall is then strengthened by the attachment of additional murein (thickening). Inhibition of the splitting process blocks further pushing outwards of the added material but triggers annularly closing of the cross wall. On completion of the cross wall, a precise cutting of the septum that may involve murosomes (lytic enzymes wrapped into vesicles) allows for cell separation. Murein turnover appears not to occur in gram-positive cocci.

In rod-shaped bacteria the sites of wall growth alternate during the cell cycle (Fig. 19.10). First, the sacculus is elongated while strictly maintaining the diameter. This process occurs as a result of incorporation of new material all over the cylindrical part of the wall, with the poles being metabolically silent. The growth mechanism is also different from that of cocci because it gives rise to an enormous amount of murein turnover. Almost 50% of the murein is released from the sacculus per generation. In the case

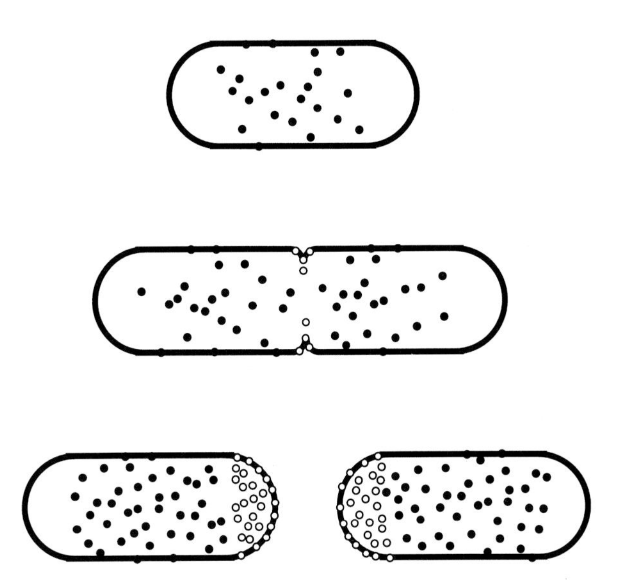

FIGURE 19.10 Schematic representation of the growth modus of rod-shaped bacteria. Solid circles represent the random insertion of new material into the cylindrical part of the cell during cell elongation. Note that the diameter of the cell does not change. Open circles represent incorporation of new material in a zonal growth zone at the site of cell division. As a result, the new polar caps formed after cell division are exclusively made of murein synthesized in the growth zone.

of gram-positive bacteria the material newly added to the cylindrical part of the rod is passed on step by step from the inner layers to the outermost layers from where it is finally removed by murein hydrolases. In the case of gram-negative bacteria, it may be the release of the docking strands that is responsible for murein turnover. Septum formation involves a switch to localized incorporation of new material at the equator of the cell, which is defined by the FtsZ protein that forms a ring structure at the site of cell division.

ACKNOWLEDGMENTS

I thank David Edwards for a critical reading of the manuscript.

BIBLIOGRAPHY

Archibald, A. R., Hancock, I. C., and Harwood, C. R. (1993). Cell envelope. *In* "*Bacillus subtilis* and Other Gram-Positive Bacteria" (A. L. Sonensheim, J. A. Hoch, and R. Losick, eds.). ASM Press, Washington, DC.

Ghuysen, J.-M., and Hakenbeck, R. (1994). Bacterial cell wall. *In* "New Comprehensive Biochemistry", Vol. 27 (A. Neuberger and L. L. M. van Deenen, eds.). Elsevier, Amsterdam.

Higgins, M. L., and Shockman, G. D. (1976). Study of a cycle of cell wall assembly in *Streptococcus faecalis* by three-dimensional reconstructions of thin sections of cells. *J. Bacteriol.* **127**, 1346–1358.

Höltje, J.-V. (1998). Growth of the stress-bearing and shape-maintaining murein sacculus of *Escherichia coli*. *Microbiol. Mol. Biol. Rev.* **62**, 181–203.

Koch, A. L. (1995). "Bacterial Growth and Form." Chapman & Hall, New York.

Nanninga, N. (1998). Morphogenesis of *Escherichia coli*. *Microbiol. Mol. Biol. Rev.* **62**, 110–129.

Park, J. T. (1996). The murein sacculus. *In* "*Escherichia coli* and *Salmonella*" (F. C. Neidhardt *et al.*, eds.), pp. 48–57. ASM Press, Washington, DC.

Rogers, H. J., Perkins, H. R., and Ward, J. B. (1980). "Microbial Cell Walls and Membranes." Chapman & Hall, London.

Schleifer, K. H., and Kandler, O. (1972). Peptidoglycan types of bacterial cell walls and their taxonomic implications. *Bacteriol. Rev.* **36**, 407–477.

Shockman, G. D., and Barret, J. F. (1983). Structure, function and assembly of cell walls of gram-positive bacteria. *Annu. Rev. Microbiol.* **37**, 501–527.

van Heijenoort, J. (2001). Formation of the glycan chains in the synthesis of bacterial peptidoglycan. *Glycobiology* **11**, 25R–36R.

Vollmer, W., and Höltje, J.-V. (2001). Morphogenesis of *Escherichia coli*. *Curr. Opin. Microbiol.* **4**, 625–633.

WEBSITE

The *E. coli* Cell Envelope Protein Data Collection
http://www.cf.ac.uk/biosi/staff/ehrmann/tools/ecce/ecce.htm

20

Chemotaxis

Jeff Stock and Sandra Da Re
Princeton University

GLOSSARY

adaptation The return to a preset behavioral state following a response to an altered environmental condition.

attractant A chemical that causes a positive chemotaxis response.

excitation A behavioral response.

information A significant perturbation in a signal transduction system.

intelligence The ability to respond successfully to a new situation.

learning The acquisition of altered sensibilities from previous experience.

memory An internal record of past experience.

receptor A protein that specifically interacts with a particular stimulus to generate a signal that leads to a cellular response.

repellent A chemical that causes a negative chemotaxis response.

sensing The acquisition of information.

Chemotaxis in microbiology refers to the migration of cells toward attractant chemicals or away from repellents. Virtually every motile organism exhibits some type of chemotaxis. The chemotaxis responses of eukaryotic microorganisms proceed by mechanisms that are shared by all cells in the eukaryotic kingdom and generally involve the regulation of microtubule- and/or microfilament-based cytoskeletal elements. In this article, we will be concerned only with bacterial chemotaxis.

All bacteria share a conserved set of just six different regulatory proteins that serve to direct cell motion toward favorable environmental conditions. The same regulatory system operates irrespective of whether motility involves one or several flagella, or whether it occurs by a mechanism such as gliding motility that does not involve flagella. The same basic system mediates responses to a wide range of different chemicals including nutrients such as amino acids, peptides, and sugars (which are usually attractants) and toxic compounds such as phenol and acid (generally repellents). The same proteins also mediate responses to oxygen (aerotaxis), temperature (thermotaxis), osmotic pressure (osmotaxis), and light (phototaxis). As a bacterium moves it continuously monitors a spectrum of sensory inputs and uses this information to direct motion toward conditions that are optimal for growth and survival. To accomplish this task, the chemotaxis system has developed molecular correlates of processes such as memory and learning that are widely associated with sensory motor regulation by higher neural systems.

I. RESPONSE STRATEGY

A. Biased random walk

In a constant environment, motile bacteria generally move in a random walk of straight runs punctuated by brief periods of reversal that serve to randomize the direction of the next run. The chemotaxis system

The Desk Encyclopedia of Microbiology
ISBN: 0-12-621361-5

functions by controlling the probability of a reversal. If, during a run, the system determines that conditions are improving, then it sends a signal to the motor that suppresses reversals so that the cell tends to keep moving in the preferred direction. If, on the other hand, the system determines that conditions are getting worse, then it sends a signal for the motor to change direction. The effect is to bias the random walk so that cells tend to migrate toward attractants and away from repellents (Fig. 20.1). Thus, bacterial chemotaxis is effected by the simple strategy of using environmental cues to modulate the probability of random changes in direction. By using this mechanism, individual cells never have to determine in which direction they want to move. Instead, they simply determine whether they want to continue on course or change direction. The biased random walk strategy is essential to bacterial chemotaxis because it provides a mechanism whereby bacteria can direct their motion despite the fact that bacterial cells are far too small to have a sense of direction.

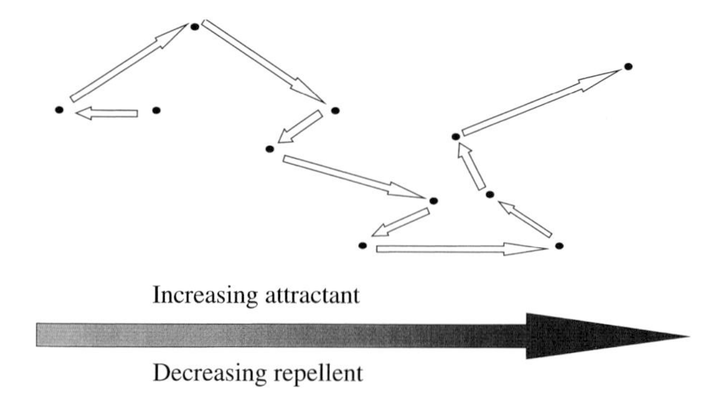

FIGURE 20.1. Chemotaxis is accomplished by a biased random walk mechanism. Bacterial swimming behavior involves a series of runs (indicated by arrows) punctuated by motor reversals that randomize the direction of the subsequent run. Cells migrate toward attractants and away from repellents by increasing their average run lengths in the preferred direction.

B. Temporal sensing and memory

Bacterial cells are generally only a few micrometers long. This is too small to possibly measure differences in attractant or repellent concentration over the length of their bodies. In the early 1970s, through the work of Macnab, Koshland, Berg, and others, it was shown that bacteria solve this problem by sensing changes in attractant and repellent concentration in time rather than in space. In other words, chemotaxis depends on a temporal rather than a spatial sensing mechanism. As a cell moves it constantly compares its current surroundings to those it has experienced previously. If the comparison is favorable, the cell tends to keep going; if not, it tends to change direction. This mechanism implies a memory function whereby the present can be compared with the past to determine whether conditions are getting better or worse as time (and movement in a given direction) proceeds.

C. Excitation and adaptation

One of the ways it was shown that bacterial chemotaxis works by a temporal rather than spatial sensing mechanism was to suddenly transfer a population of randomly moving bacteria from one spatially uniform environment into another that contained a uniform distribution of an attractant or repellent chemical. In this type of experiment there are no spatial gradients; cells are exposed only to a temporal change in their environment. As expected from a temporal sensing mechanism, when cells are suddenly exposed to attractants the entire population determines that it is on a good course and the tendency to change direction is uniformly suppressed (Fig. 20.2). The opposite effect is seen when cells are suddenly exposed to repellents. All the individuals in the population, despite the fact that they are moving many different ways, suddenly determine that they need to change the direction of their motion. This result clearly shows that bacteria must have a way of comparing the past with the present—they must have memory.

After a period of time, bacteria that have been transferred to a new environment gradually adapt so that their behavior returns precisely to the same random walk as that before they were exposed to the attractant or repellent stimulus. This occurs despite the fact that the attractant or repellent is still present. Thus, bacteria do not respond to absolute concentrations of attractant and repellent chemicals. They respond only to changes.

There is a close relationship between memory and adaptation. If one moves a population of bacteria that has adapted to an environment with an attractant back to an environment lacking attractant, the cells think they are moving in a bad direction so they all change course as if they had been exposed to a repellent. The opposite happens with cells that are adapted to a repellent. Thus, an increase in attractant concentration is equivalent to a decrease in repellent concentration and vice versa.

In bacterial chemotaxis, the sense and degree of excitation in response to a new place in time are only determined in relation to the memory of the old one, with the memory for the past environment being set by the process of adapting to it. In effect, there must be two core mechanisms at work in chemotaxis: an excitation mechanism that controls the probability of a motor reversal and an adaptation mechanism that

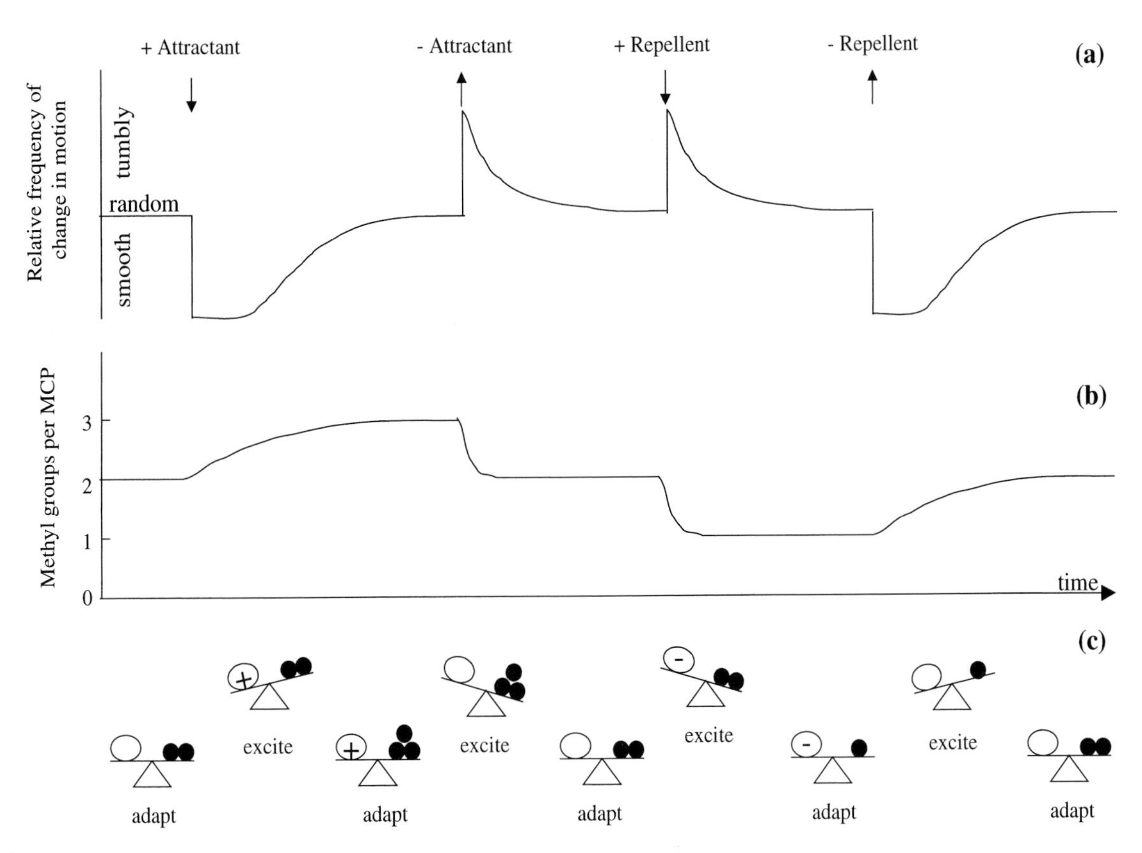

FIGURE 20.2 Excitation and adaptation in bacterial chemotaxis. (a) Addition of attractant causes cells to continue swimming smoothly without changing direction until they adapt back to their prestimulus random behavior, removal of attractant causes cells to frequently change direction or tumble until adaptation restores random behavior, adding repellent causes the same tumbling response as removing attractant, and removing repellent causes the same swimming response as adding attractant. (b) In the absence of attractants or repellents a typical MCP has approximately two of four possible methylated glutamates, addition of a saturating attractant stimulus causes addition of approximately one methyl group, removal of the attractant causes loss of this group, addition of a saturating repellent stimulus causes loss of one more methyl groups, and removal of repellent causes addition of a methyl group. (c) MCP signaling depends on the balance between stimulus (O, no stimulus; +, attractant; −, repellent) and the level of methylation (●, one methyl group; ● ●, two methyl groups, etc.). A sudden change in stimulus causes an imbalance that leads to excitation, and the level of methylation then changes to restore the balance.

modulates the sense and degree of excitation with respect to a preset default value.

II. GENETICS OF BACTERIAL BEHAVIOR

A. Chemotaxis mutants

The most common strategy that has been used to isolate mutants that are defective in chemotaxis has involved selecting for cells that cannot swarm from a colony inoculated into the center of a dish filled with semisolid nutrient agar. Chemotactic cells form a colony at the point where they are initially inoculated, and the growing cells consume nutrients in the culture media creating attractant gradients that cause them to swarm outward from the center. Mutant cells that are deficient in chemotaxis are left behind at the center. This strategy produces several different classes of mutant strains. By far the most common are strains that are not motile. In bacteria whose motility depends on flagella, these nonmotile strains can be subdivided into two classes: Fla mutants, which have lost the ability to make flagella, and Mot mutants, which make flagella that are paralyzed. Mutant strains that are unable to swarm but are fully motile are categorized into two additional subclasses: Che mutants, which are generally nonchemotactic, and blind mutants, which are unable to swarm in semisolid agar with one type of nutrient but can swarm normally in agar that contains other nutrients.

B. Genetic analysis of *Escherichia coli* chemoreceptors

It was through the selection and characterization of blind mutants from *E. coli* that Julius Adler and colleagues first demonstrated that chemosensing in bacteria is mediated by specific receptors in the cell envelope rather than by some other mechanism such as nutrient utilization. The first chemoreceptors to be identified genetically were the galactose and ribose binding proteins that had previously been shown to function in the transport of their respective sugar ligands. The selection of ribose and galactose transport mutants with normal chemosensing abilities established that uptake and metabolism were not required for chemotaxis. Further genetic analysis indicated that ribose and galactose sensing required another component, termed Trg (taxis to ribose and galactose). Early work on the ribose and galactose receptors revealed another important aspect of chemotaxis—the possibility that bacterial cells could exhibit a simple form of learning. The galactose and ribose receptors are specifically induced by growth in the presence of galactose and ribose, respectively. Thus, whereas a naive cell is unable to respond to either sugar, once it is allowed to grow on ribose or galactose the corresponding receptor is induced and the cell has now learned to respond to that sugar.

In addition to the sugar receptors, Adler's group identified two *E. coli* genes, *tar* and *tsr*, that were required for responses to amino acids and several repellents. Tar was required for sensing of the attractants aspartate, glutamate, and maltose and the repellents cobalt and nickel. The maltose response also required an inducible maltose binding protein similar to the ribose and galactose binding proteins that had been shown to be mediated by Trg. Tsr was required for chemotaxis to serine, alanine, and several other amino acids and repellents. In contrast to the genes that encode binding proteins, *tar, tsr,* and *trg* are expressed in the same regulon as the flagellar genes of *E. coli* so that as long as a cell is motile it can sense aspartate and serine.

C. Che genes

The principal task of analyzing the *E. coli* Che mutants fell to a former colleague of Adler's, John S. Parkinson, who determined that there were two major Che complementation groups in *E. coli*, designated *cheA* and *cheB*. Later studies by Parkinson and others established that the CheA locus was composed of two genes (*cheA* and *cheW*), whereas the CheB locus was composed of four genes (*cheR, cheB, cheY,* and *cheZ*). Strains defective in *cheA, cheW, cheY,* or *cheR* exhibited a smooth swimming phenotype, never changing their direction of motion. In contrast, *cheB* and *cheZ* mutants had a constantly changing, tumbly pattern of swimming behavior. Whereas the ability of *cheB* and *cheZ* mutants to change direction was still suppressed by the addition of attractants, and *cheR* mutants could still reverse in response to repellent stimuli; *cheW, cheA,* and *cheY* mutants were completely unresponsive. From these results Parkinson was able to conclude that the CheW, CheA, and CheY proteins were essential for excitation, whereas CheR, CheB, and CheZ were involved in adaptation (for a summary of the Che genes and their protein products see Table 20.1). The fact that mutants defective in excitation were invariably smooth swimming suggested that the excitation mechanism produced a signal that caused a change in the direction of motion, and that in the absence of this hypothetical signal the cell would rarely, if ever, change its direction.

TABLE 20.1 *E. coli* che genes

Gene	Protein M_r (kDa)	Function
cheR	32	Methylation of MCPs
cheB	36	Demethylation of MCPs
cheW	18	Coupling CheA to MCPs
cheA	73	Histidine kinase
cheY	14	CheY-P binds to flagellar switch to cause change in swimming direction
cheZ	24	CheY-P phosphatase

III. ROLE OF PROTEIN METHYLATION IN ADAPTATION

A. Methionine requirement for chemotaxis

One of the most important discoveries from Adler's pioneering work on the *E. coli* chemotaxis system was the serendipitous finding that methionine was required for chemotaxis. In his initial characterization of *E. coli* chemotaxis, Adler employed an assay that had first been developed in the late nineteenth century by the great German microbiologist and botanist, Pfeffer. This method, called a capillary assay, simply involves placing the tip of a glass capillary tube that contains an attractant chemical into a suspension of bacteria. As the attractant diffuses from the capillary tip, an attractant gradient is established which the cells follow up into the capillary tube. After about 1 hr the capillary is withdrawn and the bacteria inside are counted to provide a measure of the chemotaxis response. Unlike with swarming on semisolid agar, capillary assays do not require cell growth, and they are generally performed with cells suspended in a defined buffer solution. Adler observed that when he performed this type of assay with a mutant *E. coli* strain that required methionine for growth, the chemotaxis responses to attractants were generally depressed. Among all the amino acids, this effect was specific for methionine. Further analysis showed that as cells became starved for methionine they lost the ability to adapt so that addition of attractants such as aspartate or serine caused a smooth swimming behavior similar to that observed with *cheR* mutants.

B. Methylaccepting chemotaxis proteins

Subsequent studies established that the effect of methionine depletion stemmed from a requirement for the universal methyl donor, *S*-adenosylmethionine (AdoMet), which is produced from methionine and ATP through the action of AdoMet synthase:

$$\text{ATP} + \text{methionine} \rightarrow \text{AdoMet} + \text{PPi} + \text{Pi}$$

AdoMet is required for the methylation of a wide range of different macromolecules, including proteins, DNA, and RNA, as well as numerous different small molecules. It was shown that in chemotaxis the requirement for AdoMet is to methylate a set of ~60-kDa membrane proteins that were termed methylaccepting chemotaxis proteins (MCPs). These proteins were identified with the products of genes such as *tar, tsr*, and *trg* that had been implicated in chemosensing. In fact, Tar and Tsr were shown to bind aspartate and serine, respectively, and to act directly as the membrane receptors for these attractants. Trg, on the other hand, acts indireclty as a receptor for ribose and galactose through interactions with the corresponding periplasmic binding proteins. Furthermore, maltose is detected by Tar via the periplasmic maltose binding protein. Thus, each MCP can detect several different stimuli either by binding a stimulatory ligand directly or through indirect interactions that are mediated by periplasmic binding proteins.

Methylation studies, analyses with anti-MCP antibodies, and, most important, DNA sequencing have shown that the MCPs are a large and highly conserved family of proteins that are invariably associated with bacterial chemosensing. There are five different MCPs encoded in the *E. coli* genome (Fig. 20.3), including Tar, Tsr, and Trg as well as a sensor for cellular redox potential termed Aer that is responsible for aerotaxis and a receptor that mediates responses to dipeptides termed Tap. Chemotaxis systems in other species of bacteria have an equivalent or larger number of different MCPs.

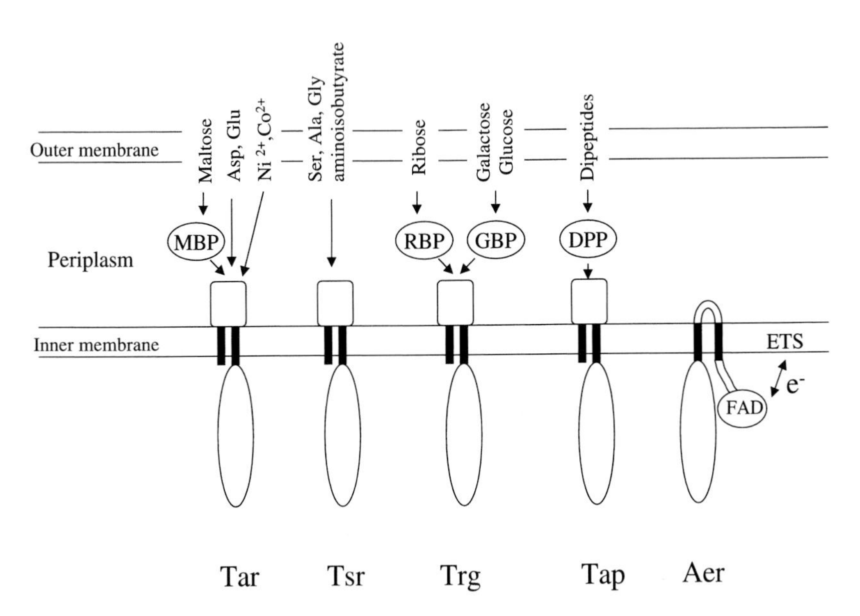

FIGURE 20.3 *Escherichia coli* chemoreceptors. There are five MCP receptors in *E. coli:* Tar, Tsr, Trg, Tap, and Aer. These transmembrane proteins either bind stimulatory ligands directly or interact with ligand-bound periplasmic binding proteins (MBP, maltose binding protein; RBP, ribose binding protein; GBP, galactose binding protein; DPP, dipeptide binding protein). Aer differs from other MCPs in that it has an intracytoplasmic sensing domain with an associated flavin cofactor, FAD. It is thought that Aer senses cellular oxidative potential through redox interactions with the electron transport system in the membrane.

C. Structural and functional organization of sensing and signaling domains of receptor MCPs

The MCPs are the principal sensory receptors of the bacterial chemotaxis system. They have a structural organization, membrane topology, and mode of function that is typical of type I receptors in all cells, including important vertebrate type I receptors such as the insulin, growth hormone, and cytokine receptors. In recent years, the Tar protein from *Salmonella typhimurium* has emerged as both the archetypal MCP and as a model to understand general principles of type I receptor function. Tar has the typical membrane topology of a type I receptor with an N-terminal extracytoplasmic sensing domain connected via a hydrophobic membrane-spanning sequence to an intracellular signaling domain. The sensing and signaling domains can function independent of the membrane and independent of one another. Most MCPs have this structural organization. Sequence comparisons indicate that, as one might expect, the extracytoplasmic sensing domains tend to be highly variable, whereas the cytoplasmic signaling domains, which interact with the Che proteins, are highly conserved. In fact, the tools of genetic engineering have been used to construct several different hybrid receptors with one MCP's sensing domain connected to another's signaling domain. In every case the hybrids exhibit sensory specificities equivalent to those of the MCP that contributed the N-terminal portion.

The sensory and signaling domains can also be produced as independent soluble protein fragments. This approach has been used with many type I receptors to produce protein fragments that are free of the membrane and can therefore be much more easily crystallized for X-ray diffraction studies. Determination of the X-ray crystal structure of the sensing domain of Tar in the presence and absence of aspartate revealed a dimer of two α-helical bundles with aspartate binding at the subunit interface.

The structure of the signaling domain has not been defined in detail, but it too is predominantly α-helical. The signaling domain is composed of a highly conserved central region that binds the CheW and CheA proteins and thereby connects the receptor to the remainder of the chemotaxis signal transduction system. This region is flanked on both sides by methylated α-helices that together contain four or more potential sites of glutamate methylation and demethylation. Attractants cause increases in the level of methylation and repellents cause decreases. These changes are responsible for adaptation to attractant and repellent stimuli (Fig. 20.2).

D. Receptor methylation enzymology

The MCP receptors are methyl esterified at several specific glutamate residues. The methylation reaction is catalyzed by an AdoMet-dependent methyltransferase encoded by the *cheR* gene. Receptor methyl groups are removed through the action of a specific methyl esterase encoded by the *cheB* gene. CheR and CheB are both soluble monomeric proteins. Their structures have recently been determined by X-ray crystallographic methods. CheR is tethered to Tar and Tsr via a tight interaction with the C-terminal four amino acids, which are identical in these two MCPs but are absent in Trg, Tap, and Aer. Tar and Tsr are present in cells at about 10-fold higher levels than Trg, Tap, and Aer, and considerable evidence suggests that the latter, so-called minor receptors, function in higher order complexes with the major receptors, Tar and Tsr. CheR tethering to major receptors puts the enzyme in position to methylate the associated minor receptors.

The active site of CheB contains a Ser–His–Asp catalytic triad that is characteristic of serine hydrolases. An N-terminal regulatory domain occludes the active site so that the enzyme is relatively inactive. CheB is activated to remove receptor methyl groups by the same signal that causes a change in the direction of cell movement in response to repellent stimuli. This provides a feedback mechanism that contributes to the adaptive phase of the chemotaxis response. Thus, repellent addition or attractant removal produce an excitatory signal to change direction. At the same time, this signal activates CheB, leading to a rapid decrease in the level of methylation that causes adaptation. The converse is true with attractant addition or repellent removal. It is as if the receptor signaling system functions as a balance between the effects of stimulatory ligands and methylation (Fig. 20.2). Addition of attractant or repellent offsets the balance to produce a positive or negative excitatory signal, and changes in methylation restore the balance to effect adaptation.

IV. MECHANISM OF SIGNAL TRANSDUCTION IN CHEMOTAXIS

A. Receptor–CheW–CheA signaling complexes

The CheA protein is a kinase that binds ATP and catalyzes the phosphorylation of one of its own histidine residues. The rate of autophosphorylation of the isolated CheA protein is very slow. The physiologically relevant form of CheA seems to be in a stable complex with CheW and the MCP receptors. The rate of CheA autophosphorylation in these receptor-signaling complexes can be elevated at least 100-fold or completely inhibited depending on the level of receptor methylation and the binding of stimulatory ligands. Attractants such as serine or aspartate have an inhibitory effect, whereas increased levels of methylation cause dramatic increases in kinase activity. Because of the dimeric nature of the receptor sensing domain and the fact that CheA is a dimer, it was assumed that the receptor–CheW–CheA signaling complex had a 2:2:2 stoichiometry. Recent results indicate a much more complex architecture, with the thousands of receptors in a cell clustering together in a higher order complex with CheW and CheA (Levit *et al.*, 1998). It has been hypothesized that packing interactions within these signaling arrays function to control kinase activity in response to the binding of stimulatory ligands or changes in the level of receptor methylation.

B. Motor regulation and feedback control

The level of phosphorylation of the CheY protein controls the probability that a cell will change its direction of motion. CheY is a 14-kDa monomeric enzyme that catalyzes the transfer of a phosphoryl group from the phosphohistidine in CheA to one of its own aspartate residues. CheY phosphorylation induces a conformational change in the protein that causes it to bind to switching proteins at the flagellar motor. Repellent-induced increases in the rate of CheA phosphorylation produce elevated levels of phospho-CheY that bind to the motor to enhance the probability of motor reversal. Phospho-CheY spontaneously dephosphorylates to terminate the response. In *E. coli* this autophosphatase reaction is dramatically enhanced by the CheZ protein. Thus, addition of attractants inhibits CheA autophosphorylation, and CheZ activity leads to a rapid decrease in the level of phospho-CheY, a reduction in the level of phospho-CheY bound to the motor, and a decrease in the probability that a cell will change direction.

The CheY protein is homologous to the regulatory domain of the CheB protein and, like CheY, the regulatory domain of CheB acts to transfer phosphoryl groups from the phosphohistidine in CheA to one of its own aspartate residues. Phosphorylation of CheB causes a dramatic increase in demethylation activity that leads to a decrease in receptor methylation and a concomitant decrease in the rate of CheA autophosphorylation. Thus, the same mechanism that acts to produce a motor response feeds back to cause adaptation. The signal transduction mechanism that mediates *E. coli* chemotaxis is summarized in Fig. 20.4.

V. PHYLOGENETIC VARIATIONS

Most of our understanding of bacterial chemotaxis has come from studies of the system in *E. coli*. Other motile bacteria that have been investigated have MCPs and all the same Che proteins as those in the *E. coli* system except for CheZ, which has only been found in enterics. There appear to be a number of variations on the *E. coli* scheme, however. Studies in *Bacillus subtilis* indicate that in this species the system is reversed so that attractants activate CheA, and CheY phosphorylation suppresses the tendency for the cell to change direction. The *B. subtilis* system also appears to have additional components that are not found in *E. coli*.

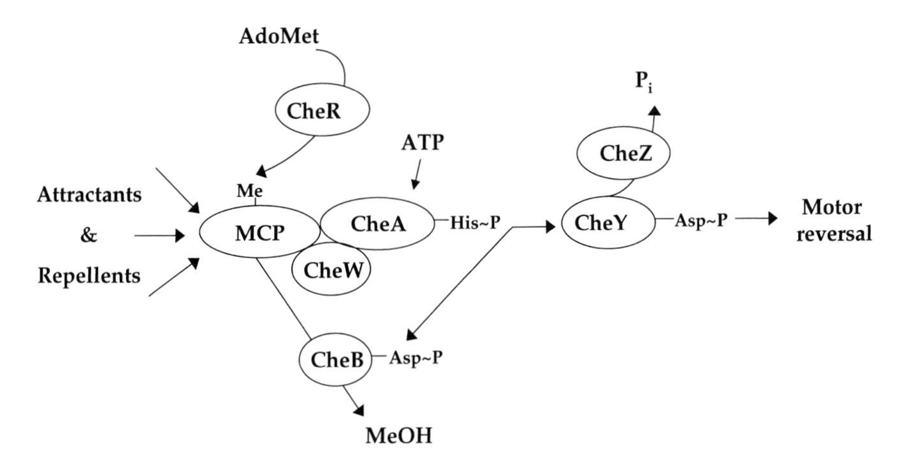

FIGURE 20.4. Biochemical interactions between the Che proteins that mediate chemotaxis responses in *E. coli*.

Many species have several copies of one or more of the chemotaxis genes. In some instances, it is apparent that there are multiple chemotaxis systems functioning in different cell types. The best example of this is provided by *Myxococcus xanthus*, in which different systems operate at different stages of development to control different types of motility. In contrast, two CheY proteins in *Rhodobacter sphaeroides* seem to supply divergent functions in one signal transduction network. One CheY interacts with the motor to control swimming behavior, whereas the other CheY functions as a CheA phosphatase to drain phosphoryl groups out of the system.

The chemotaxis system seems to be eubacterial in origin. There is no evidence for any of the chemotaxis components in eukaryotic cells. Although homologous systems are found in archaea such as *Halobacterium salinarum* and *Archaeoglobus fulgidus*, the sequences of the component proteins are so closely related to those of *B. subtilis* that one can be fairly certain that they originated by lateral transfer from a *B. subtilis* relative.

BIBLIOGRAPHY

Adler, J. (1975). Chemotaxis in bacteria. *Annu. Rev. Biochem.* **44**, 341–356.

Berg, H. C., and Brown, D. A. (1972). Chemotaxis in *Escherichia coli* analysed by three-dimensional tracking. *Nature* **239**, 500–504.

Bren, A., and Eisenbach, M. (2000). How signals are heard during bacterial chemotaxis: protein–protein interactions in sensory signal propagation. *J. Bacteriol.* **182**, 6865–6873.

Djordjevic, S., and Stock, A. M. (1998). Structural analysis of bacterial chemotaxis proteins: Components of a dynamic signaling system. *J. Struct. Biol.* **124**, 189–200.

Levit, M. N., Liu, Y., and Stock, J. B. (1998). Stimulus response coupling in bacterial chemotaxis: Receptor dimers in signalling arrays. *Mol. Microbiol.* **30**, 459–466.

Macnab, R. M., and Koshland, D. E., Jr. (1972). The gradientsensing mechanism in bacterial chemotaxis. *Proc. Natl. Acad. Sci. USA* **69**, 2509–2512.

Milburn, M. V., Prive, G. G., Milligan, D. L., Scott, W. G., Yeh, J., Jancarik, J., Koshland, D. E., Jr., and Kim, S. H. (1991). Three-dimensional structures of the ligand-binding domain of the bacterial aspartate receptor with and without a ligand. *Science* **254**, 1342–1347.

Parkinson, J. S. (1977). Behavioral genetics in bacteria. *Annu. Rev. Genet.* **11**, 397–414.

Parkinson, J. S. (1998). Protein phosphorylation in bacterial chemotaxis. *Cell* **53**, 1–2.

Silversmith, R. E., and Bourret, R. B. (1999). Throwing the switch in bacterial chemotaxis. *Trends Microbiol.* **7**, 16–22.

Springer, M. S., Goy, M. F., and Adler, J. (1979). Protein methylation in behavioural control mechanisms and in signal transduction. *Nature* **280**, 279–284.

Stock, J., and Da Re, S. (2000). Signal transduction: response regulators on and off. *Curr. Biol.* **10**(11), R420–R424.

Stock, J., and Levit, M. (2000). Signal transduction: hair brains in bacterial chemotaxis. *Curr. Biol.* **10**(1), R11–R14.

Stock, J. B., and Surette, M. G. (1996). Chemotaxis. *In "Escherichia coli* and *Salmonella:* Cellular and Molecular Biology" (F. C. Neidhardt, Ed.), pp. 1103–1129. ASM Press, Washington, DC.

WEBSITE

The *E. coli* Cell Envelope Protein Data Collection
http://www.cf.ac.uk/biosi/staff/ehrmann/tools/ecce/ecce.htm

21

Chromosome, Bacterial

Karl Drlica
Public Health Research Institute

Arnold J. Bendich
University of Washington

GLOSSARY

DNA supercoiling A phenomenon occurring in circular, covalently closed, duplex DNA molecules when the number of turns differs from the number found in DNA molecules of the same length but containing an end that can rotate. Supercoiling creates strain in closed DNA molecules. A deficiency of duplex turns generates negative supercoiling; a surplus generates positive supercoiling.

DNA topoisomerases Enzymes that change DNA topology by breaking and rejoining DNA strands. Topoisomerases introduce and remove supercoils, tie and untie knots, and catenate and decatenate circular DNA molecules.

nucleoid A term for the bacterial chromosome when it is in a compact configuration, either inside a cell or as an isolated structure.

origin of replication A location on the chromosome (*oriC*) where initiation of replication occurs. For *Escherichia coli*, *oriC* is approximately 250 nucleotides long, and during initiation it specifically interacts with several proteins to form an initiation complex.

recombination A process in which two DNA molecules are broken and rejoined in such a way that portions of the two molecules are exchanged.

replication fork The point at which duplex DNA separates into two single strands during the process of DNA replication. Associated with replication forks are DNA helicases to separate the strands and DNA polymerases to synthesize new DNA strands.

Bacterial Chromosomes are intracellular repositories of genetic information. In molecular terms, chromosomes are composed of (i) large DNA molecules that store information; (ii) RNA molecules in the process of copying information from specific genes; and (iii) proteins that repair DNA damage, duplicate the DNA, and control patterns of gene expression. Proteins also fold and bend the DNA so it fits in the cell. In many bacteria, chromosomal DNA appears to be circular, but examples of linear chromosomes have also been found. An individual cell can contain multiple identical copies of a single chromosome. Two and three distinct chromosome types have been observed in some species. When cells contain very large plasmids, the distinction between plasmid and chromosome blurs if both carry genes essential for cell growth. Chromosomal DNA has fewer duplex turns than would be found in linear, B-form DNA under the same conditions, and this deficiency of turns places torsional (superhelical) strain on the DNA. The strain lowers energy barriers for strand separation, and in this sense the chromosome is energetically activated. DNA replication begins at a precise point and continues bidirectionally until the two replication forks meet in a region 180° from the origin in circular DNA or until they reach the ends of linear DNA. As DNA is replicated, attached proteins pull the daughter chromosomes apart while other proteins untangle the DNA. Nucleotide sequence analysis of many bacterial genomes reveals that chromosomes have a rich evolutionary history of sequence rearrangements, which are due largely to fragments of DNA moving into and out of chromosomes.

The Desk Encyclopedia of Microbiology
ISBN: 0-12-621361-5

I. HISTORICAL INTRODUCTION

Bacterial chromosomes were discovered much later than their eukaryotic counterparts because they do not exhibit the striking condensation at metaphase that makes eukaryotic chromosomes so easy to visualize. By the 1940s evidence that bacteria undergo spontaneous mutation finally emerged, and this established that bacteria must have mutable genes, the functional elements of chromosomes. At about the same time, Avery and associates uncovered the chemical nature of genetic material with a transformation experiment in which a character for polysaccharide synthesis was transferred from one strain of *Pneumococcus* to another using extracted DNA. At the time, the result was not universally accepted as evidence for genetic exchange, partly because the so-called "transforming principle," DNA, exerted its effect after an unknown number of steps and partly because there was no molecular framework for explaining how DNA could function as genetic material. In 1952, Hershey and Chase announced that phage DNA, not protein, is injected into bacterial cells during infection, and a year later Watson and Crick provided the structural framework for DNA. At that point DNA became widely accepted as the carrier of hereditary information. By 1956, nucleoids, as bacterial chromosomes are sometimes called, could be seen in living cells as discrete, compact structures (a modern view is shown in Fig. 21.1). Gentle extraction methods eventually yielded large, intact DNA molecules, and by the early 1970s it became possible to isolate a compact form of the chromosome for biochemical study (Fig. 21.2). DNA supercoiling was discovered in the mid-1960s, and within a decade enzymes that introduce and remove supercoils had been characterized by Wang, Gellert, and Cozzarelli. The existence of these enzymes, DNA topoisomerases, gave credence to the idea that chromosomal DNA is under torsional tension. During the 1980s, the dynamic, regulated nature of supercoiling emerged as a major structural feature that needed to be considered whenever the activities of the chromosome were discussed. The development of rapid methods for determining nucleotide sequences led in the 1990s to complete sequences for many bacterial genomes. Information from DNA sequence analyses made it likely that all living organisms share a common ancestor, and inferences could be drawn about the nucleotide sequence history of chromosomes.

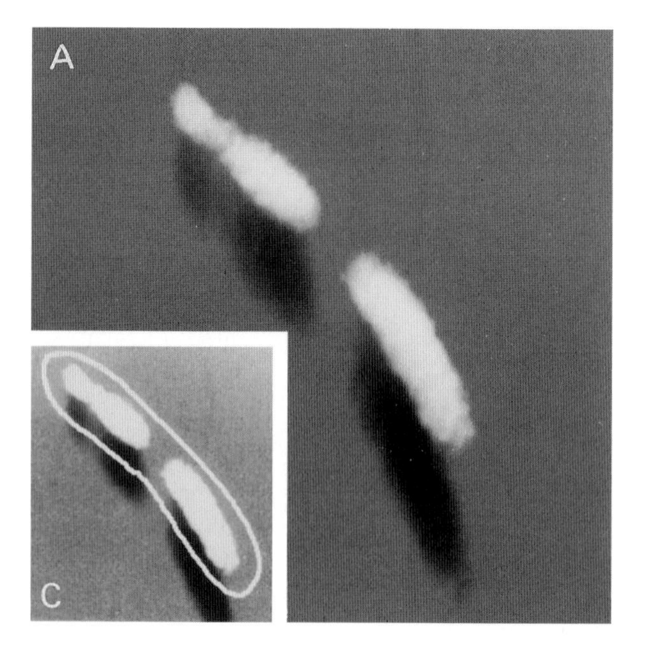

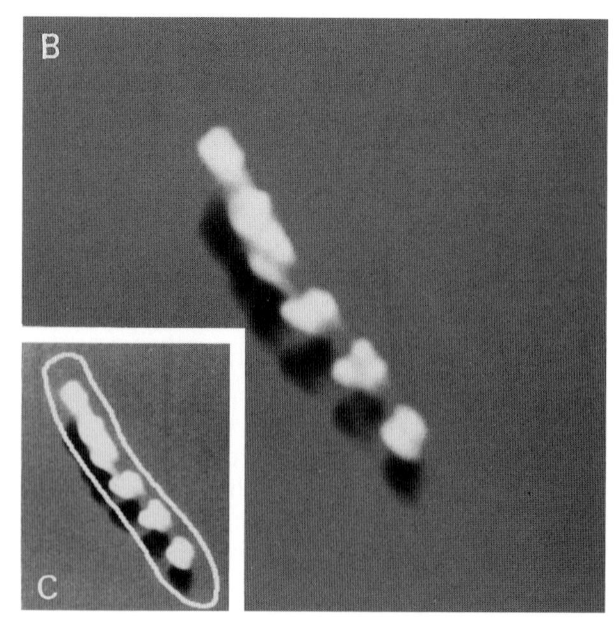

FIGURE 21.1 Bacterial nucleoids. Nucleoids of *Escherichia coli* K-12 were visualized in a confocal scanning laser microscope as developed by G. J. Brankenhoff (1985). Elongated cells were obtained by growth in broth. Then the nucleoids were stained with the DNA-specific fluorochrome DAPI (0.1 μg/ml) added to the growth medium. Under these conditions the stain had no effect on growth. The cells were observed either alive (a) or after fixation with 0.1% osmium tetroxide (b). Since the cell boundary is not easily visualized, it has been sketched in for reference (c). Multiple nucleoids were present because these fast-growing cells contain DNA in a state of multifork replication. In live cells, the nucleoid has a cloud-like appearance and a smooth boundary with the cytoplasm (protuberances, if present, would be smaller than 200 nm). Magnification = $9000\times$ (courtesy of Dr. Conrad Woldringh, Department of Molecular Cell Biology, University of Amsterdam, The Netherlands).

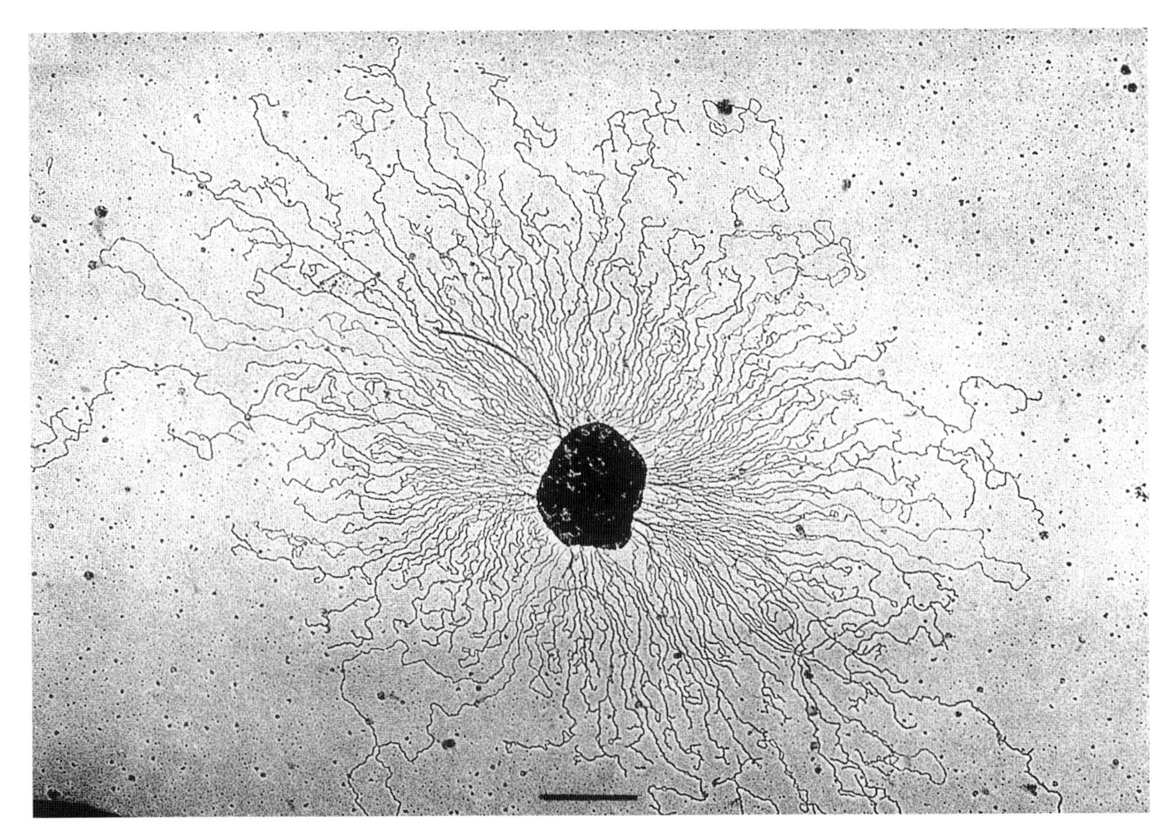

FIGURE 21.2 Isolated bacterial chromosome. Electron micrograph of a purified, surface-spread *E. coli* chromosome. Photograph prepared by R. Kavenoff and O. Ryder; reprinted from *Chromosoma* by permission of the publisher. Bar represents 1 μm.

An emerging theme is the dynamic nature of bacterial chromosomes. In terms of nucleotide sequence, massive gene shuffling has occurred over the course of evolution. With respect to three-dimensional structure, portions of the genome move to particular regions of the cell at specific times during the cell cycle, even as the bulk of the DNA threads through stationary replication forks. At the level of DNA conformation, changes can occur within minutes after alteration of cellular environment. All three types of change are influenced, and in some cases directed, by protein components of chromosomes. In this article, we discuss major concepts concerning chromosome structure and point out unanswered questions. We conclude by describing a speculative model of the chromosome that may help put some of the details in perspective.

II. CHROMOSOME FORM AND NUMBER

Bacterial DNA has been found in both circular and linear forms. For *Escherichia coli*, chromosomal circularity is supported by three lines of evidence. First, circles were observed when radioactively labeled DNA was extracted from cells and then examined by autoradiography. In these experiments almost all the molecules were so tangled that their configurations were unclear, but a few appeared as large circles more than 1 mm in length (cells are only 1 or 2 μm long). This chromosomal length is generally shorter than that expected from determination of the genomic nucleotide sequence. Second, genetic mapping studies are most easily interpreted as if the genes are arranged in a circle, although a linear interpretation is still possible (mapping can be ambiguous since a large linear bacteriophage DNA is known to have a circular genetic map). Third, two replication forks emerge from a single origin of replication and move bidirectionally toward a single terminus located 180° opposite the origin on the circular map. Replication data, which clearly reflect the whole cell population, are the strongest of the three lines of evidence. Conclusive evidence for circularity, which is still lacking, would be visualization of circular images of most of the DNA molecules present.

In 1989, chromosomal DNA molecules of *Borrelia burgdorferi* were found to have a linear form. Linear chromosomes were subsequently observed in *Streptomyces* species, *Rhodococcus fasciens*, and *Agrobacterium tumefaciens*. With *Streptomyces*, DNA ends contain repetitive sequences as well as terminal proteins that prime synthesis complementary to the 3′ end of the DNA. In *B. burgdorferi*, the ends are hairpins that facilitate complete replication. Thus, bacteria have chromosomal ends that function similarly to the telomeres of linear chromosomes in eukaryotic cells.

Many bacteria carry all their genes in a single genetic linkage group, as if they have a single type of chromosome. However, there is a growing list of species in which useful or essential genes are found on two or more chromosomes. The number of large, circular-mapping molecules is two for *Vibrio* species, *Deinococcus radiodurans, Leptospira interrogans, Rhodobacter sphaeroides*, and *Brucella* species, three for *Rhizobium meliloti*, and two to four among isolates of *Burkholderi (Pseudomonas) capecia*. Some *Agrobacterium* species contain one circular- and one linear-mapping chromosome. Thus, the idea that prokaryotes contain only one circular chromosome has been abandoned. Indeed, those with more than one chromosome (genetic linkage group) may constitute a sizable class since the vast majority of bacterial species has yet to be examined.

It is occasionally difficult to distinguish between chromosomes and plasmids since some plasmids are very large and contain genes essential for cell growth. Moreover, some large plasmids integrate into, and excise from, chromosomes. Thus, chromosome number in some species may be variable.

The existence of multiple copies of chromosomal regions, as well as entire chromosomes (multiploidy), is well-known in eukaryotes. In addition, a eukaryotic cell can contain thousands of copies of mitochondrial and chloroplast genomes. Multicopy genes and chromosomes are also common among bacteria. For example, *E. coli* contains approximately 11 genome equivalents per cell when growing rapidly in rich medium, whereas this number is between 1 and 2 during slow growth. In this organism high copy number may facilitate rapid cell doubling (see Section VII). For *Deinococcus radiodurans*, the ploidy levels are 10 during exponential growth and 4 during stationary phase; the corresponding values are about 7 and 3 for *Methanococcus janaschii*. Even slowly growing cells, such as *Borrelia hermsii* (minimum doubling time 8 h), can be multiploid. This bacterium contains 8–11 genome copies when grown *in vitro* and up to 16 copies when grown in mice. *Azotobacter vinlandii* represents a more extreme example. Its ploidy in rich medium increases from 4 to 40 and then to >100 as the culture progresses from early exponential through late exponential to stationary phase. Ploidy level then decreases at the start of a new growth cycle. This spectacular increase in chromosome copy number in *Azotobacter* is not observed with cells grown in minimal medium. For the intracellular symbiont *Buchnera* the mean copy number increases to 240 (the range is 10–600) during development of its aphid host. The current record holder is *Epulopiscium fishelsoni*. Its DNA content varies by four or five orders of magnitude among individuals at different stages of the life cycle. Perhaps the increase represents an example in which DNA is used as a stored nutrient, much as fat and starch are used by other organisms. Another use for genome copies accumulated when nutrients are plentiful could be to supply genomes by dilution to smaller progeny cells when nutrients become limited, illustrating an "engorge now, divide later" strategy of cell proliferation. Regardless of the reason for polyploidy, it is clearly not restricted to eukaryotic cells. Indeed, bacterial and eukaryotic chromosomes can no longer be considered different with respect to form (both types can be linear), ploidy levels (both types can be multiploid), and number of linkage groups (bacteria, which often have one, can have several; eukaryotes, which usually have many, can have only one, as seen with the ant *Myrmecia pilosula*).

III. GENE ARRANGEMENTS

Gene mapping in bacteria was originally based on the ability of an externally derived, genetically marked fragment of DNA to recombine with the homologous region of a recipient's DNA. The frequency with which two nearby markers recombine is approximately proportional to the distance between them. Mutations were collected for a variety of purposes, and characterization usually included determining map position on the chromosome. The resulting genetic maps revealed relationships among genes such as operon clusters, showed orientation preferences that might reflect chromosomal activities, and suggested that some chromosomal information may have been derived from plasmids and phages. The discovery of restriction endonucleases led to a quantum advance in genetic mapping since these enzymes allowed the accurate construction of maps in terms of nucleotide distances. Practical nucleotide sequencing methods, which became available in the late 1970s, are rapidly yielding the complete nucleotide sequences for many bacterial species. Data are being obtained at three levels: (i) the genetic map, with the genes and their map locations correlated with the role of the gene products in cell metabolism, structure, or regulation; (ii) the physical map in terms of locations of restriction sites; and (iii) the nucleotide and corresponding protein amino acid sequences. It is becoming clear that all living organisms probably arose from a common ancestor. Thus, information on nucleotide sequence and gene function in one organism can be applied to many other organisms.

One of the conclusions from genome studies is that large families of genes exist. For example, in *E. coli* there is a family of 80 membrane transport proteins, and half the protein-coding genes of *Mycobacterium tuberculosis* have arisen through duplication. These observations support the idea, which originated from analysis of eukaryotic genomes, that new genes arise from the duplication and modification of old ones.

The conservation of gene structure makes it possible to use nucleotide sequence information for comparison of genetic maps among bacterial species. One of the features revealed is clustering of related genes. For example, the genetic maps of *Bacillus subtilis*, *E. coli*, and *S. typhimurium* show a grouping of many genes for biosynthetic and degradative pathways. Such grouping could be for purposes of coordinated regulation since some adjacent genes produce polycistronic messages. A completely different view of the same data maintains that functionally related genes move horizontally (from one organism to another) as clusters because the products of the genes work well together, increasing the probability of successful transfer. Both ideas are likely to be accurate.

It is becoming increasingly clear that bacterial genomes can be quite malleable, even at the level of large rearrangements. For example, comparison of the genomic maps of *E. coli* and *S. typhimurium* reveals that a large inversion has occurred. Similar compari-sons of *S. typhimurium* LT2 and *S. typhi* Ty2 show that these two genomes differ by at least three inversions. In a study of 21 *P. aeruginosa* isolates, nine inversions were detected.

Many comparisons make it possible to divide bacteria into two groups based on genomic stability. Representatives of the stable genome group are *E. coli*, *S. typhimurium*, and *Halobacterium salinarum*. Highly rearranged genomes are found in *S. typhi*, *Helicobacter pylori*, and *P. aeruginosa*. The reasons for differences among organisms are not clear. Some rearrangements, such as insertions or deletions, may be more deleterious for certain bacteria, or perhaps the opportunity for rearrangements, such as the occurrence of recombinational hot spots, is greater in some organisms than others. The latter explanation appears to be more likely for chloroplast and mitochondrial chromosomes, which in many ways resemble bacterial chromosomes. After approximately 300 million years of evolution, the order of chloroplast genes is highly conserved among most land plants, including mung bean. However, chloroplast gene order is completely scrambled in pea, a plant closely related to beans. Massive rearrangement of genes is also evident when mitochondrial chromosome maps are compared among types of maize. Thus, gene order in organelles appears to have little functional significance, and it can be subject to frequent recombination if the opportunity arises.

As complete genomic nucleotide sequences become increasingly available, new questions will be raised. For example, what is the minimal number of genes required for independent life? Organisms of the genus *Buchnera* have the smallest cellular genomes, probably because many of the needs of these obligate symbionts are supplied by their hosts. Genomic comparison of the intracellular parasite, *Mycoplasma genitalium* with a free-living pathogen, *Hemophilus influenzae*, indicates that the minimal set of genes to sustain a cell may be approximately 250. Nucleotide sequences should also help identify genes involved in pathogenicity by comparison of virulent and avirulent strains of a pathogen. Such an approach has uncovered a "pathogenicity island," a collection of virulence genes, in *H. pylori*. The island is bounded by 31-base pair (bp) direct repeats, as if it had been transferred horizontally into an ancestor of *H. pylori*. In some bacteria, horizontal transfer may have been quite extensive. For example, in *E. coli* as much as 15% of the genome, (700 kbp) may have been acquired from foreign sources such as integrative bacteriophages, transposons in plasmids, and conjugative transposons (genetic elements that cannot replicate independently but cause conjugation, a form of cell-to-cell DNA transfer, to occur).

IV. RECOMBINATION

Intracellular DNA experiences a variety of perturbations that must be repaired to maintain the integrity of the genome and to allow movement of replication forks. Cells have several ways to repair DNA damage, one of which involves recombination (recombination is a process in which DNA molecules are broken and rejoined in such a way that portions of the two molecules are exchanged). Damaged sequences in one molecule can be exchanged for undamaged ones in another (Fig. 21.3). It is now thought that the raison d'être for recombination is its role in DNA repair, a process that may occur thousands of times per cell generation.

Recombination is also involved in DNA rearrangements arising from the pairing of repeated sequences. When the repeats are in direct orientation, duplications and deletions occur; inversions occur when the

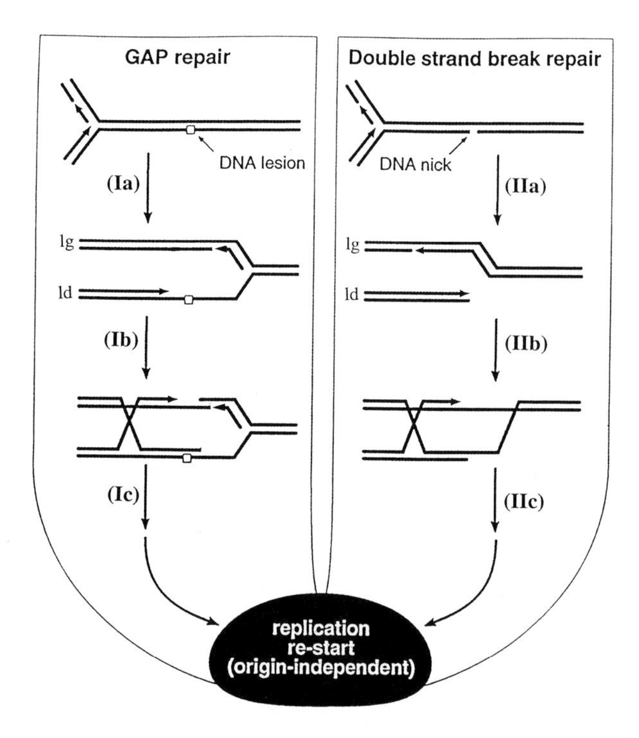

FIGURE 21.3 Major pathways for recombinational DNA repair in bacteria. The principal function of recombination in bacteria is to restart replication forks stalled at sites of DNA damage. Two pathways are depicted. First, gap repair: (1a) A DNA lesion blocks movement of the leading strand (1d), causing disassembly of the replication fork. (Ib) The RecA and RecFOR proteins replace the lesion with parental DNA from the lagging strand (1g) side of the fork by recombination. (Ic) The recombination intermediates are then resolved by the RuvABC and RecG proteins so replication can restart. Second, double-strand break repair: (IIa) A single-strand break (nick) blocks movement of the leading strand (1d), causing disassembly of the replication fork. (IIb) The RecA and RecBCD proteins regenerate the fork by recombination with parental DNA from the lagging strand (1g) side of the replication fork. (IIc) The recombination intermediates are then resolved and repaired by the RuvABC and RecG proteins so replication can restart (adapted from Cox, 1997, *Proc. Natl. Acad. Sci. USA* **94**, 11765, copyright 1997 National Academy of Sciences, USA).

repeat is inverted. In *B. subtilis* a cascade of sequential rearrangements has been identified in which large transpositions and inversions have been attributed to recombination at specific junction points in the chromosome. Several other examples are found with the *rrn* clusters, sets of similar assemblies of ribosomal and transfer RNA genes. Rearrangements at the three *rrn* clusters of *Brucella* are thought to be responsible for the differences in chromosome size and number among species of this genus. Other repeated sequences that facilitate rearrangements are the *rhs* loci (*r*ecombination *h*ot *s*pot), duplicate insertion sequences, and experimentally introduced copies of the transposon Tn10.

A third consequence of recombination is the insertion of genes from mobile elements into chromosomes. These elements, which include transposons, plasmids, bacteriophages, integrons, and pathogenicity islands, move from one cell to another and sometimes from one species to another. In a spectacular example of gene transfer and its evolutionary effects across kingdoms, the acquisition of a symbiosis island of 500 kbp converts a saprophytic *Mesorhizobium* into a symbiont of lotus plants. Because they mediate such sweeping change, mobile genetic elements may represent the most important means for generating the genetic diversity on which selection operates. For mobility, and thus generation of genetic diversity, such genetic elements require recombination activities.

V. DNA TWISTING, FOLDING, AND BENDING

A. DNA supercoiling

Circular DNA molecules extracted from bacteria that grow at moderate temperature have a deficiency of duplex turns relative to linear DNAs of the same length. The deficiency places strain on DNA, causing it to coil. This coiling is loosely referred to as negative supercoiling (an excess of duplex turns would give rise to positive supercoiling). The strain is spontaneously relieved (relaxed) by nicks or breaks in the DNA that allow strand rotation; consequently, supercoiling is found only in DNA molecules that are circular or otherwise constrained so the strands cannot rotate. Since processes that separate DNA strands relieve negative superhelical strain, they will tend to occur more readily in supercoiled than in relaxed DNA. Among these activities are initiation of DNA replication and initiation of transcription. Negative supercoiling also makes DNA more flexible, facilitating DNA looping, wrapping of DNA around proteins, and the formation of cruciforms, left-handed Z-DNA, and other non B-form structures. In a sense, negatively supercoiled DNA is energetically activated for most of the processes carried out by the chromosome.

Negative supercoils are introduced into DNA by gyrase, one of several DNA topoisomerases found in bacteria. DNA topoisomerases act through a strand breaking and rejoining process (Fig. 21.4) that allows supercoils to be introduced or removed, DNA knots to be tied or untied, and separate circles of DNA to be linked or unlinked. The effect of gyrase is modulated by the relaxing activity of topoisomerase I. Since gyrase is more active on a relaxed DNA substrate and topoisomerase I on a highly negatively supercoiled one, the two topoisomerases tend to reduce variation

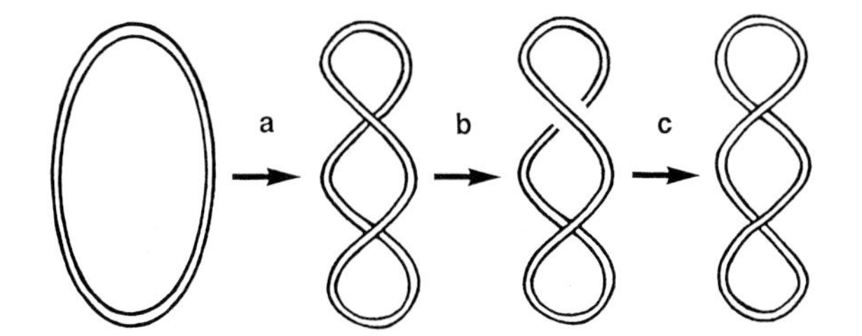

FIGURE 21.4 Negative supercoils. Gyrase generates negative supercoils by passing one duplex strand through the other. In the scheme shown, the enzyme binds to the DNA and creates a positive and a negative node (a). At the upper, positive node the duplex is broken (b). The bottom strand is then passed through the break, which is then sealed (c). [Reprinted from *Trends in Genetics* **6**, Drlica, Bacterial topoisomerases and the control of DNA supercoiling, p. 433, copyright 1990, with permission from Elsevier Science.]

in supercoiling under stable growth conditions. In addition, lowering negative supercoiling increases gyrase expression and decreases topoisomerase I expression. Thus, negative supercoiling is a controlled feature of the chromosome.

Supercoiling is influenced by the extracellular environment. For example, when bacteria such as *E. coli* are suddenly exposed to high temperature, negative supercoiling quickly drops (relaxes), and within a few minutes it recovers. The reciprocal response is seen during cold shock. Presumably these transient changes in DNA supercoiling facilitate timely induction of heat and cold shock genes important for survival. Supercoiling is also affected by the environment through changes in cellular energetics. Gyrase hydrolyzes ATP to ADP as a part of the supercoiling reaction; ADP interferes with the supercoiling activity of gyrase while allowing a competing relaxing reaction to occur. Consequently, the ratio of [ATP] to [ADP] strongly influences the level of supercoiling reached. Changes in oxygen tension and salt concentration are examples in which cellular energetics and supercoiling change coordinately. Collectively, these observations indicate that chromosome structure changes globally in response to the environment.

Supercoiling is influenced locally by transcription. During translocation of transcription complexes along DNA, RNA polymerase does not readily rotate around DNA. Consequently, transcription generates positive supercoils in front of complexes and negative supercoils behind them. Since topoisomerase I removes the negative supercoils and gyrase the positive ones, transcription and similar translocation processes have only transient effects on supercoiling. However, there are cases in which induction of very high levels of transcription results in abnormally high levels of negative supercoiling. In such situations transcription-mediated changes in supercoiling provide a way for specific regions of a DNA molecule to have levels of supercoiling that differ greatly from average values. Problems can occur because excessive negative supercoils behind a transcription complex allow nascent transcripts to form long hybrids with the coding strand of DNA. Such structures, called R-loops, probably interfere with gene expression. Problems also occur as a result of a buildup of positive supercoils in front of a transcription complex since helix tightening will slow transcription. This is probably why strong gyrase binding sites are scattered throughout the chromosome immediately downstream from active genes.

B. Chromosome folding

Multiple nicks are required to relax chromosomal supercoils. Consequently, it was concluded that the DNA must be constrained into topologically independent domains. Current estimates place the number at approximately 50 domains per genome equivalent of DNA in *E. coli*, or approximately 1 per 100 kbp of DNA. Since superhelical tension and topological domains are detected in living cells, it is unlikely that the domains are an artifact of chromosome isolation. From a functional point of view, dividing the chromosome into domains prevents a few DNA nicks or gaps from relaxing supercoils in the entire chromosome. The existence of domains also makes it possible to introduce supercoils into the chromosome before a round of replication finishes; in the absence of domains, the gaps following replication forks would relax any supercoils that gyrase might introduce. Thus, the domains allow the bulk of the chromosome to maintain a supercoiled state.

How the domains are established is unknown. They are present after cells are treated with rifampicin, an inhibitor of RNA synthesis; thus, they are not simply the by-products of DNA–RNA polymerase–RNA complexes. Even after exhaustive deproteinization, fluorescence microscopy shows that nearly every nucleoid in preparations from both exponential- and stationary-phase *E. coli* appears as a rosette or loose network of 20–50 large loops. Networks probably also occur in the linear chromosomes of *B. burgdorferi*. When examined by pulsed-field gel electrophoresis and fluorescence microscopic imaging of individual DNA molecules, some, and occasionally most, of the DNAs are in an electrophoretically immobile network larger than the size of the genome (some *B. burgdorferi* DNA also behaves as unit length linear molecules). Discovering the nature of the bonding in the apparently all-DNA networks may be central to understanding bacterial chromosome structure.

C. DNA bending

Associated with the *E. coli* chromosome are five small proteins that either bend DNA or bind to bends and stabilize them. The most abundant is HU, a protein with long flexible arms that reach around DNA and force it to make a U-turn. HU does not recognize a specific nucleotide sequence; therefore, it is considered to provide a general bending activity. Several examples have been found in which HU serves as an architectural protein, assisting in the formation of DNA–protein complexes that carry out site-specific recombination. HU also provides the DNA bending needed for certain repressors to bring distant regions of DNA together in loops that block initiation of transcription.

HU was initially classified as a histone-like protein on the basis of its amino acid composition and its ability to wrap DNA into nucleosome-like particles *in vitro*. Nucleosomes, which have long been a distinctive feature of eukaryotic nuclei, are ball-like structures in which approximately 200 bp of DNA is wrapped around histone proteins. Nucleosomes occur at regular intervals along DNA, giving nuclear chromatin a "beads-on-a-string" appearance and compacting the DNA by a factor of approximately seven. True bacteria (eubacteria) do not have true histones or nucleosomes, although some archaebacteria (archaea) do. It is unknown whether HU wraps intracellular DNA or just bends it.

Closely related to HU is a bending protein called IHF (integration host factor). It too can cause DNA to turn 180° on itself, but unlike HU, IHF recognizes specific nucleotide sequences. There are many examples in which IHF helps form a DNA loop between promoters and transcription activators bound far upstream from promoters, thereby facilitating initiation of transcription. IHF also participates as an architectural protein during the formation of site-specific DNA–protein complexes. The best known of these is the intasome generated by bacteriophage lambda during integration into the bacterial chromosome.

Two other bending proteins, factor for inversion stimulation (FIS) and leucine-responsive regulatory protein (LRP), serve as sensors and global regulatory agents. FIS senses the bacterial growth phase, and its expression is sharply elevated shortly after a culture begins growing. In older cultures the rate of FIS synthesis decreases to almost zero. FIS binds to specific nucleotide sequences, some of which are so close to a promoter that FIS acts as a repressor. In other cases, FIS acts as an upstream activator of transcription. LRP responds to the nutrient status of the cell, particularly amino acid levels. It acts as a repressor for many genes involved in catabolic (breakdown) processes and an activator for genes involved in metabolic synthesis. Both FIS and LRP also have architectural roles when they bend DNA to form protein–DNA complexes for site-specific recombination.

The fifth protein is called H-NS (histone-like nucleoid structuring protein). Unlike the other four, it does not actively bend DNA. Instead, it binds to DNA that is already bent. If an appropriate bend is near the promoter of a gene, H-NS will bind and act as a mild repressor. It is likely that 100 genes are affected in this way. H-NS action appears to be a general way to keep the expression of many genes down-regulated until their products are needed.

It is not known whether a bending protein is involved in packaging the chromosome as a whole. It is interesting to note that when H-NS is expressed at very high levels, the chromosome compacts. Although such an event is lethal, it may reflect a moderate compacting action by H-NS at normal concentration. Also relevant may be the observation that cells do not tolerate the absence of HU, IHF, and H-NS together, although cells are viable when only one of the three proteins is present. Perhaps at least one of these DNA-bending proteins is needed for chromosome compaction.

VI. CHROMOSOME INACTIVATION

In eukaryotic cells large portions of genomes are rendered transcriptionally inactive by heterochromatinization, a local DNA compaction that is readily observed by light microscopy. Bacterial chromosomes are too small for locally compacted regions to be seen; consequently, we can only guess about their existence. However, evidence is accumulating that bacteria have systems that condense entire chromosomes. In *Caulobacter crescentus* two cell types exist: swarmer cells and stalk cells. When a swarmer cell differentiates into the stalked form, the nucleoid changes from a compact to a more open structure, possibly reflecting an activation of the genome. In a second example, a histone H1-like protein in *Chlamydia trachomatis* probably causes chromosomal condensation during the conversion of the metabolically active reticulate body to the inactive, extracellular elementary body form. Another example occurs during sporulation in *Bacillus*. In this case the chromosome of the spore is bound with new proteins as its transcriptional activity ceases. Still another case is when the cells of the archaebacterium *Halobacterium salinarium* progress from early to late exponential phase of growth. In this prokaryote, the nucleoid obtained by gentle lysis procedures changes from a form containing naked DNA to one having a beads-on-a-string appearance typical of nucleosomal DNA; this change, seen by electron microscopy, is also reflected in nucleoid sedimentation properties. Finally, fluorescence measurements of DNA and RNA within the enormous cells (up to 500 μm long) of *Epulopiscium fishelsoni* suggest that decondensation and dispersion of the nucleoid are accompanied by increased transcriptional activity. As tools with sufficient resolution become available, it should be possible to determine whether the subchromosomal condensation-decondensation strategy of gene regulation in eukaryotic cells also exists in bacteria.

VII. CHROMOSOME DUPLICATION

The major features of chromosome replication have been established for many years. Semiconservative replication was demonstrated by density-shift experiments in 1958, and a few years later the autoradiograms prepared by Cairns revealed a partially replicated circle containing a large replication "eye." In the early 1970s it became clear that bidirectional replication begins at a fixed origin (*oriC*) and that the two forks proceed in opposite directions until they reach a terminus 180° from *oriC* on the circular map. Under conditions of rapid growth, bacterial chromosomes can contain more than one pair of replication forks. It is the presence of multiple forks that allows multiple chromosomes to be present, which in turn enables *E. coli* cells to double at shorter intervals than required for the forks to traverse the chromosome.

The isolation of heat-sensitive mutations in genes called *dnaA* and *dnaC* made it possible to uncouple initiation from the elongation phase of replication. Then the origin (*oriC*) was cloned by its ability to confer replication proficiency to a plasmid lacking an origin of replication. The availability of the origin on a small piece of DNA and purified initiation proteins allowed Kornberg to develop an *in vitro* initiation system. From this system it was learned that initiation involves the specific binding of the DnaA protein to *oriC* and the wrapping of origin DNA around DnaA. Local DNA strand separation then occurs at the origin, and single-stranded binding protein attaches to the separated strands. This helps stabilize what appears to be a single-stranded bubble in duplex DNA. The DnaB helicase, helped by the DnaC protein, binds to the replication bubble and enlarges it. Then DNA polymerase binds, and new strands are synthesized as one replication fork moves away from the origin in the clockwise direction and a second fork moves in the counterclockwise direction.

Several factors prevent a second round of replication from beginning immediately after the first. One is the need for negative supercoils. During the elongation phase of DNA replication nicks and gaps are present in DNA undergoing lagging strand synthesis. These interruptions in DNA are eventually sealed, but while they are present they relax any supercoils introduced by DNA gyrase. Another restriction is mediated by the SeqA protein. It binds tightly to hemimethylated DNA sites near *oriC* and is thought to prevent the DnaA protein from binding to *oriC*. Eventually these sites become fully methylated and available for DnaA binding. These processes, plus interactions with several other proteins, make initiation of replication a highly regulated process.

Termination of replication is also complex. When the replication forks reach a position about half-way around the circular map from *oriC*, they encounter nucleotide sequences called *ter*. The *ter* sequences act as binding sites for a protein that impedes the movement of the DNA helicases associated with replication. Since the replication apparatus (replisome) is likely to be a large structure, the two converging forks might not completely replicate the chromosome. That task may be left to DNA repair enzymes that can unwind the double helix and replicate short regions. The unwinding process can leave circular molecules interlinked (long linear DNA could be locally interlinked due to constraint of strand rotation by chromosomal proteins). DNA topoisomerase IV appears to be the major unlinking (decatenating) activity (bacteria that lack this enzyme probably use DNA gyrase to decatenate their chromosomes; when nicks or gaps are present, topoisomerases I and III can decatenate interlinked circles).

Sensitive probes for specific regions of the chromosome have provided support for two major cytological concepts. First, fluorescent labeling of DNA polymerase revealed that replication forks are located at the center of the cell, where they appear to remain throughout most of the cell cycle. When multifork replication occurs, two additional replication centers, each probably containing a pair of forks, can be seen situated between the mid-cell forks and the cell poles. What holds the replication apparatus in place is unknown, although circumstantial evidence implicates an attachment to the cell membrane. The fixed nature of replication forks supports the idea that replicating bacterial DNA threads through a stationary replication apparatus, a notion that has long been advocated for eukaryotic cells.

The second concept focuses on the movement of *oriC*. The region can be visualized microscopically by treating cells with fluorescent antibodies directed at proteins that bind to repeated nucleotide sequences placed near *oriC* (or any other specified region). In newly formed cells, *oriC* and the replication terminus are located at opposite poles of the nucleoid. During replication, *oriC* moves briefly toward a mid-cell position. Then two copies of *oriC* become visible at the nucleoid pole, apparently having been drawn back after replication. Later, one copy abruptly moves to the opposite edge of the nucleoid. The replication terminus then migrates to a mid-cell position, and late in the cell cycle two termini can be seen pulling apart. The septum that separates the new daughter cells forms between the termini. The localization of *oriC* and its rapid movement, which is about 10 times faster than cell elongation, indicate that bacterial chromosomes undergo a form of mitosis.

Although the details of bacterial "mitosis" are poorly understood, several proteins exhibit properties expected of mitotic proteins. For example, in *B. subtilis* the Spo0J protein appears to participate in chromosome partitioning by binding to multiple sites on the chromosome near, but not at, *oriC*. Ten related, 8-bp inverted repeat DNA sequences are scattered across approximately 800 kbp of the 4200-kbp chromosome. Eight of these sites are bound *in vivo* to Spo0J. Such distribution of bacterial centromere-like DNA elements is similar to the most common type of distribution of functional centromeric DNA in eukaryotes, the *CEN*-containing regional centromeres. Thus, Spo0J may hold the new and old copies of *oriC* near one pole of the nucleoid until one copy of *oriC* is moved to the opposite pole. Chromosome condensation, which is probably too slight to see with current methods, may involve the action of the Smc protein, a homolog of a eukaryotic protein family known to participate in DNA condensation. When *smc* is mutant in *B. subtilis*, the Spo0J protein is not found at its polar position on the nucleoid, and newly formed septa cut some daughter chromosomes in a guillotine-like manner.

The extrusion of newly replicated DNA from the centrally-located replisome may provide some organization to the chromosome: genes located near *oriC* on the genetic map are also positioned near *oriC* in the cell. Likewise, genes mapping near *ter* are spatially located near the terminus of replication.

As replicated chromosomes are pulled apart, DNA tangles must occur. The double-strand passing activities of gyrase and topoisomerase IV are well suited to resolve the tangles. Consistent with this idea, both enzymes are distributed around the *E. coli* chromosome, as judged by DNA cleavage induced by the quinolone inhibitors of topoisomerases. In *E. coli*, DNA cleavage occurs at approximately 100-kbp intervals for gyrase and 200-kbp intervals for topoisomerase IV.

VIII. SPECULATIONS ON CHROMOSOME PACKAGING DYNAMICS

In the past two decades, our understanding of bacterial chromosomes has advanced remarkably along two fronts. First, a combination of genetics and biochemistry has taught us a great deal about the proteins that manipulate DNA. Second, we now have complete genomic sequences for many bacteria. These sequences indicate that bacterial genomes are quite flexible with respect to gene order. However, few experiments bear on how the long chromosomal DNA molecule is compacted to fit inside a cell. Here, we offer a speculative scheme that may help form a framework for understanding future discoveries.

One key concept is that cytoplasmic proteins and other large cytoplasmic molecules are at such high concentration that they compact DNA through macromolecular crowding. This idea, which has recently been refined by Zimmerman, requires no specific DNA compacting proteins, and so it accommodates the apparent absence of nucleosome-like particles in eubacteria. We envision that chromosomal activities involving bulky protein complexes occur at the edges of the nucleoid (Fig. 21.5). For example, the replication apparatus, which is likely to be attached to a multienzyme complex that supplies deoxyribonucleoside triphosphates, is probably situated at the edge of the compacted portion of the chromosome, especially if replication proteins are bound to the cell membrane. Likewise, transcription, which in bacteria is coupled to translation, also probably occurs on DNA emerging from the compacted mass of nucleoid DNA because ribosomes are seen only outside the nucleoid (extrachromosomal localization is especially likely when transcription–translation complexes are bound to the cell membrane via nascent membrane proteins). Consistent with this idea, pulse-labeled nascent RNA is preferentially located at the nucleoid border, as is topoisomerase I (as noted previously, topoisomerase I may serve as a cytological marker for transcription since it is probably localized behind transcription complexes to prevent excess negative supercoils from accumulating).

If the replication and transcription–translation machinery are located on the surface of the nucleoid,

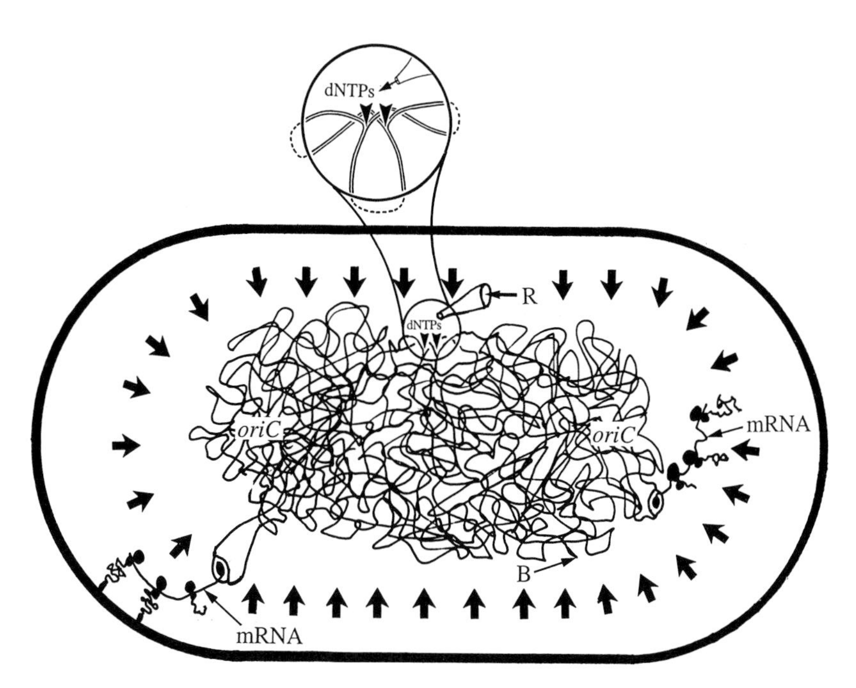

FIGURE 21.5 Overview of bacterial chromosome structure. The figure shows a schematic representation of a bacterial cell and its chromosomal DNA (nucleoid). The replication apparatus is located at the edge of the nucleoid at a mid-cell position. Two replication forks (arrowheads) are shown in close proximity. DNA is thought to thread through a stationary replication apparatus. The funnel-like structure represents a multienzyme complex responsible for synthesis of deoxyribonucleoside triphosphates (dNTPs) from ribonucleoside diphosphates (R). In the enlargement of the replication apparatus the dashed lines outside the circle represent the connections between the DNA strands masked by the large amount of DNA in the cell. The polar distribution of *oriC* regions is indicated. Macromolecular crowding (arrows) contributes to DNA condensation, with additional packing occurring from protein-induced DNA bending (B) at many points on the chromosome (only one bending point is labeled). Two examples of coupled transcription–translation are shown to occur at the edge of the nucleoid. In the lower left, nascent protein is bound to the cell membrane, drawing a region of DNA out of the compact part of the nucleoid. To provide transcriptional access to all regions of the genome, the interior and exterior regions of the nucleoid are assumed to exchange rapidly.

DNA movement must occur to allow access to all nucleotide sequences. Replication-based movement probably occurs by DNA threading through stationary replication forks. Such movement would not be sufficient for transcriptional access to the whole genome since some genes can be induced when DNA replication is not occurring. Perhaps compacted DNA is sufficiently fluid that genes frequently pass from interior to exterior. At any given moment, in some fraction of the cell population each gene may be at the surface of the nucleoid and available for transcription. Capture of a gene by the transcription–translation apparatus would hold that gene on the surface. During induction of gene expression, the fraction of cells in which a particular gene is captured would increase until most of the cells express that gene. For the chromosome as a whole, many genes would be expressing protein during active growth, and therefore many regions would be held outside the nucleoid core by the transcription–translation apparatus. Kellenberger suggested that such activity explains why the nucleoid appears more compact when protein synthesis is experimentally interrupted. The idea of gene capture for expression requires that the replication apparatus be strong enough to pull the DNA through itself even when genes are bound to ribosomes via mRNA and to the cell membrane via nascent proteins still attached to ribosomes. A fixed RNA polymerase must also pull DNA.

Capture of the *oriC* region by the replication apparatus might be similar to gene capture for transcription. With the fluid chromosome hypothesis, some replication proteins would assemble at mid-cell, whereas others would assemble with the centromere-like DNA elements and Spo0J (in *B. subtilis*) at the poles and dislodge *oriC* from its polar connection. Once liberated, a mobile *oriC* would be captured by the replication apparatus at mid-cell. As *oriC* and nearby regions are drawn through the replication forks and replicated, new binding sites for a chromosome partition protein (Spo0J) would be created. Once these sites were filled, the two daughter *oriC* regions might pair through Spo0J–Spo0J interactions and return to the polar position. Other proteins would later disrupt the Spo0J–Spo0J interactions as the new Spo0J–*oriC* complex is moved to the other pole of the nucleoid.

Movement of DNA must generate tangles, just as loops in fishing line snarl when reels are not carefully attended. The decatenating activities of the type II topoisomerases, DNA gyrase and topoisomerase IV, are available to remove the tangles, explaining in part why these two enzymes are found at many spots on the chromosome. The movement of the daughter chromosomes to opposite cell poles would provide the directionality needed by the topoisomerases to untangle the loops.

The abundant DNA bending proteins, such as HU, IHF, and H-NS, probably facilitate chromosome compaction in a dynamic manner by rapidly exchanging between DNA-bound and unbound states. Likewise, the topoisomerases probably respond rapidly to local perturbations in supercoiling to maintain the proper level of supercoiling throughout the chromosome.

Fixed topological domains are not easily accomodated by a fluid model for bacterial chromosome structure, since they require specialized, pulley-like structures to allow the DNA movement needed for replication. Transient domains could be generated by recombination intermediates, provided that recombination occurs often enough to maintain approximately 50 domains per genome.

IX. CONCLUDING REMARKS

Many of the features found in bacterial chromosomes are remarkably similar to those of eukaryotic chromosomes: linear maps, one or more dissimilar chromosomes [the number can be as high as four among bacteria and as low as one in eukaryotes ($2N = 2$)], high ploidy (copy number) levels, and a mitotic-like apparatus used in cell division. Consequently, the prevalent belief that profound differences exist between prokaryotic and eukaryotic chromosomes is gradually eroding. A major distinction, other than the presence of a nuclear membrane in eukaryotes, revolves around histones and their compaction of DNA into nucleosomes. True bacteria lack histones and nucleosomes, and so DNA compaction must occur by other means. However, some archaea have histones and stable nucleosomes, whereas some unicellular eukaryotes lack both. Thus, at the chromosome level the difference between prokaryotes and eukaryotes may not be as extensive as generally thought.

ACKNOWLEDGMENTS

We thank M. Gennaro and G. L. G. Miklos for critical comments on the manuscript. The authors' work was supported by grants from the National Science Foundation, the American Cancer Society, and the National Institutes of Health.

BIBLIOGRAPHY

Bendich, A. (2001). The form of chromosomal DNA molecules in bacterial cells. *Biochimie* **83**, 177–186.

Bendich, A., and Drlica, K. (2000). Prokaryotic and eukaryotic chromosomes: what's the difference? *BioEssays* **22**, 481–486.

Brankenhoff, G. J., van der Voort, H., van Spronsen, E., Linnemans, W., and Nanninga, N. (1985). Three-dimensional chromatin distribution in neuroblastoma nuclei shown by confocal scanning laser microscopy. *Nature* **317**, 748–749.

Bussiere, D. E., and Bastia, D. (1999). Termination of DNA replication of bacterial and plasmid chromosomes. *Mol. Microbiol.* **31**, 1611—1618.

Cox, M. (1997). Recombinational crossroads: Eukaryotic enzymes and the limits of bacterial precedents. *Proc. Natl. Acad. Sci. USA* **94**, 11764–11766.

Cozzarelli, N. R. (1980). DNA gyrase and the supercoiling of DNA. *Science* **207**, 953–960.

Cunha, S., Odijk, T., Suleymanoglu, E., and Woldringh, C. L. (2001). Isolation of the *Escherichia coli* nucleoid. *Biochimie* **83**, 149–154.

Drlica, K. (1990). Bacterial topoisomerases and the control of DNA supercoiling. *Trends Genet.* **6**, 433–437.

Drlica, K., and Zhao, X. (1997). DNA gyrase and topoisomerase IV as targets of the fluoroquinolones. *Microbiol. Mol. Biol. Rev.* **61**, 377–392.

Gordon, G. S., and Wright, A. (2000). DNA segregation in bacteria. *Annu. Rev. Microbiol.* **54**, 681–708.

Hiraga, S. (2000). Dynamic localization of bacterial and plasmid chromosomes. *Annu. Rev. Genet.* **34**, 21–59.

Kolsto, A. B. (1997). Dynamic bacterial genome organization. *Mol. Microbiol.* **24**, 241–248.

Lemon, K., and Grossman, A. (1998). Localization of a bacterial DNA-polymerase: Evidence for a factory model of replication. *Science* **282**, 1516–1519.

Moller-Jensen, J., Jensen, R., and Gerdes, K. (2000). Plasmid and chromosome segregation in prokaryotes. *Trends Microbiol.* **8**, 313–320.

Nanninga, N., Woldringh, C., and Rouviere-Yaniv, J. (2001). Bacterial nucleoid, DNA replication, segregation and cell division. *Biochimie* **83**, 147–148.

Pettijohn, D. E. (1996). The nucleoid. *In "Escherichia coli* and *Salmonella typhimurium"* (F. Neidhardt *et al.*, eds.), pp. 158–166. ASM Press, Washington, DC.

Reeve, J. N., Sandman, K., and Daniels, C. J. (1997). Archael histones, nucleosomes, and transcription initiation. *Cell* **89**, 999–1002.

Shapiro, L., and Losick, R. (2000). Dynamic spatial regulation in the bacterial cell. *Cell* **100**, 89–98.

Zimmerman, S. B., and Murphy, L. D. (1996). Macromolecular crowding and the mandatory condensation of DNA in bacteria. *FEBS Lett.* **390**, 245–248.

WEBSITE

Website for Comprehensive Microbial Resource of The Institute for Genomic Research and links to many other microbial genomic sites
http://www.tigr.org/tigr-scripts/CMR2/CMRHomePage.spl

22

Conjugation, Bacterial

Laura S. Frost
University of Alberta

GLOSSARY

plasmid An extrachromosomal DNA segment, usually circular, which is capable of autonomous replication via a segment of the plasmid called the replicon.

transconjugant A general term for a recipient cell that has successfully been converted to donor cell by conjugation.

transposon A segment of DNA that is replicated as part of a chromosome or plasmid. It encodes a mechanism, called transposition, for moving from one location to another, leaving a copy at both sites.

Bacterial Conjugation was first described by Lederberg and Tatum in 1946 as a phenomenon involving the exchange of markers between closely related strains of *Escherichia coli*. The agent responsible for this process was later found to be a site on the chromosome called the F ("fertility") factor. This finding was the basis of bacterial genetics in the 1940s and 1950s and was used extensively in mapping the *E. coli* chromosome, making it the pre-eminent prokaryotic organism at that time. It was also shown that F could excise out of the chromosome and exist as an extrachromosomal element which was capable of self-transfer to other bacteria and that mobilization of the chromosome was the serendipitous function of F integrated randomly into its host's DNA. The F sex factor of *E. coli* also imparted sensitivity to bacteriophages which required the F pilus, encoded by the F transfer region, as an attachment site during infection.

In the 1960s many other conjugative plasmids were isolated, many of which carried multiple antibiotic resistance markers. These plasmids were termed R ("resistance") factors and were found in many instances to repress pilus expression and conjugation by F, a process termed fertility inhibition (fi^+). The number of conjugative plasmids has increased tremendously in the past few decades and includes self-transmissible plasmids isolated from gram-negative and -positive bacteria as well as mobilizable plasmids. Conjugative transposons and members of the IncJ incompatibility group move between cells using a conjugative mechanism, whereby they excise and integrate into the host chromosome via a process reminiscent of lysogenic phage. An example of true conjugative phages include one from *Staphylococcus aureus* and the STX element of *Vibrio cholera*. These elements might more precisely be called conjugative genomic islands since they are conjugative, self-transmissible and are obligatory residents of the host chromosome.

In general, the transfer and replication functions of these mobile elements are often physically linked and the type of transfer system is closely aligned with the nature of the replicon which is described by incompatibility groups (Inc). An excellent summary of the properties of many conjugative plasmids is given in Shapiro (1977).

Bacterial conjugation is now known to be one of the principal conduits for horizontal gene transfer among microorganisms. The process is extremely widespread and can occur intra- and intergenerically as well as between kingdoms (bacteria to yeast, plants or animal cells). The effect of this process on evolution

The Desk Encyclopedia of Microbiology
ISBN: 0-12-621361-5

has been immense, with bacteria rapidly acquiring traits both good (hydrocarbon utilization) and bad (antibiotic resistance, toxins). The sequences of bacterial genomes has revealed many genomic islands containing homologues of conjugative systems which have been shown to be involved in protein and DNA transfer between cells, protein secretion and DNA excretion and uptake during transformation. While the mechanism for this extraordinary versatility in substrate transport remains undetermined, these homologous systems are known collectively as Type IV secretion (Christie, 2001) which is not to be confused with Type II secretion that elaborates a Type IV pilus. Type IV secretion systems also often elaborate a pilus, such as the conjugative pilus of gram-negative transfer systems, which has been classified as Type II (Ottow, 1975). Once again, bacterial conjugation is at the forefront of microbiology, but this time the emphasis is on the process rather than its utility as a geneticist's tool. Excellent summaries of related topics are provided in books by Clewell (1993) and Thomas (2000).

I. THE CONJUGATIVE PROCESS

Unlike the other processes contributing to horizontal gene transfer, transformation, and transduction, conjugation can be characterized by two important criteria. There must be close cell-to-cell contact between the donor and recipient cells and DNA transfer must begin from a specific point on the transferred DNA molecule, be it a plasmid, transposon, or chromosome (Fig. 22.1). This point is encoded within the origin of transfer (*oriT*) called *nic*. The proteins which act on these sites are encoded by *tra* (transfer) or *mob* (mobilization) regions. In general, each conjugative element encodes an array of proteins for mating pair formation (Mpf), whereas another set of proteins are involved in processing and transferring the DNA (Dtr). The Mpf genes can further be classified into the genes for pilus formation and mating pair stabilization (Mps) in gram-negative bacteria or aggregate formation in gram-positive cocci. A system to prevent close contact between equivalent donor cells is called surface exclusion. The gene products that process the DNA in preparation for transfer usually include a protein (relaxase) that cleaves the DNA in a sequence- and strand-specific manner at *nic* and remains covalently bound to the 5′ end in all cases that have been examined. This nucleoprotein complex, plus other auxiliary proteins bound to the *oriT* region, is called the relaxosome, whereas the complex formed by the transport machinery is known as the transferosome.

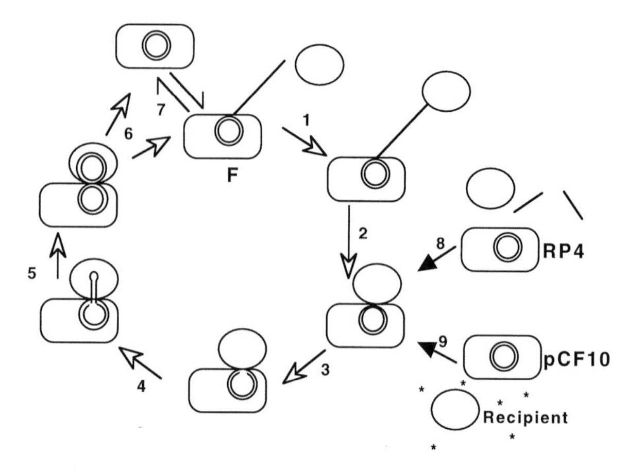

FIGURE 22.1 Summary of the mating process for universal (plasmid F) and surface-preferred (plasmid RP4) conjugation systems in gram-negative bacteria and the pheromone-activated system of *Enterococcus faecalis* (plasmid pCF10). In universal systems, the pilus attaches to a receptor on the recipient cell surface (1) and retracts to form a stable mating pair or aggregate (2). DNA transfer is initiated (3), causing transport of a single strand in the 5′ to 3′ direction (4). Transfer is associated with synthesis of a replacement DNA strand in the donor cell and a complementary strand in the recipient (5). The process is terminated by disaggregation of the cells, each carrying a copy of the plasmid (6). The transfer systems of conjugative plasmids in gram-negative bacteria can be repressed (7) or derepressed (constitutive; 8). Cells carrying RP4 and related plasmids express pili constitutively but the pili are not seen attached to the bacteria. Such cells form mating pairs by collision on a solid surface (8). In gram-positive bacteria, such as the enterococci, the donor senses the presence of pheromone (*), released by the recipient cell, which triggers mating pair formation and DNA transfer (9). Donor cells are shown as oblongs and recipient cells as ovals.

A process which prevents the transfer of DNA into the recipient cell after Mpf has occurred is called entry exclusion. Previously, surface and entry exclusion were used interchangeably; however, as the details of the process are refined, it is important to make this distinction.

In gram-negative bacteria, the process of DNA transfer is triggered upon cell contact, whereas in *Enterococcus faecalis* and T-DNA transport by *Agrobacterium tumefaciens*, contact between cells induces a complex program of gene expression leading to DNA transport. Although the sequences for many conjugative elements have been completed and comparisons have revealed information on the evolution of conjugative elements, a study of the conjugative process has only been done in some depth for IncF, IncP, Incl, and the Ti plasmid of *A. tumefaciens* in gram-negative bacteria and for the pheromone-responsive system found for some plasmids in *E. faecalis*. Some information is available for the integration/excision process of conjugative transposons as well as the role of the *mob* genes in mobilizable plasmids. In addition, conjugation in *Streptomyces* has been studied in detail but is quite different than that described and may use a DNA transport mechanism related to the process of DNA partition during cell division or septation in *Bacillus subtilis* (see Section V.B).

II. PHYSIOLOGICAL FACTORS

The level of transfer efficiency varies dramatically among the various systems. For derepressed or constitutively expressed systems such as F (IncFI) or RP4 (IncPα) maximal levels of mating (almost 100% conversion to plasmid-bearing status) are possible within 30 min. Plasmids undergoing fertility inhibition usually have a 100–1000-fold reduction in mating efficiency, whereas other plasmids, especially the smaller plasmids of gram-positive bacteria and conjugative transposons, mate at barely detectable levels in the best of circumstances. One critical factor that affects mating efficiency is temperature with very precise optimums usually being the rule. For instance, F and RP4 mate optimally at 37–42 °C and IncH plasmids and the Ti plasmid at approximately 20–30 °C. Other factors, such as oxygen levels, nutrient availability, and growth phase, contribute to mating efficiency. F^+ cells in late stationary phase are known as F^- phenocopies because they are able to accept incoming F DNA and are not subject to surface or entry exclusion. Where information is available, conjugation appears to be maximal over a short temperature range in nutrient-rich environments with good aeration for aerobic organisms.

A. Liquid versus solid support

The ability of some conjugative systems to mate equally well in liquid media or on a solid support is one of the hallmarks of conjugation. Although all conjugative elements can mate well on a solid support, usually a filter placed on the surface of a pre-warmed nutrient agar plate, many transfer systems, including those of the IncF group and the pheromone-responsive plasmids of *Enterococcus*, mate very efficiently in liquid media. This difference can be traced to the nature of the mating pair formation process, with thick, flexible pili of gram-negative bacteria associated with systems that mate well in liquid media, whereas rigid pili, not usually seen attached to the cells (e.g. IncPα), require a solid support for efficient mating. The aggregation substance of *E. faecalis* allows high levels of transfer in liquid media but other gram-positive systems and conjugative transposons mate at low levels and absolutely require a solid support. In general, it appears that mating systems requiring a solid support depend on collision between donor and recipient cells, whereas systems that mate well on either medium have a mechanism for initiating contact between freely swimming cells (thick, flexible pili, aggregation substance). Description of the media requirements for many gram-negative plasmid transfer systems is given in Bradley *et al.* (1980).

III. CONJUGATIVE ELEMENTS

Naturally occurring conjugative elements include plasmids, conjugative transposons, with either of these incorporated into the host chromosome to allow chromosome mobilization ability (CMA) resulting in high frequency of recombination (Hfr). Free plasmids can be divided into self-transmissible (Mpf plus Dtr genes) or mobilizable (Dtr or Mob genes) and can vary in size from a few kilobases to large plasmids of 100–500 kb.

A. Plasmids

In general, gram-negative transfer systems are approximately 20–35 kb in size and reside on plasmids from 60 to 500 kb, whereas mobilizable plasmids are less than 15 kb. The transfer or mobilization regions often represent half or more of the coding capability of the plasmid. Table 22.1 contains a list of selected plasmids and their characteristics, including their pilus type and mating medium preference. In non-filamentous gram-positive plasmids, the smaller plasmids (130 kb) usually have a requirement for a solid support during mating and mate at low levels, whereas the larger plasmids mate efficiently in liquid media and express genes for aggregate formation (e.g. *Enterococcus*, *Staphylococcus*, *Lactobacillus*, and *Bacillus thuriengensis*). *Streptomyces* is able to mate at high frequency and has the added property of CMA. Each large, self-transmissible plasmid can supply the needed Mpf functions for many mobilizable plasmids. These mobilizable plasmids have been used to construct vectors which are maintained in the recipient cell or deliver their cargo of DNA via mobilization but are unable to replicate in the new host (suicide vectors). This has been a boon to the genetics of otherwise recalcitrant bacteria.

B. Chromosome mobilization

F undergoes integration into the chromosome via four transposable elements (IS2, IS3a and -b, and Tn*1000*) which either mediate cointegrate formation via a transposition event or more frequently undergo

TABLE 22.1 Selected conjugative/mobilizable plasmids and conjugative transposons

Mobile element	Size (kb)	Inc. group/ pheromone	Copy no.	Mating surface/pilus type	Mating efficiency/host range
Gram-negative bacteria					
F	100	IncFI	1–2	Liquid/flexible (II)	High (derepressed)/narrow
RP4	60	IncPα	4–6	Solid/rigid (II)	High (constitutive)/broad
Co1IB-P9	93	IncI1	1–2	Liquid/rigid (II), thin (IV)	Low (repressed)/narrow
pTiC58	~200	HSL	1–2	Solid/?	Low (repressed)/narrow
vir	25 (T-DNA)	Plant exudate	1	Plants/rigid (II)	—
Gram-positive bacteria					
pAD1	60	cAD1		Liquid	High (10^{-2}/donor)/narrow
pAMβ1	26.5	—	3–5	Solid	Low (10^{-4}/donor)/broad
pIJ101	8.8	*Streptomyces*	~300	Solid	High/broad (actinomycetes)
Mobilizable plasmids					
Co1E1	6.6	—	~10	Liquid	High (IncF, -P, -I)/narrow
RSF1010	8.9	IncQ	~10	Solid	High (IncP)/broad
pMV1158	5.5	—		Solid	Low (pAMβ1)/broad
Conjugative transposons					
Tn916	18.5	—	—	Solid	Low (~10^{-8}/donor)/broad
TcREmR DOT	80	*Bacteroides*	—	Solid	Low (~10^{-5}/donor)/narrow

homologous recombination between these sequences and similar elements on the chromosome. Once incorporated into the chromosome, the F replicon is suppressed by the chromosomal replication machinery allowing stable maintenance of the Hfr strain. Like F, which was found incorporated into the host chromosome, other examples of naturally occurring Hfr strains have been reported in the literature. Hfr strains have also been constructed using homologous gene segments shared by a plasmid and its host for mapping the host's genome. Since the advent of pulsed-field gel electrophoresis for mapping chromosomes and large-scale sequencing facilities, the utility of Hfr strains has waned. The procedure for using Hfr strains for chromosome mapping requires that F integrate near a locus with a few genetically defined markers. The direction of transfer of the chromosome and the time of entry of the markers into the recipient cell are functions of the position and orientation of F in the chromosome and its distance from each marker. By laboriously measuring the time of entry of each marker (e.g. antibiotic resistance and amino acid biosynthesis), which must be able to recombine into the recipient's chromosome in such a way as to announce its presence, a map of the chromosome can be generated. The process of mobilizing the entire *E. coli* chromosome takes approximately 90 min and markers that are distal from F *oriT* are transferred much less efficiently than those more proximally located, a process called the "gradient of transmission." The last portion of the chromosome to be transferred contains the F transfer region; consequently, the recipient cells in an Hfr mating are seldom, if ever, converted to F^+ (Hfr) status.

Another related property of F is imprecise excision out of the chromosome with adjoining chromosomal sequences being incorporated into the circular F element which are often large enough to encode complete operons. These elements are known as F′ factors, such as F*lac*, F*gal*, and F*his*.

C. Conjugative transposons

The first conjugative transposon was isolated from *E. faecalis* (Tn*916*; Table 22.1) in Clewell's lab in 1981 and expressed *tet* (M) (tetracycline) resistance. This element excises from the chromosome, circularizes into a nonreplicative intermediate, and expresses functions for transfer to a wide range of recipient cell types at low frequency on solid surfaces (Salyers *et al.*, 1995). The enzymes responsible for excision and integration (Int and Xis) are related to the corresponding enzymes in lambda phage. The absence of a small repeated sequence flanking the conjugative transposon as well as a non-random site selection mechanism for integration suggests that these elements are evolutionarily more related to phage than true transposons. Excision of the element results in staggered ends which form a heteroduplex structure upon circularization. This heteroduplex is derived from flanking sequences in the chromosome ("coupling sequences"). Through mapping and sequencing the site of insertion and determining which end of the heteroduplex is inherited in the recipient, a model involving single-stranded DNA transfer has been proposed. A second important group of conjugative transposons has been found in the anaerobic gram-negative genus *Bacteroides*. These elements are usually associated with antibiotic resistance (*tetX* and *erm*). The interesting property of increased transfer proficiency in the presence of tetracycline has been noted by Salyer's group.

Conjugative transposons demonstrate amazing versatility in mobilizing DNA. They can mobilize coresident plasmids directly or form cointegrates with plasmids or the chromosome. They are also able to harbor other mobile elements such as transposons and integrons and move them between cells. In the case of *Bacteroides*, the conjugative transposons are able to excise and mobilize small nonconjugative, nonreplicative segments of DNA found in the chromosome called NBUs (non-replicating *Bacteroides* units). Although conjugative transposons have been identified in many genera of bacteria, especially in gram-positives, the details of the conjugation process remain obscure. An *oriT* region in Tn*916* and a relaxase homolog in *Bacteroides* have been identified but the nature of the proteins involved in Mpf are unknown (Bonheyo *et al.*, 2001; Hinerfeld and Churchward, 2001).

IV. GRAM-NEGATIVE CONJUGATION

With the exception of *Bacteroides*, all gram-negative transfer systems encode a conjugative pilus (Type II) which is essential for mating pair formation and DNA transport (Fig. 22.1). In thick, flexible pilus mating systems, the pili are easily found attached to the donor cell and examples of the pilus mediating contact between the donor and recipient cell are easy to visualize in the electron microscope. In the case of rigid pilus mating systems, pili are rarely seen attached to cells but are seen as bundles of pili accumulating in the medium. Although there is considerable homology between all the transfer systems examined to date within the gram-negative group of organisms, two broad classes can be identified. These include IncF-like and IncP-like plasmids, which also represent mating systems categorized by their ability to mate in liquid versus solid media or systems encoding thick, flexible pili versus rigid pili. Some of the gene products in IncI transfer systems are distantly related to those of IncP, however the remaining gene products are distinct and have homology to the Icm/Dot proteins of *Legionella pneumophila*. All conjugative systems identified to date have a relaxase ($TraI_{F,P}/Nik_I$), and a protein that energizes DNA transport ($TraD_F/TraG_P/TrbC_I$) as well as a protein that is involved in pilus assembly ($TraC_F/TrbE_P/TraU_I$). F-like systems have the characteristic Mps genes, *traG* and *traN*, whereas P-like systems all have a TrbB homologue. IncI systems also have a TrbB ($TraJ_I$) homologue but also encode a second type IV pilus (*pil*) used in Mps. The principal systems studied to date include F (Frost *et al.*, 1994); P-like systems include IncP plasmids RP4 and R751 (Pansegrau *et al.*, 1994); ColIb-P9 (IncI, Komano *et al.*, 2000), R27 (IncHI1, Lawley *et al.*, 2002) which has features of F and P systems; pCU1 and pKM101 (IncN; Pohlman *et al.*, 1994), R388 (IncW; Gomis-Ruth *et al.*, 2001), and the Ti plasmid of *A. tumefaciens* (Christie and Vogel, 2000) (Table 22.2).

TABLE 22.2 Similarities/homologies shared by pilus synthesis and transport/virulence systems[a]

Function	IncF	IncP	Ti	Pt1	IncN	Cag[b]	IncI[c]	Dot[c]
Pilin	TraA	TrbC	VirB2	PtlA	TraM		TraX	
Pore	TraL	TrbD	VirB3	PtlB	TraA			
Pore	TraC	TrbE	VirB4	PtlC	TraB	Cag544	TraU	
Pore	TraE	TrbJ	VirB5		TraC			
Pore	TraG	TrbL	VirB6	PtlD	TraD			
Lipoprotein	TraV	TrbH	VirB7	PtlI	TraN			
Pore		TrbF	VirB8	PtlE	TraE			
Pore/secretin	TraK	TrbG	VirB9	PtlF	TraO	Cag528	TraN	
Pore	TraB	TrbI	VirB10	PtlG	TraF	Cag527	TraO	DotG
Transport		TrbB	VirB11	PtlH	TraG	Cag525	TraJ	DotB
Pore?							TrbA	DotM
Transport	TraD	TraG	VirD4		TraJ	Cag524	TrbC	

[a]Based on Christie (1997). IncF (F), IncP (RP4), Ptl (*Bordetella pertussis* toxin secretion), IncN (pKM101).

[b]Cag ORFs are taken from Tomb *et al.* (1997). *Nature* **388**, 539–547.

[c]Comparison of IncI1 (R64) gene products to *dot* virulence determinants from *Legionella pneumophila* are taken from Vogel *et al.* (1998). *Science* **279**, 873–876.

A. The pilus

1. Structure

The conjugative pilus is a thin filament expressed in relatively low numbers (one to three per cell) which has no set length (usually 1–20 μm) and is randomly distributed on the surface of the cell (Fig. 22.2). The diameter of the pilus is approximately 6–11 nm, with most pili having a diameter of approximately 9 nm (Paranchych and Frost, 1988). Pili isolated from cells usually contain a "knob" at the base of the pilus which represents unassembled pilin subunits derived from the inner membrane of the cell. They also often have a pointed tip, suggesting an unusual configuration of pilin subunits at the tip, and can aggregate into large bundles which can be pelleted in an ultracentrifuge. The pilus is usually composed of a single repeating subunit of pilin arranged in a helical array with a hollow lumen clearly visible in negatively stained electron micrographs. F pili, which are the best studied, have a diameter of 8 nm with an inner lumen of 2 nm. The pilin subunits are arranged as repeating layers of five subunits with each layer rising 1.28 nm. There are 25 subunits in two turns of the helix with a crystallographic repeat of 3.2 nm.

The F pilus is expressed as a propilin of 121 amino acids (*traA*) which requires TraQ, a putative chaperone, for insertion in the inner membrane, where it is stored in a pool of approximately 100,000 subunits. The 51-amino acid leader peptide is cleaved by host

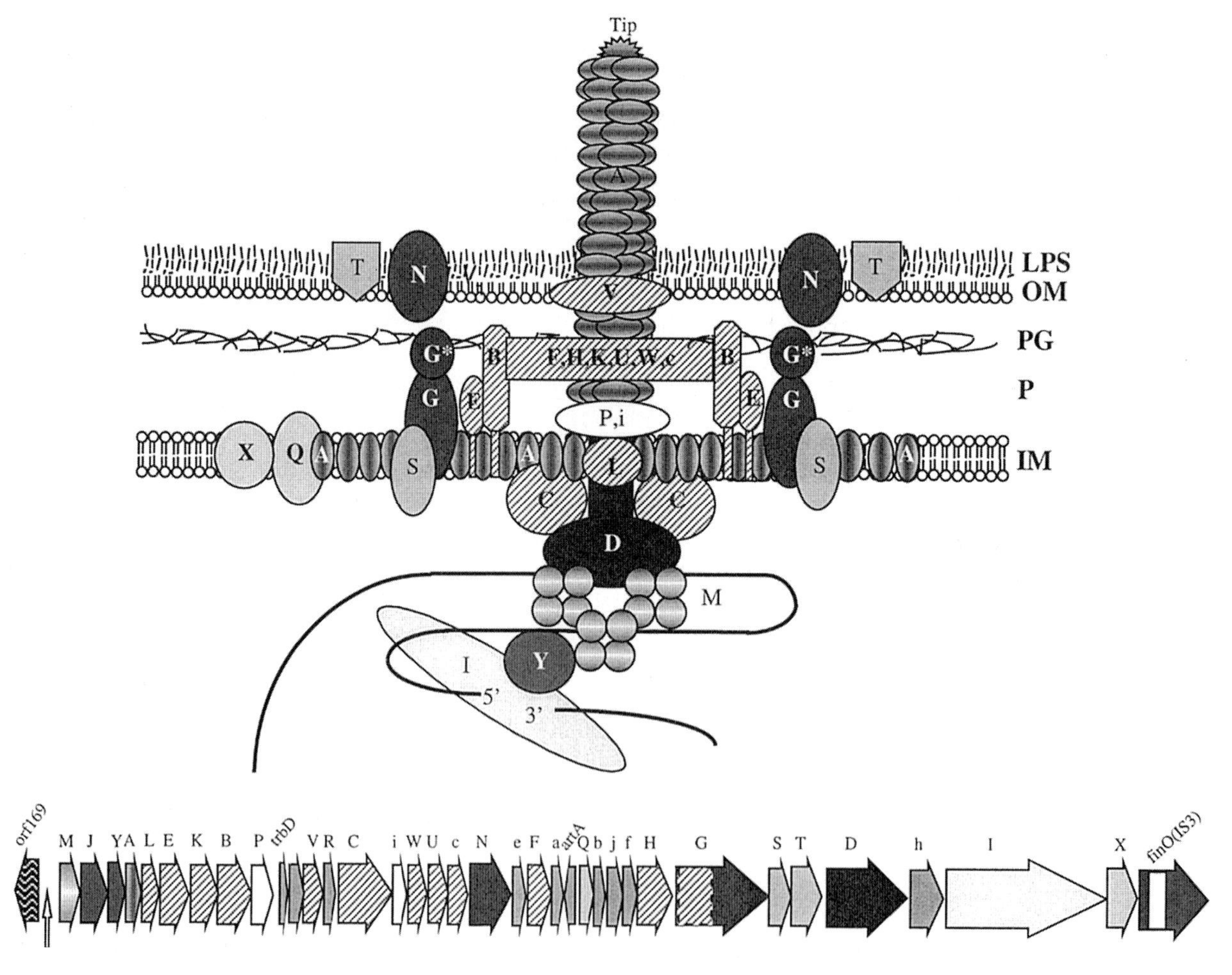

FIGURE 22.2 The F transfer region. The organization of the F transfer region is shown below the relaxosome/transferosome complex showing the cellular location of the Tra proteins. The DNA processing genes are TraD, TraI (relaxase) and TraM. TraY is a regulatory protein that is required for relaxosome formation. TraA (pilin), TraQ (pilin chaperone), TraX (N-acetylation of pilin), TraV (lipoprotein), and TraL, -E, -K, -B, -C, -W, -U, -F, -H, TrbC, and TraG (N-terminal domain) are involved in F pilus assembly. TraN and the C-terminal domain of TraG are involved in mating pair stabilization. TraS and TraT are responsible for entry and surface exclusion, respectively. The nature of the F pilus tip is unknown but is inferred to be important in establishing contact with the recipient cell. Orf169, a putative transglycosylase, and the regulatory protein TraJ are not shown. The functions of the other genes are unknown. *tra* genes are shown as capital letters and *trb* genes are shown as lower-case letters. The site of cleavage within *oriT* is shown as *nic* with *orf169* being the first gene to enter the recipient cell.

leader peptidase (LepB) and the pilin subunit is acetylated at the N terminus by TraX. Transposon insertion studies have revealed that mature pilin is oriented as two α-helical transmembrane segments within the inner membrane, with the N and C termini facing the periplasm. Assembly by the TraL, -E, -K, -B, -V, -C, -W, -U, -F, -H, and -G and TrbC proteins results in the subunits oriented within the fiber such that the acetylated N terminus is buried within the structure and the C terminus is exposed on the sides. The pilus appears to assemble at its base rather than at the tip based on evidence using the slowly assembled pili expressed by the IncHI1 plasmid R27.

The RP4 pilin subunit is expressed as a 15-kDa prepilin polypeptide (*trbC* in Tra2), which is processed three times to give a 78 aa cyclic peptide. The cleavage reactions at the N and C termini are undertaken by LepB (N terminus) and an unidentified host peptidase as well as by a pilin-specific LepB homolog, TraF, expressed by RP4, which cleaves a further four amino acids from the C terminus and cyclizes the subunit by joining the N and C termini (Eisenbrandt *et al.*, 2000). The RP4 transfer region is separated into two parts, Tra1 and -2. In addition to *traF* of Tra1, the essential genes *trbB-L* of Tra2 are required for pilus assembly, with TrbK being involved in entry exclusion. The structure of TrbB has been determined and is a soluble, double ring-like structure which utilizes ATP during pilus synthesis and presumably, DNA transfer. A homolog of TraX is present in RP4 (TraP), although its substrate and function are unknown. Cyclic pilins have been identified in the Vir system of the Ti plasmid and are suspected in a number of other systems that carry a $TraF_P$ homologue.

2. *Phage attachment*

Conjugative pili act as the primary receptor for a wide range of bacteriophages (Frost, 1993). These phages can be divided broadly into those that bind to the pilus tip and those that bind to the sides of the pilus. The structure of the phages includes the single-stranded DNA filamentous phages, the small isometric RNA phages, and the complex double-stranded DNA tailed phages which usually attach near the pilus tip. The filamentous phages specific for F-like pili (Ff phages; M13, f1, and fd) attach via a defined region of the pilin attachment protein to an unknown receptor at the pilus tip. RNA phages such as R17 and Qβ, which belong to different phage groups, bind to specific residues in F pilin exposed on the pilus sides. For RP4 (IncPα), the filamentous phage Pf3 binds to the sides of the pilus as does the RNA phage PRR1. Tailed phages such as PRD1 and PR4 bind to the pilus tip and have a broad host range, including cells bearing IncP, -W, -N, and -I plasmids.

Although the pilus is required for initial attachment, the transfer region is not necessarily required for phage penetration or growth. The Ff phages are thought to contact the cell surface via the process of pilus retraction where they interact with the TolA protein and penetrate the cell via the TolQRA pathway. RNA phages R17 and Qβ have differing requirements for the TraD protein, which energizes the transport of nucleic acid through the conjugation pore, suggesting that R17 is imported via the F transfer machinery, whereas Qβ is taken up by another pathway.

3. *Role in conjugation*

The role of the pilus in conjugation has been controversial. It clearly has a role in Mpf, but whether it has a role in initiating DNA transfer or is part of the transfer apparatus has been difficult to determine genetically. Mutations that affect pilus formation block Mpf and DNA transfer. Mutations that affect mating pair stabilization (TraN and -G) allow the initial contacts to form between cells via the pilus and also allow the initiation of DNA synthesis in the donor cell but block DNA transport into the recipient cell. Although indirect, this is the best evidence that the pilus is involved in the signaling process and the Mps genes are involved in transport possibly by forming the conjugation pore. Whether pore formation also requires an intact pilus remains unknown. Other experiments in which donor and recipient cells were not allowed to establish cell-to-cell contact suggested that mating was possible through the extended pilus, and the homology between the pilus assembly genes and protein transport systems also argues for a role for the pilus in DNA transfer.

B. Mating pair formation

The pilus is thought to identify a receptor on the recipient cell surface, which triggers retraction of the pilus into the donor cell, although the route of the pilin subunits in this process is unknown. Pilus outgrowth is thought to require energy, whereas retraction occurs by default in the absence of assembly. Thus, factors that negatively affect cell metabolism (temperature, poisons, and carbon source) cause retraction. Whether pilus outgrowth and retraction are ongoing processes or whether binding of recipient cells or phage trigger retraction is unknown.

Early studies to identify mutations in the recipient cell that affected conjugation (Con^-) revealed that various components of the heptose-containing inner core of the lipopolysaccharide (LPS) were generally important in Mpf, whereas OmpA was required for efficient conjugation by the F transfer system. The F-like systems were each affected by different mutations in the *rfa* (now *waa*) locus in the recipient cell, whereas the IncH plasmids seemed to recognize a generalized negative charge on the recipient cell surface. The requirement for OmpA by F as well as for specific side chains in the LPS appears to be a function of the outer membrane protein TraN, whereas the idea that the pilus recognizes negatively charged moieties non-specifically remains a possibility. A second protein identified in F that is involved in Mps is TraG. Mutations in *traG* can be categorized into two classes with ones near the 5′ end affecting pilus formation and others affecting mating pair stabilization (Mps) in the 3′ portion of the gene.

An interesting variation on Mps has been identified for the IncI1 transfer systems (R64 and Collb-P9) which express two types of pili: thin flexible pili which are required for Mps and thick rigid pili that are required for DNA transfer. Research on the thin pili of R64 by Komano's group has revealed that they are composed of type IV pilin (similar to the pili found in pathogens such as *Neisseria gonorrhoeae*) of 15 kDa (*pilS*). These pili have a protein at their tip (pilV) whose gene undergoes rearrangement via site-specific recombination by the *rci* gene product to form seven possible fusion proteins, each recognizing a specific LPS structure (e.g. *pilVA′*). Whether these pili retract in order to bring the donor and recipient cells together is unknown, however, this is a general feature of type IV pili. Aside from these two cases (F and R64 thin pili), little is known about Mpf in other gram-negative systems.

1. *Surface/entry exclusion*

Surface or entry exclusion reduces redundant transfer between equivalent donor cells. Such transfer is thought to be deleterious to the donor cell and is exemplified by the phenomenon of lethal zygosis which occurs when a high ratio of Hfr donor to recipient cells is used. Multiple matings with a single recipient cell result in its death because of severe membrane and peptidoglycan damage as well as induction of the SOS response resulting from the influx of a large amount of single-stranded DNA. The surface exclusion genes were first identified as the *ilz* locus (immunity to lethal zygosis).

The mechanism of entry or surface exclusion is unknown, although some sort of exclusion mechanism has usually been found to be associated with the transfer systems studied to date. One exception is the conjugative transposons, which transfer at such low frequency that redundant transfer might not be a factor. Surface exclusion in the F system involves TraT, a lipoprotein found in the outer membrane of potential recipient cells, that forms a pentameric structure and blocks mating pair stabilization. Whether it interacts with the pilus or another component of the F transfer system is unknown. The TraS protein of F is an inner-membrane protein in the recipient cell which blocks the signal that DNA transfer from the donor cell should begin and is thus associated with the property of entry exclusion. TraS has been shown to exhibit plasmid specificity for the TraG protein from the same mating system expressed in the donor cell. TrbK, a lipoprotein found in the inner membrane of RP4-containing cells, is also thought to cause entry exclusion.

C. DNA metabolism

1. *Organization of oriT*

In gram-negative transfer systems, the origin of transfer (*oriT*) can be ~40–500 bp in length and contains intrinsic bends and direct and inverted repeats which bind the proteins involved in DNA transfer. The *nic* site, which is a strand- and sequence-specific cleavage site, is cleaved and religated by the relaxase. In most cases, the relaxase requires an auxiliary protein(s) which directs the relaxase to the *nic* site and ensures the specificity of the reaction. The sequences of the *nic* sites identified to date reveal five possible sequences represented by IncF, -P, and -Q and certain gram-positive plasmids such as pMV1158. In addition, there is usually a protein that binds to multiple sites within *oriT* forming a higher order structure in the DNA which is essential for the process. This protein also appears to have a function in anchoring the relaxosome to the transport machinery (Lanka and Wilkins, 1995; Zechner *et al.*, 2000).

2. *Mechanism of DNA transfer*

After mating pair formation, a signal is generated that converts the relaxosome from the cleavage/religation mode to one in which unwinding of the DNA is coupled to transport through the conjugation pore in an ATP-dependent manner. The transfer rate is ~750 nucleotides per second, with the F plasmid (100 kb) transferred in approximately 3 min.

In IncF plasmids, TraI is the relaxase/helicase enzyme which binds to a site near *nic* and generates an equilibrium between cleavage and religation. This reaction requires supercoiled template DNA and Mg^{2+} as well as the auxiliary proteins F TraY and host integration host factor (IHF) *in vitro*. The signal that triggers the helicase activity of TraI, which is essential for DNA transfer, is unknown, as is the function of TraI* produced by a translational restart in the *traI* mRNA. TraY binds near *nic*, whereas TraM binds to multiple sites, in conjunction with IHF, to form a nucleoprotein complex required for transfer to be initiated. TraM also binds to the protein that energizes transport, TraD, which is an inner-membrane protein that utilizes ATP probably via its two NTP binding domains. The crystal structure of the soluble domain of TrwB, a $TraD_F$ and $TraG_P$ homolog, has revealed a hexameric ring reminiscent of packaging proteins of phage and the F′-ATPase.

In RP4, a similar arrangement of proteins at *oriT* exists except that there is no role for the host protein, IHF. The relaxase protein, called TraI, cleaves at *nic* as a complex with TraJ, whereas TraH stabilizes the TraI, J complex. TraK binds and bends the DNA at *oriT* to form the nucleosome-like structure thought to be needed to initiate DNA replication.

In all cases, the 5′ phosphate generated by the cleavage reaction remains covalently bound to the relaxase enzyme via a tyrosine residue which is similar to the initiation of replication by the rolling circle mechanism in some phage and plasmid replicons. The DNA is transferred in a 5′ to 3′ direction, with the first genes to enter the recipient cell called the leading region. Transfer seems to be a precise process with termination of transfer after one copy of the plasmid has been delivered to the recipient cell. A sequence in *oriT*, near *nic*, is important for termination by the relaxase in a religation reaction. The DNA is released into the recipient cell and both strands are replicated in a manner dependent on the PolIII enzyme using discontinuous synthesis in the recipient and continuous synthesis either from the free 3′ end at *nic* or from an RNA primer in the donor. Although synthesis and transport are coupled in conjugation, DNA synthesis does not drive, nor is it required, for DNA transfer.

In RP4, the transport of a primase protein, Pri (*traC* encoded in Tra1), or Sog by IncI plasmids, has been demonstrated to occur simultaneously with the transport of the DNA, with hundreds of copies being transferred to the recipient cell. This protein appears to initiate DNA synthesis in the recipient cell via primer formation, although it is not essential for conjugation. In F, no primase is transferred and DNA synthesis is thought to begin via a mechanism utilizing *ssi* sites for single-stranded initiation.

3. *Leading region expression*

The first genes to enter the recipient cell in the leading region include genes for preventing restriction of the incoming DNA as well as the SOS response (*psi*) and for plasmid maintenance via poison-antidote systems such as CcdAB and Flm in F, Hok/Sok in R1, and Kil/Kor in RP4. One gene that is highly conserved in the leading region of conjugative plasmids is *ssb*, a single-stranded DNA binding protein that is not essential to the process. These genes are transcribed from single-stranded DNA promoters that ensure expression prior to the completion of replication. Another interesting but nonessential gene is *orf169* in F or TrbN in RP4 which is related to lytic transglycosylases such as lysozyme. Perhaps this gene, which is the first to enter the recipient cell during F transfer, has a role in establishing a new transferosome in the recipient cell, especially in strains of bacteria in the natural environment which might require a transglycosylase to rearrange the peptidoglycan in preparation for pilus assembly and DNA transfer.

D. Regulation

The regulation of the genes involved in conjugation has been extensively studied in F, RP4, and Ti (see Section VII) and in gram-positive conjugation, but there is little information on other systems. The regulation of F transfer gene expression depends on both host and plasmid-encoded factors, whereas the regulation of RP4 appears to be independent of the host. Also, F is unusual in that there is no evidence for coregulation of transfer and replication, a salient feature of other conjugation systems.

In F, there are three main transcripts encoding *traM*, *traJ* (the positive regulator of transfer operon expression), and *traY-I* (33 kb). The $P_{Y\text{-}I}$ promoter is controlled by a consortium of proteins, including the essential TraJ protein; TraY, the first gene in the operon; and IHF and SfrA (also known as ArcA) encoded by the host. TraY also controls *traM* expression from two promoters which are autoregulated by TraM. The translation of the *traJ* mRNA is controlled by an antisense RNA FinP which requires an RNA binding protein FinO for activity (fertility inhibition; see Section IV.D.1). Certain mutations in *cpxA* also affect TraJ stability probably via an indirect mechanism involving upregulation of factors that respond to stress applied externally to the cell.

In RP4, transfer is tied very closely to replication with the main replication promoter for TrfA divergently oriented and overlapping with the first of two promoters for the Tra2 transfer region, P_{trbA} and P_{trbB}. In this system, the *trfA* promoter is activated first in the transconjugant, promoting plasmid replication. The P_{trbB} promoter and the promoters in Tra1, which express the genes for Dtr, are also activated in order to establish a new transferosome. Eventually, the main global regulators KorA and KorB repress expression from these promoters and allow transcription from P_{trbA} which maintains the level of Tra2 proteins in the donor cell. TrbA is a global regulator that represses P_{trbB} and the three promoters in Tra1 encoding the genes for Dtr. Thus, conjugation leads to a burst of transcription that establishes the plasmid in the new donor cell followed by transcription from either the *trfA* or the *trbA* promoters during vegetative growth.

1. *Fertility inhibition*

Fertility inhibition (Fin) is a wide-spread phenomenon which limits the transfer of competing plasmids coresident in a single cell. The fertility inhibition systems of F-like plasmids (R factors) repress F and also autoregulate the expression of their own transfer regions. These systems have two components, the antisense RNA FinP and the RNA binding protein, FinO, which together prevent translation of the *traJ* mRNA, with TraJ being the positive regulator of the $P_{Y\text{-}I}$ promoter. FinO protects FinP antisense RNA from degradation by the host ribonuclease, RNase E, allowing the FinP concentration to increase sufficiently to block *traJ* mRNA translation. Although *finP* is plasmid-specific, *finO* is not, and can be supplied from many F-like plasmids. This is the basis of the fi^+ phenotype noted in the 1960s for various R factors. F lacks FinO since *finO* is interrupted by an IS3 element and consequently is constitutively derepressed for transfer.

In F-like plasmids ($FinOP^+$), 0.1–1% of a repressed cell population expresses pili. If conjugation is initiated, the transconjugant is capable of high frequency of transfer (HFT) for approximately six generations until fertility inhibition by FinOP sets in. This phenomenon, in addition to surface/entry exclusion which ensure transfer only to recipient cells, contributes to the epidemic spread and stable maintenance of a plasmid in a natural population of bacteria.

Other Fin systems are specified by one plasmid and are directed against another. For instance, F encodes PifC, which blocks RP4 transfer, whereas RP4 encodes the Fiw system, which blocks the transfer of coresident IncW plasmids. Each system has a unique mechanism which has made the study of Fin systems more difficult and has tended to downplay the importance of this phenomenon in the control of the dissemination of plasmids in natural populations.

V. GRAM-POSITIVE CONJUGATION

Conjugative elements in non-filamentous gram-positive bacteria can be subdivided into three groups: (i) small plasmids (130 kb), usually associated with MLS resistance, exhibit moderate transfer efficiency on solid surfaces over a broad host range (pAMβ1 and pIP501); (ii) large plasmids (+60 kb), which mate efficiently in liquid media over a narrow host range and undergo clumping or cell aggregate formation (pAD1 and pCF10 in *Enterococcus*; pSK41 and PGO1 in *S. aureus*); and (iii) conjugative transposons.

There is no evidence for pili in gram-positive conjugation. In the few systems studied in detail, detection of a recipient cell results in expression of an aggregation substance (AS, Agg, or Clu) which covers the surface of the donor cell and results in the formation of mating aggregates which are visible to the naked eye. This "fuzz" is the result of a complex pattern of gene expression which has been studied in detail only rarely, mostly in the large plasmids of *E. faecalis*. This group of plasmids responds to pheromones expressed by the recipient cell, whereas the signal that triggers mating aggregate formation in other systems is poorly understood. Plasmids in *S. aureus* produce pheromones that trigger transfer by *E. faecalis*, suggesting a mechanism to broaden the host range for plasmids of the latter organism.

A. *Enterococcus faecalis*

The first conjugative plasmid identified in gram-positive bacteria was pAD1, which carried a hemolysin/bacteriocin determinant responsible for increased virulence and which caused clumping or aggregation 30 min after the addition of plasmid-free cells (Dunny and Leonard, 1997). Later it was shown that aggregation and subsequent plasmid transfer were induced by pheromones produced by the recipient cell and released into the medium. These pheromones are hydrophobic peptides of seven or eight amino acids with a single hydroxyamino acid. Each recipient cell releases several pheromones, with plasmids from a particular incompatibility group recognizing a specific pheromone in which specificity resides in the N terminus of the peptide. In addition to their role in conjugation, the function of these pheromones is unknown, as is their source. Once the plasmid has become established in the transconjugant, expression of the pheromone is repressed by preventing its synthesis in or release from, the donor cell. Picomolar amounts of pheromone added to donor cultures can induce the expression of the transfer genes. Usually the concentration is approximately 1–10 nM and a slight increase in pheromone levels is sufficient to induce clumping. Pheromones are named after the transfer system that recognizes them; thus, pAD1 recognizes cAD1 and produces iAD1, an inhibitory peptide that blocks accidental induction of transfer.

The pheromone is recognized by a plasmid-encoded protein and imported into the cell via the host Opp system (oligopeptide permease). In pAD1, it binds to a repressor, TraA, inactivating it and allowing transcription of TraE1, a positive regulator required for induction of transfer gene expression. In pCF10, the pheromone cCF10 causes anti-termination of transcription of the *prg* genes (*p*heromone *r*esponsive *g*ene) generating the 530 nt RNA Q_L and the mRNA for aggregation substance and transfer proteins. The Q_L RNA, in conjunction with the pheromone, associates with the ribosome causing preferential translation of the transfer genes. In either case, transcription of the aggregation substance (AS or Asa1 for pAD1 and Asc10 for pCF10) ensues, and this substance is deposited asymmetrically on the surface of the cell until the cell surface is covered. This binds to the binding substance (BS; lipoteichoic acid) on the recipient cell to give the mating aggregates so characteristic of conjugation. Interestingly, both AS and BS have been associated with increased virulence in a rabbit endocarditis model, with AS having RGD motifs (arginine-glycine-aspartic acid) which are known to promote binding to the integrin family of cell surface proteins.

Once the pCF10 transferosome is established in the transconjugant, a cytoplasmic membrane protein binds pheromone and prevents its release. The inhibitory peptide, iCF10, is synthesized from a shorter version of Q_L called Q_S. A surface exclusion protein, Sea1 or Sec10 (for pAD1 or pCF10, respectively), is also expressed to reduce aggregation between donor cells and provides a second level of control to prevent redundant mating between plasmid-bearing cells.

There also seems to be a close relationship between replication and transfer in these plasmids, with the replication protein PrgW embedded within a region that negatively regulates transfer in pCF10. The replication protein also appears to have a requirement for pheromone that is not understood. Similarly, the *oriT* for pAD1 is within the *rep* gene for plasmid replication initiation.

B. *Streptomyces*

The transfer systems found on conjugative plasmids in the large genus *Streptomyces* differ significantly from those of both non-filamentous, gram-positive bacteria and gram-negative systems in that there is only a single essential transfer protein and no evidence for a relaxase or *nic* site has been found. *Streptomyces* are a medically important source of antibiotics and other therapeutic compounds and are thought to be a major reservoir for antibiotic resistance mechanisms which protect these bacteria from the arsenal of antibiotics they produce.

The conjugative plasmids of *Streptomyces* range in size and structure from the circular 9-kb plasmid pIJ101 to the large linear plasmid SCP1 (350 kb). The phenomenon of conjugation in this genus was first identified because of the ability of certain integrating plasmids to mobilize chromosomes (CMA). One such plasmid, SLP1 (17 kb), excises and integrates at the 3′ end of an essential $tRNA^{Tyr}$ gene in *S. lividans*, with tRNA genes providing the loci for integration of many phage and plasmids.

Streptomycetes are soil microorganisms that undergo a complex differentiation program whereby spores germinate and form substrate mycelia that penetrate into the support surface followed by the erection of aerial hyphae which are multinucleate mycelia. These mycelia eventually septate and form spores. As the cells enter the hyphal stage, they begin to produce the array of secondary metabolites characteristic of the organism and also enter a phase in which they are competent for conjugation between substrate mycelia or mycelia and other organisms such as *E. coli*.

The intermycelial transfer of a plasmid requires one essential protein (e.g. Tra in pIJ101) which has homology to the cell division protein FtsK in *E. coli* and SpoIIIE which is located at the asymmetric septum of sporulating *B. subtilis* cells and ensures that a copy of the chromosome enters the forespore. Since no relaxase protein has been found associated with pIJ101, it is tempting to speculate that this might be a conjugation system closely related to partitioning mechanisms involving double-stranded DNA.

Once the DNA has been transferred to one compartment of a long mycelium, the plasmid is distributed to all compartments by the *tra* and the *spd* gene products. This process slows the growth rate of the cell, and areas in which plasmids are being spread via inter- and intramycelial transfer form "pocks" of more slowly growing cells which resemble plaques on a phage titer plate. This phenomenon has been useful in identifying cells containing conjugative plasmids.

VI. MOBILIZATION

Mobilization is a widespread phenomenon whereby a smaller plasmid encoding its own *nic* site and Dtr genes (*mob*) utilizes the transport machinery of a usually larger plasmid to effect its own transfer. Many plasmids can be mobilized by plasmids from many Inc groups. For instance, ColE1 can be mobilized by plasmids from IncF, -P, and -I groups and less effectively and with different requirements by IncW plasmids. The Mob proteins of ColE1 consist of MbeA (relaxase) and MbeB and -C, which aid in relaxosome formation. MbeD has an entry-exclusion function. In the vector pBR322, derived from ColE1, the *mob* genes are deleted and only an *oriT* region for IncP plasmid mobilization remains. Although ColE1 requires TraD from F for mobilization, the closely related plasmid CloDF13 supplies its own TraD-like protein, a difference which is commonly seen among mobilizable plasmids.

The most remarkable mobilizable plasmid is RSF1010 and its relatives (~8.6 kb in size) from the IncQ group. These plasmids are mobilized very efficiently by plasmids from the IncP group into an extremely broad group of recipients, including bacteria, yeast, and plants. This plasmid encodes three Mob proteins, with MobA being the relaxase. Like ColE1, it requires the TraG protein of RP4 for efficient mobilization but it is only efficiently mobilized by F TraD if a C-terminal deletion is made. The *oriT* region is a mere 38 bp in size and is homologous to *oriT* regions in plasmids from gram-positive bacteria, all of which use a rolling circle mechanism during transfer. RSF1010 can be mobilized into plants and between agrobacteria by the *vir* region (not *tra*; see Section VII) and between strains of *Legionella* using the virulence determinants encoded by the *dot* and *icm* loci involved in macrophage killing (see Section VIII).

Small gram-positive plasmids are also mobilizable by self-transmissible plasmids but less is known about them. The utility of conjugative transposons as genetic tools in gram-positives has overshadowed interest in these plasmids, which tend to be mobilized at low frequencies over long time periods. One plasmid, which replicates and transfers via the rolling circle mechanism using different origins, is pMV1158. It encodes a relaxase (MobM) which cleaves at a *nic* sequence unique to a group of mobilizable plasmids found in gram-positive bacteria representing the fourth class of *nic* sequences.

VII. TRANSFER TO PLANTS

The phenomenon of DNA transfer from the bacterium *A. tumefaciens* to plant cells has features of both gram-negative (pilus expression) and gram-positive (induction of *tra* gene expression) and has been dealt with here separately (Fig. 22.3). *Agrobacterium tumefaciens* carrying large conjugative plasmids such as Ti

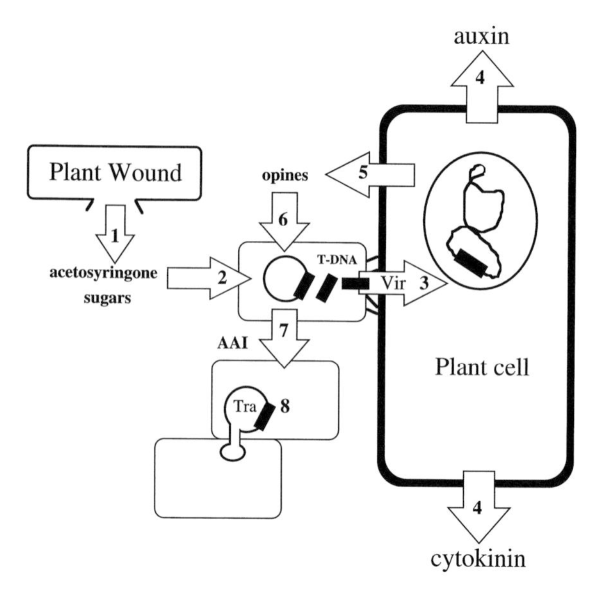

FIGURE 22.3 The signaling pathway used to stimulate T-DNA complex transfer to the plant nucleus and Ti conjugative transfer between agrobacteria. Wounded plant tissue releases phenolics (acetosyringone) and sugars (1) that are detected by the VirA, G two-component regulatory system (2). This induces expression of the *vir* genes which transports the T-DNA to the plant nucleus (3). The T-DNA is incorporated into the plant genome and produces the phytohormones auxin (indoleacetic acid) and cytokinin (4) which trigger tumorigenic growth of the plant tissue. The plant also produces opines (5), whose synthesis is encoded on the T-DNA. These unusual amino acids serve as a food source for *Agrobacterium* and also result in the induction of synthesis (6) of the conjugation factor, *N*-β-oxo-octanoyl- homoserine lactone (AAI; 7). AAI allows "quorum sensing," which determines cell density with respect to Ti-plasmid-bearing cells resulting in conjugative transfer to other agrobacteria (8).

(tumor-inducing) or Ri (root-inducing) of +200 kb in size cause crown gall disease in plants whereby they induce the formation of tumors at the site of infection. The Ti plasmid encodes a sensor-response regulator system, VirA and VirG, respectively, which in conjunction with ChvE, a chromosomally- encoded periplasmic sugar binding protein, process signals from wounded plant tissue. The phospho- transfer reaction from VirA to VirG induces gene expression from the *virA, -B, -D, -E,* and *-G* operons on the Ti plasmid. In addition, the *virC, -F,* and *-H* operons are induced but these operons express nonessential gene products that affect host range or the degree of virulence. The signals generated by the plant include phenolic compounds, simple sugars, and decreased pH or phosphate content.

The *virB* region encodes 11 proteins which are homologous to the gene products in the Tra2 region of RP4 and distantly related to the gene products of F. They encode the gene for prepropilin (VirB2), which is processed to cyclic pilin via a mechanism similar to that for RP4 pilin. A potential peptidase, homologous to TraF in RP4, has been identified (VirF) but its role has not been proven. The assembly of the VirB pilus is very temperature dependent, with an optimum of 19°C and a corresponding optimum for the transfer process. Considerable work has been done to understand the organization of the Vir proteins in the conjugative pore. In general, there appears to be a set of proteins that forms pores in the inner (VirB6, 8, and 10) and outer (VirB7, and 9) membranes with VirB2, 3, 4 and 5 involved in pilus assembly and associated with these complexes (Das and Xie, 2000).

The specific segment of single-stranded DNA that is transferred to the plant nucleus is called the T-DNA and can be characterized by the right (RB) and left (LB) borders which are direct repeats of 25 bp. The T-DNA of nopaline-producing Ti plasmids is approximately 23 kb in length and contains genes for plant hormone expression (13 kb), a central region of unknown function, and a region for opine (nopaline) biosynthesis (~7 kb). The relaxase, VirD2, in conjunction with VirD1, which is similar to RP4 TraJ, cleaves at RB and subsequently LB in a TraI-like manner and remains attached to the 5′ end. VirC1, which binds to an "overdrive" sequence near the RB, and VirC2 of certain Ti plasmids enhance T-intermediate formation. Unlike other transfer systems, many copies of the T-DNA segment accumulate in the cytoplasm suggesting replacement replication is important in this system. The accumulation of T-DNA strands has been puzzling but might represent a strategy by the bacterium to ensure infection of the larger, more complex plant cell.

The DNA in the VirD2–T-DNA complex (T-complex) is coated with the single-stranded DNA binding protein, VirE2, in preparation for transport through a conjugation pore composed of the VirB proteins (VirB2–VirB11). These proteins are very homologous to the Tra2 gene products for pilus synthesis in RP4 (Christie and Vogel, 2000), again suggesting the importance of the pilus to DNA transport. One of the Vir proteins, VirB1, which is nonessential, resembles the transglycosylase of F (Orf169) and RP4 (TrbN), whereas a truncated version of VirB1 (VirB1*) is excreted into the rhizosphere and mediates adhesion between the bacterium at the site of transfer and the plant. The process is energized by an RP4 TraG homolog, VirD4, which has NTP binding motifs that are required for its activity. Once the T-complex has entered the plant cytoplasm, the DNA is transported to the nucleus via nuclear localization signals on the VirE2 and VirD2 proteins. The T-DNA is randomly integrated into the plant genome whereupon it begins to elicit signals for plant hormone production resulting in tumor formation. The T-DNA encodes for the synthesis of auxin (indoleacetic acid) and cytokinin isopentenyl adenosine, plant hormones that elicit uncontrolled growth at the site of infection. The bacteria derive nutrients from the tumor by the devious method of opine (unusual amino acid) production encoded by the T-DNA. Opines can be classified into approximately nine different types of compounds, including octopine, nopaline, and agrocinopine, with up to three different opines being encoded by a particular T-DNA. Thus, Ti plasmids are often referred to as octopine- or nopaline-type plasmids, for instance, depending on the opine they specify. The opines are excreted from the plant and taken up by the *A. tumefaciens* bacteria encoding a region on the Ti plasmid involved in opine utilization. The genes for opine catabolism (e.g. *occ* for octopine catabolism) match the genes for the synthesis of that class of opines on the T-DNA.

An interesting aspect of Ti plasmid biology is the induction of conjugative transfer between agrobacterial cells in response to the presence of opines. The genes for this process (*tra*) are distinct from the genes for T-DNA transfer (*vir*) and encode a transfer region with homology to RP4 as well as the *vir* region.

Conjugative transfer by Ti has a narrow host range, limited to the genus of *Agrobacterium*. However, the host range can be extended to *E. coli* if an appropriate replicon is supplied, suggesting that it is plasmid maintenance and not conjugative functions that affect host range.

The process of inducing conjugative transfer in these bacteria is unique and fascinating (Fuqua *et al.*, 1996). Initially, there is a low level uptake of opines which activates the regulatory protein OccR in octopine-type plasmids and inactivates the repressor protein AccR in nopaline-type plasmids such as pTiC58. This leads to increased expression of the *tra* and opine utilization genes by activation of TraR. TraI (not to be confused with the relaxase proteins of F and RP4 plasmids) is a LuxI homolog which synthesizes a signalling compound, *N*-β-oxo-octanoyl-homoserine lactone (*Agrobacterium* autoinducer; AAI) belonging to a diverse class of homoserine lactone-like compounds involved in quorum sensing or gene activation in response to cell density (Fig. 22.3). TraR is a LuxR-like regulatory protein which detects increased levels of AAI and induces transfer gene expression to maximal levels. The result is the dissemination of the genes for opine utilization among the agrobacteria in the rhizosphere. Thus, the system demonstrates a certain degree of chauvinistic behavior since the original colonizer of the plant cell shares its good fortune with its neighbors who then out-compete other bacteria in the rhizosphere.

VIII. EVOLUTIONARY RELATIONSHIPS

With the advent of high-throughput sequencing and easily available databases, comparison of gene sequences has become routine. Considering that the mechanism for conjugation varies surprisingly little among the systems described previously, the high degree of relatedness of these systems with one another is expected (Table 22.2). However, the remarkable finding that there is homology between conjugative systems and transport mechanisms for many toxins and virulence determinants as well as tranformation systems in naturally competent bacteria has generated increased interest in conjugation resulting in its inclusion in Type IV secretion systems. These is almost gene- for-gene homology between the transport system for pertussis toxin of *Bordetella pertussis* (*ptl*) and the *virB* region of the Ti plasmid, which is in turn, homologous to the genes for pilus synthesis in IncN, -P, and -W transfer systems and is distantly related to those of F-like plasmids. Recently, five *vir* homologs (VirB4, -9, -10, and -11 and VirD4) have been found in the *cag* pathogenicity island of *Helicobacter pylori* while a separate type IV genomic island is responsible for natural transformation in this organism. Also, some of the Dot/Icm proteins involved in the pathogenesis of *Legionella pneumophila*, which can mobilize RSF1010, are homologous to genes in the Tra region of the IncI1 plasmid R64 and the Tra2 region of RP4. Other prominent type IV systems include DNA excretion in *Neisseria gonorrhoeae* which closely resembles F (Dillard, personal communication) as well as type IV-like genomic islands in *Rickettsia prowazekii* and *Brucella abortus* (Christie and Vogel, 2000). Analysis of the genome sequences in the databases reveals many type IV-like sequences, suggesting that this is a widespread transport mechanism. The array of substrates that can be transported by type IV systems is impressive, as is their ability to transport these substrates in either direction. This suggests that a primordial transport system, perhaps derived from a filamentous phage, has been adapted to many uses.

IX. CONJUGATION IN NATURAL ENVIRONMENTS

Although the process of conjugation is thought to be relevant to the adaptation of organisms to environmental conditions such as the acquisition of antibiotic resistance under continuous pressure for selection, there is much to be learned about the process in nature (Davison, 1999; Paul, 1999). Conjugation can be demonstrated in the gut of animals, biofilms, soil, aquatic environments including wastewater, on the surface of plants and animals, etc. However, the level of transfer is usually very low. Most experiments have utilized common lab strains and plasmids, which are good model systems for study, but might be irrelevant in nature. Considering the diversity of bacterial species, their vast numbers, and the time scale for their evolution, we have only scratched the surface of this phenomenon in the natural environment. However, studies on "domesticated" lab strains and plasmids have allowed predictions about the conditions that favor transfer. Most conjugative systems require actively growing cells in exponential phase and have a fairly precise temperature optimum. The majority of systems studied to date mate more efficiently on solid media. Those systems that mate efficiently in liquid media seem to be found in enteric bacteria and might be associated with diseases which are transmitted via the water supply. More information is required on the natural hosts for conjugative elements and their contribution to the evolution of these elements in an ecological niche. In addition, the most likely route of transmission, which appears to involve many intermediate organisms, is usually impossible to predict or detect because of the complexity of the system and the unknown role of nonculturable organisms in this process. Thus, we can isolate a plasmid from its environment and we can find evidence for its transfer to a new species, but we cannot, at this time, follow the plasmid as it makes its way in the world.

BIBLIOGRAPHY

Bonheyo, G., Graham, D., Shoemaker, N. B., and Salyers, A. A. (2001). Transfer region of a Bacteroides conjugative transposon, CTnDOT. *Plasmid* **45**, 41–51.

Bradley, D. E., Taylor, D. E., and Cohen, D. R. (1980). Specifications of surface mating systems among conjugative drug resistance plasmids in *Escherichia coli* K-12. *J. Bacteriol.* **143**, 1466–1470.

Christie, P. J. (2001). Type IV secretion: intercellular transfer of macromolecules by systems ancestrally related to conjugation machines. *Mol. Microbiol.* **40**, 294–305.

Christie, P. J., and Vogel, J. P. (2000). Bacterial type IV secretion: conjugation systems adapted to deliver effector molecules to host cells. *Trends Microbiol.* **8**, 354–360.

Clewell, D. B. (1993). "Bacterial Conjugation." Plenum, New York.

Das, A., and Xie, Y.-H. (2000). The *Agrobacterium* T-DNA transport pore proteins VirB8, VirB9, and VirB10 interact with one another. *J. Bacteriol.* **182**, 758–763.

Davison, J. (1999). Genetic exchange between bacteria in the environment. *Plasmid* **42**, 73–91.

Dunny, G. M., and Leonard, B. A. B. (1997). Cell–cell communication in gram-positive bacteria. *Annu. Rev. Microbiol.* **51**, 527–564.

Frost, L. S. (1993). Conjugative pili and pilus-specific phages. *In* "Bacterial Conjugation" (D. B. Clewell, ed.), pp. 189–221. Plenum, New York.

Eisenbrandt, R., Kalkum, M., Lai, E. M., Lurz, R., Kado, C. I., and Lanka, E. (1999). Conjugative pili of IncP plasmids and the Ti plasmid T pilus are composed of cyclic subunits. *J. Biol. Chem.* **274**, 22548–22555.

Frost, L. S., Ippen-Ihler, K., and Skurray, R. A. (1994). Analysis of the sequence and gene products of the transfer region of the F sex factor.

Fuqua, C., Winans, S. C., and Greenberg, E. P. (1996). Census and consensus in bacterial ecosystems: The LuxR–LuxI family of quorum-sensing transcriptional regulators. *Annu. Rev. Microbiol.* **50**, 727–751.

Gomis-Ruth, F. X., Moncalian, G., Perez-Luque, R., Gonzalez, A., Cabezon, E., de la Cruz, F., and Coll, M. (2001). The bacterial conjugation protein TrwB resembles ring helicases and F1-ATPase. *Nature* **409**, 637–641.

Hinerfeld, D., and Churchward, G. (2001). Specific binding of integrase to the origin of transfer (*oriT*) of the conjugative transposon Tn*916*. *J. Bacteriol.* **183**, 2947–2951.

Lanka, E., and Wilkins, B. M. (1995). DNA processing reactions in bacterial conjugation. *Annu. Rev. Biochem.* **64**, 141–169.

Lawley, T. D., Gilmour, M. W., Gunton, J. E., Standeven, L. J., and Taylor, D. E. (2002). Functional and mutational analysis of conjugative transfer region 1 (Tra1) from the IncHI1 plasmid R27. *J. Bacteriol.* **184**, 2173–2180.

Komano, T., Yoshida, T., Narahara, K., and Furuya, N. (2000). The transfer region of IncI1 plasmid R64: similarities between R64 *tra* and *Legionella icm/dot* genes. *Mol. Microbiol.* **35**, 1348–1359.

Ottow, J. C. G. (1975). Ecology, physiology, and genetics of fimbriae and pili. *Annu. Rev. Microbiol.* **29**, 79–108.

Pansegrau, W., Lanka, E., Barth, P. T., Figurski, D. H., Guiney, D. G., Haas, D., Helinski, D. R., Schwab, H., Stanisich, V. A., and Thomas, C. M. (1994). Complete nucleotide sequence of Birmingham IncPα plasmids. Compilation and comparative analysis. *J. Mol. Biol.* **239**, 623–663.

Paranchych, W., and Frost, L. S. (1988). The physiology and biochemistry of pili. *Adv. Microbial Physiol.* **29**, 53–114.

Paul, J. H. Microbial gene transfer:an ecological perspective. *J. Mol. Microbiol. Biotechnol.* **1**, 45–50.

Pohlman, R. F., Genetti, H. D., Winans, S. C. (1994). Common ancestry between IncN conjugal transfer genes and macromolecular export systems of plant and animal pathogens. *Mol. Microbiol.* **14**, 655–668.

Salyers, A. A., Shoemaker, N. B., Stevens, A. M., and Li, L. Y. (1995). Conjugative transposons: an unusual and diverse set of integrated gene transfer elements. *Microbiol. Rev.* **59**, 579–590.

Shapiro, J. A. (1977). Bacterial plasmids. *In* "DNA Insertion Elements, Plasmids, and Episomes" (A. I. Bukhari, J. A. Shapiro, and S. L. Adhya, eds.), pp. 601–670. Cold Spring Harbor Laboratory Press, Cold Spring Harbor, NY.

Thomas, C. M. (2000). "The Horizontal Gene Pool: Bacterial Plasmids and Gene Spread", Harwood Academic, Amsterdam.

Zechner, E. L., de la Cruz, F., Eisenbrandt, R., Grahn, A. M., Koraimann, G., Lanka, E., Muth, G., Pansegrau, W., Thomas, C. M., Wilkins, B. M., and Zatyka, M. (1999). Conjugative DNA transfer processes. *In* "The Horizontal Gene Pool: Bacterial Plasmids and Gene Spread" (C. M. Thomas, ed.), pp. 87–174. Harwood Academic Publishers, Amsterdam.

WEBSITES

Website on bacterial conjugation (M. E. Mulligan)
http://www.mun.ca/biochem/courses/4103/topics/conjugation.html

Relevant links to PubMed
http://www.ohsu.edu/cliniweb/G5/G5.386.html

23

Crystalline bacterial cell surface layers (S layers)

Uwe B. Sleytr and Paul Messner

Centre for Ultrastructure Research and Ludwig Boltzmann Institute for Molecular Nanotechnology, University of Agricultural Sciences

GLOSSARY

crystalline surface layers (S layers) In prokaryotic organisms, composed of protein or glycoprotein subunits.

S-layer (glyco)proteins Present as monomolecular arrays on surfaces of archaea and bacteria.

two-dimensional protein crystals Regular arrays of (glyco)proteins present as outermost envelope component in many prokaryotic organisms.

Surface Layers (S layers) are cell envelope components on prokaryotic cells consisting of two-dimensional crystalline arrays of (glyco)protein subunits. S layers have been observed in species of nearly every taxonomical group of walled bacteria and represent an almost universal feature of archaeal envelopes. As porous crystalline arrays covering the cell surface completely, S layers have the potential to function: (i) as protective coats, molecular sieves, and molecule and ion traps; (ii) as structures involved in cell adhesion and surface recognition, and (iii) in archaea, which possess S layers as exclusive wall components, as a framework that determines and maintains cell shape.

I. INTRODUCTION

The different cell wall structures observed in prokaryotic organism, particularly the outermost envelope layers exposed to the environment, reflect evolutionary adaptations of the organisms to a broad spectrum of selection criteria. Crystalline cell surface layers (S layers) are now recognized as common features of both bacteria and archaea.

Most of the presently known S layers are composed of a single (glyco)protein species endowed with the ability to assemble into two-dimensional arrays on the supporting envelope layer. S layers, as porous crystalline membranes completely covering the cell surface, can apparently provide the microorganisms with a selective advantage by functioning as protective coats, molecular sieves, molecule and ion traps, and as a structure involved in cell adhesion and surface recognition. In those archaea that possess S layers as exclusive envelope components outside the cytoplasmic membrane, the crystalline arrays act as a framework that determines and maintains the cell shape. They may also aid in cell division. In pathogenic organisms S layers have been identified as virulence factors.

S layers, as the most abundant of bacterial cellular proteins, are important model systems for studies of structure, synthesis, assembly, and function of proteinaceous components and evolutionary relationships within the prokaryotic world. S layers also have considerable application potential in biotechnology, biomimetics, biomedicine, and molecular nanotechnology.

II. LOCATION AND ULTRASTRUCTURE

Although considerable variation exists in the complexity and structure of bacterial cell walls, it is possible to classify cell envelope profiles into the following

The Desk Encyclopedia of Microbiology
ISBN: 0-12-621361-5

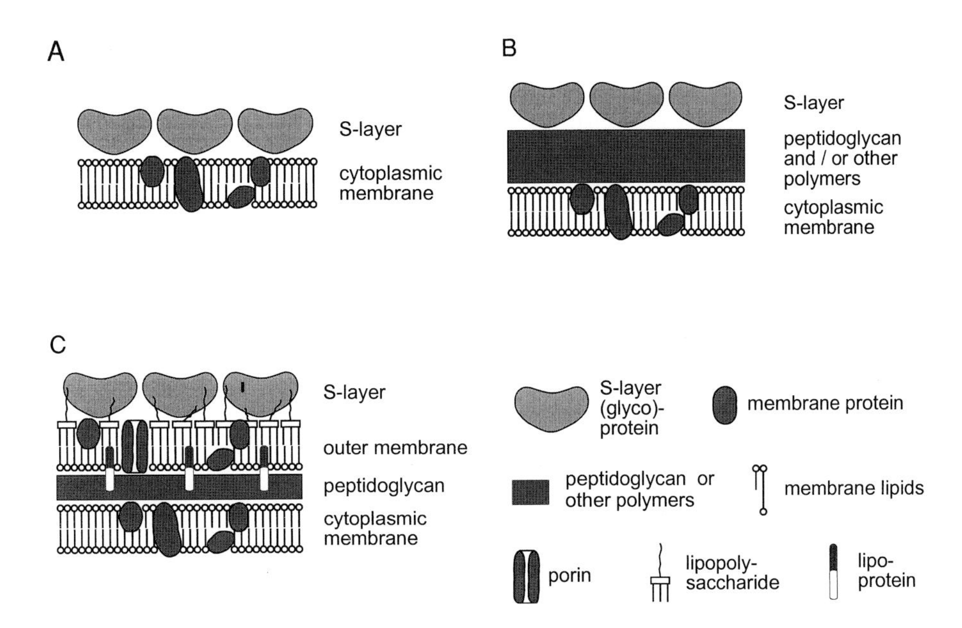

FIGURE 23.1 Schematic illustration of major classes of prokaryotic cell envelopes containing crystalline cell surface layers (S layers). (a) Cell envelope structure of gram-negative archaea with S layers as exclusive cell wall component; (b) the cell envelope as observed in gram-positive archaea and bacteria. In bacteria the rigid wall component is primarily composed of peptidoglycan. In archaea other wall polymers are found; (c) in gram-negative bacteria the S layer is closely associated with the outer membrane. [Modified from Sleytr, U.B., Messner, P., Pum, D., and Sára, M. (eds.) 1996. "Crystalline Bacterial Cell Surface Layer Proteins", R. G. Landes Company/Academic Press, Austin, TX.]

main groups on the basis of structure, biochemistry, and function (Fig. 23.1).

1. Cell envelopes formed exclusively of a crystalline S layer composed of (glyco)protein subunits external to the cytoplasmic membrane (most halophilic, thermophilic and acidophilic, alkaliphilic, barophilic, and gram-negative archaea (Fig. 23.1a).
2. Gram-positive cell envelopes of bacteria with a rigid peptidoglycan containing sacculus of variable thickness outside the cytoplasmic membrane, and gram-positive cell envelopes of archaea with a rigid sacculus composed of pseudomurein or other polymers (Fig. 23.1b).
3. Gram-negative envelopes of bacteria with a thin peptidoglycan sacculus and an outer membrane (Fig. 23.1c).

Although not a universal feature as in archaea, crystalline arrays of (glyco)proteins have been detected as outermost envelope components in organisms of most major phylogenetic branches of gram-positive and -negative bacteria.

Currently, the most useful electron microscopic preparation procedure for detecting S layers on intact cells is freeze-etching (Fig. 23.2). S layer completely cover the cell surface at all stages of cell growth and division in both archaea and bacteria. High-resolution electron microscopical and scanning force microscopical studies revealed, that S-layer lattices can have

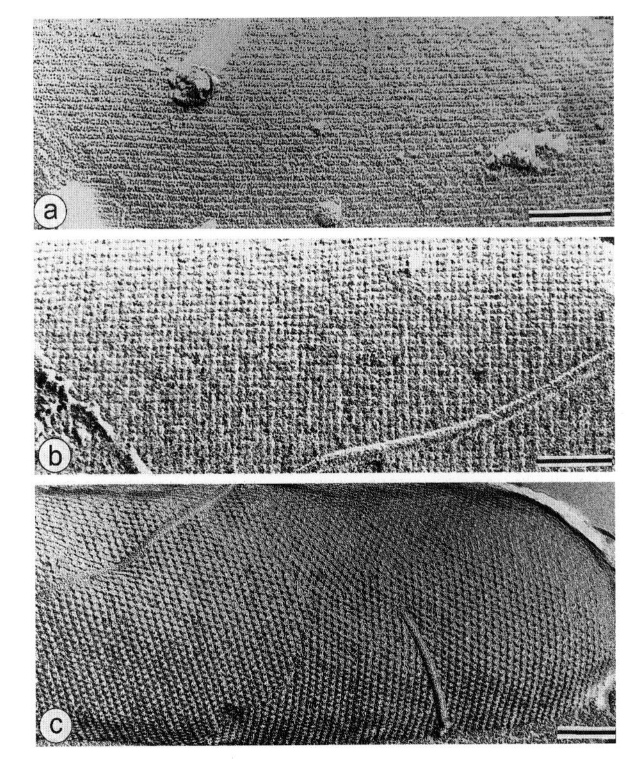

FIGURE 23.2 Electron micrographs of freeze-etched preparations of intact bacteria. (a) *Lactobacillus acidophilus* SH1; (b) *Thermoanaerobacterium thermosaccharolyticum* D120-70; (c) *Thermoanaerobacter thermohydrosulfuricus* L111-69. The oblique (a), square (b), and hexagonal (c) S layer completely covers the cell surface. Bars = 100 nm.

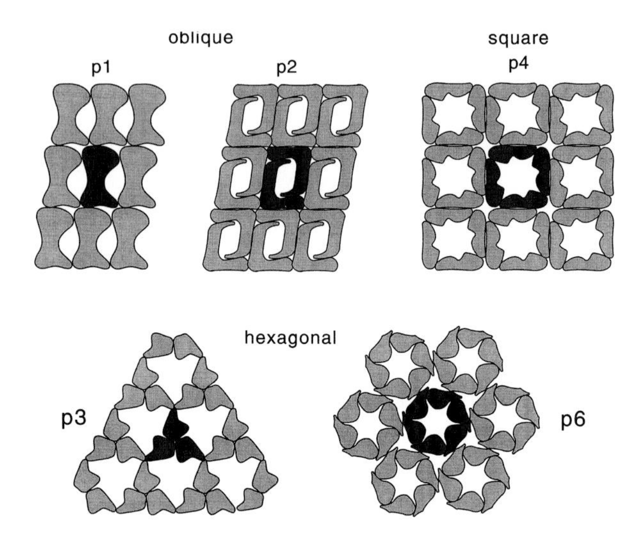

FIGURE 23.3 Schematic illustration of the space groups found for S-layer lattices. The unit cells which are the building blocks of the lattice are composed of mono-, di-, tri-, tetra-, or hexamers.

oblique (p1, p2), square (p4), or hexagonal (p3, p6) symmetry (Fig. 23.3) with a center-to-center spacing of the morphological units (composed of one, two, three, four, or six identical monomers) of approximately 5–35 nm. Among archaea, hexagonal lattices were shown to be predominant. S layers are generally 5–25 nm thick and have a smooth outer and a more corrugated inner surface. Since S-layer lattices are monomolecular assemblies of identical subunits they exhibit pores of identical size and morphology (Fig. 23.3). S layers can display more than one type of pores. From high resolution electron microscopical and permeability studies, pore sizes in the range of approximately 2–8 nm and a porosity of the protein meshwork between 30 and 70% have been estimated.

Comparative studies on the distribution and uniformity of S layers have revealed that in some species individual strains can show a remarkable diversity regarding lattice symmetry and lattice dimensions. In some organisms, two or even more superimposed S-layer lattices have been identified.

III. ISOLATION, CHEMICAL CHARACTERIZATION, AND ASSEMBLY

S layers of different bacteria may vary considerably with respect to their resistance to disruption into their monomeric subunits, and a wide range of methods have been applied for their isolation and purification. The subunits of most S layers interact with each other and with the supporting envelope layer through non-covalent forces. Additionally, secondary cell wall polymers have been recognized as components which facilitate a specific interaction between the S-layer monomers and the cell shape determining peptidoglycan sacculus. Most commonly, in gram-positive bacteria, a complete disintegration of S layers into monomers can be obtained by treatment of intact cells or cell walls with high concentrations of H-bond-breaking agents (e.g. urea or guanidine hydrochloride). S layers from gram-negative bacteria frequently disrupt upon application of metal chelating agents (e.g. EDTA and EGTA), cation substitution (e.g. Na^+ to replace Ca^{2+}), or pH changes (e.g. $pH < 4.0$). From extraction and disintegration experiments it can be concluded that the bonds holding the S-layer subunits together are stronger than those binding the crystalline array to the supporting envelope layer. There are some indications that S layers of some archaea (e.g. *Thermoproteus* species, and *Methanospirillum hungatei*) are stabilized by covalent bonds between adjacent subunits.

Most S layers are composed of a single, homogeneous protein or glycoprotein species. Sodium dodecyl sulfate polyacrylamide gel electrophoresis (SDS-PAGE) revealed apparent molecular masses for subunits in the range of approximately 40–220 kDa.

Comparison of amino acid analyses and genetic studies on S layers from both archaea and bacteria have shown that the crystalline arrays are usually composed of weakly acidic proteins. Typically they contain 40–60% hydrophobic amino acids and possess a low portion, if any, of sulfur-containing amino acids. The isoelectric points p*I* are in the range of 4–6 but for *Methanothermus fervidus* and for lactobacilli p*I* values in the range of 8–10 have been determined.

A few post-translational modifications are known to occur in S-layer proteins, including protein phosphorylation and protein glycosylation. In an S-layer (termed A layer) mutant of *Aeromonas hydrophila* the presence of phosphotyrosine residues has been detected by a comparison of the molecular masses obtained from SDS-PAGE and the DNA sequence. A more frequently observed modification of S-layer proteins from archaea as well as bacteria is their glycosylation. Whereas amongst archaea most S layers appear to be glycosylated, evidence for S-layer glycoproteins in bacteria is limited to members of the Bacillaceae. The glycan chains and linkages of these bacterial and archaeal glycoproteins are significantly different from those of eukaryotes (Fig. 23.4). Most archaeal S-layer glycoprotein glycans consist of only short heterosaccharides, usually not built of repeating units. The predominant linkage types are N-glycosidic bonds such as glucose → asparagine and

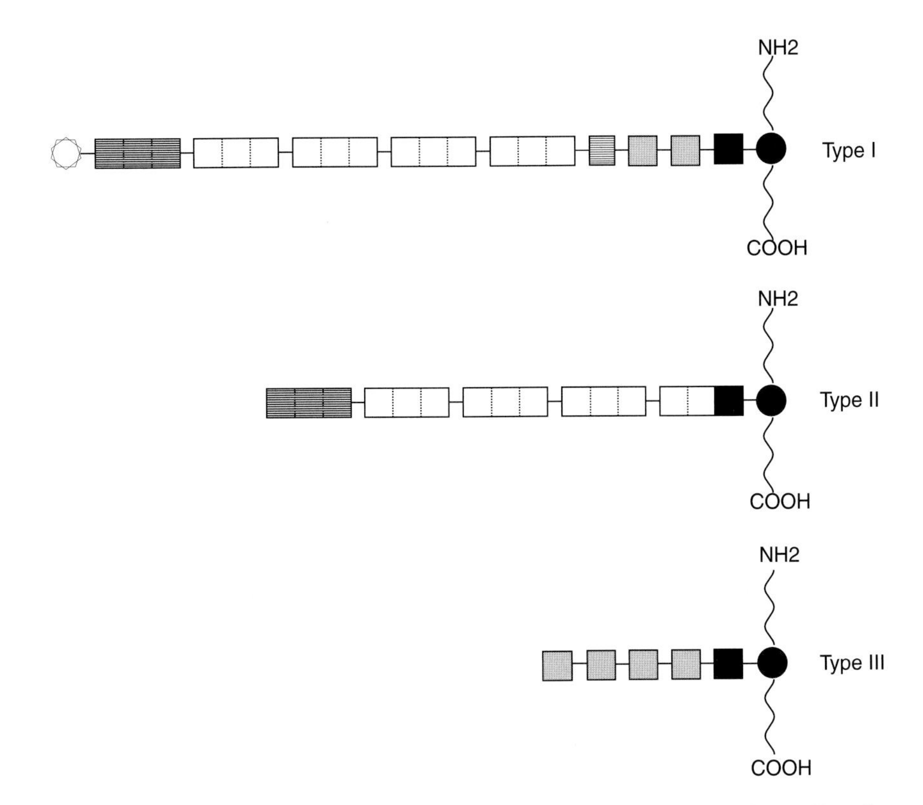

FIGURE 23.4 Schematic presentation of the architectural building plans of S-layer glycoprotein glycans. Type I, extended structure comprising a core region and an S-layer glycan chain (predominantely in bacterial S-layer glycoproteins); Type II, simplified extended structure, comprising a pseudo-core region and an S-layer glycan chain; Type III, heterosaccharide structure. On bacterial S-layer glycoproteins, this type presumably represents only the core structure without repeating units attached. In archaea, however, this type of S-layer glycan is the predominant one. *In vivo*, no repeating units are synthesized by these organisms and the heterosaccharide could consist of 10–20 sugar residues. (Repeating units, composed of two to six monosaccharides, are represented by open symbols; shaded squares indicate residues of core-oligosaccharides; possible variations in length, either in the core or the glycan chain, are indicated by hatched squares; sugar (squares) and amino acid (circles) residues participating in the glycosidic linkage are indicated by full symbols. The star-like symbol at the nonreducing end of the type I glycan represents modification of the terminal unit with O-methyl groups.) [Modified from Schäffer, C. and Messner, P. (2001). *Biochimie* **83**, 591–599.]

N-acetylgalactosamine → asparagine. The opposite situation can be found with bacterial S-layer glycoproteins. To date, only O-glycosidic linkages have been found. Among these are common linkages such as β-galactose → serine/threonine (Fig. 23.5), but also novel linkage types such as β-glucose → tyrosine and β-galactose → tyrosine have been observed. Most of the glycans are assembled of identical repeating units with up to 150 monosaccharide residues. Structurewise they are comparable to lipopolysaccharide (LPS) O-antigens of gram-negative bacteria.

Isolated S-layer subunits of gram-positive and -negative bacteria have also shown the ability to recrystallize on the cell envelope fragments from which they had been removed, on those of other organisms, or on untextured charged or uncharged inanimate surfaces.

To date, the most detailed self-assembly and reattachment experiments have been performed with S layers from Bacillaceae. S layers of these organisms reveal a high anisotropic charge distribution. The inner surface is negatively charged, whereas the outer face is charge-neutral, approximately pH 7. This characteristic of the S-layer subunits appears to be essential for the proper orientation during local insertion in the course of lattice growth.

Detailed studies have been performed to elucidate the dynamic process of assembly of S layers during cell growth. Freeze etching preparations of rod-shaped cells generally reveal a characteristic orientation of the lattice with respect to the longitudinal axis of the cylindrical part of the cell (see Fig. 23.2). For maintenance of such a good long-range order during cell growth S-layer protomers must have the ability to recrystallize on the supporting envelope layer. Labeling experiments with fluorescent antibodies and colloidal gold/antibody marker methods indicated that different patterns of S-layer growth exist for

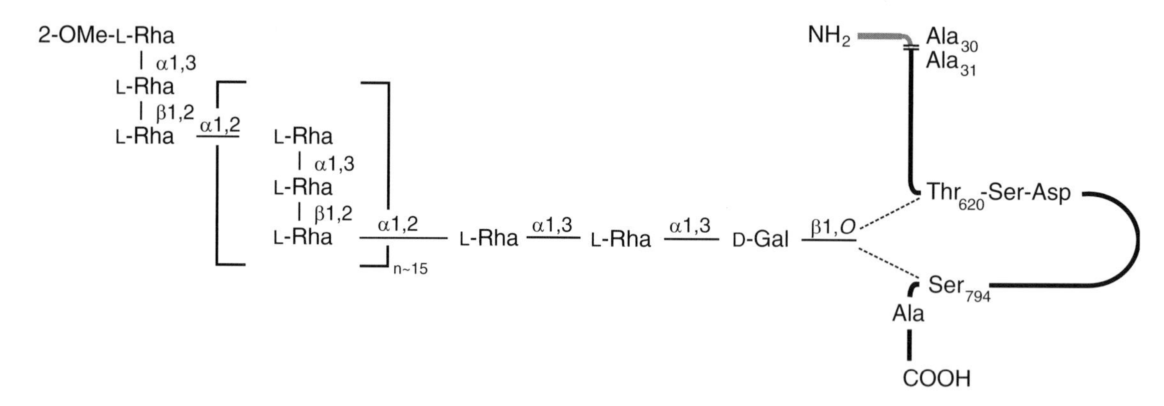

FIGURE 23.5 Complete S-layer glycoprotein structure of *Geobacillus stearothermophilus* NRS 2004/3a, showing the glycosylation sites on the S-layer protein SgsE precursor and the signal peptide. [Modified from Schäffer, C., Wugeditsch, T., Kählig, H., Scheberl, A., Zayni, S., and Messner, P. (2002). *J. Biol. Chem.* **277**, 6230–6239.]

gram-positive and -negative bacteria. In gram-positives, growth of the S-layer lattice primarily occurs on the cylindrical part of the cell by insertion at multiple bands or helically arranged bands; in gram-negatives, incorporation of new subunits occurs at random. In both types of organisms, entirely new S-layer material also appears at regions of incipient cell division and the newly formed cell poles.

IV. GENETICS AND BIOSYNTHESIS

An intact, "closed" S layer on an average-sized, rod-shaped cell consists of approximately 5×10^5 monomers. Thus, at a generation time of about 20 min, at least 500 copies of a single polypeptide species with an M_r of approximately 100,000 have to be synthesized, translocated to the cell surface, and incorporated into the S-layer lattice per second.

During the last decade, a substantial amount of information of the biosynthesis and the genetics of S layers has accumulated. Insight into the molecular organization of S-layer biosynthesis was obtained from cloning experiments and sequencing studies. Sequence comparison of S-layer genes from different archaea and bacteria have revealed that homologies between non-related organisms are low, although their amino acid compositions show no significant differences. High homology scores are usually explained by evolutionary relationships but other factors, such as growth conditions and environmental stress, may also be determinative for structural homologies of S-layer genes.

Since S layers are the predominant proteins of the bacterial cell their promoters must be very strong and efficient. For example, the promoter of the S-layer gene from *Lactobacillus acidophilus* is twice as effective as that of the gene encoding lactate dehydrogenase, which is considered as one of the strongest promoters in many bacteria.

Important for understanding of S-layer gene regulation was the observation that single bacterial strains can express different genes. Currently, the best investigated organism is *Campylobacter fetus* which interferes with reproductive function in ungulates. In this system all studies have demonstrated that only a single promoter exists and that antigenic variation is due to recombination events. *Campylobacter fetus* rearranges a single promoter strictly by a single DNA inversion event and at frequencies independent on the size of the DNA fragment. This allows expression of different S-layer gene cassettes. The variation enables the organism to circumvent the host's immune response.

Environmental factors can also induce changes in S-layer synthesis. This phenomenon has been studied in detail in *Geobacillus* (formerly *Bacillus*) *stearothermophilus* strains. When the wild-type strain was grown in continuous culture under oxygen limitation the corresponding S-layer gene was stably expressed. After relieving oxygen limitation, expression of a second S-layer gene was induced, resulting in an S-layer protein with decreased molecular mass of the monomeric unit. Both genes showed only low homology (<50%) and they possess quite different N termini. These regions recognize different secondary cell wall polymers of the bacterium as binding sites for S layers and indicate a highly coordinated change of cell envelope components due to different physiological conditions.

Concomitant with the genetic studies, knowledge was gathered about the secretion of S-layer proteins. General protein secretion pathways are known in a number of bacteria. With the exception of the S layers from *Campylobacter* and *Caulobacter* strains, all others are produced with a signal peptide, suggesting the classical route of secretion. For the S layers of *Aeromonas salmonicida* and *Caulobacter crescentus* specific secretion pathways, such as an ABC transporter system, have been described.

The remarkable diversity in glycan structures (Fig. 23.4) raised interesting questions about the biosynthesis of prokaryotic glycoproteins. Recently a model has been proposed for how the different N-linked glycans are synthesized in *Halobacterium halobium*. It includes transfer of dolichol-linked saccharides to consensus sequences for N-glycosylation on the S-layer polypeptide. Lipid-activated oligosaccharides with short-chain (C_{55}–C_{60}) dolichol species rather than undecaprenol for S-layer glycosylation were observed not only in *H. halobium* but also in *Haloferax volcanii* and *M. fervidus*. In comparison to archaeal S-layer glycoproteins, less detailed information is available regarding the biosynthesis of glycosylated S layers of bacteria. Only recently, the exact positions of glycosylation sites have been determined on the sequenced structural gene *sgsE* of *Geobacillus stearothermophilus* NRS 2004/3a (Fig. 23.5). The nucleotide sequences of three S-layer glycan biosynthesis clusters are now completed which show the molecular organization of the glycosylation machinery of S-layer glycan biosynthesis in bacteria. Fuctional expression of several nucleotide-activated sugars, which are constituents of the mature S-layer glycans, has been achieved. Currently, we are investigating the remaining genes in the clusters which code for enzyme proteins such as glycosyl transferases, transporters of the glycan chains across the cytoplasmic membrane, modification enzymes, and ligases which eventually link the S-layer glycan to the S-layer polypeptide.

V. FUNCTIONAL ASPECTS AND APPLICATION OTENTIAL

Since prokaryotic organisms possessing S layers are ubiquitous in the biosphere, it can be expected that the porous network of monomolecularly arranged (glyco)proteins· has evolved as the result of quite diverse interactions of the bacteria with specific environmental and ecological conditions.

Although relatively few data are available on specific functions of S layers, there is strong evidence that the crystalline arrays have the potential to function as: (i) cell shape-determining framework; (ii) protective coats, molecular sieves, and molecule or ion traps; and (iii) promoters of cell adhesion and surface recognition (Table 23.1).

TABLE 23.1 Functions of S layers

Determination of cell shape and cell division (in archaea that possess S layers as the exclusive wall component)
Protective coats
Prevents predation by *Bdellovibrio bacteriovorus* (in gram-negative bacteria)
Phage resistance by S-layer variation
Prevention or promotion of phagocytosis
Adhesion site for exoenzymes
Surface recognition and cell adhesion to substrates
S layers function as physicochemical and morphological well-defined matrices
Masking the net negative charge of the peptidoglycan-containing layer in Bacillaceae
Isoporous molecular sieves
Molecular sieves in the ultrafiltration range
Delineating in gram-positive bacteria a compartment (periplasm)
Preventing non-specific adsorption of macromolecules
Virulence factor in pathogenic organisms
Important role in invasion and survival within the host
Specific binding of host molecules
Protective coat against complement killing
Ability to associate with macrophages and to resist the effect of proteases
Production of S layers which do not immunologically cross-react (S-layer variation)
Fine grain mineralization

Sleytr, U. B. (1997). *FEMS Microbiol. Rev.* **20**, 5–12.

A. Cell shape determination

Because of their structural simplicity and from a morphogenetic point of view, it was suggested that S layers, like lipid membranes, could have fulfilled barrier and supporting functions as required for self-reproducing systems (progenotes) during the early period of biological evolution.

Analysis of the distribution of lattice faults of hexagonal S layers of archaea which possess an S layer as their sole cell wall constituent (Fig. 23.1a) provided strong evidence that complementary pairs of pentagons and heptagons play an important role in site for the incorporation of new subunits, in the formation and maintenance of the lobed cell structures, and in the cell fission process. The latter appears to be determined by the ratio between the increase in protoplast volume and the increase in actual S-layer surface area during cell growth.

B. S layers related to pathogenicity

Particular attention has been paid to S layers present on pathogenic organisms. Crystalline arrays present on members of the genera *Aeromonas* and *Campylobacter* comprise one of the best studied S-layer systems. In *Aeromonas salmonicida*, which causes furunculosis in salmons, generally an S layer is required for virulence. The S layer physically protects the infecting cells against proteolysis and complement and is essential for macrophage infiltration and resistance. In *Campylobacter fetus* subspecies *fetus*, the agent that causes abortion in sheep and cattle and various systemic infections or acute diarrheal illness in humans, S layers appear to make the cells resistant to phagocytosis and to bactericidal activity of serum.

Crystalline surface layers identified on other pathogens of humans and animals, including species of *Bacteroides, Brucella, Chlamydia, Cardiobacterium, Rickettsia, Wolinella, Treponema, Clostridium*, and *Bacillus*, may be of similar functional relevance as virulence factors.

C. S layers as molecular sieves and promoters for cell adhesion and surface recognition

Data obtained by high-resolution electron microscopy indicate that S layers are highly porous structures with pores of defined size and morphology (see Fig. 23.3). Permeability studies on S layers of mesophilic and thermophilic Bacillaceae provided information on their molecular sieving properties. For example, S layers of *G. stearothermophilus* revealed sharp exclusion limits for molecules larger than 30–40 kDa, indicating a limiting pore diameter in the crystalline protein meshwork of about 4.0 nm. With some mesophilic Bacillaceae, pore sizes as small as approximately 2.5 nm have been determined. These permeability studies also showed that the pores in the S-layer membranes have a low tendency for fouling, a feature regarded as essential for an unhindered exchange of nutrients and metabolites up to a defined molecular size.

S layers acting as molecular sieves have the potential to function not only as barriers preventing molecules from entering the cell (e.g. lytic enzymes, complements, antibodies, and biocides). They can also generate a functional equivalent to the periplasmic space of gram-negative bacterial envelopes in preventing the release of molecules (e.g. enzymes and toxins) from the cell. On the other hand, S layers of Bacillaceae were found to function as adhesion site for cell-associated exoenzymes.

Cells of *Aquaspirillum serpens* and other gram-negative species were shown to be resistant to predation by *Bdellovibrio bacteriovorus* when they were covered by an S layer. However, S layers of different organisms (e.g. Bacillaceae) were demonstrated to act as specific sites for phage adsorption. Analysis of phage-resistant mutants of *Bacillus sphaericus* showed that although the crystalline arrays were present on all mutants, the molecular weight of the S-layer subunits had changed.

With regard to cell adhesion and cell recognition properties of S layers, an important aspect is the frequently observed anisotropic distribution of charged groups on both faces of the (glyco)protein lattice. Most Bacillaceae possessing S layers reveal a charge-neutral outer surface, which physically masks the net negatively charged peptidoglycan sacculus. In comparison with S-layer-deficient strains, S-layer-carrying Bacillaceae have also shown a greater ability to adsorb to positively charged or hydrophobic surfaces.

D. Application potential

Based on the data obtained in fundamental studies on S layers, a considerable potential in biotechnological and nonbiological applications is evident. Applications for S layers have been found in the production of isoporous ultrafiltration membranes, as supports for a defined covalent attachment of functional molecules (e.g. enzymes, antibodies, antigens, protein A, biotin, and avidin) as required for affinity and enzyme membranes, in the development of biosensors, or in solid-phase immunoassays. S-layer membranes have also been used as support for Langmuir lipid films or liposomes, mimicking the molecular architecture of gram-negative archaea and virus envelopes (see Fig. 23.1a). S-layer fragments or self-assembly products are well suited for a geometrically well-defined covalent attachment of haptens and immunogenic or immunostimulating substances. These haptenated S-layer structures act as strong immunopotentiators. Finally, cloning and characterization of genes encoding S-layer proteins, opened new areas of applied S-layer research. Incorporation of functional domains without hindering self-assembly of S-layer subunits into regular arrays will lead to new types of recombinant vaccines, affinity matrices, diagnostics, biocompatible surfaces, and in combination with liposomes as drug-targeting and -delivery systems as well as systems for gene therapy. Recently it was demonstrated that S layers can be employed in non-life science nanostructure technologies (e.g. non-linear optics and molecular electronics) and as matrix for controlled biomineralization.

BIBLIOGRAPHY

Beveridge, T. J. (1994). Bacterial S -layers. *Curr. Opin. Struct. Biol.* **4**, 204–212.

Beveridge, T. J., and Graham, L. L. (1991). Surface layers of bacteria. *Microbiol. Rev.* **55**, 684–705.

Boot, H. J., and Pouwels, P. H. (1996). Expression, secretion and antigenic variation of bacterial S-layer proteins. *Mol. Microbiol.* **21**, 1117–1123.

Kneidinger, B., Graninger, M., Puchberger, M., Kosma, P., and Messner, P. (2001). Biosynthesis of nucleotide-activated D-*glycero*-D-*manno*-heptose. *J. Biol. Chem.* **276**, 20935–20944.

König, H., and Messner, P. (eds.) (1997). International Workshop on Structure, Biochemistry, Molecular Biology, and Applications of Microbial S-Layers (Special issue). *FEMS Microbiol. Rev.* **20**, 1–178.

Messner, P., and Schäffer, C. (2000). Surface layer glycoproteins of Bacteria and Archaea. *In* "Glycomicrobiology" (R. J. Doyle, ed.), pp. 93–125, Kluwer Academic/Plenum Publishers, New York.

Messner, P., and Schäffer, C. (2003). Prokaryotic glycoproteins. *In* "Progress in the Chemistry of Organic Natural Products", Vol. 85 (W. Herz, H. Falk, and G. W. Kirby, eds.), pp. 51–124, Springer-Verlag, Vienna.

Moll. D., Huber, C., Schlegel, B., Pum, D., Sleytr, U. B., and Sára, M. (2002). S-layer-streptavidin fusion proteins as template for nanopatterned molecular arrays. *Proc. Natl. Acad. Sci. USA* **99**, 14646–14651.

Pum, D., and Sleytr, U. B. (1999). The application of bacterial S-layers in molecular nanotechnology. *TIBTECH* **17**, 8–12.

Sára, M., and Sleytr, U. B. (2000). S-layer proteins. *J. Bacteriol.* **182**, 859–868.

Schäffer, C., and Messner, P. (2001). Glycobiology of surface layer proteins. *Biochimie* **83**, 591–599.

Schäffer, C., Wugeditsch, T., Kählig, H., Scheberl, A., Zayni, S., and Messner, P. (2002). The surface layer (S-layer) glycoprotein of *Geobacillus stearothermophilus* NRS 2004/3a. Analysis of its glycosylation. *J. Biol. Chem.* **277**, 6230–6239.

Schuster, B., and Sleytr, U. B. (2000). S-layer-supported lipid membranes. *Rev. Mol. Biotechnol.* **74**, 233–254.

Sidhu, M. S., and Olsen, I. (1997). S-layers of *Bacillus* species. *Microbiology*, **143**, 1039–1052.

Sleytr, U. B. (1997). Basic and applied S-layer research: an overview. *FEMS Microbiol. Rev.* **20**, 5–12.

Sleytr, U. B., and Beveridge, T. J. (1999). Bacterial S-layers. *Trends Microbiol.* **7**, 253–260.

Sleytr, U. B., and Sára, M. (1997). Bacterial and archaeal S-layer proteins: structure–function relationships and their biotechnological applications. *TIBTECH*, **15**, 20–26.

Sleytr, U. B., Messner, P., Pum, D., and Sára, M. (1993). Crystalline bacterial cell surface layers. *Mol. Microbiol.* **10**, 911–916.

Sleytr, U. B., Messner, P., Pum, D., and Sára, M. (eds.) (1996). "Crystalline Bacterial Cell Surface Layer Proteins." Landes/Academic Press, Austin, TX.

Sleytr, U. B., Messner, P., Pum, D., and Sára, M. (1999). Crystalline bacterial cell surface layers (S-layers): from supramolecular cell structure to biomimetics and nanotechnology. *Angew. Chem. Int. Ed.* **38**, 1034–1054.

Sleytr, U. B., Sára, M., and Pum, D. (2000). Crystalline bacterial cell surface layers (S-layers): a versatile self-assembly system. *In* "Supramolecular Polymerization" (A. Ciferri, ed.), pp. 177–213, Marcel Dekker, New York.

Sleytr, U. B., Sára, M., Pum, D., and Schuster, B. (2001). Characterization and use of crystalline bacterial cell surface layers. *Prog. Surface Sci.* **68**, 231–278.

Sleytr, U. B., Sára, M., Pum, D., and Schuster, B. (2001). Molecular nanotechnology and nanobiotechnology with two-dimensional protein crystaly (S-layers). *In* "Nano-Surface Chemistry" (M. Rosoff, ed.), pp. 333–389, Marcel Dekker, New York.

Sleytr, U. B., Sára, M., Pum, D., Schuster, B., Messner, P., and Schäffer, C. (2002). Self-assembly protein systems: microbial S-layers. *In* "Biopolymers", Vol. 7, Polyamides and Complex Proteineceous Matrices I (A. Steinbüchel, and S. R. Fahnestock, eds.), pp. 285–338, Wiley-VCH, Weinheim.

Sumper, M., and Wieland, F. T. (1995). Bacterial glycoproteins. *In* "Glycoproteins" (J. Montreuil, J. F. G. Vliegenthart, and H. Schachter, eds.), pp. 455–473, Elsevier, Amsterdam.

WEBSITES

A website of the Center for Ultrastructure Research, University of Agricultural Sciences, Vienna
http://www.boku.ac.at/zuf/res0.htm

List of bacterial names with standing in nomenclature (J. P. Euzéby)
http://www.bacterio.cict.fr/index.html

24

Culture collections and their databases

Mary K. B. Berlyn
Yale University

GLOSSARY

alkanotrophic Capable of assimilating propane, *n*-butane, and other *n*-alkane gases and liquid hydrocarbons as sole carbon and energy sources.

hyphomycetes Grouping of asexually reproducing fungi ("fungi imperfecti") in which the asexual spores (conidia) are produced on loose, cottony hyphae.

lyophilization The process of simultaneously freezing and drying materials under a vacuum.

patent depository In accordance with patent laws of most countries, novel microorganisms involved in a patent must be deposited in a recognized patent depository. Signatories to an International Budapest Treaty agreed that deposits in any approved International Depository Authority will be acceptable for meeting their patent office deposition requirements. (See, e.g., *http://www.atcc.org/ProgramsAndServices/PatentDep.cfm)*

A Stock Culture Collection is a repository for strains or varieties or species of organisms developed for the purpose of preservation and distribution of a useful range of the organisms. A distinction can be made between genetic stock collections, which are primarily mutant derivatives of one or more founder strains, along with related non-mutant strains, and germplasm and type culture collections, which have more heterogeneous holdings. This article is limited to microbial stock centers, although a few nonmicrobial centers are cited in recounting the history of genetic stock collections. The descriptions of early stock centers are presented in approximately chronological order, followed by genetic stock centers and by the later, national collections and consortia, which are grouped by country. These and more specialized collections are also listed under type of organism. Descriptions and access information are presented in a Table of Stock Centers.

I. FEDERATIONS SPONSORING DATABASES THAT IDENTIFY COLLECTION RESOURCES

The World Federation for Culture Collections (WFCC) is a Commission of the International Union of Biological Sciences (IUBS) and a federation within the International Union of Microbiological Societies (IUMS) (*http://wdcm.nig.ac.jp/wfcc/AboutWFCC.html* and *www.wfcc.info*). Its purpose is to support and promote establishment of culture collections. In collaboration with the Microbial Resources Centres (MIRCEN) and other organizations described below, it sponsors a database that is a directory of culture collections, the World Data Centre for Microorganisms, WDCM. This database began as a registry of worldwide culture collections initiated by V. B. D. Skerman at the University of Queensland in the 1960s. The registry was published in 1972 as the World Directory of Collections of Cultures of Microorganisms (WDCCM) by the WDCM, as a result of a proposal by the Japanese Federation of Culture Collections. The Directory was supported by the United Nations Educational, Scientific, and Cultural Organization

The Desk Encyclopedia of Microbiology
ISBN: 0-12-621361-5

(UNESCO), the World Health Organization (WHO), the Australian Commonwealth Scientific and Industrial Research Organization (CSIRO), and the Canadian Research Council and its second edition in 1982 also by the United Nations Environmental Program (UNEP). Its fourth edition was published in 1993. Currently, the WDCM provides an on-line database of the collections and also of databases on microbes and cell lines. The server is located at the National Institute of Genetics in Japan (*http://wdcm.nig.ac.jp;* and *http://wdcm.nig.ac.jp/DOC/menu3.xml*). It includes large and small collections, numbering nearly 500, located in public and private institutions in 55 countries. Particularly useful is an index of species and serovars held by culture collections with links to the culture collections holding the selected strain of interest. Many of these collections are cited in later sections of this article. The species list is found at *http://wdcm.nig.ac.jp/wfcc/species_list.html* and the index at *http://wdcm.nig.ac.jp/index.html*. The WDCM also allows querying by species or collection (e.g. *http://wdcm.nig.ac.jp/fsearch.html*) or by key word (*http://wdcm.nig.ac.jp/DOC/menu2.xml*) and presents a list of consulting services provided by the various stock centers (*http://wdcm.nig.ac.jp/wfcc/services_list. html*) and a listing by country (*http:// wdcm.nig.ac.jp/ hpcc.html*).

MIRCEN is a UNESCO-sponsored network of academic and research institutes to foster collaboration in environmental, applied microbiological, and biotechnological research in cooperation with the National Commissions of Member States of the United Nations (*http://www.asm.org/International/index.asp?bid=14221*). MIRCEN is also a sponsor of the WDCM database. This is acknowledged by the official designation of the database as the WFCC-MIRCEN WDCM. *Rhizobium* is a particular focus for MIRCEN, and other areas include biotechnology, aquaculture, and bioinformatics. Historically UNESCO's role in preserving microbial diversity dates to 1946 (see DaSilva, 1977). Currently, the largest members of the culture collection network include the Belgian Coordinated Collections (BCCM), the German Collection (DSMZ), Braunschweig, The Culture Collection of Hungary, the National Collection of Type Cultures (NCTC) in London, C.A.B. International Mycological Institute (IMI), and the American Type Culture Collection (ATCC), all of which are described in sections that follow. However, the network also supports collecting and coordination of collections at MIRCEN centers and groups throughout the world, including Bangkok, Dakar, Hawaii, Nairobi, and sites in African, Latin American, European, Arab, and Asian countries too numerous to mention. MIRCEN also sponsors a server located in Cairo for searching microbial collections In addition to the WDCM, the bioinformatics area of MIRCEN includes BITES, the Biotechnological Information Exchange System in Slovenia, and Bioinformatics MIRCEN at the Karolinska Institute in Stockholm (*http://home.swipnet.se/~w-2586/jacky/mircen*).

Other organizational sponsors of the WDCM are the National Institute of Genetics Center for Information Biology of Japan (CIB), the Japan Science and Technology Corporation (JST), the Committee on Data for Science and Technology (CODATA), UNESCO, and UNEP.

The Microbial Strain Data Network (MSDN) is a nonprofit organization that provides links to databases of interest to microbiologists and biotechnologists. It is sponsored by UNEP. The secretariat is in Sheffield, UK, and it is managed by an international committee. The European node was established in conjunction with CABRI, the Common Access to Biotechnological Resources and Information project. The Institute of Microbiology of China is a collaborator and has established an MSDN node in China as part of the Microbial Information Network of China, and there are nodes in India and Brazil as well. Databases for small, specialized collections as well as the large heterogeneous national and international collections are included in the MSDN Web site. It allows searches of databases in the Czech Republic, Siberia, India, Japan, Russia, Slovenia, Argentina, Bulgaria, and the UK (*http://www.im.ac.cn/msdn.html, http://www.bdt.fat.org.br/mrcs, http://wdcm.nig.ac.jp/msdn/MSDN.html*, and *http://panizzi.shef.ac.uk/msdn*).

The CABRI consortium provides the MSDN node in the UK and for Europe, providing links to MSDN, several of the large European culture collections, and cell culture and biological materials sources (*http://www.cabri.org/collections.html*).

The National Veterinary School of Toulouse has a website by J.P.M. Euziby with an extensive list of culture collections throughout the world (*www.bacterio.cict.fr/collections.html*) including links to websites (*www.bacterio.cict.fr/links.html*).

The Microbial Germplasm Database (MGD) at Oregon State University emphasizes strain collections available from individual university, industry, government, and National Science Foundation (NSF)-supported laboratories and research stations, in contrast to public collections. It includes algae, bacteria, fungi, and viruses, and provides contact information for the laboratory collections. It also publishes an on-line newsletter. The MGD is supported by the US Department of Agriculture (USDA) and NSF (*http://mgd.nacse.org/cgi-bin/mgd*). Bergey's Manual Trust (responsible for publishing new editions of the classic authority on bacterial taxonomy, *Bergey's*

Manual of Determinative Bacteriology) maintains on its Web site a list of major collections and consortia (*http://www.cme.msu.edu/bergeys/resourceinfo.html*).

There are also directories with more general purposes which include information about stock collections, for example, *http://www.seaweed.ie, http://biodiversity.uno.edu/~fungi/fcollect.html*, and MycDB. MycDB is sponsored by the WHO and Follereau Foundation, originally at the Royal Institute of Technology in Stockholm andnowat the Pasteur Institute in Paris, and is primarily a source of molecular biology information on mycobacteria, but it does provide information on availability of stocks of mycobacterial species in stock centers (*http://www.pasteur.fr/mycdb*). Databases that link or search multiple collections are also presented at the end of Table 24.1 (Section VII).

II. EARLY GERMPLASM AND TYPE CULTURE COLLECTIONS

The Collection of the Bacterial Strains of Institut Pasteur (CIP) traces its origins to Dr. Binot, who began to collect strains in 1891. It is a private, nonprofit collection that has become enlarged primarily by collaboration with the research laboratories of the Pasteur Institute. It maintains, preserves (by lyophilization since 1952), distributes, and provides information about strains. The CIP joined the World Federation of Culture Collections (WFCC) and the European Culture Collections Organization (ECCO, a collaborative organization for curators founded in 1982), and it is part of the Microbial Information Network Europe (MINE). Querying for strains can be performed at *http://www.pasteur.fr/CIP*. The individual records include links to other collections, as well as descriptions and literature references.

The Mycological Collection of the Catholic University of Louvain [now part of the Belgian Coordinated Collections of Microorganisms (BCCM)—see National Collections and Consortia in Table 24.1] was founded in 1892–1894 by Prof. Philibert Biourge, and in 1901 Biourge's *Penicillium* collection was augmented with Prof. F. Dierckx's *Penicillium* and *Aspergillus* strains. During World War I, however, much of the collection disappeared, with accessions increasing again after the war. It was managed in the 1930s through the 1960s by Prof. P. Simonart, then by Prof. G. L. Hennebert as it merged with his fungus collection, which had begun in 1956. Other significant mergers brought in hyphomycetes from Prof. J. A. Meyer and the yeast collection of Biourge. The current holdings are described in Table 24.1 as part of the BCCM consortium (*http://www.belspo.be/bccm*).

The Uppsala University Culture Collection (Mykoteket or UPSC) has its origins intermingled with that of the Botanical Museum at Uppsala University and its extensive herbarium of plants, algae, and fungi, which date back to 1785 and C. Thunberg, a disciple of Linnaeus, and more specifically to the E. Fries fungal herbarium of the mid-1800s. The current fungal culture collection, the UPSC or Mycoteket, holds over 3000 strains. Its on-line database is found at *http://www-hotel.uu.se/evolmuseum/mykotek/index.html*.

The American Museum of Natural History/Society of American Bacteriologists Collection originated with an 1899 recommendation by the Society of American Bacteriologists (now the American Society for Microbiology) calling for a central repository for bacterial cultures. Such a collection was begun at the American Museum of Natural History in New York by MIT professor C.-E. A. Winslow in 1911.

Similarly, a proposal by the Association Internationale des Botanistes led to the establishment of the Centraal Bureau voor Schimmelcultures (CBS) of the Netherlands in 1903 as a central collection of fungi, with 80 cultures available in 1906. In 1922 it moved to Delft Technical University. It was initially supported privately, but it became an institute of the Royal Netherlands Academy of Arts and Sciences, financially supported by the Dutch government, in 1968. The CBS includes as a separate unit the Phabagen Collection (see Microbial Genetic Stock Centers) and the Bacterial Collection of Kluyver Laboratory of Microbiology (LMD), which have merged to become the Netherlands Culture Collection of Bacteria (NCCB). See the section on national collections in Table 24.1. A goal of the CBS is to preserve representatives of virtually all fungal groups that can be cultured, and the cultures currently number 35,000 with annual increases of approximately 1000 (*http://www.cbs.knaw.nl/www/collection.htm* and *http://www.cbs.knaw.nl/www/cbshome.html*).

The USDA Culture Collection also had its origins in the early 1900s. Although plant germplasm is the best-known of the collections supported and administered by the US Department of Agriculture, the early history of US microbial germplasm and genetic stock collections finds the USDA in a central role. Drs. Charles Thom and Margaret Church established a USDA collection in Washington, DC, in 1913 that had begun with Dr. Thom's mold cultures in Connecticut and developed into the source of strains for both the USDA/Agricultural Research Service (ARS) Culture Collection and the American Type Culture Collection.

The USDA/ARS Culture Collection developed from this early USDA collection. In 1940, the Northern Regional Research Laboratory (NRRL) opened in

Peoria, and Dr. Kenneth Raper formally established the ARS Collection there. (The laboratory is now named the National Center for Agricultural Utilization Research, NCAUR.) The NRRL collection included approximately 2000 Thom and Church cultures of molds, deposits of citric acid-producing aspergilli, bacterial strains, and L. J. Wickerham's yeasts. Large-scale use of lyophilization (freeze-drying) for preserving cultures was pioneered at the NCAUR for these cultures. Collections contributed to the ARS collection from later research programs include Mucorales, Taphrina, filamentous fungi, yeasts, aerobic spore-forming bacteria, rhizobia, and actinomycetes. The collection accepts strains that are nonpathogenic to humans and animals and are not fastidious in their growth requirements. Plant and animal cell lines are not accepted. Active research programs of curators and other scientists at the center, as well as outside researchers, utilize and enhance the collections and their applications. A notable, historic contribution was the development of an industrial process for producing penicillin during World War II. The collections include a patent culture collection as well as the "open" public collection, totaling over 80,000 strains (*http://nrrl.ncaur.usda.gov/the_collection3.htm* and *http://nrrl.ncaur.usda.gov*). On-line databases for the Actinomycetales, bacterial, and filamentous fungal collections can be searched at *http://nrrl.ncaur.usda.gov/ searchcc.htm*.

The USDA/ARS also maintains a Collection of Entomopathogenic Fungal Cultures (ARSEF), begun in the 1970s at the University of Maine and moved to the Plant Protection Research Unit on the Cornell campus in Ithaca, New York. A fungal herbarium is also associated with this collection (*http://www.ars.usda.gov/is/np/systematics/fungibact.htm*).

A third USDA/ARS collection is the National *Rhizobium* Germplasm Resource Center. The first cultures were isolated from northern Virginia in 1913 and increased by USDA scientists in response to research demands, particularly in the 1930s and 1940s. Because of its international agricultural significance, it was initially funded by the US Agency for International Development, and in the 1980s it was designated a UNESCO Microbiological Resource Center (see MIRCEN above). Since 1990, the Agricultural Research Service has supported the collection (*http://bldg6.arsusda.gov/pberkum/Public/cc1a.html*). A searchable index can be found at the Web site for the Germplasm Resources Information Network of the USDA/ARS (*http://www.ars-grin.gov/cgi-bin/nmgp/rhy/search.pl*).

An algal collection developed by Prof. E. G. Pringsheim and colleagues at the University of Prague (Charles University) formed the basis for both the CCAP, the Culture Collection of Algae and Protozoa, now in the UK, and the CCALA, the Culture Collection of Autotrophic Organisms in the Czech Republic. The collection developed by Pringsheim, V. Czurda, and F. Mainx moved with Prof. Pringsheim to England and was later expanded and directed by Prof. E. A. George at Cambridge University, with support from the UK Natural Environment Research Council. Its two current components, the Institute of Freshwater Ecology (IFE) collection and Dunstaffnage Marine Laboratory collection, are cited in the algal collections section (see Table 24.1). The algal collection of Profs. V. Uhlir and Pringsheim in Prague was merged with the algal collection at the Institute of Microbiology, Czech Academy of Sciences at Trebon, to form the CCALA, also cited with algal collections in Table 24.1.

The American Type Culture Collection (ATCC) was founded in 1925 by representatives from the National Research Council, the Society of American Bacteriologists, the American Phytopathological Society, the American Zoological Society, and the McCormick Institute for Infectious Diseases. It was located at first at the McCormick Institute in Chicago, moving in 1937 to the Georgetown University School of Medicine and to houses in Washington, DC, in 1947 and 1956. It was incorporated as an independent, nonprofit institution in 1947. It then included microorganisms, animal and plant viruses, and cell lines. In 1964, it moved into a facility in Rockville, Maryland, and in 1998 to Manassas, Virginia. It has been supported by many grants from the US National Science Foundation (NSF) and National Institutes of Health (NIH) directed toward the various specialized collections and activities—hybridomas, bacterial collections, databases, yeasts, protozoa, probes, clones, etc.—as well as by users' fees. The holdings are extensive, with over 14,000 bacterial strains, 500 viral strains, and 26,000 strains of 6500 yeast and fungal species. It is also an official US patent repository and conducts workshops and education sessions. It is the distribution source for microbial genome clones for completely sequenced prokaryotic genomes from the Institute for Genomic Research (TIGR) in Maryland and for many cDNA clones (*http://www.atcc.org*).

The International Mycological Institute (IMI) Culture Collections are part of the institute founded in 1920 and is currently part of CAB (originally called Commonwealth Agricultural Bureaux) International, a nonprofit organization supported by 32 governments. The current collection has over 16,500 strains of filamentous fungi, yeasts, and bacteria. It is a patent depository and offers identification, preservation, testing, consultation, and training services and undertakes applied research. A database may be searched at *http://www.cabi-bioscience.org/html/Resource Centre.htm*. (See also */ReferenceCollections.htm*)

III. EARLY GENETIC STOCK CENTERS

In most cases, genetic stock centers originated when early geneticists working with the species recognized that the stocks were valuable for future research and made accommodations for preserving and disseminating these stocks. The realization of the need for the stocks was often accompanied by the realization that a means for disseminating new scientific results in a rapid and somewhat informal way was also required for advancing scientific progress with the organism. For most stock centers the information function has progressed from species-specific newsletters and strain lists to comprehensive on-line databases.

Among the earliest stock centers in the United States are the *Drosophila* Stock Center and the Maize Cooperation Genetics Stock center. The *Drosophila* center grew from the mutant collection originating with T. H. Morgan and his students at Columbia, beginning in 1913. Their stocks, maintained by Bridges, were provided to anyone requesting them. In 1928, when Morgan, Bridges, Sturtevant, and the collection moved to the California Institute of Technology, that became the site of the *Drosophila* stock center and of the stock list published in the *Drosophila* Information Service. This stock center is currently at the University of Indiana and includes approximately 6000 stocks (T. Kaufman and K. A. Matthews, *http://flybase.bio.indiana.edu/stocks*).

The Maize Genetics Cooperation Stock center had its origins in the 1928 Winter Science meetings in New York, when a group of maize geneticists discussed work on the maize linkage maps and the idea of an organized Maize Genetics Cooperation. A formalized proposal in 1932 included provision for the Maize Genetics Cooperation News Letter and the Maize Genetics Cooperation Stock Center. Marcus M. Rhoades served as first secretary of the Newsletter and first director of the stock center. Currently, the Maize Cooperation Stock Center is at the University of Illinois, USDA/ARS (M. Sachs), and the Maize Genetics Cooperation Newsletter is produced at ARS/USDA, University of Missouri (E. Coe). The collection includes nearly 80,000 pedigreed samples, including alleles of several hundred genes, combinations of such alleles, chromosome aberrations, ploidy variants, and other variations. Details about the collection, its history, available stocks, and request forms can be found at *http://w3.ag.uiuc.edu/maize-coop/mgc-info.html*. (The newsletter can be found on the Maize Genome Database site at *http://www.agron.missouri. edu/mnl.html*.)

IV. MICROBIAL GENETIC STOCK CENTERS

A. Bacterial Genetic Stock Centers

Sexuality and the ability to make genetic crosses between bacteria was discovered in *E. coli* in the 1940s. The use of microorganisms to explore the relationship between genes and biochemical pathways also became established in that decade. Individual laboratories accumulated large numbers of *E. coli* mutants, primarily derivatives of the wild-type isolate *E. coli* K-12. In the 1960s, it became apparent that a national repository would greatly aid the free exchange of strains and the advance of molecular genetics.

At that time, support from the US NSF was a critical element in developing such a resource, and the NSF supported a proposal to begin with E. A. Adelberg's Yale University collection of stocks and add important strains and sets of strains from laboratories worldwide. This became the *E. coli* Genetic Stock Center (CGSC) at Yale, curated for 25 years (until 1993) by B. Bachmann. The CGSC holds over 8000 strains and a plasmid library of cloned segments covering nearly all of the *E. coli* genome. Unlike the previous two centers, a newsletter was not associated with the founding of the stock center, but it soon assumed the functions of registering gene names and allele numbers, as set forth in the widely accepted guidelines for bacterial nomenclature by Demerec *et al.* (1966), and meeting subsequent needs for registry to avoid duplication of designations for deletions, insertions, and F-primes. It also took on responsibility for periodic publishing of the linkage map for *E. coli* K-12, and with that came a contingent role of "authenticating" canonical gene symbols. These information roles provided a natural progression to an on-line database covering gene names, functions, map locations, strain genotypes, mutation information, and supporting documentation, as well as links to information from other bacterial, genetic, and molecular biology databases. The development and maintenance of the database, as well as the Stock Center itself, have been supported by the NSF (*http://cgsc.biology.yale.edu*).

Phabagen, the Phage and Bacterial Genetics Collection, includes 5000 mutant bacterial strains (*E. coli* and *Agrobacterium tumefaciens*), 450 cloning vectors, 800 other plasmids, two plasmid-containing gene banks of *E. coli*, and over 100 phages. It was established in the early 1960s with deposits of bacterial mutants from researchers of the Working Community Phabagen. It has, since 1990, been part of CBS (see national consortia in Table 24.1) and is merging with the Kluyver Institute's LMD Collection to form the NCCB, the Netherlands Culture Collections of Bacteria (*http://www.cbs.knaw.nl/nccb/database.htm*).

The National Institute of Genetics of Japan (NIG) established a Genetic Stock Center in 1976 and developed an extensive collection of *E. coli* strains. In 1997 it was reorganized as the Genetic Strains Research Center, Microbial Genetics Center, and the Center for Genetic Resource Information. The microbial collections include approximately 4000 strains of *E. coli* genetic derivatives and 400 cloning vectors (*http://shigen.lab.nig.ac.jp/ecoli/strain* and *http://shigen.lab.nig.ac.jp/cvector/cvector.html*).

The *Salmonella* Genetic Stock Center (SGSC) at the University of Calgary, British Columbia, Canada, had its origins in the laboratory of M. Demerec at Cold Spring Harbor Laboratory and Brookhaven National Laboratory, Long Island, New York, in the 1950s and 1960s as genetic derivatives primarily of *Salmonella typhimurium* (or *Salmonella enterica*, subsp. *enterica*, serovar Typhimurium) strain LT2. It remained with the Demerec laboratory during his long career, and then it was moved and expanded at the University of Calgary by K. E. Sanderson. It currently has several thousand strains, cosmid and phage libraries, and a set of cloned genes. Many of the mutant strains are organized into special purpose kits, useful for specific genetic techniques or analyses. In addition to the mutants, it has the *Salmonella* Reference Collection (SARC) representing all subgenera of *Salmonella*. The SGSC is supported by the Natural Sciences and Engineering Research Council of Canada (*http://www.acs.ucalgary.ca/~kesander*).

Bacillus subtilis has been studied as the model spore-forming bacterium, and many mutant strains have been isolated. In 1980 the *Bacillus* Genetic Stock Center was founded at Ohio State University. It includes over 1000 genetically characterized *B. subtilis* strains, over 200 strains of other *Bacillus* species, and *E. coli* strains bearing shuttle plasmids or cloned *Bacillus* DNA. It publishes a newsletter and a genetic map for *Bacillus subtilis*. The NSF supports this stock center (*http://bacillus.biosci.ohio-state.edu*).

B. Fungal and Algal Genetic Stock Centers

The Fungal Genetic Stock Center (FGSC) was organized as a result of recommendations by the Genetics Society of America in 1960. It was originally located at Dartmouth College, directed by R. Barratt, then moved to California State University in Humboldt, and in 1985 to the University of Kansas Medical Center. Its holdings include nearly 9000 strains of filamentous fungi, mostly genetic derivatives of *Neurospora crassa*, *Aspergillus nidulans*, and *Aspergillus niger*, but include also *Neurospora tetrasperma* and isolates of other *Neurospora* and *Aspergillus* species. The collection also contains *Fusarium* species and mutants, and *Nectria* and *Sordaria* mutants and species. The stock center publishes, mails, and makes available on its website the Fungal Genetics Newsletter, meeting abstracts and announcements, and a bibliography. The website has information on genes, alleles, and maps, and the center supplies plasmids, clones, and gene libraries for *N. crassa* and *A. nidulans*, as well as the strains. The FGSC is supported by the NSF (*http://www.fgsc.net*).

The Yeast Genetic Stock Center (YGSC) originated at the University of California at Berkeley in 1960, founded and directed by R. Mortimer for genetic derivatives of *Saccharomyces cerevisiae*, primarily originating from "founder" stocks of C. Lindegren at Southern Illinois University in Carbondale. The Mortimer laboratory also published linkage maps and linkage data summaries. Maps and linkage data, but not stock descriptions, are currently provided by the *Saccharomyces* Genome Database (*http://www.yeastgenome.org*). The 1200 genetic stocks in the collection were moved in 1998 to the ATCC, and stock information is available from there (*http://www.atcc.org/SpecialCollections/YGRRC.cfm*).

The Peterhof Genetic Collection of Yeasts (PGC) is located in the Biotechnology Center at St. Petersburg State University in Russia. It has over 1000 genetically marked yeast strains, mainly lines originating from a diploid cell of an inbred strain of *Saccharomyces cerevisiae* (the Peterhof genetic lines, as distinguished from the Carbondale origin of the strains in the YGSC), other genetically marked *S. cerevisiae* strains, and segregants of crosses between the latter and the Peterhof lines. The collection is searchable via the MSDN, at *http://panizzi.shef.ac.uk/msdn/*.

The *Chlamydomonas* Genetics Center (CGC), at Duke University, collects, describes, and distributes nuclear and cytoplasmic mutant strains, and genomic and cDNA clones, of *Chlamydomonas reinhardtii*, which has served as a model organism for algae and for photosynthetic organisms. It is supported by the NSF. The Web and gopher sites provide, in addition, information on genetic and molecular maps of *Chlamydomonas reinhardtii*, plasmids, sequences, and bibliographic citations (*http://www.biology.duke.edu/chlamy*).

C. Information management at Genetic Stock Centers

The dissemination of information was often an equal partner with the dissemination of stocks in the call by the scientific community for founding of the stock centers. Some stock centers distributed printed catalogs, and others published periodic announcements summarizing their holdings. Many stock center curators or directors took responsibility for publishing frequently updated versions of the genetic map for the major organism in their collection. In 1988, a workshop on computerization of databases for genetic stock collections was held in conjunction with the 16th International Congress of Genetics, August 1988, Toronto. This meeting acknowledged the information functions of the stock centers in describing and distributing the organisms and also in producing genetic maps, standardizing allele designations, and documenting gene function and nomenclature, and it discussed and made recommendations for facilitating database development and database construction. In the US, the NSF and other funding agencies played a prominent role in encouraging stock centers to recognize the need for modernization and computerization of their valuable records.

Development of the Internet and the World Wide Web has altered and enhanced the performance of these information functions and the way that users interact with the stock centers. It has allowed information formatted in many different ways at the many stock collections sites to be searched and presented to the user and allows links among stock centers and between stock centers and databases with sequence and other types of information. It allows the centers to provide on-line direct access to the information about stocks, gene characterization, and the genetic map updates. Citations of Web addresses for all microbial stock centers and ancillary information associated with them have been given throughout this article and are included as part of the entries in Table 24.1.

V. NATIONAL COLLECTIONS AND CONSORTIA WITH BROAD HOLDINGS

During the past three decades, many countries have moved to consolidate the administration of their stock centers into a single administrative and information unit or federation. Examples of some of these consortia are presented in Table 24.1 by country. Particularly extensive consortia include those in the United Kingdom, France, Belgium, and the Netherlands, as well as the USDA collections. These and other national collection consortia in the Czech Republic, Poland, Russia, the Ukraine, China, Japan, Indonesia, Canada, and Brazil are described under the heading National consortium, comprehensive in Table 24.1. Other countries have long had centralized national collections rather than individual specialized or general collections. Such collections are listed under National, comprehensive type collections in Table 24.1. The DSMZ in Germany is one of the largest, and others are located in Belarus, Bulgaria, Hungary, India, Thailand, Taiwan, New Zealand, and Mexico. Still other comprehensive collections serve national and international communities but are not administered or entirely supported by governmental agencies. For example, the ATCC (American Type Culture Collection) in the United States is only partially supported by federal grants and is administered as an independent, nonprofit institution. Such collections are type classified as Other—comprehensive in Table 24.1. The US collections are not organized under any single coordinating administrative body, although the USDA/ARS collections are coordinated through the USDA and the ATCC represents a consolidation of many different types of collections. There is a US Federation of Culture Collections to promote cooperation and information sharing among the diverse stock centers. It cooperates with the International World Federation of Culture Collections in sponsoring international meetings among directors and curators of stock collections. The International Mycological Institute is internationally supported (CAB International), with a secretariat in the UK, and this collection and university collections in Sweden, Finland, and Australia, and Japan's Institute for Fermentation (IFO) Collection in Osaka are classified as Other—comprehensive in Table 24.1. National collections that are specialized rather than comprehensive are presented under organism-specific headings in Table 24.1.

VI. ORGANISM-SPECIFIC COLLECTIONS

The organism-specific collections are presented in Table 24.1 under the following broad headings: bacteria; fungi, including yeasts; and algae, protozoa, and cyanobacteria. Type headings provide a more specific indication of the holdings. Previously described collections, in either the national consortia part of the table or in narrative sections, are cited again in Table 24.1 in order to expedite locating the collection and its Web URL.

VII. A TABLE OF STOCK COLLECTIONS

To facilitate look-up of specific kinds of collections and especially to facilitate locating the Web address of a collection, a tabular rather than text format is used to present collections by type and country, to provide brief descriptions of the collections, and to cite the Web addresses. Clearly a totally comprehensive listing of such diverse and widespread resources is not possible, and regrettable oversights have no doubt occurred. Widely used and Web-accessible sites have been emphasized. See legend for information on searching for collections not listed in the table.

TABLE 24.1 Representative stock collections: location, description, and access

Many of the descriptions briefly cite special functions, history, and holdings, if this information has not been previously cited. Preservation and distribution functions are assumed in all cases and are not cited specifically. The table first presents comprehensive collections, with country presented in uppercase, then organism-specific collections (Bacteria; Fungi, Algae, protozoa, and cyanobacteria), and finally databases that cover multiple collections. Web URLs may change at any time, and searches may be necessary to update those shown in the table. No claim can be made that this is an exhaustive compilation of all the world's stock centers. Comprehensive database searches and links given at the end of the table can be used to search for collections not listed in the table. For example, *http://wdcm.nig.ac.jp/CCINFO Search.html* will retrieve all WDCM collections by country or organism. (See last entry in the Table.)

Type	Country	Designation	Description	Web URL
Comprehensive				
National consortium, comprehensive	United Kingdom	UK National Collections include the following:	See individual collection descriptions below	*http://www.ukncc.co.uk*
		NCTC, National Collection of Type Cultures, London	Has about 5000 cataloged strains of medical and veterinary significance and 13,000 uncataloged strains. A patent and safe depository. Jointly with the DSM (Germany) is the Resource Centre for plasmid-bearing bacteria for Europe. Supported by the Public Health Laboratory Service.	*http://www.ukncc.co.uk/html/members/nctc/nctc_info.htm* *http://www.phls.co.uk/labservices/nctc/accessions.htm*
		NCIMB Ltd., National Collection of Industrial, Marine and Food Bacteria Ltd., (formerly NCIMB and NCFB, National Collection of Food Bacteria) Aberdeen, Scotland	Includes 3800 accessions of general industrial and scientific significance, (actinomycetes and other bacteria, plasmids, and phages) and 1500 marine bacteria. Originally funded by Ministry of Agriculture, Fisheries and Food. Patent depository. Identification, service, and contract work. About 2000 bacteria of significance to food and dairying. A patent and safe depository. Originally funded by Ministry of Agriculture, Fisheries and Food, Institute of Food Research and is now affiliated with NCIMB, Ltd.	*http://www.ncimb.co.uk* and *www.uKncc.co.uk/ncimb.html*
		NCYC, National Collection of Yeast Cultures, Institute of Food Research, Norwich	Includes brewing yeast strains, genetically defined strains of *Saccharomyces cerevisiae* and *Schizosaccharomyces pombe*, and general yeast strains, totaling over 2700 nonpathogenic yeasts. A patent and safe depository. Performs yeast identification services. Member of the European Culture Collections Organization (ECCO) and the WFCC.	*http://www.ifrn.bbsrc.ac.uk/NCYC*
		NCPF, National Collection of Pathogenic Fungi, London	Has 1500 human and animal pathogenic fungi and performs identification and contract research. Supported by the Public Health Laboratory Service.	*http://www.ukncc.co.uk/html/members/Ncpf/Ncpf_info.htm*
		NCWRF, National Collection of Wood Rotting Fungi in Watford	Over 500 cultures of ~300 species of wood-rotting basidiomycetes. Funded by the Department of the Environment. Managed by CABI-Bioscience.	As above, except */ncwrf/ncwrf_info.html*
		NCPPB, National Collection of Plant Pathogenic Bacteria, Harpenden	Has 3000 phytopathogenic bacteria, representing a wide range of hosts and locations for species causing most of the known bacterial plant diseases, and has some bacteriophages as well. Funded by the Ministry of Agriculture, Fisheries and Food. Provides information, identification, and contract services.	*www.ncppb.com*
		See also CABI.Bioscience (IMI), RCR, and IACR collections	Classified in this table as Other—comprehensive, fungal—*Rhizobium*, RCR; plant pathogens, IACR.	
		CCAP, Culture Collection of Algae and Protozoa	Had its origins with the early culture collections, moving from Prague to England in the 1920s (see early collections section in text),	*http://www.ife.ac.uk/ccap* and *http://www.ukncc.co.uk/html/members/ccap/ccap_info.htm*

TABLE 24.1 *Continued*

Type	Country	Designation	Description	Web URL
			undergoing development at Cambridge University and later moving to two sites—the freshwater algae and all protozoa to the Institute of Freshwater Ecology (IFE) Windermere Laboratory at Ambleside and the marine algae to Dunstaffnage Marine Laboratory (DML) in Scotland. Approximately 2000 strains of algae and protozoa. A patent depository.	
National consortium, comprehensive	France	CNCM, National Collection of Cultures of Microorganisms	Established in 1976 by French Ministerial authorities. Housed in the Pasteur Institute. A patent and safe depository. Coordinates activities of the specialized collections of the Pasteur Institute. Member collections include the CIP (see early collections section in text and bacterial type in this table), which incorporates the *Pasteurella* laboratory collection, the Pasteur Institute collections of cyanobacteria (see PCC under collections of algae, protozoa, and cyanobacteria), *Rhodococcus, Gordonia, Lactobacillus*, coryne forms, and *Pasteurella*, as well as fungal, viral, and animal cell collections. Member of the World Federation for Culture Collections and the European Culture Collections Organization.	*http://www.pasteur.fr/Bio/RAR96/Colloq.html*
National consortium, comprehensive	Belgium	BCCM, Belgian Coordinated Collections of Microorganisms. Includes the following:	Consortium of four research-based collections, containing 50,000 documented strains of bacteria, filamentous fungi, and yeasts, and over 1500 plasmids. Supported by the Belgian Federal Office for Scientific, Technical, and Cultural Affairs. Patent and safe deposit services as well as contract research and training.	*http://www.belspo.be/bccm*, search at *http://www.belspo.be/bccm/db/index.htm*
		BCCM/MUCL, the Mycological Collection of the Catholic University of Louvain	Originated with the earliest collections (see early collections section of text). Holds 25,000 strains of fungi and yeast, including representatives of over 3300 species of zygomycetes, ascomycetes, hyphomycetes, and basidiomycetes. Focuses on agro-industrial cultures. Also holds a mycological herbarium of 40,000 fungal specimens. Undertakes research in systematic and applied mycology and provides software for yeast identification.	*http://www.belspo.be/bccm/mucl.htm#main*
		BCCM/LMBP, the Plasmid and cDNA Collection, University of Ghent	Includes plasmids that replicate in prokaryotic (*E. coli* and *Lactobacillus*) and eukaryotic (yeast, animal cell) hosts and 10 cDNA libraries.	*http://www.belspo.be/bccm/lmbp.htm#main*
		The BCCM/LMG Bacterial Collection, University of Ghent	Holds over 16,000 strains representative of 1300 species, subspecies, or pathovars, with emphasis on phytopathogenic bacteria and species of medical and veterinary significance. Also has marine *Vibrio* species and strains of some biotechnological interest.	*http://www.belspo.be/bccm/lmg.htm#main*
		The BCCM/IHEM Collection of Biomedical Fungi and Yeasts, Brussels	A collection of filamentous and yeast fungi of public health and environmental interest, containing over 6500 strains of human and animal pathogenic and allergenic species. These include specific disease isolates, e.g. 500 *Aspergillus fumigatus* and 300 *Candida albicans* isolates. Located at the Scientific Institute of Public Health—Louis Pasteur in Brussels	*http://www.belspo.be/bccm/ihem.htm#main*
National consortium, comprehensive	The Netherlands	CBS, The Centraal Bureau voor Schimmelcultures	The CBS collection and NCCB, described below, are separate units of the organizational structure of the CBS, an institute of the Royal Netherlands Academy of Arts and Sciences.	*http://www.cbs.knaw.nl*
		CBS, the Centraal Bureau voor Schimmelcultures Collection, Baarn and Delft	See early collections section of text. Currently includes separate collections and databases for filamentous fungi in Baarn (over 28,000 strains) and yeasts in Delft (4500 strains), with the IGC collection of 1250 yeast strains in Oeiras, Portugal, also included in the database, and a database for taxonomic and nomenclatural data for	*http://www.cbs.knaw.nl*, *http://www.cbs.knaw.nl/search_fdb.html*, *http://www.cbs.knaw.nl/yeast/webc.asp* *http://www.cbs.knaw.nl/fusarium/database.html*; and

TABLE 24.1 *Continued*

Type	Country	Designation	Description	Web URL
			Fusarium and the Aphyllophorales. One of the largest collections of fungi in the world, totaling over 35,000 strains, its goal is to have representatives of all fungal groups that can be cultured. Plans to integrate yeasts and fungi into one institute. Serves as the Dutch node of MINE, the Microbial Information Network of Europe, an EC-sponsored network of European collections. Performs identification and patent deposit services.	*http://www.cbs.knaw.nl/aphyllo/database.html*
		NCCB, Netherlands Culture Collections of Bacteria, Utrecht	Merger of the Bacterial Collection of Kluyver Laboratory of Microbiology at the Technical University of Delft (Kluyver Institute of Biotechnology), the LMD, containing over 4500 strains, and the Phabagen Collection (see Microbial Genetic Stock Centers) at the University of Utrecht. Participant in CABRI. Patent and safe depository, taxonomic research, identification, and consultation services.	*http://www.cbs.knaw.nl/nccb, http://www.cbs.knaw.nl/nccb/about.htm*
National consortium, comprehensive	Germany	DSMZ, Deutsche Sammlung von Mikroorganismen und Zellkulturen, Braunschweig	Founded in 1969 as the national culture collection of Germany, part of the Gesellschaft für Strahlenforschung in Munich. Moved to the Gesellschaft für Biotechnologische Forschung in Braunschweig, in 1988 became a private company, and in 1996 became an independent non-profit organization supported by the Federal Ministry of Research and Technology and the State Ministeries. Holds over 8700 bacteria and archaea, 300 plasmids, 100 phages, 2300 filamentous fungi, 500 yeasts, 700 plant viruses, as well as plant, human, and animal cell lines. Serves as a safe and patent respository, and performs teaching, service, and identification functions. On-line searchable databases for microorganisms, cell lines, and plasmids, as well as a bacterial nomenclature database.	*http://www.dsmz.de, http://www.dsmz.de/bactnom/bactname.htm*
National consortium, comprehensive	Czech and Slovak Republics	Czech Federation of Culture Collections	Includes a number of microbial collections, the larger collections listed below. Member of the WFC and ECCO.	*http://www.natur.cuni.cz/fccm*
		The Czech Collection of Microorganisms, Brno	Established at the J. E. Purkyne University in 1963, based on the bacterial collection of the university begun in 1957 and merged with the Culture Collection of the Institute of Microbiology, of the Czechoslovak Academy of Science in Prague. Includes over 2000 strains (representing more than 170 genera and 600 species) of bacteria, over 300 strains of fungi, mycoplasmas, and viruses, particularly animal pathogens and organisms of interest to food microbiology, medicine, education, agriculture, and industry. Performs identification and consultation services and serves as a safe repository. Member of the WFCC.	*http://www.sci.muni.cz/ccm/ccmang.htm* and *http://wdcm.nig.ac.jp/CCINFO/CCINFO.xml?* 65 #ch12
		The Czech Culture Collection of Fungi (CCF) at Charles University, Prague	Combined the collections of the Biological Institute of the Czechoslovak Academy of Sciences and the Charles University Botany Department research collection in 1964–1965. Approximately 1600 strains of zygomycetes and ascomycetes.	*http://www.natur.cuni.cz/fccm/#ccf* and *http://wdcm.nig.ac.jp/CCINFO/CCINFO.xml? 182*
		CCBAS, the Culture Collection of Basidiomycetes	Founded in 1959 and contains 630 strains and 253 species, including some rare species.	*http://www.biomed.cas.cz/ccbas/fungi.htm*
National consortium, comprehensive	Poland	PCM, Polish Collections of Microorganisms	Combines several individual collections at the Hirszfeld Institute of the Polish Academy of Sciences, Wroclaw. These include the DMVB and the CRS, Collection of *Rhizobium* Strains, in Pulawy, the Industrial Microorganism collection in Lodz and Warsaw, the Dairy Cultures collection in Olsztyn, the National *Salmonella* Centre in Gdynia-Redlowo, and the Research and Development Center for Biotechnology in Warsaw. Represents	*http://wdcm.nig.ac.jp/CCINFO/CCINFO.xml? 106*

(*Continued overleaf*)

TABLE 24.1 *Continued*

Type	Country	Designation	Description	Web URL
			a wide range of bacteria, yeasts, and filamentous fungi, which are searched for and described from a single Web site.	
National consortium, comprehensive	Russia	VKM, the All-Russian Collection of Microorganisms, Institute of Biochemistry chemistry and Physiology of Microorganisms	Merger in 1980 of the Institute of Microbiology's collection of yeast cultures of V. Kudryavtsev begun in the 1930s and the Institute of Biochemistry and Physiology's collections of microorganisms begun in 1968. Over 9000 strains, encompassing 3000 species of 650 genera, of bacteria, actinomycetes, filamentous fungi, and yeast. Includes genetically modified strains of *E. coli* and yeast as well as a few genetic derivatives of *Agrobacterium tumefasciens, Bradyrhizobium japonicum*, and *Streptomyces galilaeus*. Russian Academy of Sciences, Pushchino, Moscow region.	Searches and information via WDCM at *http://www.vkm.ru* and *(www.vkm.ru)* and *http://wdcm.nig.ac.jp/CCINFO/CCINFO.xml? 342*
		VKIM, All-Russian Collection of Industrial Microorganisms	Institute of Genetics and Selection of Microorganisms, Moscow.	
National, comprehensive	Ukraine	IMV, Zabolotny Institute of Microbiology and Virology Collection	Sponsored by the Academy of Sciences in Kiev.	See MSDN
National, comprehensive	Belarus	INMIB Collection of the Institute of Microbiology	Academy of Science of Belarus, Minsk.	*http://www.ac.by/organizations/institutes/inobio.html#off3578*
National, comprehensive	Bulgaria	NBIMCC, National Bank for Industrial Microorganisms and Cell Cultures, Sofia	Founded in 1983, encompasses the Bulgarian Type Culture Collection, and includes fungi and yeasts, bacteria and actinomycetes, plasmids, viruses, and cell lines.	*http://wdcm.nig.ac.jp/CCINFO/CCINFO.xml? 135*
National, comprehensive	Hungary	NCIM, National Collection of Agricultural and Industrial Microorganisms, Budapest	At the University of Horticulture and Food Industry Includes over 1600 strains of bacteria, yeasts, and fungi. It was established in 1974 and is a patent and safe depository and performs purification, identification, consultation, and training services. Member of the WFCC and ECCO.	*http://ncaim.kee.hu.* *http://wdcm.nig.ac.jp/CCINFO/CCINFO.xml? 485*
National, comprehensive	India	NCIM, National Collection of Industrial Microorganisms, Pune	A collection of over 100 bacteria, nearly 2000 fungi and yeasts, and a few algae and protozoa. National Chemical Laboratory in Pune, sponsored by the Council for Scientific and Industrial Research.	*http://kelvin.ncl.rcs.in/ncim*
National, comprehensive	India	ITCC, the Indian Type Culture Collection, New Delhi	Division of Mycology and Plant Pathology, Indian Agricultural Research Institute.	*http://wdcm.nig.ac.jp/CCINFO/CCINFO.xml? 430*
National, comprehensive	India	MTCC, Microbial Type Culture Collection and Gen Bank, Chandigarh	Established 1986 at the Institute of Microbial Technology. Actinomycetes, bacteria, yeast, fungi, plasmids	*http://imtech/ernet.in/mtcc* See also *http://wdcm.nig.ac.jp/CCINFO/CCINFO.xml? 773*
National consortium, comprehensive	China	CCCCM, the China Committee for Culture Collections of Microorganisms	Provides search capability by name or keyword for databases encompassing the following three collections as well as others. It also provides a national node of the MSDN that allows querying of some general and many specialized collections in many countries (CCCCM). Oversees the following three collections:	See Micro-Net of China *http://www.im.ac.cn.* Search collections at *http://www.im.ac.cn/database/catalogs.html*. MSDN node: *http://www.im.ac.cn/msdn.html*
		CCCM, the Center for Culture Collection of Microorganisms, Beijing	Institute of Microbiology, Chinese Academy of Science. Includes 1900 yeasts, 5500 other fungi, 2200 bacteria, and 1400 actinomycetes; also included are the type culture collection of China and patent strain repository.	*http://www.im.ac.cn/en/tcfmr.php* *http://www.micronet.cn* *http://im.ac.cn/sklomr*
		CCVCC, the China Center for Virus Culture Collection	Collects, preserves, classifies, and studies viruses, now numbering 600, including insect, plant, bacterial, and animal viruses. Chinese Academy of Sciences, Wuhan Institute of Virology, established in 1979.	*http://www.micronet.cn/institutes/ccvcc/ccvcc.html*
		ACCC, the Agricultural Culture Collection of China, Beijing	The ACCC holds more than 2000 strains of bacteria, actinomycetes, fungi, rhizobia, and edible fungi. Chinese Academy of Agricultural Sciences, Beijing, established in 1980.	*http://www.micronet.cn/institutes/accc/accc.html*

TABLE 24.1 *Continued*

Type	Country	Designation	Description	Web URL
National, comprehensive	Thailand	TISTR, the Thailand Institute of Scientific and Technological Research Collection	Linked to WDCM and can be queried at the Web site shown. The Tropical Database of Thailand is also cited by WDCM as a TISTR project. Supported by the Ministry of Science, Technology and Environment of Thailand.	*http://wdcm.nig.ac.jp/CCINFO/CCINFO.xml? 383*
National consortium, comprehensive	Japan	JFCC, Japan Federation of Culture Collections	Maintains a search capability for the various collection databases for bacteria, fungi, viruses, and algae.	*http://wdcm.nig.ac.jp/wdcm/JFCC.html*
		IAM Culture Collection, Institute of Applied Microbiology, University of Tokyo	Founded in 1953 and in 1993 reorganized in the IMCB (Institute of Molecular and Cellular Biosciences) to broaden coverage to higher organisms and their cultured cells. The microbial collection contains over 3400 strains of bacteria, yeasts, filamentous fungi, and algae.	*http://www.iam.u-tokyo.ac.jp/misyst/ColleBOX/IAMcollection.html*
		The Japan Collection of Microorganisms (JCM) in the Institute of Physical and Chemical Research (RIKEN)	Has since 1980 distributed microorganisms from a collection that includes over 3500 strains of bacteria, including actinomycetes, 2000 strains of fungi, including yeasts, and a hundred strains of archaea. Collections such as the KCC Culture Collection of Actinomycetes, Kaken Chemical Co., Tokyo, have transferred their holdings to the JCM. On-line catalog allows searching by scientific names or key words and contains information about strain history, taxonomy, reference citations, and cultivation conditions. Database also can be used to search for strains of bacteria, fungi, and yeasts, bacteriophages and other viruses, algae, and protozoa in Japan and elsewhere.	*http://www.jcm.riken.go.jp/JCM/aboutJCM.html, http://wdcm.nig.ac.jp/wdcm/JFCC.html*
		National Institute of Genetics Cloning Vector and *E. coli* Collections	Provides vectors of *E. coli* that can be stored stably as purified DNA. It also has the *E. coli* Genetic Resources collection—see bacterial collections in this table.	*http://www.shigen.nig.ac.jp/cvector/cvector.html*
National, comprehensive	Taiwan	The CCRC, Culture Collection and Research Center	Founded by the Food Industry Research and Development Institute in Hsinchu, Taiwan, in 1982. Began with food microbiology and expanded to serve the interests of industry, agriculture, medicine, environmental research, education, and biotechnology. Includes bacteria, yeasts, filamentous fungi, bacteriophages, and plasmids and phage vectors.	*http://wdcm.nig.ac.jp/CCINFO/CCINFO.xml? 59*
National consortium, comprehensive	Indonesia	Various collections, mostly governmental and university	Collections in Indonesia are listed at the WDCM Website shown. Include yeasts, algae, lichens, protozoa, fungi, bacteria, and viruses.	*http://wdcm.nig.ac.jp/indonesia.html*
Other comprehensive	Australia	UNSWCC, the University of New South Wales Culture Collection	Established in 1969. Provides catalogs of algae and protozoa, fungi and yeast, and bacteria.	*http://www.micro.unsw.edu.au/cult.html*
		The Division of Food Research, CSIRO, in New South Wales	Has a collection of 3000 fungi and 100 yeasts.	*http://wdcm.nig.ac.jp/CCINFO/CCINFO.xml? 18*
		The University of Queensland Microbial Culture Collection in Brisbane	Has over 3500 cultures of bacteria and also has algae, fungi, yeasts, and viruses. It provides identification and training and patent deposit services.	*http://wdcm.nig.ac.jp/CCINFO/CCINFO.xml? 13*
		Other collections	Thirty other culture collections in Australia are at the Web address shown.	*http://wdcm.nig.ac.jp/hpcc.html*
National, comprehensive	New Zealand	The International Collection of Microorganisms from Plants, Auckland	Includes bacteria and fungi, with primary emphasis on plant pathogens.	*http://wdcm.nig.ac.jp/CCINFO/CCINFO.xml? 589* and *http://nzfungi.landcareresearch.co.nz/icmp/search_cultures.asp*
		The New Zealand Reference Culture Collection of Microorganisms	Sponsored by the New Zealand Dairy Research Institute.	*http://wdcm.nig.ac.jp/CCINFO/CCINFO.xml? 318*

(*Continued overleaf*)

TABLE 24.1 *Continued*

Type	Country	Designation	Description	Web URL
Bacterial—lactic acid and other	Japan	NRIC, the Nodai Research Institute Culture Collections, Tokyo University of Agriculture	See also entry under Fungi in this table. Food-fermentation microorganisms. 2500 lactic acid bacterial (also 400 other bacteria, 1400 yeast cultures) Member Japan Society for Culture Collections	*http://wdcm.nig.ac.jp/CCINFO/CCINFO.xml? 747* *http://wdcm.nig.ac.jp/wdcm/JFCC.html*
Bacterial—Antarctic	Australia	ACAM, the Australian Collection of Antarctic Microorganisms	University of Tasmania, Heterotrophic bacteria collected from Antarctica and subantarctic islands and Southern Ocean. Established 1986.	*http://www.antcrc.utas.edu.au/antcrc/micropro/acaminfo.html*
Bacterial—*Escherichia* and *Klebsiella*	Denmark	Statens Serum Institute, WHO International *Escherichia* Centre	Sponsored by the World Health Organization of the UN, maintains ~400 serovars. Collaborating Centre for Reference and Research on *Escherichia* and *Klebsiella*. Performs serotype identification services as well as providing cultures. Copenhagen.	Artellerivej 5,300 Copenhagen S, Denmark Fax: 4532 683868
Bacterial—*E. coli*	United States	ECOR, the *E. coli* Reference Collection and DEC and other STEC (Shiga toxin-producing *E. coli*) reference collections	The ECOR, established by H. Ochman and R. Selander (1984) from human and animal hosts to represent variation in O and H serotypes, the DEC collection of diarrheagenic strains, and other STEC reference collections at Michigan State STEC center	*http://www.shigatox.net/stec,* *http://foodsafe.msu.edu/whittam/ecor,* and *http://shigatox.net/cgi-bin/deca*
Bacterial—*Bacillus*, genetic	United States	BGSC, *Bacillus* Genetic Stock Center, Ohio State University	See bacterial genetic stock centers section of text. Publishes genetic map. *B. subtilis* and other *Bacillus* species in addition to strain distribution.	*http://bacillus.biosci.ohio-state.edu*
Bacterial—*E. coli*, genetic	United States	CGSC, *E. coli* Genetic Stock Center, Yale University	See bacterial genetic stock centers section of text. Publishes genetic map, registers gene symbols and allele numbers in addition to strain distribution.	*http://cgsc.biology.yale.edu*
Bacterial—*E. coli*	Japan	NIG, National Institute of Genetics	See Bacterial Genetic Stock Centers section of text. Approximately 4000 strains and over 400 cloning vectors	*http://shigen.lab.nig.ac.jp/ecoli/strain* *http://www.shigen.lab.nig.ac.jp/cvector/cvector.html*
Bacterial—*Salmonella*, genetic	Canada	SGSC, *Salmonella* Genetic Stock Centre, University of Calgary	See bacterial genetic stock centers section of text. Publishes genetic map, registers gene symbols and allele numbers in addition to strain distribution.	*http://www.acs.ucalgary.ca/~kesander*
Bacterial—*Salmonella*	Poland	National *Salmonella* Center	See National, comprehensive section of this table, PCM	*http://wdcm.nig.ac.jp/CCINFO/CCINFO.xml? 784*
Bacterial—*Rhizobium*	Poland	CRS Collection of *Rhizobium* strains, Pulawy	See National, comprehensive section of the table, PCM, and other listings in WDCM.	*http://wdcm.nig.ac.jp/CCINFO/CCINFO.xml? 106,* and *http://wdcm.nig.ac.jp/hpcc.html*
Bacterial—*Rhizobium*	United Kingdom	RCR Rothamstead Collection of *Rhizobium*, Harpenden and Welsh WPBS	Collection of *Rhizobium* of the Soil Microbiology Department.	*http://www.iacr.bbsrc.ac.uk/ppi/cultures and wdcm.nig.ac.jp/CCINFO/CCINFO.xml? 607*
Bacterial—*Rhizobium*	United States	USDA/ARS National Collection of *Rhizobium*	Part of the USDA/ARS National *Rhizobium* Germplasm Research Center, combining USDA/ARS collections with the Lipha-Tech, Boyce Thompson Institute, University of Minnesota ARS Collection, the University of São Paulo, Brazil Collection, and the Centro Internacional de Agricultura Tropical (CIAT) collection in Cali, Colombia.	*http://www.ars.usda.gov/is/np/systematics/rhizobium.htm*
Bacterial—alkanotrophic	Russia	IEGM, Institute of Ecology and Genetics of Microorganisms, Moscow	Regional specialized collection of alkanotrophs (which assimilate *n*-alkanes gases and liquid hydrocarbons as carbon and energy source) with 400 species, 1000 strains. Includes mycobacteria and actinomycetes—*Rhodococcus, Gordona, Micrococcus, Brachybacteria* species. Collecting began in 1975 for studies of bio-indicators of gas and oil fields and pollution. Russian Academy of Science, Urals Branch.	*http://wdcm.nig.ac.jp/wdcm/IEGM_readme.html* and *CCINFO/CCINFO.xml? 768*
Fungi, including yeasts				
Plant pathogens	United Kingdom	The Institute of Arable Crops—Rothamstead (IACR) Plant Pathology Culture Collections	Specialized collection that holds over 600 soil-borne fungi from around the world, with particular emphasis on cereal root pathogens of the Gaeumannomyces–Phialophora complex of fungi. The GPDATA records include collecting information and pathogenicity tests, and queries	*http://www.rothamsted.bbsrc.ac.uk ppi/cultures*

TABLE 24.1 *Continued*

Type	Country	Designation	Description	Web URL
			based on name or origin of culture can be entered from the Website.	
Fungi	Czech Republic	FCCM (CCF) and CCBAS	See National, comprehensive section of this table.	*http://www.biomed.cas.cz/ccbas/fungi.htm*, and *http://www.natur.cuni.cz/fccm*
Fungi	Canada	CCFC and UAMH	See National consortium section of this table.	*http://sis.agr.gc.ca/brd/ccc/cccintro.html* and *http://www.devonian.ualberta.ca/uamh/index.html-ssi*
Filamentous fungi and yeasts	Belgium	The Mycological Collection of the Catholic University of Louvain, MUCL, and IHEM	*Penicillium*, *Aspergillus*, hyphomycetes and other fungi, and yeasts. See National Consortium section of this table, Belgium, BCCM/MUCL and BCCM/IHEM.	*http://www.belspo.be/bccm/*
Filamentous fungi and yeasts	The Netherlands	See National, comprehensive section of table, CBS	Note particularly filamentous fungi in Baarn, yeasts in Delft and Oeiras, Portugal, and taxonomic and nomenclature databases for *Fusarium* and Aphyollophorales.	*http://www.cbs.knaw.nl/search_fdb.html* and */yeast/webc.asp*
Filamentous fungi and yeasts	United Kingdom	See National consortium section, NCPF, NCYC, NCWRF, IACR	Note particularly NCPF for pathogenic fungi, NCYC for yeast cultures, NCWRF for wood-rotting fungi, and the NCPPB and IACR for plant pathogens.	*http://dtiinfo1.dti.gov.uk/bioguide/culture.htm*
Filamentous fungi and yeasts	Multinational, in UK	IMI, International Mycological Institute	See early collections section of text. Holds 16,500 strains. Part of CAB International, supported by 32 governments.	*http://www.cabi-bioscience.org/html/grc.htm*
Filamentous fungi and yeasts	Sweden	UPSC, Mycoteket, Uppsala University	Over 3000 fungal cultures. See early collections section of text.	*http://www-hotel.uu.se/evolmuseum/mykotek/index.htm*
Fungi—mycorrhizal	United States	The International Culture Collection of Vesicular–Arbuscular Mycorrhizal Fungi (INVAM) at West Virginia University	Holds and distributes strains representing more than 60 of the 154 known mycorrhizal species and has other strains yet to be identified. NSF-supported collection of vesicular and vesicular arbuscular mycorrhizal fungi, originated with N. Schenck at the University of Florida in 1985, moved to West Virginia University in 1990, combining with J. Morton's collection. Over 1100 cultures, including species in the genera *Acaulospora*, *Entrophospora*, *Gigaspora*, *Glomus*, and *Scutellospora*. Website information on species, nomenclature, methods, and services; querying by name or site characteristic.	*http://invam.caf.wvu.edu/* Links to general mycorrhizal information are also maintained (*http://mycorrhiza.ag.utk.edu and http://invam.caf.wvu.edu/otherinfo/Articles/articles.htm*)
Fungi—mycorrhizal (arbuscular)	European	BEG, the European Bank of the Glomales	European "Stock Center without Walls," with isolates representing temperate, tropical, and polar representatives of Glomales species. Database of these European endomycorrhizal collections can be queried by genus, species, continent, country, biome, or associated plant.	*http://www.kent.ac.uk/bio/beg*
Fungi—entomopathogenic	United States	ARSEF, USDA-ARS Collections of Entomopathogenic fungi, Ithaca, New York	See early collections section of the text and National collection section of this table.	*http://www.ppru.cornell.edu/Mycology/ARSEF_Culture_Collection.htm*
Fungi (including yeast)	United States	USDA/ARS/NCAUR Peoria, Illinois	Extensive holdings. See early collections section in text. Largest U.S. yeast collection (including Antarctic Marine Yeast Collection, described below). Also *Aspergillus* and *Penicillium* collection and collection of other filamentous fungi. Research, identification services, and patent depository.	*http://nrrl.ncaur.usda.gov*
Marine yeasts		The Antarctic Marine Yeasts Collection	Established in 1989 to preserve 3000 strains of yeast isolated and studied during the 1960s from the Antarctic, South Pacific, and Indian oceans. Contributed by J. Fell of the University of Miami to NCAUR Yeast Collection.	See *http://nrrl.ncaur.usda.gov*
Yeast	United States	The Phaff Collection of Yeasts and Yeast-like Microorganisms at University of California, Davis	Specialized collection emphasizing microbial diversity in natural habitats. Dedicated in 1996, but its history traces back to wine- and food-related yeasts at the College of Agriculture and Experiment Station and the Food Science Department. In 1939, H. J. Phaff began to develop the collection into a yeast taxonomy and ecology collection.	*Website http://www.phaffcollection.com/home.asp*

(*Continued overleaf*)

TABLE 24.1 *Continued*

Type	Country	Designation	Description	Web URL
			Includes over 6000 strains from North and South America, Asia, Australia, and Europe.	
Filamentous fungi and yeasts, primarily	Slovenia	MZKI, Microbial Culture Collection of the National Institute of Chemistry	Formerly the Filamentous Fungi Collection, Kernijski Institute in Ljubljana, now at the National Institute of Chemistry, Hajdrihova. Storage (including bacteria), identification, patent depository, and consultation services.	*http://wdcm.nig.ac.jp/CCINFO/CCINFO.xml? 599*
Filamentous fungi and yeasts	Australia	See National collection section of table	Comprehensive section of table—National consortium, especially CSIRO collection, New South Wales.	*http://wdcm.nig.ac.jp/hpcc.html*
Filamentous fungi and yeasts, primarily	Brazil	ESAP, Escola Superior de Agricultura collection in São Paulo	Approximately 1000 cultures of fungi and yeasts and 70 bacteria.	*http://wdcm.nig.ac.jp/CCINFO/CCINFO.xml? 720*
Filamentous fungi and yeasts	Brazil	URM, Federal University of Pernambuco	Approximately 2000 fungi and yeasts.	*http://wdcm.nig.ac.jp/CCINFO/CCINFO.xml? 604*
Filamentous fungi and yeasts	Brazil	IOC, Culture Collection of the Instituto Oswaldo Cruz in Rio de Janeiro	Approximately 1400 fungal cultures. Training and consultation services.	*http://wdcm.nig.ac.jp/CCINFO/CCINFO.xml? 720*
Fungi	Japan	NRIC, the Nodai Research Institute Culture Collection, Tokyo	Tokyo University of Agriculture. Primarily food fermentation microorganisms. Over 1400 yeast cultures and 400 molds. See also bacterial entry in table.	*http://wdcm.nig.ac.jp/CCINFO/CCINFO.xml? 747*
Fungi	India	DUM, Delhi University, Mycological Herbarium collection	Department of Botany Mycological Herbarium. 200 cultures of fungi. Performs identification services.	*http://wdcm.nig.ac.jp/CCINFO/CCINFO.xml? 40*
Fungi—Basidiomycetes	India	ITCC, India Type Culture Collection Mushroom Cultures	Indian Agricultural Institute Department of Pathology. Has *Agaricus, Auricularia, Lentinus*, and *Pleurotus* species.	*http://panizzi.shef.ac.uk/msdn/* and *http://wdcm.nig.ac.jp/CCINFO/CCINFO.xml? 430*
Fungi	Russia	RIAM, Research Institute of Applied Microbiology, Obolensk	Moscow region, Russian Fungal Collection.	*http://panizzi.shef.ac.uk/msdn*
Fungi—Basidiomycetes	Russia	Le(BIN) Basidiomycete Collection, St. Petersburg	Komarov Botanical Institute of the Russian Academy of Sciences.	*http://panizzi.shef.ac.uk/msdn*
Yeast	Russia	MSU, Moscow State University Yeast Database	Department of Soil Microbiology. See also yeast collection within VKM National collection, Comprehensive section of this table.	*http://panizzi.shef.ac.uk/msdn*
Yeast	Germany	IMET National Kulturensammlung von Microorganismen, Jena	Zentralinstitut fur Mikrobiologie und Experimentelle Therapie, ZIMET. Also yeasts within DSMZ National Collection.	Available at *http://www.dsmz.de*
Yeast (also see Comprehensive—other)	Japan	IFO, Culture Collection of the Institute for Fermentation, Osaka	A large collection of over 4400 bacteria, 7700 fungi, and 3000 yeasts as well as animal cell lines and viruses.	*http://www.ifo.or.jp/index_e.html*
Fungi and yeast	United States	Fungal Genetics Stock Center and Yeast Genetics Stock Center	Described in the text section on early genetic stock centers.	*http://www.fgsc.net and http://www.atcc.org/SpecialCollections/YGRRC.cfm* and *searchcatalogs/Fungi-Yeasts.cfm/*
Fungi and yeast	Brazil	Individual Collections	See National consortium section of this table.	
Yeast	Russia	Peterhof Genetic Collection of Yeasts	Described in the text section on early genetic stock centers.	*http://wdcm.nig.ac.jp/msdn/peteri.html*
Algae, Protozoa, and Cyanobacteria				
Algae, protozoa, and cyanobacteria	United States	UTEX, the Culture Collection of Algae at the University of Texas	Over 2100 strains of primarily freshwater algae. Began at Indiana University in 1954 and moved with Prof. R. C. Starr to Austin in 1976. Supported by the NSF and the University of Texas Organized Research Units. Lists of UTEX cultures and genera can be accessed from the Website, along with information on culture media and links to other collections.	*http://www.bio.utexas.edu/research/utex*
Algae, protozoa, and cyanobacteria	United Kingdom	CCAP, the Culture Collection of Algae and Protozoa	See early collections section of text and National Consortium—UK entries in this table. Approximately 2000 strains of algae and protozoa at IFE and DML. Patent depository.	*http://www.ife.ac.uk/ccap*

TABLE 24.1 *Continued*

Type	Country	Designation	Description	Web URL
Algae, protozoa, and cyanobacteria	Czech Republic	CCALA, the Culture Collection of Autotrophic Organisms	Began with early algal collection of Uhlir and Pringsheim in Prague, 1913, and merged in 1979 with the collection of the Institute of Microbiology, Czech Academy of Sciences at Trebon, founded in 1960. See early collections section of text. Over 200 strains of cyanobacteria and over 300 algae, and also has mosses, liverworts, ferns, and duckweed. Searchable via the MSDN.	*http://www.butbn.cas.cz/ccala/ccala.htm*
Algae, protozoa, and cyanobacteria—national center	United States	CCMP, Provasoli-Guillard National Center for Culture of Marine Phytoplankton Boothbay Harbor, Maine	Originated with the collections of Dr. Luigi Provasoli at Yale University and Dr. R. Guillard at Woods Hole Oceanographic Institution and became a national center as a result of recommendations of a 1980 workshop. Originally located at Woods Hole, but moved to the Bigelow Laboratory at Boothbay Harbor in 1981. Approximately 1450 cultures of primarily well-identified organism in pure culture. Special services include culture purification, safe deposit of private cultures, supplying mass cultures, DNA extractions, and occasional courses on culturing phytoplankton. Supported by the National Science Foundation. The culture database contains taxonomic and collection information, and the Website also presents information on culture techniques.	*http://ccmp.bigelow.org*
Algae	Germany	SVCK, the Culture Collection of Conjugatophyceae	Began at the Free University in Berlin and in 1959 moved to Marburg. Established at the Institute of General Botany of the University of Hamburg in the 1960s. Approximately 500 strains, collected world-wide. Education, training, and research functions. The Website lists culture and collection information by family.	*http://www.rrz.uni-hamburg.de/biologie/b_online/d44_1/44_1.htm*
Cyanobacteria	France	PCC, Pasteur Culture Collection of Cyanobacteria	Axenic cyanobacterial strains, one of the specialized collections of the Institut Pasteur in Paris and the Collection Nationale de Cultures de Microorganismes (CNCM) of France. See National collection section of table. Member of the WFCC and ECCO. Information includes accession number, brief description, history of isolation, cross-references with other collections with same species, medium for growth, properties, synonymous names for the species, and references. Keyword searches or specific, designated property searches are provided on their Web search pages.	*http://www.pasteur.fr/Bio/PCC/General.htm, http://www.pasteur.fr/recherche/banques/PCC*
Algae and cyanobacteria	Russia	IPPAS, the Culture Collection of Microalgae, Moscow	Institute of Plant Physiology, Russian Academy of Sciences. Established in 1958 and contains 600 cultures of cyanobacteria and algae. Patent depository for countries of the former USSR. Member of the ECCO. On-line searches can be conducted via the MSDN.	*http://wdcm.nig.ac.jp/CCINFO/CCINFO.xml? 596*
Algae	Canada	NEPCC, North East Pacific Culture Collection of Marine Microalgae at the University of British Columbia	Founded by F. J. R. Taylor in the late 1960s. Supported by the Department of Botany of the university Focuses on species of the North Pacific, but includes algae from other areas, and has many dinoflagellates. The Website lists the algae in the collection by family and gives collection and culturing information.	*http://beluga.ocgy.ubc.ca/projects/nepcc*
Algae and cyanobacteria	Canada	UTCC, University of Toronto Culture	Maintains and distributes cultures of over 350 isolates of primarily freshwater algae and cyanobacteria. Established in 1987 from several collections at the University of Toronto, with support from the Ontario Ministry of Colleges and Universities and currently from the Canadian Natural Sciences and Engineering Council and the university Department of Botany. Services include safe deposit, custom isolation of algae, identification, training, and specialized media preparation.	*http://www.botany.utoronto.ca/utcc/index.html*
Algae	Japan	Marine Biotechnology Institute (MBI) of Japan		See *http://seaweed.ucg.ie/cultures/CultureCollections.html*

(Continued overleaf)

TABLE 24.1 *Continued*

Type	Country	Designation	Description	Web URL
Algae—*Chlamydomonas*	United States	*Chlamydomonas* Genetics Center (CGC), Duke University	See genetic stock centers section of text. Mutants, cloned DNA, information on mapping and sequences.	*http://www.biology.duke.edu/chlamy*
Diatoms	United States	The Loras College Freshwater Diatom Culture Collection, Dubuque, Iowa	1200 freshwater diatoms, 64 genera, 350 species and varieties.	*http://www.bgsu.edu/Departments/biology/facilities/algae/html/DiatomCulture.html*
Algae	Australia	Murdoch University's algal collection, Perth	Approximately 400 strains of algae of commercial importance or potential. Includes *Dunaliella, Haematococcus, Spirulina, Chlorella,* and others. Also serves as a safe culture deposit for companies.	*http://seaweed.ucg.ie/cultures/CultureCollections.html#Murdoch*
Algae	France	Thallia Pharmaceuticals, SA, Ecole Centrale de Paris	About 250 strains of microalgae.	*http://seaweed.ucg.ie/cultures/CultureCollections.html#anchor480299*
Database Links				
Comprehensive database of collections	Internationally administered	MSDN, Microbial Strain Data Network	Access to large and small collections, private and public. MSDN is a nonprofit company with secretariat in Sheffield and nodes in China, India, and Brazil *et al.*	*http://panizzi.shef.ac.uk/msdn* (Europe), *http://bioinfo.ernet.in* (India), *http://www.micronet.cn/msdn.shtml* (China), *http://www.bdt.fat.org.br/bdt* (Brazil) (also *http://www.arabdecision.org/show_func_3_14_22_1_2_3797.htm* for MIRCEN—Cairo)
Comprehensive database of collections	International, server in Japan	WDCM, World Data Center for Microorganisms	Access to large and small collections, private and public. Broad searches across multiple databases and cross-linking with collection databases.	*http://wdcm.nig.ac.jp, http://wdcm.nig.ac.jp/DOC/menu3.xml, http://wdcm.nig.ac.jp/simple_search.html* and */hpcc.html*
Comprehensive database of collections	Japan—WDCM	WDCM—AHMII Agent for Hunting Microbial Information across the Internet	Queries across multiple databases.	*http://wdcm.nig.ac.jp/AHMII/ahmii.html*
Listing of collections	United States	Bergey's Manual Trust	Maintains on its Web site a list of major collections and consortia.	*http://www.cme.msu.edu/bergeys/resourceinfo.html*
Listing of collections	International	MIRCENS—Microbial Resources Centres	In addition to the on-line databases that MIRCEN sponsors, it publishes the World Directory of Collections of Cultures of Microorganisms, the World Catalogue of *Rhizobium* Collections, and the World Catalog of Algae and supports collecting and coordination of collections in MIRCEN centers and networks throughout the world. See Section I of the text.	*http://www.unesco.org/science/* and *http://www.asm.org/International/index.asp?bid=14221*
Laboratory collections	United States	MGD, Microbial Germplasm Database	Allows querying over numerous specialized and laboratory collections.	*http://mgd.nacse.org/cgi-bin/mgd*
Fungal database	United States	University of Kansas Database of Mycological Resources	Part of the University of Kansas BioDiversity and Biological Collections Web server.	*http://biodiversity.bio.uno.edu/~fungi/fcollect.html*
Algal database	Ireland	Seaweed	National University of Ireland, Galway.	*http://www.seaweed.ie*
Algal database	Japan—WDCM	WDCM—Algae	World Catalog of Algae search engine.	*http://wdcm.nig.ac.jp/simple_search.html*
Any WDCM collection	List or query by country	—	Use query interface shown under Japan WDCM above, of this list.	*http://wdcm.nig.ac.jp/hpcc.html* or */CCINFO search.html*

VIII. ORGANIZATIONS PROMOTING CULTURE COLLECTIONS

Organizations fostering stock center activity include the previously cited ECCO, European Culture Collections Organization, national federations such as the US Federation of Culture Collections and the Japan Federation of Culture Collections (which sponsor annual meetings as well as international meetings), UNESCO's MIRCEN, and the WFCC, World Federation of Culture Collections. An International Congress for Culture Collections, sponsored by the international federations, is held every 4 years.

BIBLIOGRAPHY

DaSilva, E. J. (2000). Microbial Resources Centres: Springboard for networking in the 21st century. *http://wdcm.nig.ac.jp/wdcm1999/a_MIRCEN.html.*

Demerec, M., Adelberg, E. A., Clark, A. J., and Hartman, P. E. (1966). A proposal for a uniform nomenclature in bacterial genetics. *Genetics* **54**, 61–76.

Hawksworth, D. L. (1990). "WFCC Guidelines for the Establishment and Operation of Collections of Cultures of Microorganisms." Simworth Press, Richmond, Surrey, UK, and *http://www.wdcm.nig.ac.jp/wfcc/wfcc_guidelines.html.* Revised, *http://www.wdcm.nig.ac/wfcc/GuideFinal.html.*

Hunter-Cevera, J. C. (1998). The importance of culture collections in industrial microbiology and biotechnology. *http://www.bdt.fat.org.br/oea/sib/cevera.*

Hunter-Cevera, J. C., and Belt, A. (1996). "The Importance of Culture Collections in Preservation and Maintenance of Microorganisms Used in Biotechnology." Academic Press, San Diego.

Jong, S.-C. (1989). Microbial germplasm. *In:* "Beltsville Symposia in Agricultural Research 13: Biotic Diversity and Germplasm Preservation, Global Imperatives" (L. Knutson and A. K. Stoner, eds.), pp. 241–273. Kluwer Academic Publishers, Boston.

Kurtzman, C. P. (1986). The ARS culture collection: Present status and new directions. *Enzyme Microb. Technol.* **8**, 328–333.

Nierlich, D. P., Benson, D., Berlyn, M. K. B., Blaine, L., Karp, P., and Sanderson, K. (1999). World Wide Web Resources for Microbiologists. *Biotechniques* **6**, 70–78.

Ochman, H., and Selander, R. (1984). Standard reference strains of *Escherichia coli* from natural populations. *J. Bacteriol.* **157**, 690–693.

WFCC Executive Board. (1999). WFCC Guidelines for the Establishment and Operation and Collections of Cultures of Microorganisms. 2nd edn. WFCC Newsletter No. 30. *http://www.wdcm.nig.ac.jp/wfcc/GuideFinal.html.*

WFCC Publications: see *http://wdcm.nig.ac.jp/wfcc/publications.html.*

WFCC Workshop on the Economic Value of Microbial Genetic Resources at the Eighth International Symposium on Microbial Ecology. (1998). Halifax, Canada. *http://wdcm.nig.ac.jp/wfcc/Halifax98.html#t1.*

Wackett, L. P. (2000). Web alert. Microbiol Culture Collection. *Environ. Microbiol.* **2**, 119–120.

WEBSITES

Website of Common Access to Biological Resources and Information (CABRI)
http://www.cabri.org

UNESCO//World Federation of Culture Collections Technical Information Sheets
http://wdcm.nig.ac.jp/wfcc/NEWSLETTER/Newsletter32-2.html

25

Developmental processes in bacteria

Yves V. Brun
Indiana University

GLOSSARY

chemotaxis Movement toward or away from a chemical.

hypha (pl. hyphae) A single filament of a mycelium.

mycelium A network of cellular filaments formed by branching during the growth phase of fungi and actinomycetes.

phosphorelay A signal-transduction pathway in which a phosphate group is passed along a series of proteins.

regulon A group of genes controlled by the same regulatory molecule.

septum A partition that separates a cell into two compartments.

sigma factor The subunit of the RNA polymerase holoenzyme that confers promoter specificity.

surfactant A substance that reduces the surface tension of a liquid.

TCA cycle The cyclic pathway by which the two-carbon acetyl groups of acetyl-CoA are oxidized to carbon dioxide and water.

vegetative growth Exponential growth that usually occurs by simple binary cell division and produces two identical progeny cells.

Bacterial Development generates specialized cell types that enhance the ability of bacteria to survive in their environment. In addition to changes in gene expression, developmental processes in bacteria involve changes in morphology and changes in function that play an important role in the life cycle of the organism. Eukaryotic organisms add sexual reproduction to the functions of development, but bacterial developmental processes are asexual.

Bacteria use two basic strategies to respond to changes in their environment. In the first and simplest strategy, they induce the expression of genes that enable them to deal with the environmental change. For example, starvation for inorganic phosphate, the preferred source of the essential element phosphorus, induces the Phosphate (Pho) regulon. The Pho regulon includes genes for the high-affinity phosphate-transport proteins that increase the ability of the bacterium to transport phosphate and genes that allow the bacterium to metabolize organic forms of phosphate. These responses are relatively simple in that they usually involve a two-component regulatory system that activates the transcription of a set of genes required for the response. At the other extreme of complexity are bacteria that undergo complex developmental transformations in response to stress or as part of their normal life cycle. These developmental responses involve not only changes in gene expression, but also changes in cellular morphology, metabolic chemistry, and association with cells of other species.

I. FUNCTION OF DEVELOPMENT

Bacterial development produces cells that have four basic types of functions. Representative examples of these functions are given in Table 25.1. The most common product of bacterial development is a resting cell with relatively low metabolic activity and a higher resistance to physical and chemical stress than

The Desk Encyclopedia of Microbiology
ISBN: 0-12-621361-5

TABLE 25.1 Examples of prokaryotic development[a]

Resting cells		
Resting cell	**Representative genus**	**Group**
Endospore	*Bacillus*	Gram positive
	Metabacterium	Gram positive
	Thermoactinomyces	Gram positive
Aerial spore	*Streptomyces*	Gram positive
Zoospore	*Dermatophilus*	Gram positive
Cyst	*Azotobacter*	Proteobacteria
	Methylomonas	Proteobacteria
	Bdellovibrio	Proteobacteria
Myxospore	*Myxococcus*	Proteobacteria
	Stigmatella	Proteobacteria
Exospore	*Methylosinus*	Proteobacteria
Small dense cell	*Coxiella*	Proteobacteria
Elementary body	*Chlamydia*	Chlamydia
Akinete	*Anabaena*	Cyanobacteria

Complementary cell types			
Cell	**Function**	**Representative genus**	**Group**
Heterocyst	Nitrogen fixation		
Vegetative cell	Oxygenic photo-synthesis	*Anabaena*	Cyanobacteria

Dispersal cells		
Cell	**Representative genus**	**Group**
Baeocyte	*Pleurocapsa*	Cyanobacteria
Elementary body	*Chylamidia*	Chlamydia
Gonidium	*Leucothrix*	Proteobacteria
Hormogonium	*Oscillatoria*	Cyanobacteria
Swarm cell	*Proteus*	Proteobacteria
Swarmer cell	*Caulobacter*	Proteobacteria
Zoospore	*Dermatophilus*	Gram positive

Symbiotic development			
Cell	**Representative interaction**	**Function**	**Group**
Bacteroid	*Rhizobium*–legume	N_2 fixation	Proteobacteria
	Frankia–alder	N_2 fixation	Gram positive

[a]From Shimkets, L., and Brun, Y. V. (1999). Prokaryotic development: strategies to enhance survival. *In* "Prokaryotic Development" (Y. V. Brun and L. Shimkets, Eds.), pp. 1–7. American Society for Microbiology, ASM Press.

the vegetative cell; the best-studied example is endospore formation in *Bacillus subtilis*. The second type of function of differentiated cells is dispersal. Dispersal can be propelled by flagella or can simply be aided by wind, water, or animals in the case of nonmotile cells; the dispersal swarmer cell of *Caulobacter crescentus* is the product of an asymmetric division that also produces a sessile stalked cell. The production of cells whose physiology is complementary represents the third type of function of bacterial development; this is best exemplified by formation of heterocysts that are specialized for nitrogen fixation in the Cyanobacterium *Anabaena*. Finally, bacterial development can lead to the establishment of a symbiotic relationship, as in the case of nodulation of legume roots by Rhizobium. Examples of the various developmental functions are presented here. In order to give a flavor of the research in this field, one example (*Caulobacter*) is described in slightly more detail.

II. ENDOSPORE FORMATION IN *BACILLUS SUBTILIS*

Endospore formation has been found exclusively in gram-positive bacteria and is best understood in *Bacillus subtilis*. The primary signal for the initiaton of sporulation is nutrient starvation. Cell density is also important for efficient sporulation, presumably to ensure that cells are sufficiently abundant. It may be that if starving cells are at a high density, it is better to sporulate rather than compete for nutrients. However, if cells are at a low density, the chances of finding additional nutrients is higher and sporulation less desirable. Before initiating sporulation, cells monitor many intracellular factors, such as DNA replication and the TCA cycle. The integration of the extracellular and intracellular signals is regulated through a multicomponent phosphorelay that controls the prosphorylation of the transcriptional regulator SpoOA. The initiation of sporulation by the formation of a polar septum instead of the vegetative midcell septum requires the accumulation of a sufficient concentration of SpoOA~P. The subsequent engulfment of the prespore by the mother cell compartmentalizes the prespore inside the mother cell (Fig. 25.1). The genome of the mother cell provides the components for constructing the spore exterior and the genome of the forespore provides the components for constructing the spore interior. The forespore ultimately becomes a metabolically quiescent and stress-resistant spore that can give rise to future progeny by germination when conditions improve. The mother cell is discarded by lysis after the completion of sporulation.

The regulation of events in the mother cell and the forespore is due to the presence of four different sigma factors, two in each compartment, which assures that each genome gives rise to a different set of products (Fig. 25.2). Activation of σ^F in the forespore depends on polar septation. σ^E is synthesized as an inactive precursor whose activation by proteolitic processing in the mother-cell compartment is

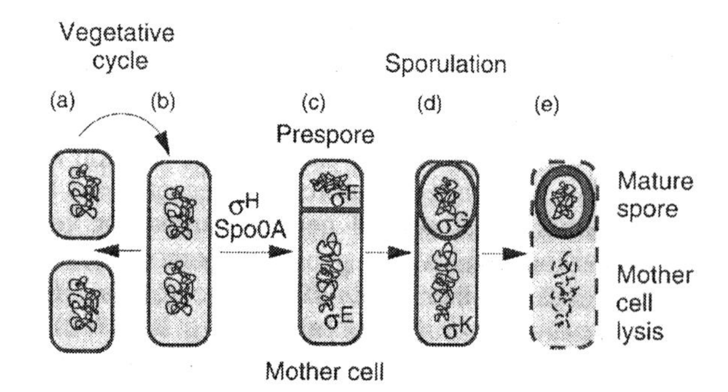

FIGURE 25.1 Life cycle of *Bacillus subtilis*. (a) Vegetative growth occurs by binary fission when nutrients are plentiful. (b) and (c) When starved, the vegetative midcell division is replaced by a highly asymmetric polar division that compartmentalizes the cell into a prespore and a mother-cell compartment. (d) The prespore is engulfed by the mother cell. (e) After formation of the spore cortex (thick circle), the mother cell lyses and releases the mature spore. [From J. Errington (1996). Determination of cell fate in bacillus subtilis, *Trends in Genetics* 12: 31–34, Copyright (1996), with permission from Elsevier Science.]

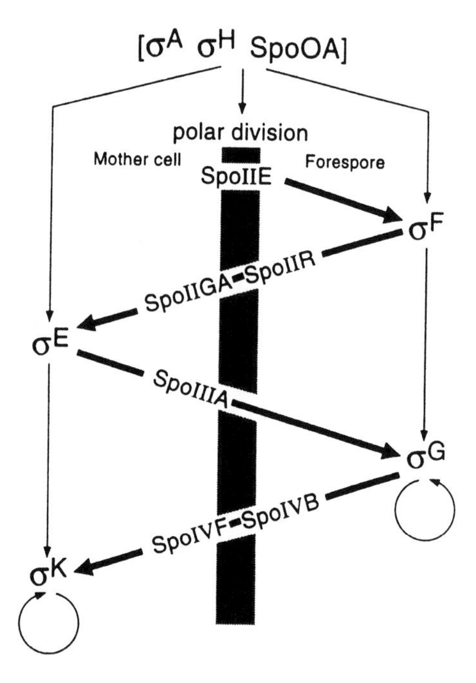

FIGURE 25.2 Criss-cross regulation of cell type-specific sigma factors. [With permission, from P. Stragier and R. Losick (1996). Molecular genetics of sporulation in bacillus subtilis, *Annual Review of Genetics* 30: 297–341, © 1996 by Annual Reviews, www.annualreviews.org.]

dependent on activation of σ^F in the forespore. The transcription of the σ^G gene requires σ^F and thus only occurs in the forespore. σ^G activation depends on proteins made in the mother cell under the control of σ^E. Finally, σ^K is only synthesized in the mother cell under the control of σ^E and is activated in σ^G-dependent manner.

FIGURE 25.3 Scanning electron micrographs of spore chain in the aerial mycelium of *Streptomyces coelicolor*. Bar, 1 μm. [From K. Chater (1998). Taking a genetic scalpel to the Streptomyces colony. *Microbiology* 144: 1465–1478.]

III. SPORULATION IN *STREPTOMYCES COELICOLOR*

The aerial mycelium of *Streptomyces coelicolor* forms by directed cell growth and differentiates into a series of spores (Fig. 25.3). The vegetative mycelium grows in the nutrient substratum by the linear growth of cell wall close to the hyphal tip (Fig. 25.4). Branching of the vegetative mycelium allows close-to-exponential increase of the mycelial mass. Septation is infrequent in the vegetative mycelium and the vegetative septa do not allow cell separation. With time, the vegetative mycelium becomes more dense, producing aerial hyphae that grow quickly. Rapid growth occurs at the expense of nutrients derived from the substrate

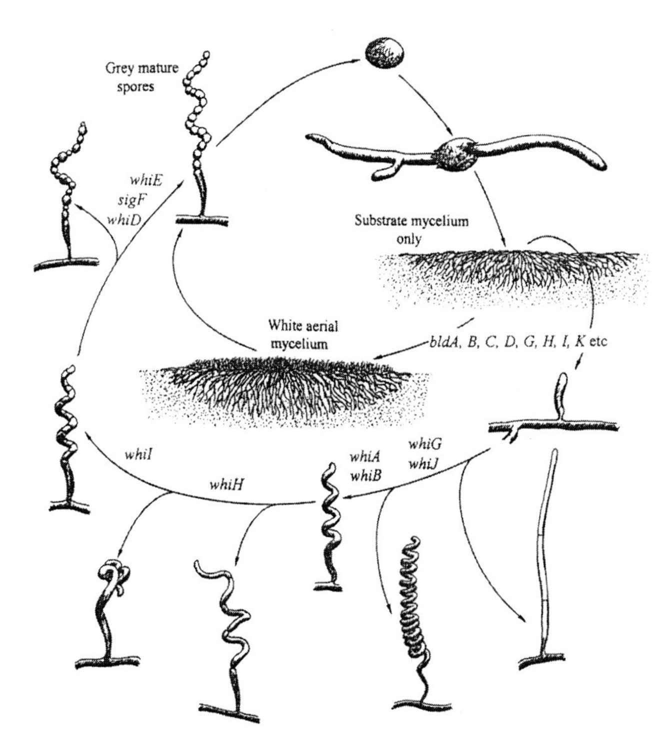

FIGURE 25.4 Life cycle of *Streptomyces coelicolor*. Genes important in the different stages of the life cycle are shown. The phenotypes of various mutants are shown by arrows diverging from the normal life cycle immediately before the gene designation. [From K. Chater (1998). Taking a genetic scalpel to the Streptomyces colony. *Microbiology* 144: 1465–1478.]

mycelium and aerial hyphae emerge from the surface of colonies. The formation of the aerial hyphae requires a set of genes called *bld* genes because mutants of these genes fail to develop a hairy surface layer (*bald*). Most of these mutants fail to produce a small extracellular surfactant protein SapB. SapB coats the surface of the aerial hyphae with a hydrophobic outer surface. This may permit growth through the surface tension barrier at the air–colony interface. Antibiotics are produced, presumably to protect the nutrients released from lysing substrate mycelium from other bacteria. All antibiotic production, as well as aerial mycelium development, is prevented in *bldA* mutants. The *bldA* gene encodes the only tRNA that efficiently recognizes the rare leucine codon UUA. Genes required for vegetative growth do not contain TTA codons. TTA codons are found in regulatory genes involved in antibiotic production. Is it thought that an increased production of mature *bldA*-encoded tRNA during development allows the efficient translation of UUA codons in regulators of antibiotic production and in genes involved in development whose identify is still unknown. Growth of the aerial hyphae eventually stops and regularly spaced sporulation septa are formed synchronously. Thus, the cell separation required for dispersal occurs by sporulation at the surface of colonies. Sporulation requires *whi* genes, identified because mutations in these genes prevent the formation of mature grey spores and the aerial mycelium remains white. Most of the early-acting *whi* genes appear to be regulatory. *whiG* encodes a sigma factor, while *whiI* and *whiH* encode transcriptional regulators.

IV. SWARMING BACTERIA

Swarming differentiation produces cells (swarm cell) capable of a specialized form of translocation on a surface and occurs in a variety of bacteria, both gram-positive and gram-negative. These include *Proteus, Bacillus, Clostridium,* and *Vibrio* species, *Serratia marcescens, Rhodospirillum centenum, E. coli,* and *S. typhimurium*. The swarm cell differentiation is triggered by growth on an appropriate solid medium, for example a petri plate. Initially, cells grow vegetatively as short rods with a small number of flagella. Differentiated swarm cells are long (20–80 μm), multinucleate, non-dividing cells with up to 50-fold more flagella per unit cell-surface area than vegetative cells. Swarm cells migrate rapidly across the plate (Fig. 25.5). Swarming differentiation is not a starvation response and is not an obligatory stage in the life cycle of these bacteria. An important signal for swarm cell differentiation is bacterial contact with a solid surface. In *P. mirabilis* and *V. parahaemolyticus,* surface-sensing is mediated by the flagella. After a certain period, the migration of swarm cells slows down and the swarm cells divide (consolidation) and revert to the vegetative-swimmer cell type (Fig. 25.6). Vegetative growth continues until a second phase of swarm cell differentiation is initiated. This differentiation–consolidation cycle continues until the surface of the plate is covered. Swarming can have a function in host–pathogen interactions. For example, *Proteus mirabilis* mutants that are deficient in swarming are unable to establish kidney infections.

The *flhDC* master regulatory operon is critical for the control of swarming differentiation. Artificial overexpression of FlhDC induces swarm cell differentiation without the need for contact with a solid surface. FlhD and FlhC form a complex that activates transcription of genes encoding flagellar export, structural, and regulatory proteins. In addition, FlhDC represses cell division.

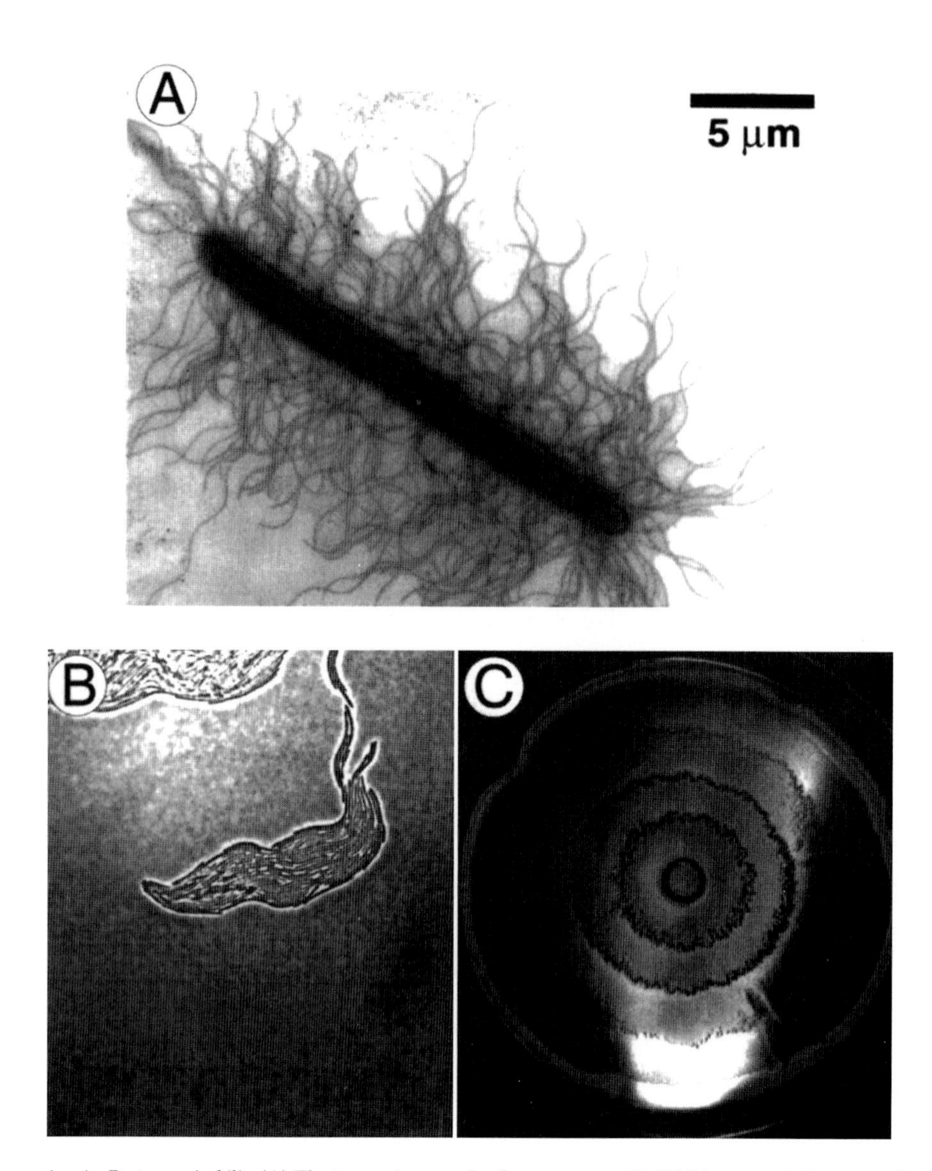

FIGURE 25.5 Swarming in *Proteus mirabilis*. (A) Electron micrograph of a swarmer cell. (B) Movement of a mass of cells at the swarming periphery. (C) Characteristic colony morphology. The pattern is produced by alternating cycles of differentiation, movement, and consolidation. [From BACTERIA AS MULTICELLULAR ORGANISMS, edited by James A. Shapiro and M. Dworkin. Copyright © 1997 by Oxford University Press, Inc. Used by permission of Oxford University Press, Inc.]

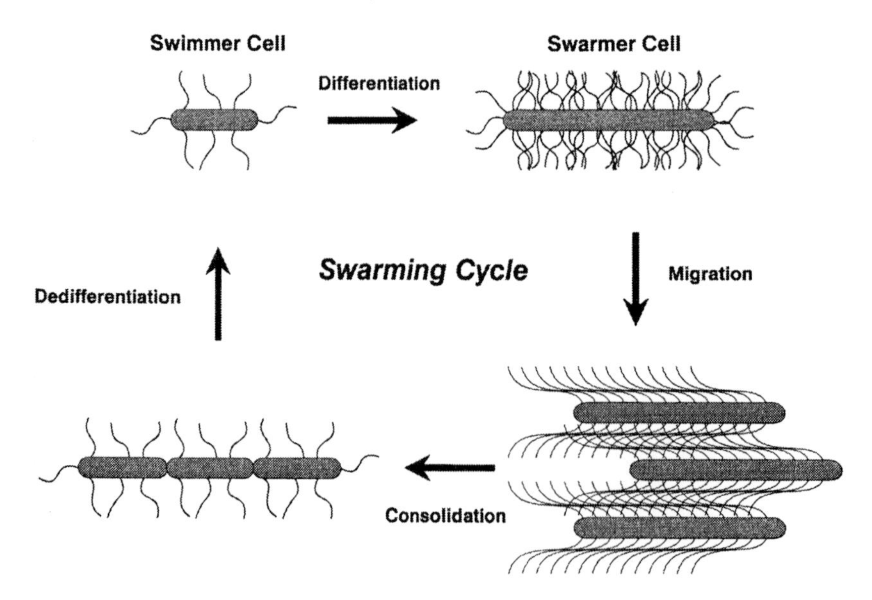

FIGURE 25.6 The *Proteus mirabilis* swarm cell-differentiation cycle. [From BACTERIA AS MULTICELLULAR ORGANISMS, edited by James A. Shapiro and M. Dworkin. Copyright © 1997 by Oxford University Press, Inc. Used by permission of Oxford University Press, Inc.]

V. DIMORPHIC LIFE CYCLE OF *CAULOBACTER CRESCENTUS*

A distinguishing feature of the development of stalked bacteria is that it is an integral part of the growth of the cell and not an alternative to it, as are the other bacterial developmental processes that occur in response to stress. The molecular mechanisms that control the developmental cycle of stalked bacteria have been studied most extensively in *Caulobacter crescentus*. Each division of *Caulobacter* cells gives rise to a swarmer cell and a stalked cell (Fig. 25.7). The swarmer cell is dedicated to dispersal and the stalked cell is dedicated to growth and the production of new swarmer cells. The obligatory time spent as a chemotactic swarmer cell presumably ensures that progeny cells will colonize a new environmental niche instead of competing with attached stalked cells. The swarmer cell has a single polar flagellum and is chemotactically competent. During this dispersal stage of their life cycle, swarmer cells do not replicate DNA and do not divide. After approximately one-third of the cell cycle, in response to an unknown internal signal, the swarmer cell sheds its flagellum, initiates DNA replication, and synthesizes a stalk at the pole that previously contained the flagellum. Located at the tip of the stalk is the holfast, the adhesion organelle that allows *Caulobacter* to attach to surfaces. The holdfast appears at the tip of nascent stalks during swarmer to stalked cell differentiation. Cell growth is accelerated at the time of swarmer-to-stalked cell differentiation and eventually leads to the formation of a predivisional cell in which a flagellum is synthesized *de novo* at the pole opposite the stalk. Unlike the swarmer cell, the progeny stalked cell is capable of initiating a new round of DNA replication immediately after cell division.

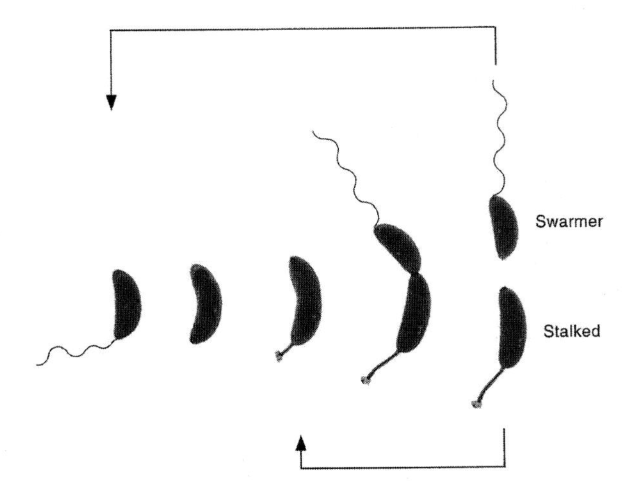

FIGURE 25.7 The *Caulobacter crescentus* cell cycle. [From cover of *Genes and Development*, Vol. 12, No. 6. Kelly, A. J., Sackett, M., Din, N., Quardokus, E., and Brun, Y. V. (1998). Cell cycle dependent transcriptional and proteolytic regulation of FtsZ in *Caulobacter Genes and Development* **12**: 880–893.]

A. Cell-cycle regulation of flagellum synthesis

The best understood event in *Caulobacter* development is the biosynthesis of the flagellum (Fig. 25.8). A new flagellum is synthesized at every cell cycle and is localized at the pole opposite the stalk. The flagellum is, for the most part, similar to that of *E. coli* and its synthesis requires more than 50 genes. The expression of these flagellar genes is temporally ordered during the progression through the cell cycle; their order of transcription approximates the order of assembly of their protein products in the flagellum. Most flagellar genes can be grouped into four classes, forming a regulatory hierarchy that dictates their order of expression. First to be transcribed, immediately after the differentiation of the swarmer cell into a stalked cell, are the gene for the MS ring that anchors the flagellar basal body in the cytoplasmic membrane, genes that encode the proteins of the switch complex, and genes for the flagellar export apparatus. In addition, Class II genes encode the regulatory proteins FlbD and σ^{54} that are required for the transcription of most Class III and IV genes. These early-expressed genes make up Class II in the flagellar regulatory hierarchy. The expression of Class II genes is required for the transcription of the next set of flagellar genes, the Class III genes. Class III genes encode proteins that make up the rest of the basal body (the rings anchored in the peptidoglycan cell wall and in the outer membrane and the rod that traverses the rings) and the proteins that compose the hook structure. The expression of Class III genes is required for the transcription of the last flagellar genes to be expressed during the cell cycle, the Class IV genes that encode the flagellins that make up the helical filament. The flagellar regulatory cascade is triggered by the response regulator CtrA, which by definition occupies Class I of the regulatory hierarchy. CtrA activates the transcription of Class II genes by binding to a conserved sequence in their promoter region. *In vitro* experiments indicate that phosphorylated CtrA (CtrA~P) is required for the transcriptional activation of these genes. The *flbD* gene is the last gene of the Class II fliF operon whose transcription depends on CtrA~P. The promoter of the *rpoN* gene contains a putative CtrA binding site. Consequently, the initiation of the

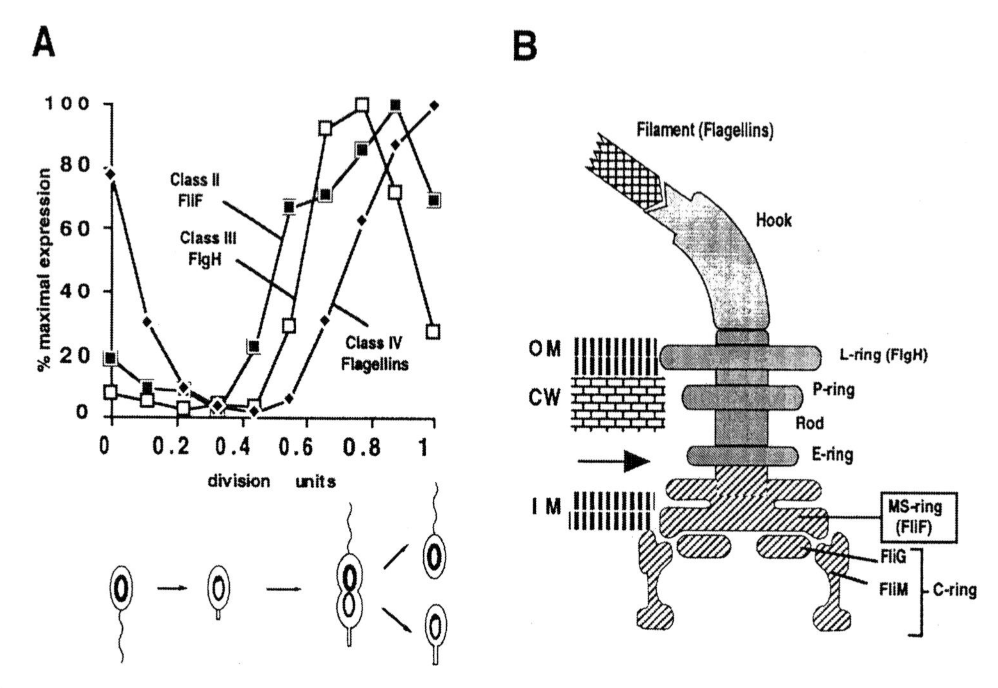

FIGURE 25.8 Cell cycle-dependent expression of *Caulobacter* flagellar proteins. (A) The expression of a representative protein from each class is shown. (B) Diagram of the *Caulobacter* flagellum. [From U. Jenal and L. Shapiro (1996). Cell cycle-controlled proteolysis of a flagellar motor protein that is asymmetrically distributed in the *Caulobacter* predivisional cell. *The EMBO Journal* **15**: 2393–2406, by permission of Oxford University Press.]

transcription of Class II genes of the flagellar regulatory cascade by CtrA~P results in the synthesis of the regulatory proteins FlbD and σ^{54}, which are required for the transcription of Class III and IV genes. CtrA also regulates DNA replication and cell division, providing a mechanism to coordinate the expression of flagellar genes with those events (see next section).

B. Regulation of cell division and DNA replication

In *Caulobacter*, different stages of development require the completion of specific stages of the replication and division cycles. The inhibition of DNA replication blocks flagellum synthesis by preventing the transcription of early flagellar genes that are at the top of the flagellar regulatory hierarchy. Cells inhibited for DNA replication are also blocked for cell division and form long smooth filamentous cells with a stalk at one pole and flagella at the opposite pole. Cells that can replicate DNA, but that are blocked in cell division, are also affected in their progression through development. The initiation of cell division plays an essential role in the establishment of differential programs of gene expression that set up the fates of the progeny cells.

In all bacteria examined, the abundance and subcellular location of the tubulin-like GTPase, FtsZ, are critical factors in the initiation of cell division. FtsZ is a highly conserved protein that polymerizes into a ring structure associated with the cytoplasmic membrane at the site of cell division. FtsZ recruits other cell-division proteins to the site of cell division and may constrict, providing mechanical force for division. In *Caulobacter*, FtsZ is subject to a tight developmental control. After cell division, only the stalked cell contains FtsZ. Transcriptional and proteolytic controls contribute to the cell cycle and developmental regulation of FtsZ (see Fig. 25.9).

The initiation of DNA replication and *ftsZ* transcription are controlled by the cell-cycle-response regulator CtrA. CtrA directly binds to five sites in the origin of replication and prevents the initiation of DNA replication. CtrA is present in swarmer cells, where it blocks DNA replication and represses *ftsZ* transcription. CtrA is degraded during swarmer cell differentiation, thus coordinating the onset of the replication and division cycles. The degradation of CtrA depends on the ClpXP protease. Late in the cell cycle, when DNA replication is complete and cell division has been initiated, CtrA is synthesized and represses *ftsZ* transcription and initiation of DNA replication. Just before cell separation, CtrA is degraded in the

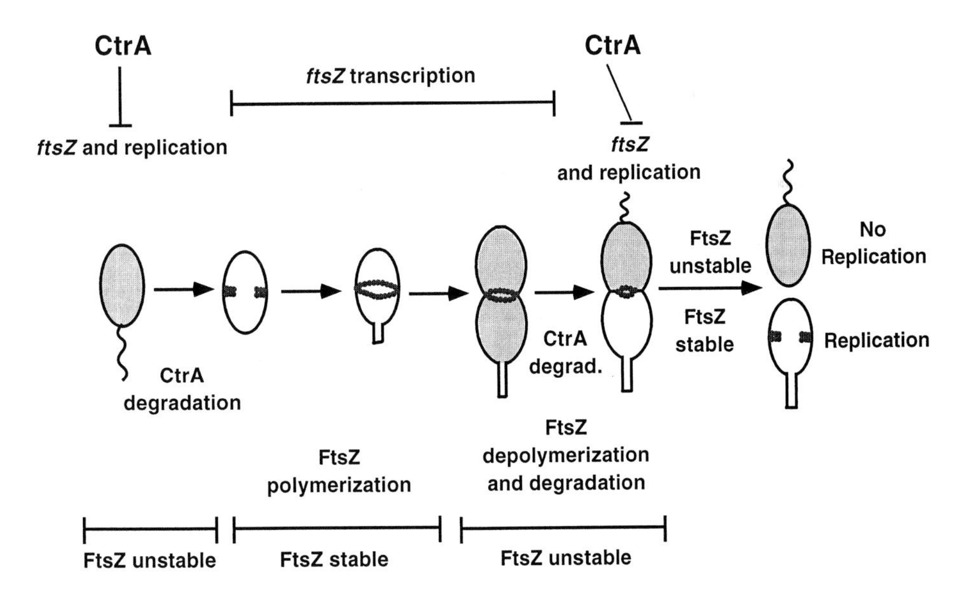

FIGURE 25.9 Model of FtsZ regulation in *Caulobacter*. CtrA (gray shading inside cells) represses *ftsZ* transcription in swarmer cells. During swarm cell differentiation, CtrA is degraded and allows *ftsZ* transcription to be turned on. FtsZ concentration increases and FtsZ polymerizes and forms a ring at the site of cell division. During this time FtsZ is stable. The reappearance of CtrA inhibits *ftsZ* transcription. FtsZ depolymerizes as the cell and the FtsZ ring constrict. FtsZ is rapidly degraded, especially in the swarmer pole. [From Kelly, A. J., Sackett, M. Din, N., Quardokus, E., and Brun, Y. V. (1998). Cell cycle dependent transcriptional and proteolytic regulation of FtsZ in *Caulobacter*. *Genes and Development* **12**: 880–893.]

stalked compartment. The absence of CtrA from stalked cells after cell division allows *ftsZ* transcription to resume and DNA replication to be initiated.

Proteolytic control of FtsZ is superimposed on the transcriptional control. FtsZ molecules are stable as they assemble into the FtsZ ring and are degraded rapidly once cells have begun to constrict. FtsZ is particularly unstable in the swarmer compartment of the predivisional cell, leading to its disappearance from swarmer cells after cell division. This two-tiered level of regulation ensures that FtsZ is only present in the cell that will initiate a new cell cycle immediately after cell division.

VI. FRUITING-BODY FORMATION IN MYXOBACTERIA

Fruiting-body formation in myxobacteria is only one example of bacterial social behavior. The entire life cycle of myxobacteria is pervaded by social behavior. Myxobacterial cells move together and feed cooperatively to maximize the efficiency of extracellular degradation. The enclosure of myxospores in the fruiting body allows then to be dispersed together and ensures that a sufficiently large population of cells will be present after germination to facilitate social interactions. The best-studied example of fruiting-body formation in myxobacteria is in *Myxococcus xanthus* (Fig. 25.10). When cells perceive a nutritional down shift, they enter the developmental pathway that leads to fruiting-body formation (Fig. 25.11). Fruiting-body formation can only occur if cells are on a solid surface, to allow gliding motility, and if the cell density is high. When these three conditions are met, cells move into aggregation centers and eventually form mounds containing approximately 10,000 cells. As many as 90% of cells lyse during aggregation. The surviving cells differentiate into resistant and metabolically quiescent myxospores during the last stages of fruiting-body formation. Fruiting-body development is regulated by a series of intercellular signals. Five categories of signals are involved: (1) the A signal is a mixture of amino acids and peptides that serves to monitor cell density; (2) the B signal acts early in development and its production depends on the Lon protease, but the nature of the signal has not been identified; (3) the C signal is associated with the cell surface and is the last of the five signals to act, controlling both aggregation and sporulation (the chemical nature of C signal is not known; all mutations that prevent C-signal formation map to the *csgA* gene and the CsgA protein itself could be the signal or it could produce the signal through an enzymatic activity); (4) the D signal requires the normal function of the *dsgA* gene that encodes the translation initiation factor 3 (IF3) but neither the manner by which *dsgA*

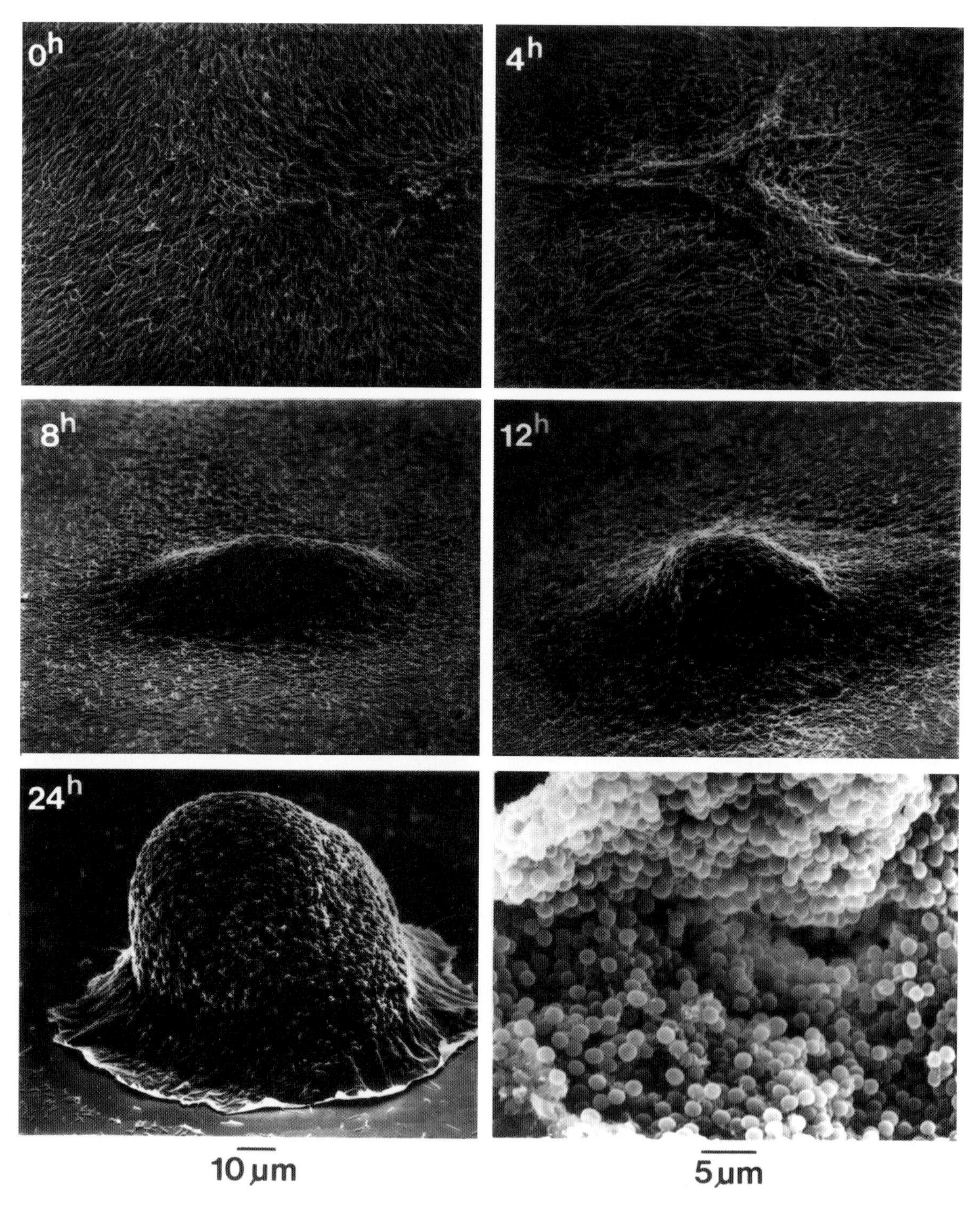

FIGURE 25.10 Scanning electron micrographs showing different stages of fruiting-body development in *Myxococcus xanthus*. The lower right panel shows spores from an open fruiting body. [From Kaiser, D., Kroos, L., and Kuspa, A. (1985). Cell interactions govern the temporal pattern of *Myxococcus* development. *Cold Spring Harbor Symposia on Quantitative Biology* **50**: 823–830.]

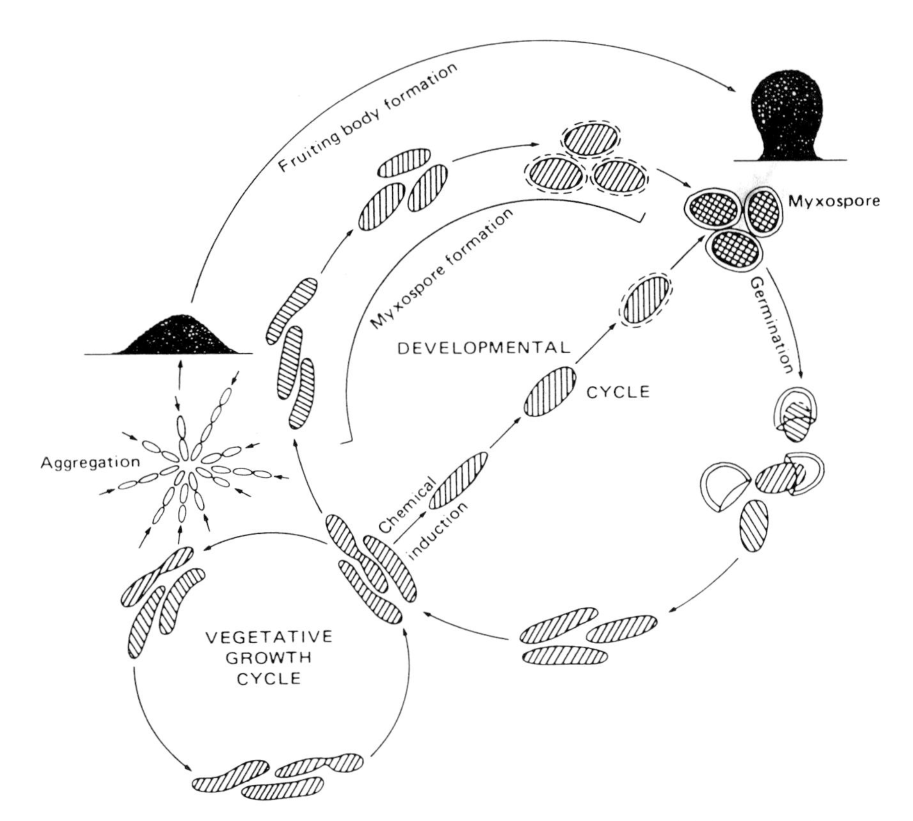

FIGURE 25.11 Life cycle of *Myxococcus xanthus*. [From Dworkin, M. (1985). Developmental Biology of the Bacteria. Benjamin/Cummings, Menlo Park, CA.]

functions in the production of the signal nor the identity of the signal are known; and (5) the E signal is thought to consist of branched-chain fatty acids liberated by a phospholipase and passed between cells to function as short-range signals.

VII. HETEROCYST DIFFERENTIATION IN CYANOBACTERIA

The purpose of heterocyst formation in cyanobacteria such as *Anabaena* is the production of a cell specialized for nitrogen fixation in order to separate two incompatible processes. The oxygen generated by the photosynthetic activity of vegetative cells is sufficient to inactivate the nitrogenase that is required to convert atmospheric N_2 to ammonium. When nitrogen fixation is required, the detrimental effect of oxygen is circumvented by sequestering nitrogenase in the anaerobic environment of the heterocyst, in an otherwise aerobic filament of vegetative cells. In the presence of ammonium or nitrate, cyanobacteria grow as undifferentiated vegetative filaments. When these cells are starved for nitrogen, heterocyst formation is induced (Fig. 25.12). The heterocyst is a terminally differentiated cell, but the differentiating cell passes through a proheterocyst stage that

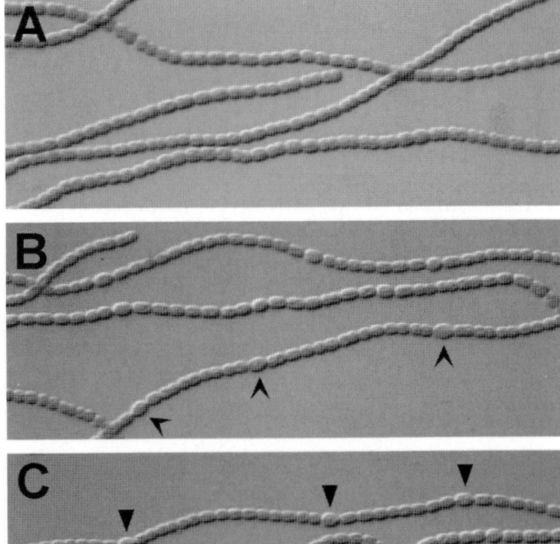

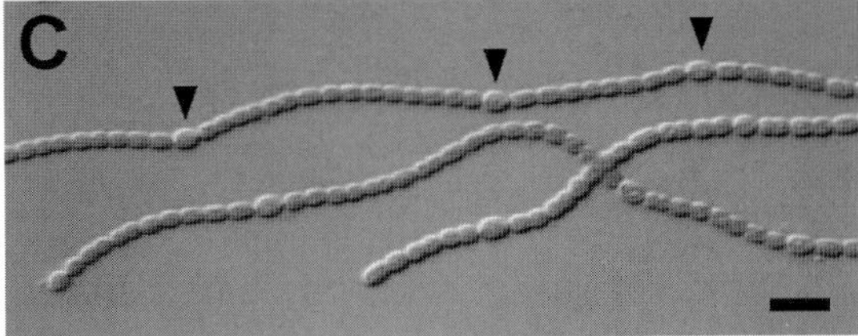

FIGURE 25.12 *Anabaena* sp. strain PCC-7120 filaments grown in nitrate medium (A) Filaments were subjected to nitrogen step-down for 18 hr (B), and 24 hr (C) to induce heterocysts. Developing proheterocysts are indicated on one filament with arrowheads in (B), and mature heterocysts are indicated with triangles in (C). The strain used in these figures contains a reporter plasmid that does not affect wild-type development of heterocysts. Scale bar, 10 μm. Photo by Ho-Sung Yoon.

can go back to vegetative growth under appropriate conditions. Single heterocysts form at approximately every 10 cells in a filament. Heterocystpattern formation is controlled in part by a diffusable signal encoded by the *PatS* gene. The PatS peptide is produced by proheterocysts and inhibits development of neighboring cells by creating a gradient of inhibitory signal.

VIII. THE PREDATORY LIFESTYLE OF *BDELLOVIBRIO*

The predation of gram-negative bacteria by *Bdellovibrio* includes a dimorphic life cycle (Fig. 25.13). During the obligatory intraperiplasmic growth phase, *Bdellovibrio* use their prey's cytoplasmic contents as their growth substrate. During the attack phase, they search for a new prey but do not grow. No DNA replication occurs during the attack phase; however, RNA and protein are synthesized. Thus, in addition to their dispersal function, the attack cells have the ability to attach to and enter bacterial prey. The function of the intraperiplasmic cells is to grow and to produce more attack phase cells. Attack cells are highly motile (100 μm/s) by virtue of a single polar flagellum. The attack phase continues until a suitable prey is encountered or until energy is exhausted. The attack phase has been described as a "race against starvation to find a susceptible prey". *Bdellovibrio* attack cells attach to prey cells and enter the prey 5–10 min after attachment. The flagellum is shed during the entry process. The prey cell is transformed into a bdelloplast by the action of a glycanase that solubilizes part of the peptidoglycan. The biochemical modification of the prey's peptidoglycan and lipopolysaccharide make the prey inaccessible to other *Bdellovibrio* cells. The *Bdellovibrio* cell then begins the systematic degradation of host macromolecules, which is complete in about 60 min. DNA replication begins during the intraperiplasmic growth phase and occurs without cell division to produce a multinucleate filament whose size is determined by the size of the prey, ranging from 4 to 100 times the length of an attack-phase cell. The inhibition of cell division during intraperiplasmic growth while DNA replication is occurring presents an interesting contrast to the usual coupling of replication and division in many bacteria. Swarming bacteria like *Proteus vulgaris* also inhibit cell division during growth as part of swarm cell differentiation. As part of the growth phase, *Bdellovibrio* incorporates some of the outer-mbrane proteins of the prey directly into its own membrane. Once growth becomes limited by the depletion of nutrients, elongation ceases and cell division is initiated simultaneously between the nucleoids. Flagella are synthesized *de novo* and the bdelloplast is lysed, releasing the attack-phase cells.

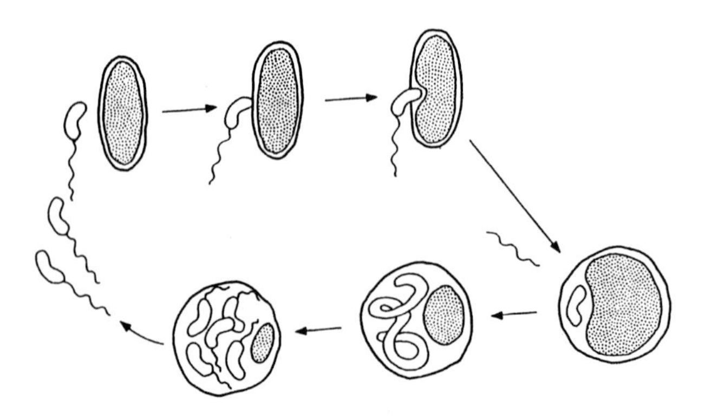

FIGURE 25.13 Life cycle of *Bdellovibrio*. [From Thomashow, M. F. and Cotter, T. W. (1992). *Bdellovibrio* host dependence: the search for signal molecules and genes that regulate the intraperiplasmic growth cycle. *J. Bacteriol.* **174**: 5767–5771.]; originally from Thomashow and Rittenberg (1979). *In* Developmental Biology of Prokaryotes, Blackwell Science Ltd.

IX. CONCLUSION

The study of bacterial development has had a major impact on our understanding of the bacterial cell. In particular, it is now clear that bacterial cells, even those that do not differentiate, are not simply bags of enzymes. Bacterial cells are highly organized at the level of protein localization. The spatial constraints of differentiating cells are combined with temporal constraints that control the ordered progression through the developmental program. A major challenge will be to determine how temporal and spatial control are integrated during bacterial development.

BIBLIOGRAPHY

Belas, R. (1997). *Proteus mirabilis* and other swarming bacteria. *In* "Bacteria as Multicellular organisms" (J. A. Shapiro and M. Dworkin, eds.), pp. 183–219. Oxford University Press, Oxford.

Brun, Y. V., and Shimkets, L. (eds.) (1999). "Prokaryotic Development." ASM Press, Washington, DC.

Chater, K. (1998). Taking a genetic scalpel to the Streptomyces colony. *Microbiology* **144**, 1465–1478.

Golden, J. W., and Yoon, H.-S. (1998). Heterocyst formation in Anabaena. *Curr. Opinion Microbiol.* **1**, 623–629.

Jelsbak, L., and Sogaard-Andersen, L. (2000). Pattern formation: fruiting body morphogenesis in *Myxococcus xanthus*. *Curr. Opin. Microbiol.* **3**, 637–642.

Jenal, U. (2000). Signal transduction mechanisms in *Caulobacter crescentus* development and cell cycle control. *FEMS Microbiol. Rev.* **24**, 177–191.

Kaiser, D. (1999). Cell fate and organogenesis in bacteria. *Trends Genet.* **15**(7), 273–277.

Kim, S., Kaiser, D., *et al.* (1992). Control of cell density and pattern by intercellular signaling in *Myxococcus* development. *Annu. Rev. Microbiol.* **46**, 117–139.

Martin M. E., and Brun Y. V. Coordinating development with the cell cycle in *Caulobacter. Curr. Opin. Microbiol.* **3**, 589–595.

Osteras M., and Jenal, U. (2000). Regulatory circuits in *Caulobacter. Curr. Opin. Microbiol.* **3**, 171–176.

Shapiro, J. A., and Dworkin, M. (eds.) (1997). "Bacteria as Multicellular Organisms." Oxford University Press, Oxford.

Shapiro, L., and Losick, R. Dynamic spatial regulation in the bacterial cell. *Cell* **100**, 89–98.

Stragier, P., and Losick, R. (1996). Molecular genetics of sporulation in *Bacillus subtilis. Annu. Rev. Genet.* **30**, 297–341.

Thomashow, M. F., and Cotter, T. W. (1992). *Bdellovibrio* host dependence: The search for signal molecules and genes that regulate the intraperiplasmic growth cycle. *J. Bacteriol.* **174**, 5767–5771.

Ward, M. J., and Zusman, D. R. (1999). Motility in *Myxococcus xanthus* and its role in developmental aggregation. *Curr. Opin. Microbiol.* **2**, 624–629.

Wolk, C. P. (1996). Heterocyst formation. *Annu. Rev. Genet.* **30**, 59–78.

Wu, J., and Newton, A. (1997). Regulation of the Caulobacter flagellar gene hierarchy; Not just for motility. *Mol. Microbiol.* **24**, 233–239.

WEBSITES

Website on developmental checkpoints (S. Cutting)
http://www.sun.rhbnc.ac.uk/~uhba009/bacillus/research_development html#checkpoint

Website on myxobacteria (Dept. of Microbiology, University of Minnesota)
http://www.microbiology.med.umn.edu/faculty/myxobacteria/index.html

Website on myxobacteria (Microbial World)
http://helios.bto.ed.ac.uk/bto/microbes/myxococc.htm#crest

Website on *Caulobacter crescentus* (L. Shapiro)
http://devbiol.stanford.edu/Caulo/index.htm

CyanoSite: A Webserver for Cyanobacterial Research (Mark A. Schneegurt)
http://www-cyanosite.bio.purdue.edu/

26

Diversity, Microbial

Charles R. Lovell
University of South Carolina

GLOSSARY

division The largest phylogenetic grouping within a domain. This grouping can be considered analogous to the kingdom and includes one or more phyla.

domain The largest phylogenetically coherent grouping of organisms. The three domains of living organisms are the Bacteria, the Archaea, and the Eucarya.

monophyletic A group of organisms descending from a common ancestor having the significant traits that define the group.

species (bacterial or archaeal) The smallest functional unit of organismal diversity consisting of a lineage evolving separately from others and with its own specific ecological niche.

subdivision The largest phylogenetic grouping within a division. This grouping can be considered analogous to a phylum, and includes one or more orders.

The Microorganisms include all organism taxa containing a preponderance of species that are not readily visible to the naked eye. As a consequence of this anthropocentric definition, the microbiota have historically included numerous types of organisms differing from each other at the most fundamental levels of cellular organization and evolutionary history. We typically consider the major groups of microorganisms to be the bacteria, archaea, fungi, algae, protozoa, and viruses. This collection spans the breadth of organismal evolutionary history and ecological function, and represents the great majority of the major phylogenetic divisions of living organisms. Although these microbial groups differ from each other in many important ways, they have in common enormous diversity and great importance in the biosphere.

I. THE SIGNIFICANCE OF MICROBIAL DIVERSITY

We are only beginning to appreciate the true extent of microbial diversity and have only recently developed methods suitable for its exploration. Interest in this area has been spurred by the crucial roles microorganisms play in global ecology and in numerous human endeavors. Major ecological roles of microorganisms and applications of microbial products and processes in human enterprises are still being uncovered, but what we currently know provides an imposing picture of the importance of the microbiota in global and human ecology.

Microorganisms are prime movers in all ecosystems. They mobilize some elements by converting them from nonvolatile to volatile forms, permitting large-scale mass transport of these elements among different geographic locations. They capture carbon dioxide (and other one-carbon compounds) through several known fixation pathways, accounting for approximately 50% of global primary productivity. Microorganisms participate in and frequently dominate the interconversions of chemical compounds, producing and regulating the great variety of materials necessary to the biosphere. Effectively all decomposition processes, including spoilage of foods, are dominated by the microbiota. Microorganisms are found in

The Desk Encyclopedia of Microbiology
ISBN: 0-12-621361-5

all environments in which liquid water and an energy source can be found. These include such extreme environments as polar sea ice (in fluid-filled channels and pockets), hydrothermal vents, hot and cold deserts, brines approaching salt saturation, extremely acidic and alkaline waters, and deep (>1 km) subsurface solid rock formations. It can be accurately stated that the Earth is a microbial world.

Microorganisms also participate in a plethora of symbiotic (literally "living together") interactions with each other and with higher organisms. Some are common commensals using the products of their hosts without causing any discernible harm. Some form mutualistic interactions in which they and their hosts both actively benefit from the symbiosis. These mutualisms range from production of essential vitamins and facilitation of digestive processes by the microflora found in the digestive tracts of animals to the elegant and species- or strain-specific nitrogen-fixing interactions that occur between many plant species and bacteria. Finally, microorganisms are infamous for their parasitic interactions with higher organisms. These range from subtle, such as the influence of *Wolbachia* infections on sex determination in arthropods, to catastrophic. The impacts of pathogenic microorganisms on human populations have on occasion been devastating and the plagues that ravaged Europe and Asia throughout the Middle Ages are no more notable in this regard than the AIDS pandemic of the present. The impacts of parasites on the structuring of higher organism communities can obviously be very significant and, as Dutch elm disease demonstrates, even a single pathogen can drive a host species to local extinction.

The exploitation of microorganisms having useful properties is central to many human enterprises and has fostered substantial bioprospecting efforts. Numerous microbial products ranging from antibiotics to foods are important commodities, but perhaps the most significant impact of beneficial microorganisms lies in agriculture. Many microorganisms that grow in mutualistic interactions with plants have been used to increase crop yields and to control pests. Nitrogen-fixing mutualisms between plants and bacteria form the basis for crop rotation and can facilitate reclamation of overexploited or marginal lands. The mycorrhizal fungi, which grow in intimate association with plant roots, enhance the ability of plants to take up mineral nutrients and water. This type of symbiosis permits plant growth in many marginal environments and most plants participate in it. Numerous bacteria and non-mycorrhizal fungi colonize the surfaces of plants and can successfully outcompete potentially pathogenic organisms, preventing infections. In addition, the highly specific insecticidal toxins produced by *Bacillus thuringiensis* have been harnessed in insect pest control strategies and offer the major advantage of not harming beneficial insect species. Agriculture is, and always has been, highly dependent on microorganisms.

The vast array of microbial activities and their importance to the biosphere and to human economies provide strong rationales for examining their diversity. Loss of biodiversity is an ongoing global crisis and loss of the microbiota indigenous to human-impacted environments or associated with higher organisms facing extinction is a very real concern. As a consequence of the growing recognition of the importance of species loss, achieving a meaningful assessment of microbial diversity has been the topic of several major workshops and meetings. This is a period of very active inquiry into microbial diversity.

A. Approaches to the examination of microbial diversity

Past studies of microbial diversity were highly dependent on our ability to isolate these organisms into pure (i.e. single species) laboratory cultures for characterization. However, in most environments only 1–5% of the microscopically visible Bacteria and Archaea can be successfully cultivated, and obtaining pure cultures of many algae, protozoa, and fungi is problematic. In addition, detailed morphological, physiological, and behavioral characterization of any microorganism can be extremely time-consuming. As a consequence, many pure culture isolates have only been partially characterized. Recent applications of molecular biological tools, chiefly the polymerase chain reaction (PCR), DNA–DNA hybridization, DNA cloning and sequencing, and efficient analysis of DNA sequences, have provided a means to catalog microorganisms without the necessity for isolation and offer useful methods for streamlining the characterization of pure cultures.

Commonly used strategies for assessing microbial diversity employ PCR to amplify specific bacterial genes extracted from samples from the environment of interest. Nucleotide sequences of the genes encoding the small subunit ribosomal RNA (16S rRNA for Bacteria and Archaea and 18S rRNA for the Eucarya), when properly aligned, provide substantial information on the relatedness of unknown organisms to known species and support the construction of phylogenetic trees for microorganisms (Fig. 26.1). Other genes of interest for characterizing microbial diversity encode key enzymes involved in microbially mediated processes [i.e. functional group genes; *nifH*

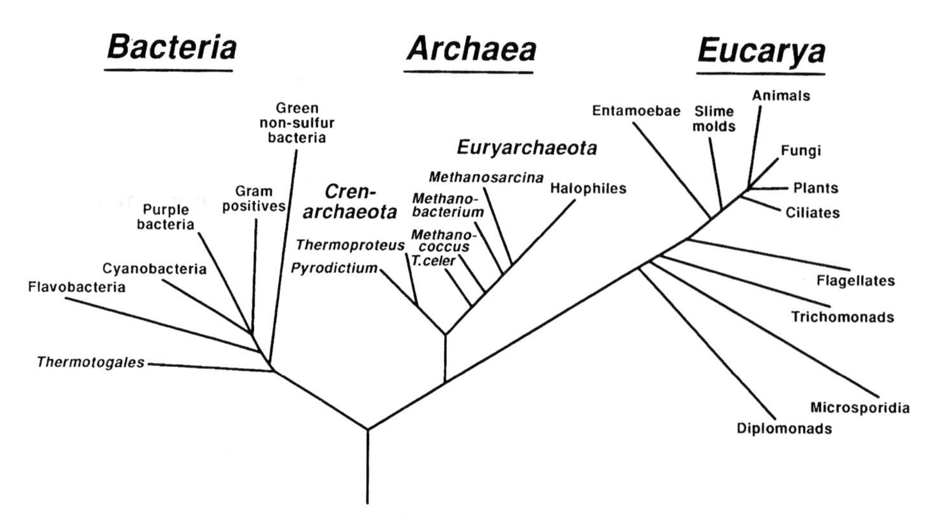

FIGURE 26.1 The universal phylogenetic tree showing the three domains of living organisms—the Archaea, the Bacteria, and the Eucarya (reproduced with permission from Woese, 1994).

(nitrogenase iron protein) for nitrogen fixation, *amoA* (ammonium monooxygenase) for ammonium oxidation, *rbcL* (ribulose bisphosphate carboxylase/oxygenase) for CO_2 fixation, etc.]. The amplified gene sequences can be resolved using various gel electrophoresis methods to provide a fingerprint of the microbial community present in the sample, or they can be cloned into an appropriate vector for propagation and nucleotide sequence analysis. The sequences of these genes are then available for phylogenetic analysis and species-specific domains within them can be used in DNA–DNA hybridization studies to quantify organisms of interest in environmental samples. Similar phylogenetic analyses of uncharacterized pure cultures allow the relatedness of these organisms to known species to be efficiently determined. Knowledge of the type of organism studied greatly facilitates selection of appropriate testing criteria for its complete identification. These technologies, although not perfect, have opened the door to the vast unknown world of microorganisms, clarifying the relationships of microorganism groups to each other and to higher organisms and facilitating the cataloging of species that have thus far resisted isolation and cultivation efforts. The picture that has emerged shows that of the three known domains of living organisms two, the Bacteria and the Archaea, consist entirely of microorganisms, and the third, the Eucarya, contains many major microbial taxonomic divisions. It is also clear from recent and ongoing studies that many new taxonomic lineages of microorganisms await discovery. The following sections summarize what is known, or at least widely accepted, about the diversities of the major groups of microorganisms and indicate some interesting areas in which information is mostly lacking. It should be noted that most current estimates of the diversity of all major microbial groups (Table 26.1) are quite conservative and may be significant underestimates of the true numbers.

TABLE 26.1 Estimated numbers of described and undescribed species in the major groups of microorganisms according to recent surveys[a]

Group	Estimated described species	Estimated undescribed species
Bacteria	5500	>1,000,000
Archaea	260	?
Fungi	72,000	>1,500,000
Algae	40,000	>200,000
Protozoa	40,000	>200,000
Viruses	6500	>1,000,000

[a]See Bibliography for sources.

II. BACTERIAL DIVERSITY

The Bacteria constitute an extremely diverse domain of prokaryotic microorganisms. They have a very simple cellular architecture, lacking the membrane-bound organelles characteristic of eukaryotic cells. They reproduce asexually, typically by simple fission, and are genetically haploid. Bacteria acquire soluble nutrients by transport across the plasma membrane but are unable to engulf colloids or particles through any endocytosis mechanisms. Virtually all known bacterial taxa have cell walls composed of an aminosugar polymer called peptidoglycan and this material is unique to this domain. Numerous distinctions can also be found between the Bacteria and other types of organisms in the details of their information processing machinery (i.e. DNA replication, RNA synthesis, and protein synthesis systems), chromosome structure, flagellar structure, and many physiological features.

Although unified by many features of cell structure and function and clearly forming a monophyletic domain, the Bacteria encompass enormous physiological diversity and can be found in almost every conceivable ecological niche. These organisms are found in extreme environments ranging from Antarctic sea ice to superheated marine hydrothermal vents, from extremely dilute aqueous systems to solutions approaching salt saturation, from highly acidic waters approaching a pH value of 0 to alkaline waters with pHs in excess of 12, and from the near vacuum of the upper atmosphere to the crushing pressures of the Challenger Deep. Bacteria can even be found in the highly radioactive environment of nuclear power plant reactor chambers. Although a few very hot environments appear to be dominated by the Archaea, the bacteria are ubiquitous inhabitants of all other environments and are important (and sometimes the only) participants in numerous ecological processes. Bacteria dominate mineral cycling and degradative processes in most environments, are important primary producers, and participate in all known types of symbiotic interactions, including parasitism on higher organisms, mutualistic interactions with other microorganisms and with higher organisms, and even active predation by some bacteria (such as *Bdellovibrio* species) on other bacterial species. The Bacteria dominate the biosphere like no other type of organism on Earth.

The bacterial domain encompasses an imposing diversity of physiological properties and ecological functions. The bacteria also clearly exemplify many of the major problems frequently encountered in assessing microbial diversity. Simply stated, nothing concerning the proper identification, classification, or phylogenetic analysis of bacteria is simple. Detailed observation of natural bacteria is complicated by their small size and their frequent growth on surfaces and in multispecies associations (biofilms, microcolonies, etc.). Even when they can be observed with a suitable degree of precision, bacteria display very few taxonomically useful morphological features. Most are simply spherical or rod shaped and a given species cannot be differentiated from organisms having similar shapes, but quite different activities, on the basis of appearance. As a consequence, bacterial identification is highly dependent on laboratory cultivation of species of interest in order to perform detailed physiological testing. However, the overwhelming majority of bacteria in nature cannot be readily cultivated on artificial laboratory growth media. Whether this is due to some incapacity of some observable cells or to our failure to formulate suitable growth media is not known, but the consequences for systematic appraisal of bacterial diversity are extremely important.

The lack of information obtainable through microscopic observations and the difficulties encountered in isolating even numerically dominant species into pure culture make bacterial diversity studies highly dependent on molecular biological approaches. Through such methods it is possible to determine phylogenetic relatedness of unknown bacterial species, including those that resist cultivation, to known species. However, complete characterization of new bacterial species still requires extensive physiological characterization of laboratory cultures. In addition, since bacteria reproduce asexually, the exact definition of a species is not completely clear. The general trend has been that intensively studied taxa are highly speciose, whereas less studied taxa tend to have few and sometimes genetically heterogeneous species. Finally, the species is not always the finest phylogenetically or ecologically significant subdivision of the Bacteria. Many (all?) species contain numerous strains that can be differentiated only through detailed molecular, physiological, and/or immunological testing. Strains are particularly important to us when they differ in their interactions with higher organism hosts. For example, different strains of a given *Rhizobium* species are nitrogen-fixing mutualists of different leguminous plant hosts. Each bacterial species, whether it associates with higher organisms or not, appears to be a radiation of closely related strains which may differ significantly in the ecological niches they exploit and which contribute to even greater bacterial diversity. Deciding the level of detail desirable in diversity studies and evaluating the feasibility of obtaining a complete catalog of bacterial diversity in all but the most species-poor environments are ongoing concerns.

Currently, there are 36 monophyletic divisions of bacteria (Fig. 26.2). These range from the *Proteobacteria*, a division which includes a great many well-characterized species, to the *Acidobacterium* division, which has only three cultivated species, and the 13 "candidate" divisions (Hugenholtz *et al.*, 1998) which currently lack characterized species altogether. Several additional division-level lineages are only

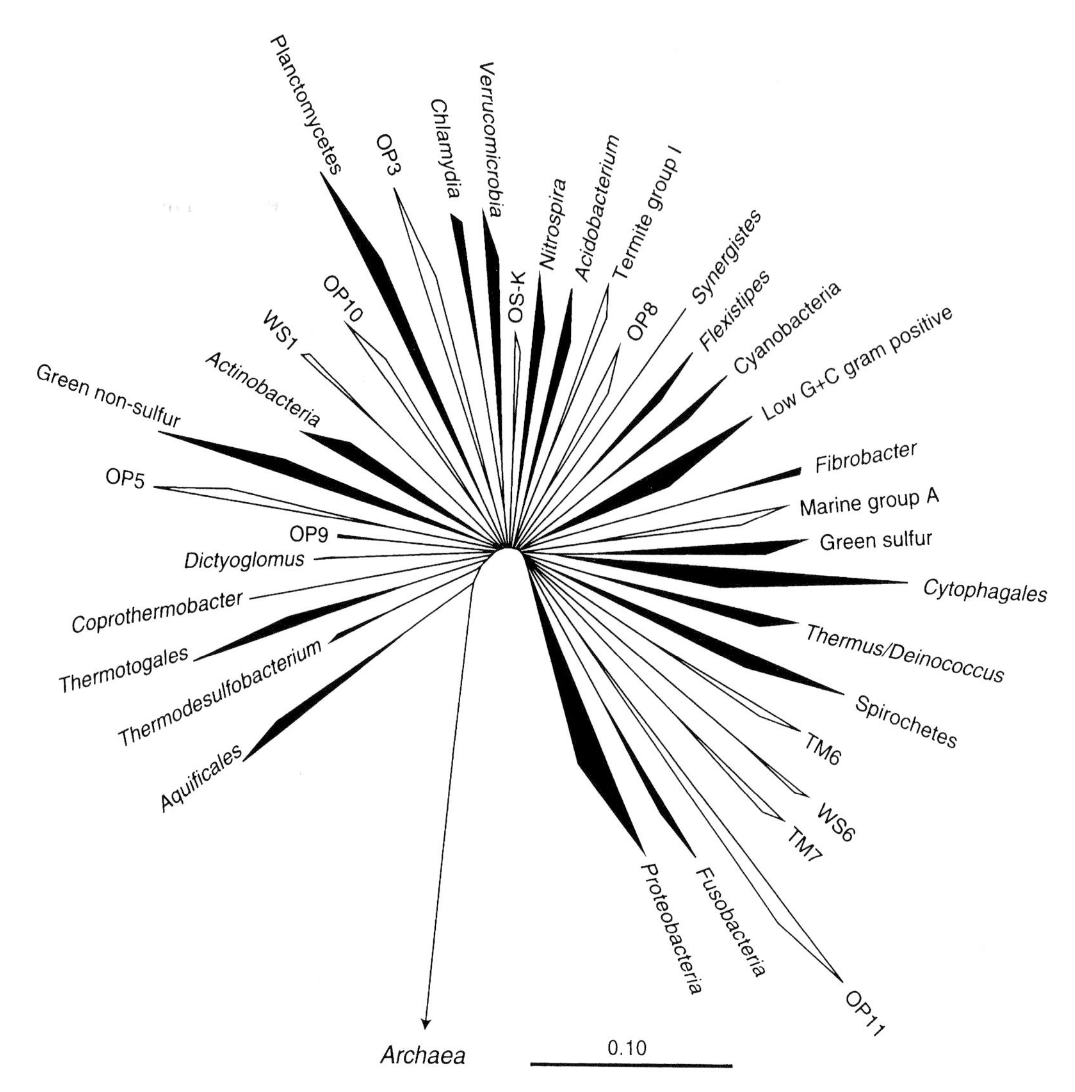

FIGURE 26.2 Phylogenetic tree showing the 36 currently recognized divisions of the Bacteria. Wedges represent division level groups of two or more sequences. Divisions containing characterized species are shown in black, and those containing only environmental sequences are shown in outline (reproduced with permission from Hugenholtz *et al.*, 1998). Scale bar represents 0.10 base changes per nucleotide in the 16S rRNA gene sequences examined.

represented by unique environmental 16S rRNA sequences. The total number of bacterial divisions is unknown but almost certainly in excess of 40. In addition to their phylogenetic coherence, some divisions are also unified in ecophysiological function (the Cyanobacteria, the Green Sulfur Bacteria, etc.), but it is important to note that this is the exception rather than the rule. For example, the *Proteobacteria* division includes numerous different physiological types ranging from aerobic heterotrophs to strictly anaerobic photoautotrophs and obligate intracellular parasites. The degree of physiological diversity within many divisions is unknown and is particularly problematic for divisions having no characterized species.

The combination of molecular phylogenetic analysis and classical physiological characterization provides a powerful approach for identifying new bacterial species and placing them into a phylogenetically coherent framework, but bacterial diversity presents an imposing challenge at all levels. At present (April, 2002), there are only about 5500 validly published bacterial species representing 993 genera. A total of only about 11,400 bacterial 16S rRNA gene sequences are currently available. It is clear that only a small fraction of bacterial diversity has been cataloged, even for most well-studied environments. The vast majority of bacteria have not been cultivated and unknown 16S rRNA gene sequences are routinely recovered. In addition, many ecosystems, particularly in the tropics, await detailed study. Even common and easily accessed environments, such as temperate terrestrial soils and freshwater sediments, have been the subjects of few systematic surveys. Currently, it seems unlikely that an accurate estimate of bacterial diversity can be presented or defended.

III. ARCHAEAL DIVERSITY

The Archaea, like the Bacteria, are prokaryotic in terms of their cellular architecture. They reproduce asexually and depend on dissolved substances for carbon and energy. There, however, the similarity between the two domains ends. The Archaea are quite different from the Bacteria in the cell wall polymers they produce, their unusual lipids, resistance to broad spectrum antibiotics that inhibit most bacteria, and numerous details of their information processing machinery. The discovery by Carl Woese that the Archaea constitute a distinct monophyletic domain divergent from both the Bacteria and the Eucarya (Fig. 26.1) not only made sense of the disparate characteristics of these organisms but also it revolutionized our understanding of the evolution and diversity of life on Earth.

Most of the Archaea that have been validly described and published are extremophiles falling into four major categories. These include the high salt-requiring extreme halophiles, the anaerobic methane-producing methanogens, various types of Archaea that grow optimally at extremely high temperatures (thermophiles or hyperthermophiles), and the cell wall-less archaeon *Thermoplasma*. The United States National Center for Biological Information (NCBI) GenBank database in 2002 listed a total of only 73 genera and 257 described species of Archaea. Several hundred sequences from unknown Archaea are also available, but we cannot make predictions about the characteristics of these organisms. Some general features of the known archaeal groups are given below.

The extreme halophiles inhabit highly saline environments and generally require oxygen and at least 1.5 M NaCl for growth. Virtually all these organisms can grow at saturating levels of salt. Although such highly saline environments are relatively rare, the extremely halophilic Archaea display a surprising diversity of cell morphologies, from rods and cocci to flattened disks, triangles, and rectangles. Given the unusual environment in which these organisms grow, the currently described nine genera may include most of the diversity of extreme halophiles.

The methanogens are restricted to anoxic habitats and all produce methane. Suitable habitats for methanogens are much more common than the high-salt environments required by the extreme halophiles and consequently the known diversity of methanogens is greater. There are seven major groups of methanogens containing at least 25 genera, including extremely halophilic methanogens, some methanogens that require highly alkaline growth conditions, and several thermophilic types. Major methanogenic habitats include soils ranging from pH neutral to acidic peat-rich soils, all types of freshwater sediments, both geothermally heated and unheated marine sediments, and the digestive tracts of higher animals, such as ruminants, and those of arthropods, including insects and crustaceans. Given this broad range of methanogenic environments, it is clear that the diversity of these organisms could greatly exceed the current rather sparse collection of species.

The hyperthermophilic Archaea include all non-methanogenic organisms that require temperatures in excess of 80°C for optimum growth. These Archaea are the most extremely thermophilic of all known organisms and several can grow at temperatures in excess of 100°C. Most hyperthermophiles are obligate anaerobes and most use reduced sulfur compounds (mineral sulfur or H_2S) in their metabolism. Ideal conditions for these organisms can be found in sulfur and sulfide-rich geothermally heated waters, such as those of hot springs and deep-sea hydrothermal vents. Although such habitats can have extremely low pH values, most hyperthermophiles have been recovered from neutral pH to mildly acidic locations. The widespread occurrence of geothermally heated waters, soils, and sediments, and the range of pH values and nutrient chemistries occurring within them, tends to support the idea of substantial diversity of hyperthermophilic Archaea. Certainly, the current catalog contains a broad range of organisms differing in physiology and growth requirements.

Thermoplasma, the cell wall-less thermophilic Archaeon, appears to be very restricted in its habitat range. All but one strain of this organism has been isolated from self-heating coal refuse piles. Although other, less transient habitats for this organism have been sought, only one strain of *Thermoplasma* has been recovered from geothermal environments. With such a limited distribution and narrow habitat range, this archaeal subdivision would be predicted to have very limited diversity.

Based on 16S rRNA gene sequence analysis, these diverse organisms can be readily grouped into two major divisions (kingdoms; Fig. 26.3). The *Euryarchaeota* include the halophilic and methanogenic Archaea

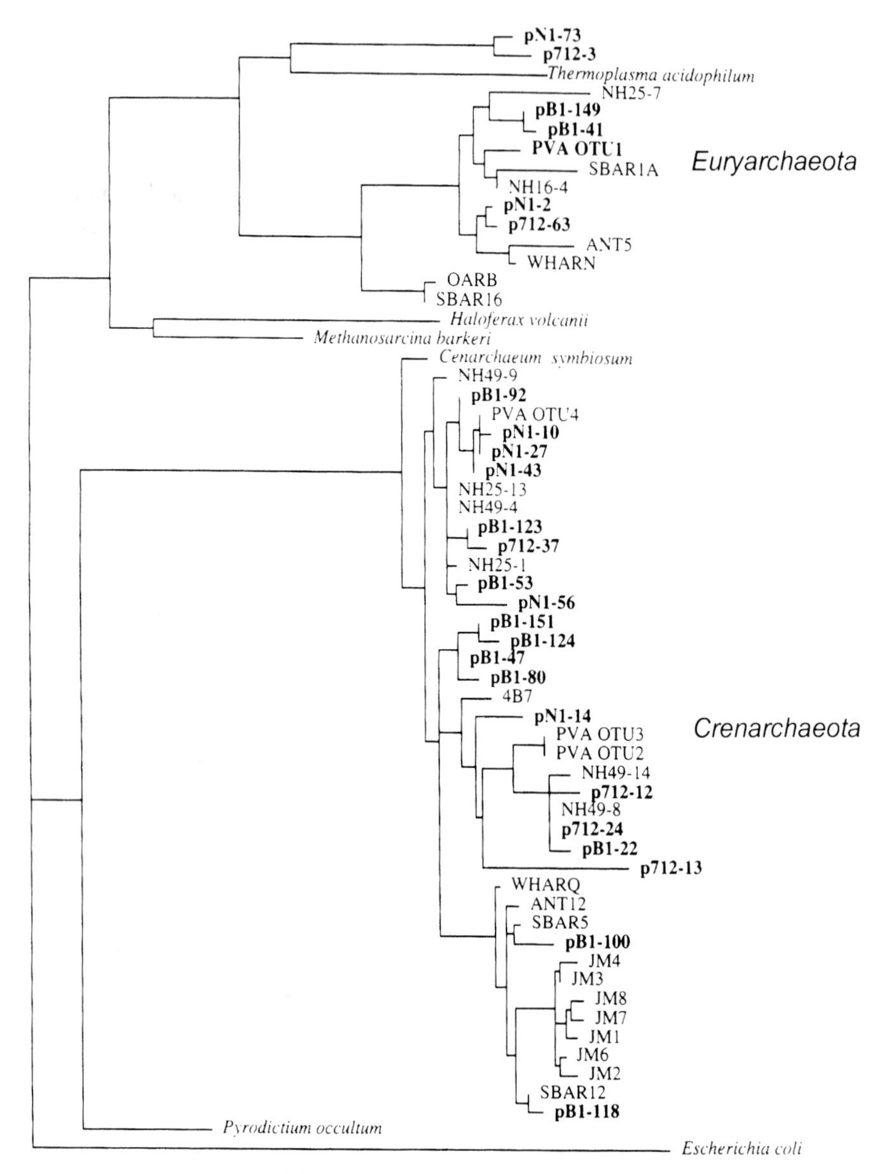

FIGURE 26.3 Phylogenetic tree for the Archaea showing the major groups of currently undescribed Archaea. Abbreviations represent cloned environmental 16S rRNA gene sequences (for published sources, see Fuhrman and Davis, 1997). The following clones were obtained from water samples: NH clones, pN1 clones, p712 clones, and 4B7 from the Pacific Ocean; WHAR clones from Woods Hole, MA; SBAR from Santa Barbara, CA; OAR from Oregon; and ANT clones from Antarctica. pGrfA4 is from freshwater lake sediment and PVA clones are from a volcanic seamount near Hawaii. JM clones are from an abyssal holothurian and *Crenarchaeum symbiosum* is from an uncharacterized marine sponge symbiont. Scale bar represents 0.10 base changes per nucleotide in the 16S rRNA gene sequences examined (adapted with permission from Fuhrman and Davis, 1997).

growing at moderate temperatures and *Thermoplasma*. The *Crenarchaeota* include all the hyperthermophilic organisms. It is easy to see why most researchers have long considered Archaea to be confined to extreme environments and thus likely to be restricted to very limited diversity. However, new findings from moderate- to low-temperature environments show that unknown Archaea are prominent members of the microbial communities found there. DNA extracted from samples of soils, marine and freshwater sediments, marine plankton, sponges, and the gut contents of holothurians (sea cucumbers) all contained 16S rRNA sequences specific to the Archaea. Archaea from these environments include a crenarchaeotal group that represents as much as 20% of the planktonic marine prokaryote assemblage (Fuhrman and

Davis, 1997) and appears to occur in both surface and deep marine waters worldwide. Sequences related to this marine group have also been recovered from freshwater sediments and terrestrial soils. Two newly recognized marine euryarchaeotal groups distantly related to *Thermoplasma* have also been discovered (Fuhrman and Davis, 1997). None of these organisms have been isolated into pure culture and little can be deduced about their physiologies and growth requirements. All that we currently know is that the Archaea are more diverse and far more abundant in nonextreme environments than was previously thought. It is worthwhile to consider that we have had the molecular biological tools required for discovery of these unknown Archaea for only a short time and that the Archaea have only been considered a separate domain for slightly longer than 20 years. We need to know a great deal more than we currently know in order to develop any meaningful estimates of archaeal diversity or any clear realization of their significance in most environments.

IV. FUNGAL DIVERSITY

The fungi are an extremely diverse division of the Eucarya that includes both unicellular species and species that grow as finely divided networks of filaments called hyphae. Unicellular species are all microscopic, but hyphal networks can support fruiting bodies that are readily visible without magnification. All are included among the microorganisms on the basis of their lack of differentiated tissues. Primary features that distinguish the fungi from other Eucarya are: (i) use of chitin as a cell wall polymer (but not in all fungi); and (ii) the absence of photopigments. Fungal nutritional requirements and core metabolic processes are neither unusual for eukaryotes nor particularly diverse. However, their morphologies and life cycles display considerable diversity, providing the basis for classical fungal taxonomy. This scheme has a serious limitation in that mature reproductive structures are required for correct identification of some organisms and these structures can be difficult to find or are produced haphazardly by some fungi. The operational solution to this problem has been inclusion of all fungi for which sexual reproductive structures have not been seen in a single polyphyletic subdivision. The seven classically defined fungal taxonomic subdivisions are given here.

The Ascomycota (Ascomycetes or sac fungi) include about 35,000 described species. Many species of the red, brown, and blue-green molds that cause food spoilage, as well as the powdery mildews that cause Dutch elm disease and chestnut blight, are Ascomycetes. The Basidiomycota (Basidiomycetes or club fungi) include about 30,000 known species. Among these are the smuts, rusts, shelf fungi, stinkhorns, puffballs, toadstools, mushrooms, and bird's nest fungi. The human pathogen *Cryptococcus neoformans*, which causes cryptococcosis, is also a Basidiomycete, as are the important plant pathogens, the smuts and rusts. The Zygomycota (Zygomycetes) include about 600 known species. Among these are the common bread molds and a few species parasitic on plants and animals. The Deuteromycota (Deuteromycetes or fungi imperfecti) include all species for which a sexual reproductive phase has not been observed—about 30,000 known species. Several human pathogens, including the organisms causing ringworm, athlete's foot, and histoplasmosis, are included in this group. The aflatoxin-producing species *Aspergillis flavus* and *A. parasiticus*, important organisms in fungal food poisonings, are also in this group. Other Deuteromycetes are responsible for production of antibiotics or are used in production of foods, such as cheeses and soy sauce. When the reproductive (perfect) stage of a Deuteromycete is characterized, it is transferred from the fungi imperfecti to the appropriate group. These four subdivisions of fungi form a monophyletic cluster of true fungi. Molecular phylogenetic analysis has revealed that the remaining three classically defined fungal groups are more closely related to other types of organisms than to the four true fungal subdivisions. The *Oomycota* (Oomycetes or water molds) resemble fungi in that they grow in a finely branched network of hyphal filaments. However, the Oomycetes have cell walls composed of cellulose, produce motile asexual zoospores, and are closely related to the algal Phaeophyta and Chrysophyta. The Oomycetes are saprotrophs and several are important plant pathogens, the most famous of which is *Phytophthora infestans*, the organism responsible for the Irish potato famine. The *Myxomycota* (plasmodial slime molds; about 700 known species) and *Acrasiomycota* (cellular slime molds; about 50 known species) lack cell walls of any kind during vegetative growth, display amoeboid motility, and are related to the protozoa. Both types are saprotrophic and feed by phagocytosis. When food resources become limiting, both types of slime molds will form fruiting structures. None of the slime molds are known to be important plant or animal parasites or to produce commercially valuable products. However, slime molds are significant participants in organic matter turnover in soils.

The major ecological functions of fungi are well understood. Soil fungi are ubiquitous saprotrophs and responsible for much of the decay of soil organic matter. These organisms are highly adapted to the saprotroph function, producing a variety of important extracellular enzymes for degradation of insoluble substrates and infiltrating the soil matrix and decaying organic matter through hyphal growth. Although primarily terrestrial, fungi are also found in aquatic ecosystems, including highly saline intertidal soils, and are particularly significant in the decay of plant materials in these environments. Fungi are also important members of the digestive tract flora of many animals and actively participate in the breakdown of plant materials, particularly in the ruminant animals and many insects.

Fungal growth at the expense of plant biomass is not confined to dead biomass. Some fungi are also very significant plant parasites responsible for economically important damage to crop plants. The complex relationship of the mycorrhizal fungi with their plant hosts, which can be characterized as mutualistic, parasitic, or fluctuating between these roles, should not be overlooked in this regard. The mycorrhizae promote plant growth in many cases and are very important agents in the structuring of plant communities, but they can also damage their hosts under stressful environmental conditions. Fungal parasitism in some cases results in spoilage of food products, particularly the seed heads of maturing grain crops. An interesting fungus important in the spoilage of grains is the Ascomycete *Claviceps purpurea*, which produces the alkaloid mycotoxin ergot. Ergot is a potent hallucinogen and consumption of ergot-contaminated rye is thought to have been the cause of medieval dancing fits, in which the populations of whole villages danced wildly until exhausted. Accusations of witchcraft and the resulting Salem witch trials and executions may also have been due to ergotism.

Although they are less frequently cited as inhabitants of extreme environments than the extremist species among the Bacteria or Archaea, the fungi can grow across a broad range of environmental conditions and include species that could certainly be considered extremophiles. In general, fungi can tolerate greater extremes of salinity, pH, and desiccation than non-extremophilic species of bacteria. The diverse array of fungi participating in the lichen symbiosis can be found from temperate to very cold and dry environments worldwide. The microbiota of the rock varnishes common in hot deserts and the endolithic microflora growing within rocks in the Antarctic dry valleys both include fungi. Fungi are found in anaerobic environments including subsurface marine sediments, water-saturated decaying organic matter, the rumens of ruminant mammals, and the hindguts of numerous insect species. Fungi inhabit soils having temperatures from below freezing up to about 60°C. Any effort to catalog total fungal diversity should certainly include such extreme environments, which are clearly more common than is usually realized.

There are between 72,000 and 100,000 known species of fungi, but total species numbers have been estimated at approximately 1.5 million, approximately six times the estimated number of vascular plant species (Hawksworth and Rossman, 1997). This estimate should certainly be considered conservative since it does not include the mostly undescribed fungi growing commensalistically, mutualistically, or parasitically in or on animals. The estimated millions of undescribed insect species in the tropics are particularly noteworthy in this regard. Since mutualistic and parasitic fungi are typically quite host specific, many of these insect species may harbor unknown fungi. Given the important ecological functions of fungi as saproptrophs in organic matter decomposition and symbionts of terrestrial plants, it is sensible to consider plants first in estimates of fungal diversity. However, the many interactions of fungi with animals are easily overlooked and a great many undescribed fungal species are likely to be involved in them. The tropics, with their abundance of plant and insect species that are found nowhere else, likely represent the largest global reservoir of undescribed fungi, and characterization of the diversity of tropical fungi has only started.

The introduction of molecular biological methods for phylogenetic characterization of fungi should greatly facilitate identification of unknown species, particularly those that do not reliably produce fruiting bodies. Classical description of new species on the basis of morphological and life cycle characteristics continues at a respectable rate, but most fungal species, including many already available in culture, herbarium, and private collections, have yet to be validly described. It is certain that discovery of new fungal species will be limited primarily by the resources available for this effort for many years to come.

V. ALGAL DIVERSITY

The algae are a diverse assemblage of photosynthetic Eucarya. Most species are unicellular and thus counted among microorganisms. Some are much larger and these macroalgae will not be discussed here. The algae all contain chlorophyll *a* as well as a variety of additional pigments, giving them colorations ranging from green to yellow, red, and brown. All algae carry out oxygenic photosynthesis and most are essentially aquatic in character, including those found at the surface of saturated soils. The algae inhabit all moderate (i.e. non-extreme) marine and freshwater environments that receive sufficient light to support photosynthesis. They are also found in a variety of unusual or extreme environments, including Antarctic sea ice, mountain snow fields, hot springs (up to maximum temperatures of about 60° C), and acidic waters (pH 4 or 5). A few species are able to tolerate very dry environments, such as dry soils, and the endolithic algae grow within rocks in the extremely cold, dry environment of Antarctic dry valleys. The halophilic *Dunaliella* can be found in many salt lakes and is often the only oxygenic phototroph present. The algae are ubiquitous, although sometimes unnoticed, members of virtually all microbial communities wherever light is available.

Algal diversity is impressive. There are at least 40,000 known algal species, with an extrapolated total of more than 200,000 species. These organisms are not monophyletic and their classical taxonomy is largely based on morphology and a few key phenotypic traits. The most important of these is their performance of oxygenic photosynthesis, their panoply of photopigments, their cell wall structure and chemistry, and the carbon reserve materials they synthesize when light is not a limiting resource. Consideration of these traits lead to the division of the algae into six major groups; the Chlorophytes, the Euglenophytes, the Crysophytes, the Rhodophytes, the Phaeophytes, and the Pyrrophytes (Table 26.2). This phenotypic classification system, although internally consistent, does not reflect the evolutionary history or the true phylogeny of the algae. The cyanobacteria, once known as the Cyanophytes, are clearly bacteria and should not be included among the algae. The Chlorophytes and Rhodophytes are closely related to the higher plants, but the Crysophytes and Phaeophytes are more closely related to the oomycetes, and the Pyrrophytes are more closely related to the ciliated protozoa.

The major ecological functions of the microalgae are reasonably well understood. Algae are important primary producers in most aquatic environments, including sediments exposed by tidal action, where they can contribute to the formation of elaborate and highly active microbial mats. The algae contribute an estimated 30–50% of global primary production and are perhaps most significant in offshore marine waters in which fixed carbon is in short supply. In coastal marine systems, a quite different role of algae is also observed. Many marine microalgae produce potent toxins. Toxin-producing dinoflagellates commonly belonging to the genera *Gymnodinium* and *Gonyaulax* can form massive blooms (red tides) resulting in severe losses to fisheries. Recently, the toxin-producing "ambush predator" dinoflagellate, *Pfiesteria piscicida*, was identified as an important pathogen responsible for fish kills along the Atlantic and Gulf coasts of North America. This organism has a very complex life cycle, including 24 distinct stages, and its toxin has potent activity against a variety of animals, including humans. Toxic algal blooms are increasing in frequency and have been linked to sewage and agricultural runoff into coastal marine waters and to transport of dinoflagellate cysts in ship ballast water. Such environmental impacts may acquaint us with other, once obscure taxa and reveal additional capacities in better known organisms.

The majority of algal diversity remains to be detected and described, and it is very likely that numerous new organisms will be recognized in the near future. This may be particularly true of bloom-producing species. However, the major algal phylogenetic lineages have

TABLE 26.2 Characteristics of the major groups of algae

Group	Common name	Morphology	Photopigments	Carbon reserves	Cell wall
Chlorophytes	Green algae	Unicellular or macroscopic	Chlorophyll *a* and *b*	Starch (α-1,4-glucan), sucrose	Cellulose
Euglenophytes	Euglenoids	Unicellular	Chlorophyll *a* and *b*	Paramylon (β-1,2-glucan)	None
Chysophytes	Golden-brown algae	Unicellular	Chlorophyll *a*, *c*, and *e*	Lipids	Silicate
Rhodophytes	Red algae	Unicellular or macroscopic	Chlorophyll *a* and *d*, phycocyanin, phycoeurythrin	Floridean starch (α-1,4-glucan and α-1,6-glucan), fluoridoside (glycerol-galactoside)	Cellulose
Phaeophytes	Brown algae	Macroscopic	Chlorophyll *a* and *c*, xanthophylls	Laminarin (β-1,3-glucan), mannitol	Cellulose
Pyrrophytes	Dinoflagellates	Unicellular	Chlorophyll *a* and *c*	Starch (α-1,4-glucan)	Cellulose

probably been identified and important improvements in methods for analyzing photopigment profiles and algal morphologies should facilitate recognition of new microalgae in field samples. Much work remains, particularly in terms of detailed phylogenetic analyses, but it is fair to say that we understand the range of important algal niches in the environment and that the major ecotypes of algae have been identified.

VI. PROTOZOAL DIVERSITY

Like the algae, the protozoa are a diverse collection of eukaryotic microorganisms defined on the basis of a small set of phenotypic traits. The protozoa lack cell walls and pigmentation, and most are motile. The absence of photopigments separates these organisms taxonomically from the algae (Euglenophytes and Pyrrophytes are currently placed in the algae) and motility distinguishes them from the true fungi. The slime molds are motile, phylogenetically related to the protozoa, and probably should be counted among them. The protozoa are very widely distributed, inhabiting all aquatic environments in which temperatures are above freezing but below about 60° C as well as soils and the digestive tracts of many animals. The major ecological function of the protozoa is as primary consumers of other microorganisms, and they are important predators on bacteria, small microalgae, and each other. Protozoa can also take up high-molecular-weight organic solutes and colloids by means of pinocytosis and low-molecular-weight organic molecules by simple diffusion. Some protozoa grow primarily as saptrotrophs, whereas others are important parasites of higher organisms. About 40,000 known species of protozoa have been documented, but this may be less than 25% of the total.

The major taxonomic groups of protozoa are the Mastigophora, the Sarcodina, the Ciliophora, and the Sporozoa. The Mastigophora are motile by means of flagella and are commonly known as flagellates. They are closely allied with the Euglenophytes, which are capable of purely heterotrophic growth and can lose their chloroplasts if maintained in the dark. Important flagellate parasites of higher organisms include the trypanosomes, such as *Trypanosoma gambiense*, the organism responsible for African sleeping sickness. The Sarcodina include the amoebas, which lack shells, and the foraminifera, which produce calcium carbonate shells during active growth. The Ciliophora are motile by means of cilia and feed primarily on particulate materials and microbial cells. Few ciliates are parasitic on higher organisms. In contrast, the Sporozoa are all obligate parasites. This large group is characterized by the lack of motile adult stages and by absorption of organic solutes as their primary means of obtaining nutrition. The Sporozoa include the plasmodia, which are the pathogens responsible for malaria.

As is the case for the algae, the protozoa are not monophyletic. Phylogenetic analysis has revealed that the Ciliophora and Pyrrophytes are more closely related to each other than to the other protozoal or algal groups. Mastigophora and Euglenophytes are also more closely related to each other than to other groups. The Sporozoa are very diverse and include some lineages that are highly divergent from all other known eukaryotic lines of descent. Data from more protozoa, particularly among the Sporozoa, will be required to fully evaluate the several known phylogenetic lineages and to properly place newly discovered lineages. Such data will also be very helpful in eliminating synonymous species from the published literature.

Considerations of the ecological functions of the protozoa, as well as the absence of solid evidence for specific geographic distributions of protozoal species, imply fairly restricted species diversity for these organisms (Finlay *et al.*, 1998). The best studied large protozoal group in terms of diversity is the Ciliophora native to marine and freshwater sediments. Only about 3000 species are known, but based on the appearance of certain species in all suitable habitats these may represent the majority of extant species. In other environments even ciliate diversity is poorly characterized. Foissner (1997) estimated that 70–80% of soil ciliates are unknown and that global diversity of these organisms is in the range of 1300–2000 species. Ciliates in the digestive tracts of animals are also very poorly characterized and may represent a substantial pool of undocumented protozoal diversity. Certainly the gut flora of an animal species would be considered to reflect the habitat, population structure, food source(s), and digestive tract architecture of that animal and consequently be somewhat characteristic of that species. The diversity of all other groups of protozoa is much more poorly characterized than that of the ciliates, and estimates of 200,000 protozoal species may not be excessive.

Although the taxonomic characterization of protozoa remains difficult and time-consuming, molecular biological methods for phylogenetic analysis should greatly facilitate identification and description of new species. Given the ubiquity of some protozoal species across suitable habitats worldwide (Finlay *et al.*, 1998), characterization of protozoal diversity in certain habitats may not be as difficult as currently thought. The largest underexplored reservoir of protozoal diversity appears to be animals, particularly arthropods, in which important protozoal digestive tract flora can be abundant and are poorly characterized. This presents a particular problem in the case of endangered animal species, whose flora may also be lost in the event of extinction.

VII. VIRAL DIVERSITY

No matter whether it is more correct to consider viruses as "living organisms" or as renegade genome fragments, the viruses have very important impacts on other types of organisms and should not be neglected in discussions of microbial diversity. These obligately parasitic entities have no independent physiological activities and require a suitable host for reproduction, which damages or destroys host cells. The importance and success of viruses as parasites in higher organisms is well known. Numerous illnesses in animals and plants have viral origins and the rapidity with which viruses spread through a susceptible host population can be alarming. The Influenza pandemic of 1918 was particularly noteworthy for its rapid propagation and lethality. Identification of such pathogenic viral strains and development of vaccines against them are foci of major national and international efforts.

In addition to the damage done by out-of-control viral reproduction, many viruses can affect their hosts more subtly through genetic modification. Temperate viruses can insert their genome into that of an appropriate host and be propagated as a stable provirus within the host for many generations. This interaction, called lysogeny in bacteria–virus systems, is a form of recombination and can confer new properties to the host. Important examples of this type of recombination-driven change in host phenotype are provided by the pathogenicity islands of *Vibrio cholerae*, enterohemorrhagic *Escherichia coli* strains, and some other enteric pathogens. These bacterial species and strains display pathogenic traits, such as adhesion to specific host receptors and toxin production, only if the correct provirus (prophage) is present. The genetic modifications brought about by lysogeny can not only dictate whether or not a host bacterium is pathogenic but also control the severity of infection by the recombinant pathogen. Many organisms, particularly microorganisms, harbor proviruses or provirus-like sequences in their genomes and the importance of most of these to the host organisms is unknown. Although we have been chiefly interested in viruses that infect humans, our domesticated animals, and crop plants, all types of organisms have viral parasites. The impacts of these less studied viruses on their host populations in nature are only beginning to be unraveled. Recent findings indicate that viruses are active participants in microbial food web processes and may exert some control over microbial population dynamics in nature.

Perhaps the diversity of no other microbial group is as poorly characterized and as widely underestimated as that of the viruses. This is partly due to difficulties in identifying and cultivating the host organisms necessary for propagating the viruses found in the environment. Viral taxonomy, which was long dependent on phenotypic characteristics, such as host range, symptoms of infection, morphology of the viral particle, and the type of nucleic acid (single-stranded DNA or RNA or double-stranded DNA or RNA) composing the genome, now also employs nucleotide sequence analysis of viral genomes. However, most types of viruses remain to be identified and no hosts for these organisms are known. There are about 6500 described species of viruses, including 2500 animal viruses, 2000 plant viruses, and 2000 bacterial viruses. However, the total number of virus species has been estimated to be 500,000, and this may be a significant underestimate of true viral diversity.

Recent findings, particularly in aquatic systems, show that viruses are extremely numerous in nature (Paul *et al.*, 1996). Viral numbers are typically on the order of 10^9–10^{10} virions per litre of water and estimates of viral production in nature are very high. From 1 to 4% of bacteria in freshwater and marine systems contain mature viruses and, since complete viruses are only visible in cells during the final 10% of the viral replication cycle, actual frequencies of infected cells could be much higher. In addition, as many as 4% of bacteria in coastal marine waters may contain stable proviruses and much higher proportions of culturable aquatic species are known to be lysogenic. Viruses are present in all the microbial environments that have been examined to date and with numbers as high as those reported for aquatic systems, and because of the diversity of morphologies observed it is clear that viruses represent a vastly underestimated pool of microbial diversity. This is particularly apparent when we consider the fact that more than 70 different viruses, representing six major viral groups, have been isolated using a single marine bacterium as the host strain (Paul *et al.*, 1996). If this pattern holds true across all types of organisms, and there is no compelling reason to think that it may not, viral diversity is much greater than currently appreciated. In addition, many viruses are quite mutable, adding genetic variation among related strains to the already large task of cataloging viral diversity. It is likely that each viral isolate represents only one of a radiation of closely related but distinguishable genotypes. This additional diversity within strains is not inconsequential, as is clear from the differences in pathogenicity among different variants of HIV and influenza A.

Although the occasional human pathogen, such as the Ebola virus, introduces itself through grim displays of lethality, most viruses will only be discovered and described through painstaking, systematic efforts. It is clear that this will require a major commitment of time and resources, but the foundations for describing natural viral diversity have been laid and this effort can be now undertaken with some expectation of long-term success. A key consideration is the types of microbial, plant, and animal hosts that should be emphasized in these studies although, as Ebola illustrates, the host range for a given virus may prove very elusive.

VIII. TOWARDS A FUNCTIONAL SURVEY OF MICROBIAL DIVERSITY

Currently, most of the earth's microbial diversity is completely unknown and current estimates of the numbers of species extant are widely considered to be significant underestimates. The microbiota likely dominate global biodiversity, and efforts to characterize microbial diversity and to determine the interplay between this diversity and ecosystem function have recently accelerated. Given the ever-expanding rate of global change brought about by human activities, two extremely important organizing foci for diversity assessment can be identified.

First, habitats that are currently threatened by human activities are clearly of immediate interest to diversity survey efforts. Most tropical regions of the world are either severely human impacted or threatened. This is of particular interest since these regions are home to an enormous diversity of plant and animal species. Each of these higher organism species is host to numerous commensal, mutualist, and parasitic microorganisms, and when a host organism faces extinction so too may some members of its microflora. Bioprospecting in the tropics for microorganisms that produce useful products has begun, but many novel species may be lost before they can be documented and preserved. Consideration of habitat loss and the potential for host organism extinction provide spatial and temporal frames of reference for microbial diversity survey efforts.

Second, some microbial activities, such as nitrogen fixation, primary production, and methane oxidation, are clearly essential to proper ecosystem function and are restricted to specific functional groups of microorganisms. The diversity of a given functional group may be limited, as seems to be the case for methane-oxidizing bacteria and for nitrogen-fixing bacteria in some habitats. Low diversity may also be an important consideration for microorganisms that grow in symbiotic relationships with higher organisms in which the degree of specificity of the microorganisms for their hosts can be great. If the active species within an essential microbial functional group are lost due to habitat modification or local extinction of a higher organism host species, key local ecosystem functions could deteriorate. Evaluation of environmentally sensitive, low-diversity functional groups provides a rationale for exploring connections between microbial diversity and ecosystem function and an ecological focus for diversity survey efforts.

Clearly, much work will have to be done to support a meaningful assessment of microbial diversity. However, the technology necessary to pursue this effort exists, and some important work has already been done. Compiling this information and systematic elimination of synonymous listings will be greatly facilitated by use of the World Wide Web, and several relevant websites have already been constructed (Table 26.3). The future will certainly produce a much greater understanding of the extent and importance of microbial diversity and most likely even more questions.

TABLE 26.3 Websites providing taxonomic or phylogenetic information on microorganisms[a]

Web sites for bacterial nomenclature
http://www-sv.cict.fr/bacterio/ (J. P. Euzeby)
http://www.dsmz.de/bactnom/bactname.htm (Website of the DSMZ, M. Kracht, database administrator)
Index of fungi (not free of charge)
http://www.cabi.org/catalog/taxonomy/indfungi.htm (CABI)
Index of viruses
http://life.anu.edu.au/viruses/Ictv/index.html
Other sites with links to culture collections, databases, and phylogenetic analysis
http://www.cme.msu.edu/RDP (Ribosome Database Project, B. L. Maidak, curator)
http://wdcm.nig.ac.jp/ (WDCM, World Data Centre for Microorganisms)
http://ftp.ccug.gu.se/ (CCUG)

[a]Information provided by Dr. Manfred Kracht, Deutsche Sammlung von Mikroorganismen und Zellkulturen GmbH, Braunschweig, Germany.

BIBLIOGRAPHY

Amann, R. I., Ludwig, W., and Schleifer, K.-H. (1995). Phylogenetic identification and *in situ* detection of individual microbial cells without cultivation. *Microbiol. Rev.* **59**, 143–169.

Caldwell, D. E., Colwell, R. R., Harayama, S., Kjelleberg, S., Givskov, M., Molin, S., Nielsen, A. T., Heydorn, A., Tolker-Nielsen, T., Sternberg, C., Nealson, K., Pace, N. R., Rainey, P., Stackebrandt, E., Tindall, B. J., and Stahl. D. A. (2000). Crystal ball: leading scientists in the field of environmental microbiology consider the technical and conceptual developments that they believe will drive innovative research during the first years of the new millennium. *Environ Microbiol.* **2**, 3–10.

Finlay, B. J., Esteban, G. F., and Fenchel, T. (1998). Protozoan diversity; Converging estimates of the global number of free-living ciliate species. *Protist* **149**, 29–37.

Foissner, W. (1997). Global soil ciliate (Protozoa, ciliophora) diversity. A probability-based approach using large sample collections from Africa, Australia, and Antarctica. *Biodiversity Conservation* **6**, 1627–1638.

Fuhrman, J. A., and Davis, A. A. (1997). Widespread Archaea and novel Bacteria from the deep sea as shown by 16S rRNA gene sequences. *Marine Ecol. Prog. Ser.* **150**, 275–285.

Hammond, P. M. (1995). Described and estimated species numbers: An objective assessment of current knowledge. *In* "Microbial Diversity and Ecosystem Function" (D. Allsopp, R. R. Colwell, and D. L. Hawksworth, eds.), pp. 29–71. Cambridge University Press, Cambridge, UK.

Hawksworth, D. B., and Rossman, A. Y. (1997). Where are all the undescribed fungi? *Phytopathology* **87**, 888–891.

Hugenholz, P. B., Goebel, B. M., and Pace, N. R. (1998). Impact of culture-independent studies on the emerging phylogenetic view of bacterial diversity. *J. Bacteriol.* **180**, 4765–4774.

Pace, N. R. (1997). A molecular view of microbial diversity and the biosphere. *Science* **276**, 734–740.

Paul, J. H., Kellogg, C. A., and Jiang, S. C. (1996). Viruses and DNA in marine environments. *In* "Microbial Diversity in Time and Space" (R. R. Colwell, U. Simidu, and K. Ohwada, eds.), pp. 115–124. Plenum, New York.

Williams, D. M., and Embley, T. M. (1996). Microbial diversity: Domains and kingdoms. *Annu. Rev. Ecol. Systematics* **27**, 569–595.

Woese, C. R. (1994). There must be a prokaryote somewhere: Microbiology's search for itself. *Microbiol. Rev.* **58**, 1–9.

Woese, C. R. (2000). Interpreting the universal phylogenetic tree. *Proc. Natl. Acad. Sci. USA.* **97**, 8392–8396.

WEBSITES

A website on Microbial Diversity Research Priorities by the American Society of Microbiology
http://www.asmusa.org/pasrc/microbia.htm

Diversity of Life Index (microbial)
http://www.geocities.com/RainForest/6243/diversity2.html

The Woods Hole Institute of Oceanography Microbial Diversity Course
http://courses.mbl.edu/MICDIV/

A Molecular View of Microbial Diversity and the Biosphere (N. R. Pace)
http://cas.bellarmine.edu/tietjen/RootWeb/a_molecular_view_of_microbial_di.htm

Comprehensive Microbial Resource of The Institute for Genomic Research (TIGR) and links to many other microbial genomic sites
http://www.tigr.org/tigr-scripts/CMR2/CMRHomePage.spl

27

DNA repair

Lawrence Grossman
The Johns Hopkins University

GLOSSARY

endonuclease Nuclease that hydrolyzes internal phosphodiester bonds.

excision Removal of damaged nucleotides from incised nucleic acids.

exonuclease Nuclease that hydrolyzes terminal phosphodiester bonds.

glycosylase Enzymes that hydrolyze N-glycosyl bonds linking purines and pyrimidines to carbohydrate components of nucleic acids.

incision Endonucleolytic break in damaged nucleic acids.

ligation Phosphodiester bond formation as the final stage in repair.

nuclease Enzyme that hydrolyzes in the internucleotide phosphodiester bonds in nucleic acids.

resynthesis Polymerization of nucleotides into excised regions of damaged nucleic acids.

transcription The synthesis of messenger and RNA from template DNA.

The Ability of Cells to survive hostile environments is due, in part, to surveillance systems which recognize damaged sites in DNA and are capable of either reversing the damage or of removing damaged bases or nucleotides, generating sites which lead to a cascade of events restoring DNA to its original structural and biological integrity.

Both endogenous and exogenous environmental agents can damage DNA. Many repair systems are regulated by the stressful effects of such damage, affecting the levels of responsible enzymes, or by modifying their specificity. Repair enzymes appear to be the most highly conserved proteins showing their important role throughout evolution. The enzyme systems can directly reverse the damage to form the normal purine or pyrimidine bases or the modified bases can be removed together with surrounding bases through a succession of events involving nucleases, DNA-polymerizing enzymes, and polynucleotide ligases which assist in restoring the biological and genetic integrity to DNA.

I. DAMAGE

As a target for damage, DNA possesses a multitude of sites which differ in their receptiveness to modification. On a stereochemical level, nucleotides in the major groove are more receptive to modification than those in the minor groove, the termini of DNA chains expose reactive groups, and some atoms of a purine or pyrimidine are more susceptible than others. As a consequence, the structure of DNA represents a heterogeneous target in which certain nucleotide sequences also contribute to the susceptibility of DNA to genotoxic agents.

A. Endogenous damage

Even at physiological pH values and temperatures in the absence of extraneous agents, the primary structure of DNA undergoes alterations (Table 27.1). Many specific reactions directly influence the informational content as well as the integrity of DNA. Although the rate constants for many reactions are inherently low, it is because of the enormous size of DNA and its

The Desk Encyclopedia of Microbiology
ISBN: 0-12-621361-5

TABLE 27.1 Hours needed for a single event at pH 7.4 at 37 °C

Event	Single-stranded DNA (2×10^6 base pairs)	Double-stranded DNA (2×10^6 base pairs)	Molecular events per genome per day
Depurination	2.5	10	24×10^3
Depyrimidination	50.8	200	12×10^2
Deamination of C	2.8	700	3.4×10^2
Deamination of A	140.0	?	?

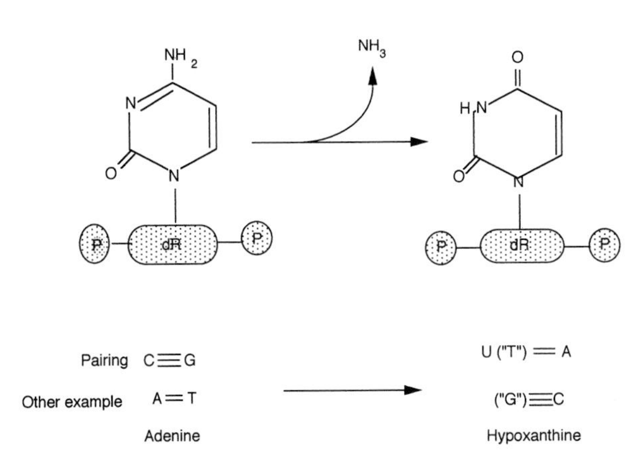

FIGURE 27.1 Deamination reactions have mutagenic consequences because the deaminated bases cause false recognition.

persistence in cellular life cycles that the accumulation of these changes can have significant long-term effects.

1. *Deamination*

The hydrolytic conversion of adenine to hypoxanthine, guanine to xanthine, and cytosine to uracilcontaining nucleotides is of sufficient magnitude to affect the informational content of DNA (Fig. 27.1).

2. *Depurination*

The glycosylic bonds linking guanine in nucleotides are especially sensitive to hydrolysis—more than the adenine and pyrimidine glycosylic links. The result is that apurinic (or apyrimidinic sites) are recognized by surveillance systems and as a consequence are repaired.

3. *Mismatched bases*

During the course of DNA replication there are noncomplementary nucleotides which are incorrectly incorporated into DNA and manage to escape the editing functions of the DNA polymerases. The proper strand and the mismatched base are recognized and repaired.

4. *Metabolic damage*

When thymine incorporation into DNA is limited either through restricted precursor UTP availability or through inhibition of the thymidylate synthetase system, dUTP is utilized as a substitute for thymidine triphosphate. The presence of uracil is identified as a damaged site and acted on by repair processes.

5. *Oxygen damage*

The production of oxygen, superoxide, or hydroxyl radicals as a metabolic consequence, as well as the oxidative reactions at inflammatory sites, causes sugar destruction, which eventually leads to strand breakage.

B. Exogenous damage

The concept of DNA repair in biological systems arose from studies by photobiologists and radiobiologists who studied the viability and mutagenicity in biological systems exposed to either ionizing or ultraviolet (UV) irradiation. Target theories, derived from the random statistical nature of photon bombardment, led to the identification of DNA as the primary target for the cytotoxicity and mutagenicity of ultraviolet light. In addition, most of the structural and regulatory genes controlling DNA repair in *Escherichia coli* were identified, facilitating the isolation and molecular characterization of the relevant enzymes and proteins.

1. *Ionizing radiation*

The primary cellular effect of ionizing radiation (Fig. 27.2) is the radiolysis of water which mainly generates hydroxyl radicals (HO•). The hydroxyl radical is capable of abstracting protons from the C4′ position of the deoxyribose moiety of DNA, thereby labilizing the phosphodiester bonds and generating single- and double-strand breaks. The pyrimidine bases are also subject to HO• addition reactions.

2. *Alkylation damage*

This modification occurs on purine ring nitrogens (cytotoxic adducts), at the O^6 position on guanine, and

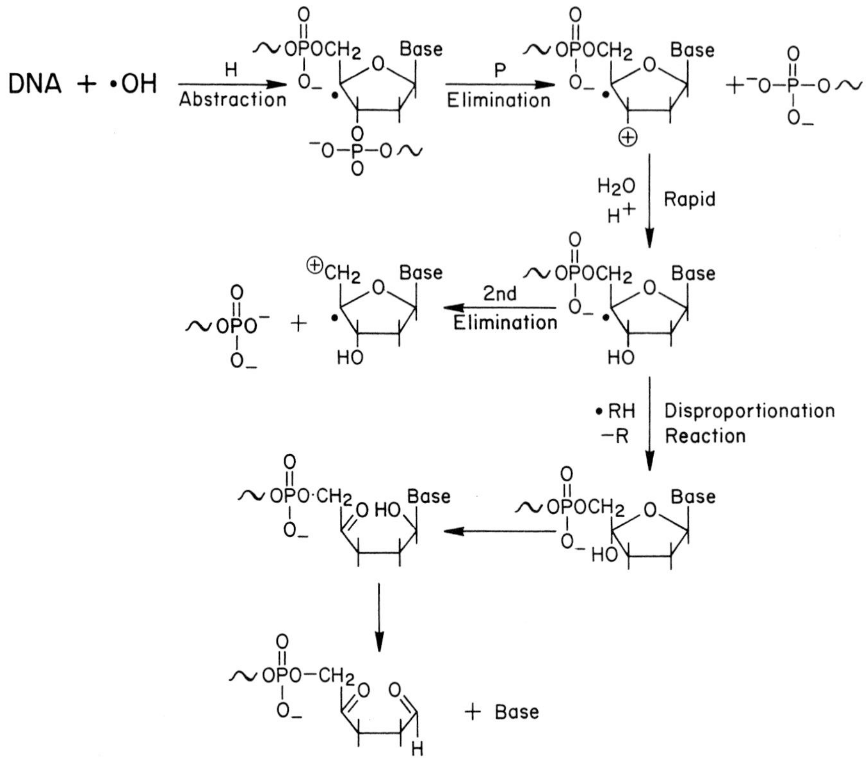

FIGURE 27.2 DNA backbone breakage by ionizing radiation.

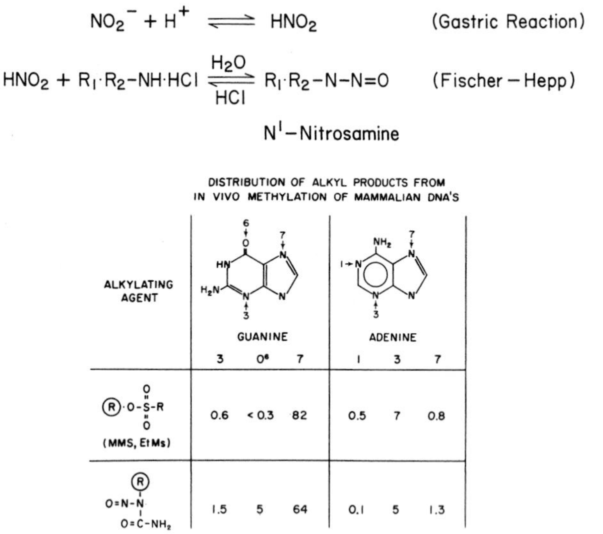

$$NO_2^- + H^+ \rightleftharpoons HNO_2 \quad \text{(Gastric Reaction)}$$

$$HNO_2 + R_1 \cdot R_2\text{-}NH \cdot HCl \underset{HCl}{\overset{H_2O}{\rightleftharpoons}} R_1 \cdot R_2\text{-}N\text{-}N{=}O \quad \text{(Fischer-Hepp)}$$

N'-Nitrosamine

DISTRIBUTION OF ALKYL PRODUCTS FROM IN VIVO METHYLATION OF MAMMALIAN DNA'S

ALKYLATING AGENT	GUANINE 3	GUANINE O6	GUANINE 7	ADENINE 1	ADENINE 3	ADENINE 7
Ⓡ-O-SO₂-R (MMS, EtMs)	0.6	<0.3	82	0.5	7	0.8
Ⓡ O=N-N-C(=O)-NH₂	1.5	5	64	0.1	5	1.3

FIGURE 27.3 The formation of alkylation sites in DNA exposed to nitrites.

at the O4 positions of the pyrimidines (mutagenic lesions) and the oxygen residues of the phosphodiester bonds of the DNA backbone (biologically silent) (Fig. 27.3). Alkylating agents are environmentally pervasive, arising indirectly from many foodstuffs and from automobile exhaust in which internal combustion of atmospheric nitrogen results in the formation of nitrate and nitrites.

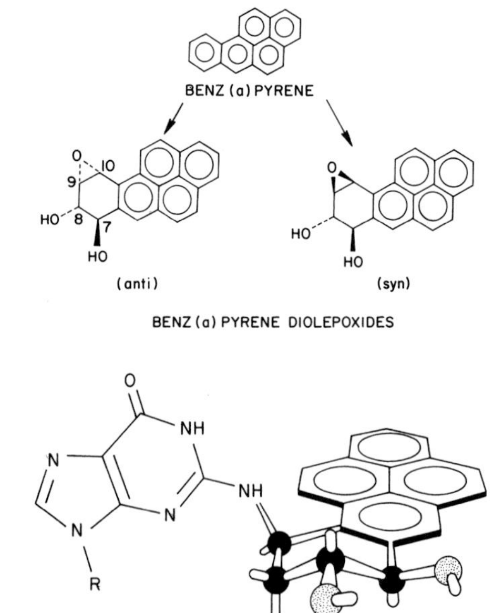

FIGURE 27.4 Bulky adducts formed in DNA exposed to benz(a)-pyrene.

3. Bulky adducts

Large, bulky polycyclic aromatic hydrocarbon modification occurs primarily on the N^2, N^7, and C^8 positions of guanine, invariably from the metabolic activation of these large hydrophobic uncharged macromolecules to their epoxide analogs (Fig. 27.4). The major source of these substances is from the combustion of tobacco, petroleum products, and foodstuffs.

4. *Ultraviolet irradiation*

Most of the UV photoproducts are chemically stable; their recognition provided direct biochemical evidence for DNA repair. The major photoproducts are 5,6-cyclobutane dimers of neighboring pyrimidines (intrastrand dimers), 6,4-pyrimidine-pyrimidone dimers (6-4 adducts), and 5,6-water-addition products of cytosine (cytosine hydrates).

II. DIRECT REMOVAL MECHANISMS

The simplest repair mechanisms involve the direct photoreversal of pyrimidine dimers to their normal homologs and the removal of O-alkyl groups from the O^6-methylguanine and from the phosphotriester backbone as a consequence of alkylation damage to DNA.

A. Photolyases (photoreversal)

The direct reversal of pyrimidine dimers to the monomeric pyrimidines is the simplest mechanism, and it is chronologically the first mechanism described for the repair of photochemically damaged DNA (Fig. 27.5). It is a unique mechanism characterized by a requirement for visible light as the sole source of energy for breaking two carbon–carbon bonds.

The enzyme protein has two associated light-absorbing molecules (chromophores) which can form an active light-dependent enzyme. One of the chromophores is $FADH_2$ and the other is either a pterin or a deazaflavin able to absorb the effective wavelengths of 365–400 nm required for photoreactivation of pyrimidine dimers. It is suggested that photoreversal involves energy transfer from the pterin molecule to $FADH_2$ with electron transfer to the pyrimidine dimer resulting in nonsynchronous cleavage of the C^5 and C^6 cyclobutane bonds. Enzymes that carry out photoreactivation have been identified in both prokaryotes and eukaryotes.

B. Alkyl group removal (methyl transferases)

Bacterial cells pretreated with less than cytotoxic or genotoxic levels of alkylating agents before lethal or mutagenic doses are more resistant (Fig. 27.6). This is an adaptive phenomenon with anti-mutagenic and anti-cytotoxic significance. During this adaptive period, a 39-kDa Ada protein is synthesized that specifically removes a methyl group from a phosphotriester bond and from an O^6-methyl group of guanine (or from O^4-methyl thymine). The O^6-methyl group of guanine is not liberated as free O^6-methyl guanine during this process, but it is transferred directly from the alkylated DNA to this protein; the Ada protein (methyl transferase) and an unmodified guanine are simultaneously generated. These alkyl groups specifically methylate cysteine 69 and cysteine 321, respectively, in the protein.

The methyl transferase is used stoichiometrically in the process (does not turnover) and is permanently inactivated in the process. Nascent enzyme, however, is generated because the mono- or dimethylated transferase activates transcription of its own "regulon" which includes, in addition to the *ada* gene, the *alk* B gene of undefined activity and the *alk* A gene which sponsors a DNA glycosylase. The latter enzyme acts on 3-methyl adenine, 3-methyl guanine, O^2-methyl cytosine, and O^2-methyl thymine. The methylated Ada protein can specifically bind to the operator of the *ada* gene acting as a positive regulator. Down regulation may be controlled by proteases acting at two hinge sites in the Ada protein.

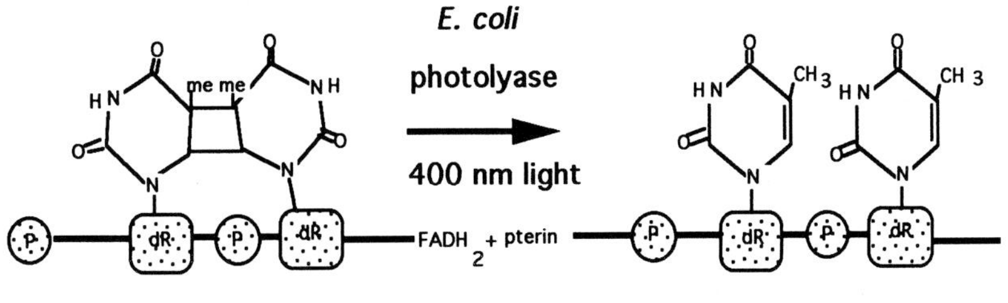

FIGURE 27.5 The direct photoreversal of pyrimidine dimers in the presence of visible light.

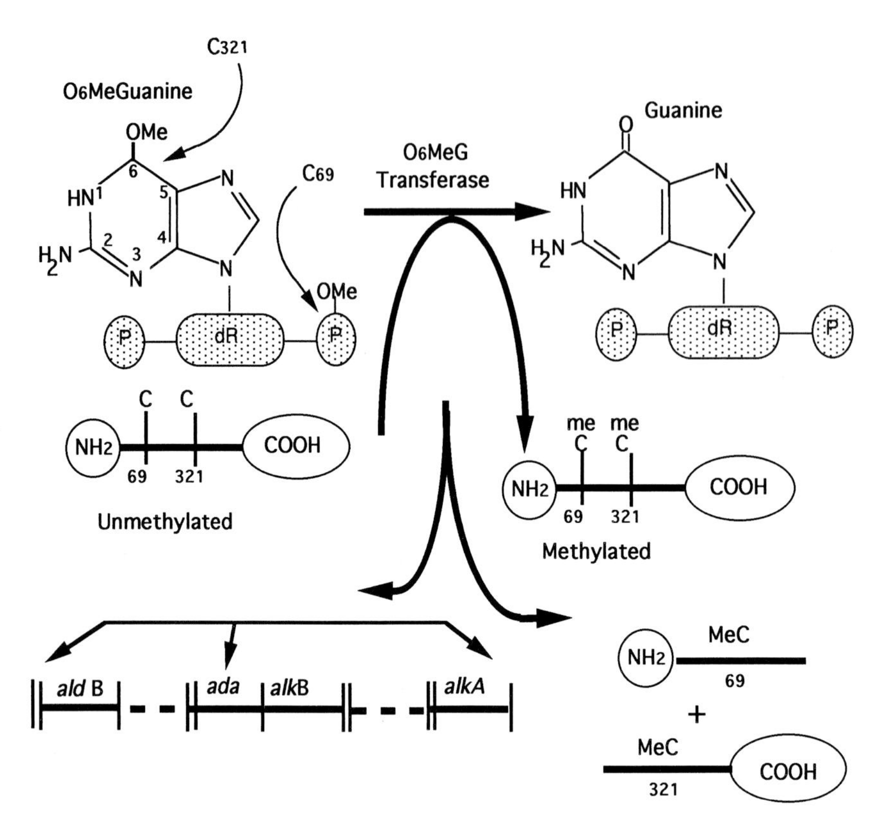

FIGURE 27.6 The direct reversal of alkylation damage removes such groups from the DNA backbone and the O^6 position of guanine. Such alkyl groups are transferred directly to specific cysteine residues on the transferase, the levels of which are influenced adaptively by the levels of the alkylating agents. The methylation of the transferase inactivates the enzyme which is used up stoichiometrically in the reaction. The alkylated transferase acts as a positive transcriptive signal turning on the synthesis of unique mRNA. Regulation of transferase levels may be influenced by a unique protease.

III. BASE EXCISION REPAIR

A. Base excision repair by glycosylases and apyrimidinic or apurinic endonucleases

Bases modified by deamination can be repaired by a group of enzymes called DNA glycosylases, which specifically hydrolyze the N-glycosyl bond of that base and the deoxyribose of the DNA backbone generating an apyrimidinic or an apurinic site (AP site) (Fig. 27.7). These are small, highly specific enzymes that require no cofactor for functioning. They are the most highly conserved proteins, attesting to the evolutionary unity both structurally and mechanistically from bacteria to man. As a consequence of DNA glycosylase action, the AP sites generated in the DNA are acted on by a phosphodiesterase (Fig. 27.8) specific for such sites that can nick the DNA 5′ and/or 3′ to such damaged sites. If there is a sequential action of a 5′ acting and a 3′ acting AP endonuclease, the AP site is excised generating a gap in the DNA strand.

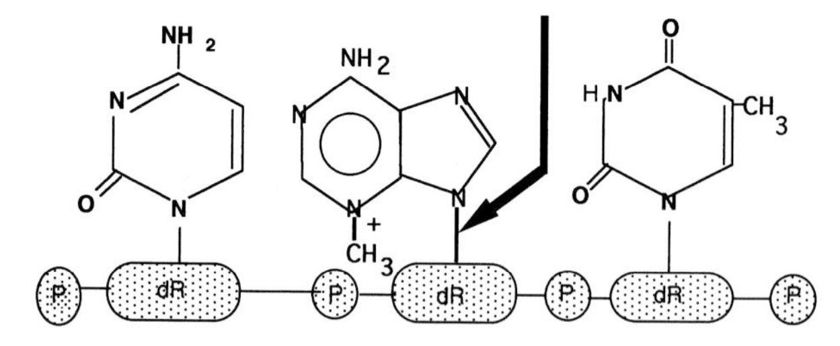

FIGURE 27.7 DNA glycosylases hydrolyze the N-glycosyl bond between damaged bases and deoxyribose generating an AP (apyrimidinic or apurinic) site (arrow).

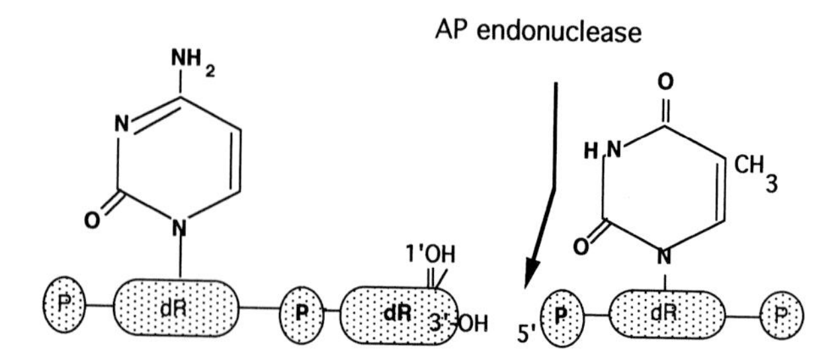

FIGURE 27.8 Endonucleases recognize AP sites and hydrolyze the phosphodiester bonds 3′, 5′, or both sides of the deoxyribose moiety in damaged DNA.

B. Glycosylase-associated AP endonucleases

An enzyme from bacteria and phage-infected bacteria, encoded in the latter case by a single gene (*den*), hydrolyzes the N-glycosyl bond of the 5′ thymine moiety of a pyrimidine dimer followed by hydrolysis of the phosphodiester bond between the two thymine residues of the dimer (Fig. 27.9). This enzyme, referred to as the pyrimidine dimer DNA glycosylase, is found in *Micrococcus luteus* and phage T4-infected *E. coli*. This small, uncomplicated enzyme does not require cofactors and is presumed to act by a series of linked β elimination reactions. An enzyme behaving in a similar glycosylase–endonuclease fashion but acting on the radiolysis product of thymine is thymine glycol, which has been isolated from *E. coli* and is referred to as endonuclease III.

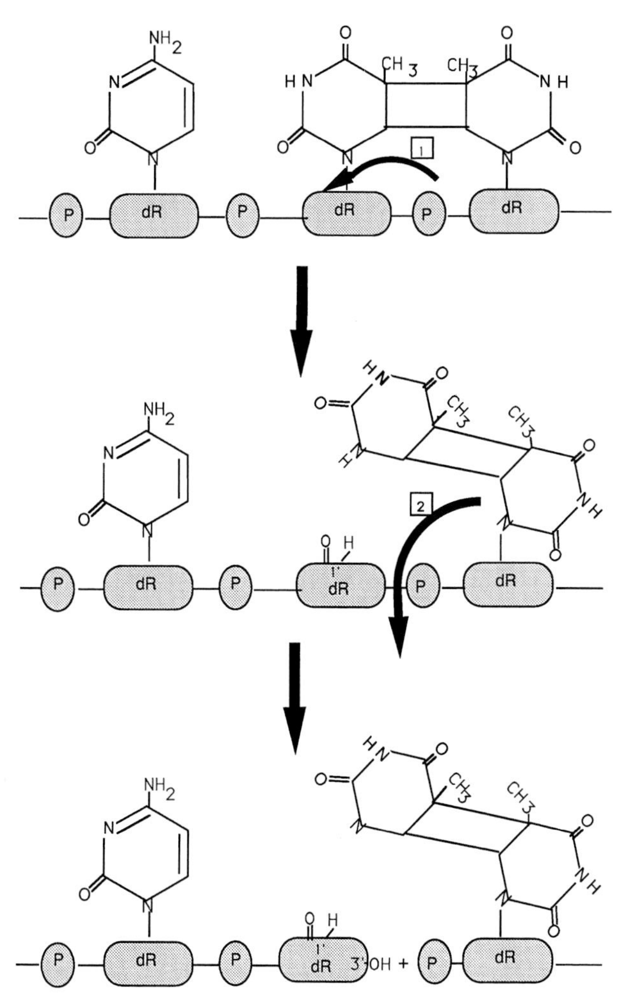

FIGURE 27.9 The same enzyme that can hydrolyze the N-glycosyl residue of a damaged nucleotide also hydrolyzes the phosphodiester bond linking the AP site generated in the first N-glycosylase reaction.

IV. NUCLEOTIDE EXCISION REPAIR

The ideal repair system is one that is somewhat indiscriminate and which can respond to virtually any kind of damage. Such a repair system has been characterized in *E. coli* in which it consists of at least six gene products of the *uvr* system. This ensemble of proteins consists of the UvrA protein that binds as a dimer to DNA in the presence of ATP, followed by the UvrB protein which cannot bind DNA by itself. Translocation of the $UvrA_2B$ complex from initial undamaged DNA sites to damaged sites is driven by a cryptic ATPase associated with UvrB which is activated by the formation of the $UvrA_2B$-undamaged DNA complex. This complex is now poised for endonucleolytic activity catalyzed by the interaction of the $UvrA_2B$-damaged DNA complex with UvrC to generate two nicks in the DNA seven nucleotides 5′ to the damaged site and the three or four nucleotides 3′ to the same site. These sites of breakage are invariant regardless of the nature of the damage. In the presence of the UvrD (helicase III), DNA polymerase I, and substrate deoxynucleoside triphosphates, the damaged fragment is released and this is accompanied by the turnover of the UvrA, UvrB, and UvrC proteins. The continuity of the DNA helix is maintained based on the sequence of the opposite strand. The final integrity of the interrupted strands is restored by the action of DNA polymerase I, which copies the other strand, and by polynucleotide ligase, which seals the gap (Fig. 27.10). The levels of the Uvr proteins are regulated in *E. coli* by an "SOS" regulon monitoring many genes, including *uvr*A, *uvr*B, possibly *uvr*C, and *uvr*D, as part of the excision repair system. It also includes the regulators of the SOS system, the uvrD and recA proteins; cell division genes *sul*A and *sul*B; recombination genes *rec*A, *rec*N, *rec*Q *uvr*D, and *ruv*; mutagenic by-pass mechanisms (*umu*DC and *rec*A); damage-inducible genes; and the lysogenic phage λ. The LexA protein negatively regulates these genes as a repressor by binding to unique operator regions. When the DNA is damaged (e.g., by UV light), a signal in the form of a DNA repair intermediate induces the synthesis of the RecA protein. When induced, the RecA protein acts as a protease assisting the LexA protein to degrade itself, activating its own synthesis and that of the RecA protein as well as approximately 20 other genes. These genes permit the survival of the cell in the face of life-threatening environmental damages. Upon repair of the damaged DNA, the level of the signal subsides, reducing the level of RecA and thus stabilizing the integrity of the intact LexA protein and its repressive properties on all

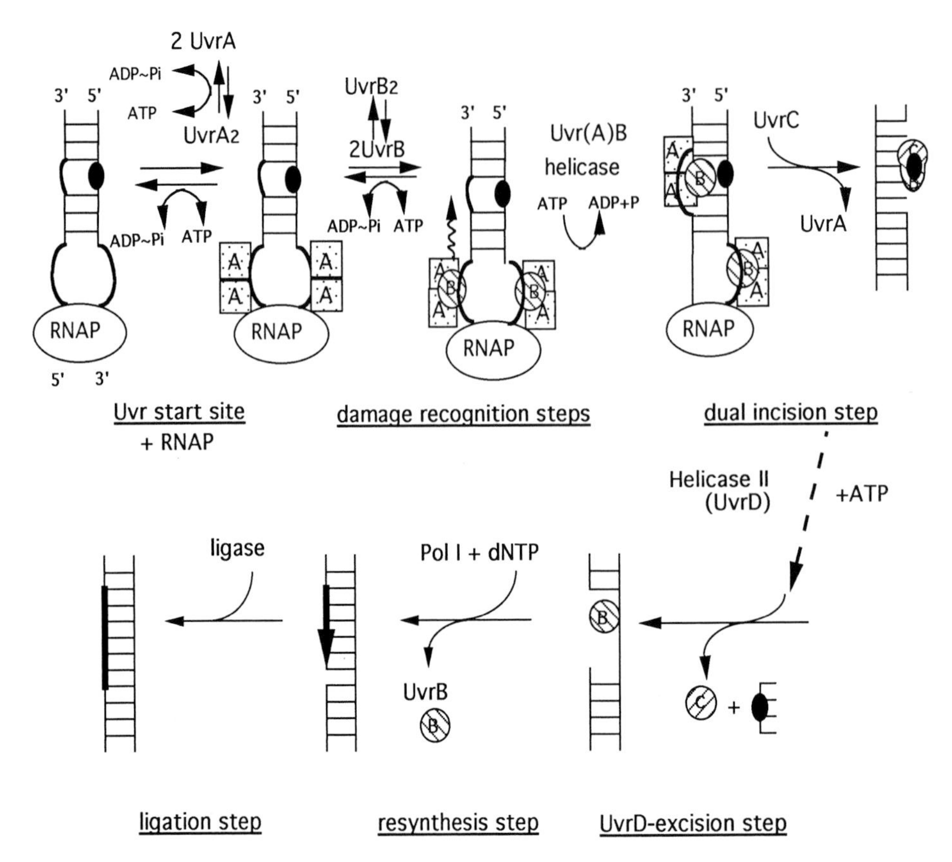

FIGURE 27.10 Nucleotide excision reactions. In this multiprotein enzyme system the UvrABC proteins catalyze a dual-incision reaction seven nucleotides 5′ and three or four nucleotides 3′ to a damaged site. The UvrA protein, as a dimer, binds to undamaged sites initially and in the presence of UvrB, whose cryptic ATPase is manifested in the presence of UvrA providing the energy necessary for translocation to a damaged site. This pre-incision complex interacts with UvrC, leading to the dual-incision reaction. The incised DNA-UvrABC does not turnover and requires the coordinated participation of the UvrD and DNA polymerase reactions for damaged fragment release and turnover of the UvrABC proteins. Ligation, the final reaction, restores to integrity the DNA strands.

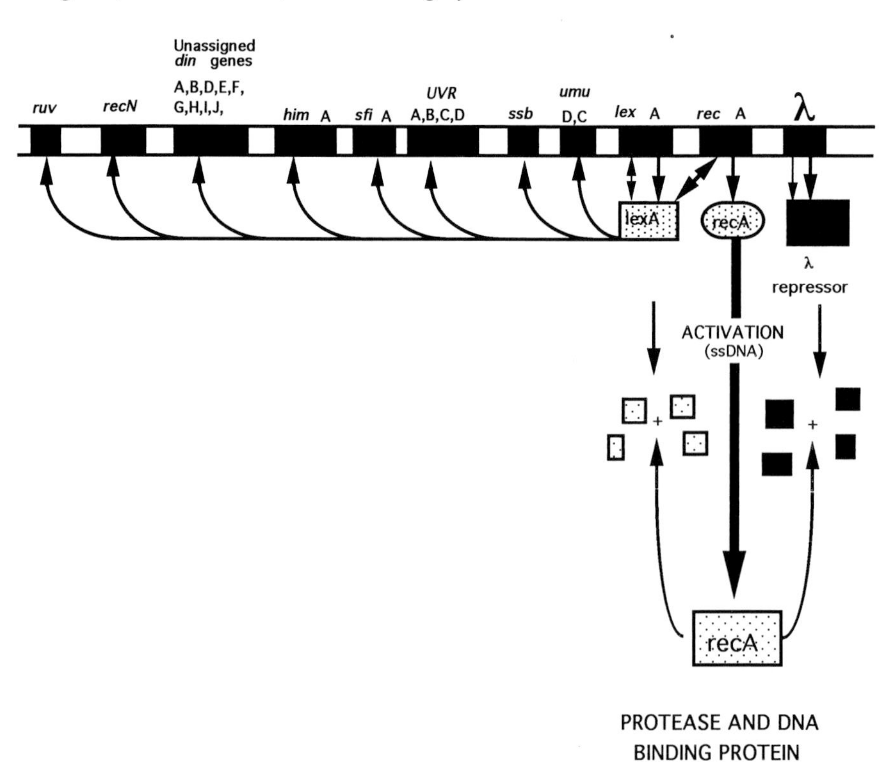

FIGURE 27.11 Regulation of the nucleotide excision pathway by the "SOS" system. The *lex* A and phage λ repressors negatively control a multitude of genes which are turned on when bacterial cells are damaged. This leads to the over-production of the recA protein which assists in the proteolysis of the LexA and phage λ proteins, thereby derepressing the controlled gene systems. When DNA is fully repaired the level of recA declines, restoring the "SOS" system to negative control.

the other genes (Fig. 27.11). Then the cell returns to its normal state.

V. TRANSCRIPTION-COUPLED NUCLEOTIDE EXCISION REPAIR

It appears that the structural and biological specificty associated with transcriptional processes limit DNA repair to those damaged regions of the chromosome undergoing transcription and that damage in those quiescent regions is persistent. Within expressed genes the repair process is selective for the transcribed DNA strand for damage such as pyrimidine dimers, and this "coupling" to transcription has been shown in *E. coli*. As a consequence, DNA repair occurs preferentially in active transcribed genes. Preferential repair occurs in the transcribed strand in actively expressed genes. Nucleotide excision repair (NER) differs in the two separate DNA strands of the lactose operon of UV-irradiated *E. coli*. The level of repair examined in the uninduced condition is about 50% after 20 min in both strands. As a consequence of a 436-fold induction of β-galactosidase, most of the dimers (70%) are removed from the transcribed strand of the induced operon within 5 min, whereas the extent of repair in nontranscribed strands is similar to that of the uninduced condition. This selective removal of pyrimidine dimers from the transcribed strand of a gene is abolished in the absence of significant levels of transcription.

As shown in Figs. 27.12 and 27.13, the RNAP when binding to its promoter site generates a defined distortion 3′ (downstream) to the RNAP binding site which provides a "landing site" for the $UvrA_2B$ complex. Strand specificity is dictated by the $5' \rightarrow 3'$ directionality of the $UvrA_2B$ helicase which can translocate only on the non-transcribed strand because RNAP interferes with the directionality on the lower strand. Nicking occurs only on the strand opposite to the strand which the $UvrA_2B$ endonuclease binds; hence, it is the transcribed strand which is initially repaired in this model.

A. Effect of pyrimidine dimers on transcription and effect of RNAP on repair

T–T photodimers in the template strand constitute an absolute block for transcription, whereas those in the complementary strand have no effect. Irradiation of cells with UV results in truncated transcripts because the pyrimidine dimers become a "stop" site. In the absence of ribonucleoside triphosphates, promoter-bound RNAP does not translocate and, hence, has no effect on repair on a T–T photodimer downstream from the transcriptional initiation site no matter whether the photodimer is in the transcribed or in the nontranscribed strand.

B. Transcription repair coupling factor

Transcription repair coupling (TRC) is achieved through the action of the transcription repair coupling

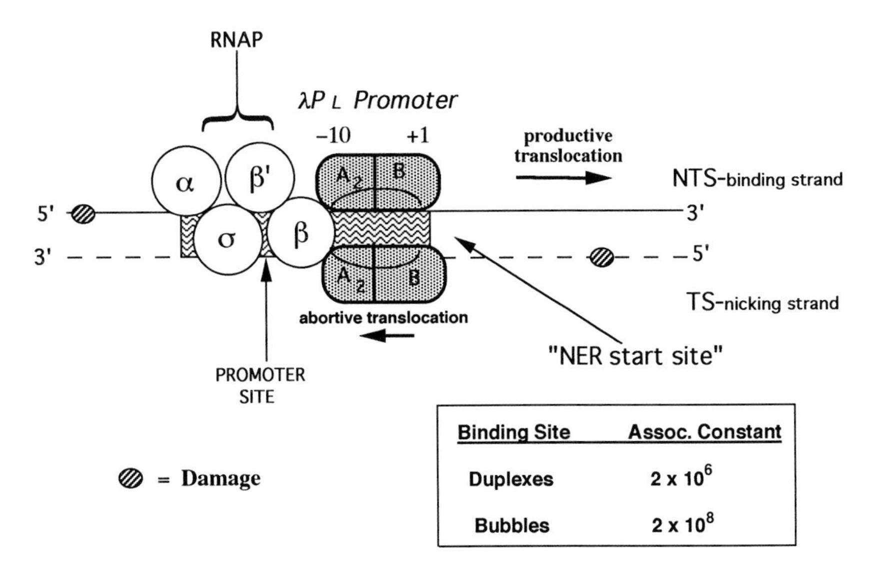

FIGURE 27.12 Strand selectivity by the *E. coli* $UvrA_2B$ helicase.

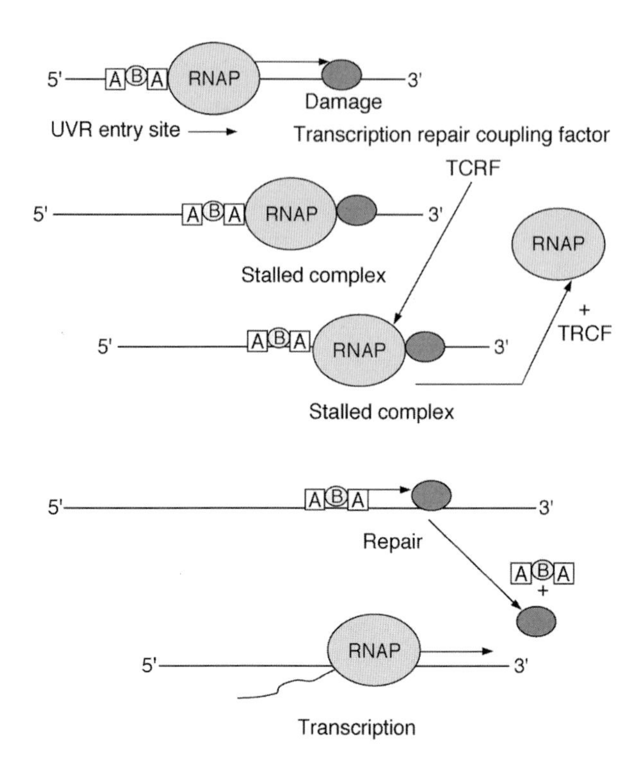

FIGURE 27.13 Relief of RNAP-damaged sites as stalled complexes by the TRCF when RNA precedes the Uvr complex.

factor (TRCF). TRCF is the product of the *mfd* gene (*m*utation *f*requency *d*ecline), which maps at 25 min on the *E. coli* chromosome. The cloned *mfd* gene is translated into a 1148-amino acid protein of ~130 kDa. The Mfd protein can nonspecifically bind dsDNA (and less efficiently ssDNA) in an ATP-binding-dependent manner, with the ATP hydrolysis promoting its dissociation. The amino acid sequence of Mfd reveals motifs which are characteristic of many DNA and RNA helicases. However, *in vitro* purified Mfd does not show either DNA or RNA helicase activity. N-terminal 1–378 residues of Mfd, have a 140-amino acid region of homology with UvrB and bind UvrA protein.

In vitro transcription is inhibited by NER damage in a transcribed strand, whereas it has no effect on the noncoding strand. This inhibition is thought to result from a stalled RNAP at the site of damage. TCRF is able to release the stalled RNAP in an ATP hydrolysis-dependent manner. Moreover, it actually stimulates NER of the transcribed strand, so that it becomes faster than the nontranscribed one. Based on all these observations, it is concluded that TRCF-Mfd carries out preferential repair of the transcribed strand by (i) releasing RNAP stalled at damaged sites and (ii) recruiting the $UvrA_2B$ complex to damaged sites through the high-affinity interaction with UvrA.

VI. MISMATCH CORRECTION

Many mechanisms do not recognize damage but do recognize mispairing errors that occur in all biological systems (Fig. 27.14). In *E. coli* mismatch correction is controlled by seven mutator genes; *dam* (methyl directed), *mut*D, *mut*H *mut*L, *mut*S, *mut*U, *uvr*D, and *mut*Y. In mismatch correction one of the two strands of the mismatches is corrected to conform with the other strand. Strand selection is one of the intrinsic problems in mismatch repair and the selection is achieved in bacterial systems by adenine methylation, which occurs at d(GATC) sequences. Since such methylation occurs after DNA has replicated, only the template strand of the nascent duplex is methylated. In mismatch repair only the unmethylated strand is repaired, thus retaining the original nucleotide sequence. The MutH, MutL, and MutS proteins appear to be involved in the incision reaction on this strand, with the remainder of the proteins plus DNA polymerase III and polynucleotide ligase participating in the excision–resynthesis reactions.

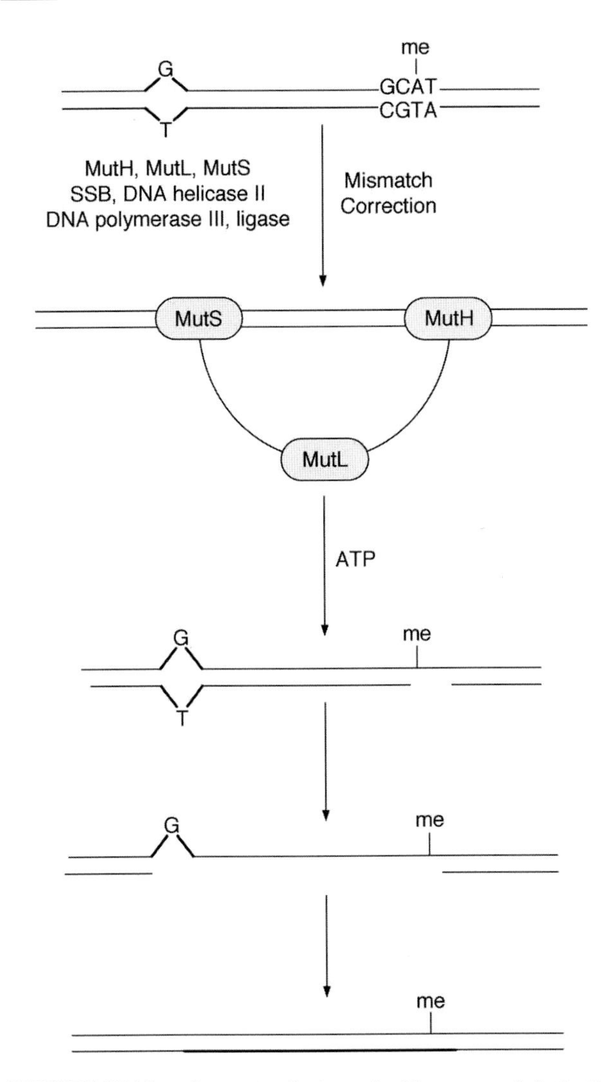

FIGURE 27.14 In the repair of mismatched bases strand distinction can be achieved by the delay in adenine methylation during replication. It is the nascent unmethylated strand which serves as a template for the incision reactions catalyzed by many proteins specifically engaged in mismatch repair processes.

BIBLIOGRAPHY

Courcelle, J., Ganesan, A. K., and Hanawalt, P. C. (2001). Therefore, what are recombination proteins there for? *Bioessays* **23**, 463–470.

Cox, M. M. (1999). Recombinational DNA repair in bacteria and the RecA protein. *Prog. Nucleic Acid. Res. Mol. Biol.* **63**, 311–366.

Friedberg, E. C. (1984). "DNA Repair." Freeman, New York.

Grossman, L., Caron, P. R., Mazur, S. J., and Oh, E. Y. (1988). Repair of DNA containing pyrimidine dimers. *FASEB J.* **2**, 2696–2701.

Hanawalt, P. C. (2001). Controlling the efficiency of excision repair. *Mutat. Res.* **485**, 3–13.

Lindahl, T., Sedgwick, B., Sekiguchi, M., and Nakabeppu, Y. (1988). Regulation and expression of the adaptive response to alkylating agents. *Annu. Rev. Biochem.* **57**, 133–157.

Makarova, K. S., Aravind, L., Wolf, Y. I., Tatusov, R. L., Minton, K. W., Koonin, E. V., and Daly, M. J. (2001). Genome of the extremely radiation-resistant bacterium *Deinococcus radiodurans* viewed from the perspective of comparative genomics. *Microbiol. Mol. Biol. Rev.* **65**, 44–79.

Modrich, P. (1989). Methyl-directed DNA mismatch correction. *J. Biol. Chem.* **264**, 6597–6600.

Rupp, W. D. (1996). DNA repair mechanisms. In "*Escherichia coli and Salmonella*. Cellular and Molecular Biology", 2nd Edition. (F. C. Neidhardt, Ed.). ASM Press, Washington, DC.

Sancar, A., and Sancar, G. B. (1988). DNA repair enzymes. *Annu. Rev. Biochem.* **57**, 29–67.

Walker, G. C. (1985). Inducible DNA repair systems. *Annu. Rev. Biochem.* **54**, 457.

Weiss, B., and Grossman, L. (1987). Phosphodiesterases involved in DNA repair. *Adv. Enzymol.* **60**, 1–34.

WEBSITE

Website for Comprehensive Microbial Resource of The Institute for Genomic Research and links to many other microbial genomic sites
http://www.tigr.org/tigr-scripts/CMR2/CMRHomePage.spl

28

DNA replication

James A. Hejna and Robb E. Moses
Oregon Health Sciences University

GLOSSARY

chromosome Package of genes representing part or all of the inherited information of the organism.

exonuclease Enzyme that degrades DNA from a terminus.

gyrase Enzyme that introduces supercoiling into the DNA duplex in an adenosine triphosphate-dependent reaction.

helicase Enzyme that unwinds duplex DNA and requires adenosine triphosphate.

polymerase Enzyme that synthesizes a nucleic acid polymer.

replication Act of duplicating the genome of a cell.

replicon Replicative unit, either part or the whole of the genome; in *Escherichia coli*, the entire genome is considered a replicon.

topoisomerase An enzyme that alters the topology of DNA, either one strand at a time (type I) or two strands at a time (type II). Gyrase and Topoisomerase IV are type II enzymes.

DNA Replication in *Escherichia coli* is a carefully regulated process involving multiple components representing more than 20 genes participating in duplication of the genome. The process is divided into distinct phases: initiation, elongation, and termination. The synthesis of a new chromosome involves an array of complex protein assemblies acting in sequential fashion in a carefully regulated and reiterated overall pattern. The scheme for DNA replication is under careful genetic control. The process is localized on the DNA structure by both DNA sequence and topology and requires specific protein–DNA interactions. DNA replication in *E. coli* is bidirectional and symmetrical.

I. DEVELOPMENT OF THE FIELD

This article will focus primarily on *Escherichia coli* as a model organism, with the assumption that what is true for *E. coli* is generally true for other prokaryotes. In gram-positive bacteria such as *Bacillus subtilis*, this assumption has been largely substantiated. Sequencing of several prokaryotic genomes has revealed homology to many of the genes involved in *E. coli* DNA replication. The development of our understanding of DNA replication in prokaryotes depends on a combination of biochemical and genetic approaches. Using several selection techniques, many laboratories isolated *E. coli* mutants that were conditionally defective (usually temperature sensitive) in DNA replication. This method of identifying genes involved in DNA replication assumed that defects of such genes resulted in the death of the cell. When a large series of mutants was assembled, they fell clearly into two broad categories: those in which DNA replication ceased abruptly following a shift to restrictive conditions and those in which DNA replication ceased slowly. The former class is called fast stop and the latter slow stop. The first category represents cells containing mutations in gene products that are required for the elongation phase of DNA replication, and the latter category contains cells with defects

The Desk Encyclopedia of Microbiology
ISBN: 0-12-621361-5

in gene products that are required for the initiation of new rounds of DNA replication.

The identification of temperature-sensitive, *dnats*, mutants was one requirement for understanding DNA replication. The second requirement was the development of systems that could be biochemically manipulated but that represented all or part of the authentic DNA replication process in *E. coli*. Several successive systems offered increasing advantages. Systems in which permeable cells allowed free access of small molecules to minimally disturbed chromosomes were the earliest. These systems allowed definition of the energy and cofactor requirements for the elongation phase of DNA replication. The limitation was that they did not permit access by macromolecules to the replication apparatus and therefore did not allow complementation of defects in proteins required for DNA replication. Also, such systems did not allow the initiation of new rounds of replication.

The development of lysate replication systems rested on the recognition that the failure to maintain the complex process of DNA replication in early studies was due to dilution of the components, resulting in disassembly of the replication structure and loss of functions required for DNA replication. The concentrated lysate systems depended on the bacterial chromosome, but it was quickly recognized that small bacterial phage chromosomes could be utilized as exogenous templates for DNA replication because they permitted the addition of proteins to allow complementation of defects in DNA replication. Lysates made from mutants defective in a specific step of DNA replication could be used to define the step at which the defect occurred and to identify the protein product complementing the defect. This allowed assignment of protein products to genes. Such systems, however, did not allow the study of initiation of new rounds of DNA synthesis on the host chromosome.

Two theoretical shortcomings of such systems are (i) that such systems might not define all the proteins required for DNA replication by the host and (ii) that such systems might require a protein for replication of the phage DNA not ordinarily required by the host chromosome. The following general point derived from these studies is worth remembering: Although *E. coli* contains numerous proteins that have overlapping or similar enzymatic function, the participation of a protein in the replication process is carefully regulated, reflecting a specific role. The DNA polymerases are the best example. Each of the three recognized DNA polymerases of *E. coli* (Table 28.1) has similar enzymatic capabilities, but ordinarily only DNA polymerase III catalyzes replication. This restriction of activity can be partially explained on the basis of protein–protein interactions. The complete basis for the regulation is not understood, and the reasons for its being advantageous to the cell are not clear.

TABLE 28.1 DNA polymerases of *E. coli*

	I	II	III
Molecular mass (kDa)	103	88	130
Synthesis	5′→3′	5′→3′	5′→3′
Initiation	No	No	No
5′-Exonuclease	Yes	No	No
3′-Exonuclease	Yes	Yes	Yes[a]
Gene	*polA*	*polB*	*polC* (*dnaE*)

[a]In a separate protein.

II. CONTROL OF DNA REPLICATION

Control of DNA replication relies on the regulation of new rounds of replication. In *E. coli*, there are two components of control: the DnaA protein and the structure of the origin of DNA replication (*oriC*). The region of the *E. coli* origin of DNA replication is at 85 min on the genetic map (based on a total of 100 min). On the sequenced *E. coli* genome, the origin is located between nucleotides 3923371 and 3923602. Thus, there is a fixed site on the *E. coli* genome that represents the appropriate place for the initiation of DNA replication. This DNA initiation is referred to as "macroinitiation" as opposed to repetitive initiation, which must occur multiple times during the "elongation" phase of DNA replication. The latter is referred to as "microinitiation." It seems that the requirements for macroinitiation at the *oriC* region include those needed for microinitiation plus additional requirements. There is a region of approximately 250 bp that must be present for DNA replication to initiate.

A. Macroinitiation

Several features regarding this region are notable. There are multiple binding sites for the DnaA protein (the DnaA boxes) (Fig. 28.1). This is a nine-nucleotide sequence that has been shown to bind the DnaA protein. There are also multiple promoter elements, suggesting the involvement of RNA polymerase in macroinitiation. Possible DNA gyrase binding sites are also present. Another notable feature is the presence of multiple Dam-methylase sites (GATC sequence).

The definition of the *oriC* region depends on cloning of this region into plasmids constructed so that replication of the plasmid depends on the function of the

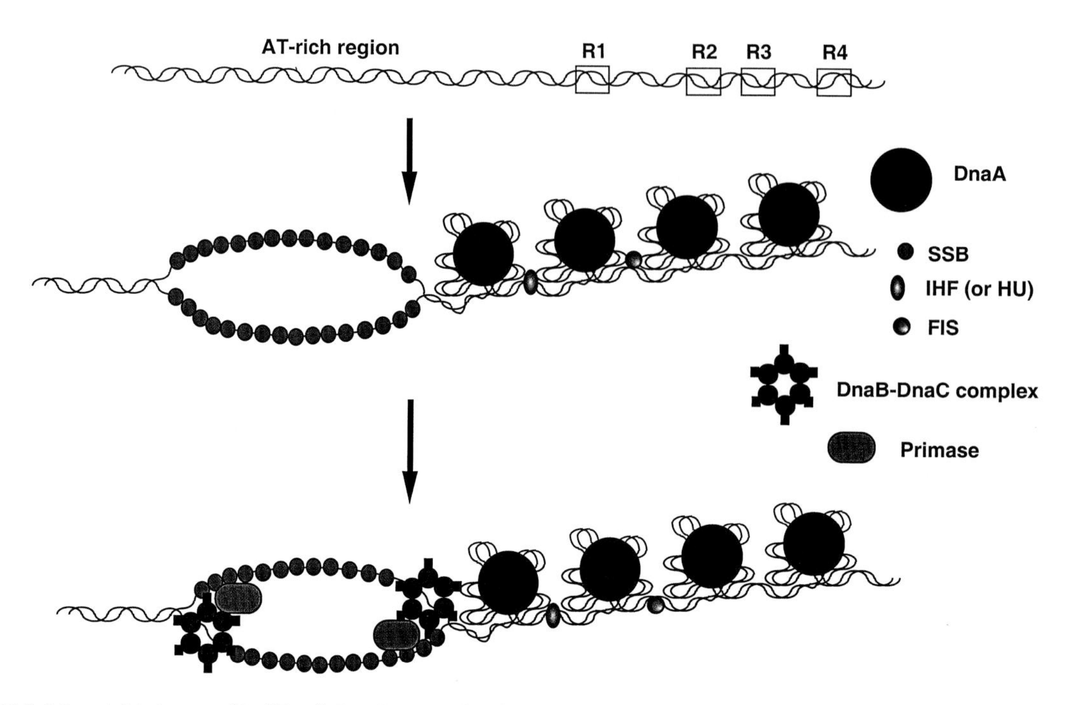

FIGURE 28.1 Macroinitiation at *oriC* of *E. coli*. DnaA protein binds to DnaA boxes R1–R4. The DNA is probably wrapped around the DnaA proteins, and a higher order nucleoprotein complex involving as many as 30–40 DnaA polypeptides is generated with the assistance of DNA-binding proteins, such as IHF, HU, and FIS. The winding of DNA into this complex leads to the compensatory unwinding of an AT-rich region adjacent to the DnaA boxes. Single-stranded DNA is then coated by Ssb protein. DnaC recruits DnaB, the helicase which drives the DNA unwinding at the replication fork, to the junction between single-stranded DNA and double-stranded DNA. Primase then associates with DnaB and synthesizes the RNA primer. The RNA primer is subsequently extended by DNA polymerase III holoenzyme as the DnaB helicase unwinds the chromosome ahead of the polymerase.

oriC sequence. This has allowed development of an *in vitro* assay system for macroinitiation of DNA replication. The cloning of *oriC* confirmed that the specificity of macroinitiation in *E. coli* resides in the origin.

The primary protein actor in macroinitiation is the DnaA protein. This protein binds at multiple sites within the *oriC* structure as noted. In addition to binding at the DnaA box consensus sequence, the DnaA protein displays a DNA-dependent adenosine triphosphatase (ATPase) activity and appears to display cooperative binding properties. This suggests that the possible role in initiation is a change of conformation of DNA by DnaA protein interactions. The DnaA protein also binds in the promoter region of the DnaA gene, suggesting autoregulation, which is supported by genetic studies. It appears that the DnaA protein must act positively to initiate DNA replication in *E. coli*.

Both protein and RNA synthesis are required for macroinitiation to occur in *E. coli*. The macroinitiation phase may be further subdivided into stages. The earliest step involves the binding of the DnaA protein to the *oriC* structure. This results in a conformational change of the origin. This complex then binds the DnaA to DnaB and DnaC proteins (which are also required for the elongation phase of DNA replication). DnaC plays a unique role in the delivery of the DnaB protein to the replication structure. The resulting complex appears to unwind the DNA strands since the DnaB protein functions as a helicase, an ATP-dependent unwinding activity. This allows the binding of single-strand binding (Ssb) protein, which allows priming such as that which occurs in the elongation of DNA replication. This stage is followed by the propagation of microinitiation and elongation phases of replication.

Thus, the proteins required for the macroinitiation of *E. coli* DNA replication appear to include DnaA and DnaC, which have specific roles, as well as DnaB, Ssb protein, gyrase, the DnaG primase protein, and the replicative apparatus of DNA polymerase III holoenzyme complex (see Section III.B). Studies also suggest a direct role for RNA polymerase in the macroinitiation of DNA replication.

The DnaA protein offers important support of the replicon hypothesis. Mutations in the DnaA protein demonstrate that all of the *E. coli* chromosome is under a unit control mechanism which defines it as a single replicon. Integration of certain low-copy number plasmids into the chromosome suppresses the phenotype in DnaA mutants that were defective in macroinitiation. This "integrative suppression" shows a general control of macroinitiation and supports the replicon hypothesis.

B. Microinitiation

Microinitiation is the hallmark of the elongation phase of DNA replication. During this phase, repeated initiation occurs along the DNA. The microinitiation step appears to be analogous to the initiation step studied in the *in vitro* lysate systems using small circular phage genomes. In the prokaryotic cell, the requirements for microinitiation appear to mimic those of the phage systems G4 and ϕX174, which do not display a requirement for the DnaA protein or the features of the *oriC* region.

Cell proteins required for microinitiation include the DnaB protein. The DnaB protein contains a nucleoside triphosphate activity that is stimulated by single-stranded DNA. It also displays DNA helicase activity. In addition, it appears to undergo protein–protein interactions with the DnaC protein. The DnaB mutants are notable for a rapid cessation of DNA synthesis at restrictive conditions. It appears that the DnaB protein is one of the "motors" that moves the replication complex along the DNA (or moves the DNA through the replication complex). The DnaB protein is typical of the proteins involved in DNA replication in that it may have more than one role.

The DnaG protein of *E. coli* is the primase. This protein is capable of synthesizing oligonucleotides utilizing nucleoside (or deoxynucleoside) triphosphates. It appears that physiologically its role is to synthesize RNA primers, which can be utilized by the DNA polymerase III holoenzyme complex to initiate DNA synthesis. As shown in Fig. 28.2, at least some portion of DNA replication in most organisms is discontinuous. That is, part of the DNA is synthesized in short pieces (termed Okazaki pieces). This is the result of the restriction for DNA synthesis in the $5' \rightarrow 3'$ direction. Since the replication fork requires apparent growth of the nascent strands in both the $5' \rightarrow 3'$ direction and the $3' \rightarrow 5'$ direction, studies were initiated that searched for precursors or enzymes that would allow growth in the $3' \rightarrow 5'$ direction. None were found. The hypothesis of discontinuous synthesis states that, on a microscopic scale, DNA is synthesized discontinuously in a $5' \rightarrow 3'$ direction in small pieces to allow an overall growth in the $3' \rightarrow 5'$ direction on one strand (the lagging strand). This hypothesis predicts the existence of a relatively uniform class of small nascent DNA strands prior to joining, and it also predicts joining activity for such DNA strands. Both of these predictions are fulfilled.

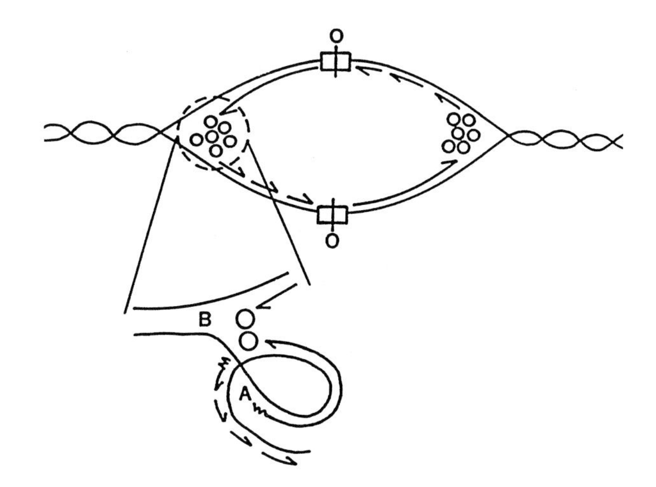

FIGURE 28.2 Model of the replication region. O, the origin; A, primed, elongating nascent strand; B, the point at which the replicative enzyme will release from the A strand and reinitiate a new discontinuous strand.

It appears that in *E. coli* DNA is synthesized on one lagging strand in pieces of approximately 1000 nucleotides, which are then covalently linked via the action of DNA ligase following synthesis. Because none of the DNA polymerases in prokaryotes have been found to initiate synthesis *de novo*, this hypothesis leads to the prediction that RNA synthesis, which can be demonstrated to initiate *de novo*, forms primers that are utilized for DNA strand synthesis. Identification of the DnaG primase activity satisfies this prediction.

The polarity restriction of DNA synthesis by DNA polymerases permits one strand to be made continuously, as indicated in the model. It appears that relatively few initiations are made in this (leading) strand and, in fact, that macroinitiation may serve to prime the whole length of the strand.

Ssb protein has analogs throughout nature. In *E. coli*, this protein is relatively small (approximately 19 kDa) and functions as a tetramer. This protein is required for DNA replication. It appears to have several roles. In the single-strand phage systems *in vitro* it confers specificity on the origin of DNA replication and there is no reason to doubt that it performs a similar role in *E. coli*. It probably maintains DNA in a more open state under physiologic conditions during DNA replication in the cell. Evidence suggests that the Ssb protein may participate in a nucleosome like structure (a nucleoprotein complex that compacts the chromosome, analogous to eukaryotic chromatin), perhaps with *E. coli* HU protein. HU, like IHF and FIS, is a small DNA-binding protein that bends the double helix, thereby facilitating the action of other DNA-binding proteins such as DnaA. It is possible that the coating of DNA strands by Ssb protein protects against nucleolytic degradation during replication. Lastly, Ssb protein appears to stimulate the rate of synthesis of DNA polymerases under particular conditions. Whether or not this is the case during DNA replication is not clear.

The 5′→3′ exonuclease of polymerase I is essential in *E. coli*, and such an enzymatic activity meets the requirement for the elimination of leftover RNA primers on the lagging strand. After the replication fork has moved on, leaving a 3′ terminus of the newly synthesized Okazaki fragment adjacent to an RNA primer, polymerase I extends the 3′ terminus of the nascent DNA strand while digesting the RNA primer in a process called "nick translation." DNA ligase is required for joining the Okazaki pieces made during discontinuous DNA synthesis. This enzyme is also required for DNA replication because mutants conditionally defective in ligase are also conditionally defective in DNA synthesis.

In addition to the previously mentioned proteins, proteins such asDNA gyrase and DNA topoisomerase I (ω protein) may play a critical role during the microinitiation phase of DNA replication.

Genes *priA*, *-B*, and *-C* (for primosome), *dnaT*, and their products play a role in the assembly of the primosome structure and are required for replication of at least some single-stranded phages and plasmids. However, the effect on the cell of a deficiency is modest.

III. DNA STRAND SYNTHESIS

The genetics of DNA strand synthesis are reflected in the DNA polymerase. *Escherichia coli* is known to contain at least three distinct and separate DNA polymerases, all possessing the following enzymatic activities: synthesis exclusively in the 5′→3′ direction, utilization of 5′ deoxynucleoside triphosphates for substrates, the copying of single-stranded DNA template, incorporation of base analogs or ribonucleoside triphosphates at low efficiency under altered conditions (such as in the presence of manganese), a 3′ editorial exonuclease that preferentially removes mismatched 3′ termini, and a rate of synthesis that does not approach that of DNA replication in the cell (Table 28.1). Despite these similarities, distinct physical differences exist, and the cell uses exclusively DNA polymerase III for DNA replication. The synthesis subunit for DNA polymerase III, the α subunit, is encoded by the *polC* (*dnaE*) gene.

A. DNA polymerases

DNA polymerase I, encoded by the *polA* gene, appears to be an auxiliary protein for DNA replication. Cells lacking this enzyme demonstrate viability, although those lacking the notable 5′→3′ exonuclease activity of this enzyme are only partially viable unless grown in high salt. DNA polymerase I is very important for survival of the cell following many types of DNA damage, and in its absence the cell has persistent single-stranded breaks that promote DNA recombination. DNA polymerase I appears to be a particularly potent effector with DNA ligase in sealing single-stranded nicks, perhaps because of its ability to catalyze nick translation in which the 5′ exonucleolytic removal of bases is coupled to the synthesis activity. Neither of the other DNA polymerases appear to possess this property.

DNA polymerase II is an enzyme without a defined role in the cell. It has been cloned and overproduced and has been found to bear a closer relationship to T4 DNA polymerase and human polymerase alpha than to either of the other two *E. coli* DNA polymerases. Nevertheless, it is capable of interacting with the subunits of the DNA polymerase III complex. Cells that completely lack the structural gene for DNA polymerase II (the *polB* gene) show normal viability and normal repair after DNA damage in many circumstances. Among suggested roles is synthesis to bypass DNA damage.

DNA polymerase III is the required replicase of *E. coli*. The fact that it plays a significant role in DNA replication is demonstrated because *dnaEts* mutants contain a temperature-sensitive DNA polymerase III. Despite having properties similar to those of DNA polymerase I and II, DNA polymerase III is specifically required for DNA replication. This is a reflection of its ability to interact with a set of subunits that confer particular properties on the complex. In the complex (termed the holoenzyme) DNA polymerase III takes on the properties of a high rate of synthesis and great processivity. Intuitively, processivity may be thought of as the ability of an enzyme catalyzing the synthesis of DNA to remain tracking on one template for a long period of time before disassociating and initiating synthesis on another template. Highly processive enzymes are capable of synthesizing thousands of nucleotides at a single stretch before releasing the template. DNA polymerase III appears to be uniquely processive among the *E. coli* DNA polymerases.

B. Holoenzyme DNA polymerase III

In addition to the α-synthesis subunit, there are at least nine constituents of the DNA polymerase III holoenzyme complex (Table 28.2). Most of them have been shown to be the products of required genes as demonstrated by the fact that mutations in that gene produce conditional cessation of DNA replication (*dnaE, dnaQ, dnaN,* and *dnaX*) or that "knockout" mutants are inviable (*holA*, encoding the δ subunit, and *holB*, encoding the "d" subunit). A knockout mutation of *holE* did not impair cell viability, implying that the θ subunit is dispensable for normal growth. The γ protein and the τ protein appear to be products of the same *dnaX* gene. This is a case in which frame-shifting termination of protein synthesis plays a role in producing different proteins from the same gene. Mutants constructed with a frameshift in the *dnaX* gene that abolish production of γ but do not affect τ are viable; however, τ has been shown to be essential. The β subunit is the product of the *dnaN* gene. This subunit appears to confer specificity for primer utilization upon the complex and to increase the processivity. The ε protein of the holoenzyme complex is known to provide a powerful 3 editorial exonuclease activity. This is manifest by the fact that in addition to lethal mutants in this gene (*mutD*), mutants that show increased error rates in DNA replication (mutators) can be isolated.

Physical studies as well as genetic studies indicate that the DNA polymerase III holoenzyme complex exists in a dimer form. The stoichiometry of the various subunits suggests that the dimer is not exactly symmetrical, but it does appear to be symmetrical for the α β, and ε subunits. The holoenzyme comprises two dimerized β subunits (β_4), a dimeric core pol III′, ($\alpha_2\varepsilon_2\theta_2\tau_2$), and a single γ complex ($\gamma_2 d_1 d_1 \chi_1 \psi_1$) that appears to be involved in loading the β processivity

TABLE 28.2 Proteins involved in *E. coli* DNA replication

Protein	Subunit	Gene	Size	Function
DnaA		*dnaA*	58	Conformational change of DNA at *oriC*, macroinitiation
DnaB		*dnaB*	52	ATP-dependent DNA helicase, unwinding DNA at the replication fork
DnaC		*dnaC*	27	Recruitment of DnaB to *oriC*–DnaA complex
Gyrase	A	*gyrA*	96	GyrA52–GyrB^2 tetramer maintains the chromosome in a negatively supercoiled state, affecting global regulation of replication, and possibly local regulation of initiation at *oriC*
	B	*gyrB*	88	ATPase subunit of gyrase
Topoisomerase IV	A	*parC*	83	Decatenation of replicated chromosomes
	B	*parE*	70	Subunit of topo IV
Topoisomerase I		*topA*	110	Relaxes supercoils, affecting global regulation of replication
Ssb		*ssb*	19	Protects single-stranded DNA from nucleolytic degradation
DNA polymerase III	α	*dnaE*	129.9	DNA synthesis
	ε	*dnaQ*	27.5	$3' \rightarrow 5'$ proofreading exonuclease
	θ	*holE*	8.6	Stimulates ε
	τ	*dnaX*	71.1	Coordinates both halves of pol III holoenzyme by linking DnaB with pol III, interacts with primase
	γ	*dnaX*	47.5	Subunit of γ complex, β clamp loading; binds ATP
	δ	*holA*	38.7	Subunit of γ complex, interacts with β
	δ'	*holB*	36.9	Subunit of γ complex, cofactor of γ ATPase
	χ	*holC*	16.6	Subunit of γ complex, interacts with Ssb
	ψ	*holD*	15.2	Subunit of γ complex; links χ and γ
	β	*dnaN*	40.6	Processivity, sliding clamp
Primosome	Pri A (N′)	*priA*	80	Recognition and binding to primosome assembly site (PAS), ATPase, helicase
	Pri B (N)	*priB*	11.4	stabilization of Pri A–PAS interaction
	Pri C (N″)	*priC*	20.3	Primosome assembly
	Dna T (I)	*dnaT*	20	Primosome assembly
	Dna G (primase)	*dnaG* (*parB*)	64	Primase, synthesizes RNA primer, interacts with DnaB
Ter (Tus, Tau)		*tus*	34	Contrahelicase, blocks Dna B by binding to Ter sites
HU	HU-α	*HupA*	10	Histone-like protein, condenses DNA
	HU-β	*HupB*	10	Forms heterodimer with HU-α
FIS		*fis*	11	Regulates initiation at *oriC*, small DNA-binding protein, modulates transcription factors and inversion
IHF	IHF-α	*himA*	11	Modulates initiation at *oriC*, involved in site-specific recombination
	IHF-β	*himD*	10	Forms heterodimer with IHF-α
Polymerase I		*polA*	103	Eliminates Okazaki primers, DNA repair polymerase
DNA Ligase		*lig*	74	Joins DNA fragments during replication

clamp onto the DNA template. The physical and genetic evidence supporting dimerization of DNA polymerase III is in accordance with a structural model for replication. This is a so-called inch-worm, or trombone, model of DNA replication (Fig. 28.3). As indicated in the model, a dimer at the growing fork would allow coupling of rates of synthesis on the leading and lagging strands, i.e., the strand made continuously and the strand made discontinuously. Because the strand made discontinuously may require frequent initiation, one might expect synthesis of this nascent DNA strand to be slower. To prevent a discrepancy in growth rate between the strands, dimerization of the synthesis units for the two strands is a method for locking the rates in step. It is possible to do this by assuming that the DNA template for the lagging strand loops out in such a way as to provide permissible polarity for the nascent strand.

A key player in organizing the replisome (that is, the multi-protein complex that replicates the chromosome) appears to be the τ subunit, which has been shown to interact with both DnaB helicase and primase. The coupling of pol III to DnaB explains the high level of processivity on the leading strand while allowing the other half of the pol III holoenzyme to cycle on and off the lagging strand during microinitiation.

IV. TERMINATION OF REPLICATION

The termination of DNA replication is complicated by topological problems created by the circular nature of the bacterial chromosome. The double-helical structure of the template DNA, given a semiconservative mode of DNA replication, results in two interwound chromosomes which must be unlinked by topoisomerases. Decatenation of the newly replicated chromosomes is accomplished primarily by topoisomerase IV, a double-stranded DNA topoisomerase that closely resembles DNA gyrase (topoisomerase II). Another problem is that bidirectional replication on a circular template must be coordinated so that the opposing replication forks meet and terminate DNA

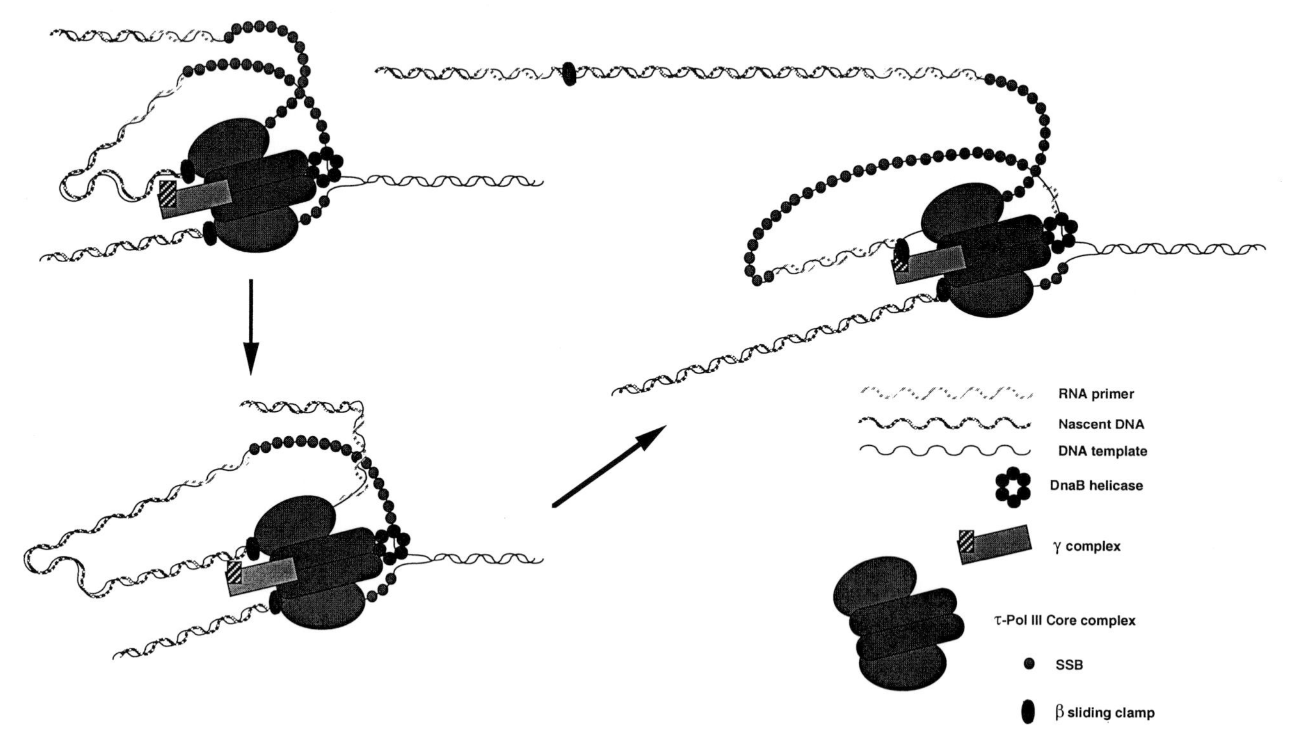

FIGURE 28.3 Fork progression in *E. coli*. The dimeric structure of the DNA polymerase III holoenzyme couples leading and lagging strand DNA synthesis during replication. Leading strand synthesis (bottom strand) is continuous and processive, but lagging strand synthesis must reinitiate many times during replication of the chromosome. This constraint is due to the antiparallel nature of DNA and the ability of DNA polymerases to synthesize DNA only in the $5' \rightarrow 3'$ direction. As the replication fork progresses, new primers are synthesized by primase on the lagging strand. The asymmetrical γ complex loads a new β dimer onto the primed DNA template, and then the β complex associates with the core polymerase to extend DNA synthesis from the $3'$ end of the primer. DNA synthesis on the lagging strand proceeds only as far as the previously replicated Okazaki fragment, at which point the β subunit, along with the nascent DNA, is released from the core polymerase, allowing the next cycle of synthesis to initiate at the next primer.

synthesis at a defined location. Otherwise, part of the genome might be overreplicated.

A specific region of the chromosome, approximately directly opposite the origin on the *E. coli* map at the 30- to 32-min region, represents the termination region. This region is particularly sparce in genetic markers. A specific terminator protein, ter (or Tus), binds to sequences called Ter (or τ). Ter sequences are arrayed in the termination region in an inverted repeat configuration such that binding of Tus protein to Ter sequences confers a polarity to blockage of the replication fork. Tus protein has been described as a polar contrahelicase because its ability to inhibit DnaB, the replication fork-specific helicase, depends on the polarity of the Ter sequence with respect to the origin of replication. A replication fork is unimpeded by Tus bound to a ter sequence in the forward orientation, but it is blocked by Tus bound to a ter sequence in the reverse orientation. Because of the inverted configuration of Ter sites in the termination region, a replication fork that progresses through a forward-oriented Ter site will be stopped at the next reverse-oriented site. The termination region thus "traps" replication forks, ensuring that part of the genome is not overreplicated. *Bacillus subtilis* has evolved a similar mechanism, with a termination protein, RTP, that binds to specific inversely-oriented repeats, also called Ter sequences. Of note is the fact that the *E. coli* Ter sites are widely separated by approximately 350 kb, whereas the innermost oppositely oriented Ter sites in *B. subtilis* are only 59 bp apart. The basic strategy for termination in the two organisms is quite similar, but the molecular mechanisms appear to be different. DNA replication termination may also be a control for cell division.

BIBLIOGRAPHY

Boye, E., Lobner-Olesen, A., and Skarstad, K. (2000). Limiting DNA replication to once and only once. *EMBO Re.* **1**, 479–483.

Bramhill, D., and Kornberg, A. (1988). A model for initiation at origins of DNA replication. *Cell* **54**, 915–918.

Davey, M. J., and O'Donnel, M. (2000). Mechanisms of DNA replication. *Curr. Opin. chem. Biol.* 4, 581–586.

Echols, H. (1986). Multiple DNA-protein interactions governing high-precision DNA transactions. *Science* **233**, 1050–1056.

Kelman, Z., and O'Donnell, M. (1995). DNA polymerase III holoenzyme: structure and function of a chromosomal replicating machine. *Annu. Rev. Biochem.* **64**, 171–200.

Manna, A. C., Karnire, S. P., Bussiere, D. E., Davies, C., White, S. W., and Bastia, D. (1996). Helicase-contrahelicase interaction and the mechanism of termination of DNA replication. *Cell* **87**, 881–891.

Marians, K. J. (1996). Replication for propagation. *In "Escherichia coli* and *Salmonella*: Cellular and Molecular Biology" (R. Curtis, III, *et al.*, Eds.), 2nd edn, pp. 749–763. ASM, Washington, DC.

Marians, K. J. (2000). Replication and recombination intersect. *Curr. Opin. Genet. Dev.* **10**, 151–156.

McHenry, C. S. (1988). DNA polymerase III holoenzyme of *Escherichia coli. Annu. Rev. Biochem.* **57**, 519–550.

Peng, H., and Marians, K. J. (1993). Decatenation activity of topoisomerase IV during *oriC* and pBR322 DNA replication *in vitro. Proc. Natl. Acad. Sci. USA* **90**, 8571–8575.

Yuzhakov, A., Turner, J., and O'Donnell, M. (1996). Replisome assembly reveals the basis for asymmetric function in leading and lagging strand replication. *Cell* **86**, 877–886.

Zavitz, K. H., and Marians, K. J., (1991). Dissecting the functional role of PriA protein-catalysed primosome assembly in *Escherichia coli* DNA replication. *Mol. Microbiol.* **5**, 2869–2873.

WEBSITE

Animation of DNA replication
http://www.microbelibrary.org/FactSheet.asp?SubmissionID=404

29

DNA restriction and modification

Noreen E. Murray
University of Edinburgh

GLOSSARY

ATP and ATP hydrolysis Adenosine triphosphate (ATP) is a primary repository of energy that is released for other catalytic activities when ATP is hydrolyzed (split) to yield adenosine diphosphate.

bacteriophages (lambda and T-even) Bacterial viruses. Phage lambda (λ) is a temperate phage and therefore on infection of a bacterial cell one of two alternative pathways may result; either the lytic pathway in which the bacterium is sacrificed and progeny phages are produced or the temperate (lysogenic) pathway in which the phage genome is repressed and, if it integrates into the host chromosome, will be stably maintained in the progeny of the surviving bacterium. Phage λ was isolated from *Escherichia coli* K-12 in which it resided in its temperate (prophage) state. T-even phages (T2, T4, and T6) are virulent coliphages, that is, infection of a sensitive strain of *E. coli* leads to the production of phages at the inevitable expense of the host. T-even phages share the unusual characteristic that their DNA includes hydroxymethylcytosine rather than cytosine.

conjugation, conjugational transfer Gene transfer by conjugation requires cell-to-cell contact. Conjugative, or self-transmissible, plasmids such as the F factor of *E. coli* encode the necessary functions to mobilize one strand of their DNA with a defined polarity from an origin of transfer determined by a specific nick. The complementary strand is then made in the recipient cell. Some plasmids are transmissible but only on provision *in trans* of the necessary functions by a conjugative plasmid. A conjugative plasmid, such as the F factor, can mobilize transfer of the bacterial genome following integration of the plasmid into the bacterial chromosome.

DNA methyltransferases Enzymes (MTases) that catalyze the transfer of a methyl group from the donor *S*-adenosylmethionine to adenine or cytosine residues in the DNA.

efficiency of plating This usually refers to the ratio of the plaque count on a test strain relative to that obtained on a standard, or reference, strain.

endonucleases Enzymes that can fragment polynucleotides by the hydrolysis of internal phosphodiester bonds.

Escherichia coli strain K-12 The strain used by Lederberg and Tatum in their discovery of recombination in *E. coli*.

glucosylation of DNA The DNA of T-even phages in addition to the pentose sugar, deoxyribose, contains glucose attached to the hydroxymethyl group of hydroxymethyl-cytosine. Glucosylation of the DNA is mediated by phage-encoded enzymes, but the host provides the glucose donor.

helicases Enzymes that separate paired strands of polynucleotides.

recombination pathway The process by which new combinations of DNA sequences are generated. The general recombination process relies on enzymes that use DNA sequence homology for the recognition of the recombining partner. In the major pathway in *E. coli*, RecBCD generates the DNA strands for transfer and RecA protein promotes synapsis. The RecBCD enzyme, also recognized as exonuclease V, enters DNA via a double-strand end.

The Desk Encyclopedia of Microbiology
ISBN: 0-12-621361-5

It tracks along the DNA, promoting unwinding of the strands but, in response to a special nucleotide sequence termed Chi, it makes a nick, following which strand separation continues and the single-stranded DNA with a 3′ end becomes available for synapsis with homologous DNA. *In vitro*, the mechanism by which RecBCD makes a DNA end is determined by the reaction conditions; its activity as a potent exonuclease is influenced by the relative concentrations of ATP and Mg ions.

SOS response DNA damage induces expression of a set of genes, the SOS genes, involved in the repair of DNA damage.

Southern transfer The transfer of denatured DNA from a gel to a solid matrix, such as a nitrocellulose filter, within which the denatured DNA can be maintained and hybridized to labeled probes (single-stranded DNA or RNA molecules). Fragments previously separated by electrophoresis through a gel may be identified by hybridization to a specific probe.

transformation The direct assimilation of DNA by a cell, which results in the recipient being changed genetically.

Awareness of the Biological Phenomenon of Restriction and Modification (R–M) grew from the observations of microbiologists that the host range of a bacterial virus (phage) was influenced by the bacterial strain in which the phage was last propagated. Although phages produced in one strain of *Escherichia coli* would readily infect a culture of the same strain, they might only rarely achieve the successful infection of cells from a different strain of *E. coli*. This finding implied that the phages carried an "imprint" that identified their immediate provenance. Simple biological tests showed that the occasional successful infection of a different strain resulted in the production of phages that had lost their previous imprint and had acquired a new one.

In the 1960s, elegant molecular experiments showed the "imprint" to be a DNA modification that was lost when the phage DNA replicated within a different bacterial strain; those phages that conserved one of their original DNA strands retained the imprint, or modification, whereas phages containing two strands of newly synthesized DNA did not. The modification was shown to provide protection against an endonuclease, the barrier that prevented the replication of incoming phage genomes. Later it was proven that the modification and restriction enzymes both recognized the same target, a specific nucleotide sequence. The modification enzyme was a DNA methyltransferase that methylated specific bases within the target sequence, and in the absence of the specific methylation the target sequence rendered the DNA sensitive to the restriction enzyme. When DNA lacking the appropriate modification imprint enters a restriction-proficient cell it is recognized as foreign and degraded by the endonuclease. The host-controlled barrier to successful infection by phages that lacked the correct modification was referred to as "restriction" and the relevant endonucleases have acquired the colloquial name of restriction enzymes. Similarly, the methyltransferases are more commonly termed modification enzymes. Classically, a restriction enzyme is accompanied by its cognate modification enzyme and the two comprise a R–M system. Most restriction systems conform to this classical pattern. There are, however, some restriction endonucleases that attack DNA only when their target sequence is modified. A restriction system that responds to its target sequence only when it is identified by modified bases does not, therefore, co-exist with a cognate modification enzyme.

Two early papers documented the phenomenon of restriction. In one, Bertani and Weigle (1953), using temperate phages (λ and P2), identified the classical restriction and modification systems characteristic of *E. coli* K-12 and *E. coli* B. In the other, Luria and Human (1952) identified a restriction system of the second, nonclassical kind. In the experiments of Luria and Human, T-even phages were used as test phages, and after their growth in a mutant *E. coli* host they were found to be restricted by wild-type *E. coli* K-12 but not by *Shigella dysenteriae*. An understanding of the restriction phenomenon observed by Luria and Human requires knowledge of the special nature of the DNA of T-even phages. During replication of T-even phages the unusual base 5-hydroxymethyl-cytosine (HMC) completely substitutes for cytosine in the T-phage DNA, and the hydroxymethyl group is subsequently glucosylated in a phage-specific pattern at the polynucleotide level. In the mutant strain of *E. coli* used by Luria and Human as host for the T-even phages, glucosylation fails and, in its absence, the nonglucosylated phage DNA becomes sensitive to an endonuclease present in *E. coli* K-12 but not in *S. dysenteriae*. Particular nucleotide sequences normally protected by glu cosylation are recognized by an endonuclease in *E. coli* K-12 when they include the modified base, HMC, rather than cytosine residues. In the T-phage experiments the modified base is hydroxymethyl-cytosine, but much later it was discovered that methylated cytosine residues can also evoke restriction by the same endonucleases.

The classical (R–M) systems and the modification-dependent restriction enzymes share the potential to attack DNA derived from different strains and thereby "restrict" DNA transfer. They differ in that in one case an associated modification enzyme is required to protect DNA from attack by the cognate restriction enzyme and in the other modification enzymes specified by different strains impart signals that provoke the degradative activity of restriction endonucleases.

I. DETECTION OF RESTRICTION SYSTEMS

A. As a barrier to gene transfer

This is exemplified by the original detection of the R–M systems of *E. coli* K-12 and *E. coli* B by Bertani and Weigle in 1953. Phage λ grown on *E. coli* strain C (λ . C), where *E. coli* C is a strain that apparently lacks an R–M system, forms plaques with poor efficiency efficiency of plating (EOP) of 2×10^{-4} on *E. coli* K-12 because the phage DNA is attacked by a restriction endonuclease (Fig. 29.1). Phage λ grown on *E. coli* K-12 (λ K) forms plaques with equal efficiency on *E. coli* K-12 and *E. coli* C since it has the modification required to protect against the restriction system of *E. coli* K-12 and *E. coli* C has no restriction system (Fig. 29.1). In contrast, λ K will form plaques with very low efficiency on a third strain, *E. coli* B, since *E. coli* B has an R–M system with different sequence specificity from that of *E. coli* K-12.

Phages often provide a useful and sensitive test for the presence of R–M systems in laboratory strains of bacteria, but they are not a suitable vehicle for the general detection of barriers to gene transfer. Many bacterial strains, even within the same species and particularly when isolated from natural habitats, are unable to support the propagation of the available test phages, and some phages (e.g. P1) have the means to antagonize at least some restriction systems (see Section IV.C). Gene transfer by conjugation can monitor restriction although some natural plasmids, but probably not the F factor of *E. coli*, are equipped with antirestriction systems. The single-stranded DNA that enters a recipient cell by conjugation, or following infection by a phage such as M13, becomes sensitive to restriction only after the synthesis of its complementary second strand, whereas the single-stranded DNA that transforms naturally competent bacteria may not become a target for restriction because it forms heteroduplex DNA with resident (and therefore modified) DNA and one modified strand is sufficient to endow protection. Transformation can be used to detect restriction systems, but only when the target DNA is the double-stranded DNA of a plasmid.

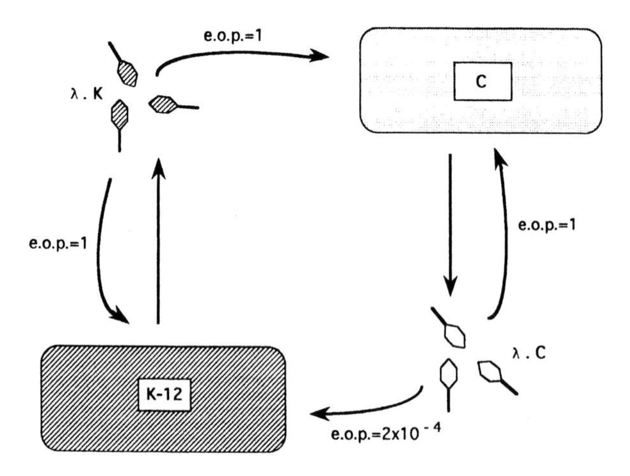

FIGURE 29.1 Host-controlled restriction of bacteriophage λ. *Escherichia coli* K-12 possesses, whereas *E. coli* C lacks, a Type I R–M system. Phage λ propagated in *E. coli* C (λ . C) is not protected from restriction by *Eco* KI and thus forms plaques with reduced efficiency of plating (EOP) on *E. coli* K-12 as compared to *E. coli* C. Phages escaping restriction are modified by the *Eco* KI methyltransferase (λ . K) and consequently form plaques with the same efficiency on *E. coli* K-12 and C. Modified DNA is indicated by hatch marks (reproduced with permission from Barcus and Murray, 1995).

B. *In vitro* assays for DNA fragmentation

Endonuclease activities yielding discrete fragments of DNA are commonly detected in crude extracts of bacterial cells. More than one DNA may be used to increase the chance of providing a substrate that includes target sequences. DNA fragments diagnostic of endonuclease activity are separated according to their size by electrophoresis through a matrix, usually an agarose gel, and are visualized by the use of a fluorescent dye, ethidium bromide, that intercalates between stacked base pairs.

Extensive screening of many bacteria, often obscure species for which there is no genetic test, has produced a wealth of endonucleases with different target sequence specificities. These endonucleases are referred to as restriction enzymes, even in the absence of biological experiments to indicate their role as a barrier to the transfer of DNA. Many of these enzymes are among the commercially available endonucleases that serve molecular biologists in the analysis of DNA (Table 29.1; see Section VII). *In vitro* screens are applicable to all organisms, but to date restriction and modification systems have not been found in eukaryotes, although some algal viruses encode them.

C. Sequence-specific screens

The identification of new R–M genes via sequence similarities is sometimes possible. Only occasionally are gene sequences sufficiently conserved that the

TABLE 29.1 Some Type II restriction endonucleases and their cleavage sites[a]

Bacterial source	Enzyme abbreviation	Sequences 5′ → 3′ 3′ ← 5′	Note[b]
Haemophilus influenzae Rd	*Hind*II	**GTPy ↓ PuAC** **CAPu ↑ PyTG**	**1, 5**
	*Hind*III	↓ **AAGCTT** **TTCGAA** ↑	**2**
Haemophilus aegyptius	*Hae*III	**GG ↓ CC** **CC ↑ GG**	**1**
Staphylococcus aureus 3A	*Sau*3AI	**↓ GATC** **CTAG ↑**	**2, 3**
Bacillus amyloliquefaciens II	*Bam*HI	↓ **GGATCC** **CCTAGG** ↑	**2, 3**
Escherichia coli RY13	*Eco*RI	↓ **GAATTC** **CTTAAG** ↑	**2**
Providencia stuartii	*Pst*I	↓ **CTGCAG** **GACGTC** ↑	**4**

[a] The cleavage site for each enzyme is shown by the arrows within the target sequence.
[b] 1, produces blunt ends; 2, produces cohesive ends with 5′ single-stranded overhangs; 3, cohesive ends of *Sau*3AI and *Bam*HI are identical; 4, produces cohesive ends with 3′ single-stranded overhangs; 5, Pu is any purine (A or G), and Py is any pyrimidine (C or T).

presence of related systems can be detected by probing Southern transfers of bacterial DNA. Generally, screening databases of predicted polypeptide sequences for relevant motifs has identified putative R–M systems in the rapidly growing list of bacteria for which the genomic sequence is available. Currently, this approach is more dependable for modification methyltransferases than for restriction endonucleases, but the genes encoding the modification and restriction enzymes are usually adjacent. Many putative R–M systems have been identified in bacterial genomic sequences.

II. NOMENCLATURE AND CLASSIFICATION

A. Nomenclature

R–M systems are designated by a three-letter acronym derived from the name of the organism in which they occur. The first letter is derived from the genus and the second and third letters from the species. The strain designation, if any, follows the acronym. Different systems in the same organism are distinguished by Roman numerals. Thus, *Hind*II and *Hind*III are two enzymes from *Hemophilus influenzae* strain Rd. Restriction endonuclease and modification methyltransferases (ENases and MTases) are sometimes distinguished by the prefixes R. *Eco*RI and M. *Eco*RI, but the prefix is commonly omitted if the context is unambiguous.

B. Classification of R–M systems

R–M systems are classified according to the composition and cofactor requirements of the enzymes, the nature of the target sequence, and the position of the site of DNA cleavage with respect to the target sequence. Currently, three distinct, well- characterized types of classical R–M systems are known (I–III), although a few do not share all the characteristics of any of these three types. In addition, there are modification-dependent systems. Type I systems were identified first, but the Type II systems are the simplest and for this reason will be described first. A summary of the properties of different types of R–M systems is given in Fig. 29.2.

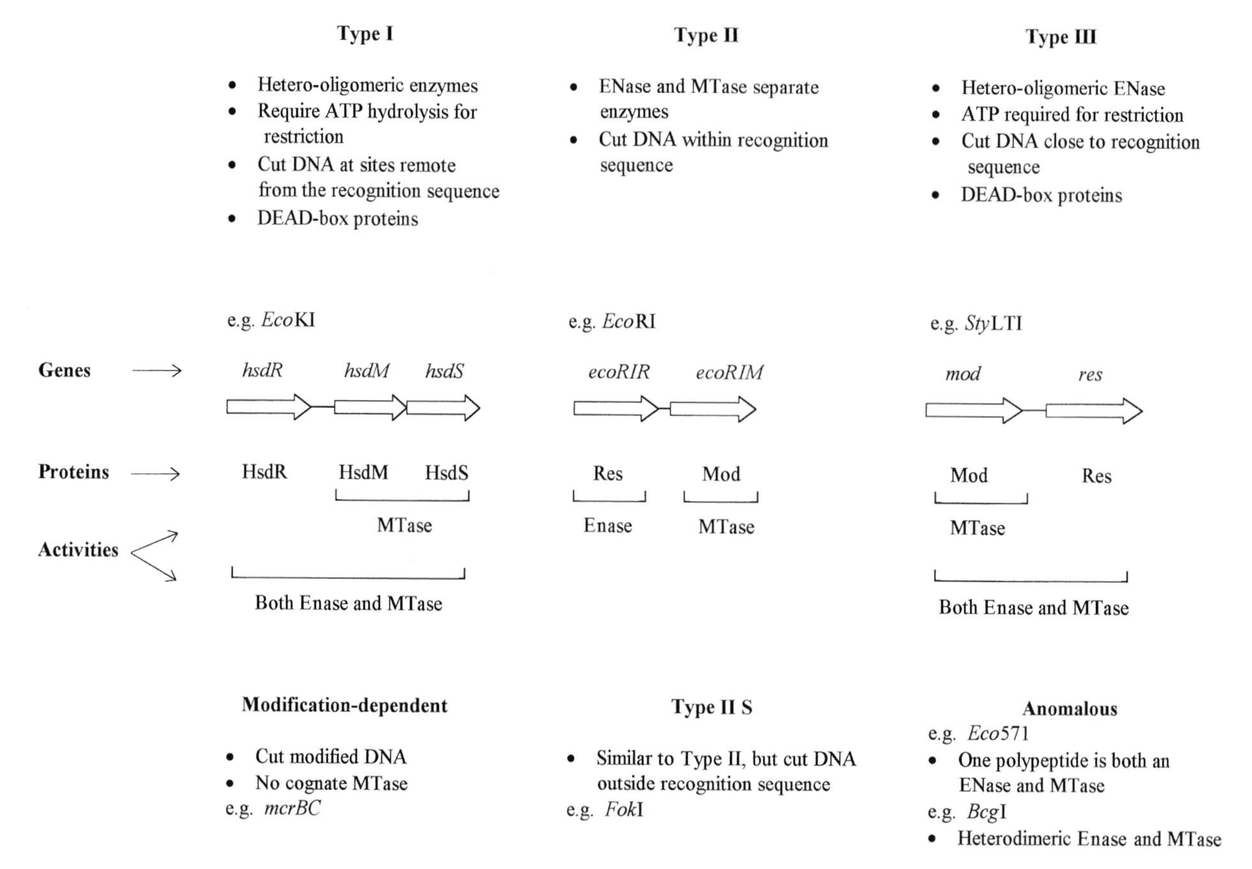

FIGURE 29.2 The distinguishing characteristics and organization of the genetic determinants and subunits of the different types of restriction and modification (R–M) systems. R–M systems are classified on the basis of their complexity, cofactor requirements, and position of DNA cleavage with respect to their DNA target sequence. The Types I–III systems are the classical R–M systems. The restriction enzymes of the Types I and III systems contain motifs characteristic of DEAD-box proteins and may therefore be helicases. Helicase activity could be associated with DNA translocation. The Type II S systems are a subgroup of the Type II systems that cleave DNA outside their recognition sequence. Some systems do not fit readily into the current classification (See Roberts *et al.*, 2003 for an updated classification of anomolous systems). The modification-dependent systems are not classical R–M systems, because they have no cognate methylase and only cut DNA that contains certain modifications. ENase, restriction endonuclease; MTase, methyltransferase (reprinted from *Trends in Microbiology* **2** (12), G. King and N. E. Murray, pp. 461–501. Copyright 1995, with permission from Elsevier Science).

C. Type II R–M systems

A Type II R–M system comprises two separate enzymes; one is the restriction ENase and the other the modification MTase. The nuclease activity requires Mg^{2+}, and DNA methylatièon requires *S*-adenosylmethionine (AdoMet) as methyl donor. The target sequence of both enzymes is the same; the modification enzyme ensures that a specific base within the target sequence, one on each strand of the duplex, is methylated and the restriction endonuclease cleaves unmodified substrates within, or close to, the target sequence. The target sequences are usually rotationally symmetrical sequences of from 4 to 8 bp; for example, a duplex of the sequence 5′ -GAA*TTC is recognized by *Eco*RI. The modification enzyme methylates the adenine residue identified by the asterisk, but in the absence of methylated adenine residues on both strands of the target sequence the restriction endonuclease breaks the phosphodiester backbones of the DNA duplex to generate ends with 3′ hydroxyl and 5′ phosphate groups. The Type II systems can be subdivided according to the nature of the modification introduced by the MTase: N6-methyladenine (m6A) and N5 and N4 methylcytosine (m5C and m4C). Irrespective of the target sequence or the nature of the modification, ENases differ in that some cut the DNA to generate ends with 5′ overhangs, some generate 3′ overhangs, and others produce ends which are "blunt" or "flush" (Table 29.1).

Type II restriction enzymes are generally active as symmetrically arranged homodimers, an association that facilitates the coordinated cleavage of both strands of the DNA. In contrast, Type II modification enzymes act as monomers, an organization consistent with their normal role in the methylation of newly replicated DNA in which one strand is already methylated.

The genes encoding Type II R–M systems derive from the name of the system. The genes specifying R. *Bam*HI and M. *Bam*HI, for example, are designated *bamHIR* and *bamHIM*. Transfer of the gene encoding a restriction enzyme in the absence of the transfer of the partner encoding the protective MTase is likely to be lethal if the recipient cell does not provide the relevant protection. Experimental evidence supports the expectation that the genes encoding the two components of R–M systems are usually closely linked so that cotransfer will be efficient.

A subgroup of Type II systems, Type IIS, recognizes asymmetric DNA sequences of 4–7 bp in length. These ENases cleave the DNA at a precise but short distance outside their recognition sequence; their name is derived from their shifted (S) position of cutting. Type IIS systems have simple cofactor requirements and comprise two separate enzymes, but they differ from classical Type II systems in the recognition of an asymmetric DNA sequence by a monomeric ENase, and require two MTase activities, but see Roberts *et al.* (2003) for the current classification of unusual Type II R–M systems.

D. Type I R–M systems

Type I R–M systems are multifunctional enzymes comprising three subunits that catalyze both restriction and modification. In addition to Mg^{2+}, endonucleolytic activity requires both AdoMet and adenosine triphosphate (ATP). The restriction activity of Type I enzymes is associated with the hydrolysis of ATP, an activity that may correlate with the peculiar characteristic of these enzymes—that of cutting DNA at non-specific nucleotide sequences at considerable distances from their target sequences. The Type I R–M enzyme binds to its target sequence and its activity as an ENase or a MTase is determined by the methylation state of the target sequence. If the target sequence is unmodified, the enzyme, while bound to its target site, is believed to translocate (move) the DNA from both sides toward itself in an ATP-dependent manner. This translocation process brings the bound enzymes closer to each other and experimental evidence suggests that DNA cleavage occurs when translocation is impeded, either by collision with another translocating complex or by the topology of the DNA substrate.

The nucleotide sequences recognized by Type I enzymes are asymmetric and comprise two components, one of 3 or 4 bp and the other of 4 or 5 bp, separated by a non-specific spacer of 6–8 bp. All known Type I enzymes methylate adenine residues, one on each strand of the target sequence.

The subunits of a Type I R–M enzyme are encoded by three genes: *hsdR*, *hsdM*, and *hsdS*. The acronym *hsd* was chosen at a time when R–M systems were referred to as host specificity systems, and *hsd* denotes *host* specificity of *D*NA. The three *hsd* genes are usually contiguous, but not always in the same order. *hsdM* and *hsdS* are transcribed from the same promoter, but *hsdR* is from a separate one. The two subunits encoded by *hsdM* and *hsdS*, sometimes referred to as M and S, are both necessary and sufficient for MTase activity. The third subunit (R) is essential only for restriction. The S (specificity) subunit includes two target recognition domains (TRDs) that impart target sequence specificity to both restriction and modification activities of the complex; the M subunits include the binding site for AdoMet and the active site for DNA methylation. Two complexes of Hsd subunits are functional in bacterial cells: one comprises all three subunits ($R_2M_2S_1$) and is an R–M system, and the other lacks R (M_2S_1) and has only Mtase activity.

E. Type III R–M systems

Type III R–M systems are less complex than Type I systems but nevertheless share some similarities with them. A single heterooligomeric complex catalyzes both restriction and modification activities. Modification requires the cofactor AdoMet and is stimulated by Mg^{2+} and ATP. Restriction requires Mg^{2+} and ATP and is stimulated by AdoMet. The recognition sequences of Type III enzymes are asymmetric sequences of 5 or 6 bp. Restriction requires two unmodified sequences in inverse orientation (Fig. 29.3a). Recent evidence indicates that Type III R–M enzymes, like those of Type I, can translocate DNA in a process dependent on ATP hydrolysis, but they hydrolyze less ATP than do Type I systems and probably only translocate DNA for a relatively short distance. Cleavage is stimulated by collision of the translocating complexes and occurs on the 3′ side of the recognition sequence at a distance of approximately 25–27 bp. This contrasts with cleavage by Type I enzymes in which cutting occurs at sites remote from the recognition sequence. Because only one strand of the recognition sequence of a Type III R–M system is a substrate for methylation, it might be anticipated that the immediate product of replication would be sensitive to restriction. It is necessary to distinguish the target for modification from that needed for restriction in order to understand why this is not so. Restriction is only elicited when two unmethylated target sequences are in inverse orientation with respect to each other and, as shown in Fig. 29.3b, replication of modified DNA leaves all unmodified targets in the

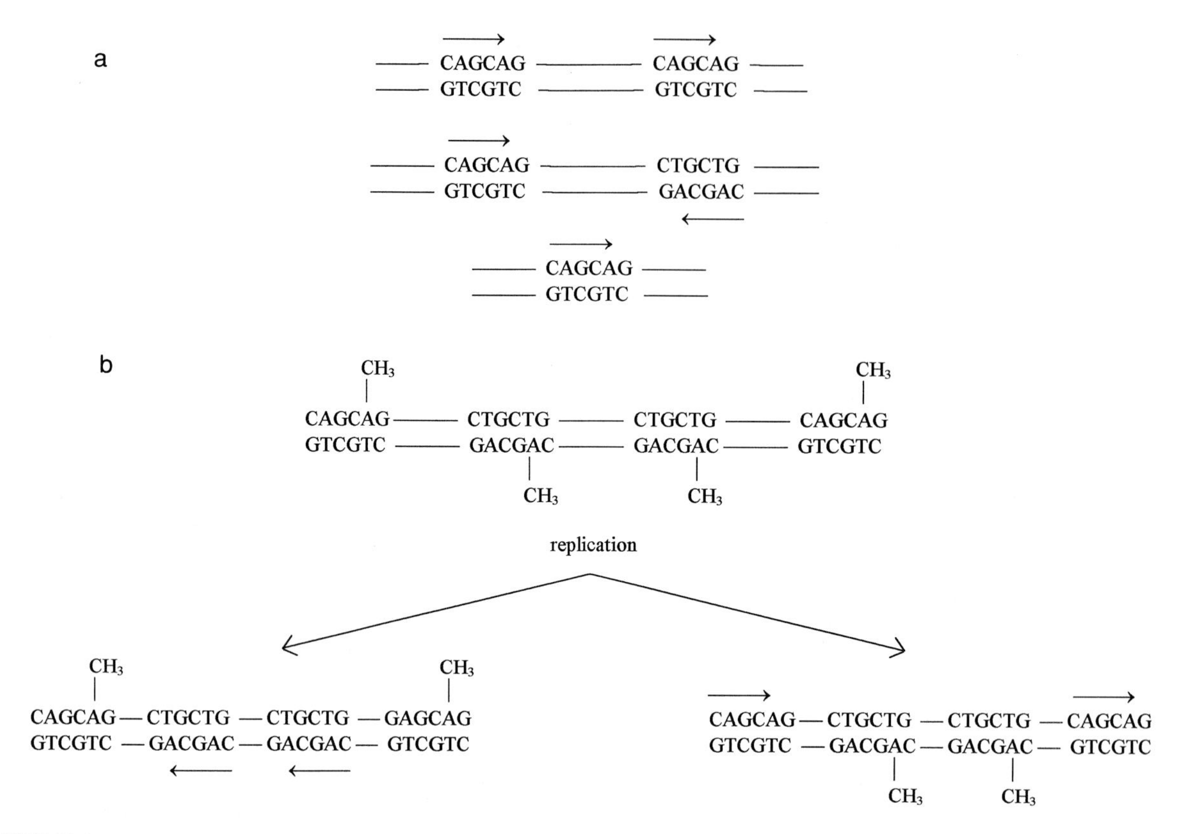

FIGURE 29.3 DNA substrates for a Type III R–M system (*Eco*P15I). The top strand of each duplex is written 5′ to 3′ the arrows identify the orientation of the target sequences. Solid lines indicate polynucleotide chains of undefined sequence. (a) Only pairs of target sequences shown in inverse orientation (line 2) are substrates for restriction. A single site in any orientation is a target for modification. (b) Replication of modified DNA leaves all unmodified target sites in the daughter molecules in the same orientation, and therefore insensitive to restriction.

same orientation. When two targets are appropriately positioned in close proximity, cutting can be elicited in the absence of DNA translocation.

The bifunctional R–M complex is made up of two subunits, the products of the *mod* and *res* genes. The Mod subunit is sufficient for modification, whereas the Res and Mod subunits together form a complex with both activities (Fig. 29.2). The Mod subunit is functionally equivalent to the MTase (M_2S) of Type I systems and, as in Type I R–M systems, imparts sequence specificity to both activities.

F. Modification-dependent restriction systems

These systems only cut modified DNA. They are variable in their complexity and requirements. The simplest is *Dpn*I from *Streptococcus pneumoniae* (previously called *Diplococcus pneumoniae*). The ENase is encoded by one gene, and the protein looks and behaves like a Type II enzyme except that it only cuts its target sequence when it includes methylated adenine residues.

Escherichia coli K-12 encodes three other distinct, sequence specific, modification-dependent systems. Two, Mrr and McrA, are specified by single genes. Mrr is distinguished by its ability to recognize DNA containing either methylated adenine or 5-methylcytosine in the context of particular, but undefined, sequences. McrA and McrBC both restrict DNA containing modified cytosines (hydroxymethylcytosine or methylcytosine). The *mcr* systems (*m*odified *c*ytosine *r*estriction) are those first recognized by Luria and Human by their ability to restrict nonglucosylated T-even phages (*rglA* and *B*; *r*estricts *gl*ucose-*l*ess phage). McrBC is a complex enzyme with a requirement for GTP rather than ATP. McrBC, like the ATP-dependent restriction enzymes, translocates DNA.

G. Other systems

As more R–M systems are identified, new enzymes with novel properties continue to be found. *Eco*571 and *Bcg*I, for example, are most similar to Type II systems, but are not ideally suited to this classification (Fig. 29.2). *Eco*571 comprises a joint ENase–MTase and a separate MTase. The former behaves as if it were a Type IIS endonuclease fused to a MTase; it cleaves to one side of the target sequence and methylates the sequence on one strand. Cleavage is stimulated by AdoMet but not by ATP. The separate MTase behaves like a Type II modification enzyme.

III. R–M ENZYMES AS MODEL SYSTEMS

A. Enzyme activities, sequence recognition, including base flipping

Structures of the crystals of several Type II restriction ENases have been determined, some in both the presence and the absence of DNA. The symmetrically arranged dimers of the Type II enzymes interact with their specific target sequences by the combined effects of different types of interactions including hydrogen bonding and electrostatic interactions of amino acid residues with the bases and the phosphate backbone of the DNA. No general structure, such as a helix-turn-helix or zinc finger (often found in proteins that interact with DNA) defines the sequence—specific interactions with the DNA, and amino acids that are widely separated in the primary sequence may be involved in interactions with the target nucleotide sequence. Three-dimensional structures of Type II restriction endonucleases show that their catalytic centers share a common structural core comprising four β-strands and an α-helix. Comparisons of the active sites of *Eco*RV, *Eco*RI, and *Pvu*II identify a conserved tripeptide sequence close to the target phosphodiester group and a conserved acidic dipeptide that represent the ligands for the catalytic cofactor Mg^{2+} essential for ENase activity. These two sequences are also relevant to the active site for Type I restriction endonucleases.

The structure of a monomeric MTase interacting with its target sequence provided an important solution to the question of how enzymes that modify a base within a DNA molecule can reach their substrate. The cocrystal structure of M. *Hha*I bound to its substrate showed that the target cytidine rotates on its sugar–phosphate bonds such that it projects out of the DNA and fits into the catalytic pocket of the enzyme. Such base flipping was confirmed for a second enzyme, M. *Hae*III, which also modifies cytosine, and circumstantial evidence supports the notion that this mechanism may be true for all MTases regardless of whether they methylate cytosine or adenine residues.

Comparative analyses of the amino acid sequences of many MTases identified a series of motifs, many of which are common to MTases irrespective of whether the target base is cytosine or adenine. These motifs enable structural predictions to be made about the catalytic site for DNA methylation in complex enzymes for which crystals are not available.

B. DNA translocation

Specific interactions of large R–M enzymes with their DNA substrates are not readily amenable to structural analysis. The relative molecular weight of *Eco*KI is in excess of 400,000 and useful crystals have not been obtained. Nevertheless, these complex enzymes have other features of mechanistic interest. Much evidence supports models in which DNA restriction involves the translocation of DNA in an ATP-dependent process prior to the cutting of the substrate. In the case of Type I R–M enzymes, the breaks in the DNA may be many kilobases remote from the target sequence. Molineux *et al.* (See Murray, 2002), using assays with phages, have shown that *Eco*KI can transfer (translocate) the entire genome (39 kb) of phage T7 from its capsid to the bacterial cell. For linear DNA, the evidence supports the idea that cutting by Type I R–M systems occurs preferentially midway between two target sequences. For Type III enzymes the breaks are close to the target sequence, but in both cases the endonuclease activity may be stimulated by the collision of two translocating protein complexes.

The most conserved features of the polypeptide sequences of Type I and Type III R–M systems are the motifs characteristic of adenine MTases and the so-called DEAD-box motifs found in RNA and DNA helicases. The latter motifs acquired their collective name because a common variant of one element is Asp-Glu-Ala-Asp, or DEAD when written in a single-letter code. The DEAD-box motifs, which include sequences diagnostic of ATP binding, are found in the subunit that is essential for restriction (HsdR or Res) but not for modification. It is not known whether an ATP-dependent helicase activity drives the translocation of DNA, although circumstantial evidence correlates ATPase activity with DNA translocation. Mutations in each DEAD-box motif have been shown to impair the ATPase and endonuclease activities of a Type I ENase.

IV. CONTROL AND ALLEVIATION OF RESTRICTION

A. Control of gene expression

The expression of genes coding for R–M systems requires careful regulation. Not only is this essential to maintain the protection of host DNA in restriction-proficient cells but also it is especially important when R–M genes enter a new host. Experiments show that many R–M genes are readily transferred from one laboratory strain to another. The protection of host DNA against the endonucleolytic activity of a newly acquired restriction system can be achieved if the functional cognate MTase is produced before the restriction enzyme. Representatives of all three types of classical R–M systems have been shown to be equipped with promoters that might permit transcriptional regulation of the two activities.

Transcriptional regulation of some of the genes encoding Type II systems has been demonstrated. Genes encoding repressor-like proteins, referred to as C-proteins for controlling proteins, have been identified in some instances. The C-protein for the *Bam*HI system has been shown to activate efficient expression of the restriction gene and modulate the expression of the modification gene. When the R–M genes are transferred to a new environment, in the absence of C-protein there will be preferential expression of the modification gene, and only after the production of the C-protein will the cells become restriction proficient.

For complex R–M systems, despite the presence of two promoters, there is no evidence for transcriptional regulation of gene expression. The hetero-oligomeric nature of these systems presents the opportunity for the regulation of the restriction and modification activities by the intracellular concentrations of the subunits and the affinities with which different subunits bind to each other. Nevertheless, efficient transmission of the functional R–M genes of some Type I systems requires ClpX and ClpP in the recipient cell. Together these proteins comprise a protease, but ClpX has chaperone activity (i.e. an ability to help other proteins to fold correctly). These proteins function to permit the acquisition of new R–M systems; they degrade the HsdR subunit of active R–M complexes before the endonuclease activity has the opportunity to cleave unmodified DNA.

B. Restriction alleviation

The efficiency with which a bacterial cell restricts unmodified DNA is influenced by a number of stimuli, all of which share the ability to damage DNA. DNA damage provokes the induction of the SOS response, and one consequence of this is a marked reduction in the efficiency with which *E. coli* K-12 restricts incoming DNA. This alleviation of restriction is usually monitored by following the EOP of phages—unmodified in the case of classical systems or modified in the case of modification-dependent restriction systems. Alleviation of restriction is characteristic of complex systems and can be induced by ultraviolet light, nalidixic-acid, 2-aminopurine, and the absence of Dam-mediated methylation. The effect can be appreciable and a variety of host systems contribute to more than one pathway of alleviation. Recent experiments have shown for *Eco*KI that ClpXP is necessary for restriction alleviation; therefore there appears to be a connection between the complex mechanisms by which restriction activity is normally controlled and its alleviation in response to DNA damage.

C. Antirestriction systems

Many phages and some conjugative plasmids specify functions that antagonize restriction. An apparent bias of functions that inhibit restriction by Type I R–M systems may reflect the genotype of the classical laboratory strain *E. coli* K-12, which is a strain with a Type I but no Type II R–M system.

The coliphages T3 and T7 include an "early" gene, *O.3*, or *ocr*, the product of which binds Type I R–M enzymes and abolishes both restriction and modification activities. *O.3* protein does not affect Type II systems. The *O.3* gene is expressed before targets in the phage genome are accessible to host restriction enzymes so that $O.3^+$ phages are protected from restriction and modification by Type I systems. Phage T3 *O.3* protein has an additional activity; it hydrolyzes AdoMet, the cofactor essential for both restriction and modification by *Eco*KI and its relatives. Bacteriophage P1 also protects its DNA from Type I restriction, but the antirestriction function Dar does not interfere with modification. The Dar proteins are coinjected with encapsidated DNA so that any DNA packaged in a P1 head is protected. This allows efficient generalized transduction to occur between strains with different Type I R–M systems.

Coliphage T5 has a well-documented system for protection against the Type II system *Eco*RI. As with the *O.3* systems of T3 and T7, the gene is expressed early when the first part of the phage genome enters the bacterium. This first segment lacks *Eco*RI targets, whereas the rest of the genome, which enters later, has targets that would be susceptible in the absence of the antirestriction protein.

Some conjugative plasmids of *E. coli*, members of the incompatibility groups I and N, also encode antirestriction functions. They are specified by the *ard* genes located close to the origin of DNA transfer by conjugation so that they are among the first genes to be expressed following DNA transfer. Like the *O.3* proteins of T3 and T7, the protein encoded by *ard* is active against Type I R–M systems.

Bacteriophage λ encodes a very specialized antirestriction function, Ral, which modulates the *in vivo* activity of some Type I R–M systems by enhancing modification and alleviating restriction. The systems influenced by Ral are those that have a modification enzyme with a strong preference for hemimethylated DNA. Ral may act by changing the MTase activity of the R–M system to one that is efficient on unmethylated target sequences. Unmodified ral^+ λ DNA is restricted on infection of a restriction-proficient bacterium, because *ral* is not normally expressed before the genome is attacked by the host R–M system, but phages that escape restriction and express *ral* are more likely to become modified as Ral serves to enhance the efficiency with which the modification enzyme methylates unmodified DNA.

Some phages are made resistant to many types of R–M systems by the presence of glucosylated hydroxymethylcytosine (HMC) in their DNA (e.g. the *E. coli* T-even phages and the *Shigella* phage DDVI). The glucosylation also identifies phage DNA and allows selective degradation of host DNA by phage-encoded nucleases. Nonglucosylated T-even phages are resistant to some classical R–M systems because their DNA contains the modified base HMC, but they are sensitive to modification-dependent systems, although T-even phages encode a protein (Arn) that protects superinfecting phages from McrBC restriction. It has been suggested that phages have evolved DNA containing HMC to counteract classical R–M systems, and that host-encoded modification-dependent endonucleases are a response to this phage adaptation. In this evolutionary story, the glucosylation of HMC would be the latest mechanism that renders T-even phages totally resistant to most R–M systems.

In some cases, a phage genome can tolerate a few targets for certain restriction enzymes. The few *Eco*RII sites in T3 and T7 DNA are not sensitive to restriction because this unusual enzyme requires at least two targets in close proximity and the targets in these genomes are not sufficiently close. For the Type III enzymes, the orientation of the target sequences is also relevant. Since the target for restriction requires two inversely oriented recognition sequences, the T7 genome remains refractory to *Eco*P151 because all 36 recognition sequences are in the same orientation. The unidirectional orientation of the target sequences is consistent with selection for a genome that will avoid restriction. Considerable evidence supports the significance of counterselection of target sequences in phage genomes, in some cases correlating the lack of target sequences for enzymes found in those hosts in which the phages can propagate.

V. DISTRIBUTION, DIVERSITY, AND EVOLUTION

A. Distribution and diversity

R–M systems are almost ubiquitous among prokaryotes. Over 3000 have been identified. The most complete documentation is that for Type II systems. This is a result of the many *in vitro* screens for sequence-specific endonucleases put to effective use in the search for enzymes with different specificities. The endonucleases identified include more than 200 different sequence specificities. Type II restriction enzymes have been detected in 12 of the 13 phyla of Bacteria and Archaea. The lack of representation in one phylum, Chlamydia, could result from a sampling bias, but it could reflect the fact that the available genomic sequences are from parasitic species that grow within eukaryotic cells.

There is no reliable biological screen for Type I systems and there has been no practical incentive to search for enzymes that cut DNA at variable distances from their recognition sequence. The apparent prevalence of Type I systems in enteric bacteria may simply reflect the common use of *E. coli* and its relatives in genetic studies. Genetic tests currently provide evidence for Type I enzymes with approximately 20 different specificities, and recent analyses of genomic sequences from diverse species indicate that Type I systems are widely distributed throughout the prokaryotic kingdom.

Janulaitis and colleagues screened natural isolates of *E. coli* for sequence-specific ENases and detected activity in 25% of nearly 1000 strains tested. Another screening experiment searched for restriction activity encoded by transmissible resistance plasmids in *E. coli*. The plasmids were transferred to *E. coli* and the EOP of phage λ was determined on the exconjugants. Approximately 10% of the transmissible antibiotic resistance plasmids were correlated with the restriction of λ and the ENases responsible were shown to be Type II. However, plasmid-borne Type I, Type II, and Type III systems are known in *E. coli*. Because of the transmissible nature of many plasmids, the frequency with which R–M systems are transferred between strains could be high and their maintenance subject to a variety of selection pressures not necessarily associated with the restriction phenotype.

The evidence, where available (i.e. for *E. coli* and *Salmonella enterica*), is consistent with intraspecific diversity irrespective of the level at which the diversity is examined. In *E. coli*, there are at least six distinct mechanistic classes of restriction enzyme (types I–III, and three modification-dependent types) and the anomalous *Eco*571 (Fig. 29.2). The Type II systems in *E. coli* currently include approximately 30 specificities, and at least 14 Type I specificities have been identified. Bacterial strains frequently have more than one active restriction system. Four systems are present in *E. coli* K-12, and *H. influenzae* Rd has at least three systems that are known to be biologically active, whereas three more are indicated in the genomic sequence.

Genome sequences now provide a wealth of information about the distrbution of coding sequences for R–M systems; these do not necessarily indicate whether the genes specify active enzymes.

B. Evolution

R–M enzymes may be dissected into modules. A Type II MTase comprises a TRD and a module that is responsible for catalyzing the transfer of the methyl group from AdoMet to the defined position on the relevant base. The catalytic domains share sequence similarities, and these are most similar when the catalytic reaction is the same, that is, yields the same product (e.g. 5 mC). Given the matching specificities of cognate ENase and MTase, it might be expected that their TRDs would be of similar amino acid sequence. This is not the case; it seems likely that the two enzymes use different strategies to recognize their target sequence. Each subunit of the dimeric ENase needs to recognize one-half of the rotationally symmetrical sequence, whereas the monomeric MTase must recognize the entire sequence. The absence of similarity between the TRDs of the ENase and its cognate MTase suggests that they may have evolved from different origins.

Restriction enzymes that recognize the same target sequence are referred to as isoschizomers. A simple expectation is that the TRDs of two such enzymes would be very similar. This is not necessarily true. Furthermore, the similarities observed do not appear to correlate with taxonomic distance. The amino acid sequences of the isoschizomers *Hae*III and *Ngo*PII, which are from bacteria in the same phylum, show little if any similarity, whereas the isoschizomers *Fnu*DI and *Ngo*PII, which are isolated from bacteria in different phyla, are very similar (59% identity).

Type I R–M systems are complex in composition and cumbersome in their mode of action, but they are well suited for the diversification of sequence specificity. A single subunit (HsdS or S) confers specificity to the entire R–M complex and to the additional smaller complex that is an MTase. Any change in specificity affects restriction and modification concomitantly. Consistent with their potential to evolve new specificities, Type I systems exist as families, within which members (e.g. *Eco*KI and *Eco*BI) are distinguished only by their S subunits. Currently, allelic genes have been identified for at least seven members of one family (IA); each member has a different specificity. It is more surprising that allelic genes in *E. coli*, and its relatives, also specify at least two more families of Type I enzymes. Although members of a family include only major sequence differences in their S polypeptides, those in different families share very limited sequence identities (usually 18–30%). Clearly, the differences between gene sequences for Type I R–M systems are no indication of the phylogenetic relatedness of the strains that encode them. Note that despite the general absence of sequence similarities between members of different families of Type I enzymes, pronounced similarities have been identified for TRDs from different families when they confer the same sequence specificity.

The information from gene sequences for both Type I and Type II systems, as stated by Raleigh and Brooks (1998), "yields a picture of a pool of genes that have circulated with few taxonomic limitations for a very long time."

Allelic variability is one of the most striking features of Type I R–M systems. Both the bipartite and asymmetrical natures of the target sequence confer more scope for diversity of sequence specificity than the symmetrical recognition sequences of Type II systems. The S subunit of Type I enzymes includes two TRDs, each specifying one component of the target sequence. This organization of domains makes the subunit well suited to the generation of new specificities as the consequences of either new combinations of TRDs or minor changes in the spacing between TRDs. In the first case, recombination merely reassorts the regions specifying the TRDs, and in the second case unequal crossing over within a short duplicated sequence leads to a change in the spacing between the TRDs. Both of these processes have occurred in the laboratory by chance and by design.

For Type I R–M systems, the swapping or repositioning of domains can create enzymes with novel specificities, but the evolution of new TRDs with different specificities has not been witnessed. In one experiment, strong selection for a change that permitted a degeneracy at one of the seven positions within the target sequence failed to yield mutants with a relaxed specificity.

VI. BIOLOGICAL SIGNIFICANCE

The wide distribution and extraordinary diversity of R–M systems, particularly the allelic diversity documented in enteric bacteria, suggest that R–M systems have an important role in bacterial communities. This role has traditionally been considered to be protection against phage. Laboratory studies following bacterial populations under conditions of phage infection indicate that R–M systems provide only a transitory advantage to bacteria. Essentially, an R–M system with a different specificity could assist bacteria in the colonization of a new habitat in which phage are present, but this advantage would be short-lived as phages that escape restriction acquire the new protective modification and bacteria acquire mutations conferring resistance to the infecting phages. It can be argued that one R–M system protects against a variety of phages, and the maintenance of one R–M system may compromise the fitness of the bacterium less than the multiple mutations required to confer resistance to a variety of phages. No direct evidence supports this expectation. It is relevant to remember that the restriction barrier is generally incomplete, irrespective of the mechanism of DNA transfer, and that the fate of phage and bacterial DNA fragmented by ENases may differ. A single cut in a phage genome is sufficient to prevent infectivity. Fragments generated from bacterial DNA will generally share homology with the host chromosome and could therefore be rescued by recombination. The rescue of viable phages by homologous recombination requires infection by more than one phage or recombination with phage genomes that reside within the host chromosome. A protective role for R–M systems in no way excludes an additional role that influences genetic recombination.

In *E. coli*, and probably bacteria in general, linear DNA fragments are vulnerable to degradation by exonucleases, particularly ExoV (RecBCD). Therefore, the products of restriction are substrates for degradation by the same enzyme that is an essential component of the major recombination pathway in *E. coli*. However, degradation by RecBCD is impeded by the special sequences, designated Chi, that stimulate recombination. It has been shown that a Chi sequence can stimulate recombination when RecBCD enters a DNA molecule at the site generated by cutting with *Eco*RI. It seems inevitable that fragmentation of DNA by restriction would reduce the opportunity for recombination to incorporate long stretches of DNA; however, given that DNA ends are recombinogenic, restriction could promote the acquisition of short segments of DNA.

Radman and colleagues have suggested that R–M systems are not required as interspecific barriers to recombination since the DNA sequence differences between *E. coli* and *Salmonella* are sufficient to hinder recombination. It is evident, however, that selection has maintained a diversity of restriction specificities within one species, and consequently restriction is presumed to play a significant role within a species, in which DNA sequence differences are less likely to affect recombination. Detailed analyses of the effects of restriction on the transfer of DNA between strains of *E. coli* are currently under way and the molecular techniques are available to monitor the sizes and distributions of the DNA fragments transferred between strains.

Kobayashi and colleagues view R–M genes as "selfish" entities on the grounds that loss of the plasmid that encodes them leads to cell death. The experimental evidence for some Type II R–M systems implies that the cells die because residual ENase activity cuts incompletely modified chromosomal DNA. The behavior of Type I systems, on the other hand, is different and is consistent with their ability to diversify sequence specificity; when new specificities are generated by recombination, old ones are readily lost without impairing cell viability.

VII. APPLICATIONS AND COMMERCIAL RELEVANCE

Initially, the opportunity to use enzymes that cut DNA molecules within specific nucleotide sequences added a new dimension to the physical analysis of small genomes. In the early 1970s, maps (restriction maps) could be made in which restriction targets were charted within viral genomes and their mutant derivatives. Within a few years, the same approach was generally applicable to larger genomes. The general extension of molecular methods to eukaryotic genomes depended on the technology that enabled the cloning of DNA fragments, that is, the generation of a population of identical copies of a DNA fragment. In short, DNA from any source could be broken into discrete fragments by restriction ENases, the fragments could be linked together covalently by the enzymatic activity of DNA ligase, and the resulting new combinations of DNA could be amplified following their recovery in *E. coli*. Of course, to achieve amplification of a DNA fragment, and hence a molecular clone, it was necessary to link the DNA fragment to a special DNA molecule capable of autonomous replication in a bacterial cell. This molecule, the vector, may be a plasmid or a virus. Importantly, it is usual for only one recombinant molecule to be amplified within a single bacterial cell. In principle, therefore, one gene can be separated from the many thousands of other genes present in a eukaryotic cell, and this gene can be isolated, amplified, and purified for analysis. The efficiency and power of molecular cloning have evolved quickly, and the new opportunities have catalyzed the rapid development of associated techniques, most notably those for determining the nucleotide sequences of DNA molecules, the chemical synthesis of DNA, and recently, the extraordinarily efficient amplification of gene sequences *in vitro* by the polymerase chain reaction (PCR). In some cases, amplification *in vitro* obviates the need for amplification *in vivo* since the nucleotide sequence of PCR products can be obtained directly.

The bacterium *E. coli* remains the usual host for the recovery, manipulation, and amplification of recombinant DNA molecules. However, for many of the commonly used experimental organisms the consequence of a mutation can be determined by returning a manipulated gene to the chromosome of the species of origin.

The recombinant DNA technology, including screens based on the detection of DNA by hybridization to a specific probe and the analysis of DNA sequence, is now basic to all fields of biology, biochemistry, and medical research as well as the "biotech industry." Tests dependent on DNA are used to identify contaminants in food, parents of children, persons at the scene of a crime, and the putative position of a specimen in a phylogenetic tree. Mutations in specific genes may be made, their nature confirmed, and their effects monitored. Gene products may be amplified for study and use as experimental or medical reagents. Hormones, cytokines, blood-clotting factors, and vaccines are amongst the medically relevant proteins that have been produced in microorganisms, obviating the need to isolate them from animal tissues.

Most of the enzymes used as reagents in the laboratory are readily available because the genes specifying them have been cloned in vectors designed to increase gene expression. This is true for the ENases used to cut DNA. It is amusing that in the 1980s the generally forgotten, nonclassical, restriction systems identified by Luria and Human (1952) were rediscovered when difficulties were encountered in cloning Type II R–M genes. It was soon appreciated that cloning the genes for particular MTases was a problem in "wild-type" *E. coli* K-12; the transformed bacteria were killed when modification of their DNA made this DNA a target for the resident Mcr restriction systems. Rare survivors were *mcr* mutants, ideal strains for recovering clones of foreign DNA rich in 5mC, as well as genes encoding MTases.

BIBLIOGRAPHY

Cheng, X., and Roberts, R. J. (2001). AdoMet-dependent methylation, DNA methyltransferases and base flipping. *Nucleic Acids Res.* **29**, 3784–3795.

Dryden, D. T., Murray, N. E., and Rao, D. N. (2001). Nucleoside triphosphate-dependent restriction enzymes. *Nucleic Acids Res.* **29**, 3728–3741.

Murray, N. E. (2002). Immigration control of DNA in bacteria: self versus non-self. *Microbiology* **148**, 3–20.

Pingoud, A., and Jeltsch, A. (2001). Structure and function of type II restriction endonucleases. *Nucleic Acids Res.* **29**, 3705–3727.

Raleigh, E. A., and Brooks, J. E. (1998). Restriction modification systems: Where they are and what they do. *In* "Bacterial Genomes: Physical Structure and Analysis" (F. J. de Bruijn, J.R. Lupski, and G.M. Weinstock, eds.), pp. 78–92. Chapman & Hall, New York.

Redaschi, N., and Bickle, T. A. (1996). DNA restriction and modification systems. *In "Escherichia coli* and *Salmonella*. Cellular and Molecular Biology" (F. C. Neidhardt, ed.), 2nd edn, ASM Press, Washington, DC.

Roberts, R. J. *et al.* (2003). A nomenclature for restriction enzymes, DNA methyl-transferases, homing endonucleases and their genes. *Nucleic Acids Res.* **31**, 1805–1812.

WEBSITES

Extensive bibliography on restriction and modification
http://www.port.ac.uk/departments/biological/restriction/r-m_refs.html

Links to PubMed on restriction and modification
http://www.ohsu.edu/cliniweb/D8/D8.586.150.html

Restriction Enzyme Database
http://rebase.neb.com/rebase/rebase.html

Pedro's Biomolecular Research Tools. A collection of links to information and services useful to molecular biologists
http://www.public.iastate.edu/~pedro/research_tools.html

30

DNA Sequencing and Genomics

Brian A. Dougherty
Bristol-Myers Squibb Pharmaceutical Research Institute

GLOSSARY

bioinformatics Biological informatics, specifically the computer-assisted analysis of genome sequence and experimental data.

DNA microarrays Miniaturized arrays containing thousands of DNA fragments representing genes in an area of a few square centimeters. These "chips" are used mostly to monitor the expression of genomes at the level of messenger RNA.

functional genomics The study of gene function at the genome level.

genomics The study of the genome, or all of the genes in an organism.

proteomics The study of the protein complement, or proteome, of an organism.

whole genome shotgun sequencing A random approach to genome sequencing based on shearing of genomic DNA, followed by cloning, sequencing, and assembly of the entire genome, and ultimately leading to finished, annotated genomic sequence. Sequencing of entire genomes can also be achieved by a sequential shotgun sequencing of a set of overlapping, large insert clones.

Advances in DNA Sequencing, computing, and automation technologies have allowed the sequencing of entire genomes from living organisms. Bacterial genomes range in size from 0.5 to at least 10 Mb (1×10^6 base pairs) and are composed of approximately 90% coding regions; therefore, high-throughput sequencing of randomly cloned DNA fragments represents the most cost-effective way to rapidly obtain genome information for these organisms.

Since the publication in 1995 of the first whole genome sequence for a bacterium, over 60 microbial genome sequences have been published (Table 30.1), and high coverage sequence data have been added to the public domain throughout this period for many other genomes. Sequencing of a microbial genome requires logistical planning for the challenges of high-throughput sequencing, bioinformatic analysis, and presentation of the data in a concise, user-friendly format, usually in the form of a World Wide Web site to supplement a journal publication; this is an intensive process that is often under-appreciated. Following the sequencing, analysis, and initial assignment of function based on similarity to previously identified genes in sequence databases, analysis of gene function at the laboratory bench is required to confirm gene function. These functional genomics studies will also require a substantial scale-up in order to keep up with the high-throughput pace of genome sequencing. Simply stated, in the pre-genomic era genes were usually identified based on a selection for function, followed by the laborious tasks of cloning, subcloning, and sequencing; genome sequencing has reversed this process today, and the challenge of the postgenomic era will be to determine how all the gene products of a genome function, interact, and allow the organism to live.

The Desk Encyclopedia of Microbiology
ISBN: 0-12-621361-5

TABLE 30.1 Published microbial genomes[a]

Genome	Mb	Publication	Genome	Mb	Publication
Bacteria			**Bacteria**		
Agrobacterium tumefaciens	5.3	*Science* **294**: 2317–2323; 2323–2328 (2001)	*Pseudomonas aeruginosa*	6.3	*Nature* **406**: 959–964 (2000)
Aquifex aeolicus	1.5	*Nature* **392**: 353 (1998)	*Rickettsia conorii*	1.27	*Science* **293**: 2093–2098 (2001)
Bacillus halodurans	4.2	Nucleic Acids Res. **28**: 4317–4331 (2000)	*Rickettsia prowazekii*	1.1	*Nature* **396**: 133–140 (1998)
Bacillus subtilis	4.2	*Nature* **390**: 249–256 (1997)	*Salmonella typhi*	4.8	*Nature* **413**: 848–852(2001)
Borrelia burgdorferi	1.44	*Nature* **390**: 580–586 (1997)	*Salmonella typhimurium*	4.8	*Nature* **413**: 852–856 (2001)
Buchnera sp.	0.64	*Nature* **407**: 81–86 (2000)	*Sinorhizobium meliloti*	6.7	*Science* **293**: 668–572 (2001)
Campylobacter jejuni	1.64	*Nature* **403**: 665–668 (2000)	*Staphylococcus aureus*	2.81	*Lancet* **357**: 1225–1240 (2001)
Caulobacter crescentus	4.01	*PNAS* **98**: 4136–4141 (2001)	*Streptococcus pneumoniae*	2.2	*Science* **293**: 498–506 (2001)
Chlamydia muridarum	1.07	*Nuc. Acids Res.* **28**: 1397–1406 (2000)	*Streptococcus pyogenes*	1.85	*PNAS* **98**: 4658–4663 (2001)
Chlamydia pneumoniae AR39	1.23	*Nuc. Acids Res.* **28**: 1397–1406 (2000)	*Streptomyces avermitilis*	8.7	*PNAS* **98**: 12215–12220 (2001)
Chlamydia pneumoniae CWL029	1.23	*Nat Genet* **21**: 385–389 (1999)	*Streptomyces coelicolor*	8.7	*Nature* **417**: 141–147 (2002)
Chlamydia pneumoniae J138	1.22	*Nucleic Acids Res.* **28**:2311–2314 (2000)	*Synechocystis* sp.	3.57	*DNA Res.* **3**: 109–136 (1996)
Chlamydia trachomatis	1.05	*Science* **282**: 754–759 (1998)	*Thermoanaerobacter tengcongensis*	2.69	*Genome Res.* **12**: 689–700 (2002)
Clostridium acetobutylicum	4.1	*J. Bacteriol.* **183**: 4823–4838 (2001)	*Thermotoga maritima*	1.8	*Nature* **399**: 323–329 (1999)
Deinococcus radiodurans	3.28	*Science* **286**: 1571–1577 (1999)	*Treponema pallidum*	1.14	*Science* **281**: 375–388 (1998)
Escherichia coli EDL933	5.5	*Nature* **409**: 529–533 (2001)	*Ureaplasma urealyticum*	0.75	*Nature* **407**: 757–762 (2000)
Escherichia coli K-12	4.6	*Science* **277**: 1453–1474 (1997)	*Vibrio cholerae*	4	*Nature* **406**: 477–483 (2000)
Escherichia coli RIMD	5.6	*DNA Res.* **8**: 11–22 (2001)	*Xanthomonas axonopodis*	5.17	*Nature* **417**: 459–463 (2002)
Haemophilus influenzae	1.83	*Science* **269**: 496–512 (1995)	*Xanthomonas campestris*	5.07	*Nature* **417**: 459–463 (2002)
Helicobacter pylori 26695	1.66	*Nature* **388**: 539–547 (1997)	*Xylella fastidiosa*	2.68	*Nature* **406**: 151–157 (2000)
Helicobacter pylori J99	1.64	*Nature* **397**: 176–180 (1999)	*Yersinia pestis*	4.65	*Nature* **413**: 523–527 (2001)
Lactococcus lactis	2.36	*Genome Res.* **11**:731–753 (2001)	**Archaea**		
Listeria innocua	3.01	*Science* **294**: 849–852 (2001)	*Aeropyrum pernix*	1.67	*DNA Res.* **6**: 83–101 (1999)
Listeria monocytogenes	2.94	*Science* **294**: 849–852 (2001)	*Archaeoglobus fulgidus*	2.18	*Nature* **390**: 364–370 (1997)
Mesorhizobium loti	7.59	*DNA Res.* **7**: 331–338 (2000)	*Halobacterium* sp.	2.57	*PNAS* **97**: 12176–12181 (2000)
Mycobacterium leprae	3.26	*Nature* **409**: 1007–1011 (2001)	*Methanobacterium thermoautotrophicum*	1.75	*J. Bacteriol.* **179**: 7135–7155 (1997)
Mycobacterium tuberculosis	4.4	*Nature* **393**: 537 (1998)	*Methanococcus jannaschii*	1.66	*Science* **273**: 1058–1073 (1996)
Mycoplasma genitalium	0.58	*Science* **270**: 397–403 (1995)	*Pyrococcus horikoshii*	1.8	*DNA Res.* 5: 55–76 (1998)
Mycoplasma pneumoniae	0.81	*Nucleic Acids Res.* **24**: 4420–4449 (1996)	*Sulfolobus solfataricus*	2.99	*PNAS* **98**: 7835–7840 (2001)
Mycoplasma pulmonis	0.96	*Nucleic Acids Res.* **29**: 2145–2153 (2001)	*Thermoplasma acidophilum*	1.56	*Nature* **407**: 508–513 (2000)
Neisseria meningitidis MC48	2.27	*Science* **287**: 1809–1815 (2000)	*Thermoplasma volcanium*	1.58	*PNAS* **97**: 14257–14262 (2000)
Neisseria meningitidis Z2491	2.18	*Nature* **404**: 502–506 (2000)	**Eukaryote**		
Pasteurella multocida	2.4	*PNAS* **98**: 3460–3465 (2001)	*Saccharomyces cerevisiae*	13	*Science* **274**: 563–567 (1996)

[a]Current as of May 2002. See Table 30.3 for websites with updated genome project lists.

I. INTRODUCTION: DNA SEQUENCING

DNA sequencing has seen many technological changes during the past two decades, evolving from slab polyacrylamide gels that could resolve 20–50 bp at a time to the 700-bp reads of today. In the late 1970s, DNA sequences were manually read from X-ray film using chemical and enzymatic methods to incorporate radiolabel. During the 1990s fluorscently labeled DNA was resolved on slab gels and read using computerized base-calling programs using sequencers such as the Applied BioSystems 377 model. In the late 1990s, capillary sequencers such as the Amersham MegaBACE and the Applied Biosystems 3700 models were introduced, and dramatically increased the throughput of automated DNA sequencing. The availability of these instruments was the catalyst that allowed the shotgun sequencing of the fruit fly genome, and greatly accelerated the schedule for completion of the Human Genome Project.

Upon publication of the 1.8-Mb chromosome of *Haemophilus influenzae* in 1995, it became apparent to the scientific community that sequencing of bacterial and even eukaryotic genomes was possible by scaling-up existing sequencing technologies. Although the utility of the whole-genome shotgun approach (Fig. 30.1) for sequencing microbial genomes is unquestionable today, the original *H. influenzae* project was in fact deemed an unproven, high-risk venture and was not funded because such a large segment of DNA had not been sequenced by a random approach. The standard paradigm for any large-scale sequencing projects was based on strategies used for sequencing the human genome and related model organisms. These "top-down" techniques consisted of two phases, an up-front mapping and cosmid-ordering phase, followed by subcloning and sequencing of a minimal tiling path (a complete set of cloned genome fragments with minimal overlap among the cosmids). Sequencing of larger segments of a microbial genome was not attempted due to a computational limitation at that time—existing software packages for the assembly of random DNA fragments were not sufficiently robust to assemble a DNA segment much larger than the size of a cosmid insert (~40 kbp). However, advances based on an unconventional approach to sequencing the human genome set the stage for whole-genome sequencing of smaller genomes. The Venter laboratory at the National Institutes of Health (NIH) and later at The Institute for Genomic Research (TIGR) developed the expressed sequence tag (EST) method for streamlining gene discovery in human genome sequencing by enriching for and sequencing messenger RNAs. By optimizing the conditions for high-throughput fluorescent sequencing and developing the computational software to assemble hundreds of thousands of random sequences into contiguous sequences (contigs), the stage was set to attempt a shotgun approach to sequencing an entire genome, which had last been done for bacteriophage λ in 1977.

The whole genome shotgun sequencing technique is described in the following section, and it consists of four phases: library construction, random sequencing, gap closure, and editing/annotation (Fig. 30.1).

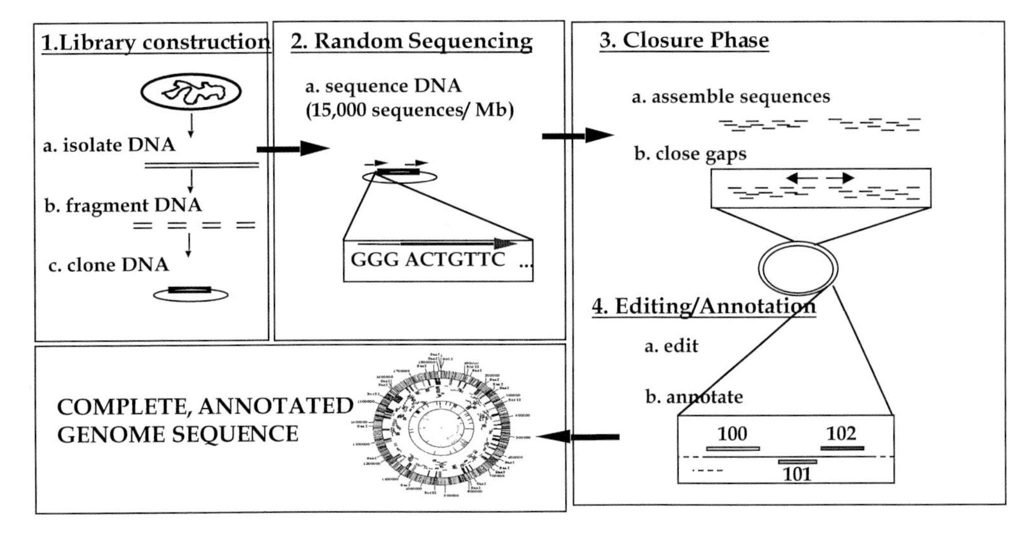

FIGURE 30.1 Strategy for a whole genome random sequencing project.

II. MICROBIAL GENOME SEQUENCING PROJECT

A. Library construction phase

Library construction is the most critical phase of a microbial genome sequencing project. Assuming a completely random library, Lander and Waterman calculated that the sequencing statisitics would follow a Poisson distribution (i.e. sequencing enough random reads to add up to one genome equivalent would leave 37% of the genome unsequenced; see Table 30.2). Sequencing of a nonrandom library will result in a significant deviation from the Lander–Waterman model for random sequencing, and the generated assemblies will have too many gaps to be efficiently closed. A library construction procedure has been developed (Fleischmann *et al.*, 1995) to achieve a high order of randomness, and it consists of the following steps:

1. Random shearing of genomic DNA and purification of ~2 kbp fragments.
2. Fragment end repair and ligation to blunt-ended, dephosphorylated vector.
3. Gel purification of the "v + i" (vector plus single insert) band from other forms.
4. Final polishing of free ends and intramolecular ligation of v + i DNA.
5. High-efficiency transformation of *Escherichia coli* strains (DH10B, SURE, etc.).
6. Direct plating onto two-layer antibiotic diffusion plates.

This optimized method results in efficient sequencing because the library is a collection of highly random, single-insert fragments that can be sequenced from both ends using universal primers. Key factors for success include random shearing of the DNA (usually by nebulization), purification of a narrow size range of insert (using minimal amounts of longwave UV light), a second gel purification of v + i to minimize plasmid clones lacking insert or containing chimeric inserts, propagation of clones in a highly restriction-deficient *E. coli* background, and outgrowth of transformed cells as individual colony-forming units on an antibiotic diffusion plate (rather than standard outgrowth as a mixture of clones in liquid medium, which may select against certain slower growing clones in the transformed population). Even with all these safeguards in place, however, gaps in the genome do occur. This is due to both the statistics of random sequencing and the inability to clone certain DNA fragments, such as those containing strong promoters (e.g. the 16S rRNA promoter in *H. influenzae*) or "toxic" genes (e.g. the complete *Hind*III restriction enzyme gene without the cognate methylase gene present in the transformed *E. coli* strain).

TABLE 30.2 Lander-Waterman calculation for random sequencing of a microbial genome (2-Mb size, 500-bp average sequence read length)

No. sequences	% genome sequenced	No. gaps	Average gap size	Fold coverage	Total bp sequenced
4000	63.21	1472	500	1×	2,000,000
8000	86.47	1083	250	2×	4,000,000
12,000	95.02	597	167	3×	6,000,000
16,000	98.17	293	125	4×	8,000,000
20,000	99.33	135	100	5×	10,000,000
24,000	99.75	59	83	6×	12,000,000
28,000	99.91	26	71	7×	14,000,000
32,000	99.97	11	63	8×	16,000,000

B. Random sequencing phase

After the library is constructed, random sequencing of the individual clones is performed. This involves: (i) the production of tens of thousands of templates in 96-well blocks; (ii) sequencing of the purified plasmid template DNA in both directions; (iii) transfer of the edited sequence reads to a database; and (iv) assembly of the sequence data into contigs.

To ensure that the expected Poisson distribution of random sequences is obtained, the assembled sequence data from the first several thousand sequences are plotted relative to the Lander–Waterman equation (Fig. 30.2). The sequencing and assembly of a verified random library continues until approximately 7- or 8× genome coverage is achieved, which should result in a manageable number of gaps to close (Table 30.2). Based on diminishing return, after sequencing to 7- or 8× random genome coverage with the 2-kbp insert library, directed sequencing strategies are employed for gap closure; for genome sequencing projects in which gap closure will not be pursued, the project goes directly to the annotation phase.

C. Closure phase

Of the three phases of a whole-genome sequencing project discussed so far, this is the most time-consuming; however, closure of a genome results in a complete list of the set of genes. If a certain gene or metabolic pathway is missing from a closed genome, its absence is not due to it being located in an unsequenced gap. Moreover, a complete genome is a linear string of nucleotides, rather than a collection of contiguous sequences of uncertain orientation relative to one

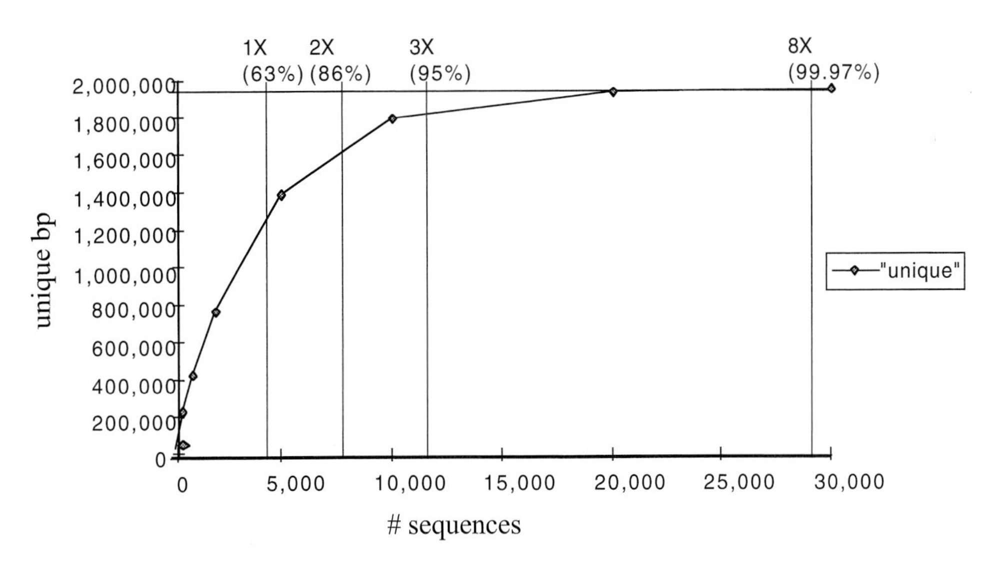

FIGURE 30.2 Plot of Lander–Waterman calculation for random sequencing for a 2-Mb genome with 500-bp average sequence read lengths.

another, and contains genes in linear progression with exact 5′ and 3′ coordinates.

The forward and reverse reads of the plasmid clones are an important tool used for genome closure. Once the assembled sequences are obtained, any template with a forward sequence read in one assembly and a reverse in the other represents a link between assemblies and is used as a template for primer walking to close the gap. Clone linkage for closure of potentially larger gaps is provided by end sequencing a large insert lambda library and mapping positions of the forward and reverse reads (separated by about 20 kbp) on the assembled contigs.

The previously mentioned gaps are called "sequencing" gaps since a DNA template is available; closure of "physical" gaps is more challenging because no template is immediately available. Depending on the number of gaps to close and tools at one's disposal, a template is generated for a physical gap by several means, including combinatorial polymerase chain reaction (PCR), screening for large insert clones that span the gap, Southern hybridization fingerprinting, and cycle sequencing using genomic DNA as a template. Once a template is obtained, it is sequenced until the physical gap is closed. The gap closure data are then incorporated into the assembly data, the assembled contigs are joined, and the process is repeated until closure of the entire genome is achieved.

D. Editing and annotation phase

Following genome closure the sequence is edited and annotated. Editing consists of proofreading the sequence and resolving ambiguities or regions of low coverage by dye terminator sequencing. In addition, possible frameshifts are identified, PCR amplified from genomic DNA, and resequenced. It should be noted, however, that some apparent base insertions, deletions, and frameshifts are authentic. Examples include polymorphisms in repetitive sequences of the *Campylobacter jejuni* genome and more than 20 authentic frameshifts in protein coding regions of the lab-propagated *H. influenzae* Rd organism (several of the counterparts of these genes in other *H. influenzae* strains have been demonstrated to be full-length, functional genes).

Annotation begins with gene finding, which is not an insignificant task in prokaryotes; although stop sites of open reading frames (ORFs) are straightforward, identifying the proper translation initiation codon can be challenging. Annotation also involves characterization of other features of the genome, including repeated nucleotide sequences, gene families, and variations in nucleotide composition. Furthermore, all genes are classified into functional role categories and analyzed as a whole for the presence of various metabolic pathways. In this way, whole-genome sequencing provides a complete picture of the metabolic potential of an organism and lays the groundwork for many follow-up studies once the genetic complement of the cell has been defined.

III. BIOINFORMATICS AND GENOME DATABASES

The need for computational tools to handle genome project data is increasing, and the URLs listed in Table 30.3 represent a fraction of the genomic websites that

TABLE 30.3 Selected genomic resources from the world wide web

Complete and in progress genome projects	
http://www.tigr.org/tigr-scripts/CMR2/CMRHomePage.spl	CMR: Comprehensive Microbial Resource (TIGR)
http://wit.integratedgenomics.com/GOLD/	GOLD: Genomes Online Database (IG)
http://www.ncbi.nlm.nih.gov:80/PMGifs/Genomes/micr.html	Entrez Microbial Genomes Listing (NCBI)
Genome sequence data	
http://www.ncbi.nlm.nih.gov/Entrez/Genome/org.html	Entrez Genome Browser
http://www.tigr.org/tdb/mdb/mdb.html	TIGR Microbial Database; links to projects
http://www.pasteur.fr/Bio/SubtiList.html	Subtilist server for *B. subtilis*
http://www.pasteur.fr/Bio/Colibri.html	Colibri server for E. coli
http://www.ncbi.nlm.nih.gov/PMGifs/Genomes/yc4.html	NCBI Yeast Genome; links to YPD, MIPS, SGD
Analysis of genome sequence data	
http://mol.genes.nig.ac.jp/gib/	Genome Information Broker
http://motif.stanford.edu/	The Brutlag Bioinformatics Group
http://www.expasy.ch/	Expasy Molecular Biology Server
http://www.ncbi.nlm.nih.gov/COG/	Clusters of Orthologous Groups (COGs) site
http://www.sanger.ac.uk/Software/Pfam/	P fam (Protein families database, HMMs)
http://lion.cabm.rutgers.edu/~bruc/microbes/index.html	SEEBUGS Microbial Genome Analysis
http://www.genome.ad.jp/kegg/	KEGG: Kyoto Encyclopedia of Genes and Genomes
http://www.ecocyc.org/	EcoCyc: Encyclopedia of E. coli Genes/Metabolism
Compilations of genome-related sites/links	
http://www.hgmp.mrc.ac.uk/GenomeWeb/	GenomeWeb List of Other Genome Sites
http://www.qiagen.com/bioinfo/	Links to Bioinformatic WWW Servers
http://www.genome.ad.jp/	GenomeNet WWW Server
http://www.public.iastate.edu/~pedro/research_tools.html	Pedro's Biomolecular Research Tools

continue to proliferate. The genome project monitoring sites given are among the most comprehensive and are continually updated as new genome projects are both initiated and completed. In fact, it is estimated that in addition to the 60-plus microbial genomes completed 7 years after the publication of *Haemophilus* (Table 30.1), hundreds more are in progress. The genome project websites listed provide extensive annotation, role categorization, search capabilities, and genome segment display and retrieval. In the near future, experimental data from functional genomics work (expression of gene under various conditions, phentoype of knockout mutant, protein–protein interaction information, etc.) will be incorporated into these existing structures and provide much added value to the basic research community.

During the course of a genome sequencing project, a variety of tools are required. At the DNA sequencing and assembly stage, software for high-throughput sequencing is needed for automated base calling (e.g. phred and ABI Base Caller) and data tracking/management. An assembly program (e.g. phrap and TIGR Assembler) is then used to find overlaps among the random sequences and build large sets of contiguous sequence. Gaps in the sequence are then closed by directed sequencing; programs to clearly display, track, and update this information are crucial. Gene finding is then performed using programs that search for ORFs among the assembled sequence data (e.g. Genemark, Critica, and Glimmer) and that use computational techniques to help discriminate the signal of real genes from the noise of potential ORFs found in a microbial genome. The annotation stage involves human inspection of the list of candidate genes and other data to help generate a catalog of genes from the organism (more like a rough draft rather than a final list). Procedures used to determine if a gene is authentic include examining similarity scores, multiple sequence alignments, and published data when similar to that of known genes and examining the sequence for gene-like characteristics (e.g. the presence and proper spacing of a promoter, ribosome binding site, and initiation codon; the presence of a similar gene in the database; the position of the gene in an operon; or nucleotide or codon usage similar to that of the rest of the genome). Functional role category can also be assigned based on similarity to genes of known function, and a picture of the metabolic capacity of the organism emerges. Finally, it should be emphasized that quality annotation is based on bioinformatic analysis beyond assigning the top BLAST similarity score. Motif searches (e.g. Prosite, Blocks, and e-motif) reveal local similarity that may be missed at the global level of a BLAST search. Also, multiple sequence alignments and phylogenetic relationships between members of a gene family can provide insight into possible function. More detailed analyses are currently providing information about relationships among predicted proteins that may be missed by solely relying on BLAST searches, and gene families are being constructed using hidden Markov modeling (Pfam) and other techniques (e.g. COGs). Finally, molecular modeling comparisons such as threading are being used to gain insight into relationships between proteins seen only at the three-dimensional level.

IV. FUNCTIONAL GENOMICS APPROACHES

A. Comparative genomics

The resources required for a microbial whole-genome sequencing project is not insignificant, with a cost of approximately $1 million and requiring 1 or 2 years to complete. However, high-throughput sequencing of microbial DNA is the most cost-effective method to obtain genomic information: a typical microbial genome would give several thousand gene sequences at a cost of several hundred dollars per gene (significantly less than the per gene cost of the pregenomic era). This cost is reduced even further by sequencing at a lower average redundancy (3–6× coverage), providing a catalog of the majority of the genes in an organism of interest. These data are sufficient for the purpose of rapidly acquiring a proprietary database of genomic data, often from multiple organisms, and performing functional genomics work with the ultimate goal of producing vaccines or antibiotics. However, the goal of the complete genome project, published in a peer-reviewed journal, is to sequence every base pair of a genome and become a foundation for downstream research by providing a contiguous list of genes, exact 5′ and 3′ coordinates for each gene, reconstruction of metabolic pathways while accounting for all genes in an organism, and other qualities not available from collections of large contigs.

There are some cases in which genome sequences from two isolates of the same species exist, but often one genome is held as proprietary information (this happens regularly for important microbial pathogens). Genome projects in which data are publicly available for two species of bacteria include those for human pathogens such as *Neisseria meningitidis Helicobacter pylori*, *Staphylococcus aureus*, and *Mycobacterium tuberculosis* to name a few. The latter case will be of special interest because CSU-93 is a highly infectious strain of *M. tuberculosis*, and it is possible that "*in silico*" comparisons of the genome sequences relative to the H37Rv strain may point to the mutation(s) or gene(s) responsible for this phenotype. Finally, comparisons by the Blattner laboratory of the completed *E. coli* K-12 strain to the enterohemorrhagic *E. coli* O157:H7 have provided some interesting findings. Although conventional wisdom would have suggested that the pathogenic O157 was a K-12 backbone plus some large segments of additional DNA, such as the 92-kb virulence plasmid or the 43-kb LEE pathogenicity island, it was determined from shotgun sequencing that the O157 genome is actually a fine mosaic of O157 and K-12 DNA and includes 1.2 Mb of DNA unique to O157 while lacking ~0.15 Mb of DNA known to be unique to the completed K-12 genome. These results provide a glimpse into the importance of comparative genomics work and the complex ways in which genomes/microbes evolve and adapt.

B. RNA-level differential gene expression

Researchers are interested in differential gene expression; that is, what genes are up- and down- regulated under a specific set of conditions (wild-type versus isogenic mutant, growth with and without an antibiotic present, and growth in the lab versus in a host for pathogens). Prior to genome sequences being widely available, differential display PCR was used to identify differentially expressed genes. Random primers are used for reverse transcription of message followed by amplification and agarose electrophoresis. Differentially labeled bands are isolated, cloned, and sequenced to determine the expressed genes. With eukarotic cells, quantitative EST sequencing or serial anaylsis of gene expression (SAGE) can also be used for differential gene expression without *a priori* knowledge of genome sequences, but the density of genes in bacteria coupled with the inability to apply mRNA-enrichment procedures based on polyadenylation has limited the use of these techniques in bacteria.

Molecular techniques previously used to study a single gene are now being applied to entire genomes. In RNA-level gene expression, researchers have applied DNA-blot hybridization and RT-PCR techniques to monitor the expression of thousands of genes at once. DNA microarrays can be thought of as high-throughput dot blots using DNA on a solid "chip" surface to report expression patterns for thousands of genes. Several different microarraying technologies are being used, and although costs continue to decrease the equipment is currently beyond the reach of the average lab. For synthetic arrays, such as the Affymetrix chip (*http://www.affymetrix.com*), photolithographic techniques are used to synthesize oligonucleotides on a 1.2-cm^2 silicon chip (hundreds of thousands of addressable positions per chip). For spotted arrays, such as those made by Stanford University (*http://cmgm.stanford.edu/pbrown*) or Molecular Dynamics (*http://www.mdyn.com*), an *x–y–z*-stage robot is used to deposit DNA spots, (usually PCR products but also ~70 base oligonucleotides), onto a treated glass microscope slide. In either case, DNA representing thousands of genes are linked to a solid surface, then hybridized with fluorescently labeled RNA from different conditions, followed by scanning to quantitate hybridized probe. In the case of spotted arrays, the use of different fluorescent dyes for each experimental condition allows for greater sensitivity and simultaneous hybridization of two or more samples on the same chip. Thus, expression data from thousands of genes (even entire genomes in the case of microbes) are then deconvoluted and analyzed.

C. Protein-level differential gene expression

Just as gene expression can be monitored at the transcription level, it can also be measured at the translation level by analyzing the entire protein complement of the genome, or proteome. At the core of these proteomics experiments is a separation technology that was developed more than two decades ago—two-dimensional polyacrylamide gel electrophoresis (2D-PAGE). Today, however, a new generation of instruments using mass spectroscopy (MS; *http://www.asms.org*) techniques afford highly accurate mass determination, giving a rapid, high-throughout path from a spot on a 2D-PAGE gel to an identified protein. Translation of whole-genome DNA sequence data gives the theoretical mass and isoelectric point for all protein gene products, and studies indicate that proteins identified from 2D-PAGE gels by electrospray MS or amino acid sequencing techniques correlate closely to these theoretical values (although approximately 20% of the spots represent isoforms, which are protein modifications that alter predicted migration positions of proteins). Variations on these techniques are being used, such as matrix-assisted laser desorption/ionization time-of-flight (MALDI-TOF) and surface-enhanced laser desorption/ionization (SELDI). The SELDI technique is an interesting twist on chip technology, giving a "protein chip" (Ciphergen, *http://www.ciphergen.com*) capable of on-chip protein enrichment and real-time estimates of protein mass.

One question that needs to be resolved in the postgenomic era is whether differential gene expression is best measured at the RNA or protein level; it is most likely, however, that researchers doing whole-genome expression studies will want to use both microarraying and proteomics technologies to assess gene expression levels.

D. Mutagenesis

With microbes, one of the classic methods for determining gene function is via mutagenesis. For many large-scale studies, transposons or antibiotic cassettes are used to disrupt genes, providing an isogenic pair of strains, a wild type, and a null mutant. By sequencing out from transposon or cassette, one can determine a "genome sequence tag" for the disrupted gene. In the pregenomic era, identification of the entire gene and surrounding genes/operons would have involved a substantial effort of cloning and sequencing, but because of the availability of genomic sequence data, the tag and sequence data allow one to rapidly map large numbers of mutants. The recent development of *in vitro* transposition systems has simplified the transposon-hopping step to a microfuge tube reaction and has great potential for use with naturally transformable organisms. Genome footprinting is another technique based on saturation mutagenesis that allows PCR-based scanning of a whole genome for gene essentiality under different experimental conditions. Finally, rather than being restricted to a random approach for mutagenesis, the availability of genome sequence data allows the directed knockout of genes in genetically amenable systems, and publicly announced programs to knockout every gene in the genome are proceeding for *Bacillus subtilis* (*http://locus.jouy.inra.fr/cgi-bin/ genmic/madbase/progs/madbase.operl*) and *Saccharomyces cerevisiae* (*http://sequence-www.stanford.edu/group/yeast_deletion_project deletions3.html*).

One advance that is invaluable for studies involving pools of mutants is oligonucleotide tag (or "bar-code") mutagenesis. In this technique, a unique oligonucleotide tag is included in the transposon/cassette and then introduced into each mutant. In the case of "signature-tagged mutagenesis" of pathogenic bacteria, pools of 96 bar-coded mutants are used to infect a host organism, the bacteria are collected, the DNA are isolated, and the bar code is amplified and then hybridized to a 96-spot array of DNA from each mutant. By comparing hybridization of the bar-code tag from *in vivo*-grown pools of mutants with that from *in vitro*-grown mutants, one can identify those mutants less successful at causing infection in a host—potential new virulence factors. Infection of mammals, as opposed to growing mutants in laboratory media, is a much more realistic environment for learning about microbial pathogenesis, and the bar-coding procedure has been very useful because it minimizes the number of experimental animals needed for infection. The whole-genome knockout program for *S. cerevisiae* incorporates bar codes into the cassettes, and preliminary experiments for monitoring pools of mutants have successfully employed microarray chips for the DNA–DNA hybridization step. Thus, the use of DNA microarrays for parallel processing of both mutant growth rates and global transcription patterns has enormous implications for changing molecular microbiology.

E. Other functional studies

Numerous functional studies in addition to those mentioned previously are being performed for sequenced bacteria. This list is not inclusive, and it should again be mentioned that the most well-coordinated functional analysis programs currently being performed are those for *B. subtilis* and *S. cerevisiae*. Methods currently being used include: (i) reporter fusion analyses, constructed for entire genomes, to provide gene expression data using whole cells; (ii) protein–protein interactions to determine what proteins interact with each other in a cell are providing important information to be used in conjunction with gene function studies; (iii) protein overexpression, using a high-throughput, brute-force protein overexpression methodology involving designing primers to the 5′ and 3′ ends of sequenced genes, cloning into T7 polymerase-driven vectors with affinity tags, and expressing and purifying the gene products using *E. coli*, with purified protein being used for numerous applications including *in vitro* screening for novel inhibitors, use as a component for vaccines, and for determining the three-dimensional structure of the expressed protein; (iv) structural genomics—in addition to modeling the theoretical structures of all the predicted proteins in a genome, determining the actual structure of expressed proteins of interest is a focal point for structural genomics, for use in rational drug design and for adding to the accumulating structure databases and helping to refine predictive programs; and (v) metabolic reconstruction studies, with initial efforts focusing on the well-studied *E. coli*, are being used as a framework for other microbes. Complete sequence data and excellent annotation are critical for all of these types of studies.

The trend for functional analysis work is increased throughput, which usually implies both miniaturization and automation. Besides the microarrays mentioned previously, miniaturization technology is progressing to the point of "lab-on-a-chip" design. These nanoreactors are capable of performing routine molecular biological techniques in tandem; for example, add a sample to the nanoreactor and DNA is extracted, purified, and separated by size for such applications as Southern hybridization or nucleotide sequencing. Automation is also apparent in any high-throughput sequencing lab, which incorporates robots for every step possible, including clone picking and arraying, plasmid purification, and numerous pipeting and reaction steps. Clearly, the automation that has provided an explosion of genome sequences will be applied to functional studies so that laboratory experimentation can keep pace with the high-throughput of genome sequencing.

V. CONCLUSIONS

The applications of high-throughput sequencing and computational methods that led to the first whole genome sequence for a living organism is revolutionizing microbiology. The present golden era of genome sequencing will soon lead to a challenging postgenomic era, in which laboratory experiments must be designed in order to determine how the thousands of genes in each microbe's genome function, interact, and allow these organisms to survive and evolve. To meet this challenge, researchers will become even more dependent on automation and computers to allow them to continually push the limits of what is achievable in genomic sequencing and functional genomics.

BIBLIOGRAPHY

Bankier, A. T. (2001). Shotgun DNA sequencing. *Methods Mol. Biol.* **167**, 89–100.

Doolittle, R. F. (2002). Biodiversity: microbial genomes multiply. *Nature* **416**, 697–700.

Dujon, B. (1998). European Functional Analysis Network (EUROFAN) and the functional analysis of the *Saccharomyces cerevisiae* genome. *Electrophoresis* **19**, 617–624.

Emilien, G., Misteli, C., Aveaux, D., and Maloteaux, J. M. (2001). Life after sequencing. *Trends Biotechnol.* **19**, 246.

Fleischmann, R. D., *et al.* (1995). Whole-genome random sequencing and assembly of Haemophilus influenzae *Rd. Science* **269**, 496–512.

Lander, E. S., and Waterman, M. S. (1988). Genomic mapping by finger printing random clones: a mathematical analysis. *Genomics* **2**, 231–239.

Marziali, A., and Akeson, M. (2001). New DNA sequencing methods. *Annu. Rev. Biomed. Eng.* **3**, 195–223.

Moszer, I. (1998). The complete genome of *Bacillus subtilis:* from sequence annotation to data management and analysis. *FEBS Lett.* **430**, 28–36.

Nelson, K. E., Paulsen, I. T., Heidelberg, J. F., and Fraser, C. M. (2000). Status of genome projects for nonpathogenic bacteria and archaea. *Nat. Biotechnol.* **18**, 1049–1054.

Nierman, W., Eisen, J. A., and Fraser, C. M. (2000). Microbial genome sequencing 2000: new insights into physiology, evolution and expression analysis. *Res. Microbiol.* **151**, 79–84.

Read, T. D., Gill, S. R., Tettelin, H. and Dougherty, B. A. (2001). Finding drug targets in microbial genomes. *Drug Discov. Today* **6**, 887–892.

Schena, M., Heller, R. A., Theriault, T. P., Konrad, K., Lachenmeier, E., and Davis, R. W. (1998). Microarrays: biotechnology's discovery platform for functional genomics. *Trends Biotechnol.* **16**, 301–306.

Searls, D. B. (2000). Bioinformatics tools for whole genomes. *Annu. Rev. Genomics Hum. Genet.* **1**, 251–279.

Tang, C. M., Hood, D. W., and Moxon, E. R. (1997). *Haemophilus* influence: the impact of whole genome sequencing on microbiology. *Trends Genet.* **13**, 399–404.

Washburn, M. P., and Yates, J. R., 3rd (2000). Analysis of the microbial proteome. *Curr. Opin. Microbiol.* **3**, 292–297.

WEBSITES

Links to Genome Centers
http://www.qiagen.com/bioinfo/

Comprehensive Microbial Resource of The Institute for Genomic Research (TIGR) and links to many other microbial genomic sites
http://www.tigr.org/tigr-scripts/CMR2/CMRHomePage.spl

Pedro's Biomolecular Research Tools. A collection of links to information and services useful to molecular biologists
http://www.public.iastate.edu/~pedro/research_tools.html

31

Ecology, Microbial

Michael J. Klug
Michigan State University

David A. Odelson
Invitrogen Corp., Carlsbad

GLOSSARY

biofilm Matrix-enclosed bacterial populations adherent to each other and/or to surfaces or interfaces.

cell sorter An instrument that uses optical or mechanical technologies which allow the separation of cells on the basis of size or cellular properties.

chemotaxis A movement response by microorganisms in which the stimulus is a chemical concentration gradient.

community An assemblage of populations of microorganisms which occur and interact within a given habitat.

ecosystem In terms of microorganisms, the totality of biotic and abiotic interactions to which the organism is exposed.

food web The interaction of communities of organisms with varying functional capabilities.

habitat A location where microorganisms occur.

interspecies hydrogen transfer The coupling or syntrophic relationship of hydrogen producers and hydrogen consumers.

Koch's postulates A concept embodied in these postulates defines the fundamental questions needed to address the function of a microbial population within a given habitat or its role in a function.

laser scanning confocal microscope An approach to microscopy which uses intense laser light beams, optics which exclude light from parts other than the specimen, and computer-assisted image enhancement to provide a nearly three-dimensional image without sample destruction or fixation.

microelectrode A micro version of pH, O_2, and specific ion electrodes which allow the exploration of microhabitats.

PCR Polymerase chain reaction; a method for increasing the number of copies of a target nucleic-acid sequence without having to culture the organism.

phylloplane The aboveground exposed surfaces of plants that are available for the colonization of microorganisms.

population A group of individuals of one species within a defined area or space.

probe A chemical or molecular technique which allows the detection or quantitation of a population or activity of microorganisms.

rhizosphere The region of soil which adheres to plant roots or is influenced by the activities of plant roots.

rumen A chamber anterior to the digestive tract of animals which harbors microorganisms that metabolize ingesta and provide metabolic intermediates to the animal, which serve as energy and biochemical precursors.

syntrophic An interpopulation interaction that involves two or more populations which provide nutritional requirements for each other.

Microorganisms are often considered to be among the first life forms on earth. Fascination with these beings was initially predisposed to their small size and simple forms. Early

The Desk Encyclopedia of Microbiology
ISBN: 0-12-621361-5

observations of microorganisms (in the seventeenth century) by Leeuwenhoek, with the aid of the first microscope, were followed by the demonstration of the role of microorganisms in the process of fermentation and spoilage by Pasteur and, eventually, the development of a means to prevent the growth of microorganisms, that is, pasteurization.

I. EVOLUTION OF MICROBIAL ECOLOGY AS A DISCIPLINE

Isolation of causative microorganisms of disease, as well as pure culture techniques, evolved in the late 1800s; Koch's postulates provided a solid basis for studying microorganisms and their roles. The era that followed provided for the continued isolation of microorganisms, the definition of their metabolic capabilities, and, in turn, their implied roles in important biogeochemical processes, that is, in the nitrogen and sulfur cycle. Attention was given to organisms from specific habitats, that is, soil, water, animal, and plants. The findings demonstrated the vast numbers of diverse microbiological forms and functions found in nearly every location that was sampled. Eventually, microbiological subdisciplines dealing with microbial associations of natural (soil, water) and manmade (food, industrial, and other) environments were established.

Within the last 30 years, it has been recognized that common microorganisms are observed in many habitats and common principles are involved in the mechanisms describing the associations of those microbes in varying habitats. The observation that individual populations of microorganisms are rarely found alone suggests potential interactions between populations and with their surrounding physicochemical environments. Additionally, the observation that they associate or align themselves within specific "strata" or gradients of physiochemical parameters points to the large number of metabolic functions of these organisms. They also have been shown to have the ability to sense (e.g. chemotaxis) and move by various means of locomotion within these gradients to maintain selected conditions for their growth.

The field of ecology is defined as a discipline of biology which deals with organisms' interactions with each other and surrounding environments. As one might expect, these aforementioned observations of microorganisms and their habitats led to the development of the subdiscipline known as microbial ecology. This development further emphasizes the need to establish a union between the examinations of the physicochemical nature of a habitat and microbiological investigations. An equally important emphasis in microbial ecology is the structure and activities associated with microbial communities rather than the earlier emphasis in causative populations of disease or specific processes.

II. MICROBIAL COMMUNITIES

Although observations of microorganisms and their associations have been made in numerous habitats, a few of these observations are felt to highlight the importance of the interactions between microorganisms which leads to community versus population responses.

The rumen ecosystem has received considerable attention, which has been driven principally by economic considerations. Research in this system has, however, defined the syntrophic relationship between members of the microbial fermentative community and the animal's growth and survival. Technically, approaches to this ecological niche present somewhat greater difficulties than those of soil or natural waters. The ecosystem is internal to the animal and the microbes are strict anaerobes. Mechanistically, however, rumen microbial ecology is relatively easier to discern, inasmuch as both input and output to and from the system are clearly defined, in much a similar relationship as an industrial bioreactor. It is interesting to note that syntrophic community-level interactions are frequently illustrated with the rumen system, in regard to hydrogen transfer and methane production.

The classic anaerobic food web first described by Hungate, and later by Wolf, which occurs in the rumen, involves the interpopulation interactions among bacterial communities capable of plant polysaccharide hydrolysis (e.g. cellulose), monomer fermenting communities, fatty acid-oxidizing communities, and, finally, terminal communities (e.g. methanogenesis), which oxidize fatty acids and reduce CO_2 to CH_4. The hydrogen is derived from previous oxidative steps. The coupling, or syntrophic, relationship between hydrogen producers and hydrogen consumers (interspecies hydrogen transfer) is now recognized as a fundamental relationship in other systems dominated by anaerobic microorganisms. In the rumen, rates and extent of metabolism are controlled by the interdependence of one community on another. The rumen system has also served as an example of a strategy for, an approach to, and methods to conduct similar investigations of animal–microbe associations, be it the crop of tropical birds or the intestinal tract of termites. Similar interactions have been observed in other anaerobic systems (such as sewage sludge, lake and marine sediments), which suggest common controls and mechanisms associated with metabolism of complex organic compounds in all of these systems. These relationships have also been shown to be involved in the metabolism of naturally occurring and manmade halogenated compounds. One such reaction involved dehalogenation by replacement of a halogen substituent of a molecule with a hydrogen atom. The hydrogen is derived from hydrogen-producing fermentative microorganisms.

Questions related to the response of these communities and the interactions among populations to disturbance of their physical–chemical environment will rely on tools which allow analyses of change in both the nature of the populations involved and changes in their functions. Do populations adapt and respond phenotypically or do they replace each other?

Traditional food web descriptions in aquatic and terrestrial ecosystems have often failed to consider the role of microbial interactions as a contribution to carbon and nitrogen cycling in these systems. Early recognition of these contributions was blurred by the examination of individual populations of organisms and by inadequately examining the interactions and controls of population size and distribution in the surrounding communities. Figure 31.1 illustrates, in its simplest form, the relationship among primary producers in aquatic/marine or terrestrial environments, heterotrophic bacteria, and phagotrophic protozoans or zooplankton.

Primary producers release soluble organic compounds (root exudates, algal metabolites), which are consumed by heterotrophic bacteria, which are subsequently grazed by phagotrophic protozoans or zooplankton. The grazers excrete nitrogen and phosphorus, which is used by the primary producers.

In sum, these observations have strikingly modified the contemporary view of the structure and mechanistic controls and regulation of growth of higher plant and animal forms in aquatic and terrestrial systems. They also point to the importance of a thorough understanding of these interactions if one is to consider management of these associations in applied applications, for example, sustainable agriculture or aquaculture.

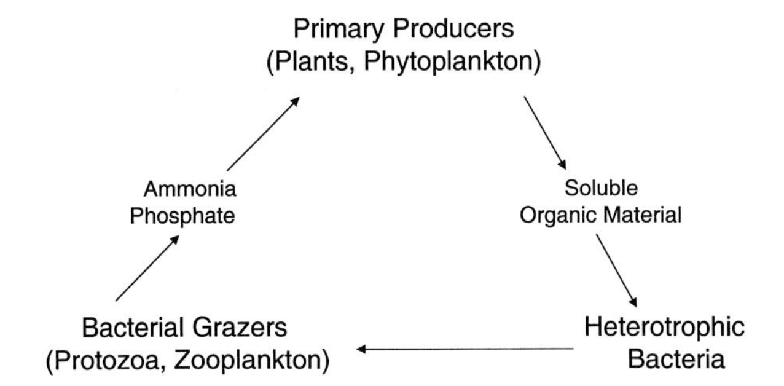

FIGURE 31.1 Relationship of primary producers, heterotrophic bacteria, and bacterial consumers in aquatic/marine and terrestrial environments.

Microbial biofilm communities were first described by Zobell and Anderson. Over the past two decades, their significance and ubiquity have been documented. It is now recognized that biofilm communities predominate numerically and metabolically in most ecosystems. Further, it is clear that biofilm cells are fundamentally phenotypically distinct from non-adhering cells. It now appears that the ability to form surface-associated structured biofilms is a common characteristic of, at least, bacteria. Remarkable intercellular and interspecies interactions, facilitated by chemicals released by these organisms, leads to complex structured communities, made up of both prokaryotic and eukaryotic organisms.

These examples, however, primarily involve interactions based on an exchange of chemical metabolites. The fact that these described interactions involve high densities of cells in close proximity to each other, for example, biofilm, can also lead to intercellular exchange of genetic information, which can lead to adaptive change in function within communities. Knowledge of naturally occurring phenotypic and genotypic viability of microorganisms has been hampered because the majority of our knowledge comes from laboratory-selected strains, grown under controlled conditions, and on a medium which has no resemblance to the environment from which they were selected. It is important to realize that approaches to microbial ecology must recognize not only isolated individual populations but, more directly, the interactions of populations within a community and the resultant effect on the overall functions of the community.

III. APPROACHES TO MICROBIAL ECOLOGY

The fundamental approach to higher plant and animal ecological studies uses quantitative observations of specific populations in various environments. For over 100 years, plant and animal ecologists have observed and detailed the frequency of occurrence of specific plant and animal populations under various environmental conditions. These observations have led to the development of models for predicting the occurrence of, as well as relationship between, the organisms and the associated environments in which they were observed. Recently, these observations have been complemented with physiological and genetic approaches that suggest specific mechanisms of selection which have led to the observed frequency or distribution of organisms. Unfortunately, our understanding of microbial systems is still embryonic in both description and prediction.

Kluyver and van Neil, in 1956, estimated that about one half the "living protoplasm" on earth is microbial. This estimate is now considered conservative, and the sheer numbers represent a daunting challenge to the microbial ecologist. Even the most modern techniques do not allow routine quantification of microbial numbers and definition of specific populations. It is principally the size of the microorganism, and, equally important, the size of the local habitat, that creates a considerable constraint on the observation of microorganisms in natural surroundings. Numbers of organisms ranging from 1 million cells per milliliter of water to 10 billion cells per gram of human fecal material. Further, the spatial distribution of these organisms within these habitats makes it difficult to recover a representative sample to examine. *In situ* observations are further complicated by our inability to observe this environment without disturbing the microorganism's natural habitat. The various surfaces and potential differences in chemistries of the habitat provide varying degrees of carbon and nitrogen resources and physicochemical environments (e.g. gaseous exchange) for colonizing microorganisms. Additionally, other organisms, micro- and macro-alike, have a potential impact on the native organisms. Grazing of microorganisms by other microbes or faunal components of a habitat have a significant impact on the numbers and types of microorganisms present.

As noted, microbial ecology evolved as a cross-discipline of standard microbiology and environmental analytical analysis. The activity of a specific population or community of microorganisms was inferred by estimating their relative numbers by direct microscope counts, viable colonial or turbidimetric determinations, or specific chemical analyses (e.g. chlorophyll for algae). Unfortunately, limitations by both the microbial and the environmental analyses have, in most cases, failed to accurately define the ecology of microbial habitats. Recently, novel analytical and microbiological methodologies have provided tools for expanding our view of microbial habitats, in a fashion that is neither constrained by the microscale of the environment, nor limited by our ability to selectively cultivate members of a community.

In situ observation of microorganisms has been expanded through use of laser confocal microscopy, an advancement which provides for a kind of "X-ray" imaging of material requiring neither disturbance nor fixation. Similarly, laser optical trapping allows the removal of individual cells from a habitat. These techniques provide new insights to our understanding of population components of communities. These increases in optical resolution also provide a means of sorting communities either by size or chemical factors through the use of cell sorters.

Advancements have also been made in developing methodology to simulate the microhabitats and physiochemical gradients which occur therein. It is interesting to note that one of the pioneering environmental microbiologists, Winogradsky simulated gradients of light, water, sulfide, and oxygen to describe relationships between sulfide-oxidizing photo-autotrophic bacteria, sulfate-reducing, and chemoautotrophic sulfur-oxidizing organisms in his Winogradsky column. Advances using gels and gradostats to simulate diffusion barriers allow us to expand our knowledge concerning relationships between the spatial heterogeneity of the physicochemical environment and the distribution of diverse groups of microorganisms. Combined with microelectrode techniques, precise analytical measurements will complement these microbial investigations. Although our abilities to observe organisms and to better understand the physical and chemical nature of their habits have improved, we still are unable to isolate a high percentage of these observable organisms. Nevertheless, within the last 30 years, the number of previously undescribed microorganisms has increased significantly.

In an effort to forego the inherent problems of the microscopic and numerical diversity of the microbial world, recent method development has approached the microbial component of the unseen habitats at the macromolecular level. The recognition that diversity is intrinsically related to the organism's genetics has provided a sound basis for separation and characterization of both microscopic and macroscopic life.

Examination of the heterogenity of DNA in soil suggests that as many as 10,000 bacterial species can be harbored in 100 g of soil. This estimate only includes the bacterial fraction and, again, reinforces the challenge presented to understanding the diversity of microorganisms in natural systems. From an ecological perspective, it also presents a challenge to understanding how so many species coexist in such small areas.

Recombinant DNA technology has allowed the identification and determination of the specific nucleotide sequence of genes. It has provided ecologists with novel methods to pursue community structure. In practice, microbial community samples can be analyzed without cultivation or microscopic observation of microorganisms. In brief, nucleic acid is isolated from the sample and utilized for hybridization studies with the corresponding gene of interest (i.e. the probe). This probe can represent either a metabolic gene or a systematic determinant, such as the 16S rRNA gene sequence. In turn, as one may infer, the sample can be evaluated in terms of population diversity at the species level or community diversity at the kingdom level (Eukaryotic versus Prokaryotic). All of these analyses have been expanded by inclusion of the polymerase chain reaction (PCR), a method allowing amplification and subsequent detection of as few as 10 microorganisms. Interestingly, these methods have, in fact, relied on previous isolation and characterization of specific microbial populations prior to isolation of a gene for use as a probe. These investigations have, however, illustrated the universal nature of certain sequences, such as from 16S rRNA genes, for random use in community analyses. An example of this technology is the use of various profiling techniques. Separate PCR-amplified 16S rRNA gene-fragment sequences are separated, and the resulting numbers provide an estimate of diversity within the community.

Areas that still remain to be explored in microbial ecology are the relationship of microbial diversity to the response of microbial communities to varying degrees of disturbance and the relationship between diversity and the function of the community. In higher plant and animal systems, the relationship of diversity within communities to their resistance to change or recovery after disturbance (e.g. fire, tillage) has been discussed and debated for decades. Various indices have been used to calculate diversity within communities. In their simplest form, these indices represent the number of species found within the community; therefore, communities with many species are described as having high diversity. Other indices relate the dominance of specific species to the total diversity, such that even communities with high diversity can have a few populations which dominate activity within the community. Our current inabilities to adequately isolate all microorganisms and the lack of distinctive morphological characteristics fail to provide us with accurate methods for measuring indices of diversity. Continued improvements in our analytical skills in identifying the macromolecular characteristics of microorganisms will provide a basis for estimating diversity at the phylogenetic level.

Profiles of cellular or phospholipid fatty acids from lipids extracted from communities or specific chemical or antigenic determinants may provide a snapshot of changes within microbial communities after disturbance. Community-level assays of the functional diversity of populations based on carbon source utilization also provide indications of changes following disturbance.

IV. FUTURE DIRECTIONS IN MICROBIAL ECOLOGY

It has often been said that advancements in microbial ecology are limited by the methods which are available to analyze microbial systems. Some of the advances made in the analytical and molecular methods over the past decade have been illustrated in this overview. Although continued advancements in the use of microelectrodes and optical methods increase our abilities to measure and observe microorganisms emphasis must increase on the refinement of techniques to discern changes in microbial community structure and the impact of these changes on the function of the community. A significant area of contemporary concern is the remediation of habitats contaminated with anthropogenic sources of organic and inorganic compounds.

The applied area of bioremediation is, and will continue to be, a timely subject in the decades to come. The use and management of the intrinsic properties of microbial communities will provide stimulus for applied microbial ecology. Although this area is yet to be accurately defined, in principle, the directive is to promote microbial dissimilation of anthropogenic compounds in a controlled manner.

In a similar manner, a reduced input of anthropogenic chemicals in agricultural systems to reduce environmental contamination will result in greater reliance on activities of the soil microbial communities.

These and other applications require further understanding of the structure and activities of microbial communities. Also required is an understanding of changes in the structure and function of these communities in relation to changes in the physiochemical environments associated with them. This, in fact, is microbial ecology.

BIBLIOGRAPHY

Amann, R. I., Binder, B. J., Olson, R. J., Chisholm, S. W., Devereux, R., and Stahl, D. A. (1990). Combination of 16S rRNA-targeted oligonucleotide probes with flow cytometry for analyzing mixed microbial populations. *Appl. Environ. Microbiol.* **56**, 1919–1925.

Arndt-Jovin, D. J., Robert-Nicoud, M., Kaufman, S. J., and Jovin, T. M. (1985). Fluorescence digital imaging microscopy in cell biology. *Science* **235**, 247–256.

Atlas, R. M., and Bartha, R. (1998). *In* "Microbial Ecology." Benjamin and Cummings, Menlo Park, CA.

Boltner, D. (2001). Ecology and industrial microbiology, Techniques, Web alert. *Curr. Opin. Microbiol.* **4**, 233–234.

Caldwell, D. E., Colwell, R. R., Harayama, S., Kjelleberg, S., Givskov, M., Molin, S., Nielsen, A. T., Heydorn, A.,

Tolker-Nielsen, T., Sternberg, C., Nealson, K., Pace, N. R., Rainey, P., Stackebrandt, E., Tindall, B. J., and Stahl, D. A. (2000). Crystal ball: leading scientists in the field of environmental microbiology consider the technical and conceptual developments that they believe will drive innovative research during the first years of the new millennium. *Environ. Microbiol.* **2**, 3–10.

Clarholm, M. (1994). The microbial loop in soil. *In* "Beyond the Biomass Compositional and Functional Analysis of Soil Microbial Communities" (K. Ritz, J. Dighton, and K. E. Giller, eds.), pp. 221–230. John Wiley and Sons.

Costerton, J. W., Lewandowski, Z., Caldwell, D. E., Korber, D. R., and Lappin-Scott, H. M. (1995). Microbial Biofilms. *Annu. Rev. Microbiol.* **49**, 711–745.

Garland, J. L. (1996). Patterns of potential C source utilization by rhizosphere communities. *Soil Biol. Biochem.* **28**, 223–230.

Geesey, G., Atlas, R., Burlage, R. S., and Stahl, D. (1998). "Techniques in Microbial Ecology." Oxford University Press, Oxford, UK.

Hedrick, D. B., Richards, B., Jewell, W., Guckert, J. B., and White, D. C. (1991). Disturbance, starvation, and overfeeding stresses detected by microbial lipid biomarkers in high-solids high-yield methanogenic reactors. *J. Indust. Microbiol.* **9**, 91–98.

Kirchman, D. L. (2000). "Microbial Ecology of the Oceans." John Wiley & Sons, New York.

Kluyver, A. J., and van Neil, C. B. (1956). "The Microbe's Contribution to Biology." Harvard University Press, Cambridge, MA.

Lander, E. S., and Waterman, M. S. (1988). Genomic mapping by fingerprinting random clones: A mathematical analysis. *Genomics.* **2**, 231–239.

Maier, R. M., Pepper, I. L., and Gerba, C. P. (2000). "Environmental Microbiology." Academic Press, San Diego, CA.

Mohn, W. W., and Tiedje, J. M. (1992). Microbial reductive dechlorination. *Microbiol. Rev.* **56**, 482–507.

Revsbech, N. P., and Jorgensen, B. B. (1986). Microelectrodes: Their use in microbial ecology. *In* "Advances in Microbial Ecology" (K. C. Marshall, ed.), pp. 293–352. Plenum Press, New York.

Santegoeds, C. M., Nold, S. C., and Ward, D. M. (1996). Denaturing gradient gel electrophoresis used to monitor the enrichment culture of aerobic chemoorganotrophic bacteria from a hot spring cyanobacterial mat. *Appl. Environ. Microbiol.* **62**, 3922–3928.

Shapiro, J. A. (1998). Thinking about bacterial populations as multicellular organisms. *Annu. Rev. Microbiol.* **52**, 81–104.

Steffan, R. J., and Atlas, E. J. (1988). DNA amplification to enhance the detection of genetically engineered bacteria in environmental samples. *Appl. Environ. Microbiol.* **54**, 2185–2191.

Torsvik, V., Goksøyr, J., and Daae, F. L. (1990). High diversity in DNA of soil bacteria. *Appl. Environ. Microbiol.* **56**, 782–787.

Whitman, W. B., Coleman, D. C., and Wiebe, W. J. (1998). Prokaryotes: The unseen majority. *Proc. Natl. Acad. Sci. USA* **95**, 6578–6583.

Wolin, M. J., and Miller, T. L. (1982). Interspecies hydrogen transfer: 15 years later. *ASM News* **48**, 561–565.

WEBSITES

Environmental Microbiology Division, ASM (American Society for Microbiology)
http://www.asmusa.org/division/n/

Environmental & General Applied Microbiology Division, ASM (American Society for Microbiology)
http://www.asmusa.org/division/q/index.html

Microbiology and Microbial Ecology Sites. Center for Microbial Ecology, Michigan Sttae University
http://www.cme.msu.edu/CME/OTHER_SITES/micro.sites.html

Website of the International Society for Microbial Ecology
http://www.microbes.org/

List of Bacterial names with Standing in Nomenclature (J. P. Euzéby)
http://www.bacterio.cict.fr/index.html

32

Emerging infections

David L. Heymann
World Health Organization

GLOSSARY

amplification of transmission The increased spread of infectious disease that occurs naturally or because of facilitating factors, such as nonsterilized needles and syringes, that can result in an increase in transmission of infections such as hepatitis.

anti-infective (drug) resistance The ability of a virus, bacterium, or parasite to defend itself against a drug that was previously effective. Drug resistance is occurring for bacterial infections such as tuberculosis and gonorrhoea, for parasitic infections such as malaria, and for the human immunodeficiency virus (HIV).

eradication The complete interruption of transmission of an infectious disease and the disappearance of the virus, bacterium, or parasite that caused that infection. The only infectious disease that has been eradicated is smallpox, which was declared eradicated in 1980.

International Health Regulations Principles for protection against infectious diseases aimed at ensuring maximum security against the international spread of infectious disease. The Regulations provide public health norms and standards for air- and seaports to prevent the entry of infectious diseases, and require reporting to the World Health Organization (WHO) the occurrence of three infectious diseases: cholera, plague, and yellow fever.

re-emerging infection A known infectious disease that had fallen to such low prevalence or incidence that it was no longer considered a public health problem, but that is presently increasing in prevalence or incidence. Re-emerging infections include tuberculosis, which has increased worldwide since the early 1980s, dengue in tropical regions, and diphtheria in eastern Europe.

Emerging Infections are newly identified and previously unknown infectious diseases. Since 1970 there have been over 30 emerging infections identified, causing diseases ranging from diarrheal disease among children, hepatitis, and AIDS to Ebola hemorrhagic fever.

I. A 25-YEAR PERSPECTIVE

In the Democratic Republic of the Congo (DRC, formerly Zaire), the decrease in smallpox vaccination coverage, poverty, and civil unrest causing humans to penetrate deep into the tropical rain forest in search of food may have resulted in breeches in the species barrier between humans and animals, causing an extended and continuing outbreak of human monkeypox. During the 1970s and 1980s, when this zoonotic disease was the subject of extensive studies, it was shown that the monkeypox virus infected humans, but that person-to-person transmission beyond three generations was rare. The outbreak of

The Desk Encyclopedia of Microbiology
ISBN: 0-12-621361-5

TABLE 32.1 Principal newly identified infectious organisms associated with diseases[a]

Year	Newly identified organism	Disease (year and place of first recognized or documented case)
Diseases primarily transmitted by food and drinking water		
1973	Rotavirus	Infantile diarrhea
1974	Parvovirus B19	Fifth disease
1976	*Cryptosporidium parvum*	Acute enterocolitis
1977	*Campylobacter jejuni*	Campylobacter enteritis
1982	*Escherichia coli* 0157:H7	Haemorrhagic colitis with haemolytic uremic syndrome
1983	*Helicobacter pylori*	Gastric ulcers
1986	*Cyclospora cayatanensis*	Persistent diarrhoea
1989	Hepatitis E virus	Enterically transmitted non-A and non-B hepatitis (1979, India)
1992	*Vibrio cholerae* 0139	New strain of epidemic cholera (1992, India)
Unclear modes of transmission, thought to be primarily transmitted by drinking water		
1985	*Enterocytozoon bieneusi*	Diarrhea
1991	*Encephalitozoon hellem*	Systemic disease with conjunctivitis, in AIDS patients
1993	*Encephalitozoon cunicali*	Parasitic disseminated disease, seizures (1959, Japan)
1993	*Septata intestinalis*	Persistent diarrhea in AIDS patients
Diseases primarily transmitted by close contact with infectious individuals, excluding sexually transmitted diseases, nosocomial infections and viral haemorrhagic fevers		
1980	HTLV-1	T-cell lymphoma leukemia
1982	HTLV II	Hairy cell leukemia
1988	HHV-6	Rosela subitum
1993	*Influenza A*/Beijing/32 virus	Influenza
1995	HHV-8	Associated with Kaposi sarcoma in AIDS patients
1995	*Influenza A*/Wuhan/359/95 virus	Influenza
2003	SARS coronavirus	Severe acute respiratory syndrome
Sexually transmitted diseases		
1983	HIV-1	AIDS (1981)
1986	HIV-2	Less pathogenic than HIV-1 infection
Nosocomial and related infections		
1981	Staphylococcus toxin	Toxic shock syndrome
1988	Hepatitis C	Parenterally transmitted non-A non-B hepatitis
1995	Hepatitis G viruses	Parenterally transmitted non-A non-B hepatitis
Human zoonoses and vector-borne diseases, including viral haemorrhagic fevers, transmitted by close contact with animals or animal products, excluding food-borne diseases		
1977	Hantaan virus	Hemorrhagic fever with renal syndrome (1951)
1990	Reston strain of Ebola virus	Human infection documented but without symptoms (1990)
1991	Guanarito virus	Venezuelan hemorrhagic fever (1989)
1992	*Bartonella henselae*	Cat-scratch disease (1950s)
1993	Sin nombre virus	Hantavirus pulmonary syndrome (1993)
1994	Sabiä virus	Brazilian hemorrhagic fever (1955)
1997	Influenza A (H5N1)	Avian influenza
1998	Nipah virus	Severe encaphalitis
Tick-borne		
1982	*Borrelia burgdorferi*	Lyme disease (1975)
1989	*Ehrlichia chaffeensis*	Human ehrlichiosis
1991	New species of *Babesia*	Atypical babesiosis
Unknown animal vector		
1977	Ebola virus	Ebola hemorrhagic fever (1976, Zaire and Sudan)
1994	Ebola virus, Ivory Coast strain	Ebola hemorrhagic fever
Soil-borne diseases, airborne diseases and diseases associated with recreational water with no evidence of direct person-to-person transmission		
1976	*Legionella pneumophilia*	Legionellosis (1947)

[a]From G. Rodier, WHO.

human monkeypox in 1996–1997, and continuing intermittently since then, is a clear example of the ability of infectious diseases to exploit weaknesses in our defenses against them.

Numerous infectious diseases have found weakened entry points into human populations and emerged or re-emerged since the 1970s (see Table 32.1). In the early to mid 1970s, for example, classic dengue fever had just begun to reappear in Latin American after it had been almost eliminated as a result of mosquito control efforts in the 1950s and 1960s. Today, dengue is hyperendemic in most of Latin America. In 2001 alone, Latin America reported over 609,000 cases of dengue, of which 15,000 were of the hemorrhagic form. These figures represent more than a doubling of cases reported for the same region in 1995. Outbreaks are also becoming more explosive. An outbreak in Brazil in 2002 caused over 500,000 cases in one of the largest outbreaks ever recorded. A pandemic in 1998, in which 1.2 million cases of dengue fever and dengue haemorrhagic fever were reported from 56 countries, was unprecedented. In 1991, cholera, which had not been reported in Latin America for over 100 years, re-emerged in Peru with over 320,000 cases and nearly 3000 deaths, and rapidly spread throughout the continent to cause well over 1 million cases in a continuing and widespread epidemic.

In North America, *Legionella* infection was first identified in 1976 in an outbreak among war veterans attending a conference in Philadelphia (US). Legionellosis is now known to occur worldwide and poses a threat to travellers exposed to poorly maintained air conditioning systems. In the Netherlands, in 1999, an outbreak of legionellosis occurred which was subsequently traced to whirlpool baths exhibited at a flower show visited by 80,000 people. Local cooling towers are considered the likely source of a large outbreak of legionellosis, involving 751 cases and two deaths, that occurred in Spain in 2001.

During this same 25-year period, a new disease in cattle, bovine spongiform encephalopathy, was identified in the United Kingdom in 1986 and, by 2002, had spread to 19 additional countries around the world. The disease is associated with the appearance in 1996 of a previously unknown variant of the invariably fatal Creutzfeldt–Jakob disease that had occurred in over 100 persons by the beginning of 2002. Within a decade, food-borne infection by *E. coli* O157, unknown in the 1970s, had become a food-safety concern in Japan, Europe, and in the Americas. Hepatitis C was first identified in 1989 and is now thought to be present in at least 3% of the world's population, while hepatitis B has reached levels exceeding 90% in populations at high risk from the tropics to eastern Europe. Tuberculosis, including multidrug-resistant forms, took advantage of weakened infrastructures in the countries of the former Soviet Union and re-emerged, with cases more than doubling in less than 7 years and with over 20% of patients in prison settings infected with multidrug-resistant strains in 2001. In the Hong Kong Special Administrative Region of China, 18 cases of human infection with influenza A virus sub-type H5N1, previously confined to birds, occurred in 1997, causing six deaths and raising considerable alarm. Human African trypanosomiasis, which had been virtually eliminated in the 1960s, resurged in an epidemic that is currently thought to infect from 300,000 to 500,000 people.

In 1976, the Ebola virus was identified for the first time as causing a disease that has come to symbolize emerging diseases and their potential impact on populations without previous immunological experience. The largest recorded outbreak, which began in Uganda in 2001, caused 425 confirmed cases and 224 deaths. Altogether, Ebola has caused just under 2000 cases and around 1250 deaths since its identification in simultaneous outbreaks in Zaire and Sudan. In 1976, at the time of the first Ebola outbreak in Zaire, HIV seroprevalence was already almost 1% in some rural parts of Zaire, as shown retrospectively in blood that had been drawn from persons living in communities around the site of the 1976 outbreak, and HIV has since become a preoccupying problem in public health worldwide.

II. MISPLACED OPTIMISM

In this same 25-year period the eradication of smallpox was achieved. This unparalleled public health accomplishment resulted in immeasurable savings in human suffering, death, and money, and stimulated other eradication initiatives. However, recent concerns about the possible deliberate use of variola virus as a biological weapon prompted some countries to consider introduction of population-wide preemptive vaccination, but after close deliberation WHO recommendations of October 2001 were accepted by most countries. The WHO is currently leading international initiatives to eradicate poliomylitis and dracunculiasis and to eliminate African trypanosomiasis, Chagas disease, leprosy, lymphatic filariasis, onchocerciasis, and blinding trauchoma. Since the global polio eradication initiative was launched in 1988, three regions have been certified as free of the disease: the Americas in 1994, the Western Pacific in 2000, and Europe in 2002. Reported cases of polio have dropped from an estimated 350,000 cases in 125 countries in 1988 to 480 cases reported to WHO in 2001 from 10 endemic countries. Reported cases of dracunculiasis have decreased from over 900,000 in 1989 to less than 64,000 in 2001, with the majority of cases in one endemic country. Of the several diseases targeted for elimination, goals have been reached for the original onchocerciasis programme in West Africa and for leprosy, where global elimination was declared in 2001. Chagas disease, likewise, continues its downward trend towards elimination.

The eradication of smallpox boosted an already growing optimism that infectious diseases were no longer a threat, at least to industrialized countries. This optimism had prevailed in many industrialized countries since the 1950s, a period that saw an unprecedented development of new vaccines and antimicrobial agents and encouraged a transfer of resources and public health specialists away from infectious disease control. Optimism is now being replaced by an understanding that the infrastructure for infectious disease surveillance and control has suffered and in some cases become ineffective. A combination of population shifts and movements with changes in environment and human behavior has created weaknesses in the defense systems against infectious diseases in both industrialized and developing countries. These weaknesses have recently come into sharp focus as countries consider preparedness plans for responding to the possible deliberate use of biological agents and recognize the importance of strong public health systems as the first line of defense for infectious disease outbreaks, no matter what their origin.

III. WEAKNESSES FACILITATING EMERGENCE AND RE-EMERGENCE

The weakening of the public health infrastructure for infectious disease control is evidenced by failures such as in mosquito control in Latin America and Asia with the re-emergence of dengue now causing major epidemics; in the vaccination programs in eastern Europe during the 1990s, which contributed to the re-emergence of epidemic diphtheria and polio; and in yellow fever vaccination, facilitating yellow fever outbreaks in Latin America and sub-Saharan Africa, including a large urban outbreak that occurred in Côte d'Ivoire in 2001. It is also clearly demonstrated by the high levels of hepatitis B and the nosocomial transmission of other pathogens such as HIV in the former USSR and Romania, and the nosocomial amplification of outbreaks of Ebola in Zaire, where syringes and failed barrier nursing drove outbreaks into major epidemics.

Population increases and rapid urbanization during this 25-year period have resulted in a breakdown of sanitation and water systems in large coastal cities in Latin America, Asia, and Africa, promoting the transmission of cholera and shigellosis. In 1950, there were only two urban areas in the world with populations greater than 7 million, but by 1990 this number had risen to 23, with increasing populations in and around all major cities, challenging the capacity of existing sanitary systems.

Anthropogenic or natural effects on the environment also contribute to the emergence and re-emergence of infectious diseases. The effects range from global warming and the consequent extension of vector-borne diseases, to ecological changes due to deforestation that increase contact between humans and animals, and also the possibility that microorganisms will breach the species barrier. These changes have occurred on almost every continent. They are exemplified by zoonotic diseases such as Lassa fever first identified in West Africa in 1969 and now known to be transmitted to humans from human food supplies contaminated with the urine of rats that were in search of food, as their natural habitat could no longer support their needs. In Latin America, Chagas disease emerged as an important human disease after mismanagement of deforested land caused triatomine populations to move from their wild natural hosts to involve human beings and domestic animals in the transmission cycle, eventually transforming the disease into an urban infection that can be transmitted by blood transfusion. Other zoonotic diseases include Lyme borreliosis in Europe and North America, transmitted to humans who come into contact with ticks that normally feed on rodents and deer, the reservoir of *Borrelia burgdorferi* in nature; and the Hantavirus pulmonary syndrome in North America. The narrow band of desert in sub-Saharan Africa, in which epidemic *Neisseria meningitidis* infections traditionally occur, has enlarged as drought spread south, so that Uganda and Tanzania experience epidemic meningitis, while outbreaks of malaria and other vector-borne diseases have been linked to the cutting of the rainforests. A 1998 outbreak of Japanese encephalitis in Papua New Guinea has been linked to an extensive drought, which led to increased mosquito breeding as rivers dried into stagnant pools. The virus is now widespread in Papua New Guinea and threatening to move farther east. Buruli ulcer, a poorly understood mycobacterial disease that has emerged dramatically over the past decade, has erupted following significant environmental disturbances, and some evidence suggests that recent explosive increases in Africa are linked to deforestation and subsequent flooding, or to the construction of dams and irrigation systems.

And finally, human behavior has played a role in the emergence and re-emergence of infectious diseases, best exemplified by the increase in gonorrhea and syphilis during the late 1970s, and the emergence and amplification of HIV worldwide, which are directly linked to unsafe sexual practices.

IV. FURTHER AMPLIFICATION

The emergence and re-emergence of infectious diseases are also amplified by two major factors—the continuing and increasing evolution of anti-infective (drug) resistance (see Table 32.2), and dramatic increases in international travel. Anti-infective agents are the basis for the management of important public health problems such as tuberculosis, malaria, sexually transmitted diseases, and lower-respiratory infections. Shortly after penicillin became widely available in 1942, Fleming sounded the first warning of the potential importance of the development of resistance. In 1946, a hospital in the United Kingdom reported that 14% of all *Staphylococcus aureus* infections were resistant to penicillin, and by 1950 this had increased to 59%. In the 1990s, penicillin-resistant *S. aureus* had attained levels greater than 80% in both hospitals and the community. Levels of resistance of *S. aureus* to other anti-infectives, and among other bacteria increased with great rapidity. By 1976, chloroquine resistant *Plasmodium falciparum* malaria was highly prevalent in southeastern Asia and 20 years later was found worldwide, as was high-level resistance to two back-up drugs, sulfadoxine-pyrimethamine and mefloquine. In the early 1970s, *Neisseria gonorrhoeae* that was resistant to usual doses of penicillin was just being introduced into Europe and the United States from Southeast Asia, where it is thought to have first emerged. By 1996, *N. gonorrhoeae* resistance to penicillin had become worldwide, and strains resistant to all major families of antibiotics had been identified wherever these antibiotics had been widely used. Countries in the western Pacific, for example, have registered quinolone resistance levels up to 69%.

TABLE 32.2 Resistance of common infectious diseases to anti-infective drugs, 1998

Disease	Anti-infective drug	Range (%)
Acute respiratory infection (*S. pneumoniae*)	Penicillin	12–55
Diarrhea (*Shigella*)	Ampicilline	10–90
	Trimethoprim	
	Sulfamethoxazole	9–95
Gonorrhea (*N. gonorrhoeal*)	Penicillin	5–98
Malaria	Chloroquine	4–97
Tuberculosis	Rifampicin	2–40
	Isonizid	

The mechanisms of resistance, a natural defense of microorganisms exposed to anti-infectives, include both spontaneous mutation and genetic transfer. The selection and spread of resistant strains are facilitated by many factors, including human behavior in over-prescribing drugs, in poor compliance, and in the unregulated sale by non-health workers. In Thailand, among 307 hospitalized patients, 36% who were treated with anti-infective drugs did not have an infectious disease. The over-prescribing of anti-infectives occurs in most other countries as well. In Canada, it has been estimated of the more than 26 million people treated with anti-infective drugs, 50% were treated inappropriately. Findings from community surveys of *Escherichia coli* in the stool samples of healthy children in China, Venezuela, and the United States suggest that although multiresistant strains were present in each country, they were more widespread in Venezuela and China, countries where less control is maintained over antibiotic prescribing. Animal husbandry and agriculture use large amounts of anti-infectives, and the selection of resistant strains in animals, which then genetically transfer the resistance factors to human pathogens or infect humans as zoonotic diseases, is a confounding factor that requires better understanding. Direct evidence exists that four multiresistant bacteria infecting humans, *Salmonella*, *Campylobacter*, *Enterococci*, and *Escherichia coli*, are directly linked to resistant organisms in animals (WHO Conference on the Medical Impact of the Use of Antimicrobial Drugs in Food Animals, Berlin, 13–17 October 1997).

Infections with resistant organisms require increased length of treatment with more expensive anti-infective drugs or drug combinations, and a doubling of mortality has been observed in some resistant infections. At the same time, fewer new antibiotics reach the market, possibly in part due to the financial risk of developing a new anti-infective drug that may itself become ineffective before the investment is recovered. There is no new class of broad-spectrum antibiotic currently on the horizon. Among the major infectious diseases, the development of resistance to drugs commonly used to treat malaria and tuberculosis is of particular public health concern, as is the emerging resistance to anti-HIV drugs. Even if the pharmaceutical industry were to accelerate efforts to develop new antimicrobials immediately, current trends suggest that some diseases may have very few and, in some cases, no effective therapies within the next 10–20 years. Many important medical and surgical procedures, including cancer chemotherapy, bone marrow and organ transplantation, and hip and other joint replacements, could no longer be undertaken out of fear that the associated compromise of immune function might place patients at risk of acquiring a difficult to treat and ultimately fatal infection.

The role of travel in the spread of infectious diseases has been known for centuries. Because a traveller can be in a European or Latin American capital one day and the next day be in the center of Africa or Asia, humans, like mosquitoes, have become important vectors of disease. During the 1990s, over 500 million people travelled by air each year (World Tourism Organization), and contributed to the growing risk of exporting or importing infection or drug-resistant organisms. In 1988, a clone of multiresistant *Streptococcus pneumoniae* first isolated in Spain was later identified in Iceland. Another clone of multiresistant *S. pneumoniae,* also first identified in Spain, was subsequently found in the United States, Mexico, Portugal, France, Croatia, Republic of Korea, and South Africa. A study conducted by the Ministry of Health of Thailand on 411 exiting tourists showed that 11% had an acute infectious disease, mostly diarrheal, but also respiratory infections, malaria, hepatitis, and gonorrhea (B. Natth, personal communication). Forced migration such as by refugees is also associated with the risk of re-emergence and spread of infectious diseases. In 1999, more than 7 million people around the world were newly uprooted. Currently, more than 35 million people are refugees or internally displaced persons. In a refugee population estimated to be between 500,000 and 800,000 in one African country in 1994, an estimated 60,000 developed cholera in the first month after the influx, and an estimated 33,000 died.

V. SOLUTIONS

Eradication and regulation may contribute to the containment of infectious diseases, but do not replace sound public health practices that prevent the weaknesses through which infectious diseases penetrate. Eradication was successful for smallpox and is advancing for poliomyelitis with virus transmission interrupted in the Americas, the Western Pacific, and Europe. Eradication or elimination applies to very few infectious diseases—those that have no reservoir other than humans, that trigger solid immunity after infection, and for which there exists an affordable and effective intervention. The development of powerful new medicines that are safe, affordable, and suitable for single-dose administration in mass campaigns has made it possible for WHO and the international community to launch recent campaigns to eliminate leprosy, lymphatic filariasis, and onchocerciasis and to consider the elimination of schistosomiasis and soil-transmitted helminthiasis in areas where control programmes have succeeded in achieving low transmission.

Attempts at regulation to prevent the spread of infectious diseases were first recorded in 1377 in quarantine legislation to protect the city of Venice from plague-carrying rats on ships from foreign ports. Similar legislation in Europe, and later the Americas and other regions, led to the first international sanitary conference in 1851, which laid down a principle for protection against the international spread of infectious diseases: maximum protection with minimum restriction. Uniform quarantine measures were determined at that time, but a full century elapsed, with multiple regional and inter-regional initiatives, before the International Sanitary Regulations were adopted in 1951. These were amended in 1969 to become the International Health Regulations (IHRs), which are implemented by the WHO.

The IHRs provide a universal code of practice, which ranges from strong national disease detection systems and measures of prevention and control including vaccination, to disinfection and de-ratting. Currently, the IHRs require the reporting of three infectious diseases—cholera, plague, and yellow fever. But when these diseases are reported, regulations are often misapplied, resulting in the disruption of international travel and trade, and huge economic losses. For example, when the cholera pandemic reached Peru in 1991, it was immediately reported to WHO. In addition to its enormous public health impact, however, misapplication of the regulations caused a severe loss in trade (due to concerns for food safety) and travel, which has been estimated as high as $770 million.

In 1994, an outbreak of plague occurred in India with approximately 1000 presumptive cases. The appearance of pneumonic plague resulted in thousands of Indians fleeing from the outbreak area, risking spread of the disease to new areas. Plague did not spread, but the outbreak led to tremendous economic disruption and concern worldwide, compounded by misinterpretation and misapplication of the IHRs. Airports were closed to airplanes arriving from India, exports of foodstuffs were blocked, and in some countries Indian guest workers were forced to return to India even though they had not been in India for several years before the plague epidemic occurred. Estimates of the cost of lost trade and travel are as high as $1700 million. Again, the country suffered negative consequences from reporting an IHR-mandated diseases due to the misapplication of the IHRs.

A further problem with the IHRs is that many infectious diseases, including those which are new or re-emerging, are not covered even though they have great potential for international spread. These range from relatively infrequent diseases such as viral hemorrhagic fevers to the more common threat of meningococcal meningitis.

Because of the problematic application and disease coverage of the IHRs, WHO has undertaken a revision and updating of the IHRs to make them more applicable to infection control in the twenty-first century. Revisions include a considerable broadening of scope to embrace all infectious diseases of international importance, especially new and re-emerging diseases. The present obligation on countries is being broadened to include a clear mechanism for confidential collaboration between the affected country and WHO to verify the presence or absence of a suspected outbreak unofficially reported by the press or electronic media. Modifications also introduce provisions for defining what constitutes a "public health emergency of international concern," thus helping to avoid inappropriate reactions to strictly localized events. It is envisaged that the revised IHRs will become a true global alert and response system to ensure maximum protection against the international spread of diseases with minimum interference with trade and travel.

Eradication and elimination cannot substitute for good public health—rebuilding of the weakened public health infrastructure and strengthening water and sanitary systems; minimizing the impact of natural and anthropogenic changes in the environment; effectively communicating information about the prevention of infectious diseases; and using antibiotics appropriately. The challenge in the twenty-first century will be to continue to provide resources to strengthen and ensure more cost-effective infectious disease control while also providing additional resources for other emerging public health problems, such as those related to smoking and aging.

BIBLIOGRAPHY

(1969). "The International Health Regulations," 3rd ed.

(2003). Global defence against the infectious disease threat. World Health Organization, Geneva.

Binder, S., Levitt, A. M., Sacks, J. J., and Hughes, J. M. (1999). Emerging infectious diseases: public health issues for the 21st century. *Science* **284**, 1311–1313.

Desselberger, U. (2000). Emerging and re-emerging infectious diseases. *J. Infect.* **40**, 3–15.

Fenner, F., and Henderson, D. A., *et al.* (1988). "Smallpox and its Eradication." World Health Organization, Geneva.

Fricker, J. (2000). Emerging infectious diseases: a global problem. *Mol. Med. Today* **6**, 334–335.

Garrett, L. (1995). "The Coming Plague: Newly Emerging Disease in a World out of Balance." Penguin Books.

Lederberg, J. (2000). Infectious history. *Science.* **288**, 287–293.

Levy, S. B. (1992). "The Antibiotic Paradox: How Miracle Drugs are Destroying the Miracle." Plenum Press, New York.

Schuchat, A. (2000). Microbes without borders: infectious disease, public health, and the journal. *Am. J. Public Health* **90**, 181–183.

WEBSITES

APEC (Asia Pacific Economic Cooperative) Emerging Infections Network
http://www.apec.org/infectious/about/index.htm

National Institutes of Allergy and Infectious Diseases NIH, USA
http://www.niaid.nih.gov/default.htm

Department of Defense Emerging Infections System, USA
http://www.geis.ha.osd.mil/

Infectious Diseases Society of America Emerging Infections Network (EIN)
http://www.idsociety.org/EIN/TOC.htm

Website of the journal Emerging Infectious Diseases
http://www.cdc.gov/ncidod/eid/

33

Energy transduction processes: from respiration to photosynthesis

Stuart J. Ferguson
University of Oxford

GLOSSARY

aerobic respiration The energetically downhill electron transfer, from a donor molecule or ion to oxygen, which is reduced to water, with concomitant coupled ion translocation and thus generation of an electrochemical gradient.

anaerobic respiration The energetically downhill electron transfer, from a donor molecule or ion to a molecule other than oxygen, or to an ionic species, with concomitant coupled ion translocation and thus generation of an electrochemical gradient. The reduction products of the acceptors can either be released from the cell or, sometimes, used as further electron acceptors.

antiport The transport of a molecule or ion up its chemical or electrochemical gradient with the concomitant movement in the opposite direction, but down its electrochemical gradient, of one or more protons or sodium ions.

bacteriorhodopsin A protein of the cytoplasmic membrane of the halophilic archaebacterium *Halobacterium salinarum* (formerly *halobium*) that has a covalently attached retinal molecule. Absorption of light by the latter pigment results in proton translocation across the membrane.

chemiosmotic mechanism The transduction of energy between two forms via an ion electrochemical gradient (*q.v.*) (usually of protons but sometimes of sodium) across a membrane. Examples of such membranes are the cytoplasmic membranes of bacteria, the inner mitochondrial membranes of eukaryotes, and the thylakoid membranes of algae.

cytochrome Hemoprotein in which one or more hemes is alternately oxidized and reduced in electron-transfer processes.

electrochemical gradient The sum of the electrical gradient or membrane potential ($\Delta\psi$) and the ion concentration gradient across a membrane (the latter is often defined as ΔpH for protons).

electron acceptor Low-molecular-weight inorganic or organic species (compound or ion) that is reduced in the final step of an electron-transfer process.

electron donor Low-molecular-weight inorganic or organic species (compound or ion) that is oxidized in the first step of an electron-transfer process.

electron transport The transfer of electrons from a donor molecule (or ion) to an acceptor molecule (or ion) via a series of components (a respiratory chain, *q.v.*), each capable of undergoing alternate oxidation and reduction. The electron transfer can either be energetically downhill, in which case it is often called respiration (*q.v.*), or energetically uphill when it is called reversed electron transfer (*q.v.*).

F_0F_1 ATP synthase The enzyme that converts the protonmotive force into the synthesis of adenosine triphosphate (ATP). Protons (or more rarely sodium ions) flow through the membrane sector of the enzyme, known for historical reasons as F_0, and thereby cause conformational changes, with concomitant ATP synthesis, in the globular F_1 part of

The Desk Encyclopedia of Microbiology
ISBN: 0-12-621361-5

the molecule. In some circumstances the enzyme can generate the proton-motive force at the expense of ATP hydrolysis.

oxidase The hemoprotein that binds and reduces oxygen, generally to water. Oxygen reductase is the function.

oxidative phosphorylation ATP synthesis coupled to a proton or sodium electrochemical gradient (*q.v.*), generated by electron transport, across an energy transducing membrane.

P/O (P/2e) ratio The number of molecules of ATP synthesized per pair of electrons reaching oxygen, or more generally any electron acceptor.

photophosphorylation ATP synthesis coupled to a proton or sodium electrochemical gradient generated by light-driven electron transport, which is often cyclic in bacteria.

proton-motive force The proton electrochemical gradient (*q.v.*) across an energy-transducing membrane in units of volts or millivolts.

quinone Lipid-soluble hydrogen (i.e. proton plus electron) carrier that mediates electron transfer between respiratory chain components.

respiration The sum of electron transfer reactions resulting in reduction of oxygen (aerobically) or other electron acceptor (anaerobically) and generation of proton-motive force.

respiratory chain Set of electron-transfer components, which may be arranged in a linear or branched fashion, that mediate electron transfer from a donor to an acceptor in aerobic or anaerobic respiration (*q.v.*).

redox potential A measure of the thermodynamic tendency of an ion or molecule to accept or donate one or more electrons. By convention, the more negative the redox potential the greater is the propensity for donating electrons and vice versa.

reversed electron transport The transfer of electrons energetically uphill toward the components of an electron-transfer chain that have more negative redox potentials. Such electron transfer can be regarded as the opposite of the respiration (*q.v.*) and is driven by the proton-motive force.

symport The transport of a molecule up its chemical or electrochemical gradient with the concomitant movement, in the same direction but down its electrochemical gradient, of one or more protons or sodium ions.

uniport The transport of an ionic species in direct response to the membrane potential across a membrane.

The Inner Mitochondrial Membrane of the microbial eukaryote and the cytoplasmic membrane of the prokaryote are the key sites where energy available from processes such as the oxidation of nutrients or from light is converted into other forms that the cell needs. Most prominent among these other forms is ATP and thus these types of membrane are concerned with oxidative phosphorylation (or photophosphorylation).

I. INTRODUCTION

Energy-transducing membranes share many common components, but most importantly they operate according to the same fundamental chemiosmotic principle. This states that energetically downhill reactions that are catalyzed by the components of these membranes are coupled to the translocation of protons (or more rarely sodium ions) across the membranes. The direction of movement is outward from the matrix of the mitochondria or the cytoplasm of bacteria. The consequence of this translocation is the establishment of a proton electrochemical gradient. This means that the matrix of the mitochondria or cytoplasm of the bacteria tends to become both relatively negatively charged (thus, called the N side) and alkaline relative to the other side of the membranes, the intermembrane space in the mitochondria and the periplasm in gram-negative bacteria (and equivalent zone in gram-positive bacteria and archae), which is thus called the P side (Fig. 33.1). This electrochemical gradient is in most circumstances dominated by the charge term, which means that there is often a substantial membrane potential across the membranes, frequently estimated to be on the order of 150–200 mV. In most circumstances, the pH gradient generated by the proton translocation is small, 0.5 unit would be an approximate average value. The membrane potential is added to the pH gradient to give the total gradient, which is usually called the proton-motive force if it is given in millivolts. The conversion factor is such that 0.5 pH unit is approximately equivalent to 30 mV. Strictly speaking, the expression of the gradient as an electrochemical potential requires that units of kJ/mol be used; in practice this is rarely done, which sometimes causes confusion. I use the term proton-motive force in this article.

II. MITOCHONDRIAL ENERGETICS

The best-known machinery for generating the proton-motive force is the mitochondrial respiratory chain. The standard mitochondrial respiratory chain is found, at least under some growth conditions, in eukarytoic microbes. The key point is that as a pair of

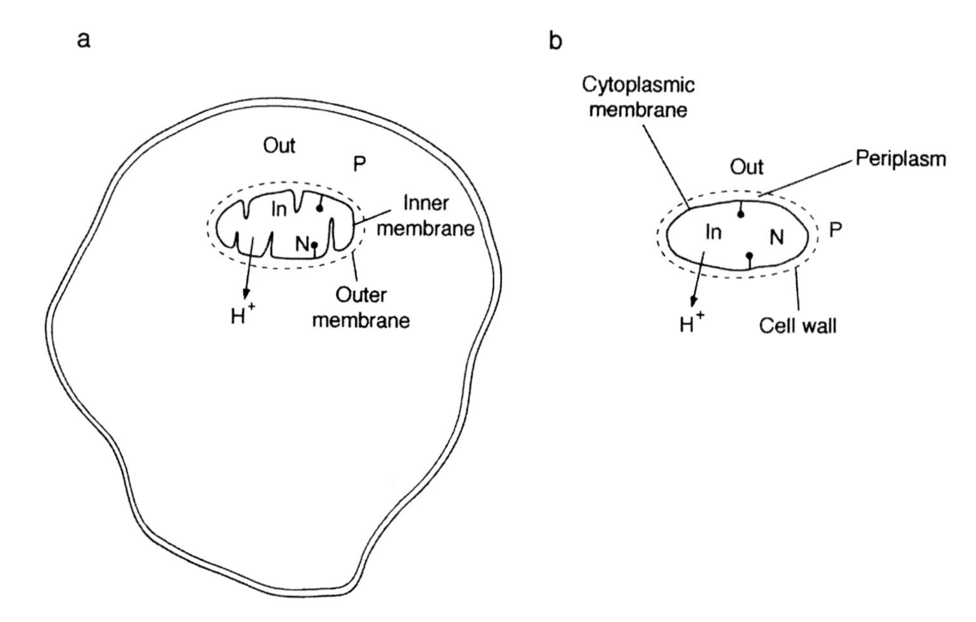

FIGURE 33.1 (a) Idealized mitochondrion in a eukaryotic cell and (b) gram-negative bacterium showing the direction of proton translocation linked to an exergonic (energetically downhill) reaction. P means a relatively positive aqueous phase (i.e. outside the inner mitochondrial membrane or the bacterial cytoplasmic membrane); N means a relatively negative aqueous phase (i.e. inside the mitochondrion (the matrix) or the cytoplasm of a bacterium. ↓ represents the ATP synthase enzyme.

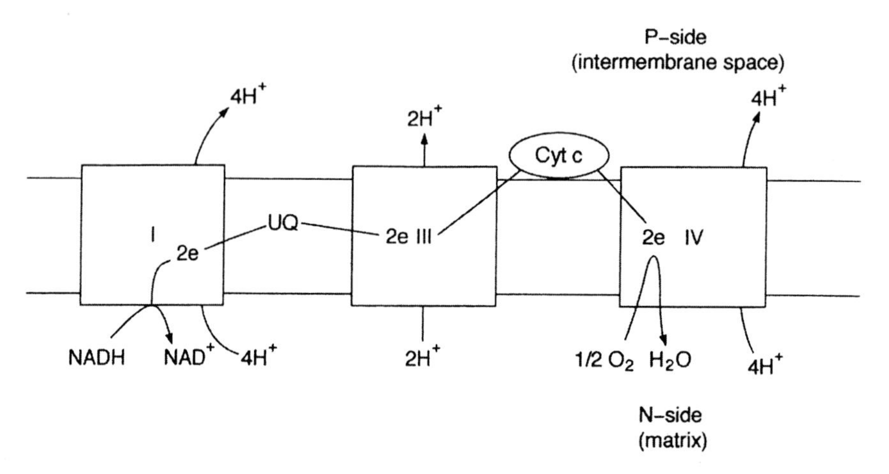

FIGURE 33.2 Considerably *simplified* representation of the proton-translocation stoichiometry, per two electrons, of the mitochondrial respiratory chain. I indicates complex I (otherwise known as NADH dehydrogenase); III, complex III (otherwise known as the ubiquinol–cytochrome *c* oxidoreductase or cytochrome bc_1 complex); IV, complex IV (otherwise known as cytochrome C oxidase or cytochrome aa_3 oxidase); P, positive aqueous phase; and N, negative aqueous phase.

electrons traverses the chain from NADH to oxygen there are three segments (formerly called sites, but this term is inappropriate because it implies equivalence and relates to a very old idea that ATP is made at three sites within the electron-transport chain) where protons can be translocated across the membrane. The first and last of these segments move four protons per two electrons, while the middle segment moves only two (Fig. 33.2) (consideration of the mechanisms of these proton translocations is beyond the scope of this article). Thus, 10 protons are moved per two electrons moving along the chain from NADH to O_2. Electrons may enter the chain such that they miss the first proton-translocating segment of the chain; succinate and the intermediates generated during fatty-acid oxidation are the most prominent examples of electron sources for this. In these cases, six protons are translocated per two electrons. The entry of electrons at the third segment would obviously give a translocation stoichiometry of four.

The proton-motive force generated can then be used to drive various uphill reactions. Most prominent is

ATP synthesis. This is achieved by protons flowing back across the membranes and through the ATP synthase enzyme, often called F_0F_1 ATP synthase. There is increasing insight into the mechanism of this enzyme; it appears to function akin to a rotary motor in which the flow of protons through the F_0 is coupled to rotation and structural changes in the F_1 part of the molecule, events that are somehow linked to ATP synthesis. It is not settled how many protons must pass through the ATP synthase to make one ATP molecule; a consensus value adopted here, even though it is not fully confirmed, is three. On the basis of "what goes one way across the membrane must come back the other," it might therefore be thought that the stoichiometry of ATP production per pair of electrons (called the P/O or P/2e ratio) flowing from NADH to oxygen would be 10/3 (i.e. 3.3) for NADH and 6/3 (i.e. 2) for succinate. However, matters are a little more complicated. The combined process of entry of ADP and Pi (phosphate) into mitochondria and the export of ATP to the cytoplasm involves the movement of one proton into the matrix (Fig. 33.3). Thus for each ATP made, the expected stoichiometry is $10/(3+1)=2.5$ for NADH and $6/(3+1)=1.5$ for succinate. These values differ from the classic textbook values of 3 and 2, respectively, but they are rapidly becoming accepted.

It was generally thought that eukaryotes were only capable of aerobic respiration. However, there is now evidence for a form of mitochondrial anaerobic respiration

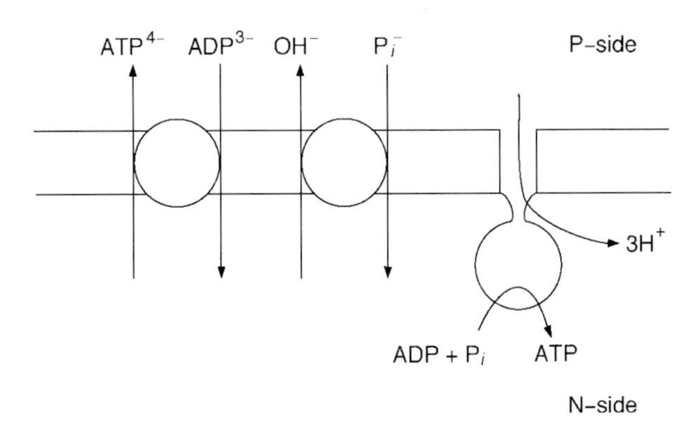

FIGURE 33.3 Charge movement, associated with ATP synthesis and translocation of adenine nucleotides and phosphate across the inner mitochondrial membrane. The stoichiometry of proton translocation through the ATP synthase is commonly taken to be 3 but this is not a definite value. The translocated protons are not believed to pass through the active site of the ATP synthase enzyme. Note that the adenine nucleotide exchange moves one positive charge into the matrix per nucleotide exchanged and the operation of the phosphate transporter effectively moves the chemical part of the proton (but not the charge) into the matrix. Thus, in combination, the two transporters move one positive charge into the mitochondrion per ATP synthesized and returned to the P phase. Note that these transporters do not operate in bacterial ATP synthesis.

in which nitrate is reduced to nitrous oxide (more typically a prokaryotic characteristic, see the following) and for a novel type of mitochondrion from the ciliate protist *Nyctotherus ovalis* that reduces protons to hydrogen. In both these examples, electrons are derived from NADH.

III. BACTERIAL ENERGETICS

Many species of bacteria employ a respiratory chain similar to that found in mitochondria in order to generate a proton-motive force. However, there are many more types of electron donor and acceptor species that can be used by bacteria (eukaryotes are restricted to the oxidative breakdown of reduced carbon compounds), and various forms of anaerobic respiration are widespread. A further general difference between bacteria and microbial eukaryotes is that in the former the protonmotive force can drive a wider range of functions and be generated in more diverse ways than in the latter. Thus, functions alongside ATP synthesis (for which the enzyme is very similar to that found in mitochondria), such as driving of many active transport processes and the motion of the flagella, are important processes that depend on the protonmotive force in many organisms (Fig. 33.4).

A common mode of active transport is known as the symport; the classic example of this is the lactose-proton symporter coded for by the *lacY gene* of the *lac* operon of *E. coli*. In this case, a transmembrane protein translocates together a proton down its electrochemical gradient and a lactose molecule up its concentration gradient (Fig. 33.5); the exact mechanism is presently unknown. There are cases known where Na is the translocated ion. A second type of transport system is the antiport (Fig. 33.5). Here the movement via a protein of the proton down its electrochemical

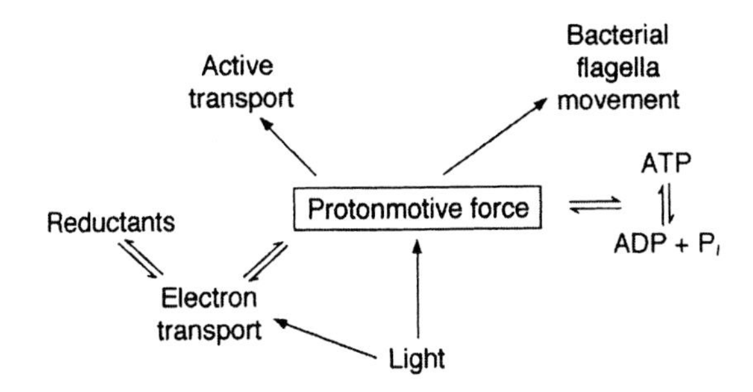

FIGURE 33.4 The central role of the proton-motive force in linking diverse reactions. Note that the direct generation of proton-motive force from light is unusual, but is exemplified by the proton-pumping bacteriorhodospin protein found in halobacteria. Normally light drives electron-transport processes, which in turn generate the proton-motive force.

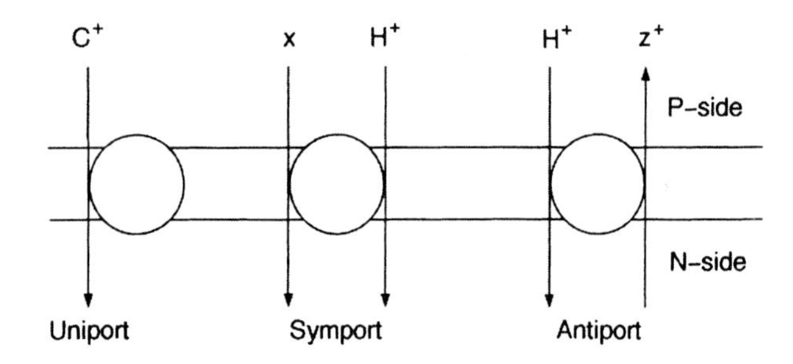

FIGURE 33.5 The three common modes of substrate transport across the bacterial cytoplasmic membrane.

TABLE 33.1 Approximate Standard Redox Potentials of Some Electron-Donor and -Acceptor Couples Used in Respiratory Processes[a]

Couple	$E^{\circ\prime}$ (mV)
N_2O/N_2	+1360
NO/N_2O	+1180
O_2/H_2O	+820
NO_3^-/NO_2^-	+430
NO_2^-/NH_4^+	+360
NO_2^-/NO	+350
fumarate/succinate	−30
methanol/formaldehyde	−180
$NAD^+/NADH$	−320
CO_2/formate	−430
CO_2/CO	−540

[a]Redox potential refers to the standard state (1 M concentrations for solutes and 1 atm pressure for gases). Conditions experienced by cells may vary significantly from these and thus the actual redox potentials of the couples should be calculated from the Nernst equation and may differ substantially from those in this table.

gradient is obligatorily linked to the movement of another species, typically an ion, in the opposite direction and up its electrochemical gradient. The third type, uniport, is the case where an ion moves in direct response to the membrane potential and is probably rarer than the other two examples in the prokaryotic world.

It is important to appreciate that not all transport processes across the bacterial cytoplasmic membrane are directly driven by the proton-motive force. Some transport reactions are driven directly by ATP. Notable among such systems are ABC (ATP binding cassette) transporters.

A more subtle aspect of prokaryotic energetics is that in some species of bacteria the proton-motive force must drive reversed electron transport under some circumstances (see later).

A common misconception when multiple functions of the proton-motive force are discussed is that this force can be divided (e.g. 100 mV for ATP synthesis and 50 mV for flagella motion). This notion is incorrect because the proton-motive force across a membrane has a single value at any one time and it is the magnitude of this force that is acting simultaneously on all energy-transducing units, be they ATP synthases, active transporters, or flagella.

IV. PRINCIPLES OF RESPIRATORY ELECTRON-TRANSPORT LINKED ATP SYNTHESIS IN BACTERIA

In principle, energy transduction on the cytoplasmic membrane is possible if any downhill reaction is coupled to proton translocation. The most familiar examples are probably those that also occur in mitochondria, for example, electron transfer from NADH to oxygen or from succinate to oxygen. In these cases, the electrons pass over a sizeable redox drop (Table 33.1). In contrast to mitochondria, various species of bacteria can use a wide variety of electron donors and acceptors. The fundamental principle is that the redox drop should be sufficient for the electron transfer to be coupled to the translocation of protons across the cytoplasmic membrane. Table 33.1 shows that such sizeable drops are associated with the aerobic oxidation of hydrogen, sulfide, carbon monoxide, and methanol, to cite just a few electron donors. Anaerobic respiration is also common with many suitable pairings of reductants and oxidants (e.g. Table 33.1). Thus NADH can be oxidized by nitrate, nitrite, nitric oxide, or nitrous oxide. The flow of electrons to these acceptors, each of which (other than nitrate) is generated by the reduction of the preceding ion or molecule, is the process known as denitrification. In *E. coli* under anaerobic conditions, formate is frequently an electron donor, and nitrate and nitrite are the acceptors, with the latter being reduced to ammonia (Table 33.1) rather than to nitric oxide, as occurs in denitrifying bacteria. A wide variety of electron-transport components, including many different types of cytochrome are involved in catalyzing these reactions. The mechanisms whereby electron transport is linked to the generation of the proton-motive force are frequently complex. However, nitrate respiration (Fig. 33.6) provides an example of one of the simplest mechanisms that corresponds to Mitchell's original redox loop mechanism.

An important point is that the consideration of the energy drop between the donor and acceptor (Table 33.1) is only a guide as to whether proton translocation, and thus ATP synthesis, can occur and,

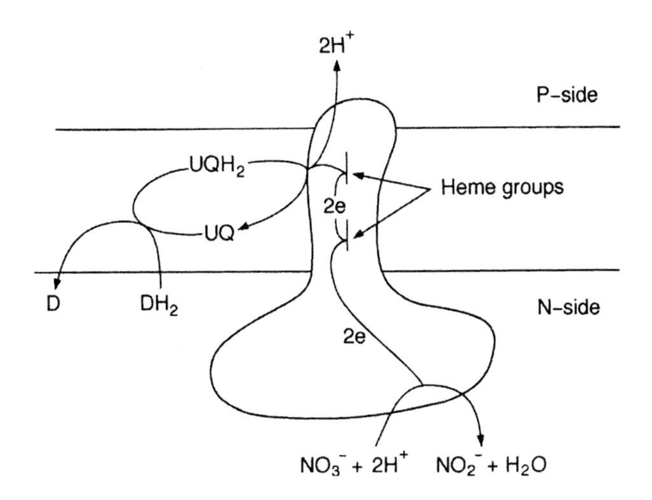

FIGURE 33.6 A simple mechanism for generating the proton-motive force, the bacterial nitrate reductase system. Oxidation of ubiquniol (UQH_2) to ubiquinone (UQ) at one side of the membrane is accompanied by release of protons to the P side and translocation of the electrons towards the N side, where they combine with protons and nitrate to produce nitrite. Overall, the process effectively translocates two protons per two electrons across the membrane. The mitochondrial electron-transport chain (see Fig. 33.2) involves more complex mechanisms for proton translocation. DH_2 is an unspecified donor to the ubiquinone, and D is the product of oxidizing DH_2.

if so, with what stoichiometry. Thus while many bacterial species can form a respiratory chain with considerable similarity to that found in mitochondria, others vary from this pattern. Notable here is *Escherichia coli*, which always lacks the cytochrome bc_1 complex, and which, following some growth conditions, has cytochrome *bo* as the terminal oxidase, but which under others has cytochrome *bd*. The consequence is that when the former proton-pumping oxidase is operating only eight protons are translocated per pair of electrons flowing from NADH to oxygen, while with the latter oxidase the stoichiometry would be six. The corresponding stoichiometry for mitochondria is ten. This example illustrates the important point that it is not just the energy drop between a donor and an acceptor that is important, but also the details of the components (or molecular machinery) in between. Another example is methanol to oxygen. Periplasmic oxidation of methanol feeds electrons into the electron-transport chain close to the terminal oxidase, yet energetic considerations alone would indicate that electrons could span more proton-translocating sites, just as they do when succinate is the electron donor (compare the redox potentials for fumarate–succinate and methanol–formaldehyde; Table 33.1). A final example to consider is the case in which both the electron donor and acceptor are in the periplasm and they are connected purely by periplasmic components. In such a case, which applies to methanol (as donor) and nitrous oxide (N_2O as acceptor), the electrons do not pass through any proton-translocating complex. Thus, no proton translocation would occur no matter what the redox drop between the two components.

It is not necessary for electrons to flow over such a large energy drop as they do when they pass from NADH to oxygen (Table 33.1) in order to generate a proton-motive force. Thus, if the driving force associated with a reaction was very small, it might still be energetically be possible for the passage of two electrons from a donor to an acceptor to cause the translocation of just one proton. If three protons are required for the synthesis of ATP, then the ATP yield stoichiometry would be 0.166 per electron flowing from electron donor to acceptor. This seemingly bizarre stoichiometry is not only energetically possible but also mechanistically possible because the chemiosmotic principle involves the delocalized proton-motive force that is generated by all the enzymes of the membrane and also consumed by them all. There is no case known that matches this extreme; nevertheless, there may well be organisms yet to be discovered that have such low stoichiometries of ATP synthesis.

One example of the lowest known stoichiometries of ATP synthesis per pair of electrons reaching the terminal electron acceptor (oxygen) occurs in *Nitrobacter*. Table 33.1 shows that the redox drop is small between nitrite and oxygen. This organism also illustrates the versatility and subtlety of the chemiosmotic mode of energy transduction. *Nitrobacter* species oxidize nitrite to nitrate at the expense of the reduction of oxygen to water in order to sustain growth. The energy available as a pair of electrons flows from nitrite to oxygen is sufficient to translocate two protons (a more detailed consideration of how this is done is outside the scope of this chapter). This means, recalling the current consensus that three protons are needed for the synthesis of one ATP molecule, that the ATP yield stoichiometry would be 0.66/2e. Nitrobacter also illustrates another important facet of energy transduction in the bacterial world. The organism is chemolithotrophic, which means that it grows on nitrite as the source not only of ATP but also of reductant (NADPH), which is required for reducing CO_2 into cellular material. Energetic considerations immediately show that nitrite cannot reduce NADP directly. What happens in the cell is that a minority of the electrons originating from nitrite are driven backward up the electron-transfer system to reduce NAD(P) to NAD(P)H. This is achieved by the inward movement of protons reversing the usual direction of electron movement (Fig. 33.7). This reversed electron-transport process is an important phenomenon in

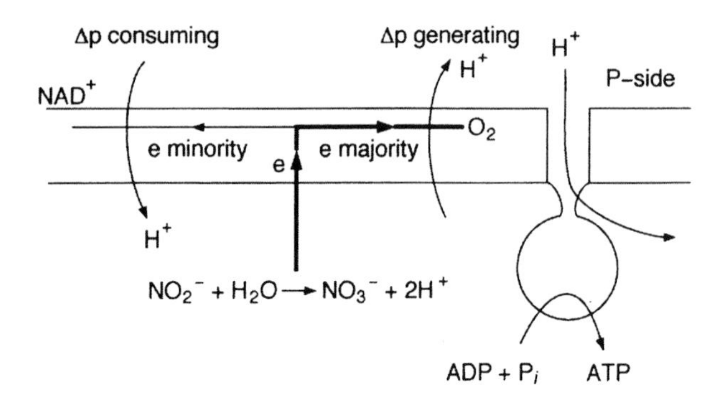

FIGURE 33.7 Reversed electron transport illustrated by the example of *Nitrobacter*. The majority of eletrons derived from nitrite flow energetically downhill to oxygen via a cytochrome oxidase, which generates a protonmotive force. A minority of electrons is driven energetically uphill by the protonmotive force so as to reduce NAD^+ to NADH. Note that in this diagram no proton stoichiometry values are implied.

a variety of bacteria, especially those growing in the chemolithotrophic mode.

Most studies of electron transport-linked ion translocation have been done with species of eubacteria. However, the same fundamental process also occurs in archaebacteria, although with some novel features that reflect some of the extreme growth modes tolerated by these organisms. For example, a key step in methane formation by methanogenic bacteria is electron transfer from hydrogen or other reductant to a small molecule contains a disulfide bond. The latter is reduced to two sulfides and the overall process is coupled to the translocation of protons across the cytoplasmic membrane. The proton-motive force thus set up can be used to drive ATP synthesis. Interestingly the ATP synthase in archaebacteria shows significant molecular differences from its counterpart in eubacteria and mitochondria, but is believed to function according to the same principle.

V. GENERATION OF THE ION ELECTROCHEMICAL GRADIENT OTHER THAN BY ELECTRON TRANSPORT

A. ATP hydrolysis

Organisms that are incapable of any form of respiration still require an ion electrochemical gradient across the cytoplasmic membrane for purposes such as nutrient uptake. One way in which this requirement can be met is for some of the ATP synthesized by fermentation to be used for ATP hydrolysis by the FoF_1 ATPase. This means that this enzyme works in the reverse of its usual direction and pumps protons out of the cell. Thus there are many organisms that can prosper in the absence of any electron-transport process, either as an option or as an obligatory aspect of their growth physiology.

B. Bacteriorhodopsin

A specialized form of light-driven generation of proton-motive force, and hence of ATP, occurs in halobacteria; these organisms are archaebacteria. The key protein is bacteriorhodopsin, which is a transmembrane protein with seven α-helices that has a covalently bound retinal. The absorption of light by this pigment initiates a complex photocycle that is linked to the translocation of one proton across the cytoplasmic membrane for each quantum absorbed. Bacteriorhodopsin is one of a family of related molecules. Another, halorhodospin, is structurally very similar and yet catalyzes the inward movement of chloride ions driven by light.

C. Methyl transferase

One step of energy transduction in methanogenic bacteria involves an electron-transfer process (see earlier). Another important process in methanogens is the transfer of a methyl group from a pterin to a thiol compound. This exergonic (energetically downhill) reaction is coupled to ion, in this case sodium, translocation across the cytoplasmic membrane.

D. Decarboxylation linked to ion translocation

In the bacterial world, the electrochemical gradients can be generated by diverse processes other than electron transport or ATP hydrolysis. For example, *Propionegenium modestum* grows on the basis of catalyzing the conversion of succinate to propionate and carbon dioxide. One of the steps in this conversion is decarboxylation of methyl-malonyl coenzyme A (CoA) to propionyl CoA. This reaction is catalyzed by a membrane-bound enzyme that pumps sodium out of the cells, thus setting up a sodium electrochemical gradient (or sodium-motive force). This gradient in turn drives the synthesis of ATP as a consequence of sodium ions reentering the cells through a sodium-translocating ATP synthase enzyme. Apart from illustrating that sodium, instead of proton circuits, can be used for energy transduction in association with the

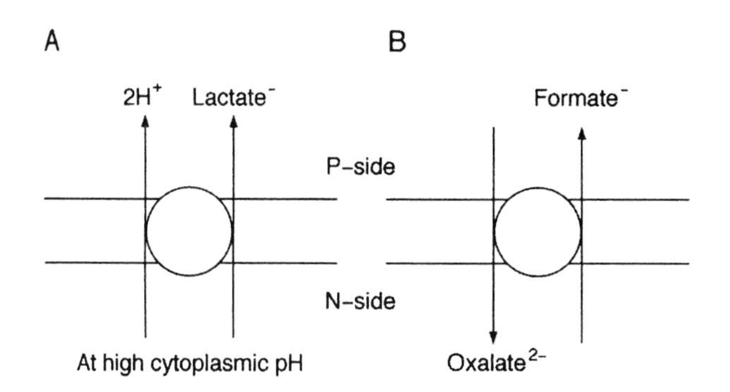

FIGURE 33.8 Two examples of generation of proton-motive force by end-product extrusion from fermenting bacteria.

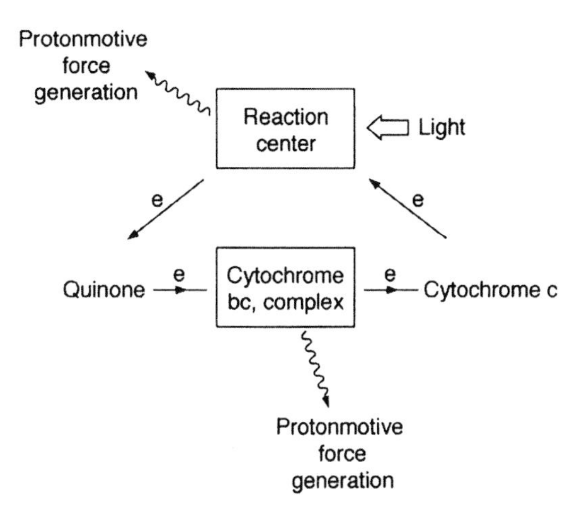

FIGURE 33.9 An oversimplified outline of the cyclic electron-transport process of photosynthetic bacteria. There are two types of reaction centers, depending on the organism. The molecular composition of the system depends on the organism. Two components contribute to the generation of the proton-motive force.

bacterial cytoplasmic membrane, this organism also illustrates that the stoichiometry of ATP synthesis can be less than one per CO_2 formed. It is believed that each decarboxylation event is associated with the translocation of two sodium ions and the synthesis of ATP with three. Thus non-integral stoichiometry is consistent with the energetics of decarboxylation and ATP synthesis. This is an important paradigm to appreciate; the underpinning growth reaction for an organism does not have to be capable of supporting the synthesis of one or more integral numbers of ATP molecules.

E. Metabolite ion-exchange mechanisms

Another example of the generation of a proton-motive force is ion exchange across the membrane. For example, in fermenting bacteria there is evidence that under some conditions an end-product of metabolism, lactic acid, leaves the cell together (i.e. in symport) with than one proton; this results in the generation of a proton-motive force (Fig. 33.8). A second example is provided by *Oxalobacter formigenes*, in which the entry of the bivalent anion oxalate is in exchange for the exit of the monovalent formate ion generated by decarboxylation of the oxalate, leading to the net generation of membrane potential (Fig. 33.8). This seems to be the principal mode of generating membrane potential in this organism.

VI. PHOTOSYNTHETIC ELECTRON TRANSPORT

Prokaryotic photosynthesis involves a cyclic electron-transport process in which a single photosystem captures light energy and uses it to drive electrons around the cycle (Fig. 33.9). The consequence of this cyclic electron flow is the generation of the proton-motive force. There are two types of photosystem found in prokaryotes. One is related to the water-splitting photosystem that is found is plants; typically this bacterial photosystem is found in organisms such as *Rhodobacter sphaeroides*. The second type of photosystem is closely related to the second photosystem of plants, the one that is concerned with the generation of NADPH. *Heliobacter* is an example of an organism carrying this type of center. Some microorganisms have both of these photosystems, arranged to operate in series as in plants. In this group are the prokaryotic blue-green algae and the eukaryotic algae.

VII. ALKALIPHILES

An interesting unresolved problem relates to energy transduction in the alkaliphilic bacteria. The problem is straightforward. These organisms can grow in an environment with a pH as high as 11 or 12. A cytoplasmic pH even as high as 9 means that the pH gradient could be as much as 3 units (equivalent to 180 mV) the wrong way around in the context of the chemiosmotic mechanism. The membrane potential always seems to be larger than 180 mV, but the total proton-motive force can be very low (e.g. around 50 mV). For some organisms that use a conventional proton-translocating respiratory chain and ATP synthase, it is not understood how they survive energetically. In other organisms, there is evidence for the role of a sodiummotive force. This would sidestep the problem of the adverse proton concentration gradient.

BIBLIOGRAPHY

Embley, T. M., and Martin, W. (1998). A hydrogen-producing mitochondrion. *Nature* **396**, 517–519.

Ferguson, S. J. (1998). Nitrogen cycle enzymology. *Curr. Opinion Chem. Biol.* **2**, 182–193.

Ferguson, S. J. (2000). ATP synthase: What determines the size of a ring? *Curr. Biol.* **10**, R804–R808.

Jormakka, M., Tornoth, S., Byrne, B., and Iwata, S. (2002). Molecular basis of protonmotive force generation: Structure of formate dehydrogenase-N. *Science* **295**, 1863–1868.

Konings, W. N., Lolkema, J. S., and Poolman, B. (1995). The generation of metabolic energy by solute transport. *Arch. Microbiol.* **164**, 235–242.

Krulwich, T. A., Ito, M., Gilmour, R., Hicks, D. B., and Guffanti, A. A. (1998). Energetics of alkaliphilic Bacillus species: Physiology and molecules. *Adv. Microbial Physiol.* **40**, 401–438.

Madigan, M. T., Martinko, J. M., and Parker, J. (2002). "Brock Biology of Microorganisms," 10th edn. Prentice-Hall, Upper Saddle River, NJ.

Neidhardt, F. C. (ed.) (1996). "*Escherichia coli* and *Salmonella*. Cellular and Molecular Biology," 2nd edn, Various articles. ASM Press, Washington, DC.

Nicholls, D. G., and Ferguson, S. J. (2002). "Bioenergetics 3." Academic Press/Elsevier, London.

Novartis Foundation Symposium 221 (1999). "Bacterial Responses to pH." John Wiley, Chichester, UK.

Unden, G., and Bongaerts, J. (1997). Alternative respiratory pathways of *Escherichia coli*: energetics and transcriptional regulation in response to electron acceptors. *Biochim. Biophys. Acta* **1320**, 217–234.

Walker, J. E. (1998). ATP synthesis by rotary catalysis (Nobel lecture). *Angew Chem. Intl. Ed.* **37**, 2308–2319.

White, D. (2000). "The Physiology and Biochemistry of Prokaryotes." 2nd edn. Oxford University Press, New York.

WEBSITE

Online Biology Book (with links). Go to ATP and Biological Energy. Go to Cellular Metabolism and Fermentation
http://www.emc.maricopa.edu/faculty/farabee/BIOBK/BioBookATP.html

34

Enteropathogenic bacteria

Farah K. Bahrani-Mougeot and Michael S. Donnenberg
University of Maryland

GLOSSARY

bacteriophage A virus that infects bacteria.

colitis Inflammation of the large intestine (colon).

cytotoxins Bacterial products that damage cells.

diarrhea Increase in the frequency of bowel movement and decrease in the consistency of stool.

dysentery Inflammatory disease of the large bowel with severe abdominal cramps, rectal urgency, and pain during stool passage and the presence of blood, pus, and mucus in stool.

enteritis Inflammation of the small intestine.

enterotoxins Bacterial products that cause fluid secretion from intestinal cells.

fimbriae (or pili) Rigid rod surface organelles with diameters of about 2–7 nm in gram-negative bacteria that often mediate bacterial adherence to host cells.

flagellae Ropelike surface organelles of 15–20 nm in diameter that provide bacteria with motility and the ability to move toward nutrients and away from toxic substances (chemotaxis).

gastritis Inflammation of the stomach.

pathogenicity island Segment of DNA that is foreign to the bacterial host and carries virulence genes.

plasmid Extrachromosomal self-replicating DNA element.

type III secretion pathway A specialized protein secretion system, responsible for export of virulence determinants by some gram-negative bacterial pathogens and symbionts of animals and plants. Some of the proteins secreted by this pathway are translocated by the bacteria into host cells.

type IV fimbriae Special type of fimbriae produced by certain pathogenic gram-negative bacteria. These fimbriae have subunits with different primary structures and often different morphologies from common fimbriae.

Enteric Infections are caused by a variety of microorganisms, including bacterial pathogens. Among these infections, diarrheal diseases are a major cause of mortality in the children of Third World countries, due to malnutrition, poor personal hygiene, and insufficient environmental sanitation. In industrialized countries, diarrhea may result from foodborne outbreaks and is common in day care centers, hospitals, and chronic care institutions, among homosexual men and immunocompromised patients. Diarrhea can result from inflammatory infections in the colon and/or small intestine caused by pathogens, such as *Shigella*, *Salmonella*, and *Campylobacter*, or noninflammatory infections in the small intestine by pathogens such as *Vibrio* and enterotoxigenic *Escherichia coli*. Diarrhea can also result from ingestion of preformed toxins, produced by bacteria such as *Clostridium perfringens* and *Bacillus cereus*. In addition, infections by enteropathogenic bacteria may cause systemic syndromes, such as typhoid fever caused by *Salmonella typhi*. Another form of enteric infection, gastritis, results exclusively from infections with *Helicobacter pylori*. We will briefly review the epidemiology, pathogenesis, and clinical features of the most important bacterial enteric pathogens of humans.

The Desk Encyclopedia of Microbiology
ISBN: 0-12-621361-5

I. BACTERIAL AGENTS OF INFLAMMATORY DIARRHEA

In the majority of cases, inflammatory diarrhea in the distal small bowel and colon occurs in response to bacterial invasion of the intestinal tissues (Table 34.1). However, in some infections, enterocolitis can be an outcome of bacterial toxicity without invasion. Causative agents of these infections include the following.

A. *Shigella* spp.

Shigella, the major etiologic agent of bacillary dysentery, is traditionally divided into four species based on biochemical and serological characteristics. These species include *S. dysenteriae*, *S. flexneri*, *S. sonnei*, and *S. boydii*. However, techniques such as multilocus enzyme electrophoresis have revealed that all *Shigella* strains are actually encompassed within the species *E. coli*. Clinical syndromes of shigellosis include a mild watery diarrhea, which is often followed by severe dysentery with blood, mucus, and inflammatory cells in feces. The incubation period ranges from 6h to 5 days. Epidemiological studies show that *Shigella* is transmitted by the fecal–oral route or by contaminated food and water. As few as 100 organisms can cause infection in an adult.

Shigella invades the colonic mucosa and this invasion involves entry and intercellular dissemination. Bacteria enter the cells by a micropinocytic process,

TABLE 34.1 Epidemiology and clinical characteristics of bacterial enteric pathogens

Pathogen	Route of transmission	Site of infection	Clinical syndrome
B. cereus	Foods such as fried rice and vanilla sauce	Small intestine	Watery diarrhea Vomiting Abdominal cramps
C. jejuni	Contaminated food and water Fecal–oral	Colon Small intestine	Watery diarrhea Dysentery
C. difficile	Environmental contamination with spores Fecal–oral	Colon	Watery diarrhea Pseudomembranous colitis
C. perfringens	Foods such as meat, turkey and chicken	Small intestine	Watery diarrhea Necrotic enteritis
EAEC[a]	?	Small intestine	Watery and mucoid diarrhea
EHEC[b]	Food contaminated with cattle feces Person–person contact	Colon	Diarrhea Hemorrhagic colitis Hemolytic uremic syndrome
EIEC[c]	Contaminated food and water Fecal–oral	Colon	Watery diarrhea Dysentery
EPEC[d]	Person–person contact	Small intestine Colon	Watery diarrhea
ETEC[e]	Contaminated food and water	Small intestine	Watery diarrhea
H. pylori	Fecal–oral (?) Person–person contact (?)	Stomach	Gastritis Peptic ulcer Gastric cancer
Salmonella (nontyphi)	Contaminated food and water Animal–person contact	Small intestine Colon	Watery diarrhea Dysentery
S. typhi	Contaminated food and water	Systemic	Typhoid fever
Shigella	Fecal–oral Contaminated food and water Person–person contact	Colon	Watery diarrhea Dysentery
S. aureus	High salt- or high sugar-containing foods	Small intestine	Watery diarrhea Vomiting
V. cholerae	Contaminated food and water Shellfish	Small intestine	Rice-water diarrhea
Y. enterocolitica	Contaminated food Person–pig contact	Small intestine Systemic	Acute diarrhea Enterocolitis Mesenteric adenitis

[a]Enteroaggregative *E. coli*.
[b]Enterohemorrhagic *E. coli*.
[c]Enteroinvasive *E. coli*.
[d]Enteropathogenic *E. coli*.
[e]Enterotoxigenic *E. coli*.

which requires polymerization of actin at the site of entry. Shortly after entry, bacteria lyse the phagocytic vacuole and move into the cytoplasm where they multiply. Within the cytoplasm, *Shigella* recruits actin microfilaments at one pole of the bacterium, which leads to the formation of a polymerized actin tail behind the bacterium and, consequently, movement of the bacteria (Fig. 34.1). The movement of the bacteria leads to the formation of cell membrane protrusions that extend from one cell into the adjacent cell and allow dissemination of bacteria without their release into the extracellular environment. The ability of *Shigella* to spread from cell to cell is measured *in vitro* by the formation of plaques on a confluent cell monolayer (plaque assay) and *in vivo* by formation of keratoconjunctivitis in guinea pigs (Sereny test).

All of the genes required for *Shigella* invasion are carried on a 200-kb virulence plasmid. A 30-kb fragment contains the genes that encode secreted proteins called IpaA, B, C, and D, the chaperones for these proteins called Ipgs, and a specialized type III secretion system for these proteins called Mxi–Spa. Several other proteins that might be important in the entry process are also secreted by the Mxi–Spa machinery. This machinery becomes activated upon contact of bacteria with the host epithelial cells. The *icsA* (*virG*) gene, which confers the ability of *Shigella* to spread from cell to cell, is located approximately 40 kb away from the *ipa–ipg–mxi–spa* region.

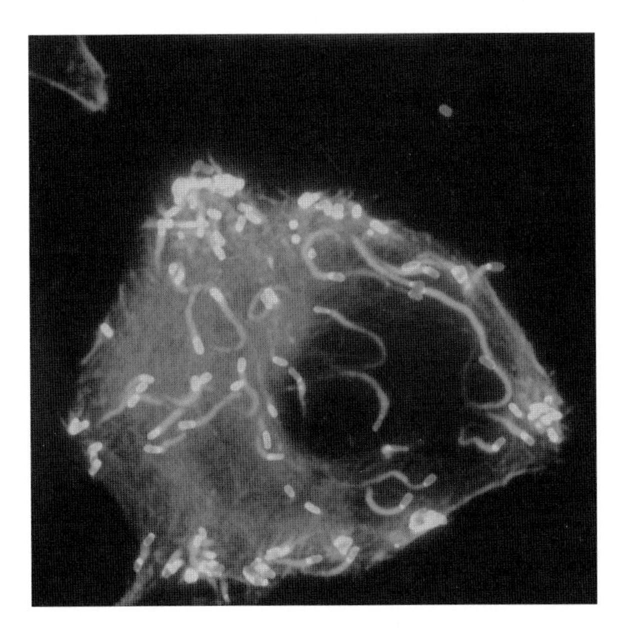

FIGURE 34.1 Actin tail formation by *Shigella flexneri*. Actin microfilaments are polymerized at one end of the bacterium and help it to move within the host cell cytoplasm. Bacteria are stained in red and actin in green. The areas where bacteria and actin colocalize appear in yellow (courtesy of Coumarin Egile and Philippe J. Sansonetti) (Plate 6).

Following secretion, IpaB and IpaC form a complex, which triggers recruitment of actin at the site of bacterial entry by a mechanism that involves the small host G protein, Rho. IpaA contributes to the entry process by interacting with other cytoskeletal proteins, such as vinculin and α-actinin. Contact with macrophages results in delivery of IpaB into the cytoplasm of macrophages where IpaB induces programmed cell death and release of interleukin-1. Release of this cytokine triggers a cascade of proinflammatory responses, which opens intercellular junctions and destabilizes the epithelia, thus facilitating bacterial invasion and ulceration of the colon.

Infections due to *S. dysenteriae* are more severe and more likely to lead to complications than are infections with other *Shigella* species. *S. dysenteriae* produces a lethal cytotoxin, called Shiga toxin. This toxin is composed of an enzymatically active A subunit and five B subunits that mediate binding of the toxin to target cell receptors. The toxin functions by cleaving 28S rRNA of eukaryotic cells and inhibiting protein synthesis. Shiga toxin may play a role in manifestations of hemolytic uremic syndrome, an occasional consequence of shigellosis.

B. Enteroinvasive *E. coli*

E. coli is commonly regarded as a harmless commensal of the intestines of humans. However, at least six varieties have been identified that possess specific pathogenic mechanisms allowing them to cause diarrhea. These pathogens include enteroinvasive *E. coli* (EIEC). EIEC is closely related to *Shigella* spp. in biochemical and serological characteristics, pathogenic mechanisms, and virulence determinants (see preceding) (Tables 34.2 and 34.3). EIEC is identified as *E. coli* based on its biochemical profile but is distinguished from other strains of *E. coli* based on genotypic and phenotypic characteristics of *Shigella* spp. Therefore, the Sereny test and plaque assay are appropriate tests for identification of EIEC.

Like *Shigella* spp., EIEC is transmitted by contaminated food and water. The infectious dose, however, is 2–3 logs higher than that for *Shigella* infection. Therefore, transmission from person to person is likely less than is the case for *Shigella*. Infection with EIEC leads to watery diarrhea, with dysentery syndrome in only some of the patients.

C. Nontyphoidal *Salmonella*

Salmonella spp. other than *Salmonella typhi* are the cause of salmonellosis in humans. There are over 2000 serotypes of *Salmonella* that infect a wide range of

TABLE 34.2 Mechanisms of pathogenicity of enteropathogenic bacteria

Pathogenicity mechanism	Organism
1. Invasion of epithelial cells	*Shigella*, EIEC[a], *Salmonella*, *Yersinia*, *C. jejuni*, *L. monocytogenes*
2. Colonization of epithelial cells	EPEC[c], EHEC[d]
A/E[b] of intestinal epithelial cells	*C. difficile*, EHEC[d] EAEC[e], non-Cholerae Vibrios, *A. hydrophilia*,
Toxin production with overt damage to epithelial cells	*P. shigelloides*, *H. pylori*, *V. cholera*, ETEC[f]
Production of an enterotoxin without overt damage to epithelial cells	
3. Release of toxins in the absence of colonization	*C. perfringens*, *B. cereus*, *S. aureus*

[a]Enteroinvasive *E. coli.*
[b]Attaching and effacing.
[c]Enteropathogenic *E. coli.*
[d]Enterohemorrhagic *E. coli.*
[e]Enteroaggregative *E. coli.*
[f]Enterotoxigenic *E. coli.*

TABLE 34.3 Mechanisms of action of toxins produced by enteropathogenic bacteria

Mode of action	Toxin	Organism	Type of toxin
Inhibition of protein synthesis by cleaving 28S rRNA	Shiga toxin	*Shigella*, EHEC[a]	Cytotoxin
ADP-ribosylation of $G_s\alpha$	Cholera toxin	*V. cholerae*	Enterotoxin
	Heat-labile toxin	ETEC[b]	Enterotoxin
Activation of guanylate cyclase	Heat-stable toxin	ETEC[b]	Enterotoxin
	EASTI[c]	EAEC[d], EPEC[e], EHEC[a]	Enterotoxin
Glucosylation of Rho	Toxin B	*C. difficile*	Cytotoxin
Pore formation	CPE[f]	*C. perfringens*	Cytotoxin
	Ace[g]	*V. cholerae*	Cytotoxin
	Aerolysin	*A. hydrophila*	Cytotoxin

[a]Enterohemorrhagic *E. coli.*
[b]Enterotoxigenic *E. coli.*
[c]Enteroaggregative heat-stable toxin.
[d]Enteroaggregative *E. coli.*
[e]Enteropathogenic *E. coli.*
[f]*C. perfringens* enterotoxin.
[g]Accessory cholerae enterotoxin.

hosts, from humans to domestic animals, birds, reptiles, and insects. Most of the *Salmonella* serotypes associated with human infections belong to subgroup 1 of the six subgroups of *Salmonella enterica.* Many serotypes are species-specific; thus, a particular serotype may be nonpathogenic in one host species and cause severe infection in another.

Gastroenteritis is the most common manifestation of *Salmonella* infections. Diarrhea begins 8–48 h after ingestion of contaminated food and lasts for 3–7 days. Infection is often food-borne and associated with consumption of foods of animal origin, such as chicken, raw milk, and undercooked eggs. The infectious dose ranges from 10^5 to 10^{10} organisms and depends on serotype, source of infection, and host factors. Food-borne salmonellosis seems to be predominantly a disease of industrialized countries. Infections can also be transmitted by the fecal–oral route, particularly among homosexual men. Immunocompromised hosts, such as HIV-infected individuals, are more prone to *Salmonella* infections. In these patients, non-typhoidal *Salmonella* usually causes bacteremia.

Salmonella spp. invade mucosal cells of the small intestine by a bacterial-mediated endocytosis process similar to *Shigella* entry. Bacteria stimulate signal transduction pathways in epithelial cells that lead to cytoskeletal rearrangements and membrane ruffling, similar to those induced by growth factors on mammalian cells. Membrane ruffling results in uptake of bacteria into cells. Unlike *Shigella, Salmonella* remains in membrane-bound vacuoles, modifies the pH of the phagolysosome, and multiplies. The organisms can also be taken up by macrophages, where they multiply and penetrate into deeper tissues (see the section following on *S. typhi*).

Most of the research on the molecular genetic basis of *Salmonella* invasion has been done on *S. typhimurium.* The genes that confer the ability of *Salmonella* to invade are located on a 40-kb pathogenicity island, called SPI-1, at the 63 min of *S. typhimurium* chromosome. SPI-1 encodes the components of a type III secretion pathway, called Inv–Spa, and the Sip proteins secreted via this machinery. Sip proteins have functions similar to *Shigella* Ipa proteins in the process of entry and induction of proinflammatory response and tissue destruction (see preceding section on *Shigella*).

Lipopolysaccharide (LPS) is another factor involved in invasiveness of *Salmonella*. Rough mutants with short O-side chains are less virulent. *Salmonella* also produces several adhesins, which might facilitate attachment to epithelial cells prior to penetration. In addition, nontyphi *Salmonella* strains produce an enterotoxin that has cytoskeleton-altering activity.

Salmonella strains induce secretion of cytokines such as IL-8 from epithelial cells. IL-8 is a chemoattractant for polymorphonuclear leukocytes (PMNs) and stimulates transmigration of PMNs through epithelial cell tight junctions into the intestinal lumen. The passage of PMNs through cellular junctions can lead to fluid leakage and consequent diarrhea.

D. *Campylobacter* spp.

Campylobacters are slender, spirally curved gram-negative rods, carrying a relatively small (1.6×10^6 bp) AT-rich genome that has been already sequenced. These organisms are microaerophilic, thermophilic, and require complex media for growth. Campylobacters are one of the most commonly reported causes of diarrhea worldwide. Infections with these organisms usually result in inflammatory dysentery-like diarrhea in adults of industrialized nations and watery diarrhea in children of developing countries. Campylobacters can also cause systemic disease. *C. jejuni* and *C. fetus* are the prototypes for diarrheal and systemic infections, respectively.

Campylobacter spp. are found in the gastrointestinal tracts of most domesticated mammals and fowls. Transmission occurs by contaminated food or water or by oral contact with feces of infected animals or humans. The organism cannot tolerate drying or freezing, thus being limited in transmission. Infections occur all year long, with a sharp peak in summer. The infectious dose can be as small as 500 or as high as 10^9 organisms, depending on the source of infection. The incubation period ranges from one to seven days and the duration of the illness is usually one week, with occasional relapses in untreated patients.

Campylobacter jejuni causes inflammation in both the colon and the small intestine. The resulting nonspecific colitis can be mistaken for acute ulcerative colitis. The inflammatory process can be extended to the appendix, mesenteric lymph nodes, and gall bladder. Bacteremia can also occur in some cases. Infections with *C. jejuni* may be followed by noninfectious complications, such as Guillain–Barre syndrome (GBS), an acute disease of peripheral nerves, and reactive arthritis.

Inflammation and bacteremia caused by *C. jejuni* suggest tissue invasion by this organism. Invasion seems to be linked to the presence of flagellae. Flagellae enable the organisms to move along the viscous environments and penetrate the intestinal mucosa. Binding of *C. jejuni* to host epithelial cells leads to bacterial uptake by a complex mechanism that remains controversial. Bacteria also penetrate the underlying lymphoid tissue and survive within the macrophages.

Campylobacter jejuni strains also produce several toxins. These toxins include a heat-labile cholera-like enterotoxin (CLT), which correlates with watery diarrhea, a cytolethal distending cytotoxin (CLDT), which alters the host cytoskeleton, and a hemolysin(s).

Campylobacter fetus causes febrile systemic illness more often than diarrheal infections. This organism has a tropism for vascular sites, thus causing bacteremia. The organism uses lipopolysaccharide (LPS) and a surface (S) layer protein, which functions as a capsule, to resist phagocytosis and serum-killing. *C. fetus* can disseminate to cause meningoencephalitis, lung abscess, septic arthritis, and urinary tract infections.

E. *Clostridium difficile*

Clostridium difficile is a gram-positive anaerobic spore-forming bacillus. This organism is widespread in the environment and is found in the intestines of several mammals, including humans. *C. difficile* is the most common recognized cause of diarrhea in hospitals and chronic care facilities in developed countries. Antibiotic therapy disrupts the normal flora of the colon and allows colonization or proliferation of *C. difficile*. Infection occurs by ingestion of spores from the environment. The spores resist the gastric acid and germinate into the vegetative form in the colon. The symptoms of disease range from mild diarrhea to severe pseudomembranous colitis. Infection with *C. difficile* is more common in the elderly, whereas neonates are resistant to colonization by this organism.

Pathogenic strains of *C. difficile* produce two very large toxins: toxin A and toxin B. Toxin A is a cytotoxin with a molecular weight of 308 kDa. This toxin stimulates cytokine production by macrophages and infiltration of PMNs, which results in the inflammation seen in pseudomembranous colitis. Toxin B is a protein of 207 kDa that also possesses cytotoxic activity. This toxin has glucosyltransferase activity that catalyzes transfer of glucose to the small GTP-binding protein, Rho. Modification of Rho leads to actin cytoskeletal disruption, which results in rounding of the cells. *C. difficile* also produces hydrolytic enzymes, such as hyaluronidase, gelatinase, and collagenase, which might contribute to destruction of connective tissue and subsequent fluid accumulation.

II. BACTERIAL AGENTS OF NONINFLAMMATORY DIARRHEA

The noninflammatory watery diarrhea caused by bacteria is usually associated with the production of an enterotoxin after bacterial colonization of the small intestine or with the presence of preformed enterotoxins in food. Occasionally, bacteria can cause a drastic effect on intestinal epithelial cells in the absence of an enterotoxin. This category of diarrhea is caused by the following.

A. *Vibrio cholerae*

Vibrio cholerae belongs to the family *Vibrionaceae* and has been the cause of seven cholera pandemics since 1817. It is transmitted by contaminated food and water. Food-borne transmission often occurs by ingestion of raw or undercooked shellfish. Since the acid-sensitive bacteria must pass through the stomach to colonize the small intestine, a high inoculum of 10^9 organisms is required to cause disease. The diarrhea can be extremely severe, with characteristic "rice water" stools, which can lead to rapid dehydration, circulatory collapse, and death.

Cholera is caused by toxigenic strains of *V. cholerae* O1 and O139 Bengal. The O1 strains can be divided into E1 Tor and classical biotypes that are epidemiologically distinct. *V. cholerae* O139 is a new strain that caused a major epidemic in 1992 in India in a population which was already immune to *V. cholerae* O1 strains. The non-O1 serogroups of *V. cholerae* cause cholera and dysentery but have not been linked to cholera epidemics.

Vibrio cholerae secretes cholera toxin (CT), which is responsible for the characteristic secretory diarrhea. CT is an enterotoxin that binds to enterocytes via five B subunits that facilitate the entry of the enzymatically active A subunit. The A subunit then catalyzes the ADP-ribosylation of the GTP binding protein, $G_s\alpha$, which results in activation of adenylate cyclase, accumulation of cAMP in enterocytes, and increase in secretion of chloride and water. Increased release of water into the intestinal lumen leads to secretory diarrhea. CT is encoded by the *ctx*AB genes carried on a filamentous bacteriophage. The receptor for the phage is a type IV fimbria, called the toxin-coregulated pilus (TCP), which is an essential colonization factor of *V. cholerae*. The gene encoding TCP is located on a 40-kb pathogenicity island (PI). This PI is associated with pandemic and epidemic strains of *V. cholerae*.

Vibrio cholerae O1 also produces two other toxins, called the zonula occludens toxin (Zot) and the accessory cholera enterotoxin (Ace). Zot affects the structure of the intercellular tight junction, zonula occludens. Ace is postulated to form ion-permeable channels in the host cellular membrane. The role of these toxins in pathogenesis is unknown.

B. Enteropathogenic *E. coli*

Enteropathogenic *E. coli* (EPEC) is an important cause of diarrhea in infants less than 2 years of age. EPEC is transmitted by the fecal–oral route by person-to-person contact. The infection occurs more frequently during the warm seasons. Infection with EPEC is often severe and leads to a high mortality rate in developing countries. The symptoms of the disease include watery diarrhea, vomiting, and fever. All EPEC strains induce a characteristic attaching and effacing (A/E) lesion on the brush border of the intestine which can be mimicked in tissue culture (Fig. 34.2). Pedestal-like structures form beneath the intimately adhering bacteria, due to the polymerization of actin. The accumulation of actin beneath the bacteria can be detected by a fluorescent-actin staining (FAS) assay. In addition, EPEC adheres to epithelial cells in tissue culture in a localized pattern, which can be detected by light microscopy.

Intimate attachment of EPEC to epithelial cells is mediated by an adhesin called intimin, which is encoded by the *eae* gene located on a 35-kb pathogenicity island, called the locus of enterocyte effacement (LEE). The LEE also encodes a type III secretion system called Esc, several proteins secreted by this secretion system called EspA, B, D, and F, and Tir. Tir becomes localized to the host cell membrane, where it serves as a receptor for intimin. The formation of the A/E lesions requires the Esc proteins, EspA, B, D, Tir, and intimin. The mechanisms that lead to diarrhea are unknown but may be related to changes in ion secretion and intestinal barrier function that have been detected *in vitro* and/or to loss of microvilli.

Localized adherence of EPEC to epithelial cells is dependent on the presence of a 90-kb plasmid, which

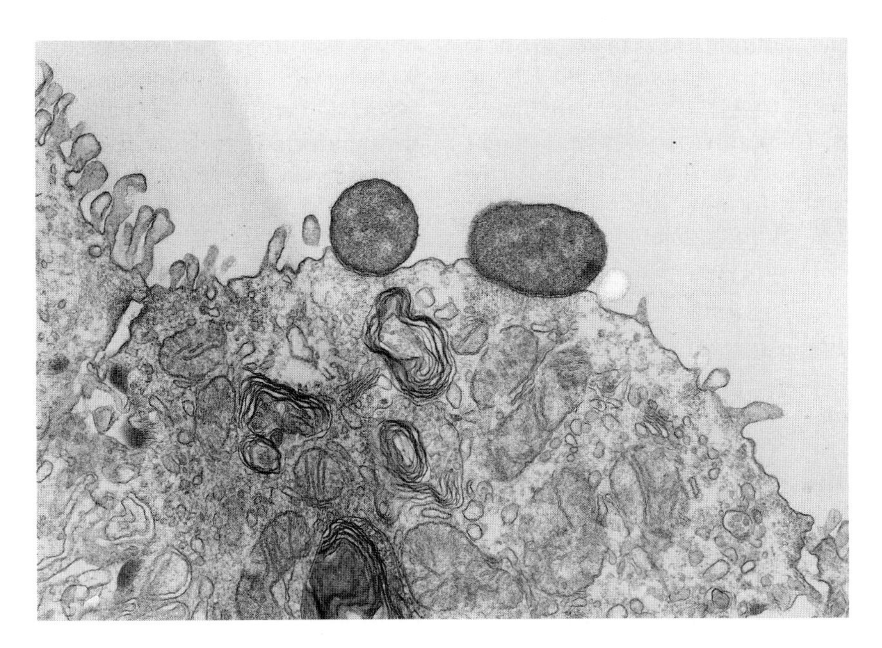

FIGURE 34.2 Typical attaching/effacing (A/E) lesions caused by enteropathogenic *E. coli* (EPEC) in an intestinal epithelial tissue culture model. Bacteria adhere intimately to the epithelial cells and induce formation of pedestals in the cell membrane (courtesy of Barry McNamara).

carries the genes required for the biogenesis of type IV fimbriae, called bundle-forming pili (BFP). BFP form ropelike structures (Fig. 34.3) that are responsible for the aggregation of EPEC bacteria to each other and for the localized adherence of EPEC to host epithelial cells. In addition to BFP, some EPEC strains produce other types of fimbriae that may also contribute to localized adherence. Some EPEC strains also produce a low molecular heat stable toxin called EAST1, similar to EAST1 of enteroaggregative *E. coli*, but the importance of this toxin in disease is unknown.

C. Enterotoxigenic *E. coli*

Enterotoxigenic *E. coli* (ETEC) is a cause of infantile and childhood diarrhea in developing countries and travelers' diarrhea in adults of industrialized countries visiting ETEC-endemic areas. Infection in travelers results in mild watery diarrhea in the majority of cases, whereas in infants from endemic areas, it can cause more severe diarrhea. ETEC has a short incubation period of 14–50 h. Epidemiological studies indicate that ETEC infection is more common in warm seasons and is transmitted through fecally contaminated food and water. The infectious dose for ETEC is approximately 10^8.

ETEC colonizes the mucosal epithelial cells of the small intestine and produces at least one of the two enterotoxins known as heat-labile toxin (LT) and heat-stable toxin (ST); both are encoded on a plasmid. LT is similar to cholera toxin (CT) in structure, function, and mode of action (see preceding). ST is a small polypeptide that activates intestinal guanylate cyclase, leading to accumulation of GMP, secretion of chloride and water, and, thus, diarrhea. ST resembles a peptide, guanilyn, normally found in the intestinal epithelium.

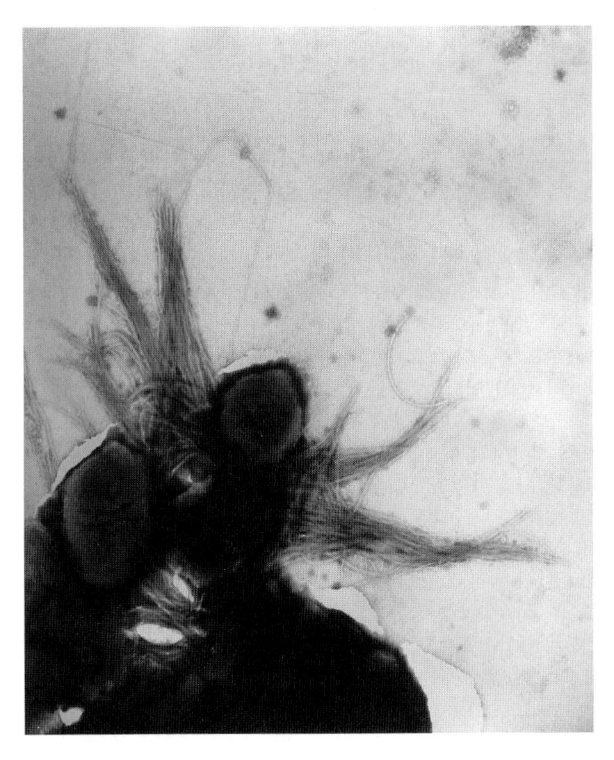

FIGURE 34.3 Bundle-forming pili produced by enteropathogenic *E. coli* (EPEC). Each bundle consists of several individual fimbrial filaments about 7 nm in diameter (reproduced with permission from Anantha *et al.*, (1998). *Infect. Immun.* **66**, 122–131).

ETEC strains produce multiple fimbriae that are host species-specific. Human ETEC fimbriae, called colonization factor antigens (CFAs), are associated with specific O serogroups. These fimbriae exhibit different morphological features, such as rigid rods similar to common fimbriae, bundle-forming flexible rods, and thin wavy filaments. In addition, many human ETEC strains produce a type IV fimbria, called Longus.

D. Enterohemorrhagic *E. coli*

Enterohemorrhagic *E. coli* (EHEC) is an emerging enteropathogen, which can cause watery diarrhea followed by bloody diarrhea, an illness designated hemorrhagic colitis (HC). EHEC is also associated with severe cases of hemolytic uremic syndrome (HUS), with a mortality rate of 5–10% The reservoir for this organism is the intestinal tract of cattle and, therefore, undercooked contaminated beef is the major source of infection. Contaminated milk, juice, lettuce, sprouts, and fast food have also caused outbreaks. Since the infectious dose is as low as 50–200 organisms, it is not surprising that EHEC is also spread by direct person-to-person contact.

The most important EHEC serotype, O157:H7, has been the cause of several food-borne outbreaks in the United States, Canada, Japan, and Europe since the 1980s. The mortality related to HUS has brought public attention to EHEC infections and has been the impetus for new regulations for handling and cooking of beef products.

The major pathogenic feature of EHEC is production of the bacteriophage-encoded Shiga toxin (or verotoxin), that is closely related to the Shiga toxin of *S. dysenteriae*. HUS is thought to be the result of hematogenous dissemination of Shiga toxin, cytotoxicity to endothelial cells, and microscopic thrombosis in the kidneys and elsewhere. Shiga toxins are also essential for development of bloody diarrhea and hemorrhagic colitis. Many EHEC strains also produce EAST1 toxin, similar to the toxin of enteroaggregative *E. coli*.

Similar to EPEC, EHEC strains also possess the LEE pathogenicity island on the chromosome (see preceding) and exhibit classic A/E histopathology. The A/E phenomenon is thought to be responsible for development of watery diarrhea, in a manner similar to EPEC diarrhea.

E. Enteroaggregative *E. coli*

Enteroaggregative *E. coli* (EAEC) is a cause of persistent diarrhea in children of developing and developed countries. EAEC infects the small bowel and causes diarrhea in less than 8 h, which may persist for >14 days. The diarrhea is usually mucoid and may be watery, with low-grade fever. EAEC has also been associated with diarrhea in human immunodeficiency virus-infected patients. More importantly, colonization with EAEC is linked to growth retardation in children, independent of the symptoms of diarrhea.

EAEC induces the formation of a mucous biofilm, which can trap and may protect the bacteria, leading to persistent colonization and diarrhea. Bacteria also elicit a cytotoxic effect on the intestinal mucosa. The cytotoxic effect is mediated by genes present on a 100-kb plasmid.

EAEC exhibits an aggregative adherence phenotype, mediated in some strains by a flexible bundle-forming fimbria called Aggregative Adherence Fimbriae I (AAF/I). AAF/I is a member of the Dr family of adhesins, present in uropathogenic *E. coli*.

EAEC strains also produce a ST-like toxin, EAST1, which is linked to the AAF/1 gene cluster on the 100-kb plasmid. The role of EAST1 in diarrhea is unknown, In addition, EAEC strains produce a 108-kDa cytotoxin that belongs to the autotransporter family of proteins. This protein also exhibits enterotoxin activity.

F. *Bacillus cereus*

Bacillus cereus is a gram-positive spore-forming rod that resides in water, soil, and as part of the normal flora in humans. This organism produces several toxins and causes two forms of toxin-mediated food poisoning, one characterized by emesis and the other by diarrhea. The production of either the emesis or diarrheal toxin is dependent on the type of the food on which the bacteria grow. The emetic toxin is a small toxin that is resistant to heat, extreme pH, and proteolytic enzymes. The toxin acts on the enteric nervous system through unknown mechanisms. The emetic toxin is associated with fried rice in the majority of cases. The emetic form of the disease has an incubation period of 2–3 h and elicits symptoms of vomiting and abdominal cramps that last 8–10 h.

The diarrheal toxin is a secretory cytotoxin consisting of a two- to three-component protein complex. This toxin induces secretion in the rabbit-ligated ileal loop and is cytotoxic in tissue culture. The mechanism of action of this toxin is unknown. The diarrheal toxin is associated with a variety of foods, such as sausage, vanilla sauce, and puddings. The incubation period for the diarrheal form ranges from 6 to 14 h. The illness is characterized by diarrhea and abdominal cramps, which may last 20–36 h.

G. *Staphylococcus aureus*

Staphylococcus aureus is a gram-positive coccus that is among the most common causes of bacterial food-borne disease. Food poisoning usually occurs by contamination of food with infected wounds on the hands, from the normal flora of skin or from the respiratory tract of food handlers. Foods such as ham and custard, which have a high concentration of salt or sugar, provide a good growth medium for *S. aureus*. One to 6 h after ingestion of the contaminated food, the symptoms of food poisoning begin, with severe vomiting and abdominal pain followed by diarrhea, which may last 24–48 h.

Food poisoning with *S. aureus* results from ingestion of the small enterotoxins A, B, C, D, or E, which are superantigens. Enterotoxin A is the most common one associated with food poisoning. These toxins act on enteric nervous system through unknown mechanisms. As little as 100–200 ng of the toxins can cause food poisoning. The *S. aureus* toxins are resistant to heat, irradiation, pH extremes, and proteolytic enzymes. Therefore, even overcooking of the contaminated food does not prevent the food poisoning. Involvement of *S. aureus* in outbreaks of food-borne disease is confirmed by detection of enterotoxins in food and by phage typing.

H. *Clostridium perfringens*

Clostridium perfringens is a gram-positive spore-forming organism that can tolerate aerobic conditions, unlike other members of Clostridia. *C. perfringens* type A exists in soil and in the intestinal tracts of most animals. This organism can cause a relatively mild food poisoning, more frequently in winter. Diarrhea and abdominal cramps develop 6–24 h following ingestion of preformed toxin and last up to 24 h. Toxigenic strains of *C. perfringens* usually grow on foods such as meat and poultry at temperatures between 15 and 50 °C, with a doubling time as short as 10 min. The spores are heat-resistant and can survive cooking and germinate after cooling. *C. perfringens* type A produces a 35-kDa cytotoxin called *C. perfringens* enterotoxin (CPE) during sporulation. CPE binds irreversibly to cells and forms ion-permeable channels in intestinal epithelial cells and acts as a superantigen that reacts with human T cells.

Clostridium perfringens can also cause non-food-poisoning diarrhea. The diarrhea is more severe, with blood and mucus in feces, and lasts longer. The disease occurs predominantly in the elderly or results from antibiotic therapy, similar to the cases with *C. difficile*.

Infections with *C. perfringens* type C can lead to necrotizing enteritis known as "pig-bel," a syndrome described in New Guinea related to consumption of large undercooked pork meals in native feasts. The symptoms of pig-bel include severe abdominal pain, bloody diarrhea, vomiting, and death due to intestinal perforation. The symptoms of pig-bel are associated with a toxin called β toxin. β Toxin is a cytotoxin with an unknown mechanism of action. This toxin is usually inactivated by proteolytic enzymes in the intestine. However, lack of proteolytic enzymes in malnourished hosts or inhibition of these enzymes by certain foods, such as sweet potato, allows the activity of β toxin and subsequent necrotizing enteritis.

III. BACTERIAL AGENTS OF ENTERIC FEVER

A. *Salmonella typhi*

Typhoid fever is a severe systemic disease caused by *S. typhi*. The disease is characterized by fever and abdominal symptoms. Infection can be transmitted by consumption of water or food contaminated with the feces of a patient or a chronic carrier. Humans are the only known reservoir for this organism, making the studies of typhoid infection difficult. However, *S. typhimurium*, which normally causes gastroenteritis in humans, causes a disease similar to typhoid in mice. Therefore, most of the studies of typhoid have focused on infection of mice and mouse macrophages with *S. typhimurium*. However, it is not clear that all of the conclusions from studies of *S. typhimurium* in mice apply to *S. typhi* in humans.

Following oral inoculation in mice, *S. typhimurium* survives the gastric acid barrier and reaches M cells, specialized epithelial cells that cover lymphoid tissues of the small intestine. Bacteria use M cells to penetrate the intestinal mucosa, whereupon they are engulfed by macrophages. Within macrophages, *Salmonella* attenuates the acidification process and multiplies. Survival within macrophages results in spread of the organisms and systemic infection. Bacteria enter the blood through the thoracic duct. Finally, bacteria are taken up by tissue macrophages in the bone marrow, liver and spleen.

The ability to multiply in macrophages and cause systemic infections is encoded by genes that reside on a second pathogenicity island, SPI-2, at 30 min on the *S. typhimurium* chromosome. (See preceding section on nontyphoidal *Salmonella* about SPI-I.) In addition to the products of these genes, intramacrophage survival is also modulated by the PhoP/PhoQ two-component regulatory system. These proteins regulate acid phosphatase synthesis and unknown genes, which are essential for survival in the acidic environment of the macrophage.

B. *Yersinia* spp.

Within the genus *Yersinia*, *Y. pestis*, *Y. pseudotuberculosis*, and *Y. enterocolitica* are pathogenic for humans. Based on DNA hybridization techniques, *Y. enterocolitica* has recently been subdivided into eight other species. *Y. pestis* is the cause of bubonic plague, the "Black Death," which claimed one-fourth of Europe's population in the fourteenth century. *Y. pseudotuberculosis* and *Y. enterocolitica* are the causes of yersiniosis, a disease more prevalent in developed countries. Yersiniosis is characterized by an enteric feverlike illness, which is accompanied by acute diarrhea. Mesenteric adenitis is a common manifestation of the disease, which causes an acute appendicitis-like syndrome, with fever and abdominal pain. Extra-intestinal manifestations can also include septicemia and nonpurulent arthritis.

Infections with *Y. pseudotuberculosis* are more common in animals and less frequent in humans. *Y. enterocolitica* is carried by healthy pigs but is pathogenic for humans. It is transmitted by ingestion of contaminated water or food, more commonly, contaminated milk. The organisms can multiply at low temperatures, such as those of refrigerated food. The infectious inoculum may be 10^9 organisms and the incubation period may last 4–7 days.

Following ingestion of *Yersinia*, bacteria adhere to and enter the intestinal epithelial cells. Infection then spreads to the mesenteric lymph nodes, where abscesses develop. Adherence of *Yersinia* to host cells is mediated by a plasmid-encoded adhesin, called YadA, and a chromosomally encoded protein, called Invasin. Invasin mediates entry into cells by interacting with β1 integrin receptors. This interaction leads to the extension of a pseudopod that forms a "zipper" around the bacterium, resulting in internalization.

At 37°C under low calcium conditions *in vitro*, *Yersinia* spp. secrete a set of proteins called Yops. Yops are virulence factors that enable bacteria to survive and multiply within lymphoid tissues of the host. The genes encoding Yops reside on a 70-kb virulence plasmid. Yops are secreted by a type III secretion system called Ysc and the secretion is regulated by temperature and contact with eukaryotic cells. Upon contact of bacteria with host cells, Yops are synthesized, secreted, and some are injected into the cytoplasm of host cells. Following injection, YopE and YopT act as cytotoxins that disrupt the actin microfilament structure. YopH is a protein tyrosine phosphatase that dephosphorylates certain proteins of macrophages. These Yops together inhibit phagocytosis by macrophages. YopP (YopJ) induces apoptosis of macrophages, which results in release of proinflammatory cytokines and subsequent inflammatory responses to infection.

Yersincia enterocolitica also produces an enterotoxin similar to heat-stable toxin of *E. coli*, called Yst. This enterotoxin might be responsible for cases of food poisoning caused by this organism.

IV. BACTERIAL AGENTS OF GASTRITIS

A. *Helicobacter pylori*

Helicobacter pylori is a spiral, microaerophilic gram-negative bacterium with two to six polar sheathed flagellae that endow the bacterium with a corkscrew mode of motility. It has a relatively small genome of 1.7 $\times 10^6$ bp which is highly AT-rich, similar to that of Campylobacters. The complete sequence of the genome has been determined for two strains. This organism is extremely prevalent, residing in large numbers (10^8–10^{10} organisms per stomach) in the stomachs of at least half of the human population. *H. pylori* probably is not found in the environment or in animals and, therefore, person-to-person contact and the fecal–oral route are the likely means of transmission. Once inoculated, the incubation period is estimated to be 3–7 days and infection can last for the lifetime of the host. *H. pylori* exhibits tissue specificity exclusively for gastric mucosal epithelial cells and does not invade beyond these tissues. Detection of the organisms is best accomplished by biopsy of stomach tissue and subsequent testing for urease activity, by serologic testing, or by culture.

Infections with *H. pylori* result in acute and chronic gastric inflammation, which, when untreated, can lead to peptic ulcers or stomach carcinoma. The majority of cases of gastric and duodenal ulcers are caused by *H. pylori*. The outcome of infection with *H. pylori* depends on a variety of bacterial, host, and environmental factors.

Gastric inflammation by *H. pylori* is mediated by several virulence factors. All *H. pylori* strains produce urease in very high amounts. Urease is a nickel-containing hexameric enzyme, which catalyzes hydrolysis of urea to ammonia. Ammonia neutralizes gastric acid of the stomach, allowing the organism to colonize. Flagellae are another important colonization factor of this organism, allowing the bacteria to move along the mucous layer of the stomach. Most *H. pylori* strains produce a vacuolating cytotoxin, called VacA, which is an autotransporter protein. This cytotoxin induces acidic vacuoles in the cytoplasm of eukaryotic cells, Also, the *H. pylori* strains that are more associated with duodenal ulcer and stomach cancer carry the cytotoxin-associated gene (*cag*) pathogenicity island. Genes on the *cag* PI are required for secretion of IL-8 and tyrosine phophorylation of host proteins.

V. MISCELLANEOUS BACTERIAL ENTERIC PATHOGENS

Thorough coverage of all enteric bacterial pathogens is beyond the scope of this article. However, we briefly describe a few other common enteropathogenic bacteria in the following section:

A. Non-cholerae vibrios

Vibrios parahaemolyticus, V. vulnificus, V. mimicus, V. fluvialis, V. furnissii, and *V. hollisae* reside in aquatic environments and prefer high salt and warm temperature habitats. These organisms have been implicated in intestinal infections following consumption of contaminated raw or undercooked shellfish. *V. parahaemolyticus* and *V. vulnificus* can also cause wound infections and septicemia. *V. parahaemolyticus* produces the thermostable direct hemolysin (TDH), which acts as an enterotoxin to stimulate intestinal secretion and subsequent diarrhea. Hemolysin may also be involved in tissue damage observed in wound infection. This hemolysin elicits the Kanagawa phenomenon on Wagatsuma agar, a phenotype specific to pathogenic strains.

Vibrios vulnificus is an invasive pathogen. Eating contaminated raw oysters leads to septicemia 24–48 h later, particularly in people with underlying liver disease. This organism produces several proteolytic enzymes and a hemolysin, which contribute to overt damage to epithelial cells in wound infections. *V. vulnificus* also produces a capsule, which has been associated with virulence.

B. *Aeromonas* spp.

Aeromonas spp. are widely distributed in marine environments and can be transmitted to humans via contaminated food, especially during summer. Three species, *A. hydrophila, A. sobria*, and *A. caviae*, cause diarrhea in humans. *A. hydrophila* produces a hemolysin called aerolysin or β-hemolysin, which, in addition to cytolytic activity, acts as an enterotoxin and induces diarrhea. It also produces other enterotoxins and cytoskeleton-altering toxins. In addition, *A. hydrophila* produces a type IV fimbria that might contribute to virulence.

C. *Plesiomonas shigelloides*

Plesiomonas shigelloides has biochemical similarities to *Aeromonas* and antigenic similarities to *Shigella* spp. It is primarily a marine microorganism, which can be transmitted to humans by raw or undercooked seafood. Diarrhea, accompanied with relatively severe abdominal cramps, occurs 24 h after ingestion of the organism. The stool may contain mucus, blood, and pus, suggesting an invasive mechanism of the disease. *P. shigelloides* produces a heat-labile enterotoxin with cytoskeleton-altering activity and a heat-stable enterotoxin with unknown mechanism of action.

Plesiomonas shigelloides also causes extraintestinal infections, such as meningitis in neonates, septicemia in immunocompromised hosts, and septic arthritis.

D. *Listeria monocytogenes*

Listeria monocytogenes is the only species of *Listeria* that is pathogenic to humans. This organism is a gram-positive rod, which is present in soil, as part of the fecal flora of many animals, and on many foods, such as raw vegetables, raw milk, cheese, fish, meat, and poultry. Diarrhea can occur following ingestion of 10^9 organisms via contaminated food. The incubation period is long and can last between 11 and 70 days. Infections with *L. monocytogenes* are uncommon in the normal population, but in immunocompromised patients, neonates, the elderly, and pregnant women, infection can lead to encephalitis, meningitis, and stillbirth.

Listeria monocytogenes invades both epithelial cells and phagocytes. Entry involves a "zippering" mechanism, similar to that described for *Yersinia* entry (see preceding). Interaction between Internalin, a bacterial protein, and E-cadherin, a receptor on epithelial cells, induces phagocytosis. Once inside the phagocytic vacuole, bacteria lyse the vacuolar membrane by a hemolysin called listeriolysin O and spread within the cytoplasm. In the cytoplasm, a bacterial protein called ActA induces assembly of actin microfilaments behind the bacterium, which results in movement of bacteria in a manner similar to movement of *Shigella* inside the cytoplasm (see preceding). Bacteria move to adjacent cells and spread. Genes involved in escape from the vacuole and in intra/intercelullar spread are carried on a pathogenicity island in the *L. monocytogenes* chromosome.

E. Enterotoxigenic *Bacteroides fragilis*

Bacteroides fragilis is a gram-negative non-spore-forming anaerobic rod that comprises a part of the normal intestinal flora of nearly all humans.

Enterotoxigenic *B. fragilis* (ETBG) strains produce a toxin that stimulates fluid secretion in the intestinal lumen and causes rounding of epithelial cells and loss of intestinal microvilli in an *in vivo* intestinal model. These strains have been linked to diarrhea in animals and in a small number of studies of humans.

F. *Clostridium botulinum*

Clostridium botulinum is the cause of food poisoning, which can occur by ingestion of preformed toxins in inadequately processed food, such as home-canned vegetables and fish. Vomiting and diarrhea occur before neurological symptoms begin. The disease is caused by neurotoxins A, B, or E. Infant botulinum is another manifestation, which is different from food poisoning in that toxins are produced after germination of spores in gut. *Clostridium botulinum* also produces a cytotoxin called C2 toxin, which can alter the cell cytoskeleton by ADP-ribosylation of G actin and preventing polymerization of G to F actin. The role of this toxin in disease is unknown.

ACKNOWLEDGMENTS

We acknowledge Rick Blank and David McGee for careful reading of this chapter and Philippe Sansonetti, Coumarin Egile, Barry McNamara, and Ravi Anantha for providing the figures. This work was supported by Public Health Services Awards AI37606, AI32074, and DK49720 from the National Institutes of Health.

BIBLIOGRAPHY

Armstrong, G. L., Hollingsworth, J., and Morris, J. G. (1998). Bacterial food-borne disease. *In* "Bacterial Infections of Humans, Epidemiology and Control", 3rd edn. (A. S. Evans and P. S. Brachman, eds.), pp. 109–138. Plenum Publishing Co., New York.

Blaser, M. J. (1995). Campylobacter and related species. *In* "Principles and Practice of Infectious Diseases," Vol. 2, 4th edn. (G. L. Mandel, J. E. Bennett, and R. Dolin, eds.), pp. 1948–1956. Churchill Livingstone, New York.

Borriello, S. P. (1998). Pathogenesis of *Clostridium difficile* infection. *J. Antimicrob. Chemo.* **41**, 13–19.

Cornelis, R. G. (1998). The *Yersinia* deadly kiss. *J. Bacteriol.* **180**, 5495–5504.

Cornelis, G. R., and Van Gijsegem, F. (2000). Assembly and function of type III secretory systems. *Annu. Rev. Microbiol.* **54**, 735–774.

Donnenberg, M. S., Zhang, H. Z., and Stone, K. D. (1997). Biogenesis of the bundle-forming pilus of enteropathogenic *Eschericia coli*: Reconstruction of fimbriae in recombinant *E. coli* and role of DsbA in pilin stability—A review. *Gene.* **192**, 33–38.

Dunn, B. E., Cohen, H., and Blaser, M. J. (1997). *Helicobacter pylori. Clin. Microbiol. Rev.* **10**, 720–741.

Frischknecht, F., and Way, M. (2001). Surfing pathogens and the lessons learned for actin polymerization. *Trends Cell Biol.* **11**, 30–38.

Guerrant, R. L. (1995). Principles and syndromes of enteric infection. *In* "Principles and Practice of Infectious Diseases," Vol. 1, 4th edn. (G. L. Mandel, J. E. Bennett, and R. Dolin, eds.), pp. 945–962. Churchill Livingstone, New York.

Isberg, R. R., and Barnes, P. (2001). Subversion of integrins by enteropathogenic **Yersinia**. *J. Cell Sci.* **114** (Pt 1), 21–28.

Kerr, J. R. (1999). Cell adhesion molecules in the pathogenesis of and host defence against microbial infection. *Mol. Pathol.* **52**, 220–230.

Lorber, B. (1997). Listeriosis. *Clin. Infect. Dis.* **24**, 1–11.

Miller, S. I., Hohmann, E. L., and Pegus, D. A. (1995). *Salmonella* (including *Salmonella typhi*). *In* "Principles and Practice of Infectious Diseases," Vol. 2, 4th edn. (G. L. Mandel, J. E. Bennett, and R. Dolin, eds.), pp. 2013–2033. Churchill Livingstone, New York.

Nataro, J. P., and Kaper, J. B. (1998). Diarrheagenic *Escherichia coli. Clin. Microbiol. Rev.* **11**, 142–201.

Parsot, C., and Sansonetti, P. J. (1996). Invasion and the pathogenesis of Shigella infections. *Curr. Top. Microbiol. Immunol.* **209**, 25–42.

Sears, C. L., Guerrant, R. L., and Kaper, J. B. (1995). Enteric bacterial toxins. *In* "Infections of the Gastrointestinal Tract" (M. J. Blaser, P. D. Smith, J. I. Radvin, H. B. Greenberg, and R. L. Guerrant, eds.), pp. 617–634. Raven Press, Ltd., New York.

Tauxe, R. V. (1998). Cholera. *In* "Bacterial Infections of Humans, Epidemiology and Control", 3rd edn. (A. S. Evans and P. S. Brachman, eds.), pp. 223–242. Plenum Publishing Co., New York.

Taux, R. V., and Hughes, J. M. (1995). Food-borne disease. *In* "Principles and Practice of Infectious Diseases." Vol. 1, 4th edn. (G. L. Mandel, J. E. Bennett, and R Dolin, eds.), pp. 1012–1024. Churchill Livingstone, New York.

Vallance, B. A., and Finlay, B. B. (2000). Exploitation of host cells by enteropathogenic *Escherichia coli. Proc. Natl. Acad. Sci. USA* **97**, 8799–8806.

WEBSITES

Comprehensive Microbial Resource of The Institute for Genomic Research and links to many other microbial genomic sites *http://www.tigr.org/tigr-scripts/CMR2/CMRHomePage.spl*

"Bad Bug Book" of the Food and Drug Administration, USA *http://vm.cfsan.fda.gov/~MOW/intro.html*

The site of salmonella.org, with many links to Salmonella related topics (S. Maloy and R. Edwards) *http://www.salmonella.org/*

Photo Gallery of Bacterial Pathogens (Neil Chamberlain) *http://www.geocities.com/CapeCanaveral/3504/gallery.htm*

Mortality and Morbidity Weekly Report (Centers for Disease Control, USA) *http://www.cdc.gov/mmwr/*

List of bacterial names with standing in nomenclature (J.P. Euzéby) *http://www.bacterio.cict.fr/index.html*

Access to online resources on Bacterial Infections and Mycoses. Karolinska Institutet *http://www.mic.ki.se/Diseases/c1.html*

35

Escherichia coli and *Salmonella*, Genetics

K. Brooks Low

Yale University

GLOSSARY

bacteriophage (*phage*) A virus that infects bacteria. Lysoganic bacteriophages sometimes insert stably into the chromosome without replicating. Lytic bacteriophages always replicate following infection, and usually kill their host bacterial cell as they produce a burst of progeny phage particles.

conjugation A process of DNA transfer between bacterial cells involving cell-to-cell contact and requiring the functions of fertility factor genes in the donor cell.

curing The elimination of a replicon such as a plasmid or prophage from a bacterial strain.

F-prime A plasmid derived by excision from a chromosomally integrated fertility factor (F), so that it carries a portion of the main chromosome, thus enabling a partial diploid (merodiploid) state.

fimbriae Hairlike proteinaceous appendages (100–1000 per cell), 2–8 nm in diameter, on the surface of bacteria that are often involved in adhesion to specific types of eukaryotic cells.

flagella Thin helical appendages on the surface of bacteria (5–10 per cell; originating at random points on the surface—peritrichous), about 20 nm in diameter, that can rotate and cause the cell to swim.

genome The total stably inherited genetic content of a cell, including chromosome and plasmids.

genotype The particular nucleotide sequence of the genome of a particular strain. A genotypic symbol (e.g., *leuB44*, a mutation in a leucine bioynthetic gene) is used to notate any change from the wild type (leu^+).

horizontal gene transfer The introduction of new genetic material into a cell's genome as a result of uptake from an outside source such as by mating (conjugation), transduction or transformation (i.e. not simply by mutation and vertical inheritance into vegetative descendants); usually used to denote gene transfer between somewhat unrelated species.

lysogenic conversion (*phage conversion*) A change in bacterial phenotype due to either phage infection or stable establishment (lysogeny) of a bacteriophage in the genome, whereby new (bacteriophage) genes are expressed, such as genes for toxin production or a new cell surface antigen.

mutation Any inherited change in nucleotide sequence in the genome.

operon A set of contiguous genes that are transcribed together into a single mRNA, together with any *cis*-acting regulatory sequences. The gene products are thus controlled coordinately. If two or more genes are controlled coordinately, but are at separate sites on the genome, they constitute a regulon.

palindrome An inverted repeated sequence. In double-stranded DNA, the ends of a palindrome are self-complementary, and can melt and reanneal to form hairpin-like structures.

pathogenicity island A cluster of contiguous genes in a genome that appear to originate from a distant

The Desk Encyclopedia of Microbiology
ISBN: 0-12-621361-5

evolutionary source and that is involved in conferring a pathogenic phenotype to a strain.

phase variation A reversible change in cell surface antigens, controlled by a genetic switch in the genome.

phenotype The observable characteristics of an organism based on its cellular or colonial morphology, biochemical structure, growth characteristics, or serological specificity. A phenotypic symbol (e.g. Leu^-, denoting a growth requirement for leucine) is used to indicate any change from the wild type (Leu^+).

plasmid An independently replicating component of the genome, usually much smaller than the main chromosome and usually not essential for cell survival.

recombination, genetic The interaction of nucleic acid elements of the genome to produce a new arrangement of base pairs; sometimes also including the addition from an outside source of new independently replicating elements such as plasmids, thus enlarging the genome.

R-prime A plasmid derived by excision from a chromosomally integrated R factor, so that it carries a portion of the main chromosome, thus enabling a partial diploid (merodiploid) state.

serotype (*serovar; bioserotype; ser*) A strain-specific immunological reactivity due to particular antigenic components on or near the cell surface of a particular strain.

transduction A gene transfer involving the injection of genetic material carried in a bacteriophage capsid.

transformation, bacterial A gene transfer by uptake of naked DNA into a cell.

Over 2300 Wild-type (naturally occurring) varieties of *Salmonella* are known to inhabit Earth and an estimated vast excess of over 10,000 (perhaps more than 100,000) wild-type varieties of *Escherichia coli* exist as well. To help explain this striking diversity, there is overwhelming evidence that a complex spectrum of vertically inherited mutational events and horizontal gene transfer events has taken place over the course of evolution, since the time these two species diverged from a common ancestor approximately 150 million year ago. The resulting population of genetic variants includes mostly innocuous inhabitants of avian, human, and other animal intestinal tracts (*E. coli* cells normally constitute about 1% of the bacterial flora in the human intestine), and also a large number of highly pathogenic variants that are able to invade the intestine, blood, and other organs, exacting a huge toll in terms of disease and death (there are 3.6 million deaths annually from *Salmonella* infections worldwide) and the associated economic burden (over a billion dollars annually in the United States alone). As is well known, certain strains of *E. coli* and *Salmonella* have been used since the 1940s to serve as laboratory workhorses for studies of mutation, physiology, gene transfer, and fundamental life processes such as mechanisms of genetic inheritance, macromolecular synthesis, biosynthetic pathways, and gene regulation. The huge families of laboratory-derived strains and seminal findings from these experiments is a crucial counterpart to the lives of *E. coli* and *Salmonella* in the wild. The genetics of these organisms is based on the study of their genomic structure and mechanisms of variation observed both in the laboratory and in nature.

I. TAXONOMIC NICHE OF *ESCHERICHIA COLI* AND *SALMONELLA*

The description of the phylogeny of *Escherichia coli* and *Salmonella* has had a particularly tortuous and confusing history. This is in part because of the medical relevance of the many pathogenic varieties of these strains and the ensuing detail of study, and in part due to the intrinsically vast spectrum of subtle antigenic variations that have evolved on the surface of these cells, and, furthermore, the ability of the cells of even one strain to alternate between antigenic types in their normal course of growth.

The larger family of bacterial genera to which *E. coli* and *Salmonella* belong is the Enterobacteriaceae (or Enterobacteria), defined in 1937 by Rahn, which has diverse natural habitats including animal intestines, as well as plants, soil, and water. The number of genera, let alone species, defined to be in this family has varied repeatedly for almost 100 years, depending on available diagnostic measures and criteria of relatedness agreed to and disagreed to by many investigators. For example, the number of genera of Enterobacteriaceae listed in Bergey's *Manual of Determinative Bacteriology*, 5th edition (1939), a widely used standard, is nine, with 67 species (not including *Salmonella*, whose taxonomic history has varied widely depending on the concept of species), whereas the 8th edition (1974) lists 12 genera and 37 species. As of the 1990s, the use of methods of DNA–DNA hybridization to determine relatedness has led to the clear resolution of over 30 genera and 100 species within the "extended" family of Enterobacteriaceae.

Within this family, certain genera such as *E. coli* and *Shigella* are very closely related (70–100% homologous); yet they are maintained in separate genera because of the practical difference in clinical diseases they cause. The next closest relatives of *E. coli* are the *Salmonella* and *Citrobacter* genera, then *Klebsiella* and *Enterobacter*. More distantly related genera include

Erwinia, Hafnia, Serratia, Morganella, Edwardsiella, Yersinia, Providencia and, most distantly, *Proteus*.

The above genera were all known before 1965 (starting with *Serratia* in 1823), mostly as a result of infection in and the fecal content of humans or animals. Since the late 1970s, 17 more genera in this family have been discovered, some from clinical origin and some from water, plants, or insects.

The major characteristics that serve to identify the Enterobacteriaceae are that they are gram-negative rod-shaped bacteria that, with minor exceptions, can grow aerobically and anaerobically, produce a catalase but not oxidase, ferment glucose and convert nitrates to nitrates, contain a common enterobacterial antigen, are usually motile by means of flagella, do not require sodium for growth, and are not spore-forming. Further subclassification depends on other subtleties in metabolic capacities, such as the ability to use various sugars. For practical reasons, many medically important strains are diagnosed also with antisera specific to components of the outer membrane. Horizontal gene transfer is believed to occur between virtually all members of the family.

E. coli was discovered in feces from breastfed infants by Theodor Escherich, in 1885, and was first named *Bacterium coli*. By the early 1900s it became clear that it differed from another similar organism known primarily as a pathogen, *Bacillus typhi* (known since 1930 as *Salmonella typhi*), and hence was renamed *Escherichia* in honor of its discoverer. *E. coli* was found to ferment lactose, whereas the (*Salmonella*) *typhi* strains could not. In hindsight, this ability to ferment lactose and the divergence of *E. coli* and *Salmonella* roughly 150 million years ago coincides rather strikingly with the assumed appearance of mammals, and the assumed first synthesis of lactose in nature, at about the same evolutionary time. The differing abilities of *E. coli* and *Salmonella* to ferment lactose, and invade mammals in a pathogenic fashion suggests a rapid adaptation to the newly emerging families of mammals on Earth.

In contrast to the discovery of *E. coli* as a common and compatible inhabitant of the human intestine, *Salmonella*, named in 1900 after Salmon, was first defined as the causative agent for hog cholera (hence the initial strain name, "choleraesuis"). Investigations of the varied and varying antigenic properties of many similar strains from diseased animals and humans led to a classification scheme (Kauffmann–White) based on three types of antigens. Further classification was based on differences in sensitivity to various bacteriophages.

The three broad classes of antigens used to categorize both *E. coli* and *Salmonella* isolates are based on three types of surface structures. The outermost are the flagella that are present in motile strains. The antigens of flagella are termed H antigens (Hauch, meaning "cloud" or "film," describing colonies of motile bacteria). Next is the group of outermost surface layers including the capsule (in some strains), envelope and fimbriae. The antigens of this group are called K (for Kapsule). Internal to the K group is a group of antigens associated with the phospholipids in the outer membrane—termed O antigens because strains that lose their motility (and are thus "ohne Hauch," that is, without cloud or film as colonies) through loss of flagella can expose the lower-lying (O) antigens. What makes this overall system feasible is that by selective treatment the outer antigens can be removed or inactivated stepwise by alcohol (for the H antigen, i.e., the flagella) and by heating to 100 °C for 2 h (which inactivates most or all of the K antigens), which thus leaves material containing only the O antigens. By preparing antibodies to the bacteria in all three states, the antisera can be subtractively purified to obtain fractions specific for each of the three antigen classes.

The results of decades of effort in this direction in the early 1900s (and continuing into the present, for more specific isolates) resulted in a well-defined spectrum of antigenic determinants that define a multitude of *E. coli* and *Salmonella* strains. In *E. coli* at least 173 distinct O antigens, 80 K antigens, and 56 H antigens are known. Although it is not likely that all conceivable combinations exist in nature, it is thought that well over 10,000 variants exist, more than for any other species. Note that the common laboratory strain, *E. coli* K-12 (also written K12 or K_{12}), does not show any K or O antigenic determinants. The name K-12 was applied in a Stanford clinic where it was isolated in 1922, long before the K, O, H system was developed.

As for *E. coli*, *Salmonella* strains from natural sources display a huge variety in O and H (and sometimes K) antigenic determinants, and over 2300 combinations have now been registered, second only to *E. coli* in complexity. Some of this complexity is due to the ability of *Salmonella* to present more than one form of H antigen (phase variation). Additional variation is introduced by the phenomenon of lysogenic conversion, whereby certain bacteriophages infect a strain, become lysogenized as a stable prophage, and express one or more new genes that alter the O surface antigens. Until the early 1940s, the names used for different isolates were taken to mean separate species, thus hundreds of *Salmonella* species are described in the early literature. In 1973, a major simplification occurred as a result of DNA-relatedness studies of

Crosa and colleagues, who found that virtually all *Salmonella* strains are closely related and can be considered a single species, divided into six subgenera (subspecies). Officially, this species is named *Salmonella choleraesuis*; however a subsequent proposal is pending to name the species *Salmonella enterica* and thus use a name not associated with any of the earlier isolates. Many investigators are already using this terminology, wherein the particular strain is defined with a serovar name that corresponds in most cases with the older "species" name, that is, *Salmonella enterica* serovar Typhimurium, or simply *Salmonella* Typhimurium.

II. THE GENOME

A. The chromosome

The genomes of *E. coli* and *Salmonella* strains consist of a single major circular chromosome of over 4 Mb of DNA (the genomes from natural isolates of *E. coli* range in size from 4.5–5.5 Mb), and commonly one or more small independently replicating (extrachromosomal) DNA plasmids, usually 100 kb or less. Thus, these organisms are haploid, although a second copy of a portion of the chromosome can be carried on a plasmid, thus creating a partial diploid (merodiploid) state. Strains of this type have been used extensively in the laboratory in genetic studies of dominance and regulation.

Though the genome is haploid, the number of copies of the main chromosome varies, ranging between approximately one per cell in the resting (stationary phase) state to sometimes four per cell in rapidly dividing cells in rich growth media. In this state there are more than one set of replication growth forks along the chromosome, which replicates divergently beginning at a site called the origin (of replication) and ending in a region called the terminus (see Fig. 35.1). Thus, in rapidly growing cultures there are at any given time more copies of genes located near the origin than near the terminus. As a result of this multiplicity of growth forks, rapidly growing cells divide and produce two daughter cells as often as every 20 min, even though the time required for any one growth fork to move from the origin to the terminus is 40 min. Genes whose products are used in high concentration during rapid growth (such as genes for components of the ribosome) tend to be located nearer the origin than the terminus, thus allowing relatively high amounts of product formation.

The chromosomes of *E. coli* and *Salmonella* are clearly related, as can be seen from Fig. 35.1, which shows the general congruence of the sequence of genes around the chromosomes for the two most commonly used laboratory strains (originally isolated from the wild), *E. coli* K-12 and *Salmonella* Typhimurium LT2. The vast majority of the genes in *E. coli* K-12 have counterparts at the corresponding regions of the *Salmonella* Typhimurium chromosome, and vice versa. At the base-sequence level, the DNA homology varies from approximately 75–99% for various protein-coding genes. The G + C to A + T DNA base ratios are also similar (50.8% G + C content for *E. coli* K-12; 52% for *Salmonella* Typhimurium). Recombination can occur between the *E. coli* and *Salmonella* chromosomes, albeit rarely (see later discussion), to form viable hybrid strains that carry portions from each chromosome, and most of the gene functions from one organism can substitute for the analogous function in the other.

Superimposed on the general congruence of these two related genomes are a number of sites in which an insertion or deletion shows a distinct difference between the two, as indicated in Fig. 35.1 (and see later discussion). Figure 35.1 also shows a region of the chromosome, in the 26–40 min region, where a large inversion has reversed the gene sequence in *E. coli* K-12 relative to *Salmonella* Typhimurium. It is believed that the *Salmonella* Typhimurium configuration of this region is the more distantly ancestral one because certain other types of Enterobacteria related to *E. coli* and *Salmonella* typhimurium carry inversions analogous to the 26–40 min inversion but differing greatly in extent, ranging from a much smaller inversion (in *Klebsiella aerogenes*) to a much larger inversion (in *Salmonella* Enteriditis). Thus, it is believed that over the course of evolution a number of independent inversion events in this region occurred, starting from a configuration similar to that of *Salmonella* Typhimurium.

A summary of types of genes and other genetic elements on the *E. coli* K-12 chromosome is listed in Table 35.1. For structural genes that encode proteins, Table 35.2 indicates the major classes based on broad functional role in the cell.

B. Extrachromosomal elements

Most natural isolates of *Salmonella*, and many of *E. coli*, carry extrachromosomal plasmids. These are broadly classified as either self-transmissable (conjugative) or not (nonconjugative). Conjugative plasmids carry a number of genes needed for conjugative DNA transfer between cells (see later discussion). Some nonconjugative plasmids can be conjugationally transferred (mobilized) when a conjugative plasmid is also

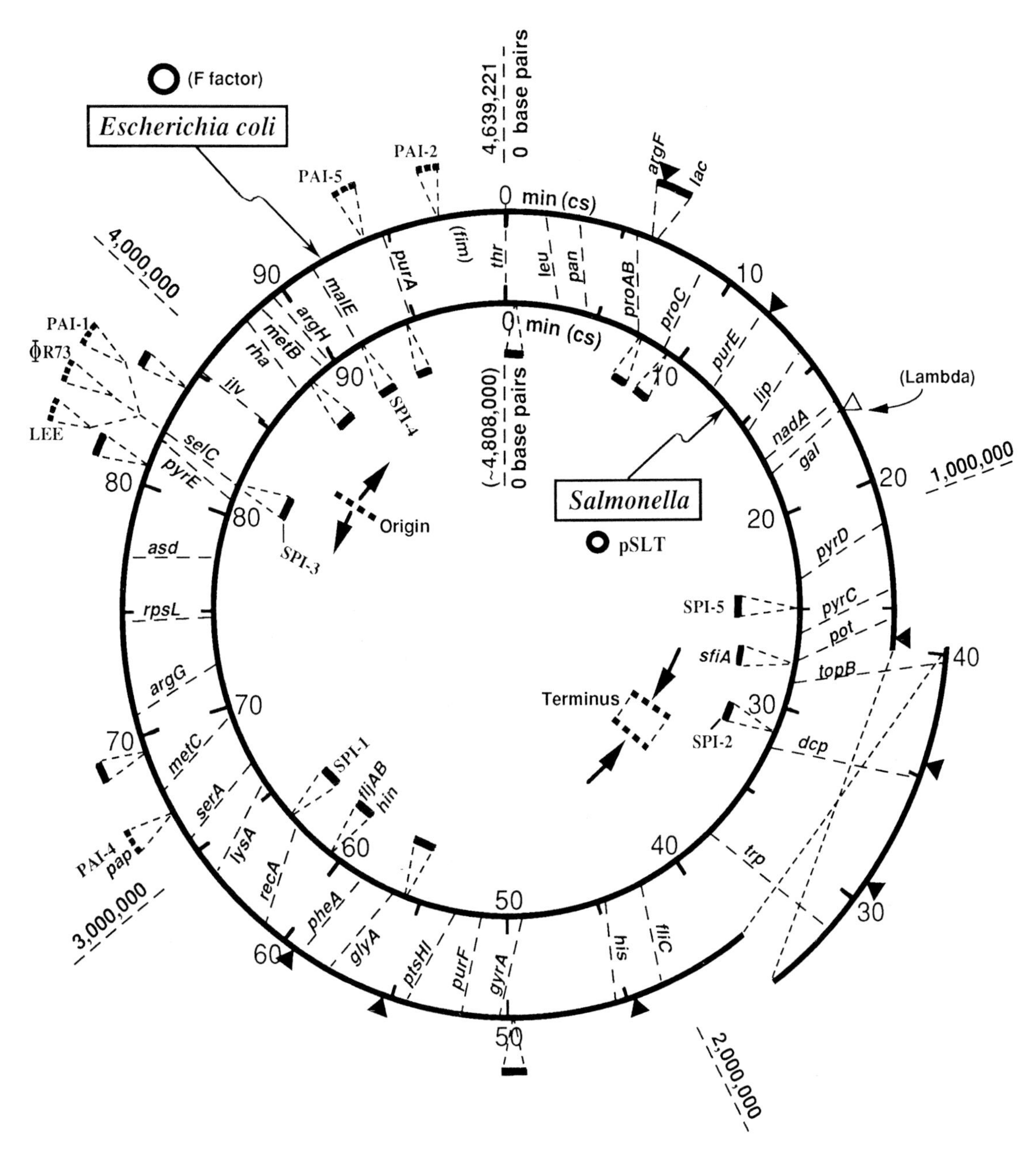

FIGURE 35.1 The two circular genetic maps of the chromosomes of representative strains of *Escherichia coli* (strain K-12 MG1655) and *Salmonella* (serotype Typhimurium, strain LT2), drawn concentrically for comparison. The maps are oriented relative to position 0 at the top, historically chosen to be at the threonine (*thr*) locus in both organisms. The outermost (radially extending) numbers (1,000,000, etc.) indicate the approximate number of nucleotide base pairs in distance clockwise along the chromosome from position 0. The numbers written along the circles (10, 20, etc.) indicate the corresponding percentage distance around the chromosome, also denoted in units of minutes (min) or, equivalently, centisomes (cs). Between the two map circles, the relative positions of representative genes or gene clusters common to both chromosomes are indicated by the gene names (*thr, leu*, etc.). Origin and Terminus indicate the positions of initiation and termination, respectively, of normal bidirectional chromosome replication during vegetative growth. The region between approximately 27 and 40 min on the *E. coli* map has an inverted gene order relative to *Salmonella*, as indicated. Representative short extra arcs of chromosome are indicated as either present on the *E. coli* chromosome but not *Salmonella* (drawn exterior to the *E. coli* map circle), or present on the *Salmonella* chromosome but not *E. coli* (drawn interior to the *Salmonella* map circle). The arcs labeled SPI-1, SPI-2, and so on denote Salmonella pathogenicity islands. Analogous islands of genes indicated by dashed arcs (PAI-1, PAI-2, etc., pathogenicity island, or LEE, locus of enterocyte effacement) on the *E. coli* map are sometimes present in certain pathogenic *E. coli* strains. The small circles represent examples of plasmids that are typically present in *E. coli* K-12 (F factor, fertility factor) or *Salmonella* Typhimurium LT2 (pSLT). The small open triangle at 17 min on the *E. coli* map indicates the chromosomal attachment site of bacteriophage lambda, which is usually present as a lysogenic prophage in *E. coli* K-12 strains. Other similar sites where other lysogenic bacteriophages sometimes are found integrated are not shown, except for ΦR73, a lysogenic phage whose chromosomal insertion site is the same one (near 82 min) used by PAI-1 and by LEE. The small filled-in triangles on the *E. coli* map indicate the positions of cryptic prophages.

TABLE 35.1 *Escherichia coli* K-12 genome, major features[a]

Elements	Number	Characteristics
Base pairs	4,639,221	Single circular chromosome
Replichores	2	Divergently replicating halves, from single origin to terminus region
G + C content	50.8%	Overabundance of G in leading replicative strands (G ~26.2%; C, A, T ~24.6%)
Protein-codable genes	4288	88% of genome; 55% transcribed in direction of replication
average size (amino acids)	317	
number with 1000–2383 amino acids	56	
number with < 100 amino acids	381	
operons (known and predicted)	2584	73% single gene (predicted), 17% two genes, 10% three or more genes; 68% single promoter (predicted), 20% two promoters, 12% three or more promoters; > 10% regulated by more than one protein
Stable RNA genes		
rRNA	7 operons	All operons transcribed in replication direction; 5 CW, 2 CCW; location biased toward origin of replication.
tRNA	86	43 operons of various sizes widely dispersed in location and orientation.
Cryptic prophages	8	All located in late-replicating halves of replichores
other phage-like remnants	33	
Integration sites (att) for various lysogenic phages	>15	Widely dispersed around genome; sites for lambda-related phages located in late replicating portions of replichores
Insertion sequences	42	Number of repeats: 1 to 11; 4 classes (gene arrangements); size range 768–1426 bp
Insertion sites for fertility factors, (F, Col V, etc.)	>20	Some are at IS sequences
Repeated sequences (or, pseudo-repeated) Chi (GCTGGTGG)	761	RecBCD-mediated recombination initiator; highly skewed (~2:1) to leading replicative strand; widely distributed
Rhs	5	5.7–9.6 kb; components and G + C content variable; function unknown; not found in most other *E. coli* strains or in *Salmonella*
REP	697	~40-bp palindromes; sometimes tandemly repeated (1–12 copies) in 355 BIME sites; possible role in mRNA stabilization, gene expression and/or replication
IRU	19	Imperfect palindromes; size ~125 bp; function unknown
RSA	6	Imperfect palindromes; size ~152 bp; function unknown
Box C	33	High G + C; size ~56 bp; function unknown
iap	23	29-bp imperfect palindromes clustered near *iap* gene on the chromosome; function unknown

[a]Strain K-12 MG1655. Abbreviations: BIME, bacterial interdispersed mosaic elements; CCW, counter-clockwise; CW, clockwise; IRU, intergenic repeat units (ERIC, enterobacterial repetitive intergenic consensis); REP, repetitive extragenic palindromic unit (PU, palindromic unit); Rhs, rearrangement hot spot.

TABLE 35.2 Distribution of major functional classes of *E. coli* protein-coding genes[a]

Functional class	Number of genes	Percent of total
DNA replication, repair, recombination	115	2.7
Nucleotide metabolism	58	1.4
RNA metabolism, transcription	55	1.3
Protein synthesis	182	4.2
Amino acid metabolism	131	3.0
Cell membranes and structure	237	5.5
Fatty-acid and lipid metabolism	48	1.1
Binding and transport	427	10.0
Carbon-compound catabolism	130	3.0
Intermediary metabolism	188	4.4
Energy metabolism	243	5.7
Regulation	178	4.2
Phage, transposons	87	2.0
Miscellaneous	577	13.5
Unknown	1632	38.0
Total	4288	

[a]Strain K-12 MG1655. Values estimated.

present to provide the necessary gene functions. Plasmids can be classified into groups based on whether they can coexist stably in the same cell. (Two different plasmids that can coexist are defined to different incompatibility, Inc, groups.) There are more than 30 Inc groups among the Enterobacterial plasmids.

About two-thirds of natural *Salmonella* Typhimurium strains carry a conjugative plasmid roughly 90 kb in size, similar to pSLT, which is present in strain LT2 as indicated in Fig. 35.1. These plasmids usually carry some of the genes that contribute to the virulence of *Salmonella* as a pathogen. Some wild-type *E. coli* strains (roughly 10% or more) also carry conjugative plasmids. The one denoted F from *E. coli* K-12 is about 100 kb in size and is unusual in that it is self-transmissable at very high frequency (see later discussion). The F factor and other F-like factors from *E. coli* strains have a number of genes involved in conjugation that are homologous to pSLT and similar plasmids from *Salmonella* Typhimurium strains.

Another type of plasmid, found in about 30% of wild-type *E. coli* strains, is the colicin factor. Colicins are toxins produced by one bacterium to kill or inhibit another bacterium. At least 20 distinct types of colicin factor, some of which are self-transmissable, have been found in *E. coli*.

III. WAYS IN WHICH THE GENOME CAN CHANGE

A. Mutation

Various changes in base sequence occur spontaneously at the rate of approximately 6×10^{-10} per base pair per generation (i.e. about 0.003 per genome) during vegetative growth. Much higher rates of mutagenesis are caused by mutagenic agents such as irradiation or exposure to certain chemicals, or even by simple prolonged starvation, as may occur in nature. The consequences of mutation can range from almost no effect (silent mutations, e.g. a base substitution in an amino acid-coding triplet in which the encoded amino acid does not change; however the change in tRNA used can lead to subtle changes in translation efficiency) to mild phenotypic changes (e.g. temperature sensitivity or leaky requirement for a growth factor; i.e. a leaky auxotroph or bradytroph, due to point mutations that alter just one or a few amino acids) to severe phenotypic changes (e.g. the inactivation of one or more genes due to deletions, frameshift mutations, or nonsense mutations). Sometimes a single mutation can result in a change in more than one phenotype or a change in production of a number of gene products, in which case the mutation is called pleiotropic. A broad range of phenotypes can be observed in *E. coli* or *Salmonella* mutants, listed in Table 35.3.

TABLE 35.3 Classes of phenotypic characteristics of mutant strains

Phenotypic class	Examples
Morphological	Colony size, texture, color; cell shape, motility, staining properties
Physiological	Dependence on temperature, oxygen, pH
Nutritional	Requirements for growth; possible carbon and energy sources
Biochemical	Breakdown products (e.g. acid) from carbohydrates, nitrates
Inhibition	Sensitivity to antibiotics, dyes, etc.
Serological	Agglutination by specific antisera
Pathogenic	Invasion capacity; virulence in variety of hosts

B. Recombination

Three distinct types of rearrangements of genetic elements (i.e. recombination) are observed throughout the living world: homologous, site-specific, and transpositional. Figure 35.2 shows diagrammatically the main topological features of these classes.

1. Homologous recombination

This refers to the interaction of two very similar DNA molecules (i.e. having almost the same base sequence

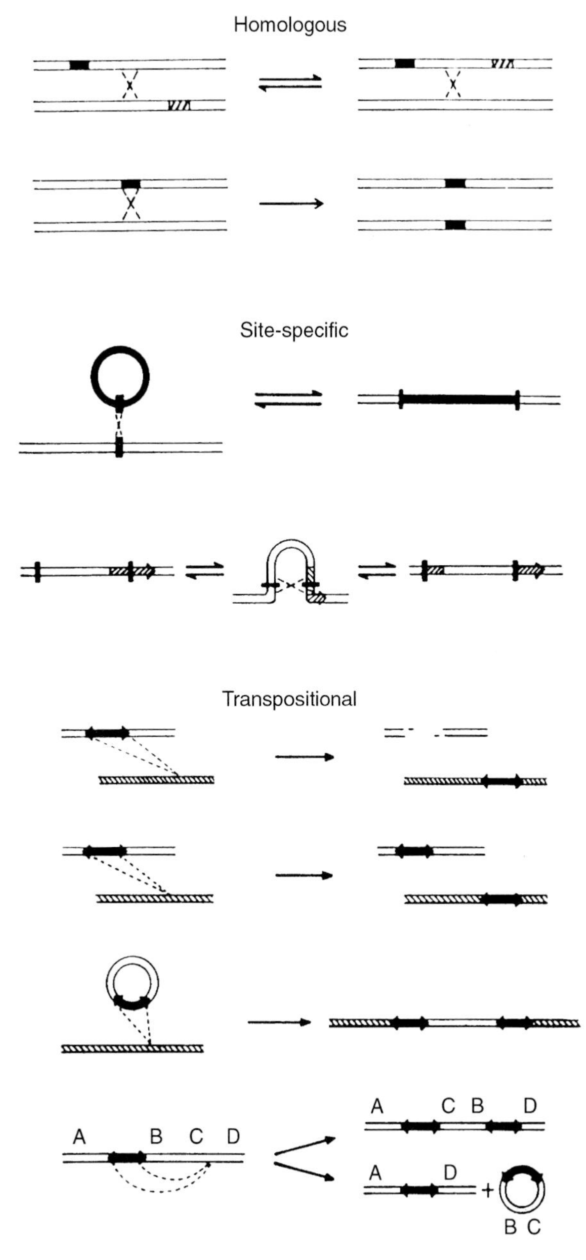

FIGURE 35.2 Schematic representation of major classes of genetic recombination. Adapted from Low, B. in Low, B. (ed.) (1988). *The Recombination of Genetic Material*. Academic press, pp. 6–7, with permission from the author and publisher.

for many hundreds of continuous bases) at equivalent sites along their DNA chains, which can result in crossover or DNA-repair events almost randomly at any site along the pair of parental DNA molecules. This can thereby result in a change in linkage, or proximity on the same chromosome, for various small (but nevertheless significant) chromosomal differences situated along the lengths of the two recombining DNA molecules. If the two recombining DNA chains are generally homologous, but contain scattered differences from one to the other (e.g. a few percent difference or more for the *E. coli* versus *Salmonella* chromosomes) recombination can still occur, but is more infrequent with increasing differences in sequences. This nearly homologous configuration is known as homeologous recombination.

Homologous recombination can thus occur following gene transfer between closely related organisms (see later discussion). However, homologous recombination also occurs, very frequently, even between elements within one cell. For instance, crossovers between some of the seven homologous copies of rRNA operons can occur during vegetative growth to produce large genetic duplications or inversions. These rearrangements are unstable and readily recombine back to the normal haploid state. Nevertheless their mere existence (in up to one-third of a growing population of cells) makes possible various changes in gene dosage or gene arrangement, which likely contributes to the evolutionary flux over long periods of time due to rare accompanying recombination events that are sometimes nonhomologous (or illegitimate).

2. *Site-specific recombination*

As the name implies, this form of recombination is a crossover event between two DNA double helices, which is catalyzed to occur at unique sites on the DNA strands, defined by the local DNA sequence involving usually 20–30 bases. A specialized enzyme recognizes these sequences, on both parental molecules, and causes a double-strand breakage and rejoining event that exchanges the partners of flanking DNA. The crossover is completely conservative, that is, no bases are added or lost at the crossover site. One configuration of site-specific recombination enables the circular form of a lysogenic phage, such as lambda (found in *E. coli* K-12) or P22 (one of a large family of related bacteriophages found in most natural *Salmonella* isolates), to integrate into the chromosome (and thus become a stable lysogen), or to be excised from it (and thus begin to replicate vegetatively). See Fig. 35.2 (site-specific, upper).

Another DNA configuration involved in site-specific recombination involves a hairpin-like intermediate such that the end result of the crossover is an inversion of the stretch of DNA located between the two crossover sites (Fig. 35.2, site-specific, lower). An example of this occurs in *Salmonella* and results in a change in gene expression for two different genes for the production of flagellin, which is assembled into filaments on the surface of the cell (see Fig. 35.3). By switching the gene expression in this way, *Salmonella* is able to change its filament antigenic structure (i.e. undergo phase variation) and thus reduce attack by host immune systems. Another mechanism for phase variation also exists in some strains, in which methylation of certain critical adenine residues will activate or deactivate gene expression, and this methylation pattern can be quasi-stably inherited for many generations, until it flip-flops back to a nonmethylated state, which reverses the antigenic phase.

3. *Transpositional recombination*

As in the case of site-specific recombination, certain specialized short DNA sequences are involved in recombination involving DNA elements called transposons. Transposons are sequences of DNA, usually in the size range of 700–10,000 bases, which can move by recombination from one location in the genome to any of a multitude of other sites (targets). The specialized sequences that promote this recombination are usually 15–30 bases long and located at the two ends of the transposon as an inverted repeat. The DNA between the two ends can encode a number of genes whose functions are either required for the transpositional recombination (e.g. transposase, resolvase) or other functions that alter cell phenotype, such as antibiotic resistance. Transposons that do not encode any internal genes except those needed for transposition are called insertion sequences (IS). The *E. coli* chromosome contains several copies of five IS, and approximately 15% of the spontaneous mutations in *E. coli* occur due to the movement of these IS to new sites. None of these particular IS are normally found in *Salmonella*, but a different IS specific to *Salmonella* strains is normally present somewhere on its chromosome.

Transposition events can either involve the movement of the entire original transposon DNA sequence to a new site (conservative transposition) or a replicative process in which the transposon is replicated to form two copies, one of which is inserted at the new (target) site and the other of which is retained at the original site (replicative transposition) (see Fig. 35.2). Even in the case of conservative transposition, a small amount of DNA synthesis is involved, in which a few of

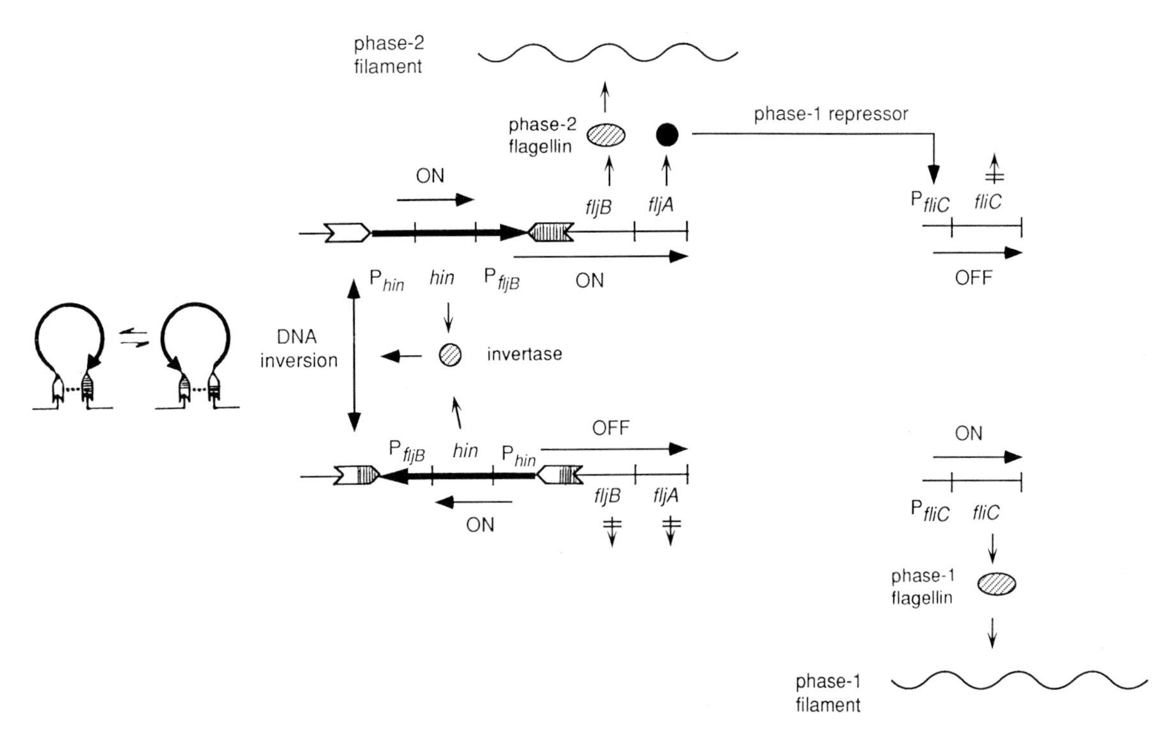

FIGURE 35.3 Mechanism for flagellar phase variation in *S. typhimurium* (i.e., the alternate expression of two structural genes for flagellin, an H antigen). The two genes, *fliC* (formerly named H1) and *fljB* (formerly named H2), are distant from each other on the chromosome at 43 and 61 min, respectively (see Fig. 35.1). The promoter for the *fljB* operon is part of a region of the chromosome that is bounded by an inverted repeat (white or shaded arrow bars), which can be inverted by site-specific recombination mediated by the Hin invertase protein, a product of gene *hin* in the invertable region. In the orientation shown at the top of the figure (phase 2), the promoter for the *fljB* operon (P_{fljB}) is correct and FljB or phase 2 flagellin is synthesized, together with the repressor of the phase 1 *fliC* operon, the *fljA* gene product (formerly named rh1). In the other orientation (phase 1), shown at the bottom of the figure, P_{fljB} is separated from the operon it promotes and neither the phase 2 flagellin nor the phase 1 repressor is synthesized. The *fliC* operon is therefore expressed, and FliC or phase 1 flagellin is synthesized. Adapted from Macnab, R. *In* Neidhardt, F., *et al.* (1996), p. 137, with permission from the author and the American Society for Microbiology.

bases at the target site are copied to create a small duplication at each end of the newly inserted transposon.

IV. GENE TRANSFER BETWEEN CELLS

A. Conjugation

In Section II, it was noted that most *Salmonella* strains and many *E. coli* strains carry fertility factors (conjugative plasmids) that can promote gene transfer from cell to cell by conjugation. Furthermore, the F-like plasmids in *E. coli* and the pSLT-like plasmids in *Salmonella* have a number of genes in common, and considerable homology between them. These facts strongly imply that over evolutionary history there has been considerable cross-talk and gene transfer among these strains. The particular F factor in *E. coli* strain K-12 is (fortunately for J. Lederberg and other early investigators in the discovery and study of conjugation) highly efficient for promoting conjugation between F^+ cells (those that carry the F factor as a plasmid) and F^- cells (those which have no F factor in any form). This is due to a mutation on the K-12 F factor in the regulatory system that, for the vast majority of fertility factors, represses the genes for conjugation. With these "normal" conjugative plasmids, this repression keeps all the cells except about 1/1000 from acting as donors. At any given time any one of the cells can, with a probability of 1/1000, become temporarily derepressed for conjugative functions and cause its own transfer into an F^- cell (thus converting it into an F^+ cell). Immediately following this transfer into the F^- cell, there is a delay before the repression of the conjugative functions builds up. Thus, such a newly formed F^+ cell is temporarily able to conjugate (act as a donor) efficiently again, with another F^- cell. In this way a rapid epidemic spread of conjugative plasmids can move into a new population of F^- cells when the opportunity arises.

On rare occasions, an F-like plasmid can recombine with the chromosome to form an Hfr (high frequency of recombination) cell (see Fig. 35.4). In the process of

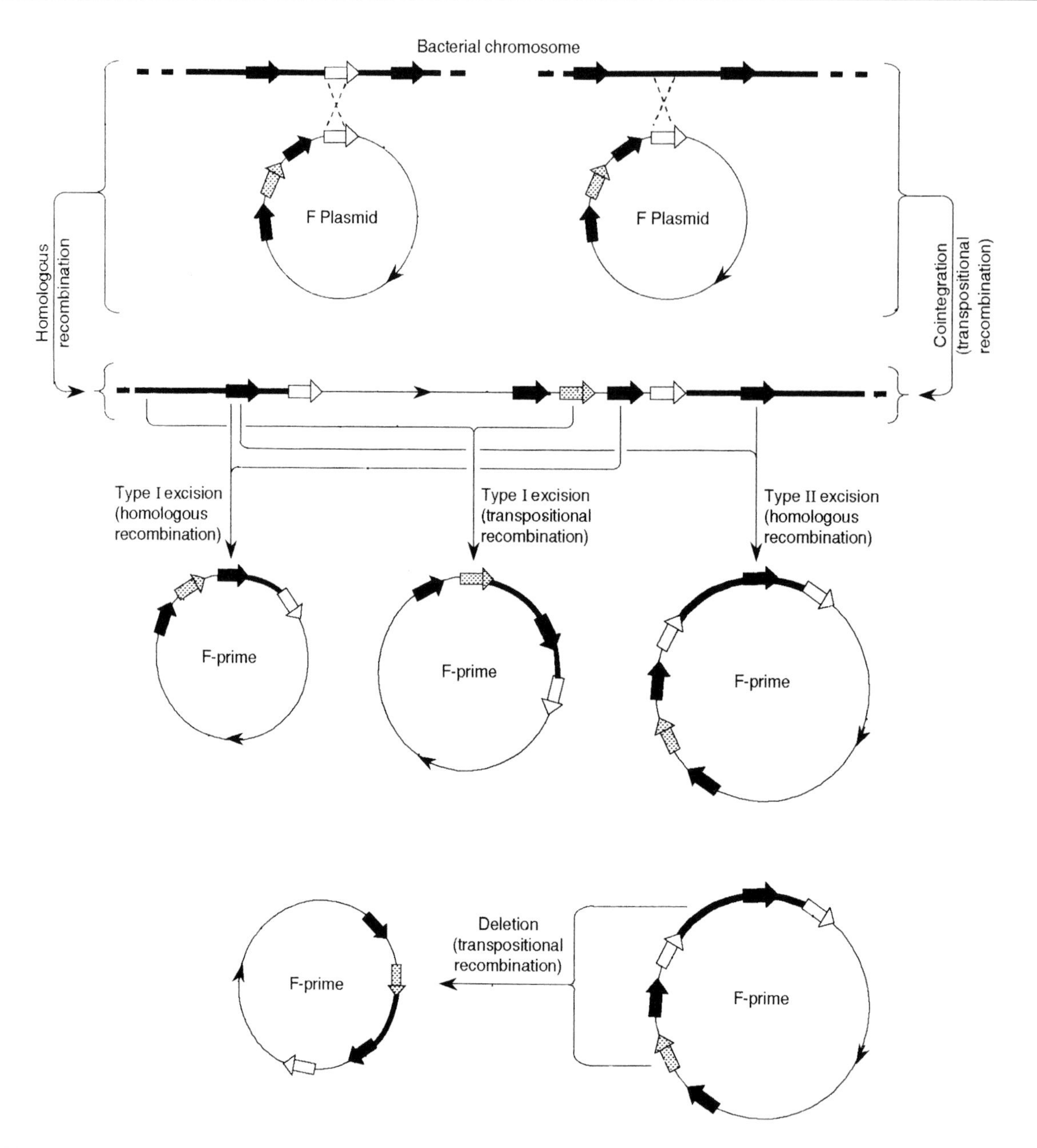

FIGURE 35.4 Interconversion between various states of the F factor in *E. coli*. The top heavy horizontal lines represent parts of the chromosome, with arrows that represent the various positions of IS. IS that are also located on the F factor (thin circular line) can recombine with the chromosome either by homologous recombination with a chromosomal IS of the same type (top left), or by transpositional recombination at a chromosomal site where there is not normally an IS (top right). The resulting chromosome with the integrated F factor (center horizontal line) is called Hfr. The Hfr chromosome can undergo subsequent recombination events, and they produce plasmids (F-primes) that carry most or all of the F factor plus a section of the chromosome. From Holloway, B. and Low, B. *In* Neidhardt, F., *et al.* (1996), p. 2414, with permission from the author and the American Society for Microbiology.

conjugation from such a cell, the transfer of the normal leading region (next to a specific site called the origin of transfer, indicated by the small arrowhead in Fig. 35.4), leads to the transfer of chromosomal genes because in the Hfr state the F factor and chromosome form one continuous DNA molecule. Considerable amounts of chromosomal genetic information can be transferred in this way, although the transfer process tends to be interrupted spontaneously so that the genes furthest around the chromosome have the least chance of being transferred.

B. Transduction

Recall that a large family of lysogenic bacteriophages related to P22 exist among wild-type *Salmonella* strains. It so happens that when phage P22 is replicating lytically in a cell and being packaged into new phage particles, occasionally (about once in every 30 times) the packaging process makes a mistake and packages a segment (about 1%) of the bacterial genome instead of the newly replicated P22 DNA. These aberrant particles are called transducing particles and can come out of the cell along with the burst of normal new phage particles. The transducing particles can travel to another cell and inject in the genomic DNA, followed in some cases by recombination into the new cell's genome. The analogous processes occur in *E. coli* strains with a phage known as P1, which can package about 2% of the genome into transducing particles. The process, discovered in 1952 by Zinder and Lederberg, is called generalized transduction and it most likely has had a major role in the horizontal transfer of genetic information among enteric bacteria. In some cases, the transduced chromosomal DNA fragment does not recombine into the chromosome but forms a stable (nonreplicating) structure that continues to be unilaterally inherited by only one of the daughter cells at each division. If the transducing fragment carries a functional gene that allows the transduced cell to grow on selective medium, only a tiny colony will grow because only one of the two daughter cells after each division will divide again. This is termed abortive transduction and has been useful in determining functional complementation between mutated genes.

Another mode of transduction, termed specialized transduction, is distinctly different from generalized transduction. Specialized transduction involves the excision of a chromosomally located lysogenic bacteriophage (i.e. in the prophage state) to form a circular replicating form. However, rarely, this excision event is imprecise and sometimes involves a crossover within the adjacent bacterial chromosomal DNA (analogous to F-prime formation, Fig. 35.4), such that the replicating phage and subsequent phage particles produced contain a small amount (a few genes) of chromosomal DNA. This DNA can thus be carried by the progeny phage particles and injected into new host cells, followed by possible recombination events. Thus, in specialized transduction only genes close to the prophage integration site can be transduced. Moreover, the altered phage particles carrying the chromosomal fragment can be grown and amplified to produce large quantities of the specific chromosomal genes for experimental purposes.

C. Transformation

The uptake of naked DNA by bacteria and the subsequent inheritance of its genetic information in the genome occur naturally in some gram-positive and gram-negative bacterial species such as *Streptococcus pneumoniae, Bacillus subtilus, Haemophilis,* and *Acinetobacter*, but only under certain growth conditions that induce competence, or the ability to take up DNA. In contrast, *E. coli* and *Salmonella* are not known to have this natural competence for DNA uptake, and it is unlikely that transformation has contributed significantly to their evolutionary genomic development. However, laboratory methods have been devised that promote the uptake of DNA into *E. coli* and *Salmonella*, thus greatly facilitating their use in molecular biological studies. These methods for artificially induced transformation competence generally involve either treatment with di- or multivalent cations (e.g., Ca^{2+}, Mg^{2+}) and heat shock, which is believed to partially disrupt the lipopolysaccharide in the outer membrane (chemical competence), or treatment with a brief pulse of high-voltage electricity (electroshock), which renders the membrane temporarily permeable to a variety of macromolecules including DNA.

BIBLIOGRAPHY

Brock, T. D. (1990). "The Emergence of Bacterial Genetics." Cold Spring Harbor Laboratory Press, Cold Spring Harbor, NY.

Charlebois, R. L. (ed.) (1999). "Organization of the Prokaryotic Genome." ASM Press, Washington, DC.

Clewell, D. B. (ed.) (1993). "Bacterial Conjugation." Plenum Press, New York.

Goodfellow, M., and O'Donnell, A. G. (eds.) (1993). "Handbook of New Bacterial Systematics." Academic Press, San Diego, CA.

Janda, J. M., and Abbott, S. L. (1998). "The Enterobacteria." Lippincott-Raven, Philadelphia.

Logan, N. A. (1994). "Bacterial Systematics." Blackwell Scientific, London.

Lund, B. M., Sussman, M., Jones, D., and Stringer, M. F. (eds.) (1988). "Enterobacteriaceae in the Environment and as Pathogens." Blackwell Scientific, London.

Neidhardt, F. C., Curtiss, R., III, Ingraham, J. L., Lin, E. C. C., Low, K. B., Magasanik, B., Reznikoff, W. S., Riley, M., Schaecter, M., and Umbarger, H. E. (eds.) (1996). "*Escherichia coli* and *Salmonella*, Cellular and Molecular Biology," 2nd edn. ASM Press, Washington, DC.

Riley, M., and Sanderson, K. E. (1990). Comparative genetics of *Escherichia coli* and *Salmonella typhimurium*. *In* "The Bacterial Chromosome" (K. Drlica and M. Riley, eds.), pp. 85–95. ASM Press, Washington, DC.

Schaechter, M., and The View From Here Group. (2001). *Escherichia coli* and *Salmonella*: The view from here. *Microbiol. Mol. Biol. Rev.* **65**, 119–130.

Sussman, M. (ed.) (1997). "*Escherichia coli*: Mechanisms of Virulence." Cambridge University Press, Cambridge.

vary in their cytotoxic potency, with the clostridial neurotoxins being the most potent exotoxins of humans. Exotoxins also vary with respect to the host that can be intoxicated. Exotoxin A of *Pseudomonas aeruginosa* can intoxicate cells from numerous species, whereas other toxins, such as diphtheria toxin are more restricted in the species that can be intoxicated. Some bacterial toxins, such as pertussis toxin, can intoxicate numerous cell types, whereas other toxins, such as the Clostridial neurotoxins, show a specific tropism and intoxicate only cells of neuronal origin. Bacterial exotoxins catalyze specific chemical modifications of hostcell components, such as the ADP-ribosylation reaction catalyzed by diphtheria toxins or the deamidation reaction catalyzed by the cytotoxic necrotizing factor produced by *Escherichia coli*. These chemical modifications may either inhibit or stimulate the normal action of the target molecule to yield a clinical pathology. Bacterial exotoxins possess an **AB** structure–function organization, in which the **A** domain represents the catalytic domain and the **B** domain comprises the receptor-binding domain and the translocation domain. The translocation domain is responsible for the delivery of the catalytic **A** domain into an intracellular compartment of the host cell.

Many bacterial exotoxins can be chemically modified to toxoids that no longer expresses cytotoxicity, but may retain immunogenicity. Studies have shown that bacterial toxins can also be genetically engineered to toxoids, which may lead to a wider range of vaccine products. Exotoxins have also been used as therapeutic agents to correct various disorders, including the treatment of muscle spasms by botulinum toxin. Nontoxic forms of exotoxins have been used as carriers for the delivery of heterologous molecules to elicit an immune response and as agents in the development of cell-specific chemotherapy. In addition, bacterial toxins have been used as research tools to assist in defining various eukaryotic metabolic pathways, such as G-protein-mediated signal transduction.

I. CLASSIFICATION OF EXOTOXINS

Exotoxins are soluble proteins produced by microorganisms that can enter a host cell and catalyze the covalent modification of a cellular component(s) to alter the host-cell physiology. The term "host cell" refers to either vertebrate cells or cells of lower eukaryotes, such as protozoa because some bacterial exotoxins intoxicate a broad range of host cells. The recognition that some pathogenic bacteria produced soluble components that were capable of producing the pathology associated with a particular disease was determined in the late nineteenth century. Roux and Yersin observed that culture filtrates of *Corynebacterium diphtheriae* were lethal in animal models and that the pathology elicited by the culture filtrate was similar to that observed during the infection by the bacterium. Subsequent studies isolated a protein, diphtheria toxin, from the toxic culture filtrates and observed that the administration of purified diphtheria toxin into animals was sufficient to elicit the pathology ascribed to diphtheria. Diphtheria toxin is a prototype exotoxin and has been used to identify many of the biochemical and molecular properties of bacterial exotoxins.

The ability of a bacterial pathogen to cause disease frequently requires the production of exotoxins, but the mere ability to produce a toxin is not sufficient to cause disease. Cholera toxin is the principal virulence factor of *Vibrio cholerae*. Administration of micrograms of purified cholera toxin to human volunteers elicits a diarrheal disease that mimics the magnitude of the natural infection. Nonetheless, nonvirulent toxin-producing strains of *V. cholerae* have been isolated and shown to lack specific biological properties, such as motility or chemotaxis. Similarly, although anthrax toxin is the principal toxic component of *Bacillus anthracis*, nonvirulent toxin-producing strains of *B. anthrasis* have been isolated and shown to lack the ability to produce a polyglutamic acid capsule. An exception to this generalization is the intoxication elicited by the botulinum neurotoxins, in which ingestion of the preformed toxin is responsible for the elicitation of disease; food poisoning by botulinum neurotoxins is an intoxication, rather than an infection by a toxin-producing strain of *Clostridium botulinum*.

Bacterial exotoxins are classified according to their mechanisms of action. The covalent modifications of host-cell components, which are catalyzed by bacterial exotoxins, include ADP-ribosylation, deamidation, depurination, endoproteolysis, and glucosylation (Table 36.1). Most cellular targets of bacterial exotoxins are proteins, although there are exceptions such as shiga toxin, which catalyzes the deadenylation of ribosomal RNA. In addition to exotoxins, there are several other classes of toxins that are produced by bacterial pathogens, including the pore-forming toxins, type III-secreted cytotoxins, heat-stable enterotoxins, and superantigens. Each of these toxins fails to perform one of the properties associated with exotoxins. The pore-forming toxins are not catalytic in their action, but instead disrupt cell physiology through the formation of pores in the host-cell plasma membrane. The type III-secreted cytotoxins can not enter host cells as soluble proteins, but instead are translocated

TABLE 36.1 Properties of bacterial exotoxins

Modification	Exotoxin	Bacterium	AB	Target	Contribution to pathogenesis
ADP-ribosylation	Diphtheria toxin	*C. diphtheriae*	**AB**	Elongation factor-2	Inhibition of protein synthesis
	Exotoxin A	*P. aeruginosa*	**AB**	Elongation factor-2	Inhibition of protein synthesis
	Cholera toxin	*V. cholerae*	**AB5**	Gsα	Inhibition of GTPase activity
	Heat-labile enterotoxin	*E. coli*	**AB5**	Gsα	Inhibition of GTPase activity
	Pertussis toxin	*B. pertussis*	**AB5**	Giα	Uncoupled signal transduction
	C2	*C. botulinum*	**A-B**	actin	Actin depolymerization
Glucosylation	Lethal toxin	*C. sordelli*	**AB**	Ras	Inhibition of effector interactions
	Toxin A and B	*C. difficile*	**AB**	RhoA	Inhibition of Rho signaling
Endoprotease	Anthrax toxin	*B. anthrasis*	**A-B**	Aedema factor A, lethal factor	Adenylate cyclase Endoprotease?
	Botulinum toxin (A–F)	*C. botulinum*	**AB**	Vesicle proteins	Inhibition of vesicle fusion
	Tetanus toxin	*C. tetani*	**AB**	Vesicle protein	Inhibition of vesicle fusion
Deamidation	Cytotoxic necrotizing factor	*E. coli*	**AB**	RhoA	Stimulation of RhoA
Deadenylation	Shiga toxin	*Shigella* spp.	**AB5**	28*S* RNA	Inhibition of protein synthesis
	Verotoxin	*E. coli*	**AB5**	28*S* RNA	Inhibition of protein synthesis

directly into the host cell by the type III secretion apparatus of the cell-bound bacterium. The heat-stable enterotoxin and superantigens do not enter the intracellular compartment of the host cell, and elicit host-cell responses by triggering signal-transduction pathways upon binding to the host-cell membrane. In this article, initial emphasis will be placed on the molecular properties of bacterial exotoxins, with a subsequent description of the general properties of pore-forming toxins, type III-secreted cytotoxins, heat-stable enterotoxins, and superantigens.

The pathology elicited by a specific exotoxin results from the catalytic covalent modification of a specific host-cell component. Although diphtheria toxin and cholera toxin are both bacterial ADP-ribosylating exotoxins, the pathogenesis elicited by each exotoxin is unique. This is due to the fact that diphtheria toxin ADP-ribosylates elongation factor-2, resulting in the inhibition of protein synthesis and subsequent cell death, whereas cholera toxin ADP-ribosylates the Gsα component of the heterotrimeric protein, which stimulates the activity of adenylate cyclase. The stimulation of adenylate cyclase elevates intracellular cAMP and the subsequent secretion of electrolytes and H2O from the cell, resulting in the clinical manifestations of cholera.

II. GENERAL PROPERTIES OF EXOTOXINS

A. Genetic organization of exotoxins

The genes encoding bacterial exotoxins may be located on the chromosome or located on an extrachromosomal element, such as a plasmid or a bacteriophage. Elegant experiments characterizing diphtheria toxin showed that the gene encoding this exotoxin was located within the genome of the lysogenic β-phage. Although both nonlysogenic and lysogenic strains of *C. diphtheriae* could establish local upper-respiratory-tract infection, only strains of *C. diphtheriae* lysogenized with a β-phage that encoded diphtheria toxin were capable of eliciting systemic disease. This established a basic property for the pathology elicited by bacteria that produce exotoxins; bacteria establish a localized infection and subsequently produce an exotoxin, which is responsible for pathology distal to the site of infection.

Most exotoxins are produced only during specific stages of growth with the molecular basis for the regulation of toxin expression varying with each bacterium. This differential expression often reflects a complex regulation of transcription, including responses to environmental conditions, such as iron. Multisubunit toxins are often organized in operons to allow the coordinate expression of their subunit components.

B. Secretion of exotoxins from the bacterium

Most bacteria secrete exotoxins across the cell membrane by the type II secretion pathway. The secretion of exotoxins by the type II secretion pathway was predicted by the determination that the amino terminus of mature exotoxins had undergone proteolysis relative to the predicted amino acid sequence. Type II secretion is also called the general secretion pathway. Type II secretion involves the coordinate translation and secretion of a nascent polypeptide across the cell membrane. During the translation of the mRNA that encodes a type II-secreted protein, the nascent polypeptide contains an amino-terminal leader sequence that is targeted to and secreted across the cell membrane. After secretion across the cell membrane, the nascent polypeptide folds into its native conformation and the leader sequence is cleaved by a periplasmic leader peptidase to yield a mature exotoxin.

Some gram-negative bacteria export the assembled exotoxin from the periplasm into the external environment via a complex export apparatus. While the heat-labile enterotoxin of *Escherichia coli* remains localized within the periplasmic space, *V. cholerae* and *Bordetella pertussis* assemble their respective exotoxins, cholera toxin and pertussis toxin, within the periplasm and then transport the mature exotoxin into the external environment. Although the multiple protein components of the export apparatus have been identified, the mechanism for export across the outer membrane remains to be resolved.

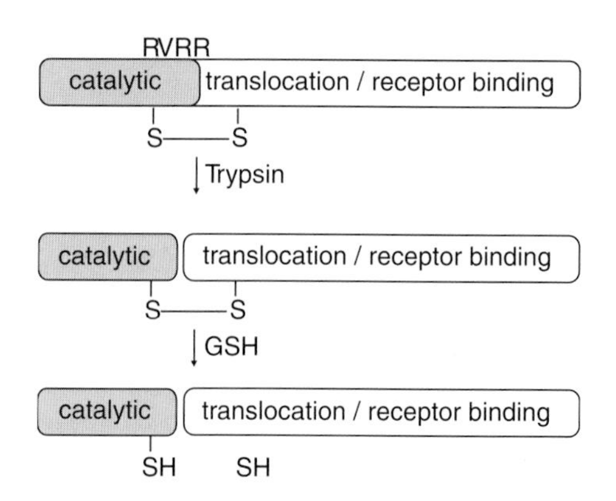

FIGURE 36.1 Bacterial exotoxins are produced as proenzymes. Most bacterial exotoxins are produced as proenzymes that undergo processing to express catalytic activity. The sequential processing of diphtheria toxin involves protein cleavage at a trypsin-sensitive site in the carboxyl terminus of the **A** domain (Arg-Val-Arg-Arg, RVRR). The **A** and **B** domains are connected by a disulfide bond, which is reduced by agents such as reduced glutathione (GSH).

C. Bacteria produce and secrete exotoxins as proenzymes

Although one property of a bacterial exotoxin is the ability to intoxicate sensitive cells, early biochemical studies observed that *in vitro* many bacterial exotoxins possessed little intrinsic catalytic activity. These perplexing observations were resolved with the determination that bacteria produce and secrete exotoxins as proenzymes, which must be activated (processed) to express catalytic activity *in vitro*. Because exotoxins intoxicate sensitive cells, the requirements for *in vitro* activation reflect the activation steps *in vivo*. Each exotoxin requires specific conditions for activation, including proteolysis, disulfide-bond reduction, or association with a nucleotide or a eukaryotic accessory protein. Some activation processes result in the release of the catalytic **A** domain from the **B** domain, whereas other activation processes appear to result in a conformational change in the catalytic **A** domain, rendering it catalytically active. Some exotoxins require sequential activation steps. Diphtheria toxin is activated by limited proteolysis, followed by disulfide-bond reduction (see Fig. 36.1).

The determination of the activation mechanism of exotoxins has also provided insight into several physiological pathways of host cells. The eukaryotic protein, ARF (ADP-ribosylation *f*actor), which activates cholera toxin *in vitro*, was subsequently shown to play a central role in vesicle fusion within the eukaryotic cell. The ability of a host-cell extract to activate cholera toxin is often used as a sign of the presence of ARF. Similarly, the characterization of the mechanisms that pertussis toxin and cholera toxin use to intoxicate eukaryotic cells has provided insight into the pathways for eukaryotic G protein-mediated signal transduction. The ability of pertusis toxin to inhibit the action of a ligand in the stimulation of a signal-transduction pathway is often used to implicate a role for G proteins in that signaling pathway.

D. AB Structure–function properties of exotoxins

Most bacterial exotoxins possess **AB** structure–function properties. (see Fig. 36.2). The **A** domain is the catalytic domain, whereas the **B** domain includes the translocation and binding domains of the exotoxin. Exotoxins are organized into one of several general types of **AB** organization. The simplest **AB** organization is represented by the diphtheria toxin, in which the **A** domain and **B** domain are contained in a single protein. Diphtheria toxin is the prototype for this class of **AB** exotoxin. Diphtheria toxin is a 535-amino-acid protein in which the amino terminus constitutes the ADP-ribosyltransferase domain and the carboxyl terminus comprises the translocation domain and receptor-binding domain. The **AB5** exotoxins are composed of six proteins that are noncovalently associated as an oligomer. Cholera toxin is the prototype for the **AB5** exotoxin. The **A** domain of cholera toxin constitutes the ADP-ribosyltransferase domain, whereas the **B5** domain is composed of five identical proteins, forming a pentamer. This is organized into a ring structure, on which the **A** domain is positioned. The five proteins that make up the **B** domain may be identical, as is the case for cholera toxin and the heat-labile enterotoxin of

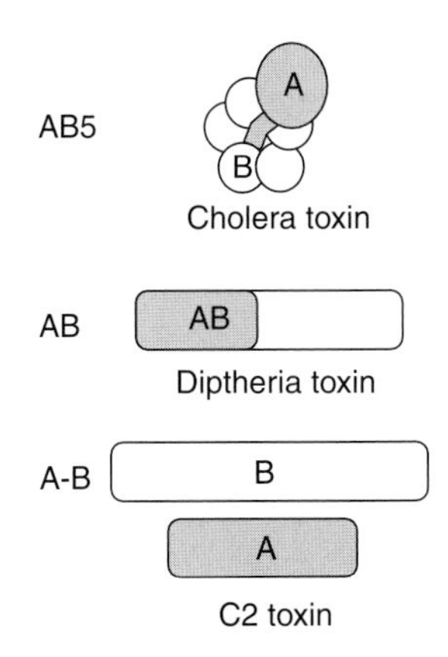

FIGURE 36.2 Bacterial exotoxins possess **AB** structure–function organization. There are three general **AB** organizations of bacterial exotoxins. The **A** domain (shaded) represents the catalytic domain whereas the **B** domain comprises the translocation and receptor-binding domains. **AB5** is represented by cholera toxin of *V. cholerae*, which is composed of six noncovalently associated proteins. **AB** is represented by diphtheria toxin of *C. diphtheriae*, in which the **A** and **B** domain are included in a single protein. **A-B** is represented by C2 toxin of *Clostridium botulinum*, which is composed of two nonassociated proteins; the two proteins associate after the binding and processing of the **B** component on the host-cell membrane.

E. coli, or may be different proteins that form a nonsymmetrical ring structure, as observed with the **B** oligomer of pertussis toxin.

The third class of **AB** exotoxin is composed of proteins that are not associated in solution, but that do associate following the binding and processing of the **B** domain to the host cell. C2 toxin is an example of this class of **A-B** exotoxin. C2 toxin is a bipartile exotoxin composed of a protein that encodes the catalytic **A** domain and a separate protein that encodes the **B** domain. The **A** domain protein of C2 toxin ADP-ribosylates actin. The **B** domain protein of C2 binds to sensitive cells and is nicked by a eukaryotic protease. The processed **B** components oligomerize and are then capable of binding either of the **A** domain proteins.

A new class of toxin organization has recently been recognized in which the **A** domain is a protein and the bacterium is directly responsible for its delivery into the cell. The bacterium binds to the eukaryotic cell and uses a type III secretion apparatus to deliver cytotoxins, also called effector proteins, into the intracellular compartment of the cell. The YopE cytotoxin of *Yersinia* is the prototype of this group of toxins. These cytotoxins do not conform to the strict definition of exotoxins because the purified cytotoxins cannot directly enter to modify host-cell physiology and are not included with the family of bacterial exotoxins. They are termed "type III secreted cytotoxins", and are described later.

Although the **A** domain possesses the catalytic activity of the exotoxin, the **B** domain possesses two specific functions, receptor binding and translocation capacity. Each exotoxin uses a unique host-cell surface component as a receptor. The cell surface receptor for each exotoxin may be specific. The cell surface receptor for cholera toxin is the ganglioside, GM1, whereas diphtheria toxin binds directly to the epidermal growth factor precursor. In contrast, the binding of pertussis toxin appears to be less specific, as pertussis toxin is able to bind numerous cell surface proteins. The ability to bind its cell surface receptor is an absolute requirement for an exotoxin to intoxicate a host cell because the deletion of the receptor-binding domain renders the exotoxin essentially noncytotoxic. After binding to the cell surface, some exotoxins are proteolytically processed or are processed during endocytic vesicle transport.

The second function of the **B** domain includes translocation capacity, which is responsible for the delivery of the **A** domain across the cell membrane. The presence of a translocation domain was predicted from early structure–function studies of diphtheria toxin, which showed that in addition to the catalytic domain and receptor-binding domain, a third function was required for the efficient expression of cytotoxicity. This third function was subsequently shown to correspond to a region of diphtheria toxin that had the propensity to interact with membranes. The crystal structure of diphtheria toxin revealed the presence of three distinct domains, representing the catalytic, translocation, and receptor-binding functions.

E. Exotoxins enter host cells via distinct pathways

Although **A** domain translocation is one of the least understood aspects of the intoxication process of exotoxins, there are several general themes that are involved in translocation of the **A** domain across the cell membrane. One translocation mechanism uses a pH gradient within the endosome to stimulate protein conformational changes in the **B** domain, making it competent to interact with the endocytic vesicle. After insertion into the endocytic membrane, the **B** domain generates a pore that is believed to be involved in the translocation of the **A** domain across the vesicle membrane in an unfolded form. After the translocation across the endocytic membrane, the **A** domain refolds to its native conformation. Subsequent to translocation of the **A** domain across the vesicle membrane, reduced glutathione may reduce the disulfide that connects the **A** domain with the **B** domain, and release the **A** domain into the cytoplasm. The potency and catalytic potential of exotoxins was demonstrated by the observation that the introduction of one molecule of the catalytic domain of diphtheria toxin into the intracellular cytoplasm was sufficient to inhibit host-cell physiology, resulting in cell death.

Other toxins, such as cholera toxin and exotoxin **A** of *Pseudomonas aeruginosa*, appear to use retrograde transport to enter the interior regions of the cell. Movement appears to occur through retrograde transport from the endosome to the Golgi apparatus and ultimately to the endoplasmic reticulum. Many exotoxins that are ultimately delivered to the endoplasmic reticulum possess a KDEL (Lys-Asp-Glu-Leu)-like retention signal sequence on their carboxyl terminus. Although the details for the actual transport pathway remain to be determined, studies with chimeric proteins have shown that the introduction of a KDEL retention sequence is sufficient to retrograde transport a protein, which is normally delivered only to the early endosome, into the endoplasmic reticulum. Thus, there is physiological precedence for the use of the KDEL sequence to retrograde transport exotoxins toward the endoplasmic reticulum. One of the basic questions concerning the intoxication process of these exotoxins is the actual mechanism of translocation and whether or not eukaryotic proteins assist in the translocation process.

F. Covalent modification of host-cell components by exotoxins

Exotoxins use several unique mechanisms to covalently modify host-cell components. The major classes of reactions are the covalent addition of a chemical group to the target protein, the cleavage of a chemical group from a target protein, or the endoproteolytic cleavage of a peptide bond of the target protein.

The ADP-ribosylation of host proteins is the prototype mechanism of action of bacterial exotoxins. Numerous bacterial exotoxins catalyze the ADP-ribosylation of specific host proteins and elicit physiological changes. In the ADP-ribosylation reaction, exotoxins use the oxidized form of nicotinamide adenine dinucleotide (NAD) as the substrate, and transfer the ADP-ribose portion of NAD to a specific amino acid via an *N*-glycosidic linkage of ADP-ribose onto the host target protein. The specific type of amino acid that is ADP-ribosylated within the target protein varies with the specific exotoxin. ADP-ribosylation may either inactivate or stimulate the activity of the target protein. Diphtheria toxin ADP-ribosylates elongation factor-2 on a post-translationally modified histidine residue called diphthamide. ADP-ribosylated elongation factor-2 is unable to perform its translocation of nascent polypeptides in the ribosome, which results in the inhibition of protein synthesis and subsequent cell death. In contrast, cholera toxin ADP-ribosylates the Gsα component of a heterotrimeric G protein. ADP-ribosylated Gsα is locked in an active conformation, which results in the stimulation of adenylate cyclase and the subsequent elevation of intracellular cAMP. Likewise, deamidation of Gln63 of RhoA by *E. coli* cytotoxic necrotizing factor (CNF) results in a constitutively active RhoA protein. Note that although most host targets for exotoxins are proteins, Shiga toxin catalyzes the deadenylation of a specific adenine on 28*S* RNA.

Recall that each exotoxin modifies a specific host-cell component, which is responsible for the specific pathology elicited by that exotoxin. Although there are no absolute rules for the types of proteins targeted for covalent modification, the most frequent targets are nucleotide-binding proteins that are involved in signal-transduction pathways, including both the heterotrimeric G-proteins and the small-molecular-weight GTP-binding proteins of the Ras superfamily. It is not clear whether this class of host protein is targeted for modification due to the presence of a common structural motif, or to its critical role in host-cell metabolism.

G. Molecular and structural properties of bacterial exotoxins

Early biochemical studies provided significant advances in defining the structure–function properties of exotoxins, resolving many of the exotoxin mechanisms of action and developing the concept that exotoxins have **AB** organization. Molecular genetics and structural biology have extended the earlier studies and provided a more detailed understanding of the biochemical and molecular relationships among the exotoxins. The biochemical characterization of diphtheria toxin and exotoxin A (ETA) of *Pseudomonas aeruginosa* showed that these two exotoxins catalyzed kinetically identical reactions during the ADP-ribosylation of elongation factor-2. In addition, both diphtheria toxin and ETA were shown to possess an active site glutamic acid, which was subsequently shown to be a signature property of exotoxins that catalyze the ADP-ribosyltransferase reaction. These observations predicted that ADP-ribosylating exotoxins would possess considerable primary amino acid homology. Thus, the determination that the genes encoding diphtheria toxin and ETA shared little primary amino acid homology was unexpected. This paradox was resolved after the analysis of the three-dimensional structures of ETA and the heat-labile enterotoxin (LT) of *E. coli*, and subsequently confirmed with diphtheria toxin. The three-dimensional structures of ETA and LT showed little similarity in their respective receptor-binding domains and translocation domains; however, the catalytic domains of ETA and LT, which are composed of seven discontinuous regions of each protein, could be superimposed on each other despite possessing homology at only three of the 43 amino acids. One of the homologous amino acids in ETA and LT was the signature active site glutamic acid. This was a remarkable finding because ETA and LT ADP-ribosylate different host target proteins and possess different **AB** organization. A common theme has evolved for describing the structure–function properties of this family of bacterial exotoxins in which the ADP-ribosylating exotoxins possess a conserved three-dimensional structure in their active sites, despite the lack of primary amino acid homology. These findings have provided a framework for the study of other classes of exotoxins produced by divergent groups of bacteria.

III. CONVERSION OF EXOTOXINS INTO TOXOIDS

A. Chemical detoxification of bacterial exotoxins

Shortly after the determination that toxic components were associated with bacterial pathogens, several studies showed that cell extracts or cell cultures of a pathogen could be treated with chemical denaturants, such as formalin, to produce nontoxic immunogenic material that could prevent the disease associated with that pathogen. In the case of diphtheria toxin and tetanus toxin, chemical modification with formalin produced toxoids that were used as acellular vaccines in large-scale immunizations. This resulted in a remarkable decrease in the incidence of both diphtheria and tetanus within the populations that were immunized. In areas where these toxoids are not administered, diphtheria and tetanus remain clinically important diseases. In addition to formalin, other chemicals have been used to detoxify bacterial exotoxins, including glutaraldehyde and hydrogen peroxide. In contrast, the chemical toxoiding of other exotoxins, such as cholera toxin and pertussis toxin, has been more difficult because the treatment of these toxins with denaturants often results in a reduction of immunogenicity. Thus, there is a need to develop alternative strategies for eliminating the cytotoxicity of certain exotoxins without compromising their immunogenicity.

B. Genetic detoxification of bacterial exotoxins

Developments in genetic engineering have provided an opportunity to produce recombinant forms of bacterial exotoxins that possess greatly reduced toxicity, but retain immunogenicity. The use of genetic engineering to develop a toxoid of pertussis toxin has been successful. The whole-cell pertussis vaccine is composed of a chemically treated preparation of *Bordetella pertussis*, which is effective in the elicitation of a protective immune response after mass immunization. However, the whole-cell pertussis vaccine is acutely reactive when administered to children. Pertussis toxin, a primary virulence determinant of *B. pertussis*, is an exotoxin that ADP-ribosylates the Giα component of heterotrimeric G proteins and effectively uncouples signal transduction between the G protein-coupled receptor and the G protein. Genetically engineered forms of pertussis toxin have been produced that possess essentially no catalytic activity or cytotoxicity, but that maintain native conformation and elicit a protective immune response when used as an immunogen. These recombinant noncytotoxic forms of pertussis toxin have been engineered with multiple mutations in their active site, virtually eliminating the risk of reversion to a cytotoxic form. Similar strategies are being applied to other bacterial exotoxins with the goal of engineering acellular vaccine candidates.

IV. THERAPEUTIC APPLICATIONS OF EXOTOXINS

One of the most exciting areas of bacterial exotoxin research has been the development of strategies to use exotoxins in therapeutic disciplines. Some therapies use the native cytotoxic form of the exotoxin. Other therapies use either the **A** or **B** domain, which is conjugated to a heterologous binding component or to effector elements, respectively, to produce a chimeric molecule with directed properties.

Botulinum toxin and tetanus toxin (BT/TT) are each a single protein that is organized as an **AB** exotoxin. The amino terminus of BT/TT expresses endopeptidase activity and constitutes the **A** domain, whereas the **B** domain possesses neuronal cell-specific receptor-binding activity. The specific association of the **B** domain with neuronal cells is responsible for the clinical manifestation of these neurotoxins. BT/TT appear to enter neuronal cells by receptor-mediated endocytosis and to deliver the **A** domain to the cytosol, where it catalyzes the endoproteolytic cleavage of host proteins that are involved in vesicle fusion. Studies have shown that botulinum toxin can be introduced into the muscles surrounding the eye to temporarily reduce muscle spasms associated with several clinical disorders.

Diphtheria toxin has been used as a carrier to stimulate an immune response against several epitopes. One epitope is polyribitolphosphate, a component of the polysaccharide capsule of *Haemophilus influenzae* type b

(Hib). Early attempts to elicit an effective immune response to purified Hib antigen resulted in the production of a T-cell-independent immune response that did not yield an effective memory. A noncatalytic mutant of diphtheria toxin, CRM197, has been used as a carrier for the Hib epitope. Immunization with the CRM197-Hib conjugate yielded a strong T-dependent immune response. Mass immunization with Hib conjugates has resulted in a dramatic reduction in the number of cases of Hib in the immunized population.

Due to their potency, the catalytic **A** domain of exotoxins have been used in the construction of chimeric immunotoxins that are designed to target cancer cells. Early studies used conjugates that were composed of the **A** domain of the diphtheria toxin coupled to an antibody that recognized a cell surface-specific antigen. The **A** chain of the diphtheria toxin was used in the first generation of immunotoxins because it had been shown to possess impressive cytotoxic potential when introduced into the cytosol of eukaryotic cells. It was estimated that the introduction of a single molecule of the **A** chain of diphtheria toxin into the cytosol was sufficient to kill that cell. In cell culture, these chimera have proven to be both potent and antigen specific. Ongoing research involves the determination of clinical situations for the use of these chimeras in a therapeutic arena.

The **B** component of anthrax toxin, protective antigen (PA) and a truncated, non-cytotoxic form of one of its **A** components (LF) has recently been used to deliver epitopes into antigen presenting cells to elicit a cytotoxic lymphocyte (CTL) response. Anthrax toxin is a tripartite toxin composed of three non-associated proteins. After binding to cells, PA is proteolytically processed and undergoes oligomerization to form a heptameric structure on the cell surface. Processed PA is able to bind either LF or edema factor (EF) and the **AB** complex undergoes receptor-mediated endocytosis in which acidification in the early endosome stimulates the translocation of the **A** domain into the cell cytosol. In the nontoxic anthrax delivery system, PA is added to antigen-presenting cells with a nontoxic LF–CTL epitope chimera used to deliver the epitope into the host cell for antigen presentation. One of the more attractive aspects of this CTL-epitope delivery system is that small amounts of PA are required to present antigen.

V. OTHER CLASSES OF BACTERIAL TOXINS

A. Pore-forming toxins

The lack of a catalytic **A** domain differentiates the pore-forming toxins from exotoxins. Thus, the pathology associated with pore-forming toxins is due solely to the generation of a pore within the membrane of the host cell. Several bacterial pathogens produce pore-forming toxins, some of which are secreted by a type I secretion pathway. Unlike type II-secreted proteins, the amino terminus of type I-secreted proteins is not processed. Type I-secreted proteins possess a polyglycine signal sequence in the carboxyl terminus of the mature toxin. There are several classes of pore-forming toxins, including members of the hemolysin family of pore-forming toxins, the aerolysin family of pore-forming toxins, and the α-toxin of *Staphylococcus aureus*. Host-cell specificity differs among pore-forming toxins. The crystal structures of several of the pore-forming toxins have been determined. The molecular events generating a pore in the membrane of a host cell have been proposed for the aerolysin family of pore-forming toxins. Aerolysin is exported by *Aeromonas hydrophilia* as a monomeric molecule, which binds to the host cell. The monomer is pro-teolytically processed and subsequently undergoes oligomerization. The oligomerized complex is inserted into the membrane and generates a pore in the center of the complex, causing the release of the cytoplasmic components of the host cell.

B. Type III-secreted cytotoxins

The lack of a **B** domain differentiates the type III-secreted cytotoxins from exotoxins. Thus, the organization of the type III-secreted cytotoxins may be represented as **A** domains that are specific effector proteins. Type III-secreted cytotoxins are transported directly into the host cells by cell surface-bound bacteria. Type III secretion of bacterial proteins is a recently defined pathway for the delivery of proteins into the cytoplasm of host cells. Type III-secreted proteins were initially recognized by the fact that the secreted mature cytotoxins were unique to proteins secreted by either the type I or II secretion pathways, whereas the amino terminus of type III-secreted proteins is not processed nor is there a polyglycine motif in their carboxyl terminus. Although it is clear that a complete type III secretion apparatus is required for the transport of type III-secreted proteins into the host cytoplasm, the mechanism for the delivery of type III-secreted proteins across the host-cell membrane remains to be resolved. Numerous bacteria have been shown to possess type III secretion pathways, including members of the genera *Escherichia, Pseudomonas, Shigella, Salmonella*, and *Yersinia*. Cytotoxicity elicited by type III-secreted cytotoxins has an absolute requirement for the type III secretion apparatus of the bacterium, as purified forms of the cytotoxins are not toxic to host cells.

The **A** domains of type III-secreted cytotoxins catalyze several unique mechanisms of action, including the depolymerization of the actin cytoskeleton, phosphatase activity, ADP-ribosyltransferase activity, and the stimulation of apoptosis. Each type III secretion apparatus appears capable of delivering numerous type III-secreted proteins into the host cell.

C. Heat-stable enterotoxins

The inability of the heat-stable enterotoxins to enter the host cell or possess catalytic activity differentiates the heat-stable enterotoxins from exotoxins. Several genera of bacteria produce heat-stable enterotoxins, including *Escherichia* and *Yersinia*. The heat-stable enterotoxin a (STa) of *E. coli* is the prototype toxin of this group. *E. coli* secretes STa into the periplasm as a 72-amino-acid precursor in which three intramolecular disulfide bonds are formed and processed into a 53-amino-acid form. The 53-amino-acid form of STa is exported into the environment, where a second proteolytic cleavage results in the production of an 18- or 19-amino-acid mature STa molecule. The mature STa binds to a protein receptor on the surface of epithelial cells, which results in an increase in the intracellular concentrations of cGMP. The intracellular increase in cGMP results in a stimulation of chloride secretion and net fluid secretion, resulting in diarrhea.

D. Superantigens

The inability of superantigens to enter the host cell or possess catalytic activity differentiates the superantigens from exotoxins. Superantigens are soluble proteins of approximately 30 kDa that are secreted by bacteria that possess mitogenic properties. Superantigens are produced by both *Streptococcus* and *Staphylococcus*. The superantigens bind to a component of the major histocompatibility complex of T lymphocytes through an antigen-independent mechanism, which stimulates proliferation of a large subset of T lymphocytes.

BIBLIOGRAPHY

Alouf J. E. (2000). Bacterial protein toxins. An overview. *Methods Mol Biol*. **145**: 1–26.

Barbieri, J. T., and Burns, D. (2003). Bacterial ADP-ribosylating exotoxins. *In*. "Bacterial Protein Toxins" (D. Burns, J. T. Barbieri, B. Iglewski, and R. Rappuoli, eds.), ASM, Washington, DC.

Collier, R. J. (1975). Diphtheria toxin: mode of action and structure. *Bacteriol. Rev*. **39(1)**, 54–85.

Dobrindt, U., Janke, B., Piechaczek, K., Nagy, G., Ziebuhr, W.. Fischer, G., Schierhorn, A., Hecker, M., Blum-Oehler, G., and Hacker, J., 2000 Oct. Toxin genes on pathogenicity islands: impact for microbial evolution. *Int J Med Microbiol*. **290**(4–5): 307–11.

Ernst, J. D. Bacterial inhibition of phagocytosis. *Cell Microbiol*. **2(5)**: 379–86.

Fivaz, M. Abrami, L. Tsitrin, Y. van der Goot FG. (2001). Aerolysin from Aeromonas hydrophila and related toxins. *Curr Top Microbiol Immunol*. **257**: 35–52.

Lesnick, M. L. Guiney, D. G. (2001). The best defense is a good offense–Salmonella deploys an ADP-ribosylating toxin. *Trends Microbiol*. Jan; **9(1)**: 2–4;

Michie. C. A., and Cohen. J. (1998). The clinical significance of T-cell superantigens. *Trends Microbiol*. **6(2)**, 61–65.

Moss, J., and Vaughan, M. (1990). "ADP-Ribosylating Toxins and G proteins: Insights into Signal Transduction." *Am. Soc. Microbiol*, Washington, D.C.

Pastan, I., Chaudhary, V., and FitzGerald, D. J. (1992). Recombinant toxins as novel therapeutic agents. *Annu. Rev. Biochem*. **61**, 331–354.

Prevost G, Mourey L, Colin DA, and Menestrina G. (2001). Staphylococcal pore-forming toxins. *Curr Top Microbiol Immunol*. **257**: 53–83

Saelinger, C. B., and Morris, R. E. (1994). Uptake and processing of toxins by mammalian cells. *Meth. Enzymol*. **235**, 705–717.

Sandvig, K., Garred, O., Holm, P. K., and van Deurs, B. (1993). Endocytosis and intracellular transport of protein toxins. *Biochem. Soc. Trans*. **21**(Pt 3), 707–711.

Sixma, T. K., Pronk, S. E., Kalk, K. H., Wartna, E. S., van Zanten, B. A., Witholt, B., and Hol, W. G. (1991). Crystal structure of a cholera toxin-related heat-labile enterotoxin from *E. coli*. *Nature* **35(6326)**, 371–377.

Stathopoulos, C., Hendrixson. D. R., Thanassi, D. G., Hultgren, S. J., St Geme, J. W., 3rd, and Curtiss, R., 3rd (2000) Secretion of virulence determinants by the general secretory pathway in gram-negative pathogens: an evolving story. *Microbes Infect. Jul*. **2(9)**: 1061–72.

Tweten, R. K., Parker, M. W., Johnson, A. E. (2001). The cholesterol-dependent cytolysins. *Curr Top Microbiol Immunol*. **257**: 15–33.

WEBSITES

"Bad Bug Book" of the U.S. Food and Drug Administration *http://vm.cfsan.fda.gov/~mow/intro.html*

Access to online resources on Bacterial Infections and Mycoses. Karolinska Institutet *http://www.mic.ki.se/Diseases/c1.html*

37

Extremophiles

Ricardo Cavicchioli and Torsten Thomas
University of New South Wales

GLOSSARY

alkaliphile An organism with optimal growth at pH values above 10.

Archaea One of the three domains (highest taxon) of life, evolutionarily distinct from the Bacteria and Eucarya (eukaryotes).

autotroph An organism able to use CO_2 as a sole source of carbon.

barophile An organism that lives optimally at high hydrostatic pressure.

chemolithotroph An organism obtaining its energy from the oxidation of inorganic compounds.

chemoorganotroph An organism obtaining its energy from the oxidation of organic compounds. Also referred to as a heterotroph.

extreme acidophile An organism with a pH optimum for growth at or below pH 3.

extremophile An organism that is isolated from an extreme environment and often requires the extreme condition for growth ("extreme" is anthropocentrically derived).

growth optimum The conditions (e.g., temperature and salinity) at which an organism grows fastest.

halophile An organism requiring at least 0.2 M salt for growth.

hyperthermophile An organism having a growth temperature optimum of 80°C or higher.

methanogen An archaeal microorganism that produces methane.

mixotroph An organism able to assimilate organic compounds as carbon sources and use inorganic compounds as electron donors.

oligotroph An organism with optimal growth in nutrient-limited conditions.

phototroph An organism that obtains energy from light.

phylogeny The ordering of species into higher taxa and the construction of evolutionary trees based on evolutionary relationships rather than on general resemblances.

prokaryote A cell or organism lacking a nucleus and other membrane-enclosed organelles, usually having its DNA in a single circular molecule. Loosely used to describe archaeal and bacterial microorganisms in contrast to eukaryotic microorganisms.

psychrophile An organism having a growth temperature optimum of 15 °C or lower, and a maximum temperature of 20 °C or lower.

Extremophiles are organisms that require extreme environments for growth. Although this is perhaps self-evident, what constitutes extreme? Extreme is a relative term, with the point of comparison being what is normal for humans. Extremophiles are therefore organisms that are "fond of" or "love" (-phile) environments such as high temperature, pH, pressure, or salt concentration; or low temperature, pH, nutrient concentration, or water availability. Extremophiles are also organisms that can tolerate other extreme conditions, including high levels of radiation or toxic compounds, or living conditions that humans consider unusual, such as living in rocks 1.5 km below the surface of Earth. In addition, extremophiles may be found in environments with a combination of extreme conditions, such as high temperature and high acidity or high pressure and low temperature.

The Desk Encyclopedia of Microbiology
ISBN: 0-12-621361-5

I. INTRODUCTION

Most extremophiles are microorganisms. For example, the upper temperature limits for archaeal, bacterial, and eukaryotic microorganisms are 113, 95, and 62 °C, respectively, in contrast to most metazoans (multicellular eukaryotes, e.g., animals and plants), which are unable to grow above temperatures of 50 °C. This example of thermal adaptation highlights an important distinction among the different classes of microorganisms (i.e., Archaea can grow at extremely high temperatures in comparison to their eukaryotic counterparts), and underscores the fundamentally different evolutionary origins of the members of the three domains of life, Archaea, Bacteria, and Eucarya. Although Archaea (formerly Archaebacteria) and Bacteria are both loosely defined as prokaryotes, they are by no means more similar to each other than Archaea are to eukaryotes. For example, archaea have a number of archaeal-specific traits (glycerol-1-phosphate–lipid backbones and methanogenesis), in addition to sharing many bacterial (metabolism, biosynthesis, energy generation, transport, and nitrogen fixation) and eukaryotic (transcription, translation, and replication) features. Due to the fact that Archaea are often found in extreme environments, the term "extremophile" is often used synonymously with "archaea", and many of the extremophiles described here are members of the Archaea. It should be noted, however, that Archaea are also found in a broad range of nonextreme marine and soil environments.

The first use of the term "extremophile" appeared in 1974 in a paper by MacElroy (*Biosystems* 6: 74–75) entitled, "Some Comments on the Evolution of Extremophiles." In the 1990s, studies on extremophiles have progressed to the extent that the First International Congress on Extremophiles was convened in Portugal, June 2–6, 1996, and the scientific journal *Extremophiles* was established in February 1997. These developments in the field have arisen due to the isolation of extremophiles from environments previously considered impossible for sustaining biological life. As a result, the appreciation of microbial biodiversity has been reinvigorated and challenging new ideas about the origin and evolution of life on Earth have been generated. In addition, the novel cellular components and pathways identified in extremophiles have provided a burgeoning new biotechnology industry.

In recognition of the importance of extremophiles to microbiology, the following sections describe the habitats, biochemistry, and physiology of a diverse range of extremophiles, followed by a concluding section on the biotechnological applications of extremophiles and their products.

II. HYPERTHERMOPHILES

A. Defining temperature classes of microorganisms

Microorganisms can generally be separated into four groups with regard to their temperature optima for growth. In order of increasing temperature, they are psychrophiles (optimum of 15 °C, maximum <20 °C), mesophiles (optimum between 20 and 45 °C), thermophiles (optimum between 45 and 80 °C), and hyperthermophiles (optimum of 80 °C or higher). On the whole, these definitions are generally accepted terms; the reader should be aware, however, that some terms may have alternative meanings in specific fields. For example, the upper temperature for the growth of yeast is about 48 °C, and the majority of yeasts can grow below 20 °C. As a result, a thermophilic yeast is defined by its inability to grow below 20 °C, but with no restrictions placed on the maximum temperature for growth.

B. Habitats and microorganisms

Hyperthermophiles have been isolated from a range of natural geothermally heated environments (Table 37.1). The terrestrial environments tend to be acidic (as low

TABLE 37.1 Examples of locations where hyperthermophiles have been isolated

Geography	Type of habitat	Location
Terrestrial	Hot springs	US, New Zealand, Iceland, Japan, Mediterranean, Indonesia, Central America, Central Africa
Subterranean	Oil reservoir (3500 m)	North Sea, North Alaska
Submarine	Shallow depth (beaches)	Italy, Sao Michel (the Azores), Djibouti (Africa)
	Sea mounts	Tahiti
	Moderate depth (120 m)	mid-Atlantic Ridge (north of Iceland)
	Deep sea (>1000 m)	Guaymas Basin (2000 m) and East Pacific Rise (2500–2700 m) (both near Mexico), mid-Atlantic Ridge (3600–3700 m), mid-Okinawa Trough (1400 m), Galapagos Rift

as pH 0.5) with low salinity (0.1–0.5% salt), whereas the marine systems are saline (3% salt) and generally less acidic (pH 5–8.5). Artificial sources of isolates include coal refuse piles and geothermal power plants.

The terrestrially based, volcanic systems include hot springs such as solfataras, which are sulfur-rich and generally acidic, or others that are boiling at a neutral pH, or that are iron-rich. Most of the heated soils and water contain elemental sulfur and sulfides, and most isolates metabolize sulfur. A by-product of oxidation tends to be sulfuric acid and as a result many of the hot springs are extremely acidic. Hyperthermophilic archaea have also been isolated from oil from geothermally heated oil reserves present in Jurassic sandstone and limestone 3500 m below the bed of the North Sea and below the Alaskan North Slope permafrost soil. As a by-product of their metabolism during oil extraction, they produce hydrogen sulfide; this condition is referred to as reservoir souring.

Numerous hyperthermophiles have been isolated from submarine hydrothermal systems, including hot springs, sediments, sea mounts (submarine volcanoes), fumaroles (steam vents with temperatures up to 150–500 °C), and deep-sea vents. The hot deep-sea vents are often referred to as "black smokers" due to the thick plume of black material that forms from mineral-rich fluids at temperatures up to 400 °C, precipitating on mixing with the cold (1–5 °C) seawater. The submarine systems are a rich source of microbial and higher eukaryotic life that spans the temperature ranges suitable for hyperthermophiles through psychrophiles. Due to the extreme depth of some of the vents (3500 m), organisms are also exposed to extreme pressure, and as a result they tend to be barophilic or barotolerant (see Section V).

The tube worms, giant clams, and mussels that inhabit the vents are dependent on the activity of the chemolithotrophic microorganisms to fix CO_2 and use a broad range of inorganic energy sources emitted from the vents. The first eukaryotic organism living at very high temperatures was identified at the Axial Summit Caldera, west of Mexico. The Pompeii worm (*Alvinella pompejana*) was observed living in a hydrothermal system in which the posterior of the worm was in 80 °C water. Interestingly, it is speculated that the ability of the worm to survive the heat may be due to the presence of microorganisms that cover its exterior and secrete heat-stable enzymes.

Most isolated hyperthermophiles are members of Archaea, and it is this characteristic that is often associated with this class of microorganism. However, it was the isolation of *Thermus aquaticus*, a bacterium, from a hot spring in Yellowstone National Park by Thomas Brock in 1969 that led to the discoveries of hyperthermophilic archaea. In 1972, Brock described *Sulfolobus*, an obligately aerobic archaeon that is able to oxidize H_2S or S^0 to H_2SO_4 and fix CO_2 as a carbon source. The hot springs that *Sulfolobus* spp. thrive in are typically as hot as 90 °C, with an acidity of pH 1–5. Some hyperthermophiles have been found in unique locations, for example, *Methanothermus* spp., which have only been isolated from one solfataric area of Iceland. In contrast, *Pyrococcus* and *Thermococcus* spp. have been found in both submarine systems and subterranean oil reserves. Of all the isolates to date, *Pyrolobus fumarii* has the highest maximum temperature (113 °C). Furthermore, it is restricted to temperatures above 90 °C. Using a microscope slide suspended in the outflow from a black smoker at 125–140 °C, microbial films have been detected on the glass surfaces. Although organisms capable of growth at such high temperatures have not been isolated, this is good evidence that microorganisms are capable of growth in this extreme temperature range.

C. Biochemistry and physiology of adaptation

A broad diversity of hyperthermophiles with varied morphologies, metabolisms, and pH requirements have been isolated, and the characteristics of a number of these are shown in Table 37.2. Although organisms such as *Sulfolobus* spp. are aerobic chemoorganotrophs, most hyperthermophiles are anaerobic and many are chemolithotrophs. The extent of metabolic diversity in the hyperthermophiles is exemplified by the strict requirement of *Methanococcus jannaschii* for CO_2 and H_2 only, in comparison to the broad spectrum of substrates used as electron donors by the sulfate-reducing archaeon, *Archaeoglobus fulgidus*, including H_2, formate, lactate, carbohydrates, starch, proteins, cell homogenates, and components of crude oil. This indicates that there are no particular carbon-use or energy-generation pathways that are exclusively linked to growth at high temperatures.

For any microorganism, lipids, nucleic acids, and proteins are all generally susceptible to heat, and there is, in fact, no single factor that enables all hyperthermophiles to grow at high temperatures.

1. *Membrane lipids and cell walls*

Archaeal lipids all contain ether-linkages (as opposed to the ester-linkages in most bacteria), which provide resistance to hydrolysis at high temperature. Some hyperthermophilic archaea contain membrane-spanning, tetra-ether lipids that provide a monolayer type of organization that gives the membranes a high

TABLE 37.2 Characteristics of selected hyperthermophiles[a]

Genus	Minimum temperature	Optimum temperature	Maximum temperature	DNA mol % G + C	pH Optimum	Aerobe (+) Anaerobe (−)	Energy yielding reactions	Morphology
Submarine isolates								
Aquifex (bacteria)	67	85	95	40	6	+	H_2 reduction	Rods
Archaeoglobus	64	83	95	46	7	−	SO_4 reduction to H_2S	Cocci
Methanococcus	45	88	91	31	6	−	$H_2 + CO_2$ reduction to CH_4	Irregular cocci
Methanopyrus	85	100	110	60	6.5	−	$H_2 + CO_2$ reduction to CH_4	Rods
Pyrobaculum	60	88	96	56	6	+/−	Organic/inorganic compounds+NO_3^- reduction	Rod
Pyrolobus	90	105	113	53	5.5	+/−	NO_3^- reduction	Lobed cocci
Thermodiscus	75	90	98	49	5.5	−	Organic compound + S^0 reduction	Discs
Thermotoga (bacteria)	55	80	90	46	7	−	Organic compounds	Sheathed rods
Terrestrial isolates								
Acidianus	65	85–90	95	31	2	+/−	Aerobic oxidation & anaerobic reduction of S^0	Sphere
Desulfurococcus	70	85	95	51	6	−	Organic compound + S^0 reduction	Sphere
Sulfolobus	55	75–85	87	37	2–3	+	Organic compound & H_2S or S^0 oxidation	Lobed sphere
Thermoproteus	60	88	96	56	6	−	Reduction of S^0	Rod

[a]All are archaeal species unless otherwise indicated. Data from Madigan, M. T., Martinko, J. M., and Parker, J. (1997). *Brock: Biology of Microorganisms*, 8th Ed. Prentice Hall; Stetter, K. O. (1998). Hyperthermophiles: Isolation, classification and properties. *In* "Extremophiles: Microbial Life in Extreme Environments" (K. Horikoshi, and W. D. Grant, Eds.), Wiley series in Ecological and Applied Microbiology, Wiley-Liss.

degree of rigidity and may confer thermal stability. The bacterium *Thermotoga maritima* also contains a novel ester lipid, which may increase stability at high growth temperatures. However, the archaeon *Methanopyrus kandleri*, which can grow at up to 110 °C, contains unsaturated diether lipids that resemble terpenoids, and it is unclear how this may affect the thermal resistance of the cell. Lipid composition also varies with the growth temperature of individual organisms. For example, in *Methanococcus jannaschii* at 45 °C, 80% of the lipid content is one lipid (archaeol), whereas at 75 °C, two lipids (caldarchaeol and macrocyclic archaeol) account for 80% of the total core.

Most archaeal species possess a paracrystalline surface layer (S-layer) consisting of protein or glycoprotein. It is likely that the S-layer functions as an external protective barrier. In *Pyrodictium* spp., the highly irregularly shaped, flagellated cells are interconnected by extracellular glycoprotein tubules that remain stable at 140 °C.

2. *Nucleic acids*

The thermal resistance of DNA could conceivably be improved by maintaining a high mol% G + C content; however, many hyperthermophiles have between 30 and 40% (Table 37.2), in comparison to, for example, the mesophile *Escherichia coli*, which has 50%. *Acidianus infernus* has a G + C content of 31%, which would rapidly lead to the melting of double-stranded DNA at its optimum growth temperature of 90 °C. Histone-like proteins that bind DNA have been identified in archaeal hyperthermophiles. It is likely that the DNA is protected by the histones and that this enables processes such as open-complex formation during transcription to occur without subsequent DNA melting. In addition, hyperthermophiles contain reverse gyrase, a type 1 DNA topoisomerase that causes positive supercoiling and therefore may stabilize the DNA.

3. *Proteins and solutes*

Heat-shock proteins, including chaperones, are likely to be important for stabilizing and refolding proteins as they begin to denature. When *Pyrodictium occultum* is heat-stressed at 108 °C, 80% of total protein accumulated is a single chaperonin, termed the "thermosome," the cells' protein-folding machine. In addition to the differential expression of certain genes throughout the growth temperature range of the organism, proteins from hyperthermophiles are inherently more stable than those from thermophiles, mesophiles, or psychrophiles. Higher stability is a result of increased rigidity (decreased flexibility) of the protein. Certain structural properties favor a more rigid protein (see psychrophilic proteins in Section IV.B.2), including a higher degree of structure in hydrophobic cores, an increased number of hydrogen bonds and salt bridges, and a higher proportion of thermophilic amino acids (e.g., proline residues that have fewer degrees of freedom). As a result, the proteins from hyperthermophiles are extremely heat stable. For example, proteases from *Pyrococcus furiosus* have half-lives of >60 h at 95 °C and an amylase from *Pyrococcus woesii* is active at 130 °C.

Protein stability may also be assisted by the accumulation of intracellular potassium and solutes, such as 2,3-diphosophoglycerate (cDPG). In *Methanopyrus kandleri* and *Methanothermus fervidus*, there is evidence that potassium is required for enzyme activity at high temperature and the potassium salt of cDPG acts as a thermal stabilizer.

D. Evolution

Earth is about 4.6 billion years old and life is believed to have evolved on Earth around 3.6–4 billion years ago. The atmosphere of early Earth was devoid of oxygen and contained gaseous H_2O, CH_4, CO_2, N_2, NH_3, HCN, trace amounts of CO and H_2, and large quantities of H_2S and FeS. For the first 0.5 billion years, the surface of the planet was probably devoid of water because the temperature was higher than 100 °C. Subsequently, the planet cooled and water liquefied. The high temperatures imply that life evolving in these conditions must have possessed thermophilic properties. Due to the high temperature and the available carbon and energy substrates, a microorganism such as H_2-oxidizing sulfur-reducing *Methanopyrus kandleri* would conceivably thrive.

Studying the phylogenetic relationship of extant (living) microorganisms by the analysis of 16*S*-ribosomal ribonucleic acid (16*S*-rRNA) sequences, reveals that the hyperthermophilic Archaea and Bacteria have short evolutionary branches that occur near the base of the tree of life. Short branches indicate a low rate of evolution and deep branches reflect a close relationship with primordial life forms. This suggests that hyperthermophiles living today may resemble some of the earliest forms of life on Earth.

There is some evidence that microbial life may have existed (or still exists) on Mars and other terrestrial bodies (e.g., on Europa, one of Jupiter's moons). In association with the characteristics of hyperthermophiles mentioned, this has led to the consideration that life on Earth may have originated from the introduction of extraterrestrial life. A possible scenario involves a meteor carrying microbial life similar to hyperthermophilic methanogens plunging into the ocean billions of years ago and initiating the use of inorganic matter to generate biological matter and the subsequent evolution of extant species. These possibilities are driving new research endeavors to discover extraterrestrial life (see Section IV.A on psychrophiles and lakes on Europa).

III. EXTREME ACIDOPHILES

A. Habitats and microorganisms

An extreme acidophile has a pH optimum for growth at or below pH 3.0. This definition excludes microorganisms that are tolerant to pH below 3, but that have pH optima closer to neutrality, including many fungi, yeast, and bacteria (e.g., the ulcer- and gastric-cancer causing gut bacterium *Helicobacter pylori*).

1. Natural environments

Extremely acidic environments occur naturally and artificially. The hyperthermophilic extreme acidophiles *Sulfolobus, Sulfurococcus, Desulfurolobus*, and *Acidianus* produce sulfuric acid in solfataras in Yellowstone National Park from the oxidation of elemental sulfur or sulfidic ores. Other members of the Archaea found in these hot environments include species of *Metallosphaera*, which oxidize sulfidic ores, and *Stygiolobus* spp., which reduce elemental sulfur. The novel, cell-wall-less archaeon, *Thermoplasma volcanium*, which grows optimally at pH 2 and 55 °C, has also been isolated from sulfotaric fields around the world. The bacterium *Thiobacillus caldus* has been isolated in hot acidic soils. *Bacillus acidocaldarius, Acidimicrobium ferrooxidans*, and *Sulfobacillus* spp. have also been isolated from warm springs and hot spring runoff.

The most extreme acidophiles known are species of Archaea, *Picrophilus oshimae* and *Picrophilus torridus*, which were isolated from two solfataric locations in northern Japan. One of the locations, which contained both organisms, is a dry soil, heated by solfataric gases to 55 °C and with a pH of less than 0.5. These remarkable species have aerobic heterotrophic growth with a temperature optimum of 60 °C and a pH optimum of 0.7 (i.e., growth in 1.2 M sulfuric acid).

In addition to Archaea, the phototrophic red alga *Galdieria sulphuraria* (*Cynadium caldarium*), isolated in cooler streams and springs in Yellowstone National

Park, has optimum growth at pH of 2–3 and 45 °C, and is able to grow at pH values around 0. The green algae *Dunaliella acidophila* is also adapted to a narrow pH range from 0 to 3.

2. *Artificial environments*

The majority of extremely acid environments are associated with the mining of metals and coal. The microbial processes that produce the environments are a result of dissimilatory oxidation of sulfide minerals, including iron, copper, lead, and zinc sulfides. This process can be written as $Me^{2+}S^{2-}$ (insoluble metal complex) $\rightarrow$ $Me^{2+} + SO_4^{2-}$; where Me represents a cationic metal. As a result of the extremely low pH in these environments, and due to the geochemistry of the mining sites, cationic metals (e.g., Fe^{2+}, Zn^{2+}, Cu^{2+}, and Al^{2+}) and metaloid elements (e.g., arsenic) are solubilized; this process is referred to as microbial ore leaching.

Most mining sites tend to have low levels of organic compounds, and as a result chemolithoautotrophs, such as the bacteria *Thiobacillus ferrooxidans, Thiobacillus thiooxidans, Leptospirillum ferrooxidans*, and *Leptospirillum thermoferrooxidans*, are prolific. In addition, mixotrophic *Thiobacillus cuprinus* and heterotrophic *Acidiphilium* spp. have been isolated from acidic coal refuse and mine drainage. Thermophilic acidophilic bacterial species include *Thiobacillus caldus* from coal refuse, *Acidimicrobium ferrooxidans* from copper-leaching dumps, and *Sulfobacillus* spp. from coal refuse and mine water. The archaeal microorganism *Thermoplasma acidophilum* is frequently isolated from coal refuse piles. Coal refuse contains coal, pyrite (an iron sulfide), and organic material extracted from coal. As a result of spontaneous combustion, the refuse piles are self-heating and provide the thermophilic environment necessary for sustaining *Thermoplasma* and other thermophilic microorganisms.

In illuminated regions (e.g., mining outflows and tailings dams), phototrophic algae, including *Euglena, Chlorella, Chlamydomonas, Ulothrix*, and *Klebsormidium* species, have been isolated. Other eukaryotes include species of yeast (*Rhodotorula, Candida*, and *Cryptococcus*), filamentous fungi (*Acontium, Trichosporon*, and *Caphalosporium*) and protozoa (*Eutreptial, Bodo, Cinetochilium*, and *Vahlkampfia*).

B. Biochemistry and physiology of adaptation

Acidophiles (and alkaliphiles; see Section VII) keep their internal pH close to neutral. Most extreme acidophiles maintain an intracellular pH above 6, and even *Picrophilus* maintains an internal pH of 4.6 when the outside pH is 0.5–4. As a result, extreme acidophiles have a large chemical proton gradient across the membrane. Proton movement into the cell is minimized by an intracellular net positive charge and as a result cells have a positive inside-membrane potential. This is caused by amino acid side chains of proteins and phosphorylated groups of nucleic acids and metabolic intermediates, acting as titratable groups. In effect, the low intracellular pH leads to protonation of titratable groups and produces a net intracellular positive charge. In addition to this passive effect, some acidophiles (e.g., *Bacillus coagulans*) produce an active proton-diffusion potential that is sensitive to agents that disrupt the membrane potential, such as ionophores.

The ability of lipids from the archaeon *Picrophilus oshimae* to form vesicles is lost when the pH is neutral, thus indicating that the membrane lipids are adapted for activity at low pH to minimize proton permeability. In *Dunaliella acidophila*, the surface charge and inside membrane potential are both positive, which is expected to reduce influx of protons into the cell. In addition, it overexpresses a potent cytoplasmic membrane H^+-ATPase to facilitate proton efflux from the cell.

The pH to which a protein is exposed affects the dissociation of functional groups in the protein and may be affected by salt and solute concentrations. Few periplasmic surface-exposed or -secreted proteins from extreme acidophiles have been studied to identify the structural features important for activity and stability. In *Thiobacillus ferrooxidans*, the acid stability of rusticyanin (acid-stable electron carrier) has been attributed to a high degree of inherent secondary structure and the hydrophobic environment in which it is located in the cell. A relatively low number of positive charges have been linked to the acid stability of secreted proteins (thermopsin, a protease from *Sulfolobus acidocaldarius*, and an α-amylase from *Alicyclobacillus acidocaldarius*), by minimizing electrostatic repulsion and protein unfolding.

IV. PSYCHROPHILES

A. Habitats and microorganisms

Over 80% of the total biosphere of Earth is at a temperature permanently below 5 °C, and it is therefore not surprising that a large number and variety of organisms have adapted to cold environments. These natural environments include cold soils; water (fresh and saline, still and flowing); in and on ice in polar or alpine regions; polar and alpine lakes, and sediments;

caves; plants, and cold-blooded animals (e.g. Antarctic fish). Artificial sources include many refrigerated appliances and equipment.

Organisms thriving in low-temperature environments (0 °C or close to 0 °C) are commonly classified as psychrophilic or psychrotolerant. Psychrophiles (cold-loving) grow fastest at a temperature of 15°C or lower, and are unable to grow over 20 °C. Psychrotolerant (also termed psychrotrophic) organisms grow well at temperatures close to the freezing point of water; however, their fastest rates of growth are above 20 °C. Psychrophilic and psychrotolerant microorganisms include bacteria, archaea, yeast, fungi, protozoa, and microalgae.

It is generally found that psychrophiles predominate in permanently cold, stable environments that have good sources of nutrition (e.g., old, consolidated forms of sea ice exposed to algal blooms). It is likely that the permanency of the cold in stable environments obviates the need for cells to be able to grow at higher temperatures. In addition, at low temperature the affinity of uptake and transport systems decrease, and as a result, psychrophiles tend to be found in environments that are rich in organic substrates, thus providing compensation for less effective uptake and transport systems.

1. *Natural environments*

Psychrophiles have been isolated from permanently cold, deep-ocean waters, as well as from ocean sediments as deep as 500 m below the ocean floor. As a consequence of the pressure in these environments, many of these psychrophiles are also barotolerant or barophiles (see Section V).

A unique source of cold-adapted microorganisms are lakes in the Vestfold Hills region in Antarctica. The Vestfold Hills lakes are only about 10,000 years old and differ in their salinity and ionic strength, oxygen content, depth, and surface ice coverage. As a result, they have proven to be a rich resource of diverse and unusual microorganisms, including a cell-wall-less Spirochaete, a coiled or "C-shape" bacterium, and the only known free-living, psychrophilic archaeal species. The archaeal species include the methanogens *Methanococcoides burtonii* and *Methangenium frigidum*, and the extreme halophile *Halobacterium lacusprofundi*. The only other low-temperature-adapted archaeal isolate that has been studied is the symbiont *Cenarchaeum symbiosum*, which was isolated from a marine sponge off the Californian coast. It is perhaps surprising that more low-temperature-adapted Archaea have not been isolated throughout the world, as 16*S*-rRNA analysis of numerous aquatic and soil samples have indicated the prevalence of Archaea in these cold habitats.

Gram-negative bacteria, including members of the genera *Pseudomonas, Achromobacter, Flavobacterium, Alcaligenes, Cytophaga, Aeromonas, Vibrio, Serratia, Escherichia, Proteus*, and *Psychrobacter* are more frequently found in cold environments than are gram-positive bacteria (e.g. *Arthrobacter, Bacillus*, and *Micrococcus*). Psychrophilic yeast are of the genera *Candida* and *Torulopsis*, and psychrotolerant members are mostly of the genera *Candida, Cryptococcus, Rhodotorula, Torulopsis, Hanseniaspora*, and *Saccharomyces*. Low-temperature-adapted fungi and molds include isolates of the genera *Penicillium, Cladosporium, Phoma*, and *Aspergillus*. The most common snow alga is *Chlamydomonas nivalis*, which produces bright red spores, marking its location clearly against the white snow background.

A remarkable, and as yet unstudied, low-temperature environment has been discovered in Antarctica. Lake Vostok is a subglacial lake found about one kilometer below Vostok Station, East Antarctica. More than sixty smaller subglacial lakes also exist in central regions of the Antarctic Ice Sheet. Due to the isolation of the lakes, they are likely to contain many novel microorganisms, some of which could be expected to have developed along a separate evolutionary path from that of currently known life. The exploration of Lake Vostoc is being considered using specialized robots based on thermal-probe technology for ice penetration, and submersible technology for lake exploration. Interestingly, the ocean on Europa (a moon of Jupiter) is also located below a kilometer-thick covering of ice. The technology developed for Lake Vostok will be a model for future exploration (in 2003–2018) of the ocean on Europa and provides the unprecedented potential to identify extraterrestrial life in an aquatic environment.

2. *Artificial environments*

The abundance of microorganisms in cold environments was realized as early as 1887, when the first low-temperature-adapted bacterium was probably isolated from a preserved fish stored at 0 °C. Since then, numerous psychrotolerant organisms have been found in artificial habitats and are frequently responsible for the spoilage of food. Members of the genera *Pseudomonas, Acinetobacter, Alcaligenes, Chromobacterium*, and *Flavobacterium* are often associated with spoilage of dairy products, whereas *Lactobacillus viridescens* and *Brochothrix thermospacta* contaminate meat products. Psychrotolerant bacterial pathogens include

Yersinia enterocolitica, Clostridium botulinum, and certain *Aeromonas* strains.

B. Biochemistry and physiology of adaptation

Unicellular organisms are unable to insulate themselves against low temperatures and, as a consequence, psychrophilic and psychrotolerant microorganisms need to adapt the structures of their cellular components.

1. *Membranes*

It is well documented that microorganisms adjust the fatty acid composition of their membrane phospholipids in response to changes in the growth temperature. Normal cell function requires membrane lipids that are largely fluid. As temperature is decreased, the fatty acids chains in membrane bilayers undergo a change of state from a fluid disordered state to a more ordered crystalline array of fatty acid chains. In order to adapt to low temperature, microorganisms decrease the transition temperature of the disordered-to-ordered state by altering the fatty acyl composition. This alteration consists of one or a combination of the following changes to the membrane lipids: an increase in unsaturation, a decrease in average chain length, an increase in methyl branching, an increase in the ratio of *anteiso*-branching compared to *iso*-branching, and an isomeric alteration of acyl chains in *sn-1* and *sn-2* positions. The most common alterations occur in fatty acid saturation and chain length. Changes in the amount and type of methyl-branching occur mostly in gram-positive bacteria. Most bacteria only have monosaturated fatty acids; however a notable exception was found in some marine psychrophilic bacteria that increase membrane fluidity at low temperatures by incorporating polyunsaturated fatty acids into their membranes.

Alterations in the membrane lipid composition can be mediated in a rapid fashion through the increase of unsaturation catalyzed by desaturases. Changes in fatty acyl chain length and the amount and type of methyl branching, however, require *de novo* fatty acid synthesis and are therefore much slower processes.

2. *Enzymes*

The thermodynamic problems associated with enzymes in an environment of low kinetic energy (i.e., low temperature) relate to the lack of sufficient energy to achieve the activation state for catalysis. Low-temperature adaptation of proteins therefore lowers the free energy of the activated state by decreasing the enthalpy-driven interactions necessary for activation. The necessity for weaker interactions leads to a less rigid, more flexible protein structure, or parts thereof. Consequently, factors that confer a more "loose" or flexible structure are expected to be important for cold-adapted proteins. Based on the comparison of proteins from low-temperature-adapted Bacteria, Archaea, and Eukarya with those from mesophiles to hyperthermophiles, including comparisons of three-dimensional structures, a number of structural differences have been identified; including the reduction of the number of salt bridges, the reduction of aromatic interactions, the reduction of hydrophobic clustering, the reduction of proline content, the addition of loop structures, and an increase in solvent interaction. It should be noted that no cold-adapted protein has been studied that exhibits all of these features. This highlights the importance of the molecular context of the changes to stability, activity, or both. As a general rule, however, enzymes from psychrophilic organisms have reduced thermostability and a lower apparent temperature optima for their activity when compared to their mesophilic or thermophilic counterparts.

3. *Cold shock and cold acclimation*

The response to a rapid decrease in temperature (cold shock) has been well studied in mesophilic laboratory microorganisms, including *Escherichia coli, Bacillus subtilis*, and *Saccharomyces cerevisiae*. In response to cold shock, the pattern of gene expression is altered (cold-shock response). In bacteria, a class of small (7–8 kDa) acidic proteins are transiently induced. These cold-shock proteins (CSPs) have been well characterized and shown, in some cases, to function as transcriptional enhancers and RNA-binding proteins. In *S. cerevisiae*, TIP1 (temperature-shock-inducible protein 1) is a major CSP; it is targeted to the outside of the plasma membrane and appears to be heavily glycosylated with *O*-mannose, therefore invoking a role for TIP1 in membrane protection during low-temperature adaptation. Following a cold shock, cells resume growth, albeit at a lower growth rate, indicating that the cold-shock response is an adaptive response aimed at maintaining growth, rather than a stress response aimed only at cell survival.

In general, the process of protein synthesis is temperature sensitive. As a result, psychrophilic microorganisms must have specially adapted ribosomes and accessory factors (e.g., initiation and elongation factors). In addition, psychrophilic and psychrotolerant microorganisnms (particularly those that are food-borne) may be exposed to sudden changes in their environmental temperature. A cold-shock response has been demonstrated in a number of psychrophilic and psychrotolerant organisms, including *Trichosporon pullulans*, *Bacillus psychrophilus*, *Aquaspirrillum arcticum*, and *Arthrobacter globiformis*; and homologs of a major *E. coli* cold-shock protein (CspA) have been identified in a number of these. The extent of the response (i.e., the number of proteins induced) in these microorganisms is dependent on the magnitude of the temperature shift. Some of the CSPs are also cold-acclimation proteins (proteins expressed continuously during growth at low temperature), suggesting that this class of CSP is important for cold-shock adaptation and growth maintenance at low temperatures.

V. BAROPHILES

A. Habitats and microorganisms

Barophiles ("weight lovers") are organisms that thrive in high-pressure habitats. An etymologically more accurate term is "piezophiles" ("pressure lovers"); however, it is less frequently used in the literature.

There is apparent confusion and contradiction in the literature concerning the defining traits of pressure-adapted organisms. This mainly stems from the complicating effects of temperature on growth rate. The effect of both temperature (T) and pressure (P) on the growth rate (k) of different organisms has been thoroughly investigated using "*PTk*-diagrams." With all other conditions held constant, there is a unique pressure, P_{kmax}, and temperature, T_{kmax}, at which the growth rate of an individual microorganism is maximal (k_{max}). These values have been used to define barophiles with $P_{\mathrm{kmax}} > 0.1$ MPa and extreme barophiles with P_{kmax} b 0.1 MPa. In addition, the values for T_{kmax} have been used to delineate psychro-, meso-, or thermo- (extreme) barophiles. Other terms used are obligate (extreme) barophiles, for organisms that cannot grow at atmospheric pressure (0.1 MPa), irrespective of temperature, and barotolerants, for those that grow best under atmospheric pressure but that can also grow at up to 40 MPa.

Barophiles were first isolated from the deep sea (deeper than 1000 m), and this environment still represents the most thoroughly studied habitat. Other high-pressure habitats include the deep-Earth and the deep-sediment layers below the ocean floor. The deep sea is a cold (<5 °C), dark, oligotrophic (low in nutrients) environment with pressure as high as 110 MPa, as is found in the Mariana Trench (almost 11,000 m). In contrast, fumarole and black-smoker hydrothermal vents, produced by the extrusion of hot subsurface water, represent a hyperthermal environment with high metabolic activity (see hyperthermophiles in Section II). The microbiota of Atlantic and Pacific vents are remarkably similar, indicating that the (hyper-) thermobarophilic microorganisms are able to survive for long periods in cold waters, thus facilitating their effective dispersal throughout the oceans. Supporting this, these microorganisms have been isolated from ocean waters far removed from hydrothermal vents.

Psychrophilic deep-sea (extreme) barophilic isolates predominantly belong to five genera of the γ-Proteobacteria—*Photobacterium*, *Shewanella*, *Colwellia*, *Moritella*, and a new group containing the strain CNPT3. One *Bacillus* species (strain DSK25) has also been isolated. Hyperthermobarophilic archaeal isolates include *Pyrococcus* spp. It is noteworthy that the difficulty in isolating microorganisms from the deep sea, and the specific enrichment and culturing techniques that are used, are likely to produce a bias in the types of barophiles that are isolated; a complete analysis of the phylogenetic distribution of barophilic microorganisms is yet to be attempted.

B. Biochemistry and physiology of adaptation

1. *General physiological adaptations*

Membrane lipids from barophilic bacteria have been well studied. In response to high pressure, the relative amount of monounsaturation and polyunsaturation in the membrane is increased. The increase in unsaturation produces a more fluid membrane and counteracts the effects of the increase in viscosity caused by high pressure. This response is analogous to adaptations caused by temperature reduction (see Section IV). Note, however, that an increase in unsaturation was not observed in two extreme barophiles that have been studied, thus indicating that alternative mechanisms of adaptation exist.

An interesting observation for many barophiles is that they are extremely sensitive to UV light and need to be grown in dark or light-reduced environments. This adaptation is not unexpected, considering the darkness that prevails in the deep sea.

There are presently no studies on barophilic microorganisms investigating the adaptation of molecular processes such as chromosomal replication, cell division, transcription, or translation. Studies of *E. coli*, however, have revealed that when cells are grown at their upper pressure limit, DNA synthesis is completely inhibited, protein synthesis is slowed down, and mRNA synthesis and decay appears to be unaffected. This demonstrates that specific cellular processes are affected by pressure, and therefore barophiles are likely to have adapted mechanisms to compensate. Pressure may stabilize proteins and retard thermal denaturation. In support of this, thermal inactivation of DNA polymerases from the hyperthermophilic microorganisms *Pyrococcus furiosus*, *Pyrococcus* strain ES4, and *Thermus aquaticus* is reduced by hydrostatic pressure.

2. *Gene and protein expression*

Pressure-regulated gene and protein expression has been observed and investigated in deep-sea bacteria. *Photobacterium* sp. SS9 expresses two outer-membrane proteins (porins) under different pressures, the OmpH protein at 28 MPa and OmpL at 0.1 MPa. The genes are regulated by a homolog of the *toxRS* system from *Vibrio cholerae*. ToxR and ToxS are cytoplasmic membrane proteins that are thought to be pressure sensors controlled by membrane fluidity.

A pressure-regulated operon has been identified in the barophilic bacterium DB6705. A complex promoter region, which is controlled by a variety of regulatory proteins expressed under different pressure conditions, was identified upstream of three open reading frames (ORFs). The function of the first two ORFs is unknown, while the third ORF encodes the CydD protein. CydD is required for the assembly of the cytochrome-bd complex in the aerobic respiratory chain. The membrane location of this protein highlights the apparent importance of membrane components in high-pressure adaptation.

VI. HALOPHILES

A. Habitats and microorganisms

The first recorded observation of organisms adapted to high salt concentrations (halophiles and halotolerants) probably dates to 2500 BC, when the Chinese noted a red coloration of saturated salterns. What they detected was most likely a bloom of extreme halophilic Archaea that possess red or orange C_{50} carotenoides.

The definition of a hypersaline environment is one that possesses a salt concentration greater than that of seawater (3.5% w/v). For water-containing environments, the salt composition depends greatly on the historical development of the habitat, and the environments are normally described as thalassohaline or athalassohaline. Thalassohaline waters are marine derived and therefore contain, at least initially, a seawater composition; however, with increasing evaporation the concentration of various salts alters depending on the thresholds for the crystallization and precipitation of different minerals (see Table 37.3). Athalassohaline water may also be influenced by the influx of seawater; however, the chemical composition is mainly determined by geological, geographical, and topographical parameters. Examples of athalassohaline environments include the Great Salt Lake in Utah and the Dead Sea.

In addition to lakes formed by evaporation in moderate climate conditions, hypersaline Antarctic lakes (e.g., Vestfold Hills; see Section IV) have been formed from the effects of frost and dryness in this environment. Antarctic and moderate temperature soils also contain salinities between 10 and 20% (w/v) and efforts have been directed at characterizing these comparatively poorly studied ecosystems. Less obvious saline habitats known to be colonized by microorganisms are animals skins, plant surfaces, and building surfaces. In addition, interest is also being focused on subterranean salt deposits as a habitat for extreme halophilic Archaea, and as a possible source of ancient prokaryotic lineages preserved in fluid inclusion bodies of salt crystals.

TABLE 37.3 Concentration of ions in thalassohaline and athalassohaline brines[a]

	Concentration (g/l)					
Ion	**Seawater**	**Seawater at onset of NaCl saturation**	**Seawater at onset of K^+-salt saturation**	**Great Salt Lake**	**Dead Sea**	**Typical soda lake**
Na^+	10.8	98.4	61.4	105.0	39.7	142
Mg^{2+}	1.3	14.5	39.3	11.1	42.4	<0.1
Ca^{2+}	0.4	0.4	0.2	0.3	17.2	<0.1
K^+	0.4	4.9	12.8	6.7	7.6	2.3
Cl^-	19.4	187.0	189	181.0	219.0	155
SO_4^{2-}	2.7	19.3	51.2	27.0	0.4	23
CO_3^{2-}/HCO_3^-	0.34	0.14	0.14	0.72	0.2	67
pH	8.2	7.3	6.8	7.7	6.3	11

[a]Data from Grant, W. D., Gemmell, R. T., and McGenity, T. J. (1998). Halophiles. *In* "Extremophiles: Microbial Life in Extreme Environments" (Horikoshi, K., and Grant, W. D., eds.), Wiley series in Ecological and Applied Microbiology, Wiley-Liss.

TABLE 37.4 Defining terms for microorganisms ranging in tolerances to salt

Category	Salt concentration (M)	
	Range	Optimum
Nonhalophilic	0–1.0	<0.2
Slightly halophilic	0.2–2.0	0.2–0.5
Moderate halophilic	0.4–3.5	0.5–2.0
Borderline extreme halophilic	1.4–4.0	2.0–3.0
Extreme halophilic	2.0–5.2	>3.0
Halotolerant	0–1.0	<0.2
Haloversatile	0–3.0	0.2–0.5

Organic compounds in hypersaline lakes are mostly produced by cyanobacteria, anoxygenic phototrophic bacteria, and by species of the green algae *Dunaliella* spp. They are found in most natural and artificial hypersaline lakes around the world. With seasonal and transient contributions from animals or plants, these environments can contain dissolved organic carbon levels up to 1 g/l. Due to the low solubility of oxygen in saline solutions, however, many habitats become anaerobic and as a result aerobic growth is often restricted to the upper layers.

Organisms living in saline habitats exhibit different levels of adaptation to salt. To account for the variety of tolerances, an extensive set of definitions exist and are summarized in Table 37.4.

A salt concentration of about 1.5 M is the upper limit for vertebrates, although some eukaryotes such as the brine shrimp (*Artemia salina*) and the brine fly (*Ephydra*) can be found in habitats with higher salinity. For salinities above 1.5 M, prokaryotes become the predominant group with moderate halophilic and haloversatile bacteria at salt concentrations between 1.5 and 3.0 M, and extreme halophilic Archaea (halobacteria) at salinities around the point of sodium chloride precipitation.

Aerobic gram-negative chemoorganotrophic bacteria are abundant in brines of medium salinity, and many strains have been isolated, including members of the genera *Vibrio, Alteromonas, Acinetobacter, Deleya, Marinomonas, Pseudomonas, Flavobacterium, Halomonas,* and *Halovibrio*. Gram-positive aerobic bacteria of the *Marinococcus, Sporosarcina, Salinococcus* and *Bacillus* species have been found in saline soils, salterns, and occasionally in solar salterns. Members of the genera *Halomonas* and *Flavobacterium* have been isolated from Antarctic lakes. Two of the most remarkable isolates from Antarctica are the extreme halophilic archaeal species *Halobacterium lacusprofundi* and a *Dunaliella* spp., which are found in association in Deep Lake. These are the only two microorganisms growing in this lake, which has 4.8 M salt and whose temperature for 8 months of the year is less than 0 °C. With the extremes of temperature, salt, and ionic strength, and with primary production rates of 10° g C/m^2/year, Deep Lake has been described as one of the most inhospitable environments on Earth.

Wherever light reaches the anoxic layer of hypersaline brines, anoxygenic phototrophs such as *Chromatium salexigens, Thiocapsa halophila, Rhodospirillum salinarum*, and *Ectothiorhodospira* are commonly found and represent the primary producers in these environments. Sulfate reducers *Desulfovibrio halophilus* and *Desulfohalobium retbaense* have been isolated from anaerobic sediments and are thought to perform dissimilatory sulfate reduction to H_2S. H_2S is subsequently used for growth by most anoxygenic phototrophs (except for members of the *Rhodospirillaceae* family). Anaerobic fermentative halophiles from the bacterial lineage of *Haloanaerobiaceae* have also been described.

Most of the bacterial species mentioned here are predominant in environments up to 2 M salt. At higher concentrations, extremely halophilic Archaea (including the confusingly named genus, *Halobacteria*) are more abundant. Most of these Archaea require at least 1.5 M salt for growth and for retaining their structural integrity. Members of the genera *Haloarcula, Halobacterium,* and *Halorubrum* are frequently found in hypersaline waters reaching the sodium chloride saturation point, and the proteolytic species *Halobacterium salinarium* is often associated with salted food. Other Archaea are also found in saline lakes, marine stromatolites, and solar ponds, and these tend to be methanogens (see Section X). In addition, the eukaryote *Dunaliella* can adapt to a wide range of salt concentrations, from less than 100 mM to saturation (5.5 M).

B. Biochemistry and physiology of adaptation

Cells exposed to an environment with a higher salinity than the one inside the cytoplasm inevitably experience a loss of water and undergo plasmolysis (Fig. 37.1). Microorganisms living in high-salt environments generally adopt one of two strategies (either salt-in-cytoplasm or compatible-solute adaptation) to prevent the loss of cytoplasmic water and to establish osmotic equilibrium across their cell membranes.

1. *Salt-in-cytoplasm adaptation*

Extremely halophilic archaea and anaerobic halophilic bacteria use a salt-in-cytoplasm strategy. This involves cations flowing through the membrane into the cytoplasm (Fig. 37.1A). Archaea accumulate intracellular potassium and exclude sodium, whereas Bacteria accumulate sodium rather than potassium. As a consequence of the high salt, intracellular components (e.g., proteins, nucleic acid and cofactors) require protection from the denaturing effects of salt. The most common protective mechanism is the presence of excess negative charges on their exterior surfaces. Malate dehydrogenase from *Haloarcula marismortui* has a 20 mol% excess of acidic residues over basic amino acid residues, compared to only 6 mol% excess for a nonhalophilic counterpart. Structural analysis shows that the acidic residues are mainly located on the surface of the protein and are either involved in the formation of stabilizing salt bridges or in attracting water and salt to form a strong hydration shell. This malate dehydrogenase is able to bind extraordinary amounts of water and salt (0.8–1.0 g water and 0.3 g salt/g protein) compared to a nonhalophilic malate dehydrogenase (0.2–0.3 g water and 0.01 g salt/g protein). Due to the adaptation mechanism in these cells, they have an obligate requirement for high concentrations of salt. In low-salt environments, the lack of salt removes the shielding effect of cations from the proteins and leads to the rapid denaturation of the three-dimensional structure.

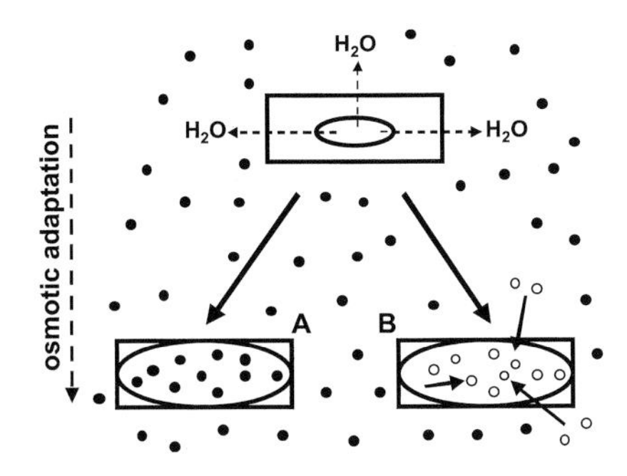

FIGURE 37.1 Effect of high salinity on intracellular water, and two strategies used for adaptation. Solid circles represent salt molecules and open circles compatible solutes.

2. *Compatible-solute adaptation*

Most halotolerant and moderately halophilic organisms, including the anoxygenic phototrophic bacteria, aerobic heterotrophic bacteria, cyanobacteria, and methanogens, maintain a salt-minimized cytoplasm. They accumulate small organic and osmotically active molecules, referred to as compatible solutes. These compounds can be synthesized *de novo* or imported from the surrounding medium (Fig. 37.1b). The latter mechanism is also used by nonhalophilic organisms; they adapt to increased salt concentrations by importing extracellular compatible solutes using transporters involved in amino acid or sugar uptake.

A large range of compatible solutes has been identified in a broad range of halophiles (Table 37.5). All these molecules are polar, highly soluble, and uncharged or zwitterionic at physiological pH values. They are strong water-structure formers and as such are probably excluded from the hydration shell of proteins (preferential exclusion), and therefore exert a stabilizing effect without interfering directly with the structure of the protein. In addition to being stabilizers against salt stress, they have also been shown to prevent the denaturation of proteins caused by heating, freezing, and drying.

3. *Membranes*

Even though the cytoplasmic interior is protected from the effects of external salt by cytoplasmic-compatible

TABLE 37.5 Compatible solutes used by halophilic microorganisms

Class of compatible solute	Specific compatible solute	Microorganism
Polyoles	Glycerol and arabite	Algae, yeast, fungi
	Glucosyl-glycerol	Cyanobacteria
Betaines	Glycine-betaine	Cyanobacteria, anoxygenic phototrophic bacteria, methanogens and *Actinopolyspora halophila*
	Dimethylglycine	*Methanohalophilus* sp.
Amino acids	Proline	*Bacillus* spp. and other *Firmicutes*
	α-glutamine	Corynebacteria
	β-glutamine	*Methanohalophilus* sp
Dimethylsulfoniopropionate		Marine cyanobacteria
Glutamine amides	*N*-α-carbomoyl-glutamine amide	*Ectothiorhodospira marismortui*
	N-α-acetyl-glutaminyl-glutamine amide	Anoxygenic phototrophic proteobacteria, *Rhizobium meliloti*, and *Pseudomonas* spp.
Acetylated diamino acids	*N*-δ-acetyl-ornithine and *N*-ε-acetyl-α-lysine	*Bacillus* spp. and *Sporosarcina halophila*
	N-ε-acetyl-α-lysine	Methanogens
Ectoines	Ectoine and hydroxyectoine	Proteobacteria, *Brevibacteria*, gram-positive cocci, and *Bacillus* spp.

solutes, the outer surface of the cytoplasmic membrane is permanently exposed to high salt concentrations. To protect the membrane in halophilic bacteria, the proportion of anionic phospholipids (often phosphatidylglycerol and glycolipids) increases with increasing salinity at the expense of neutral zwitterionic phospholipids. These alterations produce additional surface charges to the membrane and help to maintain the hydration state of the membrane.

Most halophilic Archaea posses an S-layer consisting of sulfated glycoproteins, which surrounds the cytoplasmic membrane. The sulfate groups confer a negative charge to the S-layer and possibly provides structural integrity at high ionic concentrations. In addition, archaeal ether lipids have been shown to be more stable at high salt concentrations (up to 5 M) compared to the ester lipids found in the membranes of bacteria.

VII. ALKALIPHILES

A. Habitats and microorganisms

Alkaliphiles are defined by optimal growth at pH values above 10, whereas alkalitolerant microorganisms may grow well up to pH 10, but exhibit more rapid growth below pH 9.5. Alkaliphiles are further subdivided into facultative alkaliphiles (that grow well at neutral pH) and obligate alkaliphiles (that grow only above pH 9). In addition, due to their natural habitats, many alkaliphilic organisms are adapted to high salt concentrations and are referred to as haloalkaliphiles.

Alkali environments include soils where the pH has been increased by microbial ammonification and sulfate reduction, and water derived from leached silicate minerals. These environments tend to have only a limited buffering capacity and the pH of the environments fluctuates. As a result, alkalitolerant microorganisms are more abundant in these habitats than are alkaliphiles. Artificial environments include locations of cement manufacture, mining, and paper and pulp production. Probably the best studied and most stable alkaline environments are soda lakes and soda deserts (e.g., in the East African Rift Valley or central Asia). The formation of soda lakes and deserts is similar to the formation of athalassohaline salt lakes (see Section V), with the exception that carbonate is the major anion in solution, due to the lack of divalent cations (Mg^{2+}, Ca^{2+}) in the surrounding environment. A typical soda lake composition is shown in Table 37.2.

Bacteria, Archaea, yeast, and fungi have been isolated from alkali environments. Archaeal alkaliphiles include members of the *Halobacteriaceae* (*Halorubrum*, *Natronbacterium*, and *Natronococcus* spp.) and *Methanosarcinaceae* (*Methanohalophilus* spp.). Cyanobacterial genera *Spirulina* and *Synechococcus* represent the dominant primary producers in the aerobic layers of soda lakes. Other alkaliphilic bacteria are members of the *Actinomyces*, *Bacillaceae*, *Clostridiaceae*, *Haloanaerobiales*, and γ-Proteobacteria (*Ectothiorodospira*, *Halomonadaceae*, and *Pseudomonas*). Anaerobic thermophilic alkaliphiles (alkalithermophiles) include *Clostridium* and *Thermoanaerobacter* spp., and the only representative of a new taxon, *Thermopallium natronophilum*.

B. Biochemistry and physiology of adaptation

1. pH homeostasis

Studies of alkaliphiles (particularly *Bacillus* spp.) have demonstrated that they maintain a neutral or slightly alkaline cytoplasm. This is reflected by a neutral pH optimum for intracellular enzymes compared to a high pH optimum for extracellular enzymes. The intracellular pH regulation has been shown to be dependent on the presence of sodium. The Na^+ ions are exchanged from the cytoplasm into the medium by H^+/Na^+ antiporters (Fig. 37.2). Electrogenic proton extrusion is mediated in aerobic cells by respiratory chain activity and protons are transported back into the cell via antiporters that are efficient at transporting H^+ into the cell at the expense of Na^+ export from the cell. The resulting net production and influx of protons creates a more acidic cytoplasm. In addition to controlling protons, Na^+-dependent pH homeostasis requires the reentry of Na^+ into the cell. Na^+-coupled solute symporter and sodium-driven flagella rotation ensure a net sodium balance. The combined action of the antiporters coupled with respiration provides the cell with a means to control its internal pH while maintaining sufficient Na^+ levels through symport and flagella rotation.

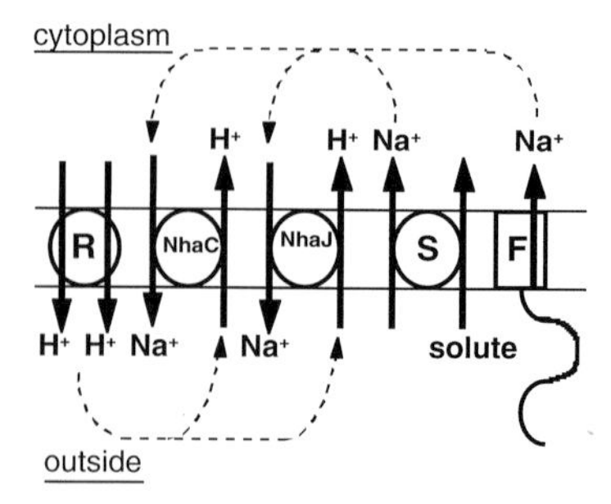

FIGURE 37.2 Schematic representation of elements involved in pH homeostasis in alkaliphilic microorganisms. R represents respiratory chain-mediated proton translocation; NhaC and NhaJ represent two separate H^+/Na^+ antiporters found in *Bacillus* spp.; S represents a Na^+-dependent solute transporter; and F represents Na^+-dependent flagella motility.

The exterior surfaces of the cell are also important for maintaining a pH differential. This is supported by evidence that the protoplasts of alkaliphilic *Bacillus* spp. are unstable in alkaline conditions. The peptidoglycan in these strains has a higher cross-linking rate at higher pH values, which may provide a shielding effect by "tightening" the cell wall. Large amounts of acidic compounds, including teichuronic acid, teichoic acid, uronic acids, and acidic amino acids, are evident in alkaliphilic cell walls compared to the cell walls of nonalkaliphilic microorganisms. The negative charge of these acidic substances may create a more neutral layer close to the outer surface of the cell.

2. *Bioenergetics*

Alkaliphiles have unique bioenergetic properties. Nonalkaliphilic respiring bacteria energize their cytoplasmic membrane with a chemiosmotic driving force (Δp) by generating an electrochemical gradient of ions that has two parameters—acidic conditions outside (caused by the extrusion of protons and described by the term ΔpH) and a positive charge outside (described by the transmembrane electrical potential, $\Delta\Psi$). The Δp is used for proton-coupled symport of solutes, proton-driven mobility (flagella) and ATP synthesis. In alkaliphilic environments, the contribution of ΔpH to Δp becomes smaller with increasing extracellular pH. However, with increasing extracellular pH, sodium ion export is increased (see Section VII.B.1) and may contribute to an increased $\Delta\Psi$. This partially compensates for the decrease of Δp by the reduction of ΔpH. Interestingly, the sodium gradient is used to energize solute transport and flagella movement, but not for ATP synthesis. No sodium-dependent ATP-synthases have been identified in alkaliphiles. In addition, ATP-synthases in alkaliphilic *Bacillus* spp. have been shown to be exclusively proton translocating.

VIII. OLIGOTROPHS

A. Habitats and microorganisms

A most important environmental factor in microbial ecology is the availability of energy, and virtually all microbial cells in nature are limited in their growth by the availability of one or more essential growth nutrients. For example, in the intestinal tract, the number of *E. coli* doubles about twice per day, whereas in ideal laboratory conditions it doubles in 20 min. Similarly, soil bacteria are estimated to grow in soil at about 1% of the maximal rate of growth observed in the laboratory. This highlights that in natural ecosystems nutrient limitation is the rule, rather than the exception.

In addition to the overall nutrient status, in aquatic and soil environments, nutrient levels are often transient; for example, a fallen leaf provides nutrients to soil microorganisms, or a dead fish provides nutrients in an aquatic environment. As a result of this apparent "feast or famine," microorganisms have adopted two main strategies for surviving in nutrient-depleted (oligotrophic) environments. Eutrophic microorganisms (also referred to as copiotrophs, saphrophytes, and heterotrophs) grow in bursts when nutrients are available and produce resting-stage cells when nutrients are in short supply (referred to as r-strategy). In contrast, oligotrophic microorganisms (also referred to as oligocarbophiles; low-nutrient, LN, bacteria; low-K_s bacteria; and dilute-nutrient-broth, DNB, organisms) grow slowly, using low concentrations of nutrients (referred to as K-strategy).

It is important to note that whereas oligotrophs grow slowly in oligotrophic environments, eutrophs are in a resting stage. In contrast, when members of these two classes are subcultured from oligotrophic environments into rich media, the eutrophs resume rapid growth, whereas the oligotrophs do not grow at all. The cellular responses that prevent growth of the oligotrophs in rich media is not understood; however, the response highlights the physiological differences between these two classes of microorganisms. In addition, the difficulty in growing oligotrophs in the laboratory has proven problematic for examining their physiology. Fortunately, some isolates have adapted (faculative oligotrophs) to growth in rich media and are amenable to laboratory studies.

Although the definition of an oligotroph and an oligotrophic environment remain the subject of debate, it is generally accepted that an oligotroph is able to grow in a medium containing 0.2–16.8 mg dissolved organic carbon/liter. The terms "obligate oligotroph" (implying the inability to grow in high concentrations of nutrients) and "facultative oligotroph" (indicating the ability to grow in low and high concentrations of nutrients) are also used to further clarify the nutritional requirements for growth.

In natural ecosystems, oligotrophs and eutrophs coexist, and the proportion of each varies depending on their individual abilities to dominate in the particular environment. For example, when the marine oligotrophic ultramicrobacterium *Sphingomonas* sp. strain RB2256 was isolated from Resurrection Bay, Alaska, it was a numerically dominant species in a place where the total bacterioplankton population was $0.2–1.07 \times 10^6$ cells/ml. In contrast, significantly lower numbers (<1%) of larger, faster-growing cells (typical of eutrophs) were able to be immediately cultured in rich media and on plates.

B. Biochemistry and physiology of adaptation

A characteristic that is often associated with oligotrophic bacteria is their ultramicro size ($<0.1\,\mu m^3$). Ultramicrobacteria (or dwarf cells) are commonly found in aquatic and soil environments. Oligotrophic ultramicrobacteria, such as *Sphingomonas* RB2256, retain their ultramicro size irrespective of growth phase, carbon source, or carbon concentration. In contrast, eutrophic microorganisms, such as *Vibrio angustum* S14 and *Vibrio* ANT-300, undergo reductive cell division during starvation, but resume their normal size ($>1\,\mu m^3$) when grown in rich media.

The characteristics that are thought to be important for oligotrophic microorganisms include a substrate uptake system that is able to acquire nutrients from its surroundings, and the capacity to use the nutrients in order to maintain its integrity and growth. As a result, oligotrophs would ideally have large surface-area-to-volume ratios, and high-affinity uptake systems with broad substrate specificities. Consistent with this, a number of microorganisms that are adapted to low-nutrient environments produce appendages to enhance their surface area. These include members of the bacterial genera, *Caulobacter, Hyphomicrobium, Prosthecomicrobium, Ancalomicrobium, Labrys*, and *Stella*. Other bacteria have tiny cell volumes (ultramicrobacteria: $<0.1\,\mu m^3$) to maximize their surface-area-to-volume ratio.

The most comprehensive physiological studies of oligotrophic marine isolates have been performed on *Sphingomonas* sp. strain RB2256. The characteristics that distinguish it from typical marine eutrophs include constant ultramicro size irrespective of growth or starvation conditions, a mechanism for avoiding predation (ultramicro size), a relatively slow maximum specific growth rate ($<0.2\,h^{-1}$), a single copy of the rRNA operon compared to 8–11 copies for *Vibrio* spp., a relatively small genome size compared to faster-growing heterotrophs, an ability to use low concentrations of nutrients, high-affinity broad-specificity uptake systems, an ability to simultaneously take up mixed substrates, an ability to immediately respond to nutrient addition without a lag in growth, and an inherent resistance to environmental stresses (e.g., heat, hydrogen peroxide, and ethanol).

The small ($0.2\,\mu m^3$) oligotrophic bacterium *Cycloclasticus oligotrophicus* RB1, which was also isolated from Resurrection Bay, shares some properties similar to *Sphingomonas* RB2256 (e.g., single copy of the rRNA operon, and relatively small cell size and genome size). Interestingly, while this chemoorganotroph is unable to grow using glucose and amino acids, it can use acetate and a few aromatic hydrocarbons such as toluene.

IX. RADIATION-RESISTANT MICROORGANISMS

A. Habitats and microorganisms

Radiation from the sun drives photosynthetic reactions, thus ensuring primary production throughout the global ecosystem. Although the visible spectrum leads to biomass production, visible light and other portions of the electromagnetic spectrum (particularly short-wavelength) also cause cellular damage. Damage to cells primarily occurs directly to nucleic acids (e.g., UV-induced thymine-dimer formation and strand breakage) or indirectly through the production of reactive oxygen species (e.g., H_2O_2, O_2^- ^0OH, and $^1O_2$), which cause damage to lipids, proteins, and nucleic acids. Due to the prevalence of natural radiation, most cells have developed a range of DNA-repair and other protective mechanisms to facilitate their survival.

In contrast to natural forms of relatively low-level radiation, microorganisms may be exposed to intense sources of radiation in the form of γ-irradiation (^{60}Co and ^{137}Cs) as a means of sterilization or by being in close proximity to nuclear reactors. It is mainly these forms of radiation that have been the sources of highly radiation-resistant extremophiles. The most well-studied of these is the bacterium *Deinococcus radiodurans*, which was first isolated in 1956 from tins of meat that had been irradiated with γ-rays. Although some microorganisms escape radiation damage by forming spores (e.g., *Clostridium botulinum*), *D. radiodurans* is resistant while in the exponential growth phase. The degree of resistance of these cells is illustrated by their ability to survive 3,000,000 rad, a dose that is sufficient to kill most spores (a lethal dose for humans is about 500 rad). As a consequence of the extreme radiation resistance of *D. radiodurans* and other microorganisms, biocides are routinely added to the cooling waters of nuclear reactors to prevent the proliferation of microorganisms. By virtue of their resistant properties, this class of extremophile can be readily isolated by exposing samples to intense UV or γ-irradiation and then plating them out on a rich medium.

Although *Deinococcus* spp. have been found in dust, processed meats, medical instruments, textiles, dried food, animal feces, and sewage, their natural habitats have not been clearly defined. Thermophilic species have, however, been isolated from hot springs in Italy, and deinococci have been identified in many soil environments. It therefore appears that this class of radiation resistant extremophiles exists in a broad range of environmental niches.

In addition to *Deinococcus* spp., some hyperthermophilic Archaea (e.g., *Thermococcus stetteri* and

Pyrococcus furiosus) are able to survive high levels of γ-irradiation.

B. Biochemistry and physiology of adaptation

Resistance to radiation could conceivably occur by two main mechanisms, prevention of damage or efficient repair of damage. In *D. radiodurans*, it has been clearly shown that the DNA is severely damaged during γ-irradiation (e.g., ~110 double-strand breaks per cell when exposed to 300,000 rad); however, within 3 h of recovery, the fragmented DNA is replaced by essentially intact, chromosomal DNA. Furthermore, even though the DNA within the cells has been extensively degraded, viability is unaffected. *P. furiosus* is also able to repair fragmented DNA after exposure to 250,000 rad when the cells are grown at 95 °C. As a result, it has been suggested that active DNA-repair mechanisms may be important determinants of survival of hyperthermophiles in their natural environments.

In addition to γ-radiation resistance, *D. radiodurans* is resistant to highly mutagenic chemicals, with the exception of those that cause DNA deletions (e.g., nitrosoguanadine, NTG). *D. radiodurans* is also extremely resistant to UV irradiation, surviving doses as high as 1000 J/m^2. The dose that is required to inactivate a single colony-forming unit of an irradiated population is for *D. radiodurans* 550–600 J/m^2 compared to just 30 J/m^2 for *E. coli*.

The DNA-repair systems in *D. radiodurans* are so effective that they have proven to be a hindrance for genetic studies, that is, it is difficult to isolate stable mutants. However, through the combined use of chemical mutagenesis and screens for mitomycin C-, UV radiation-, and ionizing radiation-sensitive strains, genes involved in nucleotide-excision repair, base-excision repair, and recombinational repair have been identified. In addition, *D. radiodurans* is multigenomic (e.g., there are 2.5–10 copies of the chromosome depending on growth rate) and a novel mechanism of interchromosomal recombination has been proposed. This process would help to circumvent problems of reassembling a complete and contiguous chromosome from the chromosome fragments that have been generated as a result of irradiation. It appears that *D. radiodurans*'s ability to use its genome multiplicity to repair DNA damage is the most fundamental reason for this species's extraordinary radioresistance.

Almost all *Deinococcus* spp. that have been isolated (even those isolated without selection for radiation resistance) are radiation resistant, thus demonstrating that the extreme resistance is not a result of selection by irradiation, but a normal characteristic of the genus. Desiccation leads to DNA damage in all cells and prevents DNA repair from occurring. For a cell to be viable when it is rehydrated, it needs to be able to repair the damaged DNA. It is therefore likely that the evolutionary process that has led to the inherent radiation resistance in *Deinococcus* is natural selection for resistance to desiccation. This is supported by evidence that *D. radiodurans* is also exceptionally resistant to desiccation.

X. OTHER EXTREMOPHILES

As a group, methanogenic Archaea (methanogens) are often considered extremophiles. They are the most thermally diverse organisms known, inhabiting environments from close to freezing in Antarctic lakes to 110 °C in hydrothermal vents. Those isolated from Antarctica include *Methanococcoides burtonii* and *Methanogenium frigidum*, and those from hydrothermal vents include *Methanopyrus kandleri*, *Methanothermus fervidus*, *Methanothermus sociabilis*, *Methanococcus jannaschii*, and *Methanococcus igneus*. Methanogens can also be isolated from a diverse range of salinities, from fresh water to saturated brines. *Methanohalobium evestigatum* was isolated from a microbial mat in Sivash Lake and grows in pH neutral, hypersaline conditions from 2.6–5.1 M. Alkaliphilic (*Methanosalsus zhilinaeae*, pH 8.2–10.3) and acidophilic (*Methanosarcina* sp., pH 4–5) methanogens have also been isolated.

Other microorganisms that may be considered to be extremophiles include toxitolerants (those tolerant to organic solvents, hydrocarbons, and heavy metals, such as *Rhodococcus* sp., which can grow on benzene as a sole carbon source), and xerophiles and xerotolerant microbes that survive very low-water activity (e.g., extremely halophilic Archaea, fungi such as *Xeromyces bisporus*, and endolithic microorganisms that live in rocks).

XI. BIOTECHNOLOGY OF EXTREMOPHILES

A major impetus driving research on extremophiles is the biotechnological potential associated with the microorganisms and their cellular products. In 1992, of the patents related to Archaea, about 60% were for methanogens, 20% for halophiles, and 20% thermophiles. Examples of "extremozymes" that are presently used commercially include alkaline proteases for detergents. This is a huge market, with 30% of the total worldwide enzyme production for detergents. In 1994, the total market for alkaline proteases in Japan alone was ~15,000 million yen. DNA polymerases have been isolated from the hyperthermophiles *Thermus aquaticus*, *Thermotoga maritima*, *Thermococcus litoralis*,

TABLE 37.6 Extremophiles and their uses in biotechnology

Source	Use
Hyperthermophiles	
DNA polymerases	DNA amplification by PCR
Alkaline phosphatase	Diagnostics
Proteases and lipases	Dairy products
Lipases, pullulanases, and proteases	Detergents
Proteases	Baking and brewing and amino acid production from keratin
Amylases, α-glucosidase, pullulanase, and xylose/glucose isomerases	Starch processing, and glucose and fructose for sweeteners
Alcohol dehydrogenase	Chemical synthesis
Xylanases	Paper bleaching
Lenthionin	Pharmaceutical
S-layer proteins and lipids	Molecular sieves
Oil degrading microorganisms	Surfactants for oil recovery
Sulfur oxidizing microorganisms	Bioleaching, coal, and waste gas desulfurization
Hyperthermophilic consortia	Waste treatment and methane production
Psychrophiles	
Alkaline phosphatase	Molecular biology
Proteases, lipases, cellulases, and amylases	Detergents
Lipases and proteases	Cheese manufacture and dairy production
Proteases	Contact-lens cleaning solutions, meat tenderizing
Polyunsaturated fatty acids	Food additives, dietary supplements
Various enzymes	Modifying flavors
β-galactosidase	Lactose hydrolysis in milk products
Ice nucleating proteins	Artificial snow, ice cream, other freezing applications in the food industry
Ice minus microorganisms	Frost protectants for sensitive plants
Various enzymes (e.g., dehydrogenases)	Biotransformations
Various enzymes (e.g., oxidases)	Bioremediation, environmental biosensors
Methanogens	Methane production
Halophiles	
Bacteriorhodopsin	Optical switches and photocurrent generators in bioelectronics
Polyhydroxyalkanoates	Medical plastics
Rheological polymers	Oil recovery
Eukaryotic homologues (e.g., *myc* oncogene product)	Cancer detection, screening antitumor drugs
Lipids	Liposomes for drug delivery and cosmetic packaging
Lipids	Heating oil
Compatible solutes	Protein and cell protectants in a variety of industrial uses (e.g., freezing, heating)
Various enzymes (e.g., nucleases, amylases, proteases)	Various industrial uses (e.g., flavoring agents)
γ-linoleic acid, β-carotene and cell extracts (e.g., *Spirulina* and *Dunaliella*)	Health foods, dietary supplements, food coloring, and feedstock
Microorganisms	Fermenting fish sauces and modifying food textures and flavors
Microorganisms	Waste transformation and degradation (e.g., hypersaline waste brines contaminated with a wide range of organics)
Membranes	Surfactants for pharmaceuticals
Alkaliphiles	
Proteases, cellulases, xylanases, lipases and pullulanases	Detergents
Proteases	Gelatin removal on X-ray film
Elastases, keritinases	Hide dehairing
Cyclodextrins	Foodstuffs, chemicals, and pharmaceuticals
Xylanases and proteases	Pulp bleaching
Pectinases	Fine papers, waste treatment, and degumming
Alkaliphilic halophiles	Oil recovery
Various microorganisms	Antibiotics
Acidophiles	
Sulfur-oxidizing microorganisms	Recovery of metals and desulfurication of coal
Microorganisms	Organic acids and solvents

Pyrococcus woesii, and *Pyrococcus furiosus* for use in the polymerase chain reaction (PCR). A eukaryotic homolog of the *myc* oncogene product from halophilic Archaea has been used to screen the sera of cancer patients. Its utility is demonstrated by the fact that the archaeal homolog produced a higher number of positive reactions than the recombinant protein expressed in *E. coli.* β-carotene is commercially produced from the green algae *Dunaliella bardawil*. The applications in industry are still limited; however, the potential applications are extensive. Some examples of their uses and potential applications are listed in Table 37.6.

The biotechnology potential is increasing exponentially with the isolation of new organisms, the identification of novel compounds and pathways, and the molecular and biochemical characterization of cellular components. Major advances are likely in the area of protein engineering. For example, the identification of the structural properties important for thermal activity and stability will enable the construction of proteins with required catalytic and thermal properties. Recently, a metalloprotease from the moderately thermophilic bacterium *Bacillus stearothermophilus* was mutated using a rational design process in an effort to increase its thermostability. The mutant protein was 340 times more stable than the wild-type protein and was able to function at 100 °C in the presence of denaturing agents, while retaining wild-type activity at 37 °C.

Advances are also likely to arise from the construction of recombinant microorganisms for specific purposes. A recombinant strain of *Deinococcus radiodurans* has been engineered to degrade organopollutants in radioactive mixed-waste environments. The recombinant *Deinococcus* expresses toluene dioxygenase, enabling it to oxidize toluene, chlorobenzene, 2,3-dichloro-1-butene, and indole in a highly irradiating environment (6000 rad/h), while remaining tolerant to the solvent effects of toluene and trichloroethylene at levels exceeding those of many radioactive waste sites. In recognition of the number of waste sites contaminated with organopollutants plus radionuclides and heavy metals around the world, and the safety hazards and cost involved in clean up using physicochemical means, the potential use of genetically engineered extremophilic microorganisms is an important and exciting prospect.

BIBLIOGRAPHY

Amend, J. P., and Shock, E. L. (2001). Energetics of overall metabolic reactions of thermophilic and hyperthermophilic Archaea and bacteria. *FEMS Microbiol. Rev.* **25**, 175–243.

Atlas, R. M., and Bartha, R. (1998). "Microbial Ecology: Fundamental and Applications," 4th edn. Benjamin/Cummings.

Battista, J. R. (1997). Against all odds: The survival strategies of *Deinococcus radiodurans*. *Annu. Rev. Microbiol.* **51**, 203–224.

Cavicchioli, R. (2002). Extremophiles and the search for extra-terrestrial life. *Astrobiology* **2**, 281–282.

Cavicchioli, R., Siddiqui, K. S., Andrews, D., and Sowers, K. R. (2002). Low-temperature extremophiles and their applications. *Curr. Opin. Biotechnol.* **13**, 253–261.

DeLong, E. F., Franks, D. G., and Yayanos, A. A. (1997). Evolutionary relationship of cultivated psychrophilic and barophilic deep-sea bacteria. *Appl. Environ. Microbiol.* **63**, 2105–2108.

Demirjian, D. C., Moris-Varas, F., and Cassidy, C. S. (2001). Enzymes from extremophiles. *Curr. Opin. Chem. Biol.* **5**, 144–151.

First International Congress on Extremophiles (1996). *FEMS Microbiol. Rev.* **18**(2–3).

Galinski, E. A. (1995). Osmoadaptation in bacteria. *Adv. Microb. Physiol.* **37**, 273–327.

Horikoshi, K., and Grant, W. D. (1998). "Extremophiles: Microbial Life in Extreme Environments". Wiley-Liss.

Kato, C., and Bartlett, D. H. (1997). The molecular biology of barophilic bacteria. *Extremophiles* **1**, 111–116.

Krulwich, T. A. (1995). Alkaliphiles: "Basic" molecular problems of pH tolerance and bioenergetics. *Mol. Microbiol.* **15**, 403–410.

Madigan, M. T., Martinko, J. M., and Parker, J. (2003). "Brock: Biology of Microorganisms," 10th edn. Prentice Hall, Upper Saddle River, NJ.

Matin, A. (1990) Keeping a neutral cytoplasm: The bioenergetics of obligate acidophiles. *FEMS Microbiol. Rev.* **75**, 307–318.

Morita, R. Y. (1997). "Bacteria in Oligotrophic Environments: Starvation-Survival Lifestyle." Chapman and Hall, New York.

Niehaus, F., Bertoldo, C., Kahler, M., and Antranikian, G. (1999). Extremophiles as a source of novel enzymes for industrial application. *Appl. Microbiol. Biotechnol.* **51**, 711–729.

Pick, U. (1998). *Dunaliella*: a model extremophilic alga. *Israel J. Plant Sci.* **46**, 131–139.

Russel, N. J. (1998). Molecular adaptations in psychrophilic bacteria: Potential for biotechnological applications. *Adv. Biochem. Eng. Biotechnol.* **61**, 1–21.

Schut, F., Prins, R. A., and Gottschal, J. C. (1997). Oligotrophy and pelagic marine bacteria: facts and fiction. *Aquatic Microb. Ecol.* **12**, 177–202.

Second International Congress on Extremophiles. (1998). *Extremophiles* **2**(2).

Sterner, R., and Liebl, W. (2001). Thermophilic adaptation of proteins. *Crit. Rev. Biochem. Mol. Biol.* **36**, 39–106.

Takami, H., and Horikoshi, K. (2000). Protein, nucleotide, genome. Analysis of the genome of an alkaliphilic *Bacillus* strain from an industrial point of view. *Extremophiles* **4**, 99–108.

WEBSITES

Comprehensive Microbial Resource of The Institute for Genomic Research and links to many other microbial genomic sites *http://www.tigr.org/tigr-scripts/CMR2/CMRHomePage.spl*

Life in extreme environments. National Space Society *http://www.astrobiology.com/extreme.html*

Life at High Temperature (T. Brock). Many photos of extremophiles, especially at Yellowstone Park *http://www.spaceref.com/redirect.html?id=0&url=www.bact.wisc.edu/bact303/b1*

List of bacterial names with standing in nomenclature (J. P. Euzéby) *http://www.bacterio.cict.fr/index.html*

The following websites provide useful information on extremophiles
http://extremophiles.org/
http://www.extremophiles2002.it/
http://link.springer.de/link/service/journals/00792/index.htm
http://www.archaea.unsw.edu.au/

38

Fimbriae, Pili

Matthew A. Mulvey, Karen W. Dodson, Gabriel E. Soto, and Scott J. Hultgren
Washington University School of Medicine

GLOSSARY

curli A class of thin, irregular, and highly aggregated adhesive surface fibers expressed by *Escherichia coli* and *Salmonella enteritidis*.

periplasmic chaperones A class of proteins localized in the periplasm of gram-negative bacteria that facilitates the folding and assembly of pilus subunits, but which are not components of the final pilus structure.

phase variation The reversible on-and-off switching of a bacterial phenotype, such as pilus expression.

pilin The individual protein subunit of a pilus organelle (also known as a fimbrin). Immature pilins, containing leader signal sequences that direct the transport of pilins across the inner membrane of gram-negative bacteria, are called propilins or prepilins.

usher Oligomeric outer-membrane proteins that serve as assembly platforms for some types of pili. Usher proteins can also form channels through which nascent pili are extruded from bacteria.

Pili, also known as fimbriae, are proteinaceous, filamentous polymeric organelles expressed on the surface of bacteria. They range from a few fractions of a micrometer to greater than 20 μm in length and vary from less than 2 to 11 nm in diameter. Pili are composed of single or multiple types of protein subunits, called pilins or fimbrins, which are typically arranged in a helical fashion.

Pilus architecture varies from thin, twisting thread-like fibers to thick, rigid rods with small axial holes. Thin pili with diameters of 2–3 nm, such as K88 and K99 pili, are often referred to as "fibrillae". Even thinner fibers (<2 nm), which tend to coil up into a fuzzy adhesive mass on the bacterial surface, are referred to as thin aggregative pili or curli. High-resolution electron microscopy of P, type 1, and S pili of *Escherichia coli* and *Haemophilus influenzae* pili has revealed that these structures are composite fibers, consisting of a thick pilus rod attached to a thin, short distally located tip fibrillum. Pili are often expressed peritrichously around individual bacteria, but some, such as type 4 pili, can be localized to one pole of the bacterium.

Pili expressed by gram-negative bacteria have been extensively characterized, and the expression of pili by gram-positive bacteria has also been reported. The numerous types of pili assembled by both gram-negative and gram-positive organisms have been ascribed diverse functions in the adaptation, survival, and spread of both pathogenic and commensal bacteria. Pili can act as receptors for bacteriophage, facilitate DNA uptake and transfer (conjugation), and, in at least type 4 pilus, function in cellular motility. The primary function of most pili, however, is to act as scaffolding for the presentation of specific adhesive moieties. Adhesive pilus subunits (adhesins) are often incorporated as minor components into the tips of pili, but major structural subunits can also function as adhesins. Adhesins can mediate the interaction of bacteria with each other, with inanimate surfaces, and with tissues and cells in susceptible host organisms. The colonization of host tissues by bacterial pathogens typically depends on a stereochemical fit between an adhesin and complementary receptor architecture. Interactions

The Desk Encyclopedia of Microbiology
ISBN: 0-12-621361-5

mediated by adhesive pili can facilitate the formation of bacterial communities such as biofilms and are often critical to the successful colonization of host organisms by both commensal and pathogenic bacteria.

I. HISTORICAL PERSPECTIVE AND CLASSIFICATION OF PILI

Pili were first noted in early electron microscopic investigations as nonflagellar, filamentous appendages of bacteria. In 1955, Duguid designated these appendages "fimbriae" (plural, from Latin for thread or fiber) and correlated their presence with the ability of *E. coli* to agglutinate red blood cells. Ten years later Brinton introduced the term "pilus" (singular, from Latin for hair) to describe the fibrous structures (the F pilus) associated with the conjugative transfer of genetic material between bacteria. Since then "pilus" has become a generic term used to describe all types of nonflagellar filamentous appendages, and it is used interchangeably with the term "fimbria."

Historically, pili have been named and grouped based on phenotypic traits such as adhesive and antigenic properties, distribution among bacterial strains, and microscopic characterizations. In the pioneering work of Duguid and co-workers, pili expressed by different *E. coli* strains were distinguished on the basis of their ability to bind to and agglutinate red blood cells (hemagglutination) in a mannose sensitive (MS) as opposed to a mannose resistant (MR) fashion. Pili mediating MS hemagglutination by *E. coli* were designated type 1 pili and these pili have since been shown to recognize mannose-containing glycoprotein receptors on host eukaryotic cells. Morphologically and functionally homologous type 1 pili are expressed by many different species of Enterobacteriaceae. Despite their similarities, however, type 1 pili expressed by the various members of the Enterobacteriaceae family are often antigenically and genetically divergent within their major structural subunits. In contrast to type 1 pili, most other pili so far identified are either nonhemagglutinating or mediate MR hemagglutination. These pili are very diverse and possess a myriad of architectures and different receptor binding specificities and functions.

Since the discovery and initial characterization of pili in the 1950s, substantial advances have been made in our understanding of the genetics, biochemistry, and structural and functional aspects of these organelles. A vast number of distinct pilus structures have been described and new types of pili continue to be identified. Pili are now known to be encoded by virtually all gram-negative organisms and are some of the best-characterized colonization and virulence factors in bacteria. Here we classify pili that are expressed by gram-negative bacteria into six groups according to the mechanisms by which they are assembled. This classification scheme is not all inclusive, but provides a convenient means for discussing the diverse types of pili, their functions, structures and assembly. Representatives of various pilus types assembled by the various pathways discussed in the following sections are listed in Table 38.1 and electron micrographs of the various pilus types are shown in Fig. 38.1.

TABLE 38.1 Pilus assembly pathways

Assembly pathway	Structure	Assembly gene products[a]	Organism	Disease(s) associated with pilus expression
Chaperone–Usher pathway				
Thick rigid pili	P pili	PapD/PapC	*E. coli*	Pyelonephritis/cystitis
	Prs pili	PrsD/PrsC	*E. coli*	Cystitis
	Type 1 pili	FimC/FimD	*E. coli* *Salmonella* sp. *K. pneumoniae*	Cystitis
	S pili	SfaE/SfaF	*E. coli*	UTI Newborn meningitis
	F1C pili	FocC/FocD	*E. coli*	Cystitis
	H. influenzae fimbriae	HifB/HifC	*H. influenzae*	Otitis media Meningitis
	H. influenzae biogroup aegyptius fimbriae	HafB/HafE	*H. influenzae*	Brazilian purpuric fever
	Type 2 and 3 pili	FimB/FimC	*B. pertussis*	Whooping cough
	MR/P pili	MrpD/MrpC	*P. mirabilis*	Nosocomial UTI
	PMF pili	PmfC/PmfD	*P. mirabilis*	Nosocomial UTI
	Long polar fimbriae	LpfB/LpfC	*S. typhimurium*	Gastroenteritis

(*Continued overleaf*)

TABLE 38.1 *Continued*

Assembly pathway	Structure	Assembly gene products[a]	Organism	Disease(s) associated with pilus expression
	Pef pili	PefD/PefC	*S. typhimurium*	Gastroenteritis
	Ambient-temperature fimbriae	AftB/AftC	*P. mirabilis*	UTI
	987P fimbriae	FasB/FasD	*E. coli*	Diarrhea in piglets
	REPEC fimbriae	RalE/RalD	*E. coli*	Diarrhea in rabbits
Thin flexible pili	K99 pili	FaeE/FaeD	*E. coli*	Neonatal diarrhea in calves, lambs, piglets
	K88 pili	FanE/FanD	*E. coli*	Neonatal diarrhea in piglets
	F17 pili	F17D/F17papC	*E. coli*	Diarrhea
	MR/K pili	MrkB/MrkC	*K. pneumoniae*	Pneumonia
Atypical structures	CS31A capsule-like protein	C1pE/C1pD	*E. coli*	Diarrhea
	Antigen CS6	CssC/CssD	*E. coli*	Diarrhea
	Myf fimbriae	MyfB/MyfC	*Y. enterolitica*	Enterocolitis
	pH 6 antigen	PsaB/PsaC	*Y. pestis*	Plague
	CS3 pili	CS3-1/CS3-2	*E. coli*	Diarrhea
	Envelope antigen F1	Caf1M/Caf1A	*Y. pestis*	Plague
	Non-fimbrial adhesins I	NfaE/NfaC	*E. coli*	UTI Newborn meningitis
	SEF14 fimbriae	SefB/SefC	*S. enteritidis*	Gastroenteritis
	Aggregative adherence fimbriae I	AggD/AggC	*E. coli*	Diarrhea
	AFA-III	AfaB/AfaC	*E. coli*	Pyelonephritis
Alternate chaperone pathway	CS1 pili	CooB/CooC	*E. coli*	Diarrhea
	CS2 pili	CotB/CotC	*E. coli*	Diarrhea
	CS4 pili		*E. coli*	Diarrhea
	CS14 pili		*E. coli*	Diarrhea
	CS17 pili		*E. coli*	Diarrhea
	CS19 pili		*E. coli*	Diarrhea
	CFA/I pili	CfaA/CfaC	*E. coli*	Diarrhea
	Cable type II pili		*B. cepacia*	Opportunistic in cystic fibrosis patients
Type II secretion pathway	Type-4A pili	General secretion apparatus (Main terminal branch)	*Neisseria* sp.	Gonorrhea, meningtitis
			P. aeruginosa	Opportunistic pathogen
			Moraxella sp.	Conjuntivitis, respiratory infections
			D. nodosus	Ovine footrot
		14 to >20 proteins	*E. corrodens*	
			Azoarcus sp.	
	Type-4B pili:	General secretion apparatus (Main terminal branch) 14 to >20 proteins		
	bundle forming pili		*E. coli*	Diarrhea
	longus		*E. coli*	Diarrhea
	CFA/III		*E. coli*	Diarrhea
	R64 pili		*E. coli*	
	toxin co-regulated pili		*V. cholera*	Cholera
Conjugative pilus assembly pathway (Type IV secretion pathway)	F pili (IncF1)	Type IV export apparatus,	*E. coli*	Antibiotic resistance
	IncN, IncP, IncW-encoded pili		*E. coli*	
	T pili	12–16 proteins	*A. tumefaciensi*	Crown gall disease
Extracellular nucleation/ precipitation pathway	Curli	CsgG/CsgE/CsgF	*E. coli* *S. enteritidis*	Sepsis
Type III secretion pathway	Hrp pili	Type III secretion apparatus,	*P. syringae*	Hypersensitive response (in resistant plants)
	EspA pilus-like structures	~20 proteins	*E. coli*	Diarrhea

[a]Chaperone/usher for chaperone–usher and alternate chaperone pathway.

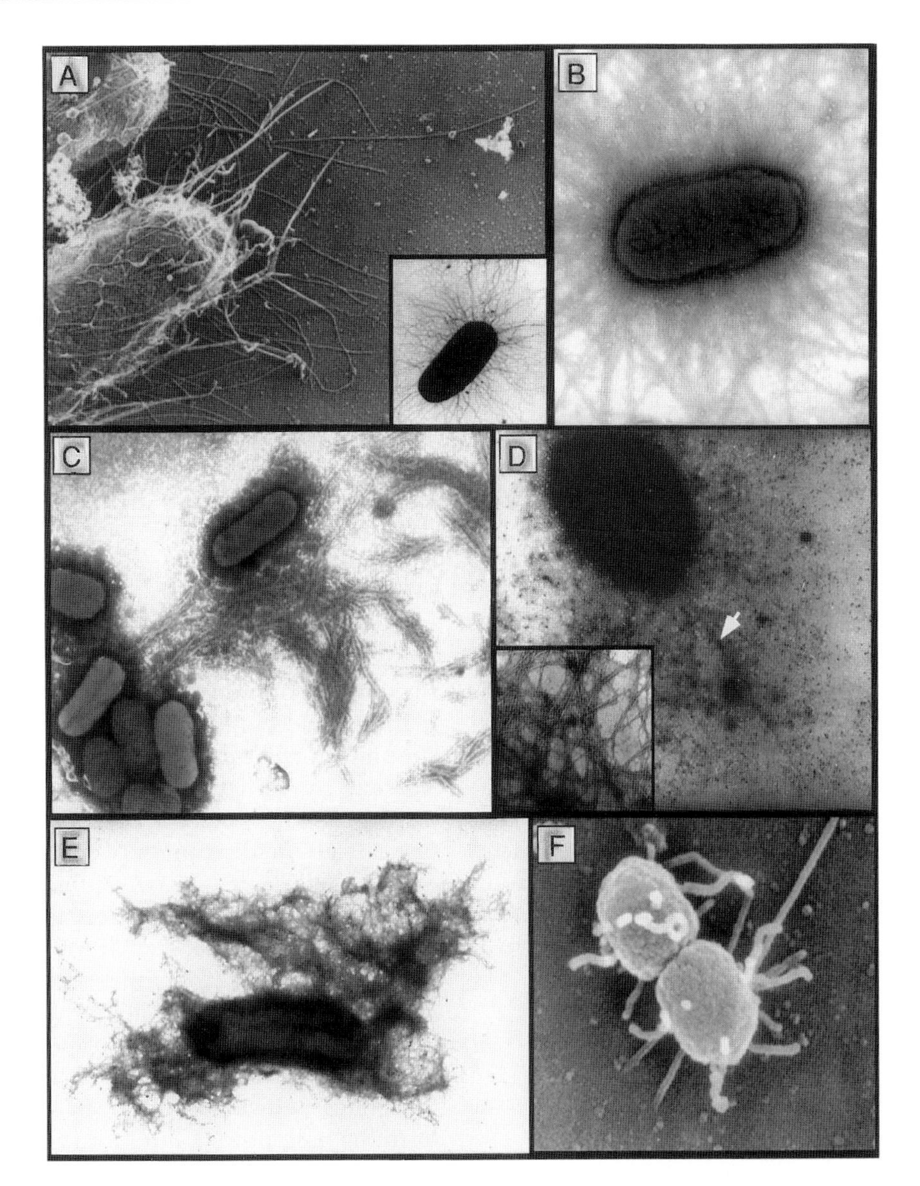

FIGURE 38.1 Montage of various pilus structures. (A) High-resolution transmission electron micrograph of *E. coli* expressing P pili (~2 nm-thick pili), the prototypical structures assembled by the chaperone or usher pathway. Inset shows a typical lower-resolution micrograph of a negatively stained bacterium expressing P pili. (B) CS2 pili (~2 nm-thick pili) assembled by the alternate chaperone pathway in *E. coli*. Photo courtesy of Harry Sakellaris and June R. Scott. (C) The polar type 4B bundle forming pili (~6 nm-wide pili) of enteropathogenic *E. coli* (EPEC). Photo courtesy of Dave Bieber and Gary Schoolnik. (D) Arrow indicates a single, polar T pilus (10 nm-thick structure), the promiscuous conjugative pilus assembled by *A. tumefaciens*. Inset shows a field of purified T-pili. Photos provided by Clarence I. Kado. (E) *E. coli* expressing curli, <2 nm-wide structures assembled by the extracellular nucleation–precipitation pathway. Photo courtesy of Stafan Normark. (F) Scanning electron micrograph of EPEC elaborating ~50 nm-thick bundles of pili containing EspA, a protein exported by a type III secretion pathway. The individual 6- to 8-nm-thick pili making up the bundles are not resolved in this micrograph. Photo provided by S. Knutton (Knutton, S., Rosenshine, I., Pallen, M. J., Nisan, I., Neves, B. C., Bain, C., Wolff, C., Dougan, G., and Frankel, G. (1998). *EMBO J.* **17**, 2166–2176) and reprinted with permission from Oxford University Press.

II. CHAPERONE–USHER PATHWAY

All pilins destined for assembly on the surface of gram-negative bacteria must be translocated across the inner membrane, through the periplasm, and across the outer membrane. To accomplish this, various adhesive organelles in many different bacteria require two specialized assembly proteins, a periplasmic chaperone and an outer membrane usher. Chaperone–usher assembly pathways are involved in the biogenesis of over 30 distinct structures, including composite pili, thin fibrillae, and nonfimbrial adhesins. Here, we focus on the structure and assembly mechanisms of the prototypical P and type 1 pilus chaperone–usher systems.

A. Molecular architecture

P and type 1 pili are both composite structures consisting of a thin fibrillar tip joined end to end to a right-handed helical rod (Fig. 38.2A and B). Chromosomally located gene clusters that are organizationally and functionally homologous encode P and type 1 pili (Fig. 38.2D and E). The P pilus tip is a 2-nm-wide structure composed of a distally located adhesin (PapG), a tip pilin (PapE), and adaptor pilins (PapF and PapK). The PapG adhesin binds to Galα(1-4)Gal moieties present in the globoseries of glycolipids found on the surface of erythrocytes and kidney cells. Consistent with this binding specificity, P pili are major virulence factors associated with pyelonephritis caused by uropathogenic *E. coli*. The minor pilin PapF is thought to join the PapG adhesin to the tip fibrillum, the bulk of which is made up of a polymer of PapE subunits. PapK is thought to terminate the growth of the PapE polymer and to join the tip structure to the rod. The pilus rod is composed of multiple PapA subunits joined end to end and then coiled into a right-handed 6.8-nm-thick helical rod having a pitch distance of 24.9 Å and 3.28 subunits per turn. The rod is terminated by a minor subunit, PapH, which may serve to anchor the pilus in the membrane.

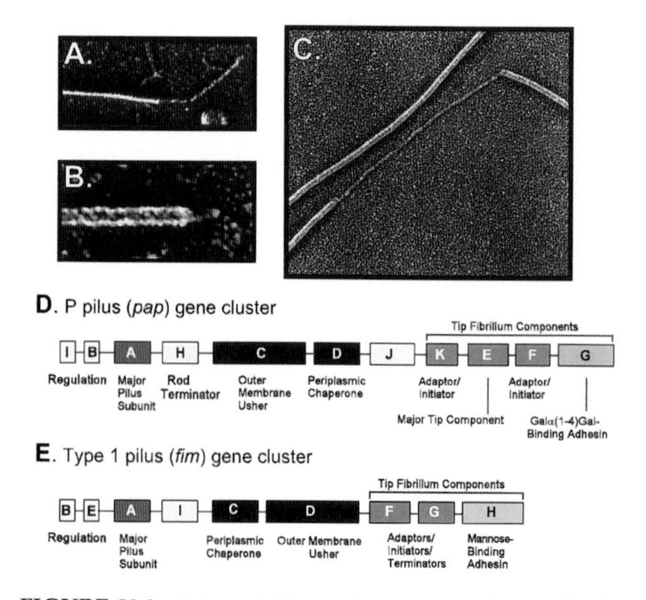

FIGURE 38.2 High-resolution electron micrographs showing the pilus rod and tip fibrillum structures of (A) P and (B) type 1 pili. Unraveling of a portion of a P pilus rod into a linear fiber is shown in (C). The images shown in (A), (B), and (C) are at different magnifications. The P (*pap*) and type 1 (*fim*) gene clusters are depicted in (D) and (E), respectively. These gene clusters share organizational as well as functional homologies.

Similar to the P pilus structure, the type 1 pilus has a short, 3-nm-wide fibrillar tip made up of the mannose-binding adhesin, FimH, and two additional pilins, FimG and FimF. The FimH adhesin mediates attachment to mannosylated receptors expressed on a wide variety of cell types and has been shown to be a significant virulence determinant for the development of cystitis. The type 1 tip fibrillum is joined to a rod composed predominantly of FimA subunits arranged in a 6- to 7-nm diameter helix with a pitch distance of 23.1 Å and 3.125 subunits per turn. Both type 1 and P pilus rods have central axial holes with diameters of 2–2.5 and 1.5 Å, respectively. Despite the architectural similarities, type 1 pili appear to be more rigid and prone to breaking than P pili. Some reports have argued that, unlike the P pilus system in which the tip subunits are thought to be located only within the tip, some of the type 1 tip subunits may also be occasionally intercalated within the rod structure.

In both P and type 1 pili, the major pilin subunits making up the rods are organized in a head-to-tail manner. Additional quaternary interactions between subunits in adjacent turns of the helical rod appear to stabilize the structure and may help drive the outward growth of the organelle during pilus assembly (see later). The disruption of these latter interactions by mechanical stress or by incubation in 50% glycerol can cause the pilus rod to reversibly unwind into a 2-nm-thick linear fiber similar in appearance to the tip fibrillum (Fig. 38.2C). Bullitt and Makowski (1995) have proposed that the ability of the pilus rods to unwind allows them to support tension over a broader range of lengths. This may help P and type 1 pili better withstand stress, such as shearing forces from the bulk flow of fluid through the urinary tract, without breaking.

In addition to composite structures exemplified by P and type 1 pili, chaperone–usher pathways also mediate the assembly of thin fibrillae such as K88 and K99 pili and nonfimbrial adhesins. K88 and K99 pili are 2- to 4-nm-thick fibers that mediate adherence to receptors on intestinal cells. They are significant virulence factors expressed by enterotoxigenic *E. coli* (ETEC) strains that cause diarrheal diseases in livestock. These pili were given the "K" designation after being mistakenly identified as K antigens in *E. coli*. In contrast to P and type 1 pili, the adhesive properties of K88 and K99 pili are associated with the major pilus subunits. The receptor-binding epitopes on the individual major pilus subunits are exposed on the pilus surface and available for multiple interactions with host tissue. In general, pili with adhesive major subunits, such as K88 and K99 pili, are thin flexible fibrillar structures. In comparison, pili with specialized adhesive tip structures, such as P and type 1 pili, are relatively rigid and rod-like.

B. Assembly model

The assembly of P pili by the chaperone–usher pathway is the best understood of any pilus assembly pathway. PapD is the periplasmic chaperone and PapC is the outer-membrane usher for the P pilus system. These proteins are prototypical representatives of the periplasmic chaperone and outer-membrane usher protein families. Figure 38.3 presents the current model for pilus assembly by the chaperone–usher pathway, as depicted for P pili.

1. *Periplasmic chaperones*

The PapD chaperone, the PapC usher, and all of the P pilus structural subunits have typical signal sequences recognized by the *sec* (general secretion) system. The signal sequences are short, mostly hydrophobic amino-terminal motifs that tag proteins for transport across the inner membrane by the *sec* system. This system includes several inner-membrane proteins (SecD to SecF, SecY), a cytoplasmic chaperone (SecB) that binds to presecretory target proteins, a cytoplasmic membrane-associated AT-Pase (SecA) that provides energy for transport, and a periplasmic signal peptidase. As the P pilus structural subunits emerge from the *sec* translocation machinery into the periplasm, PapD binds to each subunit, facilitating its release from the inner membrane. Each subunit forms an assembly competent, one-to-one complex with PapD. Proper folding of the subunits requires PapD and involves the action of the periplasmic disulfide bond isomerase DsbA. In the absence of PapD, subunits misfold, aggregate, and are subsequently degraded by the periplasmic protease DegP. The misfolding of

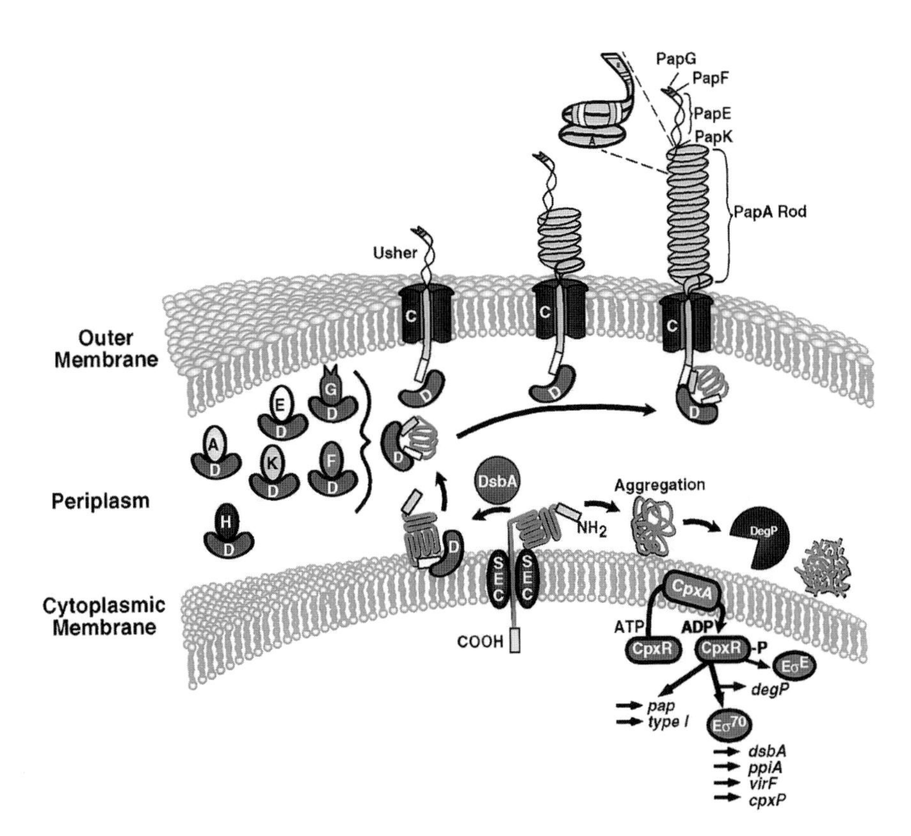

FIGURE 38.3 Model of P pilus assembly by the chaperone–usher pathway. Structural subunits for the pilus tip (PapG, PapF, PapE, and PapK) and for the pilus rod (PapA) are translocated across the inner membrane of *E. coli* via the *sec* system. On the periplasmic side of the inner membrane they interact with the chaperone PapD, which facilitates the folding and release of the subunits from the inner membrane. DsbA is required for the proper folding of both the subunits and the PapD chaperone. In the absence of the chaperone, the subunits aggregate and are degraded by the periplasmic protease DegP. The buildup of unchaperoned subunits results in the activation of the Cpx system, increasing the production of periplasmic protein-folding and -degradation proteins. Subunit–chaperone complexes are targeted to the outer membrane usher, PapC, where the chaperone is released and subunit–subunit interactions occur. The PapC usher forms a 2-nm-wide pore through which the assembled pilus structure is extruded as a linear fiber across the outer membrane. Once on the exterior of the cell, the linear PapA polymer forms a thick right-handed helical rod that is unable to slip back through the usher pore. The formation of the coiled PapA rod is thought to help drive the outward growth of the pilus.

P pilus subunits is sensed by the CpxA–CpxR two-component system in which CpxA is an inner membrane-bound sensor or kinase and CpxR is a DNA-binding response regulator (Jones, 1997). Activation of the Cpx system alters the expression of a variety of genes and may help regulate pilus biogenesis.

The three-dimensional crystal structures of both the PapD and FimC chaperones have been solved. Both chaperones consist of two Ig (immunoglobulin)-like domains oriented into a boomerang shape such that a subunit-binding cleft is created between the two domains. A conserved internal salt bridge is thought to maintain the two domains of the chaperone in the appropriate orientation. Using genetics, biochemistry, and crystallography, PapD was found to interact with pilus subunits, in part, by binding to a highly conserved motif present at the carboxyl terminus of all subunits assembled by PapD-like chaperones. The finer details of how PapD-like chaperones interact with pilus subunits were unveiled by the recent determination of the crystal structures of the PapD–PapK and the FimC–FimH chaperone–subunit complexes (Sauer, 1999, and Choudhury, 1999) (Fig. 38.4). This work demonstrated that the PapD and FimC chaperones both make similar interactions with their respective subunits. Only the PapD–PapK structure is considered here. PapK has a single domain comprised of an Ig fold that lacks the seventh (carboxy-terminal) β-strand that is present in canonical Ig folds. The absence of this strand produces a deep groove along the surface of the pilin subunit and exposes its hydrophobic core. The carboxy-terminal F strand and the A2 strand of PapK form the groove. The PapD chaperone contributes its G_1 β-strand to complete the Ig fold of the PapK subunit by occupying the groove, an interaction termed donor strand complementation. This interaction shields the hydrophobic core of PapK and stabilizes immature pilus subunits within the periplasm. Similar interactions are thought to have a central role in the maturation of subunits assembled by a variety of different chaperone–usher pathways. The residues that make up the carboxy-terminal groove formed by subunits and bound by PapD-like chaperones have been shown by mutagenesis studies to be involved in subunit–subunit interactions within the final pilus structure. Thus, in addition to stabilizing immature pilus subunits, the donor strand complementation interaction also caps one of the interactive surfaces of the

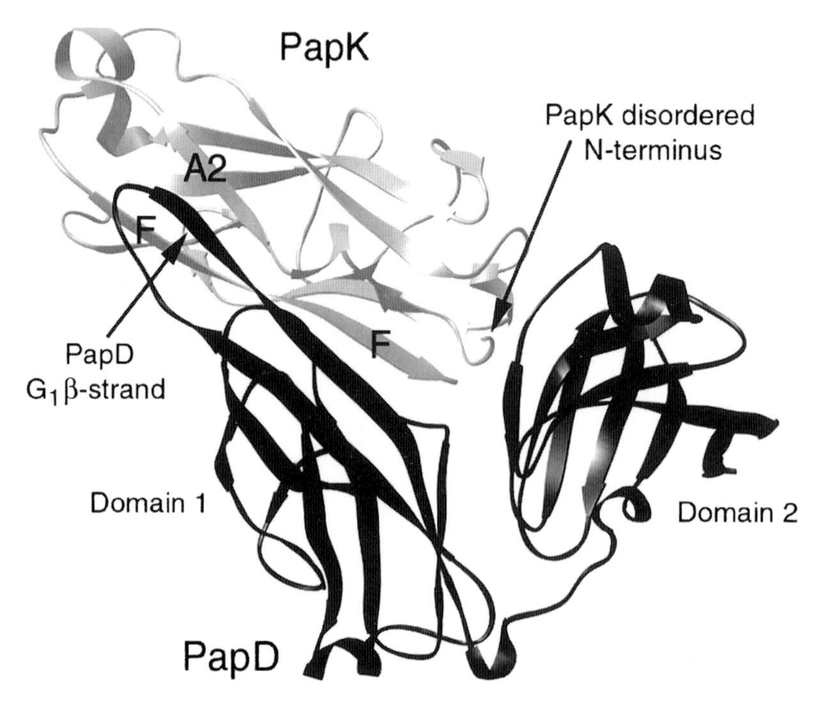

FIGURE 38.4 Ribbon model based on the crystal structure of the PapD periplasmic chaperone complexed with the PapK pilus subunit. PapD is black and the single domain of PapK is gray. The G_1 β-strand in domain 1 of PapD completes the Ig fold of the pilin, occupying the groove formed between the A2 and F strands of PapK. This interaction has been termed donor strand complementation. The eight amino-terminal amino acids of PapK are disordered in the structure. These residues have been implicated in mediating subunit–subunit interactions within the mature pilus organelle. During pilus biogenesis, it is proposed that the amino-terminal strand of a pilin can displace the G_1 β-strand of the chaperone and insert into the carboxy-terminal groove of the neighboring subunit. The mature pilus would thus consist of a linear array of canonical Ig domains, each of whose fold is completed by a strand from the neighboring subunit.

subunit and prevents premature oligomerization and aggregation of pilus subunits within the periplasm.

2. *Outer-membrane ushers*

After being formed in the periplasm, chaperone–subunit complexes are targeted to the outer-membrane usher where the chaperone is released, exposing interactive surfaces on the subunits that facilitate their assembly into the pilus. Studies in the P and type 1 pilus systems have demonstrated that the adhesin–chaperone complexes, PapDG or FimCH, bind tightest and fastest to the usher and that the adhesins are the first subunits assembled into the pilus. The binding of the chaperone–adhesin complex induces a conformational change in the usher, possibly priming it for pilus assembly. Additional subunits are incorporated into the pilus depending, in part, on the kinetics with which they are partitioned to the usher in complex with the chaperone. Conserved amino-terminal regions, in addition to the conserved carboxy-terminal motif of the pilus subunits, mediate subunit–subunit interactions within the mature pilus. Differences in the complementary surfaces in these conserved regions from one subunit to another may help dictate which of the subunits can be joined to one another during pilus assembly. Thus, the order of the subunits within the final pilus structure is determined by the specific contacts made between the various pilus subunits and also by the differential affinities of the various chaperone–subunit complexes for the usher.

In addition to acting as an assembly platform for the growing pilus, the usher protein appears to have additional roles in pilus biogenesis. High-resolution electron microscopy revealed that the PapC usher is assembled into a 15-nm diameter ring-shaped complex with a 2-nm wide central pore (Thanassi, 1998). PapC and other usher family members are thought to have a predominantly β-sheet secondary structure, typical of outer-membrane pore-forming proteins, and they are predicted to present large regions to the periplasm for interaction with chaperone–subunit complexes. After dissociating from the chaperone at the usher, subunits are incorporated into a growing pilus structure that is predicted to be extruded as a 1-subunit-thick linear fiber through the central pore of the usher complex. The packaging of the linear pilus fiber into a thicker helical rod on the outside surface of the bacterium may provide a driving force for the translocation of the pilus across the outer membrane, possibly acting as a sort of ratcheting mechanism to force the pilus to grow outward. Combined with the targeting affinities of the chaperone–subunit complexes for the usher and the binding specificities of the subunits for each other, this may provide all the energy and specificity needed for the ordered assembly and translocation of pili across the outer membrane.

III. ALTERNATE CHAPERONE PATHWAY

A variation of the chaperone–usher pilus assembly pathway has been identified in strains of ETEC. These bacteria are major pathogens associated with diarrheal diseases of travelers, infants, and young children. ETEC strains produce several types of uniquely assembled adhesive pili that are considered to be important mediators of bacterial colonization of the intestine. The best studied of these pili is CS1, which appears to be composed predominantly of a major subunit, CooA, with a distally located minor component, CooD. Several CS1-like pili have been identified and include CS2, CS4, CS14, CS17, CS19, and CFA/I pili expressed by various ETEC strains and the cable type II pili of *Burkholderia cepacia*, an opportunistic pathogen of cystic fibrosis patients. Four linked genes, *CooA*, *CooB*, *CooC*, and *CooD*, are the only specific genes required for the synthesis of functional CS1 pili. Homologous genes required for the production of CS2 and CFA/I pili have also been cloned and sequenced. Electron-microscopic examination reveals that the CS1-like pili are architecturally similar to P and type 1 pili assembled by the chaperone–usher pathway (Fig. 38.1B), although none of the proteins involved in the biogenesis of CS1-like pili have any significant sequence homologies to those of any other pilus system.

The assembly of CS1-like pili depends on a specialized set of periplasmic chaperones that are distinct from those of the chaperone–usher pathway described previously. Therefore, we refer to this mode of pilus assembly as the alternate chaperone pathway. In the case of CS1 pili, the chaperone CooB binds to and stabilizes the major and minor pilin subunits, CooA and CooD, which enter into the periplasm in a *sec*-dependent fashion (Fig. 38.5). Both CooA and CooD share a conserved sequence motif near their carboxy-termini that may function as a chaperone-recognition motif. One of the functions of CooB appears to be the delivery of the pilin subunits to an outer-membrane protein, CooC, which may function as a channel, or usher, for the assembly of pilus fibers. In addition to the pilin subunits, CooB also binds to and stabilizes CooC in the outer membrane. In the absence of the CooB chaperone, CooC and the pilin subunits are degraded. Although less well defined, the assembly of CS1 and related structures appears

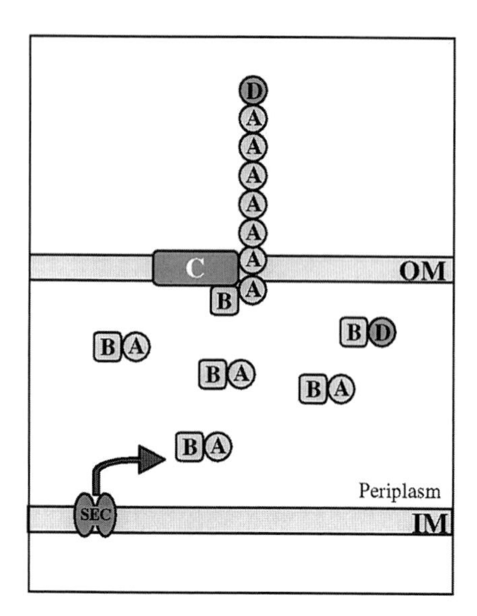

FIGURE 38.5 Assembly of CS1 pili from *E. coli* via the alternate chaperone pathway. The CooB chaperone forms periplasmic complexes with the main components of the pilus, CooA and CooD. It also appears to bind and stabilize the outer-membrane protein CooC in the absence of subunits. CooC appears to function as an outer-membrane (OM) channel for passage of the pilin fiber IM indicates innermembrane.

similar in many respects to the assembly of pili by the classic chaperone–usher pathway. Because CS1-like pili do not appear to be related to those assembled by the chaperone–usher pathway, it has been suggested that these two pilus assembly systems arose independently through convergent evolution.

IV. TYPE II SECRETION PATHWAY FOR TYPE 4 PILUS ASSEMBLY

Type 4 pili are multifunctional structures expressed by a wide diversity of bacterial pathogens. These include *Pseudomonas aeruginosa, Neisseria gonorrhoeae* and *N. meningitidis, Moraxella* species, *Azoarcus* species, *Dichelobactor nodus*, and many other species classified in these and other genera. Type 4 pili are significant colonization factors and have been shown to mediate bacterial interactions with animal, plant, and fungal cells. In addition, these pili can modulate target-cell specificity, function in DNA uptake and biofilm formation, and act as receptors for bacteriophage. Type 4 pili are also associated with a flagella-independent form of bacterial locomotion, called twitching motility, which allows for the lateral spread of bacteria across a surface.

Type 4 pili are 6-nm-wide structures typically assembled at one pole of the bacterium. They can extend up to several micrometers in length and are made up of, primarily if not completely, a small subunit usually in the range of 145–160 amino acids. These subunits have distinctive features, including a short (6–7 amino acids), positively charged leader sequence that is cleaved during assembly, *N*-methylphenylalanine as the first residue of the mature subunit, and a highly conserved, hydrophobic amino-terminal domain. The adhesive properties of type 4 pili are, in general, determined by the major pilus subunit. Additional minor components, however, may associate with these pili and alter their binding specificities. In the case of *Neisseria*, a tip-localized adhesin, PilC1, appears to mediate bacterial adherence to epithelial cells.

Recently, a second class of type 4 pili, referred to as class B or type 4B, has been defined. Type 4B pili were initially characterized in enteric pathogens and include the toxin-coregulated pilus (TCP) of *Vibrio cholera*, the bundle-forming pilus (BFP) of enteropathogenic *E. coli* (EPEC), and the longus and CFA/III pili of ETEC. Compared to the typical type 4 pilins (referred to as class A or type 4A), the known type 4B pilins are somewhat larger and have a longer (13–30 amino acids) leader sequence. Also, in place of *N*-methyl-phenylalanine as the first amino acid in the mature pilus subunit, type 4B subunits have other methylated residues such as *N*-methyl-methionine for TCP and *N*-methyl-leucine for BFP. TCP, BFP, and longus pili form large polar bundles over 15 μm in length (Fig. 38.1C). In contrast, CFA/III pili are 1–10 μm long and are peritrichously expressed. The number of pili classified as type 4B is increasing and now includes the R64 thin pilus, an organelle involved in bacterial conjugation.

Parge and coworkers solved the crystal structure of the type 4A pilin subunit (PilE) from *N. gonorrhoeae* in 1995. This work greatly advanced understanding of the structure, function, and biogenesis of type 4 pili. PilE contains 158 amino acids and was determined to have an overall ladle shape, being made up of an α–β roll with a long hydrophobic amino-terminal α-helical spine (residues 2–54) (Fig. 38.6). All type 4 pilins (types 4A and 4B) are predicted to have a fairly similar structure. The carboxy-terminal domain of type 4 pilins possesses hypervariable regions that affect the binding specificities and antigenicity of type 4 pili. In PilE, these hypervariable regions include a sugar loop (residues 55–77) with an O-linked disaccharide at position Ser-63 and a disulfide-containing region (residues 121–158), which, despite having a hypervariable nature, adopts a regular β-hairpin structure (β_5–β_6) followed by an extended carboxy-terminal tail. A disulfide-containing carboxy-terminal hypervariable region is common among the type 4 pilins. The remainder of PilE was

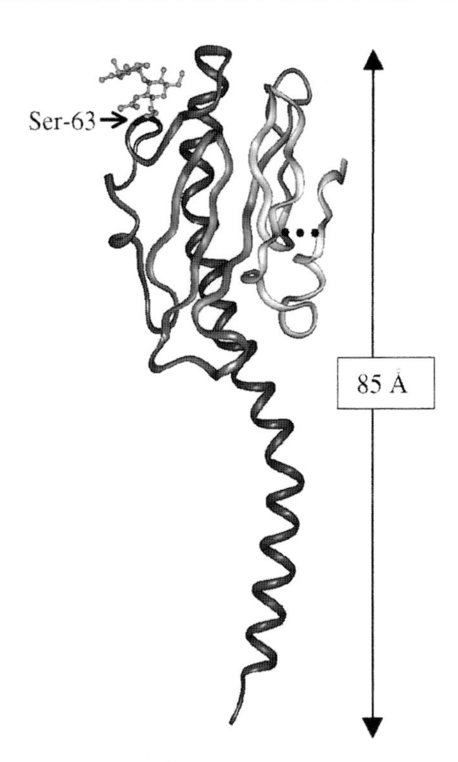

FIGURE 38.6 Ribbon model of the type 4A pilin PilE from *N. gonorrhoeae*. Secondary structural elements include a hydrophobic, conserved amino-terminal α-helical spine connected to a variable domain containing an extended sugar loop with an O-linked disaccharide at position Ser-63 and a disulfide-containing carboxy-terminal region. The disulfide bridge is indicated by a dotted line.

shown to consist of two β-hairpins forming a four-stranded antiparallel β-sheet (residues 78–93 and 103–122) with a connecting β_2–β_3 loop region (residues 94–102).

Through systematic modeling, Parge *et al.* showed that PilE was probably assembled into pili as monomers arranged in a helix with about five PilE subunits per turn and a pitch distance of approximately 41 Å. PilE subunits are predicted to be packed into pili as a three-layer assemblage consisting of an inner core of coiled conserved hydrophobic α-helices surrounded by β-sheets and an outermost layer composed of the disaccharide and hypervariable regions. Hydrophobic packing of the inner core of α-helices along with the flexibility of these helices may permit type 4 pili to bend and adopt twisted, bundled conformations, as seen in Fig. 38.1C. Hydrogen bonds throughout the middle layer of β-sheets may provide much of the mechanical stability for the pilus. The hypervariable outermost layer is not an integral part of the pilus structure and associates with the middle layer of β-sheets through only a few conserved interactions. Thus, the outermost region can be structurally pliant and accommodate extreme amino acid changes that can lead to antigenic variation and altered binding specificities without disrupting the assembly of the pilus. The antigenic characteristics of type 4 pili synthesized by *N. gonorrhoeae* can be modified extensively by a remarkable mechanism. This pathogen encodes more than 15 distinct silent pilin genes, termed PilS, that lack the invariant amino-terminal domain present in PilE. By recombination of silent PilS genes with the PilE locus, a single neisserial strain can theoretically express greater than 10 million PilE variants.

The biogenesis of type 4 pili is substantially more complicated than pilus assembly by the chaperone–usher or alternate chaperone pathways. Type 4 pilus assembly requires the expression of a myriad of genes that are usually located in various unlinked regions on the chromosome. Exceptions include TCP, BFP, and the R64 thin pilus, which are currently the only type 4 pili for which the majority of the genes required for pilus biogenesis are located within a single genetic locus. Although chromosomally located genes encode most type 4A pili, all known type 4B pili, with the exception of TCP, are encoded by plasmids. The number of genes essential for type 4 pilus biogenesis and function ranges from 14 (for pili such as BFP) to over 20 (for structures such as the type 4A pili of *N. gonorrhoeae*). In *P. aeruginosa*, it is estimated that about 0.5% of the bacterium's genome is involved in the synthesis and function of type 4 pili. Among the various bacterial species expressing type 4 pili, the genes encoding the type 4 pilus structural components are similar, whereas the regulatory components surrounding them are typically less conserved.

Several gene products are currently known to be central to the assembly of type 4 pili. These include a prepilin peptidase that cleaves off the leader peptide from nascent pilin subunits; a polytopic inner membrane protein that may act as a platform for pilus assembly; a hydrophilic nucleotide-binding protein located in the cytoplasm or associated with the cytoplasmic face of the inner membrane, which may provide energy for pilus assembly; and an outer-membrane protein complex that forms a pore for passage of the pilus to the exterior of the bacterium. Many of the components involved in type 4 pilus assembly share homology with proteins that are part of DNA uptake and protein secretion systems, collectively known as the main terminal branch of the general secretory (*sec*-dependent) pathway, or type II secretion. These secretion pathways encode proteins with type 4 pilin-like characteristics and other proteins with homology to type 4 prepilin peptidases and outer-membrane pore-forming proteins. Whether type II secretion systems can assemble pili or piluslike structures is not known.

Meyer and colleagues have described a model for type 4 pilus assembly in *N. gonorrhoea* (Fig. 38.7). The PilE propilin subunits are transported into the periplasm by the *sec* translocation machinery. Following translocation, the propilin subunits remain anchored in the inner membrane by their hydrophobic amino-terminal α-helical domains, with their hydrophilic carboxy-terminal heads oriented toward the periplasm. The removal of the positively charged PilE propilin leader sequence by the PilD signal peptidase drives the hydrophobic stems of the PilE subunits to associate to form a pilus. An inner-membrane assembly complex made up of several proteins including PilD, PilF, PilG, and PilT aids this process. The assembled pilus penetrates the outer membrane through a gated pore formed by the multimeric complex Omc. The PilC adhesin associated with the tips of type 4 pili produced by *N. gonorrhoea* may facilitate passage of the nascent pili through the Omc pore. One implication of this assembly model is that the amino-terminal region of PilE resides in a continuous hydrophobic environment during both inner-membrane transport and pilus assembly. This may allow polymerization and, interestingly, depolymerization of the pilus to proceed with only minimal energy requirements. Continued extension and retraction of type 4 pili by rounds of polymerization and depolymerization reactions are proposed as the basis of twitching motility, one of the functions of type 4 pili. This process could be controlled by nucleotide-binding proteins such as PilT, PilB, and PilU associated with the inner-membrane assembly complex. The capacity of type 4 pili to depolymerize may also provide a means for transforming DNA, which could potentially interact with the type 4 pilus fiber, to enter into the bacterial cell.

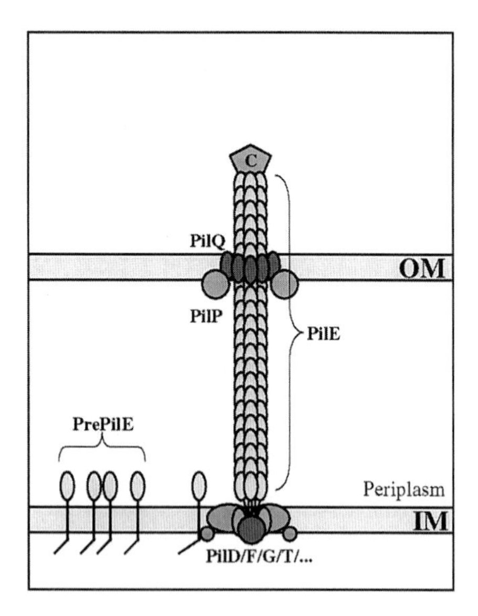

FIGURE 38.7 Model of type 4 pilus assembly by *N. gonorrhoeae*. The PilE prepilin is translocated into the periplasm aided by the *sec* machinery. PilE is processed by the PilD signal peptidase, which cleaves the positively charged leader sequence from the amino-terminus of the pilin subunit. An inner-membrane assembly complex then assembles the mature PilE subunit into a pilus fiber. PilQ mediates translocation of the pilus through the outer membrane, possibly with the assistance of other factors, such as PilP. The PilC adhesin, which appears to be incorporated at the tip of the growing pilus fiber, also seems to be required for translocation of the pilus through the outer membrane.

V. CONJUGATIVE PILUS ASSEMBLY PATHWAY

In gram-negative bacteria, certain pili, collectively known as conjugative pili, facilitate the interbacterial transfer of DNA. These pili allow donor and recipient bacteria to make specific and stable intercellular contacts before DNA transfer is initiated. In some cases, conjugative pili may also form the conduits for intercellular DNA transfer. Horizontal gene transfer, or conjugation, mediated by conjugative pili is inextricably associated with the spread of antibiotic resistance among bacterial pathogens. Conjugative pili are generally encoded by self-transmissible plasmids that are capable of passing a copy of their genes to a recipient bacterium. Closely related plasmids, with similar replication control systems, are unable to coexist in the same cell. This property has been termed incompatibility, and provides the primary basis for cataloging conjugal plasmids and the pili that they encode. Thus far, in *E. coli* alone, over 25 incompatibility groups made up of well over 100 distinct plasmids have been defined. Plasmids of a particular incompatibility group usually encode conjugative pili with similar antigenic properties, sensitivities to pilus-specific phages, and morphologies.

Among the multitude of known incompatibility groups, three morphologically and functionally distinct types of conjugative pili have been defined: rigid, thick flexible, and thin flexible pili. Rigid conjugative pili are 8- to 11-nm-wide structures that are usually specified by conjugal DNA transfer systems that function well only on solid surfaces. Thick flexible pili, on the other hand, are 8- to 11-nm-wide structures that typically, but not always, promote conjugation on solid surfaces and in liquid media equally well. Conjugal DNA transfer promoted by rigid or thick flexible pili can be enhanced, in some cases, by the presence of thin flexible pili. These pili are similar in appearance to type 4 pili and at least one member of the thin flexible pilus group (the R64 thin pilus) has been identified at the molecular level as a type 4B pilus (described previously). Thin flexible pili appear to function primarily in the stabilization of bacterial mating pairs, increasing the rate of DNA transfer. Conjugation does not occur in the presence of thin flexible pili alone, or in the absence of rigid or thick flexible pili.

The most thoroughly studied conjugative pilus is the F pilus encoded by the self-transmissible, broad host range F (fertility) plasmid, a member of the F1 incompatibility (IncF1) group of plasmids borne by *E. coli*. The F pilus system is prototypical for numerous other conjugation systems and F-pilus biogenesis is distinct from type 4 and other pilus assembly pathways. F pili are 8-nm-thick, flexible helical filaments composed primarily, if not completely, of repeating 7.2 kDa (70-amino-acid) TraA pilin subunits. Donor (F^+) cells typically express one to three F pili that are usually 1–2 μm long. Each F pilus possesses a 2-nm-wide central channel that is lined by basic hydrophilic residues, which could potentially interact with negatively charged DNA or RNA molecules during conjugation. TraA is organized into pentameric, doughnut-like disks that are stacked in the pilus such that successive disks are translated 1.28 nm along the pilus axis and rotated 28.8° with respect to the lower disk. The TraA pilin has two hydrophobic domains located toward the center and at the carboxy-terminus of the pilin. The hydrophobic domains are thought to extend as antiparallel α-helices from the central axis to the periphery of the pilus shaft. These domains are separated by a short basic region that appears to form the hydrophilic wall of the central channel of the pilus. The amino-terminal domain of TraA is predicted to face the exterior of the pilus. However, this domain is antigenically masked when the amino-terminal residue of TraA is acetylated during maturation of the pilin (see later). This modification is common among all known F-like pilins and appears to cause the amino-terminal domain to be tucked back into or along the pilus shaft. Acetylation is not essential for F-pilus assembly or function, but does help prevent aggregation of F-like pili and affects the phage-binding characteristics of these organelles. Phage are also known to recognize the carboxy-terminal hydrophobic domain of TraA. Although masked within the pilus shaft, the acetylated amino-terminal domain of TraA appears to be exposed in unassembled pilin subunits and at the distal tips of pili. The F pilus tip is believed to initiate contact between donor and recipient cells during conjugation. Alterations in the amino-terminal sequence of TraA provide the primary basis for the antigenic diversity observed among various F-like pili.

At least 16 gene products encoded by the F plasmid are involved in F pilus assembly and an additional 20 or more are needed for conjugation. Two gene products, TraQ and TraX, mediate the processing of the TraA pilin to its mature form. TraA is synthesized as a 12.8-kDa (121-amino-acid) cytoplasmic propilin that is translocated across the inner membrane where it is proteolytically processed by host signal peptidase I to yield the 7.2-kDa pilin form. TraQ, an innermembrane protein, facilitates the translocation process and may help position the TraA propilin for processing into mature pilin. In the absence of TraQ, the translocation of TraA is disrupted and most of the pilin is degraded. After processing, the amino-terminal residue (alanine) of TraA is acetylated by TraX, a polytopic inner-membrane protein. Whereas TraQ and TraX are involved in the maturation of the TraA pilin, 13 additional gene products (TraL, TraB, TraE, TraK, TraV, TraC, TraW, TraU, TraF, TraH, TraG, TrbC, and TrbI) affect the assembly of TraA into the pilus filament. Most of these proteins appear to associate with either the inner or outer bacterial membrane and may constitute a pilus assembly complex that spans the periplasmic space.

The exact mechanism by which TraA is assembled into pili is not defined. The mature TraA pilin accumulates in the inner membrane with its amino-terminus facing the periplasm. Both hydrophobic domains of TraA span the inner membrane, with the hydrophilic region of TraA connecting them on the cytoplasmic side. Small clusters of TraA also accumulate in the outer membrane and these may function as intermediates in F pilus assembly and disassembly. Large portions of the TraA sequence have the propensity to assume both β-sheet and α-helical structures, although the α-helical conformation is known to predominate in assembled pili. Frost and co-workers (1984, 1993) have suggested that a shift between β-sheet and α-helical conformations may drive pilus assembly and disassembly. F pilus assembly is energy dependent and the depletion of ATP levels by respiratory poisons such as cyanide results in F pilus depolymerization and retraction. It has been postulated that TraA is normally cycled between pili and periplasmic and inner-membrane pools by rounds of pilus outgrowth and subsequent retraction. During conjugation, F pilus retraction is thought to serve a stabilizing function by shortening the distance between bacterial mating pairs and allowing for more intimate contact.

Several components of the F pilus assembly machinery share significant homology with proteins encoded by other conjugative systems. These include proteins specified by broad host range plasmids in other incompatibility groups (such as IncN, IncP, and IncW) and many of the proteins encoded by the Ti plasmid-specific *vir* genes of the plant pathogen *Agrobacterium tumefaciens*. These bacteria elaborate 10-nm-wide promiscuous conjugative pili, called T pili (Fig. 38.1D), which direct the interkingdom transfer of a specific genetic element, known as T-DNA, into plant and yeast cells. The introduction of T-DNA into plant cells induces plant tumor formation. T pilus assembly by *A. tumefaciens* requires the expression of at least 12 *vir* gene products encoded by the Ti plasmid. VirB2 is the major, and possibly only, component of the T pilus and it is predicted to be structurally homologous to the F pilus subunit TraA.

Other than possibly stabilizing donor-recipient interactions, it is not clear how F and T pili or any pilus structures function in conjugative DNA-transfer processes. However, substantial evidence exists, at least in the case of F pili, suggesting that pilus components or the pilus itself can serve as a specialized channel for the transmission of DNA and any accompanying pilot proteins across the donor and possibly the recipient cell membranes. In light of this possibility, it is interesting to note that many components of the conjugative pilus systems encoded by the IncF, IncN, IncP, and IncW plasmids and by the *vir* genes of *A. tumefaciens* are similar to the Ptl proteins responsible for the export of the multiple subunit toxin of *Bordetella pertussis*. Furthermore, these secretion systems seem to be distantly related to transport systems used by *Legionella pneumophila* and *Helicobacter pylori* to inject virulence factors into host eukaryotic cells. Conjugative pilus systems such as those encoding F and T pili thus appear to be representative of a larger family of macromolecular transport systems. These type IV secretion systems (not to be confused with the secretion of autotransporters, such as IgA proteases, which is also known as type IV secretion) represent a major pathway for the transfer of both nucleic acid and proteins between cells (Zupan *et al.*, 1998). The understanding of how conjugative pili help mediate the intercellular transfer of macromolecules remains a significant challenge.

VI. EXTRACELLULAR NUCLEATION–PRECIPITATION PATHWAY

Many strains of *E. coli* and *Salmonella enteritidis* produce a class of thin (<2 nm), irregular, and highly aggregated surface fibers known as curli (Fig. 38.1E). These distinct organelles mediate binding to a variety of host proteins, including plasminogen, fibronectin, and human contact-phase proteins. They are also involved in bacterial colonization of inert surfaces and have been implicated in biofilm formation. Curli are highly stable structures and extreme chemical treatment is required to depolymerize them. The major component of *E. coli* curli is a 15.3-kDa protein known as CsgA, which shares over 86% primary sequence similarity to its counterpart in *S. enteritidis*, AgfA.

The formation of curli represents a departure from the other modes of pilus assembly. Whereas structures exemplified by P, CS1, type 4, and F pili are assembled from the base, curli formation occurs on the outer surface of the bacterium by the precipitation of secreted soluble pilin subunits into thin fibers (Fig. 38.8). In *E. coli*, the products of two divergently transcribed operons are required for curli assembly. The *csgBA* operon encodes the primary fiber-forming subunit, CsgA, which is secreted as a soluble protein directly into the extracellular environment. The second protein encoded by the *csgBA* operon, CsgB, is proposed as inducing polymerization of CsgA at the cell surface. In support of this model, it has been demonstrated that a $CsgA^+CsgB^-$ donor strain can secrete CsgA subunits that can be assembled into curli on the surface of a CsgA $CsgB^+$ recipient strain. Furthermore, CsgB appears to be interspersed along the length of the curli fiber, where it may initiate branching of the curli structures. In the absence of

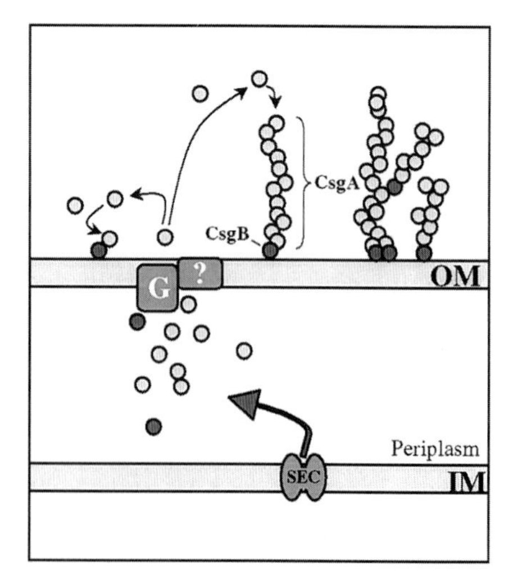

FIGURE 38.8 Model of curli assembly by the extracellular nucleation–precipitaion pathway. CsgA, the main component of curli from *E. coli*, is secreted across the outer membrane. Surface-localized CsgB serves to nucleate CsgA assembly. CsgB is also found distributed along the curli fiber, where it may serve to initiate branching of the fiber. CsgG is an outer membrane-localized lipoprotein that is required for the secretion of CsgA and CsgB, although its function is not clear.

CsgA, overexpressed CsgB is able to form short polymers on the bacterial cell surface.

The *csgDEFG* operon encodes a gene for a transcriptional activator of curli synthesis (CsgD), and three genes encoding putative assembly factors. One of these factors, CsgG, has been shown to be a lipoprotein that is localized to the outer membrane. In the absence of CsgG, curli assembly does not take place and CsgA and CsgB are rapidly degraded. The precise role of CsgG is not known. Normark and colleagues have suggested that CsgG might act as a chaperone that facilitates the secretion of the CsgA and CsgB and protects them from degradation within the periplasm. It is also possible that CsgG assembles into multimers that could function as a Csg-specific channel within the outer membrane. The roles of the *CsgE* and *CsgF* gene products are not known. It has been reported, however, that a strain lacking these two factors can export assembly-competent CsgA, suggesting that the production of CsgG alone is sufficient for functional maturation of the CsgA subunit of curli.

VII. TYPE III SECRETION PATHWAY

The various pilus assembly pathways described in the previous sections all rely on components of the *sec* machinery for the translocation of their respective pilus subunits across the inner membrane. Two new types of pili that are assembled by a *sec*-independent pathway known as type III secretion have been identified. The type III secretion system is encoded by numerous gram-negative pathogens and enables these bacteria to secrete and inject pathogenic effector molecules into the cytosol of host eukaryotic cells. About 20 gene products, most of which are inner-membrane proteins, make up the type III secretion system. The components mediating type III secretion are conserved in pathogens as diverse as *Yersinia* and *Erwinia*, but the secreted effector proteins vary significantly among species. The type III secretion apparatus, which appears to span the periplasmic space, resembles the basal body of a flagellum connected to a straight rod that extends across the outer membrane. Interestingly, all type III secretion systems encode some components with homologies to proteins involved in flagellar assembly. The secretion of proteins by the type III system is an ATP-dependent process that involves no distinct periplasmic intermediates. Type III-secreted proteins of EPEC and the plant pathogen *Pseudomonas syringae* have recently been shown to assemble into piluslike structures.

EPEC encodes four proteins, EspA, EspB, EspD, and Tir, that are secreted by a type III pathway. These proteins facilitate intimate contact between the pathogen and host intestinal cells and are required for the formation of specific (attaching and effacing) lesions. Knutton and colleagues (1998) showed that one of these proteins, EspA, can assemble into 7- to 8-nm thick peritrichously expressed pilus-like fibers that are organized into ~50-nm-wide bundles that extend up to 2 μm from the bacterial surface (Fig. 38.1F). These fibers appear to be made up of only EspA molecules. Interestingly, EspA shares substantial sequence identity with a flagellin from *Y. enterolitica*. During the infection process, the EspA fibers appear to mediate contact between EPEC and the host-cell surface prior to the establishment of more intimate bacterial attachment. The EspA fibers seem to assist the translocation of EspB effector molecules into host cells, where they can subvert host signal-transduction pathways.

In *P. syringae* and other plant pathogens, hypersensitive response and pathogenicity (*hrp*) genes control the ability of these bacteria to cause disease in susceptible plants and to elicit the hypersensitive response in resistant plants. The hypersensitive response is a phenomenon characterized by rapid localized host-cell death at the site of infection that appears to limit the spread of a pathogen in an infected plant. A subset of the *hrp* genes, recently renamed *hrc* genes, encode components of a type III secretion system. In 1997, Roine and co-workers showed that one of the proteins, HrpA, secreted by the Hrp type III secretion system is assembled into 6- to 8-nm-wide, peritrichously expressed pili. It was proposed that these pili, known as Hrp pili, are involved in mediating bacteria–plant interactions in the intercellular spaces of the host plant. In addition, Hrp pili may assist the delivery of effector proteins into host-plant cells. The exact nature and functions of the Hrp pili of *P. syringae* and the EspA-containing pili of EPEC remain to be elucidated.

VIII. REGULATION OF PILUS BIOGENESIS

Pilus biogenesis, in general, is a tightly regulated process. Ideally, the costs in energy and other resources required for pilus assembly must be balanced with any potential benefits that pilus expression might provide a bacterium. For example, by producing pili in a nutritionally poor environment a bacterium will tax its available resources, but with pili the same bacterium may be able to gain access to a more favorable location. Pathogenic and other bacteria must also control pilus expression, in some cases, to avoid attachment to unfavorable sites (tissues) within their hosts. Furthermore, pathogenic bacteria may need to modulate pilus expression to escape detection by the host immune system. Whether a bacterium expresses pili is greatly affected by environmental factors. Changes in temperature, osmolarity, pH, oxygen tension, carbon source, and nutrient availability may either increase or decrease pilus expression. The presence of iron, aliphatic amino acids, and electron acceptors other than oxygen may also influence the expression of pili. A combination of these environmental cues can stimulate (or repress) pilus synthesis and alter the expression of a variety of other factors, all of which can influence the tropism of bacteria for specific niches in the environment or in host organisms.

Environmental signals affect pilus biogenesis through global regulator proteins that can modify the transcription of pilus genes. Various global regulators have been identified and include H-NS, a DNA-binding histone-like protein that often mediates temperature regulation of pilus synthesis. H-NS appears to alter DNA topology and typically functions as a negative regulator. Regulation by carbon source can occur through the catabolite activator protein (CAP), whereas the leucine-responsive regulatory protein (Lrp) can modulate pilus expression in response to aliphatic amino acids. The CAP and Lrp regulators can control sets of pilus operons, enabling the expression of different types of pili to be coordinated and integrated with the metabolic state of the bacterial cells. In addition to these and other global regulators, specific regulator proteins encoded by genes within some pilus operons may also modulate pilus biogenesis. Multiple regulatory factors can act upon the same promoter region, switching pilus gene expression from on to off and vice versa. This on-and-off switching, known as phase variation, can also be modulated by the methylation status of a promoter region and by the inversion of sequence elements within a promoter. Two-component systems, such as the Cpx system described for P pilus assembly, also appear to be involved, at least tangentially, in the regulation of the assembly and function of a large number of pilus types.

IX. ROLE OF PILI IN DISEASE PROCESSES

The expression of pili can have substantial impact on the establishment and persistence of pathogenic bacteria in their hosts. For many bacterial pathogens, adhesive pili play a key role in the colonization of host tissues. Uropathogenic *E. coli*, for example, require type 1 pili to effectively colonize the bladder epithelium. These pili attach to conserved, mannose-containing host receptors expressed by the bladder epithelium and help prevent the bacteria from being washed from the body with the flow of urine. P pili may serve a similar function in the kidneys, inhibiting the clearance of pyelonephritic *E. coli* from the upper urinary tract. Enteric pathogens produce a wide variety of adhesive pili that facilitate bacterial colonization of the intestinal tract. These include the K88, K99, and 987P pili made by ETEC strains, the long-polar fimbriae (LPF) and plasmid-encoded fimbriae (PEF) of *S. enterica*, and the aggregative adherence fimbriae (AAF) of enteroaggregative *E. coli*. In the small intestine, TCP are essential for the attachment of *V. cholera* to gut epithelial cells. These pili also act as receptors for the cholera toxin phage (CTXΦ), a lysogenic phage that encodes the two subunits of the cholera toxin. This phage, with its encoded toxin, is transferred between *V. cholera* strains via interactions with TCP in the small intestine. Other pili also function in the acquisition of virulence factors. The uptake of DNA facilitated by type 4 pili and DNA transfer directed by conjugative pili can provide pathogens with accessory genes enabling them to synthesize a wider repertoire of virulence factors and giving them resistance to a greater number of antibiotics. Biofilm formation, which in some cases appears to require pili such as type 1, type 4, or curli, can also increase the resistance of bacteria to antibiotic treatments and may aid bacterial colonization of tissues and medical implants.

Pili are not necessarily static organelles and dynamic alterations of pilus structures during the infection process may influence the pathogenicity of piliated bacteria. For example, electron-microscopic studies (Mulvey, 1998) of mouse bladders infected with type 1-piliated uropathogenic *E. coli* showed that the pili mediating bacterial adherence to the bladder epithelial cells were 10–20 times shorter than typical type 1 pili. It is possible that the shorter type 1 pili observed are the result of pilus retraction, or breakage, during the infection process. The shortening of type 1 pili may provide a means for reeling bacteria in toward their target host cells, allowing the bacteria to make intimate contact with the bladder epithelium after the initial attachment at a distance. Within the gut, type 4B BFP promote the autoaggregation of EPEC strains, a phenomenon that probably facilitates the initial adherence of EPEC to the intestinal epithelium. Work by Bieber and co-workers (1998) suggests that, after initial attachment, an energy-dependent conformational change in the quaternary structure of BFP is needed for the further dispersal of EPEC over human intestinal cells and for the full virulence of this pathogen.

During the infection process, adhesive pili are often situated at the interface between host and pathogen where they can potentially mediate cross-talk between the two organisms. A few examples of pilus attachment inducing signal-transduction pathways in host eukaryotic cells have been reported. The binding of the type 4A pili of *Neisseria* to host receptors (probably CD46) on target epithelial cells has been shown to stimulate the release of intracellular Ca^{2+} stores, a signal known to control a multitude of eukaryotic cellular responses. Similarly, the attachment of P pili to Galα (1-4)Gal-containing host receptors on target uroepithelial cells can trigger the intracellular release of ceramides, important second-messenger molecules that are capable of activating a variety of protein kinases and phosphatases involved in signal transduction processes. The signals induced in uroepithelial cells upon the binding of P-piliated bacteria eventually result in the secretion of several immunoregulatory cytokines. The binding of type-1-piliated bacteria to mannosylated receptors on uroepithelial cells can similarly induce the release of cytokines, although apparently through different signaling pathways than those stimulated by P pilus binding. Pili can also transduce signals into bacterial cells. This was demonstrated by Zhang and Normark who, in 1996, showed that the binding of P pili to host receptors stimulated the activation of iron-acquisition machinery in uropathogenic *E. coli*. This probably increases the ability of uropathogens to obtain iron and survive in the iron-poor environment of the urinary tract. An understanding of how pili can transmit signals into bacterial cells, and the consequences of such signaling, awaits future studies. Continued research into the biogenesis, structure, and function of pili promises not only to advance our basic understanding of the role of these organelles in pathogenic processes, but may also aid the development of a new generation of antimicrobial therapeutics and vaccines.

BIBLIOGRAPHY

Abraham, S. N., Jonsson, A.-B., and Normark, S. (1998). Fimbria-mediated host-pathogen cross-talk. *Curr. Opin. Microbiol.* **1**, 75–81.

Dodson, K. W., Jacob-Dubuisson, F., Striker, R. T., and Hultgren, S. J. (1997). Assembly of adhesive virulence-associated pili in gram-negative bacteria. *In "Escherichia coli*: Mechanisms of Virulence," (M. Sussman, ed.), pp. 213–236. Cambridge University Press, Cambridge.

Edwards, R. A., and Puente, J. L. (1998). Fimbrial expression in enteric bacteria: a critical step in intestinal pathogenesis. *Trends Microbiol.* **6**, 282–287.

Forest, K. T., and Tainer, J. A. (1997). Type-4 pilus-structure: outside to inside and top to bottom—A minireview. *Gene* **192**, 165–169.

Fusseneggar, M., Rudel, T., Barten, R., Ryll, R., and Meyer, T. F. (1997). Transformation competence and type-4 pilus biogenesis in Neisseria gonorrhoeae—A review. *Gene* **192**, 125–134.

Hedlund, M., Duan, R. D., Nilsson, A., Svensson, M., Karpman, D., and Svanborg, C. (2001). Fimbriae, transmembrane signaling, and cell activation. *J. Infect. Dis.* **183** (Suppl. 1), S47–S50.

Hultgren, S. J., Jones, C. H., and Normark, S. (1996). Bacterial adhesins and their assembly. *In "Escherichia coli* and *Salmonella*" (F. C. Neidhardt, ed.), Vol. 2, pp. 2730–2756. ASM Press, Washington, D.C.

Klemm, P. (ed.) (1994). "Fimbria: Adhesion, Genetics, Biogenesis, and Vaccines." CRC Press, Boca Raton, FL.

Knight, S. D., Berglund, J., and Choudhury, D. (2000). Bacterial adhesins: structural studies reveal chaperone function and pilus biogenesis. *Curr. Opin. Chem. Biol.* **4**, 653–660.

Low, D., Braaten, B., and van der Woude, M. (1996). Fimbriae. *In "Escherichia coli* and *Salmonella*" (F. C. Neidhardt, ed.), Vol. 1, pp. 146–157. ASM Press, Washington, D.C.

Sakellaris, H., and Scott, J. R. (1998). New tools in an old trade: CS1 pilus morphogenesis. *Mol. Microbiol.* **30**, 681–687.

Sauer, F. G., Mulvey, M. A., Schilling, J. D., Martinez, J. J., and Hultgren, S. J. (2000). Bacterial pili: molecular mechanisms of pathogenesis. *Curr. Opin. Microbiol.* **3**, 65–72.

Schilling, J. D., Mulvey, M. A., and Hultgren, S. J. (2001). Structure and function of *Escherichia coli* type 1 pili: new insight into the pathogenesis of urinary tract infections. *J. Infect. Dis.* **183** (Suppl. 1), S36–S40.

Silverman, P. M. (1997). Towards a structural biology of bacterial conjugation. *Mol. Microbiol.* **23**, 423–429.

Zupan, J. R., Ward, D., and Zambryski, P. (1998). Assembly of the VirB transport complex for DNA transfer from *Agrobacterium* tumefaciens to plant cells. *Curr. Opin. Microbiol.* **1**, 649–655.

WEBSITES

The PapB protein family
http://arep.med.harvard.edu/cgi-bin/dpinteract/family?papB

The *E. coli* Cell Envelope Protein Data Collection
http://www.cf.ac.uk/biosi/staff/ehrmann/tools/ecce/ecce.htm

How bacteria crawl with links to movies (A. J. Merz)
http://www.webcom.com/alexey/moviepage.html

39

Flagella

Shin-Ichi Aizawa
Japan Science & Technology Agency

GLOSSARY

flagellar basal body The major structure of the flagellar motor, consisting of ring structures and a rod.

flagellar motor A molecular machine that converts the energy of proton flow into rotational force.

master genes Flagellar genes that regulate the expression of all the other flagellar genes, sitting at the top of the hierarchy of flagellar regulons.

polymorphic transition The interconversion of helical forms on a flagellar filament in a discrete or stepwise manner.

type III secretion system One of the protein export systems that does not use the general secretory pathway. It consists of many protein components.

The Flagellum is an organelle of bacterial motility. It consists of several substructures: the filament, the hook, the basal body, the C ring, and the C rod. The flagellar motor, an actively functional part of the flagellum, can generate torque from proton-motive force. The structural aspects of the flagellum are described here, revealed in pursuit of the identity of the flagellar motor.

I. STRUCTURE

A. Filament

The flagellum is a complex structure composed of many different kinds of proteins. However, the term flagellum, especially in earlier studies, often indicates the flagellar filament only, because the filament is the major portion of the entire flagellum. In this section, I describe the filament and occasionally call the filament just flagellum.

1. Number of flagella per cell

The number and location of flagella on a cell are readily discernible traits for the classification of bacterial species.

The number ranges from one to several hundred, depending on the species; hence the nomenclature, monotrichous (one) or multitrichous (two or more). Occasionally the term "amphitrichous" is used for two flagella.

There are three possible locations on a cell body for flagella to grow—polar (at the axial ends of the cell body), lateral (at the middle of the cell body), or peritrichous (anywhere around the cell body). In some cases, "lateral" is used as the counterpart of "polar," as in the two-flagellar systems of *Vibrio alginolyticus*, polar sheathed flagellum and lateral plain flagella (although the latter are actually peritrichous). A tuft of flagella growing from a pole is called lophotrichous. In most cases, flagella can be named by a combination of number and location, for example, polar lophotrichous flagella of *Spirillum volutans*.

Although ordinary flagella are exposed to the medium, some flagella are wrapped with a sheath derived from the outer membrane (e.g., in *Vibrio cholerae*). In an extreme case such as spirochetes, flagella are confined in a narrow space between the outer

The Desk Encyclopedia of Microbiology
ISBN: 0-12-621361-5

membrane and the cell cylinder. The flagella still can rotate; the helical cell body works as a screw, and the flagella counterbalance the torque of the cell body.

2. *Filament shape*

The filament shape is helical. In theory, there are two types of helices, right-handed and left-handed; in reality, *Salmonella* spp. have left-handed filaments and *Caulobacter crescentus* has a right-handed filament. However, it should be noted that the shapes of these two helices are not mirror images of each other.

There are several detailed filament shapes, and it will be convenient to use the names of the typical shapes found in *Salmonella* spp.—normal (left-handed), curly (right-handed), coiled (left-handed), semi-coiled (right-handed), and straight. The helical parameters of these helices are discrete and distinguishable from one another (Fig. 39.1).

Flagella can switch among a set of helical shapes under appropriate conditions; both helical pitch and helical handedness are interchangeable. The transformation of shapes can be induced by physical perturbation (torque, temperature, pH, and salt concentration of medium). Genetic changes such as point mutations in the flagellin (the component protein of the flagellar filament) gene also result in transformation of helices, but some mutant flagella, such as straight flagella, are too stiff to transform into another helix. This phenomenon, called polymorphism of flagella, is a visible example of conformational changes in proteins and therefore has evoked an idea of a functional role of flagella in motility. Could polymorphism of the flagellum by itself cause the motion? The answer is "no." Flagella are passive in terms of force generation.

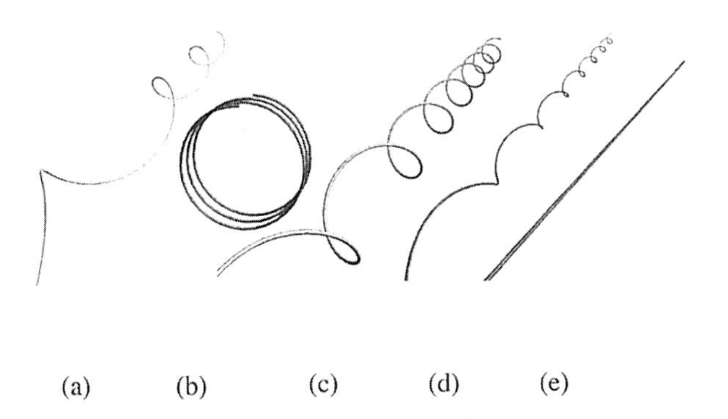

FIGURE 39.1. Helical forms of flagellar filaments. Helices are seen from a position slightly off their axial direction so that the handedness can be easily visualized. The figure shows five typical forms with their helical parameters (p, pitch; d, diameter): (a) normal ($p = 2.55\,\mu m$, $d = 0.6\,\mu m$); (b) coiled ($p = 0$, $d = 1.0\,\mu m$); (c) semi-coiled ($p = 1.26\,\mu m$, $d = 0.5\,\mu m$); (d) curly ($p = 1.20\,\mu m$, $d = 0.2\,\mu m$); and (e) straight filament ($p = \infty$, $d = 0$).

Polymorphism of flagella is observed to occur naturally on actively motile cells with peritrichous flagella. The helical transformation is necessary for untangling a jammed bundle of tangled flagella. When normal flagella in a jammed bundle are transformed into curly flagella, knots of tangled flagella run toward the free end of each flagellum to untangle the jammed bundle.

A theoretical model that explains the polymorphism successfully was presented by Dr. Chris Calladine (University of Cambridge). Twisting and bending a cylindrical rod gives rise to a helix. This model predicts 12 shapes, and eight of them have been found in existing filaments: straight with a left-handed twist, f1, normal, coiled, semi-coiled, curly I, curly II, and straight with a right-handed twist. Only a small energy barrier seems to lie between any two neighboring shapes. Polymorphic transition occurs from one shape to its neighbors; for example, in a transition from normal to curly, the filament briefly takes on coiled and semi-coiled forms.

3. *Flagellin*

The component protein of the filament is called flagellin. Although the flagellum of many bacteria is composed of one kind of flagellin, some flagella consist of more than two kinds of closely related flagellins. The molecular size of flagellin ranges from 20 to 60 kDa. Enterobacteria tend to have larger molecules, whereas species living in freshwater have smaller molecules.

One of the most characteristic features of flagellin is evident even in the primary structure of the molecule; the amino acid sequences of both terminal regions are well conserved, whereas that of the central region is highly variable even among species or subspecies of the same genera. As a matter of fact, this hypervariability of the central region gives rise to hundreds of serotypes of *Salmonella* spp.

The terminal regions are essential for binding of each molecule to another to polymerize them into a filament. Complete folding of flagellin occurs during assembly; although the terminal regions do not take on any specific secondary structure in solution, they are converted into α-helix after polymerization.

In the filament, the terminal regions are located at the innermost radius of a cylindrical structures, whereas the central region is exposed to the outside. Note that the filament is extremely stable; it does not depolymerize in water, in contrast to actin filaments or tubulin filaments, which depolymerize in the absence of salts.

This description of the flagellin molecule is not applicable to that found in archaeobacteria. Archaic flagella seem to have a system totally different from those of eubacteria; archaic flagellins have signal sequences, suggesting that the flagellum might grow from the proximal end in the outer membrane, in contrast to the distal growth of eubacterial flagellum (see Section IV).

4. *Cap protein*

Flagella have been regarded as a self-assembly system. Indeed, flagellin can polymerize into flagella under conditions that commonly promote protein crystalization *in vitro*. However, *in vivo* flagellin assembly requires another protein, without which the flagellin is secreted into the medium as monomers. The protein that helps filament formation is located at the tip and is thus called the cap protein, or HAP2 or FliD.

The cap proteins assemble in a pentamer, forming a star-shaped structure. The star hands fit in the grooves of flagellin subunits at the tip of flagellum, leaving a small gap for a nascent flagellin to insert.

B. Hook

1. *Shape*

Hook, as the name suggests, is more sharply curved (almost in a right angle) than the filament and is much shorter. The curvature indicates the flexibility of the hook, although it has to be stiff enough to transmit the torque generated at the basal structure to the filament. From these physical properties, the hook has been regarded as a universal flexible joint.

The length of the hook is 55 nm with a standard deviation of 6 nm, which is rather well controlled when compared with the length of filament. The hook length is not controlled by any molecular rulers, as might have been expected, but by a measuring cup at the base of the flagellum (see Section IV).

A polyhook is a hook of indefinite length, obtained in certain kinds of mutants. Its shape is a right-handed superhelix. The wild-type hook has the same superhelix, but consists of about one-fourth of the helical pitch.

2. *Hook protein*

The hook is a tubular polymer made of a single kind of protein, hook protein or FlgE. The molecular size of hook protein varies from 29 kDa (*Bacillus subtilis*) to 76 kDa (*Helicobacter pylori*), but it is around 42 kDa for most species.

The architecture of hook protein resembles that of flagellin—the amino acid sequence in both terminal regions is well conserved, but in the central region it is variable. Hook protein folding also completes on assembly.

3. *Scaffolding protein FlgD*

The hook does not self-assemble; it requires a helper protein, FlgD, which functions in a similar way that FliD does for filament formation. FlgD sits at the tip of the nascent hook to polymerize the hook protein coming out from the central channel. When the hook length reaches 55 nm, FlgD is replaced by HAP1 (FlgK), which remains in the mature flagellum. Because of its temporary existence, FlgD is regarded as a scaffolding protein.

4. *Hook-associated proteins*

There are two minor proteins between the hook and filament. They are called hook-associated proteins (HAPs) because they were found at the tip of the hook in several filamentless mutants. There were originally three HAPs, called HAP1, 2, and 3, in the order of their molecular size. HAP2 turned out to be located at the tip of filament as described above, leaving HAP1 and 3 between the hook and filament. They are, therefore, better termed hook-filament junction proteins.

The number of subunits of HAP1 and 3 in a filament is estimated to be 5 or 6, indicating that they form one-layer rings sitting one on another.

The roles of these two HAPs have been ambiguous. The idea of a connector to smooth the junction between the two polymers is blurred by the question, "Why are not one but two kinds necessary?" In a mutant of HAP3, filaments underwent polymorphic transitions so easily that cells cannot swim smoothly, suggesting a specific role of HAP3 as a stabilizer of filament structure.

C. *Basal structure*

Flagella have to be anchored in the cell wall. The structural entity for the anchoring was called the basal structure or basal granule, hinted at by vague images by electron microscopy. Since DePamphilis and Adler in 1974 defined the details of the basal structure, it has been called the basal body. The basal body typically consists of four rings and one rod.

The basal body does not contain everything necessary for motor function. Fragile components were detached from the basal body during purification.

In 1985, one such fragile structure was found attached to basal bodies purified by a modified method; it was named the C (cytoplasmic) ring. In 1990, another rod-like structure was found in the center of the C ring. Therefore, the basal structure (as of 2003) consists of the basal body, the C ring, and the C rod, but there may be more.

1. *Basal body*

The basal body contains rings and a rod penetrating them. The number of rings varies depending on the membrane systems; there are four rings in most gram-negatives, and two rings in gram-positives, exemplified by *B. subtilis*. Some variation in the number (such as five rings in *C. crecentus*) has been occasionally seen, but the purpose and function of a fifth ring is unclear because the cell's membranes are supposed to be the same as those of *S. typhimurium* or *E. coli*.

The structure of the basal body of *S. typhimurium* has been extensively analyzed. The physical and biochemical properties of the substructures of the basal body described later are from *S. typhimurium*, unless otherwise stated.

2. *LP-ring complex*

The outermost ring, the L ring, interacts with the lipopolysaccharide layer of the outer membrane, and the P ring just beneath the L ring may bind to the peptidoglycan layer. The LP-ring complex works as a bushing, fixed firmly enough to hold the entire flagellar structure stably in the cell surface.

The component proteins, FlgH for the L ring and FlgI for the P ring, have signal peptides, indicating that they are secreted through the general secretory pathway (GSP), which is the exception for flagellar proteins (see Section IV). FlgH undergoes lipoyl modification.

After the L and P rings have bound together to form the LP-ring complex, this complex is resistant to extremes of pH or temperature. Subjecting it to pH 11, pH 2, or boiling for 1 min does not destroy the complex, confirming that the complex serves as a rigid bushing in the outer membrane.

The essential roles of the complex are still ambiguous because mutants lacking the complex still can swim, though poorly, and because no corresponding structure has been found in gram-positive bacteria.

3. *MS-ring complex*

Earlier studies of flagellar motor function assumed that torque would be generated between the M and S rings, which face each other on the inner membrane. However, in 1990, it was shown that a single type of protein, FliF, self-assembles into a complex consisting of the M and S rings and part of the rod.

FliF is 65 kDa, the largest of the flagellar proteins. It contains no cysteine residues. It consists of several regions—terminal regions and two distinguishable central regions. Overproduction of FliF in *E. coli* gives rise to numerous MS-ring complexes packed in the inner membrane.

The S ring has been seen in the basal bodies from all the species studied so far (at least seven examples). It stays just above the inner membrane (supramembrane) and has no apparent interaction with any other structures. Besides, it is very thin (~1 nm), and the role of the S ring remains mysterious.

Although the MS-ring complex is no longer regarded as the functional center of the flagellar motor, it is still the structural center of the basal structure and plays an important role in flagellar assembly (see Section IV).

4. *Rod*

The rod is not as simple as its name suggests; it consists of at least four distinct proteins. It breaks at the midpoint when external physical force is applied to the filament, which is not expected in a structure that transmits torque to the filament.

Rod formation seems complicated because of the four component proteins. No intermediate rod structure has been observed; either there is a whole rod or no rod at all.

Because the P ring is formed on the rod, some of the rod proteins could be the target of interaction with FlgI, the subunit of the P ring.

5. *C ring*

The C ring is a fragile component of the basal structure. It is resistant to the nonionic detergent Triton X-100, but it is destroyed by the alkaline pH and high salts concentration employed by conventional purification methods. The dome shape of the C ring is easily flattened on a grid during preparation for electron microscopy.

The C ring consists of the switch proteins (FliG, FliM, and FliN) and so is sometimes called the switch complex. It is not known whether other proteins are present or not. FliG directly binds to the cytoplasmic surface of the M ring. FliM binds to FliG, and FliN to FliM. The stoichiometry of these molecules in the C ring is still controversial; 20–40 copies of FliG, 20–40 copies of FliM, and several 100 copies of FliN.

Genetic studies revealed that the switch complex plays important roles in flagellar formation, torque generation, and the switching of rotational direction. The C ring directly binds signal molecules, CheY, produced in the sensory transduction system, but the mechanism of the switching is ambiguous.

6. *Export apparatus*

Flagella have been regarded as having a self-assembly system, similar to that of bacteriophages. However, flagellar assembly is quite different from phage assembly in many ways. First, the flagellum, being an extracellular structure, assembles not in the cytoplasm but outside the cell. Second, the component proteins therefore, have to be transported from the cytoplasm to the outside. Third, assembly consequently proceeds in a one-by-one manner, at the distal end of the nascent structure.

For this kind of assembly, a protein excretion system must play an important role. As a matter of fact, among the 14 genes required in the very first step of flagellar assembly, at least seven gene products are necessary to form a protein complex, called an export apparatus. One of them, FliI, has an ATPase activity, suggesting that one step in the export process requires ATP hydrolysis as an energy source. The physical body of the export apparatus has not been identified; the C rod is a strong candidate, judging from its location in the C ring.

II. FUNCTION

The function of flagella is described here briefly so that the meaning of the structure can be understood.

Bacterial flagella rotate. There is no correlation between bacterial flagella and eukaryotic flagella, either in function or in structure; the type of movement, the energy source, and the number of component proteins differ greatly between the two. No evolutionary correlation between these two types of flagella has been shown.

Among motile bacterial species, swimming by flagellar rotation is the most common. However, several families such as myxococcus, mycoplasm, and cyanobacteria can move on a solid surface in a gliding motion; the motile organ of gliding bacteria is not known.

A. Torque

The rotational force (torque) of the flagellar motor is difficult to measure directly, but can be estimated from the rotational speed of flagella. The method most widely employed is the tethered-cell method, in which the rotation of a cell body caused by a tethered filament can be observed with an ordinary optical microscope. A more sophisticated method, which employs a laser as the light source of a dark-field microscope, allows one to measure the rotation of a flagellum on a cell stuck on the glass surface at a time resolution of millisecond.

1. *Rotational direction*

The flagella of many species (e.g., enterobacteria) can rotate both clockwise (CW) and counter-clockwise (CCW). Under ordinary circumstances, around 70% of the time is occupied by CCW rotation, which causes smooth swimming. A brief period of CW rotation causes a tumbling motion of the cell. There is no perceptible pause in switching between the two modes.

In some bacterial species such as *Rhodobacter sphaeroides*, a lateral flagellum on a cell rotates in the CW direction only, with occasional pauses. During the pauses, the filament takes on a coiled form and curls up near the cell surface. Upon application of torque to this filament, the coiled form extends to a semi-stable right-handed form that closely resembles a curly form. The CW rotation of this right-handed helix causes a forward propulsive force on the cell.

2. *Rotational speed*

Because the torque of the flagellar motor cannot be directly measured due to technical limits even today, it is estimated from the rotational speed of flagella that is believed to correlate with the torque. The highest speed of flagella observed under physiological conditions is about 200 Hz for *S. typhimurium*. High viscosity of the solution slows down the speed in a linear manner over a wide range of speeds below 200 Hz, indicating that the torque of the flagellar motor is constant in this speed regime. At the speeds higher than 200 Hz, externally forced by electro-rotation field, the torque quickly decreases and becomes zero at 300 Hz. This gives a theoretical limit of the maximal speed of the motor per se; that is, the relative speed between the rotor and stator without any load such as flagellar filament.

B. Energy source

The energy source of torque generation in the flagellar motor is not ATP but proton-motive force (PMF). PMF is the electrochemical potential of the proton, and results in the flow of protons from outside to inside the cell. PMF consists of two forms of energy, membrane potential, and entropy caused by a difference in pH between outside and inside the cell. Because these two parameters are independent and separable from each other, either one can, in principle, be abolished without affecting the other. Note that the polar flagellum of *V. alginolyticus* substitutes sodium ion (Na^+) for proton. One of the goals of flagellar research is the elucidation of the mechanism by which PMF or NaMF is converted into torque in the motor.

C. Switching of rotational direction

Switching the rotational direction of flagella is the primary basis of chemotaxis, one of the most important behaviors shown by bacteria. Damage in the switching mechanism results in a rotation biased to either CCW only or CW only.

In a strict sense, the switching mechanism will not be solved until the mechanism of rotation is solved. However, the factors involved in the mechanism are known; an effector binds to the switch complex in the flagellar motor. The effector is the phosphorylated form of CheY, a signalling protein in the sensory transduction system.

III. GENETICS

Flagellar genetics has been most extensively studied in *S. typhimurium*, especially using the enormous number of strains that Dr. Shigeru Yamaguchi (Meiji University) has collected for more than 30 years. The discussion in this section is, unless otherwise indicated, based on results obtained from these strains.

A. Flagellar genes

There are more than 50 flagellar genes, which are divided into three types, according to the null mutant phenotype.

1. *The fla genes*

Defects in the majority of the flagellar genes result in flagellar deficient (Fla^-) mutants. These genes were originally called *fla* genes. In 1985, when the number of genes exceeded the number of letters in the alphabet, a unified name system for *E. coli* and *S. typhimurium* was introduced: *flg*, *flh*, *fli*, and *flj*; one for each of the clusters of genes scattered in several regions around the chromosome (see Section III.B; Fig. 39.2).

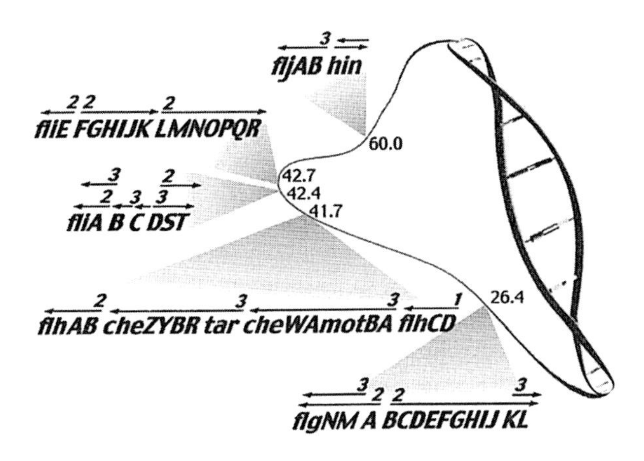

FIGURE 39.2 Genetic map of *S. typhimurium*. Flagellar genes are distributed in several clusters on the chromosome. Arrows over the genes indicate the size of operons and their transcriptional directions. The numbers on the arrows indicate classes of transcription. The regulation of classes as 2 and 3 is not simple; some operons are expressed twice, in classes as 2 and 3.

2. *The mot genes*

Mutants that produce paralyzed flagella are called motility deficient (Mot^-) mutants. There are only two *mot* genes (*motA* and *motB*) in *S. typhimurium*, but four *mot* genes (*motA*, *motB*, *motX*, and *motY*) in *R. sphaeroides* and in some other species such as *V. alginolyticus*.

3. *The che genes*

Mutants that can produce functional flagella but that cannot show a normal chemotactic behavior are called chemotaxis deficient (Che^-) mutants. These are divided into two types, general chemotaxis mutants and specific chemotaxis mutants. The former involve the proteins working in the sensory transduction (CheA, CheW, CheY, CheZ, CheB, and CheR), and the latter involve the receptor proteins (e.g., Tsr, Tar, Trg, and Tap).

B. Gene clusters in four regions

Most flagellar genes are found in gene clusters on the chromosome. They are in four regions: the *flg* genes in region I (at 26 min), the *flh* genes and *mot* and *che* genes are in region II (41.7 min), and the *fli* genes are in regions IIIa (42.4 min) and IIIb (42.7 min) (Fig. 39.2).

The *flj* operon (including *fljA* and *fljB*) at 60 min involves an alternative flagellin gene to *fliC* and is only found in *Salmonella*. Either FliC flagellin or FljB flagellin is produced at any time. The *hin* gene inverts the transcriptional direction at a certain statistical frequency; if the *flj* operon is being expressed, FljA represses *fliC*, allowing FljB flagellin alone to be produced. This alternate expression of two flagellin genes is called phase variation.

C. Transcriptional regulation

Flagellar construction requires a well-ordered expression of flagellar genes not only because there are so many genes, but also because flagellar assembly requires only one kind of component protein at a time, as described previously. There is a strict hierarchy of expression among the flagellar genes. The hierarchy is controlled or maintained by a few prominent regulatory proteins.

1. *Hierarchy: three classes*

The hierarchy of flagellar gene expression is divided into three classes; class 1 regulates class 2 gene expression, and class 2 regulates class 3. Class 1 contains only two genes in one operon, *flhD* and *flhC*.

Class 2 consists of 35 genes in eight operons. There are two regulatory genes, *fliA* and *flgM*; the rest are component proteins of the flagellum or the export apparatus.

Class 3 genes encode flagellin, *motA* and *motB*, and all the proteins involved in sensory transduction. Flagellin is one of the most abundant proteins in a cell, suggesting that the tight regulation in the hierarchy guarantees the economy of the cell.

2. *Master genes, flhDC*

Master gene products form a tetrameric complex of FlhD/FlhC, which works as a transcriptional activator of the class 2 operons.

The master operon (*flhDC*) is probably transcribed with the help of the "housekeeping" sigma factor, σ^{70}. The master operon has also been shown to be activated by a complex of cyclic AMP and catabolite activator protein (cAMP–CAP), which binds to a site upstream of the promoter.

3. *Sigma factor F (σ^F, FliA) and antisigma factor (FlgM)*

The FliA and FlgM proteins expressed from the operon competitively regulate the class 3 operons. FliA is the sigma factor that enhances the expression of the class 3 operons, whereas FlgM is an antisigma factor against FliA.

If the hook and basal body have been constructed normally, FlgM is secreted into the medium through the basal body and the complete hook, allowing free FliA proteins to work on the class 3 operons. However, if the hook and basal body construction is somehow halted in the middle of process, FlgM stays in the cytoplasm in a complex with FliA, maintaining shut off of the expression of the class 3 operons.

This intriguing regulation mechanism seems to work efficiently for peritrichous flagella. However, polar flagella which look much simpler than peritrichous ones are regulated not only by the sigma 28 (FliA) and its suppressor (FlgM) but also by sigma54 (RpoN) and its activators (FlbD, FlgR, FleQ, etc.).

4. *Global regulation versus internal regulation*

There are several external genes or factors that affect the flagellar gene expression through the master operon *flhDC*. Some of the factors show pleiotropic effects on many cellular events such as cell division, suggesting that flagellation is finely tuned with the cell division cycle due to well-organized tasks of global regulation systems.

As described, the master operon (*flhDC*) is probably transcribed using the "housekeeping" sigma factor, σ^{70}. In the 1990s, other factors regulating or modulating *flhDC* expression have been identified mainly in *E. coli*. The motility of *E. coli* cells is lost at temperatures higher than 40 °C as a result of reduced *flhDC* expression. It has been shown that some of the heat-shock proteins are involved in both class 1 and class 2 gene expression. This strongly suggests that flagellar genes are under global regulation in which the heat-shock proteins play a major role; probably the proper protein folding (or assembly) mediated by these chaperones is essential for flagellar construction.

Other adverse conditions such as high concentrations of salts, carbohydrates, or low-molecular-weight alcohols, also suppress *flhDC* expression, resulting in lack of flagella. The regulation by all these factors is independent of the cAMP–CAP pathway.

It is unknown which factors directly turn on the master *flhDC* operon in accordance with the cell cycle, and how. After the roles of global regulators on flagellar gene expression are specified, the complex regulatory network connecting flagellation and cell division will be uncovered.

IV. MORPHOLOGICAL PATHWAY

A. Steps in the morphological pathway

The order of the steps of the construction of a flagellum (the morphological pathway) has been analyzed in the same way as for bacteriophages, analyzing the

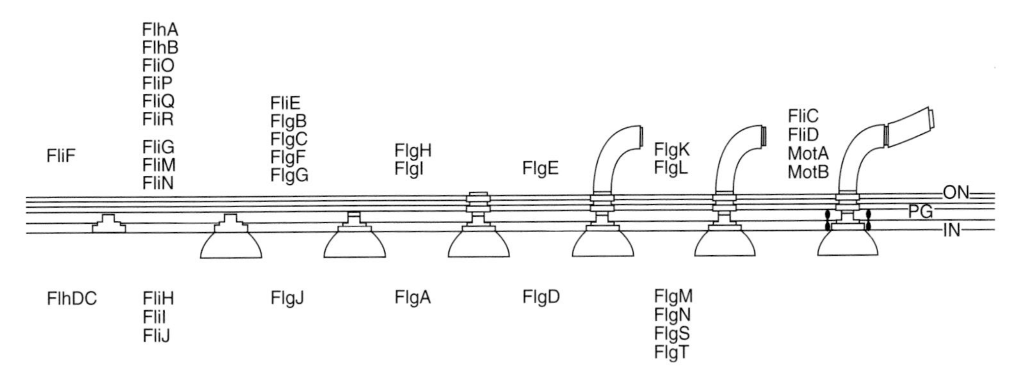

FIGURE 39.3 Morphological pathway of flagellation. Flagellar construction proceeds from left to right. Each step requires several gene products as components, shown above the membrane, and regulatory or helper proteins, as shown below the membrane, which are not included in the complete flagellar structure.

intermediate structures in various flagellar mutants and aligning them in size from small to large. Flagellar construction starts from the cytoplasm, progresses through the periplasmic space, and finally extends to the outside of the cell (Fig. 39.3).

1. *Cytoplasm*

The smallest flagellar structure recognizable by electron microscopy is the MS-ring complex; therefore, the MS-ring complex is regarded as the construction base onto which two other flagellar substructures attach, the C ring and the C rod.

2. *Periplasmic space*

The second-smallest structure consists of a rod on the MS-ring complex. When the rod has grown large enough to reach the outer membrane, the hook starts to grow. However, the outer membrane physically hampers the hook growth until the outer ring complex makes a hole in it. Among flagellar proteins, FlgH and FlgI, the component proteins of the outer ring complex, are exceptional in terms of the manner of export; these two proteins have cleavable signal peptides and are exported through the GSP.

3. *Outside the cell*

Once the physical block by the outer membrane has been removed, the hook resumes growth with the aid of FlgD until the length reaches 55 nm. Then, FlgD is replaced by HAPs, which is followed by the filament growth. The filament growth proceeds only in the presence of FliD (HAP2 or filament cap protein); without this cap, exported flagellin molecules are lost to the medium.

The number of genes necessary to proceed through each step of the construction varies; in early stages there are many genes whose roles are unidentified. Between the MS-ring complex and the rod, more than 10 genes are required. Some of those gene products form the C ring, and others are involved in the formation of the flagellar protein-specific export system.

B. Flagellar protein export as a type III secretion export system

There are several ways to export proteins outside bacterial cells. The best known pathway is the GSP. However, many flagellar proteins cannot pass through this system, because they do not have the signal sequences that are necessary for recognition by GSP.

As the number of examples of other types of protein transport has increased, the name became inappropriate. GSP is now categorized as the type II secretion system. There are now more than five secretion systems, but here I will briefly explain the type III secretion system (TTSS for short).

The flagellar protein export system is now regarded as a type III secretion system. The flagellar export apparatus consists of at least eight components. The amino acid sequences of these proteins share homology with those used for export of virulence factors from many pathogenic bacteria.

Even the structures of these two distinguishable systems resemble each other. The needle complex found first in *S. typhimurium* and then in *Shigella* looks like the basal body, consisting of several ring structures and a rod or needle.

C. The kinetics of morphogenesis

The morphological pathway of the flagellum described indicates the order of the construction steps but ignores the time consumed at each step. In order to achieve coherent cell activities, flagellar construction has to be synchronized with cell division. The most time-consuming step of flagellation seems to be the filament elongation because filaments can grow over generations.

The growth process of the filament and the hook have been carefully analyzed. By taking a closer look at elongation modes of these two polymers, we will get a glimpse of the whole kinetic process of flagellar construction.

1. *Filament growth*

In bacteria with peritrichous flagella, the number and the length of flagella are fairly well defined; there are 7–10 flagella per cell, and the average length of filament is 5–8 μm for *S. typhimurium*.

A defined number of flagella have to be supplied at each cell division. A large deviation from this number will cause disastrous results to the cell—either no flagella at all or too many flagella to swim. The number of flagella must be genetically controlled.

On the other hand, filament growth seems free from genetic control, because it continues over generations. From statistical analysis of the length distribution, the elongation rate of filaments is estimated to vary inversely to the length; thus, a filament grows rapidly in the beginning and gradually slows down to a negligible rate.

2. *Hook growth*

In contrast to the wide distribution of filament lengths, the hook length is rather well controlled at 55 nm with a deviation of 6 nm. Mutations in the *fliK* gene result in hooks with unlimited length, called polyhooks. FliK is not a molecular ruler, since a truncation in FliK also gives rise to polyhooks but not short hooks. If the ruler is short, the product measured should be short. Just recently, it was shown that mutations in switch proteins (FliG, FliM, and FliN) gave rise to short hooks with a defined length, indicating that the C ring serves as a measuring cup for hook monomers.

Statistical analysis of the length distribution of polyhooks reveals that the hook grows in a manner similar to the filament; it starts out growing at 40 nm/min and exponentially slows down to reach a length of 55 nm. After the length is 55 nm, the hook grows at a constant rate of 8 nm/min. It takes many generations for polyhooks to grow as long as several μm.

Studies of the correlation between flagellation and cell division are underway, but no definite schemes have been found.

V. CONCLUSION

The analysis of the flagellar structure is almost complete; most components of the flagellum have been identified, and the pathway of flagellar construction has been revealed. That is, the roles of ~40 flagellar genes in the flagellar construction are now known.

We will continue searching for the detailed mechanism of flagellation, including the relationship to cell division. One of the immediate goals is to answer a simple but important question, What is rotating against what? This question stems from the controversy that started at the beginning of the flagellar research. Without knowing the rotor and the stator, the mechanism of motor function will never be understood.

And then, we want to answer a more intriguing and difficult question, What is the ancestor of the flagellum? The question arose from the recent discovery of a similarity between the flagellum and the pathogenicity—not only are the gene sequences between the two systems homologous, but also their supramolecular structures resemble each other. This also leads us to the most basic question, What is the flagellum?

BIBLIOGRAPHY

Aizawa, S.-I. (1996). Flagellar assembly in *Salmonella typhimurium*. *Mol. Microbiol.* **19**, 1–5.

Aizawa, S.-I. (2001). Bacterial flagella and type III secretion systems. *FEMs Microbiol. Lett.* **202**, 157–164.

Aizawa, S.-I. (2001). Flagella, *In* "Encyclopedia of Genetics", 2nd edn., pp. 711–712, Academic Press, New York.

Aizawa, S.-I., and Kubori, T. (1998). Bacterial flagellation and cell division. *Genes Cells* **3**, 625–634.

Aizawa, S.-I., Harwood, C. S., and Kadrer, R. J. (2000). Signaling components in bacterial locomotion and sensory reception, *J. Bacterial.* **182**, 1459–1471.

Berg, H. C. (2003). The rotary motor of bacterial flagella, *Ann. Rev. Biochem.* **72**, 19–54.

Chilcott, G. S., and Hughes, K. T. (2000). Coupling of flagellar gene expression to flagellar assembly in *Salmonella enterica* serovar *typhimurium* and *Escherichia coli*. *Microbiol. Mol. Biol. Rev.* **64**, 694–708.

Kutsukake, K., Ohya, Y., and Iino, T. (1990). Transcriptional analysis of the flagellar regulon of *Salmonella typhimurium*. *J. Bacteriol.* **172**, 741–747.

Macnab, R. M. (1996). Flagella. *In "Escherichia coli* and *Salmonella typhimurium*: Cellular and Molecular Biology" (F. C. Neidhardt *et al.*, eds.), pp. 123–145. American Society for Microbiology, Washington, DC.

Macnab, R. M. (1999). The bacterial flagellum: reversible rotary propeller and type III export apparatus. *J. Bacteriol.* **181**, 7149–7153.

Namba, K., and Vonderviszt, F. (1997). Molecular architecture of bacterial flagellum. *Quart. Rev. Biophys.* **30**, 1–65.

WEBSITES

Evolution of bacterial flagella
http://minyos.its.rmit.edu.au/~e21092/flagella.htm

Bacterial motility video
http://www-micro.msb.le.ac.uk/Video/motility.html

vomiting. Fever is not seen and neither is intestinal bleeding. The organisms are ingested as cysts, which then excyst to release four sporozoites. The sporozoites then attach to and invade intestinal epithelial cells. The invasion remains superficial and *C. parvum* does not penetrate beyond the intestinal epithelial cell barrier. *C. pravum* is diagnosed by microscopy of stool to look for cysts or sporozoites, by immunofluorescene microscopy, or by enzyme immunoassay. In immunocompetent patients, the natural history of *C. parvum* infection is for the disease to be self-limiting and recovery is the rule after a week or two. The pattern is very different in immunocompromised hosts in which the infection is not cleared, and malabsorption becomes a significant and life-threatening problem. There is currently no treatment that is effective for the eradication of *C. parvum*.

B. *Giardia lamblia*

G. lamblia is probably the most frequently found enteric protozoan worldwide. This organism does not cause a dramatic enteric disease or systemic complications, yet infection with it can lead to profound malabsorption and misery in the patient. Like other enteric protozoa, it is found in fecally contaminated water and food and is yet another example of the fecal–oral transmission route. There are different *Giardia* types but only *Lamblia* are known to infect humans. The disease is initiated by ingestion of the cysts, and as few as 10–100 may be an infectious dose. Following ingestion, the cysts excyst in the proximal small intestine and release trophozoites. The trophozoites divide by binary fission and attach intimately to the intestinal epithelium, via a ventral disc on the trophozoite. Clinically, giardiasis may be extremely variable. At one extreme, there may be asymptomatic infection, and at the other, severe chronic diarrhea, leading to intestinal malabsorption. Acutely, infection usually results in watery diarrhea and abdominal discomfort. *G. lamblia* can be diagnosed by fecal microscopy, looking for either the cysts or the trophozoites. Currently, many laboratories look for the parasites using commercially available kits, that utilize either fluorescence microscopy with specific antibodies or enzyme immunoassays. In terms of therapy, metronidazole is the drug of choice.

C. *Entamoeba histolytica*

E. histolytica is one the leading causes of parasitic death in the world. It is usually spread by the fecal–oral route, either directly or via contamination of food, e.g., lettuce or water. The cyst is the infective form and cysts may survive weeks in an appropriate environment. Following ingestion, the cysts, excyst in the small bowel and form trophozoites, which then colonize the large bowel, and multiply or encyst, depending on local conditions in the intestine. There are various types of *E. histolytica*, some of which are pathogenic to humans and some of which are not. They have been differentiated based on zymodeme analysis, which is a determination of the electrophoretic mobility of certain isoenzymes. *E. histolytica* causes amebiasis, which may have various clinical manifestations. The trophozoites have the capacity to invade the host from the intestinal lumen, which in the colon results in ulceration of the mucosa, causing amebic dysentery. When the trophozoites invade further, they gain access to the portal blood vessels and are transmitted to the liver. Once in the liver, they are then able to destroy the parenchyma, resulting in a hepatic amebic abscess (this occurs in about 1% of patients with intestinal amebiasis). Intestinal infection with *E. histolytica* is diagnosed by microscopic examination of the stool, either by wet mount or trichrome stain. Serology is useful in the diagnosis of patients with invasive amibiasis. Patients with *E. histolytica* need to be treated, and metronidazole is the drug of choice. Luminal drugs that are poorly absorbed are an alternative therapy for carriers of cysts. Drugs such as paromomycin or iodoquinol are used for this purpose.

D. *Cyclospora cayetanensis*

C. cayetanensis is a recently described apicomplexan parasite that has been found in food. Most recently, it has been responsible for a number of outbreaks in North America associated with consumption of imported raspberries. It has also been associated with undercooked meat and poultry, and contaminated drinking water and swimming water. Clinically, it causes a self-limiting diarrhea, with nausea, vomiting, and abdominal pain in immunocompetent patients but may lead to a more persistent diarrhea in immunocompromised individuals. *C. cayetanenis* is diagnosed by direct stool microscopy and oocyst autofluorescence, which appears blue by Epi-illumination and a 365-nm dichroic filter and green by a 450–490-nm dichroic filter. In the laboratory, oocysts may be induced to sporulate, even in the presence of potassium dichromate used to preserve the specimens. After 1–2 weeks, approximately 40% of oocysts contain two sporocysts with two sporozoites in each. Excystation requires a number of steps and occurs when oocysts are subjected to bile salts and sodium taurcholate plus mechanical pressure. Susceptible humans are infected by ingesting sporulated oocysts and, in view of the complexity of the process and the time required, direct person-to-person spread is considered unlikely. The infection can be successfully treated with trimethoprim–sulfamethoxazole.

E. Other protozoan infections

A number of other protozoa have been associated with food- and water-borne infections in humans. These include Microsporidium, that causes watery diarrhea and malabsorption, and are becoming an increasingly recognized problem in the immunocompromised. Various microsporida, including *Enterocytozoon bieneusi* and *Septata intestinalis*, cause human disease. *Isospora belli*, another apicomplexan protozoon, is an opportunistic pathogen in immunocompromised patients. Sarcocystosis is a rare zoonotic infection in humans that can, on occasion, cause necrotizing enteritis. *Dientamoeba fragilis* was originally thought to be a commensal but now appears to be associated with a variety of gastrointestinal symptoms, including abdominal pain, nausea, diarrhea, and anorexia. *Balantidium coli* is the only ciliate known to parasitize humans. Although most infections are asymptomatic, the disease may present itself as dysentery. *Blastocystis hominis* is a strict anaerobic protozoon that infects both immunocompetent and immunocompromised hosts and results in a variety of gastrointestinal symptoms, including diarrhea, abdominal pain, nausea, vomiting, anorexia, and malaise.

V. CESTODES AND WORMS

A. *Taenia saginata*

Taenia saginata, the beef tapeworm, is highly endemic in certain areas of the world, such as parts of South America, Africa, South Asia, and Japan. Humans are the definitive host for the adult tapeworm, which is one of the largest human parasites. They may live as long as 20 years and grow up to 25 meters in length. Consumption of undercooked or raw beef containing living larval forms is how it is acquired. Cattle are the intermediate hosts, in which the hexacanth embryos emerge from the eggs and pass by blood or lymph to muscle, subcutaneous tissue, or viscera. Then, when humans eat the undercooked animal tissue, the life cycle is completed. Clinically, the worms are remarkably quiescent and nausea or a feeling of fullness may be the only symptoms. Vomiting, nausea, and diarrhea may occur. Diagnosis depends on detecting the proglottids in stool and treatment with either praziquantel or albendazole should be curative.

B. *Taenia solium*

T. solium is the pork tapeworm, and, unlike the *T. saginatum*, the larval stage can invade humans and cause infections of the central nervous system. *T. solium* is distributed worldwide and is acquired by ingesting pork meat that is infected with cysticerci. The adult worm in humans sheds proglottids, which are then eaten by pigs, after which the hexacanth embryos emerge and penetrate the pig's intestinal wall, where they migrate to muscle and other tissues. Humans then eat the larval forms, which completes the cycle. *T. solium* is usually smaller than *T. saginatum* so the clinical symptoms are even less remarkable. Infected humans may just notice the proglottids in stool. The diagnosis is dependent on identifying the proglottids and the treatment is with praziquantel or albendazole.

C. *Diphyllobothrium latum*

D. latum is a fish tapeworm and is most common in Northern Europe and Scandinavia. Eating raw fish is the greatest risk factor and the increased consumption of sushi in the United States has resulted in more cases than in the past. The life cycle is complex; after the eggs are passed in human feces, they must hatch and be eaten by copepods, which are freshwater crustaceans. They then develop into larval forms and subsequently are eaten by fish. The procecoids then invade the stomach wall of the fish and, finally, reside in the muscle of the fish. Humans then become infected by eating a fish that harbors a viable plerocercoid larva. Clinically, the infection is usually asymptomatic, but diarrhea, fatigue, and distal paresthesia are well described, as is pernicious anemia, since the tapeworm actively absorbs free vitamin B12. Diagnosis is made by identifying the ova or proglottids in stool. Treatment with praziquantel or niclosamide is effective.

D. *Hymenolepis nana*

H. nana, the dwarf tapeworm, is very common in humans and is transmitted via the fecal–oral route. This worm does not require an intermediate host. The majority of infections are asymptomatic but diarrhea, abdominal cramping, and anorexia may occur. The diagnosis is dependent on identifying the typical double lumen eggs in the stool and treatment with praziquantel should be curative.

E. Ascariasis

Ascaris is the most common intestinal helminth worldwide. *Ascaris lumbricoides* is specific for humans, who are infected by ingesting food containing the mature ova. The larvae are then released in the small intestine, enter the circulation, and then reach the pulmonary alveoli, where they develop. This pulmonary infestation may cause pneumonitis and allergic manifestations. Finally, larvae migrate up the bronchial tree and are swallowed. Humans are the definitive hosts, but soil is necessary for development of the eggs and also acts as a reservoir. Humans then ingest the developed eggs to complete the cycle. Thus, food or water that are contaminated are sources that infect humans. The diagnosis is made by finding adult worms, larvae, or eggs in the stool. Acarisis may be treated with mebendazole or pyrantel pamoate.

F. Trichuriasis

Trichuris trichiura is commonly known as whipworm and is frequently found in the same parts of the world as ascaris. Humans are the definitive host and eggs that are passed in stool mature in warm moist soil to become infective. They then contaminate food and are ingested by a new host. Clinically, the worms remain associated with the intestine and may be either asymptomatic or result in chronic diarrhea. In the context of a heavy worm burden, dysentery may develop along with malnutrition. The diagnosis is made by finding adult worms or eggs in stool. Treatment with mebendazole is usually curative.

G. *Trichinella spiralis*

This nematode begins its infection in humans following the ingestion of the first-stage larvae and its nurse cell in striated skeletal muscle tissue. The larvae are released from tissue in the stomach and pass to the small intestine, where they infect epithelial cells. The larvae then develop into adult worms and are shed in the stool. Larvae also penetrate into lymph or blood vessels and then on into muscle cells, where a nurse cell begins to form. The principal mode of transmission to humans is through the consumption of undercooked meat, usually pork. The major clinical features of this disease relate to cellular destruction secondary to the parasitic penetration of cardiac or nervous tissue. Gastrointestinal symptoms are also common and include diarrhea and vomiting. The diagnosis is dependent on the histologic identification of nurse cells containing larvae within infected muscle tissue. Serological tests are also of value. The infection may be treated with thiabendazole.

VI. NATURAL TOXINS

There are a number of naturally occurring toxins that may be present in various types of food. Many of these are associated with consumption of seafood but others are related to different and specific foods.

A. Ciguatera

Ciguatera poisoning is due to the ingestion of a neurotoxin from fish. The toxin is produced in dinoflagellates (e.g., *Gambierdiscus toxicus*). It then accumulates in the flesh of the fish. This occurs mainly in tropical and subtropical marine fin fish, including mackerel, groupers, barracudas, snappers, amberjack, and triggerfish, although not all of these types of fish are infected all of the time. In humans, the incubation period ranges from as little as 5 min up to 30 hr (with a mean of about 5 hr). There are usually gastrointestinal and neurological symptoms, including nausea, vomiting, watery diarrhea, parasthesias, ataxia, vertigo, and blurred vision. Some patients may go on to develop cranial nerye palsies or even respiratory paralysis. The symptoms may last up to a week, but then usually resolve. The initial diagnosis is usually clinical and confirming the presence of the toxin is difficult. Toxin detection can be undertaken using a mouse bioassay and enzyme immunoassays for toxin detection are being developed.

B. Scrombroid

Scrombroid poisoning typically occurs after the ingestion of spoiled fish, especially tuna and mackerel. Excess levels of histamine in the flesh of the fish are thought to be the cause of this poisoning. Histamine and other amines are formed in food by the action of decarboxylases, produced by bacteria that act on the histidine or other amino acids. Scrombroid has been associated with a number of other foods, such as Swiss cheese. However, it is more frequently associated with fish, especially if the fish has not been frozen rapidly after being caught. Clinically, symptoms may begin within 10 min to 3 hr following ingestion. Nausea, vomiting, diarrhea, flushing, and headache may all occur. Respiratory distress is a rare complication. The natural history of this disease is that it will resolve in a few hours. The diagnosis is usually clinical and can be confirmed by detecting elevated levels of histamine in the suspected food.

C. Shellfish poisoning

Four main types of shellfish poisoning have been described, including paralytic shellfish poisoning, neurotoxic shellfish poisoning, diarrheic shellfish poisoning, and toxic–encephalopathic shellfish poisoning. Shellfish poising is due to toxins made by algae (usually dinoflagellates) that accumulate in the shellfish. Paralytic shellfish poisoning is due to saxitoxin, a sodium channel toxin. Clinically, symptoms occur usually within an hour and consist of nausea, vomiting, and paralysis that may be limited to the cranial nerves or, in more severe cases, involve the respiratory muscles. Neurotoxic shellfish poisoning is due to brevitoxin, which is a lipophilic, heat-stable toxin that stimulates postgaglionic cholinergic neurons. Symptoms usually occur within 3 hr of exposure and consist of nausea, vomiting, and parasthesisa. Paralysis does usually not occur. Diarrheic shellfish poisoning, as the name implies, causes a mainly gastrointestinal disturbance with nausea, vomiting, and diarrhea. Toxic–encephalopathic shellfish poisoning (also known as amnesic shellfish poisoning) has caused outbreaks of disease in association with consumption of mussels. Symptoms include nausea, vomiting, diarrhea, server headache, and, occasionally, memory loss. With all these types of poisoning, symptoms usually occur rapidly (within 2 hr) and will usually resolve spontaneously. The exception to this is toxic–encephalopathic poisoning, when the symptoms may not occur for 24–48 hr following exposure. The diagnosis in humans is clinical. However, it may be possible to detect the presence of the toxins using either mouse bioassays or by high performance liquid chromatography (HPLC).

D. Tetrodotoxin

Tetrodotoxin is present in certain organs within puffer fish and, if ingested, can cause rapid paralysis and death. Symptoms may occur in as little as 20 minutes or after several hours. Symptoms progress from a gastrointestinal disturbance to almost total paralysis, cardiac arrhythmias, and, finally, death within 4–6 hr, after ingestion of the toxin. The diagnosis is clinical and based on history of exposure. Mouse bioassays and HPLC have been used to detect these toxins in food.

E. Mushroom toxins and aflatoxins

There are a large variety of toxins from different mushrooms that cause a wide variety of diseases in humans. These toxins can be divided into four general groups as follows: protoplasmic poisons (e.g., amatoxins, hydrazines, orellanine) that cause cellular damage and organ failure (e.g., hepatorenal syndrome); neurotoxins (e.g., muscarine, ilbotenic acid, muscimol, psilocybin) that cause coma, convulsions, and hallucinations, etc; gastrointestinal irritants that produce nausea, vomiting, and diarrhea; and disulfiramlike toxins that only cause a problem if the person ingesting the mushroom has had exposure to alcohol in the previous 48–72 hr. The diagnosis of mushroom poisoning is based largely on the clinical picture and the history of exposure. There is, however, a commercial radioimmunoassay available for the amanitins. Therapy is largely supportive but interventions to reduce toxin absorption from the intestine, such as lavage or administration of activated charcoal, may help. Plasmapheresis also helps to reduce the mortality rate.

Aflatoxins are produced by certain strains of fungi, e.g., *Aspergillus flavus* and *A. parasiticus*, that grow on various types of food and produce toxins. Nuts, especially tree nuts (Brazil nuts, pecans, pistachio nuts, and walnuts), peanuts, and other oilseeds, including corn and cottonseed, have been implicated most often. There are various types of aflatoxins (B1, B2, G1, and G2), of which B1 is the most common and the most toxic. Clinically, these toxins cause liver damage that may be in the form of cirrhosis or hepatic malignancy. Occasionally, the ingested dose of aflatoxin is so high that an acute condition develops, known as aflatoxicosis, in which there is fever, jaundice, abdominal pain, and vomiting. Diagnosis in humans is clinical, but there are assays available for the detection of the toxins in food.

F. Other natural toxins

A number of other naturally occurring toxins have been reported. These include grayanotoxin, which is from eating honey made from rhododendrons. This usually causes nausea, vomiting, and weakness soon after the honey is ingested and, typically, is self-limiting in 24 hr. Akee fruit from Jamaica contains hypoglycin A, that causes hypoglycemia and vomiting in 4–10 hr. Curcurbitacin E, from bitter cucumber, can cause cramps and diarrhea within 1–2 hr of ingestion. Hydrogen cyanide may be present in lima beans or cassava root and can lead to death within minutes. Castor beans can contain a hemagglutinin that may cause nausea and vomiting. Red kidney beans also produce a hemagglutinin, known as phytohemagglutinin. This is associated with eating raw or undercooked red kidney beans and usually causes symptoms in 1–3 hr following exposure. Patients may develop severe nausea and vomiting that can be followed by diarrhea. The toxin is heat-sensitive but needs to reach a high enough temperature to be inactivated. A number of cases have been linked to beans cooked in slow cookers, in which the temperature does not get high enough. The outcome is usually good and supportive therapy is all that is needed. Many other agents that may occur in food, such as the pyrrolizidine alkaloids causing liver damage, and a variety of chemicals and heavy metals can be included in the long list of agents that cause food-borne illness; however, it is beyond the scope of this article to discuss them.

VII. SUMMARY

As time moves on, we are finding more and more microbes of multiple types that are associated with food-borne illness. The various bacteria, protozoa, viruses, and natural chemicals discussed in this chapter include the major ones—but there may be others that we do not even know about yet. It is disconcerting that the majority of cases of gastroenteritis that may be related to food and water go undiagnosed. The majority of our epidemiological data is based on less than 5% of cases, and many laboratories do not routinely look for many of the enteric viruses that are probably causing much of the disease. In recent years, food safety has become a major issue, following a number of highly publicized outbreaks involving a variety of enteric pathogens, including *E. coli*, *Salmonella*, *Listeria*, *Cyclospora*, and Hepatitis A. The food industry has made major efforts to improve the safety of food-processing and we are beginning to see the benefits of this with lower levels of bacterial pathogens in poultry. One of the key unanswered questions in relation to food-borne illness is a determination of the outcome, in most cases. We presume that, in the vast majority of food-borne disease, the outcome is good, with no long-term problems. However, we do not know this for a fact. Occasionally however, the outcome is disastrous, with the development of conditions such as HUS, Guillain-Barré, or reactive arthropathy following infection with enteric pathogens. The story is very different in developing countries, where sanitation may be poor and where fecal contamination of food and water is frequent. In such places, food-borne illness is probably killing hundreds of thousands of children each year and resulting in nutritional deficiency in many more. In developed nations, our immediate goals are to further our understanding of how these organisms cause disease, and how to reduce the load in the animals and fresh produce that provide our food supply. In developing countries, simply instituting basic sanitation to reduce fecal–oral spread can have a huge impact on morbidity and mortality. Maintaining good personal hygiene and sanitation will reduce the chance of fecal–oral spread. Similarly, paying attention to food handling, proper cooking, and the avoidance of temperature abuse will all go a long way to reduce the burden of food-borne illness.

BIBLIOGRAPHY

Allos, B. M. (2001). *Campylobacter jejuni* infections: update on emerging issues and trends. *Clin. Infect. Dis.* Apr 15; **32**(8), 1201.

Appleton, H. (2000). Control of food-borne viruses. *Br. Med. Bull.* **56**(1), 172–183.

Balaban, N., and Rasooly, A. (2000). Staphylococcal enterotoxins. Int. J. Food Microbiol. Oct 1; **61**(1): 1–10.

Blaser, M. J., Smith, P. D., Ravdin, J. I., Greenberg, H. B., and Guerrant, R. L. (eds.) (1995). "Infections of the Gastrointestinal Tract." Raven Press, New York.

Doyle, M. P., Beuchat, L. P., and Montville, T. J. (1997). "Food Microbiology: Fundamentals and Frontiers." ASM Press, Washington.

Hohmann, E. L. (2001). Nontyphoidal salmonellosis. *Clin. Infect. Dis.* Jan 15; **32**(2), 263–269.

Karras, D. J. (2000). Incidence of foodborne illnesses: preliminary data from the foodborne diseases active surveillance network (FoodNet). *Ann. Emerg. Med.* Jan; **35**(1), 92–93.

LaMont, J. (ed.) (1997). "Gastrointestinal Infections." Marcel Dekker, Inc., New York.

Lorber, B. (1997). Listeriosis. *Clin. Infect. Dis.* **24**, 1–11.

Nataro, J. P., and Kaper, J. B. (1998). Diarrheagenic *Escherichia coli*. *Clin. Microbiol. Rev.* **11**, 142–201.

Paton, J. C., and Paton, A. W. (1998). Pathogenesis and diagnosis of Shiga toxin-producing *Escherichia coli* infections. *Clin. Microbiol. Rev.* **11**, 450–479.

Sears, C. L., and Kaper, J. B. (1996). Enteric bacterial toxins: mechanisms of action and linkage to intestinal secretion. *Microbiol Rev.* **60**, 167–215.

Smith, J. L. (1998). Foodborne illness in the elderly. *J. Food Protect.* **61**, 1229–1239.

Vazquez-Boland, J. A., Kuhn, M., Berche, P., Chakraborty, T., Dominguez-Bernal, G., Goebel, W., Gonzalez-Zorn, B., Wehland, J., and Kreft, J. 2001. *Listeria* pathogenesis and molecular virulence determinants. *Clin. Microbiol. Rev.* Jul; **14**(3), 584–640.

WEBSITES

Foodborne Illnesses. CDC
http://www.cdc.gov/health/foodill.htm

Website of the US Food and Drug Administration
http://www.dietsite.com/NutritionFacts/FoodSafetyLinkUSDA/foodborne_illnesses.htm

The Food Microbiology Division, ASM (American Society for Microbiology)
http://www.asmusa.org/division/p/index.htm

Relevant links to PubMed
http://www.ohsu.edu/cliniweb/G1/G1.273.540.html

Photo Gallery of Bacterial Pathogens (Neil Chamberlain)
http://www.geocities.com/CapeCanaveral/3504/gallery.htm

Access to online resources on Bacterial Infections and Mycoses. Karolinska Institutet
http://www.mic.ki.se/Diseases/c1.html

41

Fungal infections, Cutaneous

Peter G. Sohnle and David K. Wagner

Medical College of Wisconsin and Milwaukee VA Medical Center

GLOSSARY

candidiasis An infection caused by fungal organisms of the genus *Candida*.

dermatophytosis A condition resulting from infection of keratinized structures, including hair, nails, and stratum corneum of the skin, by three genera of fungi termed the dermatophytes.

dermis A thick layer of skin tissue below the epidermis consisting of loose connective tissue and containing blood and lymph vessels, nerves, sweat and sebaceous glands, and hair follicles.

epidermis The outer layer of the skin, consisting of dividing and maturing epidermal cells with an outermost layer of dead, keratinized cells called the stratum corneum.

keratin An epidermal cell protein that makes up the hair, nails, and stratum corneum of the skin.

mycetoma Chronic cutaneous and subcutaneous infection resulting from direct implantation of actinomycetes or true fungi.

sporotrichosis A chronic infection of the skin, subcutaneous tissue, and sometimes deep tissues with the fungus *Sporothrix schenckii*.

tinea (pityriasis) versicolor An infection of the stratum corneum of the skin with the yeastlike fungal organism *Malassezia furfur*.

Cutaneous Fungal Infections encompasses a discussion of the major cutaneous host defense mechanisms and a description of the various superficial and deeper infections of the skin. The skin has a number of physical and chemical factors which make it difficult for microorganisms to survive and grow in this location. The immune system, particularly cell-mediated immunity and the activity of phagocytic cells, appears to be important in the defense against cutaneous fungal infections. Mechanisms involved in generating cutaneous immunologic reactions are particularly complex, with epidermal Langerhans cells, other dendritic cells, lymphocytes, microvascular endothelial cells, and the keratinocytes themselves all playing important roles. Infections involving the skin include both superficial and deep mycoses. The superficial mycoses are infections of the epidermis and cutaneous appendages with a number of yeasts and filamentous fungi that are well adapted for growth at this location. The resulting diseases include cutaneous candidiasis, dermatophytosis, tinea (pityriasis) versicolor, and some related mycoses. Deeper cutaneous mycoses, such as sporotrichosis, chromoblastomycosis, and mycetoma, may begin with direct implantation of the organisms into the skin through accidental punctures involving contaminated objects. The skin may also be involved by disseminated fungal infections such as North and South American blastomycosis, cryptococcosis, and several other types of fungal infections.

I. CUTANEOUS HOST DEFENSES

A. Structure of the skin

The physical and chemical structure of the skin represents a form of defense against fungal pathogens. The skin surface is relatively inhospitable to fungal growth because of exposure to ultraviolet light, low

The Desk Encyclopedia of Microbiology
ISBN: 0-12-621361-5

TABLE 41.1 Superficial cutaneous mycoses and the common responsible pathogens[a]

Type of infection	Pathogens
A. Cutaneous candidiasis	*Candida albicans*
B. Dermatophytosis	*Trichophyton, Microsporum, Epidermophyton*
1. Tinea pedia	*T. rubrum, T. mentagrophytes, E. floccosum*
2. Tinea cruris	*E. floccosum, T. rubrum, T. mentagrophytes*
3. Tinea barbae	*T. rubrum, T. verrucosum, T. mentagrophytes*
4. Tinea unguium (onychomycosis)	*T. rubrum, T. mentagrophytes, E. floccosum*
5. Tinea capitis	*T. tonsurans, T. schoenleini (favus), T. violaceum, M. canis*
6. Tinea corporis	*T. rubrum, T. mentagrophytes, T. concentricum, T. verrucosum, T. tonsurans, M. canis, M. gypseum, E. floccosum*
C. Tinea versicolor	*Malassezia furfur* (*Pityrosporum orbiculare*)
D. Malassezia folliculitis	*M. furfur*
E. Tinea nigra	*Phaeoannellomyces werneckii* (*Exophialia werneckii*)
F. White piedra	*Trichosporon beigelii*
G. Black piedra	*Piedraia hortae*

[a]Reprinted with permission from Wagner and Sohnle, 1995, *Clin. Microbiol. Rev.* **8**, 317–335, American Society for Microbiology.

resulting in the typical swelling and redness of this type of candida infection.

In some cases, superficial *C. albicans* infections may be particularly severe and recalcitrant to treatment, producing the uncommon disorder known as chronic mucocutaneous candidiasis. This condition consists of persistent and recurrent infections of the mucous membranes, skin, and nails, along with a variety of other manifestations. The superficial infections last for years in affected patients unless they are properly treated, although deep candida infections are very rare in this situation. Oral thrush and candida vaginitis are fairly common in patients with chronic mucocutaneous candidiasis. There is often infection of the esophagus, although further extension into the viscera is unusual. Epidermal neutrophilic microabscesses, which are common in acute cutaneous candidiasis, are rare in the lesions of chronic mucocutaneous candidiasis. The oral lesions are generally tender and painful. A number of other disorders are associated with the syndrome of chronic mucocutaneous candidiasis, including endocrine dysfunction, vitiligo, dysplasia of the dental enamel, congenital thymic dysplasia, thymomas, and certain other infections. Chronic mucocutaneous candidiasis, no doubt, represents a group of syndromes with a variety of predisposing or secondary abnormalities in host defense function, most commonly deficient cell-mediated immune responses against candida antigens.

The diagnosis of superficial candidiasis is usually suspected on clinical grounds and can be confirmed in skin scrapings by demonstrating the organism using potassium hydroxide preparations and/or culture on appropriate antifugal media. Long term (3–9 months) treatment with azole antifungal drugs can produce good results in chronic mucocutaneous candidiasis, although occasional failures have occurred due to the development of resistant strains of *C. albicans*. Patients who present with chronic mucocutaneous candidiasis should be evaluated for the presence of infection with the human immunodeficiency virus and, if presenting as adults, for the possibility of thymoma.

2. *Dermatophytosis*

Dermatophytoses are infections of keratinized structures, such as the nails, hair shafts, and stratum corneum of the skin, by organisms of three genera of fungi termed the dermatophytes. Although they are not part of the normal human skin flora, these organisms are particularly well adapted to infecting this location because they can use keratin as a source of nutrients—unlike most other fungal pathogens. The different types of dermatophytosis are classified according to body site, using the word "tinea," followed by a term for the particular body site. The major types of dermatophytosis and the most frequent organisms associated with them are listed in Table 41.1. The degree of inflammation produced in the lesions appears to depend primarily on the particular organism and perhaps also to some extent on the immunological competence of the patient.

Tinea pedis (athlete's foot) is probably the most common form of dermatophytosis. This condition is a chronic toe web infection that can be scaly, vesicular, or ulcerative in form and which can sometimes produce hyperkeratosis of the sole of the foot. Tinea cruris is an expanding dermatophyte infection in the flexural areas of the groin and occurs much more frequently in males than in females. Dermatophytosis of the major surface areas of the body is termed tinea corporis. These infections frequently take the classical annular, or "ringworm," shape. Involvement of the beard area in men, a condition known as tinea barbae, is often caused by zoophilic organisms such as *T. verrucosum*. Tinea unguium is a form of onychomycosis, or fungal infection of the nails, and is most frequently caused by *T. rubrum*. Nail infections, particularly of the toenails, are among the most difficult type of dermatophytosis to treat. Infection of the hair and skin on the scalp is called tinea capitis and is more common in children than adults. Dermatophytes rarely invade the deep tissues or produce systemic infections, even in severely immunocompromised patients.

Diagnosis of dermatophytosis is made in a similar manner to that of cutaneous candidiasis, with examination of skin scrapings by potassium hydroxide preparations and culture on appropriate fungal media. The treatment of this condition has improved markedly in recent years with the development of new antifungal agents for topical application or oral administration.

However, certain kinds of dermatophytosis, including widespread infections and those of hair and nails, will often respond poorly to topical therapy and will require prolonged courses of an oral antifungal agent, such as griseofulvin, ketoconazole, itraconazole, fluconazole, or terbinafine. There is an opportunistic fungal organism, *Scytalidium dimidiatum* (*Hendersonula toruloidea*), that can produce conditions clinically mimicking those caused by the usual dermatophyte species, but which does not respond well to conventional antifungal therapy.

3. *Tinea (pityriasis) versicolor and malassezia folliculitis*

Tinea versicolor is a chronic superficial fungal infection of the skin generally affecting the trunk or proximal parts of the extremities and caused by the yeast *M. furfur* (*Pityrosporum orbiculare*). The organism is lipid-requiring and will not grow on most laboratory media. The lesions resulting from infection with *M. furfur* are macules that may coalesce into large, irregular patches characterized by fine (pityriasiform) scaling, along with hypopigmentation or hyperpigmentation. These infections can persist for years unless treated appropriately. *M. furfur* has also been postulated to play a role in certain other diseases, including atopic dermatitis, seborrheic dermatitis, psoriasis, and reticulate papillomatosis. Malassezia folliculitis is a condition that resembles several other cutaneous infections, including acne vulgaris, the macronodular lesions of disseminated candidiasis in immunosuppressed patients, the candidal papular folliculitis of heroin addicts and graft versus host disease in bone marrow transplant recipients. The papules of this condition begin as an inflammation of the hair follicles, instead of the macules typical of tinea versicolor, and may progress to frank pustules.

In tinea versicolor, potassium hydroxide preparations of skin scrapings reveal the typical grape-like clusters of yeast and tangled webs of hyphae of the causative fungus, yielding the diagnosis of this condition. The organism is not usually cultured because of the requirement for specialized media. Tinea versicolor can be treated topically with lotions or creams containing selenium or sodium thiosulfate, specific antifungal agents, or sulfur–salicylic acid shampoo. Oral azole antifungal drugs can also be used for more difficult cases. Malassezia folliculitis can be treated using topical antifungal agents or an oral azole antifungal drug.

4. *Miscellaneous superficial fungal infections*

Tinea nigra is a superficial mycosis of the palms that is most often caused by *Phaeoannellomyces werneckii* (*Exophiala werneckii*). The lesions are generally dark colored, non-scaling macules that are asymptomatic, but can be confused with melanomas and perhaps result in unnecessary surgery. Tinea nigra is most often seen in tropical or semitropical areas of Central and South America, Africa, and Asia, although some cases do occur in North America. This condition can be treated effectively with either keratinolytic agents or topical azoles. White piedra is an asymptomatic fungal infection of the hair shafts that is caused by *Trichosporon beigelii*. This infection produces light-colored, soft nodules on the hair shafts and may cause the involved hairs to break. Otherwise, this condition appears to be asymptomatic, although the causative fungus can produce serious infections in immunocompromised patients. Black piedra is similar to white piedra in that it is a nodular, generally asymptomatic, fungal infection of the hair shafts. It is caused by *Piedraia hortae* and most commonly affects the scalp hair. Black and white piedra are generally treated by clipping off the affected hairs.

B. Deeper cutaneous and subcutaneous mycoses

The dermis and subcutaneous tissues can be infected by a variety of fungal agents that are directly implanted into the skin by punctures with sharp objects contaminated by the organisms. Dissemination to the skin from other infected sites, especially the lungs, is also possible. The most common organisms causing the deep cutaneous and subcutaneous mycoses are listed in Table 41.2.

TABLE 41.2 Deep cutaneous and subcutaneous mycoses and the common responsible pathogens

Type of infection	Pathogens
A. Sporotrichosis	*Sporothrix schenckii*
B. Chromoblastomycosis	*Fonsecaea pedrosoi, F. compacta, Phialophora verrucosa, Cladophialophora carrionii, Botryomyces caespitosus, Rhinocladiella aquaspersa*
C. Eumycotic mycetoma	*Pseudallescheria boydii, Madurella mycetomatis, M. grisea, Acremonium spp., Leptosphaeria senegalensis*
D. North American blastomycosis	*Blastomyces dermatitidis*
E. South American blastomycosis	*Paracoccidioides braziliensis*
F. Histoplasmosis	*Histoplasma capsulatum*
G. Cryptococcosis	*Cryptococcus neoformans*
H. Infections with immunosuppression	*Trichosporon beigelii, Blastoschizomyces capitatus, Fusarium* spp.

1. *Sporotrichosis*

This condition is generally caused by accidental implantation of the causative fungus *Sporothrix schenckii* into the skin. The lesions most often consist of cutaneous and subcutaneous nodules extending up the limb from the site of inoculation. However, spread may occur through the lymphatics or blood vessels to the bones, joints, or other organs. It is also possible to develop lesions in the lungs by inhalation of the fungal elements. The causative organism is a dimorphic fungus that exists as either hyphae or elongated yeast cells.

The most common reservoir of the fungus in nature is on vegetation such as rose bushes, sphagnum moss, or in soil. The site of implantation may develop into a papule or pustule and cutaneous nodules may then develop proximally in a linear fashion. If the fungus is inhaled, it may cause a granulomatous pneumonitis that can cavitate and produce a clinical picture similar to tuberculosis. Immunosuppressed patients are more likely to develop disseminated disease.

Demonstration of the characteristic small, cigar-shaped yeast cells is diagnostic but often difficult. Multiple sections may have to be examined. Stellate, periodic acid-Schiff (PAS) positive eosinophilic material surrounding the organisms are known as asteroid bodies. The diagnosis of sporotrichosis is best made by culture of material from the lesions on appropriate fungal media. Isolation of this organism is usually indicative of sporotrichosis in that the fungus is not part of the normal flora of humans. Iodides may be given for cutaneous sporotrichosis, with oral itraconazole or fluconazole being used if these measures fail. For disseminated disease, either amphotericin B or itraconazole are generally effective, although relapse is common.

2. *Chromoblastomycosis*

Certain species of the dematiaceous (darkly pigmented) fungi can cause chronic cutaneous and subcutaneous infections. A number of genera can be involved, but *Fonsecaea*, *Phialophora*, *Cladophialophora*, and *Rhinocladiella* are most common. The dark pigment of these organisms is dihydroxynaphthalene melanin, which is different from the dihydroxyphenylanine melanin associated with *Cryptococcus neoformans*. Dematiaceous fungi can also cause mycetoma. Chromoblastomycosis is characterized by the presence of sclerotic (muriform) bodies in the tissues. When yeastlike cells, pseudohyphae, or hyphae of the dematiaceous fungi are present in the tissues, the term "phaeohyphomycosis" is used. Chromoblastomycosis usually results from implantation of the organisms from local trauma, usually to the feet or legs. Usually, the first lesion is an erythematous papule, followed by scaling and crusting, with eventual development into a warty structure. The pathology is characteristic of a suppurative granuloma, often with overlying pseudoepitheliomatous hyperplasia. The distribution of cases is worldwide, although most come from Central and South America.

The finding of the characteristic cross-walled, pigmented sclerotic bodies is pathognomonic of chromoblastomycosis. However, since those formed by all the relevant dematiaceous fungal species are similar, culture of the infecting organism on fungal media containing cycloheximide and antibiotics is necessary to identify it. Treatment may be difficult in that the organisms may not be sensitive to antifungal agents. Surgery or local heat may be other options in the early stages of the disease.

3. *Mycotic mycetoma*

Mycetomas are swellings with draining sinuses and grains. They usually affect the feet, legs, or hands and begin with direct implantation of the causative organisms. The latter are either actinomycetes (actinomycotic mycetoma) or the true fungi (eumycotic mycetoma). About half of the cases of mycetoma are caused by true fungi, including the genera of *Madurella*,

Leptosphaeria, Pseudoallescheria, Acremonium, and several others. Initially, pain and discomfort develop at the implantation site, followed weeks or months later by induration, abscess development, granulomas, and draining sinuses. The lesions may extend to bone and cause severe bony destruction. Eumycotic mycetoma are rare in the United States, although those caused by *Pseudallescheria boydii* are more common in the latter location.

Specimens of exudate or biopsy material should be examined by the naked eye for the presence of grains. The latter can be gram-stained and examined microscopically to differentiate actinomycotic from eumycotic mycetoma. The grains should be washed with saline containing antibacterial compounds and then cultured on Sabouraud dextrose agar containing chloramphenicol and cycloheximide, as well as on media for bacterial and actinomycotic organisms. Identification of the organisms is based on gross colonial morphology, pigmentation, and mechanism of conidiogenesis. Treatment of eumycotic mycetoma is often unsatisfactory because the causative organisms generally show poor sensitivity to available antifungal agents. Amputation of an infected limb or surgical debridement of infected tissue may be necessary. Amphotericin B or azole antifungal drugs can be used if the particular fungal strain is sensitive. If not treated effectively, mycetomas may progress for years and produce marked tissue damage, deformity, and even death.

C. Systemic mycoses with cutaneous manifestations

A number of deep fungal infections may produce cutaneous lesions as part of a disseminated disease process. In these infections, the portal of entry is usually the lung, with development of a pneumonia and spread to other organs. Dissemination is most likely to occur in patients with compromised host defenses, although blastomycosis has a very high incidence of skin lesions in noncompromised individuals. The group of infections under discussion here is different from those in the last section, where direct implantation into the skin or subcutaneous tissues is the usual mode of entry.

North American blastomycosis a systemic fungal infection well known to have cutaneous dissemination. Skin involvement occurs in approximately 60% of cases at some time during the illness, even though the pulmonary lesions may have healed at the time the skin manifestations develop. The latter generally consist of either verrucous lesions that look like squamous cell carcinomas or ulcerative lesions that begin as pustules or ulcerating nodules. The patient may either be quite ill with symptoms and signs from other manifestations of the infection or may have only skin lesions and be relatively asymptomatic otherwise. Coccidioidomycosis is a fairly similar disease which can produce papules, pustules, plaques, nodules, ulcers, abscesses, or large proliferative lesions. Dissemination to the skin is less common in coccidioidomycosis than in blastomycosis, although chronic meningitis is a more prominent feature of the former. The respective organisms of the two diseases are *Blastomyces dermatitidis* and *Coccidioides immitis;* they are endemic fungi found in different parts of the United States. A similar endemic fungal infection is histoplasmosis, caused by *Histoplasma capsulatum;* however, this disease is less likely to have cutaneous manifestations.

South American blastomycosis (paracoccidioidomycosis) is an important systemic mycosis in Latin America; this infection begins in the lungs and then frequently disseminates to the skin, mucous membranes, reticuloendothelial system, and elsewhere. It is caused by *Paracoccidioides braziliensis*. The cutaneous lesions tend to appear around the natural orifices and may be verrucous, ulcerating, crusted, or indurated and granulomatous. Cryptococcosis is an infection beginning in the lungs and caused by *Cryptococcus neoformans*. Like coccidioidomycosis, this disease tends to disseminate to the central nervous system to cause a chronic meningitis, but it can also cause skin lesions. The latter are generally papules or nodules with surrounding erythema; they are often found on the face. Severely immunosuppressed patients are at risk for disseminated disease with a number of other fungi such as *T. beigelii* (the cause of white piedra, discussed previously), *Blastoschizomyces capitatus*, and *Fusarium* spp. Each of these agents causes erythematous papules or nodules, and, in some cases, these lesions may break down to form ulcers.

Diagnosis of these various infections is usually obtained by demonstrating the organisms in specimens of sputum, blood, or biopsy material from the skin or other organs. Often, the individual fungi can be identified on appropriately stained smears or on histologic sections from biopsy material; otherwise, they may be cultured on appropriate fungal media. The latex agglutination test for cryptococcal antigen in serum or cerebrospinal fluid is very helpful for diagnosing and following cryptococcosis. Skin tests and other serological tests are generally less useful in diagnosis of the other fungal infections discussed above than is demonstration of the organisms. Treatment usually can be accomplished through the use of amphotericin B or the azole antifungal drugs. The various organisms vary somewhat in susceptibility, with

often invade and extend along blood vessels, causing tissue infarction and necrosis. A diagnosis of aspergillosis is usually made by culture and pathological examination of tissue samples in the appropriate clinical setting. A major problem with aspergillosis is the absence of good diagnostic tests for the detection of early infection. Systemic aspergillosis has high mortality despite antifungal therapy.

B. Blastomycosis

Systemic blastomycosis is caused by the dimorphic fungus *Blastomyces dermatitides*. In North America, the infection is endemic in the Mississippi and Ohio river valleys and areas adjoining the Great Lakes and the St. Lawrence River. Blastomycosis also occurs in Africa, Central America, and the Middle East. *B. dermatitides* is believed to reside in soils and decaying vegetation but its exact environmental niche remains to be defined. Initial infection is believed to occur from inhalation of conidia that germinate to yeast cells in the lung. Several outbreaks of blastomycosis have been associated with recreational activities in sites near rivers. In immunologically normal individuals, the infection is contained in the lung and humans appear to have a high level of resistance to disseminated infection. Systemic blastomycosis results when pulmonary blastomycosis disseminates to other organs. Dissemination can occur to practically any organ but the skin is the site most commonly involved. The diagnosis of blastomycosis infection is made by culture or pathological examination of involved tissue in the appropriate clinical setting. Blastomycosis is also a major fungal infection in dogs.

C. Candidiasis

Systemic candidiasis can be caused by any of several *Candida* sp. and is the most common systemic fungal infection. Systemic candidiasis is distinguished from the more common type of candidal infections, such as thrush and vaginal candidiasis, by involvement of the blood stream and internal organs. The name *Candida* is used to refer to more than one hundred fungal species, of which about a dozen are important human pathogens. *C. albicans* is the most common *Candida* sp. that causes human disease. However, other species, such as *C. glabrata, C. parapsilosis,* and *C. krusei*, commonly cause serious infections in certain patient groups. *Candida* sp. are components of the human microbial flora and systemic candidiasis differs from the other systemic fungal infections in that this infection almost always originates from the endogenous microbial flora. Candidal infections are very common in hospitalized patients and *Candida* species are frequent causes of bacteremia and deep-seated organ infections. Most cases of systemic *C. albicans* infection are iatrogenic and are related to the use of antibiotics, the presence of indwelling catheters, and surgical implantations of prosthetic devices. Antibiotics predispose to infection by suppressing the normal bacterial flora and this, presumably, allows *Candida* to proliferate and invade the gut mucosa. Indwelling catheters provide a break in the skin that allows *Candida* access to the bloodstream and the internal skin layers. Prosthetic devices provide surfaces where the fungus can attach and escape clearance by host defense mechanisms. Systemic candidal infections can affect virtually any organ and the eye, liver, spleen, and kidneys are frequently involved. The diagnosis of systemic candidiasis is usually made by culture of *Candida* sp. from a normally sterile body site. However, in many patients, making a diagnosis of systemic candidiasis is difficult because the blood cultures are negative and there are few diagnostic tests to detect deep-seated organ infections.

D. Coccidioidomycosis

The causative agent of coccidioidomycosis is *Coccidioides immitis*, a dimorphic soil fungus. This infection is found in certain areas of North and South America. In the United States, coccidioidomycosis is highly prevalent in the Southwest. This infection is believed to be acquired by the inhalation of arthroconidia that swell in tissue to become spherule-containing internal spores. Primary infection is usually either asymptomatic or reassembles a common upper respiratory infection. Individuals with occupations that involve exposure to soils in endemic areas are at higher risk. Outbreaks of coccidioidomycosis have followed archeological investigations. Systemic coccidioidomycosis occurs in less than 1% of primary infections and can involve virtually any organ. Extrapulmonary coccidioidomycosis is more common in individuals with impaired immunity. However, pregnant women and individuals of Filipino, African, and Mexican ancestry might be at increased risk for disseminated infection. Coccidioidomycosis can present as acute infection, chronic pulmonary disease, or systemic infection. The diagnosis is made by culture, histological examination of tissue, and/or serological testing.

E. Cryptococcosis

The causative agent of cryptococcosis is *Cryptococcus neoformans*. The prevalence of *C. neoformans* infection has risen dramatically in recent years in association with HIV infection. In New York City, there were more than 1200 cases of cryptococcosis in 1991, of which most occurred in patients with AIDS. Cryptococcal meningitis occurs in 5–10% of all patients with advanced HIV infection. Cryptococcal strains have been divided into two varieties known as *neoformans* and *gattii*. Variety *neoformans* is found worldwide, is associated with bird (usually pigeon) excreta, and is the predominant agent of cryptococcosis in patients with AIDS. Variety *gattii* is found in the tropics, is associated with eucalyptus trees, and can cause infections in apparently normal hosts. *C. neoformans* is unusual among fungal pathogens, in that it has a polysaccharide capsule that is important for virulence (Fig. 42.1). *C. neoformans* is acquired by inhalation, where it usually causes an asymptomatic pulmonary infection. In patients with impaired immunity, extrapulmonary dissemination can results in cryptococcal meningitis, the most common clinical presentation of cryptococcosis. Pathological examination of tissues infected with *C. neoformans* often reveals little or no inflammatory response and this phenomenon is believed to be caused, in part, by the immunosuppressive effects of the capsular polysaccharide. The diagnosis of cryptococcosis is usually made by culture from cerebrospinal fluid or blood. The capsular polysaccharide is shed into body fluid where it can be detected by serological assays. Detection of cryptococcal polysaccharide antigen is useful in diagnosis and in following the response to therapy.

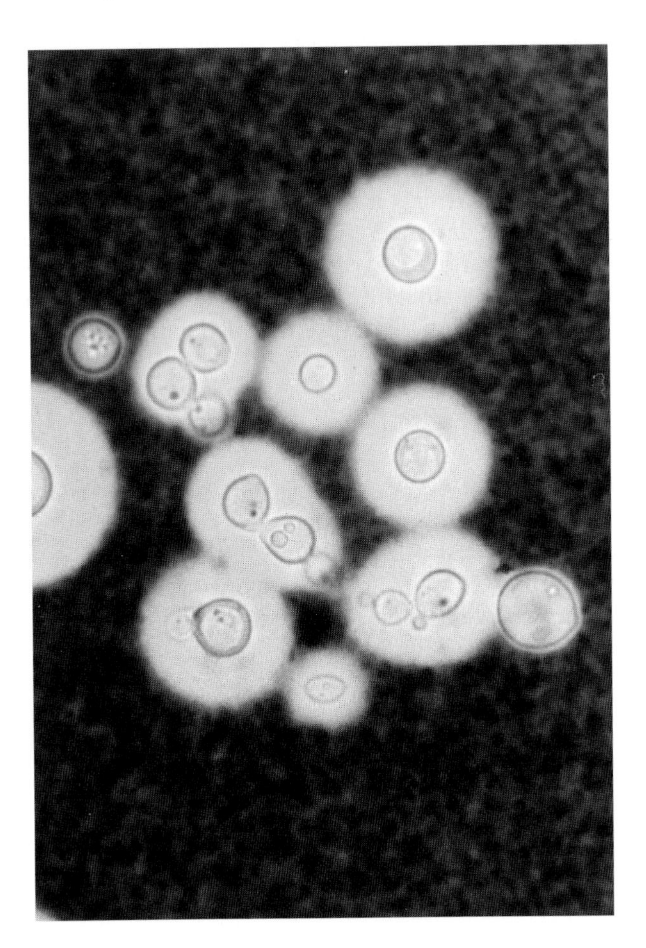

FIGURE 42.1 India-ink preparation showing *C. neoformans* cells. The cell bodies are surrounded by a polysaccharide capsule which is important for virulence. Note differences in cell size and capsule size among the various cells shown. Photograph obtained at a magnification of ×200.

F. Histoplasmosis

The causative agent of histoplasmosis is *Histoplasma capsulatum*, a soil organism that is common in the Ohio and Mississippi river valleys and in various parts of South America. *H. capsulatum* is often found in soils contaminated by bird excrement. Small epidemics have occurred because of large exposures created by disturbing contaminated sites during constructions, excavations, tree cuttings, etc. In 1978–1980, two major outbreaks of histoplasmosis occurred in the city of Indianapolis that may have been related to construction projects. *H. capsulatum* is a dimorphic fungus that grows as a mycelial form at environmental temperatures and as a yeast at mammalian body temperature. Two varieties of *H. capsulatum* are known: variety *capsulatum* is found in the Americas and variety *duboisii* is found in Africa. Infection presumably occurs by the inhalation of conidia and small mycelial fragments that convert to yeast forms in lung tissue. The clinical presentation of pulmonary histoplasmosis is similar to that of pulmonary tuberculosis, such that many cases of histoplasmosis were confused with tuberculosis until specific diagnostic methods became available. The overwhelming majority of primary pulmonary infections are asymptomatic. However, many infections become chronic and some disseminate. The probability of acquiring disseminated disease ranges from about 1 in 2000 for normal individuals to up to 27% for patients with advanced HIV infection. Disseminated disease can affect virtually any organ, with adrenal, skin, gastrointestinal, and central nervous system involvement being particularly common. The diagnosis of systemic histoplasmosis is made by culture of the fungus from blood, bone marrow, sputum, cerebrospinal fluid, or the affected body site. However, culture methods can yield false-negative results. A presumptive diagnosis can be made by visualizing yeast cells in infected tissues. In this regard, *H. capsulatum* can sometimes be detected in peripheral blood leukocytes of patients with disseminated histoplasmosis. Serological tests for histoplasma antibody and antigen can provide important clues to the presence of *H. capsulatum* infection. Antigen detection in urine samples is particularly useful in cases where the diagnosis is suspected but cultures are negative.

G. Other systemic fungal infections

Many other fungal species can cause invasive infection besides the more common systemic fungal infections listed above. In Southeast Asia, *Penicillium marneffei* is a major cause of invasive fungal infection in patients with advanced HIV infection. *Paracoccidioides brasiliensis* is a major fungal pathogen in some areas of South America. *P. brasiliensis* causes asymptomatic infection in normal individuals that can remain latent and disseminate if the immune system subsequently becomes impaired. Recently, there have been several reports of systemic fungal infection with the *Saccharomyces cerevisiae*, which is commonly known as brewer's or baker's yeast. Although rare, these cases illustrate how a usually nonpathogenic organism can cause serious infection if it colonizes a susceptible host. *Pseudallescheria boydii* is a mold that causes severe infections in patients with prolonged neutropenia or who are receiving high-dose corticosteroid therapy. *P. boydii* infections are similar to those caused by *Aspergillus* sp. *Sporothrix schenckii* is a dimorphic fungus, found in soils and plants, that can cause sporotrichosis after inoculation in the skin. Sporotrichosis has been reported throughout the world but is associated primarily with activities that result in exposure to plants, such as gardening and farming. Mucormycosis is caused by a variety of fungal species with a complex taxonomy, including *Rhizopus*, *Absidia*, and *Mucor*. In diabetic patients with poorly controlled hyperglycemia, mucormycosis is a devastating and often incurable infection. Phaeohyphomycosis is caused by a variety of fungal species that have dark cell walls. Cerebral phaeohyphomycosis is a rare but rapidly lethal brain infection that has been commonly associated with the fungus *Clasdosporium trichoides*.

IV. TREATMENT AND PREVENTION OF SYSTEMIC FUNGAL INFECTIONS

Most systemic fungal infections are fatal without antifungal therapy. Unlike the situation for bacterial pathogens, the antibiotic arsenal against the fungi is small and consists of no more than half a dozen drugs. Since the late 1950s, Amphotericin B has been the mainstay of therapy for many invasive mycoses. Amphotericin B is a powerful fungicidal agent that binds to fungal sterols and kills the fungal cell by disrupting cellular membranes. Amphotericin B may also have important immunomodulatory effects that could contribute to its therapeutic efficacy. Amphotericin B has significant toxicity that can be lessened by incorporating it into liposomal preparations but these are significantly more expensive. In recent years, other agents that target the sterol metabolic pathways have been introduced, including fluconazole and itraconazole. These agents are usually fungistatic but are much less toxic than amphotericin B and have the added advantages of being available in oral formulations. 5-Fluorocytosine is another antifungal drug that is effective when used in combination with other antifungal drugs. Several antifungal agents are in preclinical and clinical development and newer agents with enhanced efficacy and reduced toxicity may be available in the future.

A major problem in the therapy of systemic fungal infections is that antifungal chemotherapy is less effective in the setting of defective immunity. For example, antifungal therapy cannot usually eradicate *C. neoformans*, *H. capsulatum*, and *C. immitis* infections in patients with advanced HIV infection. As a result, affected individuals must be given lifelong suppressive therapy to reduce the likelihood of clinical recurrence of infection. The difficulties associated with the therapy of systemic fungal infections have stimulated interest in immunotherapy but this therapeutic strategy is still experimental.

There are ongoing efforts to develop vaccines against coccidioidomycosis, histoplasmosis, and cryptococcosis but none is currently available. At this time, the two main strategies for the prevention of systemic infection in patients at risk for infection include avoidance of infection and the use of prophylactic antifungal drugs. For many systemic fungal infections, prevention is difficult because the fungal pathogen is highly prevalent in the environment. For example, *C. neoformans* is found in high concentration in pigeon excreta in urban areas such as New York City, where many patients with advanced HIV infection live. Similarly, *H. capsulatum* and *C. immitis* are prevalent in soils of specific geographic areas of the world and avoiding exposure may be difficult for residents in those regions. Nevertheless, it is prudent for individuals with immunological disorders to avoid sites likely to contain high concentrations of aerosolized fungal pathogens such as construction sites, aviaries, chicken farms, and compost sites. Prophylactic administration of antifungal drugs has been shown to reduce the incidence of certain fungal infections, such as cryptococcosis, in patients at high risk for infection. However, there are concerns that prophylactic drug use will encourage the selection of drug-resistant fungi, and drug prophylaxis is not used routinely for the prevention of fungal infection. Another preventive strategy to reduce invasive fungal infections in patients with neutropenia is to administer colony-stimulating factors that reduce the neutropenic interval by stimulating leukocyte production.

BIBLIOGRAPHY

Calderone, R., Suzuki, S., Cannon, R., Cho, T., Boyd, D., Calera, J., Chibana, H., Herman, D., Holmes, A., Jeng, H. W., Kaminishi, H., Matsumoto, T., Mikami, T., O'Sullivan, J. M., Sudoh, M., Suzuki, M., Nakashima, Y., Tanaka, T., Tompkins, G. R., and Watanabe, T. (2000). *Candida albicans*: adherence, signaling and virulence. *Med. Mycol.* **38**(Suppl 1), 125–137.

Clemons, K. V., Calich, V. L., Burger, E., Filler, S. G., Grazziutti, M., Murphy, J., Roilides, E., Campa, A., Dias, M. R., Edwards, J. E. Jr, Fu, Y., Fernandes-Bordignon, G., Ibrahim, A., Katsifa, H., Lamaignere, C. G., Meloni-Bruneri, L. H., Rex, J., Savary, C. A., and Xidieh, C. (2000). Pathogenesis I: interactions of host cells and fungi. *Med. Mycol.* **38**(Suppl 1), 99–111.

Clemons, J. V., McCusker, J. H., Davis, R. W., and Stevens, D. A. (1994). Comparative pathogenesis of clinical and nonclinical isolates of *Saccharomyces cerevisiae*. *J. Infect. Dis.* **169**, 859–867.

Dixon, D. M., McNeil, M. M., Cohen, M. L., Gellin, B. G., and LaMontagne, J. R. (1996). Fungal infections. A growing threat. *Public Health Rep.* **111**, 226–235.

Fridkin, S. K., and W. R. Jarvis. (1996) Epidemiology of nosocomial fungal infections. *Clin. Microbiol. Rev.* **9**, 499–511.

Hazen, K. C. (1995). New and emerging yeast pathogens. *Clin. Microbiol. Rev.* **8**, 462–478.

Kwon-Chung, K. J., and Bennett, J. E. (1992), "Medical Mycology." Lea & Fabiger, Philadelphia.

Kwon-Chung, K. J., Sorrell, T. C., Dromer, F., Fung, E., and Levitz, S. M. (2000). Cryptococcosis: clinical and biological aspects. *Med. Mycol.* **38**(Suppl 1), 205–113.

Reiss, E., Obayashi, T., Orle, K., Yoshida, M., and Zancope-Oliveira, R. M. (2000). Non-culture based diagnostic tests for mycotic infections. *Med Mycol.* **38**(Suppl 1), 147–159.

Romani, L., and Howard, D. H. (1995). Mechanisms of resistance to fungal infections. *Curr. Opin. Immunol.* **7**, 517–523.

San-Blas, G., Travassos, L. R., Fries, B. C., Goldman, D. L., Casadevall, A., Carmona, A. K., Barros, T. F., Puccia, R., Hostetter, M. K., Shanks, S. G., Copping, V. M., Knox, Y., and Gow, N. A. (2000). Fungal morphogenesis and virulence. *Med. Mycol.* **38**(Suppl 1), 79–86.

Traynor, T. R., and Huffnagle, G. B. (2001 Feb). Role of chemokines in fungal infections. *Med Mycol.* **39**(1), 41–50.

Wheat, J. (1992). Histoplasmosis in Indianapolis. *Clin. Infect. Dis.* **14**(Suppl 1), S91–99.

Vartivarian, S. E. (1992). Virulence properties and nonimmune pathogenetic mechanisms of fungi. *Clin. Infect. Dis.* **14**(Suppl 1), S30–36.

WEBSITES

Website of the International Society for Human and Animal Mycology
http://www.leeds.ac.uk/isham/

Website for clinically significant mycological information
http://www.mycology.adelaide.edu.au/mycology/myco.nsf?OpenDatabase

Medical Mycology Divison, ASM (American Society for Microbiology)
http://www.asmusa.org/division/f/MainPage.html

Website of the Division of Microbiology and Infectious Diseases, NIH
http://www.niaid.nih.gov/dmid/

Access to online resources on Bacterial Infections and Mycoses. Karolinska Institutet
http://www.mic.ki.se/Diseases/c1.html

43

Gastrointestinal microbiology

T. G. Nagaraja
Kansas State University

GLOSSARY

anaerobes Microbes that are capable of generating ATP without the use of oxygen and exhibit various degrees of oxygen sensitivity.

allochthonous Nonindigenous, dormant, in transit, and not characteristic of the habitat.

autochthonous Indigenous, present during evolution of the host, and characteristic of the habitat.

cecum The proximal blind portion of the hindgut.

colon The mid-portion of the hindgut.

competitive exclusion The protective function of the normal flora of the gut to prevent entry and colonization of pathogens.

epimural bacteria Bacteria attached to the epithelial cells lining the gut.

foregut fermentation Microbial fermentation prior to the gastric or peptic digestion.

hindgut fermentation Microbial fermentation after the gastric or peptic digestion.

hydrogenosomes Cytoplasmic organelles present in anaerobic protozoa and fungi and containing enzymes that produce hydrogen from oxidation of reduced cofactors.

normal flora The population of microbes that normally reside in the host and for the most part live in harmony.

peristalsis A wave of contraction followed by relaxation that propels the digesta down the gastrointestinal tract.

rumen The largest of the four compartments of the ruminant stomach, inhabited by a myriad of microbes.

volatile fatty acids Short-chain fatty acids that are major products of microbial fermentation in the gut.

zoospore A free living and flagellated reproductive structure of fungi.

The Gastrointestinal Tract or Digestive Tract is essentially a tubular organ of varying diameter extending from the mouth to the anus. The gut is an open ecosystem, because the lumen is essentially external to the body. The gastrointestinal tract has five major regions: mouth, esophagus, stomach, small intestine (duodenum, jejunum, and ileum), and large intestine (cecum, colon, and rectum). Some of the regions, depending on the species, may be enlarged with or without sphincters or further compartmentalized. Such enlarged or compartmentalized regions slow down the transit of contents, allowing for microbial growth and fermentation.

The gastrointestinal tracts of animals, including humans, and of birds are complex microbial ecosystems. The complexity is attributable to differences in anatomical features, diet, and the health of the animal. The gastrointestinal tract contains distinct microbial populations with diverse compositions. Many of these organisms colonize and grow and are considered indigenous and, hence, are termed "normal flora," also called autochthonous microbiota. These microbes, for the most part, live in harmony with the host. Additionally, the gut flora include allochthonous microbiota that do not get established (colonization and growth) and are dormant and in passage. These are derived largely from ingested food and water and, to a small extent, from swallowed air or from another

The Desk Encyclopedia of Microbiology
ISBN: 0-12-621361-5

habitat of the host (e.g., skin, respiratory tract, or reproductive tract). The nonindigenous microbes also include a variety of gastrointestinal pathogens that may colonize and grow to establish infections. Also, some members of the normal flora could assume pathogenic roles (opportunistic pathogens) when the ecosystem is perturbed in some way or when a breach occurs in the integrity of the gut wall.

I. THE GASTROINTESTINAL ECOSYSTEM

Most of the tract offers conditions that are conducive for microbial growth. The temperature remains relatively constant (36–40°). Water and exocrine secretions (saliva and other digestive secretions) provide a moist environment. Ingested food provides the energy and other nutrients needed for microbial growth. Normal gut motility (peristalsis) helps mix the digesta, which brings microbes into contact with fresh substrate. End products of fermentation (mainly acids) are removed by absorption into the blood. Absorption, coupled with the buffering effect provided by digestive secretions (mainly saliva), helps regulate gut pH. Only in the gastric stomach and the duodenum is the low pH inhibitory to microbial growth. The gut ecosystem often is referred to as a continuous or semicontinuous culture system with more or less continuous availability of substrate, removal of end products (by absorption or passage), and passage of undigested and waste products.

The composition of the gut contents of animals is extremely complex. Physical and chemical conditions within the gut differ among species of animals but are relatively constant within species. However, within species, the composition is influenced by the diet. Among the gut ecosystems, the reticulo-rumen of cattle and the colon of humans are the most extensively investigated (Table 43.1).

The main sites of microbial fermentation differ from species to species and are broadly classified into pregastric, or foregut, and postgastric, or hindgut, fermentations (Table 43.2). Ruminants, such as cattle, sheep, goats, and buffaloes, have a capacious chamber where food is retained for a long time to allow microbial fermentation. Also, there are nonruminant foregut fermenters that have large stomachs but do not ruminate their food for secondary mastication. Hindgut fermenters rely primarily on the cecum and colon for microbial fermentation and vary considerably in their dependence on that fermentation.

TABLE 43.1 Physical, chemical, and microbiological characteristics of the rumen of cattle and the colon of humans

Characteristics	Rumen	Colon
Physical		
Capacity	30–70 liters	0.5–1.2 liters
pH	5.5–7.5	6.5–7.5
Temperature	39 °C	37 °C
Redox potential	−250–350 mV	−200–300 mV
Osmolality	250–350 mOsm/kg	300–400 mOsm/kg
Dry matter	10–18%	10–25%
Chemical		
Acids—Volatile fatty acids	Acetate, propionate, butyrate, valerate, and branched-chain fatty acids	Acetate, propionate, butyrate, and others
—Nonvolatile acids (mM)	Lactate, succinate	Lactate, succinate
Gases	CO_2, CH_4, H_2, H_2S, N_2, O_2	CO_2, CH_4, H_2, H_2S, N_2, O_2
Ammonia	Present	Present
Amino acids and peptides	Small amounts detectable 1–3 h after feeding	Trace amounts
Soluble carbohydrates	Small amounts detectable 1–3 h after feeding	Trace amounts
Complex carbohydrates		
Dietary (cellulose and hemicellulose)	Always present	Always present
Endogenous (mucopolysaccharides)	Present but minimal	Present
Minerals	Present, high sodium	Present, high sodium and chloride
Vitamins	Present and serve as major supply of B vitamins	Present but does not serve as the source of B vitamins
Microbiological		
Bacteria	10^{10}–10^{11}/g	10^{9}–10^{10}/g
Protozoa	10^{4}–10^{6}/g	Absent
Fungi	Present (not quantifiable)	Possibly present
Bacteriophages	Up to 10^{12} particles/ml	Possibly present

colonize the acidic environment because of its high ureolytic activity, which creates a microenvironment rich in ammonia. Species of *Helicobacter* also have been isolated from the stomachs of cats and dogs (*H. felis*), ferrets (*H. mustelae*), and other wild animals.

B. Ruminal bacteria

The bacterial flora of the rumen of cattle is the most extensively investigated and clearly described among the gut bacterial ecosystems. A large variety of bacterial species exist in the rumen. The numbers range from 10^8 to 10^{11}/g of ruminal contents and, with the type of diet the animal consumes, profoundly impact the numbers and species composition. Generally, the numbers are reflective of the digestibility of foodstuffs. Typically, bacterial numbers are up to tenfold higher in grain-fed cattle than in forage-fed animals. The majority of the bacteria are obligate anaerobes. Facultative bacteria comprise a very small fraction of the flora. Ruminal bacteria are predominantly gram negative and rod shaped but vary greatly in their substrate specificity and nutritional requirements (Table 43.4). Some ruminal bacteria (*Butyrivibrio*) are considered "generalists" because of their ability to ferment a range of substrates, including cellulose, hemicellulose, pectin, starch, and protein. Others are "specialists" that utilize a narrow range of substrates. Bacteria capable of utilizing cell wall (cellulose and hemicellulose) and storage polysaccharides (starch) comprise a large fraction of the flora.

Ruminal bacteria are classified into functional groups based on the substrate fermentation (e.g., cellulolytics, amylolytics, or proteolytic) and on the types of fermentation products produced (e.g., lactate producers or methanogens). In pure cultures, ruminal bacteria produce a number of fermentation products, such as lactate, succinate, formate, and hydrogen, which are normally present in low concentrations in the rumen. In addition to interspecies hydrogen transfer, which reduces the need to produce the electron-sink products, cross-feeding among ruminal bacteria allows utilization of the product of one bacterial species by another. Interspecies hydrogen transfer in the ruminal habitat involves almost exclusively methanogens as the hydrogen utilizers. Although acetogens have been shown to be present, they are unable to compete effectively with methanogens for hydrogen. Rapid utilization of hydrogen by methanogens keeps the partial pressure of hydrogen extremely low, which permits direct oxidation of reduced cofactors in nonmethanogenic bacteria. Therefore, production of hydrogen-sink products is not required. Interspecies hydrogen transfer decreases lactate, succinate, propionate, and ethanol and increases acetate.

Ruminal bacteria are nutritionally fastidious. Besides requirements for ammonia, amino acids or peptides, and B vitamins, certain ruminal bacteria, particularly fiber digesters, require one or more branched chain fatty acids (isobutyrate, isovalerate, and 2-methylbutyrate). The branched-chain fatty acids are required for the synthesis of branched amino acids or long-chain fatty acids. These acids cannot be replaced by amino acids. Therefore, culture medium employed to cultivate ruminal bacteria traditionally contains clarified ruminal fluid.

The ruminal habitat also is inhabited by certain unusual bacteria. Many of the morphological descriptions are based on electron microscopic examination of ruminal contents. The morphological types include large oval forms; crescentic cells (selenomonads); large cigar-shaped, septate, and spore-forming cells (*Oscillospira*); and large cocci in tetrads or sheets (*Sarcina* and *Lampropedia*). Relatively little is known of their physiology and ecological roles.

TABLE 43.4 Some of the predominant bacterial species of the rumen of cattle and sheep

Species	Major substrates	Major fermentation product
Anaerovibrio lipolytica	Lipid, lactate, glycerol	Acetate, propionate
Butyrivibrio fibrisolvens	Cellulose, hemicellulose, pectin, protein	Acetate, butyrate, H_2
Eubacterium ruminantium	Sugars	Acetate, butyrate, lactate
Fibrobacter succinogenes	Cellulose	Acetate, succinate
Lachnospira multiparus	Pectin, protein	Acetate, lactate, formic, H_2
Lactobacillus vitulinus	Sugars, starch	Lactate
Megasphaera elsdenii	Lactate, sugars, starch	Acetate, propionate, butyrate, H_2
Methanobrevibacter ruminantium	Hydrogen, formate	Methane
Prevotella ruminicola	Hemicellulose, pectin, protein	Acetate, succinate
Ruminobacter amylophilus	Starch, protein	Acetate, succinate
Ruminococcus albus	Cellulose, hemicellulose	Acetate, formate, ethanol, H_2
Ruminococcus flavefaciens	Cellulose, hemicellulose	Acetate, succinate, H_2

C. Human intestinal bacteria

In the human gastrointestinal tract, the esophagus, stomach and duodenum usually are not colonized by bacteria. The small intestine becomes increasingly colonized along its length, so that the bacterial flora of the ileal region of the small intestine resembles that of the colon. Obligate anaerobic counts from ileal contents have ranged from 10^6 to 10^{10}/g, whereas in the colon the numbers of anaerobes could range from 10^9 to 10^{11}/g.

Knowledge of human intestinal bacteria has been gained mainly from compositional analysis of feces because of the difficulty in obtaining *in situ* luminal samples. *In situ* samples generally are taken during surgery or from people dying from accidental injuries. Samples collected during surgery are likely to come from persons with some disease condition, and patients also are likely to have been starved and often have received antibiotics. The most representative samples are likely from persons dying from trauma, provided collection occurs soon after death. However, bacterial composition of feces generally is regarded as being fairly representative of the colonic flora.

The colonic bacterial flora in adult humans is extremely complex, and obligate anaerobes comprise 98–99% of the total flora. As many as 300 species have been isolated, but many of them have not been described fully. The dominant organisms belong to the genera *Bacteroides, Bifidobacterium, Eubacterium, Fusobacterium, Lactobacillus, Prevotella,* and *Ruminococcus* (Table 43.5). Species of *Clostridium* are only minor components. The major factor affecting the composition of the flora is the diet. Also, antibiotic administration, particularly if prolonged, can have a major effect on the bacterial flora, often resulting in adverse changes. In such instances, the protective effect of the normal flora is compromised, leading to proliferation of pathogens such as *C. difficile*.

The colonic microflora in infants is less complex. At birth, the gastrointestinal tract is sterile but rapidly becomes colonized with bacteria from the surroundings, particularly from the mother. In breast-fed infants, the colonic flora is composed largely of *Bifidobacterium* and *Lactobacillus* species, which may have the protective function of preventing colonization by enteropathogens. When infants are weaned and start to consume dry food, the flora gradually changes to resemble that of the adult human.

Despite differences in flora between the human colon and the rumen of ruminants (absence of protozoa in the colon, for example), the fermentative metabolism is very similar. Acetic, propionic, and butyric acids are the major products, with interspecies hydrogen transfer preventing the accumulation of electron-sink products like lactate or ethanol. Methanogens are present in the colon of some humans, and in their absence, sulfate-reducing bacteria are present.

The human colonic flora are not essential to the nutrition of the host, although volatile fatty acids (VFA) and vitamins may be absorbed into the blood and contribute to nutrition. The importance of gut bacteria lies in their involvement in diseases, either directly or indirectly. A significant function of the intestinal flora is to exclude enteropathogens. Use of broad spectrum antibiotics can cause overgrowth of pathogens, often resistant to the antibiotics, such as *Staphylococcus aureus, Salmonella typhimurium,* and *Candida albicans*. Another pathological condition called "small bowel syndrome" or "contaminated small bowel syndrome" results from overgrowth of

TABLE 43.5 Some bacteria frequently isolated from the intestinal contents of humans, pigs, and chickens

Human colon	Pig colon	Poultry cecum
Actinomyces naeslundii	*Bacteroides uniformis*	*Bacteroides fragilis*
Bacteroides distanosis, B. fragilis, B. thetaiotaomicron	*Butyrivibrio fibrisolvens*	*Bifidobacterium bifidum*
Bifidobacterium adolscentis, B. infantis, B. longum	*Clostridium perfringens*	*Clostridium perfringens, C. beijerinckii*
Clostridium bifementans, C. perfringens, C. ramosum	*Eubacterium aerofaciens*	*Eubacterium* spp.
Eubacterium aerofaciens, lentum	*Fibrobacter succinogenes*	*Fusobacterium* spp.
Fusobacterium necrophorum, F. mortiferum	*Lactobacillus acidophilus, L. brevis, L. cellobiosus, L. fermentans, L. salivarius*	*Gemmiger formicillis*
Lactobacillus spp.	*Peptostreptococcus productus*	*Lactobacillus acidophilus, L. fermentans, L. salivarius*
Peptostreptococcus productus, P. prevotii	*Prevotella ruminicola*	*Ruminococcus obeum*
Propionibacterium acnes	*Ruminococcus flavefaciens*	*Streptococcus faecium, S. faecalis*
Ruminococcus albus, R. bromii	*Selenomonas ruminantium*	
Streptococcus intermedius	*Streptococcus bovis, S. equinus, S. faecalis, S. salivarius*	
Veillonella spp.	*Veillonella* spp.	

bacterial population in the upper small intestine. The counts of bacteria increase 1–2 logs or more higher than in healthy humans. Anatomical defects (diverticulitis) or physiological abnormalities (e.g., lack of acid secretion in the stomach or stasis) will encourage overgrowth. Pathophysiological consequences include malabsorption and nutritional deficiencies resulting directly from bacterial activity (mucosal damage by bacterial products) and indirectly by competing for nutrients (vitamin B_{12}). Pseudomembranous colitis, caused by *C. difficile*, which almost always occurs in association with antibiotic use, has been well investigated. The proliferation of *C. difficile* is attributed mainly to the removal of competition from the normal flora. Approximately 3% of healthy adults harbor *C. difficile* in the intestinal tract. Surprisingly, healthy infants harbor both the organism and the toxin with no pathological consequences, suggesting age-related susceptibility.

Another aspect of gut flora that has been the focus of intensive research is their potential etiological role in colon cancer. The high incidence of colon cancer in North America and Western Europe is related to altered bacterial flora and production of carcinogens because of consumption of diets rich in fats and proteins of animal origin. Nitrosamines, derivatives of steroids and bile acids, and other compounds produced by the anaerobic colonic bacteria are suspected to be involved in colon cancer.

D. Intestinal bacteria of pigs

Studies concerning intestinal bacteria of pigs are of interest because of their potential role in protection against diseases. Gram positive bacteria outnumber gram negative organisms in colonic contents and feces of pigs. Species of gram positive bacteria belong to the genera *Lactobacillus, Streptococcus, Peptococcus, Peptostreptococcus,* and *Eubacterium* (Table 43.5). In contrast, the flora of the cecum is predominantly gram negative, with *Bacteroides* and *Selenomonas* being the major constituents. The significance of hindgut fermentation in pigs is difficult to assess and, undoubtedly, is related to the diet. Absorption of VFA from the pig colon has been well established, and evidence indicates that the rate of absorption for pig cecal and colonic mucosa is higher than that for equine hindgut mucosa and bovine ruminal epithelium. In pigs fed a grain diet, hindgut fermentation is of little value. However, microbial fermentation may have a significant impact on the energy metabolism of pigs eating a more natural diet that is higher in roughage. Significant methanogenic activity also occurs in the hindgut of pigs.

E. Intestinal bacteria in chickens

The gastrointestinal tracts of birds, like those of humans and animals, are inhabited by a variety of microorganisms. Because it is an important source of human food, the domestic chicken has been studied in some detail. Although crop, proventriculus, and gizzard harbor bacteria, the cecum is the region that provides a stable habitat for microorganisms. Ceca, like fermentation chambers in herbivores, have a dense population of obligate anaerobes, ranging in number from 10^9 to 10^{11}/g of cecal contents. The flora includes gram positive and gram negative cocci and rods, and some of the predominant genera include *Bacteroides, Eubacterium, Clostridium, Peptostreptococcus,* and *Propionibacterium* (Table 43.5). Two major factors known to affect cecal flora are diet and age of the chicken. Many studies have been conducted to evaluate effects of the major dietary components, carbohydrates and proteins, and also growth-promoting supplements, including antibiotics, on the cecal flora of chickens. Prior to hatching, the intestinal tract is sterile but within a few hours of hatching, streptococci, coli-aerogenes bacteria, and *Clostridium* colonize the region. However, within a few days, these facultative species are replaced by an anaerobic flora, which continues to change for several weeks and becomes increasingly complex. One consistent feature throughout the microbial development of the cecum is the presence of a large number of organisms capable of utilizing uric acid, a dominant nitrogenous compound in the urine of birds. Uric acid-degrading bacteria range from 10^8 to 10^9/g of contents. These bacteria may be important in the recycling of nitrogen in birds. The colon is quite short in chickens, and its floral composition is similar to that of the ceca.

V. PROTOZOA

The protozoa are highly specialized eukaryotic cells that are able to compete and coexist with bacteria in the gastrointestinal tract of many animals. Because of their relatively large size and active motility, a protozoan was the first microorganism to be discovered in the gastrointestinal tract. Unlike bacteria, which exist somewhere in the gastrointestinal tract of all animals, protozoa are not part of the normal flora in all animals. The occurrence of protozoa in the digestive tract is dependent on an environmentally compatible region of the tract (close to neutral pH) and a retention time for contents that exceeds the protozoan is generation time (6–24 hr). Protozoa are found in the rumens of all wild and domestic ruminants and camelids and also in the cecum and colon of some nonruminant herbivores, such as horses, elephants, and hippopotamus.

Protozoa that occur in the digestive tract of animals are grouped broadly into flagellates and ciliates. In most animals, flagellates occur in low numbers ($<10^3$/g); therefore, their contribution to overall gut fermentation is considered to be nonsignificant. However, intestinal flagellates do play a significant role in the hindgut digestion in termites and other insects, such as wood roaches, that thrive largely on cellulosic materials.

In ruminants and herbivorous nonruminants, ciliated protozoa constitute an important component of the gut flora and play a significant role in the nutrition of the host. Ciliated protozoa are classified further into holotrichs and entodiniomorphs based on certain morphological features, e.g., ciliary arrangement and shape and location of nucleus. Holotrichs have cilia covering the entire or almost the entire surface of the cell, whereas the entodiniomorphid ciliates have restricted zones of cilia. Holotrichs are generally a smaller fraction of the total ciliates and primarily ferment soluble carbohydrates. Entodiniomorphid ciliates, the dominant fraction in most animals, digest starch and structural polysaccharides (cell wall polysaccharides).

The total and generic and species compositions of ciliated protozoa are dependent on the type of host, its geographical location, nature of the diet, and frequency of feeding. Most of the studies on gut ciliated protozoa have been with the ruminal habitat. The majority of the ciliates are entodiniomorphids and many of the species are unique to ruminants and camelids. The concentration varies markedly between animals, ranging from 10^4 to 10^6/g of contents. Generally, information on hindgut protozoa is limited. Quantitative and species compositions are available from horses and elephants and show numbers ranging from 10^3 to 10^5/g of contents. As many as 70 species of ciliated protozoa have been identified in the hindgut of the horse (Table 43.6).

Protozoa are anaerobic and can ferment carbohydrates, proteins, and lipids. Fermentation products include VFA, lactic acid, CO_2, and hydrogen. Many of the anaerobic protozoa contain cytoplasmic organelles called hydrogenosomes, which are somewhat analogous to mitochondria of higher cells. These membrane-bound structures provide compartmentation to protect O_2-sensitive enzymes and produce hydrogen from the reoxidation of reduced cofactors. The hydrogenosomes also confer ciliates a certain degree of tolerance to oxygen.

TABLE 43.6 Ciliated protozoa in gastrointestinal tract of herbivores

	Number per ml or g of contents			
Location	Cattle	Sheep	Horse	Elephant
Rumen	10^4–10^6	10^4–10^6	–	–
Cecum	0	0	$2–223 \times 10^4$	401×10^4
Colon	0	0	$2–706 \times 10^4$	4.6×10^3
No. of ciliate genera	15–17	15–17	10–15	10–15
No. of ciliate species	50–55	50–55	40	17

Information on enzymes and metabolic pathways in protozoa is very limited because of the difficulty of culturing them *in vitro*. Ciliates also are colonized by bacteria on their surface, so obtaining an axenic culture of ciliates is difficult. Therefore, much of the information has been obtained from either washed cell suspensions treated with antibacterial agents or cell-free extracts for studies on enzymatic activities. Another approach frequently used to elucidate the role of protozoa in gut ecosystems is to achieve ciliate-free status. The technique is popularly called defaunation. A number of physical or chemical treatments and dietary manipulations have been described. The chemicals used are selectively toxic to ciliates. Another method to achieve defaunation is to isolate the newborn from the mother immediately after birth and raise it in complete isolation from other animals. However, in reality, achieving and maintaining ciliate-free status for a long period of time are not easy. Elimination of ciliated protozoa from the gastrointestinal ecosystem impacts profoundly on bacterial and fungal populations because of predatory activity of ciliates. Therefore, fermentative changes observed in defaunated animals are the results of absence of protozoa and increases, and possibly compositional changes, in bacterial and fungal populations.

One question that has been of interest to gut microbiologists concerned with ciliated protozoa is how essential they are to the host animal. Many of the protozoal activities do benefit the host. However, bacteria can provide equally well all the fermentative activities that ciliates can. Therefore, ciliates have no unique contribution to make to the host. Growth and digestion trials with defaunated and normal, faunated animals indicate that protozoa are not essential to the host. However, ciliated protozoa may play a significant role, either negative or positive for the host, depending on the diet.

VI. FUNGI

The existence of fungi as part of the gut microbial flora was recognized relatively recently. Early researchers had documented the existence of flagellated protozoa,

TABLE 43.7 Anaerobic fungi isolated from herbivores

Genus	Thallus type	Flagellation	Species
Anaeromyces (*Ruminomyces*)	Polycentric	Single	*A. elegans* *A. mucronatus*
Caecomyces	Monocentric or polycentric	Single	*C. communis* *C. equi*
Neocallimastix	Monocentric	Multi	*N. frontalis* *N. patriciarum* *N. hurleyensis* *N. variabilis*
Orpinomyces	Polycentric	Multi	*O. bovis* *O. joyonii*
Piromyces	Monocentric	Single	*P. communis* *P. mae* *P. dumbonica* *P. rhizinflata* *P. minutus* *P. spiralis*

but further study showed that some of the flagellates were actually fungal spores released from sporangia of the rhizoids. The presence of chitin in the cell walls of these organisms confirmed that they were true fungi. The first report of isolation of an anaerobic fungus, a species of *Neocallimaltix*, from the rumen of sheep was published in 1975. Until then, all known fungi were aerobes or facultative anaerobes, and the detection of fungus that was a strict anaerobe was a novel discovery.

Three groups of fungi occur in the gastrointestinal tract. The first group includes yeasts (*Saccharomyces*, *Candida*) and aerobic molds, which are transient, do not grow anaerobically, and are nonfunctional. However, some of the aerobic fungi could become opportunistic pathogens when conditions become favorable (immune compromised) and invade the gut wall to set up mycotic infection. The second group consists of species that parasitize certain ophryoscolecid protozoa. Two species described as chytrid fungi have been identified. Little is known about these species, and their significance remains to be elucidated. However, heavy infection of the protozoal cell with these species may result in death of the host cell. The third group includes fungi that colonize on plant material, are considered to be indigenous to the gut, and are believed to make a significant contribution to fermentation in the gastrointestinal tract. The life cycle of anaerobic fungi consists of alternating stages of motile, flagellated zoospores and vegetative, mycelial cells. The zoospores are free-living in the liquid phase of the digesta, and the vegetative stage colonizes the digesta fragments. Generally, the alternation between the stages takes about 24–32 hr.

All known anaerobic fungi are zoosporic, with the zoospores being either monoflagellated or multiflagellated. The vegetative growth, the thallus, may be either monocentric or polycentric. The genera of anaerobic fungi are defined on the basis of thallus morphology, rhizoid type (filamentous or bulbous), and number of flagella per zoospore (Table 43.7), and the species are differentiated mainly on zoospore ultrastructure. In a monocentric fungus, either the encysted zoospore retains the nucleus and enlarges into a sporangium (called endogenous zoosporangial development) or the nucleus migrates out of the zoospore, and the zoosporangium is formed in the germ tube or rhizomycelium (called exogenous zoosporangial development). In both types of monocentric development, only one zoosporangium is formed per thallus, and only the zoosporangium contains nuclei. In a polycentric fungus, the nucleus migrates out of the encysted zoospore (exogenous sporangial development) and undergoes mitosis in the rhizomycelium, which subsequently forms multiple sporangia. Thus, in polycentric fungi, both the zoosporangia and the rhizomycelium contain nuclei. Like many eukaryotes adapted to anaerobic growth conditions, anaerobic fungi lack mitochondria. They obtain energy by the fermentation of carbohydrates, which act as both electron acceptors and electron donors. They have mixed acid fermentation profiles. They ferment sugars to formate, acetate, lactate, succinate, ethanol, CO_2, and H_2. Anaerobic fungi contain hydrogenosomes, organelles containing enzymes capable of transfering reducing power from glycolytic products to hydrogen, which is then excreted to permit more ATP production via glycolysis.

Until recently, anaerobic fungi were isolated only from the gut contents or the feces of animals. However, organisms similar to *Neocallimastix* and *Orpinomyces* species were isolated from anoxic regions of a pond in a cow pasture. The significance of anaerobic fungi outside the gastrointestinal tract is not known.

Fungal biomass in the gut is difficult to quantify. Based on the chitin content, it has been estimated to be about 8 to 10% in the rumen of cattle. Zoospore counts can be obtained easily but, because of diurnal variation, counts do not provide adequate estimates of the fungal biomass. Also, colony counts are not reflective of the number, because fragments of mycelium of polycentric fungi can develop into colonies. Therefore, the term "thallus-forming units" is used for enumeration. An endpoint dilution procedure, based on most-probable number, has been developed to enumerate thallus-forming units as an estimate of the activity of the fungi in gut contents. Counts usually range from 10^3–10^7/g of ruminal contents.

The fungi produce a wide array of enzymes that can hydrolyze a range of glycosidic linkages, digest the major structural polysaccharides of plant cell walls, and allow the fungi to grow on a number of polysaccharides. Many of the polysaccharide-hydrolyzing enzymes are localized on rhizoids and rhizomycelia (vegetative stage). Anaerobic zoosporic fungi produce some of the most active polysaccharidases and phenolic acid esterases yet reported. Therefore, there is considerable interest in the biology of anaerobic fungi because of their potential application in biomass conversions.

VII. BACTERIOPHAGES

Bacteriophages have been shown to be present in the ruminal contents of cattle, sheep, and reindeer and in the cecum and colon of horses and may occur in the gut contents of all animals. Bacteriophages are the least studied microbes of the gut. Initial reports of phagelike particles were based on electron microscopic examination of gut contents. Therefore, in many instances a specific phage host was not identified. Both temperate and lytic phages have been demonstrated. More than 125 morphologically distinct phagelike particles have been documented in the ruminal contents of cattle. Some of the varieties may be degenerate forms of mature phages. It is estimated that 20–25% of ruminal bacteria may harbor temperate phages. Only a few studies have identified the specific bacterial host and have led to molecular analysis.

In a gut ecosystem, the presence of prophages in bacteria may confer a competitive advantage, outweighing the burden of additional DNA. Possible advantages could be protection against superinfection by the same phage or against infection by related phages. The possession of prophages also may enhance the genetic potential of bacteria to adapt to change in their environment.

VIII. ROLE OF GUT MICROBES

The association of microbes with the gastrointestinal tract has resulted in the development of a balanced relationship between the host and the microbes. The interaction, for the most part, is beneficial; however, the association also could result in some negative or harmful interactions. The beneficial roles include nutritional interrelationships between the gut microbes and the host and the influence of microbes in preventing the establishment of pathogens in the gastrointestinal tract.

A. Gut microbes and host nutrition

The importance of gut microbes to the host's nutrition is well documented in ruminants with foregut fermentation and in herbivorous nonruminants with hind gut fermentation. Because of microbial activity, ruminants and herbivorous nonruminants are able to utilize cell wall polysaccharides of plants as sources of energy. Another feature that makes ruminants unique is their ability to convert nonprotein nitrogen into protein. In monogastrics, the products of microbial energy-yielding metabolism, such as short-chain fatty acids, lactate, and ethanol are absorbed and utilized as carbon and energy sources by the animal tissues. Furthermore, products from lysis or digestion of microbial cells, such as proteins and vitamins, are absorbed and utilized by the animal tissues. This is supported by the observation that germ-free animals have higher requirements of B vitamins than do conventional animals.

Besides nutritional interrelationships, the normal gut flora also influences various physiological functions of the gastrointestinal tract. Evidence in support of such influences has come primarily from studies involving comparison of animals with normal gut flora and germ-free animals. Some of the intestinal functions influenced by the gut include:

1. **Transit time in the tract**. Digesta passes more rapidly in conventional animals compared to germ-free animals. The mechanism is not fully understood but possibly involves microbial products influencing smooth muscle activity.

2. **Cell turnover and enzymatic activities in the small intestine**. The cell turnover is higher and enzymatic activity is lower in conventional animals than in germ-free animals.

3. **Water, electrolyte, and nutrient absorption**. The water content of the digesta is lower and electrolyte contents (e.g., Cl^- and CHO^{-3}) are higher in the lumen of conventional animals than in that of the germ-free animals.

B. Normal flora and prevention of infections

The presence of normal flora reduces the chances for pathogens to get established in the gastrointestinal tract. The dense population of normal flora occupies the niches and space, thus making it difficult for small numbers of pathogens to compete and get established. This phenomenon is referred to as the Nurmi concept (named after the Finnish scientist who first described it); colonization resistance; bacterial antagonism; microbial interference; or, most commonly, competitive exclusion. Interest in competitive exclusion has existed for many years. However, the topic is receiving greater attention because of *C. difficile* infections in humans and potential application of the concept to reduce foodborne pathogens in chickens, pigs, and other animals. For example, *Campylobacter* or *Salmonella* infections could be prevented or minimized by feeding chicks anaerobic cecal flora from adult birds. Poultry products contaminated with *Salmonella* or *Campylobacter* are the major sources of foodborne infections in humans. The mechanisms involved in competitive exclusion are multifactorial and include regulatory forces exerted by the host, the diet, and the microbes. Some of the major microbial factors include competition for attachment sites, low pH, low redox potential, and elaboration of antimicrobial substances (e.g., VFA, lactic acid, and bacterocin).

C. Normal flora as the cause of infections

Members of the intestinal flora, which, for the most part, are beneficial and nonpathogenic as long as they reside in the intestinal tract, can cause infections in other sites in the body. Such infections often are associated with a predisposing condition that allowed the bacteria to cross the gut wall barrier (e.g., perforation of the intestines, inflamed gut wall, and surgical procedures). The infections generally are caused by mixtures of anaerobic and facultatively anaerobic bacteria and can occur in any site of the body, with abdominal cavity and liver being the common sites. Liver abscesses in cattle are classic examples of a disease condition caused by a bacterium that is a normal inhabitant of the gut but becomes a pathogen when it reaches the liver. Liver abscesses are observed frequently in beef cattle fed high-grain diets, and the primary etiologic agent is *Fusobacterium necrophorum*. The organism is a normal component of the flora of the rumen of cattle and is involved in fermentations of proteins and lactate. In cattle fed high-grain diets, the ruminal wall integrity is compromised because of high acidic conditions, and *F. necrophorum* invades and colonizes the ruminal wall, then enters the portal circulation and reaches the liver. The organism gets trapped in the capillary system of the liver, where it grows and causes abscesses in the hepatic parenchyma.

Although hundreds of species of anaerobic bacteria reside in the gastrointestinal tract, only a small fraction of these are capable of causing infections. Generally, the species isolated from the infections belong to the genera *Bacteroides*, *Fusobacterium*, *Clostridium*, and *Peptostreptococcus*. The facultative species include *Escherichia coli*, *Enterobacter aerogenes*, *Klebsiella* spp., and *Streptococcus* spp. The pathogenic mechanisms and virulence factors involved in intestinal anaerobic infections are not understood fully. All gram negative anaerobes have endotoxic lipopolysaccharide, and their biological effects are similar to those of aerobic gram negative bacteria. Also, certain fermentative products of anaerobes, such as succinic acid and amines, could contribute to the virulence. In the case of *B. fragilis*, a frequent isolate in abdominal sepsis of humans, the virulence factors include a polysaccharide capsule and an array of extracellular enzymes.

IX. CONCLUSIONS

The gastrointestinal tracts of all animals harbor a myriad of microbes performing a variety of metabolic activities that play a vital role in the health of the host. The gastrointestinal tract is a complex anaerobic microbial ecosystem. The microbial species include bacteria, protozoa, fungi, and bacteriophages. The functionally important and dominant microbes are anaerobes with fermentative metabolism. Among the gut ecosystems the rumen of domestic ruminants and the colon of the human have been investigated extensively. The interaction between the microbes and the host is complex and includes various degrees of cooperation and competition, depending on the species of animal. The cooperative interactions include nutritional interdependence and the role of normal flora in preventing the establishment of enteric pathogens. Members of the normal flora that are nonpathogenic as long as they reside in the intestinal tract can cause infections in other sites, particularly the abdominal cavity and liver, if situations allows the microbes to cross the gut wall barrier.

BIBLIOGRAPHY

Bonhomme, A. (1990). Rumen ciliates: Their metabolism and relationships with bacteria and their hosts. *Anim. Feed Sci. Technol.* **30**, 203–266.

Cheng, K. J., and Costerton, J. W. (1986). Microbial adhesion and colonization within the digestive tract. *In* "Anaerobic Bacteria in Habitats Other than Man" (E. M. Barnes and G. C. Mead, eds.), pp. 239–261. Blackwell Scientific Publ., Boston, MA.

Clarke, R. T. J., and Bauchop, T. (1977). "Microbial Ecology of the Gut." Academic Press, New York.

Dehority, B. A. (1986). Protozoa of the digestive tract of herbivorous mammals. *Insect Sci. Applic.* **7**, 279–296.

Gibson, G. R. and Roberfroid, M. B. (eds.) (1999). "Colonic Microbiota, Nutrition and Health." Kluwer Academic Publishers, Dordrech.

Hentges, D. J. (1983). "Human Intestinal Microflora in Health and Disease." Academic Press, New York.

Hill, M. J. (1986). "Microbial Metabolism in the Digestive Tract." CRC Press, Boca Raton, FL.

Hobson, P. N., and Stewart, C. S. (1997). "The Rumen Microbial Ecosystem." Blackie Academic Publication, New York.

Hobson, P. N., and Wallace, R. J. (1982). Microbial ecology and activities in the rumen: Parts I and II. *Crit. Rev. Microbiol.* **9**, 165–295.

Hooper, L. V., and Gordon, J. I. (2001). Commensal host-bacterial relationships in the gut. *Science*. May 11; **292**(5519), 1115–1118.

Gibson, G. R., and Macfarlane, G. T. (1995). "Human Colonic Bacteria." CRC Press, Boca Raton, FL.

Mackie, R. I., and White, B. A. (1997). "Gastrointestinal Microbiology. Vol. 1. Gastrointestinal Ecosystems and Fermentations." Chapman and Hall, New York.

Mackie, R. I., White, B. A., and Isaacson, R. E. (1997). "Gastrointestinal Microbiology. Vol. 2. Gastrointestinal Microbes and Interactions." Chapman and Hall, New York.

McCracken, V. J., and Lorenz, R. G. (2001). The gastrointestinal ecosystem: a precarious alliance among epithelium, immunity and microbiota. *Cell Microbiol.* Jan; **3**(1), 1–11.

Mountfort, D. O., and Orpin, C. G. (1994). "Anaerobic Fungi: Biology, Ecology, and Function." Marcel Dekker, Inc., New York.

Russell, J. B., and Rychlik, J. L. (2001). Factors that alter rumen microbial ecology. *Science*. May 11; **292**(5519), 1119–1122.

Savage, D. C. (1986). Gastrointestinal microflora in mammalian nutrition. *Ann. Rev. Nutr.* **6**, 155–78.

Trinci, A. P. J., Davies, D. R., Gull, K., Lawrence, M. I., Nielsen, B. B., Rickers, A., and Theodorou, M. K. (1994). Anaerobic fungi in herbivorous animals. *Mycol. Res.* **98**, 129–152.

Williams, A. G., and Coleman, G. S. (1991). "The Rumen Protozoa." Springer-Verlag, New York.

Wolin, M. J. (1981). Fermentation in the rumen and human large intestine. *Science* **213**, 1463–1468.

Wubah, D. A., Akin, D. A., and Borneman, W. S. (1993). Biology, fiber degradation, and enzymology of anaerobic zoosporic fungi. *Crit. Rev. Microbiol.* **19**, 99–115.

Ziemer, C. J., Sharp, R., Stern, M. D., Cotta, M. A., Whitehead, T. R., and Stahl, D. A. (2000). Comparison of microbial populations in model and natural rumens using 16S ribosomal RNA-targeted probes. *Environ Microbiol.* Dec; **2**(6): 632–643.

WEBSITES

"Bad Bug Book" of the US Food and Drug Administration
http://vm.cfsan.fda.gov/~MOW/intro.html

List of bacterial names with standing in nomenclature (J. P. Euzéby)
http://www.bacterio.cict.fr/index.html

Comprehensive Microbial Resource of The Institute for Genomic Research (TIGR) and links to many other microbial genomic sites
http://www.tigr.org/tigr-scripts/CMR2/CMRHomePage.spl

44

Genetically modified organisms: guidelines and regulations for research

Sue Tolin
Virginia Polytechnic Institute and State University
Anne Vidaver
University of Nebraska-Lincoln

GLOSSARY

confinement Procedures to keep genetically modified organisms within bounds or limits; usually, in the environment, with the result of preventing widespread dissemination.

containment Conditions or procedures that limit dissemination and exposure of humans and the environment to genetically modified organisms in laboratories, greenhouses, and some animal-holding facilities.

genetically modified organism (GMO) Any organism that acquires heritable traits not found in the parent organism; while traditional scientific techniques such as mutation can result in a GMO, the term is most frequently used to refer to modified plants, animals, and microorganisms that result from deliberate insertion, deletion, or other manipulation of deoxyribonucleic acid (DNA); also referred to as genetically engineered organisms or as organisms with modified hereditary traits.

oversight Application of appropriate laws, regulations, guidelines, or accepted standards of practice to control the use of an organism, based on the degree of risk or uncertainty associated with that organism.

recombinant DNA Broad range of techniques in which DNA, usually from different sources, is combined *in vitro* and then transferred to a living organism to assess its properties.

The Term Genetically Modified Organism (GMO) is most frequently used to refer to an organism that has been changed genetically by recombinant DNA techniques. Historically, research with GMOs has been subject to special oversight that, to this day, differs, depending on the location of the research with the organism, whether inside (contained) or outside (so-called field research), type of organism (e.g., plant, animal, microorganism) or use (e.g., medical, agricultural, environmental), and country in which one works. The oversight mechanism for contained research is principally through guidelines developed by scientists and endorsed by the private and public sector. Outside research is currently overseen by a number of federal agencies. In some countries, such as the United States, there can be overlapping jurisdictions, differing interpretations of legal statutes, and different requirements or standards for compliance by scientists who do research with GMOs in the outside environment. Scientific issues deal with differences in perspective on the risks of introductions of GMOs into the environment, the types of data required prior to the introduction to conclude the experiment is of low risk, and the types of monitoring and mitigation practices, if necessary. Nonscientific issues are also considered and include those dealing with legal and social concerns. These differences in interpretation have resulted in few introductions into the environment of microorganisms.

I. CONCERN OVER GENETICALLY MODIFIED ORGANISMS

A. The concern over safety

The new biology, dating to the 1970s and usually encompassing recombinant DNA techniques, enabled

The Desk Encyclopedia of Microbiology
ISBN: 0-12-621361-5

scientists to perform modifications of organisms with great precision and to combine DNA of organisms that can not, in current time, combine; yet these combinations are derived from components of naturally occurring organisms. The scientific community raised hypothetical questions about the safety of genetically engineered organisms, and the public questioned the potential adverse effects of organisms with the new combinations of genetic information on humans and the environment. It was argued that, as such, the organisms have not been subjected to evolutionary pressures, including dissemination and selection, and may pose a risk to humans or the environment. However, it was recognized that genetic modifications can arise by classical or molecular methods, ranging from selection of desirable combinations by farmers or bakers since antiquity to nucleotide insertion or deletion by molecular biologists.

The new biology, often called biotechnology, has generated fear that the new organisms may be unpredictable in survival and dissemination and, particularly, that transfer of the gene for the modified trait to nontarget organisms might occur. However, gene transfer occurs whether or not humans intervene. Such gene transfers are expected to have minimal consequences unless selection is imposed. Increasing evidence supports the conclusion that microorganisms, particularly bacteria, usually maintain their fundamental characteristics and their essential identities and moderate the amount of change that can be absorbed by known and unknown mechanisms. Deleterious changes can occur, and will, whether or not microorganisms are manipulated. The preliminary testing under contained conditions that is requisite and standard practice in science should, however, identify most of such gross changes. Principles for assessing risk of GMOs as developed in scientific and public forums, are based on the premise that, if one begins with a beneficial organism and imparts a neutral or beneficial trait, the probability of harm from transfer of genetic information is minimal. Some scientists are more concerned about the widespread adoption of a beneficial organism in commerce, rather than about small-scale field research, because greater exposure is likely to increase the probability of risk.

The concern over GMOs resides partly on a perceived increase in ability to survive or persist. Thus, some scientists and consumers argue that oversight of GMOs should be as stringent as that for toxic chemicals, physical disruptions such as water control projects, or exotic organism introductions. The strongest arguments for these concerns are voiced by persons who compare the risks of GMOs with that of introducing exotic organisms. The appropriateness of this analogy can be questioned because there is generally a familiarity with the organism being modified and the trait being introduced, and the fraction of the new genetic information is quite small (Table 44.1). In contrast, exotic organisms are unfamiliar and are entire genomes.

There is also a perception by some that, should there be a problem with survival or dissemination of a microorganism, nothing can be done. Essentially, the assumption is that once the gene(s) is out, it cannot be recalled. However, orderly and inadvertent movement and dissemination of microorganisms occur repeatedly because of their presence on humans, plants, and animals that are moving throughout the world at increasing rates. There are also long-standing and environmental practices that are in use to decontaminate or mitigate unwanted effects of microorganisms. Such practices are known for microorganisms associated with plants and animals, as well as for free-living microorganisms. Immediate decontamination methods include, among others, burning, chemical control, and sanitation by

TABLE 44.1 Comparison of exotic species and genetically engineered organisms

	Exotic organism[a]	Engineered organism[b]
No. of genes introduced	4000 to >20,000	1 to 10
Evolutionary tuning	All genes have evolved to work together in a single package	Organism has several genes it may never have had before. These genes will often impose a cost or burden that will make the organism less able to compete with those not carrying the new genes.
Relationship of organism to receiving environment	Foreign	Familiar, with possible exception of new genes

[a]"Exotic organism" is used here to mean one not previously found in the habitat.
[b]"Engineered organism" is used here to mean a slightly modified (usually, but not always, by recombinant DNA techniques) form of an organism already present in the habitat.
[From U.S. Congress, Office of Technology Assessment (1988). "New Developments in Biotechnology: Field Testing Engineered Organisms: Genetic and Ecological Issues." Washington, DC.]

various means. Short-term and long-term methods are abundant for plant- and animal-associated microorganisms, since a great deal of research on developing mitigation methods is conducted by scientists in the disciplines of plant pathology, veterinary medicine, and human medicine. Many of these deal with management practices and the use of genetic resistance and application of biological control organisms. Immunization of humans and animals is another type of long-term management practice to mitigate the effects of microorganisms.

B. Concerns over genetically modified domesticated organisms in agriculture and the environment

Virtually all domesticated organisms used in the production of food and fiber have been genetically modified over long periods of time, including certain live domesticated microorganisms used in making bread, beer, wine, various types of cheese, yogurt, and other foods. Selected microorganisms that have been shown to be beneficial are also widely used in the environment. These uses include, among others, microorganisms that fix nitrogen and provide nutrients for trees (mycorrhizae), as well as those used in sewage treatment plants and oil drilling. Also, naturally occurring pathogenic microorganisms are used in the testing of domesticated plants to ascertain their disease resistance. In such critical tests with known deleterious organisms, there has been no documented case of untoward effects, such as a plant disease epidemic, arising from such standard field trials.

It is widely accepted that the first step in risk assessment, whether in containment or in confined field trials, is identifying the risk by determining how much is known about the parental organism. It is also recognized that the risk can be minimized by the preferential selection of parental organisms that are generally recognized as safe because of their long history of use. In the oversight of food, such foods are categorized as GRAS, or generally recognized as safe. A similar category can be considered for microorganisms that would be introduced into the environment: GRACE, or those microorganisms that are generally recognized as compatible with the environment.

Examination of the food safety issues associated with genetic modifications has led to the conclusion that potential health risks are not expected to be any different in kind than with traditional genetic modifications. All such evaluations rest on knowledge of the food, the genetic modification, the composition, and relevant toxicological data. A recent international body concluded that rarely, if ever, would it be necessary to pursue all such evaluations exhaustively. There is reasonable agreement that a threshold should exist for regulation, or even of concern below which further evaluations on a genetically modified food product or its individual components need not be conducted. The International Food Biotechnology Council recommends flexible, voluntary procedures between food producers and processors and a regulatory agency.

II. HISTORY OF GUIDELINES AND REGULATIONS

The concern over the potential risks of GMOs led to the initiation of various mechanisms for the oversight of research conducted throughout the world. This oversight was in the form of guidelines, a set of principles and practices for scientists to follow, and regulation by laws applicable to certain processes or organisms used for certain purposes. The legal profession claims that this is the first case in which hypothetical or speculative risks have become the basis for regulation.

Codified guidelines date back to the 1970s, when the previously described concerns were raised. This led the United States National Institutes of Health (NIH), under the Department of Health and Human Services, to develop guidelines for containment of research involving recombinant DNA molecules. The first guidelines were first published in 1976 and have been updated and republished numerous times, most recently in 2002 (66 FR 57970). A public-meeting body of peers and nonscientists was assembled as the NIH's Recombinant DNA Advisory Committee (RAC) by the Office of Recombinant DNA Activities (ORDA) and was given the task to review all recombinant DNA experiments within the United States. Other countries soon followed suit.

The RAC was to assess the risk of the experiment and recommend containment conditions under which they thought the risk would be minimized for the laboratory worker and the environment. The resulting guidelines, which are now available from the NIH Office of Biotechnology Activities (OBA) website (*http://www.4.od.nih.gov/oba/*), spelled out the recommended facilities and procedures for safety to individuals and to the environment. They included such specifics as the type of pipetting one should undertake, sterilization procedures, air filtration procedures, and decontamination and mitigation procedures.

Within a short time, most of the microorganisms and experiments had been assigned a containment level, and the responsibility for overseeing such experiments was decentralized and delegated to local institutional biosafety committees (IBCs) and to other institutions in other countries. Many experiments to modify common laboratory strains of bacteria and yeast were judged to pose no risk and were exempted from the guidelines or any containment requirements. Research with other microorganisms, including viruses, required containment no greater than one would use for research with the microorganisms that did not involve genetic modification experiments. These assignments were generally consistent with the recommendations of the biomedical authorities, such as the Centers for Disease Control and Prevention in the United States.

Experiments involving introduction of GMOs into the environment were first reviewed and approved by the RAC in the early 1980s. However, they were not conducted until 1986, after investigators received approval for their research from regulatory agencies. This action signaled the beginning of oversight for such experiments by centralized regulatory authorities, rather than through guidelines that describe principles and practices for confinement of the GMO to minimize risk. In the United States, regulatory oversight currently includes research conducted by any party and is under the jurisdiction of either the United States Department of Agriculture's Animal and Plant Health Inspection Service (APHIS) or the Environmental Protection Agency (EPA). The former generally oversees plants and microorganisms that are or might be considered as plant pests, while the latter oversees research with so-called pesticidal organisms and microorganisms for other uses (industrial, manufacturing). Where there is research with microorganisms and plants, both agencies may be involved, as well as the Food and Drug Administration. More detailed descriptions of legal and jurisdictional issues can be found in publications included in the bibliography.

At the present time, there is no decentralized body in any nation for oversight of field research that is comparable to the IBC for contained research (Table 44.2). The United States Department of Agriculture (USDA) published (Federal Register, Feb. 1, 1991) a draft of guidelines for conducting research under confinement in the open environment, prepared by an advisory committee of scientists. However, these guidelines were never implemented even though they provided generalized principles for assessing and managing the risks of microorganisms, as well as plant and animals that have been genetically modified, particularly by recombinant DNA. The principles laid out in these guidelines were sufficiently generic that they were to be applicable throughout the world. Later, in 1995, this USDA committee developed performance standards specifically

TABLE 44.2 Types of oversight of genetically modified organisms[a]

Stage of development	Oversight
Laboratory, greenhouse, animal pen[b]	Decentralized oversight from federal research or regulatory agency in the form of guidelines
Small-scale field research[c]	Guidelines and regulations: Combination of decentralized and federal oversight
Scale-up or large-scale testing	Federal regulations
Commercial products	Federal, state, international regulations

[a]Reflects current practices: The degree of oversight differs in each country and among different funding and regulatory agencies.
[b]For unmodified organisms (naturally occurring, chemically altered, spontaneous or selected mutants), standards of practice apply in research, whether conducted by the public or private sector.
[c]Includes tests on land and in enclosed waters.

for research with genetically modified fish and shellfish, since no statutory authority existed for oversight by a regulatory agency. These standards can be accessed through the Information Systems for Biotechnology website (*http://www.isb.vt.edu*) and give key points relevant for safety, are useful for researchers in designing containment systems and for IBCs in evaluating these designs.

In many countries, the oversight of GMOs is essentially the same as for unmodified organisms, except for the contentious issue of planned introduction into the environment. In countries such as the United States and Canada, a sizable bureaucracy has built up to oversee both the research and product development. Even though the risks remain speculative, the fear of litigation and unknown hazards has served to minimize the actual number of introductions, particularly of microorganisms. Of the approximately 800 tests of plants, animals, and microorganisms introduced into the environment for research, approximately 10% have been microorganisms. Most of these microorganisms were modified to have marker genes that enabled them to be monitored in the environment. The first functional genes added to microorganisms and field-tested in the early 1990s were those encoding an insecticidal toxin from *Bacillus thuringiensis* added both to a pseudomonad and a coryneform bacterium. In the former case, the modified organism was killed before it was released. In the latter case, in field trials to test for insecticidal activity, plants inoculated with the GMO were successfully protected from the corn borer. But commercialization of this gene succeeded only when it was introduced into the genomes of corn (maize) and cotton.

risk alone or include other factors, such as socioeconomic considerations?
7. Should there be a "sunset" clause on termination of oversight of research with some organisms or certain types of experiments?

There has been general agreement that a centralized database for field trials would be desirable in order to compare information, including negative results that are not always published. The USDA through its Information Systems for Biotechnology program (*http://www.isb.vt.edu*) and the OECD (*http://www.olis.oecd.org/biotrack.nsf/?opendatabase*) through its Biotrack monitoring database have compiled a great deal of this information. Whether or not these activities serve the purpose of the scientific and commercial community and allay the concerns of the public remains to be seen. Thousands of tests worldwide have been conducted up to the present time and no unpredictable effects have been detected. However, it can be argued that such effects may occur in later years, and, hence, monitoring will be necessary to assess any problems that might arise. Another question that remains unanswered is how long monitoring should occur, compared with naturally occurring organisms.

Given the different views of different countries and applicable laws, global agreements on oversight likely will not be forthcoming. However, there is reasonable general agreement on standards of practice through the scientific and professional societies of the world for conducting research. There are also areas of reasonable agreement in principle, acknowledging that the process by which a genetic modification is made is not as significant as the effects of that modification, i.e., the phenotype. The same degree of oversight is not applied to unmodified organisms and organisms modified by traditional approaches. Hence, the process of modification is still the "trigger" for oversight. A second area of agreement is that familiarity or knowledge of the organism and its modification are likely to be good predictors of the characteristics of the modified organism. A third is that knowledge of the ability to confine an organism or mitigate its effect, if need be, offers a reasonable indicator of expected risk and of the potential for its management.

VI. CONCLUSIONS

Differing perspectives remain on the safety to humans and the environment of genetically modified organisms, particularly those that have been modified by recombinant DNA techniques. The concerns are particularly high with respect to the use of microorganisms in the environment. Discussions are likely to continue for several more years. It is too early to predict whether or not such tests will go forward with reasonable ease, given the stringency of the requirements to conduct the tests. The scientific concerns may not warrant the expenditure of time, effort, and money to conduct field research, since risks unique to GMOs have not yet been identified. The same question could be asked about the oversight of contained research.

Several potential oversight mechanisms would be commensurate with risk assessment and risk management and have been demonstrated as effective for laboratory research. These could include (1) categ orical exclusions, (2) only notification requirements, (3) review and approval by a local organization (e.g., institutional biosafety committees), (4) review and approval by a federal agency with an advisory group consisting of members familiar with the relevant research area, or (5) review and approval by an international agency, in cooperation with a member country.

A reasonable policy of oversight will encourage research with GMOs, especially those modified by recombinant DNA techniques. Competing perspectives may occur in different countries, and within a country, and may not be reconciled. There is no perfect oversight mechanism for any human activity, including environmental releases for research, development, or commercial purposes. There is also the recognition that various viewpoints or perspectives cannot always be accommodated or reconciled. Thus, persons of reason and broad vision will be needed to resolve some of the contentious issues dealing with planned introduction of GMOs into the environment.

BIBLIOGRAPHY

Baumgardt, B. R., and Martens, M. A. (eds.) (1991). "Agricultural Biotechnology: Issues and Choices." Purdue Univ. Agricultural Experiment Station USA.

Cordle, M. K., Payne, J. H., and Young, A. L. (1991). Regulation and oversight of biotechnological applications for agriculture and forestry. *In* "Assessing Ecological Risks of Biotechnology" (L. R. Ginzberg, ed.), pp. 289–311. Butterworth-Heinemann, Boston.

Gent, R. N. (1999). Genetically modified organisms: An analysis of the regulatory framework currently employed within the European Union. *J. Public Health Med.* **21**, 278–282.

James, C., and Krattiger, A. F. (1996). Global review of the field testing and commercialization of transgenic plants, 1986–1995: The first decade of crop biotechnology. "ISAAA Briefs No. 1." The International Service for the Acquisition of Agri-Biotech Applications, Ithaca, NY.

Levin, M., and Strauss, H. (eds.) (1991). "Risk Assessment in Genetic Engineering." McGraw-Hill, Inc., NY.

MacKenzie, D. R., and Henry, S. C. (eds.) (1991). "Biological Monitoring of Genetically Engineered Plants and Microbes." Agricultural Research Institute, MD.

Miller, H. I. (1997). "Policy Controversy in Biotechnology: An Insider's View." R. G. Landes Co. and Academic Press, Austin TX.

Miller, H. I., Burris, R. H., Vidaver, A. K., and Wivel, H. A. (1990). *Science* **250**, 490–491.

National Academy of Sciences (1987). "Introduction of Recombinant DNA-Engineered Organisms into the Environment." Committee on the Introduction of Genetically Engineered Organisms into the Environment, National Academy Press, Washington, DC.

National Research Council (1989). "Field Testing Genetically Modified Organisms: Framework for Decisions." Committee on Scientific Evaluation of the Introduction of Genetically Modified Microorganisms of Plants into the Environment. National Academy Press, Washington, DC.

Organization for Economic Cooperation and Development (1986). "Recombinant DNA Safety Considerations. Safety Considerations for Industrial, Agricultural and Environmental Applications of Organisms Derived by Recombinant DNA Techniques." OECD, Paris.

Organization for Economic Cooperation and Development (1990). "Good Development Practices for Small-Scale Field Research with Genetically Modified Plants and Micro-Organisms. A Discussion Document." OECD, Paris.

Organization for Economic Cooperation and Development (1993). "Safety Considerations for Biotechnology: Scaleup of Crop Plants." OECD, Paris.

Tiedje, J. M., Colwell, R. K., Grossman, Y. L., Hodson, R. E., Lenski, R. E., Mack, R. N., and Regal, P. J. (1989). *Ecology* **70**, 298–315.

Tolin, S. A., and Vidaver, A. K. (1989). *Annual Review of Phytopathology* **27**, 551–581.

U.S. Congress, Office of Technology Assessment (1988). "New Developments in Biotechnology—Field-Testing Engineered Organisms: Genetic and Ecological Issues." U.S. Government Printing Office, Washington, DC.

WEBSITES

NIH Guidelines for Research Involving Recombinant DNA (USA)
http://grants1.nih.gov/grants/guide/notice-files/NOT-OD-052.html

Guidelines for Recombinant DNA Experiment (Japan)
http://www.cbi.pku.edu.cn/mirror/binas/Regulations/full_regs/japan/exp/

Links to regulation in European Union member countries
http://www.biosafety.ihe.be/

Practical information on risk assessment, containment, and links to many other websites worldwide
http://www.isb.vt.edu

Information and links to regulations, focusing on Asia
http://www.isaaa.org

45

Genomes, Mapping of Bacterial

J. Guespin-Michel
Université de Rouen, France

F. Joset
LCB, CNRS, France

GLOSSARY

bacterial artificial chromosomes (BACs) Vectors based on plasmid F (from *E. coli*) origin of replication. Such vectors, which can carry 24 to 100 kbp, are stable in *E. coli*.

contigs Adjacent or partially overlapping clones.

fine mapping Ordering intragenic mutations or very tightly linked genes.

fingerprinting assembly Ordering a restriction map via comparison of the restriction patterns of overlapping clones.

gene encyclopedia (or ordered gene library) Library of cloned DNA fragments, ordered in a sequence reconstructing the gene order on the corresponding chromosome.

genetic map Linkage (or chromosomal) map showing ordered genes, independently of the method used for ordering.

linking clones Clones which, when used as probes, hybridize with two adjacent macrorestriction fragments or two clones from a library.

polarized chromosomal transfer Sequential transfer of (part of) a bacterial chromosome from a donor to a recipient cell, mediated by a conjugative plasmid.

pulsed field gel electrophoresis (PFGE) Electrophoretic device allowing separation of very long (up to Mb) DNA molecules.

physical (macrorestriction) mapping Positioning landmarks, such as restriction sites, along a bacterial chromosome (or any DNA molecule).

rare-cutter endonucleases (or rare-cutters) Endonucleases (mostly Class II restriction enzymes) that have only a limited number of recognition/cutting sites on a whole chromosome.

yeast artificial chromosomes (YACs) Shuttle yeast-*E. coli* vectors, possessing all requirements (telomeric and centromere regions), allowing their reproduction and segregation in *S. cerevisiae*. Such vectors can carry 75–2000 kbp.

Possessing Genomic (Mostly Chromosomal) Genetic Maps, i.e., ordered genes on the different elements of a genome, is an important tool for geneticists and, thus, an early aim when studying a new species. The level of achievement of this work has been closely related to the available techniques.

Until the mid-1980s, genome mapping relied on the classical concept of recombination linkage and, thus, could be achieved only in strains for which suitable natural *in vivo* DNA transfer processes were available. Possible drawbacks due to heterogeneity of recombination frequencies could not be avoided. More or less extensive chromosomal maps had thus been constructed for about 15 species. A major breakthrough has been the possibility to construct socalled physical maps, i.e., to position landmarks such as restriction sites (restriction, or physical, maps) along the DNA molecules. The first, still incomplete, physical map was published in 1987. Localization of genes along this physical map, i.e., its transformation into a genetic map, can be achieved, more or less precisely, by several methods. Extant chromosomal maps obtained by *in vivo* approaches (the case of *E. coli* is particularly enlightening) or molecular techniques

The Desk Encyclopedia of Microbiology
ISBN: 0-12-621361-5

(gene identification through partial sequencing or via hybridization with, hopefully, conserved genes from gene banks) greatly facilitate the work. Thus, each new map construction is further facilitated by comparison with all maps or genetic information available. More than 100 such maps, which widely differ in the number of genes positioned and the precision of their localization, are presently available, and, in principle, there is no cultivable strain which cannot thus be mapped. Therefore, even though the traditional genetic methods are still fully valuable for strain constructions or gene function analysis, they should no longer be useful for mapping purposes. Whole genome sequencing has been the next advance, with the first one published in 1995. Sixteen fully sequenced genomes from Bacteria and Archaea are now published, and nearly 50 more are in progress. This, however, does not render obsolete physical and genetic maps, since genome sequencing will not, for some time yet, be performed on as many species and strains as the former, which also quite often constitute useful requisites for sequencing.

I. *IN VIVO* GENETIC MAPPING OF BACTERIAL CHROMOSOMES

As soon as genetic methods became available for bacteria, they have been readily used for gene mapping, mostly applied to chromosomes, as opposed to plasmids. Various portions of chromosomes have thus been defined, depending on the techniques available. Fine mapping, used for ordering intragenic mutations or very close loci, was opposed to broad mapping, allowing the treatment of (almost) whole chromosomes. The latter is sometimes taken as the only case of genetic chromosomal mapping.

Fine mapping is based on the analysis of recombination data, assuming a direct relationship between the distance separating two markers and the recombination frequency between them (Fig. 45.1). Although it allows a reliable and precise estimation of the order of the markers, this approach is hampered due to heterogeneity of the recombination frequencies along the DNA molecule. Recombination events are initiated at particular sites (sequences) along DNA molecules, known as chi (χ) sites, which are usually not distributed regularly; thus, construction of a genetic map on recombination frequencies may be biased. Fine mapping is essentially performed via transformation (natural or artificial) or transduction. This explains why it applies only to small portions (usually a few percent) of chromosomes. It is, however, available for most species, provided sufficient effort is devoted to finding the relevant DNA transfer tools.

Chromosome mapping in gram-negative bacteria deduces gene order from the relative time required for their transfer during a sequential (polarized or oriented) transfer of the (whole) chromosome, after so-called interrupted mating experiments (Fig. 45.1). Transfer occurs between two (donor and receptor) suitably marked bacteria, the process being controlled by conjugative plasmids integrated in the chromosome and defining the polarized transfer direction. Recombination of the transferred material (i.e., marker exchange) is necessary for its stabilization, a prerequisite for its detection. However, the recombination frequencies are high enough to insure a high rate of integration, and, thus, this requirement does not bias the overall outcome. Conjugation was first discovered in *E. coli*, in which plasmid F performs the transfer (see Section VI). Plasmid F can function in only a few related gram-negative bacteria. So chromosomal mapping could be performed in other species, either via endogenous conjugative plasmids, when existing (e.g., in *Pseudomonas aeruginosa*) or via broad host-range conjugative plasmids (often of the *incP* group), engineered so as to be able to integrate more or less randomly into the desired host chromosome.

In gram-positive bacteria, the situation depends on the species. No system for polarized chromosomal transfer is available in species with low GC content, such as *Bacillus* or *Streptococcus*. Partial *in vivo* chromosomal mapping has been performed in *B. subtilis*, thanks to a very large transducing phage (genome size $\approx$ 2% of the host chromosome). In *Streptomyces*, conjugative plasmids are abundant, but there is no evidence for progressive chromosome transfer, and no kinetic mapping is possible. Therefore, final recombination frequencies are used for *in vivo* mapping. Analyses of the progeny of various four-factor crosses between doubly auxotrophic parents were first performed in *S. coelicolor* strain A3(2) (Fig. 45.1). A compiled analysis of the resultant partial maps was used to deduce the order of the markers. The resulting circular linkage map consisted of two well-marked regions, separated by two very long "silent" quadrants. Analysis of recombinants issuing from matings between appropriately marked parents showed very limited linkage between the two marked regions, thus defining the unusual length, in terms of crossover units, of these silent quadrants. Their relative lengths were later better estimated by statistical analysis of the heterozygous regions of merodiploids in a population of heteroclones (colonies arising from

A

Transformant genotype			
CitC	PolA	DnaB	%
+	+	+	33
+	+	-	21
+	-	+	39
+	-	-	7

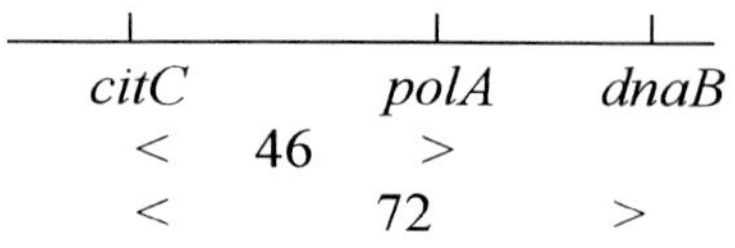

citC *polA* *dnaB*

< 46 >

< 72 >

B

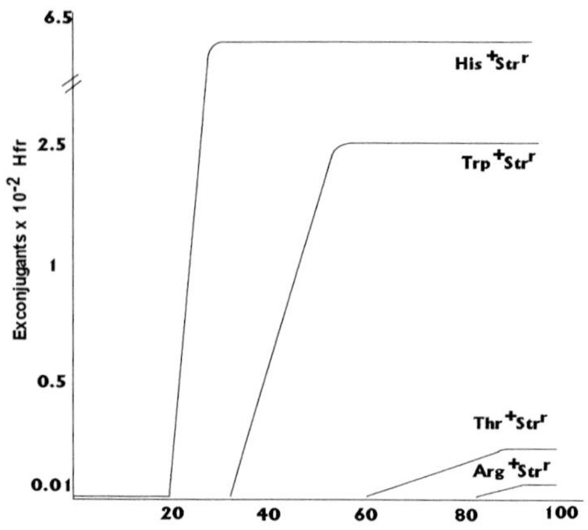

His $^{+}$Str^{r}

Trp $^{+}$Str^{r}

Thr $^{+}$Str^{r}

Arg $^{+}$Str^{r}

Exconjugants x 10^{-2} Hfr

6.5, 2.5, 1, 0.5, 0.01

20, 40, 60, 80, 100

C

1

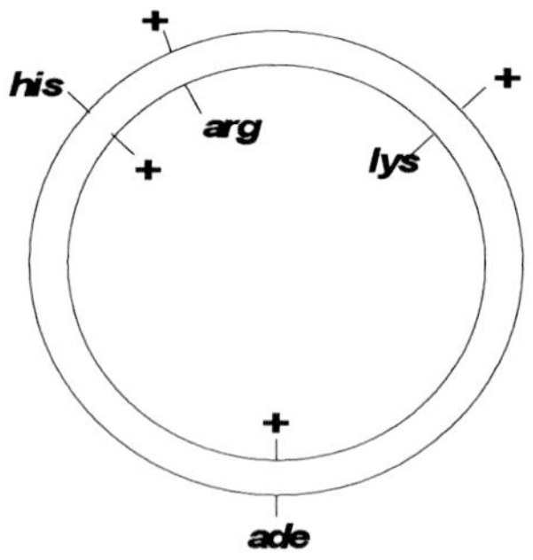

his *arg* *lys* *ade*

2

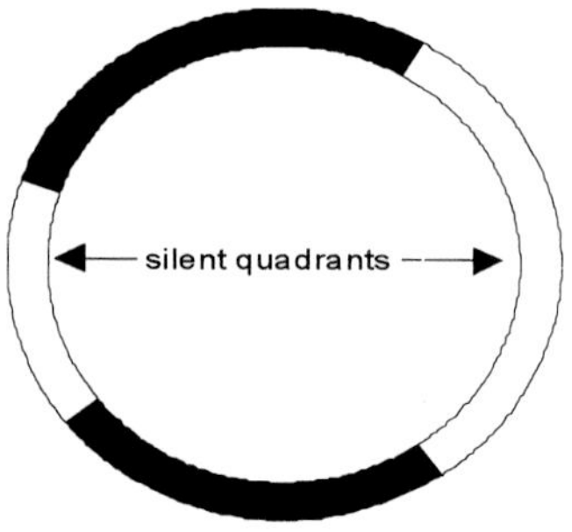

silent quadrants

Recombination between	Number of recombination events
arg – *his*	87
arg - *lys*	222
his - *lys*	249
ade – *arg*	455
ade – *his*	428
ade - *lys*	423

FIGURE 45.1 Survey of the various approaches for *in vivo* chromosome mapping in bacteria. (A) Fine mapping in *Bacillus subtilis* (the so-called 3-point test). Three loci were differently marked in a donor (*citC*$^{+}$, *dnaB*$^{+}$, and *polA*$^{-}$) and a recipient (*citC*$^{-}$, *dnaB*$^{-}$, and *polA*$^{+}$) strain; single *citC*$^{+}$ transformants were selected and their genotype determined for the other two characters (panel I); the least frequent class (7%) must correspond to a requirement for 4 crossovers, yielding the map shown, with relative distances calculated as frequencies of recombination (panel 2). (B) Chromosomal mapping by conjugation in *E. coli*. A fully prototrophic, streptomycin resistant (Str^{r}) Hfr strain was mated with a polyauxotrophic, Str^{s} F^{-} one; mating was interrupted as a function of time, and the proportion of each possible type of exconjugants determined; extrapolation towards the abscissa yields the minimal time required for transfer of each character and, thus, their relative distances (in minutes) assuming the rate of transfer is constant along the whole chromosome. (C) Mapping in *Streptomyces*. (1) Four-marker mapping in *Streptomyces coelicolor*. Doubly marked strains (parent 1: *his ade*; parent 2: *arg lys*) were crossed, recombinants selected and analyzed for their complete genotype; the recombination frequencies between pairs of markers allowed the gene arrangement to be deduced; (2) The 1992 version of the map, obtained by similar methods, carried about 130 loci (short lines outside the circle), 5 accessory elements (insertion sequences, phage insertion sites), and the two almost completely "silent" quadrants. The actual linear structure of the chromosome was later established. A circular linkage map had been arrived at because, in merodiploïds as used for genetic mapping, double crossovers are required between linear chromosomes to generate recombinants harboring complete chromosomes. From this situation ensues the presence of both ends of the chromosome from the same parent in the recombinant molecule, which leads to apparent linkage between these ends and, thus, to the erronous interpretation of a circular structure. [A and B: Adapted from Joset, F., and Guespin-Michel, J. (1993). *In* "Prokaryotic Genetics: Genome Organization, Transfer and Plasticity." Blackwell Science, Oxford. C: From Smokvina *et al.* (1988). *J. Gen. Microbio.* **134**, 395–402. D: Adapted from Kieser, H. M., Kieser, T., and Hopwood, D. A. (1992). *J. Bacteriol.* **174**, 5496–5507.]

partially diploid cells). Linkage maps were similarly established for other *Streptomyces* species.

II. PHYSICAL (MACRORESTRICTION) MAPS

Physical mapping can readily be performed for small plasmids by comparing the sizes of fragments obtained after treatment with sets of restriction enzymes, since only a limited number of such fragments are formed. To obtain the same kind of results on a larger scale, e.g., for a chromosome, one must be able to cut the DNA into a reasonable number of pieces (about 20 is optimal). This implies the formation of large fragments (most of them larger than 50 kbp), for which resolution of their size is not possible via classical agarose gel electrophoresis. Two conditions, i.e., the availability of enzymes with very few cutting sites on whole chromosomes and tools allowing size resolution for large fragments, opened the era of chromosomal physical, or macrorestriction, mapping.

A. Rarely cutting site-specific endonucleases

Physical mapping requires the formation of reproducible fragments and, thus, precludes random breaks due to shearing during extraction procedures, particularly frequent for large DNA molecules, such as bacterial chromosomes (or megaplasmids). Prevention of random breakage is achieved by trapping the cells, before any other treatment, into small plugs of agarose. Cell lysis is performed inside the agarose. Small molecules diffuse out while the DNA remains entrapped. The endonucleases can penetrate into the agarose network and the cleaved DNA fragments are electroeluted from the plug.

Several types of enzymes have only a limited number of recognition/cutting sites on whole chromosomes. Ten or so restriction enzymes (all belonging to Class II) are "rare-cutters" for many genomes, due to their 8-bp recognition sequence (Table 45.1). *Not*I is the most extensively used. But extreme GC contents or reduced representations of particular sets of nucleotides of certain genomes allow adding a few restriction enzymes with six-base recognition sequences to the list of possible rare-cutters (Table 45.1). Works are in progress to modify the recognition specificity of some frequently cutting restriction endonucleases so as to increase the size of their recognition site (this has been recently achieved for *Eco*RV). Proteins which participate in intron processing have a recognition sequence of 18–26 bp and have often proved valuable as rare-cutters. One of these, largely used for bacterial chromosome mapping, is protein I-*Ceu*I, produced by a chloroplast intron of *Chlamydomonas eugametos*. It recognizes a 26 bp sequence, usually present only in bacterial-type 23s

TABLE 45.1 Examples of endonucleases with rare-cutting frequency in bacterial genomes

Recognition particularity	Enzyme	Recognition sequence
8-nucleotide recognition sequence	*Not*I	GC/GGCCGC
	*Sse*83871	CCTGCA/GG
	*Swa*I	ATTTAAAT
	*Pac*I	TTAATTAA
	*Pme*I	GTTTAAAC
	*Sgr*AI	CACCGGCG
	*Srf*I	GCCCGGGC
	*Sgf*I	GCGATCGC
	*Fse*I	GGCCGGCC
	*Asc*I	GGCGCGCC
Hyphenated 8-nucleotide recognition sequence	*Sfi*I	GGCCNNNN/NGGCC
Rare-cutters in G + C rich genomes	*Ase*I	ATT/AAT
	*Dra*I	TTT/AAA
	*Ssp*I	AAT/ATT
Rare-cutters in T + A rich genomes	*Sac*II	CCG/CGG
	*Sma*I	CCC/GGG
	*Rsr*II	CG/GNCCG
	*Nae*I	GCC/GGC
Overlap on TAG, a rare stop codon in prokaryotes	*Spe*I	A/CTAGT
	*Xba*I	T/CTAGA
Overlap on GC/CG or CC/GG sequences, relatively rare in prokaryotes	*Avr*II	C/CTAGC
	*Nhe*I	G/CTAGC

rRNA genes. Several devices have been used to protect part of the cutting sites of frequently cutting enzymes. "Peptide nucleic acid clamps" (bis-PNAs), that bind strongly and sequence-specifically to short homopyrimidine stretches, shield overlapping methylation/restriction sites and, thus, reduce the number of accessible sites for the corresponding enzyme. A strategy called "Achilles' heel cleavage" (AC) consists in introducing into a genome a unique site (such as the phage lambda *cos* site) that can be cleaved only by a specialized enzyme.

B. Pulsed field gel electrophoresis (PFGE)

In 1982–1984, Schwartz and Cantor devised a method allowing the separation of DNA fragments ranging from 35 to 2000 kbp. The molecules were electrophoresed in an agarose gel subjected to electric fields alternately orientated at roughly right angles, hence, the name of the method, "pulsed-field gel electrophoresis," or PFGE. The rationale of the method is that the longer a DNA molecule, the slower it reorientates at each change of direction of the electric field. This, accordingly, slows down its overall migration speed along the average direction of migration. Various types of PFGE may be chosen, depending on the sizes of the fragments to separate. This also allows distinguishing circular plasmids, which migrate in a very special way in PFGE.

The reliability of this method depends on the identification of the complete set of fragments generated by the restriction treatments. Its main drawbacks are that: (i) nonambiguous resolution of the digestion fragments may be hindered if the fragments are too numerous or if two (or more) have very close lengths; (ii) fragments too small to be detected or so small that they would have eluted from the gel before it was examined, may be generated.

C. Construction of macrorestriction maps

Several strategies (a few examples will be described) can be used to reconstruct the alignment of the fragments along the chromosome. Only rarely will one strategy be sufficient.

1. The fragments obtained after digestion with one endonuclease are individually digested by a second enzyme, and reciprocally. Comparisons of the sizes of the resulting doubly digested fragments allow locating the different cleavage sites, as for plasmid mapping (Fig. 45.2). The method is optimized by 2D-electrophoresis. The main drawbacks, again, are the existence of several fragments with the same length and possible elution of small fragments from the plug during the preparation of the gel. The interest of a physical map with large intervals is limited, so further cutting with other enzymes can be pursued. The method is generally accurate if the total number of secondary fragments does not exceed 20–25, thus hampering simultaneous treatment with two enzymes.

2. Other approaches allow obtaining more precise genomic location of large numbers of fragments, i.e., to link adjacent fragments. A linking clone is a clone which contains a site for a rare-cutting enzyme and overlaps two adjacent macrorestriction fragments. Adjacent segments are called contigs, referring to their contiguous positions on the chromosome. Hybridization of a labeled restriction fragment with a whole DNA library allows the detection of the two contigs flanking the corresponding restriction site. Linking clones can be obtained from a DNA library by selecting clones which contain rare-cutter sites. The latter can be tagged for instance by insertion of a marker. Thus, the *Not*1 sites of *Listeria monocytogenes* chromosome were individually labeled with a Km^r cassette (a DNA sequence originating from a transposon, containing a gene for resistance to an antibiotic; in this case, kanamycin), the Km^r clones from the corresponding *Eco*RI libraries, selected in *E. coli*, represented the linking clones for the *Not*I restriction fragments (Fig. 45.2). Known open reading frames (ORFs) can be similarly used as markers (Fig. 45.2). Willems *et al.* (1998) have sequenced the 58 *Not*I/*Sau*3A fragments of the *Coxiella burnetii* genome. Checking in databases whether chance partial reading frames present at the *Not*I side of some fragments could correspond to registered ORFs shared by two fragments allowed linking 10 out of the 29 *Not*I fragments. Amplification by polymerase chain reaction (PCR) was then performed on the whole chromosome, using random pairs of primers directed towards the *Not*I sites of the remaining *Not*I/*Sau*3A fragments. Amplification meant that the two corresponding fragments were adjacent, i.e., were contigs.

3. A method derived from that devised by Smith and Birnstiel (1976) has been applied to *Pseudomonas aeruginosa* mapping (Heger *et al.*, 1998). Fragments formed after partial digestion by a frequent-cutter are separated by PFGE and hybridized with end probes from each of the rare-cutter fragments of the same genome (Fig. 45.2). Hybridization of a single frequent-cutter fragment with two probes identifies the rare-cutter contigs. In addition, comparison of the sizes of the partial digests can be used to establish a restriction map.

The presence of repetitive sequences (multifamily genes) or copies of mobile elements may lead to false alignments biased by erroneous apparent identity of the corresponding regions. Other methods, such as fingerprinting assembly (comparison of the restriction

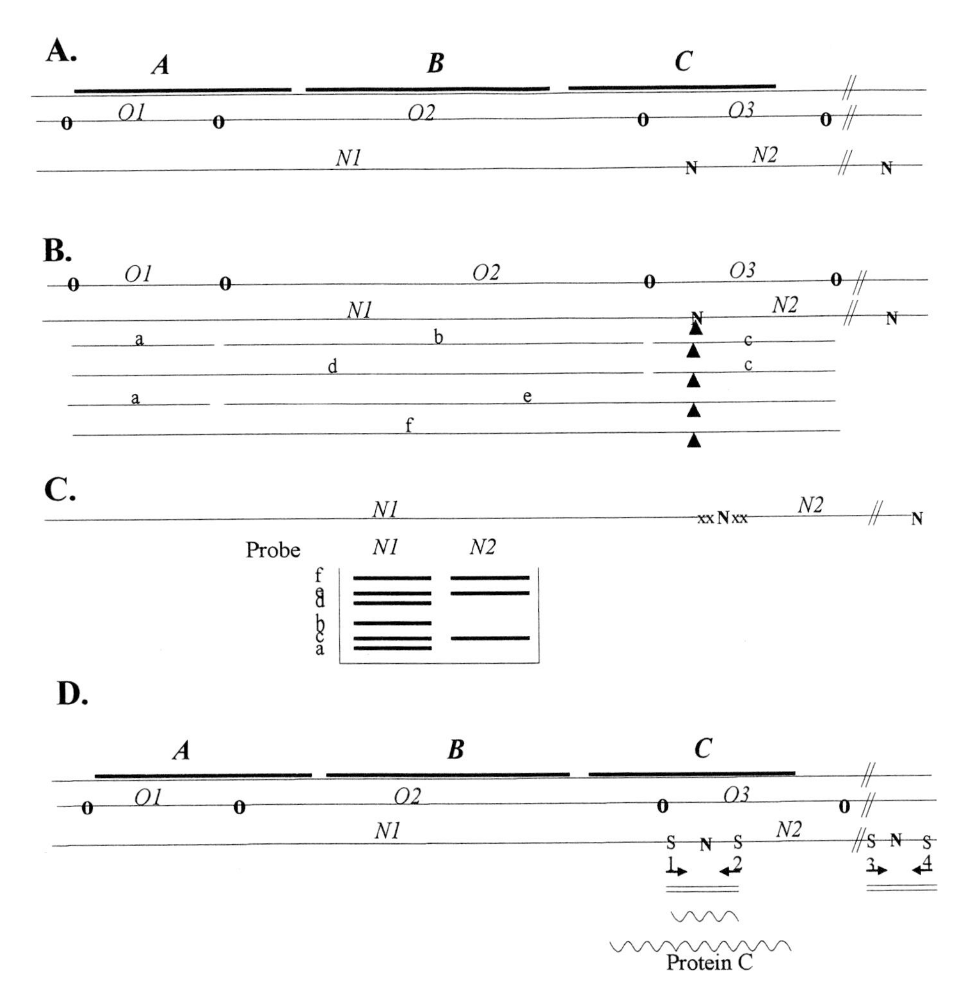

FIGURE 45.2 Construction of linking clones. Application to a portion of chromosome with genes A, B, and C. (A) Comparison of the sizes of fragments obtained after separate or double digestions by pairs of rare-cutter endonucleases (O and N) can allow positioning of the cleavage sites. *O1*, *O2*, *O3*, *N1*, and *N2* represent the fragments obtained after single digestion by enzymes O or N, respectively. (B) Insertion of a cassette (▲) in an N site and identification of the fragments obtained after partial digestion with another enzyme (O) carrying the cassette indicates that fragment *O3* links fragments *N1* and *N2* (i.e., fragments *N1* and *N2* are contigs). (C) Partial digests of the whole chromosome with enzyme O are electrophoresed and hybridized, with each of the N fragments, labeled at their extremities (xx). Only the fragments encompassing region *O3* give a signal (heavy line) with *N2*, while all are recognized by *N1*; thus *O3* overlaps (links) *N1* and *N2*. (D) If end-sequencing of N fragments indicates the presence of reading frames, primers (→) directed towards the N sites are designed. If the sequence of the fragment amplified on whole DNA from a pair of primers (here, S1 and S2) reconstitutes a continuous reading frame showing homology with a known protein, the N1 and N2 fragments are probably contigs.

patterns of supposedly overlapping clones), have been devised to check previous results or to construct contig charts. The main limit to a list of available approaches is the imagination of the workers, which should be applied to devise any technique or combination of techniques that allow overcoming a suspected problem.

The first physical chromosomal map, that of *E. coli* (strain K12), was published in 1987 (see Section VI). Since then, an exponentially growing number of macrorestriction maps have been issued. These maps, however, are not very precise, and most do not carry any genetic information. A further step was to provide actual ordered cloned libraries (so-called encyclopedias), which are useful tools for subsequent genetic mapping.

III. ORDERED CLONED DNA LIBRARIES, OR ENCYCLOPEDIAS

An encyclopedia consists of a library of DNA fragments, cloned into a vector, ordered so as to reconstruct with sufficient overlap the order on the chromosome. The first one produced, again, was for *E. coli*, in 1987 (see Section VI).

A. Choosing the vector for an ordered cloned library

The available vectors cover a large range of possible sizes of the inserts they carry: λ-based vectors (10–25 kbp), cosmids (5–50 kbp), P1-based vectors (90 kbp), yeast artificial chromosomes (YACs) (75–2000 kbp), and bacterial artificial chromosomes (BACs) (20–100 kbp). Due to their lack of stability in *E. coli*, YACs have been used for bacterial genome mapping only for *B. subtilis*, *Myxococcus xanthus*, and *Pseudomonas aeruginosa*. The BACs, based on the *E. coli* F plasmid origin of replication, are much more stable, and have been extensively used for eukaryotic gene libraries. Their first use in bacteria was for the construction of a *Mycobacterium tuberculosis* encyclopedia in 1998.

B. Assembling the encyclopedia

Constructing an encyclopedia implies finding overlapping clones, in order to define contigs. One must start with a number of clones 10- to 20-fold in excess over the number corresponding to the length of the chromosome. This ratio depends on the portion of the insert required for detection of linkage (the minimal detectable overlap, MDO). The methods used to order the clones are similar to those described for the construction of macrorestriction maps. The work can be strongly facilitated by available knowledge, such as macrorestriction or genetic maps, or sequenced regions. Then, a minimal overlapping map may be proposed. Thus 420 BAC clones (20–40 Mbp) have allowed the construction of the 4,4 Mb map of the circular chromosome of *M. tuberculosis*, but the minimal overlapping set (or miniset) requires only 68 unique BAC clones (Fig. 45.3) (Brosch *et al.*, 1998).

IV. CONVERSION OF PHYSICAL MAPS INTO GENETIC MAPS: POSITIONING GENES ALONG THE PHYSICAL MAP

Restriction maps are converted into genetic ones by locating genes with reference to the restriction sites. Localization can be achieved by DNA hybridization or by sequence comparison when available. The limiting factor of this conversion lies in the number of known genes available (i.e., cloned or sequenced) and in the precision of the physical map. In favorable cases (for instance, *E. coli*), the restriction pattern of a cloned fragment encompassing a given gene is sufficient to localize the gene on the fragment, and thus, on the chromosome. Difficulties may arise if one or several cutting sites are protected as part of an overlapping site for a different modification system in the original host, but no longer so when subcloned and amplified in the cloning host (usually *E. coli*).

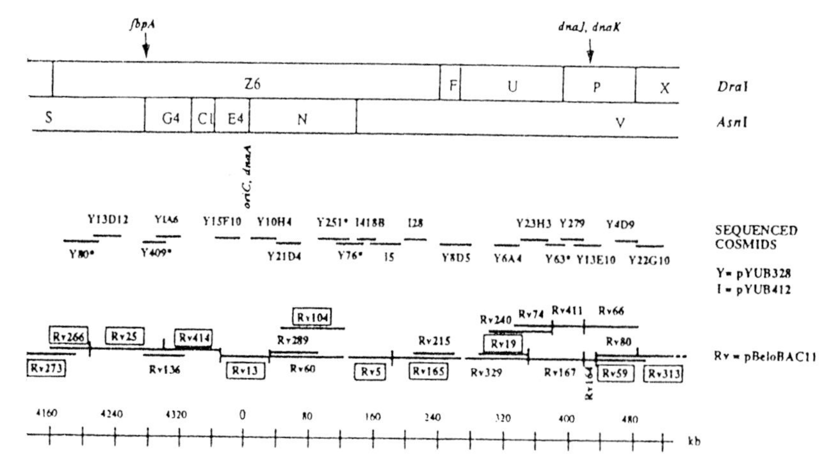

FIGURE 45.3 The BAC ordered library for *M. tuberculosis*, superimposed on the physical and genetic maps. About 1/4 of the whole chromosome is shown, starting at the arbitrary 0 (*oriC*) site. Two top lines: restriction maps for *Dra*I and *Asn*I, showing a few gene locations. Cosmids, cosmids sequenced during the sequencing project; BACs, a representative set of BAC clones positioned relatively to the cosmids by end-sequencing and restriction mapping; kb, length of the chromosome. [Adapted from Brosch, R. *et al.* (1998). *Infection and Immun.* **66**, 2221–2229.]

The genes used as probes can originate from the host itself or from a heterologous host. In the latter case, the strains should be sufficiently related and possess the same function, so that DNA sequence conservation can be expected to allow efficient annealing. PCR probes using primers covering conserved protein or DNA regions have also been thoroughly used for widely distributed genes. For instance, the cleavage site for I-*Ceu*I, which is specific of 23S rRNA genes, allows localizing these rRNA loci. A very efficient method using transposons with rare-cutting sites was also developed. When a mutation in a known gene has been obtained with such a transposon, the corresponding macrorestriction pattern displays the replacement of one fragment by two fragments, and subsequently allows a precise location of the transposon, hence, of the gene. The transposon Tn5 naturally harbors a single *Not*I site, but other transposons have been engineered so as to harbor similar single rare-cutter sites. In addition, transposon insertions can be intraspecifically transferred by conventional genetic methods, thus allowing easy comparison of chromosomal organization of related strains.

Comparisons of genetic with physical maps have shown a general good agreement with regard to gene order, but less so to distances between genes. This reflects biases introduced by nonregular distribution of recombination hotspots along the chromosome in the course of genetic mapping. Thus, restriction mapping of the chromosome of *Streptomyces* showed the estimates obtained by *in vivo* mapping to have been rather accurate. Surprisingly, however, the *Streptomyces* chromosome has turned out to be linear instead of circular.

V. GENOME SEQUENCES

Whole genome sequences are presently, and will more readily in the future be, available. Will this render the approaches via physical mapping obsolete, as the genetic methods have mostly become? This does not seem likely for several reasons. The most obvious is that, even though genome sequencing becomes cheaper, it is still time consuming. Thus, it does not seem likely that the 4000 or so presently cultivable bacterial species will have their genome sequenced, whereas a physical map is more easily feasible. It should also be recalled that a physical map, even more as an ordered library, is often a prerequisite for a genome-sequencing project, mainly for the larger bacterial genomes.

Furthermore, sequencing a genome means sequencing the chromosome of a representative strain (and possibly isolate) of a species. There is growing evidence of a large plasticity of the genomes. Thus, physical maps and/or encyclopedias will remain the easiest way to approach this problem. For instance, *Spe*I-restricted fragments of 97 strains of *Pseudomonas aeruginosa* isolated from clinical and aquatic environments have been hybridized with YACs carrying 100 kb inserts (about 3% of the chromosome) from three chromosomal regions of the well-known strain PAO. At this scale, little genomic diversity was detected in these representatives of the species. In contrast, a study of 21 strains of the same species isolated from cystic fibrosis patients, and analyzed by several rare-cutting endonucleases, revealed that blocks of up to 10% of the genome could be acquired or lost by different strains.

The problem in translating a genomic sequence into a genetic map is the identification of the genes, i.e., of the encoded functions. Sequence homology with a known element (at the DNA, RNA, or protein level), be it of prokaryotic or eukaryotic origin, allows postulating a function with some confidence. However, about 30% of the potential ORFs of all sequenced genomes do not show homology to any known element. Their identification will constitute the main challenge of what is now referred to as the "after-sequencing," or post-genomic, molecular biology.

VI. THE *E. COLI* K12 CHROMOSOMAL MAP

The building of the chromosomal map of an *E. coli* K12 strain represents, historically, a typical textbook example, since all available techniques have successively been applied until the complete molecular information was reached with the sequencing. It, thus, allows summarizing, for just one strain, all the steps in genome mapping. A genetic linkage map showing 99 genes (first edition in 1964), then 166 (1967 edition), was based on conventional mapping procedures, using recombination after conjugation or transduction by phage P1 (Fig. 45.4). This approach also provided the first proof of the circularity of a bacterial chromosome. Further accumulation of information via the same approach during the next 10–15 years led to a more complete map positioning about 1500 genes, representing 20–25% of the whole chromosome.

During the early 1980s, the introduction of molecular techniques allowed the cloning and sequencing of one-third of these genes and, thus, also provided their complete restriction maps. In 1987, Kohara and coworkers prepared a genomic library, now called an encyclopedia, using as cloning vector a modified λ phage. The whole library was contained in 1056

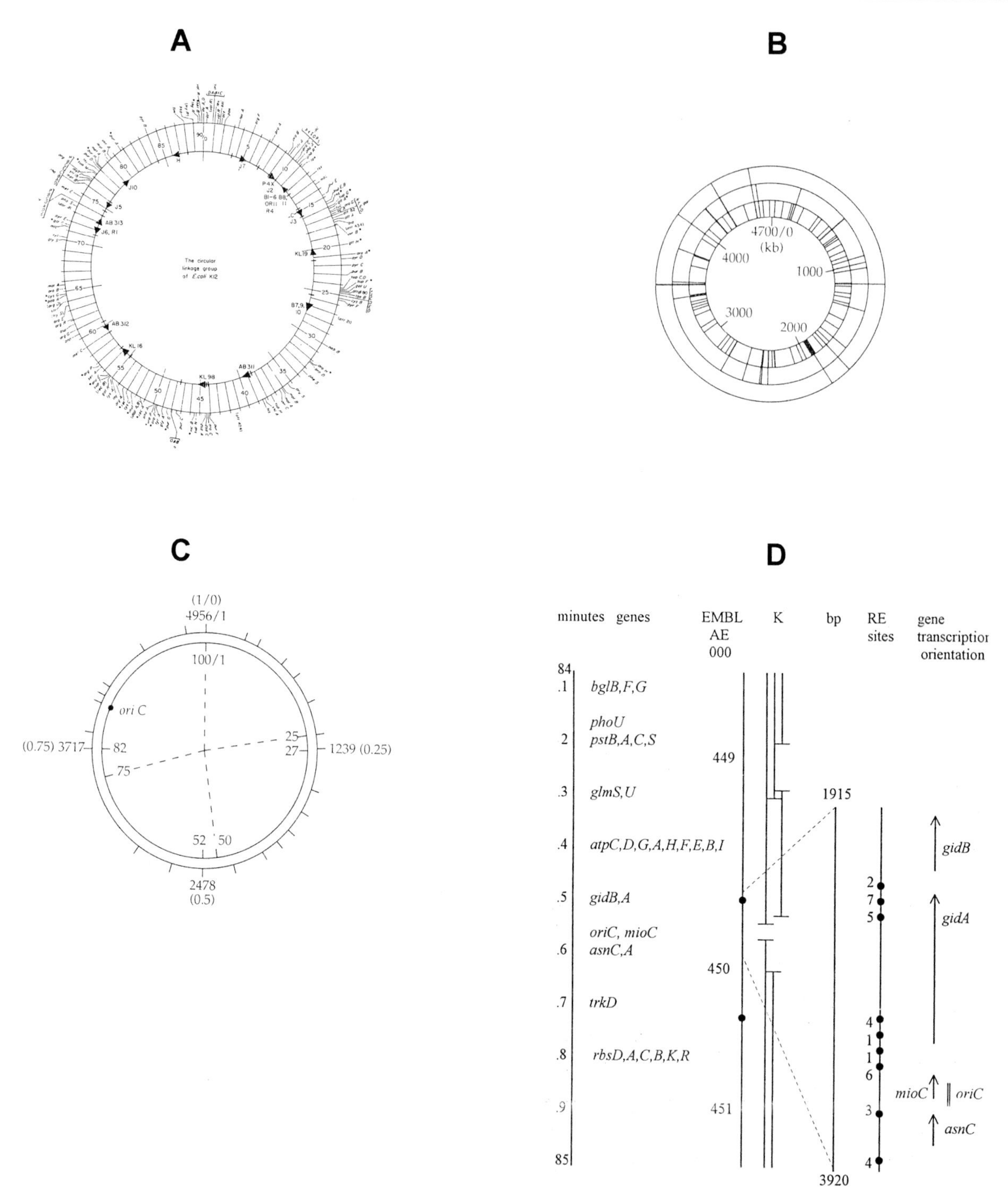

FIGURE 45.4 Successive chromosomal maps of *E. coli* K12.(A) The 1967 issue of the circular linkage map, showing 166 loci (outer circle) and Hfr origins and orientations of transfer (arrows, inner circle) used for mapping. The total length (90 minutes) had not yet been normalized to 100 units. (B) The Kohara's physical map, showing the restriction sites for only 3 endonucleases (each circle); the outer one is for the rare-cutter *Not*I. (C) Comparison of the physical (*Not*I, outer circle) and normalized genetic lengths (in minutes), with the *thr* gene arbitrarily assigned at position 100/1: several genetic distances have been distorted to match the physical map. (D) The present complete genetic map indicating the minute coordinates (only the portion around *oriC* is shown) aligned along the corresponding fully sequenced map; the latter shows, successively, the sequence entries in the GenBank/EMBL/DDBJ data libraries (EMBL), the Kohara miniset clones (K), the base position (bp), restriction segments for eight enzymes (RE), and the identified genes (genes) with their orientation. [A: Adapted from Hayes, W. (1968). "The Genetics of Bacteria and Bacteriophages." Blackwell Science, Oxford. B and C: Adapted from Joset, F., and Guespin-Michel, J. (1993). Op. cit. D: Adapted from Berlyn, M. K. B. (1998). *Microbiol. Mol. Biol. Rev.* **62**, 814–984, and Rudd, K. E. (1998). *Microbiol. Mol. Biol. Rev.* **62**, 985–1019.]

clones, each carrying 15–20 kbp-long inserts. From this library, an almost complete physical map showing restriction sites for eight endonucleases was obtained, using PFGE procedures and adapted computer programs (Fig. 45.4). It took another 2304 clones and the use of newly published information on restriction or sequencing of local regions to deal with most of the remaining ambiguities or gaps, yielding a 4700 kbp-long molecule. Simultaneously, Cantor's group, using similar approaches, proposed a 4600 kbp-long chromosome, of which all ambiguities but one were solved. This whole map was covered by 22 *Not*I fragments. Correlation between these restriction maps and the known linkage map was excellent in terms of gene order. Some distortions of the genetically estimated distances, however, were necessary for an optimal alignment with the restriction profile (Fig. 45.4). One cause for these discrepancies probably lies in unequal crossing-over frequencies along the chromosome.

The present Kohara's phage λ ordered library covering the whole genome is commercially available as a set of 476 phages, immobilized on a nylon hybridization membrane. A computer program was developed by Danchin's group to ease the experimental work of mapping a new character, by first working out its most probable location(s) through comparisons of restriction profiles. Another program (1995) allowed localizing a cloned fragment by simply determining the sizes of the hybridizing fragments obtained by the 8 restriction endonucleases used by Kohara. This allowed localizing a locus within 7 kb.

The complete sequence, published in 1997, describes a 4,639,221-bp-long chromosome, a figure very close to those reached by the previous physical data. Compilation of all data derived from genetic and molecular approaches, or predicted from sequence analyses or comparisons, has yielded what could be the complete set of information on this chromosome (see Fig. 45.4 for an example of this map). It is interesting to note that, even though location and precise length of a gene are now defined on a base level, the standard coordinate scale using a 0–100 arbitrary units (so-called minutes), derived from the times of transfer via conjugation, has been maintained. The available sequenced genome has now provided a new, easier means to localize any cloned fragment, by sequencing a small part of this fragment and aligning it by computer methods on the whole map.

The knowledge available, mostly as restriction data, from various *E. coli* strains provides growing evidence for a very low level of polymorphism in this species. Most of it may be due to movement of mobile elements (ISs, transposons, prophages, pathogenicity islands). As a consequence, the K12 map gains a wider validity than its use for the specific specimen strain from which it was constructed. This, however, is known not to be a universal situation (e.g, Streptomycetes, *Nesseria*, and to a lesser extent, *Pseudomonas*).

VII. CONCLUSION: THE INTEREST OF MAPPING BACTERIAL GENOMES

The publication of the first physical maps of bacterial genomes has started a new field in molecular biology called genomics, i.e., the study of integral genome structures. Although genomics has reached its full significance with the study of genome sequences, physical maps have provided, besides the initial tools for genome sequencing, numerous important results, such as original genomic structures (linear or multiple chromosomes, very large plasmids, and linear plasmids).

Comparison of related strains or species has been and will be of paramount importance to study the plasticity of genomes, i.e., their capacities of variations in chromosome organization, gene sequence, gene content per species, etc. Examples of this are the existence of large deletions in *Streptomyces* spp. genomes, the presence of inversions, deletions, or additions of genes or regions (e.g., in *Neisseria*, *Pseudomonas*, and *Bacillus*). To this purpose, PFGE, gene encyclopedia, or genome sequences of one or a few well-known strains serve to compare other strains of the same or related species.

BIBLIOGRAPHY

Berlyn, M. K. B. (1998). Linkage map of *Escherichia coli* K-12, edition 10: The traditional map *Microbiol. Mol. Biol. Rev.* **62**, 814–984.

Bennett, P. M. (1999). Integrons and gene cassettes: a genetic construction kit for bacteria. *J. Antimicrob. Chemother.* **43**, 1–4.

Brosch, R., Gordon, G. V., Billault, A., Garnier, T., Eiglmeier, K., Soravito, C., Barell, B. G., and Cole, S. T. (1998). Use of a Mycobacterium tuberculosis H37Rv bacterial artificial chromosome library for genome mapping, sequencing, and comparative genomics. *Infect. Immun.* **66**(5), 2221–2229.

Cole, S. T., and Saint, Giron, I. (1994). Bacterial genomics. *FEMS Reviews* **14**, 139–160.

Edwards, J. S., and Palsson, B. O. (2000). Metabolic flux balance analysis and the in silico analysis of *Escherichia coli* K-12 gene deletions. *BMC Bioinformatics* **1**, 1.

Fonstein, M., and Haselkorn, R. (1995). Physical mapping of bacterial genomes. *J. Bacteriol* **177**, 3361–3369.

Joset, F., and Guespin-Michel, J. (1993). Construction of genomic maps. *In* "Prokaryotic Genetics: Genome Organization, Transfer and Plasticity," pp. 369–387. Blackwell Science, Oxford.

Leblond, P., and Decaris, B. (1998). Chromosome geometry and intraspecific genetic polymorphism in Gram positive bacteria revealed by pulse field gel electrophoresis. *Electrophoresis* **19**, 582–588 (and, in general, this whole special issue on genome mapping).

Peterson, J. D., Umayam, L. A., Dickinson, T. M., Hickey, E. K. and White, O. (2001). The Comprehensive Microbial Resource. *Nucleic Acids Res.* **29**, 123–125.

Rudd, K. E. (1998). Linkage map of *Escherichia coli* K-12, edition 10: The physical map. *Microbiol. Mol. Biol. Rev.* **62**, 985–1019.

Schilling, C. H., Edwards, J. S., Letscher, D., Palsson, B. O. (2000–2001). Combining pathway analysis with flux balance analysis for the comprehensive study of metabolic systems. *Biotechnol. Bioeng.* 71, 286–306.

Smith, H. O., and Birnstiel, M. L. (1976). A simple method for DNA restriction site mapping. *Nucleic Acids Res.* **3**(9), 2387–2398.

Willems, H., Jagen, C., and Baljer, G. (1998). Physical and genetic map of the obligate intracellular bacterium *Coxiella burnetii J. Bacterial.* **180**(15), 3816–3822.

WEBSITES

Comprehensive Microbial Resource of The Institute for Genomic Research (TIGR) and links to many other microbial genomic sites
http://www.tigr.org/tigr-scripts/CMR2/CMRHomePage.spl

Evolutionary and Genomic Microbiology Division, ASM (American Society for Microbiology)
http://www.asmusa.org/division/r/index.html

The Electronic Journal of Biotechnology
http://www.ejb.org/

DNA Structural Analysis of Sequenced Prokaryotic Genomes (Technical University of Denmark)
http://www.cbs.dtu.dk/services/GenomeAtlas/

Pedro's Biomolecular Research Tools. A collection of Links to Information and Services Useful to Molecular Biologists
http://www.public.iastate.edu/~pedro/research_tools.html

Bacteroides website of the Virtual Museum of Bacteria

Access to online resources on Bacterial Infections and Mycoses. Karolinska Institutet
http://www.mic.ki.se/Diseases/c1.html

Comprehensive Microbial Resource of The Institute for Genomic Research (TIGR) and links to many other microbial genomic sites
http://www.tigr.org/tigr-scripts/CMR2/CMRHomePage.spl

46

Germ-free animal techniques

Bernard S. Wostmann
University of Notre Dame

GLOSSARY

chemically defined (CD) diet A diet consisting of only chemically defined small molecules that can be sterilized by ultrafiltration with a minimal loss of nutrients.

conventional Describing or referring to an animal raised under normal animal house conditions.

germfree (GF) Indicates freedom from all known microorganisms, though not always indicating freedom from virus for all species.

gnotobiotic (GN) Indicating the presence of well-defined microbial associates.

hexaflora A stable, six-member microflora often introduced to normalize specific germfree anomalies, such as the enlarged cecum. They most often consist of *Lactobacillus brevis, Streptococcus fecalis, Staphylococcus epidermidis, Bacteroides fragilis.* var. *vulgatus, Enterobacter aerogenes,* and a *Fusibacterium* sp.

Although Pasteur in 1885 had expressed the opinion that germ-free life would be impossible, 10 years later Nutthal and Thierfelder produced the first germ-free guinea pig. Their ingenuously designed but complicated glass-based equipment was soon replaced by the much more simple metal isolator systems. Basically these consisted of a main compartment equipped with gloves to handle animals, presterilized food, and other necessities, with an attached two door entry port. Germ-free animals were obtained by caesarean section and entry into the sterile compartment either directly or via a germicidal bath. During the fifties plastic started to replace metal in the isolator systems, leading to the well-known systems used not only for germfree experimentation, but nowadays for protection of prematures and patients in need of special protection against environmental contamination.

I. INTRODUCTION

As early as 1885, Louis Pasteur had mentioned to his students the desirability of being able to study metabolism without the interference of an actively metabolizing intestinal microflora, mentioning at the same time that, in his opinion, life under those conditions would be impossible (Pasteur, 1885). However, before the end of that century, Nutthal and Thierfelder (1895) had built equipment that eventually enabled a newborn cesarean-derived guinea pig to survive (Fig. 46.1). Thus, the study of germ-free (GF) life had begun.

II. GERM-FREE ISOLATORS

Soon, the above contraption was replaced by simpler two-compartment steel autoclave-type equipment, in which one compartment served as a two-way entry port for sterilizing food and bedding, the other being the sterile compartment fitted with (nowadays) neoprene gloves, long enough for handling animals, cages, and equipment (Fig. 46.2). Gloves were often fitted with thin latex hand gloves to make necessary manipulations easier. Equipment was sterilized beforehand either by steam or, later, by chemical means. In another approach, lightweight steel units

The Desk Encyclopedia of Microbiology
ISBN: 0-12-621361-5

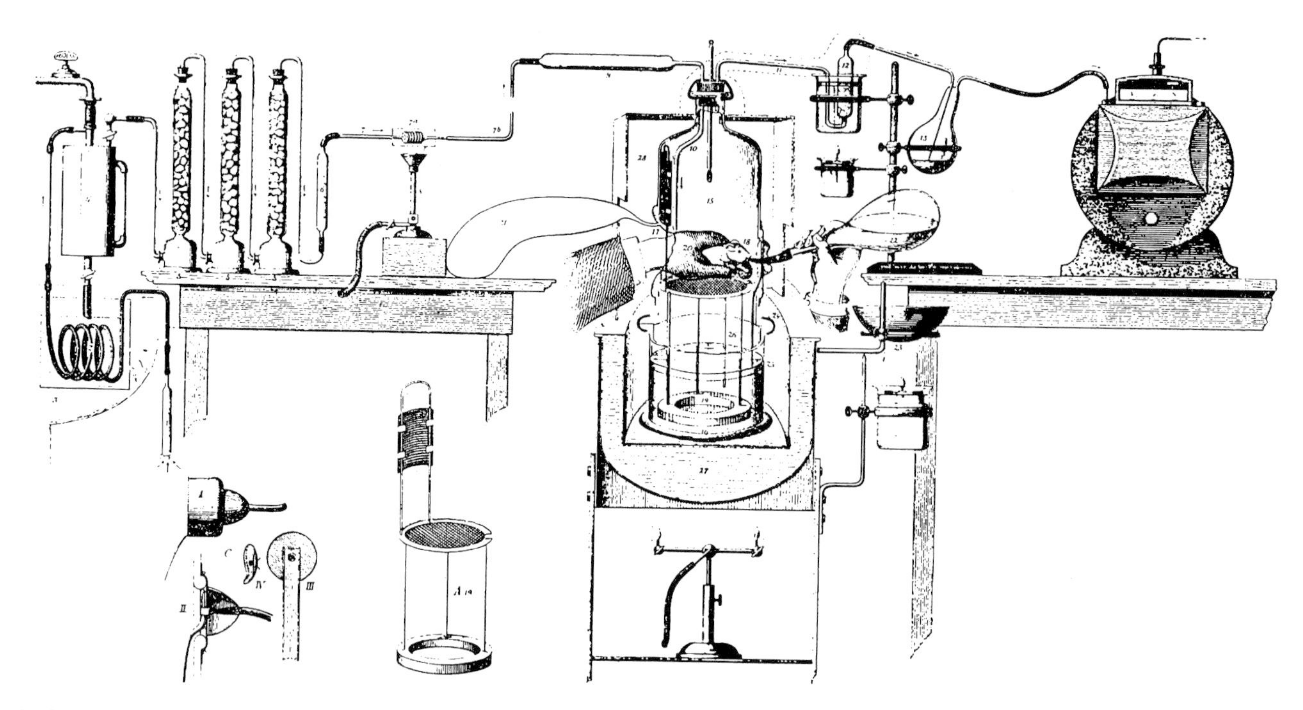

FIGURE 46.1 Equipment used by Nutthal and Thierfelder (1895) to raise Cesarean-derived germ-free guinea pigs. At left: sterilized air intake; in middle: guinea pig being fed a sterilized milk formula; at right: air exhaust system with air flow meter.

FIGURE 46.2 Steel isolator with entry port capped with plastic in preparation for chemical sterilization of materials.

with all equipment in place were sterilized in a large autoclave.

In the 1950s, plastic gradually replaced steel, although steel units are still in use in many institutions. While keeping the same two-compartment layout, the main compartment was now plastic, easily sterilized by chemical means, while food and other necessities were sterilized separately in round-bottom steel containers capped on one side with plastic. These were then connected to the main compartment via a plastic sleeve which, after chemical sterilization of the sleeve and removal of the plastic separators,

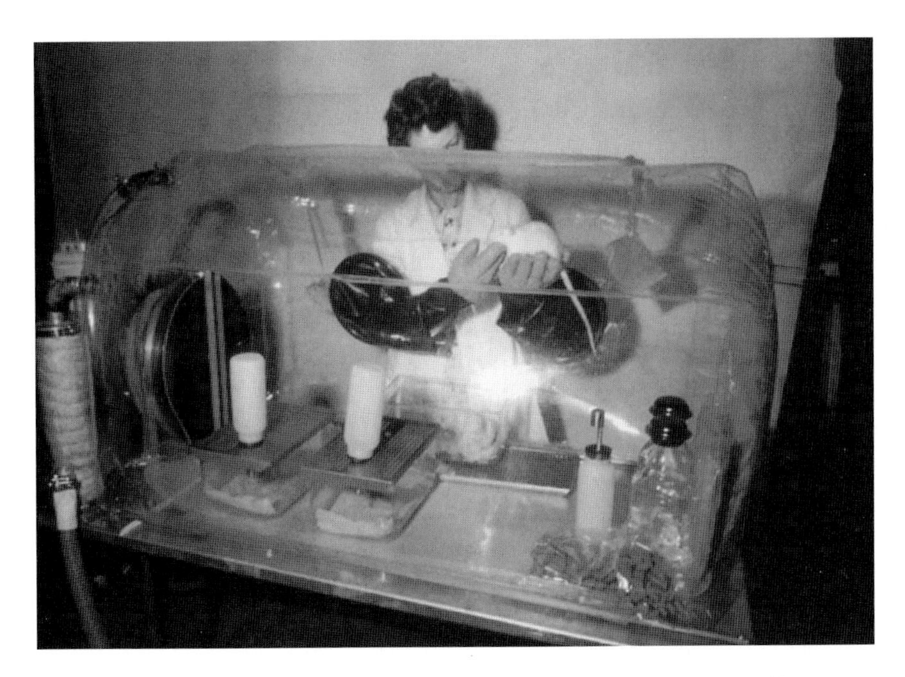

FIGURE 46.3 Plastic isolator with air filter in foreground and outlet trap partially visible behind the filter.

allowed for transfer of the sterile material. Thus, the plastic isolator was born, now in use all over the world.

Plastic isolators used for experimental purposes usually have an effective work area of 4″ × 6″ or 4″ × 8″. They resemble the well-known incubators used in hospitals to protect premature infants (Fig. 46.3). For colony breeding, much larger units are used, constructed to fit exigent requirements (Latuada, 1981). Caging requirements for GF animals generally resemble those of their CV counterparts.

The versatility of the plastic material made it possible to adjust the isolators to requirements of experimentation and for industrial animal production. Production units are often larger, although the efficiency of larger units must be weighed against the potential loss due to contamination. Some very large units have been built for specific purposes, even one with a manhole in the supporting table with an attached "half-suit," to make it possible for the caretaker to be within the unit and be able to reach any point necessary. The ultimate use has been the housing of a child with severe combined immunodeficiency (see Section V).

In a different approach, GF mice were transferred in the isolator to a standard mouse shoebox cage, which was then capped with a tight-fitting bonnet, the top consisting of HEPA-like filter material. The cage could then be removed and placed in a laminar-flow bench outfitted with HEPA filter. Properly protected personnel could now service the mice (Sedlacek *et al.*, 1981).

III. PRODUCTION OF GERM-FREE ANIMALS

Germ-free animals are obtained via cesarean section. In one approach, a specially constructed steel autoclave-type unit (as has been described) is used, which is divided into two compartments by a horizontal separator with a circular opening that is closed with a sheet of plastic. After autoclaving the units with the surgical equipment in the upper part, the lower part is opened to the environment. This (now nonsterile) part contains a small platform on which the anesthetized pregnant female can be raised so that the belly of the animal is tight against the plastic. The babies are then removed by hysterectomy. With a cauterizing knife, a cut is made through the plastic and abdominal wall, heat-sealing the abdominal cavity to the GF compartment. The babies can now be removed and placed in a separate isolator for cleaning and feeding. In this way, GF mice, rats, guinea pigs, and rabbits have been produced.

After birth, the newborn had to be fed a specifically adjusted formula, using a nipple attached to a suitable syringe-type container. Germfree guinea pigs had to be fed for only a few days because of their advanced development at birth. Other species, however, had to be fed in this way for several weeks. This turned out to present no special problems, except for the feeding of newborn GF mice. After many trials, a pair of baby C3H mice were obtained that grew to maturity, bred, and produced offspring. Thereafter, newborn mice of any strain could be foster-nursed by the GF mice already available. All of the species mentioned above reproduced well under GF conditions, rabbits only after an anomaly in Fe absorption was recognized, which needed some dietary adjustment. Germ-free gerbils have been obtained by foster nursing on GF mice that were 2–3 days postpartum. However, these animals did not reproduce because of the size of the cecum (Bartizal *et al.*, 1982).

All GF rodents and leporidae have significantly enlarged ceca. In GF rats, the distension starts about two weeks after birth. Depending on diet, the GF cecum could be from 4% to 8% of body weight, versus around 1% in the CV rat, and more in GF rabbit and guinea pigs. The enlarged cecum is caused by the absence of a microflora that normally will digest the intestinal mucus which would otherwise accumulate in the cecum. This accumulation of acidic mucus changes the local mineral balance, leaving insufficient sodium ions necessary for water removal (Gordon and Wostmann, 1973). As a result, the water content of the cecal material resembles that of the ilium. Although comparable absorption of water occurs in the colon, water content of rat GF feces is more than 20%, against about 4% in its CV counterpart. While the GF rat excretes the larger part of its water intake with the feces, the CV rats excretes most in the urine. The rather fluid feces of the GF animal may create a substantial housekeeping problem.

As mentioned earlier, GF gerbils had to be associated with a microflora consisting of six well-defined microbial species before reproduction would occur (the hexaflora). Association with this hexaflora has been used for cecal reduction, the major abnormality of the above-mentioned GF species. The hexaflora (or other multiflora) components were introduced by bringing sterile monocultures into the isolator after sterilization the outside of the usual test tubes.

Germ-free animals have also been obtained via Cesarean section in a clean, minimally contaminated environment, subsequently passing the newborn via a germicidal trap into the GF unit. This method had been originally used for rats and has later been used to produce larger GF animals like pigs, lambs, and calves (Miniats, 1984).

Depending on the requirements of the study, either noninbred or inbred strains have been used to produce the GF animal. When noninbred strains were used, regular introduction of GF animals derived from the CV stock was used to assure genetic comparability between GF and CV animals. In case inbred strains were used, it would seem that this might be unnecessary but, presumably, still advisable after prolonged periods to prevent genetic drift from occurring.

Germ-free chickens have been produced from fertilized eggs that had been incubated for at least 18 days. The eggs are then placed in the entry port of a GF isolator, chemically sterilized, and taken inside. They are spread out and temperature is kept at 37–38 °C, humidity between 70% and 80%. After hatching, the birds are kept at 37 °C for a few days. They rapidly learn to eat and drink. Thereafter, the temperature can be gradually lowered to normal room temperature (Coates, 1984).

IV. DIETS FOR GERM-FREE ANIMALS

Diets for GF animals are nowadays well established. They fall into three categories: natural ingredient diets fortified for the losses that occur during sterilization (Table 46.1); diets with minimal antigenicity, used mainly for immunological studies; and the chemically defined antigen-free diets, used for immunological and other studies where absolute definition of environment and dietary intake is required, again used mainly in immunology. For more defined work in the first category, diets based on rice starch, a well-defined source of protein like casein or soy protein, some fat, generally corn oil, and well-defined vitamin and mineral mixes have been developed (Table 46.2) and are commercially available. Diets of minimal antigenicity have been developed for the study of the very early phases of immune response in young colostrum-deprived gnotobiotic (GN) piglets (Kim and Watson, 1969).

Extreme definition was obtained with the development of the GF mouse reared on a totally chemically

TABLE 46.1 Composition of natural ingredient diet L-485

Ingredients	g/kg	Nutrient composition, %	
Ground corn	590	Protein	20.0
Soy bean meal, 50% CP[a]	300	Fat	5.3
Alfalfa meal, 17% CP	35	Fiber	3.0
Corn oil	30	Ash	5.5
Iodized NaCl	10	Moisture	11.2
Dicalcium phosphate	10	Nitrogen-free material	55.0
Calcium carbonate	5		
Lysine (feed grade)	5	Gross energy 3.9 Kcal/g	
Methionine (feed grade)	5		
Vitamin and mineral mixes	0.25[b]		
BTH	0.125		

[a]Crude protein.
[b]Kellogg, T. F., and Wostmann, B. S. (1969). Stock diet for colony production of germfree rats and mice. *Lab. Anim. Care* **17**, 589.

defined, water-soluble antigen-free diet (CD diet), consisting of dextrose, amino acids, water-soluble vitamins, minerals, and a fat supplement, sterilized by ultrafiltration (Table 46.3) (Pleasants *et al.*, 1986). This successful colony was started with inbred GF BALB/cAnN mice obtained from the GF colony maintained at the University of Wisconsin. Pregnant GF mice of this colony originally were fed natural ingredient diet L-485 (Table 46.1), then transferred to another isolator and fed CD diet. Their offspring, never having been in contact with other than the CD diet, were housed in plastic isolators in shoebox-type mouse cages, with lids modified to hold the inverted diet and water bottles. The plastic bottoms of the cages had been replaced with stainless steel wire mesh above removable drip pans. Fortified soy-derived triglycerides provided essential fatty acids plus readily available calories. Purified vitamins A, D, E, and K were added. Again, this mixture was ultra-filtered. It was then fed in stainless steel planchets that were welded to the stainless steel dividers in the cages (Fig. 46.4). Whatman Ashless filter paper provided indigestible fiber while also serving as bedding and nesting material. This was autoclaved for 25 min. at 121 °C or irradiated at 4.5 Mrad before being taken into the isolator.

Here, the original purpose had been the establishment of nutritional requirements in the absence of a metabolizing intestinal microflora. But again, without any original exposure to antigen or microflora, this proved an ideal tool for the study of the early and later development of the immune system. For these studies, the inbred GF BALB/c mouse proved to be the animal of choice, because of excellent breeding results through at least 6 generations. This was in contrast to earlier studies with inbred C3H mice, which did not reproduce beyond the second generation.

While, obviously, the GF animal maintained on a chemically defined diet is the animal of choice to study the various aspects of immunology, the GF animal maintained under less rigorous conditions will be of great value for the study of the effects of the complete

TABLE 46.2 Composition of casein–starch diet L-488F, fortified with cholesterol

	g/100 g
Casein	24.0
DL-Methionine	0.3
Starch	60.4
Cellophane spangles	5.0
Corn oil[a]	5.0
Cholesterol	0.05
Vitamin and mineral mixes[b]	5.3

[a]Contains fat-soluble vitamins.
[b]For details of the vitamin and mineral mixes, see Reddy *et al.* (1972). Studies on the mechanisms of calcium and magnesium absorption of germfree rats. *Ann. Biochem. Biophys.* **149**, 15.

TABLE 46.3 Composition of chemically defined diet L-489E14SE, per 100 g water-woluble wolids in 300 ml ultrapure H_2O

L-Leucine	1.90 g	Ca glycerophosphate	5.22 g
L-Phenylalanine	0.74 g	$CaCl_2 \cdot 2H_2O$	0.185 g
L-Isoleucine	1.08 g	Mg glycerophosphate	1.43 g
L-Methionine	1.06 g	K acetate	1.85 g
L-Tryptophan	0.37 g	NaCl	86.00 mg
L-Valine	1.23 g	$Mn(acetate)_2 \cdot 4H_2O$	55.40 mg
Glycine	0.30 g	Ferrous gluconate	55.00 mg
L-Proline	1.48 g	$ZnSO_4 \cdot H_2O$	40.60 mg
L-Serine	1.33 g	$Cu(acetate)_2 \cdot 2H_2O$	3.70 mg
L-Asparagine	1.03 g	$Cr(acetate)_3 \cdot H_2O$	2.50 mg
L-Arginine · HCl	0.81 g	NaF	2.10 mg
L-Threonine	0.74 g	KI	0.68 mg
L-Lysine.HCl	1.77 g	$NiCl_2 \cdot 3H_2O$	0.37 mg
L-Histidine · HCl · H_2O	0.74 g	$SnCl_2 \cdot 2H_2O$	0.31 mg
L-Alanine	0.59 g	$(NH_4)_6Mo_7O_{24} \cdot 4H_2O$	0.37 mg
Na L-Glutamate	3.40 g	Na_3VO_4	0.22 mg
Ethyl L-Tyrosinate · HCl	0.62 g	$Co(acetate)_2 \cdot 4H_2O$	0.11 mg
α-D-Dextrose	71.40 g	Na_2SeO_3	0.096 mg

Note: B vitamins (in mg): thiamine · HCl, 1.23; pyridoxine · HCl, 1.54; biotin, 0.25; folic acid, 0.37; vitamin B_{12} (pure), 1.44; riboflavin, 1.85; niacinamide, 9.2; i-inositol, 61.6; Ca pantothenate, 12.3; choline · HCl, 310.

Amounts of lipid nutrients in one measured daily adult dose of 0.25 mL: purified soy triglycerides, 0.22 g; retinyl palmitate, 4.3 μg (7.8 IU); cholecalciferol, 0.0192 μg (0.77 IU); 2-ambo-α-tocopherol, 2.2 mg; 2-ambo-α-tocopheryl acetate, 4.4 mg; phylloquinone, 48.0 μg. The fatty acid content is 12% palmitate, 2% stearate, 24% oleate, 55% linoleate, 8% linolenate.

FIGURE 46.4 Mouse cage used in isolator containing mice fed chemically defined diet. See text.

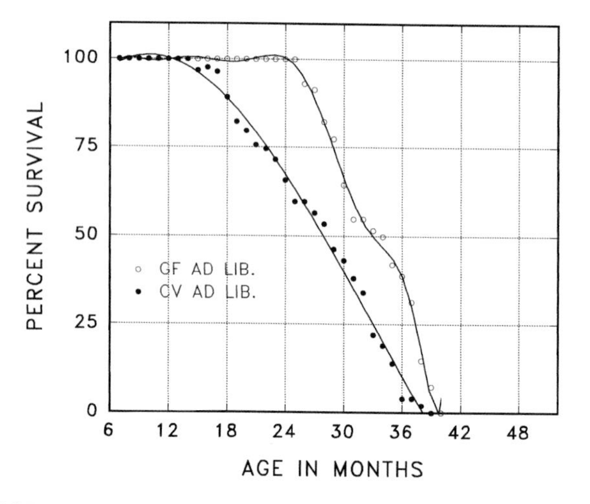

FIGURE 46.5 Life span. Percent survival of male germ-free and conventional Lobund–Wistar rats fed diet natural ingredient diet L-485.

absence of a microflora on function and metabolism. Comparing GF and CV rats, it was found that the stress of a metabolizing microflora reduces life span to an extent that otherwise can only be achieved by an involuntary reduction of dietary intake by 30% (Fig. 46.5).

V. APPLICATION OF GERM-FREE TECHNIQUES: GNOTOBIOTES

A. The boy in the bubble

In general, most controlled studies have been carried out with GF and GN rats and mice. The GF animal technique has the potential for selective association of the originally GF animal with a specific microflora element, including parasites. Mentioned earlier is the potential of bringing its distended cecum back within acceptable limits by association with a well-defined cocktail of microbial species. These are introduced by bringing monocultures grown in test tube into the main unit after chemical exterior sterilization in the entry compartment. In this way, studies have become possible of what is called the gnotobiote, the animal that harbors a desired but defined combination of microflora elements. However, the term gnotobiote per se also includes the GF animal.

Once the GF animal had been produced, the logical extension was the study of function and metabolism of this gnotobiote. Apart from anomalies mentioned earlier, the absence of a metabolizing microflora resulted in the absence of certain nutrients normally produced by the flora (e.g., vitamin K) and certain flora-produced stimuli (e.g., LPS). Because of this, diet had to be adjusted, apart from adjustment necessitated by heat or radiation sterilization. The effects of the GF state on immune function are substantial, although all GF species proved to be eventually immunologically competent. Upon antigenic stimulation, the GF animal generally shows a delayed, but eventually adequate, immune response. However, antibacterial response may be inadequate, due to slow response of macrophages, which originally are defective in chemotaxis and destructive and lytic capacity, leading to a delayed presentation of antigenic material to other elements of the immune system (Wostmann, 1996).

The possibility of the production of gnotobiotes with a stable and defined microflora opened many possibilities and can be regarded as the originally GF animal's major potential. It soon became obvious that GF rats did not develop that scourge of our society, dental caries, thereby indicating its microbial origin. GF rats were then associated with a number of different bacteria. This resulted in the recognition of *Streptococcus mutans* as the major originator of caries and put the acid-producing lactobacilli in second place (Orland *et al.*, 1955). Similarly, it was shown that GF guinea pigs inoculated with the parasite *Entomoebia histolytica*, the cause of a potentially lethal intestinal infection, do not develop any symptoms. The animals retained the amoeba for only a few days. A microflora is obviously needed to change intestinal conditions to the point where the infection could take hold (Phillips, 1964). On the other hand, it was found that a CV microflora, by its stimulation of the immune system, may affect a certain amount of protection against Schistosomiasis (Bezerra *et al.*, 1985). Studies of the various factors involved in the establishment of the nematode *Trichinella spiralis* were carried out by Przyjalkowski (Przyjalkowski *et al.*, 1983) and Despommier (Despommier, 1984). In the pork industry, GF pigs have been used to solve problems of bacterial and viral infection.

Gnotobiotes are important to the study of colonization resistance, which seeks to determine which microflora elements may be important for the flora's stability and which for its potential to resist pathogens. In a similar way, gnotobiotes enable the study of microbial translation to determine what microflora composition will enhance or inhibit certain of its members to pass the intestinal barrier and possibly cause disease.

Recently, it has become possible to establish a "normal" human microflora in originally GF mice. After the actual composition of this flora has been established, and its stability ascertained, this could open the door to a multitude of studies pertaining to human health and disease.

The plastic isolator, already in general use to protect premature infants, found its culmination in its use to protect David, the "Boy in the Bubble." Before birth diagnosed as having SCID (severe combined immune deficiency), he was born via Cesarean section and placed in an isolator, which eventually grew to a four-room apartment. Over the years, he accumulated a number of non-life-threatening organisms, but death came at the age of about $12\frac{1}{2}$ years, apparently caused by the sequela of an unsuccessful bone marrow transplant (Bealmear *et al.*, 1985).

BIBLIOGRAPHY

Bartizal, K. F., *et al.* (1982). Cholesterol metabolism in gnotobiotic gerbils. *Lipids* **17**, 791.

Bealmear, P. M., *et al.* (1985). David's story: A gift of 12 years, 5 months and 1 day. *In* "Progress in Clinical and Biological Research" Vol. 181 (B. S. Wostmann, ed.), p. 485. Alan R. Liss, Inc., New York.

Bezerra, J. A., *et al.* (1985). The life cycle of *Schistosoma mansoni* under germfree conditions. *J. Parasitol.* **71**, 519.

Coates, M. E. (1984). Production of germfree animals. Part 3. Birds. *In* "The Germfree Animal in Biomedical Research" (M. E. Coates and B. E. Gustafsson, eds.). Laboratory Animal Handbooks **9**, p. 79. Laboratory Animals Ltd., London, UK.

Coates, M. E. (ed.) (1987). "ICLAS Guidelines on the Selection and Formulation of diets for Animals in Biomedical Research" ICLAS, Institute of Biology, 20 Queensberry Place, London, UK.

Deplancke, B., and Gaskins, H. R. (2001). Microbial modulation of innate defense: goblet cells and the intestinal mucus layer. *Am. J. Clin. Nutr.* **73**, 1131S–1141S.

Despommier, D. D. (1984). Antigens from *Trigenella spiralis* that induce a protective response in the mouse. *J. Immunol.* **132**, 898.

Fitzgerald, R. F., *et al.* (1989). Cariogenicity of a lactate dehydrogenase-deficient mutant of *Streptococcus mutans* serotype c in gnotobiotic rats. *Infect. Immun.* **57**, 823.

Gordon, H. A., and Wostmann, B. S. (1973). Chronic mild diarrhea in germfree rodents: A model portraying host–flora synergism. *In* "Germfree Research, Proc. IV Internat. Symp. Germfree Research" (J. B. Heneghan, ed.), p. 593. Academic Press, New York.

Hashimoto *et al.* (ed.) (1996). Germfree life and its ramifications. "Proceedings of the XII International Symposium on Gnotobiology," Sense Printing, Japan.

Kim, Y. B., and Watson, D. W. (1969). *In* "Germfree Biology, Experimental and Clinical Aspects" (E. A. Mirand and N. Back, eds.), p. 259. Plenum Press, New York.

Latuada, C. P. (1981). Large isolators for rearing rodents. *In* "Recent Advances in Germfree Research, Proc. VII Internat. Symp. On Gnotobiology" (Sasaki *et al.*, eds.), p. 53. Tokay University, Tokyo.

Miniats, O. P. (1984). Production of germfree animals. Part 2. Farm animals. *In* "The Germfree Animal in Biomedical Research" (M. E. Coates and B. E. Gustafsson, eds.), Laboratory Animal Handbooks **9**, p. 49. Laboratory Animals Ltd., London, UK.

Nutthal, G. H. F., and Thierfelder, H. (1895). Thierisches Leben ohne Bakterien im Verdauungskanal, *Z. Physiol. Chem.* **21**, 109.

Orland, F. J., *et al.* (1955). Experimental caries in rats inoculated with *Enterococci*. *Am. Dental Assoc.* **50**, 259.

Pasteur, L. (1885). Observations relative à la note précédente de M. Duclaux. *CR Acad. Sci. Paris* **100**, 68.

Phillips, B. F. (1964). Studies of the amoeba–bacteria relationships in amebiasis III. Induced amoebic lesions in the germfree guinea pig. *Am. J. Trop. Med. Hyg.* **13**, 301.

Pleasants *et al.* (1986). Adequacy of chemically defined, water-soluble diet for germfree BALB/c mice through successive generations and litters. *J. Nutr.* **116**, 1949.

Przyjalkowski *et al.* (1983). Intestinal *Triginella spiralis* and *Triginella pseudospiralis*. *Prog. Fd. Nutr. Sci.* **7**, 117.

Sedlacek, R. S., *et al.* (1981). A flexible barrier at cage level for existing colonies. *In* "Recent Advances in Germfree Research, Proc. VII Internat. Symp. on Gnotobiology" (S. Sasaki *et al.*, eds.), p. 65. Tokay University, Tokyo.

Tzipori, S., and Widmer, G. (2000). *The biology of Cryptosporidium*. *Contrib. Microbiol.* **6**, 1–32.

Wostmann, B. S. (1981). The germfree animal in nutritional studies. *Ann. Rev. Nutr.* **1**, 257.

Wostmann, B. S. (1996). "Germfree and Gnotobiotic Animal Models 101." CRC Press, New York.

WEBSITE

Association for Gnotobiotics
http://www.gnotobiotics.org/

47

Gram-negative anaerobic pathogens

Arthur O. Tzianabos, Laurie E. Comstock, and Dennis L. Kasper
Harvard Medical School

GLOSSARY

abscess A classic host response to bacterial infection in which many anaerobic species predominate.

anaerobe A bacterium that can survive and proliferate in the absence of oxygen.

antibiotic resistance The ability of bacteria to withstand the killing effect of antimicrobial agents.

cellular immunity A T cell-dependent host response to bacteria.

intraabdominal sepsis A disease process in humans that occurs following spillage of the colonic contents within the peritoneal cavity.

Gram-Negative Anaerobic Organisms constitute an important group of pathogens that predominate in many infectious processes. These organisms generally cause disease subsequent to the breakdown of mucosal barriers and the leakage of indigenous flora into normally sterile sites of the body. The predominance of anaerobes in numerous clinical syndromes can be attributed to the elaboration of a variety of virulence factors, the ability to resist oxygenated microenvironments, synergy with other bacteria, and resistance to certain antibiotics.

I. NORMAL FLORA AND EPIDEMIOLOGY

Hundreds of species of anaerobic organisms make up the human microflora. Mucosal surfaces such as the oral cavity, gastrointestinal tract, and female genital tract are the major reservoirs for this group of organisms. Infections involving anaerobes are polymicrobial and usually result from the disruption of mucosal surfaces and the subsequent infiltration of resident flora. However, gram-negative anaerobic bacilli are the most commonly isolated anaerobes from clinical infections. The clinically important gram-negative anaerobic bacilli belong to the genera *Bacteroides, Fusobacterium, Porphyromonas, Prevotella,* and *Bilophila* (Table 47.1).

Anaerobes normally reside in abundance as part of the oral flora, with concentrations ranging from 10^9/ml in saliva to 10^{12}/ml in gingival scrapings. The indigenous oral anaerobic flora primarily comprise *Prevotella* and *Porphyromonas* species, with *Fusobacterium* and *Bacteroides* (non-*Bacteroides fragilis* group) present in fewer numbers. Aerobic or microaerophilic and anaerobic bacteria reside in approximately equal numbers in the oral cavity.

Anaerobic bacteria are present only in low numbers under the normally acidic conditions in the stomach and upper intestine. In people with decreased gastric acidity, the microflora of the stomach resemble that of the oral cavity. The upper intestine contains relatively few organisms until the distal ileum, where the flora begins to resemble that of the colon. In the colon, there are up to 10^{12} organisms/stool, with anaerobes outnumbering aerobes by approximately 1000:1. *Bacteroides* and *Fusobacterium* are the predominant gram-negative species in the colonic flora. The *Bacteroides* species present in the colon are often referred to as the *B. fragilis* group, based on their

The Desk Encyclopedia of Microbiology
ISBN: 0-12-621361-5

TABLE 47.1 Disease processes caused by gram-negative anaerobic bacteria

Anaerobic species	Colonization site	Typical disease
Bacteroides fragilis	Colon	Intra-abdominal abscesses; bacteremia, mixed soft tissue infections
Other *Bacteroides* spp.	Colon; vagina	Abscesses and necrotizing fasciitis
Fusobacterium spp.	Oral cavity; colon; vagina	Abscesses; internal jugular venous thrombophlebitis
Prevotella spp.	Oral cavity; vagina	Lung abscess; pelvic infections
Porphyromonas spp.	Oral cavity	Gingival infection
Bilophila wadsworthia	Oral cavity; vagina; colon	Abscesses; bacteremia; necrotizing fasciitis

earlier classification as subspecies of *B. fragilis*. This group encompasses several species, including *B. fragilis*, *B. thetaiotaomicron*, *B. ovatus*, *B. vulgatus*, *B. uniformis*, and *B. distasonis* (although the last organism may be reclassified as a *Porphyromonas* species). These *Bacteroides* species are all considered to be penicillin resistant, whereas the *Fusobacterium* species remain penicillin sensitive.

Prevotella, *Bacteroides*, and *Fusobacterium* are part of normal vaginal flora (10^6 organisms/g secretions). The most common isolates from clinical specimens are *P. bivia* and *P. disiens*, although *B. fragilis* is also frequently isolated from this site. *Bacteroides* species are found in approximately 50% of women, whereas *B. fragilis* is found in $<15\%$ of this population.

Bilophila species are found in normal stool specimens and are occasionally part of the normal oral or vaginal flora. *B. wadsworthia* exhibits a fairly broad resistance to β-lactam antibiotics and is of interest because of its reported occurrence in human infections.

Anaerobic gram-negative bacilli reside on mucosal surfaces and cause infection following the contamination of normally sterile sites. For example, intraabdominal infection develops after fecal spillage into the peritoneum as a result of trauma or necrosis of the intact bowel wall, and severe infections of the head and neck may arise from an abscessed tooth. It is remarkable that, despite the identification of scores of anaerobic species in normal flora, relatively few species seem to play a role in infection. After the contamination of normally sterile sites by mucosal microflora, the relatively few anaerobic bacteria that survive are those few that have resisted changes in oxidation reduction potential and host defense mechanisms. The hallmark of infection due to these gram-negative anaerobic bacteria is abscess formation, although some sepsis syndromes have been described. Typically, abscesses form at sites of direct bacterial contamination, although distant abscesses resulting from hematogenous spread are not uncommon with the more virulent anaerobes.

II. CLINICAL SYNDROMES CAUSED BY ANAEROBES

A. Anaerobic infections of the mouth, head, and neck

Anaerobes contribute to infection associated with periodontal disease and to disseminated infection arising from the oral cavity. Locally, infection of the periodontal area may extend into the mandible, causing osteomyelitis of the maxillary sinuses or infection of the submental or submandibular spaces. Other infections associated with oral anaerobes include gingivitis, acute necrotizing ulcerative mucositis, acute necrotizing infections of the pharynx, Ludwig's angina, fascial infections of the head and neck, sinusitis, and otitis.

B. Pleuropulmonary infections

Anaerobes are associated with aspiration pneumonia, lung abscesses, and empyema. In these cases, the organisms isolated generally reflect the oral flora, and more than one bacterial species is routinely involved. For example, in anaerobic lung abscesses, it is not uncommon to find up to 10 species of organisms.

C. Intra-abdominal infections

Infections originating from colonic sites, such as intra-abdominal abscesses, most frequently involve *Bacteroides* species, among which *B. fragilis* is the most common isolate. *B. fragilis* has also been associated with watery diarrhea in a few studies. Enterotoxin-producing strains are more prevalent in patients with diarrhea than in control groups. Secondary bacterial peritonitis arises when organisms from the intestine contaminate the peritoneum. The terminal ileum and colon are the most common sites of origin because of the high numbers of organisms they contain. Patients typically develop acute peritonitis following contamination, and intraperitoneal abscesses may result.

D. Pelvic infections

Anaerobes are encountered in pelvic abscess, septic abortion, endometritis, tubovarian abscess, pelvic inflammatory disease, and postoperative infections. In addition to the major anaerobic gram-negative isolates already mentioned, *P. melaninogenica*, clostridia, and peptostreptococci are commonly found in infected pelvic sites. These organisms are most often isolated when infection is not due to sexually transmitted agents.

E. Central nervous system infections

Either a single species or a mixture of anaerobic or aerobic bacteria may be found in brain abscesses; prominent among the anaerobes are *Fusobacterium*, *Bacteroides*, and anaerobic gram-positive cocci. Anaerobic brain abscesses may arise by hematogenous dissemination from an infected distant site or by direct extension from otitis, sinusitis, or tooth infection.

F. Skin and soft-tissue infections

Bacteroides species are found in necrotizing fasciitis, usually as part of a mixed anaerobic–aerobic infection. Approximately five species are typically isolated, with a 3 : 2 ratio of anaerobes to aerobes. These infections usually occur at sites that can be contaminated from oral secretions or feces; they may spread rapidly and be very destructive. Gas may be found in the infected tissues.

G. Bone and joint infections

These infections typically arise from infected adjacent soft-tissue sites. Both osteomyelitis of bone and septic arthritis are seen. *Fusobacterium* species are the most common gram-negative anaerobes isolated from infected joints, whereas infected bone may yield a wider variety of isolates.

H. Bacteremia

Anaerobic organisms, particularly *B. fragilis*, have been detected in up to 5% of blood cultures. When *B. fragilis* is isolated, patients are frequently ill with rigors and fever.

III. PATHOGENESIS

Infections involving gram-negative anaerobes are generally due to the breakdown of a mucosal barrier and the subsequent leakage of indigenous flora into closed spaces or tissue. The introduction of bacteria into otherwise sterile sites leads to a polymicrobial infection in which certain organisms predominate. These include *B. fragilis*, *Prevotella* species, *Fusobacterium* species, and *Porphyromonas* species. Although some of these organisms are numerically dominant in the normal flora, others (such as *B. fragilis*) make up a much smaller proportion. The greater ability of these organisms to cause disease more often than numerically dominant anaerobes usually indicates the possession of one or more virulence factors. These factors include the ability to evade host defenses, adhere to cell surfaces, produce toxins or enzymes, or display surface structures that contribute to pathogenic potential.

A. Synergy

The ability of different anaerobic bacteria to act synergistically during polymicrobial infection has been described but remains poorly characterized. It has been postulated that facultative organisms function in part to lower the oxidation–reduction potential in the microenvironment and that this change allows the propagation of obligate anaerobes. Conversely, studies have shown that anaerobes, including *B. fragilis*, can produce compounds such as succinic acid and short-chain fatty acids that inhibit the ability of phagocytes to clear facultative organisms. Further, it is clear that facultative and obligate anaerobes synergistically potentiate abscess formation in experimental models.

B. Role of *B. fragilis* capsular polysaccharide in abscess induction

The anaerobe most commonly isolated from clinical infections is *B. fragilis*. The most frequent sites of isolation of *B. fragilis* are the bloodstream and abscesses associated with intraabdominal sepsis. The characteristic host response to *B. fragilis* infections is the development of intraabdominal abscesses. Although the development of an abscess limits the initial spread of the organism, the host usually cannot resolve the abscess, therefore requiring surgical drainage. The high frequency of abscess formation associated with *B. fragilis* led to studies investigating this organism's pathogenic potential in relevant animal models of disease. This work identified the capsular polysaccharide as the major virulence factor of *B. fragilis* and defined the capsule's role in the induction of abscesses in an animal model of intraabdominal sepsis.

Further studies have delineated the structural attributes of the *B. fragilis* capsular polysaccharide that promotes abscess formation in animals. The capsule of strain NCTC 9343 comprises two distinct ionically linked polymers, termed PS A and PS B. Each of these purified polymers induces abscess formation when implanted with sterilized cecal contents (SCC) and barium sulfate as an adjuvant into the peritonea of rats. Historically, this adjuvant is included to simulate the spillage of colonic contents that occurs in intraabdominal sepsis. The implantation of PS A without SCC does not induce abscess formation in animals. PS A is the more potent of the two polysaccharides; less than 1 μg is required for abscess induction in 50% of challenged animals. The structures of both saccharides have been elucidated. Each polymer consists of repeating units whose possession of both positively and negatively charged groups is rare among bacterial polysaccharides. The ability of PS A and PS B to induce abscesses in animals depends on the presence of these charged groups. Numerous fecal and clinical isolates of *B. fragilis* have been examined, and this dual polysaccharide motif has been found in every strain.

The capsule of *B. fragilis* probably acts to regulate the host response within the peritoneal cavity to initiate the steps leading to abscess formation. Studies in mice have shown that the capsule mediates bacterial adherence to primary mesothelial cell cultures *in vitro*. In addition, the capsule promotes the release of the proinflammatory cytokines TNF-α and IL-1β, as well as the chemokine IL-8, from macrophages and neutrophils. The release of TNF-α from peritoneal macrophages potentiates the increase of cell-adhesion molecules such as ICAM-1 on mesothelial cell surfaces, and this potentiated response in turn leads to an increase in the binding of neutrophils to these cells. These events are the first steps leading to the accumulation of neutrophils at inflamed sites within the peritoneal cavity and are likely to lead to the formation of abscesses at these sites.

C. Abscess formation and T cells

T cells are critical in the development of intraabdominal abscesses, but little is known about the mechanisms of cell-mediated immunity underlying this host response. Attempts to define the immunologic events leading to abscess formation have been made in athymic or T cell-depleted animals. The results from these studies show that T cells are required for the formation of abscesses following bacterial challenge of animals. Studies by Sawyer *et al.* (1995) have documented a role for CD4+ T cells in the regulation of abscess formation.

B. fragilis produces a host of virulence factors that allow this organism to predominate in disease. Although the lipopolysaccharide (LPS) of *B. fragilis* possesses little biologic activity, this organism synthesizes pili, fimbriae, and hemagglutinins that aid in attachment to host-cell surfaces. In addition, *Bacteroides* species produce many enzymes and toxins that contribute to pathogenicity. Enzymes such as neuraminidase, protease, glycoside hydrolases, and superoxide dismutases are all produced by *B. fragilis*. Recent work has shown that this organism produces an enterotoxin with specific effects on host cells *in vitro*. This toxin, termed BFT, is a metalloprotease that is cytopathic to intestinal epithelial cells and induces fluid secretion and tissue damage in ligated intestinal loops of experimental animals. Strains of *B. fragilis* associated with diarrhea in children (termed enterotoxigenic *B. fragilis*, or ETBF) produce a heatlabile 20-kDa protein toxin. BFT specifically cleaves the extracellular domain of E-cadherin, a glycoprotein found on the surface of eukaryotic cells.

The pathogenesis of *P. gingivalis* relies on a broad range of virulence factors. This organism is a prominent etiologic agent in adult periodontitis. The progression of this disease is hypothesized to be related to the production of a variety of enzymes (particularly proteolytic enzymes), fimbriae, capsular polysaccharide, LPS, hemagglutinin, and hemolytic activity. *P. gingivalis* has been shown to invade and replicate host cells, a mechanism that may facilitate its spread. A class of trypsin-like cysteine proteases, termed gingipains, have recently been implicated as a major virulence factor contributing to the tissue destruction that is the hallmark of periodontal disease.

The capsular polysaccharide of *P. gingivalis* acts as a potent virulence factor facilitating a spreading infection in mice greater than that seen with unencapsulated strains. The LPS of *P. gingivalis* has been implicated in the initiation and development of periodontal disease. It has been shown that the LPS activates human gingival fibroblasts to release IL-6 via CD14 receptors on host cells. In addition, neutrophils stimulated with *P. gingivalis* LPS release IL-8.

F. necrophorum causes numerous necrotic conditions (necrobacillosis) and human oral infections. Several toxins, such as leukotoxin, endotoxin, and hemolysin, have been implicated as virulence factors. Among these, leukotoxin and endotoxin are believed to be the most important. *F. nucleatum* is a major contributor to gingival inflammation and is isolated from sites of periodontitis. This organism can co-aggregate with other oral bacteria to promote attachment to plaque; in addition, it produces several adhesins that facilitate attachment. Both *F. nucleatum* and *F. necrophorum* produce a potent LPS that is responsible for the release of numerous proinflammatory cytokines and other inflammatory mediators.

Virulence factors associated with *Prevotella* species are poorly defined. The organism's ability to interact with other anaerobes has been reported. Among their prominent virulence traits is the production of proteases and metabolic products, such as volatile fatty acids and amines. This group of organisms is particularly noted for secretion of IgA proteases. The degradation of IgA produced by mucosal surfaces allows *Prevotella* to evade this first line of host defense. A study has demonstrated that *P. intermedia* can invade oral epithelial cells and that antibody specific for fimbriae from this organism inhibits invasion.

IV. IMMUNITY

Although relatively little is known about immunity to anaerobic organisms in general, the immune response to *B. fragilis* has been studied in detail. Prior treatment of animals with PS A from this organism prevents the formation of intraabdominal abscesses after challenge with *B. fragilis* or other abscess-inducing bacteria. This protection depends on the presence of positively charged amino and negatively charged carboxyl groups associated with the saccharide's repeating unit structure. Attempts to define the immunologic events regulating abscess formation have suggested an important role for cell-mediated immunity.

Studies on rats have shown that administration of PS A shortly before or even after bacterial challenge protects against abscess formation induced by a heterologous array of organisms. This protective activity is dependent on T cells. In other words, these studies suggested that PS A elicits a rapid, broadly protective immunomodulatory response that is dependent on cell-mediated immunity.

The capsular polysaccharide of *B. fragilis* has also been shown to inhibit opsonophagocytosis, and an antibody specific for its capsule activates both the classical and alternative pathways of the complement system. Significant increases in antibody titers in patients with *B. fragilis* bacteremia have been reported, but hyperimmune globulin generated in animals specific for *B. fragilis* does not protect against abscesses. Studies have shown that specific capsular antibody does reduce the incidence of bacteremia in experimentally infected animals.

Immunity to *P. gingivalis* infections has been described and can be generated to various degrees in animals models by the capsular polysaccharide, LPS, hemagglutinin, or gingipains. The involvement of T cells in regulating the immune response to this organism has also been demonstrated. Immune T cells derived from mucosal and systemic tissues of rats given live *P. gingivalis* yielded an increase in serum and salivary responses compared to control animals. These results indicate a role for serum IgG and salivary IgA in protection against periodontal disease in which a balance between Th1 and Th2-like cells occurs in humoral immune responses to *P. gingivalis*. In a recent clinical study, patients with periodontal disease develop a significant antibody response to *P. gingivalis*, but this response does not eliminate infection.

Relatively little is known about the host immune response to *Fusobacterium* species and *Prevotella* species. *F. nucleatum* produces a protein that inhibits T cell activation *in vitro* by arresting cells in the mid-G1 phase of the cell. It is hypothesized that the suppressive effects of this protein enhance the virulence of *F. nucleatum*. In addition, *F. nucleatum* or its purified outer membrane can induce a potent humoral response in mice. Several investigators have attempted to investigate mechanisms of protective immunity against *F. necrophorum*, but potential immunogens isolated from the organism have not afforded satisfactory protection.

In studies of the host response to *P. intermedia*, a polysaccharide surface component exerted a strong mitogenic effect on splenocytes and a cytokine-inducing effect on peritoneal macrophages from both C3H/HeJ and C3H/HeN mice; this polysaccharide also stimulated human gingival fibroblasts to produce cytokines. The immunization of nonhuman primates with *P. intermedia* resulted in the production of significantly elevated levels of specific IgG. The level of serum IgA antibody also increased. Finally, coculture of *P. intermedia* with T cells significantly upregulated the expression of specific T-cell receptor-variable regions, a result suggesting that this organism has significant impact on T cells.

V. GENETICS

Knowledge of the genetic makeup of the gram-negative anaerobic pathogens lags far behind their aerobic counterparts. Sequence for only a few hundred nonredundant genes have been deposited in the databases for all these genera combined. An accurate understanding of the genetic makeup of these organisms awaits complete genome sequencing.

The G + C content of these bacteria shows some dissimilarity among genera, with variation for *Bacteroides* species reported at 41–46%, *Porphyromonas* species 41–45%, *Prevotella* species 39–51%, and *Fusobacterium* species 26–34%. Little genetic information is available for *Bilophila*. The only reported sequence for this genus is that of the 16S rDNA, which demonstrates the relatedness of this organism to other sulfur-reducing organisms such as the *Desulfovibrio* species. Similarly, few genes aside from rDNA have been sequenced from *Fusobacterium* and few factors implicated in virulence of the human *Prevotella* species have been characterized genetically.

A large proportion of the genes sequenced from *Porphyromonas* encode the various and diverse types of proteases or hemagglutinins produced by these organisms that are involved in virulence. In addition, mutational analysis of *fimA* demonstrated that fimbriae are essential for the interaction of the organism with human gingival tissue cells.

The most extensive area of genetic research in *Bacteroides* is the study of antibiotic resistance and the elements involved in the transfer of resistance genes. Aside from this area of research, the genetic analysis of other virulence factors are being performed. The gene encoding the metalloprotease toxin (*bft*) of *B. fragilis* is contained on a pathogenicity island that is present only in enterotoxigenic strains. Other products involved in aerotolerance of *B. fragilis*, such as catalase and superoxide dismutase, have also been studied at the molecular level.

VI. GENETICS OF ANTIBIOTIC RESISTANCE BY *BACTEROIDES*

Antibiotic resistance is an increasing problem in the treatment of *Bacteroides* infections: The bacteria continue to acquire genes that make them resistant to multiple antibiotics. A drastic increase in resistance to antibiotics such as tetracycline, cephalosporins, and clindamycin over the last 2 decades has necessitated the use of carbapenems, metronidazole, and β-lactamase inhibitors. Resistance to these latter agents, however, is also increasing. The genetic elements responsible for the evolution of antibiotic resistance in *Bacteroides* are the topic of this section.

A. β-Lactamases

The majority of *Bacteroides* species display some level of resistance to β-lactam antibiotics. The genes encoding the enzymes responsible for resistance vary among the *Bacteroides* species. The best studied are the β-lactamases of *B. fragilis*. Two distinct classes of β-lactamases have been described in *B. fragilis*, the active-site serine enzymes encoded by *cepA* and the metallo-β-lactamases encoded by *cfiA*. Any given *B. fragilis* strain contains only one of these β-lactamase-encoding genes, and taxonomic investigations have revealed that *cfiA*+ strains and *cepA*+ strains form two genotypically distinct groups. Although these groups cannot be differentiated phenotypically, *cfiA*+ strains exhibit a distinctive and homogeneous ribotype, can be distinguished from *cepA*+ strains by arbitrarily primed PCR, and preferentially contain most of the insertion-sequence (IS) elements described for *B. fragilis* (i.e., IS 4351, IS 942, and IS 1186).

1. cepA

The most prevalent β-lactamase of *B. fragilis* is the active-site serine enzyme encoded by the chromosomal *cepA*. This cephalosporinase does not confer protection against the carbapenems (such as imipenem) and is sensitive to the action of β-lactamase inhibitors. Not all *cepA*+ strains produce a high level of the cephalosporinase. The analysis of the sequence of *cepA* from seven *B. fragilis* strains that produce high or low levels of cephalosporinase demonstrated that the *cepA* coding regions for all strains were identical. Therefore, structural differences in the enzyme do not account for the differing levels of resistance conferred by *cepA*+ strains. Rather, the production of the enzyme is regulated at the transcriptional level by sequences upstream of *cepA*.

2. cfiA

Only approximately 3% of *B. fragilis* strains contain *cfiA*, which encodes a metallo-β-lactamase that varies in properties, substrate specificity, and activity from the *cepA* gene product. The emergence of *cfiA* has been closely monitored, as the enzyme it encodes is not affected by β-lactamase inhibitors and is active against a variety of β-lactams, including carbapenems. Only one-third of *cfiA*+ strains have been reported to produce the enzyme; the inability of other *cfiA*+ strains to do so is due to the lack of a promoter driving the transcription of *cfiA*. These silent *cfiA* genes become active when an IS element (IS 1168 or IS 1186) is inserted into the chromosome just upstream of *cfiA*. These IS elements contain outward-oriented promoters that drive the transcription of *cfiA*, leading to a 100-fold increase in the amount of β-lactamase. Given the small fraction of *B. fragilis* strains that produce the metallo-β-lactamase, carbapenems are still effective for the treatment of *Bacteroides* infections.

B. Conjugative elements involved in the transfer of antibiotic resistance

Unlike the β-lactamase genes of *B. fragilis*, the resistance of *Bacteroides* to clindamycin, tetracycline, and 5-nitroimidazole is conferred by genes carried on elements that are self-transmissible. Both conjugative transposons and conjugative plasmids are involved in the transfer of these antibiotic-resistance genes in *Bacteroides*.

1. Conjugative transposons

The conjugative transposons of *Bacteroides* are 70–80-kb elements that are normally integrated into the chromosome or on a plasmid. In addition to carrying genes for resistance to tetracycline and clindamycin, conjugative transposons contain all the necessary genes for their excision and transfer. Once the transposon has been excised from the chromosome, the element forms a covalently closed circle. Transfer of a single strand of the transposon then begins at *oriT* through a mating pore to the recipient cell. The single strand is replicated in the recipient cell and integrated into the chromosome in an orientation- and site-specific manner.

Several factors account for the ability of the conjugative transposons to propagate antibiotic-resistance genes so successfully. Their broad host range allows for their transfer between species that are only distantly related. The conjugative transposons can also mediate the mobilization of coresident plasmids and the excision and mobilization of unlinked chromosomal segments of 10–12 kb, termed nonreplicating *Bacteroides* units (NBUs). These elements contain an origin of transfer that permits their transfer by the mating pore of the conjugative transposon. Therefore, conjugative transposons allow for the spread of antibiotic-resistance genes contained on unlinked elements. Lastly, the transfer capabilities of the conjugative transposons are inducible by subinhibitory concentrations of tetracycline as described later.

2. Conjugative plasmids

Of the several conjugative plasmids described in *Bacteroides*, some contain genes conferring resistance to clindamycin or the 5-nitroimidazoles. The regions involved in transfer have been studied for many *Bacteroides* conjugative plasmids and usually involve the products of one to three *mob* genes. As are the conjugative transposons, the conjugative plasmids have been transferred across genera; however, their range is probably more restricted than that of the conjugative transposons, as recognition of the origin of replication is necessary for maintenance. Because conjugative plasmids can be mobilized by coresident conjugative transposons, the spread of 5-nitroimidazole resistance conferred by genes on conjugative plasmids can be induced by tetracycline pretreatment of cells containing conjugative transposons. Therefore, the use of tetracycline may lead to the spread of resistance to various antibiotics.

C. Antibiotic resistance genes contained on conjugative elements

1. Tetracycline resistance

Tetracycline, once effective against *Bacteroides* infections, now encounters resistance in the majority of clinical isolates. Most tetracyline-resistant *Bacteroides* contain the gene *tetQ*, which is believed to confer widespread resistance among *Bacteroides* strains. Although the *tetQ* product is most similar to the TetM and TetO classes of resistance mediated by ribosomal protection, the degree of similarity is low enough (40%) to merit a separate class of ribosomal resistance genes.

The gene *tetQ* is carried by the conjugative transposons of *Bacteroides*. What is unique about these conjugative transposons is that their transfer is increased by 100- to 1000-fold by pretreatment of the donor with subinhibitory concentrations of tetracycline. This finding led to the discovery of three regulatory genes downstream of *tetQ* on the conjugative transposon. The corresponding gene products are probably involved in the tetracycline-dependent transcriptional activation of genes involved in transfer.

Two other tetracycline-resistance determinants have been identified in *Bacteroides*. *tetX*, first cloned in 1991, encodes a product that inactivates tetracycline. An additional tetracycline-resistance determinant was found on a *Bacteroides* transposon that leads to tetracycline efflux. Neither of these products actually confers resistance to tetracycline in *Bacteroides*, and both are probably remnants of DNA transfer from other organisms.

2. Clindamycin resistance

Two reports have shown the frequency of clindamycin resistance among *Bacteroides* species at various institutions to be as high as 21.7% and 42.7%. The first clindamycin-resistance gene from *Bacteroides* (*ermF*) was sequenced in 1986. Since then, two additional genes have been sequenced and found to encode products that are 98% identical to the *ermF* product. These genes are contained on transposons, conjugative transposons, and conjugative plasmids, which accounts for their widespread distribution among *Bacteroides* strains.

The mechanism of resistance conferred by the *ermF* product involves neither inactivation nor efflux of the drug. Instead, resistance to clindamycin occurs at the level of the ribosome. The strong identity of the *ermF* product with *erm* genes from gram-positive bacteria suggests that the resistance is mediated by methylation of 23*S* rRNA, which prevents binding of the antibiotic.

The clindamycin resistance gene *ermG*, which was recently sequenced from a *Bacteroides* conjugative transposon, encodes a product that is only 46% similar to the *ermF* product. The *ermG* product is extremely similar to the *erm* products of gram-positive organisms, whose functions as 23*S* rRNA methylases have been established. Therefore, *Bacteroides* species contain two distinct genes, probably of different origins, that both confer resistance to clindamycin by the same mechanism.

3. 5-Nitroimidazole resistance

The vast majority of *Bacteroides* strains are sensitive to the 5-nitroimidazole antibiotics, and metronidazole remains the drug of choice for the treatment of *Bacteroides* infections. However, genes conferring resistance to 5-nitroimidazoles have been described in various *Bacteroides* species. Four distinct resistance genes have been sequenced (*nimA–nimD*), the products of which exhibit 67–91% similarity. *nimA, nimC*, and *nimD* are present on conjugative plasmids, and *nimB* is present in the chromosome of a *B. fragilis* clinical isolate. As are the other three *nim* genes, *nimB* is transferable by conjugation to other *B. fragilis* strains. IS elements are present just upstream of each of the *nim* genes. Some of these IS elements are highly homologous to IS 1168 and IS 1186, which control the transcription of *cfiA* of *B. fragilis*. It is likely that transcription of the *nim* genes, as with *cfiA*, is controlled by outward-oriented promoters the IS elements.

The mechanism of the resistance conferred by the *nimA* product has been studied. The gene probably encodes a 5-nitroimidazole reductase that prevents the formation of the toxic form of the drug.

BIBLIOGRAPHY

Coyne, M. J., Kalka-Moll, W., Tzianabos, A. O., Kasper, D. L., and Comstock, L. E. (2000). *Bacteroides fragilis* NCTC9343 produces at least three distinct capsular polysaccharides: cloning, characterization, and reassignment of polysaccharide B and C biosynthesis loci. *Infect. Immun.* **68**, 6176–6181.

Dorn, B. R., Leung, K. L., and Progulske-Fox, A. (1998). Invasion of human oral epithelial cells by Prevotella intermedia. *Infect. Immun.* **66**, 6054–6057.

Falagas, M. E., Siakavellas, E. (2000). *Bacteroides, Prevotella,* and *Porphyromonas* species: a review of antibiotic resistance and therapeutic options. *Int. J. Antimicrob. Agents.* **15**, 1–9.

Franco, A. A., Cheng, R. K., Chung, G. T., Wu, S., Oh, H. B., and Sears, C. L. (1999). Molecular evolution of the pathogenicity island of enterotoxigenic *Bacteroides fragilis* strains. *J. Bacteriol.* **181**, 6623–6633.

Gorbach, S. L., and Bartlett, J. G. (1974). Anaerobic infections. *N. Eng. J. Med.* **290**, 1237–1245.

Lamster, I. B., Kaluszhner-Shapira, I., Herrera-Abreu, M., Sinha, R., and Grbic, J. T. (1998). Serum IgG antibody response to Actinobacillus actinomycetemcomitans and Porphyromonas gingivalis: Implications for periodontal diagnosis. *J. Clin. Periodontol.* **25**, 510–516.

Polk, B. J., and Kasper, D. L. (1977). *Bacteriodes fragilis* subspecies in clinical isolates. *Ann. Int. Med.* **86**, 567–571.

Rasmussen, J. L., Odelson, D. A., and Macrina, F. L. (1986). Complete nucleotide sequence and transcription of ermF, a macrolide- lincosamide-streptogramin B resistance determinant from Bacteroides fragilis. *J. Bacteriol.* **168**, 523–533.

Sawyer, R. G., Adams, R. B., May, A. K., Rosenlof, L. K., and Pruett, T. L. (1995). CD4+ T cells mediate preexposure-induced increases in murine intraabdominal abscess formation. *Clin. Immunol. Immunopathol.* **77**, 82–88.

Sugita, N., Kimura, A., Matsuki, Y., Yamamoto, T., Yoshie, H., and Hara, K. (1998). Activation of transcription factors and IL-8 expression in neutrophils stimulated with lipopolysaccharide from Porphyromonas gingivalis. *Inflammation* **22**, 253–267.

Tang, Y. P., Dallas, M. M., and Malamy, M. H. (1999). Characterization of the Batl (Bacteroides aerotolerance) operon in *Bacteroides fragilis*: isolation of a *B. fragilis* mutant with reduced aerotolerance and impaired growth in in vivo model systems. *Mol. Microbiol.* **32**, 139–149.

Thadepalli, H., Gorbach, S., Broido, P., Norsen, J., and Nyhus, L. (1973). Abdominal trauma, anaerobes, and antibiotics. *Surg. Gynecol. Obstetrics* **137**, 270–276.

Tzianabos, A. O., Kasper, D. L., Cisneros, R. L., Smith, R. S., and Onderdonk, A. B. (1995). Polysaccharide-mediated protection against abscess formation in experimental intraabdominal sepsis. *J. Clin. Invest.* **96**, 2727–2731.

Tzianabos, A. O., Onderdonk, A. B., Rosner, B., Cisneros, R. L., and Kasper, D. L. (1993). Structural features of polysaccharides that induce intra-abdominal abscesses. *Science* **262**, 416–419.

Wang, Y., Kalka-Moll, W. M., Roehrl, M. H., and Kasper, D. L. (2000). Structural basis of the abscess-modulating polysaccharide A2 from *Bacteroides fragilis. Proc. Natl. Acad. Sci. USA* **97**, 13478–13483.

Wu, S., Lim, K. C., Huang, J., Saidi, R. F., and Sears, C. L. (1998). Bacteroides fragilis enterotoxin cleaves the zonula adherens protein, E-cadherin. *Proc. Natl. Acad. Sci. U.S.A.* **95**, 14979–14984.

WEBSITES

Anaerobic Gram-Negative Bacilli (S. M. Finegold)
http://gsbs.utmb.edu/microbook/ch020.htm

Photo Gallery of Bacterial Pathogens (Neil Chamberlain)
http://www.geocities.com/CapeCanaveral/3504/gallery.htm

Comprehensive Microbial Resource of The Institute for Genomic Research and links to many other microbial genomic sites
http://www.tigr.org/tigr-scripts/CMR2/CMRHomePage.spl

List of bacterial names with standing in nomenclature (J. P. Euzéby)
http://www.bacterio.cict.fr/index.html
http://www.bacteriamuseum.org/species/bacteroides.shtml

48

Gram-negative cocci, Pathogenic

Emil C. Gotschlich
The Rockefeller University

GLOSSARY

capsule An external layer, usually consisting of a complex polysaccharide, coating the surface of many species of pathogenic bacteria.

meningitis Inflammation of the meninges, which are membranes covering the brain and the spinal cord, resulting from a bacterial or viral infection.

meningococcemia An infection of the blood stream by meningococci in the absence of meningitis.

petechial skin lesions Small purplish spots on the skin caused by a minute hemorrhage.

porins Protein molecules found in outer membranes of gram-negative bacteria that serve as channels for the diffusion of water and small-molecular-weight solutes.

Gram-negative Cocci are almost invariably isolated from nasopharyngeal cultures of human beings. The majority of these are commensal species, which are classified principally by sugar-fermentation reactions performed on the organisms following their isolation from the primary throat culture. Thus, *Neisseria lactamica* derives its name from its ability to ferment lactose. On rare occasions bacteria belonging to a few of these species are able to invade the human host and cause disease. Table 48.1 lists the pathogenic gram-negative cocci.

I. INFECTION BY *NEISSERIA MENINGITIDIS* AND *NEISSERIA GONORRHOEAE*

A. Local infection

Among the *Neisseria* and related organisms listed, *Neisseria meningitidis* and *Neisseria gonorrhoeae* are the primary pathogens, that is, organisms that are able to cause disease in an otherwise healthy host. Hence, this discussion focuses on these species. DNA hybridization has indicated that *Neisseria meningitidis* and *Neisseria gonorrhoeae* are extremely closely related organisms. It is therefore not surprising that the pathogenetic strategies of the organisms are for the most part very similar. Both are able to cause infection of the mucous membranes. *Neisseria meningitidis*, transmitted by droplets from the respiratory tract of an infected individual, usually infects the mucous membranes of the nasopharynx, but has been isolated on occasion from genitourinary sites. The nasopharyngeal infection with *Neisseria meningitidis* is most often asymptomatic and is self-limited, with the infection lasting for a few weeks to months. This colonization is referred to as the carrier state and is quite common. During the winter months, the frequency of carriers is usually 10% or greater. In populations that

The Desk Encyclopedia of Microbiology
ISBN: 0-12-621361-5

TABLE 48.1 Commensal and pathogenic gram-negative cocci

Species	Distribution	Pathogenicity
Commensal species		
Neisseria lactamica	Very commonly colonizes nasopharynx of young children	Very rare instances of sepsis and meningitis
Neisseria subflava family: flava, sicca, perflava	Commonly found in nasopharyngeal cultures	
Neisseria cinerea	Nasopharynx	
Moraxella catarrhalis	50% of children are carriers during winter months	Third most common cause of acute otitis media and sinusitis in young children; sepsis very rare
Moraxella lacunata	Nasopharynx	Formerly common cause of conjunctivitis and keratitis; now rare
Moraxella bovis	Human and bovine nasopharynx	Causes outbreaks of bovine keratoconjunctivitis
Moraxella nonliqefaciens	Nasopharynx	
Moraxella osloensis	Nasopharynx	Occasionally causes invasive disease
Moraxella canis	Canine and feline oral cavity	Infections following dog and cat bites
Kingella kingi	Commonly in nasopharyngeal cultures	Occasionally causes arthritis, osteomyelitis and sepsis in children below age 2
Primary pathogens		
Neisseria meningitidis	Nasopharynx, rarely genitourinary tract	Sepsis and meningitis
Neisseria gonorrhoeae	Genitourinary tract, less commonly rectum and pharynx	Gonorrhea, sepsis, septic arthritis, and dermatitis; rarely meningitis

live in close contact, such as in boarding schools and military recruit camps, the carrier rate not infrequently will exceed 50%.

Gonococci are transmitted most often by sexual contact and infect the genitourinary tract, but are also quite frequently isolated from rectal and pharyngeal cultures. The infections of the genitourinary tract are most often symptomatic, whereas pharyngeal infections generally cause no symptoms. The conjunctivae are susceptible to gonococcal infection, particularly in neonates who acquire it by passing through an infected birth canal. Historically this infection, ophthalmia neonatorum, was extremely common and was the major cause of acquired blindness until the general acceptance of the preventive Credè procedure consisting of the instillation of drops of a 1% silver nitrate solution in the eyes of all newborns. This has been replaced by the use of less irritating antibiotic ointments.

B. Extension of local infection

Genitourinary gonococcal infection in women usually involves the endocervix and with lesser frequency the urethra, the rectum, and the pharynx. As many as 50% of infections may be asymptomatic or with insufficient symptoms to motivate the person to seek medical attention. In at least 10% of infections, there is early extension to the uterine cavity (endometritis), and subsequently ascending infection of the fallopian tubes (pelvic inflammatory disease, PID). PID is the major complication of gonorrhea. Most often it requires hospitalization for differential diagnosis and treatment, and causes tubal scarring resulting in infertility and tubal pregnancies.

Infections in men, although sometimes asymptomatic, most often cause sufficient symptoms to drive the affected individual to seek medical attention. In the days prior to availability of antibiotic therapy, epidydimitis occurred in about 10% of cases, but this complication as well as prostatitis are rarely seen today.

C. Bacteremic infection

Both *N. meningitidis* and *N. gonorrhoeae* first colonize the epithelial surface and are able to traverse epithelial cells by mechanisms to be described here. Once the organisms are in the subepithelial space, both the meningococcus and the gonococcus may invade the bloodstream. With *N. meningitidis* invasion of the bloodstream and the subsequent invasion of the meninges of the brain with resulting meningitis are very dangerous bacterial infections that can very rapidly lead to death, even with optimal antibiotic and supportive therapy. In the United States there are about 3000 reported cases each year of meningococcal meningitis, and now that meningitis due to *Haemophilus influenzae* type b is a vanishing disease as a result of the widespread acceptance of vaccination, the meningococcus is the leading cause of meningitis. In contrast with other organisms that cause meningitis, *N. meningitidis* has the ability to cause epidemic outbreaks with incidences approaching 200 or more per 100,000 per year. Epidemics of meningitis have occurred in all parts of the world, but have been a particular problem in sub-Saharan Africa; for instance in the winter of 1995–1996, there were 250,000 cases in west Africa.

In the case of the gonococcus, it is estimated that 1–2% of patients with gonorrhea do have invasion of the bloodstream, which can in rare instances be as fulminant as meningococcal infection, but usually has a much more benign course. The disease most often presents with fever, arthritis, and petechial, hemorraghic, pustular, or necrotic skin lesions. If untreated, this can progress to septic arthritis that may cause severe damage to the affected joint.

D. Treatment

Meningococcal infection is a medical emergency and the earlier effective antibiotic treatment is initiated the better the prognosis. The mortality rate for meningococcal meningitis is generally about 10%, but it is much higher in meningococcemia and shock. The drug of choice for treatment remains intravenously administered penicillin G in very high doses. However, the pneumococcus and *Haemophilus influenzae* can at times present a very similar clinical picture, and because these organisms frequently are resistant to penicillin, third-generation cephalosporins are recommended as initial therapy until the meningococcal etiology has been established. Patients need to be hospitalized because the course of the disease is unpredictable and supportive therapy is frequently needed.

The treatment of gonococcal infection has changed over the past decades due to the development of partial or complete resistance of this organism to many antibiotics. The list of recommendations proposed by the Centers for Disease Control (CDC) in 1993 take into account the need to treat this disease at the time of the clinic visit with a single dose because many patients are not compliant with regimens that require repeated administration of the medication. In addition, coinfection with *Chlamydia trachomatis* is a very common occurrence, and the treatment should also eliminate this organism.

II. MOLECULAR MECHANISMS OF INFECTION

One of the problems in the study of neisserial disease is that these organisms are restricted to human beings, and animal models have provided very little information on the pathological events occurring during the various stages of the infection. Because of the excellent response of uncomplicated gonorrhea in men to treatment with modern antibiotics, challenge studies of volunteers have been accepted as ethically justifiable. Such scientific experiments, because of their cost, can be performed only infrequently and therefore correlations have to be made with *in vitro* models that mimic the *in vivo* or natural conditions as closely as possible. The human disease has been most closely simulated by an organ culture system employing fallopian tubes that are obtained from women undergoing medically indicated hysterectomies. The epithelium lining the fallopian tubes consists principally of two kinds of cells, mucus-secreting cells, and cells that bear cilia and beat in unison to move the mucus layer. When gonococci are added to the explants *in vitro*, the first discernible interaction is that the gonococci attach by means of long hair-like projections known as pili to the surface of the mucus-secreting cells, but not to the ciliated cells. This is a distant attachment between the bacteria and the cell surface and occurs about 6 h following inoculation. Then over the next 12–18 h this distant attachment converts to a very close attachment in which the membranes of the host and the parasite come into extensive and intimate contact. The close attachment is believed to be mediated by a set of outer-membrane proteins, named opacity proteins and discussed later. These interactions initiate a signaling cascade that causes the epithelial cells to engulf the gonococci, transport the bacteria through the body of the cell in vacuoles, and egest them in an orderly fashion on the basal part of the cells onto the basement membrane. Later in the infection (24–72 h) toxic phenomena occur that result in expulsion of the ciliated cells from the epithelium. If meningococci are placed on human fallopian tubes, the same events occur, but over a shorter period of time. However, these events are not seen with commensal *Neisseria* species or with fallopian tubes that are not of human origin. The appearance of biopsies of cervical tissue taken for cancer-diagnostic purposes that were inadvertently obtained from infected patients show a picture that is quite similar.

The events transpiring in the course of the model infection have been the focus of research in order to understand meningococcal and gonococcal disease in molecular terms. Many bacterial species are able to invade epithelial cells, and in the case of Yersinia, *Salmonella*, *Shigella*, and *Listeria* quite a lot is known about this process because of the genetic tractability of these species. In the case of the *Neisseria*, this exploration is not as far advanced. Obviously the establishment of the mucosal infection depends on cross-talk between the bacterium and the host cells.

It is noteworthy that compared to the other pathogens mentioned, for which invasion occurs quite promptly, there is a long lag in the invasion by the *Neisseria* as if some slow inductive events need to occur in the host cell or the organism or both.

III. ANTIGENS

The surface antigens of the *Neisseria* have been extensively studied to gain an understanding of the molecular steps underlying the pathogenesis of these diseases as well as to identify candidate molecules for inclusion in vaccines.

A. Pili

Most peripheral on the surface are pili, which are hair-like appendages with a diameter of about 8 nm that emanate from the outer membrane of gonococci and are several bacterial diameters long. Pili consist of the helical aggregation of a single kind of protein subunit of about 18,000 MW, known as pilin. The study of pili on gonococci is enormously simplified by the fact that their presence imparts a distinctive appearance to the gonococcal colony as it grows on agar. Piliated colonies are smaller and have sharp edges when viewed with a colony microscope. On laboratory media, isolates with a different colonial appearance, namely larger colonies that are flatter and have an indistinct edge, appear; if the organisms are subcultured nonselectively, these become the predominant colonial form. This colonial form of gonococci is no longer piliated, but in some instances these strains can revert to the piliated state. This ability to turn on and off pilus expression occurs at very high frequency, on the order of 1 in a 1000 cells per generation.

Gonococci that are freshly isolated from patients invariably are piliated. It is known from challenge studies of volunteers that only piliated strains are capable of causing infection. Stable nonreverting pilus-negative strains cannot cause infections. Thus, pili seemed at first to be an ideal vaccine candidate, but it soon became evident that pili are antigenically very variable. It was noted that no two strains of gonococci appeared to have the same pili, and later it was found that the pili expressed by a single strain of gonococcus maintained in the laboratory would over time repeatedly change their antigenicity. Following the cloning of the pilin structural gene, this problem could be approached on a molecular level and the mechanism of antigenic variation is summarized in a very simplified form in Fig. 48.1. Generally, the gonococcus possesses a single genetic locus expressing the pilin, which is called *pilE*. This locus contains a complete pilin structural gene with its promoter. In addition to the expression locus, gonococci also have eight or more other loci referred to as silent loci. One of these, *pilS1*, is shown enlarged to indicate that the silent loci contain a large number of distinct pilin genes. This particular locus contains five silent pilin genes. These pilin genes are characterized by the fact that all of them are incomplete. They lack the promoter sequences and portions corresponding to the beginning of the protein coding frame. These incomplete genes are efficiently shuttled by homologous recombination into the *pilE* expression site, causing the production of a large number of serologically variant pili by a single strain of gonococcus over a period of time. The recombination events can occur in the conserved *N*-terminal portion of the coding frame and in the conserved region immediately following the termination codon. But it can also occur in several islands of conserved sequences scattered in the variable portion of the pilin genes between the minicassettes. Such antigenic changes occur at a rate of about 1 in a 1000 cells per cell division. If the new antigenic variant pilin can be assembled into intact pili, then an antigenic variation step has occurred. If the new pilin cannot be assembled into an intact pilus, the organism is pilus negative, but is able to revert to pilus positive as soon as a gene copy that can be assembled is recombined into the expression locus. Thus, the recombinational mechanism accounts for both on–off variation and for antigenic variation. This is obviously a remarkably complex genetic mechanism for varying this protein and the only conceivable evolutionary pressure to force the development of this system is, of course, the human immune system. The volunteer studies have demonstrated that the antigenic variation does occur *in vivo* and that in fact almost all of the reisolates from the infected volunteers had a different pilus type than that expressed by the infecting strain. Recent crystallographic studies have shown that pili consist of a helical aggregate of five pilin subunits per turn and that, remarkably, the exposed surface of the pilus cylinder consists of the variable domains, whereas the constant regions of the pilin molecule are buried in the cylinder.

Pili are associated with a protein of about 110,000 MW, named PilC, which is involved in the assembly of pili and also appears to be present at the tip of the pili and imparts the ability to adhere to epithelial cells. Evidence has been provided that the host antigen recognized by PilC is CD46, a widely distributed cell-surface protein that acts a complement regulatory factor. This antigen also serves as the receptor for measles virus. Meningococci also bear pili that are

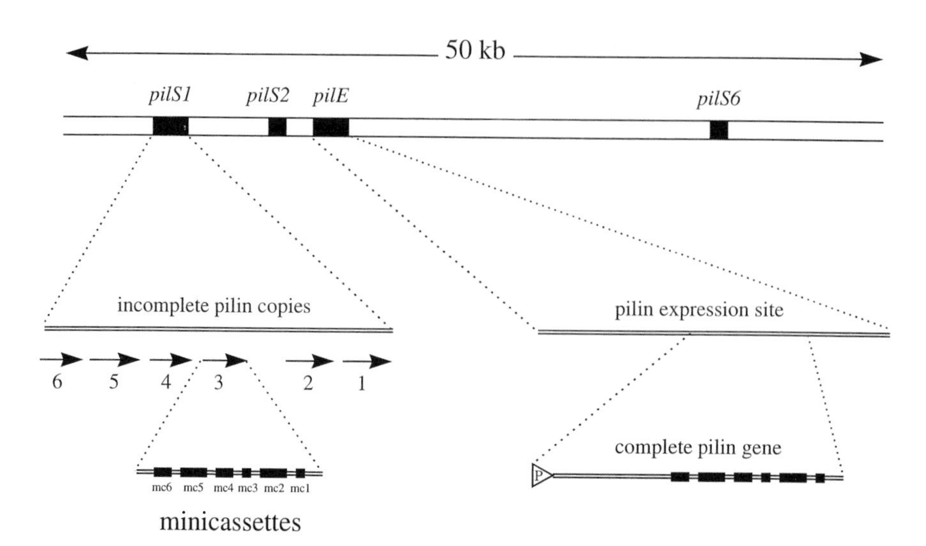

FIGURE 48.1 Genetics of pilin variation. *pilE* is the locus for the expression of the pilin protein and contains a complete pilin gene including the promoter region, indicated as a triangle. There are a number of *pilS* regions, some as indicated within 50 kb of the expression site, which contain incomplete pilin genes lacking promoter sequences and the *N*-terminal portion of the coding frame. As indicated, *pilS1* contains several incomplete pilin genes. As indicated at the bottom of the figure, the variable portions of pilin are distributed as minicassettes with intervening conserved sequences that serve as targets for homologous recombination.

very similar to gonococcal pili with a similar ability for antigenic variation.

B. Capsules

Meningococci are classified into serogroups on the basis of the chemical nature of the capsular polysaccharide they express (see Table 48.2). Epidemiologically, groups A, B, and C are the most important because they are the cause of over 90% of cases of meningitis and meningococcemia. Meningococcal disease occurs worldwide as an endemic disease principally in infants 3 months or older and in young children with a rate of two to five cases per 100,000 population. The incidence is seasonal, with winter and spring having most cases. However, meningococcal disease can also occur in epidemic form where the rate of disease can rise as high as 200–500 cases per 100,000. During epidemics, the peak incidence shifts to an older age group, children 5–7 years of age. During the first half of the twentieth century, epidemics caused by group A meningococci occurred in the United States about every 12 years. Since World War II, there has not been a major epidemic in the United States, but epidemics have occurred in many other parts of the world, notably Brazil, China, Finland, Russia, and Mongolia. However, the area of the world most severely affected is Africa in the meningitis belt that extends through all of the sub-Saharan countries from the Sahel to the rain-forest region. In this region, major epidemics affecting tens of thousands of inhabitants have occurred every 3–4 years, and the problem has worsened recently, so that in the winter season of 1995–1996 there were 250,000 cases in west Africa.

TABLE 48.2 Chemical structure of meningococcal capsular polysaccharides

Serogroup	Repeating unit in capsular polysaccharide
A	→ 6)-α-*N*-acetyl mannosamine-1-phosphate; *O*-acetyl C3
B	→ 8)-α-*N*-acetyl neuraminic acid-(2 →
C	→ 9)-α-*N*-acetyl neuraminic acid-(2 →; *O*-acetyl C7 or C8
X	→ 4)-α-*N*-acetyl glucosamine-1-phosphate
Y	→ 6)-α-glucose-(1 → 4)-*N*-acetyl neuraminic acid-(2 →; contains *O*-acetyl groups
Z	→ 3)-α-*N*-acetyl galactosamine (1 → 1)-glycerol-3-phosphate
29E	→ 3)-α-*N*-acetyl galactosamine (1 → 7)-β-KDO (2 →; *O*-acetyl C4 or C5 of KDO
W-135	→ 6)-α-galactose-(1 → 4)-α-*N*-acetyl neuraminic acid-(2 →

The presence of the capsular polysaccharides protects the organism from the natural defense of the host, such as phagocytosis by white cells and killing mediated by complement via the alternative pathway. However, the presence of antibodies to the capsular polysaccharides is protective and these antigens are the basis of the meningococcal vaccines to be discussed later. The locus for the biosynthesis of capsular polysaccharide has been characterized and is encompassed by about 25 kb of the genome. The right and left sides of the locus are conserved among serogroups and are concerned with common biosynthetic steps, such as the addition of lipid carriers and the export of the product from the cytoplasm to the exterior. The middle portion of the locus differs between serogroups and contains the enzymes responsible for the biosynthesis of the activated sugar intermediates for the particular serogroup, as well as the specific polymerase assembling the polysaccharide.

C. Outer-membrane proteins

The outer membrane of the pathogenic *Neisseria*, as is that of other gram-negative bacteria, consists of a lipid bilayer with the outer leaflet consisting principally of lipopolysaccharide (LPS). The outer membrane contains a number of integral membrane proteins, of which the porins are quantitatively predominant. The nomenclature of the neisserial outer-membrane proteins has evolved with the increasing knowledge of these proteins and is summarized in Table 48.3.

1. Porins

The porins of the pathogenic *Neisseria*, as are those of *Escherichia coli*, are postulated to consist principally of β-pleated sheets arranged perpendicularly to the membrane, with loops exposed to the cytoplasm and eight loops exposed on the surface of the organism. Each functional porin consists of a trimer of the porin subunit. The meningococcus contains two genetic loci that code for the production of outer membrane porins, originally referred to as the class 1 and class 2/3 proteins and now called *porA* and *porB*. The gonococcus, although lacking a homolog of the first porin, possesses the *porB* locus that gives rise to porins that are very similar in amino acid sequence to the class 2/3 proteins. The gonococcal porins vary antigenically to a limited extent primarily in the surface exposed loops and fall into two main classes, referred to as PIA and PIB. PIA strains predominate among gonococci isolated from the bloodstream and apparently have an increased capacity to invade the bloodstream and cause disseminated gonococcal disease. PIA strains also tend to be resistant to the bactericidal action of normal human serum. Strains that cause ascending infection of fallopian tubes are invariably of the PIB type.

TABLE 48.3 Outer-membrane proteins of pathogenic *Neisseria*

Meningococcal proteins	Gonococcal proteins	Genetic designation
Class 1	No homolog	*porA*
Class 2	Protein I, PI, PIB	*porB*
Class 3	Protein I, PI, PIA	
Class 4	PIII, rmp	*rmp*
Class 5	PII, opa	*opaA–opaJ*, or opa_{xx}–opa_{yy}
Opc	No homolog	*opc*

There are a number of indications that gonococcal porins not only serve as channels through which water and solutes of less than 1000 MW can diffuse through the outer membrane, but also play an active role in pathogenesis. Biophysical studies in artificial lipid membranes indicate that the gonococcal porins are unusual among gram-negative porins in that they are somewhat anion selective and very voltage sensitive. Voltage sensitivity means that when the protein is in a membrane, the channel will be modulated by the potential across that membrane, such that at low membrane potentials the porin molecules will be open and as the voltage is raised, the probability that the porin molecule is closed increases. In addition, it has been shown that the porins are able to bind GTP and certain other phosphate compounds and that this binding favors the closing of the porin channel.

The neisserial porins readily transfer from the outer membrane of living gonococci to foreign membranes, including human cells. It obviously becomes very interesting to ask what the functional consequences of a newly inserted voltage-dependent ion channel in the host-cell membrane may be. This has been studied in detail with human polymorphonuclear leukocytes (PMN) using purified gonococcal porin. Within seconds after the addition of porin to the leukocytes, the membrane potential of these cells becomes hyperpolarized due to chloride ion movement. Shortly thereafter, the membrane potential returns to baseline, presumably because the porin channels adjust to this hyperpolarization by closing, and the active ion pumps of these cells reestablish their baseline potential. Even though the initial effects on the cell's membrane potential by porin addition are short-lived, the consequences of a foreign voltage-regulated channel in the membrane of these cells are much longer lasting. This is seen when these cells are subsequently exposed to a stimulus such as fMLP (formyl methionyl-leucyl-proline). Normally, fMLP causes an immediate depolarization of the membrane. However, with porin channels present in the membrane, this depolarization is replaced with a prolonged hyperpolarization. Porin also markedly inhibits the aggregation of PMN. Degranulation in response to fMLP, LTB4, or the complement component C5a is also blocked, but is normal when induced with PMA. However, there is no inhibition of superoxide generation in response to these signals.

2. *Rmp*

All strains of gonococci produce an outer-membrane protein originally designated PIII. This protein migrates on a SDS-PAGE with an apparent 31,000 MW unreduced, and with an apparent 32,000 MW when exposed to reducing agents such as β-mercaptoethanol. Hence, the protein has been named *r*eduction *m*odifiable *p*rotein (Rmp). In contrast to other outer-membrane proteins, Rmp is a highly preserved antigen showing little if any variation among strains. The sequence of Rmp has substantial homology with OmpA, a protein that is universally present in all enterobacterial species. Rmp is also present in meningococci, where it was originally named class 4 protein and it is almost identical to the Rmp of gonococcus.

It has been found that complement-fixing IgG antibodies to Rmp are present in the sera of at least 15% of normal human beings with no history of prior gonococcal infection. These antibodies arise in response to the meningococcal carrier state and also by contact with the enterobacterial flora. Surprisingly, these antibodies do not mediate serum killing or opsonization of gonococci, but instead block the ability of normally bactericidal antibodies directed to other surface antigens to exert their function. These results have been substantiated with monoclonal antibodies. It was found that an anti-Rmp antibody was a powerful blocking antibody, inhibiting the activity of other bactericidal monoclonal antibodies directed to a number of different surface proteins or LPS (lipopolysaccharides). In an epidemiologic study of a population at very high risk for acquiring sexually transmitted diseases, it has been demonstrated that the presence of anti-Rmp antibodies significantly increases the risk of gonococcal infection, demonstrating the inhibitory role of blocking antibodies in the local mucosal infection. The blocking activity of anti-Rmp antibodies is not seen with meningococci, perhaps because this organism expresses quantitatively less Rmp. The molecular mechanism by which anti-Rmp antibodies act as blocking antibodies is not understood.

3. *Opa proteins*

Pathogenic *Neisseria* express another surface-exposed outer-membrane protein that in the meningococcus was called class 5 protein and in the gonococcus protein II. The presence of this class of protein leads to distinctive changes in the morphology of the colonies on agar. Gonococci that do not express protein II give rise to colonies that are transparent and resemble a beaded water droplet, whereas gonococci expressing this antigen give rise to colonies that are opaque and have a ground-glass appearance. Hence, they have been named opacity or Opa proteins. The expression of Opa can turn off and on at high frequency, and a clone of gonococcus can express at least five or six recognizably different opacity proteins over a period of time. Gonococci freshly isolated from the blood of patients with disseminated gonococcal infection do not express Opas. The same is true of isolates from pelvic inflammatory disease. Strains from males with genitourinary disease usually express Opa protein. Most remarkably, in young women not on birth control pills, the gonococci that can be isolated from the cervix vary so that at the time of ovulation the isolates express Opas, although at the time of menses they do not. So, as with pili, there is phase and antigenic variation both *in vitro* and *in vivo*. However, the mechanism is entirely different than that seen with pili. Gonococci possess 11 copies of complete *opa* genes in the gonococcal genome, and all of them have a variable number of pentameric repeats of the sequence CTCTT between the ATG initiation codon and the remainder of the protein. This repeat codes for leucine, serine, and phenylalanine, amino acids that are normally contained in the hydrophobic portion of signal sequences. The number of repeats is subject to rapid change due to slipped-strand mispairing during replication. The consequence is that with some number of repeats the beginning of the gene will be in frame with the remainder of the gene, and with a different number of repeats it will be out of frame. Thus, the expression of this class of proteins is controlled at the level of protein translation. The same pertains to meningococci, although they possess only three *opa* genes. This mechanism of genetic variation has also more recently been described in a number of other mucosal pathogens, notably *Haemophilus influenzae* and *Helicobacter pylori*.

The different *opa* genes of a few strains have been sequenced and have been distinguished either by naming them *opaA–opaJ* or by adding a numerical subscript (see Table 48.3). The loci code for mature proteins about 250 amino acids long. The genes are highly homologous, except for two regions that are very variable and a smaller region that has lesser variation. The differences among the proteins in the content of basic amino acids is noteworthy and the pI of the proteins range from about 7.0–10.0.

Since the discovery of this class of proteins, it has been noted that they increased the adhesiveness of gonococci to epithelial cells in tissue culture or to human PMN, and it was inferred that they mediated the close attachment phase in the fallopian tube model. The ligand specificity of the Opa proteins has been defined on a molecular level. It has been shown that a particular Opa protein (OpaA protein of strain MS11) recognizes heparan sulfate on the surface of epithelial cells and that heparin is able to inhibit the binding. The heparan sulfate occurs mainly on the syndecan class of molecules, of which four have been described and two of these (1 and 4) are expressed on epithelial cells. The syndecans are believed to act as receptors or coreceptors for interaction between cells and the extracellular matrix.

The other Opa proteins react with several proteins that are members of the carcinoembryonic antigen CEA family. CEA was originally described as a colon cancer associated antigen and tests for blood levels of CEA antigen are used to clinically monitor the progression of colon cancer. There are now about 20 related proteins known; their genes are clustered on human chromosome 19, they belong to the immunoglobulin (Ig) superfamily, and they have a N-terminal domain homologous to immunoglobulin variable (IgV) domain and a variable number of domains with homology to Ig constant regions. Some are transmembrane proteins with cytoplasmic tails, whereas CEA is GPI-linked. The proteins are heavily glycosylated. Several of the genes are subject to alternative splicing and various family members are expressed on a wide variety of cells, including epithelial cells. The Opa proteins of both the gonococcus and the meningococcus react with the IgV domain of the molecules irrespective of its state of glycosylation.

D. Lipopolysaccharide

As do other gram-negative bacteria, the pathogenic *Neisseria* carry lipopolysaccharide (LPS) in the external leaflet of their outer membranes. In contrast to the high-molecular-weight LPS molecules with repeating O-chains seen in many enteric bacteria, the LPS of *Neisseria* is of modest size and therefore is often referred to as lipooligosaccharide or LOS. Although the molecular size of the LPS is similar to that seen in rough LPS mutants of *Salmonella* ssp., this substance has considerable antigenic diversity. In the case of the meningococcus, a serological-typing scheme has been developed that separates strains into 12 immunotypes and the detailed structure of the majority of these has been determined. The LPS of the pathogenic *Neisseria* is heterogeneous and LPS preparations frequently contain several closely spaced bands by SDS-PAGE. Using monoclonal antibodies, it is evident that gonococci are able to change the serological characteristics of the LPS they express and that this antigenic variation occurs at a frequency of 10^{-2}–10^{-3}, indicating that some genetic mechanism must exist to achieve these high frequency variations.

The structure of the largest fully characterized gonococcal LPS molecule is shown in Fig. 48.2. To the lipid A are linked two units of keto-deoxyoctulosonic

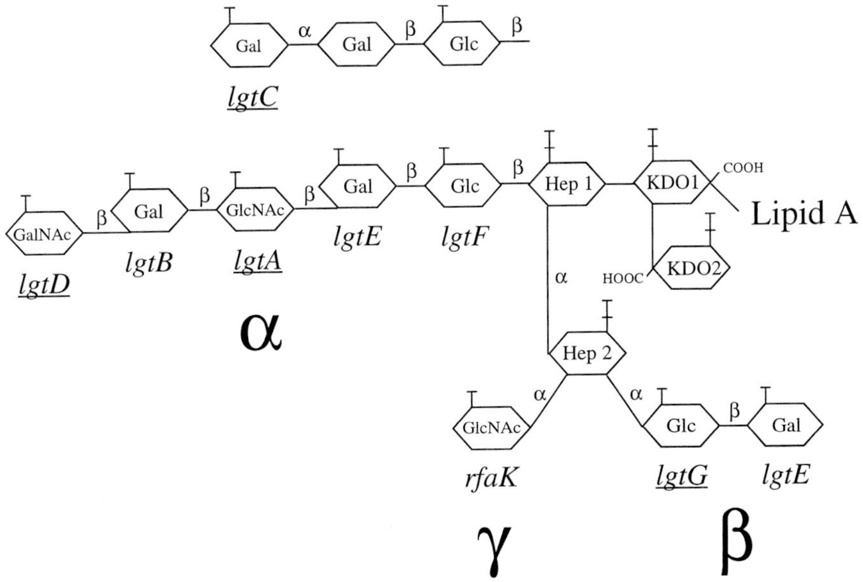

FIGURE 48.2 Genetics of gonococcal LPS synthesis. The LPS contains a lipid A portion with two residues of keto-deoxyoctulosonic acid (KDO) and linked to these are two residues of heptose (HEP) to form the inner core. This structure can bear three additional extensions, indicated as the α-, β-, and γ-chains. The largest structurally characterized α-chain is indicated in the figure and consists of glucose (Glc), galactose (Gal), *N*-acetyl glucosamine (GlcNAc), Gal, and *N*-acetyl galactosamine (GalNAc). An alternative α-chain has been characterized and is the trisaccharide shown at the top of the figure. The glycosyl-transferases responsible for the addition of each of the sugars are indicated by their genetic designation. The genes that are underlined are subject to high frequency variation. If the organism grows *in vivo* or *in vitro* in medium supplemented with cytosine monophosphate-*N*-acetyl neuraminic acid (CMP-NANA), part or all of the terminal GalNAc is replaced by NANA.

acid (KDO) and two heptoses (HEP). This inner core region as shown in Fig. 48.2 can carry three oligosaccharide extensions that have been named the α-, β-, and γ-chains. The γ-chain consisting of *N*-acetyl glucosamine (GlcNAc) appears to be always present. The β-chain, when present, consists of a lactosyl group; when it is absent, the position is substituted with ethanolamine phosphate. The α-chain in its full form consists of the pentasaccharide shown in Fig. 48.2. An alternative α-chain structure consisting of a trisaccharide is also shown. However, as indicated in Table 48.4 the sugar composition of the α-chain can vary and in every instance it is identical to human cell-surface oligosaccharides, most often part of the glycosphingolipids that in some instances are the determinants of blood-group antigens.

Gonococci possess a very unusual sialyl transferase activity, which *in vitro* is able to use exogenously supplied cytosine monophosphate-NANA (CMP-NANA) and add *N*-acetyl neuraminic acid to the LPS if the organism is expressing the lacto-*N*-neotetraose α-chain (see Table 48.4). In the human infection *in vivo*, the concentration of CMP-NANA found in various host environments is sufficient to support this reaction. The sialylation of the LPS causes gonococci to become resistant to the antibody complement-dependent bactericidal effect of serum. The resistance is to the bactericidal effect mediated by not only antibodies to LPS, but also to other surface antigens as well. Group B and C meningococci have the capacity to synthesize CMP-NANA as the precursor of their capsule biosynthesis and frequently sialylate their LPS without requiring exogenous CMP-NANA.

In the late 1990s, most of the glycosyl transferases responsible for the biosynthesis of gonococcal and meningococcal LPS have been identified, and they are shown in Fig. 48.2. This has provided an understanding of the genetic mechanism that underlies the high frequency variation in the LPS structures expressed by these organisms. Note that four of these genes (*lgtA*, *lgtC*, *lgtD*, and *lgtG*) are underlined to indicate that they contain in their coding frames homopolymeric tracts of nucleotides. In the case of *lgtA*, *lgtC*, and *lgtD*, these are stretches of 8–17 deoxyguanosines (poly-G) that can vary in size due to errors resulting from slipped-strand mispairing during replication. In *lgtG*, there is homopolymeric poly-C tract. When the number of bases in the tracts is such that the coding frames are not disrupted, the respective glycosyl transferases are produced, but, if the number changes, premature termination occurs and no functional enzyme is produced. Thus, the presence of the β-chain depends on whether functional LgtG glucosyl transferase is produced. Similarly in the instance of the α-chain synthesis, if *lgtA* is on, then the lacto-*N*-neotetraose chain will be formed and whether the terminal GalNAc is added depends on whether *lgtD* is on or off. If *lgtA* is off, then the globoside structure is synthesized if *lgtC* is on, and only the lactosyl structure is synthesized if *lgtC* is off. The gonococcus and the meningococcus have evolved a very elegant system to shift readily between a large number of different LPS structures, all of them mimics of human glycolipids. This ability to shift the expression among a number of different LPS structures is not peculiar to the pathogenic *Neisseria*, but also occurs in *Haemophilus influenzae* in which at least four genes are subject to phase variation. In this organism, the mechanism is also by slipped-strand mispairing, but occurs in repeated tetrameric sequences that can be either CAAT or GCAA. Thus, it is likely that LPS antigenic variation is important because it is an attribute of a number of mucosal pathogens.

How does this molecular mimicry, listed in Table 48.4, benefit the organism? It has been proposed that the human host may find it difficult to produce antibodies to any of these structures and that the ability to change to a different one may compound this problem. Although immune evasion is attractive as an idea, it is clear that the LOS does serve as a target for bactericidal antibodies, and, at least *in vitro*, perhaps the majority of bactericidal antibodies are directed to this antigen, rather than to other surface structures. *In vivo* this is, of course, very different because the sialylation of the LOS very effectively inhibits the bactericidal reaction and interferes with phagocytosis as well. However, only the lacto-*N*-neotetraose structure is effectively sialylated to produce the serum-resistant

TABLE 48.4 Molecular mimicry by gonococcal LPS

Human antigen mimicked	α-chain oligosaccharide
Lactosyl ceramide	Galβ1 → 4Glcβ1 → 4-R
Globoside, p^k blood group antigen	Galα1 → 4Galβ1 → 4Glcβ1 → 4-R
Lacto-*N*-neotetraose, paragloboside	Galβ1 → 4GlcNAcβ1 → 3Galβ1 → 4Glcβ1 → 4-R
Gangliosides, X_2 blood group antigen	GalNAcβ1 → 3Galβ1 → 4GlcNAcβ1 → 3Galβ1 → 4Glcβ1 → 4-R
Sialyl-gangliosides	NANAα2 → 3Galβ1 → 4GlcNAcβ1 → 3Galβ1 → 4Glcβ1 → 4-R

phenotype. Why does the organism then have a genetic mechanism to alter away from this structure? Perhaps the answer lies in the observation that sialylation of the LOS interferes with invasion of epithelial cells *in vitro*. There is also evidence that sialylated gonococci are significantly less infectious when used to challenge volunteers. It is clear that the gonococcus can circumvent LOS sialylation, either by the addition of the terminal *N*-acetyl galactosamine or by the truncation of the chain. It is also possible that the mimicry may benefit the organism by allowing it to be recognized by human carbohydrate-binding molecules such as the C-lectins, the S-lectins, and the sialoadhesins.

IV. NATURAL IMMUNITY

A. Bactericidal antibody

In the case of the meningococcus, there is clear evidence that the major predisposing factor for bloodstream invasion is the lack of biologically active antibodies to surface components and resultant failure to mediate an antibody–complement bacteriolytic reaction. This was first demonstrated in 1969 by Goldschneider and his colleagues by using two lines of evidence. The first is based on a study done in an adult population. Nearly 15,000 sera were collected from military recruits within the first week of training and stored in anticipation that a number of these would develop meningococcal meningitis during the 8-week basic training. In fact, 60 cases occurred in this cohort and in 54 of these the *Neisseria meningitidis* causing the infection could be isolated. Each of these sera, as well as 10 sera obtained from unaffected recruits serving in the same training platoons, were tested for bactericidal activity against the strain of *Neisseria meningitidis* isolated from the patient. Only 5.6% of the patients' sera were able to kill the disease causing *Neisseria meningitidis*, whereas 82% of sera obtained from unaffected recruits demonstrated bactericidal activity. The second line of evidence is the demonstration that there is an inverse relationship between the incidence of meningococcal disease and the prevalence of bactericidal antibody, and age. The disease is very rare during the first 3 months of life when maternally derived antibodies are still present. Incidence rises to a peak during between 6 and 12 months of age, when the nadir of bactericidal activity is seen. Thereafter, the incidence progressively diminishes as the prevalence of antibodies rises with age. This is the same relationship that was reported for *Haemophilus influenzae* meningitis by Fothergill and Wright in 1933. Finally, it is evident that the antibody–complement-dependent bactericidal reaction is clearly important in protection against neisserial systemic infections because deficiencies of late complement components (C6 or C8) impart a specific susceptibility to blood-borne neisserial infections, but not to other bacterial infections.

Is there natural immunity to gonococcal infection? It is established that individuals with no known immunological defective can acquire gonorrhea multiple times. In some instances, it has been documented that a single untreated consort may represent the source of the repeated infections. Thus, it has been suggested that there is no such thing as natural immunity to this disease. However, there is another side to the coin and that is the clear evidence that gonococcal infection before the days of antibiotic therapy was as a rule a self-limited disease lasting for a few weeks. This spontaneous elimination of the infection applied not only to the genitourinary disease, but also to disseminated gonococcal infection, to gonococcal arthritis (albeit with bad sequelae), and even in some instances to gonococcal endocarditis. Hence, there is ample evidence that after a period of time gonococci are killed effectively *in vivo*. In the face of this ability to self-cure, how can we explain the apparent lack of natural immunity? The most likely explanation is that gonococci are inherently so antigenically variable that the immune system requires considerable time to catch up with the repertoire of the gonococcus and eliminate the infection.

V. PREVENTION

Since the beginning of the twentieth century, attempts have been made to prepare vaccines for the prevention of meningococcal disease. Vaccines based on whole-cell preparations proved to be ineffective. In the late 1960s and 1970s, methods were developed to purify the capsular polysaccharides of the meningococcus in a form that maintained their high molecular weight. It was shown that injection of school-age children and adults with 25–50 μg of group A, C, Y, or W-135 polysaccharide resulted in a strong and long-lasting antibody response and that *in vitro* these antibodies were opsonic and bactericidal. Large-scale field trials both in the United States and overseas demonstrated that both group A and group C polysaccharide vaccines were highly effective in preventing the disease and that the protection lasted for at least 2 years. These vaccines were introduced in the US military over 20 years ago and have essentially eliminated the problem of meningococcal disease among recruits. Vaccination is employed in the military of many other countries and is required for Muslim pilgrims participating in the Hadj.

As a general rule, the immune response to purified polysaccharides is age-related, but the response varies with the antigen. Thus, responses to the group C antigen are very low at ages younger than 18 months. Children between 2 and 4 years do respond to the group C antigen, but the response is short-lived, lasting only a few months. After age 6, the responses are similar to adults. As the experience with the *Haemophilus influenzae* vaccine has demonstrated, the immune responses in this age group can be markedly improved by covalently linking the polysaccharide to a protein carrier to enhance T-cell help in the immune response. Conjugate group C vaccines are being tested.

The response to group A antigen among young children is unusually favorable. Infants who are vaccinated twice, at 3 months and again at 6 months of age will show a brisk booster immune response to the second injection that is sufficient to provide protection. This booster response has not been seen with any other polysaccharide antigen. It has been demonstrated that a protective level of group A antibodies can be maintained by immunization twice in the first year of life, then again at age 2 and upon entry to school. Unfortunately this property of the group antigen has not been taken advantage of in prevention of epidemic disease in Africa.

The group B capsular polysaccharide is a homopolymer of α(2–8)-linked *N*-acetyl neuraminic acid (see Table 48.2). This structure is present in mammalian tissues, notably on the neural cell-adhesion molecule (N-CAM), and the degree of sialylation is particularly elevated during embryonic life. Although the majority of adults have some level of antibodies to this antigen, the injection of the purified antigen generally does not raise additional antibodies. There has also been concern that engendering a strong immune response to this antigen may have deleterious effects on infants during fetal life. Therefore, group B meningococcal vaccines based on partially purified outer membranes with their LPS content reduced by detergent extraction have been prepared and have proved to be able to prevent disease under epidemic conditions. However, as noted before, there is considerable antigenic heterogeneity in meningococcal outer-membrane proteins, and a broadly effective vaccine group B is not available.

No vaccine exists for the prevention of gonorrhea, and the problem is formidable because of the extraordinary antigenic variability of this organism. Nevertheless, the experience in several European countries has demonstrated that prevention of this disease can be very effective if public education is combined with rapid treatment of infected individuals and their contacts.

VI. SUMMARY

The discrete steps that occur in the mucosal infection by the pathogenic *Neisseria* are increasingly being explained on molecular and cell biological level. It is evident that the gonococcus has developed very elaborate mechanisms to evade the immune response of human beings. With pili it has chosen the path of antigenic variation. This is an evasion mechanism that is highly developed in eukaryotic parasites such as trypanosomes, and is also seen in prokaryotes such as *Borrelia*. In the case of Rmp, the gonococcus has chosen the path of antigenic constancy as a target for blocking antibodies. With Opa proteins, the variation may be more a way to succeed in various environments in the host, rather than being an immune evasion. The biological significance of LOS variation is not yet clear, but it must be very useful because *Neisseria* and *Haemophilus influenzae* have developed, in principle, the same variation mechanism, although the specific details are quite different. In the era before ready treatment with antibiotics, self-cure of gonorrhea over a period of weeks was commonly seen, and this slow acquisition of natural immunity was probably a reflection of the time needed for the immune response to finally catch up with the variability of the particular strain infecting the human host.

BIBLIOGRAPHY

Arking, D., Tong, Y., and Stein, D. C. (2001). Analysis of lipooligosaccharide biosynthesis in the Neisseriaceae. *J. Bacteriol.* **183**, 934–941.

Blake, M. S., and Wetzler, L. M. (1995). Vaccines for gonorrhea: Where are we on the curve? *Trends Microbiol.* **3**, 469–474.

Jennings, H. J. (1990). Capsular polysaccharides as vaccine candidates. *Curr. Top. Microbiol. Immunol.* **150**, 97–128.

Merz, A. J., and So, M. (2000). Interactions of pathogenic neisseriae with epithelial cell membranes. *Annu. Rev. Cell Dev. Biol.* **16**, 423–457.

Nassif, X. (2000). Gonococcal lipooligosaccharide: an adhesin for bacterial dissemination? *Trends Microbiol.* **8**, 539–540.

Pollard, A. J., and Frasch, C. (2001). Development of natural immunity to *Neisseria meningitidis*. *Vaccine* **19**(11–12): 1327–1346.

Preston, A., Mandrell, R. E., Gibson, B. W., and Apicella, M. A. (1996). The lipooligosaccharides of pathogenic gram-negative bacteria. *Crit. Rev. Microbiol.* **22**, 139–180.

Pujol, C., Eugene, E., Morand, P., and Nassif, X. (2000). Do pathogenic neisseriae need several ways to modify the host cell cytoskeleton? *Microbes Infect.* **2**, 821–827.

Rahman, M. M., Kahler, C. M., Stephens, D. S., and Carlson, R. W. (2001). The structure of the lipooligosaccharide (LOS) from the alpha-1,2-*N*-acetyl glucosamine transferase (rfaK(NMB)) mutant strain CMK1 of *Neisseria meningitidis*: implications for LOS inner core assembly and LOS-based vaccines. *Glycobiology* **11**, 703–709.

Robbins, J. B., Schneersen, R., and Gotschlich, E. C. (2000). A rebuttal epidemic and endemic meningococcal meningitis in sub-Saharan Africa can be prevented with group A meningococcal capsular polysaccharide vaccine. *Pediatr. Infect. Dis. J.* **19**, 945–953.

Swanson, J., Belland, R. J., and Hill, S. A. (1992). Neisserial surface variation: How and why? *Curr. Opin. Genet. Dev.* **2**, 805–811.

van Putten, J. P. M., and Duensing, T. D. (1997). Infection of mucosal epithelial cells by Neisseria gonorrhoeae. *Rev. Med. Microbiol.* **8**, 51–59.

WEBSITES

Photo Gallery of Bacterial Pathogens (Neil Chamberlain)
http://www.geocities.com/CapeCanaveral/3504/gallery.htm

Identification scheme for Gram negative aerobic cocci (New York University)
http://endeavor.med.nyu.edu/courses/microbiology/courseware/infect-disease/Gram_Neg_Aerobic_Cocci3.html

List of bacterial names with standing in nomenclature (J. P. Euzéby)
http://www.bacterio.cict.fr/index.html

Access to online resources on Bacterial Infections and Mycoses. Karolinska Institutet
http://www.mic.ki.se/Diseases/c1.html

Comprehensive Microbial Resource of The Institute for Genomic Research (TIGR) and links to many other microbial genomic sites
http://www.tigr.org/tigr-scripts/CMR2/CMRHomePage.spl

49

Heat stress

Christophe Herman and Carol A. Gross
University of California

GLOSSARY

chaperone A protein that helps other proteins fold correctly by interacting preferentially with nonnative states of the protein.

protease An enzyme that hydrolyzes peptide bonds between amino acids of a protein, thus leading to the degradation of that protein.

sigma factor The subunit of transcriptase (RNA polymerase) that directs RNA polymerase to the promoter region of DNA, so that transcription originates from the appropriate point on DNA.

The Bacterial Heat Stress Response refers to the mechanism by which bacteria adapt to a sudden increase in the ambient temperature of growth. The precise components of the signal–transduction system that senses and responds to heat stress vary among bacteria. Even within a single bacterial species, several different mechanisms are used to respond to heat stress. However, the logic of the response is universal. During the induction phase, the cell senses increased temperature and produces a signal that activates a transcription factor to increase the transcription of a group of genes. Accumulation of unfolded proteins contributes to signal generation, and a majority of the proteins produced during this stress response help to restore the normal folding state of the cell. During the adaptation phase, as the folding state of the cell returns to normal, the signal is damped down and transcription of the heat shock genes declines. Thus, in general, the heat stress response is self-limiting. Only when organisms are switched to lethal temperatures does the response continue unabated for as long as the cells are able to synthesize protein. In the present article, we first consider the general inputs and outputs to emphasize the universal logic of the bacterial heat shock response. We then describe several specific responses in detail, emphasizing how different components are used to execute this logic.

I. INPUTS TO THE HEAT STRESS RESPONSE

What is the thermometer that allows the cell to sense even small changes in temperature? Unfolded or partially folded proteins are, by far, the best characterized inducers of the heat shock response, and they must comprise part of the cellular thermometer. Normally, the low levels of unfolded or partially folded proteins result either from newly synthesized proteins or those maintained in a partially folded state prior to transport across membranes. Upon heat shock, some fully folded proteins partially or completely denature, increasing the pool of unfolded proteins and the need for the cellular proteins that maintain folding state. This is a matter of crucial concern for the cell. Partially folded proteins may not simply lose their activity; they may, in fact, cause toxic outcomes for the cell through a variety of mechanisms. It is, therefore, a top cellular priority to decrease the extent of protein unfolding whenever it occurs. Additional stresses,

The Desk Encyclopedia of Microbiology
ISBN: 0-12-621361-5

such as alcohol, also increase protein unfolding, thereby activating the same response. Other consequences of heat stress, besides the general increase in level of protein unfolding, are sensed by the cell. However, most of these inputs are currently either unknown or uncharacterized, with one exception. As we shall see later, in *E. coli*, the cell directly senses the folding state of a critical RNA molecule, and this provides a secondary thermometer to sense temperature.

How is the cellular thermometer constructed? Surprisingly, the nature of the primary thermometer sensing heat stress is not known in detail for any system. However, a variety of circumstantial evidence, described later, suggests that titration of the outputs of the response, the heat shock proteins themselves, may provide a thermometer of sufficient calibration to explain the response (Fig. 49.1). Some (or many) of the heat shock proteins have dual roles in the cell: they interact with unfolded protein substrates to promote folding and they interact with the heat stress transcription factor to regulate its activity. During the induction phase of the heat stress response, an increase in the cellular concentration of unfolded proteins increases the ratio of unfolded substrates to heat shock proteins. This could titrate them away from their "homeostatic" regulatory role vis-à-vis the heat stress transcription factor. As a consequence, the amount or activity of transcription factor will increase, resulting in an increased concentration of heat shock proteins. During the adaptation phase of the heat stress response, the ratio of unfolded protein substrates to heat shock proteins normalizes, the heat shock proteins resume negative regulation of the transcription factor, and the response is damped down (Fig. 49.1).

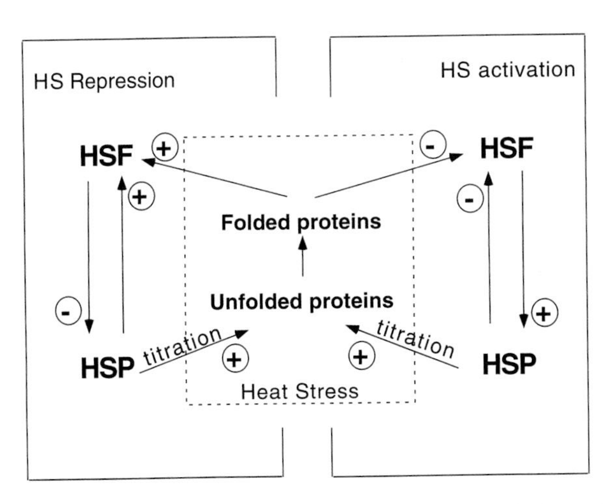

FIGURE 49.1 Mechanism of heat shock regulation in bacteria. The heat shock response can be regulated either by a repressor (left) or an activator (right). In both cases, the activity of the heat shock transcription factor is controlled by a heat shock protein; this feedback control insures homeostasis between the amount of heat shock proteins and the activity of the heat shock factor during steady-state conditions. During heat stress, the level of free heat shock proteins decreases as they are titrated by unfolded proteins. This perturbs homeostasis, resulting in increased activity of the heat shock transcription factor and more heat shock proteins. Once the unfolded proteins are repaired or degraded, homeostasis will be restored.

II. OUTPUTS OF THE HEAT STRESS RESPONSE

A. Chaperones

Chaperones, or proteins that help other proteins fold, are a major class of the proteins produced after heat stress. In the test-tube, under very dilute conditions, small proteins fold by themselves, demonstrating that the polypeptide chain itself encodes the information necessary for proper folding. However, in the cell, proteins are present at very high concentrations, and the nascent, unfolded protein has a very high potential to aggregate with other nascent chains via hydrophobic interactions, rather than proceeding on its folding pathway. By interacting with nascent and partially folded proteins, the molecular chaperones successfully thwart the tendency toward aggregation.

There are several different families of chaperones, most of which are highly conserved throughout evolution. One of the most prominent molecular chaperones is the Hsp70 family, called DnaK in bacteria, which work together with a co-chaperone Hsp40, called DnaJ in bacteria. The DnaKJ chaperone family is conserved in almost all organisms and members of each family are found in every compartment of the eukaryotic cell. Interestingly, in bacteria or in eukaryotic organelles, a third protein, GrpE, is necessary for this chaperone machine to function. GroEL, which cooperates with GroES, is a second major chaperone in bacterial cells and eukaryotic organelles, but is not found in the eukaryotic cytosol. The other major chaperones, Hsp90 and the small heat shock proteins, are conserved among most organisms; however, their role in the bacterial cell has not been completely elucidated.

Some chaperones are very large protein machines, consisting of multiple subunits of one or more proteins. For example, the GroELS chaperone machine has a central cavity consisting of two seven-membered rings of GroEL subunits. One or both ends is "capped" by a single seven-membered ring of the smaller GroES subunits. Proteins first bind to an exposed hydrophobic cavity in GroEL. Then, a combination of GroES binding and ATP hydrolysis drives conformational changes, which first expose hydrophilic residues and then drive release of the protein from the cavity. This allows at least the first critical steps of protein folding to take place within the GroEL cavity, in total isolation from other unfolded molecules in the bacterial cytoplasm.

By contrast, the DnaK chaperone machine consists of a single molecule of DnaK, which transiently interacts with its co-chaperone, a dimer of DnaJ molecules. For DnaK, its binding site for unfolded proteins is a small cleft located in the C-terminus of the molecule, which interacts preferentially with hydrophobic stretches in protein chains. The peptide-binding domain of DnaK is connected via a linker region to the N-terminal ATPase domain of the molecule. Interestingly, as was the case for the GroES co-chaperone, DnaJ both alters the rate of ATP hydrolysis and substrate release. GrpE works as a nucleotide release factor, promoting dissociation of whatever nucleotide is bound to DnaK. Despite its different construction, for both the DnaK and GroEL chaperone machines, an ATP-driven cycle of binding to and release from the chaperone underlies the action of the chaperone.

B. Proteases

Proteases, or proteins that degrade other proteins, are the second major class of proteins produced following heat stress. When proteins cannot be refolded by chaperone machines, the quality control system, composed of a variety of proteases, degrades the damaged proteins. This process serves two purposes. It releases the amino acids for reuse in making new proteins and eliminates proteins that cannot be repaired. Proteases use a variety of recognition systems, including exposure of hydrophobic C-terminal tails, internal hydrophobic stretches, and N-terminal signals.

Interestingly, many proteases are also large machines, composed of one of more types of subunit. The active sites of these proteases are contained within the central cavity, thus confining protein degradation to this cavity, where it is insulated from the cytoplasm. Such proteases also have either separate subunits or separate domains that carry out protein recognition. These portions of proteases can serve as chaperones when removed from the cleavage machinery of the protease. Many, but not all, proteases require ATP, which, most likely, drives the unfolding process.

C. Other heat shock proteins

The two slowest steps in protein folding are isomerization around the cis-trans bond of proline and the making and breaking of disulfide bonds. Protein folding catalysts are proteins that speed up these slow steps in protein folding. Some of the peptidyl prolyl isomerases (catalyzing proline isomerization) and the disulfide bond proteins are heat shock proteins.

More systematic study of the various proteins induced by heat has indicated that a number of heat shock proteins do not fall into these simple categories. The function of many such proteins is still being elucidated.

III. THE *ESCHERICHIA COLI* CYTOPLASMIC HEAT SHOCK RESPONSE: REGULATION BY σ^{32}

A. Description of the *E. coli* heat shock response

Escherichia. coli cells can rapidly sense both temperature upshifts and temperature downshifts. The induction phase begins within one minute of shift from 30 °C to the higher growth temperature of 42 °C, and the peak rate of synthesis of the 30 or more heat shock proteins is attained by 5 to 10 min after temperature upshift. At this point, the adaptation phase begins, and synthesis of the heat shock proteins declines until the new steady-state rate of synthesis is attained. At higher growth temperatures, both the maximal and steady-state rates of synthesis of the heat shock proteins are higher and the induction phase is prolonged. At temperatures so high that they are lethal to cells, the heat shock response is maintained at its maximal point for as long as the cells can synthesize protein. Indeed, at these temperatures, synthesis of heat shock proteins constitutes the major protein synthetic activity of the cell.

In the converse temperature shift from the high growth temperature of 42 °C to a lower growth temperature of 30 °C, the synthesis of heat shock proteins begins to decline within 1 to 2 min, reaching a point of minimal synthesis at about 10 min after temperature downshift. Synthesis of heat shock proteins resumes slowly, reaching the rate normal for that temperature within 50 to 100 min after downshift. Note that this response on shift to lower growth temperature is distinct from the cold shock response. The cold shock response ensues when cells are shifted to very low temperatures (20 °C or below) and involves the induction of a distinct set of proteins. The cold shock response will not be discussed here.

B. A transcriptional factor regulates the *E. coli* heat shock response

In bacterial cells, the sigma subunit of RNA polymerase directs the transcriptase to the promoter region of the gene. Most bacteria contain multiple sigmas. In addition to one or more housekeeping sigmas, responsible for the bulk of the transcription, cells have several alternative sigma factors, which allow them to respond to environmental or developmental signals. Each sigma factor recognizes a distinctive set of promoters in the cell. RNA polymerase is directed to the promoters of the heat shock genes by one such alternative sigma, called σ^{32}. During the heat shock response, the rate of transcription of the heat shock genes and, consequently, the rate of synthesis of the heat shock proteins, changes in response to alterations in the amount and/or activity of σ^{32}.

Upon temperature upshift, the amount of σ^{32} increases rapidly, reaching its peak just prior to the peak rate of synthesis of the heat shock proteins, accounting for the induction phase of the response (Fig. 49.2). The amount of σ^{32} increases both because this normally very unstable molecule (usually degraded with a half-time of only 1 min) is transiently stabilized and because the rate of translation of σ^{32} transiently increases. During the adaptation phase, σ^{32} becomes unstable again and its rate of translation decreases, resulting in a decline in the amount of σ^{32} in the cell. Eventually, σ^{32} reaches a new steady-state level characteristic of the particular growth temperature. The adaptation phase may be sharpened by some control over the activity of σ^{32}. Thus, the response of the cell to temperature upshift is primarily governed by changes in the amount of σ^{32} (Fig. 49.2). In contrast, there is little change in the amount of σ^{32} upon temperature downshift. In this case, a dramatic decrease in the activity of σ^{32} accounts for the dramatic shutoff of transcription of the heat shock genes and the consequent decrease in heat shock protein synthesis.

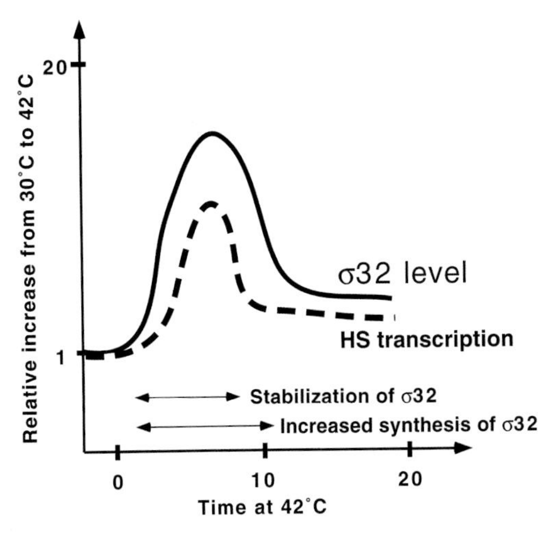

FIGURE 49.2 The time course of change in the level of the heat shock transcription factor, σ^{32}, and the transcription of heat shock genes after a shift of temperature from 30 to 42°C. The amount of σ^{32} increases both because translation of the protein increases and the protein itself becomes more stable.

C. Two (or more) thermometers control the *E. coli* heat shock response

At least two different thermometers control expression of the heat shock proteins. One thermometer controls the translation of σ^{32} mRNA in a positive way: increased physical (environmental) stress leads to increased translation (Fig. 49.3). This pathway is induced by exposure to heat but not by accumulation of unfolded proteins. A second thermometer regulates the stability of σ^{32} itself (Fig. 49.3). This pathway is induced by accumulation of unfolded proteins, as well as by exposure to heat. Elements of this thermometer may also regulate the activity of σ^{32}.

The nature of the thermometer that controls translation of σ^{32} during the induction phase has now been defined (Morita, 1999). The only player is the σ^{32} message itself. At low temperatures, the translation start site is occluded by base-pairing with other regions of the σ^{32} mRNA. Raising the temperature destabilizes this base-pairing, allowing increased translation of σ^{32}. Both mutational studies (which changed the base-pairing in the critical region) and chemical probing

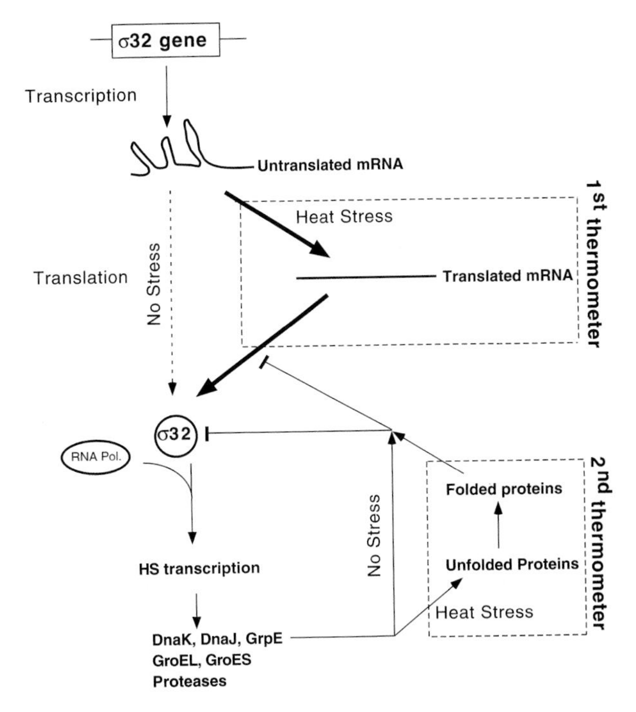

FIGURE 49.3 Model for regulation of the σ^{32} heat shock factor in *Escherichia coli*. During steady-state conditions, the mRNA of σ^{32} is poorly translated. The few σ^{32} molecules translated are subject to rapid degradation and may also be negatively regulated by the DnaK chaperone machine. During heat shock, the structure of the mRNA changes, allowing increased σ^{32} synthesis. Unfolded proteins created during the shock titrate away the DnaK machine from σ^{32}, resulting in its stabilization and increased activity. When unfolded proteins are repaired or degraded, the free DnaK machine will again promote σ^{32} degradation and inactivation.

(which assessed base-pairing as a function of temperature) are consistent with this idea. Most importantly, by using an *in vitro* assay that determines accessibility of mRNA to translation by the ribosome, it has been possible to show that no components other than σ^{32} mRNA are necessary for this regulatory system (Fig. 49.3). During the adaptation phase, translation of σ^{32} mRNA declines. The DnaK chaperone machine has been involved in this translational repression, but the mechanism by which this chaperone operates is still unclear.

The nature of the thermometer that regulates the stability and possibly the activity of σ^{32} is more complex and less defined (Fig. 49.3). Since increased production of unfolded proteins in the cell stabilizes σ^{32}, unfolded proteins are at least part of the signal that calibrates this thermometer. We also know that the DnaK, DnaJ, and GrpE heat shock proteins are required for the instability of σ^{32} and its inactivation. In other words, the DnaK chaperone machine is negatively regulating both the stability and activity of σ^{32}. A homeostatic mechanism coupling the occupancy of the DnaK chaperone machine with unfolded proteins to the amount and activity of σ^{32} has been proposed. The key concept is that the thermometer is set by competition between σ^{32} and all other unfolded or misfolded proteins that bind to the DnaK chaperone machine. During the induction phase, the increased amounts of unfolded or misfolded proteins titrate the DnaK chaperone machine away from σ^{32}, relieving their negative regulatory effects on σ^{32} stability. As a consequence, σ^{32} is stabilized and its amount will rise. This response is self-limiting. During the adaptation phase, overproduction of DnaK, DnaJ, and GrpE will restore the free pool of these chaperones to an appropriate level. This notion can be extended to account for inactivation of σ^{32} during temperature downshift. Here, a decreased amount of unfolded or misfolded proteins will allow σ^{32} to compete better for the DnaK, DnaJ, GrpE chaperone machine, with a consequence of inactivation of σ^{32}. In this model, the amount of free DnaK, DnaJ, and GrpE is a "cellular thermometer" that measures the "folding state" of the cell (Fig. 49.3). A key prediction of this model, that the amount of the DnaK chaperone machinery in the cell directly correlates with the activity of σ^{32}, has recently been verified. However, at present, this model is far from established, as many of the critical experiments to test the model have not been carried out.

D. Additional mechanisms for the cytoplasmic heat shock response in *E. coli*

Although σ^{32} controls expression of many of the genes that respond to heat stress, induction of the σ^{32} regulon is not synonymous with the cellular response to temperature upshift. Another alternative sigma factor, σ^{s}, primarily responsible for transition into stationary phase, is also induced by heat. Induction of the σ^{s} regulon after exposure to high temperature is slower than induction of the σ^{32} regulon. The signal transduction system has not been studied in detail. However, like σ^{32}, the activity and amount of σ^{s} is controlled on multiple levels, including stability, translation, and, possibly, activity. In addition, one operon (*psp*) is controlled by a dedicated activator protein that promotes transcription by yet another sigma factor, σ^{54}, after shift to very high temperature. Finally, yet another sigma factor, σ^{E}, is responsive to heat stress generated in the periplasm. That response will be described in detail in Section V.

E. Conservation and divergence of the σ^{32} paradigm for the cytoplasmic heat shock response

Homologues of σ^{32} have been identified in diverse gram-negative bacterial species. Including *E. coli*, ten gamma proteobacterial species contain σ^{32}. In addition, σ^{32} has been found in seven alpha and one beta proteobacteria. With one exception, these bacterial species have only one gene encoding σ^{32}. The exceptional species, *Bradyrhizobium japonicum*, has three genes encoding σ^{32} with distinct functional and regulatory features. Where it has been examined, the amount of σ^{32} increases with temperature, although the mechanism for accomplishing this is variable. The *E. coli* paradigm for regulating σ^{32} is most closely conserved among gamma bacteria. Here, when the temperature increases, generally, the stability and translation of σ^{32} increase as well. For some alpha bacteria, transcription of σ^{32} appears to increase with temperature. Regardless of the mechanisms for accomplishing this, the increased cellular concentration of σ^{32} gives the increased transcription of heat shock genes characteristic of the heat shock response. No homologues of σ^{32} have been identified in gram-positive bacteria.

IV. THE *BACILLUS SUBTILIS* HEAT SHOCK RESPONSE: THE CIRCE/HrcA CONTROLLED REGULON

A. The heat shock response

The *dnaK* and *groE* operons of *B. subtilis* exhibit a heat shock response similar to the one exhibited by these operons in *E. coli*. Upon shift to high temperature, transcription of the *dnaK* and *groE* operons rapidly increases, leading to a rapid increase in the rate of synthesis of the seven heat shock proteins encoded by the *B. subtilis dnaK* operon (DnaK, DnaJ, GrpE, the HrcA repressor, and three other proteins) and the two heat shock proteins encoded by the *B. subtilis groE* operon (GroEL and GroES). This response reaches its peak about 5 to 10 min after shift to high temperature. Transcription of these operons then declines, reaching a new steady-state rate by 30 to 60 min after the high temperature shift. However, the components generating this response are completely different from those responsible for the heat shock in *E. coli*.

The *dnaK* and *groE* operons of *B. subtilis* are transcribed by its housekeeping sigma (SigA). The heat shock response of these operons is controlled by a negative regulatory system, which is composed of an operator site, called CIRCE, and its cognate repressor, called HrcA (Fig. 49.4). CIRCE is a nine base inverted repeat sequence separated by nine bases (TTAGCACTC-N9-GAGTGCTAA), which is located close to or overlapping the *dnaK* and *groE* promoters. The HrcA repressor (encoded in the *dnaK* operon) binds to the CIRCE element to shut off transcription of these operons. Upon temperature upshift, repression is temporarily relieved and synthesis of the CIRCE-controlled heat shock proteins increases. The major elements of this regulatory system have been verified by genetic analysis and some *in vitro* studies. Mutating conserved bases in CIRCE results in constitutive expression of the operon downstream of the mutated element, as expected for an operator. Likewise, mutating *hrcA* results in constitutive expression of both the *groE* and *dnaK* operons. Direct binding studies, demonstrating that HrcA binds to CIRCE, have been performed with variable success. HrcA is very prone to aggregation *in vitro*, limiting the ability to characterize this reaction extensively. As will be seen in the next section, the propensity of HrcA to aggregate is likely to be for its function.

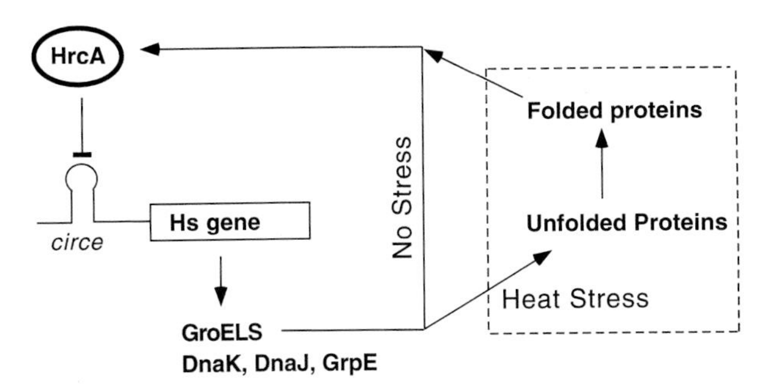

FIGURE 49.4 Model of regulation of HrcA heat shock repressor in *Bacillus subtilis*. During steady-state conditions, the GroELS chaperone machine promotes folding of the heat shock repressor, HrcA, which then binds to the CIRCE element, located upstream of the heat shock genes, and inhibits their transcription. During a heat shock, unfolded proteins titrate away the GroELS chaperone machine, preventing it from helping HrcA to fold properly and function as a repressor. Therefore, the level of heat shock proteins increases until free GroELS can again promote proper folding of the HrcA repressor.

B. The CIRCE/HrcA thermometer

Heat shock regulation in this system requires that the HrcA repressor be transiently removed from its CIRCE operator upon shift to high temperature. The mechanism for accomplishing this constitutes the cellular thermometer. A priori, two solutions seem possible. First, HrcA itself may undergo a temperature-mediated transition. Second, achieving the native state of HrcA may require the intervention of some other molecule in the cell, which is itself responsive to temperature. A variety of evidence suggests that the latter solution has been chosen.

The current idea is that HrcA requires constant interaction with a chaperone to maintain its native state. The GroEL/S chaperone is considered to play this role and, as for the *E. coli* heat shock response, a homeostatic mechanism has been proposed (Fig. 49.4). The occupancy of the *B. subtilis* GroEL/S chaperone machine with unfolded proteins is suggested to be coupled to the fraction of HrcA in the native state. During the induction phase, the increased amounts of unfolded or misfolded proteins titrate the GroEL/S chaperone machine away from HrcA, allowing it to aggregate or otherwise misfold in the cell. As a consequence, it will be unable to bind to CIRCE and repress transcription. This response is self-limiting. During the adaptation phase, overproduction of GroEL/S will restore the free pool of these chaperones to an appropriate level, HrcA will be maintained in its native state and repression from the CIRCE operator will be resumed.

Two key predictions of this model have been verified qualitatively. First, this model implies that repression by HrcA will be responsive to the amount of GroEL/S in the cell and that is true. When GroEL/S is depleted from cells, the heat shock response is induced at low temperature. Conversely, overexpressing GroEL/S represses basal synthesis at low temperature and decreases the absolute amount of induction at high temperature. Second, this model implies that overexpression of known GroEL/S substrates, but not other proteins, will induce the heat shock response. This is also true. Finally, accumulating evidence suggests that GroEL/S and HrcA interact as required by this model. GroEL/S decreases aggregation of HrcA *in vitro* and promotes binding of HrcA to the CIRCE operator, which suggests that the two proteins interact directly, rather than through intermediate molecules. However, a great deal more work remains to be done to understand precisely how this thermometer works.

Two observations make it unlikely that HrcA itself senses temperature. First, in GroEL/S depleted cells, HrcA is inactive even at low temperature, indicating that the protein is not intrinsically native at low temperature. Second, when HrcA from a thermophilic organism (*B. stereothermophilis*) is used to reconstitute the heat shock response in *B. subtilis*, heat shock occurs at the low temperature characteristic of *B. subtilis*, rather than the high temperature characteristic of *B. stereothermophilis*. Thus, it is the cellular milieu, rather than the HrcA protein per se, that senses temperature.

C. Distribution of CIRCE/HrcA regulation

The CIRCE/HrcA regulatory system is very widespread. It occurs in more than 40 different eubacteria, including gram-positive organisms, gram-negative organisms, cyanobacteria, and more distantly related eubacteria, such as chlamydia and spirochaeta. The particular genes regulated by the system vary, depending upon the organism. In gram-positive organisms with low G + C content examined thus far, CIRCE/HrcA regulates the *dnaK* and *groE* operons, but in those with high G + C content, and in α-proteobacteria, only *groE* is regulated. In addition, in some bacteria, several other genes have been reported to be regulated by this system. To add to the regulatory diversity, preliminary experiments suggest that in at least one gram-positive organism (*Lactococcus lactis*), DnaK regulates HrcA. Moreover, several cases where regulation by HrcA and σ^{32} coexist have been documented.

D. Additional mechanisms for response to heat stress in *B. subtilis*

Although understanding of the CIRCE/HrcA regulatory system is most advanced mechanistically, this system controls only nine of the many genes induced after shift to high temperature. A great many others are controlled by σ^B, an alternative sigma factor that carries out the general stress response and the starvation response in *B. subtilis*. Heat is one of the many stimuli that induce this system. σ^B is controlled by a phosphorylation cascade that affects the ability of an anti-sigma to bind to σ^B. This system is currently under intense study but, at present, there are few details about how heat stress alters this regulatory cascade. Additional heat shock proteins in *B. subtilis* are controlled by unknown mechanisms. However, recent studies in other organisms have identified several repressors in addition to HrcA that are responsive to heat stress (OrfY, HspR) and such repressors may be involved in regulating additional heat stress proteins in *B. subtilis* as well.

V. THE *ESCHERICHIA COLI* EXTRACYTOPLASMIC HEAT SHOCK RESPONSE: REGULATION BY σ^E

A. Description of the *E. coli* extracytoplasmic heat shock response

The hallmark of the gram-negative bacterial cell is the existence of two membrane layers, the inner or cytoplasmic membrane and the outer membrane, which, in turn, form the boundaries of two aqueous subcellular compartments, the cytoplasm and the periplasm. The conditions within each of these compartments differ markedly. The cytoplasm is an energy-rich, highly regulated reducing environment, in which basic cellular processes, such as transcription, DNA replication, and translation, are carried out. In contrast, the extracytoplasmic compartment is a relatively energy poor, oxidizing environment in which conditions vary with those of the external environment, due to the existence of pores in the outer membrane which allow the free exchange of small molecules and some specific substrates. Given the disparity in conditions between the two compartments, it is not surprising that the heat stress response in *E. coli* is compartmentalized. Exposure to high temperature activates both the cytoplasmic and extracytoplasmic responses, whereas conditions that specifically perturb protein folding in the periplasmic compartment activate only the extracytoplasmic response.

B. The thermometer controlling the extracytoplasmic heat shock response

The extracytoplasmic heat shock response is controlled by the alternative sigma factor σ^E, which is responsible for directing transcription of the >10 genes that comprise its regulon. Only some of these genes have been identified. These include the periplasmic protease DegP and the periplasmic peptidyl prolyl cis/trans isomerase, FkpA, which are involved, respectively, in protein degradation and folding. Interestingly, transcription of the heat shock factor σ^{32} is also under the control of σ^E, suggesting an interconnection between the two stress regulons. The induction phase of the σ^E heat stress response is slow, with maximum synthesis occurring about 20 min after shift to high temperature. The adaptation phase of this response has not been studied.

Studies on this response have advanced to the point where many of the players are known; however, the signaling mechanisms coupling the activity of σ^E to temperature are currently unknown. The central problem facing this regulatory system is to transduce a signal generated in the periplasm to the cytoplasm, where activity of σ^E is regulated. That problem has been solved by utilizing a protein chain consisting of two negative regulators. RseA, the major negative regulator of σ^E, is a membrane spanning protein whose cytoplasmic face acts as an anti-sigma factor. During steady-state growth conditions, RseA binds to σ^E, thereby preventing this sigma factor from binding to RNA polymerase. The periplasmic face of RseA binds the second negative regulator, RseB. RseB is located completely in the periplasm and is a weak negative regulator of σ^E.

The activity of σ^E is controlled by regulated proteolysis. Upon shift to high temperature, RseA is destabilized, relieving negative regulation of σ^E. At least one function of RseB is to enhance the stability of RseA. Various agents that specifically enhance unfolded proteins in the periplasmic compartment (lack of folding agents, overexpression of outer membrane proteins, interruption of lipopolysaccharide biosynthesis) all induce σ^E. How generation of unfolded proteins is coupled to destabilization of RseA is currently unknown. Moreover, additional yet to be discovered mechanisms regulate σ^E.

VI. SUMMARY AND PROSPECTS

Study of the heat stress response has led us to realize the extraordinary importance of controlling the state of protein folding in the cell. Organisms have evolved multiple, graded responses to cope with this problem. The thermometers that calibrate the response to the level of stress are composed of different materials. However, almost all of them are induced, at least in part, by unfolded proteins and respond by overexpressing a set of universally conserved heat shock proteins. Currently, many, but not all, of the types of responses to heat stress have been identified. However, on the most basic level, the way in which the thermometer controls the gradual response to temperature is not really understood for any system. Moreover, the interaction of the various heat responsive systems with each other has not yet been studied. Clearly, considerably more will be learned about this response in the next few years.

BIBLIOGRAPHY

Bukau, B., and Horwich, A. L. (1998). The Hsp70 and Hsp60 chaperone machines. *Cell* **92**(3), 351–366.

Gross, C. A. (1996). "Function and Regulation of the Heat Shock Proteins. Vol. 1, *Escherichia coli* and Salmonella: Cellular and Molecular Biology" (F. C. Neidhardt, ed.), pp. 1382–1399. American Society for Microbiology, Washington, DC.

Hengge-Aronis, R. (1999). Interplay of global regulators and cell physiology in the general stress response of *Escherichia coli*. *Curr. Opin. Microbiol.* **2**, 148–152.

Laufen, T., Mayer, M. P., Beisel, C., Klostermeier, D., Mogk, A., Reinstein, J., and Bukau, B. (1999). Mechanism of regulation of hsp70 chaperones by DnaJ cochaperones. *Proc. Natl. Acad. Sci. USA* **96**, 5452–5457.

Macario, A. J., and Conway De Macario, E. (2001). The molecular chaperone system and other anti-stress mechanisms in archaea. *Front. Biosci.* **6**, D262–D283.

Mayer, M. P., Rudiger, S., and Bukau, B. (2000). Molecular basis for interactions of the DnaK chaperone with substrates. *Biol. Chem.* **381**, 877–885.

McClellan, A. J., and Frydman, J. (2001) Molecular chaperones and the art of recognizing a lost cause. *Nat. Cell Biol.* **3**, E51–E53.

Morita, M., *et al.* (1999). Translational induction of the heat shock transcription factor sigma 32: Evidence for a built-in RNA thermosensor. *Genes Devel.* **13**, 655–665.

Narberhaus, F. (1999). Negative regulation of bacterial heat shock genes. *Molec. Microbiol.* **31**, 1–8.

Pellecchia, M., Montgomery, D. L., Stevens, S. Y., Vander Kooi, C. W., Feng, H. P., Gierasch, L. M., and Zuiderweg, E. R. (2000). Structural insights into substrate binding by the molecular chaperone DnaK. *Nat. Struct. Biol.* **7**, 298–303.

Storz, G., and Hengge-Aronis, R. (eds.) (2000). "Bacterial Stress Responses." ASM Press, Washington, DC.

Suh, W. C., Lu, C. Z., and Gross, C. A. (1999). Structural features required for the interaction of the Hsp70 molecular chaperone DnaK with its cochaperone DnaJ. *J. Biol. Chem.* **274**, 30534–30539.

Takayama, K., and Kjelleberg, S. (2000). The role of RNA stability during bacterial stress responses and starvation. *Environ. Microbiol.* **2**, 355–365.

Yura, T., and Nakahigashi, K. (1999). Regulation of the heatshock response. *Curr. Opin. Microbiol.* **2**(2), 153–158.

WEBSITES

Website for Sigma-32
http://cgsc.biology.yale.edu/cgi-bin/sybgw/cgsc/Product/17533

Website for *rpoH* gene (codes for Sigma-32)
http://arep.med.harvard.edu/cgi-bin/dpinteract/gene?rpoH

Pedro's Biomolecular Research Tools. A collection of links to information and services useful to molecular biologists
http://www.public.iastate.edu/~pedro/research_tools.html

Comprehensive Microbial Resource of The Institute for Genomic Research (TIGR) and links to many other microbial genomic sites
http://www.tigr.org/tigr-scripts/CMR2/CMRHomePage.spl

50

Horizontal transfer of genes between microorganisms

Jack A. Heinemann
Norwegian Institute of Gene Ecology

GLOSSARY

genome A collection of genetic material and genes normally dependent upon successful reproduction of the entire organism for its reproduction, for example, chromosomes.

homology Related by descent from a common ancestor. Orthologous genes are homologs diverging since organismal speciation; paralogous genes are homologs diverging since duplication; analogous genes are structurally or functionally similar but not related by descent.

horizontal gene transfer (HGT) The movement of genetic material between organisms.

horizontal gene transmission The reappearance of genetic material received by HGT in the offspring of the original recipient.

horizontal reproduction An increase in genetic material (e.g., DNA), genes or sets of genes (e.g., viruses and plasmids) due to transfer between organisms rather than organismal reproduction (vertical reproduction).

horizontally mobile element (HME) A collection of genetic material or genes that reproduce by horizontal reproduction and that may also reproduce by vertical reproduction.

host range The particular group of species in which genes transferred from another particular organism can replicate.

transfer range The particular group of species to which genes can be transferred from another organism by a particular mechanism, for example, conjugation.

vertical reproduction The concomitant reproduction of genetic material and host.

Horizontal Gene Transfer (HGT) describes the lateral movement of genes between organisms. In contrast to the vertical transmission of genes during organismal reproduction, genes transferred horizontally do not always become genes that pass to organismal offspring. HGT is also known as infectious transfer, exemplified by viruses and plasmids. Three classical mechanisms for HGT in microbes are: transformation, the uptake of nucleic acids; transduction, virus-mediated gene transfer; and conjugation, plasmid-mediated gene transfer. Conjugation alone probably occurs between all bacteria, bacteria and plants, bacteria and animals and bacteria and fungi. Possibly all organisms, microbial or not, are significantly affected by HGT.

I. THINKING ABOUT GENES, NOT GENOTYPES

It is difficult as biologists, and particularly as genetical thinkers, to consider genes separately from organisms and phenotypes. The conceptual independence of genotype and phenotype was only first introduced

The Desk Encyclopedia of Microbiology
ISBN: 0-12-621361-5

last century by W. Johannsen, who coined the term "gene." That revolution in thought is now most pertinent to those thinking about evolution. Although genes are still discovered by their effects on organismal phenotype, these effects are only indirectly related to the history and ancestry of the gene and organism.

Genes that are parts of chromosomes reproduce when the chromosome is replicated. Chromosomal replication is tightly associated with organismal reproduction. If the offspring die at any stage of the reproductive process, from incorporation of the first nucleotide at a repliction fork to their last encounter with a predator, then all the genes on the chromosomes of that organism are at an evolutionary end. Genes in chromosomes share the fate of the organism.

To the extent that gene reproduction is synchronous with organism reproduction, the evolution of particular genes and organisms is undoubtedly due to the effect of the gene on the organism. However, not all genes reproduce in this vertical fashion, that is, sychronously with organisms—at least, not all of the time. Some reproduce horizontally by transferring between organisms. When that transfer results in a gene that will be reproduced vertically, that is, inherited by the recipient organism, then the gene has been horizontally transmitted.

When genes can reproduce horizontally, then they can evolve somewhat independently of their effects on the host. If genes transfer horizontally at rates that exceed vertical reproduction, then it is possible for genes to evolve functions that cannot be determined by studying the effects of the gene on a host. This last point is especially relevant if the host–gene relationship is studied under conditions where the gene is mostly confined to reproducing at the rate of the host.

For HGT to contribute to the evolution of genes, it must occur: (a) at least infrequently, but produce strongly selected phenotypes in organisms (and probably leave records of the event through maintenance of particular DNA sequences in organismal descendants); (b) so frequently that most often the effects on organisms are unimportant to the genes reproducing horizontally (and particular DNA sequences may not accumulate in offspring); (c) frequently, but leave short nucleotide sequence records in organisms; or (d) infrequently most of the time and extremely frequently for short periods of time (e.g., during the age when mitochondria and chloroplasts first entered the ancestors of most eukaryotes). Of course, these various possibilities are not mutually exclusive.

The remainder of this article will be devoted to reviewing the mechanisms of HGT, barriers to gene transmission, estimates of transfer/transmission rates, and the difficulty of determining such rates. The article will conclude with considerations of the importance HGT has for studying evolution and assessing the risk of new biotechnologies, including the introduction of new antibiotics and genetically modified organisms.

II. THE WAY GENES REPRODUCE BETWEEN ORGANISMS

A. Mechanisms

Genes are transferred between organisms by three known routes: transformation, transduction, and conjugation.

Transduction and conjugation are conducted by viral and plasmid vectors, respectively. These vectors are themselves groups of genes that reproduce horizontally, possibly far more often than they do vertically. Transformation may have a vector of sorts, such as membrane-bound vesicles, escort proteins, or "uptake sequences." The vectors are sometimes also transmitted, but other times the vectors are only transferred with the genes. For example, transducing viruses can package chromosomal or plasmid DNA during infection or incorporate nonviral DNA into their own genome. Subsequent infections can result in a new host receiving all to none of the virus, with subsequent incorporation and inheritance of the transferred nonviral DNA. Transmission by transformation and transduction is often limited to closely related organisms because these mechanisms usually require DNA–DNA recombination and, in the case of transduction, DNA delivery is mediated by viruses that may infect a small number of species. It would be premature, however, to exclude the contribution of the growing number of broad-host range viruses being described and to equate the transfer range of viruses with the more limited range of hosts that support their infectious cycle.

Of the three mechanisms, conjugation can move the largest DNA fragments (as much as an entire bacterial genome may be moved in one conjugative encounter). Conjugation also has the broadest known transfer range, mediating exchanges between all eubacteria and transfers from prokaryotes to eukaryotes, including human cells. Conjugation, a process determined by plasmids or transposable elements, is not usually dependent on homologous recombination to achieve the formation of a recombinant.

B. Host ranges

Comparing the sequence of particular genes in different organisms has become a taxonomic tool for inferring organismal homology. Comparisons are complicated by DNA sequences that suggest a lineage different from other genes in the same organism. If that gene has been acquired by horizontal transmission, then, truly, it would be a rouge and its exceptional sequence signature could be explained. The origin of genes by HGT has often been discounted, however, when the host range of known vectors is thought to not overlap with the putative donor and recipient species involved and no other vector or ecological relationship is obvious. R. F. Doolittle (1998) calls this the "opportunity" factor. "The possibility for gene transfer is often given wider berth whenever parasitism, symbiosis or endosymbiosis is involved."

Host range determinations are generally the result of studies that require a vector or gene to be transmitted to determine retrospectively if genes had transferred. Thus, when certain plasmids or viruses do not cause demonstrable infections in an organism, they are assumed not to have transferred to that organism. The history of the host range studies using the Ti plasmid of *Agrobacterium tumefaciens* and the conjugative plasmids, like IncF and IncP, of the Gram-negative bacteria illustrate the lesson well. The transfer of DNA from *A. tumefaciens* to dicotyledonous plants is determined by the Ti plasmid. Indeed, that process, which results in tumors in certain susceptible plant species, was prematurely thought to be limited to those species. When the relevant DNA was conferred with sequences that would maintain it in other species, the host range was extended to monocots and then to fungi.

The mechanism of Ti-mediated DNA transfer is biochemically and genetically equivalent to bacterial conjugation. Since the equivocation of the mechanisms, it has been demonstrated that even mundane bacterial conjugative plasmids transfer to eukaryotes. Once again, demonstration required engineering the plasmids with a selectable marker and a strategy for replication in the eukaryotic host (either replication autonomous from the chromosomes or by integration into the chromosomes). These changes had no obvious effect on transfer. Thus, the transfer range of genes can be remarkably different from host range.

A recent finding that a DNA virus, that infects animals, evolved via recombination between a DNA virus, that infects plants, and an RNA virus, that infects animals, provides an illustration for the viruses (Gibbs and Weiller, 1999). The plant virus must have been able to transfer to animals without causing an obvious phenotype. The many transfer events preceding the evolution of the new variant virus were not detected by selecting or observing a recombinant animal (and possibly would not have been detected even with current DNA amplification technologies). The transmission event could be detected, but provides no quantitative information about the frequency of transfers of the original virus to animals. Clearly, Ti and some viruses can mediate transfer to more species than normally display the effects of that transfer because those effects are not frequently heritable or selectable. The "opportunity" factor in determining the likelihood of HGT is often less amenable to test than DNA sequence comparisons, making opportunity a very distant secondary consideration for evaluating the possibility of HGT.

III. GENE ARCHAEOLOGY

Determining which genes are most closely related, and how that reflects organismal relationships, is difficult. The underlying assumption in sequence comparisons is that sequence is the best indicator of homology. The difficulty in establishing relatedness independently of sequence information makes testing the proposition problematical. Even when sequence information is available, the identity and history of the gene are not always obvious (Fig. 50.1). Comparisons must be made between homologous genes and not just genes with homologous names—those with functional similarity (discussed by Doolittle in Horizontal Gene Transfer, 1998). For example, did two genes with similar structure and function diverge from a single sequence or converge from much different sequences? Defining genes with sufficiently similar sequences as homologous begs the question of the adequacy of the criteria for determining homology. Those who compare sequences have to beware several common problems, as will be discussed.

A. Sequence evidence of ancestry

1. *Orthologous or paralogous?*

Are the genes orthologous, diverged when the species diverged, rather than paralogous, diverged since duplication? The differences between orthologs can represent the divergence of the two organisms from a common ancestor. When gene duplication yields paralogs, each allele can change separately, reflecting the history of the genes within, instead of between, lineages.

Identifying which paralogs in different organisms reflect organismal histories is sometimes done by assuming that the most similar of the paralogs in the different species are the orthologs. However, accepting this a priori assumption can render the comparison redundant. Moreover, that test can never be independently verified because, no matter how close the

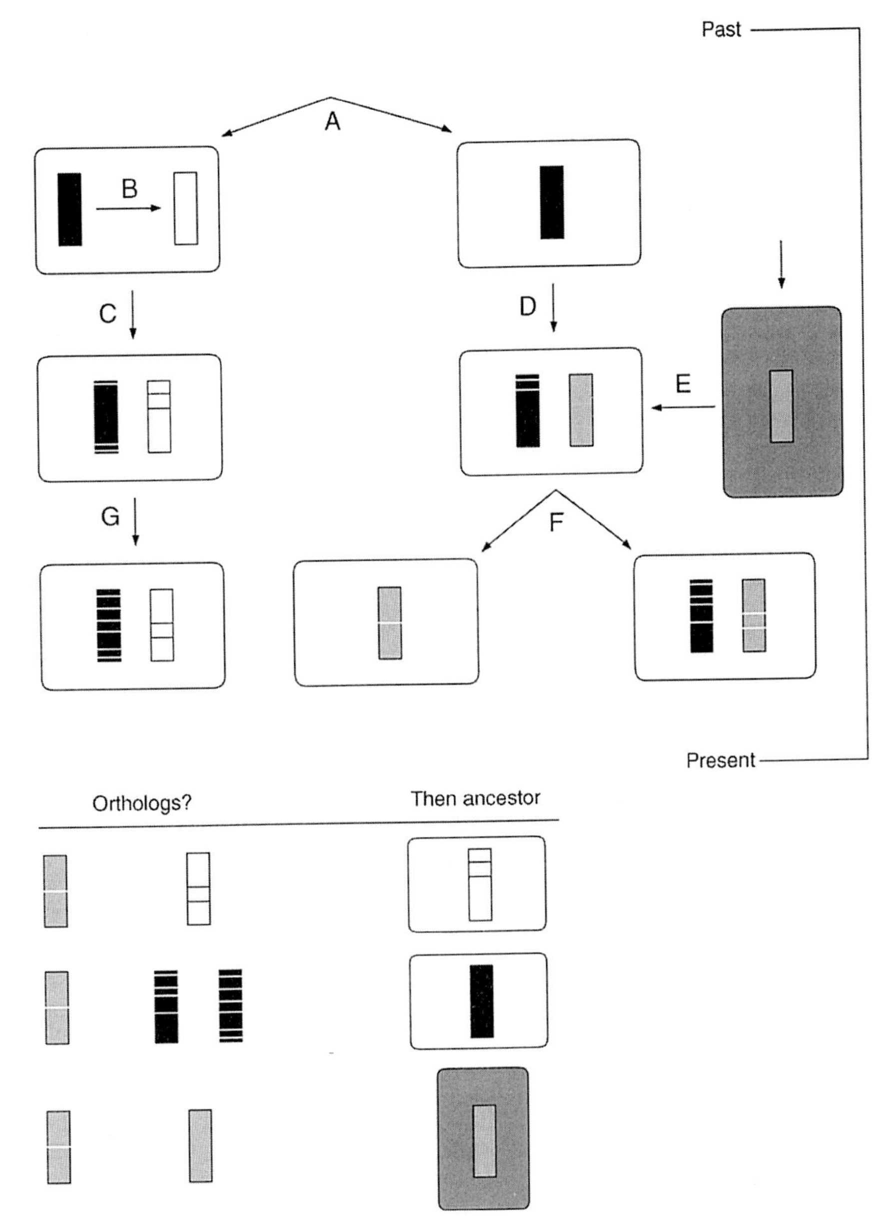

FIGURE 50.1 The complexity of inferring organismal descent from homologous genes. (Top) Illustrated is the hypothetical history of two genes (small rectangles) with similar DNA sequences and/or functions. (A) A pair of closely related organisms (large open rectangles) begins to diverge. (B) The solid black gene duplicates in one of the two, producing a paralog (small open rectangle). (C) As each organismal lineage evolves, mutations (white lines in black and gray rectangles, black lines in white rectangles) accumulate independently in the orthologous copies of the gene. The paralogs also evolve relative to each other. (D) The other lineage acquires a structurally similar, possibly homologous, copy of the same gene (small gray rectangle) by (E) horizontal transmission from another organism (large gray rectangle). (F) Perhaps as a consequence of recombination with the similar sequence, some descendants lose the original copy of the gene while others maintain both copies. (G) All genes accumulate mutations independently until the present. (Bottom) Working back from the present using only DNA sequences, which genes are orthologs and which organisms are most closely related? Illustrated are the choices for a single gene, shown left (grey rectangle) and the possible orthologs shown center. Depending on the number of nucleotides introduced into genes by horizontal recombination events, the similarity of the introduced genes, and the number of mutations in the true orthologs, the ancestry of organisms can be difficult to determine.

sequence match, the possibility that one species lost the true ortholog since the speciation event can never be excluded. Indeed, the genes may be orthologous but, because the gene was transferred horizontally, the comparison misrepresents the ancestry of the organism. Mistakes have been made by the inadvertent comparisons of paralogs (discussed by Doolittle, 1998).

2. *Mutation or recombination?*

Do the differences and similarities in sequences result from historical events other than time since divergence? Sequences can be maintained by selection or diverge rapidly through selection at rates that differ from other genes in the same lineage. Recombination can also maintain or disrupt sequences. Recombination, particularly with sequences obtained horizontally, can convert a portion of a DNA sequence. The resulting mosaic could have an overall average sequence similarity that supported the phylogeny suggested by comparing other genes between the organisms but actually be the product of genes that evolved independently.

Reports of mosaic genes are becoming increasingly common. Perhaps one of the most instructive examples of the impact of HGT on the evolution of organismal phenotypes is the story of β-lactam resistance evolution in *Neisseria* and *Streptococcus* (Fig. 50.2). Alterations in their target penicillin-binding proteins (PBP) are a common and growing means of resistance to the drugs. Not only should this type of resistance arise slowly, if at all, but it should be species-specific. The reasons for these expectations follow from the nature of the drugs and mechanisms of resistance. β-Lactams bind to several different and essential PBPs; binding to any is usually sufficient for therapy. Thus, each PBP must change to confer phenotypic resistance. Moreover, a single point mutation that together conferred resistance and maintained protein function may be impossible. Thus, up to 4 genes (as in *Streptococcus pneumoniae*) must accumulate at least two

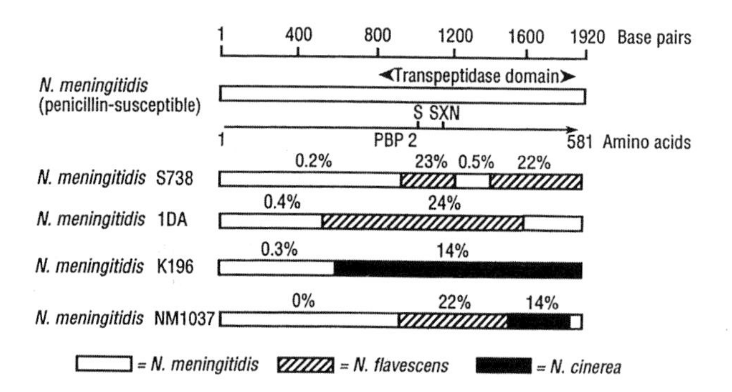

FIGURE 50.2 Mosaic PBP 2 genes in penicillin-resistant meningococci. The open rectangle represents the PBP 2 gene of a penicillin-susceptible meningococcus. The line terminating in an arrow represents PBP 2; the active-site serine residue and the SXN conserved motif are shown. The percent sequence divergence between different regions of the genes and the corresponding regions in the susceptible strain are shown for four resistant meningococci. The origins of the diverged regions are illustrated. Reprinted with permission from Spratt, B. G. (1994). Resistance to antibiotics mediated by target alterations. *Science* **264**, 390. Copyright 1994 American Association for the Advancement of Science.

changes *simultaneously* for the PBPs to retain function and for a cell to display phenotypic resistance. Based on a simple calculation using a reasonable mutation rate around 10^{-9}/base/generation, the absurd probability of a pre-existing penicillin resistant strain would be 10^{-72}/cell/generation $(10^{-9})^{4.2}$.

How did these pathogens become resistant? Individual PBPs of resistant pathogens are composed of segments of homologous PBPs from up to three different species! HGT has done what is a probabilistic impossibility for vertical evolution by mixing parts of different PBPs from various species of bacteria that have individual PBPs with low binding affinities to β-lactams.

3. *Big rare events or small common events?*

How large does a horizontally transmitted sequence have to be to reveal itself as having been acquired horizontally? Herein lies the most difficult problem in both assessing the validity of gene comparisons for determinations of ancestry and determining the rate and extent of horizontal gene transmission. A horizontally transferred gene preserved in a vertically reproducing lineage may be identified by the anomalous phylogenetic tree its sequence creates, a significant deviation from the average G + C content of the host, an unusual organization of genes or deviation from the normal codon bias of the host. Most indicators are quantitative. Thus, they are useful if large tracts of sequences with one or more large deviations from accepted norms are being analyzed. The origin of sequences becomes increasingly difficult to determine the shorter they are or the closer to accepted ranges they appear to be. The proper conclusion in those cases is uncertainty as to how much of the sequences' reproductive history has been either horizontal *or* vertical. Moreover, some of the characteristic differences between transferred and endogenous genes, like codon bias and G + C content, can change to more closely resemble organismal norms with time. Sequence comparisons limit analysis both to large tracts of DNA and to recently transmitted genes.

How many bases are retained in the average HGT event? Horizontally transferred sequences are retained through vertical reproduction only by selection or chance. Sequences that reduce fitness will likely disappear with the organism. Protein coding sequences, the source of most DNA sequences from organisms, are the least likely repository of horizontally acquired sequences. Selectable changes could be preserved in gene-regulating sequences or in sequences important for chromosome structure. Those preserved by chance are most likely to accumulate in regions already known for their high rates of change: intergenic regions and junk. Overall, only sequences much shorter than necessary to encode a protein should normally be retained in organisms.

Transferred sequences that must ultimately recombine with those in chromosomes may also be excluded by the homologous recombination and mismatch repair machinery (Table 50.1). Although such enzymes are generally necessary for efficient recombination, they also discourage the retention of dissimilar sequences. The *E. coli* RecA protein, for example, aborts recombination between sequences with less than an overall 90% identity or more than three mismatches in a row. Taking the biochemical stringency of RecA as roughly representative of most organisms, then the tracts of sequences incorporated will, in the main, be short, even if the donor and recipient are closely related. The tracts of heterologous sequences, as much as 24% diverged, in the mosaic PBPs of *Neisseria meningitidis* and *S. pneumoniae*, stretch hundreds of base pairs. Given the barriers to transmission, large mosaic structures should be created extremely rarely. Their existence is evidence of gene transfer frequencies on a scale large enough to produce such unlikely recombination events randomly.

Mutations or physiological variation in mismatch repair function (the gene products that correct mispaired bases that arise from polymerase errors or recombination between homologous but not identical sequences of DNA) dramatically increase the mutation rate of individuals and similarly decrease the barrier to interspecies recombination. For example, reducing the stringency of sequence comparisons made by mismatch repair enzymes dramatically increases the frequency of recombination in interspecies crosses of *E. coli* and *Salmonella typhimurium* (discussed in Matic *et al.*, 1996). Similar effects have been noted in eukaryotes as well as prokaryotes (Kolodner, 1996). The mismatch repair genes themselves are highly mosaic. This is to be expected because rare mutations in mismatch repair genes would make the organism more likely to retain recombination products created by horizontally transferred DNA from other organisms. When such recombinations also restored mismatch repair function, then the recombinant organism would regain its controlled mutation rate (Denamur *et al.*, 2000). HGT can increase the proportion of organisms within populations that have high mutation rates due to defects in mismatch repair functions by up to 10,000-fold in some environments (Funchain *et al.*, 2001). Thus, transient mutator phenotypes, caused by mutational or physiological cycling of mismatch repair stringency, may impact gene transfer profoundly but frequently only over short stretches of DNA and may only rarely introduce a new gene in its entirety while simultaneously restoring, or creating new, biochemical functions.

TABLE 50.1 Genes that inhibit interspecies gene transmission

Bacteria	Yeast	Humans
mutH		
mutS	*MSH1-MSH3, MSH6*	*MSH2-3, DUG1, MRP1, GTBP, p160*
mutL	*PMS1, MLH1, MLH2*	*PMS1-2, MLH1*
mutU		

4. *Homology more than sequence*

Is sequence structure *necessarily* conserved by evolution? The genomes of HMEs tend to be fluid. Over time, particular HMEs are replaced by relatives carrying different genes, such as plasmids with an expanding repertoire of antibiotic resistance genes. Tracing ancestry of HMEs by structure allows relationships to be determined over only very short periods of time. For example, the structure of infectious retroviruses can change 10^4–10^6 times the rate of other genes and defective retroviruses reproducing in synchrony with the host. The phylogeny of these viruses is confined to tracing the residues of defective viruses trapped in chromosomes of organisms or monitoring divergences on decade scales. This observation led Doolittle *et al.* (1989) to lament that "As is the case in the rest of the biological world, rapid evolutionary change appears to be associated with rapid extinction." A statement that cannot help but be true if the HME is only defined by its primary nucleic acid structure. Although contrary to their preferred conclusion, they did acknowledge that "the remarkable constellation of enzymes and structural proteins that constitute infectious retroviruses may have been assembled ... in the quite recent past ... The erratic occurrence of retrovirus-like entities in the biological world could be the result of widespread distant horizontal transfers" (Doolittle *et al.*, 1989).

B. Genome structure

How would ancestry of organisms be determined without, or with much less, reliance on nucleic acid or protein sequences? In the cases of ribonuclease H (Doolittle *et al.*, 1989) and the A8 subunit of mitochondria (Jacobs, 1991), ancestry was inferred from three-dimensional structures or other biophysical characteristics because the primary nucleic or amino-acid sequences had lost such information.

Approaches that rely indirectly on sequences are emerging. The overall composition of genes carried by organisms, their relative positions, and their regulation establish grouping patterns. As discussed above, determining the impact and extent of HGT will require looking at the effects of the process rather than at conservation of particular sequences. The recent introduction of the "Competition Model" (Cooper and Heinemann, 2000) and the "Selfish Operons Model" (Lawrence and Ochman, 1998), are producing robust tests of predictions of genome organization.

The Selfish Operon Model explains the organization of prokaryotic genes in operons as the result of HGT. The genes collected into operons are, and would need to be, nonessential for survival in at least one environment or only weakly selected, like traditional HME-borne genes, such as antibiotic resistance and novel virulence traits. Operon organization is not selected by clonal dissemination of hosts benefiting from this particular organization of its genes. Instead, the collection of genes in the operon reproduces faster by horizontal than by vertical reproduction and the genes are preserved in organisms when they transfer together.

The genes in operons are functionally codependent. Individually, they provide no selective benefit to a cell. So whereas they may be transferred horizontally as individual genes, they are lost in time from vertically reproducing lineages. As an example, imagine that the genes of the *lac* operon were once distributed around the chromosome of some ancient bacterium. That bacterium could survive occasional deletions of some intervening nonessential genes, rearrangements of genes by intrachromosomal recombination and transposition. Eventually, the genes of the *lac* operon may have come close enough together to be mobilized by a single HME or efficiently taken up by transformation. As the *lac* genes together could provide a selective benefit to a recipient cell that none of the genes could provide individually, then the recipient that maintained the cluster would be selected. Over time, the new lineage and other recipients of the transferred cluster could tolerate occasional deletions of material between the *lac* genes until the modern minimal structure of the operon emerged.

Thus, the genes were maintained vertically because of their contribution to the organismal phenotype but were organized in such a way as a result of HGT. For the model to work, HGT must occur frequently, with occasional retention of particular sequences in vertically reproducing lineages. The Competition Model, which seeks to explain HME organization, supports Selfish Operon expectations of high gene transfer frequencies. Normally, microbes are cultured under conditions that favor organismal reproduction. In tests of the Competition Model, conditions that favored HME reproduction were maintained. Sometimes gene transfer was allowed to occur as fast or faster than organismal reproduction. When multiple HMEs were mixed under such conditions, individuals with strategies to eliminate other HMEs emerged and dominated. The phenotypic expression of the genes (e.g., antibiotic resistance genes) that made HMEs successful during horizontal competition were sometimes detrimental and sometimes beneficial to the host. Studying those genes during clonal culture of the host, especially in the absence of competiting HMEs, can lead to significantly different perceptions of their function and evolutionary history.

IV. ESTIMATING RATES

The emergence of microbial cells as legitimate entities in biology was delayed by centuries because of their size. The scale of the microbial community is still largely unknown. Similarly, the recalcitrance of horizontally transferred genes to study using existing technology has led to the untenable impression that they are rare. Judging from transmission rates, HGT could be quite common. For instance, it has been estimated that 18% of the *E. coli* genome was acquired horizontally since it diverged from *S. typhimurium* 100 million years ago (Lawrence and Ochman, 1998). As much as 5% of the mammalian genome is contained between copies of the two long terminal repeats characteristic of retroviruses. As much as 30% of the mammalian genome, and 10% of the human, was created by the action of reverse transcriptase. Plant genomes may be almost half the product of reverse transcriptase. Twenty-three percent of the human major histocompatibility complex class II region is of retroviral origin, strongly suggesting that transfer alters important genetic characteristics.

A. Organism clonality

How frequently could genes be reproducing horizontally? The apparent clonality of many microogranisms amenable to the techniques used to estimate diversity would seem to support the perception of a vertical world. Evidence of clonal distribution is inappropriate evidence for conclusions about horizontal transfer, though, for several reasons. First, the number of clonal types is evidence, not of the *amount of recombination* that occurs between cells, but of the amount of recombination *and* subsequent selection of particular individuals. Second, the techniques for estimating diversity, which rely on alterations in genome sequences or protein conformations detectable by electrophoresis (e.g., restriction fragment length polymorphism (RFLP) and isozyme analysis), focus at a level of resolution that cannot detect recombination of short nucleotide sequences. Finally, the techniques are preoccupied with chromosomal characteristics. The mosaic structure of plasmids within bacteria argues for extensive interstrain recombination that is not preserved in the chromosomes. Whereas a lack of clonality, when it exists, can be concluded from such analyses, apparent clonality cannot at present preclude still enormous frequencies of HGT.

B. Viruses

Viruses in the environment provide some insight into the amount of horizontal gene flow. Since the viral life cycle can include a stable, extracellular period that other HME life cycles may not, the viruses are uniquely amenable to monitoring. The summertime free viral load of the world's oceans has been estimated at between 5×10^6 and 1.5×10^7/ml at up to 30 m depth. These viral titers were 5–100 times the estimated concentration of organisms. Up to 70% of marine prokaryotes are infected by viruses at any given time. Each of these infected hosts produce an estimated 10–100 new viruses. The absolute numbers vary somewhat between studies and season, but are consistent with counts in fresh water of 10^4 *Pseudomonas*- and chlorella-specific viruses/ml.

Finding viruses is difficult without the ability to culture each possible virus; counting the number of plasmid and transposon generations is, at present, impossible. Still, the ease of isolating plasmid-infected bacteria from natural habitats would suggest that plasmids and transposons are replicating no less than viruses. Summing the volume of water in the top 30 m of oceans, adding a factor for viral turnover on land and the contribution of nonviral HMEs, produces an, albeit crude, estimated HGT frequency of 10^{30} per day. The limited technologies available for surveying and then unambiguously culturing to purity the units that mostly reproduce horizontally leaves us with an exaggerated awareness of cellular life.

V. THE RISKS AND CONSEQUENCES

A. Effects of transfer versus transmission

The difference between transfer and transmission affects more than experimental design. The difference is the essence of the debate over the risk of gene escape from genetically modified organisms and the evolution of resistance to new generation antimicrobial agents. In both cases, we are interested in the formation of genotypes that produce organismal phenotypes better avoided. Experiments that have failed to detect the exchange of genes between organisms because no recombinants were detected have suffered from two important flaws. First, the scale of the experiments and the exposure to HME invasions were both too limited to represent the fate of genes in time and on global scales. The pace and extent of antibiotic resistance evolution is a clear indication of how extremely rare events (10^{-72}) can become certainties for populations as large as those of the HMEs and microbial cells. Second, preservation of horizontally transferred genes is the last step in the process of generating recombinant organisms. The rate limiting step in gene transmission is compromising the barriers to inheritance and expression of recombinant genes. Thus, recombinants are an inaccurate way of assaying gene transfer. Unless a risk assessment experiment can be conducted under constant selection with a flux of organisms and vectors, the effect of transfer on potentiating the creation of recombinant types cannot be estimated.

B. Potentiating conditions

Other molecules are transferred concomitantly with nucleic acids. Viruses and conjugative plasmids carry proteins into recipients. In natural vesicle-mediated transformation, proteins and DNA enter recipient cells. Except for prions, these other types of transferred molecules are genetically inert because they do not direct their own replication. Nevertheless, escort molecules, whether they be protein or polymers of nucleotides, can instigate heritable states *de novo*. For example, proteins, double-stranded RNA (e.g., RNAi, siRNA) and the methyl groups on DNA provoke heritable changes in gene expression patterns. The transfer of the *E. coli* RecA protein to *recA*$^-$ λ lysogens by conjugative plasmids can activate the latent virus, as can the transfer of damaged DNA. Once activated, the virus remains active until a new environmental signal causes its conversion to a prophage. In mammalian cells, the introduction of DNA fragments with preexisting methylation patterns different from the pattern on the homologous endogenous sequences can cause the methylation or demethylation of chromosomal sequences. In all eukaryotes so far tested, RNAi-effects can be transmitted not just horizontally from cell to cell within an organism, but across generations (Cogoni and Macino, 2000).

Thus, HGT processes are capable of more than the creation of recombinants. The impact, surely underestimated, includes the use of transferred nucleic acids for recombination with, and repair of, endogenous genes, the creation of recombinants with novel and potentially selectable phenotypes, and the potential to alter gene expression and heritable states.

ACKNOWLEDGMENTS

I thank T. Cooper and N. Gemmell for critical reading of the manuscript. This effort was supported in part by a grant from the Brian Mason Trust of Canterbury and New Zealand Lotteries Health.

BIBLIOGRAPHY

Best, S., Le Tissier, P. R., and Stoye, J. P. (1997). Endogenous retroviruses and the evolution of resistance to retroviral infection. *Trends Microbiol.* **5**, 313–318.

Cogoni, C., and Macino, G. (2000). Post-transcriptional gene silencing across kingdoms. *Curr. Opin. Genet. Develop.* **10**, 638–643.

Cooper, T. F., and Heinemann, J. A. (2000). Postsegregational killing does not increase plasmid stability but acts to mediate the exclusion of competing plasmids. *Proc. Natl. Acad. Sci. USA* **97**, 12543–12648.

Davison, J. Genetic exchange between bacteria in the environment. (1999). *Plasmid* **42**, 73–91.

Denamur, E., Lecointre, G., Darlu, P., Tenaillon, O., Acquviva, C., Sayada, C., Sunjevaric, I., Rothstein, R., Elion, J., Taddei, F., Radman, M., and Matic, I. (2000). Evolutionary implications of the frequent horizontal transfer of mismatch repair genes. *Cell* **103**, 711–721.

Doolittle, R. F. (1998). The case for gene transfers between distantly related organisms. In "Horizontal Gene Transfer" (M. Syvanen and C. I. Kado, eds.), pp. 311–320. Chapman and Hall, London and New York.

Doolittle, R. F., Feng, D.-F., Johnson, M. S., and McClure, M. A. (1989). Origins and evolutionary relationships of retroviruses. *Q. Rev. Biol.* **64**, 1–30.

Ferguson, G. C., and Heinemann, J. A. (2002). Recent history of trans-kingdom conjugation. *In* "Horizontal Gene Transfer" (M. Syranen and C. I. Kado, eds.), pp. 1–17. Academic Press, San Diego, CA.

Funchain, P., Yeung, A., Stewart, J., Clendenin, W. M., and Miller, J. H. (2001). Amplification of mutator cells in a population as a result of horizontal gene transfer. *J. Bacteriol.* **183**, 3737–3741.

Gibbs, M. J. and Weiller, G. F. (1999). Evidence that a plant virus switched hosts to infect a vertebrate and then recombined with a vertebrate-infecting virus. *Proc. Natl. Acad. Sci. USA* **96**, 8022–8027.

Gray, M. W. (1998). Mass migration of a group I intron: Promiscuity on a grand scale. *Proc. Natl. Acad. Sci. USA* **95**, 14003–14005.

Heinemann, J. A. (1999). How antibiotics cause antibiotic resistance. *Drug Discovery Today* **4**, 72–79.

Heinemann, J. A., and Roughan P. D. (2000). New hypotheses on the material nature of horizontally mobile genes. *Ann. NY Acad. Sci.* **906**, 169–186.

Huynen, M. A., and Bork, P. (1998). Measuring genome evolution. *Proc. Natl. Acad. Sci. USA* **95**, 5849–5856.

Jacobs, H. T. (1991). Structural similarities between a mitochondrially encoded polypeptide and a family of prokaryotic respiratory toxins involved in plasmid maintenance suggest a novel mechanism for the evolutionary main- tenance of mitochondrial DNA. *J. Mol. Evol.* **32**, 333–339.

Kobayashi, I. (1998). Selfishness and death: Raison d'etre of restriction, recombination and mitochondria. *Trends Genet.* **14**, 368–374.

Kolodner, R. (1996). Biochemistry and genetics of eukaryotic mismatch repair. *Genes Dev.* **10**, 1433–1442.

Lan, R., and Reeves, P. R. (2000). Intraspecies variation in bacterial genomes: the need for a species genome concept. *Trends Microbiol.* **8**, 396–401.

Lio, P., and Vannucci, M. (2000). Finding pathogenicity islands and gene transfer events in genome data. *Bioinformatics* **16**, 932–940.

Lawrence, J. G., and Ochman, H. (1998). Molecular archaeology of the Escherichia coli genome. *Proc. Natl. Acad. Sci. USA* **95**, 9413–9417.

Majewski, J. (2001). Sexual isolation in bacteria. *FEMS Microbiol. Lett.* **199**, 161–169.

Matic, I., Taddei, F., and Radman, M. (1996). Genetic barriers among bacteria. *Trends Microbiol.* **4**, 69–73.

Ochman, H., and Moran, N. A. (2001). Genes lost and genes found: evolution of bacterial pathogenesis and symbiosis. *Science* **292**, 1096–1099.

Paul, J. H. (1999). Microbial gene transfer: an ecological perspective. *J. Mol. Microbiol. Biotechnol.* **1**, 45–50.

Size Limits of Very Small Microorganisms: Proceedings of a Workshop. (1999). National Research Council.

Syvanen, M. and Kado, C. I., (eds) (2002). "Horizontal Gene Transfer." Academic Press, San Diego, CA.

Weld, R. J., and Heinemann, J. A. (2002). Horizontal transfer of proteins between species: part of the big picture or just a genetic rignette? *In* "Horitontal Gene Transfer" (M. Syranen and C. I. Kado, eds.), pp. 51–62, Academic Press, San Diego, CA.

Whitman, W. B., Coleman, D. C., and Wiebe, W. J. (1998). Prokaryotes: the unseen majority. *Proc. Natl. Acad. Sci. USA* **95**, 6578–6583.

WEBSITES

Horizontal Gene Transfer Bibliography
http://www.life.umd.edu/faculty/wilkinson/ZOOL608V/hgenes%5Chorizgene.html

Size Limits of Very Small Microorganisms: Proceedings of a Workshop. National Research Council
http://www.nap.edu/books/0309066344/html/1.html

Pedro's Biomolecular Research Tools. A collection of Links to Information and Services Useful to Molecular Biologists
http://www.public.iastate.edu/~pedro/research_tools.html

Comprehensive Microbial Resource of The Institute for Genomic Research (TIGR) and links to many other microbial genomic sites
http://www.tigr.org/tigr-scripts/CMR2/CMRHomePage.spl

Museum of Bacteria
http://www.bacteriamuseum.org

Virtue Newsletter: Science
http://www.milijolare.no/virtue/newsletter/01_05/sci-vigdis/

The Bacteriophage Ecology Group
www.mansfield.ohio-state.edu/~sabedon/

Colour Plate Section

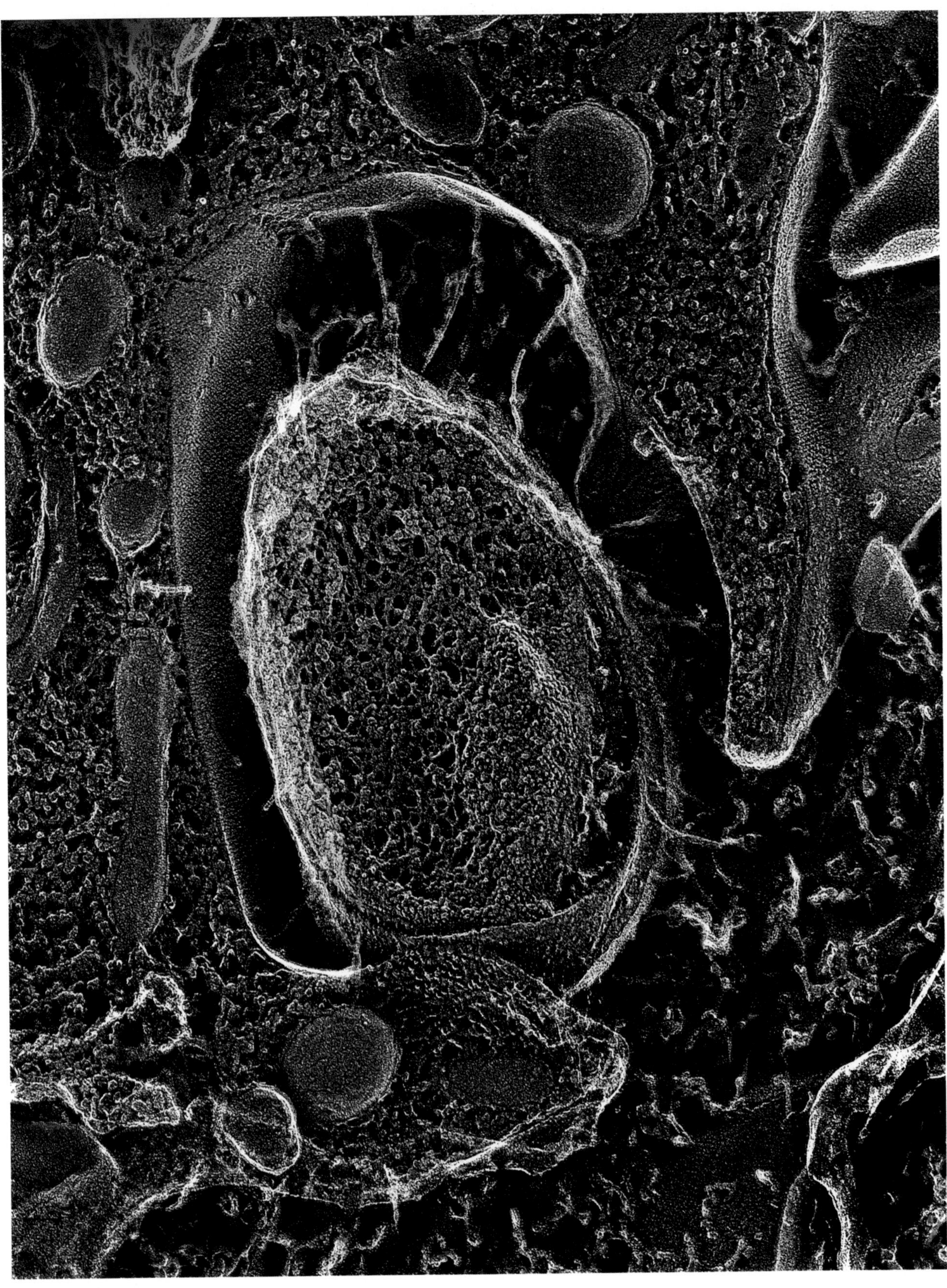

PLATE 1 The attachment of uropathogenic *Escherichia coli* to the luminal surface of the bladder epithelium by type 1 pili as visualized by high-resolution freeze-fracture, deep-etch electron microscopy. See Fig. 1.1.

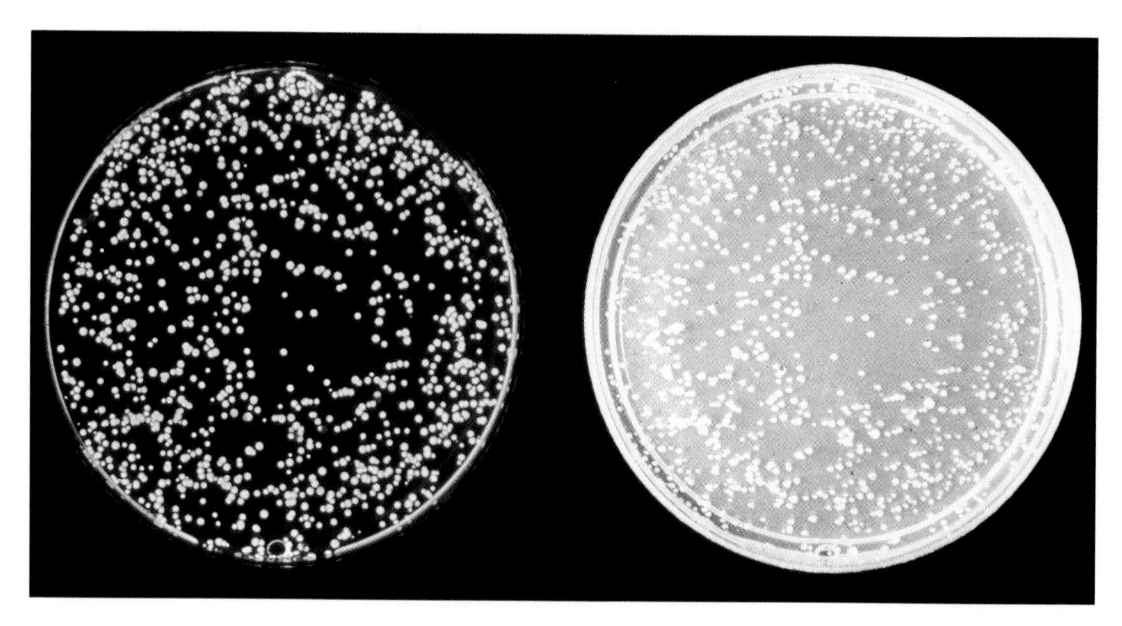

PLATE 2 Colonies of luminous bacteria photographed by their own light (left) and in room light (right). Light is emitted continuously but is controlled by a quorum-sensing mechanism (autoinducer) and is thus not proportional to growth or cell density. See Fig. 14.2.

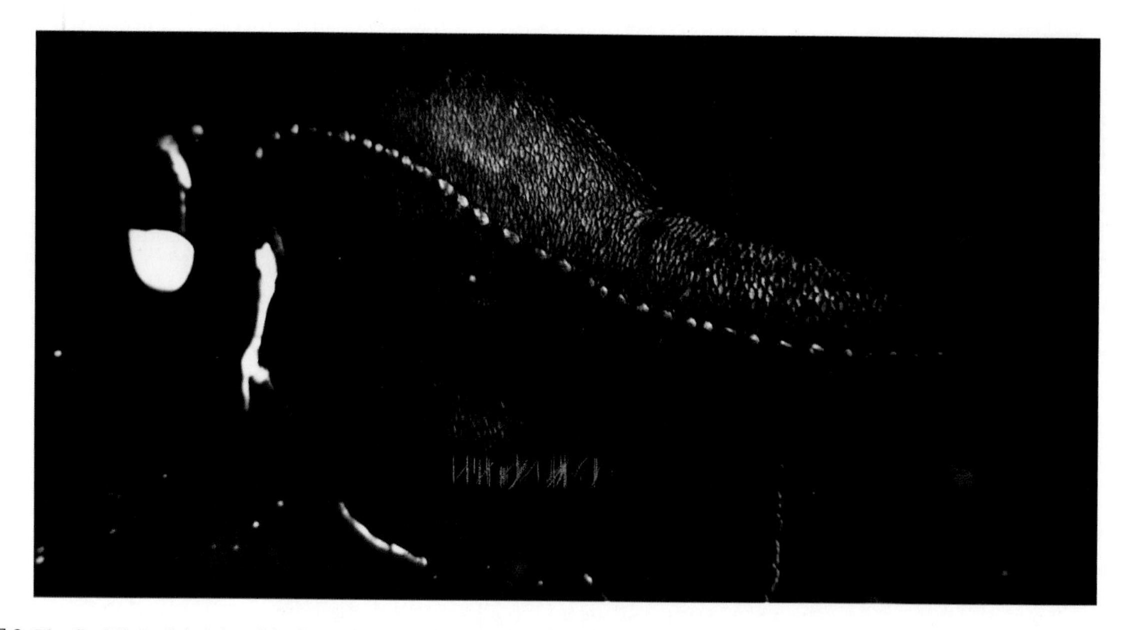

PLATE 3 The flashlight fish (photoblepharon) showing the exposed light organ, which harbors luminous bacteria and is located just below the eye. A special lid allows the fish to turn the light on and off. See Fig. 14.3.

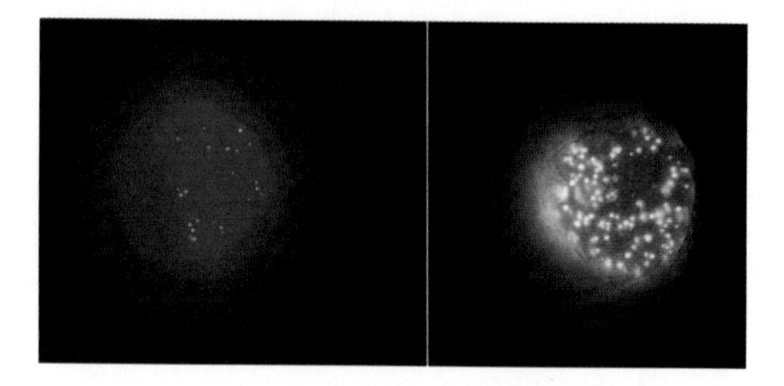

PLATE 4 Gonyaulax cells viewed by fluorescence microscopy showing scintillons (bioluminescent organelles) visualized by the fluorescence of dinoflagellate luciferin (λ_{max} of emission, 475 nm), with chlorophyll fluorescence as the red background. Scintillons are structurally formed and destroyed on a daily basis, controlled by the circadian clock. (Right) Night phase cell with many scintillons; (left) day phase cell with few scintillons. See Fig. 14.7.

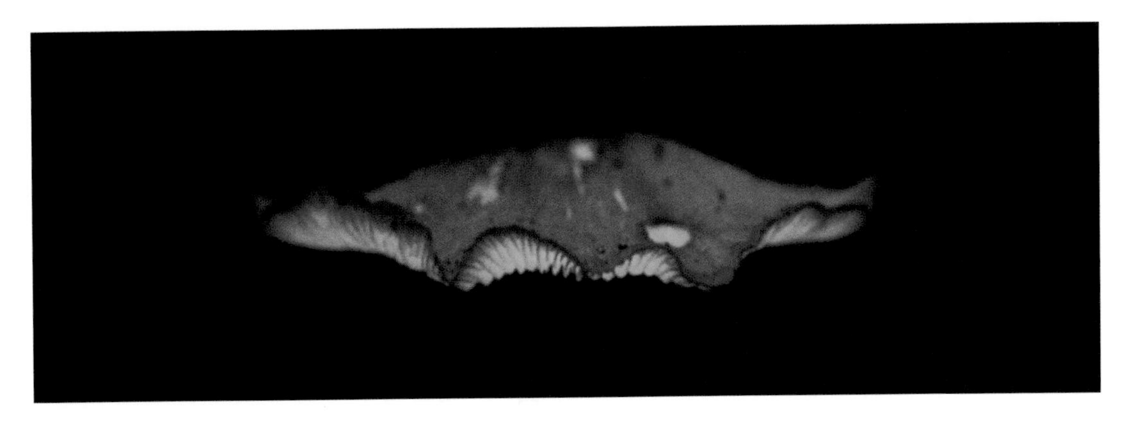

PLATE 5 Bioluminescent mushroom photographed by its own light. See Fig. 14.10.

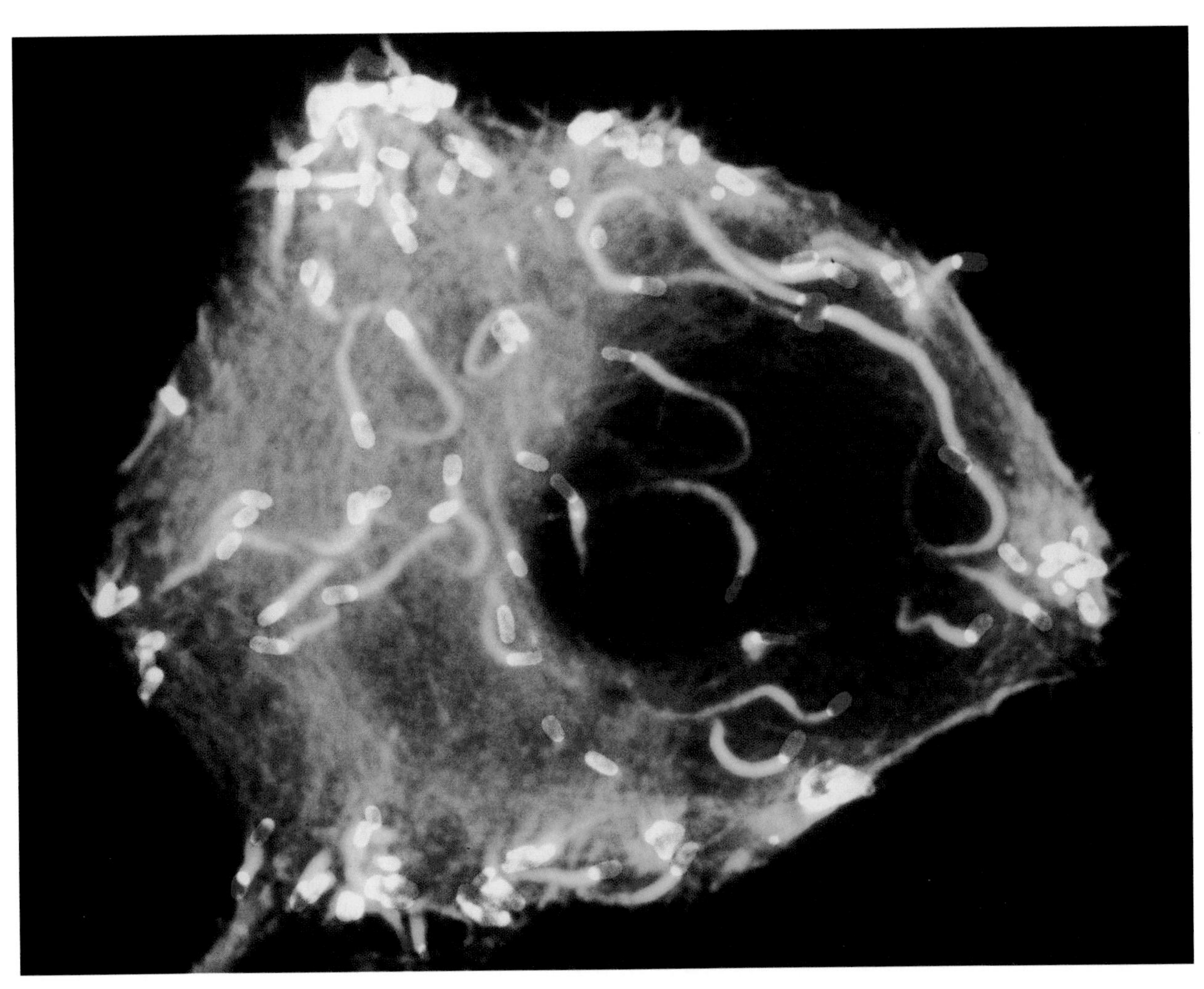

PLATE 6 Actin tail formation by *Shigella flexneri*. Actin microfilaments are polymerized at one end of the bacterium and help it to move within the host cell cytoplasm. Bacteria are stained in red and actin in green. The areas where bacteria and actin colocalize appear in yellow. See Fig. 34.1.

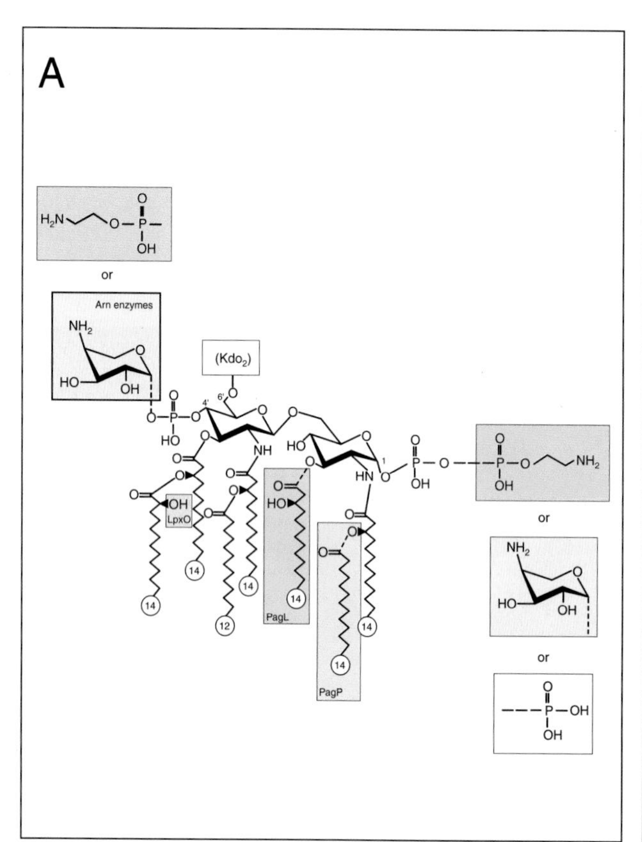

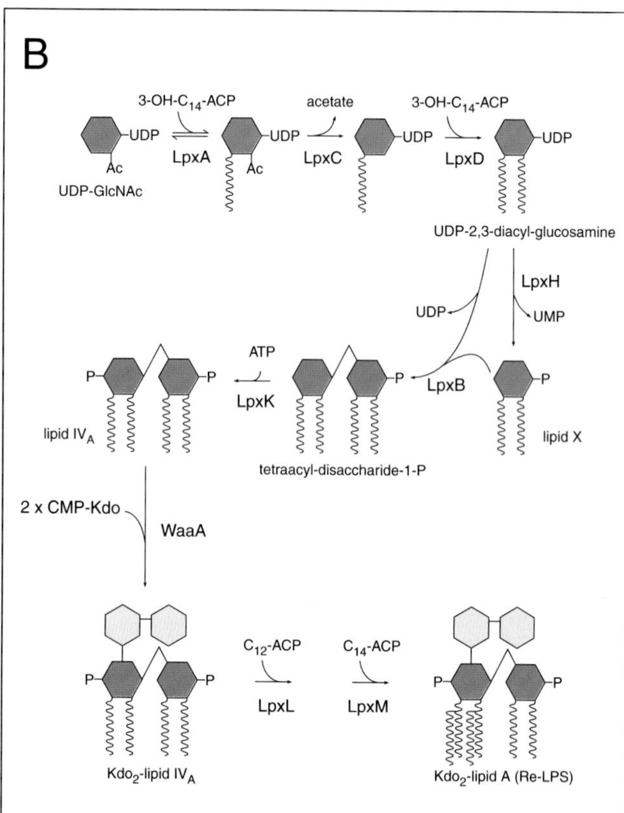

PLATE 7 Structure and biosynthesis of lipid A. Panel A shows the structure of the lipid A from *E. coli* and *Salmonella*. The basic structure is given in *black* and regulated modifications and, where known, the enzymes responsible are indicated in the *colored* boxes. 4-amino-L-arabinose (*yellow* box) is found mainly on the 4′ phosphate, whereas phosphorylethanolamine (*blue* box) is mainly located at position 1. Under some growth conditions a pyrophosphate group is found at position 1. PagP (*pink* box) is a PhoP/PhoQ-regulated palmitoyl transferase located in the outer membrane. PagL (*green* box) selectively removes a β-hydroxymyristoyl residue. LpxO (*brown* box) hydroxylates the 3′ secondary acyl chain and the process is oxygen-dependent. Panel B shows the pathway for biosynthesis of lipid A from *E. coli* K-12. The enzymes responsible for each step in the pathway are indicated. The completed Re-LPS provides an acceptor for sequential assembly of the core oligosaccharide at the cytoplasmic face of the inner membrane. The completed molecule is then exported across the inner membrane by the ABC-transporter, MsbA (not shown). See Fig. 56.2.

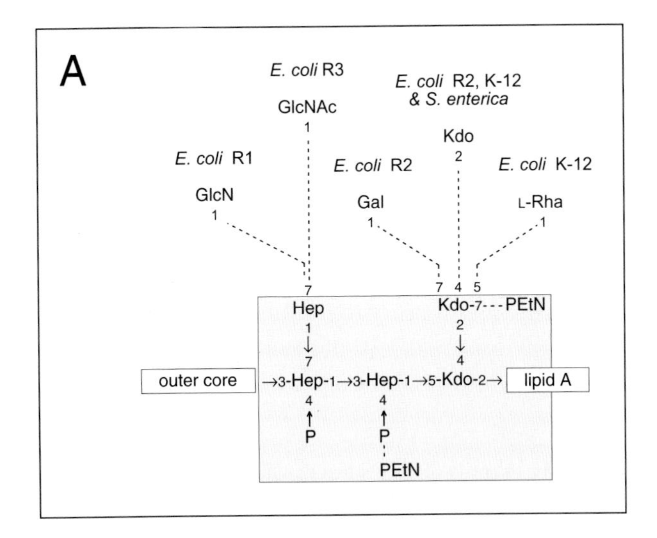

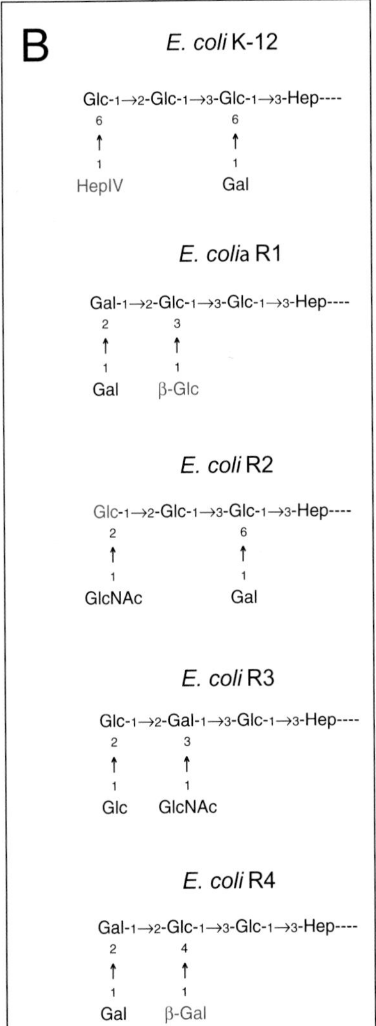

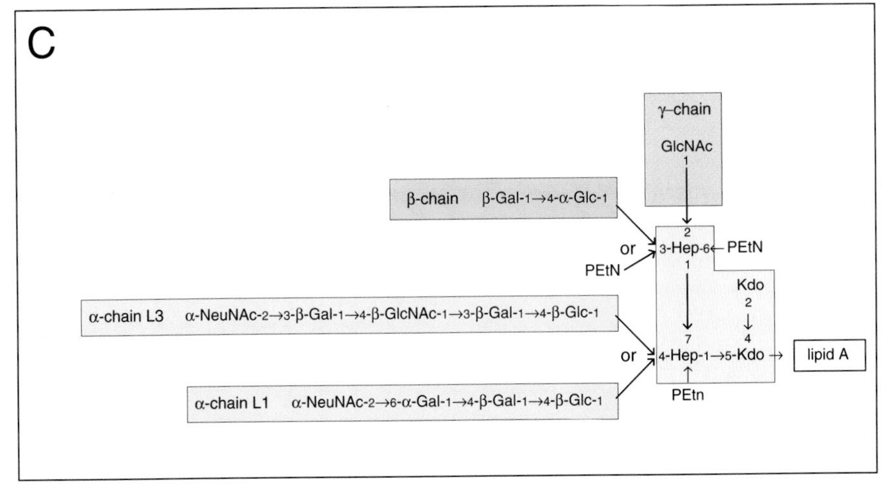

PLATE 8 Structure of the core OS region of LPS and LOS. Panel A shows the inner core region of the core OSs from *E. coli* K-12. The backbone structure (*yellow* box) is conserved but nonstoichiometric glycose modifications (dotted lines) vary in different core types. Panel B shows the known outer core structures from the five core OSs of *E. coli*. They share a common overall structure, consisting of five glycoses linked to Hep II, but they differ in glycose content, glycose sequence, and side branch position. The site of O-PS ligation is not known for all of the core OS types but, where it has been determined, the residue providing the attachment site is indicated in *red*. Panel C shows a representative LOS structure from *N. gonorrhoeae*. The LOS core (*yellow* box) is conserved among different strains but there are strain-specific differences in the attached oligosaccharide chains. Shown are 2 possible variants (L1 and L3) of α-chain (*pink* boxes) and the β- and γ-chains. Unless otherwise indicated, all residues in the various structures are linked in the α-configuration. See Fig. 56.3.

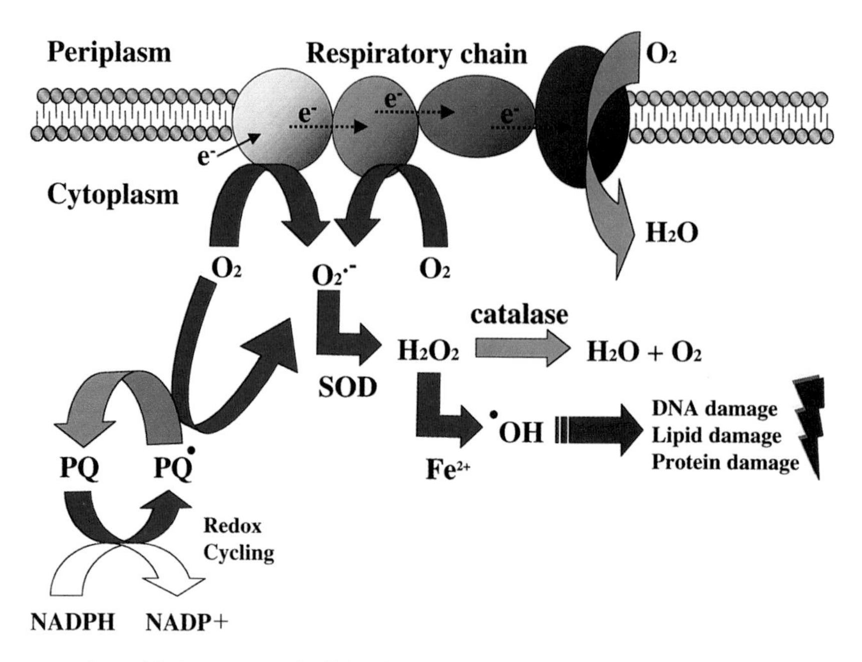

PLATE 9 Reactive oxygen species, oxidative stress, and cellular damage. The majority of the oxygen that enters the cell is reduced to water by the respiratory chain, a reaction that consumes four electrons. However, a small proportion of the oxygen molecules can be reduced in a series of one-electron reactions. Molecular oxygen forms superoxide ($O_2^{\bullet -}$) by reaction with reduced components of the respiratory chain. Superoxide can also be formed by reaction with redox-cycling drugs such as paraquat (PQ), which is enzymatically re-reduced at the expense of NADPH. Superoxide is eliminated by superoxide dismutase (SOD) to form hydrogen peroxide (H_2O_2). Hydrogen peroxide can either be detoxified by conversion into water and oxygen by catalase, or react with reduced transition metals such as iron and copper to form a hydroxyl radical ($^{\cdot}OH$). The hydroxyl radical is a highly reactive molecule than can damage virtually all the fundamental cellular components. Solid arrows indicate reactions that yield oxidants. Open arrows indicate reactions that yield innocuous products. See Fig. 66.1.

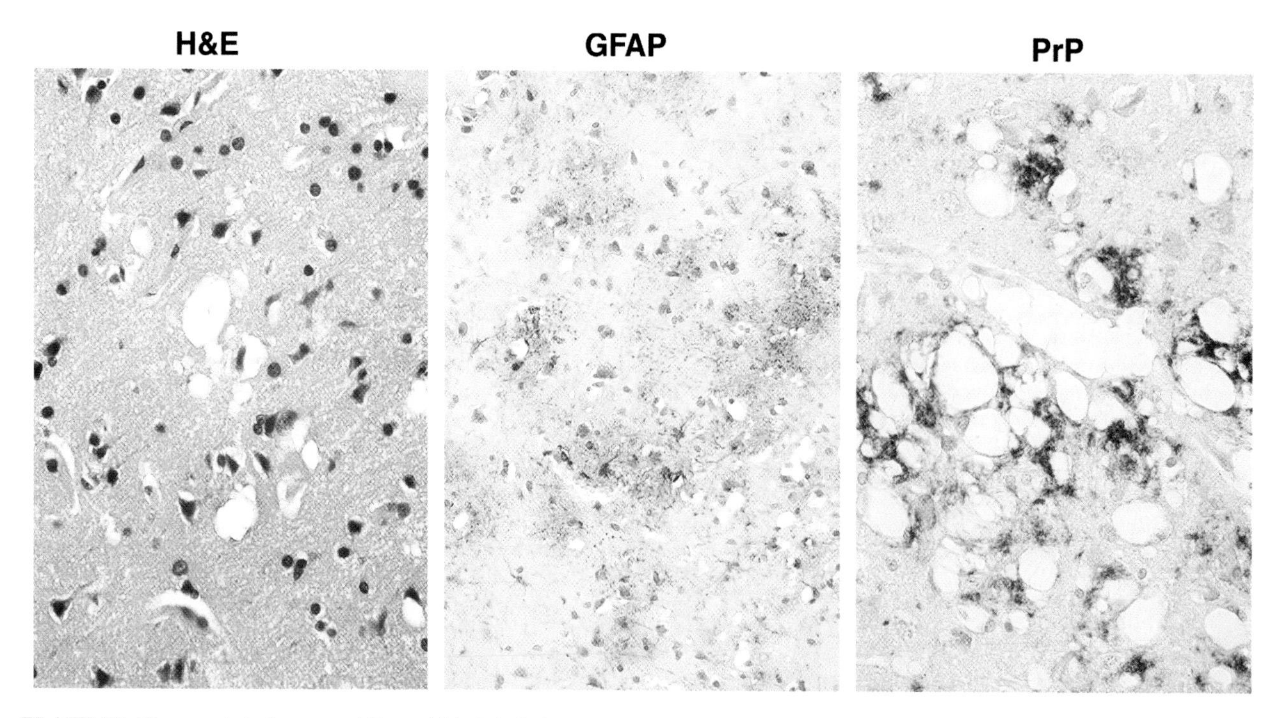

PLATE 10 Characteristic features of Creutzfeldt–Jakob disease. Hematoxylin–eosin stain (left) shows the typical vacuoles in the brain of a CJD patient, which leads to the spongiform appearance. Proliferation of reactive astrocytes is visualized by staining with antibodies against glial fibrillary acidic protein (GFAP, middle). PrP protein deposits are shown with anti-PrP immunostaining. See Fig. 71.7.

51

Human immunodeficiency virus

Luc Montagnier

Institut Pasteur, Paris, France and Queens College, New York

GLOSSARY

B cells Lymphocytes that differentiate into cells involved in antibody production.

cytokines Proteins produced by cells of the immune system that affect the immune system.

LTR (long terminal repeats) Sequences at the ends of the linear form of retroviruses.

messenger RNA processing In eukaryotes, the process whereby messenger RNA is spliced and modified at both ends.

Regulatory protein A protein that is not an enzyme but which regulates the function of other proteins.

Reverse transcriptase An enzyme that synthesizes DNA using an RNA template.

T cells Lymphocytes involved in cell-mediated immunity and activation of B cells.

Virion A mature virus particle.

I. GENERAL FEATURES

A. History

The infectious origin of acquired immunodeficiency syndrome (AIDS) was recognized soon after the identification of the disease itself in 1981. The first AIDS cases diagnosed in 1982 in hemophiliacs indicated that filtered Factor VIII concentrates could transmit the disease, suggesting that a virus or a small bacterium which could pass through bacteriological filters was the etiological agent. Indeed a retrovirus was isolated in early 1983 by L. Montagnier, F. Barre-Sinoussi, J. C. Chermann and their colleagues from a culture of activated T-lymphocytes derived from a lymph node biopsy of a homosexual patient with lymphadenopathy. Other similar isolates from full-blown AIDS patients were made by the same group in 1983, who showed, along with D. Klatzmann and J. C. Gluckman, their tropism for CD4+ lymphocytes. The viruses were found not to be antigenically related to human T cell leukemia viruses (HTLVs) but to be more closely related to the animal lentiviruses. This was despite the fact that some other studies (R. C. Gallo) produced data in favor of an HTLV-related variant. In 1984, R. C. Gallo, M. Popovic and their co-workers described, under the name of HTLV-III, a virus which proved to be identical to the virus described by the Pasteur group. The essential role of the virus, now renamed human immunodeficiency virus (HIV), in AIDS was demonstrated by epidemiological studies, particularly in cases of transmission by blood, where HIV was the only detected common factor in the donor and the receiver. A second type of virus, named HIV-2, was isolated in 1986 from West African patients with AIDS. The discovery of HIV as the etiological agent of AIDS has led to the development of rational therapeutic strategies such as reverse transcriptase inhibitors, protease inhibitors and vaccines.

B. Taxonomy and classification

By its morphology, genetic structure, nucleotide sequence, the virus belongs to the second subfamily

The Desk Encyclopedia of Microbiology
ISBN: 0-12-621361-5

of retrovirus, the lentiviruses, which includes viruses causing slow pathologies in animals, such as Visna, equine infectious anemia (EIA) and feline immunodeficiency virus.

Retrolentiviruses of primates (SIVs, HIVs) are characterized by a tropism to CD4+ lymphocytes, a property not shared by lentiviruses of ungulates.

C. Virion structure

Mature virions (diameter 100–120 nm) have a characteristic spherical morphology with a dense cone-shaped core surrounded by a bilayered phospholipid envelope in which knobs are inserted (Fig. 51.1). There are approximately 80 knobs covering the viral sphere. Each knob is made of several molecules of external glycoprotein, gp 120, possibly as trimers or tetramers linked noncovalently to preformed oligomers of the integral membrane protein, gp41.

In the case of HIV-2, dimers or tetramers of the equivalent transmembrane proteins (gp36 or gp41) are so tightly bound that they appear as such (gp80 and gp160) in electrophoretic gels under classical denaturating conditions (heating at 100 °C in 1% SDS).

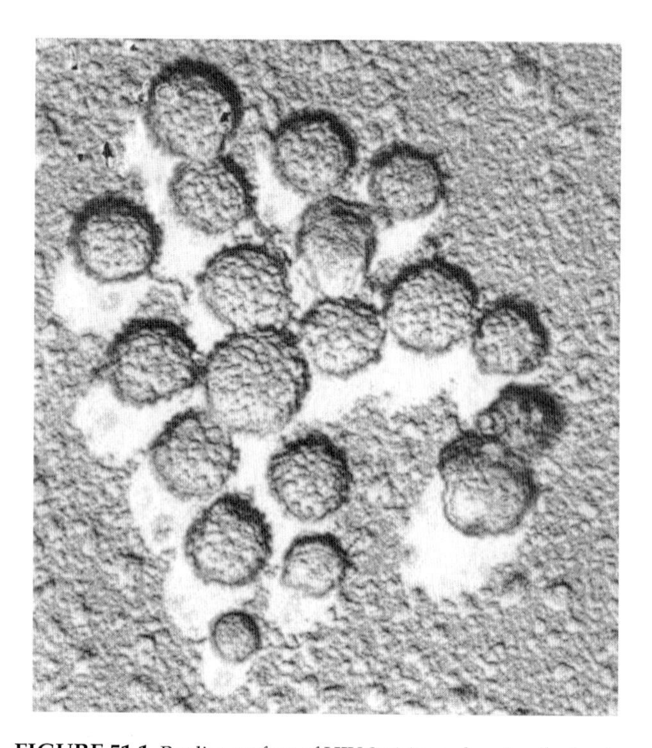

FIGURE 51.1 Replica surface of HIV-2 virions, showing the knobs formed by oligomers of the surface glycoprotein. (HIV-1 virions usually shed most of these knobs spontaneously.) (Courtesy of Drs H. Gelderblom and M. Özel.)

D. Morphogenesis

Virus assembly takes place at the surface of the plasma membrane of lymphocytes or lymphoid cells and in the membrane of intracytoplasmic vacuoles in the case of macrophages. *Gag* polypeptides are necessary to induce budding, and the subsequent morphogenesis of the virion requires the presence of the precursor glycoprotein, gp 160.

Released immature particles with uncondensed cores are transformed into mature particles with condensed cores by proteolytic cleavage of the p17/p18 protein from the gag precursor, which binds to the inner layer of the viral envelope. As this occurs, the p24 protein wraps around the nucleocapsid, formed by the two RNA molecules and the basic proteins p6 and p9.

The correct morphogenesis and infectivity of the viral particles involves cleavage by a cellular protease of the glycoprotein precursor, gp160 into gp120 and gp41.

E. Physical properties

Virions are sensitive to acidic pH (total inactivation at pH 2), ethanol (20–70%), heat (6 log inactivation at 60 °C 30 min), detergents (ionic and nonionic) and chlorine.

Virions are sensitive to radiation including UV light at 260 nm, and X- and γ-rays at a D10 value of over 500,000 rad (50 Gy).

F. Genome structure

The genome is a single-stranded RNA molecule of 9400 base pairs; there are two molecules per virion linked by noncovalent bonds. Lysine tRNA is the primer of the reverse transcriptase, which is Mg^{2+} dependent (optimal concentrations of Mg 10 mM).

The structure of the HIV genome is reflective of the complex nature of the limitations that are exerted during the replication of these viruses. Such limitations affect mainly the expression of viral genes, in an overall process that seems to optimize and synchronize expression of the viral proteins during acute replication, but could also in some instances control the expression of silent proviruses in chronically infected cells. The genome of HIV-1 and HIV-2 (Fig. 51.2), as well as of the different members of the simian immunodeficiency virus (SIV) group, contains several small genes in addition to the classical retroviral structural genes *gag*, *pol*, and *env*. Some of these genes have a critical regulatory function, others appear not to be absolutely required for *in vitro* replication. In addition to these coding elements, the HIV genome is rich in *cis*-acting sequences active at different steps of

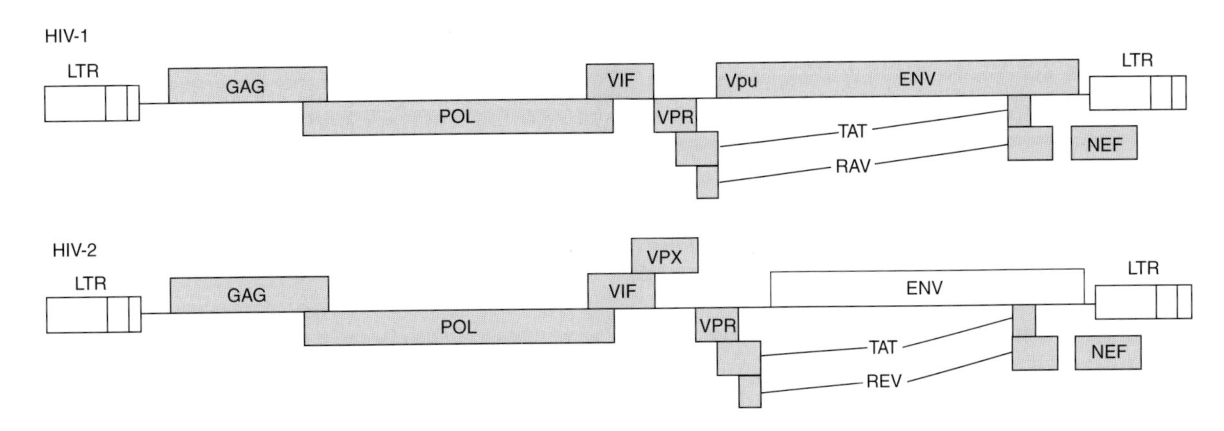

FIGURE 51.2 Genome structure of HIV-1 and HIV-2.

replication, some of which are target sites for proteins regulating viral gene expression.

G. DNA synthesis and integration

Following entry into the cell, the viral genome, represented by a dimer of two identical genomic RNA molecules, is reverse transcribed into linear, double-stranded DNA. The HIV reverse transcription process does not fundamentally differ from what is known for other retroviruses. One interesting difference, however, is that the second (plus) strand of the viral DNA has two distinct origin sites for its synthesis instead of only one. The plus-strand origin is determined, in all retroviruses, by a polypurine tract (PPT) located at the 5′ boundary of the U3 region of the long terminal repeats (LTR). HIVs and other lentiviruses have a second PPT at the center of their genome, which defines a central discontinuity in the plus strand, reflecting its use as an additional origin site. This second origin appears to improve the efficiency of the reverse transcription process, as shown by a reduced replicative capacity of HIV mutants lacking the central PPT. This feature appears to be common to all lentiviruses and spumaviruses.

During the course of an acute infection, high copy numbers of unintegrated viral DNA molecules, either linear, circular with one LTR, or circular with two LTRs, can be observed in infected cells. Such accumulation of viral DNA appears to be the result of multiple events of infection in individual cells.

Following its synthesis in the cytoplasm, viral DNA is transported to the nucleus and integrated into the host cell genome. This transport is an active process, occurring through nuclear pores, mediated by the interactions of the preintegration complex with cellular proteins. It can occur in nondividing cells, unlike the case of oncoviruses. The integration reaction, which can be reproduced *in vitro*, is mediated by the viral integrase, encoded by the C-terminal region of the *pol* gene. The integrase reacts with the termini of the linear viral DNA molecule and with apparently random sites of the host genome. Integration is regarded as being required for expression of the viral genes.

II. REGULATION OF GENE EXPRESSION

A. Transcription

The 5′ LTR of the HIV genome contains the active viral promoter. The transcriptional start site defines the boundary between the U3 and R regions of the LTR. Transcription from the HIV promoter requires the presence of cellular transcriptional activators. Upstream of the transcription start site, surrounded by the usual transcriptional complex recognition signals, two important sets of sequences have been described: these are the three copies of the SP1 transcriptional activator binding site, and two copies of an enhancer element, which reacts with the transcriptional activator NFκB. Initially described as able to control transcription of the gene coding for the kappa chain of the immunoglobulins, NFκB is a ubiquitous transactivator whose components belong to a larger gene family that include the *rel* and *dorsal* proto-oncogenes. This factor can be activated in a number of cell types of the immune system, in particular following activation of the protein kinase C pathway. Further upstream in the U3 region of the LTR, is also found a variety of potential regulatory signals; a large segment of this region, termed NRE (negative regulatory element), has been described as exerting a negative regulatory action on transcription.

B. Viral regulation of transcription: the Tat system

Transcription of viral RNAs starts at the viral LTR and is mediated by cellular RNA polymerase II. But efficient synthesis of viral mRNAs only occurs in the presence of the viral protein Tat. This protein is produced from multiply spliced RNA, in which the two coding exons of the *tat* gene are linked together. The Tat protein acts by binding to a bulged stem–loop structure present at the 5′ end of all viral mRNAs, called TAR (Tat activation region). This binding is itself mediated by a cellular protein, cyclin T. Like other cyclins, cyclin T is associated with a protein kinase (CDK9) which will phosphorylate the C-terminus of RNA-polymerase II. This phosphorylation will allow efficient elongation of viral RNAs. These Tat cellular cofactors are only present in activated cells. Their absence will allow repression of transcription of proviral DNA which could thus enter a period of latency in T lymphocytes. This is another operative regulation that can allow either high virus production or quasi latency, in addition to the upstream control operated by transcription activators (NFκB) on the LTR promoter.

There is a good crosstransactivation of the HIV-2 LTR by HIV-1 Tat, but HIV-2 Tat is a poor HIV-1 transactivator: this could be consecutive to the presence of a second stem–loop structure on the HIV-2 TAR, required for efficient HIV-2 Tat binding.

C. Regulation of mRNA processing: the Rev system

In eukaryotic cells, only fully spliced mRNAs are exported to the cytoplasm and are translated. Unspliced RNAs are retained in the nucleus and eventually degraded.

Since HIV contains a single LTR promoter element, it encodes only a single, genome-length primary transcript. Selective expression of different viral genes is determined by differential splicing of this transcript. Splice sites (recognized by cellular proteins) are relatively inefficient, so that a pool of incompletely spliced RNAs could accumulate in the nucleus.

During acute infection, three major classes of viral transcripts can be observed: the 9 kb genomic RNA molecules, producing Gag and Pol proteins, that can also be encapsidated into viral particles; the singly spliced 4.3 kb RNA molecules, coding for the env glycoproteins; and the 1.8–2 kb, multiply spliced RNA molecules, coding for the regulatory proteins. The cytoplasmic export and stability of the unspliced and singly spliced HIV transcripts, coding for the structural proteins, is controlled by the Rev protein.

In the absence of Rev, only multiply spliced messages are seen in the cytoplasm of infected cells. When subgenomic Env constructs are studied, the effect of Rev appears to involve mostly cytoplasmic transport of unspliced RNAs. This restriction, however, seems to be dependent on the presence of splice sites. It is believed that Rev is able to oppose nuclear retention of RNA molecules by elements of the splicing machinery. It has also been suggested that Rev could favor cytoplasmic stability of RNA, association to polysomes, and even translation of unspliced or singly spliced transcripts. The target site for Rev is a complex multi stem-loop structure termed RRE (Rev responsive element), located in the intron separating the *tat* and *rev* coding exon, in the *env* coding region. Rev binds to the RRE but needs to multimerize to be active, which results in a threshold effect in Rev activity. Rev binds to the RRE through its N-terminal domain, whereas its C-terminal domain behaves as a nuclear export signal (NES) which binds to cellular proteins involved in transport from nucleus to cytoplasm.

This effect has important consequences: during the early phase of a synchronous acute infection, the first transcripts to accumulate are the short multiply spliced species, followed by singly spliced and unspliced species. This shift from an early to a late phase of HIV expression appears to be under the control of accumulation of sufficient amounts of Rev. Similarly, some cells containing integrated HIV genomes bearing low basal expression levels do not produce structural RNAs; these can accumulate, however, after cell stimulation, in a process that is believed to involve accumulation of Rev. Therefore, the Rev system seems to promote synchronization of particle production during acute infection, and to delay structural protein expression in chronically infected cells, a phenomenon that could favor persistence of the infection in the context of a vigilant immune system.

III. STRUCTURAL PROTEINS

Table 51.1 summarizes virion proteins, their functions, cellular location and compares HIV-1 with HIV-2 proteins.

TABLE 51.1 Proteins of HIVs

ORF	Function	Location	HIV-1	HIV-2
gag	Precursor		p55	
	Matrix	Myristylated, outer virion shell	p17/18	p16
	Capsid	Virion: core	p24	p26
	Nucleocapsid	Virion: RNA associated	p7/p9	
	Unknown	Virion	p6	
pol	Protease	Virion	p10	
	RT (with RNase II domain)		p66	
	RT		p51	
	Integrase		p34	p34
vif	Enhances virus infectivity	Infected cell	p23	
vpr	Arrests cells in G2	Virion	p15	
vpx	Unresolved	Virion (HIV-2, siv)	Not found	p16
tat	Transcriptional transactivator	Infected cell: nucleus	p14	p20
rev	Control of RNA splicing and transport	Infected cell: nucleus and nucleolus	p19	p19
vpu	Controls env–CD4 interaction	Infected cell: membrane	p16	Not found
env	Precursor		gp160	gp140
	Surface envelope glycoprotein, receptor binding	Virion: surface	gp120	gp125
	Transmembrane glycoprotein, membrane fusion		gp41	gp36
nef	Modulates CD4 surface expression	Infected cell: membrane	p27–30	p31

A. Gag and Pol proteins

The Gag and Pol proteins are expressed from the full length, unspliced RNA molecules. In 1 : 100 of translation events, approximately, translation of *gag* is followed by a frameshift into the *pol* ORF. The Gag or Gag–Pol polyprotein precursors are then packaged into particles where the protease domain of the Gag–Pol precursor will cleave the final Gag and Pol products.

The Gag precursor is cleaved into four proteins. The first Gag protein, from the N-terminus of the precursor, is termed p17/18 for HIV-1 and p16 for HIV-2 (matrix protein); it carries a N-terminal myristic acid, which allows association of both the Gag and Gag–Pol precursors to the cell membrane. The largest Gag protein, p24 (p26 in HIV-2) is the principal constituent of the cone-shaped core characteristic of viruses of the HIV group (capsid protein). The next Gag protein, p7 or p9, is termed the nucleocapsid protein, and is tightly associated with the RNA genome: it has been found to promote both RNA dimerization and encapsidation. The most C-terminal of the Gag proteins is p6; although its function is not known, mutants lacking this protein exhibit a defect in particle budding.

The Pol region can be divided into three main functional elements. The most N-terminal is the protease, which acts as a dimer and cleaves Gag and Pol products during particle maturation. The central element is the reverse transcriptase, which acts as a heterodimer: the smaller component has polymerase activity, and the larger component has both polymerase and RNaseH activities. The action of RNaseH is required to eliminate the RNA template, with the exception of selected sequences used as primers, and to allow strand transfer events.

B. Env

The envelope glycoproteins of HIV-1 and HIV-2 are translated from a singly spliced RNA species. The glycoprotein is synthesized as a large glycosylated precursor which is then subjected to a series of biochemical rearrangements of its polysaccharide moiety and to a single cleavage event at a specific site carried out by a cellular convertase. This cleaveage separates the larger, outer portion of the envelope glycoprotein (gp 120) from the smaller, transmembrane protein (gp 41).

The gp 120 carries on a specific domain, in the form of a pocket, that binds with high affinity D1–D2 domain of the CD4 molecule present on susceptible cells. The gp 120 binding domain is conformational, involving several regions, which are conserved within HIV variants. Upon binding to CD4, the gp 120 molecule undergoes conformational changes, which exposes some other sites to cellular molecules favoring the tight association of gp 41 transmembrane protein with the plasma membrane. Unlike the CD4 binding site, these sites are located in variable regions or loops of gp 120, particularly the V3 loop. This explains why the virus can choose a variety of cell surface molecules for its entry generally in association but not always, with the CD4 main receptor. Among these surface molecules (called entry cofactors or coreceptors), two main species have been recognized, which are physiological receptors for chemokines: the CCR5 receptor for chemokines (RANTES, MIP-1, MIP-1) and the CXC4 receptor for a chemokine (SDF-1).

Very interestingly, the use of each coreceptor corresponds to viruses with different biological properties and pathogenicity. Viruses isolated at the beginning of infection use in most cases the CCR5 co-receptor (R5 viruses). They are not cytopathic *in vitro* (nonsyncytium-inducing viruses) and replicate in macrophages and activated primary T lymphocytes expressing CCR5 and CD4. In full-blown AIDS cases, new viral species appear with high power of replication *in vitro*, cytopathic effect (syncytia and single cell lysis), poor ability to replicate in macrophages and ability to grow in tumor T cell lines. They use the coreceptor CXC4 (X4 viruses). With the characterization of more viral isolates, the situation appears more complex; there are dual-tropic viruses, using both CXC4 and CCR5 coreceptors, or using alternative chemokine coreceptors (CCR3, CCR2, CCR8). There is also evidence of binding of the V3 loop to nucleolin, a nuclear protein whose one isoform is expressed at the cell surface, and with a dipeptidase present on activated T cells, CD26.

The fusion between the viral membrane and the cellular membrane involves a change in the conformation of gp 41, itself induced by the conformational changes of gp 120. It involves exposure of a hydrophobic fusion domain present at the N-terminus of gp41, which enables it to insert itself in the cellular phospholipid bilayer. The transmembrane protein also displays an unusually long intracytoplasmic domain. This domain is particularly interesting in HIV-2 and SIV; indeed, tissue culture propagation of these viruses appears to favor truncation of the intracytoplasmic tail. The correction of this truncation leads to reduced infectivity *in vitro*.

C. An important regulatory protein: Nef

This third early protein is expressed in far higher levels than Tat and Rev. Most of the protein is myristilated at its N-terminus and associated with the inner side of the plasma membrane. However, there is partial exposure of the C-terminus of the molecule at the outer cell surface and in experimental infections with Nef-expressing vectors, Nef is also in part excreted into the medium.

Expression of the intact Nef protein is not required for *in vitro* infection of peripheral blood leukocytes (PBLs) or of T cell lines. However Nef-deleted mutants of HIV and SIV are much less pathogenic *in vivo*, leading to persistent infection without disease, at least for a long period of time. Studies in the SIV/macaque model have shown that the initial virus load in lymph nodes is reduced by a factor of 10–50, and that formation of germinal centers in response to infection occurs more quickly than with the pathogenic Nef+ strains.

In the productive cycle of viral infection, several functions of Nef have been identified, which contribute to a higher or longer virus production. Nef has been shown to strongly downregulate the expression of cell surface CD4, by connecting CD4 with the cellular endocytic machinery. This prevents reinfection of the infected cell with newly produced virions, which could lead to its premature death.

Nef also downregulates to a lesser extent surface expression of MHG-1, resulting in some inhibition of recognition of the infected cells by specific cytotoxic T lymphocytes (CTL). Nef seems also, by independent, yet unclear, mechanisms, to enhance virion infectivity, perhaps by increasing the stability of the nucleocapsid at its entrance in the cell.

Perhaps the most relevant property of Nef for its role in AIDS pathogenesis is its capacity to activate CD4+ T lymphocytes by modulating signaling pathways. An extreme case is that of the PBJ14 mutant of SIV which can induce within a few weeks an acute disease in pigtail macaques, and can activate both resting CD8+ and CD4+ lymphocytes *in vitro*. These unusual properties are linked to the creation in Nef by directed mutagenesis (or selection by passage in animals) of a new SH2 (Sarc homologous region 2) binding domain (YXXL motif).

The fact that CD8 cells which are not infectable by SIV are also stimulated, suggests some paracrine effect of Nef expressed or released at the surface of infected CD4+ cells. Nef from 'normally' pathogenic strains of HIV or SIV contains a proline-rich sequence which could bind to the SH3 domain of kinases involved in the activation pathway of T lymphocytes. When mutations are introduced in this region, SIV is no longer pathogenic, unless reversion to the wild-type occurs.

In addition, recombinant Nef protein, in the 10–50 ng/ml range, can act as a costimulatory factor in T lymphocytes in which the T cell receptor was subliminally activated and permits HIV replication in such cells. Thus, Nef produced by HIV-infected cells could recruit for infection neighboring T cells which were on the verge of activation and could not otherwise be susceptible to HIV infection.

This recruitment may be important at the beginning of infection when the number of activated T cells in lymphatic tissues is low and is a limiting factor for HIV expansion.

D. Other HIV accessory proteins: Vif, Vpr, Vpu, Vpx

All these proteins are encoded by single spliced mRNAs and therefore are produced late after sufficient Rev expression.

1. *Vif*

The Vif protein is expressed at high levels in the cytoplasm of infected cells. Deletion of this gene reduces infectivity of HIV-1 virions in primary T4 cells and some tumor cell lines CH9. Some Vif proteins are also present in mature virions. It is believed that Vif can stabilize the virions upon entry in the cells and favor the formation of the reverse transcription complex.

2. *Vpr*

This small basic protein is packaged in the virion nucleocapsid in amounts equivalent to the Gag proteins and therefore can be considered as a virion structural protein. Packaging depends on its association with the p6 protein released by proteolytic cleavage from the C-terminus of the p55 Gag precursor.

Two effects of Vpr have been recognized. One, which is controversial, is to contribute to the nuclear import of the preintegration complex, by interacting with nucleoporins, and other cellular proteins, especially in non-dividing differentiated cells (macrophages).

The other function is to induce in cycling cells an arrest in G2. This effect is similar to that induced by DNA alkylating agents, and could result in the induction of the DNA repair machinery, which could be useful for integration of the proviral DNA. Rhesus macaques inoculated with a mutant SIV having a deletion in Vpr can still evolve towards AIDS-like disease, although some individuals show more resistance to disease progression.

3. *Vpu*

The *Vpu* gene is unique to HIV-1 and the related virus isolated from chimpanzees. The product, Vpu, translated from the same mRNA that produces *env*, is an integral membrane protein with a C-terminal intracytoplasmic tail.

Vpu has two functions. The first is to allow selective degradation of CD4 when the latter associates with the viral envelop precursor in the endoplasmic reticulum. This will release the Env protein from this complex and allow its subsequent incorporation into new virions.

The second function is to facilitate virion release from the cell plasma membrane. Vpu-defective mutants show accumulation of intracellular viral particles.

4. *Vpx*

The *vpx* gene is found only in HIV-2 and SIVs. Like Vpr, the protein is present in mature virions, in amounts roughly equal to that of the major Gag protein p24.

Inside the cell, Vpx is located at the inner side of the cell plasma membrane. Vpx deletion mutants have a reduced capacity to replicate in primary T lymphocytes and macrophages, although they grow normally in T cell lines.

IV. GENETICS

One striking feature of the HIVs arose as soon as the nucleotide sequence of several HIV isolates was known: HIVs display a high level of genetic variability. The most variable region of their genome is the outer region of the envelope glycoprotein, where divergence can reach 30%. The most conserved are the *gag* and *pol* genes. The leading cause of such a diversity is reverse transcriptase: this enzyme does not have any proof-correction activity, and appears to introduce errors every 10^4 nucleotides. Another cause is selection: having to replicate in the context of a vigilant immune system, HIV needs to escape the antiviral immune response. This could explain why the envelope glycoprotein is the most variable region of the genome. It has also been shown that a single infected individual does not carry a single isolate, but a heterogenous population of subisolates, sometimes referred to as "quasispecies," and that changing the conditions of viral replication (by growing the virus *in vitro*, for example) will modify the balance of that viral population. In addition the biological properties of the virus—as determined in the *in vitro* isolated—change during the course of the disease: in general, isolates made from lymphocytes of asymptomatic patients grow slower and are less cytopathic than isolates made from patients with full blown AIDS. These changes seem to be related to mutations in the envelope protein gp120, but no particular mutation can be assigned to the changes in biological properties.

Within the HIV-1 type, genetic diversity appears to be more striking when African isolates are compared with one another or with Western isolates. This is an indirect indication that African isolates may have evolved longer than their Western counterparts.

Based on nucleotide sequence of the variable regions of the envelope protein gp120 (particularly V3), a cladistic classification of HIV-1 isolates is currently being used, using capital alphabetic letters, A to J. They all belong to the group M of HIV-1s (M for major). Since the early 1990s another group of HIV-1 has been described in Central Africa (Cameroon) with some sporadic cases in Europe. This group, which differs from the **M** group by around 50% of nucleotides sequences in the *env* gene, has been termed **O** (for outlayer). All HIV-1s are highly pathogenic but display various geographic locations (see below).

More radically divergent is the HIV-2 group, which was prominent in West Africa in the 1980s, but now tends to be less prevalent than HIV-1. By serology and nucleotide sequences, HIV-2, which also displays several subtypes, is closer to SIVs than to HIV-1. In fact, it probably originates from a SIV naturally infecting sooty mangabeys living in West Africa.

HIV-2 was initially found in some West African patients with AIDS, with symptoms comparable to HIV-1 disease. However, its pathogenic potential seems to be lower than HIV-1, displaying a lower viral load in chronically infected patients and a lower capacity of transmission by sexual contacts.

V. ORIGIN AND GEOGRAPHIC DISTRIBUTION

HIV-1 seems to have originated in Central Africa. This is suggested by the larger extent of variability found in strains isolated from African patients.

Retrospective serological studies also indicate that the virus was present in the early 1970s in 0.25% of a population of young women living in Kinshasa (Zaire). However, only retrospective sporadic cases could be detected in Africa before 1980 as well as in Europe and the United States. The HIV-1 directed epidemic started at about the same time in larger cities of Central Africa and of the United States.

Molecular analysis of the virus present in a serum taken from a Zairian patient in 1959 indicates that it could be an ancestor of several identified HIV-1 clades: B, which is now prominent in North America and Western Europe, D and F present in Central Africa.

In the phylogenetic trees, two isolates made from chimpanzees are clearly closer to the M group than from the O group, the origin of which cannot be traced to any simian lentivirus. This group may have been circulating in some human populations for a longer time.

In the 1990s HIV-1 infection has become pandemic, rapidly developing in new regions such as India, Southeast Asia and South America. In Africa, the infection has spread to the East (Uganda, Kenya, Tanzania, Rwanda, Burundi) and to the North (Ivory Coast). Concerning HIV-2, its spread has been more limited from the areas where it was first detected: Guinea-Bissau, Cap Verde Islands, Senegal. HIV-2 is also present in other countries of the Western part of Africa (Ivory Coast, Burkina Faso, Cameroon, Liberia etc.) and Portuguese-speaking countries (Angola, Mozambique). An outbreak of HIV-2 infection has been detected in prostitutes of the Bombay area, India, suggesting that the virus has also reached Asia.

New rapidly spreading foci of HIV-1 infection have occurred in South Africa, Zimbabwe, Cambodia, and China. Some subtypes are becoming prominent, according to the geographic location, such as A in Ivory Coast, C and D in Central and South Africa, E in Thailand, F in Brazil. As the epidemic progresses, infection of the same individual by two variants becomes frequent, leading to the emergence of recombinants (particularly in the *env* gene) between subtypes.

As of 2002, the number of HIV-infected persons is estimated to be about 40 million worldwide.

VI. SEROLOGIC RELATIONSHIPS

Despite the large spectrum of genetic variability, the prototype strain of HIV-1, isolated at the Pasteur Institute, LAV LAI (otherwise named HTLV-IIIB) can still serve as source of antigen for the detection of HIV-1 antibodies in all geographic regions of the world, including antibodies against the envelope. By contrast, little crossreactivity exists between SIV/HIV-2 group and HIV-1 glycoproteins, whereas crossreactivity remains important in Gag and Pol proteins of the two viruses.

The lack of crossreactivity in a major epitope of the transmembrane protein (gp41, gp36) has been used to differentiate between HIV-1 and HIV-2 infection, or to detect double infection. So far a few patients have shown, by western blot using HIV-1 and HIV-2 prototypes as antigen, atypical reactivity. Such cases, in which, for instance, no antibodies against the glycoprotein of HIV-1 or HIV-2 could be detected in serum, are probably due to variants distantly related to HIV-1 or HIV-2.

With regards to animal retroviruses, SIV-infected primates do react well with all HIV-2 proteins, confirming the close relationship between SIV (particularly SIV from sooty mangabey) and HIV-2. By contrast, other animal lentiviruses (Visna, CAEV, BIV, FIV) are too distantly related to show antibody crossreactivity, with the exception of equine infectious anemia virus which shows crossreactivity with HIV-1, at the level of the Gag p24 protein.

VII. HOST RANGE AND VIRUS TRANSMISSION

HIV-1 has a very limited host range; human and chimpanzee are the only species known so far which can be chronically infected with the virus. However, no disease or deep immune depression have been observed in the hundreds of chimpanzees inoculated with human isolates of HIV-1, with one exception, whereas lack of symptoms is the exception in humans.

HIV-2 can chronically infect some macaque species (rhesus, cynomolgus); an AIDS-like syndrome has been observed in some experiments, but not in a reproducible way.

The two main routes of transmission of HIVs are blood and blood products and sexual contact. The efficiency of blood transmission (transfusion, needles, i.v. drug abuse) depends on several factors: number of virus particles, volume of blood, immune status of the receiver. Infection is particularly efficient in i.v. drug abusers.

Sexual transmission, homosexual and heterosexual, is the major mode of transmission today. All sexual practices are dangerous, but the risk is higher for anogenital intercourse, and is increased by some intercurrent genital infections (herpes, chlamydia, etc.).

Transmission from mother to child is also a major mode. In the absence of treatment, 20–30% of seropositive women give birth to an infected child. Infection can occur in the second half of pregnancy, at delivery and also by breast feeding. Severe infection of children results in death in the first year of life. Otherwise the evolution follows that seen in adults.

VIII. ROLE OF THE VIRUS IN AIDS PATHOGENESIS

Acquired Immune Deficiency Syndrome (AIDS, SIDA in French and Spanish-speaking countries) was defined in the early 1980s, biologically by a profound defect of cellular immunity associated with a deep shortage of CD4+ T lymphocytes and clinically by the occurrence of opportunistic infections and cancers. In Western countries, the most frequent infections are those of the lungs by *Pneumocystis carinii* and of the brain by toxoplasmas, followed by visceral and retinal infections by cytomegalovirus (CMV). Tuberculosis is frequent in tropical areas (Africa, Asia, South America).

Among cancers, the most frequent are disseminated and aggressive Kaposi's sarcoma caused by human herpesvirus 8 (HHV-8) and B-lymphomas, often caused by Epstein–Barr virus (EBV).

The causal relationship between HIV and the immune depression has been assessed as follows:

1. On the basis of epidemiological studies, particularly in blood donors and recipients;
2. From the tropism and cytopathic effect of HIV and viral glycoproteins on CD4+ cells;
3. On the reproduction of the disease in macaques by virus derived from molecular clones of SIV (close to HIV-2).

The natural history of HIV-1 infection has been thoroughy studied by histological and molecular techniques. Three phases can be distinguished: primary infection, a clinical latency phase, the clinical phase (full blown AIDS).

In the primary infection, entry of the virus through mucosa involves its association with dendritic cells which then transport the virus to lymphatic tissues. The virus can rapidly multiply there in activated CD4+ lymphocytes and macrophages. Since the number of activated CD4 lymphocytes is small, it becomes a limiting factor which depends on intercurring microbial infections. Biologically, this phase is characterized by a peak of viral antigens (not always present) and a large number of viral RNA copies in blood, and a peak of interferons. Specific cytotoxic lymphocytes (CTL) then appear followed by the appearance of antibodies against viral glycoproteins and internal proteins. Clinically, when the level of viral replication is high, symptoms can be detected (fever, adenopathy, headaches). However, the infection can be sometimes completely inapparent.

This episode is followed by a phase of clinical latency, with no symptoms. However, in a majority of cases, there is a slow and progressive degradation of the immune system, leading finally to clinical AIDS which will last on average 10 years, in the absence of treatment. In a minority of individuals (5–10%) (long-term nonprogressors) this degradation does not take place or is very slow. In some others, on the contrary, immune depression can occur very rapidly within 1–2 years.

These variations reflect the complex interaction of HIV with the immune system of the host. In fact, the virus continuously replicates in lymphatic tissues (lymph nodes, spleen), with a rapid clearance from the blood. A relatively small number of infected cells is involved at the beginning and their destruction (either by the virus or by cytotoxic cells) cannot solely account for the large depletion of CD4+ cells. It is likely that indirect mechanisms of cell death are involved, bearing on cells which are not infected, but are in contact with noninfectious viral particles, or gp120 shed from cells or virions. Indeed, interaction of gp120 with the CD4 receptor or coreceptors could induce a wrong signaling leading to apoptotic cell death or anergy, when the T cell receptor is stimulated.

Evidence for the preapoptotic state of a large fractional circulating CD4 cells has been obtained. Defects in antigen-presenting cells, in bone marrow renewal of precursor cells, of faster thymic involution, of high oxidative stress, have also been involved to explain the specific CD4+ cell depletion. It is also clear that a state of chronic activation bearing not only on CD4+ cells, but also on CD8, NK and B cells exists all through this phase, associated with the production of inflammatory cytokines (interferons, interleukin (IL) 6, tumor necrosis factor).

In addition, a small pool of latently infected cells exists and is probably continuously renewed. This pool will escape any kind of treatment.

In the clinical phase, with the occurrence of more pathogenic variants (CXC4 tropic) and the decrease in cell-mediated immune response, the infection from local or regional becomes systemic and precipitates the fatal evolution.

Neurological signs are also frequent, particularly a subacute encephalopathy which can develop into a dementia syndrome and brain atrophy, and seems to be due to the virus itself. Some foci of brain macrophages infected with HIV can be detected in white matter, but neither neurons nor glial cells seem to be productibly infected in the same situation. Some astrocytes express large amounts of Nef protein. Therefore, indirect mechanisms for the action of HIV (cytokines, nitric oxide?) have also to be postulated in order to explain the neuronal effects.

IX. IMMUNE RESPONSE

An immune response against HIV is mediated by T-helper cells. Antibody response against major HIV structural protein (Gag, Env) and also against Pol proteins and Nef appear within 2–3 weeks after HIV exposure, exceptionally after 6 months or more. Specific CD8+ cytotoxic T lymphocytes (CTLs) also appear, often earlier than the antibody response, directed against the same proteins. Antibodies against the viral envelope are poorly neutralizing, and are directed against variable regions of gp120, especially the V3 loop.

There is also induction of antiviral cytokines (interferons) but for unclear reasons, even large amounts of interferon are not effective to control viral expansion. There is also *in vitro* evidence for viral inhibition by other soluble factors, particularly secreted by CD8+ T lymphocytes; the increase of chemokines may inhibit by competition the entry of C5 viruses, but some other inhibitory factors produced by CD8+ cells have not yet been identified.

As the infection progresses, the specific CTL response tends to decrease and to lose its efficacy against mutant viruses, although a strong natural killer response (NK cells) can still also play a protective role.

The number and function of CD4+ T lymphocytes decreases progressively with some oscillations during the clinically latent phase. T-helper functions to recalled antigens are precociously affected, in line with a drop of IL-2 and an increase of Th2 cytokines (IL-4). The antibody response against viral proteins decreases at the clinical phase, reaching first the internal proteins. Owing to their higher titers, antibodies against gp120 and gp41 remain detectable even at late stages. These antibodies can also be detected in urine in lower titer.

X. TREATMENT

Until 1994, there was no effective treatment of HIV infection and AIDS. Monotherapy using nucleosidic or nonnucleosidic inhibitors of HIV reverse transcriptase gave disappointing results in clinical trials. It was realized that the inhibitory effect on virus replication was not strong enough to prevent the rapid occurrence of resistant mutants. Dramatic improvements appeared in 1995–1996 due to the design of a new class of potent antiviral drugs, the inhibitors of the viral protease, which prevent the maturation of infectious virions by inhibiting the cleavage of the gag-pol precursor. Another important development was the general acceptance of the concept of combination therapy, in which two, three, or even four inhibitors act synergistically and decrease the emergence of resistants. Finally the application of sensitive molecular techniques (polymerase chain reaction, branched DNA) to quantitate the viral load in plasma was an important tool to monitor the antiviral effect of the treatment.

The combined use of two inhibitors (such as azidothymidine (AZT) and 3-thiodeoxycitidine (3-TC)) of reverse transcriptase with a protease inhibitor (such as Ritonavir, Indinavir, Nelfinavir) results generally, in a strong decrease of viral load within weeks (2–4 logs), a slow but consistent increase of CD4+ cells and an improvement of the patient's condition.

Opportunistic infections are much less frequent and there is biological evidence of at least partial restoration of T helper-dependent immune functions (response to recall antigens, drop of apoptosis and cell activation markers). However, this treatment called HAART (highly active retroviral therapy) has some constraints and limitations.

1. It has to be taken daily and indefinitely.
2. Active virus multiplication is not completely suppressed, and multiresistant variants may appear, more rapidly if the patient does not adhere to strict compliance to the treatment.
3. There is a continuously renewed reservoir of cells which are latently infected and therefore have unexpressed proviral DNA which escapes the reverse transcriptase and protease inhibitors.
4. Important side effects, such as mobilization of lipids resulting in hypertriglyceridemia, insulin-resistant diabetes, can occur after long-term treatment.
5. The high cost precludes any generalization of its application for patients in developing countries.

Therefore, there is a need for intensive research to develop new types of retroviral inhibitors, to find practical tests to evaluate the spectrum of resistance and sensitivity to available drugs of the patient's virus, to find complementary treatments aimed at reducing the dosage and duration of HAART and at achieving better restoration of the immune system.

Among possible promising developments are the use of: very low doses of IL-2 (a daily subcutaneous inoculation for 6 months); hydroxyurea; and inhibitors of cell activation and a specific immunotherapy against viral proteins.

XI. PREVENTION AND VACCINE

No vaccine is yet available and the only effective prevention at present is education about the ways of transmission, systematic testing of blood donors and the use of condoms. However, an important reduction of transmission from mother to child has been achieved by treatment with AZT of the mother at the end of pregnancy and at the time of delivery and of the newborn in the first weeks of life. A lighter regimen also seems to be effective and applicable to populations of developing countries.

A complete control of the AIDS epidemic cannot be achieved without the availability of a protective vaccine. Although the use of whole virus or whole surface glycoproteins has been disappointing, there are new promising approaches based on the use of DNA, mucosal adjuvants and live vectors, internal and regulatory proteins, and conserved parts of the surface glycoproteins.

Studies of naturally resistant HIV-exposed individuals have shown that cell-mediated immunity and the secretion of soluble inhibitory factors are more important for protection than antibodies, together with a limited role of genetic factors. Individuals homozygous for mutations impairing the expression of CCR5 coreceptors are totally resistant to infections by 'R5 tropic viruses'. However, these are rare (<2%) in the caucasian population. Individuals heterozygous for these mutations can still be infected, but seem to evolve more slowly to the clinical phase of AIDS.

In fact, a large majority of exposed noninfected individuals have acquired a specific immune resistance to infection, based on CTL and specific IgAs, although the role of other genetic factors (HLA) cannot be excluded.

The efficacy trials of candidate vaccine in large populations will raise important logistic and ethical issues difficult to solve, unless an important international mobilization greater than that achieved for vaccinal eradication of poliomyelitis, is achieved.

BIBLIOGRAPHY

Coffin, J. M., Hughes, S. H., and Varmus, H. E. (1998). "Retroviruses." Cold Spring Harbor Laboratory Press, Cold Spring Harbor, NY.

Cohen, O. J., Kinter, A., and Fauci, A. S. (1997). Host factors in the pathogenesis of HIV disease. *Immunol. Rev.* **159**, 31–48.

Hirsch, M. S., and Curran, J. (1996). Human immunodeficiency viruses. *In* "Virology" (B. N. Fields, *et al.*, eds.), 3rd edn, pp. 1953–1975. Lippincott-Raven Publishers, Philadelphia, PA.

Littman, D. R. (1998). Chemokine receptors: key to AIDS pathogenesis? *Cell* **93**, 677–680.

Smith, R. A. (ed.) (1998). "Encyclopedia of AIDS." Fitzray Dearborn Publishers.

52

Identification of bacteria, Computerized

Trevor N. Bryant
University of Southampton

GLOSSARY

classification Orderly arrangement of individuals into units (taxa) on the basis of similarity, each unit (taxon) should be homogenous and different from all others.

identification Matching of an unknown against knowns in a classification, using the minimum number of diagnostic characters.

identification matrix (also known as a **probability matrix**) Rectangular table containing the percentage positive character states for a range of taxa.

identification score Means of expressing the degree of relatedness of an unknown to a known taxon.

probabilistic identification Determination of the likelihood that the observed pattern of results of tests carried out on an unknown bacterium can be attributed to the results of a known taxon within an identification matrix.

taxon General term for any taxonomic group (e.g., strain, species, genus).

Computerized Identification of Bacteria is still largely based on the determination of phenetic characters. These include morphological features, growth requirements, and physiological and biochemical activities. Other phenotypic methods include: analysis of cell wall composition, cellular fatty acids, isoprenoid quinones, whole-cell protein analysis, polyamines; pyrolysis mass spectrometry, Fourier transformation infrared spectroscopy and UV resonance Raman spectroscopy. These have not been incorporated into routine computer based identification systems although computers are used during the analytic process. Recently microbiologists have shown more interest in the use of genotypic methods to establish the taxonomic relationships between bacteria including: DNA base ratio (moles percent C+G), DNA–DNA hybridisation, rRNA homology studies and DNA-based typing methods. Computers are employed in the collection and analysis of this data. Polyphasic identification is the integration of these various techniques for identification of unknown bacteria. Few, if any, computer based polyphasic identification systems have been developed. Most computer-based identification systems use only a subset of taxonomic information available to the bacteriologist for a particular group of bacteria.

We will concentrate on identification systems where the bacteriologist can enter the results of tests carried out on an unknown, obtain a suggested identification and, where identification has not been achieved, obtain a list of additional tests that should be carried out to enable identification.

I. PRINCIPLES OF BACTERIAL IDENTIFICATION

The taxonomy of any group of organisms is based on three sequential stages: classification, nomenclature, and identification. The first two stages are the prime concern of professional taxonomists, but the end product of their studies should be an identification system that is of practical value to others. Therefore, an identification system

The Desk Encyclopedia of Microbiology
ISBN: 0-12-621361-5

is clearly dependent on the accuracy and data content of classification schemes and the predictive value of the name assigned to the defined taxa.

The ideal identification system should contain the minimum number of features required for a correct diagnosis, which is predictive of the other characters of the taxon identified. However, the minimum number of characters required is dependent on both the practical objectives of the exercise and the clarity of the taxa defined in classification. Thus many enterobacteria can be identified using relatively few physiological and biochemical tests, the numerous serotypes of *Salmonella* are recognized by their reactions to specific antisera, and the accurate identification of *Streptomyces* species requires determination of up to 50 diverse characters.

Workers at the Central Public Health Laboratory, United Kingdom demonstrated the first practical computerized identification system in the 1970s using a system for enterobacteria. During the same period, the application of computers for numerical classification was developed. This concept was subsequently applied to many bacterial groups, and these studies provided data that were ideal for the development of computerized identification schemes. Many bacterial taxonomists were slow to realize this potential, but these data now form the basis of many probabilistic identification schemes.

A wide and increasing range of computerized systems for the identification of bacteria is now reported in the scientific literature and allied to commercial kits. This reflects both the expansion of techniques used to determine characters for the classification of bacteria and the rapid developments in computer technology.

II. COMPUTER IDENTIFICATION SYSTEMS

The main approach to the identification of an unknown bacterium involves determination of its relevant characters and the matching of these with an appropriate database that defines known taxa. This database may be known as a probability matrix, or identification matrix. The ideal objective is to assign a name to the unknown that is not only correct but also predictive of some or all its natural characters. Computerized identification schemes provide a more flexible system than those of sequential systems (e.g., dichotomous keys) do.

Computerized identification can be achieved in several ways: numerical codes and probabilistic identification are the most popular approaches. Expert system and neural network have been investigated but their performance is no better than the probabilistic approach.

A. Numerical codes

These are usually based on +/− character reactions. They are applied to a relatively small set of characters that have been selected for their good diagnostic value and are applied to clearly defined taxa. Numerical codes require determination of a series of character states and the conversion of the binary results into a code number that is then accessed against the identification database. Such identification systems are particularly appropriate for the analysis of test results obtained when commercial identification kits are used.

An example is the API 20E kit that generates a unique seven-digit number from a battery of 21 tests (Table 52.1). The tests are divided into groups of three, and the results are coded 1, 2, 4 for a positive result for tests in each group. These values are then used to produce a score that reflects the test results, which can be accessed against the identification system. Organisms that generate profile numbers that are not in the identification system can be tested against appropriate computer assisted probabilistic identification systems. Numerical codes have proved to be

TABLE 52.1 Steps in the use of a system (API 20E) for the rapid computerized identification of a bacterium

			Kit test				
ABC	DEF	GHI	JKL	MNO	PQR	ST	Oxidase test
			Results				
+ − +	+ + +	− − +	− − −	+ − −	− + −	+ +	−
			Values allocated for a positive response				
124	124	124	124	124	124	12	4
			Cumulative scores for groups of three tests				
5	7	4	0	1	2	3	
			Input of scores to computer-based identification matrix				

Modified from Austin, B., and Priest, F. (1986). "Modern Bacterial Taxonomy." Van Nostrand Reinhold, United Kingdom.

convenient and effective, particularly for well-studied groups such as the Enterobacteriaceae.

B. Probabilistic identification

Probabilistic schemes are designed to assess the likelihood of an unknown strain identifying to a known taxon. In theoretical terms, the taxa are treated as hyperspheres in an attribute space (*a-space*) in which the dimensions are the characters. The center of the hypersphere (taxon) is defined by the centroid (the most typical representative), and the critical radius encompasses all the members of each taxon. Ideally, each taxon will be distinct from any others if the identification matrix has been well constructed. To obtain an identification, the diagnostic characters for an unknown strain are determined and its position in the *a-space* calculated. If it falls within the hypersphere (taxon) of a known taxon, it is identified. Thus, in essence, probabilistic identification systems allow for an acceptable number of "deviant" characters in both the known taxa and the unknown strains.

Most computer-assisted identification systems are based on Willcox's implementation of Bayes theorem.

$$P(t_i \mid R) = \frac{P(R \mid t_i)}{\Sigma P(R \mid t_i)}$$

where: $P(t_i \mid R)$ is the probability that an unknown isolate, giving a pattern of test results R, is a member of taxon (group of bacteria) t_i and $P(R \mid t_i)$ is the probability that the unknown has a pattern R given that it is a member of taxon t_i. Bayes theorem incorporates prior probabilities, these are the expected prevelance of strains included in the identification matrix. For bacterial identification most authors give all taxa an equal chance of being isolated and therefore the prior probabilities for all taxa are set to 1.0 and omitted from the equation. The above equation therefore can be re-expressed as:

$$L_i^* = \frac{L_i}{\Sigma L_i}$$

where the probabilities are now referred to as *identification scores*, or Willcox Scores. The identification scores for each taxon are normalised values and L_i^* for all taxa sums to one. Identification of an unknown isolate is achieved when L_i^* for one taxon exceeds a specified threshold value.

An example is shown below with an identification matrix consisting of three taxa for which we have the probabilities for four tests (Table 52.2). An unknown has been isolated whose results for the first three tests are positive, negative and positive, respectively. The likelihoods that the taxa a, b and c will give the pattern of results observed for the unknown is calculated by multiplying the probability of obtaining a positive result for test 1 by the probability of obtaining a negative result for test 2 by the probability of obtaining a positive result for test 3 for each taxon in turn (Table 52.3). The original identification matrix (Table 52.2) only gives the probabilities for positive results, in order to use the probability for a negative result we must subtract the matrix entries for test 2 from 1. The identification scores are expressed as normalized likelihoods in Table 52.4. In this example the unknown is not identified because a single taxon does not reach the identification threshold value. Taxa b and c are still both candidates for the identity of the unknown. Threshold values of 0.999 are typically used, for example with the Enterobacteriaceae, but with other groups of bacteria, such as the streptomycetes, values as low as 0.95 have been used. In practical terms, a value of 0.999 means that the taxon which the unknown identifies with will have at least two test differences from all other taxa in the matrix.

TABLE 52.2 Identification matrix with results of unknown

	Tests			
Taxa	**1**	**2**	**3**	**4**
a	0.01	0.20	0.99	0.90
b	0.95	0.01	0.99	0.01
c	0.99	0.10	0.85	0.99
Results of unknown	+	−	+	Missing

TABLE 52.3 Calculation of likelihood of unknown

Taxa	1		2		3		Likelihood
a	0.01	*	(1 − 0.20)	*	0.99	=	0.00792
b	0.95	*	(1 − 0.01)	*	0.99	=	0.93110
c	0.99	*	(1 − 0.10)	*	0.85	=	0.75735
					Sum	=	1.69637

TABLE 52.4 Willcox probabilities (normalized likelihoods)

Taxa			Identification score
a	0.00792/1.69637	=	0.004669
b	0.93110/1.69637	=	0.548877
c	0.75735/1.69637	=	0.446455
	Sum	=	1.000000

Whatever type of identification system is used, there are four possible outcomes:

- The unknown is identified with the correct taxon.
- The unknown is misidentified, that is, incorrectly attributed to wrong taxon.
- The unknown is not identified at all, and correctly so because the taxon to which it belongs is not present in the matrix.
- The unknown is not identified, but should have been identified with a taxon that is present in the matrix.

It is important that any system deals with these possibilities, although the last one is difficult to resolve. One problem with the identification score is that if an unknown is not represented in the matrix, but one strain within the matrix is closer to it (in *a-space*) than all others, the unknown may be identified as this strain. This is where additional criteria should be used to assist the identification process. These include, listing the differences in test results between the unknown and the strain it has been identified as, as well as the use of other numeric criteria such as taxonomic distance, the standard error of taxonomic distance measures or maximum likelihoods. Taxonomic distance is the distance of an unknown from the centroid of any taxon with which it is being compared; a low score, ideally less than 1.5, indicates relatedness. The standard error of taxonomic distance assumes that the taxa are in hyperspherical normal clusters. An acceptable score is less than 2.0–3.0, and about half the members of a taxon will have negative scores, because they are closer to the centroid than average. The maximum, or best likelihood, is the maximum probability for a taxon calculated using those tests carried out on the unknown. The calculation uses the maximum of the probabilities of a negative and positive result of a test (Table 52.5). This allows for taxa with several entries of 0.50 in a matrix. Some authors calculate the likelihood/maximum likelihood ratio, termed the modal likelihood fraction (see Table 52.6), or it's inverse and use it to decide whether to accept the identification offered by a Willcox score that has exceeded the identification threshold. In the IDENT module of the MICRO-IS program for example, the identification score is not given if the best likelihood/likelihood is greater than 100.

TABLE 52.5 Maximum possible likelihoods

Taxa	1		2		3		Best likelihood
a	(1 − 0.01)	*	(1 − 0.20)	*	0.99	=	0.78408
b	0.95	*	(1 − 0.01)	*	0.99	=	0.93110
c	0.99	*	(1 − 0.10)	*	0.85	=	0.75735

TABLE 52.6 Modal likelihood fraction

Taxa			Modal likelihood
a	0.00792/0.78408	=	0.010101
b	0.93110/0.93110	=	1.000000
c	0.75735/0.75735	=	1.000000

III. GENERATION OF IDENTIFICATION MATRICES

One of three approaches can be used to generate identification matrices.

Cluster analysis. A cluster analysis of fresh isolates and reference strains (taxa) is carried out. Phena (clusters of taxa) are selected from a dendrogram and their properties are summarised to create a starting or frequency matrix containing all characters used in the study. The identification matrix is developed by including the most useful characters and rejecting those that do not distinguish between phena. Some clusters produced by the cluster analysis are omitted from the starting matrix because they contain few members, one, two or three isolates, or comprise a group that cannot be identified.

Grouping known strains. Known strains are characterized using a range of tests and the percentage of each strain exhibiting a character calculated. The identification of each strain is assumed to be accurate and no taxonomic analysis is carried out. The problem with this approach is that isolates that have traditionally been treated as strains of one species might be a grouping of two or more species. If characterisation is carried out on strains that have been repeatedly subcultured these strains may show less metabolic activity compared to fresh isolates. This could result in a matrix that is biased against fresh isolates.

Data collected from the literature. This is the least reliable approach. Data is collected from a variety of publications and merged to create a single matrix. The authors must resolve any conflicting test results and assign probabilities for positive, negative and variable results. If the characterization methods are not adequately described in the literature, some characters may be misinterpreted and the results matrix may contain erroneous probabilities.

Whatever method is used to create the probability matrix, it is important that the characterization of unknowns is performed using the same techniques. For example, it would be inappropriate to create a matrix using miniaturized tests and use it for the identification of an unknown characterized using conventional tests.

A. Selection of characters

Whatever the size and scope of the frequency matrix, by no means all the characters used will have sufficient diagnostic value for use in an identification matrix. Therefore, the major task is to determine a minimal battery of reliable tests that will distinguish between the taxa.

The ideal diagnostic character is one that is consistently positive or negative within one taxon, and this differentiates it from most of all selected taxa. This is seldom achieved with one character, but selected groups of characters may approach this ideal. How far this is achieved depends on the consistency of the taxa studied and the objectives of the identification, which in turn influence the principles and methods used to select characters and to identify unknown strains.

Few characters can be regarded as entirely constant. Character variation may be real (e.g., strain variation) or occur from experimental error. Many tests and observations are difficult to standardize completely within or between laboratories. Therefore, any identification system should ideally take account of these sources of variation.

The first stage is to check the quality of the initial frequency matrix before proceeding further. Two criteria of particular relevance are: (a) the homogeneity of the taxa; and (b) the degree of separation or minimal overlap between them. These can be assessed using appropriate statistics calculated by the OVERMAT and OUTLIER programs.

Once it can be assumed that the frequency matrix is sound, the next step is to select the minimum number of characters from it that are required for the distinction between all the taxa included. There is some controversy about the minimum number of tests needed to separate a range of taxa effectively. One guideline is that the number of tests should at least equal the number of taxa. This may apply to relatively small, and tightly defined taxa, particularly for genera rather than species. However, for many taxa (e.g., *Bacillus*, *Clostridium*, Enterobacteriaceae, and *Streptomyces*), the large number of species would necessitate use of an excessive number of characters. Most of these problems can be solved if: (a) the aims of the identification exercise are clear; and (b) the selection of characters is approached objectively.

The ideal diagnostic character should be always positive for 50% of the taxa in the matrix and always negative for 50% of taxa in the matrix. Characters that are either always positive or always negative are clearly of no diagnostic value, as are those that have a frequency of 50% within all or most taxa. Various separation indices have been devised that can rank characters in order of their diagnostic value. The CHARSEP program incorporates several of the indices and provides a useful means for selection characters. The use of one index, the variance separation potential (VSP), is illustrated in Table 52.7; this index is based on the variance within taxa multiplied by separation potential. Values greater than 25% indicate acceptable characters.

Another program (DIACHAR) ranks characters according to their diagnostic potential. The diagnostic scores of each character for each group in a frequency matrix are ordered and the sum of scores for all selected characters in each group (Table 52.8) is also provided. The higher the score, the greater the diagnostic value of the selected characters. Sometimes, it is desirable to select a few characters that, although of

TABLE 52.7 Example of the use of the CHARSEP program to determine the most diagnostic characters of *Streptomyces* species

Characters	No. of taxa in which character is predominantly		VSP index (%)
	+ve	−ve	
Good diagnostic characters			
Resistance to phenol (0.1% w/v)	6	8	55.8
Spiral spore chains	7	6	54.9
Degredation of lecithin	13	4	48.6
Antiboisis to *Bacillus subtilis*	4	5	44.5
Poor diagnostic characters			
Production of blue pigment	22	0	0.07
Blue spore mass	22	0	0.20
Spore surface with hairy appendages	21	0	0.32
Use of l-arginine	0	19	1.32
Proteolysis	0	14	3.02

From Williams, S. T., Goodfellow, M., Wellington, E. M. H., Vickers, I. C., Alderson, G., Sneath, P. H. A., Sackin, M. J., and Mortimer, A. M. (1983). *J. Gen. Microbiol.* **129**, 1815–1830.

TABLE 52.8 Examples of the use of the DIACHAR program to evaluate diagnostic scores for characters of *Streptoverticillium* species

Species	Sum of scores
Streptoverticillium olivorteticuli	20.34
S. salmonis	19.01
S. ladakanum	18.81
S. hachijoense	18.1
S. abikoense	16.64
S. mobaraense	14.19

From Williams, S. T., Locci, R., Vickers, J., Schofield, G. G., Sneath, P. H. A., and Mortimer, A. M. (1985). *J. Gen Microbiol.* **131**, 1681–1689.

low overall separation potential, are shown by DIACHAR to be diagnostic for a particular taxon. The program BEST uses similar methods to identify useful tests and can create an identification matrix from the frequency matrix.

IV. EVALUATION OF IDENTIFICATION MATRICES

Once an identification matrix has been constructed, it is important that its diagnostic value is assessed before it is recommended for use by microbiologists who are not necessarily expert taxonomists. Matrices can be evaluated by both theoretical and practical means.

A. Theoretical evaluation

Evaluation of the matrix typically consists of determining the identification scores of the hypothetical median organism of each taxon in the matrix. This provides the best possible identification scores for each taxon included in the matrix. If any of these scores are unsatisfactory, practical identification of unknowns against such taxa will inevitably be unreliable. The MOSTTYP program does this using the Willcox probability, taxonomic distance, and the standard error of taxonomic distance as identification coefficients. The IDSC performs similar calculations and where test probabilities of 0.50 are encountered, three scores are calculated for positive, negative and missing test results.

B. Practical evaluation

Practical assessment involves entering the diagnostic character states of known taxa to the matrix. This should involve the redetermination of the diagnostic characters of a random selection of taxon representatives that have been included in the construction of both the frequency and identification matrices. It provides another assessment of experimental error in the determination of character states and its impact on the identification system. The selected representatives should then identify closely to their taxon when their identification coefficients are determined against the matrix. With a well-constructed matrix, bad identification scores are rare, but when they occur they may reflect the random choice of an atypical representative of a poorly defined taxon rather than experimental error.

The final practical evaluation of a matrix clearly involves assessment of its success in identifying unknown strains. Therefore, the appropriate characters for unknowns are determined, and their identification scores are determined and assessed by the investigator.

To date, most probabilistic identification systems have been tested against and applied to natural or "wild" isolates. However, there is an increasing use of genetic manipulation of such strains for scientific, medical, ecological, and industrial purposes. For a variety of reasons, not least patent laws, it is important to compare manipulated strains with each other and with their wild types. This is still a developing area in bacterial taxonomy, but probabilistic systems can be useful. For example, streptomycete strains that had been manipulated by various means, such as mutagens, plasmid transfer, and genetic recombination, were compared with their parent strains against an identification matrix. Most of the manipulated strains identified to the same species as their parents, indicating that most of the selected diagnostic characters were unchanged and that the identification matrix could accommodate some character state changes.

C. Computer software

Most of the programs mentioned above were developed by Prof. P.H.A. Sneath (Department of Microbiology, Leicester University, United Kingdom.) They are written in BASIC, and full details of the programs can be obtained from the following publications. Those by T.N. Bryant can be obtained from *www.som.soton.ac.uk/staff/tnb/pib.htm*

Bryant, T.N. (1987). *Computer Applications in the Biosciences* **3**, 45–48.

Bryant, T.N. (1991). *Computer Applications in the Biosciences* **7**, 189–193.

Sneath, P.H.A. (1979). MATIDEN program. *Computers & Geosciences* **5**, 195–213.

Sneath, P.H.A. (1979). CHARSEP program. *Computers & Geosciences* **5**, 349–357.

Sneath, P.H.A. (1980). DIACHAR program. *Computers & Geosciences* **6**, 21–26.

Sneath, P.H.A. (1980). MOSTTYP program. *Computers & Geosciences* **6**, 27–34.

Sneath, P.H.A. (1980). OVERMAT program. *Computers & Geosciences* **6**, 267–278.

Sneath, P.H.A., and Langham, C.D. (1989). OUTLIER program. *Computers Geosci.* **15**, 939–964.

Sneath, P.H.A., and Sackin, M.J. (1979). IDEFORM program. *Computers Geosci.* **5**, 359–367.

D. Published identification matrices

Many identification matices have been published (see Table 52.9). Most of these have been developed using

TABLE 52.9 Published identification matrices

Species/Groups covered	Taxa	Tests	Authors
Gram-negative, aerobic, non-fermenters	66	83	Holmes, B., Pinning, C.A. and Dawson, C.A. (1986) *Journal of General Microbiology* **132**, 1827–1842.
Gram-negative, aerobic rod-shaped fermenters	110	66	Holmes, B., Dawson, C.A. and Pinning, C.A. (1986) *Journal of General Microbiology* **132**, 3113–3135.
Enterobacteriaceae	41	16	Clayton, P., Feltham, R.K.A., Mitchell, C.J. and Sneath, P.H.A. (1986) *Journal of Clinical Pathology* **39**, 798–802.
Enterobacteriaceae	90	50	Holmes, B. and Costas, M. (1992). In: Board, R.G., Jones, D. and Skinner, F.A. (Eds.) *Identification Methods in Applied and Environmental Microbiology*, pp. 127–149. Oxford: Blackwell Scientific Publications.
Pseudomonas species	26	70	Costas, M., Holmes, B., On, S.L.W. and Stead, D.E. (1992) In: Board, R.G., Jones, D. and Skinner, F.A. (Eds.) *Identification Methods in Applied and Environmental Microbiology*, pp. 1–27. Oxford: Blackwell Scientific Publications.
Vibrios	31	50	Dawson, C.A. and Sneath, P.H.A. (1985) *Journal of Applied Bacteriology* **58**, 407–423.
Vibrio and related species	38	81	Bryant, T.N., Lee, J.V., West, P.A. and Colwell, R.R. (1986) *Journal of Applied Bacteriology* **61**, 469–480.
Aeromonas species	14	30	Kämpfer, P. and Altwegg, M. (1992) *Journal of Applied Bacteriology* **72**, 341–351.
Aeromonas hybridization groups	15	32	Oakley, H.J., Ellis, J.E. and Gibson, L.F. (1996) *Zentralblatt fur bakteriologie—International Journal of Medical Microbiology Virology Parasitology and Infectious Diseases* **284**, 32–46.
Campylobacteria	23	42	Holmes, B., On, S.L.W., Ganner, M. and Costas, M. (1992) In: Schindler, J. (Ed.) *Proceedings of the Conference on Taxonomy and Automated Identification of Bacteria*, pp. 6–9. Prague.
Campyylobacter, Helicobacter and others	37	67	On, S.L.W., Holmes, B. and Sackin, M.J. (1996) *Journal of Applied Bacteriology* **81**, 425–432.
Lactic acid bacteria	37	49	Cox, R.P. and Thomsen, J.K. (1990) *Letters in Applied Microbiology* **10**, 257–259.
Lactic acid bacteria	59	27	Döring, B., Ehrhardt, S., Lücke, F. and Schillinger, U. (1988) *Systematic and Applied Microbiology* **11**, 67–74.
Lactic acid bacteria	11	53	Maissin, R., Bernard, A., Duquenne, V., Baeten, S., Gerard, G. and Decallonne, J. (1987) *Belgian Journal of Food Chemistry and Biotechnology* **42**, 176–183.
Streptomyces species	23	41	Williams, S.T., Goodfellow, M., Wellington, E.M.H., Vickers, J.C., Alderson, G., Sneath, P.H.A., Sackin, M.J. and Mortimer, A.M. (1983) *Journal of General Microbiology* **129**, 1815–1830.
Streptomyces species	26	50	Langham, C.D., Williams, S.T., Sneath, P.H.A. and Mortimer, A.M.
(minor clusters)	28	39	(1989) *Journal of General Microbiology* **135**, 121–133.
Streptomyces species	52	50	Kämpfer, P. and Kroppenstedt, R.M. (1991) *Journal of General Microbiology* **137**, 1893–1902.
Streptoverticillum species	24	41	Williams, S.T., Locci, R., Vickers, J.C., Schofield, G.M., Sneath, P.H.A. and Mortimer, A.M. (1985) *Journal of General Microbiology* **131**, 1681–1689.
Actinomyces species	25	53	Seong, C.N., Park, S.K., Goodfellow, M., Kim, S.B., and Hah Y.C. (1995) *Journal of Microbiology* **33**, 95–102.
Actinomyces species	20	18	Sarkonen, N., Könnönen, E., Summanen, P., Könnönen, M., and Jousimies-Somer, H., (2001) *Journal of Clinical Microbiology* **39**, 3955–3961.
Actinoplanes species	3	19	Long, P.F. (1994) *Journal of Industrial Microbiology* **13**, 300–310.
Phytopathogenic corynebacteria (subclusters)	5	27	Firrao, G. and Locci, R. (1989) *Annuals of Microbiology* **39**, 81–92.
Coryneform bacteria	31	58	Kämpfer, P. and Seiler, H. (1993) *Journal of General and Applied Microbiology* **39**, 215–236.
Nocardioform bacteria	25	35	Kämpfer, P., Dott, W. and Kroppenstedt, R.M. (1990) *Journal of General and Applied Microbiology* **36**, 309–301.
Slowly growing mycobacteria	24	33	Wayne, L.G., Good, R.C., Krichevsky, M.I., *et al.*, (1991) *International Journal of Systematic Bacteriology* **41**, 463–472.
Non-tuberculous mycobacteria	27	23	Tortoli, E., Boddi, V. and Penati, V. (1992) *Binary—Computing in Microbiology* **4**, 200–203.
Bacillus species	44	30	Priest, F.G. and Alexander, B. (1988) *Journal of General Microbiology* **134**, 3011–3018.
Bacillus sphaericus	14	29	Alexander, B. and Priest, F.G. (1990) *Journal of General Microbiology* **136**, 367–376.
Bacillus species	36	44	Kämpfer, P. (1991) *Journal of General and Applied Microbiology* **37**, 225–247.
Aerobic gram-positive cocci catalase positive	39	60	Feltham, R.K.A. and Sneath, P.H.A. (1982) *Journal of General Microbiology* **128**, 713–120.

TABLE 52.9 *Continued*

Species/Groups covered	Taxa	Tests	Authors
Aerobic gram-positive cocci catalase negative	33	60	Feltham, R.K.A. and Sneath, P.H.A. (1982) *Journal of General Microbiology* **128**, 713–120.
Micrococcus	29	35	Alderson, G., Amadi, E.N., Pulverer, G. and Zai, S. (1991) In: Jeljaszewicz/ Ciborowski, (Ed.) *The Staphylococci, Zentralblatt für Bakteriologie Supplement 21*, pp. 103–109. Stuggart: Gustav Fisher Verlag.
Staphylococcus species	12	15	Geary, C., Stevens, M., Sneath, P.H.A. and Mitchell, C.J. (1989) *Journal of Clinical Pathology* **42**, 289–294.
Alaskan marine bacteria	86	61	Davis, A.W., Atlas, R.M. and Krichevsky, M.I. (1983) *International Journal of Systematic Bacteriology* **33**, 803–810.
Medical bacteria to genus level	60	20	Feltham, R.K.A., Wood, P.A. and Sneath, P.H.A. (1984) *Journal of Applied Bacteriology* **57**, 279–290.

the procedures described above. Success rates vary with the bacterial group under study. Probabilistic identification of gram-negative bacteria has been most effective, for example, 933 (98.2%) isolates of fermentative gram-negative bacteria and 621 (91.5%) isolates of nonfermenters were identified using a Willcox probability threshold of greater than 0.999. Of 243 vibrios isolated from freshwater, 71.6% were identified at a level greater than 0.999 and 79.4% at greater than 0.990.

When such stringent coefficient levels are applied to gram-positive bacteria, the results are often less impressive. For example, when a probability of greater than 0.999 was applied to coryneform bacteria only 50% of the unknowns identified and using a level of greater than 0.995, only 42% of streptomycete isolates were identified. If less stringent coefficients are applied to take account of the heterogeneity of such groups, a higher rate of useful identifications can be achieved. Thus, 73% of 153 streptomycete isolates were identified using a Willcox probability of greater than 0.85.

V. BACTERIAL IDENTIFICATION SOFTWARE

Several probabilistic identification programs have been published, some are available as interactive web pages although whether these pages offer any advantage over software installed on a user's machine is questionable. A list of resources is presented below. In many instances the software can be used with various identification matrices. Examples from one program, PIBWin (a Windows version of Bacterial Identifier) are shown in Figs 52.1–52.3.

BBACTID Bryant, T. N., Capey, A. G., and Berkeley, R. C. W. (1985). *Computer Applications in the Biosciences* **1**, 23–27.

BACTID Jilly, B. J. (1988). *International Journal of Biomedical Computing* **22**, 107–119.

CIBAC Döring, B., Ehrhardt, S., Lücke, F., and Schillinger, U. (1988). *Systematic and Applied Microbiology* **11**, 67–74.

Gideon *www.cyinfo.com*

IDENTIFY Jahnke, K. D. (1995). *Journal of Microbiological Methods* **21**, 133–142.

MICRO-IS Portyrata, D. A., and Krichevsky, M. I. (1992) *Binary* **4**, 31–36. *www.bioint.org/support/microis/microis.html*

MATIDEN Sneath, P. H. A. (1979). *Computers & Geosciences* **5**, 195–213.

Identmpm Maradona, M. P. (1994). *Computer Applications in the Biosciences* **10**, 71–73.

no-name Tortoli, E., Boddi, V. and Penati, V. (1992). *Binary* **4**, 200–203.

The Identifier Gibson, L. F., Clarke, C. J., and Khoury, J. T. (1992). *Binary* **4**, 25–30.

PIBWin *www.som.soton.ac.uk/staff/tnb/pib.htm*
(Bacterial Identifier)

Recognet *www.pasteur.fr/recherche/banques/recognet*

VI. OTHER APPLICATIONS OF COMPUTERS TO BACTERIAL IDENTIFICATION

We have concentrated on the use of computers to construct identification systems as well as to access them. However, Computer programs are increasingly used solely to access taxonomic data for the identification of unknown strains. Developments in computer technology can also provide a more direct link between the determination of test results and their evaluation. An example is the use of so-called "breathprints," for identification of gram-negative, aerobic bacteria. This relies on a redox dye to detect the increased respiration

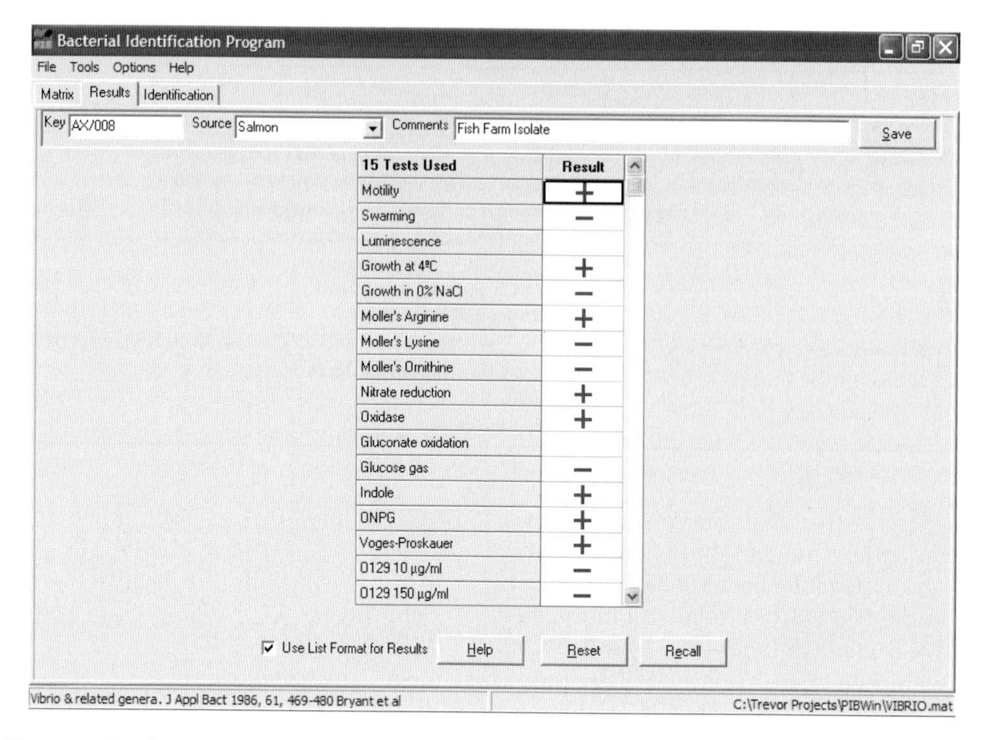

FIGURE 52.1 The screen for shows results for an isolate from Salmon reared in a fish farm entered into the PIBWin program.

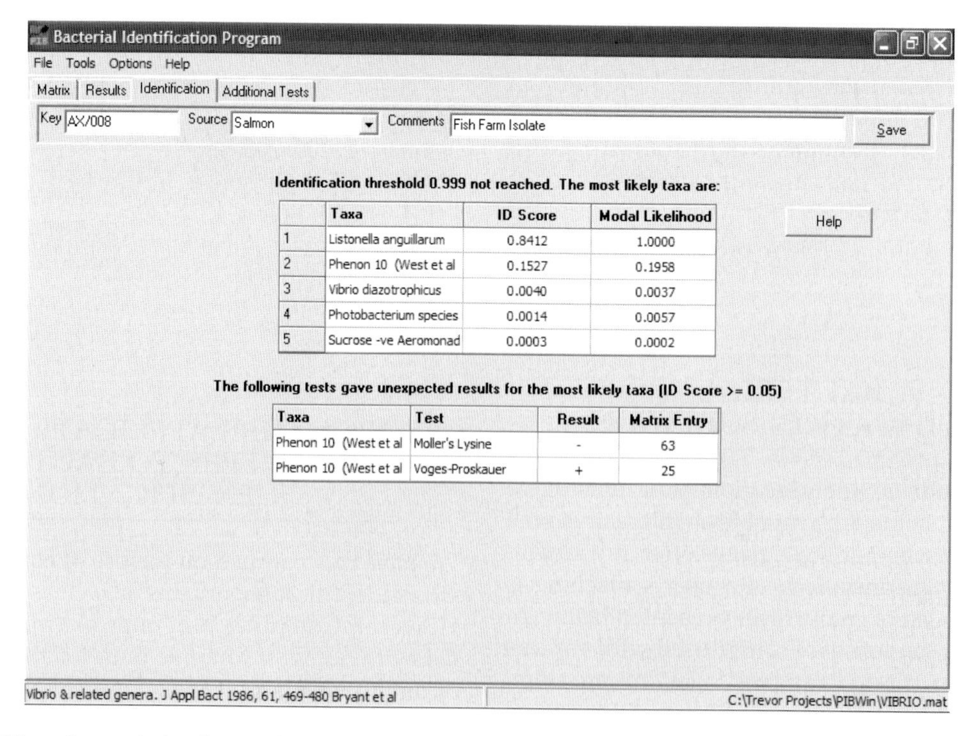

FIGURE 52.2 The unknown isolate has not been identified as no taxon exceeded the threshold of 0.999. The five possible taxa are listed by PIBWin, although three have very low identification scores (Willcox probabilities) and modal likelihood scores. Differences between test results for the unknown and likely taxa with an ID score ≥ 0.05 are listed.

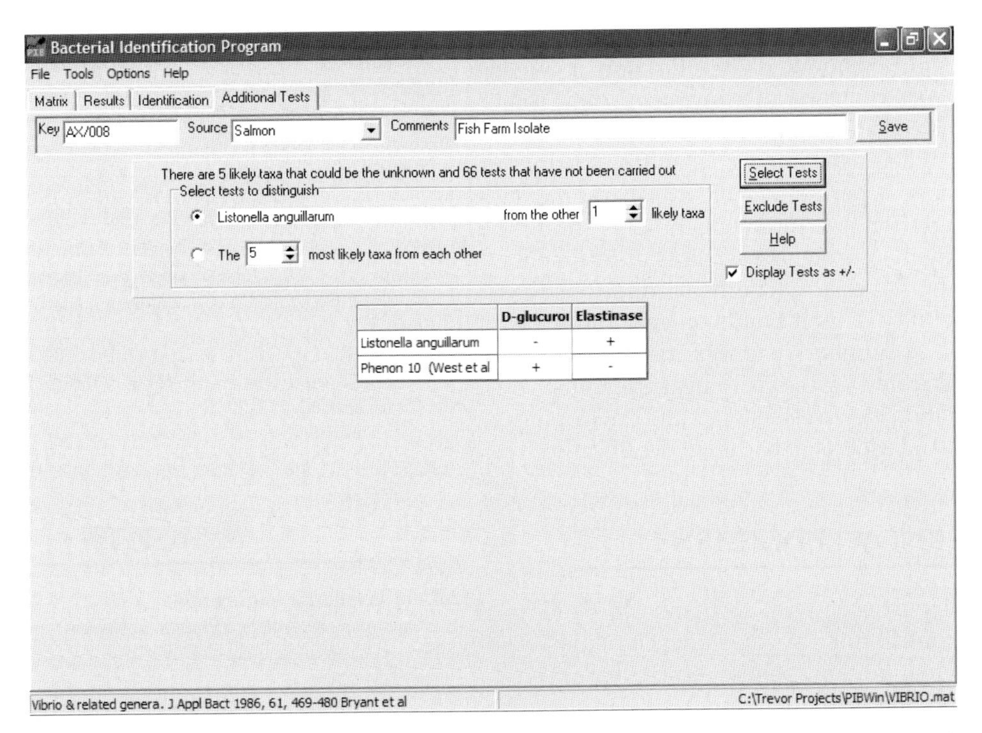

FIGURE 52.3 The PIBWin program has suggested two additional tests that will distinguish between the most likely, *Listonella anguiliarum*, and the next most likely identity, Phenon 10 (West *et al.*, 1986), for the unknown.

when a carbon source is oxidized. A range of substrates in a microtiter plate are inoculated with a strain; if a substrate is used, a pigment is formed by the redox dye, indicating a positive reaction. The pattern of these on the plate provides a breathprint, which can then be compared with those of known taxa using a system consisting of a microplate reader and a computer.

Databases for both the classification and identification of bacteria have been extended and improved by the inclusion of diagnostic characters provided by chemical analysis of cell components. These include cell wall amino acids, membrane lipids, and proteins, which have been particularly useful for the definition and identification of higher taxa such as genera or families. For example:

Analysis of bacterial lipids involves gas chromatography, which results in a printout of a set of peaks that are defined chemically but are difficult to evaluate and compare quantitatively by eye. Various programs have been used to transform the data for principal component analysis and to provide similarity and overlap coefficients for comparison of unknown strains.

The use of polyacrylamide gel electrophoresis (PAGE) to analyze protein patterns of bacteria is well established in bacterial taxonomy. The results are obtained in the form of stained bands on the gels. A variety of computer programs has been devised to facilitate their analysis. Typically, the stained gels are scanned by a computer controlled densitometer. This digitizes the continuous output of the densitometer scan, removes or corrects for background effects, and permits comparison with other stored traces. Thus, an unknown can be compared with known taxa using various similarity coefficients.

Another method of bacterial identification is pyrolysis-mass spectrometry. This involves the thermal degradation of a small sample of cells in an inert atmosphere or vacuum, leading to production of volatile fragments. Under controlled conditions, these are characteristic of a taxon, and they are separated and analyzed in a mass spectrometer. Assessment of the traces obtained requires software for performing principal components, discriminant, and cluster analysis, to allow known taxa to be distinguished and unknowns identified.

The developments in nucleic acid techniques are having a marked and exciting impact on bacterial taxonomy, where they provide a genetic assessment of taxa, which can be used to supplement or revise the existing phenetic systems. Techniques such as DNA reassociation, DNA–rRNA hybridization, and DNA and RNA sequencing are increasingly used in bacterial classification, whereas nucleic acid proves and DNA fingerprints are of great potential in identification. Computation is used in the analysis and application of such data. Despite the relative novelty of these sources of taxonomic data, ultimately a convenient and accurate means of comparing data for unknown strains with those of established taxa is still required. Thus, when determining DNA fingerprints it is useful to have a permanent record of fragment size. Precise migration measurements on gels should not be spoiled by inaccurate assessments of fragment size. Fragments have a curvilinear relationship between the mobility of their bands on gels and their molecular sizes. A number of programs have been devised to transfer and assess such measurements.

Thus, computation has an established and developing role in all stages and aspects of bacterial identification.

ACKNOWLEDGEMENT

The author and publisher wish to thank Dr. Stanley T. Williams, contributor to the Encyclopedia of Microbiology, First Edition, whose article has been revised to produce this text.

BIBLIOGRAPHY

Bryant, T. N. (1998). Probabilistic identification systems for bacteria. *In* "Information Technology, Plant Pathology and Biodiversity" (P. Bridge, P. Jeffries, D. R. Merse, and P. R. Scott, eds.), pp. 315–332. CAB International.

Canhos, V. P., Manfio, G. P., and Blaine, L. D. (1993). Software tools and databases for bacterial systematics and their dissemination via global networks. *Antonie van Leeuwenhoek, Int. J. Gen. Mol. Microbiol.* **64**, 205–229.

Langham, C. D., Williams, S. T., Sneath, P. H. A., and Mortimer, A. M. (1989). *J. Gen. Microbiol.* **135**, 121–133.

Langham, C. D., Sneath, P. H. A., Williams, S. T., and Mortimer, A. M. (1989). *J. Appl. Bacteriol.* **66**, 339–352.

Priest, F. G., and Williams, S. T. (1993) Computer-assisted identification. *In* "Handbook of New Bacterial Systematics", (M. Goodfellow and A. G. O'Donnell, eds.), 361–381. Academic Press, London.

Willcox, W. R., Lapage, S. P., and Holmes, B. (1980). A review of numerical methods in bacterial identification. *Antonie van Leeuwenhoek* **46**, 233–296.

WEBSITES

Links to Taxonomic information
www.som.soton.ac.uk/staff/tnb/microbio.htm

Approved Lists of Bacterial Names
www.dsmz.de

List of bacterial names with standing in nomenclature
www-bacterio.cict.fr

Bergey's Manual Trust
www.cme.msu.edu/bergeys/

Society for General Microbiology
www.socgenmicrobiol.org.uk/links/

Culture Collection, University of Göteborg, Sweden
www.ccug.gu.se/

A Bibliography of Bacterial Identification Matrices
www.som.soton.ac.uk/staff/tnb/pib.htm

53

Industrial fermentation processes

Thomas M. Anderson

Archer Daniels Midland BioProducts Plant

GLOSSARY

aseptic In medicine, referring to the absence of infectious organisms. In fermentation, referring to the absence of contaminating microorganisms.

auxotrophy The inability of a culture to produce an essential component or growth requirement.

bacteriophage Viruses that attack bacteria and destroy a fermentation.

bioreactor A closed vessel used for fermentation or enzyme reaction.

BOD Biological oxygen demand; the amount of oxygen required by microorganisms to carry out oxidative metabolism in water containing organic matter, such as sewage.

mass transfer The movement of a given amount of fluid or the material carried by the fluid. In bioreactors, it refers to the dispersal and solubilization of nutrients and gases in the fermentation medium.

scale up To take a process from the laboratory to full scale production, usually through intermediate steps.

sterilization The removal of all microorganisms from a material or space. Standard conditions for steam sterilization are 121 °C, 15 psig for 15 minutes.

Many Products are made by large scale fermentation today, including amino acids, enzymes, organic acids, vitamins, antibiotics, solvents and fuels (Table 53.1).

There are several advantages to making products by fermentation:

- Complex molecules that occur naturally, such as antibiotics, enzymes, and vitamins, are impossible to produce chemically.

TABLE 53.1 Fermentation products

Category	Examples	Function
Organic acids	Lactic, citric, acetic, formic, acrylic	Food acidulants, textiles, tanning, cleaners, intermediate for plastics and organic chemicals
Organic chemicals	Ethanol, glycerol, acetone, butanediol, propylene glycol	Solvents, intermediates for plastics, rubber and chemicals, antifreeze, cosmetics, explosives
Amino acids	MSG, L-Lysine, L-tryptophane, L-phenylalanine	Animal feeds, flavors, sweetners
Enzymes	Amylases, cellulases, glucose isomerase, proteases, lipases	Grain processing, cheesemaking, tanning, juice processing, high fructose corn syrup
Antibiotics	Penicillins, streptomycin, tetracycline, chloramphenicol	Human and animal health care
Bioplymers	Xanthan, dextran, poly-β-hydroxybutyrate	Stabilizers and thickeners in foods, oil well drilling, plastics
Vitamins	Vitamin B_{12}, biotin, riboflavin	Nutritional supplements, animal feed additives
Cell mass	Yeast, lactic acid bacteria	Single cell protein, baker's and brewer's yeast, starter cultures

The Desk Encyclopedia of Microbiology
ISBN: 0-12-621361-5

- Optically active compounds, such as amino acids and organic acids, are difficult and costly to prepare chemically.
- "Natural" products that can be economically derived by chemical processes, but, for food purposes, are better produced by fermentation, such as beverage ethanol and vinegar (acetic acid).
- Fermentation usually uses renewable feedstocks instead of petrochemicals.
- Reaction conditions are mild, in aqueous media, and most reaction steps occur in one vessel.
- The by-products of fermentation are usually environmentally benign compared to the organic chemicals and reaction by-products of chemical manufacturing. Often, the cell mass and other major by-products are highly nutritious and can be used in animal feeds.

There are, of course, drawbacks to fermentation processes:

- The products are made in complex solutions at low concentrations compared to chemically derived compounds.
- It is difficult and costly to purify the product.
- Microbial processes are much slower than chemical processes, increasing the fixed costs of the process.
- Microbial processes are subject to contamination by competing microbes, requiring the sterilization of the raw materials and the containment of the process to avoid contamination.
- Most microorganisms do not tolerate wide variations in temperature and pH and are also sensitive to upsets in the oxygen and nutrient levels. Such upsets not only slow the process, but are often fatal to the microorganisms. Thus, careful control of pH, nutrients, air, and agitation requires close monitoring and control.
- Although nontoxic, the waste products are high in BOD, requiring extensive sewage treatment.

Other processes considered large-scale microbial processes are industrial or municipal sewage treatment, bioremediation of contaminated soil and water, and bioleaching of metal ores. Among the largest fermentation facilities are sewage treatment plants, consisting of anaerobic primary treatment and aerobic secondary treatment facilities. Bioremediation of environmental contamination is an emerging technology. Initially, native consortia of microorganisms were used to degrade contaminants. Now, cultures are selected, developed, and grown to more rapidly rehabilitate contaminated environmental sites.

The vast majority of large-scale fermentations use bacteria, yeast and fungi, but some processes use algae, plant, and animal cells. Several cellular activities contribute to fermentation products:

- Primary metabolites: ethanol, lactic acid, and acetic acid.
- Energy storage compounds: glycerol, polymers, and polysaccharides.
- Proteins: extracellular and intracellular enzymes, single-cell proteins and foreign proteins.
- Intermediary metabolites: amino acids, citric acid, vitamins, and malic acid.
- Secondary metabolites: antibiotics.
- Whole-cell products: single-cell protein, baker's yeast and brewer's yeast, bioinsecticides.

Fermentation products can be growth associated or non-growth associated. Primary metabolites, such as ethanol and lactic acid, are generally growth associated, as are cell mass products, while secondary metabolites, energy storage compounds, and polymers are non-growth associated. Other products, such as proteins, depend on the cellular or metabolic function.

I. HISTORY

Humans have used fermentation from the beginning of recorded history to provide products for everyday use. For many centuries, most microbial processing was to preserve or alter food products for human consumption. Fermentation of grains or fruit produced bread, beer, and wine that retained much of the nutrition of the raw materials, while keeping the product from spoiling. The natural yeasts that caused the fermentation added some vitamins and other nutrients to the bread or beverage. Lactic acid bacteria fermented milk to yogurt and cheeses, extending the life of milk products. Other food products were preserved or enhanced in flavor by fermentation, such as pickled vegetables and the fermentation of tea leaves and coffee beans.

Fermentation was an art until the second half of the 19th century. A batch was begun with either a "starter," a small portion of the previous batch, or with the cultures residing in the product or vessel. The idea that microbes were responsible for fermentations was not introduced until 1857, when Louis Pasteur published a paper describing the cause of failed industrial alcohol fermentations. He also quantitatively described microbial growth and metabolism for the first time and suggested heat treatment (pasteurization) to improve the storage quality of wines. This was the first step toward sterilization of a fermentation medium to control fermentation conditions. In 1883,

Emil Christian Hansen began using pure yeast cultures for beer production in Denmark. Beer and wine were produced at relatively large scale starting in the eighteenth century to satisfy the demands of growing urban populations. In the mid-nineteenth century, the introduction of denaturation freed ethanol from the heavy beverage tax burden so that it could be used as an industrial solvent and fuel.

The first aseptic fermentation on a large scale was the acetone–butanol fermentation, which both Britain and Germany pursued in the years preceding World War I. The initial objective was to provide butanol as a precursor for butanediol, for use in synthetic rubber. After the beginning of the war, the focus of the process in Britain became acetone, which was used for munitions manufacture. Britain had previously been importing acetone from Germany. One of the people instrumental in the development of the fermentation was Chaim Weizmann, who later became the first prime minister of Israel. As the process was developed and scaled up, it was found that the producing culture, *Clostridium acetobutylicum*, would become overwhelmed by competing bacteria introduced from the raw materials. Thus, the culture medium had to be sterilized and the process run under aseptic conditions. All penetrations on the reaction vessel were steam sealed to prevent contamination. Production eventually took place in Canada and the United States, due to the availability of cheap raw materials.

In the 1920s and 1930s, the emphasis in fermentation shifted to organic acids, primarily, lactic acid and citric acid. In the United States, where prohibition had outlawed alcoholic beverages, facilities and raw materials formerly used for alcoholic beverage production became available. Lactic acid is currently used as an acidulant in foods, a biodegradable solvent in the electronic industry (as ethyl lactate), and a precursor for biodegradable plastics. Citric acid is used in soft drinks, as an acidulant in foods, and as a replacement for phosphates in detergents. As part of an effort to find uses for agricultural products during the depression, the USDA Northern Regional Research Laboratory (NRRL) pioneered the use of surplus corn products, such as corn steep liquor, and the use of submerged fungal cultures in fermentation. Previously, fungi had been grown on solid media or in the surface of liquid media. This set the stage for the large-scale production of penicillin, which was discovered in 1929 in the Britain, developed in the 1930s, and commercialized in 1942 in the United States. It was the first "miracle" drug, routinely curing bacterial infections that had previously caused serious illness or death. The demand was very high during World War II and the years following. Penicillin was initially produced as a surface culture in one-quart milk bottles. The cost, availability, and handling of bottles severely limited the expansion of production. Scientists at the NRRL discovered a new production culture on a moldy cantaloupe and developed a submerged culture fermentation. This led to significant increases in productivity per unit volume and the ability to greatly increase the scale of production by using stirred tank bioreactors. The success of penicillin inspired pharmaceutical companies to launch massive efforts to discover and develop many other antibiotics in the 1940s and 1950s. Most of these fermentations were highly aerobic, requiring high aeration and agitation. As the scale of production increased, it was found that mass transfer became limiting. The field of biochemical engineering emerged as a distinct field at this time, to study mass transfer problems in fermentation and to design large-scale fermentors capable of high transfer rates.

In the 1960s, amino acid fermentations were developed in Japan. Initially, L-glutamic acid, as monosodium glutamate, was produced as a flavor enhancer, to supplant MSG extracted from natural sources. Using cultures derived from glutamic acid bacteria, production of other amino acids followed. Amino acids are used in foods as nutrients, sweeteners, and flavor enhancers, and in animal feeds to increase the efficiency of low protein feeds. Commercial production of enzymes for use in industrial processes began on a large scale in the 1970s as well. Microbial enzymes account for 80% of all enzymes in commercial use, including grain processing, sugar production, juice and wine clarification, detergents, and high fructose corn syrup. The discovery of the tools of genetic engineering expanded the possibilities for products made by fermentation. Insulin was the first genetically engineered fermentation commercialized, developed in 1977. Since then, many genetically engineered products have been produced on a large scale.

II. CULTURE SELECTION AND DEVELOPMENT

A. Selection

Microbial processes begin with the culture used for production. Once the desired product or microbial activity is defined, the selection process begins. A culture should produce, or have the metabolic potential to produce, the desired product. Other attributes are important, such as substrate specificity, growth factor requirements, growth characteristics (pH, temperature, aeration, shear sensitivity), fermentation by-products, effect of the organism on downstream processing, and environmental and health effects. Literature searches can narrow the range of cultures to be screened, saving valuable time in culture selection. Culture collections often have cultures with some or all of the desired characteristics, but it is sometimes necessary to screen cultures from the environment. Determining the type of environment in which organisms with the desired characteristics live and how to separate them from the other cultures present takes considerable care. A thorough understanding of the physiology of the desired microorganism is necessary to design a successful isolation and selection strategy. Successful isolation of a useful culture often requires a combination of several enrichment or selective methods. Also, a system of storing and cataloging potentially useful isolates is very important so that commercially viable cultures are not lost. After an initial screening, there are usually many potentially useful isolates and secondary screening is necessary to eliminate false positives and evaluate the potential of the remaining candidates. The secondary screening is usually semi-quantitative or quantitative. The list of candidates is narrowed, as much as possible, using mass screening methods, such as agar plates with selective growth inhibitors or metabolic indicators. Large-scale screening of individual isolates in shake flask culture is very time consuming and expensive. Once a few isolates have been selected, culture and process development usually begin in parallel to condense the timeline to production.

B. Development

It is very rare that an isolate from the environment produces the desired product cost-effectively. Often, cultures make only minute quantities of the product. Increasing the productivity of the initial isolates requires a program of genetic improvement. Classical mutation and selection methods are used with most cultures selected from the environment because little is known about their genetics and whether the cultures possess cloning vectors such as plasmids, transposons, or temperate bacteriophage. Typically, a mutagen, such as ultraviolet light, ionizing radiation, or a chemical mutagen, is applied and the culture is grown in the presence of a selective growth inhibitor or toxin. The survivors are isolated and tested. This is usually an iterative process; mutant strains are screened, remutagenized, and reselected several times, often using higher concentrations or different selective agents, until a culture with commercial potential is obtained. Even after a process is successfully brought to production, culture improvement is ongoing to improve profitability and maintain a competitive advantage.

In the past two decades, the use of genetic engineering has supplemented, and sometimes replaced, classical genetic techniques. Insertion of genes into plasmids has improved the ease with which genes from one culture are transferred to another and has allowed the production of human and other mammalian proteins, such as insulin and interferons, in microbial fermentations. Plasmids with high copy numbers and strong transcriptional promoters have dramatically improved production of many proteins and enzymes. Knowledge of enzyme structure and function has led to site-specific mutagenesis, increasing the efficiency of mutagenesis programs and reducing the deleterious effects of "shotgun" mutagenesis, used in classical genetics. The introduction of polymerase chain reaction (PCR) technology has led to the ability to genetically sample environments where isolation of cultures is extremely difficult or impossible, such as cold benthic environments barely above freezing, deep sea thermal vents where microorganisms live in temperatures well above 100 °C, or acidic hot springs, such as those in Yellowstone National Park. DNA libraries from these environments can be screened for enzymes that perform under extreme conditions without having to culture the microorganisms.

III. PROCESS DEVELOPMENT AND SCALE-UP

A. Development

Process development usually overlaps culture development. The purpose of process development is the formulation of media, optimization of culture conditions, and determination of the biochemical engineering parameters used to design the full-scale bioreactors. The early stages of process development are usually performed in shake flasks, where the nutritional requirements of the culture are determined and potential media components are screened. Initial growth and production studies can be performed in shake flasks as well. The limitation of shake flask culture is that pH, oxygen content, and other environmental factors cannot be easily monitored and controlled and mass transfer studies are difficult. The next stage in process development, performed in laboratory scale fermentors, is determination of fermentation characteristics, such as pH optimum, oxygen uptake rate, growth and production rates, sensitivity to nutrients and by-products, broth viscosity, heat generation, and shear sensitivity. This information is used to determine what mode of fermentation will be used and to develop the fermentation parameters, as well as to determine the mass transfer characteristics of the fermentation used in the design of the bioreactor. In addition, the medium is developed and optimized at this stage.

A variety of biochemical engineering methods for scale-up of bioreactors have been applied over the years, including constant oxygen transfer rate, constant agitation power per volume, constant impeller tip speed, equal mixing times, or similar momentum factors or feedback control to try to maintain important environmental factors as constant as possible. Each has limitations in predicting the effect of scale-up on the process. It is more difficult to predict how a biological process will react at the commercial scale, based on laboratory and pilot plant studies, than a chemical process. This is due to the complexity of reactions and interactions that occur in a bioreactor. There are often unforseen consequences of changing the scale of operation due to the effect of heat and mass transfer on microorganisms. For example, mixing times in a laboratory or pilot scale fermentor are a few seconds, while in large-scale fermentors, mixing times can be two minutes or more. The average conditions are the same as in a smaller fermentor, but an individual microorganism encounters a variety of suboptimal conditions for a significant period of time. Also, as the fermentor size increases, the heat generation increases proportional to the volume, while the cooling capacity increases proportional to the surface area. Therefore, larger vessels require internal cooling coils to supplement the water jacket. This can aggravate mixing problems further. The properties of the fermentation broth (viscosity, osmotic pressure, substrate, and product and waste product concentrations) and gas/liquid interactions (gas lineal velocity, surface tension, pressure gradients) are also scale dependent. Thus, fermentation scale-up is often highly empirical and based on the experiences and training of the scientists and engineers involved. The cost and risk of scale-up are higher for biological systems than for other chemical systems, due to the intermediate scale pilot plant steps required to successfully predict the outcome of full-scale operations. Some of this cost can be reduced by using existing facilities or rented facilities to test the process or by using seed fermentors (typically, 5–10% as large as production fermentors) for the final pilot plant scale.

An important microbiological factor that affects the scale-up biological systems is the increase in the number of generations required for full-scale operation. Nonproducing variant strains often arise from the parent population. Most commercially used microorganisms are mutated and selected for increased product yield and rates of production. This often decreases the growth rate and hardiness of the culture. Therefore, a variant that either reverts to a previous condition or that short-circuits the selected pathways by additional mutations will have a competitive advantage. The percent of variants in the population in a given generation depends on the rate at which variants appear and the relative growth rate of the variant to the parent population, as expressed by the equation

$$\frac{X_n}{X_n + X_m} = \frac{\alpha + \lambda + 1}{(\alpha - 1) + \lambda N_g^{(\alpha + \lambda + 1)}},$$

where X_n is the number of parent culture cells, X_m is the number of variant culture cells, α is the ratio of the specific growth rate of the variant to the parent, λ is the rate of appearance of variants per genome per generation, and N_g is the number of generations. The effect of specific growth rate and the effect of appearance of variants on the parent strain population are shown graphically in Fig. 53.1 and Fig. 53.2, respectively. An unstable culture can cause serious disruptions to production in a large-scale fermentation plant. Variant strains are usually less sensitive to adverse conditions (extremes in temperature or pH, nutrient quality, anaerobic conditions) than parent strains, as well. A production strain that appears stable for the required number of generations under laboratory conditions may exhibit instability when introduced to a full-scale plant. It is important to test the stability of potential production strains under conditions that are as similar as possible to the conditions expected in the production environment. It is also important to establish a system of culture storage and management that minimizes the number of generations required to reach full-scale production and maintains the seed stock under stable conditions. Genetically engineered cultures have additional stability problems, which can be expressed in a similar mathematical expression as variant formation. Often, the genes for producing the desired product are located on plasmids that also include selectible markers, usually antibiotic resistance. In the laboratory, these cultures are maintained in media that contain the selective agent. It is not practical to use antibiotics in a large-scale plant; therefore, production cultures must have highly stable plasmids. The rate of plasmid loss can be much higher than variant formation in production cultures. Even if they aren't completely

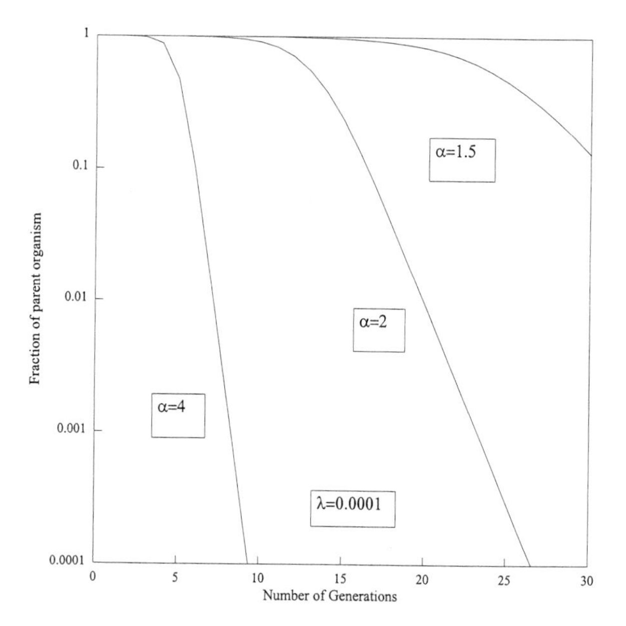

FIGURE 53.1. Effect of growth rate differences between parent and variant strain on survival of parent strain. [Reprinted from Asenjo, J. A., and Merchuk, J. C., eds. (1995). "Bioreactor System Design," Marcel Dekker, Inc., New York. Reprinted by courtesy of Marcel Dekker, Inc.]

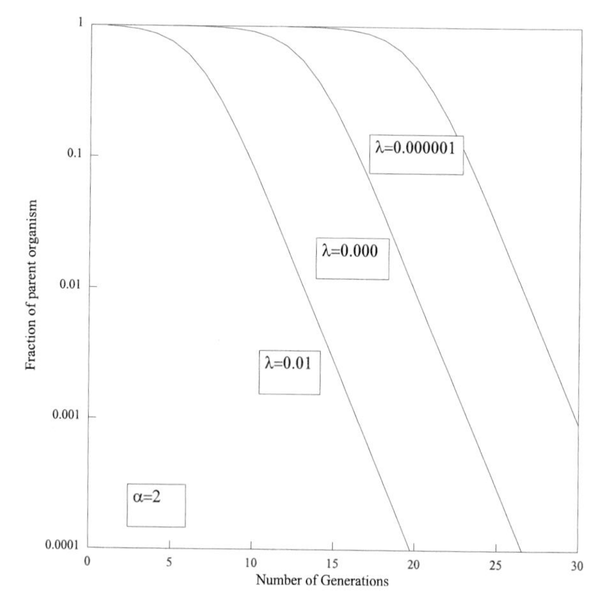

FIGURE 53.2. Effect of variant formation rate on survival of parent strain. [Reprinted from Asenjo, J. A., and Merchuk, J. C., eds. (1995). "Bioreactor System Design," Marcel Dekker, Inc., New York. Reprinted by courtesy of Marcel Dekker, Inc.]

lost, reduction of plasmid number can drastically reduce productivity per cell. Plasmid maintenance usually requires a compromise between the level expression of the cloned genes and culture growth rate.

B. Media development and optimization

The development of a suitable, economical medium is a balance between the nutritional requirements of the microorganism and the cost and availability of the medium components. The chemical constituents of the medium are determined by the composition of the cell mass and product, the stoichiometry of cell and product formation, and from yield coefficients, which can be estimated from shake flask or lab scale fermentor experiments. On a dry weight basis, 90–95% of microbial biomass consists of carbon, oxygen, hydrogen, nitrogen, sulfur, magnesium, and potassium. The remaining 5–10% is microelements (required in small amounts), primarily calcium, manganese, iron, copper, and zinc. The stoichiometry can be determined from the general equation:

$$\mathrm{CH}_a\mathrm{O}_b\ (\text{Carbohydrate}) + A(\mathrm{O}_2) + B(\mathrm{NH}_3)$$
$$\rightarrow Y_{x/s}(\mathrm{CH}_\alpha\mathrm{O}_\beta\mathrm{N}_\gamma)\ (\text{Biomass})$$
$$+\ Y_{p/s}(\mathrm{CH}_{\alpha 1}\mathrm{O}_{\beta 1}\mathrm{N}_{\gamma 1})(\text{product}) + Y_{\mathrm{CO}_{2/s}}\ (\mathrm{CO}_2) + \mathrm{D}(\mathrm{H}_2\mathrm{O})$$

The terms $Y_{x/s}$, $Y_{p/s}$, and $Y_{\mathrm{CO}_{2/s}}$ are the yield coefficients for cell mass, product and CO_2, respectively. In most large-scale fermentations, the carbon usually comes from carbohydrates, such as sugars or starches. The nitrogen is from a variety of organic and inorganic nitrogen sources and from ammonia added to control the pH. Phosphate, sulfur, and magnesium are added as salts or in complex nutrients. The micronutrients are often derived from the water or other raw materials or added as mineral salts. Some commonly used fermentation substrates are shown in Table 53.2.

Many microbes are auxotrophs for specific growth factors, such as vitamins or amino acids, and will be growth limited without adequate quantities of the compound in the medium. Other cultures may not have an absolute requirement for a growth factor, but grow slowly in the absence of the specific factor. In many cases, the exact requirements are not known, but complex substrates are required for optimal growth. It is often necessary to screen a variety of potential nutrients and combinations of nutrients to satisfy the nutritional requirements and minimize the raw materials costs. Another problem encountered with some components of culture medium is seasonal or hidden auxotrophy. This is variation in growth and production encountered due to either seasonal changes in processing or with new crops of vegetable sources of raw materials. Often, this is not seen in the laboratory, but at the large scale, due to poorer mixing in large-scale fermentors. Other considerations for selection of nutrients are regulatory approval and kosher certification for some foods.

TABLE 53.2 Fermentation raw materials

Function	Raw material
Carbon	Glucose
	Sucrose
	Lactose
	Corn syrup
	Starch
	Ethanol
	Paraffins
	Vegetable oils
Carbon, vitamins, micronutrients	Beet molasses
	Cane molasses
Simple nitrogen	Ammonia
	Urea
Simple nitrogen, sulfur	Ammonium sulfate
Amino nitrogen	Cottonseed meal
	Casein
	Soy flour
Amino nitrogen, vitamins, micronutrients	Brewer's yeast, yeast extract
	Corn steep liquor
	Distiller's dried solubles
Carbon, amino nitrogen, vitamins, micronutrients	Whey

IV. LARGE-SCALE OPERATION

A. Inoculum production

The starting place for large-scale fermentations is in the inoculum laboratory. The fermentation is doomed to failure without pure, active inoculum. This starts with the storage of culture. It is important to store the culture under conditions that retain both genetic stability and viability. A variety of methods have been used, all with advantages and disadvantages. Storing cultures in agar stabs or on agar slants is one of the oldest methods. It is simple and cultures can be stored at room temperature or under refrigeration, but viability and stability are limited and cultures must be transferred frequently. Lyophilization (freeze drying) requires no refrigeration or freezing and cultures can be stored almost indefinitely at room temperature and can serve as a long-term method of safely storing culture in the event of catastrophe. Lyophilization takes more skill and equipment than slant preparation and often has significant viability loss. Cultures are not easily revived, making it unsuitable for daily use. Cultures stored in aqueous glycerol solutions at very low temperatures, −80 °C or lower, generally have high viability and stability, are easy to produce and revive, but require cryogenic freezers or liquid nitrogen, and are, thus, susceptible to power outages and equipment failure.

Once the cultures are safely stored, they must be revived, grown in the inoculum laboratory, and transferred to growth or "seed" fermentors that feed the production fermentors. Cultures are usually grown in shake flask culture in the laboratory and transferred to a suitable container for transfer to the plant. In the laboratory, culture transfers are exposed to the air, so precautions are taken to prevent incidental contamination, such as performing transfers in rooms or hoods with HEPA-filtered air and the wearing of sterile coveralls, gloves, and masks by technicians. Once out in the plant, all transfers are made through steam-sterilized piping or hoses using differential pressure.

B. Types of bioreactors

There are a wide variety of of bioreactor designs (Figs. 53.3 and 53.4). Selection of a reactor design for a particular process depends on a variety of factors, including mass transfer considerations, mixing, shear sensitivity, broth viscosity, oxygen demand, reliability of operation, sterilization considerations, and the cost of construction and operation. Due to the complex nature of fermentation scale-up, a few basic reactor designs are used for most applications. The two most common are the stirred tank and the air lift reactors.

Stirred tank reactors use sparged air and submerged impellers to aerate and mix the broth. They are versatile and are especially adapted to highly aerobic cultures and highly viscous fermentations. The drawbacks are high energy input and the use of rotating seals on the agitator shaft, which are a contamination risk. Even within this category, there are many variations in design, such as the style, number, and placement of impellers, the height to diameter ratio, the number and placement of coils or baffles, that affect the mixing characteristics of the vessel. Due to high risk of scale-up and the high capital cost of building large-scale fermentation facilities, most plants install stirred-tank reactors.

Airlift fermentors mix the broth with air from the sparger. Some designs have an internal draft tube to direct the flow of fluid. Most airlift designs have a much greater height-to-diameter ratio than stirred tank vessels to improve oxygen transfer. The mixing is not as good as in a stirred tank but the energy input and shear forces are much lower, thus useful for shear sensitive cultures or in processes where the energy cost of agitation is a significant factor. Also at very large scale, the heat added to a stirred-tank vessel by the agitator becomes increasingly difficult to remove, adding to the cooling costs. Most large-scale airlift fermentors are used for plant effluent treatment, production of baker's yeast, or for fungal fermentations

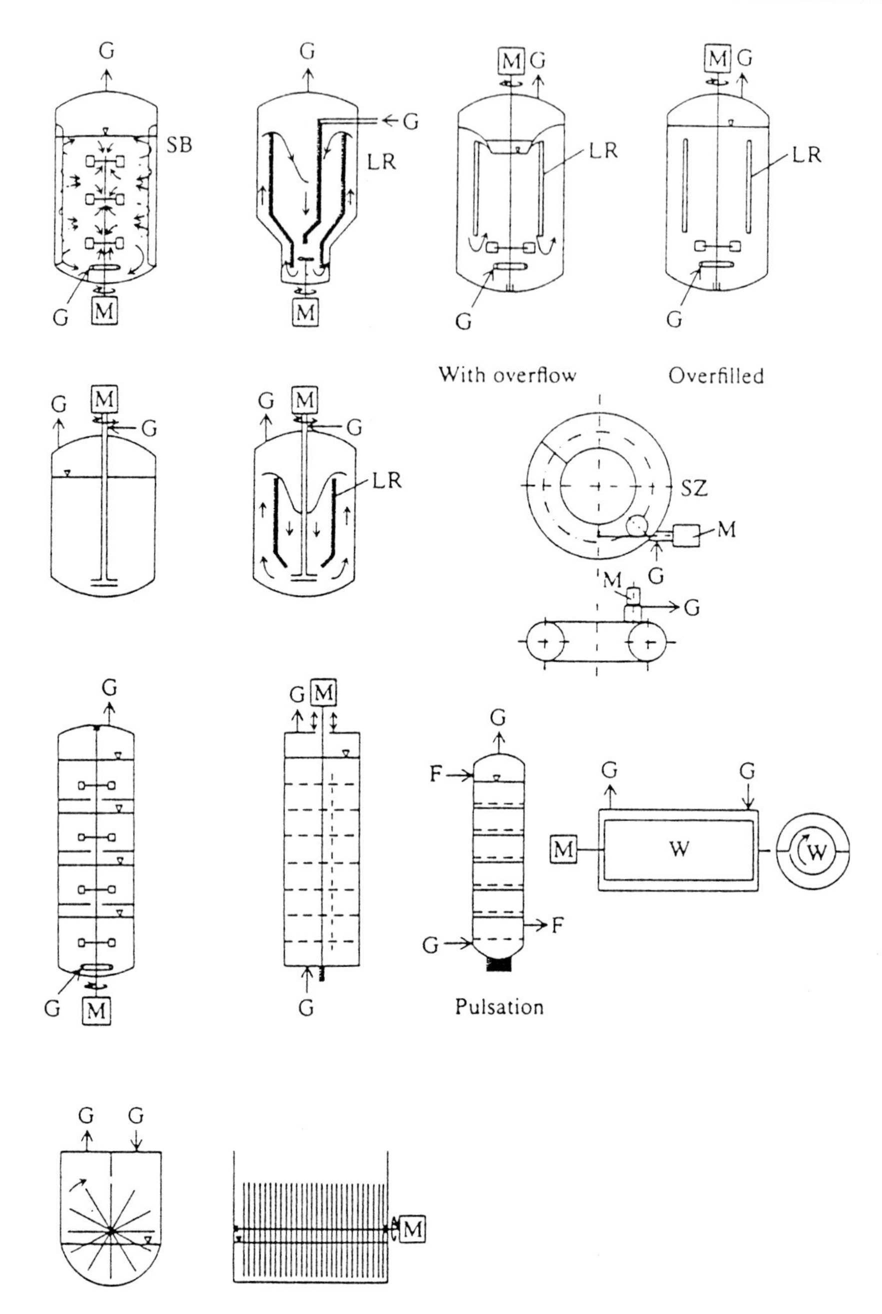

FIGURE 53.3 Bioreactors with mechanical stirring. M, motor; g, gas (air); SB, baffles; LR, conduit tube; W, roller; F, liquid. [Reprinted from Schugerl, K. (1982). New bioreactors for aerobic processes, *Int. Chem. Eng.* **22**, 591.]

where the size of the mycelial pellets is controlled by shear forces.

In addition to the mass transfer characteristics of a fermentor, other factors are important in the operation of a bioreactor, such as the ability to clean the vessel, sterile integrity of the vessel, and maintenance costs. Some reactor designs that have excellent characteristics in the pilot plant are not practical choices for large-scale operation due to mechanical complexity that causes sterility and maintenance problems on scale-up.

C. Modes of operation

There are three basic modes of operation: batch, fed batch, and continuous, with variations of these three basic modes (Fig. 53.5). In batch mode, all the ingredients required for fermentation except pH control chemicals, usually ammonia, are added to the fermentor prior to inoculation. The fermentation is run until the nutrients are exhausted, then the broth is harvested. The advantage is simplicity of operation

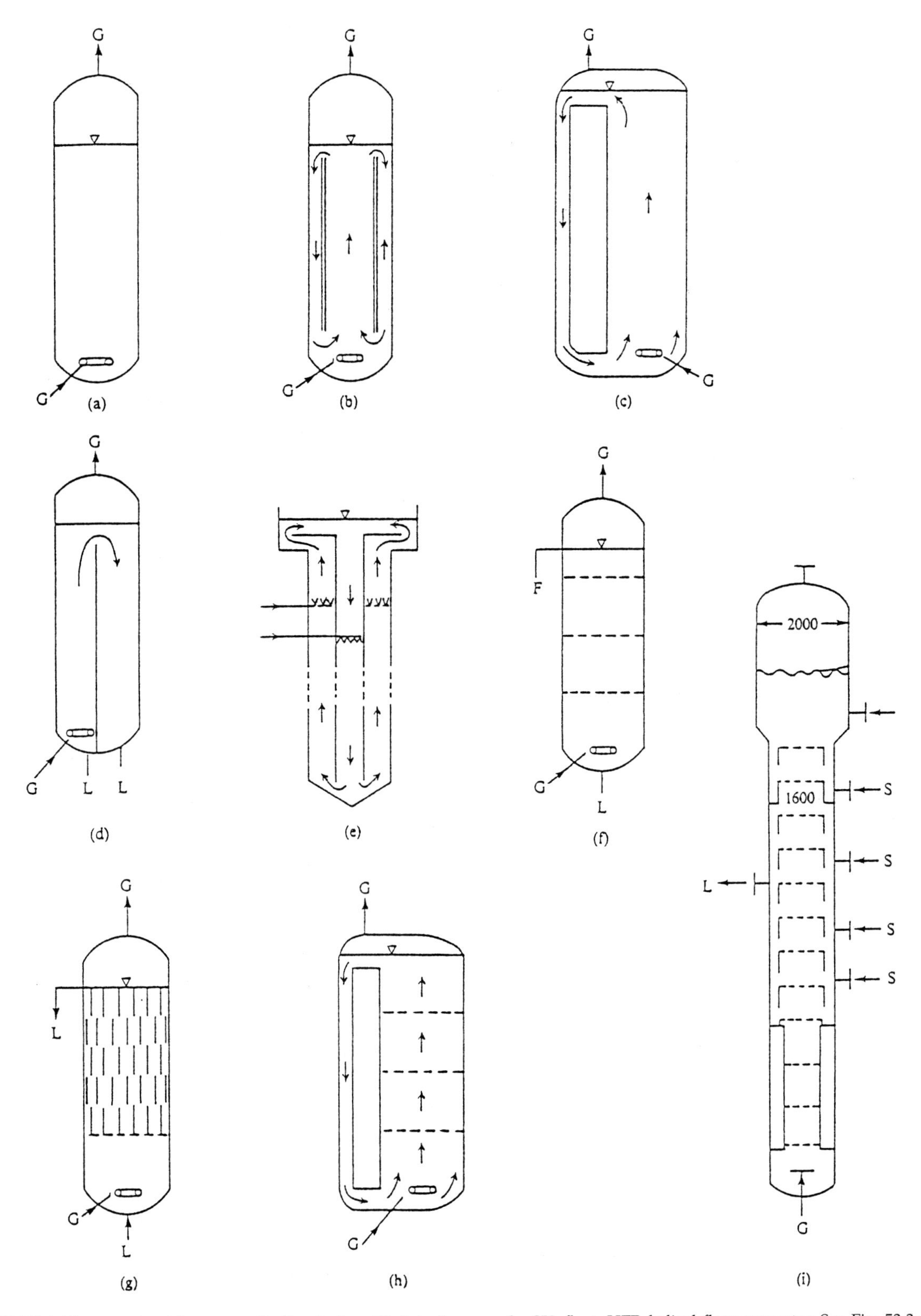

FIGURE 53.4 Bioreactors with pneumatically stirring. ID, injection nozzle; SK, float; HFP, helical flow promoter. See Fig. 53.3 for other abbreviations. [Reprinted from Schugerl, K. (1982). New bioreactors for aerobic processes, *Int. Chem. Eng.* **22**, 591.]

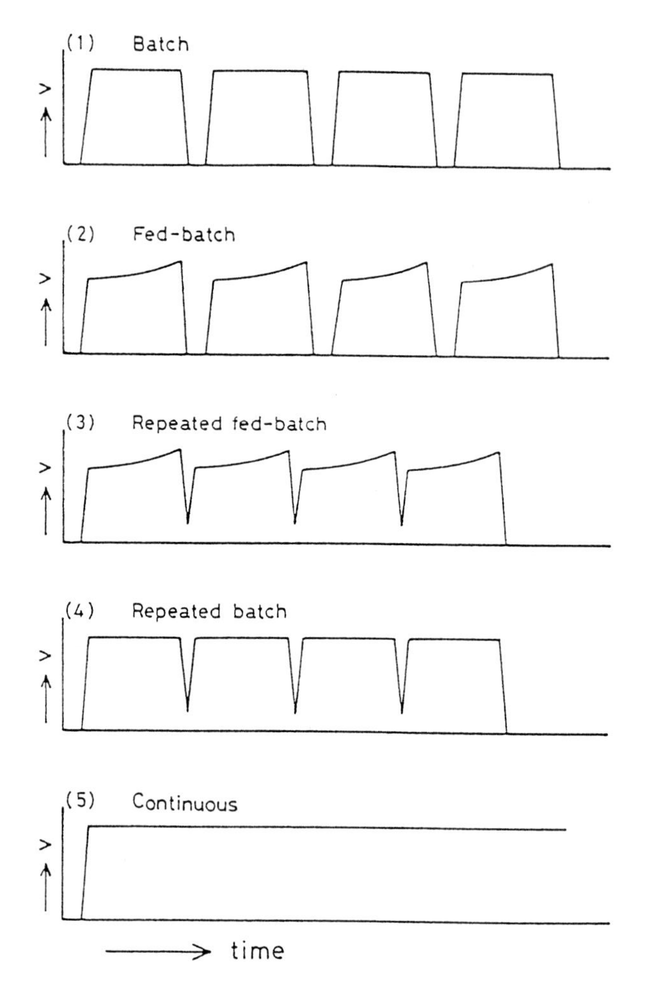

FIGURE 53.5 Changes in volume of culture broth with elapsed time for various modes of operation in bioreactors. [Reprinted from Asenjo, J. A., and Merchuk, J. C., eds. (1995). "Bioreactor System Design," Marcel Dekker, Inc., New York. Reprinted by courtesy of Marcel Dekker, Inc.]

and reduced risk of contamination. It is useful in fermentations with high yield-per-unit substrate and with cultures that can tolerate high initial substrate concentrations.

Fed batch mode starts with some of the nutrients in the fermentor before inoculation. Concentrated nutrients are added as the fermentation progresses. The advantages are the ability to add large quantities of nutrients to the fermentor by adding them gradually and the ability to control the rate of nutrient addition. This allows for high product concentrations without subjecting the culture to inhibition by high levels of nutrients. It also allows for control of culture growth rate, which is required in some fermentations to maximize productivity and yield. Overall, the use of the fermentor time is better in fed batch than straight batch fermentation, reducing fixed costs. The disadvantages are increased risk of contamination, due to the addition of nutrients through a continuous sterilizer, and increased equipment costs for continuous sterilization and flow control equipment for feed streams.

Both batch and fed batch can be run in repeated mode, with a small portion of the previous batch left in the fermentor for inoculum. The medium is then added through a continuous sterilizer. Use of the fermentor is increased by eliminating turnaround time, but the risks of contamination and genetic degradation of the culture are increased. In any case, repeated batch mode cannot be repeated indefinitely, due to maintenance and cleaning needs. Usually, repeated fermentations are run for two or three batches.

In continuous mode, the starting medium and inoculum are added to the fermentor. After the culture is grown, the fermentor is fed nutrients and broth is withdrawn at the same rate, maintaining a constant volume of broth in the fermentor. In continuous mode with cell recycle, the cell mass is returned to the fermentor using microfiltration with bacteria or screens with fungal mycelia. Continuous mode maximizes the use of the vessel and is especially good for fermentations that take a long time to reach high productivity. The disadvantages are increased risk of contamination, especially since it is difficult to keep contamination from growing through the continuous harvest line. As with repeated batch mode, continuous fermentations cannot be run indefinitely, but fermentations of several hundred hours' duration are possible.

D. Monitoring and control

Ideally, a fermentation is best controlled using online, real-time measurements of pH, oxygen, cell mass, substrate and product concentrations, metabolic state, and nutrient flow rates and applying the data to a precise model of how the culture will respond to changes in any of the parameters. Only a few of the parameters are actually measurable in real time and the models of culture behavior are often imprecise extrapolations of experimental data, although technological advances are moving in the direction of better monitoring and control.

Most physical parameters—temperature, pressure, power input, impeller speed, gas and liquid flow—can be measured accurately without invasive instrumentation. Reliable online measurement of the chemical environment has been limited to pH, dissolved oxygen, and off gas analysis. Advances in infrared analysis have recently made it possible to measure several parameters of broth composition, such as substrate, product, and ammonia concentration, simultaneously on-line. Various optical density probes, to measure cell mass, and enzyme probes, to measure nutrient or product concentration, are available, but are often subject to fouling or cannot be sterilized. Measurement of physiological characteristics, such as intracellular ATP, DNA, and RNA content, are not currently feasible and must be inferred from the physical and chemical data.

At a basic level, most fermentations are run under temperature and pH control. In other fermentations, the dissolved oxygen is controlled, usually by changing the agitator speed. In a batch fermentation, not much more is usually needed. In fed batch and continuous fermentations, it is often necessary to control the nutrient concentration at a limiting concentration in order to control the growth rate of the culture, maximize product yield, or avoid anaerobic conditions.

Substrate feed can be controlled using feedback control from pH, dissolved oxygen, or off gas analysis, either directly or by calculating physiological parameters, such as substrate utilization rate or oxygen uptake rate. In a pH-controlled feed system, the consumption of carbohydrate drives the pH down, and base (usually ammonia) is added to maintain a constant pH. When the carbohydrate becomes limiting, the pH is driven up instead. When this occurs, the feed is increased, driving the pH downward again. Sophisticated control schemes can be devised by measuring the rate of base consumption and the rate of pH increase when substrate is limiting. Without proper control loops, a pH controlled feed can bog down or overfeed, as the measurement can only tell whether there is not enough feed, not when there is too much. A dissolved oxygen (DO) controlled fermentation is maintained between two setpoints, usually between 20% and 40% dissolved oxygen, by increasing the feed when the DO rises and decreasing the feed when it drops. Without some damping of the response in DO-controlled fermentations, there is a tendency for large amplitude swings of DO and feed rate, due to delayed response and long mixing times.

E. Sterilization and contamination control

Sterilization is the process of eliminating all viable organisms from equipment and materials. For a fermentation plant, this requires:

- Presterilization of equipment (fermentor vessels, inoculation and feed piping, and air filters);
- Sterilization of feedstocks, either in the fermentor vessel or through a continuous sterilizer;
- Maintenance of aseptic conditions throughout the entire fermentation process.

Most fermentation processes require aseptic conditions for optimal productivity and yield. Some fermentations are relatively insensitive to contamination: yeast and fungal fermentations performed at low pH, fermentations with short cycle times (high growth rates), or fermentations that produce inhibitory or toxic products (alcohols, antibiotics, organic acids) tend to be resistant to contamination, but rampant, out-of-control contamination can destroy productivity altogether. Other processes are highly sensitive to contamination: repeated batch or continuous fermentations and slow-growing bacterial fermentations are more susceptible to contamination and, therefore, require more stringent conditions to maintain aseptic conditions. In addition, some contaminants can be pathogenic or produce toxins that can render products unusable or be a health risk to plant workers or consumers. Therefore, complete asepsis is the operating philosophy of most largescale plants. The philosophy of aseptic operation must be an integral part of every aspect of the operation, from design and construction to operating procedures and personnel training, for a plant to operate aseptically.

Most process equipment and medium components are sterilized by steam under pressure. The rate of cell death depends on the time and temperature of exposure to steam. The spores of the thermophile *Bacillus stearothermophilius* are the usual benchmark for determining sterilization conditions. Standard conditions for steam sterilization are 121 °C for 15 minutes, but the time required for sterilization is less at higher temperatures. Most fermentation broths and continuous feeds are sterilized by steam sterilization either *in situ* or in a continuous sterilization system, unless the medium contains heat-labile constituents.

Sterilization of the fermentor broth *in situ* requires a significant amount of time in a large vessel for heating and cooling the broth, which reduces the productive time of the fermentor. Large amounts of cooling water are required as well, and the long time required for sterilization can cause degradation of some medium components. *In situ* sterilization, however, has the advantage of being simple, thus reducing the risk of upset in the process resulting in contamination.

Continuous sterilization is sometimes used for fermentor broths but more often for nutrients added to the fermentor during the fermentation, especially for fed-batch or continuous fermentations. This is usually accomplished by either direct steam injection into the nutrient stream or the transfer of heat through a heat exchanger with no direct steam contact. After heating, a holding loop maintains the temperature for a specified time, followed by a second heat exchanger to reduce the heat of the medium before it is added to the fermentor. Continuous sterilization can be carried out at increased temperature and reduced time, to minimize the heat damage to the medium and reduce the length of holding tube required or increase the flow of fluid through the system. As a practical matter, heat from fluid returning from the holding loop is used, through a heat exchanger, to preheat broth going to the sterilizer, thus reducing the requirements for both steam to heat the incoming broth and cooling water to cool the sterilized broth. Duplicate systems are required for systems to be cleaned and maintained properly.

Filter sterilization is used for liquids with heat-sensitive components or nonaqueous liquids with low boiling points, which are added to the fermentor after sterilization and cooldown. Air and other gases are also sterilized by filtration. Two basic types of filters have been used. Depth filters consist of layers of glass wool and were mostly used to sterilize air, but can be difficult to sterilize and dry. Depth filters have been mostly replaced by absolute membrane filters, which are thin membranes with pores no larger than their rated size. Most bacteria and spores are retained by filters with 0.2 μm absolute pore size. Hydrophobic filters are used to sterilize air, usually preceded by prefilters to remove solids and water from the compressed air. Bacteriophage, which are retained on dry filters by electrostatic forces, are small enough to pass through if the air is wet, making it very important to keep the process air as dry as possible. While microbial contamination usually just reduces productivity and yield, bacteriophage can destory the production culture within a few hours of introduction into the fermentor. Fermentation plant managers live in fear of bacteriophage.

During the fermentation, the contents of the reactor must be isolated from potential contaminating sources. Besides filtering the air, all penetrations into the fermentor must be sealed from the outside. Agitators are sealed with double mechanical seals that have steam between the rotating seals. Dissolved oxygen and pH electrodes are sealed with O-rings or gaskets. Piping that penetrates the vessel, such as media fill lines, harvest lines, and inoculation lines, are isolated with steam seals, which are sections of piping between two valves that connect the fermentor to the outside "septic" environment. There is a steam source and a relief valve to vent steam condensate and maintain steam pressure above 15 psig when the line is not in use. Although valves are routinely tested for leaks, there are no valves with fine enough tolerances to prevent contamination from growing across the contact surfaces.

Of all the problems encountered in a fermentation plant, contamination is the most persistent. Contamination control is a daily, ongoing activity. The culture is monitored for contamination by microscopic examination and plating on nutrient agar, from the first shake flask through the production fermentor. This helps to prevent contaminated culture from being used in production and to trace the source of contamination whenever possible. In addition to testing the culture for contamination, other potential sources of contamination, such as sterile feed streams into the fermentor, are monitored. Unfortunately, the size of sample that is practical to test for contamination is very minute in comparison to the size of the process being tested. Contamination is usually not detected until it is a serious problem. Therefore, preventive maintenance is very important to control contamination. Fermentors, piping, valves, and filters are routinely tested for integrity. Leaky valves and cracks in cooling jackets or coils are common sites for microbial contamination to enter the process, and cooling water leaks are a common source of bacteriophage contamination. Proper procedures for sterilization, culture transfer, and fermentor operation are critical to maintaining an aseptic process. This involves careful evaluation of the effect of procedures on process integrity, writing procedures clearly, training fermentor operators properly, and monitoring the effectiveness of training.

F. Utilities

Although fermentation processes usually are performed at near ambient temperatures, the energy consumption is high due to the need for sterilization, cooling, agitation, and aeration.

1. *Steam*

Steam is required for sterilization of the medium, either in the fermentor or through continuous sterilizers. During the fermentation itself, a certain amount of steam is required to maintain the microbiological integrity of the fermentor vessel by the use of steam seals. Fermentation plants often require large quantities of steam for evaporation and drying in downstream processing, due to the low concentration of product in the broth.

2. *Cooling water*

After sterilization, fermentors need to be cooled before they can be inoculated; during the fermentation, cultures generate great quantities of metabolic heat, and agitators add some heat to the broth. This heat must be removed to maintain the proper incubation temperature. For downstream processing, cooling water is needed for certain types of crystallizers and for condensers on evaporators. Although cooling water is recycled, there is high energy input for cooling towers and chillers and evaporative water loss in cooling towers.

3. *Process water*

Much of the fermentor broth is water. The quality of the water used depends on the sensitivity of the culture to minerals in the water, effect of water quality on downstream processing, and regulatory requirements. In some cases, water must be purified by reverse osmosis for use in fermentation. Other fermentations can use water recycled from process condensers. Water is also used for some downstream process steps and for cleaning and rinsing the fermentors.

4. *Electricity*

Agitation, compressed air, and water chillers require a great amount of electricity. Where economies of scale are adequate, cogeneration of steam and electricity are very cost effective. Boilers using natural gas, oil, or coal produce steam used to feed generators for electricity. The "waste" steam produced in the generation process, which is at high enough pressure for use in fermentation, is captured and used as the process steam for the plant.

5. *Sewage treatment*

Fermentation generates waste with high BOD, mainly in the form of spent cell mass. While much of it can be used for animal feeds, some low solids streams are generated and must be treated as sewage. A lot of waste water is also generated, making the throughput for a treatment plant high. In many cases, on-site primary and secondary treatment is required to avoid paying high municipal sewage charges and fines.

G. Downstream processing

Since fermentation products are often produced in dilute aqueous solution that also includes the cell mass, metabolic by-products, and various salts, downstream processing is a major part of producing a fermentation product. The degree of purification of the product depends on the type of product and the end use. Industrial enzymes, for example, are often separated from the cell mass and concentrated by ultrafiltration to make a crude extract. On the other extreme, antibiotics for human use undergo many steps of purification. Common methods include microfiltration and ultrafiltration to remove cell mass and other debris or retain larger molecular weight proteins; concentration by evaporation or reverse osmosis; and crystallization by chilling or evaporation, pH or solvent precipitation, centrifugation, and chromatography.

BIBLIOGRAPHY

Adams, M. R., and Nout, M. R. J. (2001). "Fermentation and Food Safety." Aspen Publishers, Inc.

Alleman C. B., and Leeson, A. (eds.) (1999). "Bioreactor and Ex Situ Biological Treatment Technologies: The Fifth International In Situ and On-Site Bioremediation Symposium." San Diego, California, Battelle.

Asenjo, J. A., and Merchuk, J. C. (eds.) (1995). "Bioreactor System Design." Marcel Dekker Inc., New York.

Atkinson, B., and Mavituna, F. (1991). "Biochemical Engineering and Biotechnology Handbook." Stockton Press, New York.

Boutlon, C., and Quain, D. (2001). "Brewing Yeast and Fermentation." Iowa State University Press.

Bud, R. (1993). "The Uses of Life: A History of Biotechnology." Cambridge University Press, Cambridge.

Drioli, E., and Giorni, L. (1999). "Biocatalytic Membrane Reactors." Taylor & Francis.

Humphrey, A. (1998). Shake flask to fermentor: What have we learned. *Biotechnology Progress* **14**:1, 3–7.

Mateles, R. I. (ed.) (2001). "Directory of Toll Fermentation and Cell Culture Facilities." Candida Corp.

Pons, M.-N. (ed.) (1991). "Bioprocess Monitoring and Control." Hanseer Publishers, Munich, Germany.

Smart, K. F. (2000). "Brewing Yeast Fermentation Performance." Blackwell Science.

Wang, D. I. C., Cooney, C. L., Demain, A. L., Dunhill, P., Humphrey, A. E., and Lilly, M. D. (1979). "Fermentation and Enzyme Technology." John Wiley & Sons, New York.

WEBSITES

The Electronic Journal of Biotechnology
http://www.ejb.org/

List of Biotechnology Organizations and Research Institutes
http://www.ejb.org/feedback/borganizations.html

List of bacterial names with standing in nomenclature (J. P. Euzéby)
http://www.bacterio.cict.fr/index.html

54

Insects' symbiotic microorganisms

A. E. Douglas
University of York

GLOSSARY

bacteriocyte An insect cell harboring symbiotic microorganisms.

nitrogen recycling The metabolism of insect nitrogenous waste products (e.g., uric acid and ammonia) by symbiotic microorganisms to nitrogenous compounds (e.g., essential amino acids) that are transferred to and used by the insect.

symbiosis The intimate association between phylogenetically different organisms; often restricted to relationships from which all the organisms derive benefit.

transovarial transmission The transfer of symbiotic microorganisms to the unfertilized eggs in the ovaries of the female insect.

vertical transmission The transmission of microorganisms from a parent insect to its offspring.

Symbiotic Microorganisms are the components of the microbiota of an insect that contribute to insect survival, growth, or fecundity. They are borne by an estimated 10% of all insect species, and are located in the insect gut or tissues, often restricted to specialized cells called bacteriocytes. Most of the microorganisms are rare or unknown apart from the insect partner, and many have not been cultured *in vitro*. Historically, the microbiology of insects has been little studied and, until recently, most of the information available on symbiotic microorganisms has been derived from microscopic analysis of the insect regarding the morphology of the microorganisms, their location in the insect body, and their mode of transmission between insects. The advent of molecular techniques has transformed our understanding of symbiotic microorganisms, allowing the taxonomic identification of microorganisms and the elucidation of the molecular basis of their function. Because of the insects' dependence on their symbiotic microorganisms, these associations are of great potential value as a novel approach to insect-pest management.

I. DIVERSITY OF SYMBIOTIC MICROORGANISMS IN INSECTS

A. Distribution of symbiotic microorganisms across the microbial kingdoms

Symbiotic microorganisms include members of all microbial kingdoms (Table 54.1)—various Eubacteria, methanogens (Archaea), and protists and fungi (Eukaryota). The Eubacteria, especially members of the γ-Protobacteria, are widely represented both in the guts and cells of insects, but methanogens and protists occur in the strictly anaerobic portions of the guts of certain insects. Yeasts have been reported in the gut lumen, cells, and haemocoel (body cavity) of some species. (Basidiomycete fungi are also cultivated in the nests of some insects, e.g., fungus-gardening termites and leaf-cutting ants, but these ectosymbioses are not considered here.)

The Desk Encyclopedia of Microbiology
ISBN: 0-12-621361-5

TABLE 54.1 Survey of symbiotic microorganisms in insects

Insect	Microorganism	Location[a]	Incidence
Blattaria (cockroaches)	Flavobacteria	B in fat body	Universal
	Various bacteria	Hindgut	Universal
Isoptera (termites)	Various bacteria[b]	Hindgut	Universal
	Flagellate protists	Hindgut	Lower termites
Heteroptera	Various bacteria[b]		
Cimicidae		B in haemocoel	Universal
Coreidae		Midgut	Widespread/irregular
Lygaeidae		Midgut	Widespread/irregular
Pentatomidae		Midgut	Widespread/irregular
Pyrrhocoridae		Midgut	Widespread/irregular
Triatomidae		Midgut	Universal
Homoptera	Bacteria: including γ-Protobacteria (in aphids and whitefly) and β-Protobacteria (in mealybugs)	B in various locations; in haemocoel	Nearly universal[c]
	Pyrenomycete yeasts	Predominantly extra-cellular in fatbody/haemocoel	In delphacid planthoppers and hormaphidine aphids
Anoplura	Bacteria[b]	B, variable locations	Universal
Mallophaga	Bacteria[b]	B in haemocoel	Irregular
Diptera			
Glossinidae	γ3-Protobacteria	B in midgut epithelium	Universal
Diptera Pupiparia	Bacteria[b]	B in haemocoel	Universal
Coleoptera			
Anobiidae	Yeasts	B in midgut caeca	Universal
Bostrychidae	Yeasts	B in haemocoel	Universal
Cerambycidae	Bacteria	B in midgut caeca	Widespread
Chrysomelidae	Bacteria	B in midgut caeca	Irregular
Curculionidae	Bacteria	B in variable location	Widespread
Lucanidae	Bacteria	Midgut or hindgut	Universal
Formicidae (ants)			
Camponoti	γ3-Protobacteria		Universal
Formicini	Bacteria[b]		Irregular

[a]B indicates bacteriocytes.
[b]These bacteria have not been characterized by molecular techniques and their phylogenetic position is unknown.
[c]Absent from typhlocybine leafhoppers, phylloxerid aphids, and apoimorphine scale insects.

B. Symbiotic microorganisms in insect guts

Most insects have a substantial gut microbiota, although there are wide differences among insect taxa and among regions of the gut. Much of the literature gives misleading estimates of the microbial diversity in insect guts because the techniques commonly used are based on culturable forms, which account for only 0.1–10% of the total microbiota. An additional complication to the study of symbiotic microorganisms in insect guts is that many or all members of the microbiota are either transient (i.e., passing through the gut with the unidirectional passage of food), or commensal (i.e., resident for extended periods, but of no discernible advantage to the insect). Detailed experimental study is required to identify which, if any, of the resident gut microbiota are symbiotic microorganisms (i.e., beneficial to the insect).

The microbiology of termite guts has been studied extensively. The greatest density of microorganisms is in the anoxic proximal portion of the hindgut, known as the paunch. In all termites, this region harbors bacteria, at 10^9–10^{10} cells/ml gut volume. All the bacteria are facultative or obligate anaerobes. They comprise methanogens, spirochaetes (pillotinas, e.g., *Hollandia*, *Pillotina*, *Diplocalyx*, and *Clevelandina*, and many other unidentified forms), and other eubacteria, including species of *Enterobacter*, *Bacteroides*, *Bacillus*, *Citrobacter*, *Streptococcus*, and *Staphylococcus*. The lower termites in addition have obligately anaerobic, flagellate protists of the orders Hypermastigida and Trichomonadida and Oxymonadida at densities of up to 10^7 cells/ml (Fig. 54.1). Approximately 400 species of these protists have been reported and a few, including *Trichomitopsis termopsidus* and *Trichonympha sphaerica* (both from the termite *Zootermopsis*) have been brought into axenic culture. Higher termites lack these protists.

Apart from the termites, most studies have concerned the gut microbiota of the few insect species that are routinely reared in laboratories. As examples,

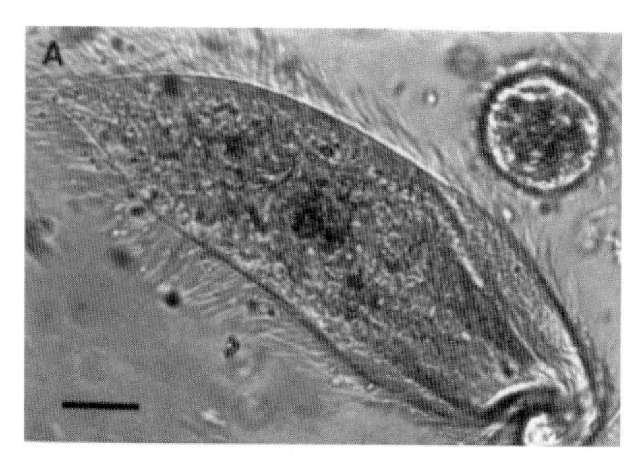

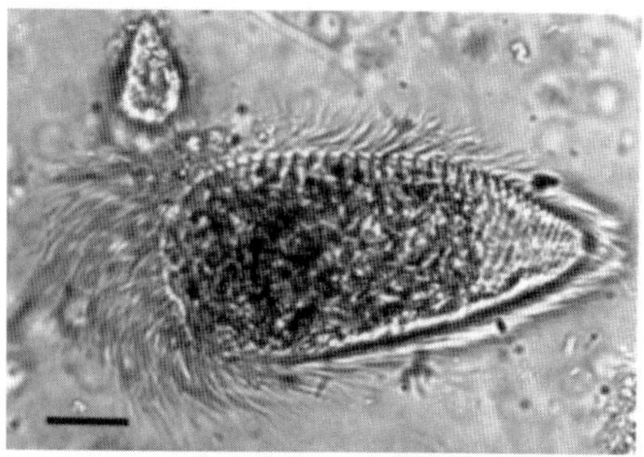

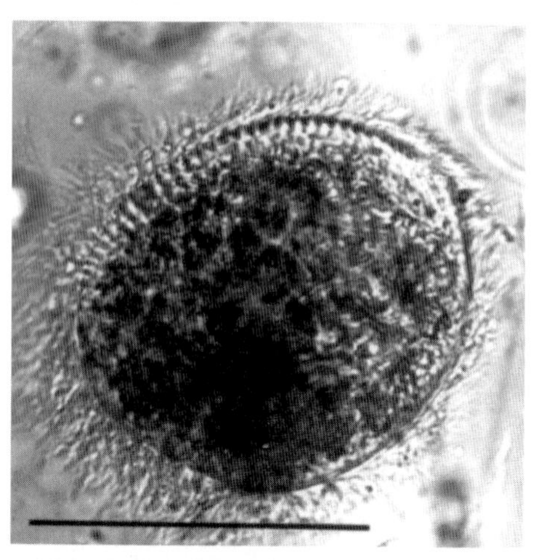

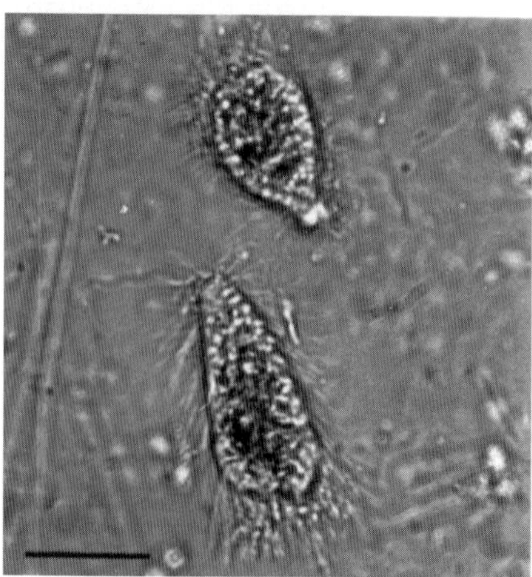

FIGURE 54.1 Symbiotic protists in the hindgut of a lower termite *Coptotermes lacteus*. Bar = 25 μm. Photographs by R. T. Czolij and M. Slaytor, reproduced from Fig. 3a of Douglas, A. E. (1992). "Encyclopedia of Microbiology," Vol. 4, pp. 165–178.

the American cockroach *Periplaneta americana* bears both obligately anaerobic bacteria (e.g., *Clostridium* and *Fusobacterium*) at densities of ~10^{10} cells/ml and facultative anaerobes (e.g., *Klebsiella, Yersinia, and Bacteroides*) at ~10^8 cells/ml; the locust *Schistocerca gregaria* has exclusively facultative anaerobes, usually at considerably lower densities than the cockroach; and the blood-feeding reduviid bug *Rhodnius prolixus* has a diversity of bacteria, including *Pseudomonas, Streptococcus, Corynebacterium,* and various actinomycetes (and not a single gut symbiont, the actinomycete *Nocardia rhodnii*, as claimed in much of the early literature).

The composition of microorganisms in the guts of insects can vary widely with environmental circumstance. The microbiota may change when insects are transferred to laboratory rearings, and specific differences in the microbiota of insects in different laboratories or under different temperature or dietary regimes have been described.

C. Symbiotic microorganisms in insect tissues and cells

Many symbiotic microorganisms are located in insect cells, where they may be well protected from the hemolymph-based defense system of the insect. In any single insect, the cells bearing intracellular symbiotic microorganisms are usually of a single morphological form (e.g., see Fig. 54.2) and location in the insect body (see Table 54.1). They are called bacteriocytes or mycetocytes, and their sole function appears to be the housing of the microorganisms.

The incidence of intracellular microorganisms in insects and, where known, taxonomic information on the microorganisms are summarized in Table 54.1. Most of the microorganisms are Eubacteria but, because they have not been brought into axenic culture, taxonomic information is available only for those forms whose 16*S* rRNA has been sequenced. Many of the bacteria are members of the γ-Protobacteria, and the microorganisms in three taxonomically disparate insect groups, aphids, tsetse flies, and ants, are particularly closely related. The bacteria in aphids and tsetse flies are known as *Buchnera* spp. and *Wigglesworthia* spp., respectively, in recognition of the research of entomologists Buchner and Wigglesworth on these systems. Other intracellular bacteria in bacteriocytes include members of the β-Protobacteria in mealybugs and flavobacteria in cockroaches. The bacteria in several insect groups, including the Anoplura and Mallophaga (the sucking and chewing lice, respectively) and various Heteroptera, have not been studied at the molecular level (see Table 54.1).

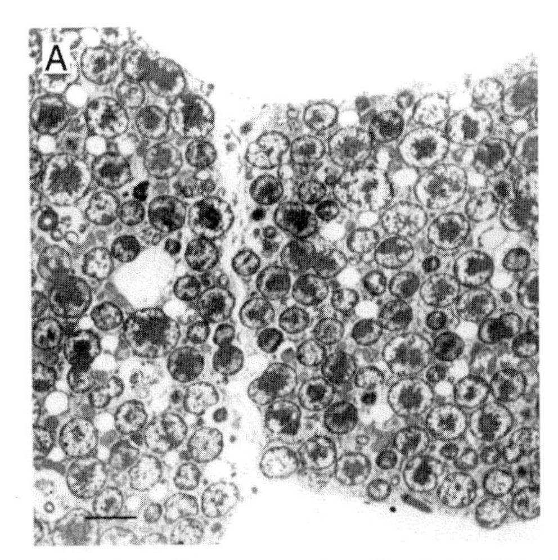

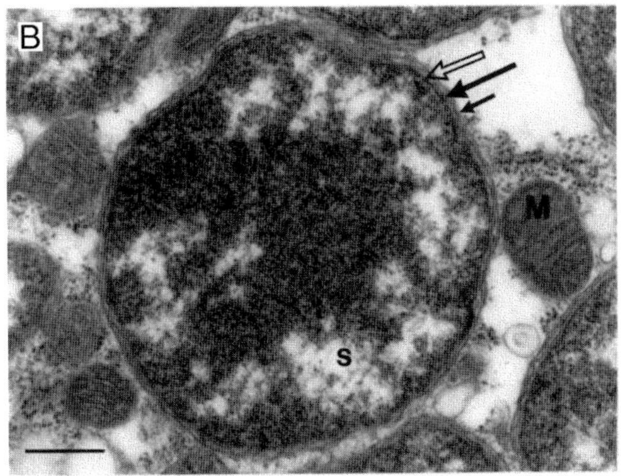

FIGURE 54.2 *Buchnera*, the intracellular symbiotic bacterium in aphids. (A) The symbiotic bacteria occupy the greater part of the bacteriocyte cytoplasm (electron micrograph of pea aphid, *Acyrthosiphom pisum*); bar = 5 μm. Reproduced from Douglas, A. E. (1997). "Symbiotic Interactions," by permission of Oxford University Press, Oxford. (B) Each symbiotic bacterium (S) is bounded by a cell membrane (↑), cell wall (↑), and insect membrane (↑). A mitochondrion (M) marks the insect cell cytoplasm (electron micrograph of black bean aphid, *Aphis fabae*); bar = 5 μm. Reproduced from Fig. 1a of Douglas, A. E. (1992). "Encyclopedia of Microbiology," Vol. 4, pp. 165–178.

Some insects bear microorganisms whose location (intracellular vs. extracellular) is variable. For example, delphacid planthoppers and hormaphidine aphids lack intracellular bacteria (unlike other planthoppers and aphids), and bear pyrenomycete yeasts in and between fat body cells. Some species of aphid and tsetse fly have secondary symbionts, in addition to *Buchnera* and *Wigglesworthia*, respectively. The secondary symbionts are variably located in cells and the haemocoel, and of uncertain significance to the insect.

II. FUNCTION OF SYMBIOTIC MICROORGANISMS

A. Symbiotic microorganisms as a source of novel metabolic capabilities

Symbiotic microorganisms are widely believed to contribute to the nutrition of insects. This was first deduced from the distribution of the associations among insects. In general, the microorganims are restricted to insects living on nutritionally poor or unbalanced diets. They are widespread or universal among insects feeding through the lifecycle on the phloem and xylem sap of plants, deficient in essential amino acids; vertebrate blood, deficient in B vitamins; and wood, which is composed principally of lignocellulose and is deficient in many essential nutrients for insects. The implication is that the symbiotic microorganisms variously degrade cellulose and synthesize essential amino acids and vitamins. They have also been implicated in the synthesis of sterols, which insects and other arthropods cannot synthesize *de novo*.

The symbiotic microorganisms can be considered to be a source of biochemically and genetically complex metabolic capabilities that the insect lacks. They have also been described as microbial brokers, mediating insect utilization of blood, plant sap, and wood.

B. Contribution of symbiotic microorganisms to the nitrogen nutrition of insects

Three routes have been identified by which symbiotic microorganisms contribute to the nitrogen nutrition of insects, the fixation of N_2, nitrogen recycling, and, the synthesis of essential amino acids.

Symbiotic microorganisms in the gut of some insects fix N_2 at appreciable rates; no intracellular N_2-fixing bacteria in bacteriocytes have been described. In particular, many termite species derive significant supplementary nitrogen from N_2-fixing bacteria (e.g., *Enterobacter agglomerans* and *Citrobacter freundii*) in the anoxic portion of their hindgut. The N_2 fixation rate varies widely among termite species, from <0.2 g N fixed/g insect weight/day in *Labiotermes* sp. and *Cubitermes* sp. to >6 g N/g/day in *Nasutitermes* species, and is also influenced by environmental conditions, including the concentration of combined nitrogen in the diet.

Nitrogen recycling refers to the microbial consumption of nitrogenous waste products of insects and the synthesis of compounds (e.g., essential amino acids) of nutritional value to the insect, which are then translocated back to the animal. Microbial utilization of insect-derived uric acid or ammonia has been demonstrated in several systems, including cockroaches, planthoppers, aphids, and termites.

For example, various bacteria, including *Streptococcus*, *Bacteroides*, and *Citrobacter* species, in the hindgut of the termite *Reticulotermes flavipes* degrade uric acid anaerobically to ammonia, carbon dioxide, and acetic acid. Experiments using ^{14}C and ^{15}N-labeled uric acid confirmed that uric acid is degraded by the hindgut microbiota in the insect and nitrogen is subsequently assimilated by the insect tissues, in the insect.

Microbial provision of essential amino acids to the insect has been studied systematically in the symbiosis between aphids and the intracellular bacteria *Buchnera*. The core evidence is nutritional, and arises from the development of chemically defined diets consisting of sucrose, amino acids, vitamins, and minerals, on which aphids can be reared. Dietary studies in which the 20 amino acids of proteins are individually omitted have revealed that many aphids have no specific requirement for most or all the amino acids that are normally dietary essentials for animals, but that aphids experimentally deprived of *Buchnera* by antibiotic treatment require all the essential amino acids. The implication, that the insect derives essential amino acids from *Buchnera*, is supported by radiotracer studies demonstrating the synthesis *de novo* of various essential amino acids by aphids bearing *Buchnera* (Fig. 54.3).

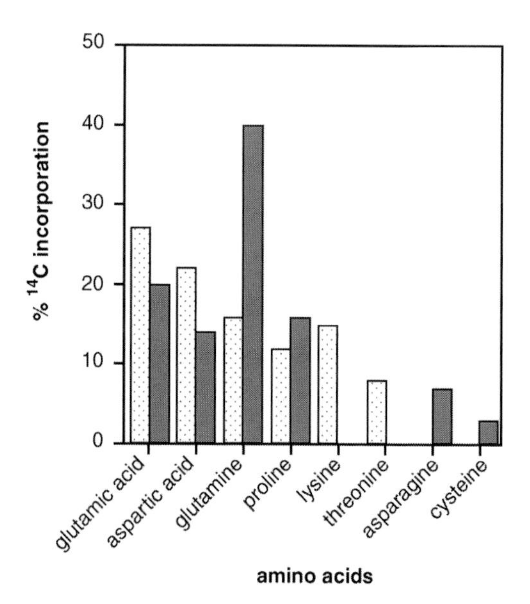

FIGURE 54.3 Essential amino acid synthesis by pea aphid-*Buchnera* symbiosis, as demonstrated by the incorporation of ^{14}C-glutamic acid into the free amino acid pool of aphids containing *Buchnera* (open-hatched) and experimentally deprived of *Buchnera* (closed-hatched). Only the aphids with the symbiotic bacteria synthesize the essential amino acids lysine and threonine. Reproduced from Fig. 1b of Wilkinson and Douglas (1996). *Entomol. Exp. Appl.* **80**, 279–282, with kind permission from Kluwer Academic Publishers.

The analysis of the plasmid profiles in *Buchnera* has revealed molecular support for the role of these bacteria in the amino acid nutrition of aphids. *Buchnera* in many aphids, including all members of the family Aphididae studied to date, bear two multicopy plasmids on which genes for the biosynthesis of tryptophan and leucine are amplified. Figure 54.4 shows the

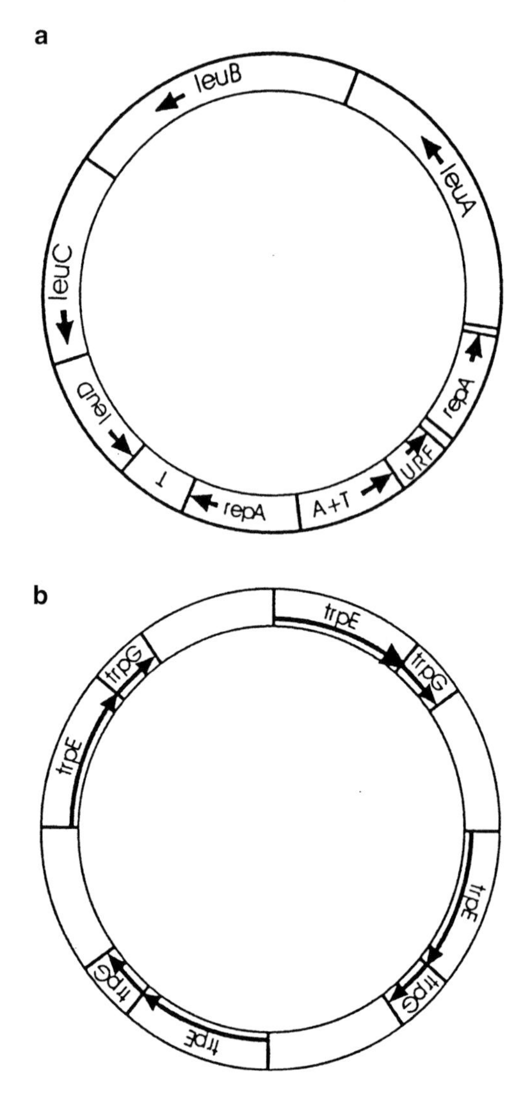

FIGURE 54.4 The plasmids of *Buchnera* from members of the family Aphididae. (a) Gene map of the leucine plasmid pRPE of *Buchnera* from *Rhopalosiphum padi* (7.8 kb) with genes *leuA–D* coding the enzymes in the dedicated leucine biosynthetic pathway. (b) Gene map of the tryptophan plasmid pBA-trpEG of *Buchnera* from *Schizaphis graminum* (14.4 kb) with four apparently identicial tandem repeats including *trpEG*, which codes for anthranilate synthase. This enzyme is the regulatory enzyme in tryptophan biosynthesis, subject to feedback inhibitition by tryptophan. It has been argued that amplification of *trpEG* results in the overproduction of anthranilate synthase and the consequent sustained tryptophan synthesis in the presence of tryptophan. Reproduced from Figs. 2 a & b of Douglas, A. E. (1997). *FEMS Microbiol. Ecol.* **24**, 1–9.

genetic organization of these plasmids in the *Buchnera* from which they were first described.

The synthesis of essential amino acid by intracellular bacteria has not been studied systematically in any insects apart from aphids. There is, however, a strong but unproven supposition that many insects, especially phloem-feeding Homoptera (e.g., whitefly and psyllids) and cockroaches, derive these nutrients from their symbiotic microorganisms. If confirmed, the essential amino acid provisioning has evolved independently in different bacterial groups.

C. Vitamin synthesis by symbiotic micro-organisms

The microbial provision of B vitamins has been proposed for the diverse insect taxa that feed on vertebrate blood and bear microorganisms in either the gut or bacteriocytes. The experimental basis for this role is exclusively nutritional, based on insect performance. Triatomid bugs deprived of their gut microbiota die as larvae, but this developmental arrest is alleviated by injection of B vitamins into the insect or vitamin supplementation of the diet. Similarly, the larvae of the louse *Pediculus* lacking symbiotic bacteria suffer high mortality, unless the blood diet is supplemented with nicotinic acid, pantothenic acid, and biotin, and B vitamins have been reported to partially restore the fecundity of tsetse fly from which the bacteria are eliminated.

Other insects feeding on nutritionally poor diets, (e.g., cockroaches and timber beetles) may also derive vitamins from symbiotic microorganisms. The anobiid beetle *Stegobium paniceum* is independent of several B vitamins, riboflavin, nicotinic acid, pyridoxine, and pantothenic acid, but insects from which the yeasts are eliminated require these vitamins for normal development.

D. Sterol synthesis by symbiotic microorganisms

Yeasts may contribute to the sterol nutrition of insects. Despite some claims in the literature, for example, that aphids derive sterols from their symbiotic bacteria *Buchnera*, bacterial provision of these nutrients is most improbable because the Eubacteria are not capable of substantial sterol synthesis.

Yeasts have been implicated in the sterol nutrition of some planthoppers and timber beetles. For example, when the yeast population in the planthopper *Laodelphax striatellus* is depleted, many of the insects die during the final molt to adulthood, but mortality was reduced from 94% to 40% by injecting the insects with either cholesterol or the plant sterol sistosterol.

E. Cellulose degradation

Many insects feeding on fiber-rich plant material, especially wood, have substantial gut microbiota. For many years, these insects have been assumed to be strictly comparable to vertebrate herbivores in which microorganisms mediate cellulose degradation. This can be illustrated by microbe-mediated cellulolysis in lower termites. Protists in the hindgut of lower termites (see Section I.B) can be eliminated by incubating the insects at elevated oxygen tensions, and these protist-free insects, commonly termed defaunated termites, cannot survive on high, cellulose diets, such as filter paper (Fig. 54.5). The protists degrade cellulose, with carbon dioxide and short-chain fatty acids, especially acetate, as the principal products of fermentation. The acetate is absorbed across the hindgut wall and metabolized as a source of energy by the aerobic tissues of the termite. Cellulase active against crystalline cellulose has been demonstrated in the protist *Trichomitopsis* and in mixed populations of protists from *Coptotermes lacteus*.

Although the experimental data on lower termites are not in serious doubt, this system cannot be generalized to all insects feeding on high-fiber diets. It is now recognized that some insects (unlike vertebrates) have intrinsic cellulases, especially endoglucanases

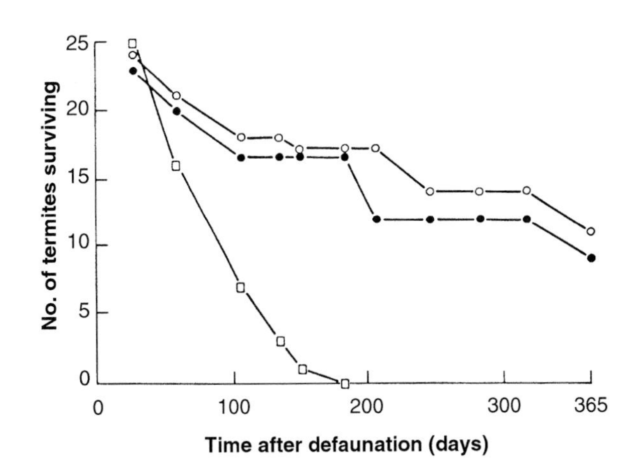

FIGURE 54.5 Survival of the lower termite *Zootermopsis* maintained on a diet of cellulose. Twenty-five days after the removal of the symbiotic protists from the hindgut defaunation by exposure to oxygen, termites were reinfected with protists from untreated termites (23 insects, filled circles), axenic culture of *Trichomitopsis termopsidis* (24 insects, open circles), and heat-killed *T. termopsidis* (25 insects, open squares), and their survival was assayed over 1 year. Reproduced from Fig. 8.8 of Smith and Douglas (1987). "The Biology of Symbiosis." Edward Arnold.

TABLE 54.2 Sites of cellulase activity in termites[a]

Gut region	Cellulase activity (% of total activity) Nasutitermes walkeri[b]	Mastotermes darwiniensis[c]
Foregut and salivary glands	5	13
Midgut	94	13
Hindgut	1	74

[a]Data from Veivers *et al.* (1982). *Insect Biochem.* **12**, 35–40 and Hogan *et al.* (1988). *J. Insect Physiol.* **34**, 891–899.
[b]A higher termite.
[c]A lower termite.

and β-glucosidases. For example, the higher termites (which lack protists) do not have cellulolytic microbiota, but instead possess high activities of intrinsic cellulases, especially in the midgut (Table 54.2).

The contribution of intrinsic and microbial cellulases to cellulose breakdown has been investigated in relatively few insects other than termites. The locust *Schistocerca gregaria* has intrinsic cellulases; the scarab beetle *Pachnoda marginata* uses the celluloytic capability of bacteria in its hindgut; and among the woodroaches (cockroaches that feed on wood), *Panesthia cribatus* uses intrinsic cellulases and has gut microbiota of noncellulolytic bacteria, whereas *Cryptocercus punctulatus* has up to 25 species of obligately anaerobic protists that degrade cellulose. There is no evidence that the efficiency of fiber digestion by the insect is influenced by the origin (intrinsic or microbial) of the cellulolytic enzymes.

Insects with microbial cellulolysis generally have a high population of methanogenic bacteria. These bacteria act as a sink for hydrogen produced by anaerobically respiring microorganisms, and so promote cellulose degradation.

III. DETERMINANTS OF THE DENSITY OF SYMBIOTIC MICROORGANISMS IN INSECTS

A. Gut microorganisms

Insect guts are physically unstable environments. The gut lumen is dominated by the unidirectional bulk flow of ingested food and many microbial cells pass directly through the gut (see also Section I.B). Microbial persistence in the gut is promoted by:

1. A higher proliferation rate than the rate of passage of the food (this is probably important in insects with cellulolytic microbiota in enlarged fermentation chambers).
2. The adhesion of microorganisms to the gut wall.
3. The sequestration of microorganisms into outpocketings of the gut (e.g., midgut caeca and Malpighian tubules).

A further aspect of the instability of the gut environment is that the cuticle and contents of the foregut and hindgut are lost at each insect molt, such that the microbiota is reestablished *de novo* multiple times through an insect's lifespan. The midgut microbiota can persist through insect molts, but is commonly lost at the metamorphosis of holometabolous insects (insects with complete metamorphosis, e.g., flies and beetles). In other respects, however, the midgut of many insects is a hostile environment for microorganisms because it is the principal site of digestive enzymes.

B. Intracellular microorganisms

The symbiotic microorganisms in the cells and tissues of insects are not generally subject to the frequent disturbance experienced by the foregut and hindgut microbiota (see Section III.A), and their populations are probably regulated by density-dependent processes, mediated by the insect.

The regulation of intracellular bacteria *Buchnera* in aphids has been well studied. In the wingless parthenogenetic morph of aphids, the bacterial population increases in parallel with aphid biomass through larval development (Fig. 54.6), and in adult aphids the bacteria occur at a density of ~10^7 cells/mg aphid fresh weight, equivalent to 10% of the total insect volume. Regulation occurs at two levels, controls over the bacterial division rate in each bacteriocyte such that they maintain a uniform density, equivalent to 60% of the cytoplasmic volume of bacteriocytes, and controls over the number of bacteriocytes through age-dependent lysis of the bacteriocytes and all enclosed bacteria. Adults of winged female aphids and male aphids tend to have a smaller *Buchnera* population than the wingless morph and these differences are mediated by both a lower rate of bacteriocyte enlargement and higher incidence of bacteriocyte death during larval development. The pos-sibility that these morph-specific differences are mediated by insect hormones has not been investigated. Reduced bacterial populations in males has also been demonstrated in leafhoppers, weevils, and cockroaches.

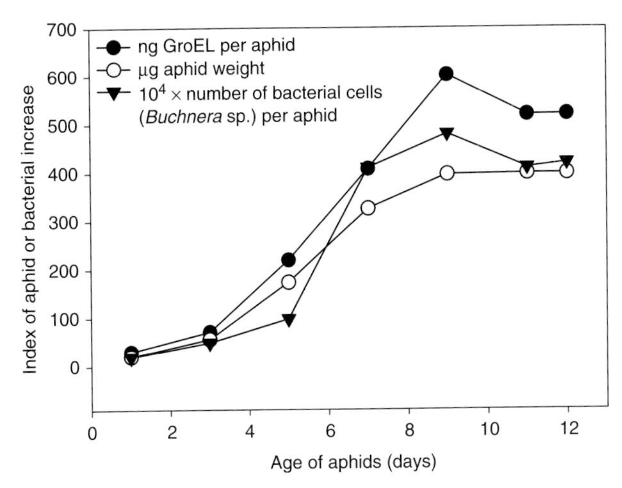

FIGURE 54.6 Controls over the population of symbiotic bacteria *Buchnera* over larval development of the aphid *Schizaphis graminum*. The increase in the bacterial population, as quantified both directly and from the increase in the bacterial protein GroEL, broadly parallels the increase in aphid weight. Reproduced from Fig. 4 of Baumann, Baumann, and Clark (1996). *Curr. Microbiol.* **32**, 279–285, copyright notice of Springer-Verlag.

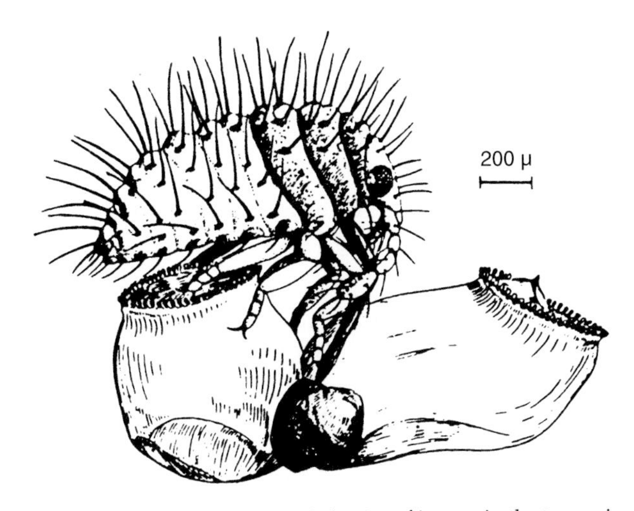

FIGURE 54.7 Role of feeding behavior of insects in the transmission of symbiotic microorganisms. The newly hatched larva of the heteropteran bug *Coptosoma scutellatum* feeds on a capsule bearing symbiotic bacteria that the mother had deposited alongside the egg. Reproduced from Fig. 94c of Buchner (1965). "Endosymbioses of Animals with Plant Microorganisms." Wiley, London.

IV. TRANSMISSION OF SYMBIOTIC MICROORGANISMS

A. Gut microbiota

All members of the gut microbiota are acquired via insect feeding. For many microorganisms, the transfer between insects is haphazard, dependent on the chance ingestion with food, but in many taxa, the insect eggs are provisioned with gut microorganisms. The transmission of some microorganisms is assured by stereotyped feeding responses of the insect. This can be illustrated by the heteropteran bug *Coptosoma scutellarum*, which bears bacteria in its midgut caeca. A capsule bearing bacteria is deposited alongside each oviposited egg. When the larva hatches, it immediately feeds on the capsule contents and acquires its complement of bacteria (Fig. 54.7), on which its growth and development depend.

Insect behavior is also implicated in the transmission of obligately anaerobic microorganisms, especially the cellulolytic protists in certain woodroaches and the lower termites. At each molt of the insect, the oxygen tension in the hindgut increases dramatically and all the protists are killed. In *Cryptocercus*, the protists initiate sexual reproduction just before each molt and are transformed into oxygen-resistant cysts. They are expelled from the hindgut into the environment, where they persist until ingested by the insect after molting. In contrast, the protists in lower termites rarely reproduce sexually and never encyst, and they are killed at each insect molt. The termites acquire a fresh inoculum of protists by feeding on a drop of hindgut contents from the anus of another colony member, a behavior known as proctodeal trophyllaxis. It has been suggested that the requirement for conspecifics as a source of protists may have been a major selection pressure for the evolution of eusociality in termites. Higher termites, which lack the protists, do not exhibit proctodeal trophyllaxis.

B. Transovarial transmission

Microorganisms located in insect tissues and cells are generally transmitted from mother to offspring via the unfertilized egg in the female ovary. As a result, the symbiotic microorganisms are present even before fertilization and, potentially, for the entire lifespan of the insect. The timing and anatomical details of transovarial transmission vary widely among insect groups, consistent with these associations having evolved independently on multiple occasions. In some insects (e.g., aphids), the bacteriocytes are closely apposed to the ovaries and the bacteria have a fleeting extracellular stage in transit from bacteriocyte to ovaries. In other insects, the bacteria have a prolonged extracellular phase, either because they are expelled from bacteriocytes at a distance from the ovaries and migrate to the ovaries (e.g., many species of lice) or because they remain on the egg surface for extended periods (e.g., cockroaches). Varying among insect taxa, the bacteria are phagocytosed directly by the egg, usually at the time of vitellogenesis (e.g., aphids and cockroaches) or taken up by insect cells at the base of each ovariole and inoculated into the posterior pole of the egg just prior to formation of the chorion (the egg shell) and ovulation.

C. Vertical transmission and its evolutionary consequences

Transovarial transmission and the more sophisticated instances of microbial transmission by egg smearing ensure that each egg is colonized by bacteria from its mother. If vertical transmission persists over many insect generations without cross-infection, the insect and microbial lineages evolve in parallel, and their phylogenies are congruent. Congruent phylogenies have been demonstrated for several insect–microbial symbioses, most notably the aphid–*Buchnera* association, in which the 16*S* rRNA sequence phylogeny of *Buchnera* is completely concordant with morphology-based phylogeny of aphids (Fig. 54.8). On the reasonable assumption that the aphids and *Buchnera* diversified in synchrony, the date of divergence between *Buchnera* and its relatives, such as *Escherichia coli*, has been estimated at 180–250 million years ago.

Vertical transmission, especially by the transovarial route, has two characteristics of significance for the evolution of symbiotic microorganisms. First, relatively small numbers of cells are transferred from parent to offspring; that is, the effective population size of the microorganisms is small. Second, the strict maternal inheritance prevents any contact between microbial populations in different insects, precluding recombination. Deleterious mutations are likely to accumulate in these small asexual populations. Supportive evidence comes from studies of sequence evolution in intracellular bacteria. First, among the *Buchnera* lineages, protein-coding genes have significantly elevated the incidence of nonsynonymous substitutions (i.e., those point mutations that alter the amino acid) and this is not paralleled by an increase in the rate of synonymous substitutions (Fig. 54.9A). Second, vertically transmitted intracellular bacteria have base substitutions in the 16*S* rRNA gene that tend to destablize the secondary structure of the rRNA molecule (Fig. 54.9B).

V. SYMBIOTIC MICROORGANISMS AND INSECT-PEST MANAGEMENT

Many insects that depend on symbiotic microorganisms for sustained growth and fecundity are pests of

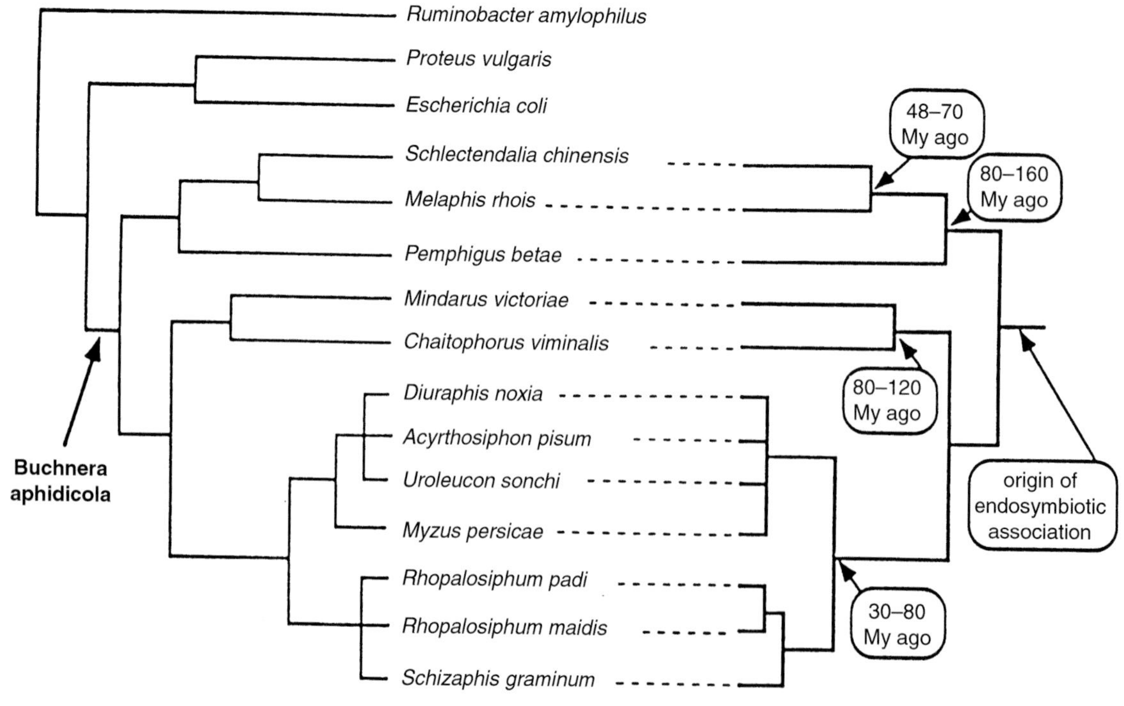

FIGURE 54.8 The congruent phylogenies of *Buchnera* and its aphid hosts. The phylogenies are based on 16*S* rDNA sequences for *Buchnera* and morphology for aphids, and the dates are estimates from aphid fossils or biogeography. Reproduced from Fig. 1 of Moran and Baumann (1994). *Trends Ecol. Evol.* **9**, 15–20, with permission from Elsevier Science.

agricultural or medical importance. They include crop pests (e.g., aphids, whitefly, grain beetles, and timber beetles) and vectors of animal and human pathogens (e.g., tsetse fly and bedbugs). In addition, certain microbial taxa have been implicated in the vector competence of their insect hosts. The transmission of luteoviruses by aphids is promoted by binding of a protein, the chaparonin GroEL derived from the intracellular bacteria *Buchnera*, to the surface of the virus particles in the insect body, and the inhibition of trypanosome infection in the tsetse fly by an insect lectin is blocked by high *N*-acetylglucosamine titers generated by the chitinase activity of a midgut bacterium.

The selective disruption of the symbiotic microorganisms in insect pests, and the consequent depression in insect performance, would be of considerable economic value but no commercial approach has been developed. There is also interest in the genetic

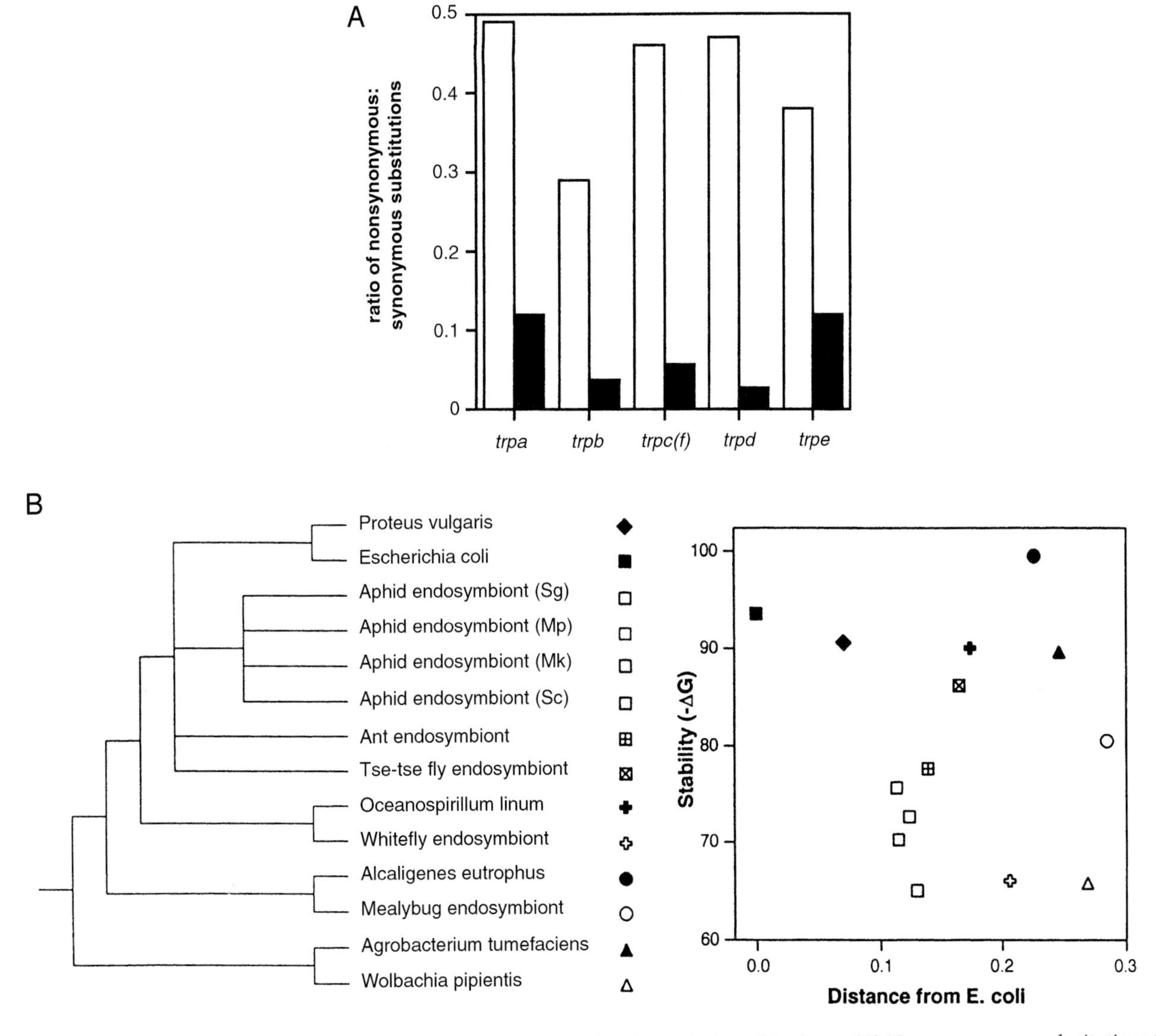

FIGURE 54.9 The consequences of vertical transmission for the molecular evolution of *Buchnera*. (A) Nonsynonymous substitutions in protein-coding genes of *Buchnera*. The number of substitutions between *Buchnera* in the aphids *Schizaphis graminum* and *Schlechtendalia chinensis* (open bars) and between the two enteric bacteria *Escherichia coli* and *Salmonella typhimurium* (solid bars) is expressed as the ratio of nonsynonymous : synonymous substitutions per nucleotide site. Data are presented for the five genes of the *trp* operon. Reproduced with permission from Fig. 1 of Hurst and McVean (1996). *Nature* **381**, 650–651. Copyright 1996 MacMillan Magazines Limited. (B) The stability of Domain I of the 16*S* rRNA in symbiotic bacteria and free-living bacteria, showing the relationships of bacterial taxa and stabilities (-Δ*G*) summed over Domain I for each organism. Reproduced from Fig. 2 of Lambert and Moran (1998). *Proc. Natl. Acad. Sci. U.S.A.* **95**, 4458–462. Copyright (1998) National Academy of Sciences, U.S.A.

manipulation of symbiotic microorganisms, for example to reduce insect-vector competence. The intracellular microorganisms are generally perceived as intractable because the methods to culture and transform them and to reintroduce the transformed bacteria into insects have not been developed. Most research has been conducted on blood-feeding insects with gut microbiota. For example, a plasmid of the actinomycete *Rhodococcus rhodnii*, which inhabits the gut of *Rhodnius prolixi*, has been genetically modified to bear the plasmid replication origins for both *R. rhodnii* and *E. coli*, and sustained infections of *Rhodnius* have been achieved with the transformed bacteria. This technology is being developed as part of a strategy to reduce the *Rhodnius*-mediated transmission of the protozoan pathogen that causes Chagas's disease in humans.

BIBLIOGRAPHY

Baumann, P., Lai, C-Y., Roubakhsh, D., Moran, N. A., and Clark, M. A. (1995). Genetics, physiology, and evolutionary relationships of the genus *Buchnera*—intracellular symbionts of aphids. *Annu. Rev. Microbiol.* **49**, 55–94.

Beard, C. B., O'Neill, S. L., Tesh, R. B., Richards, F. F., and Aksoy, S. (1993). Modification of arthropod vector competence via symbiotic bacteria. *Parasitol. Today* **9**, 179–183.

Breznak, J. A. (1982). Intestinal microbiota of termites and other xylophagous insects. *Annu. Rev. Microbiol.* **36**, 323–343.

Buchner, P. (1965). "Endosymbioses of Animals with Plant Microorganisms." Wiley, London.

Douglas, A. E. (1989). Mycetocyte symbiosis in insects. *Biol. Rev.* **64**, 409–434.

Douglas, A. E. (1998). Nutritional interactions in insectmicrobial symbioses: aphids and their symbiotic bacteria *Buchnera*. *Annu. Rev. Entomol.* **43**, 17–37.

Douglas, A. E., Minto, L. B., and Wilkinson, T. L. (2001). Quantifying nutrient production by the microbial symbionts in an aphid. *J. Exp. Biol.* **204**, 349–358.

Hurst, G. D., and Jiggins, F. M. (2000). Male-killing bacteria in insects: mechanisms, incidence, and implications. *Emerg. Infect. Dis.* **6**, 329–336.

Moran, N. A., and Telang, A. (1998). Bacteriocyte-associated symbionts of insects—A variety of insect groups harbor ancient prokaryotic endosymbionts. *BioScience* **48**, 295–304.

Moran, N. A., and Baumann, P. (2000). Bacterial endosymbionts in animals. *Curr. Opin. Microbiol.* **3**, 270–275.

O'Neill, S. L., Warren J. H., Hoffman, A. A. (Editors). (1998). "Influential Passengers: Inherited Microorganisms and Arthropod Reproduction." Oxford University Press, Oxford, UK.

Slaytor, M., and Chappell, D. J. (1994). Nitrogen metabolism in termites. *Comp. Biochem. Physiol.* **107B**, 1–10.

Thao, M. L., Moran, N. A., Abbot, P., Brennan, E. B., Burckhardt, D. H., and Baumann, P. (2000). Cospeciation of psyllids and their primary prokaryotic endosymbionts. *Appl. Environ. Microbiol.* **66**, 2898–2905.

Van Dohlen, C., Kohler, S., Alsop, S. T., and McManus, W. R. (2001). Mealybug proteobacterial endosymbionts contain proteobacteria symbionts. *Nature* **412**, 433–436.

WEBSITE

A bibliography on *Wolbachia*
http://www.wolbachia.sols.uq.edu.au/biblio.cfm

55

Iron metabolism

Charles F. Earhart
The University of Texas at Austin

GLOSSARY

bacterioferritin (BFR) A subclass of prokaryotic ferritin that contains heme.

ferritin (FTN) A large intracellular iron storage protein, present in both prokaryotic and eukaryotic cells.

haptoglobin (Hp) A serum protein that scavenges hemoglobin.

heme (Hm) A metal ion, customarily iron in the FeII state, chelated in a tetrapyrrole ring.

hemopexin A serum protein that strongly binds heme.

hemophore (Hbp) A secreted bacterial protein that shuttles environmental heme to the bacterial surface.

lactoferrin (LF) A glycoprotein present in the mucosal secretions and phagocytic cells of vertebrates that binds and transports iron.

siderophore A small molecule secreted by bacteria and fungi that chelates environmental FeIII and, as the ferrisiderophore, then binds to the microbial surface.

transferrin (TF) Iron-binding and transport protein found in the serum and lymph of vertebrates.

A Knowledge of Iron Metabolism is essential for a complete understanding of microbial growth and survival. In microbes, iron is necessary for processes such as electron transport, nitrogen fixation, removal of toxic forms of oxygen, synthesis of DNA precursors, tRNA modifications, and syntheses of certain amino acids and tricarboxylic acid cycle intermediates; some bacteria can even oxidize iron to obtain energy. Iron is such a valuable and versatile nutrilite that, of all the organisms on earth, only certain lactic acid bacteria can manage without it. The utility of iron stems primarily from the fact that its redox couple (FeII/FeIII) can have a range of potentials of -300 to $+700\,\text{mV}$, depending on the nature of the ligands and the environment surrounding the coordinated iron ions.

A major consideration in iron metabolism is that, although iron is abundant on the surface of the earth, it is relatively unavailable. At neutral pH and in an oxidizing environment, which includes most common microbial habitats, iron exists in the +3 valence state and, as such, is extremely insoluble. For a microbe inhabiting an animal host, iron availability is restricted by the presence of iron storage and transport proteins, such as ferritin, lactoferrin, and transferrin. Thus, although relatively low concentrations of iron ($5\,\mu\text{M}$) are generally sufficient for maximum growth yields, bacteria frequently find themselves in iron-deficient environments and must devote a significant amount of their resources to obtaining this metal. Also important is the fact that it is possible for bacteria to suffer from iron overload and iron assimilation must, therefore, be precisely controlled.

This article covers prokaryotic iron metabolism only. Descriptions of fungal iron assimilation are included in the reviews by Guerinot (1994) and Leong and Winkelmann (1998).

I. IRON UPTAKE

Bacteria can obtain iron from a variety of sources but, regardless of its origin, iron must be transported

The Desk Encyclopedia of Microbiology
ISBN: 0-12-621361-5

through the several microbial surface layers to reach the cytoplasm. For gram-negative bacteria, these layers minimally include an outer membrane, a monolayer of peptidoglycan, and an innermost cytoplasmic membrane. The peptidoglycan cell wall is located in the periplasm, the space between the outer and inner membranes. Gram-positive cells, in contrast, may contain only a thick, highly cross-linked peptidoglycan cell wall external to the cytoplasmic membrane.

A great variety of iron transport systems can be distinguished on the basis of the iron source and the form in which iron is mobilized, but, in general, they follow a pattern (Fig. 55.1). Thus, for iron complexed to a carrier, first, passage through the outer membrane requires an outer membrane receptor protein whose synthesis is iron-regulated. A receptor protein has specificity for a given iron–carrier complex, binds that complex only, and is sometimes synthesized in abundance only when that particular iron complex is available. In the two best-studied cases, receptor proteins have been shown to be gated pores. Second, to permit entry of complexed or free iron into the periplasm, cytoplasmic membrane proteins TonB, ExbB, and ExbD are required. These proteins function as a group to utilize the electrochemical potential of the cytoplasmic membrane to open the gated receptor pores, permitting iron and iron-chelates to pass into the periplasm. Generally, each bacterium has just one set of TonB/Exb proteins, capable of interacting with multiple receptors, but *Vibrio cholerae* and *Pseudomonas aeruginosa* have two distinct TonB systems. TonB is anchored in the cytoplasmic membrane but spans the periplasmic space, so as to be able to physically contact all receptors that require it. ExbB has three

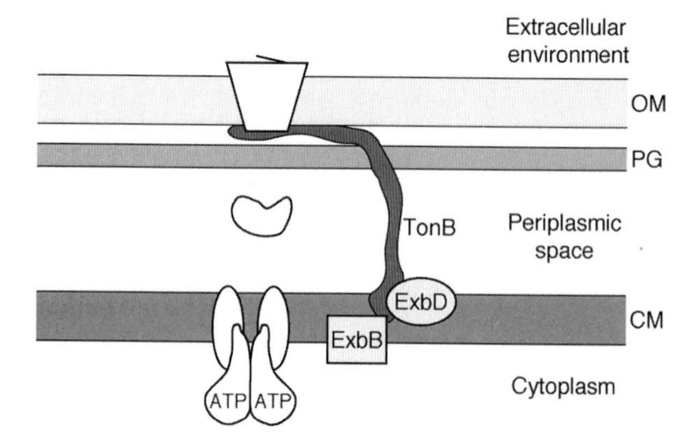

FIGURE 55.1 Generalized high affinity iron transport system of gram-negative bacteria. The three basic components are shown; they include: (i) an outer membrane receptor protein; (ii) a TonB system for energizing the receptor protein; and (iii) a periplasmic binding protein-dependent ABC transporter, located in the cytoplasmic membrane. OM, outer membrane; PG, peptidoglycan; CM, cytoplasmic membrane. See text for details.

TABLE 55.1 Consensus sequences related to iron

TonB box[a]	AspThr Hyd Val Val Thr Ala Glu
Fur binding site (Fur box, iron box)	GATAATGATAATCATTATC
DtxR binding site	TAAGGTAAGCTAACATA T G T

[a]Positions two and five are the most conserved. Position three is the most variable; Hyd indicates hydrophobic amino acid (Ile, Leu, Val, Met, Phe).

transmembrane domains and is present primarily in the cytoplasm, while ExbD has only one transmembrane segment and extends into the periplasm. Determination of the details by which TonB and its accessory proteins act is an area of active research. A straightforward mechanism would have ExbB sense the proton motive force and, with ExbD, use it to induce a conformational change in TonB, that, in turn, could bring about, by physical contact, allosteric changes in outer membrane receptors. TonB-dependent receptors have a heptapeptide (TonB box) near their amino termini that is thought to serve as a recognition sequence (Table 55.1). Third, transport across the cytoplasmic membrane employs members of the ABC supertransporter family. In this case, transport components consist of a peripheral cytoplasmic membrane ATPase, present in two copies and possessing a defining ATP binding site motif, and two hydrophobic cytoplasmic membrane proteins. These latter proteins can be similar but distinct polypeptides, one large fusion protein, or a homodimer. In addition, iron transporters have a component exterior to the cytoplasmic membrane. For gram-negative cells, the component is a periplasmic binding protein, which recognizes the iron substrate and presents it to the cytoplasmic membrane complex. In gram-positive cells, the periplasmic protein is, in some cases, replaced by a lipoprotein tethered in the cytoplasmic membrane but protruding into the peptidoglycan region.

In summary, entry of iron into the cytoplasm of gram-negative bacteria requires an outer membrane receptor protein, a TonB system, and an ABC transporter. Passage through the outer and cytoplasmic membranes depends on the proton motive force and ATP, respectively. Outer membrane receptor proteins bind just one specific iron complex, TonB systems have broad specificity, and ABC transporters often recognize several iron-complexes, provided these are structurally related.

A. Siderophore-mediated systems

A major mechanism by which bacteria obtain iron is through the production and secretion of small

(600–1000 Da), iron-chelating molecules termed siderophores. Many different siderophores have been isolated and characterized but they can generally be classified as being either of the catecholate or hydroxamate type (Fig. 55.2). Siderophores are synthesized in iron-deficient environments; after

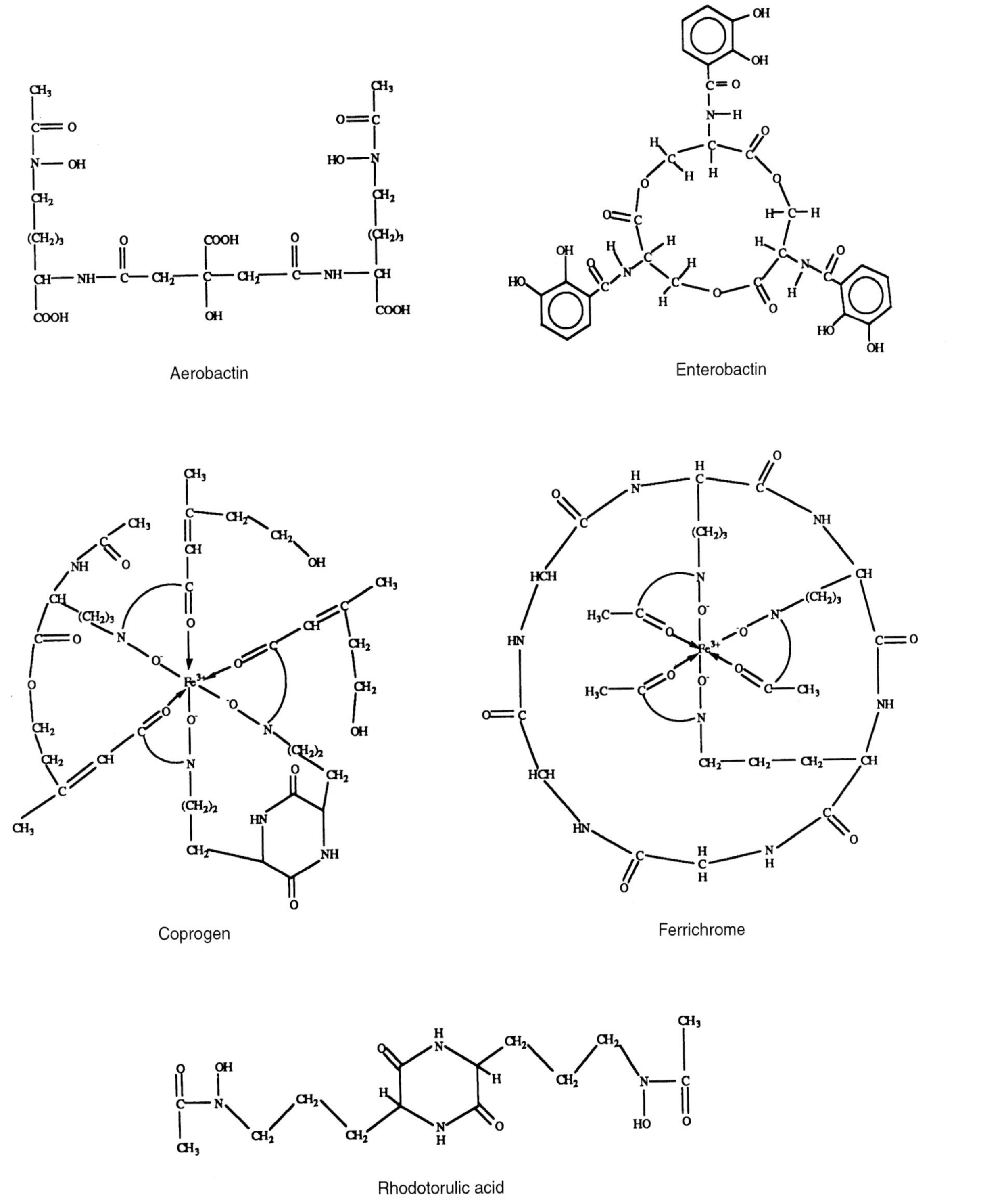

Aerobactin

Enterobactin

Coprogen

Ferrichrome

Rhodotorulic acid

FIGURE 55.2 Some representative siderophores <modified from Earhart (1996) with permission>. Aerobactin and enterobactin are bacterial products; rhodotorulic acid and the nonferrated forms of ferrichrome and coprogen are synthesized by fungi. Enterobactin is a catechol-type siderophore and the other four are hydroxamate-type siderophores. Remarkably, *Escherichia coli* can utilize all of these.

release from cells, siderophores bind iron and the siderophore– iron complex is subsequently internalized using the general type of transport system that has been described.

Siderophore biosynthesis is an interesting and growing area of research. Siderophores, generally, are built up of amino acids and hydroxy acids and, although they contain amide bonds, their synthesis does not involve ribosomes. Instead, a thiotemplate process with strong similarities to that employed for synthesis of certain peptide antibiotics is used. Nonribosomal peptide synthetases link activated precursors to the enzyme-bound cofactor 4′-phosphopantetheine (P-pant). The thioesterified precursors are then covalently linked by amide bonds in the sequence determined by their position on the synthetase. The synthetase enzymes can be huge; in some cases, molecular weights of greater than 350,000 have been reported. Also, studies on siderophore biosynthesis resulted in the discovery of a new class of enzymes, the phosphopanthetheinyl transferases, which donate the P-pant to the peptide synthetases. This essential posttranslational modification of synthetases permits them to bind activated precursors and then join the precursors together to yield the final product.

Other molecules, which do not meet the rigorous definition of siderophores, can also provide iron to cells by means of specific outer membrane receptor proteins acting in concert with binding protein-dependent ABC transporters. Citrate, when present in environmental concentrations of greater than 0.1 mM, is one such molecule, as are dihydroxybenzoic acid, a precursor of the common siderophore enterobactin, and dihydroxybenzoylserine, a breakdown product of enterobactin (Fig. 55.2).

Release of iron from siderophores is not well understood. Ferrisiderophores enter the cytoplasm and free siderophores, or modified versions, are released back into the medium. Because siderophores (i) bind FeII less avidly than FeIII and (ii) the cytoplasm is a reducing environment, enzymatic reduction of iron is thought to be the release mechanism. *In vitro* experiments have demonstrated the presence of a variety of ferrisiderophore reductase activities that employ reduced flavins to convert iron to the ferrous state. These enzymes have broad specificity and their abundance is not affected by iron availability. The Fes protein of *E. coli* (ferrienterobactin esterase) is the one case where a specific iron-regulated protein is necessary for release of siderophore-bound iron. This enzyme hydrolyzes Ent but this esterase activity may be secondary to its primary reductase role.

How newly synthesized siderophores are released is similarly unclear for most microbes. A *Mycobacterium smegmatis* mutant has been isolated that appears to be defective in siderophore secretion: the defective protein has no homologs in gene banks. For *E. coli*, the *entS* (*ybdA*) gene product encodes a protein with strong homology to proton motive force-dependent membrane efflux pumps. Secretion of the siderophore enterobactin but not its breakdown products is reduced in cells bearing *entS* mutations. The commonly held idea that siderophore secretion is facilitated by having the siderophore biosynthetic enzymes form a complex that is loosely associated with cytoplasmic membrane remains unproven. In fact, several recent studies argue that the enzymes do not form a complex *in vivo*.

B. Uptake of ferrous and ferric ions

Ferrous iron transport is useful to microbes capable of inhabiting oxygen-restricted environments, such as swamps, intestines, and marshes, or acidic locales, where reduced iron is stable and soluble. An *E. coli* ferrous iron transport system (*feo*) has been identified; two proteins, FeoA and FeoB, participate in FeII uptake. FeoB is a large cytoplasmic membrane protein with a nucleotide binding site, suggesting that ATP hydrolysis is the energy source for transport. The FeoA protein is small (less than 10,000 Da) and of unknown function. *Salmonella typhimurium*, a facultative anaerobe like *E. coli*, has *feo* genes and *Methanococcus jannischii* has a homolog of *feoB*. The latter finding is of interest in that *M. jannischii* is classified in the domain Archaea, not in the domain Bacteria. Lastly, certain aerotolerant bacteria, such as the gram-positive organism *Streptococcus mutans*, obtain iron by using a reductase exposed on the cell exterior to convert surface-bound FeIII to FeII. A ferrous ion transporter then delivers the iron to the cytoplasm.

A ferric iron acquisition system of the ABC type is present in a number of gram-negative genera including *Serratia*, where it was discovered and termed Sfu type transport, *Haemophilus*, *Yersinia*, *Actinobacillus*, and *Neisseria*. A periplasmic binding protein accepts FeIII and presents it to the cytoplasmic components of the transporter, which internalize the iron. Ferric iron transporters can also function in concert with uptake systems that have outer membrane components. In these cases, they become the terminal portion of the assimilation path. Thus, iron is removed from transferrin and lactoferrin at the outer membrane (see following), a process which requires a receptor and active TonB system, and then translocated into the cytoplasm by the ferric transporter. Also, iron can be transferred from citrate to a periplasmic binding protein for entry into the cytoplasm.

C. Uptake of iron from heme

The vast majority of iron in animals is present intracellularly as heme (Hm). Heme, in turn, serves as a prosthetic group of proteins, primarily hemoglobin (Hb), but also myoglobin and Hm-containing proteins such as cytochromes. Cell lysis is necessary for release of these proteins. Freed Hb, such as that present following hemolysis of erythrocytes, is bound by the serum glycoprotein haptoglobin (Hp). Hb not bound by Hp can oxidize, in the process releasing Hm from globin. This extracellular Hm is bound by the plasma protein hemopexin and, less specifically, by serum albumin. Therefore, extracellular Hm, available as a possible source of iron (and Hm), occurs infrequently by itself as it is usually bound to Hb, Hb-Hp complexes, and hemopexin. In no case has a siderophore been able to scavenge iron from Hm. However, as will be described, each Hm source is capable of being utilized by one or another group of bacteria. These uptake systems vary greatly. An additional complication is that there is such variability among strains of a given species that it is often not possible to list the systems present in a given species. For instance, *E. coli* lab strains cannot use either Hm or Hb as iron sources but pathogenic *E. coli* strains can often use both.

Iron assimilation pathways that recognize free Hm resemble those for iron-siderophore complexes; they require (i) a single, ligand-specific outer membrane receptor protein that is TonB-dependent and (ii) for cytoplasmic membrane passage, an ABC transporter. For several of these systems, including those of *Yersinia enterocolitica, Vibrio cholerae*, and *Shigella dysenteriae*, it has been deduced that the entire heme molecule enters the cytoplasm. (In these cases, heme supported porphyrin growth requirements, as well as providing iron.) Little is known regarding the intracellular release of iron from Hm. HemS of *Y. enterocolitica* may provide this function, and HmnO of *Corynebacterium diphtheriae*, one of only a few gram-positive species able to assimilate Hm, may be a Hm oxygenase activity.

Several genera can remove Hm from Hb. *Neisseria* and *Haemophilus* spp. have TonB-dependent Hb-binding proteins in their outer membranes. Remarkably, both of these genera have additional TonB-dependent receptors that function in iron acquisition from Hb-Hp complexes. Each of these Hb-Hp receptors may consist of two different proteins. A different mechanism for the initial steps in obtaining iron from Hm or Hb is found in *Serratia marcescens*. This organism uses an ABC transporter to secrete a small protein (HasA) that functions as a hemophore (Hbp). That is, HasA is an extracellular Hm binding protein that is necessary for uptake of Hm, either free or bound to Hb. It shuttles its bound Hm to an outer membrane receptor (HasR). *E. coli* strains harboring the virulence plasmid pColV-K30 also secrete a Hbp. Unlike HasA, Hbp is autotransported out of the cell and is bifunctional. It has protease activity, degrading Hb as well as binding and transporting the released Hm. A chromosomally encoded outer membrane receptor protein (ChuA), required for the well-known *E. coli* pathogen O157:H7 to use Hb or Hm, may be the Hbp-Hm receptor.

Only *Haemophilus* strains are known to utilize Hm associated with hemopexin. The system is not well characterized but three genes are required and one appears to encode a large secreted Hbp (HxuA). HxuA binds Hm-hemopexin, removes the Hm, and carries Hm to an outer membrane receptor. The other two genes encode proteins concerned with the secretion of HxuA. Like all *Haemophilus* Hm transport systems (free Hm, Hb:Hp, Hm:albumin), the Hm-hemopexin system requires a functional TonB protein.

D. Acquisition of iron from transferrin and lactoferrin

Transferrin (TF) and lactoferrin (LF) are extracellular iron transport molecules present in the fluids of many vertebrates; each of these related glycoproteins can bind two ferric ions. Many siderophores are capable of removing iron from these glycoproteins. *Neisseria* and *Haemophilus* produce no siderophores, however, and, instead, they have specific TonB-dependent outer membrane receptors that bind these transport proteins. An ABC transporter necessary for all non-heme iron uptake pathways (TF, LF, or iron chelates) is present in *Neisseria*; iron from these sources passes through the outer membrane and is bound by periplasmic protein FbpA, prior to entry into the cytoplasm. Outer membrane TF and LF receptors are unusual in that they appear to be bipartite; one protein has the characteristics of a typical TonB-dependent receptor, while the second is a surface-exposed lipoprotein. These receptors in *Neisseria* and *Haemophilus* show high specificity for glycoproteins of their normal hosts, such as humans.

E. Low affinity iron transport

The high affinity iron transport systems described immediately above are generally not expressed in environments containing more than 5–10 μM iron. Remarkably, the means by which bacteria assimilate iron in such iron-replete environments, oxic or anoxic, is not known. This so-called low affinity iron uptake may first require that iron be in the FeII form. Ascorbic acid, which reduces FeIII, stimulates low affinity iron uptake. FeII could be assimilated either by (i) the repressed levels of the *feo* system proteins or (ii) a cytoplasmic membrane transporter with broad specificity for divalent cations, such as the CorA protein of *E. coli* and *Salmonella typhimurium*. Also, several types of small molecules (monocatecholates, α-keto acids, and α-hydroxy acids) can, in certain genera, function as siderophores; they could mobilize extracellular ferric iron and, using a variety of transporters, provide iron.

II. IRON-DEPENDENT REGULATION

Genes encoding proteins necessary for high affinity iron uptake are regulated by iron availability. The key regulatory protein in most bacteria, gram-positive and gram-negative, is Fur. Fur, a small, histidine-rich polypeptide, is an aporepressor; in the presence of its corepressor FeII, it binds DNA. The holorepressor binds operators termed iron boxes, which are AT rich 19 bp sequences with dyad symmetry (Table 55.1). Iron boxes are located in promoters of iron-regulated genes, such that steric hindrance prevents the repressor and RNA polymerase from binding simultaneously. Adequate intracellular iron supplies thus prevent transcription of genes for transport proteins and siderophore biosynthetic enzymes. In *E. coli*, there are approximately 50 such Fur-regulated genes.

Negative control of genes by Fur does not completely explain the regulatory effects of iron. Although most iron-regulated genes are repressed by Fur in iron-replete conditions, some are positively controlled by Fur and some are induced by iron only in the absence of Fur. A subset of the *E. coli* proteins positively regulated by Fur was recently demonstrated to be controlled indirectly by a small RNA (RyhB) that could function as an antisense RNA. RyhB synthesis is negatively regulated by Fur; in the absence of holorepressor, RyhB is made and blocks synthesis of relevant proteins at a post-transcriptional initiation step. This complexity also arises because Fur can influence the expression of other regulatory systems and because some iron-regulated genes are also subject to other global control molecules. This latter observation becomes understandable when the key role of iron in metabolism is considered. For instance, iron requirements should depend in part on whether the fueling reactions being utilized require an electron transport system (respiration), with its requisite Hm and iron–sulfur proteins and propensity to generate toxic oxygen species, or not (fermentation).

Gram-positive organisms whose DNA has a high GC content, like *Corynebacterium* and *Streptomyces*, have no Fur but, instead, accomplish iron-regulated transcriptional control with DtxR proteins. The protein sequence of DtxR is unlike that of Fur but DtxR, nonetheless, functions in a Fur-like manner. FeII-DtxR binding sites are 19 bp palindromes (Table 55.1) and are positioned in the promoters of genes they regulate.

Some iron transport systems are produced only in iron-deficient environments that contain their cognate ferrisiderophores. That is, the systems are synthesized only if the intracellular iron concentration, as sensed by Fur, is low and if their specific ferrisiderophore is bound to the cell. As established in the ferridicitrate system of *E. coli*, the specific outer membrane receptor must be present, as must a functional TonB system. Binding of ferrisiderophore to the receptor transmits a signal to a cytoplasmic membrane protein that has both periplasmic and cytoplasmic domains. The signal is then passed to a cytoplasmic sigma-factorlike protein that, upon activation, associates with RNA polymerase and stimulates transcription of the necessary transport genes. Unique regulatory elements for these systems include an outer membrane receptor capable of interacting with the cytoplasmic membrane protein, the cytoplasmic membrane protein, and its partner sigma-like factor. Among Pseudomonads, this dual regulation of specific iron transport systems by Fur and surface binding of the relevant ferrisiderophore is common.

Regulation of iron assimilation not only conserves bacterial carbon and energy resources but also prevents overaccumulation of iron. Iron overload is a legitimate concern in bacteria as excess free iron can promote the formation of the toxic superoxide (O_2^-) and hydroxyl (OH·) radicals. The significance of DNA damage brought about by iron-generated hydroxyl radicals is demonstrated by the finding that *E. coli* Fur$^-$ mutants, which assimilate too much iron, that are also defective in DNA repair, cannot survive in oxic environments.

The intracellular oxidation state of iron is also used for regulation. Several proteins that control genes whose functions are related to oxygen levels contain iron–sulfur clusters; the iron present in these clusters is used as a sensing device. The *E. coli* Fnr protein, a transcriptional activator needed for anaerobic growth, is active only when its Fe–S center is reduced. In contrast, the sensor protein for superoxide stress, SoxR, is a positive regulator that is active with an oxidized Fe–S cluster.

III. INTRACELLULAR "FREE" AND STORED IRON

Two types of iron storage proteins, ferritin (FTN) and bacterioferritin (BFR), are found in bacteria. BFRs are a subfamily of FTNs, distinguished by the fact that they contain heme units. Like eukaryotic ferritins, the bacterial proteins are large (ca. 500,000 Da) and composed of 24 subunits.

The actual role of these proteins is uncertain. Unlike eukaryotic FTNs, which can accommodate over 4000 iron atoms, maximum iron contents for FTN and BFR are much less. BFR is synthesized primarily during periods of slow or no growth and may serve as an iron donor upon resumption of growth. FTN, on the other hand, is present at a constant low level throughout the growth cycle and, for at least *Escherichia coli* and *Campylobacter jejuni*, has been shown to protect against iron-catalyzed oxidative damage.

A major uncertainty regarding bacterial iron metabolism is the distribution and form of much of the intracellular iron. Even the quantity of iron in a bacterium is unclear; for the well-studied organism *E. coli*, estimates of the number of iron atoms/cell range from 100,000 to 750,000. In a variety of bacteria, Hm-containing proteins account for less than 10% of the total iron, in contrast to the situation in animals. Iron-sulfur proteins also contain 10% or less of the bacterial iron, as do FTN and BFR when cells grown under iron-deficient conditions are examined. The remainder of the bacterial iron, the majority, exists in a mobile pool about which little is known. It is this mobile iron that presumably is responsible for iron regulation and for the toxic effects of iron overload. Whether or not iron passes through this pool before being placed by ferrochelatase into protoporphyrin *IX* to form Hm, before incorporation into Fe–S clusters of proteins by unknown means, or before being converted to FeIII by the ferroxidase activity of BFR and FTN and deposited in the core of these molecules is unknown.

IV. IRON IN PRIMARY FUELING REACTIONS

Some bacteria can use the oxidation of iron compounds as their primary energy source. Bacteria capable of using inorganic, rather than organic, molecules for their fueling reactions are termed chemolithotrophs and iron-oxidizing bacteria are a major group in this nutritional category. Iron-oxidizing bacteria typically live in acidic, aerobic environments rich in both reduced iron and sulfur compounds; they grow poorly at pH values greater than 4. Among other things, low pH is critical in keeping FeII from being spontaneously oxidized to FeIII.

Thiobacillus ferrooxidans, the best studied microbe in this group, oxidizes iron using proteins in the cell envelope. An outer membrane complex oxidizes iron to the ferric form, and a periplasmic protein transfers the electrons to cytochromes in the cytoplasmic membrane; these, in turn, pass the electrons to oxygen, the ultimate electron acceptor, in the cytoplasm.

T. ferrooxidans is a major cause of water pollution, specifically, acid mine drainage. Ferrous sulfide is common in many coal and ore sites and its chemical and bacterial oxidation leads to both acidification and addition of dissolved metals to the water. Downstream, as the pH of the mine water becomes less acidic, insoluble ferric precipitates form.

In contrast to the relatively few bacteria which can oxidize iron as the initial step in generating energy, in the absence of oxygen, many bacteria, notably Shewanella species and members of the Geobacteriaceae family, can use FeIII as the terminal electron acceptor in fueling reactions. This dissimilatory reduction of iron is a form of anaerobic respiration and is of interest for several reasons. It has the geochemical significance of solubilizing iron and it is likely to represent a very early form of respiration, as all of the last common ancestors of modern organisms can reduce FeIII (using dihydrogen as the electron donor).

V. IRON AND PATHOGENICITY

A major stimulus to current studies on microbial iron metabolism is their medical significance. A brief review of the genera studied in Section I will emphasize this fact.

Bacterial pathogens find themselves in an iron-deficient environment when they invade a eukaryotic host; conditions facilitating their acquisition of iron would be expected to increase virulence. Several general observations support this idea. Humans with higher than normal iron levels show enhanced susceptibility to bacterial infections. Similarly, at least 20 bacterial pathogens are more virulent when their animal host is injected with iron compounds prior to infection. Because excess iron can have a number of effects, including some that are detrimental to host immune defenses, these data are somewhat ambiguous. They are bolstered, however, by *in vitro* experiments showing that the antibacterial effects of body fluids can be uniquely reversed by iron supplementation.

Bacteria must synthesize an appropriate iron uptake system to overcome the bacteriostatic conditions in their hosts. Pathogens which multiply extracellularly, such as in blood or on mucosal surfaces, and which do not lyse host cells must be able to obtain iron from TF or LF. This can be accomplished by siderophores or by TF- or LF-specific receptor proteins. Septicemic bacteria with defective siderophore systems are, in fact, less virulent. On the other hand, the major source of iron for intracellular pathogens is Hm and siderophore-deficient mutants of these organisms are still virulent. The required iron is taken from Hm-compounds, in this case.

For many pathogens, it has been difficult to identify specific iron uptake systems that are essential for their virulence. The multiplicity of means for assimilating iron in every species studied complicates such attempts. However, when all high affinity systems are inactivated by use of mutations in *tonB*, *Salmonella typhimurium*, *Vibrio cholerae*, and *Haemophilus influenzae* are avirulent. That a specific iron assimilation pathway can be a virulence factor now has been demonstrated for a number of pathogens. The type of transport system critical for virulence varies with the pathogen and can be species specific. Thus, *Neisseria meningitidis* lacking its outer membrane hemoglobin receptor (HmbR) is attenuated for meningococcal infection in infant rats. *Neisseria gonorrhoeae* with no functional transferrin receptor is unable to initiate urethritis in humans. High affinity ferrous iron transport (FeoB mediated) is important for *Helicobacter pylori* colonization of the gastric mucosa of mice and a functional siderophore system is required for *Yersinia pestis* virulence in mice. *Y. pestis* strains unable to either synthesize or transport the siderophore yersiniabactin are avirulent. A siderophore system is also a virulence factor in the fish pathogen *Vibrio anguillarum*. This marine microbe infects salmonid fish; approximately 10 bacteria per fish are sufficient to cause vibriosis, a fatal disease characterized by hemorrhagic septicemia. A plasmid-encoded iron-uptake system that uses the siderophore anguibactin is necessary for virulence.

Last, many pathogens utilize the low iron concentrations in the host as an environmental signal to synthesize virulence factors. Fur and DtxR-like proteins control not only iron acquisition systems but synthesis of toxins and hemolysins in these organisms.

BIBLIOGRAPHY

Braun, V., Hantke, K., and Koøster, W. (1998). Bacterial iron transport: Mechanisms, genetics, and regulation. *In* "Metal Ions in Biological Systems," Vol. 35, "Iron Transport and Storage in Microorganisms, Plants, and Animals" (A. Sigel and H. Sigel, eds.), pp. 67–145. Marcel Dekker, Inc., New York.

Bullen, D. J., and Griffiths, E. (eds.) (1999). "Iron and Infection: Molecular, Physiological, and Clinical Aspects," 2nd edn. Wiley, New York.

Byers, B. R., and Arceneaux, J. E. L. (1998). Microbial iron transport: Iron acquisition by pathogenic microorganisms. *In* "Metal Ions in Biological Systems," Vol. 35, "Iron Transport and Storage in Microorganisms, Plants, and Animals" (A. Sigel and H. Sigel, eds.), pp. 37–66. Marcel Dekker, Inc., New York.

Crosa, J. H. (1997). Signal transduction and transcriptional and posttranscriptional control of iron-regulated genes in bacteria. *Microbiol. Mol. Biol. Rev.* **61**, 319–336.

Earhart, C. F. (1996). Uptake and metabolism of iron and molybdenum. *In "Escherichia coli* and *Salmonella*: Cellular and molecular biology," Vol. 1, 2nd edn. (F. C. Neidhardt, R. Curtiss, III, J. L. Ingraham, E. C. C. Lin, K. B. Low, B. Magasanik, W. S. Reznikoff, M. Riley, M. Schaechter, and H. E. Umbarger, eds.), pp. 1075–1090. ASM Press, Washington, DC.

Expert, D., Enard, C., and Masclaux, C. (1996). The role of iron in plant host–pathogen interactions. *Trends Microbiol.* **4**, 232–237.

Genco, C. A. and Dixon, D. W. (2001). Emerging strategies in microbial haem capture. *Mol. Microbiol.* **39**, 1–11.

Guerinot, M. L. (1994). Microbial iron transport. *Annu. Rev. Microbiol.* **48**, 743–772.

Leong, S. A., and Winkelmann, G. (1998). Molecular biology of iron transport in fungi. In "Metal Ions in Biological Systems," Vol. 35, "Iron Transport and Storage in Microorganisms, Plants, and Animals" (A. Sigel and H. Sigel, eds.), pp. 147–186. Marcel Dekker, Inc., New York.

Hantke, K. (2001). Iron and metal regulation in bacteria. *Curr. Opin. Microbiol.* **4**, 172–177.

Modun, B., Morrissey, J. and Williams, P. (2000). The staphylococcal transferrin receptor: a glycolytic enzyme with novel functions. *Trends Microbiol.* **8**, 231–237

Neilands, J. B. (1995). Siderophores: structure and function of microbial iron transport compounds. *J. Biol. Chem.* **270**, 26723–26726.

Otto, B. R., Verweij-van Voght, A. M. J. J., and MacLaren, D. M. (1992). Transferrins and heme-compounds as iron sources for pathogenic bacteria. *Crit. Rev. Microbiol.* **18**, 217–233.

Payne, S. M. (1988). Iron and virulence in the family enterobacteriaceae. *Crit. Rev. Microbiol.* **16**, 81–111.

Ratledge, C., and Dover. L. G. (2000). Iron metabolism in pathogenic bacteria. *Annu. Rev. Microbiol.* **54**, 881–941.

Wandersman, C., and Stojiljkovic, I. (2000). Bacterial heme sources: the role of heme, hemoprotein receptors and hemophores. *Curr. Opin. Microbiol.* **3**, 215–220.

Winkelmann, G., van der Helm, D., and Neilands, J. B. (eds.) (1987). "Iron Transport in Microbes, Plants and Animals." VCH Press, Weinheim, Germany.

WEBSITE

List of "Iron Mavens" Labs
http://www.esb.utexas.edu/paynelab/iron.html

56

Lipopolysaccharides

Chris Whitfield
University of Guelph

GLOSSARY

core oligosaccharide (core OS) A branched and often phosphorylated oligosaccharide with varying glycan composition that is linked to lipid A. The inner core OS is more conserved and generally contains 3-deoxy-D-*manno*-octulosonic acid (Kdo) and L-*glycero*-D-*manno*-heptose (Hep) residues. The outer core OS is more variable in structure.

endotoxin In gram-negative sepsis, the lipid A component of LPS may stimulate macrophages and endothelial cells to overproduce cytokines and proinflammatory mediators. This can lead to septic shock, a syndrome involving hypotension, coagulopathy, and organ failure.

lipid A An acylated and phosphorylated di- or monosaccharide that forms the hydrophobic part of LPS.

lipooligosaccharide (LOS) A form of LPS often produced by mucosal pathogens, including members of the genera *Neisseria, Haemophilus, Bordetella* and others. LOS lacks O-PS but has oligosaccharide chains extending from the inner core OS. These chains are frequently antigenically phase-variable.

Lipopolysaccharide (LPS) An amphiphilic glycolipid found exclusively in gram-negative bacteria. LPS forms the outer leaflet of the outer membrane in the majority of gram-negative bacteria.

O polysaccharide (O-PS) A glycan chain attached to the core OS. Structures of O-PSs vary considerably and give rise to O antigens that define O-serospecificity in serological typing.

R-LPS Rough LPS, an LPS form that is truncated by the absence of O-PS and, in some cases, by parts of the core OS.

S-LPS Smooth LPS, a form of LPS common in the families *Enterobacteriaceae, Pseudomonadaceae,* and *Vibrionaceae* among others. S-LPS has a tripartite structure comprising lipid A, core oligosaccharide, and O-PS.

The Cell Envelope of gram-negative bacteria is characterized by its outer membrane. The outer membrane is an asymmetric lipid bilayer, in which the inner leaflet contains phospholipids and the outer leaflet contains the unique amphiphilic glycolipid known as lipopolysaccharide (LPS).

There are estimated to be ~10^6 LPS molecules per *Escherichia coli* cell. The distinctive structural features of LPS are crucial for the protective barrier properties of the outer membrane. In gram-negative sepsis, LPS molecules released from the bacterial surface stimulate macrophages and endothelial cells to overproduce cytokines and proinflammatory mediators, leading to the often fatal syndrome of septic shock. The involvement of LPS in this process is the reason that it is often referred to as endotoxin, and these biological effects have inspired a substantial part of LPS research. The complex structures of LPS molecules also provide fascinating research topics in the areas of synthesis and export of macromolecules, as well as in membrane biogenesis. The broad spectrum of LPS research is reflected in the activities of the International Endotoxin Society (*http://www.kumc.edu/IES/*).

The Desk Encyclopedia of Microbiology
ISBN: 0-12-621361-5

I. LIPOPOLYSACCHARIDE STRUCTURE

Early structural analyses of LPSs were driven, in part, by the need to resolve the identity of the molecule responsible for the endotoxic effect. One of the key breakthroughs in early LPS research came from the establishment of techniques for the extraction and isolation of LPS by O. Westphal and O. Lüderitz in the 1940s. Although other methods have followed, the hot aqueous phenol method they developed remains one of the most common and valuable extraction procedures in current use. This early work led to the understanding that the endotoxic phenomenon is attributable to LPS, and equally importantly, the finding that LPS molecules with similar compositions are found in different gram-negative bacteria. More recent application of analytical techniques such as nuclear magnetic resonance spectroscopy and mass spectroscopy, has led to highly refined structures for LPS molecules from diverse bacteria. It is now clear that there are general structural features or themes that are highly conserved in LPSs from different sources, but that there is significant variation in the structural fine details.

Extensive research has been performed on the LPS molecules of *Salmonella enterica* serovar Typhimurium and *E. coli*, and these LPSs form a basis for comparative analysis of other LPSs. For the purpose of discussion, the LPSs of *E. coli* and *S. enterica* sv. Typhimurium can be conveniently subdivided into three structural domains (S-LPS; Fig. 56.1). Lipid A is the hydrophobic part of the LPS molecule and is a major component of the outer leaflet of the outer membrane. Extending outward from lipid A is the branched and often phosphorylated oligosaccharide known as the core oligosaccharide (core OS). The O-antigen side chain polysaccharide (O antigen; O-PS) is a polymer of defined repeat units, attached to the core OS. The O-PS extends from the surface to form a protective layer. This complete tripartite LPS structure is known as "smooth LPS" (S-LPS), taking its name after the "smooth" colony morphology displayed by enteric bacteria that have the complete molecule on their cell surfaces. Mutants with defects in O-PS or core OS assembly produce truncated LPS molecules. For example, the widely used *E. coli* K-12 strains carry a defect in O-PS biosynthesis. The resulting colonies lack the smooth character, and the truncated LPS is, therefore, widely known as "rough LPS" (R-LPS) (Fig. 56.1). Preparations of LPS from bacteria that produce S-LPS contain a heterogeneous mixture of molecules and always have a variable amount of truncated R-LPS. Some bacteria, particularly mucosal pathogens naturally lack O-PS chains in their LPS. Their LPS contains oligosaccharide extensions attached to various points of a typical inner core OS, in a form of LPS known as lipooligosaccharide (LOS) (Fig. 56.1).

The LPS molecules from different bacteria typically show closer structural relationships in the cell-proximal lipid A and inner core OS regions and increasing diversity in the distal outer core OS and O-PS domains. The inner portions of the LPS molecule play important roles in establishing the essential barrier function of the outer membrane, and this likely places constraints on the extent of structural variation. The outer parts of the LPS molecule interact with environmental factors, such the host immune response. These selective pressures may have played a significant role in the diversification of outer LPS structures.

Some bacteria, including *Sphingomonas paucimobilis, Treponema maritime,* and *Borrelia burgorferi,* have outer membranes that lack LPS molecules entirely. In some cases genome sequences lack key genes for lipid A synthesis and therefore support compositional data. In the case of *S. paucimobilis,* glycosphingolipids probably serve to replace the lipid A, and the same may be true for the other examples. In those organisms that have a traditional LPS, it is generally thought that lipid A is essential for viability. However, the recent identification of a viable *Neisseria meningitidis* mutant with a defect in an essential step in the lipid A synthesis pathway challenges the universality of this assumption.

A. Lipid A

Free lipid A is not found on the bacterial cell surface but mild-acid treatment of most isolated LPSs releases lipid A by hydrolysis of the labile ketosidic linkage between the core OS and lipid A. Structural analysis of lipid A is hampered by its microheterogeneity, as well as its amphipathic properties. However, the lipid A components of LPSs from a variety of bacteria have now been resolved, revealing a family of structurally related glycolipids based on common architectural features. In enteric bacteria, the backbone of lipid A is formed by a disaccharide, comprising two glucosamine (GlcN) residues joined by a β1,6-linkage. The disaccharide backbone is phosphorylated at the 1 (reducing) and 4′ (nonreducing) positions and is acylated with ester and amide-linked 3-hydoxyl saturated fatty acids. In *E. coli* and *Salmonella*, the fatty acyl chains on the nonreducing GlcN residue are substituted by nonhydroxylated fatty acids, creating an asymmetric arrangement (Fig. 56.2).

Microheterogeneity is evident in lipid A preparations isolated from a given bacterium and several regulated modifications to the lipid A structure are evident in cells grown under specific conditions. Work with *Salmonella* has established that the two-component environmental

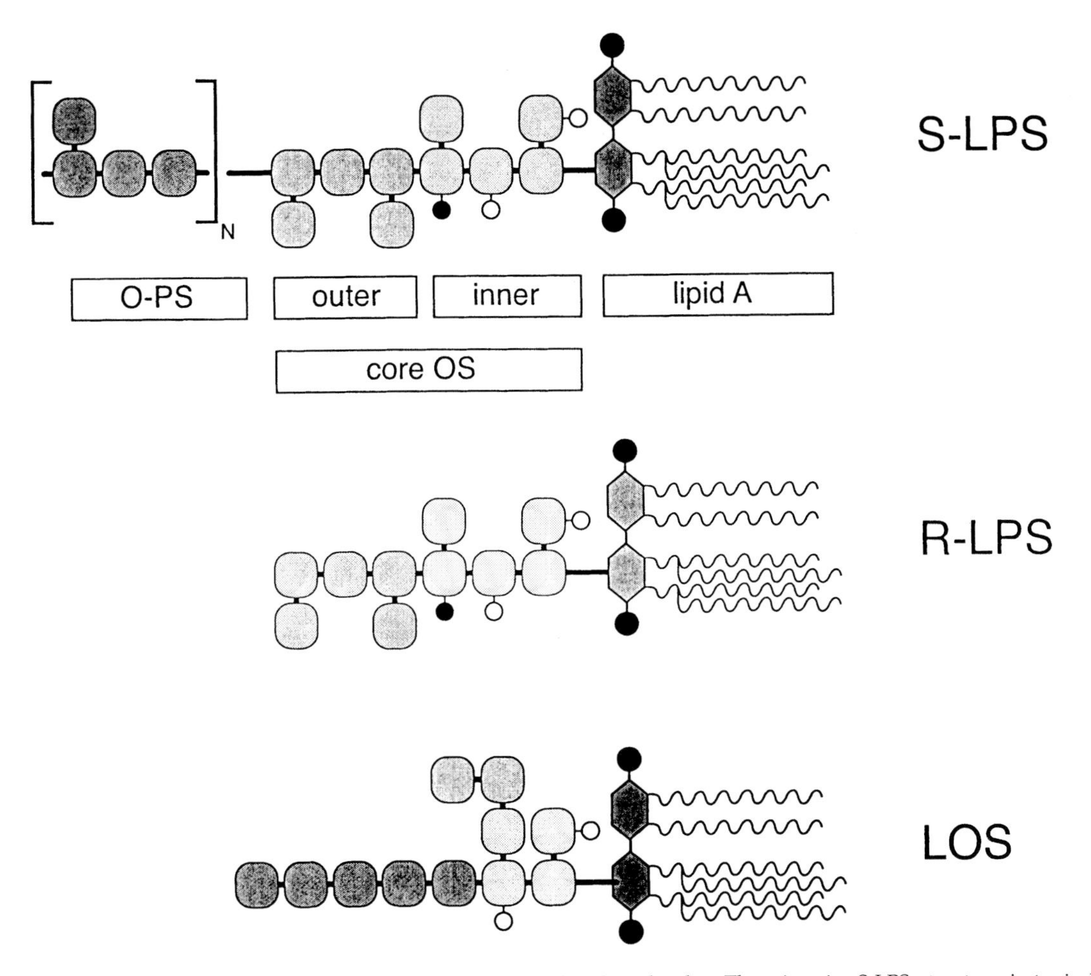

FIGURE 56.1 Schematic diagram showing three different forms of LPS molecules. The tripartite S-LPS structure is typical of LPSs produced by members of the *Enterobacteriaceae*, *Pseudomonadaceae*, and *Vibrionaceae*. These bacteria also produce a variable amount of R-LPS that lacks O-PS and, in some cases, part of the core OS. Mucosal pathogens such as *Neisseria* and *Haemophilus* spp. lack O-PS but instead may have phase-variable oligosaccharide extensions attached to the core OS, to form LOS.

sensor comprising PhoP/PhoQ is required for transcription of genes essential for virulence in mice. The sensor appears to respond to microenvironments encountered by the bacterium in the phagolysosome, with the adaptive response optimizing the bacterial physiology for intracellular growth. Among the genes whose transcription is controlled by PhoP/PhoQ are *pmrA/B*, encoding an additional two-component regulatory system. Together, these sensory systems modulate expression of enzymes that are involved in 4-amino-4-deoxy-L-arabinose (Ara4N) synthesis and its addition to lipid A, as well, as the those required for addition of an extra palmitate to form a heptaacyl lipid A in *S. enterica* sv. Typhimurium (Fig. 56.2). These alterations can have a considerable impact on the structure and function of LPS molecules. For example, *Salmonella* heptaacyl LPS shows a decreased capacity to activate release of cytokines and proinflammatory mediators (see following) and may be important for longer term survival inside the host cell. Addition of Ara4N to lipid A is an important determinant of the resistance of the bacterium to polycationic antimicrobial polypeptides, and polycationic antibiotics, such as polymyxin. The effect of this modification may be a dampening of the negative charges in lipid A, inhibiting the initial binding of polycations and their eventual perturbation of the outer membrane. Interestingly, these changes can be induced by metavanadate in *E. coli* K-12. In *P. aeruginosa*, hexaacyl lipid A modified with palmitate and Ara4N is synthesized in response to environmental cues from the cystic fibrosis lung environment. These changes confer resistance to cationic peptides and generate LPS with increased inflammatory properties. These may play a significant role in persistence of *P. aeruginosa* and tissue damage in the lungs of cystic fibrosis patients. In contrast to these

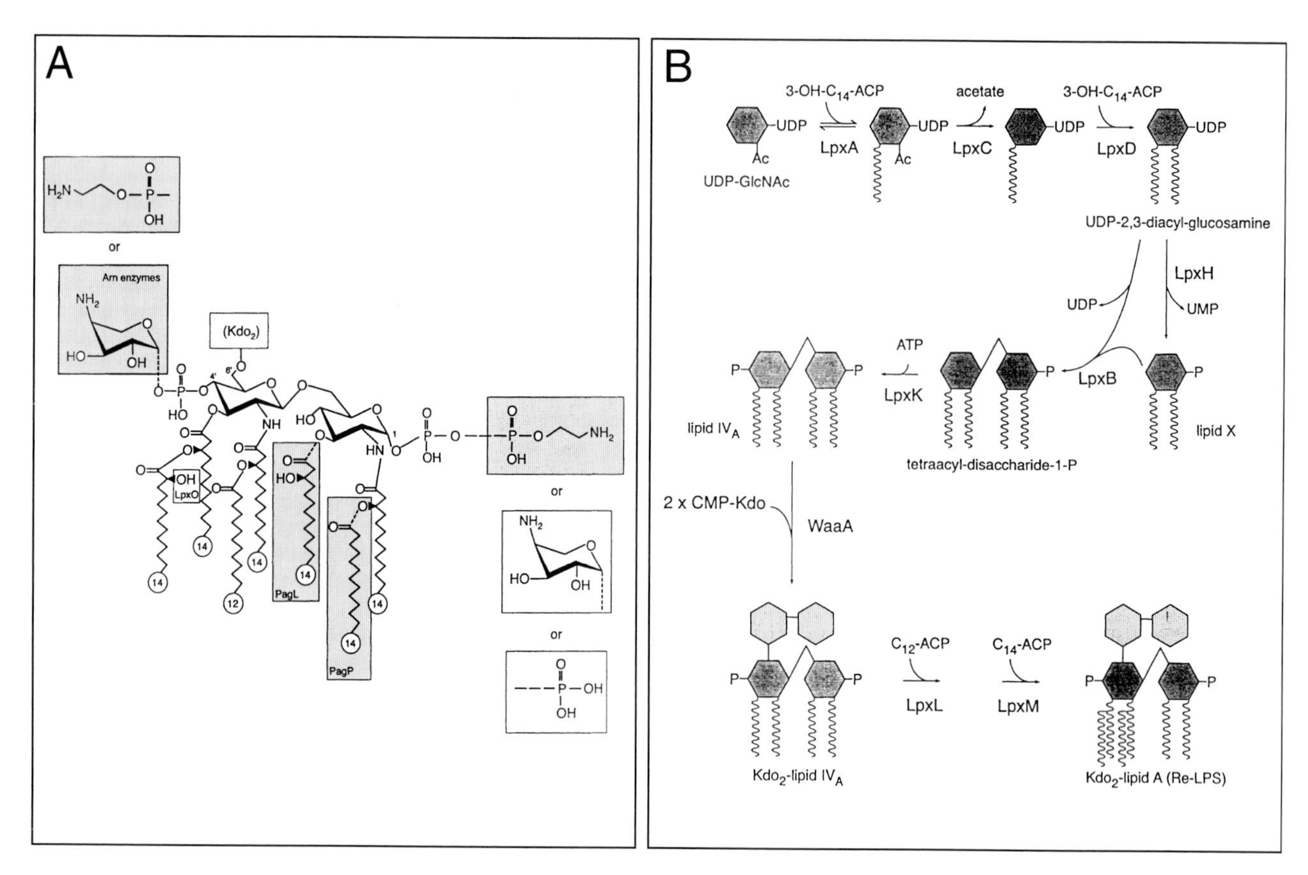

FIGURE 56.2 Structure and biosynthesis of lipid A. Panel A shows the structure of the lipid A from *E. coli* and *Salmonella*. The basic structure is given in *black* and regulated modifications and, where known, the enzymes responsible are indicated in the *colored* boxes. 4-amino-L-arabinose (*yellow* box) is found mainly on the 4′ phosphate, whereas phosphorylethanolamine (*blue* box) is mainly located at position 1. Under some growth conditions a pyrophosphate group is found at position 1. PagP (*pink* box) is a PhoP/PhoQ-regulated palmitoyl transferase located in the outer membrane. PagL (*green* box) selectively removes a β-hydroxymyristoyl residue. LpxO (*brown* box) hydroxylates the 3′ secondary acyl chain and the process is oxygen-dependent. Panel B shows the pathway for biosynthesis of lipid A from *E. coli* K-12. The enzymes responsible for each step in the pathway are indicated. The completed Re-LPS provides an acceptor for sequential assembly of the core oligosaccharide at the cytoplasmic face of the inner membrane. The completed molecule is then exported across the inner membrane by the ABC-transporter, MsbA (not shown) (Plate 7).

regulated modifications, the lipid As of *Proteus mirabilis* and *Burkholderia cepacia* are constitutively modified by Ara4N and these organisms are resistant to polymyxin under all growth conditions. It is clear from these examples that LPS should be considered to be a dynamic, rather than static, molecule, whose structure and function can be modulated in response to cues from the host.

Most lipid A variations between different species involve alterations in acylation or phosphorylation. For example, the acylation pattern is symmetrical in *N. meningitidis*, whereas *Rhodobacter sphaeroides* lipid A is distinguished by amide-linked 3-oxotetradecanoic acid and the presence of unsaturated fatty acids. The structure of the *R. sphaeroides* lipid A is of particular importance since it results in an LPS lacking the normal biological activities attributed to endotoxins. In the *Rhizobiaceae*, some lipid A molecules lack phosphate residues and the proximal glucosamine may undergo an oxidation step to form an aminogluconate residue. However, structural variations can be as extreme as the 2,3-diamino-2,3-dideoxy-D-glucose-containing monosaccharide backbone structure of lipid A molecules from *Pseudomonas diminuta* and *Rhodopseudomonas viridis*.

B. Core oligosaccharides

For the purpose of discussion of structure–function relationships, the core OS is often divided into inner and outer core regions (S-LPS; Fig. 56.1). The inner core of most known LPSs comprises characteristic residues of 3-deoxy-D-*manno*-octulosonic acid (Kdo) and L-*glycero*-D-*manno*-heptose (Hep). This region is highly conserved in enteric bacteria (Fig. 56.3A). The Kdo residues can be nonstoichiometrically modified

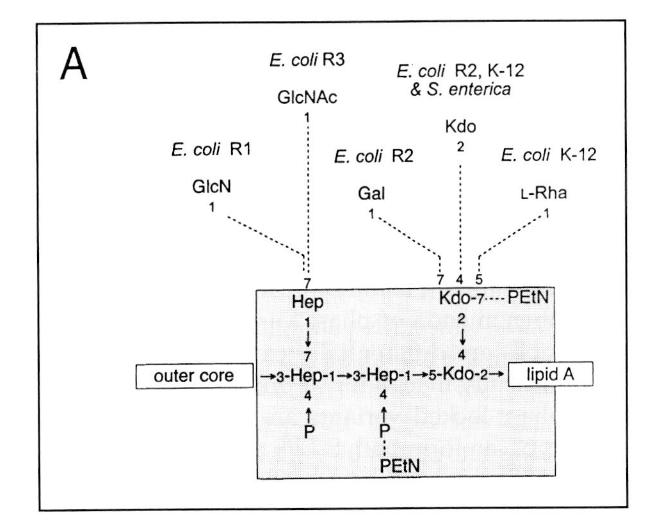

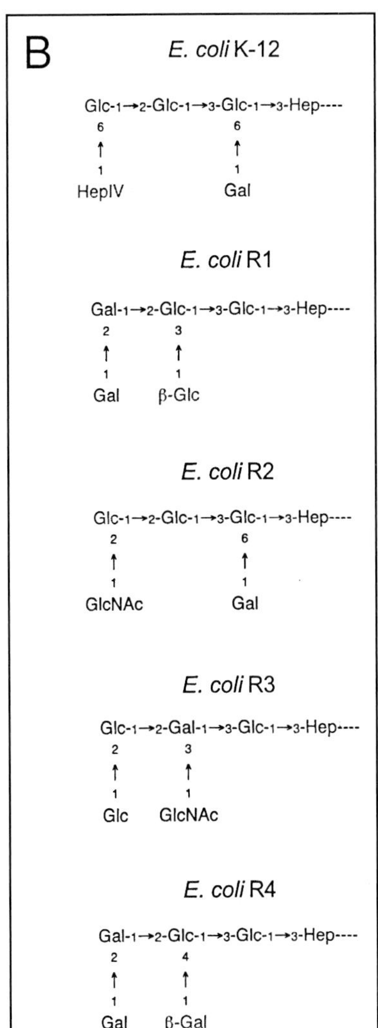

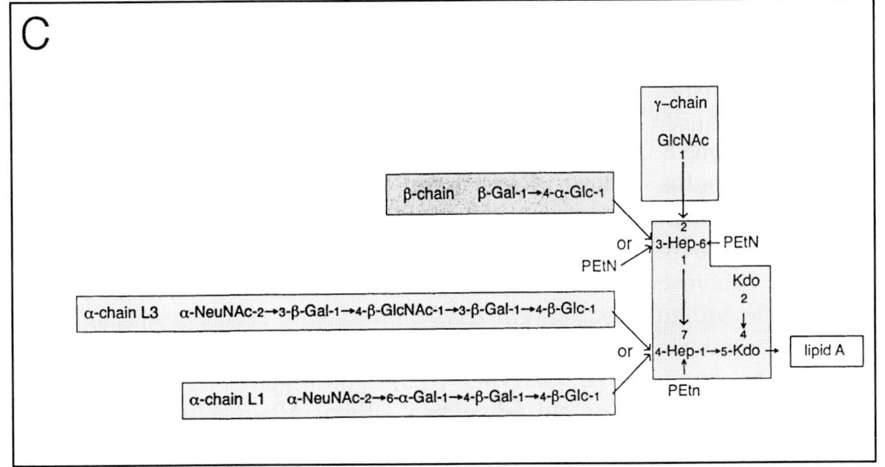

FIGURE 56.3 Structure of the core OS region of LPS and LOS. Panel A shows the inner core region of the core OSs from *E. coli* K-12. The backbone structure (*yellow* box) is conserved but nonstoichiometric glycose modifications (dotted lines) vary in different core types. Panel B shows the known outer core structures from the five core OSs of *E. coli*. They share a common overall structure, consisting of five glycoses linked to Hep II, but they differ in glycose content, glycose sequence, and side branch position. The site of O-PS ligation is not known for all of the core OS types but, where it has been determined, the residue providing the attachment site is indicated in *red*. Panel C shows a representative LOS structure from *N. gonorrhoeae*. The LOS core (*yellow* box) is conserved among different strains but there are strain-specific differences in the attached oligosaccharide chains. Shown are 2 possible variants (L1 and L3) of α-chain (*pink* boxes) and the β- and γ-chains. Unless otherwise indicated, all residues in the various structures are linked in the α-configuration (Plate 8).

by other sugars, or by 2-aminoethyl phosphate (phosphorylethanolamine; PEtN). The main chain Hep residues are also nonstoichiometrically decorated with phosphate, pyrophosphorylethanolamine (PPEtN), or a side-branch Hep. In some nonenteric bacteria, the Kdo residue proximal to lipid A is phosphorylated, or is replaced with the derivative D-*glycero*-D-*talo*-octulosonic acid (Ko). These differences may influence the lability of the linkage that is usually cleaved by mild-acid hydrolysis to release lipid A from the intact LPS molecule. In *Klebsiella pneumoniae*, Kdo is also present in the outer core OS.

The outer core OS structure is more variable among different bacteria. However, within a species, these variations are quite limited and some may have a single outer core OS type. For many years, this was thought to be the case for *Salmonella* but a second core OS structure has been reported relatively recently. *E. coli* has five known distinct outer core OS types; all contain five glycose residues but they differ in glycose content and organization (Fig. 56.3B). In addition to altering the antigenic epitopes and diagnostic bacteriophage receptors, variations in outer core OS structure give rise to altered sites for the attachment of O-PS.

chains of LOSs. These enzymes are peripheral membrane proteins that act at the cytoplasmic face of the inner membrane. In *E. coli* and *Salmonella*, the structural genes for the core OS glycosyltransferases map together with the Kdo transferase gene (*waaA*), and genes required for modification of the Heptose-region of the inner core OS (*waaPYQ*). These genes form the three separate operons in the chromosomal *waa*-region. Direct data for biochemical activities of individual core OS glycosyltransferase enzymes is unavailable in many cases. For example, studies of heptosyltransferases have generally been limited by the unavailability of the activated precursor, ADP-Hep, but the recent elucidation of the pathway for ADP-Hep biosynthesis will significantly help research in this area. Assignments of other glycosyltransferases have primarily been made by approaches where specific genes are individually mutated and the resulting LPS structure is resolved by chemical analysis. Synthesis of the core OS backbone can be carried to completion in mutants lacking the modifications that decorate the Hep-region. As a result, the precise timing of these modifications in the overall synthesis pathway is unknown.

C. Synthesis of O-polysaccharides

Despite the diversity in O-PS structures, only three mechanisms are known for the formation of O-PS (Fig. 56.5). O-PS synthesis begins at the cytoplasmic face of the inner membrane with activated precursors (sugar nucleotides; NDP-sugars) and the process terminates with a nascent O-PS at the periplasmic face. The ligation to lipid A-core then follows. The different pathways for assembly of O antigens vary in the components required for polymerization, in the cellular location of the polymerization reaction, and in the manner in which material is exported across the inner membrane. A carrier lipid, undecaprenyl phosphate (und-P), is involved in all three O-PS assembly pathways. The involvement of a carrier lipid scaffold may ensure fidelity in O-repeat unit structure, or simply provide an acceptor compatible with the membrane environment. The same three mechanisms are identified in the biosynthesis of capsular polysaccharides in gram-negative and gram-positive bacteria. In this respect, the primary distinction between the O-PSs and capsular polysaccharides is that the O-PSs are attached to lipid A-core.

The most prevalent pathways for O-PS synthesis are distinguished by the involvement (or not) of the putative "O-PS polymerase" enzyme, Wzy. The "Wzy-dependent" system (Fig. 56.5A) is the classical pathway first described in *S. enterica* serogroups A, B,

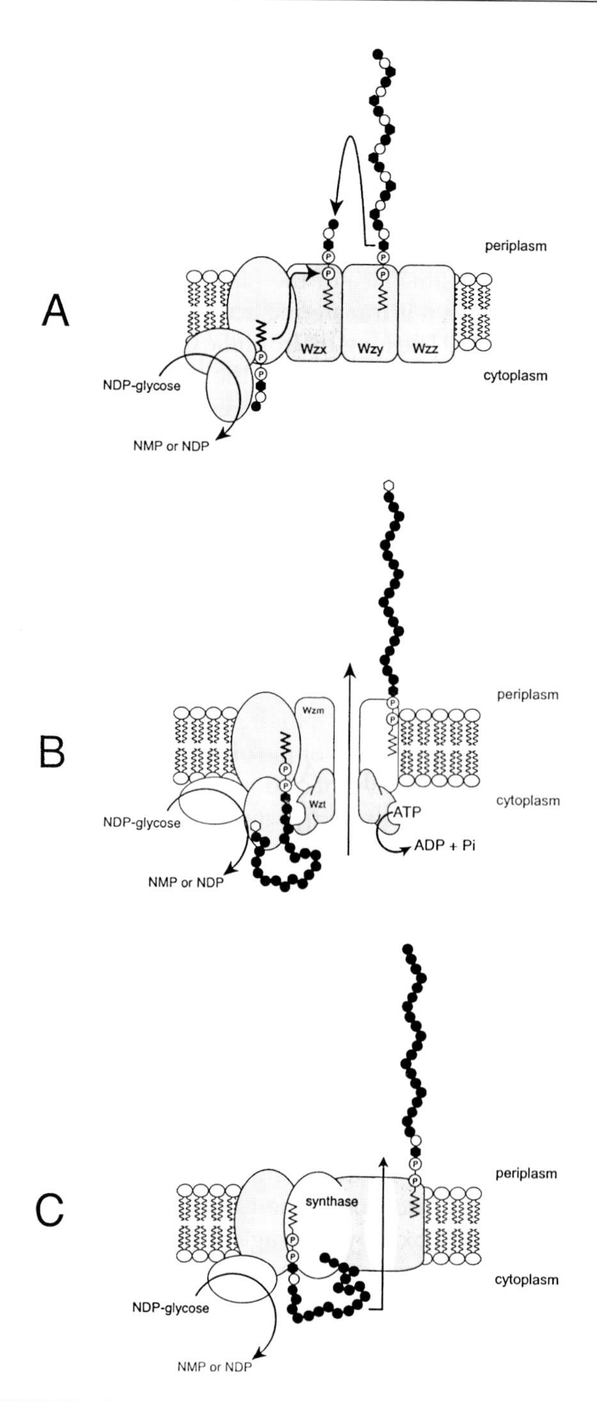

FIGURE 56.5 Models for the assembly of O-PS by the pathways termed Wzy-dependent (A), ABC-2-dependent (B), and synthase-dependent (panel C). All of the assembly systems begin at the cytoplasmic face of the plasma membrane and build the nascent O-PS in an undecaprenol pyrophosphoryl (und-PP)-linked form. Nucleotide diphospho (NDP)-sugars are the activated precursors. The enzyme complexes include integral and peripheral proteins, and key components are indicated by name. The pathways differ in the mechanism and location of O-PS polymerization and in the manner by which O-PS (or O-repeat units) are transferred across the plasma membrane. Termination of O-PS synthesis occurs with ligation of the nascent O-PS to preformed lipid A-core OS at the periplasmic face of the membrane (not shown). Modified from Raetz and Whitfield (2002).

D, and E. However, sequence and biochemical data shows the key enzymes are shared by other bacteria. In the working model for this pathway, und-PP-linked O-repeat units are assembled by glycosyltransferase enzymes at the cytoplasmic face of the plasma membrane. These reactions have been known since work in the 1960s by H. Nikaido, M. J. Osborn, P. Robbins, A. Wright, and others. The initial transferase is an integral membrane protein that transfers sugar-1-phosphate to the und-P acceptor. This is followed by sequential sugar transfers, catalyzed by additional peripheral glycosyltransferases, to form an und-PP-linked repeat unit. The polymerization reaction occurs at the periplasmic face of the membrane and utilizes und-PP-linked O-PS-repeat units as the substrate. The individual und-PP-linked O-PS-repeat units must, therefore, be exported across the inner membrane prior to polymerization, and preliminary biochemical analyses suggest that the likely candidate for this process is a multiple membrane-spanning protein, Wzx (formerly, RfbX). Polymerization of the O-PS repeat units minimally involves the putative polymerase (Wzy; formerly, Rfc) and the O-PS chain length regulator (Wzz, formerly Rol or Cld). A *wzy* mutant is unable to polymerize O-PS and its LPS comprises a single O-repeat unit attached to lipid A-core. In contrast, a *wzz* mutant makes S-LPS but loses the characteristic modal distribution of O-PS-chain lengths evident in SDS-PAGE analysis (e.g. Fig. 56.4). The O-PSs synthesized by this pathway are all heteropolymers and often have branched repeating unit structures.

In the "ABC-transporter-dependent" pathway, the O-PS is synthesized exclusively inside the cytoplasm and once complete, it is exported to the periplasm via an ABC-transporter, belonging to the ABC-2 family (Fig. 56.5B). This mechanism is, so far, confined to O-PSs with relatively linear structures and several are homopolymers. As with the Wzy-dependent pathway, synthesis is initiated at the cytoplasmic face of the inner membrane by an integral membrane glycosyltransferase enzyme, to form an und-PP-sugar. In fact, in *E. coli* the UDP-GlcNAc:undecaprenylphosphate GlcNAc-1-phosphate transferase (WecA) can initiate for either pathway. However, in the ABC-2-transporter-dependent pathway the initiating transferase acts once per O-PS chain. Additional peripheral glycosyltransferases then act sequentially and processively to elongate the und-PP-linked intermediate at the nonreducing terminus to form a fully polymerized und-PP-O-PS. The specificities of glycosyl transferases dictate the repeat-unit structure of the product. Only one und-PP acceptor is used per polymer and there is no equivalent of the polymerase (Wzy) or chain-length regulator (Wzz) enzymes. Several O-PS formed by the ABC-2-transporter-dependent pathways terminate at their non-reducing end with novel residues that are not part of the repeat-unit structure (examples include methyl groups and Kdo). These residues may act as signals for chain-termination, for initiation of export, or both. The ABC-2 family transporter comprises a transmembrane (Wzm) protein and an ATP-binding (Wzt) component. The involvement of an ABC-transporter precludes the involvement of Wzx, since there is no requirement for export of individual und-PP-O-units. It remains unknown whether the O-PS retains its attachment to und-PP during export, whether it is removed from the lipid carrier for export, or, alternatively, if an alternative carrier molecule exists.

The "synthase-dependent" pathway for O-PS biosynthesis is, so far, confined to the homopolymeric O antigen (factor 54) of *S. enterica* serovar Borreze. The model for this pathway (Fig. 56.5C) proposes that the initiating glycosyltransferase (WecA in this case) forms an und-PP-sugar acceptor that is elongated by a single multifunctional synthase enzyme, in a manner analogous to eukaryotic chitin and cellulose synthases and hyaluronan synthases from eukaryotes and bacteria. There is no dedicated ABC-2 transporter or Wzx homolog in the O:54 system and all experimental evidence points to the synthase having dual transferase-export functions. It is not known whether this system requires the nascent polymer be linked to und-PP throughout the polymerization and export processes.

In most bacteria, the genes dedicated to O-PS biosynthesis are clustered on the chromosome. The O-PS gene clusters contain a predictable spectrum of genes encoding novel sugar nucleotide synthetases, glycosyltransferases, and the characteristic enzymes such as Wzx-Wzy-Wzz, or an ABC-2 transporter, or synthase. As a result, sequence data can give an accurate first evaluation of the O-PS biosynthesis pathway involved. As might be expected from the range of O-PS structures, the O-PS biosynthesis genetic loci are highly polymorphic. Genetic recombination within and between bacterial species has played a significant role in the diversification of O-PS. In some bacteria (e.g. *Salmonella* spp., *Shigella* spp., and *P. aeruginosa*), phage-encoded genes provide additional determinants of O serotype specificity. These modifications are best characterized for Wzy-dependent systems in *Salmonella*, *Shigella*, and *P. aeruginosa*. Examples include changes in the linkage specificity of the Wzy polymerase, and acetylation or glucosylation of the polymer and all of these modifications appear to occur in the periplasm. A large number of LPS biosynthesis genes have been identified in bacteria with different lifestyles. For a current listing of known genes see the

Bacterial Polysaccharide Gene Database maintained by P. Reeves' laboratory (*www.microbio.usyd.edu.au/BPGD/default.htm*).

D. Terminal reactions in LPS assembly

Once complete, the nascent O-PS chain must be linked (ligated) to lipid A-core. All available evidence points to a periplasmic location for this reaction. Since lipid A-core is formed in the cytoplasm, it too must be exported to the ligation site at the periplasmic face of the inner membrane. An ABC-transporter, known as MsbA, appears to be responsible for this step. The mechanism underlying the ligase reaction itself remains unknown. The *waaL* gene product, often referred to as "ligase," is currently the only known protein that is essential for ligation. Its assignment as the ligase is based only on mutant phenotypes and there is no supporting mechanistic information. Interestingly, the ligase from *E. coli* K-12 will ligate structurally distinct O-PSs formed by any of the three assembly pathways, indicating that the form in which nascent O-PS is presented for ligation is conserved.

Perhaps the most interesting open questions in LPS assembly surround the process by which the completed LPS is translocated through the periplasm and then inserted into the outer membrane. In most bacteria, the translocation pathway must have relaxed specificity, since it efficiently transfers a range of LPS molecules varying from S-LPS to Re-LPS to the cell surface. Recent data has implicated an outer membrane protein, Omp85, in LPS export. There is also preliminary data that implicates the Tol system (a multiprotein complex involved in the translocation of group A colicins and filamentous bacteriophages) in surface expression of O-PS but a detailed understanding of the process is not yet available.

III. FUNCTIONS AND BIOLOGICAL ACTIVITIES OF LIPOPOLYSACCHARIDES

A. Lipopolysaccharides and outer membrane stability

From the construction of precise mutants with LPS defects it is well established that the minimal LPS molecule required for survival of *E. coli* consists of Kdo_2-lipid A (Re-LPS). Although *E. coli* can assemble an outer membrane from Re-LPS, the barrier function of the resulting outer membrane is compromised. Outer membrane integrity depends on structural elements in both lipid A and core OS. The role of Ara4N-modified lipid A in resistance to polycations has been discussed above. In *E. coli* and *Salmonella*, the phosphorylated Hep-region of the core OS is crucial for outer membrane stability by facilitating the cross-linking of adjacent LPS molecules by divalent cations or polyamines and by its interaction with positively charged groups on proteins. The inability to synthesize or incorporate Hep, or the loss of phosphoryl derivatives alone (i.e. a *waaP* mutant), gives rise to significant compositional and structural changes in the outer membrane. In *E. coli*, these mutants are known as "deep-rough" and their perturbed outer membrane structure leads to pleiotropic phenotypes, most notably hypersensitivity to hydrophobic compounds, such as detergents, dyes, and some antibiotics. *Salmonella waaP* mutants are avirulent and in *P. aeruginosa*, *waaP* is an essential gene. For such bacteria, assembly and phosphorylation of the core OS Hep-region may therefore provide further avenues for therapeutic intervention.

In most free-living bacteria, a negatively charged core OS is the important element in terms of outer membrane stability. However, phosphorylation is clearly not the only way to achieve a robust outer membrane, as some bacteria lack phosphorylation of the heptose region. For example, in the case of *Klebsiella pneumoniae*, galacturonic acid and Kdo residues in the core OS provide the only source of negative charges.

A limited number of wild-type gram-negative bacteria are viable without LPS. In the case of *S. paucimobilis*, no "typical" LPS is present in the outer membrane but, instead, the bacterium produces a glycosphingolipid–a modified ceramide derivative containing glucuronic acid and an attached trisaccharide. This lipid functionally replaces LPS in the formation of a stable outer membrane. In the case of *N. meningitidis*, mutants lacking the typical LOS-form are viable only if capsular polysaccharide is present and it is conceivable that the lipid anchor for this class of capsule partially replaces lipid A. However, there are other changes in outer membrane phospholipids and defects in expression of surface lipoproteins that may also help maintain viability. It is currently unclear whether this phenomenon extends beyond *N. meningitidis*. Perhaps the smallest wild-type LPS structure consists of only lipid A and a Kdo-trisaccharide, and is produced by members of the genus *Chlamydia*. Presumably, the intracellular growth environment for this organism, together with other features of the cell envelope, facilitate survival in the absence of a more complex LPS structure.

B. O-polysaccharides as a protective barrier

Molecular modeling of LPS structure and its organization in the outer membrane predict that the O-PS forms a significant layer on the cell surface. The O-PS partially lies flat on the cell surface, where the crossover of multiple chains forms a "felt-like" network. Since the O-PS is flexible, it can extend a significant distance from the surface of the outer membrane. It is, then, not surprising that many properties attributed to the O-PS are protective. In particular, long-chain O-PS is often essential for resistance to complement-mediated serum killing and, therefore, represents a major virulence factor in many gram-negative bacteria.

The serum proteins in the complement pathway interact to form a membrane attack complex (MAC) that can integrate into lipid bilayers to produce pores, leading to cell death. The MAC can be formed via a "classical" pathway, where surface antigen–antibody complexes initiate MAC formation, or through the "alternative" pathway, where complement component C3b interacts directly with the cell surface in the absence of antibody to facilitate MAC formation. In gram-negative bacteria with S-LPS, resistance to the alternative pathway does not result from defects in C3b deposition. Instead, C3b is preferentially deposited on the longest O-PS chains, and the resulting MAC is unable to insert into the outer membrane. In addition to O-PS chain length, complement-resistance can also be influenced by the extent of coverage of the available lipid A core with O-PS. As is often the case, there are exceptions to such generalizations. For example, there are some *E. coli* strains with S-LPS that are serum-sensitive unless an additional capsular polysaccharide layer is present. Although R-LPS variants of *E. coli* and *Salmonella* are almost invariably serum-sensitive, other bacteria (including many with LOS) use alternate strategies to achieve resistance.

The bactericidal/permeability inducing protein (BPI) is an antibacterial product found in polymorphonuclear leukocyte-rich inflammatory exudates. BPI binds LPS and may play a role in the clearance of circulating LPS but it also exhibits antimicrobial activity in the presence of serum. Resistance to BPI-mediated killing is also dependent on long chain O-PS.

C. Lipopolysaccharide and gram-negative sepsis

One potential outcome of gram-negative infections is septic shock, a syndrome manifested by hypotension, coagulopathy, and organ failure. In the United States, gram-negative sepsis accounts for 50,000–100,000 deaths each year. Septic shock results from the liberation of LPS from the bacterial cell surface, a phenomenon that naturally ensues from the growth and proliferation of bacteria. Tissue damage is not a result of direct interaction between the host tissues and an LPS "toxin," but instead results from unregulated host production of cytokines and inflammatory mediators (including tumor necrosis factor (TNF-α), and a variety of interleukins) by over-stimulated macrophages and endothelial cells. Under normal circumstances, and at regulated levels, these components have beneficial effects and lead to moderate fever, general stimulation of the immune system, and microbial killing. In sepsis, however, their overproduction leads to tissue and vascular damage and the symptoms of sepsis. Since free LPS is required to initiate the process, treatment with antibiotics and the ensuing bacterial lysis may actually exacerbate the problem.

The last decade has seen significant advances in our understanding of the manner in which LPS interacts with animal cells and stimulates their production of mediator molecules. It was suspected for some time that lipid A was the component responsible for those biological activities that LPS exhibited in sepsis. Definitive proof came from the observations that some partial LPS structures and chemically synthesized lipid A derivatives display the same biological activities as the complete molecule. Importantly, other partial structures are not only biologically inactive but can act as antagonists of LPS molecules that are active. Well-studied antagonist LPS molecules include the precursor lipid IV_A and the naturally occurring *R. sphaeroides* lipid A molecule. These structures have directed the synthesis of potent synthetic LPS antagonists that are able to negate the effects of challenge with biologically active LPS.

Circulating LPS molecules naturally form micellar aggregates and a variety of host LPS binding proteins are important in mobilizing LPS monomers from such complexes. These include BPI (see previous discussion) and LPS-binding protein (LBP), a 60 kDa acute-phase protein produced by hepatocytes. One role of these proteins is to clear and detoxify LPS. For example, LBP is known to transfer LPS to high-density lipoprotein fractions. However, LBP is also a crucial component of the signaling pathway through which animal cells are stimulated to produce cytokines and inflammatory mediators.

The central pathway by which cells recognize low concentrations of LPS (or bacterial envelope fragments containing LPS) requires the participation of a receptor protein CD14. CD14-deficient cell lines, such as 70Z/3 pre-B lymphocytes, are less sensitive to LPS (i.e. responsive to nanomolar rather than picomolar levels), unless transfected with CD14. Consistent with these results, CD14-knockout mice have been shown to be 10,000-fold less sensitive to LPS *in vivo*. In myeloid cell lines, CD14 occurs as a 55 kDa glycosylphosphatidylinositol (GPI) anchored glycoprotein, attached to the membrane (mCD14 in Fig. 56.6). However, a variety of nonmyeloid (endothelial and epithelial) cells are also responsive to LPS, via a soluble form of CD14 (sCD14). Both sCD14 and mCD14 can bind LPS to form a complex, but the kinetics of binding are slow. LBP serves to overcome this rate-limiting step by delivering LPS to mCD14 or sCD14 (Fig. 56.6) and blood taken from an LBP-knockout mouse shows a 1000-fold reduction in the ability to respond to LPS. At high LPS concentrations, CD14-independent stimulation is evident in CD14-deficient mice. For some time it was accepted that an accessory coreceptor protein was required to interact with CD14, to facilitate internalization of the lipid A signal. CD14 itself lacks a transmembrane domain to facilitate intracellular signaling, and CD14 alone is not able to discriminate between agonist and antagonist lipid A molecules; the antagonist molecules do not appear to operate by blocking the ability of agonist LPS to bind to CD14. A major development in this field was the identification of Toll-like receptor 4 (TLR4) as the lipid A-signaling coreceptor in animal cells. TLR4 recognizes CD14–lipid A complexes (or perhaps in some circumstances lipid A alone) and acts in the initiation of the signal transduction events that lead to cytokine induction. Interestingly, human TLR4 is able to discriminate between modified hexaacyl lipid A from *P. aeruginosa* and its pentaacyl form. Transmission of proinflammatory signals occurs only with the biologically active hexaacyl form. It is not yet clear which (if any) additional proteins participate in the transmembrane delivery of the lipid A signal but the soluble protein MD-2 is implicated in the process.

Once the relevant signal is transmitted across the animal cell membrane, a cascade of events leads to the release and overproduction of cytokines. The components of the latter stages of the response pathway are now beginning to be identified but they are complicated by an involvement of a multicomponent cascade. The cascade involves rapid protein phosphorylation events, and isoforms of the p38 mitogen-activated protein (MAP) kinase family play a central role. LPS antagonists block phosphorylation of p38. p38 is itself activated by the MAP kinase kinases. Downstream, p38 has a number of substrates including the myocyte enhancer factor (MEF2) family of transcription activators. The transcription factor NF-κB also plays an important regulatory role in the proinflammatory response, as well as in the development of LPS tolerance.

A variety of therapeutic approaches have been designed to interfere with specific steps in the process leading to septic shock. The numbers of mediators

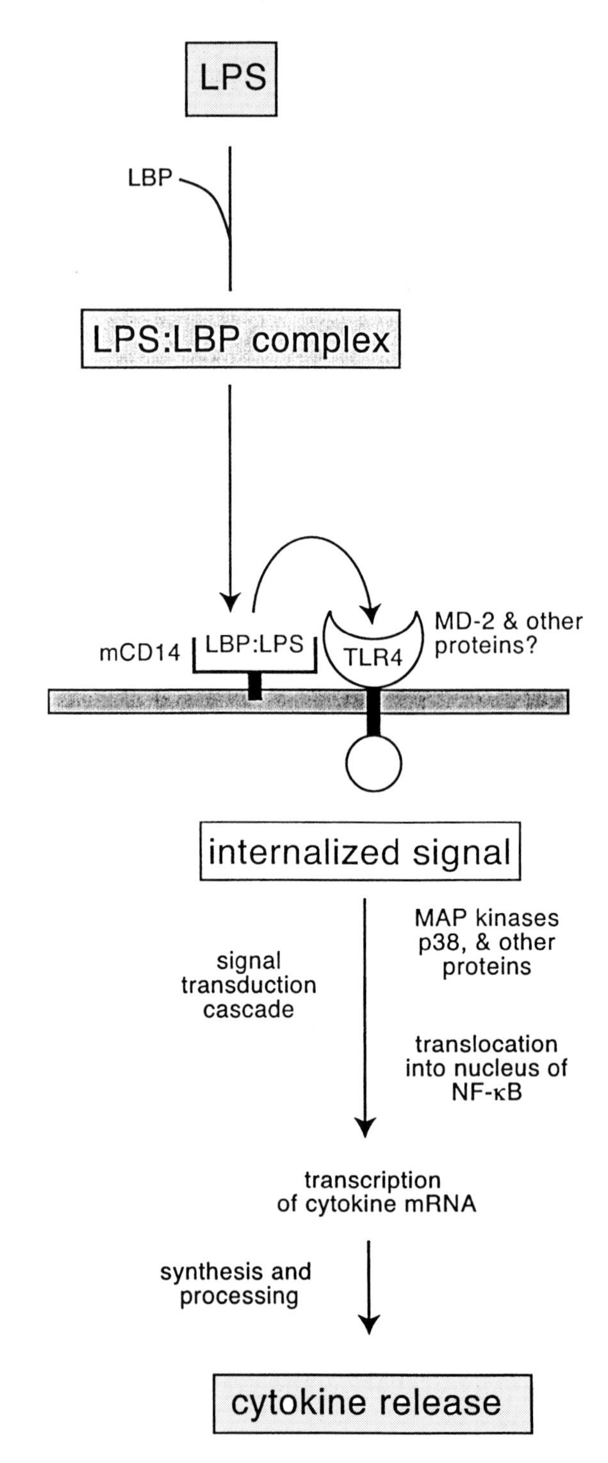

FIGURE 56.6 The pathway involved in the stimulation of macrophages and other myeloid cell lines by LPS. The LPS binding protein (LBP) delivers LPS to the GPI anchored form of CD14 (mCD14). LPS is transferred to TLR4 in complex with MD-2 and, perhaps, other proteins. This complex participates in intracellular signal generation. Once stimulated, a phosphorylation signal transduction cascade leads to transcription of genes encoding cytokines and inflammatory mediators. Elements of the pathways are also important for LPS clearing.

involved complicates strategies based on blocking cytokines themselves. Neutralizing an individual mediator would not be expected to be an effective therapy, as appears to be the case for antibody-neutralized TNF-α. Significant efforts have been directed to neutralizing the LPS signaling molecule by administering antibodies but, to date, attempts to develop therapies based on anti-endotoxin monoclonal antibodies have been disappointing. Two commercial monoclonal antibodies (E5 and HA1A) recognizing lipid A have been the subject of clinical trials but, unfortunately, these provided no compelling evidence for the protective capacity of the antibodies. However, there are antibodies that recognize the core OS of *E. coli* and *Salmonella* and that do show promise both *in vitro* and in animal models, suggesting alternate immunotherapeutic strategies. LPS neutralization could also be achieved by using proteins that bind LPS and both LBP and BPI are being pursued in this respect. Approaches that attempt to block LPS receptor pathways with synthetic LPS antagonists are proving effective in animal models. Equally promising is the application of anti-CD14 and anti-LBP monoclonal antibodies that block the formation of CD14:LPS complexes and protect against LPS exposure in animal models.

D. Molecular mimicry in LPS and LOS

As detailed structures become available for the various components in heterogeneous LPS and LOS preparations from different bacteria, it is apparent that several successful pathogens employ a strategy of molecular mimicry. In the case of *Helicobacter pylori* LOS, glycoforms have been described with structures resembling blood group antigens, including the Lewis determinants. These structures are implicated in, evasion of host immune response, autoimmune responses and in adhesion, and colonization of the bacterium. In *Campylobacter jejuni*, several LOS glycoforms contain ganglioside mimics and phase variation is common. These virulence determinants may again provide a mechanism of avoiding the host immune response but they are also suggested as a possible causative agent of autoimmune responses in the development of Guillain–Barré syndrome.

E. Biotechnological applications involving LPS

The incredible number of LPS structures provide an extensive range of oligosaccharide and polysaccharide structures with novel biological properties. These may be of value for therapeutic or other commercial applications. In one novel example, a recombinant *E. coli* strain was constructed in which the LPS core OS provided a scaffold for expression of the globotriose receptor for Shiga toxins. The *E. coli* strain efficiently adsorbs and neutralizes the toxin, affording a therapeutic approach for treating infections whose pathogenesis involves these and related toxins. Structures of glycosyltransferases, including a LOS galactosyltransferase, are now being solved at high resolution by crystallographic methods. Ultimately, this will provide general insight into the detailed structure–function relationships among glycosyltransferases, and open the possibility for engineering enzymes with novel specificities for practical applications. For information regarding known glycosyltransferases see the CAZY (Carbohydrate Active Enzymes) website maintained by B. Henrissat's laboratory (*http://afmb.cnrs-mrs.fr/CAZY/index.html*).

BIBLIOGRAPHY

Alexander, C., and Rietschel, E. T. (2001). Bacterial lipopolysaccharides and innate immunity. *J. Endotoxin Res.* **7**, 167–202.

Chang, G., and Roth, C. B. (2001). Structure of MsbA from *E. coli*: a homolog of the multidrug resistance ATP binding cassette (ABC) transporters. *Science* **293**, 1793–1800.

da Silva Correia, J., Soldau, K., Christen, U., Tobias, P., and Ulevitch, R. J. (2001). Lipopolysaccharide is in close proximity to each of the proteins in its membrane receptor complex. Transfer from CD14 to TLR4 and MD-2. *J. Biol. Chem.* **276**, 21129–21135.

Doerrler, W. T., Reedy M. C., and Raetz, C. R. (2001). An *Escherichia coli* mutant defective in lipid export. *J. Biol. Chem.* **276**, 11461–11464.

el-Samalouti, V. T., Hamann, L., Flad, H. D., and Ulmer, A. J. (2000). The biology of endotoxin. *Methods Mol. Biol.* **145**, 287–309.

Erridge, C., Stewart, J., Bennett-Guerrero, E., McIntosh, T. J., and Poxton, I. R. (2002). The biological activity of a liposomal complete core lipopolysaccharide vaccine. *J. Endotoxin Res.* **8**, 39–46.

Genevrois, S., Steeghs, L., Roholl, P., Letesson, J. J., and Van der Ley, P. (2003). The Omp85 protein of *Neisseria meningitidis* is required for lipid export to the outer membrane. *EMBO J.* **22**, 1780–1789.

Guha, M., and Mackman, N. (2001). LPS induction of gene expression in human monocytes. *Cell Signal.* **13**, 85–94.

Hajjar, A. M., Ernst, R. K., Tsai, J. H., Wilson, C. B., and Miller S. I. (2002). Human Toll-like receptor 4 recognizes host-specific LPS modifications. *Nat. Immunol.* **3**, 354–359.

Heinrichs, D. E., Yethon, J. A., and Whitfield, C. (1998). Molecular basis for structural diversity in the core regions of the lipopolysaccharides of *Escherichia coli* and *Salmonella enterica*. *Mol. Microbiol.* **30**, 221–232.

Joiner, K. A. (1988). Complement evasion by bacteria and parasites. *Ann. Rev. Microbiol.* **42**, 201–230.

Levy, O., and Elsbach, P. (2001). Bactericidal/permeability-increasing protein in host defense and its efficacy in the treatment of bacterial sepsis. *Curr. Infect. Dis. Rep.* **3**, 407–412.

Monteiro, M. A. (2001). *Helicobacter pylori*: a wolf in sheep's clothing: the glycotype families of *Helicobacter pylori* lipopolysaccharides expressing histo-blood groups: structure, biosynthesis, and role in pathogenesis. *Adv. Carbohydr. Chem. Biochem.* **57**, 99–158.

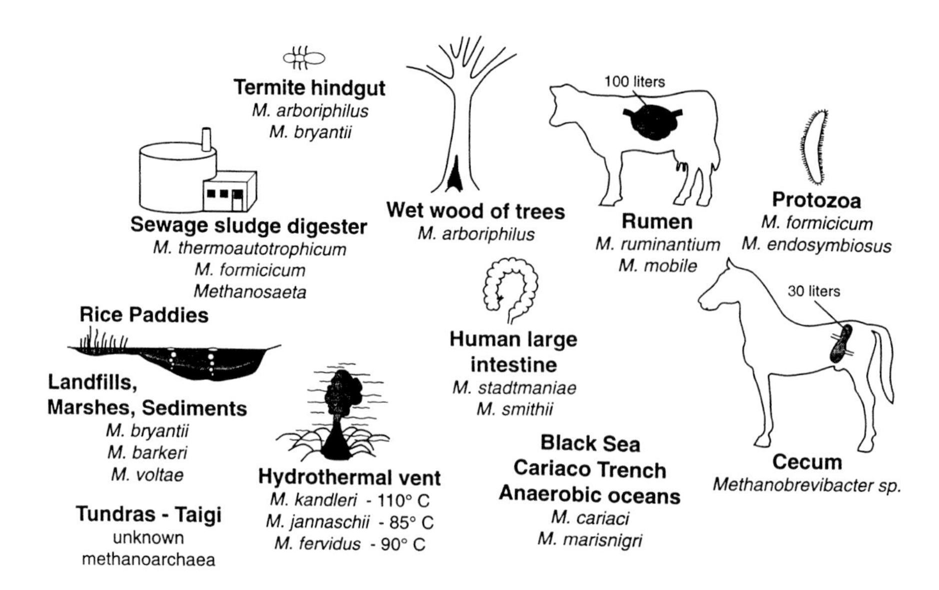

FIGURE 57.1 Habitats of the methanogenic *Archaea*. (Reproduced, with permission, from Wolfe, R. S. (1996). *ASM News* **62**, 529–534. ASM Press.)

carnivores, is eventually restored to the atmosphere as CO_2, thus completing the global carbon cycle.

I. HISTORICAL OVERVIEW

The generation of combustible gas, presumably CH_4, has been reported by Pliny as early as during the Roman Empire. Legendary manifestations of methanogenesis include the will-o-the-wisp, hypothesized to have resulted from the spontaneous combustion of marsh gas, and fire-breathing dragons, conjectured to have resulted from the accidental ignition of gas from CH_4-belching ruminants. The close association between decaying plant material and the generation of "combustible air" was first described by the Italian physicist Alessandro Volta in 1776, when he reported that gas released after disturbing marsh and lake sediments produced a blue flame when ignited by a candle. Bechamp, a student of Pasteur, was the first to establish that methanogenesis was a microbial process, which was corroborated by others throughout the remainder of the nineteenth century and early twentieth centuries. Because of the methanogens' requirement for strict anaerobic conditions, the first isolates were not reported until the 1940s. The approach used for isolation, the shake culture, involved adding microorganisms to molten agar growth medium containing a reductant, such as pyrogallol-carbonate, to prevent O_2 from diffusing into the agar. However, this approach was not suitable for isolating and maintaining methanogens in pure culture for long periods of time, as the medium was not sufficiently anaerobic. It was not until 1950 that a simple, effective technique was developed that provided the rigorous conditions required for routine isolation and culturing of methanogenic *Archaea*. The technique, referred to as the "Hungate Technique," employs gassing cannula, O_2-free gases, and cysteine-sulfide reducing buffers to prepare a highly reduced, O_2-free medium. Boiling initially deoxygenates medium and a cannula connected to an anaerobic gas line, such as N_2 or CO_2, is inserted into the vessel to displace air as the medium cools. The medium is dispensed into culture tubes or serum vials while purging with anaerobic gas and then the medium is sealed with a rubber stopper or septum. The vessel containing reduced medium and anaerobic gas effectively becomes an anaerobic chamber for culture growth. The development of the anaerobic glove box has further simplified culturing of methanogens by providing a means for colony isolation in petri plates containing anaerobic medium (Fig. 57.2). Inoculated plates are then transferred to an anaerobe jar that is purged with anaerobic gas and hydrogen sulfide to create conditions necessary for growth.

II. DIVERSITY AND PHYLOGENY

The methanogens are members of the *Archaea*, one of three domains of life proposed by C. Woese on the basis of 16S rRNA sequence, which also include the

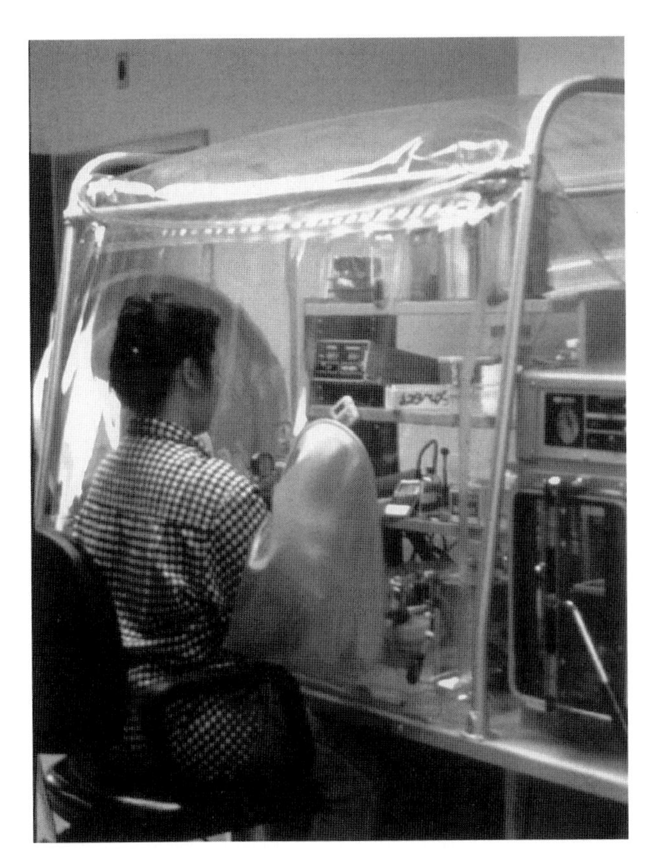

FIGURE 57.2 Anaerobic glove box used for growing methanogens. Materials are introduced into the glove box through an airlock located on the side.

Bacteria and *Eukarya* (Fig. 57.3). *Archaea* have morphological features that resemble the *Bacteria*; they are unicellular microorganisms that lack a nuclear membrane and intracellular compartmentalization. In contrast, several molecular features of the *Archaea* have similarity to the *Eukarya*; these feature include histone-like DNA proteins, a large multicomponent RNA polymerase, and eukaryal-like transcription initiation. Despite the similarities to the other domains, *Archaea* also have unique characteristics that distinguish them from the *Bacteria* and *Eukarya*. These distinguishing features include membranes composed of isoprenoids ether-linked to glycerol or carbohydrates, cell walls that lack peptidoglycan, synthesis of unique enzymes, and enzyme cofactor molecules. An additional unifying characteristic among the *Archaea* is their requirement for extreme growth conditions, such as high temperatures, extreme salinity, and, in the case of the methanogens, highly reduced, O_2-free anoxic environments.

Although the methanogens are a phylogenetically coherent group and have a limited substrate range, they are morphologically and physiologically diverse. They include psychrophilic species from Antarctica that grow at 1.7 °C to extremely thermophilic species from deep submarine vents that grow at 110 °C; acidophiles from marine vents that grow at pH 5.0 to alkaliphiles from alkaline lake sediments that grow at pH 10.3; species from freshwater lake sediments that

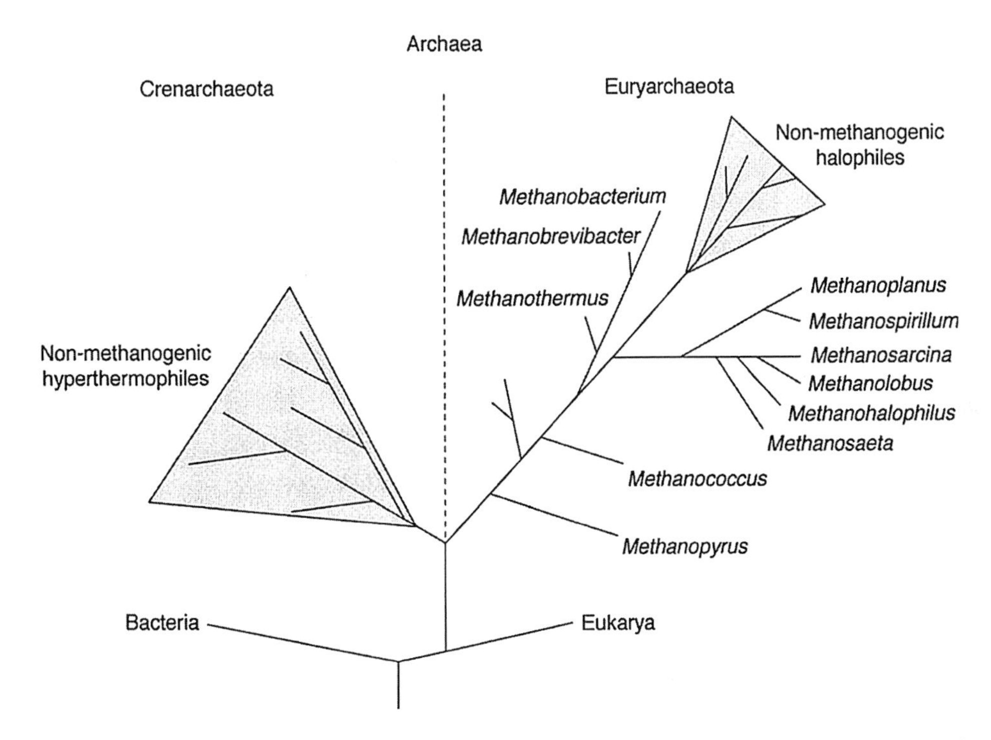

FIGURE 57.3 Phylogenetic tree based on 16S rRNA sequence showing genera of methanogens within the archaeal kingdom Euryarchaeota.

grow at saline concentrations below 0.1 M to extreme halophiles from solar salterns that grow at nearly saturated NaCl concentrations; autotrophs that use only CO_2 for cell carbon and methylotrophs that utilize methylated carbon compounds. Despite the range and diversity of growth habitats where methanogens are found, methanogens have one common attribute: they all generate CH_4 during growth.

There are currently over 60 described species of methanogens in five orders within the archaeal kingdom *Euryarchaeota* (Fig. 57.3). Characteristics of methanogenic *Archaea* are described in Table 57.1. The

TABLE 57.1 Description of methanogenic *Archaea*[a]

			Optimum growth conditions[c]		
Taxonomic epithet	**Morphology**	**Substrates**[b]	**pH**	**Temp. (°C)**	**Isolation source**
Order *Methanobacteriales*					
Family *Methanobacteriaceae*					
Genus *Methanobacterium*					
alcaliphilum	Rod	H	8.4	37	Alkaline lake sediment
bryantii	Rod	H,2P,2B	6.9–7.2	37–39	Sewage digestor
congolense	Rod	H	7.2	37–42	Cassava peel digestor
defluvii	Rod	H,F	6.5–7.0	60–65	Methacrylic waste digestor
espanolae	Rod	H	5.6–6.2	35	Kraft mill sludge
formicicum	Rod	H,F,2P,2B	6.6–7.8	37–45	Sewage digestor
ivanovii	Rod	H	7.0–7.4	45	Sewage digestor
oryzae	Rod	H,F	7.0	40	Rice field
palustre	Rod	H,F,2P,CP	7.0	37	Peat bog
subterraneum	Rod	H,F	7.8–8.8	20–40	Deep granitic groundwater
thermoaggregans	Rod	H	7.0–7.5	65	Cattle pasture
thermoflexum	Curved rod	H,F	7.9–8.2	55	Methacrylic waste digestor
thermophilum	Rod	H	8.0–8.2	62	Digestor methane tank
uliginosum	Rod	H	6.0–8.5[d]	40	Marsh sediment
Genus *Methanobrevibacter*					
acididurans	Coccobacillus	H	6.0	35	Acohol distillery waste
arboriphilicus	Coccobacillus	H,F	7.8–8.0	30–37	Cotton wood tree
curvatus	Curved rod	H	7.1–7.2	30	Termite hindgut
cuticularis	Rod	H,F	7.7	37	Termite hindgut
filiformis	Filamentus rod	H	7.0–7.2	30	Termite hindgut
oralis	Coccobacillus	H	6.9–7.4	36–38	Human subgingival plaque
ruminantium	Coccobacillus	H,F	6.3–6.8	37–39	Bovine rumen
smithii	Coccobacillus	H,F	6.9–7.4	37–39	Sewage digestor
gottschalkii	Coccobacillus	H	7	37	Horse feces
thaueri	Coccobacillus	H	7	37	Bovine feces
woesii	Coccobacillus	H,F	7	37	Goose feces
wolinii	Coccobacillus	H	7	37	Sheep feces
Genus *Methanosphaera*					
cuniculi	Coccus	H/Me	6.8	35–40	Rabbit rectum
stadtmanae	Coccus	H/Me	6.5–6.9	36–40	Human feces
Genus *Methanothermobacter*					
defluvii	Rod	H,F	6.5-7.0	60-65	Methacrylic waste digestor
marburgensis	Rod	H	6.8–7.4	65	Sewage digestor
thermoautotrophicus	Rod	H	7.2–7.6	65–70	Sewage digestor
thermoflexus	Curved rod	H,F	7.9–8.2	55	Methacrylic waste digestor
thermophilus	Rod	H	8.0–8.2	62	digestor methane tank
wolfei	Rod	H/F	7.0–7.7	55–65	Sewage/river sediment
Family *Methanothermaceae*					
Genus *Methanothermus*					
fervidus	Rod	H	6.5	83	Solfataric hot spring
sociabilis	Rod	H	6.5	88	Solfataric mud
Order *Methanococcales*					
Family *Methanococcaceae*					
Genus *Methanococcus*					
"aeolicus"	Irreg. coccus	H,F	nr	nr	nr

TABLE 57.1 *Continued*

Taxonomic epithet	Morphology	Substrates[b]	Optimum growth conditions[c] pH	Temp. (°C)	Isolation source
maripaludis	Irreg. coccus	H,F	6.8–7.2	35–39	Marine marsh sediment
vannielii	Irreg. coccus	H,F	7.0–9.0 d	36–40	Marine sediment
voltae	Irreg. coccus	H,F	6.7–7.4	32–40	Estuarine sediment
Genus *Methanothermococcus*					
thermolithotrophicus	Irreg. coccus	H,F	6.5–7.5	65	Thermal coastal sediment
Family *Methanocaldococcaceae*					
Genus *Methanocaldococcus*					
fervens	Irreg. coccus	H	6.5	85	Marine hydrothermal vent
infernus	Irreg. coccus	H	6.5	85	Marine hydrothermal vent
jannaschii	Irreg. coccus	H	6.0	85	Marine hydrothermal vent
vulcanius	Irreg. coccus	H	6.5	80	Marine hydrothermal vent
Genus *Methanotorris*					
igneus	Irreg. coccus	H	5.7	88	Marine hydrothermal vent
Order *Methanomicrobiales*					
Family *Methanomicrobiaceae*					
Genus *Methanomicrobium*					
mobile	Curved rod	H,F	6.1–6.9	40	Bovine rumen
Genus *Methanoculleus*					
bourgensis	Irreg. coccus	H,F	7.4	37	Tannery waste digestor
chikugoensis	Irreg. coccus	H,F,2P,2B,CP	6.7–7.2	25–30	Rice field
marisnigri	Irreg. coccus	H,F,2P,2B	6.2–6.6	20–25	Marine sediment
oldenburgensis	Irreg. coccus	H,F	7.5–8.0	45	River sediment
olentangyi	Irreg. coccus	H	nr	37	River sediment
palmolei	Irreg. coccus	H,F,2P,2B,CP	6.9–7.5	40	Palm oil wastewater reactor
thermophilicus	Irreg. coccus	H,F	7.0	55	Thermal marine sediment
Genus *Methanofollis*					
aquaemaris	Irreg. coccus	H,F	6.5	37	Aquaculture fish pond
liminatans	Irreg. coccus	H,F,2P,2B	7.0	40	Industrial wastewater
tationis	Irreg. coccus	H,F	7.0	37–40	Solfataric hot pool
Genus *Methanogenium*					
cariaci	Irreg. coccus	H,F	6.8–7.3	20–25	Marine sediment
frigidum	Irreg. coccus	H,F	6.5–7.9	15	Antarctic lake
frittonii	Irreg. coccus	H,F	7.0–7.5	57	Lake sediment
organophilum	Irreg. coccus	H,F,E,1P,2P,2B	6.4–7.3	30–35	Marine sediment
Genus *Methanolacinia*					
paynteri	Irreg. rod	H,F,2P,2B,CP	6.6–7.2	40	Marine sediment
Genus *Methanoplanus*					
endosymbiosus	Irreg. disk	H,F	6.6–7.1	32	Marine ciliate
limicola	Plate	H,F	7.0	40	Drilling swamp
petrolearius	Plate	H,F,1P	7.0	37	Offshore oil field
Family *Methanocorpusculaceae*					
Genus *Methanocorpusculum*					
aggregans	Irreg. coccus	H,F	6.4–7.2	35–37	Sewage digestor
bavaricum	Irreg. coccus	H,F,2P,2B,CP	7.0	37	Sugar plant wastewater
labreanum	Irreg. coccus	H,F	7.0	37	Tar pit lake
parvum	Irreg. coccus	H,F,2P,2B	6.8–7.5	37	Whey digestor
sinense	Irreg. coccus	H,F	7.0	30	Distillery wastewater
Family *Methanospirillaceae*					
Genus *Methanospirillum*					
hungateii	Sheathed spiral	H, F	6.6–7.4	30-37	Sewage sludge
Genus *Methanocalculus*[e]					
halotolerans	Irreg. coccus	H,F	6.5–7.5	35	Marine waste leachate site
pumilus	Irreg. coccus	H,F	7.6	38	Oil field
taiwanensis	Irreg. coccus	H,F	6.7	37	Estuarine sediment

(Continued Overleaf)

TABLE 57.1 *Continued*

Taxonomic epithet	Morphology	Substrates[b]	Optimum growth conditions[c]		Isolation source
			pH	Temp. (°C)	
Order *Methanosarcinales*					
Family *Methanosarcinaceae*					
Genus *Methanosarcina*					
acetivorans	Irreg. coccus, pseudosarcina	AC,ME,MA,DMS,MMP	6.5–7.5	35–40	Marine sediment
baltica	Irreg. coccus, pseudosarcina	AC,ME,MA	6.5–7.5	25	Brackish sediment
barkeri	Irreg. coccus, pseudosarcina	H,AC,ME,MA	6.5–7.5	30–40	Sewage digestor
lacustris	Irreg. coccus	H,ME,MA	nr	25	Lake sediment
mazei	Irreg. coccus, pseudosarcina	AC,ME,MA	6.5–7.2	30–40	Sewage digestor
semesiae	Irreg. coccus	ME,MA,DMS,MT	6.5–7.5	30–35	Mangrove sediment
siciliae	Irreg. coccus	AC,ME,MA,DMS,MMP	6.5–6.8	40	Marine sediment
thermophila	Irreg. coccus pseudosarcina	AC,ME,MA	6.0	45–55	Sewage digestor
vacuolata	Irreg. coccus pseudosarcina	H,AC,ME,MA	7.5	40	Methane tank sludge
Genus *Methanococcoides*					
burtonii	Irreg. coccus	ME,MA	7.7	23.4	Antarctic saline lake
methylutens	Irreg. coccus	ME,MA	7.0	30–35	Marine sediment
Genus *Methanohalobium*					
evestigatum	Irreg. coccus	ME,MA	7.4	50	Salt lagoon sediment
Genus *Methanohalophilus*					
halophilus	Irreg. coccus	ME,MA	7.4	26–36	Marine cyanobacterial mat
mahii	Irreg. coccus	ME,MA	7.4	35–37	Saline lake sediment
portucalensis	Irreg. coccus	ME,MA	6.5–7.5	40	Solar salt pond
zhilinae	Irreg. coccus	ME,MA,DMS	9.2	45	Alkaline lake sediment
Genus *Methanolobus*					
bombayensis	Irreg. coccus	ME,MA,DMS	7.2	37	Marine sediment
taylorii	Irreg. coccus	ME,MA,DMS	8.0	37	Estuarine sediment
oregonensis	Irreg. coccus	ME,MA,DMS	8.6	35	Saline alkaline aquifer
tindarius	Irreg. coccus	ME,MA	6.5	37	Lake sediment
vulcani	Irreg. coccus	ME,MA	7.2	37	Submarine fumarole
Genus *Methanosalsum* *zhilinae*	Irreg. coccus	ME,MA,DMS	8.7–9.5	35-45	Alkaline lake sediment
Genus *Methanomicrococcus*					
blatticola	Irreg. coccus	H/ME,H/MA	7.2–7.7	39	Cockroach hindgut
Family *Methanosaetaceae*					
Genus *Methanosaeta*					
concilii	Sheathed rod	AC	7.1–7.5	35–40	Pear waste digestor
thermophila	Sheathed rod	AC	7.4–7.8	35–40	Thermophilic sludge digestor
Order *Methanopyrales*					
Family *Methanopyraceae*					
Genus *Methanopyrus*					
kandleri	Sheathed rod	H	6.5	98	Geothermal marine sediment

[a]Type strain descriptions.
[b]H = hydrogen/carbon dioxide; F = formate; AC = acetate; ME = methanol; MA = methylamines; MT = methanethiol; H/ME = methanol reduction with hydrogen; H/MA = methylamine reduction with hydrogen; E = ethanol; 1P = 1-propanol; 2P = 2-propanol; 2B = 2-butanol; CP = cyclopentanol; DMS = dimethylsulfide; MMP = methylmercaptopropionate.
[c]nr = Not reported.
[d]Only a range reported.
[e]Family epithet currently uncertain.

order *Methanococcales* includes marine autotrophs that grow exclusively by CO_2 reduction with H_2. Morphologically, these species form irregularly shaped cocci. Instead of a rigid cell wall, typical of most *Bacteria*, these species form an S-layer, composed of an array of protein subunits, and are subject to osmotic lysis at NaCl concentrations below seawater. This order includes mesophilic *Methanococcus* spp., the moderate thermophile *Methanothermococcus thermolithotrophicus*, and the extreme thermophilic *Methanocaldococcus* spp.

The order *Methanobacteriales* is composed predominantly of rod-shaped cells that grow by CO_2 reduction with H_2. The exception is the genus *Methanosphaera*, which grows as cocci and uses H_2 to reduce methanol instead of CO_2. Cells have a rigid cell wall approximately 15–20 nm thick and, when stained for thin-section electron microscopy, resemble the electron dense monolayer cell wall of gram-positive bacteria. These archaeal cell walls are composed of pseudomurein, which is chemically distinguishable from bacterial murein by the substitution of *N*-talosaminuronic acid for *N*-acetylmuramic acid and substitution of $\beta(1,3)$ for $\beta(1,4)$ linkage in the glycan strands, and substitution of D-amino acids for L-amino acids in the peptide cross-linkage. *Methanothermus* species also have an additional cell-wall layer, composed of glycoprotein S-layer that surrounds the pseudomurein. *Methanobrevibacter* species are all mesophilic, *Methanobacterium* includes mesophiles and moderate thermophiles with optimal growth temperatures are high as 75 °C, *Methanothermobacter* species are exclusively moderate thermophiles, and *Methanothermus* species are extreme thermophiles, with maximum growth temperatures as high as 97 °C.

The order *Methanomicrobiales* contains genera that are diverse in morphology and physiology. Most species grow as cocci and rods. In addition, *Methanoplanus* forms flat plate-like cells with characteristically angular ends. Another species, *Methanospirillum hungateii*, forms a helical spiral. The cell walls in this order are composed of a protein S-layer and are sensitive to osmotic shock or detergents. In addition to the S-layer, *M. hungateii* also has an external sheath that is composed of concentric rings stacked together. Species are generally slightly halophilic and include mesophiles, moderate, and extreme thermophiles. Most species grow by CO_2 reduction with H_2, but some species also use formate or secondary alcohols as electron donors for CO_2 reduction.

The order *Methanosarcinales* is the most catabolically diverse species of methanogens. In addition to growth and methanogenesis by CO_2 reduction with H_2, some species grow by the dismutation, or "splitting," of acetate and by methylotrophic catabolism of methanol, methylated amines, pyruvate, and dimethylsulfide. While some species of *Methanosarcina* can grow by all three catabolic pathways, *Methanosaeta* species are obligate acetotrophs and all other genera are obligate methylotrophs. All species have a protein S-layer cell wall and most species grow as irregularly shaped cocci. However, several species of *Methanosarcina* also synthesize a heteropolysaccharide matrix external to the S-layer. This external layer can be up to 200 nm thick and is composed primarily of a nonsulfonated polymer of *N*-acetylgalactosamine and D-glucuronic or D-galacturonic acids. The matrix is called methanochondroitin because of its chemical similarity to a mammalian connective tissue component, known as chondroitin. At freshwater NaCl concentrations, *Methanosarcina* spp. that synthesize methanochondroitin grow in multicellular aggregates rather than as single cells, but when grown at marine salt concentrations or with high concentrations of divalent cations, such as Mg^{+2}, they no longer synthesize methanochondroitin and grow as single cells. *Methnaosaeta* species have an external sheath that appears similar in structure to that previously described for *M. hungateii*.

The order *Methanopyrales* is the most deeply branching methanogenic archaeon and presently includes only one species, *Methanopyrus kandleri*. This species is an obligate hydrogenotroph and grows as a rod with a pseudomurein cell wall surrounded by a protein S-layer, similar to that described for *Methanothermus*.

III. HABITATS

A. Interspecies H_2 transfer

As has been described, methanogens utilize a limited number of simple substrates. In most habitats, they depend on other anaerobes to convert complex organic matter into substrates that they can catabolize. Therefore, unlike aerobic habitats, where a single microorganism can catalyze the mineralization of a polymer by oxidation to CO_2, degradation in anaerobic habitats requires consortia of interacting microorganisms to convert polymers to CH_4. These interactions are dynamic with the methanogens affecting the pathway of electron flow, and, consequently, carbon flow, by a process called interspecies H_2 transfer. In this association, the H_2-utilizing methanogens maintain a low H_2 partial pressure that allows certain reactions to be thermodynamically favorable. One physiological group of microorganisms affected by this process is the H_2-producing acetogens. The reactions carried out by these microorganisms for growth are not thermodynamically

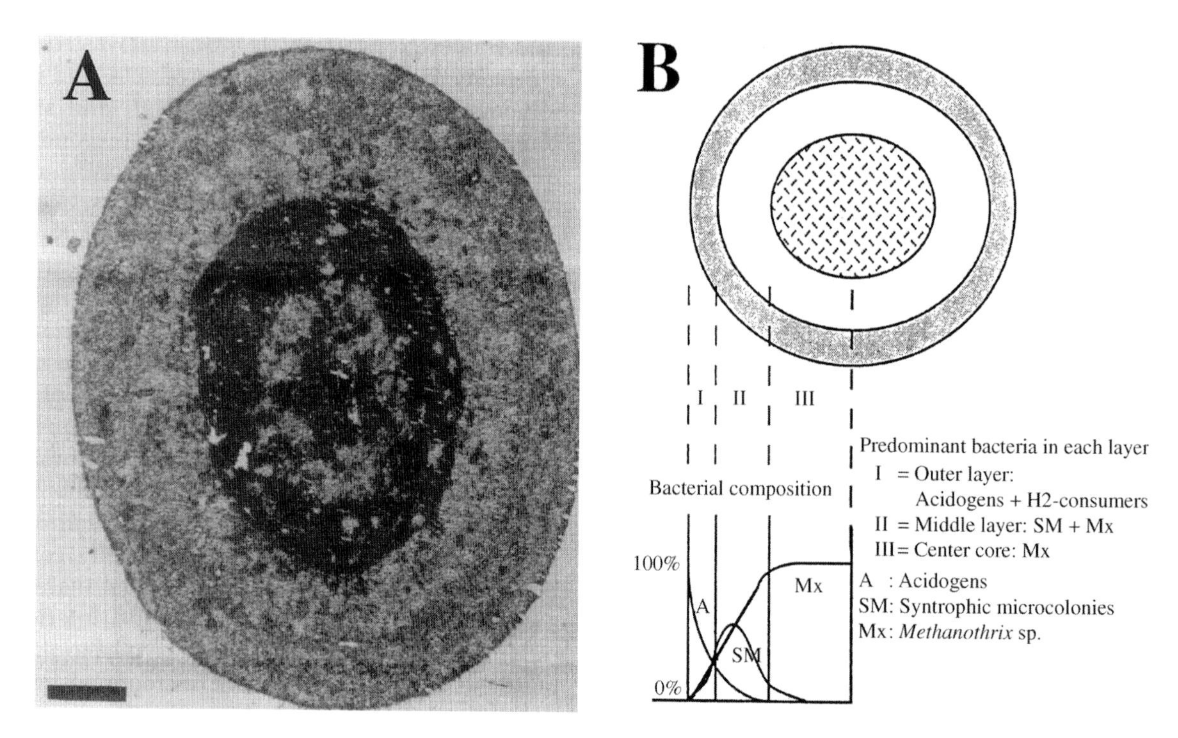

FIGURE 57.5 An anaerobic granule from a brewery wastewater digestor. Panel A shows a thin-section photomicrograph of a granule under a bright-field microscope. Bar, 0.25 mm. Panel B illustrates the bacterial composition of a granule. Methanothrix former generic epithet for Methanosaeta. (Reproduced from Fang, H. H. P., Chui, H. K., and Li, Y. Y. (1994). *Water Sci. Technol.* **30**, 87–96, and Fang, H. H. P., Chui, H. K., and Li, Y. Y. (1995). *Water Sci. Technol.* **31**, 129–135. Copyright 1994 and 1995, respectively, with permission from Elsevier Science.)

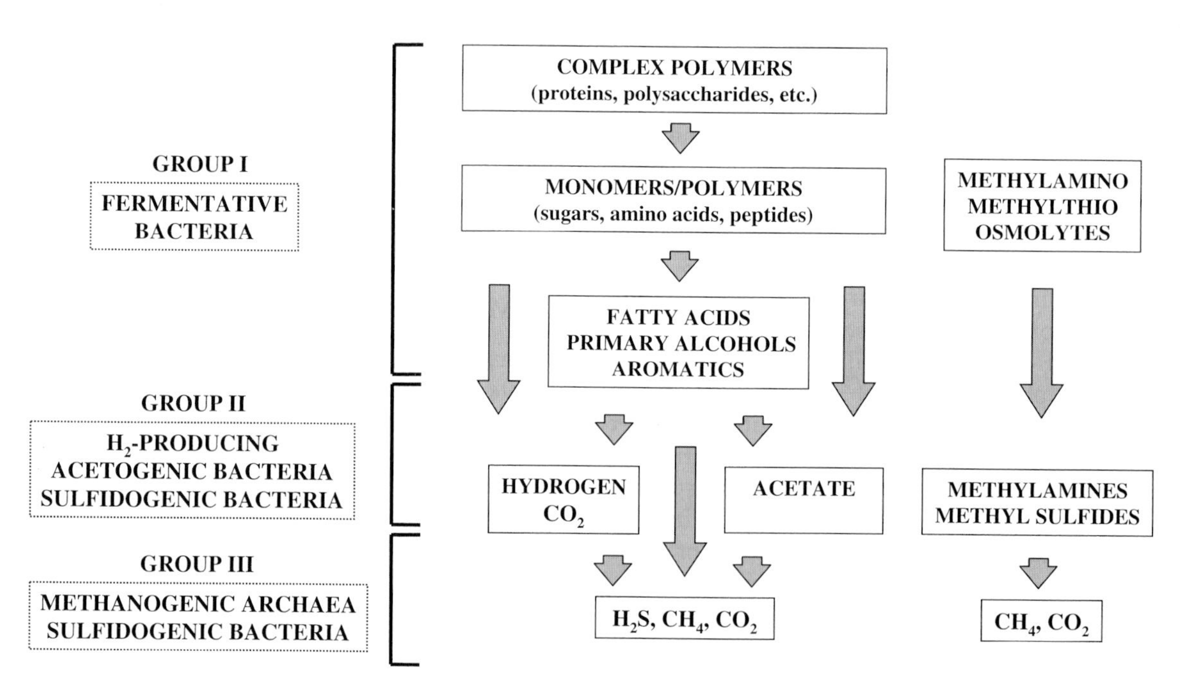

FIGURE 57.6 Carbon flow in an anaerobic microbial consortium from marine sediments.

consortium is similar to that in freshwater environments, but it is composed of halophilic and halotolerant species. Although the acetotrophic methanogens *Methanosarcina* and *Methanosaeta* have been isolated from marine methanogenic enrichments, isotope studies performed in sediment suggest that most of the acetate is oxidized by a H_2-producing syntroph rather than by splitting to CH_4. Methanogens also generate

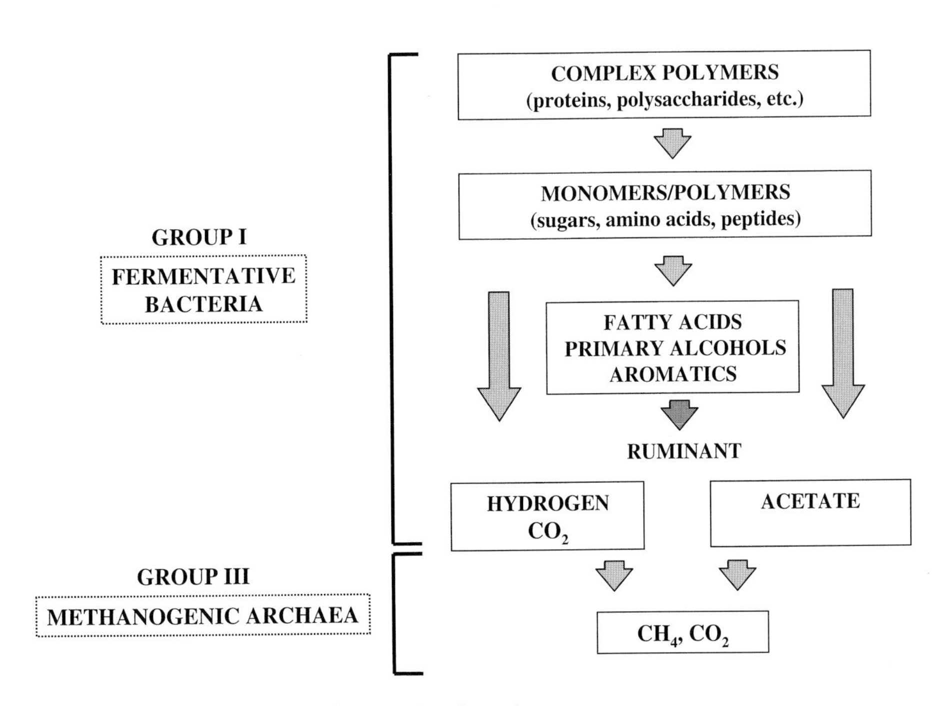

FIGURE 57.7 Carbon flow in an anaerobic microbial consortium from the rumen.

CH_4 from methylated amines and thiols, which are readily available in the marine environment as metabolic osmolytes. Since methylated amines are not used by SO_4^{-2}-reducing bacteria, this class of compound is "noncompetitive" and can be used by methanogens in habitats that contain high SO_4^{-2} concentrations.

CH_4 generated in sediments is often consumed as it diffuses through the SO_4^{-2}-reducing region of the sediments, before reaching the aerobic regions. However, the microbes that catalyze this anaerobic oxidation have not been described. Although much of the CH_4 generated in sediments is consumed in the SO_4^{-2}-reducing regions, the water column in the open ocean is supersaturated with CH_4, compared with the atmospheric concentration. This may result from a combination of unoxidized CH_4 that escapes from sediments, and methanogenic activity in the gastrointestinal tracts and fecal material of marine animals. Biologically generated CH_4 in some organic-rich buried sediments can accumulate as gas deposits. Since natural gas deposits generated abiotically are used as indicators of petroleum, these biologically generated CH_4 deposits can act as false indicators of petroleum deposits during oil exploration. Under high hydrostatic pressures generated in deep ocean sediments, biogenic CH_4 can also accumulate as solidified CH_4 hydrates.

E. Ruminant animals

Ruminant animals include both domestic (cows, sheep, camels) and wild (deer, bison, giraffes) animals. These animals have a large chamber, called the rumen, before the stomach, in which polymers, such as cellulose, are fermented by bacteria to short-chain fatty acids, H_2, CO_2, and CH_4. The rumen is similar to an anaerobic bioreactor, except that the short retention time created by swallowing saliva is less than the generation time of H_2-producing fatty-acid oxidizers and acetate-utilizing methanogens (Fig. 57.7). As a result, acetate, propionate, and butyrate are not degraded by the consortium, but are absorbed into the bloodstream of the host animal as carbon and energy sources. The volatile gases (e.g., CH_4, CO_2) are removed from the animal by belching. Acidification of the system by the acids is prevented by bicarbonate in the saliva of the animal. Carbon diverted to CH_4 and belched into the atmosphere represents a loss of energy to the animal. Ruminant nutritionists have been attempting to increase feed efficiency by adding methanogenic inhibitors, such as monensin, to feed, thereby, diverting carbon flow to metabolites that can be utilized by the animal.

F. Xylophagous termites

All known termites harbor a dense microbial community of anaerobic bacteria, and, in the case of lower termites, they also contain cellulolytic protozoa that catalyze the digestion of lignocellulose from wood. As in the rumen, these microorganisms have a synergistic relationship with the termites by converting polymers to short-chain fatty acids that are used as carbon and energy sources for their host. The carbon flow in the hindgut of soil-feeding and fungus-cultivating termites is similar to that in rumen, but in wood- and grass-eating termites, most H_2–CO_2 is converted to acetate instead of CH_4. Generally, methanogenesis outcompetes H_2–CO_2 acetogenesis and the factors that cause the predominance of acetogenesis in some termites in not known. Two species of H_2-utilizing *Methanobrevibacter* have been isolated from the hindgut of the subterranean termite that exhibit catalase activity and are particularly tolerant to oxygen exposure. Oxygen tolerance may be important for recolonization by bacterial consortia after expulsion of the hindgut contents during molting. Reinoculation is achieved by transfer of hindgut contents from other colony members, which are exposed to air during the process.

G. Human gastrointestinal tract

The human colon serves as a form of hindgut where undigested polymers and sloughed off intestinal epithelium and mucin are dewatered, fermented by bacterial consortia, treated with bile acids, and held until defecation. Fatty acids generated by fermentation are absorbed into the bloodstream and can provide approximately 10% of human nutritional needs. Methanogenesis occurs in 30–40% of the human population, with the remaining population producing H_2 and CO_2 instead. Acetogenesis and SO_4^{-2}reduction from H_2–CO_2 also occur in the human colon, but studies with colonic bacterial communities suggest that these activities are only prevalent in the absence of methanogenic activity. The level of CH_4 produced by an individual corresponds to the population levels of *Methanobrevibacter*, but factors controlling the occurrence of methanogens in the human population are not known. Diet does not appear to have a significant role in determining whether an individual harbors an active methanogenic population, but hereditary and individual physiological factors may have a role. For example, methanogens may be absent from individuals that excrete higher levels of bile acids, which are inhibitory to methanogens.

H. Protozoan endosymbionts

Methanogens are present as endosymbionts in many free-living marine and freshwater anaerobic protozoa, where they are often closely associated with hydrogenosomes, organelles that produce H_2, CO_2, and acetate from the fermentation of polymeric substrates. The products of the hydrogenosomes are substrates for methanogenesis. It is conceivable that the methanogens have a synergistic role by lowering the H_2 partial pressure to create a favorable thermodynamic shift in the protozoan's fermentation reaction. Also, evidence suggests that excretion of undefined organic compounds by the methanogen provides an advantage to the protist host. Endosymbionts are also found in flagellates and ciliates that occur in the hindgut of insects, such as termites, cockroaches, and tropical millipeds. Although rumen ciliates do not harbor endosymbionic methanogens, many have ectosymbionic methanogens that may have an analogous function.

I. Other habitats

Habitats that have a source of organic carbon and a high water content can become anaerobic as a result of respiratory depletion of oxygen, and, subsequently, support methanogenic communities. Examples include soils waterlogged by heavy rainfall, marshes, rice paddies, rotting heartwood of trees, and landfills. Most of the CH_4 generated in these habitats is released into the atmosphere. However, many landfills are now vented and collected to prevent a buildup of potentially combustible CH_4 underground or in nearby dwellings, and some communities harvest the vented biogas for heat and energy.

Methanogenic habitats are also found in geohydrothermal outsources, such as terrestrial hot springs and deep-sea hydrothermal vents. Methanogens from these environments use geothermally generated H_2-CO_2 for methanogenesis and most are hyperthermophilic, requiring growth temperatures as high as 110°C. In terrestrial sites, methanogens are usually associated with microbial mats composed of photosynthetic and heterotrophic consortia. However, in deep sea hydrothermal vents, where there is no light available for photosynthetic production of organic carbon, the methanogens and other autotrophic bacteria serve as primary producers of cell carbon for a complex community of heterotrophic microbes and animals that accumulates in the vicinity of a hydrothermal vent.

Methanogenesis also occurs in high saline environments, such as Great Salt Lake, Utah, Mono Lake, California, and solar salt ponds. Methanogens from these environments generate CH_4 from methylated amines and dimethylsulfide, which are synthesized by animals and plants as osmolytes. Methanogenesis has been detected in subsurface aquifers, where, it has been proposed, that H_2 generated by an abiotic reaction between iron-rich minerals in basalt and ground water is used as a substrate for methanogenesis. Methanogenesis has also been detected in deep subsurface sandstone, where it is hypothesized that methanogenic consortia use organic compounds that diffuse from adjacent organic-rich shale layers.

IV. PHYSIOLOGY AND BIOCHEMISTRY

A. Catabolic pathways

All methanogens generate CH_4 during growth by three basic catabolic pathways: autotrophic CO_2 reduction with H_2, formate, or secondary alcohols; acetotrophic cleavage of acetate; or methylotrophic dismutation of methanol, methylated amines, or methylthiols. The common reactions for methanogenesis are shown in Table 57.2. Six new coenzymes were discovered that serve as carbon carriers in the methanogenic pathway (Fig. 57.8). Methanofuran (MFR) is an analog of molybdopterins, which occur in enzymes that catalyze similar reactions in the *Bacteria* and *Eukarya*. Tetramethanopterin (H_4MPT) is an analog of tetrahydrofolate (H_4THF), which is also a one-carbon carrier in bacterial and eukaryal systems. Although H_4MPT was initially found to be unique to methanogens, it has since been detected in other *Archaea* and, more recently, enzymes catalyzing the methyl transfer for MFR and H_4MPT have been found to coexist with the H_4THF pathway in the CH_4-utilizing methanotrophs. Methyl coenzyme M (CoM-SH), 7-mercaptoheptanoylthreonine phosphate (HS-HTP) and cofactor F_{430} are currently unique to the methanogens.

The methanogenic sequence is initiated by a two-electron reduction of CO_2 and methanofuran by formyl-MFR dehydrogenase (a) to form formyl- MFR (Fig. 57.9). The formyl group is then transferred to H_4MPT by formyl-MFR:H_4MPT formyltransferase (b) yielding formyl-H_4MPT. A homolog of H_4MPT, tetrahydrosarcinapterin (H_4SPT), found in *Methanosarcina* spp., differs by an additional glutamyl moiety in the substituted R group. Formyl-H_4MPT cyclization to methenyl-H_4MPT is catalyzed by N5,N10-methenyl-H_4MPT cyclohydrolase (c) N5,N10-methylene-H_4MPT dehydrogenase (d) and N5,N10-methylene-H_4MPT reductase (e) catalyze the sequential reduction of methenyl-H_4MPT by the electron carrier coenzyme F_{420} to methylene-H_4MPT and methyl-H_4MPT. The methyl group is then transferred to CoM-SH by N5-methyl-H_4MPT:CoM-SH methyltransferase (f) forming methyl-S-CoM. Methyl CoM reductase (g) catalyzes the terminal reduction of methyl-S-CoM by two electrons from HS-HTP to CH_4. CoM-SS-HTP is the product of the terminal reaction, which is subsequently reduced by heterodisulfide reductase to regenerate the reduced forms CoM-SH and HTP-SH.

Methylotrophic catabolism of methanol and methylated amines requires three polypeptides. Methanol is catabolized by transfer of its methyl group to a corrinoid-binding protein, which is methylated by a substrate-specific methyltransferase, methanol:5-hydroxybenzinidazolyl (MT1). The methyl group is then transferred from the corrinoid protein to coenzyme HS-CoM by methylcobamide: CoM methyltransferase (MT2). Trimethylamine, dimethylamine, and monomethylamine each require a distinct corrinoid-binding protein, which is methylated by a substrate-specific methyltransferase. The methyl group is then transferred from the corrinoid protein to coenzyme HS-CoM by a common MT2 homolog. In contrast, catabolism of the methylthiols dimethylsulfide and methylmercaptopropionate is catalyzed by only two polypeptides: a corrinoid-binding protein tightly bound to a methylcobamide:CoM methyltransferase homolog of MT2. Methyl-S-CoM generated from methanol, methylated amines, and methylthiols is reduced to CH_4 in the methanogenic pathway, as has been described. A portion of the

TABLE 57.2 Reactions and free energy yields from methanogenic substrates

Substrate	Reaction	$\Delta G^{0\prime}$ (kJ/mol CH_4)
Hydrogen/carbon dioxide	$2CH_3CH_2OH + HCO_3^- \rightarrow 2CH_3COO^- + H^+ + CH_4 + H_2O$	−135
Hydrogen/carbon dioxide	$4H_2 + HCO_3^- \rightarrow CH_4 + 3H_2O$	−135
Formate	$4HCOO^- + 4H^+ \rightarrow CH_4 + 3CO_2 + 2H_2O$	−145
Carbon monoxide	$4CO + 5H_2O \rightarrow CH_4 + 3HCO3^- + 3H^+$	−196
Ethanol[a]	$2CH_3CH_2OH + HCO_3^- \rightarrow 2CH_3COO^- + H^+ + CH_4 + H_2O$	−116
Hydrogen/methanol	$CH_3OH + H_2 \rightarrow CH_4 + H_2O$	−113
Methanol	$4CH_3OH \rightarrow 3CH_4 + HCO_3^- + H_2O + H^+$	−105
Trimethylamine[b]	$4CH_3NH^+ + 9H_2O \rightarrow 9CH_3 + 3HCO_3^- + 4NH_4 + 3H^+$	−76
Dimethylsulfide[c]	$2(CH_3)_2S + 3H_2O \rightarrow 3CH_4 + HCO_3^- + 2H_2S + H^+$	−49
Acetate	$CH_3COO^- + H_2O \rightarrow CH_4 + HCO_3^-$	−31

[a]Other short chain alcohols are utilized.
[b]Other methylated amines are utilized.
[c]Other methylated sulfides are utilized.

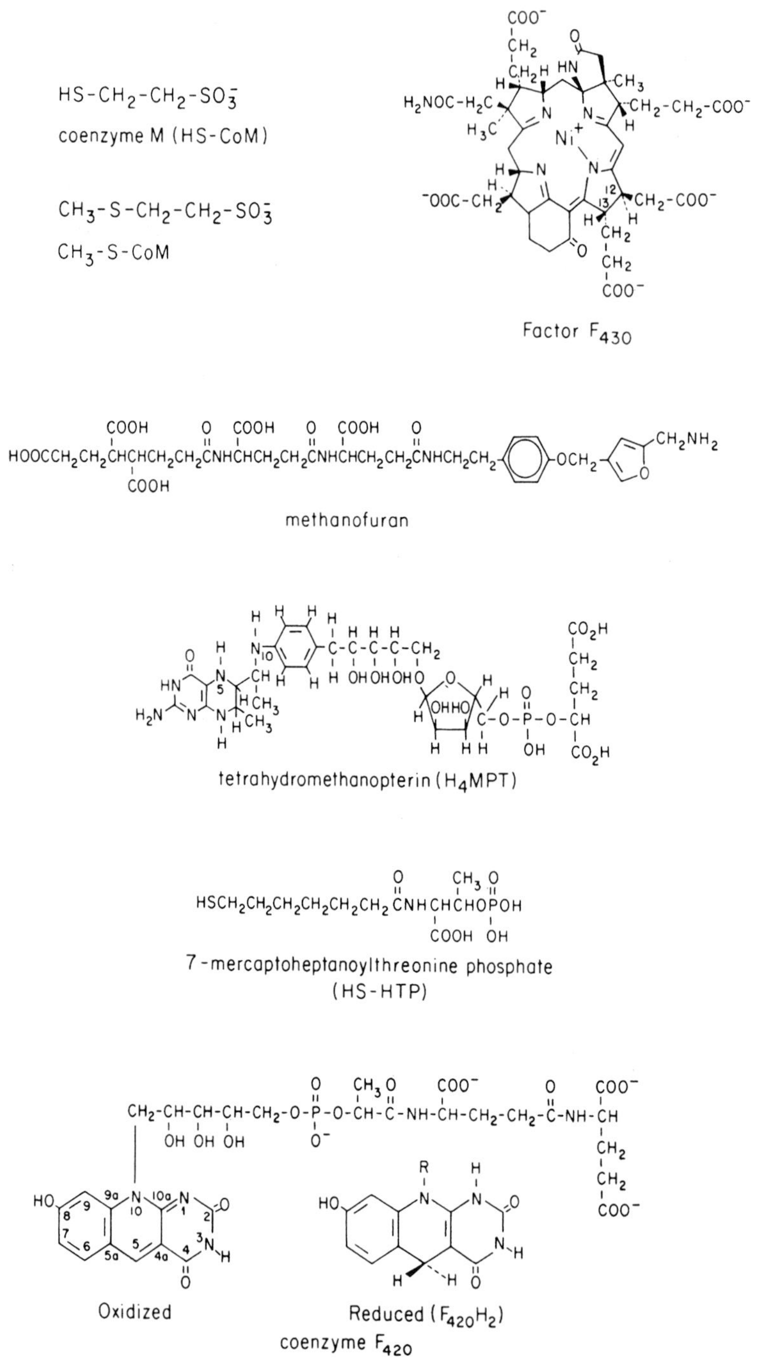

$HS\text{-}CH_2\text{-}CH_2\text{-}SO_3^-$

coenzyme M (HS-CoM)

$CH_3\text{-}S\text{-}CH_2\text{-}CH_2\text{-}SO_3^-$

CH_3-S-CoM

Factor F_{430}

methanofuran

tetrahydromethanopterin (H_4MPT)

$HSCH_2CH_2CH_2CH_2CH_2CH_2CNHCHCHOPOH$

7-mercaptoheptanoylthreonine phosphate (HS-HTP)

Oxidized

Reduced ($F_{420}H_2$)

coenzyme F_{420}

FIGURE 57.8 Structures of coenzymes that participate in the methanogenic pathway. (Reproduced, with permission, from Rouviere, P. E., and Wolfe, R. S. (1988). *J. Biol. Chem.* **263**, 7913–7916. American Society for Biochemistry and Molecular Biology, Inc.)

methyl groups generated from methylotrophic catabolism is oxidized in reverse sequence in a pathway identical to the CO_2 reduction pathway, after what appears to be a direct transfer of the methyl groups to H_4MPT. However, the mechanism of this transfer is not yet known. This oxidative sequence generates electrons for the reduction of CoM-S-S-HTP in the methyl-S-CoM reductase system.

The acetotrophic pathway for acetate catabolism proceeds by initial "activation" of acetate by formation of acetyl CoA. *Methanosarcina* spp. synthesize acetyl CoA by sequential activities of phosphotransacetylase and acetyl kinase. In contrast, activation of acetate to acetyl CoA by *Methanosaeta* spp. is catalyzed in a single step by acetyl-CoA synthase. In both genera, cleavage of the C–C and C–S bonds of acetyl-CoA is

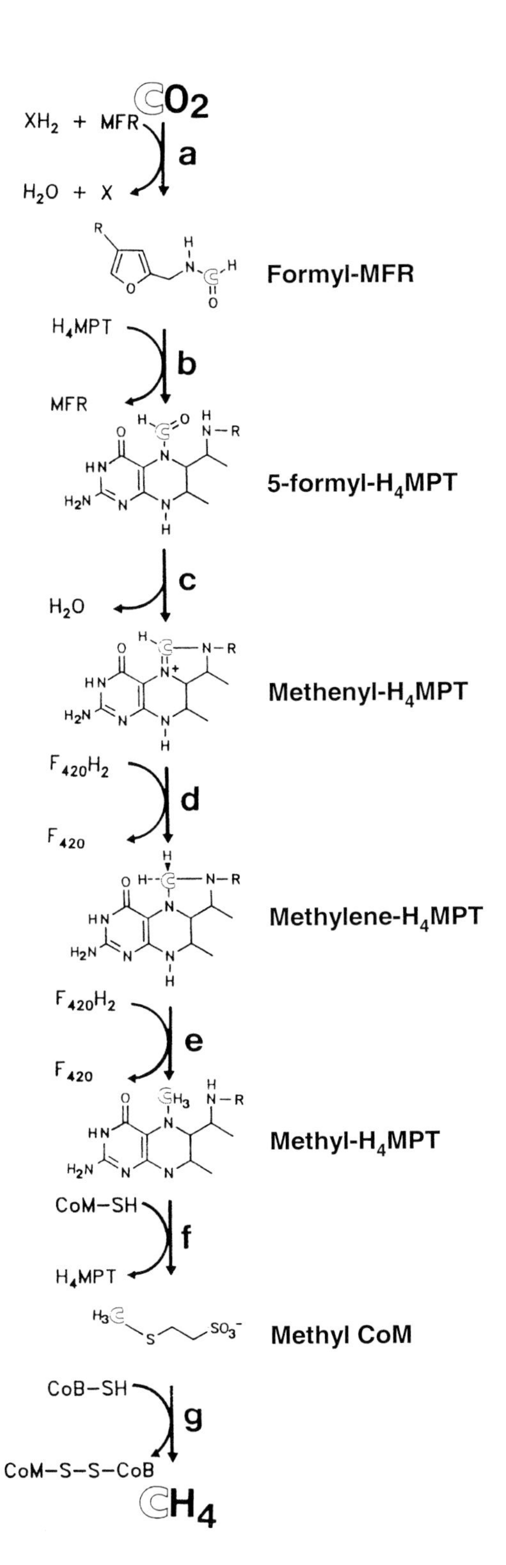

FIGURE 57.9 Methanogenic pathway in which CO_2 is sequentially reduced as coenzyme-bound intermediates to form CH_4. Details are described in the text. (Reproduced, with permission, from Weiss, D. S., and Thauer, R. K. (1993). *Cell* **72**, 819–822. Cell Press.)

then catalyzed by the acetyl-CoA decarbonylase/synthase complex, yielding enzyme-bound methyl and carbonyl groups. The complex contains CO:acceptor oxidoreductase, Co-β-methylcobamide: tetrahydropterine methyltransferase, and acetyl-CoA synthase activities. The five subunit complex consists of a two polypeptide CO-oxidizing nickel/iron-sulfur component, a two-polypeptide corrinoid/iron-sulfur component, and a single polypeptide of unknown function. The nickel/iron-sulfur component catalyzes the cleavage of acetyl-CoA, the oxidation of the bound carbonyl to CO_2, and methyl transfer to the corrinoid/iron-sulfur component. The methyl group is sequentially transferred to H_4SPT by a currently unknown process and to HS-CoM by H_4MST:CoM-SH methyltransferase. Methyl-S-CoM is reductively demethylated to CH_4 by methylreducase as previously described. The enzyme-bound carbonyl group is oxidized to CO_2. *Methanosarcina* spp. and autotrophic methanogens growing on H_2/CO_2 synthesize acetate, presumably by reversing the direction of the acetyl-CoA decarbonylase/synthase complex in a reaction analogous to acetyl CoA synthase in acetate-utilizing *Clostridia*.

B. Bioenergetics

The methanogenic *Archaea* derive their metabolic energy from autotrophic CO_2 reduction with H_2, formate or secondary alcohols, cleavage of acetate, or methylotrophic dismutation of methanol, methylated amines, or methylthiols (Table 57.2). Currently, there is little evidence to convincingly support substrate level phosphorylation. Evidence that redox reactions in the catabolic pathways are catalyzed, in part, by membrane-bound enzyme systems and are dependent upon electrochemical sodium ion or proton gradients indicates that electron transport phosphorylation is responsible for ATP synthesis. Both gradients generate ATP for metabolic energy via membrane ATP synthases. Several reactions in the methanogenic pathway are sufficiently exergonic to be coupled with energy conversion, including reduction of methyl CoM (−29 kJ/mol) and methyl transfer from H_4MPT/H_4SPT to HS-CoM (−85 kJ/mol). In addition, the oxidation of formyl-MF (−16 kJ/mol) during methyl oxidation in the methylotrophic pathway and CO oxidation (−20 kJ/mol) in the acetotrophic pathway are also exergonic.

During growth on H_2–CO_2 or H_2-methanol, the H_2:heterodisulfide oxidoreductase system catalyzes the H_2-dependent reduction of CoM-S-S-HTP to HS-CoM (Fig. 57.10). Electron translocation across the membrane generates a proton gradient for ATP synthesis via an A_1A_0 ATPase. During methylotrophic growth

TABLE 57.3 Molecular features of the three phylogeneic domains

Feature	Eukarya	Bacteria	Archaea
Genome	Multiple linear	Single circular[a]	Single circular[b]
Chromatin	Histone-mediated	No conserved mechanism	Histone-mediated
Extrachromosomal DNA	Plasmids, viruses, IS units	Plasmids, viruses, IS units	Plasmids, viruses, IS units
DNA polymerase	Families A[c], B and X	Families A, B and C	Families B and X
RNA polymerase	Three classes, complex	One class $\beta\beta'\alpha_2\sigma$	One class, complex[d]
Gene structure	Single gene	Multiple gene operon	Multiple gene operon
Transcription promoter	TATA box	−35/−10 sequence	TATA box
Transcription terminator	AAUAAA sequence	intrinsic, σ-dependent	intrinsic, oligo T
Ribosomal RNA	28S-5.8S/18S	23S-5S/16S	23S-5S/16S
Translation initiation	5′ cap	Ribosomal binding site	Ribosomal binding site
Initiator transfer RNA	Methionine	Formylmethionine	Methionine

[a]Other short chain alcohols are utilized.
[b]Some species have multiple copies of the same genome.
[c]Mitochondrial.
[d]AB′B″C based on homology to eukaryal RNA polymerase II.

site-specific recombinases and RepA-like replication initiation proteins. The functions of the remaining plasmids are cryptic at this time. The mechanism of plasmid replication in methanogens is not known, but RepA proteins are associated with plasmids and phage that replicate by a rolling circle mechanism. Lytic viruses have been detected in three strains of *Methanothermobacter* and one strain of *Methanobrevibacter smithii.* These viruses show a varying range of host specificities. A temperate virus-like particle has been isolated that can integrate into the chromosome of *Methanococcus voltae.*

In contrast to the similarities of DNA structure in methanogenic *Archaea* and *Bacteria,* DNA-modifying proteins in the methanogens share features common to both the *Bacteria* and *Eukarya* (Table 57.3). All cells must package their genomic DNA within the limited space of the cells. Although the *Bacteria* do not appear to have a conserved mechanism for DNA packaging, *Archaea* and *Eukarya* chromosomal DNAs are compacted by histones into defined structures called nucleosomes, which are further assembled to form chromatin. The small, basic histone-like proteins from *Methanothermus fervidus* have homology to *eukaryal* histones and appear to have conserved the minimal structure for creating positive superturns, required to build a nucleosome. Archaeal histone-like proteins increase the melting temperature of linear DNA *in vitro* by as much as 25°C and likely protect DNA from heat denaturation *in vivo.* Reverse gyrase, also detected in *M. fervidus,* may contribute to heat resistance of DNA by creating stable positive supercoils in inter-histone regions. Histone-like proteins isolated from mesophilic *Methanosarcina* species cause concentration-dependent inhibition or stimulation of gene transcriptions *in vitro,* which suggests that they may also have a role in gene regulation.

B. DNA replication, repair, modification, and metabolism

Although our understanding of DNA replication and repair in the *Bacteria* and *Eukarya* is well advanced, comparatively little is known about these processes in the methanogenic *Archaea.* Studies on DNA replication show that aphidicolin, a specific inhibitor of eukaryal DNA replication, inhibits DNA polymerases from *Methanococcus vannielii* and *Methanococcus voltae,* but does not inhibit DNA polymerase from *M. thermoautotrophicus.* Sequences of genes encoding both the aphidicolin-sensitive and -insensitive methanogen DNA polymerases reveal they are homologous to family B and X DNA polymerases from the *Bacteria* and *Eukarya. Methanocaldococcus jannaschii* contains a single gene with two inteins that encode a B-type DNA polymerase. In contrast, *M. thermoautotrophicus* has two polymerases: a B-type DNA polymerase composed of two polypeptides and an X-type DNA polymerase. DNA repair mechanisms have also been identified in methanogens. A photoreactivation system in *M. thermoautotrophicus* is mediated by class II photolyase with homology to metazoan photolyases. Genes encoding putative eukaryal DNA repair proteins RAD2, RAD25, and RAD51, and bacterial DNA repair proteins uvrABC, mutL and mutS have been identified in genomic sequences of methanogenic *Archaea.* Several type II restriction endonuclease-methyltransferase systems have been identified and four methanogen endonucleases are available commercially. In addition, putative type I restriction-modification enzymes have been identified by sequence annotation of the *M. thermoautotrophicus* genome. Despite their sensitivity to oxygen, a Fe-superoxide dismutase has been characterized from *M. thermoautotrophicus* and catalase activity has been detected in species of *Methanobrevibacter*. Putative genes encoding for related DNA replication and repair proteins, such as helicases, ligases, topoisimerases, endonucleases, recombinases, and replication factors, have also been identified in genomic sequences of methanogens, but their function has not yet been confirmed.

C. Gene structure and transcription

The organization of methanogenic genes in tightly linked clusters is similar to the operon configuration found in the *Bacteria*. As in the *Bacteria*, the archaeal operons are transcribed from an upstream promoter into polycistronic RNAs. However, archaeal genes are transcribed by a multicomponent RNA polymerase that is structurally homologous to eukaryal RNA polymerase and recognizes a TATA promoter with high sequence homology to the consensus motif for the *Eukarya*. Unlike the variable upstream distance of eukaryal promoters, the archaeal promoter element is located at a consistent distance, approximately 20–30 bp upstream from the transcription initiation site, a range similar to the conserved −35 bp region observed in the *Bacteria*. Site-directed deletion studies conducted with methanogenic *Archaea* by *in vitro* techniques indicate that efficient transcription and start-site selection is dependent upon a TATA promoter. This arrangement closely resembles the core structure of RNA polymerase II promoters in the *Eukarya*. Purified archaeal RNAPs from *M. thermoautotrophicus* and *M. voltae* fail to initiate site-specific transcription without the addition of TATA binding protein (TBP) and transcription factor TFIIB. Yeast and human TATA-binding proteins can substitute for TBP in a *Methanococcus*-derived archaeal cell-free transcription system, indicating that they are functionally homologous. Additional genes that have sequence similarity to the eukaryal-like transcription factors TFIIIC, TFIIE, and TFIIS have been putatively identified in the genomic sequences of methanogenic *Archaea*. Although the spatial and temporal nature of these transcription mechanisms is not yet known, the results suggest that a DNA-protein recognition site, analogous to that in the *Eukarya*, is required for site-specific RNAP recognition in the *Archaea*. This mechanism would involve sequence recognition by TFIIB-TBP and recruitment of other transcription factors to form a recognition complex that would be recognized by polymerase and initiate transcription. The mechanisms of transcriptional regulation in highly regulated genes has not yet been determined, but *in vitro* studies on *Methanococcus maripaludis* reveals that point mutations in a palindrome located downstream of the *nif*H transcription start site results in derepression of expression. This suggests that a bacterial-type repressor may mediate gene expression in some highly regulated methanogen genes. Genes encoding three distinct TBPs have been detected in the *M. acetivorans* genome, which raises the possibility that gene regulation is mediated by the formation of alternative TBP-TFB pairing. In addtion to multiple TBPs, multiple TFBs have also been detected in the genome of the extreme halophile *Halobacterium*, but there is no evidence for multiple TFBs in *M. acetivorans*. Generally, transcription is terminated following an inverted repeat sequence located downstream of methanogen genes. Transcription termination sites are similar to the ρ-independent terminators in the *Bacteria*, and likely form a stem-loop secondary structure to mediate termination. A second type of transcription terminator, which consists of a single or several tandemly arranged oligo-T sequences, is found in hyperthermophilic methanogens. The occurrence of this terminator in hyperthermophiles suggests that the stem-loop structures characteristic of σ-independent promoters may be unstable at higher growth temperatures.

D. RNA structure and translation

The stable RNAs transcribed by methanogen genes have been investigated in some detail. Methanogen ribosomes resemble bacterial ribosomes. They are composed of two protein subunits of 30S and 50S and three rRNA components of 23S, 16S, and 5S, which, assembled, yield a ribosome of 70S. Archaeal and bacterial ribosomal proteins are functionally homologous and can be interchanged to create an active ribosome *in vitro*. Genes encoding rRNA are arranged in the order 16S–23S–5S and the number of operon copies varies in number from one to four. The organization and order of genes encoding methanogen ribosomal proteins also resembles that found in the *Bacteria*. Methanogen tRNAs contain the sequence 1-methylψ CG substituted for the sequence TψCG, typically found in the arm of bacterial and eukaryal tRNAs. Introns have been detected in genes encoding tRNA. A unique feature of the archaeal genomes of *M. jannaschii* and *M. thermoautotrophicus* is the absence of a gene encoding cyteine-tRNA synthetase. This function is carried out by a dual-specificity prolyl-tRNA synthetase that recognizes and aminoacylates both $tRNA^{Pro}$ and $tRNA^{Cys}$.

Some methanogen mRNAs have poly-A^+ tails, but, as in the *Bacteria*, they average only 12 bases in length. Protein-encoding genes employ the same genetic code as *Bacteria* and *Eukarya*, and codon preferences reflect the overall base composition of the genome and the level of gene expression. The codon ATG is used frequently as an initiation codon, as well as GTG and TGG. A ribosomal-binding site located upstream of structural genes, when transcribed, is complementary to the 3′ terminal sequence of methanogen 16S rRNA. There is little information on the mechanisms of translation based on biochemical experimentation. However, genome analysis reveals that methanogens possess protein initiation factors that share homology with both bacterial and eukaryal IF proteins. Inteins have been identified by genomic sequencing of *M. jannaschii*, which suggests that protein splicing occurs in methanogens. In addition, evidence has been found for phosphorylation of proteins at a tyrosine residue, which is a mechanism of post-translational control in the *Bacteria* and *Eukarya*.

E. Genomics and gene function analysis

The genomes of the hyperthermophiles *M. jannaschii* (1.66 megabases) and *Methanopyrus kandleri* (1.7 megabases), the thermophile *M. thermotrophicus* (formerly M. thermoautotrophicum, 1.8 megabases), and the mesophiles *Methanosarcina acetivorans* (5.7 megabases) and *Methanosarcina mazei* (4.1 mega bases) have been completely sequenced and annotated. In general similar genes are found for the CO_2 reducing catabolic pathways in all four genomes. Genes encoding multiple methyltransferases and acetyl-CoA decarbonylase/synthases are only found in the genome of *M. acetivorans*, which is consistent with the ability of this species to also grow by methylotrophic and aceticlastic pathways. The large genome size of *M. acetivorans* likely reflects this species' ability to use a greater range of substrates, adapt to a broader range of environments and form complex multicellular structures compared with the more limited capabilities of the hydrogen-utilizing species. In contrast the hyperthermophiles are more "minimalist" exhibiting a paucity of genes encoding proteins for signaling and gene regulation, *grp*E-*dna*J-*dna*K heat shock operon, proteasome-chaperonin, several DNA repair proteins, DNA helicases, nitrogenase subunits, ribonucleotide reductase and proteases. Overall, the majority of archaeal open reading frames with similarity to bacterial sequences include genes for small molecule biosynthesis, intermediary metabolism, transport, nitrogen fixation, and regulatory functions. Archaeal open reading frames with similarity to eukaryal sequences include genes for DNA metabolism, transcription, and translation. The presence of Cdc6 homologs and histones suggests that DNA replication initiation and chromosome packaging is eukaryal, but detection of *fts*Z suggests that bacterial-type cell division occurs. Additional unique features include an archaeal B-type DNA polymerase with two subunits, putative RNAP A′ subunits that suggest that possibility of additional mechanisms for gene selection, and two introns in the same $tRNA^{Pro}$ (CCC) gene, which establishes a new precedent.

The availability of genome sequences of methanogens and current activities to sequence others make the development of archaeal gene-transfer systems essential for confirmation of gene function. However, advances in the genetics of methanogens have been limited to studies *in vitro* or using heterologous systems, such as *E. coli*, because of the lack of tractable gene transfer systems for these microorganisms. Two gene-transfer systems have recently been developed for species of *Methanococcus* and *Methanosarcina*. Both systems utilize hybrid shuttle vectors derived from native archaeal plasmids. Since plasmids occur in low copy number in the methanogens, the DNA to be transferred is ligated into the vector, which is then amplified in *E. coli* with ampicillin as a selectable marker. The modified plasmid is then transferred into the methanogen by polyethylene glycol-mediated (*Methanococcus* sp.) or liposome-mediated (*Methanosarcina* spp.) transformation. A second selectable marker in the plasmid, such as *pac* (puromycin acetyl transferase) controlled by a highly expressed methanogen promoter *mcr* (methyl CoM reductase), provides selection in the methanogen. Both autonomously replicating plasmids for introducing specific phenotypes and integration plasmids for disrupting genes by homologous recombination have been designed. These systems are highly efficient, yielding 10^7–10^9 transformants per μg DNA. Methods for transposon-mediated mutagenesis have also been developed. A transducing phage has been isolated from *M. thermoautotrophicus*, but it is not currently useful for gene transfer because of its limited burst size (~6 per cell). Conjugation has not been observed in methanogens. Although gene-transfer systems are somewhat limited at this time, development of new and more sophisticated systems is ongoing. Impediments that need to be overcome include a lack of understanding of mechanisms for vector replication, retention and segregation, and the limited number of selectable and phenotypic markers that are currently available for the methanogens.

VI. SUMMARY

The methanogenic *Archaea* have a pivotal role in the global carbon cycle by complementing aerobic processes that ultimately lead to the oxidation of organic carbon to CO_2. However, a steady increase in the levels of atmospheric CH_4 that has coincided with the increase in the human population is a cause for concern, since CH_4 is a greenhouse gas. Methane's contribution to global warming results from its high infrared absorbance and its role in complex chemical reactions in the stratosphere that affect the levels of ozone. Increased waste disposal activities, such as landfills, are a significant source of atmospheric CH_4. Another significant source results from agricultural activities, such as the increased use of domesticated ruminants for production of meat and dairy products and the increased development of rice paddies. Understanding the properties of methanogens and the roles they have in the global carbon cycle will have important implications in addressing the issue of global warming as the human population increases.

The application of methanogens in biotechnology has been largely limited to waste management, which is often coupled to limited biogas production. Although the petroleum crisis of the 1970s led to an interest in methanogenic biogas production by fermentation of sources ranging from agricultural products to marine kelp, cost-effective technologies were never fully developed and interest has since waned with the drop in petroleum prices. Other potential applications for methanogens include the production of novel pharmaceuticals, corrinoids, and thermo-stable enzymes. To date, methanogens have yielded only a few restriction endonucleases as commercial products. However, significant advances in our understanding of the physiology and biochemistry of methanogens over the past two decades, combined with the recent developments in gene transfer systems and advances in genome sequencing, make the application of methanogens for biotechnology more plausible in the near future.

BIBLIOGRAPHY

Amend, J. P., and Shock, E. L. (2001). Energetics of overall metabolic reactions of thermophilic and hyperthermophilic Archaea and bacteria. *FEMS Microbiol. Rev.* **25**, 175–243.

Bell, S. D., and Jackson, S. P. (2001). Mechanism and regulation of transcription in archaea. *Curr. Opin. Microbiol.* **4**, 208–213.

Bult, C. J., White, O., Olsen, G. J., Zhou, L. X., Fleischmann, R. D., Sutton, G. G., Blake, J. A., Fitzgerald, L. M., Clayton, R. A., Gocayne, J. D., *et al.* (1996). Complete genome sequence of the methanogenic archaeon, *Methanococcus jannaschii*. *Science* **273**, 1058–1073.

Daniels, L. (1992). Biotechnological potential of methanogens (Review). *Biochem. Soc. Symp.* **58**, 181–193.

Deppenmeier, U., Muller, V., and Gottschalk, G. (1996). Pathways of energy conservation in methanogenic archaea. *Arch. Microbiol.* **165**, 149–163.

Deppenmeier, U., Johann, A., Hartsch, T., Merkl, R., Schmitz, R. A., Martinez-Arias, R., Henne, A., Wiezer, A., Bäumer, S., Jacobi, C., *et al.* (2002). The genome of *Methanosarcina mazei*: Evidence for lateral gene transfer between Bacteria and Archaea. *J. Mol. Microbiol. Biotechnol.* **4**, 453–461.

Galagan, J. E., Nusbaum, C., Roy, A., Endrizzi, M. G., Macdonald, P., FitzHugh, W., Calvo, S., Engels, R., Smirnov, S., Atnoor, D., *et al.* (2002). The genome of *Methanosarcina acetivorans* reveals extensive metabolic and physiological diversity. *Genome Research* **12**, 532–542.

Ferry, J. G. (1993). Methanogenesis. *In* "Microbiology Series" (C. A. Reddy, A. M. Chakrabarty, A. L. Demain, and J. M. Tiedje, eds.), p. 536. Chapman and Hall, New York.

Forterre, P. (1997). Archaea: What can we learn from their sequences? *Curr. Opin. Genet. Dev.* **7**, 764–770.

Kates, M., Kushner, D. J., and Matheson, A. T. (1993). "The Biochemistry of Archaea (archaebacteria)," Vol. 26, (D. J. K. M. Kates, and A. T. Matheson, eds.). Elsevier, Amsterdam.

Kotelnikova, S., and Pedersen, K. (1997). Evidence for methanogenic Archaea and homoacetogenic Bacteria in deep granitic rock aquifers. *FEMS Microbiol. Rev.* **20**, 339–349.

Reysenbach, A. L., and Cady, S. L. (2001). Microbiology of ancient and modern hydrothermal systems. *Trends Microbiol.* **9**, 79–86.

Roessler, M., and Muller, V. (2001). Osmoadaptation in bacteria and archaea: common principles and differences. *Environ. Microbiol.* **3**, 743–754.

Slesarev, A. I., Mezhevaya, K. V., Makarova, K. S., Polushin, N. N., Shcherbinina, O. V., Shakhova, V. V., Belova, G. I., Aravind, L., Natale, D. A., Rogozin, I. B., *et al.* (2002). The complete genome of hyperthermophile *Methanopyrus kandleri* AV19 and monophyly of archaeal methanogens. *Proc. Natl. Acad. Sci. USA* **99**, 4644–4649.

Smith, D. R., DoucetteStamm, L. A., Deloughery, C., Lee, H. M., Dubois, J., Aldredge, T., Bashirzadeh, R., Blakely, D., Cook, R., Gilbert, K., *et al.* (1997). Complete genome sequence of *Methanobacterium thermoautotrophicum* Delta H: Functional analysis and comparative genomics. *J. Bacteriol.* **179**, 7135–7155.

Sowers, K. R., and Schreier, H. J. (1999). Gene transfer systems for the archaea. *Trends Microbiol.* **7**, 212–219.

Sowers, K. R., and Schreier, H. J. (1995). Methanogens. *In* "Archaea: A Laboratory Manual" (F. T. Robb, K. R. Sowers, S. DasSharma, A. R. Place, H. J. Schreier, and E. M. Fleischmann, eds.), p. 540. Cold Spring Harbor Laboratory Press, Cold Spring Harbor.

Thauer, R. (1997). Biodiversity and unity in biochemistry. *Anton Leeuwenhoek Int. J. Gen. M.* **71**, 21–32.

Thomm, M. (1996). Archaeal transcription factors and their role in transcription initiation. *FEMS Microbiol. Rev.* **18**, 159–171.

WEBSITES

Methanogenesis Pathway Map (University of Minnesota Biocatalysis/Biodegradation Database)
http://umbbd.ahc.umn.edu/meth/meth_map.html

Website for Comprehensive Microbial Resource of The Institute for Genomic Research and links to many other microbial genomic sites
http://www.tigr.org/tigr-scripts/CMR2/CMRHomePage.spl

List of bacterial names with standing in nomenclature (J. P. Euzéby)
http://www.bacterio.cict.fr/index.html

58

Methylotrophy

J. Colin Murrell and Ian R. McDonald
University of Warwick, Coventry, UK

GLOSSARY

c_1 compounds Compounds more reduced than carbon dioxide containing one or more carbon atoms, but no carbon–carbon bonds.

methane monooxygenase The key enzyme for the oxidation of methane to methanol in bacteria.

methanotroph A methylotrophic bacterium with the ability to grow on methane as its sole carbon and energy source (methane-oxidizing bacteria).

proteobacteria A kingdom of bacteria divided into five groups based on their 16S ribosomal RNA sequences; a group of bacteria diverse in their morphology, physiology and lifestyle.

Methylotrophy refers to the ability of microorganisms to utilize one-carbon compounds more reduced than CO_2 as sole energy sources and to assimilate carbon into cell biomass at the oxidation level of formaldehyde. Methylotrophic organisms must synthesize all cellular constituents from methylotrophic compounds, such as methane, methanol, methylated amines, halogenated methanes, and methylated sulfur species. A diverse range of both aerobic and anaerobic prokaryotes and eukaryotes can utilize methanotrophic substrates for growth.

I. HISTORICAL PERSPECTIVE

Methylotrophs were first discovered in 1892 by Leow, who described a pink bacterium growing on methanol, methylamine, formaldehyde, and also on a variety of multi-carbon compounds. This organism was called *Bacillus methylicus* and was almost certainly what is now know as the pink-pigmented facultative methylotroph (PPFM) *Methylobacterium extorquens. Bacillus methanicus* was the first methane-oxidizing bacterium, reported by Söhngen in 1906. This methanotroph was isolated in pure culture from aquatic plants. This isolate was subsequently lost but was reisolated in 1956 by Dworkin and Foster and renamed *Pseudomonas methanica*. At around that time, the PPFM *Pseudomonas* AM1 was isolated on methanol by Quayle and colleagues. This was to become the "workhorse" organism for many of the biochemical and molecular biological studies on the metabolism of methanol and has now been renamed as *Methylobacterium extorquens* AM1. In 1970, Whittenbury and colleagues isolated over 100 new strains of methane-oxidizing bacteria. The characterization of these organisms was carried out and the scheme proposed still remains the basis for current classification schemes for methanotrophs.

II. SIGNIFICANCE

A. Global carbon cycle

Methane is the most abundant organic gas in the atmosphere. It is a very potent greenhouse gas and it absorbs infrared radiation considerably more efficiently

The Desk Encyclopedia of Microbiology
ISBN: 0-12-621361-5

than CO_2 and, therefore, makes a significant contribution to global warming. Current understanding of the global methane budget suggests that methane-oxidizing bacteria play an important role in oxidizing a large proportion of the methane produced by methanogenic bacteria in environments such as wetlands, ricefields, tundra, and the marine environment. Therefore, these bacteria are a significant sink for methane in the environment in modulating net emissions of methane and may provide an important negative feedback on future methane increases in wetland and soil environments. It is, therefore, important to learn more about the role of methanotrophs in the global carbon cycle.

B. Biotechnology

Both methane and methanol are relatively cheap feedstocks for fermentation processes and methylotrophs have received considerable attention for a number of biotechnological applications. Initial work on the production of single-cell protein (SCP) was carried out with methanotrophs. Growth yields of these organisms on methane were high but two drawbacks included the high oxygen demand in the fermentation process and the explosive nature of their substrates methane and oxygen. Methanol-utilizing bacteria, e.g., *Methylophilus methylotrophus*, have been successfully used for SCP production in very large-scale fermentation processes (ICI Pruteen Process). However, due to the fall in price of agricultural protein products, such as soya protein, over recent years, SCP from methanol has not been particularly competitive on a commercial scale.

Methanol-utilizing bacteria have also been exploited for the production of vitamins, polymers, and amino acids and these processes may be more economically viable. For example, auxotrophs of thermophilic gram-positive methanol utilizers can excrete relatively large amounts of lysine and other amino acids, which can then be used for animal feedstock supplements. The possibilities of genetically engineering methylotrophs for the overproduction of amino acids is also being explored.

Methanotrophs have also been investigated for the production of bulk chemicals, such as propylene oxide. Methane monooxygenase is able to insert oxygen into a number of aliphatic and aromatic compounds other than methane. These co-oxidation properties of methanotrophs, particularly those that contain soluble methane monooxygenase (sMMO), are unique and also unusual, since these organisms appear to derive no benefit from this process. sMMO is able to co-oxidize propylene to propylene oxide, a valuable compound in organic synthesis. However, due to the toxic nature of propylene oxide, alternatives to the use of whole cells, such as immobilized enzymes, together with the problem of regenerating reductant for sMMO may need to be considered.

C. Bioremediation

Methanotrophs have received considerable attention for their potential use in bioremediation processes. The enzyme soluble methane monooxygenase not only oxidizes methane but will co-oxidize a wide variety of aliphatic, substituted aliphatic, and aromatic compounds. sMMO is able to degrade several pollutants, including vinyl chloride, trichloroethylene, and other halogenated hydrocarbons that contaminate soil and groundwater. Challenges facing the use of methanotrophs *in situ* in bioremediation processes include ensuring supply of the substrates methane and oxygen and/or reductant and also overcoming problems associated with the negative regulation of sMMO by copper ions. Methylotrophs containing specific dehalogenases may also be useful in clean-up of industrial solvent-contaminated sites. For example, dichloromethane is metabolized by some *Methylobacterium* species using a dehalogenase.

More recently, methylotrophs with the ability to degrade methyl chloride and methyl bromide have been isolated. Degradation of methyl bromide, a potent ozone-depleting gas currently used as a pesticide in agriculture, is a particularly interesting trait for methylotrophs and may be useful in mitigating methyl bromide loss to the atmosphere during soil fumigation processes.

Methylotrophic bacteria can also grow on some methylated sulfur compounds found in toxic wastes, such as paper mill effluents. Others can degrade aliphatic sulfonates and, therefore, may be important in degradation of detergents and related compounds in the environment. Research is under way to investigate the metabolism of halogenated methanes and one-carbon compounds containing sulfur, in order to be able to explore the bioremediation potential and to exploit the properties of these novel methylotrophs.

D. Expression systems

The methylotrophic yeast *Pichia pastoris* is now becoming one of the best hosts for the production of foreign proteins because of the presence of the strong methanol-inducible promoter AOX1. This allows high level expression of a large number of biotechnologically and pharmaceutically important proteins in a controlled fashion during growth of a yeast on a relatively cheap substrate.

III. HABITATS AND ECOLOGY

Methane is produced by methanogenic bacteria in a number of diverse environments in the biosphere. Methane-oxidizing bacteria, which require both methane and oxygen for growth, are generally found on the fringes of anaerobic environments and are probably responsible for oxidizing much of the methane derived from methanogens, before it escapes to the atmosphere. They appear to be ubiquitous in nature and have been isolated from many different environments, including freshwater and lake sediments, rivers, groundwater aquifers, seawater, marine sediments, rice paddies, sewage sludge, decaying plant material, acidic peat bogs, and alkaline lakes. Psychrophilic representatives may also be isolated from Arctic and Antarctic tundra and thermophilic methanotrophs growing at temperatures as high as 70 °C have recently been obtained from hot springs. They can also be isolated from polluted environments. Methanotrophs also appear to exist as symbionts, for example, in the gill tissue of marine mussels or tube worms. The carbon assimilated by these methanotrophic endosymbionts probably supplies much of the organic carbon necessary for growth of these marine organisms. Methanotrophs from certain environments, for example, the putative symbionts already described, do not respond well to conventional enrichment and isolation techniques and molecular ecology experiments employing phylogenetic and functional gene probes suggest the presence of many new, and as yet uncultivated, methanotrophs in the environment.

Methanol is also a relatively abundant substrate for the growth of methylotrophs in the environment. Methanol is released during the decomposition of plant lignins and pectins and other compounds that contain methoxy groups. Bacteria that utilize methanol are frequently found in association with methaneoxidizing bacteria, presumably growing on the methanol excreted by methanotrophs. Methanol utilizers such as *Methylobacterium* are frequently found on the leaves of aquatic and terrestrial plants and on decaying plant material. Methanol-utilizers have also been isolated from the marine environment. Bacteria that utilize methyamine, many of which also grow on methanol, are also widespread in nature. Methylated amines are the products of degradation of some pesticides, lecithin and carnitine derivatives, and of trimethylamine oxide. Methylamine-utilizing bacteria are common in both terrestrial and marine environments. Methylated sulfur compounds, such as dimethylsulfoxide (DMS), dimethylsulfide, and dimethyldisulfide, are capable of supporting the growth of certain methylotrophs, such as *Hyphomicrobium* and *Thiobacillus*. DMS is the most abundant organic sulfur gas in the environment, produced in the marine environment from the cleavage of dimethylsulfoniopropionate, an algal osmoregulator. DMS is oxidized in the upper atmosphere to sulfur dioxide and the C_1 sulfur compound methanesulfonic acid (MSA). This MSA falls to earth by wet and dry deposition and, recently, it has been demonstrated that terrestrial and marine bacteria can grow methylotrophically on this C_1 substrate. Halogenated methanes are widely used as industrial solvents in the chemical industry and some, e.g., dichloromethane, have been shown to be methylotrophic substrates for some strains of *Methylobacterium* in polluted soils. Methyl bromide and methyl chloride are natural products released into the biosphere in large amounts from marine phytoplankton, algae, and wood-decaying fungi and these are also methylotrophic substrates for newly isolated bacteria.

IV. METHANOTROPHS (METHANE UTILIZERS)

A. Physiology and biochemistry

Methanotrophs grow by oxidizing methane to methanol using the pathway shown in Fig. 58.1. Methanol is further metabolized to formaldehyde by the pyrolloquinoline quinone- (PQQ-)linked enzyme methanol dehydrogenase (MDH), an enzyme found in all gram-negative methylotrophic bacteria. Approximately half of the formaldehyde produced is further oxidized to yield carbon dioxide, resulting in generation of reducing power for biosynthesis and the initial oxidation of methane. The carbon dioxide is not fixed into cell carbon in significant amounts but is lost to the atmosphere. The remainder of the formaldehyde is assimilated into cell carbon by one of two pathways. In Type I methanotrophs, formaldehyde is condensed with ribulose phosphate into hexulose-6-phosphate in the ribulose monophosphate pathway. Type II methanotrophs utilize the serine pathway for the incorporation of formaldehyde into the cell.

The initial reaction in the RuMP pathway involves the addition of 3 mol formaldehyde to 3 mol ribulose 5-phosphate to produce 3 mol hexulose 6-phosphate. Rearrangement reactions similar to those in the Calvin cycle for carbon dioxide fixation result in the production of glyceraldehyde 3-phosphate and the regeneration of 3 mol of ribulose 5-phosphate. Glyceraldehyde 3-phosphate is used for the synthesis of cell material. The serine pathway is very different

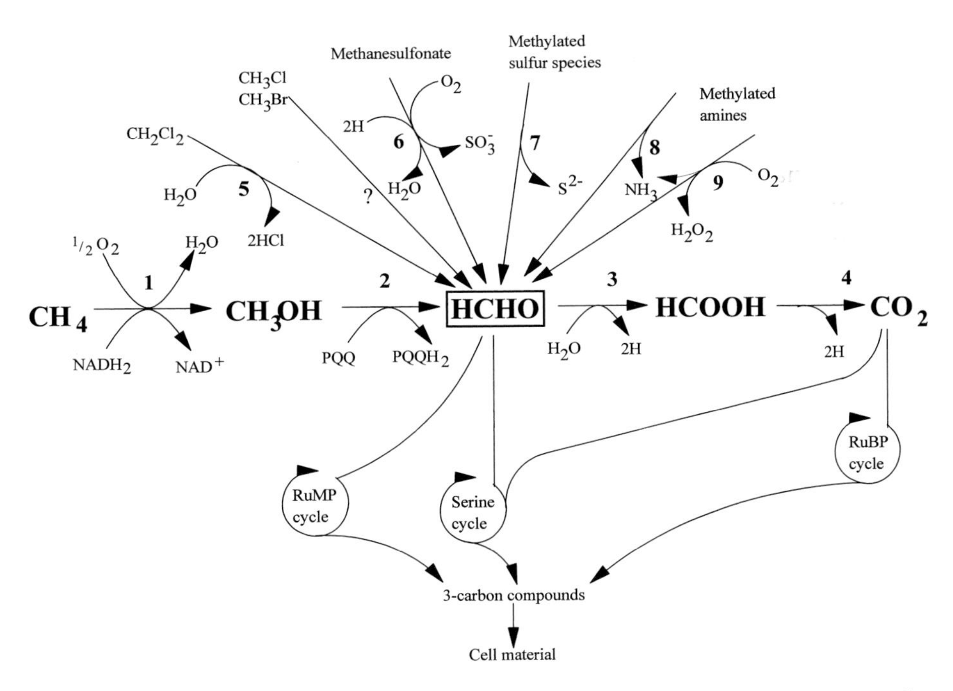

FIGURE 58.1 Metabolism of one-carbon compounds by aerobic methylotrophic bacteria. Enzymes: 1, methane monooxygenase; 2, methanol dehydrogenase; 3, formaldehyde dehydrogenase; 4, formate dehydrogenase; 5, dichloromethane dehalogenase; 6, methanesulfonic acid monooxygenase; 7, methylated sulfur dehydrogenases or oxidases; 8, methylated amine dehydrogenases; 9, methylamine oxidase.

from other formaldehyde assimilation pathways. There are no enzymatic reactions in common with the RuMP or methylotrophic yeast assimilatory pathways. In the serine pathway, 2 mol formaldehyde are condensed with 2 mol glycine to form 2 mol serine. The serine is converted to 2 mol of 2-phosphoglycerate. One phosphoglycerate is assimilated into cell material while the other is converted to phosphoenolpyruvate (PEP). The carboxylation of PEP yields oxaloacetate that is subsequently converted to 2 mol glyoxylate. Transamination of glyoxylate with serine as the amino donor regenerates the two glycine acceptor molecules. The phosphoglycerate, which is assimilated, undergoes transformations in central metabolic routes to provide carbon backbones for the synthesis of all cell materials.

Methane oxidation is carried out by the enzyme methane monooxygenase (MMO). A membranebound, particulate methane monooxygenase (pMMO) appears to be present in all methanotrophs grown in the presence of relatively high concentrations of copper ions. pMMO appears to consist of at least 3 polypeptides of approximately 46, 23, and 20 kDa, and contains a number of copper clusters. Further characterization of pMMO at the biochemical level is ongoing in a number of research groups.

In some methanotrophs, a second form of MMO, a cytoplasmic, soluble form (sMMO) is synthesized in growth conditions when the copper-to-biomass ratio is low. This sMMO is structurally and catalytically distinct from the pMMO and has a broad substrate specificity, oxidizing a wide range of aliphatic and aromatic compounds. The sMMO enzymes of *M. capsulatus* (Bath) and *M. trichosporium* OB3b both consist of three components—A, B, and C. Protein A is the hydroxylase component of the enzyme complex and contains a binuclear iron–oxo center, believed to be the reactive center for catalysis. Protein B, is a single polypeptide, contains no metal ions or cofactors, and functions as a regulatory, or coupling, protein. Protein C is a reductase containing 1 mol each of FAD and a 2Fe2S cluster which accepts electrons from NADH and transfers them to the diiron site of the hydroxylase component. The x ray crystal structure of the hydroxylase component of sMMO is now known, a fact which has further stimulated research on the mechanism of oxidation of methane by this unique enzyme.

B. Molecular biology

The genes encoding sMMO complex have been cloned from several methanotrophs. The gene cluster contains genes encoding the α, β, and γ subunits of Protein A (*mmoX*, *Y*, and *Z*), Protein B (*mmoB*), and

Protein C (*mmoC*). Derived polypeptide sequences of sMMO components from three methanotrophs showed a high degree of identity, highlighting the conserved nature of this enzyme complex. Amino acid sequences within the α subunit of Protein A align well with the four helix iron coordination bundle of the R2 protein of ribonucleotide reductase and is characteristic of a family of proteins that contains a catalytic carboxylate-bridged diiron center.

Differential expression of sMMO and pMMO is regulated by the amount of copper ions available to the cells; sMMO is expressed at low copper-biomass ratios, whereas pMMO is expressed at high copper-biomass ratios. The transcriptional regulation of the sMMO gene cluster appears to be under the control of a copper-regulated promoter. Transcription of the sMMO gene cluster is negatively regulated by copper ions. Activation of *pmo* transcription by copper ions is concomitant with repression of sMMO gene transcription in both methanotrophs, suggesting that a common regulatory pathway may be involved in the transcriptional regulation of sMMO and pMMO.

C. Molecular ecology

Difficulties in using traditional culture-based techniques to study the ecology of methanotrophs (e.g., slow growth of methanotrophs and scavenging of nonmethanotrophs on agar plates) has hampered studies. Application of molecular biology techniques to methanotrophs has been aided considerably by the sequencing of a number of 16S rRNA genes of methanotrophs and methylotrophs and the cloning of several methanotroph-specific genes. Hanson and colleagues (1996) have used 16S rRNA sequence data to examine phylogenetic relationships within genera of methanotrophs and methylotrophs.

16S rRNA data coupled with PCR technology have also been used successfully in analyzing methanotrophs in the marine environment. Seawater samples were enriched for methanotrophs by adding essential nutrients and methane. Changes in composition of the bacterial population was then monitored by analysis of 16S rDNA libraries. The dominant 16S rRNA sequence that was present in samples after enrichment on methane was found to show a close phylogenetic relationship to *Methylomonas*, indicating that novel methanotrophs related to extant *Methylomonas* spp. were present in enrichment cultures.

Genes unique to methanotrophs include those encoding sMMO and pMMO polypeptides. The high degree of identity between sMMO genes has enabled the design of PCR primers which specifically amplify sMMO genes directly from a variety of different freshwater, marine, soil, and peat samples. Results suggest that there is considerable diversity of methanotrophs in these environments.

Another functional gene probe for methanotrophs is one based on the pMMO, present in all extant methanotrophs. Sequence data on *pmo* and *amo* (ammonia monooxygenase—a related enzyme found in nitrifying bacteria) genes has allowed the design of degenerate PCR primers which will specifically amplify DNA genes encoding *pmoA* or *amoA* from many different methanotrophs and nitrifiers. Analysis of the predicted amino acid sequences of these genes from representatives of each of the phylogenetic groups of methanotrophs (α and γ Proteobacteria) and ammonia oxidizing nitrifiers (β and γ Proteobacteria) suggests that the particulate methane monooxygenase and ammonia monooxygenase may be evolutionarily related enzymes.

Another potentially useful marker is *mxaF*, encoding the large subunit of methanol dehydrogenase, which is present in virtually all gram-negative methylotrophs. *mxaF* is highly conserved and is, therefore, a good indicator of the presence of these organisms in the natural environment.

D. Phylogeny and taxonomy

Methanotrophs are all gram-negative bacteria and can be classified into two groups. Type I methanotrophs of the genera *Methylomonas, Methylobacter, Methylomicrobium, "Methylothermus," Methylococcus, Methylosphaera*, and *"Methylocaldum"* utilize the ribulose monophosphate (RuMP) pathway for the assimilation of formaldehyde into cell carbon, possess bundles of intracytoplasmic membranes, and are members of the γ-subdivision of the purple bacteria (class Proteobacteria). Type II methanotrophs, such as *Methylosinus* and *Methylocystis*, utilize the serine pathway for formaldehyde fixation, possess intracytoplasmic membranes arranged around the periphery of the cell, and fall within the γ-subdivision of the Proteobacteria. Classification schemes, based on pheno- and chemo-taxonomic studies have been strengthened as a result of the nucleotide sequencing of both 5S and 16S ribosomal RNA (rRNA) from a large number of methanotrophs and methylotrophs (Fig. 58.2). The key features of representative genera of methanotrophs are summarized in Table 58.1. Properties such as mol% G + C content of DNA, membrane fatty acid composition, nitrogen fixation, and some morphological features, can be used to discriminate between Type I and Type II methanotrophs.

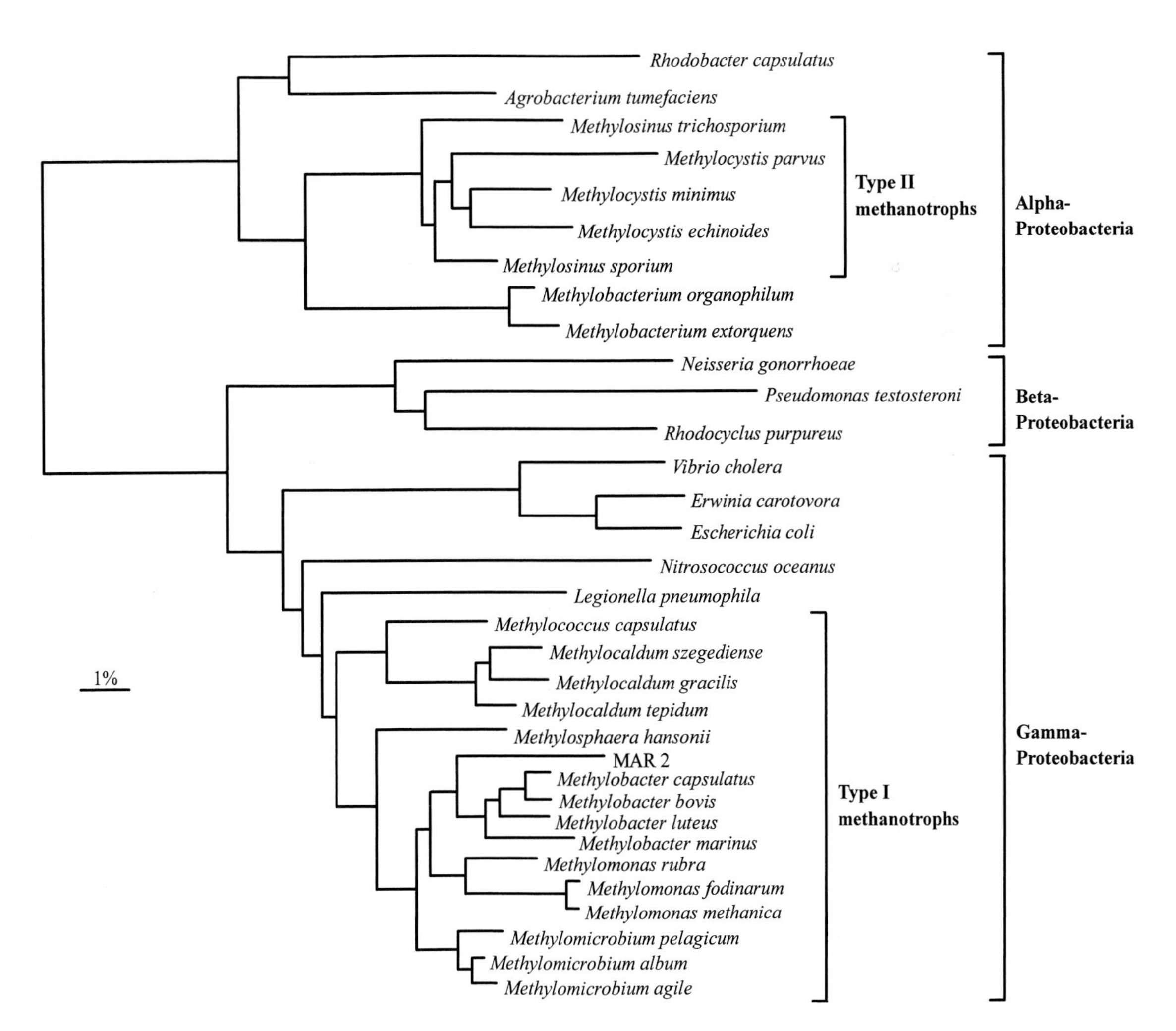

FIGURE 58.2 Phylogenetic tree of the 16S rRNA sequences from methanotrophs, methylotrophs, and other representative bacteria belonging to the α, β, and γ subdivisions of the Proteobacteria. MAR 2 is a marine methanotroph isolate.

TABLE 58.1 Characterization of methanotrophs

Genus	Phylogeny	Membrane type	Major PLFA	Formaldehyde assimilation pathway	G + C Enzyme type[a]	Mol % G + C content of DNA
Methylobacter	Gamma Proteobacteria	Type I	16:1	RuMP	pMMO	49–54
Methylocaldum	Gamma Proteobacteria	Type I	16:1	RuMP/Serine	pMMO	57–59
Methylococcus	Gamma Type I Proteobacteria		16:1	RuMP//Serine	pMMO/sMMO	59–66
Methylomicrobium	Gamma Proteobacteria	Type I	16:1	RuMP	pMMO/sMMO	49–60
Methylomonas	Gamma Proteobacteria	Type I	16:1	RuMP	pMMO/sMMO	51–59
Methylosphaera	Gamma Proteobacteria	Type I	16:1	RuMP	pMMO	43–46
Methylocystis	Alpha Proteobacteria	Type II	18:1	Serine	pMMO/sMMO	62–67
Methylosinus	Alpha Proteobacteria	Type II	18:1	Serine	pMMO/sMMO	63–67

Note: RuMP, ribulose monophosphate pathway; pMMO, particulate methane monooxygenase; sMMO, soluble methane monooxygenase. All strains grow on CH_4; some will also grow on CH_3OH.

[a]All strains possess pMMO; however, some also possess the sMMO.

V. AEROBIC METHYLOTROPHS

A. Methanol utilizers

The bacteria capable of growth on methanol are more diverse than those capable of growing on methane. They include a variety of gram-negative and gram-positive strains and include both facultative and obligate methylotrophs (Table 58.2). The methanol-utilizing bacteria can be divided according to their carbon assimilation pathway. The methanol-utilizers that contain the serine cycle for formaldehyde assimilation include *Methylobacterium* and *Hyphomicrobium* strains. Most of the gram-negative methanol utilizers that contain the RuMP cycle are obligate methylotrophs, with the exception of the facultative *Acidomonas* strains. The gram-positive methanol utilizers, which all contain the RuMP cycle, are facultative methylotrophs. All of the known gram-negative methanol- and methane-utilizing bacteria contain an enzyme for oxidizing methanol called methanol dehydrogenase. This enzyme, which oxidizes primary alcohols, contains the cofactor pyrroloquinoline quinone (PQQ). Methanol dehydrogenase is one of a family of PQQ-linked enzymes known as quinoproteins. The electrons from the oxidation of methanol are transferred from the PQQ cofactor to a specific soluble cytochrome *c* and, from there, through to other carriers to the terminal oxidase. Methanol dehydrogenases are highly conserved throughout the gram-negative methylotrophic bacteria. Studies of the molecular genetics of this system have revealed that the synthesis of a fully active methanol oxidizing pathway requires a total of at least 32 genes, among which are the *pqq* genes involved in cofactor biosynthesis and a

TABLE 58.2 Characteristics of aerobic methylotrophic bacteria

Group	Growth substrates	Major assimilation pathway	GC content (mol%)
Obligate gram-negative methylotrophs			
Methylobacillus	CH_3OH, CH_3NH_2	RuMP	50–55
Methylophaga	CH_3OH, CH_3NH_2, fructose	RuMP	38–46
Methylophilus	CH_3OH, CH_3NH_2, fructose[a]	RuMP	50–55
Facultative gram-negative methylotrophs			
Acidomonas[b]	CH_3OH, multi-carbon compounds	RuMP	63–66
Hyphomicrobium	CH_3OH, $CH_3NH_2^a$, DMSO[a], DMS[a], CH_3Cl^a, some 2-C and 4-C compounds[a]; denitrification	Serine	60–66
Methylobacterium	CH_3OH, CH_3NH_2, CH_3Cl^a, multi-carbon compounds	Serine	65–68
Methylosulfonomonas	MSA, CH_3OH, CH_3NH_2, HCOOH, poor growth on multi-carbon compounds	Serine	60–62
Microcyclus	CH_3OH, multi-carbon compounds	RuBP	65–67
Paracoccus	CH_3OH, CH_3NH_2, multi carbon compounds, denitrification	RuBP	66
Rhodopseudomonas	CH_3OH^a, multi-carbon compounds; photosynthesis	RuBP	62–72
Thiobacillus	CH_3OH^a, $CH_3NH_2^a$, H_2S, S_2O_3, multi-carbon compounds[a]	RuBP	52–69
Xanthobacter	CH_3OH, multi-carbon compounds	RuBP	67–69
Facultative gram-positive methylotrophs			
Amycolatopsis[c]	CH_3OH^a, multi-carbon compounds,	RuMP	ND
Arthrobacter	$CH_3NH_2^a$, multi-carbon compounds,	RuMP	ND
Bacillus	CH_3OH^a, $CH_3NH_2^a$, multi-carbon compounds	RuMP	60–70
Mycobacterium	CH_4^a, CH_3OH^a, multi-carbon compounds	RuMP	65–69

[a]Some strains.
[b]Formerly *Acetobacter methanolicus*.
[c]Formerly *Nocardia methanolica*.

comprehensive set of *mxa, mxb, mxc,* and *mxd* genes, some of which encode the structural enzymes and others which encode proteins involved in regulation of gene expression and protein activity.

Different types of methanol dehydrogenases occur in the gram-positive methylotrophs. A methylotrophic *Amycolatopsis* species contains an unusual quino alcohol dehydrogenase, and the methylotrophic *Bacillus* species contain a methanol dehydrogenase enzyme that is not PQQ linked, but instead is linked to NAD.

Recently, an aerobic methylotrophic bacterium, *Methylobacterium extorquens,* was found to contain a cluster of genes that are predicted to encode some of the enzymes from methanogenic and sulfate-reducing Archaea involved in C_1 transfer, thought to be unique to this group of strictly anaerobic microorganisms. Enzyme activities were also detected in *M. extorquens* and mutants defective in some of these genes were unable to grow on C_1 compounds, suggesting that the archaeal enzymes also function in aerobic C_1 metabolism. Thus, methylotrophy and methanogenesis involve common genes that cross the bacteria/archaeal boundaries.

Some bacteria, such as *Paracoccus, Thiosphaera,* and *Xanthobacter,* can grow on methanol by oxidizing this methylotrophic substrate to carbon dioxide but then fixing this carbon into cell biomass using the enzyme ribulose bisphosphate carboxylase/oxygenase (RuBISCO). These types of organisms are, therefore, considered as autotrophic methylotrophs.

B. Utilization of methylated amines

Many of the bacteria that grow on methanol are also capable of utilizing methylated amines. Some bacteria have been isolated that are capable of growth on methylated amines but do not grow on methanol or methane, such as *Pseudomonas aminovorans* and *Arthrobacter* P1. Several species of methylotrophic bacteria are able to utilize methylamine as a sole source of carbon and energy. Three different systems for the oxidation of primary amines are known. These are methylamine dehydrogenase, found in some gram-negative methylotrophs; amine oxidase, found in gram-positive methylotrophs; and indirect methylamine oxidation via *N*-methyl-glutamate dehydrogenase, found in the remaining gram-negative methylotrophs.

The methylamine dehydrogenases (MADH) are periplasmic proteins, consisting of two small and two large subunits. Each small subunit has a covalently bound prosthetic group called tryptophan tryptophylquinone (TQQ). MADHs can be divided into two groups, based on the electron acceptors that they use. The MADHs from restricted facultative methylotrophic bacteria belonging to the genus *Methylophilus,* use a *c*-type cytochrome as a electron acceptor, whereas all other MADHs use blue copper proteins called amicyanins. A group of genes called the *mau* genes are responsible for the synthesis of MADH. The *mau* gene cluster of *Paracoccus denitrificans* consists of 11 genes, 10 of which encode the structural proteins or proteins involved in cofactor biosynthesis and which are transcribed in one direction, whereas the 11th regulatory gene (*mauR*) is located upstream and is divergently transcribed.

C. Utilization of halomethanes

Certain *Methylobacterium* and *Methylophilus* species can grow on dichloromethane as sole carbon and energy source (Fig. 58.2). The key enzyme in the aerobic degradation of CH_2Cl_2 is dichloromethane dehalogenase, which catalyzes the glutathione (GSH)-dependent dehalogenation of CH_2Cl_2 to formaldehyde and chloride ions. The formaldehyde is subsequently assimilated via the serine pathway. Some *Hyphomicrobium* and *Methylobacterium* species also grow on methyl chloride, although the exact mechanism for utilization of this C_1 compound is not yet known. Methyl bromide also appears to be a C_1 substrate for certain bacteria and is currently receiving considerable attention as a potential methylotrophic substrate. Again, the bacteria responsible have not been fully characterized but it is likely that these novel bacteria are widespread in the terrestrial and marine environment and play an important role in cycling of halogenated methanes.

D. Utilization of methylated sulfur species

There are some methylotrophic bacteria which are capable of utilizing methylated sulfur compounds, such as dimethylsulfoxide (DMSO), dimethylsulfide (DMS), and dimethyldisulfide (DMDS). Most of these are of the genus *Hyphomicrobium* but certain *Thiobacillus* strains have been reported. DMSO and DMS metabolism has not been studied in great detail in methylotrophic bacteria but it is believed that in *Hyphomicrobium* DMSO is reduced to DMS, which is, in turn, converted to formaldehyde and methanethiol. The methanethiol may then be converted by an oxidase to formaldehyde and H_2S with the concomitant production of hydrogen peroxide. The formaldehyde produced would then be assimilated into cell carbon by the serine pathway in *Hyphomicrobium*. Methanesulfonic acid (MSA) is also a C_1 source for certain methylotrophs, which appear to be ubiquitous in the environment. The terrestrial strain *Methylosulfonomonas* and the marine strain *Marinosulfonomonas* oxidize MSA to formaldehyde and sulfite, using a methanesulfonic acid monooxygenase. The formaldehyde is subsequently assimilated into the cell via the serine pathway. Some strains of *Methylobacterium* and *Hyphomicrobium* can also utilize MSA. Most of the MSA-utilizers also grow well on other C_1 compounds such as methanol, methylamine, and formate, but not methane. Certain *Hyphomicrobium* species can grow on monomethylsulfate.

VI. ANAEROBIC METHYLOTROPHS

All extant methanotrophs are obligate aerobes. However, there is now good biogeochemical and biological evidence that methane oxidation occurs in marine environments, such as sulfate-rich sediments, alkaline soda lakes, and some freshwater lakes. However, to date, no anaerobic bacteria that will grow on or oxidize methane have been isolated from these environments and cultivated in the laboratory. It is not known if such bacteria are true methanotrophs or if a consortium of bacteria is involved in these processes. One hypothesis is that sulfate is the terminal electron acceptor for anaerobic methane oxidation but further experimental work is required here. Some studies indicate that, although methane is oxidized to carbon dioxide, the methane carbon is not assimilated into cell biomass.

Methanol can also be oxidized by facultatively anaerobic bacteria of the genus *Hyphomicrobium*, using nitrate as a terminal electron acceptor. Some acetogenic and methanogenic bacteria, which are strict anaerobes, are capable of growth on C_1 compounds, such as methanol and methylamines. During anaerobic growth, these bacteria convert such C_1 compounds to methane, acetate, or butyrate, rather than carbon dioxide. Carbon from the original C_1 substrate is not assimilated via formaldehyde but their methyl groups are incorporated into acetyl CoA, a precursor for cellular constituents. Therefore, following the original definition of methylotrophy, these obligate anaerobes that utilize C_1 compounds, such as methanol and methylamine, are not normally considered as methylotrophs.

VII. METHYLOTROPHIC YEASTS

The ability of some yeasts to grow on methanol as a source of carbon and energy has been discovered only relatively recently. They can be isolated from soil, rotting fruits, and vegetables, or plant material, again suggesting that methanol derived from methoxy groups in wood lignin or pectin is an important factor in the ecology of these yeasts. Methylotrophic yeasts belong to the fungi perfecti, form ascospores that are hat-shaped and homothallic. They are members of the genera *Hansenula*, *Pichia*, and *Candida* and they metabolize methanol via alcohol oxidases in peroxisomes. Assimilation of formaldehyde is accomplished by the xylulose monophosphate cycle. Yeast cultures that use methane as a sole carbon and energy source have also been described. These strains were slow growing and have received very little attention over the past 20 years.

BIBLIOGRAPHY

Anthony, C. (1982). "The Biochemistry of Methylotrophs." Academic Press, London, UK.

Conrad, R. (1996). Soil microorganisms as controllers of atmospheric trace gases (H_2, CO, CH_4, OCS, N_2O, and NO). *Microbiol. Rev.* **60**, 609–640.

Costello, A. M., and Lidstrom, M. E. (1999). Molecular characterization of functional and phylogenetic genes from natural populations of methanotrophs in lake sediments. *Appl. Environ. Microbiol.* **65**, 5066–5074.

Hanson, R. S. (1991). The obligate methanotrophic bacteria *Methylococcus, Methylomonas, Methylosinus* and related bacteria. *In* "The Prokaryotes" (A. Balows, H. G. Truper, M. Dworkin, W. Harder, and K. H. Schleifer, eds), Vol. 1, pp. 2350–2365. Springer-Verlag, New York.

Hanson, R. S., and Hanson, T. E. (1996). Methanotrophic bacteria. *Microbiol. Rev.* **60**, 439–471.

Large, P. J., and Bamforth, C. W. (1988). "Methylotrophy and Biotechnology." Longman Scientific and Technical, New York.

Leak, D. J. (1992). Biotechnological and applied aspects of methane and methanol utilizers. *In* "Methane and Methanol Utilizers" (J. C. Murrell and H. Dalton, eds), pp. 245–279. Plenum Press, New York.

Lidstrom, M. E. (1991). Aerobic methylotrophic bacteria. *In* "The Prokaryotes" (A. Balows, H. G. Truper, M. Dworkin, W. Harder, and K. H. Schleifer, eds). Vol. 1, pp. 431–445. Springer-Verlag, New York.

Lidstrom, M. E., and Tabita, F. R. (1996). "Microbial Growth on C_1 Compounds." Kluwer Academic, Dordrecht, The Netherlands.

Murrell, J. C., and Dalton, H. (1992). "Methane and Methanol Utilizers." Plenum Press, New York.

Murrell, J. C., Gilbert, B., and McDonald, I. R. (2000). Molecular biology and regulation of methane monooxygenase. *Arch. Microbiol.* **173**, 325–332.

Murrell, J. C., and Kelly, D. P. (1993). "Microbial Growth on C_1 Compounds." Intercept Press, Andover, UK.

Murrell, J. C., and Kelly, D. P. (1996). "The Microbiology of Atmospheric Trace Gases: Sources, Sinks and Global Change Processes." NATO ASI Series, Springer-Verlag.

Murrell, J. C., McDonald, I. R., and Bourne, D. G. (1998). Molecular methods for the study of methanotroph ecology. *FEMS Microbiol. Ecol.* **27**, 103–114.

Murrell, J. C., and Radajewski, S. (2000). Cultivation-independent techniques for studying methanotroph ecology. *Res. Microbiol.* **151**, 807–814.

Oremland, R. S. (1993). "Biogeochemistry of Global Change." Chapman and Hall, New York.

Valentine, D. L., and Reeburgh, W. S. (2000). New perspectives on anaerobic methane oxidation. *Environ. Microbiol.* **2**, 477–484.

WEBSITES

Website of the J. C. Murrell laboratory
http://www.bio.warwick.ac.uk/murrell/research/pathway.html

Website for Comprehensive Microbial Resource of The Institute for Genomic Research and links to many other microbial genomic sites
http://www.tigr.org/tigr-scripts/CMR2/CMRHomePage.spl

List of bacterial names with standing in nomenclature (J. P. Euzéby)
http://www.bacterio.cict.fr/index.html

59

Nitrogen cycle

Roger Knowles
McGill University

GLOSSARY

ATP Adenosine triphosphate, the energy-rich molecule synthesized in energy-yielding reactions and consumed in biosynthetic processes.

biofilm A community of microorganisms on a solid surface. They are usually embedded in polysaccharide slime produced by members of the community.

diazotroph A bacterium that can fix gaseous N_2.

heterotroph An organism deriving both its energy and its carbon from organic compounds.

lithotroph An organism deriving its energy from the oxidation of inorganic compounds.

megaplasmid A very large extrachromosomal circle of DNA present in some cells in addition to the chromosome.

periplasm The space between the inner and outer membranes of gram-negative bacteria. It may contain certain enzymes and cytochromes.

phylogeny A classification of microorganisms that represents their putative evolutionary relationships.

repression The inhibition of synthesis of an enzyme or other gene product, often brought about by an end product or other molecule that makes the enzyme reaction or the pathway in which it occurs unnecessary.

Despite the Very Large Amount of Combined Nitrogen Compounds present in the biosphere, lithosphere, and hydrosphere, over 99.9% of all global nitrogen is in the atmosphere in the form of dinitrogen. A small amount of this nitrogen gas is converted to oxides during electrical storms in the atmosphere and is washed out in rainfall. However, most transformations of nitrogen are catalyzed by microorganisms, which are thus of critical importance in the control of nitrogen availability for the growth of crop plants, in the treatment of waste, and in the conversion to forms that are leachable and result in contamination of groundwaters.

I. INTRODUCTION

Some aspects of the biochemistry and microbiology of each nitrogen cycle process are presented here, followed by ecological and environmental implications. The most reduced forms of nitrogen are the organic nitrogen compounds, such as amino acids, and the first product of their decomposition, ammonia (NH_4^+ or NH_3) (Fig. 59.1). The most oxidized form is nitrate (NO_3^-), and between ammonia and nitrate are compounds of different oxidation states: the gases dinitrogen (N_2), nitrous oxide (N_2O), nitrogen monoxide (NO or nitric oxide), and nitrogen dioxide (NO_2); and the nonvolatile hydroxylamine (NH_2OH) and nitrite (NO_2^-). Oxidative reactions (leading to the right in Fig. 59.1) tend to occur under aerobic (oxygenated) conditions, and reductive reactions leading to the left occur mostly under low-oxygen or anaerobic conditions. Environmental conditions dictate the type and abundance of microorganisms occurring in nature

The Desk Encyclopedia of Microbiology
ISBN: 0-12-621361-5

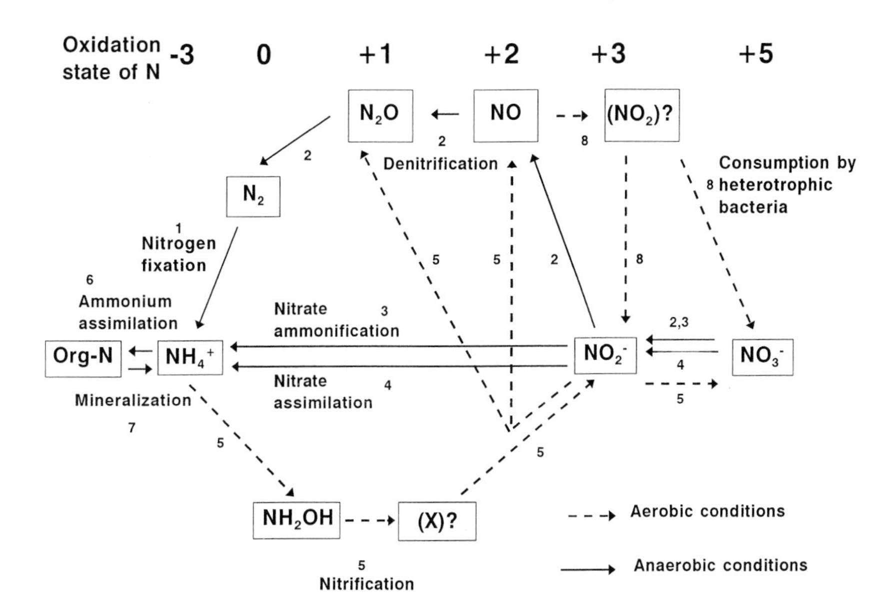

FIGURE 59.1 Relationships among the microbiological processes in the N cycle. The oxidation state of N is indicated, and aerobic and anaerobic processes are shown as dashed and solid arrows, respectively. 1, Nitrogen fixation; 2, denitrification; 3, nitrate ammonification; 4, nitrate assimilation; 5, nitrification; 6, ammonium assimilation; 7, mineralization; and 8, consumption of NO by heterotrophic bacteria.

and they thus can critically control the kinds of nitrogen transformations that are likely to occur.

II. DINITROGEN FIXATION

A. Biochemistry and microbiology

The ability to fix (assimilate) gaseous nitrogen is restricted to the archaea and the eubacteria (together, the prokaryotes) and is found in higher life forms (eukaryotes) only in association with N_2-fixing (diazotrophic) bacteria. The fixation of dinitrogen is catalyzed by an enzyme complex, nitrogenase, consisting of two subunits of dinitrogenase reductase (an iron protein), two pairs of subunits of dinitrogenase (an iron–molybdenum enzyme), and an iron–molydbdenum cofactor (FeMoCo). Dinitrogenase reductase uses ATP as the energy source to transfer energized electrons to the dinitrogenase enzyme. The very strongly triple-bonded molecule of dinitrogen is successively reduced on the surface of the FeMoCo cofactor, yielding two molecules of ammonia as the first detectable product. The ammonia is then used in the synthesis of amino acids and proteins (Section III). During dinitrogen reduction, some electrons and ATP are used to reduce protons to dihydrogen (H_2), a wasteful process. The overall stoichiometry of these reactions is

$$N_2 + 8e^- + 8H^+ + 16ATP \rightarrow 2NH_3 + 16ADP + 16P_i + H_2$$

Some (mainly aerobic) bacteria have adapted by using a recycling (or uptake) hydrogenase to reoxidize the H_2, returning electrons to the electron-transport chain and regaining valuable ATP. Nevertheless, nitrogen fixation remains a very energy-expensive process and the availability of energy and reducing power is frequently a major limiting factor.

Another limiting factor is the presence of oxygen (O_2). Because the nitrogenase complex is extremely sensitive to O_2 inactivation, it is active only in cells or environments with very low O_2 concentrations. One species of *Azotobacter* possesses a third protein in the nitrogenase complex, the FeS II or Shethna protein, which affords some protection from O_2 inactivation.

An interesting property of the nitrogenase complex is that in addition to being able to reduce dinitrogen to ammonia, it can also reduce a number of other small molecules that are somewhat similar in size and conformation to N_2. For example, it was discovered in the late 1960s that nitrogenase could reduce acetylene to ethylene (H–C≡C–H to $H_2C{=}CH_2$) and this reaction provided a very simple, inexpensive, and very sensitive assay for nitrogenase activity that could be applied to systems varying in complexity from enzyme preparations to soil samples. The sample to be assayed is exposed to C_2H_2 and the product C_2H_4 is readily detected by gas chroma-tography. In addition, nitrogenase reduces hydrogen cyanide (H–C≡N), hydrogen azide ($H{\bullet}N^-{\bullet}N^+{\bullet}N$), nitrous oxide ($N{\bullet}N^+{\bullet}O^-$), and other compounds. All of these reactions consume ATP, so the reduction of N_2O by nitrogenase is different from the reaction catalyzed by the N_2O reductase of denitrifying bacteria, which contributes to ATP synthesis (Section V.A), and which, interestingly, is inhibited by (but does not reduce) C_2H_2.

About 20 catalytic proteins, cofactors, and other components are necessary for N_2 fixation. They are coded for by genes (*nif*) that may be tightly clustered, as in *Klebsiella pneumoniae* (Fig. 59.2), or scattered throughout the bacterial chromosome, or even present on very large megaplasmids (as in some of the rhizobia that infect the roots of legumes and cause the formation of N_2-fixing root nodules). The *nif* system is tightly regulated to conserve resources for the cell, and it is repressed by both O_2 and NH_4^+.

Microorganisms possessing *nif* genes and able to fix N_2 are taxonomically and phylogenetically very diverse, but because the nitrogenase structural genes are highly conserved, some believe that the process is very ancient in the evolutionary sense, and that it may have evolved during an early geological period of decreased lightning that resulted in a nitrogen limitation. Table 59.1 shows some selected N_2-fixing genera in major phylogenetic groups, indicating the very varied ecological types represented.

It should be noted that two other, rather rare, nitrogenases are known, one containing vanadium instead of molybdenum, and the other containing only iron. They are produced only in molybdenum-deficient environments and their ecological significance is not clear.

B. Ecology of N_2 fixation

The very high ATP requirement and the high O_2 sensitivity of the nitrogenase system means that N_2-fixing organisms are active only in certain environments.

1. Environments with high C and energy availability, such as readily decomposable high-C organic matter, high-C root exudates (e.g. N_2 fixers associated with sugar cane) or photosynthate (e.g. cyanobacteria, and root nodules formed on legumes or other plants).
2. Cells or environments with low O_2 concentrations. Anaerobes such as *Clostridium* spp. naturally must avoid O_2 to grow, whether the environment is nitrogen-deficient (i.e. they need to fix N_2) or not. Facultative anaerobes such as *Klebsiella* spp. fix N_2 only when O_2 is absent, but some obligate aerobes such as *Azotobacter* can fix N_2 under aerobic conditions by using a type of "respiratory protection" in which they develop a special branch of their electron-transport chain that permits them to consume O_2 (and therefore also organic carbon) at a high rate to maintain reducing conditions in the cytoplasm where the nitrogenase is located. Indeed, *Azotobacter* has the highest respiration rate per unit cell mass of any known organism.

Many free-living diazotrophs have no specific mechanisms of O_2 protection. When they must fix N_2 they become microaerophilic and grow only in environments having O_2 concentrations less than about one-twentieth of that in the atmosphere, that is, 2.25 μM

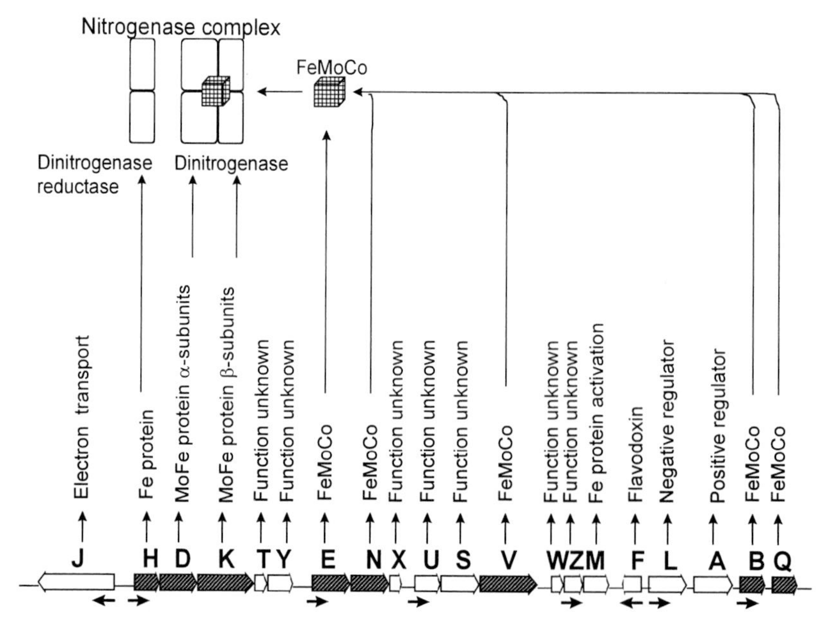

FIGURE 59.2 The *nif* genes and their products in the facultatively anaerobic N_2 fixer *Klebsiella pneumoniae*. Structural genes are shaded, and products and functions shown. The direction of transcription is shown, as is the composition of the nitrogenase complex.

TABLE 59.1 Phylogenetic distribution of selected genera of N_2-fixing organisms[a]

Groups/genera	Comment
Archaea (7)	
Methanococcus (5)	Methanogen
Proteobacteria, alpha subdivision (24)	
Azospirillum (4)	Plant-associative
Acetobacter (1)	Endophytic in rice
Bradyrhizobium (1)	Soybean nodules
Gluconacetobacter (1)	Endophytic in plants
Methylosinus (2)	CH_4 oxidizer
Rhizobium (5)	Legume nodules
Rhodobacter (5)	Photosynthetic
Proteobacteria, beta subdivision (8)	
Alcaligenes (3)	Common het'troph
Thiobacillus (1)	H_2S oxidizer
Proteobacteria, gamma subdivision (17)	
Azotobacter (6)	Aerobic het'troph
Klebsiella (4)	Env. and path.
Methylococcus (4)	CH_4 oxidizer
Pseudomonas (1)	Common het'troph
Vibrio (4)	Aquatic het'troph
Proteobacteria, delta subdivision (2)	
Desulfovibrio (7)	$SO_4^{=}$ reducer
Firmibacteria (gram positive) (3)	
Bacillus (3)	Fac. aerobes
Clostridium (16)	Anaerobes
Desulfotomaculum (3)	$SO_4^{=}$ reducer
Green sulfur bacteria (4)	
Chlorobium (3)	H_2S oxidizer
Actinomycetes (4)	
Frankia	Non-legume nodules
Streptomyces (1)	Thermophilic carboxydotroph
Heliobacteria (3)	
Heliobacterium (3)	
Cyanobacteria (>27)	Blue-green bacteria
Anabaena	
Nostoc	
Oscillatoria	
Spirulina	
Trichodesmium	Marine
Campylobacter (1)	
Campylobacter (1)	Some pathogens

[a]Numbers in parentheses after group names indicate the number of genera in the group, and those after genus names indicate the number of species in the genus with N_2-fixing ability.
Abbreviations: Env., environmental organism; Fac., facultative (grows aerobically or anaerobically); *het'troph*, heterotroph; Path., pathogenic strains exist.

dissolved O_2. Such bacteria are members of the genera *Acetobacter, Azospirillum, Aquaspirillum, Gluconacetobacter, Magnetospirillum, Pseudomonas*, and the diazotrophic methanotrophs. Some of these organisms are found inhabiting the rhizosphere of many plants, including rice. A very unusual diazotrophic *Streptomyces* actinomycete was reported to use CO as carbon and energy source, but did not reduce acetylene and its DNA did not hybridize with molecular probes for the highly conserved *nifH* and *nifDK* genes.

Cyanobacteria such as *Anabaena* produce specialized cells called heterocysts about every 10 or 12 cells in a chain of photosynthetic cells. Such heterocysts lack the O_2-producing photosystem II and support nitrogenase with carbohydrate translocated from the neighboring photosynthetic cells. Most *Anabaena* and related spp. are active in the free-living state in rice paddies and other aquatic systems. However, an *Anabaena* sp. lives symbiotically in specialized cavities in the fronds of the aquatic fern *Azolla*, which can contribute large amounts of fixed nitrogen in rice paddies.

Other cyanobacteria do not develop heterocysts. Species of *Trichodesmium* are filamentous and fix N_2 only in the center of colonial aggregates where O_2-producing photosynthesis does not occur. They are significant in, for example, the N-deficient Caribbean Sea, where, however, wave action can disrupt aggregates and allow O_2 to inhibit N_2 fixation. The single-celled *Gloeocapsa* has evolved another strategy in which it photosynthesizes during the day and fixes N_2 at night. Yet other cyanobacteria, with fungi, form symbioses called lichens that have a great capacity to colonize rocks and other inhospitable substrates. They are therefore important primary colonizers in nature, accumulating organic matter and allowing other forms of life (such as mosses, ferns, and higher plants) to become established.

Rhizobia of the genera *Rhizobium, Bradyrhizobium*, and *Sinorhizobium* are soil bacteria that can infect the roots of plants of the leguminosae (alfalfa, peas, beans, vetches, and soybean). Others can lead to the formation of stem nodules on some tropical plants such as *Sesbania*. To initiate infection, they attach to root hairs, exchange signal molecules with the plant, pass into the root-hair cell by reorientating its growth and causing an invagination of the cell wall, and promote the synthesis of an infection thread that infects cells in the root cortex. Bacteria are released from this thread and differentiate into specialized cells called bacteroids, in which the *nif* and associated *fix* genes that support nodule N_2 fixation are derepressed. The bacteroids use plant dicarboxylic acids to fuel the reduction of N_2 to ammonia, which is converted to amino acids or ureides, mainly by host enzymes, for transportation to the rest of the plant. Oxygen is required by the bacteroids and is supplied at a very low free concentration but at a high flux by leghemoglobin, a carrier having a very high affinity for O_2. The leghemoglobin comprises an apoprotein coded for by the plant and a heme moiety coded for and synthesized by the bacteroids.

Nodules formed on the roots of nonlegumes are called actinorhiza. The causal agents are species or strains of the genus *Frankia*, filamentous actinomycetes that can infect often commercially important trees and shrubs of genera such as the temperate *Alnus* (alder) and the semi-tropical *Casuarina* (horsetail pine). Such genera are important ecologically as colonizers of nutrient-poor soils and moraines, and are used for soil stabilization and windbreak purposes.

III. ASSIMILATION AND AMMONIFICATION

Most microbes in nature can not fix N_2 and must obtain their nitrogen supply in the form of ammonia (NH_4^+), nitrate (NO_3^-), or free amino acids. Ammonia is assimilated by the glutamate dehydrogenase or (in N-deficiency) the glutamine synthetase-glutamate synthase (GS-GOGAT) system. Nitrate is reduced by assimilatory nitrate and nitrite reductases to NH_4^+ (Fig. 59.1), a system that is repressed by NH_4^+, but is unaffected by O_2. Growth may be limited by C or N availability, and in agricultural soils supplied with residues having a high ratio of C to N, such as straw, the N-starved microorganisms may be serious competitors with the plant roots for available nitrogen. This suggests the advisability of reducing the C content by composting plant material before its application to soils.

Nitrogen tied up in microbial biomass components is released only on death or lysis of the cells. Other microbes then produce a variety of proteolytic enzymes and deaminases that degrade proteins, as well as other enzyme systems that attack nucleic acids and wall components. The major ultimate product of these reactions is NH_4^+ and the process is termed ammonification or mineralization (Fig. 59.1).

The ecological implication of the assimilation and ammonification reactions is that the microbial biomass in terrestrial and aquatic systems is in a state of turnover, with the two processes often being in a steady state such that there is no marked change in the concentration of NH_4^+. Perturbations of the ecosystem can upset the steady state, high C inputs resulting in net assimilation and high N inputs resulting in net ammonification. A great variety of microbes carry out these two processes and activity is likely in both aerobic and anaerobic environments.

IV. NITRIFICATION

The oxidation of NH_4^+ through NO_2^- to NO_3^- is carried out mainly by two highly specialized groups of lithotrophic bacteria that use the oxidation reactions as their sole source of energy and reducing power to fix CO_2 to the level of cellular organic C components. They can thus grow in completely inorganic environments, providing NH_4^+, CO_2, and O_2 are available. They are of great environmental importance because they convert the relatively immobile NH_4^+ to the anionic NO_3^- which is mobile; can be leached into lakes, rivers, or groundwaters; and is a major substrate for denitrification. The nitrifying bacteria are also of significance in aerobic secondary sewage treatment, in which NH_4^+ from ammonification is converted to NO_3^-. NO_3^- in drinking water obtained from contaminated lakes, rivers, or aquifers can be reduced to NO_2^- in the human gut and cause the conversion of hemoglobin to methemoglobin, with a great loss in O_2-carrying capacity. This condition is referred to as methemoglobinemia (in infants, called blue babies). The NO_2^- can also react with amines forming carcinogenic nitrosamines.

A. Biochemistry and microbiology

1. NH_4^+ oxidation

The oxidation of ammonia (NH_3 is the actual substrate) is catalyzed by an ammonia monooxygenase (AMO) in an inner membrane. O_2 is an obligate requirement because one of the atoms is incorporated into the substrate, the other is reduced to water, and hydroxylamine (NH_2OH) is produced. Reducing power is supplied from the ubiquinone pool (Fig. 59.3). The NH_2OH is oxidized to NO_2^- by means of a hydroxylamine oxidoreductase that donates the electrons released to a cytochrome 554, and then through a cytochrome 552 to the terminal oxidase that reduces O_2 to water on the inner face of the inner membrane (Fig. 59.3). The overall reactions are

$$NH_3 + O_2 + 2H^+ + 2e^- \rightarrow NH_2OH + H_2O$$
$$NH_2OH + H_2O \rightarrow NO_2^- + 5\,H^+ + 4e^-$$

The redox reactions in the membrane are associated with the extrusion of protons into the periplasm, thus

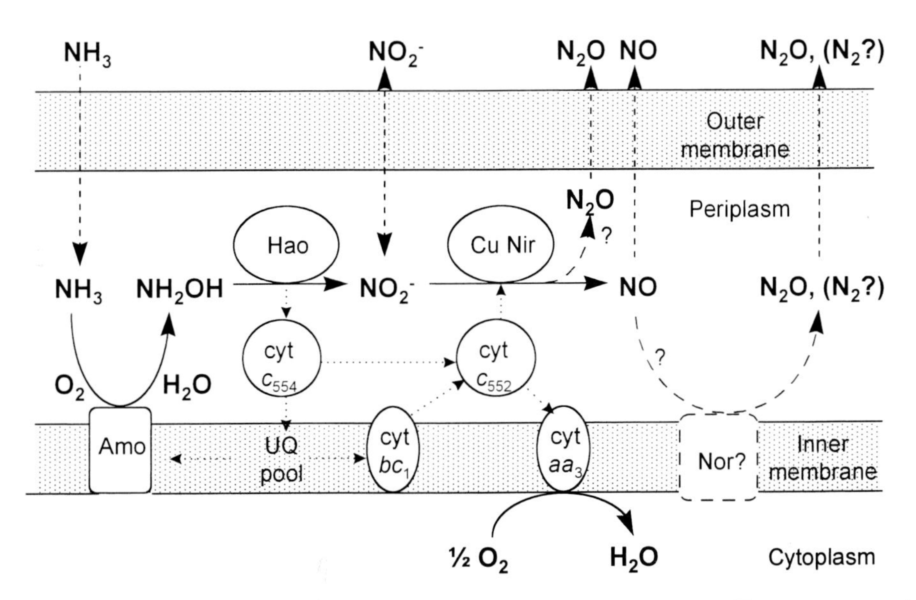

FIGURE 59.3 The cellular topography of the NH_4^+-oxidizing system in *Nitrosomonas europaea*. The components shown are ammonia monooxygenase (Amo), NH_2OH oxidoreductase (Hao), copper NO_2^- reductase (CuNir), a putative NO reductase (Nor), and several cytochromes with location and direction of electron transport (dotted lines). Solid lines indicates reactions; dashed lines, diffusion of substrates and products. The mechanism of production of N_2O is not known, but it is speculated that it could be a minor product of the CuNir or a product of an undescribed Nor. The production of N_2 is reported for one strain but the mechanism is unknown.

creating a proton-motive force and supporting ATP synthesis. ATP and reducing power are needed for growth and fixation of CO_2 by the Calvin cycle using ribulose-bisphosphate carboxylase-oxygenase (RuBisCO). The ammonia monooxygenase genes (*amo*) show significant homology with those of methane monooxygenase (pMMO), providing some evidence that the genes may be evolutionarily related.

During the oxidation of NH_4^+, small amounts (about 0.3% of the N oxidized) of NO and N_2O are produced, the amounts increasing in O_2 deficiency to 10% or more of the N oxidized. It is known that the NO is produced from NO_2^- by a copper-containing nitrite reductase (CuNir, Fig. 59.3), but the mechanism by which N_2O and, in a few cases, small amounts of N_2 are produced is not known. An NO reductase has not been isolated or characterized from these organisms. It was recently shown that NO (or NO_2) is required for effective oxidation of ammonia. Thus, air sparging of a *Nitrosomonas eutropha* culture can reduce NO concentration and inhibit growth. An NO-binding agent also inhibits growth. The mechanism of action of NO and NO_2 is not known.

In *Nitrosomonas* sp., the genes for AMO, *amo*A and *amo*B, coding for the two subunits, are duplicated, there are three copies of *hao*, and some of the cytochrome genes are present in more than one copy. There is evidence that these genes are more scattered on the chromosome than are those of the N_2 fixation and denitrification systems.

Ammonia-oxidizing bacteria are limited to six genera that are members of the beta subdivision of the Proteobacteria, except for *Nitrosococcus oceanus*, which is in the gamma subdivision (Table 59.2). Some are found commonly in soils; others in aquatic environments. Many have arrays of cytomembranes arising from infoldings of the cytoplasmic membrane and these are believed to contain important components of the NH_4^+-oxidizing system. However, much remains to be discovered about the detailed biochemistry of this system.

2. NO_2^- *oxidation*

The oxidation of NO_2^- appears to be a one-step process catalyzed by a nitrite oxidoreductase, an enzyme containing iron and molybdenum that may be located in the membrane. The reaction is

$$NO_2^- + H_2O \rightarrow NO_3^- + 2H^+ + 2e^-$$

It is not yet clear how a proton-motive force is generated for the support of ATP synthesis. The nitrite oxidoreductase can operate in the reductive direction and some strains of NO_2^- oxidizers are reported to be able to grow anaerobically. However, both the NO_2^- oxidizers and the NH_4^+ oxidizers are essentially aerobic organisms, requiring O_2 for growth and activity. The fixation of CO_2 is by means of ribulose bisphosphate carboxylase-oxygenase.

TABLE 59.2 Phylogenetic distribution of genera of nitrifying bacteria[a]

Ammonia oxidizers		*Nitrite oxidizers*	
Proteobacteria, beta subdivision		Proteobacteria, alpha subdivision	
Nitrosomonas (1)	Peripheral memb.	*Nitrobacter* (2)	Polar cap of memb.
Nitrosospira (5)	Acid soils	*Nitrococcus* (2)	Marine, tubular memb
Nitrosovibrio (2)		*Nitrospira* (1)	Marine, halophilic
Nitrosolobus (2)		*Nitrospina* (1)	Marine, no memb.
Nitrosococcus (1)	Marine		
Proteobacteria, gamma subdivision			
Nitrosococcus oceanus	Marine, equatorial stack of cytomembranes		

[a]Numbers in parentheses indicate the number of species in the genus (note that the two spp. of *Nitrosococcus* are in different subdivisions of the Proteobacteria). Memb. indicates arrays of cytomembranes that are often characteristic of the genus.

Nitrite-oxidizing bacteria are confined to four genera (Table 59.2) in the alpha subdivision of the Proteobacteria. All are lithotrophic but some can grow mixotrophically (that is, they can grow using a combination of lithotrophic and heterotrophic pathways) and one, *Nitrobacter hamburgensis*, can grow heterotrophically on organic carbon alone.

B. Ecology of nitrification

The obligatory requirement of the nitrifiers for O_2 means that nitrification occurs only in aerobic environments. Ammonia must come from ammonification, either in the same location or by diffusion from neighboring anaerobic environments where excess NH_4^+ accumulates (Fig. 59.4). Nitrification provides NO_3^-, the most important nitrogen source for plant growth, and occurs readily in most agricultural and other soils, especially at near-neutral pH values, and the production by nitrifying bacteria of NO and smaller amounts of N_2O makes such environments important global sources of these trace gases. NO plays important roles in tropospheric chemistry and is a factor in acid precipitation and ozone turnover. It was recently recognized that NO from nitrification or other sources can be consumed under aerobic conditions by many heterotrophic bacteria that convert it to NO_3^- as the major product. This process can greatly modify the net flux of NO to the atmosphere. N_2O is radiatively active, absorbing infrared radiation from Earth and thus acting as a greenhouse gas. When N_2O diffuses to the stratosphere, it is converted photochemically to NO, which then catalyzes the conversion of ozone to O_2.

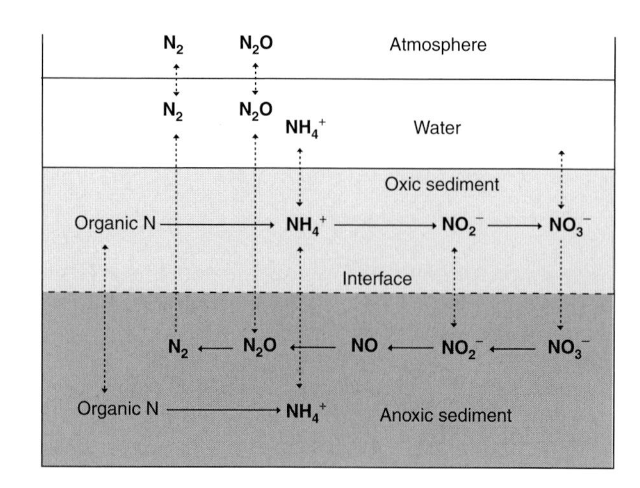

FIGURE 59.4 The stratification of N-cycle processes that occurs in aquatic sediment. Solid lines indicate reactions; dotted lines, diffusion of substrates and products. Nitrification occurs in an upper aerobic (oxic) layer and denitrification occurs in a lower anaerobic (anoxic) layer to the extent that NO_2^- or NO_3^- penetrate this layer. Mineralization of organic N occurs throughout. Ionic N oxides diffuse into the water column, and N_2O and N_2 diffuse into the atmosphere. Denitrification may be supplied with NO_3^-, either from the oxic layer nitrification or from the water column. A similar system can be established in biofilms or in water-saturated soils or soil aggregates.

Based on the behavior of nitrifying bacteria in pure culture, it was believed that nitrification would not occur in environments with pH values 2–6.5. However, it is now known that at least some of the nitrification that can occur in acidic forest soils is brought about by NH_4^+ oxidizers such as *Nitrosospira* spp. growing in colonial aggregates that are surrounded by less acid-sensitive NO_2^- oxidizers. Such aggregates are observed in aquaculture and other aquatic systems in which nitrification is an important process reducing the potential problem of ammonia toxicity and providing the denitrification substrate NO_3^-.

In agriculture, the potential losses of fertilizer nitrogen through leaching or denitrification of the nitrification product NO_3^- is considered to be a serious problem. A number of "nitrification inhibitors," such as nitrapyrin (sold as N-Serve by the Dow Chemical Company) are commercially available and usually specifically inhibit the ammonia monooxygenase enzyme, thus preventing the conversion of anhydrous ammonia (NH_3) or cationic NH_4^+ fertilizers to the leachable and denitrifyable anion NO_3^-.

Sewage treatment and certain other waste-treatment processes promote nitrification in a secondary activated sludge reactor in which biologically available organic carbon (often monitored as BOD, biochemical oxygen demand) is mineralized and converted to microbial biomass and CO_2. The NO_3^- that is produced, if fed into a river or lake as receiving water, can stimulate eutrophication and the growth of algae and macrophytes (aquatic higher plants), so it is preferably subject to tertiary or nitrogen-removal treatment in which denitrification is encouraged (Section V).

C. Heterotrophic nitrification

Although it is clear that most of the global nitrification is catalyzed by the lithotrophic nitrifiers, there are some heterotrophs that can produce NO_2^- and NO_3^- from reduced nitrogen compounds or from ammonia by mechanisms that are not yet clear.

$$RNH_2 \rightarrow RNHOH \rightarrow RNO \rightarrow RNO_2 \rightarrow NO_3^-$$
$$NH_4^+ \rightarrow NH_2OH \rightarrow NOH \rightarrow NO_2^- \rightarrow NO_3^-$$

In the heterotrophic nitrifier–denitrifier *Thiosphaera pantotropha* (recently reclassified as a strain of *Paracoccus denitrificans*), there do not appear to be genes homologous to the *amo* or *hao* of lithotrophic nitrifiers, so the nitrification mechanism would seem to be different. Other heterotrophic nitrifiers include fungi such as *Aspergillus flavus*, relatively acid insensitive and perhaps responsible for much of the nitrification in acidic forest soils, or bacteria such as *Arthrobacter* sp. However, there is no evidence that any of these organisms can conserve energy from the oxidations that they catalyze, and so the evolutionary significance of this process is not known.

It is difficult to distinguish between lithotrophic and heterotrophic nitrification activity, but in laboratory experiments, use can be made of the fact that acetylene (in low ppmv concentrations) inhibits the Amo of the former but not the oxidation system of the latter, and chlorate inhibits lithotrophic NO_2^- oxidation but apparently not heterotrophic NO_3^- production.

V. DENITRIFICATION

When oxygen becomes limiting, some microorganisms, mainly aerobic bacteria, have the ability to switch to the use of the nitrogen oxides NO_3^-, NO_2^-, NO, and N_2O as terminal acceptors of electrons in their metabolism. This process is known as denitrification, and it permits organisms to continue what is essentially a form of aerobic respiration in which the end product is dinitrogen. However, intermediates sometimes accumulate. Denitrification is of major importance because it closes the global nitrogen cycle, maintaining a balance in atmospheric dinitrogen; it is responsible for significant losses of fertilizer nitrogen in agriculture; it is a critical process in the nitrogen-removal component of modern tertiary wastewater-treatment plants; leakage of the intermediate N_2O adds to the greenhouse gas load of the troposphere and also indirectly causes catalytic destruction of stratospheric ozone; and it is possible that the long-term accumulation of nitrous oxide in closed systems may cause human or animal health problems.

A. Biochemistry and microbiology

Most of the denitrifying organisms of significance in nature are bacteria, and they have an aerobic type of electron-transport (cytochrome) chain leading to O_2 as the terminal electron acceptor. However, when O_2 is deficient or absent, a cellular regulator (Anr or Fnr) switches on the synthesis of a series of reductase enzymes that successively reduce NO_3^-, NO_2^-, NO, and N_2O to N_2. This type of respiration is less efficient than O_2 respiration, and so growth of the microorganisms is slower under anaerobic conditions. The NO_3^- reductase (Nar) is a membrane-bound molybdo-enzyme with the active site on the inner face of the membrane (Fig. 59.5). Also, a soluble NO_3^- reductase enzyme (Nap) is located in the periplasm (between the inner and outer membranes) of some gram-negative bacteria, and probably enables the organism to adapt to rapid onset of anaerobiosis. NO_2^- reductases (Nir) are of two types, copper-containing and cytochrome cd_1-containing, and are periplasmic. NO reductase (Nor) is membrane-bound, and N_2O reductases (Nos) are also mostly periplasmic Cu enzymes, except in the gliding bacterium *Flexibacter canadensis*, in which it is in the membrane.

Nitrate is the major N oxide substrate for denitrification. It is taken up through the inner membrane by means of an antiport (at least in some bacteria) that is O_2 sensitive, and thus NO_3^- uptake is inhibited by O_2. The NO_2^- product from reduction of NO_3^- exits via the NO_3^-–NO_2^- antiport and undergoes successive

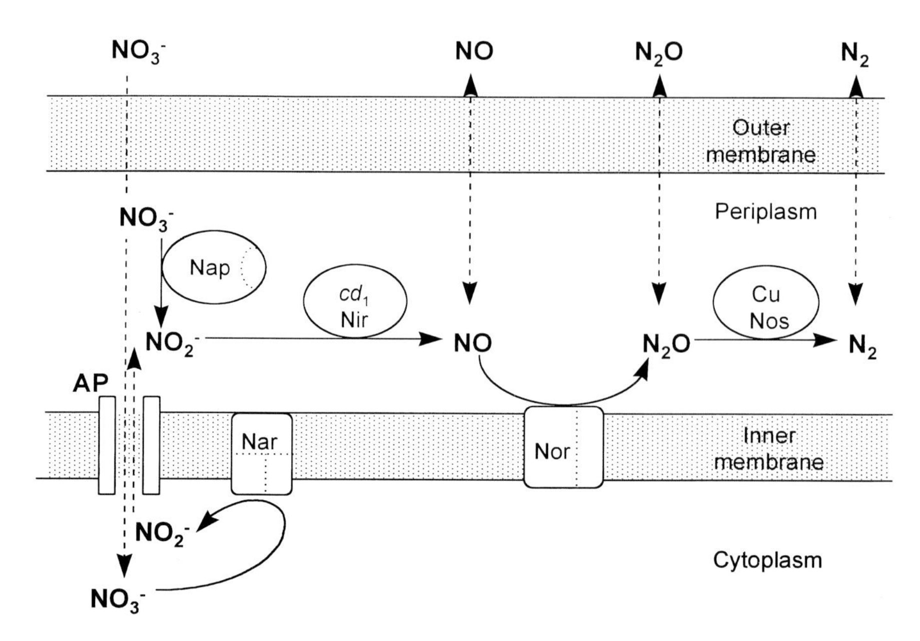

FIGURE 59.5 The cellular topography of the denitrification system in *Paracoccus denitrificans*. The components shown are a NO_3^-–NO_2^- antiport (AP); membrane-associated NO_3^- reductase (Nar) and NO reductase (Nor); and periplasmic NO_3^-, NO_2^-, and N_2O reductases (Nap, cd_1Nir, and CuNos, respectively). For simplicity, cytochromes and electron-transport pathways are not shown.

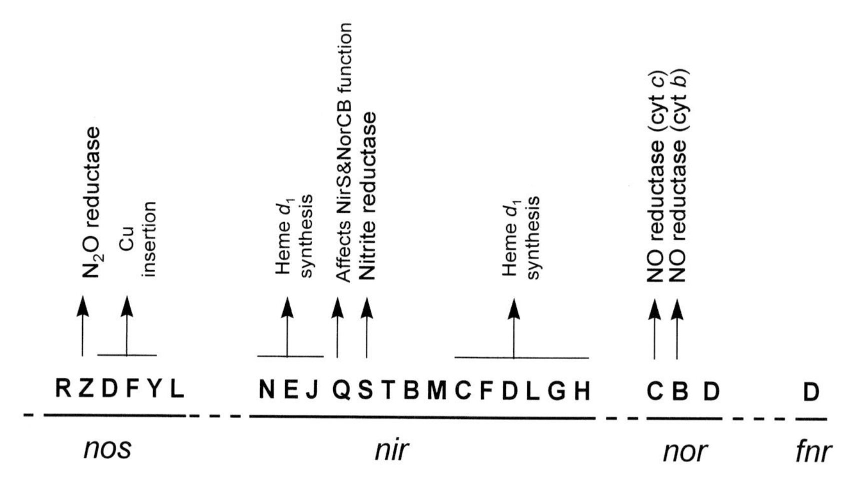

FIGURE 59.6 The chromosomal arrangement of some of the denitrification genes in *Pseudomonas stutzeri*. The *nir*, *nor*, and *nos* genes are shown to be neighboring clusters in this bacterium. In others, they may be more scattered. Products and functions are shown.

reduction by the periplasmic NO_2^- reductase (Nir), the membrane-bound NO reductase (Nor), and the periplasmic N_2O reductase (Nos) (Fig. 59.5). Electrons for these reductions are supplied via NADH dehydrogenase, the ubiquinone pool, a cytochrome bc_1 complex in the inner membrane, and other cytochromes in the periplasm (not shown in Fig. 59.5). These membrane redox reactions result in the extrusion of protons (H^+) into the periplasm and the generation of a proton motive force that allows the cell to make ATP.

The genes encoding the denitrification system are generally chromosomal and are tightly linked, as shown in Fig. 59.6 for *Pseudomonas stutzeri*. The NO_2^- and NO reductase genes, *nir* and *nor*, are coregulated, presumably to avoid the possible toxicity of NO, because NO could accumulate if Nor activity was deficient. Both low O_2 and the presence of a nitrogen oxide are necessary for the transcription of the denitrification genes. Organisms exist that possess only certain subsets of the four reductases; thus the nitrogen oxide substrates used and the final products can vary with the species (Fig. 59.7).

Phylogenetically, bacteria that are able to denitrify are found widely distributed in very diverse groups (Table 59.3), and this has led to much speculation about the evolutionary significance of this distribution.

A few fungi (for example, *Fusarium oxysporum* strains) were recently shown to denitrify, and they produce N oxide reductases with some similarities to those in bacteria. Indeed, a dissimilatory NO_3^- reductase (DNar) is found in the mitochondria of *F. oxysporum*. However, there is no evidence that such fungi are important denitrifiers in nature. Most denitrifiers are typical heterotrophs, such as *Pseudomonas, Alcaligenes,* and *Ralstonia* spp., using organic carbon as a source of reducing power.

B. Ecology of denitrification

1. *Primary controlling factors*

Denitrifiers are ubiquitous, growing best aerobically, so some potential for denitrification exists in many habitats. The actual activity depends on three primary factors—limitation of O_2, availability of N oxides, and availability of reductant (mainly as organic carbon, but some bacteria can use H_2 or sulfide).

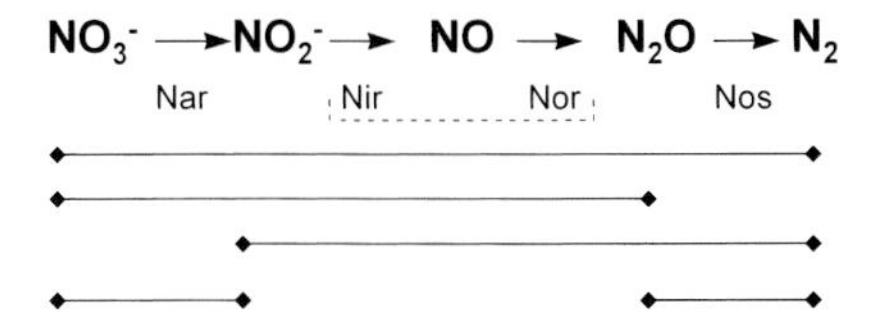

FIGURE 59.7 Schematic diagram showing that the presence of different subsets of the denitrification-pathway enzymes in different organisms can affect the substrates used and the products released.

a. Oxygen

Depletion of O_2 occurs in environments in which its consumption by biological activity is greater than its rate of supply by mass transport or diffusion. Thus, aquatic sediments, waterlogged or irrigated soils, and the centers of large water-saturated soil aggregates are some of the habitats in which the genes for N oxide reductases of denitrifiers will be derepressed and the enzymes synthesized. If respiratory activity is high and O_2 is supplied by molecular diffusion (as in biofilms), the gradients of O_2 concentration can be very steep, conditions becoming nearly anaerobic within hundreds of micrometers of an air-saturated boundary layer. Although O_2 represses the synthesis of the N oxide reductases and inhibits the activity of the enzymes, some denitrifying systems are reported to be relatively insensitive to O_2. However, studies using microelectrodes have shown significant denitrification only in the anaerobic or nearly anaerobic zone. The N_2O reductase is often reported to be the most sensitive to O_2, so exposure of a denitrifying system to O_2 may result in the release of N_2O rather than N_2. Indeed, the Nos enzyme seems to be generally more sensitive than the other reductases to unfavorable conditions. Thus, low pH values and inhibitory components in the environment can cause the release of N_2O. The marked inhibition of N_2O reduction by acetylene, coupled with the measurement of N_2O accumulation by gas chromatography, is used as a sensitive assay for denitrification.

TABLE 59.3 Phylogenetic distribution of selected genera of denitrifying organisms

Groups/genera		Groups/genera	
Archaea (4)		Proteobacteria, delta subdivision	
Halobacterium	Halophilic	*Beggiatoa*	H_2S oxidizer
Proteobacteria, alpha subdivision		*Nitrosococcus*	NH_4^+ oxidizer
Agrobacterium	Soil heterotroph	*Pseudomonas*	Soil heterotroph
Aquaspirillum	Aquatic heterotroph	Proteobacteria, epsilon subdivision	
Azoarcus	Degrades aromatics	*Campylobacter*	Some pathogens
Azospirillum	N_2 fixer	*Wolinella*	Grows on fumarate + H_2
Bradyrhizobium	Soybean nodules	Firmibacteria (gram-positive)	
Hyphomicrobium	Methylotroph	*Bacillus*	Some fac. anaerobes
Magnetospirillum	Magnetic bacterium	*Jonesia*	
Paracoccus		Cytophagales, green S bacteria	
Rhodobacter	Photosynthetic	*Cytophaga*	Gliding bacteria
Rhodopseudomonas	Photosynthetic	*Flexibacter*	Gliding bacteria
Sinorhizobium	Legume nodules		
Proteobacteria, beta subdivision			
Alcaligenes	Common heterotroph		
Neisseria	Pathogens		
Nitrosomonas	NH_4^+ oxidizer		
Ralstonia	(Alcaligenes)		
Thiobacillus	H_2S oxidizer		

b. N oxides

In environments in which O_2 depletion occurs, the denitrifiers become energy limited unless reducible N oxides are present. Nitrification is ultimately the major source of these oxides, but they also come from precipitation and fertilizers. As mentioned in Section IV.B, nitrification, the oxidation of NH_4^+ to NO_2^- and NO_3^-, is an obligately aerobic process that often occurs close to denitrification zones, on opposite sides of aerobic–anaerobic interfaces, and so these processes are frequently tightly coupled (Fig. 59.4). Such interfaces can occur in soils, but they are very important in aquatic sediments, where both the sediment and the water column may be sources of NO_3^-. The relative importance of sediment nitrification and the water column in providing NO_3^- for denitrification may depend on the depth of O_2 penetration, and thus on the steepness of the NO_3^- diffusion gradient from the surface to the zone of denitrification. Sediment nitrification is likely to be the major source of NO_3^- for denitrification in cases in which O_2 penetrates deeply, the $NO^-{}_3$ gradient from the surface to the interface is less steep, and the nitrification potential occurs close to the aerobic–anaerobic interface.

c. Reductants

In environments with low organic-matter content, the supply of electron donors (and thus energy) may be a critical factor. Denitrification capacity may be highly correlated with water-soluble soil organic carbon, and denitrifiers are reported that degrade various aromatic compounds anaerobically. Such bacteria are of interest in the bioremediation of contaminated environments. Other bacteria can use dihydrogen, elemental sulfur, or oxidizable sulfur compounds as electron donors, but it is not known whether this is of significance in nature.

2. Accumulation of intermediates and effects of plants

Transient nitrite accumulation occurs, at least in the cultures of some denitrifiers, when NO_3^- concentrations above about 0.3 mM inhibit the reduction of NO_2^-. The NO_2^- is reduced once the NO_3^- concentration decreases sufficiently. As mentioned earlier, O_2 exposure and low pH values can cause the escape of N_2O from a denitrifying system, but the escape of NO appears to be minimal because of the very high apparent affinity of NO reductase for NO. Thus, denitrification is not as significant a contributor of NO to the atmosphere as nitrification.

Active plant roots are strong sinks for O_2 and NO_3^- and are sources of oxidizable organic carbon released in exudates. The roots of aquatic macrophytes act as sources of O_2. Therefore, in either case, they have the potential to both stimulate and inhibit denitrification in the rhizosphere. The actual outcome seems to depend on the availability of NO_3^- and carbon in the O_2-depleted part of the rhizosphere soil.

3. Dissimilatory reduction of nitrate to ammonia (nitrate ammonification)

In highly reducing habitats, such as sludges and the rumen, where the ratio of available organic carbon to electron acceptor (NO_3^-) is high, the NO_3^- is reduced not by denitrification to gaseous products but via specific NO_3^- and NO_2^- reductases to NH_4^+ (Fig. 59.1). Both enzymes are soluble and cytoplasmic, and the process acts mainly as an electron sink. It is found in some fermentative organisms such as *E. coli* and some species of *Clostridium*.

VI. SUMMARY AND GLOBAL IMPLICATIONS

The microorganisms involved in the many N-cycle processes respond to changes in their environment and in the availability of their substrates. Human influences, especially since the 1930s, have greatly changed both environments and substrates. Increases in biological N_2 fixation in legume crops and in chemical production of inorganic fertilizers have roughly doubled the total N inputs to the terrestrial N cycle. The retention of N has not been sufficient to prevent losses of inorganic N and other nutrients to rivers, lakes, and oceans, and it is now known that riparian ecosystems such as wetlands or forests along shorelines are important in restricting (through assimilation and denitrification) such input of NO_3^- to rivers and lakes. The conversion of N forms has increased volatile losses of NH_3, NO, and N_2O to the atmosphere, and the long-range transport and deposition of these molecules or their oxidation products. This has resulted in the "N-saturation" of many previously N-limited natural terrestrial and aquatic ecosystems, with consequent changes in flora, decreased diversity, and increased acidity both from rainfall of nitric acid (from the oxidation of NO) and from the nitrification of $NH^+{}_4$. Other effects on the atmosphere itself include the contribution of NO to photochemical smog and of N_2O to both the greenhouse effect and the destruction of stratospheric ozone.

BIBLIOGRAPHY

Alderton, W. K., Cooper, C. E., and Knowles, R. G. (2001). Nitric oxide synthases: structure, function and inhibition. *Biochem. J.* **357** (Pt 3), 593–615.

Bédard, C., and Knowles, R. (1989). Physiology, biochemistry, and specific inhibitors of CH_4, NH_4^+, and CO oxidation my methanotrophs and nitrifiers. *Microbiol. Rev.* **53**, 68–84.

Cole, J. (1996). Nitrate reduction to ammonia by enteric bacteria: Redundancy, or a strategy for survival during oxygen starvation? *FEMS Microbiol. Lett.* **136**, 1–11.

Capone, D. G. (2001). Marine nitrogen fixation: what's the fuss? *Curr. Opin. Microbiol.* **4**, 341–348.

Colliver, B. B., and Stephenson, T. (2000). Production of nitrogen oxide and dinitrogen oxide by autotrophic nitrifers. *Biotechnol. Adv.* **18**, 219–232.

Gadkari, D., Mörsdorf, G., and Meyer, O. (1992). Chemolithoautotrohic assimilation of dinitrogen by *Streptomyces thermoautotrophicus* UBT1: Identification of an unusual N_2-fixing system. *J. Bacteriol.* **174**, 6840–6843.

Halbleib, C. M., and Ludden, P. W. (2000). Regulation of biological nitrogen fixation. *J. Nutr.* **130**, 1081–1084.

Harper, L. A., Mosier, A. R., Duxbury, J. M., and Rolston, D. E. (eds) (1993). "Agricultural Ecosystem Effects on Trace Gases and Global Climate Change." Special Publication No. 55. American Society of Agronomy, Madison, WI.

Holmes, A. J., Costell, A., Lidstrom, M. E., and Murrell, J. C. (1995). Evidence that particulate methane monooxygenase and ammonia monooxygenase may be evolutionarily related. *FEMS Microbiol. Lett.* **132**, 203–208.

Hooper, A. B., Vannelli, T., Bergmann, D. J., and Arciero, D. M. (1997). Enzymology of the oxidation of ammonia to nitrite by bacteria. *Antonie van Leeuwenhoek* **71**, 59–67.

Kobayashi, M., Matsuo, Y., Takimoto, A., Suzuki, S., Maruo, F., and Shoun, H. (1996). Denitrification, a novel type of respiratory metabolism in fungal mitochondrion. *J. Biol. Chem.* **271**, 16263–16267.

Kuenen, J. G., and Robertson, L. A. (1988). Ecology of nitrification and denitrification. *In* "The Nitrogen and Sulphur Cycles" (J. A. Cole and S. J. Ferguson, eds), pp. 161–218. Cambridge University Press, Cambridge.

Lee, S., Reth, A., Meletzus, D., Sevilla, M., and Kennedy, C. (2000). Characterization of a major cluster of nif, fix, and associated genes in a sugar-cane endophyte, *Acetobacter diazotrophicus. J. Bacteriol.* **182**, 7088–7091.

Moura, I., and Moura, J. J. (2001). Structural aspects of denitrifying enzymes. *Curr. Opin. Chem. Biol.* **5**, 168–175.

Navarro-Gonzalez, R., McKay, C. P., and Mvondo, D. N. (2001). A possible nitrogen crisis for Archaean life due to reduced nitrogen fixation by lightning. *Nature* **412**, 61–64.

Paul, E. A., and Clarke, F. E. (1989). "Soil Microbiology and Biochemistry." Academic Press, San Diego, CA.

Postgate, J. R. (1998). "Nitrogen Fixation." Cambridge: University of Cambridge.

Prosser, J. (ed.) (1986). "Nitrification." IRL Press, Oxford.

Rees, D. C., and Howard, J. B. (2001). Nitrogenase: standing at the crossroads. *Curr. Opin. Chem. Biol.* **4**, 559–566.

Revsbech, N. P., and Sørensen, J. (eds) (1990). "Denitrification in Soil and Sediment." Plenum Press, New York.

Schmidt, I., Zart, D., and Bock, E. (2001). Effect of gaseous NO_2 on cells of *Nitrosomonas eutropha* previously incapable of using ammonia as an energy source. *Antonie van Leeuwenhoek* **79**, 39–47.

Siddiqui, R. A., Warneke-Eberz, U., Hengsberger, A., Schneider, B., Kostka, S., and Friedrich, B. (1993). Structure and function of a periplasmic nitrate reductase in *Alcaligenes eutrophus* H16. *J. Bacteriol.* **175**, 5867–5876.

Sprent, J. I. (1987). "The Ecology of the Nitrogen Cycle." Cambridge University Press, Cambridge.

Stacey, G., Burris, R. H., and Evans, H. J. (eds) (1992). "Biological Nitrogen Fixation." Chapman & Hall, London.

Stevenson, F. J., and Cole, M. A. (1999). "Cycles of Soils: Carbon, Nitrogen, Phosphorus, Sulfur, Micronutrients." Wiley, New York.

Stouthamer, A. H., de Boer, A. P. N., van der Oost, J., and van Spanning, R. J. M. (1997). Emerging principles of inorganic nitrogen metabolism in *Paracoccus denitrificans* and related bacteria. *Antonie van Leeuwenhoek* **71**, 33–41.

Vitousek, P. M., Arbar, J. D., Howarth, R. W., Likens, G. E., Matson, P. A., Schindler, D. W., Schlesinger, W. H., and Tilman, D. G. (1997). Human alterations of the global nitrogen cycle: Sources and consequences. *Ecol. Appl.* **7**, 737–750.

Zumft, W. G. (1997). Cell biology and molecular basis of denitrification. *Microbiol. Mol. Biol. Rev.* **61**, 533–616.

WEBSITES

Issues in Ecology. Website of the Ecological Society of America *http://www.esa.org/sbi/sbi_issues/*

Nutrient Overload: Unbalancing the Global Nitrogen Cycle (World Resources Institute) *http://www.wri.org/wri/wr-98-99/nutrient.htm*

60

Nitrogen fixation

L. David Kuykendall
USDA

Fawzy M. Hashem
University of Maryland, Eastern Shore

Robert B. Dadson
University of Maryland, Eastern Shore

Gerald H. Elkan
North Carolina State University

GLOSSARY

cyanobacteria Photosynthetic prokaryotes of some species that fix nitrogen, either free living or symbiotically with a broad range of lower plant taxa.

diazotroph An organism capable of biological nitrogen fixation.

fixed nitrogen Chemically reduced or biologically fixed nitrogen that is in a chemical form that can be assimilated readily by most microbes or plants.

frankia Symbiotic nitrogen-fixing actinomycetes; a broad spectrum of shrubs and trees are hosts.

megaplasmid Extremely large (≥500 kb) extrachromosomal DNA molecules that encode the symbiotic nodulation and nitrogen-fixation abilities of certain legume-nodulating bacteria.

nif gene A bacterial gene coding for either nitrogenase enzyme or its subunits (i.e., *nifHDK*), or for proteins in the regulation of nitrogenase expression.

nitrogenase A highly conserved enzyme complex that catalyzes the conversion of atmospheric dinitrogen to ammonia; nitrogenase consists of two components, dinitrogenase reductase (an iron protein) and dinitrogenase (an iron molybdenum protein).

nitrogen fixation Reduction of inert atmospheric N_2 gas to a metabolizable active form by prokaryotic organisms (eubacteria and archaebacteria).

nod factors Lipochitooligosaccharides that are chemical signals produced and released by either *Rhizobium*, *Bradyrhizobium*, *Azorhizobium*, or *Mesorhizobium*. They are essential for the bacterial capacity to induce nodule organogenesis.

nod genes The genes in legume-nodulating bacteria that code for the development of nodules on the proper legume host. There are two types—a common *nod* region, which consists of a structurally and functionally conserved cluster of genes; and host-specific nodulation genes, which cannot complement nodulation defects in other species.

rhizobiophage Specific bacterial viruses (bacteriophages) virulent for *Rhizobium* and related genera.

Rhizobium, Bradyrhizobium, Azorhizobium, Sinorhizobium, Mesorhizobium, and Allorhizobium Six genera, representing four distinct families, of

The Desk Encyclopedia of Microbiology
ISBN: 0-12-621361-5

nitrogen-fixing bacteria that form nodules on the roots or stems of various legume hosts. The complex relationships they form with a limited number of host legumes vary both in host range and efficiency of symbiotic nitrogen fixation according to genetically programmed two-way communication between bacteria and host via chemical messengers.

symbiosis Mutually beneficial sharing between two distinct organisms, perhaps the best known examples of which are legumes and their bacterial microsymbionts.

Nitrogen Fixation refers to that property of some taxa of eubacteria or archaebacteria to enzymatically reduce atmospheric N_2 to ammonia. The ammonia produced can then be incorporated by means of other enzymes into cellular protoplasm. Nitrogen fixation only occurs in prokaryotes and not in higher taxa.

I. SIGNIFICANCE OF BIOLOGICAL NITROGEN FIXATION

Although nitrogen is an essential nutrient for life, little available nitrogen is present in mineral form. Above every hectare of soil at sea level, there are 78 million kg of inert N_2 (dinitrogen) gas. Plants, as eukaryotic autotrophs, need either an oxidized or reduced form of nitrogen for anabolism. Only certain prokaryotic organisms "fix" N_2 at physiological temperatures. Humans can mimic biological nitrogen fixation using the Haber–Bosch process of chemically reducing nitrogen at high temperatures. However, the process consumes precious fossil fuels, contributes to global warming, and pollutes the environment. Worldwide industrial production of NH_3 is annually about 100 million metric tons, threefourths of which is manufactured for fertilizer to grow crop plants.

Thus, humans produce about 75 million mt/year of nitrogenous fertilizer, whereas the annual contribution of reduced nitrogen from the biological process called nitrogen fixation is roughly estimated to be about two to three times that amount. As shown in Table 60.1, nitrogen fixation occurs in every natural environment, including the sea. The energy cost for nitrogen fixation is significant. Perhaps 10% of the available fossil fuel energy is used for the production of fertilizer nitrogen, whereas Hardy (1980) estimated that between 1 and 2 billion tons of plant carbohydrates derived from photosynthesis fuel the biological process of nitrogen fixation.

There are many genetically diverse nitrogen-fixing eubacteria falling into 27 families and 80 genera, and there are at least three thermophilic nitrogen-fixing genera of Archaebacteria. These nitrogen-fixing families and genera are listed in Table 60.2; the "blue-green algae" or cyanobacteria are listed in Table 60.3. With the exception of the Azotobacteraceae, there is

TABLE 60.1 Estimated annual amount of N_2 fixed biologically in various systems

System	N_2 fixed mt/year $\times 10^6$
Legumes	35
Nonlegumes	9
Permanent grassland	45
Forest and woodland	40
Unused land	10
Total land	139
Sea	36
Total	175

TABLE 60.2 Families and genera of nitrogen-fixing eubacteria, excluding cyanobacteria

Family and genus	Family and genus
Acetobacteriaceae	Methanomonadaceae
Acetobacter	*Methylobacter*
Azotobacteraceae	*Methylococcus*
Azomonas	*Methylocystis*
Azotobacter	*Methylomonas*
Azotococcus	*Methylosinus*
Beijerinckia	Pseudomonodaceae
Derxia	*Pseudomonas*
Xanthobacter	Rhizobiaceae
Bacillaceae	*Allorhizobium*
Paenibacillus (Bacillus)	*Rhizobium*
Clostridium	*Sinorhizobium*
Desulfotomaculum	Hypomicrobiaceae
Baggiatoaceae	*Azorhizobium*
Baggiatoa	Phyllobacteriaceae
Thiothrix	*Mesorhizobium*
Vitreoscilla	Bradyrhizobiaceae
Chlorobiaceae	*Bradyrhizobium*
Chlorobium	Rhodospirallaceae
Pelodictyron	*Rhodomicrobium*
Chloroflexaceae	*Rhodopseudomonas*
Chloroflexus	*Rhodospirillum*
Chromatiaceae	Spirillaceae
Amoebobactor	*Aquaspirillum*
Chromatium	*Azospirillum*
Ectothiorhodospira	*Campylobacter*
Thiocapsa	*Herbaspirillum*
Thiocystis	*Streptomycetaceae*
Corynebacteriaceae	*Frankia*
Anthrobacter	Thiobacteriaceae
Enterobacteriaceae	*Thiobacillus*
Citrobacter	Vibrionaceae
Enterobacter	*Vibrio*
Erwinia	Uncertain family
Escherichia	*Alcaligenes*
Klebsiella	*Desulfovibrio*

TABLE 60.3 Families and genera of cyanobacteria with nitrogen-fixing species

Family and genus	Family and genus
Chroococcaceae	Oscillatoriaceae
Chlorogloea	*Lyngbya*
Chroococcidiopis	*Microcoleus*
Gloeothece	Oscillatoria
Syenchococcus	*Phormidium*
Mastigocladaceae	*Plectonema*
Mastigocladus	Pleurocapsaceae
Michrochaetaceae	*Pleurocapsa*
Michrochaete	Rivulariaceae
Nistocaceae	*Calothrix*
Anabaena	*Dichothrix*
Anabaenopsis	*Gleotrichia*
Aphanizomenon	Scytonemataceae
Aulosira	*Scytonema*
Cylindrosperumum	*Tolypothrix*
Nodularia	Stigonematacaeae
Nostoc	*Fischerella*
Pseudanabaena	*Hapalosiphon*
Raphidiopsis	*Stigonema*
Richelia	*Westelliopsis*

no genus (or family) whose species are all nitrogen-fixing. The potential amount of nitrogen fixed by these bacteria depends largely on the ecosystem in which the organisms are active, as will be discussed in later sections. The amount of nitrogen fixed ranges from only trace amounts for some free-living soil bacteria to 584 kg/ha/year for the tropical tree legume *Leucaena*.

II. THE BIOLOGICAL NITROGEN FIXATION PROCESS

Dinitrogen gas is both chemically inert and very stable, requiring much energy to break the triple bond and reduce the N≡N to ammonia in an endothermic reaction: $3H_2 + N_2 \rightarrow 2\ NH_3$. This can be accomplished chemically by the Haber–Bosch process or biologically by prokaryotic organisms using adenosine triphosphate (ATP) energy to initiate the bond-breaking reaction. The prokaryotes able to fix nitrogen are extremely heterogeneous, with representatives that are autotrophic, heterotrophic, aerobic, anaerobic, photosynthetic, single-celled, filamentous, free-living, and symbiotic. Phylogenetically, these organisms are extremely diverse and yet the nitrogen-fixing process and enzyme system are similar in all of these organisms and depend on a nitrogenase enzyme complex, a high-energy requirement and availability (ATP), anaerobic conditions for nitrogenase activity, and a strong reductant.

Nitrogenase has been purified from all known nitrogen-fixing eubacteria. It consists of two components, dinitrogenase reductase (an iron protein) and dinitrogenase (an iron–molybdenum protein). Both enzymes are needed for nitrogenase activity. Nitrogenase reduces N_2 and H^+ simultaneously, using 75% of the electron flow in the reduction of nitrogen and 25% in H^+ reduction. A key characteristic of this enzyme complex is that both components are quickly and irreversibly lost on interaction with free oxygen, regardless of the oxygen requirements of the microbe. The protein usually has a molecular mass of 57,000–72,000 Da and consists of two identical subunits coded for by the *nifH* gene. It has a highly conserved amino acid sequence. The iron–molybdenum protein has a molecular mass of about 220 kDa and has four subunits, which are pairs of two different types. The α-subunit is coded for by the *nifD* gene and is 50 kDa in size. The β-subunit is coded for by the *nifK* gene and has a mass of 60 kDa. These enzymes have been sequenced and demonstrate considerable similarity. The nitrogenase complex is large and may amount to 30% of total cell protein.

Theoretically, as much as 28 moles of ATP are consumed in the reduction of 1 mole of N_2. Depending on the method used and the nature of the organism in question, *in vitro* studies show that the energy requirements vary between a minimum of 12–15 moles and 29 moles of ATP/N_2. This energy is not only required for the reduction process, but also to maintain the anaerobic conditions needed for the reaction.

Various strategies are used to exclude oxygen from the reaction in nonanaerobic bacteria. Facultative organisms fix nitrogen only under anaerobic conditions. Aerobic organisms exhibit a wide range of methods for protecting the enzyme complex from oxygen. Many grow under microaerophilic conditions accomplished by scavenging free oxygen for metabolism or sharing the ecosystem with other organisms that consume the excess O_2. Some evidence indicates that some free-living aerobes, such as *Azotobacter*, can change the conformation of the nitrogenase protein to form a less oxygen-sensitive protein. Photosynthetic aerobes such as cyanobacteria can form special cells called heterocysts in which nitrogen fixation occurs and which lack the O_2-evolving mechanism that is part of photosynthesis. The bestknown protective mechanisms are found in symbiotic nitrogen-fixing associations (i.e., *Rhizobium*–legume), in which the nitrogenase system in the endophyte is protected from excess oxygen by a nodule component, leghemogloblin. Nonleguminous nodules (with *Frankia* or cyanobacteria) are probably protected by other, as yet undescribed, oxygenrestrictive mechanisms. There are other protective mechanisms as well, but all of these require the diversion of energy from the nitrogen-fixation process itself to maintain a favorable environment for fixation.

In addition to the limitations to biological nitrogen fixation (BNF) caused by free oxygen, generally the presence in the environment of combined nitrogen, such as ammonia or nitrate, strongly inhibits both nodulation and N_2 fixation. Thus, it is unnecessary to use chemical fertilizer in large quantities for legume crop production.

III. FREE-LIVING NITROGEN-FIXING BACTERIA

There are many diverse nitrogen-fixing prokaryotes, representing 22 families and 52 genera of eubacteria (excluding the cyanobacteria), as well as three thermophilic genera of archaebacteria. Most of these genera fix nitrogen as free-living diazotrophs. These were first described by Winogradsky in 1893 (*Clostridium pasteurianum*), and Beijerinck in 1901 (*Azotobacter*). The discovery of other free-living nitrogen-fixing bacteria lagged until the availability of 15N stable isotopic and acetylene-reduction techniques became common. Thus, most of the freeliving nitrogen-fixing bacteria listed in *Bergey's Manual of Systematic Bacteriology* have been discribed using these techniques. It is generally accepted that diazotrophs obtain their carbon and energy supplies from root exudates and lysates, sloughed plant cell debris, and organic residues in soil and water. They are found "completely free-living" or in loose associations as a result of root or rhizosphere colonization (the associative bacteria are discussed separately). The quantity of nitrogen fixed is a matter of some controversy. Russian workers have estimated that *C. pasteurianum* or *Azotobacter* contribute perhaps 0.3 kg N_2/ha/year, compared with about 1000 times that amount provided by a good leguminous association. Associative organisms such as *Azospirillum* have been estimated to fix from trace amounts to 36 kg/ha. The limitations of the terrestrial BNF system is due not only to the difficulty in obtaining sufficient energy and reductant but also because of the need to divert substrate for respiratory protection of nitrogenase. The cyanobacteria can overcome the environmental constraints faced by most other nitrogen fixers. Being photosynthetically active, these prokaryotes use sunlight to fix CO_2 and are thus independent of external energy needs. The families and genera of nitrogen-fixing cyanobacteria are listed in Table 60.3.

The free-living cyanobacteria are distributed widely in humid and arid tropical surface soils. Extensive studies, especially under rice paddy conditions, have been conducted in India, Japan, and the Phillippines. Of 308 isolates from Philippine paddy soils, most were identified as *Nostoc* or *Anabaena*. Reports of fixation rates ranged from 3.2 to 10.9 kg N/ha/year. Reports of 15–20 kg N/ha/year in rice in the Ivory Coast, 44 kg N/ha/year in Lake George in Uganda, and 80 kg N/ha/year in paddy fields in India have been published. Under temperate conditions, BNF by cyanobacteria has been a major problem in the eutrophication of lakes.

About 125 strains of free-living cyanobacteria representing 10 families and 31 genera in all taxonomic groups have been shown to fix nitrogen, but the extent to which fixation occurs and conditions needed for fixation vary greatly. There are four types of nitrogen-fixing cyanobacteria—heterocystous filamentous, nonheterocystous filamentous, unicellular reproducing by binary fission or budding, and unicellular reproducing by multiple fission. Heterocysts are thickened, specialized cells occurring at regular intervals in some filamentous cyanobacteria. These cells lack the oxygen-evolving component of the photosynthetic apparatus. Heterocysts appear to be the only cells capable of fixing nitrogen aerobically as well as anaerobically. The main function of the heterocysts seems to be to compartmentalize nitrogen fixation because there is little or no photosynthesis in the heterocysts, and all of the energy translocated is available for nitrogen fixation. The other groups of cyanobacteria need to be examined as well because they appear quite active as fixers under anaerobic conditions. Nitrogen fixation in paddy rice could thus be improved.

One should not overlook the potential for the improvement of nitrogen fixation in unique environments. There are many reports of intimate associations between nitrogen-fixing bacteria and animals (e.g., termites and ruminants). Because *nif* genes can be transferred and expressed between the enteric bacteria *Klebsiella* and *Escherichia coli*, it may be possible to increase nitrogen fixation by rumen bacteria. Perhaps ultimately in ruminants, it will be practical to substitute engineered diazotrophic enterics for the plant protein now required by these animals.

IV. ASSOCIATIVE NITROGEN-FIXING BACTERIA

Since the 1960s, another type of plant–bacterial interaction has been described as "associative." This interaction was shown to result from the adhesion of the bacteria to the root surfaces of wheat, corn, sorghum, and other grasses. The major associative nitrogen-fixing systems are summarized in Table 60.4. *In vitro* studies show that many of these bacteria can achieve high rates of N_2 fixation under optimum conditions. However, in the rhizosphere, the ability to survive, grow, and colonize plant roots is a precondition limiting the potential for nitrogen-fixation. The characteristics required for an organism to flourish in the rhizosphere are the ability to withstand the changing physical and chemical soil environments, grow well and obtain all the energy needed from carbon and mineral supplies in the root zone, and compete successfully with other rhizosphere organisms for the limited energy and nutrients available. Estimates of *in vivo* fixation are extremely variable and give rise to a recurring question as to whether energy substrates in the rhizosphere are sufficient to support growth and nitrogen fixation by these associative bacteria at sufficient levels. The nitrogenase system is repressed by bound nitrogen. Therefore, in the presence of nitrogen fertilizer, BNF by free-living or associative bacteria is reduced. Thus, a mixed BNF-nitrogen fertilizer system would work better under field conditions if the microorganisms could be genetically engineered so that the expression of nitrogenase is derepressed when bound nitrogen is present.

TABLE 60.4 Major associative diazotrophs

Plant species	Principle microorganism
Rice (*Oryza sativa*)	*Achromobacter*, Enterobacteriaceae, *Azospirillum brasilense*
Sugarcane (*Saccharum* spp.)	*Azotobocter, Beijerinckia, Klebsiella, Derxia, Vibrio, Azospirillum*, Enterobacteriaceae, Paenibacillaceae
Pearl millet (*Pennisetum galucum*) and sorghum (*Sorghum bicolor*)	*Azospirillum, Paenibacillus polymyxa, P. azotofixans, P. macerans, Klebsiella, Azotobacter, Derxia, Enterobacter*
Maize (*Zea mays*)	*Azospirillum lipoferum, Azotobacter vinelandii*
Grasses	
Paspalum notatum var. *batatais*	*Azotobacter paspali*
Panicum maximum	*Azospirillurn lipoferum*
Cynodon dactylon	*Azospirillum lipoferum*
Digitaria decumbens	*Azospirillum lipoferum*
Pennisetum purpureum	*Azospirillum lipoferum*
Spartina alterniflora Loisel	*Campylobacter*
Wheat	
Triticum spp.	*Paenibacillus polymyxa*

V. SYMBIOTIC NITROGEN-FIXING BACTERIA

Only three groups of nitrogen-fixing bacteria have evolved mutually beneficial symbiotic associations with higher plants: (1) the filamentous bacterium *Frankia*, forming root nodules with a number of plants such as alder, *Purshia*, and Russian olive; (2) heterocystous cyanobacterium with a number of diverse hosts from *Cycads* to *Azolla*; and (3) *Rhizobium* and allied genera with legumes.

A. *Frankia*

The genera of dicotyledonous plants found to be nodulated by *Actinorhizae (Frankia)* are listed in Table 60.5. These include more than 180 species in more than 20 genera, representing at least eight families and seven orders of plants. There is no obvious taxonomic pattern among these hosts. Most of the species described are shrubs or trees and are found in temperate climates, but they have a wide growth range and could be grown in the tropics. In fact, *Purshia tridento* already is an important rangeland forage crop in Africa, and other *Purshia* species are harvested for firewood. *Casuarina*, a vigorous nitrogen fixer, has been planted in Thailand where it can be harvested for construction lumber after 5 years of growth. Almost all the hosts are woody, ranging from small shrubs to medium-sized trees.

TABLE 60.5 Genera of nitrogen-fixing plant genera with *Frankia* symbiosis

Genus	Number of nodulated species
Alnus	34
Casuarina	25
Ceanothus	31
Cercocarpus	4
Chamaebatia	1
Colletia	3
Coriaria	14
Cowania	1
Discaria	6
Dryas	3
Elaeagnus	17
Hippophae	1
Kentrothamnus	1
Myrica	26
Purshia	2
Rubus	1
Satisca	2
Shepherdia	3
Talguena	1
Trevoa	2

Frankia, the nitrogen-fixing endophyte found in nodules of these nonleguminous plants is a genus of prokaryotic bacteria closely related to the actinomycetes. In pure culture, these endophytes behave as microaerophilic, mesophilic, or heterotrophic organisms, usually with septate hyphae that develop sporangia. Isolates vary morphologically and nutritionally. Most strains can fix atmospheric nitrogen in pure culture. Nitrogenase genes are highly conserved, and *Frankia* nitrogenase enzymes closely resemble those of other nitrogen-fixing bacteria. *Frankia* was first reliably isolated in 1978. The organism grows slowly, requiring 4–8 weeks for visible colonies to be formed in culture. *Frankia* is similar to other aerobic actinomycetes, producing a separate filamentous mycelium that can differentiate into sporangia and vesicles. The cells are routinely grampositive, but unlike other gram-positive bacteria, *Frankia* has a discontinuous membranous layer. Molecular methods for taxonomy such as DNA–DNA hybridization have demonstrated considerable genetic diversity between isolates within this family, but only a limited number of isolates have been analyzed.

Although the majority of actinorhizal microsymbionts apparently infect their plant hosts via a root hair-mediated mechanism, some of them can infect by the direct penetration of the root by means of the intercellular spaces of the epidermis and cortex. As a result of infection, a root meristem is induced as the *Frankia* hyphae penetrate the cells, but the hyphae remain enclosed by a host-produced polysaccharide layer. *Frankia* grows in the nodule occupying a major part of the host-produced nodular tissue. Root nodule tissue usually composes 1–5% of the total dry weight of the plant. *Frankia* can form vesicles whose walls evidently adapt to O_2 in such a way that they can fix nitrogen at atmospheric O_2 levels, but the precise mechanism is obscure.

It appears that the infective abilities of each *Frankia* strain are limited to one or a few plant genera. Currently, four host-specificity groups have been identified: (1) strains infective on *Alnus*, *Comptonia*, and *Myrica*; (2) strains infective on Casuarinaceae; (3) flexible strains infective on species of *Elaeagnaceae* and the promiscuous species of *Myrica* and *Gymnostoma*; and (4) strains infective only on species of the family Elaeagnaceae. Strains in specificity groups 1 and 2 infect the host plant via the root hair, whereas group 4 isolates infect by intercellular penetration. Some strains can use both modes of infection, and these are in the "flexible" specificity group 3.

Because these organisms were only recently isolated and cultured, only in the 1990s have taxonomic, physiological, and genetic studies been started, and not all isolates have been successfully cultured. In contrast to *Rhizobium*, it is possible to routinely obtain nitrogen fixation *ex planta* in pure culture. It has been possible to inoculate the host plant with pure cultures and obtain effective nodule formation.

B. Cyanobacteria

Cyanobacteria form symbioses with the most diverse group of hosts of any other nitrogen-fixing system known. Interestingly, whereas *Rhizobium* can form associations with highly evolved plants, cyanobacteria associate with more primitive plants, such as lichens, liverworts, a pteridophyte (*Azolla*), gymnosperms (e.g., *Cycads*), and angiosperms (i.e., *Gunnera*). The cyanobacterium *Nostoc* forms symbioses with all of the taxa other than the ferns, indicating a potential for genetic manipulation to increase host range. The habitat range is also wide, from tropics to arctic, and includes freshwater, soil, saltwater, and hot springs.

As previously discussed, cyanobacteria are an important source of biologically fixed nitrogen. Whereas the free-living cyanobacteria fix up to 80 kg N_2/ha, the *Azolla–Anabaena* symbiosis can produce three times that amount. This value is based on multiple cropping, but rates in rice paddies have been reliably reported from 1.4–10.5 kg N/ha/day as daily averages over the whole growing season.

Nitrogen fixation by cyanobacteria was first reported in 1889 shortly after the *Rhizobium*–legume symbiosis was described. Until the 1970s, research with this system was intermittent, but, partly due to the recently recognized importance of these eubacteria in agriculture and the environment, research has been steadily accelerating. Cyanobacteria often have been classified by botanists using mainly morphological and anatomical criteria rather than by bacteriologists, who use molecular methods. The recognized families and genera of cyanobacteria containing nitrogen-fixing species are summarized in Table 57.3. Because morphological characteristics can vary greatly depending on growth conditions, only limited physiological, biochemical, and molecular genetic studies have been conducted. These indicate that there is considerably more diversity in this taxon than is accounted for by the traditional taxonomy.

Cyanobacteria possess the requirements for the higher plant type of photosynthesis—water is the ultimate source of reductant and oxygen is evolved with CO_2 fixation via the Calvin cycle. The photosynthetic pigments are located in the outer cell regions.

Among filamentous, heterocyst-forming cyanobacteria, the ability for BNF appears universal. In nonheterocystous filamentous forms and unicellular forms, the ability to fix nitrogen is much less common. The heterocysts, which are found spaced along the cell filaments, appear colorless. These cells cannot fix CO_2 or evolve O_2, but can generate ATP by photophosphorylation and can fix nitrogen aerobically. This is apparently due to a modified thickened cell wall, which interferes with oxygen diffusion. In these organisms then, photosynthesis and nitrogen fixation occur in different cells, which protects nitrogenase from excess oxygen stress.

As stated previously, cyanobacteria associate with almost every group of the plant kingdom, forming symbiotic nitrogen-fixing associations. These associations, however, occur with the more primitive plants. Specificity is less tight than it is with either *Rhizobium* or *Frankia*. Generally, cyanobacteria can fix nitrogen and grow independently of their plant partner. However, when they live symbiotically, photosynthesis is often diminished in favor of increased nitrogen fixation, sufficient for both partners of the symbiosis.

The cyanobacteria usually invade existing normal morphological structures in the host plant, such as leaf cavities, rather than evoking specialized structures such as the nodules caused by *Rhizobium* or *Frankia*, although in some cases, as with the roots of cycads, infection is followed by morphological change.

One reason for the renewed interest in this group of bacteria is the *Azolla–Anabaena* symbiosis. *Azolla* is a free-floating fern commonly found in still waters in temperate and tropical regions and often found in rice paddies. There are seven species. *Azolla* is a remarkable plant and under suitable conditions can double in weight in about four days. The plant forms a symbiotic relationship with the cyanobacterium *Anabaena azollae*. *Azolla* provides nutrients and a protective leaf cavity for *Anabaena*, which in turn provides nitrogen for the fern. Under optimum conditions, this symbiosis results in as much or perhaps more nitrogen fixed than does the *legume–Rhizobium* symbiosis. If inoculated into a paddy and intercropped with rice, this symbiosis can satisfy the nitrogen requirement for the rice. In Southeast Asia, where there is a wet and dry season, *Azolla is* recommended for intercropping with paddy rice, but there are problems because *Azolla* is an extremely efficient scavenger of nutrients and will compete with the rice for phosphate, so careful management is required.

Anabaena azollae symbiotically fixes about triple the amount of nitrogen fixed by free-living *Anabaena*. The high photosynthetic rate of *Azolla* no doubt supplies energy in this symbiosis. The *Anabaena* forms more heterocysts when growing in association with *Azolla* than when growing alone. Similarly, when free-living heterocystic cyanobacteria are grown in a nitrogen-starved environment, they form extra heterocysts. The hypothesis, then, is that, when sufficient energy and nutrients are present for metabolism, *Anabaena* can increase its nitrogen-fixation capacity to maximize growth. It has also been reported by several researchers that nitrogen fixation in this symbiotic system is not repressed by the presence of bound nitrogen.

Although *Azolla* has been cultivated in China and Vietnam for centuries, its use represents a new technology for most areas. Even at a relatively low technological level, there is a great potential for the *Azolla–Anabeana* system as a green manure supplement or animal feed. Little is known about the plant–bacterial interactions, but the fact that *Anabaena* can be induced to increase nitrogen fixation gives promise that the system can be optimized. The lack of repression of nitrogenase by bound nitrogen of *Anabaena* nitrogenase works in favor of a crop rotation following legumes.

C. The *Rhizobium*–legume symbioses

With worldwide distribution, the Leguminosae is one of the largest plant families. It consists of about 750 genera and an estimated 18,000 species. Members of the Leguminosae have traditionally been placed into three distinct subfamilies based on floral differences—Mimosoideae, Caesalpinoideae, and Papilionoideae. Although only about 20% of the total species have been examined for nodulation, these species are representative of all three subfamilies of legumes. Virtually all species within the Mimosoideae and Papilionoideae are nodulated, but about 70% of the species in the subfamily Caesalpinoideae are not nodulated. It is important to note that *Bradyrhizobium* strains of the cowpea type have been shown to form effective symbioses with the nonlegume *Parasponia*, which is a member of the *Ulmaceae*. This is the only verified nitrogen-fixing association between a *Rhizobium* and a nonlegumous plant.

The soil-improving properties of legumes were recognized by ancient agriculturalists. For example, Theophrastus (370–285 BC) in his "Enquiry into Plants" wrote, "Of the other leguminous plants the bean best reinvigorates the ground," and in another section, "Beans are not a burdensome crop to the ground: they even seem to manure it." However, it was only in 1888 that Hellriegel and Wilfarth established positively that atmospheric nitrogen was assimilated by root nodules. This was quickly followed by the experiments of Beijerinck, who used pureculture techniques to isolate the root-nodule bacteria and proved that they were the causative agents of dinitrogen assimilation. He initially called these organisms *Bacillus radicicola*, but they were subsequently named *Rhizobium leguminosarum* by Frank (1889).

Early researchers considered all "rhizobia" to be a single species capable of nodulating all legumes. Extensive cross-testing on various legume hosts led to a taxonomic characterization of these special bacteria based on bacteria–plant cross-inoculation groups, which were defined as "groups of plants within which the root-nodule organisms are mutually interchangeable." The concept of cross-inoculation groupings as taxonomic criteria held for a very long time and has only gradually fallen into disfavor, although much of this philosophy is retained in the taxonomic scheme. There is a wide range in the efficiency of the *Rhizobium*–legume symbiosis. Estimates for the amounts of nitrogen fixed are summarized in Tables 60.6 and 60.7.

As cells without endospores, bacteria of the family *Rhizobiaceae* are normally rod-shaped, motile, with one polar or subpolar flagellum or two to six peritrichous flagella, aerobic, and gram-negative. Considerable extracellular slime is usually produced during growth on carbohydrate-containing media with many carbohydrates used. Some *Rhizobium* and *Agrobacterium* evidently overlap; their DNA homology is very high. 16*S* RNA sequence analysis indicates very similar molecular phylogeny. Also, there is an almost complete lack of distinguishing characteristics other than those that are carried on extrachromosomal elements or plasmids. Some bacterial taxonomists are proposing the amalgamation of *Agrobacterium, Allorhizobium, Rhizobium*, and *Sinorhizobium*.

Traditionally, legume-nodulating bacteria have been recognized as falling into two major phenotypic groups according to growth rate. The term "fast growers" commonly refers to strains associated with alfalfa, clover, bean, and pea because, in culture, these organisms grow much faster (less than one-half the doubling time of slow growers, or <3 hr) than the "slow growers" exemplified by soybean and cowpea rhizobia (generation time >6 hr). Although there is phenotypic and genotypic diversity within these major groupings, and some overlap, numerous studies demonstrated the validity of this approach. *Mesorhizobium* strains, however, have intermediate growth rates between 3 and 6 hr (See Table 60.8.)

The relative fastidiousness of the slow growers has been substantiated by recent studies. Although the major biochemical pathways seem to be similar, evidence suggests that the preferred pathway may be different. 16*S* RNA analysis of the fast- and slow-growing symbionts have confirmed that these groupings

TABLE 60.6 Nitrogen fixed by pulses[a]

Plant	Average	Range
Vicia faba (faba beans)	210	45–552
Pisum sativum (peas)	65	52–77
Lupinus spp. (lupines)	176	145–208
Phaseolus aureus (green gram)	202	63–342
Phaseolus aureus (mung)	61	—
Cajanus cajan (pigeon pea)	224	168–280
Vigna sinensis (cowpea)	198	73–354
Canavalia ensiformis (jack bean)	49	—
Cicer arietinum (chickpea)	103	—
Lens culinaris (lentil)	101	88–114
Arachis hypogaea (peanut)	124	72–124
Cyamopsis tetragonolobus (guar)	130	41–220
Calopogonium mucunoides (calapo)	202	370–450

[a]In kg N/ha.

TABLE 60.7 Nitrogen fixed by tropical and subtropical forage and browse plants, green manure, and shade trees[a]

Plant	Average	Range
Centrosema pubescens	259	126–395
Desmodium intortum	897	—
Leucaena glauca	277	74–584
Lotonosis bainesii	62	—
Sesbania cannabina	542	—
Stylosanthes sp.	124	34–220
Phaseolus atropurpurea	291	—
Mikanea cordata	120	—
Pueraria phaseoloides	99	—
Enterolobium saman	150	—

[a]In kg N/ha/year.

TABLE 60.8 Differences between fast- and slow-growing rhizobia[a]

	Rhizobial type[b]	
Characteristic	Fast-growing	Slow-growing
Generation time	<3 hr	>6 hr
Carbohydrate substrate	Uses pentoses, hexoses, and mono-, di-, and trisaccharides	Uses pentoses and hexoses solely
Metabolic pathways	EMP[2], low activity Strain-specific ED, main pathway TCA, fully active PP present	EMP, low activity ED, main pathway TCA, fully active Hexose cycle present
Flagellation type	Peritrichous	Subpolar
Symbiotic gene location	Plasmids	Chromosome only
Nitrogen-fixing gene location	*nif H, D*, and *K* on same operon	*nif D, K*, and *H* on separate operons
Intrinsic antibiotic resistance	Low	High

[a]*Mesorhizobium* are intermediate phenotypically and phyllogenetically.
[b]ED, Entner–Parners pathway; EMP, Embden–Meyerhoff; PP, pentose phosphate pathway; TCA, tricarboxylic acid cycle.

indeed represent very distinct genetic phyla because the similarity coefficient (S_{AB}) of the RNA is only 0.53. Thus, with modern gene analysis, the fast-and slow-growing *Rhizobium* fall into widely separate groups. Jordan (1982) transferred the slow-growers to the new genus *Bradyrhizobium*. Recent findings using numerical taxonomy, carbohydrate metabolism, antibiotic susceptibilities, serology, DNA hybridization, and RNA analysis all demonstrate the validity of the fast- and slow-growing groupings. *Bradyrhizobium* is transferred to the new family *Bradyrhizobiaceae* in the new edition of *Bergey's Manual of Systematic Bacteriology*. A summary of some of the differences is found in Table 60.8. Thus, whereas in the first edition of *Bergey's Manual*, the slow-growing strains were placed in the new genus *Bradyrhizobium*, they now fall into their own family along with close relatives such as *Afipia* and *Nitrobacter*. The genus *Bradyrhizobium* now comprises three species, *B. japonicum*, *B. elkanii*, and *B. liaoningense*, all of which nodulate soybean (Table 60.9). Other bradyrhizobia are known to occur (e.g., the peanut bradyrhizobia) but these have not been classified. Researchers suggest that until further taxa within the genus are proposed, these should be described with the appropriate host plant given in parentheses [i.e, the peanut rhizobia–*Bradyrhizobium* sp. (*Arachis*)].

The fast-growing legume-nodulating bacteria (sometimes still called "rhizobia,") were all originally placed within the genus *Rhizobium*. There were only a few species, *R. leguminosarum*, *R. meliloti*, *R. loti*, *R. galegae*, and *R. fredii*. The first three species, *R. phaseoli*, *R. trifolii*, and *R. leguminosarum* were amalgamated into the single type species *R. leguminosarum* as biovars. The biovar *phaseoli* had tremendous genetic diversity and a number of new bean-nodulating species have been named, *R. tropici*, *R. etli*, *R. gallicum*, *R. giardinii*, or *R. mongolense*. *Rhizobium fredii* was the first of a series of species consisting of fastgrowing rhizobacteria that effectively nodulate Chinese soybean cultivars, originally thought to be nodulated only by *B. japonicum*. *R. fredii* was reassigned to a new genus, *Sinorhizobium*, in 1988. This controversial new genus also contains *S. meliloti* and several close relatives, which all share a very close phylogenetic relationship with the type species, *S. fredii*. Alfalfa plants are nodulated by *S. fredii, S. meliloti, and S. medicae*. Soybean is nodulated by *B. japonicum*, *B. elkanii*, *B. liaoningense*, *S. fredii*, and *Mesorhizobium tianshanense*. The latter new genus and species are closely related to *Rhizobium loti*, which is now appropriately named *Mesorhizobium*, along with other newly described close relatives because they are intermediate, both in growth rate and molecular phylogeny, between *Bradyrhizobium* and *Rhizobium*. Thus, they now belong to the family Phyllobacteriaceae. *Allorhizobium* is a newly proposed genus for the microsymbiont of an aquatic legume, *Neptunia natans*. The taxonomic scheme is summarized here in a list of the recognized species of legume-nodulating rhizobacteria.

TABLE 60.9 Current taxonomic classification of the rhizobia

Recognized genus	Recognized species
Allorhizobium	*Allorhizobium undicola*
Azorhizobium	*Azorhizobium caulinodans*
Bradyrhizobium	*B. japonicum, B. elkanii, B. liaoningenes*
Mesorhizobium	*M. amorphae, M. ciceri, M. huakuii, M. loti, M. mediterraneum, M. plurifarium, M. tianshanese*
Rhizobium	*R. etli, R. galegae, R. gallicum, R. giardinii, R. hainanense, R. huautlense, R. leguminosarum, R. mongolense, R. tropici*
Sinorhizobium	*S. fredii, S. medicae, S. meliloti, S. saheli, S. terangae, S. xinjiangansis*

1. *Allorhizobium undicola* fixes nitrogen with *Acacia* spp., *Faidherbia* spp., and *Lotus arabicus*. Most *A. undicola* strains are closely related to *Agrobacterium*.
2. *Azorhizobium caulinodans* forms stem nodules on *Sesbania rostrata* and readily fixes nitrogen *ex planta* when microaerobic conditions are provided. It belongs to the family Hypomicrobiaceae.
3. *Bradyrhizobium japonicum* forms root nodules on species of *Glycine* (soybean) and on *Macroptilium atropurpureum* (siratro). Some strains of *B. japonicum* express hydrogenase activity with the soybean host and are hence more efficient in symbiotic nitrogen fixation.
4. *Bradyrhizobium elkanii* normally forms root nodules on species of *Glycine* (soybean), the "non-nodulating" *rj1rj1* mutant soybean that fails to nodulate with *B. japonicum*, black-eyed peas (*Vigna*), mung bean, and *Macroptilium atropurpureum* (siratro). Unlike *B. japonicum*, *B. elkanii* often produces rhizobitoxine-induced chlorosis on sensitive soybean cultivars. Strains of *B. elkanii* often are hydrogenase positive on *Vigna* but not on *Glycine*, suggesting that they possess more symbiotic affinity or compatibility with the former than the latter.
5. *Bradyrhizobium liaoningense* is an extra-slowgrowing, soybean-nodulating *Bradyrhizobium* isolated from alkaline Chinese soils.
6. *Mesorhizbium amorphae* was isolated from nodules of *Amorpha fruticosa* growing in North China.
7. *Mesorhizobium ciceri* was isolated from chickpeas (*Cicer*) grown in uninoculated fields over a wide

geographic range, including Spain, the United States, India, Russia, Turkey, Morocco, and Syria.

8. *Mesorhizobium huakuii* was isolated from *Astragalus sinicus*, a green manure crop grown in rice fields in southern parts of China, Japan, and Korea.
9. *Mesorhizobium loti* nodulates *L. corniculatus*, *L. tenuis*, *L. japonicum*, *L. krylovii*, *L. filicalius*, and *L. schoelleri*.
10. *Mesorhizobium mediterraneum* is exclusively a *Cicer*-nodulating bacterium.
11. *Mesorhizobium tianshanense* isolates were obtained from *Glycyrrhiza pallidiflora*, *G. uralensis*, *Glycine max*, *Sophora alopecuroides*, *Swainsonia salsula*, *Caragara polourensis*, and *Halimodendron holodendron* growing in Xinjiang Region of China. Most of the host plants are wild and indigenous to that region, except *G. max*, which is of course a cultivated crop that originated in northeastern Asia.
12. *Mesorhizobium plurifarium* nodulates *Acacia senegal*, *A. tortilis*, *A. nilotica*, *A. seyal*, *Leucaena leucocephala*, and *Neptunia oleracea*, but most strains do not nodulate *Sesbania rostrata*, *S. pubescens*, *S. grandiflora*, *Ononis repens*, or *Lotus corniculatus*.
13. *Rhizobium etli* nodulates and fixes nitrogen in association with *P. vulgaris* exclusively; it includes nonsymbiotic strains.
14. *Rhizobium galegae* nodulates *Galega orientalis* and *Galega officinalis* and is specific to this plant genus.
15. *Rhizobium gallicum* nodulates and fixes nitrogen in association with *Phaseolus* spp., *Leucaena leucocephala*, *Macroptilium atropurpureum*, and *Onobrychis viciifolia*.
16. *Rhizobium giardinii* nodulates *Phaseolus* spp., *Leucaena leucocephala*, and *Macroptilium atropurpureum*.
17. *Rhizobium hainanense* is found in nodules of *Desmodium sinuatum* or *Stylosanthes guianensis*, *Centrosema pubescens*, *Desmodium triquetrum*, *D. gyroides*, *D. sinatum*, *D. heterophyllum*, *Tephrosia candida*, *Acacia sinicus*, *Arachis hypogaea*, *Zornia diphylla*, *Uraria crinita, and Macroptilium lathyroides*.
18. *Rhizobium huautlense* nodulates *Sesbania herbacea*, *S. rostrata*, and *Leucaena leucocephala*.
19. *Rhizobium leguminosarum* nodulates with some, but not necessarily all *Pisum* spp., *Lathyrus spp.*, *Vicia* spp., *Lens* spp., temperate species of *Phaseolus (P. vulgaris, P. angustifolius*, and *P. multiflorus)*, and *Trifolium* spp.
20. *Rhizobium mongolense* was recently isolated from *Medicago ruthenica*, but it also nodulates *Phaseolus vulgaris*. It is a very close relative of *Rhizobium gallicum*.
21. *Rhizobium tropici* forms nodules on *Phaseolus vulgaris* and Leucaena spp. The type strain, CFN299, nodulates *Amorpha fruticosa*.
22. *Sinorhizobium fredii* effectively nodulates *Glycine max cv.* "Peking," *Glycine soja*, *Vigna unguiculata, and Cajanus cajan*. Also nodulates alfalfa. In 1985, Dowdle and Bohlool reported new strains that are symbiotically competent with North American cultivars of soybean. Their molecular phylogeny, however, has not been determined yet.
23. *Sinorhizobium medicae* is alfalfa-nodulating, with a close phylogenetic relationship to *S. meliloti*.
24. *Sinorhizobium meliloti* forms nitrogen-fixing nodules on *Melilotus, Medicago, and Trigonella*.
25. *Sinorhizobium saheli* is found in nodules of *Sesbania* spp. growing in the Sahel and can nodulate *Acacia seyal*, *Leucaena leucocephala*, and *Neptunia oleracea*.
26. *Sinorhizobium terangae* is also found in nodules of *Acacia* spp., *Sesbania* spp., *Leucaena leucocephala*, and *Neptunia oleracea*.
27. *Sinorhizobium xinjiangense* nodulates soybean and is a close relative of *S. fredii*.

The taxonomy of the nitrogen-fixing bacteria is in a dynamic state of change. As molecular information accumulates, the cataloging of phenotypic data has not kept pace and further revision, and "reversions" will be necessary. New approaches to classification are needed because the scheme, unfortunately, does not function to allow the identification of isolates without DNA-sequence analysis.

1. *Nitrogen fixation*

The nitrogenase complex is a highly conserved enzyme system and, as stated earlier, is basically common to all of the dinitrogen-fixing prokaryotes. The evidence is conclusive that there are differences in location of nitrogenase genes of *Rhizobium* and *Bradyrhizobium*. In all of the fast-growing *Rhizobium* species, which have a chromosome size of about 3500 kb, the structural *nifH, D*, and *K* genes are localized on extremely large plasmids or megaplasmids. In *Bradyrhizobium*, *nif* genes have been mapped on the 8700-kb chromosome. In *Mesorhizobium*, on the other hand, the *nif* genes are located on the chromosome in some species and on megaplasmids in other species.

Plasmids and megaplasmids, present in a wide variety of the nitrogen-fixing bacteria, control many phenotypic and genetic characteristics of the bacterial cells. Those in *Rhizobium*, *Sinorhizobium*, and *Azorhizobium* species carry genes controlling symbiotic functions, which are clustered on a single large plasmid termed symbiotic plasmid, "pSym." Two or more plasmids that carry genes controlling symbiotic functions, *nod*, *fix* and the nitrogenase structural (*nifHDK*) genes, have been found in certain strains of the nitrogen-fixing species. In addition to symbiotic plasmid(s), rhizobia strains may carry 1–10 plasmids that range in size from 30 to more than 1000 MDa. These plasmids are highly stable and have beneficial roles in the soil environment, and plasmid profile analysis is sometimes used to discriminate among *Rhizobium* strains.

The analysis and comparison of *nif* DNA in the fast- and slow-growing "rhizobia" have established the affinity coefficient (S_{AB}) for the nucleotide sequence (*nifH*, *nifD*, and *nifK*) from the two groups. The same analysis was done comparing amino acid sequences of the nitrogenase Fe and FeMo protein polypeptides. A considerable sequence conservation reflects the structural requirements of the nitrogenase proteins for catalytic functions. The S_{AB} *nif* values (based on *nifH* sequences) between fast- and slow-growing organisms indicated that these are almost as distant from each other as they are from other gram-negative organisms. The results suggest that *nif* genes evolved in a manner similar to the bacteria that carry them rather than by a more recent lateral distribution of *nif* genes among microorganisms. Again, the phylogenetic difference between fast and slow groups is apparent. Although this general system is common to all of nitrogen-fixing prokaryotes, several concomitant alternative systems have been described; but these evidently lack biological significance and they have not been shown to be present or active in the "rhizobia."

2. *Nodule formation*

Nodule initiation and subsequent maturation is an interactive process involving the eukaryotic host legume and the prokaryotic *Rhizobium*. The process is complex, resulting in biochemical and morphological changes in both symbionts and leading to the capacity to reduce atmospheric nitrogen. Initially, the proper *Rhizobium* species proliferates in the root zone of a temperate leguminous plant and becomes attracted and attached to the root hair. A chemotactic response attracts the bacteria to the root surface. At the surface, the bacteria secrete Nod factors, which are certain chitolipooligosaccarides (CLOS) that alter the growth of epidermal root hairs so that they are deformed. CLOS molecules chemically induce nodule organogenesis in extremely low concentrations ($\sim 10^{-10}$ M). In some tropical legumes such as peanuts (*Arachis*), root hairs are not the primary invasion sites. However, the alternative invasion process, "crack entry," has been well-documented because infection occurs at the site of lateralroot emergence.

The root hair infection process consists of several events leading to nodule formation: (1) recognition by the "rhizobia" of the legume, (2) attachment to the root hair, (3) curling of the root hair, (4) roothair infection by the bacteria, (5) formation of an infection thread, (6) nodule initiation, and (7) transformation of the vegetative cells in the nodules to enlarged pleomorphic forms, called bacteroids, which fix nitrogen.

Based on morphology, there are two kinds of nodules, determinative and indeterminative. In general, indeterminative nodules are formed by fast-growing "rhizobia" and are characterized by a defined meristem during nodule growth. Determinative nodules arise from cortex tissue. Legumes nodulated by *Bradyrhizobium* form determinative nodules close to the endodermis, which is near the xylem poles in the root.

The formation of nodules on legumes is the result of a coordinated development involving many plant and bacterial genes. Studies of the nodulation *(nod)* genes of rhizobia have depended on the development of molecular genetic tools. Many of the genes involved in the nodulation process have been located and identified.

Legume roots grown axenically do not appear as morphologically distinctive from other plant roots, so that the abilities of these plants to respond to microbial signals and then alter their metabolism to form nodules are not explained by morphology alone. It has been conclusively demonstrated that genetic information from both symbionts controls nodulation and the host range of nodulation by a *Rhizobium* species. Metabolically, there are three types of nodules (often termed effective, inefficient, and ineffective). Effective nodules contain a high density of bacteria actively fixing dinitrogen. Inefficient nodules may contain a similar density of the bacteria, but only a relatively low level of fixed dinitrogen results from the symbiosis. Ineffective symbiosis occurs with bacteria not able to nodulate or fix nitrogen normally. Because the regulatory roles of the plant and bacterial genes in nitrogen fixation have not been generally elucidated, the reasons for differential nitrogen-fixing ability of nodules continues to be unclear. There is, however, an already observed compatibility in legume host–*Rhizobium* interaction that can be technologically exploited to enhance dinitrogen fixation. Such an interaction makes it possible to optimize dinitrogen fixation under field conditions in a cultivar through inoculation with an effective *Rhizobium* strain.

Nodulation genes are defined by their effect on the bacteria's ability to generate the nodulation process on the proper legume host. Because most individual species of legume-nodulating rhizobacteria can each nodulate a limited number of host legumes and plant genes also limit the symbiosis, it follows that a recognition exists between bacteria and host. Thus, there are two types of *nod* genes, a common *nod* region, which consists of a structurally and functionally conserved cluster of genes; and host-specific nodulation genes, which cannot complement nodulation defects in other genera or species. *Nod* genes have been studied in varying degrees in different species (usually in the fast-growing rhizobia). In these organisms, four genes have been identified in two transcription units (*nodD* and *nodABC*). Two additional genes, apparently on the same transcriptional unit, *nodI* and *nodJ*, have been identified. Genetic maps of the common *nod* cluster, drawing together the information from many sources, have been published. The *nodABC* appears to be functionally interchangeable among all rhizobia, and mutations in these genes cause complete nodulation failure. These genes are involved in cell division and roothair deformation. The *nodIJ* genes cause a delay in the appearance of nodules.

The second group of *nod* genes are termed "host-specific." These genes are not conserved because alleles from various rhizobia cannot substitute for each other on different hosts. Bacteria carrying mutations in these host-specific genes cause abnormal root-hair reactions. Many genetic *nod* loci have been identified in a variety of the symbiotic nitrogen-fixing species. The list includes at least 15 *nod* genes. In many cases, these have been cloned and sequenced and the gene product associated with a step in nodulation. Although the amino acid sequences of many of the nodulation-gene products have been described, the biochemical functions of these genes have not been fully determined. Possible exceptions are the *nodD* genes, which are positive gene regulators.

The centenary of this first demonstration of biological nitrogen fixation occurred during 1986. During that period, many papers were published expanding the knowledge base, both basic and applied. In 1991, the U.S. Department of Agriculture patented an improved soybean inoculant *Bradyrhizobium japonicum* strain that results in a significant increase in growth and soybean yield. Since then, this improved strain has been commercially produced. Currently, the subject of BNF is of great practical importance because of the use of fossil fuels in the manufacture of nitrogenous fertilizers. The increased scarcity and higher costs of fossil fuel have made it important to optimize biological nitrogen fixation as an alternative to chemical nitrogen. In addition, the increasing usage of nitrogen fertilizer has resulted in unacceptable levels of water pollution, which occurs only to much lesser extent when the biologically fixed forms of nitrogen are used. With the additional research capabilities resulting from the developing field of biotechnology, it is evident that interest in this field will continue and that we may reach a level of accumulated knowledge that will allow full use of BNF as an alternative to the Haber–Bosch chemical industrial process.

3. *Rhizobiophages*

Rhizobiophages occur commonly in the rhizosphere of legumes and are often associated with susceptible *Bradyrhizobium, Rhizobium,* or *Sinorhizobium* strains. They reduce rhizobial populations in soils and negatively affect the nitrogen-fixing abilities of these bacteria with the host legume plant. Rhizobiophages can be used to distinguish between rhizobial strains through "phage typing." Furthermore, rhizobiophages are potential biocontrol agents useful for reducing the number of susceptible rhizobial cells in the soils, thus decreasing nodule occupancy by the undesirable indigenous bacterial strain, and thereby increasing the nodule occupancy by superior strains used as inoculant. The use of specific bacterial viruses or bacteriophages as biocontrol agents requires the identification of symbiotically competent, rhizobiophage-resistant *Rhizobium* or *Bradyrhizobium* strains that have a demonstrated ability to promote the growth and yield of their specific legume hosts.

BIBLIOGRAPHY

Boland, G. J., and Kuykendall, L. D. (eds) (1998). "Plant-Microbe Interactions and Biological Control." Marcel-Dekker, New York.

Dilworth, M. J., and Glenn, A. R. (eds) (1991). "Biology and Biochemistry of Nitrogen Fixation." Elsevier, Amsterdam.

Elkan, G. H. (ed.) (1987). "Symbiotic Nitrogen Fixation Technology." Marcel Dekker, New York.

Elkan, G. H., and Upchurch, R. G. (eds) (1997). "Current Issues in Symbiotic Nitrogen Fixation." Kluwer, Dordrecht, The Netherlands.

Elmerich, C., Kondorsi, A, and Newton, W. E. (eds) (1998). "Biological Nitrogenase Fixation for the 21st Century." Kluwer, Dordrecht, The Netherlands.

Galibert, F. *et al.* (2001). The composite genome of the legume symbiont *Sinorhizobium meliloti*. *Science* **293**, 668–672.

Gresshof, P. M. (ed.) (1990). "Molecular Biology of Symbiotic Nitrogen Fixation." CRC Press, Roca Raton, FL.

Halbleib, C. M., and Ludden, P.W. (2000). Regulation of biological nitrogen fixation. *J. Nutr.* **130**, 1081–1084.

Hennecke, H., and Verma, D. P. S. (eds) (1990). "Advances in Molecular Genelic of Plant-Microbe Interaction." Kluwer, Dordrecht, The Netherlands.

Poole, P., and Allaway, D. (2000). Carbon and nitrogen metabolism *Rhizobium*. *Adv. Microb. Physiol.* **43**, 117–163.

Postgate, J. R. (1998). "Nitrogen Fixation." University of Cambridge, Cambridge.

Rees, D. C., and Howard J. B. (2000). Nitrogenase: standing at the crossroads. *Curr. Opin. Chem. Biol.* **4**, 559–566.

Somasegaran, P., and Hoben, H. J. (eds) (1994). "Handbook for Rhizobia." Springer-Verlag, New York.

Spaink, H. P., Kondorosi, A.; and Hooykaas, P. J. J. (eds) (1998). "The *Rhizobiaceae*: Molecular Biology of Model Plant-Associated Bacteria." Kluwer, Dordrecht, The Netherlands.

Sprent, J. I., and Sprent, P. (eds) (1990). "Nitrogen Fixing Organisms." Chapman and Hall, London.

Stacy, G., Burris, R., and Evans, H. J. (eds) (1992). "Biological Nitrogen Fixation." Chapman and Hall, New York.

WEBSITES

Website of the University of Minnesota Rhizobium Research Laboratory
http://www.rhizobium.umn.edu/

Issues in Ecology. Website of the Ecological Society of America
http://esa.sdsc.edu/issues.htm

61

Nodule formation in legumes

Peter H. Graham
University of Minnesota

GLOSSARY

bacteroids Cells of rhizobia from within the nodule that have undergone surface change, are often swollen and irregular in shape, and express nitrogenase.

cross-inoculation The ability of bacterial strains from two or more different legumes to each produce nodules on one another's host(s).

endocytosis A process releasing rhizobia from the infection thread, but surrounding them with plant-derived membrane material that shields them from the host, limiting host defense responses.

ineffective Host rhizobial combinations limited in the ability to fix nitrogen (N_2).

infection thread A plant-derived tube through which rhizobia move as they pass down the root hair or between cells in the root cortex.

inoculation The application of artificially cultured rhizobia to legume seed or soil in order to improve nodulation and N_2 fixation.

nodule A gall-like structure that develops on the root or stem of legumes following infection by compatible rhizobia.

nodulins Gene products expressed in host tissue during nodulation and N_2 fixation.

peribacteroid membrane A host-derived membrane that surrounds rhizobial cells following their release into the cells of the host.

promiscuity The ability of some rhizobia to nodulate with a range of different legume species.

symbiosis An association between two organisms in which each derives benefit.

Symbiotic N_2-Fixing Bacteria, collectively known as rhizobia, can infect leguminous plants, forming stem or root nodules. The bacteria derive energy from the host for growth and N_2 fixation and are protected from external stresses; the host has access to a form of N it could not otherwise use. Rates of N_2 fixation vary, but commonly range from 50–200 kg N_2 fixed per hectare per year reducing the plant's need for soil or fertilizer N. Collectively, nodulated crop plants fix 32–53 Tg N/year, with significant additional N_2 fixation by legumes in natural terrestrial ecosystems.

I. INTRODUCTION

Nitrogen needs in agriculture can only grow. If the emphasis is toward fertilizer N, increases to around 134 Tg N fertilizer produced year per year can be anticipated by the year 2020. At this level, N deposition and environmental N pollution in developed countries would be widespread, with fossil fuel consumed at an alarming rate. In contrast subsistence farmers in developing countries would not be able to afford or obtain fertilizer N. A more sustainable approach would be to use fertilizer N mainly for high-value crops, and to place greater dependence on N_2 fixation in other crop and pasture systems. In Brazil, an emphasis on symbiotic N_2 fixation has permitted a major increase in soybean production, while saving

The Desk Encyclopedia of Microbiology
ISBN: 0-12-621361-5

more than \$1.8 billion in fertilizer N costs annually. Field application of some recent advances in inoculation, nodulation, and N_2 fixation could benefit many crops and ecosystems.

II. EVOLUTION IN THE NODULATION OF LEGUMES

Nodulation and symbiotic N_2 fixation are restricted to a clade of plants that includes both legume and actinorhizal species. Not all legumes bear nodules; only 23% of Caesalpinioideae are nodulated, though 90% of Mimosoideae and 97% of Papilionoideae nodulate. The ability to nodulate could have arisen independently on several occasions in the evolution of legumes, including in the genus *Chamaecrista* (see Fig. 61.1). This genus has also attracted attention because some species retain the rhizobia within infection threads during symbiosis, whereas in others they are released into cells of the host. Because legumes such as bean nodulate with several distinct species of rhizobia, other studies have concluded that multiple evolution in symbiosis is unlikely. Instead, a continuum is proposed from nonnodulation, through rare or mixed nodulation capacity in each subfamily, to the abundant nodulation of most Papilionoideae and Mimosoideae. Bryan *et al.* (1996) even suggest that unnodulated legumes may fix N_2 and describe cells resembling bacteroids within the roots of *Adenanthera, Ceratonia, Bauhinia, Gleditsia*, and *Peltophorum*.

The ability of humans to manipulate both the host and the rhizobia has accelerated evolutionary change in some symbioses. Thus, trade in *Phaseolus* beans during the colonial period introduced this crop to Brazil and Eastern Africa, where acid-soil conditions limited survival of the normal bean microsymbiont *R. etli*. A result is that beans in these regions are now often associated with a more acid-tolerant *Rhizobium* spp., *R. tropici*.

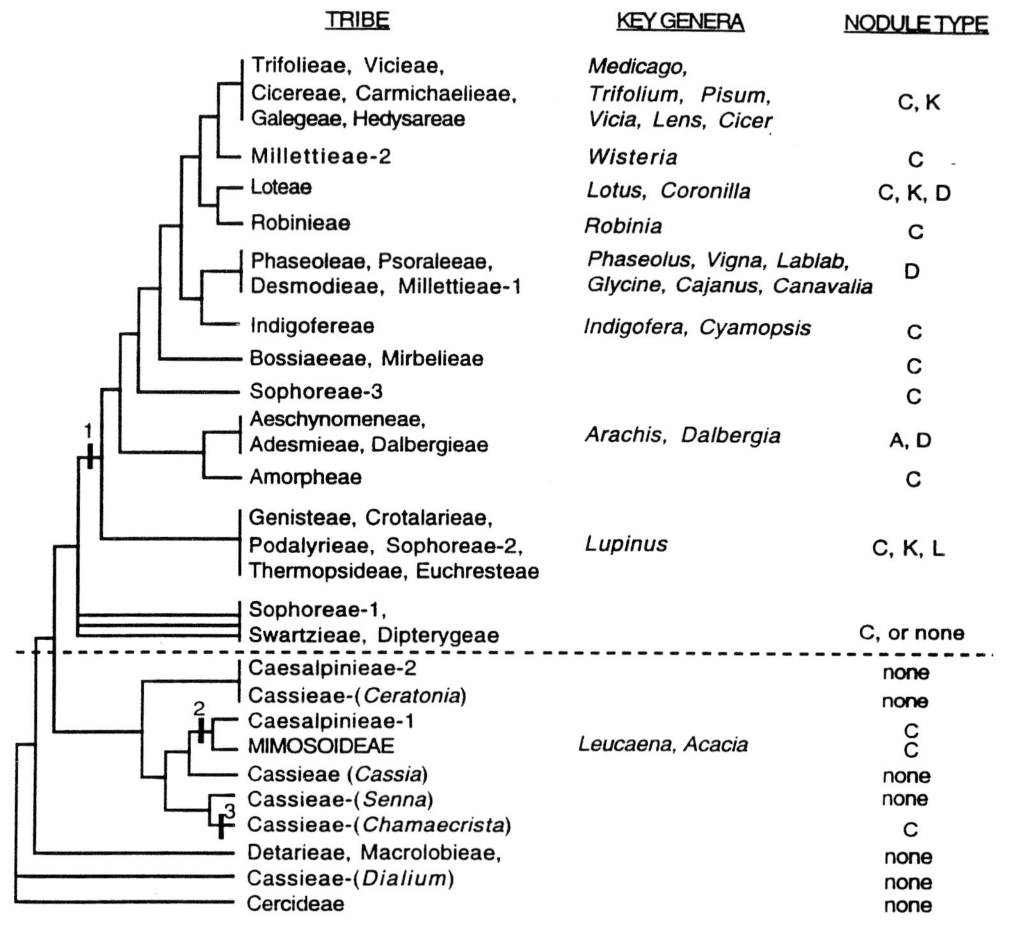

FIGURE 61.1 Phylogeny of the Leguminosae based on *rbc* L DNA-sequence data. (Modified from Doyle and Doyle (1997) and printed with permission.) Tribes appearing more than once on the tree are polyphyletic. The numbered rectangles indicate stages in evolution beyond which most species are nodulated. C and K, A and D, and L refer to indeterminate, determinate, and lupinoid (collar) nodule shapes, respectively.

Microorganisms producing root and stem nodules on legumes are divided into five genera (*Rhizobium, Azorhizobium, Mesorhizobium, Sinorhizobium,* and *Bradyrhizobium*), with separation of a sixth genus (*Allorhizobium*) possible. More than 30 species are now recognized, but rhizobial taxonomy is far from stable, and species epithets are constantly under change. The confusing taxonomic situation has not been improved by the recent discovery of methyl-oxidizing bacteria that can nodulate legumes such as *Sesbania, Crotalaria* and *Lotononis* (Moulin *et al.*, 2001). An updated list of species of root-nodule bacteria is maintained at *http://www.rhizobium.umn.edu.*

III. MECHANISMS OF INFECTION

A. Types of infection

Rhizobia can infect their hosts and induce root- or stem-nodule formation by several mechanisms, the most common of which are:

1. Root hair penetration and infection-thread formation as occurs in clovers, beans and soybeans (Hirsch, 1992).
2. Entry via wounds or sites of lateral-root emergence (Boogerd and van Rossum, 1997), as occurs in peanut and *Stylosanthes*. Rhizobia spread intercellularly or cause infected cells to collapse, colonizing the space.
3. Infection via cavities surrounding adventitious-root primordia on the stems of *Sesbania, Aeschynomene, Neptunia*, or *Discolobium* (Boivin *et al.*, 1997).

Rhizobia can also nodulate the nonlegume *Parasponia*, a process involving the formation of callus-like prenodules, intercellular spread of rhizobia via infection threads, and merging of prenodule tissue with a modified lateral root.

The same rhizobial isolate may infect one host (e.g. *Macroptilium*) via root hair penetration, but another (e.g. *Arachis*) through crack entry and intercellular spread. Similarly, one organism may produce both stem and root nodules, with the rates of N_2 fixation achieved in each not necessarily the same.

B. Visible changes during root-hair infection

The Fahraeus slide technique (Fahraeus, 1957) and the root-tip marking procedure (Bhuvaneswari *et al.*, 1981) are seminal to our knowledge of root-hair infection. One allowed repeated observation of the infection process in small-seeded legumes such as clovers; the other showed differences in the susceptibility of immature and mature root hairs to infection, and allowed research to focus on those parts of the root where infection was most common.

Compatible rhizobia begin to attach to root hairs of their host within minutes of inoculation, and attachment increases over time. Attached cells cap the root-hair tip, and are often oriented end-on to their host. Adhesion is initially mediated by the calcium (Ca)-binding protein rhicadhesin, or by plant lectins, with subsequent bonding via production of cellulose fibrils.

Rhizobia cause localized hydrolysis of the root-hair cell wall, and promote invagination of the host plasma membrane, with additional plant-cell material deposited about them as they infect. The enzymes involved in hydrolysis are cell bound and difficult to quantify, and several differ from those normally associated with plant infection. Rhizobial penetration causes root-hair growth at the point of infection to cease, and leads to root-hair curling, first visible some 6–18 h after inoculation. The proportion of root hairs infected is low, and the percentage of these giving rise to nodules, highly variable. An electron micrograph taken at this stage in infection shows the root hair curled into a shepherd's crook, penetration by several rhizobia, and the beginnings of an infection thread (see Fig. 61.2a). Rhizobia, still encased within a plant-derived infection thread, move down the root hair to the root cortex (Fig. 61.2b). Cell division in the root cortex precedes their arrival and gives rise to the nodule primordia, and in some legumes to an uninfected meristematic region. The spread of the infection thread among cells of the nodule primordium follows, with rhizobia released into their host by endocytosis. Rhizobia never

(a)

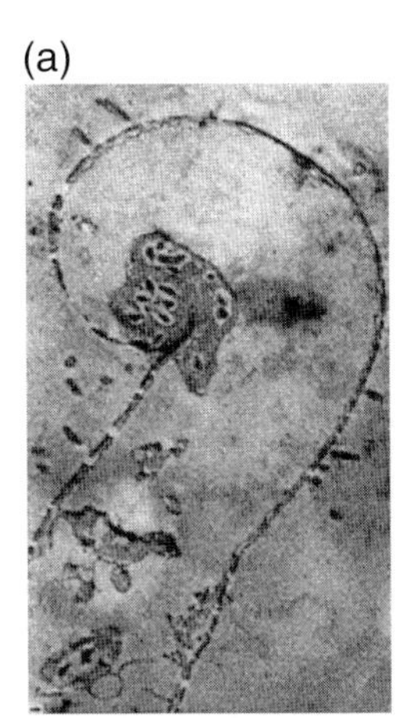

(b)

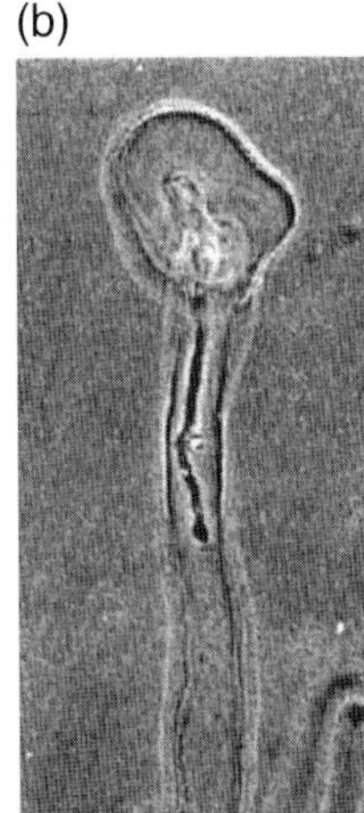

FIGURE 61.2 (a) Root hair deformation, curling, and infection—an early stage in the nodulation of clover by *R. leguminosarum bv trifolii*. (From Sahlman and Fahraeus, 1963 with permission.) (b) Rhizobia contained in the infection thread move down the root hair in the direction of the root cortex. (From Fahraeus, 1957, with permission.)

gain free intracellular access. They are initially confined by the infection thread and later surrounded by the host-derived peribacteroid membrane.

Nodulation is usually evident 6–18 days after inoculation, but this varies with the strain and cultivar used, the inoculant density, and temperature. Initially, nodulation is heaviest in the crown of the root, with secondary flushes of nodules on lateral roots as the first-formed nodules senesce.

C. The molecular basis for nodule formation

Molecular studies of infection received impetus from the demonstration that most nodulation genes in *Rhizobium* were plasmid-borne. Mapping studies followed, and have led over time to sequencing of the entire symbiotic plasmid for the promiscuous strain NGR234 (Freiberg *et al.*, 1997), and more recently to nucleotide sequence analysis for both the chromosome and plasmids of *Sinorhizobium meliloti* 1021 (*http://cmgm.stanford.edu/~mbarnett/genome.htm*). More than 50 nodulation (*nod-*) genes have now been identified among the rhizobia.

Some of these genes have been found in all rhizobia (with genes from one strain correcting mutations in another), others occur only in strains that nodulate a particular host(s). Only one nodulation gene, *nod*D, is always expressed; others are only expressed in the presence of a suitable host legume. Flavonoids excreted from the legume root are now known to induce *nod*-gene expression, with legume species varying in the mix of flavonoids produced by each, and rhizobia responding differently to specific flavonoids. Thus naringenin and genistein stimulate *nod* gene expression in *B. japonicum*, whereas luteolin is required for *S. meliloti*. Because flavonoids may potentiate rhizobial infection, but be limiting in specific cultivars or at low temperature, some flavonoid preparations are now sold commercially.

Bacterial genes involved in nodulation and N_2 fixation include genes for the synthesis and regulation of lipochitooligosaccharide nodulation factors (Long, 1996), as well as genes for exopolysaccharide, lipopolysaccharide and nitrogenase synthesis, carbon metabolism and bacteroid maturation (Oke and Long, 1999). A number of genes have regulatory functions. Nodulation (nod-) factors in the rhizobia all have the same core structure (coded for by the common nodulation genes), but vary in the side chains each carries, affecting host range (see Fig. 61.3). Strains may produce several nod factors differing slightly in composition. They act as powerful morphogens that at 10^{-9} to 10^{-11} M can deform root hairs and initiate the cortical cell division typical of nodule initiation. Table 61.1 lists some of the changes effected by these substances and the time frame in which they occur. High-affinity legume-receptor molecules for nod factors have been postulated but not identified; John *et al.* (1997) discuss the broader role of this class of compounds in plant growth regulation.

Roles for plant lectins, rhizobial extracellular polysaccharides (EPS), and lipopolysaccharides (LPS) in nodule formation continue to be explored. Lectins are nonenzymatic carbohydrate-binding proteins occurring in different forms throughout the plant. Cross-bridging between root-hair lectins and the rhizobial surface has been proposed for many years but is not fully accepted. More definitive evidence of lectin involvement in nodulation was the recent demonstration of host-range modification in white clover plants transformed with the gene for pea seed lectin. The possibility that lectins bind rhizobial nod factors has been discounted, and a more indirect effect of lectins on cell responsiveness to nod factors of all types has been proposed.

Rhizobial EPS mutants produce nodules on alfalfa devoid of rhizobia and ineffective. In contrast, hosts giving rise to determinate nodules (see Section IV.A) generally nodulate normally with such mutants. An exception is soybean, in which EPS mutant strains are delayed in nodule formation and induce a marked

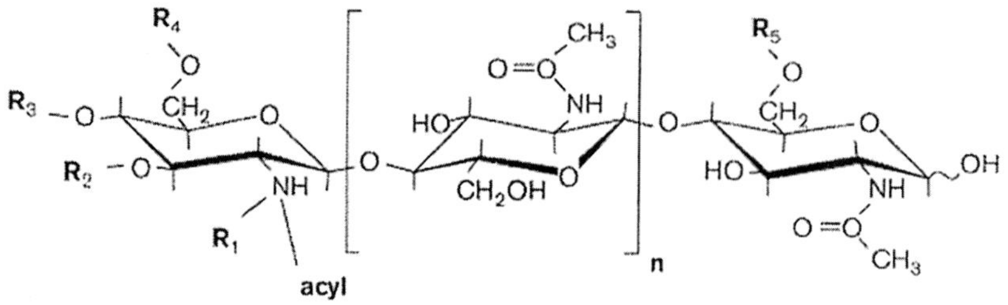

FIGURE 61.3 The general structure of Nod factors elicited by rhizobia. In this structure $n = 2$ or 3, and the substitutions possible at each indicated position are R_1 for H or methyl; R_2 and R_3 for H, carbomoyl, or 4-*O*-methyl carbomoyl; R_4 for H, acetyl, or carbomoyl; R_5 for H, sulfate, acetyl 2-*O*-methyl fucose, 4-*O*-sulfo-2-*O*-methylfucose, 3-*O*-acetyl-2-*O*-methyl-fucose, 4-*O*-acetyl-fucose, or D-arabinose; and acyl for C16–C20. Modified from Schultze and Kondorosi (1996) and used with permission.

TABLE 61.1 Effects of purified Nod factor on legumes

Plant response	Time after inoculation	Minimum concentration (M)
Depolarization of plasma membrane	15 s	10^{-11}
Increase in root hair pH	15 s	10^{-10}
Ca^{2+} spiking in root hairs	10 min	10^{-9}
Expression of early nodulin genes	6–24 h	10^{-12}
Root hair deformation	~18 h	10^{-12}
Cortical cell divisions	Days	10^{-9}
Empty nodule formation	Weeks	10^{-9}

plant defense response. For pea and alfalfa rhizobia defective in EPS synthesis, micromolar quantities of wild-type EPS, or coinoculation with a *nod*$^-$ *eps*$^+$ strain, can restore some degree of invasive ability. Because the quantity of EPS needed for nodulation is small, it has been suggested that it acts as a signal molecule for infection-thread growth beyond the epidermal cell layer.

Rhizobial LPS production is essential for normal nodule development, but reponses obtained with LPS mutants again vary with plant species. LPS mutants of *R. leguminosarum* failed to properly colonize the nodule tissue in peas, induced marked host defense responses, and were ineffective. Similar mutants of *B. japonicum* and *R. etli* were deficient in infection-thread formation, and formed nodules that were small, white, and devoid of rhizobia. In both cases, a role for rhizobial LPS in masking the elicitors of plant defense response has been proposed.

D. Nodule-specific gene expression

The interaction of host and rhizobia is accompanied by the formation of nodule-specific proteins or nodulins, originally defined as specific to infected root hairs or nodule tissue, and needed for nodule formation and function. Close to 350 such nodulins have been identified from cDNA nodule libraries in *Medicago* alone. The concept of nodulins as nodule specific has blurred over time. Many nodulins also occur in other tissues; a number have also been identified in actinorhizal or mycorrhizal symbioses. Pea mutants that neither nodulate nor form mycorrhizal associations are a further indicator of overlap in symbiosis. Some nodulins may be expressed following inoculation with either Nod factor or rhizobia; others are only expressed in the presence of rhizobia.

Nodulin expression can vary temporally and spatially. Early nodulins are involved in infection or nodule development, and may be expressed within 6 h of inoculation. Expression may be transient or, as with nodulins associated with new cell infection, may extend over the life of the nodule. Well-studied early nodulins include Enod 40, first detected in the root pericycle opposite incipient nodule primordia and postulated to function in hormone perception; Enod 12, expressed in cells adjacent to growing infection threads; and Enod 2, located in the parenchyma and postulated to have a role in regulation of oxygen supply to bacteroids.

Later nodulins are expressed just before or with the onset of N_2 fixation and are related to nodule function, carbon and nitrogen metabolism, and O_2 transport. Examples include leghemoglobin, PEP carboxylase, and enzymes involved in allantoic acid synthesis and ammonia assimilation (see Fig. 61.4). More detailed information is provided by Pawlowski (1997).

Establishment of *Medicago truncatula* and *Lotus japonicus* as model plants for genome analysis, and particularly for the molecular genetic study of root nodule and mycorrhizal symbioses (Stougaard, 2001), must accelerate progress in this area.

E. Plant defense response and symbiosis

Plants respond to invasion by rapid localized tissue necrosis, synthesis of phytoalexins or hydrolytic enzymes, and by lignification of plant tissue. Rhizobia usually avoid such responses, though enzymes involved in the flavonoid and isoflavonoid pathway leading to phytoalexin synthesis have been found in a number of legumes soon after inoculation, and glyceollin I levels in soybean show transient increases soon after infection. Chalcone synthase levels in root hairs rise soon after inoculation, but decline thereafter, whereas those in the root increase. This response appears to follow the rhizobia into the root cortex and to be a part of normal nodule development. Enhanced flavonoid levels may be needed for hormone production in the nodule, to induce cytoskeletal rearrangement, or to influence *nod* gene expression in the rhizosphere. In contrast, transient phytoalexin production during nodulation could stem from the brief exposure of the rhizobia during entry, with the host response subsequently muted as the rhizobia become enclosed within the infection thread and peribacteroid membrane. A more typical host defense response with elevated phytoalexin levels has been noted in specific soybean–*Bradyrhizobium* combinations, and following the abortion of infection threads or inoculation with EPS-defective mutants. The peribacteroid membrane could serve to protect rhizobia from host recognition, especially in organisms that spread intercellularly. Deterioration of this membrane could be a factor contributing to declining N_2 fixation in senescent nodules. Salicylic acid accumulation is an additional defense response shown in alfalfa.

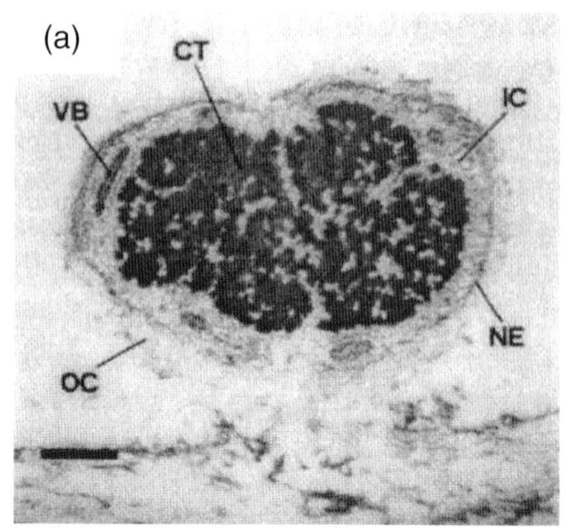

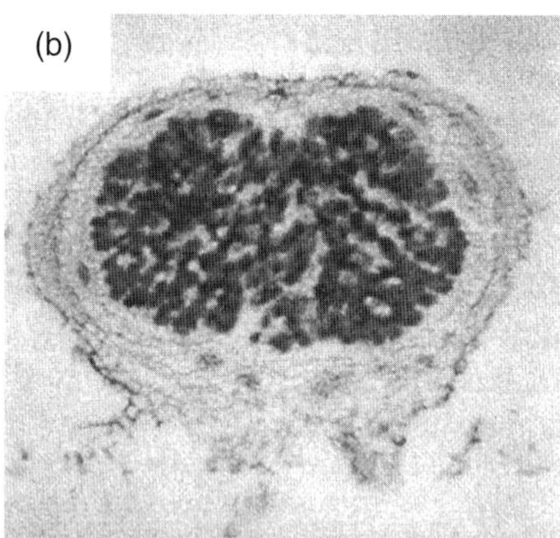

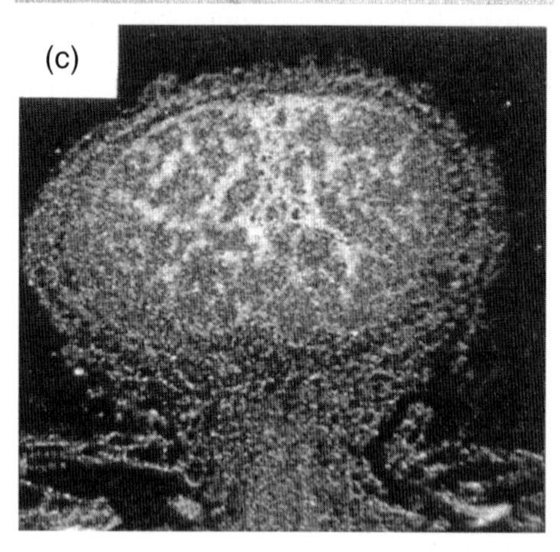

FIGURE 61.4 Sections of mature bean nodules following staining with toluidine blue and *in situ* hybridization with leghemoglobin or uricase-II probes. (a) Bright-field micrograph stained with toluidine blue, showing the determinate nodule structure, central tissue (CT), vascular bundles (VB), nodule endodermis (NE), and inner (IC) and outer (OC) cortex. Bar equals 200 μm. (b) Bright-field micrograph showing *in situ* localization of the leghemoglobin transcript. Hybridization signals are evident as dark dots in invaded cells. (c) Dark-field micrographs of the *in situ* localization of uricase-II transcripts. Silver grains are visible as white dots in uninvaded nodule cells. From Tate *et al.* (1994), with permission.

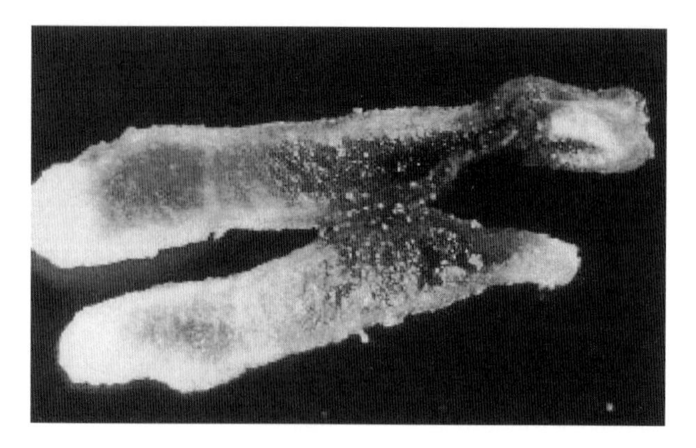

FIGURE 61.5 The internal organization of an indeterminate nodule from *Medicago sativa*, showing a white uninfected meristematic area on the left, with progression through stages of infection by *S. meliloti*, active N_2 fixation, and leghemoglobin production, to the beginnings of nodule senescence on the right. Photo courtesy of C. P. Vance, with permission.

IV. CHARACTERISTICS OF THE LEGUME NODULE

A. Nodule shape

Nodule shape is a characteristic of the host plant and is regulated by the pattern of cortical cell division following infection. Rounded, determinate nodules, such as occur on *Phaseolus* and *Glycine*, lack a persistent nodule meristem (see Fig. 61.4), whereas nodules on *Medicago* and *Trifolium*, in which the meristem is persistent, are elongate, sometimes lobed and indeterminate (see Fig. 61.5). Hirsch (1992) contrasts the points of structure and function in each nodule type. More rarely, collar nodules, as found in lupin, can completely encircle the stem.

B. Nodule number and size

The number of nodules per plant is affected by the host species and rhizobial strain, by the number and efficiency in nodulation of the inoculant rhizobia, by the presence of existing nodules, and by nutritional and soil stresses (see Section VI).

In classic studies with subterranean clover, Nutman (1967) noted that variety and strain affected both the time to nodule formation and number of nodules per plant, with the host effects usually much greater than for the strain. In crosses between lines differing in number of nodules per plant, abundant nodulation was dominant over sparse nodulation, but abundantly nodulated lines bore a higher proportion of their nodules on lateral roots.

Subsequent studies have shown a definite time limit for the infection of immature, initially susceptible root hairs. For many species, inoculation delays can mean that such root hairs mature and are not infected, and that when nodules are produced, they form further down the taproot or on lateral roots.

Autoregulation is evident in the predominantly crown nodulation of well-inoculated plants and in the irregular distribution of nodules along the root. Even young nodules that have yet to fix N_2 can limit further nodulation in the adjacent root area. Thus, when the same root-tip marked plants were inoculated twice at intervals of 15 h, each inoculation gave rise to similar numbers of new infections, but few of the infections initiated following the second inoculation developed into nodules. A very high rate of inoculation can also depress nodulation, though the basis for this response is not known. Supernodulating plants with single-gene mutations, such as have been described for soybean and pea, no longer regulate nodulation and give rise to many nodules, often at the expense of plant growth. Autoregulation of nodulation depends in some degree on the N status of the plant. Thus, a plant with many small ineffective nodules may continue to nodulate, whereas one with fewer effective nodulates will not. For this reason, the number of nodules per plant is only likely to have meaning under conditions that affect inoculant numbers or interfere with nodulation per se.

Mechanisms involved in the regulation of nodule formation have yet to be clearly defined, but involve both shoot and root factors. A role for ethylene is also possible. Legumes evolve ethylene for up to 6 days after infection, and ethylene can inhibit nodule formation. Conversely aminoethoxyvinyl glycine (AVG), an inhibitor of ethylene biosynthesis, stimulates nodulation in some plants. A hypernodulating pea mutant that is resistant to the effects of ethylene and soybean cultivars that nodulate normally following repeated treatment with ethylene have been described.

V. SPECIFICITY IN NODULATION

Given the complex signaling involved, specificity in nodulation is to be expected. Many caesalpinioid legumes never form nodules; individual accessions of normally nodulating plants may not nodulate with specific rhizobia. An example is the pea cultivar Afghanistan, which nodulates normally with strains of *R. leguminosarum bv viciae* from the Middle East, but which carries a single recessive gene preventing nodulation with most European pea rhizobia. A parallel gene overcoming resistance to nodulation has been identified in *R. leguminosarum* strain TOM. Similarly, improved soybean varieties developed in the United States do not nodulate with indigenous bradyrhizobia from African soils, though unimproved soybean lines such as Orba, Avoyelles, and Mamloxi can. This difference in inoculation response poses an interesting problem in technology transfer for

FIGURE 61.6 Response to inoculation of soybean in newly cultivated areas of Puerto Rico. Photo courtesy of R. Stewart Smith, with permission.

TABLE 61.2 Cross-inoculation groups and effectiveness subgroups

Cross-inoculation group	Effectiveness subgroups[a,b]	Common leguminous species in each group/subgroup
Bean		*Phaseolus vulgaris, P. coccineus*
Chickpea		*Cicer arietinum*
Clover	A	*Trifolium alexandrinum, T. incarnatum, T. resupinatum*
	B	*Trifolium canescans, T. dubium, T. hybridum, T. pratense, T. procumbens, T. repens*
	C	*T. affine, T. vesiculosum*
	D	*T. cherleri, T. hirtum, T. subterraneum*
	Specific	*T. ambiguum, T. fragiferum, T. glomeratum, T. rueppelianum, T. semipilosum, T. tembense, T. usambarense*
Cowpea	A	*Acacia albida, Indigofera hirsuta, Lespedeza cuneata, L. stipulacea, L. striata, Mucuna deeringiana, Phaseolus lunatus, Vigna angularis, V. mungo, V. radiata, V. unguiculata*
	B	*Arachis hypogaea, A. pintoi*
	C	*Cajanus cajan, Lablab purpureus, Macrotyloma uniflorum, M. axillare*
	D	*Macroptilium purpureum, M. lathyroides, Pueraria phaseoloides, Calopogonium caeruleum*
	E	*Centrosema* spp.
	F	*Desmodium* spp.
	G	*Stylosanthes guyanensis, S. humilis*
	Specific	*Calopogonium caeruleum, Stylosanthes hamata*
Crownvetch	A	*Coronilla varia*
	B	*Onobrychis viciaefolia*
	C	*Dalea purpurea, D. candida*
Lotus	A	*Anthyllis vulneraria, Biserrula pelecinus, Lotus corniculatus, L. pedunculatus, L. tenuis*
	Specific	*Lotus uliginosus*
Lupin	A	*Lupinus* spp.
	B	*Ornithopus* spp.
Medic	A	*Medicago sativa, M. falcata, Melilotus alba, M. indica, M. officinalis, M. sulcata*
	B	*Medicago arabica, M. hispida, M. lupulina, M. minima, M. orbicularis, M. polymorpha, M. rigidula, M. rotata, M. rugosa, M. scutellata, M. truncatula, M. tuberculata, Trigonella foenum-graecum*
	Specific	*Medicago arborea, M. laciniata*
Pea	A	*Lathyrus hirsutus, Pisum sativum*
	C	*Lens esculenta, Vicia faba, V. monantha, V. pannonica, V. villosa*
	Specific	*Lathyrus sativus, Vicia ervilla, V. sativa*
Soybean		*Glycine max, G. soja*
Others not grouped		*Astragalus, Hedysarum, Lotononis, Sesbania*

[a]Letters used to distinguish groups of legumes refer only to this table, not to inoculants made by specific manufacturers.
[b]Effectiveness subgroups determine the range of different inoculants needed for legumes. However, the recommendation of individual manufacturers may vary from the divisions shown. For example, some inoculant manufacturers provide different inoculants for peas/vetch and faba beans/lentils, while others use a single inoculant for all four species.

production areas that include both large-commercial and smallholder producers (Mpepereki *et al.*, 2000).

Each rhizobium has the ability to nodulate some, but not all legumes. Cross-inoculation groups that contain legumes nodulated by the same bacteria can be further subdivided because some host–rhizobial combinations form nodules but are ineffective in N_2 fixation, as shown in Table 61.2. As a consequence, legumes introduced into new areas of production are unlikely to find suitable indigenous rhizobia and will need inoculation to ensure good nodulation and N_2 fixation (see Fig. 61.6). More than 100 different inoculants are needed to satisfy the requirements of the major crop, tree, and pasture legumes; these are produced in a range of formats and with different cell numbers and shelf lives (*http://www.rhizobium.umn.edu*)

VI. ENVIRONMENTAL AND SOIL FACTORS AFFECTING NODULATION

A. Soil-fertility status, including level of soil N

Well-nourished plants nodulate and fix N_2 better than those that are nutrient-limited, but several elements including P, Fe, Ca, and Mo have specific functions in nodulation and nitrogen fixation. Combined forms of N can also influence nodulation.

The high P requirement of nodulated legumes reflects the energy cost of nodule formation and maintenance. Nodules are an important P sink, and plants that are deficient in this element will usually be poorly nodulated and have a low nodule mass. P also is important in attachment and signal transduction. The three-way symbiosis between legumes, mycorrhiza, and rhizobia can be critical to the N and P supply to the plant, with nodulation and N_2 fixation markedly improved by mycorrhizal colonization.

Iron is critical for symbiosis. It forms part of the active site of the nitrogenase enzymes and leghemoglobin, occurs at high levels in legume root hairs, and has an unspecified function in nodulation. Plants deficient in Fe produce nodule initials, but these may not develop until they receive additional fertilization with Fe. Both host plant and rhizobia can differ in iron-use efficiency, making for significant interaction under conditions of low Fe concentations in soil.

Calcium chelators such as EGTA ([ethylenebis(oxyethylenenitrilo)]tetra acetic acid) can inhibit the nodulation of alfalfa at concentrations as low as 0.4 mM, with the Ca requirement for nodulation of alfalfa at pH 4.8, six times greater than at pH 5.6. Because low Ca concentration limits attachment, the Ca requirement is often assumed to be for binding between rhicadhesin and rhizobia. However, Ca flux in root hairs during infection, high Ca levels in the infection thread, and Ca function in outer-membrane stability in *Rhizobium* must also be considered as contributing to the need for Ca in nodulation.

Mo is a component of nitrogenase and thus essential for N_2 fixation. The requirement for Mo is minor, but complicated at low pH by Mo adsorption in soil. Thus Mo deficiency can be overcome by liming the soil to a pH at which the element will be more available, by foliar fertilization of the growing crop, or by the use of seeds which have high Mo content.

Combined N is needed for early plant growth, and small "starter" doses of N fertilizer may improve early nodulation and N_2 fixation. However in cases in which soil N and fertilizer N application exceeds 50 kg/ha, progressive inhibition of nodule formation can be expected. Suggested mechanisms include modified carbohydrate distribution to the root, inactivation of leghemoglobin, and degradation of hormones involved in infection.

B. Environmental factors

Environmental factors including pH, temperature, and water availability can limit rhizobial survival and number in the soil, affect plant growth and development, and impact the nodulation process per se.

Soils with a pH less than 4.8 occupy more than 800×10^6 ha in Latin America alone, and are common throughout Africa, Asia, and the eastern United States. Their use in agriculture is often beset by problems of Al or Mn toxicity, or micronutrient deficiency. The reduced survival of rhizobia in acid soils is common. One study reported an average of 87,000 cells/g of the acid-sensitive *S. meliloti* in soils of near neutral pH; and only 37 cells/g in soils of less than pH 6.0. Similar data exist for *R. leguminosarum* and *R. etli*. Acid-sensitive steps in nodulation have been identified, and related to attachment and root-hair curling. Approaches used to overcome pH limitations in nodulation include the lime-pelleting of seed and inoculant to provide microenvironments of higher pH until infection is completed, and the identification of host cultivars and rhizobia more tolerant to acid soil conditions. Acid-tolerant germplasm has been important in extending soybean production in the cerrado of Brazil, and in the improved growth of annual medic species in acid-soil areas of Australia.

Rhizobia are mesophilic organisms, with an optimum temperature for growth and nodulation in the range of 20–30 °C. Exceptions include rhizobia associated with arctic or high-altitude plant species, or from high-temperature environments such as the hot dry Sahel savannah of Africa. Rhizobia from *Gliricidia*, *Lonchocarpus*, and *Leucaena* that are able to nodulate and fix significant N_2 when exposed to 40 °C for 8 h/day have also been identified. High temperatures during inoculant shipment and storage can lead to loss of the symbiotic plasmid and to a reduction in the number of cells, limiting subsequent nodulation.

BIBLIOGRAPHY

Bhuvaneswari, T. V., Bhagwhat, A. A., and Bauer, W. D. (1981). Transient susceptibility of root cells in four legumes to nodulation by rhizobia. *Plant Physiol.* **68**, 1144–1149.

Boivin, C., Ndoye, I., Molouba, F., Lajudie, P., Dupuy, N., and Dreyfus, B. (1997). Stem nodulation in legumes: Diversity, mechanisms, and unusual characteristics. *Crit. Rev. Plant Sci.* **16**, 1–30.

Boogerd, F. C., and van Rossum, D. (1997). Nodulation of groundnut by *Bradyrhizobium*–a simple infection process by crack entry. *FEMS Microbiol. Rev.* **21**, 5–27.

Broughton, W. J., Jabbouri, S., and Perret, X. (2000). Keys to symbiotic harmony. *J. Bacteriol.* **182**, 5641–5652 (Review).

Bryan, J. A., Berlyn, G. P., and Gordon, J. C. (1996). Toward a new concept of the evolution of symbiotic nitrogen fixation in the leguminosae. *Plant Soil* **186**, 151–159.

Capela, D. *et al.* (2001). Analysis of the chromosome sequence of the legume symbiont *Sinorhizobium meliloti* strain 1021. *Proc. Natl. Acad. Sci. USA* **98**, 9877–9882.

Doyle, J. J., and Doyle, J. L. (1997). Phylogenetic perspectives on the origins and evolution of nodulation in the legumes and allies. *In* "Biological Fixation of Nitrogen for Ecology and Sustainable Agriculture" (A. Legocki *et al.*, eds.), pp. 307–312. Springer, Berlin.

Fahraeus, G. (1957). The infection of clover root hairs by nodule bacteria studied by a simple glass slide technique. *J. Gen. Microbiol.* **16**, 374–381.

Freiberg, C., Fellay, R., Bairoch, A., Broughton, W. J., Rosenthal, A., and Perret, X. (1997). Molecular basis of symbiosis between *Rhizobium* and legumes. *Nature* **387**, 394–401.

Gage, D. J., and Margolin, W. (2000). Hanging by a thread: invasion of legume plants by rhizobia. *Curr. Opin. Microbiol.* **3**, 613–617.

Gualtieri, G., and Bisseling, T. (2000). The evolution of nodulation. *Plant Mol. Biol.* **42**, 181–194.

Hirsch, A. M. (1992). Developmental biology of legume nodulation. *New Phytol.* **122**, 211–237.

John, M., Rohrig, H., Schmidt, J., Walden, R., and Schell, J. (1997). Cell signalling by oligosaccharides. *Trends Plant Sci.* **2**, 111–115.

Long, S. R. (1996). *Rhizobium* symbiosis: Nod factors in perspective. *Plant Cell* **8**, 1885–1898.

Moulin, L., Munive, A., Dreyfus, B., and Boivin-Masson C. (2001) Nodulation of legumes by members of the β-subclass of Proteobacteria. *Nature* **411**, 948–950.

Mpepereki, S., Javaheri, F., Davis, P., and Giller, K. E. (2000). Soyabeans and sustainable agriculture. Promiscuous soyabeans in southern Africa. *Field Crops Res.* **65**, 137–149.

Nutman, P. S. (1967). Varietal differences in the nodulation of subterranean clover. *Aust. J. Agric. Res.* **18**, 381–425.

Oke, V., and Long, S. R. (1999). Bacteroid formation in the *Rhizobium*-legume symbiosis. *Curr. Opin. Microbiol.* **2**, 641–646.

Oke, V., and Long S. R. (1999). Bacterial genes induced within the nodule during the *Rhizobium*–legume symbiosis. *Mol. Microbiol.* **32**, 837–849.

Pawlowski, K. (1997). Nodule-specific gene expression. *Physiol. Plant.* **99**, 617–631.

Sahlman, K., and Fahraeus, G. (1963). An electron microscope study of root-hair infection by rhizobium. *J. Gen. Microbiol.* **33**, 425–427.

Schultze, M., and Kondorosi, A. (1996). The role of lipochitooligosaccharides in root nodule organogenesis and plant cell growth. *Curr. Opin. Genet. Dev.* **6**, 631–638.

Stougaard, J. (2001) Genetics and genomics of root symbiosis. *Curr. Opin. Plant Biol.* **4**, 328–335.

Tate, R., Patriaca, E. J., Riccio, A., Defez, R., and Iaccarino, M. (1994). Development of *Phaseolus vulgaris* root nodules. *MPMI* **7**, 582–589.

WEBSITES

Website of the University of Minnesota Rhizobium Research Laboratory
http://www.rhizobium.umn.edu/

Links to *Sinorhizobium meliloti* genome
http://cmgm.stanford.edu/~mbarnett/genome.htm

Comprehensive Microbial Resource of The Institute for Genomic Research (TIGR) and links to many other microbial genomic sites
http://www.tigr.org/tigr-scripts/CMR2/CMRHomePage.spl

62

Nutrition of microorganisms

Thomas Egli

Swiss Federal Institute for Environmental Science and Technology

GLOSSARY

anabolism The process of synthesis of cell components from a metabolic pool of precursor compounds.

assimilation The incorporation of a compound into biomass.

catabolism The breakdown of nutrients to precursor compounds for anabolism or for dissimilation.

chemoautotrophy The use of reduced inorganic compounds and CO_2 as the primary sources of energy and carbon for biosynthesis.

chemoheterotrophy The process in which organisms are using organic compounds as the primary sources of carbon and energy for biosynthesis.

dissimilation The oxidation of reduced (in)organic compounds to provide energy for biosynthesis and cell maintenance.

growth medium An aqueous solution (may be solidified with agar) containing all the nutrients necessary for microbial growth.

limitation of growth The restriction on microbial growth by the availability of the nutrient that is first to be consumed to completion. This growth-limiting nutrient determines the maximum amount of biomass that can be formed in this system; at low concentrations in batch culture and in the chemostat, it also determines the rate (kinetics) of growth.

nutritional categories of microorganisms Categories based on the principal carbon (CO_2 or reduced organic compounds) and energy sources (light or reduced (in)organic compounds) of microorganisms; there are four nutritional categories: photoautotrophs, photoheterotrophs, chemoautotrophs, and chemoheterotrophs.

nutrient An organic or inorganic compound that is used by microorganisms as a building block for the synthesis of new cell material. In a wider sense, also a compound not incorporated into the microorganism, but serving as a source of energy or as terminal electron acceptor. Nutrients are grouped into classes depending on the physiological purpose they serve, the quantity required, and whether or not they are essential for growth.

photoautotrophy The use of light and CO_2 as the primary sources of energy and carbon for biosynthesis.

photoheterotrophy The use of light and reduced organic compounds as the primary sources of energy and carbon for biosynthesis.

To Grow and Divide, microbial cells take up precursors and building blocks (nutrients) from the environment. In a wider sense, nutrients are also compounds that are not directly incorporated into cell material but are used by microbes to obtain the energy necessary to drive this synthesis and maintain cell integrity. Different nutritional types of microorganisms exist using different forms of carbon (CO_2 or reduced organic compounds) and energy (light or chemical energy) as the primary sources for biosynthesis. Nevertheless, the cellular composition of all microbial cells with respect to bulk components and the elemental composition is rather similar. Because of this, it is possible to estimate the general requirement of different

The Desk Encyclopedia of Microbiology
ISBN: 0-12-621361-5

nutrients for growth and to design and analyze microbial growth media. In well-designed growth media a particular, identified nutrient is growth-limiting and determines the amount of biomass that can be formed, whereas all other nutrients are present in excess (Liebig's principle). Cell metabolism and performance are strongly influenced by the nature of the growth-limiting nutrient. Therefore, many industrial fermentation processes are based on restricting the availability of a particular nutrient. This article focuses on the nutrients taken up by cells for growth and on medium design. Information on other chemico-physical factors influencing growth, such as temperature, pH, or water activity, can be found in either the classical textbook of growth by Pirt (1975) or at *http://www.bact.wisc.edu/Bact303/NutritionandGrowth/*.

I. CLASSIFICATION OF MICROORGANISMS AND NUTRIENTS

Growth and production of offspring is the ultimate goal of each microbial cell and to achieve this, cells take up nutrients from the environment for two purposes, either to serve as a source of building blocks or precursors for the synthesis of new cellular constituents or to generate energy to drive biosynthesis. Individual members of the microbial world are extremely diverse and often unique with respect to their nutritional requirements and abilities. Hence, only the main patterns of the nutritional requirements and behavior of microorganisms will be delineated here.

Two approaches are traditionally taken to describe the nutritional behavior and requirements of living cells. The two approaches do not contrast, but rather complement each other. One is to categorize organisms on the basis of the principal sources of the carbon and of the energy they are able to use for growth; the other is to categorize them on the basis of quantitative and elemental aspects of the nutrients used for growth.

A. Nutritional categories of organisms

Based on their principal carbon and energy sources, (micro)organisms are classified into four nutritional categories (Table 62.1). Most microorganisms using light as their principal source of energy are photoautotrophs (sometimes also referred to as photolithoautotrophs), whereas photoheterotrophs are a small group of specialists (certain purple and green bacteria). The ability to grow chemoautotrophically (i.e. in the dark in a medium containing only inorganic nutrients, including a reduced inorganic compound as a source of energy) is specific for bacteria and is lacking in eukaryotic microorganisms. In these three types of nutrition, the sources for carbon and for energy are clearly separated. This clear-cut distinction between carbon source and energy source is not valid for the big group of chemoheterotrophic organisms that obtain their energy from the oxidation of reduced organic compounds and at the same time use them as a source of building blocks. The terms "litho-" and "organotroph" are sometimes used to indicate in addition the source of hydrogen and electrons.

TABLE 62.1 Nutritional types of organisms based on the sources of carbon and energy used for growth

Energy source	Carbon source	Nutrition type
Light	CO_2	Photoautotroph (photolithoautotroph)
Light	Reduced organic compounds	Photoheterotroph (photolithoheterotroph)
Reduced inorganic compounds	CO_2	Chemoautotroph (chemolithoautotroph)
Reduced organic compounds	Reduced organic compounds	Chemoheterotroph (chemorganoheterotroph)

Some microbial strains are nutritionally rather flexible and could be placed in different nutritional categories. For example, the nutritional versatility of some photoautotrophic microalgae is such that they can employ equally well a chemoheterotrophic lifestyle, growing in the dark at the expense of organic carbon sources. Also it should be mentioned that, when given the chance, most autotrophs can take up and assimilate considerable amounts of reduced organic compounds (not only growth factors, as described later) and use them to feed their anabolism. The nutritional category of such microorganisms is usually based on the simplest nutritional requirements, in which phototrophy and autotrophy precede chemotrophy and heterotrophy, respectively. The degree of nutritional flexibility, in addition, is indicated by describing organisms as either obligate or facultative photo(chemo)autotrophs.

In any of the four nutritional categories there are auxotrophic strains that require low amounts of specific organic compounds, the growth factors. Unlike prototrophic strains, auxotrophs are unable to synthesize these growth factors from the principal source of carbon supplied in the medium.

B. Classes of nutrients

In everyday use, the term "nutrient" is restricted to compounds either fully or at least partly incorporated into cell material. However, biosynthesis requires

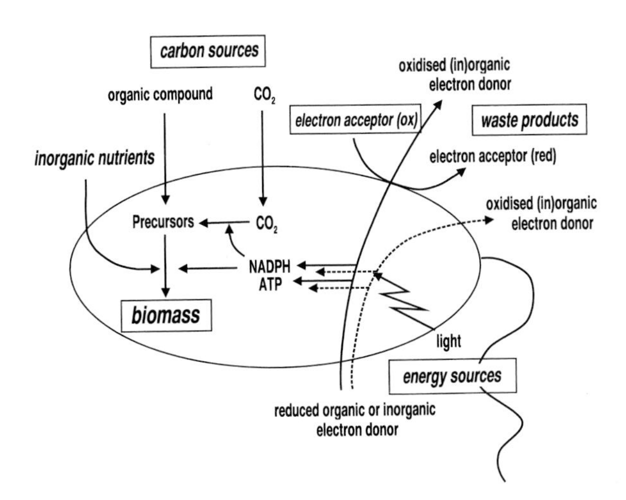

FIGURE 62.1. Simplified sketch of the physiological function of nutrients for the growth of microorganisms.

energy in addition to building blocks. Frequently, the compounds involved in the generation of energy are not incorporated into biomass, but only take part in redox processes. Hence, based on their physiological purpose, compounds essential for microbial growth can be divided into two major groups (Fig. 62.1).

1. Compounds that are either fully or partly incorporated into components of the biomass ("nutrients"), and
2. Compounds that are not incorporated into biomass but are essential for the generation of energy (electron donors or acceptors).

This distinction cannot always be made in such a clear way because there are nutritional categories of organisms in which some compounds can fulfil both functions at the same time. For example, reduced carbon sources are employed by chemoheterotrophs to obtain carbon precursors for biosynthesis as well as to generate energy, or ammonia can be used by particular chemolithoautotrophs as a source of both energy and nitrogen.

The chemical elements contained in the nutrients consumed and incorporated into new cell material can be divided into five different classes. The division is mainly based on the quantities of these elements required for growth and their occurrence in dry biomass (Table 62.2). Not considered in this table is water, which is a main constituent of all cells, making up approximately 75% of the fresh cell weight. Table 62.3 indicates that typically some 95% of the dry biomass are made up of the eight elements C, N, O, H, P, S, K, and Mg. These elements are indispensable for microbial growth. Class two elements are required in significant amounts, whereas those in class three and four are usually referred to as trace elements. For elements categorized in classes two and three, it can be demonstrated experimentally that they are essential, whereas it is difficult to prove that elements of class four are essential for growth. These elements are required in such low amounts that they are usually introduced into media as impurities of the bulk salts. Finally, a special class of nutrients are the growth factors required by auxothrophic strains. This includes a diverse group of organic compounds. The physiological role of the main nutrients is discussed in more detail in Section III.

II. ELEMENTAL COMPOSITION OF BIOMASS

The composition of a microbial cell is highly dependent on the cultivation conditions. The type of cellular constituents present (e.g. ribosomes, particular enzymes, membrane and cell wall components, and compounds in the metabolic pool) and their amount can vary enormously. Herbert (1961) has emphasized this point saying that it is useless to give the cellular composition of a microbial cell without at the same time specifying both the exact growth conditions under which this cell has been cultivated and its growth history.

Despite this diversity and variability with respect to cell constituents, the elemental composition of microbial biomass—including cell material from archaea, eubacteria, and eukaryotes—varies in a surprisingly narrow range. This is documented in the overview of the average elemental composition of microbial biomass and its variability in Table 62.3. The relative constant composition of microbial biomass with respect to the major elements results from the fact that most of the dry biomass (typically some 95%) is made up of a limited number of organic macromolecules and only a small fraction consists of monomers (metabolites and inorganic ions). Because protein, RNA, and phospholipids are the dominating components, massive changes in the content of a particular cell component are required before the overall elemental composition of the biomass is significantly affected (Table 62.4). For example, a significant increase in the carbon content of dry biomass is observed only when cells store high amounts of poly(3-hydroxyalkanoate) (PHAs) or neutral lipids, whereas it is the cellular oxygen content is primarily affected when cells accumulate glycogen. Note that an extensive incorporation of carbonaceous reserve materials results in a dilution and, hence, in a reduction of the relative content of other elements in dry biomass. A typical example is the reduced cellular nitrogen content found in cells accumulating PHA or glycogen.

TABLE 62.2 Classes of nutrients used for microbial growth based on their incorporation and occurrence in dry cell mass[a]

Class 1	Always essential	Major elements: C, H, O, N Minor elements: P, S, K(Rb), Mg
Class 2	Mostly essential	Fe, Ca, Mn, Co, Cu, Mo, Zn
Class 3	Essential in special cases	B, Na, Al, Si, Cl, V, Cr, Ni, As, Se, Sn, I
Class 4	Very rarely essential (difficult to prove)	Be, F, Sc, Ti, Ga, Ge, Br, Zr, W
Class 5	Growth factors, essential for special strains	Aminoacids, purines and pyrimidines, vitamins, hormones, etc.

[a]Based on Pirt (1975).

TABLE 62.3 Elemental composition of microbial biomass[a]

Element	% of dry weight[b] Average	Range	Typical sources used for microbial growth in the environment
Carbon	50	45[c]–58[d]	CO_2, organic compounds
Oxygen	21	18[e]–31[f]	H_2O, O_2, organic compounds
Nitrogen	12	5[g]–17[h]	NH_3, NO_3^-, organically bound N
Hydrogen	8	6[g]–8[h]	H_2O, organic compounds
Phosphorus	3	1.2[i]–10[j]	PO_4^{3-}, organically bound P
Sulphur	1	0.3–1.3	SO_4^{2-}, H_2S, organically bound S
Potassium	1	0.2[k]–5[l]	K^+ (can often be replaced by Rb^+)
Magnesium	0.5	0.1[m]–1.1	Mg^{2+}
Calcium	1	0.02–2.0	Ca^{2+}
Chlorine	0.5		Cl^-
Iron	0.5	0.01–0.5	Fe^{3+}, Fe^{2+}, organic iron complexes
Sodium	1		Na^+
Other elements , Mo, Ni, Co, Mn, Zn, W Se, etc.	0.5		Taken up as inorganic ions

[a]Data from Tempest (1969), Pirt (1975), Herbert (1976), and from results obtained in author's laboratory.
[b]Cells consist on average to 70% of their weight of water and 30% of dry matter. Average is given for gram-negative cells growing with excess of all nutrients at μ_{max} in batch culture.
[c]Carbon-limited cells containing no reserve materials.
[d]Nitrogen-limited cells storing PHA or glycogen in the presence of excess C-source.
[e]Cells grown N-limited accumulating neutral lipids.
[f]Cells grown N-limited accumulating glycogen.
[g]Nitrogen-limited cells storing PHA or glycogen in the presence of excess C-source.
[h]Cells growing at high μ containing high levels of rRNA.
[i]Cells grown P-limited.
[j]Cells accumulating the reserve material polyphosphate.
[k]Gram-positive *Bacillus* spores.
[l]Gram-positive bacilli.
[m]Magnesium-limited cells at low growth rates.

III. REQUIREMENTS AND PHYSIOLOGICAL FUNCTIONS OF PRINCIPAL ELEMENTS

A. Carbon

Dried microbial biomass consists of roughly 50% carbon, virtually all of it present as one of the many reduced organic cell constituents. Hence, as discussed (Table 62.1), the most obvious physiological function of carbon is as a source of building material for organic biomolecules. When its most oxidized form, CO_2, is used as the sole source of carbon for autotrophs, reduction to the level of organic cell material and the formation of carbon–carbon bonds is required. This process requires significant amounts of reducing equivalents (NADPH) and energy (ATP) (see Fig. 62.1). CO_2 is also employed as a terminal electron acceptor by methanogens and acetogens.

In contrast, heterotrophs use reduced carbon compounds to build their cell material and in most cases (an exception are the photoheterotrophs) the carbon compound fulfills a dual function, namely it acts as both a carbon and an energy source. In some fermenting

TABLE 62.4 Major polymeric constituents found in microbial cells and their average elemental composition[a]

	% of dry weight							
Constituent	**Average**[b]	**Range**	**%C**	**%H**	**%O**	**%N**	**%S**	**%P**
Protein	55	15[c]–75	53	7	23	16	1	—
RNA[d]	21	5[c]–30[e]	36	4	34	17	—	10
DNA[d]	3	1[c]–5[f]	36	4	34	17	—	10
Peptidoglycan	3	0[g]–20[h]	47	6	40	7	—	—
Phospholipids	9	0[i]–15	67	7	19	2	—	5
Lipopolysaccharides	3	0[h]–4[j]	55	10	30	2	—	3
Neutral lipids	—	0–45[k]	77	12	11	—	—	—
Teichoic acid[d,h]	—	0[l]–5[d]	28	5	52	—	—	15
Glycogen	3	0–50[k]	45	6	49	—	—	—
PHB	—	0–80[k]	56	7	37	—	—	—
PHA (C8)[m]	—	0–60[k]	68	9	23	—	—	—
Polyphosphate[d]	—	0–20[n]	—	—	61	—	—	39
Cyanophycin[o]	—	0–10	42	15	25	27	—	—

[a]Adapted from Herbert (1976) and extended. The figures given for the range have been collected from different organisms and, therefore, may not be applicable for particular strains.
[b]Average composition of an exponentially-growing gram-negative cell (*E. coli*), from Neidhardt *et al.* (1990).
[c]Cells storing carbonaceous reserve materials.
[d]Inclusion of the highly negatively charged polymers such as RNA, DNA, polyphosphate, or cell wall components is paralleled by the presence of appropriate amounts of counter-cations, Mg^{2+}, Ca^{2+}, or K^+.
[e]At high growth rates.
[f]Cells growing slowly.
[g]Parasitic cell wall-less species.
[h]Gram-positive bacteria.
[i]Strains replacing phospholipids under P-limited growth conditions with P-free analog.
[j]Gram-negative bacteria.
[k]Cells grown N-limited.
[l]Grown P-limited.
[m]PHA consisting of 3-hydroxyoctanoic acid.
[n]Some yeasts and bacteria.
[o]Some cyanobacteria contain the nitrogen storage material cyanophycin [(asp-arg)$_n$].

organisms, reduced carbon compounds can act as terminal electron acceptors. Typically, heterotrophic cells utilize the same carbon source for both purposes, oxidizing part of it to CO_2 (a process called dissimilation) and using the energy derived from this oxidation to synthesize cell material from the other part (assimilation). The ratio of dissimilated to assimilated carbon is essentially dependent on the degree of reduction of the carbon substrate used. The more oxidized the carbon compound, the more of it has to be dissimilated in order to provide the necessary energy to drive the synthesis processes, and the less of it can be assimilated. This is reflected in the maximum growth yield observed for various carbon sources when plotted as a function of their energy content (i.e. their degree of reduction, or heat of combustion), as shown in Fig. 62.2. Most extreme is the case of chemoheterotrophs growing at the expense of oxalate (HOOC–COOH). To generate energy, this compound is initially oxidized to CO_2, which is then assimilated in an autotrophic manner.

Heterotrophic microorganisms are extremely diverse with respect to the spectrum of carbon sources that they can use for growth. Whereas some are restricted to only a few carbon compounds (e.g. some methanotrophs appear to use only methane and methanol), others are able to metabolize and assimilate more than a hundred carbon compounds for growth. It should be added here that all heterotrophic organisms also assimilate a substantial amount of their cell carbon (typically 5–10%) from CO_2 (mainly for replenishing the tricarboxylic acid cycle when it is used as a source of building blocks for biosynthesis). Normally, this requirement for CO_2 is masked because CO_2 is produced in large amounts intracellularly from the catabolism of organic growth substrates. However, especially in freshly inoculated dilute cultures, its absence can slow down or even prevent growth on organic substrates (Pirt, 1975), and some heterotrophic mircoorganisms even require elevated concentrations of CO_2 in the culture medium.

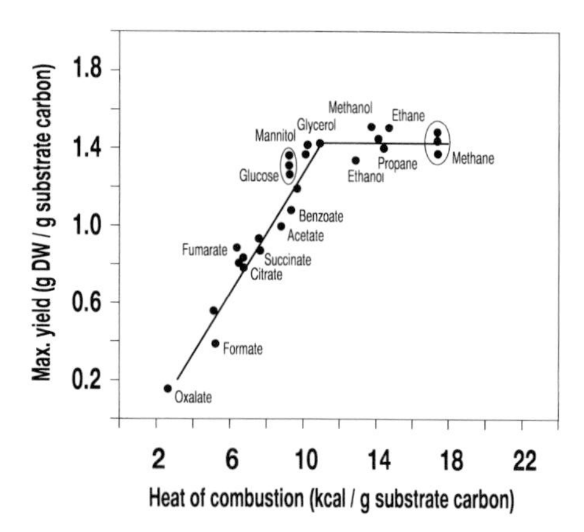

FIGURE 62.2 Maximum growth yields reported for various carbon substrates observed for heterotrophic organisms, as a function of the energy content of the carbon substrate (adapted from Linton and Stephenson, 1978).

In the case of energy excess, carbon compounds can be stored intracellularly as reserve materials in the form of polyhydroxyalkanoates, glycogen, or neutral lipids. In the case of carbon starvation, these internal carbon and energy sources are broken down to support cellular rearrangement and adaptation to the new conditions and to ensure survival.

B. Hydrogen

In cells, hydrogen is present in the form of water and as an element of all organic cell constituents. The main source of hydrogen for biosynthetic purposes is NADPH. The need for hydrogen is particularly evident for the reduction of CO_2 in autotrophs. In photo- and chemoautotrophs hydrogen equivalents used for CO_2 reduction originate from water, from the oxidation of reduced inorganic compounds. Chemoheterotrophs obtain their reducing equivalents from the oxidation of their primary carbon substrate.

C. Oxygen

As with carbon and hydrogen, oxygen is omnipresent in cells. It occurs in most of the organic components of cell material. The main sources of oxygen for the biosynthesis of particular cell components are water, molecular oxygen (but not in obligate anaerobes, where oxygen is frequently toxic) and, less obviously, CO_2. In aerobes, molecular oxygen is introduced into organic molecules with the help of mono- and dioxygenases. In addition to its function as a cell constituent, O_2 also serves as a terminal electron acceptor in aerobes.

D. Nitrogen

The cellular requirement for nitrogen is significant because it is a constituent of all major macromolecules (Tables 62.3 and 62.4). In cell components, nitrogen is mainly found in the reduced form (i.e. as primary, secondary, or tertiary amino groups). Oxidized forms (nitro- and nitroso-groups) are rarely found. Organic and inorganic forms of nitrogen in all states of oxidation, from NH_4^+ to N_2 to NO_3^-, can be used by microorganisms as sources of cell nitrogen (although some are unable to reduce oxidized forms). Note that the microbiological fixation of molecular nitrogen is of special interest to agriculture because of its availability in the air. Frequently, microbial cells exhibit nitrogen requirements in the form of special amino acids (L-forms for incorporation into proteins, or D-forms for the synthesis of cell-wall components) or peptides. Intracellularly, the assimilation of nitrogen occurs at the level of ammonia. Therefore, all more oxidized forms have to be reduced to this level before they can be used as a source of nitrogen.

Nitrogen compounds also play a major role in energy metabolism. Reduced forms (e.g. ammonia and nitrite) are used as sources of energy by nitrifying bacteria, whereas oxidized inorganic nitrogen compounds (e.g. nitrate and nitrite) are employed as terminal electron acceptors by denitrifying microbes.

E. Phosphorus

Inorganic phosphate is typically supplied in growth media as the only source of phosphorus. However, many organisms can also derive phosphorus from organic phosphates, such as glycerophosphate (organic P-sources can be used to avoid the precipitation of inorganic phosphate salts in the medium at basic pH values). Phosphate is primarily incorporated into nucleic acids, phospholipids, and cell-wall constituents. Some organisms may also store it as polymetaphosphate, which can be reused as a source of internal phosphorus or for the generation of ATP. Intracellularly, the main fraction of phosphorus is contained in ribosomal RNA, whereas ATP and other nucleic acids make up only a minor fraction of the total cellular phosphorus.

F. Sulfur

The bulk of intracellular sulfur is found in proteins (cysteine and methionine). An important function of cysteine is its involvement in the folding of proteins by the formation of disulfide bridges. Frequently, these amino acids are also found in reactive centers of enzymes (e.g. in the coordination of reactive iron centers). The sulfur-containing coenzymes and vitamins (e.g. CoA, biotin, thiamine, glutathione, and lipoic acid) are small in quantity but physiologically very important. Intracellularly, sulfur is present in a reduced form (–SH), but it is usually supplied in growth media as sulfate salt. Some organisms are not able to catalyze this reduction and therefore must be supplied with a reduced form of sulfur, for instance, cysteine or H_2S.

Many inorganic sulfur compounds are also involved in the generation of energy. Whereas reduced sulfur compounds are used as electron donors (H_2S, thiosulphate, and S^0), oxidized forms are employed as terminal electron acceptors (SO_4^{2-}, S^0).

G. Major cations

1. *Magnesium*

Magnesium is one of the major cations in cell material. Its intracellular concentration is proportional to that of RNA, which suggests that it is partly counterbalancing the negative charges of the phosphate groups in nucleic acids. Hence, its cellular concentration increases with growth rate. It is required for stabilizing the structure of ribosomes. Many enzymes are activated by or even dependent on the presence of Mg^{2+}; some important examples are enzymes catalyzing reactions dependent on ATP or chlorophylls. Magnesium is also found bound to the cell wall and the membrane, where it seems to be responsible for stabilizing the structure together with other cations.

Interestingly, the molecular ratio of Mg : K : RNA nucleotide : PO_4 in gram-negative bacteria is always approximately 1 : 4 : 5 : 8, independent of growth rate, temperature, or growth-limiting nutrient (Tempest, 1969). In gram-positive organisms, this ratio is 1 : 13 : 5 : 13, except under phosphate-limited growth conditions in continuous culture, where it is also 1 : 4 : 5 : 8. The higher K and PO_4 content of gram-positive bacteria is due to the presence of phosphate-containing cell-wall polymers (teichoic acids), which are replaced under phosphate-limited growth by nonphosphate-containing analogs (teichuronic acids).

2. *Potassium*

Potassium makes up a large part of the inorganic cations in biomass (Table 62.3). Only a small fraction of K^+ present in cells seems to be associated with binding sites of high affinity and specificity because it can be rapidly exchanged with other monovalent cations. A large fraction of K^+ is bound to RNA, for which it seems to have a stabilizing function. Therefore, as with magnesium, the requirement for K^+ increases with growth rate. Significant amounts are also found associated with the cell wall. K^+ activates a number of different enzymes, both nonspecifically (contributing to the ionic strength) or specifically (e.g. peptidyltransferase). Ions of similar size such as Rb^+ or NH_4^+ can frequently take over the function of K^+ (in contrast to magnesium, which cannot be replaced by other cations). The growth rates of many organisms are reduced when cultivated in media that is low in potassium. A variety of growth conditions affect the intracellular concentration of potassium, including the osmolarity of the medium, temperature, pH, or sodium concentration. Therefore, this cation should always be added to growth media in significant excess.

3. *Iron*

Frequently considered a trace element, iron is used in significant amounts by virtually all organisms, not only by obligate aerobes (lactobacilli seem to be the only bacteria that do not need iron for growth). Iron is the catalytic center of a number of enzymes, especially those involved in redox reactions. Most essential are the various iron-containing cytochromes in the respiratory chain, flavoproteins, or the enzymes essential for the detoxification of reactive oxygen species such as catalase or superoxide dismutase. Many of the mono- and dioxygenases initiating the breakdown of pollutants are also iron enzymes. Most bacteria require concentrations of free iron exceeding 10^{-8} M for growth. Iron (III), which is the species that prevails in aerobic environments, easily forms insoluble hydroxides and other complexes. Therefore, the acquisition of iron is a major problem for growing organisms. Many organisms react to iron limitation by excreting iron-complexing organic compounds with a high affinity to iron, the siderophores. In mineral media, iron is therefore frequently supplied complexed with an organic ligand.

A number of anaerobic bacteria (in particular nitrate- reducing strains) can use Fe^{3+} (or Mn^{4+}) as a terminal electron acceptor, reducing it to Fe^{2+} (or Mn^{2+}). On the other hand, some specialist bacteria can use Fe^{2+} (or sometimes also Mn^{2+}) as a source of energy by oxidizing it to Fe^{3+}.

4. *Calcium*

In most organisms, Ca^{2+} is present intracellularly in significantly lower amounts than Mg^{2+}, which has similar properties. The role of Ca^{2+} is not always clear; however, it seems to have important functions in stabilizing the cell wall and controlling membrane permeability. Changes in cell morphology and cell-surface properties have been reported for several microorganisms in the absence of Ca^{2+}. Ca^{2+} activates many exoenzymes such as amylase. Furthermore, a number of uptake processes are stimulated by the presence of Ca^{2+}; an example is the uptake of exogenous DNA. It appears that calcium plays extracellularly the role that magnesium plays in the cytoplasm. Often Mg^{2+} cannot replace Ca^{2+} in these extracellular functions, but strontium can. In growth media attributing a clear function to calcium is difficult because of the presence of competing divalent cations that are essential for growth, such as Mg^{2+} or Mn^{2+}.

H. Trace elements

1. *Sodium*

Microorganisms isolated from freshwater do not usually require sodium. For such organisms, it is difficult to demonstrate that this cation is essential for growth because its requirements are low and sodium is present in all bulk salts as an impurity. However, some extremely halophilic microorganisms require high concentrations of NaCl for growth. For example, in order not to disintegrate, *Halobacterium* needs more than 2.5 M NaCl in the growth medium. Sodium is also essential for certain photosynthetic bacteria and cannot be substituted for by other monovalent cations. In some marine bacteria, energy generation is even linked to the use of a Na^+-gradient, rather than an H^+-gradient. Furthermore, in many microorganisms this ion is involved in the regulation of intracellular pH using a Na^+–H^+ antiport system.

2. *Manganese*

As a substitute for the iron-containing catalase, lactobacilli produce a manganese-containing pseudocatalase for protection against molecular oxygen. A high requirement for manganese is typical for lactic acid bacteria. Many of the lignolytic peroxidases contain manganese.

3. *Cobalt*

Co^{2+}-containing coenzymes and cofactors are widespread, the best known being the coenzyme B_{12}. Cobalamines (cobalt-containing biomacrocylic compounds) are found in bacteria as well as in humans, but the highest levels are usually found in methanogenic bacteria.

4. *Nickel*

Methanogens require unusually high amounts of Ni^{2+} for growth. It was found that it is a component of the coenzyme F430 in these organisms. Furthermore, this divalent cation is a constituent of virtually all hydrogenases in both aerobic and anaerobic microorganisms that either use or produce molecular hydrogen.

5. *Copper*

Copper is the key metal in the active center of many redox reaction-catalyzing enzymes. It is present in many terminal oxidases of the respiratory chain. A number of other enzymes, such as peptidases, laccases, some nitrite reductases, or methane monooxygenase contain Cu^{2+}.

6. *Molybdenum*

A whole family of enzymes, the molybdoenzymes (also referred to as molybdenum hydroxylases) contain Mo^{2+}. This family includes the central enzyme in the reduction of nitrate to nitrite, nitrate reductase. Furthermore, the nitrogen-fixing nitrogenase contains a MoFe cofactor, which is clearly different from the molybdenum cofactor shared by other Mo-containing enzymes.

7. *Zinc*

Many bacterial (extracellular) metalloproteases contain Zn^{2+} (e.g. elastase). Many of these proteases are produced by pathogenic strains and play an important role in the pathogenesis. Other zinc-containing enzymes are alkaline phosphatase and the long-chain alcohol dehydrogenases.

IV. FEAST AND FAMINE: UNRESTRICTED VERSUS NUTRIENT-LIMITED GROWTH

In a typical laboratory shake flask culture, all the nutrients supplied in a well-designed growth medium are initially present in excess and the cells grow exponentially at the highest rate possible under these conditions. However, in every environmental and technical system, microbial growth cannot proceed unrestricted for a long time. A simple calculation makes this obvious. After 2 days of exponential growth, a single microbial cell doubling every 20 min will have produced roughly 2×10^{43} cells. Assuming an average cell weight of 10^{-12} g, this amounts to 2×10^{31} g of the biomass, or approximately 4000 times the weight of Earth. Hence, in every compartment, growth is always soon limited by the exhaustion of one or several nutrients.

A. The concept of the limiting nutrient

The term "limiting nutrient" is used with meanings, which, unfortunately, are frequently mixed up. The availability of nutrients can restrict the growth of microbial cultures in two distinct ways, namely stoichiometrically and kinetically. The stoichiometric limitation is defined by the maximum amount of biomass that can be produced from the limiting nutrient in this system ("Liebig's principle," from Justus von Liebig's agricultural fertilization studies around 1840, in which he found that the amount of a particular nutrient determined the crop on a field as long as all other nutrients were present in excess; Eq. 1). The kinetic limitation arises at low nutrient concentrations (typically in the low milligram to microgram per liter range), at which the limiting nutrient also controls the specific rate of growth of cells (μ). This kinetic control of growth rate usually follows a saturation kinetics and the Monod equation (Eq. 2) is typically used to describe the relationship between the concentration of the growth-rate-controlling nutrient and μ.

$$X = X_0 + (S_0 - s) \cdot Y_{X/S} \quad (1)$$

$$\mu = \mu_{max} \frac{s}{K_s + s}, \quad (2)$$

where S_0 is the initial and s the actual concentration of the limiting nutrient S; $X(X_0)$ is the (initial) biomass concentration; $Y_{X/S}$ is the growth yield for nutrient S, $\mu_{(max)}$ is the (maximum) specific growth rate, and K_s is the Monod apparent substrate affinity constant. This is visualized in Fig. 62.3 for growth in a closed batch culture system, in which the cells initially grow unrestricted until the consumption of the limiting nutrient leads to growth at a reduced rate, and then to growth stoppage. This determines the final concentration of biomass that can be reached. In flowthrough systems, such as a continuous culture in which fresh medium is continuously added and surplus culture is removed, the rate of addition of limiting nutrient (thought to be a single compound) controls simultaneously μ and the concentration of biomass obtained in the culture (Pirt, 1975; Kovářová and Egli, 1998).

In laboratory cultures, it is possible to cultivate cells under well-defined conditions in which the growth-limiting nutrient is known. Quantitative and practical aspects of the supply of nutrients to microbial growth

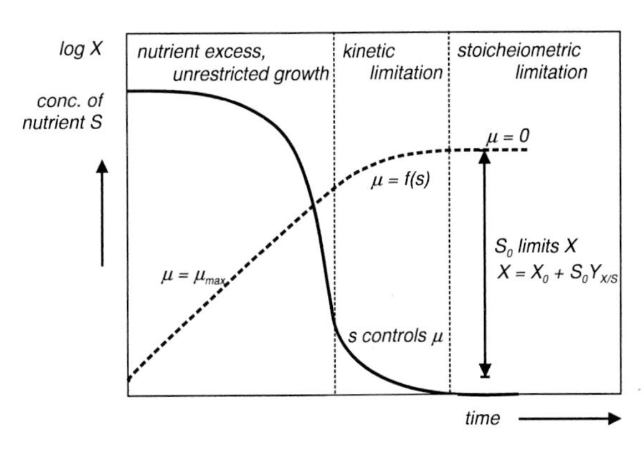

FIGURE 62.3 Kinetic and stoichiometric limitation of microbial growth in a batch culture by the concentration of the limiting nutrient (substrate)S. S_0, initial concentration of S; s, actual concentration of S; X, biomass concentration; X_0, initial biomass concentration; $Y_{X/S}$, growth yield for nutrient S.

media for controlled cultivation will be discussed in the Section V. For the cultivation of heterotrophic microorganisms for research purposes and the production of biomass, media are commonly designed with limiting carbon and energy sources, all other nutrients being supplied in excess. However, for biotechnological processes, limitation by nutrients other than carbon is frequently employed to manipulate the physiological state and metabolic performance of microbial cultures. The restriction (limitation) of specific nutrients induces or enhances the formation of many microbial metabolites and enzymes. Examples are the increased productivity in antibiotics fermentation by growth in phosphate-limited media, the production of citric acid under Fe-, Mn-, or Zn-limited batch-culture conditions, the synthesis of NAD under Zn–Mn limitation, and the accumulation of the intracellular reserve materials PHB or PHA ("bioplastic") by limiting the supply of nitrogen.

In contrast to cultivation on the laboratory and industrial scales, it is difficult to assess the nutritional regimes that govern the growth of microbial cells in environmental systems (especially the identification of the kinetic control). There are indications that microbial growth is frequently controlled not by a single nutrient, but by combinations of two or more nutrients simultaneously (Kovářová and Egli, 1998).

V. DESIGN AND ANALYSIS OF DEFINED MINIMAL GROWTH MEDIA

To grow and synthesize their own cell material, organisms must obtain all the required building blocks (or their precursors) and the necessary energy from their environment. Consequently, to cultivate microbial cells in the laboratory these nutrients must be supplied in a culture medium in adequate amounts and in a form accessible to the organism.

As a result of the physiological diversity of the microbial world, a myriad of media of different compositions have been published, for either selective enrichment or cultivation of particular microorganisms (consult, for example, LaPage *et al.*, 1970; Balows *et al.*, 1992; Atlas, 1997). All these media contain components whose nutritional function is obvious, particularly when considering their elemental or energetic function. Nevertheless, most nutritional studies made have been qualitative rather than quantitative and different nutrients have been added in more or less arbitrary amounts. Also, many of the media contain components whose reason for inclusion cannot be clearly identified because their inclusion is based more on experience or tradition than on a clear purpose.

The identification of the nutritional requirements of microbial cells usually calls for the use of defined synthetic media. The design of defined culture media is based on quantitative aspects of cell composition and it allows the influence of the growth of a microbial culture at three major levels. First, the choice of which nutrient is to limit the growth of the culture stoichiometrically and kinetically is made. Second, for nutritionally flexible microbial strains, the choice of the type of metabolism that the organism is to perform is made by the selection of the compounds that are supplied to fulfill a particular nutritional requirement, including electron donors and acceptors. Third, and often linked with the second point, the choice of the maximum specific growth rate to be achieved during unrestricted growth in batch culture is set.

A. Setting μ_{max} during unrestricted growth

In addition to physicochemical parameters such as temperature or pH, the maximum specific growth rate of a microorganism is influenced by the diversity of the nutrients supplied in the medium. This has been elegantly illustrated for the growth of *Salmonella typhimurium* by Schaechter *et al.* (1958), who used 22 media of different compositions to obtain growth of the culture at differing rates under nutrient excess conditions (a selection is given in Table 62.5). Although the four media supporting the highest specific growth rates are undefined, the other media consist of a minimal salt medium to which different carbon sources or amino acid mixtures are added. Hence, the selection of the quality of precursors supplied in the mineral medium allowed the adjustment of the growth rate of the culture in a defined and reproducible way.

B. Medium design and experimental verification of the limiting nutrient

1. *Designing a growth medium*

In the design of a defined growth medium, the initial decisions to be made are the choice of the maximum concentration of biomass the medium should allow to produce (X_{max}), and the definition of the growth-limiting nutrient (according to Liebig's principle). Typically, defined growth media for heterotrophic microbes are designed with a single carbon-energy source restricting the amount of biomass that can be produced, whereas all other nutrients (each of them usually added in the form of a single compound) are supplied in excess. Having set X_{max}, it is possible to

TABLE 62.5 Composition of a selection of media used to set the maximum specific growth rate of *Salmonella typhimurium* in batch culture[a]

No.	Medium	Comments[b]	μ_{max} (h^{-1})
1	Brain + heart infusion	Full strength	1.94
5	Nutrient broth	Diluted 1:2 with medium No. 14	1.80
6	Nutrient broth	Diluted 1:5 with medium No. 14	1.66
7	Casamino acids	1.5% + 0.01% tryptophan in medium No. 14	1.39
9	20 amino acids	20 natural amino acids + mineral salts solution	1.27
10	8 amino acids	8 natural amino acids + mineral salts solution	1.01
14	Glucose salt	0.2% glucose + mineral salts solution	0.83
15	Succinate salt	0.2% succinate + mineral salts solution	0.66
19	Methionine salt	0.06% methionine + mineral salts solution	0.56
22	Lysine salt	0.014% lysine + mineral salts solution	0.43

[a]Adapted from Schaechter *et al.* (1958).

[b]Mineral salts solution contained $MgSO_4$, Na_2HPO_4, $Na(NH_4)HPO_4$, KCl, and citric acid as chelating agent. It supported no visible growth without the addition of a carbon source.

TABLE 62.6 Design of a carbon-limited minimal medium allowing the production of 10 g/l of dry biomass[a,b]

Medium constituent	Source of, function	Growth yield assumed (g dry biomass/g element)	Excess factor assumed with respect to carbon	Mass of element (g/l)	Mass of constituent (g/l)
Glucose	C, energy	1	1	10	25.0
NH_4Cl	N	8	3	3.75	14.33
NaH_2PO_4	P	33	5	1.52	5.88
KCl	K	100	5	0.5	0.95
$NaHSO_4$	S	100	5	0.5	1.87
$MgCl_2$	Mg	200	5	0.25	0.98
$CaCl_2$	Ca	100	10	1.0	2.77
$FeCl_2$	Fe	200	10	0.5	1.13
$MnCl_2$	Mn	10^4	20	0.02	0.046
$ZnCl_2$	Zn	10^4	20	0.02	0.042
$CuCl_2$	Cu	10^5	20	0.002	0.0042
$CoCl_2$	Co	10^5	20	0.002	0.0044

[a]Based on elemental growth yields obtained from the composition of dry biomass (see Table I).
[b]Based on Pirt (1975) and Egli and Fiechter (1981). Elemental growth yields for C and the trace elements Zn, Cu, Mo, and Mn were taken from Pirt (1975). Excess factors were chosen taking into account their variation observed in dry biomass.

calculate the minimum concentration of the different elements necessary in the culture medium to produce X_{max} using the individual average elemental growth yields ($Y_{X/E}$). To ensure an excess of all the nonlimiting nutrients in the medium, their concentrations are multiplied by an excess factor (F_E). In this way, the concentrations of the nutrients required in the growth medium (E_{req}) are present in a theoretically x-fold excess with respect to the carbon source.

$$E_{req} = \frac{X_{max}}{Y_{X/E}} \cdot F_E. \tag{3}$$

An example of the design of a carbon-limited medium supporting the production of 10 g/l of dry biomass is given in Table 62.6. Note that in this medium the ingredients are chosen in such a way that it is possible to change the concentration of each of the elemental nutrients individually (e.g. by including $MgCl_2$ plus $NaHSO_4$ instead of $MgSO_4$). In addition, this medium is only weakly buffered, hence, it might be necessary to control the pH during growth.

This approach works well for the design of media for the cultivation of aerobic microorganisms at low to medium biomass concentrations. More problematic is the design of media for anaerobic cultures in which many of the medium components precipitate easily at the required redox potential, or for high cell density cultures in which solubility or toxicity problems of some of the medium ingredients are encountered (see also Section V.B.2).

An estimate for most of the elemental growth yield factors $Y_{X/E}$ can be obtained from an elemental analysis of dry biomass cultivated under unrestricted growth conditions in batch culture (compare Table 62.3). For carbon, oxygen, and hydrogen, $Y_{X/E}$ cannot be calculated directly from the elemental composition of cells because these elements are not only incorporated into the biomass, but also serve other metabolic functions. For example, carbon is not only assimilated by heterotrophs but is also oxidized to CO_2 to supply energy (see also Fig. 62.2). Also, not included in this table is the amount of electron acceptor that has to be supplied to ensure growth. Table 62.7 shows the yield coefficients for oxygen, for some of the other common electron acceptors, and for some electron donors that support chemolithoautotrophic growth.

Two points influence the choice of excess factors. First, for elements whose cellular content does not vary considerably as a function of cultivation conditions, excess factors can be set low (N, P, and S), whereas for elements that are known to vary considerably (e.g. with growth rate), they are set higher. Second, the chemical behavior of the medium component in the growth medium has to be also taken into account when choosing F_E. For example, most of the trace elements easily precipitate in growth media at neutral and basic pH and as a result their biological availability is reduced (and difficult to assess). Therefore, they are added in a 10- to 20-fold excess despite the fact that a metal complexing agent is usually added to the medium to keep them in solution (see Bridson and Brecker, 1970).

TABLE 62.7 Some growth yield factors for electron donors and electron acceptors

Electron donors	
Molecular hydrogen	$Y_{X/H_2} \approx 12\,g/mol$
Thiosulfate	$Y_{X/S_2O_3} \approx 4\,g/mol$
Fe^{2+}	$Y_{X/Fe^{2+}} \approx 0.35\,g/mol$
NH_4^+ to NO_3^-	$Y_{X/NH_4^+} \approx 1.3\text{–}2.6\,g/mol$
NO_2^- to NO_3^-	$Y_{X/NO_2^-} \approx 0.9\text{–}1.8\,g/mol$
Electron acceptors	
Molecular oxygen	$Y_{X/O_2} \approx 10^a\text{–}42^b\,g/mol$
NO_3^- to N_2	$Y_{X/NO_3} \approx 27\,g/mol^c$
NO_2^- to N_2	$Y_{X/NO_2} \approx 17\,g/mol^c$
N_2O to N_2	$Y_{X/N_2O} \approx 9\,g/mol^c$

[a]For growth with reduced substrates such as methane or *n*-alkanes.
[b]For growth with more oxidized substrates such as glucose.
[c]For growth of *Paracoccus denitrificans* with glutamate as carbon substrate.

For biotechnological purposes, for which batch and fed-batch processes are primarily used, it would be advantageous to design media that contain all the nutrients in exactly the amount required, so that all nutrients would be consumed to completion at the end of growth. This, however, is difficult to achieve due to the variability of the yield factors for the individual elements and their dependence on the cultivation conditions. Nevertheless, one of the important points in medium optimization in biotechnology is certainly to optimize the consumption of nutrients and to minimize their loss.

2. *Some practical comments on the preparation of media*

It is appropriate to add some comments on some of the most important precautions to be taken when preparing a growth medium.

Many sugars easily deteriorate during sterilization at basic pH (especially in the presence of phosphates and peptones). This leads to a browning of the medium. The products that are formed can be inhibitory for growth. This can be avoided by sterilizing the medium at a slightly acidic pH, or by sterilizing the sugars separately from the medium.

It is well known that all trace metals easily form highly insoluble phosphate salts and precipitate in growth media. This can be avoided by the addition of metal-chelating agents such as EDTA, NTA, or sometimes also carboxylic acids such as citrate or tartrate. The addition of chelating agents has a twofold effect. On one hand, it prevents the precipitation of trace metals; on the other hand, it acts as a sink for these metals, and in this way reduces their toxicity by lowering their free (for the microbes accessible) concentration.

At medium $pH > 7$, the alkaline earth metals calcium and magnesium (as the trace metals) easily precipitate in the presence of phosphate (or in the presence of carbonate ions when using a bicarbonate buffered medium, or if only hard water is available) to form highly insoluble phosphate salts. These precipitates are sometimes difficult to see with the eye, especially in shake flasks, due to the small volume of the medium. To avoid this, the medium can be sterilized either at a slightly acidic pH (which requires that pH is adjusted later) or the phosphate salts are sterilized separately from the rest of the medium and combined after cooling.

For an elaborate treatment of this subject and more detailed information, especially on some of the established media, see the review by Bridson and Brecker (1970).

3. *Experimental identification of growth-limiting nutrient*

The variability of yield factors for the various nutrients, depending on the organism used and the compounds included in a medium, requires that the nature of the growth-limiting nutrient is experimentally verified for each case. For this, the maximum concentration of biomass (X) that can be produced in such a medium is determined as a function of the initial concentration of the medium component (S_0) that is supposedly growth-limiting, with the concentration of all other medium components kept constant. This experiment can be done in either batch or continuous culture.

The typical (theoretical) relationship obtained in such an experiment is visualized in Fig. 62.4. Ideally, the relationship between S_0 of a growth-limiting nutrient and X is initially a straight line that passes through the origin; that is, when no growth-limiting nutrient is added to the medium no biomass can be produced. When the S_0 exceeds a certain concentration, a deviation from the linear relationship will be observed. It is at this concentration that another nutrient becomes growth-limiting. Deviations from this relationship can be observed when one of the bulk salts used for medium preparation contains low amounts of the limiting nutrient as an impurity, or when a certain amount of the limiting nutrient becomes inaccessible in the medium, (e.g. due to precipitation with another medium component). (Note that variations in pH due to increasing concentrations of biomass or excreted toxic products can affect cultivation conditions and influence biomass yield also.)

In practice, the linear relationship between X and S_0 is often observed, although interpretation of the data

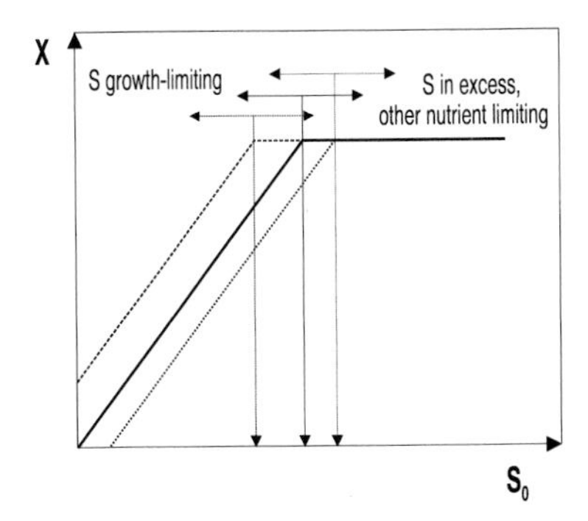

FIGURE 62.4 Concentration of dry biomass (X) that can be produced in a medium, as a function of the initial concentration of the growth-limiting substrate (S_0). (-----), X in a case where low amounts of the limiting substrate are introduced with an other medium component; (……), X in a case where part of the limiting substrate is not available for the cells, for instance, due to precipitation with another medium component.

is frequently not as straightforward as suggested by Fig. 62.4. This is demonstrated in Fig. 62.5 for *Pseudomonas oleovorans* growing in continuous culture at a fixed dilution rate with a mineral medium in which either carbon or nitrogen was limiting growth, depending on the ratio of the two nutrients. By keeping the concentration of ammonia constant and increasing the concentration of the carbon source this bacterium was cultivated at different C:N ratios. In the resulting biomass, ammonia and carbon source concentrations were measured at steady-state. When *P. oleovorans* was cultivated with citric acid as the sole source of carbon-energy the steady-state biomass concentration in the culture initially increased linearly with the concentration of the carbon source (Fig. 62.5a). Accordingly, the residual concentration of excess nitrogen decreased in the culture broth with increasing feed C:N ratios. At a C:N ratio of ~8.5 g/g, ammonia was consumed to completion. A further increase of the carbon concentration in the feed medium led to no further increase of the biomass produced. Instead, excess citric acid accumulated in the culture. Thus, one growth regime clearly limited by carbon with nitrogen in excess, and one limited by nitrogen with carbon in excess, can be recognized. When the same experiment was performed with octanoic acid as the sole carbon source, the biomass concentration also increased initially when growth was carbon-limited with excess nitrogen present in the culture (Fig. 62.5b). At a feed C:N ratio of 7.0, the residual concentration of nitrogen in the culture became undetectable. Despite this, the concentration of the biomass in the culture continued to increase linearly, with all the carbon consumed to completion when the concentration of octanoate was further increased in the feed medium. Only when the C:N ratio in the feed exceeded 14.5 did unutilized octanoate became detectable in the culture liquid. Thus, limitation based on the pattern of the biomass concentration growth became nitrogen-limited above a C:N ratio of 18.3. The residual concentration of the nitrogen source in the culture indicated the limitation of the culture by nitrogen already at C:N feed ratios higher than 7.0. The analysis of the cells showed that the effect observed was a result of channelling the surplus carbon into the formation of the reserve material polyhydroxyalkanoate (in other organisms this may be polyhydroxybutyrate, glycogen, or lipids). This dual-nutrient-(carbon–nitrogen)-limited growth regime is always observed when the organism has the ability to store carbonaceous reserve materials. Here, the extension of this zone between the two single-nutrient-limited growth regimes depends on the storage capacity of the organism for reserve material and the growth rate (Egli, 1991; for effects observed in batch culture see Wanner and Egli, 1990). Such multiple-nutrient-limited zones are not only observed for the interaction of carbon with nitrogen but also for other combinations of nutrients, such as C–P, C–Mg, or N–P. Furthermore, the extension of this zone will be

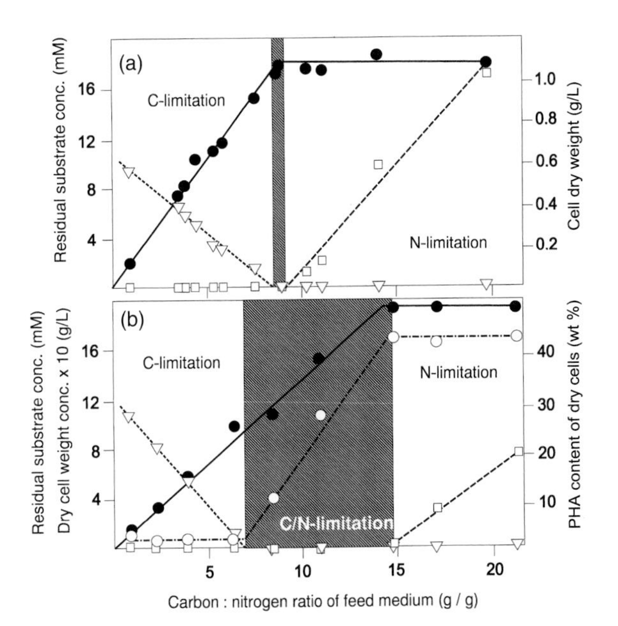

FIGURE 62.5 Growth of *Pseudomonas oleovorans* with either citrate (a) or octanoate (b) as the sole source of carbon in continuous culture at a fixed dilution rate of 0.20/hr as a function of the C:N ratio of the feed medium. The C:N ratio of the feed medium was varied by keeping the concentration of the nitrogen source (NH_4^+) constant and changing the concentration of the carbon source. (Adapted from Durner R. (1998). Ph.D. thesis No. 12591, Swiss Federal Institute of Technology, Zürich, Switzerland) ●, indicates cell dry weight; ▽, ammonium-N concentration; □, concentration of citric acid (a) or octanoic acid (b) in the culture; ○, polyhydroxyalkanoate (PHA) content of cells.

determined by the limits a microorganism exhibits with respect to its elemental composition under differently limited growth conditions.

Thus, even when a linear relationship between the biomass produced is obtained, care has to be taken in the interpretation. In such a case, it is of advantage to know which of the nutrients is the second limiting component in the medium.

C. Assessing the quality of media and some notes of caution

This approach cannot only be used to design growth media limited by a specifically selected nutrient, but it can also be used to assess the quality of various media. (Never assume that growth media reported in the literature are perfect. Many times they are not, nor are they necessarily employed for the purpose they were originally designed for!) Such an analysis gives usually a good understanding of the capacity of a growth medium with respect to the maximum biomass that it can support, the nature of the limiting nutrient, and the degree of excess of other nutrients (Egli and Fiechter, 1981).

Note that most of the classical media used before the 1960s did not include trace elements. Their addition was usually not necessary because trace elements were contained as impurities in the bulk minerals used for the preparation of the medium. Modern media are frequently prepared with "ultrapure" salts, and, not surprisingly, they fail to produce good growth unless they are supplemented with a trace-element solution. A typical example is the classic synthetic medium M9 that is used widely to grow *E. coli* in genetic studies. This medium in its original composition does not support the growth of *E. coli* for more than a few generations, after which growth slows down and finally comes to a halt.

ACKNOWLEDGMENT

The author is indebted to Peter Adriaens for his careful reading of the manuscript and for valuable comments.

BIBLIOGRAPHY

Atlas, R. M. (1997). "Handbook of Microbiological Media," 2nd edn. CRC Press, Boca Raton, FL.

Balows, A. *et al.* (1992). "The Prokaryotes: A Handbook of the Biology of Bacteria," 2nd edn. Springer Verlag, New York.

Bridson, E. Y., and Brecker, A. (1970). Design and formulation of microbial culture media. *In* "Methods in Microbiology" (J. R. Norris and D. W. Ribbons, eds), Vol. 3A, pp. 229–295. Academic Press, New York.

Egli, T. (1991). On multiple-nutrient-limited growth of microorganisms, with special reference to dual limitation by carbon and nitrogen substrates. *Antonie van Leeuwenhoek* **60**, 225–234.

Egli, T., and Fiechter, A. (1981). Theoretical analysis of media used for the growth of yeasts on methanol. *J. Gen. Microbiol.* **123**, 365–369.

Herbert, D. (1961). The chemical composition of microorganisms as a function of their environment. *Symp. Soc. Gen. Microbiol.* **11**, 391–416.

Herbert, D. (1976). Stoicheiometric aspects of microbial growth. *In* "Continuous Culture 6: Application and New Fields" (A. C. R. Dean *et al.*, eds), pp. 1–30. Ellis Horwood, Chichester, UK.

Kovářová, K., and Egli, T. (1998). Growth kinetics of suspended microbial cells: From single-substrate-controlled growth to mixed-substrate kinetics. *Microbiol. Molec. Biol. Rev.* **62**, 646–666.

LaPage, S. P., Shelton, J. E., and Mitchell, T. G. (1970). Media for the maintenance and preservation of bacteria. *In* "Methods in Microbiology" (J. R. Norris and D. W. Ribbons, eds), Vol. 3A, pp. 1–133. Academic Press, New York.

Linton, J. D., and Stephenson, R. J. (1978). A preliminary study on growth yields in relation to the carbon and energy content of various organic growth substrates. *FEMS Microbiol. Lett.* **3**, 95–98.

Neidhardt, F. C., Ingraham, J. L., and Schaechter, M. (1990). "Physiology of the Bacterial Cell: A Molecular Approach." Sinauer, Sunderland, MS.

Pirt, S. J. (1975). "Principles of Microbe and Cell Cultivation." Blackwell Scientific, Oxford, UK.

Schaechter, M., Maaløe, O., and Kjeldgaard, N. O. (1958). Dependency on medium and temperature of cell size and chemical composition during balanced growth of *Salmonella typhimurium*. *J. Gen. Microbiol.* **19**, 592–606.

Tempest, D. W. (1969). Quantitative relationship between inorganic cations and anionic polymers in growing bacteria. *Symp. Soc. Gen. Microbiol.* **19**, 87–111.

Wanner, U., and Egli, T. (1990). Dynamics of microbial growth and cell composition in batch culture. *FEMS Microbiol. Rev.* **75**, 19–44.

WEBSITE

Nutrition and Growth of Bacteria (K. Todar)
http://www.bact.wisc.edu/Bact303/NutritionandGrowth

63

Oral microbiology

Ian R. Hamilton and George H. Bowden
University of Manitoba

GLOSSARY

acquired pellicle The organic film on the surface of the tooth consisting mainly of protein and glycoproteins from saliva.
alveolar bone Bone of the jaws in which the teeth are embedded.
approximal Adjacent surfaces of teeth.
biofilm Microorganisms immobilized at a surface and frequently embedded in an organic layer (acquired pellicle) comprised of polymers of host and microbial origin.
checkerboard analysis Microbial analysis that uses whole genomic DNA probes where samples of oral bacteria are placed along one axis and the probes placed on the second axis.
community A collection of bacterial populations growing together in a defined habitat.
clone A population of cells descended from a single parental cell.
genomic species A group of organisms that share a high (>70%) DNA–DNA similarity value.
dental plaque The microbial biofilms on teeth.
gingival margin Soft tissue (gums) next to the tooth.
indigenous flora Bacteria which comprise the normal or resident flora of the ecosystem.
interproximal The region between adjacent teeth.
occlusal The surface of the tooth that is used for chewing.
periodontal pocket Pocket created adjacent to the tooth below the gingival margin.
population Bacteria of the same species, serovar, phage type or DNA fingerprint, found in a bacterial community in a given habitat.
subgingival plaque Bacteria growing below the gingival margin between the teeth and gingival tissue.
supragingival plaque Bacteria growing on the tooth surface above the gingival margin.

Oral Microbiology is the study of the bacteria that are natural inhabitants of the oral cavity. While most of the information on these bacteria relates to those genera and species that can be found in the human mouth, some information is available on the bacteria present in the mouths of a variety of animal species. Most attention has been paid to those bacteria which can be found on human tooth surfaces in microbial communities known as dental plaque, since bacteria in plaque are etiological agents of dental caries. In addition, specific bacteria growing at the margin of the tooth and the soft tissue or gingiva (gums) are responsible for the condition known as gingivitis, which results in inflammation and bleeding of the gums. Failure to remove the bacteria at the gingival margin can lead to more complex pathological conditions, collectively known as periodontal disease, which is generally characterized by the destruction of the tissue adjacent to the tooth creating "periodontal pockets." The ultimate form of periodontal disease is destructive periodontitis, a chronic condition in which pockets of up to 8–10 mm can be observed containing a wide variety of anaerobic bacteria. Associated with tissue destruction and pocket formation is the concomitant loss of the alveolar bone which anchors the teeth in the jaws and, as a consequence, advanced

The Desk Encyclopedia of Microbiology
ISBN: 0-12-621361-5

periodontal disease often leads to tooth loss. Oral bacteria are also involved in significant infections elsewhere in the body, such as lung and brain abscesses. Also, relatively recently, oral bacteria have been implicated in heart disease.

I. ORAL MICROBIAL ECOLOGY

The characteristics and properties of the microflora at various sites in the mouth, like those of any microbial ecosystem, are regulated by the nature of the habitat and the associated environment. In this respect, some discussion of the microbial ecology of the mouth is important, particularly since the two major diseases of dental caries and periodontal disease arise through alterations of the normal, or indigenous, flora rather than the invasion of foreign or alien infectious agents from outside the mouth.

A. General characteristics of the oral environment

The mouth has two major types of surface that can be colonized by bacteria: the shedding mucosal surface of the soft tissues and the non-shedding mineralized surface of the teeth. As a consequence of the unique properties of these two types of surface, both the concentration and characteristics of the flora colonizing these areas are different. In addition, each type of surface harbors many distinct microbial ecosystems that have arisen under the different microenvironments in the mouth. In general, the flora is regulated by the nutrient supply, the pH of the environment, the oxygen level, antimicrobial agents, and microbial interactions, as well as the mechanical forces of mastication (chewing) and saliva flow. Most of the bacteria in the developed ecosystems in the mouth are either facultative or obligate anaerobes. Aerobic bacteria are found in significant proportions in the early stages of microbial biofilm development, but are replaced later by the anaerobic flora and may be restricted to the surface layers of the community. Saliva plays a significant role in the development and maintenance of the oral microflora by providing nutrients and antimicrobial factors, as well as buffers and electrolytes. Adequate salivary clearance of sugar substrates and acid end-products resulting from carbohydrate metabolism by plaque bacteria is recognized as being essential for limiting the acid demineralization of the tooth enamel associated with dental caries.

1. *Mineralized tissues*

The surfaces of teeth support a variety of complex microbial ecosystems known collectively as dental plaque, which includes the plaque microflora, extracellular matrix and the associated fluid phase (plaque fluid). Depending upon an individual's oral hygiene, supragingival plaque can contain approximately 10^{11} organisms per gram wet weight with the microbial communities comprising at least 30 genera and 500 species. Early studies of dental plaque viewed the oral microflora as being similar at all sites on the tooth surface, and the properties and characteristics of plaque were studied in "pooled" samples. However, research on the microflora of plaque over the past 30 years has involved more sophisticated microbiological methods, particularly with regard to the cultivation and identification of anaerobic bacteria, and with this has come the concept that each small site on the tooth surface is comprised of a specific and unique microflora.

The development of the plaque biofilms on a clean tooth surface in the mouth begins with the deposition of an organic layer, the acquired pellicle, that is made up almost entirely of salivary proteins. Subsequent disruption of established dental plaque, however, can introduce to this organic layer proteins, polysaccharides and lipoteichoic acid of bacterial origin. The acquired pellicle serves as a nutrient source, as a homeostatic barrier to retard the loss of mineral from the tooth surface and as a promoter of plaque formation. A significant amount of research has been directed at the specific and non-specific factors involved in the adherence of oral bacteria to tooth enamel (hydroxyapatite) and to salivary-coated hydroxyapatite. This research has indicated that selected salivary proteins and glycoproteins have affinity for the hydroxyapatite surface, while certain oral bacteria, such as *Streptococcus sanguis*, *S. gordonii*, *S. mitis*, *S. oralis*, and *Actinomyces* species, have higher affinity for this surface that other members of the oral microflora. If left undisturbed, highly complex and diverse microbial communities can be established over the mineralized tooth surface and adjacent soft tissue within 3–5 days. This development involves the specific coaggregation among a wide variety of bacteria, often involving the interaction of specific cell surface components, as well as, salivary macromolecules.

Studies of the changes in oral bacterial communities during their accumulation with time have been limited to enamel surfaces (i.e. dental plaque). Early studies on the development of bacterial communities *in vivo* considered those bacteria which accumulated after 24 h and the subsequent increase in biomass complexity over periods of up to 14 days. These studies did not survey the total flora, however, the bacterial species which could be isolated most readily from enamel at 24 h were *S. sanguis* and some species of

Neisseria, Rothia, Veillonella, and *Actinomyces*. The predominantly coccal plaque persisted for 72 h after which it gradually became populated with filamentous bacteria (*Actinomyces, Leptotrichia, Eubacterium*). Concomitant with this increase in diversity was a reduction in the oxidation–reduction potential which allowed the flora to become more anaerobic and anaerobic bacteria began to compete effectively with strict aerobes which decreased in numbers with time. Importantly, species within the facultative genera, *Streptococcus* and *Actinomyces*, maintained themselves in high numbers in plaque competing effectively with other members of the community.

Dental plaque can be divided into five major types based on the location of the habitats involved: supragingival, fissure, carious lesions, gingival margin, and subgingival. Supragingival plaque is a general term including those bacteria in the many small ecosystems that reside on the tooth surface above the gingival margin. For example, unique microbial communities can be observed on smooth tooth surfaces, in the interproximal areas between the teeth and in small pits that can range over the tooth surface. Bacteria in supragingival plaque receive their nutrients both from dietary constituents and from saliva, and a significant fraction of the bacteria in these communities utilize carbohydrates as their principal energy source. Large fissures, particularly those on the biting (occlusal) surfaces of the teeth, constitute a second major and unique type of microbial ecosystem, while open (overt) carious lesions are a third type that also support unique bacterial communities.

The organisms at the gingival margin between the gum line and the tooth form a fourth general type of bacterial community. Bacteria in this region colonize a transition zone between supragingival and subgingival habitats, with the latter habitat supporting a complex community below the gingiva in subgingival plaque. In normal healthy individuals, a small pocket (sulcus), 1 or 2 mm in depth, exists below the gingival margin and the bacteria in this region receive small amounts of a protein-rich liquid (crevicular fluid) that has its origin in plasma. Thus, the bacteria in the communities at the gingival margin have three nutrient sources, diet, saliva, and crevicular fluid, the last of which contains plasma proteins and certain factors, such as hemin, that can be utilized by the more fastidious oral bacteria. Crevicular fluid also contains significant amounts of the immunoglobulins, IgG and IgM.

Failure to remove the bacteria at the gingival margin can lead to inflammation (gingivitis) and eventual destruction of the soft tissues next to the teeth, creating a pocket containing a largely Gram-negative anaerobic flora with a considerable capacity for proteolysis. Individuals with such pockets have "periodontal disease" and the ultimate stage of the disease is known as "destructive periodontitis." This condition is characterized by the presence of large pockets 8–10 mm in depth created by the destruction of the adjacent soft tissue and the periodontal ligament which anchors the tooth to the alveolar bone of the jaw. Alveolar bone destruction also occurs in advanced cases leading to the destabilization of teeth, which can often lead to tooth loss.

2. *Soft tissue*

The number of bacteria on the cheek and palatal surfaces are of the order of 5–25 per epithelial cell. The tongue, on the other hand, is a unique structure possessing numerous crypts and papillae, and epithelial cells from the dorsum of the tongue can carry more than 100 bacteria each per cell. Because the surface irregularities of the tongue allow the development of relatively high densities of bacteria, the tongue is believed to be the major source of the bacteria in saliva. Normally, the number of bacteria on oral mucosal surfaces is limited because of the continuous loss of surface epithelial cells (desquamation) and by the low growth rate of the bacteria on these surfaces; the biomass of the bacteria on oral mucosal surfaces is estimated to double only two to three times per day. Thus, epithelial cell desquamation in the mouth is an important host defense mechanism which limits the number of bacteria present on the surface at any one time. Since each surface epithelial cell has a finite life span, the growth rate of the attached bacteria is not a significant factor in determining the proportion of bacteria in the surface population. The colonization of epithelial cell surfaces appears to be determined largely by the capacity of the organism to attach and by the concentration of bacteria in the fluid phase near the surface.

3. *Saliva*

Saliva secreted into the oral cavity from the salivary glands is sterile, but soon becomes contaminated by the microflora in the existing saliva and on the various surfaces in the mouth. The bacterial concentration in saliva can approach 10^9 organisms per milliliter and fluctuates during the day, with high numbers being found following periods of sleep when the saliva flow in the mouth is significantly reduced. In addition, higher numbers are present during mastication due in part to the partial removal of bacteria from the surfaces of the mouth. The concentration of some bacteria in saliva is directly proportional to the numbers of the same bacteria in dental plaque. This observation has been used as the basis for a variety of assays of saliva to test for the presence of certain bacteria, such as species of *Streptococcus* and *Lactobacillus*, that are associated with dental caries.

B. Factors influencing oral microbial ecosystems

The oral microflora, like any natural bacterial community, is influenced by its environment. Changes in the community can be brought about by variation in the characteristics of the environment. These may be: (a) nutritional (e.g. a change in the balance of nutrients available [such as the proportion of carbohydrate in the diet]); (b) physical (e.g. loss of a surface for colonization [tooth loss]); (c) physiological (e.g. a change in the oxidation–reduction potential of the environment); or (d) competitive (e.g. certain bacterial populations will compete effectively with other populations resulting in the exclusion or reduction of these latter bacteria). Thus, if the microflora is left undisturbed, there will be a modification or a succession of the bacterial populations in the mouth until a degree of stability (climax community) is reached. Normally, stable climax plaque communities will be observed 5–7 days after the start of plaque development. The environmental factors influencing the oral flora are of two types: those determined by the host and those of an external nature.

1. *Host-associated factors*

A variety of physical–chemical factors influence the types and concentrations of the bacteria in the mouth. These include the influence of: (a) saliva and other body fluids; (b) the nature of the site or habitat; (c) the age of the host and the microbial community; and (d) microbial interactions.

a. Body fluids

Saliva interacts with all bacteria in the mouth except those found in deep fissures and carious lesions of the tooth, and in periodontal pockets. The constituents of saliva include a variety of proteins, glycoproteins, lipids, carbohydrates, and antimicrobial factors, as well as buffers and electrolytes (Table 63.1). The various polymers in saliva include enzymes, immunoglobulins, non-immune agglutinins and glycoproteins that are involved in adherence of bacteria to surfaces and in the binding of iron (lactoferrin), as well as other molecules. Although the flow rate of saliva can vary widely depending on the degree of stimulation of the salivary glands, saliva provides a ready nutrient source for bacteria on the various surfaces of the mouth. This has been shown in studies with subjects who obtained their nutrition by stomach tube. Bacteria developed on the teeth even though the subjects were not ingesting food via the mouth. Such plaque is comprised of fewer cells than that of individuals eating normally and contained fewer acid-producing (acidogenic) bacteria capable of metabolizing carbohydrate. Saliva has been shown to support the growth of a variety of bacteria that have the capacity to degrade salivary proteins and glycoproteins to their constituent amino acids and sugars, which are then utilized by bacteria for growth.

The flow rate and buffering of saliva are powerful forces for the clearance of toxic components, such as acids, from the tooth surface and for the prevention of acid demineralization of the enamel. One of the best demonstrations of the significance of the effects of saliva on dental plaque is in subjects with little or no salivary flow. Some patients with oral tumours undergo radiation therapy and, as a result, the output of the salivary glands is often seriously impaired, with salivary flow reduced by up to 95%. Absence of saliva is called "xerostomia" and patients with this condition develop caries at a rapid rate because of increases in the proportion of acidogenic and acid-tolerant (aciduric) bacteria in the plaque microflora following the reduction in saliva flow. Control of the caries can be achieved by use of fluoride mouth rinses, good oral hygiene, and by reducing the carbohydrate content of

TABLE 63.1 Components of human saliva

Enzymes	*Complex carbohydrates*
Amylase	Glycoproteins
Lysozyme	Blood group substances
Salivary peroxidase	*Immune system components*
Phosphatases	sgA, IgG, IgM
Esterases	Secretory component
β-Glucuronidase	Complement factors
Transaminases	*Small molecules*
Ribonucleases	Amino acids
Dehydrogenases	Glucose
Kallikreins	Lactate, citrate
Lipase	Ammonia, urea
Other macromolecules	Cholesterol
Albumin	Uric acid
Lactoferrin	Creatinine
Lipoproteins	*Cations/anions*
Ceruloplasmin	Na^+, K^+
Proline-rich proteins	Ca^{2+}, Mg^{2+}
Statherin	Cl^-
Cystatins	Phosphates
Histatins	Carbonates
Orosomucoid	Thiocyanate
	F^-, Br^-, I^- (trace)

Adapted from Mandel, I. D. (1974). *J. Dent. Res.* **53**, 246–266.

the diet. Saliva also contains antimicrobial agents, such as secretory IgA, salivary peroxidase/ thiocyanate, lactoferrin, and lysozyme. Since the resident or indigenous flora are resistant to these agents, these factors promote the homeostasis of the plaque microbial ecosystems by the elimination of invading microorganisms.

Another fluid of importance to bacteria near the gingival margin is gingival crevicular fluid that seeps from the sulcus at the tooth/tissue junction. This fluid contains plasma components and is, therefore, rich in proteins, vitamins, and factors that promote the growth of the more sensitive anaerobic oral bacteria in the gingival margin and sulcus area.

The microflora in the mouth is exposed to antibody from two major sources, saliva and crevicular fluid. Salivary immunoglobulins are a component of the mucosal immune system and as such comprise dimeric IgA with a secretory piece (SIgA) which is relatively resistant to proteases. However, some strains of *S. oralis* and *S. mitis* produce an SIgA protease which cleaves IgA1 at the hinge region, and it is thought that this enzyme could play a role in facilitating colonization by SIgA protease-producing bacteria. Serum immunoglobulins, including IgM and IgG, enter the mouth in crevicular fluid and consequently one might expect the diversity and specificity of antibodies of these classes to reflect those in the host's serum. In addition, there is local production of immunoglobulin by plasma cells within the gingiva.

Generally, it is thought that oral antibodies are most likely to exert their effect by influencing oral colonization by bacteria. However, examination of the specificities of either SIgA or IgG and IgM antibody in saliva shows that they react with resident oral bacteria. Despite this interaction, the presence of specific antibody does not appear to influence the survival of the oral flora. The reasons why this is so are not well understood but several mechanisms have been proposed: (a) that once the bacteria are established in the plaque biofilm, antibody is not effective; (b) that the antigens recognized are not significant in adhesion and subsequent colonization; (c) that bacteria among the resident flora are coated with host molecules from saliva and "hidden" from antibody; (d) that oral bacteria may express antigens closely similar to those of the host (molecular mimicry); (e) that the antibodies recognizing oral bacteria are natural, generated against common antigens to the host and of low avidity; and (f) that host tolerance is generated to oral bacteria during the initial colonization of infants. It seems likely that all of these mechanisms, and others that we do not understand, could influence the response of the host to the oral resident flora and also the activity of the immune system against oral organisms. There is some evidence that tolerance to *Actinomyces* and *Streptococcus* species occurs during initial colonization of infants. Host responses to the oral flora are of obvious significance when immune prophylaxis or control of oral diseases, such as caries or periodontal disease, is contemplated. In caries, active and passive immunity to *Streptococcus mutans* in animals and, in some cases, in humans, influences the extent of oral colonization by this species. One approach to the use of passive immunization in humans has been to use antibody to prevent or reduce the re-colonization by *S. mutans* after the levels of the organism have been reduced by the application of chlorhexidine.

b. Habitat

Bacteria that grow successfully on the tooth surface do not compete with those on mucosal surfaces because the two sites represent distinctly different environments. Although numerous individual unique microbial communities are found on the surface of a single tooth, there will be some bacteria common to all sites, however, the ratios and species will vary from site to site even in adjacent areas on a tooth surface. This is particularly true of the microflora in approximal sites on adjacent teeth and the approximal plaque from identical sites on the left and right sides of the same mouth. Teeth can also possess pits and fissures, and the communities present in these sites are less complex than those from smooth or approximal surfaces, primarily because the environment of a fissure is usually acidic and aciduric bacteria will survive in the deeper layers of this habitat. For this reason, the composition of fissure plaque is relatively stable and resembles in many ways the flora of a caries lesion.

c. Age of the host

The age of the host has an influence on the complexity and distribution of the oral flora. At birth, the infant's mouth is sterile, but it is colonized rapidly by bacteria from the mother. Maternal transfer has been confirmed by bacteriocin and genetic typing of isolates from mother and child. Streptococci are common among the early colonizing bacteria with *S. mitis* biovar 1, *S. oralis* and *S. salivarius* predominating, while *S. mitis* biovar 2, *S. sanguis*, *S. anginosus*, and *S. gordonii* are isolated less frequently and in lower numbers. Strains of *Actinomyces*, *Neisseria*, and *Veillonella* may also be present during the early days of life. Some organisms isolated from the infant's mouth are transients. These strains can originate from the environment and from other habitats associated with the mother. Typical of transient bacteria in the mouth's of infants immediately following birth are maternal fecal lactobacilli. These bacteria fail to survive in the mouth, presumably because of the differences between the fecal and oral environments. Eruption of the primary dentition causes significant increases in the complexity (diversity) of the oral flora. Tooth surfaces provide hard non-shedding surfaces for formation of complex biofilms and juxtaposition of the gingiva and tooth enamel provide new habitats. After tooth eruption, *Actinomyces* become established in the mouth and *Streptococcus* spp. in the plaque biofilm show different proportions to those on soft tissues. The diversity of the oral flora increases over time and within 5 years many of the bacteria commonly isolated from adults, such as *Prevotella* spp., are resident in the child's mouth.

d. Microbial interactions

In the absence of dental hygiene procedures, the accumulation of bacteria on the non-shedding surfaces of the teeth results in the formation of many complex and highly diverse plaque ecological systems. Microbial interactions are known to stimulate or inhibit microbial growth in these habitats. As plaque develops following cleaning, the initial predominantly Gram-positive flora gives way to a more Gram-negative and anaerobic flora. The reduction in the redox potential which permits the emergence of the anaerobic bacteria arises through the metabolic action of the facultative populations, which use oxygen as a terminal electron acceptor. This reduces the partial pressure of oxygen in the deeper layers of plaque and allows the facultative and obligate anaerobes to proliferate. The commensalistic relationship between facultative and obligate anaerobes was one of the first observations made in the field of "oral microbial ecology" in the early 1960s. Subsequent studies have demonstrated other beneficial elements, such as food chains, the generation of growth factors and extracellular energy polymers, as well as, much evidence of competition for nutrients and the formation of growth-inhibiting products including hydrogen peroxide, acids, and bacteriocins.

2. *External factors*

Diet and the use of antimicrobial agents are two factors having significant influences on the oral flora. The complex nature of the microflora and the natural antimicrobial agents of the host preclude the invasion of this environment by alien bacteria unless the host has been severely compromised by disease or by the application of drugs.

a. Diet

As mentioned previously, the consumption of food is not essential for the development of bacterial communities in the mouth, however, many studies have established a direct relationship between the consumption of refined sugar (sucrose) and dental caries. Although little is known of the changes in plaque on small localized areas of the tooth surface, it is known that significant changes in the biochemical properties and, to some extent, the proportions of various bacterial populations occur when the sugar content of the diet increases. In general, the flora becomes more acidogenic with lactic, acetic, and formic acids being the predominant acid end-products of sugar metabolism. Increased sugar intake by the host also results in an increase in the rate of sugar consumption and acid formation by the saccharolytic oral flora as these bacteria "adapt" to increases in this energy source. Since high-carbohydrate diets result in the formation of a more acidic plaque environment, aciduric bacteria such as *Lactobacillus* species and *S. mutans* tend to flourish and dominate the plaque ecosystems when high sugar intake is maintained for a period of time. Some studies have also reported concomitant increases in the lactic acid-utilizing *Veillonella* which are probably associated with the increased levels of lactic acid in the plaque matrix. The detection of high level of *Lactobacillus* species and *S. mutans* strains in human saliva is being used in some countries as a measure of caries activity. The conclusion to be drawn from studies on dental plaque from a variety of races (Japanese, Western European, Russian, New Guinea, and Australian Bushmen) is that, in general, plaque composition is relatively independent of diet. It seems likely that the bacterial communities can adapt to a wide range of diets and only a change in the ratios of the populations may occur. These changes in population ratios in the microbial communities may be most easily detected after the consumption of high-carbohydrate diets.

b. Antimicrobial agents

Firm evidence exists for the alteration and the partial elimination of the oral flora by applications directly to the mouth of high concentrations of chemical agents, such as fluoride, iodine, chlorhexidine, and other compounds. More controversial is the potential antimicrobial effect of low levels of water-borne fluoride (1 μg/ml). Such levels are too low to influence the growth of oral bacteria, but result in the partial inhibition of carbohydrate metabolism by the saccharolytic flora, thereby, reducing the rate of acid formation on the tooth surface. The administration of antibiotics to the host can have profound effects on large communities of bacteria, often leading to an imbalance in the microbial ecological equilibrium in the mouth. An example of this would be the overgrowth of the oral cavity by *Candida ablicans* and related species producing a condition known as candidiasis. These yeastlike fungi are normal inhabitants of the mouth and the elimination of the bacterial flora permits the transformation of these organisms from a commensal to a pathogenic role.

II. THE ORAL MICROFLORA

A. Normal oral flora

Any description of the resident flora of the oral cavity must take into consideration the variety of small ecosystems in the various oral habitats. The normal flora of such ecosystems describes those bacteria which are regularly present in all healthy members of the host population. The qualitative assessment of oral bacterial communities indicates that they are comprised of a relatively stable collection of species and strains can be found in similar habitats in the mouths of different members of the same species of animal. As mentioned previously, variations occur in the numbers and ratios of the organisms in a given mouth and can be related to diet, physiology, oral hygiene, disease, and the impact of agents such as fluoride, which are in general use by the human population.

Comparisons made among different species of animals show that while some genera are common to several mammals, different animal species show variations in the complexity of their flora and harbour different species of a given microbial genus. Among the best-studied animals, apart from humans, are those which naturally develop caries or periodontal disease, and those in which these diseases can be produced experimentally. The oral flora of the non-human primates is similar to that of humans, while dogs and cats are also known to have complex microbial communities. The flora of other animals, such as hamsters, rats and mice, is less complex. Few studies of the oral flora in herbivores, insectivores and carnivores, other than cats and dogs, have been made and nothing is known of the microflora which is resident in the oral cavities of reptiles and fishes. Even in humans and primates, relatively few studies have been directed at defining the number of microbial species and the diversity of oral bacterial communities. An overview of the flora associated with the oral cavities of humans and animals is shown in Table 63.2. This includes some examples of genera present in the oral cavities of animals along with a listing of a few important species to indicate the variation among animals. The animals examined were all in captivity and little is known of the oral microflora of animals in their natural habitats.

It is likely that given similar environments in different animal species, similar microbial genera will be present. For example, spirochetes and *Peptostreptococcus* spp. are present in dogs with periodontal disease. However, *Staphylococcus*, which is unusual in human mouths, can be isolated from the mouths of mice, reflecting a different habitat. Different species having similar niches may also be present in the mouths of different animals. One example of this is the presence of *Actinomyces denticolens*, *A. slackii*, and *A. howellii* in cattle. These three species are physiologically similar to *A. naeslundii* genomic species 1 and 2 isolated from humans and probably occupy the same niches in the oral communities in cattle as these latter organisms do in humans.

Currently, at least 30 genera have been identified in the human oral flora (Table 63.3). The approaches to defining the taxonomy of oral bacteria do not differ from those used for organisms associated with other surface habitats of humans, however, knowledge of the diversity within many genera at the species level and below is limited. This is particularly true for species and strains which have not been associated with oral disease and which might be considered members of the resident or natural flora. However, as the studies of the microflora of diseased sites have become more detailed, more precision has been used in the classification of isolates in keeping with the realization that several different species or strains of oral bacteria can be involved in oral disease. In particular, identification of bacteria by use of DNA probes and their identification from 16S rRNA sequencing has made it clear that in health and disease the oral flora is extremely complex with a diversity of species and strains that had not been appreciated previously. As examples, the use of the "checkerboard" analysis by Socransky and his colleagues has revolutionized the analysis of oral bacterial communities at the species level. The increased precision and discrimination provided by more sophisiticated tests used in analysis of larger numbers of isolates has resulted in recognition of more species diversity within some oral genera and also the revision of older classifications to define new genera and

TABLE 63.2 Bacteria common to humans and animals

Organisms	Human	Monkey	Cattle	Giraffe	Tiger	Cat	Dog	Rat	Mouse
		Animals							
Streptococcus	+	+	+	+	+	+	+	+	+
S. sanguis	+	+	–[a]	ND[b]	ND	ND	+	–	–
S. mitis/mitior[d]	+	+	–	+	ND	ND	+	–	–
S. salivarius	+	+	–	ND	ND	ND	ND	–	–
Actinomyces	+	+	+	+	+	+	+	+	+
A. viscosus[e]		(+)	(+)[c]	(+)	(+)	(+)	(+)	(+)	(+)
A. naeslundii	+	+	(+)	(+)	(+)	(+)	(+)	(+)	(+)
A. howellii	ND	ND	+	–	–	–	–	–	–
A. denticolens	ND	ND	+	–	–	–	–	–	–
A. slackii	ND	ND	+	–	–	–	–	–	–
Bacteroides[f]	+	+	+	+	+	+	+	+	–
Pigmented proteolytic	+	+	–	+	+	+	+	–	–
Pigmented saccharolytic	+	+	–	+	+	+	+	–	–
Non-pigmented	+	+	–	+	+	+	+	–	–
Porphyromonas	+	+	–	–	–	+	+	–	–
P. gingivalis	+	(+)	–	–	–	(+)	(+)	–	–
Porphyromonas spp.[g]	+	+	–	–	+	+	+	–	–
Prevotella	+	+	(+)	(+)	(+)	(+)	(+)	–	–
P. intermedia	+	+	–	–	–	–	–	–	–
P. melaninogenica	+	+	–	–	–	–	–	–	–
Prevotella spp.[h]	+	+	(+)	(+)	(+)	(+)	(+)	–	–
Fusobacterium	+	+	–	ND	+	+	+	–	–
Veillonella	+	+	–	ND	ND	–	+	–	–
Neisseria	+	+	+	+	+	+	+	–	–
Lactobacillus	+	+	–	–	–	–	–	–	+

[a]– denotes "no information."
[b]ND denotes "not detected."
[c](+), denotes the presence of bacteria which resemble the species.
[d]Strains designated *Streptococcus mitior* have been reclassified as *Streptococcus mitis*.
[e]*Actinomyces viscosus* is now to be limited to strains in animals other than humans. Human strains are now included in *A. naeslundii* genomic species 2.
[f]Several oral *Bacteroides* have been reclassified into *Porphyromonas* and *Prevotella*.
[g]*Porphyromonas* isolated from animals (see Table 63.4).
[h]*Prevotella* includes oral organisms previously classified as saccharolytic *Bacteroides*.

TABLE 63.3 Bacterial and fungal genera of the human oral flora

Gram-positive rods and filaments	Gram-negative rods and filaments	Gram-positive cocci	Gram-negative cocci
	Bacteria		
Facultative/aerobic			
Actinomyces	*Actinobacillus*	*Streptococcus*	*Neisseria*
Corynebacterium	*Capnocytophaga*	*Stomatococcus*	*Branhamella*
Lactobacillus	*Eikenella*	*Enterococcus*	
Rothia	*Hemophilus*		
	Simonsiella		
Anaerobic			
Actinomyces	*Bacteroides*	*Peptostreptococcus*	*Veillonella*
Bifidobacterium	*Campylobacter*		
Eubacterium	*Centipeda*		
Propionibacterium	*Fusobacterium*		
	Leptotrichia		
	Mitsuokella		
	Porphyromonas		
	Prevotella		
	Selenonomas		
	Treponema		
	Wolinella		
Fungi			
Candida			

species (Table 63.4). *Streptococcus* is a good example of the former, while revision of the classification of the oral "Bacteroides" to define *Prevotella* and *Porphyromonas* typifies the latter case. Also, studies of representatives of oral genera isolated from sites other than the mouth has contributed to revealing the potential diversity of species of oral genera. In particular, identification of Gram-positive rods from infections in humans and animals has extended our knowledge of *Actinomyces* with new species, such as, *A. neuii, A. radingae, A. turicensis*, and *A. hyovaginalis* being described. The origin of these organisms in infections is not known, but they may be resident in mouth like the other species of this genus.

In common with bacteria from other habitats, clonal diversity has been demonstrated among strains of species of oral bacteria. It is generally accepted that often one or relatively few clones of a pathogenic bacterial species may be responsible for causing disease. In contrast, resident or commensal species will be represented in a given habitat by several clones. The term "clone" describes a strain descended from a single parent and theoretically identical to its parent. However, parental DNA may not be preserved intact in daughter cells due to its modification during division through intragenic mutations and exchange of DNA between cells. Given this proviso, clones (genetic variants) among oral bacterial species have been demonstrated by a variety of techniques (multilocus enzyme electrophoresis, restriction fragment length polymorphism, ribotyping, arbitrary primed polymerase chain-reaction and repetitive extragenic palindromic PCR) with different degrees of discrimination. Examples include up to 61 types of *S. mutans*; 93 types of *S. mitis*; 114 types of *A. naeslundii*; 38 types of *Actinobacillus actinomycetemcomitans*; and 78 types of *P. gingivalis*. Some oral species are represented by several clones in an individual, while others exist as a single predominant clone. For example, one clone of *P. gingivalis* usually predominates in a person, while up to 20 clones of *A. naeslundii* genomic species 2 can be isolated. The biological significance, if any, of extensive clonal diversity within oral resident species is not known, although if each clone has a different phenotype that is best suited to a specific host's environment, clonal diversity among strains may enhance the survival of a species. Genetic typing methods have also proved useful to demonstrate transfer of strains of species from mothers to infants and among family members.

1. *Flora of soft tissues*

Studies on the oral flora of humans and animals have tended to concentrate on the organisms in dental plaque, while few studies have been made on the microflora of soft tissues. Some data are available on the flora of the human tongue, which includes species of *Actinomyces* (*A. naeslundii, A. odontolyticus*), *Bacteroides melaninogenicus* (*Prevotella melaninogenica*), *Neisseria, Streptococcus* (*S. salivarius, S. oralis, S. mitis*) and *Rothia mucilagenosa*, with the tongue probably representing the natural habitat of this latter organism. Although the same genera are found on other surfaces in the mouth, different species are found on soft and hard tissue surfaces. For example, *S. salivarius* regularly colonizes the soft tissues, but is relatively uncommon in dental plaque, while *S. sanguis* and *S. mutans* are found more commonly on tooth surfaces. Similarly, *A. naeslundii* genomic species 1 is thought to occur more commonly on soft tissues than *A. naeslundii* genomic species 2.

2. *Flora of dental plaque*

Dental plaque can be seperated into two types, supragingival and subgingival, based on the site of its

TABLE 63.4 Oral species within selected bacterial genera isolated from humans

Genus	Species
Streptococcus	*sanguis* (biovars 1–4), *parasanguis, gordonii* (biovar 1–3), *mitis* (biovar 1, 2), *oralis, salivarius, vestibularis, salivarius, intermedius, anginosus, constellatus* Mutans group *mutans, sobrinus, cricetus, rattus,* (*ferus, macacae, downei,* isolated from the oral cavity animals)
Actinomyces	*israelii, gerencseriae, naeslundii* (genomic species 1 and 2) *odontolyticus, meyeri, georgiae* (*bovis, viscosus, denticolens, howellii, slackii, hordeovulneris, hyovaginalis, pyogenes,* isolated from animals)
Prevotella	*melaninogenica, intermedia, nigrescens, loeschii, denticola, buccae, buccalis, heparinolytica, oralis, oris, veroralis, zoogleoformans, oulora*
Porphyromonas	*gingivalis, endontalis, assacharolytica* (*macacae, levii, cangingivalis, catoniae, crevioricans, cansulci* are similar organisms from animals)
Fusobacterium	*alocis, sulci, peridonticum, nucleatum* (*simiae,* from monkeys)
Treponema	*denticola, macrodentium, oralis, socranskii, vincentii, scoliodontium*

accumulation. Supragingival plaque is a significant reservoir for members of the resident flora and is consistently present in health or disease. In contrast, although a small gingival sulcus is present at the junction of teeth and gingiva in health, subgingival plaque accumulates in the "pockets" between the gingivae and the tooth surface formed during periodontal disease.

The most diverse supragingival plaque communities develop in the protected interproximal areas between adjacent teeth (Table 63.5), while more exposed areas, such as, occlusal surfaces support a less complex community. After a period of undisturbed development in a protected area, dental plaque may reach equilibrium with the local environment (climax community) and will be the most diverse including a wide range of bacteria. However, as plaque is normally disrupted frequently during normal mastication and oral hygiene procedures (e.g. tooth brushing and flossing), it is unlikely that climax communities develop in all habitats.

The subgingival habitat in periodontal disease provides a distinctly different environment from supragingival plaque and, consequently, supports a different flora (Table 63.5). Comparisons between the composition of the subgingival community in health and disease show that, in general, the same species of bacteria are present. It seems most likely that ecological succession takes place in the subgingival community causing changes in the populations of species leading to disease. It should be noted that although a "pocket" may still exist after tissue healing, the subgingival flora changes with increases in Gram-positive species and a return to a community similar to that of supragingival plaque.

B. Oral flora in disease

1. *Oral microbial pathogenesis*

There is little doubt that all of the bacterial diseases which cause destruction of oral tissues involve more than one type of organism and should be considered "mixed" infections. It can be proposed that different members of the infecting mixture enhance the virulence and allow the establishment of potential pathogens. Studies of mixed infections in experimental animals have shown that certain combinations of organisms are more virulent than others and that nutritional factors and immunity to various host defense mechanisms are factors in their pathogenesis. As examples, succinate produced by *Actinomyces* is a growth factor for *Prevotella* and the proteases of *Porphyromonas gingivalis* will degrade complement factors and immunoglobulins (antibodies), thereby, protecting bacteria from the host's immune system.

A significant feature of caries, and possibly periodontal diseases, is that changes in the environment of the tissue habitat can result in changes to the local oral flora encouraging the development of a potentially or overtly pathogenic bacterial community. For example, in the later stages of the succession of bacteria in dental plaque, a single species, serovar or strain, may be dominant in the community. These organisms are often those that are sufficiently virulent to produce tissue destruction in experimental animals and it can be assumed that they are playing a significant role in the natural disease process. Typical of these organisms are two of the "mutans streptococci," *S. mutans* and *S. sobrinus*, which increase in numbers in supragingival plaque in association with the decalcification of tooth enamel leading to dental caries. However, the importance of non-bacterial factors of the local environment in the development of disease must be emphasized as high concentrations of *S. mutans*, and possibly *S. sobrinus*, also occur in plaque in the absence of caries development.

2. *Dental caries*

a. *Enamel caries*

As mentioned previously, dental caries results from the destruction of teeth by acids produced by the bacteria in dental plaque metabolizing dietary carbohydrate. The consumption of sugar results in a typical plaque pH versus time curve ("Stephan Curve") which is characterized by the immediate reduction in plaque pH to a minimum value followed by a return to more neutral pH values. The extent and duration of the pH minimum, which can reach values lower than pH 4.0, is determined by the concentration of sugar in the diet, the period of food intake, the numbers of acidogenic bacteria in plaque, and the flow rate and buffering capacity of saliva at the time of eating. The reduction in plaque pH indicates that the rate of acid formation by plaque bacteria at any one time has exceeded the buffering and clearing properties of saliva. Generally, enamel demineralization occurs in a region at or below pH 5.5, the "critical pH region," while remineralization will occur above this pH because saliva is supersaturated with calcium and phosphate, the main components of enamel. Thus, small areas of enamel below dental plaque can be subjected to cycles of demineralization and remineralization during the normal course of food consumption each day. The greater amount of time that the enamel surface at a site is subjected to an acidic environment, the greater the opportunity for caries development at that location.

The plaque bacteria associated with caries generally have the capacity to degrade carbohydrate substrates

TABLE 63.5 Selected species among supragingival and subgingival dental plaque communities

Interproximal supragingival plaque	Subgingival plaque in periodontal disease
Streptococcus	
S. sanguis biovars 1–4, *S. gordonii* biovars 1–3, *S. mutans*, *S. sobrinus*, *S. anginosus*, *S. intermedius*, *S. oralis*, *S. mitis*	*S. intermedius*, *S. oralis*
Actinomyces	
A. naeslundii genomic species 1 and 2, *A. israelii*, *A. gerencseriae*, *A. odontolyticus*	*A. naeslundii*, *A. israelii*, *A. gerencseriae*, *A. meyeri*, *A. georgiae*, *A. odontolyticus*
Lactobacillus	
L. casei, *L. acidophilus*, *L. fermentum*, *I. plantarum*, *L. brevis*, *L. salivarius*	*L. uli*
Propionibacterium	
P. proprionicus, *P. acnes*	*P. proprionicus*, *P. acnes*
Corynebacterium	
C. matruchotti	
Bifidobacterium	
B. dentium	*B. dentium*
Leptotrichia	
L. buccalis	
Eubacterium	
E. alactolyticum, *E. saburreum*	*E. alactolyticum*, *E. saburreum*, *E. timidum*, *E. nodatum*, *E. brachy*
Peptostreptococcus	*P. anaerobius*, *P. micros*
Veillonella	
V. parvula, *V. dispar*, *V. atypica*	*V. parvula*
Actinobacillus	
A. actinomycetemcomitans	*A. actinomycetemcomitans*
Haemophilus	
H. segnis, *H. paraphrophilus*, *H. aphrophilus*	*H. segnis*
Campylobacter	*C. concisus*, *C. curvus*, *C. rectus*
Prevotella	
P. melaninogenica, *P. nigrescens*, *P. buccae*, *P. oris*, *P. denticola*, *P. loescheii*	*P. melaninogenica*, *P. nigrescens*, *P. denticola*, *P. oris*, *P. oralis*, *P. buccae*, *P. veroralis*, *P. tannerae*
Porphyromonas	*P. gingivalis*. *P. endontalis*.
Bacteroides	"*B. forsythus*" (closely related to *Porphyromonas*)
Fusobacterium	
F. nucleatum subsp. *polymorphum*	*F. nucleatum* subsp. *nucleatum*, subsp. *polymorphum*, subsp. *vincentii*, subsp. *fusiforme*, *F. periodontium*, *F. alocis*
Centipeda	*C. periodontii*
Selenomonas	*S. sputigena*, *S. noxia*, *S. infelix*, *S. dianae*
Treponema	*T. socranskii*, *T. vincentii*, *T. denticola*, *T. pectinovorum*
Wolinella	
W. recta	

rapidly to acid end-products, and also to grow and metabolize in acidic environments. Thus, these organisms are both acidogenic (acid-producing) and aciduric (acid-tolerant), properties that enable them to maintain a low environmental pH in association with the surfaces of teeth. Most significant among these so-called "cariogenic" bacteria are those included among the "mutans streptococci" (Table 63.4). In humans, the species most closely associated with caries are *S. mutans* and *S. sobrinus*. A second genus specifically associated with caries is *Lactobacillus*, which colonizes the tooth surface later than *S. mutans*, probably after destruction of the surface enamel. Lactobacilli are found in high numbers in carious enamel lesions which are habitats with characteristics different from those of non-decalcified enamel. Thus, caries represents a mixed infection where bacterial succession and the emergence of opportunistic pathogens is associated with a changing environment. As such, it can be used as a model for other mixed infections in the mouth. Most recently, it has been shown that early decalcification of enamel (i.e. within 4 days of plaque development in a protected area) can be associated with organisms other than *S. mutans* and *S. sobrinus*. This suggests that there are other organisms in the mouth that can initiate enamel decalcification. A potential role for other organisms in caries is suggested by studies with experimental animals where caries can be produced by different species of *Streptococcus, Actinomyces*, and *Lactobacillus*.

b. *Caries of root surfaces*

As a person ages, the gingiva may recede so that the root of the tooth becomes exposed to the oral cavity and, since the mineral of the tooth root is less acid-resistant than enamel, caries can develop readily on these surfaces. Recent studies of samples from well defined areas of root caries lesions in extracted teeth have demonstrated the complexity of the flora of lesions, which includes Gram-positive and Gram-negative anaerobes and also provided evidence for the association of different communities within lesions during their progression. The flora of some lesions is dominated by *A. naeslundii*, while in others *A. gerencseriae* and *A. israelii* form a high proportion of the Gram-positive flora. In contrast, the flora of other lesions are dominated by mutans streptococci and *Lactobacillus* spp. Given the possible variation in local environments and the dynamic nature of lesion formation, root caries can be described as a polymicrobic infection typified by the variety of different communities associated with carious tooth roots.

3. *Periodontal diseases*

The term, "periodontal disease," is used to describe a collection of diseases of the gingiva ranging from inflammation to extensive tissue destruction. Some of the diseases are well-defined clinically, while in others the signs and symptoms may vary and an accurate diagnosis can be difficult. Thus, inflammation of the gingiva (gingivitis), which can be reversed by oral hygiene procedures, can readily be diagnosed and described, as can juvenile periodontitis. Some of the other periodontal diseases may represent more than one disease and are less well defined. These include adult periodontitis, rapidly advancing periodontitis and refractory periodontitis, with the recognition of the latter being based on its lack of response to normal methods of treatment. The flora of periodontal pockets (Table 63.5) is complex with a high diversity of bacteria. Over 500 taxa have been isolated from periodontal pockets and one study of adult periodontitis identified 47 species as being most numerous in samples. Despite the complexity of the flora and difficulties with accurate diagnosis, it has been possible to associate specific bacteria with lesions of certain periodontal diseases. Although there was good evidence for succession of bacteria to dominance in caries, there was less for ecological succession in periodontal disease Recently, checkerboard analysis of periodontal pocket communities has shown that the proportion of bacterial populations in the pocket change during periodontal disease, with groups of species becoming predominant. These results would support the concept of ecological succession in periodontal disease resulting in dominance by putative periodontal pathogens.

a. *Inflammatory*

Gingivitis is unique among diseases affecting the gingiva in that it can be reversed by the use of rigorous dental hygiene procedures. Experimental gingivitis has been studied in humans and the early microbiology of gingivitis is similar to that of supragingival plaque at the gingival margin. However, as inflammation becomes more severe and bleeding of the gingiva occurs, certain species become common in the plaque. These species include, *Actinomyces israelii, Fusobacterium nucleatum, Streptococcus intermedius, Campylobacter sputagena, Prevotella nigrescens, Peptostreptococcus micros* and *Campylobacter ochracea*. "*Bacteroides intermedius*", which has been associated with pregnancy gingivitis, now includes *Prevotella intermedia* and *Prevotella nigrescens*.

b. *Destructive*

The site associated with this type of disease is the "pocket" created when there is a loss of attachment of the gingiva to the tooth. These mixed, anaerobic infections can cause extensive tissue destruction, including loss of connective tissue and the resorption of alveolar bone, which can result in tooth loss. Perhaps the best example of a relationship between a form of periodontal disease and a specific bacterium is that of localized juvenile periodontitis and *Actinobacillus actinomycetemcommitans*. This organism is often dominant in the microflora associated with the lesions of this disease and strains producing leucotoxin are particularly virulent. Certain other bacteria are also associated with areas of gingival and bone destruction, the most significant being *Porphyromonas gingivalis*,

Prevotella melaninogenica, *Prevotella intermedia*, *Fusobacterium nucleatum*, *Capnocytophaga*, *Selenomonas*, and *Treponema*. The most virulent of these is *Porphyromonas gingivalis* (formerly known as *Bacteroides gingivalis*). This organism can cause fatal infections in mice and produces potent proteases which degrade collagen, complement and immunoglobulins. The other organisms that may be significant in periodontal disease are *Bacteroides forsythus*, *Prevotella buccae*, *Fusobacterium* spp., *Peptostreptococcus* spp., and *Wolinella recta*. Although different species have been identified as potential pathogens in periodontal disease, a specific role for them in the disease process has yet to be established. Several unclassified bacteria have also been found in high numbers in individuals with periodontal disease and allocating specific roles for the range of bacteria in the subgingival flora is a major task for future research.

4. *Other infections outside the mouth*

The biofilms on teeth, tongue, and the soft tissues of the mouth are reservoirs for resident oral bacteria which are significant opportunist pathogens. Infection is generally accepted to be the result of entry of bacteria into the blood stream via the gingival crevice during eating, tooth brushing, and other simple procedures that manipulate oral tissues. These bacteremias are usually transient and a normal occurence in healthy hosts. However, in immunosuppressed subjects and those with some tissue damage or indwelling protheses, these transients can cause life-threatening infections.

Oral streptococci (*S. sanguis, S. gordonii, S. mutans*) are among the most common organisms colonizing biofilms on prostheses and damaged tissues and in particular are responsible for subacute bacterial endocarditis. *S. oralis, S. intermedius, S. anginosus*, and *S. constellatus* cause liver and brain abscesses among the general population, and also serious blood-borne infections in immunosuppressed patients. Most recently, oral bacteria have been implicated in low-birth-weight infants and contributing to atheroma in heart disease. Actinomycosis is a chronic significant mixed infection which can occur at any site in the body and it is generally accepted that the organisms associated with this condition originate from the mouth. The pathogen which gives its name to the disease is usually *A. israelii*, although *A. gerencseriae* and *Propionibacterium proprionicus*, produce a similar disease. The infection is almost invariably mixed, and includes oral bacteria. *Actinobacillus actinomycetemcomitans* is commonly found in association with *Actinomyces* and *Propionibacterium* in actinomycosis.

C. Control and prevention of oral diseases of bacterial etiology

The control of bacterial infections in the mouth other than caries and periodontal diseases follows the general principles of treatment for infections elsewhere in the body and is based primarily on antibiotic therapy. Caries and periodontal diseases are approached differently, although antibiotics are used as adjuncts to the therapy for some types of periodontal disease. Emphasis is placed on the prevention of caries and periodontal disease through an improvement of oral hygiene. Oral hygiene procedures, such a brushing and flossing, cause not only disruption of dental plaque, but serve to reduce its mass and its potential to develop into a pathogenic community.

Apart from physical disruption, other methods aimed more directly at the control of bacteria are used to prevent caries and periodontal diseases. Perhaps the most effective and useful of these methods is the application of fluoride to prevent dental caries. Fluoride reduces caries by several mechanisms, including the enhancement of re-mineralization of the carious lesion and the conversion of the normal hydroxyapatite mineral structure of tooth enamel to the less acid-sensitive fluorapatite. Fluoride also has significant inhibitory effects on bacterial metabolism and these are enhanced at low pH. The major antimicrobial effect of fluoride is the inhibition of carbohydrate metabolism by bacteria, which naturally reduces the rate and extent of acid end-product formation, the direct cause of caries. This reduces the ecological advantage afforded aciduric bacteria by making the environment in dental plaque less acidic. Consequently, fluoridation remains one of the best examples of the public health application of a preventive agent for control of a disease of bacterial etiology.

For the prevention of periodontal disease, there is no compound as effective as fluoride is in caries prevention. However, short term use of mouthrinses can be helpful for the reduction of dental plaque and for the elimination of bacteria. The effectiveness of many mouthrinses as antibacterials is questionable, although chlorhexidine (Hibitane) is recognized by most workers to be effective in reducing the oral flora. This agent is available as a mouthrinse, varying in concentration from 0.1–0.2% and is also available as a tooth gel in some countries. Chlorhexidine is used clinically for patients with extensive periodontal disease to restore the level of oral hygiene before dental treatment, for elderly and handicapped persons, and to reduce the numbers of bacteria in the mouth of immunocompromised patients prior to surgery.

A considerable body of evidence now exists establishing the association of *S. mutans* with dental caries in humans and, as a result, immunization has been proposed as a preventive measure. Considerable research has centered on isolating protective antigens and defining the optimum time and conditions for immunization. The antigens that have been tested in experimental animals include those on the surfaces of whole cells of *S. mutans*, cell-surface proteins, and enzymes, such as glucosyltransferase, which are considered significant in the etiology of caries. Two main routes of immunization, systemic and oral, have been explored as has passive immunity in rat pups from immunized mothers. Passive immunity by use of monoclonal antibodies in humans has been shown to affect the re-colonization of their mouths by *S. mutans*. Also, humans rinsing their mouths with milk from cows immunized with *S. mutans* had reduced numbers of this organism relative to controls and the strains isolated from subjects taking milk antibodies exhibited colonial variation. There is now little doubt that it is possible to influence the numbers and colonization of *S. mutans* in the human mouth through a variety of immunization procedures. The problems of cross-reactivity of whole cells of *S. mutans* to heart tissue have been overcome by the use of pure preparations of cell-surface proteins or enzymes, and recombinant proteins. To date, although human volunteers have been immunized with *S. mutans* by the oral route, there have been no clinical studies of the effectiveness of any vaccine in controlling caries in humans.

Immune prevention and control of periodontal disease has also been proposed, based on generation of active immunity against putative periodontal pathogens. Generally the antigens are cells or specific components (fimbriae) of *P. gingivalis*. Data from experimental animal infections indicates that immunization can influence oral colonization by *P. gingivalis* and bone loss in infections.

III. CURRENT AND FUTURE TRENDS

The application of molecular biological techniques in the study of the characteristics and properties of important oral bacteria is now well advanced with considerable focus now on genomics, proteomics and biofilms. The genome sequences of nine major oral organisms: *Actinobacillus actinomycetemcomitans, Candida albicans, Fusobacterium nucleatum, Porphyromonas gingivalis, Prevotella intermedia, Streptococcus mitis, Streptococcus mutans, Streptococcus sanguis*, and *Treponema denticola*, have been completed or are almost complete (see *http://www. Nidr.nih.gov/research/dbts/Pol Microbial Genome Seq Projects.asp*). Future genome sequencing will include *Streptococcus gordonii, Streptococcus sobrinus, Bacteroides forsythus, and Actinomyces* species. These advances in genomic sequencing have lead to the relatively new field of science called proteomics.

Proteome analysis is designed to use the information in a genomic sequence to describe the function of the protein complement of a cell or tissue type. At present, proteome analysis involves the separation of the proteins extracted from cells by two-dimensional electrophoresis followed by the isolation of the proteins of interest and subjecting them to mass spectrometry. The resulting mass fingerprints are then used to scan existing protein sequence databases in order to identify the proteins. Proteome analysis is now being employed to examine the regulation of the physiology of oral bacteria, with particular current interest focused on the alterations in protein synthesis initiated by *S. mutans* when exposed to acidic environments. The organism is known to induce an acid tolerance response by the up- and down-regulation of the synthesis of certain proteins as the pH of the environment decreases, a process that is known to enhance its survival under such conditions. The acquisition of this aciduric property is believed to contribute to the dominance of the organism in the acidification of small habitats on the tooth surface that can lead to the initiation of dental caries.

Although there is an extensive literature on the phenomenon of adherence of oral bacteria to surfaces, much of the early data obtained on the characteristics and properties of oral bacteria *per se* involved their growth in liquid-phase (planktonic) cultures. Considerable evidence now exists demonstrating that the physiological properties of bacteria in biofilms, such as dental plaque, are different from those of planktonic cultures, particularly with respect to resistance to adverse environmental conditions and toxic antimicrobial agents. This knowledge has resulted in an upsurge in research into alterations in cellular physiology required for, and triggered by, biofilm formation and role of various proteins, pathways and genes responsible for the enhanced resistance of oral bacteria in biofilms.

BIBLIOGRAPHY

Alam, S., Brailsford, S. R., Adams, S., Allison, C., Sheehy, E., Zoitopoulos, L., Kidd E. A., and Beighton, D. (2000). *Appl. Environ. Microbiol.* **66**, 3330–3336.

Allison, D. G., Gilbert, P., Lappin-Scott, H. M., and Wilson, M. (eds.) (2000). "Community Structure and Co-operation in Biofilms." Cambridge University Press, Cambridge.

Bowden, G. H. W. (1990). *J. Dent. Res.* **69**, 1205–1210.

Bowden, G. H. (1991). Which bacteria are cariogenic in humans ? *In* "Risk Markers for Oral Disease," Vol. 1, Dental Caries (N. W. Johnson, ed.), pp. 266–286. Cambridge University Press, Cambridge.

Bowden, G. H., and Hamilton, I. R. (1998). Survival of oral bacteria. *Crit. Rev. Oral Biol. Med.*, **9**, 54–85.

Bowden, G. H., Ellwood, D. C., and Hamilton, I. R. (1979). Microbial ecology of the oral, cavity. *In* "Advances in Microbial Ecology" (M. Alexander, ed.), pp. 135–217. Plenum, New York.

Cole, M. F., Bryan, S., Evans, M. K., Pearce, C. L., Sheridan, M. J., Sura, P. A., Wientzen, R., and Bowden, G. B. (1998). *Infect. Immun.* **66**, 4283–4289.

Ebersole, J. L., Cappelli, and Holt S. C. (2001). *Acta Odontol. Scand.* **59**, 161–166.

Frandsen, E. V., Poulsen, K., Curtis, M. A., and Kilian, M. (2001). *Infect. Immun.* **69**, 4479–4485.

Gibson, F. C. 3rd, and Genco, C. A. (2001). *Infect. Immun.* **69**, 7959–7963.

Gronroos, L., and Alaluusua, S. (2000). *Caries Res.* **34**, 474–480.

Hamilton, I. R. (1990). *J. Dent. Res.* **69**, 660–667.

Katz, J., Black, K. P., and Michalek, S. M. (1999). *Infect. Immun.* **67**, 4352–4359.

Kolenbrander, P. E., and London, J. (1992). Ecological significance of coaggregation among oral bacteria. *In* "Advances in Microbial Ecology," Vol. 12 (K.C. Marshall, ed.), pp. 183–217. Plenum Press, New York.

Kuramitsu, H. K., Ellen, R. P. (eds.) (2000). "Oral Bacterial Ecology." Horizon Sci. Press, Oxford.

Marcotte, H., and Lavoie, M. C. (1998). *Microbiol. Mol. Biol. Rev.* **62**, 71–109.

Moore, W. E. C., and Moore, L. V. H. (1994). *Periodont.* 2000, **5**, 66–77.

Newman, H. N., and Wilson, M. (eds.) (1999). "Dental plaque Revisited. Oral Biofilms in Health and Disease." Bioline, Cardiff.

Novak, M. J. (ed.) (1997). *Adv. Dent. Res.*, **11**, 1–196.

Paster, B. J., Boches, S. K., Galvin, J. L., Ericson, R. E., Lau, C. N., Levanos, V. A., Sahasrabudhe, A., and Dewhirst, F. E. (2001). *J. Bacteriol.* **183**, 3770–3783.

Scannapieco, F. A. (1994). Saliva-bacterium interactions in oral microbial ecology. *Crit. Rev. Oral Biol. Med.* **5**, 203–248.

Svensäter, G., Sjögreen, B., and Hamilton, I. R. (2000). *Microbiology* **146**, 107–117.

Svensäter, G., Welin, J., Wilkins, C., Beighton, D., and Hamilton, I. R. (2001). *FEMS Microbiol. Lett.* **205**, 139–146.

WEBSITES

List of bacterial names with standing in nomenclature (J. P. Euzéby) *http://www.bacterio.cict.fr/index.html*

Photo Gallery of Bacterial Pathogens (Neil Chamberlain) *http://www.geocities.com/CapeCanaveral/3504/gallery.htm*

Access to online resources on Bacterial Infections and Mycoses. Karolinska Institutet
http://www.mic.ki.se/Diseases/c1.html

Comprehensive Microbial Resource of The Institute for Genomic Research (TIGR) and links to many other microbial genomic sites *http://www.tigr.org/tigr-scripts/CMR2/CMRHomePage.spl*

64

Osmotic stress

Douglas H. Bartlett
University of California, San Diego

Mary F. Roberts
Boston College

GLOSSARY

compatible solutes Low-molecular-weight organic compounds that enhance the ability of organisms to survive in environments in which the extracellular solute concentration exceeds that of the cell cytoplasm.

halophile An organism that requires NaCl for growth; salt-lover.

hyperosmotic Pertaining to an increase in osmotic pressure or osmolality.

osmolality Osmotic pressure expressed in terms of osmols or milliosmols per kilogram of water.

osmophile An organism that requires high osmotic pressure; osmatic-pressure-lover.

turgor pressure The pressure exerted by the contents of a cell against the cell membrane or cell wall.

water activity The ratio of the vapor pressure of water in equilibrium with an aqueous solution divided by the vapor pressure at the same temperature of pure water. Values of a_w vary between 0 and 1.

Water is The Solvent of Life, and the degree to which it is available exerts a profound influence on all living systems. The movement of water from regions of higher to lower concentration (higher-solute concentration) across a semipermeable membrane is called osmosis. In biology, osmosis frequently involves the movement of water through a membrane phospholipid bilayer. The osmotic state of a system can be quantitated by one of several measurements. Osmotic pressure, for example, is the external pressure that is just sufficient to prevent osmosis between a solution and pure water. It is often expressed in terms of osmolality. An osmole is a gram molecular weight of osmotically active solutes in 1 kg of water and can be determined by one of several colligative properties (such as freezing-point depression). Osmotic stress results when the process of osmosis generates a cytoplasm of osmolality that is either less than or greater than optimal for a particular organism.

I. GENERAL CONSIDERATIONS

When a microbial cell is in a medium that is hyperosmotic relative to its cytoplasm, the cell tends to lose water by osmosis. Conversely, if the cell is in a medium that is hypoosmotic by comparison to its own fluids, it will accumulate water. In the absence of an appropriate response there may be growth inhibitory if not lethal consequences. The control of internal osmolality is related to other aspects of cell physiology. At low osmolality, the need to preserve internal pH, ionic strength, metabolite concentration, and the concentration of specific ions a certain range is an important aspect of the adaptational response.

The Desk Encyclopedia of Microbiology
ISBN: 0-12-621361-5

For example, K^+ is the most abundant cation in the cell, and it is kept the range of 100–150 mM in the case of *Escherichia coli*. In contrast, at high osmolality, the need to supplement the cellular milieu with solutes that are least inhibitory to cellular processes will predominate. Adaptation to osmolali ties near the upper limit of tolerance typically also confers enhanced resistance to other stressful conditions, such as heat or cold shock.

Prokaryotes and other osmoconforming organisms such as plants and invertebrates are generally considered to be close to being isoosmotic with their environment. However, in fact, because of the increased osmolality of the cytoplasm relative to the environment and the resulting inward flow of water, prokaryotic and eukaryotic microorganisms and the cells of many other osmoconforming organisms retain a positive hydrostatic pressure, termed turgor pressure, which pushes out against their cell membranes and cell walls. In response to turgor pressure, the prokaryotic cell wall expands elastically until its tension exerts an opposing force equal to the hydrostatic pressure. In this way the cell wall acts like a pressure vessel. The pressure inside gram-negative bacterial cells is on the order of 2 atmospheres (30 pounds per square inch) and can reach up to about 20 atmospheres in the case of some gram-positive bacteria. Measurements of bacterial turgor pressure indicate that it does not vary much during cell division or during the various phases of growth. It has been proposed that turgor pressure is maintained within narrow bounds in order to produce the necessary mechanical stress for cell-wall expansion. This then could be one of the reasons why it is important for cells to accumulate osmolytes as they grow and divide.

In describing the osmotic adaptations of microorganisms, another important term is water activity, abbreviated a_w. This is the ratio of the vapor pressure of water in equilibrium with an aqueous solution divided by the vapor pressure at the same temperature of pure water. The values of a_w vary between 0 and 1. Microbial life has evolved widely different preferences for a_w and the amount of change in a_w that can be tolerated. Thus, osmotic stress is different to different prokaryotes. *E. coli* typically contains 70% of its cytoplasm as water, although this value may vary by close to twofold, depending on the osmolality of its environment. Bacteria such as certain species of *Caulobacter* and *Spirillum* can grow at a_w close to 1.000 (pure water); marine bacteria are adapted to the a_w of seawater, 0.980; some members of the Archaea can grow at a_w down to 0.750; and certain fungi can still grow at a_w as low as 0.700. Similarly, endolithic lichen and bacteria of deserts must be adapted to extremely low a_w values. The bacterial species *Salinivibrio costicola* as well as bacteria of the family *Halomonadaceae* are particularly fascinating because of the tremendous range of a_w they will tolerate. They will grow over a range of water activities between 0.98 (close to freshwater) to 0.86 (close to saturated NaCl). Decreasing water activity by freeze-drying or salting is one of the principal means of preventing food spoilage by microorganisms. Water potential, a related energy term that is expressed in units of pressure, is also used to describe the concentration of water in some scientific fields.

Organisms that can grow in media of high and average solute concentrations are called osmotolerant, and those that actually prefer elevated osmolalities are termed osmophiles. In the environment, osmotic adaptations are usually correlated with the presence of particular solutes. Because of this it is important to distinguish general osmotic requirements from those regarding specific chemicals. Marine microorganisms, for example, typically have a specific need for a certain amount of sodium ions in addition to a moderately increased osmolality over their terrestrial microbial counterparts. They are thus moderate halophiles. Certain members of the Archaea domain, such as *Haloferax* and *Halobium* species, have still higher requirements for sodium ions and are described as extreme halophiles. In terms of the salt concentrations enabling growth, microorganisms may grow from close to 0 M to close to saturating concentrations at 5.2 M sodium chloride. *Salinivibrio costicola* and bacteria of the family *Halomonadaceae* are described as haloversatile.

Dessication results in extreme osmotic stress and low a_w values. Some microbes die almost instantly in air, whereas the spores of algae, fungi, and bacteria and the cysts of protozoa are very resistant to drying. The vegetative cells of many gram-positive cells and bacteria of the family *Deinococcaceae* are also quite resistant to dessication. In the former case, this is believed to stem in part from their thick cell walls, which may help protect membrane integrity. Protection of damage to chromosomal DNA during drying is also critical for cell survival. The extremely efficient DNA-repair capabilities of the *Deinococcaceae* have been found to be required for their dessication resistance.

The focus here is on the physiological processes by which microorganisms adapt to osmotic stress. However, note that one of the most important responses of motile microorganisms to changes in osmolality is to swim away into a new more favorable osmotic environment. This process is known as osmotaxis. Bacteria are repelled by pure water, attracted to their particular optimal osmolality, and repelled again by supraoptimal osmolality. By moving back into a region of optimal osmolality, the need for substantial physiological modification can be greatly diminished. How microorganisms sense osmolality and modulate their motility is largely unknown. However, it appears to be mediated by a direct effect on the rotation switch at the base of the flagellum. Chemoreceptor proteins in the cell membrane also help cells adjust their motility in response to particular solutes such as NaCl.

II. EFFECTS OF INCREASED OSMOLALITY ON PROTEINS

During exposure to hyperosmotic stress conditions, solutes will increase in concentration inside cells. Many of the deleterious effects of hyperosmotic stress can be interpreted in the context of solute effects on macromolecules. Increased osmolality tends to promote the dehydration of proteins. In a sense, proteins can be considered as behaving similar to a semipermeable membrane. The exposure of a protein to an environment in which the a_w is lower than that of the water bound up to the protein will result in the release of the bound water to the surrounding solution.

One line of evidence indicating the profound influence of osmolality on protein structure has come from the isolation of osmoremedial mutants. These microorganisms encode a mutant protein that is nonfunctional when cells are grown in media of low ionic strength, but that regains at least partial function when the cells are grown in media of higher osmolality. Osmoremedial mutations have been identified in genes encoding both soluble and membrane proteins, of monomeric or multimeric configuration. Most if not all temperature-conditional mutants are also osmoremedial.

Different solutes can have different effects on proteins and their structure and activity. K^+ and glutamate are important intracellular osmolytes at low concentration, but, as do other ions, will inhibit many enzymes when present at high concentrations. Solutes that carry a net electric charge are generally more damaging to protein stability than nonpolar or zwitterionic solutes. The charged solutes can be further subdivided according to the Hofmeister series (also called lyotropic series) according to their effects on proteins. Stabilizing ions do not bind strongly to proteins or interfere with the water molecules coating the protein surface. These ions tend to preserve the native protein structure. Thus, stabilizing solutes distribute themselves away from the protein, which is preferentially hydrated near its surface. Because the exclusion of stabilizing solute molecules from the water around the protein reduces the entropy of the system, it is accompanied by an energetic cost. This cost can be reduced by minimizing the area on the surface of the protein that is exposed to solvent, thus resulting in a more compact protein structure. Thus, even in the presence of stabilizing solutes, in the absence of protein modification, increased ionic strength will alter protein structure.

The Hofmesiter series does not apply to all proteins. An exception are the proteins of some of the extremely halophilic Archaea. Proteins isolated from these organisms are stable only under conditions of high ionic strength. Indeed the interior of Archaea of the family Halobacteriaceae may accumulate more than 5 M KCl. Their proteins have an excess of acidic over basic residues when compared to their corresponding nonhalophilic proteins. Because of this acidic character, these proteins are able to fold correctly only in the presence of sufficient counterions, such as K^+. High potassium levels may also balance the amount of water bound to proteins, and the burial of weakly hydrophobic residues such as alanine and glycine.

III. K^+ UPTAKE

Instead of passive cell-volume regulation, prokaryotes respond to osmotic increases by selectively increasing the concentration of certain solutes, thus reducing water activity but returning cell volume and turgor pressure to approximately prestress levels. The exposure of *E. coli* cells to hyperosmotic stress results in a sequential series of adaptive steps. The first line of defense is a large increase in the rate of potassium (K^+) uptake. This appears to be necessary for the resumption of growth of the plasmolyzed cells. Subsequently, the synthesis of glutamate is induced. This charged amino acid serves as the principal counterion for K^+. In some organisms (e.g., *Salmonella typhimurium*), the accumulation of glutamate is necessary for optimal growth in media of high osmolality. However, because glutamate is a substrate for a variety of cellular enzymes (specifically, a key intermediate in nitrogen metabolism and a component of proteins), its concentration in cells may be regulated.

With the uptake or synthesis of compatible solutes (see later), K^+ levels in *E. coli* decrease as a result of efflux and glutamate levels decrease as a result of turnover. The synthesis of trehalose results in this dissacharide eventually replacing K^+ (and glutamate) as the major osmolyte inside the cell. Mutants impaired in K^+ uptake produce trehalose more rapidly and mutants unable to synthesize trehalose exhibit increased intracellular K^+ concentrations. Thus, *E. coli* and presumably many other bacteria have evolved a signaling process for communicating and controlling the levels of the osmolytes K^+ and trehalose together inside the cell. K^+ is believed to act as a second messenger that influences not just trehalose levels, but many of the later steps of the cellular response to hyperosmotic stress. Consistent with this hypothesis, there is a dependency on external K^+ for osmoadaptation in the absence of external compatible solutes.

E. coli has three K^+ uptake systems, the predominant Trk, the inducible Kdp, and the minor Kup systems. Although all of these transport systems respond to osmotic upshift by increasing K^+ influx, an interesting aspect of the Kdp system is the transcriptional regulation of the genes encoding this transporter in response to osmolarity. Two proteins, the integral membrane protein KdpD and the soluble KdpE, control Kdp-transporter gene expression as members of the two-component sensor kinase–response regulator protein family in which one protein, in this case KdpD, acts as a sensor of some signal, and the other protein, in this case KdpE, is the response regulator that controls the cell's response to the perceived signal. Kdp gene expression is principally controlled by K^+ levels, but is also influenced by osmolality because increases in medium osmolality also increase the threshold K^+ concentration for induction. Because osmotic upshock transiently increases Kdp expression it was once thought that the decrease in cell turgor pressure following such a stress is the signal governing osmolality control of Kdp gene expression. Later experiments with KdpD mutants insensitive to K^+ indicated that KdpD can also control steady-state Kdp gene expression as an effective osmosensor. In order to do so, it may respond to osmotically induced changes in membrane structure.

IV. COMPATIBLE SOLUTES

A. Introduction

The response of the vast majority of organisms in all three domains of life to hyperosmotic stress is the eventual intracellular accumulation either by synthesis or transport of organic molecules (charged or neutral) for osmotic balance. The exception to this rule are all members of the family Halobacteriaceae within the Archaea domain and some groups of anaerobes within the Bacteria domain (e.g., *Halobacteroides* and *Haloanaerobium*), which have high intracellular inorganic cation concentrations. The most halophilic Archaea can accumulate very high (up to ~7 M) concentrations of inorganic ions, chiefly K^+ and Na^+. The intracellular concentration of these cations varies with external NaCl; hence they can be considered osmolytes.

However, in most microorganisms, the accumulation of inorganic cations such as K^+ is only a transient response that is followed by the accumulation of zwitterionic organic solutes such as proline or glycine betaine. An organic molecule whose intracellular concentration can vary widely (reaching molar levels in halophiles) without affecting the activity of enzymes or the integrity of cellular structures is termed a compatible solute. The distribution of commonly occurring organic osmolytes found in Bacteria, Eukarya, and Archaea, shown in Tables 64.1–64.4, is limited to only a few classes of organic compounds including sugars and polyhydric alcohols, free α- and β-amino acids and their derivatives, methylamines (with glycine betaine as perhaps the most ubiquitous osmolyte), and other simple net neutral compounds (e.g., β-dimethylsulfoniopropionate, taurine, and urea) and anionic molecules (e.g., diglycerol phosphate and cyclic 2,3-diphosphoglcyerate, cDPG). In most cases the accumulation of the organic solutes by transport from the culture medium is energetically preferred over biosynthesis of osmolytes. There is also a general sense that the chemical nature of the solute is important in determining the degree of osmotolerance of the organism.

B. Sugars and polyhydric alcohols

Sugars and polyols, such as glycerol, glucosylglycerol, sucrose, and trehalose, are commonly used for osmotic balance among diverse organisms, including prokaryotes, yeast, plants, algae, and mammalian cells (Table 64.1). Sugars and polyols are often accumulated by organisms that must deal with dessication. Sorbitol is a good example that can either be internalized from the medium (if it is present) or synthesized *de novo*. High concentrations of this polyol have been shown to protect proteins during dehydration by osmotic or thermal stress and to preserve proteins during storage. Sorbitol has been shown to function as an osmolyte in sugar-tolerant yeasts (*Hansenula anomala*), as well as in the gram-negative, strictly fermentative and ethanologenic bacterium *Zymomonas mobilis*. Other polyols, notably glycerol, mannitol, and arabitol, are accumulated by several species of yeast and algae. As specific examples, the marine algae *Dunaliella* sp. accumulates glycerol, whereas the fungi *Dendryphiella* sp. uses arabitol and mannitol as osmolytes.

TABLE 64.1 Sugars and polyols used by cells for osmotic balance

Solute	Structure	Organisms
Neutral		
glycerol	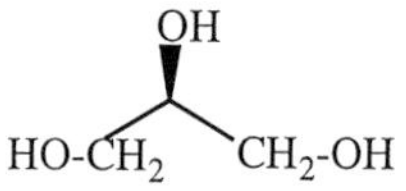	Algae (*Dunaliella* sp.), cyanobacteria, fungi, yeast
arabitol	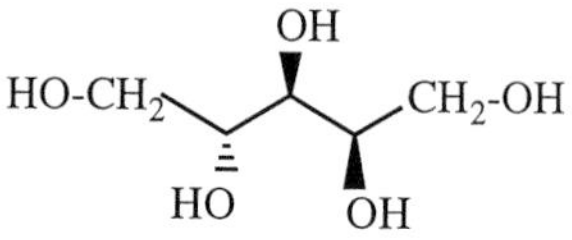	Fungi (*Dendryphiella* sp.), sugar-tolerant yeasts
mannitol	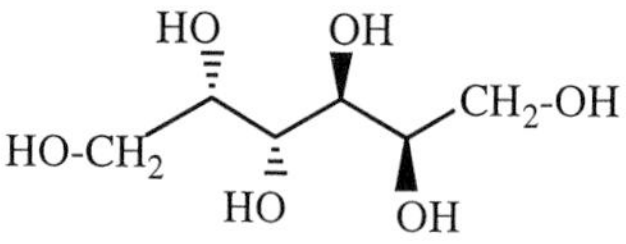	Algae (*Dunaliella* sp.), cyanobacteria, fungi, vascular plants
sorbitol	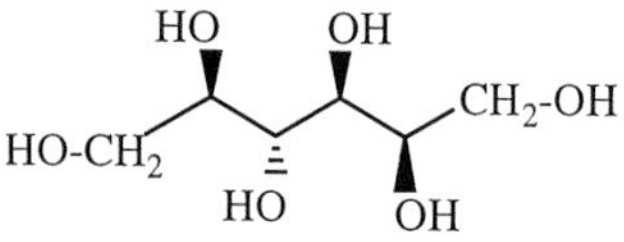	Algae, gram-negative bacteria (e.g., *Zymomonas mobilis*), animals (e.g., renal medulla cells of mammals), vascular plants, yeasts (*Hansenula anomala*)
myo-inositol	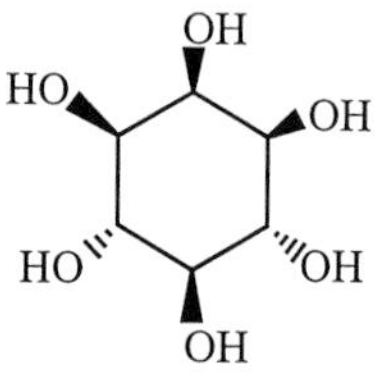	Plants, mammalian brain
fructose	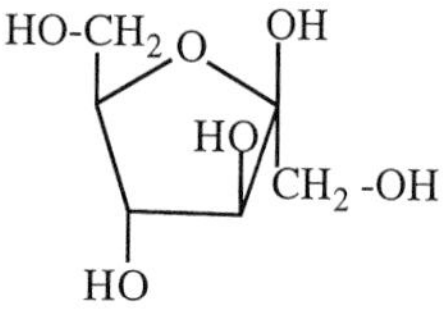	Plants
glucose	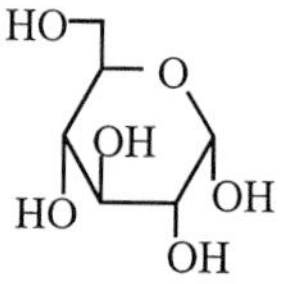	Plants
glucosylglycerol	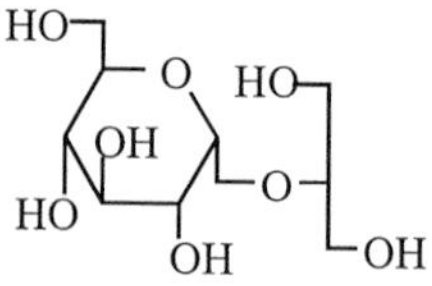	Marine cyanobacteria, freshwater cyanobacteria (*Synechocystis* sp. and *Microcystic firma*), phototrophic eubacterium (*Rhodobacter sulfidophilus*), *Pseudomonas mendocina*, *P. pseudoalcaligenes*
sucrose	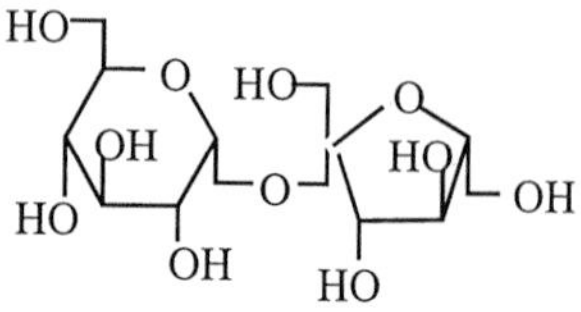	Cyanobacteria (*Anabaena* sp., *Phormidium autumnale*, and *Chroococcidiopsis* sp.), plants and animals
trehalose	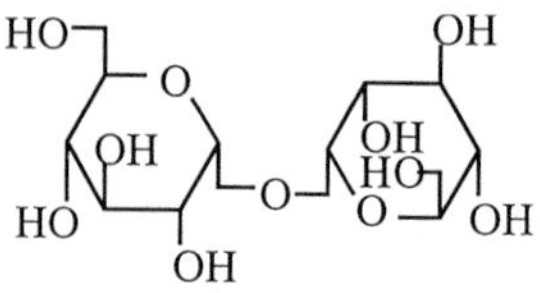	Cyanobacteria (*Anabaena* sp., *Phormidium autumnale*, *Chroococcidiopsis* sp.), plants and animals

TABLE 64.1 *Continued*

Solute	Structure	Organisms
β-fructofuranosyl-α-mannopyranoside		Soil bacterium (*Agrobacterium tumefaciens*)
Anionic		
glucosylglycerate		*Methanohalophilus portucalensis*, marine heterotroph
β-mannosylglycerate		*Archeoglobus fulgidus*, *Thermus thermophilus* (α-Isomer) *Rhodothermus marinus*
sulfotrehalose		*Natronobacterium* sp., *Natronococcus occultus*
di-*myo*-inositol-1,1′-phosphate (DIP)		Hyperthermophilic archaea: *Methanococcus igneus*, *Pyrococcus furiosus*, *Pyrococcus woesei*, *Thermatoga maritima*
diglycerol phosphate		*Archeoglobus fulgidus*

Soluble carbohydrates, including glucose and fructose, are relatively common in the osmotic adaptation of plants. Disaccharides (e.g., sucrose and trehalose) are also frequently accumulated as osmolytes by a wide range of organisms, particularly if they are available in the growth medium. *Escherichia coli* synthesizes trehalose in response to hyperosmotic stress, and mutants defective in trehalose synthesis display increased osmosensitivity (as well as heat sensitivity). Trehalose has been shown to stabilize membrane structure and lower the temperature for phase transition from gel to liquid crystalline state, which can be important during the rehydration of dessicated cells.

A more unusual disaccharide with a documented osmotic response in cells is β-fructofuranosyl-α-mannopyranoside; this solute is found in the soil bacterium *Agrobacterium tumefaciens*. Glucosylglycerol is the organic osmolyte that accumulates in cyanobacteria in response to increased salinity. Its accumulation has been documented primarily in cyanobacteria from marine habitats. This solute has also been reported in two nonphotosynthetic salt-tolerant (1.2 M NaCl) microorganisms (*P. mendocina* and *P. pseudoalcaligenes*).

Although most of the carbohydrates that function as osmolytes are neutral, there are several anionic sugars (modified with carboxylate, sulfate, or phosphate groups) that are used for osmotic balance in Archaea. The very high intracellular K^+ present in some of these organisms requires appropriate counterions. Glucosylglycerate, a negatively charged structural analog of glucosylglycerol, is synthesized and accumulated in *Methanohalophilus portucalensis*, an extreme halophile, grown under nitrogen-limiting conditions. Although trehalose is a common solute accumulated in Bacteria, a charged variant of this, sulfotrehalose (the 1→1 α-linked glucose disaccharide with a sulfate group attached to one of the glucose moieties at C-2 position) functions as an osmolyte in halophilic, alkaliphilic Archaea such as *Natronococcus* and *Natronobacterium* spp. Concentrations of the sulfotrehalose are balanced by the intracellular K^+ ions. Studies of *Natronococcus occultus* showed that these cells could transport and accumulate exogenous neutral disaccharides (sucrose, trehalose, and maltose) for osmotic balance. The accumulation of solutes from the medium suppressed sulfotrehalose biosynthesis.

There are two other types of charged polyols detected in Archaea that are symmetric phosphodiesters. Diglycerol phosphate is the major organic solute detected in *Archaeoglobus fulgidus*. Di-*myo*-inositol-1,1′-phosphate (DIP) has been found to have a role as an osmolyte in hyperthermophilic Archaea including *Methanococcus igneus*, several *Pyrococcus* species, and *Thermotoga maritima*. In both *M. igneus* and *P. furiosus*, DIP was synthesized and accumulated to high intracellular concentrations only at supraoptimal growth temperatures, suggesting the possibility that DIP has a role as a thermoprotectant in addition to its use as an osmolyte. Derivatives of DIP, notably di-2-*O*-β-mannosyl-DIP, are also observed in members of the order *Thermotogales*.

C. α-Amino acids and derivatives

Free amino acids, such as glutamate and proline, are also used for osmotic balance in a wide range of organisms (Table 64.2). Often high-affinity uptake systems exist to internalize these solutes. The neutral solute proline is accumulated from the external medium by an osmotically induced proline transport system, ProP and ProU, in enteric bacteria. The importance of proline to osmotic adaptation of *E. coli* is exemplified by the fact that proline-overproducing mutants are more resistant to osmotic stress. In *Staphylococcus aureus*, the most halotolerant, grampositive nonhalophile bacterium, proline acts as an osmoprotectant along with choline, glycine betaine, and taurine in high-osmotic strength medium. Proline- (and glycine-) containing peptides, as well as amino acid pools, resulting from exogenous peptone in the environment, also contribute to maintaining turgor and stimulating cell growth at high osmolarity in the gram-positive intracellular parasite *Listeria monocytogenes*.

Ectoine, a neutral cyclic amino acid derivative that is also a tetrahydropyrimidine (for structure see Table 64.2), is synthesized by a number of halotolerant and halotrophic bacteria. This unusual solute was first detected in the extremely halophilic phototrophic bacteria of the genus *Ectothiorhodospira* and is assumed to have a protective function similar to that of proline and glycine betaine. Hydroxyectoine, differing from ectoine by a hydroxy group in position 3 of the pyrimidine ring, has also been detected in some salt-stressed halotolerant bacteria.

In addition to being a precursor and a nitrogen donor for the biosynthesis of several amino acids, glutamic acid also serves as a compatible solute in many species. Accumulation of this amino acid can occur either by transport from the medium or from *de novo* synthesis. Because glutamate is an anionic solute, it is usually accompanied by a monovalent cation to maintain charge balance in the cell. Bacteria with lower salt tolerances, such as *E. coli*, *Enterobacter aerogenes*, and *Pseudomonas fluorescens*, accumulate glutamate in response to changes in salinity. In more halophilic Bacteria such as *Halomonas elongata*, moderate concentrations of glutamate are accumulated, but they are relatively insensitive to extracellular NaCl. However, in the most halophilic Archaea, other organic anions are accumulated.

Peptides may also be used by organisms to maintain osmotic balance. Interestingly, the constituent amino acids include those with documented roles as osmolytes (i.e., glutamate, proline, hydroxyproline, and glycine). A specific dipeptide, *N*-acetylglutaminylglutamine amide (NAGGN), was found to accumulate in *Rhizobium meliloti*, the root nodule symbiont of alfalfa, as well as *Pseudomonas fluorescens*. The more general situation is for cells to internalize peptides from complex media enriched with peptone in response to osmotic stress.

D. β-Amino acids and derivatives

For use as osmolytes, amino acids have the advantage of high solubility and ease of synthesis, but the disadvantage of they being coupled to protein (and other metabolite) biosynthesis. β-Amino acids are relatively rare in nature and they are not incorporated in

TABLE 64.2 Amino acids and derivatives used as osmolytes in cells

Solute	Structure	Organisms
α-Amino acids		
glycine	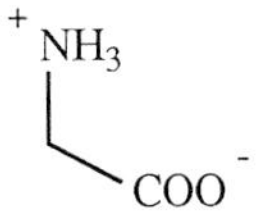 $^+NH_3$, COO^-	*Listeria monocytogenes*
proline	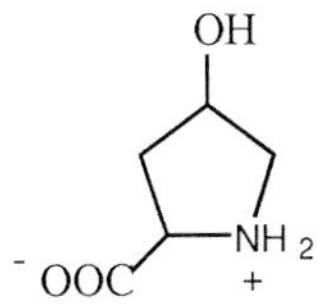 OH, ^-OOC, NH_2^+	*E. coli, Bacillus subtilis, Listeria monocytogenes, Salmonella typhimurium, Staphylococcus aureus, Streptomyces clavuligerus*, vascular plants
hydroxyproline	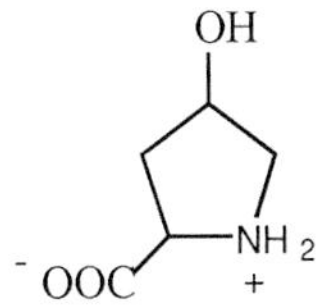 OH, ^-OOC, NH_2^+	*Listeria monocytogenes*
ectoine	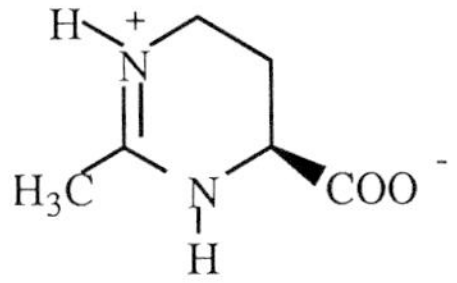 H, N^+, H_3C, N, H, COO^-	*Ectothiorhopospira, E. coli, Brevibacterium linens, Halomonas* SPC1, *Marinococcus halophilus*
hydroxyectoine	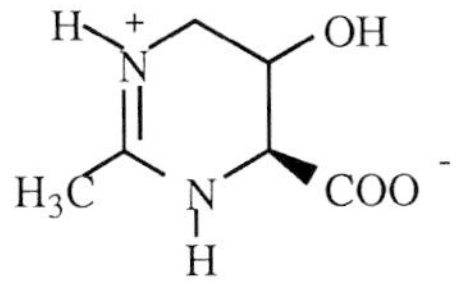 H, N^+, OH, H_3C, N, H, COO^-	Actinomycete A5-1 *Nocardiopsis* sp., *Halomonadaceae*
glutamate	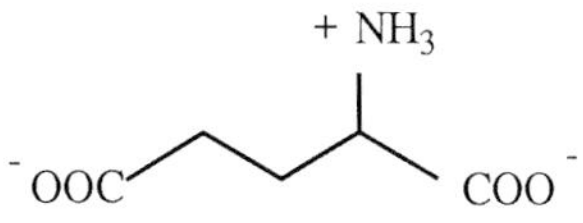 $+ NH_3$, ^-OOC, COO^-	*E. coli, Enterobacter aerogenes, Halomonas elongata, Methanococcus* sp., *Methanobacterium thermoautotrophicum, Natronococcus occultus, Pseudomonas fluorescens, Salmonella typhimurium*
β-Amino acids		
β-glutamate	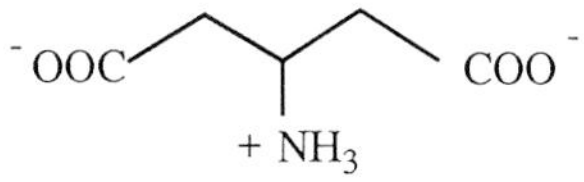 ^-OOC, COO^-, $+ NH_3$	*Methanococcus thermolithotrophicus, M. igneus, M. jannaschii*
β-glutamine	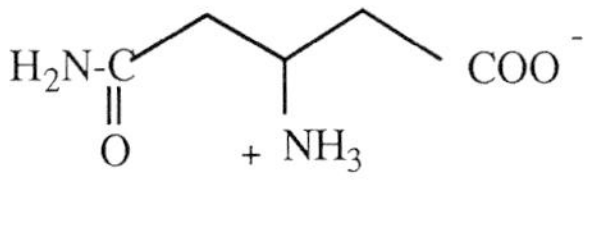 H_2N-C(=O), COO^-, $+ NH_3$	*Methanohalophilus portucalensis*
Nε-acetyl-β-lysine	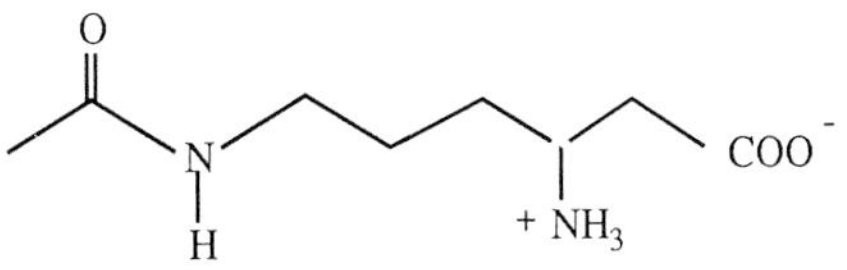 O, N, H, COO^-, $+ NH_3$	*Methanosarcina thermophila, Methangenium cariaci, Methanphalophilus* sp., *Methanococcus thermolithotrophicus*
Peptides		
N-acetylglutaminylglutamine amide (NAGGN)	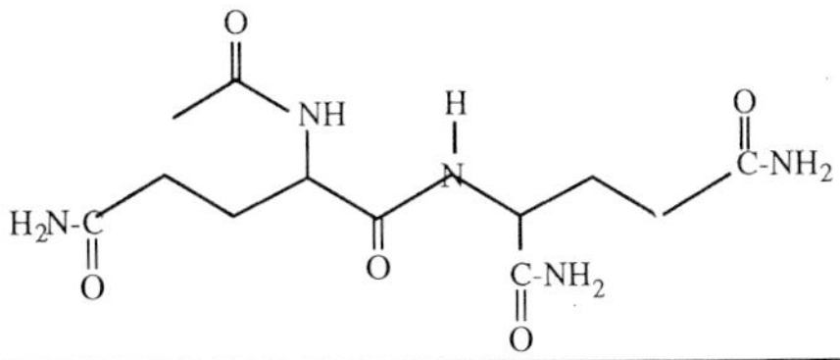 O, NH, H, N, O, C-NH_2, H_2N-C(=O), O, C-NH_2, O	*Pseudomonas fluorescens, Rhizobium meliloti*

In bacteria such as *Rhizobium meliloti* and *Agrobacterium tumefaciens*, mutants blocked in the production of these compounds are more sensitive to hypoosmotic stress.

In response to gradual decreases in osmolality, bacteria employ specific efflux systems to release turgor pressure. K^+, glutamate, proline, betaine, and trehalose can all be specifically released from cells. Sudden increases in turgor require more dramatic measures and can lead to loss of many osmolytes including even ATP, through less-specific channels. Stretch-activated channels may be activated under these conditions. Mechanosensitive ion channels have been found in all three domains of life including the extremely halophilic archaeon *Haloferax volcanii*. These channels are believed to be important in the efflux of osmolytes in response to a sudden shift to lower osmolality. These proteins exhibit gating by mechanical force on the membrane alone. MscL is one such identified protein. A figure portraying a channel, such as MscL, opening in response to membrane stretch is shown in Fig. 64.1. Such channels may also operate in the outer membrane of gram-negative bacterial cells, but have not been discovered in this membrane environment yet.

Another consequence of bacterial exposure to sudden hypoosmotic stress may be the need to expel water as well as osmolytes to relieve excess turgor pressure. Many bacterial cells have been found to possess a specific water-channel protein, designated aquaporin, the product of the *aqpZ* gene in the case of *E. coli*. Genome sequencing has suggested the presence of water channels in *Haemophilus influenzae, Mycoplasma genitalium* and *Synechocystis* PCC6803. The transport of water across cell membranes is essential to life and similar water channels have been detected in plants and animals. In prokaryotes, such channels may be less essential under isoosmotic growth conditions because the uptake of water needed for bacterial volume increase and cell growth and division can be met by simple diffusion, taking into account the water permeability of simple lipid bilayers and the high surface-to-volume ratio of microorganism. However, AqpZ may be needed for the efflux of water under conditions of hypoosmotic stress. Consistent with this hypothesis *aqpZ* gene transcription is highest at low osmolality and a*aqpZ* knockout mutant is inhibited in growth at low osmolality.

VI. ADDITIONAL OSMOSENSORS

Osmoregulation of virulence gene expression and of outer-membrane proteins has been observed in many bacteria. In these cases the regulation does not appear to be required for osmotic adaptation per se, but rather to conditions that correlate with a given osmolality, such as the presence of a possible host organism, a high- or low-nutrient environment, or the presence of toxic chemicals. Such osmoregulation in enteric bacteria is mediated by the EnvZ and OmpR proteins. This regulatory system is one of the premier paradigms of prokaryotic signal transduction. These two proteins control the expression of the outer-membrane protein genes *ompC* and *ompF* in response to medium osmolality. OmpR is a DNA-binding protein that can be phosphorylated by the inner-membrane-spanning protein EnvZ. These proteins, like the KdpD and KdpE proteins (Section III), are members of the two-component sensor kinase–response regulator protein family. In this case EnvZ acts as the sensor and OmpR is the response regulator. The induction of *ompC* expression and repression of *ompF* expression results from increased phosphorylation of OmpR at high osmolality.

Several facts indicate that EnvZ does not respond specifically to osmolality, but to one or more related factors. For example, the high osmolality signal is most effective in the case of solutes that cannot traverse the outer membrane. Also, supplementation of media with betaine inhibits high-osmolality signaling

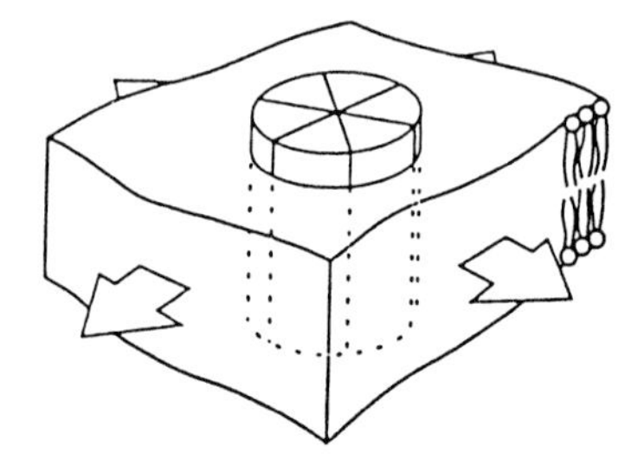

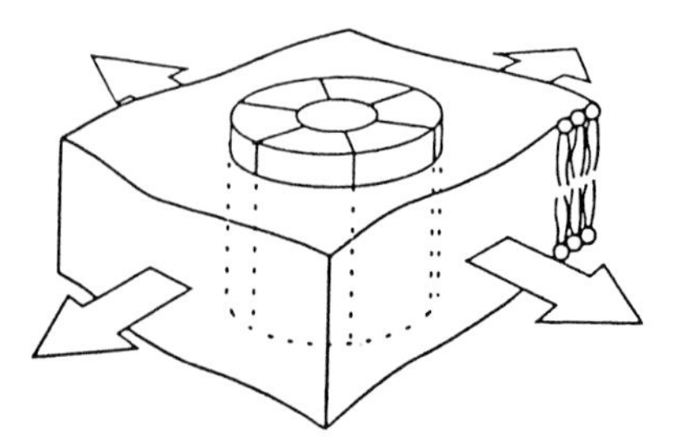

FIGURE 64.1 Model of a mechanosensitive channel. Diagram indicating a channel that is closed at low mechanical stretch force (left) and open at high mechanical stretch force (right). Taken from Sukharev *et al.* (1997) with permission.

by EnvZ. Because betaine supplementation would be predicted to affect cytoplasmic osmolyte levels, the betaine effect is most easlily interpreted as altering a cytoplasmic signal. However, because the periplasmic domain of EnvZ is known to be important for osmosensing, a periplasmic signal must also be necessary. EnvZ may be sensing osmolytes on both sides of the cytoplasmic membrane.

Another widespread osmosensor, designated ToxR, is found bacteria of the family Vibrionaceae. Many comparisons can be drawn between this protein and EnvZ and OmpR. The ToxR protein of the human pathogen *Vibrio cholerae* coordinates the activation of virulence genes, as well as the inverse expression of a pair of outer-membrane protein-encoding genes in a fashion similar to that seen in the EnvZ–OmpR system. Also, like EnvZ, ToxR appears to function as a membrane-spanning oligomer and its periplasmic domain is also critical for osmosensing. Temperature influences the activity of both the EnvZ–OmpR and the ToxR systems. Unlike the EnvZ—OmpR system, ToxR lacks the cytoplasmic portion of the EnvZ sensor and the amino-terminal domain of the OmpR regulator. As a result, osmolality signals perceived by ToxR may be directly transmitted to its DNA-binding domain the cytoplasm without the need for a two-component signaling system. Furthermore, unlike EnvZ, ToxR function is influenced by a second membrane protein, the ToxS protein, which may enhance ToxR oligomerization. To what extent ToxR is truly responsive to osmolality and not ionic effects remains to be determined. The fact that ToxR exhibits a biphasic response to medium salt concentration suggests that, like EnvZ, ToxR does not represent a true osmosensor.

VII. OSMOREGULATION MODEL

A model for bacterial osmoregulation based in large part on that presented by Csonka and Epstein (1996) for *E. coli* and *S. typhimurium* is shown in Fig. 64.2. The initial signal for a change in osmolality is proposed to

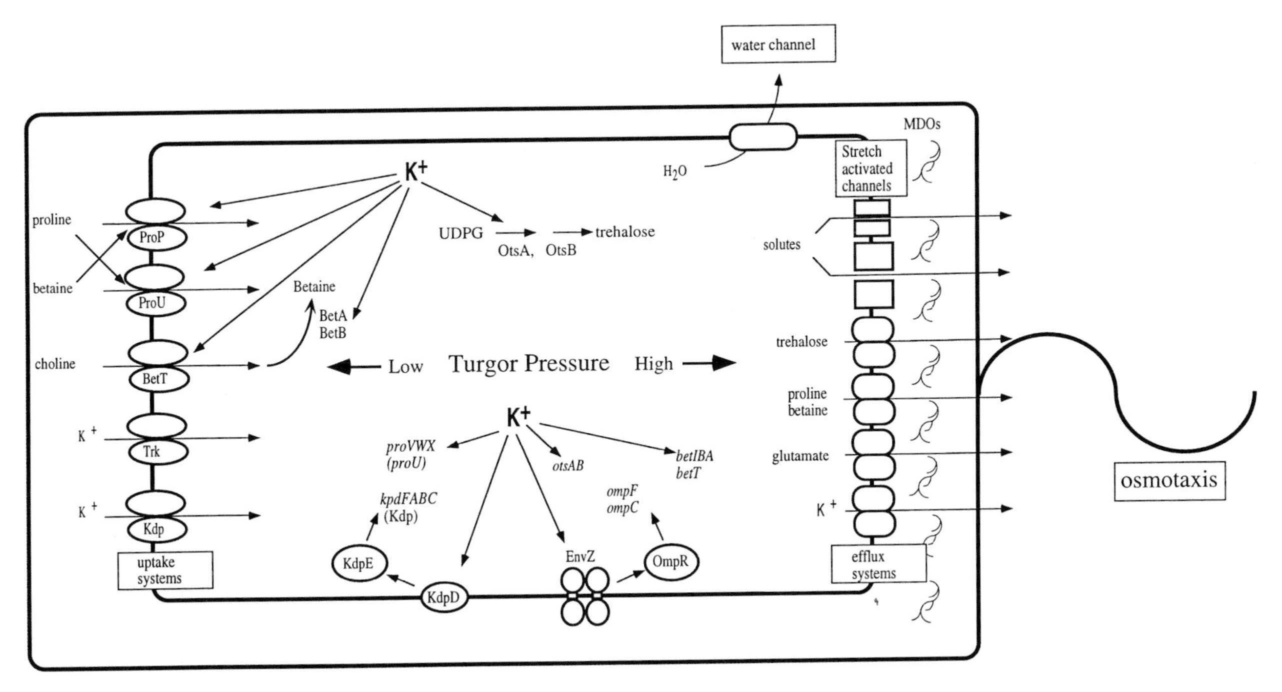

FIGURE 64.2 Model of osmoregulation in *Escherichia coli*. The outer box represents the outer membrane and the inner box the inner membrane. The lower left portion of the schematic indicates the K^+ uptake systems, which are activated by decreased turgor pressure resulting from exposure to increased medium osmolarity. K^+ uptake may be responsive to the effect of a change in turgor on membrane structure. Uptake of proline, betaine, and choline is dependent on K^+ uptake. The right portion of the model indicates the efflux systems, which may be activated at increased turgor pressure resulting from exposure to decreased medium osmolarity. This includes the transport out of the cell of specific osmolytes, such as K^+, glutamate, betaine, and trehalose, as well as the less specific release of osmolytes by stretch-activated channels and the release of water through the aquaporin water channel protein AqpZ during sudden increases in turgor. Membrane-derived oligosaccharides (MDOs) preferentially accumulate in the periplasmic space during osmotic downshifts. K^+ uptake also regulates the expression of many osmoregulated genes, as indicated in the lower portion of the figure, and enzyme activity, as indicated in the upper portion of the figure. Osmotaxis will be used to bias cell movement toward favorable osmotic environments and away from unfavorable osmotic environments. Modified from Csonka and Epstein (1996) with permission.

be a change in turgor pressure or a consequence of a change in turgor, such as membrane stretch or some other structural change in the membrane or cell wall. During hyperosmotic conditions, the decrease in cytoplasmic turgor pressure results in the net accumulation of K^+ (reflecting the sum of its uptake and efflux). This in turn will induce the synthesis or transport of the principal counter ion for K^+, glutamate. At osmolaties above 250 mosM, solutes less perturbing than K^+ are transported or synthesized. So, the increased transport of K^+ leads to increased transport and synthesis of compatible solutes such as betaine, proline, and trehalose. The accumulation of these compounds increases turgor above prestress levels, which results in the efflux of K^+ and glutamate. During osmotic downshift, the increase in turgor will stimulate specific efflux systems and the synthesis of periplasmic glucans. If the magnitude of the osmotic change is great enough, stretch-activated ion channels and possibly the water channels will provide a release valve mechanism for the rapid removal of both solutes and water. Osmotaxis is another important aspect of the osmotic response, as it provides a mechanism for cell movement into more favorable osmotic environments.

This model is largly based on information gathered from studies of enteric bacteria. Much more work is needed on microorganisms possessing adaptations for extremes of the levels and ranges of a_w that they will tolerate so that a more universal understanding of osmotic stress adaptation can be unveiled. The growing list of genetic methodologies available for use in extreme halophiles and haloversatile bacteria indicate a great opportunity for advanced studies of osmotic stress adaptation in these organisms.

BIBLIOGRAPHY

Bohin, J. P. Osmoregulated periplasmic glucans in Proteobacteria. (2000). *FEMS Microbiol Lett.* **186**, 11–19.

Calamita, G. (2000). The *Escherichia coli* aquaporin-Z water channel. *Mol. Microbiol.* **37**, 254–262.

Calamita, G., Kempf, B., Bonhivers, M., Bishai, W. R., Bremer, E., and Agre, P. (1998). Regulation of the *Escherichia coli* water channel gene aqpZ. *Proc. Natl. Acad. Sci. U.S.A.* **95**, 3627–3631.

Csonka, L. N. (1989). Physiological and genetic responses of bacteria to osmotic stress. *Microbiol. Rev.* **53**, 121–147.

Csonka, L. N., and Epstein, W. (1996). Osmoregulation. *In* "*Escherichia coli* and *Salmonella typhimurium* Cellular and Molecular Biology" (F. C. Neidhardt *et al.*, eds), pp. 1210–1223. American Society for Microbiology Press, Washington, DC.

DasSarma, S. (1995). Halophilic Archaea: An overview. *In* "Archaea, a Laboratory Manual" (F. T. Robb *et al.*, eds), pp. 3–11. Cold Spring Harbor Laboratory Press, Cold Spring Harbor.

Desmarais, Jablonski, D. P., Fedarko, N. S., and Roberts, M. F. (1997). 2-Sulfotrehalose, a novel osmolyte in haloalkaliphilic Archaea. *J. Bacteriol.* **179**, 3146–3153.

Grant, W. D., Gemmell, R. T., and McGenity, T. J. (1998). Halophiles. *In* "Extremophiles: Microbial Life in Extreme Environments" (K. Horikoshi and W. D. Grant, eds), pp. 93–132. John Wiley & Sons, New York.

Hohmann, I., Bill, R.M., Kayingo, I., and Prior, B.A. (2000). Microbial MIP channels. *Trends Microbiol.* **8**, 33–38.

Imhoff, J. F. (1986) Osmoregulation and compatible solutes in eubacteria. *FEMS Microbiol. Rev.* **39**, 57–66.

Le Rudulier, D., Strom, A. R., Dandekar, A. M., Smith, L. T., and Valentine, R. C. (1984). Molecular biology of osmoregulation. *Science* **224**, 1064–1068.

Martins, L. O., Huber, R., Stetter, K. O., Da Costa, M. S., and Santos, H. (1997). Organic solutes in hyperthermophilic Archaea. *Appl. Environ. Microbiol.* **63**, 896–902.

Roberts, M. F. (2000). Osmoadaptation and osmoregulation in archaea. *Front Biosci.* **5**, D796–D812.

Robertson, D. E., and Roberts, M. F. (1991). Organic osmolytes in methanogenic archaebacteria. *BioFactors* **3**, 1–9.

Severin, J., Wohlfarth, A., and Galinski, E. A. (1992). The predominant role of recently discovered tetrahydropyrimidines for the osmoregulation of halophilic eubacteria. *J. Gen. Microbiol.* **138**, 1629–1638.

Somero, G. N., Osmond, C. B., and Bolis, C. L., (1992). "Water and Life: Comparative Analysis of Water Relationships at the Organismic, Cellular, and Molecular Levels." Springer-Verlag, New York.

Storz, G., and Hengge-Aronis, R. (eds). (2000). "Bacterial Stress Responses." ASM Press, Washington, DC.

Sukharev, S. I., Blount, P., Martinac, B., and Kung, C. (1997). Mechanosensitive channels of *Escherichia coli*: The MscL gene, protein, and activities. *Annu. Rev. Physiol.* **59**, 633–657.

Ventosa, A., Nieto, J. J., and Oren, A. (1998). Biology of moderately halophilic aerobic bacteria. *Microbiol. Molec. Biol. Rev.* **62**, 504–544.

Vreeland, R. H. (1987). Mechanisms of halotolerance in microorganisms. *CRC Crit. Rev. Microbiol.* **14**, 311–356.

Wood, J. M. (1999). Osmosensing by bacteria: signals and membrane-based sensors. *Microbiol. Mol. Biol. Rev.* **630**, 230–262.

WEBSITE

List of bacterial names with standing in nomenclature (J. P. Euzèby) *http://www.bacterio.cict.fr/index.html*

65

Outer membrane, Gram-negative bacteria

Mary J. Osborn
University of Connecticut Health Center

GLOSSARY

***β*-barrel** A protein-folding motif in which several amino acid sequences forming *β*-sheets fold in such a way as to form a barrel-like structure.

EDTA Ethylenediaminetetracetic acid, a chelator of divalent cations.

GlcNac *N*-acetyl-D-glucosamine.

glycosyl transferase An enzyme catalyzing the transfer of a sugar residue, generally from its activated nucleotide sugar derivative to an acceptor molecule.

Kdo 2-keto-3-deoxy-D-manno-octulosonate.

lipid bilayer The fundamental structural basis of biological membranes, in which two layers of lipids are apposed, with hydrophilic groups of each facing the aqueous medium and hydrophobic fatty acyl chains forming the interior.

lipopolysaccharide A complex polymer exposed at the cell surface and responsible for major antigenic specifities and endotoxic properties.

periplasmic space The region between the cytoplasmic and outer membranes, containing the cell wall, hydrolytic enzymes, components of active transport systems, and proteins required for maturation of newly synthesized periplasmic proteins.

plasmolyzed cells Cells incubated in hypertonic medium to retract the inner membrane and expand the periplasmic space.

SDS-PAGE Polyacrylamide gel electrophoresis, incorporating the denaturant sodium dodecyl sulfate in sample and gel buffers.

The Cell Envelope of Gram-negative Bacteria consists of an inner cytoplasmic membrane, the peptidoglycan (murein) cell wall, and an outer membrane that bounds the cells. The murein cell wall underlies the outer membrane and is generally anchored to it covalently through the murein lipoprotein and by noncovalent interactions with the outer-membrane protein, OmpA. The outer membrane is the interface between the cell and its external environment. It acts as a protective barrier, mediates interactions between bacteria and between bacteria and the animal or plant host cells, and anchors externally disposed organelles such as pili. The composition and structure of the outer membrane is highly specialized, reflecting its specialized functions (See Fig. 65.1). The enteric bacteria, specifically *Escherichia coli* and *Salmonella typhimurium*, provide the paradigm for outer-membrane structure and function and will form the major basis of this review. However, it should be recognized that the specifics of structure and function may differ significantly as one moves away from the enteric family.

I. OUTER-MEMBRANE COMPOSITION AND STRUCTURE

A. Isolation of outer membrane

Two general methods have been employed to separate inner and outer membranes. The first and more rigorous takes advantage of the difference in buoyant densities between the two membranes. Cells are broken by French press or by lysozyme plus osmotic

The Desk Encyclopedia of Microbiology
ISBN: 0-12-621361-5

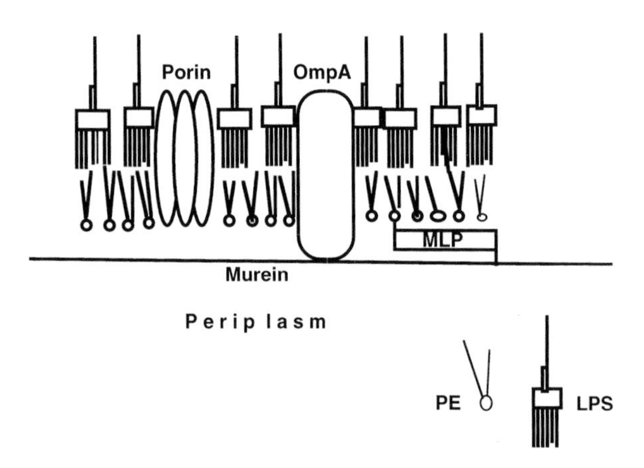

FIGURE 65.1 Schematic structure of the outer membrane. Mlp, murein lipoprotein; PE, phosphatidylethanolamine.

shock and membranes are fractionated by one of several protocols for isopycnic sucrose gradient centrifugation. The outer membrane bands at an apparent buoyant density between 1.22 and 1.27 g/cc, whereas the inner membrane band is recovered in the range of 1.12–1.15 g/cc. Unseparated material bands between the two fractions, as do several minor specialized membrane fractions that can be further purified by additional sedimentation and flotation gradient centrifugations (Ishidate *et al.*, 1986).

The second method of outer-membrane isolation takes advantage of the differential resistance of outer membrane to detergents or alkali. Detergent treatment (Triton X-100, in the absence of EDTA or Sarkosyl) is used to dissolve inner membrane, leaving the outer membrane presumably intact. Alternatively, major outer-membrane protein species and lipopolysaccharide (LPS) remain insoluble following treatment with alkali, whereas inner membrane is solubilized. These methods are far quicker and more convenient than isopycnic centrifugation, but risk the solubilization of the peripheral outer-membrane-associated proteins. The assumption that all inner-membrane proteins are solubilized is also difficult to verify.

B. Outer-membrane proteins

In contrast to the inner cytoplasmic membrane, the outer membrane is essentially metabolically inert and contains a limited number of major protein species. These include the porins, which are responsible for solute permeation across the membrane and proteins (murein lipoprotein, OmpA) that interact covalently or noncovalently with the underlying murein. In addition, a variety of less abundant species mediate interactions with the external environment (e.g., adherence), assembly of surface structures such as pili and flagella, conjugation, and secretion of proteins to the external medium.

C. Outer-membrane lipids

1. *Lipopolysaccharide structure*

Lipopolysaccharide is a unique and major constituent of the outer membrane and is responsible for the endotoxic properties of gram-negative bacteria. The LPS of enteric bacteria consists of three structural regions—Kdo-lipid A, a unique glycolipid anchoring the molecule into the outer membrane; a core oligosaccharide region (subdivided into backbone and outer core); and peripheral O-antigen chains, which are responsible for major immunological specificities of the intact organism. (See Fig. 65.2.)

Kdo-Lipid A is highly conserved among gram-negative bacteria. It consists of a fatty acylated phosphorylated glucosamine disaccharide characterized by N-linked and O-linked 3-hydroxymyristoyl residues (or shorter chain 3-hyroxy fatty acids in some nonenteric genera), as well as saturated fatty acyl substitutions on the 3-hydroxy fatty acids. Other decorations (phosphoryl ethanolamine, 4-aminoarabinose) are variably present. A Kdo disaccharide is linked to the 4′ position of the disaccharide and provides the attachment site for the core saccharide unit.

The inner-core backbone region contains two or three L-glycero-D-mannoheptose residues, and again may have additional phosphate and pyrophosphorylethanolamine decorations. The outer-core oligosaccharide is glycosidically linked to the nonreducing heptose residue, and in the enterics typically contains galactose, glucose, and *N*-acetylglucosamine. The structures are similar, but not identical, among the enteric genera. Genera such as *Hemophilus* and *Neisseria* produce only lipoligosaccharides (LOS), which are analogous to the core LPS of enterics.

O-antigen consists of long polysaccharides, generally composed of oligosaccharide repeating units. The structure is characterized by the frequent presence of unusual deoxy, dideoxy, and amino sugars; the composition, structure, and immunospecificity vary widely within and between genera, and form the basis of serological typing. Wild-type bacteria with the complete O-antigenic LPS are referred to as smooth, whereas mutants lacking O-antigen are called rough. In certain rough mutants of enteric bacteria, a second polysaccharide, enterobacterial common antigen (ECA), takes the place of the O-chains. A form of ECA that is widely distributed in the enterics is anchored in the outer membrane by attachment to a phosphoglyceride. ECA structure is conserved among the enterics.

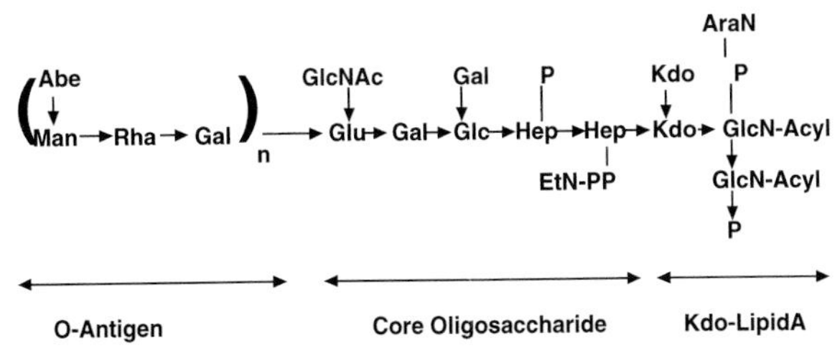

AraN
Abe GlcNAc Gal P Kdo P
Man → Rha → Gal)n → Glu → Gal → Glc → Hep → Hep → Kdo → GlcN-Acyl
EtN-PP GlcN-Acyl
P
O-Antigen Core Oligosaccharide Kdo-LipidA

FIGURE 65.2 Structure of *S. typhimurium* LPS. Abe, abequose; AraN, 4-aminoarabinose; EtN, ethanolamine; Gal, galactose; Glc, glucose; GlcN-acyl, *N,O*-di-3-hydroxymyristoyl glucosamine and *N,O* di-3-hydroxy(3-O-lauroyl or myristoyl)myristoyl glucosamine; GlcNAc, *N*-acetylglucosamine; Hep, L-glycero-D-mannoheptose; Kdo, 2-keto-3-deoxyoctulosonate; Man, mannose; Rha, rhamnose.

Both the O-antigenic LPS of enteric bacteria and the LOS of *Neisseria* and *Hemophilus* are subject to changes in structure and antigenicity due to altered patterns of gene expression. In *Salmonella*, for example, a number of temperate converting phages carry genes that encode glycosyl transferases or change the expression of host glycosyl transferases or the pattern LPS acetylation. *Neisseria* are capable of altering LOS glycosyl transferase expression by a process of antigenic variation.

2. *Enterobacterial common antigen*

Enterobacteria share a widely distributed surface-exposed antigenic specificity called. ECA is a heteropolysaccharide composed of the unusual amino sugars, *N,O*-diacetyl glucosamine, *N*-acetylaminomannuronic acid, and *N*-acetylfucosamine. It can exist in two forms in the outer membrane. A form anchored to the external leaflet of the outer membrane by linkage of the reducing terminus to a phosphatidic acid residue is generally present. The second form, in which ECA replaces the O-chains of LPS, is found in some LPS mutants that lack O-antigen.

3. *Phospholipid composition*

In addition to lipopolysaccharides, outer membranes contain conventional phospholipids. However, the phospholipid composition is typically very simple, with phosphatidylethanolamine accounting for well over 90% of the total in the enteric family.

D. Outer-membrane structure

1. *The enteric paradigm*

The lipid bilayer of the outer membrane is highly asymmetric, with LPS confined exclusively to the external leaflet of the membrane, with phosphatidyl ethanolamine restricted, probably exclusively, to the inner periplasmic leaflet. The total number of acyl chains in the two types of lipid are calculated to be roughly similar. The asymmetric distribution of lipids in the membrane is disrupted physiologically by deep rough mutations in the LPS backbone, which give a phenotype deficient in major outer-membrane proteins and result in the appearance of substantial amounts phospholipid in the outer membrane.

The major species of proteins (porins, OmpA) are integral transmembrane membrane proteins; however, certain proteins, (e.g., murein lipoprotein) are anchored specifically to the inner leaflet of the membrane, and others to the surface-exposed outer leaflet. The outer membrane is linked covalently to the underlying murein cell wall by the murein lipoprotein and noncovalently through interaction with the periplasmic domain of OmpA. Experimental treatments that disrupt the attachment of outer membrane to peptidoglycan can result in randomization of lipopolysaccharide across the bilayer, and perhaps other alterations in the normal membrane topology.

2. *Zones of adhesion*

Using light and electron microscopy of plasmolyzed cells, sites are observed at which the inner membrane fails to retract from outer membrane and the two membranes appear to be more or less firmly associated in some way. Such zones of adhesion were early postulated to function as bridges in the translocation of proteins, polysaccharides, and lipids to the outer membrane, but their biochemistry remains largely obscure. It seems likely that the contact sites observed represent several different kinds of structures. For example, annuli traversing the cell circumference, the periseptal annuli, are thought to play a role in septum formation and appear to be relatively stable. Type III secretion systems and the biogenesis of Type IV pili both include a number of periplasmically oriented proteins that are thought to function in secretion of virulence proteins or pilus subunits across the outer membrane. Other proteins, such as TonB, which is required for a number of specific transport functions, are also postulated to bridge between the inner and outer membranes.

II. OUTER-MEMBRANE PROTEINS

A. Porins

1. *General pores*

In order to reach active transport systems in the inner membrane, nutrient solutes must first be able to diffuse across the outer membrane to the periplasmic space. A family of ubiquitous major 34- to 37-kDa outer-membrane proteins, including OmpC, OmpF, and PhoE of *E. coli* and other enteric bacteria, provide fixed nonspecific diffusion pores for permeation of small hydrophilic solutes. The proteins typically exist in the outer membrane as homotrimers. Their amino acid sequences show no extended hydrophobic regions and no extensive predicted α-helical structure. Rather, the crystal structure of the OmpF porin of *E. coli* has revealed an amphipathic β-barrel structure, with the hydrophobic faces lying among the three subunits and at the external surface of the trimer. Each of the monomer units contains a transmembrane hydrophilic pore approximately 7×11 Å in size. By diffusion measurements, the OmpF channel is slightly larger than that of OmpC. The fixed pores of OmpF, OmpC, and PhoE allow diffusion-limited permeation of small hydrophilic solutes up to a limit of about 600 Da, with the rate of permeation highly dependent on the diameter of the hydrated solute. OmpC and OmpF are nonspecific, but favor neutral or cationic solutes over anions. PhoE, though part of the phosphate regulon, is not specific for phosphate and is considered a general channel with a preference for anions over cations.

The expression of OmpC and OmpF is subject to complex reciprocal regulation, at both transcriptional and posttranscriptional levels. OmpF is repressed by high temperature (37 °C), high osmolarity, and probably other factors of the host environment, and is therefore repressed in the animal host. Presumably, the repression of the porin with the larger channel in favor of OmpC, whose channel is smaller, provides additional protection against bile salts and other toxic molecules involved in host defenses. Transcriptional regulation is mediated by the EnvZ–OmpR two-component system, which responds to osmotic stress and results in the down-regulation of OmpF and activation of OmpC. In addition, a small antisense RNA, micF, is divergently transcribed from the OmpC operator and is thought to inhibit the translation of OmpF mRNA in response to environmental cues.

2. *Specific diffusion channels*

The group of nonspecific porins will not accommodate solutes larger than 600 Da, and specialized diffusion channels are required for the uptake of solutes such as maltodextrins, vitamin B12, and iron–siderophore complexes. LamB, a component of the inducible maltose regulon of *E. coli*, has a trimeric structure similar to that of OmpF, and indeed the LamB pore allows the permeation of a variety of small molecules in addition to maltose and maltodextrins. However, the LamB channel is distinguished by the presence of a binding site for maltose oligosaccharides, which confers a higher degree of specificity for this family saccharides. Diffusion of iron–siderophore complexes is mediated by a group of at least five outer-membrane channel proteins in *E. coli*, whereas B12 uptake is dependent on the Btu channel. In these cases, the channel protein exhibits high-affinity binding sites for its substrate and permeation is substrate-specific. It is of interest that most of these outer-membrane channel proteins also act as receptors for specific bacteriphages or colicins. LamB, as the name suggests, was originally identified as the receptor for phage lambda.

B. OmpA

OmpA is a highly abundant outer-membrane protein of approximately 35 kDa. The protein is monomeric in the membrane and consists of two domains; the N-terminal domain has a transmembrane β-barrel structure, whereas the globular C-terminal region is exposed at the periplasmic face of the membrane. The protein is only partially denatured in SDS at temperatures below 100 °C, the basis for its initial identification as a "heat modifiable" protein in SDS-PAGE. Two physiological functions have been identified. OmpA is required for *E. coli* to act as an efficient recipient in conjugation, a function of its surface-exposed N-terminal domain, and anchors the outer membrane noncovalently to the underlying murein cell wall through its periplasmically oriented C-terminal domain. Mutants lacking OmpA tend to form spherical cells with unstable outer membranes. It is of interest that OmpA, although monomeric, can act as a weak nonspecific porin in reconstituted systems. This is probably not physiologically significant in *E. coli*, but in *Pseudomonas* species, which lack OmpC/F homologs, the OmpA homolog, OprF, is the major nonspecific porin.

C. Outer-membrane lipoproteins

1. *Murein lipoprotein*

Murein lipoprotein is a small (7.2 kDa), highly abundant protein that provides covalent linkage between the outer membrane and murein. The protein is hydrophilic and predominantly α-helical. The protein lies at the periplasmic face of the outer membrane and is integrated into the bilayer by its N-terminal lipid modification. The C-terminal cysteine residue is modified by thioether linkage to a diglyceride residue, and, in addition, the α-amino group is fatty acylated. About one-third of the murein lipoprotein, the bound form, is covalently attached to the underlying murein by isopeptide linkage of the carboxyl group of murein diaminopimelic acid (DAP) to the ε-amino group of the C-terminal lysine residue. Murein lipoprotein mutants unable to form the attachment to murein have unstable outer membranes that leak periplasmic proteins and release outer-membrane blebs into the medium.

2. *Other outer-membrane lipoproteins*

Over 25 lipoprotein genes have been identified in *E. coli*, encoded on chromosomal, phage, or plasmid genomes. The majority have been localized to the outer membrane, although a significant number are found in inner membrane. A variety of functions have been assigned to members of the outer-membrane class, including involvement in the efflux pump for acriflavine, osmoregulation, surface exclusion, serum (complement) resistance, and release of colicins.

Lipoproteins are not limited to gram-negative bacteria. For example, in *Bacillus* species, some binding proteins required for the active transport of various substrates, as well as certain secreted proteins, are anchored to the external face of the cytoplasmic membrane as lipoproteins. Spirochetes offer a particularly dramatic example; lipoproteins apear to make up the majority of proteins in the outer membranes of treponemes and *Borrelia*. Indeed, close to 100 lipoprotein genes and pseudogenes have been identified in the *Borrelia* genome (Fraser *et al.*, 1997). Their functions are largely unknown, but presumably mediate host–pathogen interactions. Derived amino-terminal lipopeptides are highly immunogenic and induce strong inflammatory responses.

III. FUNCTIONS OF OUTER MEMBRANE

A. Barrier function

In comparison with the cytoplasmic membrane and other biological membranes, the outer membrane exhibits strikingly reduced permeability to a wide variety of lipophilic compounds, including bile salts and other detergents, lipophilic antibiotics, and dyes. Lipopolysaccharide is primarily responsible for the barrier properties of the membrane. Permeation of lipophilic compounds across membranes requires the initial intercalation of the compound into the hydrophobic interior of the lipid bilayer. Penetration into the outer membrane is hindered in the presence of LPS for two reasons. Lipid A fatty acids are saturated; therefore, they are more highly ordered and of lower mobility than those of a conventional phospholipid leaflet. In addition, each LPS molecule contains 6–7 fatty acyl residues, allowing an increased number of intermolecular hydrophobic interactions than phosphoglycerides having only two fatty acids.

The barrier function is disrupted under conditions that compromise the molecular organization of the membrane. Treatment with EDTA removes divalent cations that cross-bridge LPS phosphate groups and enhance LPS–LPS association. Mutations affecting the biosynthesis of the heptose-containing backbone region of LPS result in a deep rough phenotype, characterized by the reduced abundance of outer-membrane porins and appearance of substantial amounts of phosphatidylethanolamine in the external leaflet of the outer membrane. Polymyxin and other polycations also disrupt LPS organization and the barrier function by electrostatic interaction with the polyanionic LPS. Mutants of *Salmonella typhimurium* resistant to polymyxin have markedly increased amounts of the aminosugar 4-aminoarabinose in their lipid A, thus reducing LPS acidity and binding to added polycations.

B. Host–pathogen interactions

The outer membrane plays a crucial role in many aspects of pathogenesis, including the evasion of host defenses and the adherence of invading bacteria to host surfaces, which is important for the initial establishment of infection as well as intracellular invasion and survival. LPS O-antigen is antiphagocytic and both outer-membrane proteins and LPS have been implicated in resistance to serum complement. Outer-membrane proteins can themselves act as adhesions or invasins, mediating specific receptor–ligand interactions with host-cell surfaces (e.g., the PII (opa) protein of *N. gonorrheae* and the invasin of *Yersinia* species). Alternatively, adherence may be mediated by pili or fimbriae (e.g., the P pilus of uropathogenic *E. coli*) whose assembly requires outer-membrane usher proteins. Type III secretion systems required for intracellular invasion by many gram-negative bacteria include outer-membrane components that form multimeric structures mediating the direct injection of virulence proteins into host-cell cytosol. Other outer-membrane proteins specifically facilitate the secretion of protein toxins into the medium.

IV. ASSEMBLY OF OUTER MEMBRANE

A. Export of outer-membrane proteins

Outer-membrane proteins are synthesized as precursor proteins with classic signal sequences and are exported to the periplasm by the Sec system for polypeptide secretion. The signal sequence is removed by the leader peptidase (or, in the case of lipoproteins, by a specific prolipoprotein signal peptidase), located at the periplasmic face of the inner membrane. The interaction of soluble porin monomers in the periplasm with (presumably nascent) lipopolysaccharide is thought to promote a conformation switch allowing its assembly into mature trimers. Oligomerization is required for integration into the outer membrane. The role of LPS is consistent with the fact that deep rough LPS mutants are deficient in outer-membrane porins. Similarly, a newly synthesized OmpA monomer is found in the periplasm in an open conformation that is much more sensitive to proteolysis than is the mature protein. Interaction with nascent LPS is not required for the maturation of the periplasmic intermediate or for integration into the outer membrane, but interaction with LPS within the outer membrane is necessary for OmpA function. A periplasmic protein, Skp/OmpH, has been shown to bind both porins and OmpA, and is required for effective assembly into outer membrane. Presumably, Skp/OmpH acts as a chaperone, facilitating the maturation of the nascent periplasmic proteins.

The targeting of outer-membrane proteins from periplasm to the outer membrane appears to depend on tertiary or quaternary structures of the proteins, rather than on any obvious conserved targeting sequence. Amphipathic β-barrel structures have been identified or predicted for the major transmembrane porins and for OmpA. This structure is necessary for integration of these proteins into the membrane, and may be sufficient.

B. Biogenesis of lipopolysaccharide

1. *Biosynthetic pathways*

The modular structure of LPS—Kdo-lipid A, core, and O-antigen—is reflected in its genetics and biosynthetic pathways. In enteric organisms, most genes for biosynthesis of core and O-antigen are clustered in the *rfa* and *rfb* regions, respectively, although the genes required for synthesis of the Kdo-lipid A moiety are scattered in several locations. Core LPS and O-antigen are synthesized by two independent pathways, involving different mechanisms of polysaccharide chain elongation. The core pathway uses the classic mechanism in which single sugar residues are transferred from their respective nucleotide sugars to the nonreducing terminus of the growing chain, whereas the O-antigen pathway uses the membrane isoprenoid coenzyme bactoprenyl-P (undecaprenyl-P) for the polymerization of oligosaccharide repeating units.

The synthesis of the Kdo-Lipid A portion (Fig. 65.3) begins in the cytosol with 3-hydroxymyristoylation of UDP-GlcNAc, followed by deacetylation of the amino sugar and addition of a second, N-linked 3-hydroxymyristoyl residue. A portion of the diacyl nucleotide sugar is then hydrolyzed to diacyl glucosamine-1-P (lipid X), which acts as acceptor for transfer of the diacylglucosamine residue from the UDP derivative. An additional phosphate residue is then transferred from ATP to the 4′ position of the resulting tetraacyl disaccharide-P, followed by transfer of two Kdo residues from CMP-Kdo to the 6′ position. In the final steps, the two 3-hydroxymyristoyl residues of the nonreducing glucosamine are esterified with saturated acyl residues (laurate and myristate in *E. coli*). Lipid A also contains variable amounts of phosphyorylethanolamine and 4-aminoarabinose, but little information is available on the enzymology of their addition. The addition of Kdo is unusual in that the two residues are added in concerted fashion by a single bifunctional Kdo transferase. The number of Kdo residues varies among genera. *Chlamydia* LPS contains three, added by a trifunctional transferase, whereas in *Hemophilus influenzae*, only a single Kdo and a monofunctional transferase are present.

It is of interest that UDP-GlcNAc is the starting substrate for the synthesis of both lipidA and murein. It is clearly important for the cell to regulate entry into the two pathways in such a manner as not to compromise either because both products are essential for viability. Regulation occurs at the level of the second enzyme of the lipid A pathway, deacetylation of UDP-*O*-myristoyl-*N*-acetylglucosamine.

The addition of the heptosyl residues of the backbone and the sugars of the outer core is catalyzed by a series of peripheral membrane-associated glycosyl transferases with transfer single-sugar residues successively from the nucleotide sugars to the nonreducing terminus of the growing chain. The transferases, as

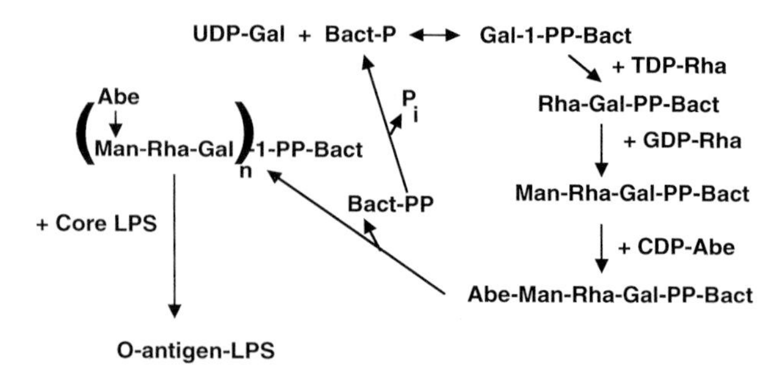

FIGURE 65.3 Lipid A biosynthetic pathway. ACP, acyl carrier protein; other abbreviations as in Fig. 65.2.

well as the enzymes required for synthesis of ADP-L-glycero-D-mannoheptose, are encoded by genes of the *rfa* locus. The expression of both *rfa* and *rfb* genes is regulated by the transcriptional activator, RfaH, which is also required for expression of the hemolysin of enteric bacteria and surface functions encoded by the F-factor.

O-antigen chains are typically composed of oligosaccharide repeating units, whose synthesis and poly-merization are determined by the *rfb* operon. The oligosaccharide unit is assembled on the membranebound coenzyme bactoprenyl (undecaprenyl)-phosphate (Fig. 65.4), and all the steps of the pathway are catalyzed by membrane-associated enzymes. The pathway begins with transfer of the reducing terminal sugar-1-P from its nucleotide sugar to the coenzyme to form bactoprenyl-pyrophosphoryl-monosaccharide. The remaining sugars are then transferred sequentially to the nonreducing terminus of the growing oligosaccharide. It should be noted that murein biosynthesis also requires bactoprenyl-P, and in certain mutants in which incomplete lipid-linked intermediates in the O-antigen pathway accumulate, the coenzyme is unavailable for murein synthesis and cell lysis ensues.

Polymerization, catalyzed by O-polymerase, takes place by a mechanism in which new oligosaccharide units are introduced at the reducing end of the growing polymer by transfer of polymer to the incoming oligosaccharide unit. The process is analogous to the mechanisms of elongation of polypeptides and fatty acids. In the final step of LPS assembly, polymeric O-antigen chains are added to the independently synthesized core LPS by *O*-ligase. O-antigen-containing LPS forms a ladder in SDS-PAGE, indicative of a high degree of heterogeneity in O-chain length. The distribution of polysaccharide chain lengths is presumably determined in part by the kinetics of competing O-polymerase and O-ligase, but is also under genetic control by the the *wzz* (*rol*) gene.

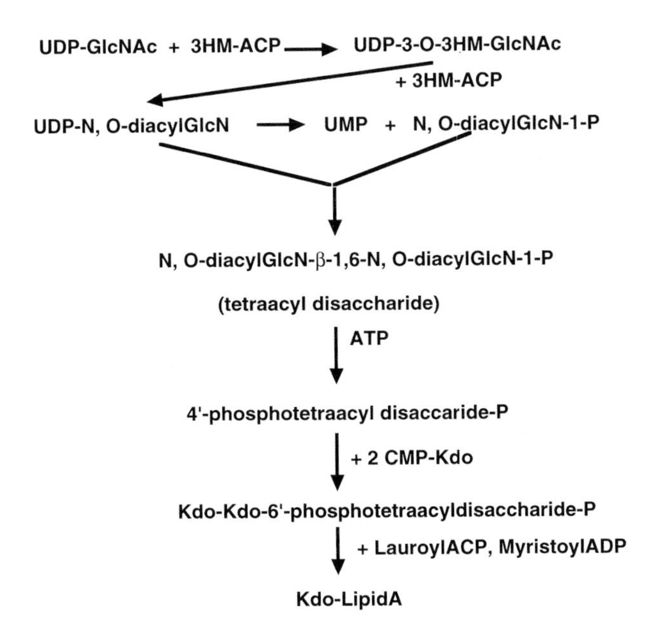

FIGURE 65.4 O-antigen biosynthetic pathway. Bact, bactoprenol; other abbreviations as in Fig. 65.1.

2. *Biosynthesis of enterobacterial common antigen*

In mutants lacking O-antigen, ECA can be attached to core LPS as though it were an *O*-polysaccharide. The ECA polymer is composed of oligosaccharide repeating units, such as O-antigen, and biosynthesis parallels that of O-antigen, using a similar bactoprenyl-P pathway. Presumably, the phosphoglyceride form of the ECA polysaccharide is formed by the transfer of the completed polymer chain to a phosphatidic acid acceptor.

3. *Topology of LPS biogenesis and export to outer membrane*

Both the core LPS and oligosaccharide intermediates of O-antigen synthesis are assembled at the cytoplasmic face of the inner membrane, yet the attachment of O-antigen to the core takes place at the periplasmic face of the inner membrane. Thus, the flip-flop of both across the inner membrane must be invoked. Similar considerations hold for the assembly of murein from its comparable lipid-linked intermediates. The mechanisms of these transmembrane translocation steps are not well understood. The presence of Kdo residues is necessary for the efficient translocation of lipid A to the outer membrane. In addition, evidence (Chu *et al.*, 1995; Zhou *et al.*, 1998) strongly suggests that the *msbA* gene of *E. coli* and the *abcA* gene of *Aeromonas salmonicida*, both members of the ATP-dependent ATP binding cassette (ABC) superfamily of transporters, are required for the translocation of cytosolically oriented core LPS or O-antigen intermediates to the periplasmic face of the inner membrane. Other ABC systems are known to facilitate the secretion of capsular polysaccharides, whose synthesis takes place in cytosol. The mechanism of LPS translocation to the outer membrane also remains unclear. The process is independent of ongoing protein synthesis, and directly or indirectly requires both ATP and a membrane potential across the inner membrane. Immunoelectron microscopy shows LPS newly incorporated into the outer membrane preferentially localized over zones of adhesion between the inner and outer membranes, but the molecular nature of such contact sites is unknown.

C. Export of phospholipid to outer membrane

The problem of the mechanism of translocation of phospholipid to outer membrane is somewhat analogous to that for LPS. Phosphatidylethanolamine and other phospholipids are believed to be synthesized in the cytoplasmic leaflet of the inner membrane and rapidly transposed to the periplasmic leaflet. The translocation of phosphatidylethanolamine to the periplasmic leaflet of the outer membrane is energy-dependent, rapid, and apparently reversible. Although phospholipid-exchange protein has been identified in the periplasm of purple bacteria, efforts to detect such activity in enteric bacteria have been unsuccessful. It has been postulated that translocation occurs at zones of adhesion, but direct evidence is lacking.

BIBLIOGRAPHY

Chu, S., Noonan, B., Cavaignac, S., and Trust, T. J. (1995). Endogenous mutagenesis by and insertion sequence element identifies Aeromonas salmonicida AbcA as an ATP-binding cassette transport protein required for biogenesis of smooth lipopolysaccharide. *Proc. Natl. Acad. Sci. U.S.A.* **92**, 5754–5758.

Driessen, A. J., Manting, E. H., and van der Does, C. (2001). The structural basis of protein targeting and translocation in bacteria. *Nat. Struct. Biol.* **8**, 492–498.

Fraser, C. M., Casjens, S., Huang, W. M., *et al.* (1997). Genomic sequence of a Lyme disease spirochaete, *Borrelia burgorferi. Nature* **390**, 580–586.

Ishidate, K., Creeger, E. S., Zrike, J., *et al.* (1986). Isolation of differentiated membrane domains from Escherichia coli and Salmonella typhimurium, including a fraction containing attachment sites between the inner and outer membranes and the murein skeleton of the cell envelope. *J. Biol. Chem.* **261**, 428–443.

Koebnik, R., Locher, K. P., and Van Gelder, P. (2000). Structure and function of bacterial outer membrane proteins: barrels in a nutshell. *Mol. Microbiol.* **37**, 239–253.

Muller, M., Koch, H. G., Beck, K., and Schafer, U. (2000). Protein traffic in bacteria: multiple routes from the ribosome to and across the membrane. *Prog. Nucleic Acid Res. Mol. Biol.* **66**, 107–157.

Nikaido, H. (1996). Outer membrane. *In* "Escherichia coli and Salmonella" (F. Neidhardt, ed.), 2nd ed, pp. 29–47. ASM Press, Washington, DC.

Raetz, C. R. H. (1996). Bacterial lipopolysaccharides: A remarkable family of bioactive macroAmphiphiles. *In* "Escherichia coli and Salmonella" (F. Neidhardt, ed.), 2nd ed, pp. 1035–1063. ASM Press, Washington, DC.

Rick, P. D., and Silver, R. P. (1996). Enterobacterial common antigen and capsular polysaccharides. *In* "Escherichia coli and Salmonella" (F. Neidhardt, ed.), 2nd ed, pp. 104–122. ASM Press, Washington, DC.

Wu, H. C. (1996). Biosynthesis of lipoproteins. *In* "Escherichia coli and Salmonella" (F. Neidhardt, ed.), 2nd ed, pp. 1005–1014. ASM Press, Washington, DC.

Zhou, Z., White, K. A., Polissi, A., Georgopolous, C., and Raetz, C. R. (1998). Function of Escherichia coli MsbA, an essential ABC family transporter, in lipid A and phospholipid biosynthesis. *J. Biol. Chem.* **273**, 12466–12475.

WEBSITES

The *E. coli* Cell Envelope Protein Data Collection
http://www.cf.ac.uk/biosi/staff/ehrmann/tools/ecce/ecce.htm

Website of an outer membrane research at the University of Tübingen, Germany
http://www.mikrobio.uni-tuebingen.de/membran/ag-braun/projekte.html

66

Oxidative stress

Pablo J. Pomposiello
University of Massachusetts

Bruce Demple
Harvard School of Public Health

GLOSSARY

oxidative stress The excess production or insufficient disposal of intracellular oxidants.

reactive oxygen species (ROS) Partially reduced oxygen derivatives, such as superoxide ($O_2^{\cdot -}$) and hydrogen peroxide (H_2O_2).

Aerobic Metabolism requires the exposure of cells to oxygen, which sometimes reacts non-enzymatically with cellular components to generate free radicals and other reactive molecules. These reactive by-products can damage all biological macromolecules and thus interrupt growth and cause mutations. This damage is limited by small molecules such as glutathione, which neutralizes some free radicals, and by enzymes such as superoxide dismutase (SOD) and catalase, which eliminate specific reactive species. Under some circumstances, the rate of free-radical production increases or cellular-defense activities are diminished, which results in oxidative stress. Oxidative stress can arise in many ways—through metabolic changes (e.g. inactivation of some components of the electron transport chain); through exposure to environmental agents that divert electron flow (e.g. the herbicide paraquat); and through immune responses to bacterial infection (e.g. superoxide and nitric oxide generated by activated macrophages). Aerobic organisms have evolved inducible defense mechanisms against various types of oxidative stress. From single proteins to complex self-regulating genetic networks, these defenses scavenge reactive oxygen species and mediate the repair of cellular damage. The knowledge of these antioxidant mechanisms is more advanced in enteric bacteria than in any other group of organisms, although progress is being made in gram-positive bacteria and yeast. It is clear that genes that originally aided microorganisms in colonizing an oxidizing atmosphere were in many cases recruited into different regulatory networks, helping the cell to fine-tune its metabolism according to a wide variety of metabolic and environmental conditions.

I. SOURCES OF OXIDATIVE STRESS

A. Aerobic metabolism

Various metabolic pathways involving the exchange of electrons between biochemical intermediates have the potential to generate oxidative damage by the anomalous transfer of single electrons. In this fashion, aerobic metabolism, photosynthesis, and denitrification are metabolic functions whose oxidant by-products have to be kept at concentrations that are compatible with cellular integrity. In a sense, the reactive by-products may be considered to be unavoidable "leaks" in these natural processes (Fig. 66.1). Oxygen

The Desk Encyclopedia of Microbiology
ISBN: 0-12-621361-5

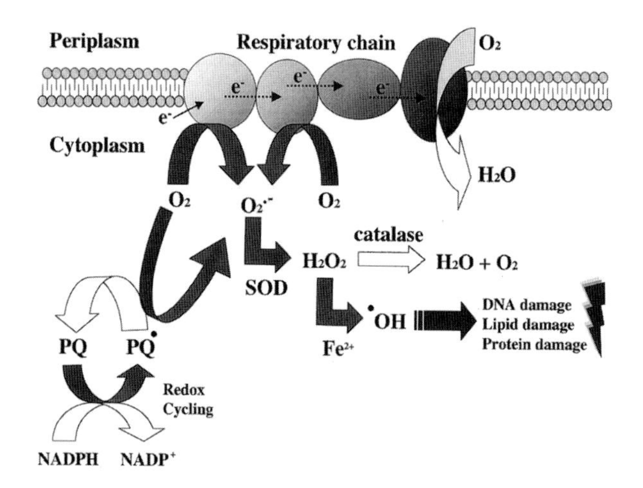

FIGURE 66.1 Reactive oxygen species, oxidative stress, and cellular damage. The majority of the oxygen that enters the cell is reduced to water by the respiratory chain, a reaction that consumes four electrons. However, a small proportion of the oxygen molecules can be reduced in a series of one-electron reactions. Molecular oxygen forms superoxide ($O_2^{\bullet -}$) by reaction with reduced components of the respiratory chain. Superoxide can also be formed by reaction with redox-cycling drugs such as paraquat (PQ), which is enzymatically re-reduced at the expense of NADPH. Superoxide is eliminated by superoxide dismutase (SOD) to form hydrogen peroxide (H_2O_2). Hydrogen peroxide can either be detoxified by conversion into water and oxygen by catalase, or react with reduced transition metals such as iron and copper to form a hydroxyl radical (·OH). The hydroxyl radical is a highly reactive molecule than can damage virtually all the fundamental cellular components. Solid arrows indicate reactions that yield oxidants. Open arrows indicate reactions that yield innocuous products (Plate 9).

competes with respiratory components to oxidize some elements of the electron-transport chain and thus yields superoxide, which is readily converted to H_2O_2 by the enzyme SOD. H_2O_2 can be safely disposed of by catalases, or it can react further, especially with reduced metals such as Fe^{2+}, to generate still more unstable products, notably the highly reactive hydroxyl radical ($^{\bullet}OH$; see Fig. 66.1). DNA is a critical target for oxidative damage, and the DNA lesions caused by oxidants can disrupt replication and lead to mutations. The metal centers of enzymes constitute another group of targets for oxidative damage, with critical sensitive activities such as ribonucleotide reductase, essential for producing DNA precursors, and aconitase, a pivotal component of the citric acid cycle. Unsaturated lipid components of the membrane react to form lipid peroxides in chain reactions with many products derived from a single free radical. The ultimate breakdown of lipid peroxides yields still another reactive compound, malondialdehyde, which can form mutagenic DNA damage. Thus, the action of ROS can exert widespread effects in the cell, both directly and indirectly.

B. Chemical and physical agents

In addition to these normal metabolic sources, environmental compounds can divert single electrons to generate oxygen radicals. These redoxcycling compounds undergo enzymatic reduction and are then reoxidized by O_2, a cyclic process that generates a flux of superoxide catalytically (Fig. 66.1). The variety of these superoxide-generating compounds is large and includes many types of quinones, naphthoquinones, and nitroquinolones, which can act as efficient sources of oxidative stress. Several physical agents can impose oxidative stress; ionizing radiation produces ROS by radiolysis of water, whereas ultraviolet light produces H_2O_2 through photochemical reactions involving various chromophores, including the amino acid tryptophan.

C. Photosynthesis

Aerobic metabolism is not the only source of oxidative stress. If photosynthetic cells are exposed to light in excess of their synthetic capabilities, light-harvesting antennae can transfer excitation energy to ground-state oxygen and yield singlet oxygen, another highly reactive species (though not actually a free radical). In addition, light-driven electron transport systems may divert electrons to oxygen instead of $NADP^+$, resulting in the same reactive derivatives as found for aerobic respiration. This light-dependent production of oxygen derivatives is termed photooxidative stress.

D. Nitric oxide

During bacterial denitrification, the aerobic conversion of nitrate to N_2, nitric oxide ($NO^{\bullet}$) is formed as a product of nitrite reduction. The accumulation of this toxic intermediate is minimized by the catalytic conversion of $NO^{\bullet}$ into nitrous oxide (N_2O) by the enzyme nitric oxide reductase. NO· is also a key cytotoxic weapon of the mammalian (and other) immune systems, and it is produced by many nonimmune cell types during inflammatory responses. Moreover, $NO^{\bullet}$ reacts very rapidly with superoxide to generate another unstable and even more reactive compound, peroxynitrite ($ONOO^-$). Thus, the ultimate effects of one reactive species may be entwined with the effects of another.

The bacteriostatic and bactericidal effects of ROS have been exploited by macrophages and other phagocytic cells to attack microbial infections. After endocytosis of bacteria, macrophages and neutrophils are activated by the bacterial lipopolysaccharide coat to produce copious amounts of superoxide and H_2O_2. Macrophages and many other cell types produce $NO^•$ during inflammatory responses.

II. PHYSIOLOGY OF OXIDATIVE STRESS

The diverse sources of oxidative stress have probably modeled the evolution of adaptations to aerobic life (Pomposiello and Demple, 2002). Consistent with this idea, aerobic organisms display both constitutive and inducible defenses against ROS. Several types of small molecules in the cytoplasm have clear roles in scavenging ROS and aiding the repair of potential damage.

A. Glutathione

Glutathione (γ-L-glutamyl-L-cysteinylglycine; GSH) is present in bacteria, fungi, animals and plants as the major low-molecular-weight thiol, typically at millimolar concentrations. GSH acts as a chemical scavenger of radicals such as $^•OH$, and also as a H-atom donor to restore macromolecules that have been attacked by free radicals. GSH in eukaryotic cells, including yeast, is also a cofactor for the H_2O_2-destroying enzyme GSH peroxidase and can be enzymatically conjugated to oxidative products to mark them for disposal from the cell.

B. NAD(P)H

Reduced GSH is maintained by the enzyme GSH reductase, which reduces the oxidized form (GSSG). In organisms where GSH is not present, usually a related peptide thiol is present, together with the respective thiol reductase. These critical reactions depend on the reducing power of NADPH, which is equilibrated in the cell with NADH (nicotinamide adenine dinucleotide). NADPH also supports other antioxidant enzymes, such as alkyl hydroperoxide reductase. Ironically, NAD(P)H is involved in redox-cycling reactions that generate high fluxes of superoxide in the cell (Fig. 66.1).

C. Thioredoxins and glutaredoxins

A family of small proteins, the thioredoxins and glutaredoxins, act as efficient thiol donors to many cellular proteins. Thioredoxins and glutaredoxins have a pair of conserved cysteine residues that become oxidized to a cystine disulfide as other proteins are reduced. Reduced thioredoxin is in turn regenerated by a reductase using NADPH; reduced glutaredoxin is regenerated using GSH as a reducing donor.

D. Superoxide dismutase

The conversion of ROS into less dangerous products is catalyzed by several enzymes that are almost universal among aerobic organisms, which reflects the universality of the biochemistry of oxidative stress. Superoxide dismutase catalyzes the dismutation of superoxide into hydrogen peroxide and oxygen and plays a central role in the protection of aerobic organisms against ROS. All aerobic organisms have at least one form of SOD. *E. coli* has three isozymes, encoded by the *sodA*, *sodB*, and *sodC* genes. The products of the *sodA* and *sodB* genes are cytoplasmic, whereas the *sodC* product is periplasmic. The three SODs of *E. coli* differ in the metals at their active sites; the SodA protein contains manganese, SodB iron, and SodC copper and zinc. Bacterial strains lacking both of the cytoplasmic SOD enzymes suffer DNA damage during aerobic growth, which results in an elevated mutation rate. The aerobic growth of SOD-deficient *E. coli* in rich media is only slightly impaired, but in minimal media growth it is abolished unless amino acids, particularly the branched-chain types, are provided. This conditional multiple auxotrophy, which is not observed during anaerobic growth, is probably due to oxidative inactivation of biosynthetic enzymes that contain iron–sulfur centers.

E. Catalases and peroxidases

Catalases and peroxidases are heme-containing enzymes that eliminate H_2O_2 by related mechanisms. Catalases redistribute electrons among H_2O_2 molecules by alternate two-electron oxidation and reduction, which generates oxygen and water. Peroxidases oxidize an organic compound while generating H_2O. *E. coli* has two catalases—HPI, encoded by the *katG* gene and predominant during exponential growth; and HPII, encoded by *katE* and present mainly in stationary phase.

III. BIOCHEMICAL BASIS OF THE RESPONSES

Many organisms exhibit adaptive responses to oxidative stress; that is, the exposure of the cells to a sublethal level of oxidative stress enhances the resistance to subsequent, higher levels of oxidative stress. These adaptive responses depend on protein synthesis, and different sets of proteins may be synthesized upon exposure to different types of oxidative agents. The induction of some key proteins in response to oxidative stress has been known for some time. For example, SOD activity increases in *E. coli* that is grown in high levels of oxygen or exposed to the redox-cycling agent paraquat. Paraquat also induces glucose-6-phosphate dehydrogenase, evidently to replenish NADPH used up in antioxidant reactions (e.g. by GSH reductase). Exposure to H_2O_2 induces catalase activity in many organisms and increases GSH reductase levels in *E. coli* and *S. typhimurium*. We now know that these inductions reflect the activation of coregulated groups of genes affecting the expression of many additional proteins.

The number of oxidative stress-inducible proteins described ranges from one in *Mycobacterium bovis*, to at least 80 in *E. coli* (Barbosa and Levy, 2000; Pomposiello *et al.*, 2001; Zeng *et al.*, 2001). The number, identity, and degree of activation of the proteins induced by oxidative stress has been studied by two-dimensional gel electrophoresis. In this technique, radioactively labeled proteins are resolved in two dimensions, first by isoelectrofocusing and then by denaturing gel electrophoresis. Two-dimensional gel analysis allows the comparison of the overall pattern of proteins synthesized during various growth conditions, and in various genetic backgrounds. Thus, this type of analysis has been applied to mutant strains that either overexpress or fail to induce sets of oxidative stress proteins, which has led to the isolation of genes coding for global regulators. Note, however, that some inductions can be overlooked by focusing solely on two-dimensional gel analysis, and the number established by this method should be taken only as a lower limit. The complexity of some responses is truly daunting; for example, the number of oxidative stress proteins in *E. coli* is probably more than 100 if evidence from other methods is combined with the two-dimensional gel approach.

IV. GENETIC BASIS OF THE RESPONSES

Bacterial gene function is regulated mainly at the transcriptional level. Cells have evolved transcriptional modulators that sense oxidative stress and activate genes whose products avert or repair the damage caused by ROS. These transcriptional modulators usually activate multiple and unlinked promoters, which as coregulated groups constitute regulons. In most cases, one or more activated genes decrease the activating stimulus in a type of negative feedback, which often makes the responses self-regulating (for reviews, see Hidalgo and Demple, 1996; Jamieson and Storz, 1998).

A. OxyR

OxyR is a redox-sensitive protein and a member of the LysR family of DNA-binding transcriptional modulators. OxyR is a homotetramer in solution and it activates as many as 10 genes in response to an increase in the intracellular concentration of H_2O_2 (Fig. 66.2). The protein exists in two forms, reduced and oxidized, and exhibits redox-regulated DNA binding. Only oxidized OxyR binds tightly to the promoters of its target genes and activates transcription, evidently through contacts with the α-subunit of RNA polymerase. Reduced OxyR binds only to the *oxyR* promoter itself, and thereby limits the expression of this regulatory protein. Active oxidized OxyR has an intramolecular disulfide bond between cysteines 199 and 208, which is evidently formed by direct reaction with H_2O_2 or another oxidant. Reduced inactive OxyR is maintained by thiol-disulfide exchange proteins, such as glutaredoxin, which is dependent on GSH. OxyR activates the transcription of *katG* (HPI catalase), *ahpFC* (alkyl hydroperoxide reductase), *dps* (general DNA-binding protein that may exclude Fe), *gorA* (GSH), *grxA* (glutaredoxin-A), and *oxyS* (small untranslated RNA that may regulate other genes post-transcriptionally). The induction of *gor* and *grxA* may

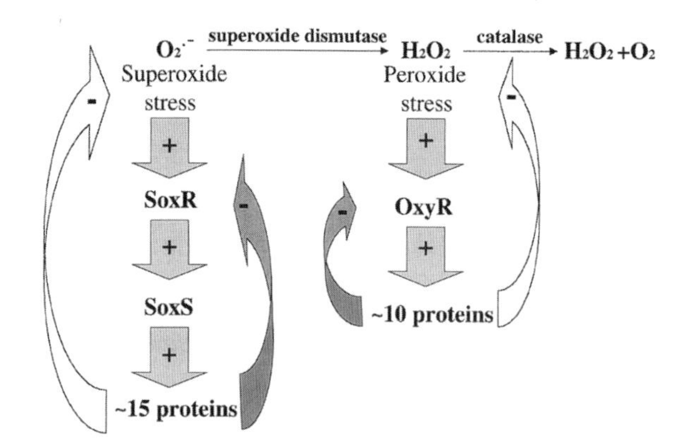

FIGURE 66.2 Regulation and homeostasis of the responses to oxidative stress in *Escherichia coli*. A rise in the intracellular concentration of superoxide ($O_2^{\cdot -}$) activates the redox-sensing protein SoxR, which controls the induction of the protein SoxS. SoxS enhances the transcription rate of at least 15 genes, including *sodA*, and this concerted response alleviates the superoxide stress through proteins that scavenge superoxide and repair oxidative damage. Superoxide is reduced to hydrogen peroxide (H_2O_2) either spontaneously or enzymatically by SOD. A rise in the concentration of hydrogen peroxide activates the protein OxyR, which in turn enhances the transcription rate of as many as 10 genes, including *katG*, and this response alleviates the hydrogen peroxide stress. Both the SoxRS and the OxyR responses are homeostatic; the induction of the effector proteins results in a decrease of the stress signal, and therefore a reduction of the response (open arrows). The induced proteins may directly switch off the regulators (solid arrows), as glutaredoxin and GSH do for OxyR.

ultimately down-regulate the response by regenerating reduced OxyR.

B. SoxR and SoxS

The SoxR protein is a redox-sensing transcriptional activator that belongs to the MerR family of DNA-binding proteins. SoxR is a homodimer of 17-kDa subunits, each one containing a redoxactive [2Fe–2S] center. The DNA-binding activity of SoxR does not depend on the iron–sulfur clusters, but only SoxR with oxidized [2Fe–2S] centers activates transcription. The activation of SoxR through the one-electron oxidation of its [2Fe–2S] centers therefore corresponds to an allosteric transition in the promoter DNA–SoxR complex. This transition does not increase binding by RNA polymerase, but rather stimulates the formation of the "open" complex essential for initiating transcription. It is not known what cellular activities operate to maintain the [2Fe–2S] centers of SoxR in the reduced state. After the exposure of bacteria to superoxide-generating agents, such as the redox-cycling compound PQ (Fig. 66.1), or to NO (either directly or following phagocytosis by murine macrophages), oxidized SoxR activates the transcription of *soxS*, a gene encoding a second transcriptional modulator (Fig. 66.2). The SoxS protein, a 13-kDa monomer homologous to the C-terminal portion of AraC protein, binds to specific sequences in 15 or more promoters and recruits RNA polymerase to stimulate their transcription. The SoxS-regulated genes in *E. coli* include *sodA* (Mn-containing SOD), *zwf* (glucose-6-phosphate dehydrogenase), *micF* (an antisense RNA that inhibits translation of the porin OmpF), *nfo* (endonuclease IV, a repair enzyme for oxidative DNA damage), *fpr* (ferredoxin reductase), *acrAB* (cellular efflux pumps), and *fumC* (redox-resistant fumarase). However, SoxS in other organisms may induce different genes: activation of *S. typhimurium soxRS* induces Mn-SOD and endonuclease IV, but not fumarase-C or glucose-6-phosphate dehydrogenase. Like many other regulatory proteins, both SoxS and SoxR limit their own expression by binding the promoters of their own structural genes. So far, SoxR and SoxS have been identified only in gram-negative bacteria.

C. Sigma S

The alternative sigma factor, sigma S (σ^S), encoded by the *rpoS* gene, is expressed during starvation or in the stationary phase in a variety of bacterial species. RNA polymerase containing σ^S activates the expression of several genes that counteract oxidative stress, and bacteria that are starved or in stationary phase are more resistant to oxidative stress than are cells growing exponentially. The σ^S-activated genes include *katE* (catalase-hydroperoxidase HPII), *dps* (protective DNA-binding protein), *xthA* (exonuclease III, another DNA-repair enzyme for oxidative damage), and *gorA* (glutathione reductase). The regulation of σ^S activity is poorly understood and seems to be under complex transcriptional, translational, and proteolytic control.

D. Fur

Antioxidant defense genes are also controlled by proteins of the Fur family. *E. coli* Fur protein is a repressor governing a system of genes involved in iron uptake, and the system is switched on when bacteria are grown in media with limiting Fe concentrations. It is unclear whether Fur senses iron levels directly or through some other signal. Fur in *E. coli* also regulates the *sodA* gene and thus Mn–SOD activity. More complex regulation involving Fur-related proteins has been described in *Bacillus subtilis*, in which the PerP protein regulates genes encoding a catalase, an alkyl hydroperoxide reductase, and a Dps-like protein. Although this collection of gene products is reminiscent of the *E. coli oxyR* system, the regulatory mechanism is quite different. PerP is a repressor, possibly with a metal corepressor, so that gene induction depends on lowered DNA binding by the regulator (i.e. derepression rather than positive control). Mutant strains lacking PerP have increased resistance to H_2O_2.

V. BIOLOGICAL IMPLICATIONS

Genetic responses to oxidative stress are widespread in microbes, although the molecular mechanisms of stress signal transduction are known in only a few cases. These responses mobilize diverse biochemical activities that operate to exclude environmental toxins, repair oxidative damage, eliminate ROS, and diminish radical production. These actions can collectively alleviate the stress, even if elevated radical production continues, and are thus homeostatic. The continuous variation of OxyR activation during the normal aerobic growth of *E. coli* provides an especially clear example.

The sensors of oxidative stress take advantage of reactions that are deleterious for most other proteins. The formation of a disulfide bond in a cytoplasmic protein is often inactivating, but this same modification is the signal that OxyR has evolved to use for its activation. Similarly, some iron–sulfur centers are damaged by oxidation, but SoxR exploits this chemistry as a signal transduction device connecting oxidation to gene activation.

For both the *oxyR* and the *soxRS* regulons, several regulon genes are involved in multiple regulatory pathways. The *sodA* gene is an excellent example, as it is controlled by at least five different regulatory systems in addition to *soxRS*. The *katG* and *dps* genes are both regulated jointly by *oxyR* and *rpoS*. Whether this theme will reemerge in other species awaits additional experimentation.

Microbial responses to oxidative stress are by no means limited to bacteria. The yeast *Saccharomyces cerevisiae* has received particular attention. Physiological and genetic analysis shows that distinct regulatory systems govern adaptive resistance to H_2O_2 compared to redox-cycling (superoxide-generating compounds). The response to H_2O_2 involves a large number of inducible proteins, including some of the same defense functions mentioned previously: catalase, SOD, glucose-6-phosphate dehydrogenase, and GSH reductase (Godon *et al.*, 1998). The modulated expression of metabolic enzymes in H_2O_2-treated yeast is proposed to remodel metabolism to increase the regeneration of NADPH at the expense of glycolysis. The expression of a portion of the H_2O_2-inducible proteins in yeast (e.g. GSH) is controlled by YAP1, a yeast homolog of the c-Jun—c-Fos transcription activators of mammalian cells. Although YAP1 itself does not seem to be the redox-sensing component of this system, the signaling mechanism that activates this transcription factor in response to oxidative stress is unknown.

What of the evolutionary pressures that have shaped the oxidative-stress responses in bacteria? Clearly, the colonization of the aerobic world required antioxidant defenses, and SOD and catalase appear to have ancient origins. One could suppose that the ability to regulate these defenses was especially useful for organisms exposed to changing levels of oxygen, as in the facultative aerobic lifestyle of *E. coli* and *S. typhimurium*. The chemical warfare conducted among organisms may also have selected for inducible genetic systems, and for the inclusion of certain other genes within the coregulated groups. The driving force for such inducible resistance is not necessarily restricted to the mammalian immune system; plants also employ H_2O_2 and NO to mediate systemic immunity (Dangl, 1998). The diversity of microbial interactions might well have elicited diverse systems for coping with oxidative stress.

BIBLIOGRAPHY

Barbosa, T. M., and Levy, S. B. (2000). Differential expression of over 60 chromosomal genes in *Escherichia coli* by constitutive expression of MarA. *J. Bacteriol.* **182**, 3467–3474.

Cabiscol, E., Tamarit, J., and Ros, J. (2000). Oxidative stress in bacteria and protein damage by reactive oxygen species. *Int. Microbiol.* **3**, 3–8.

Dangl, J. (1998). Plants just say NO to pathogens. *Nature* **394**, 525–526.

Godon, C. *et al.* (1998). The H_2O_2 Stimulon in *Saccharomyces cerevisiae*. *J. Biol. Chem.* **273**, 22480–22489.

Hengge-Aronis, R. (1999). Interplay of global regulators and cell physiology in the general stress response of *Escherichia coli*. *Curr. Opin. Microbiol.* **2**, 148–152.

Hidalgo, E., and Demple, B. (1996). Adaptive responses to oxidative stress: The *soxRS* and *oxyR* regulons. *In* "Regulation of Gene Expression in *Escherichia coli*" (E. C. Lin and A. Simon Lynch, eds.), pp. 435–452. R. G. Landes, Austin, TX.

Jamieson D. J., and Storz, G. (1997). Transcriptional regulators of oxidative stress responses. *In* "Oxidative Stress and the Molecular Biology of Antioxidant Defenses" (John G. Scandalios, ed.), pp. 91–116. Cold Spring Harbor Laboratory Press, New York.

Lynch, A. S., and Lin, E. C. C. (1996). Responses to molecular oxygen. *In* "*Escherichia coli* and *Salmonella*. Cellular and Molecular Biology", 2nd edn. (F. C. Neidhardt, ed.). ASM Press, Washington, DC.

Pomposiello, P. J., Bennik, M. H. J., and Demple, B. (2001). Genome-wide transcriptional profiling of the *E.coli* responses to superoxide stress and to sodium salicylate. *J. Bacteriol.* **183**, 3890–3902.

Pomposiello, P. J. and Demple, B. (2002). Global adjustment of microbial physiology during free radical stress. *In* "Advances in Microbial physiology", Vol. 46. (R. K. Poole, ed.), in press.

Storz, G., and Hengge-Aronis, R. (eds). (2000). "Bacterial Stress Responses." ASM Press, Washington, DC.

Touati, D. (2000). Iron and oxidative stress in bacteria. *Arch. Biochem. Biophys.* **373**, 1–6.

Zheng, M., Wang, X., Templeton, L. J., Smulski, D. R., LaRossa, R. A., and Storz, G. (2001). DNA microarray-mediated transcriptional profiling of the *Escherichia coli* response to hydrogen perodixe. *J. Bacteriol.* **183**, 4562–4570.

WEBSITES

Website for *soxR* gene and bibliography
http://cgsc.biology.yale.edu/cgi-bin/sybgw/cgsc/Site/27798

Website for *soxS* gene and bibliography
http://cgsc.biology.yale.edu/cgi-bin/sybgw/cgsc/Site/27801

Website for *marRAB* genes and bibliography
http://cgsc.biology.yale.edu/cgi-bin/sybgw/cgsc/Site/28841

Website for *rpoS* gene and bibliography
http://cgsc.biology.yale.edu/cgi-bin/sybgw/cgsc/Site/18208

67

pH Stress

Joan L. Slonczewski
Kenyon College

GLOSSARY

acid resistance, acid survival, or acid tolerance The ability of a microbial strain to exist for an extended period at pH values too acidic for growth, retaining the ability to be cultured after the acidity is neutralized.

acid shock Sudden decrease of pH of the growth medium to a level of acid at or below the limit for growth.

acidophile A species whose optimal external pH for growth is in the acidic range, generally lower than pH 6. Hyperacidophiles may grow below pH 3.

alkaliphile A species whose optimal external pH for growth is in the alkaline range, generally higher than pH 8.

base resistance The ability of a microbial strain to exist for an extended period at pH values too alkaline for growth, retaining the ability to be cultured after the alkali is neutralized.

membrane-permeant weak acid (permeant acid) A weak acid, usually an organic acid, which can permeate the cell membrane in the hydrophobic protonated form and then dissociate in the cytoplasm, producing hydronium ions and depressing intracellular pH.

membrane-permeant weak base (permeant base) A weak base, usually an organic base, which can permeate the cell membrane in the hydrophobic unprotonated form and then become protonated in the cytoplasm, producing hydroxyl ions and increasing intracellular pH.

neutrophile A microbial species which grows best in the neutral range of external pH, generally pH 6–8.

pH homeostasis The maintenance of intracellular pH within a narrow range during growth over a wider range of extracellular pH.

protonmotive force The proton potential across the cell membrane (Δp or $\Delta\mu H$), composed of the chemical gradient of protons ($Z\Delta pH$) minus the transmembrane electrical potential ($\Delta\psi$).

thermoacidophile (extreme thermophile; hyperthermophile) One of several archaean species, usually sulfur oxidizers, that grow optimally in extreme heat and extreme acid.

transmembrane pH difference (ΔpH) The difference in pH between the cytosol and the external medium. Usually the pH difference is maintained across the cellular membrane or the inner membrane of gram-negative organisms. The chemical gradient, $Z\Delta pH$, is a component of the proton potential.

The Balance between Hydronium and Hydroxyl ion concentrations in aqueous solution is most commonly represented by pH, the negative logarithm of hydronium

The Desk Encyclopedia of Microbiology
ISBN: 0-12-621361-5

concentration. pH affects microbial growth in numerous ways. In any given environment, from geothermal springs to human tissues, pH determines which species survive. Extracellular pH can be a signal for microbial behavior, whereas intracellular pH affects enzyme activity and reaction rates, protein stability, and structure of nucleic acids and many other biological molecules. In theory, every macromolecule is a "pH sensor."

Historically, pH has played many key roles in the development of microbiology. Since ancient times, fermentation has produced storable food products containing inhibitory acids (dairy products and vinegar) or bases (Japanese natto from soybeans; African dadawa from locust beans). In mining, leaching by acidophilic lithotrophs contributes to the recovery of valuable minerals; unfortunately, acidophiles also contribute to the decay of monuments. Changes in pH caused by growth on indicator media are used to identify microbial species—for example, fermentation of sugars in MacConkey media or the respiration of TCA cycle components in Simmons agar. The function of many modern therapeutic agents, including food preservatives, antibiotics, and dental fluoride, requires a transmembrane pH gradient for concentration within the cell.

I. RANGE OF PH FOR GROWTH OF MICROORGANISMS

All microbes have evolved to grow within a particular range of external pH. Acidophiles are defined approximately as growing optimally within the range pH 0.5–5, neutrophiles within pH 5–9, and alkaliphiles within pH 9–12. For example, the neutrophile *Escherichia coli* in rich broth can grow at pH 4.4–9.2, although its rate of growth is greatly decreased at the extremes. Some species have growth ranges more narrow than those previously mentioned, whereas others overlap two ranges; for example, yeast grows in the acidic range as well as the neutral range.

Among eubacteria, the acidophiles, such as *Thiobacillus ferrooxidans*, tend to be iron or sulfur metabolizers which produce H_2SO_4. Among archeans, many sulfur oxidizers live in extreme heat and acid (the "thermoacidophiles"). These species are isolated from geothermal springs as well as deep-sea hydrothermal vents. Their cells are known for unusual membrane phospholipids and glycolipids of exceptional strength, derived from isopranoid diether or tetraether. They can maintain a transmembrane pH difference (ΔpH) of several units, which is in part compensated by an inverted electrical potential (inside positive).

Neutrophiles include the majority of organisms which grow in association with human bodies. The enterobacteriaceae are neutrophiles, including *E. coli*, in which the effects of pH stress have been studied extensively. An interesting property of neutrophiles is the ability to grow either with an outwardly directed ΔpH (at low external pH) or with an inverted ΔpH (at high external pH), requiring the cell to "spend" much of its electrical potential to maintain the inverted gradient (Fig. 67.1).

The best studied alkaliphiles are the *Bacillus* species, often found in soda lakes. In some alkaliphilic *Bacillus* species, cell walls rich in negetively charged amino acids may assist the exclusion of hydroxyl ions. Krulwich has studied the ability of species such as *B. firmus* to grow faster at pH 10.5 than at pH 7.5, maintaining oxidative phosphorylation despite an inverted ΔpH which takes up most of the proton potential. Models involving sequestration of protons by the respiratory chain complex or the proton-translocating ATPase have been proposed.

During the study of pH and other stresses, a factor often neglected is the change of pH caused by growth of the culture. Microbes possess enormous capacity to change the pH of their environment, as exemplified by their well-known role in food production. Nevertheless, investigators commonly attempt to study microbial growth in media lacking adequate pH control. In any experiment, pH must be controlled either by a chemostat or by growth in buffers of appropriate pk_a for the pH range of interest (Table 67.1).

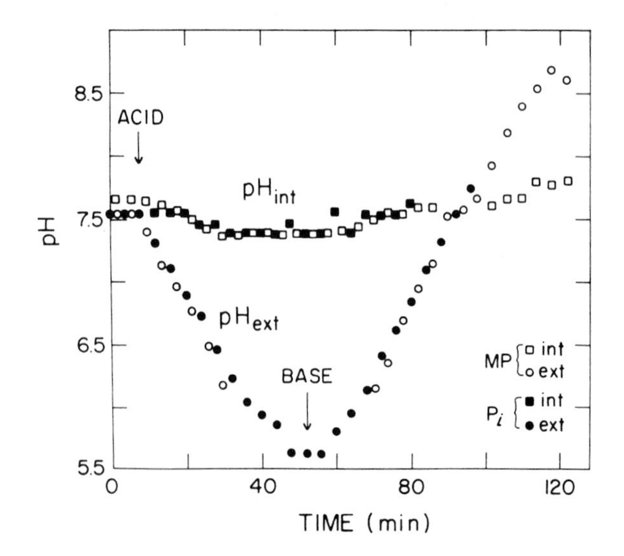

FIGURE 67.1 pH homeostasis in *Escherichia coli*. As acid or base or base are added over time, a suspension of *E. coli* cells maintain internal pH within a relatively narrow range. Both internal and external pH were measured independently by P^{31} NMR resonances of inorganic phosphate (P_i) and of methylphosphonate (MP) (reprinted with permission from Slonczewski *et al.*, 1981).

TABLE 67.1 Buffers for microbial growth media[a]

Buffer abbreviation	Chemical name	pk_a at 37°C	Useful pH range
HOMOPIPES	Homopiperazine-*N,N'*-bis-ethanesulfonic acid	4.55	3.9–5.1
MES	2-(*N*-morpholino)ethanesulfonic acid	5.96	5.5–6.7
PIPES	Piperazine-*N,N'*-bis(2-ethanesulfonic acid)	6.66	6.1–7.5
MOPS	3-(*N*-morpholino)propanesulfonic acid	7.01	6.5–7.9
TAPS	3-[*N*-tris-(hydroxymethyl)methylamino]-propanesulfonic acid	8.11	7.7–9.1
CAPSO	3-(Cyclohexylamino)-2-hydroxy-1-propanesulfonic acid	9.43	8.9–10.3
CAPS	3-(Cyclohexylamino)-1-propanesulfonic acid	10.08	9.7–11.1

[a]Examples of organic sulfonate buffers used to maintain external pH during microbial growth (data from Research Organics 1998/1999 "Catalog of Biochemicals").

The concentration of buffer typically needs to be as high as 100 m*M*, and the pH of the media must be measured both before and after microbial growth. In poorly buffered media, some genes whose expression control was initially ascribed to anaerobic regulation turn out in fact to be regulated by pH.

II. MEASUREMENT OF INTERNAL pH

Most microbial species maintain some degree of pH homeostasis; that is, the internal or cytoplasmic pH is maintained at a value different from that of the external medium, usually closer to neutrality. Study of pH stress generally requires measurement of intracellular pH; in the case of eukaryotes, measurement of mitochondrial and other organellar compartments is also required. The measurement of intracellular pH remains fraught with difficulties. Bacterial cells are too small to permit introduction of microelectrodes, but several alternative methods have been used, including equilibration of radiolabeled permeant acids and bases, nuclear magnetic resonance (NMR) of pH-titratable phosphates, and fluorescence microscopy of pH-titratable dyes.

A. Radiolabeled permeant acids

Radiolabeled permeant acids can equilibrate across the cell membrane primarily by permeation of the hydrophobic protonated form, HA. If the dissociation constant of the acid is well below the external pH, then the transmembrane concentration gradient of the acid (primarily in the deprotonated form) will approximately equal the ΔpH. This measurement, together with an independent determination of the intracellular volume, can be used to calculate the internal pH (pH_{in}), according to the formula:

$$pH_{in} = \log[(A_{in}/A_{ex})\,(10^{pk} + 10pH_{ex}) - 10^{pk}]$$

where A_{in} is the total internal concentration of radiolabeled acid, A_{ex} is the external concentration, and pH_{ex} is the external pH. Padan and others have applied this method to observe ΔpH in many bacterial species.

Considerations for use of radiolabeled weak acids are reviewed by Kashket (1985). The advantage of the use of radiolabeled acids is the relative ease of measurement of large numbers of samples. Disadvantages include the large number of potential sources of error and the loss of sensitivity that occurs at small values of ΔpH. Where internal pH is lower than external pH, sensitivity falls off rapidly, and a permeant base must be used. Some organic acids and bases are actively transported by the cell; this effect, however, is usually small compared to the equilibration time of the acid or base.

B. P^{31} NMR of titratable phosphates

A method of measurement permitting independent observation of internal and external pH is that of NMR observation of pH-titratable P^{31} shifts in phosphates. This method enables highly accurate and reproducible pH measurement on time scales as short at 10 s. Both positive and inverted pH gradients can be observed in the same experiment. Figure 67.1 shows the results of P^{31} NMR of inorganic phosphate (P_i) and methylphosphonate (MP) used to demonstrate internal pH homeostasis in *E. coli* during gradual addition of acid (HCl) or base (NaOH). A drawback of the NMR method, however, is the requirement for highly concentrated suspensions of cells so that the intracellular volume fills 5–10% of the total suspension in order for sufficient intracellular signal to be obtained.

For eukaryotic microbes such as yeast, fluorescent indicators have been developed which can distinguish the pH of different compartments. Use of this method is likely to increase, although it remains impractical for most bacteria.

III. pH HOMEOSTASIS

Most microbes are capable of protecting their cytoplasm from rapid pH change by maintaining internal pH within a range more narrow than that of the external environment. The degree or tightness of regulation differs for different organisms. *Escherichia coli* regulates internal pH very closely, even sacrificing proton potential to maintain an inverted ΔpH at high external pH (Fig. 67.1). Other organism, such as *Enterococcus faecalis*, can maintain only a positive ΔpH. Still others appear to sacrifice pH homeostasis in order to avoid a large ΔpH which can drive the uptake of toxic organic acids.

Maintenance of internal pH can be vital for cell survival even under conditions in which growth does not occur. As shown for many enteric bacteria, cells can survive exposure to extreme pH for several hours, followed by regeneration of colonies at neutral pH. During survival in extreme acid, internal pH decreases below the level found during growth; but it must remain higher than a critical level in order for viability to be maintained.

A. Mechanisms of pH homeostasis during growth and survival

The full mechanisms of bacterial pH homeostasis remain unknown, even in *E. coli*, in which the problem has long been studied. Different species have evolved different mechanisms of homeostasis. In *E. faecalis*, the proton-translocating ATPase is induced at low pH to pump out more protons, thus maintaining internal pH slightly above 7. In *E. coli*, however, ATPase mutants maintain pH normally. At low external pH, internal pH is maintained by flux of potassium ion through several transporters, as shown by Epstein and colleagues. At high alkalinity, sodium ion regulates pH through the base-inducible sodium–proton antiporter (Padan and Schuldiner, 1987). In alkaliphilic *Bacillus* species, Krulwich has shown that sodium–proton antiporters maintain an inverted Δp of up to H 2.3 units during growth at pH 11.

In dormant cells, forced to survive outside the pH range for growth, internal pH deviates from the optimum but may still be maintained at a difference from external pH. In *E. coli*, there appears to be a critical value of internal pH (approximately pH 5), below which cells lose the ability to recover and grow in neutralized media. Many different factors contributing to survival are being studied, including the σ^S (stationary-phase sigma factor) regulon and the glutamate decarboxylase system. It is interesting that two different "steps" of pH maintenance can be defined: one for growth (approximately pH 7. 5) and one for survival (approximately pH 5).

B. Avoiding uptake of permeant acids

In some species, particularly the lactococci and the streptococci, bacteria may forego pH homeostasis in order to avoid uptake of permeant acids. As shown by Russell and colleagues, the buildup of fermentation acids such as acetate eventually slows and halts the growth of all microbes, but some organisms are far more resistant to fermentation acids than others. Fermentation acid-resistant bacteria such as *Lactococcus lactis* and *Streptococcus bovis* appear to allow their intracellular pH to decrease below pH 6 in order to diminish the uptake of acids by the transmembrane pH gradient. This question remains controversial because of the difficulties of measuring the transmembrane pH difference accurately using radiolabeled permeant-acid probes.

C. Microbial modification of external

In addition to metabolic adjustment of cytoplasmic pH, another strategy of bacteria is to modulate the external pH so as to lessen pH stress. Amino acid decarboxylases and deaminases can produce organic bases or acids, respectively, to neutralize external acid or base. In pathogens such as *Helicobacter pylori* and *Proteus mirabilis*, the enzyme urease neutralizes their host environment, playing a critical role in virulence. Microbial effects on pH contribute significantly to the stability of aquatic ecosystems.

IV. pH STRESS AND ORGANIC ACID STRESS

pH stress has been defined differently by different researchers. Generally, the operational definitions of pH stress fit one of two categories: (i) adaptation to growth at the extreme acidic or alkaline end of the pH range, and (ii) survival outside the pH range of growth, with recovery of colony-forming capacity after the pH is brought within the growth range. With respect to extreme acid, the latter category is variously termed acid survival, acid tolerance, or acid resistance. The definition of an acid-resistant strain depends on the degree of acidity designated "extreme," the time of exposure, and the death rate or percentage of recoverable colonies designated as "resistant." One commonly used definition of acid resistance in *E. coli* is the recovery of 10% or more colonies following incubation of a culture for 2 h at pH 2.5. It is increasingly recognized, however, that even smaller levels of resistance could still provide a competitive edge to a pathogen.

The definition of pH stress is complicated by the presence of membrane-permeant organic acids or bases, whose uptake into the cell may be concentrated by the transmembrane pH gradient (ΔpH). Even small traces of such organic acids can halt the growth of lithotrophic acidophiles. The fermentative production of acids or bases contributes enormously to food production.

The potential effects of pH stress are manifold. Protein stability is affected; pH stress can cause unpredictable auxotrophies as the result of a single stressed enzyme in a biochemical pathway. For this reason, organisms generally show a broader pH range of growth in complex media than in minimal media. At high pH, the protonmotive force is "spent" in order to maintain an inverted ΔpH; also, DNA can be destabilized, resulting in induction of SOS response in *E. coli*. At low pH, cell membranes may become destabilized, and the cell then becomes hypersensitive to organic acids and uncouplers. It is not surprising that cells have evolved many genetic responses to alleviate pH stress and even to take advantage of it, particularly in pathogenesis.

V. GENETIC RESPONSES TO pH STRESS

Bacterial responses to pH stress include many kinds of genetic regulation. Gene products induced by acid or base may function to neutralize the internal and/or external pH, to transport protons, to avoid production of acids or bases, or to turn on a cascade of virulence factors. Genes regulated by pH are most commonly coregulated by other environmental factors, such as cations, anaerobiosis, and enzyme substrates. Certain genes respond specifically to permeant acids at low external pH; their regulation may actually respond to internal acidification. Gene products conferring acid resistance may also offer cross-protection to other environmental stresses such as starvation or osmotic shock.

A. Mechanisms to neutralize acid or base

An example of a pH response cycle in *E. coli*, dissected by Olson and by Bennett, is that of lysine decarboxylase, the *cadBA* operon (Fig. 67.2). Lysine decarboxylase (CadA) removes carbon dioxide from lysine, thus producing cadaverine, an amine which protonates to

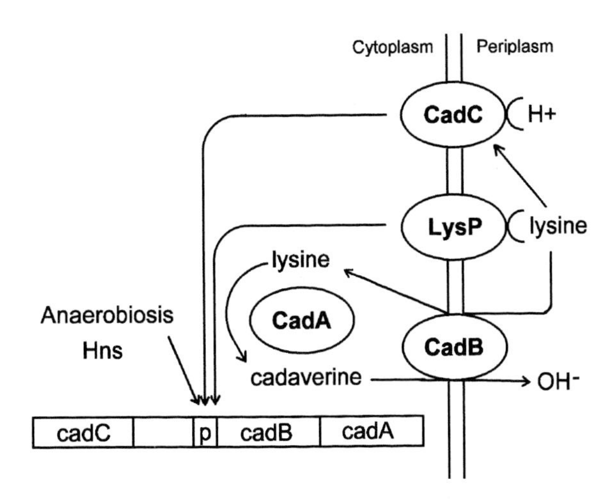

FIGURE 67.2 Acid regulation of lysine decarboxylase and cadaverine transport: A mechanism to reverse external acidification.

generate OH−. When both lysine and acidity (pH 5 or 6) occur externally, they interact with membrane-bound receptors LysP and CadC. CadC contains a pH sensor domain as well as a DNA-binding domain, which binds the promoter of *cadBA*, inducing expression of the lysine/cadaverine antiporter CadB as well as CadA. CadB brings in lysine and then exports the product cadaverine, which protonates in acid, increasing the external pH. Expression of *cadBA* is also enhanced by anaerobiosis; this is logical because anaerobic growth generally produces acid.

An example of a cation transport system regulated by pH is the sodium–proton antiporter, NhaA, whose expression is regulated by NhaR. As shown by Padan, NhaA is coinduced by sodium ion at high pH and functions to protect *E. coli* from both stresses. At high pH, Na^+ is exported in exchange for H^+, which helps reverse internal alkalinization. In alkaliphilic *Bacillus firmus*, Krulwich has shown that a more elaborate system of antiporters maintains internal pH, and pH homeostasis at high pH requires Na^+.

B. Virulence regulators

For many virulence regulators, pH provides a signal to initiate pathogenesis. In *Vibrio cholerae*, the regulator ToxR, which induces production of cholera toxin, is activated by acidification. Presumably, acid within the human stomach is the signal. In the plant pathogen *Agrobacterium tumefaciens*, the virulence factor VirG shows similar activation by acid. Reversal of pH stress can enhance pathogenesis; for example, *Proteus mirabillis* produces urease to alkalinize the urinary tract, resulting in calcification and increased colonization by the pathogen.

The multiple drug-resistance regulon *mar* is induced by permeant acids, especially benzoate derivatives, at low external pH. Thus, acidic pH stress may increase antibiotic resistance of enteric bacteria in the host environment.

VI. FOOD MICROBIOLOGY AND OTHER APPLICATIONS OF pH STRESS

Historically, the production of growth-limiting organic acids or bases has been essential to production of diverse foods in various human societies. The acids or bases produced function to prevent spoilage, improve digestabability, and add interesting flavors.

A. Acidic fermentation

In western European traditions, the best known food-making processes involve fermentation acids such as lactate or acetate. Dairy fermentations involve conversion of lactose to lactic acid, initially by streptococci (lactococci), succeeded by the more acid-resistant lactobacilli. Final pH reached in yogurt or cheese is in the range of pH 4 or 5, in combination with hundred-millimolar concentrations of lactate. The addition of lactose to sausage facilitates a similar fermentation process. Fermentation of vegetables, such as cabbage to produce sauerkraut, usually involves *Leuconostoc* species, reaching pH 3.5. Fruit juices are usually fermented by yeast to produce ethanol, but the ethanol can be fermented to acetate (final pH 3.5) to make vinegar.

The role of fermentation acids in inhibiting growth of gastrointestinal pathogens deserves further study. The concentration of fermentation acids in the human colon is comparable to concentrations which retard growth of pathogens *in vitro*, particularly in infants whose colonic pH is approximately 5. Microbial tolerance to fermentation acids may prove to be an important virulence factor; for example, *E. coli* O157:H7 is more resistant to acetate than are nonpathogenic strains of *E. coli*.

B. Alkaline fermentation

In Southeast Asian and African societies, an important group of foods are produced by alkaline fermentations. Most commonly, alkali-tolerant *Bacillus subtilis* and related species conduct fermentation with production of ammonia. In Japan, soybeans are fermented to produce natto; the breakdown of proteins to amino acids significantly enhances digestibility of the food. In West and Central Africa, the inedible locust beans are fermented to an edible product, dadawa. The fermentation process decreases the levels of toxic substances in the bean. Similarly, melon seed is fermented to ogiri, and leguminous oil beans are fermented to ugba. All of the alkaline fermentations produced provide critical sources of protein as well as culturally significant sources of flavor. Nutritional value is enhanced by hydrolysis of proteins and by increase of vitamin levels. The products have a long shelf life and often serve to make inedible foods edible.

VII. ENVIRONMENTAL ROLE OF pH STRESS

In aquatic and soil environments, pH can be the crucial factor determining which microbial species grow and what forms of metabolism are sustained. Acid stress in lakes can reduce the rate of bacterial decomposition, but bacteria can also help reverse aquatic acidity. Diatoms are so sensitive to pH that the species distribution of diatoms in sediment (diatom stratigraphy) is used to study historical rates of aquatic acidification. Figure 67.3 is a calibration curve in which a calculation based on populations of acidophilic, neutrophilic, and alkaliphilic diatoms accurately predicts aquatic pH.

In soil, bacteria generally grow best in the neutral range, whereas fungal species grow better at pH 5 or 6. Increased acidification can affect bacteria either directly, by interfering with pH homeostasis, or indirectly, by increasing the concentration of toxic metal ions.

Acidity in soil and fresh water is often increased by "acid rain," the deposition of sulfuric and nitric acids in part caused by human-made chemical pollution. The effects of acid rain on microbial processes such as decomposition are generally seen at pH levels a unit lower (pH 5) than those which interfere with aquatic invertebrates or fish (pH 6). In soil, acidification to below pH 3.5 (or liming to increase pH to 7) can inhibit mycorrhizal fungi; the significance of this inhibition for agriculture is unclear.

As in food, microbes in soil or water can dramatically alter the pH of their own environment by metabolic production or consumption of acids or bases. Consumption or production of CO_2 can shift water pH by up to 3 units before atmospheric reequilibration occurs; for example, heterotrophic metabolism beneath ice cover can decrease pH, whereas photosynthesis by algal blooms can increase pH. Other processes increasing acidity include ammonium assimilation and oxidation of HS or Fe(II); processes increasing alkalinity include nitrate assimilation, ammonification, sulfate assimilation, and Fe(III) oxidation.

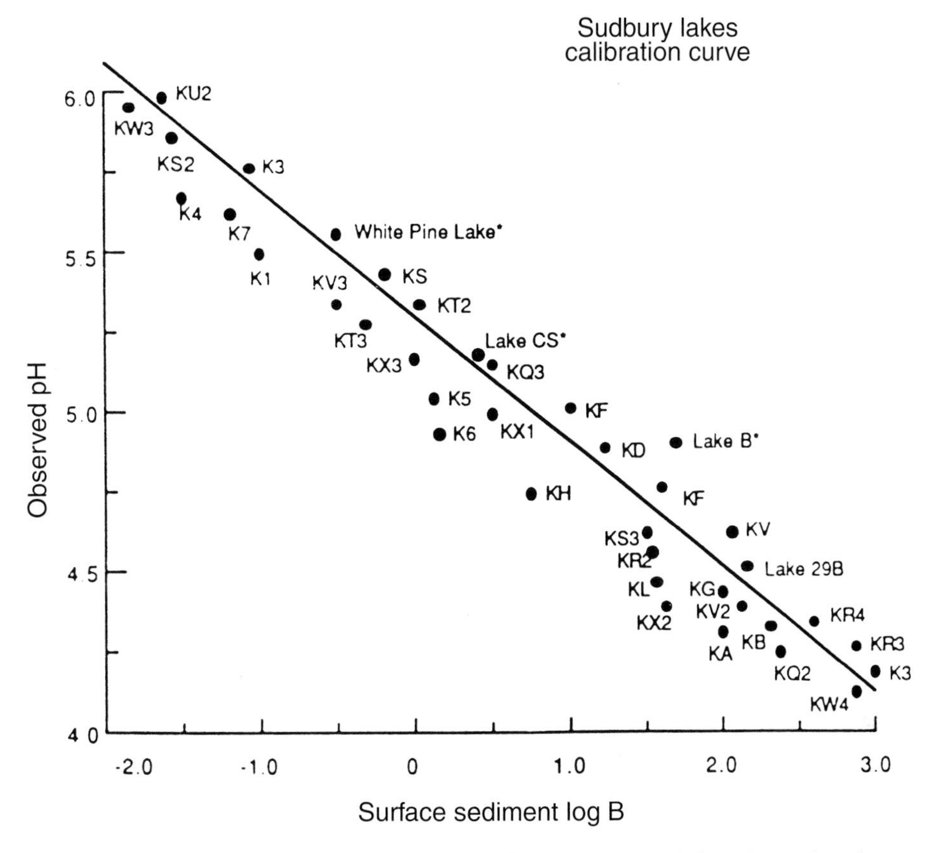

FIGURE 67.3 Diatom-inferred pH for 38 study lakes in the Sudbury and Algoma regions. Lake pH was plotted versus the diatom species index log B of Renberg and Hellberg (1982) (reprinted with permission from Dickman and Rao, 1989).

Overall, the interplay between microbes and pH is of profound significance from the standpoint of microbial physiology and genetics, virulence, ecology, and industrial applications. Stress from extreme pH has diverse effects on growth of microorganisms, which in many cases respond by altering the pH of their environment. Nonetheless, although the effects of pH stress are well documented, the mechanisms of microbial response and adaptation continue to pose many important questions for research.

BIBLIOGRAPHY

Booth, I. R. (1985). Regulation of cytoplasmic pH in bacteria. *Microbiol. Rev.* **49**, 359–378.

Chadwick, D. J., and Cardew, G. (eds.) (1999). "Bacterial Responses to pH," Novartis Foundation Symposium No. 221. Wiley, Chichester, UK.

Dickman, M., and Rao, S. S. (1989). *In* "Acid Stress and Aquatic Microbial Interactions," p. 123. CRC Press, Boca Raton, FL.

Hall, H. K., Karem, K. L., and Foster, J. W. (1995). Molecular responses of microbes to environmental pH stress. *Adv. Microbiol. Physiol.* **37**, 229–272.

Hofmann, A. F., and Mysels, K. J. (1992). Bile acid solubility and precipitation in vitro and in vivo: The role of conjugation, pH, and Ca^{2+} ion. *J. Lipid Res.* **33**, 617–626.

Kashket, E. R. (1985). The proton motive force in bacteria: A critical assessment of methods. *Annu. Rev. Microbiol.* **39**, 219–242.

Kirkpatrick, C., Maurer, L. M., Oyelakin, N. E., Yontcheva, Y., Maurer, R., and Slonczewski, J. L. (2001). Acetate and formate stress: opposite responses in the proteome of *Escherichia coli. J. Bacteriol.* **183**, 6466–6477.

Krulwich, T. A., Ito, M., Gilmour, R., Sturr, M. G., Guffanti, A. A., and Hicks, D. B. (1996). Energetic problems of extremely alkaliphilic aerobes. *Biochim. Biophys. Acta* **1275**, 21–26.

Myrold, D. D., and Nason, G. E. (1992). Effect of acid rain on soil microbial processes. *In* "Environmental Microbiology" (R. Mitchell, ed.), pp. 59–81. Wiley–Liss, New York.

Olson, E. R. (1993). Influence of pH on bacterial gene expression (MicroReview). *Molec. Microbiol.* **8**, 5–14.

Padan, E., and Schuldiner, S. (1987). Intracellular pH and membrane potential as regulators in the prokaryote cell. *J. Membrane Biol.* **95**, 189–198.

Rao, S. S. (ed.) (1989). "Acid Stress and Aquatic Microbial Interactions." CRC Press. Boca Raton, FL.

Russell, J. B., and Diez-Gonzeles, F. (1998). The effects of fermentation acids on bacterial growth. *Adv. Microbial Physiol.* **39**, 205–234.

Slonczewski, J. L., and Blankenhorn, D. (1999). Acid and base regulation in the proteome of *Escherichia coli. Novartis Found. Symp.* **221**, 75–83.

Slonczewski, J. L., and Foster, J. W. (1996). pH-regulated genes and survival at extreme pH. *In "Escherichia coli* and *Salmonella typhimurium:* Cellular and Molecular Biology" (F. C. Neidhardt, R. I. Curtiss, C. A. Gross, J. L. Ingraham, and M. Riley, eds.), 2nd edn, pp. 1539–1552. ASM Press, Washington, DC.

Slonczewski, J. L., Rosen, B. P., Alger, J. R., and Macnab, R. M. (1981). *Proc. Natl. Acad. Sci. USA* **78**, 6272.

Stancik, L. M., Stancik, D. M., Schmidt, B., Barnhart, D. M., Yoncheva, Y. N., and Slonczewski, J. L. (2002). pH-dependent expression of periplasmic proteins and amino acid catabolism in *Escherichia coli. J. Bacteriol.* **184**, 4246–4258.

Storz, G., and Hengge-Aronis, R. (eds.) (2000). "Bacterial Stress Responses." ASM Press, Washington, DC.

Wang, J., and Fung, D. Y. C. (1996). Alkaline-fermented foods: A review with emphasis on pidan fermentation. *Crit. Rev. Microbiol.* **22**, 101–138.

WEBSITE

Website on proteomic methods to study pH stress (J. L. Slonzcewski) *http://biology.kenyon.edu/slonc/labtools/2d_method.html*

68

Plant pathogens

George N. Agrios
University of Florida

GLOSSARY

disease cycle The chain of events involved in disease development, including the stages of development of the pathogen and the effects of the disease on the host.

facultative Having the ability to act in a certain way.

haustoria Feeding organs of some fungi and parasitic higher plants that absorb water and nutrients from host cells.

life cycle The stage or successive stages in the growth and development of an organism that occur between the appearance and reappearance of the same stage (e.g., spore) of the organism.

parenchyma Plant tissue composed of thin-walled cells which synthesize or store foodstuffs and usually leave intercellular spaces between them.

pathogen A living organism that can cause disease on certain other living organisms (hosts).

phloem Tube-like cells of the plant conductive system that carry sugars and other organic substances from leaves to other parts of the plant.

vector A specific insect, mite, nematode, or fungus that can acquire and transmit a specific pathogen from an infected to a healthy host.

virulence Relative ability of a pathogen to cause disease on a given host plant.

xylem Tube-like cells of the plant conductive tissue that carry water and minerals from the roots to other parts of the plant.

Plants, like humans and animals, are affected by diseases (Fig. 68.1). In all organisms, diseases are caused by internal or external abiotic factors, such as nutritional or environmental conditions; by quasi-biotic factors, such as genetic abnormalities; or by pathogens. Pathogens are living entities, mostly microorganisms, such as certain fungi, prokaryotes such as bacteria and mollicutes, viruses, protozoa, nematodes, etc. that can attack other organisms and cause disease. In addition to these pathogens, diseases in plants can also be caused by several plants that parasitize other plants and by some green algae. With the exception of the few diseases caused by parasitic plants and green algae, however, the vast majority of diseases in plants are caused by the same groups of pathogenic microorganisms as those that cause diseases in animals and humans. Plant pathogens vary considerably in size (Fig. 68.2), shape (Figs 68.2–68.4), and multiplication (Fig. 68.5). Like all pathogens, those affecting plants vary considerably in their host specificity; some are able to infect all or most plants belonging to one species, a genus, or several plant families, whereas others can infect only one or a few varieties within one plant species.

Plant pathogens cause disease in plants by disturbing the metabolism of plant cells and tissues through enzymes, toxins, growth regulators, and other substances they secrete and by absorbing foodstuffs from the host cells for their own use. Some pathogens may also cause disease by growing and multiplying in the xylem and phloem vessels of plants, thereby blocking

The Desk Encyclopedia of Microbiology
ISBN: 0-12-621361-5

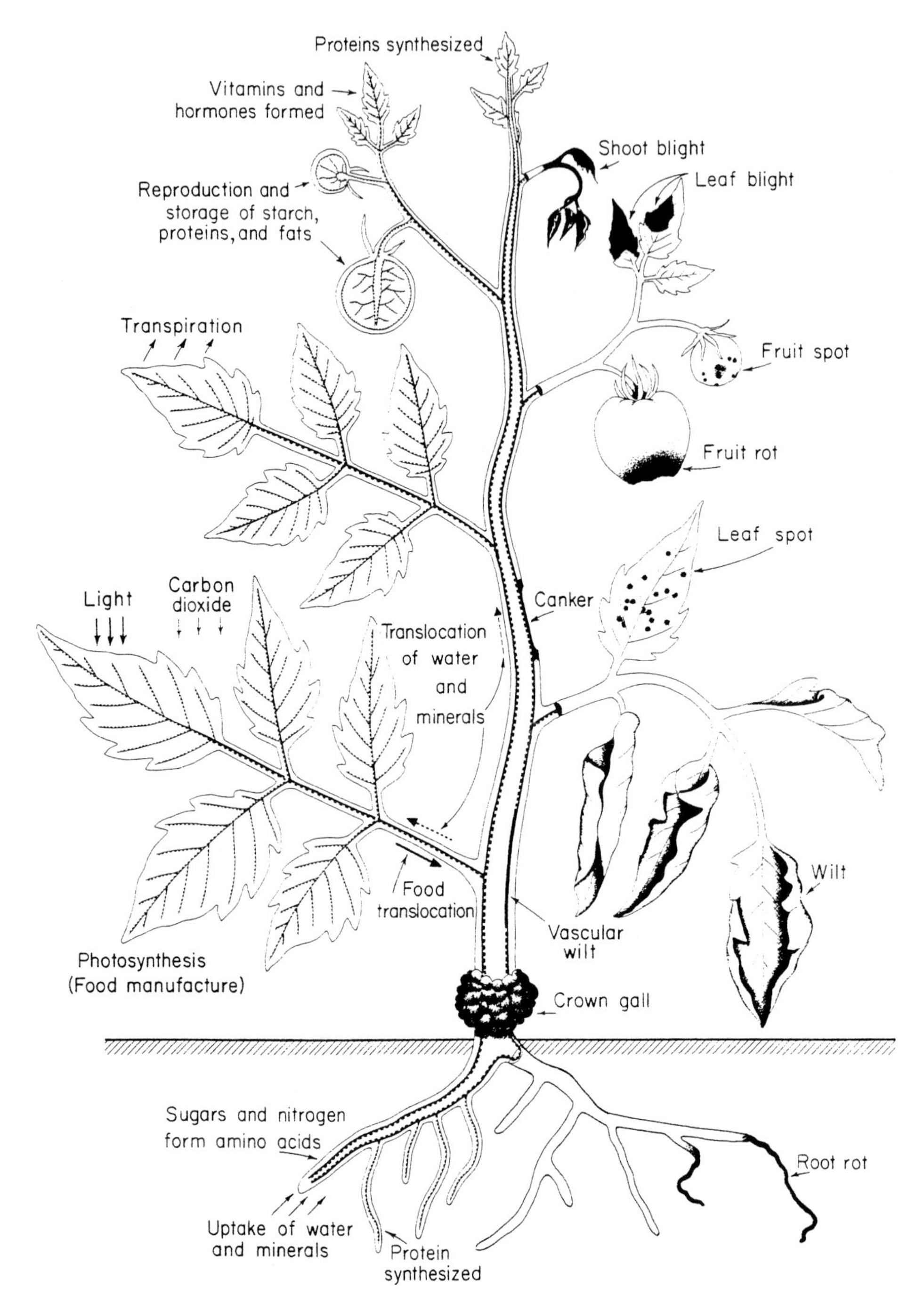

FIGURE 68.1 Schematic representation of the basic functions in a plant (left) and the interference in these functions (right) caused by some common types of plant diseases (from Agrios, 1997, p. 5).

the upward transport of water and minerals and the downward translocation of sugars. Infected plants develop a variety of symptoms (Fig. 68.1) that may vary in severity, ranging from insignificant to death of the entire plant. It is estimated that plant pathogens are responsible for a 16% loss (approximately $190 billion) of the attainable annual world crop production estimated at $1.2–1.3 trillion.

I. TYPES AND GENERAL CHARACTERISTICS OF PLANT PATHOGENS

A. Fungi

Most pathogens of plants are fungi. They cause the majority (approximately 70%) of all plant diseases.

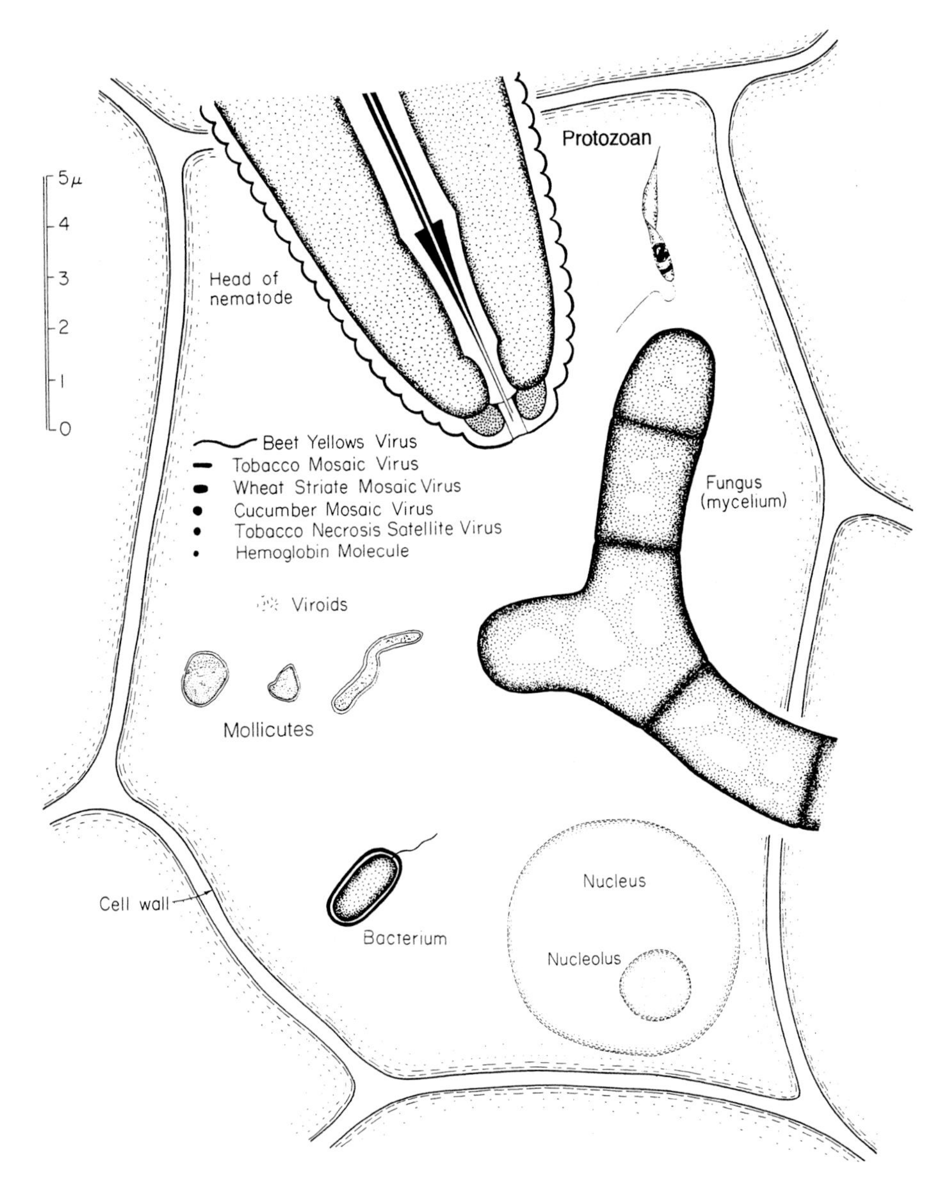

FIGURE 68.2 Schematic diagram of the shapes and sizes of certain plant pathogens in relation to a plant cell (from Agrios, 1997, p. 6).

More than 10,000 species of fungi can cause disease in plants. Some of the plant pathogenic fungi are obligate parasites (biotrophs) because they can grow and multiply only by remaining in constant association with their living host plants. Others are nonobligate parasites and either require a living host plant for part of their life cycles but are able to complete their cycles on dead organic matter or they are able to grow and multiply on dead organic matter and on living plants.

1. *Morphology*

Most fungi have a branching filamentous body called a mycelium (Figs 68.2, 68.3, and 68.5). Mycelium produces numerous branches that grow outward in a radial fashion and produce a colony (Fig. 68.3). Each branch of the mycelium is called a hypha. Hyphae are tubular and generally of uniform thickness (1–5 μm in diameter). In some fungi the mycelium is a more or less continuous tube containing many nuclei; in others, the mycelium is partitioned into cells by cross-walls, with each cell containing one or two nuclei. The length of the mycelium in some fungi is only a few millimeters, but in others it may be several to many centimeters long. In some lower fungi, now thought to be protozoa, the body consists of a mass of wall-less protoplasm containing numerous nuclei, and it is called a plasmodium.

The females of some of these nematodes become swollen at maturity and have pear-shaped or spherical bodies. Nematode bodies are transparent, unsegmented, and have no legs or other appendages. Plant parasitic nematodes have a hollow stylet or spear that is used to puncture holes in plant cells and through which they withdraw nutrients from the plants.

2. *Reproduction*

Nematodes have well-developed reproductive systems that distinguish them as female and male nematodes. The females lay eggs usually after fertilization by males but in some cases without fertilization. Many species lack males. Nematode eggs hatch into juveniles that resemble the adult nematodes but are smaller. Juveniles grow in size and each juvenile stage is terminated by a molt. All nematodes have four juvenile stages, with the first molt usually occurring in the egg. After the final molt the nematodes differentiate into males and females and the new females can produce fertile eggs in the presence or absence of males. One life cycle from egg to egg may be completed within 2–4 weeks in favorable weather, longer in cooler temperatures. In some nematodes only the second-stage juvenile can infect a host plant, whereas in others all but the first juvenile and adult stages can infect. When the infective stages are produced, they must feed on a susceptible host or they will starve to death. In some species, however, some juveniles may dry up and remain quiescent, or the eggs may remain dormant in the soil for years.

3. *Classification*

The plant pathogenic nematodes can be classified as follows:

Kingdom: Animalia
- Order: Tylenchida
  - Anguinidae—*Anguina* (seed-gall nematode), *Ditylenchus* (bulb and stem n.)
  - Belonolaimidae—*Belonolaimus* (sting n.), *Tylenchorhynchus* (stunt n.)
  - Pratylenchidae—*Pratylenchus* (lesion n.), *Radopholus* (burrowing n.)
  - Hoplolaimidae—*Hoplolaimus* (lance n.), *Rotylenchus* (spiral n.)
  - Heteroderidae—*Globodera* (round cyst n.), *Heterodera* (cyst n.), *Meloidogyne* (root knot n.)
  - Criconematidae—*Criconemella* (ring n.), *Hemicycliophora* (sheath n.)
  - Paratylenchidae—*Paratylenchus* (pin n.)
  - Tylenchulidae—*Tylenchulus* (citrus n.)
  - Aphelenchoidea—*Aphelenchoides* (foliar n.), *Bursaphelenchus* (pine wilt n.)
- Order: Dorilaimida
  - Longidoridae—*Longidorus* (needle n.), *Xiphinema* (dagger n.)
  - Trichodoridae—*Paratrichodorus* and *Trichodorus* (stubby root nematodes)

4. *Ecology*

Almost all plant pathogenic nematodes live part of their lives in the soil, especially in the top 15–30 cm, where most of the plant roots are located. Many nematodes live freely in the soil, feeding superficially on roots and underground stems. Others spend much of their lives inside their hosts, but often their eggs, preparasitic juvenile stages, and the males spend some of their lives in the soil. Soil temperature, moisture, and aeration affect survival and movement of nematodes in the soil. A few nematodes survive in the tissues of the plant they infect and in their insect vector, and they seldom, if ever, enter the soil.

5. *Dissemination*

Nematodes spread through the soil slowly under their own power, probably no more than 1 to possibly a few meters per growth season. Nematodes, however, can be easily spread by anything that moves soil particles, such as irrigation or runoff water, farm equipment, animal feet, and dust storms. A few plant pathogenic nematodes are spread from plant to plant by their specific insect vector.

6. *Infection*

When an infective nematode juvenile or adult reaches the surface of a plant, it places its mouth-parts (six lips) in contact with the plant surface while the stylet moves back and forth, exerting pressure as it tries to pierce the outer epidermal cell wall. Once that occurs, the nematode, through the stylet, secretes saliva into the cell, the contents of which are liquefied and absorbed through the stylet. Some nematodes are rapid feeders and feed only on epidermal cells, moving their stylet from one cell to the next without ever entering the plant (ectoparasitic nematodes). Others (endoparasitic nematodes) enter the plant parts and feed slower while inside the plant. Some of the latter (sedentary nematodes) become attached to one area of the plant and do not move. Some of the nematodes (migratory) enter the plant, feed internally for varying lengths of time, and then exit the plant and move about freely. Depending on where the female of a particular nematode species is feeding, she lays her eggs inside or outside the plant. When the eggs hatch, the new infective juveniles either cause further infections inside the plant or infect new plants. The mechanical injury caused by nematodes in the infected area, as well as the removal of nutrients from plants by nematodes, certainly has a detrimental effect on the plant. It is thought, however, that much greater damage by nematodes on plants is caused by the enzymes, growth regulators, and toxic compounds contained in the secretions of the nematode into the plant and by the ports of entry for other pathogens (fungi and bacteria) created by nematodes on plant roots.

7. *Epidemiology*

Plant pathogenic nematodes overseason as eggs in the soil or plant debris, as juveniles or adults in roots or stems of perennial plants, in vegetative propagative organs and some infected seeds, and as dehydrated juveniles or adults in contact with plant tissues. At the beginning of the growth season, the eggs, often stimulated by hatching factors excreted by host plants, hatch and produce juveniles which are attracted to host plants and cause infection. Dormant adults and juveniles are also activated and may cause new infections. Before or following infection, the nematode goes through its remaining juvenile stages, if any, develops into an adult, and subsequently lays eggs that can hatch; the juveniles or adults then infect and repeat the disease cycle. Nematodes are favored in their survival and spread by moderate temperatures and moisture. Too high or too low temperatures kill the nematodes, as do dry soils and flooding.

8. *Management and control*

As with other plant pathogens, nematodes are best controlled by keeping them out of a field through quarantines and use of nematode-free propagative materials, seeds, and transplants. Infected propagative materials are freed of nematodes by treating them with hot water. When nematodes are present in a field, they can be managed primarily by planting resistant plant varieties and by applying pre-plant or post-plant nematicides. Several cultural practices, such as plowing the field repeatedly before planting to expose more nematodes to the sun and dry air or flooding the field to kill nematodes by suffocation, are helpful but difficult to carry out. Planting trap crops that attract nematodes but do not allow them to reach the egg-laying stage or destroying the trap plants before the nematodes reproduce are also possible but difficult. Recently, several antagonistic bacteria and fungi have been found that provide experimental control of plant pathogenic nematodes, but these are not available commercially.

E. Protozoa

To date, only a small number of trypanosomatid flagellate protozoa have been found to parasitize and to cause disease in plants. Some flagellate protozoa parasitize the latex-bearing cells (the laticifers) of laticiferous plants, and it is not clear whether these protozoa cause disease to their hosts. Some have been isolated from fruits such as tomato, on which they cause considerable damage. Several flagellate protozoa, however, parasitize the phloem of plants. These parasites are obligate parasites that live and reproduce in the phloem sieve elements of their host plants and in their insect vector. It has been difficult to isolate, grow in pure culture, and inoculate these parasites into healthy plants to reproduce the disease; therefore, their role as causes of plant diseases was uncertain. Nevertheless, protozoa have been found associated with and are now considered as the causes of a disease of coffee trees, several diseases of palm trees, and possibly a disease of cassava plants. Plant pathogenic protozoa have been found in the tropic and semi-tropic areas of the Americas.

1. *Morphology*

The protozoa found associated with plant diseases to date are spindle-shaped single cells that have a typical nucleus and one or more long, slender flagella at some or all stages of their life cycle (Figs 68.2, 68.4, and 68.5). The size of flagellate protozoa in the phloem of infected plants varies, even for the same pathogen, depending on the stage of life cycle at which the pathogen is observed. The sizes observed are $12–18 \times 1.0–2.5$, $4–14 \times 0.3–1.0$, and 3 or $4 \times 0.1–0.2\,\mu m$.

2. *Reproduction*

Plant pathogenic flagellate protozoa reproduce by longitudinal fission. At different stages of infection, reproduction of the flagellates produces the forms and sizes listed previously.

3. *Classification*

The following is a classification of plant pathogenic protozoa:

Kingdom: Protozoa
Phylum: Euglenozoa
Order: Kinetoplastida

Family: Trypanosomatidae
Genus: *Phytomonas*

Most plant pathogenic protozoa seem to belong to one genus (*Phytomonas*), but some, especially those affecting fruits, probably belong to one or more genera other than *Phytomonas*.

4. *Ecology*

Plant pathogenic protozoa survive and multiply in the phloem elements of their host plants and in their insect vectors. Some of them, of debatable pathogenicity, survive and multiply in the latex-bearing cells (laticifers) of laticiferous plants or in tissues of fruits they colonize.

5. *Dissemination*

Plant pathogenic protozoa are disseminated from plant to plant primarily by insect vectors belonging to the families Pentatomidae, Lygaeidae, and Coreidae. Slower dissemination occurs through natural root grafts between infected and healthy trees.

6. *Management or control*

There are no effective controls against plant diseases caused by protozoa. Management practices such as planting protozoa-free nursery plants and planting them away from infected ones may help reduce the disease, as may efforts to control the insect vectors of these pathogens.

F. Parasitic higher plants

More than 2500 species of higher plants live parasitically on other higher plants. Relatively few of these plants, however, affect and cause disease on cultivated plants or on forest trees of commercial significance. The parasitic plants produce flowers and seeds like all plants (Figs 68.4 and 68.5). They belong to widely separated botanical families and vary greatly in dependence on their host plants. Some parasitic plants (e.g. mistletoes), have chlorophyll but no roots; therefore they depend on their hosts only for water and nutrients. Others (e.g. dodder) have little or no chlorophyll and no true roots, therefore, in addition to water and inorganic nutrients, they also depend on their hosts for photosynthetic products. Parasitic higher plants obtain water and nutrients from their host plants by producing and sinking into the vascular system of their host stems or roots food-absorbing organs, called haustoria. The following are the most important parasitic higher plants and the botanical families to which they belong:

Cuscutaceae—*Cuscuta* sp., the dodders
Viscaceae—*Arceuthobium*, the dwarf mistletoes of conifers
Phoradendron—the American true mistletoe of broad-leaved trees
Viscum—the European true mistletoes
Orobanchaceae—*Orobanche*, the broomrapes of tobacco
Scrophulariaceae—*Striga*, the witchweeds of many monocotyledonous plants

Parasitic higher plants vary in size from a few millimeters to several centimeters in diameter and from 1 cm tall to upright green plants more than 1 m tall. Some, however, are orange or yellow leafless vine strands that may grow to several meters in length and entwine around the stems of many adjacent host plants (Fig. 68.4). Parasitic higher plants reproduce by seeds. Seeds are disseminated to where host plants grow by wind, run-off water, birds, and cultivating equipment. Seeds of some parasitic plants are forcibly expelled to significant distances (10 m or more). Parasitic higher plants overseason on perennial hosts or as seeds on the host or on the ground. In the spring, the seeds germinate and the seedling infects a new host plant. Control or management of parasitic plants depends on removing infected plants carrying the parasites and avoiding bringing seeds of parasitic plants into new areas.

G. Parasitic green algae

Although green algae are primarily free-living organisms, many genera of green algae live as endophytes of many hydrophytes and seem to cause little or no damage to their host plants. The green alga genus *Rhodochytrium* of the family Chlorococcaceae and the genus *Phyllosiphon* of the family Phyllosiphonaceae infect numerous weeds and occasionally a few minor cultivated plants. Green algae of the genus *Cephaleuros* of the family Trentepohliaceae, however, are true parasites of many wild and cultivated terrestrial plants and cause diseases of economic significance.

Cephaleuros green algae, particularly *Cephaleuros virescens*, occurs commonly as a leaf spot and stem pathogen on more than 200 plant species growing primarily in the tropics between latitudes 32°N and 32°S. Fruit lesions are less frequent but occur on many host plants. Some of the economically most important plants affected by green algae are tea, coffee, black pepper, cacao, citrus, and mango. *Cephaleuros* green algae consist of a disk-like vegetative thallus composed of symmetrically arranged cells. Algal filaments are produced that grow mostly between the cuticle and the epidermis of host leaves and, under some conditions, between the palisade and mesophyll cells of leaves. These green algae reproduce by means of zoospores in zoosporangia, which can be disseminated by wind, rain splashes, and wind-driven rain and can infect new leaves, shoots, and fruit of plants. Most infections occur at the end of the rainy season. Upon infection, plant cells next to the invading thallus turn yellow, whereas adjacent cells enlarge and divide. In stressed plants the infecting thallus expands, whereas cells in earlier invaded tissues die and form a lesion. Lesions on leaves and shoots may be so numerous that they nearly cover the entire surface. Control of parasitic green algae, when needed, can be obtained by spraying plants likely to be affected with appropriate fungicides at the time most infections occur.

II. CONCLUSIONS

Plant pathogens belong to the same groups of organisms that include the pathogens that cause diseases in humans and animals. Most plant diseases are caused by fungi, followed by viruses, prokaryotes (bacteria and phytoplasmas), nematodes, and protozoa, respectively. The species of these organisms that infect plants do not infect humans or animals—with the possible exception of a few bacteria that may affect humans and a few storage plant organs. The nature, replication, disease-causing mechanism, etc., of plant pathogens are identical or very similar to those of their counterparts that infect humans and animals. They differ from them, however, not only in their host range but also in their ecology, dissemination, epidemiology, and control. In addition to the previously mentioned common pathogens, plants are also affected by viroids, parasitic higher plants, and parasitic green algae, none of which has been shown to affect humans or animals.

BIBLIOGRAPHY

Agrios, G. N. (1997). "Plant Pathology," 4th edn. Academic Press, San Diego, CA.

Alexopoulos, C. J., Mims, C. W., and Blackwell, M. (1996). "Introductory Mycology," 4th edn. Wiley, New York.

Brown, D. J., and MacFarlane, S. A. (2001). "Worms" that transmit viruses. *Biologist (London)* **48**, 35–40.

Dollet, M. (1984). Plant diseases caused by flagellated protozoa (*Phytomonas*). *Annu. Rev. Phytopathol.* **22**, 115–132.

Goto, M. (1992). "Fundamentals of Bacterial Plant Pathology." Academic Press, San Diego, CA.

Hooykaas, P. P. J., Hall, M. A., and Libbenga, Kr. (1999). "Biochemistry and Molecular Biology of Plant–Pathogen Interactions." Elsevier Science, Amsterdam.

Hull, R. (2002). "Matthews Plant Virology," 4th edn. Academic Press, San Diego, CA.

Joubert, J. J., and Rijkenberg, F. H. J. (1971). Parasitic green algae. *Annu. Rev. Phytopathol.* **9**, 45–64.

Kuijt, J. (1969). "The Biology of Parasitic Flowering Plants." Univ. of California Press, Berkeley, CA.

Lambais, M. R., Goldman, M. H., Camargo, L. E., and Goldman, G. H. (2000). A genomic approach to the understanding of *Xylella fastidiosa* pathogenicity. *Curr. Opin. Microbiol.* **3**, 459–462.

Maloy, O. C., and Murray, T. D. (2000). "Encyclopedia of Plant Pathology." John Wiley and Sons, NewYork.

Matthews, R. E. F. (1991). "Plant Virology," 3rd edn. Academic Press, New York.

Nelson, G. C. (2001). "Genetically Modified Organisms in Agriculture." Academic Press, San Diego, CA.

Nickle, W. E. (ed.) (1991). "Manual of Agricultural Nematology." Dekker, New York.

Olson, A., and Stenlid, J. (2001). Plant pathogens. Mitochondrial control of fungal hybrid virulence. *Nature* **411**, 438.

Prell, H. H., and Day, R. R. (2001). "Plant–Fungal Pathogen Interaction: A Classical and Molecular View." Springer-Verlag, New York

Pennisi, E. (2001). The push to pit genomics against fungal pathogens. *Science* **292**, 2273–2274.

Rogers, P., Whitby, S., and Dando, M. (1999). Biological warfare against crops. *Sci. Am.* **280**, 70–75.

Staskawicz, B. J., Mudgett, M. B., Dangl, J. L., and Galan, J. E. (2001). Common and contrasting themes of plant and animal diseases. *Science* **292**, 2285–2289.

Vance, V., and Vaucheret, H. (2001). RNA silencing in plants—defense and counterdefense. *Science* **292**, 2277–2280.

WEBSITES

Internet resources in agriculture, food and forestry. Links to plant pathogens
http://agrifor.ac.uk/browse/cabi/detail/920d83863a69a2d3a2c00e1b388cbb6e.html

List of bacterial names with standing in nomenclature (J. P. Euzéby)
http://www.bacterio.cict.fr/index.html

All the Virology on the WWW (D. M. Sander). Many links, incl. to the Big Picture Book of Viruses
http://www.tulane.edu/~dmsander/garryfavweb.html

The WWW Virtual Library: Plant Parasitic Nematodes, Many links
http://www.schwekendiek.com/axel/plantparasiticnematodes.html#potato

Links to various groups of plant parasites
http://www.microbelibrary.org/

Comprehensive Microbial Resource of The Institute for Genomic Research (TIGR) and links to many other microbial genomic sites
http://www.tigr.org/tigr-scripts/CMR2/CMRHomePage.spl

The professional society of plant pathologists with numerous links to all types of plant pathogens and books about them
http://www.apsnet.org

An excellent resource for finding anything related to plant pathogens and plant diseases. Plant Pathology Internet Guidebook
http://www.ifgb.uni-hannover.de/extern/ppigb/ppigb.htm

69

Plasmids, Bacterial

Christopher M. Thomas
The University of Birmingham

GLOSSARY

bacteriophage or phage Virus infecting a bacterium.

conjugative transfer Transfer of plasmid DNA from one bacterium to another by a process that involves physical contact between donor and recipient bacteria.

phenotype Property conferred on the host organism by a gene or group of genes.

plasmid Extranuclear genetic element that can reproduce autonomously.

transduction Carriage of nonphage DNA between bacteria protected by a phage coat and injected by the normal phage injection process.

transformation Uptake of naked DNA by a bacterium followed by recombination, normally resulting in a change in the properties of the bacterium; for some bacteria, complex transformation processes result in conversion of DNA to single-stranded linear form so that only plasmid molecules which are dimeric, containing two tandem copies, can be converted back to circular form.

transposable elements Defined segments of DNA which can move from one site to another in plasmids, chromosomes, or viruses, in some cases leaving a gap at the orginal sites but in other cases replicating the element so that one copy remains at the original site; transposition depends on recombination enzymes encoded within the element, although defective elements can sometimes be activated by another element in the same cell.

The Microbiology of Bacterial Plasmids encompasses the study of the constituent parts of these genetic elements, the phenotypes they confer on their hosts, and the genetic processes they promote within the bacteria carrying them.

The existence of a sexual system promoting genetic exchange and recombination between mutant strains of *Escherichia coli* was discovered in the late 1940s and the role of a genetic factor called F (fertility) was established a few years later. The name plasmid was coined by Lederberg in 1952 to describe all extranuclear genetic elements capable of autonomous replication, but he also included temperate bacteriophage, such as lambda, under this definition. The role of plasmids in capturing antibiotic resistance genes and promoting the spread of multiple resistance determinants as single units was recognized in the late 1950s. The physical nature of plasmids as circles of DNA was demonstrated in the late 1960s and by the mid-1970s their potential as vehicles for recombinant DNA technology employing restriction enzymes and DNA ligase was already being exploited. At the same time, plasmids carrying biodegradation determinants were identified and there were the first hints that plasmid DNA could be transferred to plant cells, resulting in "plant cancers." Molecular biology has provided a good understanding of how plasmids multiply and spread, although many questions remain. Comparison of the DNA sequences and predicted replication proteins of many plasmids has

The Desk Encyclopedia of Microbiology
ISBN: 0-12-621361-5

identified the main groups to which plasmids belong. Linear plasmids have been discovered in many species as a result of new isolation and electrophoresis procedures. A challenge for the future is to assess the importance of plasmids as the basis of a pool of genes which are available to many species, thereby promoting both diversity and adaptability in bacteria.

I. BACTERIAL PLASMIDS: DEFINITIONS

A plasmid is a genetic element that can replicate in a controlled way physically separate from the chromosome. Although plasmids are known in a few eukaryotic systems, this article is limited to bacterial plasmids. Bacterial plasmids can be minimal, consisting only of the functions needed for controlled replication, or they can be almost indistinguishable from bacterial chromosomes in their size and number of genes carried. However, a variety of properties are used to distinguish plasmids from chromosomes. The most important criterion is that the plasmid does not carry genes that are essential for the structure and growth of the bacterium. In other words it should be possible to displace the plasmid from the bacterium without loss of bacterial viability as long as the environment is rich in nutrients (containing many alternative carbon, nitrogen, and energy sources) and does not contain specific harmful compounds, such as antibiotics or toxic ions. Recent DNA sequencing of some genetic elements in the size range 100–500 kb which could be plasmids has shown them to carry many genes of primary metabolism, typical of the bacterial chromosome in *E. coli*, and they have therefore been classified as chromosomes rather than plasmids. A second important criteria is that the replication of the plasmid is not coupled to the chromosome replication cycle. Although there has been dispute regarding whether plasmids such as the sex factor F replicate in step with the chromosome, the general pattern seems to be that plasmids control replication so that it occurs at a random time during the cell cycle, whereas all chromosomal replication origins initiate at a specific cell mass and thus maintain a 1 : 1 ratio with each other. Although the first plasmids identified were double-stranded DNA molecules, some plasmids have been found to have a double-stranded linear form. The first linear plasmids were found to have proteins attached to their ends like linear phage genomes, whereas other classes of linear plasmids have telomeres that contain a hairpin joining the two strands, making the ends resistant to exonucleases. Circular plasmids are generally covalently closed on both strands and are often negatively supercoiled, which means that they are stressed by having fewer turns around the helical axis than DNA has when allowed to adopt its most stable conformation. This stress gives the DNA a tendency to melt more easily. To compensate for this stress the molecule twists up on itself and forms a more compact form which increases its resistance to shear forces, an important principle exploited in plasmid isolation.

Plasmids can range in size from approximately 1 kb to hundreds of kilobases. Their copy number can vary from 1 per chromosome to 50 or more per chromosome. They may confer no identifiable phenotype (cryptic) or they may carry multiple phenotypic determinants. Any gene or genes can be carried by a plasmid, as illustrated by the range of phenotypes listed in Table 69.1, but those that are found commonly on plasmids seem to be those that can function without the need for additional genes. For example, antibiotic resistance in many cases is conferred by a single enzyme which degrades or exports a specific antibiotic or modifies the target for the antibiotic. Other phenotypes such as heavy metal resistance or degradative pathways often involve multiple proteins, but on plasmids all necessary genes are generally found as a block, although not necessarily as a single operon. The other factor that may be important in selecting genes which are favored by a plasmid location is the selection of those which confer a phenotype that can be strengthened by an increase in gene dosage. For example, the level of resistance to penicillin conferred by β-lactamases is strongly dose dependent.

II. OVERVIEW OF PLASMID EVOLUTION

Some plasmids may be as ancient as the chromosomes of the bacteria that carry them. Others may have evolved recently. A plasmid comes into existence when a segment of DNA gains the ability to replicate autonomously. This involves the creation of a replicator region by mutation. Such a unit of replication is called a "replicon." The term originally referred to the unit of replication—that is, the segment of DNA that is replicated—but it currently is generally used to define the genetic elements needed for autonomous replication. Many strategies for replication are employed by plasmids (see Section V). The strategy which is easiest to imagine evolving is that employed by ColE1 and similar plasmids, and it involves a transcriptional unit mutating so that the transcript it produces remains associated with the DNA template so

TABLE 69.1 Examples of phenotypes conferred by bacterial plasmids

Phenotype	Element	Microbe
Antibiotic production	SCP1	*Streptomyces coelicolor*
Antibiotic resistance	RP4 (IncP)	*Pseudomonas aeruginosa*
Bacteriophage resistance	pNP40	*Lactococcus lactis*
Bacterocin	p9B4–6	*L. lactis*
Biphenyl/4-chlorobiphenyl degradation	Tn*4371*	*Ralstonia eutropha*
Capsule production	pX02	*Bacillus anthracis*
Chemotaxis/chemosensor	pNod	*Rhizobium leguminosarum*
Colicin immunity	ColE2–P9$^+$	*Escherichia coli*
Colonization antigens	pK88	*E. coli*
Conjugative transfer	F	*E. coli*
Crystal protein (insecticide)	pHD2	*Bacillus thuringiensis*
Ecological competence in soil	pRtrW14-2c	*Rhizobium leguminosarum*
Electron transport proteins	pTF5	*Thiobacillus ferroxidans*
Enterotoxins	pTP224	*Escherichia coli*
Exopolysaccharide production (galactoglucan)	pRmeSU47b	*R. meliloti*
Galactose epimerase	pSa (IncW)	*E. coli*
Gas vacuole formation	pHH1	*Halobacterium*
H_2S production	pNH223	*E. coli*
Hemolysin production	pJH1	*Streptococcus faecalis*
Herbicide degradation (2,4-D)	pJP4	*Alcaligenes eutrophus*
High rate of spontaneous mutation	pMEA300	*Amycolatopsis methanolica*
Hydrogen uptake	pIJ1008	*R. leguminosarum*
Insecticide degradation (carbofuran)	pCF01	*Spingomonas* spp.
Iron uptake	pJM1	*Vibrio angularum*
Lactose fermentation	pLM3601	*Streptococcus cremoris*
Lysine decarboxylation	pGC1070	*Proteus morgani*
Melanin production	pNod	*R. leguminosarum*
Metal resistance	pMERPH (IncJ)	*Pseudomonas*
Nitrogen fixation	pIJ1007	*R. leguminosarum*
Nodulation functions (Sym plasmid)	pPN1	*Rhizobium trifoli*
Oncogenic suppression of Ti plasmid	pSa	*Shigella*
Pigmentation	pPL376	*Erwinia herbicola*
Plant alkaloid degradation	pRme41a	*R. meliloti*
Plant tumors	Ti plasmid	*Agrobacterium*
Protease production	pLM3001	*Streptococcus lactis*
Restriction/modification	pR1eVF39b	*R. leguminosarum*
Reverse transcriptase (mitochondrial plasmid)	pFOXC1	*Fusarium oxysporum*
Sex pheromones	pAD1	*Streptococcus faecalis*
Siderophore production	pDEP10 (IncF1me)	*E. coli*
Sucrose utilization	CTnscr94	*Salmonella senftenberg*
Sulfur oxidation (dibenzothiophene)	pSOX	*Alcalignes eutrophus*
Tolerance to acidity	pRtrANU1173b	*R. leguminosarum*
Tolerance to NaCl	pRtrW14-2b	*R. leguminosarum*
Toluene degradation	Tol plasmids	*Pseudomonas putida*
UV protection	R46 (IncN)	*S. typhimurium*
UV sensitization	R391 (IncJ)	*E. coli*
Virulence	pX01	*B. anthracis*

that it is able to prime replication. Mutant replicons which replicate more efficiently are likely to proliferate up to the point that they start depressing host growth significantly because of metabolic burden or other deleterious effects. Mutants which acquire a way of autogenously regulating their replication so that initial replication is rapid but reaches a limit will then be selected. Regulation by production of antisense transcripts is one of the simplest ways in which such an autogenous circuit could appear: Transcription of the opposite strand to the pre-primer RNA creates an inhibitor that builds up as the plasmid copy number increases and thus eventually switches off replication when a critical threshold is reached.

Starting from the autonomous replicon, gene acquisition by recombination and transposition could lead to possession of auxiliary maintenance functions and transfer functions, which provide an advantage for the element, and phenotypic markers, which may aid the plasmid by giving its host a selective advantage. The first acquisition of extra DNA would occur in a location which does not inactivate the replicon. Subsequent insertions would tend to occur in the same region because this region, which can accommodate insertions, is now enlarged. There will therefore tend to be apparent clustering of such insertion events, and thus the segments that are acquired in this way will also appear clustered.

The plasmids we currently see have probably been through many rounds of recombination and transposition and deletion and insertion. Events which are deleterious would not survive, whereas events which give a fitter plasmid will displace the parent. Rearrangements which result in functionally complementary genes becoming clustered may be selected because such clustering reduces the chances of these functions becoming separated. Once such compact groups of complementary functions exist, they will tend to be inherited as modules—recombination events within the module will be disfavored compared to the ones that preserve the module. On the other hand, homologous recombination will lead to reassortment of allelic differences within a modular unit. Modules built up in this manner may be gained by a different genetic system and exploited in a different way. For example, the complex units that provide plasmids with a key part of their transfer apparatus can be shown by sequence alignments to derive from the same genes as those which drive the export of toxins from the cell. The logic is that both processes involve transport of proteins to the surface of the bacterial cell. Similar gene order over a 10 or 11 gene cluster indicates that the genes were inherited as assembled gene blocks.

Plasmids are thus mosaics of genes, each of which may have its own history. However, to understand plasmids one must consider their component parts and then examine the groupings in which they are found. Some combinations of genetic fragments may be longer lived than others, as illustrated by the backbone of the IncP-1 plasmids (Fig. 69.1). However, given the diversity of plasmids, it is only now, with DNA sequencing and computer databases, that this approach is feasible. Previously, other approaches were used to provide order from the bewildering diversity.

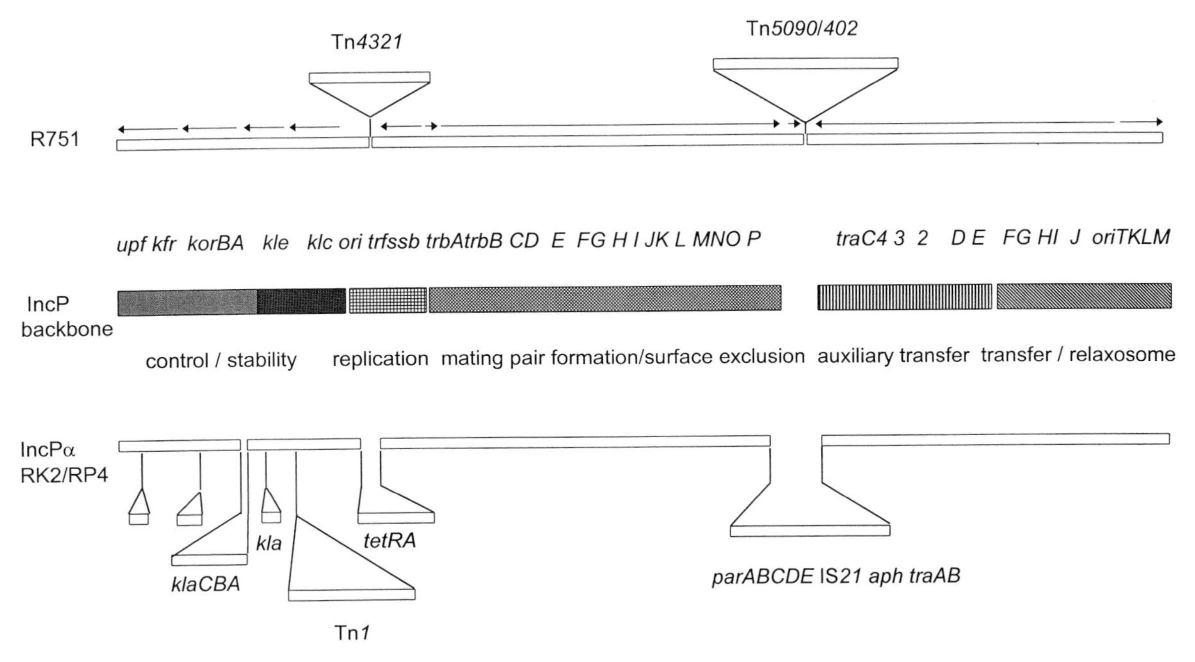

FIGURE 69.1 Organization of two related plasmids belonging to the same incompatibility group (see Section III). The IncP-1 plasmids were originally isolated in *Pseudomonas. aeruginosa* but they can transfer efficiently between and maintain themselves stably in all gram-negative bacterial species. Their transfer system can also promote transfer of DNA to gram-positive bacteria and yeast. The complete sequences of both plasmids are known. The core of replication, stable inheritance, and transfer genes can be aligned as shown. For simplicity, they are shown linearized at an arbitrary point, although these are circular plasmids. Arrows indicate the transcriptional organization of these genes. Although the two plasmids carry different phenotypic markers, the insertion of DNA carrying these genes, often but not always on transposable elements, has occurred repeatedly in the same places in the backbone. These plasmids, isolated in the late 1960s and early 1970s, confer antibiotic resistance, but related plasmids found in the environment carry biodegradation capability.

III. INCOMPATIBILITY AND PLASMID CLASSIFICATION

The number of plasmids now known is very large. Catalogs describing some of these plasmids are available from public culture collections, but many plasmid descriptions are buried in the literature. Methods for grouping plasmids are essential for the cataloging process. The most universally accepted way is to start by grouping plasmids on the basis of their replication and maintenance systems. A method of determining whether two plasmids use the same system was devised in the 1960s—the incompatibility test. Plasmids with different selectable markers were introduced into the same cell line, selecting for both plasmids. Selection for one or both of the plasmids was then removed and a culture grown. After approximately 20 generations of growth, the bacteria were spread on selective medium to determine whether both plasmids were still present in all bacteria or whether segregation had occurred to give bacteria with one plasmid or the other but not both. Plasmids that caused each other to segregate were called "incompatible" and intuitively it was concluded that this was because they depended on the same replication and maintenance functions. Therefore, incompatible plasmids were categorized into the same group, whereas plasmids that were compatible were categorized into different groups. Since these groupings depended on the use of the same host, different incompatibility groupings were established for enteric bacteria (particularly *E. coli*), *Pseudomonas* species (since most *Pseudomonas* plasmids seemed able to replicate in more than one *Pseudomonas* species), and *S. aureus*, in which much early plasmid work was done because of the prevalence of resistance to commonly used antibiotics such as penicillin. Some plasmids were never classified for a variety of reasons. Other plasmids showed anomalous behavior, displacing but not being displaced or showing apparently more than one type of incompatibility. Many of these properties are the result of the plasmids being cointegrates of plasmids of two groups so that they can use alternative replication systems. Thus, if one of the two systems is competing with that on another plasmid, the second replication system in the hybrid can continue to maintain the plasmid. This competition will therefore add a selective pressure for plasmids which are not hybrids to evolve new specificities rapidly, leading to related plasmids which have novel initiation and control specificity so that they are not subject to direct competition. Because of the complexity of real incompatibility data and because DNA sequencing is becoming easier, comparison of plasmid sequences is now a much simpler and more reliable indicator of plasmid groups. However, it does not provide information about incompatibility, which can only be determined empirically. In some cases, the real incompatibility data are very important for assessing plasmid behavior in a microcosm because they define those that coexist and those that compete.

IV. THE HORIZONTAL GENE POOL

A bacterium can carry many plasmids. These can be completely unrelated to each other or may be derived from a common ancestor whose replication/ replication control systems have diverged sufficiently so that they do not compete with each other. Low-copy-number plasmids can be hundreds of kilobases in size and may carry approximately 10% of the bacterial genome, although as noted previously an element of this size is generally referred to as a minichromosome if it carries a standard density of housekeeping genes, as has been established for certain *Rhodobacter* and *Pseudomonas* species. The presence of two or more large plasmids (as in *Rhizobium* species) may result in 10–20% of the cellular DNA being plasmid encoded. Two strains may differ simply in the plasmid(s) they carry. These plasmids may be lost and replaced by others. These plasmids may come from a closely related strain or from an unrelated species which carries broad host range plasmids. The potential number of genes that these populations of plasmids can carry into a bacterium is very extensive. This pool is called the horizontal gene pool since it is not confined to a single species. Any gene that moves onto a plasmid enters this pool as long as the plasmid is not confined to its host. Current knowledge of the bacterial kingdom is limited to a relatively small sector. Sampling of plasmids from new bacteria from uncharacterized environmental niches suggests the existence of many plasmids unrelated to plasmids that are already known. However, the sequencing studies performed to date have provided a base to which it is increasingly possible to find matches for new plasmids when some or all of their DNA sequence is determined. A major research priority for bacterial genetics is to sample the horizontal gene pool of all culturable bacteria and, if possible, the non-culturable bacteria in order to establish the diversity of this pool and provide data on the extent of exchange occurring within this pool.

V. REPLICATION STRATEGIES

A general characteristic of plasmids is their feedback control loop, which responds to bacterial growth to allow replication to occur randomly throughout the cell cycle. Plasmid molecules appear to be randomly selected for replication. Recent evidence suggests that DNA PolIII may be located primarily at the predivision site (PDS) at the cell midpoint. For some groups of plasmids there is evidence that the Rep protein functions at a membrane site. Selection of a plasmid DNA molecule for replication depends on both initiation and elongation competence. One of the problems for all plasmids is that they must move away from the PDS prior to cell division and migrate to the new zone of PolIII replication potential. This will particularly be a problem for low-copy-number plasmids. For plasmids such as F, P1, and RK2, this may explain why partitioning appears to have evolved to take their DNA to the one-quarter and three-quarter positions where the next round of replication should occur. Many mechanisms have evolved to allow initiation. These have been studied in detail for circular plasmids and are described later, followed by discussion of the mechanisms of replication control. For linear plasmids, recent studies have shown that replication initiates internally and proceeds towards the ends. How the ends are replicated is the subject of research.

A. Rolling circle replication

One of the most ubiquitous replication strategies involves the introduction of a nick into one strand of the plasmid DNA by a Rep protein encoded by the plasmid (Fig. 69.2). This double-strand replication origin (*dso*) generally consists of at least two inverted sequence repetitions (IVR) in the DNA. One IVR is the specific binding site for the Rep protein dimer, whereas the other is the target for the nicking activity of the Rep protein, where it becomes attached to the 5′ end of the nick, leaving the 3′ end to prime leading strand synthesis. Replication from this 3′ end copies the intact DNA circle, displacing the nicked strand. In phage replication this strategy can generate linear molecules consisting of many copies of the circular genome by repeated replication around the circle—hence the name rolling circle replication (RCR). There is controversy whether the nicked *ori* IVR is extruded as a hairpin from the supercoiled plasmid DNA molecule: Current data suggest that extrusion is not essential. For some groups of RCR plasmids the Rep protein has been identified in the membrane fraction, and this may help to bring the plasmid to PolIII which will copy the leading strand and this will also hold the 5′ end close to the replication fork—a vital feature to allow termination which occurs when PolIII replicates past the origin again. This allows Rep to nick the *dso* again, transferring the 5′ end which was attached to Rep to the 3′ end and thus reforming the DNA circle. The extra piece of DNA past the origin is transferred to Rep, where it remains, resulting in inactivation of Rep after one round of replication. Lagging strand synthesis occurs as a separate process starting from an origin (*sso* or single-strand origin) which forms a hairpin in the single-stranded lagging strand DNA which has been displaced by the replication fork. The hairpin is recognized by RNA polymerase in a highly species-specific way to create an RNA primer which is first elongated by DNA PolI and then by PolIII. How termination of lagging strand synthesis occurs is not known. Although most common in gram-positive bacteria, such plasmids are also found increasingly in gram-negative bacteria. Some such plasmids (e.g., the well-characterized plasmid pLS1 from *Streptococcus*) can replicate efficiently in both gram-positive and gram-negative species. The most limiting factor seems to be the efficiency of *sso* function.

B. DNA polI-dependent replicons

One of the conceptually simplest forms of replicon is illustrated by plasmids such as ColE1, whose relatives have been exploited in gram-negative bacteria as cloning vectors of the common pUC series. These plasmids will continue to replicate in bacteria after protein synthesis is inhibited but will not replicate if transcription is inhibited or if DNA PolI is lacking, as in a PolA$^-$ strain. Studies during the 1970s and 1980s showed that replication depends on a transcript produced by RNA polymerase from a promoter approximately 500 nucleotides upstream of the *ori* region (Fig. 69.3). This preprimer is folded into a complex secondary structure which allows it to remain associated with the DNA of the *ori* region, where it is processed by RNaseH to generate a primer which is used by DNA PolI to initiate leading strand synthesis. After leading strand extension for approximately 100 nucleotides, a primosome assembly site is reached which allows lagging strand synthesis to be initiated, and at this stage DNA PolIII takes over leading strand synthesis.

Many other plasmids seem to use a similar initiation process based on the processing of a transcript created by RNA polymerase. In the case of the pAMβ1 family of plasmids from the gram-positive bacteria, it seems clear that the role of the Rep protein is to promote processing of the transcript at the replication origin. A similar situation may apply to the *repFIC* replicon found in the well-studied plasmid R1, although this plasmid does not appear to use DNA PolI to initiate leading strand synthesis.

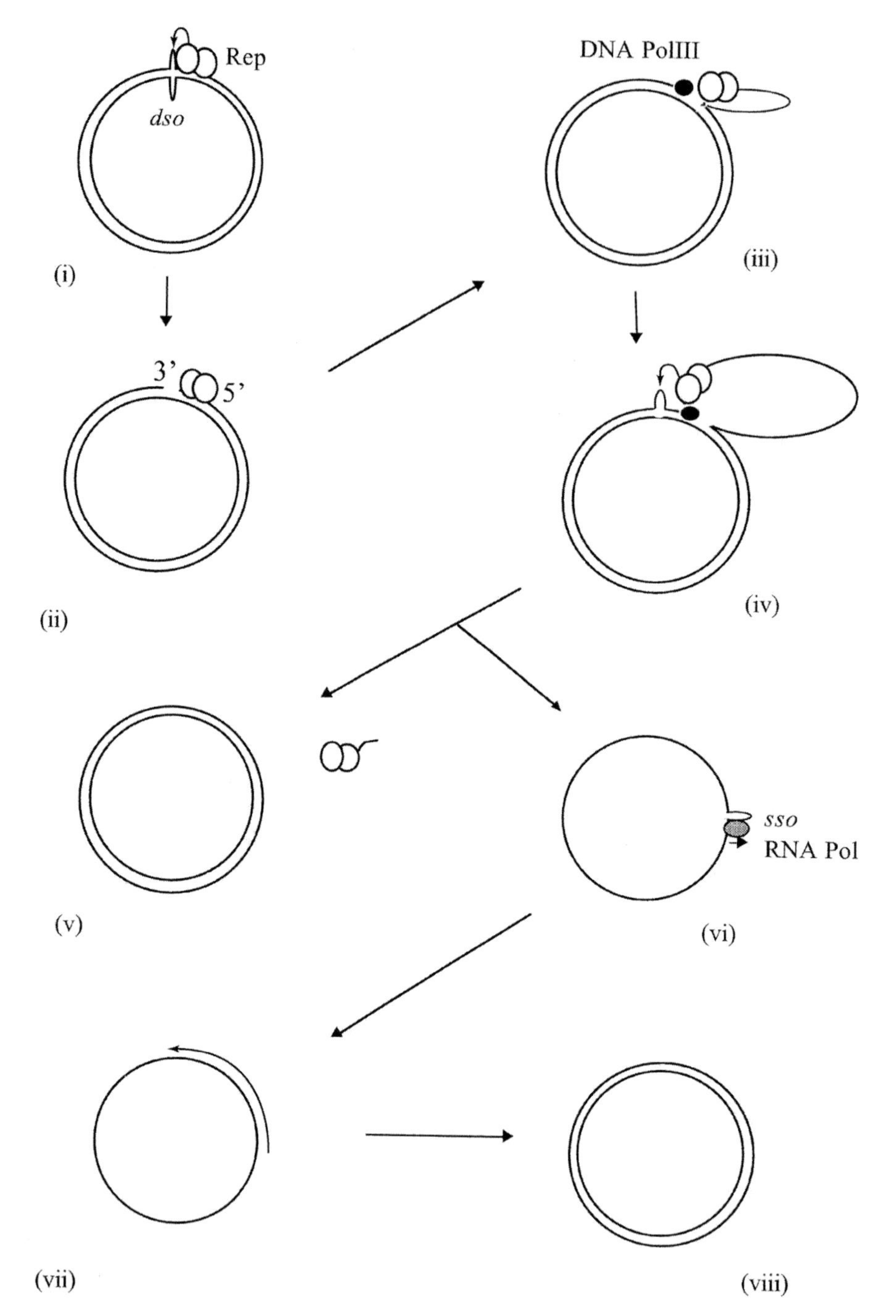

FIGURE 69.2 Rolling circle replication. Dimeric Rep protein binds to its target in the double-strand origin (*dso*) next to an inverted sequence repetition which may be extruded as hairpins in supercoiled DNA (i). Nicking of one strand occurs and Rep remains covalently attached to the 5′ end (ii). The 3′ end is extended by DNA PolIII, displacing the paired strand (iii). When replication goes past *dso* again, Rep attacks it (iv), reforming the circle (v), displacing a ssDNA circle (vi), and leaving a short ssDNA tail. The ssDNA circle is converted back to dsDNA after RNA polymerase has created a primer on a hairpin stem formed at the single-strand origin (*sso*) (vi–viii).

C. Iteron-activated replicons

The replication origins of many plasmids contain tandemly repeated sequences adjacent to an AT-rich region, where directed melting of the DNA occurs and leading strand synthesis is initiated. The repeats termed iterons (because they are iterated), each approximately 21 bp in length, are the binding sites for a plasmid-encoded Rep protein. This arrangement is similar to that in all known chromosomal replication origins (*oriCs*), in which the iterons bind DnaA protein, as well as in the origins of such bacteriophage

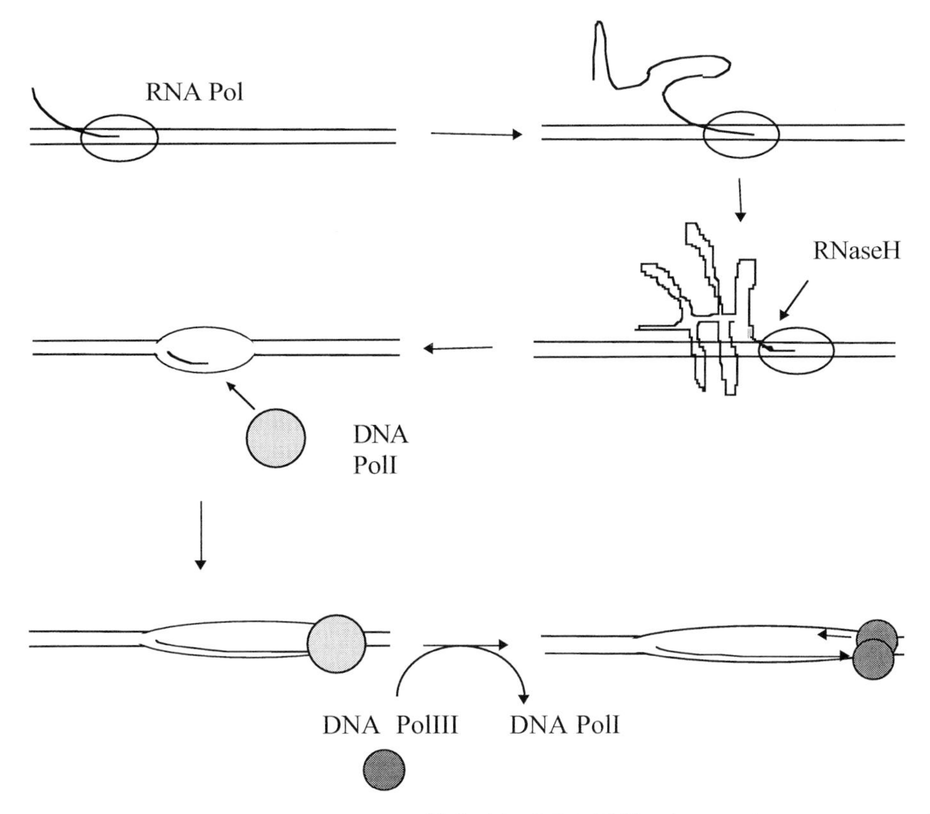

FIGURE 69.3 DNA polymerase I-dependent replication as established for ColE1. RNA polymerase generates a transcript that folds into a three-dimensional structure that promotes association of the RNA with the DNA at the replication origin. Here, it is processed by RNaseH to generate a primer which is used by DNA PolI to initiate leading strand synthesis. After approximately 100 nt the replication fork reaches a primosome assembly site which initiates lagging strand synthesis. DNA PolIII takes over from DNA PolI. In the IncFII plasmids a similar creation of an RNA primer may occur by Rep-mediated processing.

as lambda, in which the iterons bind the O protein. Many plasmid replication origins of this sort also contain binding sites for DnaA, and these plasmids display a varying degree of dependence on the activity of this host protein. Thus, in *E. coli* mutants that are temperature sensitive for replication due to a mutation in *dnaA*, the F plasmid can suppress the temperature sensitivity if it integrates into the chromosome so that chromosome replication can be initiated from within the F plasmid ("integrative suppression"). The broad host range IncP-1 plasmids have four DnaA binding sites upstream of a group of five iterons in the replication origin, and these are essential for replication in most bacterial species except *P. aeruginosa*, in which their removal causes only a partial defect in replication resulting in a lowered copy number. The role of DnaA in these plasmids may be to lead in other host proteins such as DnaBC, whose helicase activity is needed to unwind the replication origin so that DnaG primase can enter and generate the primer for leading strand synthesis. However, clearly DnaA is not essential in all circumstances; therefore, some Rep proteins can perform the recruitment process. Many Rep proteins are identified largely as dimers, but in general it appears that the active form of Rep is the monomer. Also, various host chaperone proteins (proteins that bind to and act catalytically to promote the normal folding and assembly of the active form of other proteins) are needed to convert dimers to monomers. Dimers can actually be inhibitory as described in Section V.D. The IncQ plasmids are unusual in coding for three Rep proteins. One of these proteins is an iteron-binding protein (like Rep of other plasmids), whereas the other two proteins are a helicase and a primase, making the plasmid largely independent of the host replication machinery, apart from DNA polymerase. This correlates with an extremely broad host range.

D. Replication control

An essential feature of plasmids appears to be the ability to sense their own copy number and to switch off their replication as the copy number increases to higher than the norm. It is generally accepted that copy number control systems conform in some way to the "inhibitor dilution model," in which the plasmid produces an inhibitor either constitutively so that its concentration is proportional to the copy number or in a burst after replication so that further replication is prevented until the bacterium has grown enough to dilute the inhibitor to below its critical inhibitory level. This principle was confirmed by the demonstration that in a hybrid between a low-copy-number and a high-copy-number plasmid, the high-copy-number replicon dominates, silencing the low-copy-number replicon, unless the high-copy-number replicon is inactivated. The production of an unstable inhibitor that senses copy number is illustrated best through the high-copy-number plasmids ColE1 (from *E. coli*) and pT181 (from *S. aureus*) and the low-copy-number plasmid R1 (from *E. coli*). In each case, a highly structured antisense RNA attacks an RNA target that is essential for replication. In ColE1 RNAI binds to preprimer RNA and prevents it from folding to the form that can be processed into primer essential for replication initiation. In pT181, Cop RNA induces transcription termination on the mRNA that codes for the Rep protein. In R1, Cop prevents translation of a leader peptide needed to allow translation of the rep mRNA. In each of these cases, the antisense RNAs are highly reactive, binding to their targets through single-stranded loops, and it is the speed of these reactions that is essential for their activity.

Iteron-containing replicons seem to use a different control strategy that is not fully understood. In these plasmids, it is generally found that overproduction of the Rep protein does not result in runaway replication as might be expected, and in some cases overproduction of Rep is actually inhibitory. Mutations affecting copy number show changes in the ability of Rep to form multimers and to interact with iterons in the replication origin. Since the active initiator is a monomer, it is suggested that dimers link replicated molecules together (handcuffing) and block further replication. The mechanism by which handcuffed molecules are pulled apart is not clear, although it may normally be connected to an active partitioning cycle. Some Rep proteins control their own synthesis by binding to an inverted repeat of the iteron binding sequence, suggesting that there are different multimeric forms of the Rep protein able to alternatively bind an inverted repeat or two direct repeats on different molecules. Often, there are also additional groups of iterons, adjacent to the replication origin, which are not essential for replication but which modulate initiation frequency, possibly by titration of Rep or by formation of inhibitory complexes.

VI. STABLE INHERITANCE

Plasmid copy number is controlled as described in Section V.D. Control systems have generally evolved to maintain large plasmids ($>$50 kb) at copy numbers similar to that of the bacterial chromosome, whereas small plasmids tend to be present at much higher levels (15–20 per chromosome or more). At such high copy numbers, even in slowly growing bacteria there would be 30–40 copies of the plasmid in predivisional cells. In faster growing bacteria there will be proportionately more chromosomal equivalents per cell (because the speed at which an individual chromosome can be replicate does not increase as growth rate increases so that increased replication rate is achieved by more frequent initiation of the replication cycle), and there will generally be an increase in the number of plasmids, although not necessarily in direct proportion to the chromosome. The probability of producing a plasmid-free cell is $P = 2^{-(2n-1)}$, where n is the copy number per baby cell, assuming that each copy of the plasmid is distributed randomly. For example, a plasmid that is present at 1 copy per baby cell will, on average, be present at 2 copies prior to cell division, and in half of divisions both plasmid molecules will be transmitted to only one of the two daughter cells. Thus, one-fourth of cells in each round of cell division will have lost the plasmid. For a plasmid that is present at 10 copies per baby cell, there will be 20 copies in predivisional cells and the probability of a plasmid-free cell occurring is $0.5^{19} = 1.9 \times 10^{-6}$; thus, approximately 1 in a million bacteria would be expected to lose the plasmid at each round of cell division. Accordingly, high-copy-number plasmids have generally acquired those stable inheritance functions related to ensuring that individual plasmid genome copies do segregate randomly. On the other hand, low-copy-number plasmids have also acquired genes which allow them to segregate in a nonrandom way and to combat plasmid loss in other ways.

A. Multimer resolution

All plasmids have the potential to form multimers by homologous recombination between identical DNA molecules in the bacterial cell. This should lead to great segregational instability because dimers and higher multimers tend to take over if unchecked. Also, by reducing the number of physically separate DNA molecules, they reduce the efficiency of both random and active partitioning for a given number of copies of the plasmid genome. A dimer has two origins and is therefore twice as likely to be chosen for replication as a monomer. However, once initiated it will produce two plasmid genome units and thus this effectively counts as a double-replication event. If the plasmid controls replication by producing a repressor in proportion to the number of plasmid copies present, replication of a dimer doubly reduces the potential for monomers to replicate and the cell enters a runaway situation in which dimers take over ("dimer catastrophe"). This scenario should not apply to RCR plasmids since the replication mechanism will terminate when it reaches the second origin in the dimer and will generate a monomer.

Many multimer resolution systems have been identified, the best studied of which is the *cer* system of the high-copy-number *E. coli* plasmid ColE1, whose replicon is related to the replicon found in many of the high-copy-number cloning vectors. The *cer* locus encodes a target for four host proteins: ArgR, PepA, XerC, and XerD. ArgR is the repressor that coordinates expression of operons encoding arginine metabolism genes, whereas PepA is a peptidase involved in polypeptide processing. XerC and -D are recombinases also involved in chromosomal site-specific recombination. Thus, ColE1 has recruited many host proteins, including two with apparently unrelated function, to promote its stable inheritance. Both ArgR and PepA are hexamers composed of two layers of trimers. They have the potential to stack on top of each other in alternating layers. They form the core around which the plasmid DNA wraps and holds XerC and XerD on two sides of this trimeric structure. Potentially, this recombinosome could promote recombination between sites either in the same molecule (resolution) or in different molecules (dimerization). Mutant sites can be obtained which catalyze both of these reactions, but normally the resolution pathway is favored to the exclusion of the dimerization pathway. It is thought that this is due to the need for the protein complexes at the *cer* site to be held next to each other in a supercoiled dimer for long enough to allow the slow strand exchange reaction to occur. This would not be possible in noncovalently joined molecules. The *cer* region also contains an additional stability determinant—a promoter that directs the transcription of a short region called *rcd* which causes retardation of cell division. The promoter appears not to function in monomeric plasmids but may be activated in the nucleoprotein complex at the paired *cer* sites. The transcript folds into two hairpins which occupy almost the whole length of the RNA. Although the target for this molecule is not known, it seems to serve the purpose of slowing growth of bacteria with dimers or completely halting division of bacteria which have been taken over by multimers. It therefore seems to preempt the need for postsegregational killing (*psk*) systems.

B. Killing of plasmid-free segregants

psk systems were first discovered when it was found that the *ccd* locus on plasmid F, which appeared to block cell division in bacteria with a plasmid that could not replicate, was responsible for generating two populations of the bacteria as the plasmid was lost: viable bacteria with the plasmid and nonviable bacteria which had lost the plasmid. It was subsequently shown that the CcdB protein is a poison of DNA gyrase when it is not associated with an antidote protein, CcdA. Both CcdA and CcdB are produced from the plasmid, but CcdA decays rapidly when the plasmid is lost and the action of CcdB kills these plasmid-free segregants. Plasmid loss can be mimicked by addition of rifampicin, which blocks transcription, and many of the *psk* systems have been discovered or confirmed by this test.

The best studied *psk* system is the *hok/sok* system of plasmid R1 (Fig. 69.4). Host killing is achieved by the 52-amino acid polypeptide Hok, which causes membrane depolarization. Translation of *hok* depends on activation by translation of the preceding and overlapping gene *mok*. The need for *mok* thus helps to silence *hok* when it is repressed. As it is transcribed, the primary transcript of the *mok/sok* region forms a metastable structure which buries the *mok* and *hok* translational signals in an inaccessible form. When the transcript is complete the 3′ tail interacts with the 5′ end, causing refolding and favoring an alternative structure in which the translational signals for *mok* and *sok* are now blocked by pairing with the 3′ tail. Degradation of the mRNA occurs from the 3′ end so that as removal of nucleotides proceeds the *mok* and *hok* translational signals become accessible again. In cells with plasmid, an unstable countertranscript (Sok) is constantly being produced and blocks translation of the activated *hok* transcript. However, if the plasmid has been lost, then by the time the *hok* mRNA has been activated the Sok RNA has disappeared and Hok is produced, resulting in cell death. The action of this *hok/sok* locus is sometimes referred to as programmed cell death. It certainly is one of the most interesting examples of a timed sequence of events in bacteria and shows how different cellular responses can be triggered depending on how the system is set by the state of the cell.

C. Active partitioning

Many low- or medium-copy-number plasmids appear to be lost at a lower frequency than one would expect on the basis of their copy number. Genetic regions required for this activity have been mapped. In some

role of phage in transferring plasmids in natural environments is currently under investigation.

C. Conjugative transfer

The mechanism of gene spread that is traditionally plasmid associated is conjugative transfer, a process that depends on direct contact between donor and recipient bacteria. In recent years, gene systems promoting conjugative transfer have been found as part of elements that are not plasmids, emphasizing the mosaic nature of genetic elements found in bacteria as referred to previously. For example, certain transposons can promote their own spread via conjugation, and other chromosomal segments such as some recently identified pathogenicity islands (chromosomal segments encoding virulence determinants that, based on sequence characteristics such as G + C composition and codon usage, appear to be of heterologous origin, resulting from horizontal transfer) also appear to be able to promote conjugative transfer. Conjugative transfer systems can vary in complexity. Plasmids of *Streptomyces* species do not encode a complex surface to promote fusion between donor and recipient bacteria. *Streptomyces* grow as mycelia, and when two such colonies grow into the same space fusion between different organisms may occur at reasonable frequency, allowing plasmids to move from one cytoplasm to another. The major protein needed to promote this process is an ATPase which may bind to the DNA at the pore between the two cytoplasms and help to translocate the DNA past the membrane junction. One or more additional proteins are needed to help the plasmid spread throughout the mycelium of the new host.

A more sophisticated transfer mechanism found in other gram-positive bacteria is the pheromone-induced transfer system found in Enterococci. Adhesion between donor and recipient bacteria is promoted by specific exported proteins encoded by the plasmid, although no surface appendage such as a pilus is made. Normally, these transfer genes are switched off and only switched on in the presence of potential recipients which are sensed because they produce pheromones (short hydrophobic peptides) that accumulate in the medium. Donors import the pheromone by an oligopeptide uptake system and this causes the induction of the transfer genes. A range of pheromones are produced naturally by Enterococci. Each plasmid responds to just one pheromone. On entry into a new host the plasmid inhibits production of its cognate pheromone so that it can no longer act as a recipient. The plasmid also produces an antagonistic peptide that ensures that low levels of the pheromone still inside the cell do not switch the plasmid into transfer mode. These plasmids are therefore highly specialized to transfer between strains of Enterococci.

The best studied conjugative plasmids are those from gram-negative bacteria, belonging either to the IncF group or the IncP-1 group (Fig. 69.5). The IncF plasmids include the sex factor F from *E. coli*, which was probably noticed for its ability to promote genetic exchange between different mutant strains because its transfer genes are switched on continuously as a result of a mutation that knocks out a key regulator of these genes. Approximately 40 genes are needed for conjugative transfer, and a large proportion of these are involved in producing the apparatus in the bacterial envelope that promotes formation of pairs between donor and recipient bacterium. A key part of this apparatus is a long, thin, flexible protein tube called the pilus. This makes the initial contact with a suitable recipient and then appears to disassemble from the base, bringing the two bacteria into a close contact that is sufficiently stable to maintain the pairing in unshaken liquid culture. This seems particularly suited to promote transfer in a fluid environment such as the animal gut. In contrast, the IncP-1 plasmids make a short, rigid pilus that is more prevalent in detached form in the liquid surrounding the bacteria than attached to the bacterial surface. Conjugative transfer of plasmids encoding these sort of pili is very inefficient in liquid but occurs at high frequency on solid surfaces—agar in the laboratory and probably in biofilms on soil particles or on stone surfaces in lakes or rivers.

The role of pili is not resolved, although the consensus of opinion is in favor of the pilus facilitating contact between bacteria, which then leads to localized membrane fusion, degradation of a small area of the cell wall, and formation of a pore through which the DNA can move by a process that is still not understood properly. That the apparatus is focused on assembly of this pore rather than simply with assembly of a pilus is suggested by the fact that many proteins encoded as part of plasmid mating pair apparatuses show similarity to protein export systems as well as to more obviously analogous systems, such as those that promote DNA uptake in naturally competent bacteria. Nevertheless, some researchers are convinced that the pilus is a protein tube down which DNA can travel from one bacterium to another, protected from the environment. Recently, new plastic materials have been bored with uniform diameter holes just big enough to accommodate a pilus, and these have been used to separate donor and recipient bacteria in tightly sealed containers. Although

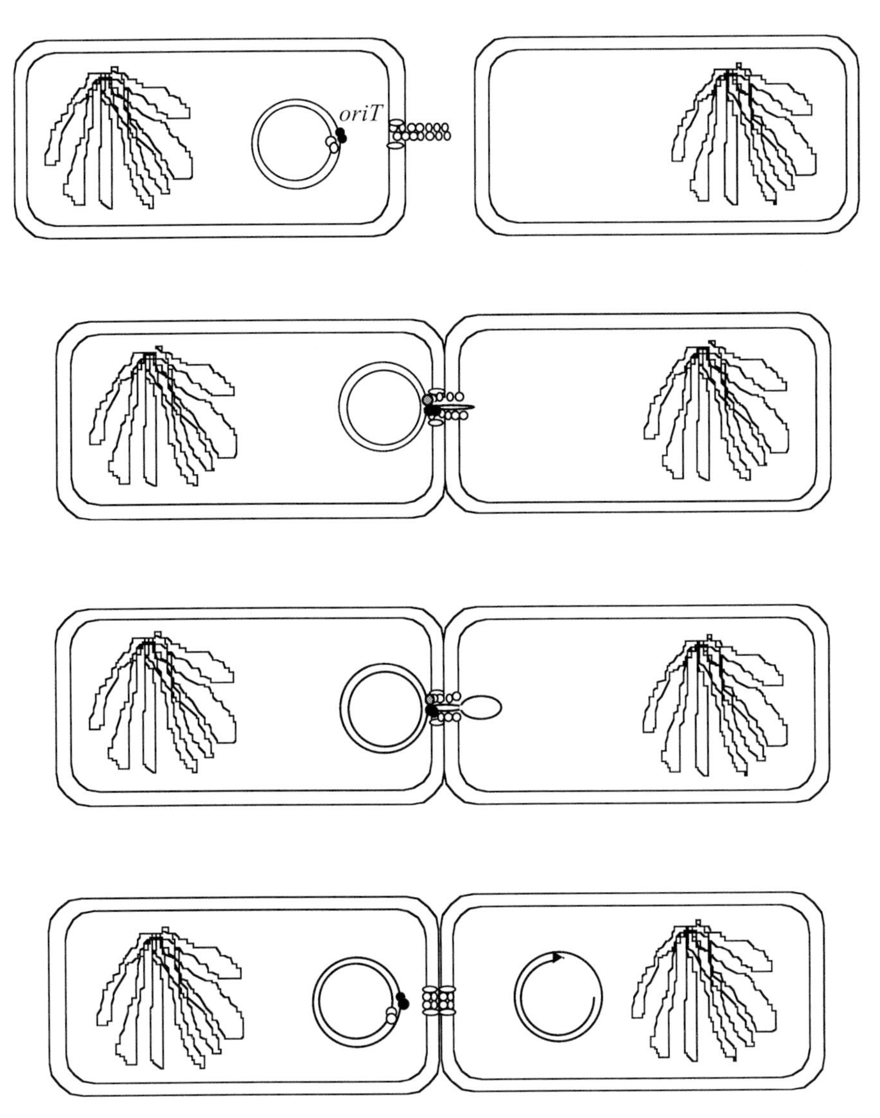

FIGURE 69.5 The transfer cycle established for conjugative plasmids of *E. coli*. A plasmid-encoded pilus (shown here consisting of two types of protein components) on the donor bacterium makes contact with the surface of the recipient and promotes contact which leads to local fusion and pore formation. Plasmid DNA carrying a relaxasome (a protein complex at *oriT*; small open and closed circles) associates with the pore, and rolling circle replication (see Fig. 69.2) is initiated, extruding a single-stranded loop into the recipient. When the replication process reaches *oriT* again, the donor molecule is reformed and a ssDNA circle is released into the recipient. Primase initiates the conversion back to dsDNA. The small circles and elipses do not imply the actual structure of any one system but rather are used to indicate a possible structure of any system.

recombinant bacteria appeared to be obtained, these experiments have failed to sway the general view that the pilus is normally just a means for selecting a mate and moving close to it. To prevent donors from trying to mate with bacteria already in possession of a related plasmid, most plasmids produce one or more proteins that are localized to the bacterial surface and which prevent the surface contacts from becoming more than a temporary association (surface exclusion).

The other essential component of these conjugative transfer apparatuses is the relaxasome, a nucleic acid–protein complex at a specific site on the plasmid which introduces a nick into one strand of the plasmid DNA and inititates the rolling circle replication process that results in the transfer of a copy of the plasmid into the recipient bacterium. This complex is called a relaxasome because plasmid DNA normally exists in a supercoiled and thus stressed state. When the complex at *oriT* introduces a nick, it dissipates this stress because if the proteins are stripped off then the strands become free to rotate about each other and the plasmid becomes relaxed. Relaxasomes on different plasmids vary in the number of proteins associated, but three is average. Probably the most important of these proteins is the relaxase, which introduces the nick into *oriT* and becomes covalently joined to the 5′ side of the gap so that the 3′ OH can serve as the primer for elongation and that the relaxase is in a position to reform a circular plasmid molecule when the replication process passes the *oriT* sequence again. In many respects this process is very similar to the replication of RCR plasmids described previously; indeed, sequence motifs in both the *oriT* region and in the relaxase protein are conserved in *dso* and Rep sequences, respectively, of RCR plasmids. There are also features in common with the transfer of DNA from bacteria to plant cells promoted by the Ti plasmid of *Agrobacterium tumefaciens*. The relaxasome is linked to the Mpf apparatus by an adapter protein (TraG in the case of IncP-1 plasmids) that determines the specificity of the system and may relay to the relaxasome the signal that the bacterium has docked with a potential recipient.

Some plasmids are unable to transfer on their own but can do so in the presence of a specific conjugative plasmid. They are said to be mobilizable. An essential requirement is that they have an *oriT* which can attract and function with the Tra proteins of the conjugative plasmid or that they have their own Mob functions and these can link to the transfer apparatus of the conjugative plasmid by the TraG-like adapter protein. The discovery of the role of the specific adapter protein has led to much greater understanding of the interactions between different transfer systems and mobilizable plasmids.

Expression of transfer genes is controlled, but in different systems this has different consequences for transfer frequency. Transfer systems related to F (but not F itself) are controlled very tightly so that they are normally switched off. A key element of this negative control is an antisense RNA (FinP), which provides the regulatory specificity, and an accessory protein FinO, which promotes the binding of the antisense RNA to its target. These control circuits respond to a variety of physiological stimuli and at low frequency generate transfer-proficient bacteria. After transfer there is a delay before FinO and FinP accumulate to repressing levels, and so new transconjugants are transfer proficient. As long as there are potential recipients available, this cycle will continue, creating a wave of transfer through the plasmid-negative population. Eventually, when there are no more potential recipients the transfer genes will get switched off. Ti and related plasmids of *Agrobacterium* species, as well as symbiosis plasmids of *Rhizobium* species, control their transfer genes in response to donor cell density. Plasmid-positive bacteria manufacture a homoserine lactone at low levels; however, when this accumulates because there are many bacteria, feedback occurs and the transfer genes are switched on. For Ti plasmids a second stimulus is needed—the presence of the complex amino acids that are manufactured by the plant tumors generated by the plasmids. The transfer of T-DNA to plant cells is controlled by a chemical signal (acetosyringone) which is synthesized in response to wounding, a necessary prerequisite for invasion by the bacterium.

An alternative general strategy adopted by some plasmids is to have one or more autogenous circuits which allow the transfer genes to be expressed continuously but kept at a low level once the transfer apparatus has been assembled so that the plasmid is as little a burden on its host as possible. In this way, the plasmid can always transfer at high frequency but the host growth rate is not burdened. The IncP-1 plasmids are the best studied example of this type of strategy.

D. Plasmid establishment in a new host

Many factors may limit the efficiency with which a plasmid establishes itself after transfer. The most obvious of these is degradation of the incoming DNA by restriction endonucleases if the DNA is not appropriately modified. Although it has been suggested that transfer by conjugation through a single-stranded DNA intermediate somehow allows a plasmid to avoid restriction systems, until recently there was no indication regarding how this might occur. Comparison of the frequency of transfer into strains with and without restriction shows very clearly that restriction systems do reduce transfer. When additional restriction sites are introduced into such a plasmid, the difference is even more dramatic. Plasmids have evolved two strategies to cope with this problem. One strategy is illustrated by the broad host range plasmids of the IncP-1 group which can transfer between almost all gram-negative bacterial species and in so doing must clearly encounter diverse restriction barriers. Examination of the complete nucleotide sequence of these plasmids showed that they are depleted for restriction targets for those restriction systems found in the *Pseudomonas* species, which are thought to be their natural hosts. This depletion could easily occur by point mutation because a base substitution at any one of six positions should abolish cutting of the DNA by the cognate restriction enzyme.

A second strategy is to allow transfer replication to deliver multiple copies of the plasmid to the recipient so that the restriction system is effectively saturated. A third, more complicated strategy exhibited by some plasmids is to encode a protein that interferes with the restriction system in the recipient. The *ardA* gene encodes such a protein which blocks the type I restriction systems encoded by enteric bacteria. Clearly, it would be important for such a protein to work before the plasmid DNA has been converted into a double-stranded form that would be sensitive to restriction. Unless the plasmid carries the protein across with it, this poses a problem because gene expression does not normally occur on single-stranded DNA. The solution is similar to that found for priming DNA replication on the lagging strand of rolling circle replication plasmids: Extended, imperfect inverted repeats in the single-stranded DNA form double-stranded hairpins that are recognized by RNA polymerase and this initiates transcription to express the genes in the leading region of the plasmid—the segment that enters the recipient bacterium first. Therefore, the protection afforded by *ardA* can only be effective if the plasmid goes through a single-stranded phase—that is, it protects during conjugative transfer but not after simple uptake of double-stranded DNA during transformation or transduction.

VIII. CHROMOSOME MOBILIZATION AND GENE ACQUISITION

One of the most important properties of plasmids is their ability to move genes from one bacterium to another without the need for recombination to allow the genes to be established in their new host. Continuous acquisition of new genes by plasmids is important to provide the variation in the plasmid population on which selection can act, as bacterial populations experience changing conditions. Transposable elements provide much of this gene movement and at the same time can promote interactions between different genomes. Transposition of an element from the chromosome to a plasmid can lead to the formation of a cointegrate in which the plasmid becomes temporarily a part of the chromosome and as such can promote conjugative movement of chromosomal DNA from one bacterium to another if the conditions are favorable for plasmid transfer. Alternatively, the presence of homologous transposable elements in the chromosome and in the plasmid can allow recombination between these repeated elements to achieve the same effect (cointegrate formation). The F plasmid contains multiple copies of various common insertion sequences which allow it to integrate into the chromosome in this way. (By chance one of these elements is inserted into the *finO* gene that is needed for repression of the transfer genes.) This results in a high frequency of transfer, and if the plasmid integrates into the chromosome then one also observes a high frequency of chromosome mobilizing ability (Cma). This property led to the discovery of F, the first plasmid identified. Although Cma of plasmids may be an unselected consequence of the conjugative transfer apparatus that has evolved for advantage of the plasmid, Cma illustrates the importance of plasmids as elements which promote the adaptability and diversity of bacteria.

BIBLIOGRAPHY

Bingle, L. E., and Thomas, C. M. (2001). Regulatory circuits for plasmid survival. *Curr. Opin. Microbiol.* **4**, 194–200.

Chattoraj, D. K. (2000). Control of plasmid DNA replication by iterons: no longer paradoxical. *Mol. Microbiol.* **37**, 467–476.

Clewell, D. B. (ed.) (1993). "Bacterial Conjugation." Plenum, New York.

del Solar, G., and Espinosa, M. (2000). Plasmid copy number control: an ever-growing story. *Mol. Microbiol.* **37**, 492–500.

Gerdes, K., Moller-Jensen, J., and Bugge Jensen, R. (2000). Plasmid and chromosome partitioning: surprises from phylogeny. *Mol. Microbiol.* **37**, 455–466.

Hiraga, S. (2000). Dynamic localization of bacterial and plasmid chromosomes. *Annu. Rev. Genet.* **34**, 21–59.

Khan, S. A. (2000). Plasmid rolling-circle replication: recent developments. *Mol. Microbiol.* **37**, 477–484.

Moller-Jensen, J., Jensen, R. B., and Gerdes, K. (2000). Plasmid and chromosome segregation in prokaryotes. *Trends Microbiol.* **8**, 313–320.

Summers, D. (1996). "Bacterial Plasmids." Blackwell, Oxford.

Thomas, C. M. (ed.) (2000). "The Horizontal Gene Pool: Bacterial Plasmids and Gene Spread." Harwood Academic. Reading, UK.

Thomas, C. M. (2000). Paradigms of plasmid organization. *Mol. Microbiol.* **37**, 485–491.

WEBSITES

List of genome plasmids NCBI
http://www.ncbi.nlm.nih.gov/PMGifs/Genomes/o.html

Pedro's Biomolecular Research Tools. A collection of Links to Information and Services Useful to Molecular Biologists
http://www.public.iastate.edu/~pedro/research_tools.html

Comprehensive Microbial Resource of The Institute for Genomic Research (TIGR) and links to many other microbial genomic sites
http://www.tigr.org/tigr-scripts/CMR2/CMRHomePage.spl

Plasmid genome database
http://molbiol.ox.ac.uk/%7Edfield/cgi-bin/Plasmid/

70

Polymerase chain reaction (PCR)

Carol J. Palmer
University of Florida

Christine Paszko-Kolva
Accelerated Technology Laboratories, Inc.

GLOSSARY

carryover In molecular biology, describing the amplification product of a previous PCR reaction that is carried to another PCR reaction, potentially leading to a false positive result.

DNA hybridization The process by which two complementary strands of DNA bind to each other in a mixture of DNA strands.

infectious agent Any bacterium, virus, protozoan, prion, or other agent that can cause infection or disease in a human, animal, or other host.

PCR (polymerase chain reaction) An *in vitro* method for the enzymatic synthesis of specific DNA sequences, using two oligonucleotide primers that hybridize to opposite strands and flank the region of interest in the target DNA. Following a series of repetitive cycles that involve template denaturation, primer annealing, and the extension of the annealed primers by *Taq* polymerase, there is an exponential accumulation of the specific target DNA.

RT-PCR (reverse transcriptase polymerase chain reaction) A technique similar to conventional PCR, except that the starting material is RNA rather than DNA. Because of this, a DNA copy must first be made from the RNA, utilizing an enzyme known as reverse transcriptase. Once the copy of the DNA is made, the PCR proceeds as usual.

Taq polymerase A thermostable DNA polymerase that was isolated from a bacteria *Thermus aquaticus*. This enzyme allowed PCR to be easily automated. Due to the heat stability of the enzyme, fresh enzyme no longer had to be added after each amplification cycle.

The Polymerase Chain Reaction has proven to be a powerful new diagnostic tool for microbiologists. The technology is based on repeated cycles of enzymatic amplification of small quantities of specific DNA or RNA sequences in target organisms until a threshold signal for detection is obtained. This technology has been applied to microbiology via direct detection of microorganisms, detection of genes that code for virulence factors, identification of the presence of genes responsible for antimicrobial resistance, and typing of bacterial isolates in epidemiological investigations.

I. INTRODUCTION

There are various methods and approaches that can be taken to identify microorganisms. Bacteria are perhaps among the easiest to identify since they can be subjected to numerous detection methods, ranging from direct staining (a variety of differential stains are available), culture on artificial media, (agar or tissue culture), serological techniques, or a range of molecular methods (PCR, DNA hybridization, or direct sequencing). Many of the same methods used to identify bacteria can also be applied to the identification of protozoans and

The Desk Encyclopedia of Microbiology
ISBN: 0-12-621361-5

viruses with minor modifications. The focus of this article will be a review of the polymerase chain reaction (PCR) and the advances that have been made in microbial detection since the invention of the polymerase chain reaction.

Current detection methods date back to the beginnings of microbiology and utilize growth on artificial media, often impregnated with differential dyes, or various broths from pre-enrichment to selective pre-enrichment. These methods often require days to weeks to obtain visible growth and suspect colonies still need to undergo further confirmatory biochemical testing. Newer methods involve the use of both polyclonal and monoclonal antibodies, either alone or in conjunction with a separation matrix, such as magnetic beads, following a pre-enrichment culture step. Although these techniques often represent significant time savings over traditional cultural methods, they may lack sensitivity and specificity. Because of such shortcomings, there is a tremendous need for rapid, accurate, and sensitive pathogen detection methods. Molecular methods provide the answer in many cases. Quite often, molecular methods are used in conjunction with classical methods to obtain results even faster. The remainder of this section will discuss the application of polymerase chain reaction (PCR) technology to two divisions of microbiology: infectious disease and environmental analysis. In addition, a recent improvement in PCR, known as the 5′ nuclease assay, as it is applied to both clinical and environmental microbial identification, will be discussed.

II. PCR TECHNOLOGY

Since its discovery in the 1980s by Kary Mullis, the PCR has been modified and enhanced into a powerful tool for the genetic detection of agents of infectious disease, sequencing of entire genomes of microorganisms (as well as the ongoing sequencing of the human genome), environmental analysis, forensic science, detection of food-borne pathogens, and identification of specific DNA markers in individuals. New applications of this exciting technology will continue to appear in the literature as the imaginations of scientists around the world continue to exploit and modify the basic PCR technique.

The PCR requires two synthetic oligonucleotide primers that are complementary to regions on opposite strands of a target piece of DNA, a target sequence in a DNA sample that occurs between the pair of primers, a thermostable DNA polymerase and the four deoxyribonucleotides. The procedure, performed in a themocycler, usually consists of three basic cycles and is completed within minutes. The first step is denaturation, which increases the temperature within the sample vial to about 94 °C, causing the double-stranded dDNA within the sample to separate into two pieces. The second step, renaturation, is completed by dropping the temperature within the thermocycler to about 55 °C, which allows the primers to anneal with their complementary sequences in the source DNA. The final step is the synthesis portion of the reaction, wherein the temperature is raised to about 74 °C, the optimum temperature for the catalytic functioning of Taq DNA polymerase. In this step, target DNA is extended, replicating to form additional copies of the target DNA. Within 30 cycles, over a million copies of the original target DNA can be reproduced.

III. DIAGNOSIS OF INFECTIOUS DISEASE

Initially, applications of PCR to infectious disease diagnostics focused on microorganisms that were impossible or slow to grow in culture, were difficult to cultivate, or posed significant health hazards using standard recovery techniques. Examples of these types of organisms include *Mycobacterium tuberculosis, Helicobacter pylori, Chlamydia trachomatis,* and HIV. The PCR has also played a key role in the identification of new or reemerging diseases, such as hantavirus and cyclospora, and has also been applied to retrospective analysis of samples from outbreaks or deaths caused by previously unidentified etiological agents.

Another area in the application of PCR to infectious disease is related to the rapid global expansion of HIV and the need to monitor new drugs in its treatment. This has led to the advent of discoveries and automation both in tracking the disease and in therapy. Patients infected with HIV can now have automated analysis of precise HIV viral loads, which can inform the clinician as to which stage of the disease the patient is in and how they are responding to antiviral therapy. This has been particularly useful with the introduction of the new protease inhibitors. Clinicians can quickly assess the impact of these new antiviral drugs on the course of HIV disease by monitoring viral load.

IV. TROPICAL INFECTIOUS DISEASES

More recently, diagnosis of tropical diseases has received significant attention in relation to PCR evaluated for the detection of *E. coli* O157:H7 in ground beef and other food samples. This represents a major advance in diagnostic microbiology and, as, in most cases, will undoubtedly be used for clinical applications first, before the technology is widely accepted for environmental applications.

The future of PCR remains bright and the various modifications will continue to lead to more user-friendly, integrated, high throughput pathogen detection systems that will translate into a better quality for life for us all. This quality improvement will come from a protected food/water supply and more rapid clinical diagnosis. This technology will be exploited in environmental and diagnostic microbiology, and numerous other fields.

BIBLIOGRAPHY

Bachoon, D. S., Chen, F., and Hodson, R. E. (2001). RNA recovery and detection of mRNA by RT-PCR from preserved prokaryotic samples. *FEMS Microbiol. Lett.*

Erlich, H. A. (ed.) (1989). "PCR Technology: Principles and Applications for DNA Amplification." Stockton Press.

Ferre, F. (1994). Polymerase chain reaction and HIV. *Clin. Lab. Med.* **14**, 313–333.

Freeman, W. M., Walker, S. J., and Vrana, K. E. (1999). Quantitative RT-PCR: pitfalls and potential. *Biotechniques* **26**, 112–122, 124–125.

Halford, W. P. (1999). The essential prerequisites for quantitative RT-PCR. *Nat. Biotechnol.* **17**, 835.

Hill, W. E. (1996). The polymerase chain reaction: Applications for the detection of foodborne pathogens. *Crit. Rev. Food Sci. Nutr.* **36**, 123–173.

Hussong, D., Colwell, R. R., O'Brien, M., Weiss, A. D., Pearson, A. D., Weiner, R. M., and Burge, W. D. (1987). Viable *Legionella pneumophila* not detectable by culture on agar media. *Biotech.* **5**, 947–950.

Innis, M. A., Gelfand, D. H., Sninsky, J. J., and White, T. J. (1990). "PCR Protocols: A Guide to Methods and Applications." Academic Press, San Diego.

Lambolez, B., and Rossier, J. (2000). Quantitative RT-PCR. *Nat. Biotechnol.* **18**, 5.

Persing, D. H., Smith, T. F., Tenover, F. C., and White, T. J. (1993). Diagnostic Molecular Microbiology: Principles and Applications. American Society for Microbiology, Washington, DC.

Pillai, S. D. (1997). Rapid molecular detection of microbial pathogens: Breakthroughs and challenges. *Arch. Virol. Suppl.* 1367–1382.

Prediger, E. A. (2001). Quantitating mRNAs with relative and competitive RT-PCR. *Methods Mol. Biol.* **160**, 49–63.

Rabinow, P. (1996). "Making PCR: A Story of Biotechnology." The University of Chicago Press, Chicago.

Stokes, N. A., and Burreson, E. M. (1995). A sensitive and specific DNA probe for the oyster pathogen *Haplosporidium nelsoni*. *J. Eukaryot. Microbiol.* **42**, 350–357.

Thaker, V. (1999). In situ RT-PCR and hybridization techniques. *Methods Mol. Biol.* **115**, 379–402.

Walliker, D. (1994). The role of molecular genetics in field studies on malaria parasites. *Int. J. Parasitol.* **24**, 799–808.

WEBSITES

PCR Guide, Weitzmann Institute—Genome and Informatics
http://bioinformatics.weizmann.ac.il/mb/bioguide/pcr/contents.html

DNA and PCR Protocols—US Department of Commerce (many links)
http://micro.nwfsc.noaa.gov/protocols/

The Polymerase Chain Reaction Jump Station
http://www.highveld.com/pcr.html

71

Prions

Christine Musahl and Adriano Aguzzi
University of Zurich

GLOSSARY

amyloid plaques Characteristic aggregated fibrils with high beta-sheet content, stainable by a dye and pH indicator called Congo red (CR). CR intercalates with a universal structure shared by all amyloids, which consists of antiparallel beta sheet extensions arranged in a quasi-crystalline fashion.

knockout mice Genetically engineered mice lacking the ability to express a specific protein because the gene of interest was ablated.

PrP^C Cellular, normal host prion protein.

PrP^{Sc} Scrapie-associated, pathological, potentially "infectious" prion protein.

PrP protein; prion Cellular protein of unknown function, primarily expressed on cells of the central and peripheral nervous system, as well as on lymphocytes. PrP is expressed in at least two different isoforms, one of which is thought to be the infectious agent of transmissible spongiform encephalopathies (TSEs).

reactive gliosis Strong proliferation of reactive astrocytes as seen by staining with antibodies against glial fibrillary acidic protein (GFAP). While not specific for TSEs, gliosis is extremely prominent in CJD and scrapie.

spongiform changes Highly characteristic hallmark of most TSEs. Formation of vacuoles (microscopic holes within cells) in the gray matter of the brain, giving it a spongelike appearance.

transgenic mice Genetically modified mouse strains, expressing at least one foreign or altered gene.

transmissible spongiform encephalopathies (TSEs) Creutzfeldt–Jakob disease (CJD), Gerstmann–Sträussler–Scheinker syndrome (GSS), fatal familial insomnia (FFI), bovine spongiform encephalopathy (BSE), and scrapie are the most common transmissible neurodegenerative diseases.

The Prion is defined as the infectious agent that causes transmissible spongiform encephalopathies (TSEs). It has been the subject of exciting discoveries and passionate controversies over the last 40 years. A large body of evidence indicates that prions do not contain informational nucleic acids. In 1996, S. B. Prusiner was awarded the Nobel Prize for Medicine for the hypothesis that prions consist of a modified form of the normal cellular protein called PrP^C.

Although the human forms of prion diseases are rare, the recent epidemic of bovine spongiform encephalopathy (BSE) in the UK has most dramatically raised the issue of transmissibility of these diseases from affected animals to humans.

I. MOLECULAR BIOLOGY OF PRIONS

A. The Prnp gene

The gene encoding the cellular isoform of the prion protein (PrP^C, or simply PrP) is located on the short arm of chromosome 20 in humans and on chromosome 2 in the mouse. It was termed *Prnp* (in the mouse) or

The Desk Encyclopedia of Microbiology
ISBN: 0-12-621361-5

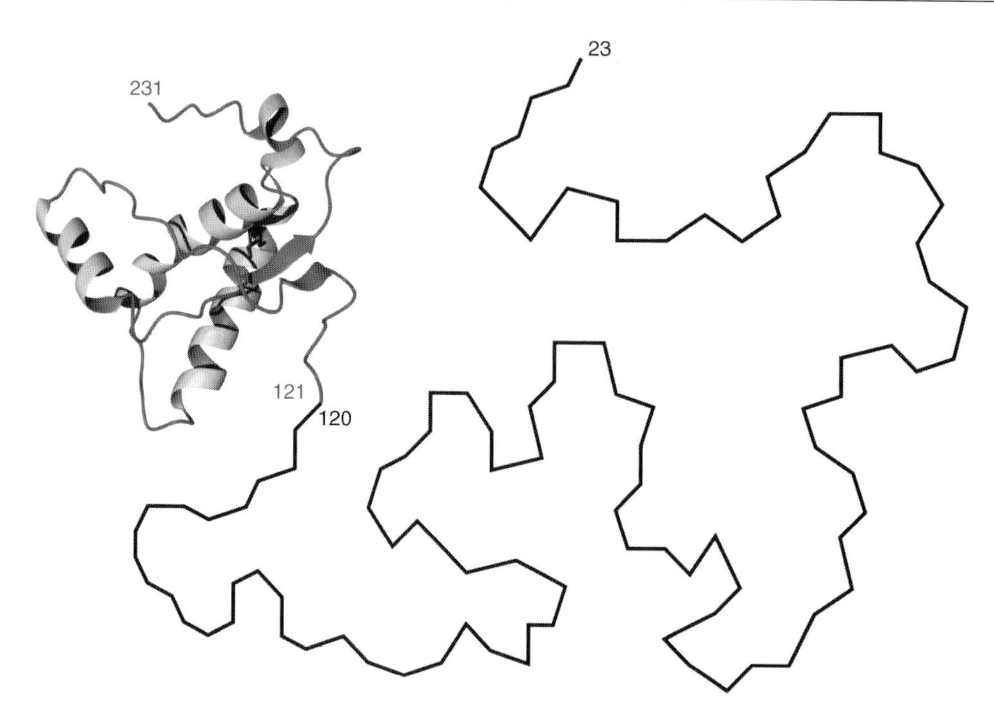

FIGURE 71.2 Three-dimensional structure of recombinant prion protein produced in *E. coli*, which corresponds to the residues 121–231 of mouse PrP^C. The structure was obtained by nuclear magnetic resonance (NMR) analysis. Analysis of a full-length recombinant mouse PrP by the groups of R. Glockshuber and K. Wüthrich shows that the N terminus (amino acids 23–120, not shown) is extremely flexible, unstructured, and devoid of α-helix or β-sheet structure.

of a normal cellular protein (Fig. 71.3). In 1982, Stanley Prusiner proposed that prions would not contain nucleic acids and are identical with PrP^{Sc}, which is the main proteinaceous constituent of the amyloid plaques found in many TSEs. Enrichment for infectivity from scrapie-infected hamster brains by Prusiner and colleagues led to the identification of PrP 27–30 (the protease resistant core of PrP^{Sc}, named according to its molecular weight of 27–30 kDa) copurifying with infectivity. In a collaboration with Prusiner, Leroy Hood, and Charles Weissmann were then able to determine parts of the sequence of the hamster prion protein in 1984. Oesch and colleagues, in 1985, cloned the cognate gene, which turned out to be encoded by the host genome. The normal cellular protein encoded by the *Prnp* gene was termed PrP^C (c_cellular). These substantial discoveries allowed the prion hypothesis to be laid on firm scientific ground.

PrP^{Sc}, the partially protease-resistant isoform of PrP^C, can be detected by digesting protein extracts derived from infected brains with proteinase K and subjecting them to Western blot analysis. Following protease treatment, PrP^{Sc} looses 67 N-terminal amino acids, giving rise to PrP 27–30, which tends to aggregate into insoluble polymers. PrP^{Sc} was shown to accumulate intracellularly in cytoplasmic organelles and in the extracellular space in form of amyloid plaques. The isolation of fractions enriched for infectious materials from the brain of scrapie-sick animals leads to a enrichment of PrP^{Sc}. Conversely, the isolation of PrP^{Sc} using affinity chromatography leads to enrichment of the infectivity. A further landmark speaking in favor of the protein only hypothesis is the demonstration by Hsiao and colleagues of linkage between familial forms of the disease GSS and mutations in the prion gene.

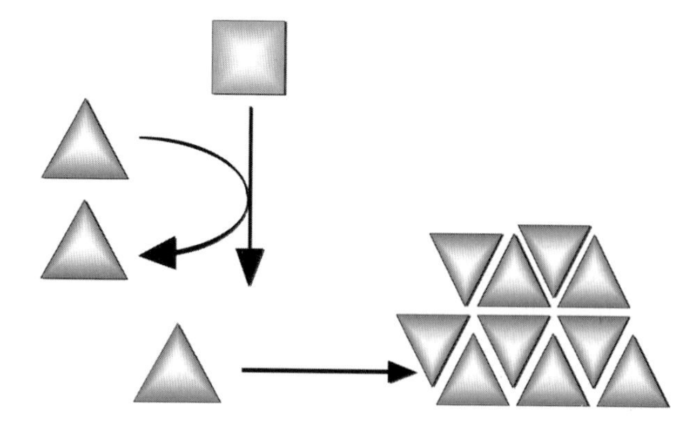

FIGURE 71.3 The Protein-only Hypothesis. According to this theory, each PrP^C molecule (rectangle) is converted into the PrP^{Sc} conformation upon interaction with PrP^{Sc} (triangle). Newly converted PrP^{Sc} is then able to convert further molecules and tends to form insoluble aggregates.

The molar ratio between the number of infectious particles and the number of PrP^{Sc} molecules in infectious brain extracts is usually in excess of $1:10^5$. This fact raises great analytical problems: it might be very difficult to purify the infectious agent if the latter consisted of subspecies of PrP^{Sc} or of a molecular modification of the PrP^C molecule which differs from what was operationally defined as PrP^{Sc}.

A. Mice devoid of PrP

The protein only hypothesis proposes that PrP^{Sc}, when introduced into a normal host, causes the conversion of PrP^C into PrP^{Sc}. Therefore, an animal devoid of PrP^C should be resistant to prion diseases.

Homozygous *Prnp*$^{o/o}$ ("PrP knockout") mice, first generated by the group of C. Weissmann in 1992, are viable and remain free of scrapie for at least 2 yrs after inoculation with prions. Also, there is no multiplication of prions in the *Prnp*$^{o/o}$ animals. Accordingly, heterozygous *Prnp*$^{o/+}$ mice, which express PrP^C at about half the normal level, show delayed development of disease when exposed to prions. Introduction of murine PrP transgenes into *Prnp*$^{o/o}$ mice restores susceptibility to mouse scrapie.

A synaptic phenomenon called *long term potentiation* (LTP), which is presumed to be important for short-term memory and learning, is apparently impaired in homozygous *Prnp*$^{o/o}$ mice. In addition, *Prnp*$^{o/o}$ mice exhibit aberrant sleep patterns and degeneration of cerebellar neurons in age, but it is unclear whether these observations are causally related to PrP^C. A different strain of *Prnp* knockout-mice (generated independently) was reported to show ataxia and Purkinje cell degeneration developing after 70 weeks of age. However, since no such phenotype was observed in other PrP-knockout mice generated by Büeler *et al.*, it remains unclear whether this observation is related to a function of PrP or, rather, absence of neighboring regulatory elements within noncoding portions of the *Prnp* locus.

B. The virus and virino hypotheses

The virino hypothesis has been proposed as an alternative to the protein-only hypothesis. It states that the infectious agent consists of viral nucleic acids complexed to host-derived PrP. This theory claims that PrP can be recruited by the viral nucleic acids as a sort of coat. The existence of many different strains of prions has been put forward as an argument in favor of this hypothesis. These strains retain their phenotypic characteristics even if they are propagated serially in one and the same inbred strain of susceptible experimental animals (mice, hamsters, or even minks).

On the other hand, no proof for the existence of a viral nucleic acid in infectious extracts was brought about yet. Finally, it has been convincingly demonstrated by Riesner and colleagues in 1991 that nucleic acid longer than 50–100 nucleotides cannot possibly be important for the infectivity of prion fractions. This finding is supported by the demonstration of the extraordinary resistance of the infectious agent toward treatments that damage or degrade nucleic acids, as shown by inactivation experiments with ultraviolet and ionizing radiation and the very low molar ratio between nucleic acids and infectious units in highly purified prion fractions.

C. Prion strains and the species barrier

The existence of prion strains is a formidable intellectual obstacle for the protein only hypothesis, and no really satisfactory explanation has been found for this phenomenon to date.

Infectious material from different sources can produce distinct and reproducible patterns of incubation time, distribution of CNS involvement, and of proteolytic cleavage of PrP^{Sc}. These properties are retained even after several passages in isogenic mice. In contrast to pathogens with a nucleic acid genome, prions have to be able to encipher strain-specific properties in the tertiary structure of PrP^{Sc}. Indeed, there seem to be conformational differences in PrP^{Sc} that correlate with such behavior. For example, the TSE strains hyper (HY) and drowsy (DY) describe the behavior of the minks affected. They are associated with characteristic incubation time, locations of neuropathology, and characteristic degrees of proteinase K susceptibility of the respective PrP^{Sc}, including partial treatment producing different NH_2 termini.

An example of strain properties in human prion diseases shows that differences in the amino-acid sequence of PrP are not necessary to produce different strains: The human disease fatal familial insomnia (FFI), associated with the mutation D178N, shows a proteinase K-resistant PrP^{Sc} species of 19 kDa after deglycosylation. In contrast, both familial and sporadic Creutzfeldt–Jakob disease (CJD), associated with the same mutation, show a 21 kDa species. Inoculation of the respective human brain homogenates into PrP-deficient mice expressing a chimeric mouse–human mouse (MHuM) PrP transgene, produced disease associated with the respectively sized PrP^{Sc}. This suggests that the two distinct PrP^{Sc} species can act as template upon a single primary MHuM PrP structure and impart onto it their own respective conformation.

The different PrP conformations could represent either different tertiary structures or different

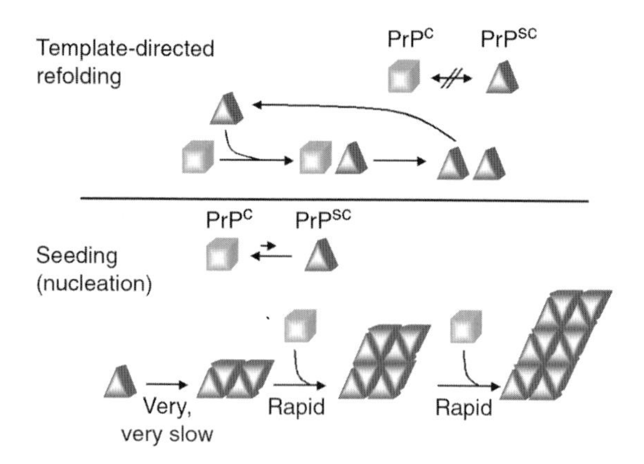

FIGURE 71.5 Two models for conversion of PrP^C to PrP^{Sc}. In the template-directed model, kinetically inaccessible PrP^{Sc} (triangle) is more stable than PrP^C (rectangle). PrP^{Sc} facilitates conversion by catalyzing the rearrangement of PrP^C. According to the seeding hypothesis, conversion between PrP^C and PrP^{Sc} is reversible, but monomeric PrP^{Sc} is less stable than PrP^C. PrP^{Sc} aggregates, once formed, can grow by recruiting further monomeric PrP^{Sc} from the solution.

molecules. According to this template assistance model, the genetically inherited and spontaneous diseases result from mutations that increase the rate at which mutant PrP^C spontaneously converts to PrP^{Sc} and/or enhance the number of intermediate molecules.

In a second proposed mechanism, formation of PrP^{Sc} is a nucleation-dependent polymerization. The conversion between PrP^C and PrP^{Sc} is reversible in the absence of a preexisting aggregate, with a PrP^{Sc} monomer being less stable than PrP^C. PrP^{Sc} polymers, however, might bind to and stabilize the PrP^{Sc} conformation, thereby promoting the conversion of PrP^C. The initial nucleation process, thus, represents the barrier to a stable conversion, since formation of low-order aggregates is not favored until a minimum size nucleus is attained.

The requirement for the formation of a stable nucleus before conversion's being accomplished predicts certain characteristics of the aggregation process, such as kinetics displaying a *lag* phase and a critical protein concentration-dependence for the initial formation of aggregates. The *in vitro* conversion process, as shown by Caughey and colleagues in 1995, seems to show such features. The observation that fractions containing high-order PrP^{Sc} aggregates (>300 kDa) can mediate the conversion to protease resistance in *in vitro* reactions, while smaller-sized fractions do not, can be explained by the relatively large size of the minimum stable nucleus, that would tend to make such a particle insoluble. Infection would, thus, circumvent the slow, thermodynamically unlikely step of nucleation by introducing a "seed" that initiates aggregation.

There are biological examples for both proposed mechanisms. In the case of the template-assisted conversion mechanism, PrP^C has a conformation that does not spontaneously form the more stable PrP^{Sc}. A number of proteins have been observed to be separated from their true free energy minima by a large barrier. These include influenza hemagglutinin and a number of proteases, including subtilisin and α-lytic protease. In the case of α-lytic protease, the conversion from a globulelike intermediate, I, to the native state, N, is extremely slow, with little or no conversion over a period of a month. But if the pro-peptide region (a naturally occurring polypeptide cleaved from the same translation product as α-lytic protease during its maturation) is bound in either cis or trans, conversion is dramatically accelerated.

There are also examples for the nucleation–polymerization. There is a resemblance, for example, to tubulin polymerization or bacterial flagellar polymerization. The soluble monomer flagellin is incorporated into the growing end of a flagellum. In liquid phase, flagellin units are unable to spontaneously polymerize even at nearly millimolar concentrations. If a seed of fragmented flagellum is placed into the mixture, polymerization rapidly takes place. The polymerizing monomers can even adopt the conformation of heterologous seed material, indicating a "templating" behavior.

It is important to mention that there could be a hybrid mechanism. The surface of an PrP^{Sc} aggregate, initially formed by a nucleation process, could catalyze the conformational change of PrP^C monomers.

III. YEAST PRIONS

Psi^+, a trait of the budding yeast, *Saccharomyces cerevisiae*, was first described by Brian Cox in 1965. The efficiency at which a certain suppressor-strain could misread UAA stop codons as sense is dependent upon the presence of a non-Mendelian factor, named [PSI^+]. Wickner suggested in 1994 that [PSI^+] is a prion form of the Sup35 protein (Sup35p). He proposed that, in [psi^-] cells, the conformation of Sup35p is fully functional ($Sup35p^{Psi^-}$) and promotes efficient termination at stop codons (we know now that Sup35p codes for the translational release factor eRF3). In [PSI^+] cells, some or all of the Sup35p is proposed to take on a new, biologically inactive conformation ($Sup35p^{PSI^+}$), leading to less efficient termination and, thus, nonsense suppression. The SUP35 aggregates appear to act as a nucleus, which, similarly to the seed of a crystal, promotes the aggregation of newly synthesized SUP35 protein and allows the propagation of the PSI+ state in a manner analogous to the propagation of prions. There is now evidence that Sup35p exists in different structural states in [*PSI*$^+$] and [*psi*$^-$] cells. Sup35p in lysates of [*PSI*$^+$], but not [*psi*$^-$] strains, shows increased protease resistance and aggregation, two characteristics typical of vertebrate prions. These results can be interpreted in terms of the seeded nucleation model; however, the initial rate of unseeded fiber nucleation is not as dependent upon the concentration of soluble Sup35p monomers as predicted by the original polymerization model.

The chaperone protein Hsp 104 is known to facilitate the folding of proteins and is required for the propagation of [*PSI*$^+$]. Both deletion or overexpression of Hsp 104 result in disappearance of the SUP35 aggregates and loss of the PSI+ state. The fact that the impairment of Sup35p, either by Mendelian mutations in *SUP35* or by the presence of [*PSI*$^+$], causes similar phenotypes is consistent with the prion model. A multicopy plasmid carrying *SUP35* efficiently induces the *de novo* appearance of [*PSI*$^+$]. This can be interpreted as evidence for the prion model, since the *SUP35* overexpression increases the probability that a Sup35p^{psi-} molecule would take on a prion shape by chance. It has also been shown that [*PSI*$^+$] can reappear, arguing against the possibility that "curing" is due to the loss of a cytoplasmic nucleic acid with no nuclear master gene. Furthermore, a dominant mutation, which causes the loss of [*PSI*$^+$], *PNM2* (*P*si-*N*o-*M*ore) is an allele of *SUP35* with a missense mutation.

Another cytoplasmically inherited genetic element in yeast, [Ure 3], appears to propagate by a similar mechanism, as shown by Wickner and Lindquist. There is reason to believe that the behavior of these unconventional genetic elements points to a molecular process that is broadly distributed and functions in a wide variety of biological contexts. Unfortunately, in the case of PrP conversion, no chaperone component (like Hsp 104) has been identified in the cellular components where conversion appears to occur. Therefore, it still remains to be verified whether the mammalian PrPSc can propagate itself through a mechanism analogous to those of the yeast prions.

IV. PRION DISEASES

A. A historical perspective

Scrapie is a naturally occurring disease which has been known to infect sheep in Great Britain for at least 250 years. In 1936, Cuille and Chelle showed that the disease could be transmitted via inoculation of a homogenate of spinal cord from sick animals into the brain of healthy sheep and goats. These findings have been confirmed by the experiments of Gordon in 1946: Inoculated animals developed a disease identical to that spontaneously occurring in sheep. In further experiments, atypical properties of the infectious agent were demonstrated, such as a very long incubation period, a remarkable resistance towards inactivation with high temperatures, treatment with alkalizing agents, aldehydes, and UV light irradiation. These properties were later on defined as some of the central features of prions.

In humans, a group of diseases with similar pathological characteristics was described to occur sporadically or in a hereditary fashion (Fig. 71.6). These diseases include Creutzfeldt–Jakob disease, GSS syndrome as well as an infectious disease called kuru, which affected the Fore population in the northern provinces of New Guinea. Already 35 years ago, William Hadlow drew the attention of scientists to the fact that these diseases may represent a particular form of scrapie in humans. These speculations of Hadlow were later confirmed in inoculation experiments performed by D. Carleton Gajdusek and colleagues. In these seminal experiments, Gajdusek accomplished transmission of kuru, and also of CJD, to chimpanzees. For these experiments Gajdusek was awarded the Nobel Prize for Medicine in 1976.

In the last several years, a new disease was found to affect cows: bovine spongiform encephalopathy (BSE). It is very likely that feeding of cows with rendered offal derived from meat and bones of scrapie-infected sheep was responsible for the transmission of scrapie and for the outbreak of BSE.

B. Transmissible spongiform encephalopathies (TSE)

The neuropathological hallmarks of TSEs, or prion diseases, are spongiform changes, astrocytic gliosis, neuronal loss, and PrP-positive plaques (Fig. 71.7). Unfortunately, until now, diagnosis of TSE could be reliably made only postmortem. Although the human forms of these diseases are rare, the epidemic proportions of the bovine spongiform encephalopathy (BSE) forced researchers worldwide to urgently reconsider the question of transmissibility to humans.

1. *Animal prion diseases*

Sheep scrapie is the prototype of the growing group of TSEs. The typical symptoms of scrapie-sick sheep include hyperexcitability, pruritus (a chronic itching of the skin, leading to the scraping behavior), and myoclonus (a brief, sudden, singular, shocklike

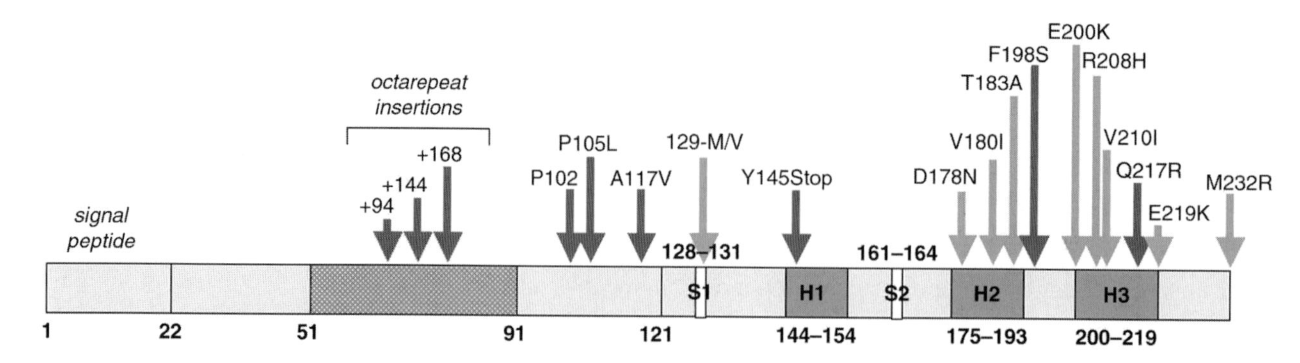

FIGURE 71.6 Mutations and polymorphisms associated with the human hereditary prion diseases. The positions of the octarepeats, α-helices (H1-3) and β-sheet (S1, S2) conformations are indicated. Mutations correlated with plaque-type deposits in GSS, spastic paraparesis, and Alzheimer-like phenotype are indicated with dark arrows. Mutations correlated with synaptic deposits found in sporadic and familial CJD, as well as FFI, are shown with grey arrows. The position of the M/V polymorphism at codon 129 is indicated. It seems that amino-proximal mutations induce a GSS-like phenotype, while most mutations close to the carboxyl terminus result in a CJD-like phenotype. Based on drawings published by S. B. Prusiner and by J. Tateishi.

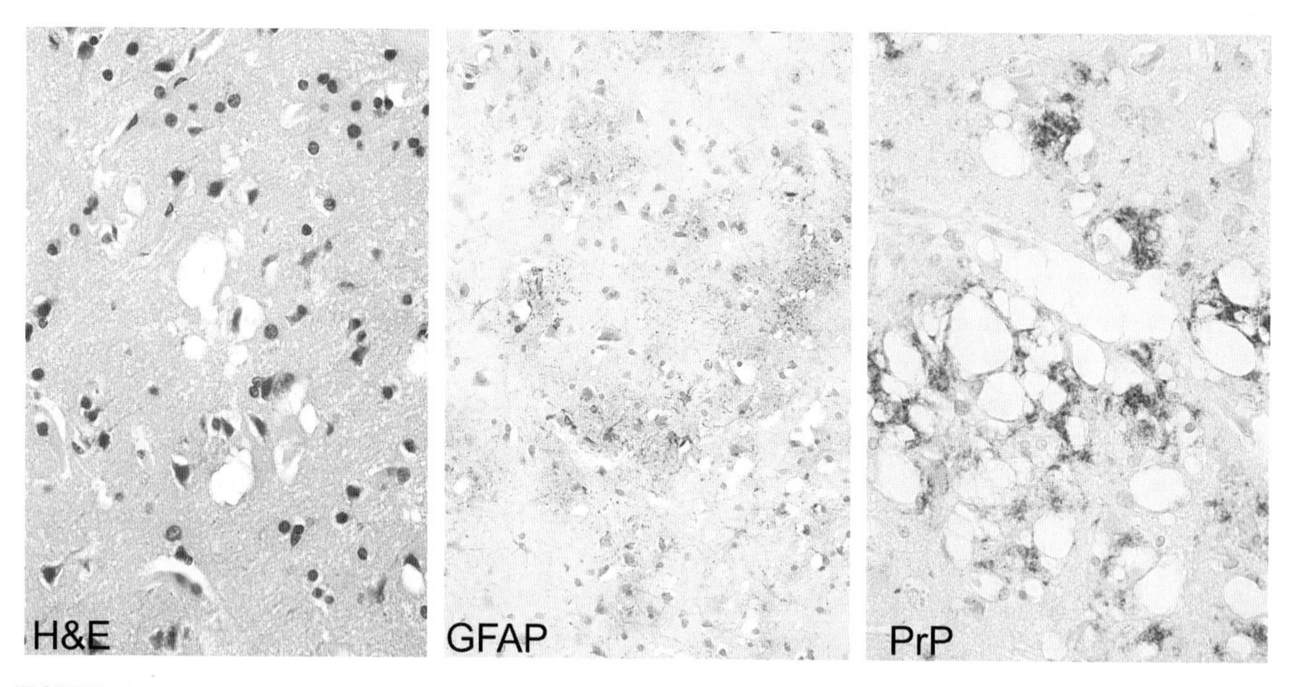

FIGURE 71.7 Characteristic features of Creutzfeldt–Jakob disease. Hematoxylin–eosin stain (left) shows the typical vacuoles in the brain of a CJD patient, which leads to the spongiform appearance. Proliferation of reactive astrocytes is visualized by staining with antibodies against glial fibrillary acidic protein (GFAP, middle). PrP protein deposits are shown with anti-PrP immunostaining (Plate 10).

muscle contraction). The disease is characterized by a rapid progression, which leads to tetraparesis (paralysis of all four legs) and, ultimately, to the death of the affected animal.

The clinical symptoms of BSE are insidious and consist of behavioral changes (including aggressive behavior, which is proverbially atypical in cows) and uncoordinated gait. Histologically, the brain exhibits lesions similar to those of TSEs in other species. It is not excluded that cases of BSE may have been seen in England as early as 1985, but probably not before that year, despite earlier anecdotal reports. The numbers of cases continued to increase, provoking a major epidemic, and peaked in summer 1992.

It has been suggested early on that one common exposure of cattle to prions came about through the use of a dietary protein supplement, meat and bone meal (MBM), that was regularly fed after weaning. No credible alternative hypothesis has been put forward on the origin of the BSE epidemic, and the incidence of new cases has been precipitously declining some 4–5 years after a more-or-less effective ban of MBM was put in place.

2. *Human prion diseases*

Since the first description by A. M. Jakob and H. G. Creutzfeldt, five human diseases have been identified as TSEs. The disease bearing the authors' names, Creutzfeldt–Jakob disease, occurs sporadically, may be transmitted, and has a genetic basis in 10–15% of all cases. Creutzfeldt–Jakob disease generally presents as a progressive dementia.

Other genetic diseases are GSS) syndrome and fatal familial insomnia (FFI). GSS is transmitted by autosomal dominant inheritance (Fig. 71.6). It is characterized by missense mutations of *PRNP*, associated with specific neuropathological lesions, and by multicentric amyloid plaques. The latter can be labeled by antibodies directed against the prion protein. This restrictive definition justifies retaining the name of GSS syndrome and excludes observations of hereditary prion diseases without multicentric amyloid plaques, as well as sporadic forms with multicentric plaques. It has been possible to transmit GSS to chimpanzees and to mice. An additional member of the group of spongiform encephalopathies, fatal familial insomnia (FFI), was also transmitted experimentally in at least three instances to mice. Kuru is a form of spongiform encephalopathies transmitted by ritual cannibalism, which affected the Fore population in the northern provinces of New Guinea.

The newest form of CJD in humans, new variant CJD (nvCJD), was first described in 1996 and has been considered evidence for a link between human TSEs and BSE. nvCJD has a distinct pathology characterized by abundant "florid plaques," decorated by a daisylike pattern of vacuolation. The age of onset in nvCJD is much lower than in sporadic CJD. The notion that nvCJD could be transmitted from cattle to primates is supported by several arguments, including the observation that intracerebral inoculation of BSE-infected brain extracts into macaque monkeys produced disease and pathology resembling that in the nvCJD patients.

C. Pathology of prion diseases

The questions revolving around the nature of the prion are important and fascinating. Less glittering, but certainly not less important, is the question of how prions induce brain damage. Very little is understood about the mechanisms of the latter phenomenon. The accumulation of some protein in brain, by itself, does not necessarily explain the dire consequences of TSE on the brain of its host. They include a highly characteristic vacuolation and, eventually, death of nerve cells, activation of astrocytes and microglial cells to an extent unparalleled by other pathological conditions, and, invariably, lethal impairment of the electrical functions of the brain.

Characteristic lesions under the light microscope consist of spongiform changes in the neuropil nerve cells and astrocytes with nerve cell degeneration and astrocytosis. These changes are often observed in the gray matter of the cerebrum and cerebellum. The distribution of the histopathologic lesions may vary according to the strain of the agent, as has been shown with inbred strains of mice. It may also vary with the site of inoculation. If the infection reaches the brain through the optic nerve, the spongy degeneration is clustered around the occipital lobe. Amyloid plaques containing PrP may be seen between cells in some TSEs (e.g. kuru, hamster scrapie) and stained with Congo red (Fig. 71.7). Electron microscopy shows twiglike structures 12–16 nm in width and 100–500 nm long, which are found only in TSE and are now called scrapie-associated fibrils (SAF).

Whether the damage is brought about by accumulation of the pathological prion protein or, rather, by the abrupt withdrawal of its normal isoform during the course of the disease is not clear to date. The fact that PrP-deficient mice are reasonably healthy only apparently refutes the latter hypothesis, since such mice may have had the time to adapt to a prionless life from early on. The crucial experiment to settle this question may be a "conditional knockout" allowing abrupt shutoff of PrP^C expression in adult life.

If PrP^{Sc} deposition was the cardinal event in pathogenesis, why is it that, in many instances of human and experimental TSE, extremely little PrP^{Sc} can be detected, even in terminal disease? And, if PrP^{Sc} deposition is an important event at all, it would seem that it can exert deleterious effect only through some PrP^C mediated processes, since PrP-deficient nerve cells are not affected, even after long-term exposure to PrP^{Sc}.

While the prime target of damage seems to be neuronal, profound neuronal loss is not always seen in TSEs. Instead, astrocytic activation occurs very early and in an extremely consistent fashion. It can be easily reproduced *in vitro* and leads to significant physiological effects, such as impairment of the blood–brain barrier. Since astrocytes belong to the few cell types identified that are capable of supporting prion replication, elucidation of the role of this cell type in TSE pathogenesis will be an exciting task for the years to come.

Growing evidence incriminates another cell type in brain damage, not only in TSE, but in diseases ranging from Alzheimer's to multiple sclerosis and even stroke: microglial cells. *In vitro* experiments seem to indicate that activation of microglia may be quite pivotal in effecting neuronal damage in TSE, and that this phenomenon is dependent on expression of PrP^{C}. The details of this pathway of cell death, however, still escape our understanding, and the proof that these phenomena occur in the brain during the course of the disease (and not only in petri dishes with explanted cells) is still missing.

V. PERIPHERAL PATHOGENESIS

A. Prions and the immune system

It has long been observed that, even following intracerebral inoculation of mice with prions, there is early acquisition of infectivity in the spleen, long preceding any appearance of infectivity in the brain. Consistent with a primary replication step in the lymphoreticular system that favors neuroinvasion, SCID mice are relatively resistant to CNS disease following intraperitoneal inoculation.

Although prions are most effective when directly administered to the brain of their hosts, this situation occurs mainly in experimental lab work and does not reflect the reality of the common routes of prion infections. In humans, most cases of iatrogenic CJD transmission were traced to intramuscular injection of growth hormone and, to a lesser extent, pituitary gonadotropins. The other example of massive, efficient human-to-human transmission is New Guinea's kuru, where oral uptake of infectivity was accomplished in the course of cannibalistic rituals.

Therefore, peripherally administered prions can reach the brain of their host. This neurotropism is remarkable, especially since no pathologies can be identified in organs other than the CNS. In addition, it may be important to identify "reservoirs" in which prions multiply silently during the incubation phase of the disease.

One such reservoir of PrP^{Sc} or infectivity is doubtlessly the immune system, and a wealth of early studies points to the importance of prion replication in lymphoid organs. The nature of the cells supporting prion replication within the LRS, however, is still uncertain. Inoculation of various genetically modified immunodeficient mice lacking different components of the immune system with scrapie prions revealed that the lack of B-cells renders mice unsusceptible to experimental scrapie. While defects affecting only T-lymphocytes had no apparent effect, all mutations affecting differentiation and responses of B-lymphocytes prevented development of clinical scrapie. Since absence of B-cells and of antibodies correlates with severe defects in follicular dendritic cells (FDCs), the lack of any of these three components may prevent clinical scrapie. Yet mice expressing immunoglobulins exclusively of the IgM subclass, without detectable specificity for PrP^{C}, developed scrapie after peripheral inoculation: Therefore, differentiated B-cells seem crucial for neuroinvasion of scrapie, regardless of B-cell receptor specificity. FDCs have been incriminated, because PrP^{Sc} accumulates in FDCs of wild-type and nude mice (which suffer from a selective T-cell defect).

Repopulation of immunodeficient mice with fetal liver cells (FLCs) from either PrP-expressing or PrP-deficient mice and from T-cell deficient mice, but not from B-cell deficient mice, is equally efficient in restoring neuroinvasion after *i.p.* inoculation of scrapie prions. This suggests that cells whose maturation depends on B-cells or their products (such as FDCs) may enhance neuroinvasion. Alternatively, B-cells may transport prions to the nervous system by a PrP-independent mechanism.

We have recently learned that "Type 4"-PrP^{Sc}, one of the hallmarks of new variant CJD, accumulates in the lymphoid tissue of tonsil in such large amounts that it can easily be detected with antibodies on histological sections. Infectivity can accumulate also in intestinal Peyer's plaques, where it replicates almost immediately following oral administration of prions.

Immune cells are unlikely to transport the agent all the way from LRS to CNS, since prion replication occurs first in the CNS segments to which the sites of peripheral inoculation project. This implies that the agent spreads through the peripheral nervous system, analogously to rabies and herpes viruses.

B. Development of drugs against prions

While common neurodegenerative conditions, such as Alzheimer's disease, have attracted a very large mind share in the pharmaceutical industry, the same has not yet been the case for TSE. One unavoidable consideration is that the exceeding rarity of TSE in humans renders them an uneconomical target for therapeutical efforts. Sadly, it is not unlikely that the BSE epidemics may change the epidemiology of human TSE considerably.

It has been argued that the possibility that we may be witnessing an incipient nvCJD epidemic may not be attractive to pharmaceutical companies, since the former will be self-limited and will subside just as

BSE is doing now, with a delay mainly determined by the (hitherto unknown) incubation time of the disease in humans. This viewpoint is not necessarily true: the time periods needed for the development cycle of new drugs are rapidly falling, as impressively demonstrated by the crop of antiretroviral agents which are now hitting the market. Given that the peak of the BSE epidemic in humans may be reached in 10–20 years, it would seem quite urgent and important to start development of therapeutic agents.

Unfortunately, by the time the first signs and symptoms of disease are recognized, significant CNS damage is consistently present. It, therefore, appears that, in the case of BSE transmission, postexposure prophylaxis may provide a more viable alternative, just as treatment of presymptomatic HIV-infected individuals is more effective than treatment of terminally ill, neurologically impaired, and often demented patients with large retroviral burdens in their CNS.

Perhaps at some point, it will be possible to treat patients even after prions have reached the brain and made themselves apparent. Two classes of substances have raised some (limited) hope to this end: amphotericins, sulfated polyanions, and anthracyclin antibiotics. While the mechanisms of action of the former is totally unknown, the latter is thought to intercalate into the highly ordered structures of prion amyloid and to disrupt it. As exciting as this approach may sound, the published data solely indicate that the apparent virulence of hamster prion preparation was only reduced after pre-incubation with iododeoxyrubicine, which procedure can hardly be described as "treatment." It seems a safe prediction that structural data on PrP^{Sc} and its aggregated forms will be extremely important for designing more effective intercalating agents.

BIBLIOGRAPHY

Abdulla, Y. H. (2001). A plausible function of the prion protein: conjectures and a hypothesis. *Bioessays* **23**, 456–462.

Aguzzi, A., Klein, M. A., Montrasio, F., Pekarik, V., Brandner, S., Furukawa, H., Kaser, P., Rockl, C., and Glatzel, M. (2000). Prions: pathogenesis and reverse genetics. *Ann. NY Acad. Sci.* **920**, 140–157.

Aguzzi, A., and Weissmann, C. (1997). Prion research: The next frontiers. *Nature* **389**, 795–798.

Haltia, M. (2000). Human prion diseases. *Ann. Med.* **32**, 493–500.

Harris, D. A. (1999). "Prions: Molecular and Cellular Biology." Horizon Scientific Press.

Hope, J. (2000). Prions and neurodegenerative diseases. *Curr. Opin. Genet. Dev.* **10**, 568–574.

Masison, D. C., Edskes, H. K., Maddelein, M. L., Taylor, K. L., and Wickner, R. B. (2000). [URE3] and [PSI] are prions of yeast and evidence for new fungal prions. *Curr. Issues Mol. Biol.* **2**, 51–59.

Prusiner, S. B. (1996). Prions prions prions. *Curr. Top. Microbiol. Immun.* **207**. Springer-Verlag.

Prusiner, S. B. (1998). Prions. *Proc. Natl. Acad. Sci. U.S.A.* **95**(23), 13363–13383.

Prusiner, S. B. (2001). Shattuck lecture–neurodegenerative diseases and prions. *N. Engl. J. Med.* **344**, 1516–1526.

Reilly, C. E. (2000). Beta-sheet breaker peptides reverse conformation of pathogenic prion proteins. *J. Neurol.* **247**, 319–320.

Soto, C., and Saborio, G. P. (2001). Prions: disease propagation and disease therapy by conformational transmission. *Trends Mol. Med.* **7**, 109–114.

Wickner, R. B., Taylor, K. L., Edskes, H. K., Maddelein, M. L., Moriyama, H., and Roberts, B. T. (2000). Prions of yeast as heritable amyloidoses. *J. Struct. Biol.* **130**, 310–322.

WEBSITES

Website on Prions (S. Heaphy)
http://www-micro.msb.le.ac.uk/3035/prions.html

BSE and CJD Information and Resources (CDC)
http://www.cdc.gov/ncidod/diseases/cjd/cjd.htm

72

Protein secretion

Donald B. Oliver
Wesleyan University

Jorge Galan
Yale University

GLOSSARY

general secretion pathway Pathway by which proteins containing signal peptides or membrane signal–anchor domains are integrated into or transported across the cytoplasmic membrane of bacteria.

protein export or secretion Transport of proteins to extracytoplasmic compartments, including the exterior of the cell or to other cells.

sec-machinery or translocon Components that promote protein export via the general secretion pathway.

signal peptide or secretion signal A region of a secreted protein that contains information for interaction with the secretion machinery.

Protein Secretion, or protein export, is the movement of proteins from one cellular compartment to another or to the exterior of the cell. In bacteria, this involves the movement of presecretory and membrane proteins from the cytoplasm, their initial site of synthesis, to a number of aqueous or membranous compartments, depending on the type of bacterium under study (see Fig. 72.1). In recent years, it has become evident that a variety of pathogenic bacteria are also capable of secreting their proteins into animal and plant cells.

Bacteria possess several different pathways for promoting protein secretion, and these are generally classified according to the secretion machinery utilized for protein export (Table 72.1). The most commonly utilized pathway is the Sec-dependent or general secretion pathway (GSP). The GSP promotes protein translocation across the cytoplasmic membrane. It is driven by a collection of cytoplasmic, peripheral, and integral membrane proteins that are collectively known as the Sec machinery, or translocon. The GSP can be thought of as the major housekeeping protein secretion pathway of the bacterial cell. It is homologous to the pathway promoting import of presecretory and membrane proteins into the endoplasmic reticulum of eukaryotic cells. Certain gram-negative bacteria possess terminal branches of the GSP for secretion of proteins across the outer membrane. Besides the GSP, there are several other specialized protein secretion pathways that are utilized by more limited sets of proteins. These include (i) the twin-arginine protein secretion pathway that promotes the export of redox-cofactor-containing enzymes, (ii) ABC protein secretion pathways, that are driven by ABC (ATP-binding cassette) exporters, and promote the secretion of toxins, proteases, lipases, and specific peptides, and (iii) the type III or contact-dependent protein secretion pathways that promote binding and import of proteins into animal and plant cells. These latter two pathways are typically associated with bacterial pathogens and specialized microbial niches.

I. GENERAL SECRETION PATHWAY

A. Targeting signals on preproteins

The hallmark of a GSP-dependent presecretory protein (preprotein) is the possession of a characteristic signal peptide that is often located at its amino

The Desk Encyclopedia of Microbiology
ISBN: 0-12-621361-5

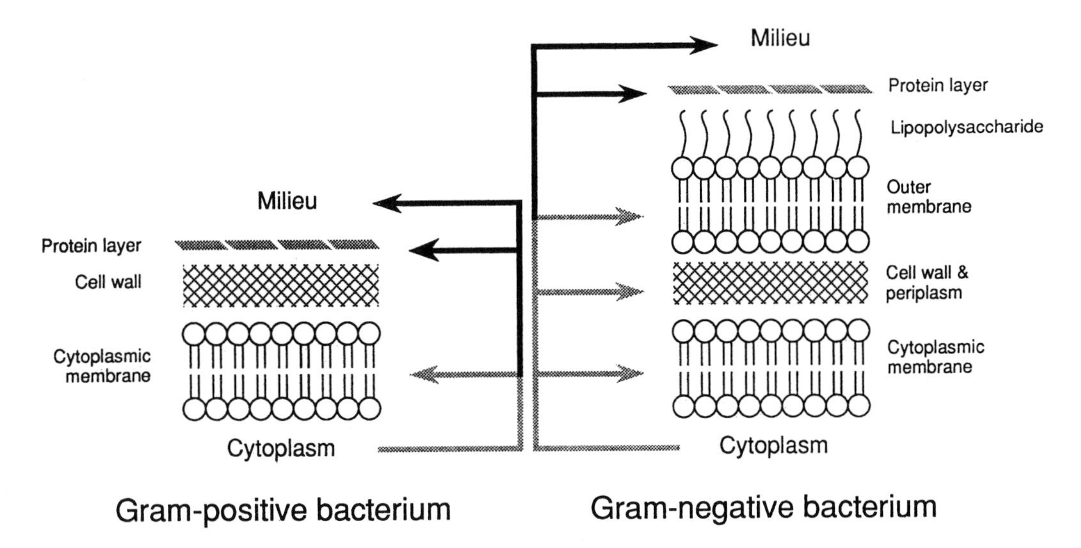

FIGURE 72.1 Destinations of secreted proteins in bacteria. Black arrows indicate destinations of extracellular secreted proteins, while shaded arrows indicate destinations of intracellular secreted proteins. [From Pugsley, A. (1993). *Microbiol. Rev.* **57**, 51.]

terminus (Fig. 72.2). Such signal peptides serve as recognition elements for the Sec-machinery, and they also delay the rate of protein folding, a potentially important feature, since protein export by the GSP must occur prior to the completion of protein folding. Secretory preproteins can interact with the Sec-machinery in either a cotranslational or posttransltional fashion. Sec-dependent signal peptides range in length from 18 to 30 amino acid residues (typically being longer in gram-positive bacteria). Although they possess little in the way of primary sequence homology, they do contain three characteristic regions: (i) a short, positively charged N-domain, (ii) a hydrophobic H-domain of 7 to 15 uncharged amino-acid residues, and (iii) a short, polar C-domain that contains a signal peptidase cleavage site. Two consensus cleavage motifs exist within the C-domain, one for general secretory proteins and one for lipoproteins. Lipoproteins need to be modified with diglyceride at a cysteine residue at the beginning of the mature region to allow signal peptide cleavage. Mutational studies show that the H-domain is the most critical for signal peptide function, and truncation or introduction of charged residues within the H-domain often prevents export of the protein. In contrast, the N-domain appears less critical, although its overall charge can affect the kinetics of protein secretion. Alteration of the C-domain may prevent signal peptidase cleavage; while the resulting preprotein is still secreted, it remains tethered to the cytoplasmic membrane by its uncleaved signal peptide. The signal peptide is thought to function as a loop, allowing insertion of an amino-terminal hairpin of the preprotein into the cytoplasmic membrane. Whether such insertion occurs directly into the lipid bilayer, by analogy to model systems, or is promoted by prior interaction with the translocon is unknown.

Additional recognition elements are contained within preproteins since signal peptide deletion still allows a low level of protein secretion to occur. Residual export can be greatly augmented by certain mutations in the Sec machinery that generally suppress signal peptide defects (see Prl following). One type of recognition element for chaperones that targets preproteins to the translocon appears to be exposed hydrophobic protein surfaces. It is thought that delayed folding kinetics of preproteins, with resultant hydrophobic surface exposure, facilitates preprotein interaction with specific chaperones for maintenance of the export-competent state. However, the structural basis for such recognition has remained largely illusory.

The final sorting of secretory proteins to the periplasm, cell wall, and outer membrane or protein layer occurs subsequent to translocation across the cytoplasmic membrane. The periplasm is probably a default pathway that requires no specific sorting signals to achieve proper targeting. By contrast, topogenic signals are required to promote association with the cell wall or outer membrane. The nature of these signals is only now being clarified for specific systems, and it is unclear, at present, how general such signals are. For outer membrane protein targeting, the possession of β-structure or the ability to bind to lipopolysaccharide appears to be important. In *E. coli* lipoproteins, the identity of the second mature

TABLE 72.1 Classification of bacterial protein secretion pathways

Pathway	Alternative name	Distribution	Translocation pathway	Protein substrates	Location of protein secretion signals	Machinery components
General secretion pathway	Sec-dependent pathway	Ubiquitous Bacteria and Archaea	Cytoplasmic membrane traversal or integration	Most periplasmic, cell wall, and outer membrane proteins, and some cytoplasmic membrane proteins	N-terminal signal peptide or signal-anchor domain, hydrophobic patches on molten globule state of preproteins?	Sec machinery or translocon
Main terminal branch of the general secretion pathway	Type II pathway	Limited Gram-negative animal and plant pathogens	Outer membrane traversal	Restricted number of hydrolytic enzymes and toxins	Patch signal(s) on tertiary or quaternary structure?	Type II exporter
Twin-arginine protein secretion pathway		Nearly ubiquitous? Bacteria and Archaea	Cytoplasmic membrane traversal or integration	Redox enzymes	N-terminal twin arginine signal peptide, patch signal(s) on tertiary structure?	TAT exporter
ABC protein secretion pathway	Type I pathway	Limited Bacteria and Archaea	Single step traversal of cytoplasmic and outer membrane (if present)	Restricted number of hydrolytic enzymes and toxins	Folded C-terminal secretion signal	ABC exporter
Type III protein secretion pathway	Contact-dependent protein secretion pathway	Limited Gram-negative animal and plant pathogens	Single step traversal of both bacterial and target cell membranes	Restricted number of agonists and antagonists of eukaryotic target cell responses	N-terminal secretion and targeting signals	Type III exporter

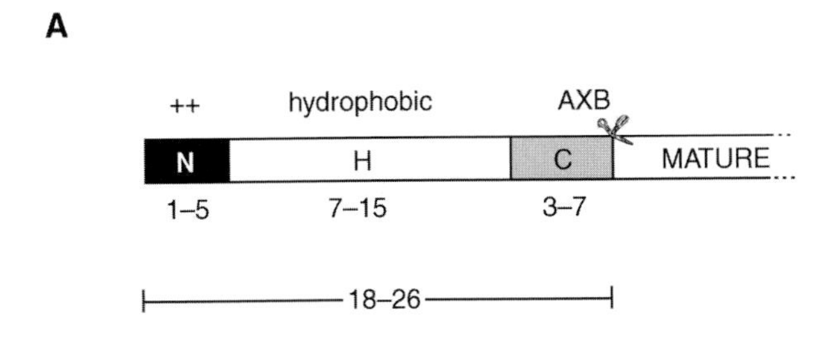

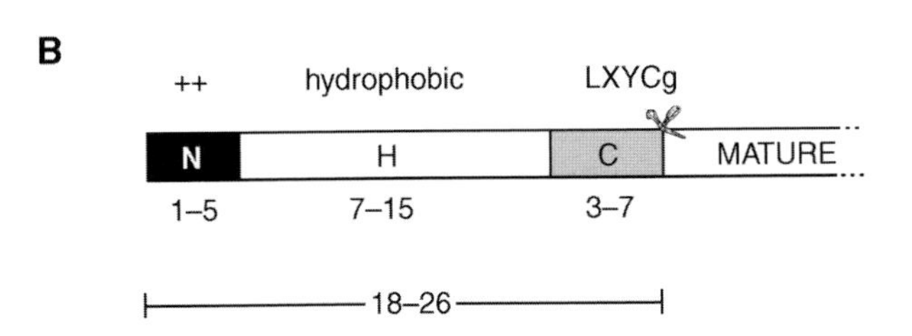

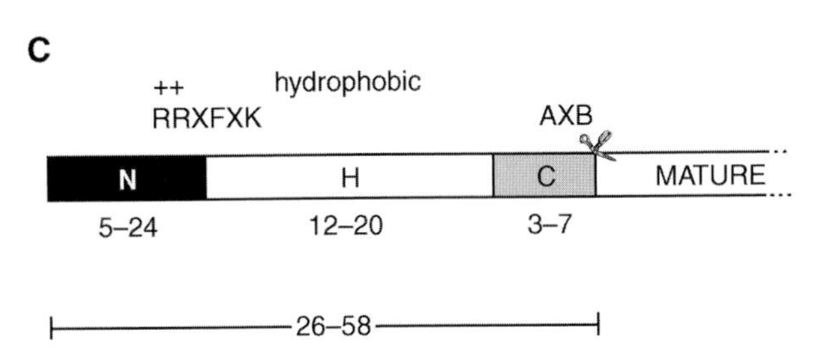

FIGURE 72.2 Structure of signal peptides. N-, H-, and C-domains for each signal peptide are indicated, along with their approximate length in amino acids. ++ indicates a net positive charge. The scissors indicate the signal peptidase cleavage site. (A) Signal peptide of a typical preprotein utilizing the GSP. The consensus cleavage motif AXB is indicated, where A indicates small uncharged or larger aliphatic amino acid residues, X indicates any residue, and B indicates small uncharged residues. (B) Signal peptide of a lipoprotein utilizing the GSP. The consensus cleavage motif LXYCg is indicated, where L indicates leucine, X and Y tend to be small uncharged amino-acid residues, and Cg indicates glyceryl-modified cysteine. (C) Signal peptide of a preprotein utilizing the twin-arginine secretion pathway. The twin-arginine motif RRXFXK is indicated, utilizing single amino-acid code, except X indicates any amino acid. The AXB cleavage motif is the same as in (A). [Adapted from Fekkes, P., and Driessen, A. (1999). *Microbiol. Mol. Biol. Rev.*]

amino-acid residue is critical for distinguishing inner from outer membrane destinations. In certain gram-positive bacteria, an LPXTGX motif (X indicates any amino acid residue), followed by a carboxy-terminal hydrophobic domain and charged tail, promotes anchoring of cell wall proteins.

Certain integral membrane proteins have been shown to utilize the GSP for their biogenesis, particularly when they contain large periplasmic domains. In this case, the signal-anchor domain (transmembrane segment) fulfills a function analogous to the signal peptide, except that no cleavage takes place. For complex polytopic membrane proteins, odd-numbered transmembrane segments serve to initiate membrane protein insertion, while even-numbered segments serve as membrane anchor domains to arrest translocation (Fig. 72.3). Insertion of an amino-terminal portion of a membrane protein in an N_{out}–C_{in} orientation (so-called amino-terminal tail insertion) has been shown to be independent of the GSP and may require a novel pathway promoted by a homolog of Oxalp, a protein responsible for amino-terminal tail insertion of mitochondrial membrane proteins.

B. Chaperones and targeting factors

As nascent or newly completed chains of presecretory and membrane proteins emerge from the ribosome, they encounter a variety of chaperones that maintain them in a unfolded state and target them to the translocon. While a number of chaperones have been shown to play at least minor roles in protein export (i.e., trigger factor, a ribosome-associated proline isomerase, and heat shock proteins GroE, DnaK, DnaJ, and GrpE), two major chaperones have been characterized in *E. coli*: SecB and signal recognition particle (SRP). SecB is present in gram-negative bacteria (its existence in gram-positive bacteria is uncertain), while SRP is conserved among all three domains.

SecB is a small tetrameric protein that binds a particular group of preproteins, either cotranslationally or posttranslationally, and targets them to the translocase through its specific interaction with SecA, a subunit of translocase. SecB is nonessential for cell viability, although its absence results in substantial secretion defects for SecB-dependent proteins. Unlike the more complex heat shock chaperones, SecB is not an ATPase, and it can only prevent protein folding, not reverse it. The structural basis of SecB-preprotein recognition is unclear, although it requires a hydrophobic patch on SecB. Of interest, this region overlaps with acidic residues that are the SecA-binding site on SecB. These results have lead to a model (Fig. 72.4) where preproteins are passed from SecB to SecA by a "hand-off" mechanism.

Bacteria possess an SRP particle and SRP receptor that is homologous to its eukaryotic counterparts and are essential for cell viability. The *E. coli* homologs of SRP54 and 7S RNA are Ffh or P48 and 4.5S RNA, respectively. FtsY is the *E. coli* homolog of eukaryotic SRP receptor α subunit, and no SRP receptor β subunit homolog has been identified yet. While there is no evidence for a protein translation arrest by bacterial SRP, the membrane targeting function of bacterial SRP appears to occur similarly to eukaryotic SRP. The interaction of Ffh with proteins must occur when they are still ribosome-associated, and it is dependent on

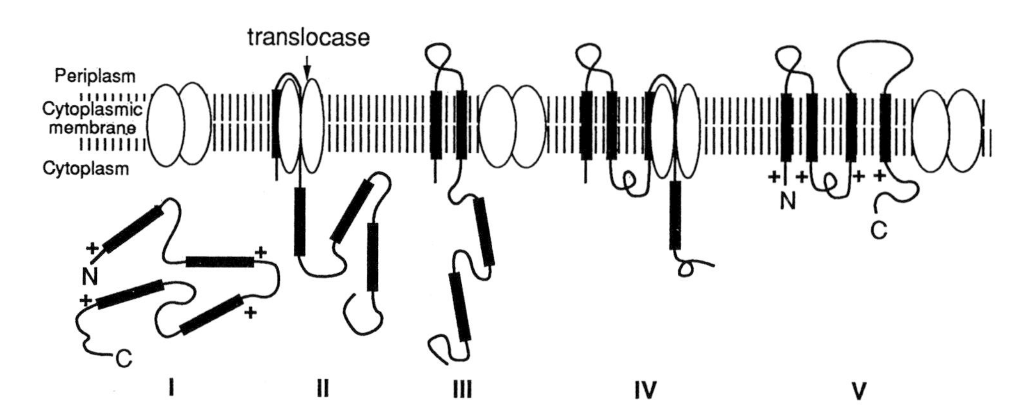

FIGURE 72.3 Sec-dependent insertion of integral membrane proteins utilizing signal-anchor domains. Signal-anchor domains are indicated by black rectangles and + indicates positively charged regions that help promote the correct membrane topology. I, II, III, IV, and V indicate successive stages of membrane protein integration. [From Pugsley, A. (1993). *Microbiol. Rev.* **57**, 70.]

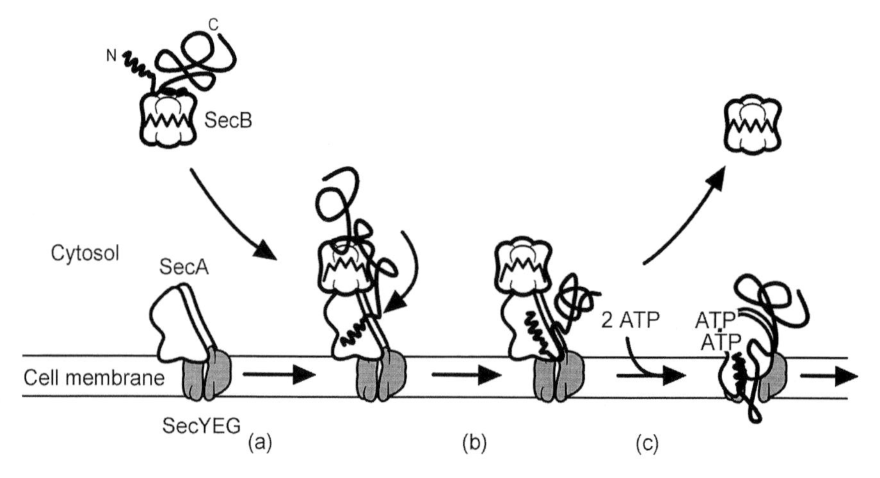

FIGURE 72.4 "Hand-off" model for preprotein transfer from SecB to SecA. (a) SecB targets the preprotein (wavy black line) to membrane bound SecA, where signal peptide binding to SecA stimulates interaction between SecA and SecB. (b) SecA–SecB interaction causes the release of the mature portion of the preprotein from SecB. The zigzag symbol represents the region of SecB that associates with both SecA and preprotein and which changes conformation during the hand-off. (c) Upon binding ATP, SecA changes conformation, resulting in initiation of protein translocation and release of SecB. [From Driessen *et al.* (1998). *Curr. Opin. Microbiol.* **1**, 218.]

the hydrophobicity of the signal peptide or signal-anchor domain. Of the three domains of Ffh (amino-terminal N domain, GTPase G domain, and methionine-rich M domain), the N and G domains appear to contain the signal peptide-binding region, while the M domain is involved in interaction with 4.5S RNA. Membrane-bound FtsY and GTP hydrolysis are required for signal peptide release from SRP. Upon release from SRP, presecretory or membrane proteins associate with the translocon. Although the details of this "hand-off" reaction are not yet known, it appears likely that the protein is targeted to SecA. While characterization of *E. coli* SRP–SRP receptor has proceeded well, considerable controversy surrounds the identity of the *in vivo* substrates of SRP. While SRP mutation or depletion affects somewhat the secretion of certain periplasmic and outer membrane proteins, recent data indicates that this system plays an important role in the biogenesis of particular integral membrane proteins. The effects of SRP depletion on periplasmic and outer membrane secretion may be indirect, since SecY protein requires SRP receptor (and presumably SRP) for its biogenesis.

C. The *E. coli* translocon

The bacterial translocon consists of core and accessory subunits (Fig. 72.5). The core structure is composed of two essential proteins: cytoplasmic membrane protein SecYE and membrane-dissociable SecA ATPase. This catalytic core of translocase has been shown to be required for both *in vivo* and *in vitro* protein translocation. Three accessory proteins have also been

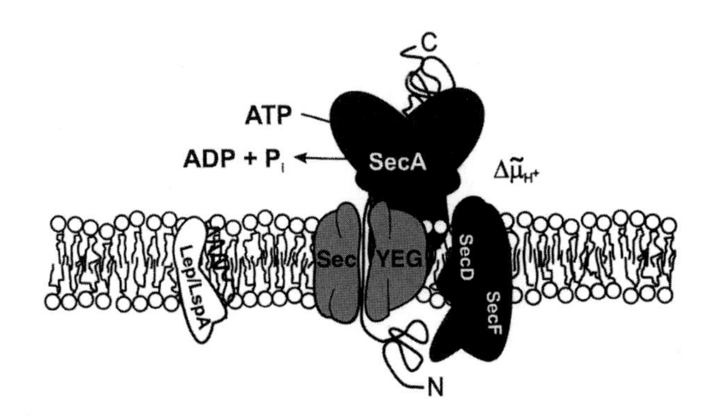

FIGURE 72.5 Structure of the translocon of *E. coli*. Core subunits SecA and SecYE are indicated, along with accessory subunits SecG, which associates with SecYE, and SecDF. A partially translocated preprotein (wavy black line) is shown as well as the two energy sources, ATP and $\Delta\mu_H^+$. Lep/LspA indicate signal peptidase I and II, respectively. [From Fekkes, P., and Driessen, A. (1998). *Microb. Rev.*]

characterized in some detail: integral membrane proteins SecD and SecF (which probably function together and are contained in a single protein in *B. subtilis* and certain other bacteria) and SecG protein. These accessory proteins are essential for cell growth only at low temperatures, and they enhance *in vivo* and *in vitro* protein translocation under a variety of conditions.

SecYE, in addition to constituting the SecA receptor, has been proposed to form a channel for protein translocation. This would be analogous to the Sec61 channel present in the endoplasmic reticular membrane of eukaryotes. SecYE/Sec61$\alpha\gamma$ homologs are present in all three domains of life. Bacterial SecY consists of 10 transmembrane domains with small cytoplasmic and periplasmic loops flanking these domains. By contrast, SecE protein is much smaller, consisting of only three transmembrane domains in *E. coli* (only one of which is essential) and a single transmembrane domain in *B. subtilis*.

SecA is a large dimeric protein present only in the bacterial domain. Most bacterial SecA proteins probably contain at least five domains: two ATP-binding domains, a preprotein-binding domain, and domains for binding SecB and SecYE proteins. This organization allows SecA to serve as an assembly factor for binding preproteins or SecB-bound preproteins to translocase. SecA interacts with preproteins by recognition of both signal peptide and mature portions, and the basic N-domain and hydrophobic H-domain of the signal peptide are important for binding. The order of assembly of the translocation complex has not been defined, although it appears likely that SecA binding to the cytoplasmic membrane and SecYE precedes the preprotein binding step.

SecA also serves as a motor protein to drive protein translocation, utilizing its translocation ATPase activity. This activity appears to correspond to successive cycles of membrane insertion and retraction by SecA protein, and it has been proposed that a mobile domain of SecA acts like a molecular sewing machine needle to drive protein translocation in a stepwise manner (Fig. 72.6). Cross-linking studies suggest that the translocating preprotein is contained in a channel that is formed by SecA and SecY proteins, which is shielded from lipid. Secretory proteins must be driven through the translocon to the periplasmic side of the membrane, while Sec-dependent membrane proteins undergo limited translocation, until a signal-anchor domain arrests further translocation and allows their release into the lipid bilayer. Little is known about how the translocon allows such lateral release of membrane proteins.

SecG is a small protein with two transmembrane domains that undergoes a cycle of membrane topology inversion during protein translocation. This cycle is coupled to SecA membrane cycling and enhances it in some way. Recent work suggests that SecG promotes a conformational change in SecA that allows it to insert into the membrane at SecYE.

It has been suggested that SecDF is important for the stabilization of membrane-inserted SecA, thereby preventing backward slippage of preproteins during the translocation cycle. However, this suggestion seems at odds with the observation that archaea lack a *secA* homolog but possess *secD* and *secF* homologs. Since SecD and SecF each possess a large periplasmic domain, it has been speculated that they may play a role late in protein secretion, such as in the folding or release of translocated proteins, their presentation to leader peptidase, or the disassembly and recycling of the Sec machinery.

Genetic studies suggest that the translocon has a proofreading activity that recognizes signal peptides on proteins. Mutations allowing export of preproteins with defective signal peptides (designated *prl* for protein localization) are alleles of *secA* (*prlD*), *secE* (*prlG*), *secG* (*prlH*), and *secY* (*prlA*). *prlA* mutants are the strongest suppressors, and they can efficiently secrete proteins that lack a signal peptide (although not cytoplasmic proteins). Genetic analysis of this system suggests that certain regions of SecY and SecE are involved in signal peptide recognition, and this step is regulated by SecA ATPase activity, which is likely to be utilized as a molecular clock to define the kinetic window for preprotein recognition.

D. Energetics and mechanism of protein translocation

Studies of protein translocation indicate that ATP hydrolysis is essential for *in vitro* protein translocation,

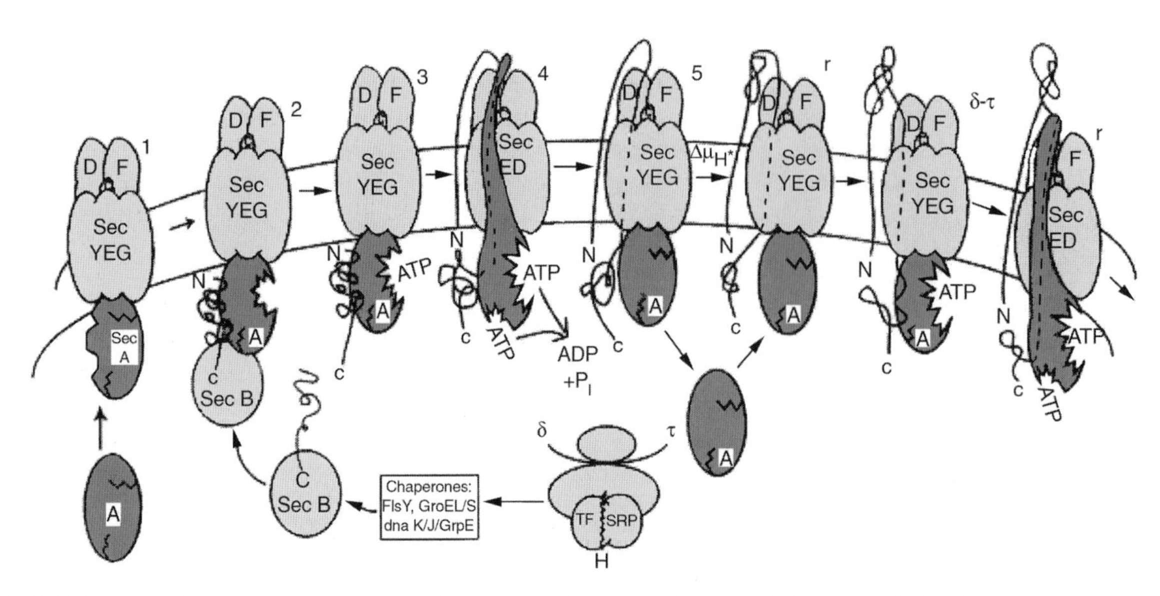

FIGURE 72.6 Proposed protein translocation cycle for *E. coli* GSP. Preprotein (wavy black line) and ATP-binding to SecA promote a cycle of SecA membrane insertion and retraction that drives translocation of the first segment of the preprotein (steps 1–4). A later step (step 5) in protein translocation is driven by PMF ($\Delta\mu_{H}^{+}$). In the lower portion of the figure, the role of SecB and other chaperones in delivering preprotein to SecA is depicted, along with the ability of SecYEG-bound SecA to exchange with cytosolic SecA. [From Duong *et al.* (1997). *Cell* **91**, 568.]

whereas PMF stimulates the rate of translocation of many preproteins. Hydrolyzable ATP serves to drive SecA membrane cycling, which is required for insertion of the initial loop of the preprotein into the translocator. Translocation of the remainder of the preprotein can occur by either SecA-generated force or PMF. The exact nature of PMF-dependent translocation is uncertain, although it is clear that PMF stimulates the activity of the translocon in some manner.

The energetics and mechanism of protein translocation in archaea is not known. Their SecY/Sec61α and SecE/Sec61γ homologs are more eukaryotic-like, and they lack any detectable SecA homolog. This suggests that protein translocation may be more similar to eukaryotes, where translation drives protein translocation via a tight junction between the ribosome and the translocon. Alternatively, PMF or an as-yet-undiscovered ATPase may drive protein secretion in this domain.

E. Distal steps during protein export

As preproteins emerge on the periplasmic side of the plasma membrane, they must undergo structural modifications that typically include signal peptide removal, protein folding, disulfide bond formation, and oligomerization. Studies reveal that there are a variety of protein catalysts that promote these events.

Signal peptidases have been characterized for general secretory proteins and lipoproteins. Signal or leader peptidase I enzymes cleave signal peptides from nonlipoproteins that utilize the GSP. These enzymes are membrane-anchored and contain a periplasmic catalytic domain that represents a novel type of serine protease, which appears to employ a serine–lysine dyad in the catalytic mechanism. Their substrate specificity is primarily determined by amino-acid residues at the -1 and -3 (relative to the cleavage site) positions of the preprotein, where small uncharged residues at -1, and small uncharged or larger aliphatic residues at -3, are required (see Fig. 72.2). Preprotein recognition is also an important element in catalysis, as the catalytic rate differs by many orders of magnitude for preprotein versus simple peptide substrates. *In vivo*, signal peptide processing occurs relatively late in protein secretion, and little is known about the interplay between the translocon and processing enzymes. Bacterial species differ in whether they possess one essential (e.g., *E. coli*) or multiple, redundant, signal peptidase I enzymes. In *B. subtilis*, for example, there are five closely related signal peptidase I enzymes, two of which are coordinately induced at the onset of maximal protein secretion.

Signal peptidase II enzymes are specific for lipoproteins that utilize the GSP. They are membrane-embedded enzymes that recognize a sequence around the processing site, Leu-X-Y * gCys or Leu-X-Y-Z * gCys (where X, Y, and Z tend to be small uncharged amino-acid residues, gCys indicates glycerylcysteine and * is the cleavage site) (see Fig. 72.2). The presence of the glyceride modification is a prerequisite for processing and, subsequent to cleavage, an additional fatty acid is added to the glycerylcysteine through an amide linkage. The antibiotic globomycin inhibits signal peptidase II enzymes. Subsequent to processing, lipoproteins may remain anchored in the cytoplasmic membrane, they may be transported to the outer membrane, or they may be released into the extracellular media, in certain instances. Transport to the outer membrane of gram-negative bacteria is accomplished by a transport pathway consisting of two components: Lo1A is a periplasmic chaperone that promotes lipoprotein release from the inner membrane, and Lo1B is an outer membrane receptor for Lo1A that binds the Lo1A–lipoprotein complex and promotes lipoprotein assembly into the outer membrane.

Secretory proteins and periplasmic domains of membrane proteins must undergo folding as they exit the translocon. Chaperones are thought to work not by direct promotion of protein folding per se, but rather, by preventing improper interactions (e.g., aggregation) that sidetrack a protein from its correct folding pathway. Bacterial periplasmic chaperones need to be dissimilar from ATP-dependent chaperones of Hsp60 and Hsp70 families present in many eukaryotic organelles or the bacterial cytoplasm, since the periplasm contains little, if any, nucleoside triphosphate.

Pilus- and fimbriae-specific chaperones represent a large and well-characterized group. Unlike cytoplasmic chaperones that recognize their substrate in an unfolded state, these periplasmic chaperones recognize their substrate in a more native state and function by capping interactive surfaces that function in oligomerization and assembly reactions. In one prototypic system from uropathogenic *E. coli*, the periplasmic chaperone–substrate complex interacts with an outer membrane "usher," which determines the order of assembly of the various pilus subunits in the outer membrane on the basis of its affinity for them. The genes for pilus- and fimbriae-specific chaperones are ordinarily located within operons that encode the structural components of these organelles, and they are typically contained on plasmids or chromosomal segments unique to pathogenic bacterial strains.

Proline isomerases (peptidyl–prolyl cis/trans isomerases or rotamases) catalyze the cis/trans isomerization of peptidyl–prolyl residues within proteins, normally a slow step. Such enzymes are found in both the bacterial cytoplasm and periplasm. *E. coli* contains at least three periplasmic proline isomerases, each belonging to a different family: PpiA (RotA) (cyclophilin family), FkpA (FKBP family), and SurA (PpiC family). *ppiA* is regulated by the Cpx two-component system that senses perturbations of the periplasmic milieu.

The bacterial periplasm contains an array of enzymes that promote the formation and correct distribution of disulfide bonds in secretory and membrane proteins. Enzymes that promote disulfide bond formation in proteins are termed disulfide oxido-reductases, while those that promote the rearrangement of disulfide bonds within proteins are termed disulfide isomerases. At the heart of the mechanism of such enzymes is a catalytic pair of cysteines, often in a Cys-X-X-Cys (where X represents a nonspecific amino-acid residue) configuration, which can undergo oxidation (to promote substrate reduction, in the case of an isomerase) or reduction (to promote substrate oxidation, in the case of an oxido-reductase). These enzymes must undergo the reverse redox reaction catalyzed by a recharging enzyme in order to be regenerated. Several different disulfide oxido-reductases/isomerases are known in *E. coli* and other bacteria (Fig. 72.7). DsbA appears to be the major enzyme catalyzing net disulfide bond formation of periplasmic and outer membrane proteins, and it is reoxidized by its partner, DsbB, a transmembrane protein. The reoxidation of DsbB itself appears to be dependent on the respiratory chain. DsbC appears to function primarily as an isomerase to correct aberrant disulfide bonds that form within proteins. Reducing potential for DsbC reduction comes ultimately from cytoplasmic thioredoxin and is transferred to the inner membrane protein DsbD before reaching DsbC. Additional Dsb proteins also exist. It is clear from these studies that a delicate redox balance is present in the periplasm that allows the correct array of disulfide bonds to be formed and maintained.

II. TERMINAL BRANCHES OF THE GENERAL SECRETION PATHWAY

Certain species of gram-negative bacteria possess pathways for transporting proteins across the outer membrane once they have reached the periplasm via the GSP. A number of such pathways are specific for a single protein, and they may require none (e.g., *N. gonorrhoeae* IgA protease) or only a single additional protein (e.g., *S. marcescens* hemolysin) to promote the export process. Space does not allow coverage of this iconoclastic group of proteins. The most important pathway for outer membrane traversal is the main terminal branch (MTB) of the GSP or type II protein secretion pathway. Hallmarks of the MTB include: (i) this pathway is present in a limited number of plant and animal pathogens (currently, *Klebsiella oxytoca, Erwinia chrysanthemi* and *carotovora, Xanthomonas campestris,*

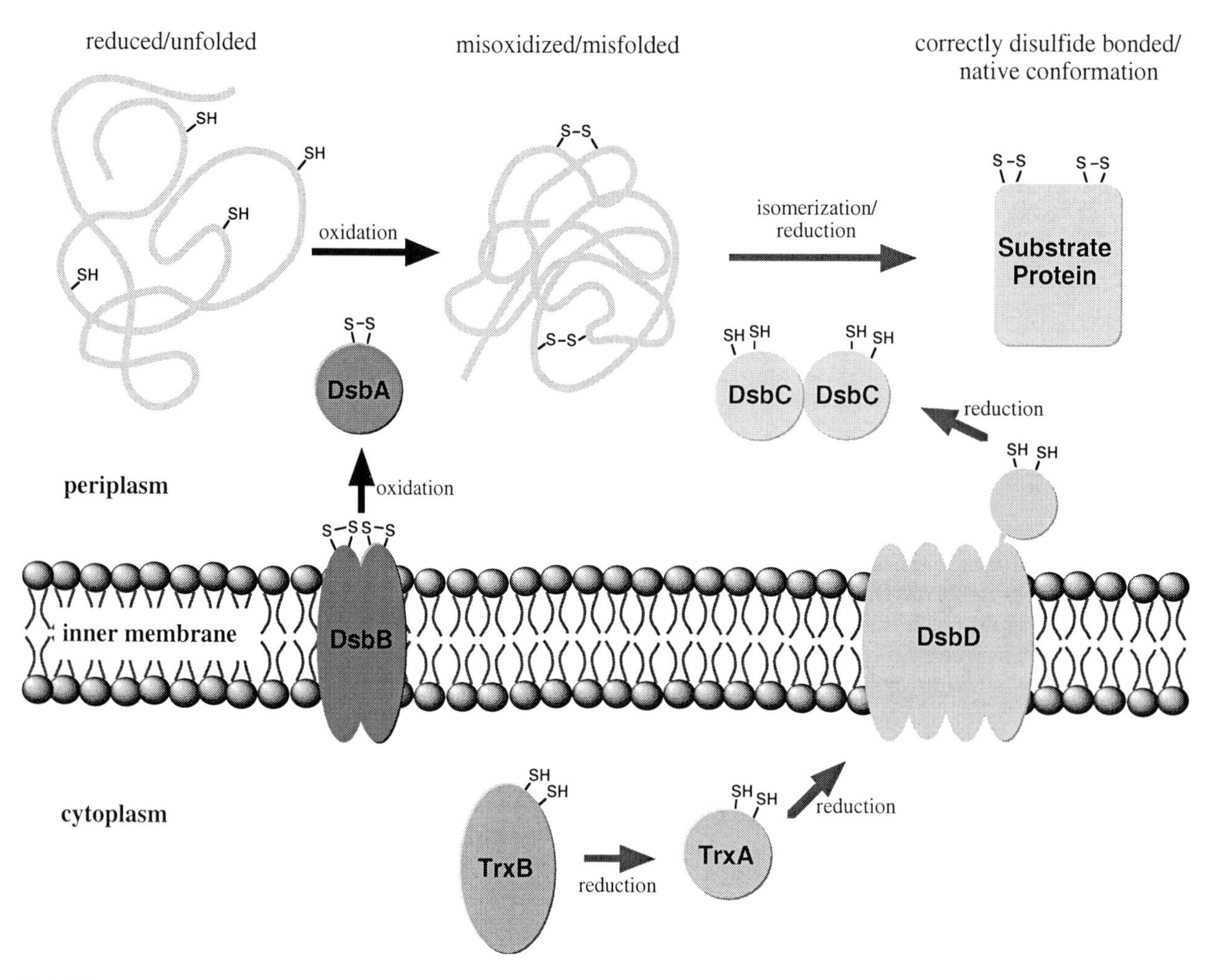

FIGURE 72.7 Model for disulfide bond formation in the *E. coli* periplasm. The roles of DsbA as an oxido-reductase and DsbC as an isomerase are depicted along with their regeneration utilizing DsbB and DsbD, respectively. TrxA and TrxB indicate thioredoxin 1 and thioredoxine reductase, respectively. [From Rietsch, A., and Beckwith, J. (1998). *Annual Rev. of Genetics* **32**, 163–184, with permission from the *Annual Review of Genetics*, Vol. 32, ©1998, by Annual Reviews.]

Pseudomonas aeruginosa, *Aeromonas hydrophila*, and *Vibrio cholerae*), (ii) it is responsible for the secretion of a limited number of proteins (often referred to as exoproteins) that include hydrolytic enzymes, such as cellulase and pectinase for plant pathogens, and proteases, lipases, and toxins for animal pathogens, (iii) secretion occurs on folded protein substrates that are present in the periplasm, (iv) a given exporter can secrete several structurally diverse exoproteins but not structurally similar exoproteins from a heterologous exporter, (v) they are complex systems that require 12–14 secretion machinery components that are located in both the inner and outer membranes, and (vi) the genes encoding these components are tightly linked in a single regulon, while the genes encoding exoproteins are often unlinked.

Correctly folded protein substrates are required for secretion by the MTB. The starch-degrading enzyme pullulanase ordinarily contains disulfide bonds prior to its secretion, although the presence of disulfide bonds is not requisite for secretion. Secretion of cholera toxin requires prior oligomerization of the A monomer and the B pentamer within the periplasm. Even the presence of single amino-acid alterations is sufficient to block secretion of particular protein substrates (e.g., *A. hydrophila* aerolysin). These examples suggest that surface elements within the final tertiary structure of exoproteins are recognized by the MTB secretion machinery (i.e., a patch signal), although particular recognition motifs have not been defined further. The existence of multiple, redundant secretion signals may hamper their definition in certain cases.

Similar recognition motifs are probably present on structurally unrelated exoproteins, since a given MTB exporter may secrete several diverse exoproteins (e.g., *P. aeruginosa* Xcp secretes elastase, exotoxin A, alkaline phosphatase, and phospholipase C). In contrast, homologous and structurally similar exoproteins from related bacteria are generally not secreted from bacteria that possess a different MTB exporter. The basis for specific recognition by these systems is yet to be clarified.

MTB exporters are complex systems, often consisting of over one dozen proteins. The best characterized systems are the *K. oxytoca* Pul and the *E. chrysanthemi* Out systems (Fig. 72.8), which have been reconstituted in *E. coli*. DNA sequence analysis indicates that all MTB exporters contain a number of similar components (identities from 30–60%) (Fig. 72.9). Several of these proteins are similar to ones that are required for the assembly of type IV pili. In particular, four pilin-like proteins (pseudopilins) are present in the Pul, Out, and other systems (Pul/Out-GHIJ). Remarkably, pseudopilin signal peptides are cleaved by a specific peptidase (Pul/OutO) that is functionally related to prepilin peptidase. In addition, there are two other components (Pul/OutEF), that are also homologous to proteins required for pilin assembly. One of these

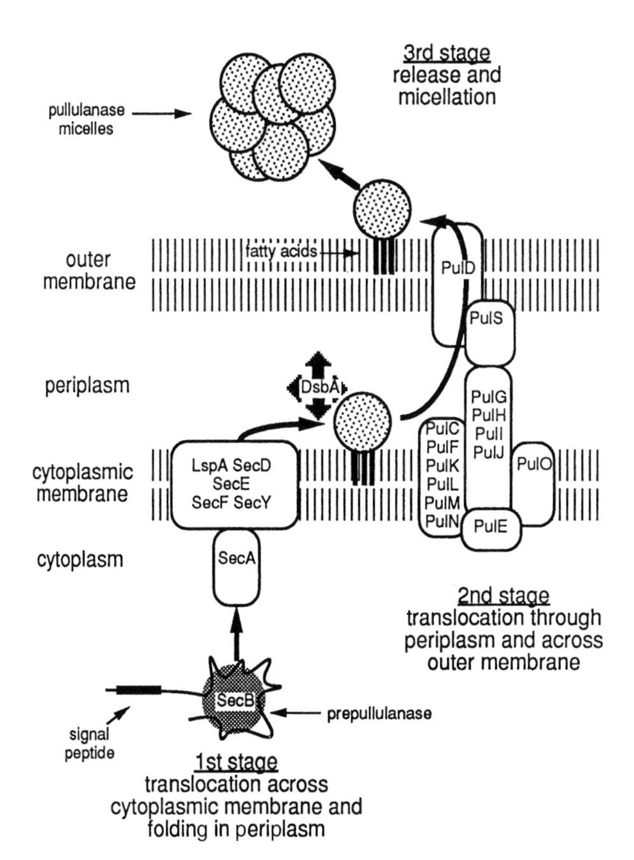

FIGURE 72.8 Model of the three stages of secretion of pullulanase by *K. oxytoca*. A probable arrangement of the Pul exporter responsible for the second stage of secretion is indicated. [From Pugsley, A. (1993). *Microbiol. Rev.* **57**, 89.]

proteins (Pul/OutE) contains a predicted ATP-binding site and is likely to function as an ATPase or kinase. However, rather than serving as an energy source for exoprotein secretion or regulating it by phosphorylation, it seems more likely that this protein functions in exporter assembly, in keeping with the role of its homolog in pilus assembly. The striking similarities between two seemingly different systems suggest that pseudopilins could be assembled into a piluslike structure, which functions as a core component of the MTB exporter. While this is an attractive hypothesis and is similar to the parallelism in structure between type III exporters and flagella (see following), it currently lacks compelling evidence in its favor.

A variety of techniques have been utilized to localize the individual proteins that comprise the MTB exporter. Remarkably, 11 out of 14 proteins in the Pul system appear to be integral cytoplasmic membrane proteins. This seems strange for a system, whose function is to translocate proteins solely across the outer membrane. Several suggestions have been proffered in this regard: (i) certain inner membrane proteins may function indirectly to promote exporter assembly, (ii) a multiprotein complex may be needed for transfer of energy between the inner and outer membranes during the export cycle (see following), (iii) exoproteins may be specifically concentrated and targeted for translocation by their association with an inner membrane protein complex, (iv) a large and elaborate inner membrane protein complex may be needed to properly gate the outer membrane protein channel, which must be large in order to accommodate folded protein substrates.

Central to the function of MTB exporters is a single integral outer membrane component (Pul/OutD) that exists in large multimeric complexes (10–14 monomers) that are presumed to be protein translocation channels. Accordingly, such proteins have been termed secretins. MTB secretins are homologous to an outer membrane protein encoded by filamentous phage (pIV), which presumably functions as a secretin for extrusion of phage particles across the bacterial outer membrane. A single pIV oligomer contains ~14 monomers arranged in a cylindrical fashion with a internal diameter of approximately 8 nM. Like trimeric outer membrane porins, pIV is rich in β-sheet structure and resistant to detergent-induced dissociation. Little is known about how secretin channels are gated.

Secretins possess an amino-terminal periplasmic domain that appears to determine exoprotein substrate specificity. Substrate-specific binding has been demonstrated for certain secretins, while domain exchanges between pIV homologs allows for substrate specificity

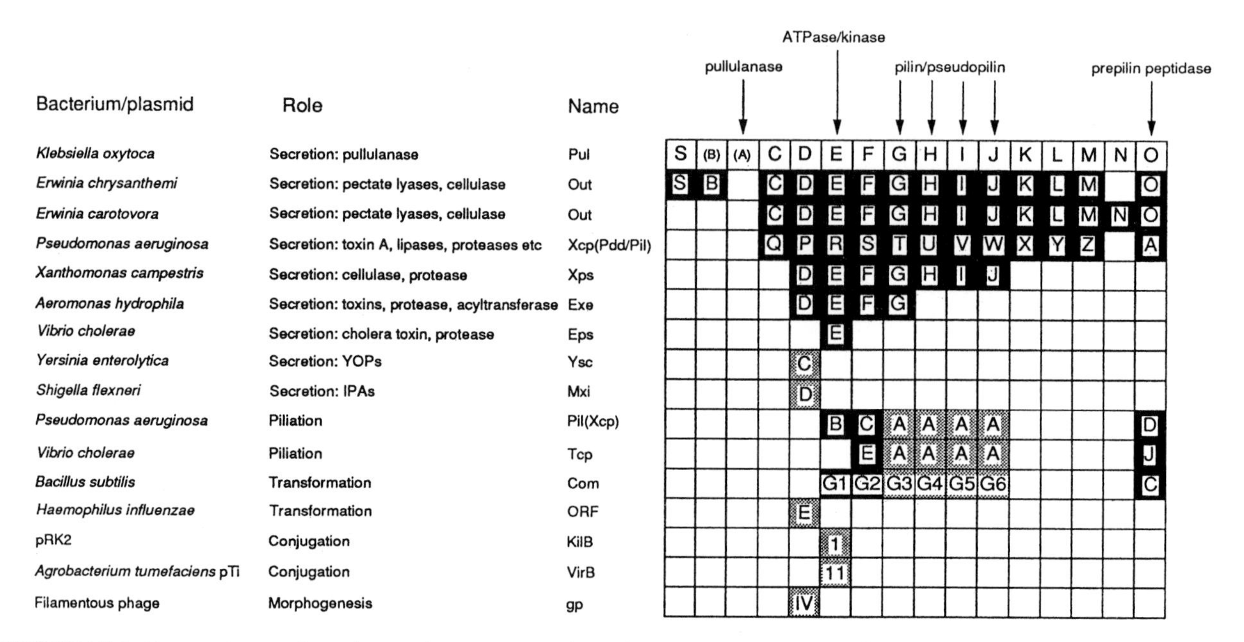

FIGURE 72.9 Chart indicating homologies of components shared by MTB exporters as well as machineries promoting movement of macromolecules across bacterial envelopes. Black boxes indicate >20% overall sequence identity to the corresponding Pul counterpart, while shaded boxes indicate significant homologies that are less than this value. Known or probable functions of these components are given at the top of the figure. [From Pugsley, A. (1993). *Microbiol. Rev.* **57**, 84.]

switching. These experiments suggest that the amino-terminal domain of secretins may be the initial docking site of exoproteins during their transit across the outer membrane.

In vitro systems are needed to define the translocation mechanism of MTB exporters. *In vivo* systems suggest a requirement for PMF and are equivocal on an ATP requirement. If ATP hydrolysis is needed only for exporter assembly, then it is possible that such systems are driven solely by PMF. Precedent exists already for PMF-dependent import of certain small molecule ligands across the outer membrane via the TonB-dependent pathway. TonB has been suggested to "shuttle" PMF-driven conformational energy to the outer membrane in order to promote transport processes. However, it is clear that considerable future work will be required to understand the function of the many components present in MTB exporters, their overall organization and assembly, the basis of exoprotein specificity, as well as the translocation mechanism.

III. TWIN-ARGININE PROTEIN SECRETION PATHWAY

A specialized pathway for the secretion of redox enzymes across the cytoplasmic membrane of most bacteria (except mycoplasmas, *Borrelia burgdorferi*, and methanogens) and, probably, archaea exists that is independent of the GSP. Redox enzymes that employ iron–sulfur clusters, molybdopterin, polynuclear copper, tryptophan tryptophylquinone, and flavin adenine dinucleotide as cofactors appear to utilize this pathway, while enzymes that have iron porphyrins, mononuclear type I or II copper centers, dinuclear Cu_A center, and pyrrolo–quinoline quinone as cofactors utilize the GSP for their secretion. The hallmark of this recently characterized pathway is the presence of a signature motif on the signal peptide of preproteins utilizing this pathway. A consensus sequence containing an invariant arginine pair, S/T-R-R-X-F-L-K (where X represents any amino acid), is present between the N- and H-domains of the signal peptide (see Fig. 72.2). Aside from this feature, such signal peptides are similar to those found for GSP-dependent preproteins, although they are often somewhat longer than their GSP counterparts (up to 58 amino-acid residues). This characteristic feature has lead to the name "*t*win *a*rginine *t*ranslocation (TAT) pathway" to describe this novel pathway.

A second characteristic of the TAT pathway is that their cognate preproteins appear to be undergo extensive folding and cofactor binding in the cytoplasm prior to insertion and translocation across the membrane. This feature is in contrast to the GSP, that requires the preprotein to be in an extended or loosely folded state for engagement by the Sec machinery.

Aside from the atypical signal peptide, precisely what targeting information is also contained within the folded preprotein is not yet known. Of interest, there are metalloprotein subunits that lack any signal peptide entirely, which are secreted by association with their complementary subunit, containing a TAT-dependent signal peptide (i.e., piggyback secretion). This suggests that the TAT machinery may possess an usually large translocation channel for accommodating its substrates.

Mutational studies in *E. coli* have identified several proteins that are likely to be components of the TAT machinery. Such mutants demonstrate that the TAT machinery is nonessential in heterotrophic bacteria under appropriate conditions (e.g., in the presence of oxygen). TatA and TatE are homologs of Hcf106, which is a component of the ΔpH-dependent thylakoid import pathway of maize. Plant thylakoids possess three protein import pathways (SRP-dependent, Sec-dependent, and ΔpH-dependent), and the ΔpH-dependent pathway catalyzes the import of preproteins containing the twin-arginine motif. Thus, the TAT pathway appears to be present in all three domains of life. An additional protein unrelated to Hcf106, TatB, also functions in the bacterial TAT pathway. All three proteins are predicted to be integral cytoplasmic membrane proteins. *tatA*, *tatB*, and *tatE* mutants accumulate enzymatically active redox enzymes in their cytoplasm, but they are normally proficient in the secretion of proteins utilizing the GSP. This phenotype suggests that a specific mechanism(s) exist(s) that excludes redox enzymes from the GSP. By contrast, blocking the Sec machinery has no effect on the secretion of redox enzymes. Taken together, there is compelling evidence that the GSP and TAT pathways are distinct and mutually exclusive.

Little is known about the energetics of protein translocation of the TAT pathway. It is unclear whether ΔpH (which is often very minimal in bacteria such as *E. coli*) could drive protein translocation in this system, by analogy to the thylakoid pathway. The only crossover between the GSP and TAT machineries is likely to be utilization of a common signal peptidase I, which should also cleave TAT-dependent preproteins, based on the similarity of the C-domains of these signal peptides.

IV. ABC PROTEIN SECRETION PATHWAYS

A number of GSP-independent protein secretion pathways are promoted by a family of exporters that contain a highly conserved protein of the ATP-binding cassette (ABC) superfamily, and consequently, they are referred to as ABC protein secretion pathways or type I systems. ABC protein exporters are members of the ubiquitous ABC superfamily of traffic ATPases, which include importers and exporters of a large variety of small and macromolecules. There are over a dozen ABC protein exporters described currently, and this number is likely to grow significantly in the future. Hallmarks of ABC protein exporters include (i) they secrete a group of homologous proteins that can include toxins, proteases, lipases, and specific peptides, depending on exporter type, (ii) their protein substrates lack GSP-dependent signal peptides, but instead usually possess uncleaved carboxy-terminal secretion signals that vary by exporter type, (ii) they are relatively simple systems that include an ABC protein and a few accessory proteins, and (iii) protein translocation occurs in a single step across one or both bacterial membranes (without any periplasmic intermediate for gram-negative bacteria). This latter feature, combined with the ability to fuse ABC-dependent secretion signals to a large number of heterologous proteins and obtain efficient secretion, make these exporters attractive for biotechnological applications.

The nature of ABC-dependent secretion signals has been studied extensively, most notably for RTX (repeats in toxin) toxin and metalloprotease families. Deletion and gene fusion studies show that the hemolysin of uropathogenic *E. coli* contains a secretion signal in its carboxy-terminal 60 amino-acid residues, while PtrG protease of *E. chrysanthemi* contains a secretion signal encompassing only its last 29 residues. Furthermore, these minimal secretion signals are effective in promoting secretion of a large variety of naive passenger proteins by gene fusion. Such heterologous protein secretion occurs exclusively via the exporter that is homologous to the secretion signal, demonstrating the specificity of the secretion signal for a given exporter. Genetic analysis indicates that secretion signals are very tolerant to single amino-acid changes, and they suggest that such signals function in a tertiary conformation (i.e., a patch signal) that is recognized by the ABC component of the exporter.

Upstream of the secretion signals of toxins, proteases, and lipases is a conserved motif consisting of a glycine-rich sequence (GGXGSD) that is repeated 4 to 36 times. This repeated motif has been shown to form a β parallel roll structure that binds calcium ions in the case of certain metalloproteases. This arrangement may promote better separation of the secretion signal from the rest of the protein, thereby affording efficient presentation of the signal to the exporter.

Typical ABC protein exporters consist of an ABC protein and one to several accessory proteins, depending on the family (Fig. 72.10). All traffic ATPases possess a highly conserved structure consisting of two membrane-embedded hydrophobic domains and two hydrophilic ATP-binding domains. This architecture probably allows the ATP-binding domains to drive substrate transport across a channel formed by the hydrophobic domains (a situation akin to SecA and SecYE proteins of the GSP). Although these domains can be contained in a single polypeptide (e.g., mammalian multidrug resistance transporter or cystic fibrosis chloride channel), the most common arrangement for bacterial protein exporters is as a homodimer.

Two additional accessory proteins are found in ABC protein exporters of gram-negative bacteria that allow protein secretion across both membranes. A protein of the membrane fusion protein family (MFP) appears to functionally link the inner and outer membranes during the translocation cycle (e.g., HlyD in Fig. 72.10). MFP proteins have an amino-terminal hydrophobic domain anchored to the inner membrane, a central periplasmic domain, and a carboxy-terminal domain with predicted β-sheet structure that interacts with the outer membrane. The second component is an outer protein that functions as a protein channel for outer membrane traversal (e.g., TolC in Fig. 72.10). This latter protein is the least specific of the three, since exporters can share a common outer membrane protein (e.g., TolC for *E. coli* hemolysin, colicin V, and *S. marcesscens* hemoprotein exporters).

The genes encoding ABC and MFP components are usually adjacent to the structural genes encoding the exoproteins, while the gene for the outer membrane protein may or may not be linked. Additional genes encoding cytoplasmic activators that posttranslationally modify exoproteins may be present within these genetic clusters. This arrangement allows functional transfer of these systems to other bacteria by horizontal genetic transmission.

Much less is known about the mechanism of protein translocation through ABC exporters compared to the GSP. Since secretion signals are usually

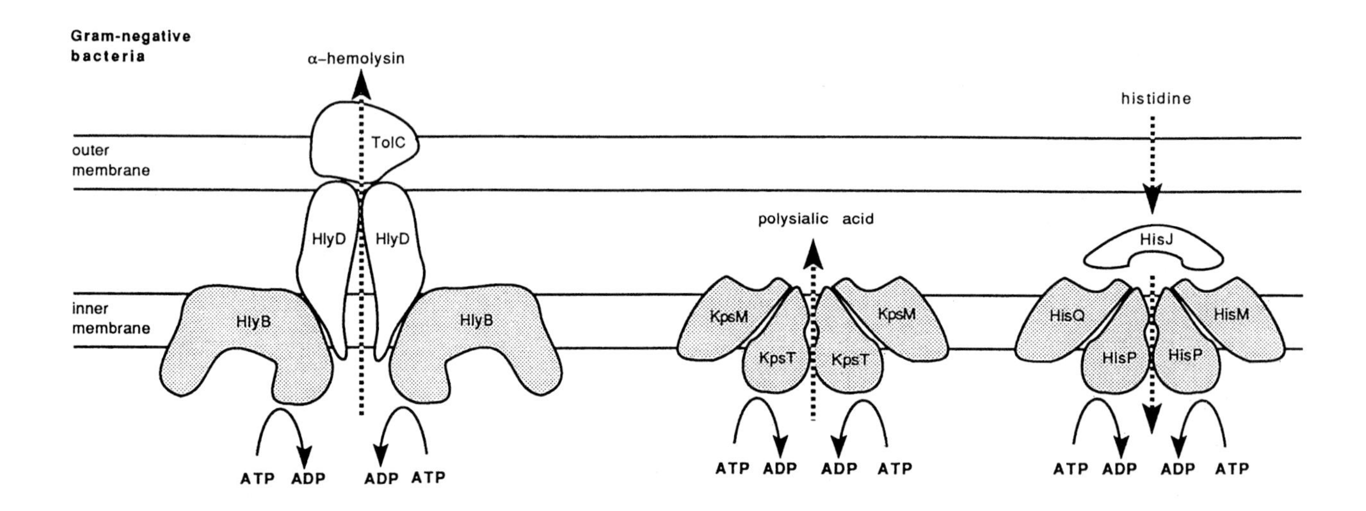

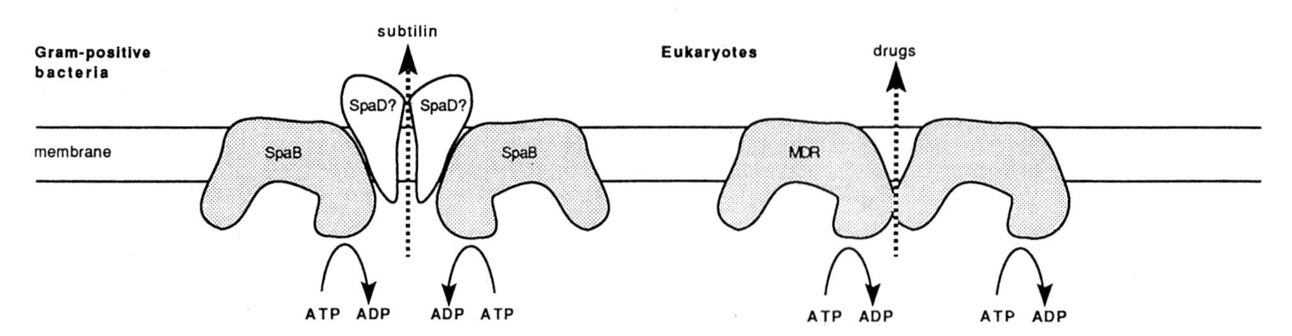

FIGURE 72.10 Models of ABC transporters. Depicted are the *E. coli* α-hemolysin and *B. subtilis* subtilin exporters (both protein exporters), along with the *E. coli* polysialic acid and mammalian multidrug exporters and the *S. typhimurium* histidine importer. [From Fath, M., and Kolter, R. (1993). *Microbiol. Rev.* **57**, 997.]

carboxy-terminal, exoprotein translocation occurs in a posttranslational manner. There is no known requirement for cytoplasmic chaperones, and whether the exoprotein is translocated in a folded, loosely folded, or unfolded state is not clear. Translocation initiates when the secretion signal interacts with the ABC protein, which defines the substrate specificity of a given system. This primary interaction leads to an ordered assembly (or stabilization) of the exporter: the ABC protein binds tightly to the MFP, which, in turn, binds to the outer membrane component. All three components must act in a concerted fashion, since the absence of the outer membrane protein leads to cytoplasmic, not periplasmic, accumulation of exoproteins. The path that the exoprotein takes through the exporter has not been defined yet.

Protein exporters require energy in the form of both ATP and PMF. PMF appears to promote the initial binding of the exoprotein to the exporter. The ABC protein drives at least a portion of the translocation cycle by ATP-dependent conformational cycling, although whether this involves membrane cycling of a mobile domain of the ABC protein akin to SecA protein is unknown. *In vitro* protein translocation systems for ABC protein exporters should clarify these and other issues relating to the translocation mechanism.

V. TYPE III PROTEIN SECRETION PATHWAYS

In recent years, it has become apparent that several gram-negative pathogenic bacteria have evolved a specialized protein secretion system, termed type III, or contact dependent, to secrete virulence determinants. This complex machinery directs the secretion and subsequent translocation into the host cell of several bacterial virulence effector proteins. These proteins, in turn, stimulate or interfere with host cell responses for the benefit of these bacterial pathogens. Remarkably, type III secretion sfiystems have been found not only in bacteria pathogenic for animals, such as *Salmonella* spp., *Yersinia* spp., *Shigella* spp., *E. coli*, *Bordetella* spp., and *Pseudomonas aeruginosa*. but also in plant pathogenic bacteria of the genera *Erwinia*, *Xanthomonas*, *Ralstonia*, and *Pseudomonas*.

A characteristic feature of this specialized secretion system is the absence of typical *sec*-dependent signal sequences in their substrate proteins, which are, therefore, secreted in a *sec*-independent manner without processing of their amino termini. Another property of these systems is the requirement of an activating signal for their efficient function. Certain growth conditions in the laboratory (e.g., low concentration of Ca^{2+}) have been shown to activate type III secretion systems in some bacteria. However, the physiologically relevant signals that activate these systems are not known, although it is widely believed that such signals must be derived from bacterial contact with host cells. Indeed, bacteria/host cell interactions lead to a marked increase in secretion through this pathway. Contact activation of protein secretion does not require *de novo* protein synthesis, implicating a posttranslational mechanism in the regulation of this process.

Type III secretion systems are most often encoded within chromosomal regions, known as pathogenicity islands, or on large virulence-associated plasmids. The nucleotide composition of the relevant coding regions is most often A + T rich and significantly different from that of the host chromosome, suggesting their acquisition by some mechanism of horizontal gene transfer. Consistent with this hypothesis is the frequent finding of sequences encoding remnants of transposable elements, bacteriophage proteins, or insertion sequences in the immediate vicinity of genes encoding type III secretion systems.

A. The secretion apparatus

A core set of at least 17 highly conserved proteins are thought to be components of the secretion machinery, based on their requirement for promotion of type III secretion. One of these components shares sequence similarity to the secretin or PulD family of protein exporters. As discussed previously, secretins are essential components of all type II secretion systems that are thought to form a channel through which the secreted substrates traverse the outer membrane. A subset of the type III secretion machinery components exhibits significant sequence similarity to components of the flagellar export apparatus. This group of proteins includes several integral membrane proteins and a membrane associated ATPase, which is thought to energize this system. This sequence homology, coupled to the architectural similarity between flagella and the type III system (see following), strongly suggests an evolutionary relationship between type III secretion and flagellar assembly. Another subset of conserved components of type III secretion systems includes several lipoproteins. The number of lipoproteins present in each system is not always the same and their primary sequence is poorly conserved also. However, some lipoproteins are highly conserved within a subset of bacterial species.

Recent electron microscopy studies have revealed the supramolecular organization of components of the type III secretion machinery of *Salmonella typhimurium*.

Although structural information on type III secretion systems from other bacteria is not available, the high degree of sequence similarity among the different components suggests a similar architecture for all type III protein exporters. The studies in *Salmonella typhimurium* revealed that the type III secretion machinery is organized in a supramolecular complex resembling a needle (needle complex) that spans both the inner and outer membranes of the bacterial envelope (Fig. 72.11). This needle complex possesses cylindrical symmetry with two clearly identifiable domains: a slender, needlelike portion, projecting outwards from the surface of the cell, and a cylindrical base, which anchors the structure to the inner and outer membranes. The base structure has some resemblance to the flagellar basal body. It is composed of two outer and two inner rings. The inner rings are 40 nm in diameter and 20 nm in thickness and interact with the cytoplasmic membrane. The outer rings interact with the outer membrane and the peptidoglycan layer and are 20 nm in diameter and 18 nm in thickness. The inner and outer rings are connected by a rod structure. The needle structure itself is 80 nm in length and 13 nm in thickness and appears hollow. The number of needle structures observed per bacterial cell varies between 10 to 100. The overall architecture of the needle complex clearly resembles that of the flagellar basal body, giving further support to the hypothesis that these two structures are evolutionary related (Fig. 72.11). Furthermore, the organization of this complex suggests a mechanism by which proteins can be secreted through the bacterial envelope without a periplasmic intermediate. Biochemical analysis of the purified needle complexes revealed that they are composed of at least three proteins: InvG (a member of the secretin family) and two lipoproteins, PrgH and PrgK. Neither of these proteins has true homologs in the flagellar system, although FliF, a component of the flagellar MS ring structure, is also a lipoprotein. Further characterization of the needle complex will most likely yield additional structural components.

B. Substrates of type III secretion systems

Despite the high degree of conservation among the components of type III secretion systems across bacteria species, there is little similarity among the substrates of the different systems, in particular, those proteins that will ultimately mediate or interfere with host cell responses. Therefore, it is evident that different pathogenic bacteria have adapted the function of the type III secretion systems for their own specific requirements. Proteins secreted through this system can be divided into at least four different categories: (i) Proteins that are required for the secretion process itself. By analogy to the flagellar system, these proteins are candidates for structural components of the needle complex; (ii) Proteins that are involved in the regulation of the secretion process. This category may include proteins that act as "plugs" of the translocation channel; (iii) Proteins that promote the process of translocation of the secreted substrates into eukaryotic host cells. These proteins have been termed translocases; (iv) Proteins that act as effectors of bacterial-induced host cellular responses. This latter category includes proteins that are agonists of cellular responses or antagonists of physiological host cell responses. Such proteins are the least conserved

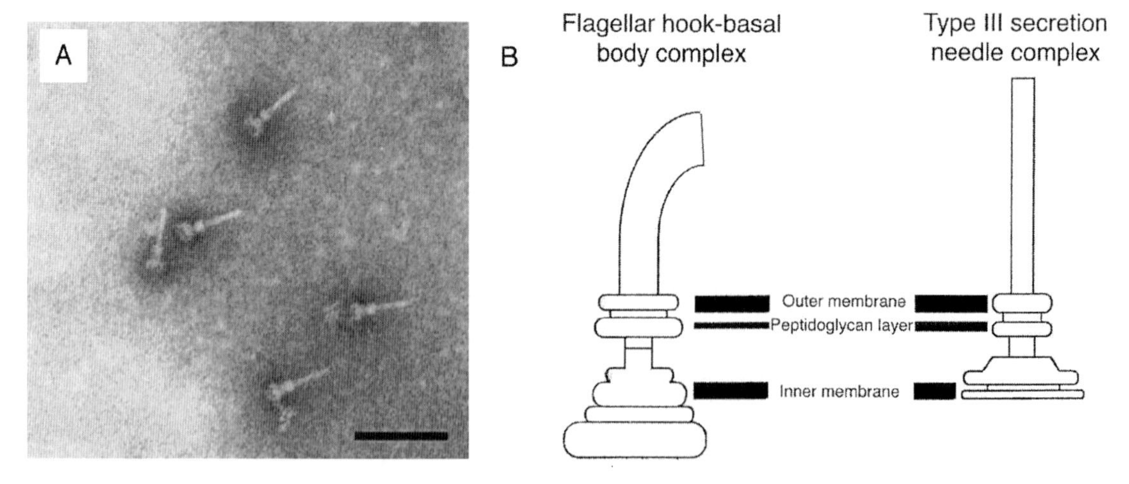

FIGURE 72.11 Type III exporter (needle complex) of *Salmonella typhimurium*. (A) Electron micrograph of purified needle complexes from *Salmonella typhimurium*. Scale bar, 100 nm. (B) Diagram of flagellar hook basal body and needle complexes aligned to show similarities. [Courtesy of Shin-Ichi Aizawa.]

among the different systems, a further indication that different bacteria have assembled a set of effectors that is best suited for their specific requirements. Effector proteins are predicted to act on eukaryotic cell target molecules and include tyrosine protein phosphatases, serine/threonine protein kinases, and exchange factors for small molecular weight GTPases.

C. Specific chaperones, export and translocation signals

A characteristic feature of type III protein secretion systems is that most of their substrates are associated with specific cytoplasmic chaperones or partitioning factors. Unlike other well-characterized chaperones, such as GroEL or Hsp70, these specialized chaperones have a rather narrow binding specificity and lack nucleotide-binding or -hydrolyzing activities. While they exhibit little primary amino-acid sequence similarity, they do share a number of biochemical properties, such as a relatively small size (15–18 kDa), a low isoelectric point, and a predominantly α-helical secondary structure. They are usually, although not always, encoded in the vicinity of their cognate target proteins. Overall, the function of the type III associated chaperones is not well understood.

Deletion analysis of type III secreted proteins has established the existence of well-defined independent domains that are involved in either their secretion or translocation into the host cell. In general, the first 10 to 20 amino-acid residues of these proteins are required for their secretion, whereas an adjacent domain of 60 to 70 residues is involved in their translocation into the host cell. Interestingly, this translocation domain overlaps with the binding site for their specific chaperone, suggesting the intriguing possibility that these chaperones are directly involved in the events that lead to the translocation of the different type III secreted proteins into the eukaryotic host cell. Recent studies carried out in *Yersinia* spp. have suggested the existence of two independent type III secretion pathways. One pathway appears to be mediated by a translationally coupled secretion signal, present in the first ~60 nucleotides of the coding mRNAs. The other pathway is chaperone-dependent and is mediated by a domain located between residues 15 and 100 of substrate proteins.

The role that this family of chaperones plays in the secretion process itself is not well understood, and it is the subject of some controversy. Most of the studies have been carried out in *Yersinia* spp. and *S. typhimurium*. In these bacteria, absence of a given chaperone results in reduced secretion of the cognate substrate protein. However, it has been difficult to establish whether reduction of secretion is due to degradation of the target protein in the absence of the chaperone, the direct involvement of the chaperone in the secretion process itself, or both. It is clear that secretion of type III substrate proteins can take place in the absence of specific chaperones. However, this process requires either removal of the chaperone binding site from the cognate protein or the absence of specific secreted substrate proteins that presumably engage in direct interaction with the cognate protein. Thus, it has remained controversial whether the function of these chaperones is to deliver the secreted target proteins to the type III secretion machinery, to maintain them in a secretion competent state, to prevent premature association with other cognate secreted products, or some combination of the above. More studies will be required to clarify the physiological role, if any, of these secretion mechanisms in the different type III secretion systems, as well as to better understand the role of specific chaperones in this process.

BIBLIOGRAPHY

Aldridge, P., and Hughes, K. T. (2001). How and when are substrates selected for type III secretion? *Trends Microbiol.* **9**, 209–214.

Bardwell, J. (1994). Building bridges: disulfide bond formation in the cell. *Mol. Microbiol.* **14**, 199–205.

Berks, B. (1996). A common export pathway for proteins binding complex redox cofactors? *Mol. Microbiol.* **22**, 393–404.

Binet, R., Letoffe, S., Ghigo, J., Delepelaire, P., and Wandersman, C. (1997). Protein secretion by gram-negative bacterial ABC exporters—a review. *Gene* **192**, 7–11.

Celli J., Deng W., and Finlay, B. B. (2000). Enteropathogenic Escherichia coli (EPEC) attachment to epithelial cells: exploiting the host cell cytoskeleton from the outside. *Cell Microbiol.* **2**, 1–9.

De Kievit, T. R., Gillis, R., Marx, S., Brown, C., Iglewski, B. H. (2001). Quorum-sensing genes in Pseudomonas aeruginosa biofilms: their role and expression patterns. *Appl. Environ. Microbiol.* **67**, 1865–1873.

Driessen, A., Fekkes, P., and van der Wolk, J. (1998). The Sec system. *Curr. Opin. Microbiol.* **1**, 216–222.

Driessen, A. J., Manting, E. H., and van der Does, C. (2001). The structural basis of protein targeting and translocation in bacteria. *Nat. Struct. Biol.* **8**, 492–498.

Duong, F., Eichler, J., Price, A., Leonard, M. R., and Wickner, W. (1997). Biogenesis of Gram-Negative bacterial envelope. *Cell* **91**, 567–573.

Fath, M., and Kolter, R. (1993). ABC transporters: bacterial exporters. *Microbiol. Rev.* **57**, 995–1017.

Fekkes, P., and Driessen, A. J. M. (1999). Protein targeting to the bacterial cytoplasmic membrane. *Microbiol. Molec. Biol. Rev.* **63**, 161–173.

Hueck, C. J. (1998). Type III secretion systems in bacterial pathogens of animals and plants. *Microbiol. Molec. Biol. Rev.* **62**, 379–433.

Kubori, T., Matsushima, Y., Nakamura, D., Uralil, J., Lara-Tejero, M., Sukhan, M., Galán, J., and Aizawa, S.-I. (1998). Supramolecular structure of the *Salmonella typhimurium* type III protein secretion system. *Science* **280**, 602–605.

Lund, P. A. (2001). Microbial molecular chaperones. *Adv. Microb. Physiol.* **44**, 93–140.

Muller, M., Koch, H. G., Beck, K., and Schafer, U. (2000). Protein traffic in bacteria: multiple routes from the ribosome to and across the membrane. *Prog. Nucleic Acid Res. Mol. Biol.* **66**, 107–157.

Plano, G. V., Day, J. B., and Ferracci, F. (2001). Type III export: new uses for an old pathway. *Mol. Microbiol.* **40**, 284–293.

Pohlschroder, M., Prinz, W., Hartmann, E., and Beckwith, J. (1997). Protein translocation in the three domains of life variations on a theme. *Cell* **91**, 563–566.

Pugsley, A. (1993). The complete general secretory pathway in gram-negative bacteria. *Microbiol. Rev.* **57**, 50–108.

Russel, M. (1998). The macromolecular assembly and secretion across the bacterial cell envelope: type II protein secretion systems. *J. Mol. Biol.* **279**, 485–499.

Sandkvist, M. (2001). Type II secretion and pathogenesis. *Infect. Immun.* **69**, 3523–3535.

Sperandio, V. V. (2000). The elusive type III secretion signal. *Trends Microbiol.* **8**, 395.

Tran Van Nhieu, G., Bourdet-Sicard, R., Dumenil, G., Blocker, A., and Sansonetti, P. J. (2000). Bacterial signals and cell responses during Shigella entry into epithelial cells. *Cell Microbiol.* **2**, 187–193.

Wandersman, C. (1996). Secretion across the bacterial outer membrane. *In "Escherichia coli* and *Salmonella* Cellular and Molecular Biology" (F. C. Neidhardt, R. Curtiss, J. L. Ingraham, E. C. Lin, K. B. Low, B. Magasanik, W. S. Reznikoff, M. Riley, M. Schaechter, and H. E. Umbarger, eds., pp. 955–966. American Society for Microbiology, Washington, DC.

WEBSITE

EcoCyc: Encyclopedia of *E. coli* Genes and Metabolism
http://ecocyc.pangeasystems.com/ecocyc/

73

Quorum sensing in gram-negative bacteria

Clay Fuqua
Indiana University

GLOSSARY

acyl HSLs Acylated homoserine lactones, diffusible signal molecules synthesized by a wide range of gram-negative bacteria.

biofilm A community of microorganisms associated with a surface.

bioluminescence The production of photons of visible light by living organisms.

homologous (homologs) Having a shared common ancestry, or referring to individuals that share a common ancestry; often determined by structural comparison.

Lux-type box Presumptive binding sites for LuxR-type proteins located directly upstream of quorum-regulated genes.

microcolony A microscopic aggregate of bacterial cells.

operon Genes within the same transcriptional unit under the control of a common promoter element.

regulon A set of genes under the control of a common regulator; not necessarily genetically linked on the DNA.

synthase A broad class of enzymes that catalyze biosynthetic reactions.

A Growing Number of bacteria are known to monitor their own population density via processes that are collectively described as quorum sensing. Most often, specific genes within the microbe are switched on at a defined population density, a bacterial "quorum," resulting in activation of functions under the control of the quorum sensor.

In almost all cases, the ability to sense a bacterial quorum involves the release of a signal molecule from the bacterium that accumulates proportionally with cell number. At a threshold concentration, the signal molecule interacts with a bacterial receptor that activates, either directly or indirectly, the expression of quorum-dependent genes. Quorum sensing is an example of multicellular behavior in bacteria, where individual cells communicate with each other to coordinate their efforts. Quorum sensing has been well documented in a number of systems including myxococcal fruiting body development, antibiotic production in several species of *Streptomyces*, and sporulation in *Bacillus subtilis*. In gram-positive bacteria, population density is often monitored using oligopeptide-based signal molecules that are actively secreted from cells under the appropriate conditions (for a review, see Kaiser and Losick, 1994; Dunny and Leonard, 1997). In contrast, many gram-negative bacteria produce acylated homoserine lactones (acyl HSLs), which act as cell density cues that freely permeate the cellular envelope and diffuse into the environment. These compounds were originally discovered in the bioluminescent marine vibrios, but have now been identified in a wide range of gram-negative bacteria. While the mechanism of acyl HSL quorum sensing is conserved in different bacteria, the functions under

The Desk Encyclopedia of Microbiology
ISBN: 0-12-621361-5

acyl HSL control are as diverse as the bacteria that produce them. This discussion will focus on the discovery and elucidation of acyl HSL-type quorum sensing in the marine vibrios, the more recent realization that these systems function in many different gram-negative bacteria, and the mechanisms by which acyl HSLs regulate bacterial behavior and influence microbial ecology.

I. DISCOVERY: AUTOINDUCTION IN MARINE VIBRIOS

A. Regulation of bioluminescence

Acyl HSL quorum sensing, originally called autoinduction, was first discovered in the luminescent marine vibrios, *Vibrio fischeri* and *Vibrio harveyi*. *V. fischeri* is a symbiont that colonizes the light organs of several marine fishes and squids, while *V. harveyi* is an enteric bacterium that resides in the intestines and on fecal matter of certain fish. In the early 1970s, it was observed that the level of bioluminescence dropped significantly following inoculation of *V. fischeri* or *V. harveyi* into broth media, increasing again in the late stage of culture growth (Fig. 73.1). However, if the conditioned cell culture fluids from dense bioluminescent cultures were added to newly inoculated cultures, bioluminescence was initiated significantly earlier. This suggested the presence of one or more inducing signals in the fluid phase of older cultures. The signal was confirmed to be bacterially derived and, hence, was described as an "autoinducer," and the phenomenon was likewise described as autoinduction. It was found that the autoinduction of bioluminescence was somewhat species-specific and that the signal was acting as an indicator of cell density, as opposed to a temporal or nutritional regulator. Initial fractionation of the cell-free culture fluids implicated a single heat-stable compound. The *V. fischeri* autoinducer was purified and chemically identified in 1981 as 3-oxo-*N*-(tetrahydro-2-oxo-3-furanyl)hexanamide, or *N*-3-(oxohexanoyl) homoserine lactone (3-oxo-C6-HSL; see Fig. 73.1). Some years later, the *Vibrio harveyi* autoinducer activity was attributed to *N*-3-(hydroxybutyryl)homoserine lactone (3-OH-C4-HSL).

The nomenclature adopted for description of acyl HSLs is as follows (i) the prefix denotes the substituent at the third carbon (3-oxo for an oxygen, 3-OH for a hydroxy, and no prefix for fully reduced), (ii) C followed by the length of the acyl chain, and (iii) HSL for homoserine lactone. Unsaturated bonds in the acyl chain are indicated as Δ, followed by the first bonded position in the chain (e.g. an unsaturated double bond between the seventh and eighth carbon is Δ7). See Figs. 73.1 and 73.2.

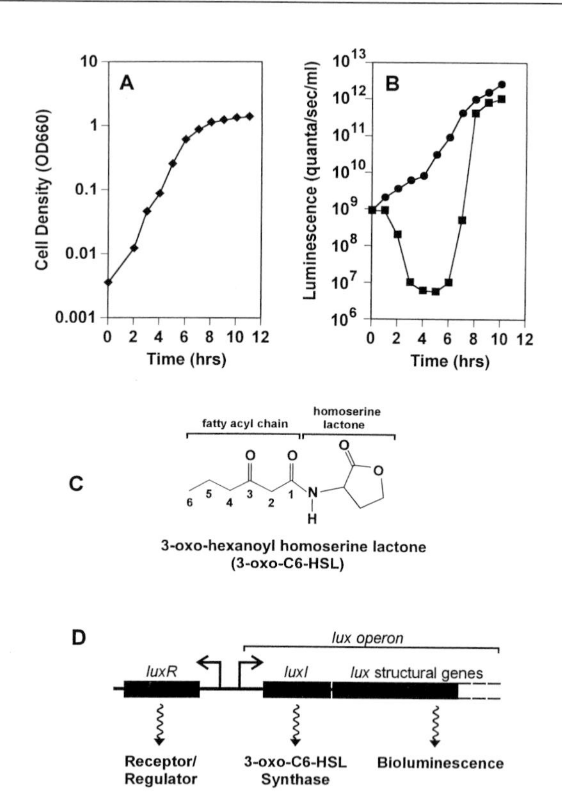

FIGURE 73.1 Autoinduction of bioluminescence. Culture of *V. fischeri* grown in seawater complete medium. (A) Culture density (◆) measure as optical density at 660 nm. (B) Luminescence as quanta per second per ml for the unsupplemented culture (■) and for an identical culture supplemented with 3-oxo-C6-HSL, the active component in conditioned *V. fischeri* culture fluids (●). (C) Chemical structure of 3-oxo-C6-HSL. Numbers identify positions on the acyl chain. (D) Operon structure for the *V. fischeri lux* genes. A and B adapted and modified from Dunlap and Greenberg, 1991, and Dunlap and Greenberg, 1985.

B. Genetics of autoinduction

In 1983, the *V. fischeri* bioluminescence (*lux*) genes were isolated and expressed in *E. coli*. *E. coli* expressing the cloned *lux* genes were not only bioluminescent, but exhibited cell density-dependent regulation. Molecular genetic analyses of the cloned *lux* locus revealed that two genes, designated *luxR* and *luxI*, were necessary and sufficient for autoinduction of bioluminescence (Fig. 73.1). The *luxR* gene is essential for response to 3-oxo-C6-HSL and encodes a transcriptional regulator that activates expression of the *lux* operon. The *luxI* gene is required for 3-oxo-C6-HSL synthesis, but is dispensable for response to the factor. The two regulatory genes are physically linked with the *luxR* gene expressed divergently from the *lux*

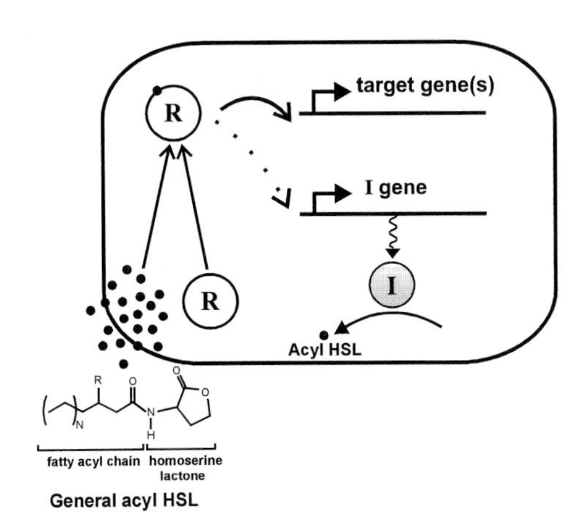

FIGURE 73.2 General model for acyl HSL quorum sensing. LuxR-type protein and LuxI-type protein are represented as spheres labeled with R and I, respectively. The acyl HSL is represented as a filled circle. Filled arrows indicate noncovalent association and catalysis. The dashed arrow implies that activation of I gene expression is variable between different bacteria. Squiggles indicate expression and translation of the I gene. On the generalized acyl HSL structure, R can be H, OH, or O, and N equals the number of carbons 0–10.

operon, of which *luxI* is the first followed by the structural genes responsible for bioluminescence (*luxICDABEG*).

C. A Model for cell density sensing

Physiological, genetic, and biochemical analysis of *lux* gene regulation in *V. fischeri* has provided a general model for acyl HSL-type quorum sensing. At low cell densities, such as in seawater, a small amount of 3-oxo-C6-HSL is synthesized by limiting pools of LuxI. The signal molecule is rapidly depleted, due to passive diffusion across the bacterial envelope, and does not accumulate. Upon colonization of the light organ, *V. fischeri* is provided a nutrient-rich environment. As cell number increases, the relative concentration of 3-oxo-C6-HSL also rises. At a specific threshold concentration, probably where levels inside and outside the cell are equivalent, the signal binds to the LuxR receptor protein, presumably causing a conformational change that results in LuxR-dependent activation of the *lux* operon promoter and a concomitant increase in bioluminescence. Elevated transcription of the *luxI* gene causes an increase in 3-oxo-C6-HSL synthesis, constituting a positive-feedback loop. While many of the details of this model are currently under investigation, this basic pattern of regulation continues to serve as a paradigm for acyl HSL quorum sensing.

II. QUORUM SENSING IN DIVERSE GRAM-NEGATIVE BACTERIA

A. Identification of acyl HSL and LuxR–LuxI-type proteins

For quite some time, the production and perception of acyl HSLs was thought to be a peculiar form of regulation restricted to the marine vibrios. Findings from several laboratories during the 1990s have radically altered this view, and it is apparent that many bacteria employ acyl HSLs as regulatory molecules (Table 73.1 and see reviews). A variety of lines of investigation have led to discovery of acyl HSL quorum sensors.

Many of these, particularly the first few, were discovered serendipitously. Examination of elastase gene regulation in *Pseudomonas aeruginosa* led to identification of the *lasR* gene, encoding a protein homologous to LuxR. Closely linked to *lasR* was the *lasI* gene, homologous to *luxI*, that directs synthesis of 3-oxo-C12-HSL. Likewise, studies on the regulation of conjugal transfer of the *Agrobacterium tumefaciens* Ti plasmid, identified the regulatory proteins TraR and TraI, homologous to LuxR and LuxI.

Contemporaneously, purification and structural analysis of a so-called "conjugation factor" identified 3-oxo-C8-HSL as the cognate acyl HSL of the *A. tumefaciens* system. Examination of mutants repressed for elaboration of degradatory exoenzymes and antibiotic production in several subspecies of *Erwinia carotovora* led to discovery of LuxR–LuxI-type quorum sensors. As it became apparent that acyl HSLs are important regulatory molecules, several quorum sensors were identified by initial detection of the pheromone itself, using acyl HSL-responsive reporter systems, followed by genetic isolation of the presumptive *luxI* and *luxR* homologs (see Table 73.1 for a listing of bacteria that produce acyl HSLs).

B. Generalities and mechanisms

The general features of acyl HSL quorum sensors are conserved in different species of bacteria. Most are comprised of homologs of the LuxR acyl HSL receptor and the LuxI acyl HSL synthase (Fig. 73.2; also see Section V for exceptions). Members of the LuxR and LuxI families of proteins share 18–25% and 28–35% identity across the length of each protein, respectively. Several bacteria, typically enteric species, synthesize 3-oxo-C6-HSL, identical to the *V. fischeri* factor. Other bacteria synthesize acyl HSLs that vary from the *V. fischeri* pheromone with respect to length and degree of saturation on the acyl chain, as well as the

TABLE 73.1 Selected LuxR–LuxI quorum sensors from gram-negative bacteria[a]

	Components identified[b]				
Bacterium	**Receptor**	**Synthase**	**Acyl HSL**	**Target function**	**Integrated with**[c]
Vibrio fischeri	LuxR	LuxI	3-oxo-C6	Bioluminescence (*lux*)	Glucose (CRP), Fe
Pseudomonas aeruginosa	LasR RhlR	LasI RhlI	3-oxo-C12 C4	Virulence factors Rhamnolipids, RpoS	RhlR, GacA, Vfr, LasR, RpoS regulon
Erwinia carotovora[d]	ExpR CarR	ExpI CarI	3-oxo-C6 3-oxo-C6	Exoenzymes Antibiotics	Carbon and sugar metabolism ?
Agrobacterium tumefaciens	TraR	TraI	3-oxo-C8	Conjugal transfer	Opines
Pseudomonas aureofaciens	PhzR	PhzI	C6	Phenazine antibiotics	LemA-GacA
Pantoea stewartii	EsaR	EsaI	3-oxo-C6	Exopolysaccharide (EPS)	EPS regulation (Rcs)
Serratia liquefaciens	?	SwrI	C6, C4	Swarming motility	amino acids, cell contact
Rhodobacter sphaeroides	CerR	CerI	Δ7-C14	Cell disaggregation	?
Escherichia coli	SdiA[e]	—	?	Cell division	Division control

[a] The list of quorum sensors is not exhaustive; see Fuqua *et al.*, 1996, and other reviews for more complete listings.
[b] The LuxR homolog, the LuxI homolog, and the primary acyl HSL are indicated. Only full-length, functional proteins are included. See text for discussion of individual systems.
[c] Indicates additional regulation that affects the function of the quorum sensor.
[d] Different LuxR–LuxI homologs have been identified in different subspecies of *Erwinia carotovora*, although several lines of evidence suggest that multiple LuxR–LuxI proteins may coexist in the same strain.
[e] The *E. coli* complete genome sequence reveals no *luxI* homolog.

substituent at the third position (Figs 73.1 and 73.2). Acyl HSLs typically have an acyl moiety with an even number of carbons, ranging from 4–14, associated with the homoserine lactone via an amide bond. Acyl HSLs and LuxR–LuxI-type regulators are thought to control their diverse target genes in response to population density via a mechanism analogous to that described for *V. fischeri* (Fig. 73.2). In environments that support only low cell density, acyl HSL, synthesized by the LuxI-type protein, dissipates by diffusion across the bacterial envelope. As conditions change to support larger numbers of cells or the flow characteristics of the environment are altered (i.e., diffusion is limited), acyl HSLs accumulate. At a specific concentration that probably varies for each microbe and each environmental niche, acyl HSLs interact with the LuxR-type receptor, elevating expression of target genes. While the *luxI* homolog is often among the target genes, as with the positive feedback described for *V. fischeri*, this is not always the case and is not essential for the function of the quorum sensor.

III. BIOSYNTHESIS OF ACYL HSLs BY LuxI-TYPE PROTEINS

A. Enzymology

It can be readily demonstrated that *luxI* and many of its homologs are required for acyl HSL synthesis. Likewise, the production of the appropriate acyl HSLs by heterologous hosts, such as *E. coli* expressing LuxI homologs, indicates that these proteins utilize common precursors and impart the biosynthetic specificity observed in the parent microbe. However, only recently has the prediction that LuxI and its homologs are acyl HSL synthases been proven. A major barrier to these findings was the identity and complexity of the substrates for biosynthesis. The general structure of acyl HSLs (Fig. 73.2) suggested that they were products of fatty acid and amino acid metabolism. Analysis of *V. fischeri* cell extracts using labeled precursors implicated methionine metabolic intermediates, and either biosynthetic or β-oxidative intermediates of fatty acid metabolism. More recent genetic studies with LuxI and the *A. tumefaciens* TraI protein have refined this view further, strongly implicating *S*-adenosyl methionine (AdoMet) and fatty acid biosynthetic intermediates as acyl HSL precursors.

Isolation of the purified acyl HSL synthases, either as native proteins or affinity-tagged fusion derivatives, has allowed *in vitro* analysis of acyl HSL synthesis. Under the appropriate reaction conditions, several LuxI-type proteins will synthesize acyl HSLs from AdoMet and acyl–acyl carrier protein (acyl ACP) conjugates, the primary intermediates of fatty-acid biosynthesis. A tentative reaction scheme has been formulated to describe the enzymatic reaction (Fig. 73.3). First, the acyl HSL synthase binds to AdoMet and an acyl–ACP carrying the correct length acyl chain with the appropriate oxidation state at the third position (all variations found at the third position on acyl HSLs are also present at the corresponding

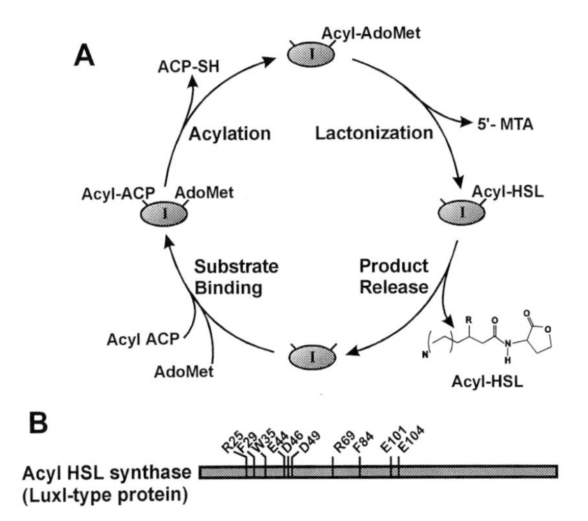

FIGURE 73.3 (A) Tentative catalytic cycle for acyl HSL synthesis. Bars extending from LuxI indicate noncovalent interaction with substrates and products. (B) LuxI protein is represented as a bar and amino acid residues conserved throughout the LuxI family are indicated (amino acids are in single letter code: D, aspartic acid; E, glutamic acid; F, phenylalanine; R, arginine; W, tryptophan). Numbering is relative to the LuxI amino acid sequence.

position during fatty-acid biosynthesis). The nucleophilic amino group on the methionine moiety of AdoMet attacks the thioester bond that conjugates the fatty acyl chain to ACP. Thus forms the amide linkage found in all acyl HSLs, displacing a reduced ACP. Lactonization of the methionine on AdoMet releases the acyl HSL and the reaction by-product 5′-methylthioadenosine (MTA), allowing the acyl HSL synthase to enter another catalytic cycle. While the current evidence suggests this is a plausible model, it remains to be rigorously tested.

B. Structure of acyl HSL synthases

Little is currently known regarding the molecular architecture of the acyl HSL synthases. They share no apparent similarity to other protein families in the sequence databases. However, the LuxI-type acyl HSL synthases now number over 15, and amino acid sequence comparisons within the family are informative. There are 10 amino acid residues, clustered in the amino terminal halves of the proteins (res. 25–104 in LuxI), conserved in all members of the family (Fig. 73.3). Of these, seven have side chains that carry charges at neutral pH. The other three carry aromatic side chains. Strikingly, loss of function mutations have been isolated at each of the seven conserved, charged amino acid residues (*luxI* and *P. aeruginosa rhlI*). Mutations in the three conserved aromatic residues severely reduce, but do not abolish, the activity of acyl HSL synthases. Many acyl transferases, enzymes that catalyze the exchange of acyl chains, utilize cysteine and serine residues at their active sites. At least for the acyl HSL synthases tested thus far, cysteine and serine residues do not appear to be required for the catalysis. Rather, it is likely that the conserved charged residues direct catalysis, probably via specific acid–base interactions with the substrates at the active site. There is currently no evidence that the acyl chain is transiently linked to the protein during catalysis, in contrast to some of the acyl transferases. A detailed view of acyl HSL synthesis and the architecture of the active site awaits additional biochemical and structural analyses.

IV. PERCEPTION OF AND RESPONSE TO ACYL HSLs

A. Entry into cells and receptor interaction

Although the prevailing dogma is that all acyl HSLs passively diffuse across the bacterial envelope, this idea has only been tested for 3-oxo-C6-HSL. Other acyl HSLs are also exchanged very rapidly, and it is, therefore, assumed that they, too, freely cross the envelope. However, it is possible that the exchange of some acyl HSLs is facilitated by transporters. Additionally, some acyl HSLs may also substantially partition into lipid bilayers, as opposed to simply crossing them.

Once inside the cell, all evidence suggests that the acyl HSLs interact directly with LuxR-type receptor proteins. This interaction alters the conformation of the receptor and fosters transcriptional regulation of target genes. Recognition of the acyl HSL involves contacts with the homoserine lactone ring as well the acyl chain portion of the molecule. Most LuxR-type proteins can distinguish between acyl HSLs whose structure differs only by the length of the acyl chain or oxidation state at the third position. In general, LuxR–LuxI regulatory pairs have evolved so that the LuxR-type protein is most sensitive to the primary product synthesized by its corresponding acyl HSL synthase. However, LuxR-type proteins will recognize acyl HSLs other than their cognate factor, albeit with reduced efficiency. The degree of specificity varies with each quorum sensor, some demonstrating high stringency and others with a broader recognition spectrum. Interestingly, although *A. tumefaciens* strains expressing wild-type levels of the TraR protein recognize only 3-oxo-C8-HSL, elevated expression causes a dramatic loss of specificity, suggesting a role for intracellular pool sizes of the LuxR-type protein in recognition of acyl HSLs.

B. Modular architecture of acyl HSL receptors

LuxR-type proteins have two distinct, physically separable activities—one is to bind acyl HSLs and the other is to bind specific DNA sequences. There is evidence that the sequences within the amino-terminal half of the LuxR protein, (amino acid residues 79–127), bind 3-oxo-C6-HSL (Fig. 73.4). Conversely, a protein consisting of only the carboxy-terminal 95 amino acids of LuxR binds to DNA and can also activate transcription, although it is not responsive to 3-oxo-C6-HSL. In the wild-type LuxR protein, these two functions are linked and interdependent. Interaction with acyl HSLs via the amino-terminal region regulates binding of the LuxR-type protein to its target sites on the DNA. At least for LuxR, this appears to be a negative regulation, where the amino-terminal half of the protein inhibits the activity of the carboxy-terminal DNA-binding region in the absence of 3-oxo-C6-HSL. Binding of the ligand relieves this inhibition and allows the protein to activate transcription.

Comparisons of amino acid sequences between members of the LuxR family reflect the binary structure that has been discussed. While the overall similarity between proteins is relatively low, the similarity within the presumptive acyl HSL-binding region and the DNA-binding region is substantially higher. The acyl HSL-binding region is a signature sequence, shared only by members of the LuxR family (Fig. 73.4). In contrast, the carboxy terminal DNA-binding region (amino acid residues 183–250) contains a helix–turn–helix (H–T–H) motif, found in many proteins that interact with DNA. Furthermore, the HTH motif and sequences that surround it firmly place these DNA-binding modules into the much larger FixJ superfamily (Fig. 73.4). Most other members of this superfamily are two-component type response regulators, with amino terminal domains that are phosphorylated at conserved aspartate residues by a sensor kinase. Interestingly, for several of these FixJ-type response regulators, phosphorylation within the regulatory region relieves inhibition of an otherwise constitutive DNA-binding activity.

Genetic analysis suggests that transcriptional activation also requires multimerization of LuxR-type proteins. Nonfunctional variants of both LuxR and TraR, lacking their DNA-binding region, act as dominant negative inhibitors of their wild-type counterparts. The nonfunctional protein, in effect, poisons the wild-type protein by forming inactive complexes. On LuxR, the region between amino acid residues 116–161, overlapping with the acyl HSL ligand recognition region, is required for multimerization (Fig. 73.4). Formation of LuxR multimers must be integrated with the intramolecular events that occur upon ligand interaction. It is unclear whether multimerization occurs subsequently or concomitantly with the acyl HSL interaction. It is possible that binding of the acyl HSL stimulates multimerization and that this shift in contacts abolishes the inhibition of DNA-binding by the amino terminal half of the protein.

C. DNA binding and transcription activation

In 1989, Devine and Shadel identified the *lux* operator, a 20 bp inverted repeat centered at −40 relative to the *lux* operon promoter. Specific mutations in this sequence caused a reduction in transcriptional activation by LuxR. As other acyl HSL-based quorum sensors were identified, it was recognized that similar sequences existed upstream of many of the genes under cell-density dependent control. These sequences not only shared the basic inverted repeat motif, and the position relative to the regulated promoter, but also primary sequence similarity with the *lux* operator. These *cis*-acting sequences are now referred to as *lux*-type boxes and they are the presumptive binding sites for LuxR-type proteins. The *lux*-type box can be 18–20 bp in length, is generally located between −40 and −44 from the transcriptional start site, and is often essential for transcriptional regulation of target genes by LuxR-type proteins. However, not all genes that are activated by LuxR-type proteins have recognizable *lux*-like boxes upstream of their promoters. Therefore, these sequences appear to be a common, although not

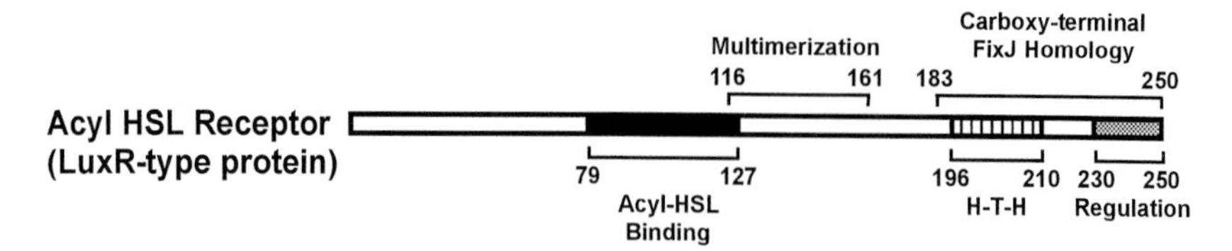

FIGURE 73.4 Modular structure of LuxR-type proteins. Functional regions are based primarily on genetic analyses of LuxR. The numbering scheme is relative to the LuxR amino acid sequence. See text for details.

obligatory, component of acyl HSL-regulated promoters.

The position of *lux*-type boxes relative to their associated promoters suggests that LuxR-type proteins make specific contacts with RNA polymerase, much the same as other prokaryotic transcriptional activators. While it is still unclear how transcription is stimulated, preliminary findings suggest that the carboxy terminal portion of the α subunit of RNA polymerase, a site of contact for many regulators, is required for transcriptional activation. A carboxy terminal fragment of LuxR with constitutive DNA binding activity (see Section IV.B) was shown by *in vitro* DNA-binding assays to associate with the *lux* promoter region. However, specific binding to the *lux* operator required the presence of both the LuxR carboxy terminal fragment and RNA polymerase. It is possible that this apparent synergy may be an artifact of using the truncated form of LuxR and the full-length protein may not exhibit this characteristic. Full-length LasR and TraR associate with their binding sites *in vitro* in the absence of RNA polymerase.

D. LuxR-type proteins that function as repressors

The majority of LuxR-type proteins activate transcription. Consequently, null mutations in most *luxR* homologs result in low level, noninducible expression of target genes. However, mutations of the *esaR* gene from *Pantoea stewartii* (formerly, *Erwinia stewartii*) that regulates capsular polysaccharide (EPS) synthesis (Table 73.1), results in elevated synthesis of EPS. This and other data indicate that EsaR is a repressor and that inducing levels of 3-oxo-C6-HSL (the cognate acyl HSL produced by EsaI) cause a derepression of transcription. A simple model would have EsaR bound to regulated promoters in the absence of inducer and dissociated in its presence. If this is correct, EsaR must adopt an active conformation when not associated with ligand and convert to an inactive conformation upon binding of the acyl HSL. There are no features in the EsaR amino acid sequence that are notably different from LuxR-type activators. However, several other LuxR-type proteins are also suspected to be repressors.

E. Modulation of responses to acyl HSLs

Although the basic components of acyl HSL quorum sensors are LuxI- and LuxR-type regulatory proteins, there are several examples where additional factors directly impinge upon, or are necessary for, proper regulation. The clearest example of this is the *A. tumefaciens* TraR–TraI quorum sensor, where several additional regulatory proteins modulate the TraR-dependent transcriptional activation. One of these is the TraM protein, identified in several strains of *A. tumefaciens* and species of *Rhizobium*. TraM acts to inhibit the function of TraR under noninducing conditions. Mutations within *traM* result in constitutive activation of TraR-dependent genes and loss of cell density-responsiveness. These results suggest that TraM is an integral component of the *A. tumefaciens* quorum sensor. In fact, the *traM* gene is itself activated by TraR and 3-oxo-C8-HSL, creating a negative feedback loop. It is unclear why the *A. tumefaciens* quorum sensor has incorporated this additional component, and the physiological role is a current topic of investigation.

There is an additional layer of regulation in at least some strains of *A. tumefaciens*. TraR is also inhibited by the activity of the Trl protein, a protein highly similar to the amino terminal half of TraR (amino acid residues 1–181) but missing the carboxy terminal DNA-binding region. Genetic analysis reveals that Trl is dominant to TraR, suggesting that it forms inactive TraR–Trl heteromultimers. The *trl* gene is itself under the control of plant-released signals, similar to signals that control expression of the *traR* gene (Table 73.1). Apparently the balance of TraR and Trl, within the cell at any given time, also integrated with the effect of TraM, dictates whether TraR will activate target gene expression.

The LuxR protein of *V. fischeri* is also modulated, not by additional regulatory proteins, but rather by C8-HSL. This factor is synthesized by the AinS acyl HSL synthase, distinct from LuxI (see Section V). In the absence of 3-oxo-C6-HSL, the AinS-produced factor is a mild inducer of the *lux* genes. However, *V. fischeri* with mutations in *ainS*, and with a functional copy of *luxI*, activate expression of the *lux* genes at significantly lower cell densities than wild-type cells. The effect of the *ainS* mutation can be masked by addition of exogenous C8-HSL, suggesting that this factor is acting as an inhibitor of LuxR, perhaps competing with 3-oxo-C6-HSL, for access to the acyl HSL-binding site.

P. aeruginosa produces two primary acyl HSLs as well, 3-oxo-C12-HSL and C4-HSL, the products of the LasI and RhlI proteins, respectively (Table 73.1). While each acyl HSL acts predominantly through its own LuxR type receptor (LasR and RhlR), there is evidence that 3-oxo-C12-HSL can reduce the responsiveness of RhlR to C4-HSL, possibly through direct competition for the acyl HSL-binding site. Although it is not clear how common it is for bacteria to utilize mutliple quorum sensors, as is the case in *P. aeruginosa* and *V. fischeri*, these observations suggest that acyl HSLs produced within the same cell can have physiologically relevant interactions *in vivo*.

V. VARIATIONS ON THE PARADIGM

A. Quorum sensing in *V. harveyi*

As discussed, the original observations regarding autoinduction of bioluminescence were made with the two marine vibrios, *V. fischeri* and *V. harveyi*. Both bacteria control bioluminescence in response to cell density, by production of diffusible autoinducers, and both produce acyl HSLs, 3-oxo-C6-HSL and 3-OH-C4-HSL, respectively. For technical reasons, studies of the *V. fischeri* system proceeded more rapidly and have led to development of the LuxR–LuxI paradigm. It was assumed, quite logically, that the *V. harveyi* autoinduction mechanism would be similar. However, the reality is, in fact, strikingly more complicated (Fig. 73.5). While *V. harveyi* does produce an acyl HSL with which it monitors its own population density, the proteins responsible for synthesis of 3-OH-C4-HSL and its effect on *lux* gene expression share no homology to LuxR and LuxI. Production of 3-OH-C4-HSL requires the activity of two genes *luxL* and *luxM*, one or both of which comprise the acyl HSL synthase (LuxL/M). As expected, this factor is freely diffusible across the bacterial envelope and accumulates with increasing cell density. Response to 3-OH-C4-HSL by induction of *lux* genes, requires a two-component type sensor kinase, called LuxN, which interacts with a cognate response regulator called LuxO. LuxO is a repressor of *lux* genes and is, presumably, inhibited in the presence of 3-OH-C4-HSL. Inactivation of the 3-OH-C4-HSL quorum sensor, by mutating, either the *luxLM* acyl HSL synthase or the *luxN* sensor, does not, however, abolish this microbe's ability to sense its own population density. There is a second quorum sensing system that also regulates *lux* genes (Fig. 73.5). This apparently redundant system relies on a signal molecule(s), called AI-2, that is released from cells and accumulates with increasing cell numbers. Response to AI-2 requires the activity of another sensor kinase, LuxQ, that also appears to interact with the LuxO response regulator to control *lux* gene expression. An additional transcriptional activator called LuxR (not related to the *V. fischeri* LuxR) also affects *lux* expression, but is not responsive to either 3-OH-C4-HSL or AI-2. The AI-2 based system is also dispensable for quorum sensing, provided the 3-OH-C4-HSL system is intact. Therefore, the *V. harveyi* quorum sensors appear to be functionally redundant and mechanistically distinct from the canonical LuxR–LuxI system. Recent findings suggest that 3-OH-C4-HSL may act as an intraspecies signal, while AI-2 may allow *V. harveyi* to monitor the presence of other bacteria. In support of this, AI-2-like signals have now been detected from *E. coli* and *Salmonella typhimurium*, as well as other bacteria, while 3-OH-C4-HSL has only been reported for *V. harveyi*.

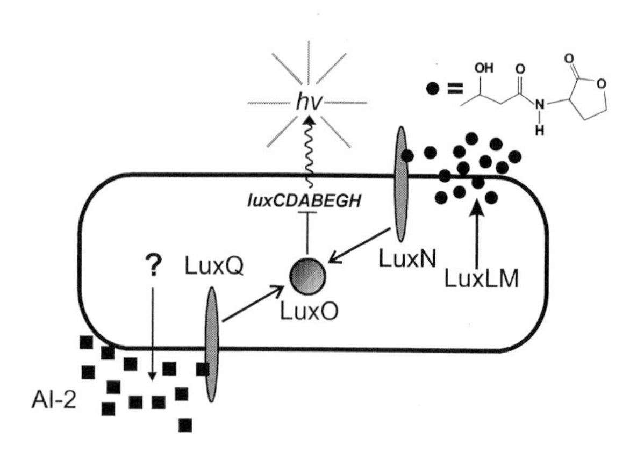

FIGURE 73.5 Regulation of bioluminescence in *V. harveyi*. Simplified model of dual quorum sensors, adopted and modified from Bassler and Silverman, 1995. LuxQ and LuxN are transmembrane, two-component type sensor kinases that presumably phosphorylate the common response regulator, LuxO, a repressor of *lux* genes. An additional regulator of *lux* genes, the *V. harveyi* LuxR protein, has been omitted for clarity and because its role in quorum-sensing has not been established. The signal molecule produced by LuxL/M is 3-OH-C4-HSL, while the structure of AI-2 signal is not yet known. See text for additional details.

B. A second acyl HSL in *V. fisheri*

Homologs of the *V. harveyi* LuxM and LuxN proteins have been identified in *V. fischeri*. The AinS protein is homologous to LuxM and is required for synthesis of C8-HSL, in addition to the 3-oxo-C6-HSL synthesized by LuxI. A gene encoding a presumptive sensor kinase, AinR, homologous to LuxN, is physically linked to *ainS*. The regulatory targets of AinR have not been identified, although AinS and C8-HSL may have a physiological role in modulating LuxR activity (see Section IV.E). The discovery of LuxL/M–LuxN in *V. harveyi* and the homologous AinS–AinR system in *V. fischeri* suggests that these proteins constitute a second family of quorum sensors, based on acyl HSLs, but mechanistically distinct from LuxR–LuxI type quorum sensors. The fact that acyl HSL quorum sensing has apparently evolved twice is remarkable and may reflect the potency of these compounds as signal molecules.

VI. REGULATION OF QUORUM SENSORS

A. Integration of acyl HSL quorum sensing into cellular physiology

As techniques for genome-scale analysis of gene expression become available, it is more and more apparent that different aspects of cellular metabolism are linked through regulatory networks. Acyl HSL quorum-sensing regulators are often embedded in or otherwise integrated with additional control circuitry (Table 73.1). For example, expression of the *luxR* gene is activated by the catabolite repressor protein (CRP), resulting in glucose repression of bioluminescence. In *A. tumefaciens*, expression of the *traR* gene occurs only in the presence of opines, plant-released compounds that tie the function of the quorum sensor to the rhizosphere. Most, if not all, acyl HSL regulators are positioned within a larger regulatory hierarchy. It appears that, for each microbial species, the regulatory context is different, probably reflecting the specific environmental niche the microbe occupies and the varying roles for quorum sensing (see Table 73.1).

A well-studied case of this regulatory integration is found in *P. aeruginosa*, where there are two distinct quorum sensors, the Las and Rhl systems (Table 73.1). LasR and 3-oxo-C12-HSL were identified as regulators of elastases and other virulence factors, while RhlR and C4-HSL were found to regulate production of rhamnolipid surfactants. LasR, in fact, controls the expression of *rhlR*, possibly facilitating a hierarchical staging of gene activation where the targets of LasR are expressed first, followed by those under RhlR control. Another regulatory target for Las, and possibly Rhl, are the *xcp* genes encoding an exoenzyme secretion system, modulating certain aspects of protein secretion in response to cell density. The Las–Rhl quorum sensing cascade is itself under control of at least two regulators: (i) the *P. aeruginosa* GacA protein, a two-component type response regulator, homologs of which control virulence in plant pathogenic pseudomonads, and (ii) the Vfr protein, a *P. aeruginosa* CRP homolog. The physiological significance of GacA and Vfr control of the Las–Rhl dual quorum sensors is not known.

VII. ECOLOGY AND FUNCTION OF QUORUM SENSING

A. Multiple roles for acyl HSL signaling?

The diversity of bacteria known to employ acyl HSLs is expanding with the discovery of each new quorum sensor. Quorum sensing may come into play under any situation where it is beneficial for a specific process to be differentially regulated as a function of population density. Acyl HSLs often regulate genes involved in host–microbe interactions, symbiotic or pathogenic (Table 73.1). The host environment provides conditions where high cell density responses are relevant and beneficial. However, acyl HSLs are clearly not restricted to regulation of interactions with host organisms, as they have been identified in microbes with no known host association, such as the free-living photosynthetic bacterium *Rhodobacter sphaeroides*. Acyl HSLs may also facilitate concerted and rapid responses by a population to external stimuli. For example, certain members of a population may respond to a specific, low-level stimulus, such as the presence of a host-released signal molecule, by stimulating production of an acyl HSL quorum sensor. Dissemination of the acyl HSL signal throughout the local environment might act to amplify the original host signal and enable a coordinated response by the bacterial population. Such signaling may occur between bacteria of the same species or different bacteria that employ similar acyl HSLs (see Section VII.C).

B. Microcolonies and biofilms

Many bacteria are readily grown in laboratory culture at cell densities sufficient to trigger acyl HSL-dependent gene expression. In the natural environment, however, such luxuriant conditions are far less prevalent, and large numbers of cells at high densities are probably not the norm. The flow characteristics and physical dimensions of any given environment will also influence if and when acyl HSLs accumulate. Squid light organs colonized with *V. fischeri* contain concentrations of 3-oxo-C6-HSL well above those required to activate LuxR and affect light production. In this case, the confined (~15 nL), nutrient-rich environment of the light organ supports a high cell density (~10^{10} cfus/ml) and provides the appropriate physical characteristics to maintain inducing concentrations of the acyl HSL. However, other environmental niches occupied by quorum sensing bacteria are less tailored to provide such conditions. A number of plant pathogens control elaboration of extracellular virulence factors by acyl HSL quorum sensing and, in several cases, the appropriate timing of this attack is crucial to the success of the infection. Clearly, where active infections occur, inducing conditions have been achieved. What is the composition of the population that leads to accumulation of the acyl HSL and subsequent induction? Although the answer is not certain, plant-associated bacteria often form homogeneous or mixed microcolonies. The relative cell densities on the interior of the microcolony are likely to be quite high and, combined with the physical barriers posed by the host as well as polymeric material (e.g., EPS) produced within the microcolony, may provide conditions that lead to acyl HSL accumulation. Bacterial cohorts analogous to plant-associated microcolonies are also thought to colonize infection sites in animal hosts.

A biofilm is another bacterial structure common to a variety of environments. Biofilms are surface-associated, organized layers of bacteria, usually bound in a matrix of extracellular material. These bacterial communities can exhibit a surprising architectural complexity, with aqueous channels, chambers, and interconnected microcolonies of cells. Biofilm formation is a major concern in medical and industrial settings, where members of the biofilm demonstrate greater recalcitrance than do their planktonic counterparts to antimicrobial strategies. It is clear that biofilm communities provide environments that foster intercellular interactions between bacteria, and acyl HSL signaling may be an important aspect of existence in a biofilm. In fact, acyl HSLs have been detected in the effluent of living biofilms from several natural environments. Furthermore, *P. aeruginosa* strains unable to synthesize 3-oxo-C12-HSL form defective biofilms, with far greater sensitivity to antimicrobial treatment. This observation suggests that not only is acyl HSL signaling facilitated by conditions in the biofilm, but that these molecules may also play a role in the development of the film. It remains to be seen how universal the role of acyl HSLs is in biofilms of different bacteria.

C. Mixed messages: do acyl HSLs mediate interspecies communication?

Many bacteria release and respond to acyl HSLs of similar structure. If these bacteria grow in close proximity to each other, such as in a microcolony or biofilm, it is possible, perhaps even unavoidable, that one species of microbe will respond to the acyl HSL of another species. Current findings suggest that cohabiting acyl HSL-producing microbes can influence each other's behavior *in situ*. *P. aeruginosa* and *Burkholderia cepacia* are common opportunistic pathogens of patients with cystic fibrosis and may form mixed species communities within the same individual. Each bacteria produces acyl HSLs that can potentiate elaboration of virulence functions by the other bacteria, perhaps collaboratively pathogenizing the host organism. A different example has been provided by examination of artificial mixed populations of the plant biocontrol agent *Pseudomonas aureofaciens*. Coinoculation of host plants (wheat) with a mixture of wild-type *P. aureofaciens* producing C6-HSL and a mutant strain (PhzI-) responsive to, but unable to synthesize, the pheromone, resulted in strong activation of target genes in the mutant strain. Findings such as these suggest that, in many environments, opportunities for interspecies signaling are not only possible, but prevalent.

D. Host responses to acyl HSLs

Interactions between microbes and their hosts are often games of detection and response, with each partner evolving the capacity to perceive and manipulate the other. Animals and plants have developed elaborate mechanisms by which they defend themselves from foreign agents. Acyl HSLs provide lines of communication between infecting bacteria, and, therefore, it may be beneficial for host organisms to develop the capacity to "eavesdrop" on the coded messages of their microbial colonizers. In addition, acyl HSLs might stimulate the immune system in a manner detrimental to the host, acting directly as classic virulence factors. Several studies have shown that a subset of acyl HSLs exhibits immunomodulatory activity in animal hosts, affecting cytokine production and immune cell proliferation. While these observations are intriguing, a much more complicated question that remains to be addressed is whether this activity is relevant *in situ*? If so, how do these responses affect the interaction between microbe and host?

At least one animal host has evolved the capacity to interfere with acyl HSL signaling. The marine sponge *Delisea pulchra* produces a pair of halogenated furanones, structural analogs of acyl HSLs, that efficiently block acyl HSL-dependent induction of swarming motility in *Serratia liquefaciens* and LuxR activation of bioluminescence. It is thought that the furanones function by directly inhibiting the interaction of acyl HSLs with LuxR-type proteins. It is unclear what role(s) the furanones play in the natural setting. Are they specific antagonists of quorum sensing or do they coincidentally inhibit the process?

VIII. CONCLUSIONS

Acyl HSL quorum sensing, for many years considered to be a specialized form of gene regulation in bioluminescent *Vibrio* species, is now known to be quite prevalent in gram-negative bacteria. The general mechanism appears to be conserved among a wide range of bacteria, although those functions regulated by the quorum sensor are extremely varied between different bacteria. As our understanding of the function and role(s) of acyl HSLs in bacterial communities improves, the prospects for utilizing these potent signal molecules in the control of bacterial behavior also improve. Likewise, studies on all forms of bacterial quorum sensing should continue to blur the distinction between unicellular and multi-cellular lifestyles, as well as to expand our appreciation of the lines of communication that underlie complex ecosystems.

BIBLIOGRAPHY

Chugani, S. A., Whiteley, M., Lee, K. M., D'Argenio, D., Manoil, C., and Greenberg, E. P. (2001). QscR, a modulator of quorum-sensing signal synthesis and virulence in Pseudomonas aeruginosa. *Proc. Natl. Acad. Sci. U.S.A* **98**, 2752.

Dunlap, P. V., and Greenberg, E. P. (1991). Role of intercellular communication in the *Vibrio fischeri*–monocentrid fish symbiosis. *In* "Microbial Cell–Cell Interaction" (M. Dworkin, ed.). ASM Press, Washington, DC.

Dunlap, P. V., and Greenberg, E. P. (1985). Control of *Vibrio fischeri* luminescence gene expression in *Escherichia coli* by cyclic AMP and cyclic AMP receptor protein. *J. Bacteriol.* **164**, 45–50.

Dunny, G. M., and Leonard, B. A. B. (1997). Pheromone-inducible conjugation in *Enterococcus faecalis*: Interbacterial and host–parasite chemical communication. *Annu. Rev. Microbiol.* **51**, 527–564.

Fuqua, C., and Greenberg, E. P. (2002). Listening in on bacteria: acyl-homosterine lactone signalling. *Nat. Rev. Molec. Cell Biol.* **3**, 685–695.

Fuqua, W. C., Winans, S. C., and Greenberg, E. P. (1994). Quorum sensing in bacteria: The LuxR/LuxI family of cell density-responsive transcriptional regulators. *J. Bacteriol.* **176**, 269–275.

Fuqua, C., Winans, S. C., and Greenberg, E. P. (1996). Census and consensus in bacterial ecosystems: The LuxR–LuxI family of quorum-sensing transcriptional regulators. *Annu. Rev. Microbiol.* **50**, 727–751.

Gray, K. M. (1997), Intercellular communication and group behavior in bacteria. *Trends Microbiol.* **5**, 184–188.

Hastings, J. W., and Greenberg, E. P. (1999). Quorum sensing: the explanation of a curious phenomenon reveals a common characteristic of bacteria. *J. Bacteriol.* **181**, 2667–2668.

Kaiser, D., and Losick, R. (1994). How and why bacteria talk to each other. *Cell* **73**, 873–885.

Lazazzera, B. A. (2000). Quorum sensing and starvation: signals for entry into stationary phase. *Curr. Opin. Microbiol.* **3**, 177–182.

Parsek, M. R., and Greenberg, E. P. (2000). Acyl-homoserine lactone quorum sensing in gram-negative bacteria: a signaling mechanism involved in associations with higher organisms. *Proc. Natl. Acad. Sci. U S A* **97**, 8789–8793.

Singh, P. K., Schaefer, A. L., Parsek, M. R., Moninger, T. O., Welsh, M. J., and Greenberg, E. P. (2000). Quorum-sensing signals indicate that cystic fibrosis lungs are infected with bacterial biofilms. *Nature* **407**, 762–764.

Ruby, E. G. (1996). Lessons from a cooperative bacterial–animal association: The *Vibrio fischeri–Euprymna scolopes* light organ symbiosis. *Annu. Rev. Microbiol.* **50**, 591–624.

Swift, S., Throup, J. P., Williams, P., Salmond, G. P. C., and Stewart, G. S. A. B. (1996). Gram-negative bacterial communication by N-acyl homoserine lactones: A universal language? *Trends Biochem. Sci.* **21**, 214–219.

WEBSITE

Quorum sensing site. University of Nottingham. With links *http://www.nottingham.ac.uk/quorum/*

74

Recombinant DNA, Basic procedures

Judith W. Zyskind
San Diego State University

GLOSSARY

α-complementation Active β-galactosidase enzyme complex formed between peptide containing N-terminal region (~15%) of β-galactosidase and β-galactosidase peptide missing the amino terminus.

autoradiography Use of x-ray film to detect the presence of radioactive material in gels and filters.

cDNA DNA copy of mRNA synthesized by reverse transcriptase, a DNA polymerase that can use either RNA or DNA as a template.

cloning Process of inserting foreign DNA into a plasmid or bacteriophage vector and reproducing this recombinant DNA in a host cell, such as *Escherichia coli*.

cos site *cos* or cohesive end site of λ phage; the site of action of terminase, the enzyme that cuts at *cos* during packaging, leaving a 12-b single-stranded sequence at the 5′ ends.

electroporation High efficiency uptake of DNA caused by artificially inducing cell permeability with a high-voltage pulse that is applied to a suspension of cells and DNA.

endonuclease Nuclease that cleaves internal phosphodiester bonds in nucleic acids.

field inversion gel electrophoresis (FIGE) Separation of high molecular weight DNA molecules by agarose gel electrophoresis with an electrical field that pulses both forward and backward, with a pause between each pulse.

ligation Phosphodiester bond formation between two nucleic acid molecules by DNA ligase.

multiple cloning site (MCS) DNA sequence containing single copies of many sites cut by different restriction endonucleases; also called polylinker.

origin of replication Site at which DNA replication is initiated.

polymerase chain reaction (PCR) Amplification of minute amounts of DNA in a test tube using primers that flank the sequence to be amplified by a thermoresistant DNA polymerase.

replica plating Transfer of an impression of colonies, made on sterile velveteen fabric held tightly over a circular form, to a fresh agar plate, or transfer of clones from a multiwell microtiter dish containing a genomic or cDNA library, using a device with steel prongs.

restriction enzyme mapping Determining the location of restriction enzyme cleavage sites in a piece of DNA without resorting to nucleotide sequence analysis.

reverse genetics Cloning of a gene after determining a portion of the amino acid sequence of its protein product and subsequent introduction of mutations into the cloned gene that are then moved to the chromosome.

Southern blotting Procedure for transferring denatured DNA either from agarose gels or from another source, for example, colonies or plaques, to nitrocellulose or nylon membrane.

The Desk Encyclopedia of Microbiology
ISBN: 0-12-621361-5

Recombinant DNA Technology is the application of techniques currently used to isolate and analyze genes which involves inserting foreign DNA into a host cell, such as *Escherichia coli*. This technology also can include determining the nucleotide sequence of this DNA in order to characterize the functions of this and its products. The introduction of PCR technology for amplifying small amounts of DNA has provided additional tools in this field.

I. CLONING STRATEGIES

A. Direct selection by complementation

Figure 74.1 illustrates a typical cloning experiment in which a selection method is available for the gene to be cloned. The gene in this cloning example is the *E. coli* *lacZ* gene encoding β-galactosidase. Plasmids containing the *lacZ* gene can be identified by complementation of a *lacZ* mutant, which is incapable of using lactose as a carbon source and, therefore, is unable to grow on minimal lactose medium. In the experiment shown in Fig. 74.1, the plasmid vector, pBR322, contains two genes, *bla* and *tet*, conferring resistance to ampicillin (Ap) and tetracycline (Tc), respectively. One of the genes can be used to select for the presence of the plasmid and the other gene can be used to screen for inserts by insertional inactivation. For example, DNA molecules inserted into the *tet* gene inactivate this gene. Plasmids carrying such inserts and transformed into bacteria can be selected on agar plates containing Ap and subsequently screened for inserts by the inability to grow on Tc-containing agar plates. The origin of replication of pBR322 is derived from a high copy number plasmid related to ColE1, so 30–50 copies are present in each cell. New genes introduced into pBR322 also will be present at this high copy number.

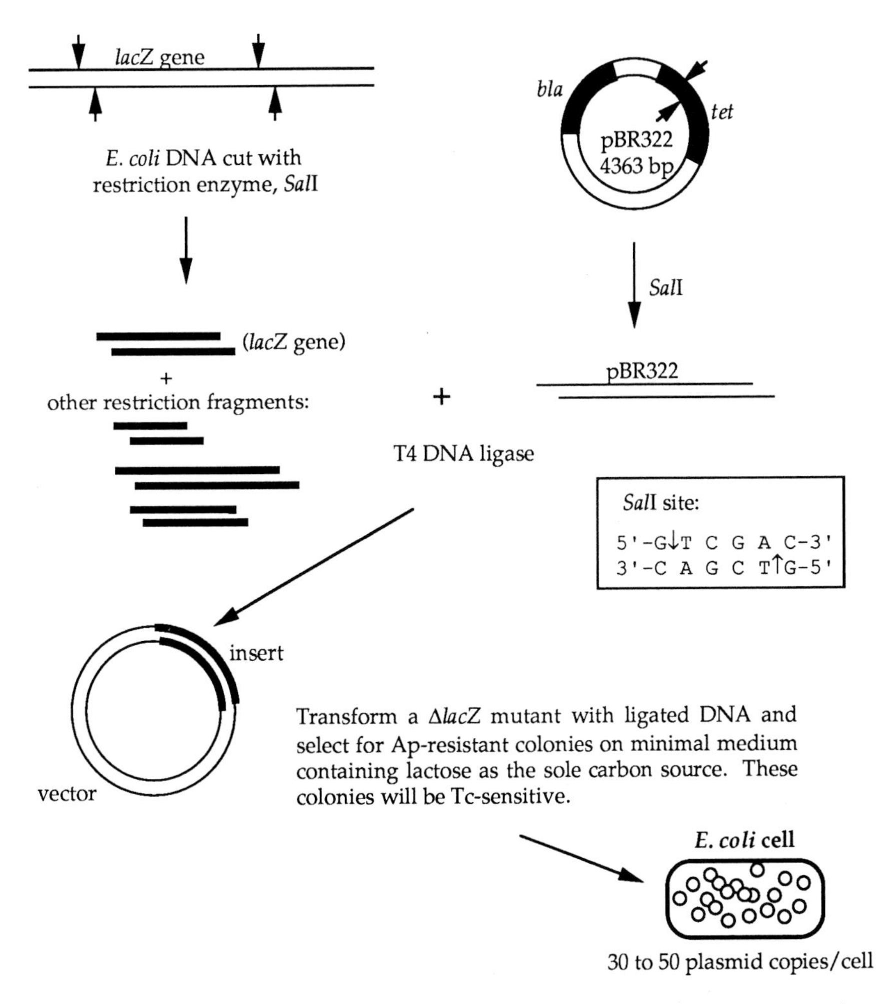

FIGURE 74.1 Cloning the *E. coli lacZ* gene. Components include plasmid vector, pBR322, restriction enzyme that cuts once in the vector, *Sal*I, *E. coli* chromosomal DNA, DNA ligase, and a *lacZ* mutant of *E. coli*. bp, Base pairs.

In the cloning experiment outlined in Fig. 74.1, chromosomal DNA is isolated from *E. coli*, then cut with the type II restriction enzyme *Sal*I, which cuts DNA at a specific sequence to give staggered or sticky ends that are complementary. The plasmid cloning vector pBR322, which has a single *Sal*I site, is also cut with *Sal*I. The two DNAs are mixed, complementary *Sal*I ends from chromosomal and vector DNA molecules hybridize to each other, and T4 DNA ligase forms phosphodiester bonds between adjacent nucleotides. The ligated DNA is introduced, either by transformation or electroporation, into a cloning host, in this case, a strain of *E. coli* deleted for the *lacZ* gene, and the cells are plated on minimal agar plates containing Ap and lactose as the sole carbon source. Colonies that develop will contain pBR322 with the *lacZ* gene inserted in the *Sal*I site.

Part of the cloning process involves plasmid replication in the host cell during colony formation. There are approximately 10^5 cells in a single colony, each with 30–50 copies of the same chimeric plasmid. This single colony and cells derived from it are frequently referred to as a clone. To analyze the cloned DNA further, plasmid DNA is isolated from clones. Depending on the scale of plasmid recovery desired, there are two main methods for isolating plasmid DNA. Miniprep procedures for plasmid isolation from small broth cultures will yield sufficient amounts of plasmid DNA to determine the size of the insert by restriction enzyme mapping and to sequence double-stranded DNA. Large-scale plasmid isolation procedures frequently rely on cesium chloride (CsCl) density gradient centrifugation in the presence of ethidium bromide to separate linear chromosomal DNA from the plasmid molecules that are circular and supercoiled. Kits for both small- and large-scale isolation of plasmid and chromosomal DNA are available from a variety of suppliers.

Ligation of the vector ends would recreate the vector in the experiment shown in Fig. 74.1. To prevent self-ligation of the vector, the phosphate groups at the 5′ ends of the vector DNA can be removed by a phosphatase such as bacterial alkaline phosphatase. Self-ligation of the vector also can be prevented by cutting both the vector and insert DNA with two different restriction enzymes that produce ends that are not compatible. This highly efficient method, called forced directional cloning, yields insertion of a restriction fragment in only one orientation and is used frequently with vectors containing a multiple cloning site (MCS).

DNA can be cloned from any organism to produce chimeric DNA molecules composed of DNA from different sources. For example, in the cloning experiment shown in Fig. 74.1, DNA from any prokaryotic organism and cDNA from any eukaryotic organism could be substituted for *E. coli* DNA, as long as the organism produces β-galactosidase and the gene is expressed in the cloning host.

B. Screening a library

If the gene to be cloned cannot be selected for directly, then identification becomes more difficult. A genomic library consisting of many different clones, each containing a different insert DNA in the plasmid vector, is screened for specific clones by looking for a particular physiological trait or by hybridization. For example, if a large number of Ap-resistant colonies was isolated on rich medium and screened for Tc-sensitivity in the experiment shown in Fig. 74.1, such a collection of Ap-resistant, Tc-sensitive clones could be considered a genomic library. The clone or clones containing the *lacZ* gene can be identified as red colonies after replica plating the genomic library onto MacConkey–lactose Ap agar plates.

The presence of the *lacZ* genes also can be identified by colony hybridization. A probe can be made from the *lacZ* gene of *E. coli* and labeled appropriately so that the probe can be detected by the presence of radioactivity, color, chemiluminescence, or fluorescence. The probe can then be used to identify the *lacZ* gene of another organism by hybridization with DNA from the genomic library that has been transferred to nitrocellulose (Southern blot). In this technique, nitrocellulose or nylon membrane cut to the size of a petri dish is placed on an agar plate, and the genomic library is transferred by replica plating, or by using toothpicks, to the surface of the membrane. After colonies form, the membrane with the colonies is moved to a lysing and denaturing solution, and the plasmid DNA is fixed to the membrane. The labeled *lacZ* probe is hybridized to the nitrocellulose or nylon membrane. Depending upon how the probe was labeled, the clone with the *lacZ* gene can be identified by detecting the presence of radioactivity, color, chemiluminescence, or fluorescence.

More commonly today, if the DNA sequence of the genome is known, the desired gene is identified by a homology search, then amplified by PCR, using primers generated from sequences flanking the gene. Restriction endonuclease sites can be included in the primers to facilitate cloning.

C. Choosing an *Escherichia coli* host for cloned DNA

Foreign DNA is degraded when introduced into *E. coli* cells because of the resident host restriction–modification system. Three genes are involved in the Type I *Eco*K restriction–modification system in *E. coli* K-12 strains, but only mutations in the *hsdR* gene confer the phenotype r^-m^+. Mutations in the other two genes; *hsdM* and *hsdS*, give an r^-m^- phenotype. Strains that have a mutation in the *hsdR* gene contain no *Eco*K restriction endonuclease activity but do have the *Eco*K methyltransferase. Such strains are good cloning hosts for foreign DNA.

There are at least three *E. coli* restriction endonucleases that cut methylated DNA. If the DNA to be cloned contains methylated DNA, additional mutations that eliminate these restriction endonucleases should be included in the cloning host. These mutations include *mcrA* (*m*odified *c*ytosine *r*estriction), *mrcB*, and *mar* (*m*odified *a*denine *r*estriction). If the gene to be cloned is homologous to sequences in the chromosome, then a *recA* mutation could be used to prevent homologous recombination between the cloned DNA and the chromosome. For example, in Fig. 74.1, a *recA* mutant host would be a preferred host for cloning the *E. coli lacZ* gene if the host strain contained a point mutation, rather than a deletion of the *lacZ* gene.

D. Reverse genetics

The peptide product of a gene can be used to identify and clone that gene. If a partial amino acid sequence is available from an isolated protein, a mixed probe can be synthesized and labeled. A mixed probe is a set of oligonucleotides corresponding to every combination of codons for each amino acid. One of the oligonucleotides will be identical in sequence to the gene sequence and will hybridize to the gene if present in a genomic library.

Replacement mutagenesis can be carried out once a gene is cloned. A total deletion (null mutation) of the cloned gene can be engineered, then introduced into the chromosome to replace the wild-type gene. Replacement requires a double reciprocal homologous recombination event between sequences flanking the gene in the clone and on the chromosome. A *recD* mutation, which eliminates the RecBCD exonuclease V but does not destroy homologous recombination, is useful for cloning linear DNA in *E. coli* when replacing a wild-type gene on the chromosome with a mutation or deletion of that gene. The phenotype of this mutation can give valuable clues to the function of the gene and its product in the cell. If the gene being replaced is required for cell growth, the deletion of the gene cannot be recovered. For required genes, the deletion is constructed in the presence of conditional expression of the gene product where, either on a plasmid or elsewhere in the chromosome, the gene is placed under the control of an inducible promoter, such as the arabinose promoter.

E. Characterizing the protein products of cloned genes

Three methods are available to examine the sizes of proteins encoded by plasmids: use of minicells, maxicells, or a coupled *in vitro* transcription–translation system. Minicells are small chromosomeless cells that are produced by *minB* mutants of *E. coli* and can be isolated from normal cells. High copy number plasmids segregate into minicells, which are able to transcribe and translate plasmid-encoded genes. Proteins synthesized in minicells can be labeled with [^{35}S]methionine, isolated, separated by SDS–polyacrylamide gel electrophoresis, and visualized by autoradiography. Proteins encoded by cloned DNA can be correlated with sizes of open reading frames (ORFs) in the cloned DNA if the nucleotide sequence is known or with the size of a known protein that the cloned DNA is suspected to encode. The maxicell method involves UV irradiation of a host cell so double-stranded breaks occur in the large chromosome but not in all λ or plasmid cloning vectors present in the cell. Chromosomal DNA is degraded and the cloned DNA in the vector is available for *in vivo* transcription and translation. Proteins synthesized by maxicells can be characterized in the same manner as those made in minicells. Plasmid-encoded proteins also can be labeled using a cellular extract, called a Zubay extract, that contains all the necessary enzymes and components for both transcription and translation. Correlation of the size of the protein or of truncated versions encoded in cloned DNA, as determined by one of these methods, with the size of the insert will suggest where in the cloned DNA the ORF for this protein is located and how long the ORF in the nucleotide sequence is expected to be.

II. CLONING VECTORS

A. Plasmid vectors

The first cloning vectors were constructed from plasmids with high copy numbers. DNA yields are important in cloning. As long as the gene to be cloned is not detrimental to the cell, the vector of choice is a high copy number cloning vector. When the gene product might prove lethal to *E. coli*, low copy number vectors, such as those derived from the F plasmid, are useful.

Unlike chromosomal DNA replication, most of the high copy number vectors do not require synthesis of a plasmid-encoded protein for DNA replication, so plasmid yields can be increased up to 50-fold by chloramphenicol amplification. When chloramphenicol is added to a culture in late log phase, plasmid DNA continues to replicate but chromosomal DNA is inhibited because chloramphenicol inhibits the initiation of chromosomal DNA replication.

Several properties that have been engineered into plasmid cloning vectors include (1) a selectable marker, which is almost always an antibiotic resistance gene; (2) a way to detect inserts by insertional inactivation, either by loss of antibiotic resistance or α-complementation; (3) a multiple cloning site (MCS), which is a cluster of unique restriction enzyme sites for cloning; and (4) SP6, T3, or T7 bacteriophage promoters flanking the MCS, making possible the synthesis of RNA complementary to either the coding or noncoding strand of DNA inserted into the MCS. These vectors also may contain the origin of replication from a filamentous bacteriophage, such as f1 or M13. When a helper phage is added to cells containing such vectors, called phagemids, the cloned DNA can be packaged into bacteriophage heads and isolated as single-stranded DNA that can be used as a template for the chain-termination method of sequencing DNA.

B. Bacteriophage λ cloning vectors

Lambda cloning vectors are useful when cloning large pieces of DNA, for example, when constructing a genomic library. DNA sequences inserted into these vectors replace λ DNA sequences not required for the phage lytic cycle. DNA of phage λ is linear, double-stranded, and 48.5 kilobase pairs (kb) long. The vector is first cut with a restriction enzyme, and the λ arms are isolated from the nonessential central region. These arms are ligated to foreign DNA with T4 DNA ligase, and the DNA is packaged with complementary extracts from two *E. coli* strains that have been infected with different λ mutants incapable of either replicating or packaging their own DNA. The requirement that DNA molecules with λ *cos* sites be separated by ~50 kb to be packaged provides a strong selection for inserts of a given size range. The advantages of producing a genomic library in phage λ are that large pieces of DNA can be cloned and that the library can be stored as 1 ml of lysate in the refrigerator.

Cosmids, plasmids containing the λ *cos* site, also can be used to clone large pieces of DNA. After ligation, cosmids containing inserted foreign DNA can be packaged in λ heads by the *in vitro* packaging extracts and transduced into an *E. coli* host. A genomic library constructed with a cosmid vector can be stored in the refrigerator after *in vitro* packaging, whereas libraries made with plasmid vectors consist of colonies stored in multiwell microtiter dishes at −70°C.

III. TYPE II DNA RESTRICTION ENDONUCLEASES

Restriction endonucleases are enzymes that cleave DNA at recognition sites (Type II) or cleave away from recognition sites (Type I and III). The only important class of restriction endonucleases for cloning DNA are Type II because DNA cleavage occurs within or adjacent to the Type II recognition sites. It is important to be aware of Type I and Type III restriction endonucleases because their presence in the cell can interfere with cloning DNA. The Type II restriction endonucleases and methyltransferases are separate peptides and can be purchased from commercial suppliers.

Examples of Type II restriction enzyme sites are shown in Fig. 74.2. The name of a restriction enzyme and its corresponding methyltransferase includes the first letter of the genus name and the first two letters

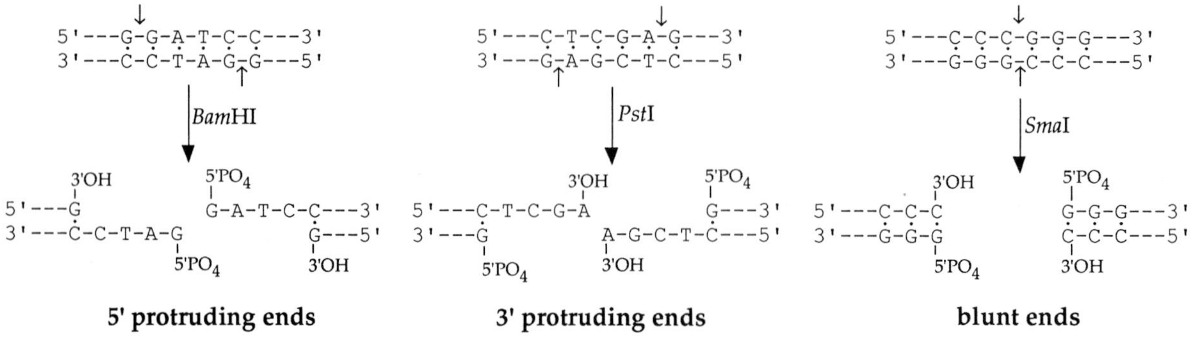

```
5'---G-G-A-T-C-C---3'        5'---C-T-G-C-A-G---3'        5'---C-C-C-G-G-G---3'
3'---C-C-T-A-G-G---5'        3'---G-A-C-G-T-C---5'        3'---G-G-G-C-C-C---5'

        BamHI                        PstI                         SmaI

3'OH         5'PO4              3'OH         5'PO4           3'OH         5'PO4
5'---G       G-A-T-C-C---3'     5'---C-T-G-C-A  G---3'       5'---C-C-C   G-G-G---3'
3'---C-C-T-A-G       G---5'     3'---G  A-G-C-T-C---5'       3'---G-G-G   C-C-C---5'
     5'PO4          3'OH            5'PO4   3'OH                 5'PO4    3'OH
```

5' protruding ends **3' protruding ends** **blunt ends**

FIGURE 74.2 Examples of the three types of ends that are generated by Type II restriction endonucleases. Dots signify hydrogen bonds between bases. Dashes between nucleotides indicate phosphodiester bonds.

of the species name of the organism from which the enzyme was isolated. This abbreviation is followed by letters or numbers referring to strain or type of bacteria or bacteriophage or plasmid that encodes the restriction enzyme. The presence of "r" or "m" before the name denotes that it is the restriction endonuclease or the methyltransferase, respectively. When "r" or "m" is missing in the name, it is assumed to be the restriction endonuclease.

Restriction methyltransferases are enzymes that methylate either cytosine or adenine in sites recognized by restriction endonucleases, thereby conferring resistance of the DNA to cutting by the endonuclease. The nucleotide methylated by the m·*Eco*RI enzyme is shown in Fig. 74.3. Many methyltransferases can be purchased commercially and have been used in cloning experiments to protect certain sites from cutting.

Type II restriction endonucleases recognize and cleave at specific sequences, usually 4–6 base pairs (bp) long, although a few recognize sites 7–8 bp long. Most sites have dyad symmetry: the double-stranded sequence is identical to the sequence after 180° rotation. Three types of DNA ends, examples of which are shown in Fig. 74.2, are produced by these enzymes: 5′ protruding ends, 3′ protruding ends, and blunt ends. All cleavage reactions leave 5′ phosphate and 3′ hydroxyl ends, both of which are required substrates for ligation.

The Type II restriction enzyme site cut by r·*Eco*RI and methylated by m·*Eco*RI is shown in Fig. 74.3. *Eco*RI leaves 5′ protruding ends; the single-stranded 5′ overhangs are identical for all *Eco*RI fragments. These single-stranded ends, called sticky ends, are cohesive, that is, when the fragments are juxtaposed the nucleotides are complementary and can form hydrogen bonds (small dots, Fig. 74.3). A recombinant DNA molecule is produced when 5′ and 3′ ends from separate fragments are joined by T4 DNA ligase.

The frequency of restriction enzyme sites in DNA varies with the size of the site and the G + C content of DNA. If the G + C content is 50%, a tetranucleotide site would appear once every 256 bp or $(1/4)^4$, whereas a hexanucleotide site would appear once every 4096 bp or $(1/4)^6$. Restriction enzymes that recognize 4-bp sites will generate fragments that have an average length of 256 bp, whereas enzymes that recognize 6-bp sites will generate fragments that have an average length of 4096 bp, although the distribution of sizes about these averages is very high. Larger DNA fragments produced by 6-bp cutters will frequently contain the sequence of a whole gene. The average gene is about 1 kb in bacteria. Small DNA fragments generated by 4-bp cutters will rarely contain the sequence of a whole gene but are used for shotgun cloning into vectors for nucleotide sequencing projects.

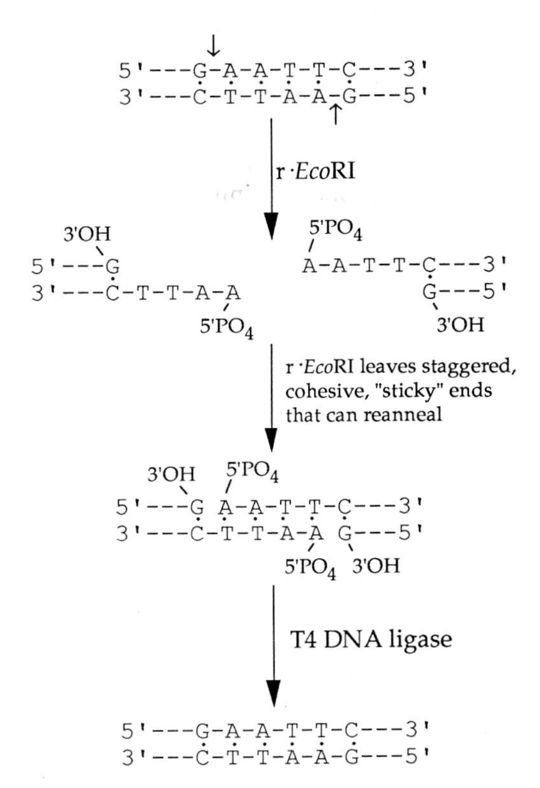

5′---G-A-A-T-T-C---3′
3′---C-T-T-A-A-G---5′

m·*Eco*RI

CH_3
5′---G-A-A-T-T-C---3′
3′---C-T-T-A-A-G---5′
CH_3

*Eco*RI site modified by m·*Eco*RI is no longer sensitive to r·*Eco*RI

FIGURE 74.3 Activities of the *Eco*RI restriction endonuclease (r·*Eco*RI) and the *Eco*RI methyltransferase (m·*Eco*RI), and the recombinant DNA product produced after ligation of two *Eco*RI fragments. Dots signify hydrogen bonds between bases. Dashes between nucleotides indicate phosphodiester bonds.

To produce overlapping fragments in a genomic library, two different restriction enzymes, a 4-bp and a 6-bp cutter, are used. In this approach, chromosomal

DNA is partially digested with a restriction enzyme that recognizes a 4-bp sequence, for example. *Sau*3A (↓ GATC). After size-fractionating the restriction fragments on agarose gels or sucrose gradients, the DNA is cloned into a vector that has been cut with a restriction enzyme that gives compatible ends, for example, *Bam*HI (G ↓ GATCC).

Once a fragment of DNA is cloned, a useful signature of that DNA, aside from its nucleotide sequence, is the location of restriction enzyme sites in the DNA. Restriction enzyme mapping of DNA involves cutting the DNA with restriction enzymes and determining the sizes of restriction enzyme fragments. Linear DNA molecules are separated according to size by agarose or acrylamide gel electrophoresis and visualized using the intercalating dye, ethidium bromide, which binds to DNA and fluoresces when irradiated with UV light (see Fig. 74.4). A restriction map can be constructed from the sizes of fragments generated with multiple enzymes, individually and in combination. Restriction enzyme mapping of large genomes, such as bacterial chromosomes, requires the use of restriction enzymes that recognize 8-bp sites [frequency is $(1/4)^8$ or approximately every 65 kb] and field inversion gel electrophoresis (FIGE), which resolves fragments of DNA up to 2000 kb in length.

Restriction fragments carrying specific genes can be identified by Southern blotting, probe hybridization, and detection of the probe. After denaturation, DNA restriction fragments can be transferred by diffusion from agarose gels to nitrocellulose or nylon membranes; this procedure is called Southern blotting. The blot is hybridized to an appropriately labeled nucleic acid probe, then washed, and the presence of the probe detected by autoradiography or chemiluminescence after exposure to x-ray film. Bands that appear on the X-ray film indicate the DNA bands that have hybridized to the radioactive probe. An example of Southern blotting is shown in Fig. 74.5.

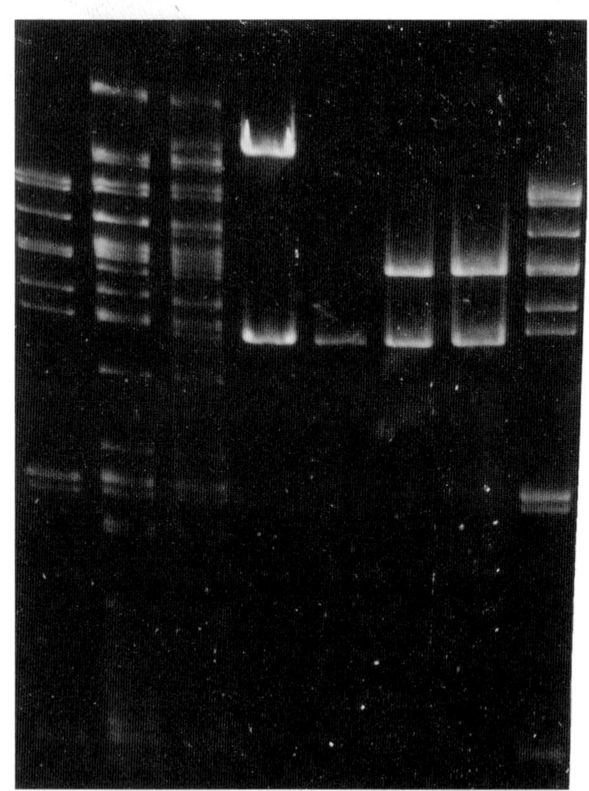

FIGURE 74.4 Agarose gel electrophoresis of *Eco*RI-digested DNA species. Lanes A and H, F plasmid; lane B, *Salmonella typhimurium* F′ plasmid FST27-D1; lane C, *S. typhimurium* F′ plasmid FST27; lane D, pJZ1; lane E, *Eco*RI fragment carrying *kan* gene; lane F, pJZ2; lane G, pML31. Plasmid pJZ1 contains the *S. typhimurium* chromosomal origin of replication. Plasmids pJZ2 and pML31 contain the F plasmid origin of replication. [Reproduced from Zyskind, J. W., Deen, L. T., and Smith, D. W. (1979). *Proc. Natl. Acad. Sci. U.S.A.* **76**, 3097–3101.]

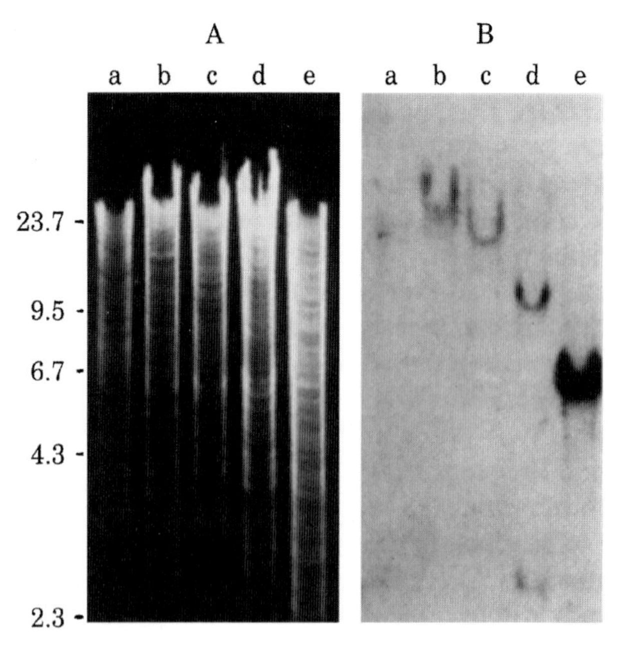

FIGURE 74.5 Southern blot analysis of *Sal*I restriction enzyme digests of chromosomal DNA isolated from marine bacteria. Chromosomal DNA was isolated from *E. coli* (a), *Vibrio fischeri* (b), *Photobacterium leiognathi* (c), *P. phosphoreum* (d), and *V. harveyi* (e). (A) Ethidium bromide stained agarose gel. (B) Autoradiogram of a Southern blot of the DNA in (A) after the blot was hybridized with a probe containing the *V. harveyi* origin of replication. [Reproduced from Zyskind, J. W., Cleary, J. M., Brusilow, W. S. A., Harding, N. E., and Smith, D. W. (1983). *Proc. Natl. Acad. Sci. U.S.A.* **80**, 1164–1168.]

IV. NUCLEOTIDE SEQUENCING

Our greatly expanded knowledge of gene structure, expression, and function, as well as of protein structure and function, is due mainly to rapid DNA sequencing methods. The two types of sequencing procedures in common use today are the chemical degradation method, developed by Maxam and Gilbert, and the dideoxynucleotide chain-termination method of Sanger and co-workers.

In the chemical degradation method, a single- or double-stranded DNA fragment is radioactively labeled at one end, the labeled DNA fragment is divided into four or five aliquots, and chemical reactions are performed so that only one or two bases are modified in each reaction. A phosphodiester bond at each modified base is cleaved with piperidine, producing a "nested set" of labeled fragments that terminates at the location of modified residues. Labeled DNA fragments from these reactions are separated by electrophoresis in adjacent lanes on a polyacrylamide gel to which the denaturant urea has been added. Autoradiography is used to detect the labeled DNA fragment ladders, and the DNA sequence can be read from the location of bands in each lane, since this corresponds to the order of the bases from the labeled end.

The dideoxynucleotide chain-termination method depends on termination of DNA synthesis after incorporation of a dideoxynucleotide. Dideoxynucleotides can be incorporated into DNA by many DNA polymerases but, because these nucleotides lack a 3′ hydroxyl group, further DNA synthesis is prevented. Template DNA can be single-stranded M13mp viral DNA, DNA amplified by asymmetric PCR, or double-stranded DNA that has been denatured. A primer is annealed to a template strand that is to be sequenced, forming hybrid molecules. The hybrid is divided evenly among four tubes, each containing a different dideoxynucleoside triphosphate and all four deoxynucleoside triphosphates, one of which is radioactively labeled. After addition of a DNA polymerase, such as modified T7 DNA polymerase (Sequenase®), the primer is elongated until incorporation of a chain-terminating dideoxynucleotide. The nested set of labeled DNA fragments in the four reactions is separated by denaturing polyacrylamide gel electrophoresis, and the sequencing ladders are detected by autoradiography (see Fig. 74.6). The dideoxynucleotide chain-termination method has been automated and is currently being used in commercially available DNA sequencers. In the most common variant of this method, each of the four dideoxynucleotides used to terminate DNA synthesis is covalently linked to a differently fluorescing dye. The chain extension reaction is carried out in a single reaction vessel, and the resulting products are separated by gel electrophoresis in a single lane. The characteristic fluorescence spectrum of each dye identifies the terminal base on each chain extension product.

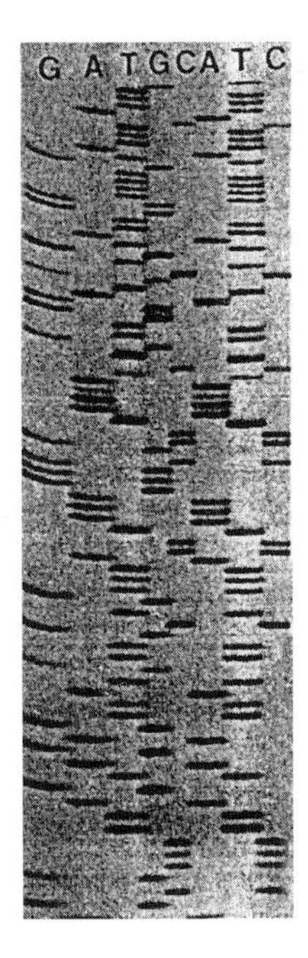

FIGURE 74.6 Autoradiograph of dideoxynucleotide chain-termination sequencing reactions. Sequence of DNA is 5′ AGCGCCCTTA GTAGAGTTAT GTTGCTGTTT ACCTAAAGGG CGCCTAAAAT CTGTTGGATG CTGTATTGGT TTTGTATTTC ATTTG3′. [Reproduced from Zyskind, J. W., and Bernstein, S. I. (1992). "Recombinant DNA Laboratory Manual." Academic Press, San Diego, California.

Once the nucleotide sequence is known, it can be analyzed using computer programs that search for ORFs and for specific sites such as promoters or binding sites for DNA binding proteins. The ORFs can be translated to determine the amino acid sequences of encoded proteins. The DNA and protein sequences can be compared with sequence libraries to determine whether the gene or protein has been isolated previously. Amino acid sequence similarities with other proteins can provide clues to protein function. For example, many bacterial protein kinases involved in signal transduction have been identified after finding sequence similarities with known protein kinases.

V. APPLICATIONS

Recombinant DNA methodology combined with genetic, biochemical, and immunological approaches has made possible rigorous gene analysis that was impossible 20 years ago. Promoters can be linked to reporter genes, such as *lacZ* or *chb*; these fusions can be used to determine the direction of transcription, promoter strength, and mechanisms regulating expression, including distinguishing transcriptional from posttranscriptional control. If the fusions are contained within high copy number plasmids, reporter enzyme activity may vary, due to differences in plasmid copy number or to titration of regulator proteins. For this reason, fusions often are integrated into the chromosome for single copy analysis of promoter activity using λ vectors or a simplified λ site-specific recombination system combining the λ *attP* site, the fusion with the reporter gene, and the λ *int* gene. Specific nucleotides in sequences of interest are easily changed (site-specific mutagenesis) using either (1) primer oligonucleotides that contain one base change and are complementary to the inserted sequence in M13 bacteriophage vectors or (2) PCR mutagenesis in which the base change is included in one of the primers. The development of *in vitro* mutagenesis has led to a new field called protein engineering, in which amino acid changes are introduced into a protein to change, and perhaps improve, its activity or properties. Theoretically, any chemical reaction could be catalyzed by an enzyme; perhaps in the near future it will be possible to predict the amino acid sequences of proteins with new enzymatic activity. If the amino acid sequence of a peptide is known, the corresponding gene can be chemically synthesized using codons that optimize expression in the cloning host. Expression vectors allow production of large amounts of a protein in bacterial cells by placing the appropriate gene under control of a strong but regulated promoter such as P*tac*. Currently, in the biotechnology industry, bacterial cells are used as factories for producing many important peptides, such as human insulin.

BIBLIOGRAPHY

Baxevanis, A., and Ouellette, B. F. F. (eds.) (1998). "Bioinformatics: A Practical Guide to Analysis of Genes and Proteins." John Wiley & Sons, New York.

Drlica, K. (1996). "Understand DNA and Gene Cloning: A Guide for the Curious," (3rd edn.). John Wiley & Sons, New York.

Eun, H. M. (1996). "Enzymology Primer for Recombinant DNA Technology." Academic Press, San Diego, CA.

Fredrickson, D. S. (2001). "The Recombinant DNA Controversy: A Memoir. Science, Politics, and the Public Interest 1974–1981." ASM Press, Washington DC.

Kalabat, D. Y., Froelich, J. M., Phuong, T. K., Forsyth, R. A., Newman, V. G., and Zyskind, J. W. (1998). Chitobiase, a new reporter enzyme. *Bio Techniques* **25**, 1030–1035.

Kreuzer, H., and Massey, A. (2000a). "Recombinant DNA and Biotechnology: A Guide for Teachers." ASM Press, Washington DC.

Kreuzer, H., and Massey A. (2000b). "Recombinant DNA and Biotechnology: A Guide for Students." ASM Press, Washington DC.

Lambrook, J., Fritsch, E. F., and Maniatis, T. (1989). "Molecular Cloning: A Laboratory Manual, (2nd edn.)." CSH Laboratory Press, Cold Spring Harbor, NY.

Lewin, B. (1997). "Genes VI." Oxford Univ. Press, Oxford, UK.

Watson, J. D., Gilman, M., Witkowski, J., and Zoller, M. (1992). "Recombinant DNA, (2nd edn.)." W. H. Freeman and Co., New York.

Wu, R., Simon, M. I., and Abelson, J. N. (eds.) (1995). "Recombinant DNA Methodology II (Selected Methods in Enzymology)." Academic Press, San Diego, CA.

Zyskind, J. W., and Bernstein, S. I. (1992). "Recombinant DNA Laboratory Manual," Academic Press, Inc., San Diego, CA.

WEBSITES

NIH Guidelines for Research Involving Recombinant DNA (USA) *http://grants1.nih.gov/grants/guide/notice-files/NOT-OD/02-052.html*

Guidelines for Recombinant DNA Experiment (Japan) *http://www.cbi.pku.edu.cn/mirror/binas/Regulations/full_regs/japan/exp/*

Links to regulations in European Union member countries *http://www.alss.be/documentation/regulations/overview.htm*

Pedro's Biomolecular Research Tools. A collection of Links to Information and Services Useful to Molecular Biologists *http://www.public.iastate.edu/~pedro/research_tools.html*

75

Sexually transmitted diseases

Jane A. Cecil and Thomas C. Quinn
The Johns Hopkins University

GLOSSARY

epidemiology The study of the relationships of various factors determining the frequency and distribution of disease in a community.

incidence The rate at which new cases of a disease occur in a community in a given time period.

prevalence The total number of cases of a given disease in a community at a certain time.

The Term "Sexually Transmitted Diseases (STDs)" refers to a number of distinct infections, which are typically transmitted through sexual contact and result in a variety of clinical manifestations. The acquisition and transmission of STDs depend primarily on sexual behavior, but some STDs such as hepatitis B and cytomegalovirus can also be acquired through nonsexual contact in childhood in areas with poor living conditions. Additionally, STDs can be transmitted from mother to child during pregnancy and childbirth. STDs are caused by a variety of organisms, including bacteria, viruses, protozoa, and ectoparasites. The clinical syndromes associated with infection by one or more of these organisms can range from asymptomatic infection to severe life-threatening illness.

I. INTRODUCTION

Over the 1980s and 1990s there have been tremendous advances in our understanding of sexually transmitted diseases (STDs). There has been an appreciation of new pathogens, new clinical syndromes, emerging antimicrobial resistance, the increasing importance of viral pathogens including the human immunodeficiency viruses (HIVs), as well as better understanding of the epidemiology of many STDs. There are approximately 50 distinct clinical syndromes associated with at least 25 different organisms. Table 75.1 summarizes some of the most important pathogens, the associated clinical syndromes, and potential complications.

Sexually transmitted diseases constitute a major public health problem worldwide, despite improvements in diagnosis, treatment, and prevention strategies. STDs are seen in virtually every society, though developing nations and underserved populations experience the greatest burden of disease. The World Health Organization (WHO) estimates that there were more than 333 million new cases worldwide of syphilis, gonorrhea, chlamydia, and trichomoniasis in adults aged 15–49 years in 1995. The global prevalence of infection with the agents of common viral STDs, such as genital herpes simplex virus, genital human papillomavirus, hepatitis B, and HIV, is estimated to be in the billions of cases, since many individuals are infected with more than one of these pathogens. In the U.S., there are approximately 12 million new cases of STDs annually, giving the U.S. the highest rate of curable STDs in the developed world. Of the top ten most frequently reported diseases to the Centers for Disease Control and Prevention (CDC) in 1995, five were STDs.

Sexually transmitted diseases are becoming recognized as a source of tremendous social, health, and

The Desk Encyclopedia of Microbiology
ISBN: 0-12-621361-5

TABLE 75.1 Major sexually transmitted agents and the diseases they cause

Agent	Disease or syndrome
Bacteria	
Neisseria gonorrhoeae	Urethritis; epididymitis; proctitis; bartholinitis; cervicitis; endometritis; salpingitis and related sequelae (infertility, ectopic pregnancy); perihepatitis; complications of pregnancy (e.g. chorioamnionitis, premature rupture of membranes, premature delivery, postpartum endometritis); conjunctivitis; disseminated gonoccal infection (DGI)
Chlamydia trachomatis	Same as *N. gonorrhoeae*, except for DGI; also, lymphogranuloma venereum,[a] Reiter's syndrome; infant pneumonia
Treponema pallidum	Syphilis[a]
Haemophilus ducreyi	Chancroid[a]
Calymmatobacterium granulomatis	Donovanosis[a]
Mycoplasma hominis	Postpartum fever; salpingitis
Ureaplasma urealyticum	Urethritis; low birth weight (?); chorioamnionitis (?)
?*Gardnerella vaginalis, Mobiluncus* sp., ?*Bacteroides* sp.	Bacterial vaginosis
?Group B β-hemolytic streptococci	Neonatal sepsis; neonatal meningitis
Viruses	
Herpes simplex virus (HSV-2, HSV-1)	Primary and recurrent genital herpes
Hepatitis B virus (HBV), hepatitis C virus (HCV)	Acute, chronic, and fulminant hepatitides B and C, with associated immune complex phenomena and sequelae including cirrhosis and hepatocellular carcinoma
Cytomegalovirus (CMV)	Congenital infection: gross birth defects and infant mortality, cognitive impairment (e.g. mental retardation, 8th nerve deafness), heterophile-negative infectious mononucleosis, protean manifestations in the immunosuppressed host
Human papilloma virus (HPV)	Condyloma acuminata; laryngeal papilloma in infants; squamous epithelial neoplasias of the cervix, anus, vagina, vulva, penis
Molluscum contagiosum virus (MCV)	Genital molluscum contagiosum
Human immunodeficiency virus (HIV-1, HIV-2)	AIDS and related conditions
Human T-lymphotropic virus type 1	T-cell leukemia/lymphoma; tropical spastic paraparesis
Protozoa	
Trichomonas vaginalis	Vaginitis; urethritis (?); balanitis (?)
Fungi	
Candida albicans	Vulvovaginitis; balanitis; balanoposthitis
Ectoparasites	
Phthirus pubis	Pubic lice infestation
Sarcoptes scabiei	Scabies

[a]These infections are responsible for the syndrome known as genital ulcer disease.

economic costs worldwide. The 1993 World Development Report ranked STDs, excluding AIDS, as the second leading cause of healthy life lost among women between the ages of 15 and 44 years in the developing world. There are many factors contributing to the disease burden exacted by STDs. Challenges to the control of STDs include limitations on the availability and quality of medical care, public education regarding safe sexual practices, access to condoms and other methods of prevention, convenient and affordable methods of diagnosis and treatment, the emergence of antibiotic resistance, and the social stigma associated with STDs. Furthermore, STDs are commonly asymptomatic in both women and men. All of these factors can result in a delay or complete lack of appropriate treatment.

Thus, the vast majority of disease burden from STDs is a result of the complications and sequelae that may follow initial infection. When left untreated, infections can migrate upward from the lower reproductive tract and result in pelvic inflammatory disease (PID), chronic pelvic pain, infertility, ectopic pregnancy and tubo-ovarian abscess in women, and prostatitis, epididymitis, or orchitis in men. It has been estimated that the economic cost of PID alone exceeds $2 billion per year in the U.S. Untreated infections in pregnant women may lead to fetal loss, stillbirths, low birth weight, and eye, lung, or neurologic damage in a newborn. The viral STDs are typically incurable and result in chronic disease. The most obvious example of this is HIV infection and AIDS.

Another example is human papillomavirus, of which some types have been found to cause cervical cancer in women.

In general, the major STDs pose a greater threat to women's health, for a number of reasons. Women are more likely to get an STD from a man than the reverse because STDs are more efficiently transmitted from men to women. Women are more likely to be asymptomatic than men, and thus women may be less likely to seek medical attention in a timely manner. STDs are often more difficult to accurately diagnose in women than in men because symptoms are less specific in women and diagnostic tests are less sensitive. Finally, the potential complications of STDs are more common and more serious in women.

Extensive studies of the epidemiology of STDs have revealed several risk factors consistently associated with the acquisition of STDs. These risk factors appear to apply to most, if not all populations. Risk factors include multiple sexual partners, urban residence, single marital status, and young age. The risk of STDs for young adults and adolescents has not been fully appreciated. STD prevention efforts in young people have remained controversial in the U.S. despite the fact that approximately 3 million teenagers acquire an STD each year. Young adults and adolescents are at greatest risk for acquiring an STD for a number of reasons, related to patterns of sexual behavior and other behaviors associated with STDs, such as substance abuse. There is some evidence that biological factors may also play a role where young females are concerned. Thus any successful STD prevention program needs to specifically include this population. Another factor associated with the acquisition of an STD is a history of prior STD infection. Similarly, simultaneous infection with more than one STD is not uncommon. In fact, several studies have found that approximately 25% of individuals with gonorrhea are also infected with *Chlamydia trachomatis*. On the basis of these data, treatment recommendations include antibiotics that are effective against both organisms.

The control of STDs has become an even higher public health priority in the setting of the HIV epidemic. STDs and HIV share many behavioral risk factors, and thus efforts at educating individuals to eliminate high-risk behaviors and adopt safer sexual practices may help reduce the incidence of both STDs and HIV. In addition, many studies have clearly documented that both ulcerative and nonulcerative STDs increase the risk of HIV transmission. Thus, an individual with both HIV and another STD is more likely to transmit HIV to a sexual partner, compared to an HIV-infected individual without another STD. Similarly, an HIV-uninfected individual with another STD is more likely to get HIV from an HIV-infected sexual partner than if he or she did not have an STD at the time of sexual intercourse. This association between STDs and HIV is further strengthened by the findings of an STD treatment study in the Mwanza region of Tanzania, by Grosskurth and colleagues. In this study, communities provided with improved STD diagnosis and treatment experienced a substantial reduction in the incidence of HIV, as a result of a decrease in sexual transmission.

Fortunately, given the renewed interest in the control of STDs, there have been many recent advances in the field. Intensive research efforts have yielded improvements in the clinical recognition, microbiologic diagnosis, and treatment of STDs. Epidemiologic data have helped to identify high-risk populations, and thus prevention programs and resources can be appropriately targeted. New clinical algorithms have aided in the early recognition of PID, urethritis, cervicitis, and vaginal infections. Newer diagnostic assays are more sensitive and more specific for quickly diagnosing diseases caused by specific pathogens. The development of nucleic acid amplification tests for some STDs have enabled health care providers to easily and accurately screen large numbers of individuals, even in the absence of symptoms, with little discomfort to the individual. This has made it possible to identify and treat those individuals who may be more likely to transmit infection or to develop complications, because they would not be prompted to seek health care given an absence of symptoms. Finally, newer antibiotics ensure the possibility of effective treatment of many STDs, including those that have become resistant to older therapeutic agents.

The STDs discussed in this article have been selected because they are among the more common STDs. In addition, they are illustrative of important variations in epidemiology, diagnosis, treatment, prevention, and pathogenesis, concepts that are central to our understanding of STDs.

II. GONORRHEA

A. Epidemiology

Despite recent declines in new cases of gonorrhea, it remains one of the most commonly reported diseases in the United States. The greatest number of cases reported to the Centers for Disease Control and Prevention (CDC) was 1,013,436 in 1978. Currently there are an estimated 600,000 new infections of *N. gonorrhoeae* each year in the United States. These numbers likely underestimate the true prevalence of gonorrhea in the

U.S., however, given that many health care workers do not comply with reporting guidelines and many patients are treated empirically without confirmatory diagnostic testing because of highly suggestive symptoms or known exposure. Estimates of disease burden attributable to gonorrhea are even less reliable in many developing countries. Available data suggest, however, that gonorrhea remains relatively common in the developing world as it does in the U.S. In contrast, gonorrhea seems to be relatively rare in Canada and much of western Europe.

As in the U.S., the incidence of gonorrhea in many developed nations has declined in recent years. The decline in cases has been felt to be due in part to the HIV epidemic and the subsequent reduction in high-risk behaviors seen in many segments of the population. Other factors believed to have contributed to the decline in gonorrhea cases include an increase in condom use, screening of asymptomatic individuals at risk for gonorrhea, and contact tracing used to identify individuals who are at risk for gonorrhea given a known exposure to an infected individual. In the U.S., the decline in gonorrhea cases has not been observed consistently throughout the population, however. The most striking decline in gonorrhea has been among homosexual men and white men and women. In contrast, smaller declines have been observed among urban, minority segments of the population, primarily African-American and Hispanic. In 1995, the reported rates of gonorrhea were 37 times higher for African-Americans than for whites. In 1996, the Centers for Disease Control reported gonorrhea rates in African-American adolescents (15–19 years olds) that were almost 25 times greater than for white adolescents. Similarly, greater declines have been observed among older age groups compared to the young.

Numerous factors have been identified that correlate with an increased risk of gonorrhea. Gonorrhea is clearly more common among African-Americans, although some of this disparity is a result of greater attendance of minority individuals at public health clinics, where reporting of cases is much better than in other health care settings. Young age is also a risk factor for gonorrhea. In 1995, 77% of all reported cases of gonorrhea occurred in individuals between the ages of 15 and 29 years. Furthermore, the reported incidence of gonorrhea was nearly twice as high for sexually active adolescent girls compared to sexually active women aged 20–24 years. Young women seem to be at particularly high risk for gonorrhea. This may be due in part to patterns of sexual behavior, but there is also a biologic basis to this observation. Young women have larger areas of cervical ectropion (columnar epithelial cells on the ectocervix) that provide a greater area of susceptible cells for infection. Low socioeconomic status, early onset of sexual activity, multiple sexual partners, unmarried marital status, and sexual activity associated with illicit drug use are also risk factors for gonorrhea. Similarly, individuals with a prior history of gonorrhea as well as prostitutes and their clients are at increased risk for gonorrhea. Although the use of oral contraceptives is not generally associated with higher rates of gonorrhea, their use may increase the risk for acquiring gonorrhea when a woman is exposed to an infected partner, when compared to her risk if she uses no contraception at all.

B. Clinical manifestations

The term gonorrhea is often thought to refer to an illness characterized by urethral discharge in men or vaginal discharge in women. The causative organism, *Neisseria gonorrhoeae*, can, however, infect several anatomic sites including the cervix, urethra, rectum, oropharynx, and conjunctiva, and is associated with a wide spectrum of disease. An individual with gonorrhea may have no symptoms at all, may have localized symptomatic disease, may have localized complicated disease, or may be very ill with disseminated infection. Infections caused by *N. gonorrhoeae* are typically limited to superficial mucosal surfaces lined by columnar or cuboidal, nonkeratinized epithelial cells, such as those sites noted above. Infection of these mucosal surfaces is usually accompanied by a marked inflammatory response that results in the production of copious, purulent discharge. As noted, however, infection may also be asymptomatic. Of men and women who report sexual exposure to partners with gonorrhea, urethral infection in men is asymptomatic approximately 10% of the time and cervical infection in these women is asymptomatic about 40–50% of the time. Pharyngeal and rectal infections secondary to gonorrhea are typically asymptomatic. Of interest, individuals who develop disseminated gonococcal infection (DGI) commonly have asymptomatic mucosal infections, detected only by screening diagnostic tests. It is unknown whether strains that cause DGI are more likely to cause asymptomatic mucosal infection or if individuals who lack symptoms simply do not seek medical treatment and therefore the infection eventually spreads.

1. *Gonococcal infections in women*

The most common form of uncomplicated gonorrhea in women is infection of the uterine cervix. The vagina is usually not infected because it is lined by squamous epithelium. The urethra is colonized in 70–90% of infected women. In women who have undergone a hysterectomy, the urethra is the most common site of infection. The Skene glands (periurethral glands) and Bartholin glands are also commonly infected in the setting of endocervical infection with *N. gonorrhoeae*. Symptoms typically appear within 10 days of infection, although the incubation period can be quite variable. The symptoms associated with endocervical infection may include vaginal discharge, genital itching, intermenstrual bleeding, unusually heavy menstrual bleeding, or painful urination. An infected woman may have all, none, or any combination of these symptoms, and symptoms may range from mild to severe. Physical examination may be normal but typically reveals a purulent discharge from the cervix, redness and swelling of the cervix, and easily induced cervical bleeding. These findings, as well as the symptoms noted, are not specific for gonorrhea, however. Thus, evaluation of sexually active women suspected of having gonorrhea must include testing for *Chlamydia trachomatis*, *Candida albicans*, *Trichomonas vaginalis*, bacterial vaginosis, as well as other infections.

Pelvic inflammatory disease (PID) is the most serious complication of *N. gonorrhoeae* infection in women, affecting approximately 10–20% of all women with acute infection. PID is a term referring to endometritis, salpingitis, tubo-ovarian abscess, peritonitis, or any combination of these findings. Gonococcal PID results from upward spread of the organism from the initial site of infection (the endocervix) into the upper genital tract. Symptoms of PID can include fever, unilateral or bilateral lower abdominal pain, pain associated with sexual intercourse, abnormal menses or intermenstrual bleeding, or other complaints associated with an intraabdominal infection. These symptoms may be severe or mild. Findings on physical exam include abdominal or uterine tenderness, adnexal tenderness, cervical motion tenderness, and occasionally an adnexal mass or fullness associated with a tubo-ovarian abscess. PID may be caused by a number of other infections, including *Chlamydia trachomatis* as well as many other bacteria. Treatment of PID includes broad-spectrum, often intravenous, antibiotics targeting chlamydia, gonorrhea, and anaerobic bacteria. PID is a significant complication not only because of the acute manifestations, but because of the long-term sequelae, including infertility, ectopic pregnancy, and chronic pelvic pain. Approximately 10% of women will become infertile after a single episode of PID. For those women that remain fertile, there is a 10-fold increase in the risk of ectopic pregnancy.

The manifestations of gonorrhea in pregnant women are similar to those in nonpregnant women, although PID is less common. Pregnant women with genital gonorrhea are at significant risk, for spontaneous abortion, premature rupture of membranes, premature delivery, and acute chorioamnionitis, as well as for transmitting gonorrhea to their newborns during delivery. Neonates infected with gonorrhea can suffer eye and pharyngeal involvement, as well as other complications. Consequently, all newborn infants are treated prophylactically with topical antibiotics to prevent ocular infection.

2. *Gonococcal infections in men*

For symptomatic men, the most common manifestation of infection with *N. gonorrhoeae* is urethritis. Infected men often develop a urethral discharge that can range from scant and clear to copious and purulent. Acute infection is usually accompanied by dysuria (painful urination), which develops after the onset of discharge, or may be the sole manifestation. Symptoms typically develop 2–5 days after exposure to an infected sexual partner, although there can be a delay of over 2 weeks. Untreated acute infection can result in complications, including acute epididymitis, acute and chronic prostatitis, and rarely cellulitis, penile lymphangitis, or periurethral abscess. The most common complication is acute epididymitis, which presents with gradually worsening unilateral scrotal pain. Examination of a man with epididymitis may reveal previously unnoticed urethral discharge as well as tenderness and swelling of the posterior of the scrotum. In young men under the age of 40 years, coinfection with other STDs, particularly *Chlamydia trachomatis*, is as common as it is for their female counterparts. Thus, treatment of gonococcal urethritis and associated complications should be targeted toward both organisms. In older men, over the age of 40 years, structural abnormalities of the urinary tract predispose to infections caused by gram-negative bacteria and other urinary tract organisms.

3. *Other gonococcal infections*

Anorectal gonorrhea is uncommon in heterosexual men, but it is seen more frequently in homosexual men and in approximately 35–50% of women with gonococcal infection of the endocervix. Receptive anal intercourse is the principal mode of transmission in homosexual men, whereas rectal infections in women more commonly arise from local contamination by cervicovaginal secretions. Most men and women with anorectal gonorrhea are asymptomatic; however, some will develop symptoms of proctitis, including severe rectal pain, tenesmus, rectal discharge, and constipation. Other symptoms may include anal itching, painless purulent discharge, or a small amount of rectal bleeding.

Pharyngeal infection rarely occurs without concomitant genital infection. Estimates suggest that of patients with gonorrhea, pharyngeal infection occurs in 10–20% of heterosexual women, 3–7% of heterosexual men, and 10–25% of homosexually active men. The vast majority of gonococcal pharyngeal infections are asymptomatic; however, these cases may rarely be associated with acute pharyngitis, tonsillitis, fever or lymph node enlargement in the neck.

Neisseria gonorrhoeae infections of the eye occur most commonly in newborns (ophthalmia neonatorum) as a result of exposure to infected maternal secretions during delivery. These infections can result in corneal scarring or perforation of the cornea, and thus prompt diagnosis and treatment are crucial. Fortunately, the use of silver nitrate or erythromycin eye drops at birth is effective in preventing gonococcal ophthalmia neonatorum. Adults can also acquire gonococcal ocular infection through autoinoculation, and develop keratoconjunctivitis as a result.

4. *Disseminated gonococcal infection*

Disseminated gonococcal infection (DGI) occurs in approximately 1–2% of individuals with untreated gonorrhea as a result of hematogenous spread of the organism from typically asymptomatic mucosal infections of the pharynx, cervix, urethra, or rectum. DGI is more common in women than in men, and bacteremia probably begins 7–30 days after initial infection. DGI is often referred to as an arthritis–dermatitis syndrome, given the usual clinical characteristics of an asymmetric polyarthritis and associated rash. The joint manifestations in DGI may begin with arthralgias or painful tenosynovitis which progress to frank arthritis later in the course of disease. Approximately 30–40% of patients with DGI will present with overt arthritis, usually involving the wrist, metacarpophalangeal, ankle, or knee joints, although any joint may be involved. Joint fluid cultures are often negative except when the fluid has a high white blood cell concentration, usually more than 40,000 white blood cells per cubic millimeter. Overall, only 20–30% of joint fluid cultures from patients with DGI will be positive for the organism. The rash associated with DGI is composed of a small number (fewer than 30) of tender, necrotic pustules on a reddened base, usually located on the distal extremities. The skin lesions may, however, appear as macules, papules, petechiae, bullae, or ecchymoses. Despite the fact that DGI represents a bloodstream infection, patients often do not clinically appear particularly ill. Blood cultures, like joint fluid cultures, are only positive in 20–30% of patients with DGI. As previously noted, the mucosal infections in DGI are usually asymptomatic. Thus, the diagnosis of DGI can be difficult to make. When DGI is suspected it is very important to obtain cultures of all potential mucosal sites, in addition to blood and joint fluid, for *N. gonorrhoeae*, to help confirm the diagnosis. Rarely, DGI can be further complicated by seeding of the meninges or the heart valves, resulting in potentially life-threatening gonococcal meningitis or endocarditis.

Disseminated gonococcal infection can also occur in newborns of infected mothers, when exposed to secretions during vaginal delivery. The illness in neonates can be similar to that in adults, with associated meningitis and arthritis. Gonococcal disease in newborns usually manifests 2–5 days after birth.

C. Diagnosis

Gram stain and culture are the most widely used methods for detection of *N. gonorrhoeae*, although some newer nonculture techniques are proving to be more convenient and equally accurate in some settings. Specimens for both Gram stain and culture must be obtained directly from the urethra or cervix with a special swab. For detection of gonococcal cervicitis or urethritis in symptomatic or high-risk individuals, a Gram stain of an appropriately obtained specimen can provide rapid and accurate results. A Gram stain is considered positive if gram-negative diplococci are seen within polymorphonuclear leukocytes. Gram stain results are more difficult to interpret on rectal and pharyngeal specimens because of the presence of other gram-negative organisms. Isolation of *N. gonorrhoeae* by culture is the standard method for diagnosis of gonococcal infections of any site, and is necessary for antibioticresistance testing. Optimal recovery of *N. gonorrhoeae* requires direct inoculation of the specimen onto selective media, followed by placement into a 35 °C–37 °C incubator containing 5–7% carbon dioxide. The most commonly used selective media include modified Thayer-Martin, Martin-Lewis, and New York City media. These media contain antimicrobial substances to suppress the overgrowth of other microorganisms that can obscure the presence of the small colonies typical of *N. gonorrhoeae*. *Neisseria gonorrhoeae* can survive for 5–7 h in special transport media, permitting clinics without the necessary laboratory capabilities to transport clinical specimens to nearby laboratories better able to process the specimens appropriately.

Nonculture techniques detect gonococcal antigens, gonococcal enzyme activity, or gonococcal genetic material. One test uses a single-stranded DNA probe to detect gonococcal ribosomal RNA, and it is fairly sensitive and highly specific for gonorrhea. In 1996, the ligase chain reaction assay was approved in the U.S. for the diagnosis of gonorrhea. Similar assays using the polymerase chain reaction and other nucleic acid amplification techniques have also been developed. These assays are as sensitive as culture (95–98%) and highly specific for detecting *N. gonorrhoeae*. They have the added advantage of being equally sensitive and specific when used on first-void urine specimens, compared to swab specimens which can be impractical to obtain and uncomfortable for the patient. Currently, there are no acceptable serologic tests for the diagnosis of gonorrhea.

D. Treatment

For several decades penicillin was the drug of choice for the treatment of gonococcal infections. By the late 1980s, however, penicillin resistance had become widespread, and penicillin was no longer a reliable choice for effective therapy. Antibiotic resistance may be plasmid-mediated or chromosomally mediated, and some strains of *N. gonorrhoeae* have become resistant to a variety of antibiotics, including tetracycline, erythromycin, and quinolones. Resistance to these antibiotics is less common. However, quinolone-resistant *N. gonorrhoeae* appears to be a growing problem, and is increasingly prevalent on the west coast of the United States and is common in Asia and the Pacific. The CDC now recommends that quinolones no longer be used to treat gonococcal infections in Hawaii and also advises against their use in California, in cases where susceptibility testing is not feasible. There are a number of effective antibiotic regimens, which are summarized in Table 75.2. Tetracyclines and quinolones should not be used in pregnant women or very young children. The choice of an antibiotic regimen depends on the severity of

TABLE 75.2 Therapy for gonococcal infections[a,b]

Uncomplicated Gonococcal Infections of the Cervix, Urethra, or Rectum

Recommended Regimens:

Ceftriaxone 125 mg intramuscularly (IM), single dose

or

Cefixime 400 mg orally (PO), single dose

or

Ofloxacin 400 mg PO, single dose

or

Ciprofloxacin 500 mg PO, single dose

or

Levofloxacin 250 mg PO, single dose

and

Therapy should include a regimen effective against *Chlamydia trachomatis*, such as

Doxycycline 100 mg PO twice daily × 7 days

or

Azithromycin 1 g PO, single dose

Epididymitis

Recommended Regimens:

Ceftriaxone 250 mg IM, single dose

and

Doxycycline 100 mg PO twice daily × 7 days

Conjunctivitis

Ceftriaxone 1 g IM, single dose, and lavage the infected eye with saline solution once. Topical therapy alone is inadequate.

Disseminated Gonococcal Infection

Patients with DGI should also be treated for *Chlamydia trachomatis*, unless coinfection has been excluded

Recommended Regimen:

Ceftriaxone 1 g IM or intravenously (IV) every 24 h

Parenteral therapy should be continued for 24–48 hr following the onset of clinical improvement, after which therapy may be changed to one of the following therapy to complete at least one week of treatment:

Cefixime 400 mg PO twice daily

or

Ciprofloxacin 500 mg PO twice daily

or

Ofloxacin 400 mg PO twice daily

Pelvic Inflammatory Disease

Recommended Parenteral Regimens for Severe Disease:

Regimen A

Cefotetan 2 g IV every 12 h

or

Cefoxitin 2 g IV every 6 h

and

Doxycycline 100 mg PO every 12 h × 14 days

Continue IV antibiotics until clinically improved, then complete course of therapy with PO doxycycline.

or

Regimen B

Clindamycin 900 mg IV every 8 h

and

Gentamicin 1.5 mg/kg every 8 h, following loading dose

Continue IV therapy until improved, then complete 14-day course with doxycycline or clindamycin 450 mg PO 4 times a day.

Recommended Oral Regimens for Milder Disease:

Ofloxacin 400 mg PO twice daily × 14 days

or

Levofloxacin 500 mg PO daily × 14 days

and

Metronidazole 500 mg PO twice daily × 14 days

[a]The regimens suggested here are for nonpregnant adults only. Recommended treatments and dosing for pregnant women and young children are different.

[b]In addition to the regimens suggested here, there are numerous alternative regimens for each of the above conditions. Please refer to the Sexually Transmitted Diseases Treatment Guidelines 2002, published by the Centers for Disease Control and Prevention for more details.

the infection, ease of administration, cost and availability, as well as effectiveness against coinfecting pathogens, such as *C. trachomatis*. This latter point is important because at least 10–20% of men and 30–40% of women with uncomplicated gonorrhea are also infected with chlamydia. Thus, in areas where the rate of coinfection is high, individuals diagnosed with gonorrhea should be treated for both gonorrhea and chlamydia. In general, individuals with one STD are at risk for having other STDs, and should be evaluated accordingly. The most common cause for treatment failure is repeated exposure to an untreated sexual partner. Given the prevalence of asymptomatic gonorrhea, treatment of all sexual partners contacted within the previous 60 days is recommended, regardless of whether they have symptoms suggestive of infection. If the patient's most recent sexual encounter was more than 60 days prior to the onset of symptoms or diagnosis, then the patient's most recent partner should be treated.

E. Organism and pathogenesis

Neisseria gonorrhoeae belongs to the family Neisseriaceae, which includes several nonpathogenic species, as well as the well-recognized pathogen *Neisseria meningitidis*. *Neisseria gonorrhoeae* is a fastidious gram-negative diplococcus, 0.6–1.0 μm in diameter, with complex growth requirements described previously. *In vivo*, gonococci grow under relatively anaerobic conditions and are capable of growing under strictly anaerobic conditions *in vitro*. *Neisseria gonorrhoeae* produces high levels of catalase and produces an oxidase. Because oxidase production is easy to detect in the laboratory, oxidase positive gram-negative diplococci that are obtained from appropriate clinical specimens, and that grow on selective media, are presumed to be *N. gonorrhoeae* in most cases. *Neisseria gonorrhoeae* can be distinguished from other *Neisseria* species on the basis of sugar fermentation patterns. *Neisseria gonorrhoeae* utilize only glucose, pyruvate, and lactate as their carbon source. Other *Neisseria* species, *Kingella dentrificans, Moraxella* species, and *Branhamella catarrhalis* can be mistaken for *N. gonorrhoeae*, particularly given their similar appearance on Gram stain. Such errors can have serious repercussions for patients and their health care providers given the social and medicolegal implications of being diagnosed with an STD.

The structure, molecular biology, and pathogenesis of *N. gonorrhoeae* have been studied extensively, yet many processes remain elusive. The cell wall is made up of pili, a capsule, several characteristic outer membrane proteins referred to as protein I, Opa protein (protein II), protein III (Rmp), iron regulating proteins, lipooligosaccharide (LOS), and H.8 (Lip), and other proteins that can vary with growth conditions. The pili are hairlike filamentous appendages that extend from the cell surface. Pili are an important virulence factor. Pili, in concert with protein II, enable gonococci to attach to the microvilli of host nonciliated columnar epithelial cells. Protein I is the most prominent outer membrane protein and functions as a porin, providing a hydrophilic channel through the outer membrane. Protein I triggers endocytosis of gonococci by the host mucosal cell. Following endocytosis, the membrane of the mucosal cell retracts around the organism, forming a membrane-bound vacuole which is transported to the base of the cell, where gonococci are subsequently released into the subepithelial tissue. Unlike most of the other outer membrane proteins, protein I is always expressed, and it is useful in identifying individual strains of gonococci as each strain produces a single antigenically stable protein I. Protein I also appears to inhibit neutrophil function, and thus may limit the ability of the host to respond to infection.

In addition to enhancing the adhesion of gonococci to mucosal cells, the Opa protein, or protein II, appears to mediate infection by enhancing gonococcal cell-to-cell adhesion, which may result in a more infectious unit. It specifically appears to enhance adhesion to conjunctival cells, epithelial cells, and neutrophils. Protein III (Rmp) appears to be able to block bactericidal antibodies produced by the host directed at other surface proteins acting as antigens, specifically protein I and LOS. Lipooligosaccharide (LOS) is similar to the lipopolysaccharide component characteristic of the cell wall of all gram-negative bacteria, except that it lacks long, hydrophilic, neutral polysaccharides. LOS appears to mediate most of the damage to host tissues. LOS is the primary target of host antibodies, and it regulates complement activation on the cell surface of gonococci in addition to prompting the release of enzymes, such as proteases and phospholipases. Evidence suggests that gonococcal LOS stimulates the production of tumor necrosis factor (TNF) in fallopian tube organ cultures, and inhibition of TNF has been shown to limit tissue damage. LOS also appears to be capable of molecular mimicry, and thus hinders the ability of the host to mount an effective host immune response. Pili, protein II, and LOS all also exhibit extensive intrastrain and interstrain antigenic variation, as well as phase variation where pili and protein II are concerned. Such variation likely represents a mechanism by which gonococci can escape the host immune response. The role of the other outer membrane proteins in the pathogenesis of gonococci is not well understood.

III. CHLAMYDIA

A. Epidemiology

Chlamydia, caused by *Chlamydia trachomatis*, is currently considered the most common STD in the U.S. There are an estimated 4 million cases in the U.S. each year, at a cost of $2.4 billion. In 1986 chlamydia became a reportable disease in several states, and for years thereafter the number of reported cases increased steadily, likely as a result of improved diagnostic techniques, screening practices, and reporting in more states. Chlamydia became the most common reportable disease in the U.S. in 1995, when there were 477,638 cases reported. Chlamydia is a problem internationally as well. In 1995, the WHO estimated that there were 89 million new cases worldwide. As with other STDs, data on the true incidence and prevalence of chlamydia are limited by a number of factors. First, chlamydial infections are often asymptomatic. In a screening study of women attending a family planning clinic, 70% of the women found to have genital infections secondary to chlamydia had neither clinical findings nor symptoms of infection. Thus, a significant number of cases likely go unrecognized by the patient or health care provider. Second, when symptoms are present, individuals are often treated empirically without diagnostic confirmation, and thus these cases of chlamydia are never reported. Third, some of the laboratory methods used to diagnose chlamydia are relatively insensitive and may yield false negative results, leading to an underestimate of disease burden.

Several studies in a variety of settings, utilizing different diagnostic screening strategies, have provided some estimates of the prevalence of genital chlamydial infections in the U.S. Approximately 3–5% of asymptomatic men and women seen in general medical settings are infected with chlamydia, whereas women and men screened in STD clinics have a prevalence of 15–20%. More recent studies using highly sensitive diagnostic assays reported even higher prevalence data. In one study, Burstein and colleagues demonstrated that the prevalence of chlamydial genital infections among adolescent females screened in family planning, STD, and schoolbased clinics in Baltimore, Maryland, was 29.1%. The majority of infected girls were asymptomatic. In a study of 13,204 new female army recruits the prevalence was 9.2%. There was wide geographic variation in the prevalence of chlamydial infection according to the state of origin of the recruit. More than 15% of women from South Carolina, Georgia, Alabama, Louisiana, and Mississippi were infected with chlamydia. In New Jersey, North Carolina, Kentucky, Texas, Oklahoma, and Arkansas the prevalence was 10–15%. In five states, including Washington and Oregon, the prevalence was less than 5%, most likely as a result of aggressive public health initiatives in place in those states since the late 1980s, aimed at reducing the number of chlamydial infections. In a similar study of 2245 male army recruits the prevalence of chlamydial urethritis was 5.3% overall, with similar geographic variation by state. Of the men infected with chlamydia, less than 15% reported having any symptoms to suggest infection.

Specific risk factors for chlamydial infections have been consistently identified in these studies. The single most important risk factor for chlamydial infections in women is young age. In fact, the CDC now recommends annual screening of all sexually active adolescent women and women ages 20–25 years, whether or not symptoms are present. In a study of adolescent females aged 12–19 years, 14-year-old females had the highest age-specific chlamydia prevalence rate. In the study of female army recruits, 95% of those infected with chlamydia were less than 25 years of age. Furthermore, studies of reinfection have shown that up to 20–30% of female adolescents had evidence of another chlamydial infection within 6 months of the initial infection. Young age, however, does not appear to be a consistent risk factor for chlamydial infections in men. In the study of male army recruits, age did not correlate with infection. Other risk factors associated with chlamydial infection mirror those associated with other STDs and include multiple sex partners, a recent new sex partner, inconsistent condom use, and nonwhite race. Among female army recruits, African-American women had more than a threefold greater risk of infection compared to white female recruits. Another important risk factor for chlamydia is the diagnosis of gonorrhea. As noted, approximately 20% of men and 40% of women infected with gonorrhea are simultaneously infected with chlamydia. In the above study of adolescent females, however, attempts to identify the majority of those infected using such risk factors, other than young age, as a screening measure were unsuccessful. The investigators recommended that all sexually active adolescent females be screened for chlamydia every 6 months, regardless of symptoms, prior infections, condom use, or partner risks.

Chlamydia is also a problem for pregnant women and their newborns. Prevalence studies in pregnant women using relatively insensitive screening techniques yielded prevalence rates of 5–7%, although prevalence rates of 21–25% have been identified in inner city and Native American populations. Of infants born to untreated infected mothers, approximately 60–70% can become infected with chlamydia, a significant proportion of which go on to develop pneumonia and ocular complications as a result.

Given the results of these studies, some areas of the U.S. have instituted broad-based screening and treatment programs, and the prevalence of chlamydial infections in these areas has declined dramatically. In the Pacific Northwest of the U.S., screening for chlamydia in patients attending family planning clinics began in 1988, and in STD clinics in 1993. The prevalence of chlamydia in this region declined from 10–12% in the late 1980s to 4–5% in 1995. One challenge to these efforts is the relatively long incubation period of chlamydia of 1–3 weeks. Infected individuals may not develop symptoms for a prolonged period of time, if at all, and thus may remain sexually active while infectious. The efficiency with which chlamydia is transmitted between sexual partners is not well understood. Studies evaluating the sexual partners of infected individuals demonstrated an infection rate of 85% among female contacts of men with chlamydial urethritis. Traditionally researchers suspected that men were more likely to transmit chlamydia to women than the converse, but more recent data from studies using highly sensitive diagnostic techniques suggest that transmission may be equal in either direction.

B. Clinical manifestations

There are several subtypes of *C. trachomatis*, the trachoma and LGV (lymphogranuloma venereum) subtypes being the only human pathogens. LGV subtypes are transmitted through sexual contact and cause ulceration, as well as painful lymphadenopathy and abscess formation. LGV infections are extremely rare in industrialized nations; they are less common than infections caused by the trachoma subtype and will not be discussed further here. Genital infections caused by the trachoma subtype of *C. trachomatis* have many of the same characteristics as those caused by *N. gonorrhoeae*. Both organisms preferentially infect columnar epithelial cells and do not appear to be capable of growth in squamous cells. Thus, like gonorrhea, chlamydia infects the urethra, with extension to the epididymis and possibly the prostrate gland in men, the squamocolumnar junction of the endocervix in women, with the potential for spread to the endometrium and upper genital tract, and the rectum. *Chlamydia trachomatis* can also cause conjunctivitis. In contrast to gonorrhea, however, chlamydial infections generally have a longer incubation time, are associated with a less robust inflammatory response, and are more likely to be asymptomatic. Unlike gonorrhea, chlamydia does not disseminate to cause a systemic infection like DGI. Unfortunately, there are no unique signs or symptoms to help make the diagnosis of *C. trachomatis* infection of the genital tract. The associated symptoms and physical findings, if present at all, can also result from other infections or complications, and therefore one has to have a high index of suspicion to diagnose a chlamydial infection.

1. Chlamydial infections in women

In women, *C. trachomatis* most commonly infects the cervix, although approximately 50% of infected women will have simultaneous infection of the urethra. Chlamydial infection of the cervix typically results in cervicitis with a mucopurulent discharge, present in approximately 20–40% of infected women. On exam, there is sometimes visible swelling and easily inducible bleeding in a zone of ectopy, an area with a relative increase in exposed columnar cells at the squamocolumnar junction. As previously noted, however, in studies in which women were screened for chlamydia, 70% or more of infected women did not have any abnormal clinical findings. Women with cervical ectopy have a higher prevalence of chlamydia than those without ectopy. Cervical ectopy is normally present in 60–80% of sexually active adolescent females, and declines with age. This finding may in part explain the higher prevalence of chlamydia among adolescent females. Oral contraceptives are also associated with cervical ectopy, and this may account for the increased risk of chlamydia, and gonorrhea, for women on oral contraceptives. Chlamydial urethritis has also been well documented in women and is estimated to occur without concomitant cervical infection about 25% of the time. The prevalence of isolated urethral infection increases with age. Chlamydial urethritis may be associated with painful and frequent urination, although the majority of women with urethritis do not have these symptoms. Nonetheless, *C. trachomatis*, as well as *N. gonorrhoeae*, should be considered as a possible diagnosis in sexually active women with urinary symptoms if there is evidence of coexisting cervicitis or if other STD risk factors are present.

The most serious form of *C. trachomatis* infection is upper genital tract disease. Approximately 10% of women with cervical chlamydia infection will develop symptomatic pelvic inflammatory disease (PID). There is likely an even larger proportion of infected women who develop asymptomatic upper tract disease. All of these women are at increased risk for infertility and ectopic pregnancy as a result of tubal inflammation and scarring. The clinical manifestations of PID secondary to chlamydia are similar to those of PID caused by gonorrhea, and are discussed more fully in a previous section. As with other sites of infection, chlamydial infection of the upper tract is clinically milder and more likely to be asymptomatic than infections caused by gonorrhea or anaerobic bacteria, despite ongoing tubal damage. Like gonorrhea, chlamydia can cause perihepatitis, with or following the development of PID, referred to as the Fitz–Hugh–Curtis syndrome. Women who are infected with chlamydia in the first trimester of pregnancy are at increased risk of postpartum endometritis following vaginal delivery.

2. *Chlamydial infections in men*

More common than gonococcal urethritis is the entity known as nongonococcal urethritis (NGU). A diagnosis of NGU is made when *N. gonorrhoeae* cannot be identified in urethral specimens from a man with urethritis. *Chlamydia trachomatis* is believed to be the causative organism in 15–55%, with a lower prevalence among older men of cases of NGU. Other organisms include *Ureaplasma urealyticum, Trichomonas vaginalis*, and herpes simplex virus. NGU is often characterized by the presence of inflammatory cells in the urine and a urethral discharge that is usually less purulent and relatively scant compared to the urethral discharge associated with gonorrhea. Infected men may experience painful or frequent urination. However, the symptoms are variable enough that one cannot reliably distinguish NGU from gonococcal urethritis clinically. If a Gram stain of the urethral specimen shows a large number of inflammatory cells (specifically, polymorphonuclear cells) without intracellular diplococci, then a presumptive diagnosis of NGU is made. A diagnosis of chlamydial urethritis requires specific diagnostic testing.

Men may also experience urethritis following treatment for gonorrhea, known as postgonococcal urethritis. These men have recurrent or persistent symptoms of urethritis despite appropriate treatment for gonorrhea. *Chlamydia trachomatis* is thought to account for 70–90% of cases of postgonococcal urethritis. Men who develop this syndrome were most likely infected with both gonorrhea and chlamydia simultaneously, but were treated with antibiotics effective against gonorrhea but not chlamydia. Because of the longer incubation period for chlamydia, symptoms of urethritis may develop even after an apparent initial response to therapy for gonococcal urethritis. Because of the high rate of coinfection, presumptive treatment for both organisms is recommended when gonorrhea is diagnosed.

Chlamydial infections of the lower genital tract in men may also become complicated. *Chlamydia trachomatis* is the most common cause of epididymitis in sexually active young men, accounting for more than 60% of cases. Epididymitis results from the spread of an untreated chlamydial urethral infection to the upper gential tract, and it is characterized by progressive unilateral scrotal pain and swelling. Chlamydial urethritis has also been associated with the subsequent development of Reiter's syndrome. Reiter's syndrome consists of urethritis, conjunctivitis, arthritis, and characteristic skin lesions. Some studies have found that 80% of men with Reiter's syndrome have evidence of prior or ongoing chlamydial infection. This syndrome is more common is men with the HLA-B27 haplotype. These patients may also develop arthritis or tenosynovitis, without the other features of Reiter's syndrome, in association with infection with *C. trachomatis*, and usually improve following treatment for chlamydia.

3. *Other chlamydial infections in adults*

Both men and women can develop proctitis as a result of rectal infection with *C. trachomatis*. Rectal infections are usually asymptomatic, but if proctitis develops patients may experience rectal pain, bleeding, and a mucous discharge. Adults can also develop an acute follicular conjunctivitis as a result of autoinoculation of the eye with infected genital secretions. Conjunctivitis can develop within 1–3 weeks and does not typically result in permanent damage or vision loss.

4. *Chlamydial infections in neonates*

Neonates exposed to *C. trachomatis* during vaginal delivery are at risk for developing conjunctivitis and pneumonia. Nasopharyngeal infection is also common. Conjunctivitis develops within 5–21 days and is characterized by copious purulent discharge and redness of the eye. As in adults, it is usually self-limited and does not result in permanent eye damage or vision loss. Pneumonia caused by chlamydia can be life-threatening, however. The incubation period for chlamydial pneumonia in neonates is 2–12 weeks. Infected infants often have a history of conjunctivitis and present with rhinitis, rapid breathing, and a characteristic cough, in the absence of fever. There is evidence to suggest that infants who have chlamydial pneumonia frequently go on to develop asthma or obstructive airway disease, even after appropriate treatment of the initial pneumonia.

C. Diagnosis

Because *C. trachomatis* is an obligate intracellular organism, cell culture has traditionally been the gold

body (EB). The EB is the stable, extracellular, infectious form of the organism and does not appear to be metabolically active. Attachment is partially charge dependent and appears to be mediated by heparan sulfate-like molecules that act as a bridge between receptors on the surface of the organism and the host cell. After attachment to the host cell, the organism, in the EB form, is endocytosed into the host cell, although the exact mechanism for uptake into the host cell is unclear. Following entry into the cell, the organism undergoes a morphological change within the first 6–8 h, which transforms the organism from an EB into a reticulate body (RB). The RB is the intracellular, metabolically active, dividing form of the organism. The RB is not infectious and cannot survive outside the host cell. The RB rely on the host cell for precursors and energy in order to synthesize their own macromolecules such as RNA, DNA, and proteins. The RB divide by binary fission, and in the first 18–24 h after entry into the host cell, some RB revert back to the EB form. The process by which chlamydiae undergo the morphological switch between the EB and RB form is not known. After the initial 24 h, there is a progressive increase in the number and proportion of EB, and ultimately, around 48–72 h after entry, the host cell ruptures, releasing the infectious EB into the extracellular environment to infect other susceptible host cells. The entire cycle takes place within the phagosome, and, interestingly, phagolysosomal fusion does not occur until cell rupture is about to take place. The inhibition of phagolysosomal fusion is attributed to a surface protein antigen on the EB that acts as an inhibitor.

The pathogenesis of *C. trachomatis* is not well under-stood. There is no direct *in vivo* evidence for a latent stage, where chlamydiae persist without ongoing replication. Since the life cycle ends with the destruction of the host cell, then presumably untreated *C. trachomatis* may cause progresive damage in infected individuals. The disease and associated symptoms of *C. trachomatis* infection are believed to be primarily the result of the inflammatory response of the host to both the organism and destroyed host cells. Available evidence, primarily from animal data, suggests that much of chlamydial disease manifestations stem from the host's immune reaction to the infection. This is supported by the observation that oftentimes second infections inflict greater damage. The responsible stimulus for the immune response appears to be a sensitizing antigen belonging to the 60-kDa heat-shock protein (HSP 60) family. This antigen is loosely bound to EB and is excreted by infected cells. Some women who are infertile or experience ectopic pregnancy because of tubal injury have been found to have high levels of antibodies to HSP 60.

IV. SYPHILIS

A. Epidemiology

Syphilis is a complex and fascinating disease, caused by infection with the bacterium *Treponema pallidum*. The disease is a systemic one, capable of affecting nearly every organ in the body and thus mimicking many other disease processes. Syphilis is characterized by several different stages, including primary, secondary, tertiary, and latent stages, and it can persist for decades if untreated. Syphilis is distributed worldwide and is particularly a problem in the developing world, where it is the principal cause of genital ulcer disease. Prior to the development of penicillin in the 1940s, syphilis was a huge public health problem in the U.S. At the beginning of World War II, the U.S. Public Health Service estimated that approximately 2.5% of all Americans were infected with syphilis. With the advent of penicillin, however, the prevalence of primary and secondary cases declined by about 93%. Fortunately, penicillin remains fully effective against syphilis, with no evidence of the emergence of penicillin resistance.

Nonetheless, syphilis remains a significant public health concern. Beginning in the mid-1980s there was a marked and progressive resurgence of primary and secondary (P&S) syphilis in the U.S. that peaked in 1990, with a total of 55,132 reported cases of P&S syphilis, representing a 20.2% increase over the previous year. The reasons for these trends are unclear. Of concern was the dramatic increase in cocaine use in the U.S. during the mid- to late 1980s. Cocaine use has been associated with high risk sexual behaviors. Also of concern have been declines in the resources allocated for public health programs involved in the control of syphilis over the previous two decades. During the 1980s, the majority of cases were initially observed in homosexual men, and subsequently in intravenous drug-using individuals. In the 1980s, more than 40% of infected men reported other men as their sexual contacts. With the onset of the AIDS epidemic and the associated changes observed in sexual behavior among homosexual men, the proportion of syphilis cases attributable to this segment of the population declined dramatically. Subsequently, mirroring trends in the HIV epidemic, a growing proportion of syphilis cases were seen in inner city, heterosexual, drug-using populations.

Since 1990, however, there has been a consistent decline in reported cases of syphilis in the U.S. In 1996, 11,387 cases of P&S syphilis were reported, the lowest number since 1959. The majority of these cases were from the South. As observed with many other

STDs, there were marked discrepancies in the proportion of cases among specific races or ethnic groups. The rate of P&S syphilis among African-Americans was nearly 50 times that among whites. The rate of P&S syphilis among Hispanics was 3 times that seen in whites. Cases of congenital syphilis have also declined, but of the 1160 reported cases in 1996 where the race or ethnicity of the mother was known, 90% of all reported cases were either African-American or Hispanic, yet these groups comprised only 23% of the female population. Certainly, biases in reporting explain some of these findings, but cannot explain them completely. Other more fundamental risk factors for STDs that are also associated with race or ethnicity, such as socioeconomic status, access to quality health care and health education, and others, may also play an important role.

B. Clinical manifestations

Syphilis is unique from other STDs in its ability to spread throughout the body in the bloodstream of infected individuals, resulting in involvement of multiple distant organs including the central nervous system, bones, arteries, and other sites. Because it is a systemic disease, with sometimes nonspecific signs and symptoms, syphilis can mimic many other disease processes, and patients may be easily misdiagnosed if syphilis is not considered and ruled out by appropriate diagnostic testing. Transmission usually occurs as a result of sexual contact with an infected individual who has the typical mucosal lesions associated with primary and secondary syphilis. Syphilis can also be transmitted via contact with other infectious lesions seen in the mouth or on the skin in secondary syphilis, or it may be transmitted transplacentally from mother to fetus, resulting in congenital syphilis. In addition, infected mothers can transmit syphilis to their newborns through breast-feeding.

Following contact with an infected individual, *T. pallidum* begins to multiply at the site of entry of the organism, resulting in the formation of a papule, after an incubation period of 10–90 days. The papule eventually converts into a painless superficial ulcer, known as a chancre, characteristic of primary syphilis. Chancres are highly contagious lesions, owing to the presence of a high concentration of *T. pallidum*. During this period, infected individuals often develop local lymph node enlargement, but often times these signs go completely unnoticed. After a period of 2–6 weeks, the chancre usually heals spontaneously.

Subsequent spread and multiplication of the organism in other organs results in findings characteristic of secondary syphilis. These signs and symptoms of secondary syphilis usually do not appear until after a 2- to 24-week period following resolution of the primary stage, during which the infected individual may be completely asymptomatic. Individuals with secondary syphilis typically develop a rash but often experience other symptoms such as low-grade fever, malaise, headache, sore throat, generalized lymph node enlargement, and muscle aches. Patients may also develop bone inflammation and hepatitis, both of which are usually asymptomatic. The rash associated with secondary syphilis usually begins with a mild, transient macular rash, which often goes unnoticed. This quickly evolves into a diffuse, symmetric, papular rash, which involves the entire trunk and extremities, classically including the palms of the hands and the soles of the feet. The papules are red or reddish-brown, discrete, and often scaly in appearance. Patients typically experience severe itching. Other typical mucocutaneous lesions, known as condylomata lata, are large, raised white or grayish patches that occur in warm, moist areas, such as the mouth, vagina, or rectum. These skin lesions contain a high concentration of treponemes and are thus highly contagious. Hair loss as well as hyper- or hypopigmentation of the skin can occur also. As seen in primary syphilis, the signs and symptoms of secondary syphilis usually resolve spontaneously after a 2- to 6-week period, even without treatment, as a result of the host's immune response.

About 25% of individuals with primary or secondary syphilis remain infected and may experience multiple relapses of the secondary stage, usually within the first year of infection. After that, approximately one-third of patients with primary or secondary syphilis clear the infection completely without treatment. The remainder enter a stage referred to as latent syphilis, for which there are no signs or symptoms, and a diagnosis of syphilis can only be made by serologic screening. Latent syphilis can persist for decades. This stage is arbitrarily divided into early latent and late latent stages. By definition, patients with early latent syphilis are asymptomatic and have been infected for less than 1 year. Those with late latent syphilis are asymptomatic and have been infected for more than 1 year. A diagnosis of early syphilis can be made if within the preceding year the individual has had a documented seroconversion, unequivocal symptoms of primary or secondary syphilis, or a sex partner with documented primary, secondary, or early latent syphilis. Individuals with syphilis are believed to be infectious and can thus transmit syphilis to their partners during the primary, secondary, and early latent stages. If the duration of latent syphilis infection cannot be documented for certain, then patients are assumed to have late latent syphilis for the purposes of treatment. One-half of patients who develop latent syphilis will continue to have serologic evidence of infection but will never develop late complications of the disease. The other half may remain asymptomatic for decades but will ultimately develop late manifestations of syphilis. Individuals with late latent syphilis are immune to reinfection with *T. pallidum*.

This latter group of individuals, if untreated, eventually experience disease progression and develop what is referred to as tertiary syphilis. Individuals with tertiary syphilis may have involvement of their skin, bones, central nervous system, heart, and arteries, as well as other sites. At this stage, syphilis may be characterized by the presence of destructive granulomatous lesions known as gummas. Gummas may be found in the bones and skin, but they may also be found in other organs, including rarely the heart and digestive tract. The pathogenesis of gumma formation is not known, but gummas usually result in extensive damage to surrounding tissues. Tertiary syphilis involvement of the central nervous system and cardiovascular system may also result from direct invasion by treponemes, without gumma formation. Cardiovascular involvement is thought to result from the multiplication of treponemes in the aorta and proximal coronary arteries. The associated arteritis can lead to severe aortic valve insufficiency, aneurysm formation, and myocardial infarction. The neurologic effects of tertiary syphilis take two forms. Meningovascular syphilis results from involvement of the meninges, the tissues surrounding the brain. The parenchymatous form affects the brain, referred to as general paresis, and spinal cord, known as tabes dorsalis, directly. Thus, the neurologic manifestations of tertiary syphilis may include meningitis, stroke, dementia, cranial nerve deficits, sensory disturbances, and paralysis. Approximately 80% of deaths from syphilis are attributable to cardiovascular involvement, and the remainder result from neurologic disease. Fortunately, tertiary syphilis has become extremely rare since the introduction of penicillin.

Congenital syphilis is a potentially devastating disease, resulting from transplacental transmission of syphilis from an infected mother to her fetus. Approximately 50% of infected fetuses are spontaneously aborted or stillborn. Those that survive to delivery may exhibit a wide range of signs and symptoms of syphilis. Newborns with symptoms before the age of 2 years have early congenital syphilis, and they develop skin lesions, mucous membrane involvement often with copious secretions, bone involvement, severe anemia, and enlargement of the liver and spleen. Children with late congenital syphilis may not have symptoms until after the age of 2 years, at which time they may develop interstitial keratitis with subsequent blindness, tooth deformities, deafness, neurosyphilis, cardiovascular involvement, and skeletal deformities. Pregnant women who receive prenatal care are routinely screened for syphilis given the terrible consequencs of infection for the newborn.

Individuals with syphilis and concurrent HIV infection appear to have a similar experience with regard to primary and secondary syphilis, with one notable exception. Concurrent HIV infection has been recognized to have a significant impact on neurologic involvement in individuals with syphilis. Numerous case series have documented rapid progression from early syphilis to neurosyphilis in patients infected with both HIV and syphilis. Many of these patients develop meningitis, cranial nerve deficits, including hearing loss and blindness, and even stroke, often in the context of therapeutic failure with conventional doses of penicillin for primary or secondary syphilis. Thus, evaluation and treatment recommendations vary somewhat for individuals with both HIV infection and syphilis.

C. Diagnosis

The most sensitive and specific test for the diagnosis of primary syphilis is the identification of treponemes on dark-field microscopy, when performed by an experienced observer on an adequate sample of fluid obtained from the surface of a chancre. *Treponema pallidum* has a characteristic corkscrew appearance on dark-field microscopy, and false positive results do not occur in experienced hands. Serologic tests can be useful in the diagnosis of primary syphillis, although false negative results occur 10–20% of the time. The RPR (rapid plasma reagin) and the VDRL (Venereal Diseases Research Laboratory) are referred to as "nontreponemal tests," because they do not measure antibodies specific to treponemal components. As a result, these assays may occasionally be positive in other disease states and are less specific than treponemal tests. These assays are also more likely than treponemal tests to be negative in primary syphilis and thus should not be used to rule out a diagnosis of primary syphilis. The RPR titer does, however, correlate with disease activity and often becomes negative following adequate treatment, although it may remain positive at a low titer throughout one's life. Thus, even when the diagnosis of syphilis is made by other means, it is useful to obtain an RPR to monitor subsequent disease activity. Tests such as the FTA-ABS (fluorescent treponemal antibody absorbed) or the MHA-TP (microhemagglutination assay for antibody to *T. pallidum*) are referred to as treponemal tests. These assays are more likely to be positive in primary syphilis than the RPR or VDRL and remain positive for life, but they are not useful in monitoring disease activity. Because these assays are specific for syphilis, but are more costly and labor intensive, they are not used for screening but rather to confirm a diagnosis of syphilis when an individual has a positive RPR or VDRL, for stages other than primary syphilis.

Secondary syphilis can be diagnosed by dark-field examination of material taken from skin lesions of individuals with suspected secondary syphilis. Serologic assays are much more reliable in diagnosing secondary syphilis compared to primary syphilis, however, and are the preferred method of diagnosis in this setting because of their relative convenience. The RPR is always positive in secondary syphilis, usually with a high titer. The treponemal tests are also always positive in secondary syphilis, and may be used to confirm that a positive RPR is in fact due to syphilis. The RPR is then used to monitor the response to therapy.

Because individuals with latent syphilis are asymptomatic, latent syphilis is usually diagnosed as a result of screening with an RPR or VDRL, followed by confirmation with a treponemal test. A diagnosis of latent syphilis is easily made if an individual recalls a recent history of a chancre or skin rash for which he or she did not seek medical care and is currently asymptomatic. Unfortunately, most asymptomatic individuals with a positive RPR are unable to provide a diagnostic medical history. For public health purposes, these individuals are assumed to have latent syphilis and are thus treated, although it may be the case that they have a positive serology despite previous adequate therapy or may have partially treated syphilis. Most local public health departments keep a record of all positive syphilis serologies and attempt to maintain records on treatment and subsequent-response to therapy. If a current RPR titer is positive at two or more dilutions less than, or fourfold lower than, previously documented (e.g. from 1:16 to 1:4), then the individual has likely had an appropriate response to prior therapy.

A diagnosis of neurosyphilis is easily made when there are neurologic findings in the setting of positive serologies and abnormal cerebrospinal fluid (CSF). However, the presence of only two of these three criteria is usually sufficient to mandate treatment for neurosyphilis. The CSF in patients with neurosyphilis may be normal or may exhibit any combination of the following findings, including an elevated opening pressure, an elevated white blood cell count (usually 10–200 mononuclear cells per cubic millimeter), an elevated protein concentration, or a positive CSF VDRL. A positive CSF VDRL is diagnostic of neurosyphilis, but this test is not sensitive and is often negative in neurosyphilis. Stroke, dementia, and other neurologic abnormalities more often than not are due to etiologies other than syphilis, but may coincidentally occur in the presence of positive serology. If, however, a patient has neurologic findings possibly due to syphilis in the presence of positive RPR or treponemal test, these individuals should be treated for neurosyphilis, even if the CSF is normal, and evaluated for other disease processes as indicated (e.g. carotid artery disease in the case of stroke). CSF examination is most useful for confirming the diagnosis of neurosyphilis when the CSF is abnormal in the context of positive serology. If an individual has a positive serology and abnormal CSF, the individual should be treated for neurosyphilis even in the absence of clinical findings of neurosyphilis. This latter scenario is most common in HIV-infected individuals. In the absence of gummas, diagnosis of cardiovascular disease from tertiary syphilis is usually made on clinical grounds in the presence of positive serology or a history of syphilis and characteristic findings on angiography. Other forms of tertiary syphilis are usually diagnosed on the basis of the identification of a gumma.

D. Treatment

Parenteral penicillin G is the preferred treatment for all stages of syphilis. The exact preparation used, the dose, and the duration of therapy depend on the stage of syphilis and certain characteristics of the infected individual. HIV may alter the response to therapy, and treatment failures have been reported in HIV-infected individuals. It is thus recommended that all individuals with syphilis be tested for HIV, and in areas of high prevalence of HIV, individuals with syphilis that are negative on initial HIV testing should be retested 3 months later. Adults with primary, secondary, or early latent syphilis should be treated with a single dose of 2.4 million units of benzathine penicillin intramuscularly. Individuals with primary or secondary syphilis who have signs or symptoms of neurologic or ophthalmic involvement should undergo CSF evaluation and slit-lamp examination to rule out neurosyphilis and ocular involvement, respectively. In the absence of neurologic findings, routine lumbar puncture is not indicated, as CSF is often abnormal in primary and secondary syphilis and does not necessarily reflect true neurosyphilis. Patients should be reevaluated clinically and serologically at 6 and 12 months. If patients have persistent clinical findings, a rising RPR titer, or failure of the titer to decline fourfold within 6 months of treatment for primary or secondary syphilis, then they most likely represent treatment failures or have been reinfected. These patients should undergo lumbar puncture to rule out neurosyphilis and should be treated with three intramuscular injections of 2.4 million units each of benzathine penicillin 1 week apart, if neurosyphilis is not present. Patients with primary, secondary, or early latent syphilis who are not pregnant and are allergic to pencillin may alternatively be treated with doxycline 100 mg orally twice a day for 2 weeks. Treatment failures are more likely with doxycycline, and these patients require especially close follow-up. Pregnant patients who are allergic to penicillin should be desensitized and treated with penicillin.

About one-third of patients with primary syphilis and two-thirds of patients with secondary syphilis develop an unusual reaction to treatment known as the Jarisch–Herxheimer reaction. This reaction typically consists of fever, chills, headache, worsening rash, and even hypotension in severe cases. The onset is usually within the first 4 h following the initiation of therapy, and the reaction resolves within 24 h. It is believed to result from the release of treponemal components, resulting in an endotoxin-like reaction. The Jarisch–Herxheimer reaction may be confused with an allergic reaction to penicillin but is not an indication for discontinuing therapy. Instead, treatment should continue and the patient should be treated with nonsteroidal anti-inflammatory medication, in addition to other supportive measures as indicated.

Patients with late latent syphilis, or latent syphilis of unknown duration, should be treated with a total of 7.2 million units of benzathine penicillin administered as three doses of 2.4 million units each, intramuscularly, 1 week apart. Evaluation should include a lumbar puncture to rule out neurosyphilis for all patients with latent syphilis if they have neurologic or ophthalmic signs or symptoms, evidence of active tertiary syphilis, evidence of treatment failure as described above, or HIV infection with late latent syphilis or syphilis of unknown duration. Patients should be reevaluated clinically and serologically at 6, 12, and 24 months. An alternative regimen for nonpregnant, penicillin-allergic patients is doxycycline 100 mg orally, twice daily for 4 weeks. As with early syphilis, penicillin-allergic pregnant patients should undergo desensitization, followed by treatment with penicillin. Treatment for tertiary syphilis other than neurosyphilis is the same as that for late latent syphilis, except that all patients with tertiary syphilis should undergo lumbar puncture to rule out neurosyphilis.

Neurosyphilis can occur at any stage of syphilis, and patients with any evidence of meningitis, ophthalmic or auditory symptoms, or cranial nerve palsies should undergo a lumbar puncture to evaluate for neurosyphilis. Because of an increased risk of neurosyphilis, HIV-infected individuals with late latent syphilis or syphilis of unknown duration, patients with evidence of treatment failure, and patients with active tertiary syphilis should also undergo a CSF examination prior to therapy, even in the absence of signs or symptoms of neurosyphilis. Those diagnosed with neurosyphilis should receive aqueous crystalline penicillin G 18–24 million units a day, administered as 3–4 million units intravenously every 4 h for 10–14 days. This course should be followed with a dose of 2.4 million units of benzathine penicillin intramuscularly, to provide a duration of therapy comparable to that for latent syphilis. There are no adequate alternatives to penicillin in the treatment of neurosyphilis. Individuals with neurosyphilis who are allergic to penicillin should undergo desensitization, followed by the recommended course of penicillin. Many experts also recommend a follow-up lumbar puncture 6 months following the completion of therapy, to ensure an adequate response.

Partner notification and treatment are as important for cases of syphilis as for other STDs. Transmission of syphilis occurs in approximately 50% of cases where there is direct sexual contact with lesions of an individual who has primary or secondary syphilis. The Centers for Disease Control and Prevention recommend that all individuals exposed within 90 days preceding the diagnosis of primary, secondary, or early latent syphilis in a sex partner be treated presumptively, given the concern for false negative serologies shortly after exposure. For individuals exposed more than 90 days prior to the diagnosis of primary, secondary, or early latent syphilis in a sex partner, the CDC recommends presumptive treatment if serologic testing is not immediately available and follow-up cannot be guaranteed. Long-term sex partners of individuals who are diagnosed with a late form of syphilis should be evaluated for syphilis clinically and by serologic testing, then treated accordingly.

E. Organism and pathogenesis

Treponema pallidum is one of several treponemal species associated with human disease. Pathogenic species belonging to the genus *Treponema* are responsible for yaws, endemic syphilis, pinta, periodontal disease, as well as venereal syphilis discussed here. *Treponema pallidum* is an obligate human parasite, with no animal or environmental reservoirs. It is a helical corkscrew-shaped gram-negative bacterium,

6–20 μm in length and 0.10–0.18 μm in diameter, placing it below the resolution of light microscopy. The unstained organism is best visualized by dark-field or phase-contrast microscopy. *Treponema pallidum* generally stains poorly with many dyes but can be visualized using silver impregnation techniques. Like other gram-negative bacteria, *T. pallidum* has an outer membrane, an inner membrane, and a thin cell wall consistig of peptidoglycan. The outer membrane is somewhat unique, however, in that it lacks lipopolysaccharide, thus rendering it more susceptible to damage resulting from physical disruption or detergent use during handling. The outer membrane also contains relatively few proteins, which may account for the limited antibody response seen with this pathogen. *Treponema pallidum* is also unique in that it has flagella, located in between the inner and outer membranes, which are responsible for its characteristic motility, consisting of rapid rotation along its longitudinal axis, as well as bending and flexing. Such motility is thought to play an important role in the organism's ability to invade and disseminate.

Treponema pallidum is a fastidious organism and has not been successfully cultured *in vitro*, and thus diagnosis depends on direct visualization and serologic testing. For study purposes, viable organisms can, however, be maintained for several weeks in tissue cell culture. The organism appears to be microaerophilic and is very sensitive to environmental conditions, being easily inactivated by mild temperature fluctuations, desiccation, and chemical agents including most disinfectants. These characteristics have limited study of the organism. Recently, the complete genome of *T. pallidum* was successfully sequenced. The genome consists of 1,138,006 base pairs containing 1041 predicted coding sequences. The information gained from sequencing has helped to elucidate systems for DNA replication, transcription, translation, and repair, as well as possible virulence factors and metabolic processes. Potential virulence factors include a family of 12 membrane proteins and several putative hemolysins. Catabolic and biosynthetic capabilities appear to be limited. Glucose is the main carbon and energy source for *T. pallidum*, although pyruvate appears to be an alternative.

The pathogenesis of *T. pallidum* is not well understood. Following penetration of mucosal surfaces, *T. pallidum* appears to adhere to host cells and then begin multiplication. Treponemes have been shown to adhere to a variety of cell types, perhaps through an interaction with fibronectin and host cell receptors. Following multiplication, the organisms disseminate via the circulation and invade distant organs, aided by their unique motility. Lesion resolution is thought to be the result of cell-mediated immunity, involving phagocytosis by macrophages activated by lymphokines and opsonic antibodies, released from antigen-specific sensitized T and B cells. Despite the destruction of millions of organisms, some persist and eventually cause the later stages seen in syphilis. There are several possible mechanisms by which treponemes escape the host immune response, probably the most important of which is the paucity of outer membrane proteins that would ordinarily provide ample targets for the immune response. In addition, treponemes may also have some surface components that inhibit complement-mediated killing. Secondary syphilis results from the dissemination and further multiplication of persistent treponemes. Some manifestations may also result from immune complex deposition, especally with regard to the skin and kidney involvement seen at this stage. Following a period of latency, persistent treponemes go on to invade the central nervous system, the cardiovascular system, and other organs in patients with tertiary syphilis. Here, much of the damage seen is likely due to the organism's invasive capabilities and the associated delayed hypersensitivity response of the host.

BIBLIOGRAPHY

Black, C. M. (1997). Current methods of laboratory diagnosis of *Chlamydia trachomatis* infection. *Clin. Microbiol. Rev.* **10**, 160–184.

Centers for Disease Control and Prevention, Division of STD Prevention. (1997). "Sexually Transmitted Disease Surveillance, 1996." U.S. Department of Health and Human Services, Public Health Service. September.

Centers for Disease Control and Prevention (2002). Sexually transmitted diseases treatment guide lines. *MMWR 2002* **51** (No. RR-6).

Cohen, M. S. (1998). Sexually transmitted diseases enhance HIV transmission: No longer a hypothesis. *Lancet* **351** (Suppl. 3), 5–7.

Drugs for sexually transmitted diseases (1995). *Med. Lett. Drugs Therap.* **37**(964), 117–122.

Erbelding, E., and Quinn, T. C. (1997). The impact of antimicrobial resistance on the treatment of sexually transmitted diseases. *Infect. Dis. Clin. North Am.* **11**(4), 889–903.

Gaydos, C. A., and Quinn, T. C. (1998). Ligase chain reaction for detecting sexually transmitted diseases. *In* "Rapid Detection of Infectious Agents" (Specter *et al.*, eds.), Plenum, New York.

Gerbase, A. C., Rowley, J. T., and Mertens, T. E. (1998). Global epidemiology of sexually transmitted diseases. *Lancet* **351** (Suppl. 3), 2–4.

Green, T., Talbot, M. D., and Morton, R. S. (2001). The control of syphilis, a contemporary problem: a historical perspective. *Sex. Transm. Infect.* **77**, 214–217.

Holmes, K. K., Mardh, P. A., Sparling, P. E., Wiesner, P. I., Cates, W., Lemon, S. M., and Stamm, W. E. (eds.) (1999). "Sexually Transmitted Diseases." McGraw-Hill, New York.

Hook, E. W. (1998). Is elimination of enzemic syphilis transmission a realistic goal for the USA? *Lancet* **351**(Suppl. 3), 19–21.

Jackson, S. I., and Soper, D. E. (1997). Sexually transmitted diseases in pregnancy. *Obstet. Gynecol. Clin. North Am.* **24**(3), 631–644.

Kalman, S., Mitchell, W., Marathe, R., Lammel, C., Fan, J., Hyman, R. W., Olinger, L., Grimwood, J., Davis, R. W., and Stephens, R. S. (1999). Comparative genomes of *Chlamydia pneumoniae* and *C. trachomatis*. *Nat. Genet.* **21**(4), 385–389.

Mayaud, P., Hawkes, S., and Mabey, D. (1998). Advances in the control of sexually transmitted diseases in developing countries. *Lancet* **351**(Suppl. 3), 29–32.

Mindel, A. (1998). Genital herpes—How much of a public health problem? *Lancet* **351**(Suppl. 3), 16–18.

Murray, C. J. L., and Lopez, A. D. (eds.) (1998). "Health dimensions of sex and reproduction: the global burden of sexually transmitted diseases, HIV, maternal conditions, perinatal disorders, and congenital anomalies. Vol 3 of Global burden of disease and injury series." Harvard School of Public Health, Boston.

Rompalo, A. M. (2001). Can syphilis be eradicated from the world? *Curr. Opin. Infect. Dis.* **14**, 41–44.

WEBSITES

The Sexually Transmitted Disease website of the National Institutes of Allergy and Infectious Diseases (USA)
http://www.niaid.nih.gov/dmid/stds/

Sexually Transmitted Diseases. A site with text, many photographs, radiographs. By RB Roberts
http://edcenter.med.cornell.edu/Pathophysiology_Cases/STDs/STD_TOC.html

Human Papillomavirus Sequence Database Website of the National Institutes of Allergy and Infectious Diseases (USA)
http://hpv-web.lanl.gov/

Website of Sexually Transmitted Diseases, a Journal of the American Sexually Transmitted Diseases Association
http://www.stdjournal.com/

Photo Gallery of Bacterial Pathogens (Neil Chamberlain)
http://www.geocities.com/CapeCanaveral/3504/gallery.htm

Access to online resources on Bacterial Infections and Mycoses. Karolinska Institutet
http://www.mic.ki.se/Diseases/cl.html

Comprehensive Microbial Resource of The Institute for Genomic Research (TIGR) and links to many other microbial genomic sites
http://www.tigr.orgtigr-scripts/CMR2/CMRHomePage. spl

76

Skin microbiology

Morton N. Swartz
Massachusetts General Hospital and Harvard Medical School

GLOSSARY

acne An inflammatory disease involving the pilosebaceous unit.

carbuncle A larger, painful, deeper, more serious, and more nodular lesion than a furuncle. It is due to *Staphylococcus aureus* and characteristically occurs on the nape of the neck, upper lip, back, or thighs and progresses to drain externally around multiple hair follicles.

cellulitis An acute edematous, suppurative, spreading inflammation of the deep subcutaneous tissues producing an area of erythema and tenderness with indistinct margins. The most common etiologic agents are group A streptococci and *Staphylococcus aureus*.

ecthyma A group A streptococcal process which begins much like impetigo at a site of minor trauma but extends through the epidermis, producing a shallow ulcer covered by a crust.

erysipelas An acute superficial form of cellulitis involving the dermal lymphatics, usually caused by group A streptococci, and characterized by a bright red, edematous, spreading process with a raised, indurated border.

furuncle A painful erythematous nodule (boil) caused by *Staphylococcus aureus* and formed by localized inflammation of the dermis and subcutaneous tissue surrounding hair follicles.

gas gangrene An acute severe infection, usually resulting from dirty penetrating wounds in which the subcutaneous tissues and muscles contain gas and a serosanguineous exudate. The process is a histotoxic infection due to *Clostridium perfringens*, *C. septicum*, or other *Clostridium* species. (Also known as *clostridial myonecrosis*.)

glabrous Bare, without hair.

impetigo A contagious pyoderma caused by direct inoculation of group A streptococci or *Staphylococcus aureus* into superficial abrasions of the skin. The lesion is confined to the epidermis and initially consists of a fragile vesicopustule with an erythematous halo progressing to a yellow-brown crust.

lymphangitis Inflammation of lymphatic vessels. Acute lymphangitis is evidenced by painful subcutaneous red streaks along the course of lymphatics.

necrotizing fasciitis A fulminating form of cellulitis that spreads to involve the superficial and deep fascia causing thrombosis of subcutaneous vessels and gangrene of overlying tissues. Type I necrotizing fasciitis is commonly due to a mixture of one or more anaerobes with one or more facultative species; Type II is due to group A streptococci.

paronychia An infection involving the folds of tissue surrounding the nail.

pilosebaceous Pertaining to hair follicles and sebaceous glands.

pyoderma Any purulent skin disease.

The Human Skin Surface is a distinct ecosystem made up of a large number of microbial species and the chemical and physical environment with which they interact. Since the cutaneous environment varies considerably from sector to sector (glabrous as opposed to hairy regions; dry areas compared to intertriginous or moist ones), the interactions

The Desk Encyclopedia of Microbiology
ISBN: 0-12-621361-5

have added complexity. The normal skin flora comprises a permanent group of microbial species, the so-called resident flora, and variable types of temporary surface colonizers, the so-called transient flora. It is a set of important reciprocal effects that determines ultimately what microorganisms will persist as permanent colonizers. These include, on the human host's part, the structural integrity of the epidermis, biochemical and immunological defenses, local anatomic features (hair, pilosebaceous apparatus, etc.), and alterations produced by physiological changes (increased sebum secretion in adolescence) and environmental factors (increased surface moisture secondary to skin occlusion or heightened environmental temperature). On the part of the colonizing microorganism they include, in addition to the foregoing, important ecological determinants such as availability of nutrients, specific ligands for attachment to host keratinocytes, and capacity to survive adverse microbial interactions (competitive colonization suppression by other species by virtue of the latter's growth advantage through use of available nutrients in this particular niche or by virtue of the latter's ability to produce antibiotics that suppress the growth of another species). For primary pathogens among the transient flora, which includes microorganisms such as group A streptococci, the intrinsic virulence of the species is a major determinant of the development of skin lesions and invasive infection.

I. HOST DETERMINANTS OF COLONIZATION AND INVASION

A. Structural integrity of the skin

The most differentiated layer (stratum corneum) of the epidermis functions primarily as a defense against inordinate water loss from the body, but it also serves as protection against invasion by microorganisms through its relative dryness, an inhospitable environment for pathogens requiring moisture for sustained growth. Also, through continuous desquamation of the outermost keratinocytes, adherent surface microorganisms are constantly shed. In contrast to the intact stratum corneum as a strong defense against invasion by normal resident flora or by transient pathogenic colonizers, cracks produced by trauma or various primary skin diseases can provide ready ingress, particularly in moist areas, for these microorganisms.

B. Biochemical defenses

Free fatty acids, particularly long-chain polyunsaturated ones such as linolenic and linoleic acid in the skin may have a role (not conclusively proven) in preventing permanent skin colonization by species such as *Staphylococcus aureus* and *Streptococcus pyogenes* (group A streptococcus). Removal of the skin surface lipids with solvents increases the duration of survival of *Staphylococcus aureus* on the skin, an effect reversed by replacement of lipid. It has been suggested that this inhibitory action of such unsaturated fatty acids accounts for the relatively poor survival of *S. aureus* on the skin surface vis-à-vis coagulase-negative staphylococci. Fatty acids released from the triglycerides of sebum (e.g., by hydrolysis effected by resident flora) may act bacteriostatically by lowering local pH as well. In general, these antimicrobial effects are limited to transient microbial pathogens and not to resident flora. A category of the resident flora, propionibacteria, is suppressed in growth to some extent by linolenic and linoleic acids. This effect may vary with species: *Propionibacterium acnes* is inhibited at higher concentrations of both fatty acids than is *P. granulosum*, perhaps accounting for lower population densities of the latter than of the former on the skin surfaces. Gram-negative and gram-positive bacteria have roughly similar susceptibilities to fatty acids. Thus, the presence of fatty acids on the skin is unlikely to be the major factor in accounting for the low incidence of gram-negative bacteria among the resident cutaneous flora.

C. Immunologic defenses

Immunoglobulins, Langerhans cells of the epidermis, and cytokines (particularly interleukin 1α, IL-lα) may play roles in immunologic defense at the cutaneous surface and alter the composition and invasive capacity of some pathogens of the transient flora. In view of the very low levels of specific immunoglobulins on the normal skin surface, they are unlikely to provide any significant antibacterial action unless a primary exudative dermatosis is present. Langerhans cells, a type of dendritic cell involved in immune surveillance for bacterial and other antigens that may breach the defenses at the stratum corneum, provide for early antigen recognition that initiates an anamnestic immune response (particularly cell-mediated). The cornified epithelial cells of the skin surface contain large quantities of IL-lα that normally are not released. However, their release is stimulated by local inflammation or bacterial action, thus setting off autocrine (and paracrine) effects with production of further cytokines and attraction of inflammatory cells to the skin surface.

D. Anatomic features

Although the skin can be considered as a single organ, its regional differences in hair, depth of cornification, numbers of sweat and sebaceous glands, and local moisture content (e.g. in contiguous surfaces) provide a variety of differing ecological niches. The majority of resident bacteria on the skin are located at two microscopic sites: within the more cornified layers of the stratum corneum and within the hair follicles.

1. *Stratum corneum*

Cocci and bacilli, often existing in microcolonies and representing the principal genera of the resident flora, have been observed on the surface of the stratum corneum in skin scrubbing experiments, cellulose tape strippings through successive layers of the epidermis, in histologic sections of skin, and by scanning electron microscopy. Large variations occur in numbers of organisms in different anatomic areas. Abnormal keratinization and hyperproliferation, as in psoriasis, have been associated with increased colonization with *S. aureus*.

2. *Hair follicles*

The greatest concentration of microorganisms on the skin is in the hair follicles. Members of the anaerobic genus *Propionibacterium* are distributed in the follicle closer to the skin surface than are coagulase-negative staphylococci, another component of the resident flora, which colonize the deeper portion of the follicle in larger numbers. *Propionibacterium acnes*, the most numerous anaerobic coryneform bacterium found on the human skin, is found in largest numbers in areas rich in sebaceous glands (e.g., forehead, alae nasi, scalp), and in much smaller numbers in the dry areas of the arm and leg. Abnormal keratinization in hair follicles causing occlusion, associated with abnormalities of sebum production leading to increased proliferation of resident *P. acnes* and subsequent inflammation, are pathogenic features in the development of acne lesions.

3. *Gross regional variations*

Resident flora vary in population densities and predominant species among various localized anatomic sites, which can be viewed as specialized ecological niches that differ from the more extensive flat, exposed surfaces. Such niches include more occluded areas such as axillae (favoring large populations of either coryneform bacteria or staphylococci), toe-web spaces (colonized commonly with large numbers of gram-negative bacilli and dermatophytic fungi), and the groin, as well as the scalp (favoring proliferation of large numbers of staphylococci, propionibacteria, and *Pityrosporum* species).

E. Physical factors affecting composition of skin flora: role of hydration

Increased water content at the skin surface, when the relative humidity approaches 100%, as when the skin of the arm is occluded with a plastic wrap, leads to rapid microbial growth to high population densities. Under such conditions, counts of coagulase-negative staphylococci, lipophilic coryneforms, and gram-negative bacilli increase by four orders of magnitude in 72 h. Similarly, densities of skin flora are highest in workers in hot, humid climates, pyodermas are more frequent during hot, humid seasons of the year, and infections by dermatophytes become more problematic under similar conditions. A normally less numerous component of the skin flora, *Acinetobacter* spp., becomes more numerous in the summer heat. The role of hydration in altering numbers of skin bacteria and lesional pathogenesis has been shown in the case of *Pseudomonas aeruginosa*. Under conditions of experimental occlusion for up to 7 days, exposure of normal skin to 10^6 *P. aeruginosa* has not produced lesions; in contrast, the addition of dressings producing hyperhydration caused development of papules and pustules. These results are congruent with the observation of the occurrence of widespread folliculitis in individuals using "hot tubs" and jacuzzis contaminated with *P. aeruginosa*.

As mentioned earlier, an occluded area such as an axilla, by virtue of its increased hydration, favors a heavy growth of staphylococci and corynebacteria. Several species of coryneform bacilli, under conditions of excessive sweating, are able to colonize hair shafts, grow extensively as visible yellow- or red-colored colonies thereon, producing the unsightly and maladorous (but unimportant) condition known as trichomycosis axillaris. The pungent axillary odor appears to result from metabolism of testosterone locally in the apocrine glands by coryneform species.

II. BACTERIAL ADHERENCE AS A FACTOR IN DETERMINING ECOLOGICAL NICHES

Bacteria tend to adhere strongly to cutaneous surfaces since vigorous washing removes less than one-half the normal flora. Thus, adhesins have been assumed to play a role, but specific components of the microorganisms adherent to corneocytes are not yet defined. It is known that *S. aureus* adhere in greater numbers to the nasal mucosa of normal nasal carriers of this species, and from the nose they may be spread as transients to the skin surface. The specific corneocyte adhesin(s) of chronic nasal carriers of *S. aureus* with which this organism's ligand interacts is not yet defined, but adherence to corneocytes can be blocked by pretreatment of these cells with staphylococcal teichoic acid. Abnormal skin, as in patients with atopic dermatitis, promotes better adherence of *S. aureus* to the cornified squamous epithelium.

Selectivity exists in adherence of bacterial cells to epithelial surfaces. *Staphylococcus epidermidis* adheres more readily to skin cells than to urinary tract cells. Strains of *S. epidermidis*, common constituents of cutaneous flora, have a capacity to adhere to vascular catheters extending through the skin surface, accounting for the prominent role of this microorganism in catheter-related bloodstream infections. Similarly, selectivity exists in regard to adherence of members of the transient skin flora. Strains of *S. pyogenes* serotypes involved in impetigo and related superficial cutaneous infections adhere better to skin cells than do strains of *S. pyogenes* isolated from the throat; likewise pharyngeal isolates adhere better to buccal epithelial cells than to skin cells.

Competition between selected strains of the same species may occur not only at the level of nutrient utilization but also at the level of adherence. Such competitive adherence was shown in the 1960s when it was noted that, during an outbreak of *S. aureus* infections, neonates whose skin had been precolonized naturally by a strain of *S. aureus* (strain 502A) acquired from a nurse were protected against the more virulent strain (type 80/81) then ambient and causing infection in the newborn. Deliberate attempts at colonization with strain 502A were subsequently shown to be effective in preventing infection with more virulent epidemic strains of *S. aureus*, but this approach was abandoned when it became evident that strain 502A could itself sometimes produce skin infections when introduced.

The skin acquires surface bacteria shortly after birth. For example, the nose is sterile in 90–100% of infants at birth. Within 72 h, 40% of infants have become colonized with *S. aureus*. *In vitro* results parallel these: binding of *S. aureus* to nasal epithelial cells obtained during the first 4 days of life is very low, reaching a level comparable to that of adult cells on the fifth day.

A relationship between pathogenicity and adherence has been suggested by the gradation in affinity of various candidal species to corneocytes: *Candida albicans* > *C. stellatoidea* > *C. parapsilosis* > *C. tropicalis* > *C. krusei* > *C. guilliermondii*.

III. RESIDENT (NORMAL) SKIN FLORA: GRAM-POSITIVE BACTERIA

The principal components of the resident cutaneous flora consist of staphylococci and coryneform bacteria (to a lesser extent *Acinetobacter* spp. and *Micrococcus* spp.) and lipophilic yeasts, primarily *Malassezia furfur* (Table 76.1).

A. Coagulase-negative staphylococci

The Baird–Parker classification of staphylococci in the 1960s identified three species (*Staphylococcus aureus, S. epidermidis, S. saprophyticus*) and *Micrococcus*. By current taxonomy *Micrococcus* and *Staphylococcus* are considered as separate and distinct genera, and about 24 species of coagulase-negative staphylococci have been distinguished utilizing phenotype characteristics and, most importantly, DNA hybridization. These species have been isolated primarily from the skin (with the exception of *S. saprophyticus* being recovered from the urine) of humans and animals (Table 76.2). Most of the human species have been isolated from normal human skin: *Staphylococcus capitis, S. cohnii, S. haemolyticus, S. hominis, S. simulans, S. warneri*, and *S. xylosus*. *Staphylococcus capitis* was identified on the scalp initially, and *S. auricularis* was first found in the human ear. *Staphylococcus schleiferi* and *S. lugdunensis* have caused bacteremias, but attempts to define a cutaneous niche for the latter have been largely unsuccessful. Certain of the coagulase-negative staphylococcal species making up the normal skin flora share the property of novobiocin resistance with the urinary tract species *S. saprophyticus*.

TABLE 76.1 Resident (normal) skin flora

Gram-positive bacteria
Coagulase-negative staphylococci (numerous species)
Staphylococcus aureus
Coryneform bacteria
Corynebacterium spp.: *C. minutissimum, C. jeikeium*, group CLC, group D2
Rhodococcus spp.
Brevibacterium spp.
Dermobacter spp.
Propionibacterium acnes
Micrococcus spp.
Gram-negative bacilli
Acinetobacter spp.: *Acinetobacter calcoaceticus–baumannii* complex, *A. johnsonii, A. lwoffii*
Pseudomonas aeruginosa (localized to toe-webs)
Fungi
Malassezia furfur (yeast forms formerly designated *Pityrosporum ovale* and *P. orbiculare*)

TABLE 76.2 Coagulase-negative staphylococci

Human origin (colonizer, pathogen)	Animal origin (colonizer, pathogen)
Common pathogens	Isolates primarily from animals
Staphylococcus epidermidis	*S. arlettae*
S. saprophyticus (principally urinary tract)	*S. caprae*
Uncommon pathogens	*S. carnosus*
S. capitis	*S. caseolyticus*
S. caprae	*S. chromogenes*
S. cohnii	*S. delphini*
S. haemolyticus	*S. equorum*
S. hominis subsp. *hominis*	*S. felis*
S. hominis subsp. *novobiosepticus*	*S. gallinarum*
S. lugdunensis	*S. hyicus*
S. saccharolyticus	*S. intermedius*
S. schleiferi	*S. kloosii*
S. simulans	*S. lentus*
S. warneri	*S. muscae*
S. xylosus	*S. pasteurii*
Rare pathogens	*S. sciuri*
S. auricularis	*S. vitulus*

Distribution studies of various coagulase-negative staphylococcal species among humans show some person-to-person variability and variability depending on anatomic areas studied; for example, striking variations have been observed in very similar populations of teenagers in the prevalence on the skin of *S. xylosus*. While *S. epidermidis* is the species of coagulase-negative staphylococci most widely distributed over the human body surface, various individual staphylococcal species predominate in individual anatomic areas, as determined in the small number of quantitative studies performed. Thus, in adults *S. epidermidis* is the dominant staphylococcal species isolated from the scalp, face, chest, and axilla, while *S. hominis* has a lesser, but still important, role in these areas. In contrast, on the dry areas of the arms and legs *S. hominis* is found about as frequently as *S. epidermidis* and roughly equals or exceeds the latter in percentage of total staphylococci found on these areas. *Staphylococcus saprophyticus* along with *S. cohnii* and *S. xylosus*, three closely related species, have been found particularly on the feet. Whereas before the age of puberty these three species account for about 5% of the staphylococcal flora on the feet of both males and females, at puberty in females they rise to account for about 45% or more of the staphyloccal flora. The more common location for *S. haemolyticus* has been the thighs.

The viable counts of individual coagulase-negative staphylococal species vary from person to person and from area to area of the body. For example, viable counts of *S. epidermidis* in the axillae of 11 carriers averaged 3×10^4 per cm^2, with a range of 8×10^2–2×10^5 per cm^2; on the forehead the mean count was 2×10^3 per cm^2, with a range from 3×10^0 to 4×10^4 per cm^2. Great variations exist in staphylococcal counts on healthy feet. The greatest population density of staphylococci on the feet are about the toes, where counts exceed 1×10^6 per cm^2.

Although the members of the "resident flora" are being considered here, their persistence in a given location may vary. Whereas 3 of 16 normal individuals were shown to carry *S. saccharolyticus* on the forehead for periods of at least 16, 27, and 38 months, another individual carried only small numbers of this microoganism and only on one occasion. In a study of staphylococcal isolates of personnel in Antarctica over a period of 42 weeks, the so-called resident staphylococcal flora was found to be made up of a mixture of "permanent" and "temporary" resident strains. Of 17 individuals studied, about one-third carried their own individual clone (defined by polyacrylamide gel electrophoresis and Western blotting) of *S. capitis* on their scalps for most or all of the prolonged period of study. To emphasize regional anatomic differences in resident floral strains (over and above distinctions at the species level), some of the individuals in this study carried a clone of *S. capitis* on their chin that was different from the one on their scalp. In contrast to the "relative permanence" of *S. capitis* on the scalps of individuals at an Antarctic base, isolates of *S. warneri*, *S. haemolyticus*, and *S. saprophyticus* recovered in the early stages of the study were not recovered subsequently, indicating that these were "temporary" resident flora constituents at these sites that had come from other reservoir sites. Dissemination or dispersal of members of the skin flora occurs, predominantly via squames or skin scales.

The coagulase-negative skin flora species colonize the newborn at somewhat differing rates. *Staphylococcus epidermidis*, *S. haemolyticus*, and *S. hominis* are present in the majority of samples obtained during the first week of life, even as early as the first day of life in some instances. Other species colonize later (10–32 weeks).

B. Coagulase-positive staphylococci

Coagulase-positive staphylococci have several characteristics by which these potential pathogens are recognized: coagulase, a heat stable nuclease, and protein A. Although several other coagulase-positive species are known (*S. intermedius*, a colonizer and pathogen of dogs; *S. hyicus*, a colonizer of pig and cattle skin and a pathogen of piglets; *S. delphini*, a cause of suppurative

IV. RESIDENT (NORMAL) SKIN FLORA: GRAM-NEGATIVE BACTERIA

Gram-negative bacilli, with the exception of *Acinetobacter* spp., are relatively rare components of the resident flora on normal skin. On the basis of DNA hydridization 17 genospecies of *Acinetobacter* have been delineated, including *A. calcoaceticus, A. baumannii, A. johnsonii,* and *A. lwoffii*. Since *A. calcoaceticus* and *A. baumannii* are closely related genotypically and phenotypically, they are often described as the *A. calcoaceticus–baumannii* complex. Carriage of *Acinetobacter* as resident flora occurs in about 25% of normal adults, primarily in axillae, groins, antecubital fossae, and toe-webs. These consist mainly of strains of *A. johnsonii, A. lwoffii,* and as yet unclassified genotypes. The carriage rate is higher in patients with eczema, particularly on lesions. Carriage of *Acinetobacter* is commoner in summer months, presumably related to increased perspiration. *Acinetobacter* skin infections, often nosocomial as are other *Acinetobacter* infections, although uncommon, include operative wound infections, cellulitis, and skin abscesses. Many of the strains of *Acinetobacter* isolated from invasive infections such as bacteremia (often vascular catheter induced), pneumonia, meningitis, and surgical wound infection belong to *A. baumannii.*

Other than *Acinetobacter* spp., gram-negative bacilli are rarely constituents on the normal skin, except as transient colonizers of exogenous origin, such as nosocomial carriage of *Klebsiella* and *Enterobacter* on the hands of health-care workers, *P. aeruginosa* colonization on use of occlusion (hyperhydration) in dermatologic therapy, and *P. aeruginosa* carriage in toe-webs. The latter probably accounts for the common involvement of the latter organism in trench foot or immersion foot of soldiers. Similarly, another pseudomonad, *P. cepacia* (now known as *Burkholderia cepacia*), has been found in the macerated web spaces of the condition known as swamp foot. Exogenously acquired *P. aeruginosa* folliculitis from use of contaminated whirlpool baths, on clearing of all lesions, may be followed, some weeks later after heavy exercise, by recurrence of the original rash, suggesting that there had been persistent carriage of *P. aeruginosa* on normal skin. *Proteus* spp. can be found normally on the nasal mucosa of about 5% of healthy individuals and in the toe-web spaces along with *Pseudomonas* spp.

V. RESIDENT (NORMAL) SKIN FLORA: FUNGI

Members of the genus *Malassezia* represent the principal resident fungal skin flora and are found on all adults, particularly around openings of sebaceous glands in the superficial layers of the stratum corneum. Although dermatophytes (*Epidermophyton* spp., *Trichophyton* spp.) may be recovered from normal-appearing skin, their role in these intances (residents or transients) is unclear as yet. *Malassezia furfur* is the name currently given for the lipophilic yeast, which assumes yeast and short hyphal forms in the superficial lesions of tinea (pityriasis) versicolor of which it is the etiology. Yeast forms of this organism, designated in the past as *Pityrosporon ovale* and *Pityrosporon orbiculare*, are those seen in carriage on normal individuals. Colonization of human skin with these fungi begins in infants, with carriage reaching a peak of 100% in early adult life. Their numbers are normally greatest on the chest, back, and scalp where sebaceous glands are plentiful, and the former two being areas where the lesions of tinea versicolor, a disease associated with overgrowth of this colonizing lipophilic yeast, is prone to develop.

VI. TRANSIENT SKIN FLORA: GRAM-POSITIVE BACTERIA

Table 76.3 lists species of transient skin flora.

A. *Staphylococcus aureus*

While persistent nasal carriage of *S. aureus* occurs in approximately 30% of normal individuals, resident carriage on normal skin occurs in about 6% of normal

TABLE 76.3 Transient skin flora

Gram-positive bacteria
Staphylococcus aureus
Streptococcus pyogenes (group A streptococcus)
Enterococcus spp.
Streptococcus agalactiae (group B streptococcus)
Peptostreptococcus spp.
Clostridium perfringens
Erysipelothrix rhusiopathiae
Corynebacterium diphtheriae
Bacillus anthracis
Gram-negative bacteria
Pseudomonas aeruginosa
Haemophilus influenzae
Escherichia coli
Klebsiella spp.
Enterobacter spp.
Proteus spp.
Serratia spp.
Citrobacter spp.
Bacteroides spp.
Aeromonas hydrophila
Halophilic noncholera vibrios

individuals and transient carriage, in another 7%. Transient skin carriage may be generated in areas previously uncolonized by distribution of the microorganism from another site of more persistent colonization on the same individual. This may occur in chronic nasal carriers and in individuals who have resident *S. aureus* colonizing or infecting preexisting skin lesions (e.g. eczema) or in individuals with other underlying conditions such as diabetes mellitus. Transmission of *S. aureus* is likely to be carried out in the aforementioned circumstances by the hands of individuals themselves. In addition, nosocomial transmission of *S. aureus* to another patient, producing transient colonization, with or without subsequent secondary infection of skin lesions, may be effected through the hands of hospital personnel. This is of particular concern with regard to the difficult-to-treat methicillin-resistant *S. aureus* (MRSA) and is the reason why "contact precautions" are employed to prevent spread of this pathogen to vulnerable hospitalized patients.

1. *Pyodermas due to* Staphylococcus aureus

Resident or transient *S. aureus* are the infecting organisms in a number of superficial staphylococcal pyodermas. Impetigo is a highly contagious superficial unilocular vesicopustular process located between the stratum corneum above and the stratum granulosum below. Although group A streptococci were most commonly isolated in the past, currently *S. aureus* (of endogenous or exogenous origin) is most frequently found in the lesions, which are usually located near the openings of hair follicles. Chronologically, after appearance in the patient's nose, the *S. aureus* strain is disseminated about 11 days subsequently to normal skin and thence is spread to skin lesions (at sites of minor abrasions or insect bites) about 11 days later, producing the characteristic lesions of impetigo. A second form of impetigo, bullous impetigo, is produced by *S. aureus* strains belonging to phage Group II and is characterized by vesicles which rapidly progress to flaccid bullae. It occurs mainly in newborn and young children. *Staphylococcus aureus* folliculitis is a pyoderma located within hair follicles, producing perifollicular inflammation or superficial small dome-shaped pustules at the openings of hair follicles. A furuncle ("boil") is a deep-seated inflammatory nodule about a hair follicle (commonly in areas subject to friction and perspiration) and usually follows a superficial folliculitis. It most often occurs without evident predisposing skin lesions but may follow preexisting abrasions, scabies, or insect bites. A carbuncle is similar to a furuncle but is a serious, more extensive, deeper placed painful lesion that develops when suppuration occurs under thick inelastic skin in areas such as the nape of the neck and back. Recurrent furunculosis is not uncommon and may result from autoinoculation from previous lesions, recurrent shedding of *S. aureus* from nasal carriage, contact sports such as wrestling, and staphylococcal infections in other family members.

B. *Streptococcus pyogenes* (group A streptococcus)

1. *Pyodermas due to group A streptococcus*

Spread of group A streptococci usually occurs by transfer of organisms from an infected individual or carrier (upper respiratory tract) through close personal contact.

a. Impetigo

Impetigo, currently, is less frequently due to group A streptococci than to *S. aureus* in Europe and the U.S., whereas formerly the reverse obtained. While streptococcal strains involved in pharyngeal disease have predominantly belonged to M serotypes 1, 3, 5, 6, or 12, those involved in group A streptococcal impetigo have newer M types such as 31, 49, 52, 53, 55–57, 60, 61, and 63. This form of impetigo occurs predominantly in preschool-age children and is highly contagious. Group A streptococci are members of the transient skin flora only during transitory carriage and while active lesions of streptococcal impetigo and other streptococcal pyodermas persist. Group A streptococci appear as transient flora on the normal skin of children approximately 10 days prior to the development of impetigo. Pharyngeal and nasal carriage of these microorganisms is not detectable until 14–20 days after skin colonization, indicating that spread has most likely been from skin contact with another individual with streptococcal impetigo or other streptococcal pyodermas. Minor trauma (abrasions, insect bites) prior to the acquisition of the group A streptococcal strain on otherwise normal skin predisposes to the appearance of impetigo. Nonsuppurative complications of group A streptococcal infections occur in the form of acute rheumatic fever and acute glomerulonephritis. While acute rheumatic fever may follow untreated group A streptococcal pharyngitis or tonsillitis in less than 2–3% of instances, it does not develop following streptococcal skin infections. Acute glomerulonephritis, on the other hand, may result from infection of either the skin or respiratory tract. Specific M serotypes are more likely to be nephritogenic: Pharyngitis-associated strains include serotype

12, 1, 4, 25; serotypes 2, 49, 55, 57, and 60 are pyoderma-associated strains. The frequency of acute glomerulonephritis following infection with a known nephritogenic strain can be as high as 10–15%.

b. Ecthyma

Ecthyma, also a group A streptococcal process, begins in a fashion much like impetigo with a superficial vesiculopurulent lesion, but it extends more deeply, penetrating through the epidermis and producing a shallow ulcer covered by a crust or eschar.

c. Erysipelas

Erysipelas, also almost always due to group A streptococci (uncommonly due to group C or group G streptococci; rarely *S. aureus*), is a superficial infection of the skin involving mainly the dermis and dermal lymphatics. The lesion has a bright red, edematous (peau d'orange) appearance and rapidly spreads peripherally. The process frequently begins with a very small break in the skin which has disappeared by the time of onset of the lesion. The group A streptococci have usually transiently colonized the skin of the involved lesion from a preceding upper respiratory tract infection, although when erysipelas has developed group A streptococci are often no longer isolated on throat culture. The disease is rapidly progressive, causes high fever, and requires prompt therapy with penicillin G or a β-lactamaseresistant penicillin.

d. Acute cellulitis

Acute cellulitis is a spreading infection involving both the skin surface and, particularly, the deeper subcutaneous tissues. It is usually due to group A (sometimes group B, C, or G) streptococci and follows transient skin colonization by the microorganism at a site of a recent puncture wound, or, on a lower extremity, at the site of a stasis ulcer or spread from a minor break in the skin in an interdigital web area, commonly due to tinea pedis.

e. Acute lymphangitis

Acute lymphangitis, most commonly caused by group A streptococci but sometimes by *S. aureus*, is an inflammatory process involving the subcutaneous lymphatics. The portal of entry is usually an acute puncture wound or abrasion that is contaminated at the time of acquisition by a group A streptococcus from the upper respiratory tract. Lymphangitis causes red linear streaks, a few millimeters to several centimeters in diameter, that progress from the local lesion toward regional lymph nodes. Prompt therapy with penicillin G or a penicillinase-resistant β-lactam drug is required to treat the frequently complicating bacteremia.

f. Streptococcal gangrene

Streptococcal gangrene, also known as necrotizing fasciitis type II or "the flesh-eating bacterial infection" in the lay press, is a gangrenous, edematous process involving the subcutaneous tissues and fascia followed by necrosis of the overlying skin. The group A streptococci involved in the process usually have entered through a puncture wound, or laceration or surgical incision, or through a nonvisible break in the skin. Transient colonization of the skin site previously or at the time of the skin injury by group A streptococci from the upper respiratory tract or from another colonized skin surface is the source of the pathogen. Extensive surgical debridement, as well as antibiotic therapy, is necessary for treatment in view of the extensive undermining of tissue in this life-threatening process.

A similar type of process, type I necrotizing fasciitis, occurs particularly on the abdominal wall, perineum, and lower extremities and is commonly due to a mixed infection with one or more anaerobes (e.g. *Peptostreptococcus* spp., *Bacteroides* spp.) along with one or more facultative species (e.g. various non-group A streptococci, such as *Enterococcus* spp. and various Enterobacteriaceae). Crepitus (due to gas in the subcutaneous tissues) is often present. The source of this type of mixed infection is commonly from dissection of an intestinal process such as an intestinal perforation or a perirectal abscess, or spread from a decubitus ulcer or surgical wound infection. As in the case of necrotizing fasciitis due to group A streptococci, treatment consists of extensive debridement of the process, drainage of the feeding focus of infection, and, because of the mixed nature of the infecting microorganisms, broader antimicrobial coverage (based on gram-stained smears of exudate and culture results) than for type II necrotizing fasciitis.

C. Non-group A streptococci

As noted earlier, group C and G streptococci can occasionally produce erysipelas and cellulitis, and group B streptococci may be responsible for erysipelas (in infants) and cellulitis. *Enterococcus* spp., group B streptococci, and *Peptostreptococcus* spp. are normally members of the lower intestinal tract flora. *Enterococcus* spp. are common causes of urinary tract infections. Thus, contamination of a skin surface with fecal or urinary flora (particularly in incontinent elderly patients) can readily cause transient colonization and secondary infection of decubitus and stasis ulcers, eczematous dermatitis, and abrasions with such microorganisms.

1. *Progressive bacterial synergistic gangrene*

A somewhat distinctive mixed infection of the skin, progressive bacterial synergistic gangrene occurs in the setting of abdominal surgery (about stay sutures or adjacent to a colostomy opening) or chronic ulceration on an extremity. The lesion consists of a central ulcer surrounded successively by a rim of gangrenous skin and an advancing zone of purplish erythema. The advancing margin contains anaerobic or microaerophilic streptococci, whereas the central ulcerated area contains *S. aureus* (rarely *Proteus* or other gram-negative bacilli); both types of organisms are necessary for this synergistic infection. Contamination (transient) of an operative wound or ulcer by such flora can lead to subsequent infection at the site of the original lesion.

2. **Streptococcus iniae** *infections*

Cellulitis and lymphangitis have occurred due to *S. iniae*, normally a fish pathogen, in individuals handling fresh aquacultured fish (tilapia) or preparing the fish for cooking in the household. Cellulitis can occur, following (or simultaneous with) transient skin colonization with the microorganism, after a percutaneous injury while handling the fresh fish.

D. Other gram-positive bacteria

1. Clostridium perfringens

Clostridium perfringens is a normal constituent of the bowel flora. Transient skin colonization from fecal or soil contamination, or direct wound contamination with fecal material or soil contents, can result either in anaerobic cellulitis due to this microorganism or in clostridial myonecrosis (gas gangrene), an incredibly toxemic infection primarily involving skeletal muscle. Gas gangrene has prominent cutaneous manifestations (edema of subcutaneous tissues, tense blebs containing dark brown fluid, patches of skin necrosis). Both anaerobic cellulitis and clostridial myonecrosis are characterized by gas in the tissues and thin serous exudate (containing numerous short, plump gram-positive rods without spores, but a paucity of polymorphonuclear leukocytes). Both processes develop as a result of a traumatic dirty wound with extensive muscle or soft-tissue damage or following wound contamination during surgery, usually involving bowel or gallbladder.

2. Bacillus anthracis

Until very recently, anthrax has been primarily a disease of domestic and wild animals, but humans have become involved accidentally through exposure to animals and their products. Anthrax still occurs in animal reservoirs in Africa, the Middle East, India, and South America while it had been virtually, eliminated in the United States by the year 2000 (~125 human cases annually in the early twentieth century, but less than 1 case per year during the past 20 years). This changed dramatically during October–November. 2001 with the outbreak in the Eastern US of bioterrorism-associated anthrax transmitted through the mail delivery system. In this outbreak there were 23 cases—11 patients with inhalation anthrax (5 fatal) and 12 patients with cutaneous anthrax (all recovered). In the past, while inhalational and gastrointestinal forms of anthrax have occurred, the most common site of human infection has involved the skin on an exposed part of the body on which an abraded area has been inoculated with the organism from contact (usually occupational) with wool, hides, and other animal products.

Bacillus anthracis, a large gram-positive aerobic rod, forms spores in the environment and on culture but not in tissues. The major virulence factors of *B. anthracis* are encoded on two plasmids—one bearing genes for the synthesis of a polyglutamyl capsule that inhibits phagocytosis of vegetative forms and the other bearing those for synthesis of the exotoxin it secretes. The exotoxin is a binary one, consisting of a single receptor-binding moiety, known as protective antigen (PA), and two enzymatic moieties designated as edema factor (EF) and lethal factor (LF). Following secretion from *B. anthrasis* as nontoxic monomers, the three proteins assemble into toxic, cell-bound complexes; but only after the PA has been proteolytically cleaved by a cell-surface protease, heptamerized, enabling it to strongly bind EF and LF. EF is an adenylate cyclase that inhibits phagocytic cell function and is responsible for the prominent edema at sites of infection. LF is a metalloprotease that inhibits intracellular signalling in macrophages and stimulates the release by macrophages of tumor necrosis factor α and interleukin-1. The latter is the likely mechanism for the sudden death from toxic effects occurring when extensive infection is complicated by high-grade bacteremia.

Cutaneous anthrax begins as a painless or pruritic papule 1–7 days after spores have been introduced and develops into a clear or hemorrhagic bulla containing large gram-positive bacilli but only rare leukocytes. The bulla ruptures, undergoes necrosis and eschar formation. The lesion is surrounded often by a striking, nonpitting, brawny, gelatinous edema and is relatively painless unless secondary infection has supervened.

3. Erysipelothrix rhusiopathiae

Erysipeloid, an uncommon acute cellulitis due to *Erysipelothrix rhusiopathiae*, occurs primarily in fishermen, butchers, and housewives who handle raw fish, poultry, and meat products. After the microorganism, a thin microaerophilic gram-positive rod, is inoculated through a break in the skin, a serpiginous violaceous lesion with sharply defined borders subsequently develops.

4. Corynebacterium diphtheriae

Cutaneous diphtheria still occurs in underdeveloped areas of the world, and outbreaks have occured among alcoholics on "skid row" and among Native Americans in the western United States. Spread of infection occurs from carriage in the pharynx of the patient or contact with another carrier, but in children spread to others may occur via direct contact with skin lesions. Most cases of cutaneous diphtheria represent transient colonization and then infection by *C. diphtheriae* of a preexisting skin lesion such as a traumatic abrasion, eczema, or ecthyma, but in occasional patients cutaneous diphtheria begins as a primary pustular lesion. Whether it originates as a primary or a secondary infection, the lesion ultimately appears as an ulcer, covered partially with a grayish membrane and a purulent exudate, surrounded by a zone of edema and erythema. In 3–5% of patients with cutaneous diphtheria, neurologic findings develop as a result of elaboration and spread of *C. diphtheria* exotoxin.

VII. TRANSIENT SKIN FLORA: GRAM-NEGATIVE BACTERIA

A. *Pseudomonas aeruginosa*

Colonization and infection of the skin with *Pseudomonas aeruginosa* has been considered earlier under the section on Resident Flora. However, this gram-negative nonfermentative, obligately aerobic bacillus is found ubiquitously in nature, particularly on fresh vegetables and in water. Exogenous contamination and transient colonization with *P. aeruginosa* producing hot tub-associated folliculitis has already been considered. *Pseudomonas aeruginosa* colonization and secondary infection of preexisting skin lesions such as decubitus ulcers and thermal burns are not uncommon occurrences and may be complicated by bacteremia. Increased colonization of the gastrointestinal tract and skin with this microorganism occurs in patients who are granulocytopenic or immunocompromised. Secondarily infected lesions often produce a purulent exudate with a greenish color and a "fruity" odor.

Pseudomonas aeruginosa is responsible for some painful paronychias, sometimes associated with green-blue discoloration of the fingernail, in individuals who have their hands chronically in water and are thus exposed to transient colonization by this microorganism. External otitis, also known as "swimmer's ear," is frequently caused by *P. aeruginosa* which colonizes the external ear transiently from exposure to water. Subsequent mild trauma allows ingress of the microorganism into the area of the pinna, producing a characteristic macerated, swollen appearance.

B. *Haemophilus influenzae*

Most invasive *H. influenzae* infections (meningitis, bacteremia, epiglottitis, cellulitis) are caused by encapsulated type b strains and occur in children. Widespread use of the polysaccharide–protein conjugate vaccine has reduced the incidence of such infections by 95%. *Haemophilus influenzae* cellulitis characteristically occurs in younger children (aged 6–24 months) and commonly involves the face, neck, or upper extremities. Although the exact pathogenesis of the cellulitis is uncertain, in most instances it has been preceded by an upper respiratory infection. The association of otitis media, a previously common infection caused by *H. influenzae* type b, with cellulitis of the face has led to the suggestion of possible spread of infection from the ear via lymphatics in the pathogenesis of buccal cellulitis. Involvement of the face, neck, and upper extremities is suggestive of "fall-out" from the respiratory tract onto the skin of the aforementioned areas. Subsequent transient colonization and invasion locally through a small break in the skin is the likely pathogenesis of cellulitis with this striking anatomic distribution. Typically, *H. influenzae* type b cellulitis follows coryza or pharyngitis in an infant and is characterized by an increase in fever and a purple-red tender area of edema. Unlike the distinct margins of erysipelas, the border of *H. influenzae* cellulitis is indistinct.

C. Enteric gram-negative bacilli

Members of the Enterobacteriaceae (*Escherichia coli, Klebsiella, Proteus, Enterobacter, Citrobacter, Serratia*) as well as *Bacteroides* spp. and other members of the anaerobic lower intestinal flora occasionally can be the etiology in acute cellulitis, particularly when it occurs in the elderly, in diabetics, or following skin trauma, surgery, or subcutaneous dissection of infection from the colon or perineum. Granulocytopenia, prior extensive antimicrobial use, and chronic illness are additional predisposing factors. Such mixed infections in the form of acute cellulitis or necrotizing fasciitis may also follow introduction of microorganisms from unhygienic skin surfaces as well as via contaminated narcotics delivered by "skin popping." Such mixed infections, in any of the foregoing settings, may exhibit subcutaneous gas formation.

D. *Aeromonas hydrophila*

Aeromonas spp. are commonly found in fresh and brackish waters associated with fish or aquatic animals. *Aeromonas* can produce diarrheal disease (most common infection), can be transiently carried in the intestinal tract, occasionally can cause soft tissue infections, and is rarely responsible for bacteremia and sepsis in immunocompromised hosts. Aeromonads are nonsporulating facultatively anaerobic bacilli. Of the over a dozen species, *A. hydrophila* is the one most commonly producing soft tissue infection. Exposure of the skin to contaminated water provides an opportunity for transient colonization. Invasion occurs through previously sustained sites of skin trauma in the colonized area or at the time of trauma coincident with the exposure to freshwater. The increasing clinical use of leeches following reimplantation or flap surgery has been an additional source of soft tissue infections due to *A. hydrophila*. This microorganism normally inhabits the foregut of leeches, and potentially contaminates the wound area. The high frequency (7–20%) occurrence of *Aeromonas* infection in patients treated with leeches has led to the prophylactic use of antibiotics with this procedure.

Cellulitis due to *A. hydrophila* may develop rapidly after an acute trauma and may be complicated by the development of bacteremia and a systemic sepsis syndrome. Another, much rarer, form of infection, *A. hydrophila* myonecrosis, can follow penetrating trauma and muscle injury occurring in a freshwater environment. It occurs within 24–48 h of sustaining trauma, progresses rapidly, and is characterized by prominent pain, systemic toxicity, marked edema, and the presence of gas in fascial planes and muscle. In all these respects this process resembles clostridial gas gangrene. Bacteremia frequently is present. Extensive surgical debridement and prompt initiation of antimicrobial therapy are required. Most strains are susceptible to ciprofloxacin, trimethoprim–sulfamethoxazole, aminoglycosides (except streptomycin), third-generation cephalosporins, and carbapenems.

E. Halophilic noncholera vibrios and non-o1 *Vibrio cholerae*

Four *Vibrio* species in addition to non-O1 *Vibrio cholerae* (strains not belonging to serogroup O1 and only occasionally responsible for sporadic cases and outbreaks of diarrhea) are capable of producing cellulitis or traumatic wound infections: *V. vulnificus, V. alginolyticus, V. damsela*, and *V. parahaemolyticus*. The *Vibrio* spp. are halophilic but will grow on blood agar plates, however. These microorganisms are commonly present in saltwater and estuarine sediments as well as on fish and shellfish, particularly along the coast of the Gulf of Mexico (and to a lesser extent along the Atlantic and Pacific coasts) in the summer months. *Vibrio vulnificus* is the most pathogenic of these species and is capable of causing severe wound infections and a "primary septicemia" syndrome, which may follow 24–48 h after ingestion of raw oysters or other uncooked seafood. The more typical *V. vulnificus* infection occurs as cellulitis several days following a laceration sustained in seawater or brackish inland lakes. Sometimes cellulitis develops without discernible antecedent skin trauma following exposure to seawater. Here the sequence presumably has been transient skin colonization with subsequent penetration through a small break in the skin. Primary septicemia or severe cellulitis complicated by septicemia are more likely to occur in patients with underlying cirrhosis of the liver, hemochromatosis (with its high serum iron level), diabetes mellitus, renal failure, leukemia, and processes requiring corticosteroid therapy. It is important to recognize the entity of "primary septicemia" since bullous skin lesions resembling those of a primary *V. vulnificus* cellulitis may be a feature of the "primary septicemia" syndrome, usually of gastrointestinal origin.

Traumatic wound infections due to *V. vulnificus* commonly involve the lower extremity and usually consist of cellulitis with large hemorrhagic bullae, but may consist of pustular lesions with associated lymphangitis and lymphadenitis. The cellulitis may be very painful, cause high fever, and progress rapidly, producing local necrotizing vasculitis with extensive skin necrosis and ulcer formation, myositis, and gangrene requiring amputation. Treatment of infected traumatic wounds due to *V. vulnificus* and similar *Vibrio* species requires debridement of necrotic lesions and antimicrobial therapy with a third-generation cephalosporin plus an aminoglycoside or with a combination of a tetracycline (e.g., doxycycline) or a fluoroquinolone plus an aminoglycoside (gentamicin or tobramycin).

Non-O1 *V. cholerae*, although primarily a gastrointestinal pathogen, may cause cellulitis and necrotizing fasciitis following skin trauma and contact with seawater in subtropical areas. As in the case of *V. vulnificus* infections, patients with chronic liver disease are particularly vulnerable, and "primary septicemia" can result in cellulitis with hemorrhagic bullae.

VIII. OTHER FORMS OF MICROBIOLOGICAL INVOLVEMENT IN INFECTIOUS PROCESSES OF THE SKIN

Other routes for ingress into the skin for infecting microorganisms not parts of the "normal resident" or "transient" flora include spread by direct contact, spread by vector inoculation, spread via bloodstream dissemination, or spread from infection of contiguous anatomic structures. Owing to space limitation examples will be cited here in tabular form with the aim to provide illustrative examples without any attempt at completeness (see Tables 76.4–76.6).

TABLE 76.4 Infections of the skin caused by direct contact with an individual or environmental source

Disease process	Causative microorganism
Fungal diseases	
Epidermophytosis ("athlete's foot")	*Microsporum* spp., *Trichophyton* spp., *Epidermophyton* spp.
Common yeast infections	*Candida* spp., *Torulopsis* spp.
Mycetoma	*Pseudoallescheria boydii*, *Madurella* spp., *Exophiala*, etc.
Sporotrichosis	*Sporothrix schenckii*
Sexually transmitted diseases	
Syphilis	*Treponema pallidum*
Endemic (nonvenereal) syphilis (Bejel)	*T. pallidum*
Yaws	*T. pertenue*
Pinta	*T. carateum*
Chancroid	*Haemophilus ducreyi*
Lymphogranuloma venereum	*Chlamydia trachomatis* (serovars L1–L3)
Granuloma inguinale	*Calymmatobacterium granulomatis*
Mycobacterial diseases	
Primary inoculation tuberculosis (susceptible host)	*Mycobacterium tuberculosis*
Tuberculosis verrucosa cutis (inoculation in person with +PPD)[a]	*M. tuberculosis*
Fishtank granuloma	*M. marinum*
Buruli ulcer	*M. ulcerans*
Pyogenic skin abscesses and ulcers	*M. chelonae* subsp. *chelonae*, *M. chelonae* subsp. *abscessus*
Leprosy	*M. leprae*
Bacterial diseases	
Ulceroglandular tularemia	*Francisella tularensis*
Melioidosis	*Burkholderia pseudomallei*
Glanders	*B. mallei*
Mycetoma	*Nocardia brasiliensis*
Viral diseases	
Herpes labialis	HSV type 1
Herpes progenitalis	HSV type 2
Orf	Orf parapoxvirus
Milker's nodule	Paravaccinia virus
Molluscum contagiosum	Molluscum contagiosum virus
Warts	Human papillomavirus

[a] +PPD, positive tuberculin skin test.

TABLE 76.5 Infections involving skin caused by insect vector or animal bite

Disease process	Causative microorganism
Cat scratch disease (primary lesion)	*Bartonella henselae*
Bacillary angiomatosis	*B. henselae*
Lyme disease (erythema chronicum migrans)	*Borrelia burgdorferi*
Boutonneuse fever, South African tick-bite fever (1° lesion)	Rickettsia conorii
Rickettsialpox (1° lesion)	*Rickettsia akari*
Scrub typhus (1° lesion)	*Rickettsia tsutsugamushi*
Animal bites: cellulitis and infected wounds	*Pasteurella multocida, P. canis, Capnocytophaga canimorsus, Haemophilus felis, H. aphrophilus, Neisseria canis, N. weaveri, Weeksella zoohelcum, Prevotella melaninogenica, P. denticola, Fusobacterium nucleatum, Porphyromonas canoris, P. gingivalis, Veillonella parvula*

TABLE 76.6 Infections involving the skin due to hematogenous dissemination

Bacterial infections (bacteremic)
- Gram-positive bacteria
  - *Staphylococcus aureus*
  - *Streptococcus pyogenes* (group A streptococcus)
  - *Enterococcus* spp. (acute bacterial endocarditis)
  - Histotoxic clostridia (primarily *Clostridium septicum*)
- Gram-negative bacteria
  - *Pseudomonas aeruginosa*
  - *Burkholderia pseudomallei* (melioidosis)
  - *Salmonella typhi*
  - *Neisseria gonorrhoeae*
  - *Neisseria meningitidis*
  - *Vibrio vulnificus*
  - *Bartonella bacilliformis*

Spirochaetes
- *Treponema pallidum* (secondary syphilis)
- *Borrelia burgdorferi*

Fungi
- *Histoplasma capsulatum*
- *Blastomyces dermatitidis*
- *Coccidioides immitis*
- *Paracoccidioides brasiliensis*
- *Candida* spp.
- *Cryptococcus neoformans*

Mycobacteria
- *Mycobacterium tuberculosis* (acute miliary tuberculosis)
- *M. leprae* (leprosy)

Rickettsia
- *Rickettsia rickettsii* (Rocky Mountain spotted fever)
- *R. typhi* (endemic typhus)
- *R. prowazekii* (epidemic typhus)

Viruses
- Coxsackie virus A16 (hand-foot-and-mouth disease)
- Enterovirus 71 (hand-foot-and-mouth disease)
- Herpes simplex type 1
- Herpes simplex type 2
- Varicella-zoster
- Vaccinia
- Variola (smallpox)
- Human immunodeficiency virus (acute retroviral syndrome)

BIBLIOGRAPHY

Kloos, W. E., and Bannerman, T. L. (1994). Update on clinical significance of coagulase-negative staphylococci. *Clin. Microbiol. Rev.* **7**, 117–140.

Lesher, J. L. (2000). "An Atlas of Microbiology of the Skin." Parthenon Publ., Lancaster, UK.

Noble, W. C. (1983). "Microbial Skin Disease: Its Epidemiology." Arnold, London.

Noble, W. C. (ed.) (1992). "The Skin Microflora and Microbial Skin Disease." Cambridge University Press, Cambridge.

Noble, W. C. (1993). Ecology and host resistance in relation to skin disease. *In* "Dermatology in General Medicine" (T. B. Fitzpatrick, A. Z. Eisen, K. Wolff, J. M. Freedberg, and K. F. Austen, eds.), 4th edn., Chap. 17. McGraw-Hill, New York.

Noble, W. C., and Somerville, D. A. (1974). "Microbiology of Human Skin." Saunders, Philadelphia.

Rupp, M. E., and Archer, G. L. (1994). Coagulase-negative staphylococci: Pathogens associated with medical progress. *Clin. Infect. Dis.* **19**, 231–245.

Swartz, M. N. (2001). Recognition and management of anthrax—an update. *N. Engl. J. Med.* **345**, 1621–1626.

Swartz, M. N., and Weinberg, A. N. (1993). Infections due to gram-positive bacteria. *In* "Dermatology in General Medicine." (T. B. Fitzpatrick, A. Z. Eisen, K. Wolff, J. M. Freedberg, and K. F. Austen, eds.), 4th edn., Chap. 187. McGraw-Hill, New York.

Weinberg, A. N., and Swartz, M. N. (1993). General considerations of bacterial diseases. *In* "Dermatology in General Medicine" (T. B. Fitzpatrick, A. Z. Eisen, K. Wolff, J. M. Freedberg, and K. F. Austen, eds.), 4th edn., Chap. 186. McGraw-Hill, New York.

Weinberg, A. N., and Swartz, M. N. (1993). Gram-negative coccal and bacillary infections. *In* "Dermatology in General Medicine" (T. B. Fitzpatrick, A. Z. Eisen, K. Wolff, J. M. Freedberg, and K. F. Austen, eds.), 4th edn., Chap. 188. McGraw-Hill, New York.

WEBSITES

Photo Gallery of Bacterial Pathogenes (Neil Chamberlain)
http://www.geocites.com/CapeCanaveral/3504/gallery.htm

Access to online resources on Bacterial Infections and Mycoses. Karolinska Intitutet
http://www.mic.ki.se/Diseases/cl.html

National Institutes of Allergy and Infectious Disease NIH, USA
http://www.niaid.nih.gov/default.htm

77

Soil microbiology

Kate M. Scow
University of California

GLOSSARY

bulk soil The portion of the soil not under the influence of plant roots.

decomposition Breakdown of a compound into simpler compounds, often by microorganisms.

denitrification The biochemical reduction of nitrate or nitrite to gaseous nitrogen either as molecular nitrogen or as an oxide of nitrogen.

humus Dark-colored organic by-products consisting of microbial cell walls and other resistant molecules formed from free-radical reactions of sugars, amino acids, and products of lignin decomposition.

mineralization The conversion by microorganisms of an element from an organic form to an inorganic form.

nitrification The biochemical oxidation of ammonium to nitrite and nitrate by microorganisms.

rhizosphere The portion of soil in the immediate vicinity of plant roots in which the microbial communities are influenced by the presence of the roots.

soil The dynamic natural body comprising Earth's surface layer, composed of mineral and organic materials and living organisms.

Soil Microbiology is concerned with microorganisms, primarily bacteria and fungi, that spend at least part of their life cycles in soil. Soil is defined as the dynamic natural body comprising Earth's surface layer, composed of mineral and organic materials and living organisms. Much of soil microbiology is also relevant to microbiology in deeper subsurface environments (e.g., groundwater). The chemical and biological balance of the planet is dependent on processes carried out by soil microbial communities. Global scale processes in which soil microorganisms are active participants include energy flow, organic matter decomposition, and biogeochemical cycling. The difficulty of separating soil organisms from their environment has fostered a strong tradition in soil microbiology of studying processes, with less emphasis on studying individual organisms than is the case in other areas of microbiology.

This article is an introduction to the microorganisms found in soils, describes the soil habitat and environmental factors affecting microbial communities, considers several microbial processes in soils, discusses interactions among soil organisms (including symbioses between microorganisms and plants), and concludes with a discussion of emerging topics and research areas in soil microbiology.

I. OVERVIEW OF THE SOIL MICROORGANISMS

A. Classification

The organisms inhabiting the soil span a wide range of taxonomic and functional groups. As many as 1000–5000 different genotypes may be present in a gram of soil based on the reannealing kinetics of DNA

The Desk Encyclopedia of Microbiology
ISBN: 0-12-621361-5

extracted directly from soil and then denatured. Soil microorganisms include representatives of the three phylogenetic domains of Bacteria, Archaea, and Eukarya, domains that are differentiated by sequence patterns of the small subunit ribosomal RNA Bacteria and fungi are the most abundant groups in soil with respect to biomass and numbers. Archaea, including methanogens, extreme halophiles, and extreme thermophiles, are present, but their significance in soils that are neither flooded nor exposed to environmental extremes is not well understood. Phototrophic bacteria and green algae are abundant only when soils are flooded (e.g., in wetlands, rice paddies, poorly drained soils). Viruses of bacteria, fungi, fauna, and plants are common and exist both in and outside of their soil-dwelling hosts. Many organisms in soil are not metabolically active but appear to be in wait for an opportunity when environmental conditions favor them.

Soil microorganisms are also categorized on the basis of the functions (usually metabolic) they perform. All life-forms require a source of electrons (electron donor), an electron sink, and a source of carbon. Metabolic requirements in these three categories form the basis for a useful functional classification system that targets processes more than individual organisms. Organisms are classified as phototrophs, lithotrophs, or organotrophs according to whether they obtain energy from light, inorganic chemicals, or organic chemicals, respectively. A broad array of organic compounds, both natural and human-made, can be metabolized by organotrophs. Reduced forms of numerous inorganic elements, including nitrogen, iron, manganese, sulfur, selenium, and others, provide energy for a diverse group of lithotrophic bacteria. On the basis of whether they obtain their carbon from carbon dioxide or organic chemicals, organisms are classified as autotrophs or heterotrophs, respectively. Organisms that respire are categorized as aerobic if they use oxygen as an electron acceptor and anaerobic if they use other compounds as electron acceptors. These alternate electron acceptors include oxidized forms of nitrogen, iron, manganese, sulfur, selenium, and carbon dioxide. The terms strict and facultative are used to designate whether an organism is exclusively aerobic or anaerobic, or able to live under both conditions, respectively.

B. Major groups

Viruses, though not technically classified as forms of life, are often considered in soil microbiology. They range in diameter from 0.02 to 0.25 μm. Although viruses are obligate parasites of the bacteria, fungi, plants, and fauna species that inhabit the soil, many can persist outside their hosts, in some cases for years. Viruses may play a role in regulating microbial populations, though parasitism, and are believed to be important vectors for the exchange of genetic material among prokaryotes.

Bacteria are a very diverse group of organisms in soil, and most major taxonomic groups are represented in most soils. As many as 13,000 bacterial species are estimated to be in soils based on analysis of DNA. Common genera found in soil include *Acinetobacter, Agrobacterium, Alcaligenes, Arthrobacter, Bacillus, Caulobacter, Cellulomonas, Clostridium, Corynebacterium, Flavobacterium, Micrococcus, Mycobacterium, Nocardia, Pseudomonas, Streptomyces*, and *Xanthomonas*. Bacteria capable of most types of metabolism (e.g., aerobes and anaerobes, organotrophs and lithotrophs, autotrophs and heterotrophs) can be found, at least at low densities, in most soils. Actinomycetes, many of which have filamentous growth forms, are common soil inhabitants. They are relatively resistant to desiccation and are abundant in desert soils.Actinomycetes are more abundant at higher or neutral pH than under acidic conditions. Previously, specific groups of bacteria that were thought to be most abundant in soil corresponded only to those organisms that could be cultured on laboratory media (e.g., gram-negative bacteria, spore-forming bacteria, and actinomycetes). More recent methods that describe microbial communities based on extraction and characterization of DNA (e.g., cloning libraries) contradict some of these previous assumptions about soil organisms and indicate that some of the most abundant species have never been described.

Bacteria range in size from approximately 0.3–1.0 μm in diameter and exhibit a variety of morphologies, including rods, cocci, and filamentous forms. Bacteria are usually smaller when living in soil, where nutrient conditions are poor, than when grown in nutrient-rich laboratory media. Although many soil bacteria are potentially motile, their movement is constrained by the low moisture conditions usually present in soil. Numbers of bacteria usually range from 10^8 to 10^9 cells in a gram of soil from the surface horizon, and the total biomass of bacteria usually ranges from 400 to 5000 kg in a hectare of agricultural soil (Table 77.1). Bacteria commonly grow in small colonies of 2–20 cells. Many species form spores or other types of resistant bodies that permit them to endure harsh conditions in soil.

Soil fungi are a diverse group with a broad range of morphologies and life cycles. In many ecosystems, such as forests, they constitute the largest biomass (500–5000 kg/ha) of all the soil organisms. In fact, the

TABLE 77.1 Relative numbers and biomass of microbial and faunal populations in surface Soils[a,b]

	Number		Biomass[c]	
Organisms	**per m²**	**per gram**	**kg/ha**	**g/m²**
Microflora				
Bacteria	10^{13}–10^{14}	10^{8}–10^{9}	400–5000	40–500
Actinomycetes	10^{12}–10^{13}	10^{7}–10^{8}	400–5000	40–500
Fungi	10^{10}–10^{11}	10^{5}–10^{6}	1000–20,000	100–2,000
Algae	10^{9}–10^{10}	10^{4}–10^{5}	10–500	1–50
Fauna				
Protozoa	10^{9}–10^{10}	10^{4}–10^{5}	20–200	2–20
Nematodes	10^{6}–10^{7}	10–10^{2}	10–150	1–15
Mites	10^{3}–10^{6}	1–10	5–150	0.5–1.5
Collembola	10^{3}–10^{6}	1–10	5–150	0.5–1.5
Earthworms	10–10^{3}		100–1700	10–170
Other fauna	10^{2}–10^{4}		10–100	1–10

[a]From THE NATURE AND PROPERTIES OF SOIL by Brady/Weil, © 1996. Reprinted by permission of Prentice-Hall, Inc., Upper Saddle River, NJ.
[b]Surface soils are generally considered 15 cm (6 in.) deep, but in some cases (e.g., earthworms) a greater depth is used.
[c]Biomass values are on a live weight basis. Dry weights are about 20–25% of these values.

largest individual organism known, inhabiting a total of 2.5 square miles, is a soil fungus inhabiting a forest soil. Fungi are far less diverse metabolically than are the bacteria; most fungi are organoheterotrophs. At least 70,000 fungal species have been identified. These include members of Oomycota, Chytridomycota, Zygomycota, Ascomycota, and Basidiomycota. The oomycetes, chytrids, and zygomycetes (including "sugar fungi") are rapid initial colonizers of organic materials added to soil. The chytrids tend to be found in very wet soils and include animal and plant parasites. The zygomycetes include the endomycorrhizal fungi. The basidiomycetes are slow growers and include the ectomycorrhizal fungi and lignin degraders such as white rot fungi.

The most common fungal growth form in soil is mycelial growth. The mycelium is defined as the filamentous network of hyphae that grows by apical extension. Filamentous organisms, particularly fungi, are uniquely suited to colonize large volumes of soil containing heterogeneously distributed food sources. Some fungi can endure harsh environmental conditions or predation through having resistant structures such as spores or because their hyphae are made up of recalcitrant organic molecules. Reproduction can be through dissemination of sexual and asexual spores. Many plant pathogens are fungi, many of which spend at least part of their life cycle in soil, outside of plants. Beneficial and, in some cases, necessary symbiotic relationships occur between mycorrhizal fungi and most plant species. Fungi are usually more abundant than bacteria in soils that are not physically disturbed, that receive large inputs of complex organic matter, and where organic materials remain on the soil surface.

Green algae and cyanobacteria, both phototrophic unicellular organisms, can be cultured out of most soils. Because these phototrophs are sensitive to desiccation and because sunlight cannot penetrate far into the soil matrix, algal and cyanobacterial population densities are usually low except in flooded or poorly drained soils. Algae and cyanobacteria may be the major primary producers in ecosystems where plant growth is limited by low temperature or moisture. In deserts, for example, blooms following rainfall events provide the main sources of carbon and nitrogen to these systems.

II. SOIL HABITAT AND DISTRIBUTION OF MICROORGANISMS

The soil habitat exhibits substantial heterogeneity which, in turn, fosters and maintains an enormous biological diversity. The spatial distribution of microorganisms is far from uniform as evidenced by the fact that microbial populations occupy only a few percent of the total mass or volume of soil. The heterogeneity in the distribution of microbial populations is manifested at the scale of the soil particle and aggregate, within the soil profile, and at the field scale.

A. Particle and aggregate scale

At the scale of micrometers to millimeters, soil consists of mineral particles complexed with organic material and of the voids (pores) created by their arrangements in space. The organic matter fraction of soil ranges from less than 0.1–80% on a volume basis, with most surface soils having organic matter contents between 1 and 7%. Microbial biomass usually makes up 3–5% of the total organic matter in soils. The mineral particles (clay, silt, and sand) are components of larger structures, called aggregates, in which the particles are held together by microbial polysaccharides and hyphae. Soil texture is defined by the relative proportions of clay, silt, and sand. Soil pores, between and within aggregates, contain either gas or solution, often in an approximately equal ratio. The pore size distribution depends on the relative proportions of clay, silt, and sand-sized particles, and it strongly influences which types of soil organisms are active. Numerous microenvironments, differing in redox status, pH, nutrient levels, pore size, and other factors can coexist within microns of one another.

Microbial distribution at the microscale is a function of soil pore size and access of the organisms to nutrients and electron acceptors. Figure 77.1 shows the distribution of microorganisms on a soil aggregate. The size of a microorganism strongly influences its location. Bacterial cells range from 0.3 to 1 μm in diameter and 1–2 μm long for nonfilamentous forms and up to 15 μm long for actinomycetes. The diameters are similar in size to large clay particles and are small enough to permit bacteria to inhabit the small pores found within soil aggregates. Fungal hyphae range in diameter from 2 to 10 μm, and their lengths vary considerably. Being too large to penetrate into most soil aggregates, fungal hyphae grow on the surfaces of and between aggregates. Fungal hyphae are important in binding small aggregates into macroaggregates. Microbial population densities decrease as one moves toward the center of an aggregate. This is due to the fact that nutrients and oxygen are consumed by organisms living at the surface of the aggregate. Organisms capable of anaerobic respiration, such as denitrifiers, are often active within the anoxic centers of soil aggregates. Most methods for measuring soil properties are for relatively large volumes of soil and lump together phenomena occurring in numerous microenvironments. Highly sensitive techniques capable of making measurements at spatial scales relevant to microorganisms are usually not possible to employ in complex systems such as soil.

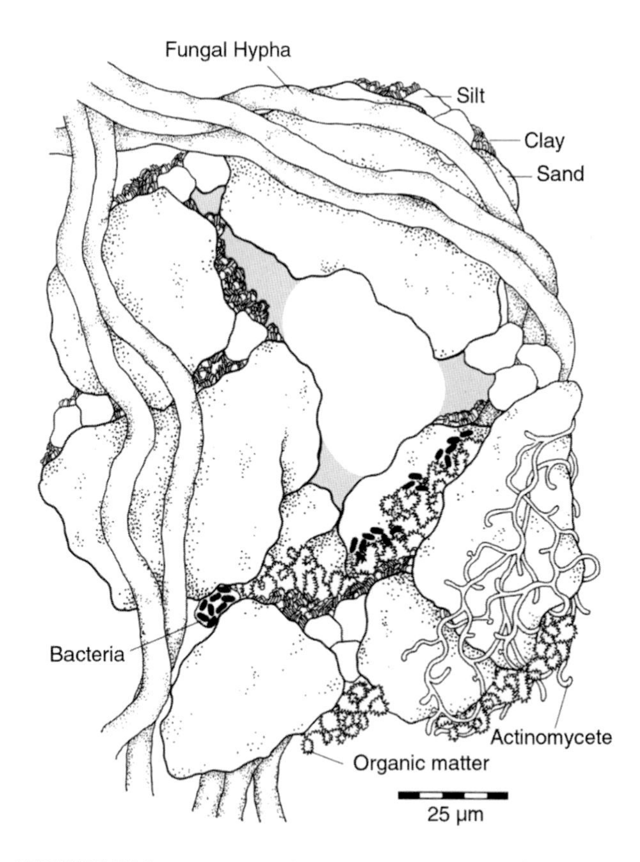

FIGURE 77.1 Distribution of microorganisms on a soil aggregate. From PRINCIPLES AND APPLICATIONS OF SOIL MICROBIOLOGY by Sylvia, D. *et al.*, © 1991. Reprinted by permission of Prentice-Hall, Inc., Upper Saddle River, NJ.

B. Soil profile scale

At the scale of millimeters to meters, there is a horizontal pattern in a soil's composition, with a higher organic matter content in the surface, which declines with depth, and higher amounts of clay deeper in the soil profile. Figure 77.2 shows the distinct horizons that can be found in most soils. These horizons are referred to as the organic (O), eluvial (A), illuvial (B), and unconsolidated parent material (C) horizons. Generally the density of microbial populations declines with soil depth, primarily because of corresponding decreases in organic carbon content, and, to a lesser degree, in oxygen concentration.

There is also a lateral and horizontal pattern in the composition of soil that is created by the spatial distribution of plants. Plants, particularly their roots, are essential members of soil communities and an important source of carbon and other nutrients to microorganisms. The rhizosphere is defined as the zone of soil under the influence of plant roots (Fig. 77.3). The soil not contained within the rhizosphere is commonly referred to as bulk soil. The rhizoplane is defined as

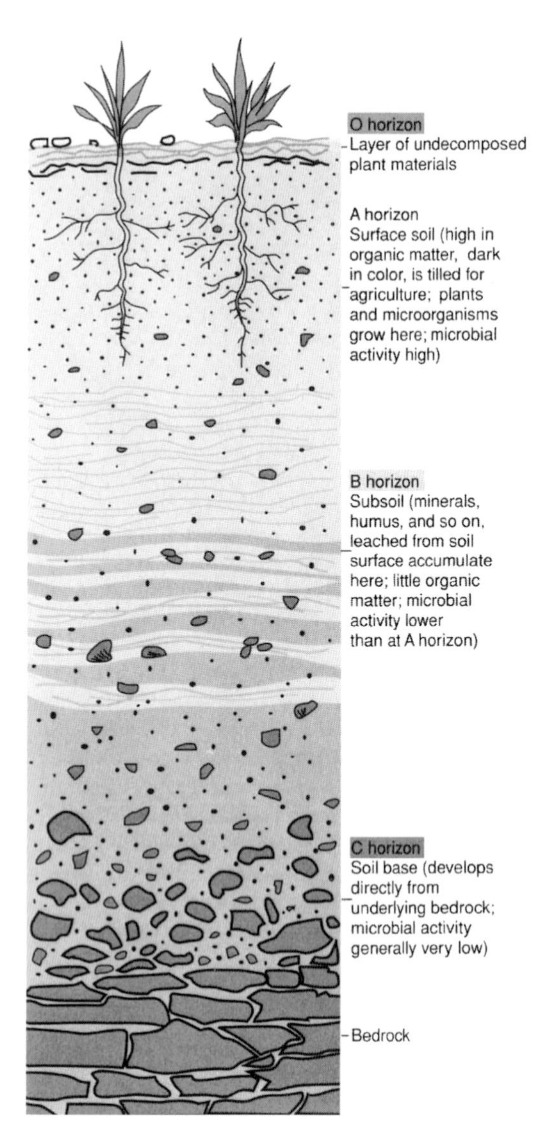

FIGURE 77.2 Profile of a well-developed soil. From BROCK BIOLOGY OF MICROORGANISMS 8/E by Madigan/Martinko/Parker, © 1997. Reprinted by permission of Prentice-Hall, Inc., Upper Saddle River, NJ.

the environment at the surface of the plant root. Rhizosphere soil often experiences greater fluctuations in water content, a higher or lower pH, and different nutrient and oxygen concentrations than does bulk soil. Plant roots release a variety of carbon compounds that support higher microbial populations and activity in the rhizosphere than in bulk soil. These compounds include low molecular weight compounds, called secretions and exudates, and larger molecular weight compounds in the form of mucilages and sloughed off plant cells (Fig. 77.3). Mucigel, which provides carbon to microorganisms, is a complex of plant and microbial cells, polymeric substances, and soil particles and is found at the interface between the plant root and soil.

Rhizosphere communities contain larger population densities, greater microbial biomass, and higher levels of certain types of microbial activities than do adjacent communities in the bulk soil. Gram-negative bacteria and denitrifiers are cultured in greater numbers from rhizosphere than bulk soil samples. Fungi, particularly mycorrhizal and plant pathogenic species, are also abundant in the rhizosphere. Conflicting data prevent making conclusions about whether microbial communities are more diverse in the rhizosphere or bulk soil. As tools for characterizing microbial communities continue to improve, we will have the ability to answer such questions.

Heterogeneity in soil chemistry, largely driven by microbial reactions, develops in soil when a source of reductant (e.g., organic substance) is made available or oxygen availability is greatly reduced, for example, after flooding of the soil. The redox and available electron acceptors available at a particular location determine which types of microorganisms can live there. With an increasing distance from the source of the reductant, there is a change in the dominating electron acceptors utilized, creating a spatial gradient in microbial communities. Changes in metabolism that occur along the gradient result from both changes in the dominant members of the community as well as from changes in the electron acceptors being used by the same organisms (e.g., facultative anaerobes). Soil physical structure can also create spatial gradients in electron acceptor utilization and thus microbial communities. For example, from the surface to interior of a large soil aggregate there is a change from the utilization of oxygen to that of nitrate as the major electron acceptor. Terminal electron acceptor process (TEAP) analysis is the study of the chemistry of electron acceptors and their products. This information can provide proof that biodegradation of a pollutant is occurring at a particular site.

C. Field and landscape scale

At the scale of meters to kilometers (the landscape scale), soil physical and chemical properties vary depending on topography, parent material, vegetation and other biota present, climate, and time since its initial formation. Figure 77.4 depicts the major forces at work in soil formation. Microbial activities play an important role in soil development through organic matter turnover and mineral weathering. An example of weathering is the production of carbon dioxide which is converted to carbonic acid which, in turn, dissolves limestone and other minerals.

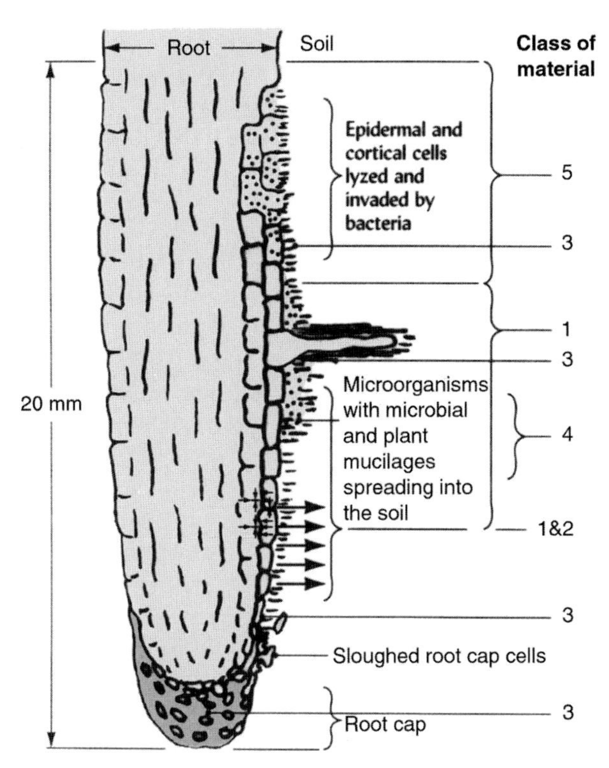

FIGURE 77.3 Substances released by plant roots into the rhizosphere. Adapted from Rovira, Foster, and Martin (1979).

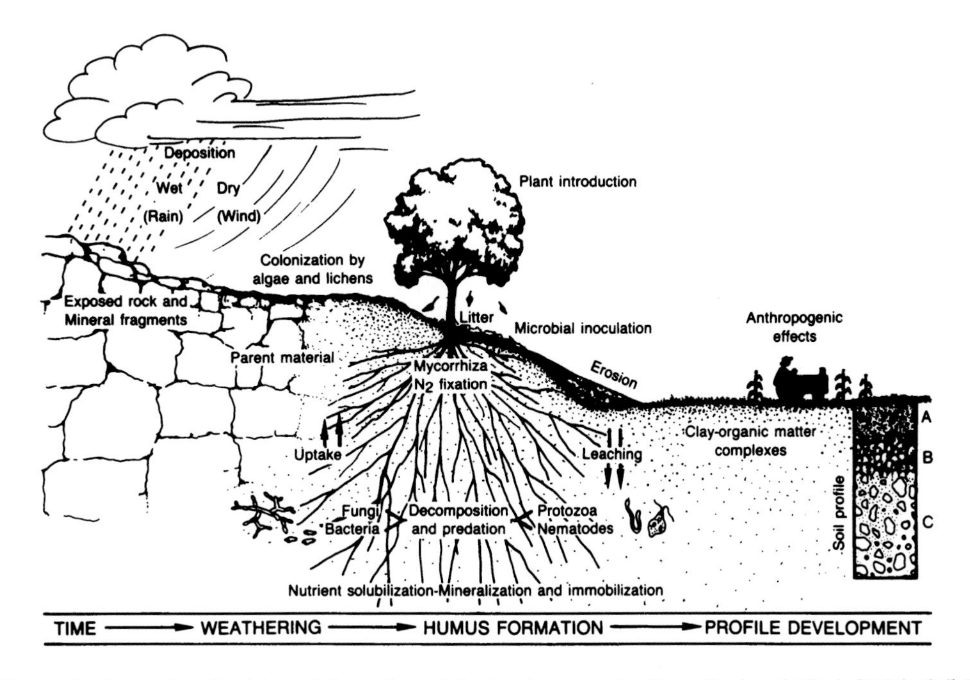

FIGURE 77.4 The major factors involved in soil formation at the landscape scale. From Paul and Clark (1996), *Soil Microbiology and Biochemistry*, p. 21, Academic Press, San Diego.

Systematic and predictable relationships between soil taxonomy and microbial community composition have not been made on a broad scale. Certain factors associated with a feature of the landscape, such as flooding in a low-lying area, have large effects on community composition. The effects of various environmental factors on microbial communities are discussed below.

III. FACTORS AFFECTING MICROBIAL COMMUNITIES

Soil microorganisms are essentially aquatic organisms and require liquid medium or at least a thin film of moisture on the surfaces of solid media. Water is held in soil by adsorption onto surfaces or as free water in the pores between soil particles. Soil water is usually measured as the work required to remove the water from the soil and is expressed in terms of suction or potential.

Soil biological activity is at a maximum when moisture is abundant, but with abundant enough water-free pores to permit adequate oxygen diffusion from the atmosphere, that is, at −10 to −100 kPa. Soil moisture also buffers temperature fluctuation and influences the diffusion, mass flow, and concentrations of nutrients and gases. At low moisture contents, the distribution of the soil's water becomes discontinuous, and the solution remaining is held in small pores. Changing the soil moisture content alters the composition of microbial communities. Fungi, actinomycetes, and spore-forming bacteria are favored when moisture is low. The filamentous growth form of fungi permits these organisms to bridge air-filled pores and thus utilize nutrients in other locations. Algae, protozoa, and facultative and anaerobic bacteria are favored by abundant moisture.

The oxygen concentration at a particular location in soil is affected by the rate of diffusion of oxygen and the biological demand for oxygen at that site, which in turn is determined by available carbon, available electron acceptors, and the abundance of organisms. The oxidation–reduction (redox) potential is a measurement of the likelihood that a substance will gain or lose electrons. When measured directly in soil, the redox potential actually represents numerous redox reactions. Table 77.2 summarizes the major categories of reactions, and their redox potentials, involved in the reduction part of redox reactions. For each of these half-reactions, there is a corresponding half-reaction in which a carbon compound is oxidized. The redox potential in soil ranges from below −240 to 820 mV at pH7. Microbial activity usually lowers the redox potential of soil as oxygen and then the other electron acceptors are consumed. Aerobic and anaerobic organisms can coexist within a soil. Anoxic environments often exist as microsites in what are largely aerobic soils; the anoxia is caused by depletion of oxygen around pockets of high carbon concentrations. Long-term flooding of soil can lead to significant changes in microbial communities. Decomposition of organic compounds is usually most rapid when coupled with the utilization of oxygen, rather than other compounds, as the electron acceptor.

Soil organisms live in an environment of constantly changing osmotic pressure, even in nonsaline soils, because of large daily and seasonal fluctuations in soil moisture content. Soil solution ranges from nearly pure water, for example, in areas of high rainfall, to high salt concentrations, such as in saline soils. Under ideal conditions, the protoplasm of cells has a slightly higher solute concentration than the surrounding soil solution, so water tends to enter cell and turgor is maintained. At low soil osmotic potentials, water molecules move into cells, which may cause expansion and rupturing. At high soil osmotic potentials, water moves out of cells and results in dehydration and shriveling. Organisms are protected from osmotic shock by rigid cell walls, and some species generate intracellular chemicals for osmotic balance.

The largest variations, both diurnal and seasonal, in soil temperature occur near the surface where soil is directly exposed to the atmosphere. Microorganisms, as a group, exist over a temperature range from 0° to 80°C, but they are most active between 20° and 40°C. Every organism has a temperature range within which it is active. Within the temperature range specific to each organism, most microbial reaction rates increase by 1.5–3 times for every 10°C increase

TABLE 77.2 Temporal sequence of terminal electron acceptors used when carbon is available[a]

Terminal electron acceptor and ultimate reduced product	Environmental process	Redox potential at pH 7 (mV)	Soil biota involved
$O_2 + e^- \rightarrow H_2O$	Aerobic respiration	+820	Plant roots, aerobic microbes, animals
$NO_3^- + e^- \rightarrow N_2$	Denitrification	+420	*Pseudomonas*
$Mn^{4+} + e^- \rightarrow Mn^{3+}$	Manganese reduction	+410	*Bacillus* etc.
Organic matter + $e^- \rightarrow$ organic acids	Fermentation	+400	*Clostridium* etc.
$Fe^{3+} + e^- \rightarrow Fe^{2+}$	Iron reduction	−180	*Pseudomonas*
$NO_3^- + e^- \rightarrow NH_4^+$	Dissimilatory nitrate reduction	−200	*Achromobacter*
$SO_4^{2-} + e^- \rightarrow H_2S$	Sulfate reduction	−220	*Desulfovibrio*
$CO_2 + e^- \rightarrow CH_4$	Methanogenesis	−240	*Methanobacterium*

[a]From Kilham (1994). Reprinted with the permission of Cambridge University Press.

in temperature. Psychrophilic, mesophilic, and thermophilic microorganisms can all be isolated from many types of soils, even in temperate climates. Temperature extremes usually only suppress microbial activity rather than kill off large portions of the community. Storage of soils at low temperatures (e.g., at 4°C), often utilized in experiments to suppress biological activity, fails to completely shut down many microbial processes and can lead to misleading information.

Soil pH, which exhibits a broad range across soils, has both direct and indirect effects on microorganisms. Direct effects of low pH include denaturation of proteins and alteration of pH-sensitive enzyme activity. Indirect effects include impacts of pH on the availability and/or chemical forms of toxic (e.g., aluminum) or essential (e.g., phosphate) ions and of organic acids and bases. Bacteria are active across a wide range of pH values from 1 to 9; however, many have optimal activity at a neutral pH. The sulfur oxidizing bacteria are among the most acid loving of soil organisms. Fungi are competitive at pH values ranging from 2 to 7, whereas most actinomycetes and cyanobacteria are active only at pH 6 and higher. The composition of microbial communities can be modified by altering soil pH through acidification or liming, and these practices are sometimes used to control pathogens of plants. Microbial activity lowers the pH of soil through production of organic acids during fermentation and lignin degradation, through production of inorganic acids during oxidation of reduced forms of nitrogen and sulfur, and from carbon dioxide production during aerobic respiration. Microorganisms can increase soil pH via reduction reactions that consume protons, such as occur in anoxic soils. The pH of a previously acidic soil will reach neutrality within weeks of flooding due to reduction reactions.

A large fraction of many compounds in soil are not in the soil solution but instead are associated with surfaces, are partitioned into organic matter, or exist in precipitate form. Microorganisms take up undissolved compounds, if at all, at slower rates than the same compounds in solution. Therefore, a large fraction of many elements and molecules exist in forms that are not biologically available. This is the case with mineral ions that exist as precipitates or are held on exchange complexes on clays, oxides, and organic matter. Similarly, many organic compounds, both naturally occurring and human generated, exceed their water solubility limit in soil or are strongly associated with the hydrophobic portion of organic matter. The reduced bioavailability of environmental pollutants greatly decreases their rates of biodegradation or transformation, but also lowers the toxicity of these pollutants to sensitive organisms.

Soils contain numerous substances that may be toxic to some organisms. These include heavy metals, hydrogen sulfide, organic acids excreted by plant roots or generated by microorganisms (e.g., acetic, butyric, lactic acids), antibiotics, carbon dioxide (when transfer of gases is impeded), and humanmade substances such as pesticides and industrial wastes. Many compounds that are toxic to one group of organisms can be utilized as carbon and energy sources, or in other ways, by another portion of the community.

IV. MAJOR MICROBIAL PROCESSES IN SOIL

A. Gas exchange

Soil microorganisms are important regulators of Earth's atmosphere through the gases they emit and consume. Soil microorganisms are involved in the cycling of all major elements, and many of these elements have gaseous forms. Low molecular weight, volatile compounds produced by microorganisms include carbon dioxide, from respiration; methane, from methanogenic processes in anaerobic environments; and N_2, NO, and N_2O from denitrification and hydrogen sulfide from sulfate reduction in anoxic environments. Gases emitted by microorganisms are, in turn, consumed by other types of microorganisms. Thus carbon dioxide is consumed by autotrophic bacteria, methane by methanotrophic bacteria, nitrogen gas by nitrogen-fixing bacteria, and hydrogen sulfide by sulfur oxidizing bacteria. Through these activities, microorganisms affect the composition, as well as chemistry, of the atmosphere. Long-term transport of volatile compounds emitted by microorganisms can be lead to nutrient deposition into other ecosystems. Microorganisms can also produce volatile forms of heavier elements, such as selenium and mercury, through methylation reactions. These methylation reactions can lead to the transfer of toxic chemicals both form, as in the case of selenium, or into, as for mercury, environments where sensitive organisms can be exposed. Certain gases are directly, or indirectly, destructive of the ozone layer or contribute to global warming. As Earth's atmosphere is changing, there is a growing interest in better understanding soil microbial processes both to reduce the production of greenhouse gases and to enhance potential microbial sinks for greenhouse gases.

B. Elemental cycling

The oxidation states, and in some cases the physical forms, of elements change when used by microorganisms as electron donors or acceptors. A biogeochemical cycle is the result of a series of changes in the chemistry and physical locations of a specific element; thus, biological processes are driving forces of biogeochemical cycles. In some cases, abiotic reactions are faster than biological reactions (e.g., iron oxidation at high pH) and will also contribute to the cycling of elements.

The nitrogen cycle is one of the most well-studied biogeochemical cycles (Fig. 77.5). Nitrogen-fixing bacteria, both free-living in the soil and symbiotic with plants (see below), are largely responsible for the transfer of nitrogen (N_2) from the atmosphere to the soil. The organic forms of nitrogen resulting from N fixation enter the soil with the death of microorganisms and plants and are eventually mineralized to ammonium by most species of bacteria and fungi. In the process called nitrification, the reduced form of N as ammonium is oxidized to nitrite and nitrate primarily by the lithoautotrophic bacterial genera *Nitrosomonas*, *Nitrosospira*, and *Nitrobacter*. Nitrate is also used as an electron acceptor by a variety of bacteria and transformed to dinitrogen gas and, to a lesser extent, nitric and nitrous oxides. In addition, both nitrate and ammonium are incorporated into cellular material (thus into organic form) by most microorganisms in the process of immobilization (see below). All of the above reactions form a closed loop in the cycling of nitrogen in the environment.

The significance of microbial transformations of certain elements is evident throughout the terrestrial environment. Some of the more striking colors of soil are microbial by-products. Many of the red and orange pigments in soil are from oxidized iron, some of it of microbial origin. In waterlogged soils, various shades of gray, green, and black result from reduced sulfur and iron and from their interactions with each other and other elements. Many of the characteristic odors of anoxic environments are microbial in origin. The typical rotten egg smell of anoxic soils is reduced

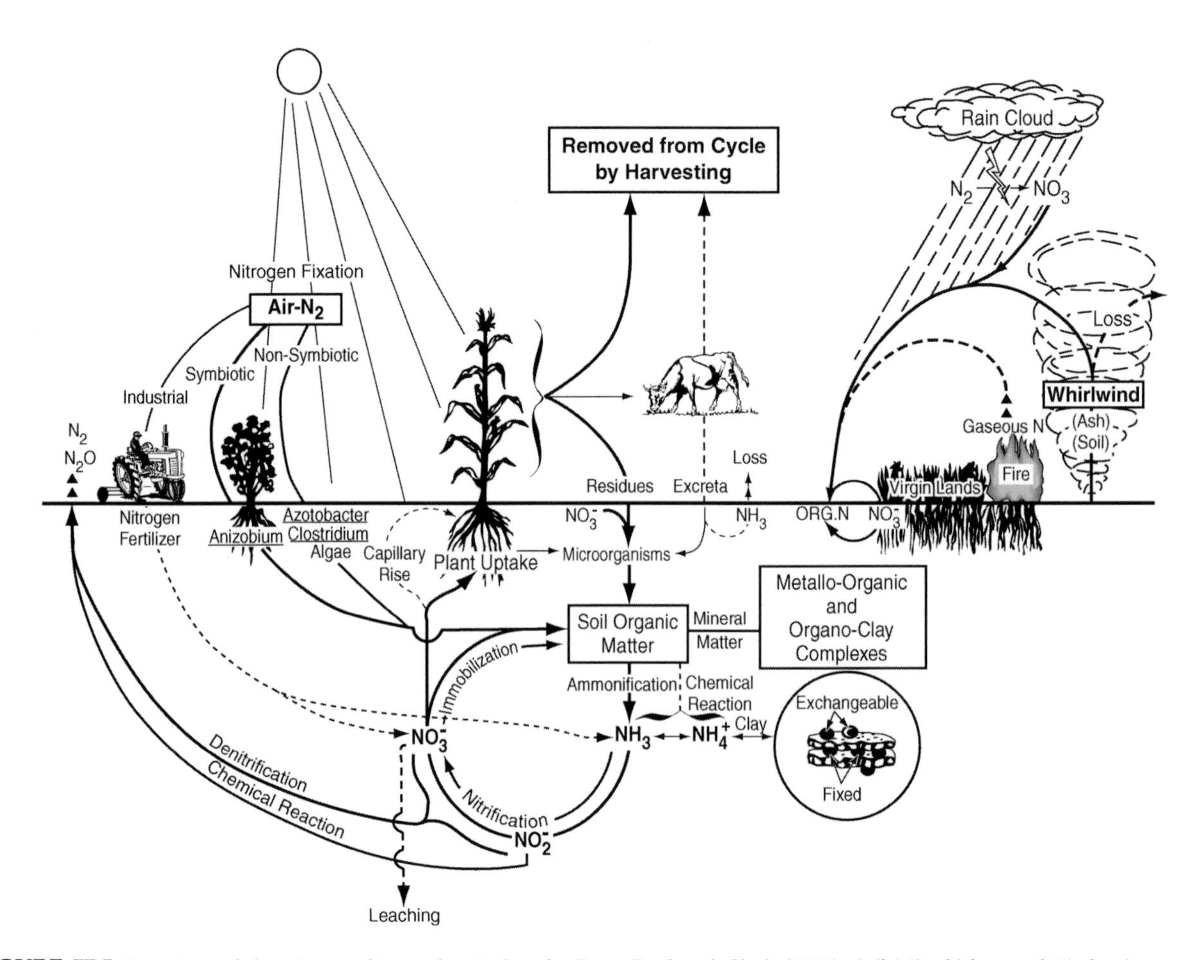

FIGURE 77.5 Overview of the nitrogen biogeochemical cycle. From Paul and Clark (1996), *Soil Microbiology and Biochemistry*, p. 186, Academic Press, San Diego.

sulfur resulting from the use of sulfate as an electron acceptor. Putrescence is associated with organic sulfur compounds resulting from decomposition or fermentation products produced by anaerobic processes. Microbial redox reactions alter the oxidation states of many inorganic pollutants (e.g., selenium and chromium) and thus can decrease (and sometimes increase) the toxicity of these elements.

C. Decomposition, mineralization, and immobilization

Organic compounds produced by plants via photosynthesis continually enter the soil as root exudates when the plant is living and as debris when the plant dies. These complex organic polymers are broken down into smaller organic and mineral components by the combined forces of soil fauna and microorganisms in a process called decomposition. The nutrients released by this process are thus made available for uptake and incorporation into plants and other organisms. Most carbon fixed by terrestrial photosynthesis is returned to the atmosphere within 1 year by microbial decomposition. Plant residues make up the largest fraction of carbon that enters the soil, and microorganisms and animals contribute the remainder. On entering the soil, approximately 60–70% of organic residues are decomposed in the first year, and the remainder decays much more slowly. Many hydrolytic and oxidative extracellular enzymes are essential in the early phases of decomposition because they produce smaller molecules from polymers (e.g., cellulose, hemicellulose, lignin) that are otherwise too large to be absorbed by cells. Nutrients are released from organic residues by the process of mineralization. Mineralization is defined as the conversion of organic chemicals to their inorganic constituents, primarily carbon dioxide, water, and/or ammonium.

During anabolism, soil microorganisms assimilate inorganic forms of elements into their cells in a process referred to as immobilization. Depletion of soil inorganic nutrient pools by microbial immobilization can temporarily limit plant growth. Although both mineralization and immobilization are always occurring simultaneously, the carbon to nitrogen ratio of available organic substrates determines which process will dominate. Thus net immobilization of inorganic nitrogen is expected during decomposition of compounds with high carbon to nitrogen ratios. A large portion of immobilized elements eventually becomes available for plant uptake when microorganisms are preyed on by protozoa or lysed by environmental conditions.

An important product of decomposition is soil humus. Humus is defined as dark-colored organic byproducts consisting of microbial cell walls and other resistant molecules formed from free-radical reactions of sugars, amino acids, and products of lignin decomposition. Thus, microorganisms contribute to humus formation in several ways. They form or release smaller molecules from larger organic polymers. These smaller molecules are condensed into humic substances by reactions catalyzed by primarily fungal-derived extracellular enzymes. Fungi create complex and recalcitrant molecules such as melanin and tannins to withstand predation and other stresses, which in turn become humus precursors when the fungi die. Another type of soil organic material important to soil structure are polysaccharides. Many microorganisms in soil live in a mesh of extracellular polymeric substances, usually polysaccharides. Production of these substances, formed by both bacteria and fungi, occurs under conditions of unbalanced growth (e.g., when carbon is more abundant than nitrogen) or under moisture stress. Cross sections of bacterial colonies in soil, viewed by electron microscopy, reveal that colonies of bacteria consist of a few individual cells enmeshed in a nest of polysaccharides which, in turn, are often surrounded by a coating of clay particles.

D. Biodegradation and transformation of pollutants

Many organic and inorganic pollutants end up in soil intentionally, through their use as pesticides, and unintentionally when they migrate from contaminated waste sites or are deposited on soil from the atmosphere. Biodegradation of organic pollutants can be considered under the general category of decomposition. In its most general form, biodegradation is defined as an alteration in the chemical composition of a molecule mediated by a biological process. Microorganisms are able to utilize many organic pollutants as sources of energy and carbon and use some of the highly chlorinated pollutants as electron acceptors under anoxic conditions. Bioremediation is defined as the decontamination of polluted environments via biological activity. Different types of bioremediation include simple monitoring of naturally occurring biological and abiotic processes leading to pollutant containment or removal (intrinsic remediation), stimulation of natural processes through amendment with nutrients and electron acceptors (biostimulation), or inoculation with microorganisms (bioaugmentation). In soil, biodegradation rates may be substantially reduced by association of pollutants with the solid phase of soil, a process which decreases the availability of the pollutant to microorganisms.

V. INTERACTIONS AMONG SOIL ORGANISMS

A single-species population of microorganisms rarely occurs in soil; instead microorganisms are members of complex communities. The soil food web describes all organisms in soil and their interrelationships. Figure 77.6 is a simplified depiction of a soil web. The composition of a community is governed by the biological steady state, which is a function of the associations and interactions of the members of the community. With environmental change, this steady state may be upset and may shift to a new set of relationships. These relationships can be antagonistic, positive, or neutral.

Antagonistic relationships include predation and parasitism, ammensalism, and competition. Examples of predation and parasitism among soil microorganisms include the bacterium *Bdellovibrio* infesting other bacteria and viral infections of bacteria and fungi. Examples of ammensalism include the alteration of the immediate environment through an organism's

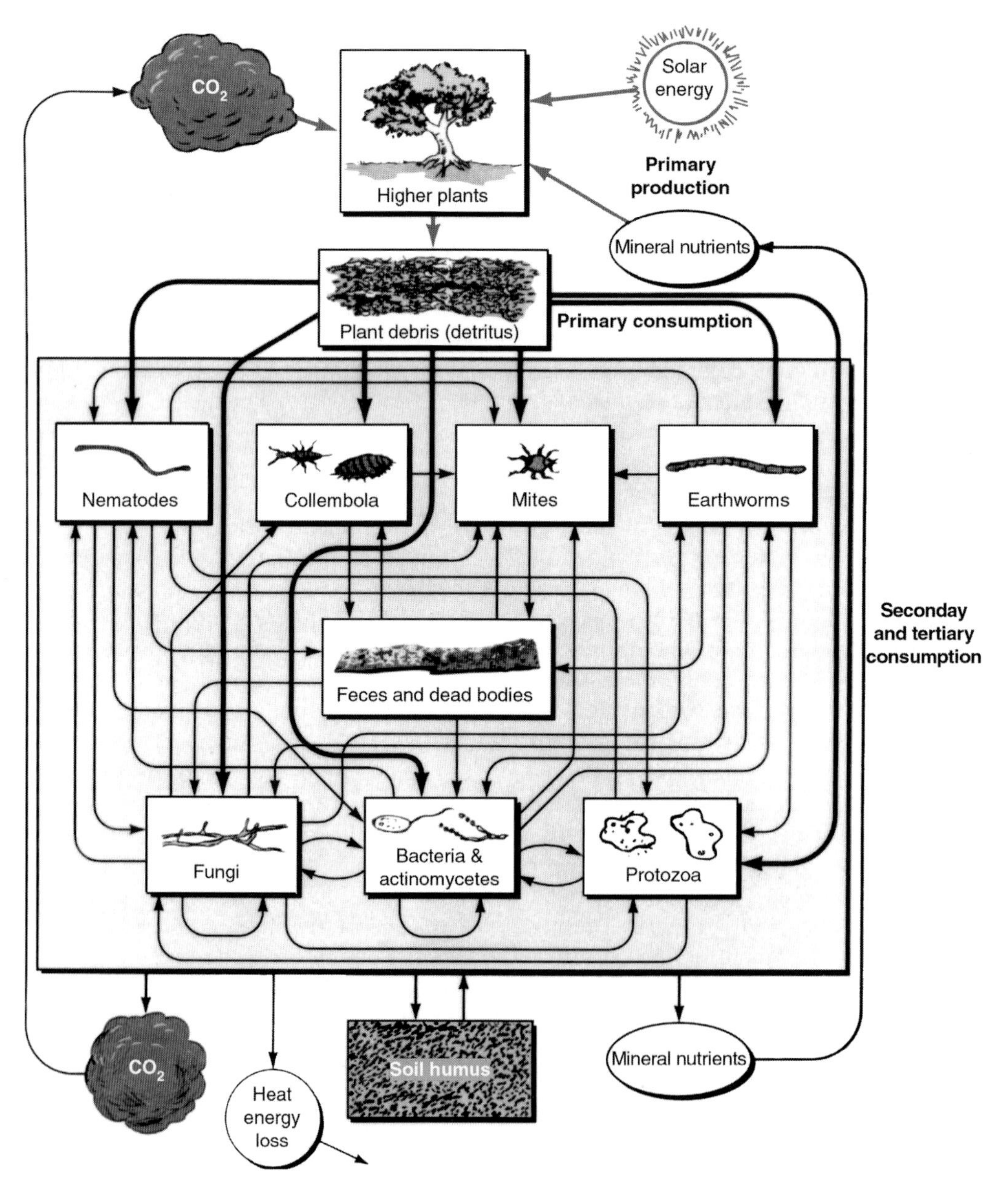

FIGURE 77.6 Overview of the soil food web. From THE NATURE AND PROPERTIES OF SOIL by Brady/Weil, © 1996. Reprinted by permission of Prentice-Hall, Inc., Upper Saddle River, NJ.

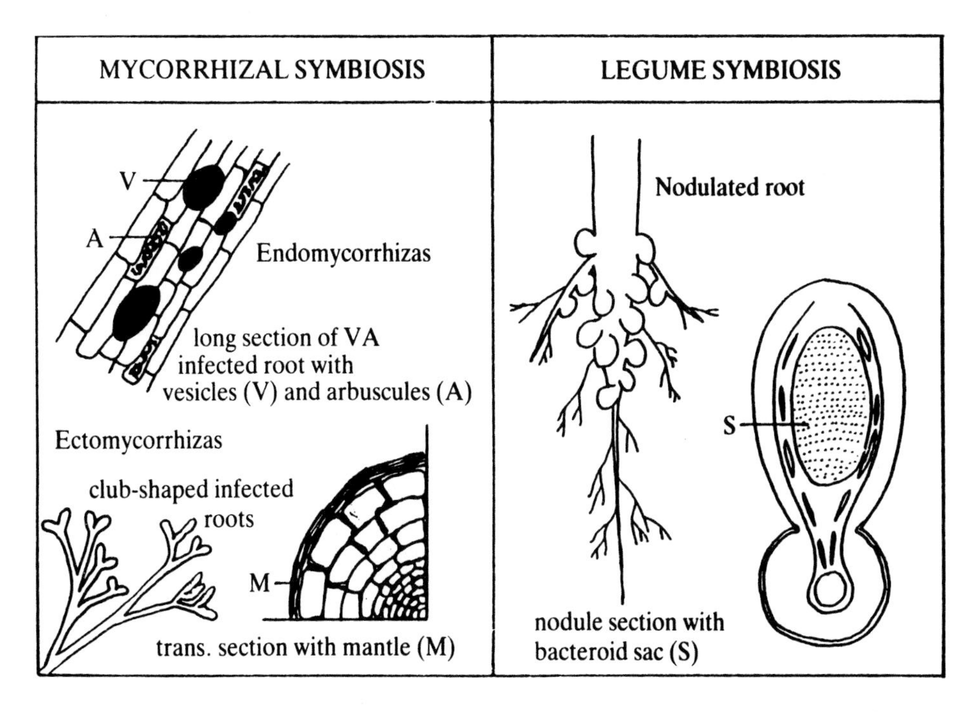

FIGURE 77.7 Plant root/microbial relationships: the mycorrhizal and legume symbioses. From Kilham (1994). Reprinted with the permission of Cambridge University Press.

normal metabolic activities (e.g., through production of acid). Other soil microorganisms, most notably the actinomycetes, produce specific toxic compounds, such as antibiotics, that inhibit or eliminate bacteria, yeast, and other fungi. Most of the earliest antibiotics used for medical purposes were discovered to be produced by the common soil actinomycete *Streptomyces*. These antibiotics include chloramphenicol, streptomycin, neomycin, erythromycin, cycloheximide, and tetracycline. The soil microbiologist Selman Waksman was awarded the Nobel Prize for his pioneering research on antibiotics. Competition for resources, particularly carbon, is considered one of the most important controls of microbial populations in soil. Competitive organisms may have faster growth rates or higher enzyme affinities for the resource in demand than do the less competitive species. Some organisms produce specific compounds, such as siderophores, that increase their access to limiting nutrients, in this case iron.

Positive relationships include mutualism (including symbiosis), commensalism, and synergism. Mutualistic reactions are involved in nitrification and lignin decomposition, both multistep processes in which different organisms are responsible for different metabolic steps. For example, ammonium oxidizing bacteria such as *Nitrosomonas* or *Nitrosospira* produce nitrite which, in turn, is oxidized by other bacterial genera, such as *Nitrobacter*. Without *Nitrobacter*, buildup of nitrite would eventually become toxic to the ammonium oxidizers. Some of the most important symbiotic relationships involving soil microorganisms are relationships with plants. These symbioses between plants and microorganisms are essential to the survival or competitiveness of many plant species. One example of symbiosis is the relationship between leguminous plants (e.g., clover, alfalfa, soybeans, vetch) and root-nodule bacteria, primarily of the genera *Rhizobium* and *Bradyrhizobium* (Fig. 77.7). Another example is the relationship between the actinomycete *Frankia* and actinorrhizal plants, including members of the families Betulaceae, Casuarinaceae, Elaeagnaceae, Myricaceae, Rosaceae, and Rhamnaceae. In both types of mutualism, nodules formed on the plant root provide an environment in which the bacterial symbiont can convert atmospheric nitrogen to a form that is usable by the plant. Atmospheric dinitrogen is reduced to ammonium and then transported into the plant. The plant, in turn, provides the bacteria with sugars necessary for the high energy-requiring demands of N_2 fixation. Amounts of nitrogen fixed by symbiotic bacteria far exceed amounts fixed by free-living bacteria potentially able to fix nitrogen.

The other major type of mutualism in soil involves most plant species and primarily two kinds of mycorrhizal ("fungus root") fungi: ectomycorrhizal and endomycorrhizal fungi (Fig. 77.7). The fungus benefits from the association by having direct access to sugars provided by the plant. The most obvious benefit to the plant is an increase in the volume of soil the plant can exploit. Mycorrhizal hyphae can extend from 5 to 15 cm into the soil from the plant root and penetrate soil pores that are too small for plant roots to enter. The fungal mycelia can increase the plant root surface area by as much as 10 times. Mycorrhizae substantially increase plant uptake of immobile nutrients, such as phosphate. Other benefits to the plant, depending on the kind of mycorrhizal fungus, include increased uptake of nitrogen and other elements, protection against plant pathogens and heavy metals, increased aggregation of soil particles, and possibly drought resistance.

The ectomycorrhizae include hundreds of species of fungi (primarily Basidiomycetes) that are associated primarily with tree species (e.g., pine, birch, oak, spruce, fir) in temperate and arid ecosystems. The plant rootlet is bound by layer of fungal material, called the mantle. The inner portion of mantle is connected to hyphae extending between plant root cells of the epidermis and the outer cortex, forming a mycelial network called the Hartig net. Many ectomycorrhizal fungi can be cultivated on laboratory media and thus have been well studied. The endomycorrhizal fungi (mostly Zygomycetes), also known as vesicular–arbuscular (VA) mycorrhizal fungi, are associated with more than 80% of plant species, including nearly all cultivated plants, forest and shade trees, shrubs, and herbaceous species. These fungi penetrate the cell wall of plant cortical cells and form highly branched hyphal structures, known as arbuscules, that are sites of nutrient exchange between plant and fungus. Because endomycorrhizal fungi cannot grow without their plant hosts, less is known about the physiology of endo- than ectomycorrhizal fungi. Plant biologists are now recognizing that if studies of plant nutrition are to be realistic, they must consider the contribution of mycorrhizae to the uptake of nutrients by plants.

VI. EMERGING RESEARCH AREAS

Soil supports an extraordinary diversity of microorganisms; however, surveys of soil indicate that a substantial number of organisms have not been identified and characterized. The development of molecular tools has permitted the recent discoveries of many new, previously unidentified bacterial genotypes. Considerable effort is directed at characterizing this diversity and particularly using methods that do not initially require traditional enrichment and isolation of organisms. The challenge of the next decade will be to link new information on the composition and structure of soil microbial communities back to the soil processes that have long been the focus of soil microbiology. In addition, the evolution of soil microbial communities; in particular the role of horizontal gene transfer in their evolution, is an area of intense study.

With the growing interest in sustainable management of agriculture and forests, and in reducing synthetic chemical inputs to these ecosystems, new research is being conducted on the role of microorganisms in soil fertility and biological control. Biotechnology is being used to manipulate microorganisms to enhance processes they already perform, or to perform new functions. The role of microorganisms in accelerating and ameliorating global climate change is also an area of active study.

BIBLIOGRAPHY

Coyne, M. S. (1999). "Soil Microbiology: An Exploratory Approach". Chapman and Hall, International Thomson Pub., New York.
Killham, K. (1994). "Soil Ecology." Cambridge Univ. Press, Cambridge, UK.
Paul, F. A., and Clark, F. E. (1996). "Soil Microbiology and Biochemistry," 2nd Ed. Academic Press, San Diego.
Silvia, D. M., Fuhrmann, J. J., Hartel, P. G., and Zuberer, D. A. (1998). "Principles and Applications of Soil Microbiology." Prentice-Hall, Upper Saddle River, NJ.
Tate, R. L. (1994). "Soil Microbiology". Wiley, New York.
Tate, R. L. (2000). "Soil Microbiology". Wiley, New York.
Wackett, L. P. (2001). Soil DNA and the microbial metagenome. An annotated selection of World Wide Web sites relevant to the topics in Environmental Microbiology Web alert. *Environ. Microbiol.* **3**, 352–353.

WEBSITES

A glossary of soil microbiology terms
http://dmsylvia.ifas.ufl.edu/glossary.htm
Website on the Significance of Soil Microbiology (IMPACT Project of the European Union)
http://www.ucc.ie/impact/agrisf.html
CliniWeb relevant links in PubMed
http://www.ohsu.edu/cliniweb/G1/G1.273.540.274.html
List of bacterial names with standing in nomenclature (J. P. Euzéby)
http://www.bacterio.cict.fr/index.html
Soil biological communities (Idaho Bureau of Land Management)
http://www.blm.gov/nstc/soil/index.html

78

SOS response

Kevin W. Winterling
Emory & Henry College

GLOSSARY

constitutive Refers to the expression of a set of genes/operons in the absence of the inducer for expression.

din genes Genes that are induced in the presence of DNA damage.

lytic phase One potential pathway of bacteriophage or prophage that leads to the lysis of the host cell and the release of new phage into the surrounding medium.

prophage A bacteriophage that has had its DNA incorporated into the bacterial chromosome.

regulon A set of genes and/or operons that are all regulated by the product(s) of the same regulatory gene(s).

reporter gene A gene that is fused to another gene so that the expression of that gene may be assayed. The product of the reporter gene is typically more stable and easier to detect than the gene to which it is fused.

−10 and −35 sites Short sequences of DNA that lie approximately 10 and 35 base pairs upstream of the transcription start site of that particular gene. These two sites are necessary for the binding of RNA polymerase.

The SOS Response refers to the expression of a global regulon in response to DNA damage. It involves the coordinated induction of a number of un-linked *D*amage *In*ducible or din genes that are involved in DNA repair, inhibition of cell division, and enhanced survival and mutagenesis of the bacterial population, as well as the induction of bacteriophage.

I. *ESCHERICHIA COLI* AS THE PARADIGM

Throughout their life cycle, all living cells will be exposed to a variety of stressful and ever changing conditions to which they must respond appropriately in order to continue to survive. One of the most significant and common traumas encountered by a living cell is the alteration of the structure of its DNA. These alterations or changes to DNA may occur as a result of exposure to a variety of physical and chemical agents or as a result of imperfect DNA replication. If left uncorrected, these changes may lead to mutations, which may ultimately be lethal. In addition, the products of DNA damage often block DNA replication and thus pose an immediate threat to the survival of the cell. Obviously, for these reasons, it would be beneficial for living cells to have a way of correcting and/or accommodating DNA damage in order to survive. Many prokaryotes have evolved such a system in which a set of unlinked genes is coordinately expressed in response to such DNA damage. This global response has come to be known as the "SOS response."

Since the SOS phenomenon was first suggested by Miroslav Radman nearly 25 years ago, the induction of the SOS response in *Escherichia coli* has been studied extensively and has become the paradigm for the

The Desk Encyclopedia of Microbiology
ISBN: 0-12-621361-5

prokaryotic response to DNA damage. The goal of this chapter is to provide an overview of the characteristics of the SOS response as they have come to be known through the extensive study of *E. coli*. In addition, the properties of the SOS response in the gram positive bacterium, *Bacillus subtilis* will be discussed.

II. REGULATION OF THE SOS RESPONSE

The SOS system is regulated by two proteins which are themselves expressed as part of the SOS response. One protein, RecA, has several functions (among them is the facilitation of genetic recombinational events) in addition to its role as the positive regulator of this global response. The other protein, LexA, serves as the transcriptional repressor of the system. As DNA damage accumulates in a bacterial cell, specific enzymatic or coprotease activities of RecA become activated (designated as RecA*). RecA* forms a long polymer filament by complexing with other RecA* molecules and DNA. This nucleoprotein filament, as it is called, is recognized and bound by free LexA protein. This interaction of the nucleoprotein filament and LexA leads to the cleavage of LexA, thus diminishing the pool of intact protein that is available to bind to the operators of various SOS genes and operons. This releases the SOS genes from the negative transcriptional repression of LexA and allows them to be expressed.

The first observations that a DNA damage induced survival mechanism existed in bacteria was reported by Jean Weigle. He demonstrated that when ultraviolet-irradiated bacteriophage λ was plated on *E. coli* cells that were previously irradiated with ultraviolet light, the bacteriophage survival and the rate of bacteriophage mutagenesis were greater than if the bacteriophage had been plated on unirradiated bacteria. These phenomena are referred to as W (Weigle) reactivation and W (Weigle) mutagenesis, respectively. It was not until the mid-1970s, however, that significant insights into the regulatory pathway of the SOS system in *E. coli* came from the further examination of the bacteriophage, λ. It was observed that as a consequence of exposure to DNA damage, λ cI repressor (cI controls/represses the lytic phase) was being cleaved and, subsequently, the lytic phase of λ was derepressed. Interestingly, this process of cI cleavage and subsequent λ induction was not seen in *recA* (Def) strains (strains of *E. coli* in which the *recA* gene had been mutated), suggesting that RecA was in some way regulating the induction of λ. The hypothesis that RecA was involved in λ induction was further reinforced by *in vitro* assays that showed λ cI was indeed cleaved when incubated in the presence of purified RecA protein.

About this same time, Gudas and colleagues noticed that after cells had been exposed to DNA damaging agents, there was a dramatic increase in the cellular level of a specific but unknown protein, which they called protein X. Further analysis showed that the synthesis of protein X was also similarly elevated in various *recA* and *lexA* mutant strains. Subsequent purification and characterization of the unknown protein revealed its identity to be RecA. On the basis of these observations, it was suggested that LexA repressed the expression of the *recA* gene while RecA played a role in the inactivation of the LexA protein itself. Furthermore, the connection was made that like λ repressor, LexA was most likely cleaved by RecA at the onset of SOS induction. Indeed, the purification and characterization of LexA, and the isolation of LexA (Def) mutants that allowed for the constitutive expression of the SOS system, proved this hypothesis to be correct.

Additional study of the regulation of other SOS genes has revealed that the SOS phenomenon in *E. coli* ubiquitously requires the interaction of the products of the *recA* gene and the *lexA* gene. The LexA protein is the cellular repressor for the many genes that are expressed as part of the SOS response (Table 78.1), including *recA* and *lexA* itself. LexA functions as the cellular repressor of the SOS system by binding to a 16 base pair region of DNA that displays dyad symmetry [5′-CTGT-$(AT)_4$-ACAG-3′]. This conserved sequence of DNA is known as the "SOS box" and is located upstream of most SOS genes and SOS operons. Even in the repressed state, there is, however, a basal level of expression of genes that comprise the SOS regulon. Specifically, there is a sufficient amount of LexA protein to act as the SOS repressor and enough RecA present to fulfill the cell's need for recombinational repair and to induce the SOS response.

III. LexA AS A REPRESSOR

Differential expression of SOS genes is seen in *E. coli*. That is, a certain level of DNA damage does not lead to the equal expression of all of the SOS genes. Such regulation occurs, in part, because LexA binds to the operators of various SOS genes with differing affinities. As a result, certain damage inducible (din) genes are fully expressed at low concentrations of a particular inducing agent, whereas other din genes may require more time and/or an increased dosage to be significantly induced. This graded response allows cells

TABLE 78.1 DNA damage inducible genes in *Escherichia coli* that appear to be regulated by LexA

Gene	Function
dinA	DNA polymerase II
dinB	λ mutagenesis
dinD	Cold-sensitive mutant
dinF	Unknown
dinG	Helicase
dinH	Unknown
dinI	Unknown
dinJ	Unknown
dinK	Unknown
dinL	Unknown
dinM	Unknown
dinN	Unknown
dinO	Unknown
dinP	Identical to *dinB*
lexA	Transcriptional repressor of SOS
recA	Recombination, SOS regulator, SOS mutagenesis
recN	Recombinational repair
ruvAB	Recombinational repair
sbmC	Resistance to Microcin B17
ssb	Single-stranded binding protein
sulA	Inhibitor of cell division
umuDC	SOS mutagenesis
uvrA	Nucleotide excision repair
uvrB	Nucleotide excision repair
uvrD	DNA helicase

with only minor DNA damage to induce error-free repair processes (nucleotide excision repair) without inducing other more dramatic pathways that may ultimately be error-prone (SOS mutagenesis). In fact, substantially more DNA damage must accumulate in the cell in order for the UmuDC proteins (which are required for error-prone repair of DNA and SOS mutagenesis) to be expressed. The continued accumulation of unrepaired DNA lesions ultimately leads to the arrest of cellular division (*sulA*) and the eventual induction of temperate bacteriophage that may exist as prophage in the cell. As cells begin to recover from the damage inducing conditions, the inducing signal diminishes and repression of the SOS regulon returns to preinduction levels.

As briefly mentioned above, there are minor differences among the operators of the SOS genes in *E. coli* that result in different levels of expression. The extent of repression of the individual genes of the SOS regulon depends on at least four factors. The first is operator strength: The actual sequences of individual SOS boxes will vary slightly from the consensus sequence listed above. These slight differences in individual SOS boxes may vary their relative ability to bind LexA by up to a factor of 17. Indeed, there are large differences in the experimentally determined dissociation constants (K_d) of LexA for the SOS box of one gene versus another. LexA shows the highest affinity for the *sulA* gene and a much weaker affinity for the *lexA* SOS box. LexA seems to have the weakest affinity for the SOS box of the *uvrD* gene. Intuitively, this makes sense. The prolonged presence of SulA will lead to the death of the cell. LexA must be readily available to appropriately regulate the SOS response.

The second factor influencing the extent of repression is the location of the operator with respect to the promoter: Some SOS boxes overlap the −35 site, some are located between the −35 and −10 sites, some overlap the −10 site, while still others are located downstream of the −10 site. In the case of the *uvrA* gene, the SOS box overlaps the −35 region of the promoter, which appears to allow LexA to interfere with the action of RNA polymerase at an early stage of transcription initiation. The diverse locations of other SOS boxes with respect to the promoter elements suggest that LexA may inhibit other stages of transcription as well. In those cases where LexA is bound downstream of the transcription start site (*uvrD*, *umuDC*, and *sulA*), it is possible that both LexA and RNA polymerase are bound simultaneously, as is the case with the classic example of the *lac* repressor. In this situation, the presence of the repressor hinders the formation of a competent transcription complex.

Promoter strength is also a factor, as evidenced by the variable levels of expression of SOS genes in *lexA* (Def) and *lexA* temperature sensitive (ts) strains. The induced level of expression of *sulA* is 110 times greater than the uninduced level of expression of *sulA*. Whereas the induction ratio for *sulA* is 110-fold, the induction ratio for another SOS gene, *uvrB*, is only 3.6-fold. If regulation of the SOS genes simply relied on LexA binding affinity, all SOS genes would presumably have similar levels of expression in *E. coli* strains that contain a nonfunctional repressor.

Finally, the presence of additional operators, upstream of certain SOS genes most likely plays a role in SOS regulation. While most SOS genes have only one SOS box associated with them, some have multiple operators (*recN* has three operators, *lexA* has two). Unfortunately, the effect of these additional SOS boxes has not yet been fully determined, but the relationship is believed to be cooperative in nature.

The fact that the SOS box displays dyad symmetry [5′-CTGT-$(AT)_4$-ACAG-3′] led many researchers in the field to suggest that LexA would bind target DNA as a dimer (two monomers of protein, one monomer to each half of the dyad). In fact, protein–DNA binding assays have shown that LexA does indeed interact with target DNA as a dimer. Many DNA binding proteins that are functionally active as dimers or higher order oligomers are readily capable of forming these structures in solution. Both CAP (dimer) and the *lac* repressor (tetramer) exhibit the ability to form stable multimers in solution, and λ repressor has been shown to form dimers in solution with an association constant (K_a) of $5.9 \times 10^{-7}\,M^{-1}$. Despite this precedent, researchers have shown that in all likelihood, LexA does not readily form dimers in solution. The K_a for the monomer–dimer equilibrium of LexA is a relatively low $2.1 \times 10^{-4}\,M^{-1}$. At the *in vivo* concentration of LexA (1300 molecules of LexA monomer per cell) one would therefore predict that the predominant portion of LexA should be in the monomeric state. Indeed, extensive sedimentation analysis revealed that LexA is predominantly a monomer in solution.

The X-ray crystallographic studies of several protein–DNA complexes show that one subunit of each protein monomer contacts each half-site of the operator. For many of these proteins, these protein–DNA complexes form by binding of preformed dimers to their target sites. Since it was evident that LexA did not form dimers in solution, it was suggested that LexA bound target DNA as a monomer, then dimerized while in contact with the target DNA. Since the amino-terminal domain was found to bind an SOS box half-site with the same affinity as the intact LexA protein, it was determined that this region, alone, is responsible for DNA binding. In similar experiments, the K_{dimer} (dimerization constant) of the carboxyl terminus was determined to be the same as the K_{dimer} for intact LexA, thus leading to the conclusion that this region alone is accountable for LexA dimerization. Finally, by demonstrating that intact LexA has a much higher affinity for the intact SOS operator than it does for a half-operator, it was concluded that LexA dimerizes on the DNA in a cooperative manner. It seems that once the first LexA monomer binds to an operator half-site, the second monomer binds much more quickly via a combination of protein–protein and protein–DNA interactions.

Like many other proteins that bind to DNA, the DNA binding domain of LexA displays a helix–turn–helix motif. The work performed by Knetgel *et al.* delineates some of the specific interactions of LexA with its binding site, the SOS box. These authors demonstrated that there are three α-helices in the amino terminus, with helices II and III constituting the primary structure of the helix– turn–helix motif. It is helix III, which comprises amino acid residues 40–52, that protrudes into the major groove of the DNA and is responsible for the affinity of LexA to DNA. Ser^{39}, Asn^{41}, and Ala^{42} contribute hydrophobic interactions in addition to Asn^{41}, Glu^{44}, and Glu^{45}, which form the direct hydrogen bonds to the CTGT half-site of the SOS box. There are many other nonspecific protein–DNA contacts that have proved important for binding. Their role is to enhance the stabilization of the LexA monomer to the operator half-site and consequently promote dimerization.

Not all of the base pairs within the consensus sequence of the SOS box are of equal importance. On the basis of the distribution and the degree to which proteins are expressed by operator constitutive mutants, the four base pairs CTGT seem to be the most important for LexA recognition. Within these four base pairs, the two central base pairs (TG) seem to be absolutely required, since these are conserved in all known operator half-sites. The importance of the $(AT)_4$ region is very difficult to assess, owing to the high degree of variability present. This region may not be involved in LexA contact since it is not always protected in assays that methylate bases that are not specifically in contact with protein. This sequence most likely favors LexA binding indirectly by providing the proper spacing between the two half-sites of the SOS box, as has been suggested in the case of the repressor of bacteriophage 434.

IV. THE SOS INDUCING SIGNAL

As discussed above, the activated form of RecA, RecA*, is an important physiological requirement for the induction of the SOS system. It is not the DNA damage itself that leads to the activation of RecA, but most likely the accumulation of single-stranded DNA (ssDNA) that occurs when DNA replication is stalled due to bulky lesions and mutations. There is definitive evidence supporting this hypothesis. First, RecA may be activated *in vitro* by the addition of ssDNA and ATP to the reaction mixture. Additionally, infection of *E. coli* with the filamentous ssDNA phage f1 does not typically induce the SOS response. However, similar infection with an f1 mutant that is defective in the initiation of minus-strand DNA synthesis routinely does lead to the induction of the SOS response.

V. MECHANISM OF LexA CLEAVAGE

Following DNA damage, large stretches of single-stranded DNA are generated when DNA polymerase III dissociates at a lesion and then reassociates approximately 1 kb downstream from the lesion. RecA protein binds to this single-stranded DNA to form spiral nucleoprotein filaments on DNA. Free LexA recognizes this structure and binds within the deep helical groove of the RecA nucleoprotein filament. This interaction with RecA then facilitates the cleavage of LexA, at a scissile peptide bond located between residues Ala^{84} and Gly^{85}, approximately the center of the protein. This Ala^{84}–Gly^{85} bond connects the amino-terminal domain of the protein (which is responsible for DNA binding) to the carboxyl-terminal domain (responsible for dimerization). On cleavage, the level of functionally active LexA protein available to bind the SOS boxes located in the operator/promoter regions of SOS genes decreases dramatically. The end result is to release SOS genes from the negative transcriptional regulation of LexA. Once the DNA is repaired and the inducing signal subsides, RecA returns to an inactive state and the subsequent autocatalytic activity of LexA ceases. Consequently, the cellular pool of functional LexA increases, and expression of the SOS regulon returns to preinduction levels.

It was originally believed that RecA was responsible for the enzymatic cleavage of LexA via a classic protease mechanism. This hypothesis was disproved by Little and colleagues when they showed that the mechanism for LexA cleavage involves both intramolecular and intermolecular reactions that occur independently of RecA. The role of RecA appears to be that of a coprotease that increases the rate of LexA autodigestion under physiological conditions. This autocatalytic cleavage also occurs when LexA is incubated at an alkaline pH, thus proving LexA cleavage is capable of occurring in a manner independent of RecA. This type of cleavage is observed in a number of other functionally and/or structurally related proteins such as the repressors of bacteriophages λ, 434, P22, and ϕ80 and the mutagenesis proteins UmuD, MucA, and $RumA_{(R391)}$. A small number of amino acid residues have been found to be highly conserved among these proteins (the previously mentioned alanine–glycine cleavage site and appropriately spaced serine and lysine residues), and these conserved amino acids presumably play principal roles in the autocatalytic cleavage process. Earlier studies proposed that the LexA cleavage reaction was very similar to that of serine proteases or other signal peptidases, but more recent studies have shown that the reaction is more like that of the TEM1 β-lactamase. The proposed mechanism of autocatalytic cleavage suggests that the lysine residue in the carboxyl domain removes a proton from the carboxyl domain serine residue, which in turn acts as a nucleophile to attack the alanine–glycine bond. In addition, the lysine residue may donate a proton to the α-amino group when the bond is broken.

VI. SOS MUTAGENESIS

The SOS response is often thought of as being a system that is responsible for repairing DNA damage, when in reality, it is a mechanism that enhances the survival of the cell. Often, after an organism has been exposed to a DNA damaging agent such as ultraviolet light or mutagenic chemicals, all of the mutations and lesions that result cannot be repaired by the methods to which the cell has access. Typically, this DNA damage would block the replication of DNA and lead to the death of the cell. SOS mutagenesis allows the replication machinery to bypass these DNA lesions. The end result is mutated DNA, but the cell has survived. Because mutations themselves may have extremely negative effects on the viability of the cell, SOS mutagenesis is tightly regulated (discussed above).

Three of the gene products that are produced as part of the SOS response are required for SOS mutagenesis: RecA, UmuD, and UmuC. In addition to these three LexA regulated genes, DNA polymerase is also required for this process. Given the appropriately intense inducing dose, the SOS system produces the UmuD and UmuC proteins via transcription of the *umuDC* operon. In its initial transcribed state, the UmuD protein is not mutagenically active. In much the same way that the LexA protein undergoes an autocatalytic cleavage, UmuD is cleaved. UmuD is structurally related to the aforementioned proteins, LexA and λ cI. A comparison of the amino acid sequences of the three proteins reveals a conserved alanine–glycine cleavage site as well as the appropriately spaced serine and lysine residues. The cleavage of UmuD yields a modified protein that is referred to as UmuD′. UmuD′ then forms a homodimer and complexes with a UmuC monomer to form a complex ($UmuD'_2C$) that is essential for SOS mutagenesis. This complex, together with RecA, seems to allow the DNA replication enzyme, DNA polymerase III, to bypass specific types of DNA damage.

Although it is generally accepted that the primary components that are required for SOS mutagenesis are the proteins mentioned above, the mechanism by which these proteins allow DNA replication to continue past sites of DNA damage is still under intense study. The importance of SOS mutagenesis is supported by the fact that homologs of UmuD and UmuC are found in a number of other bacteria, as well as bacteriophage P1. The presence of UmuDC homologs even transcends into the eukaryotic kingdom in such organisms as the yeast *Saccharomyces cerevisiae* and the nematode *Caenorhabditis elegans*.

VII. THE SOS RESPONSE IN BACILLUS SUBTILIS

The SOS response has long been considered an integral part of the ability of a bacterial cell to survive environmental insults and faulty metabolic processes, but it has since become evident that a DNA repair mechanism, such as the SOS system, is capable of more than enhancing cell survival and regulating the rate of mutagenesis. DNA repair systems also play important roles in viral activation, DNA replication, genetic recombination, metabolism, and cancer. Unfortunately, *E. coli* lacks a readily identifiable developmental cycle, and therefore is not an appropriate model for studying the relationship between DNA repair systems and these other phenomena. On the other hand, the gram-positive soil bacterium *B. subtilis* appears to be an ideal paragon for studying the relationship between DNA repair mechanisms and other developmental cycles. In addition to having a defined SOS system, *B. subtilis* maintains the ability to differentially sporulate, develop motility, produce degradative enzymes, express antibiotics, and become naturally competent in response to environmental stimuli. In addition to the SOS regulon of *E. coli*, similar systems and the cognate regulatory proteins exist in many other gram-negative bacteria. The apparent importance of this stress response mechanism suggests that it is of ancient origin, and its evolution might have even preceded the divergence of the gram-negative and gram-positive eubacteria. Using a set of isogenic strains that differed only in the presence of specific mutations in genes that were thought to be involved in DNA repair and recombination, it was demonstrated that an SOS system did exist in the gram-positive, spore forming bacterium *B. subtilis*. Furthermore, the SOS systems of both *E. coli* and *B. subtilis* are induced in a RecA-dependent manner and have similar phenotypic characterizations of DNA repair, enhanced survival and mutagenesis, prophage induction, and the inhibition of cell division. A number of din genes (including *recA*) were isolated, mapped, cloned, and sequenced using reporter gene technology. This information allowed for the identification of a consensus sequence, GAAC-N_4-GTTC, that was found in the promoter region of each identified din gene. Deletion analysis of this consensus sequence allowed investigators to conclude that the sequence and/or adjacent regions are essential to the proper SOS regulation of each din gene. Since this sequence was found in the promoter regions of *recA* genes in othergram-positive organisms, it was hypothesized to be the gram-positive version of the SOS box, although there is no sequence homology present between it and the SOS box found in *E. coli*.

Owing to the sequence difference between the gram-negative SOS box and the putative gram-positive SOS box, the site in *B. subtilis* was delineated the "Cheo box." Also, since the exact role of the Cheo box had not been fully elucidated, it seemed premature to actually label it as the SOS box. At times however, editorial constraints have, in fact, led to the Cheo box being designated as the gram-positive SOS consensus sequence or SOS box.

The *B. subtilis recA* gene and its gene product have been extensively characterized; however, the identity and role of the repressor of the *B. subtilis* SOS regulon remained somewhat enigmatic. A number of studies by several researchers eventually identified the repressor as the product of a previously isolated din gene, *dinR*. The DinR protein is the same approximate size as LexA (~23 kDa) and also exhibits 34% identity and 47% similarity to LexA. These similarities include regions thought essential for autocatalytic proteolysis. The further characterization of DinR and the operator regions of SOS genes in *B. subtilis* has led to the redefinition of the SOS repressor binding site. This site is now referred to as the "DinR box" and has a consensus sequence of 5′-CGAAC-N_4-GTTCG-3′.

Unlike the *E. coli* SOS response, in which there is only one mechanism for induction, the SOS phenomena of *B. subtilis* are currently classified into four distinct types. The increased complexity of the *B. subtilis* SOS response is primarily attributable to its relationship with the development of natural competence. Again, using reporter gene technology, it was determined that a number of DNA damage inducible genes were also induced following the onset of competence. As noted above, RecA protein is activated by the presence of single-stranded regions of chromosomal DNA, which have been demonstrated to accumulate as cells become competent. It was hypothesized that activated RecA then stimulates the autocatalytic activity of a LexA homolog, and SOS genes are subsequently derepressed. Once it was noted that several *din* genes were induced by competence in *B. subtilis* strains that lacked functional RecA protein, it became evident that SOS induction is more complex in *B. subtilis* than in *E. coli*.

Four types of SOS phenomena in *B. subtilis* have been subsequently described. These types have been classified according to the nature of their mechanisms of DNA damage induction and competence induction. The type of SOS induction that appears to be the most similar to the induction of the *E. coli* SOS system is known as the Type I phenomenon. This includes the following events: expression of the genes *dinB, dinC,* and *uvrB,* induction of certain prophage (ϕ105, SPO2), error-prone repair, and W reactivation. Type I events are induced by DNA damage as well as by the development of the competent state with both being RecA dependent.

The Type II SOS phenomenon of *B. subtilis* has only one characterized phenotype, filamentous growth. Filamentation in *B. subtilis* is induced by DNA damaging agents, but not by the development of competence. Surprisingly, the phenomenon of filamentation in *B. subtilis* is a RecA-independent event.

The genes (*recA* and *dinA*) that have been classified as part of the type III phenomenon are regulated in the most complex manner yet observed. Expression of *recA* and *dinA* is induced by the presence of DNA damaging agents, but only in the presence of RecA itself. The expression of *recA* and *dinA* is also induced by the development of competence; however, under these conditions, RecA protein need not be present. It appears that damage induction of these genes occurs in the prototypical manner that requires activated RecA to stimulate the autocatalytic activity of the LexA homolog. However, the competence induction of these genes appears to occur via a separate competence specific pathway, since several genes that are specific to the competence pathway (*spo0A, spo0H, degU, comK*) are required for the competence induction of *recA*.

The Type IV phenomenon involves the induction of the prophage of the SPβ family, such as ϕ3T. There are two types of temperate bacteriophage that infect *B. subtilis:* "smart" phage and "naive" phage. The smart phage, which include SPβ and ϕ3T, are capable of differentiating between SOS induction by DNA damage and SOS induction by the development of competence. Therefore, the smart bacteriophage are induced on DNA damage only, and not by the development of competence. The induction of smart bacteriophage is a RecA-dependent event.

VIII. THE SOS SYSTEM IN OTHER PROKARYOTES

Escherichia coli has served as the paradigm for study of the SOS system in gram-negative bacteria, and *B. subtilis* has served as the model for study in gram-positive populations. However, the SOS regulon has been found to exist in many other bacteria based on a variety of observations suggesting that the two key regulatory elements, RecA and LexA, have been conserved throughout evolution.

More than 60 highly conserved homologs of RecA have been characterized in a wide variety of bacteria. In addition, there is evidence for LexA-like regulation in 30 species of gram-negative bacteria. This was demonstrated by the introduction of a plasmid containing a *recA* operator/promoter that was fused to a reporter gene (*recA-lacZ*), which was capable of being both induced and repressed in relation to the presence or absence of DNA damaging agents, respectively. More direct evidence for the existence of SOS regulatory networks in other bacteria has been provided by the successful cloning of the *lexA* genes from several organisms as well as genome sequencing projects that have identified *lexA*-like genes in other organisms.

BIBLIOGRAPHY

Bridges, B. (2000). DNA polymerases and SOS mutagenesis: can one reconcile the biochemical and genetic data? *Bioessays* **22**(10), 933–937.

Dubnau, D. (1991). *Microbiol. Rev.* **55**.

Friedberg, E. C., Walker, G. C., and Siede, W. (1995). "DNA Repair and Mutagenesis," 2nd edn. American Society for Microbiology, Washington, DC.

Hanawalt, P. C. (1989). Concepts and Models for DNA Repair: From *E. coli* to mammalian cells. *In* "Environmental Molecular Mutagenesis," Vol. 14.

Koch, W. H., and Woodgate, R. (1998). "DNA Damage and Repair: DNA Repair in Prokaryotes and Lower Eukaryotes" (J. A. Nickoloff and M. F. Hoekstra, eds.), Humana, Totowa, NJ.

Maor-Shoshani, A., Reuven, N. B., Tomer, G., and Livneh, Z. (2000). Highly mutagenic replication by DNA polymerase V (UmuC) provides a mechanistic basis for SOS untargeted mutagenesis. *Proc. Natl. Acad. Sci. USA* **97**(2), 565–570.

Napolitano, R., Janel-Bintz, R., Wagner, J., and Fuchs, R. P. (2000). All three SOS-inducible DNA polymerases (Pol II, Pol IV and Pol V) are involved in induced mutagenesis. *EMBO J.* **19**(22), 6259–6265.

Pham, P., Bertram, J. G., O'Donnell, M., Woodgate, R., and Goodman, M. F. (2001), A model for SOS-lesion-targeted mutations in *Escherichia coli. Nature* **409**(6818), 366–370.

Smith, B. T., and Walker, G. C. (1998). Mutagenesis and more: umuDC and the *Escherichia coli* SOS response. *Genetics* **148**(4), 1599–1610.

Sutton, M. D., and Walker, G. C. (2001). Managing DNA polymerases: Coordinating DNA replication, DNA repair, and DNA recombination. *Proc. Natl. Acad. Sci. USA* **98**(15), 8342–8349.

Walker, G. C. (1996). The SOS response in *Escherichia coli.* In *"Escherichia coli and Salmonella. Cellular and Molecular Biology,"* 2nd edn. (F.C. Neidhardt, ed.) ASM Press, Washington.

Yasbin, R. E., Cheo, D. L., and Bol, D. (1993). *"Bacillus subtilis* and Other Gram-Positive Bacteria, Biochemistry, Physiology, and Molecular Genetics" (J. A. Hoch, A. L. Soneshein, and R. Losick, eds.), American Society for Microbiology, Washington, DC.

WEBSITES

The SOS response in *E. coli*
http://info.bio.cmu.edu/courses/03441/termpapers/97TermPapers/SOS-response/index.htm

Links to SOS in PubMed
http://www.ohsu.edu/cliniweb/G5/G5.386.html

79

Space flight, effects on microorganisms

D. L. Pierson and S. K. Mishra
NASA/Johnson Space Center

GLOSSARY

commensalism Symbiotic relationship in which one species benefits and the other is unharmed.

endogenous Originating or produced within an organism or one of its parts.

exogenous Originating outside an organism; infections can be of exogenous or endogenous origin.

microgravity The condition of an environment in which acceleration due to gravity is approximately zero; also termed weightlessness.

pedicel Slender stalk that supports the fruiting or spore-bearing organ in some fungi.

solar particle event Sudden eruption on the surface of the sun that results in an increased flux of high-energy particles, which in turn increases the exposure of spacecraft to ionizing radiation.

Spacelab Manned laboratory developed by the European Space Agency for flights aboard the U.S. Space Shuttle; pressurized habitable modules and pallets adapted to specific missions are carried in the Shuttle's payload bay.

The Effect of Space Flight on microbial function has been of concern to microbiologists since humans first began to explore space. Because microorganisms will be present on board manned and unmanned spacecraft, the potential exists for colonization of the vehicle itself as well as its inhabitants. The combination of the closed nature of spacecraft and the stressful nature of space flight (e.g., acceleration, weightlessness, radiation) increases the possibility of microbially induced allergic reactions and infections among space crew. Space flight is also suspected of altering human immune function as well as bringing new environmental selection pressures to bear on endogenous and exogenous microbiota. The combined effect of these processes may render normally harmless commensal or environmental microbes pathogenic to humans. Furthermore, colonization of the vehicle itself may result in system fouling, biodegradation of sealants, and perhaps the production of toxic metabolites and environmental pollutants. As the number and duration of manned flights increase, it has become imperative to characterize the effects of space flight and related factors on microbial growth, physiology, virulence, and susceptibility to antibiotic agents to protect the health of the crew and the integrity of the spacecraft.

At present, very few research data are available to address these concerns. Although many microorganisms have been used as models to study the effects of cosmic radiation, microgravity, vibration, and hypervelocity on living systems during a number of missions over the past 40 years (Tables 79.1 and 79.2), the severe constraints involved in performing experiments in space have largely precluded exhaustive studies. The absence of gravity, which mandates the development of specialized equipment and procedures, the restrictions imposed on power, weight, and volume, and intense competition for the crew's time during space flights require that experiments be simple and easily performed with little or no crew involvement. Thus, many basic questions concerning the effects of space on

The Desk Encyclopedia of Microbiology
ISBN: 0-12-621361-5

TABLE 79.1 U.S. and Soviet missions carrying microbiology experiments

Flight	Country	Launched	Manned duration
Sputnik	USSR	1957–1961	Unmanned
Vostok	USSR	1961–1965	
Gemini	U.S.	1964–1966	10 Manned flights
Cosmos 110	USSR	Feb 1966	
Apollo	U.S.	1967–1972	
Biosatellite II	U.S.	Sept 1967	2 days
Zond 5	USSR	Sept 1968	Unmanned
Zond 7	USSR	Aug 1969	
Cosmos 368	USSR	Oct 1970	
Salyut 1	USSR	April 1971	23 days
Skylab 1	U.S.	May 1973	Unmanned
Skylab 2	U.S.	May 1973	28 days
Skylab 3	U.S.	July 1973	59 days
Soyuz 12	USSR	Sept 1973	
Skylab 4	U.S.	Nov 1973	84 days
Salyut 3	USSR	June 1974	14 days
Cosmos 690	USSR	Oct 1974	
Salyut 4	USSR	Dec 1974	41 days
Apollo–Soyuz	U.S.–USSR	July 1975	9 days
Salyut 5	USSR	June 1976	33 days
Salyut 6	USSR	Sept 1977	1,192 days
Cosmos 1129	USSR	Sept 1979	
Salyut 7	USSR	April 1982	1,805 days
Spacelab 1	U.S.	Nov 1983	10 days
Spacelab 3	U.S.	July 1985	7 days
Spacelab 2	U.S.	July 1985	8 days
Spacelab D-1	U.S.	Oct 1985	7 days
Mir	USSR	Feb 1986	366 days (max)

TABLE 79.2 Organisms used in space flight experiments

Prokaryotes	Eukaryotes
Bacteria	Protozoa
Actinomyces aureofaciens	*Euglena gracilis*
Actinomyces erythreus	*Paramecium aurelia*
Actinomyces levoris[a]	*Paramecium tetraurelia*
Actinomyces streptomycin	*Pelomyxa carolinensis*
Aerobacteria aerogenes	*Tetrahymena periformis*
Aeromonas proteolytica	*Tetrahymena pyriformis*
Bacillus brevis	Fungi
Bacillus subtilis	Molds
Bacillus thuringiensis	*Aspergillus niger*
Clostridium butyricum	*Chaetomium globosum*
Clostridium sporogenes	*Neurospora crassa*
Escherichia coli	*Penicillium roquefortii*
Hydrogenomonas eutropha	*Phycomyces blakesleeanus*
Methylobacterium organophilum	*Polyporus brumalis*
Methylomonas methanica	*Trichoderma viride*
Methylosinus sp.	*Trichophyton terrestre*
Proteus vulgaris	Yeasts
Pseudomonas aeruginosa	*Candida tropicalis*
Staphylococcus aureus	*Rhodotorula rubra*
Streptomyces levoris[b]	*Saccharomyces cerevisiae*
Bacteriophage	*Saccharomyces vivi*
Aerobacteria aerogenes phage 1321	*Zygosaccharomyces bailii*
	Slime mold
Escherichia coli phage T1, T2, T4, T7, and λ	*Physarum polycephalum*
	Algae
Salmonella typhimurium phage P-22	*Chlamydomonas reinhardtii*
	Chlorella ellipsoidea
	Chlorella pyrenoidosa
	Chlorella sorokiniana
	Chlorella vulgaris
	Scenedesmus obliquus

[a,b]Synonyms: *Actinomyces* in Russia, *Streptomyces* in the U.S.

microbial structure and function have yet to be resolved. This article presents an overview of information collected to date on the effects of the space flight environment on microbial physiology and function.

I. MICROBIAL SURVIVAL AND GROWTH IN SPACE

Studies on the effect of extreme conditions on microorganismal growth and survival began as early as 1935, when high-altitude balloons were used to investigate the effects of low temperatures, decreased pressures, and increased radiation. From 1954 to 1960, experiments on nearly 30 high-altitude balloons and sounding-rocket flights revealed that *Neurospora* spores and vegetative bodies could survive direct exposure to the environment at 35–150 km above Earth's surface. In the 1960s, viable organisms from cultures of *Penicillium roquefortii* and *Bacillus subtilis* carried on the Gemini 9A and 12 missions were recovered after nearly 17 h of direct exposure to space. Parallel attempts to detect microorganisms in the extraterrestrial environment in analyses of micrometeorites collected during the Gemini missions, and lunar samples collected during the Apollo flights, revealed no evidence of viable microorganisms nor any identifiable biological compounds. Therefore, the potential for contaminating Earth with extraterrestrial life forms, an early concern, was judged to be extremely unlikely.

Studies of microbial behavior in space performed to date, although numerous, have produced inconclusive, occasionally contradictory results (Tables 79.3 and 79.4). The first experiments on the unmanned Sputnik orbital satellites (1957–1961) used microorganisms to identify the gross effects of galactic radiation, weightlessness, and other related factors on biological systems. Studies of bacteriophage induction have been many (see later) and have dated back to the second Soviet satellite mission in August 1960, which included flight experiments with *Clostridium butyricum*, *Streptomyces* spp., *Aerobacteria aerogenes* 1321 bacteriophage, and T-2 coliphage. Viability

80

Sporulation

Patrick J. Piggot
Temple University School of Medicine

GLOSSARY

cortex A peptidoglycan that is unique to spores and sporulating organisms. There are very few peptide bridges, and many of the muramic residues in the glycan chain are in the form of a lactam with no attached peptide side chain.

engulfment The process by which the developing prespore is completely surrounded by the mother cell.

mother cell One of the two cells that is formed by the sporulation division. It is required for spore formation, but lyses when the spore is formed.

prespore One of the two cells that is formed by the sporulation division. It develops into the mature spore. It is sometimes called the forespore.

σ factor A protein that binds to RNA polymerase core enzyme, designated E, to form RNA polymerase holoenzyme, E–σ. The σ factor determines the specificity of the binding of the holoenzyme to promoter sequences in DNA.

vegetative cell A bacterial cell from a culture that is growing exponentially.

Spores are a dormant form of bacteria. They are resistant to a variety of environmental stresses that would kill the vegetative (growing) form of the bacteria. The stresses include heat, desiccation, irradiation, and chemicals such as ethanol and chloroform. Sporulation is the process by which the spores are formed from vegetatively growing bacteria, and is a response to nutrient depletion. The best studied examples of sporulation are of members of the genera *Bacillus* and *Clostridia*. The description is generally also valid for sporulation of members of related genera such as *Sporosarcina* and *Thermoactinomyces*.

Sporulation has been most extensively studied with *Bacillus subtilis*. Formation of heat-resistant spores from vegetative cells of *B. subtilis* takes about 7 h at 37 °C. This species has been favored because it has very good systems of genetic analysis. Efficient systems of genetic exchange by transformation and transduction mean that it is easy to transfer genes and mutations from one strain to another. It is also easy to identify mutants of *B. subtilis* that cannot sporulate, because, on appropriate media, the mutant colonies are poorly pigmented compared to colonies of the sporulating parental strains. The complete sequence of the *B. subtilis* genome (chromosome) has recently been determined, further facilitating studies of sporulation of this species.

I. STAGES OF SPORULATION

The morphological changes during sporulation have been characterized for a number of species by electron microscopy. The basic sequence of changes is similar for all species of *Bacillus* and *Clostridium* that have been studied. It is illustrated in Fig. 80.1. Identification of successive stages by Roman numerals follows the

The Desk Encyclopedia of Microbiology
ISBN: 0-12-621361-5

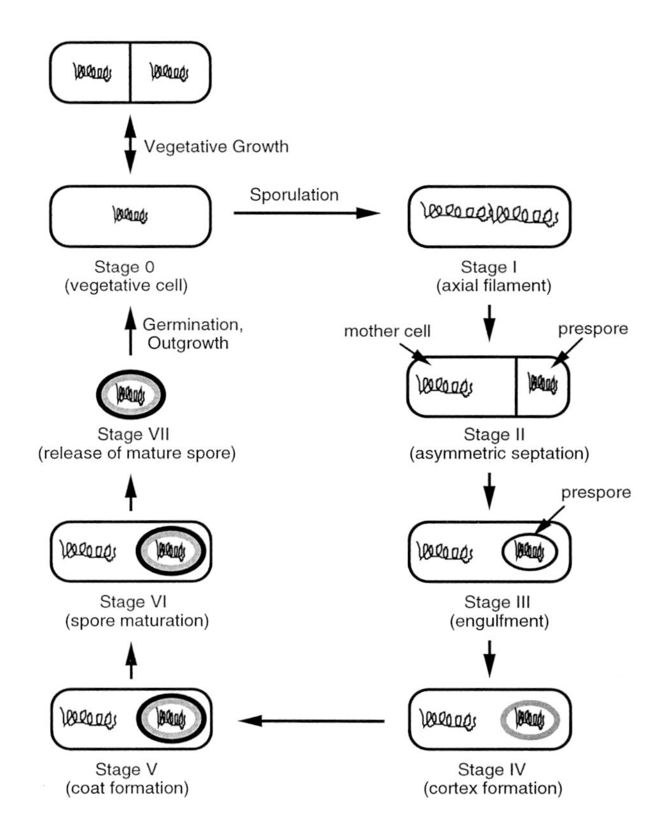

FIGURE 80.1 Schematic representation of the stages of sporulation. (I thank Jun Yu and Edward Amaya for help in the preparation of this and the other figures.)

convention introduced by Ryter and now generally used. Vegetative (growing) cells are rod shaped. They are defined as at stage O. Stage I is the formation of an axial filament of chromatin. Stage II defines the completion of a division septum at one pole of the cell. This division produces two cell types, the smaller prespore (sometimes called the forespore) and the larger mother cell. Stage III is defined as the completion of engulfment of the prespore by the mother cell. Stage IV is the deposition of two layers of cell-wall material, the cortex and the primordial germ-cell wall, between the opposed membranes that surround the engulfed prespore. Deposition of layers of coat material around the outside of the prespore defines stage V. By stage VI the prespore has matured into the heat-resistant spore, but the spore is still located within the mother cell. Lysis of the mother cell to release the mature spore is defined as stage VII.

It should be noted that development of the spore takes place inside the mother cell from stages III to VII. For this reason, the process is often called *endo*spore formation. Bacteria of some less-studied genera, for example, *Thermoactinomyces* and *Sporosarcina*, also form endospores by a process very similar to that described above, although they do not have rod-shaped vegetative cells: *Thermoactinomyces* has vegetative mycelium and *Sporosarcina* cocci; interestingly the sporulation division for *Sporosarcina* appears to be symmetrically located. Sporulation within other genera such as *Streptomyces* and *Myxococcus* does not involve endospores and is also in other ways significantly different from the process described above; it is not discussed here.

A. Spo$^-$ mutants and *spo* loci

Sporulation is not an obligatory part of the life cycle of endospore formers. Consistent with this, many different mutants have been described that cannot sporulate but can grow perfectly well vegetatively. Such Spo$^-$ mutants have been most extensively analyzed for *B. subtilis*. Operationally, Spo$^-$ mutants which have been tentatively identified because of the poor pigmentation of the colonies, can be positively identified by phase-contrast microscopy through their failure to produce phase-bright spores. They can also be recognized by their sensitivity to heat—typically spores survive 20 min at 80°C, but Spo$^-$ mutants and vegetative cells of Spo$^+$ strains are killed by this treatment. (The heat resistance of spores is species dependent. In general, species that grow at higher temperatures give spores that resist higher temperatures.) Analysis by electron microscopy of thin sections has shown that, in general, particular mutants are blocked at a recognizable stage of sporulation and are designated accordingly. For example: SpoII mutants are blocked at stage II—the sporulation septum is formed, but engulfment is not completed; SpoIII mutants are blocked at stage III—the prespore is completely engulfed by the mother cell, but none of the protective layers, such as cortex and coat, have yet been formed.

The mutations that give rise to the Spo$^-$ mutant phenotype map, by definition, in *spo* loci. More than 40 *spo* loci have been identified in *B. subtilis*. These loci were defined initially by groups of *spo* mutations that mapped close together. In almost all cases they have now been located on the completed sequence of the *B. subtilis* genome, and have been shown to correspond to individual genes or to operons, which are groups of adjacent genes expressed as a unit (Table 80.1).

II. INITIATION OF SPORULATION

Sporulation is favored by high cell density. It can be initiated by depletion of the source of carbon, of nitrogen or, in some conditions, of phosphorous. It can

TABLE 80.1 Guide to selected sporulation loci discussed in text

Locus	Encoded function
cotE	Scaffold protein for spore coat assembly
cotG	Tyrosine-rich coat protein
gerA	Germination response to L-alanine
gerB	Germination response to GFAK
kinA	Kinase that phosphorylates SpoOF
kinB	Membrane-associated kinase that phosphorylates SpoOF
phrA	Peptide inhibitor of RapA
rapA	SpoOF-PO_4 phosphatase
rapB	SpoOF-PO_4 phosphatase
sigK	Pro-σ^K
spoOA	Response regulator; pivotal transcription regulator
spoOB	Phosphotransferase
spoOE	SpoOA-PO_4 phosphatase
spoOF	Response regulator
spoIIA	Operon encoding σ^F, SpoIIAB (anti-sigma factor) and SpoIIAA (anti-anti-sigma factor that is inactivated upon phosphorylation by SpoIIAB)
spoIIE	Membrane protein; SpoIIAA-PO4 phosphatase
spoIIG	Pro-σ^E and SpoIIGA (likely processing enzyme)
spoIIR	E-σ^F-transcribed gene required for pro-σ^E processing
spoIIIE	DNA translocase
spoIIIG	σ^G
spoIVA	Scaffold protein for coat assembly
spoIVB	E-σ^G-transcribed gene required for pro-σ^K processing
spoVA	DPA transport to prespore
spoVF	DPA synthesis
spoVID	Scaffold protein for coat assembly

only be initiated in cells that are actively replicating their genome, and only during part of the vegetative cell-division cycle. It is known that a fall in concentration of the nucleotides GTP and GDP always occurs at the start of sporulation. Further, artificial depletion of GTP and GDP can trigger sporulation in some media that ordinarily would not support sporulation. These nucleotides, or something derived metabolically from them, are plausible candidates as mediators of the starvation signal. However, it is not known how a fall in the concentration of these nucleotides leads to the complex pattern of gene expression that is triggered after the start of sporulation and is necessary for spore formation.

The critical early event that sets in motion this complex pattern of gene expression is the activation of the transcription regulator SpoOA by phosphorylation. Phosphorylation of the SpoOA protein is achieved through a series of reactions collectively referred to as the phosphorelay (Fig. 80.2). In the phosphorelay a protein kinase, KinA or KinB, phosphorylates the SpoOF protein. The phosphate group is then transferred to SpoOA through the action of the SpoOB phosphotransferase.

Countering this sequence that leads to SpoOA phosphorylation are several phosphatases: SpoOE specifically dephosphorylates SpoOA-PO_4; RapA and RapB specifically dephosphorylate SpoOF-PO_4. The balance of all these reactions determines if SpoOA is phosphorylated sufficiently to activate a series of genes that must be expressed for successful sporulation.

The phosphorelay is the conduit for the external and internal signals that trigger spore formation. Thus, the target of the cell-density signal appears to be the RapA phosphatase. This phosphatase is inhibited by a pentapeptide derived from processing of a small protein called PhrA. The PhrA protein is secreted, probably as a 19-residue peptide. It is processed and then reimported by an oligopeptide permease. At high cell density it is thought that the concentration of the 5-residue peptide derived from PhrA is sufficient to inhibit the RapA phosphatase, thus increasing the concentration of SpoOF-PO_4 and fueling the phosphorylation of SpoOA; strains deleted for the *phrA* gene or for the oligopeptide permease genes sporulate very poorly. Cell-cycle, DNA-replication and tricarboxylic-acid-cycle signals also feed through the phosphorelay, as do nutritional-starvation signals; the molecular details of these various signal mechanisms have not been worked out.

SpoOA-PO_4 activates or represses expression of a large number of genetic loci. Crucial to subsequent events during sporulation, it activates transcription into mRNA, and hence translation into protein, of the *spoIIA, spoIIE,* and *spoIIG* loci, which will be discussed below; and it is necessary for the asymmetrically located sporulation division.

III. THE SPORULATION DIVISION

The asymmetrically located sporulation division is often considered the defining early morphological event in sporulation of *B. subtilis*. The machinery for division is similar to that for vegetative division—for example, both processes require the tubulin-like FtsZ protein. However, there are several clear differences between the two types of division: (i) The sporulation division septum is much thinner than the vegetative division septum. (ii) The two cells that result from the sporulation division do not separate from each other as occurs following vegetative division. Rather, the mother cell engulfs the prespore. (iii) Autolysis of the wall material (peptidoglycan) within the sporulation septum begins in the center of the septum and ultimately there is apparently complete loss of wall material. In contrast, autolysis of the wall material of the vegetative septum begins at the periphery of the septum and proceeds inwards. Moreover, there is

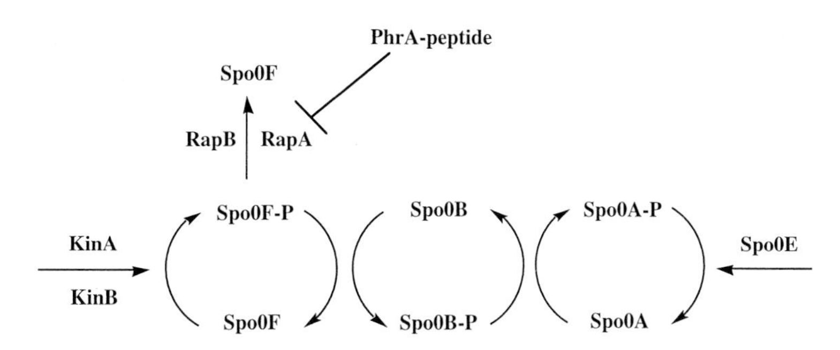

FIGURE 80.2 Representation of the reactions of the phosphorelay. (Adapted from Perego and Hoch, 1996, and Piggot, 1996). Two kinases, KinA and KinB, are activated by unknown signals to phosphorylate SpoOF. In the figure, phosphorylated forms of proteins are indicated by -P, for example, SpoOF-P. The phosphate group is transferred from SpoOF to the key transcription regulator SpoOA by the phosphotransferase SpoOB. SpoOA is activated by phosphorylation. The protein-phosphatase SpoOE specifically dephosphorylates SpoOA-P. RapA and RapB are protein phosphatases that specifically dephosphorylate SpoOF-P; their action also reduces the phosphorylation of SpoOA because of the action of the phosphorelay. A pentapeptide derived from PhrA inhibits the action of RapA. It is likely that there are additional regulators of the phosphatases.

little loss of wall material—the split septum provides the wall for the poles of the nascent cells. (iv) The sporulation septum is formed before a chromosome is completely packaged into the prespore. The prespore initially contains only the origin-proximal one-third of a chromosome. A DNA translocase, SpoIIIE, is located in the center of the septum and is required for transfer of a complete chromosome into the prespore. (v) The septum is asymmetrically located, with respect to the cell poles, during sporulation but not vegetative growth. The *spoIIE* locus mediates this SpoOA-directed switch in septum location. (vi) Gene expression becomes compartmentalized after the sporulation division, with some genes expressed in the prespore and other genes expressed in the mother cell.

These comments about the structure of the sporulation septum apply to all species of endospore former that have been studied, with the exception that the sporulation septum is symmetrically located in *S. ureae*. Comparatively little is known about sporulation genes in species other than *B. subtilis*. However, homologs of *spoOA* were identified in all of a wide range of endospore-forming species that were analyzed. The core components of the regulatory systems described in succeeding sections are conserved in all the species whose sequence has been determined: *B. anthracis*, *B. halodurans*, *B. stearothemophilus*, *Clostridium acetobutylicum* and *C. difficile*.

IV. COMPARTMENTALIZATION OF GENE EXPRESSION, AND DIFFERENT σ FACTORS

As soon as the spore septum is completed, different genes start to be expressed in the prespore and in the mother cell. This compartmentalized gene expression is a consequence of different sporulation-specific RNA polymerase σ factors becoming active, σ^F in the prespore and σ^E in the mother cell. The σ factors direct the RNA polymerase to transcribe particular genes into mRNA by recognizing DNA sequences called promoters. Different σ factors direct recognition of different promoters, and so σ^F directs transcription of a different set of genes from σ^E. These two σ factors are synthesized before the spore septum is formed, but are not active. Why they only become active upon septation and why their activities are compartmentalized are questions that have not been fully answered. These questions are presently exciting a lot of research activity. Because σ^F needs to be active in order for σ^E to be activated, and not vice versa, activation of σ^F is considered first.

The *spoIIA* locus is a three-gene operon. Its transcription is activated by SpoOA-PO_4 soon after the start of sporulation. σ^F is encoded by the third gene of the *spoIIA* operon, *spoIIAC*. The second gene, *spoIIAB*, encodes a protein, SpoIIAB, that binds to σ^F and inhibits its action. The SpoIIAB protein is thus an anti-sigma factor. The first gene of the operon, *spoIIAA*, encodes an anti-anti-sigma factor that binds to SpoIIAB and releases σ^F from σ^F–SpoIIAB complexes (Fig. 80.3). SpoIIAB is also a protein kinase that can specifically phosphorylate SpoIIAA; SpoIIAA-PO_4 is not able to bind to SpoIIAB. To complete the list of known participants in this system, the SpoIIE protein is a phosphatase that specifically dephosphorylates SpoIIAA-PO_4. The balance of these various reactions determines if σ^F is active or not.

The critical step in activation of σ^F is thought to be the activation of SpoIIE phosphatase activity. In the sporulating cell before septation, this phosphatase is not active, SpoIIAA is completely phosphorylated and SpoIIAB is bound to σ^F, thus blocking σ^F activity.

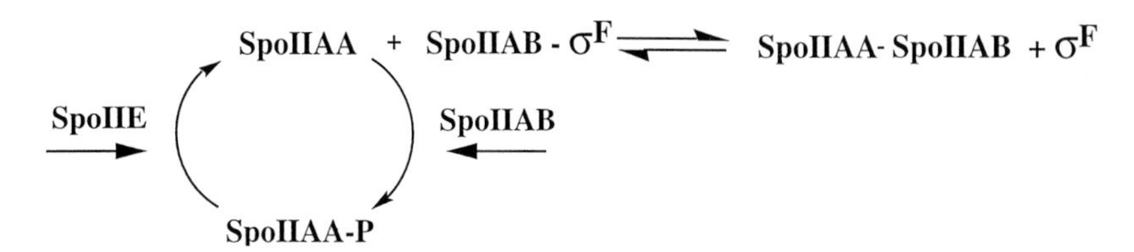

FIGURE 80.3 Outline of the activation of σ^F during sporulation (modified from Piggot, 1996). SpoIIAB is an anti-sigma factor that binds noncovalently to σ^F and inactivates it. SpoIIAA can bind noncovalently to SpoIIAB, thereby releasing and activating σ^F. SpoIIAB also functions as a kinase to phosphorylate SpoIIAA; SpoIIAA-P cannot bind to SpoIIAB and reverse the inhibition of σ^F activity. SpoIIE is a septum-located phosphatase that specifically dephosphorylates SpoIIAA-P, thus activating SpoIIAA and hence σ^F. Activation of the SpoIIE phosphatase is thought to be the critical step leading to σ^F activation during sporulation.

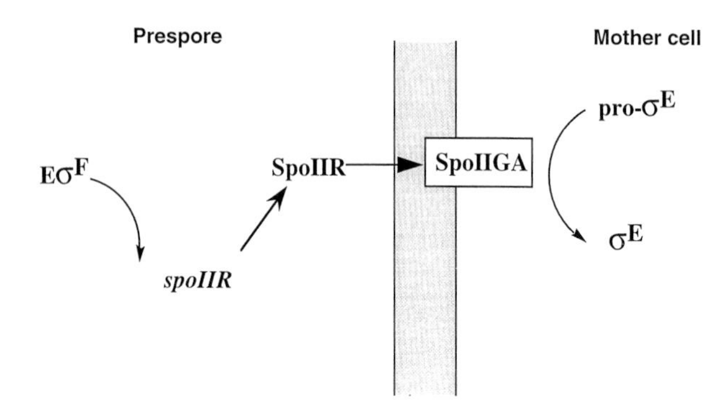

FIGURE 80.4 Outline of the activation of σ^E during sporulation (modified from Piggot, 1996). Formation of SpoIIR in the prespore causes activation in the mother cell of pro-σ^E by its proteolysis to σ^E. This action is mediated by the septally located SpoIIGA, which is thought to be the protease that catalyzes the reaction.

Something about septation activates the SpoIIE phosphatase; its activity releases unphosphorylated SpoIIAA which binds SpoIIAB, thus freeing and hence activating σ^F. It is known that SpoIIE is a membrane protein that is found predominantly in the spore septum, but what activates SpoIIE is not known. σ^F becomes active on the prespore side of the sporulation septum, but not the mother cell side of the septum. It is not certain how its activity is compartmentalized.

The mother cell-specific σ^E is activated in a different way from σ^F. σ^E is encoded by the second gene of the *spoIIG* operon as an inactive precursor, pro-σ^E. An N-terminal pro sequence of 27 residues must be proteolytically cleaved for σ^E to become active. The likely processing enzyme is SpoIIGA, which is encoded by the first gene of the *spoIIG* operon. SpoIIGA is a membrane protein that is located in the spore septum. Activation of SpoIIGA, and hence σ^E, requires the activity of the SpoIIR protein (Fig. 80.4). The SpoIIR protein is produced exclusively in the prespore from a gene that is exclusively under the control of σ^F. Thus, activation of σ^F in the prespore leads very rapidly to activation of σ^E in the mother cell. These two σ factors initiate the programs of compartmentalized gene expression, but neither directs the process that leads to compartmentalization. The activation of σ^F and σ^E is required for progression to the next stage of sporulation, engulfment. Mutations in the structural genes for σ^F and σ^E and in the genes for their activation all block sporulation at stage II, septation.

V. ENGULFMENT

The process of engulfment starts soon after the sporulation division septum is formed. Autolysis of septal wall material starts from the center of the septum and proceeds to its periphery, resulting in removal of essentially all wall material. At this time the septum is very flexible. The middle of the septum appears, in electron micrographs, to bulge into the mother cell. The annulus of attachment of the septum to the cylindrical wall of the organism moves to the prespore pole and then fuses to itself. The consequence of this fusion is that the prespore is completely surrounded by membranes of opposed polarity, and is further shielded from the external medium because it is completely surrounded by the mother cell. Completion of engulfment is defined as stage III. By this stage, the organism is "committed" to form spores: addition of rich medium does not reverse the sporulation process. This contrasts with the organism at the start of stage II, where both mother cell and prespore can revert to vegetative growth upon addition of rich medium.

Sporulation mutants have been found that are blocked at different stages of the engulfment process. Some of the corresponding genes are under σ^F control, others under σ^E control. Thus, both prespore and mother cell contribute to engulfment. Still other genes are under the control of the main vegetative σ factor, σ^A. Although a number of such engulfment genes have been identified, the biochemical mechanisms involved in engulfment are not well understood.

VI. POST-ENGULFMENT DEVELOPMENT

By stage III, the mother cell and the prespore are very different from each other. Notably, the prespore is now completely surrounded by the mother cell and so is not in contact with the medium. The prespore has not, however, acquired any of the resistance or dormancy properties that characterize the mature spore. These are developed in the succeeding stages of sporulation. Completion of engulfment activates two additional σ factors, σ^G in the prespore and σ^K in the mother cell. Activation of these σ factors is required for the subsequent stages of sporulation, including the development of spore resistances. Transcription of the structural gene for σ^G, *spoIIIG*, is directed by σ^F and consequently is confined to the prespore. It also requires expression of an unidentified mother cell-specific gene (or genes). This latter control is one of several controls found throughout post-septation development, where interaction between mother cell and prespore is required for further development. σ^G is initially inactive, and is only activated upon the completion of engulfment. Its activation also requires expression of mother cell-specific genes.

The molecular mechanisms of σ^G activation have not been worked out. σ^G and σ^F have overlapping promoter specificity, and once activated, σ^G directs transcription of *spoIIIG* and hence its own synthesis. A number of prespore genes transcribed before engulfment by E-σ^F continue to be transcribed after engulfment by E-σ^G, because of this overlapping specificity. E-σ^F activity ceases as E-σ^G becomes active. Several new genes become active after engulfment whose promoters are recognized by σ^G but not by σ^F. Of particular importance in this last category are the *ssp* genes which encode small acid-soluble proteins (SASPs) that are critical to ultraviolet resistance (see below).

Transcription of the structural gene for σ^K, *sigK*, is directed by σ^E, and consequently is confined to the mother cell. σ^K is initially formed as an inactive precursor, pro-σ^K. Activation of pro-σ^K, by proteolytic removal of the N-terminal pro sequence (which shows no similarity to the pro sequence for σ^E), requires the product of a particular σ^G-directed gene, *spoIVB*. Thus σ^K activation in the mother cell is tied to prior events in the prespore and, indirectly, requires completion of engulfment. In an interesting parallel to events in the prespore, σ^K and σ^E have overlapping specificity, and when activated, σ^K directs transcription of *sigK* and hence its own synthesis. Several other mother cell-specific genes are also transcribed both before and after engulfment, because of this overlapping specificity. E-σ^E activity ceases as E-σ^K becomes active. A number of genes whose promoters are recognized exclusively by σ^K are transcribed only after engulfment; notable among the latter are several *cot* genes which encode coat proteins (see below).

In some strains of *B. subtilis* an additional, very striking control is exercised on *sigK*. In the genome of the germ line (i.e. vegetative cells, prespores and spores) the 5′ and 3′ portions of the *sigK* gene are separated by a 48-kbp DNA sequence known as SKIN. There is a site-specific recombination specifically in the mother cell (directed by σ^E) that excises SKIN and forms an intact copy of *sigK*. In other strains of *B. subtilis* and in other species of *Bacillus*, *sigK* is intact in the germ line and there is no need for this site-specific recombination.

VII. DEVELOPMENT OF SPORE RESISTANCES

Spore resistance to heat, to irradiation and to chemicals, such as chloroform and octanol, develops in the succeeding stages of sporulation. Several spore constituents formed during this period are thought to contribute, to varying degrees, to these resistances. Formation of two of the constituents, the cortex and the coat, are associated with stages IV and V, respectively. The possible roles of these structures are discussed below. The roles of two other major spore constituents are then considered: SASPs; and dipicolinic acid (DPA). During this period the spore core (i.e. the spore cytoplasm) decreases in volume, so that the final volume of the spore core is about half that of the prespore cytoplasm immediately following engulfment.

Two types of peptidoglycan, the cortex and the primordial germ-cell wall are assembled around the prespore following the completion of engulfment. The cortex has a very different structure from other peptidoglycans and is only found in spores and sporulating organisms. It is degraded upon germination. The cortex shares with other peptidoglycans a glycan chain of alternating *N*-acetylglucosamine and muramic acid residues. It is unique in that approximately 50% of the muramic residues are in the form of a lactam and have no attached peptide side chain. The other muramic residues have attached to them either L-alanine or a peptide side chain. There are strikingly few peptide crosslinks in the cortex peptidoglycan. These differences are thought to make the cortex a more flexible structure than vegetative cell-wall peptidoglycan. It was long thought that this distinctive structure might be critical to the dehydration that is central to spore heat resistance. However, recent results indicate that normal spore dehydration can be obtained in mutants containing no muramic acid lactam, and in mutants with a fourfold increase in crosslinking. The results suggest that these aspects of cortex structure are not important for the process of dehydration, but may still have a role in the maintenance of dehydration that has been achieved by other means. There remains a correlation between the extent of dehydration and the heat resistance of spores. Mutants with little or no muramic acid lactam are dramatically impaired in germination; it seems likely that the muramic acid lactam is a major specificity determinant of germination-lytic enzymes. The primordial germ-cell wall (PGCW), as its name implies, becomes the wall of the cell that is formed upon germination of the spore; this contrasts with the cortex which is degraded during germination. PGCW structure, so far as it has been determined, is that of the vegetative cell wall. Both cortex and PGCW are assembled between the opposed membranes that surround the engulfed prespore. The cortex is the outer of the two peptidoglycans. Where they have been distinguished, enzymes specific to the cortex are found to be synthesized in the mother cell and enzymes specific to the PGCW in the prespore.

The spore coat appears in electron micrographs as electron-dense layers that surround the prespore. There is an inner coat that is lamellar and lighter staining and an outer coat that is darker staining. The coat is composed of a number of proteins which together make up 50% or more of the total protein of the mature spore. Coat proteins tend to be insoluble and difficult to work with biochemically. However, for *B. subtilis* some 20 *cot* genes that encode coat proteins have been identified. All the known *cot* genes are transcribed only in the mother cell, most of them exclusively by E-σ^K, although a few have promoters recognized by σ^E. Assembly of coat layers in the mother cell only loosely follows the pattern of transcription of *cot* genes. The layers are assembled on a scaffold that includes the SpoIVA, SpoVID and CotE proteins. Each of these proteins has been shown to localize to the outer surface of the prespore. Of the three, SpoIVA is required first. *spoIVA* mutants assemble coats in whorls within the cytoplasm of the mother cell, and there are no coat layers around the prespore; in such mutants, SpoVID and CotE do not assemble around the prespore. *spoVID* and *cotE* mutants are blocked at later stages of the coat assembly process. The genes for the scaffold proteins, *spoIVA, spoVID* and *cotE*, are all transcribed by RNA polymerase containing the earlier-expressed σ^E.

The coat layers contain several types of covalent crosslinkage between protein subunits. These include disulfide bonds, (ϵ–γ) glutamyl–lysyl isopeptide bonds, which are found in keratins, and *o,o*-dityrosyl linkages, which are found in insect cuticle. The structures of the individual coat proteins are varied. For example, the CotG protein is tyrosine rich. There is a superoxide dismutase located in the coat, and, in combination with a peroxidase, it may help assemble CotG into the insoluble coat matrix through the formation of the dityrosyl protein–protein bridges. Other coat proteins resemble β-keratin. The effects of mutation in the various *cot* genes vary considerably. The overall picture that emerges is that the coat layers contribute to the resistance of spores to organic solvents, such as chloroform and octanol, and to the action of lysozyme; also that the coat contributes to the germination properties of spores.

SASPs constitute up to 20% of the protein of the dormant spore. They are degraded very rapidly on germination, and the amino acids released contribute significantly to the outgrowth of the germinated spore. There are two main types of SASP, of which one, known as the α/β type, is important to spore resistance to UV irradiation. This resistance is achieved because the α/β-type SASPs bind to DNA and change the properties of the DNA. They convert the DNA from its normal B conformation to the A conformation. UV irradiation of DNA of spores or of DNA solutions containing α/β-type SASP produces a product known as spore photoproduct. This spore photoproduct is very efficiently repaired upon germination. Consequently, the spores are more resistant to UV irradiation than are vegetative cells in which pyrimidine dimers (the typical product of UV irradiation of almost all living cells) are formed and are less efficiently repaired. The repair of the spore photoproduct during germination is critical to the UV resistance. The α/β-type SASPs have also been shown to confer resistance to dry heat. These last experiments indicate that the major cause of the lethality of dry heat to spores is damage to DNA. The mechanism of resistance of spores to dry heat (e.g. resistance of a freeze-dried spore preparation) is thus different from that for wet heat (e.g. resistance of an aqueous suspension of spores) where SASPs have a relatively minor role. (Papers discussing spore heat resistance usually mean resistance to wet heat unless they specifically mention dry heat; this usage is retained here.)

Dipicolinic acid (pyridine-2,6-dicarboxylic acid; DPA) is another major constituent of the spore core, accounting for as much as 10% of dry weight. Most DPA in the spore is thought to be complexed with divalent cations, predominantly Ca^{2+}. Its role in spore dormancy and possibly resistance (see below) may reflect its association with these cations which are present in a correspondingly high concentration. DPA is synthesized from dihydrodipicolinate, an intermediate in lysine biosynthesis. Its synthesis is catalyzed by the product of the *spoVF* locus. Although DPA accumulates in the spore core, transcription of *spoVF* is directed by σ^K in the mother cell. The uptake and/or maintenance of DPA in the prespore depends, at least in part, on the σ^G-directed *spoVA* locus. DPA enhances significantly the SASP-dependent formation of spore photoproduct upon UV irradiation, and so sensitizes spores to UV irradiation. DPA is thought to be required for the maintenance of spore dormancy. The role of DPA in spore (wet) heat resistance has been the subject of controversy. The controversy has centered on spores of DPA^- mutants: some groups have reported that the spores have normal heat resistance, whereas other groups studying different mutants (including *spoVF* mutants of *B. subtilis*) have found the spores to be heat sensitive. It is clear that production of DPA is not sufficient to confer heat resistance. It is also clear that spores can have a greatly reduced level of DPA and still be heat resistant. However, there is a low spontaneous level of formation of DPA so that *spoVF* mutants (or other "DPA^-" mutants) are not completely free of DPA. This reviewer leans to the view that a low level of DPA may be required for full heat resistance. In support of this interpretation are experiments with the *spoVF* mutants of *B. subtilis* in which exogenous addition of a low amount of DPA significantly increased the heat resistance of populations of spores formed by the mutants.

It is likely that resistance to wet heat is multifactorial. As discussed above, cortex, SASP and DPA are factors that have been shown to have or are suspected of having a role. None has been shown to be the sole determinant of heat resistance. The one strong correlation with heat resistance that has not been shaken over the years is of extensive dehydration of the spore core. This dehydration suggests that a high turgor pressure must somehow be maintained by the structures of the spore. The dehydration changes the density of the developing spore, and mature spores have an unusually high density for living cells of about 1.3 g/ml. The increase in density changes the optical property of the spore, which becomes bright when viewed by phase-contrast microscopy, as opposed to the dark vegetative cell and mother cell. Operationally, phase-bright spores provide a very easy way to identify spores. For the sporulation aficionado, phase-bright spores are a joy to behold—except when a Spo^- mutant is expected!

VIII. GERMINATION

Not only are mature spores highly resistant, but they are dormant. Within the limitations of the techniques used, no metabolism has been detected in dormant spores. Spores have been reported to survive for decades, centuries and even millennia! Nevertheless, when a germinant is provided, spores germinate and return to vegetative growth. Usually, this process is divided into two stages: first, germination, which is rapid (a few minutes) and results in the loss of resistances and the resumption of active metabolism; second, outgrowth, which is slower (about 2 h) and involves the resumption of macromolecule synthesis and the development into normal, vegetatively growing bacilli.

Most known germinants are nutrients, and a wide range have been described. They include amino acids, nucleosides, and sugars. Often they are species-specific. L-alanine probably works on the widest range of species. For *B. subtilis*, a number of *ger* loci have been defined by mutations that impair germination. In some cases the mutants are impaired in the response to particular germinants. For example, *gerA* mutants are defective in their response to L-alanine, and *gerB* mutants are defective in their response to glucose, fructose, L-asparagine and KCl (which together function as a single germinant, GFAK). The two loci are transcribed in the prespore by E-σ^G. The encoded proteins are thought to provide membrane-located receptors for their respective germinants; the mechanism of signal transmission is not known. Other *ger* loci encode proteins that involved in the response to both L-alanine and GFAK; germination is thought to be a multistep process.

The events of germination happen very rapidly, and it is difficult to fit a precise chronology to them. This task is complicated by the heterogeneity of the response within the spore population. The heterogeneity can be reduced, but not eliminated, by first "activating" spores (a pre-germination treatment) with a treatment such as mild heat. The earliest germination events include the release of DPA into the medium and the loss of heat resistance. Hydrolysis of the cortex is rapidly initiated by germination-specific lytic enzymes. All this occurs in the absence of macromolecular synthesis—thus it is not dependent on transcription. There is little ATP in the dormant spore, and energy is initially provided by phosphoglyceric acid; ATP synthesis is one of the earliest events that is initiated upon addition of germinant. The germinated spore is phase dark. During outgrowth, the germinated spore swells considerably; protein, RNA, and DNA synthesis is initiated, and the first vegetative division usually occurs about two hours after the initiation of germination, given that a growth medium is provided. The organism has now returned to vegetative growth.

BIBLIOGRAPHY

Henriques, A. O., and Moran, C. P., Jr. (2000). Structure and assembly of the bacterial endospore coat. Methods. **20**, 95–110.

Perego, M., and Hoch, J. A. (1996). Protein aspartate phosphatases control the output of two-component signal transduction systems. *Trends Genet.* **12**, 97–101.

Piggot, P. J. (1996). Spore development in *Bacillus subtilis. Curr. Opin. Genet. Dev.* **6**, 531–537.

Piggot, P. J., Moran, C. P., Jr., and Youngman, P. (eds) (1993). "Regulation of Bacterial Differentiation". American Society for Microbiology, Washington, DC.

Popham, D. L., Helin, J., Costello, C. E., and Setlow, P. (1996). Muramic lactam in peptidoglycan of *Bacillus subtilis* spores is required for spore outgrowth but not for spore dehydration or heat resistance. *Proc. Natl. Acad. Sci. USA* **93**, 15405–15410.

Sonenshein, A. L. (2000). Control of sporulation initiation in *Bacillus subtilis. Curr. Opin. Microbiol.* **3**, 561–566.

Sonenshein, A. L., Hoch, J. A., and Losick, R., (eds). (2002). "*Bacillus subtilis* and Its Closest Relatives: From Genes to Cells". ASM Press, Washington, DC.

Stragier, P., and Losick, R. (1996). Molecular genetics of sporulation in *Bacillus subtilis. Annu. Rev. Genet.* **30**, 297–341.

WEBSITES

SubtiList World-Wide Web Server
http://genolist.pasteur.fr/SubtiList/

The Non-redundant *Bacillus subtilis* data base
http://pbil.univ-lyon1.fr/nrsub/nrsub.html

Bacillus subtilis Japan Functional Analysis Network
http://bacillus.genome.ad.jp/

Bacillus subtilis Genetics at the University of London
http://web.rhul.ac.uk/Biological-Sciences/cutting/index.html

Micado (formerly Mad Base). A relational database on *B. subtilis* genetics
http://locus.jouy.inra.fr/cgi-bin/genmic/madbase_home.pl

Website on myxobacteria (Dept. of Microbiology, University of Minnesota)
http://www.microbiology.med.umn.edu/faculty/myxobacteria/index.html

Website on myxobacteria (Microbial World)
http://helios.bto.ed.ac.uk/bto/microbes/myxococc.htm#crest

81

Starvation, Bacterial

A. C. Matin
Stanford University

GLOSSARY

ancillary factors Proteins that influence promoter recognition by an RNA polymerase.

chaperones Proteins that are required for correct folding of newly synthesized proteins. They also prevent protein denaturation during stresses, and can renature damaged proteins.

chemostat An apparatus that makes it possible to grow bacteria under steady-state conditions at submaximal growth rates due to low concentrations of an essential nutrient.

promoters Sequences upstream of the transcriptional start site of a gene (usually −10 and −35 nucleotides upstream of the start site) recognized by individual species of RNA polymerase holoenzymes.

semistarvation Conditions under which bacteria grow at a rate less than their maximal potential due to the limitation of an essential nutrient. Existence under complete or semistarvation is the norm for bacteria in nature.

sigma factors Small proteins that combine with the RNA polymerase core enzyme. The resulting RNA polymerase holoenzyme can transcribe various genes. Each species of RNA polymerase holoenzyme recognizes specific promoter sequences.

starvation An environmental condition in which bacteria do not grow at all due to the lack of an essential nutrient.

starvation promoters Promoters that are selectively switched on during starvation or semistarvation conditions. Their sequences are recognized by starvation specific RNA polymerases.

starvation proteins Proteins whose levels either go up or that are uniquely synthesized in starving bacteria.

vegetative cells Nonstarved, actively growing bacterial cells.

virulence Cellular features that enable a bacterium to cause disease.

Starvation which is frequently experienced by bacteria, causes them to differentiate into forms that are much more resistant to killing. This results from the synthesis of special proteins, called the starvation proteins. These proteins strengthen the bacterial cell envelope, prevent damage to vital cell constituents, and enhance the cell's capacity to repair DNA and essential proteins. A secondary sigma factor, called σ^S, becomes stabilized in starving cells. Its concentration increases, leading to the formation of a new species of RNA polymerase. The latter recognizes the regulatory region of the starvation genes, increasing their transcription. Starved cells are probably also enhanced in their disease-causing ability.

I. IMPORTANCE OF STARVATION

Bacteria in nature mostly exist in a state of semi- or complete starvation because most natural environments are deficient in nutrients. The available food for bacteria in the oceans is only a fraction of a milligram

The Desk Encyclopedia of Microbiology
ISBN: 0-12-621361-5

per liter (6–10 μg/l in freshwater, 0.4 g/100 g of soil), and although the quantitative aspects are not known, it is likely that disease-causing bacteria also experience nutrient deprivation while colonizing their host. In nearly all of these environments, bacteria on average grow only at a fraction of the rate of which they are genetically capable—indeed, estimated growth rates in many natural environments can range from close to zero to a generation time of hundreds of days.

Within the constraints of their genetic endowment, bacterial characteristics can very enormously depending on their growth conditions. To understand what bacterial physiology is like under natural conditions, there is increasing interest in studying bacterial characteristics under partial or complete starvation conditions. An elucidation of these characteristics is a prerequisite for purposeful manipulation of bacteria toward beneficial ends. Indeed, insights gained from such studies promise to provide radically new approaches for environmental cleanup, as well as microbial containment. Completely starving bacteria are studied in the laboratory utilizing flask (batch) bacterial cultures that enter the stationary phase due to exhaustion of an essential nutrient. For studies on semistarving bacteria, the chemostat is the instrument of choice.

II. APPEARANCE OF STARVING BACTERIA

With respect to changes in shape produced by starvation, bacteria can be divided into two major groups, those exhibiting a very marked morphological differentiation and those showing only a minimal alteration. The former group is exemplified by species of the genus *Bacillus* which, on starvation, form structures called endospores (Fig. 81.1a) that appear markedly different from their actively growing counterparts. Another example is provided by the myxobacteria, which form fruiting bodies (Fig. 81.1b) that differ strikingly from their vegetative cells. In the second group, which encompasses a majority of bacteria, starvation-induced changes are confined to diminution of cell size, protoplast shrinkage with consequent periplasm enlargement, and nucleoid condensation (Fig. 81.1c). In this article, I consider only the second group. Because of lack of pronounced morphological changes in the bacteria, it was recognized only relatively recently that this group, too, undergoes a profound alteration in its gene expression pattern on starvation. Studies involving this group of bacteria have concentrated mainly on *Escherichia coli*, although many other bacteria have also been examined in this respect, namely, *Vibrio, Salmonella, Pseudomonas*, and others.

III. THE STARVATION PROTEINS

Use primarily of two-dimensional gel electrophoresis showed that when bacteria such as *E. coli* enter the stationary phase, they increase the synthesis of a number of proteins, many of which are unique to the starvation state. These proteins are called starvation proteins. They are synthesized mainly in the first few hours of starvation and fall into early, middle, and late temporal groups. Their synthesis is accompanied by a progressive increase in the resistance of the cells to a variety of stresses. Examples of the stresses to which the starved cell becomes more resistant include starvation itself, hostile temperature, oxidative or osmotic state of the environment, and deleterious chemicals such as chlorine. In other words, fully starved cells of *E. coli* (i.e. those starved for about 4 h) exhibit a marked general resistance.

Starvation is a composite stress. One of its effects is to diminish the cellular redox status, generating oxidative stress. The dearth of ATP and lowered protonmotive force undermine ion transport. Maintenance of pH homeostasis is thereby rendered more difficult, exposing the cell to acid or alkaline stress; and the inability to concentrate ions produces osmotic stress. Survival in the face of these assaults necessitates induction of chaperones (see below) to prevent cellular damage and of repair mechanisms capable of operating in energy-depleted cells. Survival during starvation thus requires mechanisms to resist along with starvation, also its constituent stresses; it is therefore logical that a starved cell exhibits enhanced resistance to multiple stresses.

IV. HOW STARVATION PROTEINS PROTECT

A. Metabolic amplification

The starvation proteins help the cell survive a dearth of nutrients in two broad ways. A number of these proteins are concerned with increasing the scavenging capacity of the cell for the missing nutrient. For example, under the scarcity of glycerol as carbon substrate, *E. coli* synthesizes a different pathway for its uptake and metabolism. The key enzyme of this pathway is glycerol kinase, which utilizes an ATP molecule for glycerol uptake and has a high affinity for this

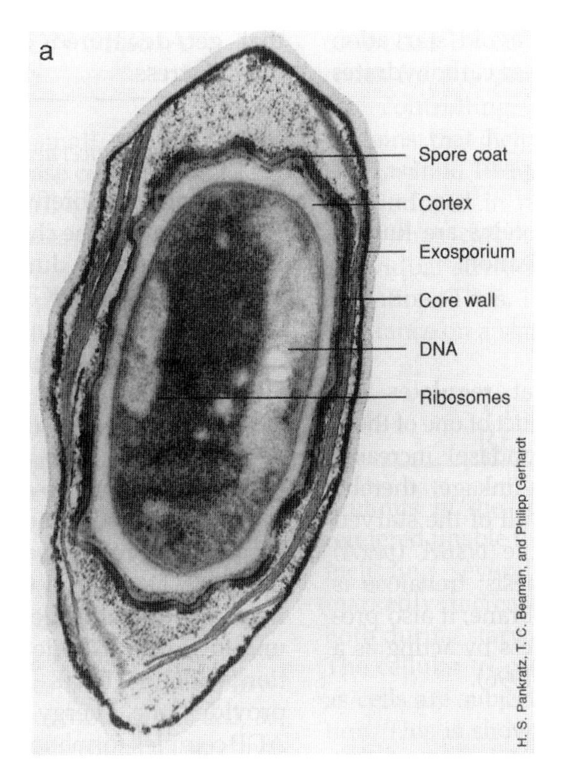

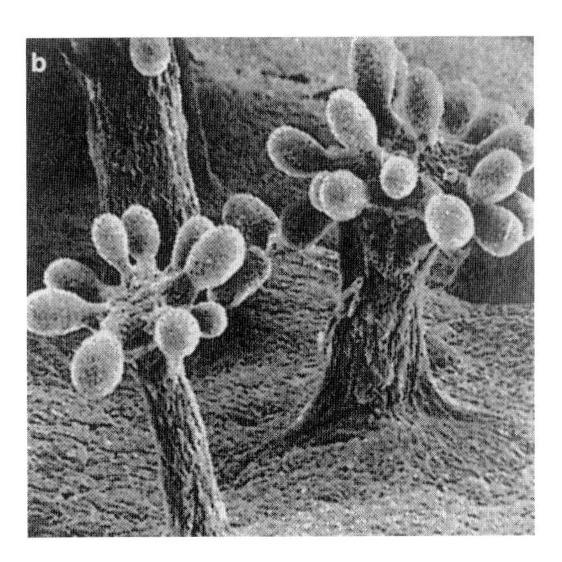

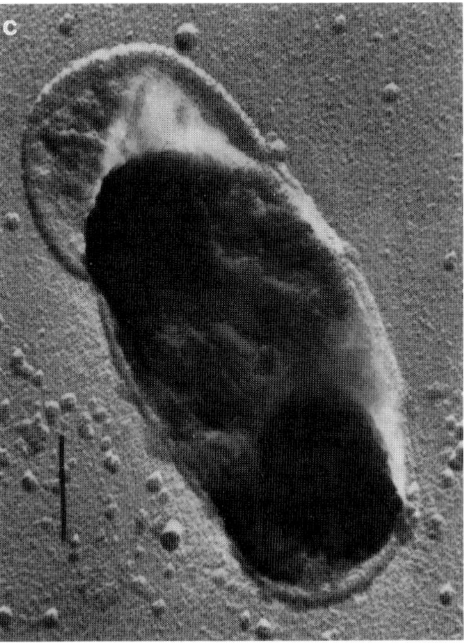

FIGURE 81.1 (a) A bacterial endospore. Reproduced with permission from Brock *et al.* (1997). Prentice-Hall, Upper Saddle River, NJ. (b) A bacterial fruiting body. Reproduced with permission from Brock *et al.* (1997). Prentice-Hall, Upper Saddle River, NJ. (c) A starved *E. coli* cell. Reproduced with permission from Reeve *et al.* (1984). *J. Bacteriol.* 160, 1041–1046. American Society for Microbiology, Washington, DC.

compound. Similarly, under phosphate starvation, a high affinity phosphate transport system is induced (PST), and when potassium is scarce the high affinity KDP system for K^+ uptake is utilized. Another strategy to escape starvation is synthesis of proteins that enable the cell to utilize substrates other than the one which became limiting. An example is the induction of the CstA protein in *E. coli* when glucose runs out. This protein appears to be concerned with the utilization of peptides, which are abundant in the gut,

on *rpoS* mRNA and σ^S protein levels and their halflives reveal that both the transcription of the *rpoS* gene and the translational efficiency of the *rpoS* mRNA decrease during starvation, resulting in up to an 80% decrease in σ^S synthesis rate (from ~55 in the exponential phase to 13 pmol/mg protein/min in the stationary phase). But a concurrent increase in σ^S stability more than offsets the decreased synthesis, resulting in the observed increased levels of the sigma factor. In starved cells this protein exhibits a 7- to 16-fold greater half-life (Fig. 81.2) (Matin *et al.*, 1999).

B. The biochemical basis of σ^S stability

The biochemical basis of the σ^S protein degradation is partly understood, but little information is available on its differential stability in exponential versus starved cells. σ^S is degraded by a protease called the ClpXP protease. The protease is made up of two different proteins, ClpP and ClpX. Neither alone has significant proteolytic activity; indeed ClpX by itself acts as a chaperone to rescue certain denatured proteins. Once combined with ClpP, however, ClpX acquires proteolytic activity against specific proteins. ClpP, incidentally, can also pair up with other chaperones, such ClpA, and ClpY to become a protease, but each combination targets different proteins (Gottesman, 1996). Thus, σ^S is degraded by ClpXP but not ClpAP protease, and many targets of the latter are not affected by the former. σ^S protein is also not affected by other proteases of *E. coli*, such as the Lon protease and very likely also the FtsH (HflB) protease. These conclusions are based on a study of *E. coli* mutants. Those devoid of either the *clpP* or *clpX* gene show a stable σ^S and increased levels of this sigma factor in all phases of growth, whereas *lon* or *clpA* mutants behave like the wild type in this respect.

The ClpXP protease levels do not change during transition to the stationary phase, nor does its activity toward another of its target proteins show any alteration during this transition. Thus, it is highly probable that σ^S protein becomes resistant to the protease in the stationary phase. Certain proteins have been found to associate with σ^S only in exponential phase cells, and it is possible that their association puts σ^S in a configuration that is vulnerable to the ClpXP-mediated proteolysis. The situation may be analogous to the postulated mechanism of σ^{32} instability in normal, unstressed cells. It is thought that the association of this sigma factor with certain chaperones (Dnak, DnaJ, and GrpE) in nonstressed cells is responsible for its sensitivity to proteases such as Lon, FtsH, and ClpQY. On heat stress, the chaperones dissociate from the sigma factor, possibly altering its configuration to a form that is resistant to proteolysis (Yura *et al.*, 1993). What proteins may be concerned with altering σ^S stability are not known; they probably do not include the key chaperones like GroEL, DnaJ, GrpE, or CbpA, as mutants devoid of these proteins show a wild-type stability pattern for σ^S. A lack of DnaK, however, is known to destabilize σ^S in the stationary phase.

The ClpXP protease appears to target amino acids 173–188 of the σ^S protein. If these amino acids are deleted from the σ^S protein, it becomes resistant to the protease. This region approximates the site at which σ^S interacts with the RNA polymerase core enzyme. This aspect may have a role in the different sensitivity of σ^S to proteolysis in different phases. Another possibility is that the 173–188 amino acid region of the sigma protein serves as the site recognized by the ClpXP protease. This protease appears to need two sites for recognizing its target protein, one of which (LDA/L) is probably common to all of its substrates, while the other is not; the 173–188 amino acid patch may be the second site needed to target σ^S.

C. Potential regulation at the translational level

The 173–188 amino acid patch of σ^S (the "turnover" site) is encoded by nucleotides 519–564 of the coding region of the *rpoS* gene. Computer-assisted analysis showed that this region is complementary to the translational apparatus (i.e. the Shine–Dalgarno sequence, the start codon, and the surrounding sequences) of *rpoS* mRNA. Bonding of this "antisense element" with the translational apparatus can cause the mRNA to loop on itself, blocking or diminishing translation. Indeed, it was first thought that the major mechanism of changes in σ^S levels during the exponential to stationary phase transition is increased translational efficiency of the *rpoS* mRNA in the stationary phase due to a relaxation in this phase of the mRNA secondary structure.

As the direct measurements of the *rpoS* mRNA translational efficiency during stationary phase transition showed decreased rather than increased efficiency (see above), this is obviously not the mechanism of increased σ^S levels during glucose starvation. But it remains an intriguing possibility that this coincidence—the antisense element of the *rpoS* mRNA encoding the turnover element of the RpoS (σ^S) protein—exists to permit regulation at both the translational and posttranslational levels. σ^S has diverse roles in cell physiology, requiring a strict control of its levels. In situations where, due to conflicting requirements, σ^S sensitivity to ClpXP protease cannot be enhanced, and yet σ^S levels need to be lowered,

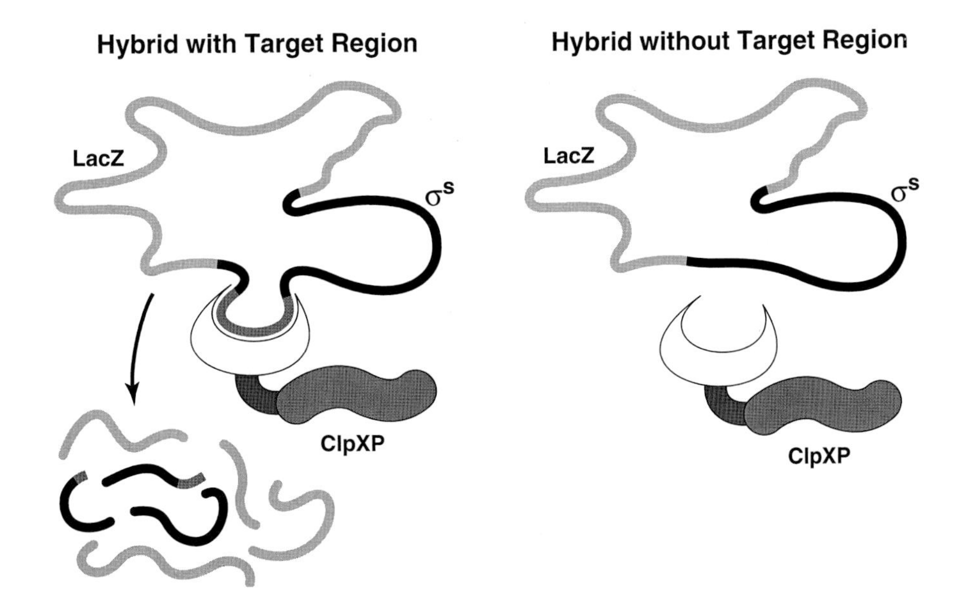

FIGURE 81.3 In fusion strains making the RpoS–LacZ hybrid protein containing the ClpXP target region (left), the hybrid is attacked by the ClpXP protease, accounting for low levels of β-galactosidase in that fusion. Fusion strains making the RpoS–LacZ hybrids without the target region (right) show high levels of β-galactosidase because the hybrids are not attacked by the protease. Reproduced with permission from Matin *et al.* (1999). Survival strategies in the stationary phase. *In* "Microbial Ecology and Infectious Disease" (E. Rosenberg, ed.). American Society for Microbiology, Washington, DC.

decreased translational efficiency through the use of the antisense element could indeed be involved. The proteins H-NS and HF-1 may have a role in influencing rpoS mRNA translation. H-NS is a histonelike protein whose synthesis increases in the stationary phase, whereas HF-1 apparently has the capacity to relax helical regions around the mRNA ribosomal binding sites (Matin *et al.*, 1999; Hengge-Aronis, 1996). Similarly, the untranslated RNAs *dsrA* and *oxyS* have been postulated to influence *rpoS* mRNA translation. The stringent response regulator ppGpp also regulates σ^S synthesis, possibly at the transcriptional level.

That translational efficiency changes are responsible for the increased σ^S levels in starving cells was inferred from the behavior of protein fusions of *rpoS* to the *lacZ* gene. Unless a translational fusion contained a certain minimum coding region of the *rpoS* gene, it produced high levels of β-galactosidase in all phases of growth. Only the fusion containing the nucleotides comprising the antisense element (519–564 nucleotides) mimicked the induction pattern of σ^S during the exponential to stationary phase transition of the wild type bacteria; fusions devoid of this element showed high levels of β-galactosidase in all growth phases. The initial interpretation of these results was that the presence of the antisense element prevented efficient translation in the exponential phase. However, the findings that the antisense element codes for the turnover element of σ^S and that the translational efficiency actually decreases in the stationary phase have led to the conclusion that the low levels of β-galactosidase in the fusions with the antisense element are due to posttranslational regulation. The hybrid RpoS–LacZ proteins formed by these fusions would contain the σ^S turnover region, leading to its degradation by the ClpXP protease. The hybrid protein generated in the fusions devoid of the antisense element would, on the other hand, not be targeted by the protease, and such fusions show high β-galactosidase levels in all phases (Fig. 81.3) (Matin *et al.*, 1999).

VII. THE σ^S-DEPENDENT PROMOTERS

RNA polymerase holoenzymes exercise their transcriptional selectivity by recognizing different promoter sequences. The −10 and −35 promoter sequences recognized by $E\sigma^{70}$, for example, are quite different from those recognized by $E\sigma^{32}$. Given that $E\sigma^S$ is involved in the expression of genes with a role in starvation and stresses, it was expected that the promoter sequence recognized by $E\sigma^S$ would be unique. It turns out, however, that the genes whose expression is dependent on $E\sigma^S$ possess a consensus −10 sequence that is very similar to that recognized by

Eσ^{70} (CTATACT versus TATAAT); the latter holoenzyme is the major agent of gene transcription in rapidly growing cells. In agreement with this finding, it has been shown by *in vitro* experiments that several promoters can be recognized by both Eσ^S as well as Eσ^{70}.

What then determines the ability of Eσ^S to affect differential gene transcription? Only tentative answers can be given to this question at this time. For one thing, there are some differences in the promoter region recognized by Eσ^S and Eσ^{70}. Thus, sequences upstream of the −17 position appear crucial for recognition by the Eσ^S holoenzyme. In addition, at approximately the −35 region, the Eσ^S-recognized promoters possess an AT-rich region rather than the Eσ^{70} consensus −35 sequence. Such a sequence produces curved DNA, and it appears that such a configuration facilitates promoter recognition by Eσ^S. Nor can it be discounted on the present evidence that the −35 sequence of the Eσ^S-recognized promoters is wholly unimportant. Conserved cytosines followed by a −33 guanine appear to be essential for recognition by Eσ^S. More important, however, in determining whether a gene is transcribed by Eσ^S may be factors other than the promoter sequence. These include the ionic composition of the cell (which is likely to be significantly influenced by different stresses), the presence of ancillary factors, stress-induced changes in a core component of RNA polymerase, and the proportion at a given time of the Eσ^{70} and Eσ^S holoenzymes.

The diverse stresses against which Eσ^S-transcribed genes afford protection are probably an ever present and recurrent threat in nature. But while the product of these genes is protective, it is probably inimical to growth. Rapid growth, when circumstances are conducive, is as important to survival as the ability to withstand stresses. Hence there is the need to constantly, and at short notice, shift between expression of Eσ^{70}- and Eσ^S-transcribed genes. This is probably why the two proteins, σ^S and σ^{70}, chemically resemble each other, and why the promoters their holoenzymes recognize differ not so much in their sequence, but rather in such factors as changes in cellular ionic, ancillary factor, and RNA polymerase core component composition.

VIII. STARVATION AND VIRULENCE

It is likely, though not proven, that one of the stresses encountered in the host by an invading bacterium is starvation. Inside the host, bacteria generally grow at a slower rate than their genetic potential permits, and they express many starvation genes. Both of these effects could be due to stresses other than starvation. However, the host mobilizes high affinity ligands to capture essential nutrients, and rapid exponential growth of the invader would soon deplete any environment with finite replenishment capacity, regardless of its initial plenitude. Starvation therefore is likely to be a contributing stress for a pathogenic invader.

The regulation of the expression of virulence genes in many pathogenic bacteria appears to be consistent with this premise. An essential gene cluster for *Salmonella typhimurium* virulence is the *spvABCD* operon, which is borne on a virulence plasmid. The expression of this operon is mediated by σ^S. Further, σ^S is involved in the expression of the *spvR* gene, which also regulates expression of the *spvABCD* operon. Thus, this operon is maximally expressed in starving cells of this bacterium. σ^S levels increase in *S. typhimurium* when it is phagocytosed by macrophages, and mutants deficient in σ^S are avirulent—indeed, they can serve as effective vaccines.

slyA is another gene of *S. typhimurium* that is relevant to its virulence. Transcriptional fusions to this gene show strong induction in the stationary phase as well as when phagocytosed by the macrophages. Mutants deficient in this gene synthesize fewer proteins in stationary phase and inside the macrophages, and they show reduced virulence. Similarly, the stress protein HtrA is important for the ability of *S. typhimurium* to replicate in the macrophages. HtrA is a serine protease that is believed to dispose of denatured periplasmic proteins. This protease is also important for the virulence of *Yersinia enterocolitica* and *Brucella abortus*. The starvation protein ClpC facilitates replication of *Listeria monocytogenes* in macrophages; *clpC* mutants of this bacterium are less virulent. Likewise, σ^E, which is also probably a starvation protein, controls the mucoidy phenotype of *Pseudomonas aeruginosa* that enables this bacterium to colonize the lung of cystic fibrosis patients.

IX. FUTURE PERSPECTIVES

A sharp focus on the molecular physiology of starving bacteria is only a relatively recent development. Among the questions that are presently being investigated is the nature of the growth phase-dependent sensitivity of σ^S to the ClpXP protease. This question impinges on a broader question in biology, namely, what determines the ability of different proteases to selectively target specific proteins. Whether the antisense element does have a role in translational control of σ^S is another important question. Little is known about the way in which starvation is sensed by a bacterium and linked to changes in σ^S levels. A protein called SprE or RssB, which bears homology to the response-regulator class of proteins, has been implicated in this process. However, the nature of the sensor remains unknown. Also important is further elucidation of the apparent promiscuity of the $E\sigma^S$-regulated promoters.

Since starvation promoters are maximally induced during slow growth, they have been utilized to dissociate expression in bacteria of useful activities from a need for rapid growth. This is potentially an area of considerable importance, for instance, in improving *in situ* bioremediation and vaccine effectiveness, but awaits further research to be effectively exploited. The role of stress proteins in enhancing cellular resistance and bacterial virulence needs to be understood at a deeper biochemical and molecular levels. The resulting knowledge can provide novel means of controlling bacterial populations, both when their destruction is desired, as in disease, or when their perpetuation is the objective, as in many beneficial processes carried out by bacteria.

BIBLIOGRAPHY

Fu, J. C., Ding, L., and Clarke, S. (1991). Purification, gene cloning and sequence analysis of an L-isoaspartyl protein carboxyl methyltransferase from *Escherichia coli*. *J. Biol. Chem.* **266**, 14452–14572.

Gottesman, S. (1996). Proteases and their targets in *Escherichia coli*. *Annu. Rev. Genet.* **30**, 465–506.

Hengge-Aronis, R. (1996). Regulation of gene expression during entry into the stationary phase. *In* "*Escherichia coli* and *Salmonella typhimurium:* Cellular and Molecular Biology" (F. C. Neidhardt, R. Curtiss III, J. L. Ingraham, E. C. C. Lin, K. B. Low, Jr., B. Magasanik, W. S. Reznikoff, M. Riley, M. Schaechter, and H. E. Umbarger, eds.), pp. 1497–1512. American Society for Microbiology, Washington, DC.

Magasanik, B. (1996). Regulation of nitrogen utilization. *In* "*Escherichia coli* and *Salmonella typhimurium:* Cellular and Molecular Biology" (F. C. Neidhardt, R. Curtiss III, J. L. Ingraham, E. C. C. Lin, K. B. Low, Jr., B. Magasanik, W. S. Reznikoff, M. Riley, M. Schaechter, and H. E. Umbarger, eds.), pp. 1344–1356. American Society for Microbiology, Washington, DC.

Matin, A. (1996). Role of alternate sigma factors in starvation protein synthesis—Novel mechanisms of catabolite repression. *Res. Microbiol.* **147**, 494–504.

Matin, A. (2001). Stress response in bacteria. *In* "Encyclopedia of Environmental Microbiology", vol. 6, pp. 3034–3046, John Wiley and Sons, New York.

Matin, A., Baetens, M., Pandza, S., Park, C. H., and Waggoner, S. (1999). Survival strategies in the stationary phase. *In* "Microbial Ecology and Infectious Disease" (E. Rosenberg, ed.), American Society for Microbiology, Washington, DC.

Munster, U., and Chrost, R. J. (1990). Advanced biochemical and molecular approaches to aquatic microbial ecology. *In* "Brock/Springer Series in Contemporary Biosciences" (J. Overbeck and R. J. Crost, eds.), pp. 8–46. Springer-Verlag, New York.

Record, M. T., Jr., Reznikoff, W. S., Craig, M. L., McQuade, K. L., and Schlax, P. J. (1996). *Escherichia coli* RNA polymerase ($E\sigma^{70}$), promoters, and the kinetics of steps of transcription initiation. *In* "*Escherichia coli* and *Salmonella typhimurium:* Cellular and Molecular Biology" (F. C. Neidhardt, R. Curtiss III, J. L. Ingraham, E. C. C. Lin, K. B. Low, Jr., B. Magasanik, W. S. Reznikoff, M. Riley, M. Schaechter, and H. E. Umbarger, eds.), pp. 792–820. American Society for Microbiology, Washington, DC.

Takayama, K., and Kjelleberg, S. (2000). The role of RNA stability during bacterial stress responses and starvation. *Environ. Microbiol.* **2**, 355–365.

Yura, T., Nagai, H., and Mori, H. (1993). Regulation of heat shock response in bacteria. *Annu. Rev. Microbiol.* **47**, 321–350.

WEBSITES

Pedro's Biomolecular Research Tools. A collection of Links to Information and Services Useful to Molecular Biologists
http://www.public.iastate.edu/~pedro/research_tools.html

Comprehensive Microbial Resource of The Institute for Genomic Research (TIGR) and Links to many other Microbial Genomic Sites
http://www.tigr.org/tigr-scripts/CMR2/CMRHomePage.spl

A Website on Protein Folding *in vivo* by the Semmelweis University of Medical Sciences, Budapest
http://pps9900.cryst.bbk.ac.uk/projects/attila/Examples.html

82

Strain improvement

Sarad Parekh

Dow AgroSciences

GLOSSARY

DNA recombination A laboratory method in which DNA segments from different sources are combined into a single unit and manipulated to create a new sequence of DNA.

fermentation A metabolic process whereby microbes gain energy from the breakdown and assimilation of organic and inorganic nutrients.

gene Physical unit of heredity. Structural genes, which make up the majority, consist of DNA segments that determine the sequence of amino acids in specific polypeptides. Other kinds of genes exist. Regulatory genes code for synthesis of proteins that control expression of the structural genes, turning them off and on according to circumstances within the microbe.

gene cloning Procedure employed where specific segments of DNA (genes) are isolated and replicated in another organism.

genetic code The linear sequence of the DNA bases (adenine, thymine, guanine, and cytosine) that ultimately determines the sequence of amino acids in proteins. The genetic code is first "transcribed" into complementary base sequences in the messenger RNA molecule, which in turn is "translated" by the ribosomes during protein biosynthesis.

genetic recombination When two different DNA molecules are paired, those regions having homologous nucleotide sequences can exchange genetic information by a process of natural crossover to generate a new DNA molecule with a new nucleotide sequence.

interspecific protoplast fusion Method for recombining genetic information from closely related but nonmating cultures by removing the walls from the cells.

metabolic engineering A scientific discipline that integrates the principles of biochemistry, chemical engineering, and physiology to enhance the activity of a particular metabolic pathway.

mutation Genetic lesion or aberration in DNA sequence that results in permanent inheritable changes in the organism. The strains that acquire these alterations are called mutant strains.

plasmid An autonomous DNA molecule capable of replicating itself independently from the rest of the genetic information.

primary metabolites Simple molecules and precursor compounds such as amino acids and organic acids that are involved in pathways that are essential for life processes and the reproduction of cells.

secondary metabolites Complex molecules derived from primary metabolites and assembled in a coordinated fashion. Secondary metabolites are usually not essential for the organism's growth.

The Science and Technology of designing, breeding, manipulating, and continuously improving the performance of microbial strains for biotechnological applications

The Desk Encyclopedia of Microbiology
ISBN: 0-12-621361-5

is referred to as "strain improvement." The science behind developing improved cultures has been enhanced recently by a greater understanding of microbial biochemistry and physiology, coupled with advances in fermentation reactor technology and genetic engineering. In addition, the availability and application of user-friendly analytical equipment such as high pressure liquid chromatography (HPLC) and mass spectroscopy, which raised the detection limits of metabolites, have also played a critical role in screening improved strains.

I. INTRODUCTION

The use of microbes for industrial processes is not new. Improving the commercial and technical capability of microbial strains has been practiced for centuries through selective breeding of microbes. In making specialty foods and fermented beverages (such as alcohol, sake, beer, wine, vinegar, bread, tofu, yogurt, and cheese), specific strains of bacteria and fungi isolated by chance have been employed to obtain desirable and palatable characteristics. Now, with integrated knowledge of biochemistry, chemical engineering, and physiology, microbiologists have taken a more scientific approach to the identification of microbial strains with desired traits.

Later spectacular successes observed in improvement of the industrial strains by mutation and genetic manipulations in the production of penicillin and other antibiotics led to strain development as a driving force in the manufacture of pharmaceuticals and biochemicals. Microbes are now routinely used in large-scale processes for the production of lactic acid, ethanol fuel, acetone–butanol, and riboflavin as well as for the commercial production of enzymes such as amylases, proteases, and invertase. Efforts were also made by chemical engineers to improve fermenter designs on the basis of understanding the importance of culture media components, sterile operations, aeration, and agitation. Today, production of hormones, steroids, vaccines, monoclonal antibodies, amino acids, and antibiotics are testimonies to the important role of strain improvement in the pharmaceutical industry.

The intent of this article is to briefly describe strategies employed in strain improvement, the practical aspects of screening procedures, and the overall impact that strain improvement has on the economics of fermentation processes. Readers are also urged to review the additional articles described in the bibliography, especially the basic concepts as well as the theoretical basis of genetic mutations and screening improved strains.

II. ATTRIBUTES OF IMPROVED STRAINS

Microbial strain improvement cannot be defined simply in terms of modifying the strain for overproduction of bioactive compounds. Strain improvement should also be viewed as making the fermentation process more cost-effective. Some of the traits unique to fermentation process that make a strain "improved" are the ability to (a) assimilate inexpensive and complex raw materials efficiently; (b) alter product ratios and eliminate impurities or by-products problematic in downstream processing; (c) reduce demand on utilities during fermentation (air, cooling water, or power draws); (d) excrete the product to facilitate product recovery; (e) provide cellular morphology in a form suitable for product separation; (f) create tolerance to high product concentrations; (g) shorten fermentation times; and (h) overproduce natural products or bioactive molecules not synthesized naturally, for example, insulin.

A. Need for strain improvement

Microbes (fungi, bacteria, actinomyces) that live freely in soil or water and produce novel compounds of commercial interest, when isolated from their natural surroundings, are not ideal for industrial use. In general, wild strains cannot make the product of commercial interest at high enough yields to be economically viable. In nature, metabolism is carefully controlled to avoid wasteful expenditure of energy and the accumulation of intermediates and enzymes needed for their biosynthesis. This tight metabolic and genetic regulation, and synthesis of biologically active compounds, is ultimately controlled by the sequence of genes in the DNA that program the biological activity. To improve microbial strains, the sequence of these genes must be altered and manipulated. In essence, microbial strain improvement requires alteration and reprogramming of the DNA (or the genes) in a desired fashion to shift or bypass the regulatory controls and checkpoints. Such DNA alterations enable the microbe to devote its metabolic machinery to producing the key biosynthetic enzymes and increasing product yields. In some cases, simple alteration in DNA can also lead to structural changes in a specific enzyme that increases its ability to bind to the substrate, enhance its catabolic activity, or make itself less sensitive to the inhibitory effects of a metabolite. On the other hand, when the changes are made in the regulatory region of the gene (such as the promoter site), it can lead to deregulation of gene expression and overproduction of the metabolite. A typical example is

combinations of desired characteristics in microbes. Genetic engineering is usually employed to create targeted mutations on the genes, unlike other methods of mutation that are random. It must be emphasized that this technology is not a way of constructing new forms of life. Even the genetic materials of the simplest organisms are highly complex, and insertion of a few genes from an unrelated organism will not create a new microbe.

On the basis of the method of screening and selection chosen, there are basically two methods of improving microbial strains through random mutation: (1) random selection and (2) rationalized selection.

1. *Random selection*

Random mutagenesis and selection is also referred to as the classic approach or nonrecombinant strain improvement procedure. Improved mutants are normally identified by screening a large population of mutated organisms, since the mutant phenotype may not be easy to recognize against a large background. After inducing mutations in the culture, the survivors from the population are randomly picked and tested for their ability to produce the metabolite of interest. This approach has the advantage of being simple and reliable. Moreover, it offers a significant advantage over the genetic engineering route alone by yielding gains with minimal start-up and sustaining such gains over years despite a lack of scientific knowledge of the biosynthetic pathway, physiology, or genetics of the producing microbe (Lein, 1986). This empirical approach has been widely adapted by the fermentation industry, following the successful improvement in penicillin titers since World War II. One drawback to the random selection approach is that it relies on nontargeted, nonspecific gene mutations, so many strains need to be screened to isolate the improved mutant in the mixed population. In addition when the culture is mutagenized, multiple mutations may be introduced in the strain. This may result in the enfeeblement of the organism lacking the properties of interest.

This process of strain improvement involves repeated applications of three basic principles: (1) mutagenesis of the population to induce genetic variability, (2) random selection and screening from the surviving population of improved strains by smallscale model fermentation, and (3) assaying of fermentation broth/agar for products and scoring for improved strains. It must be emphasized that the action of the mutagenic agent on DNA can not only cause genetic alteration but can also induce cell death, owing to irreversible damage to the DNA or formation of lethal mutations. Hence, after the mutagenic treatment, mutants are sought among the surviving population with the anticipation that each of the surviving cells harbors one or several mutations. Each time an improved strain is derived through mutation, it is used again as the parent strain in a new cycle of mutation, screening by fermentation (liquid or solid), and assay (Fig. 82.1). This random procedure of mutant selection is continued until a strain is derived that is statistically superior in performance compared to the control strain prior to the mutagenic treatment. The objective of mutagenesis is to maximize the frequency of desired mutations in a population while minimizing the lethality of the treatment. For this purpose, nitrosoguanidine (NTG) has been the mutagen of choice because it offers the highest possible frequency of mutants per survivor (Baltz, 1986). The efficiency of the random selection process is dependent on several factors: the type of culture used (such as spores or conidia), mutagen dose and exposure time, the type and damage to DNA, conditions of treatment and posttreatment, frequency of mutagen treatment, and the extent of yield increase detectable.

In addition to the mutation conditions, the test or quantitative and analytical screening procedures employed (bioassays, radioimmunoassays, chromatography, HPLC) also play a critical role in successful isolation of superior mutants. The ability to detect a gain mutant among the randomly selected mutants is greatly influenced by the process as well as the variability within the process and the actual titer differences between the improved strain and the control (Rowlands, 1984). The screening procedure therefore is usually designed to maximize the precision and selectivity of improved cultures (gain per sample tested) and to minimize the variability (measured as the coefficient of variation) when treating the unmutagenized control and reference samples. All the strains tested, including the control, are normally worked up all the way head-to-head from the initial cell clone stage to the final screening stage. This is essentially the test of significance. As such, if all the treatment conditions were the same, a successful test should show a statistical difference between the means of the control and improved cultures.

Putative mutants isolated after a primary run are subjected to secondary and tertiary confirmations (replications and repetitions) to raise the level of confidence and observe the anticipated titer differences on data collections (Fig. 82.1). A desired improvement in the strain is typically obtained with less testing if the selection system is less susceptible to variability and if the coefficient of variation is lower. Considerable efforts are therefore directed at troubleshooting

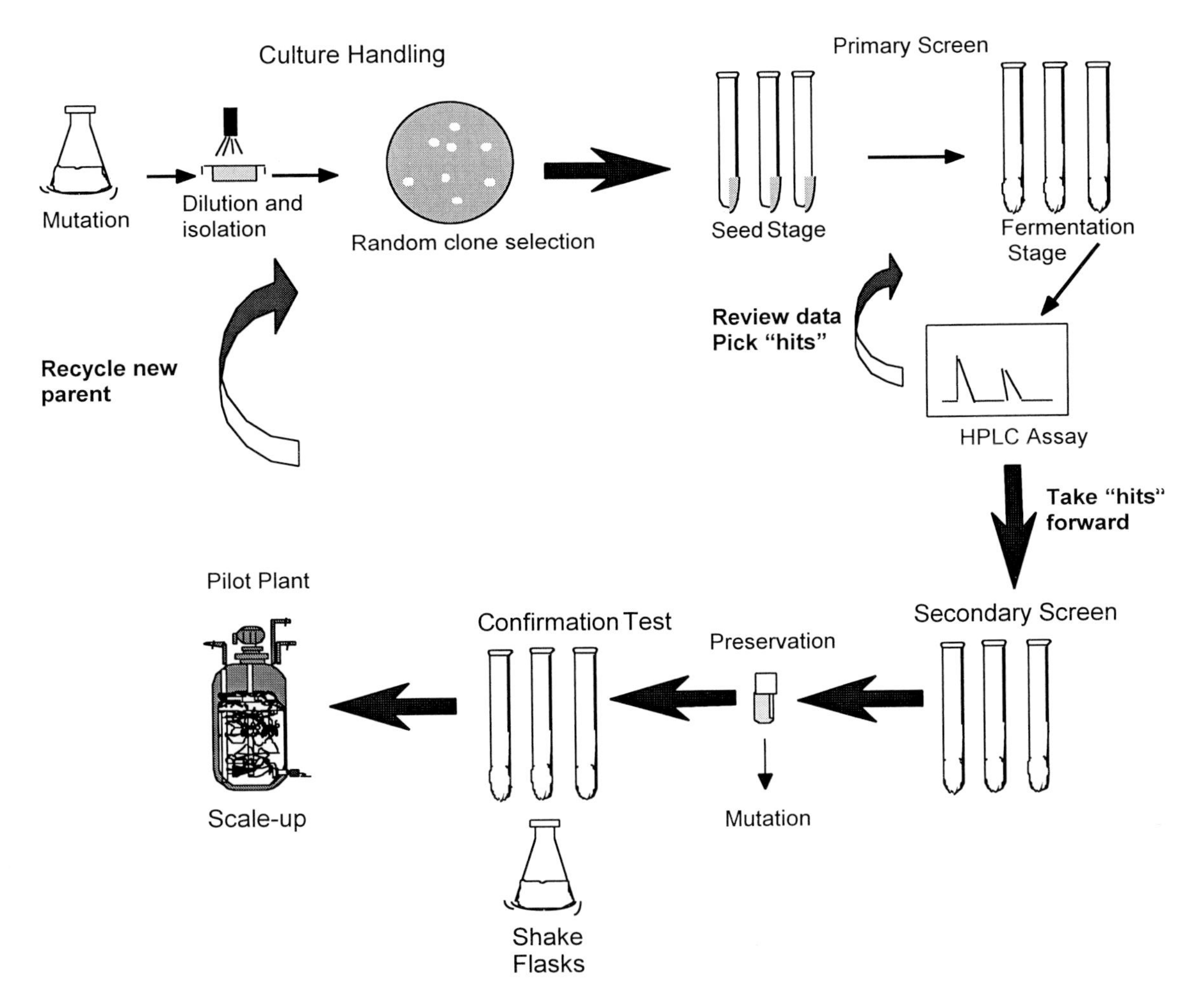

FIGURE 82.1 Typical steps in mutation and random strain selection process.

important handling procedures to identify and eliminate key contributors to the errors in the process. Furthermore, the screening strategy is carefully chosen so that the medium and fermentation parameters mimic large-scale production. This increases the possibility that the improved performance of the mutant will be achieved at scale-up. The random approach of strain selection relies on delivering small incremental improvements in culture performance. Although the procedure is repetitive and labor intensive, this empirical approach has a long history of success and has given dramatic increases in titer improvement, as best exemplified by the improvements achieved for penicillin production in which titers over 50 g/liter are reported—a 4000-fold improvement over the original parent strain (Crueger and Crueger, 1984). These authors have also cited certain actinomyces or fungal strains capable of overproducing metabolites in quantities as high as 80 g/liter. Not surprisingly, therefore, pharmaceutical and other fermentation industries typically adopt this technique for selection of improved mutants for many of their processes. The historic successes (e.g., production of antibiotics and other secondary metabolites, enzymes, and amino acids) bear testimony to creating superior strains through this procedure (Queener and Lively, 1986).

The procedure of mutation followed by random selection is laborious and requires screening a large number of strains to obtain desired mutants. This is because in random screening procedures a high percentage of mutants examined will be carried over as survivors from the mutagenesis and will exhibit the same or lower yields than the parent strain. Factors that impact the success of the random program and accelerate strain improvement are the following: the extent of yield improvement, the frequency of induced mutations, the amount of time for turnaround of the mutation selection cycle, and the testing capacity. In addition, the success of a strain improvement program also depends on resource allocation.

The key labor-intensive steps in classic strain improvement programs include the isolation of individual mutated cultures, preparation and distribution of sterile media, transfer of clones and their inoculum to initiate fermentation, assays of fermentation broth, and repeated confirmations. Furthermore, the more complex the regulation and biosynthesis of a desired compound, the greater is the number of strains that need to be evaluated and replicated. As a rule, mutants with very high yields are rarer than those with subtle improvements. So, to increase the odds, a larger number of improved strains are examined, raising the probability of detecting improved strains. Thus, if the strain improvement program is operated manually, successful improvements detected will roughly be proportional to the number of personnel allocated (Rowlands, 1984). The advantage of manual screening is in the trained eyes of the microbiologists. They can visually detect the alteration in the morphology, pigmentation, and growth characteristics of mutants during the selection process. In fact, isolation of pelleted strains of filamentous organisms is commonly based on morphological criteria.

To increase the efficiency of random selection, ways by which the key steps in the process can drive the throughput higher without adding labor are typically sought. In some instances high throughput screens have been automated with robotics technology. This allows screening of large populations with minimum resources by miniaturizing wherever possible the equipment required and constructing an automated integrated system. In an industrial system, sterile media are robotically dispensed in custom designed sterilizable and cleanable/disposable modules, each having over 100 tubes or bottles. Individual clones are detected by an optical system and plugged from an agar based medium into liquid seed medium. The inoculations of seed stage culture to fermentation vials are also accomplished by robots. The extraction and HPLC analysis of the fermentation broth is also automated to match the throughput of the screening stage. The advantage of such automation is that it facilitates the capture and downloading of process data and allows statistical process control approaches to be implemented where refinement of the process is required. The success of automated programs requires skillful microbiologists, and constant monitoring and evaluation of the screening system to ensure that all aspects of the automation are functioning efficiently without introducing variability. The significant disadvantage of robotic systems is the initial high capital investment and continued maintenance of equipment and software (Nolan, 1986).

Although other sophisticated techniques are being developed to generate improved strains, random high throughput selection and mutation will continue to be an integral part of any strain improvement program. The random approach is least useful for microbes that are less susceptible to mutagenesis, such as some fungi (owing to their diploid or polyploid genome structure) and bacteria with very efficient repair systems. In a typical manually operated strain improvement project, the expected frequency of gain could be of the order 1 in 10,000, where about 10,000 mutants may pass through the primary screen before a higher producing candidate is identified. However, as the titer increases, depending on the pathway, the organism, the product, and the history of the production strain, significantly larger gains are required to detect an improved mutant. The use of prescreening and rational selection allows for a significant improvement in the efficiency of the selection process.

2. *Rationalized selection*

An alternate approach to random screening requires a basic understanding of product formation and the fermentation pathway; this can be acquired through radioisotope feeding studies and isolation of mutants blocked in various pathways (Queener *et al.*, 1978). These observations can shed light on the metabolic checkpoints, and suggest ways to isolate specific mutants. For example, environmental conditions (pH, temperature, aeration) can be manipulated, or chemicals can be incorporated in the culture media to select mutants with desired traits. This approach is used in many instances as a prescreen, since selecting for a particular mutant is unlikely to guarantee a hyper-yielding mutant. In some instances, by adding toxic substances to the media, the sensitive parent strains are prevented from growing, and only the resistant mutant clones propagate. Such an enrichment procedure has been used to isolate mutants with increased biosynthetic capacity through a change in a regulatory mechanism (leading to either an enzyme resistant to inhibition or an enzyme that is expressed constitutively) or mutants that are modified in the transport or degradation of compounds, which ultimately leads to higher product formation. This rationalized technique is powerful, and when the logic behind the mutant selection criteria is sound, the effectiveness of mutant gains is much greater than with random selection.

Rationalized selection for strain improvement does not generally require a sophisticated understanding of molecular biology to manipulate environmental or cultural conditions. It does, however, require some understanding of cellular metabolism and product synthesis to design the right media or environmental conditions. The procedure is useful in selecting strains overproducing metabolites, antibiotics, simple molecules, amino acids, or enzymes (Queener and Lively, 1986). Some of the mutants derived through rationalized selections are described below.

a. Auxotrophic mutants

Many metabolic processes have branched pathways, and isolating mutants blocked in one branch of the metabolic pathway can cause accumulation of simple products such as amino acids, nucleotides, and vitamins made by other branches. Auxotrophic strains are blocked at some point in a pathway vital for growth, and unless the specific nutrients or products of the pathway are supplied in the media, the auxotrophs do not survive. Auxotrophs are primarily isolated by plating the mutagenized population on a complete medium that has all the nutrients needed for growth. The clones are then replica plated to minimal medium lacking some specific nutrients, and auxotrophs that fail to grow on minimal media are identified. Most auxotrophic strains give poor antibiotic yields; however, some prototrophic revertants have demonstrated improved antibiotic production such as in tetracycline production (Queener *et al.*, 1978).

b. Regulatory mutants

Since anabolism and catabolism in any organism are tightly regulated, selection and screening of microbes with less efficient regulation, and optimizing culture conditions, can lead to relaxed regulation and overproduction of microbial products. A broad understanding of metabolic pathway bottlenecks is necessary for a rational approach to developing improved regulatory mutants. Isolating strains relaxed in regulation can usually be accomplished by selecting strains desensitized to feedback inhibition (enzyme activities) or feedback repression (enzyme synthesis) involved in the pathway. One difficulty in applying analog-resistant mutants to strain improvement is that many analogs of primary metabolites need to be tested and some either do not inhibit growth or inhibit growth only at very high concentrations.

(1) *Mutants Resistant to Feedback Inhibition* In many microbes, the end products of metabolism, when accumulated in the microbial cell, inhibit the enzyme activities of many pathways. The end product causes conformational changes by binding to a specific (allosteric) site on the enzyme, and inhibits activity. The binding is usually noncompetitive. Mutation in the structural gene, however, can alter the enzyme binding site and prevent these inhibitory effects. By studying the interaction of various analogs of end products and their resistance, improved strains can be selected that lack feedback inhibition and thus overproduce metabolites of interest. For example, some analogs (acting through these regulatory controls) prevent the synthesis of compounds required for growth and thus cause cell death. Supplementing the screening medium with these analogs selects only mutants with altered enzyme structure and desensitized to inhibition effects to grow. Such procedures have led to development of superior mutant strains of *Arthrobacter, Bacillus, Streptomyces, Aspergillus*, and *Corynebacterium* that overproduce amino acids, nucleotides, and vitamins (Demain, 1983). In some cases, the rational selective agent is biological rather than chemical. Resistance to actinophage has been used to isolate superior vancomycin-producing strains of *Streptomyces orientalis* (Soviet Union Patent 235–244-A, 1969).

(2) *Mutants Resistant to Repression* Here intermediates, products of catabolism (derived from breakdown of compounds containing carbon, nitrogen, or phosphorus), or end products regulate the amount of biosynthetic enzymes synthesized and, therefore, the amount of final product formed. However, mutations at the operator site or other regulatory sites on the gene relieve such end-product repression and allow overproduction of the biosynthetic enzyme. For example, it is well known that antibiotics inhibit their own biosynthesis (e.g., penicillin, chloroamphenicol, puromycin, streptomycin), where key enzymes required for the architecture of these complex molecules are repressed. Mutant strains less sensitive to antibiotic production are therefore isolated to provide higher yields (Elander and Voumakis, 1986). In a similar context, constitutive mutants have been selected that form enzymes (amylase, glucoamylase, lipase, protease) independent of cultural conditions or the presence of inducing compounds.

Resistance to an antimetabolite is not the sole means of selecting product-excreting mutants resulting from a desensitized enzyme system. Removing enzymes sensitive to feedback repression or the end product that causes inhibition during fermentation also accelerates productivity. Elimination of end-product inhibition or repression effects have been demonstrated by adding chemicals during the fermentation process to trap the end product or the inhibitor. *In situ* end-product extraction, or adding a mechanical device during fermentation (such as specific membrane modules with a particular molecular weight cutoff), allows the percolation of the final product from accumulating in the broth. Increasing permeability of the cell membrane is another method of controlling intracellular product accumulation, enhancing the extracellular metabolite flux. This approach has been exploited to improve the titers of monosodium glutamate (MSG) from *Corynebacterium*, *Micrococcus*, and *Brevibacterium* (Demain, 1983).

Last, ways of stabilizing the activity of enzymes involved in the assembly of molecules have been reported to augment product formation and strain performance. For example, in gramicidin biosynthesis amino acids are added to stabilize *in vivo* gramicidin S synthetase enzymes and prolong the longevity of biosynthetic activity (Demain, 1983).

c. *Other procedures*

Mutant strains suspected of metabolic impairment can also be assayed visually for the presence or absence of specific enzyme activities by plating and spraying on the "diagnostic" solid or liquid culture medium with selective reagents, dyes, or an indicator organism. For example, the agar plug method has been used to detect production of an antibiotic by measuring the extent of growth inhibition of an organism sensitive to the antibiotic. The diameter of the resulting zone of inhibition serves as a measure of antibiotic production. Other procedures rely on the use of chromogenic agents, which are normally converted to a visible product by a specific biochemical reaction or reorganization of the redox level in the media. This leads to visual detection from the large background population of a specific strain having the biochemical activity of interest. Examples of these detection substrates are phenol red for acid-base reactions, 2-nitrophenyl-β-D-galactopyranoside (ONPG) for galactosidase, 6-nitro-3-phenylacetamidobenzoic acid (NIPAB) for penicillin amidase, 4-nitrophenylphosphate (PNPP) for phosphatase, nitrocefin for β-lactamase inhibition, and azocasein for proteinases (Elander and Vournakis, 1986; Queener and Lively, 1986). Tetrazolium and methylene blue (EMB, eosin methylene blue agar for *Escherichia coli* and other coliforms) are commonly employed for detection of oxidation–reduction reaction complexes exhibited by strains of interest. These reactions can be coupled into high throughput screening systems, giving the possibility of targeting whole cells or isolated enzymes for strain selection. For example, carotenoids have been demonstrated to protect *Phaffia rhodozyma* against singlet oxygen damage. A combination of Rose Bengal and thymol in visible light has been employed to select carotenoid-overproducing strains (Schroeder and Johnson, 1995). Enrichment with a singlet oxygen system led to development of mutants with increases in certain carotenoids but a decrease in astaxanthin.

B. Genetic recombination

In addition to the manipulation of microorganisms by mutation, the techniques of genetic recombination can be employed to get new strains containing novel combinations of mutations and superior microbial strains. Generally, genetic recombination methods include those techniques that combine two DNA molecules having similar sequences (homologs). Through the special event of crossing-over, they are reunited to give a new series of nucleotide sequences along the DNA that are stable, expressible genetic traits. This mechanism of gene alteration and strain modification is called genetic recombination. This definition includes the techniques of protoplast fusion, transformation, and conjugation. Most recently, recombinant DNA technology has been employed to assemble new combinations of DNA *in vitro*, which are then reinserted into the genome of the microbe, creating new varieties of microbe not attainable through traditional mutation and rationalized selection approaches. This approach overlaps the other methods to some extent in that it involves transformation of microbes with laboratory-engineered specific recombinant molecules via plasmid or phage vectors (Hamer, 1980).

1. *Protoplast fusion*

Fusing two closely related protoplasts (cells whose walls are removed by enzyme treatment) is a versatile technique that combines the entire genetic material from two cells to generate recombinants with desired traits that cannot be obtained through a single mutation. The technique has the advantage of producing hybrids from cells that are sexually incompatible. The procedure of forced mating allows mingling of DNA that is not dependent on appropriate sex factors and is not influenced by barriers of genetic incompatibility.

The procedure relies on stripping the cell wall of the microbes with lytic enzymes, stabilizing the fragile protoplasts with osmotic stabilizing agents, and using a chemical agent or an electric pulse (electrofusion) to induce membrane fusion and to form a transient hybrid. In the hybrid, the genes align at homologous regions, and crossing-over of genes creates recombination within the fused cells. After recombination, the protoplasts are propagated under specific conditions that favor regeneration of cell walls. The unwanted parents are discriminated against by incorporating selective markers in the screening process (e.g., auxotrophy, extracellular enzymes, morphological differences, levels of antibiotic produced) so that only recombinants grow and form viable cells. The efficiency of the technique is influenced by the fragility of cells, the types of genetic markers, the fusing agent used, and the protoplast regeneration capability.

The use of protoplast fusion has been reported (Matsushima and Baltz, 1986) to improve a wide range of industrial strains of bacteria and fungi including *Streptomyces, Nocardia, Penicillium, Aspergillus,* and *Saccharomyces.* This technique is frequently employed in the brewing industry for improving yield and incorporating traits such as flocculation to aid beer filtering, efficient utilization of starch, contamination control, and minimizing of flavors during brewing. Many of these traits are not easily achievable through simple mutation. One advantage of protoplast fusion is the high frequencies at which recombinants are produced under nonselective conditions in the absence of sex factors and without need of specific mating types. In *Streptomyces coelicolor* frequencies as high as 20% have been observed. Another interesting feature is that more than two strains can be combined in one fusion. In some instances, four strains have been fused to yield recombinants containing genes derived from all four parents. This approach can be extremely useful in accelerating strain development. However, because of the absence of control over the amount of genetic material from any one strain retained in the recombinant, protoplast fusion may not improve the strain in the desired fashion. The big disadvantage to this approach has been the genetic instability of the fused strains and the lack of control over which genetic alterations occur. A detailed description of protocols and considerations for the application of protoplast fusion to a variety of industrial microbes is available (Matsushima and Baltz, 1986).

2. *Transformation*

Transformation is the process involving the direct uptake of purified, exogenously supplied DNA by recipient cells or protoplasts. When this occurs, the donor DNA may either combine with the recipient DNA or exist independently in the cell. This leads to changes in the amount and organization of the recipient microbe DNA, hopefully improving it with some of the characteristics coded by the donor DNA. Transformation can be mediated by total genomic DNA or cloned sequences in plasmid or phage DNA. Essentially, the cultures to be transformed are cultivated in a specific physiological manner to develop the competency to make them readily accept foreign DNA. Having selectable markers on the donor DNA allows easy identification of transformants. This procedure allows the transfer of genetic material between unrelated organisms. Certain microorganisms have a well-established gene cloning system that provides a great potential for improving strains by transformation. Transformation methods for strain improvement pertaining to primary or secondary metabolites have been demonstrated in *Streptomyces, Bacillus, Saccharomyces, Neurospora,* and *Aspergillus* (Elander and Vournakis, 1986).

3. *Conjugation*

Conjugation introduces mutational changes in microbes through unidirectional transfer of genetic material from one strain to the other; it is mediated by plasmid sex factors. Conjugation requires cell-to-cell contact and DNA replication. This mode of genetic exchange can achieve transfer of chromosomal DNA or plasmid DNA. Several strains have been modified by this procedure to make them resistant to specific antibiotics and microbial contamination.

The application of conjugate plasmid recombination technology is employed for strain improvement of *Lactococcus* starter cultures in the dairy industry, which is often plagued by problems with phage. Phage infection can lead to slow acid production, which can economically impact a cheese factory. Furthermore, owing to the nonaseptic nature of the dairy starter culture (open vat in cheese making, the presence of mixed flora in milk), phage-resistant strains are desired as starter cultures for fermentation. Various naturally occurring phage-resistant strains have a number of resistance mechanisms (e.g., abortive infection, restriction/modifications, and absorption inhibitions) that in many instances are carried on conjugative plasmids. Several lactic acid bacteria have therefore been modified by conjugation and transformation procedures to acquire phage resistance in dairy starter cultures. Typical methods for conjugative transfer involve mating by donor and recipient cells on milk agar, followed by harvesting of cells and further isolation on selective medium. This procedure has been applied successfully to construct nisin-producing *Lactococcus* strains (Broadbent and Kondo, 1991).

C. Cloning and genetic engineering

1. In vitro *recombinant DNA technology*

By employing restriction endonucleases and ligases, investigators can cut and splice DNA at specific sites. Some endonucleases have the ability to cut precisely and generate what are known as "sticky ends." When different DNA molecules are cut by the same restriction enzyme, they possess similar sticky ends. Through a form of biological "cut and paste" processes, the lower parts of the one DNA is made to stick well onto the upper part of another DNA. These DNA molecules are later ligated to make hybrid molecules. The ability to cut and paste the DNA molecule is the basis of "genetic engineering." A useful aspect of this cut and paste process involves the use of plasmid, phage, and other small fragments of DNA (vectors) that are capable of carrying genetic material and inserting it into a host microbe such that the foreign DNA is replicated and expressed in the host. A wide array of techniques can now be combined to isolate, sequence, synthesize, modify, and join fragments of DNA. It is therefore possible to obtain nearly any combination of DNA sequence. The challenges lie in designing sequences that will be functional and useful (Hamer, 1980).

The protocol to modify and improve strains involves the following steps:

a. Isolate the desired gene (DNA fragment) from the donor cells.
b. Isolate the vector (a plasmid or a phage).
c. Cleave the vector, align the donor DNA with the vector, and insert the gene into the vector.
d. Introduce the new plasmid into the host cell by transformation or, if a viral vector is used, by infection.
e. Select the new recombinant strains that express the desired characteristics.

For successful transfer of a plasmid/phage vector, it must contain at least three elements: (1) an origin of replication conferring the ability to replicate in the host cell, (2) a promoter site recognized by the host DNA polymerase, and (3) a functional gene that can serve as a genetic marker. A great deal of literature exists on the theoretical overviews, and laboratory manuals on the use of recombinant DNA for strain modification and improvements are available.

2. *Site-directed mutagenesis for strain improvement*

So far the mutations and the modifications of the strains discussed have been randomly directed at the level of the genome of the culture. The application of recombinant technology and the use of synthetic DNA now make it possible to induce specific mutations in specific genes. This procedure of carrying out mutagenesis at a targeted site in the genome is called site-directed mutagenesis. It involves the isolation of the DNA of the specific gene and the determination of the DNA sequence. It is then possible to construct a modified version of this gene in which specific bases or a series of bases are changed. The modified DNA can now be reinserted into the recipient cells and the mutants selected. Site-directed mutagenesis has found valuable application in improving strains (Crueger and Crueger, 1984), by enhancing the catalytic activity and stability of commercial enzymes, for example, penicillin G amidase.

Since the mid-1970s the synergistic use of classic techniques along with rational selection and recombinant DNA has made a significant impact in developing improved strains. Fermentation processes for products as diverse as human proteins and antibiotics and other therapeutic agents (chymosin, lactoferrin) have benefited from these combinatorial approaches. Transcription, translation, and protein secretion, activation, and folding are one or more of the rate-limiting steps critical for the overproduction of such therapeutic proteins. Achieving overproduction of active therapeutic proteins in bacterial or fungal heterologous gene expression systems has been made amenable due to the mix of classic and rational selection procedures. Genetic engineering along with classic methods has been used on numerous occasions to improve the performance of yeast and bacteria in alcohol fermentation, expand the substrate range, enhance the efficiency of the fermentation process, lower by-product formation, design yeast immune to contamination, and develop novel microbes that detoxify industrial effluents. Cost-effective production by fermentation of alcohols (ethanol, butanol) that can be used as substitutes for fossil fuels has been aided by this technology. Bacterial manufacturing of large quantities of hormones, antibodies, interferons, antigens, amino acids, enzymes, and other therapeutic agents to combat diseases has also become possible by recombinant DNA technology and strain improvement programs. Through increased gene dosage, improved efficiencies of antibiotic production have been achieved to relieve one or more rate-limiting steps. Novel and hybrid antibiotics and bioactive compounds have also been produced by combining different biosynthetic pathways in one organism that would have been difficult or impossible to manufacture through synthetic chemistry (e.g., *Cephalosporium acremonium* and *Claviceps purpurea*). Moreover, using recombinant DNA techniques, entire sets of genes for antibiotic biosynthesis have been cloned into a heterologous host in a single step. By cloning portions of the biosynthetic genes from one producer to another strain, hybrid compounds have also been synthesized, with novel spectra of activities and pharmacological applications. An example of this is the production by *Streptomyces peucetius* Subsp. *caesius* of adriamycin (14-hydroxydaunomycin), an antitumor antibiotic (Crueger and Crueger, 1984).

Occasionally it has been found that certain improved mutants produce extremely high levels of a specific enzyme. When analyzed, these mutants had multiple copies of a structural gene coding for the specific enzyme of interest. Increasing the number of gene copies in the cell (through gene cloning) has therefore been employed to overproduce enzyme precursors and their end product. In addition, mutations at the promoter or regulatory site have been demonstrated to alter secondary metabolite productivity (Hamer, 1980). For example, in *Saccharopolyspora erythraea*, specific mutations at a ribosomal RNA operon terminator site altered the transcription and expression of the erythromycin gene cluster, and strains harboring these mutations overproduced enzymes involved in the later steps of erythromycin biosynthesis (Queener *et al.*, 1978).

Once an improved strain is confirmed through bench-work studies, additional efforts are necessary to validate its performance. It is normally purified by reisolation, and the reisolates are verified for strain variability, homogeneity, and performance. They are preserved in large lots for examination under pilot plant conditions before being introduced for large-scale production.

V. IMPROVED STRAIN PERFORMANCE THROUGH ENGINEERING OPTIMIZATION

Major improvements in fermentation are no doubt attributed to superior strains created through mutation or genetic alterations. Further improvement in culture performance can also be achieved by giving a strain the optimum environmental and physical conditions. During the strain improvement process, it is important to keep in mind that the ultimate success also depends on optimization of fermentation design factors. The use of batch or fed batch, continuous or draw-and-fill operation, the extent of shear, broth rheological properties, and oxygen and heat transfer characteristics all contribute to improvement in strain performance. The application of biochemical engineering principles can be used to design environmental parameters that shift kinetics of metabolic routes toward the desired product.

A. Improving strain performance through optimizing nutritional needs

The environment in which the altered strain is grown is known to influence higher product yield and get the best performance out of the culture. Since the media commonly used for production are different from the ones in which mutants are screened, media optimization is requisite to achieving the best response from the improved strains when scaled up to production. The media for production are reformulated so that they meet all growth requirements and supply the required energy for growth and product synthesis. Early bench work is typically performed with biochemically defined media to elucidate metabolite flux and regulation (inhibition or repression) by specific nutrients and physical variables. Later research is done to develop complex media that are more cost-effective to support cultural conditions of improved strains and maximize product synthesis without producing additional impurities that may impact isolation of the product. Additional issues, such as inoculum media and transfer criteria, media sterilization, pH, cultivation variables, and the sensitivity of the culture to different batches of raw material, are addressed during media optimization.

Statistical computer-based methods and response surface modeling are available for the study of many variables at the same time. A full search is normally made of every possible combination of independent variables to determine appropriate levels that give the optimum response in strain performance. Success in this area can be enhanced if additional physiological data are available, such as the role of precursors, the steps in the biosynthetic pathway, carbon flux through the pathway, and the regulation of primary and secondary metabolism by carbon and nitrogen. Controlling the levels of metabolites and precursors during fermentation aids in controlling lag and repression or toxicity effects. The removal of inhibiting products has been practiced where increasing the concentration up to economical levels demonstrates poor process kinetics. Adding chelating agents has been beneficial if the fermentation is found to be sensitive to substrate-specific repression. Further improvement in strain performance and productivity gain has been observed when the key enzymes participating in product formation are stabilized. For example, biosynthesis of the antibiotic gramicidin has been improved greatly by adding precursor amino acids that are substrates for the key enzymes (Demain, 1983).

B. Influence of bioengineering in improving strain performance

The ultimate destination of an improved strain is a large fermenter in which the desired product is made for commercial use. Conversion of laboratory processes to an industrial operation is called scale-up. It is not a straightforward process, requiring the use of methods of chemical engineering, physiology, and microbiology for success. The goal of the scale-up team is to cultivate improved strains under optimum production conditions. Open communication, data feedback, and synergy between engineers and scientists are vital to facilitate successful launch and scaling up of new and improved strains. Factors such as media sterilization for culture seed and production, methods of aeration and agitation, power input, control of viscosity, and evaporation rates are considered when moving new strains into production. In addition, sterility factors, heat transfer, impeller types, baffle types and positioning, the geometry and symmetry of the fermenter, mixing times, oxygen transfer rates, respiratory quotient and metabolic flux, disengagement of gases, and culture stability are all important in bringing the improved mutants from the laboratory to industrial scale production. Furthermore, metabolic feeds and the impact of the addition of feeds subsurface or surface, as well as timing of additions, are also optimized for directing ways toward the desired product. In some cases, process control and parameter optimization are facilitated using near-infrared spectroscopy and Fourier transform infrared spectroscopy when integrated in the fermentation process. This allows fermentation broth analysis and the ability to assay *in situ*, avoiding sample preparation and permitting timely adjustment of environmental parameters. However, on-line mass spectroscopy analysis is helpful mostly in a fermentation that is less sensitive to specific variables, as it cuts down routine assay work, sample preparation time, and the need for expensive equipment.

After the successful introduction of the improved strain in fermenters, production processes are validated and designed to run automatically for comparative and consistent operations. In some cases mathematical modeling of the physiological state and microbiological process are elucidated for maximizing strain performance. Typically this is done in three stages: (a) qualitatively analyzing relationships among growth, substrate consumption, and production (usually based on the assumption of metabolic pathway and biogenesis of product); (b) establishing mathematical formulations and kinetic equations of the model, emphasizing the role of operator functions and technical operation associated with overproduction; and (c) estimation of parameters and simulation of the model on the basis of experimental data. During these scale-up and modeling studies, emphasis is also placed on the capital and operating costs as well as the reliability of the process (Hemker, 1972; Flynn, 1983).

Among the various strains of microbes that have been scaled up, a few problems have invariably been noticed when scaling up filamentous organisms. The viscous nature of the culture creates heterogeneity, uneven mixing and distribution of bubbles, and failure to disperse micelle and floc formation. Several of these factors can be addressed up front during strain selection. The development of morphological mutants with short mycelia (higher surface area per unit volume) has been beneficial. This change can also influence the release of heat and spent gas without causing gradients in the fermenters. In addition, methods of mixing, nutrient feeding, and pH control also play critical roles in successful scale-up of improved mycelial strains. Finally, data on performance of the broth in pilot and large-scale purifications are also crucial for approval of improved strains for market production.

C. Metabolic engineering for strain development

In the broadest sense, metabolic engineering is a new technology in strain improvement that optimizes, in a coordinated fashion, the biochemical network and metabolic flux within the fermenters, with inputs from chemical engineering, cell physiology, biochemistry, and genetics. By systematically analyzing individual enzymatic reactions and pathways (their kinetics and regulation), methods are designed to eliminate bottlenecks in the flow of precursors and to balance stoichiometrically the distribution of metabolites for optimum product formation. Nuclear magnetic resonance studies of metabolic flux analysis and kinetic measurements are further combined with thermodynamic analysis of the biological process to predict better strain performance. The principles governing a biosynthetic pathway, including genetic controls, interaction with complex raw material sources, and bioreactor operations and mathematical modeling strategy, are exploited to exceed the microbe's capability and improve its productivity. Metabolic engineering applications in strain improvement have found a special niche as a result of their previously observed successes in the production of amino acids and biopolymers from strains of *Brevibacterium*, *Corynebacterium*, and *Xanthomonas* (Rowlands, 1984).

VI. SUMMARY

Strain development has been the icon of the fermentation industries. Discoveries in mutation, protoplast fusion, genetic manipulations, and recombinant DNA technology and the experience gained from modern reactor design and operation of fermenters have revolutionized the concept of microbial strain development. Although greater improvements in overproduction of metabolites and antibiotics of specific microbes have resulted from essentially random empirical approaches to mutation and strain development, future strain development technology will be supplemented by more knowledge-based scientific methods. With the advances in understanding biosynthetic pathways, elucidation of regulatory mechanisms related to induction and repression of genes, as well as bioengineering design, it will be possible to apply new strategies and limitless combinations for isolating improved strains. Furthermore, tailoring genes through the avenue of *in vitro* DNA recombination techniques in both bacteria and fungi has been shown to be feasible. Perhaps these areas will facilitate new strategies and have higher impact on industrial strain improvement.

BIBLIOGRAPHY

Baltz, R. H. (1986). Mutagenesis in *Streptomyces* spp. *In* "Manual of Industrial Microbiology and Biotechnology" (A. Demain and N. A. Solomon, eds.), pp. 184–190. American Society of Microbiology, Washington, D.C.

Broadbent, J. F., and Kondo, J. K. (1991). Genetic construction of nisin-producing *Lactococcus cremoris* and analysis of a rapid method for conjugation. *Appl. Environ. Microbiol.* **57**, 517–524.

Crueger, W., and Crueger, A. (1984). Antibiotics. *In* "Biotechnology: A Textbook of Industrial Microbiology," pp. 197–233. Akademische Verlagsgesellschaft, Wiesbaden, Germany. English translation copyright 1984 by Science Tech, Madison, WI.

Demain, A. L. (1983). New applications of microbial products. *Science* **219**, 709–714.

Elander, R., and Vournakis, J. (1986). Genetics aspects of overproduction of antibiotics and other secondary metabolites. *In* "Overproduction of Microbial Metabolites" (Z. Vanek and Z. Hostalek, eds.), pp. 63–82. Butterworth, London.

Flynn, D. (1983). Instrumentation for fermentation processes. *In* "IFAC Workshop, 1st, Helsinki, Finland, 1982. Modeling and Control of Biotechnical Processes: Proceedings," pp. 5–6. Pergamon, Oxford.

Hamer, D. H. (1980). DNA cloning in mammalian cells with SV40 vectors. *In* "Genetic Engineering, Principles and Methods" (J. K. Setlow and A. Hollander, eds.), Vol. 2, pp. 83–102. Plenum, New York and London.

Hemker, P. W. (1972). *In* "Analysis and Simulation of Biochemical Systems," Proc. 8th FEBS Meeting, pp. 59–80. Elsevier North-Holland, Amsterdam.

Lein, J. (1986). Random thoughts on strain development. *SIM News* **36**, 8–9.

Matsushima, P., and Baltz, R. (1986). Protoplast fusion. *In* "Manual of Industrial Microbiology and Biotechnology" (A. Demain and N. Solomon, eds.), pp. 170–183. American Society of Microbiology, Washington, DC.

Nolan, R. (1986). Automation system in strain improvement. *In* "Overproduction of Microbial Metabolites" (Z. Vanek and Z. Hostalek, eds.), pp. 215–230. Butterworth, London.

Queener, S., and Lively, D. (1986). Screening and selection for strain improvement. *In* "Manual of Industrial Microbiology and Biotechnology" (A. Demain and N. Solomon, eds.), pp. 155–169. American Society of Microbiology, Washington, DC.

Queener, S., Sebek, K., and Veznia, C. (1978). Mutants blocked in antibiotic synthesis. *Annu. Rev. Microbiol.* **32**, 593–636.

Rowlands, R. T. (1984). Industrial strain improvement: Mutagenesis and random screening procedures. *Enzyme Microbial Technol.* **6**, 3–10.

Schroeder, W., and Johnson, E. (1995). Carotenoids protect *Phaffia rhodozyma* against singlet oxygen damage. *J. Indust. Microbiol.* **14**, 502–507.

WEBSITE

The Electronic of Biotechnology
http://www.ejb.org/

83

Sulfur cycle

Piet Lens, Marcus Vallero, and Look Hulshoff Pol
Wageningen Agricultural University

GLOSSARY

acidophiles Bacteria that prefer acidic conditions, that is, microbes with a low pH optimum, typically below pH 4.

assimilatory reduction Reduction of a compound for the purpose of introducing its building elements into cellular material.

dissimilatory reduction Reduction of a compound in an energy-yielding reaction. The element does not become incorporated into cellular material.

sulfate reducing bacteria (SRB) Name of a group of bacteria belonging to a diversity of genera that gain their metabolic energy from the reduction of sulfate to sulfide.

sulfide oxidizing bacteria Name of a group of bacteria belonging to a diversity of genera that gain their metabolic energy from the oxidation of sulfide to sulfate. Under certain conditions, the oxidation is incomplete and stops at elemental sulfur, thiosulfate, or sulfite.

sulfuretum Habitat with a complete sulfur cycle (plural: sulfureta).

sulfur reducing bacteria Name of a group of bacteria belonging to a diversity of genera that gain metabolic energy from the reduction of elemental sulfur to sulfide.

Sulfur Cycling is a natural environmental process in which sequential transformation reactions interconvert sulfur atoms in different valence states ranging from −2 to +6. The reactions of the sulfur cycle alter the chemical, physical, and biological status of sulfur and its compounds so that sulfur cycling can occur. Many of the reactions of the sulfur cycle are mediated by microorganisms. The transformation reactions involved represent a continuous flow of sulfur-containing compounds among Earth's various compartments (soil, water, air, and biomass). Besides cycling of sulfur on a macroscale, internal sulfur cycles within one compartment also exist. These internal cycles depend on gradients of oxygen and the sulfur compounds, which can occur on a large scale (e.g., in stratified lakes) or on a micrometer scale (e.g., in laminated mats and wastewater treatment biofilms). Disrupting of the sulfur cycle can lead to several serious environmental problems. On the other hand, sulfur biotransformations are the basis of a whole set of environmental bioremediation technologies.

I. PLANETARY SULFUR FLUXES

Sulfur is the eighth most abundant element in the solar atmosphere and the fourteenth most abundant element in Earth's crust. Sulfur is present in several of the large environmental compartments present on Earth (Table 83.1). The lithospheric compartment is the largest and contains roughly 95% of this element. The second largest compartment is the hydrosphere, and Earth's oceans contain approximately 5% of the total sulfur. Sulfate is the second most abundant anion

The Desk Encyclopedia of Microbiology
ISBN: 0-12-621361-5

in seawater. The other compartments in which sulfur is found together comprise $< 0.001\%$ of the remaining sulfur.

Within the frame of global biogeochemical cycling (Fig. 83.1), sulfur is transformed with respect to its oxidation state, formation of organic and inorganic compounds, and its physical status (gas, liquid, or solid; soluble or insoluble). In oxidizing conditions, the most stable sulfur species is sulfate. In reducing environments, sulfide is formed. Sulfide and some organic compounds containing reduced sulfur, for example, dimethyl sulfide (DMS), carbonyl sulfide (COS), and carbonyl disulfide (CS_2), are volatile and can escape to the atmosphere. From there, sulfur compounds are redeposited into the litho-, hydro-, and pedospheres, either directly or after conversion to sulfate, via the gaseous intermediate SO_2.

TABLE 83.1 Amount of sulfur contained in various components of earth

Component	Amount of sulfur (kg)
Atmosphere	4.8×10^{9}
Lithosphere	2.4×10^{19}
Hydrosphere	
Sea	1.3×10^{18}
Freshwater	3.0×10^{12}
Marine organisms	2.4×10^{11}
Pedosphere	
Soil	2.6×10^{14}
Soil organic matter	1.0×10^{13}
Biosphere	8.0×10^{12}

Sulfide has a high chemical reactivity with some metal cations, leading to the formation of poorly soluble metal sulfides. This results in an accumulation of solid-state reduced sulfur stocks within global planetary sulfur cycling, owing to the formation of insoluble metal sulfides in most anaerobic environments, such as marshes, wetlands, freshwater, and sea sediments all over the globe. The formation of solid-state reduced sulfur proceeded even in the early history of planetary biogeochemical cycling. Accumulation of sulfur in anaerobic, highly organic-rich deposits of biomass resulted in contamination of coal by metal sulfides as well as organically bound sulfur compounds (sulfur content ranging between 0.05 and 15.0%). Similarly, mineral oils and petroleum can contain substantial amounts of sulfur compounds (0.025–5%). Under certain conditions, large—commercially exploitable—quantities of elemental sulfur can accumulate in petroleum reservoirs (see Section III). Other stocks of accumulated solid-state reduced

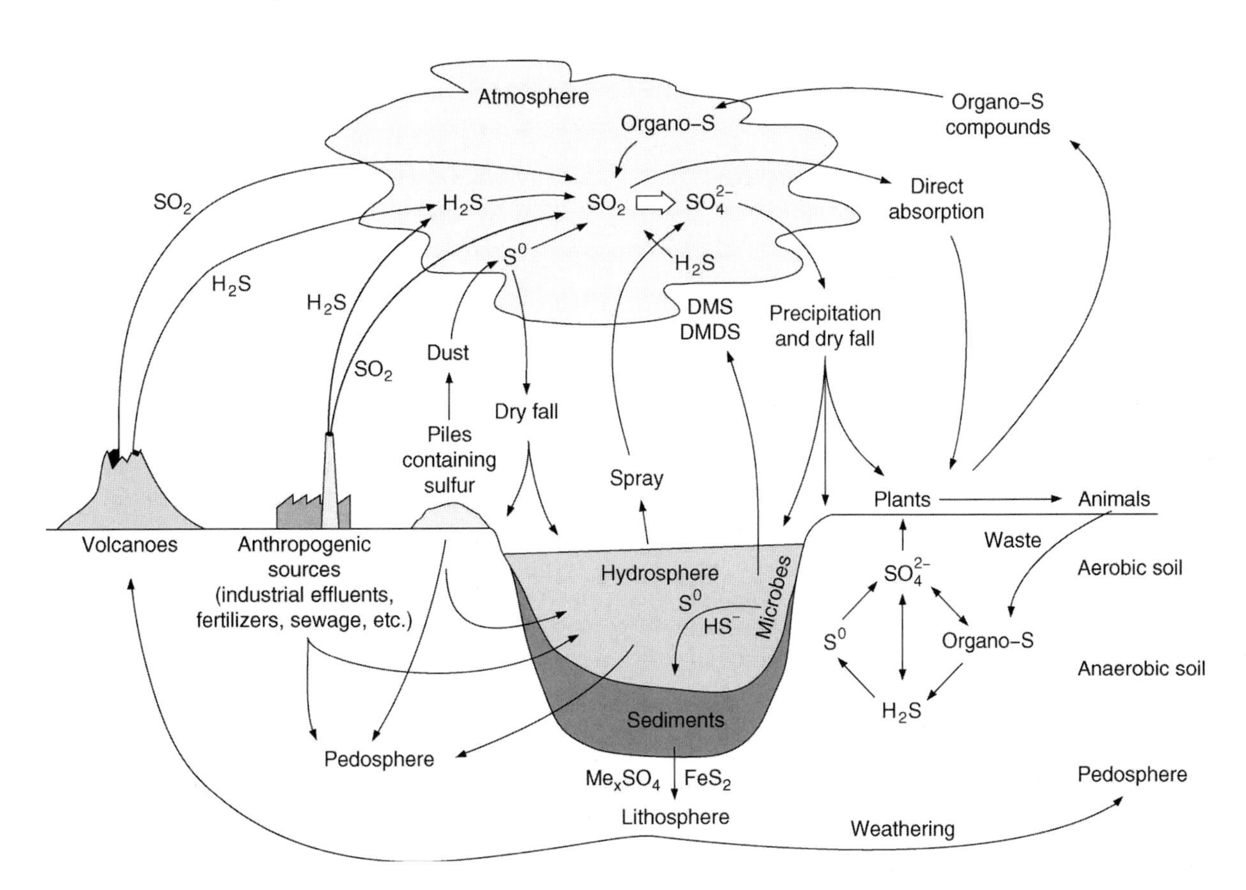

FIGURE 83.1 Simplified version of the overall sulfur cycle in nature.

sulfur are the sulfidic ores, for example, pyrite (FeS2), covellite (CuS), chalcopyrite (CuFeS2), galena (PbS), and sphalerite (ZnS).

Increasing anthropogenic extraction of sulfur-containing compounds from the lithosphere considerably perturbs the global sulfur cycle. The most important anthropogenic flux is the emission of sulfur into the atmosphere (113 Tg S/year, i.e., 113×10^{12} g S/year). In the continental part of the sulfur cycle, this flux is comparable only with weathering (114.1 Tg S/year) and river runoff to world oceans (108.9 Tg S/year). The second major anthropogenic flux of sulfur is the pollution of rivers, and subsequent runoff of the river waters (104 Tg S/year). The anthropogenic sulfur inputs on the globe result in an acceleration of sulfur cycling. This is manifested in elevated levels of sulfate in runoff waters, buildup of sulfide in anaerobic environments, and, after exposure of reduced-sulfur stocks to air, acidification of the environment and leaching of toxic metals (see Section IV). At the same time, the increased anthropogenic emissions to the atmosphere bring about other adverse environmental effects, for example, acid deposition. The negative consequences of acid rain are well known and include, for example, extensive damage to forests and wildlife and detrimental effects on buildings, constructions, and artifacts such as art works, statues, etc.

II. THE MICROBIAL SULFUR CYCLE

The behavior of sulfur compounds in the environment is highly influenced by the activity of living organisms, particularly microbes. In Fig. 83.2, the stocks of sulfur with different oxidation status (marked by squares) are given: S^{2-} is the sulfide form, S^0 is elemental sulfur, SO_4^{2-} is sulfate, and C-SH represents the stock of organic sulfur compounds. Arrows indicate the trophic status of microbes in each process, distinguishing autotrophic (using inorganic CO_2) and heterotrophic (using organic carbon compounds, C_{Org}).

Since the 1980s, the ecology of bacteria with a role in the sulfur cycle has received considerable attention from different scientific fields (e.g., microbial mats and sediments, wastewater treatment biofilms, and corrosion). This involved various advanced analytical techniques, for example, quantification of reaction products at the micrometer scale using microelectrodes for oxygen, sulfide, pH, and glucose; determination of metabolic and transport processes using ^{1}H, ^{13}C, and ^{31}P nuclear magnetic resonance (NMR) techniques, X-ray Absorption Nean Edge Spectroscopy (XANES),

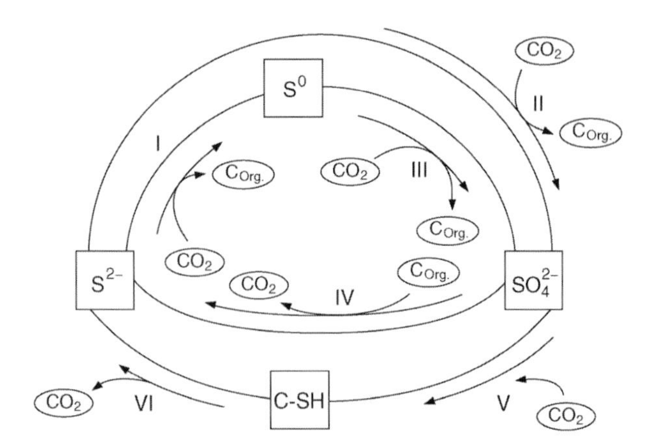

FIGURE 83.2 Schematic representation of the different pathways occurring in the microbial sulfur cycle. After Tichý *et al.* (1998).

and studies of population dynamics using 16S ribosomal RNA (rRNA) based detection methods.

A. Microbial sulfate reduction

Sulfate salts are the major stock of mobile sulfur compounds. They are mostly highly soluble in water, and considerable amounts can be transported in the environment. In the microbial sulfur cycle, sulfate is converted to sulfide by sulfate reducing bacteria (SRB) via dissimilatory sulfate reduction (pathway I in Fig. 83.2). This process of bacterial respiration occurs under strictly anaerobic conditions and uses sulfate as terminal electron acceptor. Electron donors are usually organic compounds, eventually, hydrogen:

$$8H_2 + 2SO_4^{2-} \rightarrow H_2S + HS^- + 5H_2O + 3OH^- \quad (1)$$

SRB include the traditional sulfate reducing genera *Desulfovibrio* and *Desulfotomaculum* in addition to the morphologically and physiologically different genera *Desulfobacter, Desulfobulbus, Desulfococcus, Desulfonema,* and *Desulfosarcina* (Widdel, 1988). In the presence of sulfate, SRB are able to use several intermediates of the anaerobic mineralization process. Besides the direct methanogenic substrates molecular hydrogen (H_2), formate, acetate, and methanol, they can also use propionate, butyrate, higher and branched fatty acids, lactate, ethanol and higher alcohols, fumarate, succinate, malate, and aromatic compounds. In sulfidogenic breakdown of volatile fatty acids, two oxidation patterns can be distinguished. Some SRB are able to completely oxidize volatile fatty acids to CO_2 and sulfide as end products. Other SRB lack the tricarboxylic acid cycle and carry out an incomplete oxidation of volatile fatty acids, with acetate and sulfide as end products.

In addition to the reduction of sulfate, reduction of sulfite and thiosulfate is also very common among

SRB. *Desulfovibrio* strains have been reported to be able to reduce di-, tri-, and tetrathionate. A unique ability of some SRB, for example, *Desulfovibrio dismutans* and *Desulfobacter curvatus*, is the dismutation of sulfite or thiosulfate:

$$4SO_3^{2-} + H^+ \rightarrow 3SO_4^{2-} + HS^- \quad (2)$$

$$S_2O_3^{2-} + H_2O \rightarrow SO_4^{2-} + HS^- + H^+ \quad (3)$$

Some SRB were found to be able to respire oxygen, despite being classified as strictly anaerobic bacteria. Thus far, however, aerobic growth of pure cultures of SRB has not been demonstrated. The ability of SRB to carry out sulfate reduction under aerobic conditions remains nevertheless intriguing and could be of significance for microscale sulfur cycles (see Section III).

In the absence of an electron acceptor, SRB are able to grow through a fermentative or acetogenic reaction (Widdel, 1988). Pyruvate, lactate, and ethanol are easily fermented by many SRB. An interesting feature of SRB is their ability to perform acetogenic oxidation in syntrophy with hydrogenotrophic methanogenic bacteria, as described for cocultures of hydrogenotrophic methanogenic bacteria with *Desulfovibrio* sp. using lactate and ethanol or with *Desulfobulbus*-like bacteria using propionate. In the presence of sulfate, however, these bacteria behave as true SRB and metabolize propionate as the electron donor for the reduction of sulfate.

B. Microbial sulfur oxidation

Different bacteria can oxidize various reduced sulfur compounds, for example, sulfide, elemental sulfur, or thiosulfate. Oxidation of sulfide to elemental sulfur (pathway II in Fig. 83.2) is performed by autotrophic bacteria. Equation (4) gives the stoichiometry of the chemoautotrophic process, which proceeds aerobically or microaerobically. In addition, photoautotrophic sulfide oxidation can also occur under anaerobic conditions. Photosynthetic sulfur bacteria are capable of photoreducing CO_2 while oxidizing H_2S to S^0 [Eq. (5)], in a striking analogy to the photosynthesis of eukaryotes [Eq. (6)].

$$2H_2S + O_2 \rightarrow 2S^0 + 2H_2O \quad (4)$$

$$2H_2S + CO_2 + h\nu \rightarrow 2S^0 + [CH_2O] + H_2O \quad (5)$$

$$2H_2O + CO_2 + h\nu \rightarrow 2O_2 + [CH_2O] \quad (6)$$

Sulfide can also be completely oxidized to sulfate. Equation (7) gives a formula for the chemoautotrophic process, although photoautotrophic oxidation of sulfide to sulfate can also occur.

$$H_2S + 2O_2 \rightarrow SO_4^{2-} + 2H^+ \quad (7)$$

This oxidation reaction, catalyzed by, for example, *Thiobacillus*, involves a series of intermediates, including sulfide, elemental sulfur, thiosulfate, tetrathionate, and sulfate (Kelly *et al.*, 1997):

$$SH^- \rightarrow S^0 \rightarrow S_2O_3^{2-} \rightarrow S_4O_6^{2-} \rightarrow SO_4^{2-} \quad (8)$$

Eventually, oxidation of sulfide may proceed in oxygen-free conditions, using nitrate as the electron acceptor [Eq. (9)]. Oxidation of sulfide in anoxic conditions is mostly carried out by bacteria from the genus *Thiobacillus*, such as *T. albertis* and *T. neapolitanus.*

$$\begin{aligned} 0.422\,H_2S + 0.422\,HS^- + NO_3^- + 0.437\,CO_2 \\ + 0.0865\,HCO_3^- \\ + 0.0865\,NH_4^+ \rightarrow 1.114\,SO_4^- \\ + 0.5\,N_2 + 0.0842\,C_5H_7O_2N\ \text{(biomass)} \\ + 1.228\,H^+ \end{aligned} \quad (9)$$

Elemental sulfur can be oxidized to sulfate via chemoautotrophic or photoautotrophic microorganisms (Fig. 83.2, pathway IV). The stoichiometry of the chemoautotrophic process is given in Eq. (10).

$$2\,S^0 + 3O_2 + 2H_2O \rightarrow 2SO_4^{2-} + 4\,H^+ \quad (10)$$

The biological oxidation of reduced sulfur is mediated by a diverse range of bacterial species (Table 83.2). They can be divided into two main groups: the aerobic and microaerobic chemotrophic sulfur oxidizers (sometimes called the colorless sulfur bacteria) and the anaerobic phototrophic sulfur oxidizers (sometimes called the purple and green sulfur bacteria).

The most common microorganisms associated with sulfide oxidation are *Thiobacillus* spp. These non-spore-forming bacteria belong to the colorless sulfur bacteria. Thiobacilli are gram-negative rods about 0.3 μm in diameter and 1–3 μm long. Most *Thiobacillus* species are motile by polar flagella. All *Thiobacillus* species grow aerobically, although anaerobic growth [see Eq. (9)] has been observed for some species as well. Elemental sulfur accumulates on the cell surface, in contrast to filamentous colorless sulfur bacteria, for example, *Thiothrix* sp., *Beggiatoa* sp., and *Thioploca* sp., which accumulate elemental sulfur intracellularly. *Thiothrix* spp. are commonly found in flowing sulfidic water containing oxygen in both marine and freshwater environments such as outlets of sulfidic springs and wastewater treatment plants. They differentiate ecologically from other filamentous colorless sulfur bacteria in that *Thiothrix* spp. prefer hard substrates to which they attach with a holdfast, whereas *Beggiatoa* spp. and *Thiploca* spp. do not attach and prefer soft bottom sediments. The diameter of the filaments is variable in all three genera and is used to define species.

TABLE 83.2 Sulfide and sulfur oxidizing bacteria

Genus or group	Habitat	Comments
Chemotrophs		
Thiobacillus	Soil, water, marine	Mostly lithotrophs, one Fe(II) oxidizer, some thermophiles, deposit S^0 outside cell
Sulfobacillus	Mine tips	Lithotroph, spore-former, Fe(II) oxidizer, thermophilic
Thiomicrospira	Marine	Lithotroph
Beggiatoa	Water, soil, marine	Gliding bacteria, deposit S^0 inside cell, difficult to isolate
Thiothrix	Water, soil, marine	
Thioploca	Water, soil	
Achromatium	Water	
Thiobacterium	Water	Deposit S^0 inside cell, not grown in pure cultures
Macromonas	Water	
Thiovulum	Water	
Thiospira	Water	
Sulfolobus	Geothermal springs	Archaebacterium, lithotroph, Fe(II) oxidizer, thermophilic
Phototrophs		
Chlorobiaceae (green S bacteria)	Water, marine	Lithotrophs, deposit S^0 outside cell
Chromatiaceae (purple S bacteria)	Water, marine	Mixotrophs, deposit S^0 inside cell, except for *Ectothiorhodospira* spp. which deposit S^0 outside cell
Chloroflexaceae	Geothermal springs	Mixotrophs, deposit S^0 outside cell, thermophiles
Oscillatoria (blue-green algae/cyanobacteria)	Water	S^0 deposited outside cell under anoxic conditions

The phototrophic bacteria make up the Chlorobiaceae (green sulfur bacteria), the Chromatiaceae (purple sulfur bacteria), and the filamentous thermophilic flexibacteria, examplified by *Chloroflexus aurantiacus*. Phototrophic bacteria contribute substantially to both the sulfur cycle and the primary productivity of shallow aquatic environments, but they are less significant in the mid-ocean where they are limited by light availability.

In addition to the specialized bacteria listed in Table 83.2, a considerable number of common bacteria (e.g., *Bacillus* spp., *Pseudomonas* spp., and *Arthrobacter* spp.) and fungi (e.g., *Aspergillus* spp.) have also been shown to oxidize significant amounts of reduced sulfur compounds when grown in pure culture. Even if such transformations are incidental, these organisms are present in large numbers in soils and the aquatic environment compared to *Thiobacillus* spp., suggesting that nonlithotrophic sulfur oxidation may be quantitatively important.

Species of *Thiobacillus*, as the main representatives of the acidophiles, play a key role in the degradation of sulfidic materials (Johnson *et al.*, 1993). In addition to the oxidation of reduced sulfur compounds as their energy source [see Eqs. (7) and (10)], they can also gain energy from the conversion of ferrous (Fe^{2+}) to ferric (Fe^{3+}) iron. The best known acidophile is *Thiobacillus ferrooxidans*, which combines the ability to oxidize both sulfur compounds and ferrous iron. The bacterium is able to oxidize, at low pH values (pH 1–4), sulfidic minerals such as pyrite according to the following reaction [Eq. (11)]:

$$4FeS_2 + 15O_2 + 2H_2O \rightarrow 2Fe_2(SO_4)_3 + 2H_2SO_4 \quad (11)$$

Thiobacillus ferrooxidans was isolated from acid mine drainage water in the late 1940s, together with *T. thiooxidans*. The latter species can only oxidize sulfur and reduced sulfur compounds and lacks the ferrous iron-oxidizing capacity. In contrast to *T. ferrooxidans, T. thiooxidans* cannot attack sulfidic minerals on its own, but it can contribute to their solubilization in a syntrophic relation with ferrous iron oxidizers such as *Leptospirillum ferrooxidans*. The latter microorganism can only convert ferrous iron and has no sulfur-oxidizing capacity.

The group of acidophiles also comprises facultative autotrophs or even obligate heterotrophs (Fortin *et al.*, 1996). A representative of this subgroup is *T. acidophilus*, a facultatively autotrophic sulfur compound oxidizer that can also grow on, for example, sugars. Apart from solubilizing sulfidic minerals, facultatively autotrophic acidophiles are also crucial in removing low molecular weight organic acids, which are toxic to the obligate autotrophs even at very low concentrations (5–10 mg/l). These toxic effects are due to the uptake of organic compounds with carboxylic groups, which are undissociated in the outer medium with low pH but are dissociated in bacterial cytoplasm with circumneutral pH.

C. Transformations of organic sulfur compounds

The formation and degradation of organic sulfur compounds (C–SH) are not solely microbial processes, and numerous other organisms participate in them. Particularly, the formation of organic sulfur (Fig. 83.2, pathway V) is accomplished by all photosynthesizing organisms, including algae and green plants. Dimethyl sulfide (DMS) is the most common product of oceanic green algae sulfur conversions. Green plants and many microorganisms assimilate sulfate as their sole sulfur source. Therefore, they reduce sulfate via a reductive process, assimilatory sulfate reduction, in which the formed sulfide is incorporated into organic matter via a condensation reaction with serine derivatives to generate the amino acid cysteine.

Conversion of organic sulfur to sulfide occurs during the decomposition of organic matter (Fig. 83.2, pathway VI). Considerable environmental risks are encountered in these processes, especially regarding the volatilization of organic sulfur compounds and associated odor pollution (Smet *et al.*, 1998).

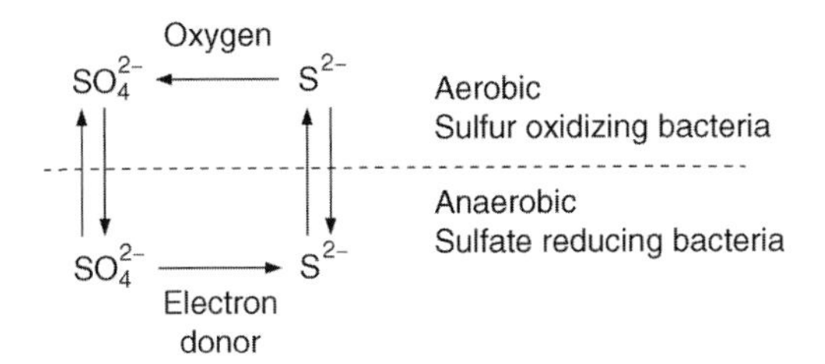

FIGURE 83.3 Schematic representation of the cyclic reactions prevailing in a microbial sulfur cycle between sulfate reducing and sulfide oxidizing bacteria.

III. SULFUR CYCLING WITHIN ECOSYSTEMS

The sulfur cycle, together with the carbon and nitrogen cycle, is distinguished from most other mineral cycles (e.g., P, Fe, Si) by exhibiting transformations from gaseous to ionic (aqueous or solid) forms. Each of the transformations (see Section II) is reliant on appropriate cellular and ambient oxygen (or redox) conditions. Thus, cycling of these elements depends on the presence of oxygen (or redox) gradients. These gradients vary in size from the smallest, which can be generated over distances of only a few micrometers, for example, biofilms and sediments, to larger zones (a few millimeters) such as in soil crumbs and microbial mats. In exceptional circumstances, these gradients may extend over several tens of meters, for example, in geothermal springs, stable stratified lakes, or sewer outfalls. Of particular interest for the sulfur cycle are the steep gradients present in microenvironmental or microzonal conditions, as they allow interactions between cells that normally cannot coexist, namely, between anaerobic sulfate reducers and aerobic sulfide oxidizers.

In some cases, very neat cyclic reactions are possible. The clearest example is the sulfur cycle as illustrated in Fig. 83.3. Habitats with a complete sulfur cycle are

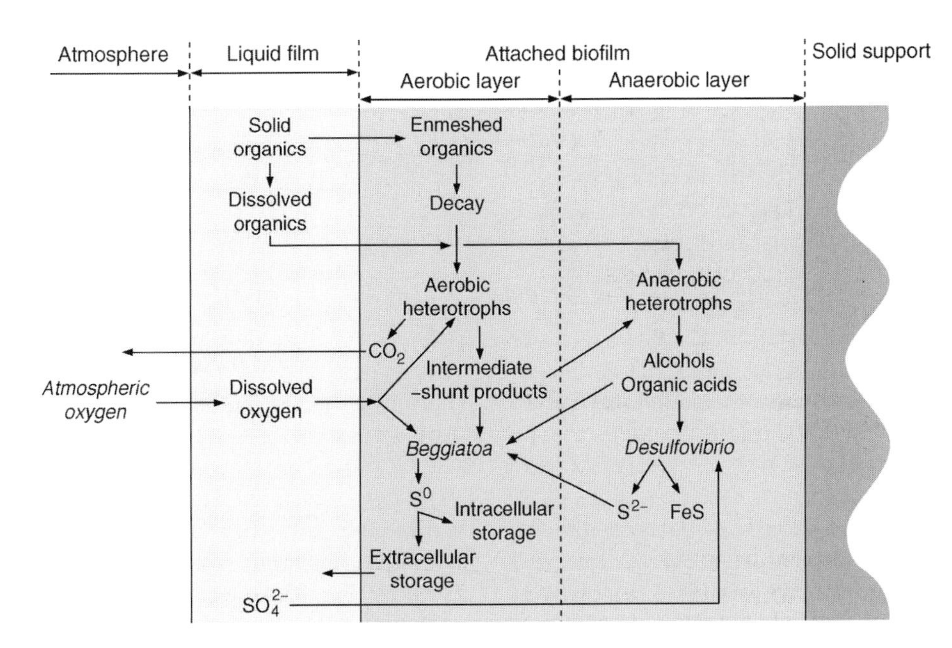

FIGURE 83.4 Reactions involved in a sulfur cycle in a biofilm of a rotating biological contactor type wastewater-treatment system. After Alleman *et al.* (1982).

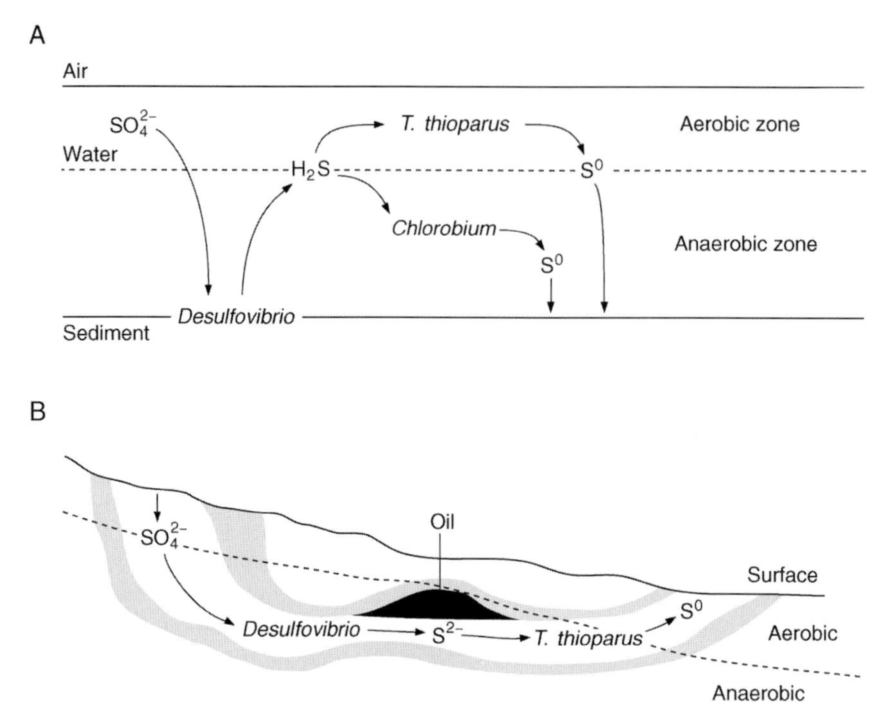

FIGURE 83.5 The biological deposition of sulfur (A) in a lake and (B) in geological strata mediated by the disrupting of sulfur cycling activities.

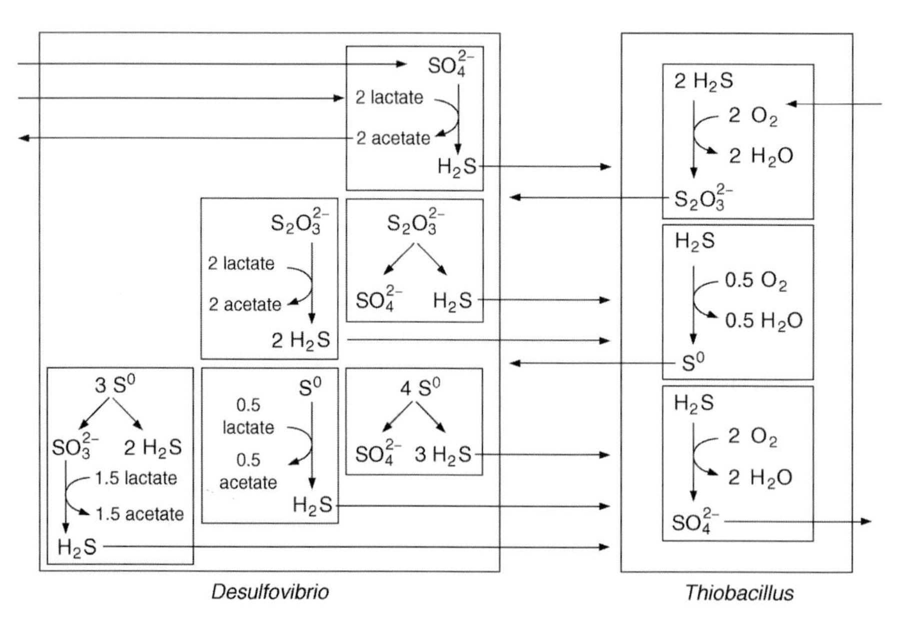

FIGURE 83.6 Possible pathways of sulfur transformations in mixed cultures of *Desulfovibrio* and *Thiobacillus* supplied with lactate, sulfate, and oxygen. After Van den Ende *et al.* (1997).

known as sulfureta. Both groups of bacteria may be found close to the border between the aerobic and anaerobic habitats, and neither organism can grow in the other's space. Yet, they are totally dependent on sulfur compounds that diffuse between them. Such cycles are common in marine or estuarine sediments, and they are also evident in stratified water bodies and fixed film wastewater-treatment systems (Fig. 83.4).

Figure 83.4 depicts possible blocks of the cyclic activity, for example, sulfide precipitating with heavy metals (e.g., FeS) or accumulation of elemental sulfur in sulfide oxidizing bacteria. The latter also prevails in natural, light exposed environments, as illustrated in Fig. 83.5, which depicts some scenarios for the biogenic deposition of elemental sulfur in a lake and in the vicinity of petroleum reservoirs. Formation of colloid/solid elemental sulfur does not necessarily imply a complete blocking of sulfur cycling activity. The sulfur can be reduced to sulfide by, for example, *Desulfovibrio desulfuricans*, thus allowing the sulfur cycle to proceed (Fig. 83.6).

IV. ENVIRONMENTAL CONSEQUENCES AND TECHNOLOGICAL APPLICATIONS

Disruption of sulfur cycling can also have serious environmental consequences. Below, an overview is given of the major environmental effects when the sulfur cycle is unbalanced. This imbalance can be a result of both natural and anthropogenic processes. A number of technological applications that utilize bacteria of the sulfur cycle are also presented.

A. Environmental consequences

Microbial transformations of solid-state reduced sulfur compounds that are used or affected by anthropogenic activities represent major environmental risks. Materials such as fossil fuels, ores, anaerobic sediments, or solid waste may undergo oxidative changes, resulting in a solubilization of sulfur from the solid phase. Thus, large amounts of sulfuric acid are formed [see Eq. (11)], which are transported off site in so-called acid mine drainage (Dvorak *et al.*, 1992). The acid inhibits plant growth and aquatic life. Its major environmental consequence is, however, the solubilization of cationic heavy metals.

In tropical regions of the world, periodic flooding and draining of maritime estuarine soils leads to the accumulation of reduced sulfur and its subsequent oxidation (Begheijn *et al.*, 1978). This results in production of sulfuric acid and appearance of acid sulfate soils, sometimes called "cat clays." In many cases, treatment (neutralization) of these sites is required, for example, before farming may be allowed.

Sulfate-rich wastewater is generated by many industrial processes that use sulfuric acid or sulfate-rich feed stocks (e.g., fermentation or seafood-processing industries). Also, the use of reduced sulfur compounds in industrial processes, namely, sulfide (tanneries, Kraft pulping), sulfite (sulfite pulping), thiosulfate (fixing of photographs), or dithionite (pulp bleaching), contaminates wastewater with sulfurous compounds.

Pollution by humans has greatly increased the sulfur dioxide levels in the atmosphere. When dissolved in water in the clouds, SO_2 acidifies the rainwater. Acidic rainfall has been implicated in the reduction of growth of forests in North America and Europe. Acid rain also has a detrimental effect on buildings, constructions, and other artifacts such as ancient art works, causing etching and destroying their original beauty.

Many of the volatile organic and inorganic sulfur compounds are extremely odorous. Farmyard feedlots and improperly managed rendering and composting facilities are examples of places where offensive sulfur gases are generated.

B. Technological applications

Hydrometallurgical processes play an important role in the extraction of metals from certain low-grade ores, for example, those with $<1\%$ Cu by weight. Hydrometallurgy consists of the dissolution of metals from minerals, usually by constant percolation of a leaching solution through beds of ore (Bailey and Hansford, 1993). The process is most effective in ores that contain substantial amounts of pyrite and is based on the rapid oxidation of S^{2-} and Fe^{2+} promoted by acidophiles. In the literature, two mechanisms by which acidophiles attack the insoluble metal sulfides are proposed, namely, indirect and direct leaching (Evangelou and Zhang, 1995). In the indirect mechanism, the ferric iron acts as a chemical oxidizer of the sulfidic minerals. The ferric iron is a product of bacterial oxidation. In this mechanism, the biological sulfur-oxidizing capacity of the acidophiles is of no relevance. In the direct mechanism, sulfidic mineral solubilization involves both the biological ferrous iron and sulfur compound oxidation. Pyrite oxidation by *T. ferrooxidans* was shown to be mediated by both the biological oxidation of ferrous iron and sulfide.

Microbial leaching (bioleaching) is also proposed for decontamination of polluted solid wastes, soils, or sediments (Couillard and Mercier, 1992). Successful demonstration of the bioleaching of toxic metals from an anaerobically digested sewage sludge was demonstrated on a technological scale. Experimental attempts to use bioleaching for heavy metal removal from freshwater sediments are reported as well. The process may use the autoacidification potential of the sediment, driven by the presence of reduced sulfur and ferrous compounds. If the latter is insufficient

TABLE 83.3 Environmental biotechnological applications using processes of the microbial sulfur cycle

Application	Sulfur conversion utilized	Waste stream
Wastewater treatment		
Sulfate removal	Sulfate and/or sulfite reduction + partial sulfide oxidation to S^0	Industrial wastewaters, acid mine drainage, and spent sulfuric acid
Sulfide removal	Partial sulfide oxidation to S^0	Industrial wastewaters
Heavy metal removal	Sulfate reduction	Extensive treatment in wetlands or an-aerobic ponds High rate reactors for process water, acid mine drainage, and groundwater
Microaerobic treatment	Internal sulfur cycle in the biofim	Domestic sewage
Off-gas treatment		
Biofiltration of gases	Oxidation of sulfide and organosulfur compounds	Biogas, malodorous gases from composting and farming
Treatment of scrubbing waters	Sulfate and/or sulfite reduction + partial sulfide oxidation to S^0	Scrubbing waters of SO_2-rich gases
Solid waste treatment		
Bioleaching of metals	Sulfide oxidation	Sewage sludge, compost
Desulfurization	Oxidation	Rubber
Gypsum processing	Sulfate reduction	
Treatment of soils and sediments		
Bioleaching of metals	Sulfide oxidation	Dredged sediments and spoils
Degradation xenobiotics	Sulfate reduction	Polychlorinated bifenyl (PCB)-contaminated soil slurries

relative to the sediment's buffering capacity, additional sulfur can be added to achieve satisfactory extraction yields.

Reduced sulfur compounds are commonly added to alkaline soils so that the sulfuric acid produced decreases the soil pH to a level acceptable for plant growth. Moreover, sulfur compounds can be used as fertilizers to improve crop production in those areas of the world where sulfur is a limiting nutrient. This applies, for example, in Western European countries, where atmospheric deposition of sulfur to farmland decreased as a result of the stringent control of sulfur dioxide emissions.

Better insight into sulfur transformations enabled the development of a whole spectrum of new biotechnological applications for the bioremediation of polluted waters, gases, soils, and solid wastes (Table 83.3). These technologies rely on both bacterial sulfate reduction and sulfide/sulfur oxidation (Lens *et al.*, 2002). They allow the removal of sulfur and organic compounds as well as heavy metals and nitrogen. Until recently, biological treatment of sulfur-polluted wastestreams was rather unpopular because of the production of hydrogen sulfide under anaerobic conditions. Gaseous and dissolved sulfides cause physiochemical (corrosion, odor, increased effluent chemical oxygen demand) or biological (toxicity) constraints, which may lead to process failure. However, anaerobic treatment of sulfate-rich wastewater can be applied successfully provided a proper treatment strategy is selected (Lens *et al.*, 1998), which depends on the aim of the treatment: (i) removal of sulfur compounds, (ii) removal of organic matter, or (iii) removal of both.

Microbial sulfur transformations can provide a unique tool to control pollution by both sulfur compounds and heavy metals. In some cases, rather extensive techniques using natural processes are applied, for example, the use of wetlands or anaerobic ponds to treat voluminous aqueous streams like acid mine drainage or surface runoff waters. These systems require low maintenance and can sustain their function for prolonged time intervals. Subsequent regeneration of these systems using bioleaching and treatment of spent extraction liquor by SRB may be a logical step. In general, technological applications using various processes of the microbial sulfur cycle may provide many beneficial effects in the future.

BIBLIOGRAPHY

Alleman, J. E., Veil, J. A., and Canaday, J. T. (1982). Scanning electron microscope evaluation of rotating biological bio-film. *Wat. Res.* **16**, 543–550.

Bailey, A. D., and Hansford, G. S. (1993). Factors affecting biooxidation of sulfide minerals at high concentrations of solids—A review. *Biotech. Bioeng.* **42**, 1164–1174.

Begheijn, L. T., van Breemen, N., and Velthorst, E. J. (1978). Analysis of sulfur compounds in acid sulfate soils and other marine soils. *Commun. Soil Sci. Plant Anal.* **9**, 873–882.

Couillard, D., and Mercier, G. (1992). Metallurgical residue for solubilization of metals from sewage sludge. *J. Environ. Eng.* **118**, 808–813.

Dvorak, D. H., Hedin, R. S., Edenborn, H. M., and McIntire, P. E. (1992). Treatment of metal-contaminated water using bacterial sulfate reduction: Results from pilot-scale reactors. *Biotech. Bioeng.* **40**, 609–616.

Evangelou, V. P., and Zhang, Y. L. (1995). Pyrite oxidation mechanisms and acid mine drainage prevention. *Crit. Rev. Environ Sci. Technol.* **25**, 141–199.

Fortin, D., Davis, B., and Beveridge, T. J. (1996). Role of *Thiobacillus* and sulfate-reducing bacteria in iron biocycling in oxic and acidic mine tailings. *FEMS Microbiol. Ecol.* **21**, 11–24.

Johnson, D. B., McGinness, S., and Ghauri, M. A. (1993). Biogeochemical cycling of iron and sulfur in leaching environments. *FEMS Microbiol. Rev.* **11**, 63–70.

Kelly, D. P., Shergill, J. K., Lu, W.-P., and Wood, A. P. (1997). Oxidative metabolism of inorganic sulfur compounds by bacteria. *Antonie van Leeuwenhoek* **71**, 95–107.

Lens, P. N. L., Visser, A., Janssen, A. J. H., Hulshoff Pol, L. W., and Lettinga, G. (1998). Biotechnological treatment of sulfate rich wastewaters. *Crit. Rev. Environ. Sci. Technol.* **28**, 41–88.

Lens, P. N. L., Vallero, M., Esposito, G., and Zandvoort, M. (2002). Perspectives of sulphate reducing bioreactors in environmental biotechnology. *Rev. Environ. Sci. Bio/Technol.* **1**, 311–325.

Smet, E., Lens, P., and van Langenhove, H. (1998). Treatment of waste gases contaminated with odorous sulfur compounds. *Crit. Rev. Environ. Sci. Technol.* **28**, 89–116.

Tichý, R., Lens, P., Grotenhuis, J. T. C., and Bos, P. (1998). Solid-state reduced sulfur compounds: Environmental aspects and bioremediation. *Crit. Rev. Environ. Sci. Technol.* **28**, 1–40.

Van den Ende, F. P., Meier, J., and Van Gemerden, H. (1997). Syntrophic growth of sulfate-reducing bacteria and colorless sulfur bacteria during oxygen limitation. *FEMS Microbiol. Ecol.* **23**, 65–80.

Widdel, F. (1988). Microbiology and ecology of sulfate- and sulfur reducing bacteria. *In* "Biology of Anaerobic Microorganisms" (A. J. B. Zehnder, ed.), pp. 469–586. Wiley, New York.

WEBSITES

Animation of sulfur cycle (T. M. Terry)
http://www.microbelibrary.org

Links for bacteria involved in sulfur cycle (C. Hagedorn and N. Lowe)
http://soils1.cses.vt.edu/ch/biol_4684/Cycles/Soxidat.html

PowerPoint version of the sulfur cycle
http://www.ppi-far.org/ppiweb/ppibase.nsf/$webindex/article=A56B59A0852569B5005D14CC92A2158B

General Information about anaerobic reactors and sulfur cycle
http://www.uasb.org

Web Portal on Microorganisms
http://www.microbes.info

84

Transcriptional regulation in prokaryotes

Orna Amster-Choder
The Hebrew University Medical School

GLOSSARY

−10 element A consensus sequence centered about 10 bp before the start point of transcription, which is involved in the initial melting of DNA by RNA polymerase.

−35 element A consensus sequence centered about 35 bp before the start point of transcription, which is involved in the initial recognition by RNA polymerase.

promoter A sequence of DNA whose function is to be recognized by RNA polymerase in order to initiate transcription. A typical *E. coli* promoter contains two conserved elements, a −10 element and a −35 element (see above).

RNA polymerase The enzyme that synthesizes RNA using a DNA template (also termed DNA-dependent RNA polymerase).

sigma (σ) factor A subunit of bacterial RNA polymerase needed for initiation. The σ factor has major influence on the selection of promoters.

start point The position on the DNA that corresponds to the first base transcribed into RNA.

terminator A DNA sequence that causes RNA polymerase to terminate transcription and to dissociate from the DNA template.

transcription The synthesis of RNA on a DNA template.

transcription unit The DNA sequence that extends from the promoter to the terminator; it may include more than one gene.

The First Step in Gene Expression is the transcription of the coding DNA sequences to discrete RNA molecules. Specific DNA regions, defined as promoters, are recognized by the transcribing enzyme, a DNA-dependent RNA polymerase. The RNA polymerase binds to the promoter and initiates the synthesis of the RNA transcript. The enzyme catalyzes the sequential addition of ribonucleotides to the growing RNA chain in a template-dependent manner until it comes to a termination signal (terminator). The DNA sequence between the start point and the termination point defines a transcription unit. An RNA transcript can include one gene or more. Its sequence is identical to one strand of the DNA, the coding strand, and complementary to the other, which provides the template. The base at the start point is defined as +1 and the one before that as −1. Positive numbers are increased going downstream (into the transcribed region), whereas negative numbers increase going upstream. The immediate product of transcription, termed primary transcript, which extends from the promoter to the terminator, is almost always unstable. In prokaryotes, messenger RNA (mRNA) is usually translated concomitantly with being transcribed, and is rapidly degraded when not protected by the ribosomes, whereas ribosomal RNA (rRNA) and transfer RNA (tRNA) are cleaved to give mature products.

Transcription is the principal step at which gene expression is controlled. DNA signals and regulatory proteins determine whether the polymerase will choose to transcribe a certain gene and whether the

The Desk Encyclopedia of Microbiology
ISBN: 0-12-621361-5

whole process of transcription will be accomplished successfully. The timing of transcription of specific genes is influenced by environmental conditions and by the growth cycle phase.

The molecular picture of how genes are transcribed and the nature of the regulatory mechanisms that control transcription are far from being complete, but lots of progress has been made. Work on relatively simple organisms, bacteriophages and bacteria, has provided new insights into the mechanisms that are involved in the regulation of gene expression, transcriptional-control mechanisms being among them. Although there are significant differences in the organization of individual genes and in the details and the complexity of the regulatory mechanisms among prokaryotes and eukaryotes, it is clear that basic principles are shared among all organisms. Due to the relative simplicity of prokaryotic biochemical pathways, their easy manipulation in the laboratory, and the advanced tools that are available for changing their genotype and for testing the resulting phenotype, it is easier to infer these basic principles by studying prokaryotes.

I. THE TRANSCRIPTION MACHINERY

A. RNA polymerase catalyzes transcription

RNA synthesis is catalyzed by the enzyme RNA polymerase. The reaction involves the incorporation of ribonucleoside 5′ triphosphate precursors into an oligoribonucleotide transcript, based on their complementarity to bases on a DNA template. RNA polymerase catalyses the formation of phosphodiester bonds between the ribonucleotides. The formation of a phosphodiester bond involves a hydrophilic attack by the 3′-OH group of the last ribonucleotide in the chain on the 5′ triphosphate of the incoming ribonucleotide. The incoming ribonucleotide loses its terminal two phosphate groups, which are released in the form of pyrophosphate. In this manner, the RNA chain is synthesized from the 5′-end toward the 3′-end.

The best characterized RNA polymerases are those of eubacteria, for which *E. coli* is a prototype. Unlike eukaryotic cells, in which various types of polymerases are dedicated to the synthesis of the certain types of RNA, in eubacteria a single type of polymerase appears to be responsible for the synthesis of mRNA, rRNA, and tRNA.

The dimensions of bacterial RNA polymerase are approximately 90 × 95 × 160 Å. The molecular weight of the complete *E. coli* enzyme is approximately 465 kDa. About 7000 molecules of RNA polymerase are present in an *E. coli* cell, but the number of molecules engaged in transcription at any given time varies from 2000 to 5000, depending on the growth conditions.

The DNA sequence that is being transcribed by RNA polymerase is transiently separated into its single strands, with one of the strands serving as a template for the synthesis of the RNA strand. This region is therefore defined as the *transcription bubble*. As the RNA polymerase moves along the DNA, it unwinds the duplex at the front of the bubble and rewinds the DNA at the back, so that the duplex behind the transcription bubble reforms. Thus, the bubble moves with the RNA polymerase, and the RNA chain is elongated. The length of the transcription bubble varies from 12–20 bp. The length of the transient hybrid between the DNA and the newly synthesized RNA sequence within the transcription bubble is a matter of controversy and the estimates range from 2–12 nt. Beyond the growing point, the newly synthesized RNA chain enters a high-affinity binding site within the RNA polymerase.

B. Bacterial RNA polymerase consists of multiple subunits

The RNA polymerases of certain phages consist of single polypeptide chains. These polymerases recognize a very limited number of promoters and they lack the ability to change the set of promoters from which they initiate transcription. In contrast, bacterial RNA polymerases consist of several subunits. The most studied bacterial RNA polymerase, the one in *E. coli*, exists in two forms, an holoenzyme and a core enzyme. The holoenzyme is capable of selective initiation at promoter regions, whereas the core RNA polymerase is capable of elongation and termination, but not selective initiation. The two forms of the polymerase consist of two identical α-subunits, one β-subunit and one β'-subunit, but only the holoenzyme contains an additional subunit, one of several σ proteins. The subunit composition of the holoenzyme is summarized in Fig. 84.1. The α-subunit plays an important role in RNA polymerase assembly, which proceeds in the pathway $\alpha \rightarrow \alpha_2 \rightarrow \alpha_2\beta\beta' \rightarrow \alpha_2\beta\beta'\sigma$. The α-subunit plays some role in promoter recognition and in the interaction of RNA polymerase with transcriptional activators (see later). The β- and β'-subunits together constitute the catalytic center of RNA polymerase. The β-subunit was demonstrated to contact the template DNA, the newly synthesized RNA, and the substrate ribonucleotides. The β'-subunits contacts the RNA chain as well. Mutations in the genes that encode β and β' show that both subunits are

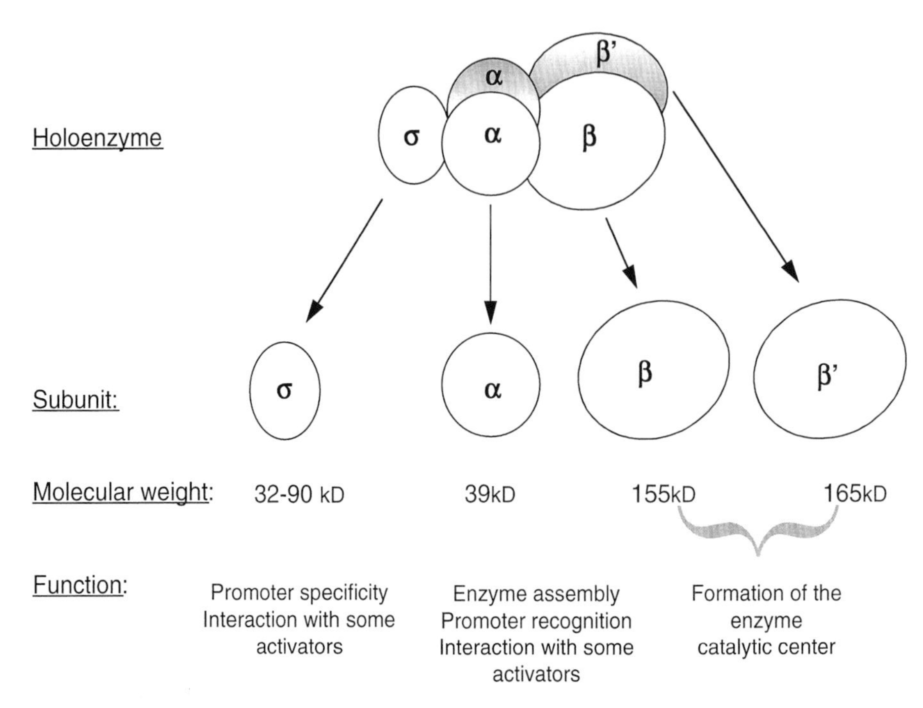

FIGURE 84.1 *E. coli* RNA polymerase holoenzyme consists of four types of subunits.

involved in all stages of transcription. The sequences of β and β' show homology to the sequences of the largest subunits of eukaryotic RNA polymerases. This conservation through the evolution hints that the mechanisms by which all RNA polymerases catalyze transcription share common features. The assignment of individual functions to the different subunits of the core polymerase are only a rough estimation because most likely each subunit contributes to the activity of the enzyme as a whole.

The σ factor is involved only in transcriptional initiation. Its function is to ensure that the polymerase binds in a stable manner to DNA only at promoters, not at other sites. There are several types of σ factors in the bacterial cell (see later). The major σ factor in *E. coli,* which is required for most transcription reactions, is σ^{70}. Although σ^{70} has domains that recognize the promoter DNA sequence, as an independent protein it does not bind to DNA, perhaps because its DNA-binding domain is sequestered by another domain of the σ^{70} molecule (when σ^{70} is shortened from its N-terminus, it is able to bind to DNA, suggesting that the N-terminal region has an inhibitory effect on the ability of σ^{70} to bind to DNA). However, upon binding to the core enzyme and the formation of the holoenzyme complex, σ^{70} undergoes a conformational change and it now contacts the region upstream of the start point. The dogma is that the σ factor discharges from the core enzyme when abortive initiation is concluded and RNA synthesis is successfully initiated (although findings suggest that, at least in some cases, σ can remain associated with the polymerase at postinitiation steps). The released σ factor becomes immediately available for use by another core enzyme. *E. coli* cells contain about 3000 molecules of σ^{70}, enough to bind about one-third of the intracellular core RNA polymerase.

C. The ability of RNA polymerase to selectively initiate transcription is dependent on the presence of σ factor

Bacterial polymerases have to recognize a wide range of promoters and to transcribe different genes on different occasions. Specificity of gene expression is in part modulated by substituting one species of σ for another, each specific for a different class of promoters. In *Bacillus subtilis,* σ factors are implicated in the temporal regulation of sporulation. In *E. coli,* alternative σ factors are used to respond to general environmental changes. The σ factors are named either by their molecular weight (e.g., σ^{70}), or after their genes, which are usually termed *rpo* (e.g., RpoD for σ^{70}).

When cells are shifted from low to high temperature, the synthesis of a small number of proteins, the heat-shock proteins, transiently increases. The σ^{32} protein (RpoH) is responsible for the transcription of the heat-shock genes. The basic signal that induces the production of σ^{32} is the accumulation of unfolded proteins that results from the increase in temperature.

The heat-shock proteins play a role in protecting the cell against environmental stress. Several of them act like chaperones, preventing the unfolding (denaturation) of proteins. The heat-shock proteins are synthesized in response to other conditions of stress, implying that the production of σ^{32} is induced by conditions other than elevation in temperature. The σ^{32} protein is unstable and it is rapidly degraded when it is not needed. Another σ factor, σ^{E}, appears to respond to more extreme temperature shifts that lead to the accumulation of unfolded proteins, which are usually found in the periplasmic space or outer membrane. Less is known about this σ factor and about the genes it controls. When *E. coli* and other enteric bacteria are nitrogen limited, the synthesis of a number of proteins is dramatically induced. The increased production of these proteins enlarges the capacity of cells to produce nitrogen-containing compounds, and to use nitrogen sources other than ammonia. The transcription of the genes that encode these proteins is dependent on σ^{54} (also known as σ^{N}). The expression of the flagellar genes under normal conditions depends on σ^{28} (also known as σ^{F}).

When *E. coli* cells are starved, they shift from the exponential-growth phase to the stationary phase. The ability of the cells to cope with starvation depends on the production of many proteins. This is enabled due to the synthesis of σ^{S} (RpoS), which transcribe the relevant genes. Unlike *E. coli*, when *B. subtilis* cells are starved, they can form spores. Sporulation involves the differentiation of a vegetative bacterium into a mother cell and a forespore; the mother cell lyses and a spore is released. The process of sporulation involves a drastic change in the biosynthetic activities of the bacterium, in which many genes are involved. This complex process is concerted at the level of transcription. The principle is that in each compartment (the mother cell and the forespore) the existing σ factor is successively displaced by a new σ factor that causes the transcription of a new set of genes. Communication between the compartments occurs in order to coordinate the timing of the changes in the mother cell and the forespore.

The promoters identified by the various σ factors are organized similarly, having their important elements centered around 10 and 35 nt upstream from the start point (see later), except for σ^{54}, whose promoters have slightly different characteristics. However, the ability of different σ factors to cause RNA polymerase to initiate at different sets of promoters stems from the fact that each type of σ factor recognizes promoter elements with unique sequences. σ^{54} differs from the other σ factors also in its ability to bind to DNA independently, and by the influence of sites that are distant from the promoter on its activity. In many aspects, σ^{54} resembles eukaryotic regulators.

A comparison of the sequences of the different σ factors identifies four regions that have been conserved. Several sequences in these regions were identified with individual functions, such as interaction with core RNA polymerase or contacting the various promoter elements.

II. TEMPLATE RECOGNITION: PROMOTERS

A. Promoter recognition depends on conserved elements

Template recognition begins with the binding of RNA polymerase to the promoter. Promoter is a sequence of DNA whose function is to be recognized by RNA polymerase in order to initiate transcription. The information for promoter function is provided directly by the DNA sequence, unlike expressed regions, which require that the information be transferred into RNA or protein to exercise their function. Two main approaches were used to identify the DNA features that characterize a promoter. The first is comparative sequence analysis and the second is the identification of mutations that alter the recognition of promoters by RNA polymerase. The sequence of more than 100 *E. coli* promoters recognized by the major RNA polymerase species, σ^{70} holoenzyme, has been determined. Their comparison revealed an overall lack of extensive conservation of sequence over the 60 bp associated with RNA polymerase. Nevertheless, statistical analysis revealed some commonalities. A typical *E. coli* promoter, which is recognized by σ^{70}, contains four conserved features: the start point, the −10 region, the −35 region, and the distance between the −10 and −35 regions. The start point is usually a purine. It is often the central base in the sequence CAT, but this is not a mandatory rule. The −10 region is a hexanucleotide that centers approximately 10 bp before the start point, although this distance is somewhat variable. Its consensus sequence is TATAAT (in the antisense strand). The conservation is $T_{80}\ A_{95}\ T_{45}\ A_{60}\ A_{50}\ T_{96}$, where the numbers refer to the percent occurrence of the most frequently found base at each position. The −35 region is a hexanucleotide sequence that centers approximately 35 bp upstream of the start point. Its consensus is TTGACA and the conservation is $T_{82}\ T_{84}\ G_{78}\ A_{65}\ C_{54}\ A_{45}$. The favored spacing between the −10 and the −35 sequences is 17 bp.

For most promoters, there is a good correlation between promoter strength and the degree to which the −10 and −35 elements agree with the consensus sequences. The significance of the conserved promoter features was further emphasized by the finding that most mutations that alter promoter activity (i.e., affect the level of expression of the gene(s) under the control of this promoter) change the sequence of the particular promoter in an expected fashion. Mutations that increase the similarity to the proposed conserved −10 and −35 sequences or bring the spacing between them closer to 17 bp, usually enhance the promoter activity (*up mutations*); mutations that decrease the similarity to the conserved sequences or bring the spacing between them more distant from 17 bp, usually reduce the promoter activity (*down mutations*). The nature of down mutations in the −35 and −10 regions of various promoters led to the conclusion that the −35 region is implicated in the recognition of the promoter by RNA polymerase and the formation of a closed transcription complex, whereas the −10 region is implicated in the shift of the closed complex to the open form (see later). The fact that the −10 region is composed of AT base pairs, which require low energy for melting, makes its suitable to assist in unwinding and, thus, in converting the transcription complex to its open form.

There are several exceptions to the proposed generalized pattern. For example, some promoters lack one of the conserved sequences, the −10 or the −35 region, without a corresponding effect on promoter activity. In some cases, it was proposed that another sequence compensates for the lack of a consensus sequence. In still other cases, it was concluded that the promoter cannot be recognized by RNA polymerase alone, and the involvement of additional proteins, which overcome the deficiency in intrinsic interaction between RNA polymerase and the promoter, is required. Other exceptional promoters were discovered due to the isolation of promoter mutations that do not affect any of the conserved promoter features described so far. Rather, they are the outcome of base substitutions in other sequences in the vicinity of the conserved sequences or the start point. One explanation for these findings is that the analysis that generated the sequence characteristics of a typical promoter may have missed some sites that contribute to the transcription initiation process, possibly because it included too many promoters, both weak and strong. Alternatively, other base pairs in the promoter could be recognized by RNA polymerase, but might become significant only if canonical recognition sites are absent.

The isolation of deletions that progressively approach specific promoters from the upstream region demonstrated the involvement of specific upstream sites in the recognition of RNA polymerase in some cases. This led to the discovery that some *E. coli* promoters contain a third important element in addition to the −10 and −35 sequences. This element was named the upstream element, or *UP element*, because it is located approximately 20 bp upstream of the −35 region. Its sequence is AT-rich and it was first identified in the strong promoters of the *rrn* genes, which encode ribosomal RNA. It is believed now that promoter strength is a function of all three elements, −10, −35, and UP, with very strong promoters, such as the *rrn* promoters, having all three elements with near-consensus sequences and with weaker promoters having one, two, or three nonconsensus promoter elements. It has been found that, whereas σ^{70} is responsible for the recognition of the −10 and −35 regions, the UP element interacts with the α-subunit of RNA polymerase.

Finally, the activity of some promoters is affected by sequences downstream to the −10 region, or even downstream to the transcription start point. The sequences immediately around the start point seem to influence the initiation event. The effect of the initial transcribed region (from +1 to +30) on promoter strength is explained by the influence of this region on the rate at which RNA polymerase clears the promoter.

Unlike the promoters described so far, which are recognized by holoenzymes that contain σ^{70} or a close homolog, a minority of the cellular holoenzymes use σ^{54} and have different basal elements located at −12 and −24. Transcription initiation from these promoters relies also on enhancer-like elements that are remote (upstream) from the promoters (see later).

B. Possible mechanisms for promoter recognition

How does RNA polymerase find the promoter sequences? How does it identify a stretch of ~60 bp which defines a promoter in the context of 4×10^6 bp that make up the *E. coli* genome? Three models were suggested to explain the ability of RNA polymerase to find promoters. The first model assumes that RNA polymerase moves in the cell by random diffusion. It associates and dissociates from loose binding sites on the DNA until, by chance, it encounters a promoter sequence that allows tight binding to occur. According to this model, movement of an RNA polymerase molecule from one site on the DNA to another is limited by the speed of diffusion through the medium. However, this parameter might be too low to account for the rate at which RNA polymerase finds promoters. The second model addresses this problem by assuming that once an RNA polymerase molecule binds to a DNA sequence, the bound sequence is directly displaced by another sequence; the enzyme exchanges sequences very rapidly, until it binds to a promoter that allows an open complex to form and transcription initiation to occur. According to this model, the association of the polymerase with DNA sequences and their dissociation are essentially simultaneous. Thus, the time spent on site exchange is minimal and the search process is much faster in comparison to the speed calculated based on the first model. This model fits the accepted notion that core polymerases that are not busy in transcription are stored by binding to loose sites on the DNA. The third model assumes that RNA polymerase binds to a random site on the DNA and starts sliding along the DNA molecule until it encounters a promoter. The actual mechanism by which RNA polymerase finds promoters might combine features of the various models.

III. TRANSCRIPTION INITIATION

A. Stages of the transcription-initiation process

Transcription initiation is the phase during which the first nucleotides in the RNA chain are synthesized. It is a multistep process that starts when the RNA polymerase holoenzyme binds to the DNA template and ends when the core polymerase escapes from the promoter after the synthesis of approximately the first nine nucleotides.

The stages of the transcription initiation process, which are summarized in Fig. 84.2, can be described in terms of the types of interaction between the RNA polymerase and the nucleic acids that are involved. The first stage in transcription initiation is the formation of a complex between the holoenzyme and the DNA sequence at the promoter, which is in the form of a double-stranded DNA. This complex is termed a closed binary complex or *closed complex*. The second stage is the unwinding of a short region of DNA within the sequence that is bound to the RNA polymerase. The complex between the polymerase and the partially melted DNA is termed an open binary complex or *open complex*. The conversion of the closed complex into the open complex leads to the establishment of tight binding between the RNA polymerase and the promoter sequence. For strong promoters, the conversion into an open complex is irreversible. The third stage is the incorporation of two ribonucleotides, and the formation of a phosphodiester bond between them. Because the complex at that stage contains an RNA as well as DNA, it is called an initiation ternary complex. Up to seven additional ribonucleotides can be added to the RNA chain without any movement of the polymerase. After the addition of each base, there is a certain probability that the enzyme will release the short (up to nine bases long) RNA chain. Such an unsuccessful initiation event is termed an abortive initiation. Following an abortive initiation, RNA polymerase begins again to incorporate the first base. Several rounds of abortive initiations usually occur and the result is the formation of short RNA chains that are 2–7 bases long. When initiation succeeds, that is, a nine-base-long RNA chain is formed and is not released, the last stage in transcription initiation occurs. At that stage the σ factor is released from the polymerase. As a consequence, a complex containing core polymerase, DNA, and RNA is formed. This complex is called an elongation ternary complex. The departure of the polymerase from the promoter to resume elongation is termed promoter escape or promoter clearance.

The activities of genes are frequently regulated at the initiation step of transcription. Therefore, the initiation of transcription is a very precise event that is tightly controlled by a variety of regulatory mechanisms.

B. Repression of transcription initiation

Gene expression is sometimes negatively regulated by a repressor protein that, when bound to DNA, inhibits transcription initiation. The ability of the repressor to bind to DNA is in turn modulated by the binding of an effector molecule to the repressor. The regulation of the *lac* operon expression in *E. coli* is a paradigm for this type of transcriptional control. LacI is a repressor protein that blocks the initiation of transcription from the promoter of the *lac* operon. LacI binds to a site on the DNA, termed an *operator*, that overlaps with the promoter. Because of this overlap, the binding of the LacI repressor and of RNA polymerase are competitive events. That is, RNA polymerase cannot bind to the promoter until the repressor is removed from the operator. The binding of one of several β-galactoside compounds to the repressor destablizes the repressor–operator complex and allows RNA polymerase to bind to the promoter to initiate transcription. Interestingly, *lac* has two additional binding sites for LacI, an upstream site and a downstream site located in the first gene of the operon. Compared to the operator, the additional sites have a lower affinity for the repressor protein and it was suggested that they do

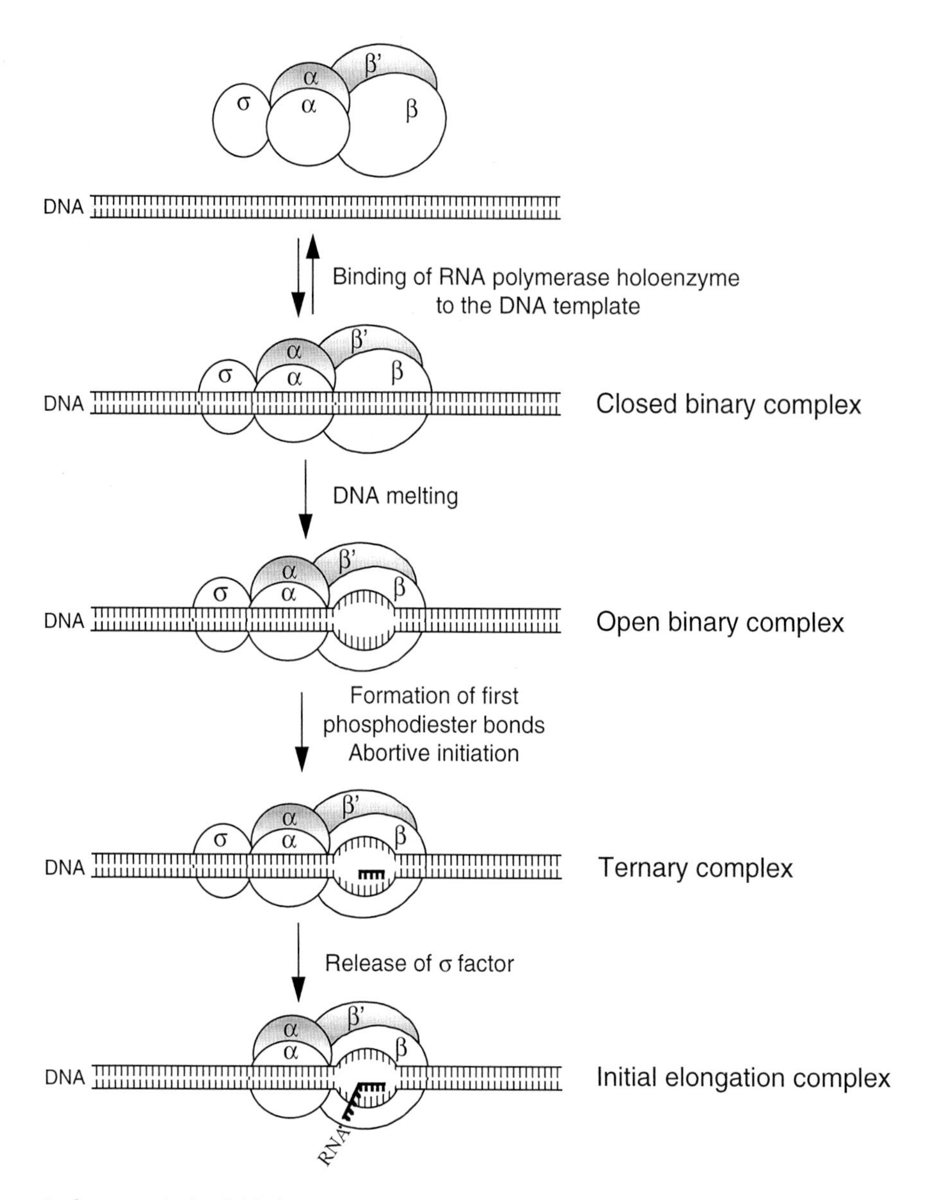

FIGURE 84.2 Stages in the transcription initiation process.

not directly participate in the inhibition of transcription initiation. Rather, the secondary binding sites seem to stabilize the repressor–operator complex.

There are important exceptions to the *lac* repression paradigm. An example is the repression of the *gal*-operon transcription initiation by a different and less understood mechanism. The *gal* operon contains two repressor binding sites, both required for maximum efficacy of the *gal* repressor, yet neither of these operators overlap with the promoter sequences. Thus, the binding of the *gal* repressor to its operators does not seem to compete directly with binding of RNA polymerase to the *gal* promoter. It was suggested that the *gal* repressor, when bound to both binding sites, holds the DNA in a conformation that is unfavorable for binding to RNA polymerase.

C. Activation of transcription initiation

The frequency of transcription initiation from many promoters is enhanced by activator proteins. In most cases, these proteins bind within or upstream from the promoter and seem to act by making a direct contact with RNA polymerase. The activators that interact with the most abundant form of RNA polymerase involved in transcription initiation in *E. coli*, the σ^{70} holoenzyme, can be roughly divided into two groups, those that interact with the σ-subunit of RNA polymerase and those that interact with the σ^{70} subunit.

The best characterized activator from the first group is the catabolite gene-activator protein (CAP). The CAP target site on the DNA was determined in a number of systems. Comparative sequence analyses led to the definition of a consensus sequence for CAP binding. CAP-binding sites are found at various location relative to the transcription start point in different systems. The most studied case is the activation of transcription initiation from the *lac* promoter by CAP. The regions that are required for the activation on both CAP and the α-subunit of RNA polymerase were defined. CAP acts as a dimer, and although the activating region is present in both subunits of the CAP dimer, transcription activation at the *lac* promoter requires only the activating region of the promoter-proximal subunit. CAP interacts with the C-terminal domain of the α-subunit (αCTD). The αCTD constitutes an independently folded domain, which is connected to the remainder of α by a flexible linker. This allows αCTD to make different interactions in different promoters. The simplest model for transcription activation by CAP is that CAP binds to the DNA and recruits the αCTD, and thus the RNA polymerase holoenzyme, to the promoter.

The best characterized activator from the second group is the cI protein of bacteriophage λ (λcI). λcI binds to a site on the DNA that overlaps the -35 element of the λP_{RM} promoter. The activating region in λcI was defined and was demonstrated to directly contact a specific region in σ^{70}.

The existence of at least two groups of activators that bind to separate targets on the DNA and to different components of RNA polymerase raised the possibility that, at some promoters, RNA polymerase might be contacted simultaneously by two or more activators. This was demonstrated in an artificial system that was engineered to contain both CAP and cI binding sites upstream from the λP_{RM} promoter. Both activators were shown to make contact with RNA polymerase, most likely with CAP contacting αCTD and with λcI contacting σ^{70}, and their effect on transcription activation was synergistic. Transcription activation by two activators was demonstrated also in natural promoters. One conclusion from studies with various types of activators is that many activators seem to function by helping recruit DNA-binding domains of RNA polymerase to DNA, thus supplementing suboptimal RNA polymerase–DNA interactions with protein–RNA polymerase interactions.

In contrast to the copious σ^{70} promoters, the rare σ^{54} promoters, which contain -12 and -24 basal elements instead of the well-known -10 and -35 elements, seem to be regulated solely by activation rather than by repression. σ^{54} activators (the most studied is NtrC) bind to enhancer-like sites on the DNA; that is, the sites are remote from the promoters (upstream) and their precise location is not critical for transcriptional activation. In fact, these sites can be moved kilobases in *cis* and retain their residual function. Thus, unlike σ^{70} activators that bind to sites that enable direct communication with RNA polymerase, σ^{54} activators, once bound to their DNA target sites, cannot touch the polymerase without looping out the intervening DNA. This seems to be the reason why σ^{54} promoters frequently require the help of integration host factor (IHF), which enhances the bending of the DNA, as a cofactor. It is accepted that σ^{54} polymerase can bind to its promoters to form a closed complex. However, this polymerase cannot transcribe because it cannot melt the DNA. Once the upstream activator binds to its target site upstream of the promoter, it loops out of the sequence between its binding site and the promoter and touches the complex. This interaction triggers the melting of DNA (with the help of a helicase activity) and the creation of a transcription bubble. Thus, σ^{54} activators catalyze the conversion of the polymerase–promoter complex from a closed state to a transcription-ready open state, rather then tethering the RNA polymerase to the promoter.

D. Regulation of transcription initiation via changes in DNA topology

The template for transcription is a negatively supercoiled DNA. Because the formation of an open transcription complex requires DNA melting, and because the degree of superhelicity affects the energy needed for the melting, it was anticipated that the superhelical character of a template would affect the properties of this template. Indeed, the efficiency of some promoters is influenced by the degree of supercoiling. Most of these promoters are stimulated by negative supercoiling, although few are inhibited. The effects of superhelicity on the process of transcription initiation have been shown *in vitro* in numerous studies and *in vivo* by the use of inhibitors of gyrase, which introduces negative supercoils. The reason why some promoters are sensitive to the degree of supercoiling, whereas others are not, might have to do with the fact that the sequence of some promoters is easier to melt and is therefore less dependent on supercoiling. Alternatively, because various regions on the bacterial chromosome are believed to have different degrees of supercoiling, the location of the promoter might determine whether it is sensitive to changes in superhelicity.

IV. TRANSCRIPTION ELONGATION

The initiation phase ends when RNA polymerase succeeds in extending the RNA chain beyond the first nine nucleotides and escapes from the promoter. At that stage the elongation process begins and the enzyme starts moving along the DNA, extending the growing RNA chain. During the transition from initiation to elongation, the size and shape of the RNA polymerase undergoes successive changes. The first change is the loss of the σ factor. Whereas the holoenzyme covers approximately 75 bp (from −55 to +20), after the loss of σ, the polymerase covers approximately 55 bp (from −35 to +20). At that stage the polymerase is displaced from the promoter (promoter clearance or escape) and undergoes a further transition to form the elongation complex, which covers only 35–40 bp, depending on the stage during elongation. The polymerase now becomes tightly bound to both the nascent transcript and the DNA template, making it very stable.

The average rate of transcript elogation by the various RNA polymerases is 40 nt/s. However, this rate varies dramatically among RNA polymerase and loosely correlates with the subunit complexity of the enzyme. Thus, the simple single-subunit bacteriophage RNA polymerases are the most rapid of all DNA-dependent RNA polymerases (several hundred nt/s), bacterial RNA polymerases transcribe at an intermediate rate (50–100 nt/s), and eukaryotic polymerases, although diversified, appear to be the slowest (20–30 nt/s).

RNA polymerases are not as accurate as DNA polymerases. The reason for the difference in fidelity between RNA and DNA polymerases might be that RNA polymerases do not have the robust proofreading mechanism that characterizes DNA polymerases. Of course, the fidelity of DNA replication is of greater importance than that of transcription because unlike replication errors, misincorporation during transcription does not result in permanent and inherited genetic changes.

A. Blocks to transcription elongation

Transcript elongation does not occur at a constant rate. Throughout the elongation phase, RNA polymerase can be paused, arrested, or terminated. During a *transcriptional pause*, the polymerase temporarily stops RNA synthesis for a certain amount of time, after which it can resume the elongation process. Thus, pausing can be described as transcriptional hesitation. In contrast, during a *transcriptional arrest* the polymerase stops RNA synthesis and cannot resume it without the aid of accessory proteins. Throughout both pauses and arrests, RNA polymerase remains stably bound to the DNA template and to the nascent transcript. These features distinguish paused and arrested polymerases from those that have terminated and thus detached from the DNA. Pausing and termination are sometimes related because pausing is a prerequisite for termination, at least in the case of ρ-dependent termination (transcription termination is discussed in Section V). However, not all pauses are termination precursors. The time it takes a stalled polymerase to resume elongation varies among pause sites from very short periods of time, which cannot be accurately measured, to several minutes. The fraction of the polymerases that respond to an elongation block is also variable because ternary complexes differ in their ability to recognize pausing signals.

Transcriptional pause and arrest signals can be intrinsic; that is, sequences in the nascent transcript or in the DNA template whose interaction with RNA polymerase can inhibit the progression of the ternary complex, such as RNA regions, which have the propensity to form a stable secondary structure. In addition, extrinsic factors may obstruct the progress of RNA polymerase during transript elongation. There are numerous examples of RNA polymerases from various organisms being physically blocked by DNA-binding proteins during RNA synthesis in natural and artificial systems. An example is the purine repressor, which binds well downstream from the *purB* operon transcriptional start point and blocks the polymerase during elongation (it should be noted that in many cases RNA polymerase is able to transcribe beyond the DNA-binding proteins in its path by either displacing them or bypassing them). In addition to DNA-binding proteins, which are the most obvious obstacles for RNA polymerase, factors that perturb the structure of DNA can also inhibit the progression of RNA polymerase and thus interfere with transcript elongation, for instance, extreme positive or negative supercoiling, unusual DNA structures such as Z-DNA, and DNA lesions. The efficiency of such potential impediments to block RNA polymerase from elongating depends on various local factors and on the type of the RNA polymerase. For example, T7 RNA polymerase can efficiently bypass gaps in the DNA template strand that are 1–5 nt, and less efficiently gaps as large as 24 nt.

Transcriptional pausing is involved in various regulatory mechanisms. An example is the attenuation of amino acid biosynthetic operons, which depend on a transcriptional pause that leads to the precise coordination of the position of RNA polymerase with the ribosome translating the nascent RNA. Although transcriptional arrest has been characterized *in vitro* and the evidence for its occurrence in the cell is only circumstantial, it is also believed to be implicated in the regulation of many genes. These predictions are based on the recognition that if an arrest occurs within the coding region of a gene, the arrested complex would block subsequently initiated RNA polymerases, thereby effectively repressing RNA synthesis from the affected gene.

B. Transcript cleavage during elongation

RNA polymerases in ternary complexes are able to endonucleolytically cleave the nascent transcript near the 3′-end. The resulting 5′-fragment remains stably bound to the RNA polymerase, whereas the 3′-fragment is released from the enzyme. Transcript cleavage does not result in transcript termination, but rather in the creation of a new 3′-terminus for chain elongation. Following the cleavage, RNA polymerase can correctly resynthesize the discarded RNA segment and continue with the elongation. The size of the released 3′-fragment varies. There are reports of cleavage products ranging from 1–17 nt. The cleavage rate also varies, depending on the particular ternary complex. The mechanisms that cause the variations in both cleavage-product size and cleavage rate are poorly understood. It is likely that the conformational changes that occur within the ternary complex as it moves along the DNA template play a crucial role in determining the rate of the cleavage reaction and the size of the cleaved fragment.

The actual cleavage is carried out by the catalytic site of RNA polymerase, which catalyzes polymerization. However, the cleavage reaction seems to involve accessory proteins in addition to RNA polymerase. In *E. coli*, the GreA and GreB proteins stimulate the polymerase to cleave and release the 3′-fragment and to resume transcription elongation. Thus, GreA and GreB can be defined as cleavage-stimulatory factors. GreA and GreB also affect the size of the released 3′-fragment. Other factors, such as NusA, can regulate the cleavage properties induced by GreA and GreB.

The physiological role of transcript cleavage reaction catalyzed by RNA polymerase has not been determined. However, one accepted hypothesis is that transcript cleavage serves to rescue RNA polymerases that are arrested during elongation. It has been shown that arrested RNA polymerase complexes can be reactivated by transcript cleavage *in vitro*. The cleavage-stimulatory factors, GreA and GreB, were shown to release the RNA polymerase from the elongation arrest. GreA can suppress the formation of arrested complexes *in vitro*; it can act only if it is present before the polymerase arrests. GreB can act after the polymerase has stopped; its action is triggered by the appearance of an arrested complex. It has been suggested that GreA and GreB stimulate promoter escape at some promoters *in vivo*.

Another role that has been presumed for transcript cleavage is to increase the fidelity of transcription by facilitating the removal of misincorporated nucleotides. To accomplish transcript cleavage and elongation resumption, the RNA polymerase, after it has stalled, is assumed to either move backwards along the DNA template or undergo structural changes so that the catalytic site is repositioned to the new 3′-end to which bases should be added. The ability of the polymerase to withdraw, to cleave, and discard the 3′-end of the nascent transcript; to replace the removed sequence with a newly synthesized one; and to continue with the elongation in a template-directed manner is what makes this process so suitable for proofreading.

C. The Inchworm model for transcription elongation

The old-fashioned view of the elongation process as a smooth forward motion, during which RNA polymerase moves 1 bp along the DNA template for every base added to the newly synthesized RNA chain, might still hold for some regions of DNA. However, evidence has accumulated for a different type of movement of the polymerase during elongation. Thus, a new model for RNA polymerase translocation has evolved: the *inchworm model*. This model, which is illustrated in Fig. 84.3, describes the movement of RNA polymerase on the DNA template in a discontinuous inchworm-like, fashion. The model predicts that the process of RNA chain elongation is a cyclic process that consists of discrete translocation cycles. Each cycle involves the steady compression of the RNA polymerase on the DNA template followed by a sudden expansion. According to this model, the upstream (back) boundary of the enzyme moves steadily during elongation, as the RNA chain is extended. However, the downstream (front) boundary of the enzyme does not move while several nucleotides are added; it then "jumps", that is, it moves 7–8 bp along the DNA. At the beginning of each cycle, RNA polymerase stretches across ~35 nt of template DNA; it gradually compresses from the back end till it covers only ~28 nt; it then releases from the front end and stretches back to cover ~35 nt. As the RNA polymerase compresses, the nascent RNA chain

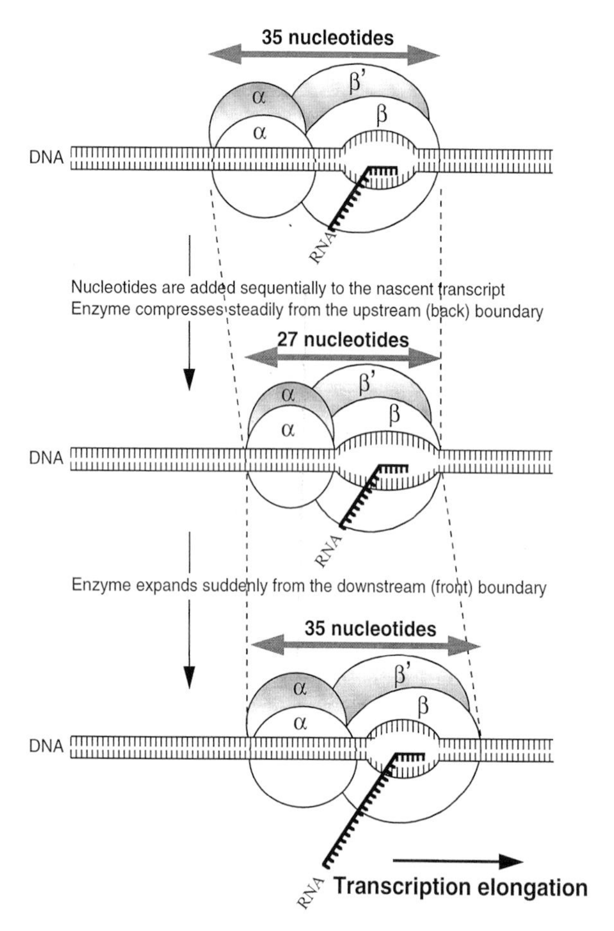

FIGURE 84.3 Discontinuous inchworm-like movement of RNA polymerase during transcription elongation.

the complex becomes longer and the single-stranded transcription bubble enlarges as well. The internal tension that these changes probably create in the enzyme is released when the front end expands discontinuously.

The inchworm model for transcription elongation postulates that RNA polymerase binds to the DNA template at two separate sites, one downstream in the direction of transcription and one upstream, that can move independently of each other. This permits the polymerase to move in an inchworm-like manner, so one DNA binding site on the polymerase remains fixed to the DNA, whereas the other moves along the DNA. There is now both direct and indirect evidence that validate this assumption. The downstream DNA binding site in the *E. coli* polymerase was found to be double-strand-specific, whereas the upstream is single-strand-specific and interacts with the template strand. The model also assumes that the catalytic site of RNA polymerase is linked to the movement of the upstream DNA site, but can move independently of the downstream DNA-binding site. The inchworm model also makes predictions about the existence of more than one RNA-binding site on the ternary complex. It has been shown that nascent RNAs interact with at least three sites on the *E. coli* polymerase, two on the β-subunit and one on the β'-subunit. It is believed that together the DNA and RNA sites account for the remarkable stability and flexibility of ternary complexes. However, the precise size, placement, and strand specificity of these nucleic acid-binding sites are currently being elucidated.

D. Transcriptional slippage

RNA polymerase usually synthesizes RNA transcripts that are precisely complementary to the DNA template. However, in rare circumstances, RNA polymerase can undergo transcriptional slippage that results in the synthesis of a transcript that is either longer or shorter than the sequence encoded by the DNA template. Such a slippage appears to occur only when the polymerase transcribes homopolymeric runs. It has been proposed that the generation of transcripts that are shorter than the encoding template is due to translocation of RNA polymerase without the incorporation of nucleotides, whereas the longer products are due to RNA polymerase incorporating nucleotides without translocation. Transcriptional slippage can occur during both the initiation and the elongation phases. However, the minimal length of the consecutive template nucleotides that can promote slippage in the two phases in different. During initiation, homopolymeric runs as short as 2 or 3 nt can be reiteratively transcribed by RNA polymerase. During elongation, RNA polymerase tends to slip only on longer runs, but the precise requirements have not been elucidated. In one case, slippage by the *E. coli* RNA polymerase during elongation was reported to require runs of at least 10 dA or dT nucleotides, whereas runs of dG at the same length did not result in slippage. In some cases, the ability to slip seems to require a transcriptional pause in addition to the homopolymeric run. Transcriptional slippage is sometimes an important means of regulating transcription. For example, slippage has been reported to play an important role in the regulation of transcription initiation at several bacterial operons, including *pyrBI*.

E. Implications of DNA topology on transcription elongation

Because DNA has a helical secondary structure, a rotation about its axis is necessary to accomplish transcription elongation. This requires that either the entire transcription complex rotates about the DNA or that the DNA itself rotates about its helical axis. Under conditions in which the RNA polymerase rotation is constrained, for example, due to the presence of ribosomes attached to the nascent RNA chain (which is often the case in bacteria), the DNA will rotate through the enzyme. Consequently, the process of transcription will tend to generate positive supercoils in the DNA ahead of the advancing RNA polymerase and negative supercoils behind it. Excessive torsional stress in the DNA will arise if the DNA is anchored at various points (as is the case for circular DNA, such as the bacterial chromosome), or from the movement of RNA polymerase in opposite directions along the DNA. DNA topoisomerases are the natural candidates to remove this tension. It was suggested that gyrase, which can relieve positive supercoils, and topoisomerase I, which removes negative supercoils, amend the situation in front of and behind the RNA polymerase, respectively. This model is supported by the finding that when the activities of gyrase and topoisomerase I are inhibited or otherwise defective, transcription causes major changes in DNA supercoiling. A possible implication of this is that transcription, in addition to having a significant effect on the local structure of DNA, is responsible for generating a significant proportion of supercoiling that occurs in the cell.

F. Implications of DNA replication on transcription

Transcription regulation is carefully coordinated with DNA replication and chromosome segregation. In *E. coli* and in other bacteria and bacteriophages, heavily transcribed genes are oriented such that replication and transcription occur in the same direction. Despite this arrangement, because DNA replication occurs 10–20 times faster than transcription, RNA polymerase and DNA polymerases do collide. The outcome of such an encounter is hard to predict. In *E. coli* there is evidence suggesting that the replication fork can displace the elongation complex. However, in bacteriophage T4, the movement of the replication apparatus does not seem to disrupt the elongation complexes, regardless of the direction of their motion relative to the replication fork. Interestingly, when direct collisions occur between the DNA and RNA polymerases of T4, the RNA polymerase switches from the original DNA template strand to the newly synthesized daughter strand. The mechanism that allows the strand exchange without the dissociation of the elongation complex is not known, but probably relies on the various contacts with the DNA and RNA. Whatever the mechanism, the cell needs to coordinate the replication and transcription processes carefully.

V. TRANSCRIPTION TERMINATION

Transcriptional elongation is highly processive and can lead to the production of RNA transcripts that are thousands of nucleotides long. The processivity is due to the high stability of the complex between the RNA polymerase and the nucleic acids during elongation. It is this stability that necessitates the involvement of specific signals and factors to implement the termination of transcription. To enable efficient termination, the termination signals or factors should cause drastic alterations of the interactions that are responsible for the stable elongation. At termination, RNA polymerase stops adding nucleotides to the RNA chain, all the hydrogen bonds that hold the RNA–DNA hybrid together break leading to the release of the transcript, the DNA duplex reforms, and the enzyme dissociates from the DNA template. The sequence of these events is still not clear because attempts to determine whether the release of the RNA polymerase is simultaneous with the transcript release or occurs subsequently have given ambiguous results. Once the transcript is released from the complex, it is unable to reattach in a way that allows transcriptional elongation to resume. Therefore, the transcript release is the commitment step that makes the termination process irreversible. On one hand, this mechanism ensures the termination at the end of genes and prevents the expression of adjacent distinct genetic units; on the other hand, this mechanism provides an opportunity to control gene expression.

The exact point at which termination of an RNA molecule occurs in the living cell is difficult to define. The 3′-end of an RNA transcript looks the same whether it is generated by termination or by cleavage of the primary transcript. Therefore, the best identification of termination sites is provided by systems in which RNA polymerase terminates *in vitro*. An authentic 3′-end can be identified when the same end is generated *in vitro* and *in vivo*. In *E. coli*, two types of terminators were discovered, intrinsic terminators that do not require ancillary proteins, and terminators that require the involvement of termination factors.

A. Intrinsic terminators

Intrinsic terminators are sites at which core polymerase can terminate transcription *in vitro* in the absence of any other factor. The best characterized intrinsic terminators are the ones recognized by *E. coli* RNA polymerase. Intrinsic terminators are characterized by a GC-rich sequence with an interrupted dyad symmetry followed by a run of about 6–8 dA residues on the template strand.

The transcription of the GC-rich sequence with the interrupted inverted repeats will give rise to an RNA segment that has the potential to fold into a stable stem-and-loop secondary structure (sometimes described as a hairpin structure). There is much indirect evidence that this structure is indeed formed in the nascent RNA. For example, mutations that interrupt the pairing in the stem part decrease the efficiency of termination, and compensatory mutations that restore the pairing recover the efficiency. There is also a strong correlation between the predicted stability of the structure and the termination efficiency. DNA oligonucleotides that are complementary to one arm of the stem in the stem-loop structure effectively reduce the efficiency of termination, presumably by annealing to the RNA sequence, and thus prevent the RNA from folding into the stem-loop structure. The sequence of the loop in the stem-loop structure also influences the stability of the RNA secondary structure, but the rules for contributing to loop stability have not been fully elucidated. How does the stem-loop structure contribute to termination? It is suggested that the formation of this structure in the newly synthesized RNA sequence, which is still in contact with the polymerase, causes the polymerase to pause and destabilizes the ternary complex.

The other structural feature of an intrinsic terminator, the run of the dA residues (which is sometimes interrupted) in the template strand, is located at the very end of the transcription unit. The transcription of this sequence will generate a run of rU residues at the 3′-end of the RNA transcript. The hybrid between the dA and the rU residues is significantly less stable than most other hybrids, due to weak base-pairing, and it thus requires the least energy to break the association between the strands. This poor base-pairing is assumed to unwind the DNA–RNA hybrid and destablize the interaction of the nucleic acids with the paused polymerase. The importance of the dA run has been established by mutational analysis. The importance of the length of the dA stretch was confirmed by introducing deletions that shortened this element; although the polymerase could still pause at the stem-loop, it no longer terminated. Interestingly, the actual termination can occur at any one of several positions towards the end of the dA run.

The DNA sequence within 30 bp downstream to the transcription stop point, which does not reveal an obvious consensus sequence, is also important for termination in certain cases. For example, changes in the sequence 3–5 bp downstream to the stop point of T7 early-gene terminator can reduce the efficiency of termination from 65% to 10%. Although these sequences are not transcribed, they are near or within the contact point between the RNA polymerase and the DNA in the transcription complex. The way these sequences can affect transcription is by influencing the unwinding of the DNA or the progression of the polymerase along the DNA. Alternatively, the stability of the binding of the polymerase could vary depending on the sequence at the contact points.

B. Rho-dependent termination

The best characterized termination factor is the bacterial ρ (rho) protein. ρ is a classic termination factor in the sense that it provides a mechanism for dissociating nascent transcripts at sites that lack intrinsic terminators. The sequences that form a ρ-dependent terminator extend from at least 60 bp upstream to about 20 bp downstream of the actual stop point. A sequence comparison of several ρ-dependent terminators did not reveal a consensus. ρ binds to the nascent RNA chain and a common feature of the RNA sequence to which it binds is a relatively high C and low G content. In addition, ρ has a strong preference for sufficiently long segments of unstructured RNA (lacking base-pairing).

ρ causes RNA polymerase to terminate preferentially at points that are natural pause sites. There is no evidence that ρ affects the elongation-pausing specificity of the polymerase. In addition to ρ being an RNA-binding protein, ρ contains an RNA–DNA helicase activity; it hydrolyzes ATP to energize the separation of an RNA–DNA hybrid. Thus, ρ is acting primarily as an RNA-release factor. The current model for ρ action is that it binds to the RNA transcript at sites that are unstructured and rich in C residues; it then translocates along the RNA until it catches up with the polymerase at sites where the enzyme pauses; ρ unwinds the RNA–DNA hybrid in the transcription bubble; and termination is completed by the release of ρ and RNA polymerase from the nucleic acids. Some ρ mutations can be suppressed by mutations in the genes that encode the β- and β'-subunits of RNA polymerase. This implies that in addition to interacting with the nascent RNA chain, ρ also interacts with the polymerase.

The lack of stringent sequence requirements for a ρ-dependent transcripton terminator raises the possibility that such terminators might be fairly frequent in DNA sequences, not only at the ends of operons but also within genes. What prevents ρ from terminating within genes? Because transcription and translation are coupled in prokaryotes, the mRNA chain that emerges from the transcription complex is protected by ribosomes, probably preventing ρ from gaining access to the RNA. The phenomenon of polarity (a nonsense mutation in one gene prevents the expression of subsequent genes in the operon) can be explained by the release of ribosomes from the transcript at the nonsense-mutation site, so that ρ is free to attach to and move along the mRNA; when it catches up with RNA polymerase, it terminates transcription, thus preventing the expression of distal parts of the transcription unit. What prevents ρ from acting on transcripts that are not translated, such as rRNAs and tRNAs? One reason seems to be the lack of ρ-binding sites on these RNAs because they are highly structured. Ribosomal RNA molecules are further protected by the binding of ribosomal proteins. Another mechanism that protects rRNA operons against ρ-dependent termination relies on sequences near the start of the rRNA genes that dictate antitermination. It was suggested that this mechanism increases the rate of transcriptional elongation of rRNA operons by preventing pausing.

The Psu protein encoded by bacteriophage T4 antagonizes ρ-dependent transcription termination. Although the mechanism that enables Psu to oppose ρ is unknown, its general anti-ρ action suggests that it acts directly on ρ, either by binding to it or by modifying it.

C. Auxiliary termination factors

Although some polymerases can spontaneously terminate transcription at intrinsic terminators, the efficiency of termination *in vitro* is often enhanced significantly by the presence of additional factors. It therefore seems that the DNA signals that characterize intrinsic terminators are necessary, but sometimes not sufficient. The best characterized of these auxiliary termination proteins is NusA. A less-studied factor named τ (tau) is known to enhance and modify recognition of some strong intrinsic terminators for *E. coli* RNA polymerase. ρ-dependent termination can also be enhanced by an auxiliary factor, the NusG protein.

The ability of NusA to increase the efficiency of termination at some intrinsic terminators might be attributed to its capability to increase the rate of pausing at certain sites. Enhanced pausing would allow more time for the conformational change that leads to the release of the nascent transcript. Some intrinsic terminators predicted to form not a very stable secondary structure, such as the one in the ribosomal protein S10 operon leader, depend on NusA for their operation, and can therefore be defined as NusA-dependent terminators. In the case of the S10 leader terminator, the effect of NusA is enhanced substantially by the product of one of the genes in this operon, the ribosomal protein L4. Because L4 is an RNA-binding protein, one assumption is that it binds to the S10 leader RNA and enhances its folding into a stem-loop structure. Alternatively, L4 bound to the nascent RNA can affect the elongation properties of RNA polymerase. In the unusual case of the intrinsic terminator in the attenuator region preceding the gene for the β-subunit of *E. coli* RNA polymerase, NusA, rather than enhancing termination, reduces termination efficiency. The efficiency of termination at ρ-dependent sites is not increased by NusA.

NusG has a minor effect on termination at some intrinsic terminators. However, NusG plays a significant role in the function of some ρ-dependent terminators. This role was deduced from the strong effect of NusG loss on the activity of some ρ-dependent terminators *in vivo*. NusG might be acting directly to enhance ρ activity. Alternatively, it might be acting indirectly by slowing the dissociation of ribosomes from mRNA, thus preventing ρ from binding to the transcript.

Interestingly, the presence of the σ factor in excess significantly increases the rate of RNA polymerase recycling. Hence, the σ factor can also be considered a termination factor. This activity supports the notion that σ can remain associated with the polymerase at postinitiation steps.

D. Antitermination

Antitermination is used as a control mechanism in phages to regulate the progression from one stage of gene expression to the next, and in bacteria to regulate expression of some operons. Antitermination occurs when RNA polymerase reads through a terminator into the genes lying beyond. The terminators that are bypassed can therefore be defined as conditional terminators. Antitermination is not a general mechanism that can occur in all terminators, but is, rather, dependent on the recognition of specific sites in the nucleic acids. Many of the antitermination mechanisms rely on the modification of the polymerase that makes it bypass certain terminators or on changes in the structure of the transcript caused by RNA-binding proteins.

The N protein of bacteriophage λ mediates antitermination necessary to allow RNA polymerase to read through the terminators located at the end of the immediate early genes in order to express the delayed early genes. The recognition site needed for antitermination by N, called *nut* (for N utilization), lies upstream from the terminator at which the action is eventually accomplished. The *nut* site consists of two sequence elements, a conserved 9-nt sequence called boxA, which is also a part of the bacterial *rrn* operon antiterminator signals, and a 15-nt sequence called boxB, which encodes an RNA that would form a short stem structure with an A-rich loop. A number of host proteins, including NusA, NusG, ribosomal protein S10 (NusE), and NusB, participate in the N-mediated antitermination process (Nus stands for N utilization substance). According to the current model for N-mediated antitermination at ρ-dependent terminators, N recognizes and binds to the boxB stem-loop structure formed on the nascent transcript, whereas NusB and S10 bind to the boxA sequence on the RNA. These proteins are held together through interactions with core RNA polymerase that are stabilized by NusA and NusG. Hence, a ribonucleoprotein complex is formed at the *nut* site and stays attached to t elongating RNA polymerase. This complex prevents RNA polymerase from pausing, thus denying the ρ factor the opportunity to cause termination, and the polymerase continues past the terminator. N also suppresses termination at intrinsic terminators; however, NusA suffices for N to prevent termination at these sites. Other phages related to λ have different N proteins and different antitermination specificities. Each phage has a characteristic *nut* site recognized specifically by its N-like protein. All these N-like proteins seem to have the same general ability to interact with the transcription apparatus in an antitermination capacity.

The Q protein is required later in bacteriophage λ infection. It allows RNA polymerase to read through the terminators located at the end of the immediate early genes, in order to express the late genes of bacteriophage λ. Q has a different mode of action than N. It recognizes and binds to a site on the DNA, called *qut*. The upstream part of *qut* lies within the λ-promoter P_R, whereas the downstream part lies at the beginning of the transcribed region. Thus, Q antitermination is specific for RNA polymerase molecules that have initiated at the P_R promoter. The part of *qut* that lies within the transcribed region includes a signal that causes RNA polymerase to pause just after initiation. This pause apparently allows Q to interact with the polymerase. Once bound, the Q-modified enzyme is released from the pause and is able to read through most transcription terminators, both intrinsic and ρ-dependent. It seems that the modification of the polymerase by Q increases the overall rate of transcription elongation and permits the polymerase to hurry past the terminators. Interestingly, the pause of the polymerase early in the transcription unit, which is a prerequisite for Q-mediated antitermination, involves the binding of the σ subunit of the RNA polymerase holoenzyme to the nontemplate stand of DNA in the transcription bubble up to 15 nt downstream from the start point of transcription. Thus, an initiation factor acts in concert with a DNA-binding termination factor to modify the elongation properties of RNA polymerase. Once again, it is shown σ can remain associated with the polymerase and play a role in postinitiation steps.

RNA polymerase molecules that are engaged in transcribing the ribosomal RNA (*rrn*) operons are modified in a way that makes them bypass certain terminators within the rRNA genes. The modification is established by the recognition of a sequence signal that is nearly identical to the boxA sequence involved in N-mediated antitermination. It has been shown that a heterodimer of NusB and S10 protein binds to the boxA sequence on the RNA of one of the *rrn* operon. It was therefore proposed that the mechanism of transcriptional antitermination in the *rrn* operons, similar to the mechanism mediated by N, involves formation of a ribonucleoprotein complex on the boxA complex that is carried along with the elongating polymerase. The probable purpose of this mechanism is to ensure that transcription of the rRNAs is immune from ρ action.

Transcription of the *bgl* operon in *E. coli*, which codes for proteins required for the use of β-glucoside, is also controlled by antitermination. One of the operon products, the BglG protein, prevents the termination of transcription at two intrinsic terminators. The first terminator is in the 5′ untranslated leader of the transcript and the second is in the intergenic region between the first and second genes of the operon. BglG is an RNA-binding protein that recognizes and binds to a specific sequence partially overlapping the sequence of both terminators. By binding to its RNA target site, BglG stabilizes a secondary structure, which is an alternative to the terminator structure. Thus, BglG binding to the *bgl* transcript prevents the formation of the terminators and the polymerase can read through them. BglG exerts its effect as a transcriptional antiterminator only when the expression of the operon is required, that is, when β-glucosides are present in the growth medium. In the absence of β-glucosides, BglG is kept as an inactive monomer due to its phosphorylation by BglF, the β-glucoside phosphotransferase. Upon addition of β-glucosides, BglF dephosphorylates BglG, allowing BglG to dimerize and act as an antiterminator. The effect of BglG is reminiscent of the effect of the ribosomes in the *E. coli trp* operon transcriptional attenuation.

BIBLIOGRAPHY

Busby, S., and Ebright, R. H. (1994). *Cell* **79**, 743–746.

Chamberlin, M. J., and Hsu, L. M. (1996). *In* "Regulation of Gene Expression in *Escherichia coli*" (E. C. C. Lin and A. S. Lynch, eds.), pp. 7–25. Landes, Austin, TX.

Choy, H., and Adhya, S. (1996). *In* "*Escherichia coli* and *Salmonella*: Cellular and Molecular Biology" (F. C. Neidhardt *et al.*, eds.), 2nd edn., pp. 1287–1299. American Society for Microbiology, Washington, DC.

Gralla, J. D. (1991). *Cell* **66**, 415–418.

Gralla, J. D., and Collado-Vides, J. (1996). *In* "*Escherichia coli* and *Salmonella*: Cellular and Molecular Biology" (F. C. Neidhardt *et al.*, eds.), 2nd edn., pp. 1232–1245. American Society for Microbiology, Washington, DC.

Landick, R., Turnbough, C. L., Jr., and Yanofsky, C. (1996). *In* "*Escherichia coli* and *Salmonella*: Cellular and Molecular Biology," (F. C. Neidhardt *et al.*, eds.), 2nd edn., pp. 1263–1286. American Society for Microbiology, Washington, DC.

Reznikoff, W. S., Siegele, D. A., Cowing, D. W., and Gross, C. A. (1985). *Annu. Rev. Genet.* **19**, 355–387.

Richardson, J. P. (1993). *Crit. Rev. Biochem. Mol. Biol.* **28**, 1–30.

Roberts, J. W. (1996). *In* "Regulation of Gene Expression in *Escherichia coli*" (E. C. C. Lin and A. S. Lynch, eds.), pp. 27–45. Landes, Austin, TX.

Uptain, S. M., Kane, C. M., and Chamberlin, M. J. (1997). *Annu. Rev. Biochem.* **66**, 117–172.

WEBSITE

Analysis of *E. coli* Promoters
http://www.cbs.dtu.dk/dave/MScourse/promoters.html

Links to Genome Centers
http://www.qiagen.com/bioinfo/

Comprehensive Microbial Resource of The Institute for Genomic Research (TIGR) and links to many other microbial genomic sites
http://www.tigr.org/tigr-scripts/CMR2/CMRHomePage.spl

Pedro's Biomolecular Research Tools. A Collection of Links to Information and Services Useful to Molecular Biologists
http://www.public.iastate.edu/~pedro/research_tools.html

85

Transduction: host DNA transfer by bacteriophages

Millicent Masters
Edinburgh University

GLOSSARY

abortive transductant A bacterium that has received and expresses a transduced DNA fragment in generalized transduction, but that fails to integrate it into its genome; abortively transduced DNA is not replicated, but can be unilinearly inherited.

attachment site A site on a phage (attP) or chromosome (e.g., attB, attλ, or attphi80) at which a site-specific integrase can join circular phage DNA and host chromosome.

bacteriophage capsid The protein coat that protects and transfers the phage chromosome.

concatemer A single molecule of DNA consisting of more than two phage genomes joined head to tail.

cotransduction The delivery of two genetic markers to a recipient cell by a single phage; cotransducible chromosomal markers must normally be close enough to be included on a single piece of DNA small enough to fit in a phage head.

headful packaging Incorporating DNA into a maturing bacteriophage particle using a single site-specific cut that initiates the entry of DNA into the phage head and, when the head is filled, cutting the DNA again to create a chromosome with a terminal repeat, that is longer than a genome.

induction The destruction of phage-replication repressor in a lysogen, leading to initiation of the phage lytic cycle.

lysogeny A quiescent state in which a temperate bacteriophage is replicated as part of the host genome, either passively, if integrated into the chromosome, or by maintenance as a plasmid.

package To actively incorporate DNA into a maturing bacteriophage particle; the initiation of phage DNA packaging is site-specific and starts from a *pac* (P1, P22) or *cos* (λ) site.

site-specific packaging Incorporating DNA into a maturing bacteriophage particle by cutting specific sites; the length of DNA packaged is determined solely by the distance between the specific cutting sites.

stable transductant A bacterium that has received DNA via a viral vector and incorporated that DNA, or a part of it, into its genome.

temperate Being able either to infect and lyse a host with the production of further phages in an infectious cycle, or to enter the lysogenic state and be stably inherited as part of the host genome.

terminal redundancy The DNA sequence repeated at the ends of a phage chromosome; generally indicative of headful packaging.

virulent Not being able to lysogenize and, on infection, being able to enter only the lytic cycle.

Genetic transduction is the transfer from a donor to a recipient cell of nonviral genetic material in a viral coat. Transduction is one of the three ways by which genetic material can be moved from one bacterium to another. The

The Desk Encyclopedia of Microbiology
ISBN: 0-12-621361-5

others are transformation, in which naked DNA is actively or passively taken up by recipient cells, and conjugation, in which cell–cell contact is initiated via structures encoded principally by plasmids. DNA is either transferred through these structures or protected by them during transfer. Both transformation and conjugation are evolved systems. Naturally tranformable bacteria, such as *Bacillus subtilis*, *Haemophilus influenzae*, and *Streptococcus pneumoniae* have sizable sets of genes concerned with the transformation process. Conjugational systems are essential to the plasmid way of life, allowing plasmids to move between hosts, not necessarily of the same species, more or less promiscuously. Transduction, on the other hand, may well be simply an accidental concomitant of the way in which bacteriophages multiply and package their DNA.

I. HISTORY AND CONCEPTS

A. Discovery of transduction in bacteria

Transduction was discovered by Zinder and Lederberg in the early 1950s, at a time when the existence of "the gene" was not yet universally accepted. Conjugation had been discovered in *Escherichia coli* by Lederberg and Tatum in 1946 and it was conjugational transfer of DNA between two differently marked strains of *Salmonella typhimurium* that Zinder and Lederberg were expecting to demonstrate when they discovered tranduction. Conjugation had been shown to require cell–cell contact because it failed to occur when donor and prospective recipient were grown in the same vessel, but separated by a filter that allowed macromolecules and medium, but not cells, to pass through. On the contrary, Zinder and Lederberg were to find that a filter did not impede gene transfer between *Salmonella* strains.

Because it was expected that conjugational gene transfer would allow the coinheritance of any pair of donor genes, however far apart, the strains tested were all doubly marked (i.e., contained mutations in two different genes) to avoid mistaking revertants or contaminants for recombinants. Although many strain pairings were tried, transduction was discovered only when a particular pair of differently marked strains of *S. typhimurium* was tested. This single success was the consequence of two happy accidents. We know now that in transduction only linked markers can be coinherited. By good fortune, one of the strains used, LT-22, was both marked with two closely linked genetic markers and was lysogenic for a bacteriophage, P22, that is able to transduce. Phages produced by spontaneous lysis of LT-22 crossed the filter, infected the other, sensitive, strain and carried transducing DNA home. Transformation was excluded as a mechanism by showing that transducing activity was resistant to DNase treatment and further experimentation showed a phage to be responsible for the gene transfer. A few years later, bacteriophage P1, a lysogenizing phage of *E. coli*, was also shown to mediate transduction. Because any single genetic marker can be transferred by P1 or P22, the process is called generalized transduction. Transduction by these two phages continues to be an important genetic tool. Figure 85.1 shows an overview of generalized transduction based on the behavior of the *E. coli*-P1 system.

In 1956, Morse, Lederberg, and Lederberg described another sort of transductional system. They found that the bacteriophage λ was able to transduce the galactose (*gal*) genes, which are located next to the site on the chromosome at which the lysogenized phage is integrated. Because they could find no other genes that λ would transduce, they named the process specialized transduction. It was later shown that the biotin (*bio*) genes, which with the galactose genes flank the λ integration site, can also be transduced. Specialized transduction in a modern guise has since achieved new importance because genetic engineering techniques permit the artifical construction of specialized transducing particles containing any desired gene.

B. Specialized and generalized transduction compared

Specialized and generalized transduction are processes that are quite different in detail, but which have in common the fact that a bacteriophage acts as a vector to transfer host DNA from donor to recipient. The two processes are compared in Table 85.1. They are described separately in the sections that follow.

II. SPECIALIZED TRANSDUCTION

A. Phage λ and the generation of low-frequency transducing lysates

Specialized transduction by coliphage λ follows naturally from its temperate lifestyle. λ particles contain linear phage DNA with cohesive ends that result from staggered cuts within *cos* sites, made during phage DNA packaging. These cohesive (or sticky) ends permit the DNA to circularize after infection. After circularization, a choice is made between two possible outcomes of the infection. In the lytic cycle, the λ circle replicates rapidly, initiating a process that will result in the lysis of the host and release of progeny phage particles about an hour later. An intermediate in lytic growth is a DNA concatemer consisting of several

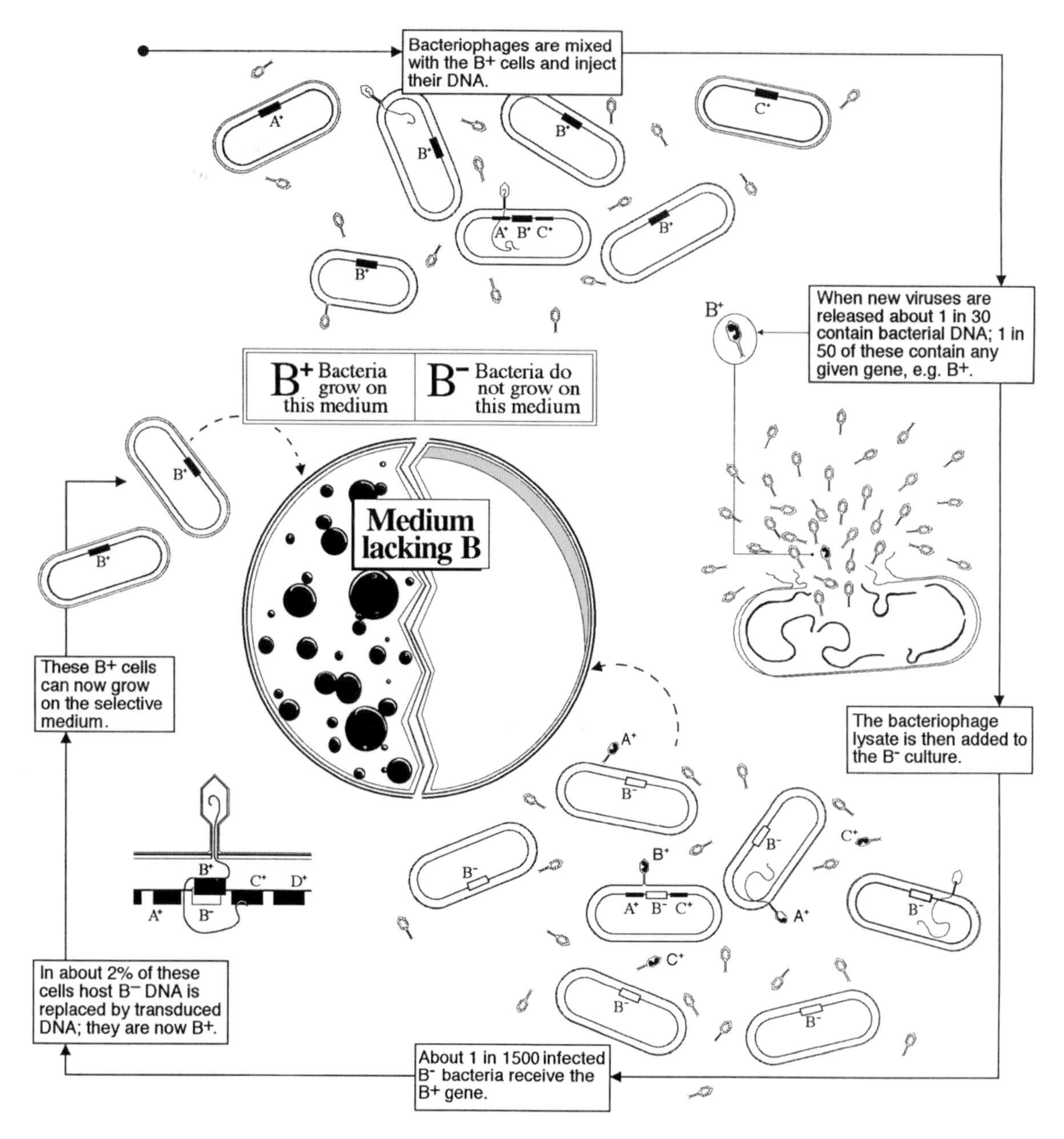

FIGURE 85.1 Overview of the generalized transduction process (from Masters, 1996, with permission). The picture, which proceeds clockwise from the upper left, applies in outline to all generalized transducing systems, but the numerical information is based on the *E. coli*-P1 system. It shows the production of a transducing lysate from a prototrophic strain and the transduction of a gene, B, from donor to recipient. The donation of B^+ to the B^- recipient confers a selectable phenotype (growth on B-deficient medium). Although A^+ and C^+ are also donated and genetic exchange can occur, these exchanges will not lead to an alteration in recipient genotype because the recipient is already A^+C^+.

TABLE 85.1 Comparison of specialized and generalized transduction

	Specialized	Generalized
Genes transduced	Adjacent to chromosomal insertion site	Any
Process creating transducing lysates	Induction of lysogen	Induction or infection
Transducing particles	Contain both phage and host DNA covalently linked on a single molecule	Host DNA only
Process creating transducing particles	Aberrant excision of lysogenic phage DNA	Mischoice of packaging substrate
Transduced progeny	Unstable, partially diploid lysogens	Stable, haploid recombinants
Best-known enterobacterial transducing phages	λ and related phages	P1, P22, T4

full-length genomes arranged head to tail. This is the packaging substrate, which a phage enzyme cleaves at successive *cos* sites to create the cohesive-ended, genome-length fragments of fixed sequence that are packaged into phage heads.

In the lysogenic response, the repression of replication is established and a chain of events is initiated that lead, via a site-specific recombinational event between *attP* on the phage and *attB* on the host DNA (*attB* is located between the *gal* and *bio* genes), to the integration of the phage into the continuity of the bacterial chromosome. Once integrated, it is passively replicated, once per division cycle, by the host's replicative machinery. It remains quiescent, except for the continual production of a repressor that prevents entry into the lytic cycle. If repression fails or is interfered with deliberately by induction, the phage is cut out of the chromosome to reverse the process that occurred at lysogenization. If all goes smoothly, excision will be accurate and a normal phage is generated.

Specialized transducing phages arise when excision occurs inaccurately to form a circle of about the correct length, but with the wrong DNA. Provided that the replication origin is retained, such a DNA molecule can go through the replication cycle including concatemer formation, *cos* cutting, and packaging. Phages that have acquired *gal* (λd*gal*, where "d" signifies defective) will have lost phage capsid genes and be unable to undergo further lytic growth without help. However, because the cell in which the particle originates probably also contains complete phage genomes (excised perhaps from a sister chromosome), the requisite enzymes and structural components will be available to allow the assembly and liberation of defective transducing particles. λ*bio* phages, in contrast, contain replication and capsid genes and so are not defective. They will, however, lack some of the genes needed to attain lysogeny. Aberrant excisions that lead to recoverable transducing particles are rare; only 10^{-5}–10^{-6} phages are of this type. Figure 85.2 illustrates the events that lead to the formation of a specialized transducing particle.

B. Transduction by specialized transducing phages

If a lysate such as the one described is prepared on a *gal*$^+$ donor and used to transduce a *gal*$^-$ recipient, gal$^+$ progeny will be found at low frequency. These progeny will frequently be diploid for the *gal* genes as a result of inheriting *gal* DNA from both phage and host, although haploid progeny can arise by the replacement of *gal*$^-$ by *gal*$^+$ with loss of the phage. If the ratio of phage to recipient is low, cells receiving λd*gal* will not be coninfected with an intact phage. Diploid *gal*$^+$ progeny will either be λd*gal* lysogens with the phage integrated at the attachment site, or, more often, cointegrates of phage and chromosome generated by a single homologous recombination event between *gal* sequences. These cointegrated Gal$^+$ progeny will be unstable because the integration is readily reversible. Both types of lysogen will be immune to λ infection, but will not produce any phage on induction because they lack coat protein genes. Stable, haploid, *gal*$^+$, λ-sensitive progeny can also arise by the resolution of the cointegrate with allele exchange and the subsequent loss of phage DNA.

If the ratio of phage to recipient is high, cells receiving λd*gal* are likely to also be infected with a normal phage. In this case, dilysogens can form with the defective and normal phages in tandem array. The Gal$^+$ phenotype of such progeny will be unstable because homologous recombination between the tandemly repeated genomes will lead to frequent phage loss. Dilysogens are inducible because the normal phage encodes all necessary proteins for phage maturation; the product of induction will be a mixed lysate that can contain equal numbers of transducing and infectious particles. These high-transducing lysates are not only convenient for further gene transfer, but provided a precloning, pre-PCR method for obtaining purified *gal* sequences. In *E. coli* strains from which *attB* is missing, λ can lysogenize at low frequency at alternative sites. This property was used to generate transducing phages for a variety of markers around the chromosome before it became possible to use restriction endonucleases to create genomic libraries *in vitro*.

C. Other specialized transducing systems

Any temperate bacteriophage that integrates into the chromosome of its host should be capable of specialized transduction. This supposition has not been extensively tested, but high-transducing lines of the lambdoid phages P22 (*Salmonella*) and phi80 (*E. coli*), the corynephage γ, and *Bacillus* phages SPβ and H2 have been reported (for references see Weisberg, 1996).

In vitro packaging allows specialized transducing phages to be made for any gene *in vitro*, and specially constructed λ vectors are widely used for making genomic libraries or for cloning particular genes. A λ cloning vector is a phage that has been modified so as to provide unique restriction endonuclease cutting sites at the designated cloning site. Many vectors have had inessential λ DNA deleted to provide more "space" for cloned DNA and have lost the DNA needed to lysogenize. However, lysogenizing vectors are available in which DNA fragments several genes in length can be cloned, allowing such genes to be stably maintained in

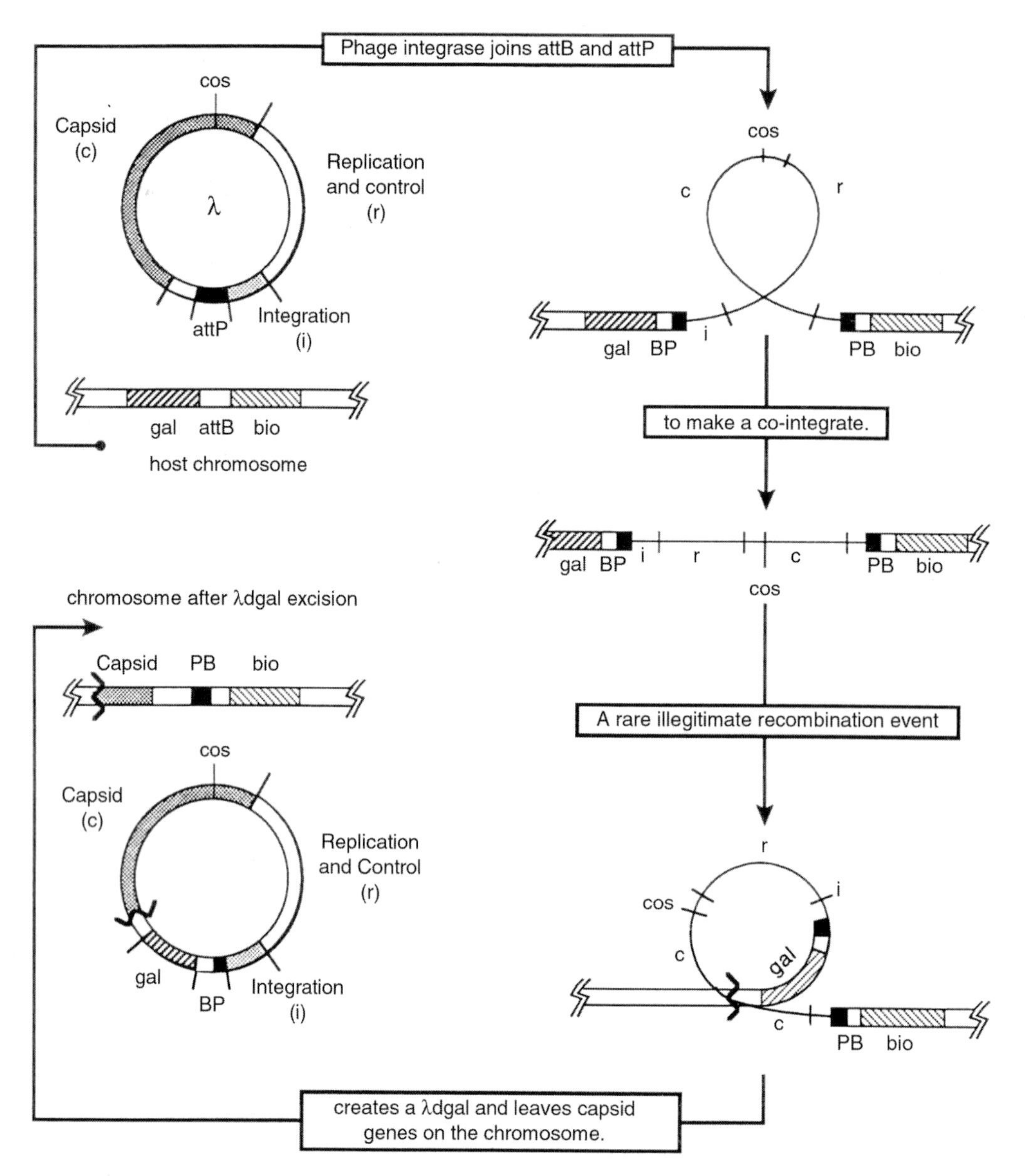

FIGURE 85.2 The creation of a specialized transducing phage. The figure reads clockwise from the upper left; phage λ is used as an example. Site-specific recombination between *attP* and *attB* leads to the formation of a cointegrate between λ and the chromosome to create a normal λ lysogen. After induction, a further site-specific event (not shown) reverses the process in most cases. Rarely, however, an illegitimate recombination event takes place that leaves phage DNA on the chromosome and a similar amount of chromosomal DNA (*gal* in this case) as part of a defective phage. The bold zig-zags mark the site of illegitimate recombination and the consequent boundary between phage and chromosomal DNA.

single copy, as part of a phage, at the phage-attachment site. Special-purpose transducing phages of this sort have been made that permit the controlled expression of foreign genes in *E. coli*, such as, for example, phage T7 RNA polymerase, used to amplify the expression of selected proteins engineered to be transcribed by the polymerase. Other phages allow the linking of gene promoters or translation–initiation regions to reporter genes, so that expression can be quantified. Because λ lysogens contain single copies of cloned genes, problems encountered when gene products are overexpressed, as they usually are from plasmid clones, are avoided. The use of specialized transducing phages as mutagens is mentioned in Section VII.A. For a fuller discussion of cloning in λ see the chapter by N. Murray in "Bacterial Genetic Systems" (Miller, 1991).

III. GENERALIZED TRANSDUCTION: BASIC FACTS

The most extensively studied and used generalized transducing systems are the *Salmonella*-P22 and

E. coli-P1 systems. In both systems, transduction is a rare event with perhaps $1/10^5$ phage particles able to transduce a particular selected marker. Because transduction is infrequent, no single recipient cell will receive more than one transduced DNA fragment. Therefore, cotransduction will occur only if the markers are close enough together to be included on the same piece of DNA. A P1 head contains a linear DNA molecule about 100 kb, or 2% of the chromosome, in length. P22 is smaller and holds about 44 kb of linear DNA, about 1% of the chromosome. These lengths define the maximum distance between cotransducible markers. Because transduced DNA is linear, it can become incorporated into the recipient chromosome only via a double crossover event. Because homologous pairing must be achieved, the requisite crossovers seldom occur very close to the ends of the transduced fragment and, in practice, the closer two markers are, the more likely they are to be cotransduced. Figure 85.3 shows the frequency of cotransduction of two markers as a function of the distance between them. Cotransductional mapping has been extremely valuable in placing closely linked markers, not separable by conjugational mapping, in the correct relative positions on genetic maps.

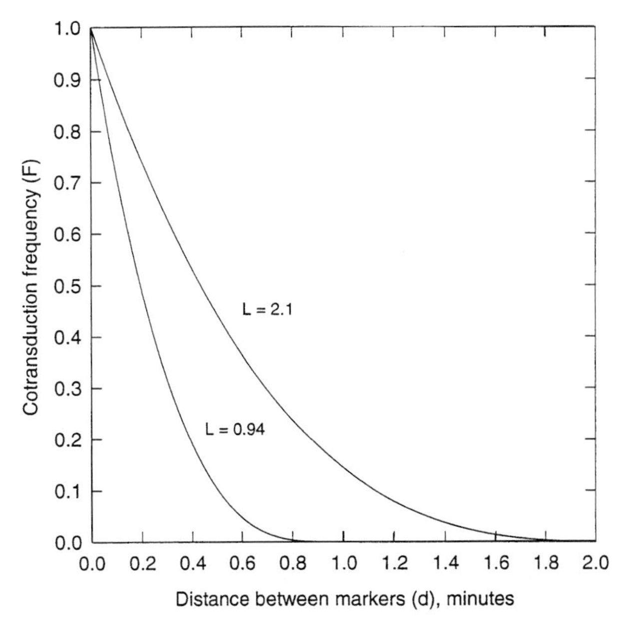

FIGURE 85.3 Mapping by cotransduction (from Masters, 1996, with permission). Cotransduction frequencies (*F*) are plotted as a function of the distance between markers (*d*) according to the formula $F = (1 - d/L)^3 \cdot L$ is the length, in map units, of the packaged DNA, and is 2.1 for P1 and 0.94 for P22. The formula assumes that all genes are equally likely to be transduced and so is more applicable to transduction with P1 or with high-transducing mutants of P22 than to transduction with wild-type P22, which packages DNA selectively.

IV. GENERALIZED TRANSDUCTION: HOW TRANSDUCING PARTICLES ARE FORMED

A. How generalized transducing phages package their DNA

Generalized transducing particles are formed when the phage DNA-packaging mechanism seizes on host DNA, instead of phage DNA, as a packaging substrate. In order to understand how transducing particles are formed, it is thus necessary to understand how the phages that transduce package their own DNA. P1, P22, and most other naturally transducing phages that have been studied package their DNA from concatemers by a headful packaging mechanism. Concatemers are polymers of phage genomes that arise during DNA replication. In cells infected with P1 or P22, a phage-specific packaging endonuclease recognizes a particular site on the concatemeric substrate, termed the *pac* site. Cutting at this site creates an end that then enters a waiting phage head. DNA packaging proceeds until the head is full and the DNA is then cut again (this time not at a specific site). An empty phage head replaces the full one and the process is repeated. An important point is that the second and subsequent cuts are not site-specific; the terminal cut occurs when the phage head is filled. Each phage head holds more than a phage genome's worth of DNA, leading to what is termed terminal redundancy, which describes the fact that each new phage DNA molecule has a rather long sequence repeated at both ends. Successive daughter DNA molecules cut from a single concatemer will begin and end with different sequences (Fig. 85.4). This permuted terminal redundancy is crucial for replication during the next cycle of infection, as it permits the replication of linear molecules without the loss of unique genetic material and provides a means for subsequent concatemer formation.

B. Packaging host DNA

How does this packaging mechanism foster the packaging of host DNA? Both P1 and P22 *pac* sites have been added to host chromosomes experimentally. In each case, the transduction of markers to one side of the introduced site is increased dramatically, by as much as several orders of magnitude. The DNA that is packaged with increased frequency is not only that which contains the *pac* site, but also that of subsequent "headfuls." In a *Salmonella* system in which *in situ* replication of P22, which proceeds into the

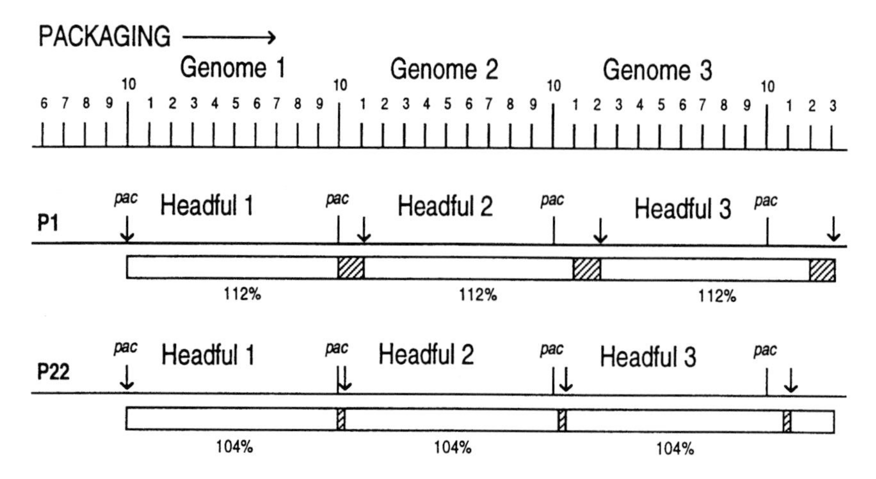

FIGURE 85.4 Headful packaging of phage DNA results in terminal redundancy. The figure shows the packaging of phage DNA, startng at a *pac* site, from a notional concatemer in which each genome is divided into 10 units. A P1 head holds 1.12 genome equivalents, leading to a terminal redundancy of 12%. The capacity of a P22 head is only slightly larger than a genome equivalent; as a result, the terminal redundancy is only 4%. Terminally redundant DNA is shown as crosshatched, whereas vertical arrows indicate the sites from which successive encapsidations begin.

neighboring chromosome, can be induced, even the 12th headful was transduced at 1000 times the normal frequency. One hundredfold increases in P1 transduction were observed when P1-*pac* was inserted into the chromosome. These observations suggest that the chromosomes of *E. coli* and *Salmonella* lack exact matches to P1-*pac*, or P22-*pac*. However, packaging of transduced DNA might nonetheless proceed from sites sufficiently similar to *pac* to allow the initiation of packaging at low efficiency. It appears that this is so for the *Salmonella*-P22 but not for the *E. coli*-P1 system.

1. *P22 packages chromosomal DNA from pac-like sites*

If the chromosome had *pac*-like sites, their numbers would be unlikely to be great, and markers closer to and downstream from *pac* might be expected to be packaged with higher frequencies. Consistent with this idea, different genes are transduced with quite different frequencies, varying over a range of ~1000-fold. A further expectation, if *pac*-like sites are few, is that processive head-filling would mean that packaged fragments would tend to be in register, and cotransduction frequencies would be skewed to reflect this. A test confirmed this expectation. Making a deletion of chromosomal material altered the cotransduction frequencies of downstream markers in a way that suggested that the packaging register had shifted. Finally, P22 mutants able to transduce at greatly increased frequency (up to 1000-fold) were isolated and shown to have a mutated packaging endonuclease. The range in transduction frequencies was greatly reduced in these mutants, consistent with a loss in packaging specificity.

2. *P1 packaging of host DNA is unlikely to proceed from pac-like sites*

The range of frequencies with which P1 transduces different markers is not very great; about 20-fold is the maximum. The measurement of the relative frequencies of host genes in transducing lysates by quantitative hybridization showed that they varied over a quite narrow range, about threefold. This suggests that the packaging of transducing DNA by P1 is initiated non-specifically, either from a large variety of possible sequences or from accessible structures, such as free DNA ends. Despite this, each transducing particle does not appear to originate from a separate cell; rather cells that produce one are much more likely to produce a second, carrying a different host DNA fragment, than do infected sister cells. This suggests that the packaging of transducing DNA, once initiated, may well be processive (continue along the same molecule). The hybridization studies described allowed the fraction of host DNA in the lysate to be estimated at 5%. Because each transducing phage carries 2% of the chromosome the frequency of transduction for an individual marker would be ~10^{-3} if all transducing particles yielded recombinants, rather than the 10^{-5} observed. Clearly, the delivery to a recipient cell is not sufficient to ensure stable inheritence (see Section V.B).

V. GENERALIZED TRANSDUCTION: TRANSDUCED DNA IN THE RECIPIENT CELL

Once packaged within phage particles, transducing DNA is delivered to recipient cells in the same manner as is phage DNA. Linear DNA molecules are injected into the cell through the wall and cytoplasmic membranes. Three possible fates await them there—degradation by nucleases, recombination with homologous recipient DNA to form a stable haploid transductant, or persistence within the cytoplasm in a form refractory both to recombination and degradation. For both the P1 and P22 systems, only about 1% of the transferred DNA becomes part of the recipient chromosome.

A. Some transduced DNA is degraded

Escherichia coli contains several exonucleases able to destroy unprotected incoming linear DNA. Thus successful transformation of *E. coli* with linear DNA requires the use of nuclease-deficient strains, and, even so, the recovery of transformants is poor. Transduced DNA, in contrast, resists degradation; successful transduction does not require nuclease-deficient hosts. If the fate of radioactively labeled donor DNA is monitored, the fraction of DNA that is degraded and recycled can be measured. In recipients unable to undergo genetic exchange, no more than 15% of the label entering cells can be recovered from the host chromosomal DNA. This material is not in continuous stretches, consistent with the expectation that it is likely to consist of recycled nucleotides. Surprisingly, however, recoverable chromosomal label was not increased much in recombination-proficient cells. Thus, there appears to be significant recycling of nucleotides, but very little physically detectable recombination. Most transduced DNA must have some other fate.

B. Abortive transduction

Most of the transduced DNA, up to 90% of label in some cases, remains in the cytoplasm in a stable form. It neither replicates nor is degraded, and can be physically recovered for at least 5 h after infection. This DNA is referred to as abortively transduced DNA, and the cells that harbor it are termed abortive transductants. Abortive transduction was first noted by Lederberg in the mid-1950s and was proposed to explain a curious phenomenon that accompanied the transduction of nonmotile *Salmonella* to motility. Motility is scored on semisolid agar, on which motile bacteria swim and form diffuse patches rather than discrete colonies. After transduction, transduced progeny were observed instead to form trails, consisting of strings of colonies of nonmotile cells, each apparently arising from a single nonmotile descendent of a motile cell. This could be understood if the parent motile cell were an abortive transductant harboring a stable nonreplicating, but transcriptionally active fragment of DNA encoding the synthesis of flagella. At division, only one daughter would inherit this DNA; this cell would continue to make flagella and would swim off. Its sessile sister would found a colony. The abortively transduced piece of DNA would continue to be unilinearly inherited as long as the original recipient continued to divide. Later, abortive transductants for nutritional markers were found; these are barely visible colonies, each containing about 10^5 cells. Each such colony contains one cell able to found a similar colony. If that cell alone were able to make the enzyme encoded by the transduced gene, the colony would grow linearly, each daughter inheriting enough gene product to grow for a bit, but only the original cell continuing to divide to produce a visible microcolony.

Thus it appears likely that the bulk of P1- and P22-transduced DNA assumes the abortive configuration in the recipient cell. DNA in this configuration is neither degraded nor recombined into the host chromosome. What is the abortive configuration? Increased numbers of complete transductants can be obtained at the expense of abortive transductants by treating the phage with UV to produce gaps in its DNA. Recombinational processes can be initiated at gaps, and the effect of UV on transduction suggests that, not surprisingly, abortively transduced DNA lacks gaps. However, recombination is most often initiated at the ends of linear molecules, and the DNA injected during phage infection is linear. Are the ends of abortively transduced DNA somehow protected? The answer appears to be a definite yes. If host DNA is made heavy by density-labeling before infection, and the label removed at the time phage are added, all heavy particles in the eventual lysate will contain transducing DNA; these can be separated from the light infective phages and used to infect cells. The heavy transducing DNA can be extracted from transduced cells soon after transduction and its physical state examined. This experiment was performed using P1 with startling results: Abortively transduced DNA appears to consist of circular molecules, the ends of which are held together by protein. Unfortunately the identity and origin of this protein have not been established. An intriguing suggestion made by Yarmolinsky and Sternberg is that the protecting protein is in fact the packaging endonuclease. They suggest that when bound to a genuine *pac* site, the protein is transferred along the concatemer, initiating headful cuts at intervals and finally being released. When it is initially bound to bacterial DNA, perhaps at an end, this does not occur and it is instead packaged bound to the end of the transducing DNA. In the recipient, it could attach to the other end, as well, creating abortive circles. In contrast, for P22, an injected internal head protein is implicated in abortive transduction. P22 mutants deficient in this protein yield an increased number of complete transductants at the expense of abortives.

C. Stable transduction

Only about 2% of transduced DNA can be recovered from the recipient chromosome in continuous segments long enough (>500 bp) to suggest that they were introduced into the chromosome by recombination. DNA of this length is only found in the chromosomes of recombination-proficient recipients, consistent with the idea that it is a recombination product. Because abortive circles appear stable and do not recombine once they are formed, the recombining DNA molecules that are the source of stable transductants presumably do not achieve the abortive state. This may be because they lack the protective protein at one or both ends, because circularization fails to occur, or because the DNA molecule is internally damaged in a way that makes it a suitable substrate for recombination.

In the *E. coli*-P1 system, in which the packaging frequency reflects gene frequency in the donor population, the range in frequencies of stable transduction is about fivefold greater than the variation in packaging frequency. Thus, there are factors in the recipient cell that favor the integrative recombination of particular markers. It seems plausible that the success of transductional recombination is affected by the proximity of the "recombinator" sequence chi, a 9-bp sequence that stimulates recombination in its vicinity. The stimulation of transductional recombination by chi has been demonstrated in specific crosses, but it would be interesting to correlate the known locations of chi with measured transduction frequencies, in order to determine whether the observed variations in transduction can be attributed to the presence of the element.

The introduction of recombinogenic structures into transduced DNA can stimulate integrative recombination at the expense of abortive transduction. Lysates treated with UV, which should contain what will become gapped DNA, give a greater proportion of complete transductants, as do transduced fragments containing origins of replication. The initiation of replication on such fragments in the recipient cell would be expected to give increased numbers of recombinogenic DNA ends.

Stable transductants are haploid cells in which the recipient allele has been replaced with the donor allele. The initial product of recombination, however, will be a heterogenote, most likely because only a single strand of recipient DNA is replaced. Two cycles of DNA replication will thus be required before two identical donor-type chromosomes are available for segregation to daughter cells. There is therefore a lag before the number of transduced cells can begin to increase; the earliest divisions will produce one transduced and one parental-type cell. Because of this, colonies formed by transduced mixtures that have been plated immediately after phage addition can contain cells with both donor and recipient genotypes. Further purification of such colonies is required to score the cotransduction of unselected recessive markers.

VI. OTHER TRANSDUCING SYSTEMS

A variety of other coliphages are capable of carrying out generalized transduction. These include the virulent headful-packaging phages T1 and T4, for which special mutants must be used to avoid donor DNA degradation and excessive killing of recipients, and the temperate headful-packaging phage Mu. Certain mutants of λ are capable of generalized transduction, but the packaging pathway is not understood and the production of usable generalized transducing lysates require special manipulations. Of the aforementioned, only transduction by T4 has found a niche. Because T4 packages 172 kb, or 3.5% of the chromosome's DNA per particle, it can transduce fragments about 50% longer than those transduced by P1; this has been useful. It is also less fussy about its host and can package DNA from mutants, such as those deficient in recombination, which are poor hosts for P1.

Generalized transduction has been described for a great many species of bacteria, suggesting that a generalized transducing vector can probably be found for any species susceptible to bacteriophage infection. There are 11 main branches on the eubacterial evolutionary tree; generalized transducing phages have been described for organisms on at least three of these, the spirochetes, the gram-positive eubacteria, and the proteobacteria. These latter two branches include most of the familiar and extensively studied bacterial species, whereas three of the others consist exclusively of thermophiles or intracellular parasites for which phages might be expected to be rarer. Generalized transduction has been used to study both the high and low G + C gram-positive bacteria (*Bacilli* and *Actinomycetes*). Generalized transducing phages have been described for species in at least three of the five subdivisions of the proteobacteria, the γ subdivision, which includes the coliforms and a variety of pseudomonads, the δ subdivision (*Myxococcus* and *Desulfovibrio*), and the ε subdivision (*Campylobacter*). Gene transfer in the α subgroup species, *Rhodobacter capsulatus*, is mediated by small particles, resembling phage heads, that package and deliver 4.6-kb fragments of host DNA, but the production of the particles has not been associated with presence of a phage. Interestingly, bacteriophages have been described for halophilic *Archaeobacteria*, but there is, as yet, no report of a transductional system.

The discussion has so far has centered on the transduction of chromosomal markers. There have, however, been many reports of transduction of plasmids by a variety of phages. Although the integration of the plasmid into the chromosome or the introduction of a *pac* site into a plasmid greatly facilitates plasmid transduction in several experimental systems, transduction can also occur in the absence of these facilitating factors. In at least some cases, plasmid–concatemer formation seems to accompany infection, providing a mechanism for the formation of a long-enough substrate to constitute a phage headful. It is not clear whether a long substrate is required in all cases or whether single circular molecules can sometimes be transduced.

VII. LABORATORY USES OF TRANSDUCTION

The understanding of the processes and structures that constitute a cell can be sought either with a gene-to-phenotype, or a phenotype-to-gene approach. Both strategies require the generation and analysis of mutants. Although many genome sequences have now been completed and methods have been devised for readily producing the protein products of selected genes the understanding of the roles of particular gene products in cells will continue to require the phenotypic analysis of specific mutants. Conversely, generating and analyzing groups of mutants sharing a particular or related phenotype permits the establishment of functional connections between the products of different genes. The study of mutations requires their generation, mapping, and transfer between strains for specialized phenotypic analyses. Transduction continues to have an important role in all three of these stages in the functional analysis of genes.

A. Mutagenesis

Because a transducing phage contains a relatively small fragment of DNA, transduction–cotransduction can be used in one of several ways to make sure that a mutation generated by random chemical means is confined to a limited segment of the genome. If a mutation is sought that is linked to a selectable marker, either a donor strain (UV or radiomimetic agents are popular for this) or an already prepared transducing lysate (hydroxylamine can be used to mutate DNA already in phage capsids) can be treated with a mutagen and transductants inheriting the linked marker can be analyzed for their mutational status. Happily, collections of *E. coli* strains exist, each with a mapped transposon insertion, such that a transductionally linked transposon is available for any chromosomal gene.

If mutation is to be by transposon mutagenesis, transducing phages can be used to deliver the transposon. A group of specialized transducing λ phages, which cannot replicate except in special (permissive) hosts, have been engineered to contain mini-transposons (which have the advantage of being unable, once transposed, to independently transpose again). If a nonpermissive strain is infected with these phages and a transposon encoded drug-resistance marker selected, surviving progeny will be the products of transposition. P1 can inject its DNA into, but not productively infect, a wide range of nonpermissive species, e.g., P1::Tn5 has been used many times for the mutagenesis of *Myxococcus xanthus*.

B. Mapping

Transduction has been used both for relatively rough and for fine-structure genetic mapping. Approximate gene distances can be calculated from cotransduction frequencies by using the curves in Fig. 85.3. These curves were derived by assuming that transducing DNA is packaged at random from donor DNA and that inheritence in the recipient follows a pair of recombinational exchanges equally likely to occur at any point on the transduced fragment. Each of these assumptions is only approximately true, as described earlier. For P1, the first is reasonably correct, but recombination is favored at certain sites more than others. For P22, packaging is not random, although high-transducing P22 mutants package less selectively. Because cotransduction frequency is only a good measure of distance if each of the markers is transduced with the same frequency, marker-dependent differences in transduction frequencies can limit its usefulness as a measure of distance. The availability of the set of transposon insertion strains described previously has made it possible to roughly map, relative to a transposon insertion site, any mutation whose location is approximately known.

Fine-structure genetic mapping using multifactorial transductional crosses was an important technique in its day, but is unlikely to be done again. The molecular technologies of cloning or PCR amplification and sequencing can order closely linked markers more straightforwardly (if not more cheaply). Fine-structure mapping required the tedious construction of multiply marked, reciprocally mutant, strains that could then be crossed. Markers were ordered on the assumption that because two crossovers are more common than four, rare recombinant classes are likely to have required multiple exchanges. Anomalies have occurred that led to incorrect conclusions.

The use of these techniques requires that the approximate location of the mutated gene of interest be known. If a new mutation has been isolated on the basis of its phenotypic consequences, its position on the chromosome may be entirely unknown. There are several choices for mapping such a mutation. One genetic approach is to use conjugational analysis (rough mapping) followed by transduction (fine mapping). An alternative for *Salmonella* is to use the Benson and Goldman (1992) set of P22 prophages, which are located at intervals on the chromosome, to directly map mutations by transduction. Lysates prepared on these lysogens transduce at such high frequency that they can be used to localize a mutation with spot tests. A fully molecular approach, such as finding a complementing clone in a genomic library and analyzing its DNA directly, can of course bypass genetic procedures. For *E. coli*, with its fully sequenced genome, the affected gene can be quickly identified in this way.

C. Strain construction

Transduction really comes into its own as the simplest available method for moving genes between strains. Most sophisticated studies require either the combination of mutations in a strain to detail their interactions, the transfer of a mutation to a standard strain for comparability of behavior, or the transfer of a mutation to a specilized strain to determine its effect, for instance, on the expression of particular reporter genes. Linked drug-resistance markers, which can, if necessary, first be transduced to the mutant strain, can be used to facilitate the necessary transductions.

Transduction can also be used in multistep procedures to introduce mutations or deletions constructed *in vitro* onto the chromosome. Plasmids containing the desired constructions may, in a minority of cells, integrate into the chromosome by a single-crossover event. The unstable cointegrate thus formed can resolve with an exchange of alleles. If the mutation incorporates a selectable marker, transduction can be used to separate the chromosomal DNA with the desired allele from the plasmid, by transducing it to a new plasmid-free strain.

VIII. TRANSDUCTION OUTSIDE THE LABORATORY

In this article, I have discussed the mechanism of transduction and the uses to which it has been put in the laboratory. We know from the epidemic spread of drug-resistance plasmids that conjugation occurs in natural environments. Does transduction occur in nature as well? This question has not been extensively investigated, but a number of reports suggest that it very likely does. The principal organism for which there are reports of transduction in natural environments is *Pseudomonas aeruginosa. P. aeruginosa* transduction has been demonstrated on leaves and even between bacteria on neighboring plants. Both chromosomal and plasmid markers were transferred between strains in lake water in which phages are present. There is probably also marine transfer. The transductional transfer of drug resistance between *P. aeruginosa* strains in hospitals has also been recorded. Transduction has also been reported to occur between strains of *Streptococcus thermophilis* in its natural environment, yogurt! It thus appears likely that transduction does occur in the wild and might be responsible for a significant fraction of natural gene transfer. The ability of P1 to transfer DNA between *E. coli* and the distantly related *Myxococcus xanthus* suggests that transduction plays a role in horizontal gene transfer between distantly related species in addition to a role in the transfer of genes between more closely related species.

ACKNOWLEDGMENT

I thank David Donachie for the preparation of Figures 1, 2, and 4.

BIBLIOGRAPHY

Benson, N. R., and Goldman, B. S. (1992). Rapid mapping in *Salmonella typhimurium* with Mud-P22 prophages *J. Bacterial.* **132**, 1673–1681.

Masters, M. (1996). Generalized transduction. *In "Escherichia coli* and *Salmonella* Cellular and Molecular Biology" (F. C. Neidhardt, *et al.* eds.), pp. 2421–2441. ASM Press, Washington, DC.

Miller, J. H.(ed.) (1991). Bacterial genetic systems. *Meth. in Enzymol.* **204**.

Weisberg, R. A. (1996). Specialized transduction. *In "Escherichia coli* and *Salmonella* Cellular and Molecular Biology" (F. C. Neidhardt, *et al.* eds.), pp. 2442–2448. ASM Press, Washington, DC.

Yarmolinsky, M. B., and Sternberg, N. (1988). Bacteriophage P1. *In* "The Bacteriophages" (R. Calendar, ed.), Vol. 1, pp. 291–438, Plenum Press, NY.

Zinder, N. D. (1992). Forty years ago: The discovery of bacterial transduction. *Genetics* **132**, 291–294.

86

Transformation, Genetic

Brian M. Wilkins[†] *and Peter A. Meacock*
University of Leicester

GLOSSARY

competence Physiological state of cells allowing binding and uptake of exogenous DNA.

homologous recombination Physical exchange between DNA molecules having the same or a very similar nucleotide sequence.

heteroduplex DNA Duplex DNA in which the two strands have different genetic origins. A heteroduplex may contain mismatched bases.

pheromone A species-specific chemical produced by one organism to alter gene expression or behavior in a second organism.

transformation The uptake of naked DNA from the extracellular environment with production of genetically different progeny cells.

Genetic Transformation is the process by which bacteria and microbial eukaryotes take up fragments of naked DNA from the extracellular medium to produce genetically different progeny cells called transformants. The process is manifested by genetically and ecologically diverse organisms in which it is encoded by chromosomal genes. Natural transformation systems require the development of competence by the recipient cell as a specialized physiological state necessary for the binding and uptake of exogenous DNA as linear fragments. Considerable diversity exists among the transformation apparatuses of various bacteria, with some showing specificity for homologous DNA, but a unifying theme of natural systems is the uptake of random fragments of single-stranded (ss) DNA. Transforming DNA that is related at the nucleotide-sequence level to the chromosome of the recipient cell can be incorporated into the resident genome by homologous recombination to give a region of heteroduplex DNA. Mismatched bases may block formation of the heteroduplex or be removed from the recombinant structure by excision from the donor strand. Plasmid DNA is also taken up by natural systems as single-stranded fragments; these may interact to regenerate circular plasmid molecules, albeit with low efficiency. Natural transformation is viewed as a gene-exchange process that enhances the genetic diversity of a bacterial population, as well as maintaining the constancy of the bacterial genome by potentiating recombinational repair of damaged DNA. Artificial transformation procedures have been devised for many microbial species, allowing the uptake of duplex and circular forms of DNA. Such procedures provide a cornerstone of recombinant DNA technology.

[†]Deceased.

I. HISTORY OF TRANSFORMATION AND DIVERSITY OF SYSTEMS

Transformation was discovered by Frederick Griffith in 1928 through work with *Streptococcus pneumoniae*, which causes pneumonia in humans and lethal infections in mice. His experimental system used pathogenic wild-type pneumococci, which had a smooth colony morphology due to the production of a polysaccharide

The Desk Encyclopedia of Microbiology
ISBN: 0-12-621361-5

capsule, and an avirulent rough mutant that lacked a capsule. Griffith found that the rough strain acquired virulence and the ability to produce a capsule when mixed with heat-killed wild-type cells and injected into the peritoneal cavity of mice. Some 16 years later, Oswald Avery, Colin MacLeod, and Maclyn McCarty reported a systematic chemical analysis of the transforming principle, showing it to possess the properties of DNA. This was a historic landmark not only in paving the way for the identification of DNA as the genetic material, but also in establishing transformation as the first genetic transfer process to be described for bacteria.

Natural transformation is encoded by genes in the bacterial chromosome and involves the uptake of naked DNA from the extracellular environment. Hence, there is no requirement for a living donor cell. The process supports the uptake of random small fragments of DNA. These are rendered single-stranded during their transport into the recipient cell and are recoverable as material of about 10^4 nucleotides in size. Transformation differs significantly from the two other quasisexual processes for gene transfer between bacteria, namely bacterial conjugation and transduction. Transduction is a bacteriophage-mediated process. It involves the accidental packaging of a double-stranded fragment of bacterial or plasmid DNA of the donor cell into a phage particle, which then acts as the transfer vector to the recipient cell. The size of a transduced DNA fragment is generally greater than that of transforming DNA; it may exceed 100 kb, depending on the DNA packaging capacity of the phage particle. Bacterial conjugation is the dedicated transfer process of self-transmissible plasmids and a class of mobile genetic elements called conjugative transposons. According to the general model of conjugation based on systems studied in *Escherichia coli*, the process requires a specialized contact between the donor and recipient bacterium to form the transport pore for a specific DNA strand. This is transferred progressively in the 5′ to 3′ direction from a genetically defined origin. As a secondary property, conjugative plasmids can cause the transfer of very large sectors of chromosomal DNA, which may exceed 2000 kb in size.

Many genetically and ecologically distinct bacteria are known to be naturally transformable. The list includes gram-positive and gram-negative (eu) bacteria, including cyanobacteria, and archaea. Mechanistic studies of transformation have focussed on the similar systems determined by *S. pneumoniae* and the soil bacterium *Bacillus subtilis*. These are viewed as gram-positive models. Gram-negative paradigms are the systems specified by *Haemophilus influenzae* and *Neisseria gonorrhoeae*—the aetiological agents of spinal meningitis and gonorrhoea, respectively. The *Haemophilus* and *Neisseria* systems differ significantly from each other and from the gram-positive models.

Transformation is likely to contribute importantly to gene flux between bacteria in natural environments. The process has also been used for mapping the relative positions of chromosomal genes of some naturally competent bacteria. Mapping involves the selection of transformants containing a specific donor gene and scoring the fraction containing one or more unselected donor markers. The derived values allow the ordering of genes on the basis that cotransformation frequency is inversely proportional to genetic distance. The usefulness of the technique is limited by the small fraction of the genome that is transferred on any one DNA fragment and by the fastidious growth requirements of a number of transformable bacteria, which make it difficult to isolate mutations in sufficient density per unit of chromosomal DNA. Such problems are bypassed by recombinant DNA technologies that allow mapping with reference to restriction fragment maps or, in the case of *B. subtilis* and *H. influenzae*, to the known nucleotide sequence of the entire chromosome.

The need in molecular biology to introduce DNA modified *in vitro* back into cells has led to the development of artificial procedures for the transformation of prokaryotic and eukaryotic microorganisms that do not develop natural competence. Methods used to promote the uptake of DNA include the treatment of whole cells with metal cations in conjunction with a heat shock, use of a brief high-voltage electrical discharge, and polyethylene glycol (PEG) treatment of osmotically stabilized protoplasts.

II. REGULATION OF COMPETENCE DEVELOPMENT

Transformation systems require the acquisition of competence as a physiological state enabling the bacteria to take up high-molecular-weight fragments of exogenous DNA. In many systems, competence is a transient condition associated with the induced synthesis of a small number of specific proteins. Growth conditions favoring the development of competence by different bacteria vary considerably. For example, competence may occur throughout the exponential growth phase to decline at the stationary phase, as found for *Deinococcus radiodurans* and *Mycobacterium smegmatis*, whereas the process develops in the early- to mid-exponential phase in *S. pneumoniae* and at the entry to the stationary phase in *B. subtilis*. Competence is observed to be a constitutive property in *N. gonorrhoeae*. Yet another variation is apparent in the fraction of cells that acquires competence under optimal inducing conditions.

Competence development has been characterized most extensively for *B. subtilis*. Here, it is a postexponential response involving a complex network of signals that sense the nutritional state of the environment and cell density. Many of the regulatory proteins in the cascade participate in other postexponential-phase responses such as sporulation, motility, and the production of extracellular degradative enzymes and antibiotics. Competence optimally develops late in exponential growth in a glucose-minimal medium, or early in stationary phase, but occurs in only a subfraction of cells making up 10% or less of the population. The state is associated with a cell type that is more buoyant and relatively dormant in macromolecular synthesis. Twenty or more genes have been identified for the development of competence, which collectively function in an elaborate complex of regulatory networks involving cell density-sensing systems, protein phosphorylation, and protein–protein and protein–DNA interactions (Fig. 86.1). The various regulatory signals converge to control the expression and activation of the key competence transcription factor, ComK. This is required for the expression of late genes in the *comC, comE, comF*, and *comG* loci, which determine the biosynthesis of the DNA-binding and -uptake apparatus.

The activity of ComK is negatively controlled by two proteins, ClpC and MecA. ClpC, in the presence of ATP, is thought to function as a molecular chaperone to increase the affinity of MecA for ComK, thereby forming a ternary complex that blocks the transcriptional function of ComK. The protein is released from the inhibitory complex via convergent pathways initiated by two cell-density-sensing pheromones. One pheromone, ComX, is a 9- or 10-amino-acid peptide that is processed posttranslationally and secreted from *B. subtilis* by a dedicated exporter system. ComX is detected by the ComP–ComA by two-component regulatory system; the pheromone is sensed by the membrane-associated ComP histidine protein kinase to trigger its autophosphorylation and the subsequent transfer of the phosphate group to the ComA response regulator. The phosphorylated form of the regulator, ComA~P, functions as a transcription factor to activate the expression of an operon specifying the ComS protein. In turn, ComS releases ComK from the ternary complex with MecA and ClpC.

The second pheromone, competence-stimulating factor (CSF), is a five-amino-acid peptide. It is probably

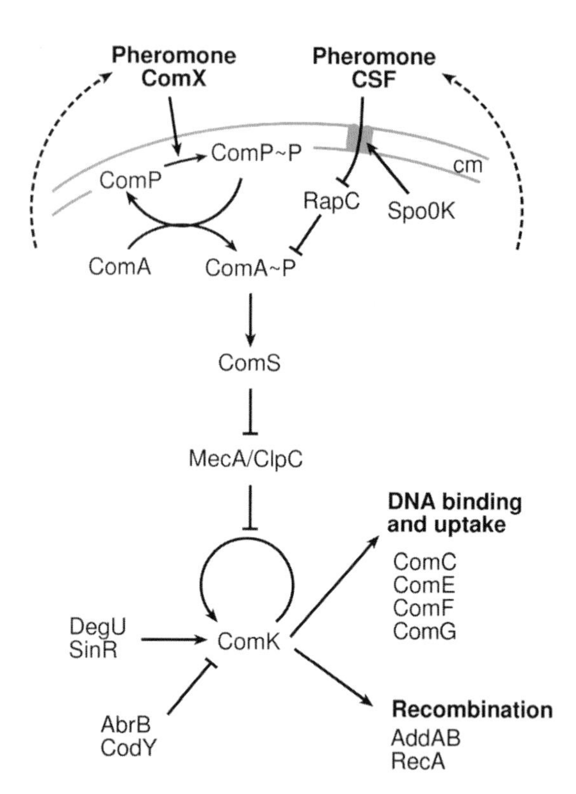

FIGURE 86.1 Model for pheromone regulation in *B. subtilis*, showing control of the late competence genes, which function in DNA binding and uptake, and recombination genes. Blocked lines indicate negative action.

taken up into the cell by the Spo0K oligopeptide permease transporter to modulate the phosphorylation level of ComA by inhibiting a phosphatase (RapC) active on ComA~P. The involvement of extracellular pheromones provides a quorum-sensing mechanism for monitoring cell density and triggering competence development at high cell concentrations. Under such circumstances, other cells of the same species might be releasing DNA through autolysis or, more intriguingly, by extrusion as an active donor function.

ComK protein is singularly required as a promoter-binding protein to stimulate the transcription of the late *com* loci, as well as for competence-related expression of recombination loci such as the *recA* and *add* genes. ComK protein in addition enhances transcription of its own gene via autoregulation. Transcription of *comK* is influenced by other DNA-binding proteins that integrate information on nutrient availability. Among these, AbrB and CodY are negative regulators, whereas DegU and SinR have stimulatory roles.

In contrast to the *B. subtilis* system, competence arises in *S. pneumoniae* as a transient state in the early- to mid-exponential phase of growth to affect nearly every cell in the culture. The state is marked by a shutdown of general protein synthesis and production of competence-specific proteins. As described for *B. subtilis*, competence development in *S. pneumoniae* depends on the accumulation of an extracellular pheromone. This is a 17-residue peptide, known as competence-stimulating peptide (CSP), which is sensed by a two-component signal transduction system to stimulate the expression of the competence genes. Interestingly, there are two promoters for the gene encoding the CSP precursor; one is constitutive and the other is autoinducible. The arrangement allows CSP that has accumulated to a basal level by expression of the constitutive promoter to trigger a cascade of pheromone production from the autoregulated promoter.

The regulation of competence in *H. influenzae* is quite different in that almost every cell can develop the state as a stable internally regulated condition. The same applies to a number of other gram-negative bacteria. The response is triggered by a transition into unbalanced growth following a shift-down from a rich to a nutrient-limited medium, with the cyclic AMP-catabolite activator receptor protein complex playing a regulatory role. *H. influenzae* also achieves low levels of competence in the late-exponential phase of growth.

III. DNA BINDING AND UPTAKE

A. Overview of DNA transport across cell membranes

A particularly challenging aspect of gene transfer is the nature of the specialized channels that allow DNA as a polyanion to pass through the hydrophobic environment of cell membranes and enter the cytoplasm of the recipient cell. Considerable variation exists in the protein complexes making up such channels. As discussed later, some of the DNA transporters show intriguing relationships to systems dedicated to protein secretion. These include systems for the secretion of exoenzymes and for biosynthesis of type IV pili. The latter are extracellular appendages that mediate the adherence of virulent bacteria to host eukaryotic cells, as well as a flagellum-independent form of translocation across surfaces called twitching motility.

A common feature of transformation systems that have been examined at the molecular level is that DNA enters the cytoplasm of the recipient cell in single-stranded form. Considerable diversity exists between systems with regard to the DNA uptake apparatus and whether or not it shows species specificity in using homologous DNA. In the *B. subtilis* and *S. pneumoniae* paradigms, DNA transport across the single-cell membrane of the cell envelope occurs through a surface-exposed complex of proteins. The transformation of gram-negative bacteria is complicated by the fact that the DNA must traverse an outer membrane, which characteristically contains an outer leaflet of lipopolysaccharide (LPS), as well as a cytoplasmic or inner membrane. In *H. influenzae*, DNA uptake proceeds through a specialized membrane vesicle, whereas passage of transforming DNA across the outer membrane of *N. gonorrhoeae* requires the type IV pilus.

B. *Bacillus subtilis* and *Streptococcus pneumoniae*

The inhibitory effects of nonhomologous competitor DNAs indicate that *B. subtilis* and *S. pneumoniae* take up DNA without sequence specificity. The first stage in DNA uptake involves the noncovalent binding of double-stranded (ds) DNA to about 50 competence-specific receptor sites on the surface of the cell. Duplex DNA is bound much more efficiently than ssDNA, ssRNA, or dsRNA. Binding is initially loose in that the DNA is released by washes of high ionic strength or by competitor DNA. Loose binding is followed by tight binding in which the DNA is resistant to release by washing but remains exposed to the external medium, as shown by its susceptibility to shear and exogenous DNase. The DNA then undergoes double-stranded cleavage by one or more surface-exposed endonucleases, which may be localized at the binding sites. The size distribution of the resulting fragments is in the range of ~5–15 kb.

The uptake of DNA, defined by its acquisition of resistance to exogenous DNase, commences 1–2 min after binding and occurs at a rate of ~100 nt/s at 28°C. Uptake culminates in the entry of ssDNA into the cytoplasm of the recipient cell. The uptake channel is thought to be an aqueous pore in a surface-exposed complex of proteins. The complex includes an entry nuclease that potentiates intracellular transport of ssDNA fragments with the associated degradation of the second strand and release of oligonucleotides into the medium. The entry nuclease of *S. pneumoniae* is the 30-kDa membrane-bound EndA endonuclease; the polarity of its activity is such that the nonhydrolyzed strand enters the cell with a leading 3′-terminus.

Further evidence for a single-stranded intermediate comes from the eclipse phase. This is a transient stage in the transformation process in which DNA that has been taken up into the recipient cell—as defined by its resistance to extracellular DNase—cannot be isolated in a form that transforms a second cell. The explanation is that entrant DNA is single-stranded in the eclipse phase and therefore an unsuitable substrate for uptake by a second cell. Another feature of the eclipse is the coating of the entrant strand with a ssDNA-binding protein. The eclipse protein is thought to protect the DNA from nucleases and aid the next step in the transformation process, namely homologous recombination with the resident genome.

Several models have been proposed to explain the bioenergetics of DNA transport. The classic proposal, made in 1962 by Sanford Lacks, that DNA uptake by *S. pneumoniae* is powered by the hydrolysis of the second strand by the membrane-localized EndA endonuclease remains controversial because the nuclease might be required to generate the single-stranded substrate of a transporter. Other mechanisms may include one or both components of the proton-motive force, an anion-exchange reaction dependent on proton cycling, an ATP-dependent system acting on ssDNA, and a transporter consisting of a complex of poly-β-hydroxybutyrate and calcium salts of inorganic polyphosphates, as discussed further in Section VI. Such a complex accumulates in the membrane fraction of competent bacteria and may serve to transport polyanionic salts, such as ssDNA, through the cytoplasmic membrane.

The DNA uptake apparatus of *B. subtilis*, as described by David Dubnau and his colleagues, is a complex of proteins encoded by some 15 genes belonging to the dispersed *comC*, *comE*, *comF*, and *comG* loci (Fig. 86.2). Homologs of some of these proteins have been described for other gram-positive and gram-negative bacterial-transformation systems. Similarities have also been detected at the predicted amino-acid-sequence level between ComG polypeptides and proteins involved in both the secretion of exoenzymes and the biosynthesis of type IV pili.

The seven ComG proteins of *B. subtilis* are membrane-associated and essential for DNA binding. ComGC, ComGD, ComGE, and ComGG have prepilin-like signal sequences, and together with ComGF, may be assembled in a complex on the outer face of the cell membrane. The processing of the proteins is mediated by ComC, which resembles the prepilin protease family. ComGA and ComGB are located in the cell membrane and may play morphogenetic roles, as judged from properties of apparent equivalents active in type IV pilus biogenesis. The ComG apparatus might function to aid transfer of transforming DNA across the cell wall to receptors on the outer surface of the membrane or to remodel the cell wall to allow DNA access to the transmembrane transporter.

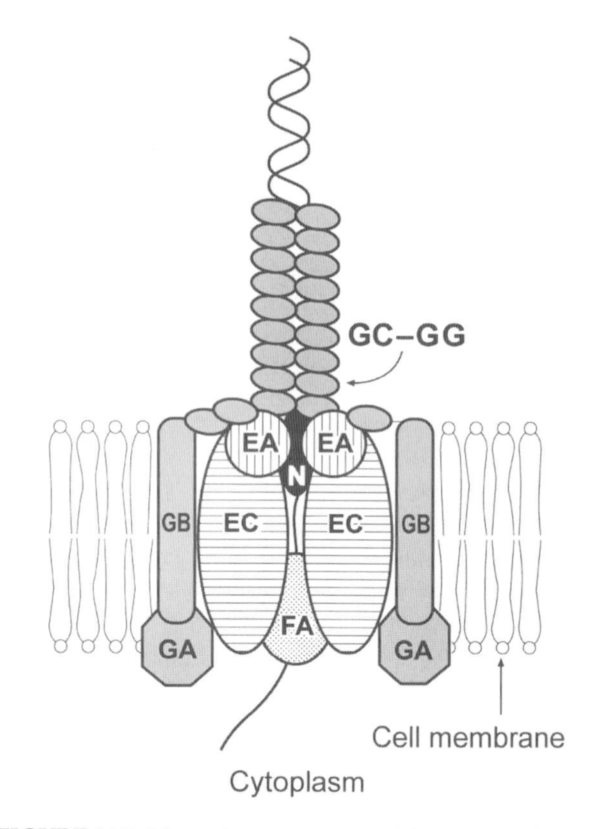

FIGURE 86.2 Schematic representation of the DNA uptake apparatus of *B. subtilis*. Extracellular duplex DNA enters the protein complex at the top. Transport of one strand into the cytoplasm of the cell is associated with the degradation of the second strand by an entry nuclease (N). Other labeled structures in the complex are late Com proteins.

ComEA and ComEC are integral membrane proteins. Mutational studies suggest that ComEA function to couple DNA binding to uptake, whereas ComEC contributes to the DNA-uptake channel. ComFA is another membrane protein required for DNA uptake; it resembles by its amino acid sequence a family of ATP-dependent RNA–DNA helicases, in particular the subfamily containing the DEAD (Asp-Glu-Ala-Asp) motif. However, ComFA shows the highest similarity to *E. coli* replication protein PriA. This protein has the unusual property of translocating along ssDNA in the 3′ to 5′ direction via an ATP-dependent process. Thus, ComFA may use ATP hydrolysis to translocate ssDNA through an import channel consisting of ComEC subunits with ComEA localized at the entry port.

C. Gram-negative systems showing specificity of DNA uptake

The transformation systems of *H. influenzae* and *N. gonorrhoeae* differ from the *B. subtilis* paradigm in showing specificity for the uptake of homologous DNA. Such selectivity is manifest through the lack of competitive inhibition by other types of DNA. Hamilton Smith and colleagues discovered in 1979 that the specificity of DNA uptake by *H. influenzae* is conferred by a specific oligonucleotide uptake sequence (US), subsequently identified as 5′-AAGTGCGGT-3′ and present at 1465 copies in the 1.8×10^3-kb genome. The US motif is widely distributed around the genome, but tends to be localized in inverted repeats downstream of genes. Here it might contribute to transcription termination. A different US, 5′-GCCGTCTGAA-3′, is responsible for the specificity of DNA uptake by *N. gonorrhoeae*, but again the sequence is found in transcriptional terminators. Such sequences may have acquired a role in transformation in response to their dispersal in the genome, which makes them suitable as tags for identifying diverse sectors of homologous DNA following its fragmentation.

DNA uptake by *H. influenzae* is effected by a membrane-protected structure called a transformasome. There are several of these structures per competent cell; each has a diameter of ~20 nm and consists of a membrane vesicle that emanates from a zone of adhesion of the inner and outer membranes to protrude beyond the cell surface. Duplex DNAis inferred to bind to a protein receptor on the outside of the transformasome and then to enter this structure irreversibly, where it is protected from exogenous DNase. Single-stranded linear DNA is released from the transformasome into the cytoplasm in a polar manner with a leading 3′-terminus.

In marked contrast, the uptake of transforming DNA by *N. gonorrhoeae* requires the type IV pilus of the organism (Fig. 86.3). The complete pilus is required, including PilC, which is associated with the cell surface and tip of the organelle. The pilus system might be required to present the US receptor at the cell surface or to translocate transforming DNA across the outer membrane. Transport might be achieved by the depolymerization of the pilus fiber, as is thought to occur in twitching motility. Passage of the transforming DNA across the remainder of the cell envelope involves competence proteins apparently unassociated with the pilus. Of these, ComL and Tpc are thought to cause localized restructuring of the polysaccharide–peptide polymers making up the peptidoglycan cell wall. Tpc appears to be loosely associated with the peptidoglycan, but ComL, a lipoprotein, is covalently bound to it. ComA as an integral membrane protein aids DNA transport across the inner membrane; interestingly, ComA is related at the amino-acid-sequence level to the ComEC DNA uptake protein of *B. subtilis*.

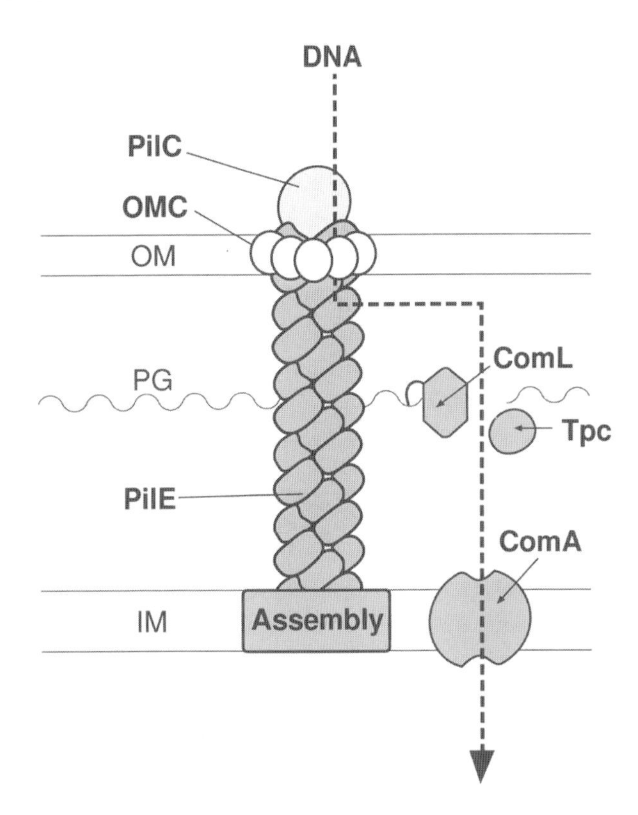

FIGURE 86.3 Model of the route of DNA uptake through the cell envelope of *N. gonorrhoeae*. The structure to the left is the type IV pilus, consisting of the assembly center in the inner membrane, PilE subunits, and the terminal PilC protein. OMC is the multisubunit outer-membrane pore complex that is thought to act as the gated pore for export of the pilus. Cell envelope components are OM, outer membrane; PG, peptidoglycan; IM, inner membrane.

IV. INTEGRATION OF TRANSFORMING DNA INTO THE RECIPIENT GENOME

A. Homologous recombination of chromosomal genes

Single-stranded fragments of chromosomal DNA entering the cell are integrated into the recipient genome by homologous recombination. The process has a central requirement for an equivalent of the *E. coli* K-12 *recA* gene product. This protein binds ssDNA to form long nucleoprotein filaments essential for strand pairing with the homologous duplex and subsequent exchange reactions. The expression of *recA*, as well as some other recombination genes, is induced as part of the competence response (Fig. 86.1). The product of strand exchange is a region of heteroduplex DNA (Fig. 86.4a). Such a structure may range in size from about 40 bp, which corresponds to the minimal segment that can be processed effectively, up to an average of ~12 kb. If heteroduplex DNA is formed from nonidentical DNA strands

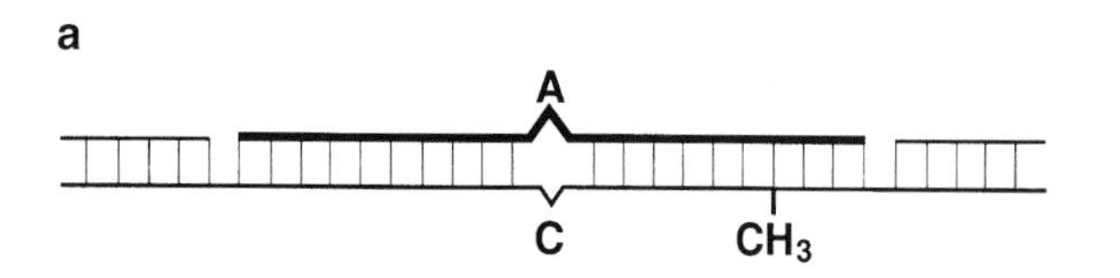

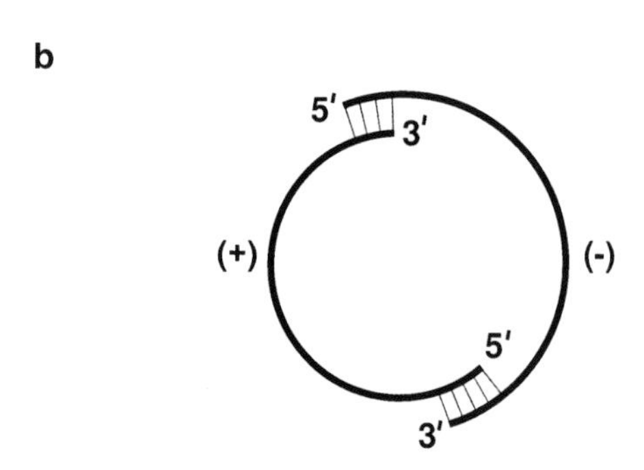

FIGURE 86.4 Establishment of transforming DNA in the recipient cell. (a) Heteroduplex DNA formed by homologous recombination of an entrant strand of more than ~40 nt (bold line) and the complementary strand of the resident chromosome of the recipient cell (light lines). Gaps are sealed by DNA ligase. A and C represent mismatched nucleotides due to sequence divergence of the donor and recipient genomes. CH_3 indicates modification methylation of a restriction-enzyme recognition site in the recipient DNA; the entrant strand is unmodified. (b) Regeneration of a plasmid by the annealing of overlapping fragments of complementary (+) and (−) strands. A duplex circle is generated by DNA synthesis initiated at the 3′-OH termini.

derived from genetically different organisms, the duplex will contain mismatched bases. These may influence recombination frequencies, as discussed later.

B. Processing of plasmid DNA

As is the case for chromosomal DNA, plasmids are taken up as linear single-stranded molecules that are subject to extensive fragmentation. Plasmid DNA gives rise to very few transformants in natural systems, presumably due to the difficulty of forming circular unit-length molecules from ssDNA fragments. The frequency of transformation by plasmid monomers is dependent on the square of the DNA concentration, suggesting that intact plasmids can be reassembled in the recipient cell from sections of different monomers taken up as separate events. The mechanism may involve the annealing of overlapping fragments of complementary (+) and (−) strands, followed by infilling of single-stranded regions by repair synthesis to give an intact duplex (Fig. 86.4b).

Plasmids also exist as multimers. Transformation frequencies by multimeric DNA show a linear dependence on DNA concentration, implying that a single multimer can give rise to a transformant. Monomeric circles might be generated by the annealing of overlapping parts of (+) and (−) strand fragments derived from one multimer. Another plausible explanation is that single fragments are sufficiently large to contain more than a complete set of plasmid genes; in effect one entrant (+) strand would possess a terminally redundant repeat. A unit-size circle would be produced if synthesis of the nascent (−) strand proceeded to copy one terminal (+) repeat, to be followed by the slippage of the newly synthesized end to pair with the other (+) terminus. The completion of the cycle of replication would give a circular (−) strand.

C. Sexual isolation: Mismatch repair and DNA restriction

Nucleotide-sequence divergence is a barrier to the effective recombination of genes from closely related species, which is described as sexual isolation. The barrier is manifest by an exponential relationship between the reduction in transformation frequency and the sequence divergence of donor and recipient DNAs. One contributory factor to sexual isolation is the activity of mismatch-repair systems of the type that function to remove mispaired nucleotides inserted as replication errors during vegetative growth. An example is the Hex mismatch-repair system of *S. pneumoniae*; this operates preferentially to correct the donor contribution in a mismatched hetero-duplex, thereby reducing the transformation efficiency of some markers by as much as 20-fold. Transition mismatches, such as A-C and G-T, are eliminated more effectively than transversions (e.g., A-G) and are inherited inefficiently as low-efficiency markers.

The Hex system is thought to recognize the mismatch in the newly formed heteroduplex and then initiate a bidirectional search for unsealed gaps flanking the donor-strand insertion. Thus, repair is targeted to the incoming strand, which may be removed entirely to be replaced by infilling through DNA synthesis. In this way, joints with imperfectly paired nucleotides are aborted. It is also possible that mismatch-repair enzymes inhibit the formation of mispaired heteroduplex joints in transformation. A precedent for this suggestion comes from properties of the *E. coli* counterpart of the Hex system, namely the Mut system. Purified Mut proteins have been shown to block RecA-mediated exchanges *in vitro* between nonidentical DNA molecules.

The mismatch-repair system is only partly responsible for impaired recombination of sequence diverged DNA in transformation. One line of evidence for this conclusion is the considerable sexual isolation detected when a repair-defective mutant is used as the transformation recipient; a second is the ready saturation of the Hex system by multiple mismatches in a single heteroduplex region. Thus, the main factor contributing to sexual isolation may be the physical obstacle that mismatches impose on RecA-mediated exchanges between nonidentical DNA strands.

DNA-restriction systems impose another barrier to productive gene exchange between different strains and species. Classic restriction endonucleases preferentially recognize dsDNA and cleave the DNA when the enzyme specificity sites lack modification methylation of specific bases in the target sequence. The prevalence and diversity of restriction-modification systems imply that restriction is a commonly encountered obstacle to productive gene exchange. Although restriction systems severely disrupt patterns of inheritance of unmodified chromosomal DNA transferred in conjugational and transductional crosses, they have little effect on the inheritance of unmodified chromosomal genes transferred by natural transformation systems. An explanation is that transforming DNA is refractory to cleavage because it is single-stranded and acquires protection following recombinational insertion into the recipient chromosome by the preexisting methylation of the recipient strand in the heteroduplex (Fig. 86.4a). Such hemimethylated DNA is the preferred substrate of modification enzymes, which primarily function as maintenance methyltransferases to effect the rapid conversion of hemimethylated DNA into the fully methylated state. In contrast to chromosomal DNA, unmodified plasmid DNA is sensitive to restriction during transformation. The explanation is that plasmids are regenerated from single-stranded fragments of entrant DNA and DNA synthesis on these strands gives a duplex molecule that is unmodified. Unmodified double-stranded plasmids transferred by artificial transformation procedures are likewise sensitive to restriction in the recipient cell.

V. ECOLOGICAL IMPLICATIONS OF BACTERIAL TRANSFORMATION

Transformation in natural environments has been examined for only a few organisms, but the elaborate nature of DNA uptake mechanisms and the ever-increasing list of species known to develop natural competence points to the conclusion that the process is important in the wild. Transformation is potentiated by the significant amounts of extracellular DNA—much being of microbial origin—that are present in a variety of environments, including marine water, freshwater, sediments, and soil. DNA concentrations in excess of 20 μg/l water and up to 1 μg/g sediment have been recorded. In order to persist outside the cell, released DNA must acquire protection from ubiquitous extracellular DNases. Protection is conferred by the capacity of DNA to adsorb to the surfaces of clay and sand particles in soils and to humic acids. Binding to clay minerals is effected by H-bonds and electrostatic interactions between negative charges on DNA and positive charges on the edges of the clays. These interactions apparently change the conformation of the DNA in some way to confer protection from nucleases while, at the same time, allowing the DNA to interact with binding sites on competent cells. Evidence also exists that transformation occurs in marine water as well as in river epilithon—defined as the community of organisms in the slimy layer on the surfaces of stones in water—but not in sediments.

The notion that transformation is an evolutionary strategy for potentiating gene exchange raises the intriguing possibility that some cells in a population become genetic donors and release DNA under the conditions that favor competence development in recipients. In the well-studied *B. subtilis* system, DNA release has been detected at spore generation, competence development, and cell death. The release of DNA may reflect the activity of autolytic enzymes that hydrolyse moieties of the cell wall, as observed at the end of exponential growth of batch cultures. DNA may also be released by some active export process compatible with cellular survival. An interesting example of the coupling of donor and recipient activities is seen in the phenomenon of cell contact transformation. The process has been observed for several gram-positive and gram-negative organisms; it is bidirectional and requires cell-to-cell contact, yet is DNase sensitive and independent of detectable plasmids and phages.

In addition to promoting genetic diversity, gene transfer by bacterial transformation may contribute to maintaining the integrity of genomes by allowing recombinational repair of DNA damaged by various chemical and physical agents in the environment. DNA damage induces the expression of a network of dispersed genes, which collectively make up the SOS regulon. The repair scenario proposes that damaged sectors of the recipient genome are replaced by intact sequences acquired from other cells by transformation and processed by the recombination pathway of the SOS response. It should be pointed out that whereas classic SOS functions, including the derepression of the *recA* gene, are expressed spontaneously at competence development in *B. subtilis*,

DNA-damaging treatments do not induce the organism to develop competence. The same applies to *H. influenzae*. Irradiation with ultraviolet light increases the rate of homologous recombination, but the response is apparently a consequence of the recombinogenic lesions that are generated in the recipient genome, rather than a reflection of stimulated DNA uptake.

A further idea is that transformation is a mechanism for scavenging the considerable amount of DNA present in natural habitats. According to this hypothesis, competence is an adaptation to starvation conditions that causes DNA uptake to potentiate alternate metabolic processes. Thus, nucleosides and bases salvaged from the DNA uptake process are channeled into the nucleotide pools or catabolized further to provide alternative sources of carbon, nitrogen, phosphate, and energy. The genetic complexity of the DNA uptake process, coupled with the specificity of some systems for homologous DNA, would seem to argue against the hypothesis that nutrient scavenging is the driving force in the evolution of transformation processes.

VI. ARTIFICIAL TRANSFORMATION SYSTEMS

The transformation of bacteria that do not develop natural competence was first reported in 1970 for the paradigm organism of bacterial genetics, the gram-negative *E. coli* strain K-12. During a study of bacteriophage transfection, M. Mandel and A. Higa discovered that *E. coli* could take up purified DNA of bacteriophage λ without the need for coinfection with helper virus if the cells were treated with high concentrations of $CaCl_2$ solution at 0°C; the entry of the transfecting DNA was achieved by exposing the mixture to a brief heat pulse after allowing time for adsorption of the molecules to the cell surface. Subsequently it was found that, once induced to the competent state by the ice-cold calcium treatment, amino acid-requiring mutants of *E. coli* could be transformed back to the wild-type state with linear fragments of dsDNA. However, such transformants, which arose by homologous recombination, could only be established in strains lacking exonuclease V, an enzyme that degrades linear DNA. Stanley Cohen and coworkers went on to report that both *E. coli* and its close relative *Salmonella typhimurium* could also take up plasmids as circular dsDNA molecules; their protocol used a 2 min heat pulse at 42°C followed by the incubation of the cells in growth medium to allow the phenotypic expression of the plasmid-borne drug-resistance genes prior to plating on antibiotic medium selective for the growth of transformants. This finding provided the mechanism whereby recombinant DNA molecules generated *in vitro* could be returned to bacterial cells for *in vivo* study, and this has since become a cornerstone of modern molecular biology. At present, this process is purely a laboratory phenomenon and whether *E. coli* uses something equivalent as a mechanism for genetic exchange in its natural habitat remains unresolved. From these initial observations of reported frequencies ranging across 10^3–10^6 transformants/μg input plasmid DNA, the method has been improved through the systematic manipulation of inducing conditions and strain genotypes, most notably by Douglas Hanahan, such that 100- to 1000-fold-higher frequencies can now be achieved. Using similar approaches, it is now possible to introduce DNA molecules into a diverse range of bacteria previously thought to be untransformable.

A number of parameters have been identified that contribute to the efficiency of artificial transformation of *E. coli*. First, the genetic composition of the strain can have a marked effect, most likely accounted for by differences in the cell-surface lipopolysaccharide composition. Thus, variant rough strains with shortened, or completely lacking in, LPS O side chains are more easily transformed than wild strains. The peripheral long O side chains probably restrict access of the DNA to the uptake sites on the cell surface through either physical hindrance or ionic-charge interaction and are therefore dispensable, whereas the LPS core appears to be an essential component of the cell envelope for optimal uptake. Secondly, it is clear that although an artificially induced state, chemical competence is influenced by the physiological state of cells at the time of treatment. In general, the highest transformation frequencies are achieved with competent cells made from rapidly proliferating, late-exponential-phase cultures growing in rich medium. Moreover, growth at reduced temperatures (25–30°C) can also improve competence, and such cells can be induced to take up DNA by a less severe heat pulse. It has long been known that lipid composition is markedly affected by growth conditions, with membranes of low-temperature cells having a higher proportion of unsaturated fatty acids, which makes them more fluid; such membranes might allow easier transit of macro-molecules. In addition, the growth of cells in the presence of Mg^{2+} ions improves competence, probably by modifying the structure of the LPS layer through the substitution of protein–LPS interactions with divalent-cation ionic bonds, thereby facilitating the removal or loosening of the LPS layer during the competence-inducing procedure.

Treatment with multivalent cations at 0°C is an absolute requirement for the induction of competence. However, other divalent ions can be substituted for calcium; barium, hexamine-cobalt, manganese, and rubidium have all been shown to work, and in some cases more effectively. Moreover, the addition of other chemicals, in particular dimethyl sulfoxide and the sulfydryl-reducing agent dithio-threitol, and certain physical procedures such as freeze–thaw can also be beneficial with particular strains. All of these probably influence cell-surface structure and membrane fluidity and so enable DNA molecules to access and transit the cell envelope. Thus, simple and complex conditions have been developed that can be applied to all strains of *E. coli* with the expectation of one or other giving a high transformation yield; the reader is directed to the excellent review by Hanahan and Bloom (1996).

Chemical transformation may kill a large proportion of the cell population and, even under optimal conditions with excess DNA, only around 10% of the total viable cell population becomes transformed. The use of equimolar mixtures of two compatible and distinguishable plasmids has shown that co-transformation occurs at a relatively high frequency; under conditions of DNA saturation, 70–90% of cells may receive both plasmid species. Thus, although only a small fraction of the cells may become induced to a competent state, those that do so are able to take up multiple DNA molecules very efficiently. However, when DNA is limiting, still less than 1% of the input plasmid molecules give rise to a transformant.

What then are the sites of DNA entry and what is the actual mechanism of DNA uptake in chemically induced cells? The cell envelope of a gram-negative bacterium like *E. coli* comprises outer and inner membranes separated by a peptidoglycan wall layer. The two membranes are fused to each other through holes in the rigid wall. It is thought that these zones of adhesion, estimated at 400/cell, constitute the channels through which DNA and other molecules are transported into the cell. Consistent with this, titration of transformation frequency against increasing amounts of transforming DNA suggests that there are probably a few hundred uptake sites per cell. Furthermore, there is a linear relationship between the increase in plasmid size and the decrease in transformation probability per plasmid molecule; and both supercoiled and relaxed DNAs transform with comparable frequencies across the size range of 2–66 kb. It seems most likely that uptake occurs by an active process; if it were simply passive diffusion through pores of a fixed size, there would be a dramatic drop in transformation efficiency with molecules larger than the pore size and compact supercoiled forms of large molecules would transform more efficiently than their relaxed isomers. There is also no evidence from competition experiments for the requirement of specific uptake sequences in the transforming DNA. Finally, the ability of competent *E. coli* cells to excise pyrimidine dimers introduced into transforming DNA molecules by prior UV irradiation indicates that it is dsDNA that is taken up, rather than ssDNA as found for natural transformation processes.

The uptake process itself can be divided into. stages. First, DNA molecules must bind to the cell surface at the sites of uptake. Chemical treatment probably exposes these sites via modification of the LPS structure, while the 0°C environment causes the membrances to adopt a suitable phase state for DNA association. Because both DNA and the phopholipid-based LPS are anionic polymers, it is likely that another function of the divalent cations is to shield and neutralize the phosphate backbone of the DNA, so overcoming repellent-charge interactions with the cell surface.

Two possibilities exist for the molecular channel that provides the route of DNA uptake. One suggestion is the cobalamin (vitamin B_{12}) transporter, which is known to be located at zones of adhesion; evidence comes from the observation that increasing amounts of cobalamin reduce transformation frequency. Thus the reason that inclusion of hexamine-cobalt salts during cell preparation boosts the transformation efficiency may be through their activation of the cobalamin transport system. Alternatively, it has been proposed by Rosetta Reusch and colleagues that the uptake channel is composed of poly-β-hydroxybu-tyrate (PHB) complexed with polyphosphate (polyP) and calcium ions. PHB has been isolated from the cytoplasmic membrances of many bacteria and euk-aryotic cells, and found to accumulate to significant levels in *E. coli* during the induction of competence. Accumulation coincides with the appearance of a new high-temperature phase transition of the membrances, indicating that the lipid bilayer has undergone structural modification. The view is that the PHB–polyP–Ca^{2+} complex forms a cylindrical channel that spans the cytoplasmic membrane, with the PHB constituting the lipophilic exterior and the polyP–Ca^{2+} within the hydrophilic interior. After passage across the outer membrane, facilitated by disruption of the LPS layer by divalent cations, the introduced DNA displaces the polyP from the PHB pore and so enters the cell. Although transformants can be obtained at low frequencies without the heat pulse, this almost certainly increases membrance fluidity and so aids the release of DNA into the bacterial cytoplasm. Interestingly PHB accumulation has been correlated with the development of competence in the naturally transformable organisms *H. influenzae*, *B. subtilis*, and *Azotobacter vinlandii*; in the latter two cases an altered membrane phase transition has also been detected. Therefore, there may be similarities between natural and chemically induced competence at the level of the DNA-transport mechanism.

Electroporation is an alternative method for the introduction of naked DNA molecules into microbial cells and has been applied successfully to many organisms previously intractable to even chemical induction of competence. DNA and cells suspended in a solution of very low ionic strength are placed in a cuvette with two closely spaced electrodes and subjected to a brief high-voltage electrical pulse. The discharge causes a transient reversible depolarization and permeabilization of the cell membranes, so inducing the formation of pores, which allow entry of DNA molecules. The procedure can be used to introduce a variety of molecules including proteins and RNA, as well as DNA of various conformations. The yield of transformants is optimized by the alteration of the parameters of the electrical field discharged through the cells, and frequencies of 10^{10} transformants/μg plasmid DNA are achievable in *E. coli*. In contrast to the chemical method, electroporation can achieve the transformation of almost every viable cell if the DNA is in excess. Only 10% of the input molecules are taken up when the DNA is the limiting factor. Studies using mixtures of plasmid DNAs have shown that, as in the chemical method, uptake is efficient and multiple plasmid molecules can be transferred in a single process. Although this procedure has no requirement for treatment with divalent cations, *E. coli* mutant strains lacking LPS O side chains generally give higher yields of transformants, suggesting that, here again, accessibility of the DNA through the cell surface is important.

Once introduced by either chemical transformation or electroporation, covalently closed plasmids possessing a functional replication origin will be propagated as autonomous replicons. In contrast, genes carried on linear DNA fragments can only be established through homologous recombination. However, linear DNA fragments are subject to degradation by the exonuclease V component of RecBCD enzyme. Thus, efficient transformation with chromosomal genes can only be achieved in *E. coli* strains that lack exonuclease V, but remain recombination proficient. Appropriate strains are *recB recC* of *sbcB* triple mutants or the more robust *recD* mutants. The use of such strains in conjunction with linearized transforming DNA is a valuable procedure for targeting gene replacements and insertions to the bacterial chromosome and large plasmids.

The establishment of transforming duplex DNA is subject to the action of classic restriction systems that target molecules lacking modification methylation. There also exist methylation-dependent systems (Mcr) that recognize and cleave DNA methylated at certain cytosine residues. These systems may be particularly obstructive to the cloning of DNA from higher eukaryotes because this is often methylated extensively at CpG motifs. Although restriction systems reduce the overall transformation frequency, none constitutes an absolute barrier because escaping molecules acquire the modification status of the host cell and thus become immune to further attack.

A third method for the introduction of plasmid DNA, protoplast transformation, was developed for the antibiotic-producing streptomycetes by Mervyn Bibb, Judith Ward, and David Hopwood, and was subsequently applied by other workers to several other commercially important organisms, such as the corynebacteria. Osmotically stabilized protoplasts were treated with PEG, which precipitated the transforming DNA onto the exposed membrane surfaces, to promote uptake, and then were allowed to regenerate cell walls. Although the procedure was highly effective with plasmid and phage dsDNAs that under-went autonomous replication, giving frequencies approaching 10^7 transformants/μg input DNA, it worked only poorly with circularized and linear fragments of chromosomal dsDNA that required homologous recombination in order to be inherited. Subsequently, it was discovered that the latter type of transformation could be stimulated several hundred-fold by the denaturation of the transforming DNA, which presumably generates ssDNA for strand invasion of the resident chromosome during homologous recombination. The procedure has significant practical application in streptomycetes because it facilitates targeted genetic manipulations of the chromosome such as gene disruption and replacement, mutational cloning, and complementation screening with ordered gene libraries. The stimulation of homologous recombination by transformation with denatured linear chromosomal DNA may also be applicable to genetic manipulation of other organisms for which transformation with dsDNA normally gives a high background of illegitimate integration events.

VII. TRANSFORMATION OF MICROBIAL EUKARYOTES

Although the initial claims for the transformation of yeast cells were recorded in 1960, these early studies failed to eliminate other explanations such as genetic reversion. Thus, the first convincing demonstration that a microbial eukaryote could be transformed with naked DNA was that of Albert Hinnen and colleagues who, in 1978, succeeded in converting a nonreverting *leu2* double mutant of bakers' yeast, *Saccharomyces cerevisiae,* to wild type using a cloned copy of the *LEU2* gene carried on a bacterial plasmid vector. The integration of the transforming DNA into the host chromosome was confirmed via Southern blot hybridization. Simultaneously, Jean Beggs showed that yeast could also be transformed with *in vitro* constructed recombinant plasmids capable of autonomous replication. In both cases, transformation involved the conversion of the yeast cells to spheroplasts by enzymatic degradation of the cell-wall β-glucan in the presence of sorbitol as an osmotic stabilizer. Polyethylene glycol, a known inducer of membrane fusion, and calcium were added to precipitate the DNA onto the spheroplasts, which were then plated in agar containing sorbitol in order to stabilize the transformants during their regeneration of a cell wall. Although technically difficult and time consuming, this procedure is capable of giving transformation frequencies up to $10^4/\mu g$ input DNA. Similar spheroplast transformation methods exist for several microbial eukaryotes, including the fission yeast *Schizosaccharomyces pombe*, and filamentous fungi, such as *Neurospora crassa, Aspergillus nidulans*, and *Trichoderma reesei.*

Most yeast researchers now use one of two procedures that allow the direct transformation of intact cells. The first, initially developed by H. Ito and colleagues and later optimized by R. H. Schiestl and R. D. Geitz, relies on the treatment of yeast cells with lith-iumions. As with chemical transformation of bacteria, the lithium ions seem to permeabilize the yeast cell wall and facilitate the access of the DNA to the cytoplasmic membrane. To achieve maximal transformation frequencies, an excess of carrier DNA is added along with the transforming DNA and the mixture precipitated onto the cells with PEG at 30°C; DNA entry is effected by a 42°C heat pulse. Single-stranded DNA has a greater stimulatory effect as a carrier than dsDNA. Possibly ssDNA, which is thought to be a less-suitable substrate for uptake than dsDNA, saturates non-productive DNA binding sites. The net result is that ssDNA increases the effective concentration of transforming duplex DNA. The nature of the uptake channel and the molecular mechanism by which the DNA is taken into the cell nucleus are unknown. The second method is electroporation, which is performed in a manner very similar to that used for bacterial cells. In both methods, yeast cells are plated directly onto medium selective for growth of transformants and transformation frequencies as high as $10^7/\mu g$ DNA have been reported, although frequencies in the range 10^3–10^5 are more common. Again, related methods have been developed that can be applied to a range of yeasts, including several of commercial interest such as *Kluyveromyces spp., Hansenula polymorpha, Pichia pastoris,* and *Yar-rowia lipolytica.*

All of these methods allow the introduction of both linear and circular dsDNA molecules and synthetic oligonucleotides. Because the frequency of mitotic recombination is high, recombinants of nuclear genes can be recovered easily. Indeed, homologous recombination between transforming DNA and the corresponding chromosomal region is stimulated several hundredfold if the DNA is linearized before introduction. Thus, targeted integration, gene replacement, and rescue of mutant chromosomal alleles into low-copy centromere-containing plasmids are all possible, as is the standard introduction of genes on extrachromosomal vectors. Such attributes make yeasts the most versatile eukaryotic organisms for molecular genetic studies.

A development for transforming microbial eukaryotes involves the use of biolistic devices. This process, originally developed for the transformation of plant material, literally involves firing microscopic tungsten or gold particles, previously loaded with DNA, at cells or tissues; the explosive force is generated either by the percussion of a gunpowder cartridge or rapid discharge of highly compressed gas, the net effect being to propel the microprojectiles at supersonic speeds into, and often through, the target material. This technology has been used to transform yeast cells not only with nuclear chromosomal genes, but also with mitochondrial DNA; thus, nonrespiring mitochondrial mutants could be transformed to respiratory proficiency and rho-zero (ρ^0) mutants lacking all mitochondrial DNA could be transformed to a stable rho-minus (ρ^-) state with a recombinant plasmid carrying a section of yeast mitochondrial DNA. The direct selection of mitochondrial transformants was not possible and they were recovered as cotransformants of nuclear gene transformants at a frequency of 10^{-3}–10^{-4}. However, despite this limitation, the discovery offers real possibilities for the molecular manipulation and study of eukaryotic organelle genomes.

BIBLIOGRAPHY

Bibb, M. J., Ward, J. M., and Hopwood, D. A. (1978). Transformation of plasmid DNA into *Streptomyces* at high frequency. *Nature* **274**, 398–400.

Dreiseikelmann, B. (1994). Translocation of DNA across bacterial membranes. *Microbiol. Rev.* **58**, 293–316.

Dubnau, D. (1993). Genetic exchange and homologous re-combination. *In:* "*Bacillus subtilis* and Other Gram-Positive Bacteria: Biochemistry, Physiology, and Molecular Genetics" (A. L. Sonenshein *et al.*, eds), pp. 555–584. American Society for Microbiology, Washington, DC.

Dubnau, D. (1997). Binding and transport of transforming DNA by *Bacillus subtilis*: The role of type-IV pilin-like proteins—a review. *Gene* **192**, 191–198.

Dubnau, D. (1999). DNA uptake in bacteria. *Annu. Rev. Microbiol.* **53**, 217–244.

Dubnau, D., and Provvedi, R. (2000). Internalizing DNA. *Res. Microbiol.* **151**(6), 475–80.

Fussenegger, M., Rudel, T., Barten, R., Ryll, R., and Meyer, T. F. (1997). Transformation competence and type-4 pilus biosynthesis in *Neisseria gonorrhoeae*—a review. *Gene* **192**, 125–134.

Gietz, R. D., Schiestl, R. H., Willems, A. R., and Woods, R. A. (1995). Studies on the transformation of intact yeast cells by the LiAc/ssDNA/PEG procedure. *Yeast* **11**, 355–360.

Grossman, A. D. (1995). Genetic networks controlling the initation of sporulation and the development of genetic competence in *Bacillus subtilis*. *Annu. Rev. Genet.* **29**, 477–508.

Hanahan, D., and Bloom, F. R. (1996). Mechanisms of DNA transformation. *In* "*Escherichia coli* and *Salmonella:* Cellular and Molecular Biology" (F. C. Neidhardt *et al.*, eds), 2nd ed., pp. 2449–2459. American Society for Microbiology, Washington, DC.

Kleerebezem, M., Quadri, L. E., Kuipers, O. P., and de Vos, W. M. (1997). Quorum sensing by peptide pheromones and two-component signal-transduction systems in Gram-positive bacteria. *Mol. Microbiol.* **24**, 895–904.

Lorenz, M. G., and Wackernagel, W. (1994). Bacterial gene transfer by natural genetic transformation in the environment. *Microbiol. Rev.* **58**, 563–602.

Majewski, J., and Cohan, F. M. (1998). The effect of mismatch repair and heteroduplex formation on sexual isolation in Bacillus. *Genetics* **148**, 13–18.

Se-Hoon, O., and Chater, K. F. (1997). Denaturation of circular or linear DNA facilitates targeted integrative transformation of *Streptomyces coelicolor* A3(2): possible relevance to other organisms. *J. Bacteriol.* **179**, 122–127.

Sherman, F. (1997). An introduction to the genetics and molecular biology of the yeast *Saccharomyces cerevisiae*. *In:* "The Encyclopedia of Molecular Biology and Molecular Medicine" (R. A. Meyers, ed.), Vol. 6, pp. 302–325. VCH, Weinheim, Germany.

Tortosa, P., and Dubnau, D. (1999). Competence for transformation: a matter of taste. *Curr. Opin. Microbiol.* **2**(6), 588–592.

Yin, X., and Stotzky, G. (1997). Gene transfer among bacteria in natural environments. *Adv. Appl. Microbiol.* **45**, 153–212.

WEBSITE

Access to relevant links in PubMed
http://www.ohsu.edu/cliniweb/G5/G5.386.html

87

Transposable elements

Peter M. Bennett
University of Bristol

GLOSSARY

composite transposon A modular structure comprising an intrinsically nontransposing sequence flanked by copies of the same IS element.

conjugative transposon A DNA element that mediates both its own transposition and conjugal transfer from one bacterial cell to another.

insertion sequence A small cryptic DNA element that can migrate from one genetic location to another completely unrelated location.

inverted repeat A sequence that defines both ends of a transposable element and that is found as an inverted duplication.

target site A site at which a transposition event occurs (occurred).

transposing bacteriophage A bacteriophage that replicates using a form of transposition.

transposition immunity The inhibition of transposition of one copy of a transposable element by a second copy of the same element on the target DNA.

transposon A DNA element that can migrate from one genetic site to another unrelated site and that encodes functions other than that required for transposition.

Transposable Elements are discrete DNA sequences that can move from one location on a DNA molecule to another location on the same or on a different DNA molecule. The process of transposition is, *ipso facto*, a recombination event in that it involves breaking and reforming phosphodiester bonds. It requires no extended homology between the element and its sites of insertion; accordingly, it does not depend on the homologous recombination system of the host cell. In practice, this means that transposition is *recA*-independent. Several types of transposable elements are known and transposition can occur by one of several different mechanisms. Hence, although the term "transposable element" implies a defined genetic entity, the term "transposition" indicates simply that genetic rearrangement by DNA relocation has occurred, by any one of a number of mechanisms. Transposable elements have been discovered in gram-negative and gram-positive bacteria, in archaebacteria, and in yeasts and are probably responsible for much of the macromolecular rearrangements of microbial genomes.

I. INTRODUCTION

A. General structure of a transposable element

Two points of caution should be stated at the outset. First, although transposition results in the insertion of one DNA sequence into another, in some instances, but not all, the transposed sequence is also retained at its original location (i.e. the event is not only recombinational but also replicative; see later). Second, not all transposition events involve the movement of a defined DNA element (e.g. IS*91*-like elements can effect cotransposition of varying lengths of DNA adjacent to one side of the element; see later).

Most reported transposition events in microbes involve transposable elements alone. Each element is

The Desk Encyclopedia of Microbiology
ISBN: 0-12-621361-5

a defined structure that is preserved precisely from one transposition event to the next. Most, but not all, elements terminate in short, perfect or near perfect inverted repeats (IRs), which function as recognition signals for the transposition enzymes. Terminal IR sequences are usually 15–40 bp long (but may be much longer, e.g. the imperfect IRs of Tn7 and of bacteriophage Mu) and differ from one transposable element to another. Related elements have related IRs. Each fully functional element encodes at least one protein that is needed for its own transposition, normally termed a transposase. In most cases, the transposase processes the ends of the element in readiness for recombination; in addition, functions unrelated to transposition may be encoded.

B. Insertion sequences and transposons

The smallest elements capable of self-promoted transposition are the prokaryotic insertion sequences (IS elements), which range in size from about 0.7 to 2 kb. Each encodes only one or possibly two functions necessary for the control and execution of its own transposition. Transposons are larger structures that, in addition to a transposition system, accommodate functions unrelated to transposition, which confer a predictable phenotype on the host cell (e.g. resistance to an antibiotic), unlike IS elements, which may alter the cell phenotype at random by mutation. Indeed, it was this mutational activity that first drew attention to the existence of transposable elements in bacteria in the mid-1960s. However, transposable DNA sequences were first discovered in maize by Nobel laureate Barbara McClintock in the early 1950s; she suggested, correctly, that they were the cause of the variations in pigmentation of maize kernels. This conclusion was based on a detailed and elegant genetic analysis. Physical evidence to support the genetic case for transposable elements was first obtained from studies of bacterial systems in the late 1960s, culminating in the descriptions of three bacterial insertion sequences, IS*1*, IS*2*, and IS*3*. Modern molecular genetics has revolutionized and considerably simplified the task of detecting, isolating and characterizing transposable elements in both prokaryotic and eukaryotic systems.

Transposons fall into two general categories, sometimes called composite transposons and complex transposons. However, a better description would be composite and noncomposite, for reasons that will become clear. A composite transposon has a modular structure in which a central unique sequence, accommodating genes not involved in transposition, is flanked by two copies of an IS element in direct or, more commonly, in inverted repeat. One or both of these terminal elements provide the unique functions necessary for the transposition of the composite structure and often retain the ability to transpose as independent IS elements (see later).

Complex or noncomposite transposons have no obvious modular structure. Functions unconcerned with transposition (i.e. those equivalent to the central unique section of composite elements) are encoded by genes that are integrated into the basic transposable element, alongside transposition functions. In most cases, the whole gene ensemble is flanked by short IRs, as seen for the majority of IS elements, and no component of a noncomposite transposon can transpose independently of the rest of the structure. In crude comparative terms, a noncomposite transposon is more analogous to an IS element than to a composite transposon. Indeed, some noncomposite transposons (e.g. Tn3-like elements), as do IS elements, can operate in pairs to transpose the DNA sequences flanked by them. This cooperative behavior seems to be a feature of many prokaryotic transposable elements and has been exploited in the use of mini-Mu, a truncated derivative of the transposing bacteriophage Mu, to clone chromosomal genes *in vivo*.

C. Transposing bacteriophages

Transposing bacteriophages are typified by bacteriophage Mu. This element was identified by J. Taylor in the early 1960s in a culture of *Escherichia coli* and was the first bacterial transposable element to be discovered, although this was not fully appreciated at the time. The bacteriophage was so named because Taylor deduced, correctly, that the increased number of auxotrophic mutants he found among populations of *E. coli* lysogenized by this temperate phage had been generated by phage integration into genes encoding biosynthetic enzymes. Hence, the phage was described as a mutator phage, now abbreviated to Mu. He concluded that, unlike bacteriophage λ, Mu inserts at many different sites in the *E. coli* chromosome. Transposing bacteriophages replicate by transposition. They differ from other transposable elements in that they can exist independently of other DNA molecules and the host cell as bacteriophage particles.

D. Distribution of transposable elements

Transposable elements are widely distributed among eubacteria and archeae and are found in lower and higher eukaryotes. Many IS elements have been identified in both gram-positive and gram-negative bacteria and several have been found in Archaebacteria such as *Halobacterium* sp. and *Methanobrevibacter smithii*. Indeed, wherever a serious search has been made for such elements, they have usually been found. In addition, as the complete sequences of various bacterial genomes have been determined, so a number of putative, previously unknown elements have been revealed. However, the confirmation of the identities of these elements must await the demonstration that they can transpose, and do not simply represent antique transposition events of obsolete elements. The failure to detect a transposable element is more likely to indicate a deficiency in the detection system than the absence of transposable elements.

Transposons, both composite and complex, have been found in many gram-positive and gram-negative bacteria, often in bacteria of clinical or veterinary origin. These sources account for the preponderance of antibiotic resistance transposons; the relative paucity of other markers found on transposons probably reflects no more than the intense interest in aspects of bacteria that impinge directly on human welfare, rather than that antibiotic-resistance genes have a special relationship with transposons. Most bacterial transposons have been found on plasmids rather than on bacterial chromosomes, although there are a few notable exceptions. This finding probably indicates no more than that once a gene is located on a plasmid its ability to spread horizontally to other bacteria is markedly enhanced and, hence, the likelihood of detection is also significantly increased. It also seems that, for some transposons, transposition to a plasmid location occurs at a significantly higher frequency than to bacterial chromosomes, despite the greater capacities of the latter and despite the fact that insertions are not site-specific. The reason for this is unknown.

E. Nomenclature

The designations given to transposable elements are not usually assigned at random or at the whim of the researchers who discovered them. To avoid accidental assignment of the same code to two different elements, it has been widely accepted that numbers are allocated from a central directory. This resource is administered by Dr. Esther Lederberg at the Department of Medical Microbiology, Stanford University, California. Periodically, lists of new assignments are published. The convention originally adopted is simple. Insertion sequences are designed IS followed by one of a set of numbers allocated to the research worker or laboratory from the central directory, on request. The digits are written in italics. Likewise, transposons are designated Tn followed by an appropriate number (in italics). The designation is intended as no more than a simple, unique identifier. More recently, additional letters are added to indicate the bacterial source of the element and the numbering system has been restarted (e.g. IS*Rm2*, an IS element from *Rhizobium meliloti*, and IS*M1*, an IS element from *Methanobrevibacter smithii*, are quite distinct from IS*2* and IS*1*, respectively, both of which originated in *E. coli*, but which are also found in other members of the Enterobacteriaceae). These simple rules have, however, not been followed when naming some conjugative transposons, particularly those found in *Bacteroides* sp., because they fail to indicate the nature of the element.

II. IS ELEMENTS AND COMPOSITE TRANSPOSONS

A. History

The first transposable elements to be recognized as such in bacteria were three small cryptic elements, appropriately designated IS*1*, IS*2*, and IS*3*, discovered as a consequence of the analysis of strongly polar mutations in the *gal* and *lac* operons of *E. coli* (i.e. mutations that substantially eliminated the expression not only of the gene in which the mutations were located but also the expression of the other genes in the operon, located promoter distal to the mutant gene). Genetically, the mutations were found to behave like point mutations, but, although they reverted spontaneously, the frequencies of reversion were unaffected by chemical mutagens known to induce base-substitution and frameshift mutations. Hence, it was suggested that the genetic lesions resulted from the insertion of additional DNA sequences at the sites of mutation. A series of elegant physical studies using, initially, λ.*gal*-transducing phages and then electron-microscope analysis of heteroduplex DNA structures provided physical evidence for mutation by DNA insertion.

B. Structure

Bacterial IS elements are, structurally, the simplest form of transposable element. To date, approximately 500 have been reported. Their sizes (Table 87.1), rarely larger than 2 kb, preclude them encoding more than the one or two functions that are necessary for transposition. Of those that have been examined in sufficient detail, many have one large open reading frame (ORF) that uses the majority of the sequence. In a few cases, notably IS*1*, IS*3*, IS*10*, IS*50*, IS*903* and IS*911*, the

TABLE 87.1 Some prokaryotic is elements[a]

Insertion element	Family	Size (bp)	Terminal IRs (bp)	Target site duplication (bp)
Gram-negative bacteria				
IS*1A*	IS*1*	768	18/23	9
IS*2*	IS*3*	1331	32/41	5
IS*3*	IS*3*	1258	29/40	3,4
IS*4*	IS*4*	1426	16/18	11–13
IS*5*	IS*5*	1195	15/16	4
IS*6*	IS*6*	820	14/14	8
IS*10R*	IS*4*	1329	17/22	9
IS*21*	IS*21*	2131	30/42	4
IS*50R*	IS*4*	1534	8/9	8–10
IS*91*	IS*91*	1830	0	0
IS*150*	IS*3*	1443	22/31	3
IS*492*	IS*110*	1202	0	5
Gram-positive bacteria				
IS*110*	IS*110*	1558	0	0
IS*231A*	IS*4*	1656	20	10–12
IS*431R*	IS*6*	790	17/20	NR
IS*904*	IS*3*	1241	31/39	4
IS*L1*	IS*3*	1257	21/40	3
IS*S1S*	IS*6*	808	18	8
Archaebacteria				
IS*H1*	IS*5*	1118	8/9	8
IS*H2*	IS*4*	521	19	10–12
IS*H23*	IS*NCY*	1000	23/29	9
IS*H51-1*	IS*4*	1371	15/16	5
IS*M1*	IS*NCY*	1381	33/34	8

[a]NR, not recorded; IS*NCY*, family not assigned.

products of these predicted genes have been shown to encode proteins necessary for transposition. For the remainder, it is assumed the presumptive genes encode transposition functions. Some elements have smaller ORFs, which overlap or are contained within the large ORF. Whether these are real genes, the products of which have a role in transposition of the element, is largely unknown, although in one or two cases (e.g. IS*10* and IS*50*), the products have been shown to have regulatory functions (i.e. to control the frequency of transposition).

Most of the IS elements characterized to sequence level appear to belong to one of a relatively small number of extended families. Seventeen independent familial groups have been identified, and what has become increasingly clear is that members of most individual families can be distributed across many different eubacterial and archaebacterial genera. There are exceptions. Members of the IS*1* family appears to be restricted to the enterobacteria, whereas IS*66* and similar elements have only been found in bacteria of the rhizosphere. The largest group is the IS3 family, which now comprises more than 40 members distributed among both the eubacteria and the archeae, suggestive, perhaps, of a somewhat ancient lineage. One current database contains sequence information on approximately 500 IS elements isolated from 73 bacterial genera representing 159 bacterial species (see Mahillion and Chandler, 1998).

One variation on gene arrangement is seen on IS*1*, one of the smallest transposable elements known. It is 768 bp long, delimited by near-perfect 23-bp IRs, and has two short adjacent ORFs, *insA* and *insB*, which have been shown by mutation analysis to be necessary for transposition. The two sequences, in different reading frames, are cotranscribed. However, *insB* is not a true gene, in the sense that it does not encode a specific protein. Rather, the ORF- designated *insB* encodes the second half (C terminus) of a fusion peptide, InsAB, that has the bulk of InsA as its N-terminal half. This apparent violation of the rules governing the termination of peptide synthesis (i.e. the failure to terminate protein synthesis at end of the *insA* transcript) is accomplished by a proportion of the ribosomes that initiate the translation of *insA* making a −1 frameshift toward the end of the *insA* section, before the termination codon is reached. The frameshift, which is directed by the nucleotide sequence, puts these ribosomes into the reading frame that accommodates *insB* and translation continues until the termination codon signaling the completion of InsB synthesis is encountered.

The InsAB protein is the IS*1* transposase, whereas the InsA peptide functions as a transposition regulator. Both proteins compete to bind to the IRs of IS*1*. However, whereas InsAB can mediate transposition, InsA cannot, so, by occluding InsAB from the IRs, InsA can inhibit IS*1* transposition. Interestingly, the transposition frequency is determined not by the absolute amount of either protein, but by their ratio. Elements of the IS3 group also encode fusion peptides that are transposases (see later).

Several well-studied IS elements were discovered as terminal repeats of composite-resistance transposons. Notable among these are IS*10*, IS*50*, and IS*903*. The first constitutes the flanking elements, in inverted repeat, of Tn*10*, which encodes resistance to tetracycline; the second forms the terminal IRs of Tn*5*, and element that confers resistance to kanamycin (Km), bleomycin (Bl), and streptomycin (Sm), although the last determinant is silent in some hosts, including *E. coli*. IS*903* forms the terminal IRs of Tn*903*, another transposon encoding resistance to kanamycin. Although IS*10* (Fig. 87.1) and IS*50* (Fig. 87.2) both belong to the IS*4* group of elements, it is interesting to note that they differ in the way that each controls its frequency of transposition. IS*903* controls its frequency of transposition in yet another fashion. IS*903* is an IS*5*-like element of 1057 bp with perfect 18-bp

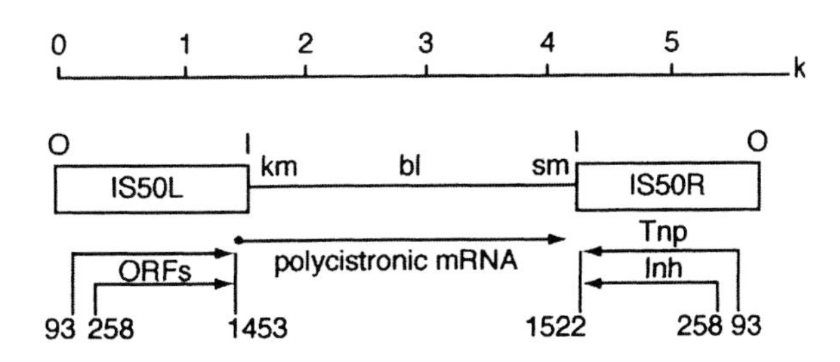

FIGURE 87.1 Tn*5* was detected on the resistance plasmid JR67, which originated in a strain of *Klebsiella pneumoniae*, when it transposed to phage λ. Tn*5* is 5.7 kb, with a central module of 2.6 kb encoding resistance to kanamycin (Km), bleomycin (Bl), and streptomycin (Sm), flanked by inverted copies of IS*50*, designated IS*50L,R*. The resistance genes are transcribed as a single operon from a promoter within IS*50L*. Deletion analysis has indicated that IS*50R* alone is responsible for Tn*5*-IS*50* transposition. IS*50R* is 1534 bp, has imperfect 19-bp IRs (seven mismatches) and encodes two peptides designated Inh and Tnp, of 421 and 476 amino acids, respectively. Tnp is the IS*50* transposase; Inh is an inhibitor of Tn*5*/IS*50* transposition. (It reduces the frequency of transposition after Tn*5*/IS*50* has established in a new host. On moving to another cell, the transposition of IS*50*/Tn*5* is subject to zygotic induction.) The Tnp and Inh proteins are virtually identical; they differ in that Inh lacks the first 55 amino acids of Tnp. The shortening is not a result of posttranslational modification; rather, Inh is produced from its own transcript. IS*50R* and IS*50L* differ at only one position, 1453. The change has profound effects. It simultaneously creates the promoter for the resistance operon and a premature translation termination signal for the two IS*50* peptides, both of which are shortened by 23 amino acids at their C termini. The loss of these amino acids inactivates both proteins. I, junctions of IS*50* elements with central resistance module; mRNA, messenger RNA; O, junctions of Tn*5* and carrier molecule. Numbers indicate translation initiation and termination points (relative to nucleotide sequence). Arrows indicate direction of transcription and translation.

terminal IRs. It encodes a single peptide of 306 amino acids, the transposase, which is unstable with a half-life of approximately 3 min. The decay of transposase prevents a build-up of transposase activity in the cell and ensures a low transposition frequency.

C. Mechanisms of transposition

What is known about the mechanisms of transposition of IS elements comes primarily from studies using IS*1*, IS*10*, IS*50*, IS*903*, and IS*911*. One early important finding was that, in most cases, IS sequences at new insertion sites are flanked by direct repeats (DRs) of short sequences (generally 2–12 bp) that are found as single copies at the insertion target sites. It was realized that this arrangement would result automatically if, in preparing the target site for insertion, both DNA strands at the target site are cut, with the individual cleavage sites on the different strands slightly staggered. Then, when the IS element is joined to the short single-strand extensions created by the staggered cuts and the single-strand gaps on

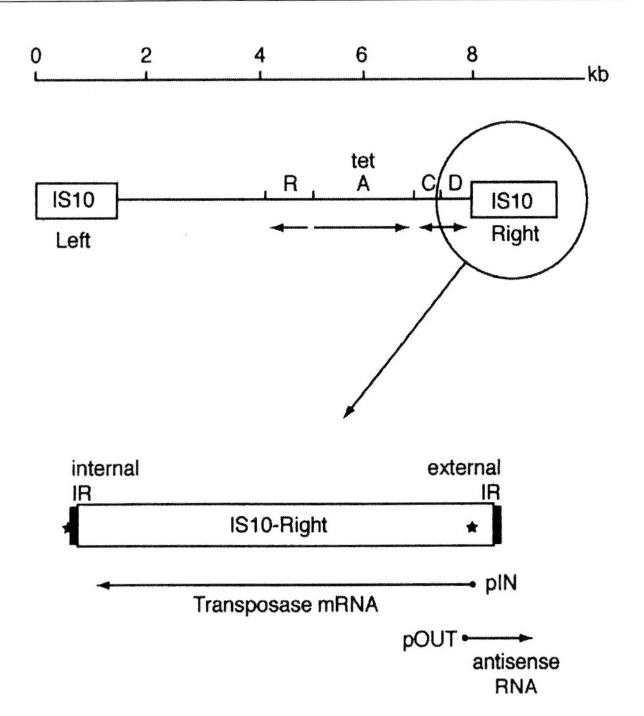

FIGURE 87.2 Tn*10* originated on the resistance plasmid [112.34] 222 (otherwise called NR1 and R100). It is 9.3 kb with an approximate 6.7 kb central module encoding inducible tetracycline resistance, flanked by inverted copies of IS*10*, designated IS*10R,L*. Tetracycline resistance is conferred by the TetA protein, which forms a tetracycline antiport that pumps tetracycline out of the cell. Expression of *tetA* is repressed by TetR and induced by tetracycline. The expression of the divergently transcribed *tetC* and *tetD* (functions unknown) is also induced by tetracycline. The sequences of the two copies of IS*10* differ at several positions; IS*10R* is responsible for transposition. The IS*10* transposase is encoded by a 1206-nt open reading frame transcribed towards the *tet* genes from promoter pIN. A second promoter, pOUT, located 35 bp downstream from pIN, directs transcription in the opposite direction to pIN to produce a short RNA molecule (69 nt) that is partially complementary to the *tnp* mRNA. This small RNA molecule is practically stable, with a half-life of approximately 70 min and a level of about five copies per copy IS*10*. In contrast, the *tnp* transcript is unstable (half-life, 40 s) and of low abundance (approximately 0.25 copy per copy IS*10*). When the 5′-ends of these RNA molecules anneal, the translation of tnp mRNA is blocked (an example of antisense RNA translation inhibition). The transposition activity of IS*10* is also regulated by *dam* methylation of DNA. Two *dam* sites (*) are involved. One overlaps the −10 box of the *tnp* promoter (pIN); the second is at the other end of IS*10* where transposase binds (to mediate the transposition of IS*10*). Methylation of the pIN *dam* sequence reduces the efficiency of pIN, whereas methylation of the second site reduces the binding of transposase to that end of the element. Both effects act to damp down transposition activity of IS*10*; however, only methylation at pIN will affect the transposition of Tn*10*, a differential affect that enhances the coherence of the composite element.

both sides of the element are filled in (by gap-repair DNA synthesis), short DRs will be created at the junctions of the element and the target. This is essentially what happens with many transposable elements. Extensive sequence analysis has also shown that, although the sequences of the DRs vary from one insertion to another and the sizes of the DRs vary from one element to another, for any given element, the size of the DRs is usually constant, suggesting that target site cleavage is element, rather than host, directed. These findings apply to most transposable elements, including phage Mu. Hence, if, during the course of determining the nucleotide sequence of a length of DNA, a structure is discovered that is delimited by IRs and flanked by short DRs, then it is reasonable to conclude that a transposable element has been discovered, even if the particular copy is no longer active. Examples of such serendipitous discoveries are readily found in the total genome sequences of a number of bacterial species. With one or two elements, transposition does not result in target-site duplications (i.e. following a transposition event the element is not flanked by short DRs). In these cases, insertion may be into flush-cut target sites.

The transposition mechanism of many IS elements, including IS*10* and IS*50*, is conservative; the element is disconnected from the donor molecule as a double stranded sequence that is then moved to a new site (note that this is not true of all transpositions, as seen later). The simplest conservative model is that described as the cut-and-paste mechanism (Fig. 87.3). In this process, the two ends of the element are brought together (synapsed) and the cognate transposase cleaves both DNA strands at both its ends, disconnecting it completely from the donor DNA molecule. Formally, this generates a free form of the element, in the sense that the element is no longer covalently joined to a carrier DNA. However, free transposable elements are not normally seen in cells, although IS circles, in which the ends are held together by protein, have been reported for IS*10* when the transposase is greatly over-expressed. Evidence from experiments with IS*10* and other elements suggests that these

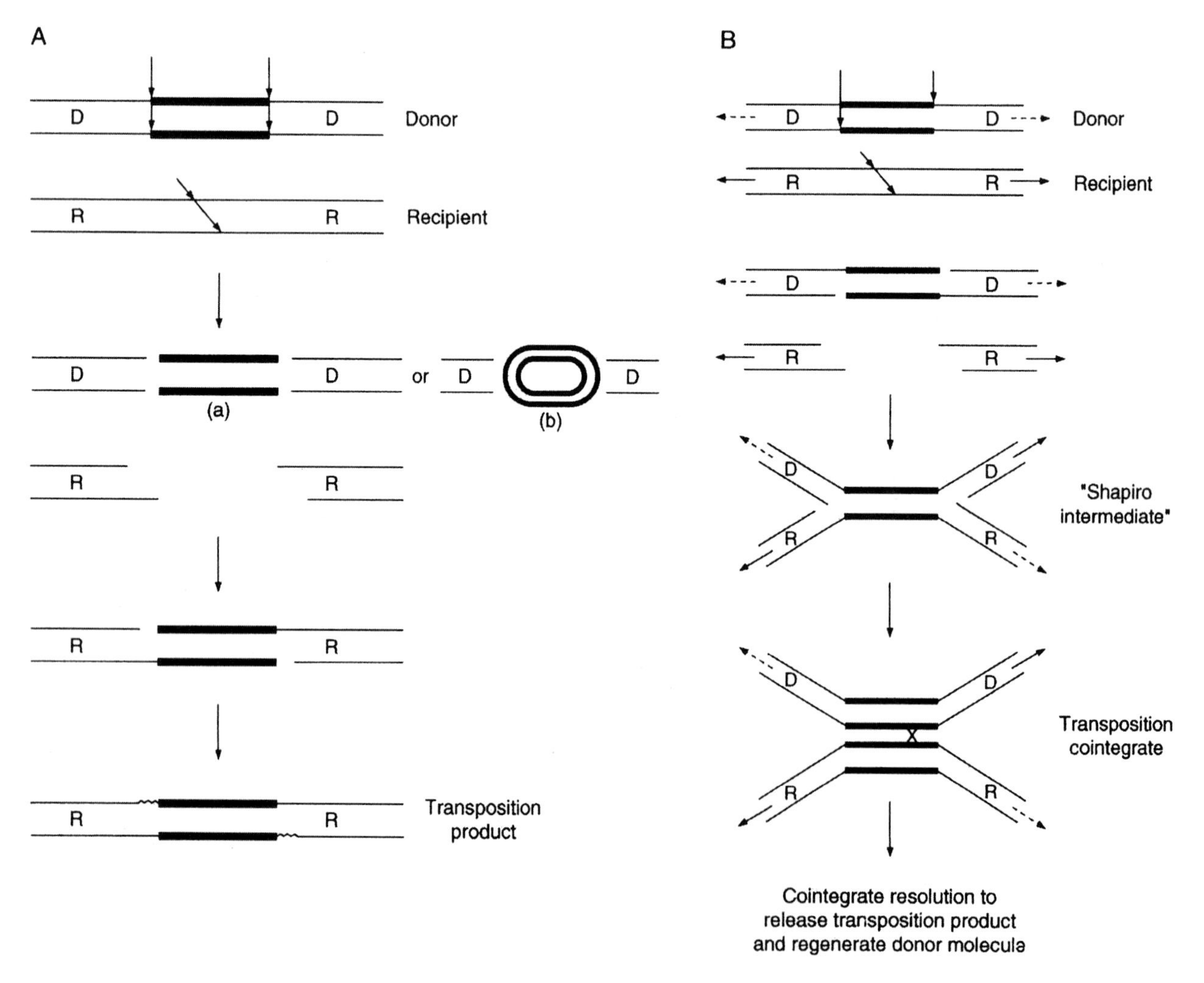

FIGURE 87.3 Models of transposition. (A) Conservative, cut-and-paste transposition with (a) linear DNA intermediate (although the ends may be held together by transposase), and (b) circular DNA intermediate; (B) replicative transposition.

circular DNA–transposase protein complexes represent a transposition intermediate that is normally part of a transposition complex called a transpososome. The synapsing of the ends of IS*10*, prior to cleavage, is believed not to be a directed reaction but, rather, one that relies on the random collision of IR–transposase complexes that assemble at the ends of the element. The transpososome then attacks the target site directed by transposase. The free 3′ OH groups at both ends of the element act as nucleophiles, attacking 5′ phosphate groups in the DNA backbone at the target site to effect transesterification reactions. Synapsing the ends of the element holds them in the correct alignment to execute what is probably normally a coordinated, but slightly staggered attack by the element on the two strands of DNA. These reactions, which require no additional energy input, insert the transposon into its new site, in the process creating short single-strand gaps, one on each strand of the DNA, that flank the element. Gap-repair DNA synthesis completes the insertion by restoring strand continuities and generating the flanking DRs. The fate of the donor molecule, which has been cleaved at the original IS insertion site, is unknown, but it is thought normally to be lost (degraded).

Some IS elements not only transpose conservatively in this manner, but their transposition can also, on occasion, give rise to what are called transposition cointegrates. These structures result from transposition mechanisms that generate what is called a Shapiro intermediate, which allows the semiconservative replication of the element (Fig. 87.3; also, see later). Transposition cointegrates carry two copies of the element as direct repeats. IS*1* can generate transposition cointegrates as well as simple insertions. IS-generated cointegrates are relatively stable entities, particularly if generated and maintained in a *recA* background (i.e. in a host that is deficient for homologous recombination). Cointegrate formation has been exploited to identify and isolate new transposable elements.

The transposition of some IS elements involves the production of a true circular intermediate (i.e. the element is disconnected from its donor site and its ends are joined to produce a circular DNA molecule) (Fig. 87.3). IS*3* and IS*911*, both of which are IS*3*-like, give rise to circular intermediary forms. The production of IS circles is dependent on two element-encoded proteins, OrfA and OrfAB. As in the case of IS*1*, the latter protein is a fusion protein, the product of *orfA* and *orfB*, two *orfs* that have relative reading frames 0 and −1 and that overlap slightly. As is IS*1* transposase, IS*3* and IS*911* fusion proteins are the results of translational frameshifting. Both OrfA and OrfAB proteins bind to the terminal IRs of their respective elements. The increased production of OrfA and OrfAB stimulates the production of IS circles and both proteins are needed for high-frequency transposition. This is in contrast to IS*1*, in which the smaller protein InsA inhibits the activity of the fusion protein InsAB. IS*911* circles have been shown to be efficient substrates for intermolecular transposition *in vitro*. When the element is circularized, the terminal IRs are separated by 3 bp, which were originally linked to one end of the donor molecule. This spacing segment is lost when the element is inserted into the target site. One interesting consequence of circularizing both IS*3* and IS*911* is that strong promoters are created that can drive the expression of *orfA* and *orfAB*. These new promoters are significantly stronger that the indigenous promoters on the linear structures. The −35 hexamer box of the new promoter is provided by one IR and the −10 hexamer box by the other, which are side by side in the circular form of the elements. The formation of these strong promoters by IS circularization may be a device to ensure that there are adequate levels of transposition proteins to drive the second stage of the transposition process (i.e. the insertion of the element into target sites) and to ensure that the transposase level is boosted only when there is a suitable substrate for it.

Although transposition of many IS elements is essentially a cut-and-paste exercise, some elements use a replication strategy. IS*91* and related elements are distinctive in that they have no IRs, insert into a specific tetranucleotide sequence, and do not generate target-site duplications.

The sequence analysis of IS*91*-like elements, each approximately 1.7 kb in size, has revealed that at one end of each element is a highly conserved sequence that resembles the leading-strand replication origins (ls-origin) of a number of gram-positive plasmids that replicate by rolling circle (RC) replication to generate single-stranded (ss) DNA copies, that are then converted to double-stranded DNA. The other end of each IS element is also highly conserved with a region of dyad symmetry followed by a tetranucleotide sequence that matches the specific insertion site. Each IS*91*-like element possesses a single large ORF that encodes a protein with striking similarities to the RC-plasmid replication proteins (REPs). These findings suggest that the transposition of IS*91* and similar elements involves RC replication and a ssDNA intermediate, possibly circular.

D. Transposition of composite transposons

Many composite transposons, originating in a broad spectrum of bacteria, including gram-positive and

TABLE 87.2 Some prokaryotic composite transposons[a]

Transposon	Size (kb)	Terminal elements	Target duplication (bp)	Marker(s)[b]
Gram-negative bacteria				
Tn*5*	5.7	IS*50* (IR)	9	KmBlSm
Tn*9*	2.5	IS*1* (DR)	9	Cm
Tn*10*	9.3	IS*10* (IR)	9	Tc
Tn*903*	3.1	IS*903* (IR)	9	Km
Tn*1525*	4.4	IS*15* (DR)	8	Km
Tn*1681*	4.7	IS*1* (IR)	9	HST
Tn*2350*	10.4	IS*1* (DR)	9	Km
Tn*2680*	5.0	IS*26*[c] (DR)	8	Km
Gram-positive bacteria				
Tn*3851*	5.2	NR	NR	GmTbKm
Tn*4001*	4.7	IS*256* (IR)	8	GmTbKm
Tn*4003*	3.6	IS*257* (DR)	8	Tm

[a]DR, direct repeat; HST, heat-stable enterotoxin; IR, inverted repeat; NR, not recorded.
[b]Resistance genes: Bl, bleomycin; Cm, chloramphenicol; Gm, gentamicin; Km, kanamycin; Sm, streptomycin; Tb, tobramycin; Tc, tetracycline.
[c]IS*26* and IS*6* are synonyms.

gram-negative organisms, encode resistance to antibiotics; a few encode functions other than drug resistance (Table 87.2). In general, these elements appear not to be phylogenetically related, although different transposons carrying essentially the same drug-resistance gene are known, as are transposons in which completely different central sequences are bracketed by copies of the same IS element (Table 87.2). Composite structures transpose because each is treated simply as an extended version of the element that forms the terminal repeats. This is possible because each terminal element is delimited by short IRs, so the composite structure also is delimited by the same IRs. An interesting quirk of these systems is that, because all that is needed for transposition, apart from the cognate transposase, is a pair of IRs, what is called "inside-out" transposition can also occur, provided that the donor molecule is circular. In this case, the terminal elements form a composite structure with the carrier DNA molecule, abandoning the functions that characterized the original transposon. The new structure is delimited by the pair of IRs that were located at the inside junctions of the IS elements and their original central module, rather than the outside pair that delimit the original transposon.

III. Tn3 AND RELATED TRANSPOSONS

A. Complex transposons—general

Although it is convenient to consider composite transposons as a group because of their structural similarities, few appear to be phylogenetically related and individual elements may transpose by different mechanisms. Complex transposons show a more complicated genetic arrangement than either IS elements or composite transposons, in that genes that do not encode transposition functions have been recruited into and become part of the basic transposable element. However, just as the term "composite transposon" denotes only that such an element has a modular structure with terminal IS repeats, so the term "complex transposon" means no more than that the transposon does not possess a modular structure with terminal IS repeats and that it is not a bacteriophage.

B. Tn3

The archetype complex transposon is Tn*3*, originating on the Resistance (R) plasmid, R1. It is virtually identical to the first drug-resistance transposon discovered, Tn*1*, found on another R plasmid, RP4. Both elements were originally called TnA and the transposition functions are fully interchangeable. The transposons encode TEM β-lactamases that confer resistance to ampicillin and carbenicillin and a few other β-lactam antibiotics. The two enzymes, designated TEM 1 and TEM 2 (encoded by Tn*3* and Tn*1*, respectively) differ by only one amino acid (a Gln-to-Lys substitution at position 37) and have essentially identical substrate specificities. Both Tn*1* and Tn*3* are widely distributed among gram-negative bacteria from clinical and veterinary sources, with Tn*3* being somewhat more common. In recent years, these enzymes have evolved to create a large family of more than 100 members that are collectively referred to as extended-spectrum β-lactamases. These variants have expanded substrate spectra that incorporate some of the newer cephalosporins (the third- generation cephalosporins, e.g. ceftazidime and cefotaxime). A second set of variants are insensitive to β-lactamase inhibitors, such as clavulanic acid (inhibitor resistant β-lactamases). Undoubtedly, the speed at which these mutants have arisen and have become established owes much to the widespread distribution of the parent genes on Tn*1* and Tn*3*.

C. Tn3-related transposons

Tn*3* is typical of an extended family of phylogenetically related transposons, members of which have been found in both gram-negative and gram-positive bacteria. Between them they encode resistance to several antibiotics and to mercuric ions; a couple specify catabolic functions, whereas several are cryptic (Table 87.3). The evolutionary relationships among these elements were first revealed when it was discovered that apparently unrelated transposons (Tn*3*, Tn*501*, Tn*551*, and Tn*1722*)

TABLE 87.3 Some Tn*3*-like transposons[a]

Transposon	Size (kb)	Terminal IRs (bp)	Target (bp)	Marker(s)[b]
Gram-negative bacteria				
Tn*1*	5	38/38	5	Ap
Tn*3*	4.957	38/38	5	Ap
Tn*21*	19.6	35/38	5	HgSmSu
Tn*501*	8.2	35/38	5	Hg
Tn*1000*	5.8	36/37	5	None
Tn*1721*[c]	11.4	35/38	5	Tc
Tn*1722*	5.6	35/38	5	None
Tn*2501*	6.3	45/48	5	None
Tn*3926*	7.8	36/38	5	Hg
Tn*4651*	56	32/38	5	xyl
Gram-positive bacteria				
Tn*551*	5.3	35	5	Ery
Tn*917*	5.3	38	5	Ery
Tn*4430*	4.2	38	5	None
Tn*4451*	6.2	12	NR	Cm
Tn*4556*	6.8	38	5	None

[a]NR, not recorded.
[b]Ap, ampicillin; Cm, chloramphenicol; Ery, erythromycin; Hg, mercuric ions; Sm, streptomycin; Su, sulphonamide; Tc, tetracycline; xyl, xylose catabolism.
[c]Tn*1721* is a composite structure that uses Tn*1722* as its basis for transposition.

have different, but clearly similar, short (35–40 bp) terminal IR sequences. When the elements were fully sequenced, it became apparent that the similarities extend to the transposition genes as well (Fig. 87.4).

D. Transposition functions

Each Tn*3*-related element encodes a transposase of approximately 1000 amino acids. The genes, approximately 3 kb long and designated *tnpA*, are clearly ancestrally related. Most of these transposons also encode a second recombination enzyme, called a resolvase. These enzymes, encoded by genes designated *tnpR*, are site-specific recombinases. Each acts at a particular site, designated *res*, located on the transposon adjacent to *tnpR*. These enzymes mediate the second stage of a two-step transposition mechanism.

E. Family branches

The Tn*3*-related transposons split naturally into two main branches of the family. On each of the main branches, the transposition functions of the elements are closely related. Tn*3* is the type element for one

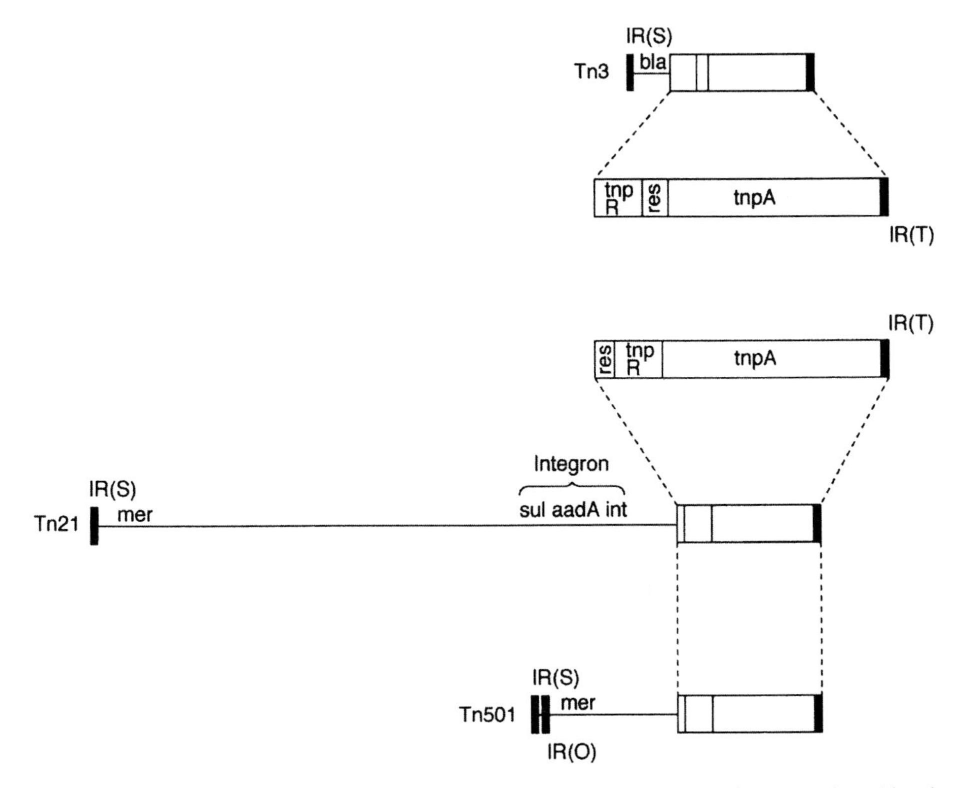

FIGURE 87.4 Schematic representation of Tn*3* and some related elements. Genes depicted: *aadA*, aminoglycoside adenylyltransferase A; *bla*, TEM β-lactamase; *int*, integrase. IR(T), inverted repeat adjacent to *tnpA*, and IR(S), second IR, are represented by thick vertical bars; *mer*, operon conferring resistance to mercuric ions; *res*, resolution site; *sul*, sulfonamide-resistant dihydropteroate synthetase gene; *tnpA*, transposase gene; and *tnpR*, resolvase gene.

branch (which includes Tn*1000* and Tn*1331*) and Tn*21* for the other (Table 87.3). Tn*21* is larger than Tn*3* (20 versus 5 kb) and confers resistance to streptomycin, spectinomycin, and sulfonamide, as well as to mercuric ions. The antibiotic-resistance genes on Tn*21* and related elements have been acquired as gene cassettes inserted into integrons located on the elements (see later). Notwithstanding their different sizes, both Tn*3* and Tn*21* devote approximately the same genetic capacity, 4 kb, to transposition. In addition to the two main groups, there are one or two elements that are clearly Tn*3*-like, but that are not closely related to either Tn*3* or Tn*21* or, indeed, to each other.

The Tn*21* branch of the family contains many elements that differ principally in the number of resistance determinants carried (Table 87.3). These elements have the same or almost the same transposition functions, which are functionally interchangeable. Much of the diversification seen in the group is thought to be of relatively recent origin, reflecting the activities of integrons and the acquisition and loss of gene cassettes (see later). Some elements, such as Tn*501* and Tn*1721*, are less closely related to Tn*21* than others, such as Tn*2424* and Tn*2603*. Sequence analysis of the transposition functions of Tn*501* and Tn*1721* indicates significant degrees of divergence from the type element, Tn*21*, but they are much more closely related to Tn*21* than to Tn*3*. The Tn*21*-like transposons present what is possibly the most successful diversification of a single transposable element discovered. However, the identity of the parent element (i.e. the one that lacks accessory genes) is unknown.

The structures of Tn*3* and some related elements are depicted in Fig. 87.4, which illustrates the two arrangements of the transposition functions seen in the extended family. On Tn*21* and its close relatives, both transposition genes have the same orientation and the gene arrangement—*res, tnpR, tnpA*. On Tn*3* and closely related elements, *tnpR* and *tnpA* are adjacent but opposed. They are transcribed divergently from a short common promoter region, in which is also located the *res* site. One consequence of this gene arrangement is that TnpR acts not only as a site-specific recombinase, but also as a repressor for the transposase gene, *tnpA*. Knockout mutations of Tn*3* *tnpR* generate Tn*3* derivatives that transpose at elevated frequencies to produce transposition cointegrates that are easily recovered in a *recA* host.

F. Recruitment of antibiotic resistance genes to Tn*21*-like transposons: integrons and gene cassettes

Molecular analysis of several Tn*21*-like elements revealed that many of the antibiotic resistance genes carried by these elements have not been inserted at random. Rather, they are located, as single genes, or sets of genes, at essentially the same place on the transposon backbone, specifically on one side or the other of the *aad* gene, if this is carried, or as a replacement for it if the *aad* gene is absent. This finding, given the variety of genes involved (approximately 40 resistance-gene cassettes have been reported, as well as a handful of cassettes with genes of unknown function) and the size of Tn*21* (20 kb, of which only 4 kb is needed for transposition), suggested that the diverse resistance genes had been captured by a site-specific recombination mechanism. This is now known to involve elements termed integrons and mobile DNA sequences called gene cassettes.

An integron is defined as a genetic element that has a site, *attI*, at which DNA, in the form of a gene cassette, can be integrated by site-specific recombination, and that encodes an enzyme, integrase (product of *int*), that mediates the site-specific recombinations. A number of integrons also accommodate a gene for sulfonamide resistance, *sul*, at one end. A gene cassette is a discrete genetic element that may exist as a free circular nonreplicating DNA molecule when moving from one site to another, but that is normally found in linear form as part of another DNA element, such as a transposon, plasmid, or bacterial chromosome. Gene cassettes normally accommodate only one gene and a short sequence called a 59-base element that serves as the specific cassette recombination site. Accordingly, gene cassettes are small, normally about 500–1000 bp. The genes on cassettes, in general, lack promoters and are expressed from a specific promoter on the integron located beside *attI*. In a few cases, cassettes carry two genes; these are likely to have arisen from the fusion of two gene cassettes which at one time were side by side. Because the expression of the genes on the majority of cassettes is from the integron promoter adjacent to *attI*, integration of cassettes is not only site-specific, but also orientation-specific. Several cassettes can be integrated sequentially at *attI*, which is regenerated with each insertion. When this occurs, the genes on the cassettes are expressed as an operon. Hence, a distinctive genetic array for an integron can be envisaged as *int attI* gene cassette(s) *sul* (Fig. 87.5).

Three distinctive classes of integrons have been reported. Class 1 accommodates the majority of integrons found, including those on Tn*21*-like transposons. The basic integron of class 1, designated In0, lacks gene cassettes; that is, it has the genetic array *int attI sul*. New integrons are generated by the insertion of one or more cassettes at *attI*, or the deletion of one or more cassettes from an existing integron. The order of cassettes, from *attI*, can also be changed by site-specific

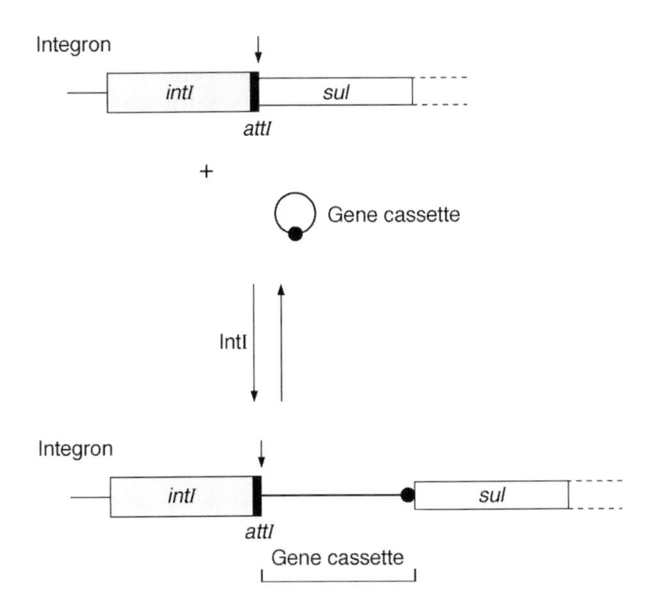

FIGURE 87.5 Integrons and gene cassette integration and excision. The recipient integron shown represents the basic class 1 element, In0; that is, it has no gene cassette between the integrase gene *intI* and the sulfonamide resistance gene, *sul*. IntI first mediates the excision of a gene cassette from one integron array, releasing it as a free circular form comprising a single gene followed by a 59-base element (solid circle). The gene cassette is then inserted, by site specific recombination mediated also by IntI, into *attI* (represented as the solid vertical bar) on the recipient integron.

excision of a promoter-distal cassette and its reinsertion at *attI*.

The movement of gene cassettes in and out of integrons is a random process, similar to most other genetic rearrangements in bacteria. Which rearrangements survive and which are lost is a matter of circumstance and natural selection; if the gene combination is of benefit to the host, it will be selected and established in the population by clone amplification. Most of the genes discovered on cassettes are antibiotic- or disinfectant resistance genes; however, this is probably not a true picture of the diversity of genes on these elements, but, again, only a reflection of the intense interest of medical microbiologists in resistance genes for the past two to three decades. The discovery, in a number of bacterial species, of superintegrons encoding a wide variety of functions supports this view.

G. Mechanism of transposition

The transposition of members of the Tn*3* family of transposons involves replication and recombination (Fig. 87.3). Each transposition event generates a target site duplication, usually one of 5 bp and, as in the previously described mechanisms, the duplication arises because the cleavage of the target to allow transposon insertion involves slightly staggered cuts on the two DNA strands. The mechanism of transposition of Tn*3*-like elements differs from those described previously in that the element is not wholly disconnected from the donor site. Rather, single-strand cleavages are introduced by the cognate transposase at both ends of the transposon on opposite strands to expose 3′ OH groups. These are then used as nucleophiles to attack two 5′ phosphate groups at the target site, one on each DNA strand, separated by 5 bp. Two transesterification reactions join the transposon to the target DNA, generating what is termed a "Shapiro intermediate," after J. A. Shapiro, who first proposed its existence (the structure was also proposed independently, at about the same time, by A. Arthur and D. Sherratt). The transesterification reactions not only join the transposon to the target, but also generate free 3′-OH groups on both sides of the transposon that can be used to prime the replication of the element. Whether one or both is used *in vivo* is not known. The replication of the transposon is executed by host-cell enzymes.

When the element has been replicated and double strand continuity has been restored, the product generated is a transposition cointegrate. This molecule contains the transposon donor and target replicons in their entireties, joined by directly repeated copies of the transposon, one at each replicon junction. The final product, a copy of the target replicon with a single copy of the transposon, is released by cointegrate resolution. Normally this is achieved by a site-specific recombination reaction, mediated by the *tnpR* gene product resolvase, using the two *res* sites present on the cointegrate (one on each copy of the transposon). Should either *tnpR* or *res* be damaged, then host mediated *recA*-dependent recombination across the duplicate transposon sequences can also resolve the cointegrate to the final transposition product and a molecule that is indistinguishable from the original transposon donor (Fig. 87.3).

H. One-ended transposition

The replicative mechanism of transposition outlined in Section III.G explicitly involves both ends of the element in a nondiscriminatory manner (i.e. both ends are needed, but are treated identically). This explains the need for perfect or near-perfect terminal IRs because both ends must bind transposase. Surprisingly, however, when one IR sequence is deleted TnpA-mediated transposition still occurs, albeit at much-reduced frequency and necessarily modified. Such events are referred to as one-ended transpositions. A one-ended transposition is a replicative recombination mediated by some (Tn*3*, Tn*21*, and Tn*1721*), but not all Tn3-like elements. It involves one IR sequence (instead of two) and the cognate transposase. The products resemble replicon fusions in which the donor replicon, rather than a discrete part of it (i.e. the transposon), has been inserted into the recipient replicon. One end of the inserted DNA is always the solitary IR sequence, and the inserted sequence is flanked by 5-bp DRs, as for a normal transposition event, consistent with one-ended transpositions being true transpositions. Unlike normal transpositions, however, the products are not uniform in size and insertions both less than and greater than the unit length of the donor replicon have been reported. The different inserts form a nested set of DNA fragments with a copy of the IR sequence defining an anchored end. These data are not accommodated by the model described (Fig. 87.3B). However, a nested set of insertions, that all start from the same point on the donor molecule are indicative of a mechanism that involves RC replication. Whereas the two-ended mechanism precisely determines the sequence to be transposed, rolling circle transposition determines precisely only one end of the insert. How the other end is determined is not known.

I. Transposition immunity

Tn*3* and some of its relatives, but not Tn*21*, display a characteristic termed transposition immunity. Elements that show this behavior will not normally transpose on to another DNA molecule that already carries a copy of the transposon. The effector for inhibition has been identified as the element's IR sequence and a single copy affords significant protection, but the mechanism is not fully understood. In the case of Tn*3*, transposition immunity can be largely overcome by increasing the level of transposase (by the use of *tnpR* knockout mutants). Transposition immunity has also been demonstrated for Tn*7* and bacteriophage Mu, in which the mechanism is somewhat better understood.

IV. CONJUGATIVE TRANSPOSONS

A. Origin and properties

Conjugative transposons, first described in gram-positive bacteria, have also been found in gram-negative anaerobes such as *Bacteroides* sp. The type element Tn*916*, which encodes resistance to tetracycline, was discovered in *Enterococcus faecalis*. As the collective name for the elements implies, they not only transpose but also promote their own transfer from one cell to another. They differ from plasmids in that they do not replicate autonomously, but are replicated passively, as parts of the replicons into which they insert.

Tn*916* is 16.4 kb and indistinguishable from Tn*918*, Tn*919*, and Tn*925*, which are independent isolations of essentially the same element. A related conjugative transposon, Tn*1545*, is 23.5 kb and encodes resistance to kanamycin and erythromycin in addition to tetracycline. These elements promote their own conjugal transfer at frequencies of 10^{-9}–10^{-5}.

B. Transposition

Tn*916* is typical of the group and is the best studied. These elements can mediate conjugal transfer followed by transposition into the chromosome of the recipient cell, chromosome-to-plasmid transposition, and transposon excision. The last of these activities is often tested in *E. coli* rather than in *Enterococcus faecalis* because the frequency of excision is considerably higher in the former microbe. More than 50% of the Tn*916* sequence encodes conjugation functions, located as a block of genes at one end of the element. These functions are not strictly necessary for transposition, although mutations in some interfere with chromosome-to-plasmid transposition. Mutations that block transposition map to the other end of the element. Such mutations also inhibit transposon excision. Two transposition genes, designated *xis-Tn* and *int-Tn*, have been identified on Tn*1545*.

The transposition of Tn*916* and related elements is a two-stage process. The first step is the excision of the element as a circular intermediate, similar to that involved in transposition of IS*911* (Fig. 87.3). Staggered double-strand cuts at both ends of the element generate a linear DNA molecule with 5′ single-stranded hexanucleotide overhangs. End-to-end joining creates a circular DNA molecule with a 6-bp heteroduplex spacer separating the ends of the element. The circular DNA species may then undergo one of two productive fates: (a) the insertion of the transposon into one of the DNA molecules in the cell; or (b) the conjugal transfer to another cell followed by insertion of the transposon into a suitable DNA molecule in the new host (e.g. the bacterial chromosome). The mechanism appears to be similar to the integration–excision systems of lambdoid phages, and involves functions related to the Int and Xis proteins that mediate integration and excision of bacteriophage λ into the *E. coli* chromosome.

V. SITE-SPECIFIC TRANSPOSONS

A. General comments

Most transposable elements show little or no target site specificity, although bias toward insertion into certain regions of DNA molecules has been reported for some elements (e.g. IS*1* has been reported to favor insertion in AT-rich tracts, as has Tn*3*; and although Tn*5* usually displays low target-site selectivity, a preference for some "hotspots" has been observed); but these biases are preferences, not requirements. In contrast, a few elements display a marked, if not exclusive, specificity of insertion.

B. Tn7

Tn*7* was first identified on the R plasmid R483. It carries two resistance genes, *dfrI* and *aadA*, the former encoding a trimethoprim-resistant dihydrofolate reductase, whereas the latter codes for an aminoglycoside adenylyltransferase, which mediates resistance to streptomycin and spectinomycin. Tn*7* is 14 kb long, generates 5-bp target-site duplications on transposition, and inserts in one orientation at high frequency into single sites on the chromosomes of a number of gram-negative bacteria, including *E. coli*, *Klebsiella pneumoniae*, *Pseudomonas* sp., and *Vibrio* sp. At a lower frequency (10^{-4}), it will insert, with little target-site selectivity, into many different DNA molecules (e.g. plasmids) that lack the specific insertion site. However, if the specific insertion site is engineered into a plasmid, Tn*7* will insert into that plasmid at high frequency in a site-specific orientation-specific manner.

C. Chromosomal site of Tn7 insertion

The locus of Tn*7* insertion on the *E. coli* chromosome has been located at 84′, between *phoS* (encodes a periplasmic phosphate-binding protein) and *glmS* (encodes glucosamine phosphate isomerase), and designated *att*Tn*7*. The 5-bp target site is part of the *glmS* transcriptional terminator, which is located about 30 bp from the end of the translational reading frame. Unexpectedly, the target site (i.e. the site of insertion) and the attachment site (*att*Tn*7*, the sequence that determines the locus specificity) are distinct. The attachment site is a sequence of approximately 50 bp, located 12 bp from the 5-bp target site and extending into *glmS*. This relationship between *att*Tn*7* and *glmS* is also seen in other bacteria, such as *Klebsiella pneumoniae* and *Serratia marcescens*, in which chromosomal insertion is specific. Although the various attachment sites in the different bacteria are homologous, the target sites differ completely, indicating that the sequence of the actual site of insertion of Tn*7* has little or no role to play in the specificity, but is determined solely by its distance from *att*Tn*7*. Hence, Tn*7* exploits sequence data held in a highly conserved host gene, *glmS*, to select an insertion site, but is then inserted beside rather than in *glmS*, ensuring that the insertion does not harm the host.

D. Transposition functions of Tn7

Tn*7* is unusual in the number of transposition functions it requires. Five have been identified, sequenced, and designated *tnsA*, *B*, *C*, *D*, and *E*. The genes encode peptides of 31, 78, 63, 59, and 61 kDa, respectively. The first three are required for all transpositions. High frequency site-specific transposition requires, in addition, the product of *tnsD*, whereas low-frequency random-site transposition requires instead the *tnsE* product.

E. Mechanism of transposition of Tn7

A Tn*7* *in vitro* transposition system, with a plasmid carrying *att*Tn*7* as target, has been developed. Using this system, it has been shown that the transposon is first disconnected from the donor molecule by two double-strand staggered cuts, each of which generates 5′ overhangs of three bases at the ends of the element. The DNA of these overhangs comes from the carrier DNA molecule, not Tn*7*. The uncoupled linear Tn*7* sequence is then joined to the target (i.e. the mechanism is cut-and-paste) (Fig. 87.3). This is thought to involve nucleophilic attacks by the 3′-OH groups at the ends of the uncoupled transposon on phosphate groups at the target site. Two transesterification reactions, one targeted to each strand of the DNA duplex and separated by 5 bp, cleave the recipient DNA at the target site and fuse it to Tn*7* (which is then joined at both ends to the target DNA by short ssDNA sequences). It is assumed that repair processes then remove the short ssDNA extensions at the 5′-ends of the transposon (i.e. the remnants of the previous insertion site, which would be mismatched with the short stretches of ssDNA to which the transposon is attached) and fill in the resulting single-strand gaps, generating the 5-bp DRs that flank the Tn*7* insertions.

It has been found *in vitro* that the transposition of Tn*7* proceeds via a DNA-protein complex that contains a transposon donor and target DNA molecules, four transposition proteins (TnsA, B, C, and D) and ATP. ATP may be directly involved in the reaction because TnsC is an ATP-binding protein, but it may also be needed to ensure that the DNA substrates have an appropriate degree of supercoiling, maintained by DNA gyrase.

The cell-free system faithfully reproduces several of the features of Tn*7* transposition that are characteristic of the element. Transposition *in vitro* is site- and orientation-specific with respect to *att*Tn*7*, as it is *in vivo*, and shows transposition immunity: When the target molecule carries a copy of Tn*7*, transposition of a second copy of Tn*7* into it is blocked.

F. IRs of Tn7

Tn*7* is unusual in the length of the terminal sequences needed for transposition; 75 bp are needed at one end of the element and 150 bp at the other, but these sequences form IRs of only 30 bp. Artificial elements flanked by IRs of the shorter sequence will transpose when transposition functions are provided. In contrast, a DNA sequence flanked by IRs of the 150-bp sequence does not transpose. The basis for this differentiation has not been determined. Each terminal sequence contains several analogs of a 22-bp consensus sequence; these have been shown to bind TnsB and are thought to serve as nucleation centers for the assembly of the DNA–protein complex required for transposition.

G. Tn554

Tn*554* is a 6.7-kb transposon encoding resistance to erythromycin and spectinomycin in *Staphylococcus aureus*, where it transposes into the chromosome at high frequency, primarily into a single site, designated *att*Tn*554*, in one orientation only. From its sequence, six Tn*554* ORFs were identified, of which five have been demonstrated to be genes—the three gene *tnsABC* transposition operon accounts for approximately half the transposon's coding capacity, whereas *ery* and *spc* code for resistance to erythromycin and spectinomycin, respectively. Tn*554* also displays transposition immunity.

The mechanism of transposition of Tn*554* differs from that of most other elements in the sense that not only is the element transposed, but so also are a few base pairs from the carrier molecule. This short additional sequence is located on one side of the element in the donor, but is transferred to the opposite side in the transposition products. Hence, the sequence changes with each sequential transposition. The transposition mechanism is not known, but current data suggest a cut-and-paste mechanism involving a circular intermediate.

H. Tn502

Tn*502* is a poorly characterized gram-negative transposon that encodes resistance to mercuric ions. It appears not to be related to Tn*21*, Tn*501*, or other Hg transposons of the Tn*3* family. It is notable because it displays site-specific insertion into a plasmid, rather than into a chromosome. Tn*502* is 9.6 kb and inserts at high frequency in one orientation into a single site on the IncP plasmid RP1 (RP4). If this site is deleted, insertion then occurs into many sites at a much lower frequency. In this respect, its behavior resembles that of Tn*7*, high-frequency site-specific transposition versus low-frequency random-site transposition. There is insufficient information to determine whether Tn*502* is related to other site-specific elements.

VI. TRANSPOSING BACTERIOPHAGES

A. Structure of Mu

Transposing bacteriophages use a transposition strategy to replicate. The type element of the group is bacteriophage Mu, discovered by L. Taylor in 1963. Taylor recognized that Mu was a temperate phage and concluded that because lysogenization often created auxotrophs, Mu must be able to integrate into the *E. coli* chromosome at many different sites, in some cases damaging genes responsible for biosynthetic functions. It is now well established that Mu integrates into the chromosome using a conservative (cut-and-paste) transposition strategy and that, subsequently, Mu replicates by multiple rounds of transposition via Shapiro intermediates (Fig. 87.5).

The linear genome of Mu is 37 kb, but each phage particle carries a DNA molecule of approximately 39 kb. This is constructed from the Mu genome with about 150 bp of host DNA on one side and 1–2 kb of host DNA on the other. These flanking sequences are different for individual copies of the phage genome and reflect the site of insertion of that copy of Mu, prior to the assembly of the phage particle. Genome packaging proceeds by a headful mechanism that disconnects the phage from the carrier molecule by first cutting the carrier DNA approximately 150 bp beyond one end of Mu (the end nearest the replication functions) and then again approximately 39 kb away, on the other side of the phage genome, so generating a linear copy of the genome flanked by host DNA.

In its genome structure, Mu is unremarkable. It has the usual arrays of genes needed for phage assembly, that is, head and tail production. It has two replication (transposition) genes, designated A and B, which are located close to one end of the genome (traditionally depicted on the left in linear maps) and it has a 3-kb segment that can invert by site-specific recombination. This region is homologous to an invertible section of the phage P1 genome; inversion of these regions changes the host specificity of the phages by promoting the expression of alternative tail fiber genes. Mu and P1 are not related in any other respect.

B. Mu replication and integration

More than a decade after its discovery, it was realized that Mu replicates by transposition and the study of Mu replication *in vitro* has been an invaluable model transposition system. Phage replication involves formation of a Shapiro intermediate, which establishes a replication origin(s) from which the replication of the phage genome proceeds. Because the transposition or replication is primarily intramolecular, at least initially, this results in gross molecular rearrangements (i.e. deletions and inversions) of the bacterial chromosome, which is also progressively fragmented into many small circular DNA molecules, each carrying one or two copies of the Mu genome (an inevitable consequence of rounds of intramolecular transposition). Unlike Tn3 and related transposon cointegrates, which are resolved by a site-specific resolution system (*res*/resolvase), Mu cointegrates are not specifically resolved. If resolution does occur, as it may, it is the result of host-mediated, *recA*-dependent recombination. The resolution of cointegrates is not important to the phage and lack of resolution does not inhibit further rounds of transposition or replication. Multiple cycles of Mu transposition do not require the presence of a second replicon in the cell; the chromosome suffices, although if a second replicon is present in the cell, it suffers the same fate as the chromosome.

In addition to replicating by transposition, Mu is established in a new host also by transposition, from the extended linear sequence carried in the phage particle into a replicon carried by the host, usually the bacterial chromosome. The transposition initially generates a Shapiro-type intermediate, but one in which the DNA molecule carrying the transposon (Mu) is linear instead of circular. Again, the intermediate is formed by transesterification reactions involving 3′-OH groups at the ends of the phage DNA, created by single-strand cleavage by the phage A protein. Then, instead of the inserted DNA (i.e. the phage genome) being replicated, a second pair of cleavages, this time at the 5′-ends of the genome, disconnects it completely from the terminal remnants of previous host DNA, which are discarded, and double-strand continuity is restored to seal the phage genome into a new site, flanked by 5-bp DRs.

That Mu lysogeny is established by a conservative integration was elegantly demonstrated as follows. Phage particles produced in a *dam*$^+$ host were used to infect a *dam* mutant of *E. coli* (which cannot methylate its DNA). The newly integrated phage DNA was found to be fully methylated, not hemimethylated, indicating that no replication had occurred prior to or as a consequence of insertion in the new host. The conservative transposition used to establish Mu in a new host and phage replication both require the A gene product, which is the Mu transposase. The role of the B gene product, which is also needed for efficient integration and replication, is largely one of target-site selection and facilitating the assembly of the transposition complex.

C. D108 and other transposing bacteriophages

Mu was the first transposing bacteriophage to be discovered. Only one other transposing coliphage is known, D108, which is closely related to Mu. The two genomes display 90% sequence homology, with the main divergence being at the ends of the genomes. One such region includes the 5′-ends of the A genes, the consequence of which is that the two A gene products, although they do complement each other, do so only poorly. Transposing phages have also been identified in *Pseudomonas* sp. The sizes of their genomes (~40 kb) and the structures of the DNA packaged to form the phage particles are strikingly similar to those of Mu and D108, but no gross similarities in DNA sequence with Mu have been found. Possible transposing phages have also been found in *Vibrio* sp., identified on the basis that lysogenization can result in auxotrophy, which was how Mu was originally detected.

VII. YEAST TRANSPOSONS

A. Types

Finally, mention must be made of transposons found in yeast. Several types have been identified in *Saccharomyces cerevisiae*. The majority are components of the nuclear DNA, but one, designated Ω, is found in mitochondrial DNA. No one element has been found in both compartments.

Several of the yeast transposons (e.g. Ty1, Ty2, Ty3) are what are known as retrotransposons because the way in which they transpose is clearly akin to the mechanism of retroviral replication. Ty1 has been shown to transpose by a process involving a reverse transcriptase step. Ty2 and Ty3 closely resemble Ty1, and also require reverse transcriptase for transposition.

B. Structure of Ty elements

The Ty elements are 5–6 kb long and typically have long terminal direct repeats (LTRs), which can themselves transpose (called δ in the cases of Ty1 and Ty2, and σ for the LTRs of Ty3). The Ty elements and their LTRs have no IRs, as such, although, as do many retroviruses, the transposable sequences terminate 5′-TG·CA. In keeping with the view that these elements can be regarded as transposons is the finding that intact elements are nearly always flanked by 5-bp duplications of host sequence (i.e. target-site duplication). Both Ty1 and Ty2 accommodate two ORFs, designated TYA and TYB. The equivalent ORFs of the two elements indicate a considerable degree of similarity between the proteins of the two transposons. Several domains of the Ty1 TYB protein show significant similarities to retroviral *pol* functions.

C. Transposition of yeast transposons

Yeast cells engaged in high-frequency transposition of Ty1 or TY2 contain large numbers of virus-like particles that contain Ty-RNA, reverse transcriptase, and capsid proteins encoded by the Ty element. It is thought that these particles are transposition intermediates. Transposition requires the conversion of the RNA to DNA and integration of this into the nuclear DNA. That the mechanism of retrotransposon transposition was likely to be analogous to the replication of retroviruses was originally inferred from the structural analysis of the elements and their transcripts, which resemble the structures of retroviral proviruses and viral RNA, respectively.

In contrast to the retrotransposons, the mitochondrial element Ω appears to transpose directly as DNA. In this sense, it more closely resembles transposable elements in bacterial cells.

A common type of transposable element in eukaryotes are what are called LINE elements (long interspersed nucleotide elements), after the first characterized L1 of mammals. Nucleotide-sèquence analysis of the elements revealed ORFs with striking similarities to the *gag* and *pol* genes of retroviruses. These elements are also known as the non-LTR retrotransposons, to emphasize their structural difference from LTR retrotransposons, such Ty1, and to indicate that transposition involves an RNA intermediate. Although mainly identified in higher eukaryotes, a LINE-like element, TAD, has been identified in the fungus *Neurospora crassa*. These elements represent a second class of retrotransposon.

VIII. CONSEQUENCES OF TRANSPOSITION IN BACTERIA

A. Genome rearrangements

The most obvious consequence of transposition is the insertion of one DNA sequence into another, an event that may disrupt a gene and cause a mutation. This property has been widely exploited in genetic analysis, and insertional inactivation by transposons is commonly used to locate genes of interest. Several ingenious "suicide systems" have been devised to deliver the transposon into the cell, where it becomes established only if it is transposed on to a resident replicon because the delivery vehicle cannot itself be replicated.

Insertional inactivation can also be exploited to capture transposable elements, including IS elements that encode no predictable phenotype. Vectors that express products that are lethal when the cells are exposed to particular culture conditions have been developed; insertions into the gene that encodes the lethal protein prevent the potentially fatal expression; that is, cells that contain a damaged copy of the gene survive and form colonies on the selective agar, whereas cells that contain an undamaged copy of the gene express it and die. The vector pKGR, carrying the *sacRB* cassette from *Bacillus subtilis*, is one such capture system. The expression of *sacB* is lethal to *E. coli* when cultured on medium containing sucrose, so carrying pKGR is lethal under these growth conditions; such cells fail to form colonies on sucrose containing agar. In contrast, cells with pKGR derivatives with *sacB* inactivated readily form colonies. The system allows, therefore, positive selection for loss of function, hence, loss of expression of *sacB*.

Transposable elements mediate DNA rearrangements other than simple insertions. The sites at which transposable elements have been inserted are often hot spots for deletions and inversions. These rearrangements involve the sequences on one side or other of the element, but not the element itself, which remains in place and intact. Many of these events are the results of intramolecular transpositions; that is, the element and its prospective target site are on the same DNA molecule. For elements such as Tn*3* and Mu that transposase via Shapiro intermediates, whether intramolecular transposition results in a deletion or an inversion of the adjacent sequence depends only on how the 3′-OH ends of the element attack the target. If each attacks a phosphate group on the same DNA strand as itself, then the transposition event automatically results in DNA fragmentation. The two sections on the starting molecule separated by the element (at its original site) and the target-site end become separate circular DNA molecules, each with a copy of the transposable element. However, only one of the fragments will carry the origin of replication and so can be replicated and survive; the result is a deletion adjacent to the original copy of the transposon. Conversely, if each 3′-OH attacks a phosphate group on the opposite DNA strand, then the transposition will generate a duplicate but inverted copy of the element at the new site, and the two sections of the carrier DNA separated by the old and new copies of the element are inverted with respect to each other.

Many other transposable elements also promote the formation of adjacent deletions and inversions, including elements thought to transpose by conservative cut-and-paste mechanisms. Precisely how these rearrangements are achieved is largely unknown.

B. Replicon fusion and conduction

When plasmid-to-plasmid transposition gives rise to cointegrates, these may offer the opportunity for conduction, that is, conjugal transfer of a nonconjugative plasmid by a conjugative one fused to it. The plasmids are cotransferred to the recipient cell as a single cointegrate DNA molecule, after which resolution may occur to release the donor replicon carrying one copy of the element and the target replicon carrying a second copy of the transposable element. The transposable element may initially be on either of the participating plasmids. Conduction systems have also been used to study transposable elements that are cryptic (e.g. IS elements) exploiting the element's ability to create replicon fusions and using plasmid markers to follow the movement of the cointegrate.

IX. CONCLUSION

The ability of transposable elements to insert and to generate deletions and inversions accounts for much of the macromolecular rearrangement that is observed among related bacterial plasmids. That bacterial chromosomes are subject to the same array of mutational events can readily be demonstrated in the laboratory. Nonetheless, chromosomal rearrangement seems to occur less often than might be expected, given the sizes of the DNA molecules and the apparent abundance of transposable elements. Whether this is because many chromosomal rearrangements prove, immediately or in the longer term, to be deleterious to host survival, particularly in competitive situations, and so are never established in the population or because chromosomal rearrangements occur less frequently than we might expect remains to be seen.

BIBLIOGRAPHY

Bennett, P. M. (1999). Integrons and gene cassettes: A genetic construction kit for bacteria. *J. Antimicrob. Chemother.* **43**, 1–4.

Bennett, P. M., and Hawkey, P. M. (1991). The future contribution of transposition to antimicrobial resistance. *J. Hos. Infect.* **18** (Suppl. A), 211–221.

Berg, D. E., and Howe, M. M. (eds) (1989). "Mobile DNA." American Society for Microbiology Press, Washington, DC.

Craig, N. L. (1995). Unity in transposition reactions. *Science* **270**, 253–254.

Craig, N. L. (1997). Target site selection in transposition. *Annu. Rev. Biochem.* **66**, 437–474.

Craig, N. L., Craigie, R., Gellert, M., and Lambowitz, A. M. (eds) (2002). "Mobile DNA II." American Society for Microbiology Press, Washington, DC.

Eickbush, T. H. (1992). Transposing without ends: The non-LTR retrotransposable elements. *New Biol.* **4**, 430–440.

Goryshin, I. Y., and Reznikoff, W. S. (1998). Tn5 in vitro transposition. *J. Biol. Chem.* **273**, 7367–7374.

Hallet, B., and Sherratt, D. J. (1997). Transposition and site-specific recombination: Adapting DNA cut-and-paste mechanisms to a variety of genetic rearrangements. *FEMS Microbiol. Rev.* **21**, 157–178.

Kleckner, N., Chalmers, R. M., Kwon, D., Sakai, J., and Bolland, S. (1996). Tn10 and IS10 transposition and chromosome rearrangements; mechanism and regulation *in vivo* and *in vitro*. *Curr. Top. Microbiol. Immunol.* **204**, 49–82.

Mahillon, J., and Chandler, M. (1998). Insertion sequences. *Microbiol. Mol. Biol. Rev.* **62**, 725–774.

Mizuuchi, K. (1992). Transpositional recombination: Mechanistic insights from studies of Mu and other elements. *Annu. Rev. Biochem.* **61**, 1011–1051.

Mizuuchi, M., Baker, T. A., and Mizuuchi, K. (1995). Assembly of phage Mu transpososomes—cooperative transitions assisted by protein and DNA scaffolds. *Cell* **83**, 375–385.

Reznikoff, W. S. (1993). The Tn5 transposon. *Annu. Rev. Microbiol.* **47**, 945–963.

Rice, L. B. (1998). Tn916 family conjugative transposons and dissemination of antimicrobial resistance determinants. *Antimicrob. Agents Chemother.* **42**, 1871–1877.

Rowe-Magnus, D. A., Gueroot, A. M., Ploncard, P., Dychinco, D., Davies, J., and Mazel, D. (2001). The evolutionary history of chromosomal super-integrons provides an ancestry for multiresistant integrons. *Proc. Natl. Acad. Sci. U.S.A.* **98**, 652–657.

Salyers, A. A., Shoemaker, N. B., Stevens, A. M., and Li, L. Y. (1995). Conjugative transposons: An unusual and diverse set of integrated gene transfer elements. *Microbiol. Rev.* **59**, 579–590.

Sarnovsky, R. J., May, E. W., and Craig, N. L. (1996). The Tn7 transposase is a heteromeric complex in which DNA breakage and joining activities are distributed between different gene products. *EMBO J.* **15**, 6348–6361.

Ton-Hoang, B., Polard, P., and Chandler, M. (1998). Efficient transposition of IS911 circles *in vitro*. *EMBO J.* **17**, 1169–1181.

WEBSITES

Lectures on transposable elements, University of Leeds
http://www.bmb.leeds.ac.uk/mbiology/ug/ugteach/gene2020/transpos.html

The Barbara McClintock papers at the American Philosophical Society
http://www.amphilsoc.org/library/mole/m.htm/

88

Two-component systems

Alexander J. Ninfa and Mariette R. Atkinson
University of Michigan Medical School

GLOSSARY

autophosphatase reaction The catalysis of the dephosphorylation of a phosphorylated residue found on itself by an enzyme.

autophosphorylation reaction The catalysis of the phosphorylation of a residue found on itself by an enzyme.

cross-regulation The convergence of parallel signal transduction pathways permitting the stimulation of one pathway to affect the output of the parallel pathways. The phosphorylation of a receiver domain by multiple, independently regulated, transmitter domains is an example of cross-regulation.

phosphotransfer reaction Reaction in which a phosphoryl group is transferred from a site on a protein to a different site on the same or a different protein, or to a small molecule.

response regulator A protein containing a receiver domain, which brings about the final step in a signal trandsuction pathway resulting in regulation of the target of the system. For example, many response regulators are transcription factors that activate or repress gene expression on phosphorylation of their receiver domain.

The Two-component Regulatory Systems are a related family of signal transduction systems that use the transfer of phosphoryl groups to control gene transcription and enzyme activity in response to various stimuli.

I. OVERVIEW

The key components of these systems are two distinct protein domains, referred to as the transmitter domain (T) and the receiver domain (R). Each of these domains has enzymatic activities (Fig. 88.1). As examples, we shall use the NRI–NRII two-component system regulating nitrogen assimilation in *Escherichia coli* and the KinA–Spo0F–Spo0B–Spo0A two-component system regulating sporulation in *Bacillus subtilis.* In these systems, NRII and KinA are proteins containing the T domain. The common feature of the T domain, which is always dimeric, is that these proteins bind ATP and phosphorylate themselves on a conserved histidine residue (Fig. 88.1A). The R domain catalyzes the transfer of phosphoryl groups from the phosphorylated histidine of the T domain to a highly conserved aspartate residue on itself (Fig. 88.1A). In the NRI–NRII system, the NRI protein contains an R domain, which transfers phosphoryl groups from NRII~P to itself, whereas in the sporulation system, the Spo0F protein consists of an R domain that catalyzes the transfer of phosphoryl groups from KinA~P to itself. From the phosphorylated R domain, phosphoryl groups may be subsequently transferred to water, as in the NRI–NRII system, or in some cases to another histidine residue within a domain referred to as the phosphotransfer (PT) domain (Fig. 88.1B). In the sporulation system, phosphoryl groups are transferred from Spo0F~P to Spo0B. From the

The Desk Encyclopedia of Microbiology
ISBN: 0-12-621361-5

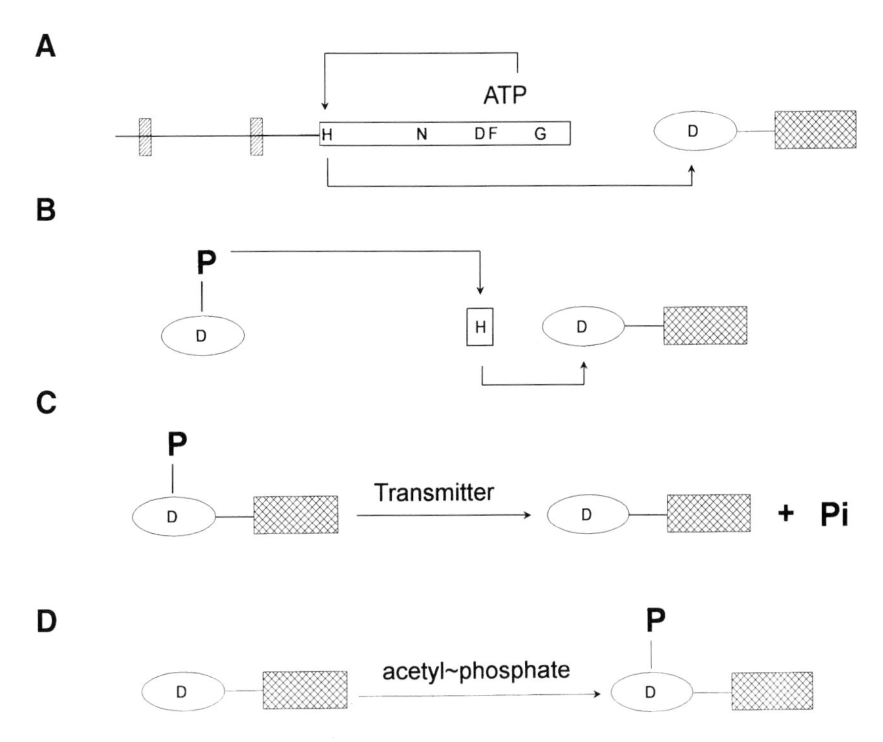

FIGURE 88.1 Enzymatic activities of transmitter, receiver, and phosphotransfer domains. Transmitter domains are depicted as rectangles containing H, N, D, F, and G motifs. Receiver domains are depicted as ovals with the conserved aspartate represented with D. Phosphotransfer domains are depicted by small rectangles with the conserved histidine represented with H. (A) Flow of phosporyl groups to the response regulator in a typical two-component system. The transmitter protein is depicted as containing an N-terminal domain (thin line) with two transmembrane segments (depicted as hatched rectangles). The response regulator is depicted as containing an N-terminal receiver domain and a C-terminal transcriptional activation domain (crosshatched). (B) Flow of phosphoryl groups from a phosphorylated receiver domain to a phosphotransfer domain, to the receiver domain of a response regulator. (C) Dephosphorylation of a phosphorylated response regulator by a transmitter domain. (D) Phosphorylation of a response regulator receiver domain by acetyl phosphate. For details, see text.

phosphorylated PT domain, phosphoryl groups may be transferred to another R domain (Fig. 88.1B). In the sporulation system, the Spo0A protein contains an R domain that catalyzes the transfer of phosphoryl groups from Spo0B to itself. Thus, in this latter class of systems, a three-site phosphorelay ($T \rightarrow R \rightarrow PT$) is used to deliver phosphoryl groups to an R domain.

For each two-component system, the phosphorylation of a key R domain is used to control the activities of various other protein domains, which may be either directly associated with the R domain or part of a distinct protein or macromolecular complex. The protein bearing this key R domain, which directly affects a cellular activity, has been referred to as the response regulator. For example, NRI and Spo0A, but not Spo0F, are considered to be response regulators. Phosphorylation of the R domain of the response regulator brings about a conformational change in the R domain that may be propagated to associated domains or alter the interactions of the R domain with other proteins. Information is transduced by mechanisms that alter the phosphorylation state of the R domain of the response regulator and by so doing alter the associated or interacting proteins.

Two-component systems form the core of a wide variety of different cellular signal-transduction systems, and not surprisingly the domains noted above are used in different ways in different systems. In many cases the T domain has another activity; it catalyzes the dephosphorylation of the phosphorylated R domain (Fig. 88.1C). This is true of the NRI–NRII system, where NRII catalyzes the dephosphorylation of NRI~P. In some signal transduction systems, multiple T and R domains, each with distinct functions, are employed. Finally, in almost all systems, additional regulatory proteins and/or protein domains are present. These permit the perception of stimuli, and they may affect the autokinase or phosphatase activities of the T domain in response to stimuli, or in cases where a phosphorelay is used, may affect the flow of phosphoryl groups through the relay or the phosphorylation state of the R domain of the response regulator at the end of the chain. In the *B. subtilis* sporulation system, distinct phosphatases act at each step to regulate the flow of phosphoryl groups to the response regulator Spo0A.

Yet another important feature of the R domain is that in many cases this domain may catalyze its own phosphorylation directly from small molecule phosphorylated metabolic intermediates, such as acetyl phosphate (Fig. 88.1D). Acetyl phosphate seems to be a key intracellular stress messenger; it is formed from acetyl~CoA and may signal a perturbation in one or more of the metabolic pathways using acetyl~CoA. When acetyl phosphate accumulates, the phosphorylation of the R domain of certain response regulators may participate in the adaptation to stress. Since a number of different response regulators are phosphorylated by acetyl phosphate, factors affecting this common source of phosphoryl groups may simultaneously affect many cellular processes. As will be discussed later in the article, the common use of acetyl phosphate may serve as a mechanism for communication between different two-component signal transduction systems.

II. STRUCTURE AND FUNCTION RELATIONSHIPS IN THE TRANSMITTER, RECEIVER, AND PHOSPHOTRANSFER DOMAINS

The distinguishing feature of the two-component signal transduction systems is the presence of the two protein domains corresponding to the T domain and the R domain. In addition, some two-component systems also contain the PT domain. The primary amino acid sequence of these domains is sufficiently conserved to permit the identification of homologous domains from the conceptual translation of DNA sequences.

A. Transmitter domains

The T domain (Fig. 88.2A) is about 250 amino acids in length, and is less well conserved than the R domain. However, within this domain are several short segments that are highly conserved (Fig. 88.2A). Near the N terminus of the domain is a conserved segment known as the H-box, which contains the highly conserved histidine residue that is the site of autophosphorylation (see below). Approximately 100 amino acids away, another highly conserved segment known as the N-box is found, followed by the D-box, F-box, and G-box. Although more information will be required for a definitive functional assignment, some evidence suggests that the D-, F-, and G-boxes are involved in the binding of ATP and catalysis (see below). These segments of the T domain are shared with other classes of proteins that bind ATP, such as serine kinases (Fig. 88.3G), topoisomerases, and chaperones.

Essentially none of the conserved residues depicted in Fig. 88.2A are completely conserved. However, the

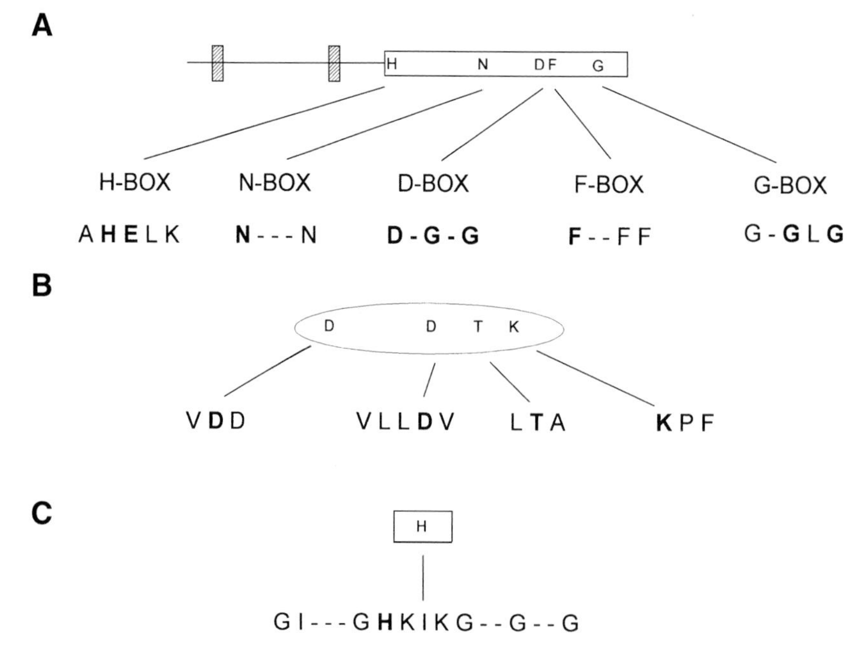

FIGURE 88.2 Conserved motifs within transmitter, receiver, and phosphotransfer domains. Amino acids are depicted using the standard single-letter code; the most highly conserved residues are depicted in bold. (A) The transmitter domain. (B) The receiver domain. (C) The phosphotransfer domain.

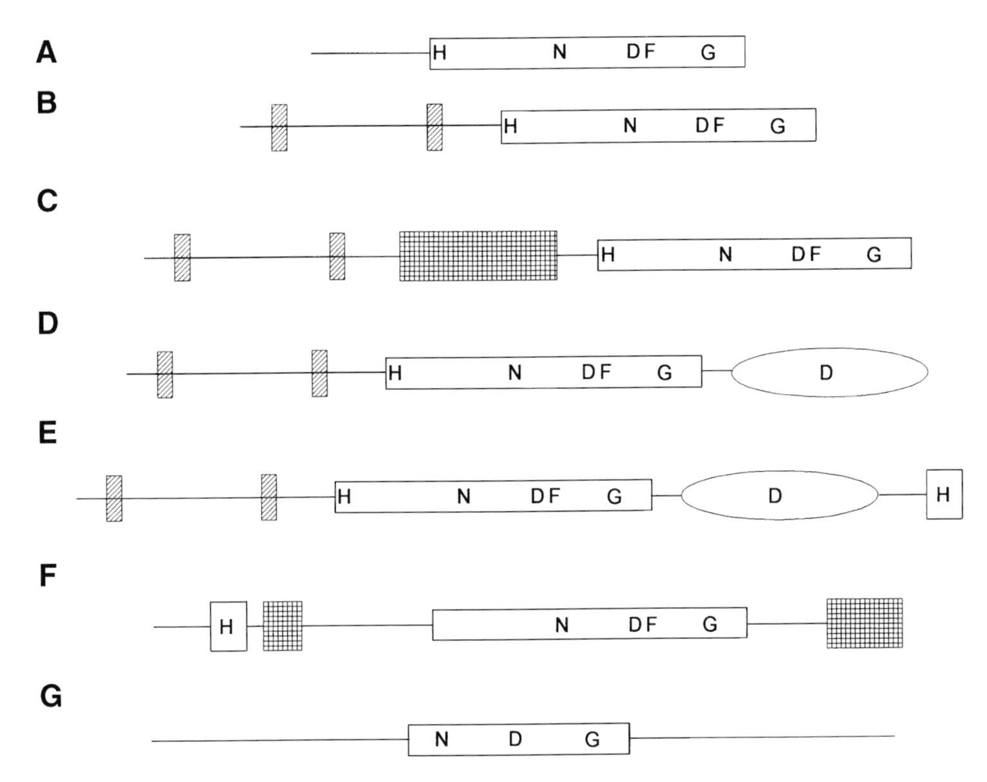

FIGURE 88.3 Association of transmitter domains with other protein domains. Symbols are as in Fig. 88.1, and as below. For C, an additional sensory domain is depicted with checkerboard hatching; for F, domains for the association with the receiver and transmembrane receptors are depicted with checkerboard hatching. For details, see text.

D- and G- boxes are the best conserved parts of the T domain, and as noted above, these features are also found in other classes of proteins that bind ATP. The F-box is the most variable, and is not recognizable in several bona fide T domains. The H-box is invariant with the exception of the CheA and FrzE proteins, which are orthologous. CheA and FrzE are involved in the regulation of bacterial chemotaxis toward attractants and away from repellants. In these two proteins, an H-box is completely absent and is replaced functionally by a PT domain found near the N-terminus of the protein (Fig. 88.3F).

To examine the role of these conserved regions, each of the highly conserved residues found within T domains was mutagenized by site-specific mutagenesis of the *E. coli glnL* gene encoding the paradigmatic transmitter, NRII (NtrB). NRII is involved in the regulation of nitrogen assimilation in response to signals of nitrogen and carbon status. As already noted, NRII is among those transmitter proteins that has both kinase and response regulator phosphatase activity. Alteration of the highly conserved residues affected the ability of NRII to become autophosphorylated, but did not diminish the phosphatase activity of NRII. Indeed, some of the mutations appeared to increase the phosphatase activity of NRII. These results indicate that the conserved sequences are directly involved in the kinase activity. Also, alteration of the conserved H-box histidine in NRII did not diminish the NRII phosphatase activity, indicating that this phosphatase activity is not the reversal of the phosphotransfer reaction by which the R domain becomes phosphorylated.

The site of NRII phosphorylation was directly determined by proteolysis and peptide mapping after labeling the protein with radioactive ^{32}P. This analysis directly confirmed that the H-box histidine is the site of NRII autophosphorylation. Similarly, the site of CheA autophosphorylation was mapped to the conserved histidine within the N-terminal PT domain of this protein.

The role of the G-box residues in NRII was also examined by characterization of altered proteins with conservative replacements in this region. This analysis suggested that the conserved glycine residues are important for the binding of ATP. Whereas the wild-type protein has very high affinity for ATP and can be readily cross-linked to ATP by UV irradiation, the proteins with G-box mutations were defective in autophosphorylation at low ATP concentrations, and could not be readily cross-linked to ATP by UV irradiation.

In addition to the conserved regions discussed so far, several recent studies suggest that some T domains may also share a conserved helical structural motif adjacent to the H-box, known as the X-box. This helical segment forms part of the dimer interface in these proteins. In NRII this region is predicted to lie immediately downstream from the H-box, whereas in several other T domains, this region is predicted to map immediately upstream of the H-box. The H- and X-box regions of transmitter proteins are similar to the region surrounding the site of phosphorylation in the Spo0B PT protein.

The autophosphorylation of the dimeric T domain occurs by a trans-intramolecular mechanism in which ATP bound to one subunit is used to phosphorylate the H-box of the other subunit within the dimer. This was demonstrated conclusively for the NRII T domain. Although this domain forms a stable dimer, the subunits may be reversibly dissociated and reassociated by treatment with low concentrations of urea, followed by dialysis. This has permitted the formation of heterodimers containing subunits with different mutations *in vitro*. Heterodimers containing one subunit lacking the histidine site of autophosphorylation and one subunit defective in the ATP-binding G-box are capable of autophosphorylation, because the intact ATP-binding portion of the H-box mutant can bring about the phosphorylation of the intact H-box site of the G-box mutant subunit.

Several lines of evidence have suggested that the T domain is actually two protein domains, not one. First, as already noted, a number of other ATP-binding proteins have the N-, D-, and G-boxes, but lack the H-box. Second, the CheA kinase has been cleaved into two pieces bearing the PT domain and the N-, D-, F-, and G-boxes. The latter of these peptides is able to bring about the phosphorylation of the disconnected PT domain. Finally, the T domain of the *E. coli* EnvZ protein (depicted in Fig. 88.3B) has been cleaved into two peptides by scission between the H- and N-boxes. The peptide containing the N-, D-, F-, and G-boxes is able to phosphorylate the disconnected H-box. The EnvZ transmitter is involved in the regulation of porin gene expression in response to the osmolarity of the growth medium.

As already noted, some of the T domains catalyze not only the phosphorylation of the H-box histidine, but also the dephosphorylation of the phosphorylated response regulator R domain. Among this class of T domains are those found in the NRII and EnvZ proteins of *E. coli*. Mutations affecting the phosphatase activity of these proteins have been identified. Some of these mutations map in other domains of these proteins, and may affect signal perception (see below). However, some of the mutations mapped to the H-box (but not to the conserved histidine residue) and nearby flanking portions of the T-domain. One interpretation of these findings is that the nonphosphorylated form of the H-box must interact with the phosphorylated receiver domain to bring about its dephosphorylation. This conclusion is consistent with the observation, already noted, that mutations reducing the binding of ATP appear to increase the phosphatase activity of NRII.

How are stimuli perceived in two-component regulatory systems? One mechanism seems to involve control of the autophosphorylation and/or phosphatase activates of the T domain by associated sensory domains. The T domain is always found associated with other unrelated domains in proteins. The most common arrangement is shown in Fig. 88.3A and B; in these proteins the T domain is found at the C-terminal end of the protein, and the N-terminal end is involved in signal perception. In many cases, the N-terminal signal-perception domain contains two transmembrane segments, with the portion of the protein between these forming an extracellular domain. The NarX and NarQ transmitters of *E. coli* control respiratory gene expression in response to the presence of the alternative electron acceptors nitrate and nitrite. In these transmitters, the ligands nitrate and nitrite have been shown to bind to the extracellular sensory domain. This binding appears to reciprocally regulate the autophosphorylation and response regulator phosphatase activities of the transmitter domains. In some other *E. coli* systems, a transmembrane transmitter protein interacts not with small molecules, but with other membrane proteins that are apparently responsible for the detection of small molecule signals.

Another transmitter of interest is the FixL protein, which regulates nitrogen fixation in rhizobia. This protein (Fig. 88.3C) contains an extracellular sensory domain and an additional cytoplasmic sensory domain that is bound to heme and functions in oxygen sensation. Signals from the oxygen-sensing domain appear to reciprocally regulate the kinase and phosphatase activities of the transmitter domain.

In the case of *E. coli* NRII (Fig. 88.3A), the transmitter protein is soluble, and information on the availability of carbon and nitrogen is transmitted by interaction of NRII with another signal-transduction protein called PII. When PII binds to NRII, the kinase activity of NRII is inhibited and the phosphatase activity of NRII is activated. Signals of carbon and nitrogen status regulate the ability of PII to bind to NRII.

The chemotaxis system of enteric bacteria is an example of the integration of a two-component regulatory system with a complex signal-perception mechanism. The CheA transmitter protein binds transmembrane receptor proteins that detect ligands, and this binding requires an adapter protein, CheW. CheA that is not coupled to the receptors has very low autophosphorylation activity, whereas CheA in the complex with receptors and CheW is very active. This activity is greatly reduced on interaction of the receptors with their ligands. Thus, in this system regulation appears to be mainly due to the regulation of the rate of transmitter autophosphorylation.

B. Receiver domains

The receiver domain is about 125 amino acids in length, and the entire domain is generally highly conserved (Fig. 88.2). Within this domain there are four segments that are essentially invariant. Near the N-terminus, one or more acidic residues are found, which play a role in the chelation of Mg^{2+}. The phosphorylated aspartate residue occurs near position 55 in the R domain. A highly conserved threonine residue, which is occasionally replaced by serine, is found near position 87. Finally, a very highly conserved lysine residue is found at approximately position 109.

In some cases, the receiver domain alone comprises the whole protein (Fig. 88.4A). For two such cases, the SpoOF protein controlling sporulation in *Bacillus subtilis* and the CheY protein controlling chemotaxis in enteric gram-negative bacteria, the structure of the domain has been solved by nuclear magnetic resonance (NMR) or X-ray cyrstallographic methods. In the *E. coli* NRI (NtrC) protein, a receiver domain is found at the N terminus of the protein, and the associated domains are involved in transcriptional activation. The structure of the isolated N terminal receiver domain of NRI has been solved by NMR methods. Finally, the NarL transcription factor, like NRI, contains a receiver domain at its N terminus, and a transcriptional activation domain at its C terminus. The complete structure of NarL has been solved by X-ray diffraction methods. In all of these cases, the receiver domain consists of a five-stranded β-sheet surrounded by α-helices. At one end of the receiver domain, the highly conserved acidic residues form an

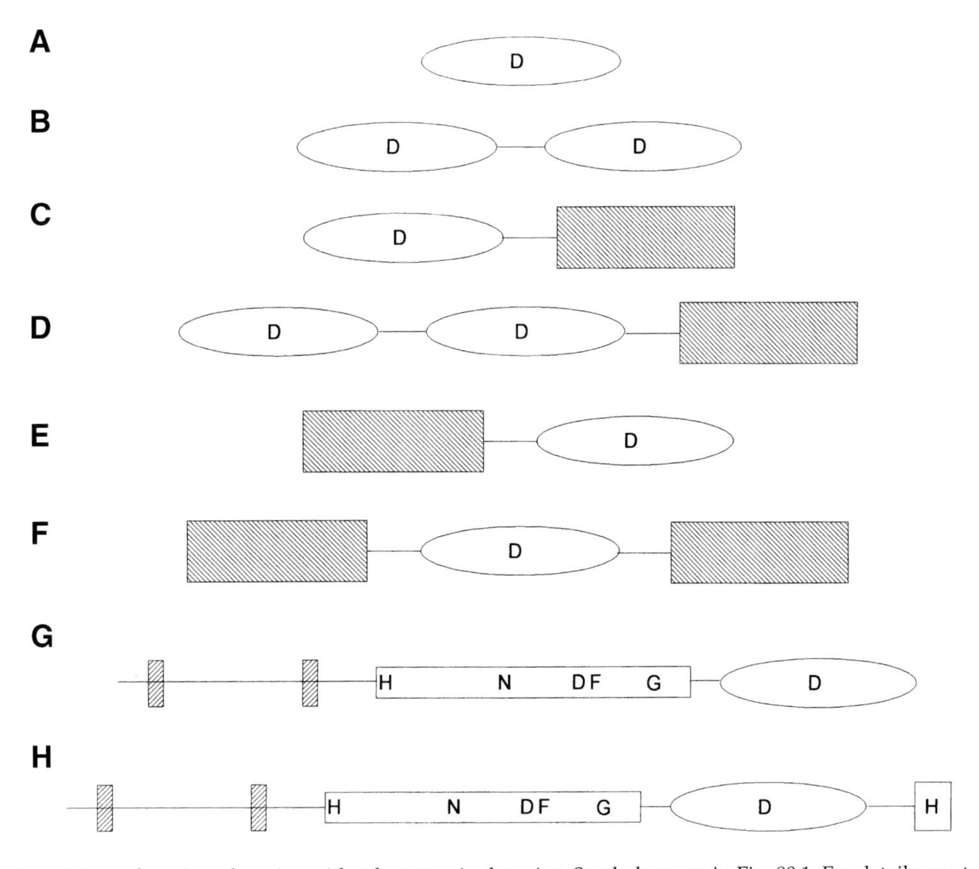

FIGURE 88.4 Association of receiver domains with other protein domains. Symbols are as in Fig. 88.1. For details, see text.

acidic pocket. The aspartate at approximately position 55 is the site of phosphorylation, while the aspartate near position 10 plays a key role in chelation of a Mg^{2+} metal ion necessary for catalysis. Given the high sequence conservation among response regulator domains, they probably all share a similar structure.

The site of phosphorylation of the receiver domains of *E. coli* CheY and NRI were determined by reduction of the acyl phosphate with tritiated borohydride, followed by mapping of the site. These experiments confirmed the identity of the modified residue. The site of phosphorylation of a number of different receiver domains has been altered by site-specific mutagenesis, and either the resulting mutant proteins were not modified or the modification was vastly decreased. Among the latter group of proteins, the residual phosphorylation was observed to occur on a serine residue, or the modified protein had properties suggesting that a serine or threonine was modified. This modification may represent an artifact that is due to the absence of the usual site of phosphorylation. Interestingly, in some cases receiver domains have been mutated so that the aspartate at the site of phosphorylation is converted to glutamate. This mutation results in a protein with properties similar to the phosphorylated form of the receiver, suggesting that the larger glutamate residue at this position may mimic the effect of phosphorylating the natural aspartate residue.

A number of protein domains have been identified that are remarkably similar to the receiver domain but lack the aspartate residue that is phosphorylated in bona fide receivers. One of these proteins, the FlbD transcription factor of *Caulobacter crescentus*, is active when unphosphorylated and was not able to be phosphorylated using several different transmitter domains. Apparently, the pseudoreceiver domain of this protein and its relatives serves some other regulatory function that does not require phosphorylation.

Aspartyl phosphates in proteins are relatively unstable, especially at basic pH. At neutral pH, studies of model compounds have suggested an intrinsic stability of acyl phosphates of several hours in aqueous solutions. A similar stability is observed with most phosphorylated receiver proteins when they are denatured with sodium dodecyl sulfate (SDS). When proteins are not denatured, however, a wide range of stabilities is observed for the receiver domains, ranging from a few seconds to the predicted several hours. Thus, many of the phosphorylated receiver domains are considerably more unstable than predicted. Since denaturation of the phosphorylated proteins results in the expected phosphoryl-group stability, it is clear that in the native structure, the phosphorylated aspartate is rendered unstable by nearby structural features of the protein. One of the factors affecting the stability of the phosphorylated receiver is the conserved lysine found at position 109, since alteration of this residue has been observed in several cases to stabilize the phosphorylated form of the protein. It has been proposed that the receiver domain "catalyzes" its own dephosphorylation, and this activity has been referred to as the "autophosphatase" activity. The relationship between this autophosphatase activity and the phosphatase activities of transmitter domains is unclear at the present time. It has been proposed that the transmitter domains may not actually be phosphatases, but rather may be activators of the autophosphatase activity intrinsic to the receiver domain.

In addition to the receiver domain autophosphatase activity and phosphatase activity that may be associated with a transmitter domain, the phosphorylated receiver domains are frequently subject to dephosphorylation by additional, unrelated phosphatases. The mechanism of action of these phosphatases is currently not well understood. They may act by stimulating the intrinsic autophosphatase activity. Regulation of the appearance or activity of these unrelated phosphatases serves to permit additional stimuli to affect the phosphorylation state of the receiver module.

How does the phosphorylation of the receiver domain bring about signal transduction? The response regulator proteins contain receivers that regulate other protein domains, either in the same or different proteins. The phosphorylation of the receiver domain in these proteins apparently results in a conformational change that affects the target domains that are regulated. A typical response regulator organization is shown in Fig. 88.4C. In these proteins, an N-terminal receiver domain is associated with a C-terminal domain. Various types of C-terminal domains are found, such as transcriptional activation domains and domains with various enzyme activities. In the case of the Spo0A response regulator, which is a transcription factor regulating the initiation of sporulation in *Bacillus subtilis*, the unphosphorylated N-terminal receiver domain inhibits transcriptional activation activity. Phosphorylation of this domain results in an altered conformation in which the receiver no longer is able to inhibit the binding of DNA and the activation of transcription at the regulated promoter elements.

A similar mechanism is observed with the *E. coli* NRI (NtrC) response regulator protein, which is an enhancer-binding transcriptional activator that works in concert with a specialized form of RNA polymerase containing σ^{54} to activate genes in response to nitrogen starvation. Unphosphorylated NRI is a dimer of identical subunits, which is essentially unable to activate transcription. Phosphorylation of the receiver domain of NRI results in the formation of an oligomer, likely a tetramer or octamer of NRI subunits (i.e. a dimer or tetramer of dimers), that is able to activate transcription. Thus, the unphosphorylated receiver domain of NRI appears to act by preventing the oligomerization necessary for the formation of the active form of the protein.

The mechanism of signal transduction by the *E. coli* CheY chemotaxis regulator is apparently somewhat different. This protein is a response regulator that consists entirely of the receiver domain (Fig. 88.4A). CheY interacts with the switch proteins of the flagellar motor to bring about reversals or pauses in the rotation of the flagella. This in turn causes a jittery swimming motion known as "tumbling," which serves to randomize the direction of the bacteria. Phosphorylated CheY, but not unphosphorylated CheY, is able to interact with the motor switch proteins. Thus, activity of the motors is controlled by the extent of CheY phosphorylation. Both unphosphorylated CheY and CheY~P are monomeric, and the distinction made by the motor switch proteins must therefore be due to conformational differences between the two forms of CheY.

C. Phosphotransfer domains

As the phosphotransfer domain was only recently discovered, much less is known about it. The domain is approximately 60 amino acids in length, and it contains a highly conserved histidine and a pattern of conserved glycine residues (Fig. 88.2C). The conserved histidine is apparently the site of phosphorylation, as deduced from the properties of the phosphorylated domain and the result of alteration of the histidine to a nonphosphorylatable residue.

As noted above, the CheA transmitter lacks an H-box and is instead directly phosphorylated within its N-terminal PT domain. This observation indicates that the PT domain may serve as a target for direct phosphorylation by the kinasedomain of transmitters. However, it is clear from experiments with the ArcB transmitterof *E. coli* and several other transmitters that the PT domain usually requires phosphotransfer from an R domain to become phosphorylated.

Interestingly, the first protein observed to have the function of a PT domain is not related by homology to the PT domains found in other systems. This protein was the Spo0B protein of *B. subtilis*, which, as already noted, forms part of the phosphorelay that delivers phosphoryl groups to the receiver domain of the response regulator Spo0A. The sporulation phosphorelay consists of several transmitters (KinA, KinB, perhaps others) that phosphorylate the Spo0F receiver. Spo0F~P then transfers its phosphoryl groups to Spo0B, which in turn transfers its phosphoryl group to Spo0A. Currently, there are no known homologs of Spo0B in the GenBank database. However, since the function of Spo0B has been clearly demonstrated, it raises the possibility that there exist additional proteins with the functions of the PT domain, which are not related to either Spo0B or the PT domain by homology.

Recently, an *E. coli* protein has been identified that is a phosphatase specific for phosphorylated PT domains. This phosphatase, the SixA protein, contains a conserved arginine-histidine-glycine (RHG) motif found in many other phosphatases, and becomes transiently phosphorylated on the RHG histidine motif as part of the phosphatase catalytic mechanism.

III. INCORPORATION OF T, R, AND PT DOMAINS IN SIGNAL TRANSDUCTION SYSTEMS

The arrangements in which the T, R, and PT domains are found in signal transduction systems are summarized in Figs 88.3 and 88.4. As already noted, the T domain is typically found associated with an N-terminal sensory domain (Fig. 88.3A and B) or multiple N-terminal sensory domains (Fig. 88.3C). In numerous cases, T and R domains are found in the same protein, termed hybrid transmitters (Fig. 88.3D). In yet other cases, T, R, and PT domains are contained within a single protein, which we call compound transmitters (Fig. 88.3E). The CheA and FrzE proteins represent a special class in that they lack the H-box and contain instead an N-terminal PT domain (Fig. 88.3F). Finally, as already noted, there are numerous cases of proteins that are not bona fide transmitters but contain the N-, D-, and G-boxes characteristic of the transmitter domain (Fig. 88.3G). This plethora of domain arrangements indicates the modular nature of the T, R, and PT domains and suggests that additional arrangements are likely to be discovered.

Receiver domains may be found unassociated with other protein domains, as in the Spo0F and CheY proteins (Fig. 88.4A), or associated with another receiver domain (Fig. 88.4B). These proteins may serve as response regulators (CheY) or as part of phosphorelay systems. The typical response regulator contains a receiver domain associated with the domains it controls (Fig. 88.4C). Many different types of associated domains have been observed in this arrangement. The PleD protein of *C. crescentus* represents a variation in which two receiver domains are associated with a domain that is controlled (Fig. 88.4D). The receiver domain of response regulators is less often found downstream from associated domains under its control (Fig. 88.4E) or sandwiched between unrelated domains (Fig. 88.4F). Finally, as already noted, receiver domains may be found associated with T domains in hybrid transmitters (Fig. 88.4G) and with T and PT domains in compound transmitters (Fig. 88.4H). These proteins are almost certainly part of phosphorelay systems.

IV. FACTORS AFFECTING THE PHOSPHORYLATION STATE OF THE RESPONSE REGULATOR

A. Flow of phosphoryl groups

The flow of phosphoryl groups to the response regulator is depicted in Fig. 88.5 for some of the arrangements of T, R, and PT domains. In the typical bacterial two-component system, phosphoryl groups are transferred from ATP to the H-box histidine, followed by transfer to the aspartate residue of the R domain, from which they are transferred to water (Fig. 88.5A). In the various phosphorelay systems depicted in Fig. 88.5B–D, the flow of phosphoryl groups is from the H-box histidine to an R domain, then to the PT domain, and finally to the receiver domain of the response regulator. This basic theme seems to occur regardless of whether the various domains are on separate proteins (Fig. 88.5B and D) or are part of a compound transmitter (Fig. 88.5C).

B. Role of acetyl phosphate

As already noted, some receiver domains are efficiently phosphorylated by acetyl phosphate. In one case, the RssB response regulator of *E. coli* that controls the stability of σ^S, phosphorylation by acetyl phosphate seems to be the main mechanism for activating the response regulator (i.e. there is no cognate transmitter, it is a one-component system). In other cases, a transmitter and acetyl phosphate both contribute to response regulator phosphorylation. These systems usually have a transmitter protein with phosphatase activity toward the phosphorylated receiver.

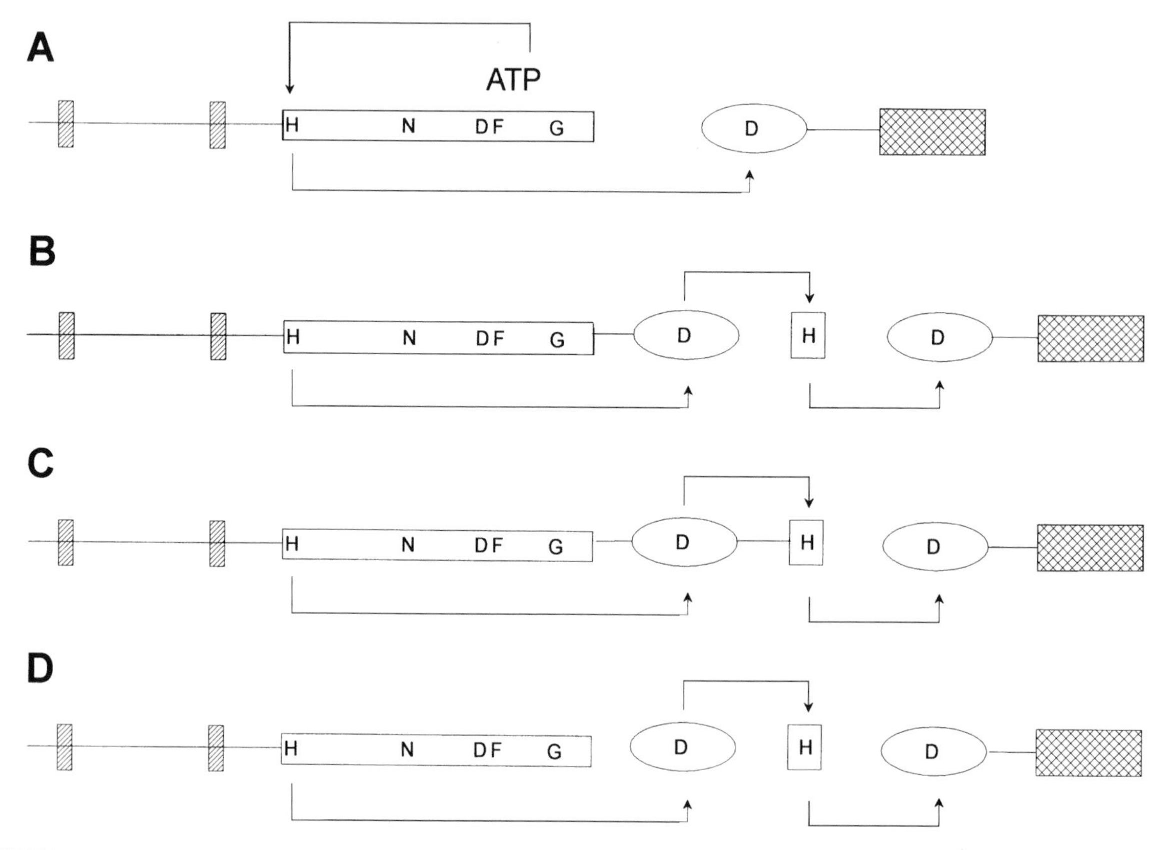

FIGURE 88.5 Flow of phosporyl groups in a typical two-component system and in phosphorelay systems. Symbols are as in Fig. 88.1. (A) Flow of phosphoryl groups in a typical two-component system. (B, C, and D) Flow of phosphoryl groups in phosphorelay systems with differing domain associations. For details, see text.

In vivo, this phosphatase activity is required to prevent the inappropriate activation of the system by acetyl phosphate. Indeed, these phosphatase activities have a role in regulating the intracellular concentration of acetyl phosphate, since the result of their activity is the conversion of acetyl phosphate to acetate and inorganic phosphate (P_i). Since multiple two-component systems are affected by acetyl phosphate, this serves as a means for efficient communication between regulatory systems.

Various roles have been proposed for acetyl phosphate activation of receiver domains. Phosphorylation of these domains by acetyl phosphate may provide a basal extent of phosphorylation that can be quickly increased above the regulatory threshold on activation of the transmitter kinase activity. Alternatively, acetyl phosphate may be required to achieve the full phosphorylation of the receiver domain, even when the transmitter kinase activity is fully stimulated.

C. Cross-regulation

In numerous cases, multiple transmitter domains have been shown to bring about the phosphorylation of a receiver domain. When each receiver domain has an apparent "cognate" transmitter as well as the ability to be phosphorylated by other transmitters, this phenomenon is an example of cross-regulation. In every case where cross-regulation has been shown to occur, only one of the transmitter domains, the "cognate" transmitter, has the ability to bring about the dephosphorylation of the receiver domain. Cross-regulation permits the activation of receivers by stimuli affecting different transmitters, while the phosphatase activity of the cognate transmitter may prevent inappropriate activation of the system. Cross-regulation may permit more rapid phosphorylation of the receiver under certain conditions than would be possible if only a single transmitter could phosphorylate the receiver.

D. Multiple, independently regulated phosphatases

The obvious advantage of signal transduction phosphorelay systems is that each phosphorelay step may serve as a point for control of the system output. This has been shown to be true for the phosphorelay system regulating *B. subtilis* sporulation. The decision to initiate sporulation in this organism is a "consensus" decision affected by numerous stimuli, such as the availability of nutrients and the density of the bacterial culture. A multiplicity of phosphatases have been found, which affect the various steps of the phosphorelay. Each of these phosphatases imposes a requirement for the appropriate stimuli before the initiation of sporulation may proceed.

V. OCCURRENCE OF T, R, AND PT DOMAINS

The two-component systems constitute the most common type of signal transduction system found in bacteria, with over 100 different systems known. In many cases, a given system has been identified in numerous bacteria, such that about 300 examples of bacterial transmitter proteins are currently known. At the time this article was prepared (July 1998) new bacterial transmitter proteins were appearing in the GenBank database at the rate of 15–20 per month, largely because of rapid progress in the sequencing of bacterial genomes. The biological functions of many of the newly discovered systems are not known.

In the course of preparing this article, a search was made of the completed *Escherichia coli* genome sequence for homologs of the paradigmatic transmitter protein, nitrogen regulator II (NRII or NtrB). Twenty-three clearly homologous sequences were identified, as well as a number of proteins that aligned to part of the conserved T domain. Since some transmitter proteins may have diverged from NRII sufficiently to remain unrecognized in the homology search (performed using advanced BLAST at the National Center for Biotechnology Information Web site: *http://www.ncbi.nlm.nih.gov/cgi-bin/BLAST*), this represents the minimal number of transmitter proteins in *E. coli*.

A somewhat higher number of receiver domains are known in bacteria, as might be expected since multiple receiver domains may be functionally linked to a single transmitter protein. An advanced BLAST search of GenBank using the paradigmatic *E. coli* CheY response regulator protein sequence as the query sequence revealed 506 sequence entries displaying high similarity; the vast majority of these were bacterial receiver proteins. A similar search of the *E. coli* genome sequence revealed 32 receiver domains highly homologous to CheY, as well as several proteins that shared homology to parts of CheY. Thus, *E. coli* has 32 receiver domains. At the time this article was prepared, the number of receiver domain sequences in GenBank was growing at the rate of about 20 entries per month.

Until recently, it was thought that the two-component systems were uniquely prokaryotic in distribution; however, it is now clear that this is not the case.

The budding yeast *Saccharomyces cerevisiae* is known to contain a hybrid protein with both T and R domains, Sln1. In addition, *S. cerevisiae* contains a PT protein, Ypd1, and the response regulator SSK1, which work in concert with Sln1 to regulate gene expression in response to medium osmolarity, as well as another protein with a receiver domain, SKN7. During the preparation of this article, it was observed that ORF YLR006c (accession number e245767) from the yeast genome sequence is homologous to the receiver domain. The fission yeast *Schizosaccharomyces pombe* contains the receiver protein Mcs4 and two transmitter proteins, Mak1 and Mak2. Mcs4 is involved in cell cycle control. *Neurospora crassa* contains a hybrid transmitter, Nik1, and a second transmitter protein, Nik2. *Aspergillus nidulans* also contains a Nik1 protein similar to the *N. crassa* Nik1 protein. Finally, *Candida albicans* has been shown to contain orthologs of the Sln1 and Nik1 proteins. Thus, the yeasts contain a limited number of two-component systems. *Dictyostelium discoideum* contains at least three transmitter proteins, two of which are hybrid transmitters with associated R domains. In addition, *D. discoideum* contains a response regulator protein, RegA, which acts in concert with the DhkA transmitter protein. The *D. discoideum* two-component systems are involved in various aspects of sporulation and germination control.

The higher plant *Arabidopsis thaliana* contains three transmitter proteins, ETR1, ERS, and CKI1; of these the ETR1 and CKI1 proteins, regulating responses to ethylene and cytokines, respectively, are hybrids containing R domains as well. Tomato contains an ortholog of the ETR1 protein, NR (neverripe). In addition, *A. thaliana* contains five additional proteins with receiver domains and four PT proteins, as deduced by PCR amplification using primers to the conserved regions of these domains. Thus, two-component systems are clearly present in plants.

In addition to the proteins listed above, a large number of proteins are present in GenBank that are clearly related to the transmitter proteins, yet lack the conserved H-box motif containing the phosphorylated histidine residue of bona fide transmitter proteins. These proteins are probably not histidine kinases, and in two cases, *B. subtilis* anti-sigma factor SpoIIAB protein and mitochondrial branched chain keto acid dehydrogenase kinase, the proteins have been shown to be serine kinases. The phytochromes of higher plants are also likely to be in this category. The ATP-binding site of the bona fide transmitter proteins is conserved in these proteins, as well as in a number of other proteins such as the HS90 family of chaperone proteins and type II DNA topoisomerases. Included among the latter group are many putative proteins found in human sequences from the high throughput genome sequencing projects, as well as the human disease gene associated with hereditary nonpolyposis colon cancer (accession U07418). During the course of preparing this article, many such putative proteins were identified within the human genome (e.g. accession numbers AF029367, AC002353, AC003035). However, bona fide transmitter proteins, with an intact H-box, seem to be lacking in the human sequences available so far (as of July 1998). A simple BLAST search of human genomic and expressed sequences using the NRII sequence as query readily identifies many such sequences, but further analysis revealed that they are the result of bacterial sequence contamination of the human sequence libraries. Similarly, the rice expressed sequence library within GenBank is extensively contaminated with bacterial sequences, and therefore is essentially useless for the identification of bona fide transmitter proteins in this organism.

BIBLIOGRAPHY

Aizawa, S. I., Harwood, C. S., and Kadner, R. J. (2000). Signaling components in bacterial locomotion and sensory reception. *J. Bacteriol.* **182**, 1459–1471.

Bourret, R. B., Borkavitch, K. A., and Simon, M. I. (1991). Signal transduction pathways involving protein phosphorylation in prokaryotes. *Annu. Rev. Biochem.* **60**, 401–441.

Hoch, J. A., and Silhavy, T. J. (eds.) (1995). "Two-Component Signal Transduction." American Society for Microbiology Press, Washington, DC.

Loomis, W. F., Shaulsky, G., and Wang, N. (1997). Histidine kinases in signal transduction pathways of eukaryotes. *J. Cell Sci.* **110**, 1141–1145.

Ninfa, A. J. (1996). Regulation of gene expression by extracellular stimuli. *In "Escherichia coli* and *Salmonella typhimurium*: Cellular and Molecular Biology." (F. C. Neidhardt, ed.), 2nd edn., Chap. 39. American Society for Microbiology Press, Washington, DC.

Parkinson, J. S. (1993). Signal transduction schemes of bacteria. *Cell* **73**, 857–871.

WEBSITE

Comprehensive Microbial Resource of The Institute for Genomic Research (TIGR) and links to many other microbial genomic sites
http://www.tigr.org/tigr-scripts/CMR2/CMRHomePage.spl

89

Vaccines, Bacterial

Susan K. Hoiseth
Wyeth Vaccines Research

GLOSSARY

antibody A serum protein produced in response to an antigen and which has the property of combining specifically with that antigen. Different classes of antibody (lgG, lgM, lgA, etc.) have different functional properties and may be opsonic, bactericidal, or neutralizing, or interfere with the attachment of pathogens to the host cell.

antigen A molecule which is able to react specifically with antibody.

capsule A loose, gel-like structure on the surface of some bacteria. Capsules are usually, but not always, polysaccharide in nature.

cell-mediated immune response A specific immune response mediated by T cells and activated macrophages rather than by antibody.

endotoxin The lipopolysaccharide component of the outer membrane of gram-negative bacteria. The lipid A portion triggers a host inflammatory response via tumor necrosis factor and interleukin-1.

T-independent antigen An antigen capable of stimulating B cells to produce antibody directly without requiring T cell help.

toxin (exotoxin) A protein, present in cell-free extracts of pathogenic bacteria, capable of causing a toxic effect on the host.

toxoid A toxin which has been treated so as to inactivate its toxicity but which is still capable of inducing immunity to the active toxin.

Vaccination is the practice of using modified (killed or attenuated) microorganisms, or portions thereof, to induce immunity to a particular disease without actually causing the disease. This may be through preventing infection or by limiting the effects of infection.

I. BRIEF HISTORY

Jenner's work using live cowpox virus to vaccinate humans against smallpox dates back to 1796. The practice of variolation (the introduction of dried pus from smallpox pustules into healthy individuals) dates back to even earlier times. Variolation is known to have been practiced during the sixteenth century in India and China and may have originated in central Asia as early as the tenth or eleventh centuries. Although often effective, variolation was quite risky and resulted in death in 2–3% of those receiving the treatment. It is Jenner, however, who is generally credited with the first scientific study and the development of a true "vaccine" safe enough for widespread use. Jenner's vaccine utilized live cowpox virus, which caused only a mild disease in humans but which was sufficiently related to the human virus to confer cross-protection against smallpox. The first bacterial vaccines for human use were not developed until 100 years after Jenner's first viral vaccine. These were killed, whole cell vaccines for typhoid, cholera, and plague, developed in the late 1890s.

The Desk Encyclopedia of Microbiology
ISBN: 0-12-621361-5

TABLE 89.1 Effect of vaccination on bacterial diseases for which universal infant immunization is recommended (United States)[a]

Disease	Baseline (prevaccine) annual cases	1998 Provisional cases	% Decrease
Diphtheria	175,885[b]	1	100[c]
Pertussis	147,271[d]	6279	95.7
Tetanus	1314[e]	34	97.4
Haemophilus influenzae type b	20,000[f]	54[g]	99.7

[a]Data from the Centers for Disease Control (1999).
[b]Average annual number of cases, 1920–1922.
[c]Rounded to nearest tenth.
[d]Average annual number of cases reported during 1922–1925.
[e]Estimated number of cases based on reported number of deaths during 1922–1926 assuming a case fatality rate of 90%.
[f]Estimated number of cases from population-based surveillance studies before vaccine licensure in 1985.
[g]In children younger than 5 years of age. Excludes 71 cases of *Haemophilus influenzae* disease of unknown serotype.

TABLE 89.2 Bacterial vaccines licensed for use in the United States[a]

Vaccine	Reference for ACIP recommendation[b]
Diphtheria, tetanus, pertussis (DTP)	*MMWR* **40** (RR-10), 1991
Acellular pertussis (DTaP)	*MMWR* **46** (RR-7), 1997
Haemophilus influenzae type b	*MMWR* **40** (RR-1), 1991
Pneumococcal polysaccharide	*MMWR* **46** (RR-8), 1997
Preumococcal conjugate	*MMWR* **49** (RR-9), 2000
Meningococcal	*MMWR* **49** (RR-7), 2000
Typhoid	*MMWR* **43** (RR-14), 1994
Cholera	*MMWR* **37** (40), 1988
Anthrax	*MMWR* **49** (RR-15), 2000
Plague	*MMWR* **45** (RR-14), 1996
Tuberculosis (BCG)	*MMWR* **45** (RR-4), 1996
Lyme	*MMWR* **48** (RR-7), 1999[c]

[a]Licensed for use by the Food and Drug Administration, Center for Biologics Evaluation and Research.
[b]Recommendations for vaccine use are provided by the Advisory Committee on Immunization Practices, administered through the US Centers for Disease Control (CDC). They are published as Recommendations and Reports (RR) in *Morbidity and Mortality Weekly Report* (MMWR) and are available on-line at *www.cdc.gov/mmwr*. Other sources for vaccine recommendations include those provided by the Committee on Infectious Diseases of the American Academy of Pediatrics (published in their *Red Book*, with periodic updates published in the journal *Pediatrics*) and *The Guide for Adult Immunization* published by the American College of Physicians.
[c]This product was withdrawn from the market in February 2002.

Vaccines are one of the greatest achievements of biomedical science and public health. Since the introduction of wide-spread vaccination for many bacterial diseases, there has been a dramatic decline in the morbidity and mortality associated with these diseases (Table 89.1). A good example is diphtheria vaccine, which became available in 1923. In 1920, there were 147,991 cases of diphtheria reported in the United States, with 13,170 deaths; in 1998 there was only a single case reported. This decline is not seen in countries in which diphtheria vaccination rates are low. A striking example occurred in the newly independent states of the former Soviet Union following the breakup of the union. Vaccine coverage decreased, and the incidence increased dramatically starting in 1990. The incidence peaked in 1995, after which renewed efforts in vaccinating helped lower the incidence.

II. BACTERIAL PATHOGENESIS: TOXINS, CAPSULES, AND OTHER VIRULENCE FACTORS

Although the early killed whole cell typhoid, cholera, and plague vaccines were moderately effective, later vaccines relied on an understanding of some of the mechanisms by which various bacteria cause disease and the factors they elaborate in order to circumvent host defense mechanisms. The recognition of tetanus and diphtheria as toxin-mediated diseases led to the production of inactivated toxins (toxoids) as vaccines for these two diseases. Similarly, the recognition that bacterial capsules impart antiphagocytic properties to bacteria, and that antibodies to the capsule are often protective, led to the development of purified capsular polysaccharide or polysaccharide–protein conjugate vaccines against pneumococcal, meningococcal, and *Haemophilus influenzae* infections (see Sections III. B–D).

Diseases caused by bacteria that are either facultative or obligate intracellular pathogens have been more difficult to vaccinate against. These organisms will require identification of novel antigens and/or vaccine delivery methods. One approach currently being tested is the use of DNA vaccines. This approach involves injecting nonreplicating plasmid DNA directly into an animal or human. The antigen of interest is expressed under the control of a mammalian promoter and is synthesized in the tissues of the vaccine recipient (usually muscle). DNA vaccines generally elicit a strong cell-mediated immune response, which is thought to be necessary to protect against intracellular pathogens. Other approaches currently being investigated include the use of novel adjuvants (immune-enhancing agents) and/or use of oral or intranasal delivery of vaccines for mucosal pathogens. These approaches are still mainly experimental (although oral administration is used for some existing vaccines). The reader is referred to Levine *et al.* (1997) and Plotkin and Orenstein (1999) for further discussion of new vaccines that are still being tested in animal models or which are in preliminary stages of testing in humans. The remainder of this article will focus on bacterial vaccines currently licensed in the United States for use in humans. These vaccines are listed in Table 89.2, along with references for the current US Advisory Committee for Immunization Practices (ACIP) recommendations for use of each vaccine.

III. LICENSED VACCINES

A. DTP/DTaP (diphtheria, tetanus, and pertussis/DT plus acellular pertussis)

1. *Tetanus*

Tetanus is an acute, often fatal disease caused by release of a potent neurotoxin from the anaerobic bacterium, *Clostridium tetani.* The organism is acquired through environmental exposure, often contamination of a wound with soil. It is also found in animal feces. Neonatal tetanus, due to infection of the umbilical cord stump, is an enormous problem in developing countries, with an estimated 1.2 million deaths per year worldwide.

Vaccines for tetanus followed from von Behring and Kitasatos's purification of the toxin in 1890 and their finding that injection of minute amounts of the toxin into animals generated antibodies in the survivors that neutralized the toxin. This work led to the production of antitoxin in horses that was then used to treat humans. Active immunity in humans was not produced until the 1920s, when it was demonstrated that the toxin could be inactivated by treatment with formaldehyde but was still capable of generating neutralizing antibodies. The resulting tetanus "toxoids" became commercially available in the United States in 1938, but they were not widely used until the military began routine prophylactic use in 1941. Tetanus vaccination has been part of routine infant immunization in the United States since 1944 and has been highly successful in reducing the morbidity and mortality associated with the disease (Table 89.1). Tetanus toxin is one of the more potent toxins known, and the disease has a high fatality rate (up to 90% without treatment; ~20% with antibiotics and good intensive care). By the time the patient is symptomatic, a significant amount of toxin has already been released. Hence, pre-existing immunity is crucial. Current U.S. guidelines call for routine immunization of infants at 2, 4, 6, and 15–18 months of age, followed by a booster dose prior to school entry. An additional dose is given at 11 or 12 years of age and then every 10 years thereafter. The first five doses are usually combined with diphtheria and pertussis vaccines. There are additional guidelines for wound prophylaxis, and for dirty wounds tetanus immune globulin may also be given.

2. *Diphtheria*

Diphtheria is a respiratory infection caused by the gram-positive bacterium, *Corynebacterium diphtheriae.* Like tetanus, diphtheria is a toxin-mediated disease. The hallmark of the disease is a membranous inflammation of the upper respiratory tract, with widespread damage to major organ systems from the toxin. The toxin is a protein synthesis inhibitor and causes damage mainly to the cardiovascular and nervous systems and sometimes the kidneys. The bacterium generally remains localized in the respiratory tract, and deep tissue invasion and bacteremia are extremely rare. The major manifestations of disease are due to systemic spread of the toxin.

Diphtheria antitoxin was first given in 1891 and was produced commercially in Germany in 1892. A combination of toxin/antitoxin was used in the United States beginning in 1914. By 1923, formalin-inactivated toxoids became available, and in 1948 diphtheria toxoid was combined with tetanus toxoid and pertussis as DTP. Current US guidelines call for doses at 2, 4, 6, and 15–18 months of age, followed by a booster dose prior to school entry. The amount of diphtheria toxoid given to older children and adults is usually less than that given to infants in order to prevent local injection site reactogenicity in people who may already have significant antibody titers. This reduced dose is often given combined with a tetanus booster and is referred to as Td, with the lower case "d" indicating the reduced amount of diphtheria toxoid. As indicated in Table I, the use of diphtheria toxoid has been highly effective in controlling disease in the United States.

3. *Pertussis*

Pertussis (whooping cough) is caused by the gram-negative bacterium, *Bordetella pertussis.* The hallmark of the disease is a protracted cough, often lasting many weeks. Pertussis is spread by the respiratory route, and the initial symptoms are indistinguishable from those of other upper respiratory infections. The disease is most severe in infants, in whom intense coughing leads to a forced inspiratory "whoop." Major complications include bronchopneumonia, encephalopathy, and prolonged vomiting (eating may trigger coughing episodes, followed by vomiting). Prior to the introduction of a pertussis vaccine, an average of 147,271 cases per year were reported in the United States (Table 89.1), with 5000–10,000 deaths. Epidemic peaks of disease occur at 2- to 5-year intervals, and as many as 270,000 cases were reported in a single peak year. By the late 1930s, there was evidence from controlled field trials that several whole cell pertussis vaccines provided significant protection against disease. Following introduction of widespread vaccination in the 1940s, the incidence of pertussis in the United States decreased to a low of 1010 cases in 1976. The incidence has risen slightly since 1976, with 7796 cases reported in 1996 and 6279 in 1998. Some of this increase reflects increased disease in older children and adults, perhaps due to waning immunity. The number of cases remained at about 6000–7000 cases per year through the end of the 1990s and into 2000.

The original pertussis vaccines were inactivated whole cell preparations (typically heat, thimerosal, or formalin inactivated). Like other gram-negative whole cell vaccines, they contained considerable amounts of endotoxin (lipopolysaccharide from the gram-negative outer membrane) and were commonly associated with local injection site pain and erythema, fever, drowsiness, and fretfulness. Other rare but more serious events (seizures and encephalopathy) have also been reported in temporal association with whole cell pertussis vaccines. Cause and effect of these rare events has been difficult to prove, however, because these vaccines are given to infants at an age when previously unrecognized underlying neurological and developmental disorders first become manifest. Because of concerns regarding the use of whole cell pertussis vaccines, however, several countries (Japan, Sweden, England, and Wales) curtailed or suspended pertussis vaccination, and epidemic pertussis recurred. They subsequently resumed routine childhood vaccination with either whole cell vaccine or acellular vaccines. Many studies have shown that if there was a relationship between whole cell pertussis vaccine and neurological problems, it was sufficiently rare that the benefits of vaccination far exceeded the risks. In the absence of vaccination, the risk of death or encephalopathy from disease was much greater than the risks (real or perceived) from vaccination. This issue should be largely put aside because new, more purified (and less pyrogenic) acellular pertussis vaccines have recently been licensed.

Although the correlates of protective immunity for pertussis are still not fully defined, several acellular pertussis vaccines have been tested in clinical efficacy studies and have been shown to be effective. Acellular vaccines were first licensed in Japan in 1981 and were tested extensively in large efficacy trials in Sweden, Italy, Germany, and Senegal in the late 1980s and early 1990s. They were licensed in the United States in 1991 for use as the fourth and fifth doses of the DTP series and for use as the primary series in infants in 1996. These vaccines contain inactivated pertussis toxin, either singly or with one or more of the following: filamentous hemagglutinin (FHA), pertactin (a 69-kDa nonfimbrial, outer membrane agglutinogen), and fimbrial agglutinogens (attachment factors that allow binding of the organism to ciliated epithelial cells of the upper respiratory tract). Pertussis toxin (previously known as lymphocytosis-promoting factor, histamine-sensitizing factor, or islet cell-activating factor) plays a major role in the systemic manifestations of disease and contributes to protective immunity. However, other toxins (adenylate cyclase, tracheal cytotoxin, and heat-labile toxin) and colonization factors may also contribute to pathogenesis and to protective immunity (based on animal models). The correlation of antibody titers to the various acellular vaccine components and clinical efficacy has not been clearcut. However, several different acellular vaccines gave overall efficacies approximately comparable to the whole cell vaccines, and local reactions, fever, and other systemic events occur significantly less often with acellular vaccines than with whole cell vaccines. Acellular vaccines, in combination with diphtheria and tetanus toxoids (DTaP), are currently recommended by the ACIP for all five doses.

B. *Haemophilus influenzae*

Haemophilus influenzae type b (Hib) causes serious invasive diseases (meningitis, septicemia, cellulitis, and epiglottitis) primarily in young children. Before the availability of vaccines, Hib was the most common cause of bacterial meningitis in children younger than 5 years of age. Other capsular serotypes of *H. influenzae* (a, c, d, e, and f) only rarely cause disease.

The first-generation Hib vaccine, licensed in 1985, consisted of purified type b capsular polysaccharide. The capsule is a major virulence determinant for the organism, allowing it to avoid phagocytosis. Antibodies directed to the capsule promote phagocytosis and clearance of the organism. The first-generation polysaccharide vaccine was licensed only for use in children 18 months of age and older. Infants posed a greater challenge and required new technologies for development of a successful vaccine. This is due to the fact that the immature infant immune system does not respond well to T-independent antigens. Whereas infants respond to T-dependent protein antigens such as diphtheria and tetanus toxoids, they do not respond well to T-independent polysaccharide antigens (they lack the subset of B cells capable of responding directly to antigens without T cell help). This problem was overcome by chemically conjugating the polysaccharide to a protein carrier and converting the response to a T-dependent response. The first Hib conjugate vaccine was licensed for use in infants in 1990. Prior to introduction of Hib vaccines, there were an estimated 20,000 cases of invasive Hib disease per year in the United States, with approximately 1000 deaths and considerable long-term sequelae in survivors (hearing loss, learning disabilities, or mental retardation). After less than 10 years of routine infant immunization, the number of cases in children younger than 5 years of age decreased to 54 in 1998 (excluding 71 cases of unknown serotype). The reduction in disease incidence has exceeded that predicted based on the number of completely immunized infants and suggests an element of "herd immunity" (through reduction in transmission of the organism from vaccinated individuals to unvaccinated infants).

The type b vaccine will not protect against unencapsulated ("nontypeable") *H. influenzae*. These latter strains are a frequent cause of otitis media but rarely cause systemic disease. Efforts are under way to develop protein-based vaccines to protect against otitis media caused by nontypeable strains.

C. Pneumococcal vaccine

Infections with *Streptococcus pneumoniae* (pneumococci) are most common in the very young (<2 years of age) and in the elderly (>65 years of age) and in individuals with other underlying medical conditions. Predisposing conditions include immunocompromized individuals, those with functional or anatomic asplenia, chronic cardiovascular diseases (congestive heart failure or cardiomyopathy), chronic pulmonary diseases (emphysema and chronic obstructive pulmonary disease), and chronic liver disease. The organism colonizes the upper respiratory tract, and many people carry the organism without developing disease. Damage to respiratory mucosa by a viral respiratory infection, cigarette smoke, air pollutants, opiates, or aging may predispose to infection. Disease can occur with the following manifestations: (i) disseminated infection (bacteremia or meningitis); (ii) pneumonia or other lower respiratory tract infection; or (iii) upper respiratory tract infection, including otitis media and sinusitis. It is estimated that each year in the United States pneumococcal disease accounts for 3000 cases of meningitis, 50,000 cases of bacteremia, 500,000 cases of pneumonia, and 7 million cases of otitis media (*Morbid. Mortal. Weekly Rep.* 46, RR-8, 1997).

More than 90 immunologically distinct serotypes of pneumococci have been described, varying in structure of the capsular polysaccharide. The capsule is a major virulence determinant for the pneumococcus, and antibodies to the capsule are protective. Protection, however, is type specific, necessitating the inclusion of multiple serotypes in the vaccine. Fortunately, some types are more frequent causes of disease than others, and vaccines are designed to include the most common types. Hexavalent pneumococcal polysaccharide vaccines were available for use in the United States starting in 1946 but were withdrawn after only a few years because of the apparent success of treating pneumococcal disease with penicillin. It was not until the 1960s that it became apparent that there was still considerable morbidity and mortality from pneumococcal disease, despite antimicrobial therapy. This is perhaps more true today, with the emergence of penicillin-resistant pneumococci. Extensive field trials were conducted in South African gold miners in the 1970s, and a 14-valent capsular polysaccharide vaccine was licensed in 1977, followed by a 23-valent vaccine in 1983. As was the case for *H. influenzae*, these polysaccharide vaccines do not work well in infants. However, unlike *Haemophilus*, for which the vast majority of disease occurs in infants and young children, pneumococcal disease is also a significant problem in the elderly. Pneumococcal polysaccharide vaccine is recommended for those aged 65 or older as well as individuals 2 years of age or older who have certain chronic diseases, asplenia, or who are immunocompromised. The vaccine is currently underutilized in the elderly population, with only approximately 30% of those aged 65 or older having received the vaccine. The ACIP issued a revised recommendation in 1997 (*Morbid. Mortal. Weekly Rep.* 46, RR-08) that the vaccine be used more extensively. The serotypes in 23-valent vaccine cover at least 85–90% of the serotypes causing invasive disease in the United States.

Pneumococcal polysaccharide conjugate vaccines have recently been developed and a 7-valent vaccine was licensed for use in infants in 2000. A large-scale efficacy trial of this 7-valent conjugate vaccine in infants showed 97% efficacy against invasive pneumococcal disease caused by vaccine serotypes. Because the conjugates are considerably more complex to make than the polysaccharide vaccine, the conjugate vaccines will probably not contain all 23 serotypes currently available in the polysaccharide vaccine. Fortunately, the seven most common serotypes account for ~80–90% of invasive disease in young children in the United States and Canada, so even a 7-valent vaccine is having a significant impact on invasive disease. Serotype prevalence varies in different parts of the world, and higher-valent conjugate vaccines are being developed for global use.

D. Meningococcal vaccines

Neisseria meningitidis causes both endemic and epidemic disease, mainly in the form of meningitis or meningococcemia (fulminant sepsis). In the United States, the disease is mainly endemic, with small outbreaks occurring in localized settings. Epidemic disease occurs mainly in the developing world, especially sub-Saharan Africa and Asia. There are approximately 2600 cases of meningococcal disease per year in the United States with a case fatality rate of 12%. These fatalities occur despite appropriate antimicrobial therapy. Particularly for menigococcemia, disease progresses very rapidly, and 60% of patients have experienced symptoms for less than 24 h prior to hospital admission; death may occur within hours. Death from meningococcemia is due to disseminated intravascular coagulation and hypovolemic shock (gram-negative endotoxic shock); purpura is frequently present, and if the patient survives, skin graft or amputation may be required. Neurological complications may occur in up to 20% of survivors of meningococcemia or meningitis. Although the incidence of meningococcal disease is not as high as that of pneumococcal disease (or *Haemophilus* before introduction of Hib conjugate vaccines), the severity of the disease and the rapidity with which it strikes previously healthy individuals make it a target for vaccination.

Like pneumococcal and *Haemophilus* infections, the ability of the meningococcus to survive in the bloodstream is related to the presence of a polysaccharide capsule, and antibodies to the capsule are protective. Although there are at least 13 different capsular serogroups of meningococci, the majority of disease is caused by groups A, B, C, Y and W-135. In the United States, most cases are B, C, or Y; the number of cases of Y has increased during the past several years. Epidemics in the developing world are mainly serogroup A, especially in the "meningitis belt" of sub-Saharan Africa.

Previously, military recruits experienced high rates of meningococcal disease. Since the introduction of routine vaccination with an A/C polysaccharide vaccine in 1971 (and later a quadrivalent A, C, Y, and W-135 vaccine), rates in recruits have decreased substantially. Routine vaccination of civilians is not recommended, however, because, like other polysaccharide vaccines, meningococcal polysaccharide vaccine is relatively ineffective in children younger than 2 years of age (among whom risk for disease is highest). The A and C vaccines have estimated efficacies of 85–100% in older children and adults and have been useful for controlling serogroup C outbreaks (provided a large segment of population at risk is vaccinated). In Quebec, Canada, 1.6 million doses of polysaccharide vaccine were administered to 6-month-olds through 20-year-olds during a serogroup C outbreak in 1993. The overall efficacy was estimated to be 79%, with greater efficacies in teenagers and lower efficacies in children younger than 5 years of age. In addition to use in outbreak settings, the A, C, Y, and W-135 polysaccharide vaccine licensed for use in the United States is recommended for use in people with terminal complement component deficiencies, people with asplenia, and for travelers to parts of the world where there is epidemic disease. Several conjugate vaccines have recently been developed, and these result in better efficacies in the younger age groups. Although not yet available in the United States, serogroup C conjugates have been licensed in the United Kingdom and a number of other countries where they had previously experienced a high incidence of serogroup C disease.

There is no serogroup B vaccine currently available in the United States. The serogroup B capsule contains polysialic acid and is poorly immunogenic in humans. This is likely related to the fact that human tissues also contain sialic acid determinants, and vaccine-induced cross-reactivity with human tissue may be undesirable from a safety standpoint. Efforts to develop a serogroup B vaccine have therefore focused mainly on surface proteins. This has presented a challenge since there is considerable strain to strain variability in the major surface proteins. A protein-based vaccine is available in Cuba and some Latin American countries, but it is specific for the predominant strain in Cuba. This vaccine can provide protection against epidemics caused by strains homologous to the vaccine but not against endemic disease caused by strains of heterologous subtypes. This vaccine would not be expected to have much of an impact in countries such as the United States or Canada, where endemic disease is caused by a diverse number of strains of differing surface protein subtypes. Efforts are ongoing to identify conserved, protective surface proteins and/or to combine several of the less conserved proteins into a multivalent vaccine.

E. Special-use vaccines: typhoid, cholera, plague, and anthrax

These are all diseases of very low incidence in the United States and for which vaccination is indicated for only limited numbers of individuals (travelers to parts of the world where disease is a problem, military personnel, etc.).

1. *Typhoid*

Typhoid fever is an enteric fever caused by *Salmonella typhi*. It is an acute, systemic illness resulting from ingestion of food or water contaminated with *S. typhi*. The organism causes a severe, generalized infection of the reticuloendothelial system, intestinal lymphoid tissue, and gallbladder and results in high fever, headache, malaise, and abdominal discomfort. Without treatment, typhoid fever has a case fatality rate of 10–20%; with appropriate antibiotic treatment, fatality rates decrease to <1%. Antibiotic resistance, however, is a continuing problem.

Prior to the introduction of water treatment at the beginning of the twentieth century, typhoid fever was a major problem in large cites in Europe and the United States. Once water filtration and chlorination were introduced, typhoid rates plummeted. There are currently only a few hundred cases per year reported in the United States, and the majority of these occur in people who have traveled to parts of the world where typhoid is still endemic or who are immigrants from these countries. Most other cases can be traced to contaminated food, handled by food service workers who can be classified into one of the two previous categories. Worldwide, typhoid remains an enormous problem, with an estimated 33 million cases and 500,000 deaths per year. These cases occur largely in countries that lack proper sewer and water systems.

There have been a variety of typhoid vaccines available since the late 1890s, and they have been evaluated in numerous efficacy trials. These were mainly heat-killed, phenol-preserved whole cell vaccines, or acetone-killed and dried whole cell vaccines. Efficacies are in the 70% range, with some higher and some lower. Efficacy is dependent on the number of organisms ingested, and this will vary from country to country depending on the level of contamination of the water supply, etc. The killed whole cell vaccines, although providing considerable protection, do have unpleasant side effects. These are due mainly to the endotoxin component (lipopolysaccharide from the gram-negative outer membrane). Side effects include fever, malaise, local injection site erythema, induration, and pain; some vaccine recipients experience general disability for 1 or 2 days. Considerable efforts have gone into developing less reactogenic typhoid vaccines, and two new vaccines have been licensed in the United States during the past 10 years. The live attenuated strain, Ty21a, was generated in the 1970s by chemical mutagenesis. It has a mutation in the *galE* (galactose epimerase) gene, and in the absence of exogenous galactose it is unable to synthesize full-length lipopolysaccharide (LPS); it is also Vi [virulence antigen (capsule)] negative. For many years it was believed that the *galE* and *via* (Vi) mutations were the major attenuating lesions in Ty21a. However, recent work has shown that a strain with defined deletions in *galE* and *via* is still capable of causing typhoid in humans. The precise mechanism of attenuation in Ty21a is still not understood, although a mutation in *rpoS* (stress-induced sigma factor) may also contribute. Despite the undefined nature of attenuation, Ty21a has been found to be extremely safe. Ty21a is given orally and was tested extensively in efficacy studies in Alexandria, Egypt, in the late 1970s and in Chile and Indonesia during the 1980s. It gave efficacies of 96% in Egypt with a liquid formulation, but gave lower efficacies for several different formulations in Chile and Indonesia. The liquid formulation used in Egypt was not suitable for mass production, and a variety of enteric (acid-resistant) and gelatin capsule formulations were tested in Chile along with a sachet formulation containing lyophilized organisms and buffer to be reconstituted with water just prior to drinking the vaccine. The enteric capsules were superior to the gelatin capsules plus bicarbonate and gave an efficacy of 67%. The sachet formulation provided 78% protection. Efficacy in Indonesia was 42% for three doses of enteric capsule and 53% for the sachet formulation. A four-dose regimen of the enteric capsule formulation was licensed for use in the United States in late 1989. Efficacy in persons from nonendemic areas who travel to endemic areas, however, has not been tested.

An alternative approach is the Vi capsule. Purified Vi capsular polysaccharide vaccines were tested in field trials in Nepal and South Africa in the 1980s and conferred protection rates of 55–72%. A Vi typhoid vaccine was licensed in the United States in 1994 and is given intramuscularly.

2. *Cholera*

Like typhoid, cholera is a disease related to poor sanitation. The causative agent, *Vibrio cholerae*, excretes a potent enterotoxin (cholera toxin) that leads to severe, dehydrating diarrhea, frequently with vomiting. Fluid loss can reach 15–20 l per day, resulting in death from hypovolemic shock and/or electrolyte imbalance. Fluid and electrolyte replacement is essential, and such treatment dramatically reduces the fatality rates. Disease in the United States is rare (~10–20 cases per year), but it is endemic in parts of Asia, Africa, South America, and the Middle East.

Parenteral, killed whole cell vaccines for cholera have been available in many parts of the world since the late 1800s. These vaccines confer only approximately 50% protection, and booster doses are needed every 6 months. They have many of the side effects of other whole cell gram-negative vaccines (fever, headache, and general malaise). Efforts have been under way to make an improved cholera vaccine, and an oral formulation consisting of killed whole cells plus the B subunit (non-toxic, binding subunit) of cholera toxin has been licensed in Sweden. The vaccine is less reactogenic when given orally and gave efficacies in the 60% range for vaccinees older than 5 years of age. Efforts have also focused on live attenuated strains, and a genetically engineered strain (CVD103 HgR) is licensed in several countries (but not in the United States). This strain is deleted for the A (active) subunit of cholera toxin but makes the B subunit; therefore, it is capable of generating antitoxic (B subunit) antibodies and antibody to the bacterial cell surface (LPS, etc.). Both types of antibodies are thought to contribute to protection. However, results from a recent efficacy trial of a single oral dose of CVD103 HgR in Indonesia are disappointing.

Current ACIP guidelines do not call for routine vaccination of US residents traveling to parts of the world where cholera is endemic, but some countries may require proof of vaccination for entry. The vaccine available in the United States is the parenteral killed whole cell product.

3. *Anthrax*

Anthrax is extremely rare in the United States and in most parts of the world, with an average of 0.25 cases per year reported in the United States between 1988 and 1996. Disease is acquired by contact with *Bacillus anthracis* spores, either by cutaneous exposure or through inhalation. In the industrialized world, disease is associated with processing of animal materials such as wool, hair, hides, and bones. In agricultural settings in Africa and Asia, meat and animal carcasses may also be a source of infection. Although disease in the United States is rare, interest in anthrax vaccines was renewed following a 1979 epidemic in Sverdlovsk, Russia, that was associated with accidental release of spores from a military laboratory and because Iraq produced weapons containing anthrax spores during the 1991 Gulf War. Interest was further accelerated by the 2001 bioterrorist attacks through the United States mail.

Anthrax produces two main toxins, lethal factor and edema factor, which are unusual in that they share the same binding (B) subunit, which is referred to as "protective antigen." The vaccine available in the United States is composed of a cell-free culture filtrate, adsorbed onto aluminum hydroxide, plus low concentrations of formaldehyde and benzethonium chloride as preservatives. Vaccine potency is tested by guinea pig challenge, and although the protective antigen is believed to contribute to immunity the importance of the various components is not completely defined. The vaccine is recommended for industrial workers who handle wool, goat hair, hides, and bones imported from countries in which animal anthrax occurs (mainly from Asia, Africa, and parts of South America and the Caribbean) and for laboratory researchers working with virulent strains. The US military also vaccinates against anthrax. A live, attenuated anthrax vaccine is available in the former Soviet Union, and a similar attenuated vaccine is used for vaccinating domestic animals in many countries throughout the world.

4. *Plague*

Plague is a natural (zoonotic) infection of rodents and their fleas. During the Middle Ages, plague killed approximately 25 million people in Europe and was known as the "Black Death." The causative organism, *Yersinia pestis*, is a gram-negative coccobacillus belonging to the Enterobacteriaceae. Although currently disease is rare, the potential for epidemics still exists. In the United States, an average of 13 cases per year were reported to the Centers for Disease Control during the years 1970–1995. Disease in the United States occurs mainly in the southwestern states (New Mexico, Arizona, and Colorado, with smaller numbers of cases reported from California and nine other western states). On a worldwide basis, an average of 1087 cases per year were reported to the World Health Organization during the years 1980–1994 (although disease may be under-reported from countries in which surveillance and laboratory capabilities are inadequate). Epidemics are most likely to occur in countries with poor sanitary conditions, where large populations of rats live in close proximity to humans.

The plague vaccine currently available in the United States is a formaldehyde-inactivated whole cell preparation. It is recommended for use in laboratory researchers working with virulent strains and for field-workers, agricultural consultants, and military personnel working in areas where they may be exposed to enzootic or epizootic disease.

F. Tuberculosis

Mycobacterium tuberculosis can cause disease in any organ of the body, but pulmonary tuberculosis is by far the most common form of disease. The organism is spread via the respiratory route, and infection at sites other than the lung is usually the result of dissemination from a primary lung lesion. The study of tuberculosis and tuberculosis vaccines is complicated by the fact that only a small percentage of those infected with the organism develop clinical disease; in those that do develop disease, the onset of symptoms may occur from several weeks to many decades following initial infection.

In the mid-1800s, mortality from tuberculosis in the larger eastern cities of the United States averaged 400 per 100,000 people per year. It had long been recognized that crowded living and working conditions favored the spread of tuberculosis, and improvement of these conditions reduced the incidence of tuberculosis, even before the availability of antimicrobial therapy or vaccination. In the United States, the mainstay of tuberculosis control has been early detection and treatment of patients with active disease and contact identification to provide preventive therapy for persons infected but not yet symptomatic. This strategy worked reasonably well until the 1980s, when the incidence of disease began to increase and outbreaks of multi-drug-resistant strains began to appear (mainly in HIV-positive individuals and their health care workers and in correctional facilities and among the homeless). Although a tuberculosis vaccine was developed in the 1920s, and has been given to billions of people throughout the world, it has never been widely used in the United States mainly because it converts people to skin test positive and thereby eliminates an important screening test used to identify infected but asymptomatic people. Additionally, vaccine efficacy estimates from different trials have varied widely from 0 to 80%.

The current vaccine is the live, attenuated bacille Calmette–Guerin (BCG) strain, developed by Calmette and Guerin at the Pasteur Institute by subculturing a strain of *Mycobacterium bovis* on artificial media every 3 weeks for 13 years. Unfortunately, different sublines of BCG have resulted from propagation and production of vaccine in different laboratories throughout the world, and this may partially account for some of the discrepancies in efficacy reported for the various trials. Despite the widespread use of BCG vaccine in most countries of the world, there are still an estimated 8 million cases of active disease, and 2 or 3 million deaths per year worldwide. Although overall efficacy may be less than ideal, BCG vaccination is believed to reduce the more serious meningeal and disseminated forms of tuberculosis in children.

In the United States, the risk of tuberculosis in the general population is relatively low, and routine BCG vaccination is not recommended. Vaccination should be considered for an infant or child continually exposed to an untreated or ineffectively treated patient who has pulmonary tuberculosis, where the child cannot be removed from the infectious patient or given long-term preventive antimicrobial therapy. Efforts are ongoing to develop a better tuberculosis vaccine.

G. Lyme disease

Lyme disease vaccine was licensed in the United States in December 1998, but in February 2002, it was withdrawn from the market by the manufacturer. This vaccine contains lipidated outer surface protein A (OspA) from *Borrelia burgdorferi*, produced recombinantly in *Escherichia coli*. Lyme disease is named for the community in Connecticut in which disease was first recognized to occur in epidemic fashion. Clinical descriptions matching that of Lyme disease had been described earlier, but it was not until 1975 in Lyme, Connecticut, that the infectious nature was recognized. The causative agent was first isolated in 1982 by Burgdorfer and colleagues.

Lyme disease is transmitted through the bite of an infected tick, and disease generally occurs in stages. In the initial stage there is usually (~85% of cases) a characteristic skin lesion (erythema migrans) emanating from the site of the tick bite, accompanied by fever, headache, malaise, or stiff neck. Approximately 5–15% of infected individuals develop neurologic or cardiac symptoms within a few months of the initial infection, and if untreated this frequently leads to late-stage infection and Lyme arthritis. Most patients respond to antibiotic therapy although refractory cases do occur. Treatment is more likely to be delayed in patients who do not develop the initial skin lesion.

The majority of Lyme disease in the United States (~88%) occurs in the northeast, the upper Midwest, and northern California. It is also prevalent in the temperate European countries, including Germany, Sweden, and Austria, and the central regions of the former Soviet Union. More than 12,000 cases of Lyme disease were reported in the United States in 1998. Control efforts have focused on tick avoidance, the wearing of clothing that covers the arms and legs when working or playing in tick-infested areas, and the use of insect repellents. Routine use of the vaccine, even in areas with the highest incidence of disease, was not believed to be cost-effective. However, vaccination was recommended for individuals in areas of high or moderate risk who engaged in recreational, property maintenance, or occupational activities that resulted in frequent or prolonged exposure to tick-infested habitats.

It was not recommended for persons in high-to moderate-risk areas who had minimal or no exposure to tick-infested habitats or for persons who resided, worked, or recreated in areas of low to no risk.

The mechanism of action of this Lyme disease vaccine is unlike that of other vaccines. The bacteria express OspA in the infected tick but downregulate OspA and express mainly OspC in humans. Antibodies to OspA are apparently ingested by the tick when it takes a blood meal on the human host, and these kill the *Borrelia* within the tick gut. This suggests that relatively high levels of antibody would need to be maintained over time. Prelicensure efficacy studies showed 76% efficacy 1 year after the third dose of vaccine, but the long-term efficacy was not evaluated.

IV. CONCLUDING REMARKS: DISEASE ERADICATION

Global eradication of smallpox through vaccination was a milestone in history. Polio eradication is on the horizon, and elimination of measles and rubella from the United States, Canada, and several European countries is also targeted. Whether bacterial diseases can be similarly eradicated will depend on several factors: (i) Is there an animal or environmental reservoir for the pathogen? (ii) Does immunity prevent actual infection/colonization (as opposed to protection against the systemic effects of the disease)? (iii) Is there a long-term carrier state? and (iv) Can sustained global vaccination efforts be implemented? *Haemophilus influenzae* type b vaccine has had the unanticipated benefit of helping reduce transmission of the organism to unvaccinated infants (a form of "herd immunity"). Although Hib disease is still a problem globally, the dramatic reduction in Hib disease in the United States following introduction of conjugate vaccines, combined with the fact that there is no animal reservoir for the organism, suggests that global eradication of Hib might be achievable through vaccination. Global eradication, however, requires tremendous effort and resources. Major bacterial diseases are more likely to be controlled by vaccination but may not be completely eradicated.

BIBLIOGRAPHY

Centers for Disease Control (1999). Achievements in public health, 1900–1999. Impact of vaccines universally recommended for children—United States, 1990–1998. *Morbid. Mortal. Weekly Rep.* **48** (12), 243–248.

Centers for Disease Control. *Morbid. Mortal. Weekly Rep.* Annual Summaries.

Centers for Disease Control. *Morbid. Mortal. Weekly Rep.* Recommendations and Reports. (Specific references for each vaccine are listed in Table 89.2)

Institute of Medicine. Vaccine-related reports available online at *www.search.nationalacademies.org/* (search "vaccines").

Levine, M. M., Woodrow, G. C., Kaper, J. B., and Cobon, G. S. (eds) (1997). "New Generation Vaccines," 2nd edn. Dekker, New York.

Plotkin, S. A., and Orenstein, W. A. (eds) (1999). "Vaccines," 3rd edn. Saunders, Philadelphia, PA.

Thole, J. E., van Dalen, P. J., Havenith, C. E., Pouwels, P. H., Seegers, J. F., Tielen, F. D., van der Zee, M. D., Zegers, N. D., and Shaw, M. (2000). Live bacterial delivery systems for development of mucosal vaccines. *Curr. Opin. Mol. Ther.* **2**, 94–99.

Vaccine supplement (1998, May). *Nature Med.* **4**(5).

WEBSITES

Emerging Infectious Diseases (an NIH journal)
http://www.cdc.gov/ncidod/eid/about.htm

Center for Vaccine Development, University of Maryland
http://medschoo.umaryland.edu/CVD/SOM.HTML

Division of Microbiology and Infectious Diseases, NIH
http://www.niaid.nih.gov/dmid/

90

Vaccines, Viral

Ann M. Arvin
Stanford University School of Medicine

GLOSSARY

adaptive immunity B- and T-lymphocyte-mediated, memory immune responses that control viral infections through specific interactions with virus-infected cells and virions.

attenuation Genetic alteration of infectious viruses to reduce their potential to cause disease.

tropism Pattern of infectivity for cells and organs that is characteristic of the viral pathogen.

immunogenicity Capacity to elicit adaptive immunity to proteins of the virus.

protective efficacy Capacity to protect against disease usually caused by the virus.

I. GENERAL PRINCIPLES

The fundamental objective of vaccination against viral pathogens is to induce adaptive immunity in the naive host, which protects from disease upon any subsequent exposures to the infectious agent. In the absence of vaccine-induced immunity, the initial control of a viral infection depends on mechanisms that comprise the innate immmune system, such as production of interferon-alpha (IFNα) or lysis of virus-infected cells by natural killer cells. Innate immunity limits viral spread but these defenses are often not sufficient to block symptoms of illness during the interval necessary to elicit adaptive immunity against the virus. In the extreme circumstance, life-threatening complications may result in the interim. Adaptive antiviral immunity consists of the clonal expansion of T lymphocytes and B lymphocytes that have the functional capacity to recognize specific viral proteins and to interfere with viral replication and transfer of virions from infected to uninfected cells within the host.

In order to induce adaptive immunity, viral proteins must be processed by dendritic cells or macrophages, which are specialized antigen-presenting cells that mediate the cell surface expression of viral peptides in combination with the class I or class II major histocompatibility complex (MHC) proteins. MHC-restricted antigen presentation creates populations of "memory" T lymphocytes within the CD4 and CD8 subsets that are primed to synthesize cytokines, such as interleukin 2 (IL2) or interferon gamma (IFNγ), when exposed to the same viral peptide–MHC class I or class II protein complex. Cytokines modulate the inflammatory response, expanding and recruiting antigen-specific, cytotoxic T lymphocytes (CTL) to the site of viral infection, and inducing B lymphocytes to produce antibodies of the IgM, IgG, and IgA subclasses that can bind to proteins made by the pathogen or mediate antibody-dependent cellular cytotoxicity.

Adaptive immunity that protects against viral pathogens can be achieved by inoculation of the naive host with infectious virus which has been attenuated for its capacity to cause disease, or by exposure of the

The Desk Encyclopedia of Microbiology
ISBN: 0-12-621361-5

host to viral proteins administered in a noninfectious formulation. Effective priming of adaptive T lymphocyte and B lymphocyte responses by a viral vaccine is expected to block most or all symptoms of infection when the host is exposed to the pathogen. The immunogenicity of a vaccine is defined as its capacity to elicit adaptive immunity, whereas protective efficacy refers to the prevention of disease, which is a consequence of the effective induction of virus-specific immunity. Vaccine-induced immunity may not prevent asymptomatic or abortive infection during these encounters, but memory, or "recall", responses to the viral proteins should eliminate any serious morbidity or risk of mortality known to be associated with the infection in a susceptible, non-immunized individual. Antiviral responses elicited by vaccination provide "active", as distinguished from "passive" immunity. Passive antiviral immunity is provided by virus-specific immunoglobulin class G (IgG) antibodies, which may be acquired transplacentally by infants, or by administration of immunoglobulins, such as rabies or varicella-zoster immune globulin. Active immunity as elicited by effective vaccines, mimics the memory immunity that follows natural infection and is persistent, whereas passively acquired antibodies are metabolized over a half-life of about four weeks and protection is transient.

The challenge of designing viral vaccines that elicit adaptive immunity that is sustained and protective can be addressed using several different strategies, which are often dictated by characteristics of the pathogen and the target population requiring protection. Historically, variolation against smallpox was the first attempt to induce active immunity against a virus by inoculation, as described in early texts from China. Nevertheless, variolation differs from vaccination because the unaltered variola virus was given, with disease modification presumed to result from administering a low infectious inoculum by a cutaneous route. The first success of viral vaccination is attributed to Benjamin Jesty, an English farmer, who used cowpox to prevent smallpox in 1774, as recounted by Edward Jenner, who published his own experience with vaccination in "Variolae Vaccinae", 1798. The remarkable achievement of the global eradication of smallpox was accomplished two hundred years later, using vaccinia vaccine. Many viral diseases can now be prevented by immunization and efforts are in progress to make new vaccines that will provide effective prophylaxis against many other human viral pathogens, or in some cases, when given as 'therapeutic' vaccines, which are intended to control the progression of chronic viral infections, such as human immunodeficiency virus (HIV) (Table 90.1).

TABLE 90.1 Live viral vaccines for prevention of human disease

Current	Under development
Measles	Influenza A and B
Mumps	Respiratory syncytial virus
Rubella	Parainfluenza viruses, 1, 2, and 3
Varicella	Herpes simplex viruses 1 and 2
Polioviruses 1, 2, and 3	Cytomegalovirus
Yellow fever	Rotavirus
Adenovirus	
Vaccinia	

Vaccination is also used to control viral diseases in non-human species.

II. LIVE VIRAL VACCINES

Live attenuated viral vaccines are now licensed in the United States and elsewhere for the prevention of measles, mumps, rubella, varicella, and polioviruses 1, 2, and 3 (Table 90.2). When conditions of special risk for exposure exist, live attenuated yellow fever vaccine, live adenovirus, and vaccinia are given as prophylaxis. Live viral vaccines contain an infectious virus as the primary component, which has been attenuated in order to reduce or eliminate its potential to cause disease in the naive host. Vaccine strains are made from RNA viruses, including measles, mumps, rubella, poliovirus, rotavirus, and yellow fever, as well as from DNA viruses, such as varicella, adenovirus, and vaccinia. Attenuation of virulence is accomplished by laboratory manipulations of the naturally occurring, "wild type" virus, which is referred to as the parental strain of the vaccine virus. The parental strains of live attenuated viral vaccines are obtained from an individual experiencing the typical disease caused by the virus. Alternatively, attenuation is achieved by taking advantage of host range differences in virulence between human and closely related animal viruses. Proteins made by the animal virus are similar enough to those encoded by the human pathogen to elicit protective, adaptive immune responses. When a suitable animal model is available, a reduction in the capacity of the vaccine virus is demonstrated and an alteration in the potential to cause disease may be documented. Although strain selection and characterization can be done *in vitro* and in animal models to predict safety, sequential evaluation of vaccine strains in individuals who have natural immunity, followed by gradual dose escalation studies in susceptible individuals is required to prove safety for human use.

TABLE 90.2 Non-infectious vaccines for prevention of human viral diseases

Current	Under development
Polioviruses 1, 2, and 3	Human papillomavirus
Influenza A and B	Human immunodeficiency virus
Hepatitis A	Herpes simplex virus 1 and 2
Hepatitis B	Respiratory syncytial virus
Japanese encephalitis virus	Hepatitis C
Tick-borne encephalitis	

The attenuation of a live attenuated vaccine strain is defined clinically by its loss of the potential to cause disease. The attenuated virus should retain infectivity at the site of inoculation, which may be by subcutaneous injection or by oral or intranasal delivery to mucosal cells. In order to be attenuated, the tropisms of the parent virus that would otherwise allow it to produce damage to the host must be incapacitated. For example, the attenuation of polioviruses requires that the vaccine strains be incapable of infecting cells of the central nervous system. Some recipients may have reactions to live attenuated vaccines, including low fever, but the incidence of reactogenicity, such as fever, associated with the administration of effectively attenuated vaccine strains is low, and other manifestations, such as rash, are mild.

The attenuation of RNA and DNA viruses to produce vaccine strains used in most licensed vaccines is accomplished by traditional approaches in which the parent virus undergoes passage *in vitro*, using human or non-human cells, or sequentially in both human and non-human cells, or by growth in chick or duck embryo cells in eggs. An additional strategy for achieving attenuation is to modify environmental conditions, such as adapting the virus to grow at low temperatures. Cold-adapted viruses are less able to replicate at human body temperature. Measles vaccine is made by passage in chick embryo cells, with further attenuation achieved by cold passage at 32 °C. Rubella vaccine, RA27/3 strain, is derived by passage in human cells only, including growth at 30 °C. Varicella vaccine was attenuated by passage in guinea pig embryo cells and cold passage. In the case of rotavirus, attenuation is achieved by reassortment of genes from strains infectious for humans with genes from related, non-human virus. As a consequence of these manipulations, the vaccine virus remains infectious but its ability to replicate in the human host is limited and the cycles of viral replication that occur in the vaccine recipient do not result in a reversion to virulence.

By definition, the vaccine strain must retain genetic stability in order to preserve its attenuation. Sequence differences from the parent strain have been implicated in the attenuation of poliovirus strains 1, 2, and 3 that are used to make live poliovirus vaccines. However, the genetic basis for the attenuation of most traditional vaccine strains is not known. The evidence that these strains are genetically stable is inferred from the preservation of the attenuation phenotype when the vaccine is given to susceptible individuals. Even when sequence information is available for vaccine strains, it is difficult to determine which sequence differences from the parent strain are most essential for the biologically observed modification of virulence. The definition of genetic markers of attenuation is complex because the traditional procedures for making live vaccines typically yield many variations in the genome sequence of the vaccine strain and genetic stability can be predicted to be multifactorial. In general, genetic stability is enhanced as the number of mutations in the vaccine strain increases. For example, the vaccine poliovirus type 1 has 56 mutations in 7441 nucleotides compared to only 10 of 7429 differences in the type 3 strain. Live attenuated vaccines may also contain mixed populations of the vaccine virus that have different genetic alterations, as has been described for the rubella vaccine. In most cases, biological attenuation means that the vaccine virus also loses its transmissibility to other susceptible individuals who are in close contact with the vaccine recipient. However, when vaccine strains are transmissible, the genetic stability of the vaccine virus must also be preserved after replication in secondary contacts.

The immunogenicity of live attenuated viral vaccines depends upon the selection of an appropriate infectious dose for inoculation, whether the vaccine is given by systemic or mucosal routes. The identification of the proper dosage regimen for administration is also important. Many live attenuated viral vaccines must be given as several doses in order to establish persistent adaptive immunity in the majority of naive recipients. Some live virus vaccines consist of mixtures of vaccine strains because protection must be conferred against disease caused by different subgroups of the wild type virus, as illustrated by the trivalent oral poliovirus vaccine. In other cases, several live attenuated virus strains are combined to facilitate simultaneous immunization of the susceptible host against unrelated viruses, as exemplified by the measles–mumps–rubella vaccine. The challenge of designing these multivalent vaccines is to assure that each attenuated vaccine strain is present in a high enough inoculum to allow it to replicate adequately at the site of inoculation in the presence of the other strains. A balance of the components must be achieved to prevent interference by more potent vaccine viruses which might impair the immunogenicity of the other vaccine strains. The establishment of adaptive immunity to all components may depend upon a multiple dose regimen, as is recommended for live attenuated polio vaccine. The timing between doses of live attenuated viral vaccines is also important because interference can occur when one live virus vaccine is given too soon after another. The required interval is usually at least 4 weeks, to avoid reducing the infectivity of the second vaccine strain as a result of replication of the first vaccine strain, or antiviral immune responses elicited by the first vaccine, such as interferon production.

Healthy young children constitute the primary target population for the live attenuated vaccines to prevent measles, mumps, rubella, varicella, and polioviruses 1, 2, and 3. In contrast, the live yellow fever and adenovirus vaccines are used in individuals who are considered to have a particular risk of disease. Yellow fever vaccines are used to prevent disease in local populations and visitors to endemic areas. These vaccines are made from a strain first developed in the 1930s, which was attenuated by passage in monkeys and then prolonged tissue culture passage. The vaccine strain causes a low level of viremia, which is also characteristic of infection with wild type virus, but multiple sequence changes from the parent strain have been demonstrated and clinical experience demonstrates that its pathogenic potential to cause life-threatening dissemination is eliminated. Adenovirus vaccines against serotypes 4 and 7 have been used to control outbreaks among military recruits. Prevention of respiratory tract infection is achieved by the oral administration of live adenovirus, in tablets which are coated to prevent acid inactivation in the upper gastrointestinal tract. In this instance, attenuation results from the route of inoculation without any molecular alteration of the viral genome.

III. NON-INFECTIOUS VACCINES

Non-infectious vaccines are licensed for influenza, polio, hepatitis A, hepatitis B, rabies, Japanese encephalitis virus, and tick-borne encephalitis (Table 90.2). The vaccines are made by inactivating the infectious virus after growth in tissue culture or eggs, or by using protein components of the virus only. These vaccines are referred to as "killed" or "inactivated" vaccines, or as "subunit" vaccines. This approach to vaccine design has the advantage of eliminating concerns about the infectious component of attenuated live viruses. While attenuation of virulence is the major issue in making live viral vaccines, immunogenicity is the primary concern in designing inactivated vaccines. Alum is used as an adjuvant to provide the amplification of adaptive immunity that is achieved by viral replication in the case of live attenuated vaccines. The induction of a balanced host response against viral proteins is of critical importance in the production of inactivated vaccines, as illustrated by the occurrence of atypical measles disease in children who were immunized with a formalin-inactivated, alum-precipitated measles vaccine. Formalin-inactivated respiratory syncytial virus vaccine was associated with severe lower respiratory tract infection in immunized infants who were infected with the wild type virus. Although formalin inactivation creates safe, inactivated vaccines for other viral pathogens, cross-linking by formaldehyde may have changed the conformation of viral proteins, inducing antibodies against amino acid epitopes that were not elicited in the normal host response to viral proteins made during replication in host cells. Immunization with the inactivated measles vaccine appears to have resulted in formation of antigen–antibody immune complexes when viral infection occurred. This misdirection of the adaptive immune response resulted in immune-mediated disease, instead of protective immunity in some vaccine recipients. Since inactivated vaccines are not as immunogenic for inducing memory host responses as natural infection or live viral vaccines, most dose regimens for inactivated or subunit vaccines incorporate 'booster' doses to assure the long-term persistence of virus-specific immunity.

Inactivated influenza vaccine is used to protect individuals who are at risk for life-threatening disease during the annual epidemics of influenza A and B. The target populations for this vaccine are elderly adults, immunocompromised patients, and those with chronic pulmonary or cardiac diseases. Because influenza viruses undergo rapid antigenic changes, it is necessary to formulate the vaccine annually to contain the hemagglutinin and neuraminidase proteins from the two predominant circulating strains of influenza A and the major influenza B strain, which are identified through a global surveillance network. The component viruses are grown in embryonated eggs, inactivated by formalin and combined in a trivalent vaccine. Subunit preparations of influenza vaccine are made by detergent treatment to increase relative concentrations of HA and NA proteins. In the case of influenza vaccine, the need for repeated immunization is dictated by the genetic capacity of influenza viruses to undergo antigenic drift and shift, requiring adminstration of the new vaccine to high risk populations before each winter epidemic begins. Inactivated polio vaccine is also a trivalent vaccine made from formalin-inactivated strains of polioviruses 1, 2, and 3. The manufacture of inactivated polio vaccine is complicated by the need to achieve complete inactivation of these non-attenuated viruses while maintaining immunogenicity that is protective against paralytic disease caused by each of the three polio serotypes. The current enhanced potency vaccine given as five doses, beginning in infancy, with later booster doses, is now recommended as an alternative to live attenuated polio vaccine in developed countries. Regimens combining initial immunization with inactivated vaccine followed by doses of live attenuated vaccine are also effective. Inactivated polio vaccine must be given by injection, which is a practical limitation to their use in developing countries.

Whereas most viral vaccines are designed to prevent the disease caused by acute, primary infection, the benefit of hepatitis B vaccine results from preventing chronic active infection and the late sequelae of hepatic failure and hepatocellular carcinoma. In contrast to influenza and polio, hepatitis B virus does not replicate in tissue culture. The vaccine for hepatitis B consists of the surface antigen of the virus, which is a glycoprotein that forms the outer envelope of the virion. When expressed by introducing the gene sequence into yeast or mammalian cells, the hepatitis B surface antigen self-assembles into a particle structure, which contributes to its immunogenicity as a single viral protein, and allows it to be used as an effective single protein subunit vaccine. Recombinant DNA vaccines have replaced vaccines in which surface antigen particles were extracted from the plasma of chronic carriers. The success of the hepatitis B vaccine depends in particular upon the timely delivery of the vaccine to infants. Vaccination beginning at birth blocks the transmission of hepatitis B virus from carrier mothers to their infants who are otherwise at high risk for chronic infection. Although both viruses cause hepatitis, hepatitis B is a DNA virus while hepatitis A is an enterovirus belonging to the same family as polioviruses. New vaccines for hepatitis A have been licensed, which contain formalin inactivated virus grown in tissue culture and administered with alum or liposomal adjuvants. Hepatitis A vaccine is useful for susceptible adults, who may be exposed by occupation, travel, and other risk factors, as well as during community outbreaks. Hepatitis A vaccine may be recommended for universal administration to children in developed countries where the cost of vaccination is acceptable even though the risk of serious disease in young children is low.

Inactivated vaccines are licensed to prevent three viral causes of central nervous system disease, including rabies, Japanese encephalitis virus, and tick-borne encephalitis. In 1885, Louis Pasteur inoculated Joseph Meister with spinal cord material from infected rabbits that was inactivated by drying but contained some infectious virus. This work initiated a vaccine method based on the use of nervous tissue from infected animals, which continued to be used during the first half of the twentieth century, with later modifications made to improve viral inactivation. The current rabies vaccine is made from virus grown in human cells in tissue culture and inactivated with β-propiolactone, which eliminates the risks of adverse reactions to myelinated animal tissues. The administration of the vaccine to exposed individuals is simplified to a five or six dose regimen instead of the 14–23 doses required for earlier rabies vaccines. Whereas most viral vaccines protect against infections acquired by human to human transmission, immunization of domestic animals with inactivated rabies vaccine is critical for disease prevention. The Japanese encephalitis virus is a flavivirus, related to the St. Louis encephalitis virus and other members of this family, which is maintained as a mosquito-borne pathogen in Asia. Although most infections are asymptomatic, some individuals develop encephalitis that is fatal or causes severe, permanent neurologic damage. The licensed vaccine for Japanese encephalitis is an inactivated preparation purified from infected mouse brain although inactivated and live attenuated vaccines made in tissue culture are used in China. The need for immunization is restricted to populations in endemic areas and for travelers who are visiting rural areas in these countries during the peak season for transmission in summer and fall. Tick-borne encephalitis virus is also a flavivirus, with subgroups called Far Eastern and Western virus types. The distribution of endemic areas includes parts of Europe and Russia. The vaccine used in Europe is made from formalin-inactivated virus grown in chick embryo cells. Immunization is recommended for populations in endemic areas and for travelers to these areas who may have increased risk of exposure to ticks.

IV. VACCINE IMMUNOLOGY

Because the development of new viral vaccines takes years and is very costly, immunologic criteria are used to judge the probable efficacy of candidate vaccines. Laboratory assays for assessing vaccine immunogencity measure the production of antibodies directed against viral proteins, including IgG and secretory

IgA antibodies, as well as their functional capacity to neutralize the virus *in vitro* or to mediate antibody-dependent cellular cytotoxicity. Because of the importance of cell-mediated immunity for defense against viral infections, assays for cytokine production by T cells stimulated with viral antigens in vitro and for T-cell-mediated cytotoxicity are useful measures of the establishment of virus-specific memory immunity. Establishing accurate correlates of protection must be done in field trials during which large cohorts of vaccinees are exposed to wild type virus. In most instances, only simple laboratory assays can be performed when so many individuals must be tested. Serologic assays are used for this purpose even though protection is likely to require adequate T-cell-mediated immunity. It is necessary to have a reliable expected attack rate for transmission of the viral pathogen whereas under field conditions, rates of transmission are affected by many variables, such as the proximity and duration of contact with the index case. Therefore, most clinical vaccine trials require large populations of subjects. Whether protective immunity is induced by viral vaccines is often proved conclusively only after widespread implementation of immunization programs, as illustrated by the impact of vaccines against childhood diseases, such as measles, mumps, and rubella.

The specific goals of vaccine immunology are to demonstrate that vaccination induces adaptive immunity against relevant viral antigens in the naive host and, when possible, to identify immunologic responses that are associated particularly with protection of vaccine recipients against the usual consequences of infection with the wild type virus. In practice, the definition of immunologic correlates of vaccine protection is rarely straightforward. The immunogenicity of vaccines is usually assessed by comparison with immune responses that follow natural infection with the same virus but in most instances, specific correlates of protection are not known for naturally acquired immunity. The redundancy of the mammalian immune system means that a broad range of adaptive immune responses to the pathogen can be measured in the healthy immune individual. For example, individuals who have antibodies to viral proteins can also be expected to have antigen-specific CD4+ and CD8+ T cells. Vaccine-induced antibodies, especially those with neutralizing activity against the virus, have been considered the first line of defense against infection when the immunized host encounters the wild type virus. These antibodies may limit initial replication at the site of viral inoculation. Primary vaccine failure is defined as a failure of the initial doses of the vaccine regimen to induce virus-specific antibodies. Effective vaccines are expected to elicit seroconversion in most vaccine recipients, which often requires the administration of several doses. Nevertheless, seroconversion is not invariably a predictable marker of protective immunity. Immunization usually induces a range of antigen-specific antibody responses in different individuals. In some cases, detection of any antibodies to viral proteins correlates with protection whereas in other cases, a "protective" titer is defined as greater than or equal to a particular concentration of antibodies. The appropriate laboratory marker of protection may also differ depending upon the nature of the vaccine that is being evaluated. For example, inactivated vaccines often elicit high titers of virus-specific antibodies that correlate with protection while live attenuated vaccines are more likely to induce cellular immunity and lower antibody titers. Despite these differences, the inactivated and live attenuated forms of vaccine may be equally effective against the same pathogen. Live attenuated viral vaccines often induce a more persistent cell-mediated immune response, which affords protection even when antibody titers fall below the threshold of detection in standard serologic assays. In the case of live viral vaccines, rates of vaccine virus shedding after the inoculation of naive subjects may be as reliable a marker of protection as immunologic assays. Depending upon the assessment of risk, correlates of vaccine protection may be defined by deliberate, direct challenge of immunized volunteers with the wild type virus.

Whether or not precise correlates of protection can be defined, immunologic assays are useful for demonstrating effects of vaccine composition and host factors on the response to viral vaccines. These analyses provide valuable insights into the effect of age. For example, immunologic studies of live attenuated varicella vaccine revealed that adolescents and young adults require a two dose regimen to achieve humoral and cell-mediated immune responses that are equivalent to those induced by a single dose in young children. The immunogenicity of vaccines given to young infants may be diminished by transplacentally acquired antibodies, as is observed in measles immunization. Immunologic assays are also useful to determine whether different viral vaccines are compatible when administered concurrently. Since the immunogenicity of viral vaccines in infants is influenced by nutritional status and factors such as the prevalence of intercurrent infections with gastrointestinal pathogens, vaccine formulations that are effective in developed countries may not be appropriate in other circumstances. A need to adjust vaccine dosage or regimen may be evident from comparative immunogenicity studies. Assessing the interval over which adaptive immune responses remain detectable is necessary because waning immunity may indicate the need for booster doses of the vaccine. In addition to primary vaccine failure in which the initial immunogenicity is inadequate to prevent disease caused by wild type virus, secondary vaccine failures occur when immunity declines over time to non-protective levels. Finally, the control of vaccine-preventable disease often depends not just upon immunogenicity in individual vaccinees but upon achieving adequate levels of herd immunity. Immunologic assays provide information necessary to predict whether the local introduction of the virus is likely to be sustained through secondary transmissions and result in a community outbreak.

V. MOLECULAR APPROACHES TO VIRAL VACCINE DESIGN

The advances in molecular biology that have occurred during the past several decades have generated new opportunities for making viral vaccines. Molecular approaches to the design of human viral vaccines will have a major impact in helping to address deficiencies of existing vaccines and allowing the invention of vaccines against infectious diseases that are not preventable by immunization at this time (Table 90.3). Molecular techniques are already being implemented to improve licensed vaccines, as illustrated by the use of cDNA clones to reduce the frequency of poliovirus mutations during vaccine manufacture and the use of reassortant methods to incorporate new influenza viral antigens into available virus strains that replicate to the levels required for production of the inactivated influenza vaccine. Hepatitis B vaccine is now made from recombinant surface antigen protein and the live attenuated rotavirus vaccine contains genetic reassortants that incorporate genes from human and primate viruses.

Molecular approaches that are being developed to create new or improved vaccines include the genetic engineering of live attenuated viral vaccines, in which targeted mutations or deletions are made in genes that are determinants of virulence or tissue tropism, the synthesis of replication defective viruses as vaccines, the expression of recombinant viral proteins and peptides from plamids and in constitutively expressing mammalian cells, the synthesis of virus-like particles from viral proteins made in the absence of the viral genome, the administration of "naked DNA" corresponding to viral genome sequences that are immunogenic, and the use of attenuated human viruses or host range mutants as vectors for expressing genes from unrelated viruses. These strategies have the potential to allow vaccination against viral pathogens such as human papilloma virus, which cannot be grown in tissue culture. Molecular methods will also be valuable to redesign current vaccines; for example, the use of DNA vaccines may diminish the interference with measles vaccine immunogenicity associated with transplacentally acquired maternal antibodies. Progress is also being made in the creation of novel adjuvants such as cytokines or immunostimulatory DNA sequences that modulate the host response to enhance antiviral immunity.

TABLE 90.3 Molecular approaches to the design of human viral vaccines

Genetically engineered attenuation
Genome reassortants
Host range variants
Replication defective viruses
Recombinant viral vectors
Recombinant proteins and peptides
Virus-like particles
DNA vaccines

VI. PUBLIC HEALTH IMPACT

The ultimate success of a viral vaccine is realized when the implementation of vaccine delivery programs results in a global reduction of the disease burden caused by the pathogen. The standard set by the smallpox vaccine campaign provides a challenge to eradicate other viral diseases that continue to cause serious disease and death. The worldwide control of measles and polio are the current priorites for eradication initiatives. Even when effective vaccines are available, the need to vaccinate very high percentages of the susceptible population in order to block transmission presents an obstacle to disease control. Viral vaccines are often labile unless frozen, necessitating an intact "cold chain" during transport to remote areas, and sterile needles and syringes must be available. Mass vaccine campaigns supported by funds from international agencies provide a practical response to these problems through "vaccination days", as was demonstrated by the successful administration of polio vaccine to millions of children in India in a single day.

The opportunity for reducing the disease burden by immunization depends on the viral pathogen, how it is transmitted and the pathogenic mechanisms by which it causes disease. Viral pathogens have evolved concurrently with the human host so that persistence in human populations is assured. Smallpox eradication succeeded by case identification and vaccination of close contacts but polioviruses circulate by causing asymptomatic infection in most individuals. The cycle of transmission of these viruses may be broken by achieving high levels of vaccine immunity through several summer–fall seasons. Measles is expected to be difficult to eradicate because it is highly contagious, requiring only a few susceptibles in the population to cause an outbreak. Some viruses, most notably the herpesviruses and human immunodeficiency virus, cause a life-long persistent infection, associated with intermittent or chronic viral shedding. Control of these viruses differs from those that cause acute infection because re-introduction of the virus into the population can occur readily. Human immunodeficiency virus presents the exceptionally difficult problem of marked antigenic diversity and rapid emergence of virus subpopulations that can escape adaptive immune responses.

Despite these obstacles, vaccine strategies are essential to reduce the impact of viral diseases because of the limited availablity of effective antiviral drugs for most viruses, their short-term efficacy in many circumstances, and the relative cost of antiviral drugs compared with vaccines. The global impact of viral vaccines on public health is recognized as the most important intervention provided by modern medicine.

BIBLIOGRAPHY

Ada, G., and Ramsay, A. (1997). "Vaccines, Vaccination and the Immune Response", Lippincott-Raven, Philadelphia, PA.

Bloom, B. R., and Widdus, R. (1998). Vaccine visions and their global impact. *Nature Med.*, **4** (5 Suppl), 480–484.

Fields, B. N., Knipe, D. M., and Howley, P. M. (1996). "Fields Virology", 3rd edn., Lippincott-Raven Press, Philadelphia, PA.

Long, S., and Prober, C. G. (1997). "Textbook of Pediatric Infectious Diseases", Lippincott Raven Press, Philadelphia, PA.

Plotkin, S. A., and Orenstein, W.A. (1999). Vaccines, 3rd edn., W.B. Saunders Co., Philadelphia, PA.

WEBSITES

All the Virology on the WWW
http://www.virology.net/garryfavweb.html

Emerging Infectious Diseases (an NIH journal)
http://www.cdc.gov/ncidod/eid/about.htm

Center for Vaccine Development, University of Maryland
http://medschool.umaryland.edu/CVD/SOM.HTML

Division of Microbiology and Infectious Diseases, NIH
http://www.niaid.nih.gov/dmid/

91

Viruses

Sondra Schlesinger and Milton J. Schlesinger
Washington University School of Medicine

GLOSSARY

genome replication The synthesis of multiple copies of a virus genome transcribed from a complementary strand of RNA or DNA.

plaques Clear areas in a lawn of bacteria or in a monolayer of cultured cells caused by the cytopathic effects of a virus infection. The plaque arises from infection of a single cell. The spread of the virus from the original cell to neighboring cells is limited by the addition of a solid medium (such as agar) to the dish which confines the spread to a small area. The number of plaques is used as a measure of the number (titer) of infectious particles.

transformation A change in the properties of a cell usually associated with a loss in growth control. Transformed cells are more tumorigenic than their normal counterparts.

virion Extracellular virus particle. These particles may be only nucleoproteins or the nucleoprotein may be surrounded by an envelope consisting of lipid and protein.

virus replication Synthesis of new virus particles includes both replication of the genome and assembly of virus particles.

In 1953, Luria Defined Viruses as "submicroscopic entities, composed of nucleic acid and protein, capable of being introduced into specific living cells and of reproducing inside such cells only." Any definition of a virus, however, should emphasize the unique mechanism of their replication: Viral nucleic acid genomes and virus-encoded proteins are synthesized as separate components followed by their assembly into new particles within the host cell. Organized growth followed by binary division, characteristic of all other microorganisms and eukaryotic cells, does not occur during virus replication. A second distinctive characteristic of viruses is that they are obligate parasites and depend totally on metabolic activities of the host cell for progeny formation. Most viruses are much smaller in size than bacteria; this factor led to the recognition that they were distinct entities from bacteria as causes of transmissible diseases.

The term "virus" is derived from the Latin word for poison and stems from the original observation that a virus infection caused detectable damage to the host organism, often leading to lethality.

I. STRUCTURE, COMPOSITION, AND CLASSIFICATION

Virus particles are mostly in the shape of helical rods or icosahedral spheres in which organized arrays of protein subunits surround and protect the nucleic acid genome. Several viruses are pleomorphic in shape and others, such as some of the bacteriophages, have complex structures composed of a head which contains the packaged genome and a tail which

The Desk Encyclopedia of Microbiology
ISBN: 0-12-621361-5

functions in the attachment to the host cell and in the injection of the genome into the cell. In general, cell-free viruses, referred to as virions, are structured to be relatively stable to most environmental conditions but also capable of conversion to a labile form for rapid disassembly in their host cell.

Many viruses have highly uniform and symmetrical structures and several have been crystallized. X-ray diffraction patterns of these crystals have led to the determination of the virus structure at the atomic level. The first high-resolution structure based on X-ray diffraction was tomato bushy stunt virus, an icosahedral plant virus. Structures of the small icosahedral RNA animal viruses, poliovirus and rhinovirus, quickly followed and enhanced technologies soon led to structures of the small DNA viruses, polyoma and SV40. More than two dozen complete viral capsids have been solved by X-ray crystallography. These have revealed a common structure among the small icosahedral viruses consisting of a protein subunit core composed of a wedge-shaped, eight-stranded anti-parallel β-barrel motif. Atomic structures of several virus subcomponents, such as the hemagglutinin and neuraminidase of influenza virus, the hexon protein of adenovirus, and the protease and reverse transcriptase of human immunodeficiency virus (HIV), have been determined.

Electron microscopy is required for visualization of viruses and was utilized initially to also interpret structures of viruses. Isolated, purified virus particles and those associated with intracellular structures can be detected after fixation, dehydration, and staining of samples (Fig. 91.1). Technical advances in electron microscopes and computer analysis of images have

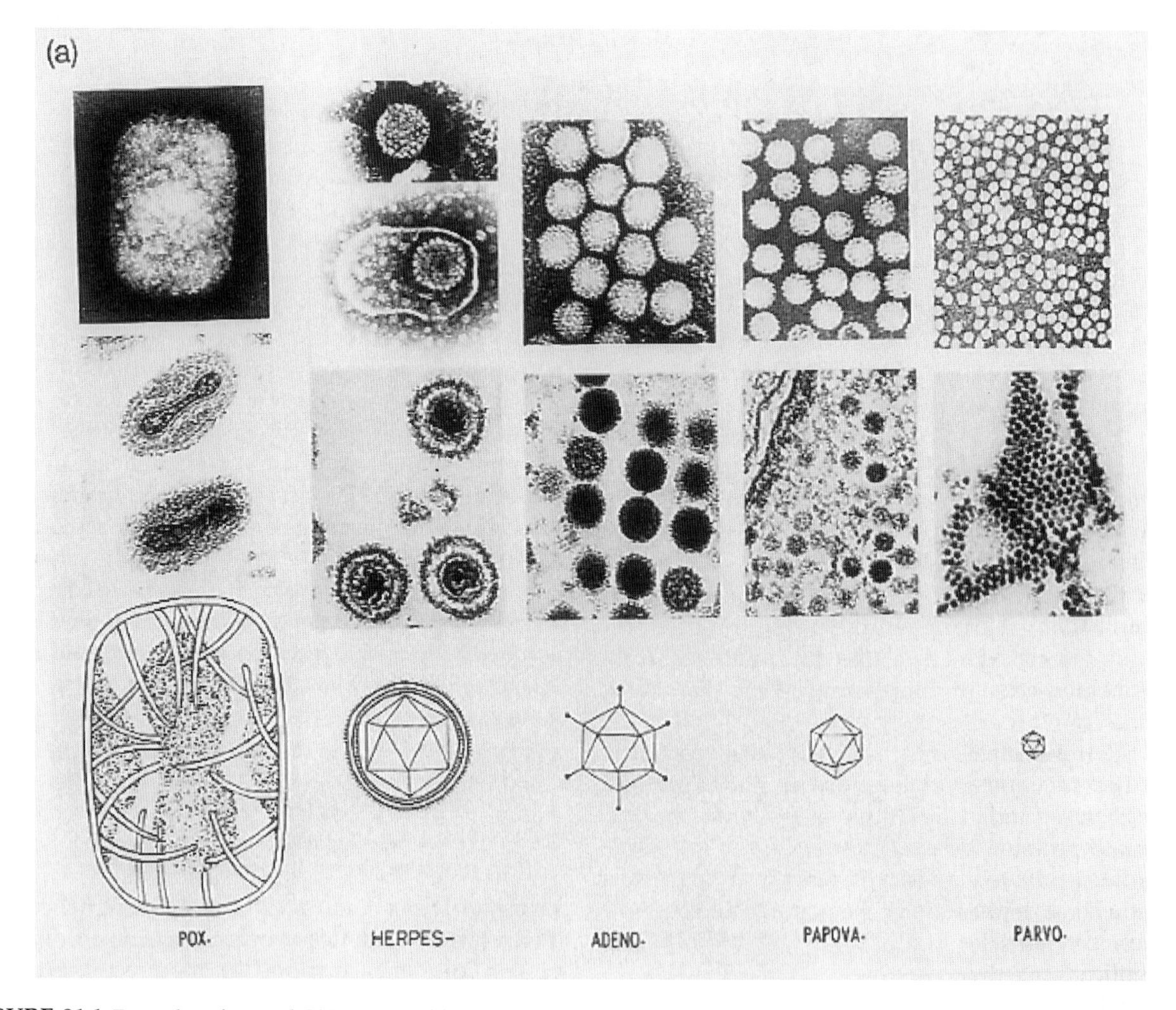

FIGURE 91.1 Examples of several DNA viruses (a) and RNA viruses (b). The top rows show negatively stained preparations of extracellular virus; the middle rows show virus particles in thin-sectioned cells. Magnification × 50,000. The bottom row depicts the viruses and indicates relative sizes. Myxo refers to the Orthomyxoviridae family (reproduced with permission from Landry and Hsiung, 1996, copyright Yale University Press).

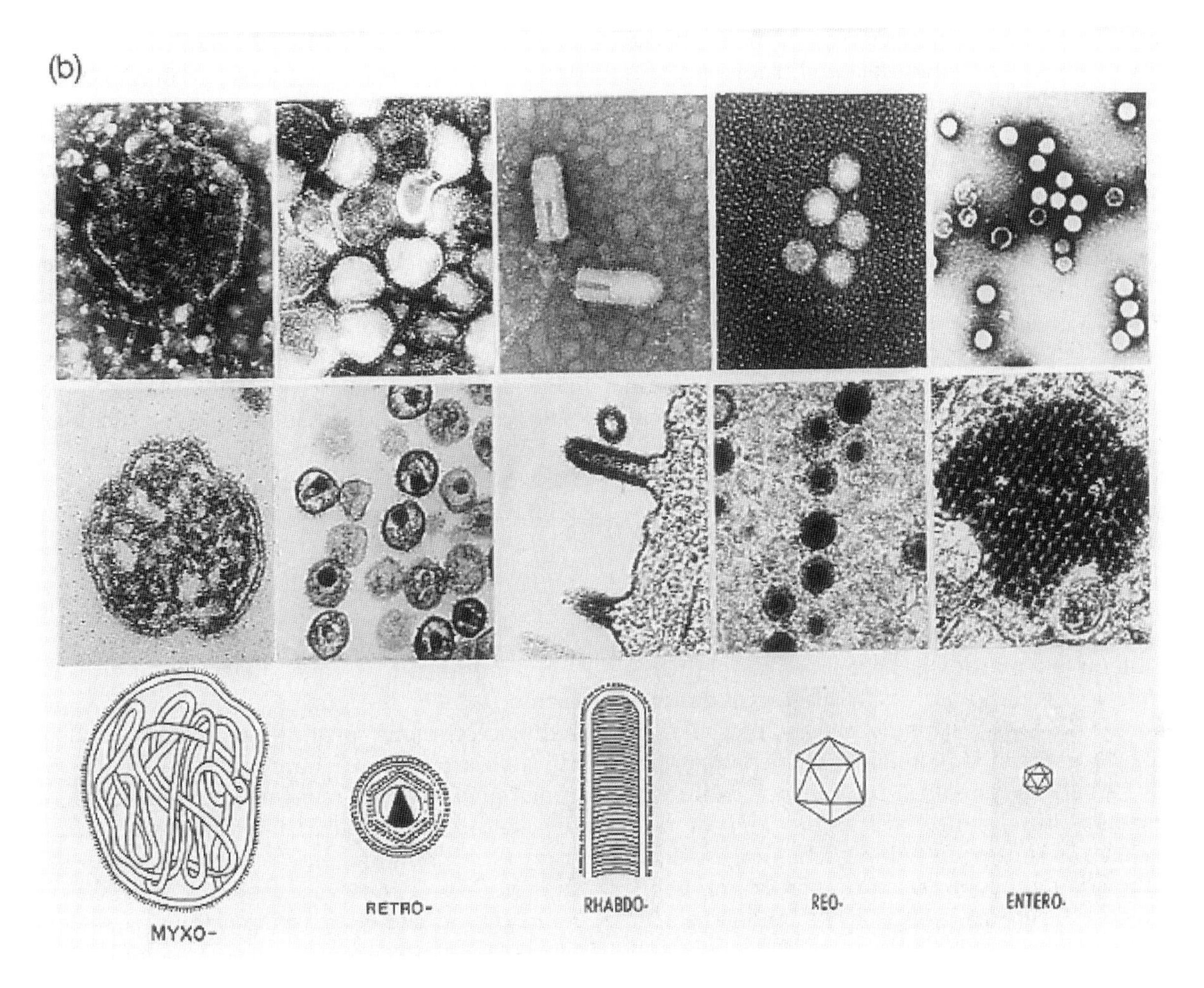

FIGURE 91.1 *continued.*

allowed virions to be examined in vitreous ice (cryoelectron microscopy) and have led to a determination of virus structures with resolution on the order of 9–10 Å.

In addition to protein and nucleic acid, many viruses are enveloped by a lipid bilayer which contains transmembranal proteins encoded by virus genomes, but the lipids and lipid components are derived from the host cell. They are selectively incorporated into the virus particle during the stages of virus assembly and budding from host cell membranes. The proteins are arrayed as oligomeric spikes protruding outward from the membrane. Many viruses also contain enzymes, encoded by virus genomes, that are required for the early stages of virus replication.

The major groups of viruses are distinguished by their nucleic acid genome, which consists of either DNA or RNA. Except for the retroviruses, which are diploid, all viruses contain only a single copy of their genome. The nucleic acid can either be single or double stranded and arranged as a single molecule or in separate segments. The size of the nucleic acid ranges from 1000 or 2000 to 300,000 bases or base pairs, which is equivalent to 3 or 4 genes to more than 100 genes. The genome may be circular or linear and covalently linked or modified at its termini by various chemical structures.

Viruses with DNA as their genome are classified based on whether they are single or double stranded. Viruses with RNA as their genome are classified into three groups. One contains single-strand RNA that has been designated as positive sense because the genomic RNA serves as an mRNA and the isolated genomic nucleic acid is infectious by itself. The second group is composed of RNA genomes that have been designated as negative sense because the genomic RNA is complementary to the viral mRNAs. The third group contains double-stranded RNA. Negative-sense RNA and the double-strand RNA viruses require the initial transcription of the input genome RNA by virion-associated transcriptases upon entry to produce mRNAs and initiate the replication cycle. Additional viruses, that is, the retroviruses and hepadnaviruses, are classified as RNA and DNA reverse-transcribing viruses.

Within each of the groups, the viruses are further subdivided based on the host used for their replication, that is, bacteria, protozoa, fungi, plants, invertebrates, and vertebrates. Most viruses are restricted to a single species of host and within a multicellular host grow only in certain cells. Some viruses use more than one species as a host during their natural life cycle. One such group was originally called arboviruses (arthropod-borne viruses) because they are transmitted by mosquitoes or other arthropods to vertebrates, replicating in both hosts.

Early efforts to classify viruses were based on common pathogenesis, common organ tropism, or common ecological niches or modes of transmission. These attempts broke down when it became clear that the same virus could cause vastly different symptoms in different hosts. For example, the monkey B virus (a herpes virus) is asymptomatic in these animals but usually lethal in humans. In addition, different viruses may cause similar symptoms; for example, rhinoviruses, coronaviruses, and adenoviruses all can cause the common cold. As more information about isolated and purified viruses was obtained, it became more rational to group viruses based on common structures and composition. Table 91.1 illustrates some examples of the current classification system. A complete taxonomic list contains one order, 71 families, and 164 genera. Approximately 4000 different species of virus are placed in this classification. Virus orders are groupings of families that share common characteristics but are distinct from other virus families and orders. Most virus families have distinct particle morphology (Fig. 91.1); similar physicochemical properties of the genome, such as particle mass, density, and pH stability; similar genome structure with regard to type, size, and segmentation of the nucleic acid as well as strategies of replication; similar kinds of proteins and lipids; and antigenic properties. Subfamilies allow for separation of members that have more complex relations within the family. Genera group viruses share common characteristics which distinguish them from viruses of other genera. A virus species has been defined by the International Committee on Taxonomy of Viruses as a polythetic class of viruses that constitutes a replicating lineage and occupies a particular ecological niche.

Supergroups represent a new kind of virus classification which recognizes similarities in genome sequences among positive-sense RNA viruses that cut across traditional families. Most of these are in genes encoding enzymes involved with virus replication and this suggests that these viruses evolved from a common ancestor. One superfamily includes the enveloped positive strand RNA alphaviruses, a member of the Togaviridae family, the human hepatitis E virus, a nonenveloped RNA virus that has not been classified, and the plant virus family Bromoviridae.

TABLE 91.1 Examples of virus taxonomy[a]

Family	Genus	Host category type species
Double-stranded DNA viruses		
Siphoviridae	Unnamed, the λ-like phages	Bacteria: coliphage λ
Baculoviridae	Nucleopolyhedrosis	Invertebrates: *Autographa californica* nuclear polyhedrosis virus
	Granulovirus	Invertebrates: *Plodia interpunctella* virus
Adenoviridae	Mastadenovirus	Vertebrates: human adenovirus 2
DNA and RNA reverse transcribing viruses		
Retroviridae	"Unnamed mammalian type B retroviruses"	Vertebrates: mouse mammary tumor virus
	Lentiviruses	Vertebrates: human immunodeficiency virus
Negative-sense, segmented RNA viruses		
Orthomyxoviridae	Influenza virus A, B	Vertebrates: influenza A A/PR/8/34(H1N1)
Bunyaviridae	Hantavirus	Vertebrates: Hantaan virus
Positive-sense, single-stranded RNA viruses		
Picornaviridae	Enterovirus	Vertebrates: polio virus 1
	Rhinovirus	Vertebrates: human rhinovirus 1
Togaviridae	Alphavirus	Vertebrates: Sindbis virus

[a]From Murphy (1996), in *Fields Virology* (B. N. Fields, D. M. Knipe, and P. M. Howley, eds.), copyright Lippincott–Raven Press, 1996.

Some viruses are referred to as satellites in that they require other closely related virus gene expression in order to complete their replication cycle. These include hepatitis delta virus, which grows only in the presence of coinfection with hepatitis B virus, and the adenoassociated virus, a member of the Parvoviridae family, which requires coinfection with adenovirus to replicate.

II. HISTORICAL DEVELOPMENT

The first virus discovered was tobacco mosaic virus (TMV), the agent responsible for a disease in tobacco plants. That this disease was transmissible by an agent produced in the sick plant was noted by Adolph Mayer in Holland in 1879. The identification of the agent as nonbacterial was made in 1892 in St. Petersburg by Ivanovsky, who recovered the agent in a filtrate from a filter that blocked passage of bacteria. The term virus had been used somewhat generally in the late nineteenth and early twentieth centuries to signify a toxic preparation of microorganisms made by biological material but distinct from other known microorganisms. In 1898, Beijerinck repeated Ivanovsky's experiments but proposed that the transmissible agent was a fluid rather than a virus particle. The first animal virus reported was the filterable agent that caused foot and mouth disease in cattle which was discovered by Loffler and Frosch in 1898; the first human virus was identified as the filterable agent that caused yellow fever and was discovered by Walter Reed and associates of the US Army Commission in 1900. Viruses causing tumors were first noted in chickens in 1911. They were discovered by Peyton Rous (at the Rockefeller Institute), who also worked on the rabbit papilloma virus identified by Shope in the late 1920s. In all these cases, it was the ability to cause disease in its host that actually defined and led to the name of the agent, and its designation as a virus was based on passage through filters that retained bacteria.

In 1915, F. W. Twort noted that viruses could kill bacteria. Also in 1915, F d'Herelle observed clear areas in lawns of shigella bacilli spread on agar culture dishes and referred to them as *taches vierges* or plaques. He showed that the killer agent was a filterable virus and called it bacteriophage in a 1917 publication that described the plaque assay, which involved a limiting dilution procedure of the toxic agent and led him to conclude that the agent was particulate. d'Herelle first described the binding of virus to a host cell as the initial stage of the infectious cycle, noting that the initial adsorption step was a major determinant of host cell specificity. He described cell lysis as the cytopathic effect of virus growth in the bacteria, which led to the release of newly replicated virus. Bacteriophages were discovered for many other bacteria and were initially tested as potential therapeutic agents for curing bacterial diseases.

The first evidence that viruses contained nucleic acids as well as proteins was obtained from experiments with bacteriophage by M. Schlesinger in Germany in the early 1930s. He also determined the mass for the phage particle based on diffusion studies, and his studies led to the hypothesis that viruses were essentially small, dense nucleoprotein particles.

The morphology of viruses had to await the development of the electron microscope because they were too small to be resolved by optical instruments. TMV was the first virus visualized in the electron microscope, which was developed in Germany in the late 1930s. A T-even bacteriophage was the first virus specimen observed with the electron microscope in the United States by Tom Anderson in 1942.

Crystals of TMV were obtained by Wendell Stanley in 1935, and their chemical analysis revealed that 5% of their mass was RNA. An X-ray diffraction pattern of these crystals was reported by Bernal in 1941 and showed the repeating unit structure of the virus rod. Experiments in the 1950s demonstrated that TMV could be reversibly dissociated into its component RNA and protein and that the RNA contained the genetic information for this virus. Prior to these studies, Hershey and Chase had reported in 1952 that the DNA of bacteriophage contained the genetic information for these viruses. This result provided one of the important landmarks in establishing that the genetic information was carried by nucleic acid.

A major concept which helped lead to an initial understanding of virus structure was the proposal by Watson and Crick in the early 1950s that the small size of viral genomes contained limited amounts of genetic information; thus, the virus capsid coats must be organized in repeating subunits of single or small numbers of protein species and a self-assembly mechanism should lead to their final structure. Caspar and Klug laid down the ground rules for this type of self-assembly that could lead to isometric structures in a classic paper published in 1962.

The report in 1926 that variants of TMV bred true and were revertible suggested that this virus contained mutable genetic information, but virus genetics did not advance significantly until studies of the bacteriophages in the 1940s. Emory Ellis and Max Delbruck were primarily responsible for initiating studies with the bacteriophages of *Escherichia coli*, and they elaborated the one-step growth curve experiment originally devised by d'Herelle and elaborated on by Burnet in the mid-1930s, who also noted distinctive phage variants in host range and in antigenic responsiveness. The ready isolation of mutants and their analysis by recombination led to major insights into the molecular nature of genetic events.

The breakthrough that led to advances in identifying and characterizing animal viruses can be attributed to the development of tissue culture systems in the 1950s. The various cells and culture conditions allowed for development of plaque assays for animal viruses and for analysis of animal virus growth under controlled and reproducible conditions. Identification of biochemical events in the replication of these viruses became tractable in these cell cultures. Importantly, such cultures could be used for growing virus in amounts useful for vaccine production, as exemplified by the work of Enders and colleagues at Harvard Medical School using monkey kidney cells to produce the human poliovirus. Thus, in the 1960s and 1970s, molecular components of many different kinds of viruses were identified and viral genomes began to be mapped initially by recombination, then by DNA restriction enzymes, and, ultimately, by determination of the sequence of the nucleotides.

A surprise that came from these initial studies was the observation that some viruses contained, within the particle, enzymes that were essential for the initial stages of virus replication; the first identified was a DNA-dependent RNA polymerase carried in the poxvirus. Some RNA viruses were found to contain RNA-dependent RNA transcriptases tightly bound to the ribonucleoprotein structure. Another major breakthrough occurred in 1970 with the independent discoveries by Baltimore and Temin that a class of RNA-enveloped viruses responsible for cell transformation contained an enzyme that copied DNA from RNA. Temin proposed several years earlier the existence of a DNA copy of the Rous sarcoma RNA virus based on the requirement for cell replication to obtain progeny of the replicating virus and the inhibition of virus replication by a drug which blocked DNA transcription to RNA. In addition to the essential role of this enzyme in the life cycle of the RNA viruses, now known as retroviruses, it became possible to convert any RNA molecule into DNA. In the 1980s and 1990s, with the introduction of genetic engineering, the genomes of both DNA- and RNA-containing viruses could be cloned and modified. The ability to clone and amplify viral genes has played an essential role in identifying new viruses and developing diagnostic assays for viruses such as HIV and hepatitis C virus.

Undoubtedly, HIV is the most important virus currently circulating in the human population. This retrovirus, the etiologic agent for adult immunodeficiency syndrome (AIDS), was first recognized in the early 1980s. Transmission of this virus—primarily through sexual contact and, initially, blood transfusions—has led to an ongoing worldwide pandemic in which an estimated 30 million individuals have been infected.

III. REPLICATION

A. Attachment

The first step in the virus replication cycle is the attachment of the virus to its host cell. This interaction is relatively specific and, in general, accounts for the selectivity of a virus for its particular host and, within multicellular hosts, for a particular type of cell. There are unique domains within proteins on the outer surfaces of viruses which function as sites for attachment to host cell components, referred to as virus receptors. The latter are frequently molecules on the outer surface of the host cell which function as normal cellular components but have been coopted by a virus in order to promulgate its entry into the cell. The type of specificity that appears to play such an important role in the tropism of animal and bacterial viruses has not been observed for plant viruses.

Virus receptors are highly diverse and range from simple subcomponents of surface proteins, such as sialic acid, to more complex polypeptides, such as the members of the Ig superfamily of membrane embedded surface proteins. Measles virus utilizes the complement receptor CD46. HIV recognizes CD4 on T cells. The bacteriophage lambda uses the bacteria surface protein that normally functions to bind maltose for utilization of this sugar by the bacteria. Closely related viruses may use quite distinct cell receptors; for example, the intercellular adhesion molecule-1 is the receptor for one group of rhinoviruses, whereas another group uses a receptor that recognizes the low-density lipoproteins. Different viruses can use similar receptors; for example, sialic acid is recognized by orthomyxo viruses, papovaviruses, reoviruses, and coronaviruses. Heparin sulfate is recognized by herpesviruses and by some alphaviruses. Many viruses use more than a single receptor for host-cell binding, which can extend their host range. For example, HIV binds to CD4 on T cells and also to the glycolipid, galactosyl ceramide, on neuronal cells and intestinal epithelium.

Binding affinities of viruses to cells are not exceptional, and effective attachment is the result of multiple interactions due to the number of repeating domains on the surface of a virus that can bind receptors. In several cases, binding may require a coreceptor which most likely functions in the second stage of virus entry. Another mechanism of virus binding is one in which the virus is complexed with an antibody, and the Fc region of the antibody interacts with cells, such as macrophages, that contain Fc receptors on their surface. This type of interaction can lead to enhanced infection in cases in which an individual has been previously exposed to the virus.

B. Entry and uncoating

The binding of virus to a host cell is followed rapidly by uptake and delivery of the virus genome into the cytoplasm or nucleus of the cell. Two major pathways are known for accomplishing this step in animal cells. One is receptor-mediated endocytosis, a general mechanism for uptake which consists of the formation of a membrane pit coated with clathrin, a protein capable of forming a protein-coated lipid vesicle. The receptor-bound virus enters the cell incorporated in the clathrin-coated vesicle. The clathrin coat dissociates, leaving an endocytic vesicle that takes up protons via an ATP-driven pump which lowers the pH within the vesicle. Proteins on the surface of the virus respond to this lower pH by changes in their conformation such that they can form pores within the endosome membrane that act as entry channels for translocation of the virus genome sequestered within the particles.

For many viruses, particularly RNA viruses such as influenza virus, alphaviruses, and vesicular stomatitis virus, which contain a lipid envelope surrounding their nucleocapsid, channel formation requires a fusion of the virus with cell lipids. This fusion is driven by virus-encoded glycoproteins that localize to the surface of the virus particle, but their fusogenic potential is cryptic and nonfunctional in the virus particle outside the host cell. The low pH of the endosome provides one kind of signal for activation of the fusogenic structure. The conformation change in the influenza glycoprotein, the hemagglutinin, leads to an elongated, hydrophobic coiled coil structure.

The second entry pathway delivers virus genetic material directly from the cell surface and is generally believed to require coreceptors for pore formation. Their role would presumably be to trigger conformational changes in virus surface proteins that mimic the low pH-driven systems. For HIV, two specific chemokines have been identified as coreceptors.

For picornaviruses such as poliovirus and rhinovirus and some plant icosahedral viruses, crystallographic analyses coupled with cryo-EM provide information about how the binding of these viruses to host cell receptors may trigger them for disassembly. Rhinovirus has on its surface a crevice or canyon that contains sites for binding the host cell receptor and the floor of the canyon has a pocket occupied by a long-chain fatty acid or sphingosine. The latter stabilizes virus against heat and low pH. In a model for virus penetration into the host cell the binding of receptor releases the pocket factor and leads to conformational changes in protein capsid subunits VP1 and VP3 and movement of VP4 from an internal position in the capsid shell to an external location. This creates a pore in the membrane that allows the threading of the RNA from its compact form in the virus to an extended configuration in the cell cytoplasm.

C. Genome expression and replication

Uptake of a virus into a cell initiates the eclipse phase of replication, which is defined as that period in which infectious particles are not detected within the cell. This illustrates one of the unique characteristics of viruses: the synthesis of the different components followed by their assembly into infectious particles. Virus genome expression generally is initiated very soon after entry into the host cell. For DNA viruses that infect higher eukaryotic cells, with the exception of the Poxviridae, the genomic material must be transported to the nucleus. Poxviruses replicate entirely in the cytoplasm and their genes code for the enzymes required for their DNA replication and transcription. DNA viruses that replicate their genomes in the nucleus (and bacterial DNA viruses) may use host enzymes to carry out these reactions. For most DNA viruses, however, genomic replication includes reactions that are distinct from the ones carried out by the host cell and are often dissimilar for different viruses. Those activities that are unique to the replication of a specific DNA viral genome would require the introduction of new proteins encoded by the viral genome. Some enzymes that seem to duplicate host cell functions may also be encoded by a virus genome; this redundancy may serve as a means to increase or to differentially regulate some of the reactions required for viral genome replication.

Most RNA viruses, with some exceptions such as members of the Orthomyxoviridae and Retroviridae families, replicate entirely in the cytoplasm and no nuclear functions are required for their replication. Most retroviruses require insertion of their genome into the host cell chromosome for their expression, and for many such viruses cell division is required for the DNA copy of the virus to be integrated. Lentiviruses, which include HIV and are members of the Retroviridae family, possess nuclear localization signals in their nucleoproteins and other virus gene products that enable virus genome integration to occur in the absence of host cell division.

An essential and common feature for all viruses is their need to synthesize mRNAs that will be translated into the proteins required for replication and assembly. For positive-sense RNA viruses, the incoming genomic RNA is a messenger RNA and can be translated directly to produce proteins required for further transcription and replication of the RNA. Cells do not contain the enzymes for transcribing the genomes of negative-sense RNA viruses or the DNA of poxviruses into mRNA in their cytoplasm, and for these viruses, transcriptases carried within the particles transcribe the genome into mRNA as the first stage of intracellular replication. Viral mRNAs are translated using the host cell ribosomes and translation factors. For many positive-sense RNA viruses, the mRNA is translated from a single initiation site near the 5′ end of the genome. A polyprotein is produced which is proteolytically processed to smaller, individual polypeptides that function in genome replication and in assembly of progeny virus. Much of this proteolytic processing is carried out by proteases encoded in the virus genome. The genes of most DNA viruses and both the double-stranded and negative-sense RNA viruses are transcribed as mRNAs encoding single polypeptides in the manner similar to those of the host cell. Viruses can utilize a variety of means to expand the number of proteins encoded by their genome. These include translation of proteins from more than one initiation site in a gene, alternative splicing, frameshifting, and posttranscriptional editing which involves the insertion of additional nucleotides into mRNAs causing a change in the reading frame.

A separation of virus replication into early and late stages can be readily distinguished for the large DNA bacteriophages and for DNA viruses infecting animals. The early stages are those that occur before the genome is replicated and include transcription of the genome into mRNAs and, for the large DNA viruses, synthesis of proteins involved in regulation of transcription and enzymes required for viral DNA replication. Genome replication—the synthesis of many copies of the genome which will be incorporated into newly formed progeny—is required to initiate the late steps, which involve the synthesis of the structural proteins of the virus and the interaction of the genome with those proteins leading to assembly of infectious particles and, for many viruses, their release from the cell. Viral mutants blocked in early steps are unable to carry out the late steps, but mutants with defects in late steps can complete the early steps.

The division into early and late steps can also be made for some RNA viruses but not for all. Members of the Picornaviridae and Flaviviridae families are examples of viruses that initiate translation at a single site on the RNA genome. The viral proteins required for genome replication and virus structure are synthesized at essentially the same time. There is still a kinetic division, as there is for all viruses, into the eclipse phase in which the viral proteins are synthesized and the genome is replicated and a phase of assembly and release. However, the steps cannot be separated genetically as described previously. Replication of these RNA genomes requires the synthesis of complementary copies of the genome which then serve as templates for the synthesis of new copies of genomic RNA. The viral proteins essential for these steps have been termed the nonstructural proteins; they are encoded in the virus genome but are not part of the virus particle. Host proteins may also be involved in genome replication, but for most viruses the proteins and how they function have not been determined.

Some positive-strand RNA viruses, such as members of the Togaviridae, the Coronaviridae, and the plant viruses Bromoviridae, translate nonstructural proteins required for transcription and replication from genomic RNA and other proteins, including those required for virus structure and assembly from subgenomic mRNAs. This division permits different proteins to be produced in different amounts. Those proteins that function as enzymes in transcription and replication are needed at lower concentrations than structural proteins that are required in large amounts for assembly. Early steps are those involved in translation of the incoming genomic RNA to produce the proteins for transcription and replication followed by their interaction with genomic RNA to synthesize complementary strands. Later steps include the production of more genomic RNA and the subgenomic RNAs, translation of these RNAs, and assembly, but these are difficult to distinguish except in the cartoons shown in textbooks.

The first step in expression of the genome of negative-sense RNA viruses is transcription of the genome into mRNAs. This step can be distinguished from subsequent steps because it can occur in the absence of any viral protein synthesis and does not require the host cell. Host cell factors are required, however, for the subsequent steps to generate the positive-strand complement of the genome that serves as a template for the replication of more genomic RNA.

D. Assembly

The final stages in virus production involve selective interactions between the newly formed virus-encoded structural proteins and the newly replicated genomic nucleic acid to form nucleoprotein particles, often referred to as nucleocapsids. These structures may have helical or icosahedral symmetry. For some viruses, assembly initiates at a nucleation site in which specific regions of the virus nucleic acid bind the protein subunit followed by interactions with additional polypeptides. In general, it is the sequential addition of closely related or identical protein subunits that drives a self-assembly pathway leading to the formation of the nucleocapsids. Alternatively, in other viruses, oligomeric substructures of proteins form prior to binding nucleic acid, and for some of the larger DNA viruses the nucleic acid is incorporated into the particle after the capsid has been almost completely formed. For these more complex viruses, substructures form prior to the final assembly process, and additional scaffolding proteins and chaperones as well as enzymes that cleave protein and nucleic acids are required transiently to effect the assembly. There does not appear to be a specific mechanism by which nonenveloped viruses are released from cells. The cytopathic effects of infection can cause sufficient disintegration of the cells to allow for virus release and spread to neighboring cells.

The final stages of assembly and release of newly replicated viruses that are enveloped by lipid bilayers occurs by a process called budding. In this activity, the assembled nucleocapsids bind to the cytoplasmic face of a cellular membrane. In most cases, the clustering of a viral-encoded, membrane-associated protein acts as a nucleation site and binding is followed by additional protein–lipid interactions which drive the lipid bilayer to fold around the particle, ultimately fusing and pinching off the virus segment from the host cell but keeping the host cell sealed. Figure 91.2 shows electron micrographs illustrating the assembly of the alphavirus, Sindbis virus. The virus-encoded proteins at the budding sites are either transmembranal glycoproteins

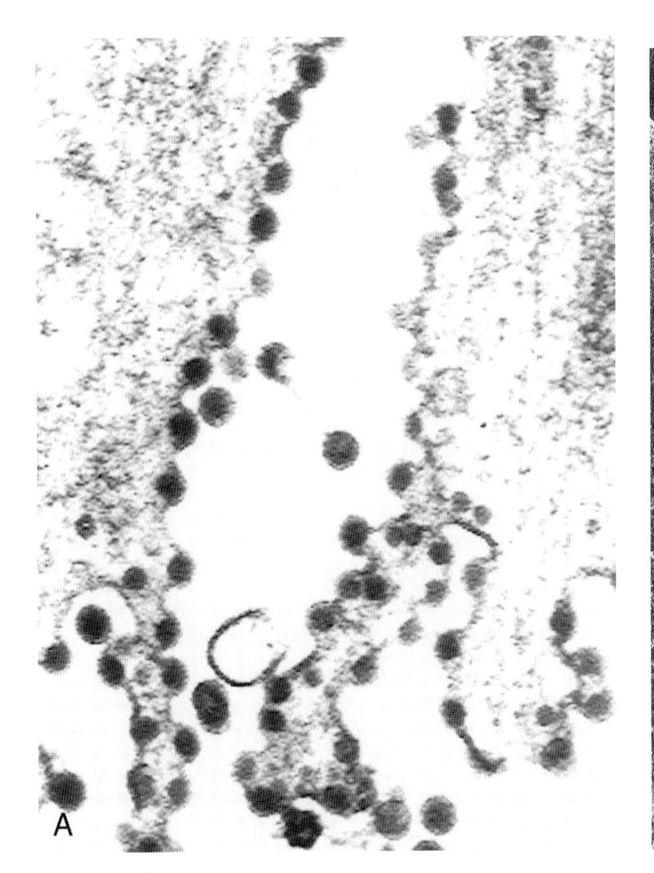

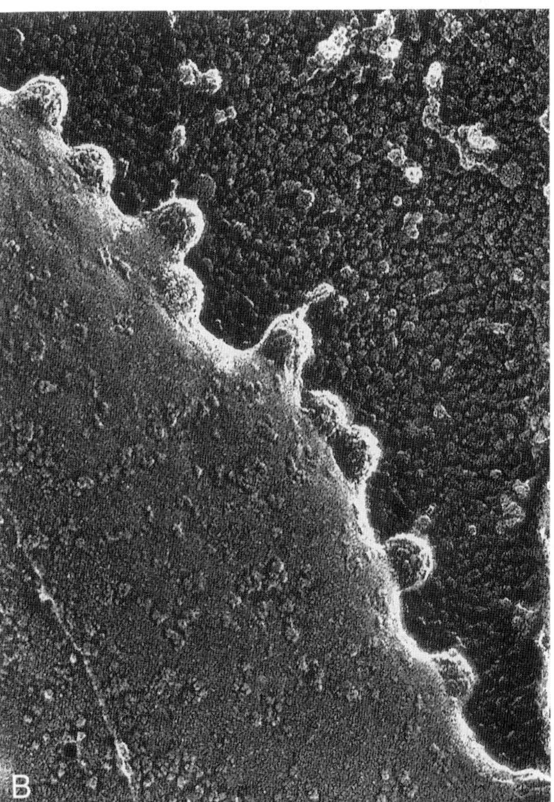

FIGURE 91.2 Electron micrographs of Sindbis virus, a member of the Togaviridae family, budding from the plasma membrane of infected cells. (A) Fixed and stained thin sections. Darkly stained, icosahedral nucleocapsids of the virus are shown at the cytoplasmic face of the plasma membrane prior to release. Magnification × 100,000 (courtesy of M. Aach-Levy, Department of Cell Biology and Physiology, Washington University School of Medicine). (B) Samples were quick frozen, deep etched, and freeze fractured and surface replicas were prepared. Structures protruding from the cell surface represent membrane envelopment of virus nucleocapsids. Magnification × 150,000 (courtesy of Professor John Heuser, Department of Cell Biology and Physiology. Washington University School of Medicine).

synthesized and transported to a specific membrane or proteins with lipophilic domains attracted to the cytoplasmic face of membranes. The different membranes of a eukaryotic cell (i.e. the endoplasmic reticulum, the Golgi stacks, the nuclear membrane, or the plasma membrane) can be utilized for this budding process, and the choice of membrane is determined by where the virus-encoded membrane proteins are localized. In the process, host cell membrane proteins are for the most part excluded from the virus membrane, but there is a selection or enrichment of specific lipids at the budding sites.

IV. GENETICS

A low level of mistakes in base pairing can occur during replication of the genome of a virus and may result in the formation of mutants detectable as phenotypic variants. For DNA-containing viruses, corrective repair or proofreading decreases the rate of mutation to about 10^{-6}–10^{-8}. However, with RNA-containing viruses or those utilizing RNA in genome replication (e.g. retroviruses), these values are much higher—on the order of 10^{-3}–10^{-5}. A stock of an RNA virus is essentially a population that almost always contains variants, and for this reason the population has been termed "quasispecies." Selective pressures during the replication cycle can lead to the emergence of different virus species. For example, antiviral drugs can give rise to new variants that have become drug resistant. Variants can also be isolated by specifically inserting nucleotide base changes into viral genomes by site-directed mutagenesis utilizing recombinant DNA methodologies.

A special set of deletion mutants may arise during continuing virus replication and act to limit virus growth. These mutants retain the set of sequences required for genome replication and for packaging of the genome into cell-free particles but otherwise have extensive deletions and rearrangements of the virus genome. They are defective and require the nondefective virus as helper to provide the missing enzymes and factors required for replication. These mutants are termed defective interfering or DI particles. The replication of their genomes competes effectively with that of the standard virus genome and can damp down the severity of infection.

Genetic recombination between viral DNA genomes may occur at homologous regions and has been analyzed in detail, especially for several of the DNA bacteriophage. DNA viruses also recombine with and become incorporated into the host cell genome, leading to what has been termed a provirus or latent state. Recombination between DNA molecules that occurs by a mechanism of break and rejoin of molecules is a well-established phenomenon; recombination between nonsegmented RNA virus genomes is less well understood. It appears to occur by a copy choice or template switching mechanism during nucleic acid synthesis. RNA viruses with segmented genomes, such as the influenza and reoviruses, can participate in a reassortment of genetic information in cells that are multiply infected with different strains of the same virus. Some progeny will then contain gene segments from both parents. For reassortment to be detected, some of the gene segments in the different strains must differ genetically so that the progeny that had undergone reassortment is phenotypically different from the parents. The ability of different strains of influenza virus to undergo reassortment in nature has led to some of the pandemics caused by this virus. This type of reassortment can introduce new variants of the viral surface proteins into an influenza virus, producing a strain to which the human population has not been exposed and would not be immune.

V. VIRUS–HOST INTERACTIONS

A. Interactions between viruses and the host cell

Viruses were first recognized by their ability to kill cells and cause disease, but interaction of a virus with a host cell does not always lead to death. Perhaps the most detailed study of how a cell survives viral infection is that of the bacteriophage lambda. Lambda is a DNA-containing bacteriophage that can follow one of two pathways when it infects its bacterial host, *E. coli*. In one pathway the virus synthesizes viralspecific proteins, replicates its DNA, assembles new particles, and kills the cell. Under some conditions, however, infection follows a different pathway. Replication of the viral DNA as an independent entity is repressed by the presence of a specific viral-encoded protein termed the repressor, the lambda genome becomes integrated into the chromosome of the bacteria, and the cell survives. This is known as the lysogenic state. Lambda DNA can be excised from the chromosome under conditions that are adverse for bacterial growth leading to a lytic state and the production of high levels of new bacteriophage. The molecular mechanism by which lambda bacteriophage becomes latent and is reactivated is very well understood. There are many examples of viral latency among animal DNA viruses, in particular the herpesviruses, but the mechanisms of latency and activation for these viruses are not as well understood.

In recent years, mechanisms by which viruses are able to kill cells have come under further scrutiny, particularly in studies with cultured mammalian cells. In many cases, infection causes the cell to undergo a programmed series of reactions, known as apoptosis, that result in death. Some viruses (e.g. adenovirus and Epstein–Barr virus) code for proteins that interfere with the steps leading to apoptosis, and cells infected with these viruses may survive. In these and in many other examples in which cells survive, virus replication is suppressed. Cells that are infected with adenovirus and survive may be altered in growth control (transformed) and are able to cause tumors in animals. Epstein–Barr virus infects human B lymphocytes and does not complete its replication cycle in most cells. Although the infection is aborted, it can lead to the immortalization of the cells, allowing them to survive in culture.

Several DNA viruses of animals are able to transform cells, making them tumorigenic, and the viruses can cause tumors in animals. These viruses encode proteins such as the T antigen of polyoma and SV40, the E6 and E7 gene products of papillomavirus, and the E1a and E1b proteins of adenovirus that interact with and alter tumor suppressor proteins of normal mammalian cells. Transformation is usually observed only in cells in which the complete virus replication cycle is blocked.

The only RNA viruses associated with transformation and tumorigenesis are members of the Retroviridae family. In the first decade of the twentieth century, viruses (later identified as retroviruses) were shown to cause leukemia in chickens and a different retrovirus designated as Rous sarcoma virus was determined to cause solid tumors in chickens. More than 60 years later, the ability of retroviruses to cause tumors could be explained by the incorporation of a modified host gene into the viral genome. Infection of normal cells by a virus carrying one of these modified host genes (such as the *src* gene or the *ras* gene, which code for proteins that are involved in normal growth control) can transform that cell into one that has become cancerous. In other cases, retrovirus sequences were integrated into a position in a host chromosome that altered the level of transcription of a particular regulatory gene.

Many RNA viruses are cytopathic to the cell that they infect. These viruses may usurp the translational machinery of the cell so that viral proteins are synthesized at the expense of host cell proteins. It has been possible to isolate variants of some of these cytopathic viruses that can establish persistent infections. In most cases that have been studied in detail, cells infected with these variants produce lower levels of viral products and this may be an important factor allowing survival of both the host cell and the virus. In this type of persistent infection essentially all the cells remain infected. In other cases, infected cells synthesize cytokines such as interferons, which are proteins that induce many cellular pathways that inhibit viral replication. Cells that are producing interferons remain for the most part refractory to virus infection, but a small fraction of the cells may remain permissive for a productive infection. In this type of persistent infection most of the cells remain uninfected.

B. Interaction of viruses with the host organism

Virus infection of humans most often conjures up diseases such as polio, influenza, hepatitis, or AIDS. Mechanisms by which viruses cause disease are complex and depend on multiple factors, but for many viruses disease does not always follow infection. An important finding in the studies that led to the development of vaccines to prevent the disease of poliomyelitis was that infection with this virus was a common event and only a small fraction (approximately 1%) of those infected developed a paralytic disease. Poliovirus enters the body through the mouth and replicates in the intestine. It is only when poliovirus infects the neurons of the gray matter of the brain (medulla oblongata) and spinal cord (anterior horn cells) that the paralytic disease is likely to develop. In contrast to infection by poliovirus, infection by viruses such as measles, rabies, influenza, and HIV almost always lead to disease.

There are several different routes by which a virus can enter an individual to cause disease. In addition to the gastrointestinal route taken by enteric viruses such as poliovirus, viruses such as rhinoviruses and influenza virus enter through the respiratory tract and viruses such as HIV, papillomaviruses, and herpesvirus type 2 are most likely to enter via the genitourinary tract or the rectal mucosa, although they can also be introduced by direct inoculation (e.g. with a needle). Some viruses are usually not transmitted from person to person but are carried by insect vectors or infected animals. Viruses, particularly those in the Togaviridae and Flaviviridae families, replicate in insects and are then transmitted to humans by an insect bite. These viruses are among those originally grouped as arboviruses. A few of them, such as the virus that causes yellow fever, have been controlled in some parts of the world by elimination of their mosquito host.

Both rabies virus, a member of the Rhabdoviridae family, and hantaviruses, a member of the Bunyaviridae family, are introduced into humans through infected animals. Humans are then considered "dead-end hosts" because the virus usually is not spread from the infected person to other people.

For some viruses, the site of entry is also the site of replication and disease manifestation. This is clearly illustrated by viruses such as rhinoviruses and influenza virus which cause symptoms in the respiratory tract. This is not always the case, however, as mentioned previously for poliovirus. Viruses such as measles virus and varicella virus (the cause of chicken pox) also enter through the respiratory tract, but the symptoms of the disease (rashes) appear when the virus replicates in the skin. The most common route for a virus to spread from one site to another is through the blood stream, but some viruses such as HIV, rubella virus, and Epstein–Barr virus infect lymphocytes and can be transported within those cells to other sites in the body.

Most virus infections of humans that have been studied in detail involve acute disease, meaning that the symptoms of disease appear within a limited and defined period after infection. Although viruses are thought to also be associated with chronic diseases, it has been much more difficult to establish this connection. AIDS, however, represents a clear example in which the disease symptoms may not appear until years after the initial infection. The onset of AIDS is attributed to the ability of HIV to destroy those T cells expressing CD4 that are required in the immune response. In most cases, this leads to an inability of the infected individual to mount an immune response against opportunistic infections. The reasons why the length of time between HIV infection and disease can vary so dramatically remain unclear. For most chronic diseases, a connection between the disease and a virus infection has been much more difficult to establish, and the evidence in support of an association is derived mainly from animal models. There are several animal models in which the disease, initiated by a specific virus infection, shows a striking similarity to a human chronic disease, such as multiple sclerosis or some of the arthritic diseases. In some of these cases, the disease can be attributed to an aberrant immune response. In particular, extensive studies with animals have demonstrated that virus infection can lead to autoimmunity. One explanation for this is that virus proteins may share antigenic determinants with host proteins (molecular mimicry) and the immune response against the virus proteins could subsequently lead to an autoimmune response.

C. Protection against virus infection

1. *Vaccines*

The first virus used to inoculate humans as a way of protecting them against infection was a poxvirus isolated from cows. In 1798, Jenner performed his classic experiment of inoculation with cowpox to protect against the dreaded disease of smallpox, and the term vaccination derives from that time (*vacca* is Latin for cow). The mechanism of protection was not understood at that time, but we now know that infection with an attenuated virus will induce an immune response that can protect against subsequent infections with more virulent strains of that or very closely related viruses. The use of an attenuated strain does pose some risks, particularly of reversion to virulence, and it has not always been possible to isolate a strain that is both nonvirulent and able to induce a protective immune response. Another means of vaccination has been to use inactivated virus as the immunizing agent. The most successful example was the inactivated polio virus developed by Jonas Salk in the 1950s. Immunization with killed virus or with purified components of a virus is safer because reversion is not a risk, but protection usually does not last as long. In the past few decades vaccination against major diseases, including measles, mumps, and hepatitis B, has had a major impact on decreasing the frequency of these diseases.

2. *Eradication of virus diseases*

One of the major advances in public health in the last half of the twentieth century has been the eradication of smallpox. The elimination of this disease was made possible by the very great effort put forth by the World Health Organization (WHO) in providing mass vaccinations accompanied by extensive surveillance and containment. As a result, the world was declared free of smallpox in 1979.

Many factors made eradication of smallpox a reality, but perhaps the most important was that humans are the only natural host. There are no pockets of this virus existing in animal populations and so there would not be a natural source for reintroducing the virus into the human population. Two other important factors were the availability of a stable vaccine and the ability to recognize all cases of smallpox by its clear symptoms.

WHO has targeted polio for eradication. The problems in carrying out this program are more complex than those involved with smallpox, in part because infection by this virus does not always produce recognizable symptoms. Due to the extensive work of the Pan American Health Organization, eradication of wild-type poliovirus was achieved in the Americas in 1991. The only cases reported in these areas have been those associated with the live vaccine strains. An important part of the polio eradication effort has been the use of sophisticated molecular techniques which made it possible to identify the source of any strain of this virus and to trace its origins.

VI. VIRUSES AS VECTORS FOR GENE EXPRESSION

Traditionally, viruses have been considered as agents of disease, but in the past decade recombinant DNA technology has provided a new impetus for using viruses in beneficial ways. The genomes of many DNA and RNA viruses have been cloned as cDNAs. These cloned molecules, when introduced into cells either as DNA plasmids or after transcription into RNA, can produce infectious virus. It is becoming possible to identify each viral gene and to determine which genes are essential and at which step in the replication cycle they function. Moreover, other viral or cellular genes can be inserted into a viral genome, either in addition to or in place of native viral genes.

There are a variety of ways in which virus vectors may be used. The most obvious is in the large-scale production of a particular cellular or viral protein. Many viruses take over the protein synthetic machinery and cause the infected cell to synthesize large quantities of viral-encoded proteins. A heterologous gene inserted into a viral genome would be produced in amounts comparable to those of the virus proteins. A second area that holds much promise is that of vaccination. Attenuated viruses are now being used as vaccines. These vaccine strains could also contain genes of other viruses or other infectious agents for which there is no attenuated strain—the result being that a person would be vaccinated against several different infectious agents simultaneously.

The most valuable use of viruses as vectors would be in gene therapy. This type of therapy is being considered in the treatment of genetic diseases such as cystic fibrosis or hemophilia. Viruses provide an extremely effective means for introducing new genetic information into cells and several are currently being studied for possible use in genetic diseases and also in the treatment of cancer and cardiovascular and neurological diseases. Those viruses being tested include the DNA-containing poxviruses and adenoviruses, the RNA alphaviruses, and lentiviruses such as HIV. HIV could have special advantages for use in gene therapy, particularly because of its ability to be incorporated into the genome and express its genes in nondividing cells. At the end of the twentieth century, HIV is considered one of the most devastating infectious agents that is rampant in the human population. A hope for the future is that we will learn how this virus causes disease and delete or disarm those genes involved in pathogenesis while harnessing other components that allow the virus to be used for curing diseases.

BIBLIOGRAPHY

Baltimore, D. (1971). Expression of animal virus genomes. *Bacteriol. Rev.* **35**, 235–241.

Caspar, D. L. D., and Klug, A. (1962). Physical principles in the construction of regular viruses. *Cold Spring Harbor Symp. Quant. Biol.* **27**, 1–24.

Eigen, M., and Biebricher, C. K., (1988). Sequence space and quasispecies distribution. *In* "RNA Genetics" (E. Domingo, J. J. Holland, and P. Ahlquist, eds.), Vol. 3, pp. 211–245. CRC Press, Boca Raton, FL.

Knipe, D. M., Lamb, R., Howley, P. M., and Griffin, D. E. (eds.) (2001). "Fields' Virology", 4th edn. Lippincott Williams & Wilkins, Baltimore, MD.

Landry, M. L., and Hsiung, G. D. (1996). Diagnostic virology. *In* "Encyclopedia of Virology" (R. G. Webster and A. Granoff, eds.). Academic Press, London.

Murphy, F. A. (1996). Virus taxonomy. *In* "Fields Virology" (B. N. Fields, D. M. Knipe, and P. M. Howley, eds.), pp. 15–57. Lippincott–Raven, Philadelphia.

Nevins, J. R., and Vogt, P. K. (1996). Cell transformation by viruses. *In* "Fields Virology" (B. N. Fields, D. M. Knipe, and P. M. Howley, eds.), pp. 301–343. Lippincott–Raven, Philadelphia.

Ptashne, M. (1992). "A Genetic Switch." Cell Press/Blackwell, Cambridge, MA.

WEBSITES

American Society for Virology
http://www.mcw.edu/asv/

All the Virology on the WWW (D. M. Sander). Many links
http://www.tulane.edu/~dmsander/garryfavweb.html

The Big Picture Book of Viruses (D. M. Sanders)
http://www.tulane.edu/~dmsander/Big_Virology/BVHomePage.html

Science Org. Virology Laboratory. Many links
http://www.vir.gla.ac.uk/virlink.shtml

American Society for Virology History
http://medicine.wustl.edu/~virology/

92

Viruses, Emerging

Stephen S. Morse
Columbia University

GLOSSARY

arbovirus Arthropod-borne virus; it replicates in arthropods (e.g., mosquitoes, biting files, and ticks) and is transmitted by bite to a vertebrate host. Arbovirus is an ecological rather than a taxonomic definition; various arboviruses belong to a variety of virus families, including the Flavidiridae, the Bunyaviridae, and the genus *Alphavirus* of the family Togaviridae.

emerging viruses Viruses for which their incidence has recently increased and appears likely to continue increasing.

endemic Occurring naturally and constantly in a particular area (as opposed to epidemic).

epidemic The appearance of a disease in a population at a prevalence greater than expected.

hemorrhagic fever An infection manifested by acute onset of fever and hemorrhagic signs (blood vessel damage as indicated by petechiae on skin, internal bleeding, and, in severe cases, shock); typical of many zoonotic viruses.

incidence The number of people (in a given population) developing a specified disease or becoming infected at a specified period of time.

pandemic From Greek *pan* ("= all"); an epidemic so widespread that it covers virtually the entire world.

prevalence The frequency of a disease in a population. An epidemiologic measurement, it is similar to but not technically synonymous with incidence.

re-emerging infection An infection that had formerly been controlled but is now increasing again; usually an indication of a breakdown in public health measures.

vector An agent, usually animate, that serves to transmit an infection, e.g., mosquitoes.

xenotransplantation The transplantation of organs or cells from another species in order to replace a nonfunctioning organ (e.g., heart, liver, kidney, or pancreas) or cellular component (e.g., a bone marrow transplant).

zoonosis An infection or infectious disease of vertebrate animals that is transmissible under natural conditions to humans.

Emerging Viruses are viruses that have recently increased their incidence and appear likely to continue to do so. In less formal terms, they are viruses that have newly appeared in the human population or are rapidly expanding their range, with a corresponding increase in cases of disease. In recent years, many viral diseases have been identified for the first time. Some, such as AIDS, have made their debut alarmingly and dramatically. Other viruses, such as influenza, have long been known for their tendency to reappear periodically to cause major epidemics or pandemics. The reasons for these sudden manifestations of new viral diseases have in general been poorly understood, making it difficult to determine whether anything can be done to anticipate and prevent disease emergence. In many cases, the causes of viral emergence, although

The Desk Encyclopedia of Microbiology
ISBN: 0-12-621361-5

complex, are often less random than they seem. Because of the large number of viruses, this article is selective, focusing on common features shared by emerging viruses; no attempt has been made to include all possible candidates. For additional information on specific viruses or particular aspects, the reader is referred to the appropriate articles in this encyclopedia and to the Bibliography.

I. DEFINING EMERGING VIRUSES: "NEW" VERSUS NEWLY RECOGNIZED

A. Categories of emerging viruses: new or newly recognized?

The most noticeable category of emerging viruses includes those that seem genuinely new in some important respect, such as a first appearance in an epidemic of dramatic disease, a sudden increase in distribution, or novel mechanisms of pathogenesis. Many of the emerging viruses that immediately come to mind, such as HIV, are classified into this group. Most of this article will consider viruses in this category and what is known about the causes of their emergence.

A second group of potentially emergent viruses consists of viruses that, although not new in the human population, are newly recognized. A recent example is human herpesvirus 6 (HHV-6). Although identified only in the mid-1980s, HHV-6 appears to be almost universal in distribution and has recently been implicated as the cause of roseola (exanthem subitum), a very common childhood disease. Since roseola has been known since at least 1910, HHV-6 is likely to have been widespread for at least decades if not longer. Suggested roles of HHV-6 in chronic fatigue syndrome or as a cofactor in AIDS are still under study. Other viruses discussed later, such as parvovirus B19, are also likely to be in this category.

B. The newly recognized

The significance of the newly recognized but common viruses is still debatable. Some have been implicated in various chronic diseases, although much of the evidence is inconclusive. However, as chronic diseases have become increasingly important in industrialized societies, the impact of agents responsible for chronic diseases could be considerable. Otherwise, epidemiologically, newly recognized but common viruses generally do not represent an apparent threat because they are already widespread and may have reached an equilibrium in the population. Recognition of the agent can even be advantageous, offering new promise of more accurate diagnosis and possibly control.

Conceivably, with any type of agent, a change in the agent or (more frequently) in host condition might result in a new or more serious disease. This is evident with "opportunistic pathogens," agents that generally have limited ability to cause human infections and disease but may do so in suitable circumstances. Immunosuppressed or immunocompromised individuals, such as people with AIDS, are particularly prone to such infections. With many pathogens, expression of disease may also be altered by such host factors as nutritional or immune status or age at first infection. The effects of nutritional status are often obvious, but there may be more subtle effects as well. Recent experiments have demonstrated that a virulent variant could occur during infection of selenium-deficient mice with a normally avirulent (mild) Coxsackievirus B3 isolate. The mechanism is unknown. The virulent virus that appeared closely resembled other known virulent genotypes of the same virus. The genotype was stable and could lethally infect healthy mice. Although no other examples of this remarkable phenomenon have been found, it would be suprising if this was an isolated case.

C. The significance of new technologies in identifying new viruses

The importance of technological advances in identifying new viruses should be mentioned. Viruses such as HHV-6 and hepatitis C became apparent because the means were developed to demonstrate their existence. The recognition of HIV was dependent on the previous development of methods for growing T lymphocytes in culture. In a more general sense, the introduction of tissue culture methods, in the 1940s, was a major breakthrough in the study and characterization of viruses. It can be expected that new tools for detection will uncover new viruses. In particular, many new avenues have been opened by the recent development of such techniques as the polymerase chain reaction (PCR), which is capable of detecting 1 HIV-infected cell in 100,000. Because PCR can detect and amplify DNA in minuscule amounts of sample and is comparatively undemanding, it is rapidly finding favor in many applications. PCR has great potential for disease archeology and the study of evolution. Using PCR, many otherwise intractable samples can now be tested, even those from mummified human bodies 7000 years old and files preserved in amber. Recently, PCR was used, with generic primers to relatively well-conserved influenza virus sequences, to obtain sequence data on the influenza virus that caused the devastating pandemic of 1918 and 1919 from nucleic acid extracted from autopsy specimens preserved from victims of the pandemic.

The recent discovery of the Kaposi's sarcoma-associated herpesvirus (human herpesvirus 8) demonstrates the power of molecular diagnostic technologies, guided by epidemiological evidence, for pathogen discovery. Kaposi's sarcoma has long been known as a slowly progressing cancer in elderly Mediterranean men (essentially a chronic disease), but the disease, in a rapidly progressive form, achieved prominence in the 1980s as one of the first AIDS-associated diseases. (Interestingly, since then Kaposi's sarcoma as an AIDS-associated disease appears to have declined in incidence for unknown reasons.) Despite suspicions that Kaposi's sarcoma might be of infectious etiology, the cause remained elusive. A recently developed molecular technique, representational difference analysis, has been used to compare the DNA in tissues from Kaposi's sarcoma patients and controls. The sequence data were used to develop additional specific probes and made it possible to identify a herpesvirus in the Kaposi's patients. The virus, which has since been more extensively characterized, appears to be a gamma-herpes-virus related to such primate viruses as herpesvirus saimiri and (more distantly) Epstein–Barr virus.

II. EXAMPLES OF POTENTIALLY EMERGING VIRUSES

Predicting the greatest threats is a difficult task which is made more difficult by significant gaps in our knowledge. Although specifics differ, many viruses seem likely to merit inclusion on a world list of emerging viruses (Table 92.1). It must be cautioned that predicting the future is an endeavor notoriously fraught with pitfalls. HIV was arguably underappreciated when AIDS was first identified, and attempts to predict the next influenza pandemic are still largely guesswork. Unanticipated effects of changes in environmental or other conditions can bring an as yet undescribed or currently obscure zoonotic virus to world prominence, as occurred with HIV a few years ago or Lassa fever. Conversely, a prominent virus may become submerged by environmental changes or even driven to extinction, as was smallpox. Therefore, consideration of emerging viruses should emphasize the principles of viral emergence, and listings of specific viruses are offered only as examples.

Many viruses are currently prominent as emerging viruses or have become prominent in the recent past, including influenza; several members of the Bunyaviridae, including the hantaviruses (Hantaan, Seoul, and related viruses classified in the genus *Hantavirus*) and Rift Valley fever (in a separate genus, *Phlebovirus*, of the Bunyavirdae); yellow fever and dengue (in the Flaviviridae); possibly the arenavirus hemorrhagic fevers, including Junin (Argentine hemorrhagic fever) and Lassa fever, probably the Filoviridae (Marburg and Ebola); human retroviruses, especially HIV but also (human T-lymphotropic virus (HTLV); and various arthropod-borne encephalitides, including St. Louis encephalitis and Japanese encephalitis (both in the Flaviviridae) and Venezuelan equine encephalomyelitis, Eastern equine encephalomyelitis, Eastern equine encephalomyelitis, and (in Australia) Ross River (all three in the *Alphavirus* genus of the Togaviridae; the latter causes a dengue-like disease with arthritis as a frequent complication). These encephalitides are mosquito-borne but also have natural (non-human) vertebrate hosts.

Regarding the health of people in the United States, some viruses are of a greater concern than others. Of the viruses listed, and excluding viruses already established in the United States, influenza, HIV-2, dengue, and possibly some hantaviruses are of greatest immediate potential importance. Proximity makes dengue, which is spreading over the Caribbean basin, a special concern. Some native mosquito-borne encephalitides, such as LaCrosse and California encephalitis (both in the California group of the Bunyaviridae), St. Louis encephalitis, and Eastern equine encephalomyelitis, are generally sporadic. Venezuelan equine encephalomyelitis has been almost entirely eliminated from the United States, but reintroduction is possible.

In considering emerging viruses, it is essential to take a global view. A virus emerging anywhere in the world could, under favorable conditions, reach any other part of the globe within days. Accelerating environmental change and increased immigration also expose new populations to microbes that were once buried in the depths of rain forests or confined to remote villages. For example, in 1989, a man in an Illinois hospital died of Lassa fever. The virus is normally endemic to west Africa; the patient contracted the virus while visiting family in Nigeria but fell ill only after returning home a few days later. The ability of a virus to disseminate throughout the world in a short time is clearly demonstrated by new influenza pandemics, which blanket the globe within a few months of their inception (which usually occurs in China). Additionally, many diseases, such as Korean hemorrhagic fever (caused by Hantaan, the prototype hantavirus, with an estimated 100,000–200,000 cases a

TABLE 92.1 Examples of emerging viruses[a]

Virus	Signs/symptoms	Distribution	Natural host	Transmission
Orthomyxoviridae (RNA, 8 segments)				
Influenza[b]	Respiratory	Worldwide (from China?)	Fowl, pigs	Respiratory
Bunyaviridae (RNA, 3 segments)				
Hantaan, Seoul	Hemorrhagic fever + renal syndrome	Asia, Europe, United States	Rodent (e.g., *Apodemus*)	Contact with infected secretions
Rift Valley fever[c]	Fever ± hemorrhage	Africa	Mosquito, ungulates	Vector: *Aedes* mosquitoes
Flaviviridae (RNA)				
Yellow fever[c]	Fever, jaundice	Africa, South America	Mosquito, monkey	Vector: *Aedes aegyptia* (urban), other *Aedes* species (sylvan)
Dengue[b]	Fever ± hemorrhage	Asia, Africa, Caribbean	Mosquito, human/monkey	Vector: *Aedes aegyptia* (Asia: also *Aedes albopictus*)
Arenaviridae (RNA, 2 segments)				
Junin (Argen. HF)	Fever, hemorrhage	South America	Rodent (*Calomys musculinus*)	Contact with infected secretions
Machupo (Boliv.)	Fever, hemorrhage	South America	Rodent (*Calomys callosus*)	Contact with infected secretions
Lassa fever	Fever, hemorrhage	West Africa	Rodent (*Mastomys natalensis*)	Contact with infected secretions
Filoviridae (RNA)				
Marburg, Ebola	Fever, hemorrhage	Africa	Unknown	Contact; nosocomial through contaminated needles
Retroviridae (RNA + reverse transcriptase)				
HIV[b]	AIDS, etc.	Worldwide	?Primate	Blood transfusion; nosocomial through contaminated needles; sexual transmission

[a]Modified from Morse and Schluederberg (1990). © 1990 by The University of Chicago Press.
[b]Viruses of special concern for the near future.
[c]Transmitted by arthropod vector.

year in China), may be serious causes of illness or death in certain regions of the world, even if they are not an imminent threat to the health of people in the United States. The following sections provide a brief description of some of these viruses. Additional information on specific viruses or viral families can be found under the appropriate cross-references in this encyclopedia or in the sources cited in the Bibliography.

A. Arenaviruses

Members of the Arenaviridae cause hemorrhagic fevers in humans; their natural hosts are rodents. Both Old World and New World representatives are known. New World representatives include Junin (Argentine hemorrhagic fever), whose natural hosts include the rodent *Calomys musculinus*, and Machupo (Bolivian hemorrhagic fever), whose host is the rodent *Calomys callosus*. Several additional New World arenaviruses have been identified throughout South America in the past few years. For example, one arenavirus, provisionally named Guaranito, was described in 1991 during a dengue epidemic in Venezuela. The major Old World arenavirus, Lassa fever of west Africa, has as its natural host the rodent *Mastomys natalensis*. Lassa became infamous for its high mortality rate in Western medical missionaries who first came in contact with the virus in the late 1960s and early 1970s. Infected rodent hosts usually shed virus asymptomatically in their urine, and primary infection in humans is generally by contact with infected excreta from the rodent, probably through inhalation of aerosolized virus in the excreta. Secondary cases of Lassa fever in health care providers or family members can occur through contact with patients' blood or infected secretions.

B. Filoviruses (Ebola and Marburg)

The Filoviruses (Ebola and Marburg) are among the least understood of all viruses. Their natural hosts are unknown; some believe they originated in primates, but recent evidence favors bats as the most likely natural hosts. Human disease is typically fever with hemorrhage, and mortality can be high. Ebola has caused several epidemics in Africa. An epidemic in Zaire (now Congo) in 1976 involved 278 known cases, almost all of which were hospital acquired through contaminated hypodermic equipment or contact with patients; mortality was approximately 90%. In the same year, a separate epidemic in the Sudan involved almost 300 cases, with more than 50% mortality. Outbreaks since then have included the Cote d'Ivoire (1 case in a field naturalist, who survived) and Gabon, both associated with handling chimpanzee carcasses, and a well-publicized 1995 outbreak in Kikwit, Zaire (Congo), involving 315 known cases and 77% mortality. Filoviruses in imported Old World monkeys have also been of recent concern. Marburg virus was first identified in 1967 when 25 laboratory technicians in Marburg, Germany, and in Belgrade, Yugoslavia (Serbia), became sick after handling tissues from African green monkeys (7 died). Six medical workers and family contacts subsequently became infected but recovered. Since then, there have been 3 documented primary cases (all fatal) of Marburg which was acquired by travelers in Kenya or (in one instance) Zimbabwe; three individuals who became infected by contact with the primary cases all survived. In 1989 and 1990, monkeys (Asian macaques) in a facility in Reston, Virginia, died suddenly. A filovirus was isolated, although whether this was the primary cause of death in the monkeys is still not certain. The virus, now variously known as Reston filovirus or Reston strain of Ebola, appears less virulent than classic Ebola. Several animal handlers who apparently became infected did not develop acute disease. This outbreak was the subject of the best-selling book, *The Hot Zone*, by Richard Preston.

C. Flaviviruses (Dengue and Yellow Fever)

Yellow fever, a mosquito-borne disease characterized by fever and jaundice, remains a significant world health problem. Historically, the impact of yellow fever virus was tremendous. Yellow fever was so devastating to workers in the region that the Panama Canal could be completed only after yellow fever was controlled. Historians have documented the high mortality caused by yellow fever in European settlers coming to Africa in the nineteenth century. The development of a yellow fever vaccine was one of the first successes of the Rockefeller Foundation; the same vaccine is still in use today and remains effective. Despite the availability of effective vaccines, yellow fever virus is still unvanquished and is widespread in Africa and South America. Its origins as a human pathogen date back several hundred years at least, but human infection is only incidental for yellow fever. The natural cycle of infection is the sylvatic ("jungle") cycle in monkeys in tropical areas of Africa and South America (where it was probably introduced from Africa). Most human cases occur by incidental infection of people in areas where sylvatic yellow fever is well established. The virus is mosquito-borne, but different mosquito species are involved in different settings. In the sylvatic form, the virus is carried by local forest mosquitoes, which can also transmit infections among humans. The *Aedes aegypti* mosquito, a species well adapted to living within human habitations (the genus name is Latin for "house"), is generally the vector in "urban" yellow fever. It is generally believed that transport and movement of people in the slave trade disseminated both the yellow fever virus and the *A. aegypti* mosquito from Africa to other tropical areas. Although *A. aegypti* can be found in some portions of the southeastern United States, the last yellow fever epidemic in the United States was in New Orleans in 1905.

Dengue, another flavivirus carried by many of the same mosquito species as yellow fever, is in tropical areas worldwide (Africa, Asia, the South Pacific, South America, and the Caribbean) and continues to spread. Dengue is now widespread in the Caribbean basin. Cuba had more than 300,000 cases in a 1981 epidemic, and there was an epidemic in Venezuela in the winter of 1990 and in Brazil in 1991. Travelers returning to the United States from the tropics occasionally return with dengue: The federal Centers for Disease Control reported at least 27 confirmed cases of imported dengue (in 17 states) in the United States in 1988, which is a typical yearly rate. A more severe form, known as dengue hemorrhagic fever, occurs in many areas where dengue is hyperendemic and has been postulated to result from sequential infection with different dengue viruses that now overlap geographically in many tropical areas. In the New World, dengue hemorrhagic fever was first seen during the Cuban dengue epidemic of 1981. The frequency of dengue hemorrhagic fever is increasing as several types of dengue virus extend their range.

D. Hantaviruses

In the family Bunyaviridae, members of the Hantavius genus (Hantaan, Seoul, and related viruses such as Puumala and others in Europe) classically cause hemorrhagic fevers with renal syndrome (fever, bleeding, kidney damage, and shock). Hantavirus pulmonary syndrome (HPS), caused by some New World hantaviruses, was first recognized in 1993. Various hantaviruses are found in Asia, Europe, and the United States as naturally occurring viruses of rodents. The most prominent member of this family is Hantaan virus, the cause of Korean hemorrhagic fever. Named for a river in Korea, the disease first came to Western attention during the Korean War. At least 3000 U.S. and UN troops developed Korean hemorrhagic fever, more than 300 of which died. The disease has long been known in Asia, and it has been suggested that there is a description of Korean hemorrhagic fever in a Chinese medical text dating to the tenth century. Currently, approximately 100,000–200,000 cases of Hantaan are diagnosed annually in China compared with 471 in 1955. The major natural host of Hantaan virus in Asia is the striped field mouse, *Apodemus agrarius*. This rodent is not native to the United States; consequently, Hantaan virus is not likely to become established in the United States (of course, the evolution of a host range variant capable of infecting native rodents cannot be excluded as a theoretical possibility). There are many New World hantaviruses with native rodent hosts that have not been clearly associated with human disease (as well as other New World hantaviruses that do cause disease). Because any new variant would probably have to compete with these existing viruses or would have to be geographically distinct or have a different host range, new introductions of related viruses may be limited (however, many hantaviruses are already present). This does not necessarily apply to the rodent hosts, and a new rodent-borne virus could be introduced into the United States if a suitable rodent host established itself or if the host were a cosmopolitan rodent such as the domestic rat. For example, another Hantavirus, Seoul virus, is found in rats. The virus was originally identified in rats in Korea and has since been identified in urban rats living in American cities. It has been suggested that rats carried on ships from Asia may have introduced Seoul virus into the United States. In Korea, Seoul virus has caused hemorrhagic fever with renal syndrome similar to Hantaan virus but usually considerably milder. In the United States, acute disease has not been identified, although seropositive individuals have been found in some inner-city areas. There is some evidence, although inconclusive, for a possible association with chronic renal disease, including renal hypertension.

Pathogenic hantaviruses native to the Americas were not recognized before 1993. Then, in the late spring and summer of 1993, several patients, mostly young and previously healthy adults, were admitted to hospitals in New Mexico, as well as Arizona and Colorado, with fever and acute respiratory distress; many subsequently died from pulmonary edema and respiratory failure. By the time the outbreak ended in late summer, there had been more than two dozen patients, with 60% mortality. Serology and detection of genetic sequences by PCR provided evidence that a previously unrecognized hantavirus was the cause of the outbreak, and the condition was subsequently called HPS. The major reservoir was identified as *Peromyscus maniculatus*, the deer mouse, which was also the rodent most commonly trapped near houses in the area. A high percentage (20–30%) of captured *P. maniculatus* proved positive by serology or PCR. Using PCR, the same sequences were identified in tissues taken at autopsy from several of the patients and in tissue samples from local rodents. As of January 1999, 205 cases of HPS (mostly sporadic; that is, as isolated individual cases) have been identified in 30 states in the United States and 30 cases in three western Canadian provinces. On the basis of identification of hantaviruses in stored samples and other evidence, it appears very likely that most of these viruses have long been present in their natural hosts but are newly recognized as a cause of human disease.

Many related hantaviruses have been identified throughout the Americas. In North America, other HPS-associated hantaviruses have been identified in the rodents *Peromyscus leucopus, Sigmodon hispidus* (the cotton rat), and *Oryzomys palustris*. Several hantaviruses have also been identified in South America in the past few years. One virus, dubbed Andes, caused cases of HPS in Argentina as well as several outbreaks in Chile (one in 1997 involved 25 cases). Like most other zoonotic viruses, most hantaviruses do not spread readily from person to person, but there was evidence of person-to-person transmission of Andes virus during the outbreak in Argentina.

E. Hendra and nipah viruses

In 1994, an outbreak of respiratory illness claimed the lives of 13 horses (of 20 that became ill during the outbreak) and a horse trainer (one other worker was affected but recovered) in Hendra (a suburb of Brisbane), Queensland, Australia. A virus was isolated and identified as a paramyxovirus by electron microscopy and partial gene sequencing. Because the virus appeared most closely related to the morbilliviruses (the genus within the Paramyxovirus family that includes measles and canine distemper viruses), it was originally named "equine morbillivirus." However, gene sequencing data support its classification as a member of the Paramyxovirus family that, although more closely related to the morbilliviruses than to other genera, is distinct and probably deserving of a separate genus. Other studies determined that the natural host of the virus appeared to be fruit bats (particularly *Pteropus* species). Since the virus was neither "equine" nor a true morbillivirus, it was proposed to rename the virus "Hendra." In another outbreak (in a different part of Queensland and believed to have occurred in August 1994), 2 horses died and one human subsequently developed encephalitis.

A related virus, now called Nipah, was identified in Malaysia in 1999 when, beginning in September 1998 and continuing into April 1999, cases of encephalitis appeared. The cases were originally ascribed to Japanese encephalitis, but laboratory results proved negative. Subsequent laboratory work demonstrated that a different virus, related to but distinct from Hendra, was responsible. Pigs were infected as well as humans, and most of the human cases were occupationally related. As of April 4, 1999, there were 229 cases reported, with 111 deaths. Several cases (11 cases with 1 death) also occurred among slaughterhouse workers in Singapore who worked with imported pigs. Control measures in Malaysia included the slaughter of more than 900,000 pigs by April 1999. Studies are under way to determine natural host (quite possibly bats, by analogy with Hendra) and occurrence in other domestic animals.

F. Hepatitis

Five viruses, all unrelated, are now known as possible causes of human viral hepatitis, and several others have been suggested based on molecular evidence. In addition to the familiar hepatitis A and B viruses, there are hepatitis C virus, the transfusion non-A non-B hepatitis Virus characterized in the late 1980s hepatitis E, a water-borne virus from Asia identified several years ago; and delta agent, or hepatitis D.

Hepatitis C is common in the United States and may be responsible for 98% of current post-transfusion hepatitis, especially since routine testing of blood has greatly reduced hepatitis B in transfusions. Its identification, led rapidly to development of serologic and molecular tests for the virus (which it is hoped will lead to its reduction or elimination in the blood supply) as well as to trials of therapeutic interferon. The agent is an RNA virus which appears "flavivirus-like" (resembling; but not closely related to yellow fever and dengue viruses). In addition to post-transfusion non-A non-B hepatitis, hepatitis C may be an important cause of community-acquired hepatitis. In a recent study, more than half of 59 patients with "community-acquired non-A non-B hepatitis" (no blood transfusion) in one U.S. county were seropositive, suggesting additional modes of transmission for hepatitis C. In the past few years, other putative blood-borne hepatitis viruses (GB, TT, and others) have been identified by molecular methods, but their natural history and public health significance are still unclear. Hepatitis E was identified several years ago. It is a water-borne disease widespread in Asia and South America and is caused by an RNA virus that has not been fully characterized but appears to belong to a different viral family from that of the other known hepatitis viruses.

Delta hepatitis, discovered in 1977 in Italy, causes an acute fulminant hepatitis in hepatitis B carriers. Delta is a defective agent, consisting of a small RNA and a distinctive protein known as delta antigen, wrapped in a covering of hepatitis B surface antigen. Delta therefore requires hepatitis B as a helper virus, in essence parasitizing the parasite. Luckily, delta is still not common in the United States, except in some groups that are also at high risk for hepatitis B, but it is endemic in Italy and parts of South America. Also, delta is comparatively rare in Asia, where hepatitis B is very common. Because it "borrows" the hepatitis B surface antigen and requires co-infection with hepatitis B virus, increasing use of the vaccine for hepatitis B should reduce the occurrence of disease from delta. In many parts of the world, where hepatitis B immunization may not be widely practiced, delta remains a great potential menace. Delta is unique among known animal viruses due to its small size and the fact that portions of the agent appear to be related to viroids, which are very small infectious RNA agents of plants. These ultimate parasites are tiny pieces of nucleic acid with an independent agenda. Other small RNA agents similar to delta probably exist in animals but have not been identified.

G. Influenza

Influenza is one of the most familiar viruses, but it remains an important threat. Annual or biennial epidemics of influenza A are due to antigenic drift, a mutational change in the hemagglutin (H) Surface protein of the virus so that the host's immune system no longer recognizes the antigen and the new virus can reinfect until host immune responses are mounted to the new variant antigen. Pandemics, very large epidemics that occur periodically and involve virtually the entire world, are the result of antigenic shift, a reassortment of viral genes usually involving the acquisition by a mammalian influenza virus of a new hemagglutinin gene from an avian influenza virus. There are approximately 13 subtypes of the hemagglutin gene, although only a few H subtypes have been associated with human infection. These pandemic strains have generally originated in China. In June 1991, it was reported that an influenza virus with the H3 hemagglutinin but distinct from currently circulating varieties and believed to be of avian origin had appeared in horses in north-eastern China in 1989 and 1990. In 1997, several cases of human disease associated with H5 influenza were reported from Hong Kong, eventually involving 18 confirmed human cases with 6 deaths. Although H5 influenza has long been associated with highly lethal outbreaks in poultry, human infection with H5 influenza had not previously been noted. Control measures instituted by Hong Kong officials included the slaughter of all 1.6 million chickens in Hong Kong. The virus did not appear to spread beyond Hong Kong in this outbreak.

H. Parvovirus B19

Parvoviruses are the smallest DNA viruses of vertebrates (*parvo* is from the Latin word for small). The virus particle measures approximately 20–22 nm in diameter, and the viral DNA genome is approximately 5000 bases long. Parvoviruses typically replicate in rapidly dividing cells, such as cells lining the intestine or blood cell precursors in bone marrow. Canine parvovirus is another member of this family. Another human parvovirus, adenoassociated virus, has not been associated with human disease. Parvovirus B19 was discovered fortuitously in sera from healthy blood donors with false-positive reactions for hepatitis B antigen. After the virus was characterized, additionally studies indicated that approximately 60% of adults are seropositive. The virus is worldwide in distribution. Evidence implicates B19 as the cause of erythema infection, or "fifth disease," which is a mild, self-limited febrile disease of childhood. B19 has also been associated with aplastic crises in chronic hemolytic anemias (the sudden disappearance of blood cell precursors in bone marrow). B19 and other parvoviruses have also been suggested as causes of joint disease, but this is not clearly resolved.

I. Retroviruses

Human retroviruses have become the subject of intense scientific interest largely because of HIV. The HIV pandemic has become one of the defining conditions of the late twentieth century. Two major types, HIV-1 and -2 are known. HIV-1, which has several subtypes, is the cause of the main AIDS pandemic. HIV-2 has not yet disseminated as far as HIV-1, although the potential exists. Although HIV-2 appears to cause a more slowly progressing disease than HIV-1, AIDS caused by HIV-2 has been documented. The origin of HIV-1 remains obscure, but recent evidence support the hypothesis that HIV was introduced from a subspecies of chimpanzee, *Pan troglodytes troglodytes*, in Africa. Molecular evidence from currently circulating HIV-1 strains suggests that there have been at least three successful introductions of HIV-1 viruses into the human population from this chimpanzee subspecies.

Considerable work has also been done recently on HTLV types I and II. HTLV is widespread in some populations in Asia, the Pacific, and parts of the Caribbean, and HTLV II has been reported from an isolated aboriginal group in South America, possibly suggesting relative antiquity as a human virus. HTLV II is also spreading rapidly within certain populations in the United States, such as intravenous drug users. The spectrum of disease due to HTLV is still being defined. HTLV was originally identified in adult T cell leukemia–lymphoma. In tropical areas, HTLV I has been associated with a neurological disease, tropical spastic paraparesis, and a related condition, HTLV I-associated myelopathy, has been defined. It is not known whether HTLV II is responsible for human disease, although some atypical T cell leukemias have been advanced as possibilities. Because of serologic cross-reactivity, older studies did not always distinguish the two viruses, and it is possible that some disease attributed to HTLV I might have been caused by HTLV II.

More speculatively, various researchers have recently suggested that other neurological diseases or autoimmune diseases may be caused by human retroviruses, either HTLV or currently unknown types. There is only limited evidence in human disease; however, the recent demonstration in mice of viral superantigens, homologs of cellular genes that are carried by a mouse retrovirus and that can induce abnormalities in T cell development and possibly autoimmunity, suggests that similar roles for human viruses are possible.

Speculation about endogenous retroviruses is possible, but there are no data to decide the question. Humans, like virtually all eukaryotes, contain endogenous retroviral elements (retroviral-like sequences) in cellular DNA. In certain circumstances, an endogenous retrovirus, silent for generations in the host

DNA, can regain independent existence and become capable of infecting other hosts. Murine retroviruses have demonstrated this property in the past, and Robert Huebner suggested years ago that feline leukemia virus, a common and sometimes fatal infection of cats, probably originated in this way from an endogenous rodent retrovirus. Despite this potential, a similar event has not been identified in humans.

J. Rift valley fever

Rift Valley fever virus (RVF), a mosquito-borne virus found in Africa, was first recognized in 1931 as a livestock disease in European breed sheep and cattle introduced into Africa. The apparent recipe, as expressed by Karl M. Johnson, was "foreign animals, local virus, new disease." For unknown reasons, the severity of human disease has increased since the virus was first identified. The virus is found naturally in Africa in a number of ungulates (including camels), and RVF caused epizootics in several parts of Africa, with occassional disease in occupationally exposed animal handlers, veterinarians and butchers, but no human deaths. This changed in 1975 when South Africa experienced an epizootic with some human deaths. In 1977, RVF suddenly emerged in a dramatic zoonotic outbreak in Egypt that resulted in thousands of human cases and 598 reported human deaths. Rapid action (quarantine and immunization of livestock in Israel) prevented the virus from spreading over the Mediterranean basin. An outbreak in Mauritania in 1987 followed the Egyptian pattern but on a smaller scale, with an estimate of 1264 human cases and 224 deaths. The human infection is characterized by a fever, usually with hemorrhaging; retinitis is frequently seen.

K. Spongiform encephalopathies

Bovine spongiform encephalopathy (BSE), currently a cause of great concern in the United Kingdom, is another recently emerged disease. BSE belongs to a family of diseases which also includes scrapie in sheep and whose human counterparts include Creutzfeldt–Jakob disease and kuru. All are uniformly fatal. The term spongiform encephalopathy refers to characteristic histopathological lesions in the brain. Attempts to implicate conventional viruses or viroids have been unsuccessful, and Stanley Prusiner, in work that earned him the 1997 Nobel prize in medicine or physiology, proposed that the spongiform encephalopathy agents are a novel form of infectious material which he terms "prions," infectious self-replicating proteins. Creutzfeldt–Jakob disease is a sporadic presenile dementia (occurring in people younger than the age expected for senility) with an estimated prevalence of approximately 1 case per 1 million population. Familial forms, due to mutation in a gene coding for a specific protein (the prion protein, also known as amyloid A), have recently been identified. Kuru was extensively studied by D. Carleton Gajdusek (a 1976 Nobel laureate in medicine or physiology), who determined the now well known epidemiology of kuru, which was sustained within an aboriginal New Guinea population by ritual cannibalism.

At least in crude extracts, strains of the spongiform encephalopathy agents are among the most heat-stable infectious materials known, BSE appears to be an example of interspecies transfer, in this case possibly scarpie from sheep introduced into a new host species. Incompletely rendered sheep byproducts fed to cattle have been suggested as the vehicle, although the possibility cannot be excluded that BSE existed earlier but was unrecognized. It has been proposed that changes in rendering processes in the late 1970s and early 1980s, using lower temperatures and (perhaps more critically) lesser amounts of solvents, may have permitted some infectious material to survive the process. A similar interspecies transfer seems to have previously occurred (in the 1940s), resulting in the disease now known as transmissible mink encephalopathy. The recent identification of spongiform encephalopathy in felines that were fed BSE-contaminated meat by-products seems to have a similar history. Although the risk to humans is unknown, the relative rarity of Creutzfeldt–Jakob disease, the human equivalent, suggests that the risk with scrapie may be comparatively low. Creutzfeldt–Jakob disease is not noticeably more prevalent among sheep or cattle farmers, even though scrapie has been known in sheep for at least two centuries and there have presumably been many opportunities for human exposure comparable to the acquisition of BSE by cattle. However, BSE may possibly be more transmissible. In the United Kingdom, many cases of "variant CJD" in people younger than would normally be expected to develop classic Creutzfeldt–Jacob disease have been described and are believed to be linked to BSE.

Interestingly, except for the spongiform encephalopathy agents and parvoviruses, most of these emerging viruses contain RNA genomes. Although there is no compelling reason for this, one might speculate that this might be related to the diversity and mutability of RNA viruses. At least superficially, RNA viruses represent a great diversity of viruses and replication strategies. This diversity may be due at least in part to the high mutation rates shown by many RNA viruses. This in turn has been attributed to the error-prone nature of RNA replication and especially to the lack of a "proofreading" function in this process.

III. WHAT ARE THE ORIGINS OF NEW VIRUSES?

A. Possible sources of new viruses

Although it is not possible to attribute underlying causes or precipitating factors to all episodes of viral emergence, for many episodes at least some of the causes can be identified. Of course, we cannot be certain that these are the only causes, and the known examples probably show some unintentional ascertainment bias because the most explainable are the most likely to be included. Nevertheless, it is noteworthy that many episodes are explainable.

An examination of emerging viruses might appropriately question how a new pathogen might originate. If we assume the constraints of organic evolution, which essentially require that new organisms must descend from an existing ancestor (evolutionary constraints will be briefly considered later, there are fundamentally three sources (which are not necessarily mutually exclusive): (i) the evolution of a new viral variant (*de novo* evolution), (ii) the introduction of an existing virus from another species, and (iii) dissemination of an agent from a smaller human population in which the agent may have arisen or been introduced originally. The term "viral traffic" was coined to represent processes involving the access, introduction, or dissemination of viruses to their hosts, as distinct from *de novo evolution*.

B. Evolution of new viral variants

Considerable debate has centered around the relative importance of viral evolution versus transfer and dissemination of viruses to new host populations (viral traffic) in the emergence of "new" viral diseases. Evaluating the significance of *de novo* evolution is complicated by the difficulty of demonstrating that a new isolate is truly newly evolved and not merely a new introduction of an organism that has long existed in nature but was previously unrecognized. There appear to be a few documented examples of this "*de novo* evolution" in nature. Antigenic drift in influenza is probably the best known example of viral emergence due to the evolution of a new variant. Some additional examples of possibly or apparently newly evolved viruses are listed in Table 92.2. The list is not exhaustive, although attempts have been made to make it as inclusive as possible. Most viruses on this list cause diseases typical of their viral families or similar to the parental virus from which the new variant evolved. In both humans and horses, the recombinant Western equine encephalomyelitis (WEE), for example, causes a disease similar to Eastern equine encephalomyelitis but somewhat milder (its apparent evolutionary advantage is that WEE generally has different insect and bird hosts). A polio-like syndrome described in China and South America in the early 1990s is not included because the responsible virus is still not characterized and it is not clear whether this is newly evolved or the result of viral traffic.

TABLE 92.2 Known or suggested newly evolved viruses

Virus	Virus family	Remarks	Disease
Rocio encephalitis (Brazil)	Flaviviridae	Recombinant	H[b]
Western equine encephalomyelitis (United States)	*Alphavirus* genus (Togaviridae)	Recombinant	H
Influenza H5 mutant (chickens: Pennsylvania, 1983)	Myxoviridae	New variant	Severe respiratory infection in chickens
Influenza H7 (seals: United States, 1980)	Myxoviridae		H: Conjunctivitis
Enterovirus 70	Picornaviridae	?New strain	H: Conjunctivitis
Rev-T (strain of avian reticulo-endotheliosis virus)	Retroviridae	Avian	Fulminant lymphoma in fowl
Friend virus, spleen focus-forming strains	Retroviridae	Mouse	
Canine parvovirus 2	Parvoviridae	Dogs	Enteritis, cardiomyopathy (similar to parvoviruses infecting other species)

[a]From Morse (1994).
[b]H, associated with human disease.

In addition, there have been a few examples of variants with altered biological properties. Howard Temin offers the example of an avain retrovirus [reticuloendotheliosis virus strain T (REV T), which derived from the considerably less virulent REV-A strain] that became more virulent as a result of several accumulated genetic changes. Recent evidence demonstrates that human chronic hepatitis B infection was associated with a mutation in a viral gene for precore protein. Vaccine escape mutants of hepatitis B were also recently described in a few infected individuals, although some dispute the interpretation. Despite many demonstrations of this phenomenon *in vitro*, this is one of the few documented vaccine escape mutants isolated from natural infection in the field.

C. Role of viral traffic

The previously discussed examples notwithstanding, critical examination of known examples of viral emergence indicates that the overwhelming majority of such instances can be accounted for by viral traffic. At least over the admittedly limited time span of human history, most emerging pathogens have probably not been newly evolved. Rather, they are existing agents conquering new territory. The overwhelming majority probably already exist in nature and simply gain access to new host populations. The most novel of these emerging pathogens are zoonotic (naturally occurring agents of other animal species); rodents are among the particularly important natural reservoirs. Many instances of emergence can be attributed to precipitating factors which facilitate the introduction of viruses from the environment into human hosts or aid their dissemination or expansion from a smaller human population. The following section discusses some of these causes.

IV. FACTORS PRECIPITATING VIRAL EMERGENCE

A. Causes of viral emergence: The role of viral traffic

The previous analysis suggests that viral emergence can be viewed as a two-step process, involving the introduction of a virus into a human population followed by dissemination. Emphasis should therefore be placed on understanding the conditions that affect each of these steps. Although this discussion concentrates on viruses, most of the considerations are also applicable to most other emerging pathogens.

Many of the known examples of viral emergence share common features. They are usually precipitated by environmental or social changes, often induced by human activities (Table 92.3). The significance of the zoonotic pool is most apparent for viruses, which generally require a host in order to be maintained in nature. Considering that the total number and variety of viruses in animal species are probably very large, this offers a large pool of potential new virus introductions. In such cases, introduction of viruses into the human population is often the result of human activities, such as agriculture, that cause changes in natural environments. Often, these changes place humans in contact with previously inaccessible agents or increase the density of a natural host or vector, thereby increasing the chances of human infection. Examples among the viruses reviewed here include hantaviruses, Lassa fever, and Argentine hemorrhagic fever, all of which are natural infections of rodents, and probably Marburg, Ebola, and Rift Valley fever.

In the 1993 outbreak of HPS (the outbreak in which the disease first came to the attention of medical science), unusual climatic conditions in the winter and spring preceding the outbreak may have caused a massive increase in the rodent population, increasing human exposure to the virus. Reports from the Four Corners area suggested that the winter had been unusually wet, resulting in a large crop of nuts and other rodent food and, in turn, an exceptionally large rodent population and thereby offering more opportunities for people to come in contact with infected rodents (and hence the virus). This has apparently been confirmed by the ecological research station at Sevilleta, New Mexico, which documented a 10-fold increase in area rodent populations beginning in the spring 1993. The abnormally high precipitation has been attributed to climatic events caused by the ocean current known as El Niño.

Although this outbreak was precipitated by natural environmental changes that favored an increased rodent population, the responsible environmental changes are often human induced. For example, Argentine hemorrhagic fever has spread as the pampas were cleared for maize planting. The natural host of this virus, the mouse *Calomys musculinus*, flourishes in this environment, propagating its virus in the process. Cases of Argentine hemorrhagic fever have increased proportionately. Hantaan, although unrelated to the Argentine hemorrhagic fever virus, is acquired by humans in a similar way: Increased rice planting encouraged the little field mouse *A. agrarius*, the natural host of Hantaan virus. Infected mice shed virus in secretions such as urine. Humans normally

TABLE 92.3 Probable factors in the emergence of some emerging viruses[a]

Virus family, virus	Probable factors in emergence
Arenaviridae	
Junin (Argentine HF[b])	Changes in agriculture (maize, changed conditions favoring *Calomys musculinus*, rodent host for virus)
Lassa fever	Human settlement, favoring *Mastomys natalensis*
Bunyaviridae	
Hantaan	Agriculture (contact with mouse *Apodemus agrarius* during rice harvest)
Seoul	Increasing population density of urban rats in contact with humans
Rift Valley Fever	Dams, irrigation
Oropouche	Agriculture (cacao hulls encourage breeding of *Culicoides* vector)
Filoviridae	
Marburg, Ebola	Unknown; in Europe and United States, importation of monkeys
Flaviviridae	
Dengue	Increasing population density in cities and other factors causing increased open-water storage, favoring increased population of mosquito vectors
Orthomyxoviridae	
Influenza (pandemic)	Integrated pig–duck agriculture
Retroviridae	
HIV	Medical technology (transfusion, contaminated hypodermic needles); sexual transmission; other social factors
HTLV	Medical technology (transfusion, contaminated hypodermic needles); sexual transmission; other social factors

[a]Modified from Morse (1992).
[b]HF, hemorrhagic fever.

become infected during the rice harvest by contact with infected secretions in the rice fields.

There are numerous examples of this situation in which the emergence of the new virus or pathogen is associated with changing environmental conditions that favor contact of humans with a natural host for an existing virus. Although Lyme disease is bacterial rather than viral, it appears likely that similar environmental conditions are responsible for its recent emergence. Another example is monkey pox. The name seems misleading, because various arboreal mammals in the rain forest, mostly squirrels and probably not including monkeys, appear to be the natural reservoir hosts. Human monkey pox exposures appear most likely to orginate by contact with infected arboreal rodents as a result of hunting the animals for meat of exposure while foraging. It is still uncertain whether monkey pox is emerging. There is no indication that this is occurring, and deforestation in Africa is reducing human exposure to the virus.

Agriculture provides some other unexpected examples. Viroids, infectious agents of plants that consist entirely of small RNA without a protein coat, as far as is known are spread entirely by mechanical transmission on agricultural implements such as pruning knives and harvesters. The evolution of viroids could very likely have been shaped, unbeknownst to its human agents, by these human activities.

B. Pandemic influenza

Remarkably, the same principles as those discussed previously also seem to apply in certain circumstances to viruses in which there is an essential role for viral evolution in the success of new viral variants. Influenza is perhaps the most interesting example. Robert Webster calls influenza the oldest emerging virus that is still emerging; influenza A virus is one of the few known examples (aside from some arguable cases, such as HIV, it may be the only example) of an emerging virus whose emergence (actually, for influenza, periodic reemergence) can clearly be ascribed to viral evolution. Approximately every 20 years, influenza A undergoes a major antigenic shift in one key protein, known as the hemag-glutinin (H) protein, and a pandemic results. Although most changes in influenza virus H proteins occur by so-called antigenic drift involving the accumulation of random mutations (this drift can lead to the smaller, but still medically important, influenza epidemics seen every few years), new pandemic influenza viruses occur by a different route—that of major antigenic shifts. These invariably seem to involve a reassortment of viral genes belonging to different influenza strains. Thus, the important event in generating new pandemic influenza strains has not been mutational evolution but rather reshuffling of existing genes. Where do the genes come from? It has recently been found that most influenza genes are maintained in wildflow; every known subtype of the H protein can be found in waterfowl, such as ducks. Many virologists believe that pigs are an important mixing vessel allowing influenza virus to make a transition from birds to humans. Every known major influenza epidemic has originated in south China, where a traditional and unique form of integrated pig–duck farming has been long practiced. Christoph Scholtissek and Ernest Naylor suggested that this form of agriculture may facilitate the development of new influenza reassortants by placing ducks (the reservoir of a variety of influenza strains) and pigs (thought to be "mixing vessels" for mammalian influenza strains) in close proximity. Agriculture may therefore play the leading role in emergence of this virus. Also viral traffic, reassortant viruses from the mixing of animal influenza strains and the transmission of the resulting virus to humans, appears more important than new viral evolution for human disease.

C. Arboviruses

Water is an essential factor for arthropod-borne viruses, which include many important diseases worldwide, because many of the insect vectors breed in water. Japanese encephalitis accounts for almost 30,000 human cases annually in Asia, with approximately 7000 deaths (although immunization programs in several countries promise to control the disease in humans). The incidence of the virus is closely associated with flooding of fields for rice growing. In the outbreaks of RVF in Mauritania, the human cases occurred in villages near dams on the Senegal River.

Rapid urbanization has been blamed for the high prevalence of dengue in Asia. Profusion of water storage containers in cities, necessary to supply the dense and rapidly expanding human population, has caused a mosquito population boom, with a concomitant increase in the transmission of dengue.

Many types of human activities may also disseminate mosquito vectors or reservoir hosts for viruses. It was mentioned previously that both yellow fever virus and its principal vector, the *A. aegypti* mosquito, are believed to have been spread from Africa via the slave trade. The mosquitoes were apparently stowaways in the large open-water containers that were kept on the ships to provide water during the voyage. In a more modern repetition, an aggressive vector of dengue virus, *Aedes albopictus* (the Asiatic tiger mosquito), was recently introduced into the United States in shipments of used tires imported from Asia. From its entry in Houston, Texas, the mosquito has established itself in at least 17 states. In this century, many other mosquito species have been introduced to new areas in war material being returned from foreign theaters of action.

The recent rapid spread of raccoon rabies in the United States has a similar cause. In the past few years, rabies in raccoons has moved from the southeast, where it has been localized for some time, to the northeast, and it appear that raccoons will become a major wildlife source of rabies in the northeastern United States. After identifying its first rabid raccoon in October 1989, New Jersey reported 37 rabid raccoons during the first third of 1990. The Centers for Disease Control has implicated sport hunting as the main factor in this explosive spread of raccoon rabies. In order to ensure an adequate supply of raccoons for hunting, a group of hunters imported Florida raccoons, some of which were apparently rabid, to the area of western Virginia, from whence it spread further north.

D. Gateways for viral traffic: The expansion of human viruses

Highways and human migration to cities, especially in tropical areas, can introduce remote viruses to a larger population. On a global, scale, similar opportunities are provided by rapid air travel. For example, HIV probably traveled along the Mombasa–Kinshasa highway and came to the United States presumably through travel. Health officials have also linked the movement of young men from villages to the cities, and resulting freedom from local restraints on behavior, with dissemination of HIV in Africa.

Once introduced into a human host, the success of a newly introduced pathogen depends on its ability to spread within the human population. A similar situation applies to agents already present in a limited or isolated human population because the agents best adapted to human transmission are likely to be those that already infect people. Here, too, human intervention is providing increasing opportunities for dissemination of previously localized viruses. The example of HIV demonstrates that human activities can be especially important in disseminating newly introduced pathogens that are not yet well adapted to the human host and do not spread efficiently from person to person.

Finally, as a highway for viral traffic, the possibilities for iatrogenic disease should also be mentioned. Cases of Lassa fever, Ebola, and Crimean–Congo hemorrhagic fever, in addition to more familiar viruses, have been acquired by health care workers caring for infected individuals or spread in hospitals. As demonstrated by well known examples such as HIV and hepatitis B, many viruses that might not otherwise transmit easily from person to person may be transmitted, through transfusions, organ transplants, or contaminated hypodermic needles, allowing the donor's viruses direct access to new hosts. For many viruses that were not able to spread efficiently from person to person, including HIV, this circumvents their lack of effective means of transmission. As these life-saving procedures become more widely used, and as the scarcity of donors forces medical centers to search farther afield, it is reasonable to expect more instances of disease. In the past few years, similar concerns have been raised about xenotransplantation as a possible route for introducing new zoonotic infections into the human population.

E. Animal viruses as models of interspecies transfer

In order to supplement our knowledge of interspecies transfer of viruses, examples of emerging viruses in animals might also be considered as useful models for interspecies transfer and mechanisms of viral emergence.

Two of the best studied examples are canine parvovirus and seal plague. Canine parvovirus [officially called canine parvovirus type 2 (CPV-2)] first emerged in approximately 1978 as an epidemic disease of dogs. By 1980, the virus spread globally in the dog population, and the virus is now endemic in every country that has been tested. Slight variants, designated as CPV-2a and CPV-2b, appear to have subsequently displaced the original CPV-2 in the dog population. The virus appears to be closely related to two other parvoviruses, mink enteritis virus and feline panleukopenia virus (the cause of feline distemper). Typical signs are enteritis and cardiomyopathy, often with leukopenia, in animals infected with these viruses. The ability of CPV-2 to infect domestic dogs may have been conferred by a mutation in the capsid (coat protein) gene. Consistent with this hypothesis, it is known that the host range for parvoviruses is at least partly determined by the capsid.

Seal plague, a newly recognized paramyxovirus, is related to measles and canine distemper viruses, but it is distinct. The viruses may have been transferred directly from another species of seal; it has also been speculated that an outbreak of canine distemper might have been related, although this is not clear and seems less probable. Outbreaks of canine distemper, probably from infected dogs, have been associated with die-offs of lions in the Serengeti park in Africa in recent years.

In addition to their possible utility as model systems, emerging viruses of other vertebrates are worth further consideration because serious diseases of live-stock or of food plants can have great economic impact and, in the worst case, can cause widespread starvation. Occasionally, some might be potential zoonotic viruses, as was RVF.

V. WHAT RESTRAINS VIRAL EMERGENCE?

Understanding the factors restraining emergence is of great importance, but data are limited. This section, summarizing current knowledge, is necessarily speculative in nature. It is clear that we are still discovering many of the applicable rules, as indicated by the Hong Kong example of H5 influenza in humans—a surprise when it was identified because human infection with H5 had not been known previously.

Although many viruses have high mutation rates and thus potentially may be evolving rapidly, few have shown striking changes in pathogenesis. In order to survive, viruses must be maintained in nature in a living host. There are constraints imposed by the requirements for a means of transmission and the relatively few routes by which a virus can infect a host. Such requirements must impose strong selective pressures on a virus. Therefore, although variants are continually being generated, presumably a stabilizing influence is exerted by natural selection as the virus replicates in its natural hosts. Even if viral mutation is currently unpredictable, genotypic variation and phenotypic change are not equivalent, and there remain evolutionary constraints at the phenotypic level at least. This may be the reason that ecological and demographic factors appear to be at least as important in influencing viral emergence as viral mutation or evolution. Unfortunately, when changes have occurred, they have generally not been predictable. Occasionally, an apparent recombinant virus will emerge, such as Rocio encephalitis or WEE. Why most of them are unsuccessful is not known.

In addition to emerging, viruses may also disappear or be displaced by new variants for reasons that are often poorly understood. Influenza A H7N7, once frequent in horses, has almost disappeared, having apparently been displaced by the H3 subtype. Some have suggested that a similar fate may befall H2N2 influenza in humans. The original strain of CPV-2 is being displaced by a variant, CPV-2a, which in turn will be displaced by CPV-2b. However, although smallpox was eradicated, it is apparently not being replaced by monkey pox, despite the latter's ability to cause sporadic human cases. Several recent viral emergences, such as Rocio encephalitis, Lassa fever, and Marburg disease, have fortunately remained limited or (for Rocio) apparently have resubmerged.

As Edwin Kilbourne pointed out, many zoonotic introductions fail to become self-sustaining in the human population. The restricted range of disease, often severe acute hemorrhagic fevers, and the often limited ability of the virus to spread from person to person are evidence that most of these viruses are not well adapted to human infection. Viruses already adapted to humans—many of which may have been zoonotic introductions in the past that did evolve successfully—and that are present in an isolated human population are more likely than newly introduced zoonotic viruses to disseminate if conditions favor their transmission. HIV is a case in point.

Mathematical ecologists have suggested from mathematical analysis that self-sustaining infection involves a trade-off between transmissibility and virulence (Paul Ewald, taking this analysis a step further, suggested that rate of transmission can affect virulence so that factors that cause more rapid transmission of the virus favor increased virulence and, conversely, slowing down transmission can select for decreased virulence). As in the cause of smallpox, a highly virulent virus can sustain itself if it is also easily transmitted. Many zoonotic introductions cause severe primary disease, (hence they are highly virulent), but appear to have low transmissibility and therefore have not become established in the human population. This is not an evolutionary imperative for them because they survive in nature in their natural hosts, to which they are better adapted. Nevertheless, the impact of zoonotic introductions can be severe, as shown by the Ebola outbreaks in Africa. In addition, as discussed previously, some can now be transmitted inadvertently by artificial means, such as blood transfusions or contaminated needles, circumventing the requirement to evolve greater transmissibility.

Therefore, understanding restrictions on viral variation and emergence would appear to be of prime importance. Some of this will hinge on improved understanding of viral evolution, especially in the ecological context. The relatively low rate of emergence of viral disease may be due in part to the limited entry points whereby viruses gain access to new hosts.

In addition to the considerations mentioned previously, successful viruses also need to evade or subvert host defenses. The macrophage is likely to be one important cell in this process. Other factors restraining the emergence of new viruses, such as inability to spread from person to person, are still not understood at the molecular level. Moreover, an improved understanding of receptors, tissue-specific transcription factors, and tissue-specific mechanisms of cell killing will be essential for defining target cell specificity and host selective pressures restraining viral variation.

VI. PROSPECTS FOR PREDICTION, CONTROL, AND ERADICATION

A. Strategies for anticipating viral emergence

Although many of the questions about emerging viruses may have scientific foundations, allocating resources and setting priorities are often more affected by social, economic, and political factors. This is nowhere more apparent than in considering strategies for anticipating and controlling emerging infections. Environmental and social factors can be major determinants of viral emergence. In principle, this means that emergence can be better anticipated and potentially controlled. In practice, this requires making political decisions such as defining appropriate and workable precautions for development programs, balancing programs so that both development needs and health protection requirements can be met, and mobilizing resources to accomplish these objectives on a world-wide basis. However, although molecular technologies such as sensitive immunoassays and PCR are providing powerful tools for identifying and tracking viruses, resources for global surveillance and control are currently inadequate. Research and clinical facilities are in dangerously short supply, and a critical dearth of trained researchers is expected by the next generation.

From historical experience, new viruses appear most likely to emerge from tropical areas undergoing agricultural and demographic changes and in the periphery of cities in these areas. Surveillance of such areas for the appearance of disease outbreaks or novel diseases would therefore seem advisable. In 1989, D.A. Henderson proposed an international network of surveillance centers located in tropical areas, especially on the edges of expanding tropical cities. Each center would include clinical facilities, diagnostic and research laboratories, an epidemiological unit which could include disease investigation and local response capability, and a professional training unit and would be linked to an international network for data analysis and instant response to emergencies.

One challenge has long been how to develop surveillance capabilities that could provide warning at the early stages of disease emergence, wherever this may occur, to complement the coverage provided by established centers in urban areas. The rise of the internet, and the global connectivity it offers, provides unprecedented opportunities for better disease surveillance worldwide. An example is the e-mail list ProMED-mail. In order to provide a vehicle for bringing together scientists and health officials from around the world to develop and promote coordinated plans for global disease surveillance, ProMED, the Program for Monitoring Emerging Diseases, was formed in 1993 under the auspices of the Federation of American Scientists. Because the need for effective communications was repeatedly mentioned in our discussions, in 1994 we sought the help of SatelLife, a nonprofit organization in Boston, to connect participants worldwide by e-mail. The e-mail list, dubbed ProMED-mail, rapidly evolved into a prototype system for real-time reporting of disease events and discussion of emerging diseases. The technology also makes it possible to cross-correlate similar events that may be occurring in different places. Open to all interested persons at no charge, ProMED-mail now has more than 1000 subscribers worldwide.

If we are often the engineers of viral traffic, we need better traffic engineering. Currently, analytic knowledge of viral traffic is not advanced enough to allow predictions and long-term advance planning based on its principles. However, it is conceivable that such knowledge could be made more systematic in the future, allowing better anticipation of viral and microbial traffic, and better predictive ability.

B. Existing methods for control of viral diseases

Human viruses that have been substantially controlled in industrialized countries by immunization include polio, rubella, and measles. In some populations (primarily American and European travelers from areas without yellow fever who are immunized before exposure if they are traveling to endemic areas), yellow fever is prevented by immunization. Potential for immunization exists for several other viruses, such as hepatitis B, mumps, and perhaps cytomegalovirus; vaccines of reasonable efficacy are available for these viruses. Rabies immunization in domestic animals has greatly reduced the occurrence of human rabies in the United States to one to three cases a year, although wildlife remain a source of potential exposure for both humans and domestic animals.

Public health measures have traditionally been directed at combating transmission or protecting potential susceptibles through immunization. In addition to immunization, traditional control measures include improved sanitation, mosquito control programs, health certification of travelers, and health inspection of imported livestock. Traditional public health programs have been instrumental in containing many potential threats but also have several drawbacks. Their success with the targeted diseases depends on vigilance and assiduity. Efforts may fall victim to their own success, being prematurely relaxed or abandoned, usually to save money, and allowing the conditions that precipitated the program in the first place to be reestablished. Many mosquito control programs have met with this fate after initial partial success.

In many cases, relatively simple solutions are possible, if there is sufficient global resolve to implement them. For example, rebuilding water supply systems in tropical cities to reduce or eliminate open-water storage could have a real impact on dengue. In other cases, there may be efficacious vaccines or other preventive measures, but problems in deployment allow old agents, placed under control by improved environmental conditions and public health measures, to regain a foothold or expand. Epidemics of yellow fever in some areas (Nigeria reported 600 deaths in a July 1991 outbreak) are an example. Adding yellow fever immunization to the World Health Organization (WHO) worldwide Expanded Program on Immunization, might well prevent further epidemics such as the one in Nigeria. Several U.S. cities experienced epidemics of childhood measles in 1990 and 1991. A recent government report attributes much of the increase in measles cases to cutbacks in childhood vaccination programs, with the result that some children were inadequately immunized or immunized too late. Most programs also cannot contain viruses that can spread efficiently from person to person, such as influenza. The current strategy used for influenza is to attempt to track emerging new strains and to immunize when feasible.

C. Prospects and requirements for eradication

Given that viruses that had been major scourges in the past have been controlled or (rarely) even eliminated by immunization or public health measures, it is logical to consider the prospects of eradication for some of the diseases discussed here. The greatest victory so far has been smallpox, now officially extinct. One of the most feared of all viral diseases, as well as reputedly one of the most easily transmissible, smallpox has had a long history. The historian William McNeill suggested that the Spanish conquest of Mexico may have been aided by the effects of smallpox, which the

Europeans brought with them. Jenner's work with vaccination made smallpox one of the first examples of immunization; a form of immunization, using material from smallpox lesions, may also have been practiced earlier in China. In the twentieth century, universal childhood vaccination and health control at borders succeeded in bringing the virus under control in Western industrialized countries, except for occasional imported cases and their contacts. Smallpox was last seen in the United States in 1949. As control measures continued to reduce the geographic distribution of small-pox, eradication was adopted as a feasible goal by WHO. An intensive world-wide campaign of vaccination and surveillance in smallpox-endemic areas eventually succeeded in eliminating the disease. In 1979, with an official certification by WHO that was confirmed in May 1980 by the World Health Assembly, smallpox became the first virus to be declared extinct.

This success with a previously feared virus encouraged public health agencies to consider additional candidates for eradication. (Ironically, this success has led to the dilemma of whether to destroy the existing strain collections of smallpox—currently centralized at two laboratories—or to retain them for future research. Some have expressed concern that smallpox might be obtained by terrorists or by countries interested in developing biowarfare capabilities, and there is a lack of effective antiviral treatment. In May 1999, in part because of these concerns, the World Health Assembly agreed to defer destruction for at least another 3 years.) For successful eradication, self-sustaining infection within the human population must be eliminated, and there must be no natural non-human reservoir of infection from which the virus could be reintroduced. In practice, eliminating self-sustaining infection in humans usually involves immunizing a large proportion of the susceptible population and requires effective vaccines. These characteristics—the ability to prevent infection through immunization and the lack of an additional reservoir of the virus—made smallpox vulnerable to eradication. On the other hand, yellow fever does not meet these criteria. There is a suitable vaccine so transmission to people could theoretically be eliminated, but its maintenance in the sylvatic cycle would prevent eradication of yellow fever from the environment. Many of the other viruses discussed here, including by definition all of the zoonotic and arthropod-borne viruses, have natural reservoirs that would render eradication impracticable. Viruses that have no additional reservoirs but that have long periods of infection or transmission, such as human herpesviruses and HIV, would also be difficult to eradicate unless effective lifelong immunity could be established in susceptibles. On the other hand, measles and polio meet the criteria, and are likely targets for the future. They have no known reservoirs outside humans and effective vaccines are available. Polio eradication, a world-wide goal for the beginning of the twenty-first century, will probably not be achieved but natural polio has essentially been eliminated from the Western Hemisphere. For measles, there is currently no definitive timetable for world-wide eradication, but health agencies hope to eradicate it from the Western Hemisphere by late 2000. Thus, even when eradication is possible in principle, the goal may prove elusive or impracticable.

The successes have been encouraging, but the examples of viral emergence and reemergence or resurgence should caution us against complacency. Continuing improvement in nutrition and sanitation has been responsible for reducing the overall impact of infectious disease in industrialized countries. However, most of the world has still to benefit from these simple improvements, with predictable consequences. With respect to viral emergence, human intervention is providing increasing opportunities for dissemination of previously localized viruses, as in the case of HIV and dengue. Because human activities are a key factor in emergence, anticipating and limiting viral emergence is more feasible than previously believed but requires mobilizing effort and funds, especially on behalf of the Third World. One speculation is that episodes of disease emergence may be-come more frequent as environmental and demographic change accelerate. Evidence suggests that both the scientific and the social challenges of viral emergence are likely to continue for the foreseeable future.

ACKNOWLEDGMENTS

I thank Joshua Lederberg, Mirko Grmek, Howard Temin, John Holland, Frank Fenner, Walter Fitch, Gerald Myers, Edwin D. Kilbourne, Robert E. Shope, Thomas P. Monath, Karl M. Johnson, Peter Palese, Hugh Robertson, Baruch Blumberg, D.A. Henderson, Seth Berkley, Barry Bloom, James M. Hughes, James LeDuc, Patrick S. Moore, and other colleagues for invaluable comments and discussions. Special acknowledgments for ProMED to Jack Woodall, Barbara Hatch Rosenberg, Dorothy Preslar and the Federation of American Scientists, and the members of the steering committee. In modified form, some of the views and interpretations in portions of this article are based on those expressed in earlier articles in other publications, including "The Origins of 'New' Viral Diseases" (*Environ. Carcinogen. Ecotoxicol. Rev.* **C**9(2), 1991).

I was supported by the National Institutes of Health, U.S. Department of Health and Human Services (RR 03121 and RR 01180) and by the Arts & Letters Foundation (Milford Gerton Memorial Fund). Work on emerging viruses was supported by the Division of Microbiology and Infectious Diseases, National Institute of Allergy and Infectious Diseases (NIAID), National Institutes of Health (NIH), and by the Fogarty International Center of NIH. I am grateful to Dr. John R. La Montagne, Deputy Director, NIAID, and Dr. Ann Schluederberg, former virology branch chief, for their support and encouragement.

BIBLIOGRAPHY

Anderson, R. M., and May, R. M. (1991). "Infectious Diseases of Humans. Dynamics and Control." Oxford Univ. Press, Oxford.

Benenson, A. S. (Ed.) (1995). "Control of Communicable Diseases Manual," 16th ed. American Public Health Association, Washington, DC.

Binder, S., Levitt, A. M., Sacks, J. J., and Hughes, J. M. (1999). Emerging infectious diseases: public health issues for the 21st century. *Science* **284**(5418), 1311–1313.

Desselberger, U. (2000). Emerging and re-emerging infectious diseases. *J. Infect.* **40**(1), 3–15.

Emerging Infectious Diseases [journal published by the Centers for Disease Control and Prevention (CDC)]: Current and past issues available on-line at *http://www.cdc.gov/EID/* or from the CDC homepage (*http://www.cdc.gov*), which also provides links to additional sources of information.

Fields, B. N., Knipe, D. M., and Howley, P. M. (Eds.) (1996). "Fields Virology," 3rd ed. Lippincott–Raven, Philadelphia.

Fricker, J. (2000). Emerging infectious diseases: a global problem. *Mol. Med. Today* **6**(9), 334–335.

Lederberg, J. (2000). Infectious history. *Science* **288**(5464), 287–293.

LeDuc, J. W. (1989). Epidemiology of hemorrhagic fever viruses. *Rev. Infect. Dis.* **11** (Suppl. 4), S730–S735.

McNeill, W. (1976). "Plagues and Peoples." Doubleday, New York.

Morse, S. S. (1991). Emerging Viruses: Defining the rules for viral traffic. *Perspect. Biol. Med.* **34**, 387–409.

Morse, S. S. (Ed.) (1992). "Emerging Viruses." Oxford Univ. Press, New York.

Morse, S. S. (1994). Toward an evolutionary biology of viruses. *In* "The Evolutionary Biology of Viruses" (S. S. Morse. Ed.). Lippincott–Raven, Philadelphia.

Morse, S. S. (1995). Factors in the emergence of infectious diseases. *Emerg. Infect. Dis.* **1**, 7–15. [Also available on the Internet from the CDC website.]

Morse, S. S., and Schluederberg, A. (1990). Emerging viruses: The evolution of viruses and viral diseases. *J. Infect. Dis.* **162**, 1–7.

ProMED-mail (e-mail list for reporting and discussion of infectious disease events): Available on-line (from SatelLife website) at *http://www.healthnet.org/programs/promed.html* or (from Federation of American Scientists website) at *http://www.fas.org/promed/index.html/*.

Schuchat, A. (2000). Microbes without borders: infectious disease, public health, and the journal. *Am. J. Public Health* **90**(2), 181–183.

Wilson, M. E. (1991). "A World Guide to Infections. Diseases, Distribution, Diagnosis." Oxford Univ. Press, New York.

WEBSITES

APEC (Asia Pacific Economic Cooperative) Emerging Infections Network
http://www.apec.org/infectious/about/index.htm

National Institutes of Allergy and Infectious Diseases NIH, USA
http://www.niaid.nih.gov/default.htm

Department of Defense Emerging Infections System, USA
http://www.geis.ha.osd.mil/

Infectious Diseases Society of America Emerging Infections Network (EIN)
http://www.idsociety.org/EIN/TOC.htm

Website of the journal Emerging Infectious Diseases
http://www.cdc.gov/ncidod/eid/

Emerging Infectious Diseases (an NIH journal)
http://www.cdc.gov/ncidod/eid/about.htm

93

Yeasts

Graeme M. Walker
University of Abertay Dundee

GLOSSARY

bioethanol Ethyl alcohol (ethanol) produced by yeast fermentation for use as a fuel or industrial commodity.

birth scar Concave indentations that remain on the surface of daughter cells following budding.

budding A mode of vegetative reproduction in many yeast species in which a small outgrowth, the daughter bud, appears and grows from the surface of a mother cell and eventually separates to form a new cell at cell division.

bud scar The chitin-rich, convex, ringed protrusions that remain on the mother-cell surface of budding yeasts following the birth of daughter cells.

Crabtree effect The suppression of yeast respiration by high levels of glucose. This phenomenon is found in *Saccharomyces cerevisiae* cells which continue to ferment irrespective of oxygen availability due to glucose repressing or inactivating the respiratory enzymes or due to the inherent limited capacity of cells to respire.

fission A mode of vegetative reproduction found in the yeast genus *Schizosaccharomyces*. Fission yeasts grow lengthwise and divide by forming a cell septum that constricts mother cells into two equal-size daughters.

Pasteur effect The suppression of yeast sugar-consumption rate by oxygen; alternatively, under anaerobic conditions glycolysis proceeds faster than it does under aerobic conditions. In *Saccharomyces cerevisiae*, the Pasteur effect is only observable when the glucose concentration is low (below around 5 mM) or in resting or starved cells.

respirofermentation Yeast fermentative metabolism in the presence of oxygen.

Saccharomyces cerevisiae Baker's or brewer's yeast species, which is used widely in the food and fermentation industries and is also being exploited in modern biotechnology (e.g. in the production of recombinant proteins) and as a model eukaryotic cell in fundamental biological research.

sporulation The production of haploid spores when sexually reproductive yeasts conjugate and undergo meiosis.

Yeasts are eukaryotic unicellular microfungi that are widely distributed in the natural environment. Around 800 yeast species are known, but this represents only a fraction of yeast biodiversity on Earth. The fermentative activities of yeasts have been exploited by humans for millennia in the production of beer, wine, and bread. The most widely exploited and studied yeast species is *Saccharomyces cerevisiae*, commonly referred to as "baker's yeast." This species reproduces asexually by budding and sexually following the conjugation of cells of the opposite mating type. Other yeasts reproduce by fission (e.g. *Schizosaccharomyces pombe*) and by formation of pseudohyphae as in dimorphic yeasts, such as the opportunistic human pathogen *Candida albicans*. In addition to their wide exploitation in the

The Desk Encyclopedia of Microbiology
ISBN: 0-12-621361-5

production of foods, beverages, and pharmaceuticals, yeasts play significant roles as model eukaryotic cells in furthering our knowledge in the biological and biomedical sciences. The complete genome of *S. cerevisiae* was sequenced in 1996 and that of *Schiz pombe* in 2002, and research is now underway to assign physiological functions to each of these sequenced genes. Work with yeasts is not only leading to insights into how a simple eukaryote works, but also insights into human genetics and an understanding of human heritable disorders.

I. DEFINITION AND CLASSIFICATION OF YEASTS

A. Definition and characterization of yeasts

Yeasts are recognized as unicellular fungi that reproduce primarily by budding, and occasionally by fission, and that do not form their sexual states in or on a fruiting body. Yeast species may be identified and characterized according to various criteria based on cell morphology (e.g. mode of cell division and spore shape), physiology (e.g. sugar fermentation tests), immunology (e.g. antibody agglutination and immunofluorescence), and molecular biology (e.g. ribosomal DNA phylogeny, DNA base composition and hybridization, karyotyping, and random amplification of polymorphic DNA). Molecular-sequence analyses are being increasingly used by yeast taxonomists to categorize new species.

B. Yeast taxonomy

The most commonly exploited yeast species, *Saccharomyces cerevisiae* (baker's yeast), belongs to the fungal kingdom subdivision Ascomycotina. Other yeast genera are represented in Basidiomycotina (e.g. *Cryptococcus* spp.) and *Rhodotorula* spp. and Deuteromycotina (e.g. *Candida* spp. and *Brettanomyces* spp.). There are around 100 recognized yeast genera (see Table 93.1).

C. Yeast biodiversity

Around 900 species of yeast have been described, but new species are being characterized on a regular basis and there is considerable untapped yeast biodiversity on Earth. Several molecular biological techniques are used to assist in the detection of new yeast species in the natural environment, and together with input from cell physiologists, provide ways to conserve and exploit yeast biodiversity. *S. cerevisiae* is the most studied and exploited of all the yeasts, but the biotechnological potential of non-*Saccharomyces*

TABLE 93.1 Alphabetical listing of 100 yeast genera

Aciculoconidium	*Dipodascopsis*	*Malassezia*	*Sporobolomyces*
Agaricostilbium	*Dipodascus*	*Metschnikowia*	*Sporopachydermia*
Ambrosiozyma	*Endomyces*	*Moniliella*	*Stephanoascus*
Arxiozyma	*Eremothecium*	*Mrakia*	*Sterigmatomyces*
Arxula	*Erythrobasidium*	*Myxozyma*	*Sterigmatosporidium*
Ascoidea	*Fellomyces*	*Nadsonia*	*Sympodiomyces*
Aureobasidium	*Fibulobasidium*	*Oosporidium*	*Sympodiomycopsis*
Babjevia	*Filobasidiella*	*Pachysolen*	*Tilletiaria*
Besingtonia	*Filobasidium*	*Phaffia*	*Tilletiopsis*
Blastobotrys	*Galactomyces*	*Pichia*	*Torulaspora*
Botryozyma	*Geotrichum*	*Protomyces*	*Tremella*
Brettanomyces	*Guilliermondella*	*Prototheca*	*Trichosporon*
Bullera	*Hanseniaspora*	*Pseudozyma*	*Trichosporonoides*
Bulleromyces	*Hansenula*	*Reniforma*	*Trigonopsis*
Candida	*Holtermannia*	*Rhodosporidium*	*Trimorphomyces*
Cephaloascus	*Hyalodendron*	*Rhodotorula*	*Tsuchiyaea*
Chionosphaera	*Issatchenkia*	*Saccharomyces*	*Ustilago*
Citeromyces	*Itersonila*	*Saccharomycodes*	*Wickerhamia*
Clavispora	*Kloeckera*	*Saccharomycopsis*	*Wickerhamiella*
Coccidiascus	*Kluyveromyces*	*Saitoella*	*Williopsis*
Cryptococcus	*Kockovaella*	*Saturnispora*	*Xanthophyllomyces*
Cyniclomyces	*Kurtzmanomyces*	*Schizoblastosporion*	*Yarrowia*
Cystofilobasidium	*Leucosporidium*	*Schizosaccharomyces*	*Zygoascus*
Debaryomyces	*Lipomyces*	*Sirobasidium*	*Zygosaccharomyces*
Dekkera	*Lodderomyces*	*Sporidiobolus*	*Zygozyma*

yeasts is gradually being realized, particularly with regard to recombinant DNA technology.

II. YEAST ECOLOGY

A. Natural habitats of yeast communities

Yeasts are not as ubiquitous as bacteria in the natural environment, but nevertheless yeasts can be isolated from terrestrial, aquatic, and aerial samples. Yeast communities are also found in association with plants, animals, and insects. Preferred yeast habitats are plant tissues, but a few species are found in commensal or parasitic relationships with animals. Some yeasts, most notably *Candida albicans*, are opportunistic human pathogens. Several species of yeast may be isolated from specialized or extreme environments, such as those with low water potential (i.e. high sugar or salt concentrations), low temperature (e.g. some psychrophilic yeasts have been isolated from polar regions), and low oxygen availability (e.g. intestinal tracts of animals). Table 93.2 summarizes the main yeast habitats.

B. Yeasts in the food chain

Yeasts play important roles in the food chain. Numerous insect species, notably *Drosophila* spp., feed on yeasts that colonize plant material. As insect foods, ascomycetous yeasts convert low-molecular-weight nitrogenous compounds into proteins beneficial to insect nutrition. In addition to providing a food source, yeasts may also affect the physiology and sexual reproduction of drosophilids. In marine environments, yeasts may serve as food for filter feeders.

C. Microbial ecology of yeasts

In microbial ecology, yeasts are not involved in biogeochemical cycling as much as bacteria or filamentous fungi. Nevertheless, yeasts can use a wide range of carbon sources and thus play an important role as saprophytes in the carbon cycle. In the cycling of nitrogen, some yeasts can reduce nitrate or ammonify nitrite, although most yeasts assimilate ammonium ions or amino acids into organic nitrogen. Most yeasts can reduce sulfate, although some are sulfur auxotrophs.

III. YEAST CELL STRUCTURE

A. General cellular characteristics

Yeasts are unicellular eukaryotes that portray the ultrastructural features of higher eukaryotic cells. This, together with their ease of growth, and amenability to biochemical, genetic, and molecular biological analyses, make yeasts model organisms in studies of eukaryotic cell biology. Yeast-cell size can vary widely, depending on the species and conditions of growth. Some yeasts may be only 2–3 μm in length, whereas others may attain lengths of 20–50 μm. Cell width appears less variable, between 1 and 10 μm. Table 93.3 summarizes the diversity of yeast-cell shapes.

Several yeast species are pigmented and various colors may be visualized in surface-grown colonies, for example: cream (e.g. *S. cerevisiae*), white (e.g. *Geotrichum* spp.), black (e.g. *Aureobasidium pullulans*), pink (e.g. *Phaffia rhodozyma*), red (e.g. *Rhodotorula* spp.), orange (e.g. *Rhodosporidium* spp.), and yellow

TABLE 93.2 Natural yeast habitats

Habitat	Comments
Soil	Soil may only be a reservoir for the long-term survival of many yeasts, rather than a habitat for growth. However, some genera are isolated exclusively from soil (e.g. *Lipomyces* and *Schwanniomyces*)
Water	*Debaryomyces hansenii* is a halotolerant yeast that can grow in nearly saturated brine solutions
Atmosphere	A few viable yeast cells may be expected per cubic meter of air. From layers above soil surfaces, *Cryptococcus, Rhodotorula, Sporobolomyces,* and *Debaryomyces* spp. are dispersed by air currents
Plants	The interface between soluble nutrients of plants (sugars) and the septic world are common niches for yeasts (e.g. the surface of grapes); the spread of yeasts on the phyllosphere is aided by insects (e.g. *Drosophila* spp.); a few yeasts are plant pathogens
Animals	Several nonpathogenic yeasts are associated with the intestinal tract and skin of warm-blood animals; several yeasts (e.g. *Candida albicans*) are opportunistically pathogenic toward humans and animals; numerous yeasts are commensally associated with insects, which act as important vectors in the natural distribution of yeasts
Built environment	Yeasts are fairly ubiquitous in buildings; for example, *Aureobasidium pullulans* (black yeast) is common on damp household wallpaper and *S. cerevisiae* is readily isolated from surfaces (pipework, vessels) in wineries

TABLE 93.3 Diversity of yeast-cell shapes

Gross cellular morphology	Description	Typical genera
Ellipsoidal	Oval- or 'rugby-ball'-shaped	*Saccharomyces*
Spherical	Complete spheres	*Debaryomyces*
Cylindrical	Cylinders with hemispherical ends	*Schizosaccharomyces*
Apiculate	Lemon-shaped	*Hanseniaspora, Saccharomycodes*
Ogival	Elongated cell rounded at one end and pointed at other	*Dekkera, Brettanomyces*
Bud-fission	Flask-shaped cells with septum at neck of bud	*Pityrosporum*
Filamentous	Pseudohyphal (chains of elongated budding cells) or hyphal (branched or unbranched filamentous cells that may show septa)	*Candida albicans, Yarrowia lipolytica*
Triangular	Buds are restricted to the three apices of triangular cells	*Trigonopsis*
Curved	Crescent-shaped	*Cryptococcus cereanus*
Stalked	Buds formed on short denticles or long stalks	*Sterigmatomyces*

(e.g. *Bullera* spp.). Some pigmented yeasts have uses in biotechnology. For example, the astaxanthin pigments of *Phaffia rhodozyma* have applications as fish-feed colorants for farmed salmonids, which have no means of synthesizing these red compounds.

B. Methods in yeast cytology

By using various cytochemical and cytofluorescent dyes and phase-contrast microscopy, it is possible to visualize several subcellular structures in yeasts (e.g. cell walls; capsules, if present; nuclei; vacuoles; mitochondria; and several cytoplasmic inclusion bodies). The *GFP* gene from the jellyfish (*Aequorea victoria*) encodes the green fluorescent protein (which fluoresces in blue light) and can be used to follow the subcellular destiny of certain expressed proteins when GFP is fused with the genes of interest. Immunofluorescense can also be used to visualize yeast cellular features when dyes such as fluorescein isothiocyanate and rhodamine B are conjugated with monospecific antibodies raised against yeast structural proteins. Confocal scanning laser immunofluorescence microscopy can also be used to detect the intracellular localization of proteins within yeast cells and to give three-dimensional ultrastructural information. Fluorescence-activated cell sorting (FACS) has proved very useful in studies of the yeast cell cycle and in monitoring changes in organelle (e.g. mitochondrial) biogenesis. Scanning-electron microscopy is useful in revealing the cell-surface topology of yeasts, as is atomic force microscopy, which has achieved high-contrast nanometer resolution of yeast-cell walls. Transmission-electron microscopy, however, is essential for visualizing the intracellular fine structure of ultrathin yeast-cell sections.

C. Subcellular yeast architecture and function

Transmission-electron microscopy of a yeast cell would typically reveal the cell wall, nucleus, mitochondria, endoplasmic reticulum, Golgi aparatus, vacuoles, microbodies, and secretory vesicles. Figure 93.1 shows an electron micrograph of a yeast cell.

Several of these organelles are not completely independent of each other and derive from an extended intramembranous system. For example, the movement and positioning of organelles depends on the cytoskeleton and the trafficking of proteins in and out of cells relies on vesicular communication between the endoplasmic reticulum (ER), Golgi, vacuole, and plasma membrane. Yeast organelles can be readily isolated for further studies by physical, chemical, or enzymic disruption of the cell wall, and the purity of organelle preparations can be evaluated using specific marker enzyme assays.

In the yeast cytoplasm, ribosomes and occasionally plasmids are found and the structural organization of the cell is maintained by a cytoskeleton of microtubules and actin microfilaments. The yeast-cell envelope, which encases the cytoplasm, comprises (from the inside looking out) the plasma membrane, periplasm, cell wall, and, in certain yeasts, a capsule and a fibrillar layer. Spores encased in an ascus may be revealed in those yeasts that undergo differentiation following sexual conjugation and meiosis. Table 93.4 provides a summary of the physiological function of the various structural components to be found in yeast cells.

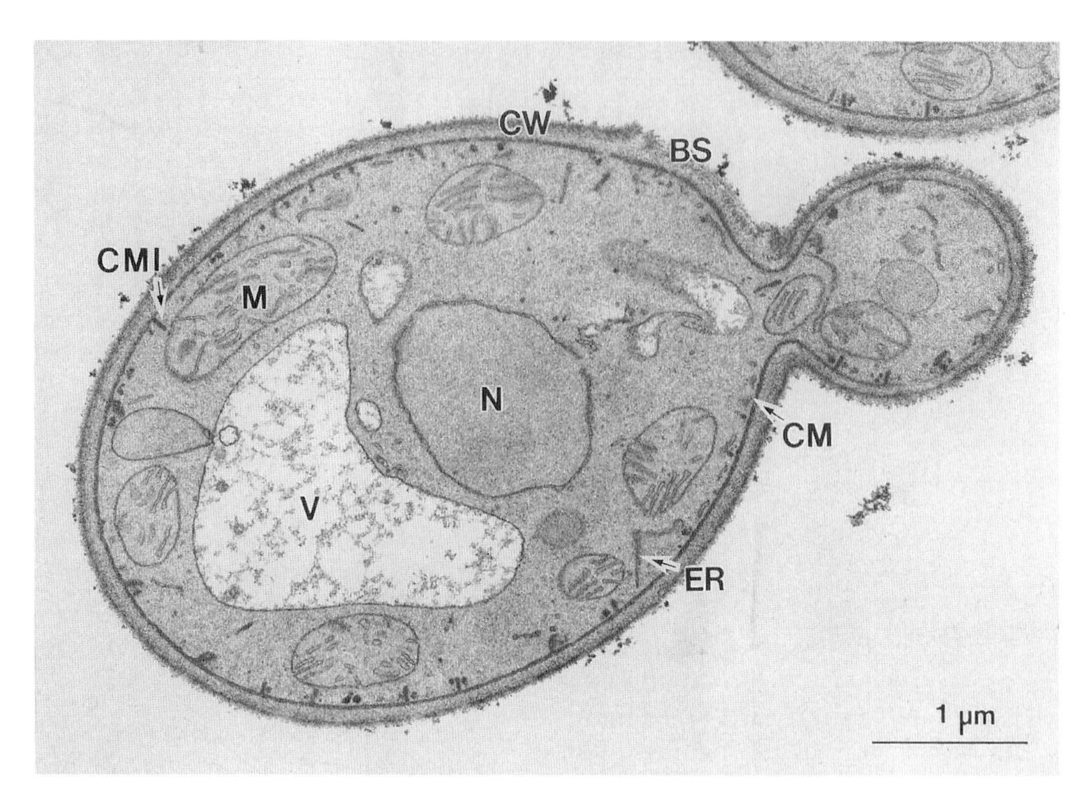

FIGURE 93.1 Ultrastructural features of a yeast cell. The transmission electron micrograph is of a *Candida albicans* cell. BS, bud scar; CM, cell membrane; CMI, cell membrane invagination; CW, cell wall; ER, endoplasmic reticulum; M, mitochondrion; N, nucleus; V, vacuole. (Courtesy of M. Osumi, Japan Women's University, Tokyo.)

TABLE 93.4 Functional components of an idealized yeast cell

Organelle or subcellular compartment	Function
Plasma membrane	Primary barrier for the passage of hydrophilic molecules in and out of yeast cells; selective permeability is mediated by specialized membrane proteins (e.g. proton-pumping ATPase)
Periplasm	This cell wall-associated region external to the membrane and internal to the wall is mainly made up of secreted mannoproteins (e.g. invertase and acid phosphatase) that are unable to permeate the cell wall
Cell wall	Involved in cell protection, cell-shape maintenance, cell–cell interactions, signal reception, surface attachment, and specialized enzymatic activities
Fimbriae	Proteinaceous protrusions emanating from surface of several basidiomycetous and ascomycetous yeasts that are mainly involved in cell–cell interactions before sexual conjugation
Capsules	Slimy extramural layers, prevalent in basidiomycetous yeasts (e.g. *Cryptococcus*), which serve to protect cells from physical (e.g., dehydration) and biological (e.g. phagocytosis) stresses
Peroxisomes	Membrane-bound organelles found in some yeasts that serve oxidative functions (e.g. use of methanol by methylotrophic yeasts)
Nucleus	The nucleoplasm contains DNA, RNA, and proteins (e.g. protamines and histones). DNA–histone complexes (chromatin) are organized in chromosomes that pass genetic information to daughter cells at cell division
Nucleolus	Crescent-shaped region within the nucleus, which is the site of ribosomal RNA transcription and processing
ER, Golgi, secretory vesicles	Secretory system for import (via endocytosis) and export (via exocytosis) of proteins
Vacuole	Membrane-bound organelle involved in intracellular protein trafficking in yeasts; also responsible for nonspecific intracellular proteolysis and as a site for storage of basic amino acids, polyphosphate, and metal cations
Mitochondria	Respiratory metabolism (aerobic conditions); fatty acid, sterol, and amino acid metabolism (anaerobic conditions)

IV. NUTRITION, METABOLISM, AND GROWTH OF YEASTS

A. Nutritional and physical requirements for yeast growth

1. *Yeast nutritional requirements*

Yeast cells require macronutrients (sources of carbon, nitrogen, oxygen, sulfur, phosphorus, potassium, and magnesium) at the millimolar level in growth media, and they require trace elements (e.g. Ca, Cu, Fe, Mn, Zn) at the micromolar level. Most yeasts grow quite well in simple nutritional media, which supplies carbon and nitrogen-backbone compounds together with inorganic ions and a few growth factors. Growth factors are organic compounds required in very low concentrations for specific catalytic or structural roles in yeast, but are not used as energy sources. Yeast growth factors include vitamins, which serve vital functions as components of coenzymes; purines and pyrimidines; nucleosides and nucleotides; amino acids; fatty acids; sterols; and other miscellaneous compounds (e.g. polyamines and choline). Growth-factor requirements vary among yeasts, but when a yeast species is said to have a growth-factor requirement, this indicates that it cannot synthesize the particular factor, resulting in the curtailment of growth without its addition to the culture medium.

2. *Yeast culture media*

It is quite easy to grow yeasts in the laboratory on a variety of complex and synthetic media. Malt extract or yeast extract supplemented with peptone and glucose (as in YEPG) is commonly employed for the maintenance and growth of most yeasts. Yeast Nitrogen Base (YNB) is a commercially available chemically defined medium that contains ammonium sulfate and asparagine as nitrogen sources, together with mineral salts, vitamins, and trace elements. The carbon source of choice (e.g. glucose) is usually added to a final concentration of 1% (w/v). For the continuous cultivation of yeasts in chemostats, media are usually designed that ensure that all the nutrients for growth are present in excess except one (the growth-limiting nutrient). Chemostats can therefore facilitate studies into the influence of a single nutrient (e.g. glucose, in carbon-limited chemostats) on yeast-cell physiology, with all other factors being kept constant. In industry, yeasts are grown in a variety of fermentation feedstocks including malt wort, molasses, grape juice, cheese whey, glucose syrups, and sufite liquor.

3. *Physical requirements for yeast growth*

Most species thrive in warm, dilute, sugary, acidic, and aerobic environments. Most laboratory and industrial yeasts (e.g. *S. cerevisiae* strains) grow best from 20 to 30°C. The lowest maximum temperature for growth of yeasts is around 20°C, whereas the highest is around 50°C.

Yeast need water in high concentration for growth and metabolism. Several food-spoilage yeasts (e.g. *Zygosaccharomyces* spp.) are able to withstand conditions of low water potential (i.e. high sugar or salt concentrations), and such yeasts are referred to as osmotolerant or xerotolerant.

Most yeasts grow very well between pH 4.5 and 6.5. Media acidified with organic acids (e.g. acetic and lactic) are more inhibitory to yeast growth than are media acidified with mineral acids (e.g. hydrochloric). This is because undissociated organic acids can lower intracellular pH following their translocation across the yeast-cell membrane. This forms the basis of action of weak-acid preservatives in inhibiting food-spoilage yeast growth. Actively growing yeasts acidify their growth environment through a combination of differential ion uptake, proton secretion during nutrient transport (see later), direct secretion of organic acids (e.g. succinate and acetate), and CO_2 evolution and dissolution. Intracellular pH is regulated within relatively narrow ranges in growing yeast cells (e.g. around pH 5 in *S. cerevisiae*), mainly through the action of the plasma-membrane proton-pumping ATPase.

Most yeasts are aerobes. Yeasts are generally unable to grow well under completely anaerobic conditions because, in addition to providing the terminal electron acceptor in respiration, oxygen is needed as a growth factor for membrane fatty acid (e.g. oleic acid) and sterol (e.g. ergosterol) biosynthesis. In fact, *S. cerevisiae* is auxotrophic for oleic acid and ergosterol under anaerobic conditions and this yeast is not, strictly speaking, a facultative anaerobe. Table 93.5 categorizes yeasts based on their fermentative properties and growth responses to oxygen availability.

B. Carbon metabolism by yeasts

1. *Carbon sources for yeast growth*

As chemorganotrophic organisms, yeasts obtain carbon and energy in the form of organic compounds. Sugars are widely used by yeasts. *Saccharomyces cerevisiae* can grow well on glucose, fructose, mannose, galactose, sucrose, and maltose. These sugars are also readily fermented to ethanol and carbon dioxide by this yeast, but other carbon substrates such as ethanol, glycerol, and acetate can only be respired by *S. cerevisiae* in

TABLE 93.5 Classification of yeasts based on fermentative capacity

Class	Examples	Comments
Obligately fermentative	*Candida pintolopesis*	Naturally occurring respiratory-deficient yeasts; only ferment, even in presence of oxygen
Facultatively fermentative		
Crabtree positive	*Saccharomyces cerevisiae*	Predominantly ferment high sugar-containing media in the presence of oxygen
Crabtree negative	*Candida utilis*	Do not form ethanol under aerobic conditions and cannot grow anaerobically
Nonfermentative	*Rhodotorula rubra*	Do not produce ethanol, either in the presence or absence of oxygen

the presence of oxygen. Some yeasts (e.g. *Pichia stipitis* and *Candida shehatae*) can use five-carbon pentose sugars such as D-xylose and L-arabinose as growth and fermentation substrates. A few amylolytic yeasts exist (e.g. *Saccharomyces diastaticus*) that can use starch, and several oleaginous yeasts (e.g. *Candida tropicalis*) can grow on hydrocarbons such as straight-chain alkanes in the C_{10}–C_{20} range. Several methylotrophic yeasts (e.g. *Hansenula polymorpha* and *Pichia pastoris*) can grow very well on methanol as the sole carbon and energy source, and these yeasts have industrial potential in the production of recombinant proteins for pharmaceutical use.

2. *Yeast sugar transport*

Sugars are transported into yeast cells across the plasma membrane by a variety of mechanisms from simple net diffusion (a passive or free mechanism), facilitated (catalyzed) diffusion, and active (energy-dependent) transport. The precise mode of sugar translocation will depend on the sugar, yeast species, and growth conditions. For example, *S. cerevisiae* takes up glucose by facilitated diffusion and maltose by active transport. Active transport means that the plasma-membrane ATPases act as directional proton pumps in accordance with chemiosmotic principles. The pH gradients thus drive nutrient transport either via proton symporters (as is the case with certain sugars and amino acids) or via proton antiporters (as is the case with potassium ions).

3. *Yeast sugar metabolism*

The principal metabolic fates of sugars in yeasts are the dissimilatory pathways of fermentation and respiration (summarized in Fig. 93.2), and the assimilatory pathways of gluconeogenesis and carbohydrate biosynthesis.

Yeasts described as fermentative are able to use organic substrates (sugars) anaerobically as electron donor, electron acceptor, and carbon source. During alcoholic fermentation of sugars, *S. cerevisiae* and other fermentative yeasts reoxidize the reduced coenzyme NADH to NAD (nicotinamide adenine dinucleotide) in terminal step reactions from pyruvate.

$$\text{Glucose} \xrightarrow{2\text{ATP}} 2\text{Pyruvate} \xrightarrow[\text{pyruvate decarboxylase}]{2CO_2}$$

$$2\ \text{Acetaldehyde} \xrightarrow[\text{alcohol dehydrogenase}]{2\text{NADH} + 2H^+ \rightarrow 2\text{NAD}^+} 2\ \text{Ethanol}$$

$$\text{Sum: Glucose} + 2\text{Pi} + 2\text{ADP} \longrightarrow 2\ \text{Ethanol} + 2Co_2 + 2\text{ATP} + 2H_2O$$

In the first of these terminal reactions, catalyzed by pyruvate decarboxylase, pyruvate is decarboxylated to acetaldehyde, which is finally reduced by alcohol dehydrogenase to ethanol. The regeneration of NAD is necessary to maintain the redox balance and prevent the stalling of glycolysis. In alcoholic-beverage fermentations (e.g. of beer, wine, and distilled spirits), other fermentation metabolites, in addition to ethanol and carbon dioxide, are produced by yeast that are very important in the development of flavor. These metabolites include fusel alcohols (e.g. isoamyl alcohol), polyols (e.g. glycerol), esters (e.g. ethyl acetate), organic acids (e.g. succinate), vicinyl diketones (e.g. diacetyl), and aldehydes (e.g. acetaldehyde). The production of glycerol (an important industrial commodity) can be enhanced in yeast fermentations by the addition of sulfite, which chemically traps acetaldehyde.

$$\text{Glucose} + HSO_3^- \rightarrow \text{Glycerol} + \text{acetaldehyde-}HSO_3^- + CO_2$$

Of the environmental factors that regulate respiration and fermentation in yeast cells, the availability by glucose and oxygen are the best understood and are linked to the expression of regulatory phenomena, referred to as the Pasteur effect and the Crabtree effect. A summary of the description of these phenomena is provided in Table 93.6.

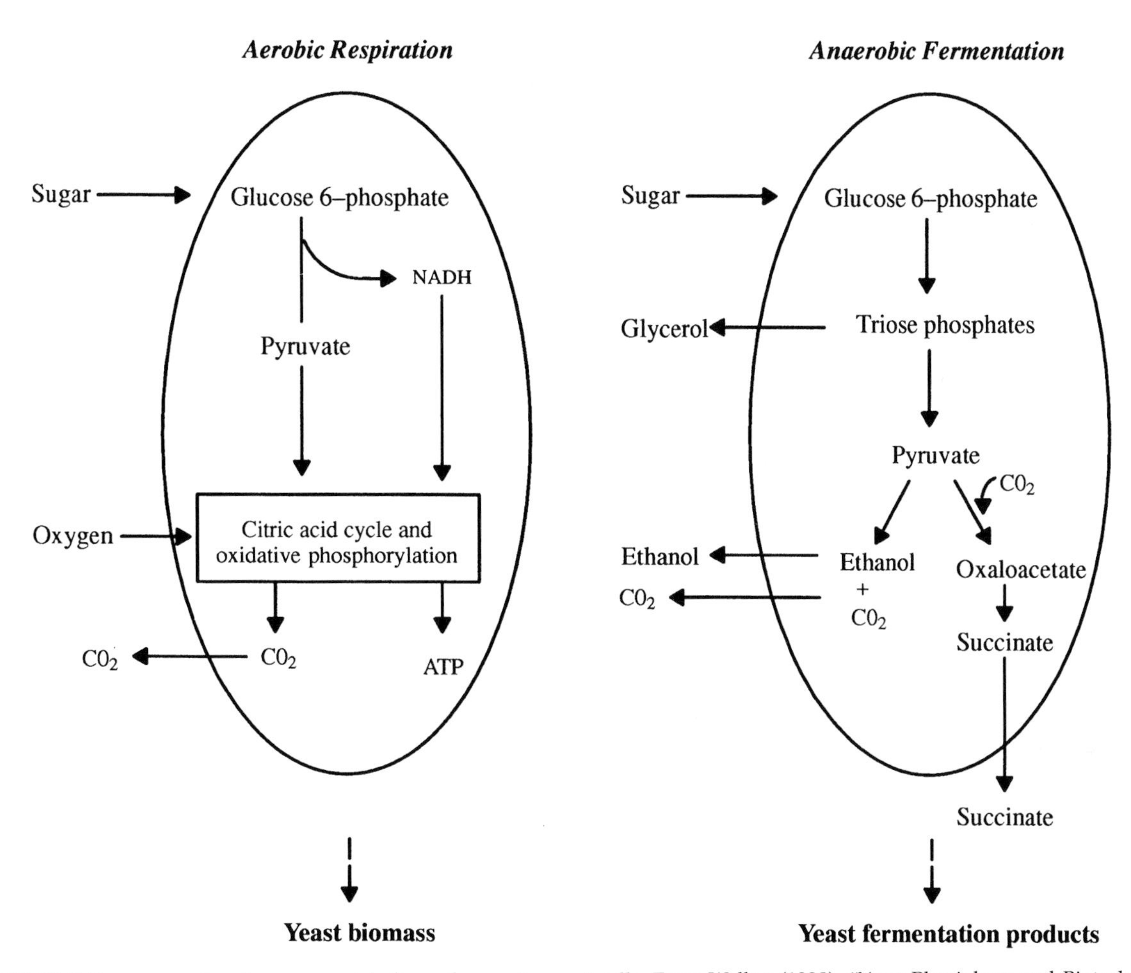

FIGURE 93.2 Summary of major sugar catabolic pathways in yeast cells. From Walker (1998). "Yeast Physiology and Biotechnology." Copyright John Wiley & Sons Limited. Reproduced with permission.

TABLE 93.6 Summary of Pasteur and Crabtree effects in yeast sugar metabolism

Regulatory phenomenon	Description	Comments
Pasteur effect	Activation of sugar consumption rate by anaerobiosis	Only observable in resting or nutrient-starved cells (e.g. *S. cerevisiae*)
Crabtree effect	Suppression of respiration by high glucose levels	Cells (e.g. *S. cerevisiae, Schiz. pombe*) continue to ferment irrespective of oxygen availability due to glucose repressing or inactivating respiratory enzymes or due to the inherent respiratory capacity of cells

C. Nitrogen metabolism by yeasts

1. *Nitrogen sources for yeast growth*

Although yeasts cannot fix molecular nitrogen, simple inorganic nitrogen sources such as ammonium salts are widely used. Ammonium sulfate is a commonly used nutrient in yeast growth media because it provides a source of both assimilable nitrogen and sulfur. Some yeasts can also grow on nitrate as a source of nitrogen, and, if able to do so, may also use subtoxic concentrations of nitrite. A variety of organic nitrogen compounds (amino acids, peptides, purines, pyrimidines, and amines) can also provide the nitrogenous requirements of the yeast cell. Glutamine and aspartic acids are readily deaminated by yeasts and therefore act as good nitrogen sources.

2. *Yeast transport of nitrogenous compounds*

Ammonium ions are transported is *S. cerevisiae* by both high-affinity and low-affinity carrier-mediated transport systems. Two classes of amino acid uptake systems operate in yeast cells. One is broadly specific, the general amino acid permease (GAP), and effects the uptake of all naturally occuring amino acids. The other system includes a variety of tranporters that display specificity for one or a small number of related amino acids. Both the general and the specific transport systems are energy dependent.

3. *Yeast metabolism of nitrogenous compounds*

Yeasts can either incorporate ammonium ions or amino acids into cellular protein, or these nitrogen sources can be intracellularly catabolized to serve as nitrogen sources (see Fig. 93.3). Yeasts also store relatively large pools of endogenous amino acids in the vacuole, most notably arginine. Ammonium ions can be directly assimilated into glutamate and glutamine, which serve as precursors for the biosynthesis of other amino acids. The precise mode of ammonium assimilation adopted by yeasts will depend mainly on the concentration of available ammonium ions and the intracellular amino acid pools. Amino acids may be dissimilated (by decarboxylation, transamination, or fermentation) to yield ammonium and glutamate, or they may be directly assimilated into proteins.

D. Yeast growth

The growth of yeasts is concerned with how cells transport and assimilate nutrients and then integrate numerous component functions in the cell in order to increase in mass and eventually divide. Yeasts have proved invaluable in unravelling the major control elements of the eukaryotic cell cycle and research with the budding yeast, *Saccharomyces cerevisiae*, and the fission yeast, *Schizosaccharomyces pombe*, has significantly advanced our understanding of cell-cycle regulation, which is particularly important in the field of human cancer.

1. *Vegetative reproduction in yeasts*

Budding in the most common mode of vegetative reproduction in yeasts and is typical in ascomycetous yeasts such as *S. cerevisiae*. Yeast buds (see Fig. 93.4) are initiated when mother cells attain a critical cell size at a time that coincides with the onset of DNA synthesis. This is followed by localized weakening of the cell wall and this, together with tension exerted by turgor pressure, allows the extrusion of the cytoplasm in an area bounded by new cell-wall material. The mother and daughter bud cell walls are contiguous during bud development. In *S. cerevisiae*, cell size at division is asymmetrical, with buds being smaller than mother cells when they separate. Scar tissue on the yeast cell wall, the bud and birth scars, remain on the daughter bud and mother cells, respectively (see Fig. 93.4).

Fission is a mode of vegetative reproduction typified by species of *Schizosaccharomyces*, which divide exclusively by forming a cell septum that constricts

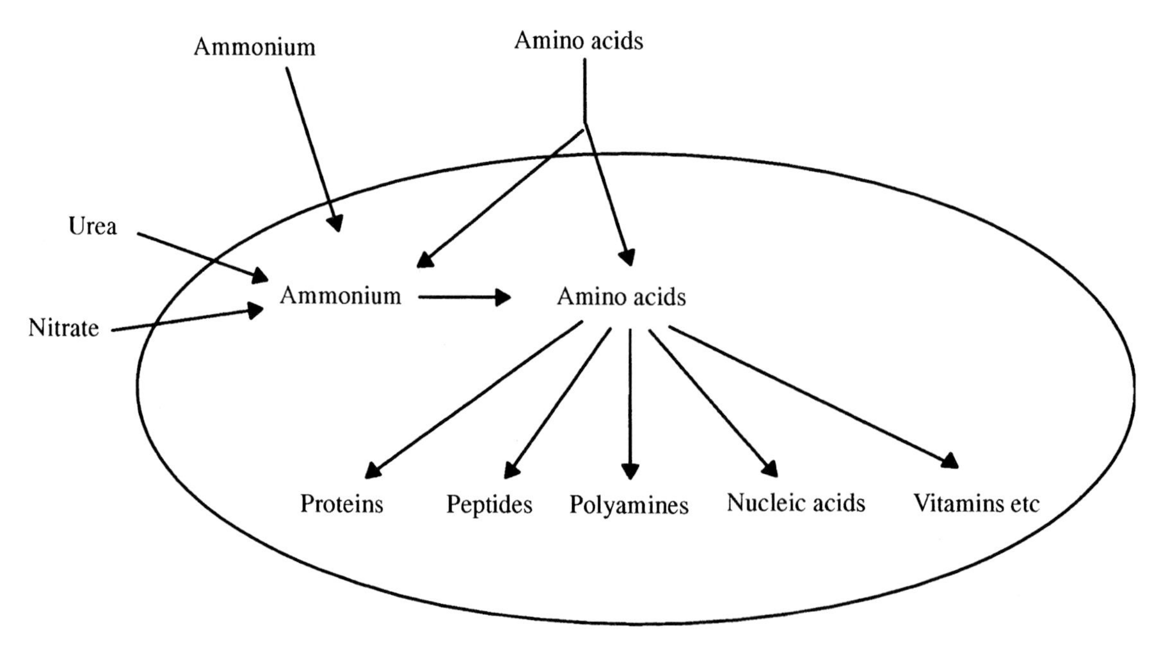

FIGURE 93.3 Overview of nitrogen assimilation in yeasts. From Walker (1998). "Yeast Physiology and Biotechnology." Copyright John Wiley & Sons Limited. Reproduction with permission.

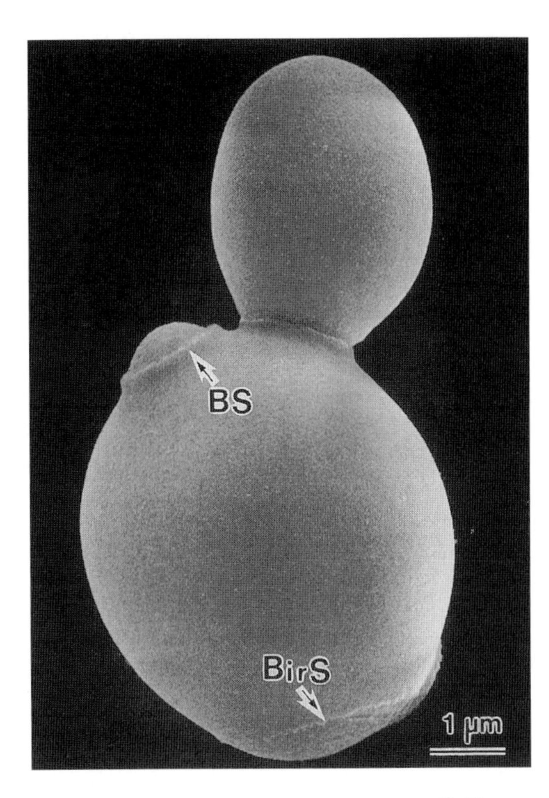

FIGURE 93.4 Bud and birth scars in a yeast cell. The scanning electron micrograph shows a bud scar (BS) and a birth scar (BirS) on the surface of a *Saccharomyces cerevisiae* cell. (Courtesy of M. Osumi, Japan Women's University, Tokyo.)

the cell into two equal-size daughters. In *Sch. pombe*, which has been used extensively in eukaryotic cell cycle studies, newly divided daughter cells grow lengthways in a monopolar fashion for about one-third of their new cell cycle. Cells then switch to bipolar growth for about three-quarters of the cell cycle until mitosis is initiated at a constant cell-length stage.

Filamentous growth occurs in numerous yeast species and may be regarded as a mode of vegetative growth alternative to budding or fission. Some yeasts exhibit a propensity to grow with true hyphae initiated from germ tubes (e.g. *Candida albicans*), but others (including *S. cerevisiae*) may grow in a pseudohyphal fashion when induced to do so by unfavorable conditions. Hyphal and pseudohyphal growth represent different developmental pathways in yeasts, but cells can revert to unicellular growth upon return to more conducive growth conditions. Filamentation may therefore represent an adaptation by yeasts to foraging when nutrients are scarce.

2. *Population growth of yeasts*

As in most microorganisms, when yeast cells are inoculated into a liquid nutrient medium and incubated under optimal physical growth conditions, a typical batch-growth curve will result when the viable cell population is plotted against time. This growth curve is made up of a lag phase (period of no growth, but physiological adaptation of cells to their new environment), exponential phase (limited period of logarithmic cell doublings), and stationary phase (resting period with zero growth rate).

Diauxic growth is characterized by two exponential phases and occurs when yeasts are exposed to two carbon growth substrates that are used sequentially. This occurs during aerobic growth of *S. cerevisiae* on glucose (the second substrate being ethanol formed from glucose fermentation).

In addition to batch cultivation of yeasts, cells can also be propagated in continuous culture in which exponential growth is prolonged without lag or stationary phases. Chemostats are continuous cultures that are based on the controlled feeding of a sole growth-limiting nutrient into an open culture vessel, which permits the outflow of cells and spent medium. The feeding rate is referred to as the dilution rate, which is employed to govern the yeast growth rate under the steady-state conditions that prevail in a chemostat.

Specialized yeast culture systems include immobilized bioreactors. Yeast cells can be readily immobilized or entrapped in a variety of natural and synthetic materials (e.g. calcium aliginate gel or microporous glass beads) and such systems have applications in the food and fermentation industries.

V. YEAST GENETICS

A. Life cycle of yeasts

Many yeasts have the ability to reproduce sexually, but the processes involved are best understood in the budding yeast, *S. cerevisiae*, and the fission yeast, *Sch. pombe*. Both species have the ability to mate, undergo meiosis, and sporulate. The development of spores by yeasts represents a process of morphological, physiological, and biochemical differentiation of sexually reproductive cells.

Mating in *S. cerevisiae* involves the conjugation of two haploid cells of opposite mating types, designated a and α. These cells synchronize one another's cell cycles in response to peptide-mating pheromones, known as a factor and α factor (see Fig. 93.5). The conjugation of mating cells occurs by cell-wall surface contact followed by plasma-membrane fusion to form a common cytoplasm. Karyogamy (nuclear fusion) then follows, resulting in a diploid nucleus. The stable

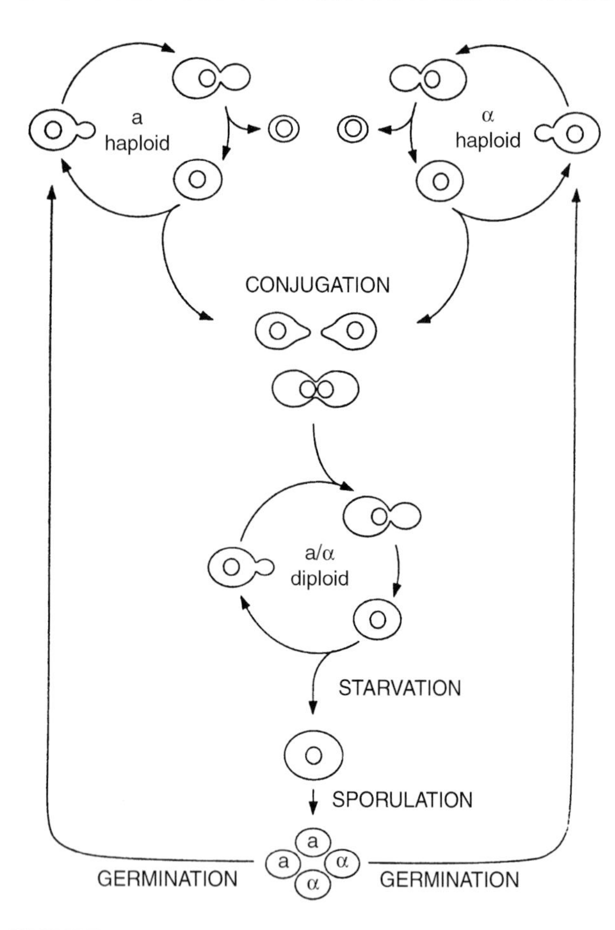

FIGURE 93.5 Life cycle of *Saccharomyces cerevisae*. From Murray, A., and Hunt, T. (1993). "The Cell Cycle, an Introduction." Copyright © 1993 Oxford University Press, Inc. Used by permission of Oxford University Press, Inc.

diploid zygote will continue mitotic cell cycles in rich growth media, but, if starved of nitrogen, the diploid cells will sporulate to yield four haploid spores. These germinate in rich media to form haploid budding cells that can mate with each other to restore the diploid state.

In *Sch. pombe*, haploid cells of the opposite mating types (designated *h*+ and *h*−) secrete mating pheromones and, when starved of nitrogen, undergo conjugation to form diploids. In *Sch. pombe*, however, such diploidization is transient under starvation conditions and cells soon enter meiosis and sporulate to produce four haploid spores.

B. Genetic manipulation of yeasts

There are several ways of genetically manipulating yeast cells including hybridization, mutation, rare mating, cytoduction, spheroplast fusion, single chromosome transfer, and transformation using recombinant DNA technology.

Classic genetic approaches in *S. cerevisiae* involve mating of haploids of opposite mating type. Subsequent meiosis and sporulation results in the production of a *tetrad ascus* with four spores, which can be isolated, propagated, and genetically analyzed (i.e. tetrad analysis). This process forms the basis of genetic breeding programs for laboratory-reference strains of *S. cerevisiae*. However, industrial (e.g. brewing) strains of this yeast are polyploid, are reticent to mate, and exhibit poor sporulation with low spore viability. It is, therefore, generally fruitless to perform tetrad analysis and breeding with brewer's yeasts. Genetic manipulation strategies for circumventing the sexual reproductive deficiencies of brewer's yeast include spheroplast fusion and recombinant DNA technology.

Intergeneric and intrageneric yeast hybrids may be obtained using the technique of spheroplast fusion. This involves the removal of yeast-cell walls using lytic enzymes (e.g. glucanases from snail gut juice or microbial sources), followed by the fusion of the resulting spheroplasts in the presence of polyethylene glycol and calcium ions.

Recombinant DNA technology (genetic engineering) of yeast is summarized in Fig. 93.6. Yeast cells possess particular attributes for expressing foreign genes and have now become the preferred hosts, over bacteria, for producing human proteins for pharmaceutical use. Although the majority of research and development into recombinant protein synthesis in yeasts has been conducted using *S. cerevisiae*, several non-*Saccharomyces* species are being studied and exploited in biotechnology. For example, *Hansenula polymorpha* and *Pichia pastoris* (both methylotrophic yeasts) exhibit particular advantages over *S. cerevisiae* in cloning technology.

C. Yeast genome and proteome projects

A landmark in biotechnology was reached in 1996 with completion of the sequencing of the entire genome of *S. cerevisiae*. The *Sch. pombe* genome project was completed in 2002. The functional analysis of the many orphan genes of *S. cerevisiae*, for which no function has yet been assigned, is underway through international research collaborations. Elucidation by cell physiologists of the biological function of all *S. cerevisiae* genes, that is, the complete analysis of the yeast proteome, will not only lead to an understanding of how a simple eukaryotic cell works, but will also provide insight into molecular biological aspects of heritable human disorders.

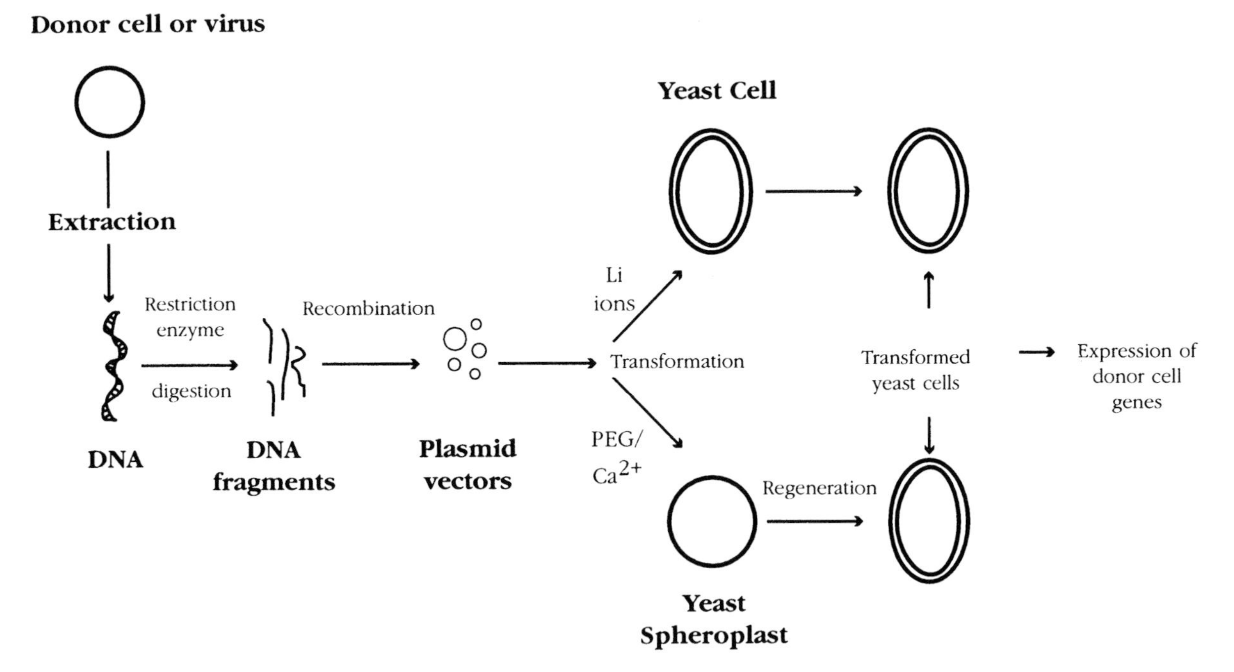

FIGURE 93.6 Basic procedures in yeast genetic engineering. From Walker (1998). "Yeast Physiology and Biotechnology." Copyright John Wiley & Sons Limited. Reproduced with permission.

VI. INDUSTRIAL, AGRICULTURAL, AND MEDICAL IMPORTANCE OF YEASTS

A. Industrial significance of yeasts

Yeasts have been exploited for thousands of years in traditional fermentation processes to produce beer, wine, and bread. The products of modern yeast biotechnologies impinge on many commercially important sectors, including food, beverages, chemicals, industrial enzymes, pharmaceuticals, agriculture, and the environment. Table 93.7 lists some of the principal industrial commodities from yeasts. *S. cerevisiae* is the most exploited microorganism known and is the yeast responsible for producing potable and industrial ethanol, which is the world's premier biotechnological commodity.

Some yeasts play detrimental roles in industry, particularly as spoilage yeasts in food and beverage production (see Table 93.8).

B. Yeasts of environmental and agricultural significance

Although a few yeast species are known to be plant pathogens (e.g. *Ophiostoma ulmi*, the causative agent of Dutch Elm disease), several have been shown to be beneficial to plants in preventing fungal disease. For example, *S. cerevisiae* has potential as a phytoallexin elicitor in stimulating cereal plant defenses against fungal pathogens, and several yeasts (e.g. *Debaryomyces hansenii* and *Metschnikowia pulcherrima*) may be used in the biocontrol of fungal fruit diseases. Other environmental benefits of yeasts are to be found in aspects of pollution control. For example, yeasts can effectively biosorb heavy metals and detoxify chemical pollutants from industrial effluents. Some yeasts (e.g. *Candida utilis*) can effectively remove carbon and nitrogen from organic wastewater.

In agriculture, live cultures of *S. cerevisiae* have been shown to stabilize the rumen environment of ruminant animals (e.g. cattle) and improve the nutrient availability to increase animal growth or milk yields. The yeasts may be acting to scavenge oxygen and prevent oxidative stress to rumen bacteria, or they may provide malic and other dicarboxylic acids to stimulate rumen bacterial growth.

C. Medical significance of yeasts

The vast majority of yeasts are beneficial to human life. However, some yeasts are opportunistically pathogenic towards humans. Mycoses caused by *Candida albicans*, collectively referred to as candidosis (candidiasis), are the most common opportunistic yeast infections. There are many predisposing factors to yeast

TABLE 93.7 Industrial commodities produced by yeasts

Commodity	Examples
Beverages	Potable alcoholic beverages: beer, wine, cider, saké, distilled spirits (whisky, rum, gin, vodka, cognac)
Food and animal feed	Baker's yeast, yeast extracts, fodder yeast and livestock growth factor, feed pigments
Chemicals	Fuel ethanol, carbon dioxide, glycerol, citric acid vitamins; yeasts are also used as bioreductive catalysts in organic chemistry
Enzymes	Invertase, inulinase, pectinase, lactase, lipase
Recombinant proteins	Hormones (e.g. insulin), viral vaccines (e.g. hepatitis B vaccine), antibodies (e.g. IgE receptor), growth factors (e.g. tumor necrosis factor), interferons (e.g. leucocyte interferon-α), blood proteins (e.g. human serum albumin), enzymes (e.g. gastric lipase)

TABLE 93.8 Some food and beverage-spoilage yeasts

Yeast species	Food spoiled
Cryptococcus laurentii, Candida zeylanoides	Frozen poultry carcasses
Zygosaccharomyces bailii, Z. rouxii	Fruits, fruit juices, vegetables
Kluyveromyces, Rhodotorula, and *Candida* spp.	Milk, yogurt, cheeses
Torulopsis, Pichia, Candida, Hansenula spp., and 'wild' species of *Saccharomyces*	Beer
Zygosaccharomyces bailii and many other yeasts	Wine

infections, but immunocompromised individuals appear particularly susceptible to candidosis. *C. albicans* infections in AIDS patients are frequently life-threatening.

The beneficial medical aspects of yeasts are apparent in the provision of novel human therapeutic agents through yeast recombinant DNA technology (see Table 93.7). Yeasts are also extremely valuable as experimental models in biomedical research, particulalry in the fields of oncology, pharmacology, toxicology, virology, and human genetics.

BIBLIOGRAPHY

Brown, A. J. P., and Tuite, M. (1999). "Yeast bene Analysis". Academic Press, London.

Burke, D., Sterans, T., and Dawson, D. (2000). "Methods in Yeast Genetics." Cold Spring Harbor Laboratory.

Dickinson, J. R., and Schweizer. M. (1998). "The Metabolism and Molecular Physiology of *Saccharomyces cerevisiae*." Taylor & Francis, New York.

Fantes, P., and Beggs, J. (Eds.) (2000). "The Yeast Nucleus." Oxford University Press, Oxford.

Johnston, J. R. (1994). "Molecular Genetics of Yeast, a Practical Approach." Oxford University Press, New York.

Kurtzman, C. P., and Fell, J. W. (1998). "The Yeasts. A Taxonomic Study." Elsevier Science, Amsterdam.

Kocková-Kratochvilová, A. (1990). "Yeasts and Yeast-like Organisms." VCH, New York.

Oliver, S. G., Winson, M. K., Kell, D. B., and Baganz, F. (1998). "Systematic functional analysis of the yeast genome". *Trends Biotechnol.* **16**, 373–378.

Panchal, C. J. (1990). "Yeast Strain Selection." Marcel Dekker, New York.

Pringle, J. R., Broach, J. R., and Jones, E. W. (1997). "The Molecular and Cellular Biology of the Yeast Saccharomyces." Cold Spring Harbor Laboratory Press, Cold Spring Harbor, NY.

Rose, A. H., and Harrison, J. S. (1989–1995). "The Yeasts." Academic Press, London.

Spencer, J. F. T., and Spencer, D. M. (1997). "Yeasts in Natural and Artifical Habitats." Springer-Verlag, Berlin.

Walker, G. M. (1998). "Yeast Physiology and Biotechnology." John Wiley & Sons, Chichester, UK.

Wolf, K. (1996). "Nonconventional Yeasts in Biotechnology, A Handbook." Springer-Verlag, Berlin.

Zimmerman, F. K., and Entian, K.-D. (1997). "Yeast Sugar Metabolism, Biochemistry, Genetics, Biotechnology and Applications." Technomic Publishing, Basel, Switzerland.

WEBSITES

Access to budding yeast, fission yeast, and Candida
http://genome-www.stanford.edu/Saccharomyces/VL-yeast.html

Candida albicans information
http://alces.med.umn.edu/Candida.html

Saccharomyces genome data base
http://genome-www.stanford.edu/Saccharomyces/

Cryptococcal Working Group, Univ. of Calgary
http://www.ucalgary.ca/~cmody/

Schizosaccharomyces pombe website (F. Hochstenbach)
http://mc11.mcri.ac.uk/tony/pombelinks/pombelink.html

Fungi & Yeasts Collection Catalog, American Type Culture Collection
http://www.atcc.org/SearchCatalogs/Fungi_Yeasts.cfm

International Society for Human and Animal Mycology
http://www.leeds.ac.uk/isham/

Medical Mycology Divison, ASM (American Society for Microbiology)
http://www.asmusa.org/division/f/MainPage.html

Fungi and Human Disease (N. Chamberlain). Links to photographs
http://www.kcom.edu/faculty/chamberlain/Website/fungi.htm

Yeast genome site
http://www.Sanger.ac.uk/yeast/home.html

Yeast molecular biology tools
http://www.fmi.ch/biology/research_tools.html

Yeast Newsgroup
http://www.bio.net/hypermail/YEAST/news:bionet.molbio.yeast

Index